Lecture Notes in Computer Science 3124

Commenced Publication in 1973
Founding and Former Series Editors:
Gerhard Goos, Juris Hartmanis, and Jan van Leeuwen

José Neuman de Souza   Petre Dini
Pascal Lorenz (Eds.)

Telecommunications and Networking – ICT 2004

11th International Conference on Telecommunications
Fortaleza, Brazil, August 1-6, 2004
Proceedings

 Springer

Volume Editors

José Neuman de Souza
Universida de Federal do Ceará, Departamento de Computação
Campus do Pici - Bloco 910, 60455-760 Fortaleza, Ceará, Brazil
E-mail: neuman@ufc.br

Petre Dini
Cisco Systems, Inc., USA and Concordia University, Canada
170 West Tasman Drive, San Jose, CA 95134, USA
E-mail: pdini@cisco.com

Pascal Lorenz
University of Haute Alsace
34 rue du Grillenbreit, 68008 Colmar, France
E-mail: lorenz@ieee.org

Library of Congress Control Number: 2004109602

CR Subject Classification (1998): C.2, H.4, H.3, H.5.1, K.4.4, K.8.1, D.2

ISSN 0302-9743

ISBN 978-3-540-22571-3   ISBN 978-3-540-27824-5 (eBook)
DOI 10.1007/978-3-540-27824-5

springeronline.com

© Springer-Verlag Berlin Heidelberg 2004

Originally published by Springer-Verlag Berlin Heidelberg New York in 2004.

Typesetting: Camera-ready by author, data conversion by PTP-Berlin, Protago-TeX-Production GmbH
Printed on acid-free paper   SPIN: 11306153   06/3142   5 4 3 2 1 0

Preface

Welcome to the 11th International Conference on Telecommunications (ICT 2004) hosted by the city of Fortaleza (Brazil).

As with other ICT events in the past, this professional meeting continues to be highly competitive and very well perceived by the international networking community, attracting excellent contributions and active participation. This year, a total of 430 papers from 36 countries were submitted, from which 188 were accepted. Each paper was reviewed by several members of the ICT 2004 Technical Program Committee. We were very pleased to receive a large percentage of top-quality contributions.

The topics of submitted papers covered a wide spectrum from photonic techniques, signal processing, cellular networks, and wireless networks, to ad hoc networks. We believe the ICT 2004 papers offer a wide range of solutions to key problems in telecommunications, and describe challenging avenues for industrial research and development.

In addition to the conference regular sessions, seven tutorials and a workshop were organized. The tutorials focused on special topics dealing with next-generation networks. The workshop focused on particular problems and solutions in heavily distributed and shareable environments.

We would like to thank the ICT 2004 Technical Program Committee members and referees. Without their support, the creation of such a broad conference program would not be possible. We also thank all the authors who made a particular effort to contribute to ICT 2004. We truly believe that due to all these efforts the final conference program consisted of top-quality contributions.

We are also indebted to many individuals and organizations that made this conference possible. In particular, we would like to thank the members of the ICT 2004 Organizing Committee for their help in all aspects of the organization of this professional meeting.

The 11th International Conference on Telecommunications was an outstanding international forum for the exchange of ideas and results between academia and industry, and provided a baseline for further progress in the telecommunications area.

We hope that the attendees enjoyed their stay in Brazil and were able to visit various points of interest in this lovely country.

June 2004

José Neuman De Souza
Petre Dini
Pascal Lorenz

IEEE

ICT 2004 General Chair

José Neuman de Souza, UFC, Brazil

ICT Steering Committee

Hamid Aghvami, King's College, London, UK
Tulin Atmaca, Institut National des Télécommunications, France
Petre Dini, Cisco Systems, Inc., USA, and Concordia University, Canada
Pascal Lorenz, University of Haute Alsace, France
Farokh Marvasti, King's College, London, UK

ICT 2004 Technical Program Co-chairs

Petre Dini, Cisco Systems, Inc., USA and Concordia University, Canada
Pascal Lorenz, University of Haute Alsace, France

ICT 2004 International Scientific Committee

Aghvami, Hamid, King's College, London, UK
Aguiar, Rui L., Universidade de Aveiro, Portugal
Agoulmine, Nazim, Université d'Évry, France
Aissa, Sonia, INRS-Telecommunications, Canada
Alencar, Marcelo Sampaio, UFCG, Brazil
Alfa, Attahiru Sule, University of Manitoba, Canada
Andrade, Rossana, UFC, Brazil
Atmaca, Tulin, Institut National des Télécommunications, France
Atwood, Bill, Concordia University, Montreal, Canada
Baier, Paul Walter, University of Kaiserslautern, Germany
Barbeau, Michel, Carleton University, Canada
Brito, Jose Marcos Câmara, National Institute of Telecommunications, Brazil
Casaca, Augusto, INESC, Portugal
Celestino, Júnior Joaquim, UECE, Brazil
Cherkaoui, Soumaya, University of Sherbrooke, Canada
Chemouil, Prosper, France Telecom R&D, France
Chen, Xiaodong, University of London, UK
Cooklev, Todor, San Francisco State University, USA
Correia, Luis M., Technical University of Lisbon, Portugal
Dini, Petre, Cisco Systems, Inc., USA and Concordia University, Canada
Ferreira, Afonso, CNRS-INRIA, France
Galis, Alex, University College London, UK
Kabacinski, Wojciech, Poznan University of Technology, Poland

Kahng, Hyun-Kook, Korea University, Korea
Lee, Mike Myung-Ok, Dongshin University, Republic of Korea
Logrippo, Luigi, University of Ottawa, Canada
Lorenz, Pascal, University of Haute Alsace, France
Marshall, Alan, Queen's University of Belfast, UK
Marvasti, Farokh, King's College, London, UK
Mohr, Werner, Siemens, Germany
Molinaro, Antonella, University of Messina, Italy
Mota, João César Moura, UFC, Brazil
Mynbaev, Djafar, City University of New York, USA
Nefedov, Nikolai, Nokia Research Center, Finland
Oliveira, Mauro, CEFET-CE, Brazil
Pujolle, Guy, LIP6, France
Sestini, Fabrizio, European Commission DG Information Society, Belgium
Sezer, Sakir, Queen's University Belfast, UK
Soulhi, Said, Ericsson Research, Canada
Souza, José Neuman de, UFC, Brazil
Stasiak, Maciej, Poznan University of Technology, Poland
Takasaki, Yoshitaka, Tokyo University, Japan
Wenbo, Wang, Beijing University of Posts and Telecommunications, China
Westphall, Carlos Becker, UFSC, Brazil

Reviewer List

Aghvami, Hamid, King's College, London, UK
Agoulmine, Nazim, LIP6, France
Aguiar, Rui Luis, Universidade de Aveiro, Portugal
Aissa, Sonia, INRS-EMT, Canada
Alencar, Marcelo Sampaio de, UFCG, Brazil
Alfa, Attahiru Sule, University of Manitoba, Canada
Andrade Rossana, Maria de Castro, UFC, Brazil
Atmaca, Tulin, INT-Evry, France
Atwood, John William, Concordia University, Montreal, Canada
Baier, Paul Walter, University of Kaiserslautern, Germany
Barbeau, Michel, School of Computer Science, Carleton University, Canada
Barreto, Guilherme de Alencar, UFC, Brazil
Brito, Jose Marcos Camara, INATEL, Brazil
Casaca, Augusto, INESC, Portugal
Cavalcanti, Francisco Rodrigo P., UFC, Brazil
Celestino, Júnior Joaquim, UECE, Brazil
Chemouil, Prosper, France
Chen, Xiaodong, University of London, UK
Cherkaoui, Soumaya, Université de Sherbrooke, Canada
Coelho, Pedro Henrique Gouvêa, Brazil

Cooklev, Todor, San Francisco State University, USA
Correia, Luis M., Technical University of Lisbon, Portugal
Costa, Max, UNICAMP, Brazil
Dini, Petre, Cisco Systems, Inc., USA and Concordia University, Canada
Farrell, Paddy, UK
Ferreira, Afonso, CNRS/INRIA, France
Galdino, Juraci Ferreira, Brazil
Galis, Alex, University College London, UK
Gameiro, Atílio, Portugal
Godoy, Júnior Walter, CEFET-PR, Brazil
Granville, Lisandro Zambenedetti, UFRGS, Brazil
Hauck, Franz J., Universität Ulm, Germany
Kabacinski, Wojciech, Poznan University of Technology, Poland
Kahng, Hyun-Kook, Korea University, South Korea
Kassler, Andreas J., SCE, NTU, Singapore
Lee, Mike Myung-Ok, Dongshin University, South Korea
Logrippo, Luigi, Université du Québec en Outaouais, Canada
Lorenz, Pascal, University of Haute Alsace, France
Madeiro, Francisco, UNICAP, Brazil
Marshall, Alan, Queen's University Belfast, UK
Martins, Filho Joaquim F., UFPE, Brazil
Marvasti, Farokh, King's College, London, UK
Medeiros, Álvaro Augusto Machado de, UNICAMP, Brazil
Mobilon, Eduardo, CPqD, Brazil
Mohr, Werner, Siemens, Germany
Molinaro, Antonella, University of Messina, Italy
Monteiro, Edmundo, Portugal
Mota, João César Moura, UFC, Brazil
Mynbaev, Djafar, City University of New York, USA
Nefedov, Nikolai, Nokia Research Center, Finland
Oliveira, Hélio Magalhães de, UFPE, Brazil
Oliveira, Mauro, CEFET-CE, Brazil
Olveira, Antonio Jeronimo Belfort de, UFPE, Brazil
Peixeiro, Custodio, Portugal
Pelaes, Evaldo, UFPA, Brazil
Pereira, Jorge, European Commission, Belgium
Pimentel, Cecilio, UFPE, Brazil
Pinto, Ernesto Leite, IME, Brazil
Pujolle, Guy, LIP6, France
Queiroz, Wamberto José Lira de, UFPB, Brazil
Ramos, Rubens Viana, UFC, Brazil
Rocha Jr., Valdemar C. da, UFPE, Brazil
Rocha, Mônica de Lacerda, CPqD, Brazil
Sabbaghian, Maryam, Canada
Salvador, Marcos Rogério, CPqD, Brazil

Sargento, Susana, Portugal
Seara, Rui, UFSC, Brazil
Sezer, Sakir, Queen's University Belfast, UK
Soulhi, Said, Ericsson Research, Canada
Souza, José Neuman de, UFC, Brazil
Stasiak, Maciej, Poznan University of Technology, Poland
Takasaki, Yoshitaka, Tokyo University, Japan
Wenbo, Wang, Beijing University of Posts and Telecommunications, China
Westphall, Carlos Becker, UFSC, Brazil
Yacoub, Michel Daoud, UNICAMP, Brazil

Table of Contents

Multimedia Services

Enhancing 3G Cellular Systems With User Profiling Techniques
for Multimedia Application QoS Adaptation 1
  G. Araniti, P. De Meo, A. Iera, D. Ursino

Delivery of Streaming Video in Next-Generation Mobile Networks 7
  S. Bates

Use of the 2500-2690 MHz Band
for Multimedia Broadcast Multicast Service 16
  A.R. Oliveira, A. Correia

Throughput Fairness and Capacity Improvement
of Packet-Switched Networks Through Space-Time Scheduling 22
  E.B. Silva, L.S. Cardoso, T.F. Maciel, F.R.P. Cavalcanti,
  Y.C.B. Silva

Antennas

Performance of MIMO Antenna Systems With Hybrids of Transmit
Diversity and Spatial Multiplexing Using Soft-Output Decoding 28
  W.C. Freitas Jr., A.L.F. de Almeida, J.C.M. Mota,
  F.R.P. Cavalcanti, R.L. de Lacerda Neto

Simple Model to Determine the Gap Effect on the Input Impedance
of a Cylindrical Dipole. Application to Standard Antennas 38
  A. Kazemipour, X. Begaud, D. Allal

An Interference Avoidance Technique
for Ad Hoc Networks Employing Array Antennas 43
  T. Hunziker, J.L. Bordim, T. Ohira, S. Tanaka

Analysis of a Cylindrical Microstrip Antenna
by the Nonorthogonal FDTD Method and Parallel Processing 53
  R.O. dos Santos, R.M.S. de Oliveira, F.J.B. Barros,
  C.L. da S.S. Sobrinho

Transmission Technologies and Wireless Networks

Unified Analysis of the Local Scatters Channel Model 60
  W.J.L. Queiroz, F.G.S. Silva, M.S. Alencar

Spacial Analysis of the Local Scatters Channel Model 67
    W.J.L. Queiroz, F.G.S. Silva, M.S. Alencar

Propagation Prediction Based on Measurement at 5.8GHz
for Fixed Wireless Access 74
    T.M. Keen, T.A. Rahman

Dimensioning of Wireless Links Sharing Voice and Data 82
    M. de Oliveira Marques, I.S. Bonatti

Communication Theory (I)

An Iterative Matrix-Based Procedure
for Finding the Shannon Cover for Constrained Sequences 88
    D.P.B. Chaves, C. Pimentel, B.F. Uchôa-Filho

Truncated Importance Sampling Simulation
Applied to a Turbo Coded Communication System 94
    B.B. Albert, F.M. de Assis

Zero-Error Capacity of a Quantum Channel 100
    R.A.C. Medeiros, F.M. de Assis

Communication Theory (II)

Communication System Recognition by Modulation Recognition 106
    A.R. Attar, A. Sheikhi, A. Zamani

An Extension to Quasi-orthogonal Sequences 114
    F. Vanhaverbeke, M. Moeneclaey

Trellis Code Construction for the 2-User Binary Adder Channel 122
    V.C. da Rocha Jr., M.L.M.G. Alcoforado

Rayleigh Fading Multiple Access Channel
Without Channel State Information 128
    N. Marina

Communication Theory (III)

Adaptive Decision Feedback Multiuser Detectors
With Recurrent Neural Networks for DS-CDMA in Fading Channels 134
    R.C. de Lamare, R. Sampaio-Neto

Comparison of Two Filterbank Optimization Criteria
for VDSL Transmission 142
    J. Louveaux, C. Siclet, D. Pinchon, P. Siohan

Phase Estimation and Phase Ambiguity Resolution by Message Passing 150
    J. Dauwels, H. Wymeersch, H.-A. Loeliger, M. Moeneclaey

An Expurgated Union Bound for Space-Time Code Systems 156
    V.-D. Ngo, H.-W. Choi, S.-C. Park

Achieving Channel Capacity With Low Complexity RS-BTC
Using QPSK over AWGN Channel 163
    R. Zhou, A. Picart, R. Pyndiah, A. Goalic

Telecommunication Pricing and Billing

Design of Sub-session Based Accounting System
for Different Service Level of Mobile IP Roaming User 171
    B. Lee, H. Kim, K. Chung

Priority Telephony System With Pricing Alternatives 183
    S. Yaipairoj, F. Harmantzis

Mobile Telephony Industry: Price Discrimination Strategies
for Interconnected Networks 192
    L. Cricelli, F. Di Pillo, N. Levialdi, M. Gastaldi

Network Performance and Telecommunication Services

A Novel ECN-Based Congestion Control and Avoidance Algorithm
With Forecasting and Verifying 199
    H.-s. Liu, K. Xu, M.-w. Xu

Competitive Neural Networks for Fault Detection and Diagnosis
in 3G Cellular Systems 207
    *G.A. Barreto, J.C.M. Mota, L.G.M. Souza, R.A. Frota, L. Aguayo,
    J.S. Yamamoto, P.E.O. Macedo*

Performance Analysis of an Optical MAN Ring
for Asynchronous Variable Length Packets 214
    H. Castel, G. Hébuterne

Enabling Value-Generating Telecommunications Services
upon Integrated Multiservice Networks 221
    D.X. Adamopoulos, C.A. Papandreou

Active Networks and Mobile Agents

An Architecture for Publishing and Distributing Service Components
in Active Networks 227
    N. Dragios, C. Harbilas, K.P. Tsoukatos, G. Karetsos

Performance Comparison of Active Network-Based and Non Active
Network-Based Single-Rate Multicast Congestion Control Protocols 234
    Y. Darmaputra, R.F. Sari

MAMI: Mobile Agent Based System for Mobile Internet 241
  M.A. Haq, M. Matsumoto

Design and Implementation of an ANTS-Based Test Bed
for Collecting Data in Active Framework 251
  V. Damasceno Matos, J.L. de Castro e Silva, J.C. Machado,
  R.M. de Castro Andrade, J.N. de Souza

Adaptive QoS Management for Regulation
of Fairness Among Internet Applications 257
  M.F. de Castro, D. Gaïti, A. M'hamed

Optical Photonic Technologies (I)

Black Box Model of Erbium-Doped Fiber Amplifiers in C and L Bands 267
  A. Teixeira, D. Pereira, S. Junior, M. Lima, P. André, R. Nogueira,
  J. da Rocha, H. Fernandes

Optical Packet Switching Access Networks
Using Time and Space Contention Resolution Schemes 272
  L.H. Bonani, F.J.L. Pádua, E. Moschim, F.R. Barbosa

Full-Optical Spectrum Analyzer Design
Using EIT Based Fabry-Perot Interferometer 282
  A. Rostami

A Powerful Tool Based on Finite Element Method
for Designing Photonic Crystal Devices 287
  A. Cerqueira Jr., K.Z. Nobrega, F. Di Pasquale,
  H.E. Hernandez-Figueroa

Optical Photonic Technologies (II)

Analysis of Quantum Light Memory in Atomic Systems 296
  A. Rostami

Wavelength Conversion With 2R-Regeneration
by UL-SOA Induced Chirp Filtering 304
  C. de Mello Gallep, E. Conforti

Performance Analysis of Multidimensional PPM Codes
for Optical CDMA Systems in the Presence of Noise 312
  J.S.G. Panaro, C. de Almeida

Smart Strategies for Single-Photon Detection 322
  J.B.R. Silva, R.V. Ramos, E.C. Giraudo

Optical Networks (I)

Inter-arrival Planning for Sub-graph Routing Protection
in WDM Networks   328
    D.A.A. Mello, J.U. Pelegrini, M.S. Savasini, G.S. Pavani,
    H. Waldman

An Adaptive Routing Algorithm
for Intelligent and Transparent Optical Networks   336
    R.G. Dante, F. Pádua, E. Moschim, J.F. Martins-Filho

Novel Wavelength Assignment Algorithm for Intelligent Optical Networks
Based on Hops Counts and Relative Capacity Loss   342
    R.G. Dante, J.F. Martins-Filho, E. Moschim

Ad Hoc Networks (I)

Assigning Codes in a Random Wireless Network   348
    F. Djerourou, C. Lavault, G. Paillard, V. Ravelomanana

Flexible QoS Routing Protocol for Mobile Ad Hoc Network   354
    L. Khoukhi, S. Cherkaoui

Integrating Mobile Ad Hoc Network into Mobile IPv6 Network   363
    A. Ali, L.A. Latiff, N. Fisal

Network Architecture for Scalable Ad Hoc Networks   369
    J. Costa-Requena, J. Gutiérrez, R. Kantola, J. Creado, N. Beijar

Ad Hoc Networks (II)

Dual Code-Sensing Spread Spectrum Random Access (DCSSRA) Protocol
for Ad Hoc Networks   381
    C.-J. Wang, W. Quan, Z.-H. Deng, Y.-A. Liu, W.-B. Wang, J.-C. Gao

Exploiting the Small-World Effect to Increase Connectivity
in Wireless Ad Hoc Networks   388
    D. Cavalcanti, D. Agrawal, J. Kelner, D. Sadok

Adaptive Resource Management in Mobile Wireless Cellular Networks   394
    M. Hossain, M. Hassan, H.R. Sirisena

Comparative Analysis of Ad Hoc Multi-path Update Techniques
for Multimedia Applications   400
    M. Kwan, K. Doğançay

Ad Hoc Networks (III)

PRDS: A Priority Based Route Discovery Strategy
for Mobile Ad Hoc Networks 410
    B. Zhou, A. Marshall, J. Wu, T.-H. Lee, J. Liu

Performance Evaluation of AODV Protocol over E-TDMA MAC Protocol
for Wireless Ad Hoc Networks 417
    V. Loscrì, F. De Rango, S. Marano

Mitigating the Hidden Terminal Problem
and Improving Spatial Reuse in Wireless Ad Hoc Network 425
    J.-W. Bang, S.-K. Youm, C.-H. Kang, S.-J. Seok

A New Distributed Power Control Algorithm
Based on a Simple Prediction Method 431
    R.A.O. Neto, F.S. Chaves, F.R.P. Cavalcanti, T.F. Maciel

Ad Hoc Networks (IV)

A Heuristic Approach to Energy Saving in Ad Hoc Networks 437
    R.I. da Silva, J.C.B. Leite, M.P. Fernandez

Calculating the Maximum Throughput in Multihop Ad Hoc Networks 443
    B.A.M. Villela, O.C.M.B. Duarte

Throughput of Distributed-MIMO Multi-stage Communication Networks
over Ergodic Channels 450
    M. Dohler, A. Gkelias, A.H. Aghvami

Delivering the Benefits of Directional Communications
for Ad Hoc Networks Through an Efficient Directional MAC Protocol 461
    J.L. Bordim, T. Ueda, S. Tanaka

Signal Processing (I)

Synchronization Errors Resistant Detector for OFDM Systems 471
    Y. Bar-Ness, R. Solá

The Discrete Cosine Transform over Prime Finite Fields 482
    M.M.C. de Souza, H.M. de Oliveira, R.M.C. de Souza,
    M.M. Vasconcelos

A Lattice Version of the Multichannel Fast QRD Algorithm
Based on A Posteriori Backward Errors 488
    A.L.L. Ramos, J.A. Apolinário Jr.

New Blind Algorithms Based on Modified "Constant Modulus" Criteria
for QAM Constellations 498
    C.A.R. Fernandes, J.C.M. Mota

Signal Processing (II)

Channel Estimation Methods for Space-Time Block Transmission
in Frequency-Flat Channels 504
   N. Nefedov

Faster DTMF Decoding 510
   J.B. Lima, R.M. Campello de Souza, H.M. de Oliveira,
   M.M. Campello de Souza

Adaptive Echo Cancellation for Packet-Based Networks 516
   V. Stewart, C.F.N. Cowan, S. Sezer

The Genetic Code Revisited: Inner-to-Outer Map, 2D-Gray Map,
and World-Map Genetic Representations 526
   H.M. de Oliveira, N.S. Santos-Magalhães

Signal Processing (III)

On Subspace Channel Estimation in Multipath SIMO
and MIMO Channels.. 532
   N. Nefedov

A Novel Demodulation Technique for Recovering PPM Signals
in the Presence of Additive Noise 541
   P. Azmi, F. Marvasti

A Novel Iterative Decoding Method for DFT Codes in Erasure Channels....... 548
   P. Azmi, F. Marvasti

Comparison Between Several Methods of PPM Demodulation
Based on Iterative Techniques 554
   M. Shariat, M. Ferdosizadeh, M.J. Abdoli, B. Makouei,
   A. Yazdanpanah, F. Marvasti

Signal Processing (IV)

A Network Echo Canceler Based on a SRF QRD-LSL Adaptive
Algorithm Implemented on Motorola StarCore SC140 DSP 560
   C. Paleologu, A.A. Enescu, S. Ciochina

Detection of Equalization Errors in Time-Varying Channels 568
   J.F. Galdino, E.L. Pinto, M.S. Alencar, E.S. Sousa

A Design Technique for Oversampled Modulated Filter Banks
and OFDM/QAM Modulations 578
   D. Pinchon, C. Siclet, P. Siohan

Nonlinear Bitwise Equalization 589
   F.-J. González Serrano, M. Martínez Ramón

Network Performance and MPLS

Modelling and Performance Evaluation of Wireless Networks 595
  G.H.S. Carvalho, R.M. Rodrigues, C.R.L. Francês, J.C.W.A. Costa,
  S.V. Carvalho

Improving End-System Performance With the Use of the QoS Management 601
  M.A. Teixeira, J.S. Barbar

Throughput Maximization of ARQ Transmission Protocol
Employing Adaptive Modulation and Coding 607
  C. González, L. Szczeciński, S. Aïssa

CR-LSP Dynamic Rerouting Implementation
for MPLS Network Simulations in Network Simulator 616
  T.A. Moura Oliveira, E. Guimarães Nobre, J. Celestino Jr.

Traffic Engineering (I)

FuDyLBA: A Traffic Engineering Load Balance Scheme for MPLS
Networks Based on Fuzzy Logic 622
  J. Celestino Jr., P.R.X. Ponte, A.C.F. Tomaz, A.L.B.P.B. Diniz

Service Differentiation over GMPLS 628
  H. Moungla, F. Krief

Sliding Mode Queue Management in TCP/AQM Networks 638
  M. Jalili-Kharaajoo

A Distributed Algorithm for Weighted Max-Min Fairness in MPLS Networks ... 644
  F. Skivée, G. Leduc

Traffic Engineering (II) and Internet (I)

Applying Artificial Neural Networks for Fault Prediction
in Optical Network Links 654
  C.H.R. Gonçalves, M. Oliveira, R.M.C. Andrade, M.F. de Castro

Capacity Allocation for Voice over IP Networks
Using Maximum Waiting Time Models 660
  S. Sharafeddine, N. Kongtong, Z. Dawy

SCTP Mobility Highly Coupled With Mobile IP 671
  J.-W. Jung, Y.-K. Kim, H.-K. Kahng

Migration to the New Internet – Supporting Inter Operability
Between IPv4 and IPv6 Networks 678
  T. Vazão, L. Raposo, J. Santos

SIP, QoS (I), and Switches

Smart Profile: A New Method for Minimising SIP Messages 688
A. Meddahi, G. Vanwormhoudt, H. Afifi

A QoS Provisioned CIOQ ATM Switch With m Internal Links 698
C.R. dos Santos, S. Motoyama

Multi-constrained Least Cost QoS Routing Algorithm 704
H. Jiang, P.-l. Yan, J.-g. Zhou, L.-j. Chen, M. Wu

The New Packet Scheduling Algorithms for VOQ Switches 711
A. Baranowska, W. Kabaciński

Optical Networks (II)

An Algorithm for the Parameters Extraction
of a Semiconductor Optical Amplifier 717
R.P. Vivacqua, C.M. Gallep, E. Conforti

All-Optical Gain Controlled EDFA: Design and System Impact 727
J.C.R.F. Oliveira, J.B. Rosolem, A.C. Bordonalli

Iterative Optimization in VTD to Maximize the Open Capacity
of WDM Networks 735
K.D.R. Assis, M.S. Savasini, H. Waldman

Link Management Protocol (LMP) Evaluation for SDH/Sonet 743
P. Uria Recio, P. Rauch, K. Espinosa

Optical Networks (III) and Network Operation and Management (I)

Experimental Investigation of WDM Transmission Properties
of Optical Labeled Signals Using Orthogonal IM/FSK Modulation Format 753
P.V. Holm-Nielsen, J. Zhang, J.J. Vegas Olmos, I. Tafur Monroy,
C. Peucheret, V. Polo, P. Jeppesen, A.M.J. Koonen, J. Prat

Performance Assessment of Optical Burst Switched Degree-Four
Chordal Ring Networks 760
J.J.P.C. Rodrigues, M.M. Freire, P. Lorenz

All-Optical Routing Limitations Due to Semiconductor Optical
Amplifiers Dynamics 766
A. Teixeira, P. André, R. Nogueira, P. Monteiro, J. da Rocha

The Hurst Parameter for Digital Signature of Network Segment 772
M. Lemes Proença Jr., C. Coppelmans, M. Bottoli, A. Alberti,
L.S. Mendes

Network Operation and Management (II)

High Performance Cluster Management Based on SNMP: Experiences on
Integration Between Network Patterns and Cluster Management Concepts 782
  R. Sanger Alves, C. Cassales Marquezan, L. Zambenedetti Granville,
  P.O.A. Navaux

A Web-Based Pro-active Fault
and Performance Network Management Architecture 792
  A.S. Ramos, A. Salles Garcia, R. da Silva Villaça, R.B. Drago

A CIM Extension for Peer-to-Peer Network and Service Management 801
  G. Doyen, O. Festor, E. Nataf

A Generic Event-Driven System for Managing SNMP-Enabled
Communication Networks 811
  A.P. Braga, R. Rios, R. Andrade, J.C. Machado, J.N. de Souza

Network Management Theory and Telecommunications Networks

Algorithms for Distributed Fault Management
in Telecommunications Networks................................... 820
  E. Fabre, A. Benveniste, S. Haar, C. Jard, A. Aghasaryan

Fault Identification by Passive Testing 826
  X.H. Guo, B.H. Zhao, L. Qian

Script MIB Extension for Resource Limitation
in SNMP Distributed Management Environments 835
  A. da Rocha, C.A. da Rocha, J.N. de Souza

UML Specification of a Generic Model for Fault Diagnosis
of Telecommunication Networks 841
  A. Aghasaryan, C. Jard, J. Thomas

Mobility and Broadband Wireless

IEEE 802.11 Inter-WLAN Mobility Control With Broadband Supported
Distribution System Integrating WLAN and WAN 848
  M. Rahman, F. Harmantzis

A Cost-Effective Local Positioning System Architecture Based on TDoA 858
  R.M. Abreu, M.J.A. de Sousa, M.R. Santos

Indoor Geolocation With Received Signal Strength Fingerprinting Technique
and Neural Networks 866
  C. Nerguizian, C. Despins, S. Affès

Implementation of a Novel Credit Based SCFQ Scheduler
for Broadband Wireless Access 876
   E. Garcia-Palacios, S. Sezer, C. Toal, S. Dawson

Cellular System Evolution (I)

BER for CMOS Analog Decoder With Different Working Points 885
   K. Ruttik

BiCMOS Variable Gain LNA at C-Band
With Ultra Low Power onsumption for WLAN 891
   *F. Ellinger, C. Carta, L. Rodoni, G. von Büren, D. Barras,
   M. Schmatz, H. Jäckel*

Joint MIMO and MAI Suppression for the HSDPA 900
   M. Marques da Silva, A. Correia

A Joint Precoding Scheme and Selective Transmit Diversity 908
   M. Marques da Silva, A. Correia

Cellular System Evolution (II)

Application of a Joint Source-Channel Decoding Technique
to UMTS Channel Codes and OFDM Modulation 914
   M. Jeanne, I. Siaud, O. Seller, P. Siohan

Transmit Selection Diversity for TDD-CDMA
With Asymmetric Modulation in Duplex Channel 924
   I. Jeong, M. Nakagawa

Planning the Base Station Layout in UMTS Urban Scenarios:
A Simulation Approach to Coverage and Capacity Estimation 932
   E. Zola, F. Barceló

A General Traffic and Queueing Delay Model
for 3G Wireless Packet Networks 942
   G. Aniba, S. Aïssa

Personal Communication, Terrestrial Radio Systems, and Satellites

MobiS: A Solution for the Development of Secure Applications
for Mobile Devices 950
   *W. Viana, B. Filho, K. Magalhães, C. Giovano, J. de Castro,
   R. Andrade*

Compensating Nonlinear Amplifier Effects on a DVB-T Link 958
   V. Vale do Nascimento, J.E.P. de Farias

Controlled Load Services in IP QoS Geostationary Satellite Networks 964
    F. De Rango, M. Tropea, S. Marano

Mobility Management

Fast Authentication for Inter-domain Handover 973
    H. Wang, A.R. Prasad

Session and Service Mobility in Service Specific Label Switched
Wireless Networks 983
    P. Maruthi, G. Sridhar, V. Sridhar

Mobility Management for the Next Generation of Mobile Cellular Systems 991
    M. Carli, F. Cappabianca, A. Tenca, A. Neri

Handoff Delay Performance Comparisons
of IP Mobility Management Schemes Using SIP and MIP 997
    H.-s. Kim, C.H. Kim, B.-h. Roh, S.W. Yoo

Multimedia Information, Network Reliability, EMC in Communications, and Multicast

A Video Compression Tools Comparison for Videoconferencing
and Streaming Applications for 3.5G Environments 1007
    N. Martins, A. Marquet, A. Correia

Fairness and Protection Behavior of Resilient Packet Ring Nodes
Using Network Processors 1013
    A. Kirstädter, A. Hof, W. Meyer, E. Wolf

Markov Chain Simulation of Biological Effects
on Nerve Cells Ionic Channels Due to Electromagnetic Fields
Used in Communication Systems 1023
    D.C. Uchôa, F.M. de Assis, F.A.F. Tejo

Multiobjective Multicast Routing Algorithm 1029
    J. Crichigno, B. Barán

Image Processing, ATM, and Web Services

A Comparison of Filters for Ultrasound Images 1035
    P.B. Calíope, F.N.S. Medeiros, R.C.P. Marques, R.C.S. Costa

Filtering Effects on SAR Images Segmentation 1041
    R.C.P. Marques, E.A. Carvalho, R.C.S. Costa, F.N.S. Medeiros

The Implementation of Scalable ATM Frame Delineation Circuits 1047
    C. Toal, S. Sezer

Web Based Service Provision
A Case Study: Electronic DesignAutomation 1057
   S. Dawson, S. Sezer

Communication, Security, and QoS (II)

Identification of LOS/NLOS States Using TOA Filtered Estimates 1067
   A.G. Guimarães, M.A. Grivet

A Hybrid Protocol for Quantum Authentication of Classical Messages 1077
   R.A.C. Medeiros, F.M. de Assis

Attack Evidence Detection, Recovery, and Signature Extraction
With ADENOIDS 1083
   F.S. de Paula, P.L. de Geus

QoS-Differentiated Secure Charging in Ad-Hoc Environments 1093
   J. Girão, J.P. Barraca, B. Lamparter, D. Westhoff, R. Aguiar

Switching and Routing

Multi-rate Model of the Group of Separated Transmission Links
of Various Capacities 1101
   M. Głąbowski, M. Stasiak

A Multicast Routing Algorithm Using Multiobjective Optimization 1107
   J. Crichigno, B. Barán

A Transsignaling Strategy for QoS Support in Heterogeneous Networks 1114
   D. Gomes, P. Gonçalves, R.L. Aguiar

Evaluation of Fixed Thresholds for Allocation and Management
of Dedicated Channels Transmission Power in WCDMA Networks 1122
   *C.H.M. de Lima, E.B. Rodrigues, V.A. de Sousa Jr.,
   F.R.P. Cavalcanti, A.R. Braga*

Next Generation Systems (I)

Efficient Alternatives to Bi-directional Tunnelling for Moving Networks 1128
   *L. Burness, P. Eardley, J. Eisl, R. Hancock, E. Hepworth,
   A. Mihailovic*

Group Messaging in IP Multimedia Subsystem of UMTS 1136
   I. Miladinovic

The Effect of a Realistic Urban Scenario on the Performance
of Algorithms for Handover and Call Management
in Hierarchical Cellular Systems 1143
   E. Natalizio, A. Molinaro, S. Marano

Providing Quality of Service for Clock Synchronization 1151
    A.C. Callado, J. Kelner, A.C. Frery, D.F.H. Sadok

Next Generation Systems (II) and Traffic Management (I)

Application of Predictive Control Algorithm to Congestion Control
in Differentiated Service Networks...................................... 1157
    M. Jalili-Kharaajoo, B.N. Araabi

Design of a Manageable WLAN Access Point 1163
    T. Vanhatupa, A. Koivisto, J. Sikiö, M. Hännikäinen,
    T.D. Hämäläinen

Performance Evaluation of Circuit Emulation Service
in a Metropolitan Optical Ring Architecture 1173
    V.H. Nguyen, M. Ben Mamoun, T. Atmaca, D. Popa, N. Le Sauze,
    L. Ciavaglia

Expedited Forwarding End to End Delay Variations 1183
    H. Alshaer, E. Horlait

Traffic Management (II)

Internet Quality of Service Measurement Tool
for Both Users and Providers 1195
    A. Ferro, F. Liberal, E. Ibarrola, A. Muñoz, C. Perfecto

A XML Policy-Based Approach for RSVP 1204
    E. Toktar, E. Jamhour, C. Maziero

SRBQ and RSVPRAgg: A Comparative Study 1210
    R. Prior, S. Sargento, P. Brandão, S. Crisóstomo

TCP Based Layered Multicast Network Congestion Control 1218
    A.A. Al Naamani, A.M. Al Naamany, H. Bourdoucen

A Novel Detection Methodology of Network Attack Symptoms
at Aggregate Traffic Level on Highspeed Internet Backbone Links 1226
    B.-h. Roh, S.W. Yoo

Wireless Access (I)

Precision Time Protocol Prototype on Wireless LAN 1236
    J. Kannisto, T. Vanhatupa, M. Hännikäinen, T.D. Hämäläinen

Moving Telecom Outside Plant Systems
Towards a Standard Interoperable Environment 1246
    G.M. Weiss, E.Z.V. Dias

Analysis and Contrast Between STC and Spatial Diversity Techniques
for OFDM WLAN With Channel Estimation 1252
  E.R. de Lima, S.J. Flores, V. Almenar, M.J. Canet

Fair Time Sharing Protocol: A Solution for IEEE 802.11b Hot Spots 1261
  A. Munaretto, M. Fonseca, K. Al Agha, G. Pujolle

Wireless LANs (II)

An Algorithm for Dynamic Priority Assignment
in 802.11e WLAN MAC Protocols 1267
  A. Iera, G. Ruggeri, D. Tripodi

DHCP-Based Authentication for Mobile Users/Terminals
in a Wireless Access Network 1274
  L. Veltri, A. Molinaro, O. Marullo

Performance Analysis of Multi-pattern Frequency Hopping Wireless LANs 1282
  D. Chen, A.K. Elhakeem, X. Wang

A Delayed-ACK Scheme for MAC-Level Performance Enhancement
of Wireless LANs 1289
  D. Kliazovich, F. Granelli

Wireless LANs (III) and Network Planning and Optimization

The VDSL Deployments and Their Associated Crosstalk
With xDSL Systems 1296
  K.-A. Han, J.-J. Lee, J.-C. Ryou

Design of PON Using VQ-Based Fiber Optimization 1303
  P. Shah, N. Roy, A. Roy, K. Basu, S.K. Das

IPv6 Deployment Support Using an IPv6 Transitioning Architecture
– The Site Transitioning Architecture (STA) 1310
  M. Mackay, C. Edwards

An Intelligent Network Simulation Platform Embedded
With Multi-agents Systems for Next Generation Internet 1317
  M.F. de Castro, H. Lecarpentier, L. Merghem, D. Gaïti

Internet (II)

Reduced-State SARSA With Channel Reassignment
for Dynamic Channel Allocation in Cellular Mobile Networks 1327
  N. Lilith, K. Doğançay

Optimal Bus and Buffer Allocation for a Set
of Leaky-Bucket-Controlled Streams 1337
  E. den Boef, J. Korst, W.F.J. Verhaegh

Scalable End-to-End Multicast Tree Fault Isolation 1347
    T. Friedman, D. Towsley, J. Kurose

Internet Service Pricing Based on User and Service Profiles 1359
    Y. Khene, M. Fonseca, N. Agoulmine, G. Pujolle

Design and Performance of Asymmetric Turbo Coded Hybrid-ARQ 1369
    K. Oteng-Amoako, S. Nooshabadi, J. Yuan

ICT 2004 Poster Papers 1382

Author Index 1385

Enhancing 3G Cellular Systems with User Profiling Techniques for Multimedia Application QoS Adaptation

Giuseppe Araniti, Pasquale De Meo, Antonio Iera, and Domenico Ursino

DIMET, Università Mediterranea di Reggio Calabria, Via Graziella, Località Feo di Vito, 89060 Reggio Calabria, Italy
{araniti,demeo,iera,ursino}@ing.unirc.it

Abstract. In this paper the authors' attention is on a dynamic QoS adaptation mechanism whose functional behavior is driven by underlying system conditions, user exigencies and preferences. The proposed mechanism differs from analogous systems as it dynamically estimates the user preferences, through a "user profiling" activity, and uses the estimated profile to dynamically adapt the resources devoted to the traffic flow. The procedure is completely automated and quite fast. In this paper the proposed mechanism is introduced into a 3G system and its effectiveness tested under different conditions.

1 Introduction

Effectively supporting multimedia applications at guaranteed and user-tailored QoS levels is an exigency strongly felt in advanced 3G and 4G Personal Communication scenarios. The highly fluctuating radio channel conditions, jointly with heavy resource requests deriving from multimedia users, force to think in terms of "adaptive QoS", or "soft QoS" [4], in contraposition to the traditional "hard QoS" or "static QoS" idea. The present paper focuses on the possibility of allowing both "system level" measurements and "user level" exigencies and preferences to drive the dynamic QoS adaptation process. In the authors' opinion, the best way to implement the highlighted ideas into an evolutionary multimedia wireless scenario is to define a distributed mechanism for QoS control. It could be implemented between the application-related and the sub-network related functions.

A significant activity in this field is carried on by many consortia and forums (IETF, PARLAY, EURESCOM, etc) and this testifies the high interest for QoS related issues in advanced wired-wireless telecommunications platforms. In our previous work we propose an adaptive QoS management mechanism [1] and we presented just simple behavior evaluations. In this paper we propose an experimental evaluation of this mechanism with reference to an UMTS environment. This choice is fully justified because of in a cellular network, resources are typically scarce and quickly fluctuating.

J.N. de Souza et al. (Eds.): ICT 2004, LNCS 3124, pp. 1–6, 2004.
© Springer-Verlag Berlin Heidelberg 2004

This paper is organized as follows: Section 2 provides an architecture for QoS management in generic multimedia environment. Section 3 presents our user-profile based mechanism for QoS management. Section 4 provides the description of the outputs of a detailed test campaign aiming at measuring the performances of our system in an UMTS environment. Finally, in Section 5, we draw our conclusions.

2   Distributed QoS Control Functionality

Our QoS-aware distributed functionality is strongly based on the knowledge of the user profile describing her/his behavior and preferences. It dynamically translates user preferences into underlying network parameters, in order to properly define the best QoS degree the end user can be provided with. Its design is inspired by the following criteria: *Subjectivity, Portability, Dynamicity*, and *Automaticity*.

Our QoS control architecture consists of five groups of modules. They handle *(i)* parameter mappings, *(ii)* resource distribution, *(iii)* network status verification, *(iv)* QoS negotiation and *(v)* users' profile management. The modules composing the architecture of our dynamic QoS control mechanism behave as follows.

The Mapper. It transforms the application level parameters (*high level parameters*) setted by the user into metrics relevant to the communications network (*low level parameters*) which the requested service operates on.

The Resource Manager. It carries out the following tasks: *bandwidth management* and *transmission of high-level parameters* to the network.

The Controller and the Monitor. These are always active components. The first handles the possible presence of network faults, while the second periodically extracts information describing the network status.

The Impulsive Monitor. It is activated whenever unexpected variations of the network status occur.

The Re-negotiator. It redefines the QoS by taking into account both user desires and network constraints.

The User Profile Manager. This is an always active component that dynamically records the behavior of a user while she/he accesses different types of services and suitably modifies her/his profile.

3   User Profile, Mapping, and Renegotiation

The User Profile has been constructed by referring to the theory of *User Modelling* [2]. Particularly, our techniques for constructing and handling the user profile are based on the assumption that, in a long period, user interests coincide with concepts stored in the information sources which she/he frequently visited in the past [3]. A further, more interesting, characteristic of our approach consists in the exploitation of XML for storing and handling the User Profile; the exploitation of XML favors data exchange over heterogeneous platforms.

The User Profile $P(u)$, associated with a user u, can be represented as a set of concepts of interest for u and a set of relationships among concepts. More details on the parameters associated to the applications are given in [1].

As previously pointed out, the *Mapper* is activated whenever high-level parameters have to be translated into low-level ones. The overall bandwidth the system requires for supporting a multimedia transmission, namely $B_{overall} = B_v + B_a + B_d$, is an interesting index which put network layer indexes into relationship with higher layers' parameters. B_v is the bandwidth the system requires for enabling the real time vision of a video sequence; it is computed, according to [5], in terms of: R_x and R_y, the video image Horizontal and Vertical Resolution, C, the Chrominance, L, the Luminance, and F, the video Frame Rate. B_a is the required bandwidth for supporting an audio transmission; it can be computed in terms of Ch (current required channel number), Fr (current Sampling Frequency), and Co (current Required Audio Codex). Finally, B_d is the bandwidth to support a data transmission (i.e. the data bit rate)

The *Re-negotiator* has to compute the values of B_v, B_a and B_d that maximize the user perceived QoS , compatibly with the network conditions.

Therefore, we need to quantify the *user satisfaction*. This is achieved by defining the so called *Satisfaction Degree* D_S of a user u as: $D_S = \frac{QoS^{net} - QoS^{R-min}}{QoS^{R-max} - QoS^{R-min}}$.

In this equation, QoS^{net} is the Quality of Service of the overall multimedia application. It is a combination of QoS contributions from each multimedia traffic component, suitably weighted by means of current values of some attributes defined within the dynamic Profile. QoS^{R-max} and QoS^{R-min} represents the maximum and the minimum overall QoS values the user is disposed to accept. These two values are derived from the signed user SLA(Service Level Agreement). For the exact calculation of these indexes the reader can refer to [1].

D_S numerator represents the distance between the actual QoS provided by the network and the minimum QoS the user might accept; the denominator denotes the distance between the maximum and the minimum QoS specified by the user. In the formula if $QoS^{net} \leq QoS^{R-min}$, then D_S is negative, this implying that the connection must be interrupted. It is worth pointing out that if $QoS^{R-max} = QoS^{R-min}$, then D_S is not defined. In this case the re-negotiation activity makes no sense since the user requirements are given in terms of hard-QoS constraints.

In our framework, it is compulsory to manage resource distribution for obtaining the highest values is possible for D_S. This task is handled as a two-steps optimization problem: *(i)* the first step consists in distributing the available bandwidth among each multimedia component. This bandwidth assignment problem is formulated as a linear programming problem. *(ii)* The second step consists in adjusting the *high level* QoS parameters of each multimedia component in order to fit the network constraints and maximize the user QoS expectation; it is worth observing that high level QoS parameters optimization is treated as a non linear optimization problem.

Interested reader can refer to [1] for additional analytical details.

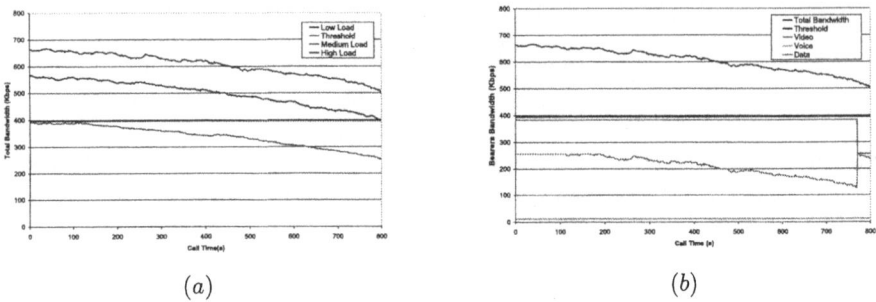

(a) (b)

Fig. 1. *(a)* Bandwidth supplied by the network in three load conditions: low (calls/second= 0.1), medium (calls/second = 0.2) and high (calls/second = 0.4); *(b)* Available video, audio and data bandwidth (0.1 calls/second).

4   Prototype and Experimental Results

In this section we will supply simulation results that demonstrate the correctness and efficiency of our algorithm. In order to obtain such results we use an UMTS system simulator. The UMTS reference scenario adopted during our simulation campaign consists of repeated modules of 4 macrocells (900m cell radius), each one overlapping the area of 4 circular microcells (300m cell radius). We consider two classes of users: *pedestrian* and *vehicular*, respectively moving at 3 Km/h and 60 Km/h nominal speed and assume that: *(i)* users are originated and enter the network following a Poisson process; *(ii)* the users freely roam and establish multimedia calls in the cells *(iii)* the sojourn time of a user in a cell is exponentially distributed. Moreover, we assume that handover events are distributed as a decreasing exponential with mean equal to the sojourn time; finally, we assume a perfect power control. The traffic generated by each user in the system can be of the following types: *(i)* *Traditional Telephone calls, (ii) Videoconference calls* characterized by voice, video, and data bearers and *(iii) Web Browsing.* A multi-tier coverage structure, supported by UMTS, has been utilized in our simulations. We have carried out a large variety of experiments by taking into consideration various user behaviors and QoS expectations. In particular, the shown results refer to the case in which the traffic distribution, keeping the percentage of vehicular (fast) users fixed at 50%, consists of 40% of video-conference connections (with average duration of 15 minutes), 40% of telephone calls (with average duration of 3 minutes), and 20% of Web-browsing activities. The reference user accesses a multimedia application exploiting all communication components (video,audio, and data). In more details, we assume that the user exhibits a special interest in the video component, she/he gives less importance to the audio and is not much interested in the data component. The Figure 1*(a)* shows the bandwidth supplied by the network to the user as a function of the time. The curves are expressed as function of call inter-arrival time, corresponding respectively to 0.1, 0.2 and 0.4 calls/second.

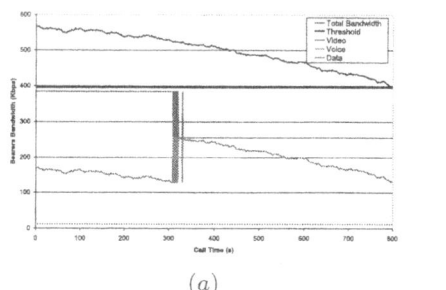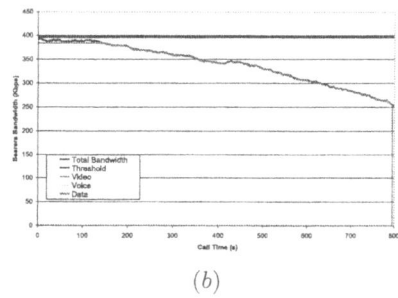

$$(a) \qquad\qquad (b)$$

Fig. 2. *(a)* Available video, audio and data bandwidth (0.2 calls/second). *(b)* Available video, audio and data bandwidth (0.4 calls/second).

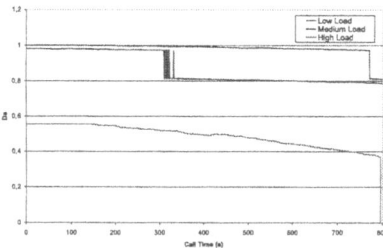

Fig. 3. D_S under low (0.1 calls/second), medium (0.2 calls/second) and high (0.4 calls/second) traffic conditions.

From Figure 1 *(a)* it is possible to note that the quality threshold under which the customer rejects the multimedia service is equal to 396.2 Kbps. If the bandwidth is less than this threshold, a re-negotiation becomes necessary. If we assume 0.1 calls/second, the total bandwidth never assumes values under the threshold. Therefore, a user will be able to contemporary use all components. In particular, during the call, the most important components for the user (video and audio) are at their maximum QoS levels, while the QoS of the data component decreases (Figure 1 *(b)*). D_S (see Figure 3) assumes always the maximum value, or differs just a little from it, even if the data bearer (the less important component for the user) continues to degrade due to its poor weight in the computation of D_S. Differently, a rate reduction of the video component (which has a greater weight) will cause a sudden lowering of D_S, as clearly emerges from Figure 3.

Figure 2 *(a)* shows the resource distribution among bearers under *medium* loading conditions. The total bandwidth assigned to the user is very close to the threshold. Initially, video obtains the maximum level of resources that can be assigned to it. Then, due to bandwidth reduction, video component are characterized by a strong variability. This is an effect of the behavior of our algorithm,which constantly monitors the instantaneous total bandwidth available for the user, and redistributes it in order to obtain always an "optimal" configuration, while satisfying the user requests. When the user total bandwidth decreases to a determined value, the video bearer cannot be maximized any more.

In Figure 3 we show D_S. As expected, it mainly depends on video bearer variations while it does not seem to be influenced by data degradation. Contemporary, D_S is generally greater than 0.5.
As a further step, we consider a highly loaded network (0.4 calls/second). In these conditions, we have a resources scarcity such that our system is forced to shoot down both data and audio components. Differently, video component curve has the same shape of the total bandwidth curve, until it reaches the minimum acceptable value; under this value the call will be interrupted (see Figure 2(b)).

It is worth pointing out that although we have a heavy load in the network, our algorithm automatically avoids to transmit audio and data components and privileges video bearer from heavy degradation. Its behaviour implies the possibility of still maintaining the user satisfaction degree at acceptable levels, this preventing the multimedia connection from being necessarily terminated.

5   Conclusions

In this paper we contributed to the research issue of QoS management in new generation wireless multimedia systems. In particular, we proposed a dynamic QoS control mechanism which exploits user profiling techniques and suitable QoS mapping functions to introduce the soft-QoS idea into *a beyond-3G system*. The exploitation of information deriving from the user and the system level, jointly with dynamic resource optimization procedures, enables the effective bandwidth exploitation and guarantees a QoS always matching the user expectations.

Acknowledgment. The authors heartily thank Dr. Francesco Azzará for his valuable support during the evaluation tool development phase.

References

1. G. Araniti, P. De Meo, A. Iera, and D. Ursino. Adaptively controlling the QoS of multimedia wireless applications through user profiling techniques. *IEEE Journal in Selected Areas of Communications*, 21(10):1536–1546, 2003.
2. A. Kobsa. Generic user modeling systems. *User Modeling and User-Adapted Interaction*, 11:49–63, 2001.
3. J. Konstan, B. Miller, D. Maltz, J. Herlocker, L. Gordon, and J. Riedl. Grouplens: Applying collaborative filtering to usenet news. *Comunications of ACM (CACM)*, 40(3):77–87, 1997.
4. D. Reiniger, R. Izmailov, B. Rajagopalan, M. Ott, and D. Raychaudhuri. Soft Qos control in the WATMnet broadband wireless system. *IEEE Personal Communications Magazine*, Feb. 99:34–43, 1999.
5. R. Steinmetz and K. Nahrstedt. *Multimedia: Computing, Communication and Applications*. Prentice Hall, Inc., Upper Saddle River,NJ 07458,USA, 1995.

Delivery of Streaming Video in Next-Generation Mobile Networks

Stephen Bates*

The University of Alberta
stephen.bates@ece.ualberta.ca

Abstract. In this paper a family of novel streaming video delivery schemes is presented. These schemes are designed for next-generation mobile communication networks and attempt to minimize uplink communications whilst smoothing downlink capacity requirements.

These schemes are based on the concept of a Feedforward Constant Delivery System. For a variety of streaming video datasets and appropriate model data we show how these systems can achieve excellent QoS with minimal uplink packet transmission. We go on to show how the scheme can be improved upon by introducing a Mean Deviation Ratio Threshold and a Buffer Occupancy Cost Function.

1   Introduction

Streaming video has already become a major traffic type on broadband wireline networks and has received considerable attention in this context [1], [2]. As the ability of mobile networks to carry digital information increases we expect to see streaming video services introduced on them also [3].

These streaming video services will range from Real-Time (RT) video conferencing to Non-Real-Time (NRT) delivery of DVD quality video. The QoS requirements of these services are very different and developing one delivery scheme for all classes of streaming video may not be the optimal approach.

In addition the handsets in next generation mobile networks have only a small amount of power and complexity to offer to such services. Any scheme which reduces the transmission requirements and complexity of the handset increases its battery life and reduces its cost. In this paper we begin by focusing on developing a delivery scheme for NRT streaming video which ensures a good QoS whilst minimizing uplink capacity. We do produce some simulation results which suggest that the technique may also be applicable to RT services.

Existing delivery techniques involve differentiation between one video stream and another and between particular frames within a video stream [4], opportunistic scheduling [5] and dynamically allocating bandwidth [6]. All of these schemes tend to be quite complex at either the base-station, handset or both. This places a strain on the network and hardware resources.

* This work was funded in part by NSERC grant G121210952 and a University of Alberta, Faculty of Engineering Startup Grant.

2 The Mobile Network System Model

Consider a mobile handset which initiates a request for a streaming video at time t_0. We assume the base-station has the resources to deliver this video as required with negligible delay.

Now divide time into slots of duration T_s. This is the time the base station takes to transmit a packet. Each packet has a fixed header size P_h and a variable data size $P_d(\hat{\gamma}_j(sT_s))$ where $\hat{\gamma}_j(t)$ is the base-station's estimate of receiver js Signal to Interference Ratio (SIR) at time t.

The transmitter must select a transmission method that is appropriate for the quality of the channel between it and receiver j. This SIR is time-varying but in our case we are only concerned with the impact this has on the choice of P_d. In our analysis we assume the SIR is used to select one of several modulation schemes. We assume the SIR to be drawn from a Markov Process with a transition matrix Γ. We can derive the average data capacity of a packet, $\overline{P}_d$, from Γ using eigenvalue decomposition.

Use $\delta_j(sT_s)$ as a kronecker delta function to indicate those time-slots where the scheduling algorithm in the base-station determines a packet should be transmitted to handset j and $\epsilon(sT_s)$ to identify when the packet at that timeslot was in error. The total data received by receiver j at the end of timeslot s can be given as

$$R_j^d(sT_s) = \sum_{i=0}^{s} P_d(\hat{\gamma}_j(iT_s))\delta_j(iT_s)(1 - \epsilon(iT_s). \tag{1}$$

Now within the handset we assume a memory buffer to store unprocessed frames and a service capacity (CPU) to process these frames. We will assume the CPU processing time is linear with frame size and a function of the CPU clock speed

$$p_i = \rho T_{cpu} F_i. \tag{2}$$

We also assume a frame can only be processed once it has been received in its entirety. We can form an expression for the arrival time of the frame (which is also the earliest time we can begin processing it) to be

$$a_i = s_i T_s : \min_{s_i}(R_j^d(s_i T_s) \geq \sum_{k=0}^{i} F_k). \tag{3}$$

Where F_i is the size of the ith video frame in bits. Therefore we can set a lower bound on the display time of each frame.

$$d_i \geq a_i + p_i. \tag{4}$$

This is a lower bound since we may not wish to display the frame at this time due to it being too close to the previous frame display time d_{i-1}. It is the set $\{d_i\}$ which determine the quality of the video stream seen by the user.

Now consider the case where the user is willing to accept some delay between t_0 and d_0. This will be the case in Non-Real-Time (NRT) services. However in this case the user does insist upon low jitter and no dropping of frames. We can investigate the relationship between the size of $d_0 - t_0$ and the impact this has on maximum required buffer size and required transmission characteristics.

Since the handset will not begin displaying the streaming video until time d_0 the buffer will fill in a monotonic fashion during this period. i.e. until this time

$$B(t) = R_j^d(t). \tag{5}$$

Also during this time the actual arrival times of these frames is irrelevant to the QoS. However from d_0 onwards we require $d_i = d_0 + iT_F$, where T_F is the frame period, so we must ensure

$$a_i \leq d_0 + iT_F - p_i. \tag{6}$$

An initial trivial result is if $d_o \geq a_{N_F-1}$ because in this case all the frames have been delivered to the handset before the playback of the video commences. In this case the buffer requirement is $\sum_i F_i$ bits and we can ensure perfect playback as long as the CPU requirements are met. In this case since the condition for (5) holds we simply require the average transmission speed (in packets/second) to be

$$c = \frac{\sum_{i=0}^{N_F-1} F_i}{T_F(N_F - 1)\overline{P}_d(1 - P_\epsilon)}. \tag{7}$$

As long as we assume these times are long enough for the mean SIR and probability of error P_ϵ to be valid. Given the length of the streaming video datasets this is a reasonable assumption. However this trivial case is not of great use unless the user decides a considerable period in advance that they wish to observe the video stream. It also places the largest demand on the handset buffer size. Of considerable more interest to us is when $d_0 << a_{N_F-1}$. Note that as $d_0 \to t_0$ we approach the cases that may be applicable to Real-Time (RT) streaming video services.

3   The Streaming Video Datasets

The streaming video sequences used in this analysis were obtained from the Technical University of Berlin. The techniques used to encode these traces and some analysis of them is given in [2]. Each sequence of the original video was encoded a number of times to achieve different image qualities (and bit rates). The subject matter of these traces seems appropriate for the type of services we wish to consider.

3.1   Data Model: A Marginal Distribution Model with I Frame Matching

It is well know that MPEG encoded video possesses an inherent periodicity due to the larger I frames which occur every 12th frame. In addition, the marginal distribution of a data stream tends to be heavy tailed but hard limited to some maximum value.

One way of producing a model with an excellent match to the data's marginal distribution is to use a marginal distribution mapping. In this case we split the real data sequence into the set of I frames and the set of not I frames and perform a match for each. We then recombine these to form our model data. Hence the model possesses the periodicity of the I frames but none to the other correlation properties that have been reported upon [7]. We refer to this model as a Marginal Model with I Frames (MMIF).

4   A Feedforward Constant Delivery System

There are several advantages to using a Feedforward Constant Delivery System (FCDS) at the base-station. These include simplifying the scheduling, reducing the amount of uplink packets and reducing the power requirements of the handset. We can derive a service model for a streaming video dataset based on the discussions in Section 2.

Consider the two time periods prior to and post the initial display time d_0. Assuming a constant service rate for each of these two periods and that at time d_0 we want to have delivered N_{d_0} frames to the handset. These two service rates $\underline{c}$ and $\overline{c}$ can be given (in packets/second) as

$$\underline{c} = \frac{\sum_{i=0}^{N_{d_0}-1} F_i}{(d_0 - t_0)\overline{P}_d(1 - P_\epsilon)} \tag{8}$$

and

$$\overline{c} = \frac{\sum_{i=N_{d_0}}^{N_F-1} F_i}{T_F(N_F - 1)\overline{P}_d(1 - P_\epsilon)}. \tag{9}$$

4.1   Simulation Results for FCDS

The frame loss rate was plotted for a variety of d_0 and N_{d_0} using suitably determined $\underline{c}$ and $\overline{c}$. These results are given for real data and a MMIF model in Figure 1.

One interesting result we can draw from this analysis is that there is no benefit to pre-stocking the buffer. In fact, in the case where the pre-stocking is large there is more packet loss for a given d_0 and convergence to zero packet loss requires a larger d_0.

Another conclusion we can draw is that in the case of a large amount of pre-stocking it is the marginal distribution of the streaming video and not its correlation properties that contribute most to the performance loss. The difference between the real data and the marginal model are closer for large N_{d_0}.

There is much less difference in performance when $N_{d_0} < 10000$ for both real data and the model. It appears that if d_0 reaches a certain threshold then lossless

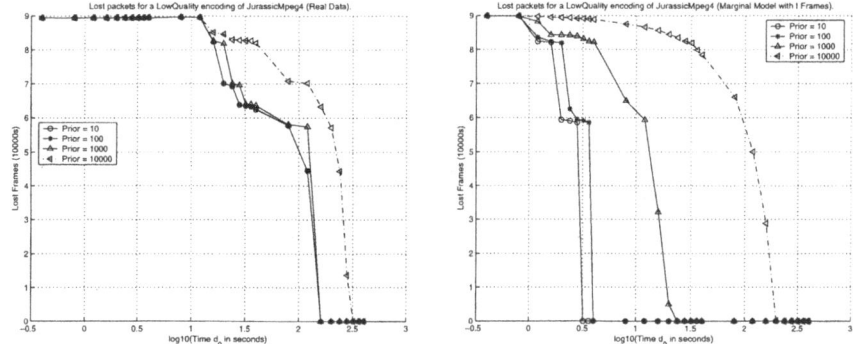

Fig. 1. The packet loss using a dual mode constant service model for the Real Data (left) and Fitted I Frame Marginal Model (right). The legend term Prior equates to N_{d_0}.

transmission can be achieved. This threshold is of the order of $\frac{1}{10}$ of the entire video sequence. Frame loss occurs due to the fact $\bar{c}$ is too small to service the periods when a number of frames larger than the frame mean occur together. The likelihood of this happening in the real data is higher than for the model since the real data is self-similar and displays long range dependence [2], [7]

5   Feedforward Constant Delivery System with Mean Deviation Ratio Adjustment

In the case of NRT streaming video we know the frame characteristics in advance. This allows us to determine where the running average of the sequence exceeds the overall mean and hence where the delivery system will be strained. By allocating extra resources at this time we can achieve improved performance.

We define the Mean Deviation Ratio (MDR), $\overline{F}_j^N$, as

$$\overline{F}_j^N = \frac{1}{\overline{F}} \sum_{i=j}^{i=j+N-1} (F_i - \overline{F}). \tag{10}$$

In Figure 2 we plot the CDF of the MDR for a variety of N for both real data and the MMIF. Note the graphs are very different and hence the MDR is related more to the correlation of the data than its marginal distribution.

Also note that as N is increased the variance of $\overline{F}_j^N$ reduces for both the MMIF and the real data. The MDR has much larger extreme values for the real data than the model. The MDR can be used to improve the performance of the FCDS by altering the service rate $c(sT_s)$. However if we do this we must adjust subsequent transmissions to ensure than over the entire post d_0 transmission $\bar{c}$ is preserved.

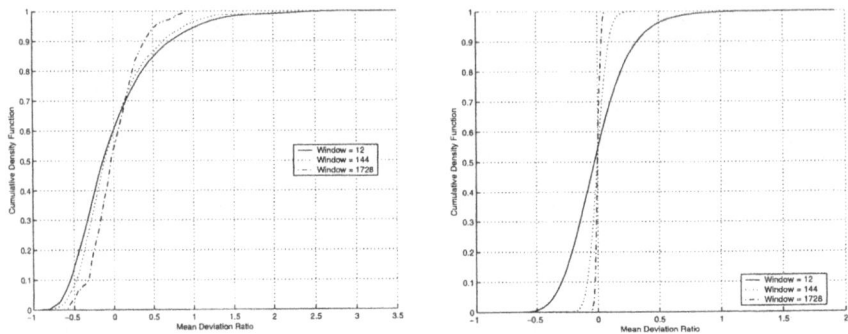

Fig. 2. The CDF of the MDR for a variety of window lengths for the Real Data (left) and the Fitted I Frame Marginal Model (right).

To do this for each timeslot in $(d_0, d_0 + (N_F - 1)T_F)$ calculate $\overline{F}_{sT_s}^{N(sT_s)}$ and

$$c(sT_s) = \begin{cases} \overline{c}'(1 + \overline{F}_{sT_s}^{N(sT_s)}) & \text{if } \overline{F}_{sT_s}^{N(sT_s)} > \Pi(sT_s) \\ \overline{c}' & \text{otherwise.} \end{cases} \tag{11}$$

Where

$$\overline{c}' = \begin{cases} \overline{c} & \text{if } sT_s = d_0 \\ \overline{c}' - \frac{c(sT_s)}{N_F - 1} & \text{if } \overline{F}_{sT_s}^{N(sT_s)} > \Pi(sT_s) \\ \overline{c}' & \text{otherwise.} \end{cases} \tag{12}$$

Where we define $\Pi(sT_s)$ to be the MDR Threshold (MDRT).

5.1   Simulation Results

The FCDS was simulated using the handset buffer size (as estimated by the base-station) to determine $N(sT_s)$ and the loss, service and buffer characteristics were plotted against $\Pi(sT_s)$ which was held constant for the duration of a given simulation.

The results for $d_0 = 1$ second is given in Figures 3 - 5. Simulations were also run for $d_0 = 10$, 60 and 300 seconds with similar results. Packet loss occurs when the MDRT is increased to unity and beyond. However from Figure 2 we note that the probability of exceeding an MDRT of this size is very small, especially when the buffer occupancy is large. Packet loss probability increases monotonically with MDRT. It appears that when the MDRT is too high the scheme does not get enough opportunities to respond to the MDR to prevent packet loss.

For all MDRT the mean service rate was close (to within 5%) of $\overline{c}$. However the standard deviation and maximum were minimized when the MDRT was close to zero. When packet loss occurred the maximum service rate increased dramatically.

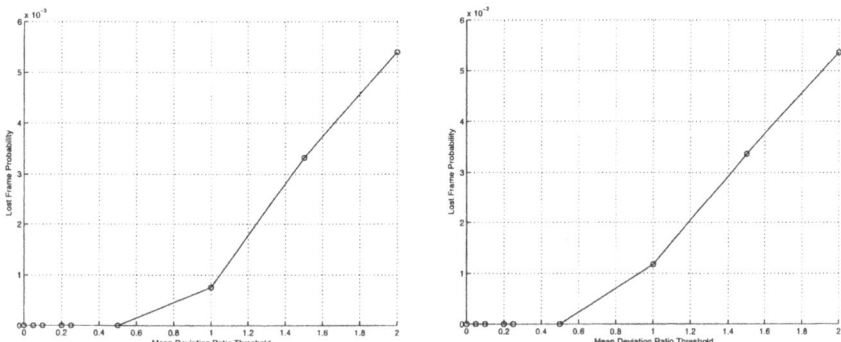

Fig. 3. The frame loss probability for the delivery of streaming video versus the mean deviation threshold without (left) and with (right) the Buffer Occupancy Cost Function (d_0 is one second).

The buffer occupancy rate is highest when the MDRT is low. This is due to the delivery scheme being overly sensitive to the MDRT and not the buffer occupancy. We address this below.

In order to reduce the occupancy of the handset buffer we introduce a Buffer Occupancy Cost Function (BOCF), $\Omega(sT_s)$, which we use to modify $c(sT_s)$ when $\overline{F}_{sT_s}^{N(sT_s)} > \Pi(sT_s)$. We do this because a large MDR may not be a problem when the buffer occupancy is large. However the MDR is an accurate indication of the upcoming stress upon the communication link and hence any attempt to reduce $c(sT_s)$ may result in performance degradation. The BOCF is a multiplicative term which alters (12) to

$$
\bar{c}' = \begin{cases}
\bar{c} & \text{if } sT_s = d_0 \\
\bar{c}' - \Omega(sT_s)\frac{c(sT_s)}{N_F - 1} & \text{if } \overline{F}_{sT_s}^{N(sT_s)} > \Pi(sT_s) \\
\bar{c}' & \text{otherwise.}
\end{cases} \tag{13}
$$

We considered

$$
\Omega(sT_s) = \frac{B_{\max} - N(sT_s)}{B_{\max}} \tag{14}
$$

in a second series of simulations. In these we set $B_{\max}$ to be 1000 frames and hard limited to be non-negative. The results are given in Figures 8 to 5. The simulations with BOCF displayed the same packet loss performance. Losses occur when the buffer is empty and the effect of the BOCF is minimal at these times. It is possible losses could be avoided by increasing the BOCF to greater than unity in cases of low buffer occupancy. The BOCF reduces the maximum and average buffer occupancy values by up to 400% at low MDRT with no impact on frame loss. However whilst the average service rate is maintained, the maximum is increased by up to 50%. Overall the impact of the BOCF is very positive.

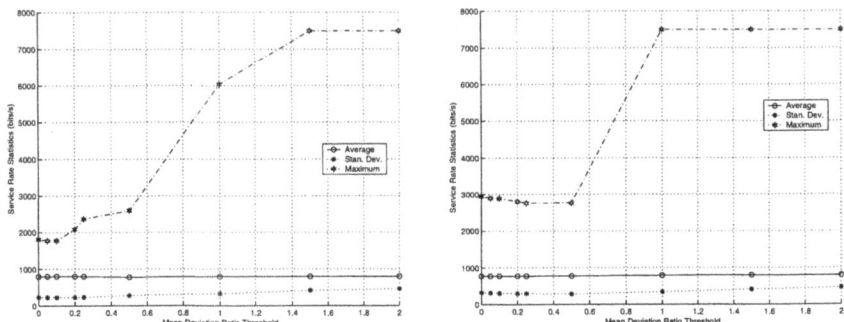

Fig. 4. The statistics of the $\bar{c}$ service rate for the delivery of streaming video versus the mean deviation threshold without (left) and with (right) the Buffer Occupancy Cost Function (d_0 is one second).

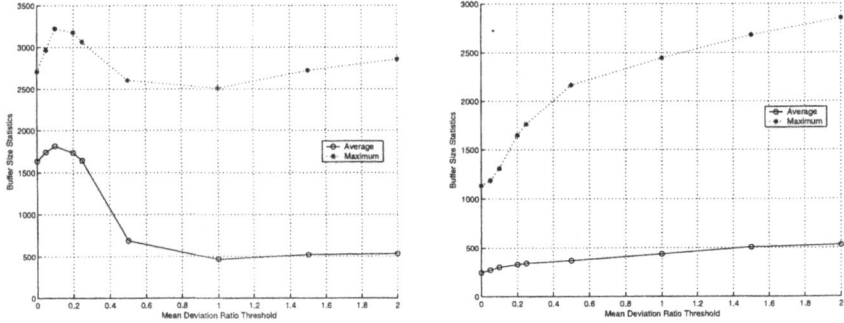

Fig. 5. The statistics of the handset buffer for the delivery of streaming video versus the mean deviation threshold without (left) and with (right) the Buffer Occupancy Cost Function (d_0 is one second).

6   Conclusions and Future Work

In this paper we have presented a novel family of streaming video delivery schemes which are suitable for NRT streaming services but show potential for RT streaming services also. These schemes minimize the uplink transmission requirements by using a Feedforward approach based on constant service capacities. We showed that good QoS could be achieved especially when the mean deviation ratio and buffer occupancy were taken into account.

We presented simulation results for both real streams and model streams which matched the data in the marginal distribution sense. Performance simulations for the model were omitted since we identified that it is the correlation structure of the data which has most impact on the MDR. We plan to investigate this further by using fractional modeling theory [8] to attempt to obtain better models with respect to MDR properties.

We also showed how the performance of the delivery scheme was not a function of d_0. Results for larger d_0 were similar to those presented in this paper. This suggests that the scheme may be developed for RT streaming services. This is an area of on-going investigation.

References

1. Loguinov, D., Radha, H.: End-to-End Internet video traffic dynamics: Statistical study and analysis. In: Proceedings of IEEE INFOCOM. (2002)
2. Fitzek, F., Reisslein, M.: MPEG–4 and H.263 video traces for network performance evaluation. IEEE Network **15** (2001) 40–54
3. Zheng, B., Atiquzzaman, M.: A novel scheme for streaming multimedia to personal wireless handheld devices. Consumer Electronics, IEEE Transactions on **49** (2003) 32–40
4. A. Ziviani, J. F. Rezende, O.C.M.B.D., Fdida, S.: Improving the delivery quality of MPEG video streams by using differentiated services. In: 2nd European Conference on Universal Multiservice Networks – ECUMN02. (2002)
5. R. Tupelly, J.Z., Chong, E.: Opportunistic scheduling for streaming video in wireless networks. In: Information Sciences and Systems, 2003 Conference on. (2003) 752–754
6. Sivaradje, G., Dananjayan, P.: Dynamic resource allocation for next generation wireless multimedia services. In: Communication Systems, 2002. ICCS 2002. The 8th International Conference on. (2002) 752–754
7. Bates, S., McLaughlin, S.: Is VBR video non-stationary or self-similar? Implications for ATM traffic characterisation. In: Proceedings of the European Simulation Muilticonference. (1996)
8. Bates, S.: The estimation of stable distribution parameters from teletraffic data. Signal Processing, IEEE Transactions on **48** (2000) 865–871

Use of the 2500–2690 MHz Band for Multimedia Broadcast Multicast Service

Ana Rita Oliveira and Américo Correia

Associação para o Desenvolvimento das Telecomunicações e
Técnicas de Informática (ADETTI)
Av. das Forças Armadas, Edifício ISCTE
1600-082 Lisboa Portugal
{rita.oliveira,americo.correia}@iscte.pt

Abstract. The viable deployment of both UMTS Terrestrial radio access network (UTRAN) modes, Frequency Division Duplex (FDD) and Time Division Duplex (TDD) in additional and diverse spectrum arrangement will be investigated. Simulations are performed to study the interference between UTRA-TDD and the UTRA-FDD having adjacent channel carriers. This expansion spectrum will be predominantly used for expanding capacity, by means of macro, micro and pico cells in "hotspot" areas. In some scenarios, the TDD and FDD cells can use the same frequency, which will increase the capacity and utilize the underused UTRA-FDD UL resources. This study is one of the tasks of the B-BONE project (Broadcasting and multicasting over enhanced UMTS mobile broadband networks). In the 2500-2690 MHz band, we investigate the feasibility of future Enhanced UMTS FDD and TDD modes with multi-carrier and MIMO to carry digital broadcast services such as Multimedia Broadcast Multicast Service (MBMS).

1 Introduction

The two major radio technologies for deployment in terrestrial networks, UTRA FDD for wide area access, and UTRA TDD for high user density local area access, are likely to be employed in different scenarios, considering the additional spectrum arrangements. Operating together seamlessly, operators can make the most efficient use of 3G spectrum with full mobility and service roaming for the broadest number of voice and data customers, maximizing the profitability of rich voice, data applications and multimedia services.

In order to, efficiently utilize both Radio Access Network (RAN) and the core network part of enhanced UMTS FDD and TDD modes for provision of broadcasting/multicasting services, specific functionalities have to be added to the current UMTS standard. The first step where already made in Release 6, to incorporate the MBMS. The main goal of this work is to provide conditions to incorporate this service in the 2500-2690 MHz band.

J.N. de Souza et al. (Eds.): ICT 2004, LNCS 3124, pp. 16–21, 2004.
© Springer-Verlag Berlin Heidelberg 2004

The TDD-CDMA for mobile radio communication is extremely efficient to handle asymmetric data and benefits resulting from the reciprocal nature of the channel. This technology optimizes the use of available spectral resources, reduces delays for the user and provides the highest level of user satisfaction, provides great flexibility for network deployment because its radio air interface gives optimum coverage in data intensive where private intranets, extranets, and virtual private networks are resident. The FDD technology is designed to operate in paired radio frequency bands in which relatively equal amounts of data flow in two directions, such as a voice call where the traffic is predictable and symmetrical with relatively constant bandwidth requirements in both the UL and DL channels. Combining FDD and TDD modes we can maximize the number of users on a system and thereby maximize average revenue per user, total revenues, and return on investment. This would be particularly feasible in the indoor scenario, where available spectrum is often needed to serve the wanted wide area traffic. For the indoor BSs, each operator will use at least one of the Wideband CDMA (WCDMA) carrier pairs that are used by the outdoor macro system, thus providing soft handover between the macro and indoor cells on this carrier. Indoor systems are expected to be mainly deployed in urban areas, but could be introduced also in hot spots surrounded by medium and low traffic areas. In UTRA, the paired bands of 1920-1980 MHz and 2110-2170 MHz are allocated for FDD operation in the UL and DL, respectively. Whereas the TDD mode is operated in the remaining un-paired bands of 1900-1920 MHz and 2010-2025 MHz [1]. For the purpose of ex-panding capacity, the 2500-2690 MHz band will be using exclusively either UTRA FDD or UTRA TDD. It is however, acknowledged that the combination of both radio technology in new bands is a valid option, as already discussed in ITU-R WP 8F.

In this paper, the interference and co-existence between the UTRA-TDD and FDD system is discussed. The interference occurs within the FDD and TDD air interfaces and additionally between the systems.

The content of this paper is organized and summarized as follows. Section 2 illus-trates the proposed scenarios for 2500-2690 MHz band. Section 3 describes the inter-ferences between FDD and TDD systems. Section 4 provides a discussion and a brief conclusion of the subjects discussed in the previous sections.

2   Proposed Scenarios for 2500–2690 MHz

The feasibility analysis of enhanced UMTS FDD and TDD to carry digital broadcast services available and trade-off analysis with other DVB solutions is being investi-gated. Compiling and publishing network protocols for the enhanced UMTS (FDD and TDD) to transport digital broadcasting/multicasting services will contribute to the dissemination of the B-BONE project results.

Co-existence between UTRA FDD and UTRA TDD within 2500 – 2690 MHz will be considered in this study but is however acknowledged that the combination of FDD and TDD in new bands is a valid option. In the table below, seven scenarios are shown from the several different possible scenarios [3].

Table 1. Spectrum Arrangement Scenarios for 2500-2690 MHz band allocation according to ITU-R.[4].

MHz 2500				2690
Portions	**A**	**B**	**C**	**D**
Scenario1	FDD UL (internal)	TDD		FDD DL (internal)
Scenario2	FDD UL (internal)	FDD DL (external)		FDD DL (internal)
Scenario3	FDD UL (internal)	TDD	FDD DL (ext)	FDD DL (internal)
Scenario4	FDD DL (external)		TDD	
Scenario5	TDD		FDD DL (external)	
Scenario6	TDD			
Scenario7	FDD DL (external)			

In this work we consider the scenarios 2, 4, 5 and 7, in order to find out witch ones performs the requirements of B-BONE project. The objective is to Enhanced UMTS (FDD/TDD) to provide via single network broadcasting/ multicasting/unicasting services in a flexible, efficient and low cost manner. This is the approach of 3GPP and some initial work was already done and included in the UMTS Release 6 called MBMS. Scenario 2 shows a graphical representation for utilizing the additional frequencies from 2500 - 2690 MHz for UTRA FDD. Thus, two distinct frequency arrangements for FDD operation within the new 2.5 GHz band would exist here which we will call "*FDD internal*", respectively "*FDD external*" throughout this Section[4]. With this scenario we could made the following observations:

- Provision of a wide range of asymmetric capacity, in particular, the UL and DL bands of the "FDD internal" system can be asymmetric;
- Provision of additional UL/DL spectrum to support new and existing operators;
- Provision of a DL capacity extension for existing Band I operators;
- Propagation loss in 2.5 GHz is higher than the actual band;
- From UE roaming, it would be beneficial if the partitioning A / B+C / D of the 2.5 GHz band could be made fixed on an as global basis as possible;
- Implementing UEs or Node B's according to the scenario 2 frequency arrangement does not require development of any new or risky implementation concepts;
- 30 MHz carrier separation between FDD UL internal and FDD DL external bands is desirable to achieve the present interference protection levels.

Scenarios 4 and 5 illustrate a graphical representation for utilizing the additional frequencies from 2500 - 2690 MHz for UTRA FDD and TDD. With these scenarios we could made the following observations:

- Provision of a DL capacity extension for existing Band I operators;
- TDD allows the autonomous frequency allocation for new operators, which do not have a frequency block in the core bands;
- The FDD DL is located next to the TDD band, at least one TDD and one FDD DL channel will experience interference caused by the other system due to the imperfect transmitter and receiver characteristics.

Scenario 7 aims to support DL optimized utilization of the 2500 – 2690 MHz band with the goal to obtain a capacity enhancing complement for UTRA operating in the Band I. With this scenario we could made the following observations:

- The introduction of each additional DL 2.5 GHz carrier adds the same DL capacity as a corresponding Band I carrier would do and increases the achievable DL/UL throughput asymmetry of the system;
- There appears to be no need for power compensating the additional 3 dB Path Loss on the 2.5 GHz carrier for *coverage reasons* as there is ample margin for DL coverage available;
- Use of Variable Duplex Technology (VDT) is an essential technological element in providing this solution.

3 Interference Model in the FDD/TDD

According to Shannon's theorem, the limiting factor of capacity in a WCDMA system is interference. In this section, we describe the interference between FDD and TDD systems. The goal is to select the best technology to allow asymmetric services such as broadcast/multicast data reception in parallel with other mobile services. The asymmetry between DL and UL capacity is necessary in order to have an efficient use of the radio resources. By then, only one of the FDD bands is required and hence the more flexible TDD link could potentially double the link's capacity by allocating most of the time-slots in one direction. FDD and TDD can share the same bandwidth, as long as the amount of interference is not excessive. Due to imperfect transmitter and receiver characteristics a certain amount of the signal power transmitted on the dedicated channel is experienced as interference on the adjacent channels. The impact of adjacent channel interference (ACI) between two operators on adjacent frequency was studied. The ACI needs to be considered, because it will affect WCDMA system where large guard bands are not possible, witch would mean a waste of frequency bands. However, ACI could be avoided by putting tight requirements on transmit and receive filter masks of base and mobile stations. There are two critical scenarios of WCDMA interference. One happens when MS from operator 1 is coming close to BS of operator 2 (located at the cell edge of operator 1) and is blocking this BS because it is transmitting with full power. Second happens when BS from operator 2 is transmitting with high power and therefore is blocking all of MSs of operator 1 in a certain area around it, caused by dead zones because of the excessive power, and/or blocking because of the exceeded input power at the MS receiver. The influence of adjacent channels on each other can be identified as Adjacent Channel Leakage power Ratio (ACLR), Adjacent Channel Selectivity (ACS) and Adjacent Channel Interference Ratio (ACIR) [2]. ACLR is defined as the ratio of the transmitted power to the power measured in an adjacent RF channel. Both the transmitted power and the received power are measured with a filter that has a Root-Raised Cosine (RRC) filter response with roll-of factor of $\alpha=0.22$, with a noise power bandwidth equal to the chip rate. ACS is defined as the measure of a receiver's ability to receive a WCDMA signal at its assigned channel frequency in the presence of an adjacent channel signal at a given frequency offset from central frequency of the assigned channel. ACIR is a measure of over all system performance. It is defined as the ratio of the transmission power to the power measured after a receiver filter in the adjacent channel(s). ACIR will be considered in our simulations, because the total amount of interference is the primary concern, this is what we called Adjacent Channel Protection Ratio. ACIR is given by:

$$ACIR = \cfrac{1}{\cfrac{1}{ACLR} + \cfrac{1}{ACS}}$$

4 Discussion and Conclusions

In our simulations the chosen topology is the university campus in the urban area of Lisbon. We considered two UMTS Portuguese operators that are operating in adjacent channels within the same area. The user profile used in the simulations is mobile users travelling in two avenues the bit rate is 144 kbps, the mobile speed is 50Km/h and three-sectored antennas are used. The applied path loss model is based on the *Walfisch-Ikegami-Model*. WCDMA internal interferences are estimated, taking into account ACLR. The requirements of ACLR defined in the table 2, are valid when the adjacent channel power is greater than -50 dBm. The scenario 2 of table 1 is the chosen reference scenario.

Table 2. Adjacent channel performance requirements.

Adjacency	Channel Separation	Max. Allowed ACLR	
		MS	BS
1rst Adjacent Carrier	5 MHz	33 dB	45 dB
2nd Adjacent Carrier	10 MHz	43 dB	55 dB
Band Separation	30 MHz	53 dB	57 dB

Other parameters are taken into account as show in table below.

Table 3. Parameter used in simulations.

Parameters	Values
Chip Rate	3.84 Mcps
MCL micro	53 dB
BS (max power)	43 dBm
BS (min power)	27 dBm
MS (max power)	21 dBm
MS (min power)	-50 dBm
Standard deviation for the shadow fading	7 dB
Max allowed noise rise in micro cell	20 dB
Max allowed noise rise in macro cell	6 dB
MS/BS noise figures	8 dB/ 5 dB
BS antennas	17.5 dBi
MS antennas	0 dBi
Frequency	2595 MHz
Processing Gain	14,3 dB

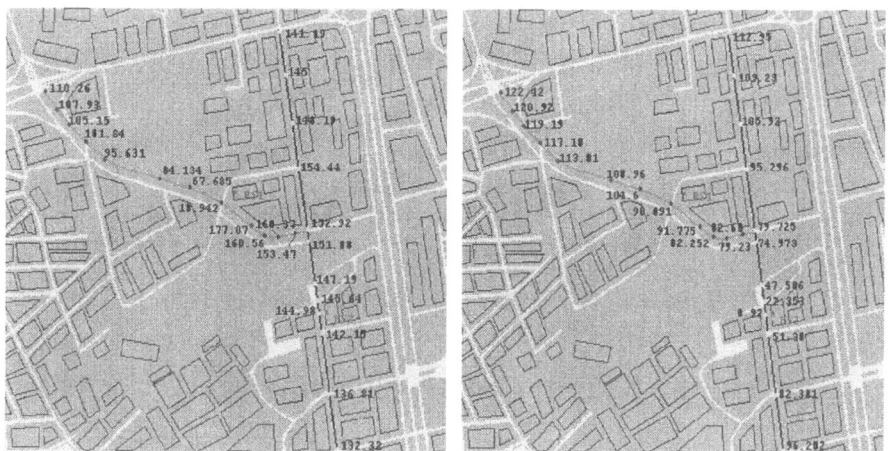

Fig. 1. a) Right map, required transmitted powers (dBm) of BS1 to block MSs; b) Left map required transmitted powers (dBm) of BS2 to block MSs.

We have considered 2 different situations. In figure 1. a) BS1 (operator 1) is transmitting to the mobile users of operator 2 and the received interference at MSs in the adjacent channel is checked against ACLR values of table 2. In figure 1. b) BS2 (operator 2) is transmitting to MSs of operator 1. We can conclude that in this case when MSs are coming close to the transmitting BS2, the required transmitted powers can block the MS connection. Considering one of the goals of B-BONE project, we have concluded that from the seven proposed scenarios for band allocation according to ITU-R, the most suitable scenarios are 2, 4, 5 and 7, in order to accommodate digital broadcasting/multicasting services on Enhanced UMTS network, such as MBMS. We have considered scenario 2. The *"FDD external"* with an extra band of 40MHz could provide to a number of existing operators an asymmetric capacity extension, with no impact on existing frequency arrangements. So that, a DL capacity extension for existing Band I operators, could support MBMS services. The *"FDD internal"* with paired bands of 60MHz each could also support MBMS services.
Nevertheless, the use of TDD is viable and allows the frequency allocation for new operators, which do not have a frequency block in the core bands.

References

1. 3GPP TS 25.101: "UE Radio Transmission and Reception (FDD)".
2. Wacker, Laiho, Novosad, Radio Network Planning and Optimisation for UMTS, Wiley.
3. 3GPP TR 25.889 V6.0.0, "3rd Generation Partnership Project; Technical Specification Group Radio Access Network; Feasibility Study considering the viable deployment of UTRA in additional and diverse spectrum arrangements (Release 6).
4. ITU-R8/112-E; 17 October 2002; Frequency arrangements for implementation of the terrestrial component of International Mobile Telecommunications-2000 (IMT-2000) in the bands 806-960 MHz, 1 710-2 025 MHz, 2 110-2 200 MHz and 2 500-2 690 MHz

Throughput Fairness and Capacity Improvement of Packet-Switched Networks Through Space-Time Scheduling

Emanuela B. Silva, Leonardo S. Cardoso,
Tarcisio F. Maciel, Francisco R.P. Cavalcanti, and Yuri C.B. Silva

Wireless Telecommunications Research Group - GTEL
Federal University of Ceará, Fortaleza, Brazil
{emanuela, leosam, maciel, rod, yuri}@gtel.ufc.br
http://www.gtel.ufc.br

Abstract. This article discusses some performance issues about space-time scheduling applied to packet data shared channels. Beamforming adaptive antennas (AA) and Spatial Division Multiple Access (SDMA) strategies were evaluated through system-level simulations considering different scheduling algorithms. Throughput fairness among users is discussed and the best antenna and scheduling algorithm configurations considering both capacity and fairness are shown.

1   Introduction

In this article we discuss some issues regarding space-time scheduling. The spatial filtering and interference reduction provided by adaptive antennas may allow the tightening of the reuse pattern without sacrificing link quality. Increased capacity can also be obtained while maintaining the same frequency reuse, by employing Spatial Division Multiple Access (SDMA) and reusing the radio channels several times within the cell. In both cases, the gains provided by adaptive antennas are exchanged for capacity either by tightening the frequency reuse or by increasing the intra-cell interference. All these strategies are evaluated within this work in order to determine which space-time configuration provides the best performance in terms of spectral efficiency and system fairness.

2   Simulation Model

A dynamic discrete-time system-level simulation tool, which simulates downlink data communication, was used in order to evaluate space-time scheduling. This simulation tool is based on the EGPRS specifications and more details about this tool can be found in [2]

In this work, system capacity is defined as the highest offered load, expressed in terms of spectral efficiency in bps/Hz/site, at which a minimum average packet throughput per user, herein of 10kbps per time slot, is ensured to at least 90% of the users in the system.

J.N. de Souza et al. (Eds.): ICT 2004, LNCS 3124, pp. 22–27, 2004.

In the simulations, a macrocellular environment was considered. Link-level results corresponding to a TU50 scenario without frequency hopping were used [3]. Perfect knowledge of the Signal-to-Interference plus Noise Ratio (SINR) was assumed, and an ideal link quality control, i.e., no delay between measurement and Modulation and Coding Scheme (MCS) selection, was considered. Pure link adaptation mode (no incremental redundancy) was also assumed.

A World Wide Web (WWW) traffic source was simulated [4] using a single 200kHz GSM/EDGE carrier per sector without frequency hopping.

2.1 Scheduling Algorithms, Beamforming Adaptive Antennas, and Spatial Division Multiple Access

In this work, four scheduling disciplines [5] with distinct properties have been considered: First-In-First-Served (FIFS), Least Bits Left First Served (LBFS), Maximum Signal-to-Interference plus Noise Ratio (MAX), and Round Robin (RR).

The Beamforming Adaptive Antenna (AA) technique used was the spatial matched filter. The antenna system consisted of an 8-element Uniform Linear Array (ULA) implementing the spatial matched filter and assuming an azimuth spread of 5° (five degrees), which is enough to completely fill the antenna nulls. Moreover, a digital filter (Kaiser window) is used to suppress side lobe levels at the expense of a broader main lobe and reduced directivity gains. The estimate of the direction of arrival is considered ideal, i.e., it is assumed that the base station knows the exact position of the mobile station. More details about the antenna model used in the simulations can be found in [2].

SDMA uses the interference reduction ability of adaptive antennas to improve capacity by reusing the spectrum within the own cell, allowing for several mobile stations to share simultaneously a same channel.

The approach consists of establishing groups of users (compatibility group) sufficiently distant of each other over the azimuthal plane, and to steer a narrow beam towards each one of them. If the steered beams are sufficiently narrow and the angular separation large enough, the intracellular co-channel interference should remain at acceptable levels. The compatibility test consists of verifying whether the angular distance between users is larger than a chosen value, thus, guaranteeing reasonable intra-cell interference levels.

Here we considered a *space-time* scheduling approach, where the time dimension is prioritized, i.e., users are sorted according to the time-domain criterion and then they are sequentially tested for compatibility with the head-of-line user. The scheduler determines the user which will be given priority, and after that other users are checked in order to build a compatibility group, if possible. In order to enter a compatibility group, users must be kept at an angular distance of 20 degrees (4 times the angular spread). It is assumed that each group can accommodate a maximum of 2 users.

3   Results and Analysis

3.1   Reference Scenario and Multiuser Diversity Gain

In an interference-limited scenario the data throughput verified by the users sharing a channel queue is limited by the high error rates and reduced channel reliability due to interference. In such high error scenarios, scheduling algorithms cannot provide significative multiuser diversity gains, since all channels have similar performances.

This can be seen in figure 1, which confirms that there are no substantial gains when changing the scheduling algorithm in the sectored scenario.

The same figure shows the improved system performance when beamforming adaptive antennas are employed. Since AA highly reduces interference, the system becomes queueing limited. The multiuser diversity gains provided by scheduling algorithms can then be better exploited, resulting in an increase of the load that can be offered to the system and higher diversity gains.

3.2   Beamforming Adaptive Antenna

Figure 2 compares the impact of AA over a WWW service running on shared channels of a GSM/EDGE system employing sectored antennas, a 1/3 frequency reuse pattern and FIFS. As it can be seen in the referred figure, a spectral efficiency gain superior to 250% can be achieved when using AA and keeping a 1/3 reuse pattern. As AA extremely reduces interference, a tighter reuse pattern (1/1) can be implemented, which results in a capacity gain of 636%.

Figure 2 also shows the extra capacity improvement that can be obtained by using a suitable scheduling algorithm.

In the case of the 1/1 reuse pattern, the interference increase leads to a superiority of MAX over the other algorithms, since it is the only channel-state dependent algorithm selecting always the best channel to transmit.

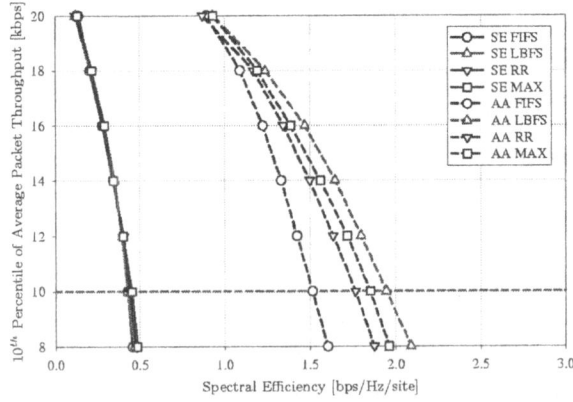

Fig. 1. Performance of different scheduling algorithms considering a 1/3 reuse pattern with and without adaptive antennas.

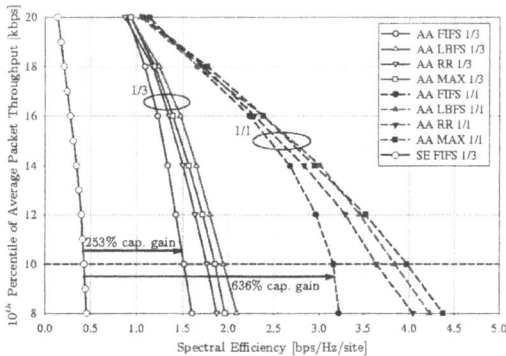

Fig. 2. Capacity gain using AA and scheduling algorithms for 1/3 and 1/1 reuses.

On the other hand, in the 1/3 reuse, due to AA interference reduction, SIR values stay mostly within an interval where the throughput gains for higher SINR values are marginal. In this case, prioritizing the delay is more beneficial in terms of multiuser diversity than prioritizing the best link quality, which causes the LBFS to be the best algorithm in this scenario.

3.3   Spatial Division Multiple Access

The performance of SDMA compared to a 1/3 system with sectored cells and using FIFS is shown in figures 3(a) and 3(b). It can be seen that the use of SDMA before modifying the frequency reuse pattern improves the spectral efficiency in 336%.

It can also be observed that changing the frequency reuse pattern to 1/1 with the FIFS algorithm (figure 3(b)), the system capacity is improved if compared to the 1/3 case using the same SDMA scheme. Indeed, using SDMA plus FIFS and assuming a 1/1 reuse, a capacity gain of 387% is provided with regard to the reference scenario.

Since the use of a 1/1 reuse pattern incurs in higher interference radiation, specially intracell interference, changing the reuse from 1/3 (figure 3(a)) to 1/1 (figure 3(b)) does not improve system capacity. This fact can be also observed in figure 3(b) where, for the 1/1 pattern, there is no difference between the scheduling algorithms. On the other hand in figure 3(a) the MAX algorithm improves system capacity in 20% due the multi-user diversity gain offered by this scenario.

One must, however, pay attention to the fact that all users belong to the same service class. Therefore, they should get almost the same QoS level. This issue refers to throughput fairness among users and is discussed in the next section.

3.4   Quality of Service Fairness

This section shows how the degree of fairness may significantly change depending on the scheduling algorithm. Here, fairness is evaluated in terms of the through-

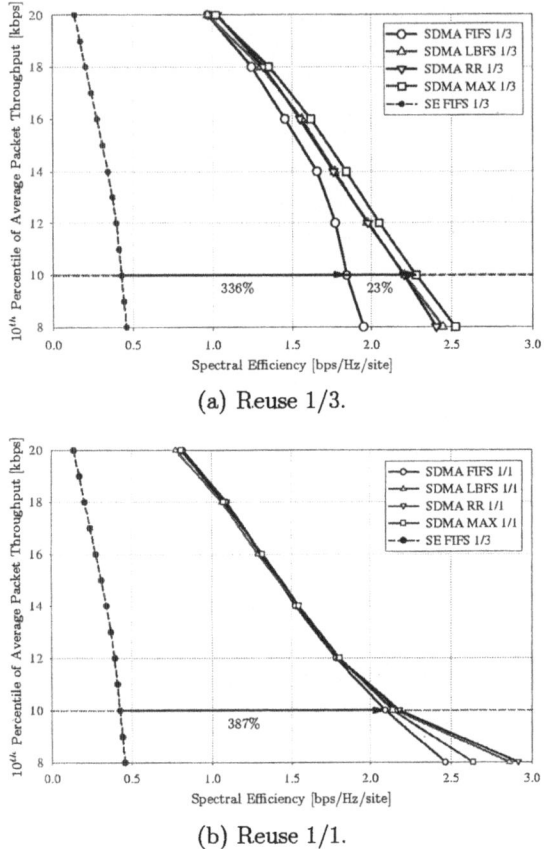

(a) Reuse 1/3.

(b) Reuse 1/1.

Fig. 3. Capacity gain using SDMA with different scheduling algorithms.

put experienced by the users. Thus, in an ideally fair configuration, all users should verify the same throughput and the difference between the 10th and the 90th percentiles would be zero.

In this work the 10th percentile of the average throughput per user is equal to 10 kbps. Table 1 shows the 90th percentile of the average throughput per user thus, the higher the 90th percentile, the more unfair the algorithm.

FIFS and MAX are the less fair algorithms, which is a natural consequence of their formulation. FIFS is not a channel-state-dependent scheduling algorithm and does not exploit channel diversity. With the MAX algorithm, the user with the best channel is always scheduled to transmit, while users with poor channels, mainly at the cell border, are severely penalized.

In the SDMA scenario, RR is the most fair among the evaluated algorithms since it gives the users equal chance to transmit. In the AA scenario FIFS and MAX still are the most unfair algorithms and LBFS is the most fair, since in this

Table 1. 90th Percentile of Average Packet Throughput per User and System Capacity for all Algorithms and Scenarios

Alg.	SE 1/3 T_{90th} [kbps]	SE 1/3 Spec. Eff. [bps/Hz/site]	SDMA 1/3 T_{90th} [kbps]	SDMA 1/3 Spec. Eff. [bps/Hz/site]	AA 1/1 T_{90th} [kbps]	AA 1/1 Spec. Eff. [bps/Hz/site]
FIFS	28.4	0.43	30.0	1.85	30.5	3.16
LBFS	28.9	0.43	26.8	2.21	**27.4**	3.84
RR	28.6	0.44	**26.7**	2.20	29.0	3.64
MAX	28.6	0.45	29.7	**2.28**	30.9	**3.96**

scenario the queue lengths become longer and the transmission delay imposed to small packets become a limiting factor.

4   Conclusions

We have seen that smart antennas can substantially improve the capacity of cellular systems, when compared to the reference scenario (SE with FIFS). It was also shown that smart antennas in SDMA mode provide the best capacity when considering a 1/3 frequency reuse pattern, and that a tightening in the reuse to 1/1 brings the system back to its interference limited characteristic, limiting the scheduling algorithm gains. In the 1/1 frequency reuse pattern, the single beam AA solution had a much better performance, providing capacity gains of up to 826%. MAX stands as the most unfair, but with the highest capacity improvement. LBFS and RR appeared as the most fair, but with slightly lower capacity gains.

References

1. R. H. Roy, "An overview of smart antennas technology and its application to wireless communication systems," *IEEE International Conference on Personal Wireless Communications.*, pp. 234 – 238, 1997.
2. W. M. de Sousa Jr., F. R. P. Cavalcanti, Y. C. B. Silva, and T. F. Maciel, "System-level performance of space-time scheduling for non real-time traffic in a 3G network," *IEEE Proceedings Vehicular Technology Conference, VTC 2002-Fall.*, vol. 2, pp. 1207 – 1211, September 2002.
3. D. Molkdar and S. Lambotharan, "Link performance evaluation of egprs in la and ir modes," *IEEE Vehicular Technology Conference*, 2000.
4. C. Johansson and L. C. De Verdier and F. Khan , "Performance of different scheduling strategies in a packet radio system," *IEEE International Conference on Universal Personal Communications, ICUPC*, vel. 1, pp. 267 – 2715, October 1998.
5. H. Fattah and C. Leung, "An overview of scheduling algorithms in wireless multimedia networks," *IEEE Wireless Communications*, vol. 9, pp. 76–83, October 2002.

Performance of MIMO Antenna Systems with Hybrids of Transmit Diversity and Spatial Multiplexing Using Soft-Output Decoding

W.C. Freitas Jr., A.L.F. de Almeida, J.C.M. Mota,
F.R.P. Cavalcanti, and R.L. de Lacerda Neto

Wireless Telecom Research Group, Federal University of Ceará, Fortaleza, Brazil
PHONE/FAX:+55-85-2889470
{walter, andre, mota, rod, raul}@gtel.ufc.br
http://www.gtel.ufc.br

Abstract. In this work we present some MIMO transmission schemes that combine transmit diversity and spatial multiplexing using four transmit antennas. Then, we show that the Bit-Error-Rate (BER) performance of these schemes can be considerably improved with the joint use of channel coding (at the transmitter) and soft-output detection (at the receiver). The SOVA approach is used to enhance performance of some detection layers that are not space-time coded. Both parallel and successive detection strategies are considered.

1 Introduction

Multiple-Input Multiple-Output (MIMO) wireless channels are known to offer unprecedent spectral efficiency and diversity gains, which can be exploited by employing antenna arrays at both ends of the wireless link [1]. Two well-known applications of MIMO channels are the use of spatial multiplexing, e.g., V-BLAST (Vertical Bell Labs Layered Space-Time) [2,3], and Space-Time Block Coding (STBC) [4,5]. The V-BLAST scheme tries to maximize spectral efficiency of the overall system as much as possible by transmitting multiple co-channel signals that are distinguished at the receiver with array processing. On the other hand, STBC schemes improve signal quality at the receiver (bit error performance at low signal-to-noise ratio) with simple linear processing, allowing even a single-element receiver to distinguish between the signals on different paths.

With the objective of providing the two gains that can be perceived in a MIMO channel, i.e. diversity and spatial-multiplexing gains, some hybrid MIMO receivers have been recently proposed [6]. Instead of their remarkable performance in terms of Bit-Error-Rate (BER) and spectral efficiency, in some of these schemes the BER performance can be further improved as long as some channel coding strategy is applied at the transmitter in order to protect the layers that are not space-time coded. Moreover, if soft-output decoding (e.g. soft-output Viterbi algorithm) is used for such layers at the receiver side, better results are expected.

J.N. de Souza et al. (Eds.): ICT 2004, LNCS 3124, pp. 28–37, 2004.

In this work we focus on the performance enhancement of some hybrid MIMO receivers due to the joint use of channel coding and Soft-Output Viterbi Algorithm (SOVA). The considered hybrid schemes employ four transmit antennas. In these schemes, some layers are space-time coded across two, three or four antennas. For the non-space-time coded layers, convolutional encoders are employed to protect data prior to transmission. At the receiver side, the SOVA detection technique is used to improve detection performance of such non-space-time coded layers. During the detection process, both parallel and successive detection strategies are considered.

The remainder of this paper is organized as follows. In Section II, we briefly describe the multiplexing-diversity trade-off. Section III is dedicated to the channel model. In Section IV, we present the Coded Hybrid MIMO Structures with Successive Interference Cancellation (CHS-SIC). Section V describes our interference cancellation algorithm. Section VI is dedicated to an overview of SOVA metrics that are considered in the decoding algorithm, while Section VII contains our simulation results. Finally, in section VIII we conclude this paper and draw some perspectives.

2   Diversity and Multiplexing Trade-off

Currently, it has been shown that an important approach to increase the data rate over wireless channels is the use of multiple antennas at both ends of the wireless link. When MIMO antenna systems are used, it can be possible to create multiple parallel channels for transmission of independent information or to add diversity by transmitting/receiving the same information in different antennas. Provided that the antennas are sufficiently spaced at both transmitter and receiver, the transmitted signals experiment independent fading, which implies in a low probability of simultaneous deep fading in all channels. Compared to traditional antenna systems, where antenna arrays are employed either at the transmitter or at the receiver, the capacity due to the use of antenna arrays at both link-ends is enormously incremented.

Up to the present moment, most of MIMO antenna schemes are designed to achieve just one of the two aforementioned gains, i.e. multiplexing gain *or* diversity gain, and is well-known that the focus in a particular gain implies a sacrifice of the other one. Therefore, hybrid MIMO transmission schemes arises as a solution to jointly achieve spatial multiplexing and transmit diversity gains. In other words, with hybrid MIMO schemes it can be possible to considerably increase the data rate while keeping a satisfactory link quality in terms of BER.

3   Channel Model

In order to formulate the channel and received signal model, let the receiver be equipped with an N-element antenna array. The transmitted signals are assumed to undergo independent fading such that the signal at the output of each element of the receive antenna array is a superposition of flat-faded versions

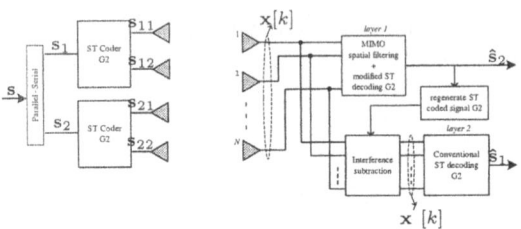

Fig. 1. G2+G2(SIC) transmitter-receiver structure

of the transmitted signals plus white Gaussian noise. Fading is also assumed quasi-static over every sub-sequence and independent between two consecutive sub-sequences. For all the MIMO transmission schemes, we assume that the total transmission power is fixed (normalized to 1) and equally divided across the transmit antennas. Ideal symbol timing is assumed at the receiver. At any time-instant k, the received signal vector can be expressed in a general form as

$$\mathbf{x}[k] = \mathbf{H} \cdot \mathbf{s}[k] + \mathbf{n}[k], \tag{1}$$

where

$$\mathbf{H} = \begin{bmatrix} \mathbf{h}_{n1} \ \mathbf{h}_{n2} \ \mathbf{h}_{n3} \ \mathbf{h}_{n4} \end{bmatrix}, \tag{2}$$

denotes the MIMO channel matrix of dimension $N \times 4$. The column vector $\mathbf{h}_{nm}$, with $1 \le n \le N$ and $1 \le m \le 4$ is the complex vector channel that links the nth receive antenna and the mth transmit antenna. The envelope of each element in the vector $\mathbf{h}_{nm}$ follows a Rayleigh distribution. The 4×1 vector $\mathbf{s}[k]$ contains the symbols transmitted from all antennas at time-instant k. The composition of vector $\mathbf{s}[k]$ depends on the specific hybrid scheme considered. The $N \times 1$ vector $\mathbf{n}[k]$ denotes the temporally and spatially Additive White Gaussian Noise (AWGN).

4   Coded Hybrid MIMO Structure (CHS)

In this section we present the hybrid MIMO transmission schemes. Figures 1 up to 3 show the architecture of the Coded Hybrid Structures with Successive Interference Cancellation (CHS-SIC) and four transmit antennas. The search for hybrid schemes that are limited to four antenna-elements is motivated by the practical feasibility of utilizing this number of antenna elements in nowaday's base station-to-mobile transmissions.

4.1   G2+G2 Scheme

The first hybrid scheme, called G2+G2, is shown in Fig. 1. It employs a four-element transmit antenna array with two vertically-layered G2 space-time coding schemes. The input sequence $\mathbf{s}$ is split into two parallel sub-sequences $\mathbf{s}_1$ and $\mathbf{s}_2$.

For every two consecutive time-instants k and $k+1$, the first multiplexed sub-sequence $\mathbf{s}_1$, constituted by symbols $s_1[k]$ and $s_1[k+1]$, enters in a parallel form in the first ST coder G2. Similarly, the second multiplexed sub-sequence $\mathbf{s}_2$ enters the second ST coder G2. At each ST coders, the sub-sequences are space-time coded using the standard G2 (Alamouti's) code [4]. The transmitted signals can be organized in a equivalent space-time coding matrix as described below

$$\mathbf{\Omega}_{G2+G2} = \begin{bmatrix} s_1[k] & s_1[k+1] & s_2[k] & s_2[k+1] \\ -s_1[k+1]^* & s_1[k]^* & -s_2[k+1]^* & s_2[k]^* \end{bmatrix}. \tag{3}$$

As all transmit antennas are co-channel, i.e. they share the same frequency band, the sub-sequence associated to a given ST coder G2 appears as an interferer to that sub-sequence associated to other ST coder G2. This type of co-channel interference is defined here as Multiple Access Interference (MAI). Considering the G2+G2 scheme, we can expand the general received signal vector (1) as the sum of a MIMO desired signal and a MIMO interferer signal as follows

$$\mathbf{x}_{G2+G2}[k] = \mathbf{H}_d^{G2}\mathbf{s}_1[k] + \mathbf{H}_I^{G2}\mathbf{s}_2[k] + \mathbf{n}[k], \tag{4}$$

where $\mathbf{H}_d^{G2}$ and $\mathbf{H}_I^{G2}$ are MIMO channel matrices of dimension $N \times 2$ and $\mathbf{s}_1[k]$ and $\mathbf{s}_2[k]$ are multiplexing sub-sequences.

Due to the presence of MAI at the receiver, a MIMO-MMSE spatial filter of the first (second) multiplexing layer is optimized to cancel interference from the second (first) multiplexing layer. Following the spatial filters, modified STBC decoders extract the diversity gains from each spatially multiplexed sub-stream and perform signal detection. Then, the two detected sub-sequences $\hat{\mathbf{s}}_1[k]$ and $\hat{\mathbf{s}}_2[k]$ are re-ordered and converted to the serial unique stream that constitutes the estimation of the overall transmitted data. When Successive Interference Cancellation (SIC) detection is used and assuming perfect interference subtraction, a diversity order of $2(N-2)$ is achieved at the first detection layer while the second one achieves a diversity order of $(N-1)$. In the G2+G2 scheme, neither channel coding nor SOVA are used, since the two spatial multiplexing layers are space-time coded. This scheme is considered in this work as a reference system for comparison with the next ones.

4.2   G3+1 Scheme

In the G3+1 hybrid scheme, two spatial multiplexing layers are present, see Fig. 2. A ST coder G3 is used at the first layer, while the other layer are non-space-time-coded, see [5]. A convolutional encoder is thus employed at the non-space-time-coded layer with the objective of provide some reliability for the data transmitted in this layer. In the traditional G3+1 hybrid scheme, the data of the non-space-time coded layer is transmitted with no protection. The transmitted signals can be organized in a equivalent space-time coding matrix as described below

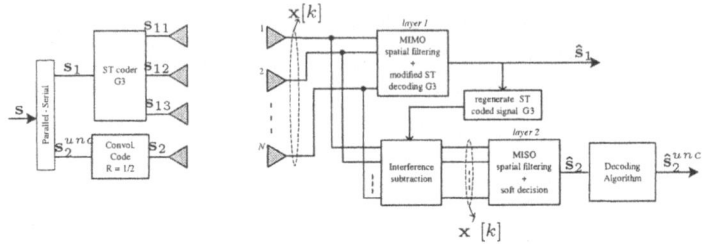

Fig. 2. G3+1(SIC) transmitter-receiver structure

$$\Omega_{G3+1} = \begin{bmatrix} s_1[k] & s_1[k+1] & s_1[k+2] & s_2[k] \\ -s_1[k+1] & s_1[k] & -s_1[k+3] & s_2[k+1] \\ -s_1[k+2] & s_1[k+3] & s_1[k] & s_2[k+2] \\ -s_1[k+3] & -s_1[k+2] & s_1[k+1] & s_2[k+3] \\ s_1[k]^* & s_1[k+1]^* & s_1[k+2]^* & s_2[k+4] \\ -s_1[k+1]^* & s_1[k]^* & -s_1[k+3]^* & s_2[k+5] \\ -s_1[k+2]^* & s_1[k+3]^* & s_1[k]^* & s_2[k+6] \\ -s_1[k+3]^* & -s_1[k+2]^* & s_1[k+1]^* & s_2[k+7] \end{bmatrix}. \tag{5}$$

In (5) the sub-sequence $\mathbf{s}_2$ represents the output of the rate-1/2 convolutional code, which means that $\mathbf{s}_2$ already is coded, thus redundancy is now present.

For this scheme, we can expand (1) to represent the received signal for the scheme G3+1 as a MIMO desired signal and a SIMO interferer signal followed by a sample noise vector

$$\mathbf{x}_{G3+1}[k] = \mathbf{H}_d^{G3}\mathbf{s}_1[k] + \mathbf{h}_I\mathbf{s}_2[k] + \mathbf{n}[k], \tag{6}$$

where $\mathbf{H}_d^{G3}$ and $\mathbf{h}_I$ are MIMO and SIMO channel matrices of dimension $N \times 3$ and $N \times 1$, respectively. In this case $\mathbf{s}_1[k]$ and $\mathbf{s}_2[k]$ are multiplexing sub-sequences.

At the receiver, since MIMO and SIMO multiplexed signals should be detected, we make use of the MIMO and MISO spatial filters at the first and second detection layers, respectively, for MAI mitigation. The MISO spatial filter provides soft-output decision to make possible the SOVA to treat the received data. Here, the diversity gain is $3(N-1)$ for the first layer and N for the second one.

4.3   G2+1+1 Scheme

The G2+1+1 hybrid scheme with channel coding is drawn in Fig. 3. Again, four transmit antennas are employed. As can be seen from the figure, this scheme consists of three spatial-multiplexing layers, where the first layer is space-time coded using ST coder G2. For the other two layers a rate-1/2 convolutional code is employed at each one. The equivalent space-time coding matrix is described as follows

$$\Omega_{G2+1+1} = \begin{bmatrix} s_1[k] & s_1[k+1] & s_2[k] & s_3[k] \\ -s_1[k+1]^* & s_1[k]^* & s_2[k+1] & s_3[k+1] \end{bmatrix}. \tag{7}$$

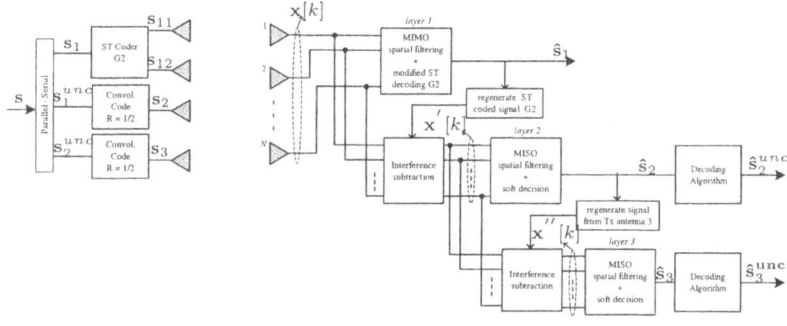

Fig. 3. G2+1+1(SIC) transmitter-receiver structure

For this scheme, the received signal $\mathbf{x}[k]$ can be expressed as the sum of a MIMO desired signal and two SIMO interference signals

$$\mathbf{x}_{G2+1+1}[k] = \mathbf{H}_d^{G2}\mathbf{s}_1[k] + \mathbf{h}_{I1}\mathbf{s}_2[k] + \mathbf{h}_{I2}\mathbf{s}_3[k] + \mathbf{n}[k], \qquad (8)$$

where $\mathbf{h}_{I1}$ and $\mathbf{h}_{I2}$ are $N \times 1$ interferer channel vectors.

For the CHS-SIC receiver architecture to cope with this hybrid scheme, a MIMO-MMSE spatial filter is employed at the first detection layer, where a space-time coded signal $\mathbf{s}_1[k]$ should be detected. At the second and third layers standard array processing is employed for MAI cancellation and detection of $\mathbf{s}_2[k]$ and $\mathbf{s}_3[k]$, respectively.

Compared to the standard ST coder G2, the traditional (non-channel coded) G2+1+1 hybrid scheme aims at achieving three times its data rate. With the use of the rate-1/2 convolutional code, the data rate of this hybrid scheme is reduced by a factor of two due to redundancy that now is added by the channel coder. Here, the diversity order of the first layer is $2(N-2)$, while the second and third layers have a diversity order of $2(N-1)$. Comparing the coded G2+1+1 scheme with the traditional G2+G2 scheme, the first offers the same spectral efficiency as the latter.

5   Interference Cancellation Algorithm

Following the signal-plus-interference models described in the previous section for the schemes, G2+G2, G3+2 and G2+1+1, we can state that signal processing algorithm at the receiver to detect a MIMO desired signal should perform the following tasks:

1. estimate the overall MIMO channel matrix $\mathbf{H}$;
2. cancel multiple access interference from channel estimation;
3. perform space-time decoding after interference cancellation.

In this section an interference cancellation algorithm is proposed to accomplish the detection of MIMO (space-time coded) signals in the presence of multiple access (co-channel) interference. According to the developed signal-plus-interference models of (4), (6) and (8), this algorithm will be applied at the two detection layers of the G2+G2 scheme as well as at the first detection layer of the G2+1+1 and G3+1 schemes, i.e., where the signal(s) of interest is(are) space-time coded.

The proposed algorithm optimizes the coefficients of a Minimum Mean Square Error (MMSE) spatial filter in such a way that the orthogonality of the space-time code is preserved as much as possible in its output signal. Stated in a different way, the spatial filter combines the received signal such that interference signals add destructively at its output signal while the MIMO structure of the desired signal are preserved as much as possible to add constructively during the subsequent processing stage, consisting of space-time decoding. For this reason, we call this interference cancellation algorithm as a two-stage algorithm, see [7].

At any time-instant k, the output signal vector of the $N \times N$ MIMO-MMSE spatial filter is given by

$$y[k] = \mathbf{W}x[k], \tag{9}$$

where

$$\mathbf{W} = \begin{bmatrix} w_{11} & w_{12} & \cdots & w_{1N} \\ w_{21} & w_{22} & \cdots & w_{2N} \\ \vdots & \vdots & \ddots & \vdots \\ w_{N1} & w_{N2} & \cdots & w_{NN} \end{bmatrix}. \tag{10}$$

We obtain the error vector at the output of the MIMO-MMSE spatial filter as

$$e[k] = \mathbf{W}x[k] - \mathbf{H}_d s_1[k] = \mathbf{W}x[k] - x_d[k], \tag{11}$$

where $x_d[k] = \mathbf{H}_d s_1[k]$ is the desired space-time coded signal associated to the first multiplexing layer of a particular hybrid transmission scheme. Contrarily to the classical MIMO-MMSE spatial filter (where the desired signal is $s_1[k]$), here the desired signal consists of the original transmitted signal modified by desired MIMO channel impulse response $\mathbf{H}_d$.

The MMSE cost function be formalized as follows

$$J_{MMSE} = E\{\|\mathbf{W}x[k] - x_d[k]\|^2\}. \tag{12}$$

Solving this unconstraint optimization problem, the obtained solution with respect to $\mathbf{W}$ is given by

$$\mathbf{W} = \mathbf{R}_{x_dx} \cdot (\mathbf{R}_{xx})^{-1}, \tag{13}$$

where $\mathbf{R}_{xx} = E\{x[k]x^H[k]\}$ and $\mathbf{R}_{x_dx} = E\{x_d[k]x^H[k]\}$ are the input covariance matrix and a cross-correlation matrix, respectively. The superscript H denotes conjugate transpose.

The coefficients of the MIMO-MMSE spatial filter can be computed after direct Least Square (LS) estimate of the MIMO channel impulse response, more details about this section could be found in [7].

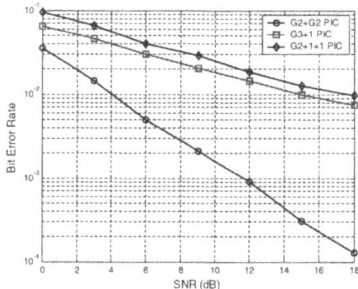 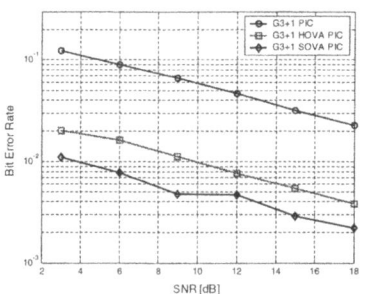

Fig. 4. Performance of uncoded hybrid schemes with parallel interference cancellation (PIC) and N=4.

Fig. 5. Influence of the SOVA on the performance of the second layer for G3+1 PIC and N=4.

6   Simulation Results

As channel code we consider the rate-1/2 memory-2 Recursive Systematic Convolutional (RSC) code that is defined by the generator (1,5/7) in octal form [8]. In the SOVA procedure, the trellis computations are done in two directions: a forward and a backward one. The first part of the algorithm, trellis is run in the forward direction and SOVA behaves like a traditional Viterbi Algorithm (VA). In the second part, the trellis is visited in the backward direction. In this part, the metric for each state are stored by the algorithm and soft-output information of bits is computed as a Likelihood Ratio (LLR) in the form described below

$$\Lambda(u_i) \triangleq \log\left(\frac{P\{u_i = 1|\mathbf{r}\}}{P\{u_i = 0|\mathbf{r}\}}\right). \tag{14}$$

where u_i is the transmitted codeword and $\mathbf{r}$ is the received sequence.

The BER performance of CHS receivers are evaluated here by means of numerical results from Monte-Carlo simulations with Parallel and Successive Interference Cancellation (PIC) and (SIC), respectively. Perfect channel estimation is assumed. The transmitted symbols are modulated with Binary-Phase Shift-Keying (BPSK). The BER curves are plotted according to the average SNR per receive antenna. Whenever notation $M \times N$ is used it refers to a MIMO scheme with M transmit and N receive antennas.

Figure 4 first shows the BER results for traditional G2+G2, G3+1 and G2+1+1 hybrid schemes without channel coding and SOVA detection. The PIC are employed. We consider $N = 4$ receive antennas for all schemes. In terms of BER performance, the G2+G2 scheme achieves the best results. This is expected since all spatial multiplexing layers are space-time coded, resulting in a higher diversity gain at the receiver.

Figure 5 shows the results for the coded hybrid scheme G3+1 with PIC detection and $N = 4$ receive antennas. As the decoding algorithm we consider two approaches: Hard Output Viterbi Algorithm (HOVA) and SOVA. We can

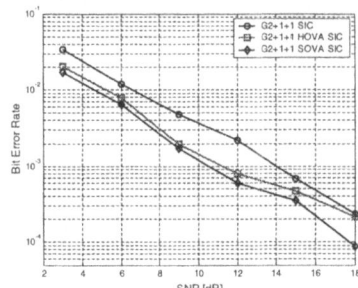

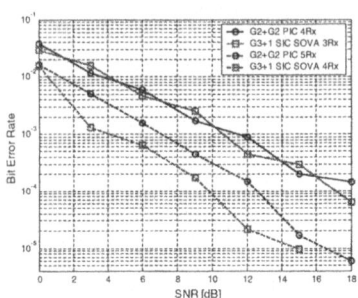

Fig. 6. Influence of the SOVA on the performance of the second layer for G2+1+1 SIC and N=4.

Fig. 7. Comparison of the influence of the number of receive antennas and the decoding algorithms for the hybrid MIMO receivers.

see that the use of a channel code in the non-space-time coded layer of G3+1 results in a huge increase in performance when compared with the scheme with no channel coding. On the other hand, in the channel coded case redundancy is introduced and therefore the effective spectral efficiency decrease from 1.5 symbols/s/Hz to 1.0 symbol/s/Hz. Its worthnoting that SOVA offers an increase in performance of about 3dB over the HOVA as expected and already demonstrated in the theory of channel coding.

In Fig. 6, the effect of the channel coding is now evaluated for the G2+1+1 scheme. Here, we employ the SIC technique at the receiver with the assumption of perfect interference subtraction. Again, $N = 4$ receive antennas are used. We observe that the gain achieved by the use of channel code and SOVA is not as evident as the gain perceived for the PIC case. When the SIC technique is used at the receiver, the spatial diversity gain due to interference subtraction plays a more important role in BER performance than the time diversity gain provided by the channel code and the use of SOVA. This observation explains the smaller improvement in performance as compared to the PIC case.

In Figure 7, we compare the receiver performance of two hybrid schemes. The first is the PIC G2+G2 with no channel coding and no SOVA. The second is the proposed coded SIC G3+1 scheme with SOVA detection. The results are shown for different number of receive antennas. It can be seen that similar performance in terms of BER achieved with the traditional PIC G2+G2 scheme and $N = 4$ is also achieved with the coded SIC G3+1 with SOVA and $N = 3$. On the other hand, the coded SIC G3+1 with SOVA and $N = 4$ outperforms the traditional PIC G2+G2 scheme and $N = 5$. These results indicate that the joint use of channel coding and SOVA in hybrid schemes could cope with a receiver with less receive antennas.

It is important to note that the performance gains in terms of BER provided by the use of channel coding at the transmitter and SOVA at the receiver come at the cost of some complexity increase while the same spectral efficiency is maintained comparing the schemes G2+G2 and G3+1. On the other hand, they can represent an attractive solution in scenarios with spatially correlated channels, where the lack of spatial diversity could be compensated by an increase in the time diversity offered by the channel coding.

7 Conclusion and Perspectives

In this work we have proposed some hybrid MIMO transmission schemes that arise as a solution for the inherent diversity-multiplexing trade-off of MIMO channels. Receiver structures based on a two-stage interference cancellation algorithm the SIC detection strategy were presented. Then, we have evaluated the performance of two hybrid MIMO receivers with to the joint the use of channel coding and soft-output detection in layers that are not space-time coded. Simulation results have shown that channel code and SOVA improve performance of G3+1 and G2+1+1 schemes, mainly when no SIC is used.

The perspectives of this work include the implementation of other MIMO hybrid schemes (considering space-frequency coding and OFDM transmissions) as well as the adaptation of the interference cancellation architectures to the multiuser MIMO context.

References

1. G. J. Foschini and M. J. Gans, "On Limits of Wireless Communications in a Fading Environment When Using Multiple Antennas", *Wireless Personal Communications*, v. 6, n. 3, Mar 1998, pp. 311-335.
2. G. J. Foschini, "Layered Space-Time Architecture for Wireless Communications in a Fading Environment when using Multiple Antennas", *Bell Labs Tech. J.*, v.1, n.2, 1996, pp.41-59.
3. G. D. Golden, G.J. Foschini, R.A. Valenzuela, P.W. Wolniansky, "Detection Algorithm and Initial Laboratory Results using the V-BLAST Space-Time Communications Architecture", *Electronics Letters*, v.35, n.7, Jan, 1999, pp. 14-15.
4. S. Alamouti, "A simple transmit diversity technique for wireless communications," *IEEE Journal of Selected Areas in Communications*, vol. 16, pp. 1451–1458, Oct 1998.
5. V. Tarokh, H. Jafarkhani, and A. R. Calderbank, "Space-time block codes from orthogonal designs," *IEEE Transactions on Information Theory*, vol. 5, pp. 1456–1467, Jul 1999.
6. A. L. F. de Almeida, W. C. Freitas Jr., F. R. P. Cavalcanti, J. C. M. Mota, and R. L. de Lacerda Neto, "Performance of a MIMO Systems with a Hybrid of Transmit Diversity and Spatial Multiplexing," *XX Simpósio Brasileiro de Telecomunicações (SBrT)*, 2003, Rio de Janeiro-RJ, Brasil.
7. A. L. F. de Almeida, W. C. Freitas Jr., F. R. P. Cavalcanti, J. C. M. Mota, and R. L. de Lacerda Neto, " A Two-Stage Receiver for Co-Channel Interference Cancellation in Space-Time Block-Coded Systems over Frequency Selective Channels," *XX Simpósio Brasileiro de Telecomunicações (SBrT)*, 2003, Rio de Janeiro-RJ, Brasil.
8. Robert H. Morelos- Zaragosa, " Ther Art of Error Correcting Coding," *John Wiley and Sons, Inc.*, New York, NY, USA.

Simple Model to Determine the Gap Effect on the Input Impedance of a Cylindrical Dipole. Application to Standard Antennas

Alireza Kazemipour[1], Xavier Begaud[2], and Djamel Allal[1]

[1] Bureau National de Métrologie, Laboratoire National d'Essais BNM-LNE,
29 Ave. Roger Hennequin, 78197 Trappes cedex, France
alireza.kazemipour@lne.fr
[2] Department COMELEC, TELECOM Paris,
46 rue Barrault, 75634 Paris cedex 13, France
xavier.begaud@enst.fr

Abstract. The effect of the Gap width on the impedance of a center-fed dipole is studied. We propose here an equivalent capacitance to model the discontinuity of the Gap region, in a closed-form formula. This model explains successfully the dependence of the input impedance to the gap characteristics as well as the current density along the dipole. The model has been used for a calculable standard dipole and the results show a good agreement with the experimental tests.

1 Introduction

The EM emissions are measured by the standard antennas. The AF is the essential parameter of a standard antenna and is defined as the ratio of the electric field strength E of a plane wave incident to the antenna to the detected voltage V_r at its output:

$$AF \ (dB/m) = E \ / \ V_r$$

Therefore the AF should be correctly determined for the precise EMC measurements. The AF could be determined by experimental methods in a Standard Site [1] or by the theoretical treatments. If the AF is evaluated directly by analytical or numerical methods, the antenna is a "calculable" one. The AF depends on all the characteristics of the antenna especially its input impedance.

The commercial softwares for the antenna simulation do not generally take into account the gap effect on the dipole input impedance. Nevertheless, for each real case we need a center-placed gap on the dipole to feed this instrument.

The problem of a dipole antenna excited by a finite gap could be treated by supposing the integral equations for a non-continuous symmetrical body [2], [3]. These solutions are not simple and generally proposed for the infinite dipoles.

Here the gap discontinuity is simulated by a simple equivalent capacitance C_{eq} which is parallel to the dipole input impedance Z_{ant} (Fig. 1).

J.N. de Souza et al. (Eds.): ICT 2004, LNCS 3124, pp. 38–42, 2004.

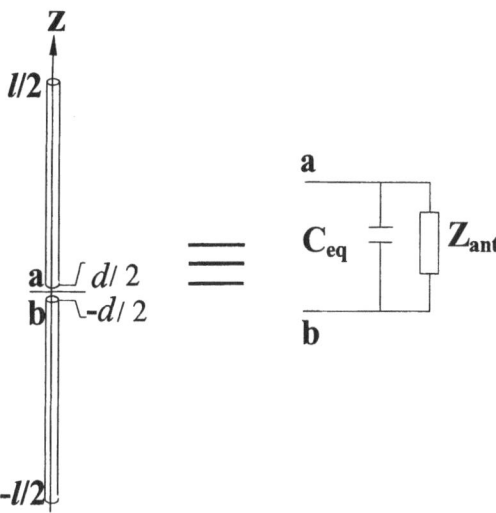

Fig. 1. Real dipole and its equivalent model

2 Theory

A general schema of a center-fed cylindrical dipole antenna is presented in Fig. 1. The total length is l with a homogenous radius a. The hollow gap is situated symmetrically between two metallic halves of the dipole and its width is d. This cylindrical hollow volume could be filled with air or a dielectric material labeled by ε_r.

The equivalent capacitance C_{eq} of the gap discontinuity could be deduced by evaluating the induced charge on this region i.e. the open end of the halves. The charge distribution along the dipole could be evaluated by the "charge continuity principle" [4]:

$$dI(z)/dz + j\omega Q(z) = 0 \qquad (1)$$

By considering the symmetrical charge distribution in figure 1:

$$Q(a+d/2) = -Q(-a-d/2) = (j/\omega)\ dI(z)/dz \ \text{ for } \ z = a + d/2 \qquad (2)$$

where $I(z)$, $Q(z)$ are the linear current density (A/m) and charge density (Q/m) along the dipole and : $\omega = 2\pi f$. The equivalent capacitance C_{eq} can be directly deduced by using the definition of the capacitance :

$$C = |Q| / |\delta V| \qquad (3)$$

$$\Rightarrow C_{eq} = |aQ(a+d/2)| / |V(d/2) - V(-d/2)| \qquad (4)$$

The term $|aQ(a+d/2)|$ represents the induced charge on each side of the gap where the factor a (radius) is the equivalent length of the open-end halve in the gap region

[4]. The term $|V(d/2) - V(-d/2)|$ is the potential difference between two halves of the dipole and could be explained by the input impedance of the dipole Z_{ant} and the current on the dipole center:

$$|V(d/2) - V(-d/2)| \approx |Zan\ I(z = 0)| \tag{5}$$

Let apply the relations (2), (4) and (5):

$$C_{eq} = \varepsilon_r\ [a\ /\ \omega|Z_{ant}\ I(0)|\ ]\ [dI(z)/dz\ ;\ z = a + d/2\ ] \tag{6}$$

The relation (6) shows the dependence of the equivalent capacitance to the radius of the dipole and the current density along the dipole as well. Here, ε_r is for indicating the effective wavelength in the dielectric.

Now, it is interesting to apply the formula (6) for a half wavelength dipole, because it is highly used as a standard antenna. In this case the longitudinal current has a quasi sinusoidal distribution:

$$I(z) = I_0\ Cos(2\pi z/\lambda) \tag{7}$$

(2), (7) $\Rightarrow$

$$|Q(d/2)| = |(1/\omega)\ dI(z)/dz\ ;\ z = a + d/2| = I_0(2\pi/\lambda\omega)\ Sin[\pi(d + 2a)/\lambda] \tag{8}$$

(6), (8) $\Rightarrow$

$$C_{eq} = \varepsilon_r\ [a\ /\ |Z_{ant}\ I_0|\ ]\ [\ I_0(2\pi/\lambda\omega)\ Sin(\pi(d + 2a)\ /\ \lambda)\ ]= \tag{9}$$

$$\varepsilon_r\ [2a\pi\ Sin(\pi(d + 2a)\ /\lambda)]\ /\ (\lambda\omega\ |Z_{ant}|\ )$$

The equivalent impedance related to capacitance C_{eq} is therefore :

$$|Z_{eq}| = 1\ /\ |C_{eq}|\omega = \lambda\ |Z_{ant}|\ /\ [\ 2\varepsilon r\ a\pi\ Sin(\pi(d + 2a)\ /\lambda)\ ] \tag{10}$$

Referring to figure 1, Z_{eq} should be used in parallel to the theoretical input impedance of dipole Z_{ant}.

For a typical example, let calculate the equivalent gap parasite impedance for a half wavelength dipole at 2 GHz: f = 2GHz; $a = d = 2.5mm$ and $\varepsilon_r = 1$;

(10) $\Rightarrow Z_{eq} = 60\ |Z_{ant}|\ \Rightarrow Z_{in} = Z_{eq}\ ||\ Z_{ant} \approx 0.98\ Z_{ant}$

This simple example shows that the perturbation of the gap on the input impedance is negligible if the gap dimensions are limited. However, by increasing the gap width and radius, C_{eq} increases directly. The gap effect is more important if the gap volume is filled with a high ε_r dielectric.

3   Experimental Results

A real system (dipole + balun) has been made to test the obtained formula (10). The Balance to Unbalance transformer (balun) is here a wide-band printed one [5], [6]. This balun permits to install directly two symmetrical halves of the dipole on its output lecher lines and to study the gap effects for a very wide-band 300-2100 MHz

(Fig. 2). This balun is printed on a dielectric substrate with ε_r equal to 5 and a thickness about 2mm. By this high dielectric permitivity, the gap effects are important.

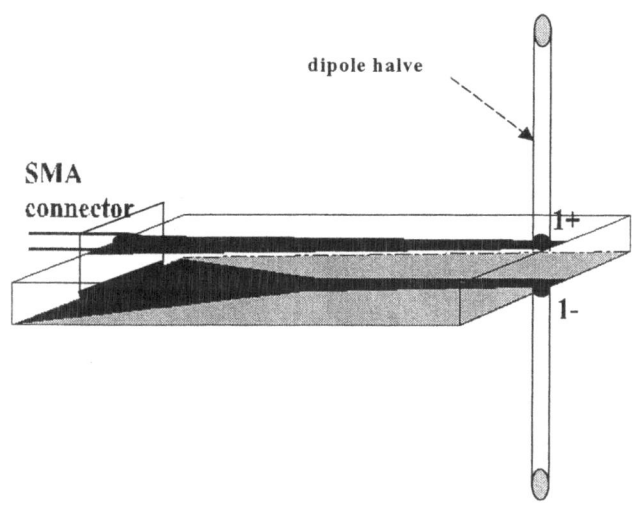

Fig. 2. Experimental set-up to study the gap effects

The input impedance of this experimental set-up is measured on SMA connector (Fig. 2). This impedance depends on the dipole impedance as well as the S parameters of the balun. The S parameters of the balun are measured [6] and by knowing them, the input impedance of the real dipole (containing the gap effect) is deduced.

The input impedance of the real dipole is presented in table 1. The indirect measured impedance of dipole is compared with the analytical results of the formula (10). Furthermore, the simulated input impedance by NEC software is presented, where the gap effect is traditionally ignored. So, the table 1 shows first of all the gap effect evaluated by the simple formula (10) comparing to an ideal case (without gap)

Table 1. Input impedance (module and phase) of the half wavelength thick dipoles; $a = 0.010$ $\lambda - 0.014\lambda$, and the gap width $d = 3$mm

f	Z_{in} (10)	Z_{in} (meas.)	Z_{ant} (NEC)
MHz	Ω < deg.	Ω < deg.	Ω < deg.
700	117 < 19°	119 < 15°	117 < 20°
900	123 < 10°	120 < 6°	117 < 20°
1100	130 < 8°	134 < 5°	125 < 18°
1400	128 < 0°	140 < -6°	120 < 15°
2100	130 < -5°	125 < -18°	120 < 15°

and secondly to the experimental results. This comparison shows the performance of the proposed model.

As indicated in table 1, the gap effects are very important for the high frequency where the gap dimensions are relatively comparable to the wavelength. For an important gap appearance (f = 2.1 GHz), the input impedance of a real half wavelength dipole could be completely changed and be capacitive !

4 Perspectives for Standard Calculable Antennas

This simple model could be used for the analytical treatment of a calculable standard dipole. The traditional dipole analysis without gap effect can impose some error on the AF up to 0.5 dB/m for a real thick dipole. The numerical simulation can cause the same error too, if the gap discontinuity is not considered. However, a numerical simulation based on FDTD can also take into account the gap effect and it could be a good challenge to compare its results with this simple analytic formula.

Anyway, the presented compact formulation can present the gap effect as un estimable capacitance which depends clearly on the gap dimensions and the current distribution along the dipole.

References

1. Salter, M. J., Alexander, M. J.: EMC Antenna Calibration and the Design of an Open-Field Site. MEAS. Sci. Technol. 2 (1991) 510 – 519
2. Do-nhat, T.: A model for feed gap field of a dipole. IEEE AP-35, Nov. (1987) 1237 – 1280
3. Do-nhat, T.: On the effect of gap width on the admittance. IEEE AP-37, Dec. (1989) 1545 – 1553
4. King, R. W.: The theory of linear antennas. Harvard University Press. Cambridge, Massachusetts (1956)
5. Begaud, X.: Analyse d'antennes et de réseaux d'antennes large bande et bipolarisation par une méthode d'éléments finis de surface. PhD Thesis, Université de Rennes – France, Dec. (1996)
6. Kazemipour, A., Begaud, X.: Numerical and analytical model of standard antennas from 30MHz to 2GHz. 2001 IEEE EMC Symposium, Aug. (2001), Montreal – Canada

An Interference Avoidance Technique for Ad Hoc Networks Employing Array Antennas

Thomas Hunziker, Jacir L. Bordim, Takashi Ohira, and Shinsuke Tanaka

ATR Adaptive Communications Research Laboratories,
2-2-2 Hikaridai, Keihanna Science City, Kyoto, 619-0288 Japan

Abstract. Array antennas have the potential to increase the capacity of wireless networks, but a distributed beamforming algorithm for maximizing the capacity in asynchronous, decentralized mobile ad hoc networks is yet to be found. In this paper we pursue an interference avoidance policy, based upon channel reciprocity in time-division duplexing, and arrive at a practical signaling technique for array antenna enhanced ad hoc networks in multi-path environments with fading. The beamforming scheme may be incorporated in an RTS-CTS-data-ACK based medium access control protocol without necessitating any exchange of channel state information. In computer simulations we find that the interference avoidance effectively bounds the packet transmission disruption probability, and that the resulting single hop transport capacity increases almost linearly with the number of antennas under certain conditions.

1 Introduction

Wireless ad hoc networks have been receiving increasing attention in the recent years, both from the academia and the private sector. Since ad hoc networks do not rely on a fixed network infrastructure or centralized administration, the cost in deploying such networks is greatly reduced. For this reason, ad hoc networks can be easily set up and torn down, which makes them suitable to support communications in urgent and temporary tasks, such as business meetings, disaster-and-relief, search-and-rescue, law enforcement, among other special applications.

The IEEE 802.11 standard for wireless local area networks includes a medium access control (MAC) protocol for ad hoc networking, letting a community of terminals in a neighborhood exchange information over a common channel. Adhering to a CSMA/CA (carrier sensing multiple access / collision avoidance) policy, a terminal listens for signals from other transmitters, and refrains from sending a signal if any interference is detected. If deployed over a large area, multiple terminals can communicate simultaneously as long as they are sufficiently apart and the resulting cochannel interference (CCI) within certain limits.

Directional antennas offer CCI mitigation capabilities and transmission range extension, thereby enhancing the spatial channel reuse. A number of MAC protocols which aim to exploit the benefits of directional antennas have been proposed for ad hoc networks, e.g. [1] [2] [3] [4] [5] [6]. In these works, simple antenna models with switchable beam directions and line-of-sight (LOS) transmission scenarios are assumed. Wireless networks in indoor or urban outdoor environments,

J.N. de Souza et al. (Eds.): ICT 2004, LNCS 3124, pp. 43–52, 2004.

however, encounter multi-path propagation, and direct LOS paths between the antennas of two terminals exist only exceptionally. More often, multiple replicas of a desired signal arrive at the receiver antenna via reflections from different directions.

Array antennas along with proper digital array signal processing techniques can effectively suppress CCI in *non*-LOS scenarios [7]. Employing array antennas, the signals become multi-dimensional from the digital transceiver point of view, and the channels between the terminals become multiple-input multiple-output (MIMO) channels. Besides of facilitating CCI suppression, this multi-dimensionality also opens new possibilities for the MAC. Sending out a signal needs not necessarily be prohibited in the presence of a nearby transmitter. Rather, with an appropriate choice of the vector signals, two or more pairs of terminals can simultaneously communicate in a small area.

The problem of jointly optimizing the beamforming at multiple transceivers is addressed in [8] for the case of a stationary, synchronous networks with multiple base stations and mobile terminals. In this paper, we focus on decentralized peer-to-peer networks with uncoordinated channel access. Furthermore, a MIMO channel between any two terminals is assumed time-invariant only for the short duration of a data packet transmission. We extend the principle of CSMA/CA to multi-dimensional signals, arriving at a MAC scheme for networks in which every terminal is equipped with an array antenna. The beamforming, including power control, is accomplished on the basis of the observed signals during the idle periods, while the frame exchange pattern for a data packed transmission accords with the IEEE 802.11 MAC protocol.

The rest of the paper is organized as follows. In Sect. 2, we describe the transmitter, receiver, and MIMO channel models. The actual beamforming schemes are defined in Sect. 3. A respective MAC protocol, based on RTS (request to send), CTS (clear to send), data and ACK (acknowledge) frames, is described in Sect. 4. The achieved gains of the enhanced MAC in terms of the transport capacity are analyzed by means of computer simulations in Sect. 5. Finally, conclusions are drawn in Sect. 6.

In the following we use boldface characters for vectors and matrices, and let a T, a H, and an asterisk in the superscript denote transposition, Hermitian transposition, and complex conjugation (element-wise in the case of vectors), respectively. Furthermore, $\|\mathbf{x}\| = \sqrt{\mathbf{x}^H \mathbf{x}}$ represents the vector norm.

2   Signal Model

We consider an ad hoc network in which all users share a common frequency band. The transmission of a data packet from one terminal to another involves the exchange of a number of frames in time-division duplexing (TDD). The total duration of a data packet transmission is T_P seconds, and multiple terminal pairs may communicate simultaneously as long as they are sufficiently apart.

The baseband equivalent vector signal transmitted from a terminal A has the form $\mathbf{s}_A y(t)$, where $y(t)$ is the normalized, modulated signal, and the $N \times 1$-vector

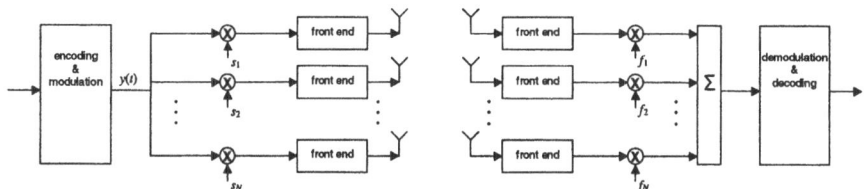

Fig. 1. Left: transmitter including a pre-filtering by $\mathbf{s} = (s_1, \ldots, s_N)^{\mathrm{T}}$; right: receiver including a post-filtering by $\mathbf{f} = (f_1, \ldots, f_N)^{\mathrm{T}}$.

$\mathbf{s}_A$ represents the linear *pre-filter*, which is also sometimes termed *beamformer*. The pre-filter assigns the signal stream for each antenna a certain complex amplitude, as shown on the left hand side in Fig. 1, where the total vector signal power $\varepsilon_A = \|\mathbf{s}_A\|^2$ is limited by $\varepsilon_A \le \varepsilon_{\max}$. Finally, the signals are digital/analog converted in the front ends and transferred to the radio frequency.

The reverse operations are performed in the receive mode, as sketched on the right hand side in Fig. 1. At the output of the front ends in the terminal B, the vector signal from the terminal A is observed as $\mathbf{r}_{B \leftarrow A} y(t)$. We restrict our attention to narrow-band transmission scenarios where $\mathbf{r}_{B \leftarrow A}$ can be written as

$$\mathbf{r}_{B \leftarrow A} = \mathbf{H}_{A \rightarrow B} \mathbf{s}_A, \tag{1}$$

with the $N \times N$-marix $\mathbf{H}_{A \rightarrow B}$ representing the MIMO channel between the digital transmitter in the terminal A and the digital receiver in the terminal B. The signals from the N antennas are in the following linearly combined by the *post-filter* $\mathbf{f}_B$, with $\|\mathbf{f}_B\| = 1$, and the signal $\mathbf{f}_B^{\mathrm{T}} \mathbf{r}_{B \leftarrow A} y(t)$ is finally demodulated and decoded.

In the rest of the paper we represent every transmitted or received vector signal simply by its spatial signature, i.e., as $\mathbf{s}_A$ and $\mathbf{r}_{B \leftarrow A}$, respectively. Furthermore, all terminals in the network are assumed similar, with identical maximal transmit power and noise figure.

Given the Euclidean distance $d_{A \rightarrow B}$ between the two terminals A and B, the elements in the MIMO channel matrix $\mathbf{H}_{A \rightarrow B}$ are modeled as independent and complex normally distributed with zero mean and variance $C d_{A \rightarrow B}^{-\kappa}$, where C and κ are positive constants. This stochastic model for the MIMO channel accounts for multi-path propagation, resulting in Rayleigh fading, and for the decay of the mean signal strength with the distance by a certain path loss exponent κ. Independent fading of the $N \cdot N$ channels between the transmitter and the receiver antennas is a reasonable assumption in environments with many scatterers and an antenna spacing in the order of at least half of the wavelength.

We assume that the coherence time of the radio channels by far exceeds T_{P}, making channel variations during a packet transmission negligible. A MIMO channel is modeled as time-invariant, but only applies for one data packet transmission. Furthermore, in a TDD scenario, the channels may be assumed reciprocal, i.e. [9]

$$\mathbf{H}_{A \rightarrow B} = \mathbf{H}_{B \rightarrow A}^{\mathrm{T}}. \tag{2}$$

3   Beamforming

Using an array antenna, causing excess CCI at a nearby terminal can be avoided by a favorable pre-filtering in the transmit mode. Moreover, CCI can be mitigated during a frame reception by an appropriate linear combining of the signals from the N antennas. Both are usually depicted as beamforming, though in multi-path scenarios beamforming is a somewhat more subtle procedure than just steering the antenna beam towards the direction of the terminal of interest.

Beamforming during signal transmission and reception are dual problems in many networking scenarios involving reciprocal channels. For instance, the same beam pattern that minimizes the transmitted signal power towards a certain neighbor also limits the interference originating from that terminal after the two terminals swap their roles as interferer and receiver. In the below devised data packet transmission scheme, we let every terminal use the pre-filter for the transmit mode also as the post-filter for the receive mode, except of a scalar factor, i.e.

$$\mathbf{s}_A = \sqrt{\varepsilon_A}\mathbf{f}_A \tag{3}$$

in the case of terminal A. Furthermore, a terminal chooses a filter when sending (or receiving) the first frame and keeps it fixed throughout the packet transmission. As we will see in the following, these rules enable a terminal to control the amount of interference it causes at active neighboring terminals.

Receiving a frame from terminal A at the time t, for instance, the signal-to-interference-plus-noise ratio (SINR) after the signal combining in the terminal B may be expressed as

$$\gamma_B(t) = \frac{|\mathbf{f}_B^T\mathbf{r}_{B\leftarrow A}|^2}{\sum_{x\in\mathcal{T}_t\backslash\{A\}}|\mathbf{f}_B^T\mathbf{r}_{B\leftarrow x}|^2 + N_0} = \frac{|\mathbf{f}_B^T\mathbf{r}_{B\leftarrow A}|^2/N_0}{\sum_{x\in\mathcal{T}_t\backslash\{A\}}|\mathbf{f}_B^T\mathbf{r}_{B\leftarrow x}|^2/N_0 + 1}. \tag{4}$$

The numerators in (4) represent the gain of the signal of interest, the sum expressions in the denominators represent the CCI, with $\mathcal{T}_t$ denoting the set of all active transmitters in the network at the time t, and N_0 defines the power of the additive thermal noise.

Two terminals successfully establish a packet transmission if the SINR's during the initial exchange of the RTS and CTS frames meet some criteria. Even if this is the case, however, success of the subsequent data frame transmission is not guaranteed, since in this phase the arrival of new interferers may lead to a sudden deterioration of the SINR. Note that a CCI term in the expression on the right hand side in (4) may be transformed as

$$\frac{|\mathbf{f}_B^T\mathbf{r}_{B\leftarrow x}|^2}{N_0} = \frac{|\mathbf{s}_B^T\mathbf{H}_{x\to B}\mathbf{s}_x|^2}{\varepsilon_B N_0} = \frac{|(\mathbf{H}_{B\to x}\mathbf{s}_B)^T\mathbf{s}_x|^2}{\varepsilon_B N_0} = \frac{\varepsilon_{\max}}{\varepsilon_B}\left|\left(\frac{\mathbf{r}_{x\leftarrow B}}{\sqrt{N_0}}\right)^T\left(\frac{\mathbf{s}_x}{\sqrt{\varepsilon_{\max}}}\right)\right|^2, \tag{5}$$

using first $\mathbf{s}_B = \sqrt{\varepsilon_B}\mathbf{f}_B$ and second the MIMO channel reciprocity (2). That is, the CCI from the terminal x is a function of ε_B and the pre-filter chosen locally by the terminal x. Our idea is to bound the expression following the factor

$\varepsilon_{\max}/\varepsilon_B$ on the right hand side in (5) by imposing constraints on the pre-filter selection. This limits the impact of a single interference source, reducing the risk that an ongoing packet transmission is disrupted.

Supposing the terminals A and B aim to establish a packet transmission at the time t_0, we constraint the choice of the pre-filters $\mathbf{s}_A$ and $\mathbf{s}_B$ by

$$\left| \left(\frac{\mathbf{r}_{A \leftarrow x}}{\sqrt{N_0}} \right)^{\mathrm{T}} \left(\frac{\mathbf{s}_A}{\sqrt{\varepsilon_{\max}}} \right) \right|^2 \leq I_0 \quad \forall x \in \mathcal{A}_{t_0} \backslash \{A, B\} \tag{6}$$

and

$$\left| \left(\frac{\mathbf{r}_{B \leftarrow x}}{\sqrt{N_0}} \right)^{\mathrm{T}} \left(\frac{\mathbf{s}_B}{\sqrt{\varepsilon_{\max}}} \right) \right|^2 \leq I_0 \quad \forall x \in \mathcal{A}_{t_0} \backslash \{A, B\}, \tag{7}$$

respectively. In these two expressions, $\mathcal{A}_t$ denotes the set of all terminals in the network which are involved in a packet transmission at the time t, and I_0 is a positive constant. To be able to comply with (6-7), we let all idle terminals listen to signals from neighbors and store the spatial signature of every detected signal in a local database. Because of $\|\mathbf{s}_A\|^2 \leq \varepsilon_{\max}$, it is actually sufficient for e.g. the terminal A to be aware of all $\mathbf{r}_{A \leftarrow x}$ with $\|\mathbf{r}_{A \leftarrow x}\|^2 > I_0 N_0$. The particular pre-/post filers are chosen as follows.

3.1 Beamforming at Data Packet Sender

The terminal A, sending the first (RTS) frame within the packet transmission, has no information about the state of the MIMO channel $\mathbf{H}_{A \to B}$ and chooses $\mathbf{s}_A$, $\mathbf{f}_A$ with the objective of minimizing CCI at the terminals $\mathcal{A}_{t_0} \backslash \{A, B\}$ under the constraints (6) and $\|\mathbf{s}_A\|^2 \leq \varepsilon_{\max}$. Hence,

$$\mathbf{f}_A = \arg \min_{\|\mathbf{f}\|=1} \max_{x \in \mathcal{A}_{t_0} \backslash \{A, B\}} \left| \mathbf{r}_{A \leftarrow x}^{\mathrm{T}} \mathbf{f} \right| \tag{8}$$

and

$$\mathbf{s}_A = \min \left\{ \frac{\sqrt{I_0 N_0 \varepsilon_{\max}}}{\max\limits_{x \in \mathcal{A}_{t_0} \backslash \{A, B\}} \left| \mathbf{r}_{A \leftarrow x}^{\mathrm{T}} \mathbf{f}_A \right|}, \sqrt{\varepsilon_{\max}} \right\} \cdot \mathbf{f}_A. \tag{9}$$

Equation (8) may be viewed as finding the optimal subspace under the objective of minimal CCI, whereas (9) represents the power control part, fulfilling the given constraints.

Unfortunately, (8) is hard to solve if many interferers are present. As a practical substitute for $\mathbf{f}_A$, the eigenvector associated with the smallest eigenvalue of the Hermitian *interference-plus-noise correlation matrix* [7]

$$\mathbf{R}_{\mathrm{i+n}} = \sum_{x \in \mathcal{A}_{t_0} \backslash \{A, B\}} \mathbf{r}_{A \leftarrow x} \mathbf{r}_{A \leftarrow x}^{\mathrm{H}} + N_0 \mathbf{I}_N \tag{10}$$

is used in the computer simulations in Sect. 5.

3.2 Beamforming at Data Packet Receiver

Upon receiving the initial frame, the terminal B chooses its filters on the basis of $\mathbf{r}_{B\leftarrow A}$. By similar algebra as in (5), we find that the gain $|\mathbf{f}_A^T \mathbf{r}_{A\leftarrow B}|^2$ of the signal from terminal B at the terminal A linearly depends on $|\mathbf{r}_{B\leftarrow A}^T \mathbf{s}_B|^2$. This motivates

$$\mathbf{f}_B = \frac{\mathbf{r}_{B\leftarrow A}^*}{\|\mathbf{r}_{B\leftarrow A}\|} \tag{11}$$

and

$$\mathbf{s}_B = \min\left\{ \frac{\sqrt{I_0 N_0 \varepsilon_{\max}}}{\max\limits_{x \in \mathcal{A}_{t_0}\backslash\{A,B\}} |\mathbf{r}_{B\leftarrow x}^T \mathbf{f}_B|}, \sqrt{\varepsilon_{\max}} \right\} \cdot \mathbf{f}_B. \tag{12}$$

4 MAC Protocol

The following RTS-CTS-data-ACK cycle accommodates the interference avoidance policy (6-7) and the ensuing beamformers (8-12). As part of the protocol, all idle terminals continuously listen to RTS and CTS frames from neighboring terminals and keep a copy of the spatial signature of every detected signal in a repository.

1. To initiate a packet transmission, the terminal A sends an RTS frame choosing a pre-filter $\mathbf{s}_A$ according to (9). The recipient (terminal B) successfully receives the frame if, using the post-filter (11), the SINR $\gamma_B(t)$ achieves a certain minimum SINR γ_{RTS}. Otherwise, the packet transmission fails at this point.
2. The terminal B replies by a CTS frame, choosing a pre-filter according to (12). The CTS frame reception is successful if, after the post-filtering by (8), the SINR $\gamma_A(t)$ achieves a certain minimum SINR γ_{CTS}. Otherwise, the packet transmission fails at this point.
3. The terminal A transmits a frame with the data, employing the same pre-filter $\mathbf{s}_A$ as for the RTS frame. The terminal B successfully receives the data if, using the same post-filter as for the RTS frame reception, the SINR $\gamma_B(t)$ never falls below γ_{DATA} during the data frame reception. Otherwise, the packet transmission fails at this point.
4. With a final ACK frame the terminal B acknowledges a successful data reception, employing the same pre-filter as for the CTS frame. The entire packet transmission is successful if, using the same post-filter as for the CTS frame reception, the SINR $\gamma_A(t)$ achieves a certain minimum SINR γ_{ACK}.

After the data packet transmission, both terminals refrain from transmitting for at least T_{P} seconds. In this time, they can bring their databases up-to-date in order to ensure a proper signaling in a following packet transmission.

5   Numerical Results

The achievable end-to-end capacity of wireless networks with omnidirectional antennas is studied in [10] by Gupta and Kumar, and extending their results to some directional antenna models, under LOS conditions, is aimed at in [11] and [12]. In the following, we analyze the performance of the proposed interference avoiding channel access scheme, in environments with multi-path propagation and Rayleigh fading. We restrict our attention, however, to the amount of information that can be transported over single hops, without paying attention to the problem of finding end-to-end routes.

5.1   Network Simulation Set-Up

In our computer simulations, two new terminals are created for every data packet transmission and randomly placed in the planar network area, and they immediately disappear after the transmission, whether successful or not. This type of network model may also be viewed as having an *infinite set of nodes* [13]. The position of the packet sender is generated in a uniform fashion, whereas the associated packet receiver is uniformly placed within a radius of $D_{\max}$, representing the maximal single hop distance. Data packet transmissions are initiated at random points in time, and the interval between the appearance of two packet senders per any area of 1 m^2 exhibits an exponential distribution with mean μ_t. The entire network covers a much lager area than $D_{\max}^2$ such that the lower "local" interferer density at the boundary does not essentially change the results.

The MIMO channels between two new terminals and all other active terminals in the network are randomly generated according to the model in Sect. 2, with a path loss exponent of $\kappa = 3$ as commonly assumed for urban environments [14]. Following the generation of the positions and channels, the packet transmission proceeds as described in Sect. 4. The RTS and CTS frames of a packet transmission starting at $t = t_0$ are assumed infinitely short and the active terminals during this phase are given by $\mathcal{A}_{t_0}$. The two frame transmissions are successful if the SINR's at the respective receivers achieve γ_{RTS} and γ_{CTS} of 6 dB each. For the data transmission we also demand a minimum SINR γ_{DATA} of 6 dB, which needs to be maintained for every $t \in [t_0, t_0 + T_{\mathrm{P}})$. The ACK frame is again infinitely short and requires an SINR of 6 dB at $t = t_0 + T_{\mathrm{P}}$.

As for the thermal noise power N_0 we assume that a signal, being sent omnidirectionally with maximal power from a transmitter at the maximal distance $D_{\max}$, is received at a mean signal-to-noise ratio of 15 dB. Hence, packet transmissions are much more likely to fail because of excess CCI than because of the thermal noise, unless the network usage is very low.

We compute the *transport capacity* [10] of the single-hop network, which we define here as the achieved bit-meters/s/Hz per network area of 1 m^2. Only a single data rate of one bit per second and Hertz is available, and every packet transmission from a terminal A to a terminal B with successful ACK frame reception is accounted for $d_{A \to B}$ bit-meters/s/Hz.

5.2 Results

Increasing the rate of the transmission attempts by decreasing μ_t lets the transport capacity grow linearly with μ_t^{-1} at first, and only marginally for μ_t close to zero. In the same time the *success ratio*, defined as the number of successful data packet transmissions over the number of attempts, approaches zero. In real networks the success ratio determines the latency, since an unsuccessful attempt implicates retransmission after some delay. As there are normally some limits on the latency, it is reasonable to compare the transport capacities of different channel access techniques at some fixed success ratio. We consider two scenarios: In scenario 1 we adjust μ_t in each simulation such that a success ratio of 50% results, whereas in scenario 2 we aim at a success ratio of 10%.

Fig. 2 shows the achievable transport capacities in these two scenarios, with the number of antennas per terminal ranging between one and six. The given transport capacities include a factor $1/D_{\max}$. This is due to the fact that the encountered packet transmit distances grow linearly with $D_{\max}$, the traffic density decreases with $1/D_{\max}^2$, and, consequently, the transport capacities scale with $1/D_{\max}$. The figure contrasts the performance with the MAC protocol in Sect. 4, with $I_0 = 2$, against the performance of two respective procedures without collision avoidance: In the first, the data sending terminal A transmits and receives omnidirectionally, utilizing $\mathbf{s}_A = (\sqrt{\varepsilon_{\max}}, 0, \ldots, 0)^{\mathrm{T}}$ and $\mathbf{f}_A = (1, 0, \ldots, 0)^{\mathrm{T}}$, whereas the data packet receiving terminal B chooses $\mathbf{f}_B$ according to (11) and $\mathbf{s}_B = \sqrt{\varepsilon_{\max}}\mathbf{f}_B$. In the second trivial channel access scheme, all terminals employ maximal transmit power and omnidirectional beam patterns.

Clearly, employing multiple antennas does not have any effect with omnidirectional transmission and reception. A beamforming at the data packet receiver based on the vector signal from the packet sender yields some valuable gain if $N > 1$. More significant improvements result from accomplishing an interference avoidance. In the case of one antenna, the superiority of a collision avoidance scheme over a random channel access scheme is well known. With multiple antennas, however, we find that an interference avoidance achieves even greater gains. We note in particular that in scenario 1, employing the proposed interference avoidance technique lets the transport capacity grow almost linearly with the number of antennas.

A data packet transmission may fail either at the beginning because the SINR requirements for the RTS and CTS frames are not attained, or at a later stage due to the sudden arrival of new interferers. We name the breakdown during the data frame or the ACK frame a disruption of a packet transmission. It is interesting to analyze the incidence rate of the latter. We consider the *disruption ratio*, which we define as the number of disrupted packed transmission over the transmission attempts, analogous to the success ratio.

We find disruption ratios around 20% in scenario 1 and 16% in scenario 2 for both techniques without collision avoidance, no matter how many antennas are employed. Hence, in scenario 2, many more packet transmissions are disrupted than successfully completed. Our interference avoidance scheme limits the disruption ratios to fractions of the success ratios, as shown in Tab. 1. With six

Table 1. Percentages of disrupted packet transmissions.

scenario	$N=1$	$N=2$	$N=3$	$N=4$	$N=5$	$N=6$
1	5.0%	2.6%	1.5%	0.9%	0.65%	0.45%
2	2.5%	1.2%	0.6%	0.3%	0.15%	0.09%

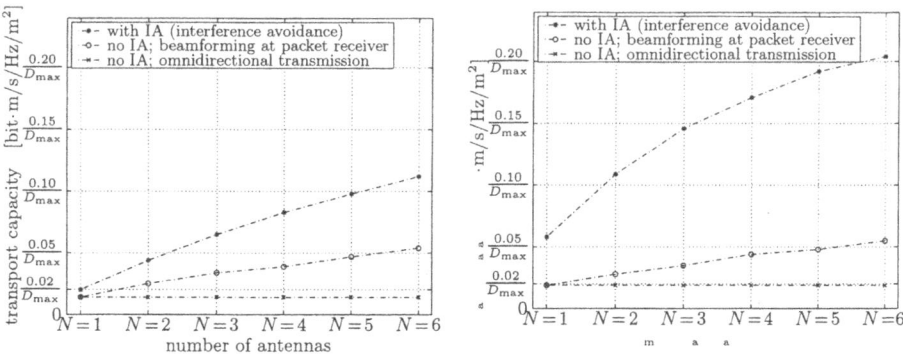

Fig. 2. Transport capacities with/without interference avoidance (IA), with different numbers of antennas; left: scenario 1 (50% success ratio), right: scenario 2 (10% success ratio).

antennas per terminal, for example, the disruption rate is less than a hundredth of the packet transmission success rate in both scenarios.

6   Conclusions

A practical beamforming technique has been proposed with the objective of avoiding CCI in decentralized asynchronous wireless networks. The beamforming scheme, incorporated in a RTS-CTS-data-ACK based MAC protocol, may be viewed as an extension of the CSMA/CA principle to the multidimensional signal case. Every channel access is preceded by a sensing for vector signals from active terminals in the neighborhood, and the spatial signature and power of a transmitted signal vector are selected under the criterion of limiting the CCI caused at the perceived neighbors.

Computer simulations have shown that the proposed beamforming can greatly reduce the collision probability and increase the transport capacity compared to channel access schemes without interference avoidance. Moreover, at a moderate traffic load with a 50% data packet transmission success ratio, the transport capacity grows almost linearly when increasing the number of antennas per terminal from one up to six.

Acknowledgements. The authors wish to thank Dr. B. Komiyama for encouraging this work. This research was supported in part by the Telecommunications Advancement Organization of Japan.

References

1. T. Korakis, G. Jakllari, and L. Tassiulas, "A MAC protocol for full exploitation of directional antennas in ad-hoc wireless networks," in *Proc. ACM MobiHoc 2003*, (Annapolis, MD), pp. 98–107, June 2003.
2. Y.-B. Ko, V. Shankarkumar, and N. H. Vaidya, "Medium access control protocols using directional antennas in ad hoc networks," in *Proc. IEEE INFOCOM 2000*, (Tel Aviv, Israel), pp. 13–21, Mar. 2000.
3. A. Nasipuri, S. Ye, J. You, and R. E. Hiromoto, "A MAC protocol for mobile ad hoc networks using directional antennas," in *Proc. IEEE Wireless Communications and Networking Conference (WCNC) 2000*, (Chicago, IL), pp. 1214–1219, Sept. 2000.
4. S. Bandyopadhyay, K. Hasuike, S. Horisawa, and S. Tawara, "An adaptive MAC protocol for wireless ad hoc community network (WACNet) using electronically steerable passive array radiator antenna," in *IEEE Proc. GLOBECOM 2001*, (San Antonio, TX), pp. 2896–2900, Nov. 2001.
5. R. R. Choudhury, X. Yang, N. Vaidya, and R. Ramanathan, "Using directional antennas for medium access control in ad hoc networks," in *Proc. ACM MobiCom 2002*, (Atlanta, GA), pp. 59–70, Sept. 2002.
6. J. L. Bordim, T. Ueda, and S. Tanaka, "A directional MAC protocol to support directional communications in ad hoc networks," in *Proc. First Int. Conf. on Mobile Computing and Ubiquitous Networking (ICMU-2004)*, (Tokyo, Japan), pp. 52–57, Jan. 2004.
7. J. H. Winters, "Optimum combining in digital mobile radio with cochannel interference," *IEEE J. Select. Areas Commun.*, vol. 2, pp. 528–539, July 1984.
8. J.-H. Chang, L. Tassiulas, and F. Rashid-Farrokhi, "Joint transmitter receiver diversity for efficient space division multiaccess," *IEEE Trans. Wireless Commun.*, vol. 1, pp. 16–27, Jan. 2002.
9. G. Lebrun, T. Ying, and M. Faulkner, "MIMO transmission over a time-varying channel using SVD," in *IEEE Proc. GLOBECOM 2002*, (Taipei, Taiwan), pp. 414–418, Dec. 2002.
10. P. Gupta and P. R. Kumar, "The capacity of wireless networks," *IEEE Trans. Inform. Theory*, vol. 46, pp. 388–404, Mar. 2000.
11. S. Yi, Y. Pei, and S. Kalyanaraman, "On the capacity improvement of ad hoc wireless networks using directional antennas," in *Proc. ACM MobiHoc 2003*, (Annapolis, MD), pp. 108–116, June 2003.
12. A. Spyropoulos and C. S. Raghavendra, "Asymptotic capacity bounds for ad-hoc networks revisited: The directional and smart antenna cases," in *IEEE Proc. GLOBECOM 2003*, (San Francisco, CA), pp. 1216–1220, Dec. 2003.
13. D. Bertsekas and R. Gallager, *Data Networks*. Englewood Cliffs, NJ: Prentice Hall, 2nd ed., 1992.
14. T. S. Rappaport, *Wireless Communications*. Upper Saddle River, NJ: Prentice-Hall, 1996.

Analysis of a Cylindrical Microstrip Antenna by the Nonorthogonal FDTD Method and Parallel Processing

Ronaldo O. dos Santos[1], Rodrigo M.S. de Oliveira[2], Fabricio Jose B. Barros[2], and Carlos Leonidas da S.S. Sobrinho[2]

[1] Instituto de Estudos Superiores da Amazonia (IESAM)
Av. Governador Jose Malcher, 1175 - Belem, Para, Brazil
ronaldo@ufpa.br
[2] Federal University of Para (UFPA)
P.O.Box 8619, 66075-900 Belem, Para, Brazil
rodrigo@deec.ufpa.br
fabricio@cetuc.puc-rio.br
leonidas@ufpa.br

Abstract. A computational code has been developed for analyzing 3-D radiation problems in curved geometries. To accomplish this, a suitable formulation is developed considering Maxwell equations in a general coordinate system and numerically solved by the use of a parallel curvilinear finite-difference time-domain (FDTD) method. In order to validate the computational code, the method analyzes a microstrip antenna mounted on a curved surface. The results obtained by the developed code are then compared to those generated by the conventional (orthogonal) FDTD method and by experimental measurements.

1 Introduction

Along with the new generation of wireless communication systems, precise analysis of new radiating elements (antennas) of diversified geometries becomes necessary to certify (and improve) the performance of these systems. An uncountable number of techniques of analysis and synthesis are employed in order to analyze several types of antennas with complex geometries [1]-[2]. General consensus, what is aimed to be achieved with these geometries, somewhat complex, is to allow that the radiator has its dimensions reduced, being able to operate in two or more bands of frequency (multi-band antennas), and to have its radiation characteristics (directivity, gain, efficiency, etc ...) optimized. The techniques used to analyze such structures usually are limited by the Cartesian coordinate system.

This way, the idea of developing a code able to analyze wide-bands for these types of radiators in curved surfaces or with its spatial orientations not coincident to the coordinates of the orthogonal system has motivated us to research the FDTD method in general coordinates.

J.N. de Souza et al. (Eds.): ICT 2004, LNCS 3124, pp. 53–59, 2004.

The nonorthogonal grid incorporation in the FDTD method was initially proposed by Holland [3], who used a nonorthogonal FDTD model based on a system of global curvilinear coordinates. This technique is named nonorthogonal FDTD or GN-FDTD. In this method, Stratton's formulation [4] was used to solve the differential Maxwell's equations numerically. However, the GN-FDTD technique has limited applications because it is necessary to have a system of coordinates that is analytically described by a global base [5]. Subsequently, Fusco in [6] developed discrete Maxwell's equations for a system of local nonorthogonal curvilinear coordinates. This method is more versatile than the GN-FDTD, but it requires a grid-generation software to discretize the analysis region. Lee, et al. [7], expanded Fusco's work to three dimensions and demonstrated the stability criterion for the method. This method (LN-FDTD),truncated by the UPML formulation proposed by Roden [5], was employed to analyze the antennas considered here.

The disadvantages associated with the LN-FDTD method, especially for the analysis of 3-D structures, are: the large memory and the long CPU time required for the calculations. A solution for these problems would be more powerful computers. Another one, financially more accessible, would be the PCs clustering technique [8]. Here, LAM/MPI library has been used along with the LN-FDTD method to make a more precise analysis of the antennas and also to show the efficiency of the parallel processing in face to the sequential one.

2   Parallel Implementation of the Nonorthogonal FDTD Code

The main idea behind the parallel implementation of the LN-FDTD algorithm is the division of the analysis domain into sub-domains. This technique is known as Data Decomposition or Domain Decomposition. Data are portioned among the processors and processed simultaneously by each processor, which execute essentially the same code, but on different boundary conditions. This is a typical implementation of the SPMD model (Single Program Multiple Data) [9]. The distribution and sharing of data is manually made in such way that continuity is ensured. The library chosen to exchange the messages is LAM-MPI[10]. Fig.1 shows how the field components are exchanged at the interfaces between sub-domains for both LN-FDTD and FDTD. Thus, different processors work simultaneously executing a part of the program, but about several data, i.e., the processor calculates all the components of the fields of its domain, passing on only those located at the interfaces.

For the microstrip antennas analyzed with a cluster with twelve machines [11], the domain of analysis was divided in equal sub-domains resulting in arrays with equal dimensions, as shown by Fig.2. Each processor calculates the electric and magnetic components corresponding to the its assigned region (sub-domains).

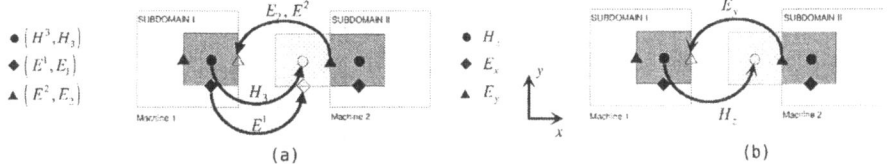

Fig. 1. Comparison between field passing schemes: (a) LN-FDTD - components located at an interface between two regions on the surface u^3; (b) FDTD - components located at an interface parallel to the y-z plane.

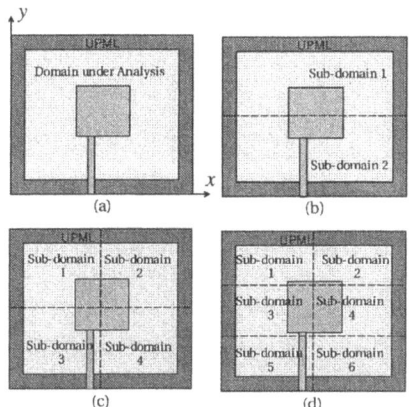

Fig. 2. Division of the numerical domain into sub-domains: (a) Numerical domain under analysis; (b) Numerical domain divided into two sub-domains; (c) Numerical domain divided into four sub-domains; (d) Numerical domain divided into six sub-domains.

3 Results

Fig.3(a) shows the microstrip antenna over a curved (cylindrical) substrate analyzed by this paper. It is basically the same regular microstrip antenna analyzed by Sheen in [12], but its patch is curved over the y-z plane, which curvature radius is $20mm$. It should be noticed, however, that just the patch is located at the curved region of the substrate, in such a way that the feeding plane is identical to that of the regular antenna.

To analyze a microstrip antenna over a curved substrate by the orthogonal FDTD method is a complicated task because the method employs regular (orthogonal) cells to model the physical boundary of this curved antenna. This way, the LN-FDTD method has been implemented and applied to perform such analysis. A curved-linear grid has been projected in order to properly model the device, as shown by Fig.3(b). This is a $61 \times 99 \times 36$ grid $(\hat{x}, \hat{y}, \hat{z})$ and the spatial increments are, approximately $\Delta x = 0.389mm$, $\Delta y = 0.4mm$, $\Delta z = 0.265mm$.

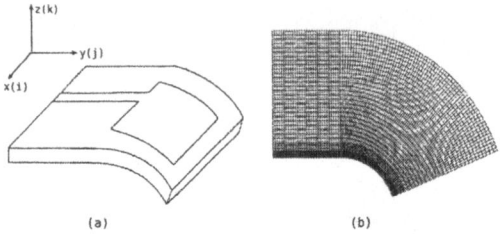

Fig. 3. (a) Rectangular microstrip antenna on a curved substrate; (b) mesh configuration on y-z cross-section.

The boundary condition used to truncate the domain is the UPML implemented in the curvilinear coordinate system 10 layers depth. The excitation source is a Gaussian pulse described by

$$E_z = exp\left[-\frac{(t - t_o)^2}{T^2}\right] \tag{1}$$

where $T = 15ps$ and $t_o = 45ps$. The time increment t used in this analysis is $0.488ps$ and the period necessary to reach the steady state is about $5000t$.

The parallel LN-FDTD implementation is convenient as long as the LN-FDTD method requires approximately three times the computational effort required by the FDTD method. In such way, the mesh illustrated by Fig.3(b) has been analyzed by employing six, four,two processors and one processor as well.

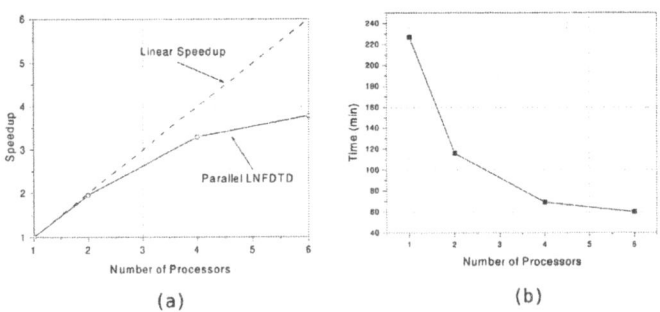

Fig. 4. (a) Comparison between the ideal speedup with the obtained by the LN-FDTD method and (b) Processing time required to analyze the microstrip antenna.

Fig.4(a) shows the speedup curve for the grid under analysis. In order to have a comparison reference, the ideal linear speedup is included. Here, the used definition of speedup is: $S = T_S/T_N$, where T_S is the time for the sequential processing and T_N is the processing time required for N processors. It is noticed that the speedup for the proposed algorithm, considering two processors, is very close to the ideal (linear). From three processors on, the obtained speedup still increases, but under the linear pattern. This behavior is basically due to the network traffic saturation, as long as the LN-FDTD involves more interface components that need to be sent and received among the domains. Fig.4(b) shows the processing time versus processors. As it can be seen there is a considerable time reduction when 6 machines are used.

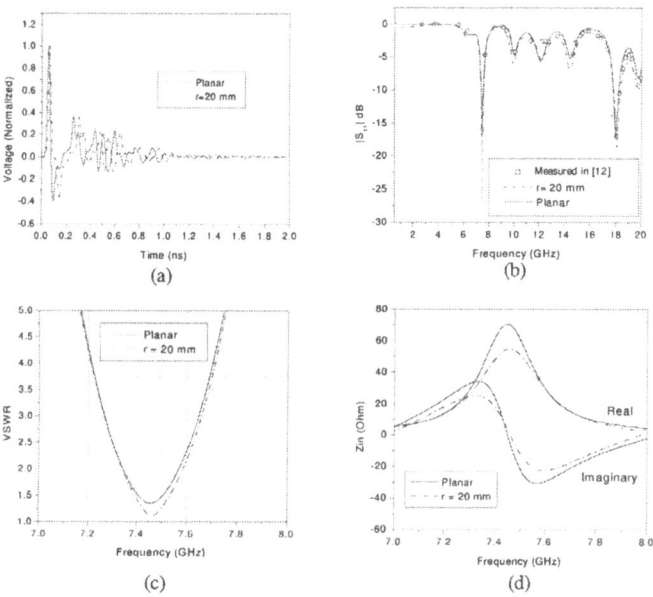

Fig. 5. (a) Voltage obtained between the terminals of the planar microstrip antenna and between the terminals of the curved-substrate microstrip antenna ($r = 20mm$); (b) Comparison between the reflection coefficients of the planar and curved substrate microstrip antennas; (c) Comparison between the V_{SWR} of the planar and cylindrical microstrip antennas; (d) Comparison between the input impedances of the planar and curved substrate microstrip antennas.

The voltage obtained between the feeding line terminals for the planar antenna and for the curved-substrate microstrip antenna is shown by Fig.5(a). Fig.5(b) shows the return loss as a function of frequency considering the simulated results for planar and curved microstrip antennas and the measured results for planar microstrip antenna [12]. From the evaluation of the return loss, the (voltage) stationary wave rate (V_{SWR}) can be obtained. Fig.5(c) compares the calculated V_{SWR} for the planar and cylindrical microstrip antennas versus frequency. Fig.5(d) shows the variation of impedance with the frequency. In this figure, the real part of the input impedance, for the curved microstrip antenna, is approximately 50Ω at the resonance frequency.

It is important to emphasize that the obtained results for the microstrip antenna by the LN-FDTD method are identical to those obtained in [13] where the GN-FDTD method is applied.

4   Conclusions

A computational code has been developed to solve 3-D radiation problems. This code has been implemented to run in the sequential and parallel computational environments, because the LN-FDTD method requires approximately three times the computational power required by the conventional FDTD method. This way, the microstrip antenna mounted on a curved substrate has been analyzed by employing six, four, two and one processor. The methodology permits the analysis of structures whose geometries are not coincident to the Cartesian coordinate system and, in such way, the physical boundary condition can be applied with no approximations.

The reflection coefficient, V_{SWR} and input impedance comparisons, revels that the curved microstrip antenna has a lower level of reflection at the neighborhood of the resonant frequency. Thus, this device matches better than the planar one to the feeding line. It should be noticed that the analyzed antenna's feeding line is not included on the curved surface, although it could be easily included on that region with the patch.

Acknowledgments. The authors are very grate to Professor D. M. Sheen for providing us his planar microstrip measured data. Thanks to CNPq, CAPES, IESAM and to contract UFPA01 (Ericsson-UFPA-Amazonia Celular) for the technical and financial supports thought these last years.

References

1. Esa, M., Nawawi, N. A. A., and Subahir, S., "Investigation of Lowering the Operating Frequency of a Microwave Device Using Simulation", *X Simposio Brasileiro de Microondas e Optoeletronica (2002), pp 121-125;*
2. Yang, F., Zhang, X. X., Ye X., R-Sammi, Y., "Wide-band E-shaped patch antennas for wireless communications", *IEEE Trans. On Antennas and prop. (2001), Vol. Ap-49, pp.1094-1100;*
3. Holland, R., "Finite-Difference Solutions of Maxwell's Equations in Generalized Nonorthogonal Coordinates", *IEEE Trans. on Nuclear Science (1983), vol. NS-30, No.6 pp. 4589-4591;*
4. Stratton, J. A., "Electromagnetic Theory", *New York: McGraw-Hill (1941);*
5. Roden, J. A., "Broadband Electromagnetic Analysis of Complex Structures with the Finite-Difference Time-domain Technique in General Curvilinear Coordinates", *Ph.D. Dissertation (1997), University of Kentucky;*
6. Fusco, M., "FDTD algorithm in curvilinear coordinates", *IEEE Trans. Antennas Propagat. (1990), AP-38, pp.76-88;*

7. Lee, J. F., Palendech, R., Mittra, R., "Modeling Three-Dimensional Discontinuities in Waveguides Using Non-orthogonal FDTD Algorithm", *IEEE Transactions on Microwave Theory and Techniques (1992), Vol. MTT-40, No.2;*

8. Almasi, G.S, Gohlied A., "Hihgly Parallel Computing", *Benjamim/Cummings Publishing (1994), 2nd edition;*

9. Andrews, G., "Foundations of Multithreaded: Parallel and Distributed Programming", *Addison wesley (2000);*

10. Message Passing Interface (MPI) on the Web, *http://www.mpi.nd.edu/lam* ;

11. Araujo, J. S., Santos, R. O., Sobrinho, C. L. S. S., Rocha, J. M., Guesdes, L. A., Kawasaki, R. Y. , "Analysis of Antennas by FDTD Method Using Parallel Processing with MPI" *International Microwave and Optoeletronics Conference (2003), Iguazu Falls, 2nd Vol., pp.1049-1054;*

12. Sheen, D. M., Ali, S. M., Abouzahra, M. D., Kong, J. A., "Application of the three-dimensional finite-difference time-domain method to the analysis of planar microstrip circuits", *IEEE Trans. Microwave Theory Tech.(1990), Vol. MTT-38, No. 7, pp. 849-857;*

13. Kashiwa, T., Onishi, T., Fukai, I., "Analysis of Miscrostrip Antennas on a Curved Surface Using the Conformal Grids FDTD Method, *IEEE Trans. Antennas Propagat. (1994), vol. AP-42, No. 3, pp. 423-427.*

Unified Analysis of the Local Scatters Channel Model

Wamberto J.L. Queiroz, Fabricio G.S. Silva, and Marcelo S. Alencar*

Federal University of Campina Grande - UFCG
Av. Aprígio Veloso, 882, 58109-970, Campina Grande, PB, Brasil
{wamberto,fabricio,malencar}@ufcg.dee.ufpb.br
Laboratory of Communications, Department of Electrical Engineering

Abstract. This article shows a simple mathematical treatment to model and simulate the local scatters channel using a Gaussian probability density for arrival angle distributions, time correlation functions and power spectral density. The correlation envelope is related to the channel parameters and is used in many applications where second order statistics are needed. The channel temporal analysis uses the WSS-US (Wide-Sense Stationary and Uncorrelated Scattering) model for modeling the directional components, considering parameters as angular spread, channel directional component and Doppler frequency.

1 Introduction

Channel modeling is an important issue for mobile communications systems performance assessment. From a description of the channel, efficient processing techniques can be devised and the system performance can be analyzed. The first channel model that included a directional component and an angular distribution for incoming signals was proposed by Lee [1] in 1974. For that channel, the outcoming signals, leaving the mobile station antenna, are reflected by scatters uniformly distributed in a circular geographic area around the mobile station and form a reflected signal cluster that reach the base station antenna array within a certain angular interval at a ϕ_0 angle. The directional channel model can be classified as Low-rank and High-rank models [2], [3]. The most known low-rank channel is the local scatters model. Each channel directional component is modeled by a stochastic process $\tilde{h}_l(t, \tau_l)$ and the directional channel model can be written as

$$\mathbf{h}(t,\tau) = \sum_{l=1}^{L} p_l \mathbf{a}(\phi_l) \tilde{h}_l(t,\tau) = \sum_{l=1}^{L} p_l \mathbf{a}(\phi_l) \left[\frac{1}{\sqrt{N}} \sum_{n=1}^{N} \alpha_{n,l} e^{-j\omega_{n,l}t} \delta(t-\tau_l) \right]. \quad (1)$$

The sum in (1) provides a way of simulating the stochastic process $\tilde{h}_l(t, \tau_l)$ by adding a sufficient number of random variables. In accordance with the Central

* The authors would like to express their thanks to Brazilian Council for Scientific and Technological Development (CNPq) for the financial support to this research

J.N. de Souza et al. (Eds.): ICT 2004, LNCS 3124, pp. 60–66, 2004.
© Springer-Verlag Berlin Heidelberg 2004

Limit Theorem, as N increases, $\tilde{h}(t, \tau_l)$ tends to a complex Gaussian process. Written in polar form the process modulus has Rayleigh distribution and a phase that is uniformly distributed in $[0, 2\pi)$. The vector elements $\mathbf{a}(\phi_l)$ are the phase of each arrival signal that reaches the base station antenna array, τ_l represents the path delay of each reflected component, p_l is used as a normalization component in order to $\sum_{l=1}^{L} p_l^2 = 1$, the coefficients α_l model the variations of the incoming signal power and $f_{n,l}$ represent the Doppler frequency in the n-th component of the l-th directional path.

Generally the classical Jake's method is used to simulate the stochastic process $\tilde{h}(t, \tau_l)$ [4]. If a linear array with M elements along the y-axis is considered, the steering vector can be written as

$$\mathbf{a}(\phi_l) = [\, 1 \; e^{-j\beta d \mathrm{sen}(\phi_l)} \; \cdots \; e^{-j(M-1)\beta d \sin(\phi_l)} \,], \tag{2}$$

where $\beta = \frac{2\pi}{\lambda}$. This vector specifies the transmitted signal phase, sampled at each array element. The arrival angle ϕ_l is a random variable that depends on the environment around the mobile station. Depending on the spatial scatters distribution, different pdfs for azimuthal angular distribution are found in the literature [1], [5]. Studies have shown, using field measurement, that when scatters have uniform distribution, a Gaussian pdf is more appropriate to model the distribution of arrival angles. Reference [6] proposed the formula

$$p(\phi) = \frac{Q}{\sqrt{2\pi\sigma_\phi^2}} e^{-\frac{(\phi-\phi_0)^2}{2\sigma_\phi^2}}, \quad -\frac{\pi}{2} + \phi_0 \le \phi \le \frac{\pi}{2} + \phi_0, \tag{3}$$

where σ_ϕ is the standard deviation of the spread-out DOA (Direction of Arrival) in the angular domain and Q is a normalization factor used in order to adjust the area of $p(\phi)$ to unity.

2   Channel Correlation Function

In order to obtain the correlation functions, a linear array structure is assumed. It can be shown that the output of this reception structure can be written in a vectorial form as $y(t) = \mathbf{w}^H [\mathbf{x}(t) + \eta(t)]$, where $\mathbf{x}(t) = [x_0(t) \; x_1(t) \; \cdots \; x_{M-1}(t)]^T$,

$$\mathbf{w} = [\, w_0 \; w_1 \; \cdots \; w_{M-1} \,]^T \quad \text{and} \quad \eta(t) = [\, \eta_0(t) \; \cdots \; \eta_{M-1}(t) \,]^T. \tag{4}$$

The components of $\mathbf{x}(t)$ represent the signal samples that can be detected at each element array, whose values are given by the convolution between the channel response $\mathbf{h}(t, \tau)$ and $s(t)$. Therefore, $\mathbf{x}(t)$ can be written as

$$\mathbf{x}(t) = \sum_{l=1}^{L} p_l \mathbf{a}(\phi_l) \left[\frac{1}{\sqrt{N}} \sum_{n=1}^{N} \alpha_{n,l} e^{-j2\pi f_{n,l} t} \right] s(t - \tau_l) = \sum_{l=1}^{L} p_l \mathbf{a}(\phi_l) g_l(t) s(t - \tau_l). \tag{5}$$

The term between brackets from this point it will be denoted by $g_l(t)$. Using this result, the array output $y(t)$ can be written as

$$y(t) = \sum_{l=1}^{L} p_l g_l(t) s(t - \tau_l) \sum_{m=0}^{M-1} w_m^* a_m(\phi_l) + \sum_{m=0}^{M-1} w_m^* \eta_m(t). \tag{6}$$

As it is known, the array factor of an antenna array can be defined as the product between the steering vector and an associated weights vector. The term $\sum_{m=0}^{M-1} w_m^* a_m(\phi_l)$ can be seen as the array factor value in ϕ_l. This array factor is defined as $F(\phi_l) = \sum_{m=0}^{M-1} w_m^* a_m(\phi_l)$.

The last sum in (6) can be seen as a Gaussian noise sample weighted by the excitation amplitude of the array elements and will be denoted by $\eta_w(t)$. Therefore $y(t)$ can be written as

$$y(t) = \sum_{l=1}^{L} p_l g_l(t) s(t - \tau_l) F(\phi_l) + \eta_w(t). \tag{7}$$

The channel response, in terms of the array factor, can be given by

$$h(t, \tau) = \sum_{l=1}^{L} p_l g_l(t) \delta(t - \tau_l) F(\phi_l). \tag{8}$$

The correlation function is given by $R_h(\xi) = E[h(t, \tau) h^*(t + \xi, \tau)]$, which gives

$$R_h(\xi) = E\left[\left(\sum_{l=1}^{L} p_l g_l(t) \delta(t - \tau_l) F(\phi_l) \right) \left(\sum_{k=1}^{L} p_k g_k^*(t + \xi) \delta(t - \tau_k + \xi) F^*(\phi_k) \right) \right].$$

Assuming that the L multipath components are uncorrelated, as assumed in the WSS-US channel model, one obtains

$$R_h(\xi) = \sum_{l=1}^{L} p_l^2 |F(\phi_l)|^2 E[g_l(t - \tau_l) g_l^*(t - \tau_l + \xi)] = \sum_{l=1}^{L} p_l^2 |F(\phi_l)|^2 R g_l(\xi), \tag{9}$$

where $g_l(t)$ is the term between brackets shown in (5) and

$$R g_l(\xi) = \frac{1}{N} \sum_{n=1}^{N} E[\alpha_{n,l} \alpha_{n,l}^*] E[\exp(-j2\pi f_{n,l}\xi)]. \tag{10}$$

Considering $E[\alpha_{n,l} \alpha_{n,l}^*] = 1$ and using the Doppler frequency concept, one has that

$$R g_l(\xi) = \frac{1}{N} \sum_{n=1}^{N} E[\exp(-j2\pi f_m \cos(\phi_{n,l})\xi)]. \tag{11}$$

Equation (11) represents the medium value of $Rg_l(\xi)$ for N channel elementary realizations. From this result $R_h(\xi)$ can be rewritten as

$$R_h(\xi) = \sum_{l=1}^{L} p_l^2 G(\phi_l) E[\exp(-j2\pi f_m \cos(\phi_l)\xi)], \tag{12}$$

where $G(\phi_l) = |F(\phi_l)|^2$.

The expected value in (12) is given by

$$\overline{Rg_l}(\xi) = \int_{-\infty}^{\infty} p(\phi_l) \exp(-j2\pi f_m \cos(\phi_l)\xi) d\phi_l. \tag{13}$$

Considering the Gaussian pdf for the arrival angle distribution and writing the result in terms of Bessel functions, one has

$$\Re\{\overline{Rg_l}(\xi)\} = J_0(2\pi f_m \xi) + Q \sum_{k=1}^{\infty} J_{2k}(2\pi f_m \xi)(-1)^k e^{-2k^2 S_\phi^2} A_k(\phi_0, S_\phi)$$

$$\Im\{\overline{Rg_l}(\xi)\} = Q \sum_{k=0}^{\infty} J_{2k+1}(2\pi f_m \xi)(-1)^k e^{-\frac{((2k+1)S_\phi)^2}{2}} B_k(\phi_0, S_\phi),$$

where $A_k(a,b)$ and $B_k(a,b)$ are defined as

$$\begin{aligned}
A_k(a,b) &= \cos(2ka)\mathcal{A}(2k,b) - \sin(2ka)\mathcal{B}(2k,b) \\
B_k(a,b) &= \cos((2k+1)a)\mathcal{A}(2k+1,b) - \sin((2k+1)a)\mathcal{B}(2k+1,b)
\end{aligned} \tag{14}$$

and the expressions $\mathcal{A}(a,b)$ and $\mathcal{B}(a,b)$ are given in [3]. Finally, the channel correlation function can be written as

$$R_h(\xi) = \sum_{l=1}^{L} p_l^2 G(\phi_l)|\overline{Rg_l}(\xi)|. \tag{15}$$

Figures 1(a) and 1(b) show the channel temporal correlation behavior for different values of angular dispersion, considering two values of the main channel directional component and a Doppler frequency equal to 100 Hz. The channel observation time in both cases is 40 ms. At first, it is possible to conclude that the channel dynamics do not depend on the Doppler frequency, but on the spatial features including the angular dispersion S_ϕ and main component direction ϕ_0. As ϕ_0 tends to 0° or 180°, the channel presents a slower behavior. On the other hand, the channel dynamics increases when the angular dispersion increases.

3   Power Spectral Density

The power spectra of the received signal in a multipath channel model can be obtained assuming that there is a Doppler frequency shift equal to $\beta v \cos(\phi)$ when the mobile station or unit receiver moves at a constant speed v. Therefore,

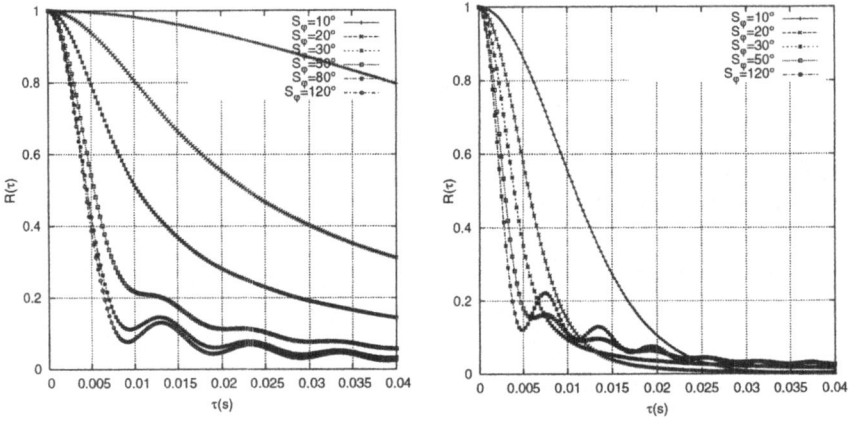

(a) Main component direction $\phi_0 = 0°$

(b) Main component direction $\phi_0 = 45°$

Fig. 1. Channel temporal correlation for different values of angular dispersion, with Doppler frequency equal to 100 Hz.

if ϕ is the incidence angle of a radio wave arriving at an antenna array, the instantaneous angular frequency, $\omega(\phi) = \omega_c + \omega_m \cos(\phi)$.

If one takes a base station directional antenna with gain $G(\phi)$ and considers that ϕ has a probability density function $p(\phi)$, then within a differential angle $d\phi$, the total power is $P_0 G(\phi) p(\phi) d\phi$, where P_0 is the power that would be received by an isotropic antenna. The total power equals the differential variations of the received power spectral density with frequency $S(f)df$. Considering that $\omega(\phi) = \omega(-\phi)$, one has

$$S(\omega)|d\omega| = 2\pi P_0 [G(\phi)p(\phi) + G(-\phi)p(-\phi)]|d\phi|. \tag{16}$$

Furthermore, $|d\omega| = |-\omega_m \sin(\phi)||d\phi| = [\omega_m^2 - (\omega - \omega_c)^2]^{\frac{1}{2}}|d\phi|$. Therefore, considering the Gaussian probability density function for the arrival angles, $S(\omega)$ can written as

$$S(\omega) = \frac{2\pi P_0 Q}{\sqrt{2\pi(\omega_m^2 - (\omega - \omega_c)^2)}} \left[\frac{G(\phi)}{\sigma_\phi} e^{-\frac{(\phi - \phi_0)^2}{2\sigma_\phi^2}} + \frac{G(-\phi)}{\sigma_\phi} e^{-\frac{(\phi + \phi_0)^2}{2\sigma_\phi^2}} \right] \tag{17}$$

and the Doppler frequency pdf can be expressed as

$$p(f) = \frac{Q}{\sqrt{2\pi(f_m^2 - f^2)}S_\phi} \exp\left[-\left(\frac{\cos^{-1}\left(\frac{f}{f_m}\right) - \phi_0}{\sqrt{2}S_\phi} \right)^2 \right] \tag{18}$$

Comparing (18) with (17), it is easy to show that the spectral power density can be rewritten as

$$S(\omega) = 2\pi P_0[G(\phi)p(\omega) + G(-\phi)p(-\omega)] \tag{19}$$

where $p(\omega)$ is the pdf of the Doppler frequency given in (18). Figure 2 shows, for example, the behavior of $p(f)$ for different values of angular dispersion S_ϕ, the main component direction of the channel ϕ_0 equal to $0°$ and maximum Doppler frequency f_m equal to 100 Hz.

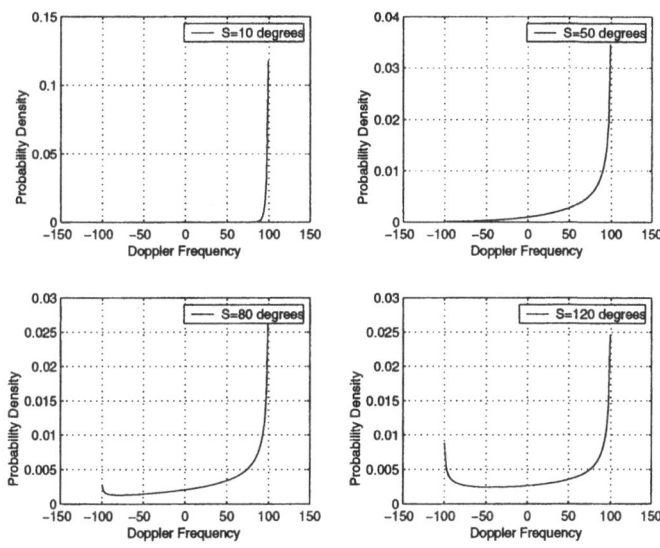

Fig. 2. Probability density behavior of the random variable f, Doppler frequency, considering four values of angular dispersion, $\phi_0 = 0°$ and $f_m = 100$ Hz.

Changes on the channel behavior depend, aside from the maximum Doppler frequency, on the angular dispersion S_ϕ and on the main component direction ϕ_0. An appropriate choice of angular dispersion depends on the environment, antenna height and distance between mobile station and radio base station.

Conclusion

This paper presented a study of the Local Scatters Channel Model using a Gaussian probability density to model the arrival angles of the received signal at the base station. An analysis for both the temporal and frequency domain was presented, using parameters of the spatial channel model. The classical Jake's method was proposed to model the multipath channel components. It was verified that the correlation envelope and the Doppler frequency strongly depend on certain parameters, such as channel main component direction and angular spread.

References

1. Lee, W.C.Y.: Effects on Correlation Between Two Mobile Radio Base Station Antennas. Proc. IEEE Trans. on Communications **21** (1973) 1214–1224
2. Ottersten, B.: Array Processing for Wireless Communications. Proc. Workshop on Statistical and Array Processing, Corfu, Greece (1996) 466–473
3. Fuhl, J.: Smart Antennas for Second and Third Generation Mobile Communications Systems . PhD thesis, Technische Universitat Wien, A-2803 Schwarzenbach, Eggenbuch 17 (1997)
4. Luiz M. Correia, Instituto Superior T cnico, P.: Wireless Flexible Personalised Communications, COST 259: European Co-operation in Mobile Radio Research. John Wiley & Sons (2001)
5. Salz, J., Winters, J.H.: Effect of Fading Correlation on Adaptive Arrays in Digital Mobile Radio. IEEE Trans. on Vehicular Technology **43** (1994) 1049–1057
6. Trump, T., Ottersten, B.: Estimation of nominal direction of arrival and angular spread using an array of sensors. Signal Processing, Elsevier (1996)

Spacial Analysis of the Local Scatters Channel Model

Wamberto J.L. Queiroz, Fabricio G.S. Silva, and Marcelo S. Alencar*

Federal University of Campina Grande - UFCG
Av. Aprígio Veloso, 882, 58109-970, Campina Grande, PB, Brasil
{wamberto,fabricio,malencar}@ufcg.dee.ufpb.br
Laboratory of Communications, Department of Electrical Engineering

Abstract. This paper presents a detailed study concerning the fading correlation functions of a circular array compared to a linear array, considering probability density functions (pdf's) related to the azimuthal arrival angle. It is considered that signals that reach each element of the array arrive respectively with uniform, Gaussian and cosine-shaped distributions. Mathematical and numerical results for these configurations are presented.

1 Introduction

In wireless communications systems the channel model has generally an important role. It would be difficult to evaluate a new method to improve a transmission link without a good channel model. In an attempt to include most of the features found in practical transmission environments, the mathematical models become more and more complex. To reduce the propagation impairments, many applications of antenna arrays, including linear and circular arrays have been studied. Therefore, it is important to know the behavior of the fading correlation functions when different types of azimuthal arrival angle distributions are considered [1], [2], [3]. In reference [4] the correlation functions are analyzed for a linear array concerning the uniform distribution and the Gaussian distribution. In this work a comparison between two usual array configurations is presented, taking into account three different distributions.

2 Problem Overview

Imagine a propagation environment in an urban area where there are many threes, buildings and other types of construction distributed around an user with a portable cell phone. Therefore, one can assume that all these elements found in this area can be modeled as a set of scatterers circularly distributed around the user. In this model, the local scattering around the mobile user generates

---

* The authors would like to express their thanks to Brazilian Council for Scientific and Technological Development (CNPq) for the financial support to this research.

J.N. de Souza et al. (Eds.): ICT 2004, LNCS 3124, pp. 67–73, 2004.
© Springer-Verlag Berlin Heidelberg 2004

signals that reach the base station antenna within a range of angles. Moreover, the spatial distribution function $p(r)$ of the scatterers around the mobile unity can be given by [5], [1], [6]

$$p(r) = \begin{cases} \frac{1}{\pi R^2}, & ||r - r_{MS}|| \leq R \\ 0, & \text{elsewhere} \end{cases}, \tag{1}$$

where r is a radial distance measured from the mobile station position, R is the radius of the scatterers circle, typically in the order of 100m to 200m, and r_{MS} is the distance between the base station and the mobile station.

Depending on the spatial distribution of the scatterers, different pdfs for azimuthal angular distributions can be found in the literature. The first one is the uniform distribution, $p_u(\phi)$, written as [2]

$$p_u(\phi) = \begin{cases} \frac{1}{2\Delta}, & -\Delta + \phi_o \leq \phi \leq \Delta + \phi_o \\ 0, & \text{elsewhere} \end{cases}. \tag{2}$$

Although this distribution can provide a closed form expression for the envelope correlation coefficient ρ at different antenna positions, it has been shown that this uniform assumption is not valid for uniform scatters distribution. The second distribution, $p_c(\phi)$, known as cosine-shaped distribution, was proposed in [1] and can be written as

$$p_c(\phi) = \begin{cases} \frac{k_2}{\pi} \cos^k(\phi - \phi_o), & -\frac{\pi}{2} + \phi_o \leq \phi \leq \frac{\pi}{2} + \phi_o \\ 0, & \text{elsewhere} \end{cases} \tag{3}$$

and the third case is the Gaussian distribution, $p_g(\phi)$, that is given by [7], [3]

$$p_g(\phi) = \frac{k_3}{\sqrt{2\pi\sigma_\phi^2}} e^{-\frac{(\phi - \phi_o)^2}{2\sigma_\phi^2}}, \quad -\frac{\pi}{2} + \phi_o \leq \phi \leq \frac{\pi}{2} + \phi_o, \tag{4}$$

where σ_ϕ is the angular standard deviation and the parameters k_2 and k_3 are chosen in order to adjust $p_c(\phi)$ and $p_g(\phi)$ areas to unity.

3   Spacial Correlation Study

After presenting the channel model and azimuthal arrival angle distributions, one can introduce the signal samples taken at the array elements. These samples, denoted by x_l and x_c respectively for a linear array and a circular array, can be put in a vector notation written as

$$\mathbf{x}_l = \begin{bmatrix} e^{-ji_0\beta d\sin(\phi)} \\ e^{-ji_1\beta d\sin(\phi)} \\ \vdots \\ e^{-ji_{N-1}\beta d\sin(\phi)} \end{bmatrix} \quad \mathbf{x}_c = \begin{bmatrix} e^{j\beta a\cos(\phi-\theta_1)} \\ e^{j\beta a\cos(\phi-\theta_2)} \\ \vdots \\ e^{j\beta a\cos(\phi-\theta_N)} \end{bmatrix} \tag{5}$$

where $i_0 = 0$, $i_1 = 1$, $\cdots$, $i_{N-1} = N - 1$ and the angles θ_n stand for the angular position of the circular array elements. For the case of uniform circular arrays $\theta_n = 2\pi\frac{n}{N}$, where N is the number of elements in each array configuration.

3.1   Linear Array Correlation Functions

Consider the correlation between two signal samples, x_m and x_n, taken from two array elements when the arrival angle distribution is uniform. In this case the spatial correlation, noted by $\rho_u(m,n)$, can be written as

$$\rho_u(m,n) = E[x_m x_n^*], \tag{6}$$

where $E[x]$ represents the expected value of x.

Applying the pdf described in (2) and expanding the result in terms of Bessel series the real and imaginary terms of $\rho_u(m,n)$, respectively, denoted by $\mathcal{R}\rho_u(m,n)$ and $\mathcal{I}\rho_u(m,n)$ can be written as

$$\mathcal{R}\rho_u(m,n) = J_0((i_m - i_n)\beta d) + 2\sum_{k=1}^{\infty} J_{2k}((i_m - i_n)\beta d)\text{Sa}(2k\Delta)\cos(2k\phi_0),$$

$$\mathcal{I}\rho_u(m,n) = 2\sum_{k=0}^{\infty} J_{2k+1}((i_m - i_n)\beta d)\text{Sa}((2k+1)\Delta)\sin((2k+1)\phi_0). \tag{7}$$

In the second case, concerning the linear array, the arrival angle distribution has a cosine shape, as given by (3). In this case, the intergrals that arise in decorrence of applying the expected value considering (3) can not be solved in general for any exponent k. Following, the autocorrelation functions for $k = 1$ and $k = 3$ are shown. In each case one obtains a different k_2 value. When $k = 1$, the adjusting parameter $k_2 = \pi/2$. Using the Bessel's series, the functions $\mathcal{R}\rho_c$ and $\mathcal{I}\rho_c$ can be written as

$$\mathcal{R}\rho_c(m,n) = J_0((i_m - i_n\beta d) - 2\sum_{k=1}^{\infty}(-1)^k J_{2k}((i_m - i_n)\beta d)\frac{\cos(2k\phi_0)}{4k^2 - 1},$$

$$\mathcal{I}\rho_c(m,n) = \frac{\pi}{2}J_1((i_m - i_n)\beta d)\sin(\phi_0). \tag{8}$$

For $k = 3$, the adjusting parameter $k_2 = 3\pi/4$. In this case the correlation functions are given by

$$\mathcal{R}\rho_c(m,n) = J_0((i_m - i_n)\beta d) + 18\sum_{k=1}^{\infty} J_{2k}((i_m - i_n)\beta d)(-1)^k\frac{\cos(2\phi_0 k)}{(4k^2 - 9)(4k^2 - 1)},$$

$$\mathcal{I}\rho_c(m,n) = \frac{3\pi}{16}\left(3J_1((i_m - i_n)\beta d)\sin\phi_0 + J_3((i_m - i_n)\beta d)\sin(3\phi_0)\right). \tag{9}$$

In the last case, considering also linear arrays, the Gaussian distribution shown in (4) is used for the azimuthal arrival angle. In this case the correlation functions obtained from Bessel's series have the following expressions for real and imaginary parts, denoted by $\mathcal{R}\rho_g$ and $\mathcal{I}\rho_g$, respectively

$$\mathcal{R}\rho_g(m,n) = \mathrm{J}_0((i_m - i_n)\beta d) + \frac{2k_3}{\sqrt{\pi}} \sum_{k=1}^{\infty} \mathrm{J}_{2k}((i_m - i_n)\beta d)\mathrm{I}_{ck}(\sigma_\phi, \phi_0)$$

$$\mathcal{I}\rho_g(m,n) = \frac{2k_3}{\sqrt{\pi}} \sum_{k=0}^{\infty} \mathrm{J}_{2k+1}((i_m - i_n)\beta d)\mathrm{I}_{sk}(\sigma_\phi, \phi_0),$$

(10)

$$\mathrm{I}_{ck}(\sigma_\phi, \phi_0) = \frac{\sqrt{\pi}}{2} \left[\cos(2k\phi_0)\mathcal{A}(2k, \sigma_\phi) - \sin(2k\phi_0)\mathcal{B}(2k, \sigma_\phi)\right] e^{-2k^2\sigma_\phi^2},$$

$$\mathrm{I}_{sk}(\sigma_\phi, \phi_0) = \frac{\sqrt{\pi}}{2} \left[\sin((2k+1)\phi_0)\mathcal{A}(2k+1, \sigma_\phi)\right.$$

$$\left. + \cos((2k+1)\phi_0)\mathcal{B}(2k+1, \sigma_\phi)\right] e^{-\frac{(2k+1)^2\sigma_\phi^2}{2}},$$

where $\mathcal{A}(a,b)$ and $\mathcal{B}(a,b)$ are defined in [4].

3.2   Circular Array Correlation Functions

For the case of circular array, the signal samples taken from the array elements are given as shown in the second vector expression of (5). Considering the uniform arrival angle distribution given in (2), the correlation function will be given by

$$\rho_u(m,n) = E\left[\exp(j\beta aC_{m,n}\cos(\phi - \phi_{m,n}))\right],$$

(11)

where $C_{m,n}$ and $\phi_{m,n}$ are, respectively, given by

$$\phi_{m,n} = \mathrm{tg}^{-1}\left(\frac{\sin\theta_m - \sin\theta_n}{\cos\theta_m - \cos\theta_n}\right) \quad \text{and} \quad C_{m,n} = \sqrt{2(1 - \cos(\theta_m - \theta_n))}. \quad (12)$$

As can be seen in Equation (11), terms as $\sin(x\cos(\theta))$ and $\cos(x\cos(\theta))$ will arise. Therefore, using Bessel's series, one obtains the real and imaginary parts of $\rho_u(m,n)$, which are given by

$$\mathcal{R}\rho_u(m,n) = \mathrm{J}_0(\beta aC_{m,n}) + 2\sum_{k=1}^{\infty}(-1)^k \mathrm{J}_{2k}(\beta aC_{m,n})\mathrm{Sa}(2k\Delta)\cos(2k(\phi_0 - \phi_{m,n}))$$

$$\mathcal{I}\rho_u = 2\sum_{k=0}^{\infty} \mathrm{J}_{2k+1}(\beta aC_{m,n})(-1)^k \mathrm{Sa}((2k+1)\Delta)\cos((2k+1)(\phi_0 - \phi_{m,n})).$$

Considering now the cosine-shaped angular distribution, the general expression for the correlation functions, when for example $k = 3$ is set, can be written as follows

$$\mathcal{R}\rho_c(m,n) = \mathrm{J}_0(\beta aC_{m,n}) + 18\sum_{k=1}^{\infty} \mathrm{J}_{2k}(\beta aC_{m,n})\frac{\cos(2k(\phi_0 - \phi_{m,n}))}{(4k^2 - 1)(4k^2 - 9)},$$

$$\mathcal{I}\rho_c(m,n) = \frac{3\pi}{16}\left(-3\mathrm{J}_1(\beta aC_{m,n})\cos(\phi_0 - \phi_{m,n}) + \mathrm{J}_3(\beta aC_{m,n})\cos(3(\phi_0 - \phi_{m,n}))\right),$$

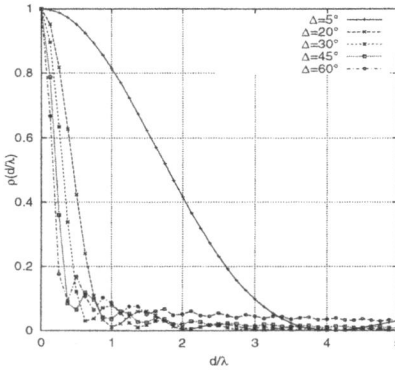

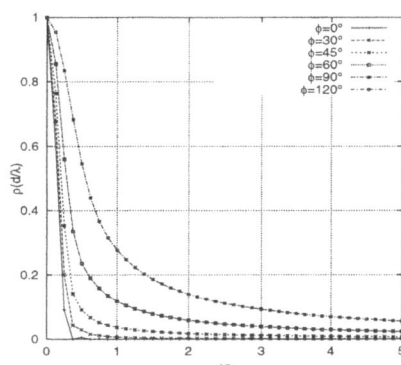

(a) Uniform distribution and main cluster direction $\phi_0 = 45°$

(b) Cosine-shaped distribution with $k = 3$ for different values ϕ_0

Fig. 1. Spatial correlatial plots between the first and third elements in a linear antenna array with eight elements.

The last case concerning circular array arises when a Gaussian distribution is used. In this case the correlation function between two samples taken from the circular array can be obtained by using Bessel's series. Therefore it is possible to write the real and imaginary parts of $\rho_g(m, n)$ as

$$\mathcal{R}\rho_g(m, n) = J_0(\beta a C_{m,n}) + k_3 \frac{2}{\sqrt{\pi}} \sum_{k=1}^{\infty} (-1)^k J_{2k}(\beta a C_{m,n}) I_{ck}(\sigma_\phi, \phi_0, \phi_{m,n})$$

$$\mathcal{I}\rho_g(m, n) = k_3 \frac{2}{\sqrt{\pi}} \sum_{k=0}^{\infty} (-1)^k J_{2k+1}(\beta a C_{m,n}) I_{sk}(\sigma_\phi, \phi_0, \phi_{m,n}),$$

(13)

$$I_{ck} = \frac{\sqrt{\pi}}{2} e^{-2k^2 \sigma_\phi^2} (\cos(2k(\phi_0 - \phi_{m,n})) \mathcal{A}(2k, \sigma_\phi) - \sin(2k(\phi_0 - \phi_{m,n})) \mathcal{B}(2k, \sigma_\phi)),$$

$$I_{sk} = \frac{\sqrt{\pi}}{2} (\cos((2k+1)(\phi_0 - \phi_{m,n})) \mathcal{A}(2k+1, \sigma_\phi)$$

$$- \sin((2k+1)(\phi_0 - \phi_{m,n})) \mathcal{B}(2k+1, \sigma_\phi)) e^{-\frac{((2k+1)\sigma_\phi)^2}{2}}.$$

4 Results

After obtaining the correlation functions for the three arrival angle distributions, for both linear and circular arrays, it is possible to show some numerical results, using the correlation envelope $\rho(m, n) = \mathcal{R}^2(\rho(m, n)) + \mathcal{I}^2(\rho(m, n))$. The first results in this section were obtained for uniform e cosine-shaped distributions. A

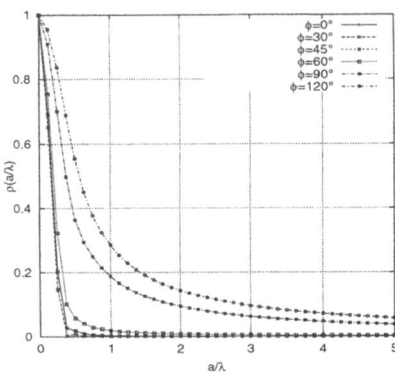 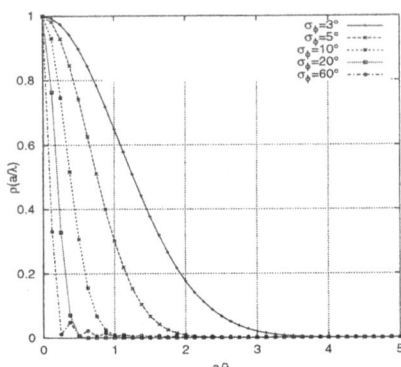

(a) Cosine-shaped distribution with $k = 3$ for different values of ϕ_0, considering the fourth and seventh elements

(b) Gaussian distribution with $\phi_0 = 90°$ for different values of σ_ϕ, considering the second and fifth elements

Fig. 2. Spatial correlation plots between elements in a circular antenna array with eight elements.

linear array with 8 elements with distance d between each element was considered. The curves obtained are shown in Fig.1(a) e Fig.1(b). The second results were obtained for Gaussian and cosine-shaped distributions, considering a circular array with 8 elements and radius a. The plots obtained can be seen in Fig.2(a) and Fig.2(b) As can be seen in Fig. 2(b) the spatial correlation depends on the array structure and main cluster angular direction. When $\phi_0 = 90°$ and Gaussian distribution is used, the circular array presents total spatial uncorrelation. Moreover, as it is known the cosine-shaped distribution approaches the Gaussian distribution when the parameter k increases. So, when a circular array is used, the spatial correlation decreases faster with smaller values of radius a. When a Gaussian distribution is used, the correlation envelope tends to zero faster than for the other cases.

5   Conclusion

This paper presented a comparative study between two array configurations concerning three types of arrival angular distribution: uniform, cosine-shaped and Gaussian distribution. Two commonly used array configurations were analyzed, the linear array and the circular array. Closed forms expressions were obtained for the correlation functions of the cosine-shaped distribution for three values of the parameter k and, as shown for the case of circular array, $k = 3$ is sufficient to obtain good results, while with linear array this value was insufficient.

References

1. Lee, W.C.Y.: Effects on Correlation Between Two Mobile Radio Base Station Antennas. Proc. IEEE Trans. on Communications **21** (1973) 1214–1224
2. Salz, J., Winters, J.H.: Effect of Fading Correlation on Adaptive Arrays in Digital Mobile Radio. IEEE Trans. on Vehicular Technology **43** (1994) 1049–1057
3. Trump, T., Ottersten, B.: Estimation of nominal direction of arrival and angular spread using an array of sensors. Signal Processing, Elsevier (1996)
4. Fuhl, J.: Smart Antennas for Second and Third Generation Mobile Communications Systems . PhD thesis, Technische Universitat Wien, A-2803 Schwarzenbach, Eggenbuch 17 (1997)
5. Ottersten, B.: Array Processing for Wireless Communications. Proc. Workshop on Statistical and Array Processing, Corfu, Greece (1996) 466–473
6. Lee, W.C.Y.: Applying the Inteligent Cell Concept to PCS. IEEE Trans. on Vehicular Technology **43** (1994) 672–679
7. Adachi, F., Feeney, M.T., Williamson, A.G., Parsons, J.D.: Crosscorrelation between the envelopes of 900 MHz signals received at a mobile radio base station site. IEE Proceedings **133** (1986) 506–512

Propagation Prediction Based on Measurement at 5.8GHz for Fixed Wireless Access

Tang Min Keen and Tharek Abdul Rahman

Wireless Communication Centre, Faculty of Electrical Engineering,
Universiti Teknologi Malaysia,
81310 UTM Skudai, Johor, Malaysia.
minkeen@yahoo.com, tharek@fke.utm.my

Abstract. This paper reviews most commonly used ray tracing techniques and applies a ray tracing technique which incorporates site specific environmental data to predict path loss in newly constructed hostels in Universiti Teknologi Malaysia (UTM) for 5.8 GHz Industrial, Scientific and Medical (ISM) band in Malaysia. Radio propagation path loss has been measured in order to verify the predicted results. As the prediction model excluding the vegetation affects that appeared in the fresnel zone clearance in the real site environment, corrections has been done on the predicted total loss in taking account the obstruction loss. It indicates a good agreement between measurements and predicted result with a deviation range of 0.01 dB to 2.82 dB.

1 Introduction

Radio propagation is heavily site-specific and can vary significantly depending on the terrain, frequency of operation, velocity of the mobile terminal, interference sources and other dynamic factors. To achieve high radio system availability, apart from the radio equipment design, good location of radio antenna sites, good radio path planning and choice of an interference-free radio channel are most important. [1] Hence, accurate prediction of radio wave propagation in a communication channel is essential before installation of any wireless system. Site specific analysis tools have proliferated for this purpose. The used of these tools has also been boosted by the availability of detailed city or building maps in electronic format. One of the basic site-specific analysis tools is ray tracing software module that applied in this paper. This technique of prediction is then verified and enhanced with actual RF measurements in the possible installation scenario.

There are two main options that are available for the implementation of a ray tracing software module known as ray launching, and point-to point ray tracing approach. [2]-[5] Both of them have their individual pros and cons. Ray tracing computes all rays receiver point individually but require an extremely high computation times. Thus, to make this technique computationally feasible, many acceleration techniques have been proposed to be implemented in this approach. On the other hand, ray

J.N. de Souza et al. (Eds.): ICT 2004, LNCS 3124, pp. 74–81, 2004.
© Springer-Verlag Berlin Heidelberg 2004

launching [6]-[8] is an option that the casting of rays from transmitter is in a limited set directions in space causing inaccuracy for those rays traveling long distance. A small constant angle separation between launched rays needs to be specified to produce reliable results. Though, this technique is very efficient computationally.

In conjunction with the two options available, there are authors that mixed the two techniques by splitting the three dimensional (3D) into two successive two-dimensional stages, without loss of generality compared with the full 3D techniques. [3], [9]-[11]

All the options in ray tracing software models have been widely used as simulation tools for the design and planning of wireless system in mobile and personal communication environments: outdoor macro cells, street micro cells, and indoor pico cells. All the available modeled ray tracing techniques approximate electromagnetic waves as discrete propagating rays that undergo attenuation, reflection, diffractions and diffuse scattering phenomena if available due to the presence of buildings, walls, and other obstructions. The total received electric field at a point is the summation of the electric fields of each multipath component that illuminates the receiver. These models have the advantage of taking 3D environments into account, and are thus theoretically more precise. In addition, they are adaptable to environment changes such as transmitter location, antenna position and frequency and predict wideband behavior as well as the waves' direction of arrival.

This paper is organized as follow: Firstly, a brief description of a ray tracing model that applied in this paper. This is followed by experimental setup and results. Then, predicted results are corrected to take accounts the vegetation obstructions in the fresnel clearance and presented in Section IV. Finally, discussions and comparisons of experimental results and predicted result are presented.

2   Propagation Prediction Model

The model applied in this paper is based on a 3D Vertical Plane Launch (VPL) ray tracing technique, developed in [12]. The VPL approach accounts for specular reflections from vertical surfaces and diffraction at vertical edges and approximates diffractions at a horizontal edge by restricting the diffracted rays to lie in the plane incidence. Some limitations and simplifications arise from this software to obtain a computational efficient model. This model neglects diffuse scattering from the walls, rays that travel under a structure and also reflections from the rooftop that travel upward and hence away from the buildings and receivers. It is believed that the rays do not contribute to the total received power in a microcellular environment, or that they occur very infrequently.

To save computer time, we restricted the number of reflections to six for each branch between vertical diffractions. The number of diffractions on horizontal wedges is not limited in any of the cases. Due to diffraction by vertical wedges is very time consuming, a limitation on the number of diffraction at vertical edge is done where any given ray path to at most two. Besides, $\varepsilon_r = 6$ is used for the reflection coefficient at walls [11] because the use of reflection coefficient for a dielectric half space with $\varepsilon_r \approx 5 - 7$ give the least error with measurements. Vegetation effects did not considered in this model due to the irregularity of the plantations along the paths. Nevertheless, corrections will be done in the end of the prediction to obtain an accurate prediction.

2.1   Site Survey

A visit to the related site is carried out. There are mainly seven blocks three-wings with eight to ten floors buildings in first hostel and two u-shaped with five floors buildings in another hostel. Transmit site is at Wireless Communication Centre (WCC), which is located at least 30 meters higher than the hostels' building. The terrain between WCC and the hostels is a small oil palm plantation. Hence, the site overlooked a terrain of light rolling hills with moderate tree densities. From the highest floor of WCC, we can clearly see these buildings and the oil palm plantation. Figure 1 (a) and (b) show the photos that captured from WCC to both the hostels, whereas figure 1(c) shows the photo that is captured from one of the receiver site in the hostel to WCC.

(a)   (b)   (c)

Fig. 1. Photographs Illustrating the Site Related

2.2   Geometry Databases

The first step of site specific propagation prediction is characterizing the geometrical and electrical attributes of the site. The obtained building plans and contour maps are digitized into databases. This prediction area covers 720 X 1280meter^2. The same building database and terrain database will be used in the simulation to predict and analyze the result on different placements of the receiver point. Figure 2 shows the visualization of the building and terrain databases for the software.

2.3   Antenna Parameters

Besides geometry databases, antenna radiation pattern and gain are also important inputs in the software. Market available planar array directional antennas are used in the wireless measurement for both the transmitting and receiving sites. Unlike omni-directional antenna that used in mobile systems, the received power can be higher as much as the antenna's gain, but only if the arriving rays lie in the angular range of main lobe. Hence, before wireless prediction and measurement being carried out, is a need to verify the antenna radiation pattern and the gain of the antenna. Measurement of radiation pattern and gain of the antenna at 5.77 GHz has been carried out in an anechoic chamber followed the procedure which printed in [13]-[15]. Two 2D patterns have been measured. They are the x-z plane (elevation plane; $\varphi = 0$), which represents the principal E-plane and the x-y plane (azimuthal plane; $\theta = \pi/2$), which represents the principal H-plane. Figure 3(a) and (b) show the principal E- and H-plane radiation pattern in polar-logarithmic form.

Fig. 2. Databases Visualization

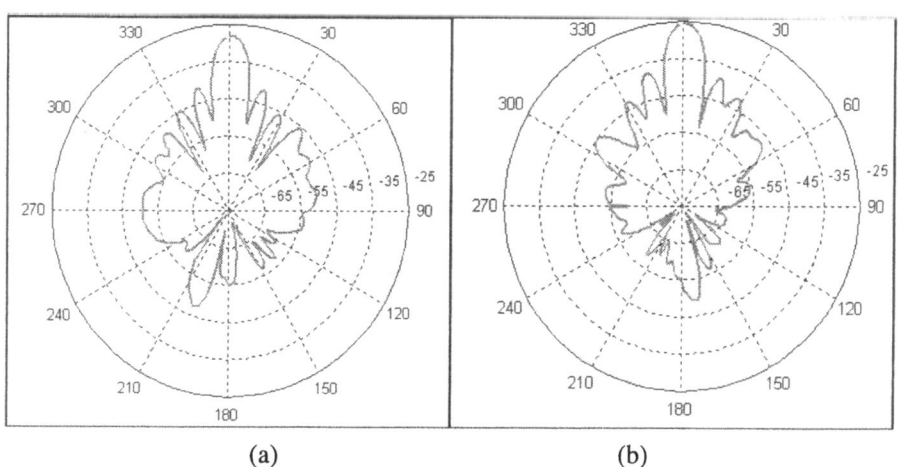

(a)	(b)

Fig. 3. (a) Principle E-Plane. (b) Principle H-Plane.

Two-antenna method of absolute-gain measurement has been carried out to obtain the gain of the antenna. This measurement was carried out with both the antennas were polarization matched and the separation between the antennas prevail far-field conditions. From the measurement, the computed gain of the antenna is 23dBi.

3   Experimental Setup

This measurement campaign was carried out in nine blocks of two hostels' buildings that include 35 local area path loss measurements, with transmitter antenna was placed at the rooftop level of Wireless Communication Centre, UTM. After determining the locations for the transmitter and receiver, the antennas are mounted onto a pole with a proper polarization. The antennas and the mounting brackets used are able to withstand strong winds to avoid any movement that could introduce misalignment. The outdoor unit is then mounted to the mounting bracket, connected to the antenna via a short RF cable and connected to indoor terminal via an IF cable. At every site, the antenna is aligned in both the azimuth and elevation planes until maximum received signal level is obtained. As all operator communications with the measurement system is achieved over the Ethernet port using hypertext transfer protocol (HTTP), it eases the access and control the terminal remotely from any geographical location.

The 35 radio paths propagated from the transmitter passing through an oil palm plantation before reaching the receivers. The distances for these links were ranges from 360 to 605 meter with the transmitter was higher than the oil palm plantation and the receivers. To assure that propagation channel were stationary in time, the measured data was averaged over 30 instantaneously sampled values in 15 minutes.

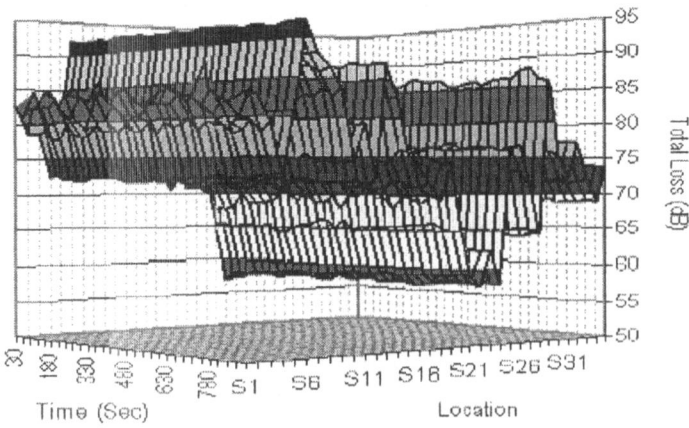

Fig. 4. Measured Loss

4   Prediction and Measurement Comparison

The discussed ray tracing wave propagation prediction software tool is used to compute the path loss value with respect to the slow fading process at the given receiver locations. Visualizations were done on every receiver location in order to understand the different ray components contained in every channel.

The 35 set experimental data obtained from field measurement is shown in Figure 4. The instantaneously sampled values are found having small fluctuation within 3.4 dB maximum ranges and with standard deviation less than 1.1 dB. Mean of the 30 sampled values is computed for comparison with prediction then.

From the 35 local area path loss predictions, there are 11 locations matched closely with measurement loss, with error range from 0.09 dB to 2.78 dB. These locations are classified with links that have 100% fresnel zone clearance with LOS or NLOS conditions. The predictions loss for another 24 locations has range from 3.61 to 33.03 dB less than measurement data. The severe difference is due to the irregular vegetation effects in fresnel zone which is hard to model and is not considered in VPL ray tracing prediction. All of these 24 locations are in LOS conditions. Hence to consider the obstruction loss of vegetation in fresnel zone, the deviation within free space loss and measurement loss will be used. The free space loss model is used to predict received signal strength when the transmitter and receiver have a clear, unobstructed LOS path between them. [16] The equation of the free space loss with R distance at frequency 5.775 GHz is

$$PL_a = 107.671 + 20\log_{10} R_{bm} \tag{1}$$

Figure 5 shows comparisons between predicted path loss and mean measured path loss, and also the computed free space loss. After VPL predicted path loss added with the computed vegetation obstruction loss, the difference between the modified predicted data and measured data is displayed in Figure 6 From this figure, is found that the modified predicted loss have a very good agreement with the measured loss. Both the computed mean and standard deviation for the error are 0.95 dB only.

5   Conclusion

The VPL ray tracing is based on an efficient 3D ray construction algorithm, taking into account a sufficient number of reflections and diffractions. Radio propagation path loss has been measured in order to verify the predicted results. As the prediction lack of important vegetation lines that appeared between the transmitter and receiver, the difference between experimental data and free space loss had been used to represent the vegetation obstruction loss in fresnel zone clearance and to correct the predicted data. The corrections were done only to those predicted ray in LOS condition with vegetation obstructions in fresnel zone and had severe difference with experimental data. Out of the 35 locations, 24 predicted were corrected. And finally, the 24 corrected data and 11 uncorrected predicted data showed good agreement with the measurement data, which the mean error and standard deviation lower than 1 dB. These small errors might be contributed by the experimental configurations such as terrain and building data inaccuracies such as wall orientation error, wrong earth levels and missing information likes construction material characteristics.

Overall, for links with LOS conditions, the obtained predicted path loss and power delay profiles that without consideration of vegetation effect can be corrected or modified accordingly in such a way described above before employed as a tool for aiding in the planning and design of wireless system, due to their accuracy and efficiency. Other LOS, OLOS or NLOS links that without vegetation obstruction in frenel zone clearance, this VPL ray tracing tool is able to provide a good prediction. For future work, the analysis of radio propagation with this method finally provides a complete set of output magnitudes that can be used to analyze BER for digital modulations and to characterize the channel where strategies like diversity, equalization, or adaptive antennas are used.

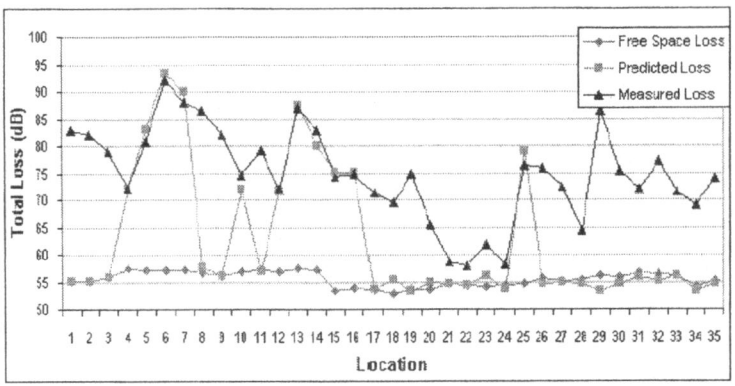

Fig. 5. Comparisons between Free Space Loss, Predicted Loss, and Measured Loss

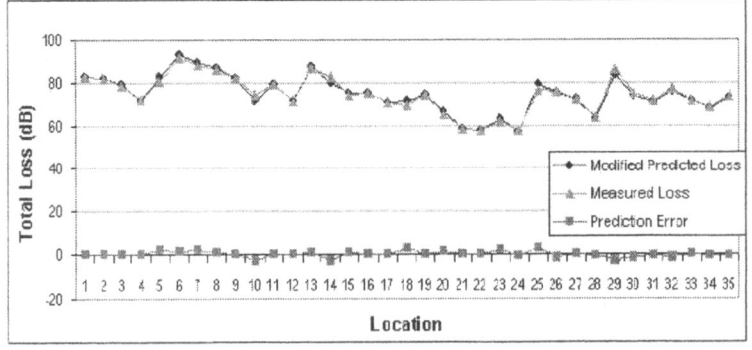

Fig. 6. Comparisons Between Modified Predicted Loss and Measured Loss

References

1. M. P. Clark: Wireless Access Network. John Wiley & Sons. West Sussex P019 1UD, UK (2000) 139-142
2. F. A. Agelet, A. Formella, J.M.H. Rabanos, F. Isasi de Vicente and F. P. Fontan: Efficient ray-tracing acceleration techniques for radio propagation modeling. IEEE Trans.Vehicular Tech., Vol. 49. No. 6. (2000) 2089-2104
3. A. Toscano, F. Bilotti and L. Vegni: Fast ray-tracing technique for electromagnetic field prediction in mobile communications. IEEE Trans. Magnetics. Vol.39. No.3. (2003) 1238-1241
4. C. F. Yang, B. C. Wu and C.J. Ko: A ray-tracing method for modeling indoor wave propagation and penetration. IEEE Trans. Ant. Propagation. Vol. 46. No.6. (1998) 907-919
5. S. H. Chen and S. K. Jeng: An SBR/image approach for radio wave propagation in indoor environments with metallic furniture. IEEE Trans. Ant. Propagation. Vol. 45. No.1. (1997) 98-106
6. G. Durgin, N. Patwari and T. S. Rappaport: Improved 3D ray launching method for wireless propagation prediction. Electronics Letters. Vol. 33. No.16. (1997) 1412-1413
7. G. Durgin, N. Patwari and T. S. Rappaport: An advanced 3D ray launching method for wireless propagation prediction. IEEE 47th Vehicular Technology Conference. Vol. 2. (1997) 785-789
8. E. Costa: Ray tracing based on the method of images for propagation simulation in cellular environments. 10th International Conference Antennas and Propagation. Vol. 2. 1 (1997) 204-209
9. J.P Rossi, J.C Bic, A.J. Levy, Y. Gabillett and M. Rosen: A ray launching method for radio-mobile propagation in urban area. Antennas and Propagation Society International Symposium, AP-S. Digest. Vol.3. (1991) 1540-1543
10. J. P. Rossi and Y. Gabillet: A mixed ray launching/tracing method for full 3-D UHF propagation modeling and comparison with wide-band measurements. IEEE Trans. Ant. Propagation. Vol.50. No. 4 (2002) 517-523
11. G. Liang and H.L.Bertoni: A new approach to 3-D ray tracing for propagation prediction in cities. IEEE Trans. Ant. Propagation. Vol. 46. No. 6 (1998) 853-863
12. George Liang, User's Manual for Site Specific Outdoor/Indoor Propagation Prediction Code, Site Ware Technologies, Inc. (1999)
13. ANSI/IEEE Std. 149-1979: Test Procedures for Antennas.
14. C. A. Balanis: Antenna Theory: Analysis And Design. 2nd edn. John Wiley & Sons, Inc. 605 Third Avenue New York (1997) 839-844
15. W. L. Stutzman and G. A. Thiele: Antenna Theory and Design. 2nd edn. John Wiley & Sons, Inc. 605 Third Avenue New York (1998) 409-415
16. Theodore S. Rappaport: Wireless Communications Principles and Practice. Prentice Hall. New Jersey, USA (1996) 70-71

Dimensioning of Wireless Links Sharing Voice and Data

Marcelo de Oliveira Marques[1] and Ivanil Sebastião Bonatti[2]

[1]National Institute of Telecommunications
P.O. Box 05 – 37.540-000 – Santa Rita do Sapucaí – MG – Brazil
marcelo@inatel.br
[2]School of Electrical & Computer Engineering – University of Campinas
P.O. Box 6101 – 13.083-970 – Campinas – SP – Brazil
ivanil@dt.fee.unicamp.br

Abstract. An algorithm for wireless link dimensioning using an approximation technique that allows the decomposition of a two-dimensional Markov chain in two one-dimensional chains is presented. The algorithm was applied to GSM/GPRS links considering as service quality parameters, the voice blocking probability, the data blocking probability and the data waiting average time. The number of channels and the preemption threshold of voice channel over the data channel have been determined, illustrating the adequacy and efficiency of the considered algorithm.

1 Introduction

The introduction of the multiple access service GPRS (*General Packet Radio Service*) on GSM networks (*Global System for Mobile Communications*) increases the use efficiency of radio resource through the sharing of data and voice users on the available channels in GSM networks [1]. The GPRS uses the GSM frequency canalization whose carrier supports a TDMA frame (*Time Division Multiple Access*) composed of eight time-slots. Users can share the transmission of voice, PDUs (*Protocol Data Unit*) and signaling over several frequency carriers.

When a mobile station wants to transmit a PDU (*Protocol Data Unit*), it requests the creation of a TBF (*Temporary Block Flow*) through the random access channel PRACH (*Packet Random Access Channel*). The mechanism of TBF creation can be viewed as the establishment of a virtual circuit defining the connection between the mobile station and the base station. This virtual circuit remains *connected* and busy during the time necessary to the complete transmission of the PDU.

The integration of voice and data users can be implemented on packet switching networks treating both in the same way, but giving priority to the voice users over the data users, in order to guarantee a maximum time delay to the voice channels [2], [3]. The multiple access protocols with reserve [4] allow the allocation of channels repeating cyclically in the appropriate cadence to synchronous users. The remaining system capacity is used to send the data signals that are sensitive to losses, but not to delays. Such integration can be implemented, for example, by the combination of the protocol TCP/IP (*Transmission Control Protocol/Internet Protocol*) for data with the protocol UDP (*User Datagram Protocol*) for voice. *Chen et al.* have investigated the

J.N. de Souza et al. (Eds.): ICT 2004, LNCS 3124, pp. 82–87, 2004.
© Springer-Verlag Berlin Heidelberg 2004

voice and data sharing with priority for the voice service and buffering the preempted data packets [5]. *Meo et al.* have investigated a dynamic sharing strategy using the information of the buffer occupation as reference for allocation of exclusive channels for data packets [6].

In this paper, an algorithm for the dimensioning of wireless links is considered using an approximation technique [7] that allows the decomposition of the two-dimensional Markov chain, in two one-dimensional chains. The assumptions and the system model are presented in Section 2, determining the capacity of the wireless link. Numerical results and discussions are presented in Section 3, and the Conclusions are presented in Section 4.

2 System Model

Each mobile station generates PDUs having Poisson distribution with average rate λ_d /n, n being the number of mobile stations in the system. The lengths of the PDUs have exponential distribution with average size $E\{R\}$. Voice calls are generated with average rate λ_v / n having Poisson distribution, with service average time $1/\mu_v$ and are exponentially distributed. The blocked voice calls are cleared of the system.

The transmission system is modeled with $C = C_v + C_d$ channels, each one has a transmission rate of 13.4 kbps (CS2 codification scheme used in GSM).

A channel, when busy due a voice call, remains unavailable during the conversation time, and is set free in the end of the communication. When busy for a PDU transmission (a mobile station is sending a PDU), the channel remains unavailable during the transmission of the PDU, unless it suffers preemption from a voice call. In that case, the remaining not transmitted bytes of the PDU are buffered and are priority sent, in a next successful access. Each station has a buffer to store the PDUs when the access is denied, being b the limit for the maximum number of PDUs in the system. If a PDU is generated and the buffer is full, the PDU is blocked and cleared of the system.

Both the voice calls and PDUs can access C_v channels, from the total of C channels, with preemption for the voice calls over the PDUs. The $C_d = C - C_v$ channels can be exclusively taken by PDUs. Each PDU uses only one channel at a time. Therefore, the resulting channel service time is equal to the division of the PDU length by the channel transmission rate (13.4 kbps). The packet retransmissions due to the error control scheme increase the PDUs average length and the parameter $1/\mu_d$ takes into account these retransmissions.

The complexity of the Markovian chain significantly grows when the number of channels or the buffer length of the system increases, making its use impractical for dimensioning purposes.

Ghani and Schwartz has shown that the probability of the number of voice channels and the conditional probability of the number of data channels could be approximated by independent queues [8]. In general, the time to transmit a PDU is significantly lesser than the duration of a voice call, and the quality of the approximation improves as the difference between the service time of the voice and data channels increases.

The number of channels allocated to voice users, Y, is described by the $M/M/C_v/C_v$ queue since, thanks to the preemption mechanism, all C_v channels are available to the voice users

The conditional number of data users (PDUs), N for a given y, can be approximated by the $M/M/C_d+x/b$ queue.

The maximum current number of channels that are used by PDUs is $C_d + x$, with $x = C_v - y$, and the preempted PDUs are not considered in the buffer of the conditional queue approximation.

The traffic parameters $\rho = \lambda_d / \mu_d$ and $\rho_v = \lambda_v / \mu_v$ and the structural parameters C, C_d and b are considered known when performing analysis is computed. The main purpose is to determine the service quality parameters: P_v, voice channels blocking probability; P_d, data blocking probability and W_d, PDUs average waiting time.

The QoS parameters P_{vspec}, P_{dspec} and W_{dspec} are specified as goals to be achieved with a minimum number of channels in the system (C and the threshold for data C_d) when dimensioning the wireless link. The voice and PDU demand traffic forecasts, characterized by ρ_v and ρ parameters, are assumed to be known.

From the $M/M/C_v/C_v$ queue, the voice channels blocking probability is given by the Erlang Formula [10].

The number of channels available for PDU transmission is dependent of the number of channels occupied by voice users. Therefore, the conditional probability of $C_d + x$ channels available for data is given by

$$q_x = \left(\sum_{k=0}^{C_v} \frac{\rho_v^{\,k}}{k!} \right)^{-1} \frac{\rho_v^{\,C_v - x}}{(C_v - x)!} \,, \qquad x = 0,1,...,C_v . \qquad (1)$$

The data channels can be modeled by the $M/M/m = C_d + x/b$ queue, where b is the limit for the maximum number of PDUs in the system.

The probability of transmitting k PDUs is given by [9].

$$p_d(k,m) = \begin{cases} p_d(0,m) \dfrac{\rho^k}{k!} & \text{for } 1 \le k < m = C_d + x. \\[3mm] p_d(0,m) \dfrac{\rho^k}{m!\, m^{k-m}} & \text{for } m \le k \le b. \end{cases} \qquad (2)$$

with

$$p_d(0,m) = \left[\sum_{k=0}^{m-1} \frac{\rho^k}{k!} + \frac{\rho^m \left(1 - \left(\dfrac{\rho}{m} \right)^{b-m+1} \right)}{m! \left(1 - \left(\dfrac{\rho}{m} \right) \right)} \right]^{-1} . \qquad (3)$$

The data channel probability distribution function is given by

$$P_{da}(d) = \sum_{x=0}^{C_v} q_x \cdot p_d(d,m), \qquad d = 0,1,...,b. \qquad (4)$$

A new TBF requisition will be blocked if $k = b$ PDUs are in the system. Thus, the conditional blocking probability is given by $p_d(b,m)$ and the data blocking probability, P_{da}, is

$$P_{da} = \sum_{x=0}^{C_v} q_x \cdot p_d(b,m). \qquad (5)$$

The average waiting time, W_d, is given by (Little Theorem)

$$W_d = \frac{1}{\lambda_d(1-P_{da})} \cdot \sum_{x=0}^{C_v} q_x L(m) - \frac{1}{\mu_d}. \qquad (6)$$

where $L(m)$ is the average conditional number of PDUs given by

$$L(m) = p_d(0,m)\left(\frac{\rho^m \frac{\rho}{m}\left(1-\left(\frac{\rho}{m}\right)^{k-m+1} - \left(1-\frac{\rho}{m}\right)(k-m+1)\left(\frac{\rho}{m}\right)^{k-m}\right)}{m!\left(1-\frac{\rho}{m}\right)^2}\right) + \rho(1-p_d(b,m)). \qquad (7)$$

2.1 Algorithm for Synthesis

The proposed algorithm has two phases. First, the number of channels C_v, which are shared by the voice and data users, is determined guarantying the bound for the voice-user blocking probability specification, P_{vspec}. Second, the number of channels, C_d, which are exclusively allocated to data users, is determined to attain the specified bounds W_{dspec} (average waiting time) and P_{dspec} (data blocking probability) values.
The determination of C_v implies the repetitive computation of the Erlang Formula, which is computed by an iterative procedure to overcome numerical difficulties [10]. The Erlang formula $E(\rho_v,n)$ gives the blocking probability of a Poissonian traffic, with intensity ρ_v, offered to a link with capacity n. In the dimensioning process the inverse of the Erlang Formula $n(\rho_v,P_v)$ is needed, which produces the minimum number of channels having blocking probability lesser than P_v.
The bisection algorithm [12] can be used to solve the numerical inequality that produces the inverse of Erlang function $n(\rho_v,Pv)$. The upper bound is given by n_s and the average number of occupied channels gives the lower bound n_l [11].
The threshold C_d can be also computed by the bisection algorithm taking the initial lower bound as zero and the initial upper bound as one. The stop condition is achieved when the data average waiting time and the data blocking probability are both lesser than the specified values.

3 Numerical Results and Discussions

The number of channels C_v and C_d ($C = C_v + C_d$) are shown in Table 1, for several values of the traffic ρ_v and ρ with parameter values: $E\{R\} = 1675$ bytes (13400 bits), $P_{vspec} = $ 1e-2, $W_{dspec} = 1$ s, $P_{dspec} = $1e-3, $\Delta = \{0,30\}$ and $b = (C_v + C_d + \Delta)$.

Note that, for the $\Delta = 0$, the number of channels exclusively dedicated to the data users (C_d) is quite small due the effectiveness of sharing the channels (C_v) used by the voice users. Note also that the number of C_d channels is inferior of the one that would be necessary, $C_{d\_no\ sharing}$, if the strategy of sharing the C_v channel was not used.

If the buffer total length is enough only to store the PDUs that are pre-empted by the voice users ($b = C_v + C_d$), i.e. ($\Delta = 0$), then the number of channels exclusively dedicated to the data users (C_d) significantly increases for small values of the voice traffic ρ_v. As the voice traffic ρ_v increases, the number of channels dedicated to data users C_d becomes very near of the values for $\Delta = 30$, indicating the effectiveness of the sharing mechanism. The number of channels C_v and the threshold C_d were also computed for buffers with $\Delta > 30$ giving pratically the same results as for $\Delta = 30$. Meaning that this value of Δ is big enough when computing the number of channels.

Table 1. C_v and C_d as function of ρ_v and ρ for $P_{vspec} = $ 1e-2; $W_{dspec} = 1$ s; $P_{dspec} = $ 1e-3; $\Delta = 30$ and $\Delta = 0$

				$\Delta = 30$			$\Delta = 0$			
ρ_v	C_v	ρ	C_d	$C_{d\ no\ sharing}$	P_d/P_{dspec}	W_d/W_{dspec}	C_d	$C_{d\_no\ sharing}$	P_d/P_{dspec}	W_d/W_{dspec}
1	5	1	1	2	0.08	0.062	2	6	0.123	2.2e-3
10	18	1	1	2	0.14	0.174	1	6	0.357	0.067
30	42	1	1	2	0.10	0.267	1	6	0.168	0.156
1	5	10	9	12	0.92	0.159	16	21	0.966	2e-4
10	18	10	9	12	0.98	0.090	10	21	0.684	0.023
30	42	10	9	12	0.88	0.100	9	21	0.993	0.062
1	5	30	31	35	0.55	0.062	42	48	0.996	6e-5
10	18	30	29	35	0.80	0.065	33	48	0.774	0.007
30	42	30	29	35	0.45	0.051	29	48	0.758	0.036

In order to analyze the robustness of the values obtained for the number of channels in the synthesis process some simulations have been made. The behavior of the P_d blocking probability and the waiting time W_d in function of the normalized voice traffic were computed.

There was a small performance degradation of the data users due the sharing mechanism. The sensitivity of the P_d data blocking probability and of the W_d waiting time to the variation of the voice traffic is larger for bigger values of the nominal voice traffic. The relative percentage of busy voice channels increases as the nominal value of the traffic is augmented. Therefore, the mechanism of sharing the channels C_v can take better profity when the nominal voice traffics are small.

4 Conclusions

The dimensioning of wireless links has been considered in this paper. The minimum number of channels that produces limited voice and data blocking probabilities and limited waiting time for the data traffic has been obtained. The voice users have preemption over data users in a parcel of channels available in the system. The threshold on the number of channels that can suffer preemption is also determined by the proposed synthesis method. The method is based on an approximation that allows decomposing the Markovian chain describing the system into two queues with analytical solutions. The proposed algorithm is computationally simple and efficient. Its application in GPRS networks illustrates the adequacy of the method to wireless links dimensioning.

References

1. ETSI TS 101 350, „General Packet Radio Service (GPRS), Overall Description of the GPRS Radio Interface, Stage 2" (1999).
2. Zahedi, A. and Pahlavan, K., „Capacity of a Wireless LAN with Voice and Data Services", IEEE Transactions on Communications, Vol. 48, No. 7, pp. 1160-1170 (2000).
3. Kalden, R., Meirick, I. and Meyer, M., „Wireless Internet Access Based on GPRS", IEEE Personal Communications, April, pp 8 – 18, (2000).
4. Kwok, Y. K. and Lau, V. K. N., „Performance Evaluation of Multiple Access Control Schemes for Wireless Multimedia Services", IEEE Proc. Commun., Vol. 148, No.2, pp 86-94 (2001).
5. Chen, W., Wu, J.C. and Liu, H., „Performance Analysis of Radio Resource Allocation in GSM/GPRS Networks", Vehicular Technology Conference, Proceedings. VTC 2002-Fall. 2002 IEEE 56th, Vol. 3, pp 1461-1465 (2002).
6. Meo, M., Marsan, M. A. and Batetta, C., „Resource Management Policies in GPRS Wireless Internet Access Systems", Proceedings of DNS'02 (2002).
7. Ni, S. and Haggman, S., „GPRS performance estimation in GSM circuit switched services and GPRS shared resource systems", Proceedings of IEEE WCNC'99, Vol. 3, pp 1417-1421 (1999).
8. Ghani, S. and Schwartz, M., „A Decomposition Approximation for the Analysis of Voice/Data Integration", IEEE Transactions on Communications, Vol. 42, No. 7, pp. 2441-2452 (1994).
9. Gross, Donald and Harris, C.M., „Queueing Theory", John Wiley & Sons, Inc (1998).
10. Bear, D., „Principles of Telecommunications – Traffic Engineering", Peter Peregrinus LTD (1976).
11. Berezner, S.A., Krzesinski, A. E. Taylor, P. G. Taylor, „On the Inverse of Erlang's Function", Journal of Applied Probability, Vol. 35, pp. 246-252 (1998).
12. Lasdon, L.S., „Optimization Theory for Large Systems", Macmillan Series in Operations Research (1970).

An Iterative Matrix-Based Procedure for Finding the Shannon Cover for Constrained Sequences

Daniel P.B. Chaves[1], Cecilio Pimentel[1], and Bartolomeu F. Uchôa-Filho[2*]

[1] Communications Research Group - CODEC
Department of Electronics and Systems
Federal University of Pernambuco
P.O. Box 7800 - 50711-970 Recife - PE - BRAZIL
cecilio@ufpe.br
[2] Communications Research Group - GPqCom
Department of Electrical Engineering
Federal University of Santa Catarina
88040-900 - Florianópolis - SC - BRAZIL
uchoa@eel.ufsc.br

Abstract. In many applications such as coding for magnetic and optical recording, the determination of the graph with the fewest vertices (i.e., the Shannon cover) presenting a given set of constrained sequences (i.e., shift of finite type) is very important. The main contribution of this paper is an efficient iterative vertex-minimization algorithm, which manipulates the symbolic adjacency matrix associated with an initial graph presenting a shift of finite type. A characterization of this initial graph is given. By using the matrix representation, the minimization procedure to finding the Shannon cover becomes easy to implement using a symbolic manipulation program, such as Maple.

1 Introduction

The discrete sequences used to transmit or store digital information often have to satisfy constraints on the occurrence of consecutive symbols [1]. Examples of constraints adopted in commercial systems are found in [2]. The set of all bi-infinite sequences satisfying a certain constraint is referred to in the symbolic dynamics literature as a *shift space*. A shift space is called *shift of finite type* if it may be specified in terms of a finite list of forbidden strings. Such set of constrained sequences can also be specified by a labeled directed graph, called the *presentation* of the shift space [3, Theorem 3.1.5]. It is often desirable to find a right-resolving presentation (the outgoing edges of each vertex are labeled distinctly) with the smallest number of vertices. The Shannon cover is this

---

* The authors acknowledge partial support of this research by the Brazilian National Council for Scientific and Technological Development (CNPq) under Grants 302402/2002-0 and 302568/2002-6.

J.N. de Souza et al. (Eds.): ICT 2004, LNCS 3124, pp. 88–93, 2004.

minimal presentation. From a practical viewpoint, the determination of such minimal right-resolving presentation is very important because many codes for constrained systems, such as finite-state codes for magnetic and optical disks [4], are constructed from this graph.

The fundamental concept used to construct the Shannon cover is to identify all classes of equivalent vertices of an arbitrary initial presentation. This identification can be carried out using the algorithm presented in [3, p. 92]. In this paper, we propose an efficient alternative algorithm for constructing the Shannon cover for a shift of finite type which iteratively manipulates binary vectors that define the entries of the symbolic adjacent matrix of the graph. The paper is organized in five sections. Section 2 contains background material on symbolic dynamics. Section 3 describes the construction of an initial presentation. We give a characterization of the initial presentation which yields a simpler procedure to identify equivalent vertices. The graph minimization is also discussed in this section. Section 4 summarizes the conclusions of this work.

2   Background on Symbolic Dynamics

This section summarizes relevant background material from symbolic dynamics. We refer the reader to [3] for further definitions and concepts from this mathematical discipline. Let $\mathcal{A}^{\mathbb{Z}}$ be the set of all bi-infinite sequences (called points) $x = \cdots x_{-2}x_{-1}x_0x_1 \cdots$ of symbols drawn from an alphabet $\mathcal{A}$ of finite size, namely $|\mathcal{A}|$. A finite sequence of consecutive symbols from $\mathcal{A}$ is called a *block*. We use the notation $x_{[i,j]} = x_ix_{i+1}\cdots x_j$ to specify a block which occurs in the point x, starting from position i to position j. We say that a point $x \in \mathcal{A}^{\mathbb{Z}}$ *contains* a block w of length $M > 0$ if there exists an index i such that $w = x_{[i,i+M-1]}$. Let $\mathcal{F}$ be a set of blocks over $\mathcal{A}$, called *set of forbidden blocks*. A shift space X is the subset of $\mathcal{A}^{\mathbb{Z}}$ consisting of all points not containing any block from $\mathcal{F}$. A shift space is called an *M-step shift of finite type* if the length of the longest block in $\mathcal{F}$ is $M + 1$. The *language* of X, denoted by $\mathcal{B}(X)$, is the set of finite blocks of symbols which occur in points of X. We say that X is *irreducible* if for every pair of blocks $u,v \in \mathcal{B}(X)$, there is a block $w \in \mathcal{B}(X)$ such that the concatenation uwv is also in $\mathcal{B}(X)$.

A finite directed graph (or simply a *graph*) G consists of a finite set of vertices $\mathcal{V}(G)$, and a finite set of directed edges $\mathcal{E}(G)$ connecting the vertices. A labeled graph is a pair $\mathcal{G} = (G, \mathcal{L})$, where G is the underlying graph of $\mathcal{G}$, and the labeling $\mathcal{L} : \mathcal{E}(G) \to \mathcal{A}$ assigns to each edge a symbol from the alphabet $\mathcal{A}$. We write $I \xrightarrow{a} J$ if there is an edge in G from I to J labeled a. A *path* in G is a block of edges $\pi = e_1e_2 \cdots e_n$ such that the terminal vertex of e_i is the initial vertex of e_{i+1}. The label of π is the block $\mathcal{L}(\pi) = \mathcal{L}(e_1)\mathcal{L}(e_2)\cdots\mathcal{L}(e_n)$. A graph G is irreducible if for every pair of vertices $I,J \in \mathcal{V}(G)$ there is a path from I to J. A vertex $I \in \mathcal{V}(G)$ is *stranded* if either no edges start at I or no edges terminate at I. A graph is *essential* if it has no stranded vertices. A labeled graph is irreducible (resp., essential) if its underlying graph is irreducible (resp., or essential).

A *sofic shift* $X_{\mathcal{G}}$ is the set of sequences obtained by reading off the labels of bi-infinite paths on G. We say that $\mathcal{G}$ presents $X_{\mathcal{G}}$, or $\mathcal{G}$ is a presentation of $X_{\mathcal{G}}$. A block $w \in \mathcal{B}(X_{\mathcal{G}})$ is said to be generated by a path π in G if $w = \mathcal{L}(\pi)$. The incoming or outgoing edges of a stranded vertex cannot possibly occur in any walk on the graph. Then, if $\mathcal{G}$ presents $X_{\mathcal{G}}$, $\mathcal{G}$ has a unique essential labeled subgraph that presents $X_{\mathcal{G}}$. The *follower set* of a vertex $I \in \mathcal{G}$, denoted by $F_{\mathcal{G}}(I)$, is the set of all blocks of all lengths that can be generated by paths in G starting from I. The set of all such blocks of length n is denoted by $F_{\mathcal{G}}^n(I)$. Two vertices I and J are called follower-set equivalent if they have the same follower set. A presentation is follower-set separated if $F_{\mathcal{G}}(I) = F_{\mathcal{G}}(J)$ implies that $I = J$. A block $w \in \mathcal{B}(X_{\mathcal{G}})$ is a *synchronizing block* for $\mathcal{G}$ if all paths in G that generates w terminate at the same vertex, say I.

3   Graph Minimization

We first describe a method to generating an initial presentation, namely $\mathcal{G}_{X[M+1]}$, of a shift of finite type [3, Theorem 3.1.5]. Without loss of generality, we will assume that the forbidden blocks in $\mathcal{F}$ all have the same length $M + 1$.

Let $\mathcal{G}_{X[M+1]}$ be a graph with vertex set $\mathcal{A}^M$, the set of all M-blocks over $\mathcal{A}$. For any two vertices $I = a_1 a_2 \cdots a_{M-1} a_M$ and $J = b_1 b_2 \cdots b_{M-1} b_M$ in $\mathcal{G}_{X[M+1]}$, there is one edge from I to J if and only if $a_2 \ldots a_M = b_1 \cdots b_{M-1}$ and $a_1 a_2 \ldots a_M b_M$ is not in $\mathcal{F}$. The label of this edge is b_M. The labeled graph $\mathcal{G}_{X[M+1]}$ that presents the shift space is right-resolving. If the shift space is irreducible, the graph $\mathcal{G}_{X[M+1]}$ is also irreducible [3, Theorem 2.2.14]. The presentation $\mathcal{G}_{X[M+1]}$ allows a very simple recursive identification of the equivalent classes.

Definition 1. *Let $\mathcal{G} = (G, \mathcal{V})$ be a labeled graph over an alphabet $\mathcal{A}$. Let $\mathcal{I}$ be a set of all vertices in $\mathcal{V}(G)$ possessing the following property: If there is an edge $I \xrightarrow{a} J$ for some $I \in \mathcal{I}$ and for some $a \in \mathcal{A}$, then there are edges from all other vertices in $\mathcal{I}$ to the same terminal vertex J labeled a. The vertices in $\mathcal{I}$ are called* out-edge equivalent, *and $\mathcal{I}$ is an* out-edge equivalence class *of $\mathcal{G}$.*

If two vertices are in different out-edge equivalence classes, then they are called out-edge separated vertices, and if all out-edge equivalence classes of a labeled graph $\mathcal{G}$ have only one element, then $\mathcal{G}$ is called out-edge separated graph. Hereafter, we refer to out-edge equivalent classes with more than one element. A set of out-edge equivalent vertices can be merged by an operation called *in-amalgamation*, defined as follows.

Definition 2. *Let $\mathcal{I} = \{I_1, I_2, \cdots, I_m\}$ be an out-edge equivalence class of a labeled graph $\mathcal{G}$. The goal is to create a new labeled graph $\mathcal{H}$ from $\mathcal{G}$ by means of an operation called in-amalgamation that replaces all vertices within an out-edge equivalent class with one representative of this set, say I_1. This operation redirects into I_1 all edges incoming to $\{I_2, \cdots, I_m\}$, and eliminates the vertices $\{I_2, \cdots, I_m\}$.*

An operation called a *round* of in-amalgamations of out-edge equivalence classes of $\mathcal{G}$ produces a new graph, say $\mathcal{H}_1$, that presents the same sofic shift by successively applying the procedure described in Definition 2 to each equivalent class. We may repeat this process p times until we end up with an out-edge separated graph $\mathcal{H}_p$.

An important property of the initial presentation $\mathcal{G}_{X^{[M+1]}}$ is that all M-blocks in $\mathcal{B}_M(X)$ are synchronizing blocks. Consequently, if $\mathcal{H}_p$ is the out-edge separated graph obtained from $\mathcal{G}_{X^{[M+1]}}$, then $\mathcal{H}_p$ is follower-set separated [5]. We will apply in the next section rounds of in-amalgamations to find the Shannon cover.

3.1   A Matrix-Based Implementation

We develop in this section a systematic procedure to simplify the labeled graph $\mathcal{G}_{X^{[M+1]}}$. Without loss of generality, we assume that the alphabet $\mathcal{A}$ is composed of integer numbers $\mathcal{A} = \{0, 1, \cdots, q-1\}$. We also consider that the vertex set of a labeled graph $\mathcal{G}$ is $\mathcal{V}(G) = \{0, 1, \cdots, |\mathcal{V}(G)| - 1\}$.

Definition 3. *Let $\mathbf{A}(\mathcal{G})$ denote the symbolic adjacency matrix of the right-resolving labeled graph $\mathcal{G} = (G, \mathcal{L})$. The (I, J)th entry of $\mathbf{A}(\mathcal{G})$, denoted by $[\mathbf{A}(\mathcal{G})]_{I,J}$, is the binary sequence $b_0 b_1 \cdots b_{q-1}$, where $b_k = 1$ if there is an edge $I \xrightarrow{k} J$ in $\mathcal{G}$, otherwise, $b_k = 0$, where $k = 0, 1, \cdots q-1$. We say that the entry $[\mathbf{A}(\mathcal{G})]_{I,J}$ is zero if it is a sequence of q zeros (no edge from I to J). Both the row and the column indices of $\mathbf{A}(\mathcal{G})$ vary from 0 to $|\mathcal{V}(G)| - 1$ (rows and columns of $\mathbf{A}(\mathcal{G})$ are indexed by vertices).*

If a labeled graph $\mathcal{G} = (G, \mathcal{L})$ is essential, all rows and columns of $\mathbf{A}(\mathcal{G})$ are nonzero. A symbolic adjacency matrix $\mathbf{A}(\mathcal{G})$ with this property is called essential. Let α, β be subsets of $\mathcal{V}(G)$. The submatrix of $\mathbf{A}(\mathcal{G})$ constructed by selecting the elements of $\mathbf{A}(\mathcal{G})$ with row indices in α and column indices in β is denoted by $\mathbf{A}(\mathcal{G})[\alpha \mid \beta]$. The Ith row and the Jth column of $\mathbf{A}(\mathcal{G})$ are denoted, respectively, by $\mathbf{A}(\mathcal{G})[I, \mathcal{V}]$ and $\mathbf{A}(\mathcal{G})[\mathcal{V}, J]$. It is of interest to define an operation called addition of two columns, in the sense defined below.

Definition 4. *Let $\mathbf{A}(\mathcal{G})[\mathcal{V}, J_1]$ and $\mathbf{A}(\mathcal{G})[\mathcal{V}, J_2]$ be two columns of $\mathbf{A}(\mathcal{G})$. If we regard each column as a vector of length $q \times |\mathcal{V}(G)|$, $\mathbf{A}(\mathcal{G})[\mathcal{V}, J_1] + \mathbf{A}(\mathcal{G})[\mathcal{V}, J_2]$ denotes a symbol by symbol componentwise logic "or" addition.*

It is worth noting that if the matrix $\mathbf{A}(\mathcal{G})$ has a set of m equal rows, say $\mathcal{I} = \{I_1, I_2, \cdots, I_m\}$, the vertices of $\mathcal{G}$ associated with these rows form an out-edge equivalence class.

The symbolic adjacency matrix $\mathbf{A}(\mathcal{H})$ of the labeled graph $\mathcal{H}$ obtained from $\mathcal{G}$ by merging an out-edge equivalence class $\mathcal{I} = \{I_1, I_2, \cdots, I_m\}$ (as described in Definition 2) is easily obtained from $\mathbf{A}(\mathcal{G})$. The first step is to add (according to Definition 4) all columns of $\mathbf{A}(\mathcal{G})$ with indices in $\mathcal{I}$. The result of this operation is $\mathbf{A}(\mathcal{H})[\mathcal{V}, I_1]$. Next, we must zero all rows and columns of $\mathcal{H}$ with indices in $\{I_2, \cdots, I_m\}$. The remaining entries of $\mathbf{A}(\mathcal{H})$ are the same as that of $\mathbf{A}(\mathcal{G})$.

Rounds of in-amalgamations of out-edge equivalence classes produce a sequence of matrices $\mathbf{A}(\mathcal{H}_0) = \mathbf{A}(\mathcal{G}_{X[M+1]})$ to $\mathbf{A}(\mathcal{H}_p)$, where all nonzero rows of $\mathbf{A}(\mathcal{H}_p)$ are distinct. Then, the essential submatrix is obtained from $\mathbf{A}(\mathcal{H}_p)$ by eliminating all rows and columns associated with stranded vertices. The resulting essential submatrix is the Shannon cover.

The entries of the symbolic adjacency matrix $\mathbf{A}(\mathcal{G}_{X[M+1]})$ are calculated as follows. The rows and columns of $\mathbf{A}(\mathcal{G}_{X[M+1]})$ are indexed by elements of the ordered set $\mathcal{P}$ of strings of length M over $\mathcal{A}$. For example, the set $\mathcal{P}$ for $q = 2$ and $M = 3$ is $\mathcal{P} = \{000, 001, 010, 011, 100, 101, 110, 111\}$. The entries of the matrix $\mathbf{A}(\mathcal{G}_{X[M+1]})$ are zero, except when the row $r_1 r_2 \cdots r_M$ overlap by M symbols with the column $c_1 c_2 \cdots c_M$. In this case, the unique nonzero symbol associated with this entry is $b_{c_M} = 1$. The final step to obtain the matrix $\mathbf{A}(\mathcal{G}_{X[M+1]})$ is zeroing exactly $|\mathcal{F}|$ entries of $\mathbf{A}(\mathcal{G}_{A[M+1]})$, those for which $r_1 r_2 \cdots r_M c_M$ is in $\mathcal{F}$. If the matrix $\mathbf{A}(\mathcal{G}_{X[M+1]})$ has zero rows or columns, these are associated with stranded vertices. So, we have to generate an essential submatrix. To simplify the notation, we also call the resulting submatrix by $\mathbf{A}(\mathcal{G}_{X[M+1]})$.

Example 1. Let X be a 4-step shift of finite type over $\mathcal{A} = \{0,1\}$, specified by the set $\mathcal{F} = \{000,\ 1010,\ 1011\}$. Alternatively, $\mathcal{F} = \{0000,\ 0001,\ 1000,\ 1010,\ 1011\}$. The essential submatrix of $\mathbf{A}(\mathcal{G}_{X[4]})$ is:

$$\mathbf{A}(\mathcal{H}_0) = \mathbf{A}(\mathcal{G}_{X[4]}) = \begin{bmatrix} 00 & 10 & 01 & 00 & 00 & 00 \\ 00 & 00 & 00 & 10 & 00 & 00 \\ 00 & 00 & 00 & 00 & 10 & 01 \\ 01 & 00 & 00 & 00 & 00 & 00 \\ 00 & 00 & 00 & 10 & 00 & 00 \\ 00 & 00 & 00 & 00 & 10 & 01 \end{bmatrix}.$$

First round: Merge two out-edge equivalence classes of $\mathbf{A}(\mathcal{H}_0)$, $\mathcal{I}_1 = \{1,4\}$, $\mathcal{I}_2 = \{2,5\}$. The new matrix is $\mathbf{A}(\mathcal{H}_1)$:

$$\mathbf{A}(\mathcal{H}_1) = \begin{bmatrix} 00 & 10 & 01 & 00 & 00 & 00 \\ 00 & 00 & 00 & 10 & 00 & 00 \\ 00 & 10 & 01 & 00 & 00 & 00 \\ 01 & 00 & 00 & 00 & 00 & 00 \\ 00 & 00 & 00 & 00 & 00 & 00 \\ 00 & 00 & 00 & 00 & 00 & 00 \end{bmatrix}.$$

Second round: Merge one out-edge equivalence class of $\mathbf{A}(\mathcal{H}_1)$, $\mathcal{I}_1 = \{0,2\}$. The new matrix is $\mathbf{A}(\mathcal{H}_2)$:

$$\mathbf{A}(\mathcal{H}_2) = \begin{bmatrix} 01 & 10 & 00 & 00 & 00 & 00 \\ 00 & 00 & 00 & 10 & 00 & 00 \\ 00 & 00 & 00 & 00 & 00 & 00 \\ 01 & 00 & 00 & 00 & 00 & 00 \\ 00 & 00 & 00 & 00 & 00 & 00 \\ 00 & 00 & 00 & 00 & 00 & 00 \end{bmatrix}.$$

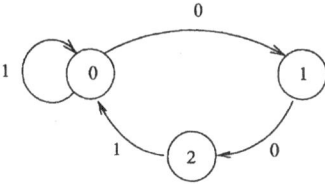

Fig. 1. Shannon cover for the forbidden set $\mathcal{F} = \{000, \ 1010, \ 1011\}$.

All nonzero rows of $\mathbf{A}(\mathcal{H}_2)$ are distinct. So, the minimization procedure has finished. The essential submatrix of $\mathbf{A}(\mathcal{H}_2)$ is:

$$\mathbf{A}(\mathcal{H}_2) = \begin{bmatrix} 01 & 10 & 00 \\ 00 & 00 & 10 \\ 01 & 00 & 00 \end{bmatrix}.$$

Figure 1 illustrates the labeled graph $\mathcal{H}_2$.

4 Conclusions

In this paper, an efficient iterative vertex-minimization algorithm to finding the Shannon cover of a shift of finite type is presented. It is based on a symbolic representation of the adjacency matrix associated with an initial graph presenting the shift. The proposed algorithm is easy to implement using a symbolic manipulation program, such as Maple. It thus constitutes an important tool for the design of codes for practical constrained systems.

References

1. B. H. Marcus and P. H. Siegel and J. K. Wolf, "Finite-state modulation codes for data storage," *IEEE J. on Select. Areas Commun.*, vol.10, pp. 5-37, January 1992.
2. K. A. S. Immink, *Coding Techniques for Digital Recorders*. Englewoods Cliffs, NJ:Prentice-Hall, 1991.
3. D. Lind and B. Marcus, *An Introduction to Symbolic Dynamics and Coding*. New Jersey, Cambridge University Press, 1995.
4. P. H. Siegel and J. K. Wolf, "Modulation and coding for information storage," *IEEE Communications Magazine*, vol. 29, no. 12, pp. 68-86, Dec. 1991.
5. Daniel P. B. Chaves and Cecilio Pimentel and Bartolomeu F. Uchôa-Filho, "On the Shannon cover of shifts of finite type," in *Proceedings of the Brazilian Telecommunication Symposium*, Rio de Janeiro, Brazil, Oct. 2003.

Truncated Importance Sampling Simulation Applied to a Turbo Coded Communication System

Bruno B. Albert and Francisco M. de Assis*

Universidade Federal de Campina Grande
Av. Aprigio Veloso 885, CEP 58109-970
Campina Grande PB Brasil
{albert,fmarcos}@dee.ufcg.edu.br

Abstract. The authors present and demonstrate an efficient method to estimate the bit error rate (BER) of turbo codes via a truncated importance sampling (TIS). The TIS method is based on the relevant code words that effectively contribute to the total BER. In order to find out these code words, a fast algorithm is proposed. Two examples are presented to describe the TIS method. The method is particularly effective at high signal to noise ratio (SNR), where Monte Carlo (MC) simulation takes too much time to obtain an estimative that is statistically sound.

1 Introduction

The main contribution of this work is to apply the error center principle [1] [2] to estimate the performance of a classical turbo code scheme. We demonstrate the viability and the accuracy of the TIS method by two examples. Two questions are posed here: 1. Which error centers are relevant to compute the BER? 2. How to find out them? The first question is answered in Section 5. The last question is answered by an algorithm to find the relevant error centers based on input bits pattern. In order to answer the second question, an algorithm based on input bits pattern is proposed.

This paper is organized as follows. In Section 2 we restate the theory required to apply the TIS method. In Section 3 we describe the simulation model and the simulation method by two examples. A fast algorithm to find out error centers is presented in Section 4. An upper bound for the error caused by the truncation of the importance sampling (IS) method is shown in Section 5. Finally we summarize our investigation in Section 6.

* This work was supported by Brazilian National Research Council - CNPq - Grant 300902/94-8, and CAPES/PROCAD program.

J.N. de Souza et al. (Eds.): ICT 2004, LNCS 3124, pp. 94–99, 2004.

2   Specific Decoding Error Probability via IS

The error bit probability when a code word $\mathbf{c}$ is transmitted $P_b(\mathbf{c})$ is defined as

$$P_b(\mathbf{c}) = \frac{1}{m} \sum_{\mathbf{c}' \in C} n_b(\mathbf{c}, \mathbf{c}') P(\mathbf{c}'|\mathbf{c}) \tag{1}$$

where $P(\mathbf{c}'|\mathbf{c})$ is the probability of decoding $\mathbf{c}'$ when $\mathbf{c}$ is transmitted and is called specific decoding error probability, $n_b(\mathbf{c}, \mathbf{c}')$ is the number of postdecoding bit errors caused by decoding $\mathbf{c}'$ instead of $\mathbf{c}$, and m is the number of information bits provided by the source. The idea is to estimate the bit error probability $P_b(\mathbf{c})$ by the specific decoding error probabilities $P(\mathbf{c}'|\mathbf{c})$, calculated for all error center $\mathbf{c}'$, and apply these results to (1). Each $P(\mathbf{c}'|\mathbf{c})$ is computed independently using the IS method, and may be estimated by

$$\hat{P}^*(\mathbf{c}'|\mathbf{c}) = \frac{1}{N_{IS}} \sum_{i=1}^{N_{IS}} w(\mathbf{Y}^{(i)}|\mathbf{c}) I_{\mathbf{c}'}(\mathbf{Y}^{(i)}) \tag{2}$$

where $\hat{P}^*(\mathbf{c}'|\mathbf{c})$ is the estimator of $P(\mathbf{c}'|\mathbf{c})$, N_{IS} is the number of simulation runs, and $\mathbf{Y}^{(i)}$ represents the random channel output sequence for the ith simulation run. It may be proved that the IS estimator (2) employing the weighting function $w(\mathbf{y}|\mathbf{c}) = f(\mathbf{y}|\mathbf{c})/f^*(\mathbf{y}|\mathbf{c})$ is unbiased [1], where $f(\mathbf{y}|\mathbf{c})$ is the channel joint density function, and $f^*(\mathbf{y}|\mathbf{c})$ is the IS simulation density.

3   System Model Description and Examples

The encoder is built using a parallel concatenation of two recursive systematic convolutional (RSC) codes with an interleaver between them, and code rate 1/3. Each component encoder is a four state RSC code with generator matrix represented by $G(D) = [1, (1+D^2)/(1+D+D^2)]$. The decoder is made up of two elementary decoders in serial concatenation scheme using a feedback loop. Each decoder uses the soft-output Viterbi algorithm (SOVA). A binary symmetric channel (BSC) with crossover probability λ is used to illustrate the principles discussed in the above Section. The biasing channel model is characterized by a time-varying crossover probability λ^* nonstationary, as in [1], and depends on the error center $\lambda^* = \lambda$ if $c_k = c'_k$, and $\lambda^* = 1/2$ otherwise, where c_k is one information bit of the transmitted code word at time t_k, and c'_k represents one bit of the error center at the same time. Thus, this biased channel model directs the transmitted code word toward the specific error center.

Example 1 As a first example, we choose a very short interleaving length to demonstrate the viability and the accuracy of the method. We use an interleaver with 8 bits length and permutation $\Pi(0,1,2,3,4,5,6,7) = (6,3,7,1,2,4,5,0)$. We take the information bits to be the all-zero code word. If the channel is memoryless and the encoder is linear with maximum likelihood (ML) decoding, than the bit error probability does not depend of the code word transmitted, thus $P_b = P_b(\mathbf{c})$ in (1). Although the iterative decoding process is not ML, we verify by simulation that our assumption is still valid. All the $2^8 - 1 = 255$ error centers were evaluated to compute the bit error probability. Figure 1(a) shows a comparison between the MC method and IS method with one decoding iteration. Instead of plotting BER versus λ, we plot BER versus E_b/N_0 the signal-to-noise ratio, where E_b is the transmitted energy per information bit and $N_0/2$ is the double sided noise power spectral density. Each point has at least 10% relative precision. We may observe the accuracy of the IS method. the point at 14dB is only estimated with the IS method. Figure 1(b) shows a comparison

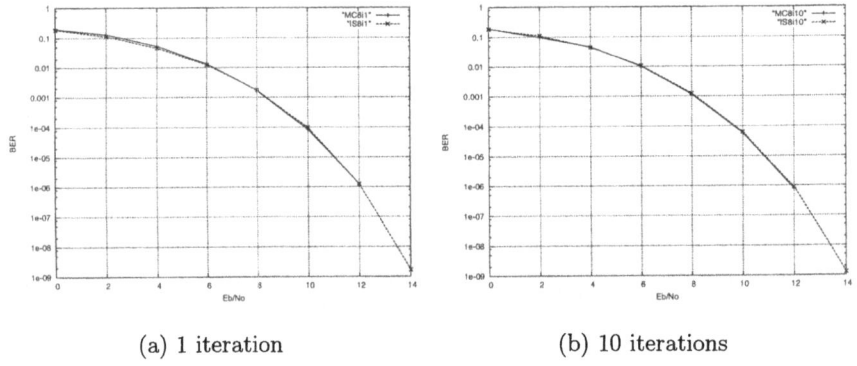

(a) 1 iteration   (b) 10 iterations

Fig. 1. MC and IS BER simulation comparison, interleaver 8 bits

between the MC method and IS method with ten iterations. We still note the accuracy of the IS method. Again, the 14dB point is only estimated with the IS method. The gain of IS over MC simulation is measured by the computing time ratio $\eta = T_{MC}/T_{IS}$, where T_{MC} is the MC simulation time, and T_{IS} is the IS simulation time. MC simulation is more efficient than IS simulation for $E_b/N_0 \leq 8$dB, $\eta = 0.40$. At 14dB we estimated the gain in 10^3 times. The IS simulation takes about 20 minutes and MC simulation would take over 13 days on a 1.7GHz PC for at least 10% relative precision.

Example 2 Now, we use a 64 bit length interleaver. In this case, we have $(2^{64}-1)$ error centers, so we can not cover all of them to compute the BER. Certainly, MC simulation would be more efficient than IS simulation in this case. However, for high SNRs few error centers dominate the summation in (1) and we can concentrate our simulation on these error centers. In [3] it was shown that for $E_b/N_0 \geq 6$dB codewords with weights ω less than 10 determine the code performance. We find out 7 error centers with this code weight restriction. Figure 2(a) shows a comparison between the MC simulation and the IS simulation with one iteration. The IS curve converges to the MC curve at 9dB. The point at 12dB is only computed with IS simulation method. For ten iterations the IS curve also converges at 9dB point, as shown in Figure 2(b). Points 10 and 12dB are only computed with IS method. The gains are measured at the same way as in example 1. MC simulation is more efficient than IS simulation for $E_b/N_0 \leq 6$dB, $\eta = 0.70$ We are not considering the time wasted to find out the error centers. Note that the IS estimated BER value at 6dB is high biased, we must take into account error centers with higher weights, that would waste more simulation time. Thus, MC gain would be even better, in this case. The gain at 12dB is estimate in 1.7×10^4, and would take over 70 days to be MC simulated with at least 10%, we take about 6 minutes with IS method on a 1.7GHz PC.

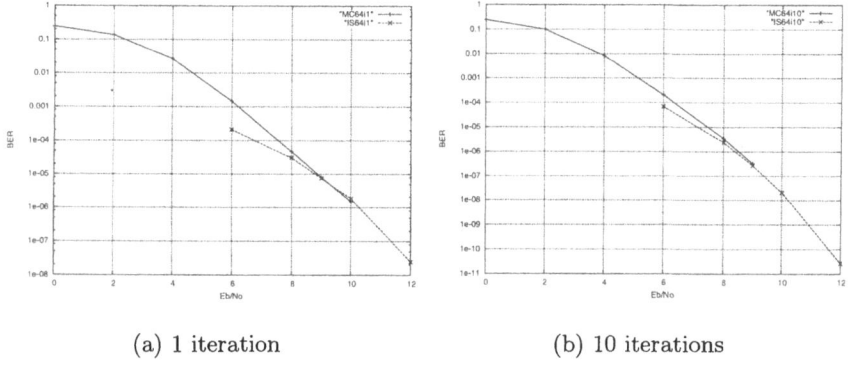

(a) 1 iteration   (b) 10 iterations

Fig. 2. MC and IS BER simulation comparison, interleaver 64 bits

4 Error Centers Searching Algorithm

The error centers searching algorithm (ECSA) is a fast algorithm based on the turbo encoder input bit patterns to find out error centers (code words) with weight less or equal than a weight upper bound W_b. Before we show the ECSA, we present a direct approach algorithm. We use the Example 2 to describe the algorithms. From [3], the codewords with weights $\omega \leq 10$ are generated by information words with weights $2, 3, 4, 5,$ and 6. We can generate all combinations of information words with these weights and select only the ones that has codewords with $\omega \leq 10$. So we must investigate $\binom{64}{2} + \binom{64}{3} + \binom{64}{4} + \binom{64}{5} + \binom{64}{6} = 83,277,936$ information words. We find out 7 words that satisfies the condition. For large interleaver lengths the number of information words to investigate may be unacceptable again. So, a better algorithm is required. The second approach is based on the information word bit patterns that produce allowed weights on the first component encoder of the turbo code. Therefore, we need to evaluate some structural properties of each RSC component encoder. Assuming the encoder start and finish in the zero state. The weight enumerator function

(WEF) of a convolutional code is defined as $T_1(X, Y) = \sum_{i=1}^{\infty} \sum_{j=1}^{\infty} a_{i,j} X^i Y^j$, where X is the indeterminate associated with the weight of the code word, Y is the indeterminate associated with the weight of the information word, and $a_{i,j}$ is the number of code words with weight i that correspond to information word of weight j. The WEF of the first RSC component encoder is $T_1(X, Y) = X^5 Y^3 + X^6 (Y^2 + Y^4) + X^7 (3Y^3 + Y^5) + X^8 (Y^2 + 6Y^4 + Y^6) + \dots$, and tell us that there is one code word of weight 5 generated by an information word pattern of weight 3. There are then two code word of weight 6 generated by two information word patterns, one of weight 2, and another of weight 4. However, the WEF do not say which information word generate the correspondent code word, this is precisely what we are looking for. Since the second RSC component encoder has only parity bits as output, we define a new WEF to reflect this fact, $T_2(W, Y) = \sum_{k=1}^{\infty} \sum_{j=1}^{\infty} b_{k,j} W^k Y^j$, where W is associated with the weight of the parity bits of the code word, Y has the same meaning as in $T_1(.)$, and $b_{k,j}$ is the number of code words with parity bits of weight k generated by information words of weight j. The WEF of the second RSC component encoder is $T_2(W, Y) = W^2 (Y^3 + Y^4 + Y^5 + Y^6 + \dots) + W^4 (Y^2 + 3Y^3 + 6Y^4 + 10Y^5 + 15Y^6 + \dots) + W^6 (Y^2 + 5Y^3 + 15Y^4 + \dots) + \dots$. We are interested on code words with weights less or equal than 10. We start combining $T_1(.)$ and $T_2(.)$ to produce code words with exactly 10 weight. Note that the allowed terms must satisfies $i + k = 10$, where $i \in \{5, 6, 7, 8, \dots\}$ and $k \in \{2, 4, 6, \dots\}$. Let (i, k) be an allowed pair, then, to satisfies $i + k = 10$ we have two possibilities, $(6, 4)$ and $(8, 2)$. Considering the later pair $(8, 2)$, the corresponding terms are $X^8 (Y^2 + 6Y^4 + Y^6), W^2 (Y^3 + Y^4 + Y^5 + Y^6 + \dots)$. Recall that the interleaver does not change the input word weight to the second encoder. Therefore, only information words with weights 4 and 6 can generate code words with 10 weight. There are 7 input patterns with this characteristic, six with information word weight of 4, term $6Y^4$, and one with information word weight of 6, term Y^6. We can search all patterns using the RSC encoder signal flow graph. Once we find out the patterns we proceed as follows. We apply a shifted version of each pattern to the turbo encoder input, and we check the code word weight, if the weight is less or equal to 10 we label the code word as an error center. For example, suppose we find out the following pattern, (1100101). The turbo code information words are shifted versions of this pattern, so we have the following inputs to check, $(11001010 \dots 0), (01100101 \dots 0), \dots, (0 \dots 01100101)$. The same procedure is done to produce code words with exactly 9, 8, and, 7 weights. We found 14 patterns and considering the worst case, when the pattern length is three, (111), we investigate at most $14 \times (64 - 3) = 854$ information words instead of $83, 277, 936$ as in the brute force case.

5 Analysis of the Effect of Truncating IS

When we truncate the summation in (1), we make an error when evaluate $P_b(\mathbf{c})$. Assuming the all-zero code word is transmitted. We define the truncated error as $\Delta = \hat{P}_b^*(0) - \hat{P}_b^{*T}(0)$, where $\hat{P}_b^*(0)$ is the complete IS simulation result, and

$\hat{P}_b^{*T}(0)$ is the TIS simulation result. We note that Δ stands for a biasing deviation of $\hat{P}_b^{*T}(0)$ given $\hat{P}_b^*(0)$ is an unbiased estimator. Δ is a random variable, and its expected value has an upper bound defined by[1]

$$E\Delta < \frac{1}{2R} \exp\left[-(W_b+1)RE_b/N_0\right] \sum_{d>W_b}^{m/R} A_d, \qquad (3)$$

where E is the expectation operator, m is the number of information bits, W_b is the allowed code word weight upper bound, R is the code rate, E_b/N_0 is signal to noise ratio, N is the interleaver length, and A_d is the number of code words with weight d. The Equation (3) shows that if we fix the code word weight upper bound W_b value, the truncated error expected value $E\Delta$ converges exponentially to zero as E_b/N_0 increases. On the other hand, if we fix a value to E_b/N_0, than $E\Delta$ also converges exponentially to zero as W_b increases.

6   Conclusions

We demonstrate the viability and accuracy of the application of the IS method to compute the BER of turbo code schemes with convolutional encoders, specially for high SNRs values. In these cases IS simulation method provides expressive gains in relation to MC simulation method. One difficult with this approach is to find out the error centers that contribute to the BER. A fast algorithm based on the turbo encoder input bits pattern is proposed to solve this problem. Although a BSC was chosen, others more complex channels may be employed. In Example 1 the IS simulation could be faster if we do not consider all 255 error centers as in Example 2. Note that the IS curves (Fig. 2(a) and Fig. 2(b)) in Example 2 can be more accurate by introducing error centers with high weights, as can be seen in Equation (3), the price is an increased simulation time. Results obtained put forward the using of MC with low SNRs, and truncated IS for high SNR values.

References

1. Sadowsky, J.S.: A new method for Viterbi decoder simulation using importance sampling. IEEE Transactions on Communications **38** (1990) 1341–1351
2. Ferrari, M., Belline, S.: Importance sampling simulation of turbo product codes. ICC2001, The IEEE International Conference on Communications **9** (2001) 2773 – 2777
3. Wen Feng, J.Y., Vucetic, B.S.: A code-matched interleaver design for turbo codes. IEEE Transactions on Communications **50** (2002) 926 – 937
4. Wicker, S.B.: Error Control Systems for Digital Communication and Storage. Prentice Hall, Inc. (1995)

---

[1] The details of this analysis is omitted here for the sake of space, and will be presented in another paper.

Zero-Error Capacity of a Quantum Channel

Rex A.C. Medeiros* and Francisco M. de Assis

Federal University of Campina Grande
Department of Electrical Engineering
Av. Aprígio Veloso, 882, Bodocongó,
58109-970 Campina Grande-PB, Brazil
{rex,fmarcos}@dee.ufcg.edu.br

Abstract. The zero-error capacity of a discrete classical channel was first defined by Shannon as the least upper bound of rates for which one transmits information with zero probability of error. Here, we extend the concept of zero-error capacity for a noisy quantum channel. The necessary requirement for a quantum channel have zero-error capacity greater than zero is given. Finally, we give some directions on how to calculate the zero-error capacity of such channels.

1 Introduction

Given a noisy classical communication channel, how much information per channel use one can transmit reliably over the channel? This problem was first studied by Shannon [1]. If X and Y are random variables representing, respectively, the input and the output of the channel, then its capacity is given by the maximum of the mutual information between X and Y

$$C = \max_{p(x)} I(X; Y), \tag{1}$$

where the maximum is taken over all input distributions $p(x)$ for X. The capacity defined above allows a small probability of error for rates approaching the channel capacity, even when the best coding scheme is used. In some situations, it may be of interest to consider codes for which the probability of error is zero, rather than codes with probability of error approaching zero. In a remarkable paper [2], Shannon defined the zero-error capacity of a noisy discrete memoryless channel as the least upper bound of rates at which it is possible to transmit information with probability of error equal to zero.

In this paper, we extend the concept of zero-error capacity for quantum channels. We establish the condition for which a quantum channel has zero-error capacity greater then zero. Then, we give some directions for finding zero-error capacity of quantum channels. This paper is organized as follows: Section 2 presents the main aspects related to the Shannon's zero-error capacity theory of a noisy classical channel. In the next section, we define zero-error capacity in a quantum scenario. Section 3.1 relates fixed points of a quantum channel to zero-error capacity. Then, we present the conclusions.

---

* Graduate Program in Electrical Engineering - Master's degree student.

J.N. de Souza et al. (Eds.): ICT 2004, LNCS 3124, pp. 100–105, 2004.

2   Classical Zero-Error Capacity Theory

Shannon defined zero-error capacity for discrete memoryless classical channels, specified by a finite transition matrix $||p(j|i)||$, where $p(j|i)$ is the probability of input letter i being received as output letter j $(i = 1, 2, \ldots, a; \; j = 1, 2, \ldots, b)$. A sequence of input letters is called an input word, and a sequence of output letters an output word. A block code of length n is a mapping of M messages onto a subset of input words of length n. So, $R = \frac{1}{n} \log M$ will be the input rate for this code [2].

A decoding scheme for a block code of length n in a zero-error context is a function from output words of length n to integers 1 to M. The probability of error for a code is the probability that the noise and the decoding scheme lead to an input message different from the one that actually occurred, when M input messages are used each one with probability $1/M$.

In a discrete channel, two input letters are adjacent if there is an output letter which can be caused by either of these two. A discrete channel may be represented by diagrams as illustrated in Fig. 1(a). This channel has at least two non-adjacent input letters, and thus the zero-error capacity is greater than zero.

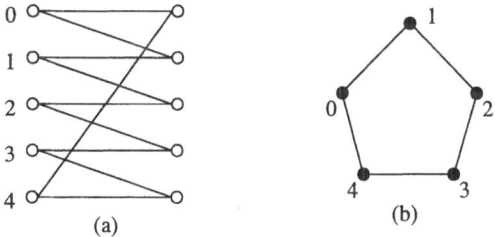

Fig. 1. (a) A discrete channel. (b) The corresponding adjacency graph.

Definition 1 (Classical zero-error capacity [2]). *The zero-error capacity of a discrete memoryless classical channel, denoted by C_0, is the least upper bound of all rates which can be achieved with zero probability of error.*

This definition means that if $M_0(n)$ is the largest number of words in a code of length n, no two of which are adjacent, the capacity C_0 is the least upper bound of the numbers $\frac{1}{n} \log M_0(n)$, when n varies through all positive integers. That is,

$$C_0 = \sup_n \frac{1}{n} \log M_0(n). \tag{2}$$

We may represent a discrete channel using a graph, called adjacency graph. This is done by taking as many vertices as there are input letters and connecting two distinct vertices with a line or branch of the graph if corresponding input letters are adjacent. The graph corresponding to the channels in Fig. 1(a) is shown in Fig. 1(b).

The problem of finding C_0 is equivalent to the following problem in graph theory [3]. Let G be the characteristic graph associated with a discrete memoryless channel D. The clique number $\omega(G)$ of a graph G is the largest cardinality of a set of vertices every two of which are connected in G. Then,

$$C_0 = \sup_n \frac{1}{n} \log \omega(G^n), \tag{3}$$

where G^n stands for the characteristic graph representing the product channel D^n [4].

Most important results concerning zero-error information theory can be found in a survey paper due to Köner and Orlitsky [4].

3   Zero-Error Capacity of a Noisy Quantum Channel

We extend the definition of zero-error capacity for a quantum channel $\mathcal{C}$. Here, a quantum channel is represented by a trace-preserving quantum map $\mathcal{E}(\cdot)$ [6].

Let $\mathcal{X}$ be the set of possible input states to the quantum channel $\mathcal{C}$, belonging to a d-dimensional Hilbert space $\mathcal{H}$, and let $\bar{\rho} \in \mathcal{X}$. We denote $\bar{\sigma} = \mathcal{E}(\bar{\rho})$ the received quantum state when $\bar{\rho}$ is transmitted through the quantum channel. Because knowledge of post-measurement states is not important, measurements are performed by means of a POVM (Positive Operator-Valued Measurements) $\{E_j\}$, where $\sum_j E_j = I$. Define $\mathcal{S}$ a finite subset of $\mathcal{X}$ and $\bar{\rho}_i \in \mathcal{S}$. Let $p(j|i)$ be the probability of Bob measures j given that Alice sent the state $\bar{\rho}_i$. Then, $p(j|i) = \text{tr}(\bar{\sigma}_i E_j)$.

We define the zero-error capacity of a quantum channel for product states. A product of any n input states will be called an input quantum codeword, $\rho_i = \bar{\rho}_{i_1} \otimes \cdots \otimes \bar{\rho}_{i_n}$, belonging to a d^n-dimensional Hilbert space $\mathcal{H}^n$. A mapping of K classical messages (which we may take to be the integers $1, \ldots, K$) into a subset of input quantum codewords will be called a quantum block code of length n. Thus, $\frac{1}{n} \log K$ will be the rate for this code. A piece of n output indices obtained from measurements performed by means of a POVM $\{E_j\}$ will be called an output word, $w = \{1, \ldots, N\}^n$.

A decoding scheme for a quantum block code of length n is a function that univocally associates each output word with integers 1 to K representing classical messages. The probability of error for this code is greater than zero if the system identifies a different message from the message sent.

Definition 2 (Zero-error capacity of a quantum channel). *Let $\mathcal{E}(\cdot)$ be a trace-preserving quantum map representing a noisy quantum channel $\mathcal{C}$. The zero-error capacity of $\mathcal{C}$, denoted by $C^{(0)}(\mathcal{E})$, is the least upper bound of achievable rates with probability of error equal to zero. That is,*

$$C^{(0)}(\mathcal{E}) = \sup_n \frac{1}{n} \log K(n), \tag{4}$$

where $K(n)$ stands for the number of classical messages that the system can transmit without error, when a quantum block code of length n is used.

According to this definition, we establish the condition for which the quantum channel C has zero-error capacity greater than zero.

Proposition 1. *Let $S = \{\bar{\rho}_i\}$, $S \subset \mathcal{X}$ be a set of $M \leq d$ quantum states, and let $\mathcal{P} = \{E_j\}$ be a POVM having $N \geq M$ elements such that $\sum_j E_j = I$. Consider the subsets*

$$A_k = \{j \in \{1, \ldots, N\}; \; tr(\bar{\sigma}_k E_j) > 0\}; \; k \in \{1, \ldots, M\}. \tag{5}$$

Then, the quantum channel C has zero-error capacity greater than zero iff there exists a set S and a POVM $\mathcal{P}$ for which at least one pair $(a, b) \in \{1, \ldots, M\}^2$, $a \neq b$, the subsets A_a and A_b are disjoints, i.e., $A_a \cap A_b = \emptyset$.

The respective quantum states $\bar{\rho}_a$ and $\bar{\rho}_b$ are said to be non-adjacent in C for the POVM $\mathcal{P}$.

Proof. Suppose $\bar{\rho}_a, \bar{\rho}_b \in S$ obeying $A_a \cap A_b = \emptyset$ for a POVM $\mathcal{P}$. We can construct a quantum block code only mapping two classical messages into the states $\bar{\rho}_a$ and $\bar{\rho}_b$. This code will be a rate equal to one and $C^{(0)}(\mathcal{E}) \geq 1 > 0$.

Conversely, if $C^{(0)}(\mathcal{E})$ is greater then zero then, and according to the Shannon's definition of zero-error capacity, at least two input states from some set S, say $\bar{\rho}_a$ and $\bar{\rho}_b$, are non-adjacent for a POVM $\mathcal{P}$, i.e., $A_a \cap A_b = \emptyset$.

Definition 3 (Optimum $(S, \mathcal{P})$ for C). *The optimum $(S, \mathcal{P})$ for a quantum channel C is composed of a set $S = \{\bar{\rho}_i\}$ and a POVM $\mathcal{P} = \{E_j\}$ for which the zero-error capacity is reached.*

As an example, we consider the well known bit flip channel. Such channel leaves a quantum state ρ intact with probability p and flips the qubit with probability $1 - p$. That is, $\mathcal{E}(\rho) = p\rho + (1 - p)X\rho X$. It is easy to see that the optimum $(S, \mathcal{P})$ for the bit flip channel is composed of the following map and POVM:

$$1 \to \bar{\rho}_1 \to |\psi_1\rangle = \frac{1}{\sqrt{2}}(|0\rangle + |1\rangle) \qquad E_1 = |\psi_1\rangle\langle\psi_1|. \tag{6}$$

$$2 \to \bar{\rho}_2 \to |\psi_2\rangle = \frac{1}{\sqrt{2}}(|0\rangle - |1\rangle) \qquad E_2 = |\psi_2\rangle\langle\psi_2|. \tag{7}$$

Then, the error zero capacity of the bit flip channel is given by

$$C^{(0)}(\mathcal{E}) = \frac{1}{1}\log(2) = 1. \tag{8}$$

For this simple channel, we see that the capacity is found for $n = 1$. This means that one can transmit one bit per use of the channel with probability of error equal to zero!

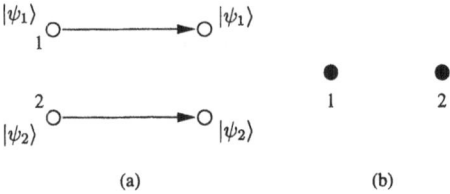

Fig. 2. (a) Discrete channel for the optimum $(\mathcal{S}, \mathcal{P})$. (b) Adjacency graph.

Given a quantum channel, a set $\mathcal{S}$, and a POVM $\mathcal{P}$, it is easy to construct a representation for the quantum channel using a discrete classical channel. For the bit flip channel presented above, the discrete channel and the adjacency graph are shown in Fig. 2. We can interpret the zero-error capacity of a quantum channel as the Shannon's zero-error capacity of the equivalent classical channel $(C_0 = 1)$.

3.1   Zero-Error Capacity and Fixed Points

Let $\mathcal{E}(\cdot)$ be a trace-preserving quantum map representing a quantum channel $\mathcal{C}$. It is easy to show that such transformation is a continuous map on a convex, compact subset of a Hilbert space (see [6, pp. 408]). The Schauder's fixed point theorem guarantees the existence of at least one quantum state ρ such that $\mathcal{E}(\rho) = \rho$. Let $D(\rho_a, \rho_b) = \frac{1}{2}\mathrm{tr}|\rho_a - \rho_b|$ be the trace distance between the quantum states ρ_a and ρ_b. We say that a quantum channel $\mathcal{C}$ is contractive if for any two quantum states ρ_a and ρ_b, $D(\mathcal{E}(\rho_a), \mathcal{E}(\rho_b)) \leq D(\rho_a, \rho_b)$. Such channel may have more than one fixed points. This suggests an interesting relation between zero-error capacity and fixed points.

Proposition 2. *Let $\mathcal{E}(\cdot)$ be a contractive trace-preserving quantum map representing a quantum channel $\mathcal{C}$, having N_f fixed points. The zero-error capacity of $\mathcal{C}$ is at least $\log N_f$.*

Proof. We can construct a quantum block code for this channel by taking N_f classical symbols, each one associated with fixed points $|\psi_1\rangle, \ldots, |\psi_{N_f}\rangle$ in a way that $\overline{\rho}_1 = |\psi_1\rangle\langle\psi_1|, \ldots, \overline{\rho}_{N_f} = |\psi_{N_f}\rangle\langle\psi_{N_f}|$. Then, we define a POVM given by $\{E_i = |\psi_i\rangle\langle\psi_i|\}$ and possibly one more element E_0 such that $\sum_{i=0}^{N_f} E_i = I$.

The graph for the discrete classical channel representing the system $(\mathcal{C} + \mathcal{S} + \mathcal{P})$ is composed of N_f non-interlinked points. The Shannon's capacity of this graph is clearly $\log N_f$.

Another interesting relation between zero-error capacity of a channel represented by a trace-preserving quantum map $\mathcal{E}(\cdot)$ and fixed points is due to the entanglement fidelity, $F(\rho, \mathcal{E})$, who measures how well a quantum channel preserves entanglement. The entanglement fidelity is given by

$$F(\rho, \mathcal{E}) = \sum_i |\mathrm{tr}\,(\rho B_i)|^2, \tag{9}$$

where $\{B_i\}$ are operation elements of $\mathcal{E}(\cdot)$. It is easy to show that $F(\rho, \mathcal{E}) = 1$ if and only if all pure states $|\psi_i\rangle$ on the support of ρ are fixed points, i.e., $\mathcal{E}(|\psi_i\rangle\langle\psi_i|) = |\psi_i\rangle\langle\psi_i|$ [6, pp. 423]. This suggest a searching for all density operators ρ having unitary entanglement fidelity. For qubit channels, it is generally a soft task. Remember that a density operator ρ obeys two rules: (1) tr $(\rho) = 1$ and (2) ρ is a positive operator. We illustrate this technique for the bit flip channel with $p = \frac{1}{2}$, $\mathcal{E}(\rho) = \frac{1}{2}\rho + \frac{1}{2}X\rho X$. Here $B_1 = \frac{1}{\sqrt{2}}I$ and $B_2 = \frac{1}{\sqrt{2}}X$. Then, making $F(\rho, \mathcal{E}) = 1$, the only vectors on the support of ρ are $|\psi_{1,2}\rangle = \frac{1}{\sqrt{2}}(|0\rangle \pm |1\rangle)$. These are the same quantum states presented in Eqs. (6) and (7). The corollary below follows directly from Proposition 2.

Corollary 1. *Let $\mathcal{E}(\cdot)$ be a contractive trace-preserving quantum map representing a quantum channel, and let ρ be quantum states such that $F(\rho, \mathcal{E}) = 1$. If d is the dimension of the Hilbert space spanned by quantum states $|\psi_i\rangle$ in the support of ρ, then the zero-error capacity of the channel is at least $C^{(0)}(\mathcal{E}) = \log d$.*

In the previous example, quantum states on the support of ρ generate the Hilbert space of dimension 2. Consequently, we conclude that $C^{(0)}(\mathcal{E}) = \log 2 = 1$. Note that, at the same time we found the zero-error capacity, we found in addition the code archiving the capacity.

4   Conclusions

We extended the Shannon's theory of zero-error capacity for a quantum channel represented by a trace-preserving quantum map. Moreover, we showed a closed connection between zero-error capacity and fixed points and we gave some directions for calculating lower bounds using trace distance and entanglement fidelity. The results were illustrated for the bit flip quantum channel.

Acknowledgments. The authors thank the Brazilian National Council for Scientific and Technological Development (CNPq) for support (CT-INFO Quanta, grants # 552254/02-9).

References

1. C. E. Shannon. A mathematical theory of communication. *The Bell System Tech. J.*, 27:379–423, 623–656, 1948.
2. C. E. Shannon. The zero error capaciy of a noisy channel. *IRE Trans. Inform. Theory*, IT-2(3):8–19, 1956.
3. L. Lovász. On the Shannon capacity of a graph. *IEEE Trans. Inform. Theory*, 25(1):1–7, 1979.
4. J. Köner and A. Orlitsky. Zero-error information theory. *IEEE Trans. Inform. Theory*, 44(6):2207–2229, 1998.
5. T. M. Cover and J. A. Thomas. *Elements of Information Theory*. John Wiley & Sons Inc., New York, 1991.
6. M. A. Nielsen and I. L. Chuang. *Quantum Computation and Quantum Information*. Cambridge University Press, Cambridge, 2000.

Communication System Recognition by Modulation Recognition

Ali Reza Attar, Abbas Sheikhi, and Ali Zamani

Department of Electrical and Electronic Engineering, Shiraz University,
P.O.BOX : 71345-1457, Shiraz, Iran
ar_attar@hotmail.com
http://www.shirazu.ac.ir

Abstract. Communication system recognition is major part of some civilian and military applications. The recognition of the system is done by inspecting the received signal properties like modulation type, carrier frequency, baud rate and so on. Therefore we need Automatic Modulation Recognition (AMR) methods plus carrier and baud rate estimation methods. In this paper we introduce a new AMR method based on time-domain and spectral features of the received signal. We have used neural network as the classifier. Some analog and digital modulations including AM, LSSB, USSB, FM, ASK2, ASK4, ASK8, PSK2, PSK4, PSK8, FSK2, FSK4, FSK8 and MSK are considered. Then using information from the received signal like baud rate, carrier frequency and modulating scheme the protocol used for signal transmission is detected.

1 Introduction

Automatic Modulation Recognition (AMR) can be used in many applications. Electronic warfare [1], Electronic support measure [2], Spectrum surveillance and management [3], Universal demodulator [4] and Counter channel jamming [5] are some of its applications to mention a few. Here, we are going to use AMR as a part of a communication system recognition method. Different methods of AMR can be categorized in two broad fields: Pattern recognition and Decision theoretic. In the past, decision making was the main method used by researchers like [6], [7], [8], [9] and [10]; but in the previous years pattern recognition methods are dominant especially using neural networks. Some of the papers dealing with this situation are [3], [4], [5] and [11]. Consider the case of receiving a previously unknown signal and trying to decode its data content. In order to decode the data correctly we need some information from the received signal as input to be able to deduce some information as output. Some authors consider a three step method for decoding data from an unknown received signal and the input/output data relationship like Fig. 1.a [12].

We claim that by using an extra step named protocol recognition like Fig. 1.b, data decoding is done more easily.

J.N. de Souza et al. (Eds.): ICT 2004, LNCS 3124, pp. 106–113, 2004.

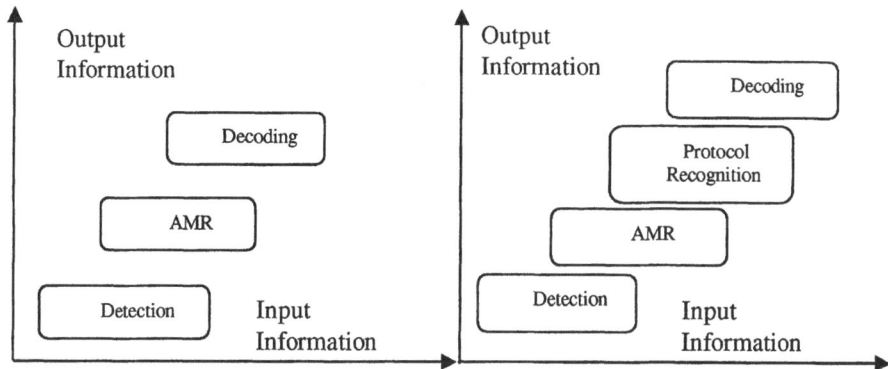

Fig. 1. a) The common input/output information relationship b) Our proposed method

In this paper we will introduce a new modulation classifier using features of received signal in both time and spectral domain and then using this information plus baud rate and carrier frequency estimation the communication system used, can be deduced using a database of protocol features. There are a number of different methods for AMR introduced by researchers of the field. Most of these methods are proper for a limited number of modulations. Each method uses its own assumptions about the known parameters of signal like carrier frequency, SNR, Baud rate and so on. So, combining two methods together can be a difficult job. The modulation set we choose includes AM, LSSB, USSB, FM, MASK, MPSK, MFSK all for M=2, 4, 8 and MSK (minimum shift keying). There is no unique way to recognize all of these modulations. The above modulation schemes are selected form nine communication systems that we try to recognize them. These systems are ACARS, ALE, ATIS, FMS-BOS, PACTOR-II, PSK31, DGPS, GOLAY and ERMES. Successful recognition of these sample systems makes the way easier to expand our method to other communication systems. In part 2 the problem definition and introduction of the features used in AMR will be presented. Part 3 contains description of the neural network structure. The carrier and baud rate estimation method is introduced in Part 4. Part 5 is the result of our simulations and finally the conclusions will be presented in part 6.

2 Problem Definition

We need some features to show the amount of variation in amplitude, phase and frequency of the signal. As we have used neural networks it is possible to use hierarchical method for classification [11]. So we grouped the following modulation schemes as metagroup1: AM, LSSB, USSB, FM, ASK2, ASK4, PSK2, PSK4, FSK2 and FSK4 and the remaining ones as metagroup2: ASK8, PSK8, FSK8 and MSK.For the first metagroup we can use the nine features presented by Nandi and Azzouz [15]. They are as follows:

$$1\text{-}\gamma_{max} = \max\left|DFT(a_{cn}(i)\right|^2 / Ns \tag{6}$$

Ns: No. of samples per block, a_{cn}: Normalized-centered instantaneous amplitude.

$$2\text{-}\sigma_{ap} = \sqrt{1/c(\sum_{an(i)>at}\phi_{NL}^2(i)) - (1/c\sum_{an(i)>at}|\phi_{NL}(i)|)^2} \tag{7}$$

ϕ_{NL}: Centered-nonlinear component of instantaneous phase, C: No. of samples in ϕ_{NL} for which $a_n(i) > a_t$, the threshold value of $a_n(i) = \dfrac{a(i)}{mean(a(i))}$ And a(i) is the instantaneous amplitude.

$$3\text{-}\sigma_{dp} = \sqrt{1/c(\sum_{an(i)>at}\phi_{NL}^2(i)) - (1/c\sum_{an(i)>at}\phi_{NL}(i))^2} \tag{8}$$

$$4\text{-}P = \frac{P_L - P_U}{P_L + P_U} \text{ where } P_L = \sum_{i=1}^{fcn}|X_c(i)|^2, \quad P_U = \sum_{i=1}^{fcn}|X_c(i+f_{cn}+1)|^2 \tag{9}$$

$X_c(i)$: Fourier transform of RF signal.

$$5\text{-}\sigma_{aa} = \sqrt{1/Ns(\sum_{i=1}^{Ns}a_{cn}^2(i)) - (1/Ns\sum_{i=1}^{Ns}|a_{cn}(i)|)^2} \tag{10}$$

$$6\text{-}\sigma_{af} = \sqrt{1/c(\sum_{an(i)>at}f_N^2(i)) - (1/c\sum_{an(i)>at}|f_N(i)|)^2} \tag{11}$$

$f_N(i)$: Normalized-centered instantaneous frequency.

$$7\text{-}\sigma_a = \sqrt{1/c(\sum_{an(i)>at}a_{cn}^2(i)) - (1/c\sum_{an(i)>at}a_{cn}(i))^2} \tag{12}$$

$$8\text{-}\mu_{42}^a = \frac{E\{a_{cn}^4(i)\}}{\{E\{a_{cn}^2(i)\}\}^2} \text{ where E\{\} means Expected value.} \tag{13}$$

$$9\text{-}\mu_{42}^f = \frac{E\{f_N^4(i)\}}{\{E\{f_N^2(i)\}\}^2}, f_N \text{ :Normalised-centred instantaneous fre-} \tag{14}$$

quency

For the second group we use the spectrum of the signal as the feature as proposed in reference [2]. In this neural network Welsh periodogram of signal is used as feature for classification. To reduce the dimension of input data, the main lobe of periodogram containing most of information is used and the remaining parts are discarded. Choosing the proper interval of periodogram is done after checking all the modulation set spectrums. Finally we need methods to estimate baud rate and carrier frequency of signal. For carrier recovery we have used the zero crossing method [1].

3 Neural Network Structure

We have used the concept of hierarchical neural networks described by [11]. In this method classification can be done in successive steps. The outputs can be classified in groups called metagroup and the neural network classifies these matagroups first. Then classification can be done within each metagroup in the same manner till getting the final result. The input data is classified as one of the first metagroup members or just as metagroup2. So we do not need an extra neural network for classification of first metagroup subsets. However signals belonging to the second metagroup are classified using another network structure to the final output result.

Neural network1 is a feedforward neural network with two hidden layers. The structure is chosen after an extensive simulation tests. The number of nodes is 9 in the input layer, 75 in the first hidden layer, 75 in the second hidden layer and 11 in the output layer. The network is trained using backpropagataion method. For better result we have used a variable rate backpropagation learning algorithm. We have used 240 block of data each containing 2048 samples of the signal in SNRs varying from 0 to 55 dB for training the networks and another set of the same size but different from those used for training as the test set. The second neural network is a one hidden layer feedforward neural network that is trained using the same method as the first one. It has only one input node. There are 80 nodes in the its hidden layer.

4 Carrier and Baud Rate Estimation

Hsue and Soliman[1] introduced a method based on zero crossing to estimate the carrier frequency. The received waveform is modeled as:

$$r(t)=s(t)+v(t) \qquad 0 \leq t \leq T_s \tag{15}$$

s(t) is the modulated signal and v(t) is an additive band limited Gaussian noise. If the signal is sampled using a zero-crossing sampler, the time tags of zero-crossing points are recorded as a sequence $\{x(i), i=1,2\ldots,N\}$. Then the zero-crossing interval sequence $\{y(i)\}$ is defined as

$$y(i)=x(i+1)-x(i) \qquad i=1,2,\ldots,N-1 \tag{16}$$

And zero-crossing interval difference sequence $\{z(i)\}$ is

$$z(i)=y(i+1)-y(i) \qquad i=1,2,\ldots,N-2 \tag{17}$$

Since Intersymbol transition (IST) affects the accuracy of estimation, we should ignore IST samples. After discarding them we obtain $\{y_a(i)\}$. The length of the resultant sequence is N_y. The carrier frequency is then estimated by averaging $y_a(i)$ samples as:

$$f_c = \frac{N_y}{2\sum_{i=1}^{N_y} y_a(i)} \tag{18}$$

For baud rate estimation we have used Wegener method [16]. The method used is as follows. We can compute the "baud length" instead of "baud rate" and they are simply the reciprocal of each other. For baud length estimation we can use the transition of data to compute a sequence of integers $\{L1, L2, ..., L_N\}$ where L_j represents the number of samples between baud transitions j-1 and j (baud length) and N is the number of transitions. If some minimum SNR is not achieved the estimation will not be accurate or even useless. If we define a false transition due to demodulator error a "glitch ", then a de-glitch process can be used to make the result immune of noise effect. The simplest method is to sort the transition lengths in an increasing order and the first (shortest) element should be discarded as a glitch. An initial estimation of baud length is assumed. A good choices is an average of a few neighboring elements. Then the following algorithm can be used:

Sum=0, For each L_j in the sorted list {#bauds=round (L_j / est_{init}),

Sum=sum+(L_j /#bauds) }, est_{final} =sum/N

Where est_{final} is the final estimation result. A very useful feature of this method is that if we have prior information about the expected signals e.g. a table of possible baud rates, we can use it effectively to make the estimation process faster and more accurate.

5   Simulation Results

We have used voice signal as the modulating signal instead of binary random data and we used 495120 samples of voice in total. In test period we used another set of 491520 to check the performance of the neural network. The results are presented in table 1 and 2 at the end of paper. In table 1 the performance result of the first neural network is included. In this case, ASK8, PSK8, FSK8 and MSK signals are classified all as class 2 data. However, the classification probability is not the same for all of these signals. Therefore we included them separately and the reader is aware that thisdoes not mean the ability of recognition of them by the first neural network. In the second table the performance result of both the first and second neural network for

Table 1. The percent of correct decision probability of the first neural network

SN	0 dB	5 dB	15	25	35	45	55
AM	87.51	90	90	100	100	100	100
LSSB	15	55.5	90	100	100	100	100
USSB	17	85.7	95	100	100	100	100
FM	90	85	90	100	100	100	100
ASK2	90	100	100	100	100	100	100
ASK4	80	100	100	100	100	100	100
ASK8	75	95	100	100	100	100	100
PSK2	10	35	85	100	100	100	100
PSK4	15	65	95	100	100	100	100
PSK8	90	95	100	100	100	100	100
FSK2	85	75	100	100	100	100	100
FSK4	75	100	100	100	100	100	100
FSK8	95	100	100	100	100	100	100
MSK	70	85	95	100	100	100	100

Table 2. The correct decision probability of the second class

0SNR	10 DB	25 Db	315 Db	425 Db	535 Db	645 Db	755 Db
8ASK8	915	1060	1180	12100	13100	14100	15100
16PSK8	1745	1870	1990	2095	21100	22100	23100
24FSK8	2530	2675	2787.5	2897	29100	30100	31100
32MSK	3320	3460	3595	36100	37100	38100	39100

Table 3. Comparison of our method with Nandi and Azzouz at 15 dB SNR

	AM	LSSB	USSB	FM	ASK2	ASK4	PSK2	PSK4	FSK2	FSK
NANDI	88.	99.8	98.5	90.1	96.8	86.5	99.5	96.8	99	99.5
OUR	90	90	95	90	100	100	85	95	100	100

the class 2 signals are introduced. Although the modulation subset considered is different from reference [2] and [15], we compare the performance of our method with Nandi and Azzouz method. The performance of Ghani method versus SNR is not presented in his paper, so we could not compare its performance with our method. As it is clear from table 3 the combination of the two methods has been done. The next step is to extract carrier frequency and baud rate. Using the methods mentioned the job can be done effectively in SNRs above 15 dB. Then using the modulation scheme information and estimated baud rate and carrier frequency and comparing this information with the known counterparts of them in the nine systems mentioned earlier we can deduce the protocol. The method is robust and fast enough to be used in on-line applications and expanding the method to recognition of more communication systems is straight forward.

6 Conclusions

We have developed a method for recognition of communication systems based on modulation recognition and baud rate and carrier frequency estimation. For automatic modulation the classification procedure is done by the hierarchical neural network method. As it is clear from the results, the overall performance of the AMR method used in this paper is above 75% even in SNR as low as 15 dB. For SNR above 35 dB the performance reaches 100%.

References

1. Hsue S. Z. and Soliman S. S.: Automatic modulation recognition of digitally modulated signals, Lecture Notes in MILCOM '89, vol. 3, (1989) 645–649.
2. Ghani N. and Lamontagne R.: Neural networks applied to the classification of spectral features for automatic modulation recognition, Lecture Notes in MILCOM '93, vol. 1, (1993) 111-115.
3. Mobasseri B.: Digital modulation classification using constellation shape, Signal Processing, (2000) 251-277.
4. Taira S.: Automatic classification of QAM signals by neural networks Lecture Notes in ICASSP 2001, vol. 2, (2001) 1309–1312.
5. Wong M. L. D. and Nandi A. K.: Automatic modulation recognition using spectral and statistical features with multi layer perceptrons, Lecture Notes in Sixth international symposium on Signal processing and its application, vol. 2, (2001) 390-393.
6. Hung C. Y. and Polydoros A.: Likelihood methods for MPSK modulation classification, IEEE Trans. On Communication, (1995) 1493–1503.
7. Polydoros A. and Kim K.: On detection and classification of quadrature digital modulations in broad-band noise, IEEE Trans. On Communication, (1990) 1199–1211.
8. Boudreau D., Dubuc C. and Patenaude F..: A fast automatic modulation recognition algorithm and its implementation in a spectrum monitoring application, Lecture Notes in MILCOM '00, vol. 2, (2000) 732-736.
9. Hong L. and Ho K. C.: BPSK and QPSK modulation classification with unknown signal levels, Lecture Notes in MILCOM '00, vol. 2, (2000) 976-980.
10. Lichun L.: Comments on signal classification using statistical moments, IEEE Trans. On Communication, (2002) 1199–1211.
11. Louis C. and Sehier P.: Automatic modulation recognition with hierarchical neural networks, Lecture Notes in MILCOM '94, vol. 3, (1994) 713-717.

12. Delgosh F.: Digital modulation recognition, Master's Thesis, Sharif Industrial university, (1998)
13. Carlson A. B. , Crilly P. B. and Rutledge J. C., communication systems, Fourth edition, McGraw Hill, (2001)
14. Proakis J. G., Digital communication, Second edition, McGraw Hill international edition, (1989)
15. Azzouz E. G. and Nandi A. K., Automatic modulation recognition of communication signals, Kluwer Academic Publishers, Boston, (1996)
16. Wegener A. W.: Practical techniques for baud rate estimation, Lecture Notes in ICASSP-92, vol. 4, (1992) 681-684.

An Extension to Quasi-orthogonal Sequences

Frederik Vanhaverbeke and Marc Moeneclaey

TELIN/DIGCOM, Ghent University, Sint-Pietersnieuwstraat 41
B-9000 Gent, Belgium
fv@telin.ugent.be

Abstract. An improved overloading scheme is presented for single user detection in the downlink of multiple access systems based on random OCDMA/OCDMA (O/O). By displacing in time the orthogonal signatures of the two user sets, that make up the overloaded system, the multiuser interference between the users of the two sets can be reduced by up to 50% (depending on the chip pulse bandwidth) as compared to Quasi-Orthogonal Sequences (QOS), that are presently part of the downlink standard of cdma2000. This reduction of the multiuser interference gives rise to an increase of the feasible Signal to Noise Ratio (SNR) for a particular channel load.

1 Introduction

In any synchronized multiple access system based on code-division multiple access (CDMA), the maximum number of *orthogonal* users equals the spreading factor N. In order to be able to cope with oversaturation of a synchronized CDMA system (i.e. with a number of users $K = N + M$: $N < K$ ($\leq 2N$), a popular approach consists of the OCDMA/OCDMA (O/O) systems [1-5], where a complete set of orthogonal signature sequences are assigned to N users ('set 1 users'), while the remaining M users ('set 2 users') are assigned another set of orthogonal sequences. The motivation behind this proposal is that the interference levels of the users are decreased considerably as compared to other signature sequence sets (e.g., random spreading), since each user suffers from interference caused by the users belonging to the other set only.

In [1-4], the potential of various O/O types with multiuser detection [6] was investigated, while [5] evaluates the downlink potential with single user detection of a particular type of O/O: 'Quasi Orthogonal Sequences (QOS)'. Especially the latter application is of practical interest, since alignment of the different user signals is easy to achieve in the downlink (as opposed to the uplink), while single user detection is the obvious choice for detection at the mobile stations. The QOS, discussed in [5], are obtained by assigning orthogonal Walsh-Hadamard sequences [7] to the set 1 users, while each set 2 user is assigned a Walsh-Hadamard sequence, overlaid by a common bent sequence with the window property [5]. These QOS minimize the maximum correlation between the set 1 and set 2 users, which was the incentive to add these QOS to the cdma2000 standard, so that overloaded systems can be dealt with [8].

Up to now, the chip pulses of *all* users are perfectly aligned in time in all considered O/O systems. However, an additional degree of freedom between the set 1 and the set 2 users has been overlooked: one can actually displace the set 2 signatures with respect to the set 1 signatures, without destruction of the orthogonality within each set. In this contribution we investigate the impact of this displacement on the

J.N. de Souza et al. (Eds.): ICT 2004, LNCS 3124, pp. 114–121, 2004.
© Springer-Verlag Berlin Heidelberg 2004

crosscorrelation between the set 1 and set 2 users, and the resulting favorable influence on the downlink performance. In section 2, we present the system model, along with the conventional QOS system. In section 3, we introduce the new type of O/O with the displaced signature sets, and we compute the crosscorrelation among the user signals. In section 4, we assess the downlink performance in terms of maximum achievable Signal-to-Noise and Interference Ratio (SNIR) as a function of channel overload and required transmit power. Finally, conclusions are drawn in section 5.

2   Conventional OCDMA/OCDMA

Consider the downlink of a perfectly synchronized single-cell CDMA system with spreading factor N and K users. Since all signals are generated and transmitted at the same base station, this signal alignment is easy to achieve, and the total transmitted signal $S_1(t)$ is simply the sum of the signals $s_k(t)$ of all users k (k = 1, ..., K):

$$S_1(t) = \sum_{k=1}^{K} s_k(t) = \sum_{k=1}^{K} \left\{ \sum_{i=-\infty}^{+\infty} A_k(i).a_k(i) \left[\sum_{j=0}^{N-1} \beta_k^{(i)}(j).p_c\left(t - i.NT_c - j.T_c\right) \right] \right\} \tag{1}$$

In this expression,

* $\beta_k^{(i)} = (\beta_k^{(i)}(1) \ ... \ \beta_k^{(i)}(N)) \in \{1/\sqrt{N}, -1/\sqrt{N}\}^N$, $A_k(i)$ and $a_k(i) \in \{1,-1\}$ are the signature sequence, the amplitude and the data symbol of user k in symbol interval i, respectively.

* The unit-energy chip pulse $p_c(t)$ is a square-root cosine rolloff pulse with rolloff α, and chip duration T_c [9]. The associated pulse, obtained after matched filtering of $p_c(t)$, is the cosine rolloff pulse $\phi_c(t)$ with rolloff α. Note that $\phi_c(t)$ is a Nyquist pulse, i.e., $\phi_c(jT_c) = \delta_j$.

We denote the channel gain from the base station to user k in symbol interval i by $\gamma_k(i)$. In order to obtain a decision statistic $z_k(i)$ for the detection of databit $a_k(i)$, the received signal is applied to a matched filter $p_c(-t)$, followed by a sampling on the instants $(iN+j)T_c$ (j = 0, ..., N-1). The resulting samples are correlated with the signature sequence $\beta_k^{(i)}$, and finally a normalization is carried out by multiplying the result by $\gamma_k^*(i)/|\gamma_k(i)|^2$. Since $\phi_c(t)$ is a Nyquist pulse, and due to the perfect alignment of the user signals, the observable $z_k(i)$ is given by (k = 1, ..., K):

$$z_k(i) = A_k(i).a_k(i) + \sum_{j \neq k} A_j(i).a_j(i).\rho_{k,j}(i) + n_k(i) \tag{2}$$

In (2), $\rho_{k,j}(i) = (\beta_k^{(i)})^T.\beta_j^{(i)}$ is the correlation between $\beta_k^{(i)}$ and $\beta_j^{(i)}$, and $n_k(i)$ is a real-valued noise sample with variance $\sigma_k^2/|\gamma_k(i)|^2$, where σ_k^2 is the power spectral density of the noise at the receiver input.

In order to assure that all users meet a predefined quality of service constraint, power control is applied in the downlink. If LNT_c is smaller than the minimum coherence time of the channels between the base station and the users[1], power control can be achieved by updating the amplitudes of the users once every L symbol intervals, based on (an estimate of) the variance of the sum of noise and interference over a time span of L symbol intervals. To meet the quality of service constraint for user k, it is necessary that this variance remains lower than the predefined threshold of user k. Since the channel gain is constant over these L symbol intervals, the variance of interest is given by ($k = 1, ..., K$):

$$\tilde{\mu}_k^2 = \sum_{j \neq k} \tilde{A}_j^2 . \tilde{R}_{k,j} + \tilde{\delta}_k \quad \text{with} \quad \tilde{R}_{k,j} = \frac{1}{L} \sum_{l=1}^{L} \left| \rho_{k,j}(i_1 + l) \right|^2 \qquad (3)$$

where $\tilde{x}$ denotes the (constant) value of x over the considered time span of L symbols, that starts at time index i_1, $\tilde{\delta}_k = \sigma_k^2 / |\tilde{\gamma}_k|^2$, and $\tilde{R}_{k,j}$ is the crossinterference between user k and user j over the considered time interval. The variance $\tilde{\mu}_k^2$ is related to the signal to interference plus noise ratio $SINR_k$ of user k by $SINR_k = \tilde{A}_k^2 / \tilde{\mu}_k^2$.

From expression (3), it is obvious that the squared crosscorrelations between the signatures of the users should be as small as possible, in order to restrict the MUltiuser Interference (MUI). As long as K remains smaller than N, taking orthogonal signatures, as is done in IS-95 [12] and cdma2000 [8], is optimum because they yield $\rho_{k,j} = 0$. If K exceeds N (i.e., an oversaturated system), the O/O system tries to eliminate as much intra-cell interference as possible, by taking orthogonal signatures for N users ('set 1 users'), and by taking another set of orthogonal signatures for the remaining M users ('set 2 users'). This eliminates the interference of (N-1) and (M-1) users on the detection of the set 1 and set 2 users, respectively. Indexing the set 1 users as the first N users and the set 2 users as the last M users, (3) turns into:

$$\tilde{\mu}_k^2 = \sum_{j=N+1}^{K} \tilde{A}_j^2 . \tilde{R}_{k,j} + \tilde{\delta}_k (k = 1,...,N) \quad ; \quad \tilde{\mu}_k^2 = \sum_{j=1}^{N} \tilde{A}_j^2 . \tilde{R}_{k,j} + \tilde{\delta}_k (k = N+1,...,K) \quad (4)$$

The signatures of the set 1 users span the complete vector space of dimension N, implying that $\sum_{j=1}^{N} \tilde{R}_{k,j} = 1$ (k = N+1, ..., K). From this expression, it is immediately seen that the maximal crossinterference between the set 1 and the set 2 users will be minimized if and only if *all* of these crossinterferences $\tilde{R}_{k,j}$ are equal to 1/N. This condition is met by means of QOS for $N = 4^n$, where the signatures of the users are fixed over all symbol intervals, so that the signatures β_k^{QOS} of the set 1 users (k = 1,

---

[1] Most channels suffer from slow fading, so that L can take on large values, since the *minimum* coherence time is typically on the order of 5 ms [10,11]. In IS-95, for instance, $N.T_c$ is about 52 μs, and the power is updated every 1.25 ms (L = 24) [12].

..., N) are the Walsh Hadamard sequences $\mathbf{WH}_N^{(k)}$ (k = 1, ..., N) of order N, and the signatures of the set 2 users are obtained by overlaying the same Walsh-Hadamard sequences by means of a bent sequence $\mathbf{Q} \in \{1,-1\}^N$ [5]:

$$\beta_k^{QOS} = diag\{Q_1, Q_2, ..., Q_N\}.\mathbf{WH}_N^{(k-N)} \quad (k = N+1, ..., K).$$

3 O/O with Displaced Signature Sets

In the conventional O/O systems, there is an additional degree of freedom that has not been exploited. Indeed, in order to make sure that the first N user signals are orthogonal, they have to be perfectly aligned in time, and the same is true for the set 2 user signals. However, the set 1 user signals do not need to be aligned with the set 2 user signals to provide this property. Hence, the displacement τ ($\tau \in [0, NT_c)$) of the set 2 users with respect to the set 1 users is an additional degree of freedom. Adopting the same notation $s_k(t)$ as in (1) for the signal of user k, the total transmitted signal $S_2(t)$ is now given by

$$S_2(t) = \sum_{i=1}^{N} s_i(t) + \sum_{i=N+1}^{K} s_i(t - \tau) \tag{5}$$

We focus on random O/O [1], where the signatures of the set 1 users and the set 2 users are obtained by overlaying the Walsh Hadamard vectors in every symbol interval i by the respective scrambling sequences $\mathbf{P}_1^{(i)}$ and $\mathbf{P}_2^{(i)}$, that are chosen completely at random and independently out of $\{1,-1\}^N$ in every symbol interval:

$$\beta_k^{(i)} = \begin{cases} diag\{P_1^{(i)}(0), ..., P_1^{(i)}(N-1)\}.\mathbf{WH}_N^{(k)} & k = 1, ..., N \\ diag\{P_2^{(i)}(0), ..., P_2^{(i)}(N-1)\}.\mathbf{WH}_N^{(k-N)} & k = N+1, ..., K \end{cases} \tag{6}$$

Let us consider the contribution $z_k^j(i)$ of set 2 user j to the decision variable $z_k(i)$ of set 1 user k in symbol interval i:

$$z_k^j(i) = \sum_{s=-\infty}^{+\infty} A_j(s).a_j(s).\rho_{k,j}^{(i,s)}(\tau) \tag{7}$$

with $\rho_{k,j}^{(i,s)}(\tau) = \sum_{m=0}^{N-1}\sum_{n=0}^{N-1} \beta_k^{(i)}(m).\beta_j^{(s)}(n).\phi_c[((i-s).N+(m-n))T_c - \tau]$.   Hence, the crossinter-ference $\tilde{R}_{k,j}$ between set 1 user k and set 2 user j is now given as

$$\tilde{R}_{k,j} = \frac{1}{L}\sum_{l=1}^{L}\left\{\sum_{s=-\infty}^{+\infty}\left|\rho_{k,j}^{(i_1+l,s)}(\tau)\right|^2\right\} \tag{8}$$

and is an (approximately) Gaussian random variable. However, its expected value is reduced by a factor $1/\lambda \geq 1$ as compared to QOS (see appendix):

$$E\left[\tilde{R}_{k,j}\right] = \frac{\lambda(\alpha,\Delta)}{N} = \frac{1}{N}\sum_{s=-\infty}^{+\infty}\phi^2\left(s.T_c - \Delta\right) \quad ; \quad \Delta = \tau - \left\lfloor\frac{\tau}{T_c}\right\rfloor.T_c \qquad (9)$$

So, as compared to the original QOS system, the expected value of the MUI of all set 1 and set 2 users is decreased by a factor $1/\lambda$, that is dependent on the rolloff α and on Δ. For $\Delta = 0$, the chip pulses of all the users are perfectly aligned, and $\lambda = 1$, whether the symbol boundaries of the set 1 and the set 2 users are aligned or not.

The function $\lambda(\alpha,\Delta)$ is the interference function of the square-root cosine rolloff pulse $p_c(t)$. This interference function was introduced in another context (PN-spread asynchronous communication) in [13], where it was shown that it can be written as a function of the Fourier transform $P_c(f)$ of $p_c(t)$:

$$\lambda(\alpha,\Delta) = F(0) + (1 - F(0)).\cos(2\pi.\Delta T_c) \quad with \quad F(0) = \frac{1}{T_c}\int_{-\infty}^{+\infty}|P_c(f)|^4 df \qquad (10)$$

According to this expression, λ is minimal for $\Delta = T_c/2$. Numerical calculations yield the following relationship between the optimal value of λ and the rolloff α of the chip pulse: $\lambda(\alpha,T_c/2) = 1 - \alpha/2$. So, it is obvious that we can obtain important decreases in MUI, by displacing the chip pulses of the set 2 users by half a chip period as compared to the chip pulses of the set 1 users. This decrease can be up to 50% for $\alpha = 1$, and amounts to 15 % for a practical rolloff value of 0.3.

4   Network Capacity of Improved O/O

If we impose on the system a common quality of service constraint, so that the $SINR_k$ for all users k has to be at least κ, then, as long as the problem is feasible, the minimum (optimum) power solution A_{min} corresponds to the case where all $SINR_k$ are exactly κ [14,15]:

$$A = \kappa(R.A + \delta) \qquad (11)$$

with $(A)_k = \tilde{A}_k^2$, $(\delta)_k = \tilde{\delta}_k$ for k = 1, ..., K, $(R)_{i,j} = \tilde{R}_{i,j}$ if i and j are from a different set, while $(R)_{i,j} = 0$ if i and j are from the same set. It is well known [14,15,16] that (11) has a positive solution A, if and only if the largest eigenvalue λ_R^{max} of R is smaller than $1/\kappa$. Hence, the maximum achievable SNIR target level $\kappa_{max}(R)$ for the system is given by $1/\lambda_R^{max}$.

For random O/O, R is a random matrix with components $\tilde{R}_{k,j}$ that have a Gaussian distribution for any (k,j), $(j,k) \in \{1,...,N\}x\{1,...,M\}$. As a consequence, λ_R^{max} is also a random variable, as well as the corresponding $\kappa_{max}(R)$. Nevertheless, for a typical value L = 20, simulations point out that the distribution of κ_{max} is strongly peaked around the value $\kappa_{max}(E[R]) = 1/\lambda_{E[R]}^{max}$, where $E[R]$ is the expected value of R (cf. (9)). This is illustrated in figure 1, where the *maximum* deviation ($\kappa_{max}(E[R]) - \kappa_{max}(R)$) over a wide range of matrices R is shown for N = 16 and 32, and $\alpha = 0.25$ and 1. It can be observed that this maximum deviation is highest at low channel overloads, but decreases sharply as the channel load increases. Moreover, the maximum deviation is significantly lower for N = 32 as compared to N = 16. The *average* deviation was found to be less than 0.05dB for channel loads in excess of 1.12 and 1.2 for N = 32 and 16 respectively. Simulations indicate that for N = 64, even lower maximum and average deviations are found, allowing us to conclude that the deviation decreases significantly when N increases. In addition to this, the deviation increases as α increases.

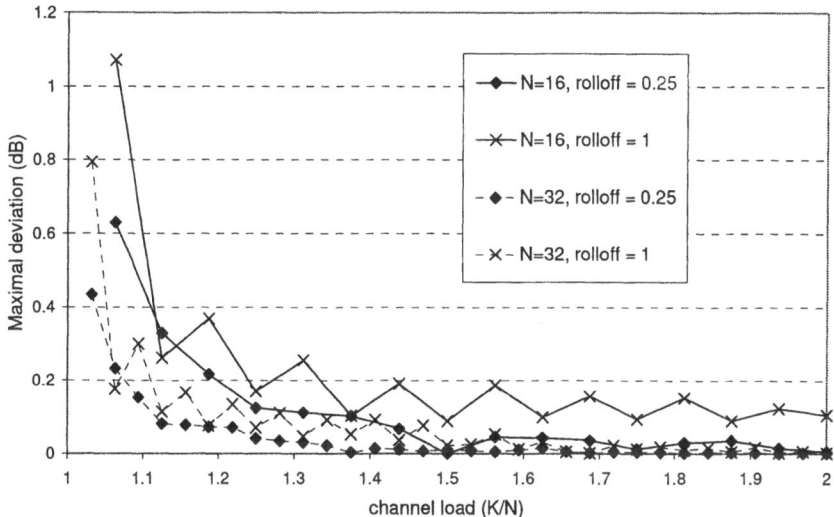

Fig. 1. Maximum deviation (dB) as a function of channel load for N = 16, 32 and α = 0.25 and 1.

Figure 2 shows κ_{max} (E[**R**]) for the improved O/O systems. As the previous discussion brought to light, this performance is representative for the actual performance of improved O/O, with very small deviations at the low values of κ_{max}, especially for moderate to high spreading factors. In figure 2, we also added the performance of QOS and the upper bound for chip-synchronous systems [17]. It is found that κ_{max} (E[**R**]) for improved O/O is about 0.58, 0.67, 0.79 and 0.97 dB above the achievable SNIR of QOS for α = 0.25, 0.5, 0.75 and 1 respectively. We also see from figure 2 that the performance of improved O/O with α = 1, 0.75, 0.5 and 0.25, exceeds the (chip-synchronous) upper bound for channel loads in excess of 1.25, 1.4, 1.56 and 1.77 respectively.

5 Conclusion

In this article, we presented a new type of random O/O that allows for a higher network capacity in the downlink as compared to the Quasi-Orthogonal Sequences, that are presently implemented in the cdma2000 standard. This improved O/O system is obtained by displacing in time (over half a chip period) the signatures of the orthogonal sets of the O/O system. Depending on the rolloff of the chip pulses, the achievable common SNIR target level can be increased by up to about 1 dB as compared to QOS. Moreover, the performance of improved O/O can exceed the chip-synchronous upper bound for channel loads in excess of 1.25.

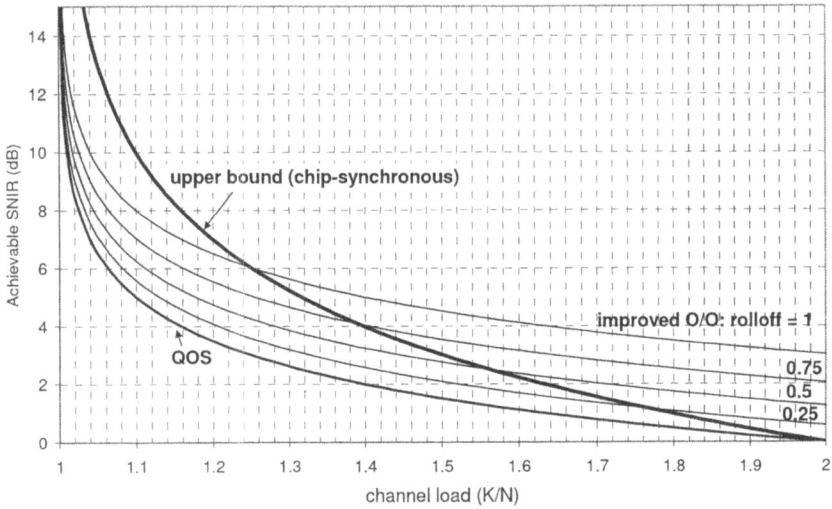

Fig. 2. κ_{max} (E[**R**]) for improved O/O

References

1. F. Vanhaverbeke, M. Moeneclaey and H. Sari, "DS/CDMA with Two Sets of Orthogonal Sequences and Iterative Detection," *IEEE Communication Letters*, vol. 4, pp. 289-291, Sept. 2000.
2. Sari H, Vanhaverbeke F and M. Moeneclaey, "Multiple access using two sets of orthogonal signal waveforms," *IEEE Communication Letters*, vol. 4, pp. 4-6, January 2000.
3. H. Sari, F. Vanhaverbeke and M. Moeneclaey, "Extending the Capacity of Multiple Access Channels," *IEEE Communications Magazine.*, pp. 74-82, Jan. 2000.
4. F. Vanhaverbeke and M. Moeneclaey, "Sum Capacity of the OCDMA/OCDMA Signature Sequence Set," *IEEE Communications Letters*, vol. 6, pp. 340-342, August 2002.
5. K. Yang, Y.-K. Kim and P. V. Kumar, "Quasi-orthogonal Sequences for Code-Division Multiple-Access Systems," *IEEE Trans. Info.*, vol. 46, pp. 982-993, May 2000.
6. S. Verdu, *Multiuser detection*, Cambridge University Press, New York 1998.
7. E. H. Dinan and B. Jabbari, "Spreading Codes for Direct Sequence CDMA and Wideband CDMA Cellular Networks," *IEEE Commun. Mag.*, pp 48-54, Sept. 1998.
8. TIA/EIA 3GPP2 C.S0002-B, "Physical Layer Standard for cdma2000 Spread Spectrum Systems, Release B," Jan. 16, 2001.
9. J. G. Proakis, *Digital Communications*, Mc. Graw-Hill Science, 2000.
10. B. Sklar, "Rayleigh Fading Channels in Mobile Digital Communication Systems Part I: Characterization," *IEEE Commun. Mag.*, pp. 136-146, Sept. 1997.
11. B. Sklar, "Rayleigh Fading Channels in Mobile Digital Communication Systems Part II: Mitigation," *IEEE Commun. Mag.*, pp. 148-155, Sept. 1997.

12. Telecommunication Industry Association, TIA/EIA, Washington, DC. *Mobile Station-base Station Compatibility Standard for Dual-mode Wideband Spread Spectrum IS-95 A*, 1995.
13. J. H. Cho and J. S. Lehnert, "An Optimal Signal Design for Band-Limited Asynchronous DS-CDMA Communications", *IEEE Trans. Inform. Theory*, pp. 1172-1185, May 2002.
14. S. V. Hanly and D. N. Tse, "Power Control and Capacity of Spread Spectrum Wireless Networks," *Automatica*, vol. 35, pp. 1987-2012, December 1999.
15. G.J. Foschini and Z. Miljanic, "A simple distributed autonomous power control algorithm and its convergence", *IEEE Trans. Vehicular Technology*, vol. 42, pp. 641-646, November 1993.
16. S. Ulukus, *Power control, Multiuser Detection and Interference Avoidance in CDMA Systems*, Phd., Rutgers University, NJ, October 1998.
17. S. Ulukus and R. D. Yates, "User Capacity of Asynchronous CDMA Systems with Matched Filter Receivers and Optimum Signature Sequences," *IEEE Trans. on Inform. Theory*, to appear.

Appendix

In order to determine $E\left(\tilde{R}_{k,j}\right)$, we take a closer look at

$$I(l) = E\left[\sum_{s=-\infty}^{+\infty}\left|\rho_{k,j}^{(i_1+l,s)}(\tau)\right|^2\right] = \sum_{s=-\infty}^{+\infty} E\left[\left|\sum_{m=0}^{N-1}\sum_{n=0}^{N-1}\beta_k^{(i_1+l)}(m).\beta_j^{(s)}(n).\phi_c\left[((i_1+l-s).N+(m-n))T_c-\tau\right]\right|^2\right]$$

Since the scrambling sequences of the different users sets are random, the chips of different users are uncorrelated, and the same is true for different chips of the same signature. This, combined with the fact that all chips belong to the set $\{1/\sqrt{N}, -1/\sqrt{N}\}$, reduces $I(l)$ to:

$$I(l) = \frac{1}{N^2}\sum_{s=-\infty}^{+\infty}\sum_{m=0}^{N-1}\sum_{n=0}^{N-1}\phi_c^2\left[((i_1+l-s).N+(m-n))T_c-\tau\right]$$

$$= \frac{1}{N^2}\sum_{s=-\infty}^{+\infty}\sum_{m=0}^{N-1}\sum_{n=0}^{N-1}\phi_c^2\left[(s.N+(m-n))T_c-\tau\right]$$

An observation of the terms in the last expression of $I(l)$ learns that each term $\phi_c^2(k.T_c-\Delta)$ occurs N times in the summation. As a consequence:

$$I(l) = \frac{1}{N^2}\sum_{s=-\infty}^{+\infty}N.\phi_c^2(s.T_c-\Delta) = \frac{1}{N}\sum_{s=-\infty}^{+\infty}\phi_c^2(s.T_c-\Delta)$$

and

$$E\left(\tilde{R}_{k,j}\right) = \frac{1}{L}\sum_{l=1}^{L}I(l) = \frac{1}{N}\sum_{s=-\infty}^{+\infty}\phi_c^2(s.T_c-\Delta)$$

Trellis Code Construction for the 2-User Binary Adder Channel

Valdemar C. da Rocha Jr. and Maria de Lourdes M.G. Alcoforado

Communications Research Group - CODEC,
Department of Electronics and Systems, P.O. Box 7800,
Federal University of Pernambuco,
50711-970 Recife PE, Brazil,
vcr@ufpe.br, mlmga@ee.ufpe.br

Abstract. This paper introduces a construction technique for trellis codes on the two-user binary adder channel (2-BAC). A condition for unique decodability for convolutional codes on the 2-BAC is derived, with an immediate extension for trellis codes. A pair of block codes (uniquely decodable on the 2-BAC) is used as a filter to eliminate just those paths through the trellis that would lead to ambiguity at the decoder. This approach in principle does not limit the sum rate of the resulting code pair.

1 Introduction

The subject of multiple-access channels is currently a topic of great practical interest in telecommunications. The simplest multiple-access channel is the two-user binary adder channel (2-BAC) [1]. This channel is memoryless and accepts at each unit of time two binary inputs, one from each user, both users assumed to produce equally likely binary digits, and emits a single output from the alphabet $\{0, 1, 2\}$. In the noiseless case, the channel output y is just the arithmetic sum of the inputs x_1 and x_2, respectively, while in the noisy case the channel output behavior is described by the conditional probability $P(y|x_1 x_2)$.

The 2-BAC channel capacity region was determined independently by Liao [2], Ahlswede [3] and Slepian and Wolf [4], and since then many papers have appeared on code construction for the 2-BAC [5]-[9], however exclusively on block codes. We remark that, except for the paper by Peterson and Costello [10], there are virtually no papers available on the subject of code construction for the 2-BAC channel using convolutional codes. One possible reason for that is the fact that the authors published a (in retrospect) pessimistic result, that the sum rate of a linear code pair on the 2-BAC can not exceed 1. Since convolutional codes are linear codes they did not seem in principle interesting for use on the 2-BAC. An approach using distance properties of linear block codes for constructing 2-BAC codes was put forward in [5] and [6]. The decoding problem, however, is usually left aside and the main goal in code construction for the 2-BAC has been in most cases just to achieve high rates. Perhaps the most serious difficulty for the practical use of 2-BAC codes has been precisely the decoding problem.

J.N. de Souza et al. (Eds.): ICT 2004, LNCS 3124, pp. 122–127, 2004.

In this paper we establish a condition for unique decodability of trellis codes on the 2-BAC in terms of output sequences and states of a 2-user trellis. We present also a code construction for the 2-BAC based on a single convolutional code and on a pair $(\mathbf{C}_1, \mathbf{C}_2)$ of short block codes uniquely decodable on the 2-BAC. A significant advantage of the resulting two-user trellis codes is that the decoding techniques used in the single-user case, like Viterbi decoding for example, apply directly to the two-user case [10]. Furthermore, we show that this new technique achieves capacity in the 2-BAC.

2   Trellis Construction for the 2-BAC

Consider a pair of trellis codes allocated one for each user. Because the channel is defined in terms of input pairs, the decoder must consider pairs of paths through the single-user trellises. That is, the a posteriori probabilities of single paths are not defined, however, the a posteriori probabilities of path pairs are defined. This leads immediately to the concept of a two-user trellis. The two-user trellis is defined such that, at any given time slot, each distinct pair of paths through the two single-user trellises corresponds to a unique path through the two-user trellis. At any given time slot each branch of the two-user trellis corresponds to a pair of branches in the single-user trellises, and each state of the two-user trellis corresponds to a pair of states in the single-user trellises. The decoder task is to discover which path along the two-user trellis is the most likely. If each single trellis has L_1 and L_2 states, respectively, the two-user trellis will have $L_1 L_2$ states.

3   Condition for Unique Decodability

In this section we characterize, in terms of states and branch labels of a trellis, a condition for unique decodability of a convolutional code pair on the 2-BAC. Let $(\mathcal{C}_1, \mathcal{C}_2)$ be a convolutional code [11, p.287-314] pair of blocklength n, the encoder of which have L_1 and L_2 states respectively. Usually a convolutional encoder is started in the all-zero state and, after N sub-blocks are generated, the sequence of sub-blocks is either truncated or terminated. In either case the resulting code can be treated as a block code. Let $a(s_i, s_p) \in \mathcal{C}_1$ denote a binary sequence of N sub-blocks starting at state s_i, $i = 1, 2, \ldots, L_1$ and terminating at state s_p, $p = 1, \ldots, L_1$, consisting of a concatenation of N sub-blocks of $\mathcal{C}_1$. Similarly, let $b(s_r, s_v) \in \mathcal{C}_2$ denote a binary sequence of N sub-blocks starting at state s_r, $r = 1, 2, \ldots, L_2$ and terminating at state s_v, $v = 1, \ldots, L_2$, consisting of a concatenation of N sub-blocks of $\mathcal{C}_2$.

We introduce next an array of rows and columns (see Fig. 1 for an example) that we call the *trellis array* $\mathcal{A}$ of the given convolutional code pair $(\mathcal{C}_1, \mathcal{C}_2)$. Each cell element in $\mathcal{A}$ is a ternary sequence $\mathbf{z}$ representing the arithmetic sum of two binary sequences $a(s_i, s_p)$ and $b(s_r, s_v)$, i.e., $\mathbf{z} = a(s_i, s_p) + b(s_r, s_v)$, where $a(s_i, s_p)$ and $b(s_r, s_v)$ indicate, respectively, the row index and the column index, and where $i = 1, 2, \ldots L_1$, $p = 1, 2, \ldots, L_1$, $r = 1, 2, \ldots, L_2$, and $v = 1, 2, \ldots, L_2$.

If we fix i in $a(s_i, s_p)$ (the row index of A) and fix r in $b(s_r, s_v)$ (the column index of A), and let $p = 1, 2, \ldots, L_1$ and $v = 1, 2, \ldots, L_2$, we obtain a sub-array denoted by $S(s_i, s_r)$.

Proposition 1. *A convolutional code pair is uniquely decodable on the 2-BAC if and only if in each sub-array $S(s_i, s_r), i = 1, 2, \ldots, L_1$ and $r = 1, 2, \ldots, L_2$, there is no repeated ternary sequences.*

Proof. Contrary to the hypothesis, suppose that the ternary sequence $\mathbf{z}$ occurs more than once in $S(s_i, s_r)$. It then follows that $\mathbf{z} = a(s_i, s_p) + b(s_r, s_v) = a(s_i, s_{p'}) + b(s_r, s_{v'}), p \neq p', v \neq v'$. Because the pair of initial states of the component binary codes, i.e., the pair s_i from $\mathcal{C}_1$ and s_r from $\mathcal{C}_2$, is common to both sub-array cells, it is not possible for the decoder to resolve the ambiguity of whether to deliver $a(s_i, s_p)$ associated to user 1 and $b(s_r, s_v)$ associated to user 2, or to deliver $a(s_i, s_{p'})$ associated to user 1 and $b(s_r, s_{v'})$ associated to user 2.

Now suppose that the ternary sequence $\mathbf{z}$ occurs only once per sub-array but may appear more than once in the trellis array. It then follows that $\mathbf{z} \in S(s_i, s_t)$ and $\mathbf{z} \in S(s_{i'}, s_{t'})$, $i \neq i'$ or/and $t \neq t'$. In this situation, however, the decoder knows in advance the pair of initial states of the component binary codes, for example the pair s_i from $\mathcal{C}_1$ and s_t from $\mathcal{C}_2$, and unambiguously separates the two user codewords. □

We remark that although *Proposition 1* specifies a necessary and sufficient condition for uniquely decodability on the 2-BAC, the use of convolutional codes will limit the maximum sum rate to 1. However, the approach that led to *Proposition 1* suggests that we change $\mathcal{C}_1$ and $\mathcal{C}_2$ and use instead trellis codes, i.e., codes not restricted to be linear. This line of reasoning is pursued in the next section.

4   Code Construction

Let $\mathcal{C}$ denote a (n, k) systematic convolutional code [11, p.303-308] and let $(\mathbf{C}_1, \mathbf{C}_2)$ denote a pair of block codes uniquely decodable on the 2-BAC. An encoder for $\mathcal{C}$, having L states, is made available to user 1 and, similarly, to user 2. User 1 feeds his messages to the encoder for $\mathbf{C}_1$, and the resulting codewords from $\mathbf{C}_1$ are fed as messages for the encoder for $\mathcal{C}$. User 2 proceeds in a similar manner, using an encoder for $\mathbf{C}_2$ followed by an encoder for $\mathcal{C}$. Essentially, the encoding operation performed by each user is a concatenation of his respective block code with code $\mathcal{C}$. Since $\mathcal{C}$ is systematic it follows that the arithmetic sum of the codewords from $\mathcal{C}$ produced by user 1 and user 2, respectively, must be uniquely decodable. This follows because the pair $(\mathbf{C}_1, \mathbf{C}_2)$ is uniquely decodable on the 2-BAC and arithmetic sums of codewords from $\mathbf{C}_1$ and $\mathbf{C}_2$ appear in the information section of the arithmetic sum of codewords from $\mathcal{C}$. The use of codes $\mathbf{C}_1$ and $\mathbf{C}_2$ causes the elimination of some rows and some columns in the original trellis array (using code C in the two array dimensions), giving origin to a *punctured trellis array*.

In terms of our derivation in the previous section, with a little abuse of notation, we denote by $(\mathcal{C}_1, \mathcal{C}_2)$ the trellis code pair generated above and state this result more formally as follows.

Proposition 2. *A trellis code pair is uniquely decodable on the 2-BAC if and only if in each sub-array $S(s_i, s_r), i = 1, 2, \ldots, L$ and $r = 1, 2, \ldots, L$, there is no repeated ternary sequences.*

If the rate of the pair $(\mathbf{C}_1, \mathbf{C}_2)$ is R it follows from our construction that $R_C = R(k/n)$ is the rate of the code consisting of the arithmetic sum of codewords from $\mathcal{C}$. Therefore, by taking $\mathcal{C}$ to be a code with rate k/n approaching 1, R_C will be very close to R. This means that if R for the pair $(\mathbf{C}_1, \mathbf{C}_2)$ achieves capacity on the 2-BAC then R_C will also achieve capacity.

Example

Let $\mathcal{C}$ be a rate $1/2$ systematic recursive convolutional code with transfer function matrix:

$$G(D) = \begin{bmatrix} 1 & \dfrac{1 + D^2}{1 + D + D^2} \end{bmatrix} \tag{1}$$

We use the code generation with $N = 2$ sub-blocks and the corresponding trellis array is shown in Fig. 1. There are ternary sequences $\mathbf{z}$ that occur more than once in each sub-array. We can observe for example that at $S(s_2, s_1)$ we have $a(s_2, s_1) + b(s_1, s_3) = a(s_2, s_2) + b(s_1, s_4) = a(s_2, s_3) + b(s_1, s_1) = a(s_2, s_4) + b(s_1, s_2) = 1111$. The use of codes $\mathbf{C}_1 = \{00, 11\}$ and $\mathbf{C}_2 = \{00, 01, 10\}$, as described earlier, causes the elimination of some rows and some columns in the table shown in Fig. 1. We thus obtain the punctured trellis array shown in Fig. 2, corresponding to a pair of uniquely decodable trellis codes.

	$b(s_1,s_1)$	$b(s_1,s_2)$	$b(s_1,s_3)$	$b(s_1,s_4)$	$b(s_2,s_1)$	$b(s_2,s_2)$	$b(s_2,s_3)$	$b(s_2,s_4)$	$b(s_3,s_1)$	$b(s_3,s_2)$	$b(s_3,s_3)$	$b(s_3,s_4)$	$b(s_4,s_1)$	$b(s_4,s_2)$	$b(s_4,s_3)$	$b(s_4,s_4)$
	0000	1110	0011	1101	1100	0010	1111	0001	1011	0101	1000	0110	0111	1001	0100	1010
$a(s_1,s_1)$ 0000	0000	1110	0011	1101	1100	0010	1111	0001	1011	0101	1000	0110	0111	1001	0100	1010
$a(s_1,s_2)$ 1110	1110	2220	1121	2211	2210	1120	2221	1111	2121	1211	2110	1220	1221	2111	1210	2120
$a(s_1,s_3)$ 0011	0011	1121	0022	1112	1111	0021	1122	0012	1022	0112	1011	0121	0122	1012	0111	1021
$a(s_1,s_4)$ 1101	1101	2211	1112	2202	2201	1111	2212	1102	2112	1202	2101	1211	1212	2102	1201	2111
$a(s_2,s_1)$ 1100	1100	2210	1111	2201	2200	1110	2211	1101	2111	1201	2100	1210	1211	2101	1200	2110
$a(s_2,s_2)$ 0010	0010	1120	0021	1111	1110	0020	1121	0011	1021	0111	1010	0120	0121	1011	0110	1020
$a(s_2,s_3)$ 1111	1111	2221	1122	2212	2211	1121	2222	1112	2122	1212	2111	1221	1222	2112	1211	2121
$a(s_2,s_4)$ 0001	0001	1111	0012	1102	1101	0011	1112	0002	1012	0102	1001	0111	0112	1002	0101	1011
$a(s_3,s_1)$ 1011	1011	2121	1022	2112	2111	1021	2122	1012	2022	1112	2011	1121	1122	2012	1111	2021
$a(s_3,s_2)$ 0101	0101	1211	0112	1202	1201	0111	1212	0102	1112	0202	1101	0211	0212	1102	0201	1111
$a(s_3,s_3)$ 1000	1000	2110	1011	2101	2100	1010	2111	1001	2011	1101	2000	1110	1111	2001	1100	2010
$a(s_3,s_4)$ 0110	0110	1220	0121	1211	1210	0120	1221	0111	1121	0211	1110	0220	0221	1111	0210	1120
$a(s_4,s_1)$ 0111	0111	1221	0122	1212	1211	0121	1222	0112	1122	0212	1111	0221	0222	1112	0211	1121
$a(s_4,s_2)$ 1001	1001	2111	1012	2102	2101	1011	2112	1002	2012	1102	2001	1111	1112	2002	1101	2011
$a(s_4,s_3)$ 0100	0100	1210	0111	1201	1200	0110	1211	0101	1111	0201	1100	0210	0211	1101	0200	1110
$a(s_4,s_4)$ 1010	1010	2120	1021	2111	2110	1020	2121	1011	2021	1111	2010	1120	1121	2011	1110	2020

Fig. 1. Trellis array, $N = 2$.

		$b(s_1,s_1)$	$b(s_1,s_3)$	$b(s_1,s_4)$	$b(s_2,s_1)$	$b(s_2,s_2)$	$b(s_2,s_4)$	$b(s_3,s_2)$	$b(s_3,s_3)$	$b(s_3,s_4)$	$b(s_4,s_1)$	$b(s_4,s_2)$	$b(s_4,s_3)$
		0000	0011	1101	1100	0010	0001	0101	1000	0110	0111	1001	0100
$a(s_1,s_1)$	0000	0000	0011	1101	1100	0010	0001	0101	1000	0110	0111	1001	0100
$a(s_1,s_2)$	1110	1110	1121	2211	2210	1120	1111	1211	2110	1220	1221	2111	1210
$a(s_2,s_3)$	1111	1111	1122	2212	2211	1121	1112	1212	2111	1221	1222	2112	1211
$a(s_2,s_4)$	0001	0001	0012	1102	1101	0011	0002	0102	1001	0111	0112	1002	0101
$a(s_3,s_1)$	1011	1011	1022	2112	2111	1021	1012	1112	2011	1121	1122	2012	1111
$a(s_3,s_2)$	0101	0101	0112	1202	1201	0111	0102	0202	1101	0211	0212	1102	0201
$a(s_4,s_3)$	0100	0100	0111	1201	1200	0110	0101	0201	1100	0210	0211	1101	0200
$a(s_4,s_4)$	1010	1010	1021	2111	2110	1020	1011	1111	2010	1120	1121	2011	1110

Fig. 2. Punctured trellis array.

5 Comments

A condition for unique decodability for convolutional codes on the 2-BAC was derived, with an immediate extension for trellis codes. The use of a pair of block codes (uniquely decodable on the 2-BAC) plays the role of a filter to eliminate just those paths through the trellis that would lead to ambiguity at the decoder. This approach in principle does not limit the sum rate of the resulting code pair. In conclusion, we have shown in this paper how to construct trellis codes for the 2-BAC.

Acknowledgements. The research of the first author received partial support from the Brazilian National Council for Scientific and Technological Development - CNPq, Grant 304214/77-9 and the research of the second author received partial support from the Brazilian Ministry of Education - CAPES. The authors thank José S. Lemos Neto for helpful discussions.

References

1. T. Kasami and Shu Lin, "Coding for a multiple-access channel", *IEEE Trans. on Inform. Theory*, Vol. IT-22, Number 2, March 1976, pp.129-137.

2. H.H.J. Liao, "Multiple access channels", Ph.D. Dissertation, Dep. Elec. Eng., Univ. of Hawaii, Honolulu, 1972.

3. R. Ahlswede,"Multi-way communications channels", 2nd Int. Symp. on Information Transmission, USSR, 1971.

4. D. Slepian and J. K. Wolf, "A coding theorem for multiple-acces channels", Bell Syst. Tech. J., vol. 51, pp. 1037-1076. 1973.

5. J.L. Massey, "On codes for the two-user binary adder channel", Oberwolfach Information Theory Workshop, Germany, April 1992.

6. V.C. da Rocha, Jr. and J. L. Massey, "A new approach to the design of codes for the binary adder channel", in *Cryptography and Coding III*, (Ed. M.J. Ganley), IMA Conf. Series, New Series Number 45. Oxford: Clarendon Press, 1993, pp. 179-185.

7. H.A. Cabral and V. C. da Rocha, Jr., "Linear code construction for the 2-user binary adder channel", *IEEE Int. Symp. on Info. Theory*, Whistler, Canada, 1995, pp. 497.

8. R. Ahlswede and V. B. Balakirsky, "Construction of uniquely decodable codes for the two-user binary adder channel", *IEEE Trans. on Inform. Theory*, Vol. 45, Number 1, January 1999, pp.326-330.

9. G. H. Khachatrian, "A survey of coding methods for the adder channel", in *Numbers, Information and Complexity* (dedicated to Rudolph Ahlswede on the occasion of his 60th birthday), Ed. I. Althofer, Kluwer, 2000.

10. R. Peterson and D.J. Costello, Jr., "Binary convolutional codes for a multiple-access channel", *IEEE Trans. on Info. Theory*, vol.25, no.1, pp.101-105, January 1979.

11. S. Lin and D. Costello Jr., *Error Control Coding: Fundamentals and Applications*, Prentice-Hall Inc., Englewood Cliffs, New Jersey, USA, 1983.

12. R. Urbanke and Quinn Li, "The zero-error capacity region of the 2-user synchronous BAC is strictly smaller than its Shannon capacity", http://lthcwww.epfl.ch/~ruediger/publications.html .

Rayleigh Fading Multiple Access Channel Without Channel State Information

Ninoslav Marina

Mobile Communications Laboratory (LCM)
School of Computer and Communication Sciences
Swiss Federal Institute of Technology (EPFL)
CH-1015 Lausanne, Switzerland
ninoslav.marina@epfl.ch

Abstract. In this paper we determine bounds of the capacity region of
a two-user multiple-access channel with Rayleigh fading when neither
the transmitters nor the receiver has channel state information (CSI).
We assume that the fading coefficients as well as the additive noise are
zero-mean complex Gaussian and there is an average power constraint
at the channel input for both senders. Results that we get show that the
lower (inner) and the upper (outer) bound of the capacity region are
quite close for low and high signal-to-noise ratio (SNR). Surprisingly,
the boundary of the capacity region is achieved by time sharing among
users, which is not the case for fading channels with perfect CSI at the
receiver. As an additional result we derive a closed form expression for
the mutual information if the input is on-off binary.

Keywords: Multiple-access channel, capacity region, Rayleigh fading,
channel state information, volume of the capacity region.

1 Introduction

Wireless communication systems are currently becoming more and more important. A challenging task for operators of mobile communication systems and
researchers is the need to constantly improve spectral efficiency, maintain a desirable quality of service, minimize the consumption of transmit power in order
to lower electromagnetic radiation and prolong the battery life. In the same time
the number of base stations has to be minimized, whilst accommodating as many
users as possible. In fulfilling these requirements, the greatest obstacle is the nature of the mobile communication channel, which is time-varying, due to rapid
changes in the environment and mobility of users. Signal strength may drop by
several orders of magnitude due to an increase in distance between transmitter
and receiver and superposition phenomena in scattering environments. This phenomenon is commonly known as fading and such channels as fading channels.
Many modern wireless systems send a training sequence inserted in the data
stream in order to provide the receiver with information about the channel. On
the other hand, some systems provide a feedback channel from the receiver to

J.N. de Souza et al. (Eds.): ICT 2004, LNCS 3124, pp. 128–133, 2004.

the transmitter and this information can help the transmitter to choose an appropriate signal to access the channel. Knowledge of the channel is known as *channel state information* (CSI). Many papers have been written on channels with perfect CSI at the receiver, at the transmitter, at both and at neither of them. In practical wireless communication systems, whenever there is a large number of independent scatterers and no line-of-sight path between the transmitter and the receiver, the radio link may be modelled as a Rayleigh fading channel. In a multi-user environment the uplink channel is typically modelled as multiple access channel (MAC). The performance of the channel strongly depends on the fact whether the state of the channel is available at the receiver and(or) the transmitter(s). In this paper we are interested in deriving the *capacity region* of the two-user Rayleigh fading channel without CSI at the receiver or the transmitter. The capacity region of a multiple access channel is the closure of achievable rates for all users [5]. This channel is of interest since in some cases the channel can vary very quickly and it will be not possible to send any information about the channel.

The case without channel state information for the single user channel has been studied in [9,7,3,6] and for the multi-user channel in [8,2]. In [3], authors show that without channel state information, the optimal input is discrete with a mass point at zero. In [9] it is shown that without channel state information, the capacity at high SNR depends double-logarithmically on the SNR. A more general result on the double logarithmic behavior at high SNR is given in [6].

In Section 2 we establish a closed form solution for the mutual information when the input is binary on-off. In Section 3 we find lower and upper bounds of the capacity region of a two-user Rayleigh fading channel. We compare the bounds in Section 4 and give conclusions in Section 5.

2   Rayleigh Fading Channels Without CSI

The capacity $C_{su}(\rho)$ of the single user channel $Y' = AX' + Z$, where $A, Z \sim \mathcal{N}_\mathbb{C}(0,1)$ and $E[|X|^2] \leq \rho$ has been derived in [3]. Having $\rho = \sigma_A^2 P/\sigma_Z^2$, the capacity of this channel is the same as the capacity of the channel with $A \sim \mathcal{N}_\mathbb{C}(0, \sigma_A^2)$, $Z \sim \mathcal{N}_\mathbb{C}(0, \sigma_Z^2)$ and $E[|X|^2] \leq P$. According to [3], the capacity achieving input distribution is discrete. For low SNR, the mutual information for binary inputs is not far from the capacity. For extremely high SNR, higher than the fading number, defined in [6], the capacity behaves as $\log(\log(\text{SNR}))$.

Next we give a closed form expression for the mutual information between the input and the output, if the input is on-off binary, namely $(0, b)$. The binary on-off input is interesting since for low SNR, it is optimal. At the end of the section we give a numerical result for the capacity.

Proposition 1. (Closed form expression for the mutual information for a particular on-off input probability p and SNR ρ): For the channel $Y = AX + Z$, when the input is binary on-off with $\Pr\{X = 0\} = 1 - p$ and power constraint

$E[|X|^2] \leq \rho$, the mutual information between the input X and the output Y, is

$$I(X;Y) = h(p) + pJ\left(\frac{p+\rho}{p^2(1-p)^{-1}}, \frac{\rho}{p}\right) + (1-p)J\left(\frac{p^2(1-p)^{-1}}{p+\rho}, \frac{\rho}{p+\rho}\right) \quad (1)$$

where $h(\cdot)$ is the binary entropy function, $J(c,d) = -\ln(1+c) + \frac{cd}{1+d} \cdot {}_2F_1(1,1+d^{-1};2+d^{-1};-c)$, and ${}_2F_1(u,v;w;z) = \frac{\Gamma(w)}{\Gamma(u)\Gamma(v)}\sum_{k=0}^{\infty}\frac{\Gamma(u+k)\Gamma(v+k)}{\Gamma(w+k)} \cdot \frac{z^k}{k!}$ is the Gaussian hypergeometric function defined in [4]. $\Gamma(q) = \int_0^{\infty}x^{q-1}e^{-x}dx$ is the Euler gamma function.

The detailed proof is given in [1].

To compute the capacity of the binary input Rayleigh fading channel without channel state information, denoted by C_b, one has to find the maximum of $I_{p,\rho}(X;Y)$ over p for different ρ. Unfortunately $dI_{p,\rho}(X;Y)/dp = 0$ is a transcendental equation and cannot be solved explicitly. The capacity and the optimizing p^* as functions of ρ are shown in Fig. 1. Note that as the power of the input signal increases the information rate of this channel goes to its limit of $\ln 2$ nats and p^* goes to 0.5. This happens since if ρ goes to ∞, we get the channel $Y = AX$.

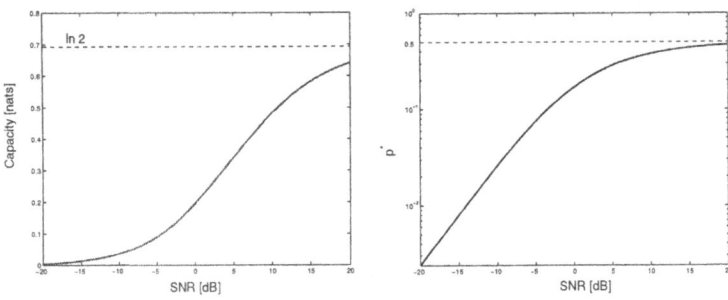

Fig. 1. Capacity C_b and the optimal probability p^* as function of ρ

3   Two-User Rayleigh Fading Channel

In this section, we give a lower and an upper bound of the capacity region of a two-user Rayleigh fading channel in the case where the channel is not known either at the transmitters or at the receiver, but all of them know the statistics of the channel exactly. The ratio of volumes [1] of the lower and the upper bound of the capacity region will serve us as a measure for the proximity of these bounds. The channel is

$$\tilde{Y} = \tilde{A}_1\tilde{X}_1 + \tilde{A}_2\tilde{X}_2 + \tilde{Z}, \quad (2)$$

where $\tilde{A}_1$, $\tilde{A}_2$ and $\tilde{Z}$ are independent and identically distributed (i.i.d.), zero-mean, complex Gaussian random variables of variances $\sigma_{A_1}^2, \sigma_{A_2}^2$ and σ_Z^2 respectively. There are input constraints, $E[|\tilde{X}_1|]^2 \leq P_1$ and $E[|\tilde{X}_2|]^2 \leq P_2$. No channel information is provided to the transmitters and the receiver. However, they perfectly know the statistics of the channel. To find "good" bounds of the capacity region of this channel, we use the results for the single user memoryless Rayleigh fading channel, studied in [3]. Channel (2) has the same capacity as the channel $\frac{\tilde{Y}}{\sigma_Z} = \frac{\tilde{A}_1}{\sigma_{A_1}} \frac{\sigma_{A_1}}{\sigma_Z} \tilde{X}_1 + \frac{\tilde{A}_2}{\sigma_{A_2}} \frac{\sigma_{A_2}}{\sigma_Z} \tilde{X}_2 + \frac{\tilde{Z}}{\sigma_Z}$. Letting $Y = \tilde{Y}/\sigma_Z$, $X_1 = \sigma_{A_1} \tilde{X}_1/\sigma_Z$ and $X_2 = \sigma_{A_2} \tilde{X}_2/\sigma_Z$, we get

$$Y = A_1 X_1 + A_2 X_2 + Z, \tag{3}$$

with $A_1 = \tilde{A}_1/\sigma_{A_1}$, $A_2 = \tilde{A}_2/\sigma_{A_2}$ and $Z = \tilde{Z}/\sigma_Z$ all being $\mathcal{N}_{\mathbb{C}}(0, 1)$. Power constraints on the new inputs become $E[\|X_1\|]^2 \leq \sigma_{A_1}^2 P_1/\sigma_Z^2 = \rho_1$ and $E[\|X_2\|]^2 \leq \sigma_{A_2}^2 P_2/\sigma_Z^2 = \rho_2$. It is clear that doing these transforms, all mutual information in the new channel remain the same. Thus, the capacity region of the channel (2) is the same as the capacity region of the channel (3). For a particular input distribution, the region of achievable rates for the channel (3) is $\mathcal{R}(p_{X_1}, p_{X_2}) = \{(R_1, R_2) \in \mathbb{R}_+^2 : R_1 \leq I(Y; X_1|X_2); R_2 \leq I(Y; X_2|X_1); R_1 + R_2 \leq I(Y; X_1, X_2)\}$. The capacity region is a closure of the convex hull of the union over all possible product input distributions $p_{X_1}(x)p_{X_2}(x)$ of all such regions $\mathcal{R}(p_{X_1}, p_{X_2})$. To compute the maximum mutual information in the capacity region for user 1 and user 2 separately, we need to analyze the single user fading channel, similarly as it is done in [3]. Given $X_2 = x_2$, the equivalent channel is $Y = A_1 X_1 + (A_2 x_2 + Z)$. This channel is the same as the single user fading channel $Y = A_1 X_1 + Z$, with larger variance of the additive noise, that is, $1 + |x_2|^2$. Thus, it behaves as the channel $Y = A_1 X_1 + Z$, with different SNR constraint, that is, $\rho' = \rho/(1 + |x|^2)$. It is shown in [3] that the capacity achieving input distribution for this channel has to be discrete with a mass point at the origin. Moreover, it is shown in the same paper that for low SNR, the maximizing input distribution is binary. Thus, the rate of user 1 is bounded by $R_1 \leq \sum_{x_2} p_{X_2}(x_2)I(X_1; Y|X_2 = x_2) \leq \sum_{x \in \mathcal{X}_2} p_{X_2}(x)C_{su}\left(\frac{\rho_1}{1+|x|^2}\right) \leq \sum_{x \in \mathcal{X}_2} p_{X_2}(x)C_{su}(\rho_1) = C_{su}(\rho_1)$, where the last inequality is achieved with equality if $p_{X_2}(0) = 1$, i.e. if user 2 is silent. By $C_{su}(\rho)$ we denote the capacity of the single user fading channel with no channel state information, for a particular SNR= ρ. Thence, the point $C_{su}(\rho_1)$ is achievable and it is the highest rate that can be achieved by user 1, using the channel while user 2 is silent. That is one point on the boundary of the capacity region, namely the extreme point on the R_1-axis. From symmetry, the same is true for user 2, i.e. the extreme point on the R_2-axis is $C_{su}(\rho_2)$.

After finding both extreme points, let us find the maximum sum rate. It is shown in [8] that if the propagation coefficients take on new independent values for every symbol (i.i.d.), then the total throughput capacity for any number of users larger than 1, is equal to the capacity if there is only one user. Hence, time division multiple access (TDMA) is an optimal scheme for multiple users. In that case the sum rate is given by $\Theta = aC_{su}(\rho_1/a) + (1 - a)C_{su}(\rho_2/(1 - a)) \leq C_{su}(\rho_1 + \rho_2)$, with $a \in [0, 1]$. Note that the maximum throughput cannot be larger than $C_{su}(\rho_1 + \rho_2)$, the capacity which is achieved if both users fully cooperate, and is equivalent to the single user capacity for SNR= $\rho_1 + \rho_2$. The latest is achieved with equality for $a = \rho_1/(\rho_1 + \rho_2)$. This is an upper bound of the

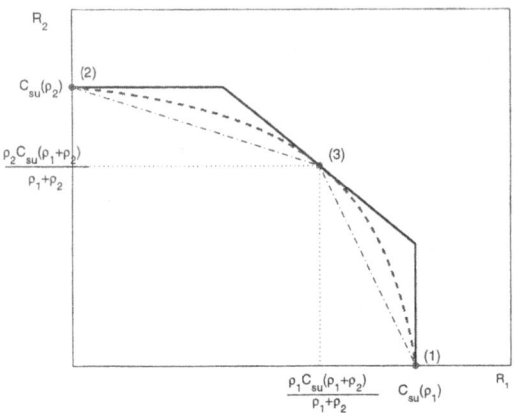

Fig. 2. Lower and upper bounds of the capacity region, for given ρ_1 and ρ_2.

capacity region, namely the pentagon $\{(R_1, R_2) \in \mathbb{R}_+^2 : R_1 \le C_{su}(\rho_1), R_2 \le C_{su}(\rho_2), R_1 + R_2 \le C_{su}(\rho_1 + \rho_2)$ (Fig. 2). A straightforward lower bound is the region obtained by connecting the points that are achievable (dash-dot line in Fig. 2). Better lower bound is the time-sharing region (dashed line in Fig. 2). It is obtained by allowing user 1 to use the channel aT seconds and user 2, $(1-a)T$ seconds. Because of the average power constraint the power used during the active period is normalized. Hence, the proposed lower bound of the capacity region parameterized by $a \in [0, 1]$ is given by

$$R_1(a) = a \cdot C_{su}(\rho_1/a)$$
$$R_2(a) = (1 - a) \cdot C_{su}(\rho_2/(1 - a)). \tag{4}$$

The capacity region touches the upper bound in the following three points $(C_{su}(\rho_1), 0)$, $(0, C_{su}(\rho_2))$ and $(\frac{\rho_1 C_{su}(\rho_1+\rho_2)}{\rho_1+\rho_2}, \frac{\rho_2 C_{su}(\rho_1+\rho_2)}{\rho_1+\rho_2})$. Note that a trivial upper bound that is much looser is the capacity region of the same channel, with perfect channel state information at the receiver.

4   Comparison

Comparing the bounds for different SNRs [1], it can be seen that for low and high SNR they are closer then for the some medium SNR. How to measure the "tightness" of the bounds? We propose comparing the *volumes* of the corresponding regions. For the two user case the volume is $V_2 = \int_{R_1} R_2 dR_1$. It is easy to compute the volume of the upper bound in terms of the single-user capacities. For $\rho_1 = \rho_2 = \rho$, $V_{UB}(\rho) = \frac{1}{2} \cdot [C_{su}(2\rho)]^2 - [C_{su}(2\rho) - C_{su}(\rho)]^2$. The volume of the lower bound is $V_{LB}(\rho) = \int_0^{C_{su}(\rho)} R_2 dR_1 = \int_0^1 R_2(a)\dot{R}_1(a) da$, where $R_1(a)$ and $R_2(a)$ are given by (4), and $\dot{R}_1(a)$ is the first derivative of R_1 with respect to a. It can be seen that for low SNR the lower and the upper bound are very close and as the SNR increases they diverge up to some SNR (~ 3 dB), where the lower and the upper bound are at maximum "distance". In this case the ratio $\mathcal{V}_{LB} \simeq 0.925\mathcal{V}_{UB}$. As SNR increases above 3 dB, the bounds approach again, i.e. the ratio of $\mathcal{V}_{LB}$ and $\mathcal{V}_{UB}$ increases and tends to 1. The results can be easily extended for an $M-$user case.

5 Conclusions

The single user Rayleigh fading channel with no side information has attracted some attention since it is useful for modelling different wireless channels. In this paper we get some insight for the multiple access Rayleigh fading channel with no CSI. We give bounds of the capacity region of the two-user Rayleigh multiple access channel. We see that the sum rate is maximized by time-sharing among users and in that case we achieve the boundary of the capacity region by giving to each user an amount of time that is proportional to its input average power constraint multiplied by the variance of the fading. This is not the case with multiple access channels with perfect CSI at the receiver only. However it is the case with the Gaussian MAC. We also see that the inner bound is always within 92.6 % (in terms of the volume of the capacity region) of the outer bound. As an open problem for future research we leave the improvement of the inner and the outer bound.

References

1. N. Marina, "Successive Decoding," Ph.D. thesis, School of Computer and Communication Sciences, École Polytechnique Fédérale de Lausanne (EPFL), Lausanne 2004.
2. N. Marina, "On successive decoding without channel state information," in *Proc. IEEE Int. Symp. Telecommunications*, Tehran, 2001.
3. I.C. Abou-Faycal, M.D. Trott and S. Shamai (Shitz), "The capacity of discrete-time memoryless Rayleigh-fading channels," *IEEE Trans. on Information Theory*, Vol. 47, No.4, May 2001.
4. M. Abramowitz and I.A. Stegun, *Handbook of Mathematical Functions*. New York: Dover, 1965.
5. T. M. Cover and J.A. Thomas, *Elements of Information Theory*. New York: Wiley, 1991.
6. A. Lapidoth and S.M. Moser, "Capacity bounds via duality with applications to multi-antenna systems on flat fading channels," *IEEE Trans. on Information Theory*, Vol. 49, No.10, Oct. 2003..
7. T. Marzetta and B. Hochwald, "Capacity of a mobile multiple-antenna communication link in Rayleigh flat fading," *IEEE Trans. on Information Theory*, Vol. 45, No.1, Jan. 1999.
8. S. Shamai (Shitz) and T.L. Marzetta, "Multiuser capacity in block fading with no channel state information," *IEEE Trans. on Information Theory*, Vol. 48, No.4, Apr. 2002.
9. G. Taricco and M. Elia, "Capacity of fading channel with no side information," *Electronic Letters*, vol. 33, no. 16, Jul. 1997.

Adaptive Decision Feedback Multiuser Detectors with Recurrent Neural Networks for DS-CDMA in Fading Channels

Rodrigo C. de Lamare and Raimundo Sampaio-Neto

CETUC/PUC-RIO, 22453-900, Rio de Janeiro - Brazil
delamare@infolink.com.br, raimundo@cetuc.puc-rio.br

Abstract. In this work we propose adaptive decision feedback (DF) multiuser detectors (MUDs) for DS-CDMA systems using recurrent neural networks (RNN). A DF CDMA receiver structure is presented with dynamically driven RNNs in the feedforward section and finite impulse response (FIR) linear filters in the feedback section for performing interference cancellation. A stochastic gradient (SG) algorithm is developed for estimation of the parameters of the proposed receiver structure. A comparative analysis of adaptive minimum mean squared error (MMSE) receivers operating with SG algorithms is carried out for linear and DF receivers with FIR filters and neural receiver structures with and without DF. Simulation experiments including fading channels show that the DF neural MUD outperforms DF MUDs with linear FIR filters, linear receivers and the neural receiver without interference cancellation.

1 Introduction

In third generation wideband direct-sequence code-division multiple access (DS-CDMA) systems high data rate users can be accomodated by reducing the processing gain N and using a low spreading factor [1]. In these situations, the multiacess interference (MAI) is relatively low (small number of users), but the intersymbol interference (ISI) can cause significant performance degradation. The deployment of non-linear structures, such as neural networks and decision feedback (DF), can mitigate more effectively ISI, caused by the multipath effect of radio signals, and MAI, which arises due to the non-orthogonality between user signals. Despite the increased complexity over conventional multiuser receivers with FIR linear filters, the deployment of neural structures is feasible for situations where the spreading factor is low and the number of high data rate users is small. In this case, the trade-off between computational complexity and superior performance is quite attractive. Neural networks have recently been used in the design of DS-CDMA multiuser receivers [2]-[5]. Neural MUDs employing the minimum MMSE [2]-[5] criterion usually show good performance and have simple adaptive implementation, at the expense of a higher computational complexity. In the last few years, different artificial neural networks structures have been used in the design of MUDs: multilayer perceptrons (MLP) [2], radial-basis functions (RBF) [3], and RNNs [4,5]. These neural MUDs employ non-linear functions to create decision boundaries for the detection of transmitted symbols, whilst conventional MUDs use linear functions to form such

J.N. de Souza et al. (Eds.): ICT 2004, LNCS 3124, pp. 134–141, 2004.

decision regions. In this work, we present an adaptive DF MUD, using dynamically driven RNNs in the feedforward section and FIR linear filters in the feedback section, and develop stochastic gradient (SG) algorithms for the proposed DF receiver structure. Adaptive DF and linear MMSE MUDs are examined with the LMS algorithm and compared to the DF and non-DF neural MUDs operating with SG algorithms. Computer simulation experiments with fading channels show that the DF neural MUDs outperforms linear and DF receivers with the LMS and the conventional single user detector (SUD), which corresponds to the matched filter.

2 DS-CDMA System Model

Let us consider the uplink of a symbol synchronous DS-CDMA system with K users, N chips per symbol and L_p propagation paths. The baseband signal transmitted by the k-th active user to the base station is given by:

$$x_k(t) = A_k \sum_{i=-\infty}^{\infty} b_k(i) s_k(t - iT) \tag{1}$$

where $b_k(i) \in \{\pm 1\}$ denotes the i-th symbol for user k, the real valued spreading waveform and the amplitude associated with user k are $s_k(t)$ and A_k, respectively. The spreading waveforms are expressed by $s_k(t) = \sum_{n=1}^{N} a_k(i)\phi(t - nT_c)$, where $a_k(i) \in \{\pm 1/\sqrt{N}\}$, $\phi(t)$ is the chip waveform, T_c is the chip duration and $N = T/T_c$ is the processing gain. Assuming that the receiver is synchronised with the main path, the coherently demodulated composite received signal is

$$r(t) = \sum_{k=1}^{K} \sum_{l=0}^{L_p-1} h_{k,l}(t) x_k(t - \tau_{k,l}) \tag{2}$$

where $h_{k,l}(t)$ and $\tau_{k,l}$ are, respectively, the channel coefficient and the delay associated with the l-th path and the k-th user. Assuming that $\tau_{k,l} = lT_c$ and that the channel is constant during each symbol interval, the received signal $r(t)$ after filtering by a chip-pulse matched filter and sampled at chip rate yields the N dimensional received vector

$$\mathbf{r}(i) = \sum_{k=1}^{K} A_k \mathbf{H}_k(i) \mathbf{C}_k \mathbf{b}_k(i) + \mathbf{n}(i) \tag{3}$$

where the Gaussian noise vector is $\mathbf{n}(i) = [n_1(i) \ \ldots \ n_N(i)]^T$ with $E[\mathbf{n}(k)\mathbf{n}^T(i)] = \sigma^2 \mathbf{I}$, where $(.)^T$ denotes matrix transpose and $E[.]$ stands for expected value, the k-th user symbol vector is given by $\mathbf{b}_k(i) = [b_k(i) \ \ldots \ b_k(i - L_s + 1)]^T$, where L_s is the ISI span. The $(L_s \times N) \times L_s$ user k code matrix $\mathbf{C}_k$ is described by

$$\mathbf{C}_k = \begin{bmatrix} \mathbf{s}_k & 0 & \ldots & 0 \\ 0 & \mathbf{s}_k & \ddots & 0 \\ \vdots & \vdots & \ddots & \vdots \\ 0 & \ldots & \ldots & \mathbf{s}_k \end{bmatrix} \tag{4}$$

where $s_k = [a_k(1) \ldots a_k(N)]^T$ is the signature sequence for the k-th user, and the $N \times (L_s \times N)$ channel matrix $\mathbf{H}_k(i)$ for the user k is expressed by

$$\mathbf{H}_k(i) = \begin{bmatrix} h_{k,0}(i) \ldots h_{k,L_p-1}(i) & \ldots & 0 & 0 \\ \vdots & \ddots & \ddots & \ddots & \ddots & \vdots \\ 0 & 0 & \ldots & h_{k,0}(i) \ldots h_{k,L_p-1}(i) \end{bmatrix} \tag{5}$$

The MAI comes from the non-orthogonality between the received signature sequences, whereas the ISI span L_s depends on the length of the channel response, which is related to the length of the chip sequence. For $L_p = 1$, $L_s = 1$ (no ISI), for $1 < L_p \leq N, L_s = 2$, for $N < L_p \leq 2N, L_s = 3$.

3   Conventional Decision Feedback MUD

Consider a one shot DF MUD (the receiver observes and detects only one symbol at each time instant), whose observation vector $\mathbf{u}(i)$ is formed from the outputs of a bank of matched filters , where $\mathbf{u}(i) = \mathbf{S}^T \mathbf{r}(i)$ with $\mathbf{S} = [s_1 \ldots s_K]$. The observation vector is represented by:

$$\mathbf{u}(i) = [u_1 \ldots u_K]^T \tag{6}$$

The detected symbols for an one shot DF multiuser receiver using FIR linear filters are given by :

$$\hat{b}_k(i) = sgn(\mathbf{w}_k^T(i)\mathbf{u}(i) - \mathbf{f}_k^T \hat{\mathbf{b}}(i)) \tag{7}$$

where $\mathbf{w}_k(i) = [w_1 \ldots w_K]^T$ and $\mathbf{f}_k(i) = [f_1 \ldots f_K]^T$ are, respectively the feedforward and feedback weight vectors for user k for the i-th symbol in a system with K users. The feedforward matrix $\mathbf{w}(k)$ is $K \times K$, the feedback matrix $\mathbf{f}(k)$ is $K \times K$ and is constrained to have zeros along the diagonal to avoid cancelling the desired symbols. In this work, we employ a full matrix $\mathbf{f}(k)$, except for the diagonal, which corresponds to parallel DF [6]. The MMSE solution for this MUD can be obtained via an adaptive algorithm, such as the LMS algorithm [7], which uses the error signal $e_k(i) = b_k(i) - \mathbf{w}_k^T(i)\mathbf{u}(i) + \mathbf{f}_k^T(i)\hat{\mathbf{b}}(i)$, and is described by:

$$\mathbf{w}_k(i+1) = \mathbf{w}_k(i) + \mu_w e_k(i)\mathbf{u}(i) \tag{8}$$

$$\mathbf{f}_k(i+1) = \mathbf{f}_k(i) - \mu_f e_k(i)\hat{\mathbf{b}}(i) \tag{9}$$

where $b_k(i)$ is the desired signal for the k-th user taken from a training sequence, $\mathbf{u}(i)$ is the observation vector, $\hat{\mathbf{b}}(i) = [\hat{b}_1(i) \ldots \hat{b}_K(i)]$ is the vector with the decisions, μ_w and μ_f are the algorithm step sizes.

4   Proposed DF Neural Receiver

In this section we present a decision feedback CDMA receiver that employs recurrent neural networks as its feedforward section and linear FIR filters, similar to the DF MUD described in the preceeding section, as its feedback section. The decision feedback

CDMA receiver structure, depicted in Fig. 1, employs dynamically driven RNNs in the feedforward section for suppressing MAI and ISI and FIR linear filters in the feedback section for cancelling the associated users in the system. With respect to the structure, RNNs have one or more feedback connections, where each artificial neuron is connected to the others. These neural networks are suitable to channel equalisation and multiuser detection applications, since they are able to cope with channel transfer functions that exhibit deep spectral nulls, forming optimal decision boundaries [8].

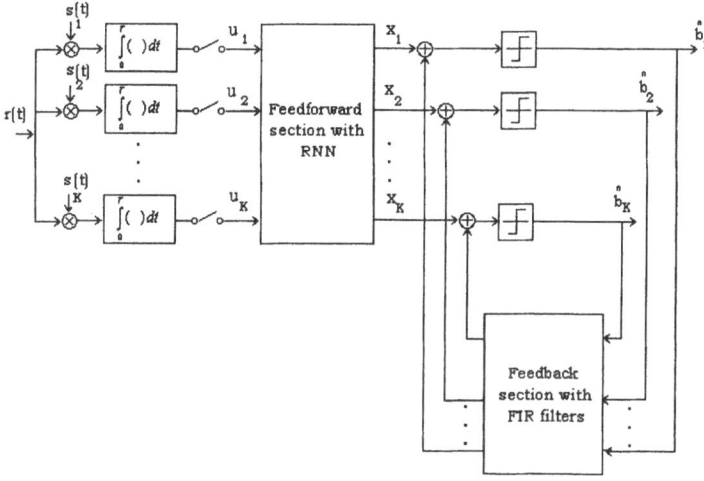

Fig. 1. Proposed DF neural receiver: RNNs are employed in the feedforward section for MAI and ISI rejection and FIR filters are used in the feedback section for cancelling associated interferers.

To describe the proposed neural DF system we use a state-space approach, where the $K \times 1$ vector $\mathbf{x}_k(i)$ corresponds to the K states of the artificial neural network for user k, the $K \times 1$ vector $\mathbf{u}(i)$ to the channel K user symbols output observation vector and the output of the DF neural MUD $\hat{b}_k(i)$ is given by:

$$\xi_k(i) = \left[\mathbf{x}_k^T(i-1)\ \mathbf{u}^T(i)\right]^T \tag{10}$$

$$\mathbf{x}_k(i) = \varphi(\mathbf{W}_k^T(i)\xi_k(i)) \tag{11}$$

$$\hat{b}_k(i) = sgn(\mathbf{D}\mathbf{x}_k(i) - \mathbf{f}_k^T(i)\hat{\mathbf{b}}(i)) \tag{12}$$

where the matrix $\mathbf{W}_k(i) = [\mathbf{w}_{k,1}(i)\ \cdots\ \mathbf{w}_{k,j}\ \cdots\ \mathbf{w}_{k,K}]$ has dimension $2K \times K$ and whose K columns $\mathbf{w}_{k,j}(i)$, with $j = 1, 2, \ldots, K$ have dimension $K \times 1$ and contain the coefficients of the RNN receiver for user k, $\mathbf{D} = \begin{bmatrix} 1\ 0\ \ldots\ 0 \end{bmatrix}$ is the $1 \times K$ matrix that defines the number of outputs of the network, $\varphi(.)$ is the activation function of the neural network, the feedback matrix $\mathbf{f}(k) = [\mathbf{f}_1(i)\ \ldots\ \mathbf{f}_K(i)]$ is $K \times K$ and as in the conventional MUD case is constrained to have zeros along the diagonal to avoid cancelling the desired symbols. Note that this type of interference cancellation provides uniform performance over the user population. In particular, we have only one output $\hat{b}_k(i)$ per observation vector $\mathbf{u}(i)$, which corresponds to the one shot approach.

5   Adaptive Algorithms for the Neural DF Receiver

In order to derive an SG adaptive algorithm that minimizes the mean squared error (MSE) for the proposed DF receiver structure, we consider the following cost function:

$$J(\mathbf{W}_k(i), \mathbf{f}_k(i)) = E[e_k^2(i)] = E[(b_k(i) - (\mathbf{D}\mathbf{x}_k(i) - \mathbf{f}_k^T(i)\hat{\mathbf{b}}(i)))^2] \qquad (13)$$

where $e_k(i) = b_k(i) - (\mathbf{D}\mathbf{x}_k(i) - \mathbf{f}_k^T(i)\hat{\mathbf{b}}(i))$. A stochastic gradient algorithm can be developed by computing the gradient terms with respect to $\mathbf{w}_{k,j}$, $j = 1, 2, \ldots, K$, and $\mathbf{f}_k$ and using their instantaneous values. Firstly, we consider the first partial derivative of $J(\mathbf{W}_k(i), \mathbf{f}_k(i))$ with respect to the parameter vector $\mathbf{w}_{k,j}(i)$ with dimension $2K \times 1$, which forms the matrix $\mathbf{W}_k$:

$$\frac{\partial J(\mathbf{W}_k(i), \mathbf{f}_k(i))}{\partial \mathbf{w}_{k,j}(i)} = \left(\frac{\partial e_k(i)}{\partial \mathbf{w}_{k,j}(i)} \right) e_k(i) = -\mathbf{D} \left(\frac{\partial \mathbf{x}_k(i)}{\partial \mathbf{w}_{k,j}(i)} \right) e_k(i) = -\mathbf{D} \Lambda_{k,j}(i) e_k(i) \qquad (14)$$

where the $K \times 2K$ matrix $\Lambda_{k,j}(i)$ contains the partial derivatives of the state vector $\mathbf{x}_k(i)$ with respect to $\mathbf{w}_{k,j}(i)$. To obtain the expressions for the updating of the matrix $\Lambda_{k,j}(i)$, we consider the update equations for the state vector $\mathbf{x}_k(i)$ given through (10) and (11). Using the chain rule of calculus in (11), we obtain the following recursion that describes the dynamics of the learning process of the neural receiver:

$$\Lambda_{k,j}(i+1) = \Phi_k(i) \left(\mathbf{W}_k^{1:K}(i) \Lambda_{k,j}(i) + \mathbf{U}_{k,j}(i) \right), \quad j = 1, 2, \ldots, K \qquad (15)$$

where the $K \times K$ matrix $\mathbf{W}_k^{1:K}$ denotes the submatrix of $\mathbf{W}_k$ formed by the first K rows of $\mathbf{W}_k$, the $K \times K$ matrix $\Phi_k(i)$ for user k has a diagonal structure where the elements correspond to the partial derivative of the activation function $\varphi(.)$ with respect to the argument in $\mathbf{w}_{k,j}^T(i)\xi_k(i)$ as expressed by:

$$\Phi_k(i) = diag \left(\varphi'(\mathbf{w}_{k,1}^T(i)\xi_k(i)), \ldots, \varphi'(\mathbf{w}_{k,j}^T(i)\xi_k(i)), \ldots, \varphi'(\mathbf{w}_{k,K}^T(i)\xi_k(i)) \right) \qquad (16)$$

and the $K \times 2K$ matrix $\mathbf{U}_{k,j}(i)$ has all the rows with zero elements, except for the j-th row that is equal to the vector $\xi_k(i)$):

$$\mathbf{U}_{k,j}(i) = \begin{bmatrix} \mathbf{0}^T \\ \xi_k(i) \\ \mathbf{0}^T \end{bmatrix}, \quad j = 1, 2, \ldots, K \qquad (17)$$

The update equation for the feedforward parameter vector $\mathbf{w}_{k,j}$ of the decision feedback receiver is obtained via a stochastic gradient optimisation that uses the expression obtained in (14) to update the parameters using the gradient rule $\mathbf{w}_{k,j}(i+1) =$

$\mathbf{w}_{k,j}(i) - \mu_n \frac{\partial J(\mathbf{W}_k(i), \mathbf{f}_k(i))}{\partial \mathbf{w}_{k,j}(i)}$ which yields the recursion for the neural section of the receiver:

$$\mathbf{w}_{k,j}(i+1) = \mathbf{w}_{k,j}(i) + \mu_n \mathbf{D}\boldsymbol{\Lambda}_{k,j}(i)e_k(i) \tag{18}$$

where μ_n is the step size for the algorithm that adjusts the feedforward section of the proposed MUD. To compute the update rule for the feedback parameter vector $\mathbf{f}_k$ of the decision feedback receiver, we compute the gradient of $J(\mathbf{W}_k(i), \mathbf{f}_k(i))$ with respect to $\mathbf{f}_k$ and obtain the following gradient-type recursion:

$$\mathbf{f}_k(i+1) = \mathbf{f}_k(i) - \mu_f e_k(i)\hat{\mathbf{b}}(i) \tag{19}$$

where μ_f is the step size of the algorithm that updates the feedback section.

6 Simulation Experiments

In this section we assess the BER and the convergence performance of the adaptive receivers. The DS-CDMA system employs Gold sequences of length $N = 15$. The carrier frequency of the system was chosen to be 1900 MHz. It is assumed here that the channels experienced by different users are statistically independent and identically distributed. The channel coefficients for each user k ($k = 1, \dots, K$) are $h_{k,l}(i) = p_l|\alpha_{k,l}(i)|$, where $\alpha_{k,l}(i)$ ($l = 0, 1, \dots, L_p - 1$) is a complex Gaussian random sequence obtained by passing complex white Gaussian noise through a filter with approximate transfer function $\beta/\sqrt{1 - (f/f_d)^2}$ where β is a normalization constant, $f_d = v/\lambda$ is the maximum Doppler shift, λ is the wavelength of the carrier frequency, and v is the speed of the mobile [9]. For each user, say user k, this procedure corresponds to the generation of L_p independent sequences of correlated, unit power ($E[|\alpha_{k,l}(i)|^2 = 1$), Rayleigh random variables and has a bandwidth of 4.84 MHz, which corresponds to the data rate of 312.2 kbps. The simulations assess and compare the BER performance of the DF and linear receivers operating with the LMS, DF and non-DF neural MUDs operating with the algorithms of Section V, the SUD and the single user bound (SU-Bound), which corresponds to the SUD in a system with a single user (no MAI). Note that for the neural receiver without DF we make $\mathbf{f} = 0$ in the structure and algorithms of Sections IV and V. The parameters of the algorithms are optimised for each situation and we assume perfect power control in the DS-CDMA system. The activation function $\varphi(.)$ for the neural receiver is the hyperbolic tangent (i.e. $tanh(.)$) in all simulations. The receivers process 10^3 data symbols, averaged over 100 independent experiments in a scenario where the mobile terminals move at 80km/h. The algorithms are adjusted with 200 training data symbols during the training period and then switch to decision directed mode in all experiments. We remark that the BER performance shown in the results refers to the average BER amongst the K users.

6.1 Flat Rayleigh Fading Channel Performance

The BER and the BER convergence performance of the receivers were evaluated in a flat Rayleigh fading channel ($L_p = 1$, $p_0 = 1$) with additive white gaussian noise (AWGN). The BER convergence performance of the MUDs is shown in Fig. 2, where the proposed

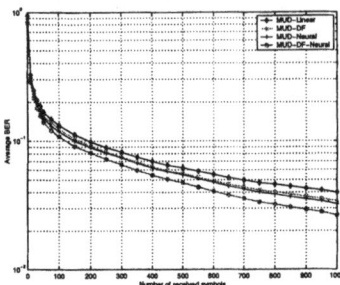

Fig. 2. BER convergence performance of the receivers in a flat Rayleigh fading channel with AWGN, $E_b/N_0 = 8$ dB and with $K = 4$ users. The parameters of the algorithms are $\mu_w = 0.005$, $\mu_n = 0.005$ and $\mu_f = 0.0015$.

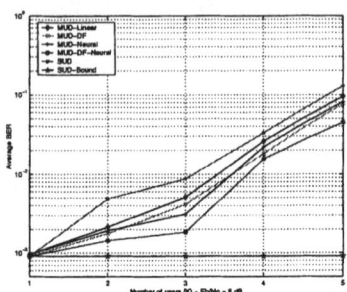

Fig. 3. BER performance of the receivers in a flat Rayleigh fading channel with AWGN, $E_b/N_0 = 8$ dB and with a varying number of users. The parameters of the algorithms are $\mu_w = 0.005$, $\mu_n = 0.005$ and $\mu_f = 0.0015$.

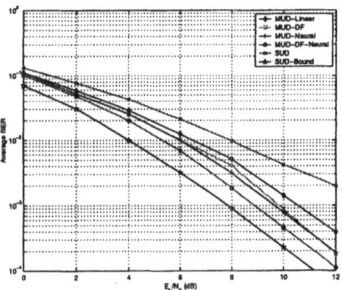

Fig. 4. BER performance versus E_b/N_0 for the receivers in a flat Rayleigh fading channel with AWGN and $K = 3$ users. The parameters of the algorithms are $\mu_w = 0.005$, $\mu_n = 0.005$ and $\mu_f = 0.0015$.

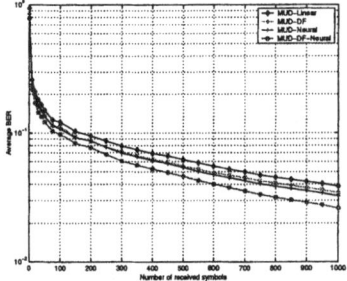

Fig. 5. BER convergence performance of the receivers in a two-path Rayleigh fading channel with AWGN, $E_b/N_0 = 10$ dB and with $K = 4$ users. The parameters are $\mu_w = 0.0025$, $\mu_n = 0.0025$ and $\mu_f = 0.0015$.

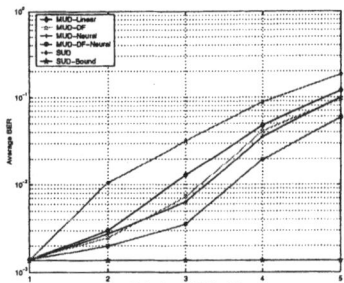

Fig. 6. BER performance versus number of users for the receivers in a two-path Rayleigh fading channel with AWGN, $E_b/N_0 = 8$ dB. The parameters of the algorithms are $\mu_w = 0.0025$, $\mu_n = 0.0025$ and $\mu_f = 0.0015$.

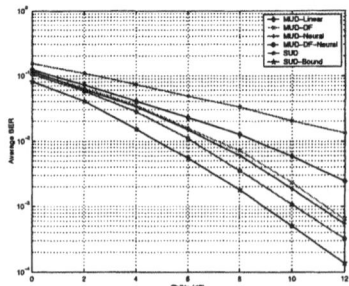

Fig. 7. BER performance versus E_b/N_0 for the receivers in a two-path Rayleigh fading channel with AWGN and $K = 3$ users. The parameters of the algorithms are $\mu_w = 0.0025$, $\mu_n = 0.0025$ and $\mu_f = 0.0015$.

DF neural MUD achieves the best performance, followed by the neural receiver without DF, the DF MUD with linear FIR filters, the linear receiver and the SUD. In Figs. 3 and 4 the BER performance versus the number of users (K) and versus E_b/N_0, respectively, is illustrated. The results show that the novel DF neural MUD achieves the best BER performance, outperforming the neural MUD, the DF MUD, the linear MUD and the SUD.

6.2 Multipath Rayleigh Fading Channel Performance

The BER and the BER convergence performance of the receivers are now assessed in a frequency selective Rayleigh fading channel with AWGN. The channel is modeled as a two-path ($L_p = 2$) and the parameters are $p_0 = 0.895$ and $p_1 = 0.447$ for each user. The BER convergence performance is shown in Fig. 5, whereas the BER performance versus the number of users (K) and versus E_b/N_0 is depicted in Figs. 6 and 7, respectively. The results show that the new DF neural MUD has the best BER performance, outperforming the neural MUD, the DF MUD, the linear MUD and the SUD.

7   Concluding Remarks

In this paper we proposed adaptive DF multiuser receivers for DS-CDMA systems using recurrent neural networks. We developed SG adaptive algorithms for estimating the parameters of the feedforward and feedback sections of the new receiver. A comparative analysis through computer simulation experiments was carried out for evaluating the BER performance of the proposed receiver and algorithms. The results have shown that the novel DF neural MUD receiver outperforms the other analysed structures.

References

1. "Physical Layer Standard for CDMA2000 Spread Spectrum Systems (Release C)," Telecommunications Industry Association, TIA/EIA/IS- 2002.2-C, May 2002.
2. B. Aazhang, B. P. Paris and G. C. Orsak,"Neural Networks for Multiuser Detection in Code-Division-Multiple-Access Communications," *IEEE Transactions on Communications*, vol. 40, No. 7, July 1992.
3. U. Mitra and H. V. Poor, "Neural Network Techniques for Adaptive Multi-user Demodulation," *IEEE Journal on Selected Areas of Communications*, Vol. 12, No. 9, December 1994.
4. W. G. Teich, M. Seidl and M. Nold," Multiuser Detection for DS-CDMA Communication Systems based on Recurrent Neural Networks Structures", *Proc. IEEE ISSSTA '98*, 02-04 September, Sun City, South Africa, pp. 863-867, 1998.
5. R. C. de Lamare and R. Sampaio-Neto, "Analysis of Adaptive Multiuser Receivers for DS-CDMA using Recurrent Neural Networks", *Proc. SBrT/IEEE International Telecommunications Symposium*, September 2002
6. G. Woodward, R. Ratasuk, M. Honig. and P. Rapajic, "Multistage decision-feedback detection for DS-CDMA," *Proc. IEEE Int. Conf. Communications (ICC)*, June 1999.
7. S. Haykin, *Adaptive Filter Theory*, 3rd edition, Prentice-Hall, Englewood Cliffs, NJ, 1996.
8. S. Haykin, *Neural Networks: A Comprehensive Foundation*, 2nd Edition, Prentice-Hall, 1999.
9. T. S. Rappaport, *Wireless Communications*, Prentice-Hall, Englewood Cliffs, NJ, 1996.

Comparison of Two Filterbank Optimization Criteria for VDSL Transmission

Jérôme Louveaux[1], Cyrille Siclet[2], Didier Pinchon[3], and Pierre Siohan[4]

[1] Université catholique de Louvain, Louvain-la-neuve, Belgium,
louveaux@tele.ucl.ac.be
[2] IUT de Saint-Dié, UHP Nancy I - CRAN, Saint-Dié des Vosges, France,
cyrille.siclet@iutsd.uhp-nancy.fr
[3] Université Paul Sabatier, MIP Laboratory, Toulouse, France,
pinchon@mip.ups-tlse.fr
[4] France Télécom R&D, DMR/DDH Laboratory, Cesson-Sévigné, France,
pierre.siohan@francetelecom.com

Abstract. In this paper, we consider the use of filterbank based multi-carrier modulation for VDSL transmission. We compare two filterbank optimization criteria, i.e. the maximization of the time-frequency localization and the minimization of the out-of-band energy and we show that filterbank optimized according to the time-frequency criterion achieve better rates and allow the use of reduced size equalizers.

1 Introduction

Filterbanks have been extensively considered in the last decade for multicarrier communications [1,2] in a way to improve the frequency separation between the subcarriers. Different kind of filters have been proposed [3,4]. Most of these families of filters leave room for some optimization of the actual filters to be used. The optimization may use different criteria (such as synchronization properties or spectral selectivity for better narrowband interference rejection) but the most usual objective is to increase the achievable bit rate. However a direct optimization of the bit rate is not feasible, and would be channel dependent. So classically the filters are optimized according to a simple criterion like the out-of-band energy, which is supposed to provide interesting features for various system aspects [5].

In this contribution, we compare two possible criteria of filter optimization and analyze how the achievable bit rate and the required receiver complexity are influenced by the choice of this criterion. The family of filters considered here is limited to the cosine modulated orthogonal filters [6], and their performance are compared in a VDSL (very high bit rate digital subscriber lines) environment. The receiver is a SIMO (single input multiple output) equalizer [5,7]. Within these specifications, it is shown that the time-frequency localization criterion provides better results than the minimum out-of-band energy criterion. The improvement is observed both on the achievable bit rate and on the reduction of the required equalization complexity.

J.N. de Souza et al. (Eds.): ICT 2004, LNCS 3124, pp. 142–149, 2004.

2 Filterbank Based Multicarrier Modulation for VDSL Transmission

We denote by $I_p(m)$ the symbols of the N_t parallel input data streams with baud rate $1/N_tT$. These N_t symbol streams enter synthesis filters with impulse responses denoted by $g_p(n)$. These filterbank output samples are shaped by the transmitter filter with impulse response $p(t)$ and sent over the channel with impulse response $c(t)$ at the rate $1/T$. The signal is corrupted by an additive white Gaussian noise (AWGN) denoted $n(t)$. After the receiver band-limiting filter $f(t)$, the received signal is sampled at the rate M/T (M stands for a possible fractionally spaced processing). We denote by $h(t) = p(t) \star c(t) \star f(t)$ (where $\star$ stands for convolution) the overall channel impulse response. The discrete-time transmitted signal $s(n)$ is given by

$$s(n) = \sum_{p=0}^{N_t-1} \sum_{m=-\infty}^{\infty} I_p(m) g_{p,f,eq}(n\frac{T}{M} - mN_tT + \epsilon) + n_f(nT/M + \epsilon) \quad (1)$$

where $n_f(t)$ is the AWGN filtered by $f(t)$, $g_{p,f,eq}(t)$ are the waveforms resulting from the cascade of the band-specific synthesis filters $g_p(n)$ and $h(t)$, i. e.

$$g_{p,f,eq}(t) = \sum_{n=-\infty}^{\infty} g_p(n) h(t - nT). \quad (2)$$

The receiver structure for which timing errors are investigated has been proposed in [7]. It is a single input multiple output (SIMO) fractionally spaced linear equalizer (FSLE) possibly modified into a SIMO fractionally spaced decision feedback (FSDF) equalizer. The SIMO equalizer computes estimates of symbols transmitted over the different branches. For band p', this estimate is computed by

$$\hat{I}_{p'}(m') = \sum_{n'=-K_1}^{K_2} c_{p'}(n') s(m'N_tM - n') \quad (3)$$

where $c_{p'}(n)$ denotes the impulse response of the p'th branch of the SIMO equalizer, and it is assumed that each impulse response is made of $K_1 + K_2 + 1$ taps. The computations were achieved in a VDSL (very high speed digital subscriber lines) context which means high bit rate applications on copper lines. ANSI document [8] specifies the line models and the noise environment to be used for performance evaluation of modulation schemes. In this paper, we focus on the VDSL1 line model, with a 24-gauge loop of 1.0 km. The used bandwidth is assumed to be 10 MHz (starting around 375 kHz). According to standard specifications, the transmit power spectral density (PSD) is constrained to be lower than -60 dBm/Hz. The additive white noise has a constant PSD of -140 dBm/Hz.

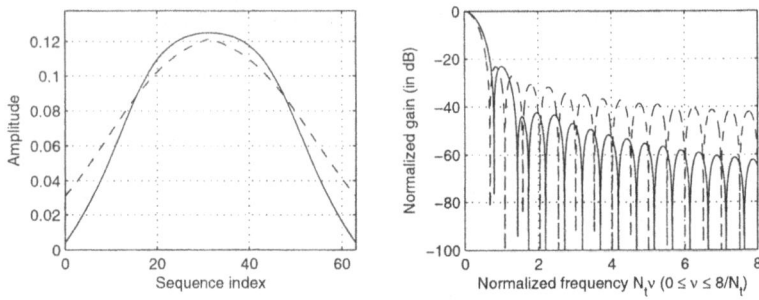

Fig. 1. Orthogonal prototype filters maximizing the time-frequency localization (solid line, $\xi = 0.91$, $J = 3.8 \times 10^{-2}$) and minimizing the out-of-band energy (dashed line, $\xi = 0.55$, $J = 1.9 \times 10^{-2}$), for $N_t = 32$ carriers and $L = 64$ coefficients.

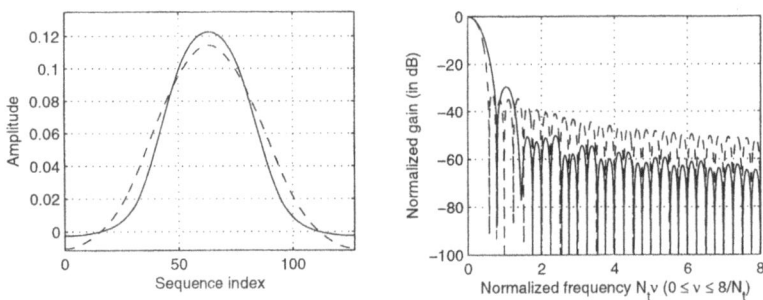

Fig. 2. Orthogonal prototype filter maximizing the time-frequency localization (solid line, $\xi = 0.97$, $J = 1.6 \times 10^{-2}$), and minimizing the out-of-band energy (dashed line, $\xi = 0.81$, $J = 2.0 \times 10^{-3}$), for $N_t = 32$ carriers and $L = 128$ coefficients.

3   Design Criteria and Filter Examples

The most common criterion for the design of prototype filters is the minimization of the out-of-band energy [6], [9]. More recently, it was also noticed that the maximization of the time-frequency localization could be an adequate criterion in the case of multicarrier transmission through time-frequency dispersive channels, as for instance the wireless mobile channel [10,11,12]. In this paper, both criteria are applied to a family of orthogonal filters[1] and the results obtained by using the corresponding filters are compared. The filterbank considered here is the set of cosine modulated filterbanks [6]. They are based on a prototype filter $p(n)$ of length $L = 2mN_t$ for some parameter m. The filters are then derived from that prototype and have the same length.

---

[1] It is usual [9] to impose the orthogonality property in such systems although it can be shown that only biorthogonality is required [13,14].

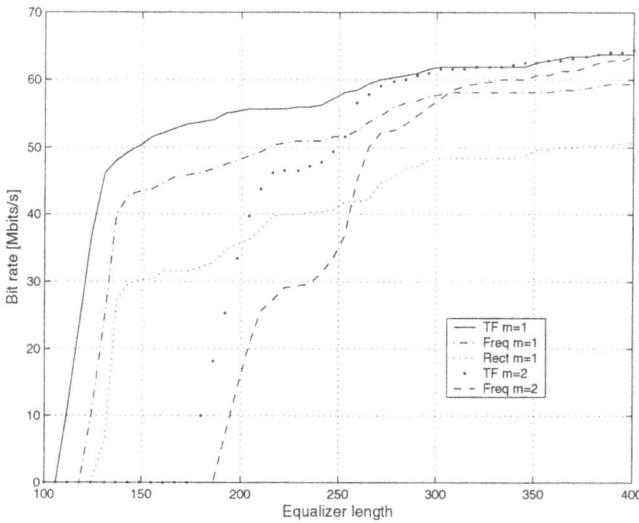

Fig. 3. Influence of the optimization criterion and of the filter length ($N_t = 32$, $m = 1$ and $m = 2$). The solid line is for the time-frequency criterion with short filters ($m = 1$), dash-dotted line for the out-of-band energy criterion with $m = 1$, small-dotted line for the rectangular prototype ($m = 1$), big-dotted line for the time-frequency criterion with long filters $m = 2$, and dashed line for the out-of-band energy criterion with $m = 2$.

3.1   Minimization of the Out-of-Band Energy

The out-of-band energy can be expressed as a function of the prototype filter p

$$J(p) = \frac{E(f_c)}{E(0)} \text{ with } E(x) = \int_x^{\frac{1}{2}} |P(e^{j2\pi\nu})|^2 d\nu, \tag{4}$$

where f_c is the cutoff frequency, ν is the normalized frequency and $0 \leq J(p) \leq 1$. It is worthwhile noting that, as mentioned in [15], and due to the power complementarity property of the orthogonal filterbank, the minimization of $J(p)$, also guarantee that $|P(e^{j2\pi\nu})|$ is nearly constant in its passband, *i.e.* in the range of frequency $\nu \in [0, f_c]$. In the case of a filterbank based multicarrier VDSL system, we use $f_c = \frac{1}{N_t}$, which represents the double of the band that would be allocated if the filters were perfectly separated in frequency. This large cuttoff frequency is used to allow a slow decrease of the frequency response and hence make it possible to get a good out-of-band rejection.

3.2   Maximization of the Time-Frequency Localization

The so-called time-frequency localization for discrete-time signal can be directly derived, as in [16], from the expressions for continuous-time signals of the second

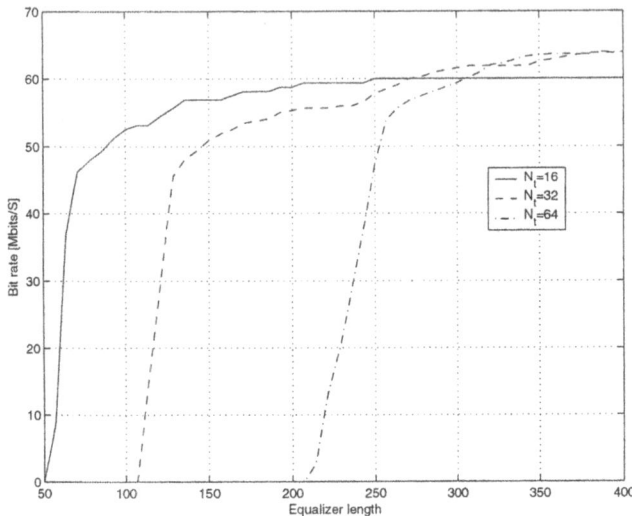

Fig. 4. Influence of the number of carriers ($m = 1$ and time-frequency criterion). The solid line is for $N_t = 16$ subcarriers, dashed line for $N_t = 32$ and dash-dotted line for $N_t = 64$.

order time and frequency moments. Here we do not explicitly refer to continuous-time functions, we prefer to use a localization measure specifically dedicated to discrete-time signals [17]. Let x be a discrete-time real-valued signal, with a norm denoted $\|.\|$, defined by $\|\mathbf{x}\| = \left(\sum_{k=-\infty}^{+\infty} x^2[k]\right)^{\frac{1}{2}}$. Following Doroslovački [17], the gravity center in time is defined by

$$T(x) = \frac{\sum_{k=-\infty}^{+\infty} (k - 1/2)\,(x[k] + x[k-1])^2}{\sum_{k=-\infty}^{+\infty} (x[k] + x[k-1])^2},$$

and the second order moments in frequency and in time are defined by

$$M_2(x) = \frac{1}{\|\mathbf{x}\|^2} \sum_{k=-\infty}^{+\infty} (x[k] - x[k-1])^2, \tag{5}$$

and

$$m_2(x) = \frac{1}{\|\mathbf{x}\|^2} \sum_{k=-\infty}^{+\infty} \left(k - \frac{1}{2} - T(x)\right)^2 \left(\frac{x[k] + x[k-1]}{2}\right)^2, \tag{6}$$

respectively. Then, the time-frequency localization measure of the prototype filter p can be defined as

$$\xi(p) = \frac{1}{\sqrt{4m_2(p)M_2(p)}}. \tag{7}$$

Similarly to [10], the normalization used here implies that $0 \leq \xi(p) \leq 1$, $\xi(p) = 1$ being the optimum.

3.3   Filter Examples

Using the optimization procedure described in [18], a large variety of filters satisfying the orthogonality property can be designed with a nearly optimal time-frequency localization or out-of-band energy. When using a filterbank based multicarrier modulation, a small number of carriers is acceptable (cf. [5]) contrary to wireless applications like DVB-T (Digital Video Broadcasting - Terrestrial) for example. That is why we have restricted our simulations to a maximum of 64 carriers. Figures 1 and 2 give the time and frequency responses of optimized prototype filters for $N_t = 32$ carriers and $L = 64$ coefficients, and $L = 128$, respectively. It is worthwhile noting that prototypes obtained for a different number of carriers (e.g. 16 or 64) have very similar time and frequency responses. In order to emphasize the frequency response around the center of the band, we have restricted ourself to $0 \leq \nu \leq \frac{8}{N_t}$. One can notice that time-frequency optimized filters seem to have a smallest out-of-band energy than frequency optimized ones. But it is not the case in fact because frequency-optimized filters better match the allocated band: they have a smaller cut-off frequency even if the decrease is slower afterwards.

4   Simulation Results

The performance of the filters presented above are now compared in a VDSL environment. The setup can be found in [7]. The performance measure is the achievable bit rate of the system as a function of the equalizer length. Fig. 3 shows the achievable bit rates with filters designed with the two criteria, for $m = 1$ and $m = 2$. They are also compared with a simple rectangular prototype. This figure clearly shows the improvement brought by the time-frequency criterion. The achievable bit rate is higher whatever the complexity used in the equalizer.

   This figure also shows the influence of the length of the prototype on the system. The achievable bit rate at very long equalizers gets higher with long prototypes (and thus long filters) but the minimal equalizer length necessary to obtain an acceptable result is much higher. Besides, for reasonable equalizer lengths, the shorter filters provide much better results. Hence it appears that the use of long filters is practically never a good choice, except perhaps for systems aiming at very high performance and not restricted in complexity. Note that, for this family of filters, it is not possible to get a smaller length than

$L = 2N_t$ ($m = 1$). Fig. 4 shows the effect of changing the number of carriers in the system. Again, the criterion used here is the time-frequency criterion, and the filters are of length $L = 2N_t$ ($m = 1$). Similarly to the effect of the filter length, it appears that larger number of carriers can reach higher performance at very high complexity, but the smaller number of carriers is usually better at reasonable equalizer lengths.

5   Conclusion

Two criteria of optimization have been compared for the choice of filters in multi-carrier modulation in a VDSL environment. It has been shown that the criterion based on the time-frequency localization provides better results than the out-of-band energy criterion (frequency localization) in every analyzed situation. Other important aspects in the design of the system have also been analyzed such as the number of carriers and the length of the filters.

References

1. Fliege, N.J.: Orthogonal multiple carrier data transmission. European Trans. on Telecommunications **3** (1992) 255–264
2. Cherubini, G., Eleftheriou, E., l er, S., Cioffi, J.M.: Filter bank modulation techniques for very high-speed digital subscriber lines. IEEE Comm. Magazine (2000) 98–104
3. Vahlin, A., Holte, N.: Optimal finite duration pulses for OFDM. IEEE Trans. on Comm. **44** (1996) 10–14
4. B lcskei, H., Duhamel, P., Hleiss, R.: Design of pulse shaping OFDM/OQAM systems for high data-rate transmission over wireless channels. In: IEEE ICC'99, Vancouver, Canada (1999)
5. Louveaux, J.: Filter Bank Based Multicarrier Modulation for xDSL Transmission. PhD thesis, Universit catholique de Louvain, Louvain-la-Neuve, Belgium (2000) http://www.tele.ucl.ac.be/~jlouveau/PhDlouveaux.pdf.
6. Vaidyanathan, P.P.: Multirate systems and filter banks. Prentice Hall, Englewood Cliffs (1993) ISBN 0-13-605718-7.
7. Vandendorpe, L., Cuvelier, L., Deryck, F., Louveaux, J., van de Wiel, O.: Fractionally spaced linear and decision feedback detectors for transmultiplexers. IEEE Trans. on Signal Processing **46** (1998) 996–1011
8. ANSI T1E1.4: Very-high-speed digital subscriber lines". Technical report, (ANSI, Secreteriat for Telecommunications Industry Solutions) draft, Rev. 7A.
9. Nguyen, T.Q., Koilpillai, R.D.: The theory and design of arbitrary-length cosine-modulated filter banks and wavelets, satisfying perfect reconstruction. IEEE Trans. on Signal Processing **44** (1996) 473–483
10. Le Floch, B., Alard, M., Berrou, C.: Coded orthogonal frequency division multiplex. Proceedings of the IEEE **83** (1995) 982–996
11. Haas, R., Belfiore, J.C.: A time-frequency well-localized pulse for multiple carrier transmission. Wireless Personal Comm. **5** (1997) 1–18
12. Lacroix, D., Goudard, N., Alard, M.: OFDM with guard interval versus OFDM/OffsetQAM for high data rate UMTS downlink transmission. In: IEEE VTC Fall'01, Atlantic City, USA (2001)

13. Kozek, W., Molisch, A.F.: Nonorthogonal pulseshapes for multicarrier comm. in doubly dispersive channels. IEEE Journal on Selected Areas in Comm. **16** (1998) 1579–1589
14. Siclet, C.: Application de la th orie des bancs de filtres l'analyse et la conception de modulations multiporteuses orthogonales et biorthogonales. PhD thesis, Universit de Rennes I, Rennes (2002)
15. Vaidyanathan, P.P.: Theory and design of M-channel maximally decimated quadrature mirror filters with arbitrary M, having the perfect reconstruction property. IEEE Trans. on Acoustics, Speech, and Signal Processing **35** (1987) 476–492
16. Siohan, P., Roche, C.: Cosine-modulated filterbanks based on extended Gaussian functions. IEEE Trans. on Signal Processing **48** (2000) 3052–3061
17. Doroslovački, M.I.: Product of second moments in time and frequency for discrete-time signals and the uncertainty limit. Signal Processing **67** (1998) 59–76
18. Pinchon, D., Siohan, P., Siclet, C.: Design techniques for orthogonal modulated filter banks based on a compact representation. To appear in IEEE Trans. on Signal Processing (2004)

Phase Estimation and Phase Ambiguity
Resolution by Message Passing

Justin Dauwels[1], Henk Wymeersch[2], Hans-Andrea Loeliger[1], and
Marc Moeneclaey[2]

[1] Dept. of Information Technology and Electrical Engineering, ETH, CH-8092
Zürich, Switzerland.
{dauwels, loeliger}@isi.ee.ethz.ch,
[2] DIGCOM Research Group, TELIN Dept., Ghent University,
Sint-Pietersnieuwstraat 41, B-9000 Gent, Belgium
{hwymeersch, mm}@telin.ugent.be

Abstract. Several code-aided algorithms for phase estimation have recently been proposed. While some of them are ad-hoc, others are derived
in a systematic way. The latter can be divided into two different classes:
phase estimators derived from the expectation-maximization (EM) principle and estimators that are approximations of the sum-product message
passing algorithm. In this paper, the main differences and similarities between these two classes of phase estimation algorithms are outlined and
their performance and complexity is compared. Furthermore, an alternative criterion for phase ambiguity resolution is presented and compared
to an EM based approach proposed earlier.

1 Introduction

This paper deals with iterative code-aided algorithms for phase estimation and
phase ambiguity resolution in a communications receiver. Coded channel input
symbols are transmitted in frames of L symbols. We consider a channel model
of the form

$$Y_k = X_k e^{j\Theta} + N_k, \tag{1}$$

where X_k is the coded channel input symbol at time $k \in \{1, \ldots, L\}$, Y_k is the
corresponding received symbol, Θ is the unknown (constant) phase, and N_k is
white complex Gaussian noise with (known) variance $2\sigma_N^2$, i.e., σ_N^2 per dimension. For the sake of definiteness, we assume that the channel input symbols X_k
are M-PSK symbols and are protected by a low-density parity check (LDPC)
code. It is convenient to break Θ into two contributions:

$$\Theta = \Phi + Q\frac{2\pi}{M} \tag{2}$$

with $0 \leq \Phi < 2\pi/M$ and $Q \in \{0, \ldots, M-1\}$. Accordingly, the problem of
estimating Θ can be decomposed in two subproblems: the problem of estimating

J.N. de Souza et al. (Eds.): ICT 2004, LNCS 3124, pp. 150–155, 2004.
© Springer-Verlag Berlin Heidelberg 2004

Φ, referred to as "phase estimation", and the problem of determining Q, which is called "phase ambiguity resolution".

Several turbo-synchronization algorithms have recently appeared that deal with constant phase rotations [1]–[5]. In [1], phase estimators are derived from the expectation-maximization (EM) principle. In [2], [3] and [4], phase estimation algorithms are presented that are approximations of the sum-product message passing algorithm applied to factor graphs of the channel model. An EM based algorithm for phase ambiguity resolution is proposed in [5].

In this paper, we compare the sum-product based phase estimators of [3] to the EM based phase estimator [1]. To this end, we formulate also the EM based estimator as a message passing algorithm. We refer to [6] for a classical exposition of the EM algorithm. We then compare the EM based phase estimator to several sum-product based algorithms in terms of performance and complexity. We also propose an alternative criterion for phase ambiguity resolution and compare it to the one proposed by Wymeersch et al. [5].

This paper is structured as follows. In Section 2, we briefly explain the factor graph we use to represent channel (1). In Section 3, we review both types of phase estimators and elaborate on the main differences between them. Section 4 considers the problem of phase ambiguity resolution. In Section 5, we investigate the computational complexity of the various estimation algorithms. In Section 6, we present simulation results.

2   Factor Graphs of the Channel Model

We use Forney-style factor graphs (FFG), where nodes (boxes) represent factors and edges represent variables. A tutorial introduction to such graphs is given in [7]. The system described in Section 1 is easily translated into the FFG of Fig. 1, which represents the factorization of the joint probability function of all variables. The upper part of the graph is the indicator function of the LDPC code, with parity check nodes in the top row that are "randomly" connected to equality constraint nodes ("bit nodes"). The nodes below the bit nodes represent the deterministic mapping $f : \left(B_k^{(1)}, \ldots, B_k^{(\log_2 M)} \right) \mapsto X_k$ of the bits $B_k^{(1)}, \ldots, B_k^{(\log_2 M)}$ to the symbol X_k. These nodes correspond to the factors

$$\delta_f \left(b_k^{(1)}, \ldots, b_k^{(\log_2 M)}, x_k \right) \triangleq \begin{cases} 1, & \text{if } f \left(b_k^{(1)}, \ldots, b_k^{(\log_2 M)} \right) = x_k; \\ 0, & \text{otherwise.} \end{cases} \tag{3}$$

In Fig. 1, the variable S is defined as $S \triangleq e^{j\Theta}$. The equality constraint node imposes the constraint $S_k = S$, $\forall k$. Furthermore, Z_k is defined as $Z_k \triangleq X_k S_k$. The row of "multiply nodes" represents the factors $\delta(z_k - x_k s_k)$. The bottom row of the graph represents the factors $p(y_k | z_k) \triangleq (2\pi\sigma_N^2)^{-1} e^{-\|y_k - z_k\|^2 / 2\sigma_N^2}$.

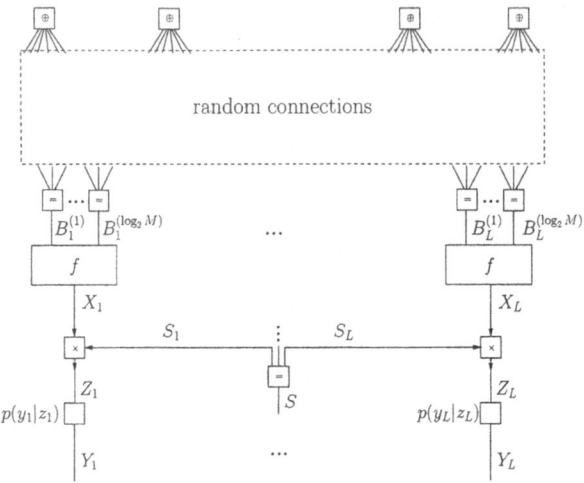

Fig. 1. FFG of LDPC code and the channel model

3 Phase Estimation Through Message Passing

Both in the sum-product based algorithms and in the EM based algorithm (which we will also view as message passing in the factor graph), the messages are updated according to the following schedule. First, the bottom row in Fig. 1 is updated, i.e., messages are sent from the received symbols Y_k towards the phase model. Then, one alternates between

1. An update of the messages along the S_k edges, which can be scheduled as a horizontal (left-to-right) sweep.
2. A subsequent horizontal (left-to-right) sweep for updating both the messages $\mu_{\boxtimes \to x_k}(X_k)$ out of the multiply nodes along the X_k edges and the messages $\mu_{f \to b_k}(B_k)$ out of the mapper nodes along the B_k edges.
3. Several iterations of the LDPC decoder.

In both algorithms, the messages out of the mapper nodes and inside the graph of the LDPC code are computed according to the standard sum-product rule [7]; we will therefore only consider the computation of the messages related to the phase Θ. First, we briefly elaborate on how the messages are computed according to the sum-product rule, then we consider the EM rule.

Straightforward application of the sum-product rule to compute the messages along the S_k edges leads to (intractable) integrals. Several methods to approximate these integral (or, equivalently, to represent the messages along the S_k edges) are proposed in [3] and [4]: numerical integration, particle methods, canonical distributions (Fourier, Gaussian and Tikhonov) and gradient methods ("steepest descent"); each of these methods leads to a different phase estimator.

The EM based phase estimator of [1] may be viewed as a message passing algorithm that is similar to the steepest descent sum-product based estimator.

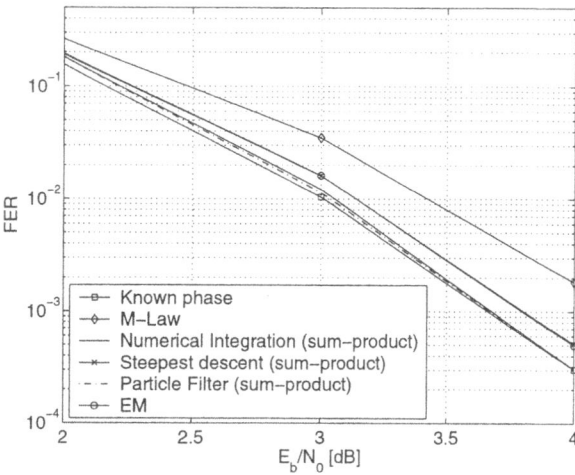

Fig. 2. FER of several phase estimators

In both algorithms, the message $\mu_{S_k \to \boxtimes}(s_k)$ is represented by an estimate $\hat{S}$. In the EM based phase estimator, this estimate is computed according to the rule [1]:

$$\hat{S} = \arg\max_s \sum_i \left(\sum_x \mu_{X_i \to \boxtimes}(x)\, \mu_{\boxtimes \to X_i}(x) \log \mu_{z_i \to \boxtimes}(xs) \right). \qquad (4)$$

The rule (4) is similar to the sum-product rule [3]

$$\hat{S} = \arg\max_s \sum_i \log \left(\sum_x \mu_{X_i \to \boxtimes}(x)\, \mu_{Z_i \to \boxtimes}(xs) \right). \qquad (5)$$

Note the difference in the position of the logarithm. In addition, the update rule (4) involves both the incoming messages $\mu_{X_i \to \boxtimes}(x)$ and the outgoing messages $\mu_{\boxtimes \to X_i}(x)$; in the expression (5), only the incoming messages $\mu_{X_i \to \boxtimes}(x)$ occur. In this particular case, the EM rule can be evaluated analytically [1]. In more complicated problems, this is not the case. One then typically uses steepest descent to compute a hard estimate. In this case, the similarity between EM based and steepest descent sum-product based synchronizers is even stronger.

4 Phase Ambiguity Resolution

The algorithms presented in the previous section make estimates of Φ (see Eq. (2)); they are not able to resolve the phase ambiguity, i.e., to detect Q. We determine Q by hypothesis testing. For each possible value of Q, we apply the algorithms presented in the previous section. We restrict the domain of the S_k messages to the set $\{\exp j\theta : \theta \in [Q2\pi/M, (Q+1)2\pi/M]\}$. We subsequently select the most likely hypothesis. Wymeersch et al. [5] derived the following EM rule for selecting Q:

$$\hat{Q} = \arg\max_q \sum_i \left(\sum_x \mu_{X_i \to \boxtimes}(x)\, \mu_{\boxtimes \to X_i}(x) \log \mu_{Z_i \to \boxtimes}(x\hat{s}_k(q)) \right), \quad (6)$$

where $\hat{s}_k(q)$ is the estimate of S_k under hypothesis Q. The rule (6) is an approximation of the MAP detector

$$\hat{Q} = \arg\max_q \log \sum_\mathbf{x} \int_\phi p(\mathbf{x}, \mathbf{y}, \phi, q) d\phi. \quad (7)$$

We propose as alternative approximation

$$\hat{Q} = \arg\max_q \sum_i \log \sum_x \mu_{X_i \to \boxtimes}(x) \mu_{\boxtimes \to X_i}(x). \quad (8)$$

If the messages $\mu_{S_k \to \boxtimes}(s_k)$ are represented as single values, then the rule (8) reduces to

$$\hat{Q} = \arg\max_q \sum_i \log \sum_x \mu_{X_i \to \boxtimes}(x) \mu_{Z_i \to \boxtimes}(x\hat{s}_k(q)), \quad (9)$$

which is very similar to the expression (6). One observes the same differences and similarities between the expressions (6) and (9) as between the expressions (5) and (4). The criterion (8) is more general than (6): the phase messages need not to be represented as single values.

5   Computational Complexity

The computational complexity of the phase estimators is directly related to the way the phase messages are represented. Both in the EM algorithm and in the steepest descent sum-product algorithm, the phase messages are single real numbers. As a consequence, the complexity of both algorithms is similar and very low. The complexity of the approach based on numerical integration and particle filtering is much larger: it is proportional to the number of quantization levels and particles respectively, which is typically choosen between 100 and 1000. It is well known that numerical integration becomes infeasable in higher dimensions. Particle filtering on the other hand scales much better with the dimension of the system.

6   Simulation Results and Discussion

We performed simulations of the sum-product based as well as the EM based algorithms for joint phase estimation and phase ambiguity resolution. We used a fixed rate 1/2 LDPC code of length 100 that was randomly generated; we did not optimize the code for the channel at hand. The symbol constellation was

Gray-encoded 4-PSK. We iterated three times between the LDPC decoder and the phase estimator, each time with hundred iterations inside the LDPC decoder. We did not iterate between the LDPC decoder and the mapper. In the particle filtering (sum-product based) algorithm, the messages were represented as lists of 100 samples; in the numerical integration (sum-product based) algorithm, the phase is uniformly quantized over 100 levels.

Fig. 2 presents the FER (frame error rate) of the presented algorithms for joint phase estimation and ambiguity resolution. We include the FER for perfect synchronization, i.e., for the case in which the phase Θ is known. Moreover, we show the FER resulting from the M-law phase estimator assuming perfect phase ambiguity resolution. We observe that the M-law estimator gives rise to a degradation of up to 0.5 dB compared to perfect synchronization. Both the EM estimator as well as the sum-product based estimators are able to reduce most of this degradation. The particle filtering and numerical integration (sum-product based) estimators have slightly lower FER than the EM algorithm and steepest descent sum-product based algorithm, but their complexity is higher.

Acknowledgements. We heavily used the collection of C programs for LDPC codes by Radford M. Neal from
http://www.cs.toronto.edu/~radford/ldpc.software.html.
This project was in part supported by the Swiss National Science Foundation grant 200021-101955 and the Interuniversity Attraction Pole Program P5/11 - Belgian Science Policy.

References

1. N. Noels et al., "Turbo synchronization : an EM algorithm interpretation", *International Conference on Communications 2003*, Anchorage, Alaska, May 11–15, 2003, pp. 2933–2937.
2. R. Nuriyev and A. Anastasopoulos, "Analysis of joint iterative decoding and phase estimation for the noncoherent AWGN channel, using density evolution," *Proc. 2002 IEEE Information Theory Workshop,* Lausanne, Switzerland, June 30 – July 5, 2002, p. 168.
3. J. Dauwels and H. -A. Loeliger, "Phase Estimation by Message Passing," *IEEE International Conference on Communications*, ICC 2004, Paris, France, June 20–24, 2004, to appear.
4. G. Colavolpe and G. Caire, "Iterative Decoding in the Presence of Strong Phase Noise," *IEEE Journal on Selected Areas in Communications,* Differential and Noncoherent Wireless Communications, to appear.
5. H. Wymeersch and M. Moeneclaey, "Code-aided phase and timing ambiguity resolution for AWGN channels", *The IASTED International Conference, Signal and Image Processing (SIP-03),* Honolulu, Hawaii, Aug. 2003.
6. A. P. Dempster, N. M. Laird and D. B. Rubin, "Maximum likelihood from incomplete data via the EM algorithm," *Journal of the Royal Statistical Society,* 39(1):1-38, 1977, Series B.
7. H. -A. Loeliger, "An introduction to factor graphs," *IEEE Signal Processing Magazine*, Jan. 2004, pp. 28–41.

An Expurgated Union Bound for Space-Time Code Systems

Vu-Duc Ngo, Hae-Wook Choi, and Sin-Chong Park

Systems VLSI Lab Laboratory, SITI Research Center, School of Engineering
Information Communication University (ICU)
Yusong P.O. Box 77, Taejon 305-714, Korea
{duc75,hwchoi,scpark}@icu.ac.kr

Abstract. The performance of multiple-input-multiple-output (MIMO) systems over flat fading channels by calculating the Union Bound (UB) was analyzed. Based on the original idea of Verdu's theorem, a new tighter UB by excluding the redundant code-matrices is introduced. The new tighter bound is calculated by the summation of the Pairwise Error Probabilities (PEPs) of code-matrices that belong to an irreducible set.

1 Introduction

The increasing requirement for faster and more reliable wireless communication links has brought systems with multiple antennas at the transmitter and the receiver, these systems are so called multiple-input multiple-output (MIMO) systems, to be considered ([1],[2]). Several papers analyze the performance of MIMO systems with different type of designs such as BLAST [1] and coded space time systems ([4], [6]). The Union Bound with simplified form is the popular method that we can use to theoretically evaluate the performance of MIMO systems but it shows the useless results for the low Eb/No . In this paper, we derive new tight bound for MIMO systems [1] over flat fading channels based on Verdu's theorem [3], this theorem introduces how to refine the Union Bound by excluding the redundant terms then makes it tighter. Our proposed bound can be archived with less complexity of computation in comparison with the conventional Union Bound due to the computations are done over the irreducible set that is the subset of the codebook. The paper is organized as follows: in Section 2, we study ML detection criteria for MIMO channels and analyze the performance of MIMO systems over the flat fading channels, adhering to Bieliegi in [4]. In Section 3 we introduce the way to tighten the Union Bound for MIMO systems based on Verdu's theorem [3]. Our summary and conclusions are presented in section 4.

2 ML Detection and Union Bound

2.1 ML Detection

Consider a MIMO system has M_t transmitting and M_r receiving antennas, respectively. Assume that a space-time system, adhering to [1], is used. The

J.N. de Souza et al. (Eds.): ICT 2004, LNCS 3124, pp. 156–162, 2004.
© Springer-Verlag Berlin Heidelberg 2004

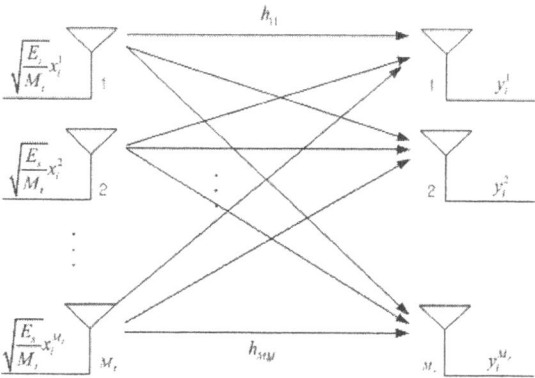

Fig. 1. Channel model.

transmitted signal with a block length N is represented as $[\mathbf{x}_1, ..., \mathbf{x}_N]$, where $\mathbf{x}_i = \left[x_i^1, ..., x_i^{M_t} \right]^T$, and transmitted over the flat fading channel as the model shown by Fig. 1. The received signal is expressed as the matrix $\mathbf{Y}_{M_r \times N}$

$$
\mathbf{Y}_{M_r \times N} = \sqrt{\frac{E_s}{M_t}} \mathbf{H}_{M_r \times M_t} \mathbf{X}_{M_t \times N} + \mathbf{E}_{M_r \times N}
$$

$$
= \sqrt{\frac{E_s}{M_t}} \begin{pmatrix} h_{1,1} & \cdots & h_{1,M_t} \\ \vdots & \ddots & \vdots \\ h_{M_r,1} & \cdots & h_{M_r,M_t} \end{pmatrix} \left(\mathbf{x}_1 \ldots \mathbf{x}_N \right) + \left(\mathbf{e}_1 \ldots \mathbf{e}_N \right), \quad (1)
$$

where $\mathbf{E}_{M_r \times N} = [\mathbf{e}_1, ..., \mathbf{e}_N]$ is the matrix of zero-mean complex Gaussian random variables with zero mean and independent real and imaginary parts that having the same variance $N_0/2$ (i.e., circular Gaussian noise). The channel is defined as the matrix $\mathbf{E}_{M_r \times M_t}$ whose entries h_{ij} are independent complex Gaussian variables, and E_s denotes the transmitted symbol energy. Under the assumption of the Channel State Information (CSI) is perfectly known to the receiver, and of AWGN, the ML detection scheme is applied, the decoded code-matrice is chosen to meet the following criteria

$$
\overline{\mathbf{X}}_{M_t \times N} = arg \min_{\mathbf{X}_{M_t \times N}} \left\| \mathbf{Y}_{M_r \times N} - \sqrt{\frac{E_s}{M_t}} \mathbf{H}_{M_r \times M_t} \mathbf{X}_{M_t \times N} \right\|_F^2, \quad (2)
$$

where we define the square of Frobenius norm of a matrix $\mathbf{A}_{m \times n}$ with the elements $a_{i}j$ as

$$
\|\mathbf{A}\|_F^2 = \sum_{i=1}^m \sum_{j=1}^n |a_{ij}|^2 = Tr\left(\mathbf{A}\mathbf{A}^H \right) = Tr\left(\mathbf{A}^H \mathbf{A} \right). \quad (3)
$$

Suppose that $\mathbf{X}^{(i)}_{M_t \times N}$ and $\mathbf{X}^{(j)}_{M_t \times N}$ are the transmitted and decoded code-matrices respectively, an incorrectly decoding decision in favor of $\mathbf{X}^{(i)}_{M_t \times N} \neq$

$\mathbf{X}^{(j)}_{M_t \times N}$ is made if only if

$$\left\| \mathbf{Y}_{M_r \times N} - \sqrt{\frac{E_s}{M_t}} \mathbf{H} \mathbf{X}^{(j)}_{M_t \times N} \right\|_F^2 < \left\| \mathbf{Y}_{M_r \times N} - \sqrt{\frac{E_s}{M_t}} \mathbf{H} \mathbf{X}^{(i)}_{M_t \times N} \right\|_F^2. \quad (4)$$

Let us define $\overline{\mathbf{X}} = \mathbf{X}^{(j)}_{M_t \times N} - \mathbf{X}^{(i)}_{M_t \times N}$, after some steps of manipulation, the inequality (4) is equivalent to

$$2\mathrm{Re}\left\{ \mathbf{E}^H \sqrt{\frac{E_s}{M_t}} \mathbf{H} \overline{\mathbf{X}} \right\} < -\frac{E_s}{M_t} \mathbf{H} \overline{\mathbf{X}} \overline{\mathbf{X}}^H \mathbf{H}^H. \quad (5)$$

Moreover, on the other hand we can easily obtain the mean and variance values of the left hand side of the inequality (5) as follows

$$\begin{cases} \varepsilon\left[2\mathrm{Re}\left\{ \mathbf{E}^H \sqrt{\frac{E_s}{M_t}} \mathbf{H} \overline{\mathbf{X}} \right\} \right] = 0 \\ \mathrm{Var}\left[2\mathrm{Re}\left\{ \mathbf{E}^H \sqrt{\frac{E_s}{M_t}} \mathbf{H} \overline{\mathbf{X}} \right\} \right] = 2N_0 \frac{E_s}{M_t} \mathbf{H} \overline{\mathbf{X}} \overline{\mathbf{X}}^H \mathbf{H}^H. \end{cases} \quad (6)$$

It follows that $2\mathrm{Re}\left\{ \mathbf{E}^H \sqrt{\frac{E_s}{M_t}} \mathbf{H} \overline{\mathbf{X}} \right\}$ is a Gaussian random variable as

$$2\mathrm{Re}\left\{ \mathbf{E}^H \sqrt{\frac{E_s}{M_t}} \mathbf{H} \overline{\mathbf{X}} \right\} = N\left(0, 2N_0 \frac{E_s}{M_t} \mathbf{H} \overline{\mathbf{X}} \overline{\mathbf{X}}^H \mathbf{H}^H \right), \quad (7)$$

and hence, by simplified $\mathbf{X}^{(a)}_{M_t \times N} = \mathbf{X}^{(a)}$ (a is an arbitrary index), the pairwise error probability (PEP) with respect to the case of fixed channel $\mathbf{H}$ can be calculated by

$$P\left(\mathbf{X}^{(i)} \to \mathbf{X}^{(j)} | \mathbf{H} \right) = Q\left(\sqrt{\frac{E_s}{M_t} \frac{\left\| \mathbf{H}(\mathbf{X}^{(i)} - \mathbf{X}^{(j)}) \right\|_F^2}{2N_0}} \right). \quad (8)$$

When the channel $\mathbf{H}$ is regarded as random channel, the probability of incorrect decision [4] (taken as the average over $\mathbf{H}$) can be bounded as follows

$$P\left(\mathbf{X}^{(i)} \to \mathbf{X}^{(j)} | \mathbf{H} \right) = \varepsilon_{\mathbf{H}}\left[Q\left(\sqrt{\frac{E_s}{M_t} \frac{\left\| \mathbf{H} \overline{\mathbf{X}} \right\|_F^2}{2N_0}} \right) \right]$$

$$= \left[\frac{1}{\det\left(I + \frac{E_s}{4M_t N_0} \overline{\mathbf{X}}^H \mathbf{H}^H \mathbf{H} \overline{\mathbf{X}} \right)} \right]^{M_r}. \quad (9)$$

To obtain the final result in the inequality (9), during manipulation we apply the fact that $\det(\mathbf{I} + \mathbf{AB}) = \det(\mathbf{AB} + \mathbf{I})$.

2.2 Union Bound

Let us define the codebook that contains all valid code-matrices as the set $\{\Omega | \mathbf{X}^{(i)} \in \Omega\}$ for any i, $0 \leq i \leq |\Omega| - 1$, where $|\Omega|$ presents the cardinality of the codebook. Assuming that all code-matrices are equally likely to be transmitted over the fixed channel $\mathbf{H}$, the total error probability can be obtained via the Union Bound and then upper bounded as

$$P(e) = \sum_{\mathbf{x}^{(i)}, \mathbf{x}^{(j)} \in \Omega; i \neq j} P\left(\mathbf{X}^{(i)} \rightarrow \mathbf{X}^{(j)} | \mathbf{H}\right)$$

$$\leq \frac{1}{|\Omega|} \sum_{\mathbf{x}^{(i)} \in \Omega} \sum_{\mathbf{x}^{(j)} \in \left\{\Omega | \{\mathbf{x}^{(i)}\}\right\}} Q\left(\sqrt{\frac{E_s}{M_t} \frac{\left\|\mathbf{H}\overline{\mathbf{X}}\right\|_F^2}{2N_0}}\right). \tag{10}$$

In the equations (9) and (10), we have the term $\left\|\mathbf{H}\overline{\mathbf{X}}\right\|_F$ standing for Euclidean distance between two code-matrices, and also can derive the average error probability of a MIMO system over the random channel in favor of Union Bound as follows

$$P(e) \leq \varepsilon_{\mathbf{H}} \left[\frac{1}{|\Omega|} \sum_{\mathbf{x}^{(i)} \in \Omega} \sum_{\mathbf{x}^{(j)} \in \left\{\Omega | \{\mathbf{x}^{(i)}\}\right\}} P\left(\mathbf{X}^{(i)} \rightarrow \mathbf{X}^{(j)}\right) | \mathbf{H} \right]$$

$$= \frac{1}{|\Omega|} \sum_{\mathbf{x}^{(i)} \in \Omega} \sum_{\mathbf{x}^{(j)} \in \left\{\Omega | \{\mathbf{x}^{(i)}\}\right\}} \left[\frac{1}{\det\left(I + \frac{E_s}{4M_t N_0} \overline{\mathbf{X}}^H \mathbf{H}^H \mathbf{H} \overline{\mathbf{X}}\right)} \right]^{M_r}. \tag{11}$$

Let consider the all zero code-matrice $\mathbf{X}^0 = [0]_{M_t \times N}$ as the transmitted code-matrice, the Union Bounds in inequalities (10) and (11) can be simplified as following equations in terms of fixed and random channels, respectively

$$P(e) \leq \sum_{\mathbf{x}^{(j)} \in \Omega; j \neq 0} Q\left(\sqrt{\frac{E_s}{M_t} \frac{\left\|\mathbf{H}\mathbf{X}^{(j)}\right\|_F^2}{2N_0}}\right), \tag{12}$$

and

$$P(e) \leq \sum_{\mathbf{x}^{(j)} \in \Omega; j \neq 0} \left[\frac{1}{\det\left(I + \frac{E_s}{4M_t N_0} \left(\mathbf{X}^{(j)}\right)^H \mathbf{H}^H \mathbf{H} \mathbf{X}^{(j)}\right)} \right]^{M_r}. \tag{13}$$

From the inequality (13) we clearly see that the performance of MIMO systems in terms of error probability critically is depended on the number of received antenna which is regarded as diversity order. We also observe that the error probability of MIMO systems is closely related to the Euclidean distance between each pair of transmitted and received code-matrices. Equations (12) and (13) show that if the transmitted code-matrice is all zero matrix then the error performance depends on the distance spectrum of code-matrices belonging to codebook when CSI is known perfectly to the receiver.

3   Tighten Union Bound by Applying Verdu's Theorem

3.1   Generalization Verdu's Theorem for MIMO Systems

In the inequalities (12) and (13), the Union Bound are bounded by using the sum of the probabilities of all the individual error events instead of a union of all the probability of error events. If one of the pairwise error region (PER) caused by the decoded code-matrice, let say $\mathbf{X} = \{\mathbf{X}^{(j)} \in \Omega | (\|\mathbf{Y}_{M_r \times N} - \sqrt{\frac{E_s}{M_t}}\mathbf{H}\mathbf{X}^{(j)}\|_F^2 <$ $\|\mathbf{Y} - \sqrt{\frac{E_s}{M_t}}\mathbf{H}\mathbf{X}^{(0)}\|_F^2)\}$, is included in the union of some of the remaining PERs then it can be excluded out of the union or the sum in the right hand side of the inequalities (12) and (13). Therefore, in principle, we can exclude all the terms that do not correspond to the hyperplanes that form the faces of the Voronoi region of the transmitted all zero code-matrice. The method toward the elimination of these terms - let say the redundant terms - out of the Union Bound was derived by Verdu [3] in the case of binary antipodal transmission over the ISI Gaussian channel. A generalization of the Verdu's theorem for coding theory was presented by Biglieri et al. [5], it shows a sufficient condition for excluding a redundant PEP out of the Union Bound. By applying a similar manner as in [5], we propose the method to exclude all the redundant code-matrices out of the codebook to form the irreducible set for calculating the tightened Union Bound of MIMO systems. Given codebook $\Omega = \{\mathbf{X}^{(0)}, ..., \mathbf{X}^{(|\Omega|-1)}\}$ of all the valid code-matrices that can be the outputs of MIMO systems, then there exists the irreducible set which belongs to Ω denoted by $\chi = \{\chi^0, ..., \chi^{p-1}\}; \chi \in \Omega$, where p denotes the size of this set and any $\mathbf{X}^{(k)} \in \Omega$ that satisfies the conditions

$$\begin{cases} \mathbf{X}^{(k)} \notin \chi \\ \mathbf{X}^{(k)} = \mathbf{X}^{(0)} + \alpha_j \sum_{j=0}^{p-1} (\chi^{(j)} - \mathbf{X}^{(0)}), \quad \alpha_j = 0, 1; \quad \chi^{(j)} \in \chi \\ \mathbf{X}^{(0)} \quad : \text{transmitted code-matrice (all zero code-matrice)} \end{cases} \quad (14)$$

or any code-matrice $\mathbf{X}^{(k)}$ can be represented as the sum of some elements of χ, all the elements of χ can not be decomposed further. Then the PEP $P(\mathbf{X}^{(i)} \to \mathbf{X}^{(j)})$ can be expurgated out of the Union Bound defined in inequalities (12) and (13), respectively.

3.2   Tightened Union Bound for MIMO Systems

Consequently, the code-matrices of the codebook except the cardinality of the irreducible set will be excluded. Simply put, the Union Bound in cases of fixed and random channels in the inequalities (12) and (13) can be tightened respectively as follows

$$P(e) \leq \sum_{\chi^{(j)} \in \chi} P\left(\chi^{(j)} \to \mathbf{X}^{(0)} | \mathbf{H}\right) = \sum_{\chi^{(j)} \in \chi} Q\left(\sqrt{\frac{E_s}{M_t} \frac{\|\mathbf{H}(\chi^{(j)} - \mathbf{X}^{(0)})\|_F^2}{2N_0}}\right)$$

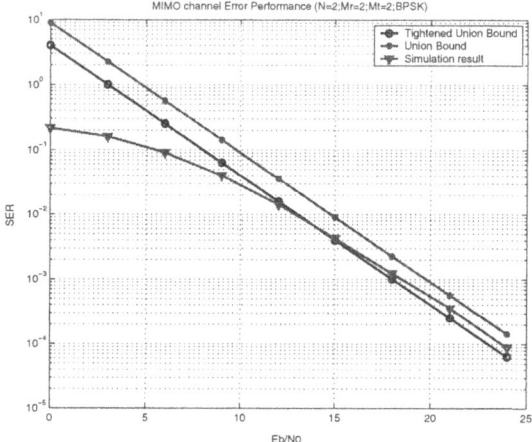

Fig. 2. Simulation result and analytical bounds.

$$= Q\left(\sqrt{\frac{E_s}{M_t}\frac{\left\|\mathbf{H}\chi^{(0)}\right\|_F^2}{2N_0}}\right) + \dots + Q\left(\sqrt{\frac{E_s}{M_t}\frac{\left\|\mathbf{H}\chi^{(p-1)}\right\|_F^2}{2N_0}}\right), \tag{15}$$

and

$$P(e) \leq \sum_{\chi^{(j)} \in \chi} \left[\frac{1}{\det\left(I + \frac{E_s}{4M_t N_0}\left(\chi^{(j)}\right)^H \mathbf{H}^H \mathbf{H}\chi^{(j)}\right)}\right]^{M_r}. \tag{16}$$

It is easy to realize the above bounds yield the tighter approximation in comparison with the conventional Union Bound in the inequalities (12) and (13), respectively. To demonstrate this conclusion, we show the simulation results of MIMO systems ($N = 2$, $M_r = 2$, $M_t = 2$ and BPSK modulation scheme) and the Tightened Union Bound (as inequality (16)) as well as the conventional Union Bound (Union Bound is approximately calculated by the exponential function of inequality (13)) for random channel on Fig. 1. It is clear that the Tightened Union Bound is tighter than the Union Bound. In the area of E_b/N_0 greater than 15dB, the new tighter bound is better than the simulation results due to using of exponential function (as the Chernoff bound)

4   Conclusions

In this paper we studied the BER performance of MIMO systems over the flat fading channels using Union Bound. We also derived the error probabilities of MIMO systems based on ML criteria for both cases of fixed and random MIMO channels. The significant step is that we base on the theorem derived by Verdu [3] and the irreducible set of MIMO code-matrices to calculate the new tighter bound. The bound is not only tighter than Union Bound but also simpler in terms of the computation complexity.

References

1. G. Foschini, "Layered space-time architecture for wireless communications in a fading environment when using multiple antennas", Bell Labs Tech. J., 41-59, 1996.
2. G. Foschini and M. Gans, "On limits of wireless communications in a fading environment when using multiple antennas", Wireless Pers. Comm., 6(3), 311-335, March 1998.
3. S. Verdu, "Maximum likelihood sequence detection for intersymbol interference channel : A new upper bound on error probability," IEEE Trans. Inform. Theory, vol. IT-33, no. 1, pp. 62-68, January. 1987.
4. E. Biglieri et al., "Performance of space-time codes for a large number of antennas", IEEE Trans. on Information Theory, vol. 48, no. 7, pp. 1794 - 1803, July 2002.
5. E. Biglieri, G. Caire, and G. Taricco, "Expurgating the union bound to error probability: A generalizeation of the Verdu-Shields theorem". ISIT 97, Ulm, Germany, p. 373, June 1997
6. V. Tarokh et al., "Space-time block coding for wireless communications: Performance results ", IEEE J. Sel. Areas Comm, vol. 17, no. 5, pp. 451-460, March. 1999.
7. A. Paulraj et al., "Introduction to Space-time wireless communications", Cambridge University Press 2003

Achieving Channel Capacity with Low Complexity RS-BTC Using QPSK over AWGN Channel

Rong Zhou, Annie Picart, Ramesh Pyndiah, and André Goalic

GET-ENST de Bretagne, Département SC, TAMCIC (CNRS-FRE 2658),
Technopôle de Brest Iroise, CS 83818 - 29238 Brest Cedex, France
rong.zhou@enst-bretagne.fr

Abstract. High code rate Block Turbo Codes (BTC) using Bose-Chaudhuri-Hocquenghem (BCH) codes have already demonstrated near Shannon performances for Quadrature Phase-Shift Keying (QPSK) over Additive White Gaussian Noise (AWGN) channel. We show here that reliable transmission can be achieved at less than 1 dB from Shannon limit with very low complexity Reed-Solomon (RS) BTC under the same transmission condition. This is due to a proper choice of RS component codes used to construct RS product codes. Furthermore the size of the coded blocks required for RS-BTC to achieve a given code rate is much smaller than for BCH-BTC which is very attractive for practical considerations.

1 Introduction

The introduction of Convolutional Turbo Codes (CTC) in 1993 [1] has considerably modified our approach to channel coding in the last ten years. The general concept of iterative Soft-Input-Soft-Output (SISO) decoding has been extended to Block Turbo Codes (BTC) [2] and also to Low Density Parity Check (LDPC) [3] codes. BTC are very efficient for high code rate (R>0.7) applications. BTC using concatenated Bose-Chaudhuri-Hocquenghem (BCH) codes can achieve reliable transmission using Quadrature Phase-Shift Keying (QPSK) over AWGN channel at less than 1 dB from Shannon limit with very high decoding speed and reasonable decoding complexity [4]. The main limitation for BCH-BTC is that very large coded blocks (>65,000 bits) are required to achieve high code rates (R>0.9).

Our motivation here was to investigate the performance of non-binary BTC. The most widely used non-binary code is Reed-Solomon (RS) code which has Maximum Distance Seperable (MDS) property. RS-BTC has already been investigated with QPSK modulation over AWGN channel in [5] and the results were declared not as good as BCH-BTC with respect to Shannon limit. After a theoretical analysis, we give here a proper choice of RS component codes used to construct RS product codes. The so constructed RS-BTC can achieve near

J.N. de Souza et al. (Eds.): ICT 2004, LNCS 3124, pp. 163–170, 2004.

Shannon performances with QPSK modulation over AWGN channel using a very low complexity turbo decoder. In addition, the block size is almost three times smaller than BCH-BTC of similar high code rate.

The paper is organized as follows. In Section II we recall the basic concepts of RS product code and turbo decoding. Section III is dedicated to a theoretical performance analysis of product codes based on different BCH and RS component codes. Different RS Product codes are simulated in Section IV to verify the theoretical conclusion in Section III. After taking the simulation results into account, a good trade-off between complexity and performance is obtained. Section V draws some general conclusions and presents some future work on RS-BTC.

2   Iterative Decoding of RS-BTC

The concept of product code introduced by P. Elias [6] is a simple and efficient method to construct powerful codes. Product codes are obtained by means of serial concatenation of two (or more) linear block codes e^1 having parameters (n_1, k_1, δ_1) and e^2 having parameters (n_2, k_2, δ_2). n_i, k_i and δ_i (i=1,2) stand respectively for code length, code dimension and Minimum Hamming Distance (MHD) of each component code. The parameters of the constructed product code P are given by $n = n_1 \times n_2$, $k = k_1 \times k_2$, $\delta = \delta_1 \times \delta_2$. The code rate of P is $R = R_1 \times R_2$ where R_i ($i = 1, 2$) is the code rate of its component code. It is shown [7] that all n_1 rows of the product code are code words of e^2 just as all n_2 columns are code words of e^1 by construction.

RS codes are a subclass of non-binary BCH codes. RS(n, k, δ) is a Maximum Distance Seperable (MDS) code which has the highest code rate for a given error correcting capability t and code length n. RS product codes are constructed from $k_1 \times k_2$ Q-ary information symbols ($k_1 \times k_2 \times q$ data bits) where $Q = 2^q$. Serial concatenation is realised by coding Q-ary symbols along rows and columns successively. Our RS product code P is constructed by using two identical component codes $e^1 = e^2 = RS(n, k, \delta)$. Each element of the Galois Field GF(Q) can be represented by q bits with the help of the Galois field generator polynomial [8]. Thus the so constructed RS product code can be described by a matrix $[E]$ having n rows and $n \times q$ columns after a Q-ary to binary decomposition, as shown in Fig. 1.

The transmission with QPSK modulation over AWGN channel can be regarded as the superposition of two independent Binary Phase-Shift Keying (BPSK) applied respectively on the in-phase and orthogonal carrier. Each bit of the RS product code matrix $[E]$ is associated with a binary symbol according to the mapping rule ($0 - > $ -1, $1 - > $ +1). At channel output, the bit level Log Likelihood Ratio (LLR) is computed and fed to the following turbo decoder.

The turbo decoder is made up of several cascaded Soft-Input-Soft-Output (SISO) decoders. Each SISO decoder is used for decoding rows and columns of the received matrix $[R]$. A SISO decoder can be divided into a Soft-Input-Hard-Output (SIHO) decoder and a reliability estimator of each decoded bit at the output of SIHO decoder. The Chase algorithm [9] is extended to construct the SIHO

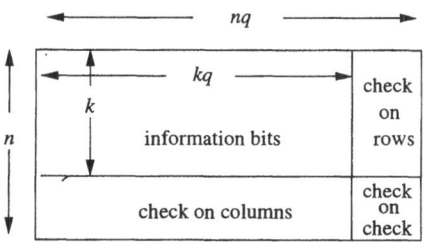

Fig. 1. RS product codes with Q-ary symbol concatenation

decoder for RS(n, k, δ) over GF(Q). Let the vector $R_v = (r_1, r_2, ..., r_i, .., r_{qn})$ be the normalized bit level LLR sequence after soft demodulation for one recieved word and $Y = (y_1, y_2, ..., y_i, .., y_{qn})$ be the hard-decision estimation of the transmitted sequence with the relationship $y_i = 0.5 \times [sign(r_i) + 1]$ $(1 \leq i \leq q \times n)$. First s least reliable positions corresponding to components with the smallest absolute values in vector R_v are determined. Test patterns of length qn of weight 0 to s with the '1' restricted to the least reliable positions are added to the sequence Y to obtain new binary sequences. For each new binary sequence, a corresponding Q-ary sequence $H = (h_1, ..., h_i, ..., h_n)$ is then obtained after the binary to Q-ary conversion. The algebraic RS decoder processes these Q-ary sequences and produces a code word set U containing at most 2^s different code words at its output. Each code word in U has its binary representation after converting every Q-ary symbol into q bits. For each code word, its binary representation is used to calculate its corresponding Euclidean Distance (ED) with respect to R_v. The Maximum-Likelihood (ML) RS code word D which has the minimum ED among all code words in U is selected as the output of SIHO decoder. At each bit position j $(1 \leq j \leq qn)$, the code words in U, whose binary representations have an opposite bit with respect to d_j, will be considered. Among these code words, the code word C which has the minimum ED to R_v is of interest. This code word C is called as the concurrent code word of D. If the search of such a concurrent code word C is successful, the soft output r'_j for bit d_j $(1 \leq j \leq qn)$ is given by [2]:

$$r'_j = (\frac{M^C - M^D}{4}) \times d_j \tag{1}$$

where M^D is ED from D to R_v and M^C is ED from C to R_v. Else we use the predefined or optimised value β and the following relation :

$$r'_j = r_j + \beta \times d_j \tag{2}$$

The difference between the soft output value r'_j and soft input value r_j is the extrinsic information w_j for bit d_j $(1 \leq j \leq qn)$. The extrinsic information w_j as well as the channel output value r_j are used to obtain the input to the next SISO decoder at position j by means of equation (3) :

$$r_j = r_j + \alpha \times w_j \qquad (3)$$

where α is the coefficient used to reduce the influence of extrinsic information in the first iterations.

In practice, a very simple and efficient stopping criterion based on syndrome check is integrated in turbo decoding process to reduce the average number of iterations [10]. If all the rows (or columns) before row (or column) decoding at a given iteration are codewords, then the decoding algorithm has converged and the decoding is stopped. However, if the algorithm has converged to a wrong code word, these errors will be taken into account in the determination of the number of errors in the subsequent iterations. A counter is used to cumulate the syndromes of Y during row (or column) decoding. Iterative decoding is stopped when counter is null at the end of row (or column) decoding. Thus the additional complexity of this stopping criterion is very low.

3   Shannon Limit of BCH and RS Product Codes

Prior to turbo codes, the Asymptotic Coding Gain (ACG) was regarded as the major parameter to evaluate the error correction capability of a code. It is defined as follows [11]:

$$G_a = 10 * log(R\delta) \qquad (4)$$

where R is the code rate and δ is MHD of the code. Although this parameter is important, it has the disadvantage of depending only on the code itself and shows only the asymptotic performance. Shannon limit is a more general parameter. It takes code length, code dimension, target error rate and the used modulation, which is independent of the code [12], into consideration. Thus we shall consider Shannon limit to compare the theoretical performance of different BCH and RS product codes.

Shannon limit of different BCH and RS product codes are given in Table 1. We consider BCH and RS product codes using classical single-error-correcting codes and their extended and expurgated versions as component codes. In order to compare the performance of codes with similar code rate, BCH and RS product codes are constructed over different Galois Fields. Differences in Shannon limit are calculated by taking product code based on classical single-error-correcting component codes as reference.

We observe that when replacing classical RS(t=1) component codes by their extended or expurgated versions, there is a significant reduction in both code rate and Shannon limit at product code level. As for BCH product codes of similar code rates, the reductions of these two parameters are negligeable. Thus the adoption of extended or expurgated BCH codes as component codes to construct product codes is not attractive for RS product code case. This is because the penalty in terms of code rate and Shannon limit is too high in the latter case.

Table 1. Shannon Limit of Different BCH Product Codes

	n^2(bits)	k^2(bits)	R	SH(dB)	$\triangle$(dB)
$BCH(31,26)^2$	961	676	0.7034	2.379	0
$BCH(32,26)^2$	1024	676	0.6602	2.089	-0.29
$BCH(31,25)^2$	961	625	0.6504	2.069	-0.31
$RS(15,13)^2$	900	676	0.7511	2.739	0
$RS(16,13)^2$	1024	676	0.6602	2.089	-0.65
$RS(15,12)^2$	900	576	0.64	2.049	-0.69
$BCH(255,247)^2$	65025	61009	0.9382	4.079	0
$BCH(256,247)^2$	65536	61009	0.9309	3.929	-0.15
$BCH(255,246)^2$	65025	60516	0.9307	3.919	-0.16
$RS(63,61)^2$	23814	22326	0.9375	4.139	0
$RS(64,61)^2$	24576	22326	0.9084	3.569	-0.57
$RS(63,60)^2$	23814	21600	0.9070	3.549	-0.59

Although the ACG of product codes based on modified RS component codes can be increased, the optimum decoding complexity for these codes increases significantly. Let us consider the SISO decoder using Chase-II algorithm. In order to achieve reliable transmission with the extended (or expurgated) RS product code, a much larger number of test patterns will be required. As a result, it will reduce its practical interest. To summarize, the best trade off between complexity and performance is to use single-error-correcting RS component codes in the construction of RS product codes.

4   Simulation Results

Following the above conclusion, four different single-error-correcting classical RS codes based on four different Galois Field $GF(8)$, $GF(16)$, $GF(32)$, $GF(64)$ have been considered for constructing RS product codes. These four product codes are noted as $RS(7,5)^2$, $RS(15,13)^2$, $RS(31,29)^2$ and $RS(63,61)^2$ and are transmitted with QPSK modulation over AWGN channel.

The Bit Error Rate (BER) is evaluated by using Monte Carlo simulation technique for different E_b/N_o values. The number of test patterns in the simulation is limited to 16 in order to keep the turbo decoder at a very low complexity level. These perfromance curves are shown in Fig. 2 for 8 iterations with stopping criterion. The BER is given for 400 matrices having at least one erroneous Q-ary symbol after decoding down to a BER of 10^{-5}. As a comparison, we consider RS product codes based on double-error-correcting RS component codes. These RS product codes are $RS(7,3)^2$, $RS(15,11)^2$, $RS(31,27)^2$ and $RS(63,59)^2$. The same simulation conditions are applied to these codes. These performance curves are shown in Fig. 3. The influence of the test pattern number on the performance of RS-BTC is also evaluated. We increased the number of test patterns from 16 to 32 for all above codes while keeping other simulation conditions unchanged. The curves are also shown in Fig. 2 and Fig. 3.

Table 2. △SH for RS-BTC over AWGN Channel

	16 test patterns	32 test patterns
$RS(7,5)^2$	0.9 dB	0.90 dB
$RS(15,13)^2$	0.8 dB	0.80 dB
$RS(31,29)^2$	0.8 dB	0.75 dB
$RS(63,61)^2$	0.8 dB	0.70 dB
$RS(7,3)^2$	2.05 dB	1.76 dB
$RS(15,11)^2$	1.85 dB	1.52 dB
$RS(31,27)^2$	1.77 dB	1.47 dB
$RS(63,59)^2$	1.64 dB	1.37 dB

The gaps to Shannon limit with 16 and 32 test patterns in the SISO decoder are computed and shown in Table 2. We can see that the improvements are negligible for $RS(t=1)$ based product codes when the number of test patterns is doubled. Thus the number of test patterns equal to 16 can achieve a good trade-off between complexity and performance for these RS-BTC. For $RS(t=2)$ based product codes, there are significant performance improvements when the number of test patterns is increased to 32. But even in this case, the gaps to Shannon limit are still greater than 1.35 dB. We thus need a more sophiscated turbo decoder with a larger number of test patterns to achieve reliable transmission with these codes.

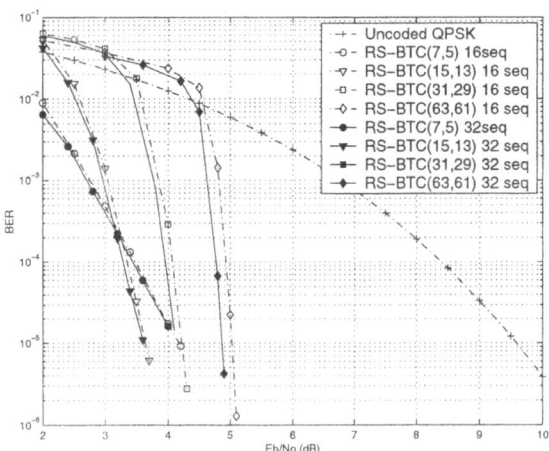

Fig. 2. RS(t=1) based RS-BTC with QPSK over AWGN channel

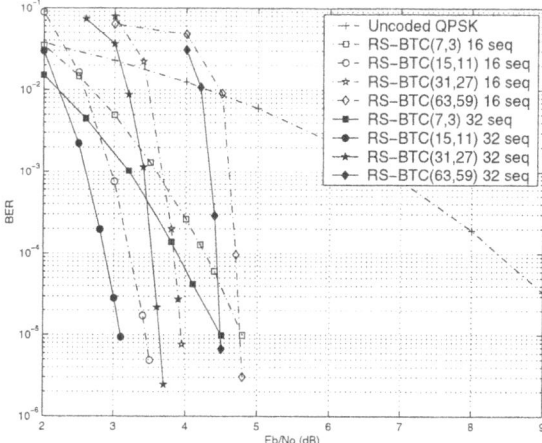

Fig. 3. RS(t=2) based RS-BTC with QPSK over AWGN channel

Table 3. BCH-BTC vs. RS-BTC when using QPSK

	K(bits)	R	$\triangle$(dB)
$BCH(64, 57)^2$	3249	0.7932	0.88
$RS(15, 13)^2$	676	0.7511	0.80
$BCH(128, 120)^2$	14400	0.8789	0.83
$RS(31, 29)^2$	4205	0.8751	0.80
$BCH(256, 239)^2$	57121	0.8716	0.90
$BCH(256, 247)^2$	61009	0.9309	0.80
$RS(63, 61)^2$	22326	0.9375	0.80

The performances of RS-BTC are compared to those of BCH-BTC with similar code rates. Three parameters: block information size K, code rate R and the gap to Shannon limit $\triangle$SH are used as comparison guidelines. They are given in Table 3 for codes with different block size. It shows that both BCH-BTC and RS-BTC can exhibit near Shannon performance with QPSK modulation over AWGN channel. When we consider the codes of similar code rates, RS-BTC have data blocks K in the order of three times smaller than those of BCH-BTC. This observation is valid for codes with different block sizes and code rates.

5   Conclusion

In this paper we show that product codes using single-error-correcting RS component codes can achieve reliable transmission at less than 1 dB from theoretical Shannon limit. Previous contribution [5] on RS-BTC considered the extended (or expurgated) RS component codes which introduce a significant penalty in terms of code rate and require the use of very large number of test patterns in the SISO decoder to achieve optimum performance. We thus propose to use classical RS($t = 1$) code as component code which yields to a product code with MHD equal to 9. With this choice of RS component codes, a very low complexity turbo decoder using only 16 test patterns can guarantee a good trade-off between complexity and performance.

The proposed RS product codes exhibit a much smaller block size than BCH product codes of similar code. As the encoding/decoding delay and memory size in hardware implementation is proportional to the size of data block, this gives a serious advantage to the newly proposed RS-BTC over BCH-BTC for high code rate applications. In addition, the RS encoder/decoder is based on Q-ary symbol processing instead of bit processing. Thus coding/decoding speed can be greatly accelerated. Future work will consider practical applications for these RS-BTC.

References

1. C. Berrou, A. Glavieux and P. Thitimajshima, "Near Shannon limit error-correcting coding and decoding: Turbo-codes",IEEE Int. Conf. on Comm. ICC'93, vol.2/3, May 1993, pp.1064-1071.
2. R. Pyndiah, A. Glavieux, A. Picart and S. Jacq, "Near optimum decoding of product codes", in Proc. of IEEE Globecom '94, vol.1/3, Nov.-Dec.1994, pp.339-343.
3. R.G. Gallagar, "Low Density Parity Check Codes", in IRE Trans. on Information Theory, Jan. 1962, pp.21-28.
4. J. Cuevas, P. Adde, S. Kerouedan and R. Pyndiah, "New architecture for high data rate turbo decoding of product code", Globecom '02, vol.2, Nov. 2001, pp.339-343.
5. O. Aitsab and R. Pyndiah, "Performance of Reed-Solomon block turbo code", Globecom'96, vol.1, Nov. 1972, pp.121-125.
6. P. Elias, "Error-free coding", IRE Trans. on Information Theory, vol. IT-4, Sept. 1954, pp.29-37.
7. F.J. Macwilliams and N.J.A. Sloane, "The theory of error correcting codes", North-Holland publishing company, 1978, pp.567-580.
8. E.R. Berlekamp, "Algebraic coding theory", McGraw-Hill Book Company, New York, 1968
9. D. Chase, "A class of algorithms for decoding block codes with channel measurement information", IEEE Trans. Inform. Theory, vol IT-8, Jan. 1972.
10. R. Pyndiah and P. Adde, "Performance of high code rate BTC for non-traditional applications", 3rd International symposium on turbo codes & related topics, Sep. 2003.
11. G.C. Clark, Jr. and J. Bibb Cain, "Error-Correction coding for digital communications", Plenum Press,New York, 1981.
12. R.G. Gallagar, "Information Theory and Reliable Communication", John Wiley & Sons, New-York 1968.

Design of Sub-session Based Accounting System for Different Service Level of Mobile IP Roaming User

ByungGil Lee, HyunGon Kim, and KyoIl Chung

Electronics and Telecommunications Research Institute
161 Gajeong-dong, Yuseong-gu, Daejeon, KOREA,
{Bglee, Hyungon, kichung}@etri.re.kr,
http://www.etri.re.kr

Abstract. An authentication, authorization and accounting (AAA) system is one of the most important components for internet service providers (ISPs) and wireless service providers (WSPs). For QoS management of accounting data in a Diameter-based AAA system, the current paper proposes an updated accounting protocol based on a sub-session state machine, instead of the original (RFC) stateless accounting framework on the server side. As such, the proposed design and implementation strategy can effectively apply the Diameter AAA system to a Mobile IP roaming user with different service levels. For co-operation between authentication, the authorization state machine, and the base protocol state machine on the AAA client side, the current paper proposes an updated accounting state machine (current standard is not clearly defined). For the interface between the Mobile IP protocol part and the AAA client of the accounting state machine, an new designed sub-session based accounting state machine is also proposed on the client side and a sub-session-based scenario demonstrated using the accounting protocol. This paper enhance new IP based accounting techniques to support soft-guaranteed QoS in a wireless Internet architecture.

1 Introduction

With the rapid growth of access network technologies and dramatic increase in subscribers, Internet service providers(ISPs) and communication service vendors are facing some difficult challenges in providing and managing network access security. To provide better services, vendors must be able to verify and keep account of mobile service subscribers' service levels. Furthermore, vendors need to be able to measure the connection time to the network for billing and resource planning purposes. One solution that meets these requirements is authentication, authorization, and accounting (AAA).

In wired network, needs of the QoS for multimedia service has been increased and algorithms of prioritizing among packet classes and delivering time sensitive packets in existing bandwidth has been studied[1].

J.N. de Souza et al. (Eds.): ICT 2004, LNCS 3124, pp. 171–182, 2004.

In addition, QoS(Quality of Service) via wireless networks is expected to become a crucial part of the mobile service provider (MSP)'s service area[2]. As an AAA protocol, IETF AAA WG was developed to overcome the shortcomings and deficiencies of the RADIUS protocol. Distinct from RADIUS, Diameter adopts a base protocol that is a kind of engine that works with Diameter applications. However, the Diameter base and application protocol [3-4], as defined by current accounting standards, which constitute a fully operable protocol suite in a real environment, still have some implementation problems for accounting that need to be solved and require enhancements for various wireless applications. For co-operation between authentication and authorization in the AAA framework, i.e. real-time accounting transfers should adopt a state machine based accounting protocol. In addition, the accounting mechanism of Diameter should able to specify a service quality for maintaining and transporting policy-based accounting records until the mobile user disconnects[4].

Thus, the current paper proposes and implements an accounting protocol that has an new sub-session based accounting state machine on both the server side and the client side. The remainder of this paper consists of a brief introduction to the AAA protocol in Chapter 2, the architecture of the proposed AAA in Chapter 3, and some final conclusions in Chapter 4.

2   Diameter AAA Protocol

The Diameter's Mobile IP application allows an AAA server to authenticate, authorize, and collect accounting information for a Mobile IP service rendered to a mobile node. The major differences between Diameter and RADIUS include:

- Peer-to-peer nature
- Explicit support for intermediaries
- Extensibility
- Built-in failover support
- Larger attribute space
- Bit to indicate mandatory status of data
- Application-layer ACKs and error messages
- Unsolicited server messages
- Peer discovery
- Negotiation capabilities

The AAA infrastructure verifies the user's credentials and provides a service policy to the serving network for which the user is authorized. The AAA infrastructure can also provide a reconciliation of charges between the serving and home domains. In the model in Fig. 1, the AAA server authenticates, i.e. authorizes the mobile node(once its identity has been established) to use Mobile IP and certain specific services in a foreign network, and takes account of the usage information.

The basic concept in the Diameter AAA standard is to provide a base protocol framework that can be extended to provide AAA application services to

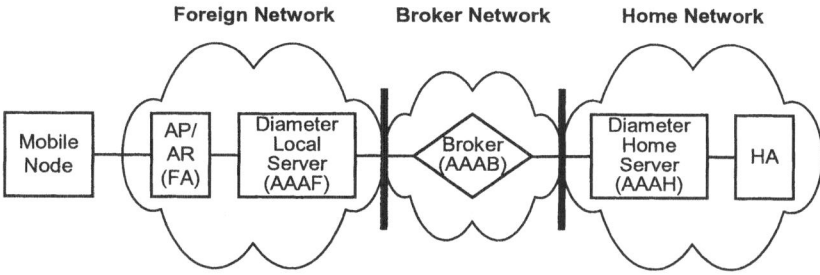

Fig. 1. AAA with Mobile IP Model

various access networks. The base protocol provides a common functionality to a set of applications, such as the capability for negotiation, delivery of command codes(Messages), peer link management and fault recovery(Watch dog), user session management, basic accounting functions, and others. The application protocols of the AAA standard have been built with massive scaling, flexibility of comprehensive attributes, and incorporated applicable level accounting support for each application service. The application of NASREQ(EAP), Mobile IPv4, Mobile IPv6, and IP-based multimedia and others is shown in Fig. 2.

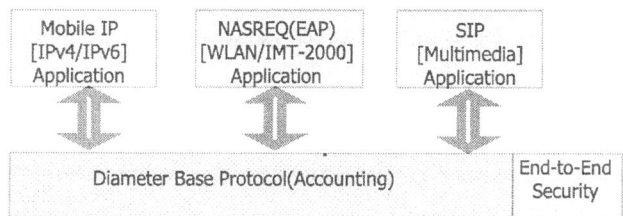

Fig. 2. AAA Application and Protocol

The Foreign(serving) authentication server(AAAF) requests proof from the external home authentication server(AAAH) that the external mobile node has acceptable credentials for the Mobile IP registration process. The Mobile IP registration message(RRQ) from the mobile node is sent by FA through the AAAF(and through the AAAB) to the AAAH server. The FA translates the Mobile IP registration protocol into the Diameter protocol and transports the message to the AAA server. If the user is able to access the required network, the Diameter Client sends an Acct start request message to the home Diameter Server. On receipt of the Acct start request message, the home Diameter Server answers the request via an Acct start answer message. As an option, during an Acct session, the Diameter Client can send snapshot accounting data in an interim record to the home Diameter Server. There is a disparate session identifier between a Foreign Agent and the home Diameter Server, and between the home Diameter Server and a Foreign Agent[3]. When the home Diameter Server that performs the Auth service identifies a new session, it creates an Accounting-Multi-Session-Id. Yet, when the Home Agent decides that the new session is caused by the movement of a mobile node(Foreign Agent Handoff), it overwrites the value of the Accounting-Multi-Session-Id received from home the Diameter Server in a Home-Agent-MIP-Request to the old value used before the movement. The Home Agent then sends the changed Accounting-Multi-Session-Id in a Home-Agent-MIP-Answer to the home Diameter Server, which sends the Accounting-Multi-Session-Id to the Foreign Agent in an AA-Mobile-Node-Answer. The AAA servers must perform authentication and authorization during the initial Mobile IP registration step. Thereafter, accounting as a next step must performed. An all accounting message must including the Accounting-Multi-Session-Id and the AVP is used to merge the accounting data from multiple accounting sessions.

3   Proposed Accounting Protocol for Mobile IP Service User

3.1   Protocol Descriptions

This section introduces the redesigned sub-session based accounting protocol and the test environment for specified accounting in a Mobile IP service based on QoS. In the Diameter accounting protocol, the accounting protocol is based on a server-directed model with capabilities for the real-time delivery of accounting information. One of the requirements for the AAA service is to support a detailed accounting service for the Mobile IP. The accounting function must maintain and transport the session-based interim and realtime accounting records until the mobile user disconnects[3-4]. Interim accounting provides protection against the loss of session summary data by providing checkpoint information that can be used to reconstruct the session record in the event that the session summary information is lost.

Therefore accounting is a crucial factor in the protocol design and implementation of the AAA. As such, the following accounting functions can be placed on the AAA service:

- Real-time Accounting
- Accounting-related services(Prepaid cards, Postpaid cards) Support (Billing Applications)
- End-to-End Security
- QoS or flow-based Accounting using sub-session scheme
- Package or usage-based accounting
- Fault resilience accounting
- Periodic transporting protocol based on accounting interim interval

Basically diameter protocol provide two different types of services to applications. The first involves authentication and authorization, and can optionally make use of accounting. The second only makes use of accounting. So, accounting state machine was none in the first case and independent accounting state machine was defined for second case.

In this paper, we defined co-operating scenario and state machine for flexible and coupled accounting. Fig. 3 describes the co-operating scenario for authentication, authorization and accounting in a case initiated state of an authentication session connection, where an open state and connecting accounting protocol is established and comes to the accounting state machine. This scenario is a connection scenario with a Mobile IP application process. The state machine starts when it receives a request from a user application. The Diameter Client sends an Auth request to the home Diameter Server. On receipt of the request, the home Diameter Server responds to the Diameter Client. Then an Auth session is started, and the state is Pending. If a successful service-specific authentication and authorization answer is received with a non zero auth-lifetime, the next state is Open. To obtain interaction between the authentication and authorization and the accounting part, the proposed accounting architecture must support the same session-based connection and a procedure for exchanging the state information and session information for a series of Authentication, Authorization and Accounting (AAA) services. The accounting state machine can also reach to the Open state of authentication and authorization.

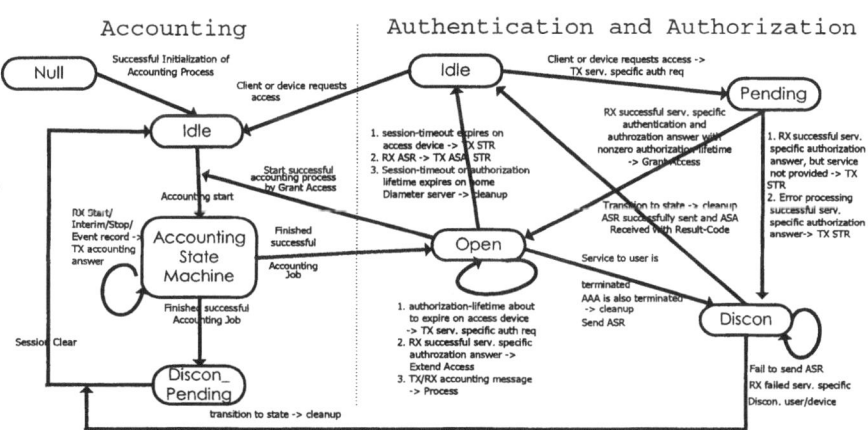

Fig. 3. Co-operative State Machine

In this paper, we present a new accounting mechanism of AAA providing mobile networks. In standard, QoS or different service level based accounting is not clearly defined up to date. So, another new sub-session mechanism can be used for specific service of different service level based on user session. We introduce another some bit for different service level in sub-session ID field. The fields of sub-session ID Avp are consist of 3 part, Diameter Avp head, Service level and Session ID.

$$< \quad Sub \quad - \quad SessionIDAvp \quad > \quad ::= < \quad Diameter \quad - \quad Avp \quad - \quad Header \quad :$$
$$Code, Flag, length >$$
$$< Sub - Session - Id >$$
$$< QoS - Code >$$

The QoS-Code field in Sub-session ID avp is type unsigned64 and contains the accounting sub-session's QoS code. It is used for the different service level as user select. If a specified interim interval exists, the Diameter client must produce additional records between the Start Record and Stop Record, Interim Record and an sub-Start Record for sub-session, sub-Stop Record for sub-session and sub-Interim Record for sub-session, and the AAA server must store all kinds of marked sub session records periodically for different service levels, as shown in Fig. 4. The subsequent procedures of session-based accounting and sub-session-based accounting are shown in Fig. 4.

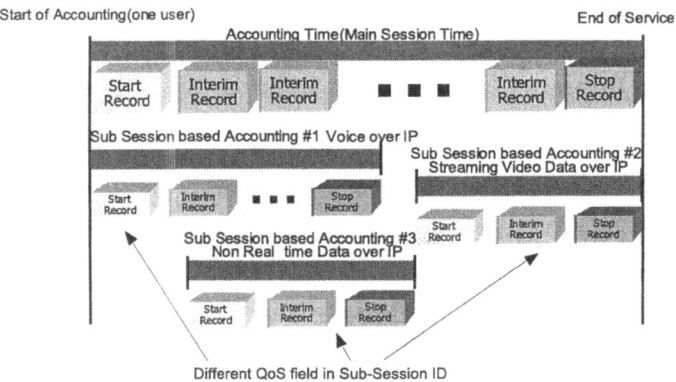

Fig. 4. Accounting process for different service level

Figs. 5 and 6 show the re-design of the client state machine and server state machine based on the original AAA protocol that is not clearly defined. Fig 5 shows the design of a session-based accounting state machine as well as a sub-session-based accounting state machine. A sub-session represents a distinctly different service that has a different level, plus it can be included in a given session. These services can occur concurrently or serially.

As shown in Fig. 6, the Diameter Client sends an Accounting raw data request message to the Mobile IP protocol part. The Mobile IP protocol part responds(sends counted volume of the used traffic data of the user), then the procedure of making the ACR Accounting packet is started. While performing the user application service, the ACR accounting messages continue to be sent. When a session is moved to the Idle state, any resources that were allocated in authentication, authorization, and accounting for the particular session must be released. The states PendingS, PendingI, PendingL, PendingE and PendingB stand for pending states to wait for an answer to an accounting request related to a Start, Interim, Stop, Event or buffered record, respectively.

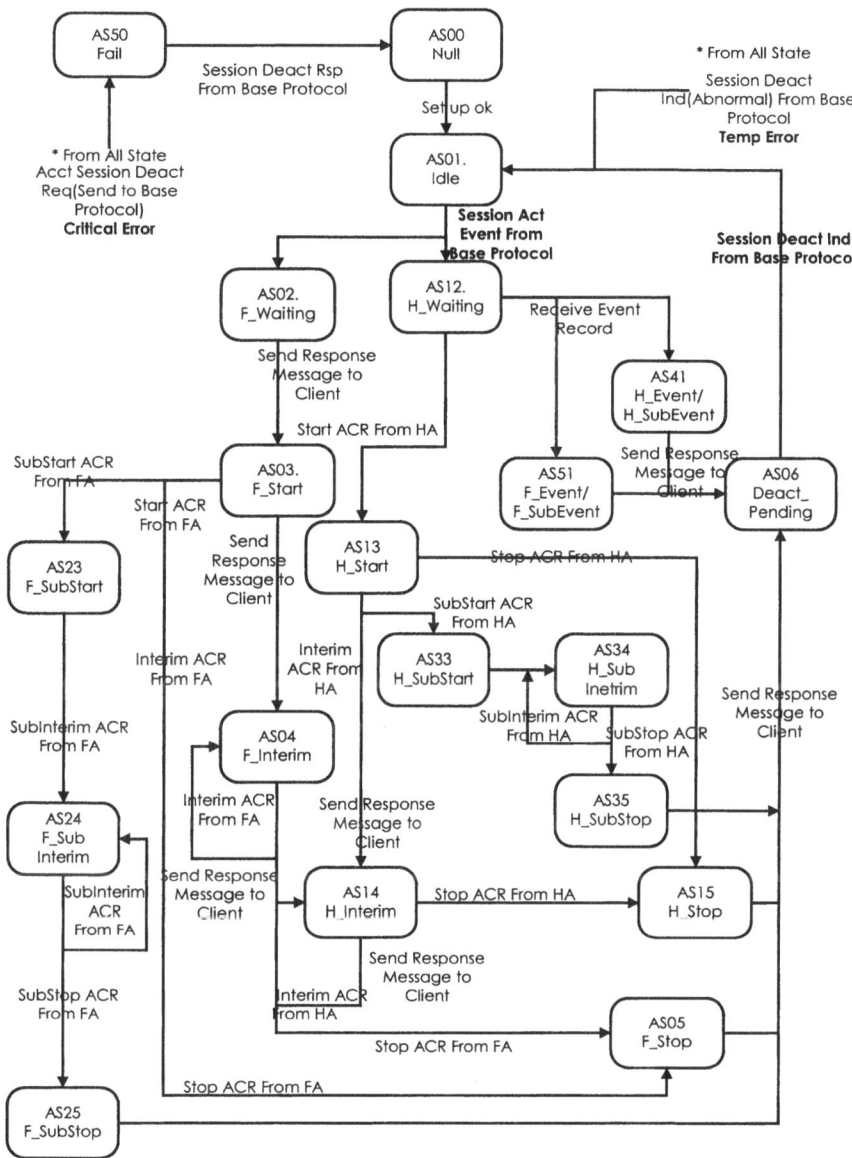

Fig. 5. Designed State Machine of AAA Accounting(Server)

3.2 System Architecture for Implementation

Our system model can be applied to the wireless environments including 3G like CDMA-2000 and UMTS, and 802.1x networks. As shown in Fig. 7, AAA in mobile environment is based on a set of client and server. The implementation protocol stacks of AAA servers are shown in Fig. 7. To address the privacy problem in wireless mobile computing, we introduce TLS protocol and to endure

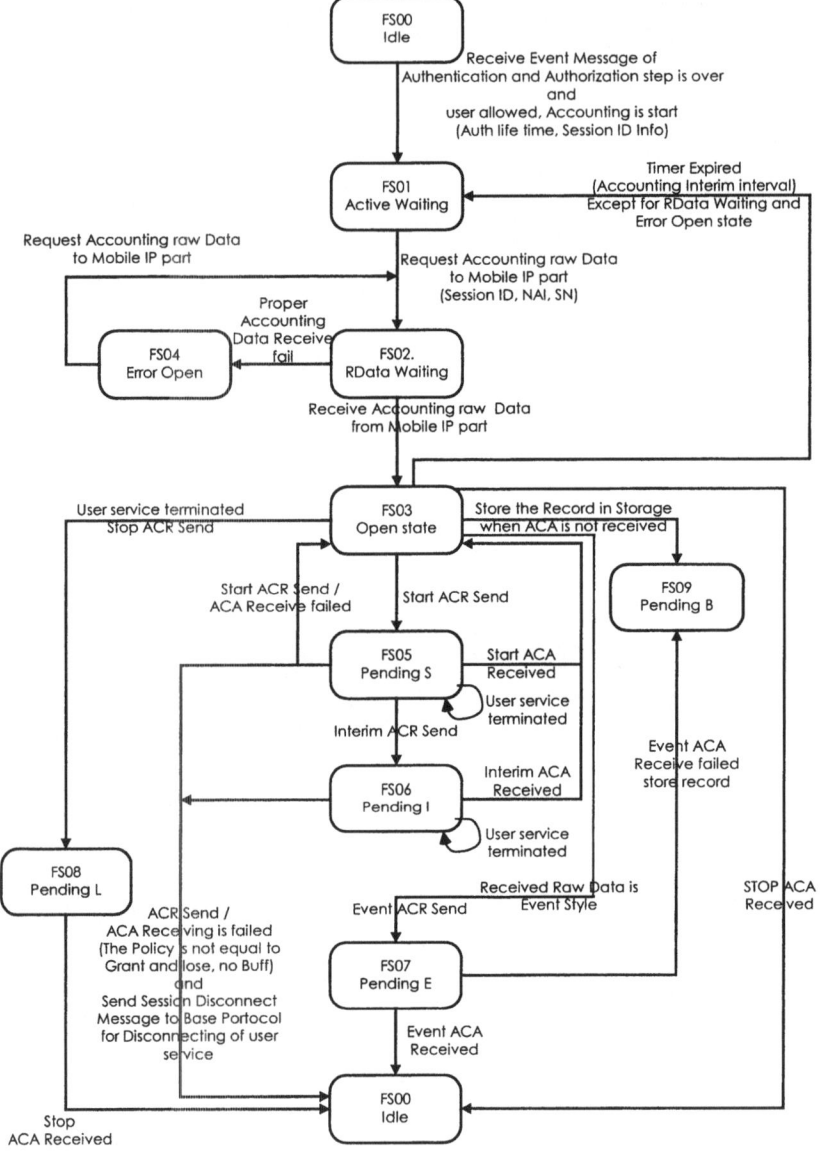

Fig. 6. Updated State Machine of AAA Accounting(FA/HA): sub-session state is omitted for want of space

the robust communication also adapt SCTP transport protocol in low layer protocol section.

Table 1 shows system description of the test environment of Diameter based AAA for Mobile IP. Fig. 8 shows the real implementation environment of the MIP-based AAA System.

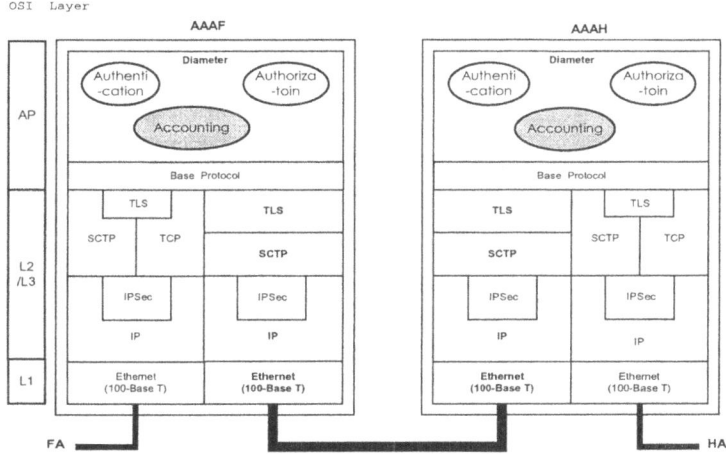

Fig. 7. Protocol Stacks of AAA servers for Implementation

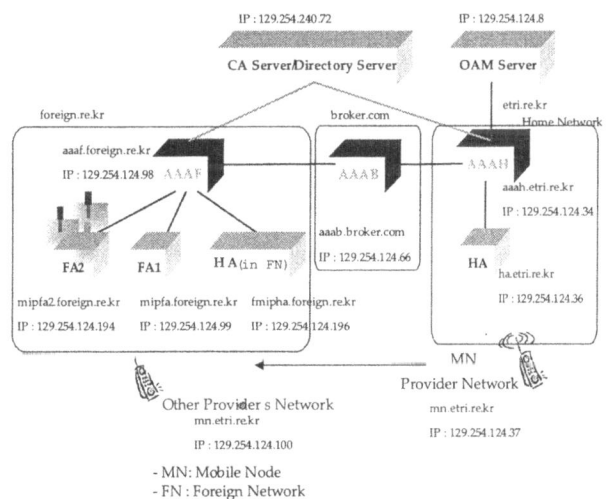

- MN: Mobile Node
- FN : Foreign Network

Fig. 8. Implementation Environment

Table 1. System Configuration

System	Description
AAA Server and Client platform	ZION LINUX SYSTEM(2CPU, Intel Zeon 700MHz)
AAA Server and Client OS	Red Hot Linux 6.2(Kernel v. 2.2.19)
Mobile Node platform	Embedded Linux for MN

4   Results of Implementation

The proposed accounting system has five sub modules (blocks) used for storing accounting records, processing accounting messages, controlling sessions(sub-

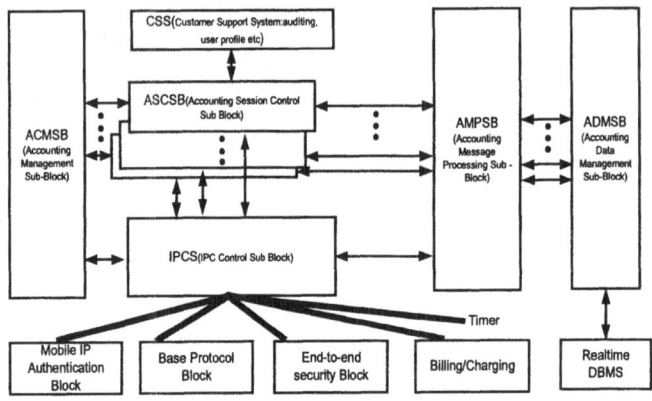

Fig. 9. Block level design of the Accounting protocol

sessions), and monitoring in/out inter-block messages, as shown in Fig. 9. To address the QoS specified accounting in wireless mobile computing, we introduce QoS code point table to transport and to endure the QoS based inter-network communication also adapt accounting mechanism of different service level for different administration-domain.

Following shows information description of the QoS based accounting of Diameter based AAA for Mobile IP in inter-network communication.

- Service Type
- Sub-Session-ID
- QoS-Level of Providing Service
- Accounting Information for Specified Service

We measured the time(processing time) spent for the authentication step and the time spent for the accounting step at AAAH, and the CPU load average of sub-session based accounting. Clearly, the accounting took more time than duration of the authentication and authorization and 3 sub-session based accounting from FA and HA has more sever system load than simple accounting's. Table 2 also shows the specified processing time of the authentication, authorization and accounting in AAAH. Table 3 also shows the averaged count of the CPU load in sub-sessions based accounting and simple accounting. The value of the CPU load average means averaged waiting job count in Linux based system.

Table 2. Comparison of processing-time in Authentication, Authorization and Authentication

	Authentication and Authorization	Accounting
Processing Time in AAAH(sec)	1.10 - 1.36	3.7 - 4.6

Table 3. Comparison of System Load in the Sub-session based Accounting

	Simple Accounting	Sub-Session based Accounting
Load Average	3.7 - 4.7	6.5 - 8.53

Proposing this solution for AAA and QoS Mobility Management issues, this study aims to contribute to the technical innovation of the 3G systems by exploiting the potential of IP-based wireless mobile multimedia networking in evolution towards 4G systems The sub-session-based accounting had a significant effect on the accounting system load. If lots of sub-sessions exist, the Diameter client must produce more additional records between the Start Record and Stop Record, Interim Record for sub-session, and the AAA server must endure sever system load for different service levels. To support a robust AAA network for a nationwide ISP(Internet Service Provider), a cautious accounting server-model design would seem to be very important. Furthermore, the accounting process must be robust to the system load, if QoS-based accounting or flow-based accounting is needed. So, we should solve this problem as a batch accounting method for QoS-based and flow-based accounting.

1. Initially, a parent accounting session is established
2. User want to start multimedia service with different service level
3. QoS-based inner sub-session is also established as marked different service level
4. Each accounting Data is counted for specified processing
5. Each accounting Request Message is send to peer
6. The sub-session is over as user's multimedia service is finished
7. Parent session is also finished

The authentication and authorization functions only occur during the network access time, then the system load of AAA (Authentication, Authorization, and Accounting) usually determines the accounting process thereafter. Thus, management of the accounting function allows a Diameter server, which maintains a robust AAA session, to provide a stable service. And in this research, we can state that everything was designed to take a more advanced approach in QoS base accounting. From a macro prospective, we get confirmation of some divided sub-session using level of different services, so, a QoS accounting method.

5   Conclusion

The current paper proposed and implemented a new accounting protocol for different service levels in a Diameter-based AAA system. We also showed the newly designed sub-session based accounting protocol. For different service level, we have introduced a QoS bit in sub-session ID and state machine with sub-session. In addition, we also tested accounting system with sub-session for Mobile IP user and presented the result that showed robust AAA system can perform the QoS based Accounting service due to system load. And we can state that our research was designed to take a more advanced approach in QoS base accounting.

References

1. S. I. Choi and J. D. Huh., "Dynamic Bandwidth Allocation Algorithm for Multimedia Services over Ethernet PONs," ETRI Journal, Volume 24, Number 6, Dec. 2002
2. T. B. Zahariadis, K. G. Vaxevanakis, C. P. Tsantilas and et al., "Global Roaming in Next-Generation Networks," IEEE Communication Magazine, Feb. 2002
3. P. R. Calhoun, J. Arkko, and E. Guttman, "Diameter Base Protocol," RFC 3588, Sep. 2003
4. P. R. Calhoun, and C. E. Perkins, "Diameter Mobile IPv4 Application," draft-ietf-aaa-diameter-mobileip-14.txt, http://www.ietf.org, Oct. 2003. IETF work in progress.

Priority Telephony System with Pricing Alternatives

Saravut Yaipairoj and Fotios Harmantzis

Stevens Institute of Technology, Castle Point on the Hudson, Hoboken, NJ 07030, USA
{syaipair, fharmant}@stevens.edu

Abstract. Dynamic pricing schemes in telecommunication networks were traditionally employed to create users' incentives in such a way that the overall utilization is improved and profits are maximized. However, such schemes create frustration to users, since there is no guarantee that they would get services at the anticipated prices. In this paper, we propose a pricing scheme for priority telephony systems that provides alternatives to users. Users can choose between a) a dynamic price scheme that provides a superior quality of service or b) a fixed low price with acceptable performance degradation. Our results verify that the proposed pricing scheme improves the overall system utilization and yet guarantees users' satisfaction.

1 Introduction

In long-haul communications, network resources are critical commodities that require an efficient allocation mechanism to users. In recent years, researchers have been focusing on resource allocation schemes, so that network resources can be utilized efficiently and total profit from resource usage is maximized. However, it is well known that network users act independently and sometimes "selfishly", regardless of the current network traffic conditions. Therefore, even with advanced resource allocation schemes in place, it is hard to avoid congestion. As a result, congestion reduces the total system utilization. Mechanisms that give users incentives to behave in ways that improve the overall utilization and performance of the network are needed. In commercial networks, pricing had been proved an effective mean to resolve the problem of scarce resource allocation.

Network users are inherently price sensitive. Via prices, the network could send signals to the users, providing them with incentives that influence their behavior [1]. Pricing thus becomes an effective mean to perform traffic management and congestion control. Such schemes are known as *dynamic pricing schemes*. In a dynamic pricing scheme, call prices change as demand fluctuates [2]. It rises in accord with demand, deterring additional users from accessing the network or holding network resources for long periods during congestion time. Therefore, such schemes create users' incentive for efficient network utilization. In addition, during the off-peak hours, the price drops from its nominal level; this will serve as an incentive to generate more traffic to an otherwise under-utilized network [3].

J.N. de Souza et al. (Eds.): ICT 2004, LNCS 3124, pp. 183–191, 2004.

However, despite the beneficiary of dynamic pricing, it has major drawbacks. Dynamic pricing schemes create frustration to users. Since price fluctuates according to demand, there is no guarantee to final charges. Users with low price expectations would risk being blocked, during congestion periods.

In this paper, we propose a middle ground where users have choices in the way they are priced: they can either accept a) a dynamic pricing scheme, where prices changes according to the system congestion levels, or b) a fixed pricing scheme, where the provider charges a low price, with users experiencing an acceptable performance degradation.

The paper is organized as follows: In Section 2, we describe the traditional dynamic pricing used in a Priority Telephony System. In Section 3, we present our priority queuing system, where the appropriate parameters are defined. Section 4 shows numerical results and how our proposed pricing scheme can improve the call admission control mechanism of the network. Discussion on the results is also taking place in that section. In Section 5, we draw the conclusion of our work.

2 Dynamic Pricing in Priority Telephony System

In telephony networks, whenever congestion occurs, the incoming telephone calls can either be blocked from the system or placed into a buffer (queue), waiting to be served whenever the telephone trunks are free. In the latter case, the Quality of Service (QoS) metric used for measuring the performance of the system is the delay that users experience in the queue. The shorter the time users spend in the queue, the better for them.

In queuing networks, users experience delay according to their priority agreement with the system, which can be described by a Priority Queuing Model. In a priority queuing system, users who require more attention are distinguished from those who can endure the quality of conventional services. Usually, the QoS required by priority users is higher and therefore should be served faster than the average (conventional) users. The price charged to priority users is clearly higher.

Currently, the service charge for telephone users is either fixed per call or flat. One of the advantages of these schemes is the simple billing and accounting processes [4, 5]. However, since users act independently and sometimes in a "selfish" manner, they utilize the system regardless of its traffic condition. Such pricing schemes do not provide incentives for users to avoid congestion during peak hours and cannot react effectively to the dynamics of the network. With dynamic pricing schemes, prices change depending on the network conditions. Users who require access to the network during peak hours and are able to afford higher prices, will be admitted to the network, while users who are not able to afford such prices are blocked. However, we argue that blocked users during congestion time (even though their pricing requirements for being prioritized are not met) would result in reality to highly dissatisfied users. By using a queue to delay, instead of block, call requests during time of congestion, it is likely that users would be more satisfied and it can potentially yield to a better network operation.

3 Model for Priority Queue with Priority Call Admission Control

Call admission control (CAC) is widely used as an effective mean to prevent overloading in telecommunication networks. According to our model, during system congestion time, admitted calls are required to meet a certain pricing requirement. In this context, a new type of CAC is introduced here, namely, the *Priority Call Admission control* (PCAC). PCAC's main function is to control the amount of incoming calls, based on users' priorities that are regulated by pricing criteria.

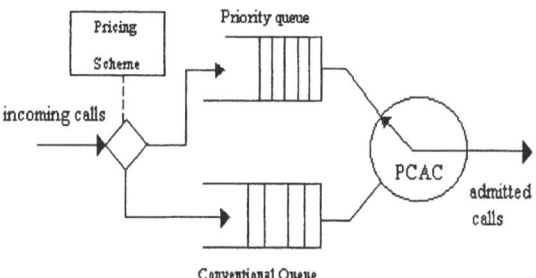

Fig. 1. Priority Queue with Pricing Alternatives

As shown in Fig 1, the system consists of two types of queues: one for the priority users and the other for the conventional ones. Both queues store incoming call requests and feed them to the telephone trunk group. The key elements in this system model are two functional blocks: the pricing block and the PCAC block. The pricing block acts as a price broadcasting point to incoming users. In this particular system, users who refuse to pay a higher price are placed in the queue for conventional users and wait until get served. Priority users who meet the higher price requirement, are placed into the priority queue, which would be served faster than the conventional one. The above procedure will only take place when the system experiences congestion. If the congestion level is not met, all calls will be placed in the conventional queue and served as soon as the system is ready.

The PCAC block can be characterized as a QoS controller that allocates system resources according to call requests coming from both queues. Since priority users require more attention than conventional users, the priority queue is served by the PCAC block in such a way that certain QoS is met. At the same time, conventional users are also served by PCAC with a QoS that is obviously inferior to the priority users. The objective of this system is to adjust system resources in such a way that we can meet the QoS constraints of both queues and maximize the number of calls being served by the system.

During peak hours, users who attempt to access the system will find themselves facing two choices: One is to accept a high price according to dynamic pricing theory, as they will enjoy the higher QoS of priority callers. The second choice is to deny the high price and be charged by a fixed low pricing scheme. As a result, in the second case, users will experience longer delays before being served, depending on the existing traffic conditions. The call procedure of the system can be described as follows:

Call Procedure

1.   Users dial in numbers and wait for a system response.
2.   The current status of the system is identified. If the system is not congested, the call requests will be placed in the conventional queue waiting for available trunks.
3.   If the system is congested, the system will notify users the approximated time they have to wait for service. Then, it announces the price for those users who consider priority status and ask for their choice (be prioritized or stay on the line).
4.   If the answer is positive, the users' requests for call are placed in the priority queue where users are served with superior QoS.
5.   For those who stay on the line, their call requests will be placed in conventional queues.

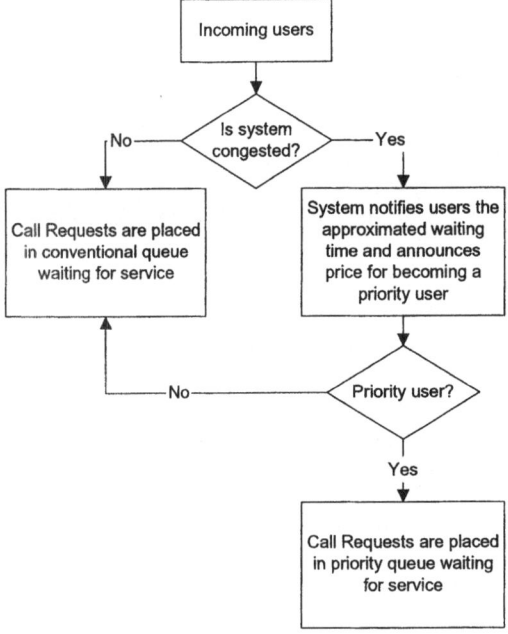

Fig. 2. Call procedure in the proposed pricing model

3.1 Directing Traffic

An important parameter used by the PCAC block is the *priority factor* (P_s), i.e., the portion of total resources that is assigned to the priority queue. With a higher P_s, more resources would be dedicated to priority queues. P_s can be adjusted regularly based on the users' QoS constraints and the incoming traffic for both user types. We assume that the average holding time of priority users is shorter than that of the conventional users. Since they are willing to pay extra, it is unlikely to spend much time in the system. Both types of users share the same resources (telephone trunks in this case). Therefore, the average holding time that the system experiences from both call types would be the weighted sum of the holding time by the priority users and conventional users, i.e.,

$$T_{avg} = P_s T_{avg\_p} + (1 - P_s) T_{avg\_c} \tag{1}$$

where T_{avg_p} is the average holding time of priority users and T_{avg_c} is the average holding time of conventional users. The P_s parameter controls the amount of network resource assigned to priority users; the remaining resources are assigned to conventional users. With this model in mind, we can assume that the trunk group is logically divided into two groups. One group is assigned for priority users and the other group is assigned for conventional users. Both of them can be studied independently. Hence, the number of trunks assigned to priority users is $N_p = P_s \cdot N$ whereas the number of trunks assigned to conventional users is $N_c = (1 - P_s) \cdot N$, with N being the total number of telephone trunks.

3.2 Dynamic Prices

The price that is broadcasted to users when the system experiences congestion, can be derived from the demand function. The demand function describes the users' reaction to the price changes. We use the demand function that appears in [7] since it is used for different priority users, which fits our model. The demand function is as follows:

$$q = e^{-(\frac{P_h}{P_0} - 1)^2} \qquad p_h \geq p_0 \tag{2}$$

where p_0 is the price charged to conventional users, p_h is the price charged to priority users, and q is the percentage of priority users who are willing to pay this higher price. From (2),

$$p_h = p_0 + \frac{p_0 \sqrt{-4\ln(q)}}{2} \tag{3}$$

The percentage of incoming users (q) gives us information regarding the number of users who are placed in each queue. We assume that the performance of each type of queue can be considered independently. The model consists of two basic queuing models, with average holding times given by (1), and arrivals dictated by the demand function (2) at a given price. Basically, these basic queuing systems form a M/M/m system which can be studied by the Erlang-C formula, i.e.

$$C^{-1}(N, a) = B^{-1}(N, a) - B^{-1}(N-1, a) \tag{4}$$

where
$$B(N, a) = \frac{\dfrac{a^N}{N!}}{\displaystyle\sum_{i=0}^{N} \dfrac{a^i}{i!}} \qquad \text{(Erlang-B formula)}$$

The QoS can be characterized by the user delay experienced in the queues. More specifically, by the tail of a delay distribution, i.e. P[user delay > R seconds] is less than a QoS requirement, e.g., 1%. Therefore, using the Erlang-C formula and the fact that we can consider both queues independently, we can identify the QoS requirement as follows:

$$P[W > t_p] = C(N_p, a_p)e^{-N_p\mu(1-\rho_p)t_p} \tag{5}$$

$$P[W > t_c] = C(N_c, a_c)e^{-N_c\mu(1-\rho_c)t_c} \tag{6}$$

where $a_p = q \cdot a_{total}$, $\rho_p = a_p / N_p$
$a_c = (1-q) \cdot a_{total}$, $\rho_c = a_c / N_c$
$a_{total} = \lambda(t) \cdot T_{avg}$

$C(N,a)$ is given by the erlang-C formula, W is the user delay (time in queue), $a_{()}$ is the load   imposed by each type of users, $N_{()}$ is the number of trunks logically assigned to each type of users, μ is average departure rate of users (1/Tavg), $\rho_{()}$ is the load per server for each type of users, and t_p and t_c are delay constraints for the priority and conventional queues respectively . We assume that t_p should be a lot less than t_c, when the system experiences congestion. Utilization of the overall system is given by

$$Utilization = \frac{\lambda(t) * T_{avg}}{N} \tag{7}$$

where $\lambda(t)$ is the arrival rate in the telephone system at time t.

3.3 Optimal Call Arrival Rate

As the system operates, the system resources are shared in a way that the QoS requirements for each user type can be achieved. An important parameter here is the maximum number of users that the network can accommodate. The number of users need to conform to the QoS constraints of both queues. This parameter is influenced by the optimal call arrival rate (λ_{opt}), which is the maximized overall arrival rate of the system. We know that λ_{opt} is embedded in (5) and (6). For different percentages of priority users (q) and priority factors (P_s), we can achieve a certain arrival rate. The λ_{opt} can be obtained when we find that arrival rate that maximizes the utilization of the system.

To obtain λ_{opt}, we need to consider equations (5) and (6). The QoS constraint in (5) and (6) can be set at a certain probability level, depending upon the user requirement. λ_{opt} can be found by setting probabilities in (5) and (6) as the QoS constraint (1% in this case). Here, we obtain the maximum arrival rate (λ) numerically, by changing P_s and q. Therefore, for a certain value of q, we can find that P_s that yields maximum call arrival rate or λ_{opt}. That is

$$\lambda_{opt} \text{ for certain } (q) = f(P_s^*, q) \tag{8}$$

where P_s^* satisfies the condition $\dfrac{df(P_s, q)}{dP_s} = 0$

4 Performance Analysis

In section 4.1, we describe the basic assumptions and parameters used in the priority queuing system. The results of our analysis are shown in section 4.2.

4.1 Assumptions and Parameters

We assume for the shake of simplicity that the considering network is a circuit-switched telephone network. The arrivals are modeled using the Poisson law (exponentially distributed inter-arrival times). The system queues are first-come first-serve (FCFS). The parameters used throughout our analysis are as follows:

1. The number of telephone trunks equal 30. Trunks assigned to each queue are regulated by the parameter P_s.
2. The average call holding time for priority users (T_{avg_p}) and conventional users (T_{avg_c}) are exponentially distributed with mean 120 seconds and 300 seconds respectively.
3. Regarding the QoS parameters:
    a. Probability of priority users kept waiting in the queue for more than 1 minutes (t_p) is less than 1%; and
    b. Probability of conventional users kept waiting in the queue for more than 10 minutes (t_c) is less than 1%.
4. The normal charging rate for conventional users using the trunks (P_o) is 8 cents per minutes. The charging rate for priority users (P_h) depends on the demand function and it is broadcasted upon arrival.

4.2. Numerical Results

Figure 3 shows the relationship between the arrival rate and priority factor (P_s). For a certain percentage of priority users, there is apparently an optimal call arrival rate (the peak of the curve) that maximizes the number of calls and still maintains a QoS requirement of both queues. As q increases, the optimal call arrival rate is increased. Figure 3 also shows the improvement in call accommodation. Without our pricing scheme, the optimal arrival rate is 5.6 calls/sec. When our scheme is used, the optimal arrival rate increases to 8.5 calls/sec, with a q of 80%. We can achieve a higher optimal call arrival rate, by degrading the QoS requirements of either type of users.

In terms of utilization, Figure 4 shows the utilization of the system. Apparently, the utilization of the system is roughly the same regardless of q. This is because the utilization of the system depends only on arrival rate $\lambda(t)$ and priority factor (P_s). Therefore, we are able to achieve high utilization of the system with minimum

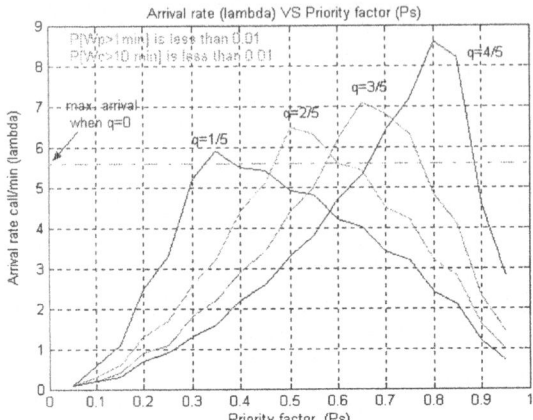

Fig. 3. Optimal arrival rate for certain percentage of priority users

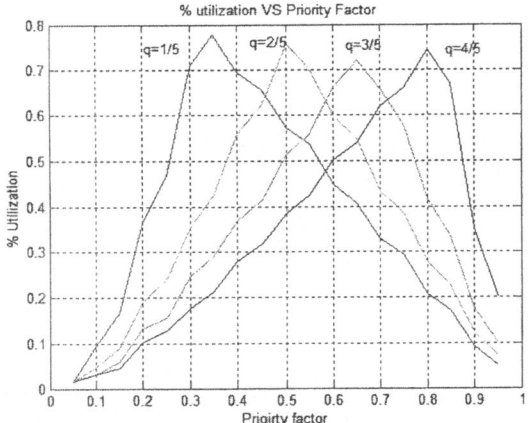

Fig. 4. The total utilization VS priority factor and percentage of priority users

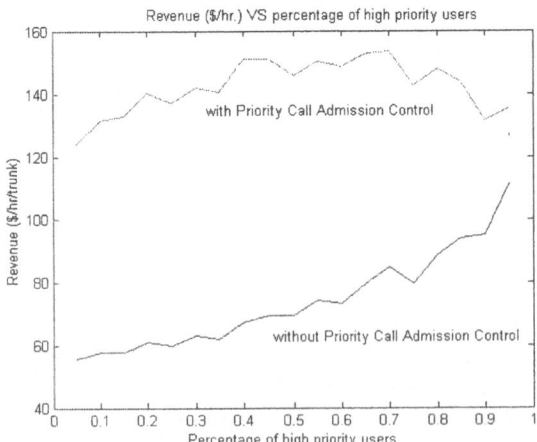

Fig. 5. Revenue with Priority Call Admission Control ($ per hour per trunk)

amount of user who is willing to be priority users. In addition, without the proposed pricing scheme, the utilization required to achieve optimal call arrival rate is 93% (from (7)). However, The system utilizes only 75% of its resources with our pricing scheme.

Figure 5 compares the revenue generated from each trunk by using the proposed scheme vs. the revenue under the traditional fixed pricing scheme, for the same amount of traffic. The revenue stream consists of the sum of the revenue created by the priority users and conventional users with their respective price factor. We observe that there is a significant revenue increase due to our way of pricing. However, the effect of the percentage of priority users (q) to the revenue is not significant. Therefore, we can operate at a level of low percentage of priority users, and still generate higher revenue.

5 Conclusion

Our proposed pricing model provides incentive for users to use system resource more efficiently. Furthermore, users will be satisfied with the fact that they can choose their pricing schemes based on their expected quality of service. The proposed system is also flexible enough to adapt to the fluctuating traffic by adjusting the pricing factor and the users' QoS requirements. In addition, our pricing model is general enough and has been proposed for voice services in mobile networks [8].

References

1. Falkner, M.: A user's perspective on Connection Admission Control: Integrating Traffic Shaping, Effective Bandwidths and Pricing. Doctoral thesis at Carleton University, Ottawa, Canada, May 12, 2000.
2. Fitkov-Norris, E.D., Khanifar, A.: Dynamic Pricing in Mobile Communication Systems. In: First International Conference on 3G Mobile Communication Technologies, 2000, 416 –420
3. MacKie-Mason, J.K., Varian, H.R.: Pricing Congestible Network Resources. In: IEEE Journal on Selected Areas in Communications,Vol. 13, Issue. 7, (1995), 1141 –1149
4. Viterbo, E., Chiasserini, C.F.: Dynamic Pricing for connection-oriented services in wireless networks. In: 12th IEEE International Symposium on Personal, Indoor and Mobile Radio Communications (2001), Vol.1, A-68 -A-72
5. Patek, S.D., Campos-Nanez, E.: Pricing of dialup services: an example of congestion-dependent pricing in the Internet. In: Proc. of the 39th IEEE Conference on Decision and Control (2000), Vol.3, 2296 -2301
6. Odlyzko, A.M.: Paris Metro Pricing for the Internet. In: Proc. ACM Conference on Electronic Commerce (1999),140-147
7. Hou, J., Yang, J., Papavassiliou, S.: Integration of Pricing with Call Admission Control for Wireless Networks. In: 54th IEEE Vehicular Technology Conference (2001), Vol. 3, 1344 -1348
8. Yaipairoj, S., Harmantzis, F.: Dynamic Pricing with "Alternatives" for Mobile Networks. In: IEEE Wireless Communication and Networking Conference (2004)
9. Peha, J.M.: Dynamic Pricing as Congestion Control in ATM Networks. In: Global Telecommunications Conference (1997), Vol.3, 1367-137

Mobile Telephony Industry: Price Discrimination Strategies for Interconnected Networks

Livio Cricelli[1], Francesca Di Pillo[2], Nathan Levialdi[2], and Massimo Gastaldi[3]

[1] Dipartimento di Meccanica, Strutture, Ambiente e Territorio, Università degli Studi di Cassino, Via G. Di Biasio 43 - 03043 Cassino (FR)
Tel. +39 0776 2993481 – Fax +39 0776 2993711 – cricelli@unicas.it
[2] Dipartimento di Ingegneria dell'Impresa, Università degli Studi di Roma "Tor Vergata", Via del Politecnico 1 - 00133 Roma
Tel. +39 06 72597359 – Fax +39 06 72597305
{dipillo, levialdi}@disp.uniroma2.it
[3] Dipartimento di Ingegneria Elettrica, Università degli Studi di L'Aquila, Monteluco di Roio - 67100 L'Aquila
Tel. +39 0862 434411 – Fax +36 0862 434403 –gastaldi@ing.univaq.it

Abstract. The changes occurring in the mobile telephony industry, both at legislative and technological level, have affected the competitive scenario in various national contexts. The main aim of this paper is to highlights the results of price discrimination strategy in terms of market shares and profits in order to identify incumbent behaviours able to inhibit competition.

1 Introduction

The increasing volume of mobile phone calls and the relative growing number of carriers are transforming mobile telecommunications from a traditional monopoly into a market with growing competitive pressure [1]. In our paper we analyse the relationship between market share and profit resulting from a strategy of price discrimination. In particular, the policy of price discrimination in the mobile telephone market is based on the possibility for the carriers to set different prices for calls originating on the carrier's own network, finish on the same network (on net) or on the network of the rival carrier (off net) [2], [3], [4]. The carriers price discrimination strategy has the aim to induce customer migration towards their own networks. This goal assumes a relevant role in the Italian mobile telecommunications industry, where the penetration rate is very high and very close to a context of full participation of customers. In particular, it is possible to observe two phases: the first in which the market was in full expansion and whereby carriers had the goal of attracting customers that still had not decided to participate in the market. The second phase, in which the market is saturated and the competitive strategy is based on attracting customers from rival networks. The decision by customers to migrate on other carriers is based on two considerations. The first one is related to the effective price level: customers search to migrate towards those carriers that have adopted a system of lower prices. The second

J.N. de Souza et al. (Eds.): ICT 2004, LNCS 3124, pp. 192–198, 2004.
© Springer-Verlag Berlin Heidelberg 2004

refers to the relative dimensions of market shares. For example, even if a carrier fixes its their on net price relatively low, but has a market share of small dimension, migration would not therefore be convenient for customers that would have a very low probability of carrying out calls on net. The focus of this paper is the analysis, with respect to the Italian market, of the convenience by mobile telephone carriers to choose a price discrimination strategy. The goal of such analysis is that of verifying if the price discrimination strategy causes a different impact in terms of profits and market shares according to relative dimensions, improving a previous analysis [5]. In particular, an extension to three carriers and interconnection between mobile and fixed is here presented. The paper is organized as follows. In the next section, we describe the competitive scenario in which operate three mobile carriers, through the analisys of traffic flow and market variables (market shares and profits). In the third section we analyse the convenience of carriers to adopt a price discrimination strategy, in terms of profits and market shares. Finally, in the last section, we present the main conclusions of our work.

2 The Model

In order to represent the Italian mobile telephony market, we consider only three operators, i, j, k with the followings traffic flows:
   1   Mobile-mobile on net. 2 Mobile-mobile off net. 3 Mobile-fixed.
The net surplus w_i of generic customer, belonging to the network i, is given by:

$$w_i = \left[u - p^i_{on-net} S_i q\Delta - p^i_{off-net}(1 - S_i)q\Delta - p^i_{mf} q(1 - \Delta) \right] \tag{1}$$

and the market shares, obtained through an extension of Hotelling model, have the following expressions:

$$S_i = \frac{1 + 2\beta + \alpha}{3} + \frac{2\sigma}{3}\left(2w_i - w_j - w_k\right); \; S_j = \frac{1 - \beta + \alpha}{3} + \frac{2\sigma}{3}\left(2w_j - w_i - w_k\right) \tag{2}$$

$$S_k = \frac{1 - \beta - 2\alpha}{3} + \frac{2\sigma}{3}\left(2w_k - w_i - w_j\right)$$

where:
- u is the utility function of subscriber;
- P_{on-net} is the price of on-net call;
- $P_{off-net}$ is the price of off-net call;
- P_{mf} is the price to connect mobile with fixed network;
- Δ is the share of calls that terminate on mobile network;
- q is the quantity of calls per customer q = Q / N where Q is the total calls volume and N is the number of customers.

The parameters β and α measure the degree of asymmetry among networks, in terms of brand loyalty. Particularly, β refers to the market leader and depends on the brand loyalty; α is related to the follower j and it depends on the competitive advantage towards the carrier k. The variable σ is the degree of substitutability among the networks, and therefore the switching cost of customer. When β and α are equal to 0 and σ is equal to 1, operators have no brand loyalty and there are not customer barriers to exit. In this case, market shares depend only on the prices. The call volume is split in on net and off net volumes as a function of market shares. The use of the market share as a proxy of termination calls probability is suitable when the aspect of the relative dimensions of operators represents a key variable of the analysis. Other modelling approaches [6] [7] split the calls traffic in on net and off net as a function of a generic probability since the focus of the analysis is not the relative dimensions among carriers. In fact, in such models, the market shares of operators are considered symmetric. For the mobile-mobile termination is necessary to distinguish among on-net and off-net calls. Particularly, the marginal cost of off net calls includes the interconnection charge, the price a carrier must pay to destination carrier to terminate its call. The interconnection charge of off-net traffic, denoted with t, in the current settlement agreements assumes a common value for the three competitors. The termination cost of the mobile-fix traffic, the parameter f of the model, is imposed by the Italian Communications Authority. Moreover c_o is the marginal cost for originating the call, c_t is the marginal cost for terminating the call, T is the interconnection charge paid by fixed operators to terminate the call on mobile network and Q_f is the total fixed to mobile calls volume. Following the above assumptions, the profit function of carrier i is given by:

$$\Pi_i = \Delta Q\, S_i S_i\, (\, p^i_{on\text{-}net} - c_o^i - c_t^i\,) + \Delta Q\, S_i S_j\, (\, p^i_{off\text{-}net} - c_o^i - t\,) + \Delta Q\, S_i S_k\, (\, p^i_{off\text{-}net} - c_o^i - t$$
$$) + \Delta\, Q\, S_i S_j\, (\, t - c_t^i\,) + \Delta\, Q\, S_i S_k\, (\, t - c_t^i\,) + (\, 1 - \Delta\,)\, Q\, S_i (\, p^i_{mf} - c_o^i - f\,) + Q_f\, S_i\, (\, T_i - c_t^i$$
$$) \qquad \text{with } i \neq j \neq k \qquad\qquad (3)$$

- $\Delta Q\, S_i S_i\, (\, p^i_{on\text{-}net} - c_o^i - c_t^i\,)$ is the profit deriving from the on-net calls;
- $\Delta Q\, S_i S_j\, (\, p^i_{off\text{-}net} - c_o^i - t\,)$ and $\Delta Q\, S_i S_k\, (\, p^i_{off\text{-}net} - c_o^i - t\,)$ are the profits deriving from the off-net calls;
- $\Delta Q\, S_i S_j\, (\, t - c_t^i\,)$ and $\Delta Q\, S_i S_k\, (\, t - c_t^i\,)$ are the profits deriving from the interconnection charge paid from carriers j and k;
- $(\, 1 - \Delta\,)\, Q\, S_i (\, p^i_{mf} - c_o^i - f\,)$ is the profit originated by mobile-fix calls;
- $Q_f\, S_i\, (\, T_i - c_t^i\,)$ is the profit deriving from the traffic volume originated on fixed network.

3 Simulations

The starting simulation point is represented by the current carriers prices. In order to analyse the different impact of a price discrimination strategy, a price variation is simulated. Particularly, the aim of the analysis is to verify if a price discrimination strategy causes an improvement on profit or on market share or on both, with respect to the relative dimensions of carriers. The bold lines of tables 1-2 are the incumbent starting simulation point, the former monopolist, that fixes a discriminatory prices, with 49% of the market. The price reduction of on net calls from the current level causes an increase of the leader market share as shown in table 1. Such increase is greater in comparison to that generated by the price reduction of off net calls, as we can observe in table 2. By decreasing the price of on net calls, the leader increases its own market share.

Table 1. Carrier market shares and profits variations with leader price discrimination strategy.

P^j_{on-net} €- cent	S_i	S_j	S_k	Π_i(000.000€)	Π_j(000.000€)	Π_k(000.000€)
10	57%	30%	13%	1.207	791	349
11	55%	31%	14%	1.217	808	369
12	54%	31,7%	14,3%	1.223	825	389
13	52%	32,5%	15,5%	1.225	841	409
14	50,6%	33,3%	16,1%	1.224	858	429
15	**49%**	**34%**	**17%**	**1.223**	**877**	**452**
16	47,4%	34,8%	17,8%	1.211	890	469
17	46%	35,6%	18,4%	1.200	906	489
18	44,4%	36,4%	19,2%	1.186	921	509
19	43%	37%	20%	1.169	937	528
20	41%	38%	21%	1.149	952	547

Table 2. Carrier market shares and profits variations related to carrier i off-net prices.

$P^j_{off-net}$ €- cent	S_i	S_j	S_k	Π_i(000.000€)	Π_j(000.000€)	Π_k(000.000€)
40	51%	33%	16%	1.232	855	425
41	51%	33%	16%	1.230	859	430
42	50%	33%	16%	1.229	864	436
43	50%	34%	16%	1.227	868	441
44	49%	34%	17%	1.225	873	446
45	**49%**	**34%**	**17%**	**1.223**	**877**	**452**
46	49%	34%	17%	1.221	881	457
47	48%	35%	17%	1.219	886	463
48	48%	35%	18%	1.217	890	468
49	47%	35%	18%	1.215	895	473
50	47%	35%	18%	1.213	899	479

In fact, already having an elevated catchment area, the leader can capture many customers from the other operators, attracted by the great probability that the phone call originates and terminates on the same network. Notice that the impact of the discrimination strategy carried out by the leader is more intensive for the smallest carrier in terms of market share (ΔS_k = -28.6%), whose customers have the lowest brand loyalty. The follower j also suffers a market share loss (ΔS_j = -11.8%). An efficient strategy for the operator j is to reduce the price of off net calls, as we can observe in the tables 3 and 4. The reduction of the price of off net calls causes an increase of the market share greater than that in obtained by reducing the on net price. Therefore, for the follower j is not convenient to choose a discrimination strategy as a function of the network destination. Moreover, the profit increase is greater in the case of off net

Table 3. Carrier market shares and profits variations with carrier j price discrimination strategy.

$P^j_{on\text{-}net}$ €-cent	S_i	S_j	S_k	$\Pi_i(000.000€)$	$\Pi_j(000.000€)$	$\Pi_k(000.000€)$
10	46%	41%	13%	1.168	937	363
11	46,2%	39,8%	14%	1.178	928	378
12	46,7%	38,7%	14,6%	1.187	920	393
13	47,3%	37,5%	15,2%	1.196	911	408
14	47,9%	36,4%	15,7%	1.206	900	423
15	48,4%	35,2%	16,3%	1.214	888	437
16	**49%**	**34%**	**17%**	**1.223**	**877**	**452**
17	49,6%	32,9%	17,5%	1.232	858	466
18	50,2%	31,8%	18%	1.241	841	481
19	50,8%	30,6%	18,6%	1.249	822	495
20	51,3%	29,5%	19,2%	1.257	802	510

Table 4. Carrier market shares and profits variations related to carrier j off-net prices.

$P^j_{off\text{-}net}$ €-cent	S_i	S_j	S_k	$\Pi_i(000.000€)$	$\Pi_j(000.000€)$	$\Pi_k(000.000€)$
35	43,7%	44,8%	11,5%	1.132	983	312
36	44,8%	42,6%	12,6%	1.150	965	340
37	45,8%	40,5%	13,7%	1.168	945	368
38	46,9%	38,4%	14,7%	1.186	923	395
39	50%	36,2%	15,8%	1.203	900	422
40	**49%**	**34%**	**17%**	**1.223**	**877**	**452**
41	50%	32%	18%	1.235	846	476
42	51,2%	29,8%	19%	1.251	815	503
43	52,2%	27,7%	20,1%	1.266	781	529
44	53,3%	25,5%	21,2%	1.281	744	555
45	54,4%	23,4%	22,2%	1.295	704	582

Table 5. Carrier market shares and profits variations with carrier k price discrimination strategy.

$P^k_{on\text{-}net}$ €-cent	S_i	S_j	S_k	$\Pi_i(000.000€)$	$\Pi_j(000.000€)$	$\Pi_k(000.000€)$
15	47,6%	32,7%	19,7%	1.197	845	505
16	47,9%	33%	19,1%	1.202	851	494
17	48,2%	33,2%	18,6%	1.206	857	484
18	48,5%	33,5%	18%	1.211	862	473
19	48,7%	33,8%	17,5%	1.215	868	461
20	**49%**	**34%**	**17%**	**1.223**	**877**	**452**
21	49,3%	34,4%	16,3%	1.225	880	437
22	49,6%	34,6%	15,7%	1.228	885	425
23	49,9%	34,9%	15,2%	1.232	891	412
24	50,1%	35,2%	14,7%	1.236	897	399
25	50,4%	35,5%	14,1%	1.240	902	386

Table 6. Carrier market shares and profits variations related to carrier k off-net prices.

$P^k_{off-net}$ €-cent	S_i	S_j	S_k	$\Pi_i(000.000€)$	$\Pi_j(000.000€)$	$\Pi_k(000.000€)$
30	42,2%	27,3%	30,5%	1.106	727	699
31	43,6%	28,6%	27,8%	1.130	758	655
32	44,9%	30%	25,1%	1.153	788	608
33	46,3%	31,4%	22,3%	1.176	817	559
34	47,6%	32,7%	19,6%	1.198	846	506
35	**49%**	**34%**	**17%**	**1.223**	**877**	**452**
36	50,4%	35,4%	14,2%	1.240	901	389
37	51,8%	36,8%	11,4%	1.259	929	324
38	53,1%	38,2%	8,7%	1.278	955	255
39	54,5%	39,5%	6%	1.296	981	181
40	55,9%	40,9%	3,2%	1.314	1.006	101

prices reduction then the on net one. This result is due to the great sensibility of network j subscribers to the variations in off-net price. In fact, the carrier j customers send the off net traffic volume higher than on net one. Carrier j off net prices reduction impact is particularly intensive on carrier k (the smallest one); in fact the carrier j market share increase is due to the consumers migrating from carrier k. On the contrary, this strategy has a low impact on carrier i because of its high level of brand loyalty. In tables 5 and 6 is shown that this strategy is absolutely not profitable for the smallest carrier k; his best competitive behaviour is to set the same price for off-net and on-net calls not differentiating at all.

4 Conclusions

The analysis of competition of mobile telecommunications market shows that the price discrimination strategy may have a different impact on market shares and profits depending on the carrier relative market dimensions. In fact, for the leader, the discriminating prices choice with respect to the destination network, has the aim to increase it own market share; in such way, as a consequence on its own market power, it imposes exit barriers for consumers and entry barrier for the potential new entrant. Thus, this price discrimination strategy represent a threat for the other carriers, especially for the smallest ones. For these reasons, the intervention of Regulation Authority is particularly requested. The price discrimination strategy is inefficient for followers, both in terms of profits and in terms of market shares.

References

1. Cricelli, L., Gastaldi, M. and N. Levialdi. 2002. "The impact of Competition in the Italian Mobile Telecommunications Market". *Networks & Spatial Economics 2*, 239-253.

2. Laffont, J-J. and J.Tirole. 1998. "Network Competition: II. Discriminatory Pricing". *Rand Journal of Economics 29*, 38-56.
3. Wright J. 2000. "Competition and termination in cellular networks." *Review of Industrial Organization 17*, 30-45.
4. Carter, M. and J. Wright. 1999. "Interconnection in Network Industries." *Review of Industrial Organization 14*, 1-25.
5. Cricelli, L., Di Pillo, F., Ferragine, C. and N. Levialdi. 2003. "Simulation in the Mobile Telephony Market: Price Discrimination Strategy Under Asymmetrical Condition" *Proceedings of SCSC 2003*, 671-676.
6. Dessein W. 2003. "Network competition in nonlinear pricing". *RAND Journal of Economics 34*, 1-19.
7. Gabrielsen, T. and S. Vagstad. 2002. "Why is on-net traffic cheaper than off-net traffic?" *mimeo*, University of Bergen.

A Novel ECN-Based Congestion Control and Avoidance Algorithm with Forecasting and Verifying*

Hui-shan Liu, Ke Xu, and Ming-wie Xu

Department of Computer Science and Technology, Tsinghua University,
Beijing 100084, China
{liuhs, xuke, xmw}@csnet1.cs.tsinghua.edu.cn

Abstract. The congestion control and avoidance algorithms can be divided into source and link algorithms. The source algorithms are based on ' pipe' model; which includes TCP Tahoe, TCP Reno, TCP NewReno, SACK and TCP Vegas. In this paper, we present a novel congestion control source algorithm-FAV(Forecast And Verify), which measures packets delay to forecast whether there will be congestion in network, and uses ECN to verify whether the forecast is precise or not. Experiments are performed by ns2 to compare FAV with TCP Reno and TCP Vegas, and the results imply that FAV algorithm can decrease loss rate and link delay efficiently, and keep link utilization high. We think FAV can control the congestion better than other source algorithms.

1 Introduction

Controlling network traffic and decreasing congestion are the efficient methods to increase network performance. The early TCP congestion control and avoidance algorithms [1],[2] regard the network as 'black box'. The data source increases the traffic by enlarging the sending window size continually till some packets are dropped by router, and the source uses this method to judge congestion. This type of algorithm has to suffer relatively high end-to-end delay and loss rate.

ECN[3] is presented to notify congestion explicitly. It uses the CE bit defined in TOS field of IP packet head to imply that congestion has taken place in the network, instead of using packet dropping simply. However, because of link delay, ECN can't notify currently network congestion condition.

In order to perceive congestion earlier and control sending rate efficiently, we have designed an algorithm to notify congestion ahead in which destination uses forecast mechanism to judge the queue length change trend of congestion node.

The paper is organized as follows. Section II introduces related works. Section III describes FAV algorithm. Section IV presents our experiments using ns2 [4] and result analysis. At last, we draw a conclusion and introduce future works.

---

* This work was supported by the Natural Science Foundation of China (No 90104002, 60203025, 60303006, 60373010) and National 973 Project Fund of China (No 2003CB314801).

J.N. de Souza et al. (Eds.): ICT 2004, LNCS 3124, pp. 199–206, 2004.

2   Related Works

Wang and Crowcroft's DUAL algorithm [5] is based on reacting to this increase of the round-trip delay. The congestion window normally increases as in Reno, but every two round-trip delays the algorithm checks to see if current RTT is larger than the average of maximum and minimum RTTs. If it is, then the algorithm decreases the congestion window by one-eighth.

Jain's CARD (Congestion Avoidance using Round-trip Delay) algorithm [6] is based on an analytic derivation of an optimum window size for a deterministic network.

Some of other algorithms are based on the flattening of the sending rate. In Wang and Crowcroft's Tri-S scheme[7], they increase the window size by one segment and compare the throughput with the last one. If the difference is less than one-half of the throughput achieved when the first segment is in transit, they decrease the window by one segment.

TCP Vegas algorithm [8],[9] is a better approach with good performance. It is based on controlling the amount of extra data, not only on dropped segments. Vegas does the judgment at the source, so that the judgment based on the measurement of RTT can't imply the congestion link direction correctly.

3   FAV Algorithm

3.1   Basic Model

Let us suppose that it is DropTail queue in gateways of network. The source A sends packets by average rate λ, and the length of packets is L_p. The middle of the network is the congestion node B, and C is the destination. We suppose that the forwarding rate of node B is R_f, and let R_f be μ (normally R_f can be looked as a static value). (see Fig 1).

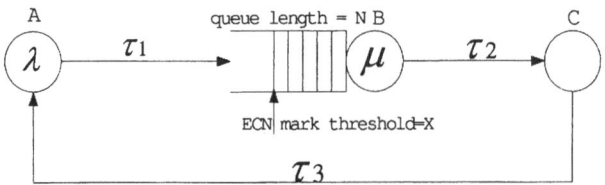

Fig. 1. Basic model

In this model, we suppose the congestion takes place at bottleneck node B, and the rest of the network will never be congested. $\tau 1$ and $\tau 2$ are link delay, and $\tau 3$ is ACK packets delay. The maximum of node B queue length is N. After the queue length exceeds threshold X, the packets will be marked CE. In this paper, we use several terms: the **queue length of packet k** means the queue length of congestion node when the packet k enters queue. **Average queue length of forecast group** means the average queue length of the forecast group.

3.2 Ingroup Forecast Formula

Because we suppose $R_f = \mu$ (it is a static variable), the forward time of a packet is $\frac{L_p}{R_f}$. The queue length when the packet enters the queue is in direct proportion to the queuing delay in the congestion node.

At the same time, the clock of the source shows Cs and the clock of the destination shows Cr; The time-stamp of packet is Ts; The time at which the destination receives the packet is T_r; When the packet n arrives at congestion node B, the length of queue in node B is ℓ_n; The length of every packet is L_p; If the clock of source and destination is completely synchronous, the packets that have been sent from node A will be delayed by Td_n and arrive at C.

$$Td_n = \tau 1 + \tau 2 + \ell_n \times \frac{L_p}{R_f}$$

Because the clock of A and C may not be synchronous, the delay result we measure is

$$Td'_n = Td_n + \zeta = \tau 1 + \tau 2 + \ell_n \times \frac{L_p}{R_f} + \zeta \quad (\zeta = Cr - Cs)$$

In order to eliminate the clock error ζ and link delay $\tau 1$ and $\tau 2$, we make difference with nearby packets delay.

$$Td'_n - Td'_{n-1} = (\ell_n - \ell_{n-1}) \times \frac{L_p}{R_f}, \text{ therefore,}$$

we can get recursion to dispel parameters R_f and L_p.

$$K_n = \frac{Td'_n - Td'_{n-1}}{Td'_{n-1} - Td'_{n-2}} \text{ and } \ell_n - \ell_{n-1} = K_n \times (\ell_{n-1} - \ell_{n-2})$$

We define the forecast formula within group as:

$$\ell_n = (K_n + 1) \times \ell_{n-1} - K_n \times \ell_{n-2} \quad (\ell_0 = 0, \ell_1 = 1)$$

We use the formula to forecast the queue length of congestion node when the packet n enters the queue.

3.3 Intergroup Forecast Formula

To judge whether the average queue length of next forecast group will be over threshold X, we calculate the average queue length of forecast group n by this formula.

$$E[\ell]_n = \frac{\sum_{k=1}^{S} \ell_k}{S}$$, S is the number of packets received within forecast group n, and

ℓ_k is the forecast queue length of packet n. We define the intergroup forecast formula as $E[\ell]_{n+1} = E[\ell]_n$. It is based on the calculation of average queue length of current group to forecast the network traffic within next RTT.

3.4 FAV Algorithm

FAV algorithm runs in the destination and controls the source-sending rate by marking ACK. The goal of FAV algorithm is to keep average queue length within a forecast group nearby the ECN mark threshold, but we can't know the threshold X. So we use $\overline{X}$ to approximate X.

When the destination receives data packets, it computes K_n and ℓ_n by measuring Td'_n. If $\overline{X}$ is larger than ℓ_n and P_{CE}, we let $\overline{X}$ be ℓ_{n-1}. If $\overline{X}$ is smaller than ℓ_n and $P_{\overline{CE}}$, we let $\overline{X}$ be ℓ_{n+1}. At the beginning of the algorithm we initialize $\overline{X}$ with 0. We modify $\overline{X}$ by comparing the forecast result and CE-marked packet. We use $P_{\overline{CE}}$ to denote that the packet isn't CE-marked, and use P_{CE} to denote that the packet is CE-marked.

Before the destination sends ACK, it computes $E[\ell]_n$ and lets *Diff* be the difference of $E[\ell]_n$ and $\overline{X}$. If *Diff* is larger than $\alpha \times E[\ell]_n$, the destination sends P_{DEC}. If *Diff* is smaller than $-\alpha \times E[\ell]_n$, the destination sends P_{NOR}. If *Diff* is between $\alpha \times E[\ell]_n$ and $-\alpha \times E[\ell]_n$, the destination sends P_{HOLD}. The ACK includes: (1) P_{NOR}, which is a normal ACK; (2) P_{DEC}, which is DEC-marked ACK; (3) P_{HOLD}, which is HOLD-marked ACK. The DEC and HOLD can be set in a bit in TOS fields of IP packet head.

When the source receives ACK, if receiving P_{NOR}, the source increases the window by one segment; if receiving P_{DEC}, the source reduces the window by one-eighth; if receiving P_{HOLD}, the source keeps the window unchanged.

4   NS Simulation and Result Analysis

The model we used to simulate the FAV algorithm is shown in Fig 2. The S_1 to S_n are the data sources and R_1 to R_n are the destinations. We create the TCP connection from every S_i to R_i ($1 \leq i \leq n$). Those connections share the congestion link that is from router1 to router2. We simulate and compare FAV algorithm with TCP Reno, TCP Vegas, in which all the gateways are DropTail. When we simulate, we set these parameters: α =1/8; the queue length of Router1 is 150; all the sources start at 0s and stop at 60s.

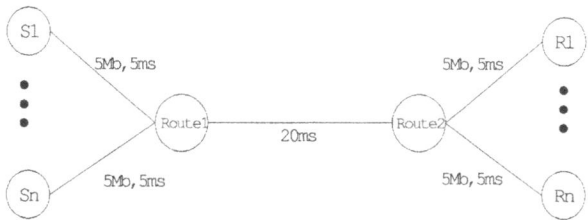

Fig. 2. Simulation topology

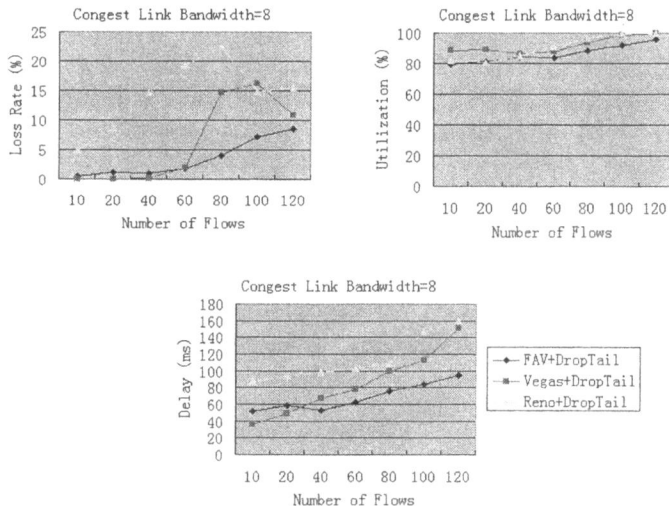

Fig. 3. Simulation Result (8Mbps)

Table 1. Statistic of Experiment 1

Index (average)	FAV	Ve-gas	Reno	FAV/Vegas	FAV/Reno
Loss Rate (%)	3.5	6.3	14.3	55%	24%
Utilization (%)	86.5	92.2	90.8	94%	95%
Delay (ms)	68	85	114	80%	60%

Experiment 1. When congestion link bandwidth is 8Mbps, we measure the Loss Rate, Utilization and Delay in bottleneck link. (See Fig 3 and table 1).

When the congestion link bandwidth is 8Mbps and the number of flows is less than 60, the loss rate of TCP Vegas algorithm is the lowest. But when the number of flows of network increases continually, the loss rate of TCP Vegas and TCP Reno increases very fast. At the same time, the loss rate of FAV still keeps low, and it is half of TCP Vegas's and quarter of TCP Reno's. Because FAV flattens the data traffic, the utilization is a little lower than TCP Vegas and TCP Reno. When the number of flows is over 30, the packets delay of FAV will be the lowest.

Experiment 2. When congestion link bandwidth is 16Mbps, we measure the Loss Rate, Utilization and Delay in bottleneck link. (See Fig 4 and Table 2).

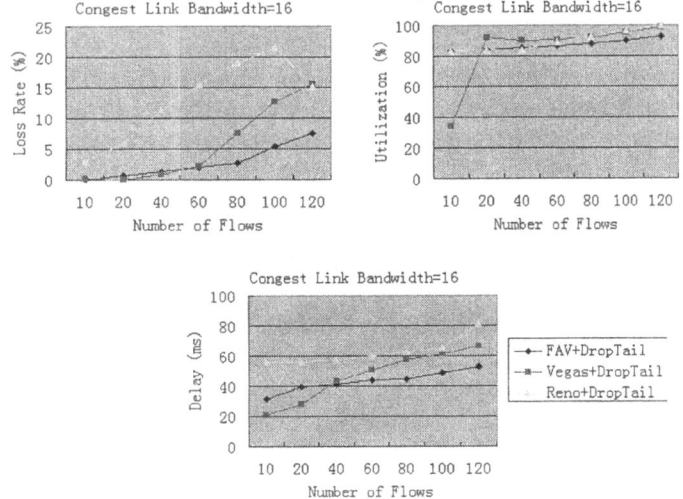

Fig. 4. Simulation Result (16Mbps)

Table 2. Statistic of Experiment 2

Index (average)	FAV	Ve-gas	Reno	FAV/Vegas	FAV/Reno
Loss Rate (%)	2.8	5.5	13.1	51%	21%
Utilization (%)	87.1	84.6	90.0	103%	97%
Delay (ms)	43	47	63	91%	68%

As we can see, when the congestion link bandwidth is 16Mb and the number of flows is less than 60, the loss rate of TCP Vegas algorithm is still the lowest. But when the number of flows of network increases continually, the loss rate of FAV still keeps low, and it is half of TCP Vegas's and one-fifth of TCP Reno's. When the congestion link bandwidth increases, the utilization of FAV remains steady, and is close to the performance of TCP Vegas and TCP Reno. At the same time the packets delay of FAV will be the lowest.

We can conclude from these experiments that the loss rate of FAV decreases obviously compared with TCP Vegas and TCP Reno. Therefore, the FAV algorithm reduces the retransmission packets and network load efficiently. With the loss rate decreasing, the end-to-end packets delay decreases too. When the congestion bandwidth increases, FAV can achieve relatively high and steady link utilization.

Experiment 3. Fairness analyzing

We use the Fairness Index [10] to analyze the algorithm fairness.

$$F(x) = \frac{(\sum x_i)^2}{n(\sum x_i^2)}$$

We do the experiments when congestion bandwidth is 8Mbps, 16Mbps and 32Mbps. (see Fig. 5)

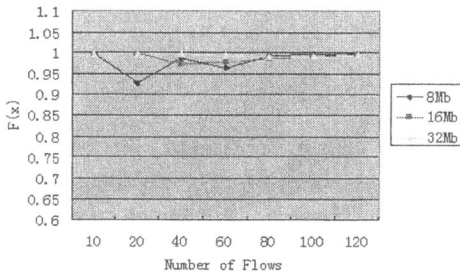

Fig. 5. Fairness Analyzing Simulation Result

The fairness parameters of FAV algorithm all exceed 0.92, and most of them are larger than 0.95. When the number of flows is over 80, the fairness parameters reach 0.99. So we believe that the fairness of FAV is well.

5   Conclusion and Future Work

Usually, the classical congestion control source algorithms apperceive packet loss or measure packets delay at the source to judge whether the network is congested or not. Because of existing link delay, it is difficult to judge the network congestion condition precisely and control the congestion efficiently. FAV algorithm forecasts the queue length of congestion node by measuring packets delay at destination, and it can forecast whether there will be congestion in the network and uses ECN to verify whether the forecast is precise or not. We believe FAV can control the congestion better. First, FAV reduces the loss rate efficiently using intergroup forecast to control data sending rate; Second, FAV decreases packets delay too; Third, FAV can adapt to the change of network traffic by using ingroup forecast and ECN-based verification mechanism, and it can keep network utilization high; Fourth, FAV can judge whether the congestion takes place in data sending link precisely by using receiver-drive mechanism.

Our paper is based on forecast idea to study the congestion control. We still need to solve a lot of problems in future work. First, when the lengths of packets are variable, we cannot presuppose the forwarding delay of every packet in the basic model. We must find a novel forecast formula. Second, since we suppose the network only has a single congestion link, in the future we should consider the condition in which the network has multi congestion links; Third, FAV is the source algorithm. If we combine the FAV with link algorithm (e.g. Random Early Detection, RED [11]), it is likely to achieve better performance.

References

1. V. Jacobson, Congestion Avoidance and Control, Proc, ACM SIGCOMM '88, pp. 314-329.
2. V. Jacobson, Modified TCP Congestion Avoidance Algorithm, Message to end2end-interest mailing list, April 1990.
3. K. Ramakrishnan, The Addition of Explicit Congestion Notification (ECN) to IP, RFC 3168, September 2001.
4. Network Simulator - ns (version 2), http://www.isi.edu/nsnam/ns/
5. Z. Wang, J. Crowcroft, Eliminating Periodic Packet Losses in 4.3-Tahoe BSD TCP Congestion Control Algorithm. ACM Computer Communication Review, 22(2):9–16, April 1992.
6. R. Jain. A Delay-Based Approach for Congestion Avoidance in Interconnected Heterogeneous Computer Networks. ACM Computer Communication Review, 19(5):56–71, October 1989.
7. Z. Wang and J. Crowcroft, A New Congestion Control Scheme: Slow Start and Search (Tri-S). ACM Computer Communication Review, 21(1):32–43, Jan. 1991.
8. L. Brakmo, S. O'Malley, L. Peterson, TCP Vegas: New techniques for congestion detection and avoidance, In Proceedings of the ACM SIGCOMM '94, pp 24-35, August 1994.
9. L. Brakmo and L. Peterson, TCP Vegas: End to End Congestion Avoidance on a Global Internet. IEEE Journal on Selected Areas in Communication, 13(8):1465-1480, October 1995.
10. Chiu, Dah-Ming, R.Jain, Analysis of the Increase and Decrease Algorithms for Congestion Avoidance in Computer Networks, Computer Networks and ISDN Systems, 17(1): 1-14, 1989.
11. S. Floyd, V. Jacobson, Random Early Detection Gateways for Congestion Avoidance. IEEE/ACM Transactions on Networking, 1(4): 397-413, 1993.

Competitive Neural Networks for Fault Detection and Diagnosis in 3G Cellular Systems

Guilherme A. Barreto[1], João C.M. Mota, Luís G.M. Souza,
Rewbenio A. Frota, Leonardo Aguayo, José S. Yamamoto[2], and
Pedro E.O. Macedo

[1] Department of Teleinformatics Engineering,
Federal University of Ceará, Fortaleza-CE, Brazil
guilherme@deti.ufc.br,
http://www.deti.ufc.br/~guilherme
[2] CPqD Telecom & IT Solutions, Campinas-SP, Brazil
sindi@cpqd.com.br,
http://www.cpqd.com.br

Abstract. We propose a new approach to fault detection and diagnosis in third-generation (3G) cellular networks using competitive neural algorithms. For density estimation purposes, a given neural model is trained with data vectors representing normal behavior of a CDMA2000 cellular system. After training, a normality profile is built from the sample distribution of the quantization errors of the training vectors. Then, we find empirical confidence intervals for testing hypotheses of normal/abnormal functioning of the cellular network. The trained network is also used to generate inference rules that identify the causes of the faults. We compare the performance of four neural algorithms and the results suggest that the proposed approaches outperform current methods.

1 Introduction

The third generation (3G) of wireless systems promise to provide mobile users with ubiquitous access to multimedia information services, providing higher data rates by means of new radio access technologies, such as UMTS, WCDMA and CDMA2000 [1]. This multi-service aspect brings totally new requirements into network optimization process and radio resource management algorithms, which differ significantly from traditional speech-dominated second generation (2G) approach. Because of these requirements, operation and maintenance of 3G cellular networks will be challenging.

The goal of this paper is to propose straightforward methods to deal with fault detection and diagnosis (FDD) of 3G cellular systems using competitive learning algorithms [2]. We formalize our approach within the context of statistical hypothesis testing, comparing the performance of four neural algorithms (WTA, FSCL, SOM and Neural-Gas). We show through simulations that the proposed methods outperform current standard approaches for FDD tasks. We

J.N. de Souza et al. (Eds.): ICT 2004, LNCS 3124, pp. 207–213, 2004.
© Springer-Verlag Berlin Heidelberg 2004

also evaluate the sensitivity of the proposed approaches to changes in the training parameters of the neural models, such as number of neurons and the number of training epochs.

The remainder of the paper is organized as follows. In Section 2, we describe the neural models and the data for training them. In Section 3, we introduce a general approach for the fault detection task and a method to generating inference rules from a trained neural model. Computer simulations for several scenarios of the cellular system are presented in Section 4. The paper is concluded in Section 5.

2   Competitive Neural Models

Competitive learning models are based on the concept of *winning neuron*, defined as the one whose weight vector is the closest to the current input vector. During the learning phase, the weights of the winning neurons are modified incrementally in order to extract *average features* from the input patterns. Using Euclidean distance, the simplest strategy to find the winning neuron, $i^*(t)$, is given by:

$$i^*(t) = \arg \min_{\forall i} \|\mathbf{x}(t) - \mathbf{w}_i(t)\| \tag{1}$$

where $\mathbf{x}(t) \in \Re^n$ denotes the current input vector, $\mathbf{w}_i(t) \in \Re^n$ is the weight vector of neuron i, and t symbolizes the iterations of the algorithm. Then, the weight vector of the winning neuron is modified as follows:

$$\mathbf{w}_{i^*}(t+1) = \mathbf{w}_{i^*}(t) + \eta(t)[\mathbf{x}(t) - \mathbf{w}_{i^*}(t)] \tag{2}$$

where $0 < \eta(t) < 1$ is the learning rate, which should decay with time to guarantee convergence of the weight vectors to stable states. The competitive learning strategy in (1) and (2) are referred to as *Winner-Take-All* (WTA), since only the winning neuron has its weight vector modified per iteration of the algorithm. In addition to the plain WTA, we also simulate three simple variants of it, namely: (1) the *Frequency-Sensitive Competitive Learning* (FSCL) [3], the well-known *Self-Organizing Map* (SOM) [4], and the *Neural-Gas algorithm* (NGA) [5].

To evaluate the performance of these competitive models on FDD tasks we need to define a set of KPIs (Key Parameter Indicators), which consist of a number of variables responsible for monitoring the QoS of a cellular system. These KPIs are gathered, for example, from the cellular system's operator, drive tests, customer complaints or protocol analyzers, and put together in a pattern vector $\mathbf{x}(t)$, which summarizes the state of the system at time t:

$$\mathbf{x}(t) = [KPI_1(t) \ KPI_2(t) \ \cdots \ KPI_n(t)]^T \tag{3}$$

where n is the number of KPIs chosen. Among the huge amount of KPIs available for selection, we have chosen the Number of Users, the Downlink Throughput (in Kb/s), the Noise Rise (in dB), and the Other-Cells Interference (in dBm). The data to train the neural models were generated by a static simulation tool. In addition, each component x_j is normalized to get zero mean and unity variance.

3   Fault Detection and Diagnosis via Competitive Models

Once we choose one of the neural models presented in Section 2 and train it with state vectors $\mathbf{x}(t)$ collected during normal functioning of the cellular network (i.e. no examples of abnormal features are available for training). After the training phase is completed, we compute the quantization error associated to each state vector $\mathbf{x}(t)$, $t = 1, \ldots, N$, used during training, as follows:

$$e(t) = \|\mathbf{E}(t)\| = \|\mathbf{x}(t) - \mathbf{w}_{i^*}(t)\| \tag{4}$$

where $\mathbf{E}(t)$ denotes the quantization error vector and i^* is the winning neuron for the state vector $\mathbf{x}(t)$. In other words, the quantization error is simply the distance from the state vector $\mathbf{x}(t)$ to the weight vector $\mathbf{w}_{i^*}(t)$ of its winning neuron. We refer to the distribution of N samples of quantization errors resulting from the training vectors as the *normality profile* of the cellular system.

Using the normality profile we can then define a numerical interval representing normal behavior of the system by computing a lower and upper limits via percentiles. In this paper, we are interested in an interval within which we can find a given percentage $p = 1 - \alpha$ (e.g. $p = 0.95$) of normal values of the variable. In Statistics jargon, the probability p defines the confidence level and, hence, the *normality interval* $[e_p^-, \ e_p^+]$ is then called (empirical) confidence interval. This interval can then be used to classifying a new state vector into normal/abnormal by means of a simple hypothesis test:

$$\begin{aligned}
\text{IF} \quad & e^{new} \in [e_p^-, \ e_p^+] \\
\text{THEN} \quad & \mathbf{x}^{new} \text{ is \textbf{NORMAL}} \\
\text{ELSE} \quad & \mathbf{x}^{new} \text{ is \textbf{ABNORMAL}}
\end{aligned} \tag{5}$$

The *null-hypothesis*, H_0, and the *alternative hypothesis*, H_1, are defined as:

- H_0: The vector $\mathbf{x}^{new}$ reflects the NORMAL activity of the cellular system.
- H_1: The vector $\mathbf{x}^{new}$ reflects the ABNORMAL activity of the cellular system.

Once a fault has been detected, it is necessary to investigate which of the attributes (KPIs) of the problematic input vector are responsible for the fault. From the weight vectors of a trained competitive neural model it is possible to extract inference rules that can determine the faulty KPIs in order to invoke the cellular network supervisor system to take any corrective action.

All the previous works generate inference rules through the analysis of the clusters formed by a subset of the NORMAL/ABNORMAL state vectors [6]. This approach is not adequate for our purposes, since the state vectors reflect only the normal functioning of the cellular network. We propose instead to evaluate the absolute values of the quantization errors of each KPI, computed for each training state vector:

$$ABS\left(\mathbf{E}(t)\right) = \begin{pmatrix} |E_1(t)| \\ |E_2(t)| \\ \vdots \\ |E_n(t)| \end{pmatrix} = \begin{pmatrix} |x_1(t) - w_{i^*1}(t)| \\ |x_2(t) - w_{i^*2}(t)| \\ \vdots \\ |x_n(t) - w_{i^*n}(t)| \end{pmatrix} \tag{6}$$

This approach is similar to that used in the fault detection task, but now we built n sample distributions using the absolute values of each component of the quantization error vector, $\mathbf{E}$. For the detection task we used only one sample distribution built from the *norm* of the quantization error vector, as described in (4). Then, for all the sample distributions, $\{|E_j(t)|\}$, $t = 1, \ldots, N$ and $j = 1, \ldots, n$, we compute the corresponding confidence intervals $[|E_j^-|, |E_j^+|]$, where $|E_j^-|$ and $|E_j^+|$ are the lower and upper bounds of the j-th interval.

Thus, whenever an incoming state vector $\mathbf{x}^{new}$ is signalized as abnormal by the fault detection stage, we take the absolute value of each component E_j^{new} of the corresponding quantization error vector and execute the following test:

IF $|E_j^{new}| \in [|E_j^-|, |E_j^+|]$,

THEN x_j is normal.

ELSE x_j is one (possible) cause of the fault.

In words, if the quantization error computed for the KPI x_j is within the range defined by the interval $[|E_j^-|, |E_j^+|]$, then it is not responsible for the fault previously detected, otherwise it will be indicated as a possible cause of the detected fault. If none of the KPIs are found to be faulty, then a *false alarm* will be discovered and then corrected. Confidence levels of 95% and 99% are used.

4   Computer Simulations

The 3G cellular environment used for system simulations is macrocellular, with two rings of interfering cells around the central one, resulting in a total of 19 cells. Other configurations are possible, with 1, 7 or 37 cells. All base stations use omnidirectional antennas at 30 meters above ground level, and the RF propagation model is the classic Okumura-Hata for 900MHz carrier frequency. Quality parameters, such as E_b/N_t^{Target} and maximum Noise Rise level are set to 5dB and 6dB, respectively. The number of initial mobile users is 60, which can be removed from the system by a power control algorithm. For each Monte Carlo simulation (drop) of the celular environment, a set of KPIs is stored and used for ANN training/testing procedures.

The first set of simulations evaluates the performance of the neural models, by quantifying the occurrence of false alarms after training them. The chosen network scenario corresponds to 100 mobile stations initially trying to connect to 7 base stations. No shadow fading is considered, and only voice services are allowed. Each data set corresponding to a specific network scenario is formed by 500 state vectors (collected from 500 drops of the static simulation tool), from which 400 vectors are selected randomly for training and the remaining 100 vectors are used for testing the neural models.

The results (in percentage) are organized in Table 1, where we show the intervals found for two confidence levels (95% and 99%). For comparison purposes, we show the results obtained for the single threshold approach. The error rates were averaged for 100 independent training runs. For all neural models,

Table 1. False alarm (FA) rates and confidence intervals for the various neural models.

Model	Proposed Approach		Approach by [7]	
	CI, FA (95%)	CI, FA (99%)	CI, FA (95%)	CI, FA (99%)
WTA	[0.366, 1.534], 12.43	[0.074, 1.836], 5.41	[0.000, 0.465], 17.91	[0.000, 1.018], 7.13
FSCL	[0.214, 1.923], 10.20	[0.136, 4.584], 1.80	[0.000, 1.126], 12.20	[0.000, 0.385], 3.00
NGA	[0.277, 1.944], 9.50	[0.1651, 4.218], 2.10	[0.000, 1.329], 10.10	[0.000, 0.941], 2.30
SOM	[0.361, 1.815], 8.75	[0.187, 2.710], 1.43	[0.000, 1.122], 13.28	[0.000, 1.191], 2.71

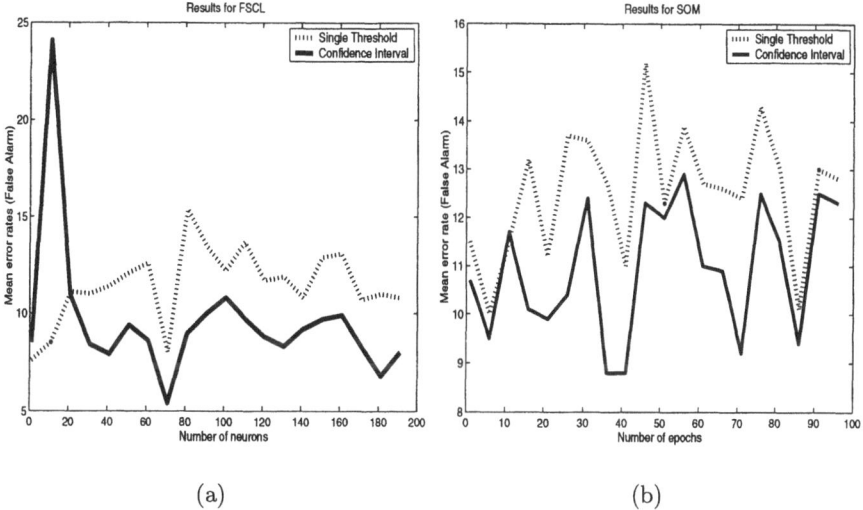

(a)   (b)

Fig. 1. Error rates of false alarms versus (a) the number of neurons (FSCL) and (b) the number of training epochs (SOM).

the number of neurons and the number of training epochs were set to 20 and 50, respectively. As expected, the NGA/SOM models performed much better than the WTA/FSCL models.

The second set of simulations evaluates the sensitivity of the neural models to changes in their training parameters. The goal is to understand how the number of neurons and the number of training epochs affect the occurrence of false alarms after training the neural models. The results are shown in Figure 1. For each case, we compare the interval-based approach proposed in this paper with the single threshold presented in [7]. The chosen network scenario corresponds to 120 mobile stations initially trying to connect to 7 base stations, for which fast and shadow fading are considered this time. Voice and data services are allowed. For the sake of simplicity, results are shown for one neural model only, since similar patterns are observed for the others.

For a given value of a parameter (e.g. the number of neurons), the neural model is trained 100 times with different initial weights. For each training

Table 2. Results (in percentage) for the joint fault detection and diagnosis tasks. FA=false alarm, AA=Absence of alarm.

Neural	$p = 95\%$			$p = 99\ \%$		
Model	**FA**	**AA**	**PERF**	**FA**	**AA**	**PERF**
WTA	5.54	0.00	91.50	2.20	0.04	95.80
FSCL	4.30	0.00	92.83	1.10	0.00	98.17
NGA	5.54	0.00	91.50	0.65	0.00	99.00
SOM	4.67	0.00	92.80	0.98	0.00	98.50

run, state vectors are selected randomly for the training and testing data sets. Also, the ordering of presentation of the state vectors for each training epoch is changed randomly. Then, the final value of the false alarm error rate is averaged for 100 testing runs. These independent training and testing runs are necessary to avoid biased estimates of the error rate.

In Figure 1a, the number of neurons is varied from 1 to 200, and each training run lasts 50 epochs. In Figure 1b, the number of epochs is varied from 1 to 100, while the number of neurons is fixed at 30. As a general conclusion, we can infer that in average the proposed approach produces better results than the single threshold method.

The last set of simulations evaluate the proposed methods for generating inference rules from competitive ANNs. Table 2 depicts the obtained results, averaged over 100 Monte Carlo simulations. ERROR I refers to the false-alarm rate, while ERROR II refers to the absence-of-alarm rate. The indicator $PERF$ (%) denotes the mean accuracy of the FDD system and it is computed as $PERF = 100 \cdot (1 - ERRORS/S)$, where S is the total number of state vectors used for testing. For each simulation, there were 8 state vectors corresponding to ABNORMAL conditions of the cellular network, and 52 state vectors reflecting NORMAL conditions. Thus, we have $S = 60$. One can infer that the maximum possible value of $ERRORS$ is S, reached only in the case of a very unreliable FDD system.

Two faulty state vectors per KPI were simulated by adding or subtracting random values obtained from Gaussian distributions with standard deviations greater than 1. The underlying idea of this procedure is to generate random values outside the range of normality of each KPI, and then, to test the sensitivity of the FDD system. It is worth emphasizing that all neural models performed very well, irrespective to their performances in the fault detection task. The only remaining error is the false alarm, which is the less crucial in a cellular network. Even this type of error has presented a very low rate of occurrence. All the ABNORMAL vectors have been found and his causes correctly assigned, i.e., all the faulty KPIs inserted in each ABNORMAL state vector have been detected.

5   Conclusion

In this paper we proposed general methods for fault detection and diagnosis in 3G cellular networks using competitive neural models. Unlike the available qualitative methods [8,9,10], the approach we took is focused on *quantitative* (numerical) results, more adequate for online performance analysis, being based on a statistically-oriented and widely accepted method of computing confidence intervals. Our methods outperformed current available single-threshold methods for FDD tasks.

Acknowledgements. The authors thank CPqD Telecom&IT Solutions/Instituto Atlântico for the financial support. The first author also thanks CNPq (DCR: 305275/2002-0).

References

1. Prasad, R., Mohr, W., Konäuser, W.: Third Generation Mobile Communication Systems - Universal Personal Communications. Artech House Publishers (2000)
2. Principe, J.C., Euliano, N.R., Lefebvre, W.C.: Neural and Adaptive Systems: Fundamentals through Simulations. John Wiley & Sons (2000)
3. Ahalt, S., Krishnamurthy, A., Cheen, P., Melton, D.: Competitive learning algorithms for vector quantization. Neural Networks **3** (1990) 277–290
4. Kohonen, T.: The self-organizing map. Proceedings of the IEEE **78** (1990) 1464–1480
5. Martinetz, T.M., Schulten, K.J.: A 'neural-gas' network learns topologies. In Kohonen, T., Makisara, K., Simula, O., Kangas, J., eds.: Artificial Neural Networks. North-Holland, Amsterdam (1991) 397–402
6. Hammer, B., Rechtien, A , Strickert, M., Villmann, T.: Rule extraction from self-organizing networks. Lecture Notes in Computer Science **2415** (2002) 877–882
7. Laiho, J., Kylväjä, M., Höglund, A.: Utilisation of advanced analysis methods in UMTS networks. In: Proceedings of the IEEE Vehicular Technology Conference (VTS/spring), Birmingham, Alabama (2002) 726–730
8. Binzer, T., Landstorfer, F.M.: Radio network planning with neural networks. In: Proceedings of the IEEE Vehicular Technology Conference (VTS/fall), Boston, MA (2000) 811–817
9. Raivio, K., Simula, O., Laiho, J.: Neural analysis of mobile radio access network. In: Proceedings of the IEEE International Conference on Data Mining (ICDM), San Jose, California (2001) 457–464
10. Raivio, K., Simula, O., Laiho, J., Lehtimäki, P.: Analysis of mobile radio access network using the Self-Organizing Map. In: Proceedings of the IPIP/IEEE International Symposium on Integrated Network Management, Colorado Springs, Colorado (2003) 439–451

Performance Analysis of an Optical MAN Ring for Asynchronous Variable Length Packets

Hind Castel and Gérard Hébuterne

INT, Institut National des Télécommunications
9, rue Charles Fourier, 91011 Evry, France
{hind.castel, gerard.hebuterne}@int-evry.fr

Abstract. We propose to study the performance of an unslotted optical MAN ring operating with asynchronous variable length optical packets. Source stations are connected to the optical ring. They take electronic packets coming from client layers and convert them into optical packets. The MAC protocol is based on the synchronous *"empty slot"* procedure. The performance model of the ring is based on an M/G/1 queue whith multiple priority classes of packets, and a preemptive Repeat-Identical service policy. Using the busy-period analysis technique, joint queue-length probability distribution is computed and exact packet delays at each source are given for different packet length distributions. Practical applications of the model are considered in order to see the impact of input parameters (packet length distributions, rank of the station) on performance measures.

1 Introduction

Packet format is one of the most important issues in the network concept since it has a major influence on the overall Quality of Qervice (QoS) the network is able to deliver to users. Packet format is of prime importance in a network performance in terms of end to end delay and of bandwidth loss. In this paper, we focus on the optical MAN which consists of several unidirectionnal optical physical rings interconnected by a centralised Hub. A ring within the MAN interconnects several nodes and consists of one or more fibres each operated in DWDM regime. We restrict to the case where all the stations transmit their packets in one wavelength, so for us the ring represents a wavelength. The multi-server case will be addressed in a future work. Sources generate packets (for example IP, or Ethernet packets) and an electronical/optical interface performs several operations which transform a variable length packet from the client domain into an optical packet. An optical packet transport network may work with either synchronous or asynchronous operation, being also called slotted or unslotted network respectively. Of prime importance are the questions of the optical packet format. Structuring the information (payload) into optical packets may be done according to different schemes : fixed length packets or variable length packets. Fixed length packets have been extensively considered with the ATM technology. This packet format is a frequent choice in slotted networks

J.N. de Souza et al. (Eds.): ICT 2004, LNCS 3124, pp. 214–220, 2004.

[1], all packets have the same size and are transmitted in a single slot. In the case the client packets are too long to be transmitted by a single slot, they are segmented into several packets which are treated as independant entities each with its own header. The advantage of fixed length packets are numerous since all the telecom techniques (traffic shaping, load balancing,...) can be adopted to really support different classes of service . But the problem is the choice of the packet size, and the guard time repetition in consecutive slots which can lead to inefficient bandwitdth utilization. Variable length packets can be used in slotted or unslotted networks. In the case of slotted networks, they are called slotted variable length packets or "train of slots" [5]. Packets may have variable sizes as long as they are multiples of a basic time slot. The packet fit a sequence of slots, no slot belonging to other packets can be inserted between them. This choice allows to use a single header for the whole packet, which reduces the waste of bandwidth. In unslotted networks, very few papers have been written about the issues related to the optical packet format. In this case, the packet format can be fixed or variable. For variable length packets, each packet has an arbitrary dimension not necessarily a multiple of a time-unit as slotted variable length packets. So no padding is required as in fixed length packet (slotted or unslotted networks), and slotted variable length packets. This results in higher bandwidth utilisation than previous solutions. Another advantage of asynchronous variable length packets is that the optical network don't need optical synchronisation. The MAC protocol studied is based on an *empty slot* procedure in order to avoid collisions at source nodes. The upstream station has the highest priority, so packets can be transmitted on the wavelength at any time. The priority of a station for transmission decreases when we go downstream on the wavelength. A station can transmit a packet if it finds a gap (corresponding to the idle period of previous stations on the ring) large enough to accomodate the packet. This gives advantage to upstream nodes, so a given node can be starved by transmissions of previous nodes. In addition, gaps can be too short to accomodate any packet. It results a waste in bandwidth and an increase of the waiting times. This phenomenon depends closely on packet length distributions, and imposes a deeper analysis. To the best of our knowledge, there are no published works trying to give modeling tools for this problem. We use mathematical models to solve this problem. We define a queueing model based on the M/G/1 queue with multiple priority classes of packets and preemptive repeat- identical service policy which has been completely solved in [2]. We have stated the stability conditions required in order to have finite queue lengths. Thus queue-length distributions are computed and mean delays at each source are given under various configurations. This paper is organized as follows. In section 2, we present the queueing model which allow us to study the performance of the optical ring. In section 3, we give mathematical equations required to the performance analysis. In section 4, the queueing model is tested and numerical results of the performance measures are showed. Mean packet delays curves of variable length optical packets are plotted and allow us to see the impact of input parameters. Finally, some concluding remarks and prospects for future work are given.

2   Queueing Model

We define an analytical model in order to compute mean packet delays at any source station. The model will provide performance measures for asynchronous variable length packet. It is based on an M/G/1 with multiple priority classes of customers. The service policy is preemptive repeat-identical. Let n be the number of stations connected to the ring. Each station is identified by an index $i(\ 1 \leq i \leq n)$ which increases when we go downstream in the ring. We suppose that the arrival rate in each station i follows a Poisson process with rate λ_i. Let S_i represents the service time of optical packets of station i. Queueing model is based on the M/G/1 with n priority classes of customers corresponding to the n source stations. Each station (so its packets) has a level of priority defined by the empty slot protocol. Station 1 contains packets of class 1, it has the highest priority because it is the upstream station in the ring. Any source station i contains packets of priority i which is lower than priorities of station $1, \ldots, i-1$ (corresponding to packets of stations $1 \ldots, i-1$) but higher than priorities $i+1 \ldots, n$ (packets of stations $i+1 \ldots, n$) .

3   Performance Evaluation Analysis

In the real system, a station i begins sending when it finds a gap *larger than the size of the packet to send*. This is equivalent to assume that station i begins sending *as soon as a gap is detected*, but it interrupts emission if the gap in not large enough, i.e. as soon as a packet arrives on the ring coming from stations $1, \ldots, i-1$, and tries again at the next gap: this amounts to represent (without any approximation) a station i as a M/G/1 system with preemptive repeat-identical service [2] receiving two kinds of customers : packets of class i and the flow of packets from upstream stations having the highest priority (classes 1, ..., i-1). The M/G/1 queue with multiple priority classes and preemptive repeat-identical service policy has been already studied, see e.g. [2]. For each station i, we analyze the busy period of the M/G/1 generated by the flows from stations $1, 2, \ldots, i-1$ (flows $i+1, \ldots, n$ have a lower priority and so have no influence). This allows to compute the joint queue-length probability distributions as presented in [2]. According to the model description, the queue-length state of station i depends especially on the busy period generated by stations $1 \ldots, i-1$ and the service time policy. So let for each station i the following random variables : H_i is the busy period generated by the station, C_i is the completion time (the elapsed time from the instant that the class i packet first receives service until the time it completes service), it is the real service time of the packets, M_i : is the queue-length and D_i is the packet delay. Queue-length process of any source station i is completely defined by the busy period density generated by sources $1, \ldots, i-1$ (denoted H_{i-1}) and the completion time density C_i. The service policy is preemptive repeat-identical, because an optical packet repeats its service until it finds an uninterrupted time duration of length equal to the service time S_i.

3.1   Mean Marginal Queue-Lengths and Mean Delays

The case of station 1 is easy because it has the higher priority, so it doesn't depend on the other stations. It amounts to study an M/G/1 with one input flow λ_1, and completion time as service time S_1. In [3], probability distributions of the queue length and the busy period have been presented. $E(M_1)$ is :

$$E(M_1) = \lambda_1 E(S_1) + \frac{\lambda_1^2 E(S_1^2)}{1 - \lambda_1 E(S_1)} \tag{1}$$

We study the case of a station i where $2 \leq i \leq n$. In [2] queue-length probability distributions are computed for any station i. A station i can be represented by an M/G/1 with arrivals λ_i and $\Lambda_{i-1} = \sum_{j=1}^{i-1} \lambda_j$, where Λ_{i-1} represents the rate arrival of packets which interrupts packet services of station i. As Λ_{i-1} generates a busy period H_{i-1}, so probability distribution of C_i depends on the probability distribution of H_{i-1} [2]. $E(C_i)$ is :

$$E(C_i) = [\frac{1}{\Lambda_{i-1}} + E(H_{i-1})][E(e^{\Lambda_{i-1}S_i}) - 1] \tag{2}$$

The busy period H_i is either initiated by the arrival of a packet of station i, or by packets of stations $1, \ldots, i-1$:

$$E(H_i) = \frac{\lambda_i}{\Lambda_i} \frac{E(C_i)}{1 - \lambda_i E(C_i)} + \frac{\Lambda_{i-1}}{\Lambda_i} \frac{E(H_{i-1})}{1 - \lambda_i E(C_i)} \tag{3}$$

Probability distributions of completion-time C_i and busy period H_{i-1} lead to the probability distribution equations of M_i [2]. $E(M_i)$ for $i = 2 \ldots, n$ is given as follows :

$$E(M_i) = \lambda_i E(C_i) + \Lambda_{i-1}\lambda_i \frac{E(H_{i-1}^2)}{2(1 + \Lambda_{i-1}E(H_{i-1}))} + \lambda_i^2 \frac{E(C_i^2)}{2(1 - \lambda_i E(C_i))} \tag{4}$$

where $E(C_i^2)$ and $E(H_i^2)$ are obtained from their probability distributions given in [2]. Using Little theorem, we can derive the delay at each station i : $E(D_i) = \lambda_i E(M_i)$.

3.2   Stability Condition

We give now necessary and sufficient conditions for all the queues to be stable. As explained in [2], the condition $\lambda_i E(C_i) < 1$, $i = 1 \ldots, n$ is necessary but not always sufficient in order to have a stable system. The goal of stability condition is to have $E(M_i) < \infty, \forall 1 \leq i \leq n$. So we must have $E(C_i^2)$ and $E(H_{i-1}^2)$ to be finite (we will see that there is no problem with others terms). We present in the following two cases of packet length distributions : deterministic, and general.
• deterministic : stability condition implies that $E(M_1) < \infty$, $E(H_1) < \infty$, and $E(H_1^2) < \infty$, $E(C_2^2) < \infty$, $E(M_2) < \infty$, $E(H_2) < \infty$, $E(H_2^2) < \infty$, $E(C_3^2) < \infty$, $E(M_3) < \infty$, ..., and $\forall i \leq n : E(H_{i-1}^2) < \infty$, $E(H_{i-1}) < \infty$, $E(C_i^2) < \infty$,

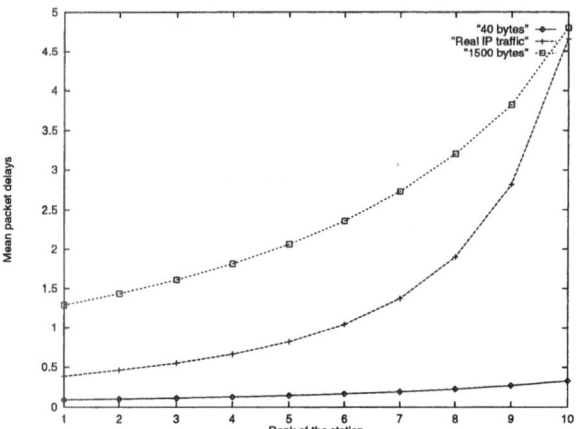

Fig. 1. Mean packet delays (in μs) versus the rank of the station for deterministic packet length distribution

$E(M_i) < \infty$. So the stability condition is sufficient. It is also necessary because if it is not verified, then the system is instable.

• general case: stability condition implies that : $E(M_1) < \infty$, $E(H_1) < \infty$ and $E(H_1^2) < \infty$, and $\forall i = 2, \ldots, n$ $E(H_i) < \infty$, but not necessarily $E(C_i^2) < \infty$ and $E(H_i^2) < \infty$. Because exponential terms in $E(C_i^2)$ formula (see [2]) are not always finite. For example, if S_i is an exponential random variable with mean $1/\mu_i$, we have to add the condition : $\mu_i > 2\Lambda_{i-1}$.

We explain now what the stability condition gives as conditions to the input parameters. So we replace $E(C_i)$ by its formula. $E(H_i)$ (i=2, ..., n) can be written in the case of preemptive repeat identical service policy as [2] :

$$E(H_i) = \frac{\rho_1 + \sum_{k=2}^{i} \lambda_k \omega_k}{\Lambda_i [1 - (\rho_1 + \sum_{k=2}^{i} \lambda_k \omega_k)]}, \quad \omega_k = \frac{1}{\Lambda_{k-1}} E(e^{\Lambda_{k-1} S_k} - 1) \qquad (5)$$

and $\rho_1 = \lambda_1 E(S_1)$. Then stability condition will give :

$$\lambda_i < \frac{\Lambda_{i-1}}{E(e^{\Lambda_{i-1} S_i} - 1)} (1 - (\rho_1 + \sum_{j=2}^{i-1} \lambda_j \omega_j)) \forall i \in [2, \ldots n], \; \rho_1 < 1 \qquad (6)$$

As explained before, for a deterministic packet length distribution, this condition is necessary and sufficient. But in the general case, it is necessary but not sufficient in order to have finite queue lengths. For example, for exponential distributions of packet lengths we have to add the condition $\mu_i > 2\Lambda_{i-1}$ $\forall i \in 2, \ldots, n$.

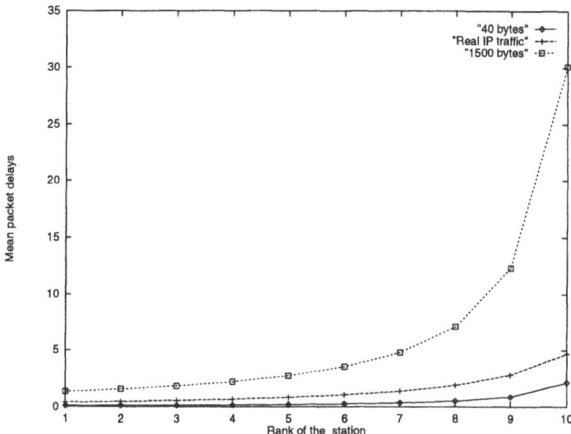

Fig. 2. Mean packet delays (in μs) versus the rank of the station for exponential packet length distribution

4   Numerical Results

We present in this section mean packet delays for different values of input parameters. For all figures, we have supposed 10 source stations and a ring transmission rate of 10 Gbit/s. The packet length distribution is the same for each station. Packet delays values are given in μs. The goal is to see the main parameters that influence packet delays. The rank of the station in the ring has certainly an impact on the delays (rank 1 corresponds to the priority 1, ... rank 10 to the priority 10). Another parameter to see is the packet length distribution. We have supposed different optical packet lengths obtained from different electronical packet lengths. According to real IP traffic measurements [4], three lengths of electronical packets have been considered : 40 bytes, 552 bytes and 1500 bytes. Optical packets are then defined according to the VLP optical packet format. The optical packet is represented by a payload (filled by an electronical packet), a guard time (50 ns), a packet header of 5 bytes, and synchronisation preamble of 4 bytes. In Fig. 1, mean packet delays versus the rank of the station are showed for deterministic packet length distribution. Each station has an input bitrate of 500 Mbits/s. Three curves are plotted for both figures. The lowest one (resp. the highest one) corresponds to IP packets of length equal to 40 bytes (resp. 1500 bytes). The middle curve shows the real IP packet length distribution [4] defined by : 40 bytes (0.6) , 552 bytes (0.25) and 1500 bytes (0.15). We can easily see the impact of the packet length on packet delays. When the size of the packet increases it may be difficult to find a gap large enough to insert the packet. So packet delays increase. Also, when the rank of the station increases (which means that its priority decreases), packet delay increases too. In Fig. 2, assumptions are the same than in Fig. 1, except the packet length distribution which is exponential. We can see that packet delays are higher than in the de-

terministic case especially when the packet length increases (for 1500 bytes).
When the input bitrate is 1 Gbits/s, we have remark that mean packet delays
are largely higher, and stability conditions were rapidly not verified (for a rank
upper than 5).

5   Conclusion

The main contribution of this paper is the definition of a queueing model which
provide the performance analysis of an asynchronous optical MAN ring for VLP
packets. Results have shown clearly the impact of the packet length distribution
on packet delays.

References

1. C. Guillemot et al. Transparent Optical Packet Switching : the European ACTS
   KEOPS project approach. IEEE/OSA Journal on Lightwave Technology, Vol.16,
   No 12, 1998, pp. 2117-2134.
2. N. K. Jaiswal, Priority Queues. New York : Academic 1966.
3. L. Kleinrock, Queueing Systems, Vol. 1 : Theory. New York, John Wiley and Sons,
   1975.
4. K.Thompson, G.J.Miller, R.Wilder, *Wide-area internet traffic patterns and characteristics,* IEEE Network, vol 11, No 6,pp10-23, 1997
5. S. Wittewrongel, Discrete-time buffers with variable-length train arrivals. IEEE
   Electronic Letters, Vol. 34, No 18, pp. 1719-1721, 3 Sept. 1998.

Enabling Value-Generating Telecommunications Services upon Integrated Multiservice Networks

Dionisis X. Adamopoulos[1] and Constantine A. Papandreou[2]

[1] Department of Technology Education & Digital Systems, University of Piraeus, Greece
dxa@unipi.gr
[2] Hellenic Telecommunications Organisation (OTE), Greece
kospap@ote.gr

Abstract. In the highly competitive telecommunications environment of today a service provider must continually evolve its network and enable new revenue-generating services faster and more cost effectively than the competition. These services should create value for the end users by satisfying all their functional requirements in an efficient manner, and they should be delivered with simplicity and ease of provisioning. This paper focuses on the transition from a requirements capture and analysis phase to a service analysis phase in the framework of an object-oriented service creation methodology, by considering the internal structure and the functionality of a telematic service. After the examination of the main activities that take place in the service analysis phase and the identification of the main artifacts that are produced during them, the overall service development process is highlighted, and the paper attempts to validate, through several service creation experiments, not only the service analysis phase, but also the overall service creation methodology. Finally, the validation results are discussed and further evaluation actions are briefly outlined.

1 Introduction

Increased competition, regulatory changes and the convergence of network technologies are causing service providers to look for innovative telecommunications services (telematic services) to differentiate their offerings and gain a competitive advantage. Telecommunication networks are gradually evolving towards integrated services networks or Multi-Service Networks (MSNs), which are networks capable of supporting a wide range of services. In MSNs, services are viewed as value adding distributed applications, composed using more elementary facilities available underneath through specialised Application Programming Interfaces (APIs) and operating on top of a general purpose communications subsystem.

For this reason, the emphasis is placed on the telecommunication software and on the rapid development of services upon open, programmable networks. Therefore, this paper proposes a structured service creation approach, emphasizing service analysis activities, that offers a viable service paradigm inside an open deregulated multi-provider telecommunications market place and is compatible with and influenced by the state of the art service creation technologies of Open Service Access (OSA), Parlay and Java APIs for Integrated Networks (JAIN) [6], and conformant to the open

J.N. de Souza et al. (Eds.): ICT 2004, LNCS 3124, pp. 221–226, 2004.
© Springer-Verlag Berlin Heidelberg 2004

service architectural framework specified by the Telecommunications Information Networking Architecture Consortium (TINA-C) [2][8]. The practical usefulness and efficiency of the proposed approach is ensured by extensive validation attempts with various services, involving comparative examination and experimentation with the use of different development environments.

2 Analysing Service Requirements

Service analysis activities constitute the service analysis phase in the service creation methodology, which is proposed to have an iterative and incremental, use case driven character. An iterative service creation life cycle is adopted, which is based on successive enlargement and refinement of a telematic service through multiple service development cycles within each one the service grows as it is enriched with new functions. More specifically, after the requirements capture and analysis phase, service development proceeds in a service formation phase, through a series of service development cycles. Each cycle tackles a relatively small set of service requirements, proceeding through service analysis, service design, service implementation and validation, and service testing. The service grows incrementally as each cycle is completed.

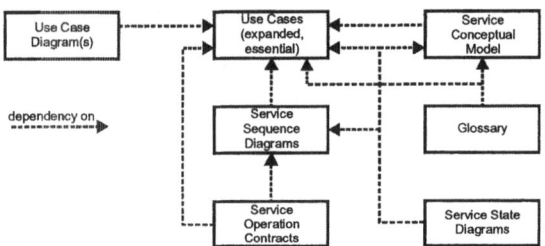

Fig. 1. Service analysis phase artifact dependencies

The aim of the service analysis phase is to determine the functionality needed for satisfying the service requirements and to define the software architecture of the service implementation. For this reason, the focal point shifts from the service boundary to the internal service structure [1]. The most important activities of this phase are examined in the following sections. The dependencies between the artifacts produced can be seen in Fig. 1.

2.1 Definition of Service Conceptual Models

The service analysis phase is the first phase of the service creation process where the telematic service is decomposed into its constituent parts, with the appropriate relationships among them, in an attempt to gain an overall understanding of the service. The resulting (main) service conceptual model involves identifying a rich set of service concepts (Information Objects, IOs) regarding the service under examination by investigating the service domain and by analysing the essential use cases [7].

The main service conceptual model is accompanied by a set of ancillary service conceptual models. These models are derived by (and correspond to) a number of generic information models deduced from the TINA-C service architecture [8] and complement semantically the main service conceptual model with useful session related concepts and structures [3].

The following steps specify the main service conceptual model:

Step 1: Identify the service concepts.

A central task when creating a service conceptual model is the identification of the service concepts. Two techniques are proposed for the identification of service concepts. The first is based on the use of a service concept category list, which contains categories that are usually worth considering, though not in any particular order of importance. Another useful technique is to consider the noun phrases in the text of the expanded use cases as candidate service concepts or attributes.

Step 2: Identify associations between the service concepts.

After identifying the service concepts, it is also necessary to identify those associations of the service concepts that are needed to satisfy the information requirements of the current use case(s) under development and which aid the comprehension of the service conceptual model. The associations that should be considered in order to be included in a service conceptual model are the associations for which the service requirements suggest or imply that knowledge of the relationship that they present needs to be preserved for some duration ("need-to-know" associations) or are otherwise strongly suggested in the service developer's perception of the problem domain.

Step 3: Identify attributes of the service concepts.

A service conceptual model should include all the attributes of the identified service concepts for which the service requirements suggest or imply a need to remember information. These attributes should preferably be simple attributes or pure data values. Caution is needed to avoid modelling a (complex) service concept as an attribute or relating two service concepts with an attribute instead of an association.

Step 4: Draw the main service conceptual model.

Adding the identified type hierarchies, associations and attributes to the initial service conceptual model, forms the main service conceptual model. It has to be noted that a verb phrase should be used for naming an association, in such a way that the association's name together with the names of the service concepts that it relates create a sequence that is readable and meaningful.

The proposed methodology considers the TINA-C service architecture (which has a direct and significant influence to subsequent service creation technologies) in a critical manner with the intention to extract from it useful concepts and guidelines / techniques. Taking into account the generic TINA-C session related information models [1] and the different types of sessions that can be established between business administrative domains, access sessions can be classified according to the specialisation hierarchy shown in Fig. 2(a). The access session related service IOs and their relationships are depicted in the information model of Fig. 2(b). In this figure, the Domain Access Session (D_AS) service IO is associated with a particular domain and represents the generic information required to establish and support access interactions between two domains. Furthermore, it is specialised into UD_AS (managed by the user), PD_AS (managed by the provider) and PeerD_AS service IOs, as each D_AS is associated with a particular access role.

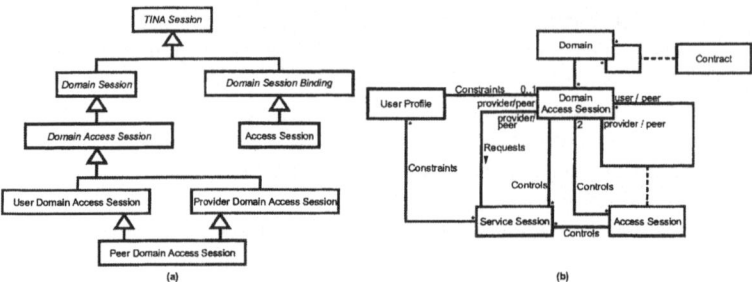

Fig. 2. Important access session related information models: (a) Classification of the access session, (b) The access session information model

Service sessions can be classified according to the specialisation hierarchy shown in Figure 3(a). The service session related service IOs and their relationships are depicted in the information model of Figure 3(b). Every service session consists of usage and provider service sessions. Each member of a session, i.e. an end-user, a resource or another session, is associated with a usage service session.

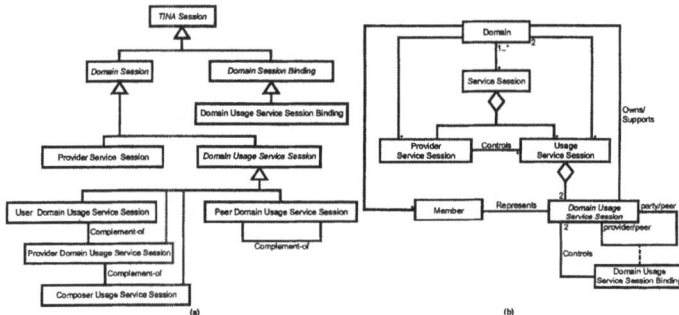

Fig. 3. Important service session related information models: (a) Classification of the service session, (b) The service session information model

2.2 Definition of Service Sequence Diagrams and Service Operation Contracts

A service sequence diagram should be done for the typical course of events of each use case and sometimes for the most important alternative courses. It depicts, for a particular course of events within a use case, the external actors that interact directly with the service, the service (as a black box), and the service events that the actors generate. Service events (and their associated service operations) should be expressed in an abstract way, emphasising their intention, and not in an implementation specific manner [7].

The behaviour of a service is further defined by service operation contracts, as they describe the effect of service operations upon the service. Their creation is dependent on the prior development of use cases and service sequence diagrams, and on the identification of service operations. UML contains support for defining service contracts by allowing the definition of pre- and post-conditions of service operations [4].

3 The Validation Approach

The proposed service creation methodology (and thus its service analysis phase) was validated and its true practical value and applicability was ensured as it was applied to the design and development of a real complex representative telematic service (a MultiMedia Conferencing Service for Education and Training, MMCS-ET). More specifically, a variety of scenarios / use cases were considered involving the support of session management requirements (session establishment, modification, suspension, resumption, and shutdown), interaction requirements (audio / video, text, and file communication), and collaboration support requirements (chat facility, file exchange facility, and voting). Due to the incremental and iterative nature of the proposed methodology these use cases were examined in nine (9) service development cycles covering a time period of almost two (2) years.

Further validation attempts of the proposed methodology and examination of its usefulness, correctness, consistency, flexibility, effectiveness and efficiency, involved a variety of service creation activities for different telecommunications services (actually service scenarios) using different development approaches. More specifically, the following telematic services were considered: Distributed collaborative design, distributed case handling, remote monitoring, remote database access, remote database utilisation, remote access to expertise, remote application running, entertainment on demand (pay-per-view), remote consultation, social conversation.

These service scenarios were developed in a small scale (with a maximum of five end-users) in both a business and an academic environment, by three (different) teams of two service developers in three different (object-oriented) manners; namely using the proposed service creation methodology, using an ad hoc approach and using a widely accepted general purpose object-oriented software development methodology (the Unified Process, UP). All service developers had similar knowledge and experience, and a manager monitored the service requirements and the reuse of the specifications and code in all cases in order to keep them at comparable levels during the application of the different service creation approaches for each service scenario.

The parameters that were examined each time a telematic service was developed (some of which are explicitly related to service analysis activities) are the following:

- The total time needed for the development of the service from the beginning of the service project until service deployment.
- The number of problems that were reported by the users of the service (e.g. not supported functions, unsatisfactory operation, unpredictable behaviour, etc.) after using it in a daily basis for two months after its deployment.
- The number of service concepts in the service conceptual models, the number of essential use cases, the number of service sequence diagrams, the number of service operation contracts, and the number of service state diagrams in which changes were performed during the first two months of service operation.
- The number of objects in the service code that was necessary to change during the first two months of service operation.
- The number of lines of service code that were added during the first two months of service operation.
- The total time needed for service maintenance activities during the first two months of service operation.

- The total time that the service was necessary to be inactive due to service maintenance during the first two months of service operation.

From the results of the service creation experiments is evident that the ad hoc approach, although it seems to be fast in some cases, is the less flexible approach with
the highest risk in misinterpreting the requirements and the highest possibility to
cause maintenance problems. The most important drawback of the UP is its difficulty
to be applied in a service creation context, which is reflected by the greater total development time, by the more changes that are needed to service analysis artifacts
when performing corrections or extensions and by the relatively verbose service code
that it produces. Therefore, the proposed process is the most flexible and cost effective approach (in terms of total development time and maintenance problems), satisfies the requirements of the users (slightly better than UP), and (when considering all
the parameters) improves the productivity of the service developers and increases the
possibility for the successful completion of a service creation project.

4 Concluding Remarks

The validation attempts described in Section 3 provided strong evidence that the proposed methodology and the corresponding service analysis activities can be used efficiently for the development of new telecommunications services in open, programmable service-driven next-generation networks and that they are correct and effective
as they can lead to the desired outcome, i.e. a successful telematic service that satisfies the requirements of its users. However, the proposed methodology, apart from
being specialised for service creation purposes in the area of telecommunications
service engineering, remains a methodology for the development of systems. Therefore, its evaluation using the Normative Information Model-based Systems Analysis
and Design (NIMSAD) framework is suggested [5]. Such an attempt will provide additional confidence on the capabilities and the quality of the proposed service creation
process and will support the increased expectations regarding its value and impact.

References

1. Adamopoulos, D.X., Pavlou, G., Papandreou, C.A.: Advanced Service Creation Using Distributed Object Technology. IEEE Communications Magazine 3 (2002) 146-154
2. Berndt, H., Hamada, T., Graubmann, P.: TINA: Its Achievements and its Future Directions.
   IEEE Communications Surveys & Tutorials 1 (2000)
3. Choi, S., Turner, J. Wolf, T.: Configuring Sessions in Programmable Networks. Computer
   Networks 2 (2003) 269-284
4. Evits, P.: A UML Pattern Language. Macmillan Technology Series (2000)
5. Jayaratna, N.: Understanding and Evaluating Methodologies, NIMSAD: A Systemic
   Framework. McGraw-Hill (1994)
6. Jormakka, J, Jormakka, H.: State of the Art of Service Creation Technologies in IP and Mobile Environments. Proceedings of IFIP World Computer Congress (2002) 147-166
7. Larman, C.: Applying UML and Patterns: An Introduction to Object-Oriented Analysis and
   Design and the Unified Process. Prentice Hall (2002)
8. TINA-C: Service Architecture. Version 5.0 (1997)

An Architecture for Publishing and Distributing Service Components in Active Networks

N. Dragios, C. Harbilas, K.P. Tsoukatos, and G. Karetsos

School of Electrical and Computer Engineering
National Technical University of Athens –GREECE
{ndragios, ktsouk}@telecom.ntua.gr

Abstract. Application level active networks provide a way of transforming the current network infrastructure into one where new services and protocols are more easily adopted without the need for standardization. In this paper we deal with an application layer active networking system, and address the problem of publishing and distributing software components, provided by trusted service providers, throughout the active network. We propose an network architecture of dedicated servers providing Content Distribution Network (CDN) functionality; this leads to smaller response times to client requests and decreased network traffic. Experimental results from a test network configuration illustrate the benefits obtained from exploiting CDN capabilities, and support the viability of our approach.

1 Introduction

The evolving technology of Active Networking (AN) aims at changing the way computer networks behave. Activeness in networks is an idea proposed in [10], and seeks to address the problem of slow adoption of new technologies and standards, as well as the slow evolution of network services. Activeness entails injecting programmability into the network infrastructure, so as to allow the rapid introduction of many new services inside the network [2].

In this paper we focus on an application level active network. Here, processing of various data flows within the network takes place at the application layer (not at the network layer). The Application Level Active Networking system (ALAN), presented in [5][6][7], has been implemented to provide users with a flexible framework for supporting active services within traditional network boundaries. Software components implementing these services, called proxylets, can be loaded onto service nodes, called active servers, where they are executed on demand. This approach is similar in spirit to the Active Services framework of [1] and the Application Level Active network of [8][9]. In this context, the distribution of service components, provided by trusted third parties, throughout the active network, is of considerable interest, for it directly affects response times to client requests and the volume of network traffic. We herein attempt to address this issue. Consequently, our focus is not on the AN per se, but on the network architecture used for publishing and distributing the service components throughout the active network, so as to better support the desired active functionality. In particular, we exploit the idea of building Content Distribution Net-

J.N. de Souza et al. (Eds.): ICT 2004, LNCS 3124, pp. 227–233, 2004.
© Springer-Verlag Berlin Heidelberg 2004

works (CDNs) for enabling speedy, uninterrupted, and reliable access to web content; this may well fit in an application layer active networking environment, where service components are requested from specific web servers.

The paper is organized as follows: In Section 2 we describe the structure of the particular application level active network under consideration. In Section 3 we present the architecture for distributing and publishing service components; this is based on forming a Proxylet Distribution Network of servers. In Section 4 we report experimental results from a test network configuration which illustrate the benefits obtained from applying CDN technology and support the viability of this approach.

2   An Application Level Active Network

The application level active networking system presented in this section was developed in the European Commission IST project ANDROID (Active Network Distributed Open Infrastructure Development) and borrows from the ALAN system discussed in [5][6][7], while the CDN service is part of ongoing work within the context of IST project mCDN.

In ALAN, clients can place proxylets onto service nodes, and execute them on demand. These nodes provide an Execution Environment for Proxylets (EEP), allowing these proxylets to be run. Requests can be sent to the EEP to load a proxylet, referenced by a URL, and the node will load the proxylet subject to checking the permissions, validity and security. Proxylets can be downloaded from a number of different sources, known as Independent Software Vendors (ISVs), and run on an EEP. Proxylets are Jar files, containing Java classes, placed on a web server and can be referenced via their URL. They are self-describing, so that they can be used effectively in a dynamic environment. This is achieved by specifying appropriate proxylet metadata, expressed in XML, which include the proxylet functional characteristics, their facilities for communicating with other components and the corresponding security policies.

3   Publishing and Distribution of Proxylets

3.1   Content Distribution Network (CDN)

A CDN is an independent network of dedicated servers that web publishers can use to distribute their contents at the edge of the Internet [4]. CDN supports caching, transparent routing of user requests and securing of the contents from modification. CDN infrastructure consists of three main sub-systems: (a) the redirection, (b) the content delivery and (c) the distribution [3]. Redirection sub-system redirects the client request towards a closer, than the content provider, server that can serve the request. Content delivery sub-system transparently delivers content from that closer location. Distribution sub-system moves contents from the origin server to the servers of the content delivery system.

Our effort is focused on building an independent network of servers that implements part of the functionality of a CDN. We call this network a Proxylet Distribution

Network (PDN). The PDN distributes the metadata of the available services, along with the actual services, i.e., the software components implementing those services. Those services are offered by ISVs, through web servers providing the proxylets. The hosts comprising the network, which provides CDN-like functionality, are called Proxylet-Brokers (PBs). One PB is located in each administrative domain, where one or more active servers may exist. PBs could also be placed anywhere in the network, since their services are quite independent from any other procedure. The gathering and distribution of information about services could be seen as an off-line process. PBs do not depend on any other component of the AN. Their network is built independently and its purpose is to serve the clients of the AN. However, PBs need to be close to active servers, where the proxylets are loaded and executed, and close to clients, who need to have quick access to information about the services available on the network.

3.2 Building the Proxylet Distribution Network

The Proxylet Distribution Network (PDN) is built in a step-by-step manner. A new PB contacts a member PB of the PDN and follows a join process. This takes place as follows: (a) the member of the PDN accepts the join request of the new PB and notifies it for all member PBs of the network, (b) the member of the PDN also sends all information it has, if any, about services provided by ISVs in this network, in the form of proxylet metadata. As a result of the join process, all PBs in the PDN have the same view of the services available. Finally, new PB contacts all other members of the PDN and identifies itself. The end result is a completely interconnected network of PBs, the Proxylet Distribution Network. PBs join or leave the PDN as needed. Whenever a new PB comes in, it informs the rest of the PDN about his arrival, whereas if some PB's failure occurs, it is removed from the list of members of the PDN. Thus, any number of PBs can access the PDN at any time, making the network of brokers a dynamic set that shrinks and stretches in an autonomous manner.

3.3 Proxylet Metadata Publishing

PDN is populated with proxylet metadata provided by ISVs. During this phase an ISV contacts a PB in order to publish the proxylets it provides. ISV sends all proxylet metadata to the PB as a number of XML files, each one representing one proxylet. The PB in turn contacts every other PB of the PDN and distributes the metadata received from the ISV. After that, all PBs know all the available services (proxylets) and associated metadata describing them. Given that one PB resides in every administrative domain, where clients have access to the network and request services from, the gathering of proxylet metadata information close to the client results in a quick browsing of the available services. The benefits of the PDN approach are quantified in Section 4, by comparing response times to client request with and without the PDN. Although the replication of all XML files at all nodes of the PDN seems to be inefficient and consuming both of bandwidth and disk space, this is not really the case. First, most of these file transfers are performed off-line, without aggravating the traffic of the actual active network, and, second, the size of these files is quite small.

3.4 Service Component Request

A client contacts the PDN through an interface to his closest PB and is presented with a list of all available proxylets, and their main characteristics contained in XML metadata files. He then selects the most appropriate service for his application and requests the downloading of the service. PB establishes a URL connection to the appropriate ISV and fetches the proxylet. The proxylet is cached locally and is available to any application. Then, PB informs the rest of the PDN about the fetching of the proxylet so that all other PBs know that the proxylet is available not only from the ISV, but also elsewhere in the PDN. Smaller response times, reduced network traffic and reliability are gained by this approach. Considerable reduction of traffic at the ISV web server is also achieved, as well as continuous reliable provision of the service, even when the ISV is inaccessible because of a possible network failure.

3.5 Update of Distributed Metadata

One of the most important issues is keeping up-to-date all proxylet associated metadata. When a new proxylet appears in an ISV site, or an existing proxylet is modified, ISV announces this change to a PB. ISV sends the new XML file to the PB, who in turn distributes this XML file to all nodes of the PDN. Upon receiving the new XML file, each PB updates its structures where proxylet associated information is stored. It may also be the case that the previous version of the software component represented by the replaced metadata has been downloaded recently. Thus, in order for the new metadata to be compatible with the corresponding Jar file, the PB should re-fetch the proxylet from the appropriate ISV.

4 Performance Evaluation

4.1 Network Topologies and Service Establishment

In order to quantify the benefits obtained by the proposed PDN approach, we measure the performance of two experimental network topologies. In both topologies a client is assumed to issue requests for locating and fetching certain proxylets. A PDN infrastructure is available only in the second case. Both topologies consist of 4 web servers, which act as ISVs, and are located at certain addresses shown in figure 1. Each ISV provides a number of proxylets of different sizes and their metadata. The selected proxylet sizes are between 20KB and 1.5MB. The client requesting the proxylets is located in the same domain as ISV2 and SV4. Fig. 1 depicts the two experimental topologies.

In the non-PDN topology, client maintains a list of ISVs, and queries them in order to locate the desired one proxylets. Client selects an ISV from his list, either based on some policy, or in random. In our experiment the client scans the ISVs sequentially, in the order ISV 1, 2, 3, 4. Client issues HTTP requests to get all metadata (XML files) available from ISV 1. After checking the metadata, if he finds the desired proxylet, he issues one more HTTP request to ISV 1, to retrieve the proxylet. In

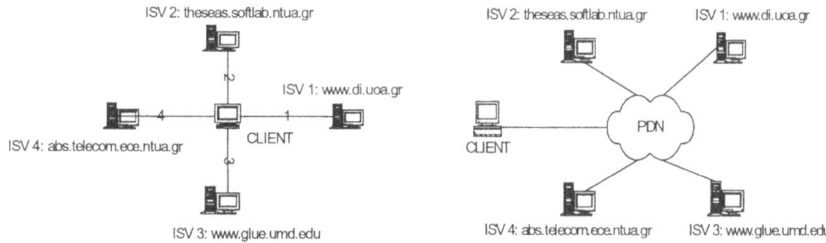

Fig. 1. Non - PDN and PDN topology

case the client does not find the desired proxylet in ISV 1, he proceeds similarly with ISV 2, 3 and 4. We measure the total response time, defined as the time that elapses from the moment the client issues the requests for the XML files to ISV 1, until the moment the desired proxylet is downloaded. We repeat our measurements for all cases where the desired proxylet is each one of all proxylets provided by all ISVs.

In the PDN topology, the client searching relies on the functionality provided by the PDN. The client contacts a PB, who resides in his domain, and hosts all metadata from all ISVs. The connection between the client and the PB is based on socket communication; hence it is faster than HTTP. After examining the XML metadata, client requests a proxylet. The PB subsequently makes an HTTP request to the appropriate ISV providing the proxylet and fetches it locally. We measure the resulting total response time until the proxylet is downloaded. We also measure the response time for the situation where the requested proxylet is already cached in the PB, in which case the ISV does not need to be contacted at all. This process is repeated for all available proxylets.

4.2 Experimental Results and Discussion

We collect the response times obtained from the experiments described above in Fig. 2. Each plot shows the response times for proxylets located in each one of the ISVs vs. the proxylet file size. The three lines in each plot represent the response times for downloading a proxylet, under the following three scenarios: (a) non-PDN topology as described in section 4.1(scenario Direct-ISV), (b) PDN topology as described as described in section 4.1 (scenario PB-ISV), (c) PDN topology with cached proxylets. The client requests a proxylet that has already been cached by the PB. (scenario PB-cache). We observe that if the desired proxylet is found in either ISV 1 or ISV 2, the response times for Direct-ISV scenario are only slightly lower than those of scenario PB-ISV. That is, the resulting PDN performance is very much comparable even with the case where the client would need to directly query only one or two ISVs in order to retrieve the proxylet. Thus, the top two plots indicate that the overhead associated with the use of the PDN is rather small. Next, for the proxylets located in ISV 3, in which case scenario Direct-ISV requires sequential scanning of ISVs 1, 2, and 3, we see that the use of the PDN already leads to smaller response times than those of scenario Direct-ISV. This is due to the fact that scenario PB-ISV requires only one HTTP connection from the PB to the ISV where the proxylet is located, regardless of the number of ISVs, as opposed to multiple HTTP requests that may be necessary without the PDN.

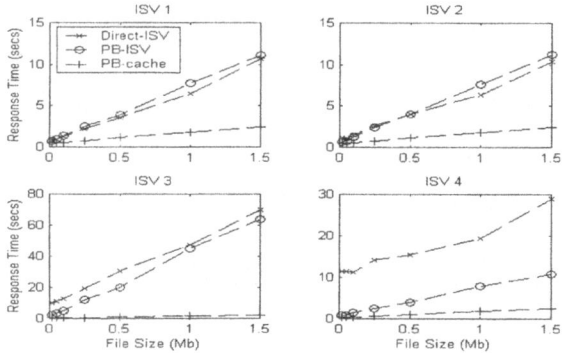

Fig. 2. Response times vs. file sizes when proxylet is located at ISV 1, 2, 3, 4, respectively.

This difference between the response times of the Direct-ISV and PB-ISV scenarios becomes more pronounced when the proxylet is located in ISV 4, in which case the client needs to contact all four ISVs if no PDN is available. It is also worth noting that the smallest response times are achieved when the desired proxylet is found at the PB cache, as shown by the PB-cache lines in all four plots. The % response time improvement under the PB-ISV scenario, as compared to the Direct-ISV scenario, is summarised in Table 1.

Table 1. Response time reduction achieved by PDN

ISV 1	ISV 2	ISV 3	ISV 4
-8.43 %	2.82 %	20.21 %	74.91 %

A further benefit of the PDN approach is that finding the desired proxylet in the PB cache yields a 70,71% response time reduction over the case where the proxylet is fetched using HTTP from ISV 1, i.e., the first ISV contacted by the client under scenario ISV.

In short, the PDN approach offers better performance than direct HTTP when the existing ISVs are three or more. Clearly, the performance gains of the PDN would become far more apparent in an active network with a large number of ISVs. This state of affairs should be typical of a realistic environment, where an increasing number of vendors provide their own proxylets to support add-ons or pure new network services.

5 Conclusions

This paper stresses the need for a content distribution mechanism to spread out the services available on an active networking system. To achieve this, we propose and implement a dedicated network of servers, providing part of a CDN's functionality. Experiments with the proposed content distribution architecture indicate that it may well offer a promising solution to the problem of distribution of services in an active networking environment. Information on services and the services themselves are distributed across a network of nodes placed close to users and the active servers, where they can be easily accessed. Reduced network traffic, quick response to applications that use those services and caching of those services are some of the benefits gained by this approach.

References

1. E. Amir, S. McCanne and R. Katz, "An Active Service Framework and its Application to Real Time Multimedia Transcoding," *Proc. SIGCOMM'98*, pp. 178-189, September 1998.
2. K. Calvert, S. Bhattacharjee, E. Zegura and J. Sterbenz, "Directions in Active Networks" *IEEE Communications Magazine*, 1998.
3. M. Day, B. Cain and G. Tomlinson, "A Model For CDN Peering," IETF Internet Draft, http://www.alternic.org/drafts/drafts-d-e/draft-day-cdnp-model-01.html.
4. M. Day and D. Gilletti, "Content Distribution Network Peering Scenarios," IETF Internet Draft, http://www.alternic.org/drafts/drafts-d-e/draft-day-cdnp-scenarios-00,01,02.html.
5. M. Fry and A. Ghosh, "Application Level Active Networking," *Fourth International Workshop on High Performance Protocol Architectures (HIPPARCH '98)*, June 1998.
6. M. Fry and A. Ghosh, "Application Layer Active Networking," *Computer Networks*, Vol. 31, No 7, pp. 655-667, 1999.
7. G. MacLarty and M. Fry, "Policy-based Content Delivery: An Active Network Approach," *Fifth International Web Caching and Content Delivery Workshop*, Lisbon, Portugal, 22-24, May 2000.
8. I. W. Marshall and M. Banfield, "An Architecture For Application Layer Active Networking," *IEEE Workshop on Application Level Active Networks: Techniques and Deployment*, November 2000.
9. I. W. Marshall, J. Crowcroft, M. Fry, A. Ghosh, D. Hutchison, D. J. Parish, I. W. Phillips, N. G. Pryce, M. Sloman and D. Waddington, "Application-Level Programmable Internetwork Environment," *BT Technology Journal* Vol. 17, No. 2, April 1999.
10. D. Tennenhouse and D. Wetherall, "Towards an Active Network Architecture," *Computer Communication Review*, Vol. 26, No. 2, pp. 5-18, April 1996.

Performance Comparison of Active Network-Based and Non Active Network-Based Single-Rate Multicast Congestion Control Protocols

Yansen Darmaputra and Riri Fitri Sari

Centre for Information and Communications Engineering Research
Electrical Engineering Department, University of Indonesia
{yansen00, riri }@eng.ui.ac.id

Abstract. This paper presents the comparisons of two single-rate multicast protocol, Active Error Recovery/Nominee Congestion Avoidance (AER/ NCA) and Pragmatic General Multicast Congestion Control (PGMCC). Both protocols use worst receiver mechanism, NACK-based feedback, and window-based transmission rate adjustment. AER/NCA is implemented using the new Active Networks technology, while PGMCC is implemented using the traditional passive network.
We discovered from the simulation results that both protocols are TCP-friendly. The implementation of active networks causes many computations to be done in the network so that it would degrade network performance. However, in high loss rates network, active networks technology could provide fast recovery to loss packets.

1 Introduction

Multicast is one-to-many communication that is suitable for group communications. Using multicast, data packet is only sent once through each link. The packet will be duplicated at intermediate routers then will be passed to the next hop router until it reaches its destination. Multicast data forwarding is done using multicast distribution trees which will exploit the hierarchy of upstream and downstream nodes.

Reliable multicast needs feedback mechanism to know the status of the packet being sent. Both AER/NCA [1] and PGMCC [3] protocols use NACK as its feedback. Single-rate multicast sends data in one transmission rate to the whole group. The transmission rate is adjusted to the receiving capacity of the worst receiver. Therefore, the existence of one slow receiver could drag down the throughput of the whole group. This will cause the drop to zero problem in which the sender estimates the loss rate much higher than the actual loss experienced by the receivers.

Active network is a new concept of network which allows packet to be processed inside the network. Active networks can be implemented in two approaches, integrated approach and discrete approach [7]. AER/NCA uses discrete approach which uses programmable router/switch that already has standard methods in it.

J.N. de Souza et al. (Eds.): ICT 2004, LNCS 3124, pp. 234–240, 2004.

This paper is structured as follows. Section 2 consists of brief description of AER/NCA and PGMCC. Section 3 contains the simulation result for AER/NCA simulations. Section 4 presents the performance comparison of AER/NCA and PGMCC protocols. Section 5 concludes the paper.

2   AER/NCA and PGMCC

The architecture of AER/NCA [1] uses discrete approach in which server is put collocated with router. AER/NCA does not require all nodes in the multicast distribution tree to be active. Active nodes are only put at specific locations in the distribution tree where it is considered useful. Figure 1 shows multicast distribution tree with active nodes support.

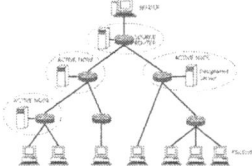

Fig. 1. Multicast Distribution Tree with Active Nodes

PGMCC [3] uses PGM-enabled router which is also called Network Elements (NE). NE provides services such as eliminating NACK and sending NACK Confirmation (NCF) downstream. NE will also help in loss recovery by maintaining retransmit state which lists interfaces from which it received NACK.

Both AER/NCA and PGMCC use the following packet: positive acknowledgement (ACK), negative acknowledgement (NACK), and Source Path Messages (SPM). SPM packet is used to inform receivers about the multicast hierarchy. AER/NCA uses another packet, called Congestion Status Message (CSM) packet which is used to inform sender about the Round Trip Time (RTT) and loss rate estimation from each receiver. For loss recovery, designated server in active node is called repair server (RS). Any receiver that detects loss will wait for a random amount of time and then sends NACK to its upstream. RTT and loss rate information is needed to choose worst receiver, AER/NCA gets it from CSM packet that contains RTT and loss rate estimation from the receiver. PGMCC gets it from field rxw_lead and rx_loss in every ACK/NACK packet. Function to compute throughput estimation is as follows:

$$T(p, RTT) = C / (RTT. \quad p) \tag{1}$$

Where p is loss rate, RTT is round trip time, and C is constant. The worst receiver will be the receiver that has the highest value of function G (G=RTT. $\sqrt{p}$). Throughput estimation will be computed every time sender receives new information of RTT and loss rate.

3 Simulation Experiment Results and Discussion

Simulation is only done for AER/NCA using the same parameter as PGMCC simulation in [3]. PGMCC simulation is not re-simulated here. Simulation is performed using NS simulator [10] that have been modified to support active networks for AER/NCA.

Unless stated, each simulation is executed 10 times and the simulation time is 500 s. We used generated data between 100-500 s of simulation time. Traffic for AER/NCA is Constant Bit Rate (CBR). Traffic for TCP NewReno is File Transfer Protocol (FTP). All queue use Droptail (FIFO) with the queue buffer of 50 unless stated. Both AER/NCA and TCP use 1000 bytes data packet. CSM and SPM packet are 76 bytes long. An active node is shown as polygon in Figure 2.

The first three sets of simulation use network topology and parameter as shown in Figure 2. These simulations use only AER/NCA session, in which 15 AER/NCA receivers behind node 2 join the session. The propagation delay on L2 is varied from 10 ms to 400 ms with 30 ms interval. The simulation result is shown on Figure 3, together with the result of simulation set 2. As expected, throughput is proportionally inversed with the increase in propagation delay. Propagation delay determines RTT from sender to receiver. The higher the RTT, the longer it takes for the sender to receive ACK. For the second set of simulation, 15 AER/NCA receivers behind node 3 also join the session. Three TCP sessions, TCP1, TCP 2, TCP 3 also travel through L1. TCP sessions start at random time from 1 s to 10 s (distributed uniformly). Figure 3 shows the simulation result.

The graph shows that at 10 ms and 40 ms, AER/NCA will equally share the bottleneck link bandwidth with the three competing TCP sessions. This behavior shows the TCP-friendliness of AER/NCA. For higher propagation delay, AER/NCA throughput will degrade depending on the worst receiver's throughput. For propagation delay below 160 ms, worst receiver will be behind node 3. For propagation delay above 160 ms, worst receiver will be behind node 2. This shows that in the presence of different links, AER/NCA will choose the worst receiver and adjust its transmission rate according to the receiving capacity of the worst receiver.

The third set of simulation activates another TCP session, TCP 4. The simulation is performed for 100 s, with fairness index computed at 1, 2, 5, 10, 15, 20, 50, and 100 s. There are three variations for propagation delay on L2, that is 10, 100, and 130 ms. In this simulation, fairness will be shown by computing fairness index using Jain's formula. Then, *fairness* F_τ can be computed as:

$$F_\tau(t) = \frac{\left(\sum_{i=0}^{n} X_i^\tau(t)\right)^2}{n \cdot \sum_{i=0}^{n} \left(X_i^\tau(t)\right)^2} \tag{2}$$

Where X is the ratio between the transmission rate and bandwidth of the link, while n is the number of sessions that runs through the same bottleneck link. Fairness index will be between 0 and 1 where 1 reflects fair division of throughput of sessions.

Figure 4 shows that, for three variation of propagation delay, fairness index will degrades between t=1s and t=2s. This is due to the initialization process that is done by AER/NCA when it starts the session. When session starts, sender must inform receivers about the multicast hierarchy. This is done by sending SPM to all of the receivers. SPM will be processed by the intermediate nodes (active nodes) and by receivers along the multicast tree.

The next three sets of simulation will use network topology and parameter as shown in Figure 5. The fourth set of simulation is performed to simulate web-like traffic, which is a bursty traffic that could generate high loss rates in the link it travels. To simulate web-like traffic, ON/OFF UDP sources are used with Mean ON and Mean OFF time taken from Pareto distribution. The simulation lasts for 200 s. All the UDP sources and receivers are placed in single node since AER/NCA module in NS has limitations in the number of nodes. The maximum nodes allowed is 128 nodes which would not permit each UDP sources and receivers to be placed separately. Figure 6 shows the simulation results, in which the more UDP sources activated, the higher the loss rate on L1. Loss packets happened due to buffer overflow. Although buffer queue on L1 is set to 100, many sources which flow through L1 will definitely overflow the buffer and cause packets to be dropped.

The fifth set of simulation use AER/NCA, TCP 2, and UDP. Only AER/NCA receivers behind node 3 join the group. This simulation is intended to examine the session's throughput in the presence of high loss rates. It is shown in Figure 7 that high loss rate makes both AER/NCA and TCP experience degradation in throughput. For loss rate under 13% (for 105 UDP sources based on previous simulation result), TCP gains higher throughput than AER/NCA, but for loss rate above 13%, AER/NCA gains higher throughput than TCP. In high loss rates, the loss recovery mechanism in AER/NCA is much better than TCP. AER/NCA utilizes active node to detect losses, buffer packets, and perform packet loss recovery

The sixth set of simulation use AER/NCA, TCP 1, TCP 2, and UDP. AER/NCA receivers behind node 2 and node 3 join the group. Link L2 is a link with very low bandwidth, therefore RTT for receivers behind L2 will be mostly determined by the transmission time. On L1, as before, RTT will be mainly determined by queuing delay. The simulation result is shown on Figure 8. In this result it could be noted that the behavior is similar as that of the previous simulation. AER/NCA throughput oscillates throughout the simulation. This happens because throughput equation used is only relevant for loss rate below 5% [2].

The last simulation uses network topology and parameter shown in Figure 9. This simulations sets deal with uncorrelated losses. The link between node 2 and receivers is set to have 1% loss rate. There are two sessions used, AER/NCA and TCP. At t = 0 s, 10 AER/NCA receivers join the group. At t=300s, 10 other AER/NCA receivers join the group. The simulation result is shown on Figure 10.

4 Comparison Between AER/NCA and PGMCC

This section presents the results comparison between AER/NCA and PGMCC. Figure 10 shows the result of the first two simulations. AER/NCA shows that it has better worst receiver election mechanism than PGMCC. At propagation delay 10 ms, throughput of PGMCC 2 is about 300 Kbps. For simulation 2 where there are 4

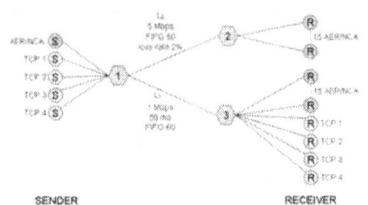

Fig. 2. Topology 1

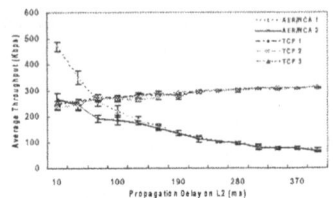

Fig. 3. Throughput of AER/NCA and 3 TCP sessions on bottleneck link L1

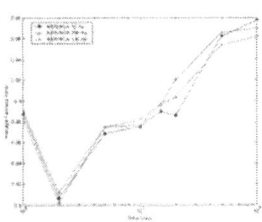

Fig. 4. Fairness Index for AER/NCA and 4 TCP sessions competing on bottleneck link L1

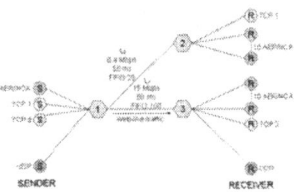

Fig. 5. Topology 2

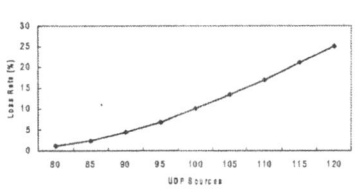

Fig. 6. Loss Rate as the function of UDP sources

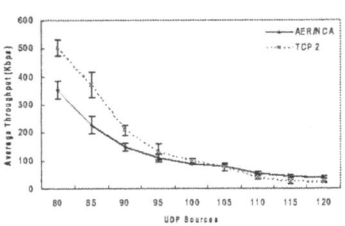

Fig. 7. Throughput of AER/NCA and TCP on bottleneck link L1 in the presence of web-like traffic

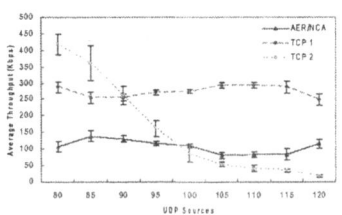

Fig. 8. Throughput AER/NCA and 2 TCP sessions in presence of web-like traffic

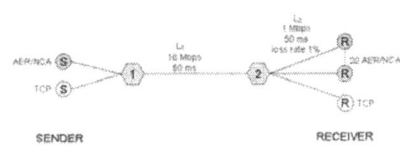

Fig. 9. Topology 3

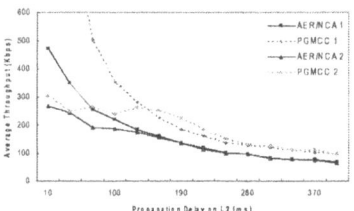

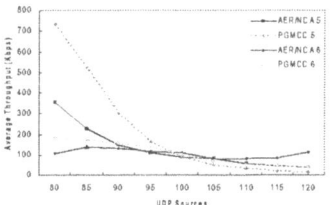

Fig. 10. Throughput of AER/NCA and PGMCC

Fig. 11. Throughput of AER/NCA and PGMCC in the presence of ON/OFF UDP Sources on bottleneck link L1

sessions (1 PGMCC, 3 TCP) competing for bandwidth on bottleneck link L1, the fair bandwidth allocation is 250 Kbps for each session. In [3], it is stated that due to the presence of a very fast link, any worst receiver switch will cause PGMCC to take more bandwidth than it should. Therefore in the presence of heterogeneous receivers, PGMCC worst receiver election mechanism is not functioning well. Figure 11 shows the comparison of throughput between AER/NCA and PGMCC in the presence of ON/OFF UDP Sources on bottleneck link L1 for simulation 5 and 6.

The fairness index of PGMCC will increase along with time. For 10 ms propagation delay, the fairness index curve of PGMCC is below the other 2 curves. This shows that in heterogeneous receivers, PGMCC will act quite unfair to other sessions. However, the value of fairness index is still acceptable; hence both AER/NCA and PGMCC have shown that they are TCP-friendly. At low loss rate (UDP Sources<100) PGMCC have more throughput than AER/NCA. But in higher loss rate, the inverse happens. In low loss rate, computations in PGMCC is less than AER/NCA, therefore its throughput is higher. In high loss rates, AER/NCA's loss recovery mechanism works more efficient than PGMCC.

5 Conclusion

In active networks, too many computations in network will decrease network performance because computations will slow down packet transmission. Therefore, it is suggested that active components (programmable router/switch, designated server) are not put at backbone/core layer networks. Backbone network is a very high speed network, in which computations in the middle of the backbone network will degrades network performance. Besides that, most packet losses occur at the edge of network, not at the backbone network [8]. Active components are also suitable to be put behind lossy due to its capability to cache packets and provide quick local retransmission.

In loss recovery mechanism, active networks has shown that it has better mechanism including best-effort caching packets at active nodes to provide local retransmissions, immediate loss detection by source router, and decentralize the burden of retransmission to repair servers. These mechanisms will increase protocol's scalability. However in high loss rate (loss rate > 5%), more precise formula is needed for worst receiver election mechanism to avoid frequent worst receiver changes in

AER/NCA. Currently passive network has been proven to be effective in providing fast packet transmission. Therefore, by combining the strengths and eliminating the weaknesses for active and passive networks and creating a new type of network which is called hybrid network, it is expected that more reliable networks could be created.

References

1. S. Kasera, S. Bhattachayya, M. Keaton, D. Kiwior, S. Zabele, J. Kurose and D. Towsley, "Scalable Fair Reliable Multicast Using Active Services," *IEEE Network*, Vol.14, No.1, pp.48-57, January/February 2000.
2. L. Rizzo, "pgmcc: a TCP-friendly Single-rate Multicast Congestion Control Scheme," *Proc. ACM SIGCOMM 2000*, pp.17-28, Stockholm, Swedia, August 2000.
3. G. Iannaccone, L. Rizzo, "Fairness of a single-rate multicast congestion control scheme," *Proceedings of International Workshop on Digital Communications*, Lecture Notes in Computer Science, vol.2170, Springer, September 2001.
4. S. Ortiz Jr, "Active Networks: The Programmable Pipeline," *IEEE Computer Technology News*, pp.19-21, August 1998.
5. A. Patel. "Active Network Technology: A Thorough Overview of Its Application and Its Future," *IEEE Potentials*, February/March 2001.
6. T. Wolf dan Sumi Y. Choi, "Aggregated Hierarchical Multicast for Active Networks," *In Proc. of the 2001 IEEE Conference on Military Communications (MILCOM)*, McLean, VA, October 2001.
7. D. Tennenhouse and David Wetherall, "Towards an Active Network Architecture," *Computer Communication Review*, Vol.26, No.2, pp.5-18, April 1996.
8. L. Lehman, S.Garland, D.Tennenhouse, "Active Reliable Multicast," *Proc. IEEE INFOCOM'98*, pp.581-589, San Fransisco, USA, November 2000.
9. G. Mark, "Marc Greis Tutorial for the UCB-LBNL-VINT Network Simulator", at http://www.isi.edu/nsnam/ns/tutorial/index.html   9 September 2003.
10. The Network Simulator NS-2. http://www.isi.edu/ nsnam/ns/
11. K. Fall and S. Floyd, "Simulation-based Comparisons of Tahoe, Reno, and SACK TCP," *ACM Computer Communications Review*, vol.26, no.3, pp.5-21, July 1996.

MAMI: Mobile Agent Based System for Mobile Internet

Mohammad Aminul Haq and Mitsuji Matsumoto

GITS, Waseda University, 1-3-10 Nishi Waseda, Shinjuku, Tokyo 169-0051, Japan,
mdaminul@asagi.waseda.jp, mmatsumoto@waseda.jp

Abstract. In this paper we propose a mobile agent based system for wireless mobile internet. We show that our agent based system has some desirable properties which are effective in compensating the drawbacks of wireless environment, mobile device limitations and TCP, HTTP protocol limitations in wireless environment. We describe the system architecture and operation scenario. We clearly define what to mobilize and how to achieve this. We discuss the benefits of using agent in this scenario and show how different optimization techniques can easily be incorporated in such a system due to its flexibility.

1   Introduction

Connecting to the Internet on the move is not what was thought during the origin of the Internet and an underlying wired network was assumed in the design. So modifying the scenario for the mobile environment without major change in the operational structure is not a trivial one. Moreover, mobile environment is hostile; it has low bandwidth, high rate of disconnection, high error rate etc. During the past few years several efforts were made to realize Internet access from mobile devices. IBM eNetwork Web Express [8], the MobiScape [1], and the Mowgli Project [14] are some of them. All of these models are based on proxies. Two proxies are set at the two end of the wireless link to intercept the data that flows on the wireless link. The client side proxy (on the mobile device) and the server side proxy (On the wired network) interact to provide wireless access to WWW. The two proxies come to know the device and browser capabilities and shape the traffic accordingly. Moreover in [1] both of them maintain cache of recently visited web pages. That means in the course of communication the server side accumulates information that are necessary for better performance of wireless Internet. But in these systems there is not provision to mobilize this information according to user movement. As a result whenever user changes his point of attachment to the fixed part of the mobile network he has to start as if he has no previous session or history which can greatly improve his performance. In this paper we propose the use of mobile agents to mobilize this information. Moreover the future mobile networks will be characterized by service variety and multi-provider scenario. So any system aimed to work successfully under this scenario should not only be able to overcome the difficulties of the wireless

J.N. de Souza et al. (Eds.): ICT 2004, LNCS 3124, pp. 241–250, 2004.

environment but also make provisions for better service flexibility. With these things in mind we will be providing a mobile agent based system for mobile Internet: MAMI. MAMI uses both static and mobile agent to provide the necessary flexibility and optimization techniques as explained in Section 3.

The rest of the document is organized as follows. Section 2 defines mobile agents and differentiates between provider and service agent. Section 3 illustrates the proposed system architecture, Section 4 discusses the system's operation, Section 5 points out the benefits achievable from such a system, Section 6 depicts related works and finally Section 7 concludes the paper.

2   Mobile Agents

A mobile agent is a software component having unique capability to transport its code and state of execution with it, in part or as a whole at run-time, migrating among the components of a network infrastructure where they resume execution. The term "state" typically means the attribute values of the agent that help it determine what to do when it resumes execution at its new environment. The "code" in an object oriented context means the class code necessary for an agent to execute. In MAMI there are basically two types of agents: provider agent and service agent.

2.1   Provider Agent and Service Agent

Before going into the detail of a provider agent and a service agent, let us A provider agent [7] is permanently available at a fixed location and offers access to local resources. On the other hand service agent [7] migrates among network components to accomplish its mission and will be using the services of a provider agent. The mission of the service agent is a user-defined set of rules in which the agent should act. The provider agent of a support station creates the service agent for the user which follows user's movement. In creating the service agent, provider agent follows user's profile so that the user can access only registered services. Mobile terminal is assisted by a terminal agent [6]. These concepts are elaborated in Section 3 and 4.

3   System Architecture

MAMI is designed to work in a mobile network with the architecture of Fig. 1. As evident, this is the general mobile network architecture. Here the mobile nodes are connected to base station via wireless link. Several of these base stations are connected to one Support Station. The concept of support station is present in all major mobile network architecture available at present but in a different term like Mobile Switching Data Point in UMTS, or Packet Processing Module (PPM) along with Message Packet Gateway (M-PGW) in i-mode [5]. We extend these components to include support for mobile agents and call it support station.

There are several support stations in the mobile network. The support stations are connected among them selves by the core network which is fixed line network. The support stations are connected to Internet where information providers and service providers publish their information and service. Another feature of support station is that it provides necessary storage to maintain caches of recently requested WebPages by the users. If the requested web page is fresh in the support station's cache, it can be returned right away, reducing web page response time in mobile Internet scenario. So the system is a hierarchy of components as appear from Fig. 1. As we have just said that support stations include necessary environment for provider and service agents. The environment should provide necessary support for the creation of agents, messaging among the agents, agent migration facility, collaboration, control, protection and destruction of mobile agents. Several mobile agent frame works have been proposed to provide such facilities like Aglets [16], and Voyager [17].

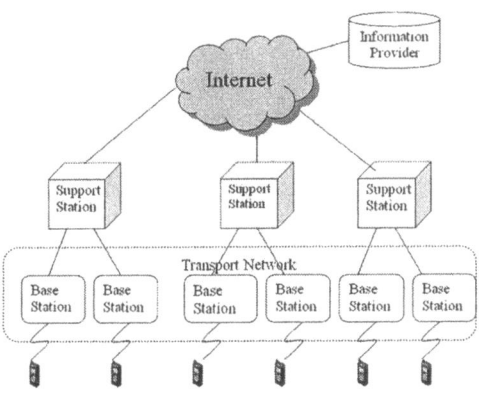

Fig. 1. System architecture

4   System Operation

This section explains how the total system works. The overall system operation is shown in Fig. 2. As can be seen, the mobile devices are supported by one program, Terminal Agent. This is a static agent and runs in the mobile devices environment. Requests for web content from browsers are all directed to this agent who is responsible to return the content. Terminal agent appears as the ultimate web content provider to all programs. Terminal agent knows the capabilities of the device very well so as to return web content in a format suitable for device's capabilities and resources. Moreover, terminal agent maintains a cache locally which contains all the recent web content fetched from outside on a LRU (Least Recently Used) basis. As mobile terminals are resource constrained devices, the size of this local cache is dynamically determined by terminal agent.

Terminal agent checks for the availability of a requested content in the local cache first. If it finds an up-to-date page in the cache then it returns it from there. This saves a good amount of wireless traffic which is not only costly but also time consuming. This activity is similar to the popular off-line browsing. Moreover terminal agent keeps track of user's interests and follows user's regular activity to predict [9] and prefetch [15] content. Again terminal agent keeps track of user's movement and helps in path prediction [10] [13]. These features are explained in Section 5 in detail.

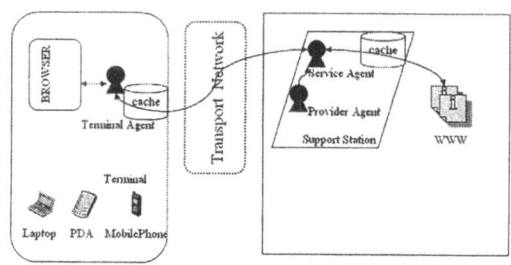

Fig. 2. System operation

In support station we have Provider Agent. This is also a static agent and it is always up and running in a specific support station. Provider agent creates Service Agent for each user. So there is one provider agent per support station but there may be several service agent active in any support station. Provider agent is well aware of the system architecture and underlying environment details as such to create the best suitable service agent for the user. Provider agent is also aware of user's registered services so that the created service agent is capable of handling only the registered service. Provider agent also involved in mobilizing service agent as explained in Section 4.1. Service Agent is the mobile agent. The mobility of service agent is discussed in Section 4.1. Service agent is the counter part of terminal agent. Service agent and terminal agent are involved in the traffic movement from Internet to terminal and vice versa. Service agent serves terminal agent with the requested information from the Internet. Since web page cache is maintained in support station service agent first looks at this cache. If the information is still up-to-date in the cache, the cache content is returned instead of going to the real web server. This speeds up the complete system. So in MAMI, two agents are at the two end of the transport link. One is in the device and one in the support station. Having two programs at the two ends give some optimization scope. For example, as long as the service agent and terminal agent can talk a common language, the underlying protocol can be anything suitable for the wireless environment, like LTP (Lightweight Transport Protocol) [5] or anything else which is able to reduce TCP/IP connection overhead [8] and HTTP headers [8]. Apart from this, since two programs are at the two ends of transport network and since both terminal agent and service agent are supported by cache, if both of them have cache entries and both the cache entries have timed out then they use differencing [8]. In this process the terminal agent asks fresh content from service agent with the information that its cache entry has timed out. Since the requested web page was available in terminal's cache, it will also be available in support station's cache. But both the entries are not up-to-date. In this case service agent will fetch fresh content from Internet and will compute the difference between the fresh content and cache content. Service agent will then send the difference to the terminal agent as a reply to the requested web page. Terminal agent will then compute the fresh content from this difference and his cache content and will serve it to the browser. To realize this technique the timeout value of the caches of both terminal agent and service agent need to be the same. This technique is useful with the observation that replies of form based queries to the same server generate responses which vary very little among themselves. This also drastically reduces the information that is exchanged over the wireless network. Moreover in MAMI, terminal agent and service agent both apply compression and decompression techniques on the exchanged information to even reduce the volume of data transfer over the wireless link.

4.1   Agent Migration

As we have already said in previous section that service agent is a mobile agent. What this means is that, service agent is active in a particular support station as long as his corresponding mobile user is attached to the support station. Whenever the user moves to a base station which is attached to a different support station, service agent migrates to that support station. This is shown in Fig. 3. Now we discuss about the movement of user and agents. Whenever the user reaches a base station that is served by a different support station than previous, the process of agent movement takes place. Now there are mainly two types of movement that can be implemented. One is where the agent program, along with its state and cache migrates to the new support station and another one is where only the state and cache of service agent is migrated and a new agent is created in the new support station by the provider agent of the new support station based on these. In MAMI we propose to use the second type of migration, i.e., only service agent's state and cache are migrated. This is because migrating a program is a costly task. Costly in the sense that it will increase the volume of information to be migrated. As a result the load in the fixed network will increase and its performance will be reduced. Moreover, if we transfer the program then we need to have the same program environment in all the base stations. Otherwise the program will not be able to execute. Although Java based mobile agents promise to execute on heterogeneous environments, for the shake of traffic reduction and performance we opt for not to migrate the code [3]. That is, whenever the user is handed over to a base station under a new support station, before the hand off process ends, the provider agent sends the corresponding service agent's status and cache to the new provider agent which then creates a new service agent for the user.

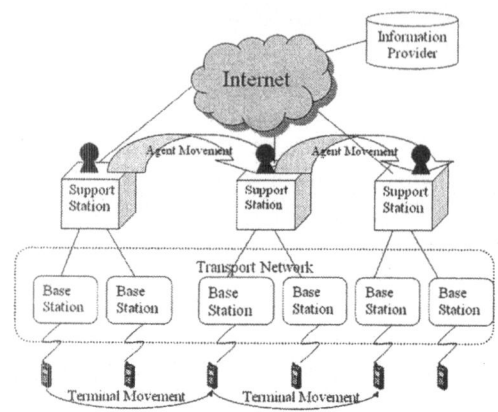

Fig. 3. Agent movement with respect to user movement

This is how base station level handoff and support station level handoff operates together. The terminal agent listens to the Location Area Identifier LAI broadcasted by base stations. The LAI changes with the support station change. So whenever the terminal agent finds out that he is in a base station which is broadcasting a new LAI than previous, he informs provider agent of current support station about this. The provider agent then collaborates with the provider agent of new support station to mobilize his service agent. Fig. 4 shows this migration procedure. It should be made clear that base station level handover does not necessary mean agent migration among support stations. The 5 steps indicated in Fig. 4 are as follows:

Step 1: Terminal agent informs provider agent about the new Location Area Identifier.

Step 2: Provider agent of current support station communicates with the provider agent of new support station. Provider agent of the new support station allocates resources for the new service agent and its cache which will be migrating soon.

Step 3: Provider agent of current base station informs current service agent to migrate.

Step 4: Service agent migrates

Step 5: Relocation of Cache.

In MAMI, since provider agents are involved in migrating the agents, the system is secured from any unwanted programs to be pushed in support stations by any malicious person. Again, since both the provider agents are programs, they can employ any suitable security algorithm to trust each other. Moreover since a new service agent is created on the new support station, MAMI works on heterogeneous support stations and there is no need for homogeneous environments in the support stations.

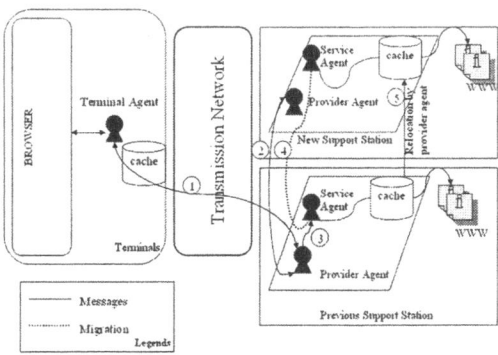

Fig. 4. Agent migration process

5   Benefits of MAMI

In the past few years agent technology has spread to many areas [7] including user interface, personal assistance, mobile computing, information retrieval, telecommunication services and service/network management. In this section we take a look at the benefits of deploying mobile agents, especially in mobile web access.

5.1   Support for Disconnected Operation

Wireless links are characterized by frequent disconnection which poses great difficulty in the synchronous nature of present Internet. If a disconnection occurs during the response of a request then the request should be made again. Mobile agents provide a good solution to this problem. In MAMI, the request from terminal agent is handed over to service agent which then fetches the information from Internet. Meanwhile the there may be no connection between terminal agent and service agent. When service agent is finished with its work it can send the reply or terminal agent may suggest not to send the reply right on but in a later time, suitable for the user. Results can be eventually gathered by the user upon reconnection [4] [7]. So there is no need to maintain a continuous open connection between the mobile node and the internet access point over the wireless network [2].

5.2   Support for Reduced Network Load

Wireless links are not only of limited bandwidth but also expensive. So reducing communication over this link has been a desired feature for any mobile internet architecture. MAMI reduces wireless data transport by several measures as discussed in Section 4. In MAMI, the service agent works on behalf of user to collect the requested information. Once the service agent has received the request from terminal agent, the mobile node need not communicate with service agent. Instead service agent will communicate with mobile node when it has something to return. Moreover since the service agent carries its state and cache with it, there is no need to resend requests or start from scratch in a new support station. So seamless mobile internet access is feasible. The service agent is well aware of the capability of the mobile terminal as such to represent data properly for the terminal. It can perform HTTP header reduction [8], differencing [8] or compression on the data. All these reduce the communication between the mobile node and the internet access point, which not only reduces the network load but also make less use of the expensive wireless link [2] [4].

5.3   Support for Enhanced Flexibility

It is difficult to change dynamically the interface of services offered by a server through which clients access the resources of a server [2]. In MAMI as long as we can keep service agent and terminal agent understand each other, we can change dynamically the interface on the client and/or server side. So in case of introduction of new services or new interface there is not need to change the terminal side. This introduces the flexibility of service introduction or modification.

5.4   Web Prefetch

Web prefetch is the technique to fetch information from the Internet which are likely to be requested in near future before the request is actually made. In MAMI, service agent along with terminal agent follows user's interests, behavior and daily routine to build intelligence in it which will guide it in requesting web pages even before the user requests it [15] [9]. Terminal agent can request to fetch content from Web into the cache of Support station well before user has made the request. So that when user really makes the request service agent does not need to go to Internet. Instead service agent sends the already fetched content from cache. In this scenario, prefetched content comes upto cache of support station. It does not go to terminal device until user really makes the request. So user do not have to pay for any prefetched content. But service providers will be interested to incorporate this feature because this will increase system's performance. This will improve response time as well which is critical in wireless WWW access. Again, since the cache of support station contains all the recent pages fetched by the users under this support station, it is more likely that a great portion of the requests are satisfiable from the local cache.

5.5   Web Cache Maintenance and Relocation

Since the speed of Internet browsing is dominated by the time to response required by the remote server, it is always a good idea to maintain a cache of recently visited sites so that further requests can be satisfied from the cache. In MAMI both the terminal agent and service agent maintain cache. As the user moves from place to place, service agent can help by bringing the user specific cache to the support station under which the user is currently residing. That is a mobile agent can follow the user's physical mobility [4] and create the current network node prepared in well advance for the user's convenience and satisfaction.

5.6 Path Prediction

MAMI applies path prediction [10][13] to improve the performance even more. Path prediction algorithm finds out the user's next cell based on current cell location, user's history and movement. Terminal agent can learn from user's movement pattern, history or daily routine and can build up knowledgebase from which it can predict the next cell where the user is going to be in near future, then it can tell provider agent to mobilize service agent's cache so that the handoff is more efficient and system's response time does no hamper with user's movement. Several path prediction algorithms [10] [13] have been proposed. In MAMI the path prediction algorithm applies intelligence along with these algorithms. The terminal agent keeps track of user's daily routine and learns about users movement pattern from the user and provider agent and terminal agent inter operates to find the most accurate next cell. Based on the result of this algorithm provider agent can mobilize the mobile agents along with the required cache to the right place.

6 Related Works

Major works related to the realization of mobile access to WWW are [8] and [1]. Both are proxy based system for mobile Internet. But none of them issues the movement of users in such a system. [12] discusses how mobile agents can be applied in mobile Internet but a complete system architecture and operation are not provided. Java based mobile agent systems performance in an ATM based testbed is presented in [12]. Mobile agent based service provisioning is discussed in [6]. How mobile agents can be used to deliver registered services to user who is moving and changing his devices dynamically. How proxy cache can be relocated so that it follows the movement of the user is discussed in [11][11]. How mobile agents can operate in heterogeneous environments is discussed in [3].

7 Conclusion

In this paper we have provided a complete system employing mobile agents. Several techniques to optimize the system performance and security are integrated with the system. The benefits or advantages or such a system is taken into full consideration. A complete mobile agent based system for mobile Internet, its different aspects, applicability and benefits were the main topics discussed in this paper.

References

1. C. Baquero et al.: MobiScape: WWW browsing under disconnected and semi-connected operation, in Proceedings of the First Portuguese WWW National Conference, Braga, Portugal, (July 1995).
2. D. B. Lange, and M. Oshima: Seven good reasons for mobile agents, Communications of the ACM, vol. 42, Issue 3, pp. 88-89, March 1999.
3. F.M.T. Brazier et al.: Agent factory: generative migration of mobile agents in heterogeneous environments.
4. G. P. Picco: Mobile Agents: An introduction, Journal of Microprocessors and Microsystems, vol.25, no.2, pp. 65-74, April 2001.
5. NTT DoCoMo R&D , i-Mode Network System,
   http://www.nttdocomo.co.jp/corporate/rd/tech_e/imod01_e.html,   accessed   on
   12th April, 2004.
6. P. Farjami et al.: Advanced service provisioning based on mobile agents, Computer Communications 23 (8) (April 2000).
7. P. Farjami et al.: A mobile agent-based approach for the UMTS/VHE concept, in proceedings of Smartnet'99-the fifth IFIP conference on intelligence in networks, Thailand, November 1999.
8. R. Floyd et at.: Mobile web access using eNetwork web express, IEEE Personal Communications Magazine (October 1998).
9. R. W. Hill et at.: Anticipating where to look: predicting the movements of mobile agents in complex terrain, Proceedings of the first international joint conference on autonomous agents and multiagent systems: part 2, 2002, Bologna, Italy.
10. S. Hadjiefthymiades et at.: ESW4: Enhanced scheme for WWW computing in wireless communication environments, ACM SIGCOMM Computer Comunication Review 29(4), 1999.
11. S. Hadjiefthymiades et at.: Using proxy cache relocation to accelerate web browsing in wireless/mobile communications, Proceedings of the tenth international conference on World Wide Web Conference, 2001, Hong Kong, Hong Kong.
12. S. Hadjiefthymiades et at.: Supporting the WWW in Wireless Communications Through Mobile Agents, Mobile Networks and Applications Vol. 7, 2002, pp.305-313, Kluwer Academic Publishers.
13. T.Liu et at.: Mobility modeling, location tracking, and trajectory prediction in wireless ATM networks, IEEE Journal on selected areas in communications 16(6), August 1998.
14. The MOWGLI Project:
    http://www.tml.hut.fi/Studies/Tik-110.300/1999/Wireless/mowgli_1.html,   last
    accessed 6th June, 2003.
15. Z. Iang et at.: Web prefetching in a mobile environment, IEEE Personal Communications, pp 25-34, Vol.5, Issue 5, 1998.
16. AGLETS: http://www.trl.ibm.co.jp/aglets last accessed on 5th July 2003 3:00 PM JST.
17. VOYAGER: http://www.recursionsw.com/products/voyager/ Last accessed on 1 July 2003 9:30AM JST.

Design and Implementation of an ANTS-Based Test Bed for Collecting Data in Active Framework

Victória Damasceno Matos[1], Jorge Luiz de Castro e Silva[2], Javam C. Machado[1], Rossana Maria de Castro Andrade[1], and José Neuman de Souza[1]

[1] Universidade Federal do Ceará (UFC)
{victoria, rossana}@lia.ufc.br, {javam, neuman}@ufc.br
[2] Universidade Federal de Pernambuco (UFPE)
jlcs@cin.ufpe.br

Abstract. The active networks approach has been presented as an alternative technology for solving several problems in conventional networks, mainly in the network management area. This work describes a data retrieval mechanism test bed, including design and implementation, which can be used for managing SNMP enabled networks, based on the ANTS (Active Node Transfer System) toolkit framework developed at the Massachusetts Institute of Technology (MIT). The results obtained from the experiments have shown that the chosen approach is a good alternative for data retrieving operations.

1   Introduction

The motivation for the proposed model came from the fact that traditional management approaches like SNMP (Simple Network Management Protocol) and CMIP (Common Management Information Protocol) generate excessive network traffic, as they have mechanisms based on the client-server model, where the manager centralizes the information and provides scheduling for execution of corrective actions, while the agent interacts with the MIB (Management Information Base) and executes the requests issued by the manager.

Another important concern is the network management applications that may benefit from the use of the active network paradigm. Regarding active network studies [1, 2, 3, 4, 5], it has been verified that one of the biggest platforms embedding this technology is the ANTS (Active Node Transfer System).

In the proposed model, some procedures were implemented within the ANTS platform, enabling data acquisition from an initialization process within an active node. In the process, an active node sends capsules (corresponding to traditional network packets) that execute locally the collection task.

With this model, which provides dynamic network traffic parameters, it is possible to assist performance management methods, mainly those that use active technologies.

J.N. de Souza et al. (Eds.): ICT 2004, LNCS 3124, pp. 251–256, 2004.

2   Active Networks

Traditional networks guarantee data sharing in the sense that packets can be efficiently carried among connected systems. They only make the processing needed to forward the packets to the destination. This kind of network is insensitive to the packets it carries. The packets are transferred without modification [6]. Computation role in these kinds of networks is extremely limited.

Active networks are a new approach to network architecture that allow their users to inject customized programs into the nodes, enabling these devices to perform customized processing on the messages flowing through them [1, 6]. This approach can be applied both to users and applications.

The results obtained from the active networks in network management applications are of great interest to this work.

2.1   Active Node Transfer System (ANTS)

Developed by the Massachusetts Institute of Technology (MIT), ANTS (Active Node Transfer System) is a toolkit, written in Java where new protocols are automatically deployed at both intermediate nodes and end systems by using mobile code techniques [2].

ANTS views the network as a distributed programming system and provides a programming language-like model for expressing new protocols in terms of operations at nodes. The tool combines the flexibility of a programming language with the convenience of dynamic deployment.

In the ANTS architecture the capsules which refer to the processing to be performed on their behalf replace the packets in traditional networks include a reference to the forwarding routines to be used for processing at each active node. There are forwarding routines that will be found in all the nodes. Others are specific to the application and will not reside at every node, but must be transferred to a node by means of code distribution before the capsules of that type can be processed for the first time. The code group is a collection of related capsule types whose forwarding routines are transferred as a unit by the code distribution system. Finally, the protocol is a collection of related code groups that are treaded as a single unit of protection by an active node. The active nodes have the responsibility to execute protocols within restricted environments that limits their access to shared resources.

3   Proposed Retrieval Mechanism

The module, which was designed and implemented, is composed of a few scripts that are responsible for the configuration of the notes to be monitored. The module also contains forwarding application capsules and Java classes that are responsible for structuring the capsules and the data collection process.

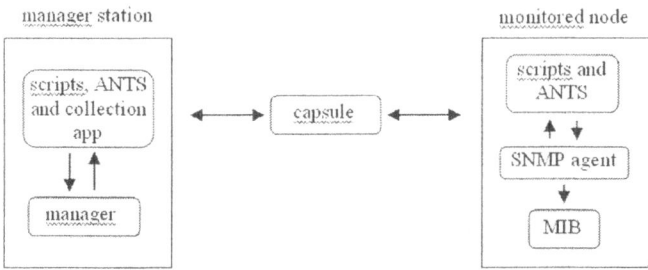

Fig. 1. Data retrieval process

In other words there is an interaction between the application and the SNMP agents in order to get MIB information from the analyzed nodes. Figure 1 shows how the mechanism works. The manager station sends capsules with routines that must be executed in the monitored node, thereby obtaining the desired information.

The first step towards the implementation was to establish ANTS active environment where an active application is due to execute the routines desired in a node. To activate a node it is necessary to associate a virtual IP and an entry to a determined active application in the node, where such configuration is passed as a parameter through a script (*coleta.config*). Thus, when the capsule of a certain application arrives at an active node, it executes there the desired routines.

Another important factor is the routes to be followed by the capsules. The forwarding must be pre-established in one script called *coleta.routes*, which can be generated by the ANTS.

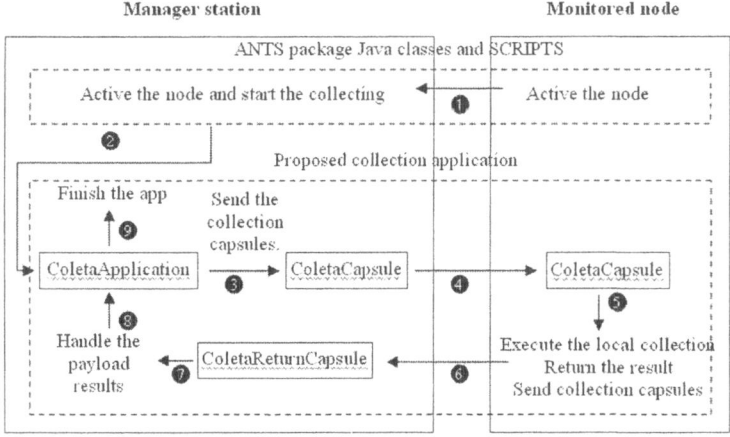

Fig. 2. Execution model of proposed retrieval mechanism.

After that (scripts and ANTS package), the application is ready to work. It is composed essentially of three distinct classes: *ColetaApplication*, *ColetaCapsule* and *ColetaReturnCapsule*.

The *ColetaApplication* class starts the data capture, manages the sending and receiving of capsules and finishes all the process in the manager node. The *ColetaCapsule* creates the capsules that embed the local collection code that will be sent to the monitored nodes. This way, it is possible to send new collection capsules to the monitored nodes, which implies a distributed model. Finally, the *ColetaReturnCapsule* creates the capsules that will return the results and treats the values returned to the manager node. Figure 2 shows the steps of this execution model.

3.1 Advantages of the Active Retrieval Mechanism

An analysis was made that compares the proposed mechanism with the native model of the management protocol using SNMP (Figure 3) in order to show the performance of both approaches.

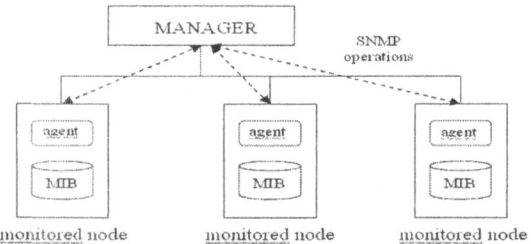

Fig. 3. Traditional network framework with SNMP protocol.

In this environment, which was based on the client-server model, there is a great number of messages exchange. The manager, thus, is interacting with all the managed devices to get the management information from the MIB agents. This can result in decreasing performance, as the management centralization will overload the manager and the tremendous number of messages exchanges between the manager and the agent will eventually increase the network traffic.

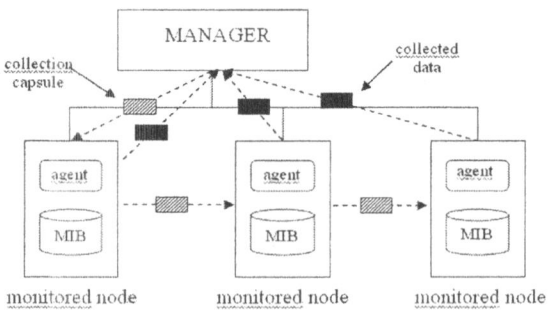

Fig. 4. Distributed framework of data collection

As previously seen, the traditional form of getting MIB data is centralized, thereby forcing the manager to capture information from all the devices that it manages. In the proposed method, any active node, not necessarily the manager, can send a capsule to run the collection in another active node and one single capsule can run the collection operation in some devices, as represented in Figure 4.

When the process is started, only one capsule that is sent to the agents can acquire information from all monitored node. This is possible because an agent is able to send collection capsules to other agents. The obtained results will be returned to the management station through capsules too.

4   Results

A comparison was made between the time spent by the proposed collection model and the traditional model of SNMP, using the ucd-SNMP tool [7]. It was discovered that, by employing the proposed mechanism, the capture of the entailed values of some OIDs, in only one node, results in a longer time. But when some nodes were added to the collection process and the distribution mechanism was used, the performance was significantly improved, as explained in Figure 4. This conclusion was arrived to by looking at the average of the obtained results that are graphically presented in Figure 5. In one experiment, two nodes (the manager and the agent) were used. In another experiment, three nodes (one manager and two agents) were employed. The benchmarks ware made with and without distributed forwarding and the varied collected OIDs numbers.

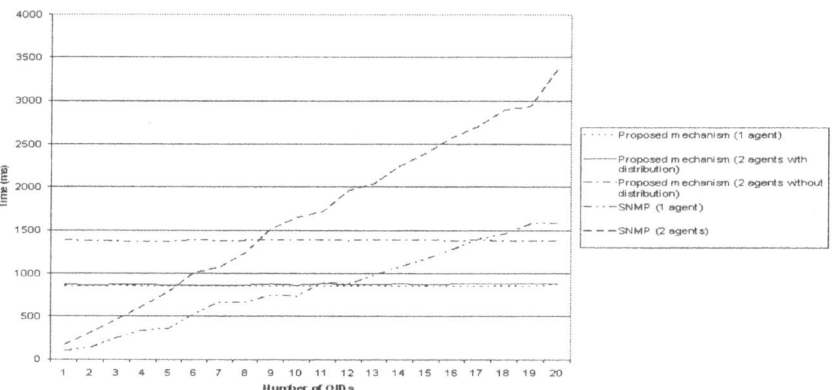

Fig. 5. Benchmarks.

A more significant improvement has been found when the number of OIDs to be consulted was modified. Using the experiment with three machines, one can notice that the increase on the number of OIDs to be obtained has more impact on the elapsed time when the environment is the traditional SNMP. This is explained by the fact that, in the proposed method, the manager sends only one capsule with all the OIDs to be collected; while in the traditional SNMP, some PDUs (Protocol Data Unit) are sent to get data.

In the practical study, it was verified that the configuration of the routes is also an excellent factor to obtain better times. This way, in an environment with many machines, it is not enough to have a distributed management without evaluating that points would be advantageous to make the decentralization. It was noted that in a monitoring of few machines, the collection system must be totally distributed, or either, a monitored node to pass the collection capsule one another node to be monitored. However, as the number of nodes increases, there can be a delay in the collection with this decentralization model. Therefore, it would be of great value the study and the implementation of an optimal solution in the establishment of the routes.

5   Conclusions

The motivation of this work is the deficiencies presented in the traditional management protocols, based on the client-server model. They are centralized management models that may generate a great number of exchanges of messages between an agent and a manager, wasting bandwidth and resulting in delays in the network monitoring.

In the analysis that were performed, it was noted that one of the most important active platforms is the ANTS, a toolkit written in Java developed by the Massachusetts Institute of Technology (MIT) that allows a dynamic construction and implementation of network protocols. For this reason, this framework was used in the development of the proposed module.

The aim of optimizing the use of the broadband and diminishing the time spent on collecting data has been reached, because the proposed model is a data collection mechanism based on distributed management and by the fact that only the capsules that will indicate the start of the local collection processing are sent though the network. This is achieved through local SNMP operations, preventing the excess of messages exchange between the client and the server. Benchmarks were conducted that prove the benefits on active data collection claimed in this work.

References

1. Tennenhouse, D. L. A Survey of Active Network Research. *IEEE Communications Magazine*, v.35, n.1, p.80-86, Jan., 1997.
2. Wetherall, D.J.; Guttag, J.V.; Tennenhouse, D.L. Ants: „A Toolkit for Building and Dynamically Deploying Network Protocols. In: IEEE OPENARCH'98, San Francisco, CA, Apr. 1998.
3. Schwartz, B.; Zhou, W.; Jackson, A.W. Smart Packets for Active Networks. *BBN Technologies*, Jan. 1998.
4. Wetherall, D.J.; Tennenhouse, D.L. The ACTIVE_IP option. In: The 7th ACM SIGOPS European Workshop, 7, Sep., 1996.
5. Gunter, C.A.; Nettles, S.M.; Smith, J.M. The Switch Ware Active Network Architecture. *IEEE Netwo*rk, special issue on Active and Pro-grammable Networks, v.12, n.3. , May/Jun, 1998.
6. Tennenhouse, D. L.; Wetherall D.J. Towards an Active Network Architecture. *Computer Communication Review*, v.26, n.2, p.5-18, Apr., 1996.
7. Net-SNMP website. Available in: <http://www.net-snmp.org>. Access in: 20 out. 2002.

Adaptive QoS Management for Regulation of Fairness Among Internet Applications

Miguel F. de Castro[1]*, Dominique Gaïti[2], and Abdallah M'hamed[1]

[1] GET/Institut National des Télécommunications
9 rue Charles Fourier - 91011 Evry Cedex, France
{miguel.castro;abdallah.mhamed}@int-evry.fr
[2] LM2S - Université de Technologie de Troyes
12 rue Marie Curie - 10010 Troyes Cedex, France
dominique.gaiti@utt.fr

Abstract. Emerging multimedia applications in the Internet characterizes the trend of service convergence in one only infrastructure. These new types of media can, however, be harmful to other flows sharing the same infrastructure, such as TCP-based applications. The most common problem is fairness, due to the unresponsive behavior of UDP flows. This article presents a fairness regulation scheme based on an adaptive QoS management architecture, where nodes are responsible for certain level of management operations, like surveillance and parameter tuning. An adaptive meta-behavior is defined in order to adjust scheduling parameters according to UDP/TCP traffic ratio.

1 Introduction

The coexistence of the new generation multimedia services with ubiquitous TCP-based applications, like WWW, FTP and Remote Database Access can be complicated. This is due to the fact that most multimedia applications are based on UDP transport protocol, which does not offer *a priori* congestion control mechanisms (*unresponsive* behavior). Therefore, when TCP and UDP applications compete for scarce resources, TCP applications take disadvantage because of their intrinsic congestion control mechanisms, which reduces throughput in order to adapt to congestion. Moreover, sharing resources among multiple TCP flows characterized by different RTT (Round-Trip Times), for example, is also subject of unfair resource distribution, as shown in [1,2]. These situations illustrate the *fairness problem* among these TCP and UDP flows.

This article proposes an adaptive approach based on manipulating scheduling parameters in order to reduce fairness problems among TCP and UDP flows. This solution is deployed over an Adaptive QoS Management Architecture based on Active Networks [3]. Simulations have shown that this adaptive approach can indeed reduce fairness problem by controlling scheduling parameters.

---

* Supported by CAPES Foundation/Brasil.

J.N. de Souza et al. (Eds.): ICT 2004, LNCS 3124, pp. 257–266, 2004.
© Springer-Verlag Berlin Heidelberg 2004

The remaining of this article is organized as follows. In Section 2, we outline the fairness problem in the Internet. Section 3 presents the Adaptive QoS Management Architecture employed to this work. A case study based on simulation is shown in Section 4. Obtained results are outlined and evaluated in Section 5, while Section 6 presents our conclusions and outlooks.

2   Fairness Among Internet Applications

Network management role in the new generation of Internet is still more important than ever. After the first big collapse threat, the Internet was enhanced by TCP's congestion avoidance and flow control. With service convergence, the challenge is being to create management mechanisms to accommodate several service behaviors under a common infrastructure.

Moreover, as emerging streaming media applications in the Internet primarily use UDP (User Datagram Protocol) transport, it is still harder to exert more strict control in order to avoid network congestions. These new UDP applications generate large volumes of traffic which are not always responsive to network congestion avoidance mechanisms, causing serious problems on fairness [4]. Hence, if no control is taken, such unresponsive flows could lead to a new congestion collapse [5].

While a definitive solution to this problem is not yet deployed, standardized and broadly employed, network management must take this problem into account and try to solve it with a core network point of view. We argue that the effects of this fairness problem among TCP and UDP applications can be minimized by performing an adaptive management on packet scheduling parameters, in order to regulate packet loss rate. To issue such an adaptive control, we employ an architecture for adaptive QoS management as described on next section.

3   An Architecture for Adaptive QoS Management

With the aim to achieve the capabilities of an adaptive and programmable QoS management described above, we propose a new architecture where a management plane is inserted into the Internet layer architecture. This kind of layer structure in three dimensions is inspired from the B-ISDN (Broadband Integrated Services Digital Network) [6] reference model, which has a separate plane where all the management functions are implemented. However, our management plane is based on Active and Programmable Networks in order to enable adaptive management. This new plane will store legacy management mechanisms and it will also be able to receive new mechanisms by offering a Management API (MAPI) and an Execution Kernel, which constitute the *Adaptive Management Execution Environment*. The new architecture for the Adaptive Management is shown in Figure 1.

The Adaptive Management Plane is composed by the system hardware (dedicated to management functions), which is shared between two entities: the set of

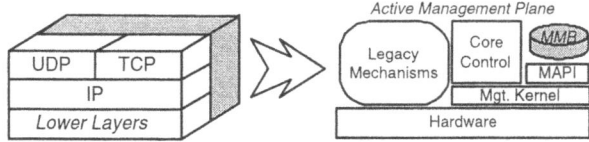

Fig. 1. Adaptive Management Architecture.

Legacy Mechanisms and the Management Kernel. On the top of the Management Kernel, there are the Core Control, the Management Application Programming Interface (MAPI) and the Management Module Base (MMB).

The Legacy Mechanisms are built-in mechanisms that are intrinsic to the network architecture chosen. Examples are Drop-Tail, ECN, RED, etc. These mechanisms were chosen to be stored separately from customized mechanisms because as they are strictly inherent to the network architecture employed (*e.g.* IPv6, TCP, UDP, etc), they are already standardized and have no need to be taken away. As they are unchangeable, they can profit of a hardware optimized implementation and represent the core management mechanisms available. Nevertheless, adaptive management is able to control them (enable and disable) and to deal with their parameters and configurations.

The MMB is a persistent memory that stores the implementations of the customized management mechanisms, which were injected into the node by management designer. MMB updates are completely controlled by the Core Control element and, consequently, by central carrier management entities.

The Management Kernel is an Operating System responsible for the creation and the processing of Execution Environments to the Core Control and to all management mechanisms that are built on the top of the MAPI and stored in the MMB. The Kernel controls local resources like queues, memory and processing power.

The role of the Core Control is to manage the execution of running mechanisms, to monitor network status and decide, based on policies installed by the carrier network manager, to swap running mechanisms with available idle mechanisms according to the observed network status. The Core Control is also responsible to serve management information to carrier management system. The MAPI contains functions and procedures needed for management mechanism designers to have access to low level functions and resources.

We identify two different development levels for programmability to adaptive management. In the first level, which is simpler to implement, the role of adaptive management is network monitoring and management control by means of adaptive reconfiguration and notifications. In this case, forwarding and management functions are mutually independent, and management processing overhead does not influence forwarding performance. On the other hand, with the second level, the MMB is able to receive new implementations of mechanisms in addition to reconfiguration capabilities. These new implementations can be, for instance, a new enhancement for Random Early Detection (RED) or a new scheduling algorithm. As this new phase imposes the execution of software-based control to forwarding functions, its feasibility depends on the development of high performance execution capabilities. So far, we are already able to see reasonable performance in software-based network elements deployed over specialized hardware, such as Intel^{TM} IXP1200 Network Processor [7]. Hence, we believe that the required capabilities for this second phase of adaptive management will be available shortly.

The architecture described in this article supports programmability of a number of modules, such as behavior, information, notification, configuration, etc. The relationship among these modules performs the desired Adaptive Management procedure. A set of implemented modules describes a *meta-configuration* or *meta-behavior*. Inside the architecture, a *meta-configuration* is seen as an instance of the Adaptive Management Execution Environment. A meta-configuration is expressed in a new Domain Specific Language (DSL) called CLAM (Compact Language for Adaptive Management).

Security in this architecture is issued by protecting the entrance of unknown active management packets inside the network cloud, *i.e.* border routers must filter this kind of packets, unless if it comes from a trusted port. Moreover, employment of limited syntax language protects the architecture from safety threats, for instance, infinite loops.

As the architecture is intended to manage a limited carrier domain, scalability is not a problem, once management functions inside the network do not increase as the number of end users increases.

Performance is also one of the main concerns on this architecture. The idea of keeping a management plane separate from operational functions is intended to render network management functions as independent as possible from transport functions. Although management plane can control configuration of transport, there is no interdependence between these two planes, *i.e.* management plane does not block transport plane while management actions are taken. Statistic gathering, for instance, is done by analyzing clones from intercepted packet samples.. Thus, while management plane analyses packet contents, original package continues its way. Hence, we consider that this architecture is compliant with ForCES (Forwarding and Control Element Separation) idea [8].

4   Case Study: Adaptive Fairness Regulation

In order to better understand the problem we want to deal with, we have performed some preliminary experiments. First of all, we reproduce the fairness problem described earlier on a simple network topology illustrated in Figure 2. Although very simple, the congested interface of the server-side IP router in this topology can represent the overall behavior of a general congested interface from a network node in more complex topologies. Hence, results obtained for the node in this simple topology can be abstracted to overall interfaces in more complex networks. This assumption is valid because we do test an isolated behavior of a network interface, without direct interaction with other network elements, thus not subject to scalability threats. Further, there is no per-flow control, which would be subject to scalability problems.

In our simulated topology, a given number of clients have access to TCP-based services (Web, FTP, Mail, DB query and Telnet) and UDP-based services (Video on Demand, Voice over IP and Videoconferencing). Across a 20 Mbps serial link, we have a set of servers in enough quantity to avoid server-side bottlenecks (servers' loads have been watched). Two IP routers enabled with a common set of QoS management mechanisms (queue management and scheduling) are disposed at the two borders of the serial link. We use OPNET^{TM} Modeler tool to perform simulations.

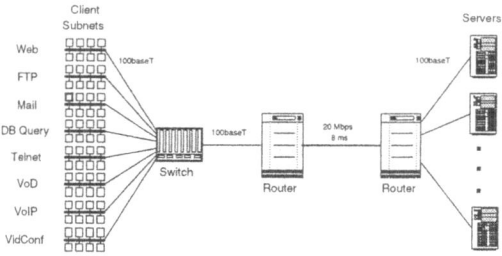

Fig. 2. Simulated Topology.

We can easily observe the phenomenon of fairness problem by a simple experimentation, where 300 TCP clients compete with a variable number of UDP clients for the 20 Mbps link. Each UDP client generates a 910 Kbps CBR (Constant Bit Rate) flow. In the router, we have an output interface with one only queue, with capacity of 100 packets, enhanced with RED (Random Early Detection) [9], with the following parameters: $(Max_{th} = 100; Min_{th} = 50; p_{max} = 10)$. We then calculate the fairness index for this scenario as proposed in [10]:

$$F(x_1, x_2, \cdots, x_n) = \frac{\left(\sum_{i=1}^{n} x_i\right)^2}{n \times \sum_{i=1}^{n} x_i^2}; \qquad x_i = \frac{T_i}{O_i} \qquad (1)$$

where T_i and O_i represent the measured and the fair throughput for flow i.

For this experiment, we consider the fair throughput for an UDP flow as the fixed CBR generation throughput, while for TCP flows we consider the fair throughput as the maximum throughput achieved by a TCP flow when there is no competing UDP flow. Figure 3(a) shows the results of this experiment. In this Figure, we can see that the fairness index calculated with Equation 1 accentually decreases as the number of UDP flows increases. Numerically, we observed along the UDP load increase that TCP flows have to reduce their throughput in average 62%, while UDP flows have their throughputs decreased in average only 11%, due to packet drops.

One possible solution to this fairness problem is to compensate the absence of congestion control mechanisms in UDP by dropping their packets. Consequently, the first step is to have a minimum classification scheme in order to identify UDP traffic and affect to it a higher discard probability. This classification may be restricted to assigning the ToS (Type of Service) field from IP header on UDP packets, which can be easily done in border nodes of the network, for example.

After this first differentiation, we must find a way to treat differently these two types of traffic. We employ the Custom Queuing (CQ) [11] from Cisco. Initially, we want to evaluate the impact in fairness of simply separating the traffic into two different queues with equivalent priorities, without – for the moment – adapting parameters. We defined two queues with capacity of 50 packets each, both of them with RED tuned as follows: ($Max_{th} = 50; Min_{th} = 30; p_{max} = 10$). Results are shown in Figure 3(b). As expected, we can observe that until the number of UDP clients reach 10, fairness index decreases, and after it begins to increase. This is due to the equality of priority that we have given to both queues. 10 UDP clients represent in average 10 Mbps of submitted throughput, which corresponds to 50% of output link capacity. As, since 10 UDP sources, there is no more bandwidth share available to UDP traffic, and the two queues are served in the same fashion, UDP packets begin to be more often rejected due to buffer overflows, and we begin to see fairness index increases again.

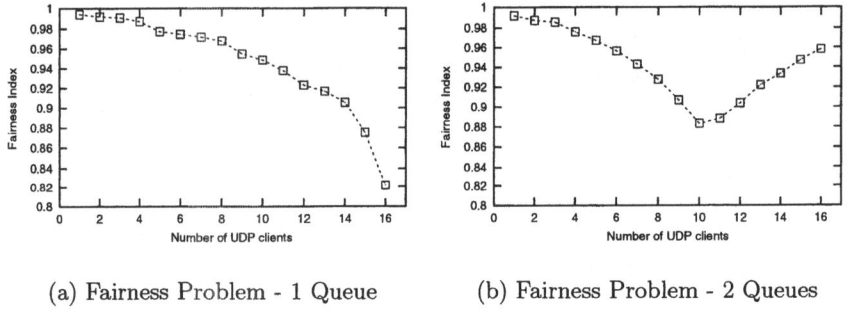

(a) Fairness Problem - 1 Queue   (b) Fairness Problem - 2 Queues

Fig. 3. Fairness Problem Observation Experiments with One and Two Queues.

Although we were able to observe a certain level of enhancement in fairness index, it is still not fair to share bandwidth in such a static fashion, giving 50% resources for each traffic type.

Therefore, it may be desirable for controlling fairness to adapt UDP packets drop according to congestion situation and to UDP/TCP traffic ratio. This adaptation can be issued by a customized meta-behavior designed over the above presented architecture.

4.1   Meta-configuration

The fairness control meta-behavior we want to build will be based on a simple heuristic: we must observe a sample of incoming packets by its marking (TCP- or

UDP-based), and then determine the appropriate "Byte Count" parameter for Custom Queuing on each of the two queues. The incoming packet sample gathering is issued by configuring the management core control to collect information from one intercepted packet at each $\Delta\tau$ ms. After collecting information from N packets, a decision must be made about changing CQ Byte Counts. Hence, one decision (change) is taken at each $\Delta T_d = \Delta\tau \times N$ ms, based on the UDP count ratio among the N collected values. Therefore, this new configuration will be activated for the next ΔT_d ms. After a decision, all counters are reset and counting process restarts.

In order to determine the best mapping between UDP counter ratio and good CQ configuration, we have performed several simulations, where the UDP load was varied. For each traffic situation, different CQ configurations were tested. Analyzing the obtained results, we have found the average expected CQ configuration profile for enhancing fairness according to UDP traffic ratio. From these experimental results, we were able, by means of a simple linear regression, to find the CQ Byte Count coefficients γ_{tcp} and γ_{udp} that characterize the ratios of total Byte Count (BC_{all}) given to TCP and UDP queues, respectively. These coefficients, in their turn, are function of the UDP and TCP traffic ratios (R_{tcp} and R_{udp} in [0;1]). Hence, the obtained coefficients are well suited to our network traffic behavior:

$$\gamma_{tcp} + \gamma_{udp} = 1$$
$$BC_{tcp} = \gamma_{tcp} \times BC_{all} \therefore BC_{udp} = \gamma_{udp} \times BC_{all}$$

In order to protect from queue starvation due to mistaken decisions, we decided to determine a floor of 0.025 for each coefficient, $i.e.$ each queue will have at least 2.5% of the total Byte Count.

Table 1. Simulation parameters.

parameter	name	value
Total Simulation Time (s)	T_{sim}	430
Transient Sim. Time (s)	T_{trans}	130
Total Byte Count	BC_{all}	20,000
Number of Collected Packets	N	1,000
Time between collects (ms)	$\Delta\tau$	(1, 2, 4, 8, 16, 32, 64)
Number of UDP Clients	N_{udp}	variable (1 $\rightarrow$ 20)
Number of TCP Clients	N_{tcp}	300

Table 1 shows all simulation parameters defined for our experiments. The simulation transient time was calculated following techniques found in [10]. Several replications of the same scenario were simulated in order to obtain a reasonable confidence interval.

From the preliminary results, we determined the good values for γ_{tcp} and γ_{udp} in order to calculate CQ parameters each ΔT_d ms. Thus, the management core control was configured with a meta-behavior where incoming traffic is observed and CQ configuration is adapted in real time. In order to test this meta-behavior, we have created a generic traffic scenario where the number of active clients is varied along time. Management core control does not have direct access to information such as the instantaneous number of active clients. Thus, UDP traffic ratio must be estimated from collected packet samples. Results are shown in the next section.

5   Results and Discussion

This section shows the results obtained from enhancing the simulated routers in our topology with the meta-behavior described in the last section. In Figure 5(a), we can see the mean UDP load inferred by the meta-behavior and the corresponding values of γ_{udp} for a scenario where $\Delta\tau = 4ms$, *i.e.* one packet examined at each $4ms$. This values of γ_{udp} were used to adapt in real-time values for BC_{tcp} and BC_{udp}. We can observe that, as stated earlier, γ_{udp} is directly proportional to the UDP traffic ratio.

In order to determine the best value to ΔT_d, we have tried different values for $\Delta\tau$, holding the number of packets examined before decision $N = 1000$. Hence, we tried $\Delta T_d = (1s; 2s; 4s; 8s; 16s; 32s; 64s)$. Figure 4(a) shows the average fairness index value calculated for each one of these scenarios (Active t=1s ... 64s). We also tested the two former scenarios where there is no adaptive management: 1 Queue (single queuing with no traffic differentiation) and 2 Queues (one queue to UDP and the other to TCP, keeping the same fix value to CQ Byte Count to each queue).

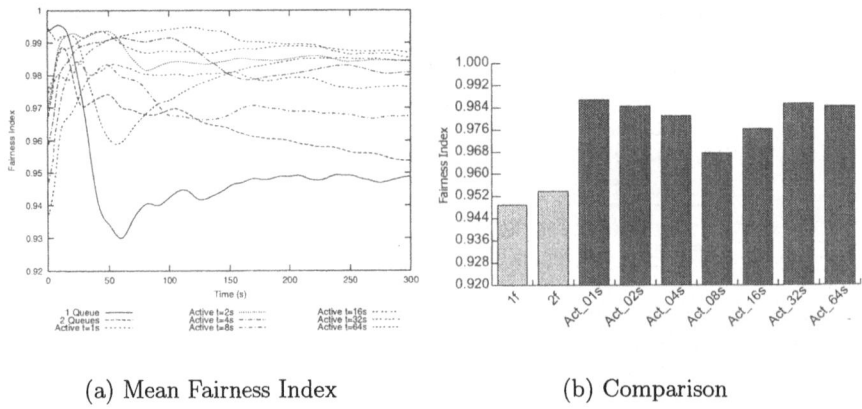

(a) Mean Fairness Index   (b) Comparison

Fig. 4. Comparison among mean fairness index values.

As we can observe from this figure, all tested scenarios with active meta-behaviors have presented fairness enhancements when compared to the two fixed configurations (1 Queue and 2 Queues).

Figure 4(b) compares the mean fairness index values for all scenarios. We can observe that when we isolate all active scenarios, we had the lower fairness index with $\Delta T_d = 8s$. Nevertheless, fairness obtained with this value of ΔT_d is still better than the static scenarios. That means that 8s was not enough to infer an accurate value for the actual UDP traffic ratio or either the decision taken based on the eight last seconds is not the best choice for the next 8 seconds. In fact, this decision depends on overall traffic behavior, more specifically its variability. On the other hand, the best values are found for $\Delta T_d = 1s$ and $\Delta T_d = 32s$. Therefore, we conclude that the decision granularity of $1s$ gets good results because it keeps in track of every minor change in traffic behavior, and the decision taken is often updates. On the other hand, $\Delta T_d = 32s$ is the choice that better fit the variability behavior of the traffic modeled in this simulations. However, $\Delta T_d = 1s$ is not interesting because it states that information must be gathered from one packet at each $1ms$, which in the worst case would mean collect the equivalent of 12 Mbps of data (considering $MTU = 1500$ bytes). Hence, this would generate an unacceptable overhead. Thus, the best choice for our network behavior is $\Delta T_d = 32s$.

If we observe the differences in fairness index shown in Figure 4(b), the difference between the best value obtained with active meta-behavior and the base fix configuration (1 Queue) is 0.038. Although this may seem a small difference, the impact over application-level performance is very significant. For instance, we have calculated the mean web page download time for a few scenarios. The results are illustrated in Figure 5(b). From these results, we have found an application-level performance recovery of average 20% when comparing $\Delta T_d = 32s$ and "1 Queue". In this case, mean page download time dropped from $6.81s$ to $5.47s$.

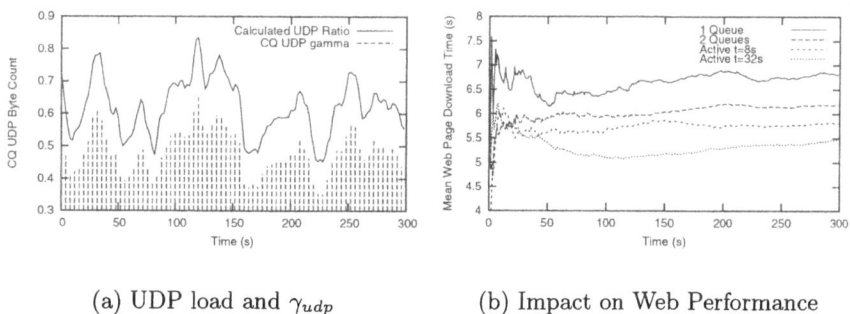

(a) UDP load and γ_{udp} (b) Impact on Web Performance

Fig. 5. Experiments Results.

6   Conclusions and Outlook

The results presented in the case study encourage the employment of adaptive management inside the network. Just as for the fairness problem, this architecture may permit the development and deployment of new heuristics to manage IP networks – especially with QoS constraints – in a flexible and efficient fashion.

In the case study presented in this article, we are able to see that fairness was indeed enhanced by means of active control. Although the best parameters found to these experiments may be specific for the applications and the traffic behavior presented in this topology, we argue that the architecture is flexible enough to allow adaptation to other network realities.

However, the problem presented in this case study does not employ all potentials of the proposed architecture. Indeed, most features offered by the adaptive management architecture was not necessary, thus not employed in this case study. As future work, we want to extend the gamma of management problems solved by means of adaptive meta-behaviors, also making use of further features of this architecture, such as customized notifications (communication) among network elements.

References

1. Brown, P.: Resource sharing of TCP connections with different round trip times. In: Proc. of IEEE INFOCOM 2000. Volume 3., Tel Aviv, Israel (2000) 1734–41
2. Zabir, S.M.S., Ashir, A., Kitagata, G., Suganuma, T., Shiratori, N.: Ensuring fairness among ECN and non-ECN TCP over the Internet. Intl. Journal of Network Management **13** (2003) 337–48
3. de Castro, M.F., M'hamed, A., Gaïti, D.: Providing core-generated feedback to adaptive media on actively managed networks. In: Proc. of IEEE/SBrT International Telecommunications Symposium (ITS 2002), Natal, Brazil (2002)
4. Hong, D.P., Albuquerque, C., Oliveira, C., Suda, T.: Evaluating the impact of emerging streaming media applications on TCP/IP performance. IEEE Communications Magazine (2001) 76–82
5. Braden, B., Clark, D.D., Crowcroft, J., Davie, B.S., Deering, S.E., Estrin, D., Floyd, S., Jacobson, V.: Recommendations on queue management and congestion avoidance in the Internet. RFC 2309, Internet Engineering Task Force (1998)
6. ITU-T: BISDN Reference Model. ITU-T Recommendation I.321. (1992)
7. Intel IXP1200: (Intel IXP1200 Network Processor Homepage) http://www.intel.com/design/network/products/npfamily/ixp1200.htm.
8. Khosravi, H., Anderson, T.W., Eds.: Requirements for separation of IP control and forwarding. RFC 3654, Internet Engineering Task Force (2003)
9. Floyd, S., Jacobson, V.: Random early detection gateways for congestion avoidance. IEEE/ACM Transactions on Networking **1** (1993) 397–413
10. Jain, R.: The Art of Computer Systems Performance Analysis. John Wiley & Sons (1991)
11. Vegesna, S.: IP Quality of Service. Cisco Press (2001)

Black Box Model of Erbium-Doped Fiber Amplifiers in C and L Bands

António Teixeira[1,2], Davide Pereira[4], Sérgio Junior[1,2], Mário Lima[1,2], Paulo André[2,3], Rogério Nogueira[2,3], José da Rocha[1,2], and Humberto Fernandes[5]

[1] Departamento de Telecomunicações, Universidade de Aveiro,
3810-193 Aveiro, Portugal
teixeira@ua.pt, http://www.ua.pt
[2] Instituto de Telecomunicações, Campus Universitário Santiago,
3810-193 Aveiro, Portugal,
{sjunior,mlima,pandre}@av.it.pt, frocha@ieee.org,
[3] Departamento de Física da Universidade de Aveiro,
3810-193 Aveiro, Portugal, rnogueira@fis.ua.pt,
[4] IC/WOW, Siemens S.A., Rua Irmãos Siemens,
2720-093 Alfragide, Portugal, davide.pereira-ext@siemens.com,
[5] Universidade Federal do Rio Grande do Norte,
Natal, Brasil, humbeccf@ct.ufrn.br

Abstract. In this paper we analyze Gain and Noise Figure of a C and an L band EDFA verifying experimentally the behavior of the involved parameters as a function of the power and the wavelength.

1 Introduction

In recent years, the EDFA has revolutionized the field of long haul transmission [1]. Typical long distance links installed today contain a booster amplifier at the transmitter, a series of in-line amplifiers along the transmission line and a preamplifier at the receiver. In contrast, the number of 3-R regenerators along the transmission line has dropped substantially in comparison with the traditionally equipped links. The predictability of the amplifiers figures, particularly the spectral characteristics for the gain and noise, plays an important role for the layout of such transmission systems; especially if the number and lengths of amplifier spans vary and/or if multivendor equipment is installed. With the advent of commercial long distance systems operating in Wavelength-Division Multiplexing (WDM), the predictability of the EDFA behavior has become crucial since the operating point of the EDFA is now affected by the power and the spectral characteristics of each WDM. In addition, optical amplifiers for WDM transmission are partly based on an increasingly complex internal circuitry.

The spectral gain and NF can be completely described by solving numerically (numerical models), for specified amplifier parameters, the propagation and rate

J.N. de Souza et al. (Eds.): ICT 2004, LNCS 3124, pp. 267–271, 2004.
© Springer-Verlag Berlin Heidelberg 2004

equations that model the interaction between the optical field with Erbium ions [2]. This methodology is heavy to solve, since longitudinal and transverse integrations of the equations are necessary, however it is the most approximated and used in EDFA simulations [3].

Another valid approach is to use the so-called analytical models; these meant that the resulting expressions providing basic EDFA characteristics (gain, noise, output spectra, etc.) turn into closed forms, i.e., no longitudinal integration being necessary. On a reduced-complexity end, the analytical models provide straightforward predictions on the basic EDFA characteristics, at the expense of some simplifying assumptions concerning the regime of operation.

In the class of analytical models we find the Black Box Model (BBM), based upon input-output measurements, which is useful when some of the internal parameters or construction details of the amplifier are not available. This will be the model used in this work. A simple and accurate BBM has been proposed, where gain and NF are described by empirical formulas [4].This approach will be applied experimentally to two EDFA one in the C-band (1530-1565nm) and the other in the L-band (1570nm – 1610nm).

2 The Model

The spectral properties of gain saturation, for a given wavelength, can be described by an empirical equation [4], which depends on the signal input power and small signal gain, given by:

$$G = \frac{G_o}{1 + \left(\dfrac{P_{in}}{P_{sat}}\right)^{\alpha}} = \frac{G_o}{1 + \left(\dfrac{G_o P_{in}}{P_{max}}\right)^{\alpha}} \tag{1}$$

In equation (1), G_0 and G are the small signal gain and the saturated gain, respectively, for a given input signal power P_{in}. The unknown terms are α and P_{sat}, or P_{max}, if we consider the maximum output power, which is given by $P_{max} = G_o P_{sat}$.

The main idea is to determine experimentally the small signal gain for the bandwidth required, and then a minimum of two sets of saturation gains for each signal wavelength are sufficient to determine the two parameters, α and P_{max}. With these parameters obtained, the saturated gain can be accurately modeled for a given signal wavelength. Using the same parameters for different signal wavelengths, the accuracy is not guaranteed, as we will see next.

In Fig.1 the measured gains for 1590nm are represented by squares, with small signal gain (Go) for $P_{in} \cong$ -45dBm equal to 26.95dB. Fitting the points results in $P_{max} =$ 26.619, and $\alpha = 0.789$. Substituting these parameters in (1), a plot of the predicted gain is made by the model, which is also shown in Fig.1 (continuous line). As we can see, there is good agreement between the experimental and modeled gains.

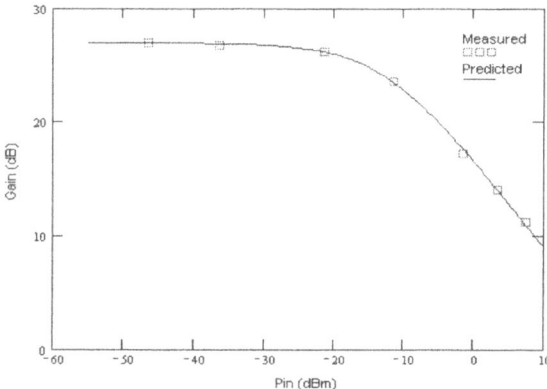

Fig. 1. Modeled and experimental gains for different input powers at 1590nm. The modeled gain is a function of P_{max} and α, previously determined using experimental gains

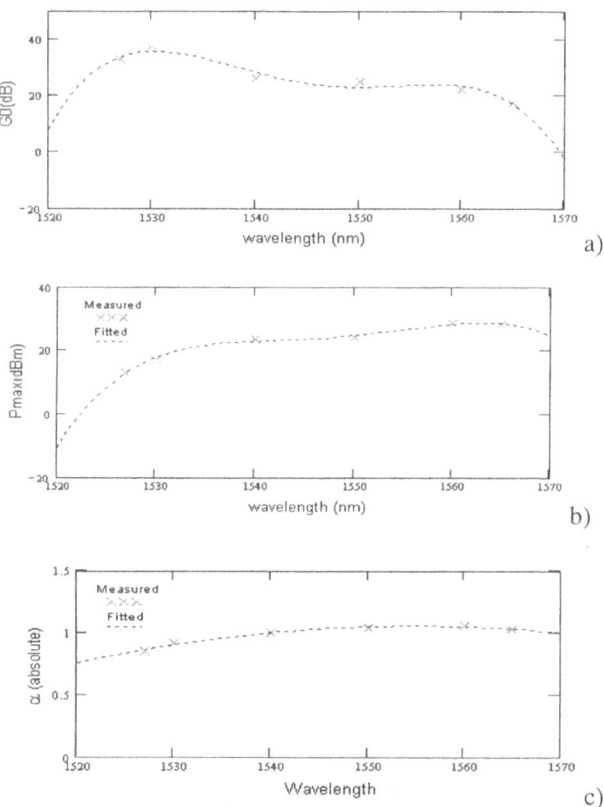

Fig. 2. Experimental and approximated values for a)G_o, b) P_{max}, and c) α.

Following a similar procedure for the rest of the wavelengths, good accuracy is kept and the parameters were extracted. Fig.2 a, b and c present G_0, P_{max} and α for C-Band, where the measured values are represented by 'x's.

Observing Figs 2 a, b and c we see that the parameters are not constant for the wavelengths considered. Reasonable approximations of these curves can be estimated by using polynomial regression, fitting regression values to the obtained by the model (G_0, P_{max}, or α), by means of a polynomial of any order.

Using polynomial regression, the coefficients of the polynomials are determined. For G_o and P_{max}, the polynomials were of fourth order, and for α, a second order polynomial was enough.

Using polynomial regression, the coefficients of the polynomials are determined. For G_o and P_{max}, the polynomials were of fourth order, and the coefficients are given by (2) and (3) respectively. To approximate α, a second order polynomial was enough, and its coefficients are given by (4).

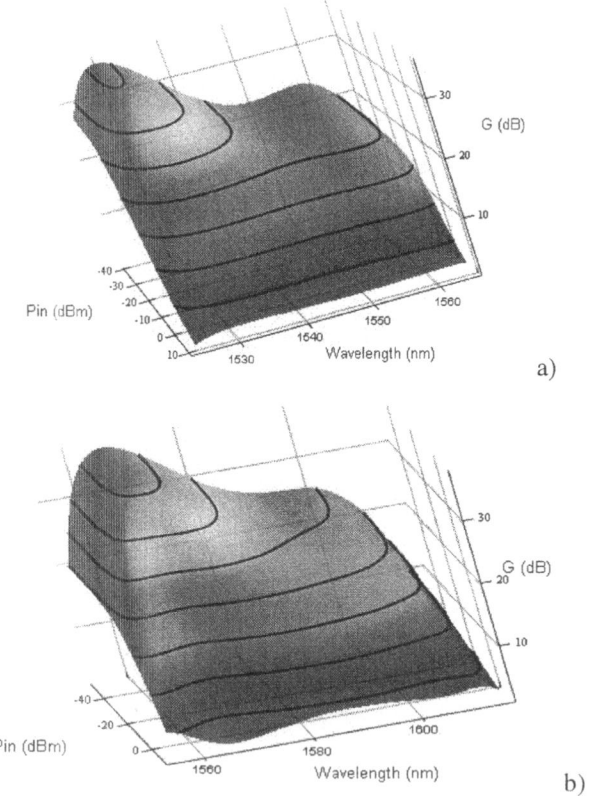

Fig. 3. 3D model representation for *C band* (a) *L-band* (b).

$$G_o(\lambda) \approx 30.22 - 0.62 \cdot (\lambda - 1580) + 0.02 \cdot (\lambda - 1580)^2 + 1.12 \cdot 10^{-3} \cdot (\lambda - 1580)^3 \\ - 4.57 \cdot 10^{-5} \cdot (\lambda - 1580)^4 \tag{2}$$

$$G_o(\lambda) \approx 30.22 - 0.62 \cdot (\lambda - 1580) + 0.02 \cdot (\lambda - 1580)^2 + 1.12 \cdot 10^{-3} \cdot (\lambda - 1580)^3 \cdot \\ - 4.57 \cdot 10^{-5} \cdot (\lambda - 1580)^4 \tag{3}$$

$$\alpha(\lambda) \approx 0.78 - 2.29 \cdot 10^{-3} \cdot (\lambda - 1580) - 2.44 \cdot 10^{-4} \cdot (\lambda - 1580)^2 \tag{4}$$

Applying (2), (3), (4) in (1), we obtain a completely defined gain model for C-Band. For the L band similar procedure was followed. In Fig. 3 we present the 3D models for the gain variation with input power and wavelength dependent parameters for the two amplifiers, C and L Band.

3 Conclusions

We have experimentally determined the parameters of BBM Model for the gain of EDFA's in C and L-bands. There is good agreement on the measured and modeled obtained values. The characterization of the BBM with the wavelength was made, and some analytical expressions for the parameters (G_o, P_{max}, α) were obtained for both amplifiers.

Acknowledgments. We would like to acknowledge the projects Wo-Net (POSI/2002/CPS40009), WIDCOM (POSI/35574/CPS/2000), FEDER and Siemens SA for the support through the ADI CONVOCO project.

References

1. E. Desurvire, D. Bayart, B. Desthieux, S. Bigo, Erbium-Doped Fiber Amplifiers, New York: John Wiley and Sons, (2002), 3-169.
2. C. R. Giles, E. Desurvire, "Modeling Erbium-Doped Fiber Amplifiers", J. Lightwave Technol., vol. IX, February (1991), 271-283.
3. VPI Component Maker™ Fiber Amplifier, Virtual Photonics Incorporated,(2002).
4. X. Zhang, A. Mitchell, "A Simple Black Box Model for Erbium-Doped Fiber Amplifiers", IEEE Photon. Technol. Lett., vol. XII, July (1998), 28-30

Optical Packet Switching Access Networks Using Time and Space Contention Resolution Schemes

Luiz Henrique Bonani[1], Fabiano João. L. Pádua[1], Edson Moschim[1],
and Felipe Rudge Barbosa[2]

[1] DSIF – School of Electrical and Computer Engineering – Unicamp
Cidade Universitária "Zeferino Vaz" – 13081-970 – Campinas-SP – Brazil
{bonani,padua,moschim}@dsif.fee.unicamp.br
[2] Fundação CPqD
Rodovia Campinas Mogi-Mirim, km 118,5 – 13088-902 – Campinas-SP – Brazil
rudge@cpqd.com.br

Abstract. Optical packet switching offers the possibility of increased granularity and more effective use of bandwidth in large capacity systems, on a scale of Tb/s. The natural integration of optical packet switching in photonic technology opens up the possibility of packet switching in transparent optical channels, where the packets remain from end-to-end in the optical domain, without the necessity of optoelectronic conversion. Therefore, the optical switching must be robust enough in order to provide conditions to solve the contention between optical packets in access networks, where the traffic is less aggregated. This work presents alternative schemes of contention resolution between optical packets to be used in access networks, without using the already established, but more expensive, wavelength conversion of WDM systems.

1 Introduction

The extraordinary increase of the Internet traffic has caused a strong traffic demand due to the provision of broadband services in optical as well as wireless networks, such as data, voice and video communications. The advent of WDM technology has already allowed significant advances in the increase of the available capacity in point-to-point optical links, but the processing capacity of the switches and electronic routers must represent serious limitations in the future optical networks. The fact that the Internet Protocol (IP) has emerged as the dominant protocol for data communications and that WDM technology utilizes the optical bandwidth more efficiently and incrementally, has motivated the use of IP over WDM in the optical layer. One of the main challenges of such approach is the existent mismatch between the transmission capacity offered by the WDM technology and the processing capacity of IP routers. In addition, there is also a concern about the performance of very large size routing tables, usually encountered in network backbones, which can easily exceed $6{\times}10^5$ entries [1]. The most critical switching functionality arises at the forwarding level.

J.N. de Souza et al. (Eds.): ICT 2004, LNCS 3124, pp. 272–281, 2004.
© Springer-Verlag Berlin Heidelberg 2004

Optical packet switching technology emerges as an alternative to surpass the ultimate limitations of the processing capacity of electronic devices for higher speed data rate transmission and throughput, by keeping a transparent payload content untouched, and processing only simple headers. The optical packets consist of a header and payload, where the former contains the routing information, which must be processed at each switching node. Thus, the header may be processed at a lower rate suitable for the available electronic switching technology, while the payload remains in the optical domain and might have higher bit rates [2]. Such approach is more suitable for optical access networks (OANs) [3], where traffic demands are expected to increase rapidly due to the introduction of new broadband services.

In access networks, traffic is less aggregated and presents somehow a more burst like behavior. The optical packet switching technology seems to be economically feasible in OANs, if the cost of implementing optical transparent nodes is compatible with the bandwidth requirements and demand of access from end users. In such context, a regular multihop 2×2 mesh network topology, such as Manhattan Street [4],[5], provides flexibility, scalability and finer granularity to such optical access network. The redundancy of optical paths prevents unnecessary demands for extra bandwidth, and avoids waste of resources as generally occurs in conventional broadcast and select topologies. In this work we study two approaches in order to develop criteria for the contention resolution between optical packets, which are the temporal and spatial contention resolution schemes. In other words, we present the temporal criterion, with optical buffers that comprise of fiber delay lines, and the spatial criterion, with deflection of optical packets. The traffic analysis is performed in a mesh network topology with a 2x2 configuration, with a new module developed for the Network Simulator (NS) that will be also used in the future to analyze that optical packet switching networks in the context of differentiated voice and data traffic, and QoS.

This work is organized as follows. In Section 2, we present, in a summarized form, the problem of contention between optical packets and a brief description of temporal and spatial contention resolution criteria. In Section 3, we analyze these two contention resolution criteria through the discrete events simulation method, using a Manhattan Street (MS) topology, and thus, acquiring the base for the analysis between these contention resolution schemes. Finally, in Section 4, we present the conclusions and final considerations for this work.

2 Techniques for Contention Resolution Without Wavelength Conversion

We consider optical networks with a topology comprising switching nodes with two input and output ports (2x2 nodes). These optical networks also include header electronic processing for the optical addressing recognition, using, for example, a frequency header labeling technique [6],[7].

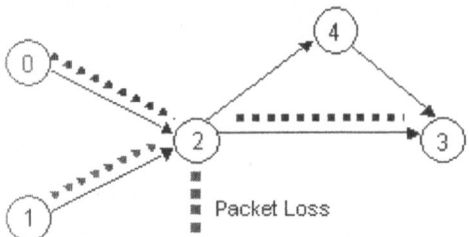

Fig. 1. Optical network without contention resolution criterion.

Hence, analyzing the traffic distribution in a simple topology, as shown in the Fig. 1, we can see that there are situations where in the two input ports of a given switch node arrive two different optical packets at the same time, having the same preferred output port. In this example, we have two traffic flows arriving at the node 2 from the nodes 0 and 1 (different input ports), and they must be redirected to the node 3 (the same output port). We can observe that, as the packets arrive to the node 2 at the same time, and considering no contention resolution method, the node is not able to perform the simultaneous processing. Consequently, one of the traffic flows is discarded, since the synchronism in the packets generation (time interval between packets of the same flow) is kept the same. Therefore, an approach to avoid the packet loss of the traffic flow from node 1 is the adoption of a contention resolution method. As seen previously, one of the solutions is the wavelength conversion in the switch node (WDM techniques). However, we analyze other techniques due to the greater simplicity of a node without wavelength conversion, and because of the smaller technological costs for the implementation of contention resolution methods.

2.1 Buffering Optical Packets

The optical buffers are also used to synchronization and flow control. They can be allocated in several parts of the routing node, such as in the input ports, in the output ports or be shared between the input and output ports[8]. We can have optical buffers in a switch, using for this, optical fiber delay lines (FDL). One of the existent models is classified as recirculating configuration, and it consists of multiple optical fiber delay lines, where each forms a loop with one circulation time equal to one packet duration. Table 1 presents a compilation for comparison of the sizes of different packets and frames in order to determine the optical fiber length of a FDL. The buffer can store multiple packets with the constraint that one packet enters and leaves the buffer at a time.

The recirculating buffer presents greater flexibility. This because the packet storage time can be adjusted by changing the circulation number, and offers additionally the capability of random access with a storage time, which depends on the number of recirculations. Nevertheless, the signal has to be amplified during each circulation to compensate for the power loss, which results in amplified spontaneous emission noise and limits somewhat the maximum buffering time.

Table 1. Estimated packet time durations and corresponding fiber buffer lengths at 2.5 Gb/s

Packet Type	Packet Size (bytes)	Packet Size (bits)	Bit Rate 2.5 Gb/s	
			Time (ns)	Fiber Length
IP Datagram	< 65500	< 524000	210000	42 km
Ethernet Frame	< 1500	< 12000	4800	980 m
ATM Cell	53	424	170	34 m

2.2 Deflection Routing for Packets

In conventional electronic networks the packet buffers in the intermediate nodes is usually implemented by a store-and-forward routing. This approach can not be simply applied to optical packet networks using just FDL. Thus, the deflection routing scheme, also called hot potato routing, represents an attractive alternative to become possible the implementation of an optical packet network without buffer, with mesh topologies, and with a contention resolution criterion.

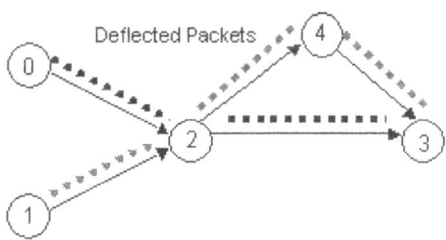

Fig. 2. Optical network with deflection routing.

According to the network presented in Fig. 1, we can see now the behavior shown in Fig. 2, where we can observe that none of the packets is lost in the circumstances presented there. Two important parameters must be considered, when we work with deflection routing: the packets delay between their source and final destination and the order of packet arrivals at the destination node. These parameters are related with the issues of quality of service (QoS). The packets delay between their source and final destination is extremely important due to the fact that the packets can remain in the network a greater time than the usual one, before they arrive to their final destinations. For example, Fig. 2 can be considered again, where initially the two packet flows have two links to cover before arriving at the final destination. However, because of the deflection routing, the flow originated at node 1 is deflected in the node 2 (redirected to the node 4), which follows its route to node 3, only at node 4, covering one link more than the initially expected.

3   Analysis of Contention Resolution Techniques Without Wavelength Conversion

In the analysis of the contention resolution techniques without wavelength conversion, we have used a methodology already used in other works [9], which follows the analysis of uniform traffic distribution [10].

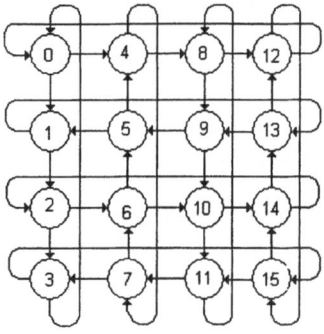

Fig. 3. Manhattan Street Topology with 16 nodes.

The network capacity C is the product of the total number of links by the bit rate, divided by the average number of hops (links) that a packet takes to travel to its destination. Therefore, in the case there are N nodes with two links, a bandwidth of S bits/s and an average number of hops $\overline{H}$, the total network capacity is given by Equation (1) [10].

$$C = \frac{2 \cdot N \cdot S}{\overline{H}} \tag{1}$$

Under uniform traffic distribution, all links are equally loaded and the expected average number of hops in a MS-16, shown in Fig. 3, is *2.933*, leading us to an aggregate capacity C of *27.27 Gb/s*. In order to represent the same traffic conditions with the NS-2 simulator [11], we have followed the methodology where packets are generated in all network nodes, with destination addresses to all other ones. As a result, the maximum network capacity will be reached when the bit rate R generated by each user (or traffic flow) is the same and given by Equation (2)

$$R = \frac{C}{N \cdot (N-1)} \tag{2}$$

Each user can admit a fraction R_e, that is understood as the bit rate corresponding to a given network load L, which can vary from 0 to 100%. So that, R_e is equal to R (the maximum bit rate) when $L = 100\%$, as we can see in Equation (3).

$$R_e = R \cdot L \tag{3}$$

For each value of R_e, varying L in intervals of 10%, we evaluate the Packet Loss Fraction (*PLF*) to the MS-16, which is our performance metric of each one of the studied contention resolution techniques. Other architectures are considered elsewhere

[12]. The *PLF* is given by Equation (4), where p is the total number of lost packets, and r is the total number of received packets.

$$PLF = \frac{p}{p + r} \qquad (4)$$

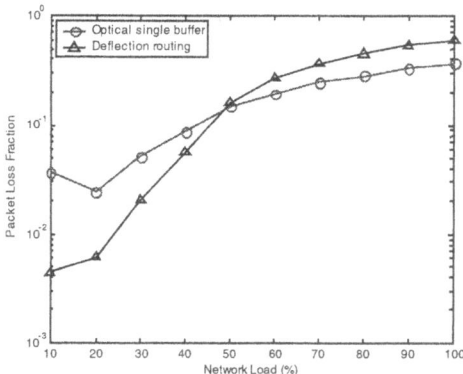

Fig. 4. Performance of MS-16 with deflection routing and optical buffer techniques using PLF.

Hence, using the discrete event simulation tool NS-2 [13], we have achieved the evaluation results for the MS-16 presented in Fig. 3, with temporal and spatial contention resolution criterion. Our scenario has considered packets size (*ps*) of 650 Bytes, links of 10 μs (2 km) with bandwidth 2.5 Gb/s, simulation time (*st*) of 2 ms and using UDP (User Datagram Protocol) in the transport layer. When using temporal contention resolution, we have adopted a recirculating optical buffer in each node output port. Also, we have considered a CPVI (Constant Packet with Variable Intervals) traffic, which comprises packets with constant size and interval between them following an exponential distribution. The network throughput (*NT*) is defined by Equation (5), where *NT* is the fraction of network resource that successfully delivers data.

$$NT = \frac{8 \cdot r \cdot ps}{C \cdot st} \qquad (5)$$

In Fig. 4 we show the performance comparison between temporal and spatial contention resolution techniques, considering an optical single buffer. Through we can note that the behavior of deflection routing is quite better, when compared with the single optical buffer (optical buffer with size of one packet), up to 50% of the network load.

This can be explained by the adopted methodology, where there are 256 users increasing their bit rates R_e linearly. Them, the network still can experience packet loss, even with the adoption of a contention resolution method, because there is the bottleneck of the link capacities, and can have network links loader than other ones. Thus, when a link reaches locally its maximum capacity S we observe packet losses. Besides, we have not considered a method for the blocking application. So, we take

conditions of traffic, even with a small network load, because all the applications are attended, and they can congest some links even with a geographical uniform distribution of traffic, as adopted in this work.

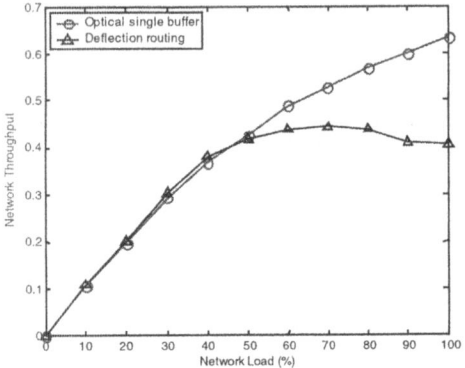

Fig. 5. Network throughput.

We have also analyzed the evolution of the network throughput for the two contention resolution schemes. The results, based on Equation (5), are shown in Fig. 5. We can see that we have almost the same performance of throughput, when considering an optical single buffer and the deflection routing, up to 50% of network load. After that, the optical buffer presents a better performance of throughput by the same reasons presented when discussing the lower PLF of the optical single buffer above the threshold of 50% of network load. Therefore, one of our studies is to build a block mechanism to avoid the loss of packets when the application is about to enter the network and the two output ports of node are busy, which occurs frequently with higher network loads, and where the problem is effectively realized. However, the simulations presented here are enough to understand that when the networks are submitted to low loads, case of access networks, the contention resolution methods without wavelength conversion are sufficient to deliver data with a satisfactory level of performance.

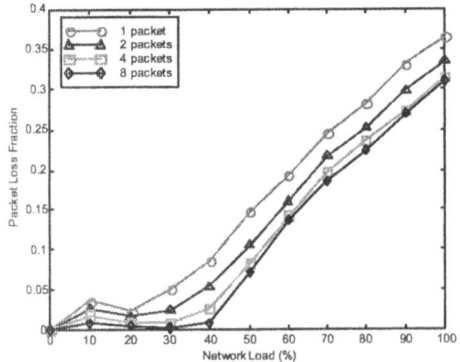

Fig. 6. Performance of MS-16 using optical buffers.

In Fig. 6 we see the performance of MS-16 with temporal contention resolution. Here we have considered a recirculating optical buffer, which size can store 1, 2, 4 or 8 packets. We can observe that the performance difference is very small when we increase the optical buffer size. It can be explained by the network load, that determines the packet losses in a given links even with a great buffer size.

In Fig. 7 we estimate, with simulation results, the average delay (AD) and its respective standard deviation (SD) when considering all packets that have traveled in the networks, as well as the average life time (LT) and its SD when considering the packets that were discarded. In this figure we consider the MS-16 topology, and additionally, three other network topologies with 4, 6 and 8 nodes, comprising 2x2 nodes, and using the deflection routing method for contention resolution. We can observe in Fig. 7 (a) that a network with a small size can present better performances of AD, when submitted to low network loads, but with a higher load, as viewed in Fig. 7 (b), the number of collisions between packets increases, taking us to a SD greater because of the deflection routing. Moreover, we can also see that the SD related to the LT is greater than the own LT, because of the packets that have been lost in the first links that they should travel, and other ones traveling a excessive number of links, limiting their possibilities of reach the final destination node.

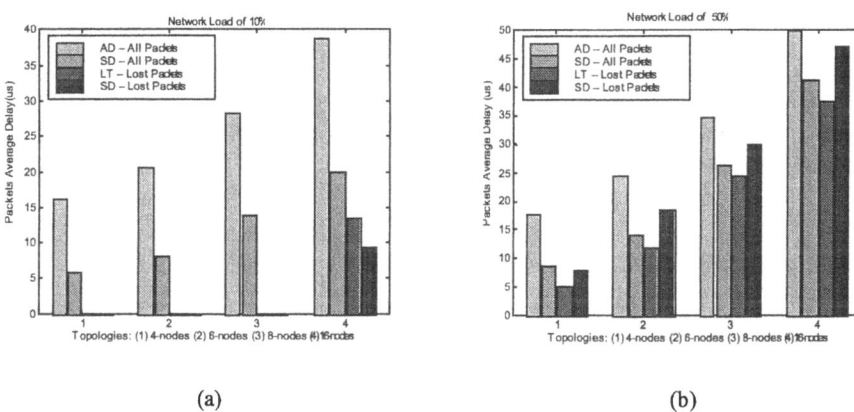

(a)   (b)

Fig. 7. Average Delay and Life Time of packets for 4 network topologies, including the MS-16.

In network loads of 50%, for the MS-16, we observe AD of 50 μs, with SD of 43 μs. It is too large when compared with the same parameters with network load of 10%. Besides, the LT is much more significant when the network load is 50%, evidencing the great number of packets, which are lost without reach their final destination. It has visible impacts in the network performance, as we can see in Fig. 4, when we analyze the two contention resolution methods for network loads greater than 50%, and the deflection routing presents a worse performance when compared with optical buffers. The other parameter related to QoS, besides the AD, is the order in the packet arrivals at the destination node. It is easy to realize that we have a significant impact for the QoS, due to the final order in the destination node, because the great value of SD, relative to AD, appoint to the need of a reordering mechanism to recover the order of packets as in the original flow.

4 Conclusions

Recent advances in optical technology and the expectation of multi-functional communications networks, with greater capacities of switching, have stimulated the research in optical networks with functionality of optical packet switching.

From the techniques of time and space contention resolution presented here, it is evident that the use of one or another is essential to a good performance of the networks. In general, this performance was satisfactory in the region of low network loads for the two studied techniques. Comparing the two schemes we realize a better performance bufferless deflection routing for lower traffic and a better performance of optical buffering for higher loads. However, to the choice of one or another, we should elect some priorities, which can take into account some economics, technological and application criteria. For example, the possibility to guarantee the QoS, the nature of traffic flows (if it is a real time or not), the kind of network where the system will work and the available technology. It is clear that we do not expect that the two techniques presented here can substitute the already established WDM systems, due to their great robustness and large use in the core networks. However, the WDM systems need a large initial investment in technologies, which have high aggregated costs. Therefore, the two techniques presented can be viewed as an alternative in some networks where we have not high traffic loads, as for example, the access networks, which should be able to switch optical packets in a near future.

References

1. Bannister, J., Touch J., Willner, A., Suryaputra, S.: How many wavelengths do we really need? A study of the performance limits of packet over wavelengths, Optical Magazine Networks, (2000), 17-28.
2. Barbosa, F.R., Maia, D., Pezzolo L., Sachs, A.C., Salvador, M.R.: Optical Pachet Switching Node for Metro-Access Networks, Proc. of European Conference on Optical Communications – ECOC'2003, (2003).
3. Bonani, L.H., Furtado, M.T., Moschim, E., Barbosa, F.R.: Analysis of Optical Packet Switching Performance with Spatial Contention Resolution for Optical Access Networks, Proc. of IASTED CSN'2003, (2003) 418-423.
4. Choudhury, A.K., Li, V.O.K.: Performance analysis of deflection routing in the Manhattan Street network, Proc. IEEE ICC, (1991), 1659-1665.
5. Chamltac, I., Fumagali, A.: An Optical Switch Architecture for Manhattan Networks, IEEE J.Select. Areas Communic. Vol.11, no. 4, (1993), p.550.
6. Barbosa, F. R., Sachs, A. C., Furtado, M. T., Rosolem, J.B.: Optical Packet Switching: a transmission and recovery demonstration using an SCM header, XIX Simpósio Brasileiro de Telecomunicações – SBT'2001, (2001).
7. Bonani, L. H., Sachs, A. C., Furtado, M. T., Moschim, E., Yamakami, A.: Non-Uniformly Distributed Traffic Analysis of Optical Networks with Optical Packet Switching Functionalities, Proc. of SBMO'2002, paper ST7.2, (2002), 163-167.
8. Tucker, R.S., Zhong W.D.: Photonic packet switching: an overview, IEICE Transactions on Communications vol.E82-B, no.2, (1999), 254-264.
9. Bonani, L. H., Sachs, A. C., Furtado, M. T., Moschim, E., Yamakami, A.: Optical Network Analysis under Non-uniform Traffic Distribution, Proc. of ITS'2002, (2002).

10. Acampora, A.S., Shah, S.I.A.: Multihop lightwave network: a comparison of store-and forward and hot potato routing, IEEE Transac. on Communications, vol.40, no.6, (1992), 1082-1090.
11. Fall, K., Varadhan K.: NS Manual, UC Berkeley, August 2001.
    http://www.isi.edu/nsnam/ns/doc/index.html, last access in 02/10/2004.
12. Bonani, L.H., Barbosa, F.R., Moschim, E.: Performance and Dimensioning Analysis of Optical Packet Switching Access Networks. Proceedings of ICT'2004, (2004).
13. Breslau, L., et. All.: Advances in Network Simulation, IEEE Computer Magazine, Vol. 33, (2000), 59-67.

Full-Optical Spectrum Analyzer Design Using EIT Based Fabry-Perot Interferometer

A. Rostami

OIC Design Lab., Faculty of Electrical Engineering, Tabriz University, Tabriz 51664, Iran
Tel: +98 411 3393724
Fax: +98 411 3300819
rostami@tabrizu.ac.ir

Abstract. In this paper, we will present the full-optical realization of spectrum analyzer system based on Electromagnetic Induced Transparency (EIT) and Fabry-Perot interferometer. In this system, the index of refraction variation due to variation of the coupling field intensity in the EIT process can be used for detection of the incident light components in the frequency domain. For extracting the frequency component, the transmission coefficient of the Fabry-Perot interferometer is used. Our proposed system can monitor the incident spectrum content at least for 5-THz bandwidth.

1 Introduction

Nowadays, high speed signal processing is basic demand from science and technology point of view. Photonics based technologies was a important alternative. But, optical to electrical and electrical to optical conversions have large delay time. So, for removing this delay time the full-optical systems were proposed in the last few years. In this direction, in this paper, we will propose a new structure for full-optical spectrum analyzer as an important block in the signal-processing domain. There are many previously published works in the spectrum analyzer area [1-6]. In [1], Fiber Bragg Grating is used for implementation of optical spectrum analyzer. In this work 14 nm bandwidth with 0.12 nm resolution were reported and the opto-electronical techniques were used. In [2] the authors proposed a high-resolution optical spectrum analyzer based on Heterodyne detection. In [3] the electro-optically tunable Er-doped Ti:LiNbo3 wave-guide Fabry-Perot cavity was used. In [4], the authors will present a spectrum analyzer using a UV-induced chirped grating on a planar wave-guide and a linear detector array. This structure has a bandwidth of 7.8 nm. In [5] a new type of optical spectrum analyzer is proposed that features the cascade connection of different arrayed wave-guide grating with different channel spacing through optical switches. In all the presented work, the mixed system (Photonics) is used for implementation of spectrum analyzer. But, in this work, we will present a novel and suitable full-optical method based on EIT and Fabry-Perot interferometer [6]. In our case, we will use the potential properties of EIT phenomenon.

The organization of this paper is as follows.

In section II, EIT based all-optical spectrum analyzer is investigated. In this section we try to present the EIT phenomenon based susceptibility control by coupling field.

J.N. de Souza et al. (Eds.): ICT 2004, LNCS 3124, pp. 282–286, 2004.
© Springer-Verlag Berlin Heidelberg 2004

Then using this alternative, we obtain the transmission coefficient of Fabry-Perot cavity included by EIT medium. Finally, the obtained results for our proposed system are reported. Finally, the paper ends with a conclusion.

2 EIT Based All-Optical Spectrum Analyzer

In this section we will propose a new structure for implementation of full-optical spectrum analyzer based on Fabry-perot interferometer. In this new structure, the quantum optical phenomenon named EIT is used. EIT describes the phenomena whereby a medium that is normally opaque to a input signal (probe or incident laser) tuned to a resonant transition can be made transparent when a coupling field is also applied simultaneously. EIT can be performed in atomic, solids and semiconductor mediums. For this case, we consider the Hetero-structure semiconductors as an EIT medium with a three-level system such as shown in Fig. (1). Application of a light beam at frequency ω_{ac} (pump beam) will, under certain conditions, allow for a signal beam at frequency $\omega = \omega_{ab}$ to experience extremely low losses whereas in absence of the former, high absorption is observed. Signal (probe) and control (pump) electric fields amplitudes are respectively related to the Rabi frequencies Ω_{ab} and Ω_{ac} depicted in Fig. (1). At EIT condition [7], the system is driven by the optical beams in to a time-invariant electronic state (dark state), which arises from a coherent combination of the eigenstates $|c\rangle$ and $|b\rangle$. The density matrix formalism can be used for obtaining the system susceptibility. Now, using quantum mechanical tools one can obtain the following relations for the susceptibility of the medium from the incident signal point of view [7].

$$\chi = \chi' + i\chi'' \tag{1}$$

$$\chi = \frac{N_a \cdot |P_{ab}|^2 \cdot [\gamma_3 + i(\Delta_{ab} - \Delta_{ac})]}{\varepsilon_0 \hbar \left[(\gamma_3 + i(\Delta_{ab} - \Delta_{ac}))(\gamma_1 + i\Delta_{ab}) + \dfrac{\Omega_{ac}^2}{4} \right]} \tag{2}$$

where Ω_{ab}, Ω_{ac} are the related Rabi frequencies which are corresponds to signal and coupling field amplitudes, $\Delta_{ab} = \omega_{ab} - \omega_p$, $\Delta_{ac} = \omega_{ac} - \omega_c$ are incident signal detuning (ω_p is the incident signal frequency) and coupling field detuning (ω_c is the coupling

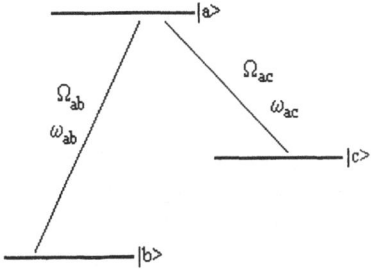

Fig. 1. 3-level Λ type EIT medium

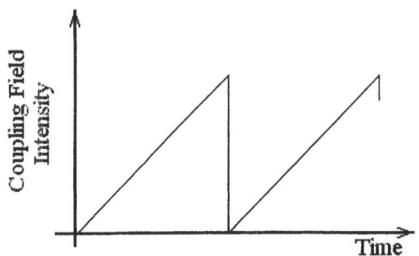

Fig. 2. Coupling Field Variation Profile

field frequency), γ_1 and γ_3 are the decay rates for off diagonal the density matrix elements relating to the transitions a-b and b-c respectively, N_a is the density of 3-level states and P_{ab} is the dipole matrix element between a and b. Now, using Eq. (2), we can obtain the index of refraction as

$$n = \sqrt{n_0^2 + \chi} \tag{3}$$

where n_0 is background index of refraction. Using smooth variation of the coupling field as Fig. (2), we can obtain the susceptibility variation and consequently index of refraction variation.

Fig. (3) shows our idea for realization of full-optical spectrum analyzer. In this structure, the transmitted intensity is used for detection of frequency contents of the incident light. The Fabry-Perot cell including EIT medium is used for our proposal. According to the light transmission in layered media [6], we obtain the

$$t = \frac{t_{12}t_{23}e^{-i\phi}}{1 + r_{12}r_{23}e^{-i2\phi}} \tag{4}$$

where t_{12} and t_{23} are the transmission coefficients from layer 1 to 2 and layer 2 to 3 respectively. Also, r_{12} and r_{23} are reflectivity from layer 1 to 2 and layer 2 to 3. The ϕ in Eq. (4) is defined as

$$\phi = \frac{2\pi}{\lambda}n_2 d \cos(\theta) \tag{5}$$

where λ is the incident light component wavelength, n_2 is the index of refraction for EIT medium, d is thickness of EIT medium and θ is the incident light angle with respect to normal axis to EIT surface. The intensity Transmission coefficient for this structure can be obtained as

$$T = \frac{(1-R)^2}{(1-R)^2 + 4R\sin^2(\phi)} \tag{6}$$

where $R = |r_{12}|^2 = |r_{23}|^2$ is reflectivity from interfaces.

Figs. (4.1,2) shows the transmission and reflection coefficients for our proposed structure. As we see, changing the incident wavelength will change the transmission coefficient peak. So, with measuring the transmission coefficient peaks, one can determine the incident frequency components. Also, the reflection coefficient can be used for spectrum analyzer purpose.

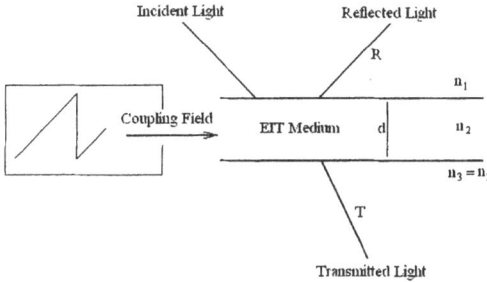

Fig. 3. Schematics of Full-optical Spectrum Analyzer

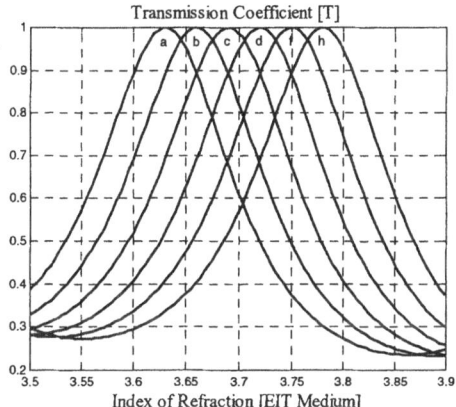

Fig. 4.1. Transmission Coefficient Vs. Index of Refraction Variation

$a)\lambda = 0.9075\mu m, b)0.9150\mu m, c)0.9225\mu m, d)0.930\mu m$, $f)0.9375\mu m, h)0.9450\mu m$

$$n_1 = n_3 = 1, d = 1\mu m$$

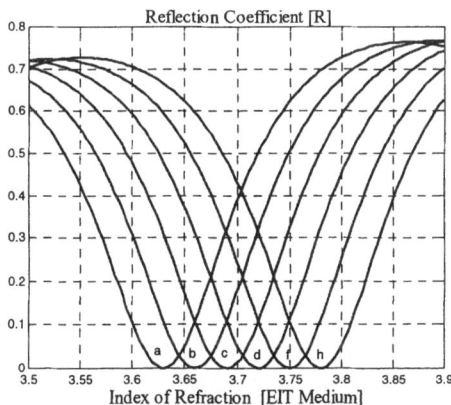

Fig. 4.2. Transmission Coefficient Vs. Index of Refraction Variation

$a)\lambda = 0.9075\mu m, b)0.9150\mu m, c)0.9225\mu m, d)0.930\mu m,$,

$f)0.9375\mu m, h)0.9450\mu m, n_1 = n_3 = 1, d = 1\mu m$

Fig. 5 shows the output-transmitted intensities for two incident components with different amplitudes.

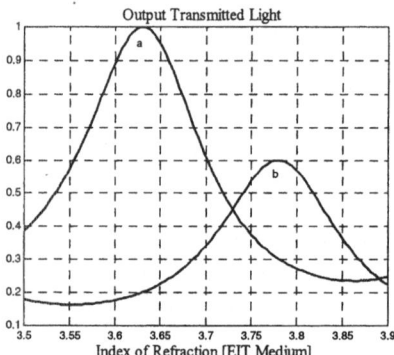

Fig. 5. Transmitted Intensities Vs. Index of Refraction

$a)\lambda = 0.9075\mu m, b)0.9450\mu m$, $n_1 = n_3 = 1, d = 1\mu m$,a: Input Amplitude=1, h: Input Amplitude=0.6

As Fig. (5) shows that the incident wavelength and amplitude can be obtained from transmitted intensities.

Conclusion

In this paper, we propose a new structure for full-optical spectrum analyzer based on EIT phenomenon. In this topology, Fabry-Perot interferometer is used for detection of frequency component of incident light. Our proposed system has 5-THz bandwidth because of EIT limitation, which is very large for communication purpose. Our structure is analog and easy for implementation with GaAs based structures.

References

1. Wagner, J. L. et al: Fiber Bragg Grating Optical Spectrum Analyzer Tap. ECOC 97, No. 48, Sept. 1997.
2. Kataoka, T. and Hagimoto, K.: High Resolution Optical Spectrum Analyzer Using a Heterodyne Detection Techniques. Proceeding of IMTC'94.
3. Suche, H. et al.: Integrated Optical Spectrum Analyzer with Internal Gain. IEEE Photonics Technology Letters, Vol. 7, No. 5, May 1995.
4. Madsen, K. et al.: Planar Waveguide Optical Spectrum Analyzer Using a UV-induced Grating. IEEE J. Selected Topics in Quantum Electronics, Vol. 4, No. 6, Nov. 1998.
5. Takada, K. et al.: Optical Spectrum Analyzer Using Cascade AWG's with different channel spacing. IEEE Photonics Technology Letters, Vol. 11, No. 7, July 1999.
6. Yeh, P.: Optical Waves in Layered Media. John Wiley, 1988.
7. Scully, M. O. and Zubairy, M. S.: Quantum Optics. Cambridge University Press, 1997.

A Powerful Tool Based on Finite Element Method for Designing Photonic Crystal Devices*

Arismar Cerqueira Jr.[1], K.Z. Nobrega[2], F. Di Pasquale[1], and
H.E. Hernandez-Figueroa[2]

[1] Scuola Superiore Sant'Anna di Studi Universitari e di Perfezionamento, Pisa, Italy.
{arismar.cerqueira,fabrizio.dipasquale}@cnit.it.
[2] Microwave and Optics Department, DMO, Electrical and Computer Engineering
Faculty, State University of Campinas, UNICAMP, Brazil.
{bzuza,hugo}@dmo.fee.unicamp.br.

Abstract. Photonic crystal devices have been regarded as the key solution for overcoming the limitations of conventional optical devices. In
recent years, their novel and unique features have been investigated and
exploited by many researches. This work presents a powerful design tool
based on Finite Element Method, FEM, to be applied in the design and
optimization of these devices, such as Photonic Crystal Fibers (PCFs)
and integrated optical waveguides with photonic crystals. The full vectorial FEM code which has been implemented can be used for accurate
modal and dispersion analysis. Some examples are presented concerning
a novel approach for designing large-mode-area endlessly single-mode
PCF and optical waveguides with photonic crystals with reduced critical
power of the optical bistability phenomenon.

1 Introduction

Photonic Crystal Fibers, also known as Holey Fibers, were proposed by Russell and his co-authors, from University of Bath, in 1992. The first working
example was produced in 1995 and reported in 1996 [1]. Since then this new
kind of fiber has been extensively studied and investigated. There are many
kinds of PCFs with different geometries and propagation mechanisms, which
have been proposed in accordance to the desired application, such as endlessly
single-mode [2], birefringent, ultrahigh nonlinear, large mode-area, dispersion
flattened or hollow-core PCFs [3] [4] [5]. Their applications are not limited to
optical communications; it is also possible to improve optical coherence tomography, frequency metrology and spectroscopy with their remarkable properties.

PCFs have multiple air holes periodically arranged around the core. This
periodic structure is called photonic crystal and it is similar to normal crystals
present in semiconductor materials. In other words, photons in a PCF have
similar behavior as electrons in a semiconductor crystal. Initially, the photonic
bandgap, PBG, had been considered the only guiding mechanism for this kind

---

* This work was partially supported by FAPESP, Brazil.

J.N. de Souza et al. (Eds.): ICT 2004, LNCS 3124, pp. 287–295, 2004.
© Springer-Verlag Berlin Heidelberg 2004

of optic fiber. Later, researches have discovered that these fibers could similarly provide innovative propagation characteristics also using the simplest and most conventional principle of total internal reflection.

Integrated optical waveguides with photonic crystals are also attracting a great deal of attention for the possibility they offer in terms of innovative features, compactness and cost reduction in telecommunication applications; possible applications include electrically pumped optical amplifiers, dispersion compensation devices, optical memories, add-drop devices, tunable lasers and filters.

FEM has become recognized as a general method widely applicable to engineering problems [6] and full-vectorial finite element formulations provide efficient and robust techniques for analyzing propagation and dispersion properties of optical fibers and integrated waveguides [7]. The main properties of these approaches are:

- No spurious solutions.
- The ability to treat a wide range of dielectric waveguides which may possess arbitrarily shaped cross-sections, inhomogeneity, transverse-anisotropy and significant loss or gain.
- Direct solution for the complex propagation constant at a specified frequency.
- Possibility to compute leaky modes.
- It is possible to take into account the polarization effects.
- Accurate expression of the sharp discontinuities in dielectric constant.
- Adaptive mesh refinement allows improvement of the solution in specific areas of high relative error.
- The global matrices comprising the eigenvalue problem are sparse and symmetric.

2   Finite Element Analysis

The large index contrast and complex structure in PCFs make them difficult to treat mathematically. Standard optical fiber analyses do not help and, in addition, in the majority of PCF cases is practically impossible to solve analytically, so Maxwell's equations must be solved numerically. The main idea consists in transforming this complicated problem, which can be described by the curl-curl equation 1:

$$\nabla \times (\epsilon_r^{-1} \nabla \times H) - k_0^2 \mu_r H = 0 \tag{1}$$

into an eigensystem, in which its eigenvalues are β/k_0, the effective indexes, and its eigenvectors are the magnetic field components. In (1) H is the magnetic field, ϵ_r and μ_r are the dielectric permittivity and magnetic permeability tensors, respectively, and k_0 is the wave number in vacuum.

FEM codes are basically divided in four basic steps [8]. The domain discretization is the first and perhaps the most important step in any finite element analysis. It consists on dividing the solution domain Ω into a finite number of subdomains, denoted as Ω^e (e=1,2,3,...M), where M is the total number of subdomains. They form a patchwork of basic elements that can have different sizes, shapes and physical properties. In our specific case the solution domain is the transverse cross-section of the optical waveguide which is divided in triangular finite elements, instead of rectangular ones because they are more suitable for irregular regions.

The second step consists in selecting interpolation functions, which provide an approximation of the unknown solution within each element. The interpolation is usually selected to be a polynomial of first(linear), second(quadratic), or higher order. We follow a nodal approach using second order polynomials as interpolating functions in each finite element (each triangle is characterized by 6 nodes,three vertices and the other three at the middles of its three sides, and the unknown function ϕ is the magnetic field components). Within the element, the unknown function ϕ is expressed as a quadratic function:

$$\phi^e(x,y) = a^e + b^e x + c^e y + d^e x^2 + e^e xy + f^e y^2 \qquad (2)$$

whose six coefficients $a^e, a^e,...,a^e$ can be determined by imposing 2 at the six nodes [8].

In the next step the curl-curl equation is transformed into a generalized eigenvalue problem by applying a variational formulation. The formulation we are considering is based on a penalty function to suppress spurious modes, as shown in (3), [8].

$$F(H) = \frac{1}{2} \int\int_\Omega [\frac{1}{\epsilon_r}(\nabla \times H)\cdot(\nabla \times H)^* + \frac{s}{\mu_r^2\epsilon_r}|\nabla\cdot(\mu_r H)|^2 - k_0^2\mu_r H\cdot H^*]d\Omega \qquad (3)$$

where s is the penalty function.

By finding the stationary solutions of functional $F(H)$, specially in the complex case, for which maximum or minimum values are meaningless, we obtain the following generalized eigenvalue system:

$$[A]\{\phi\} - \lambda[B]\{\phi\} = \{0\} \qquad (4)$$

where the eigenvalue λ is the mode effective refractive index (n_{eff}) and the eigenvector ϕ is the full vectorial magnetic distribution $(H_x, H_y$ and $H_z)$.

Finally, solving the eigenvalue system is the fourth and final step in a finite element analysis. The matrices $[A]$ and $[B]$ are sparse and symmetric, so the computational time is effectively minimized using a sparse matrix solver. They are described in [8].

The frequency dependent refractive index $n(w)$, which allows one to account for the material dispersion, has been described by using the Sellmeir equation:

$$n^2(w) = 1 + \sum_{j=1}^{M} \frac{B_j w_j^2}{w_j^2 - w^2} \qquad (5)$$

where w_j is the resonance frequency and B_j is the oscillator strength, found empirically for each material.

The dispersion parameter $D(\lambda)$ can be directly calculated from the modal effective index $n_{eff}(\lambda)$:

$$D(\lambda) = -\frac{\lambda}{c}\frac{d^2 n_{eff}(\lambda)}{d\lambda^2} \qquad (6)$$

where λ is the operating wavelength and c is the light velocity in vacuum.

3   Design Tool Based on FEM

The design tool we have developed, is essentially split in four parts, as shown in figure 1: the Photonic Crystal Fiber Topology Design Appliance, a commercial mesh generator, the FEM solver and a program for results visualization (Visual).

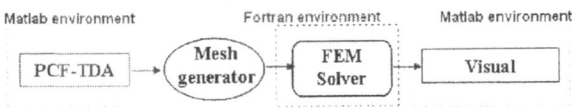

Fig. 1. Design tool scheme.

The Photonic Crystral Fiber Topology Design Appliance (PCF-TDA) is an efficient program, written in matlab, that can easily draw extremely complex PCFs in seconds. Otherwise it would be necessary to spend a quite long time to draw and optimize PCF topologies. It allows to design PCF with different air hole diameter, periodicities and also d/λ rates, taking advantage of CAD, Computer-Aided Design, properties of mesh generator.

The commercial mesh generator provides very accurate meshes, especially in areas of high relative error. This feature ensures improved solutions, through mesh refinement in such areas of the transverse waveguide cross-section where high refractive index differences are present; this is extremely important for designing photonic crystal devices, in which several refractive index discontinuities are present.

The FEM solver has been written in Fortran language and its implementation follows the steps described in the previous section to obtain waveguide modal analysis and dispersion properties.

Visual is a program, written in matlab, which allows easy and efficient visualization of PCF design results, such as magnetic field distributions and dispersion characteristics.

The main advantage of this design tool is that it is possible to select the number of eigenvalues and eigenvectors to be analyzed, or in other words, it is able to compute not only the fundamental mode, but also higher-order modes. This feature is extremely useful in PCF design, as it allows one to investigate important characteristics which affect the fiber modal properties.

4   Results

4.1   Large-Mode-Area Endlessly Single-Mode PCF

Standard fiber technology has been shown to be limited by the fundamental structure of conventional fibers which is based on a simple concentric core-cladding geometry that confines the light in the core by total internal reflection.

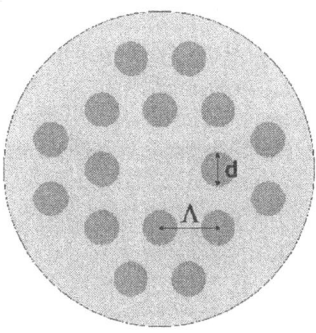
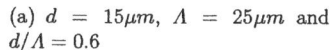
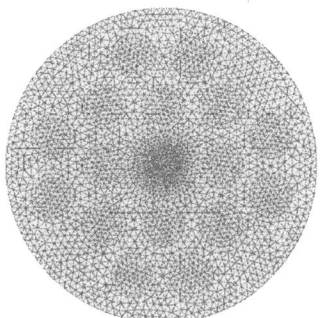

(a) $d = 15\mu m$, $\Lambda = 25\mu m$ and $d/\Lambda = 0.6$

(b) Mesh: 5083 triangular elements and 10286 nodes.

Fig. 2. Large-mode-area endlessly single-mode PCF.

The small refractive index contrasts achievable between the concentric core and cladding imposes a fundamental constraint on the design properties. Applications which require high-power in single-mode fibers with large effective area, or small highly nonlinear cores with precisely controlled dispersion properties are good examples in which the unique waveguide properties provided by PCF can offer great advantages. In particular, single-mode operation over a large wavelength range can be achieved in well-designed PCFs; their endlessly single-mode features combined with extremely large effective areas [9] make them very attractive for high power applications. As the number of modes in a PCF only depends on the ratio of air hole diameter d to the spacing between holes Λ [10], no changes in the refractive index profile are required to fabricate endlessly single-mode PCF with different core diameters. These fibers also offer interesting features due to the low nonlinearity achievable with single-mode operation and large effective area [11]. High power optical amplifiers can be realized based on this kind of PCFs: in particular, low cost booster amplifiers operating in linear conditions and pumped by broad area lasers or high power fiber Raman lasers can be very very attractive for broad-band local area and metro applications. They can be either based on Raman technology or standard EDFA technology including Er-Yb co-doped materials [12].

In this section we will point out how our FEM code can be used for designing endlessly single-mode PCFs based on a new original geometry. In traditional endlessly single-mode PCFs, the air holes are arranged in a hexagonal honeycomb pattern across their cross section [9] [10] [2]. We are proposing here a novel approach to design this kind of solid-core PCF, based on a different geometry, as shown in figure 2(a). The design parameters of this fiber are: air hole diameter $d = 15\mu m$ and inter-hole spacing $\Lambda = 25\mu m$, which ensures that this PCF is endlessly single-mode. The solid-core diameter, defined as the diameter of the ring formed by the innermost air holes, is $D = 35\mu m$, that is very large when compared to step-index single-mode fibers. Note that we have applied a mesh strategy which improves the accuracy in PCF core, as shown in figure 2(b). This mesh has 5083 triangular elements and 10286 nodes.

Endlessly single-mode PCFs have been recently presented operating in the wavelength range from 337 to 1550nm [2]. The PCF structure here proposed and shown in figure 2(b) is monomode at all wavelengths $\lambda \geq 100$nm, which represents a remarkable extension in the endlessly single-mode behavior. Figure 4 shows the magnetic field components of the fundamental mode at wavelength $\lambda = 100$nm. It is interesting to note that the magnetic field is concentrated in the PCF core, confirming that the PCF parameters (d, Λ and d/Λ) have been well designed, ensuring total internal reflection.

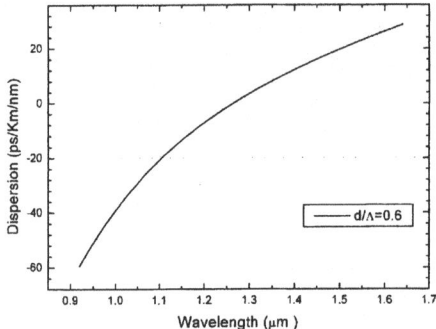

Fig. 3. PCF dispersion curve.

Also note that PCFs can be very useful for realizing dispersion compensating devices and, in general, devices in which it is possible to tailor the chromatic dispersion characteristics. On the other hand, the modal characteristics of conventional dispersion compensating fibers, DCFs, cannot be significantly changed to realize effective broad-band dispersion compensation with very short devices, due to the small index variation achievable over the transverse cross section. This limitation may be overcome by using photonic crystal fibers, which have a high-index contrast [13]. PCF can exhibit unique and useful modal characteristics obtained by optimizing design parameters such as d, Λ and d/Λ. Figure 3 shows the dispersion properties of this PCF example.

4.2   Optical Waveguide with Photonic Crystals

Circular photonic crystals, embedded in nonlinear channel optical waveguides, have been proposed to reduce the critical power of the optical bistability phenomenon [14]. This work has proved that the waveguide of figure 5(a) can reduce up to three times the critical power of the bistable behavior when compared to conventional waveguides.

In this section we will show how our FEM code can be effectively applied for designing such kind of integrated optical waveguides. Numerical results will also be compared with [14]. Considering the dimensions of the air holes in

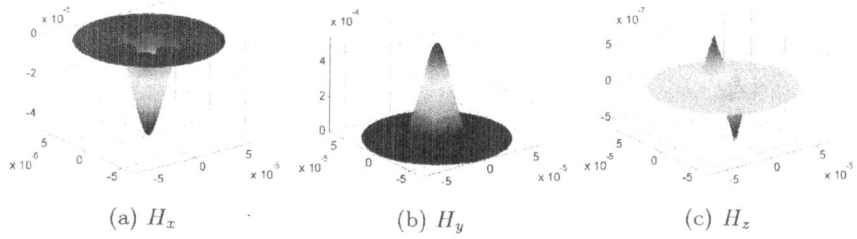

(a) H_x (b) H_y (c) H_z

Fig. 4. Magnetic field components at wavelength $\lambda = 100$nm.

figure 5(a) and comparing them with the PCF described in the previous section, one could easily understand the need of a powerful method able to deal with such imperfections without lack of numerical accuracy. Once again, a mesh strategy has been applied in order to improve the accuracy in the core, as shown in figure 5(b). This mesh has 5364 triangular elements and 10775 nodes.

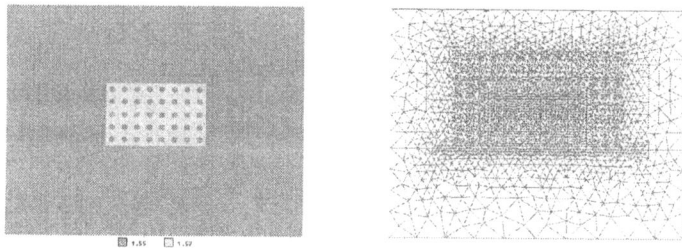

Fig. 5. Waveguide with photonic crystals and its mesh

This is an example of a multimode waveguide at $\lambda = 0.5\mu m$, as shown by the magnetic field components of figure 6. In this case the relation $d/\Lambda = 0.5$ ($d = 0.125\mu m$ and $\Lambda = 0.25\mu m$) was not large enough to trap the higher-order modes. In other words, the lobe dimensions, or the transverse effective wavelengths, can slip between the gaps, as shown in figures 6(b) and 6(c).

We have used this optical waveguide to investigate the accuracy of our FEM design tool for photonic crystals; we have computed waveguide modes and propagation constants at different wavelengths and then computed the waveguide dispersion properties. Table 1 shows a comparison between effective indexes, at different wavelengths, calculated by the proposed tool and the one described in [14]. These results show a very good agreement, with a maximum discrepancy $\eta = 0.001\%$, that proves this tool is very accurate for designing photonic crystal devices.

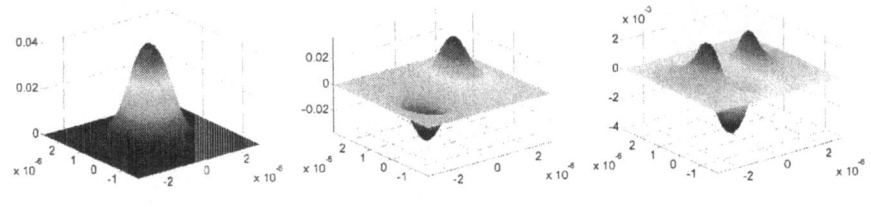

(a) Fundamental mode.   (b) Second-order mode.   (c) Third-order mode.

Fig. 6. Modes of waveguide with photonic crystals.

Table 1. Effective index comparison between the two approaches.

Wavelength (nm)	Proposed design tool	Tool described in [14]
500	1.55988	1.55986
1000	1.55175	1.55173
1500	1.54294	1.54293

5   Conclusion

It has been proposed a powerful design tool based on a full-vectorial finite-element method for designing photonic crystal devices. It is essentially divided in four parts: Photonic Crystal Fiber Topology Design Appliance, a commercial mesh generator, the FEM solver and Visual for output data visualitation. Its main advantages are: an easy and efficient interface to draw complex PCFs in seconds; improved solutions, through mesh refinement; high accuracy of the numerical solver, which is able to compute not only the fundamental mode, but also higher-order modes; an easy an useful visualization of the PCF design results, such as field distributions and dispersion characteristics.

The FEM analysis allows one to transform problems which are practically impossible to solve analytically into eigensystems, in which the eingenvalues are the mode effective indexes and eigenvectors are the magnetic field components.

Based on this powerful tool, we have presented a novel structure for endlessly single-mode PCFs, based on an original symmetry of two-dimensional photonic crystal around its solid-core. This new structure should improve the performances of traditional endlessly single-mode PCFs, since it extends its monomode operation conditions for all wavelengths $\lambda \geq 100$nm; this represents a remarkable extension in the endlessly single-mode behavior.

These PCFs with large-mode area may be used to construct low cost, high power amplifiers either based on Raman or EDFA technologies. These amplifiers can operate in linear regime since they are characterized by very low nonlinearities.

Finally, we have shown that this design tool is suitable not only for PCFs, but also for integrated optical photonic crystal devices. Comparing with published results concerning an integrated waveguide with photonic crystals for optical spatial switching, we have shown a maximum discrepancy between effective indexes lower than $\eta = 0.001\%$, proving that our techniques provides high numerical accuracy.

References

1. J.C. Knight, T.A. Birks, P.S.J. Russell, and D.M. Atkin, "All-silica single-mode fiber with photonic crystal cladding," *Opt. Letters*, vol. 21, pp. 1547–1549, 1996.
2. E. Silvestre, P.S.J. Russell, T.A. Birks, and J.C. Knight, "Endlessly single-mode heat-sink waveguide," *CLEO*, p. 428, 1998.
3. A. Peyrilloux, T. Chartier, A. Hideur, L. Berthelot, G. Mlin, S. Lempereur, D. Pagnoux, and P. Roy, "Theoretical and experimental study of the birefringence of a photonic crystal fiber," *Journal of lightwave technology*, vol. 21, pp. 536–539, 2003.
4. T. A. Birks J. C. Knight an, P. St. J. Russell, and J. P. de Sandro, "Properties of photonic crystal fiber and the effective index model," *Journal of Optical Society of America A*, vol. 15, pp. 748–752, 1998.
5. P. St. J. Russell, "Photonic crystal fibers," *Science*, vol. 229, pp. 358–362, 2003.
6. F. Di Pasquale, M. Zoboli, M. Federighi, and I. Massarek, "Finite-element modeling of silica waveguide amplifiers with high erbium concentration, *Journal of Quantum Electronics*, vol. 30, pp. 1277–1282, 1994.
7. H. E. Hernandez-Figueroa, F. A. Fernandez, Y. Lu, and J. B. Davies, "Vectorial finite element modeling of 2d leaky waveguides," *Transactions on magnetics*, vol. 31, pp. 1710–1713, 1995.
8. Jianming Jin, *The Finite Element Method in electromagnetics*, John Wiley and Sons, second edition, 2002.
9. J.C. Knight, T.A. Birks, R.F. Cregan, P.S.J. Russell, and P.D de Sandro, "Large mode area photonic crystal fibre," *Electronics Letters*, vol. 34, pp. 1347–1348, 1998.
10. T.A. Birks, J.C. Knight, and P.S.J. Russell, "Endlessly single-mode photonic crystal fiber," *Optics letters*, vol. 22, pp. 961–963, 1997.
11. G. P. Agrawal, *Nonlinear fiber optics*, Academic, 1995.
12. K. Furusawa, A. Malinowski, J., H. V. Price, T. M. Monro, J. K. Sahu, J. Nilsson, and D. J. Richardson, "Highly efficient all-glass double-clad ytterbium doped holey fiber laser," *CLEO*, pp. 46–47, 2002.
13. L. P. Shen, W.-P. Huang, G. X. Chen, and S. S. Jian, "Design and optimization of photonic crystal fibers for broad-band dispersion compensation," *Transactions on magnetics*, vol. 15, pp. 540–542, 2003.
14. K. Z. Nobrega and H. E. Hernandez-Figueroa, "Optical bistability in nonlinear waveguides with photonic crystals," to appear in *Microwave and Optical Technology Letters*, 2004.

Analysis of Quantum Light Memory in Atomic Systems

A. Rostami

E. Eng. Dept., Faculty of Engineering, Islamic Azad University of Tabriz, Tabriz, Iran.
OIC Research Lab., Faculty of Electrical Engineering, Tabriz University, Tabriz 51664, Iran.
Tel: +98 411 3393724
Fax: +98 411 3300819
rostami@tabrizu.ac.ir

Abstract. We extend the theory to describe the quantum light memory in Λ type atoms with considering γ_{bc} (lower levels coherency decay rate) and detuning for the probe and the control fields.

1 Introduction

Atomic coherence and related phenomena such as electromagnetically induced transparency (EIT) and slow light have been studied extensively in recent years [1-10]. Many application are proposed to this phenomena such as nonlinear optics (SBS, FWM and etc.), Lasing without Inversion, Laser cooling and Sagnac Interferometer [11-14]. One of important and promising applications in this field is light storage and quantum light memory that is investigated by some research groups [15-25]. The most common mechanism in this application is that the light pulse is trapped and stored in atomic excitations in the EIT medium by turning off the control field and then is released by turning on the control field. However, most of these works do not present clearly and general theory to analyze the propagation and storage of light. In addition, most of the works in EIT and slow light treat the light classically that is not proper to extend to quantum memory in which quantum state of light is to be stored. The most general theory for quantum memory was developed by M. Fleischhauer, et. al. [16,17]. They consider the light, quantum mechanically and present an excellent theory to describe the case. However, their work is not general, from our point of view, in some cases. The most deficient aspect of their work is that they do not consider the decay rate of lower levels coherency and the detuning from resonances which have important effects on the propagation and storage of light in the atomic media. It has caused their theory to be ideal and inexact. In this paper, we try to extend the theory of quantum light memory which is developed previously by M. Fleischhauer et al. to a more general and clear quantum mechanically for slow light and light storage in atomic ensemble with considering all decay rates and detuning.

The organization of this paper is as follows. In section 2, quantum mechanical model to describe slow light and light storage is presented. In this section after introducing the mathematical model, two subsection including low intensity limit and small variations and adiabatic passage limit are discussed. The result and discussion is presented in section 3. Finally, the paper is ended with a short conclusion.

J.N. de Souza et al. (Eds.): ICT 2004, LNCS 3124, pp. 296–303, 2004.

2 Mathematical Model for Light Storage

Here we further develop the theory presented in [16]. In this case, Λ type three level atoms are considered which is demonstrated in Fig. (1). Probe field couples the two $|a\rangle$ and $|b\rangle$ atomic levels together and the respected detune is defined as $\omega_{ab} - v_p = \Delta + \Delta_p$. Also the control (coupling) field couples the two $|a\rangle$ and $|c\rangle$ levels with a detuning from resonance ($\omega_{ac} - v_c = \Delta$), where v_μ are related to the probe and the control field carrier frequencies and $\omega_{\alpha\beta}$ are the resonance frequencies of corresponding levels. Δ and Δ_p are defined as one and two photon detuning respectively.

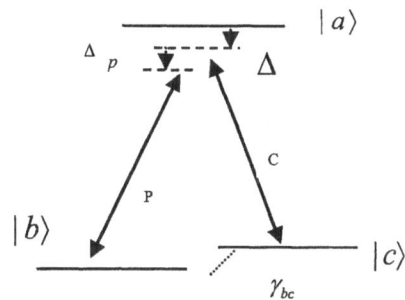

Fig. 1. Schematics of Λ type three level atoms

The probe field $\hat{E}(z,t)$ can be defined as follows [16]

$$\hat{E}(z,t) = \sqrt{\frac{\hbar v}{2\epsilon_0 V}} \hat{\mathcal{E}}(z,t) \times e^{i\frac{v_p}{c}(z-ct)} \tag{1}$$

In this relation $\hat{\mathcal{E}}_{(z,t)}$ is the slowly varying annihilation operator (dimensionless field operator) that corresponds to the envelope of probe field. V is the quantization volume of field that can be chosen to be equal to the volume of memory cell. The atomic operator for the atom j is defined as

$$\hat{\sigma}_{\alpha\beta}^{j} \overset{\Delta}{=} |\alpha_j\rangle\langle\beta_j| \tag{2}$$

In this relation $|\alpha_j\rangle$ and $|\beta_j\rangle$ are the Heisenberg Picture base atomic Kets (States) for atom number j. We can divide the memory cell to sections so that atomic operator does not change on them and every division is characterized by coordinate z. By this means one can define the collective (continuum) atomic operators as [16,26]

$$\hat{\sigma}_{\alpha\beta}(z,t) \overset{\Delta}{=} \frac{1}{N_z} \sum_{j=1}^{N_z} \hat{\sigma}_{\alpha\beta}^{j} , \tag{3}$$

where N_z is the number of atoms in the division z. For our purposes, it is easier to work with slowly varying collective atomic operators that are defined as

$$\hat{\tilde{\sigma}}_{\alpha\beta}(z,t) = \hat{\sigma}_{\alpha\beta}(z,t) e^{i\frac{\omega_{\alpha\beta}}{c}(z-ct)} . \tag{4}$$

With considering these operators the interaction Hamiltonian in interaction picture can be written as [27]

$$\hat{H}_I = -N\int_L \frac{dz}{L}(\hbar g \hat{\mathcal{E}}(z,t)\hat{\sigma}_{ab}(z,t)e^{i(\Delta+\Delta_p)t} +\frac{\hbar}{2}\Omega\hat{\sigma}_{ac}(z,t)e^{i\Delta t})+h.c.,\tag{5}$$

where N is the total number of atoms in the memory cell, L is the length of the cell, and g is the vacuum Rabi frequency that is given by $g = \dfrac{\vec{\wp}_{ij}\cdot\vec{\epsilon}\sqrt{\frac{\hbar v_k}{2\epsilon_0 V}}}{\hbar}$ and is related to atom field coupling strength in a given interaction system. Also, $\vec{\wp}_{ij}$ is the electric dipole moment corresponding to the two levels i and j, and $\vec{\epsilon}$ is the field polarization. One can find equations of motion for the atomic and field operators by substituting the above Hamiltonian in the Heisenberg-Langevin equations [26-29] as

$$[\frac{\partial}{\partial t}+c\frac{\partial}{\partial z}]\hat{\mathcal{E}}(z,t) = igN\hat{\sigma}_{ba}(z,t)\tag{6}$$

$$\frac{\partial}{\partial t}\hat{\sigma}_{bc}(z,t) = -(i\Delta_p+\gamma_{bc})\hat{\sigma}_{bc}-ig\hat{\mathcal{E}}(z,t)\hat{\sigma}_{ac}+i\Omega^*\hat{\sigma}_{ba}+\hat{F}_{bc}(z,t)\tag{7a}$$

$$\frac{\partial}{\partial t}\hat{\sigma}_{ba}(z,t) = -(i(\Delta+\Delta_p)+\gamma_{ba})\hat{\sigma}_{ba}+ig\hat{\mathcal{E}}(z,t)(\hat{\sigma}_{bb}-\hat{\sigma}_{aa})+i\Omega\hat{\sigma}_{bc}+\hat{F}_{ba}(z,t)\tag{7b}$$

$$\frac{\partial}{\partial t}\hat{\sigma}_{ca}(z,t) = -(i\Delta+\gamma_{ca})\hat{\sigma}_{ca}+i\Omega(\hat{\sigma}_{cc}-\hat{\sigma}_{aa})+ig\hat{\mathcal{E}}(z,t)\hat{\sigma}_{cb}+\hat{F}_{ca}(z,t)\tag{7c}$$

$$\frac{\partial}{\partial t}\hat{\sigma}_{aa}(z,t) = -\gamma_a\hat{\sigma}_{aa}-ig[\hat{\mathcal{E}}^+(z,t)\hat{\sigma}_{ba}-H.a.]-i[\Omega^*\hat{\sigma}_{ca}-H.a.]+\hat{F}_a(z,t)\tag{7d}$$

$$\frac{\partial}{\partial t}\hat{\sigma}_{bb}(z,t) = \gamma\hat{\sigma}_{aa}+\gamma'\hat{\sigma}_{cc}+ig[\hat{\mathcal{E}}^+(z,t)\hat{\sigma}_{ba}-H.a.]+\hat{F}_b(z,t)\tag{7e}$$

$$\frac{\partial}{\partial t}\hat{\sigma}_{cc}(z,t) = \gamma\hat{\sigma}_{aa}-\gamma'\hat{\sigma}_{cc}+i[\Omega^*\hat{\sigma}_{ca}-H.a.]+\hat{F}_c(z,t)\tag{7f}$$

The sign (+) on operators is the Dagger sign that correspond to Hermitian conjugate of the operators. $\gamma_{\alpha\alpha}$ and $\gamma_{\alpha\beta}$ are the population decay rate of level α and the coherency decay rate of levels α and β respectively. Ω is defined the Rabi frequency of control field that is given by $\Omega = \vec{\wp}_{ac}\cdot\vec{E}_c/\hbar$, where $\vec{E}_c$ is amplitude of the control field. $\hat{F}_\alpha$, $\hat{F}_{\alpha\beta}$ are δ correlated Langevin noise operators that are caused by reservoir noisy fluctuations (Vacuum Modes) [16,26,28]. In the above equations we see the $\gamma_{bc}, \Delta, \Delta_p$ terms in which in the main reference [16] they ignored. These parameters, as we will show, especially γ_{bc} have considerable effects on the memory behavior. The present equations are a set of coupled differential equations and solving them are difficult. Therefore, we use some approximations to minimize these equations.

3   Results and Discussion

Now, our simulated result are presented in the following figures.

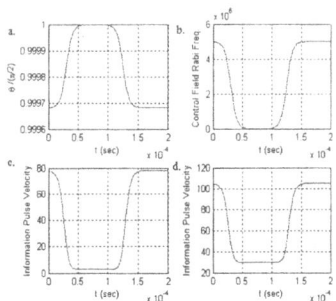

Fig. 2. a) $\theta/\!\!/_{\pi/2}$ as a function of time. b) Rabi frequency of control field as a function of time. c) Group velocity of information pulse for $\gamma_{ba} = 10^8$ $rad/\!\!\sec$. d) Group velocity of information pulse for $\gamma_{ba} = 10^9$ $rad/\!\!\sec$. ($\gamma_{bc} = 10^4$ $rad/\!\!\sec$, $\Delta, \Delta_p = 0$.)

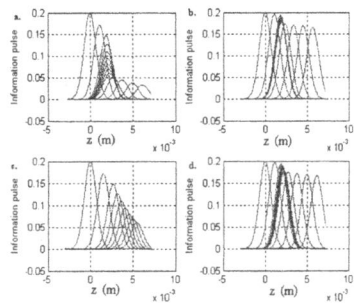

Fig. 3. Information pulse propagation for time steps of

15μ sec .a) $\gamma_{bu} = 10^8$, $\gamma_{bc} = 10^4$ ($rad/\!\!\sec$). b) $\gamma_{ba} - 10^8$, $\gamma_{bc} = 10^3$ ($rad/\!\!\sec$).

c) $\gamma_{ba} = 10^9$, $\gamma_{bc} = 10^4$ ($rad/\!\!\sec$). d) $\gamma_{ba} = 10^9$, $\gamma_{bc} = 10^3$ ($rad/\!\!\sec$). ($\Delta, \Delta_p = 0$).

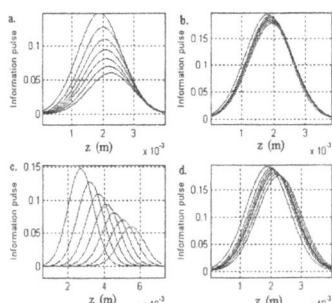

Fig. 4. Information pulse for time steps of 15μ sec in the storage region where the control field

is off. a) $\gamma_{ba} = 10^8$, $\gamma_{bc} = 10^4$ ($rad/\!\!\sec$). b) $\gamma_{ba} = 10^8$, $\gamma_{bc} = 10^3$ ($rad/\!\!\sec$).

c) $\gamma_{ba} = 10^9$, $\gamma_{bc} = 10^4$ ($rad/\!\!\sec$). d) $\gamma_{ba} = 10^9$, $\gamma_{bc} = 10^3$ ($rad/\!\!\sec$). ($\Delta, \Delta_p = 0$).

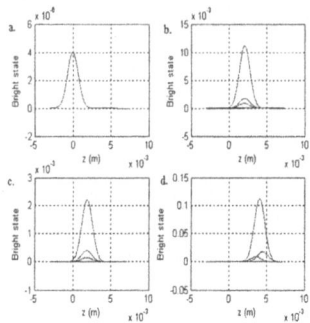

Fig. 5. a) Initial bright state for input pulse with $\gamma_{ba} = 10^8$, $\gamma_{bc} = 10^4$ rad/sec. b) Bright state for time steps of 15μ sec for $\gamma_{ba} = 10^8$, $\gamma_{bc} = 10^4$ rad/sec. c) Bright state for time steps of 15μ sec and for $\gamma_{ba} = 10^8$, $\gamma_{bc} = 10^3$ rad/sec. d) Bright state for time steps of 15μ sec and for $\gamma_{ba} = 10^9$, $\gamma_{bc} = 10^4$ rad/sec. $(\Delta, \Delta_p = 0)$.

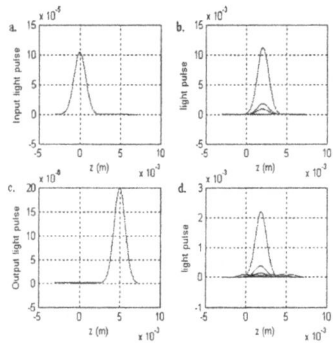

Fig. 6. a) Input light pulse inside medium for $\gamma_{ba} = 10^8$, $\gamma_{bc} = 10^4$ rad/sec. b) light pulse for time steps of 15μ sec for $\gamma_{ba} = 10^8$, $\gamma_{bc} = 10^4$ rad/sec. c) Output light pulse inside medium for $\gamma_{ba} = 10^8$, $\gamma_{bc} = 10^4$ rad/sec. d) light pulse for time steps of 15μ sec and for $\gamma_{ba} = 10^8$, $\gamma_{bc} = 10^3$ rad/sec. $(\Delta, \Delta_p = 0)$

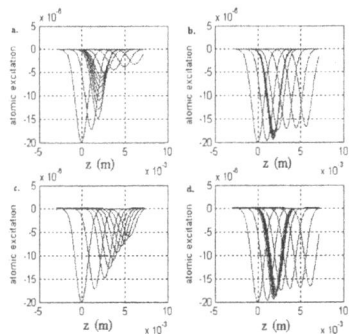

Fig. 7. a) σ_{bc} (atomic excitation) propagation for time steps of $15\,\mu\sec$. a) $\gamma_{ba}=10^{8}$, $\gamma_{bc}=10^{4}$ ($^{rad}/_{sec}$). b) $\gamma_{ba}=10^{8}$, $\gamma_{bc}=10^{3}$ ($^{rad}/_{sec}$). c) $\gamma_{ba}=10^{9}$, $\gamma_{bc}=10^{4}$ ($^{rad}/_{sec}$). d) $\gamma_{ba}=10^{9}$, $\gamma_{bc}=10^{3}$ ($^{rad}/_{sec}$). ($\Delta,\Delta_{p}=0$).

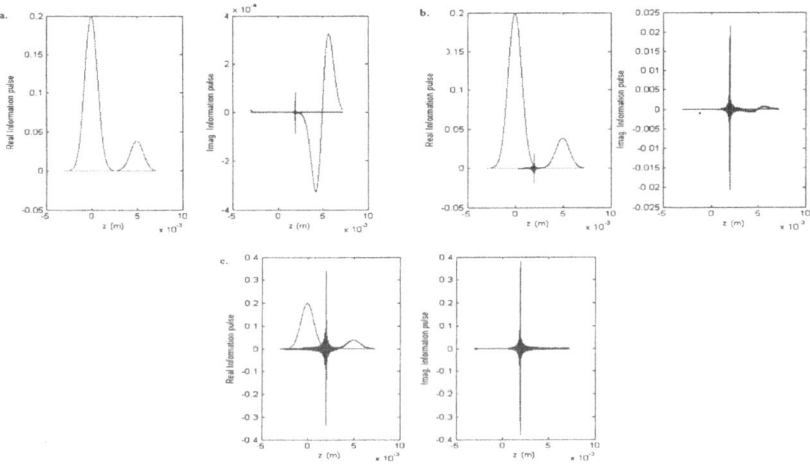

Fig. 8. Real and Imaginary parts of the information pulse at the times $t=0$ (input) and $t=165\mu\sec$ (output) plotted with each other for various values of One-Photon detuning.

a) $\Delta=2\times10^{6}$. $^{rad}/_{sec}$ b) $\Delta=4\times10^{6}$ $^{rad}/_{sec}$ .c) $\Delta=5\times10^{6}$ $^{rad}/_{sec}$.
($\Delta_{p}=0$, $\gamma_{ba}=10^{8}$, $\gamma_{bc}=10^{4\,rad}/_{sec}$).

4 Conclusion

In this paper, we further developed the quantum mechanical theory for quantum light memory in the Low intensity limit and small variations and adiabatic passage limit, primarily developed by [16]. We entered the parameters $\gamma_{bc},\Delta,\Delta_{p}$ into the formulations. We obtained and explained their effects in a clear form.

302   A. Rostami

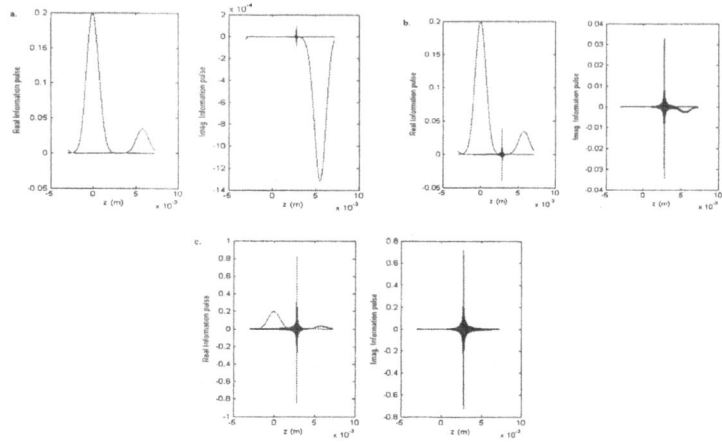

Fig. 9. Real and Imaginary parts of the information pulse at the times $t = 0$ (input) and $t = 165\mu$ sec (output) plotted with each other for various values of Two-Photon detuning.

a) $\Delta_p = 2 \times 10^2 \ ^{rad}/_{sec}$. b) $\Delta_p = 4 \times 10^2 \ ^{rad}/_{sec}$. c) $\Delta_p = 5 \times 10^2 \ ^{rad}/_{sec}$.

$$(\Delta = 0, \gamma_{ba} = 10^8, \gamma_{bc} = 10^4 \ ^{rad}/_{sec}).$$

References

1. Harris, S. E., Field J. E. and Kasapi, A. , Phys. Rev. A 46 No.1 p.46 (1992); S. E. Harris, J. E .Field and A. Imamoglu, Phys. Rev. Lett. 64 1107(1990).
2. Arimondo, E. , Progress in Optics, E. Wolf Ed. Vol: XXXV (1996) (Amsterdam: Elsevier Science).
3. Taichenachev, A. V., Tumaikin, A. M. and Yudin, V. I. , Phys. Rev. A. 61 011802 (2000).
4. Harris, S. E. , Hau, L. V. , Phys. Rev. Lett. 82 4611 (1999).
5. Advances In Atm. Mol. Opt. Physics, Ed.:B.Bederson Vol:46 Academic Press (2001).
6. Boyd, R. W. , Gauthier, D. J. , Progress in optics, E.Wolf Ed. (2002) (Amsterdam: Elsevier Science).
7. Kash, M. M. , et.al., Phys. Rev. Lett. 82 5229 (1999).
8. Hau, L. V. , et al., Nature 397, 594, (1999).
9. Ye C. Y. , and Zibrov, A. S. , Phys. Rev. A 65, 023806, (2002).
10. Kien F. L., and Hakuta, K. , Can. J. Phys. 78, 548 (2000).
11. Matsko, A. ,et al., Phys. Rev. Lett. 86, 2006, (2001).
12. Johnsson M. T. , and Fleischhauer, M. , Phys. Rev. A 66, 043808 (2002).
13. Mompart J., and Corbalan, R. , J. Opt. B.:Quantum Semiclass. Opt.2, R7 (2000).
14. Zimmer, F. , Fleischhauer, M. , (quant-ph/0312187v1 Dec.2003).
15. Fleischhauer, M. , Yelin S. F. , and Lukin, M.D., Opt. Comm. 179, 395 (2000).
16. Fleischhauer, M. , Lukin, M. D. , Phys. Rev. A 65, 022314, (2002); M. Fleischhauer, M. D. Lukin, Phys. Rev. Lett. 84, 5094, (2000); M. Fleishhauer, C. Mewes, Proceeding of the Enrico-Fermi school on quantum computation, Varenna, Italy (2001).
17. Mewes, C. , Fleischhauer, M. , Phys. Rev. A 66,   033820 (2002).
18. Juzeliunas, M. Masalas, M. Flaischhauer, Preprint:quant-ph/0210123v1 Otc.2002).
19. Kozhekin, A. E. , Molmer K. , andPolzik, E., Rev. A 62, 033809, (2000).
20. Kai-Ge, W. , Shi-Yao, Z. , Chin. Phys. Lett. 19, No.1, 56, (2002).

21. Agarwal G. S., and Dey, T. N. , (Preprint: quant- ph/0112059v2 Sep.2002).
22. Dantan A. , and Pinard, M. , (Preprint: quant-ph/0312189v1 Dec.2003).
23. C. Liu, Z. Dutton, C. H. Behroozi, L.V. Hau, Nature, 409, 490(2001).
24. Philips, D. F. , Fleischhauer, A. , Mair A. , and Walsworth, L. , Phys. Rev. Lett. 86, 783, (2001).
25. Zibrov, A. S., et al., Phys. Rev. Lett. 88, 103601, (2002).
26. Fleischhauer, M., Richter, T., Phys. Rev. A, 51, 2430, (1995).
27. Quantum Optics, Scully, M.O., Zubairy, M.S. (1997), (Cambridge University Press).
28. Fleischhauer M., and Scully, M.O., Phys. Rev. A, 49, 1973, (1994).
29. Raymer, M.G., et al., Phys. Rev. A, 32, 332, (1985); M.G.Raymer, J.Mostowski, Phys. Rev. A, 24, 1980, (1981).
30. Fleischhauer, M., Manka, A.S., Phys. Rev. A, 54, 794, (1996).
31. Vitanov, N.V., Stenholm, S., Phys. Rev. A, 56, 1463, (1997).

Wavelength Conversion with 2R-Regeneration by UL-SOA Induced Chirp Filtering

Cristiano de Mello Gallep[1] and Evandro Conforti[2]

[1] *Div. Tecnologia em Telecomunicações* – CESET/Unicamp, R. Paschoal Marmo, 1888 – Jd. Nova Itália, 13484-370, Limeira/SP – Brasil. gallep@ceset.unicamp.br.
[2] *Dep. Microondas e Óptica*, FEEC/ Unicamp, Av. Albert Einstein, 400 – Barão Geraldo, 13083-970, Campinas/SP - Brazil, conforti@dmo.fee.unicamp.br

Abstract. Simulations of ultra-long (10 mm) semiconductor optical amplifier's induced optical chirp (cross-phase modulation), with posterior conversion to amplitude modulation by proper filtering, show promising results for wavelength conversion with pulse re-amplifying and re-shaping in 10 Gbps and 20 Gbps. The dependences on filter bandwidth and optical input power (high, medium and low levels) are analyzed for conversion of originally clean and dirty eye-diagrams at the input gate within their reshaping capability.

1 Introduction

The semiconductor optical amplifier (SOA) is a prominent key-tool device to act as the nonlinear element in all-optical signal processing for WDM (wavelength division multiplexing) networks [1]. Due to the maturity of this device technology, the costs to the market are decreasing and the integration with other active and passive devices is possible. SOA-based sub-systems are being developed to give feasible alternatives for optical signal processing as wavelength conversion [1], switching [2] and pulse regeneration [3], among other important functionalities.

To implement such all-optical processing features, the phenomenon mostly used is the SOA's nonlinear gain behavior: the cross-gain modulation (XGM) and the cross-phase modulation (XPM). Another important non-linear phenomenon, the four-wave mixing (FWM), is less used due its complexity, both in hardware setup and operation, although presenting high speed response [1]. All of them - XGM, XPM and FWM - are reached under saturated optical gain conditions during SOA amplification, obtained by injecting either high input optical powers or electronic bias current into the SOA active cavity, or by doing both simultaneously.

The first nonlinear process, XGM, is the easiest to implement but the one with the slowest response, and so the worst converted eye-opening performance. It can be used to cover bit-rates not higher than 5 Gbps and with a clear input pulse's shape format, since XGM depends on very deep carrier density modulation. The second, XPM, is faster than the previous one and presents better eye-opening conversion, operating based on the fast cavity index response due to smaller carrier modulation, enabling pulse regeneration in the 10-40 Gbps data rate range. But most XPM implementations

J.N. de Souza et al. (Eds.): ICT 2004, LNCS 3124, pp. 304–311, 2004.

need expensive integrated structures to divide properly, to guide and to interfere the original and the desired output optical channels after the amplification [4].

Recently, a novel XPM scheme was proposed by Lucent Labs (U.S.A.), using discrete SOA and XPM induced red-shift chirp filtering to convert and regenerate optical pulses, without using any integrated structure [5]. That scheme however needs a very complex filtering tool, not so easily implemented. Here, the authors propose an even easier alternative, based on filtering of the SOA blue-shift induced chirp by a WDM short-band filter, to transform the frequency chirp in amplitude modulation. In this way, it is possible to implement wavelength conversion with pulse re-amplifying/re-shaping (2R regeneration) based just on discrete devices: an ultra-long (UL) SOA and a WDM filter.

2 The XPM Induced Blue-Shift Chirp Filtering

The basic scheme for the XPM induced blue-shift chirp filtering sub-system is based on an UL-SOA ($L_z \cong 10$ mm, see Table 1 for other device's simulation parameters) and a common WDM short-band filter (free-spectral range of 25, 50 or 100 GHz, with a super-gaussian like or a sin–square like frequency response), as illustrated in Fig. 1 with its operation scheme.

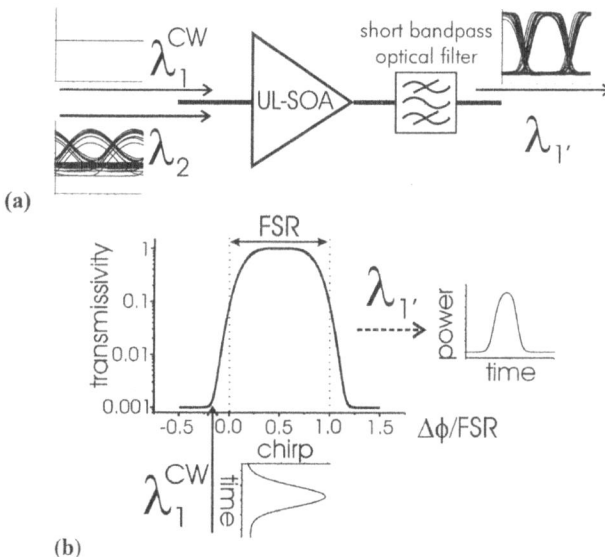

Fig. 1. Optical channel blue-shift chirp filtering (induced by XPM in UL-SOA amplification): (a) basic set-up and (b) operation scheme for 2R wavelength conversion (super-gaussian like filter response)

As a modulated channel λ_2 propagates inside the very long SOA cavity, under high optical gain (just below initial lasing oscillations), it induces itself gain-saturation and phase modulation (SPM) due to charged carrier's modulation. Simultaneously, the amplification induces XGM and XPM in a co-propagating CW channel - λ_1 (Fig.1). Due to the long cavity, the induced chirp correspondent to the XPM can be as high as 10 GHz to 20 GHz (positive), showing also residual Fabry-Perot oscillations and bit-pattern dependence.

By passing the composite output optical signal (the amplified channels λ_1 and λ_2 with also the ASE (amplified spontaneous emission) noise through an abrupt optical filter, the converted signal is obtained when the λ_1 wavelength is rightly positioned at the filter's low-frequency edge, and the modulated channel (λ_2) is positioned out of the bypass window. With this scheme, the second channel and the ASE are totally rejected and the first channel is also highly attenuated but in such time periods it presents a huge positive optical chirp induced by the modulation code in λ_2. So, the SOA XPM induced chirp (frequency modulation) in the CW channel is transformed in amplitude modulation (FM-AM conversion). By this way, a λ_1. (λ_1 + ~FSR/2) output channel is obtained with pulses varying from Gaussian like format (input pulse's shape, low SOA gain) to super-Gaussian like format (filter response's shape, high SOA gain). Due to the abrupt S-like profile in the edge of the filter band, the scheme enables wavelength conversion with simultaneous pulse amplification and re-shaping. The bit pattern dependence of SOA gain, due to average input power oscillations (depending on how big are level "1"-bit repetition in the pulse train) is clearly noted but does not degrade the eye-opening, as shown in the simulated eyes next.

2.1 Simulations of Eye-Opening in Wavelength Conversion

The SOA simulator used to study the scheme presented here is based on a modified transfer matrix method (TMM) and was presented elsewhere [6,7]. Its main operational parameters are listed in Table 1. Basically, "n_z" discrete sections are used to calculate a simultaneous propagation of two optical channels inside the SOA active cavity, in both positive and negative directions, and so for the total ASE power, with internal optical reflections considered just in the cavity edges. Passing from one discrete section "j" to the next, each channel "k" (co and counter propagating) suffers alterations in amplitude, in each discrete time "i", given by the gain [8,9]:

$$dG_{k\,i,j} = \Gamma\,[a_1\,(N_{i,j} - N_{tr}) - a_2\,(\lambda_k - \lambda_{sh})^2 + a_3\,(\lambda_k - \lambda_{sh})^3] - (\alpha_{abs} + \alpha_{scat})\,. \tag{1}$$

Here, $\lambda_{sh} = \lambda_0 - a_4\,(N_{i,j} - N_{tr})$ is the central frequency gain shift (see Table 1), and also alterations in the optical phase, defined by [9]:

$$d\theta_{k\,i,j} = \frac{2\pi \cdot dz}{\lambda_k} \cdot \left[n_{ef} + \Gamma\frac{dn}{dN}(N_{th} + (N_{i,j} - N_{tr})) \right]. \tag{2}$$

The total output phase alterations thus obtained, as the optical channels propagates for the entire SOA length, are the resultant XPM (or SPM) and its time first derivative is used to determine the induced frequency chirp for each optical channel.

The simulation of optical filtering was added to the program by properly correlation of the output power and chirp data. Two different filter shapes, sin-square like and super-gaussian like, were implemented with four parameters to describe their basic features: 1) insertion loss (due to internal absorption and coupling, ideally neglected here, [dB]); 2) response (filter function amplitude range, i.e., the transmissivity in a logarithmic scale, [dB]); 3) free-spectral range – FSR (filter bypass window, [GHz]); and 4) central frequency location – CFL (relative CW channel position to the filter response window, being the FSR range equivalent to 180°, [°] ????). To calculate the filter response, in each discrete-time point i, the absolute optical frequency location $\Delta\phi$ is determined by:

$$\Delta\phi(i) = chirp(i) + CFL * FSR / 180^\circ . \tag{3}$$

Table 1. UL-SOA simulation parameters.

Symbol	Definition	Value (unit)
n_t	# time points	6000
n_z	# space points	115
dt	time discretization	1 ps
Lx	cavity width	1.4 μm
Ly	cavity height	0.2 μm
Lz	cavity length	10147 μm
Ntr	carrier density at transparency	$2\ 10^{24}\ m^{-3}$
Nth	carrier density at threshold	$4\ 10^{24}\ m^{-3}$
α_{abs}	absorption loss	$2000\ m^{-1}$
α_{scat}	scattering loss	$100\ m^{-1}$
α_{ins}	insertion loss	2 dB
nef	effective index	3.4
a_1	gain coefficient	$2.5\ 10^{-20}\ m^2$
a_2	gain coefficient	$7.4\ 10^{18}\ m^{-3}$
a_3	gain coefficient	$3\ 10^{25}\ m^{-4}$
a_4	gain coefficient	$3\ 10^{-32}\ m^4$
dn/dN	differential refractive index	$-1.2\ 10^{-26}\ m^3$
Γ	Confinement factor	0.4
β	spont. emission coupling factor	0.0002

3 Simulation Results

For the optimized results presented in this article, the simulated SOA received a bias current of 1100 mA (+/- 10 mA, depending on input optical power). The power of the input CW optical channel - λ_1 (1550 nm) - is fixed in 100 μW and the modulated channel - λ_2 (1530 nm) – has average levels set to high. medium and low input optical powers. This features are used for a clean, a dirty and a very dirty input eye-opening. Some cases are shown in the next figures, for 10 Gbps and 20 Gbps data rates. Figure 2 shows the results for 10 Gbps and super-gaussian like filter, for a clean and a very dirty input eye-opening (high input power), and for an optimized CFL=-20° (fixed) with FSR equal to 25 GHz, 50 GHz and 100 GHz (filter response of 30 dB).

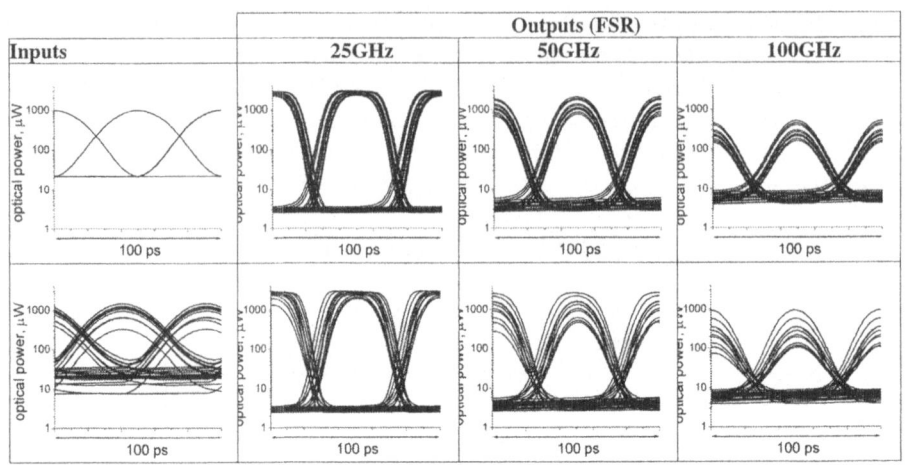

Fig. 2. Eye diagrams of input (clean and very-dirty signals with high power) and output (converted) signals in 10 Gbps, for super-gaussian like filter with FSR of 25 GHz, 50 GHz and 100 GHz and fixed CFL=-20° and response = 30 dB

As can be seen in the upper line of Fig.2 (clean input), the converted signals present wide eye-opening, with shorter and more abrupt pulses profile than the input signal, but with some fluctuations on pulse width and amplitude due to pattern dependence effects on the chirp magnitude: a long sequence of "1"-bits decreases the average optical gain and so the induced chirp. The best results are obtained for the shortest filter band (25GHz) and the highest input power, with the output eye-opening decreasing as the filter band increases or the optical input power decreases. In order to have a merit number to quantify the eye-opening regeneration, the relative eye amplitude ratio - A_{eye} is calculated, in logarithmic scale, defined by:

$$A_{eye} = 10 \cdot \log \frac{bit_1^{min} - bit_0^{max}}{bit_0^{max}} , \tag{4}$$

i.e., the dB relation of the lower "1"-bit level (bit_1^{min}) with the higher "0"-bit level (bit_0^{max}). So, for the clean input eye with A_{eye}= 16.5 dB a converted output was found with A_{eye}=28.3 dB (FSR=25GHz, high input power), showing more than 10 dB in reshaping gain (2R-gain).

Moving to corrupted input signals (lower lines in Fig.2), quite the same output forms are obtained but with higher variations in the "1"-bit levels. Although, for a dirty input eye-opening with A_{eye}= 6.5 dB an output with A_{eye}=26.2 dB was obtained (FSR=25GHz, high input power): almost 20 dB in reshaping gain.

Figure 3 presents all A_{eye} for the simulated data in 10 Gbps with super-gaussian like filter, including also the low input power level results for clean (A_{eye}=16.5 dB), dirty (A_{eye} ~12.2 dB to 13.2 dB) and very dirty (A_{eye}=6.5 dB) input eyes. Fig. 3 has also a back-to-back line (reference of 2R-gain = 0 dB) to situate the positive (region above the line) and the negative (region below the line) 2R-gain, showing that low input powers cannot be used is such lambda-conversion scheme without serious eye closing: low optical input powers cannot induce enough frequency chirp in the CW channel (due small carrier consumption) which is necessary for a good FM-to-AM conversion. This feature can be overcome by pre-amplifying the channel before passing through the UL-SOA.

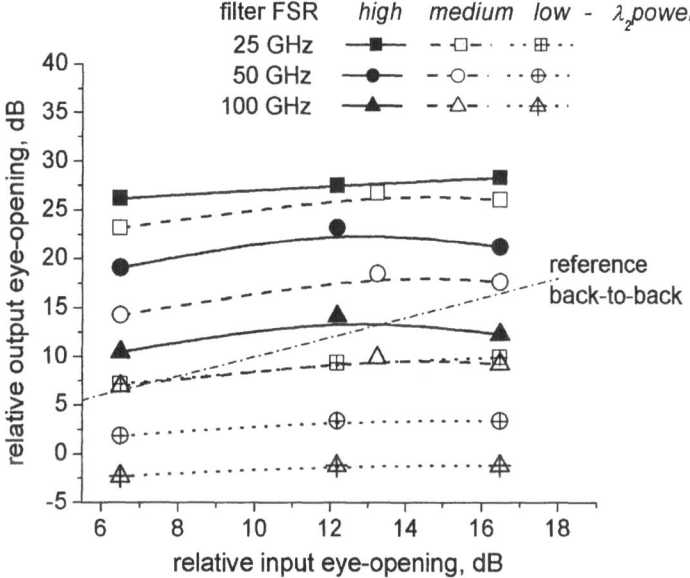

Fig. 3. 2R gain in UL-SOA with filtering scheme for wavelength conversion, 10Gbps case (CFL= -20°)

Fig. 3 also shows that with medium or high input powers and filters with FSR<50GHz, the scheme can present good 2R-gain and also that the reshaping action is higher for input eyes with bad A_{eye}.

The best results obtained for 20 Gbps data rate are presented in Figure 4: clean and dirty inputs, with high/medium power, and outputs for sin–square like filter (CFL= 0°) and super-gaussian like filter (CFL= -10°), both with response of 30 dB and FSR = 50 GHz. Each box in Fig. 4 also shows its calculated A_{eye}, in order to compare the 2R-gain from case to case. Both four cases show positive 2R-gain, with better performance for the sin-square like filter, which can equalize the "0"-bit levels due its more abrupt frequency response. Fig. 4 shows 2R-gains of 9.7 dB (sin-square like) and 6.2 dB (super-gaussian like) for the clean input eye case, and 7.4 dB (sin-square like) and 2.5 dB (super-gaussian like) for the dirty input eye case. The pattern dependence can be easily seen in Fig.4 too, feature which is also enlarged for the dirty-input case. One alternative to minimize such behavior, besides pre-amplifying the input signal, is to optimize the CFL within the FSR. With lower FSR and little positive CFL in the sin-square like filter, the bit levels can be more equalized but some penalties appear in the correspondent A_{eye}.

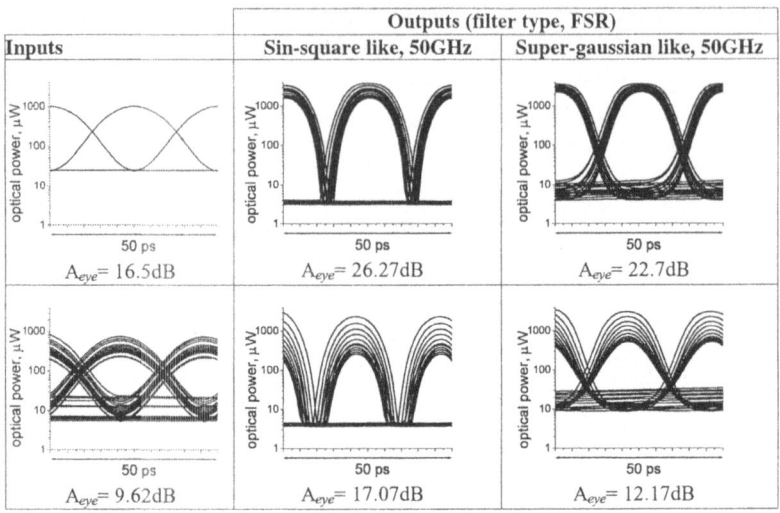

Fig. 4. Eye diagrams of input and output (converted) signals in 20 Gbps, for clean and dirty inputs, high power, and outputs for filter (sin–square like (CFL=0°) and super-gaussian like (CFL= -10°), both with response = 30 dB and with FSR = 50 GHz. Each box also presents its relative eye-opening (dB)

4   Conclusion

Some preliminary simulation results for a λ-conversion sub-system based on UL-SOA XPM frequency chirp filtering are presented for 10 Gbps and 20 Gbps data rate. Good 2R-gains are obtained (~ 20 dB) for closed input eyes with high input optical power. Although, some fluctuations on bit-"1" level occurred due bit-pattern dependence in the SOA gain. This behavior can be minimized by properly pre-amplifying the optical channel.

The set-up, totally based on discrete devices, should be experimentally implemented soon using UL-SOA provided by the HHI Labs (*HHI- Fraunhofer Institute for Telecommunications* [10], Germany) and commercial Bragg-gratings based filters. Closer eye diagrams should be tested, under practical operation conditions with BER (bit-error rate) measurements.

Acknowledgements. This work was supported in part by FAPESP (*Fundação de Amparo a Pesquisa do Estado de São Paulo*, by the CEPOF – Research Center in Photonics) and MCT (*Ministério de Ciência e Tecnologia*).

References

1. Stubkjaer, K.: Semiconductor Optical Amplifier-based All-Optical Gates for High-Speed Optical Processing. IEEE J. Selected Top.Quantum Electr., Vol.6 (2000) 1428-1435
2. Conforti, E., Gallep, C.M., Bordonalli, A.C.: Decreasing Electro-Optic Switching Time in Semiconductor Optical Amplifiers by Using Pre-Pulse Induced Chirp Filtering. In: Mørk, J., Srivastava, A. (eds.): TOPS 92, Optical Amplifiers and Their Applications 2003, Optical Society of America, Vol. 92 (2003) 111-116.
3. Leuthold, J., Mikkelsen, B., Behringer, R.E., Raybon, G., Joyner, C.H., Besse, P.A.: Novel 3R regenerator based on semiconductor optical amplifier delayed-interference configuration. IEEE Photon. Techn. Letters, Vol.13 (2001) 860-862
4. Manning, R.J., Ellis, A.D., Poustie, A.J., Blew, K.J.: Semiconductor Laser Amplifier for Ultrafast All-Optical Signal Processing. J. Optical Society of America B, Vol.14 (1997) 3204-3216
5. J. Leuthold *et. al.*: Nonblocking All-Optical Cross Connect Based on Regenerative All Optical Wavelength Converter in a Transparent Demonstration over 42 nodes and 16800 km. J. Lightwave Technol., Vol. 21 (2003) 2863- 2870
6. Gallep, C. M., Conforti, E.: Simulations on Picosecond Non-Linear Electro-Optic Switching Using an ASE-calibrated Optical Amplifier Model, Optics Communication, Vol 236/1-3 (2004) 131-139.
7. C. M. Gallep: Reduction of Electro-Optic Switching Time in Semiconductor Optical Amplifiers, PhD thesis (in Portuguese), State University of Campinas - Unicamp (2003). Available at *www.ifi.unicamp.br/photon*
8. Lee, H., Yoon, H., Kim, Y., Jeong, J.: Theoretical Study of Frequency Chirping and Extintion Ratio of Wavelength-Converted Optical Signals by XGM and XPM using SOA´s. IEEE J. of Quantum Electr., Vol.35 (1999), 1213 –1219
9. Yu, J., Jeppensen, P.: Improvement of Cascaded Semiconductor Optical Amplifier Gates by Using Holding Light Injection. IEEE J. Lightwave Tech., Vol.19 (2001) 614-623
10. *www.hhi.fraunhofer.de*

Performance Analysis of Multidimensional PPM Codes for Optical CDMA Systems in the Presence of Noise

José Santo G. Panaro[1] and Celso de Almeida[2]

[1] National Institute of Telecommunications, Inatel – S. R. do Sapucaí - MG - Brazil
panaro@inatel.br
[2] State University of Campinas, Unicamp - Campinas - SP - Brazil
celso@decom.fee.unicamp.br

Abstract. This paper analyzes the asynchronous error performance of optical code-division multiple access systems utilizing one-dimensional and multidimensional PPM codes in the presence of white Gaussian noise. The effects of the multiple access interference combined with the noise are statistically characterized by continuous probability density functions and, therefore, allowing the investigation on the error performance of the O-CDMA systems employing on-off keying (OOK) and on-off orthogonal (OOO) signaling. For OOK systems, the optimum decision threshold is determined and the error performance degradation due to deviation in the threshold is considered. Furthermore, the performances of the O-CDMA systems using OOK and OOO signaling are compared.

1 Introduction

Optical code-division multiple-access (O-CDMA) is an optical channel-encoding scheme, which utilizes characteristics of the channel to create multi-dimensional representations of the bits being transmitted. O-CDMA has many desirable features including flexibility, intrinsic security, and simplified network control and management. These features make the technology appealing for optical access networks [1].

Perhaps the main attractive feature of the O-CDMA is the possibility of multiple subscribers – each with different bit rates and protocols – share the channel simultaneously and asynchronously. The limiting factor to the traffic that the system can support is a combination of multiple-access interference (MAI) and noise. MAI is caused by the joint action of the cross-correlation between the sequences of N_u asynchronous users. In a previous work we have considered only the effect of the MAI in O-CDMA systems using PPM codes without synchronism constraint [2]. Besides the multiple access interference, in this work we take into account the receiver input noise, which is modeled as an additive white Gaussian random process.

There are some alternatives to implement PPM-coded O-CDMA systems. The simplest and first envisioned system consists in assigning a sequence of binary word, with length L and unitary weight, for each user. If the bit duration is T_b, then the chip time interval is $T_c = T_b / L$, and each code word has only one chip position different of zero. The transmission of a bit *one* is done by sending the code word assigned to the user, while the all-zero word signalizes a bit *zero* [3], [4]. Therefore, this simple scheme configures an on-off keying (OOK) unipolar signaling technique.

J.N. de Souza et al. (Eds.): ICT 2004, LNCS 3124, pp. 312–321, 2004.

More sophisticated systems can be devised by using two-dimensional PPM codes [5] – [8]. In this case, the codes are constructed on multiple wavelengths as well various time chips. Thus, every bit *one* is encoded by R PPM sequences of length L and transmitted simultaneously on distinct sub-channels (wavelengths). Therefore, a two-dimensional PPM code word can be pictured as an R by L binary matrix with a single one per row. In this case, the code weight W is identical to R. Bit *zero* is symbolized by the R by L null matrix. An example of a 4×5-PPM code word is shown in Figure 1.

An alternative signaling scheme makes use of R sets consisting of two OOK sub-channels, one set for each PPM sequence. Only one sub-channel in the set is activated during each bit interval, while the other remains in the zero state, and this scheme constitutes an on-off orthogonal (OOO) unipolar signaling technique. In this case the code weight W remains equal to R.

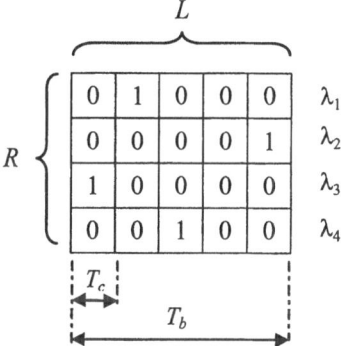

Fig. 1. Example of a two-dimensional PPM code word with $R = 4$ and $L = 5$. Each row is a PPM sequence that can be transmitted temporally on a distinct wavelength.

2   Interference and Noise Analysis

2.1 OOK Signaling

For an O-CDMA/OOK system, the intensity of the interference X_i, occasioned by the ith user over a chip time interval of the sequence of a reference user, is a random variable in the range [0, 1]. For OOK signaling, the asynchronous interference level is characterized by the probability density function given by

$$f_i(x_i) = (1-p)\,\delta(x_i) + p\,[u(x_i)-u(x_i -1)],\tag{1}$$

where $\delta(\cdot)$ is the *Dirac delta* function, $u(\cdot)$ is the *unit step* function, and p is the probability of partial superposition of the pulse of the ith user sequence on a chip of the reference sequence. Furthermore, if the data source is equiprobable then we have $P(\vec{s}_0) = P(\vec{s}_1) = 1/2$. Thus, the partial interference probability can be expressed as

$$p = \frac{2}{L}P(\vec{s}_1) = \frac{1}{L}\tag{2}$$

We have shown in [2] that the probability density function (p.d.f.) and the moment generating function (m.g.f.) of the multiple access interference for OOK signaling due to n active users are given, respectively, by

$$f_X(x) = (1-p)^n \delta(x) + \sum_{j=1}^{n} \sum_{k=0}^{j} \binom{n}{j}\binom{j}{k} p^j (1-p)^{n-j} \frac{(x-k)^{j-1}}{(-1)^k (j-1)!} u(x-k) \tag{3}$$

and

$$\Phi_X(s) = E\left[e^{sX}\right] = \sum_{j=0}^{n} \sum_{k=0}^{j} \binom{n}{j}\binom{j}{k} p^j (1-p)^{n-j} \frac{(-1)^k e^{sk}}{(-s)^j}. \tag{4}$$

The noise can be modeled as a zero-mean white Gaussian random process. Consequently, the p.d.f. of the noise can be written as

$$f_W(w) = \frac{1}{\sigma\sqrt{2\pi}} e^{-w^2/2\sigma^2} \tag{5}$$

and, accordingly, the noise m.g.f. is given by

$$\Phi_W(w) = E\left[e^{sW}\right] = e^{\sigma^2 s^2/2}. \tag{6}$$

The combined effect of the interference and the white noise over any transmitted chip results in a new random process, which can be described as

$$Y = X + W \tag{7}$$

and the distribution of the new random variable can be obtained by convoluting their interference and noise probability density functions, that is

$$f_Y(y) = f_X(x) * f_W(w). \tag{8}$$

The result of the operation in (8) can be obtained more conveniently from the m.g.f. of the random variable Y [9], calculated as

$$\Phi_Y(s) = \Phi_X(s) \cdot \Phi_W(s) = \sum_{j=0}^{n} \sum_{k=0}^{j} \binom{n}{j}\binom{j}{k} p^j (1-p)^{n-j} \frac{(-1)^k}{(-s)^j} e^{sk} e^{\sigma^2 s^2/2} \tag{9}$$

and taking its inverse transform, in order to obtain the desired p.d.f., as follows:

$$f_Y(y) = \sum_{j=0}^{n} \sum_{k=0}^{j} \binom{n}{j}\binom{j}{k} p^j (1-p)^{n-j} \frac{(-1)^k}{\sigma\sqrt{2\pi}} \underbrace{\int \cdots \int_{-\infty}^{y}}_{j} \exp\left(-\frac{(t-k)^2}{2\sigma^2}\right) dt \tag{10}$$

Although the result just obtained is expressed in an analytical form, the multiple integral operations in (10) make difficult its numerical computing. However, a more adequate expression can be evaluated with the aid of the generalized complementary error function [10], defined as

$$\text{erfc}_m(u) = \underbrace{\int \cdots \int}_{m} \text{erfc}(u)\,du = \frac{2}{\sqrt{\pi}} \int_u^\infty \frac{(t-u)^m}{m!} e^{-t^2}\,dt, \quad m \geq 0 \tag{11}$$

Therefore, Equation (10) can be rewritten as

$$f_Y(y) = (1-p)^n \frac{1}{\sigma\sqrt{2\pi}} \exp\left(-\frac{y^2}{2\sigma^2}\right)$$
$$+ \frac{1}{2} \sum_{j=1}^n \sum_{k=0}^j \binom{n}{j}\binom{j}{k} p^j (1-p)^{n-j} \frac{\left(\sigma\sqrt{2}\right)^{j-1}}{(-1)^k} \text{erfc}_{j-1}\left(\frac{k-y}{\sigma\sqrt{2}}\right) \tag{12}$$

Figure 2-a illustrates the shape of the p.d.f. of the joint action of the interference and noise for an OOK system. The figure also shows the p.d.f. of the MAI alone, computed by using (3). The asymmetrical nature of the random process is evident.

2.2 OOO Signaling

On-off orthogonal signaling utilizes two complementary channels for each PPM sequence and, consequently, the receiver employs two identical detectors. The binary decision can be taken subtracting their outputs and deciding accordingly the polarity of this new variable. Considering this scheme, the intensity of the interference X_i, caused by the ith user on the sequence of a reference user during a chip time interval, is a random variable normalized in the range [-1, 1]. Thus, the asynchronous interference intensity is characterized by the probability density function given by

$$f_i(x_i) = (1-2p)\delta(x_i) + p[u(x_i+1) - u(x_i-1)], \tag{13}$$

where p is the partial interference probability expressed in (2).

It has shown in [2] that, for OOO signaling, the p.d.f and m.g.f. of the MAI produced by n active users on any chip of the sequence of a reference user, can be expressed, respectively, as

$$f_X(x) = (1-2p)^n \delta(x) + \sum_{j=1}^n \sum_{k=0}^j \binom{n}{j}\binom{j}{k} p^j (1-2p)^{n-j} \frac{(x+j-2k)^{j-1}}{(-1)^k (j-1)!} u(x+j-2k) \tag{14}$$

and

$$\Phi_X(s) = \sum_{j=0}^n \sum_{k=0}^j \binom{n}{j}\binom{j}{k} p^j (1-2p)^{n-j} \frac{(-1)^k e^{(2k-j)s}}{(-s)^j} \tag{15}$$

The combined effect of the interference and noise over any transmitted chip results in a new random variable Y, which can be described again by (7) and (8). Thus, as the input noise power is equal to $2\sigma^2$ in this case, the resulting m.g.f. is given by

$$\Phi_Y(s) = \sum_{j=0}^n \sum_{k=0}^j \binom{n}{j}\binom{j}{k} p^j (1-2p)^{n-j} \frac{(-1)^k}{(-s)^j} e^{(2k-j)s} e^{\sigma^2 s^2} \tag{16}$$

and, taking the inverse transform of (16) and using (11), we obtain that

$$f_Y(y) = (1-2p)^n \frac{1}{2\sigma\sqrt{\pi}} \exp\left(-\frac{y^2}{4\sigma^2}\right)$$

$$+\frac{1}{2}\sum_{j=1}^{n}\sum_{k=0}^{j}\binom{n}{j}\binom{j}{k}p^j(1-2p)^{n-j}\frac{(2\sigma)^{j-1}}{(-1)^k}\mathrm{erfc}_{j-1}\left(\frac{2k-j-y}{2\sigma}\right). \tag{17}$$

Figure 2-b illustrates the shape of the resulting p.d.f. for an example of O-CDMA/OOO system. In this case the process has zero mean and is symmetric around the origin.

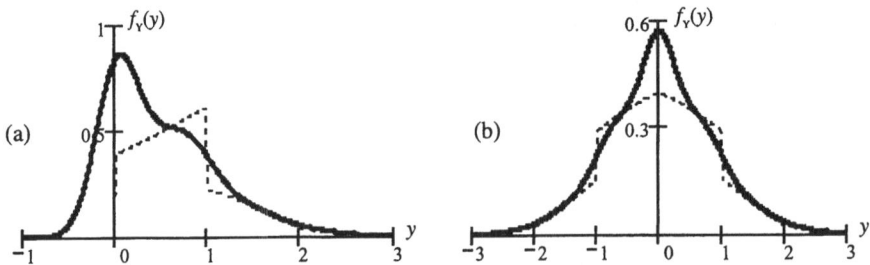

Fig. 2. Multiple access interference combined with white Gaussian noise p.d.f. for O-CDMA systems with parameters $p = 0.1$, $n = 10$ and $\sigma^2 = 0.05$. The p.d.f. of the interference alone is delineated in dotted lines, except the delta at the origin. (a) OOK signaling; (b) OOO signaling.

3   Performance Analysis

3.1 OOK Signaling

For O-CDMA systems using multidimensional OOK signaling, each one of the N_u users employs W PPM sequences that are transmitted simultaneously on distinct sub-channels. We assume that the receiver consists of a SUM detector followed by a binary decision device. In order to rearrange the individual pulses of the received sequences, the SUM detector applies a variable delay in each input signal and adds them up to produce a single resultant pulse. Thus, the ideal output signal has normalized amplitude equal to W if $\bar{s}_1$ is transmitted and zero otherwise. Moreover, the output noise remains a zero-mean white Gaussian noise with variance equal to $W\sigma^2$, where σ^2 is the input noise power of each (identical) sub-channel. Thus, the signal-to-noise ratio at the output of the SUM detector is improved and is given by

$$SNR_o = W \cdot SNR \tag{18}$$

where $SNR = 1/2\sigma^2$ is the normalized input signal-to-noise ratio of each sub-channel.

The junction of W independent interference signals described in (3) produces a new random process of identical p.d.f., except that $n = W(N_u - 1)$. Therefore, Equation (12) can be modified, resulting in the following p.d.f. to the new random variable Z:

$$f_Z(z) = (1-p)^n \frac{1}{\sigma\sqrt{2\pi W}} \exp\left(-\frac{z^2}{2\sigma^2 W}\right)$$

$$+ \frac{1}{2}\sum_{j=1}^{n}\sum_{k=0}^{j}\binom{n}{j}\binom{j}{k}p^j(1-p)^{n-j}\frac{\left(\sigma\sqrt{2W}\right)^{j-1}}{(-1)^k}\operatorname{erfc}_{j-1}\left(\frac{k-z}{\sigma\sqrt{2W}}\right)$$

(19)

where $n = W(N_u - 1)$.

If the symbols $\vec{s}_0$ and $\vec{s}_1$ are equiprobable, then the bit error probability for an O-CDMA/OOK system is given by

$$P_e = \frac{1}{2}\left[P(\text{error}\,|\,\vec{s}_0) + P(\text{error}\,|\,\vec{s}_1)\right].$$

(20)

Each conditional probability present in (20) can be expressed as

$$P(\text{error}\,|\,\vec{s}_0) = \int_\gamma^\infty f_z(z\,|\,\vec{s}_0)\,dz = \int_\gamma^\infty f_z(z)\,dz$$

$$= 1 - \frac{1}{2}\sum_{j=0}^{n}\sum_{k=0}^{j}\binom{n}{j}\binom{j}{k}p^j(1-p)^{n-j}\frac{\left(\sigma\sqrt{2W}\right)^j}{(-1)^k}\operatorname{erfc}_j\left(\frac{k-\gamma}{\sigma\sqrt{2W}}\right)$$

(21)

and

$$P(\text{error}\,|\,\vec{s}_1) = \int_{-\infty}^\gamma f_z(z\,|\,\vec{s}_1)\,dz = \int_{-\infty}^\gamma f_z(z-W)\,dz$$

$$= \frac{1}{2}\sum_{j=0}^{n}\sum_{k=0}^{j}\binom{n}{j}\binom{j}{k}p^j(1-p)^{n-j}\frac{\left(\sigma\sqrt{2W}\right)^j}{(-1)^k}\operatorname{erfc}_j\left(\frac{k-\gamma+W}{\sigma\sqrt{2W}}\right)$$

(22)

where γ is decision threshold level of the decoder.

The optimum decision threshold level, γ_o, is achieved when the sum of the end tail areas delimited by γ, under the conditional probability density functions $f_Z(z\,|\,\vec{s}_0)$ and $f_Z(z\,|\,\vec{s}_1)$, is minimized. This condition is satisfied by

$$f_Z(\gamma_o\,|\,\vec{s}_0) = f_Z(\gamma_o\,|\,\vec{s}_1).$$

(23)

The asymmetry in the distribution of the interference plus noise has two consequences. First the conditional error probabilities given by (21) and (22) result, generally, in distinct values and the user perceives an asymmetric binary channel. Moreover, γ_o is not a fixed value, changing with the signal-to-noise ratio and the traffic intensity in the system.

Figure 3 shows the optimum decision threshold and the resulting bit error probability for a reference system using $W = 3$ PPM sequences, as function of the number of users. It can be noticed that optimum threshold varies in the range $W/2 \le \gamma_o \le W$. As expected, the lower limit occurs in the absence of interference ($N_u = 1$), when only the zero-mean Gaussian noise disturbs the system.

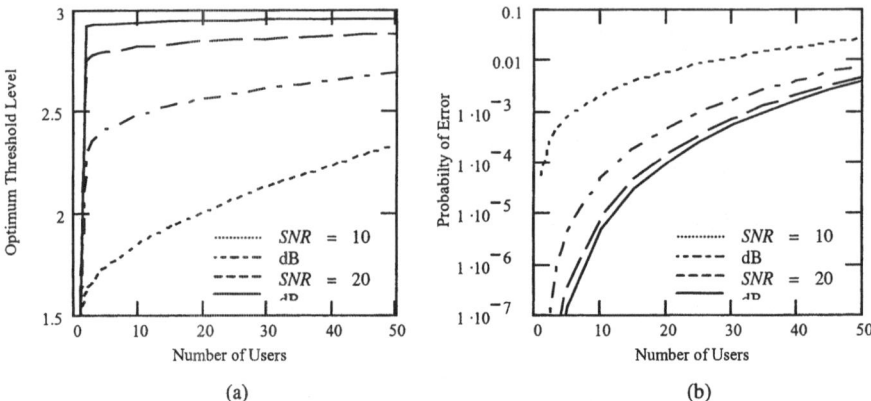

Fig. 3. Performance of an O-CDMA/OOK system using $W=3$ PPM sequences of length $L=100$, as function of the number of users: (a) optimum decision threshold; (b) minimum bit error probability.

Figure 4 reveals that the upper limit for γ_o takes place when $SNR \rightarrow \infty$. Further, when the signal-to-noise ratio is large enough, the interference dominates the error performance and the bit error probability depends almost exclusively on the number of users in the system. Under this circumstance the value of γ_o is reasonably constant, approaching W.

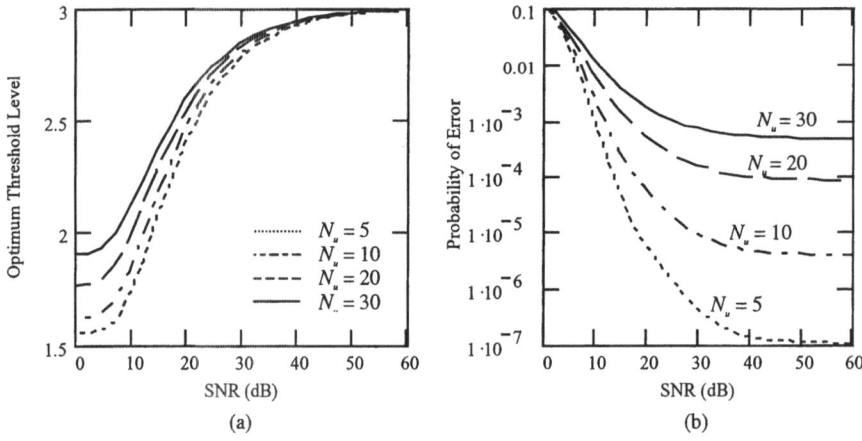

Fig. 4. Performance of an O-CDMA/OOK system using $W=3$ PPM sequences of length $L=100$, as function of the input signal-to-noise ratio: (a) optimum decision threshold; (b) minimum bit error probability.

The ideal receiver requires the dynamic estimation of the optimum decision threshold, which depends on the instantaneous traffic in the system. However, if the SNR is reasonably large, it is feasible to establish a sub-optimal threshold such as the impact on the performance should be small. Figure 5 shows that is possible to minimize the performance degradation for light, medium or heavy traffic conditions, by choosing a fixed threshold value that is optimum for a certain number of users in the desired operation range. The threshold values were fixed such as $\gamma = \gamma_o(N_u)$ for $N_u = 2$, 5 and 10. Also, the ideal case is shown for comparison.

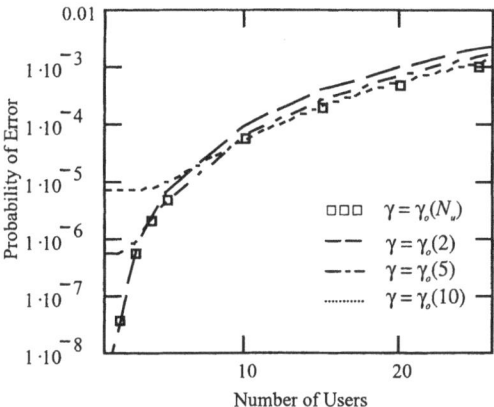

Fig. 5. Degradation of the bit error probability for some fixed decision threshold levels in O-CDMA/OOK system using $W = 3$ PPM sequences of length $L = 100$ and $SNR = 20$ dB.

3.2 OOO Signaling

Multidimensional OOO signaling employs W sets of two-reciprocal OOK channels. The receiver requires two identical SUM detectors and the decoding can be done by observing the outputs of both differentially. Thus, Equation (17) can be immediately extended for multidimensional O-CDMA/OOO systems, permitting to write that

$$
\begin{aligned}
f_Z(z) = &(1-2p)^n \frac{1}{2\sigma\sqrt{\pi W}} \exp\left(-\frac{z^2}{4\sigma^2 W}\right) \\
&+ \frac{1}{2}\sum_{j=1}^{n}\sum_{k=0}^{j}\binom{n}{j}\binom{j}{k} p^j (1-2p)^{n-j} \frac{\left(2\sigma\sqrt{W}\right)^{j-1}}{(-1)^k} \operatorname{erfc}_{j-1}\left(\frac{2k-j-z}{2\sigma\sqrt{W}}\right)
\end{aligned}
\tag{24}
$$

where $n = W(N_u - 1)$.

The ideal output signal of the differential SUM detector is bipolar with normalized amplitude equal to $\pm W$. Also, as $f_Z(z)$ is symmetrical and has zero mean, then $\gamma_o = 0$ in this case. Therefore, if the data source is equiprobable, the error probability is given by

$$
P_e = P(\text{error} \mid \vec{s}_0) = P(\text{error} \mid \vec{s}_1)
\tag{25}
$$

and, particularly, we can obtain that

$$P_e = \int_{-\infty}^{\gamma_0} f_z(z \mid \vec{s}_1)\, dz = \int_{-\infty}^{0} f_z(z - W)\, dz$$

$$= \frac{1}{2} \sum_{j=0}^{n} \sum_{k=0}^{j} \binom{n}{j}\binom{j}{k} p^j (1-2p)^{n-j} \frac{\left(2\sigma\sqrt{W}\right)^j}{(-1)^k}\, \mathrm{erfc}_j\left(\frac{2k-j+W}{2\sigma\sqrt{W}}\right) \qquad (26)$$

where $n = W(N_u - 1)$.

Figure 6 shows comparative performance plots between OOK and OOO signaling O-CDMA systems as function of the signal-to-noise ratio. For high and moderate noise channels, OOO performs better than OOK and, as can be seen, these systems usually approach their maximum attainable performance at much lower SNR levels. Only in low-noise environments OOK signaling can be slight better than OOO. However, considering that the OOK receiver is more prone to exhibit some threshold inaccuracy, this small benefit tends to be negligible in real systems.

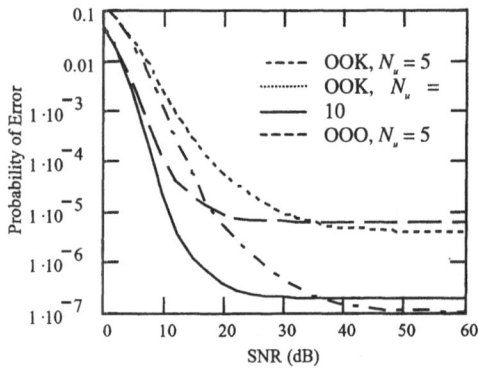

Fig. 6. Performance comparison between OOK and OOO signaling O-CDMA systems as function of the signal-to-noise ratio ($W = 3$ and $L = 100$).

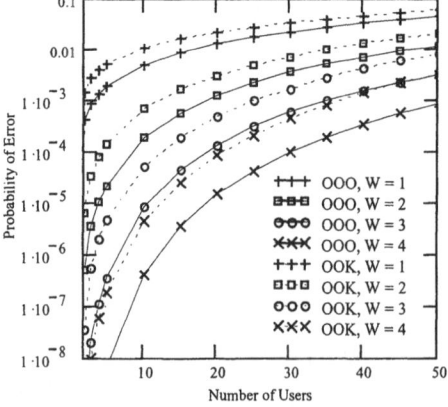

Fig. 7. Bit error probability for OOK and OOO O-CDMA for an increasing number of PPM sequences of length $L = 100$. The input signal-to-noise ratio in all sub-channels is $SNR = 20$ dB.

Finally, Figure 7 illustrates the performance evolution of OOK and OOO signaling O-CDMA systems by increasing the number of transmitted PPM sequences, as function of the number of users and for a fixed signal-to-noise ratio. The performance gain obtained by augmenting the number of PPM sequences of the code is substantial and proportionally larger for greater W values. As predicted, the OOO systems perform significantly better than the OOK systems for the established input SNR.

4   Conclusion

We have shown that the decision threshold γ for decoding OOK signals depend on the multiple access interference (MAI) and signal-to-noise ratio (SNR). However, the performance degradation of these systems can be kept relatively small by estimating the SNR alone and computing an adequate value for γ.

Differently, OOO signaling requires twice as much sub-channels than OOK, but in compensation it does not need threshold estimation and, over a large SNR range, provides better error performance than OOK systems.

References

1. Stok and E. H. Sargent, „The role of optical CDMA in access networks." *IEEE Commun. Mag., 40*(9), 2002, pp. 83-87.
2. J. S. G. Panaro and C. de Almeida, „Asynchronous performance of multidimensional PPM codes for optical CDMA systems," in *Proc. 2nd IASTED Int. Conf. Communication Systems and Networks,* Benalmádena, Spain, 2003, pp. 424-429.
3. P. A. Davies and A. A. Shaar, „Asynchronous multiplexing for an optical-fiber local area network," *Electron. Lett., 19*(10), 1983, pp. 390-392.
4. A. Shaar and P. A. Davies, „Prime sequences: quasi-optimal sequences for OR channel code division multiplexing," *Electron. Lett., 19*(21), 1983, pp. 888-889.
5. E. Park, A. J. Mendez, and E. M. Gasmeiere, „Temporal/spatial optical CDMA networks: Design, demonstration and comparison with temporal network," *IEEE Photon. Tech. Lett., 4,* 1992, pp. 1160-1162.
6. L. Tancevski, I. Andonovic, M. Tur, and J. Budin, „Hybrid wavelength hopping/time spreading code division multiple access systems," *Proc. IEE – Optoelectron., 143*(3), 1996, pp. 161-166.
7. E. S. Shivaleela, K. N. Sivarajan, and A. Selvarajan, „Design of a new family of two-dimensional codes for fiber-optic CDMA networks," *J. Light. Tech., 16*(4), 1998, pp. 501-508.
8. T. W. F. Chang, and E. H. Sargent, „Optical CDMA using 2-D codes: The optimal single-user detector," *IEEE Comm. Lett., 5*(4), 2001, pp. 169-171.
9. Papoulis, *Probability, Random Variables, and Stochastic Processes.* Singapore: McGraw-Hill, 1991, ch. 5.
10. M. Abramowitz and I. A. Stegun, *Handbook of Mathematical Functions with Formulas, Graphs, and Mathematical Tables.* New York: Dover, 1972, pp. 299-300.

Smart Strategies for Single-Photon Detection

João B.R. Silva, Rubens V. Ramos, and Elvio C. Giraudo

Federal University of Ceara
Campus do PICI, Bloco 705, C. P. 6007
60455-760 – Fortaleza, CE, Brasil
{joaobrs, rubens, elvio}@deti.ufc.br
http://www.gcq.deti.ufc.br/index.html

Abstract. The key device in single-photon detectors is the avalanche photodiode. When used in quantum key distribution, a protocol for key distribution with security guaranteed by quantum physics laws, the afterpulsing, false counts caused by electrons trapped in the gain region, limits strongly the maximal transmission rate of the system since we have, after an avalanche have occurred, to turn off the photodiode for a fixed time before enabling the avalanche photodiode to receive another photon. In this paper, aiming to overcome the afterpulsing, three smart strategies for single-photon detection are discussed using analytical and numerical procedures.

1 Introduction

Single-photon detectors have become important with the arising of setups for communication, in 1550 nm, using single-photon pulses, as quantum teleportation and quantum key distribution (QKD) [1-3]. This last has attracted much attention due to its highly desirable property of inviolability assured by fundamental laws of physics. It provides a secure change of bits, which will constitute the key for a cryptographic scheme, and it can reveal, through a higher error rate, the presence of an eavesdropper in the channel. The security of the quantum protocols is guaranteed by three factors: a suitable codification of the information in a quantum property (phase of a light wave, for example) and the fact that a quantum cannot be either split or cloned. For practical implementation of QKD's protocols, optical interferometers and single-photon detectors [4,5] are used. One of the main drawbacks that limit the transmission rate in QKD systems employing single-photon detectors is the afterpulsing. Although the afterpulsing depends on the semiconductor structure, it is possible to employ smart strategies to minimize it. In the simplest case the afterpulsing can be eliminated making the time between two consecutive gate pulses (that enables the APD to have an avalanche) $T_R = T-\tau$, where T is the period and τ the pulse width, large enough to permit all traps to become empty. This strategy clearly limits the transmission rate since, even when there is not detection during a gate pulse the hold off time, T_R, will have to be waited before enabling the APD to have an avalanche.

J.N. de Souza et al. (Eds.): ICT 2004, LNCS 3124, pp. 322–327, 2004.

2 Single-Photon Detectors

The main device in a single-photon detector is the avalanche photodiode, APD. In 1550 nm InP/InGaAs-APDs are used. The avalanche in an APD can be fired only when the reverse voltage applied to the APD (V_{APD}) is larger than a threshold value named the breakdown voltage (V_b). On the other hand, after the avalanche has finished, some carries remain kept in traps in the high field region. After some (average) time those carries become free and they can start a new avalanche. When the APD is used in the Geiger mode ($V_{APD} > V_b$), after the avalanche has been started, it must be quenched in order to do not damage the APD. This can be achieved by a circuit that senses the avalanche and decreases V_{APD} to a value lower than V_b. In QKD systems the gated mode is the most used method to quench the avalanche. The voltage through the APD remains larger than V_b only during a short time window, τ. Between two consecutive gate pulses $V_{APD} < V_b$ and an avalanche cannot be fired. The gate pulses and gated quenching circuit are shown in Fig. 1 [4,6,7,8].

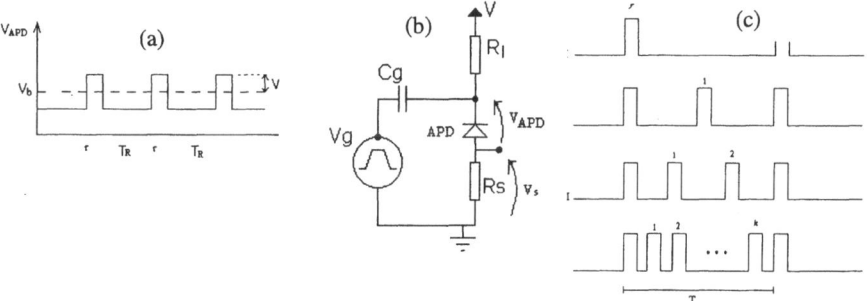

Fig. 1. (a) Gate pulses, (b) gated quenching circuit and (c) new gate pulses insert in the interval $T_R = T - \tau$.

3 Smart Strategies for Single-Photon Detection

A smarter strategy of detection that increases the transmission rate of QKD systems was proposed and experimentally tested by NTNU group [9]. In this strategy the period of hold off of the APD, T_R, takes place only after a detection. In this way, we can increase the gate pulse rate and only when detection occurs the period T_R without any pulse takes place. In Fig. 1.c it can be seen the traditional situation (I) in which the distance between two neighbors gate pulses is T_R, and the situations in which one (II), two (III) and k (IV) pulses are introduced between the two pulses of the traditional situation. The mathematical analysis of this strategy is very complicated since the probability of detection in a defined moment is dependent of the past history. Here we limit ourselves to find analytically an upper bound for the probability distribution of the number of detection for long sequences of gate pulses, and to use numerical simulations to analyze the behavior of different optical setups. Let us start considering that the sequence of gate pulses is shared in small sequences, frames, of

(k+1) pulses having duration of T, as shown in Fig. 1.c (IV). If p is the probability of detection during a gate pulse, the probability of detection during the interval T, P_T, is given by:

$$P_T = p + (1-p)p + ... + (1-p)^k p = 1 - (1-p)^{k+1}$$ (1)

Equation (1) works only for the first frame of gate pulses. For the other frames the correct equation would be $P_T = 1-(1-p)^q$, $0 \leq q \leq k$ and each possible value of q occurs with some probability that depends on what happened in the earlier frames. Hence, for simplification, let us assume that for all frames of the gate pulse sequence the probability of detection is constant and given by (1). Then, the probability distribution for the number of counts (or detections) in a sequence of gate pulses having ξ frames is the binomial distribution:

$$P(n) = \binom{\xi}{n} P_T^n (1-P_T)^{\xi-n}$$ (2)

and hence, the average number of counts is $<C>=P_T\xi$. This is the upper bound for the average value. For a numerical simulation of a sequence having 10^5 pulses, p=0.1, considering different values of k, the following results, shown in Table 1, were found:

Table 1. Average number of counts for 10^5 gate pulses, with p=0.1 and different values of k.

k	0	4	9	19	49	99	499
ξ	10^5	2.10^4	10^4	5.10^3	2.10^3	10^3	200
P_T	0.1	0.4095	0.6513	0.8784	0.9948	1	1
$<C>= P_T\xi$	10^4	8190	6513	4392	1989.6	10^3	200
$<C>_N$	10^4	7142.8	5263.5	3448.4	1695	917.9	197
$<C>_N/(\xi T)$	0.1	0.357	0.526	0.689	0.847	0.917	0.985

Observing the lines 1 and 3 of Table 1, we note that, the larger the number of pulses inserted (value of k) the larger the probability of having detection in the interval T, as expected. We can also compare the upper bound for the average number of counts, $<C>$, to the numerical value achieved, $<C>_N$. The error is in the interval [0%,21.48%]. Moreover, the larger the value of k, the shorter the time of duration of 10^5 gate pulses, hence, the interesting parameter for performance analysis is the average number of detection by the interval of duration of the 10^5 pulses. This can be seen in the last line of Table 1. Based on the results shown in Table 1, we conclude that the Norwegian strategy permits a higher transmission rate for QKD systems, and the larger the values of k the larger the transmission rate; however, there is saturation when k is close to 99. The maximal value of k for pulses with width τ and duty-cycle of 50% is $k_{max} = T/(2\tau)-1$. Considering T=0.1 ms e τ=10 ns, we have k_{max}=4999. However, as shown in the second line of Table 1, for p=0.1, k=99 it is already good enough in order to have $P_T \approx 1$. However, the larger the value of k, the closer we stay from the upper bound given by (1)-(2). At last, is useful to know that for low values of k the distribution of probability of the number of counts is like Gaussian, but when the value of k becomes large enough, the curve change its form towards a delta function. This happens because for high values of k the probability of a count in a frame tends to one and, hence, the number of counts in the in the whole sequence tends to ξ, the number of frames in the whole sequence of gate pulses.

A second strategy that can increase the transmission rate in QKD systems consists in using two APDs instead of only one. When detection happens in one APD, the next incoming photon may be guided to the other APD. Two situations are possible: APD riffled (randomly chosen) (I) and switched (II), as shown in Fig. 2.

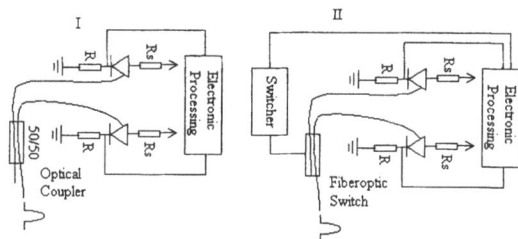

Fig. 2. Strategy of detection employing two APDs, riffled (I) and switched (II).

The difference between the two setups is because the first one (I) uses a balanced beam splitter to guide the incoming photon to the APD, hence, we have not control about which path the photon will take. On the other hand, using a fiber optic switch and an extra electronic circuit for switching, the second setup (II) permits us to realize an optimization since we can always guide the incoming photon to the next APD able to realize detection. In order to analyze the performance of setups in Fig. 2 (I) and (II), we use numerical simulations. In Fig. 3 the probabilities distributions of the number of counts are shown using the following parameters values: $p=0.075$, 10^5 pulses and k assuming the values 0 and 49, respectively.

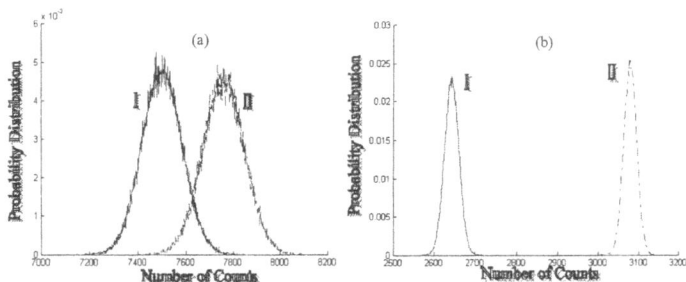

Fig. 3. Distributions of probabilities of the number of counts for (a) $k=0$ and (b) $k=49$, $p=0.075$ and 10^5 pulses. (I) Riffled. (II) Switched.

As can be observed in Fig. 3, the larger the value of k the larger the advantage of the setup employing fiber optic switch (II) over the setup employing optical coupler (I). We can also easily notice that both setups are better than the strategy using only one

APD. The average values of the number of counts for different values of k are shown in Table 2.

Table 2. Average number of counts for the setups (I) and (II), using $p=0.075$ and 10^5 pulses.

k	0	1	4	9	19	49	99	499
$<C>_I$	7501	7234.85	6524.15	5616.04	4379.88	2643.86	1592.26	381.38
$<C>_{II}$	7760	7749.80	7436.42	6734.82	5354.47	3079.47	1758.26	390.90

For low values of k, the system employing only one APD is better because the probability of detection per gate pulse, when an optical coupler or fiber optic switch are present is lower due to loss in those devices. The loss decreases the average number of photons arriving in the APD. In the simulations realized the probability of detection used $p=0.075$, corresponds to loss of approximately 1.3 dB. In order to compare the three strategies presented, in Fig. 4 it is shown the average number of counts per time for the three strategies.

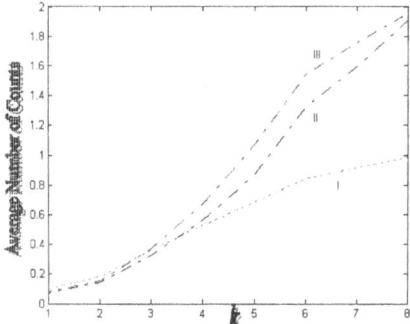

Fig. 4. Average number of counts per time, in 10^5 gate pulses for $k = 0, 1, 4, 9, 19, 49, 99$ e 499. I) One APD, $p=0.1$. II) APDs riffled, $p=0.075$. III)APDs switched, $p=0.075$.

Observing once more the Tables 1 and 2, we see that only for large values of k the system employing two APDs have transmission rate almost twice of the one reachable by the system employing only one APD.

4 Conclusions

It was discussed three possible smart strategies for single-photon detection aiming to improve the transmission rate in QKD setups. The first one, Norwegian strategy, was analyzed through numerical simulations and an upper bound was found analytically. The comparison between the upper bound and the results from the numerical simulations showed us that, the larger the value of k (number of pulses inserted) the closer to the upper bound the system works. After, two other strategies, employing two APDs were proposed. While one APD is emptying its traps, the other is able to receive a photon and start an avalanche. The two strategies are APD riffled and APD switched. In the first one an optical coupler is used and the incoming photon has 50% of chance to go to the correct APD. In the second one we have total control about which APD will receive the photon, since an optical switch is used. Both techniques were simulated numerically and, as expected, the switched scheme showed a better performance than the riffled scheme. Even with the additional loss introduced by the optical coupler or switch, the schemes employing two APDs can have a larger transmission rate than the scheme using only one APD, if the value of k is large enough.

Acknowledgment. The authors are grateful to CNPq for financial help.

References

1. Bennett, C. H., Bessette, F., Brassard, G., Salvail, L. and Smolin J.: Experimental Quantum Cryptography. Journal of Cryptology, Vol. 5. (1992) 3-28
2. Phoenix, S. J. D., Townsend, P. D.: Quantum cryptography: How to beat the code breakers using quantum mechanics. Contemporary Physics, Vol. 36 (1995) 165-195
3. Gisin, N., Ribordy, G., Tittel, W., Zbinden, H.: Quantum cryptography. quant-ph/0101098, http://xxx.lanl.gov, (2001)
4. Ribordy, G., Gautier, J. D., Zbinden, H., Gisin, N.: Performance of InGaAs/InP avalanche photodiodes as gated-mode photon counters. Applied Optics, Vol. 37. (1998) 2272-2277.
5. Stefanov, A., Gisin, N., Guinnard, O., Guinnard, L., Zbinden, ,H.: Optical Quantum Number Generator. quant-ph/9907006, http://xxx.lanl.gov, (1999)
6. Gibson, F.: Experimental evaluation of quantum cryptography system for 1550nm. Master of Science thesis, IMIT-QEO, Kungl Tekniska Högskolan, Sweden, (1998)
7. Ramos, R. V., Thé, G. A. P.: Single-photon detectors for quantum key distribution in 1550 nm: simulations and experimental results. Microwave and Optical Technology Letters, 32-2, (2003) 136-139
8. Bourennane, M., Ljunggren, D., Karlsson, A., Jonsson, P., Hening, A., Ciscar, J. P.: Experimental long wavelength quantum cryptography: from single-photon transmission to key extraction protocols. Journal of Modern Optics, 47 (2/3), (2000) 563-579
9. Vylegjanine, K.: High-speed single-photon detectors for quantum cryptosystems. Master of Science thesis, Dept. of Physical Electronic, NTNU, Norway, (2000)

Inter-arrival Planning for Sub-graph Routing Protection in WDM Networks

Darli A.A. Mello, Jefferson U. Pelegrini, Marcio S. Savasini,
Gustavo S. Pavani, and Helio Waldman

Optical Networking Laboratory
OptiNet/DECOM/FEEC - State University of Campinas - UNICAMP
Caixa Postal 6101 CEP: 13083-852, Campinas - SP, Brazil
darli@decom.fee.unicamp.br
http://www.optinet.fee.unicamp.br

Abstract. We propose an inter-arrival planning strategy that strongly reduces path and wavelength reassignment in WDM networks protected by sub-graph routing. It has been recently demonstrated that Sub-Graph Routing Protection outperforms Backup Multiplexing in terms of blocking probability and link utilization. However, a major drawback is that upon occurrence of a link failure, even connections that do not traverse the faulty link may have to reassign their path or wavelength to accommodate others, interrupting the service. In this paper we show that offline inter-arrival planning of protection resources can efficiently reduce reassignment while preserving blocking performance.

1 Introduction

Reliability is a crucial challenge for the deployment of next generation optical networks [1]. In WDM networks a link failure may lead to serious impairments to a large number of end users. High reliability standards can only be achieved by protection and restoration. Protection mechanisms can be classified into two main classes: dedicated protection or shared protection. In the first case there are two disjoint paths connecting source and destination nodes: a working path and a dedicated protection path with the same capacity. In the second case different connections share protection resources, thus saving allocated spare capacity. A common variety of shared protection is Backup Multiplexing [2], where backup paths of disjoint working paths are allocated to the same resources, thus "multiplexing" them.

In [3] Frederick and Somani introduced a novel shared protection approach to protect traffic against single link failures using sub-graph routing. In the proposed strategy although protection paths are pre-planned, no explicit spare capacity has to be allocated in the network. The new approach exhibited promising performance in terms of network utilization and request blocking when compared to Backup Multiplexing. However, a major drawback of Sub-Graph Routing Protection is that upon the occurrence of a link failure, even connections that do not traverse the faulty link may have to reassign their path or wavelength to

J.N. de Souza et al. (Eds.): ICT 2004, LNCS 3124, pp. 328–335, 2004.

accommodate connections directly harmed by the failure, causing inconvenient service interruption. This will be called *altruistic reassignment*. Datta, Frederick and Somani [4] have recently addressed this problem introducing constraints for connection allocation in sub-graphs that eliminate altruistic reassignment. Nevertheless, this approach strongly increased blocking probability.

Unlike connections physically routed in the network, sub-graphs are logical topologies stored in the system. Therefore the protection resources allocated in the sub-graphs can be optimized offline in the interval between two network events, i.e., connection arrivals or departures. Using this approach, we introduce in this paper an inter-arrival planning strategy that reduces reassignment in sub-graph routing protected networks while preserving blocking performance.

The remainder of this paper is divided into five sections. Section 2 explains Sub-Graph Routing Protection. Section 3 shows our proposal to improve reassignment probability using inter-arrival planning. Section 4 presents the methods used to evaluate the performance of the new scheme. Section 5 shows the simulation results, and Sect. 6 concludes the paper.

2   Sub-graph Routing Protection

Sub-Graph Routing Protection is a new and clever way of protecting traffic. Initially conceived to support single link failures, it has been recently extended to support multiple and Shared-Risk Link Group failures [4]. In this paper we simulate only single link failures, but the developed concepts can be extrapolated to other scenarios.

The main idea of Sub-Graph Routing Protection is rather simple. A network topology can be represented by a undirected graph $G(V, E)$ with a vertex set V and an edge set E. The set V represents nodes and the set E represents bidirectional connections. We call $G(V, E)$ the base network. A single link failure can be represented by a sub-graph G_i, which is the original graph G without an edge e_i, such that $G_i = G - e_i$. In this way all possible single-link failures in the network can be represented by L sub-graphs, where L is the cardinality of the edge set E.

In Sub-Graph Routing Protection a connection request is only accepted if it can be successfully routed in each of the L sub-graphs and in the base network. In case of a link failure the network immediately incorporates the state represented by the corresponding sub-graph.

As an illustrative example let us consider the base network $G(V, E)$ shown in Fig. 1. Sub-graphs G_1 to G_7 are formed after the base network by removing a different link, so that all link failures are represented.

A connection request must be able to be routed in the base network and in all sub-graphs to be accepted. Consider connection requests A-C, E-A and D-C arriving sequentially in the example network with only one wavelength (W=1), and that the routing algorithm chooses the shortest path available:

1. As connection A-C can be routed in the base network and all sub-graphs, it is accepted;

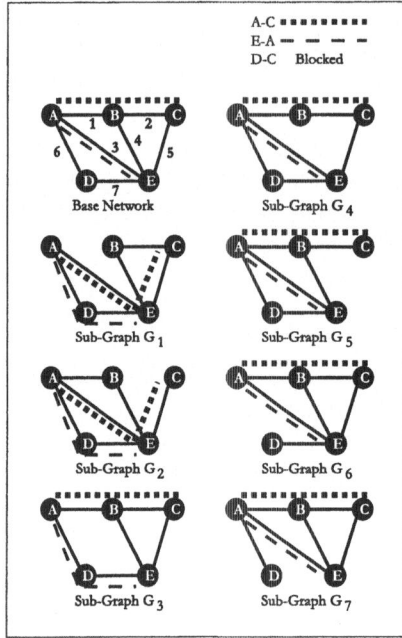

Fig. 1. Sub-Graph Routing Protection

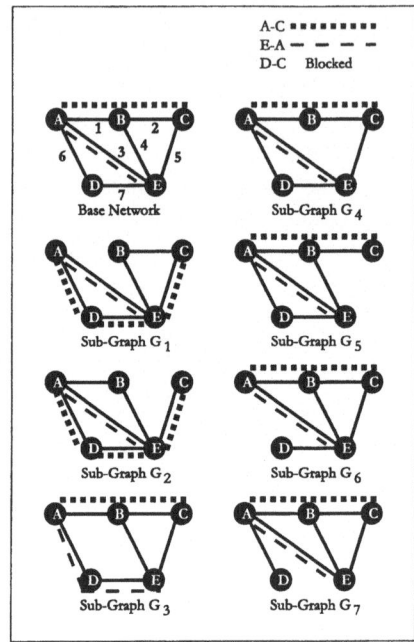

Fig. 2. Inter-Arrival Planning

2. As connection E-A can be routed in the base network and all sub-graphs, it is also accepted;
3. As connection D-C can not be routed in sub-graphs "1, 2, 3, 5 and 7", it is blocked.

The two first connections are accepted and protected against single link failures. However, one should note that if link A-B or B-C fails, connection E-A, which neither traverses links A-B nor B-C, is reassigned to path E-D-A for altruistic reasons.

3 Inter-arrival Planning

Inter-arrival strategies improve system performance at the cost of computational processing [5]. Since data is processed offline between network events, i.e., a connection arrival or departure, gains in performance can be achieved without wasting processor time. Inter-arrival planning of connections routed in sub-graphs is a feasible strategy because sub-graphs are not physical entities, but states to be incorporated by the network in case of a link failure.

A requirement for a successful inter-arrival planning is that the time interval needed for the planning - the planning time - must be shorter than the time interval between network events - the inter-arrival time -, otherwise the planning would be aborted. Nowadays the inter-arrival time in optical networks is huge, traffic can be considered incremental and provisioning a connection can take hours or even days. The simulations carried out in this paper agree with the expectation that in a near future traffic will evolve towards dynamicity, but electronic planning time will stay much shorter than inter-arrival time.

The strategy we propose optimizes the routing and wavelength assignment of connections in each sub-graph to mitigate altruistic reassignments. The idea is to prioritize in sub-graph G_i the routing of connections that do not traverse edge i in the base network, allocating them in the same path and wavelength. Connections that traverse edge i in the base network, which would be anyway reassigned in sub-graph G_i, are then allocated in the remaining capacity. If it is not possible, the inter-arrival planning is aborted and the initial state of the sub-graph restored.

The inter-arrival planning is carried out in each sub-graph G_i by the following steps:

1. Initial state of G_i is saved;
2. All connections are cleared from G_i;
3. Two connection lists are generated: C_F – connections that traverse edge i in the base network, C_N – connections that do not traverse edge i in the base network;
4. Connections of list C_N are routed in G_i as in the base network;
5. Connections of list C_F are routed using the shortest path among all wavelength planes, in order of arrival;
6. If not possible to route all connections, restore initial state.

To illustrate the inter-arrival planning we applied it to the example depicted in Fig. 1. The optimized configuration can be found in Fig. 2. Connection request A-C is accepted and routed in the base network and in all sub-graphs, as in the previous example. When connection request E-A arrives it is accepted and routed in the base network and in all sub-graphs. After this network event (the arrival of connection request E-A) lists of Table 1 are generated and the inter-arrival planning takes place, clearing the sub-graphs and filling them according to the steps described above.

First connections of list C_N are routed, succeeded by connections of list C_F. After the inter-arrival planning, connection E-A is routed in sub-graphs G_1 and G_2 just like in the base network, avoiding the unnecessary path reassignment in case of failure in links A-B or in B-C. Connection D-C is blocked for the same reason as explained before.

4   Performance Evaluation

We evaluate the performance of the inter-arrival planning strategy through simulations performed by the optical networks simulator developed at the OptiNet, Optical Networking Laboratory at the State University of Campinas - UNICAMP. The simulator was written in the Java programming language. All connections are bidirectional with 16 wavelengths. Uniform traffic is assumed. The

Table 1. Connection Lists

Sub-Graph	C_N	C_F
G_1	E-A	A-C
G_2	E-A	A-C
G_3	A-C	E-A
G_4	A-C;E-A	-
G_5	A-C;E-A	-
G_6	A-C;E-A	-
G_7	A-C;E-A	-

arrival of connection requests follows a Poisson distribution, and the holding time is exponentially distributed. We simulated three network topologies as in [4]: the 14-node, 23-link NSFNet; 9-node, 18-link 3x3 Mesh-Torus; and the 11-node, 22-link NJLATA. Two performance metrics were investigated: Blocking Probability and Probability of Reassignment. The second is the probability that a connection has to change its path or wavelength to survive a single failure in any link of the network. Although in [4] the probability of only path reassignment was investigated, we believe that wavelength reassignment is essential to evaluate the system performance since it also causes service interruption. The curves are calculated by averaging the results of the 10 last rounds of a series of 11 with 1000 connection requests each, to simulate a steady state network occupancy.

4.1 RWA in the Base Network

Connections are routed in the base network using Dijkstra's shortest path algorithm, in terms of hop count, applied to the physical network topology. If there are more than one shortest path, one is randomly chosen. Wavelength selection follows the random fit scheme.

4.2 RWA in Sub-graphs

Unconstrained RWA. refers to the seminal paper on sub-graph routing [3], which used in sub-graphs the same RWA scheme as in the base network.

Constrained RWA. refers to [4]. Here connections are constrained to follow the same path and wavelength as in the base network in those sub-graphs which contain all the links traversed by the connection in the base network. In the remaining sub-graphs, connections are routed using the same RWA strategy as in the base network.

Unconstrained RWA + inter-arrival planning. first a connection request is accepted or blocked according to the "Unconstrained RWA" policy. Then the inter-arrival planning takes place following the steps described in Sect. 3. The RWA of connections in list C_F, step "5" of the inter-arrival planning, chooses the wavelength with the shortest path, which is determined by applying Dijkstra's shortest path algorithm to all wavelength planes. The inter-arrival planning also takes place when connections depart.

5   Results

Each of the three RWA strategies exhibited the same behavior for the simulated topologies: 3x3 Mesh-Torus in Fig. 3, NJLATA in Fig. 4 and NSFNet in Fig. 5. The results can be summarized as follows:

Unconstrained RWA. leads to low blocking probability, but to an extremely high reassignment probability, which is a consequence of the random selection of the shortest path and wavelength in sub-graphs. Maybe the use of the First Fit wavelength assignment policy would already result in gains.

Constrained RWA. leads to no altruistic reassignment and consequently low reassignment probability at the cost of a pronounced raise in blocking probability, limiting the use of this policy.

Unconstrained RWA + inter-arrival planning. leads to the same low blocking probability as in the pure "Unconstrained RWA" and low reassignment probability as in the "Constrained RWA". This can be explained by the fact that here the request acceptance policy is the same as in the pure "Unconstrained RWA", while the inter-arrival planning almost eliminates altruistic reassignment.

6   Conclusion

Sub-Graph Routing Protection is a new and powerful protection scheme. However, because of its novelty there are still several ways to improve performance. An important weak point of the original proposal was the high probability of path or wavelength reassignment upon the occurrence of a failure in any link of the network. This drawback was overcome in previous works by constraining path and wavelength before accepting connection requests, which however strongly increased blocking. In this paper we introduce an inter-arrival planning strategy that resulted in low reassignment probability and blocking probability similar to the original proposal. This improvement is reached at the cost of computational capacity, but since data is processed between network events, there is no loss in system performance. We conclude that inter-arrival planning is an efficient resource to improve the performance of Sub-Graph Routing Protection.

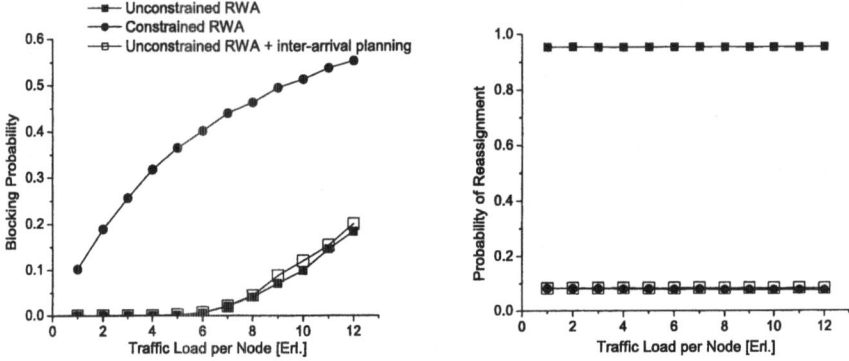

Fig. 3. 3x3 Mesh Torus

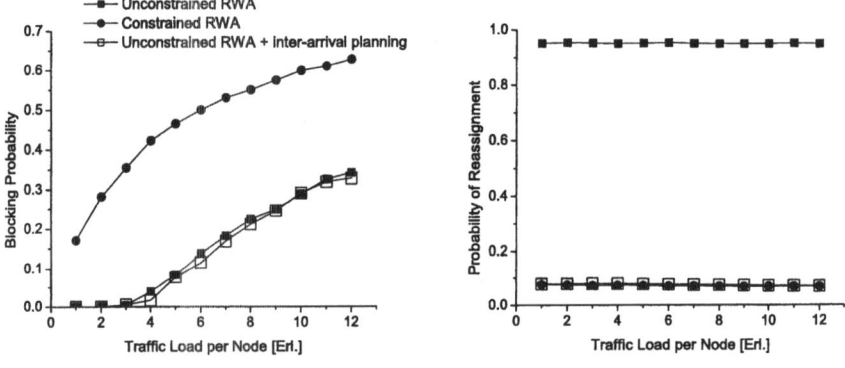

Fig. 4. NJLATA

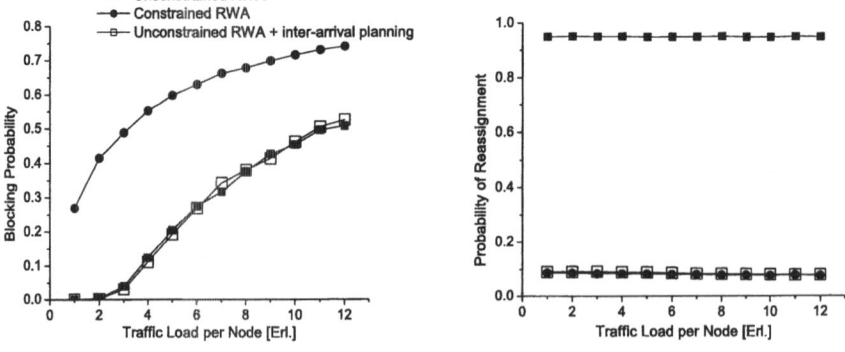

Fig. 5. NSFNet

Acknowledgment. This work has been supported by Fundação de Amparo à Pesquisa do Estado de São Paulo (FAPESP) and Ericsson Telecomunicações S.A., Brazil.

References

1. Ramaswami, R., Sivarajan, K.N.: Optical Networks: a Practical Perspective. Morgan Kaufmann Publishers (2002)
2. Mohan, G., Murthy, C.S.R., Somani, A.K.: Efficient algorithms for routing dependable connections in WDM optical networks. IEEE/ACM Trans. Networking **9** (2001) 553–566
3. Frederick, M.T., Somani, A.K.: A single-fault recovery strategy for optical networks using subgraph routing. In: Proc. ONDM'03. (2003) 549–568
4. Datta, P., Frederick, M.T., Somani, A.K.: Sub-graph routing: a novel fault-tolerant architecture for shared-risk link group failures in WDM optical networks. In: Proc. DRCN'03. (2003)
5. Ho, P.H., Mouftah, H.T.: A novel distributed protocol for path selection in dynamic wavelength-routed WDM networks. Kluwer Photonic Network Communications **5** (2003) 23–32

An Adaptive Routing Algorithm for Intelligent and Transparent Optical Networks

R.G. Dante[1,2], F. Pádua[1], E. Moschim[1], and J.F. Martins-Filho[2]

[1] Laboratório de Tecnologia Fotônica, Departamento de Semicondutores, Instrumentos e Fotônica, Faculdade de Engenharia Elétrica e de Computação, UNICAMP – SP, Brasil.
{rgdante, moschim}@dsif.fee.unicamp.br
Tel: +55 19 37883766, Fax: +55 19 32891395.

[2] Grupo de Fotônica, Departamento de Eletrônica e Sistemas, Universidade Federal de Pernambuco (UFPE), 50740-530 Recife – PE, Brazil.
{reinaldo, jfmf}@ufpe.br
Tel: +55 81 21268995 , Fax: +55 81 21268995.
http://www.fotonica.ufpe.br

Abstract. In this paper we propose an adaptive routing strategy, called *Weighted Link Capacity* (WLC), as a candidate solution for the routing problem. The WLC algorithm consists on choosing the path that minimizes the cost of network resources, whilst, in parallel, maintaining the network traffic as balanced as possible. Its strategy is based on the number of hops, link length and free link capacity. We assume two constraints to block a connection: first, when there is no free path available; second, when the distance of the path exceeds the optical transmission reach of optical equipments without 3R regeneration, which results in a bit error rate above an acceptable limit (e.g., 10^{-12}). The simulation results of dynamic traffic in two hypothetical meshed networks are shown in terms of blocking probability as a function of network load. We demonstrate that WLC outperforms the traditional routing algorithms.

1 Introduction

Intelligent and transparent optical networks (ITON) [1]-[3] require an efficient and fast wavelength routing mechanism to set up and release the connections dynamically. Automatically switched optical networks (ASON), proposed in ITU-Tb [4],[5], is a candidate for future transport technology in ITON. In ITON, the routing and wavelength assignment (RWA) problem is defined as a problem of setting up lightpaths by routing and assigning a wavelength to each connection.

We examine the RWA problem through some traditional fixed routing algorithms such as shortest-distance-path (SDP) and shortest-hop-path (SHP) [6], and some traditional adaptive routing algorithms such as link-state algorithm based on distance and network state (LS-d), and link-state algorithm based on number of hops and network state (LS-h) [7], [8]. All of those algorithms are based on Dijkstra's algorithm to

J.N. de Souza et al. (Eds.): ICT 2004, LNCS 3124, pp. 336–341, 2004.
© Springer-Verlag Berlin Heidelberg 2004

select the minimum cost routing. However, the metrics of cost for each routing algorithm mentioned before is different.

In this article we propose and demonstrate a novel adaptive routing algorithm, called *Weighted Link Capacity* (WLC), with a new routing strategy based on the number of hops, link length and free link capacity. In addition, our WLC algorithm takes into account the optical transmission reach constraint without 3R (i.e., re-amplifier, re-shape, re-time) regeneration for optical WDM equipments. We demonstrate that WLC outperforms those traditional routing algorithms.

2 Algorithm Description and Simulation Models

The WLC algorithm is proposed for single-fibre networks, however it can be extended for multi-fibre networks. We assume no wavelength conversion and a circuit-switched ITON. Exploiting the concepts of link capacity and load balancing, WLC consists of choosing the path that minimizes the cost of network resources whilst, in parallel, maintaining the network traffic as balanced as possible.

Let ψ be a network state which specifies the existing lightpaths (routes and wavelength assignments) in the network. Consider a set of wavelengths W per link, which the bit-rate of each wavelength is B, in a single-fibre network. In WLC, the link capacity on link l, $C(W,B)_l$, is defined as the throughput in bits per second, i.e.,

$$C(W,B)_l = W\,B\,. \tag{1}$$

Similar to (1), the free link capacity on link l in the state ψ, $C(u,B,\psi)_l$, is defined as

$$C(u,B,\psi)_l = W\,B\,(1-u)\,, \tag{2}$$

where u is the number of used wavelengths divided by W.

Observing the Eq. (2), when u increases $C(u,B,\psi)_l$ decreases and, consequently, that link l will eventually be congested. It means the routing strategy should avoid to select that link l as part of a chosen path for a given connection request. Therefore, the cost of choosing that link l will be high. WLC also considers the number of hops and distance as part of its routing strategy. Either the number of hops or distance can increase the cost to choose a path in case it is large or long, respectively. For example, selecting a path with large number of hops increases the blocking probability on that chosen wavelength for next connection requests. Furthermore, selecting a long path should increase the blocking probability because of the degradation of optical signal quality, which results in a high bit error rate (BER).

Thus, the routing strategy of WLC algorithm is formulated, as follows:

$$Minimize : M \tag{3}$$

where

$$M = \begin{cases} (H^K + 1) \sum_{i=0}^{n-1} \dfrac{D_i}{C(\overline{u}, B, \psi)_i^Q} & \text{if } u \neq 1 \\ \infty & \text{if } u = 1 \end{cases} \tag{4}$$

and

$$\sum_{i=0}^{n-1} D_i \leq R_{\max} \qquad \forall \lambda \in W \tag{5}$$

M is the metric of cost, H is the number of hops along the path, D_i is the length of link i, n is the number of links along the path, K and Q are constants, $R_{\max}$ is the maximum optical signal transmission reach without 3R regeneration. In this paper we assume $R_{\max} = 600$ km.

In (4), M represents the result of the influences of number of hops, link distance and free link capacity on choosing the path for a given connection request. Both influences of number of hops and link capacity on choosing the path can be as sensitive as possible through the constants K and Q. When u is equal to 1 on link l it indicates that link l is congested and so M is set as infinite.

In (5), $R_{\max}$ is 600 kilometres for each wavelength λ, maintaining the BER below an acceptable limit (e.g., 10^{-12}). It is an approach to simplify the BER calculation, once the maximum optical transmission reach depends on the number of wavelengths and several optical impairments along the path [9]. The advantage of this approach is that it saves processing time of routing calculation at each OXC. That approach could be extended in case there is a curve of maximum optical transmission reach as a function of the number of wavelengths for an acceptable limit of BER, e.g. 10^{-12}, measured from a point-to-point optical system. Finally, WLC selects the shortest-cost-path that satisfies the routing strategy in (3)-(5).

Once the WLC algorithm selects that shortest-cost-path, it invokes a wavelength assignment algorithm, e.g. First Fit (FF), for choosing a wavelength to establish a lightpath for that (s,d) pair. We assume two conditions to block a connection request: First, if there is no route which distance is less or equal to 600 kilometres; Second, if there is no available wavelength on that selected path.

Our WLC algorithm is implemented in C++ codes. The blocking probability is obtained upon the simulation of a set of bidirectional calling requests (5×10^5 calls). We use Poisson distribution for call arrivals and for call holding time. The source-destination node pair for each call follows a uniform distribution. The simulation of 5×10^5 calls takes approximately 1 minute. For comparisons we also determine the performance of some traditional routing algorithms such as shortest-distance-path (SDP), shortest-hop-path (SHP), link-state algorithm based on distance and network state (LS-d) and link-state algorithm based on number of hops and network state (LS-h) under the limit of optical transmission reach without 3R regeneration as 600 km for each lightpath. WLC is easy to implement and performs well in distributed environments.

3 Simulation Results and Discussion

This paper presents the simulation results of two hypothetical single-fibre networks: (1) metro-access meshed double rings as illustrated in Fig. 1-a (denoted as *network 1*); (2) long-haul meshed double rings as illustrated in Fig. 1-b (*network 2*). In Fig. 1-a, all possible paths satisfy (5) for any connection request (denoted as *scenario 1*); however (5) is a constraint in network 2 as shown in Fig. 1-b (*scenario 2*). Based on both cases, we examine the performance of WLC compared to SDP, SHP, LS-d and LS-h. In all simulation, we considered the following values for WLC parameters: K = 1.0; B = 10 Gbps; and Q = 2.50. For all routing algorithms applied to network 1 and 2, we assumed W = 8 wavelengths and R_{max} = 600 km.

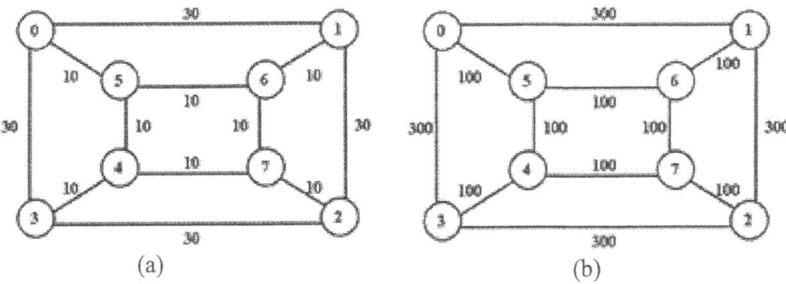

Fig. 1. Simulated networks: (a) metro-access meshed double rings; (b) long-haul meshed double rings.

Fig. 2 shows the blocking probability of all routing algorithms mentioned before as a function of network load for 8-wavelengths in network 1. We observe that the hop-based routing algorithms (e.g., SHP and LS-h) perform better than the distance-based routing algorithms (e.g., SDP and LS-d). Moreover, seeking the paths that minimize the number of hops leads to lower blocking probability than those that minimize the distance under no reach constraint (Eq. (5)). However, LS-h performs equal than WLC under no reach constraint as shown in Fig. 2.

On the other hand, Fig. 3 shows the different behaviour of hop-based routing algorithms in terms of blocking probability versus load under the reach constraint given by (5). Yet the distance-based routing algorithms and WLC maintain the same behaviour. Examining the range of load from 5 Erlangs up to 15 Erlangs we observe that SDP performs better than SHP. It is because SDP seeks the shortest-distance-paths that are equal or less than 600 kilometres and so the reach constraint does not affect them, whereas SHP seeks the shortest-hop-paths and so they are more sensitive to the reach constraint. Above 15 Erlangs, SHP performs better than SDP because of its SHP routing strategy. Similarly, LS-d performs better than LS-h for loads from 5 Erlangs up to 13 Erlangs. Above 13 Erlangs, LS-h performs better than LS-d. Attending carefully to e.g. LS-h when it performs better than LS-d, we can see that the blocking probability of LS-h rises softly compared to LS-d. Below 25 Erlangs, the predominance of blocking connection for hop-based algorithms is due to unavailable routes less or equal to 600 kilometres as shown in Fig. 3. Above 25 Erlangs, the predominance of blocking connection for hop-based algorithms is due to unavailable wavelengths on valid selected paths.

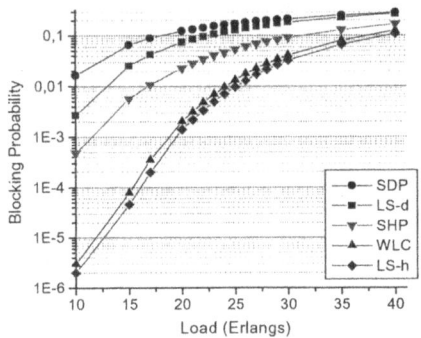

Fig. 2. Blocking probability versus load for 8-wavelengths in network 1.

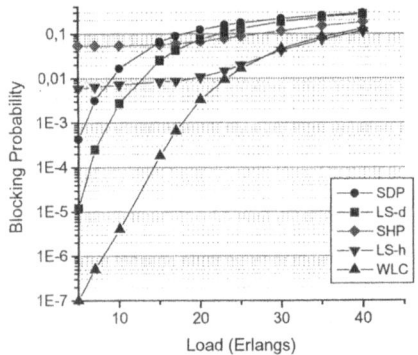

Fig. 3. Blocking probability versus load for 8-wavelengths in network 2.

Fig. 3 shows that the WLC algorithm outperforms the traditional routing algorithms SDP, SHP, LS-d and LS-h, for blocking probabilities below 1% in network 2. The overall behaviour of WLC algorithm is similar for both scenarios, in spite of its blocking probability is slightly higher for the network 1. Similar to SDP and LS-d, the blocked connections occur in WLC algorithm due to unavailable wavelengths on selected paths.

Thus, for network topologies where the path lengths can affect the optical signal quality and consequently, increase the BER above an acceptable limit, we should minimize the distance instead of number of hops.

Scenarios 1 and 2 show that WLC routing strategy adapts itself to different characteristics of network, where either it is necessary to minimize the number of hops (scenario 1) or it is necessary to minimize the distance (scenario 2), but always maximizing the link capacity. Therefore, our proposed WLC routing strategy combines harmoniously hop-count, distance and free link capacity as routing metrics and achieves low blocking rates.

4 Conclusions

In conclusion, we have proposed and demonstrated a novel adaptive routing algorithm for intelligent and transparent optical networks based on hops counts, distance and free link capacity. WLC algorithm leads to less or equal blocking probability than adaptive hop-based routing algorithm, and performs better than adaptive distance-based algorithm. It is easy to implement the WLC algorithm due to a unique calculation of routing metric from the source node of an arriving connection request to the destination node. The results in this paper clearly show that minimizing the hop-counts and distances whilst simultaneously maximizing the link capacity has a significant impact on blocking probability of optical networks.

References

1. B. Mukherjee, "WDM Optical Communication Networks: Progress and Challenges", *IEEE Journal of Selected Areas in Communications*, vol. 18, no. 10, pp. 1810-1824, Oct 2000.
2. Paul Green, "Progress in Optical Networking", *IEEE Commun.Mag.*, pp. 54-61, Jan. 2001.
3. Z. Zhang, J. Fu, D. Guo, L. Zhang, "Lightpath Routing for Intelligent Optical Networks", *IEEE Network*, pp. 28-35, Jul. 2001.
4. ITU-T Recommendation G.8080, "Architecture for the Automatically Switched Optical Networks", 2001.
5. ITU-T Recommendation G.7715, "Architecture and Requirements for Routing in the Automatically Switched Optical Networks", 2002.
6. A. Birman, "Computing Approximate Blocking Probabilities for a Class of All-Optical Networks", *IEEE Journal on Selected Areas in Commun.*, vol.14, pp.852-857, Jun.1996.
7. H. Zang, J. Jue and B. Mukherjee, "A Review of Routing and Wavelength Assignment Approaches for Wavelength-Routed Optical WDM Networks", *Optical Networks Magazine*, pp.47-60, Jan.2000.
8. C. Hsu, T. Liu, N. Huang, "On Adaptive Routing in Wavelength-Routed Networks", *Optical Networks Magazine*, pp.15-24, Jan.2002.
9. J. F. Martins-Filho, C. J. A. Bastos-Filho, S. C. Oliveira, E. A. J. Arantes, E. Fontana and F. D. Nunes, "Novel Routing Algorithm for Optical Networks Based on Noise Figure and Physical Impairments", *Proc. of the European Conference on Optical Communications – ECOC 2003*, Rimini., Italy, Sept. 2003, vol. 3, pp. 856-857, Paper We4.P.150.

Novel Wavelength Assignment Algorithm for Intelligent Optical Networks Based on Hops Counts and Relative Capacity Loss

Reinaldo G. Dante[1, 2], Joaquim F. Martins-Filho[2], and Edson Moschim[1]

[1] Laboratório de Tecnologia Fotônica, Departamento de Semicondutores, Instrumentos e Fotônica, Faculdade de Engenharia Elétrica e de Computação, UNICAMP – SP, Brasil.
{rgdante, moschim}@dsif.fee.unicamp.br
Tel: +55 19 37883766, Fax: +55 19 32891395.

[2] Grupo de Fotônica, Departamento de Eletrônica e Sistemas, Universidade Federal de Pernambuco (UFPE), 50740-530 Recife – PE, Brazil. {reinaldo, jfmf}@ufpe.br
Tel: +55 81 21268995 , Fax: +55 81 21268995. http://www.fotonica.ufpe.br

Abstract. This paper focuses on the routing and wavelength assignment (RWA) problem in intelligent and transparent optical networks operating under no wavelength conversion constraint for end-to-end connections in distributed environments. We propose and demonstrate a novel wavelength assignment algorithm based on hops counts and relative capacity loss, called *Modified Distributed Relative Capacity Loss* (M-DRCL). It consists of grouping end-to-end routes with the same number of hops in M-DRCL tables. Dissimilar to the DRCL algorithm, M-DRCL shows a new strategy of wavelength assignment, including the destination node on its analysis and assuming one, more than one, or even all potential routes from source node to destination nodes combined by the same number of hops on its tables. We present simulation results of dynamic traffic in a hypothetical meshed network in terms of blocking probabilities as a function of network load. We show that our M-DRCL algorithm outperforms the traditional WA algorithms.

1 Introduction

Dense wavelength division multiplexing, used in conjunction with wavelength-routing, is a promising mechanism for information transport in intelligent and transparent optical networks (ITON) [1]-[2]. Automatically switched optical networks (ASON) [3], [4] seems a strong candidate for future transport technology in ITON.

For such networks the routing and wavelength assignment (RWA) problem is of paramount importance. We examine some traditional wavelength assignment (WA) algorithms such as First Fit (FF), Relative Capacity Loss (RCL) [6] and Distributed Relative Capacity Loss (DRCL) [6]. The FF algorithm enumerates all wavelengths from a given set in a list and assigns the first available indexed wavelength. This scheme requires no global information. The RCL algorithm is based on MAX-SUM

J.N. de Souza et al. (Eds.): ICT 2004, LNCS 3124, pp. 342–347, 2004.
© Springer-Verlag Berlin Heidelberg 2004

[7] and it requires global information such as network state and topology. RCL requires also the connection request matrix and for that reason it is difficult and expensive to implement in a distributed environment. The DRCL algorithm is based on RCL, but the connection request matrix is unknown. DRCL depends strongly on a routing table, which is implemented using the Bellman-Ford algorithm. Given a connection request from source node (s) to destination node (d) and a routing table, DRCL considers all paths from (s) to every other node in the network, excluding (d).

In this article we propose and demonstrate a novel wavelength assignment algorithm, called *Modified Distributed Relative Capacity Loss* (M-DRCL), with a new strategy based on hops counts and RCL. Dissimilar to the DRCL algorithm, M-DRCL may consider all potential paths from (s) to (d), grouping them to the same hop-based M-DRCL table. We present simulation results of dynamic traffic in terms of blocking probability as a function of network load. We demonstrate that our algorithm outperforms the traditional WA-heuristic algorithms.

2 Algorithm Description

Let a sampling amount K of potential routes be selected and combined by the criteria of same number of hops in M-DRCL tables. Each M-DRCL table contains one or more routing with the same number of hops. On each hop-based M-DRCL table the proposed WA algorithm calculates the total and relative capacity losses (TRCL) for each wavelength. If there is no available wavelength, the connection request is blocked. On the other hand, if there is a set of available wavelengths, M-DRCL will select the wavelength that minimizes the total relative capacity loss on each hop-based M-DRCL table. Assuming that there are two or more routings for that selected wavelength on hop-based M-DRCL table, the proposed WA algorithm selects the routing that produces the minimum relative capacity loss.

If there is a drawn of minimum total relative capacities through the calculation of two or more wavelength on different hop-based M-DRCL tables, the M-DRCL will select that wavelength on minimum hop number M-DRCL table. M-DRCL is implemented in C++ codes. The blocking probability is obtained upon the simulation of a set of bidirectional calling requests (5×10^5 calls). We use Poisson distribution for call arrivals and for call holding time. The source-destination node pair for each call follows a uniform distribution. The simulation of 5×10^5 calls takes approximately 1 minute. For comparisons we also determine the performance of some traditional WA-heuristic algorithms such as First-Fit (FF), Relative Capacity Loss (RCL) and Distributed Capacity Loss (DRCL) for some fixed and adaptive routing algorithms. M-DRCL is easy to implement and performs well in distributed environments.

3 Simulation Models

Let φ be a network state which specifies the existing lightpaths (routes and wavelength assignments) in the network. The *link capacity* on link l and wavelength j in the state φ, $r(\varphi,l,j)$, is defined as the number of fibres on which wavelength j is unused on link l in Equation (1), as follows:

$$r(\varphi,l,j)=M_l-D(\varphi)_{lj} \,,\tag{1}$$

where M_l is the number of fibres on link l and $D(\varphi)$ is the number of assigned fibres on link l and wavelength j in the state φ. The path capacity, $r(\varphi,p,j)$ on wavelength j is the number of fibres on which wavelength j is available on the most-congested link along the path as

$$r(\varphi, p, j)=\min_{l\in\pi(p)} r(\varphi,l, j) \,.\tag{2}$$

The path capacity of p in state φ, $R(\varphi,p)$, is the sum of path capacities on all wavelength as

$$R(\varphi,p)=\sum_{j=1}^{W}\min_{l\in\pi(p)} r(\varphi,l, j) \,.\tag{3}$$

MAX-SUM [8] chooses wavelength j to minimize the *total capacity loss*, which can be computed as

$$\sum_{p\in P}\left(r(\varphi,p,j)-r(\varphi'(j),p,j)\right) \,.\tag{4}$$

On the other hand, RCL [6] chooses wavelength j to minimize the *relative capacity loss*, which can be computed as

$$\sum_{p\in P}\left(\frac{r(\varphi,p,j)-r(\varphi'(j),p,j)}{r(\varphi,p,j)}\right) \,.\tag{4}$$

RCL needs previously to know the traffic matrix and determines the total relative capacity loss for all connection requests in a single RCL table [7].

M-DRCL uses the RCL scheme to determine the relative capacity loss along the path from (s) to (d). However, for a given (s,d) connection request, M-DRCL groups each potential path from s to d with the equal number of hops in the same M-DRCL table. The amount of routes is given from an adaptive routing algorithm or, depending on the number of nodes in the network, all routes could be considered. We understand that two or more paths can be compared in terms of relative capacity loss when they have got the same number of hops because (1) routes with a large number of hops can result naturally in a higher capacity losses than routes with small number of hops and so they cannot be compared; (2) the minimum sum of relative capacity loss for all paths, independently on the number of hops, in a single table does not lead to the best choice of wavelength.

4 Simulation Results and Discussion

This paper presents the simulation results of two hypothetical single-fibre networks: (a) proposed by Zang in [7] as shown in Fig. 1-a (called *network 1*); (b) meshed double rings as illustrated in Fig. 1-b (called *network 2*).

Fig. 2 shows the blocking probability versus load for four different RWA algorithms on up to down sequence: (1) fixed routing algorithm + DRCL (static routing table); (2) the same fixed routing algorithm + First Fit (FF); (3) adaptive routing algorithm + DRCL (dynamic routing table, updated before each new connection request); (4) all potential routes + M-DRCL. Because of the topology of network 1 is small the blocking probability of M-DRCL is equal to DRCL.

Fig. 3 shows the blocking probability as function of load for an up to down sequence of RWA algorithms: (1) 10% of better potential routes from adaptive routing (AR) algorithm + M-DRCL (*10%M-DRCL*); (2) AR + DRCL (dynamic routing table, updated before each new connection request); (3) 90% of better potential routes from AR + M-DRCL (*90%M-DRCL*); (4) 70% of better potential routes from AR + M-DRCL (*70%M-DRCL*); (5) all potential routes + M-DRCL (*all M-DRCL*); (6) 40% of better potential routes from AR + M-DRCL (*40%M-DRCL*); (7) 20% of better potential routes from AR + M-DRCL (*20%M-DRCL*); and (8) 25% of better potential routes from AR + M-DRCL (*25%M-DRCL*) for 16-wavelengths in network 2. In Fig. 3, we observe the performance of all WA-heuristic algorithms on decreasing sequence, as follows: (1) 25%M-DRCL; (2) 20%M-DRCL; (3) 40%M-DRCL; (4) all M-DRCL; (5) 70%M-DRCL; (6) 90%M-DRCL; (7) DRCL; and (8) 10%M-DRCL.

Fig. 4 illustrates the blocking probability as function of load for 16 wavelengths in network 2, for different routing algorithms. For a given connection request, if there are two or more shortest-distance-paths to be selected in network 2, the dynamic Dijkstra's algorithm will choose that shortest-distance-path with maximum number of free wavelengths. For the blocking probability less than or equal to 1%, M-DRCL outperforms DRCL significantly, and M-DRCL supports 20% more load than DRCL.

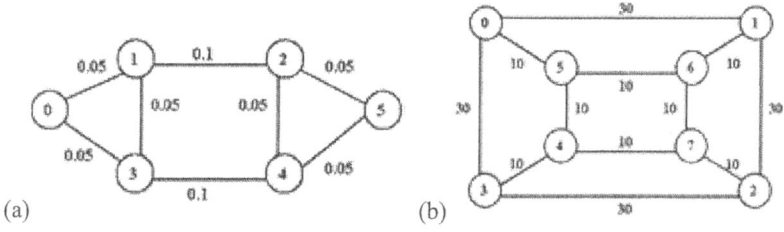

(a)   (b)

Fig. 1. Examples of network topologies: (a) proposed in [7]; (b) meshed double rings.

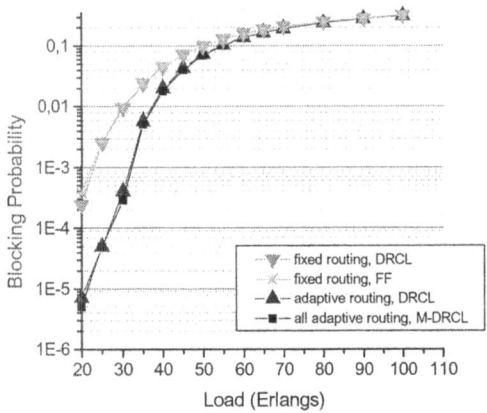

Fig. 2. Blocking probability versus load in network 1.

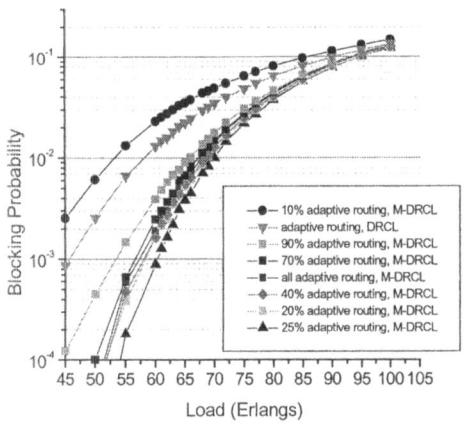

Fig. 3. Blocking probability versus load for 16 wavelengths in network 2.

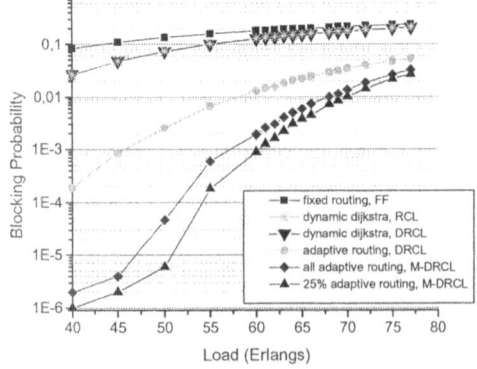

Fig. 4. Blocking probability versus load for 16-wavelengths in network 2.

We observe from Fig. 3 and Fig. 4 that M-DRCL results in lowest blocking probability at 25% of better routes from AR. It means that it is enough to select a set of few better routes from AR instead of selecting all possible routes. Under that observation, we verify that M-DRCL saves significantly the processing time of routing and wavelength assigning calculation at each OXC. If all possible routes are considered then M-DRCL will not depend on the routing algorithms, and even that M-DRCL could result in lower blocking probability than DRCL.

5   Conclusions

In conclusion, we have proposed and demonstrated a novel WA-heuristic algorithm for intelligent and transparent optical networks based on hops counts and relative capacity loss. It leads to lower or equal blocking probability than DRCL algorithm. Other advantage of our M-DRCL over DRCL consists of computing either a set of few paths or all of the paths, whereas DRCL always computes all of them from a routing table, excluding the destination node. Moreover, M-DRCL dismisses the routing table, which contains a selected routing from a node to another, to choose a wavelength on the connection request. Since each node knows the network state and topology, M-DRCL can choose the wavelength from the off-online calculation of a set of few possible routes. Thereby, it is not necessary to update the routing tables. M-DRCL analyses the influence of assigning the wavelength on the current call among the possible future calls and neighbour nodes in distributed environments.

References

1. B. Mukherjee, "WDM Optical Communication Networks: Progress and Challenges", *IEEE Journal of Selected Areas in Communications*, vol. 18, no. 10, pp. 1810-1824, Oct 2000.
2. Paul Green, "Progress in Optical Networking", *IEEE Commun.Mag.*, pp. 54-61, Jan. 2001.
3. Z. Zhang, J. Fu, D. Guo, L. Zhang, "Lightpath Routing for Intelligent Optical Networks", *IEEE Network*, pp. 28-35, Jul. 2001.
4. ITU-T Recommendation G.8080, "Architecture for the Automatically Switched Optical Networks", 2001.
5. ITU-T Recommendation G.7715, "Architecture and Requirements for Routing in the Automatically Switched Optical Networks", 2002.
6. X. Zhang and C. Qiao, "Wavelength assignment for dynamic traffic in multi-fiber WDM networks", Proc. 7th ICCCN, pp.479-485, Oct. 1998.
7. H. Zang, J. Jue and B. Mukherjee, "A Review of Routing and Wavelength Assignment Approaches for Wavelength-Routed Optical WDM Networks", *Optical Networks Magazine*, pp.47-60, Jan.2000.
8. R. A. Barry and S. Subramaniam, "The MAX-SUM Wavelength Assignment Algorithm for WDM Ring Networks", *Proc. OFC'97*, Feb.1997.

Assigning Codes in a Random Wireless Network

Fayçal Djerourou, Christian Lavault,
Gabriel Paillard*, and Vlady Ravelomanana

LIPN, UMR CNRS 7030
Université Paris XIII, 93430 Villetaneuse, France
{fd, lavault, gap, vlad}@lipn.univ-paris13.fr

Abstract. In this paper we present an algorithm that can assign codes in the Code Division Multiple Access (CDMA) framework for multihop ad hoc wireless networks. In CDMA framework, collisions are eliminated by assigning orthogonal codes to the stations such that the spectrum of frequency can be used by all transmitters of the network at the same time. In our setting, a large number n of distinguishable stations (e.g. sensors) are randomly deployed in a given area of size $|S|$. We propose an efficient and fully distributed algorithm, which assigns codes to the nodes of our network so that, for any $\ell > 0$, any two stations at distance at most $\sqrt{(1 + \ell)\,|S|\,\log n / \pi n}$ from each other are assigned two distinct codes.

1 Introduction

Multihop ad hoc wireless networks, such as sensor networks are becoming a more and more important subject of research [7]. In this paper a network is a collection of transmitter-receiver devices, referred to as *stations* (*processors* or *nodes*). It is also assumed that each such station knows only its own identity (Id). Multihop wireless networks consist in a group of stations that can communicate with each other by messages over one wireless channel. Besides, messages may go through intermediate stations before reaching their final destination. At any given time t, the network may be modeled with its *reachability graph*: for any station u and v, there exists one arc $u \to v$ *iff* v can be reached from u.

The time is assumed to be slotted and in each time slot (round) every node can act either as a *transmitter*, or as *receiver*, but *not both*. At any given time slot, a station u acting as a receiver gets a message if and only if exactly one of its neighbors transmits within the same round. If more than two neighbors of u transmit simultaneously, u is assumed to receive no message (collision). More precisely, the considered networks has no ability to distinguish between the lack of message and the occurrence of some collisions or conflicts. Therefore, it is highly desirable to design protocols working independently of the existence/absence of any collision detection mechanism. In this paper, we consider that a set of n stations are initially *randomly scattered* (following the uniform distribution) in

---

* Supported by grant 1522/00-0 from CAPES, Ministry of Education, Brazil.

a surface $\mathcal{S}$ of size $|\mathcal{S}|$. Typically, a global model for a mobile computing environment is a graph $G_t = (V, E_t)$ where V is the set of the stations and E_t is the set of links, which are present at time slot t. In our paper, the problem under consideration consists in minimizing the number of codes of the CDMA (Code Division Multiple Access), which is equivalent to coloring the graph G_t [2]. In the CDMA problem, collisions are removed by assigning orthogonal codes to the stations in such a way that the whole spectrum of frequency can be used by transmitters in the network at the same time. Each code designs the bit specification of each station in the network [1]. To deal with the CDMA, our algorithm solves the graph coloring problem.

2   Coloring an Euclidean Random Network

Our algorithm can be split in two steps as described in the following:

• First, each station has to discover its proper neighborhood. This is done using the randomized algorithm EXCHANGEID. This protocol needs $O\left((\log{(n)})^2\right)$ steps.

• Next, once the station nodes know their neighboorhood, we run ASSIGN-CODE which is a randomized (greedy) algorithm to assign codes to the current reachability graph.

2.1   Discovering the Neighborhood

Algorithm 1: EXCHANGEID

> **begin**
>> **for** $i := 1$ *to* $C(\ell) \log{(n)}^2$ **do**
>>> With probability $\frac{1}{\log n}$, each node u sends a message containing its own identity;
>
> **end**

Theorem 1. *For any fixed constant $\ell > 0$, there exists a constant $C(\ell)$ such that if the transmission radius of each station is set to $r^2 = \dfrac{(1+\ell)\,|\mathcal{S}|}{\pi n}\log n$, then, with probability tending to 1 as n tends to ∞, every node receives all the identities of all its neighbors after an execution of EXCHANGEID.*

Proof. The proof of Theorem 1 relies on the two following facts, viz., (i) the main properties of the random Euclidian network, and (ii) the number of iterations $T = C(\ell)(\log{(n)})^2$ in the loop of EXCHANGEID is sufficient for each node to send its ID *at least once* to all its neighbors. For (i), we refer to the results of [8], which can be briefly stated as follows. Denote by r the transmission range of the n nodes randomly distributed in the surface $\mathcal{S}$ of size $|\mathcal{S}| = O(n)$. Then, in the following regimes, the graph is connected with high probability and we have

(a) For fixed values of k, that is $k = O(1)$, if $\pi \frac{n}{|\mathcal{S}|} r^2 = \log n + k \log \log n + \omega(n)$, then the graph has asymptotically almost surely[1] a minimum degree $\delta = k$.

---

[1] With probability tending to 1 as $n \to \infty$. For short, a.a.s. Here and throughout this paper $\omega(n)$ is a function tending to ∞ arbitrarily slowly.

(b) Let $k = k(n)$, but $1 \ll k(n) \ll \log n / \log \log n$.

If $\pi \frac{n}{|S|} r^2 = \log n + k(n) \log \log n$, then the minimum degree (resp. maximum degree) is a.a.s. $\delta = k(n)$ (resp. $\Delta = e \log n$).

(c) If $\pi \frac{n}{|S|} r^2 = (1 + \ell) \log n$ with $\ell > 0$, then each node v of the graph has a.a.s. d_v neighbors, with

$$- \frac{\ell \log n}{W_{-1}\left(-\frac{\ell}{e(1+\ell)}\right)} + o(\log n) \leq d_v \quad \text{and} \quad d_v \leq - \frac{\ell \log n}{W_0\left(-\frac{\ell}{e(1+\ell)}\right)} + o(\log n),$$

$$(1)$$

where W_{-1} and W_0 denote the two branches of the Lambert W function. We refer here to the paper of Corless *et al.* [3] for more precision on the Lambert W function, which is now a special mathematical function of its own (Note that it is implemented in almost all computer algebra tools such as MAPLE.)

Therefore, by inequality (1), within the regime considered in the assumptions of Theorem 1, the maximum degree of the graph is (with high probability) bounded by $c_\ell \log n$ (where c_ℓ satisfies, e.g. $c_\ell = 2 W_0 \left(-\ell/e(1+\ell)\right)$ [3].

Using this latter remark, let us complete the proof of our Theorem. For any distinct pair (i, j) of connected nodes, define $X_{i \to j}^{(t)}$ as follows: $X_{i \to j}^{(t)} = 1$ iff the node j does not receive the ID of i at time $t \in \left[1, (\log(n))^2 C(\ell)\right]$ and 0 **otherwise**.

In other terms, the set $\left\{X_{i \to j}^{(t)} \mid i, j \neq i, t \in \left[1, (\log(n))^2 C(\ell)\right]\right\}$ is a set of random variables that counts the number of arcs $i \to j$, such that j never received the ID of i.

Denote by X the r.v. $X = \sum_{i \neq j} X_{i \to j}$, where $X_{i \to j} = 1$ iff $X_{i \to j}^{(t)} = 1$ for all $t \in \left[1, (\log(n))^2 C(\ell)\right]$. Now, we have the probability that i does not succeed to send its ID to j at time t:

$$\mathbf{Pr}\left[X_{i \to j}^{(t)} = 1\right] = \left(1 - \frac{1}{\log(n)}\right) + \frac{1}{\log n} \times \left(1 - \left(1 - \frac{1}{\log n}\right)^{d_j(t)}\right),$$

where $d_j(t)$ denotes the degree of j at time t.

Therefore, considering the whole range $\left[1, (\log(n))^2 C(\ell)\right]$, after a bit of algebra we obtain $\mathbf{Pr}[X_{i \to j} = 1] \leq \left(1 - \frac{e^{-c_\ell}}{\log(n)}\right)^{\log(n)^2 C(\ell)} \leq \exp\left(-\ln(n) e^{-c_\ell} C(\ell)\right)$, which bounds the probability that, for all $t \in \left[1, (\log(n))^2 C(\ell)\right]$, i never sent its ID to j.

By linearity of expectations and since from **(c)** the number of edges is of order $O(n \log n)$, we then have $\mathbb{E}[X] \leq O(n \log(n) \exp(-\ln(n) e^{-c_\ell} C(\ell)))$. Thus, $\mathbb{E}[X] \ll 1$ as $n \to \infty$ for a certain constant $C(\ell)$ such that, say $C(\ell) \geq 2 e^{c_\ell}$. (Note that this constant can be computed for any given ℓ by using, e.g. $c_\ell = 2 W_0\left(-\ell/e(1+\ell)\right)$ and this completes the proof of Theorem 1.)

Algorithm 2: ASSIGNCODE

---

begin

    Each vertex u has an initial *palette*:

    $p(u) = \{c_1, c_2, \cdots, c_{deg(u)+1}\}$; $\Delta := -\ell \log n / W_0 \left(-\frac{\ell}{e\,(1+\ell)} \right)$;

    for $i := 1$ **to** $D(\ell) \log{(n)}^2$ **do**

        For each vertex u do

        • Pick a color c from $p(u)$;

        • Send a message contaning c **with probability** $\frac{1}{\Delta+|p(u)|}$;

        if *no collision* **then**

            Every station v in $\Gamma(u)$ gets the message properly ;

            One by one (in order) every member of $\Gamma(u)$ send a message ;

            ($\star$ This step is synchronized by always allowing Δ time slots. $\star$)

        if *u receives all the $|\Gamma(u)|$ messages* **then**

            u send a message of **confirmation** and goes to sleep ;

            every station in $\Gamma(u)$ removes c from its palette ;

end

2.2   Assigning Codes

To assign one code to each node of the network, we use the following protocol which we call ASSIGNCODE. Each vertex u has an initial list of colors (palette) of size $deg(u)+1$ and starts uncoloured. It is important to remark here that the stations know their neighbors (or at least a part of them) by using the previous algorithm, viz. EXCHANGEID. Then, the protocol ASSIGNCODE proceeds in rounds. In each round, each *uncoloured vertex* u, simultaneously and independently picks a color, say c, from its list. Next, the station u attempts to send this information to his neighborhood denoted by $\Gamma(u)$. Trivially, this attempt succeeds iff there is no collision. Before its neighbors eventually assign the color c to u, they has to send one by one a message of reception. Note that this can be done deterministically in time $O(\log n)$, since u can attribute to its active neighbors in $\Gamma(u)$ a predefined ranking ranging from 1 to $|\Gamma(u)|$. Therefore, u sends an *aknowledgement* message, and all its neighbors perform *updates* of their proper palettes and of their active neighbors. Hence, at the end of such a round the new colored vertex u can quit the protocol.

Theorem 2. *For any constant $\ell > 0$, there exists a constant $D(\ell)$ such that if the transmission radius of each station is set to $r^2 = \dfrac{(1+\ell)\,|\mathcal{S}|}{\pi n} \log n$, then, with probability tending to 1 as n tends to ∞, any pair of nodes at distance at most r from each other receives two different codes after invoking the protocol* ASSIGNCODE.

Proof. Although more complicated, the proof of Theorem 2 is very similar to the one of Theorem 1. For any distinct node u, recall that $\Gamma(u)$ represents the set of its neighbors and denote by p_u the size of its current palette. Now, define the random variable Y_u as follows: $Y_u = 1$ **iff** the node u remains uncoloured after the $D(\ell)(\log n)^2$ steps of ASSIGNCODE and 0 **otherwise**.

Denote by $\Gamma_u^{(t)}$ the set of *active neighbors* of u at any given time t during the execution of the algorithm. Suppose that we are in such time slot t. Independently of its previous attemps, u remains uncoloured with probability

$$p_{u,t} = \left(1 - \frac{1}{(\Delta + p_u)}\right) + \frac{1}{(\Delta + p_u)} \times \left(1 - \left(1 - \frac{1}{(\Delta + p_v)}\right)^{|\Gamma_u^{(t)}|}\right).$$

There is at least a collision due to one neighbor $v \in \Gamma_u^{(t)}$.

Since $\forall t$, $\Gamma_u^{(t)} \le \Delta$ and $\forall v$, $1 \le p_v \le \Delta + 1$, we have

$$p_{u,t} \le \left(1 - \frac{1}{(\Delta + p_u)}\right) + \frac{1}{(\Delta + p_u)}\left(1 - \left(1 - \frac{1}{\Delta}\right)^{|\Gamma_u^{(t)}|}\right)$$

$$\le \left(1 - \frac{1}{(\Delta + p_u)}\right) + \frac{1}{(\Delta + p_u)} \times \left(1 - \left(1 - \frac{1}{\Delta}\right)^{\Delta}\right)$$

$$\le 1 - \frac{1}{e(\Delta + p_u)} \le 1 - \frac{1}{2e\,\Delta} \le 1 - \frac{1}{6\Delta}.$$

Therefore, with probability at most $\left(1 - \frac{1}{6\Delta}\right)^{D(\ell)\log n^2} \le \exp\left(-\frac{D(\ell)\log n^2}{6\Delta}\right)$, u remains uncoloured during the whole algorithm. Thus, the expected number of uncoloured vertices at the end of the protocol ASSIGNCODE is less than $\mathbb{E}[Y] = \sum_u \mathbb{E}[Y_u] \le n \exp\left(-\frac{D(\ell)\log n^2}{6\Delta}\right)$.

Since by (1), we have $\Delta = \Delta(\ell) \le -2\frac{\ell\log n}{W_0\left(-\frac{\ell}{e(1+\ell)}\right)}$, it is now easy to choose a constant $D(\ell)$ such that $D(\ell) > -\frac{12\ell}{W_0(-\ell/e(1+\ell))}$, and $\mathbb{E}[Y] \ll 1$ as $n \to \infty$. After using the well known Markov's inequality, the proof of our Theorem is completed.

2.3   Efficiency of the Algorithms

Both protocols use local competitions, which means that the "coin flipping" games to access the shared wireless channel take place only between neighbors. First, we note that the lower bound for broadcasting in a network of diameter D is given by $\Omega\left(D\log(n/D)\right)$ [5]. A node u in the network needs at least $O\left(|\Gamma(u)|\right)$ "local broadcasts". By "local broadcast" we mean the sending of information to nodes of distance at most $2r$ where r is the transmission range. From the main result of [5], it takes at least $\Omega\left(\log(deg(u))\right)$ time slots to get all the IDs of the neighbors of u. By (1), $|\Gamma(u)| = \Theta(\log n)$, an algorithm needs at least $\Omega(\log n \times \log\log n)$ time slots to exchange the IDs of all the *connected* nodes. Therefore, our protocol EXCHANGEID, which needs $O\left((\log n)^2\right)$ time slots, is at most a $O(\log n/\log\log n)$ factor away from the optimal, and ASSIGNCODE is at most a $O(\log n/\log\log n)$ factor away from the optimal.

3   Final Remarks

This paper solves the problem of assigning different codes to stations randomly deployed in any given region $\mathcal{S}$. Our results make sense and can be useful for many reasons, including:

(i) Our settings is examplified for a large number n of fire sensors dropped by planes in some large area $\mathcal{S}$. In our paper, the areas where such sensors are thrown need only be bounded. By contrast with numbers of existing papers, our results reflect real-life situations where the areas under consideration are far to be as regular as squares, rectangles or circles.

(ii) Our analysis yields key insights for the coloring problem in a rigorous framework, whereas a majority of results are based on empirical results.

(iii) Assume each deployed node to be *"active"* with constant probability p ($0 < p \leq 1$). All our results can be extended by taking the intensity of the process $n/|\mathcal{S}|$ as $p\,n/|\mathcal{S}|$ (or by simply using $n' = p\,n$ in the analysis). An *"active"* node is neither faulty nor asleep. This is especially well suited for "energy efficient" settings when some node are inative and saving their batteries, which increases the network's lifetime.

(iv) We considered herein uniform distributions, which can be approximated by Poisson point processes [4, pp 39-40]. Since Poisson processes are **invariant** [6] if their points are independently translated, the translations being identically distributed from some bivariate distribution (direction and distance), all our results remain valid. Therefore, the present results can serve to cope with **mobile networks** whenever the mobility model corresponds to the same translation distribution and whenever the $O\left((\log n)^3\right)$ time slots that are needed to achieve the coloring of the nodes can be neglected. Thus, the results of this paper can help both researchers and designers to face many realistic situations.

References

1. Battiti, R., Bertossi, A. A., Bonuccelli M. A.: Assigning Codes in Wireless Networks: Bounds and Scaling Properties. Wireless Networks, 5 (1999) 195–209
2. Bollobàs, B.: Modern graph theory. Graduate Texts in Mathematics. Springer, Berlin (1998)
3. Corless, R. M., Gonnet G. H., Hare D. E. G., Jeffrey D. J., Knuth D. E.: On the Lambert W Function. Advances in Computational Mathematics 5 (1996) 329–359
4. Hall, P.: Introduction to the Theory of Coverage Processes. Birkhäuser (1988)
5. Kushilevitz, E., Mansour, Y.: An $\Omega(Dlog(N/D))$ Lower Bound for Broadcast in Radio Networks. SIAM Journal on Comput 27(3) (1998) 702–712
6. Miles, R. E.: On the Homogenous Planar Poisson Point Process. Math. Biosciences 6 (1970) 85–127
7. Perkins, C. E.: Ad Hoc Networking. Addison-Wesley (2001)
8. Ravelomanana, V.: Asymptotic Critical Ranges for Coverage Properties in Wireless Sensor Networks. In submitted, Available upon request (2003)

Flexible QoS Routing Protocol for Mobile Ad Hoc Network

Lyes Khoukhi and Soumaya Cherkaoui

Department of Electrical and Computer Engineering, Université de Sherbrooke
J1K 2R1, QC, Canada
{Lyes.Khoukhi, Soumaya.Cherkaoui}@USherbrooke.ca

Abstract. We introduce a flexible QoS routing protocol for mobile ad hoc network called AQOPC "Ad hoc QoS Optimal Paths based on metric Classes". It provides end-to-end quality of service (QoS) support in terms of various metrics and offers accurate information about the state of bandwidth, end-to-end delay and hop count in the network. It performs accurate admission control and a good use of network resources by calculating multiple paths based on different metrics, and by generating the needed service classes. To regulate traffic, a flexible priority queuing mechanism is integrated. QoS violation detection and adaptive recovery are assured by a mechanism based on the prediction of the arrival time of data packets. The results of simulations show that AQOPC provides QoS support with a high reliability and a low overhead, and it produces lower delay than its best effort counterpart at lower mobility rates.

1 Introduction

The next generation of wireless networks will carry diverse media such as data, voice, and video. Therefore, it is necessary to provide quality delivery with regard to some parameters such as bandwidth and delay for sensitive applications using voice and video media for example. In the case of mobile ad hoc networks (MANET), the dynamic topology and the unpredicted state of the network add other dimensions to the problem of the satisfaction of Quality of Service (QoS).

Some protocols that have been proposed for routing in ad hoc networks, such as DSDV [1], AODV [2], DSR [3] and ZRP [4], do take into account the specific conditions of MANETs, but were designed without explicitly considering QoS. Therefore, they are not eligible to support multimedia enriched services such as voice applications. Several works have been proposed recently concerning QoS Routing in MANETs [6-10]. Some works [7] proposed table-driven routing approaches for QoS support. However, their performances are degraded comparatively to reactive approaches due to the problem of stale route information. In [7], the authors address the problem of supporting real time communications in a multihop mobile network and they propose a protocol for QoS routing. A ticket-based QoS routing protocol was proposed in [9]; it is based on the model assuming that the bandwidth of a link can be determined independently of its neighboring links. These proposed QoS protocols can

J.N. de Souza et al. (Eds.): ICT 2004, LNCS 3124, pp. 354–362, 2004.

also be classified into two schemes: source routing and distributed routing. Source routing schemes such as [8] suffer from problems of scalability and frequent updates of the state of the network. On the other hand, most existing distributed algorithms (exp. [13]) require the maintaining of a global network state at every node, which may also cause the problem of the scalability. Other researches focus on MAC layer: in [7, 11] a CDMA over TDMA channel model is introduced by using the notion of a time slot on a link to calculate the end-to-end path bandwidth.

The aim of QoS routing is to find routes that deliver a guaranteed end-to-end QoS. However, assuring QoS in ad hoc networks is more difficult than in others. This is due to the frequent changes in topology and it is also due to the scarcity of resources, some of which, like bandwidth, are shared between adjacent nodes. The optimization of resources usage and a good cooperation between nodes are vital to the satisfaction of some QoS in MANETs. These factors were not always considered in the works cited above. In this paper we present a new QoS routing protocol for mobile ad hoc network, named "Ad hoc QoS Optimal Paths based on metric Classes" (AQOPC). The objective is to make a good use of network resources when transmitting users data packets. For that, AQOPC searches for multiple parallel paths that satisfy QoS requirement constraints. To perform adequate path selections, the information on intermediate nodes is exploited. Two service classes, similar to the ones used in the third generation wireless telecommunication systems (3G) are introduced in AQOPC. These classes allow a wide range of requirements as defined by 3G: interactive, conversational, streaming, etc. Primary Service Classes (PSC) try to provide critical guaranties (for real-time applications, for example), whereas Secondary Service Classes (SSC) provide less critical guaranties and give other feasible paths and permit also alternative routes recovery in the case of PSC QoS violation or error detection. To regulate traffic, a flexible priority queuing mechanism, was developed.

The rest of the paper is organized as follows. In Section 2 we introduce the proposed AQOPC protocol. Protocol performances and simulation results are shown in section 3, and section 4 concludes the paper.

2   AQOPC Protocol

2.1   Route Discovery

AQOPC conforms to a pure on-demand rule. It neither maintains a global routing table nor exchanges it periodically. The protocol relies on dynamically establishing routing table entries in network nodes.

Paths Discovery. When a source node S is ready to communicate with a destination for which it has no routing information, it broadcasts a route request (RREQ) short message to its neighbors. The RREQ format is: *<packet_type, source_addr, dest_addr, sequence, previous_addr, time_to_live, hop_req, del_req, band_req>*.

The node will add a route entry in its routing table, if the RREQ is accepted, with status *checked* and rebroadcast the request to the next hop. This status remains valid for a short period $2T_D$, where T_D is equal to the requested delay *del_req*. If no reply packet arrives at this node in time, the route entry will be deleted at the node and reply packets coming after this time will be ignored.

Paths Reservation. For each RREQ received by the destination, one route replay (RREP) short message will be generated and unicasted back to the source along the reverse route. An intermediate node, which receives an RREP, checks its resources availability and applies a multi criteria admission control policy. A node will update its route status to *confirmed* if the packet is accepted. The RREP format is : *<packet_type, source_addr, dest_addr, sequence, hop_rep, del_rep, band_rep>*.

Upon receiving an RREP, a node may have to update its routing table. The main fields of each entry in a node routing table are: *<dest_addr, seq, next_hop, statuts, hop_node, band_node, del_node>*.

The *dest_addr* field corresponds to the destination address, *seq* corresponds to the sequence number of the node in the current route, and next_*hop* corresponds to the address of the next node. The field *Status* indicates the state of a node route (set to *checked* in the paths discovery phase and set to *confirmed* in the paths reservation phase). The fields *del_node, hop_node,* and *band_node* are useful for QoS control as will be shown in section 2.3. The sequence number is used in the protocol to avoid possible loops during path discovery.

2.2 Multi-criteria Admission Control

The admission control policy should guarantee QoS requirements for each accepted flow. For the source QoS requirements $[D_{max}, H_{max}, B_{min}]$ associated respectively to the maximum end-to-end delay, maximum hop count, and minimum bandwidth, the policy of multi-criteria admission control assures for each flow the 3 guaranty intervals $[I_D, I_H, I_B]$. The policy then selects the optimal path corresponding to the source requirements from the convenient guaranty interval. The values α, β for a guaranty interval $I=[\alpha, \beta]$ are calculated by the route discovery process discussed above. Note that AQOPC operates on a hop-by-hop basis.

Bandwidth Control. Bandwidth usually represents a critical requirement for real-time applications. In our study, the value of bandwidth β at node i is a function of the node bandwidth capacity BCi and the traffic at this node Ti. We do not consider external interferences; therefore, we do not deal with the reduced capacity through interference from neighbouring nodes.

We call *bandwidth domain* $\Phi(i)$ of node i that covers the first and second vicinities of i, the set defined as follow:

$$\Phi(i) = \{ \ Tij \ / \ \eta(i,j), \text{ or } \eta(\eta(i,k),j) \ \} .$$

where $(\eta(i,j), \text{ or } \eta(\eta(i,k),j))$ is the neighbouring relation defined as:
$\eta(i,j) = \{i \text{ has link with } j\}$. The bandwidth β_i of node i is then:

$$\beta_i = BCi - \sum T_{ij} , \ \forall \ T_{ij} \in \ \Phi(i) . \tag{1}$$

The bandwidth available at node i depends on the traffic generated at this node, the traffic travelling through this node, and the traffic generated at the neighboring nodes. We use β_i in (1) as the node bandwidth to compute the path offering a maximum bandwidth among n paths. Assume that $N_l \beta_k$ is the bandwidth at node k in path l, we search MAX $(\chi_b^{(n,1)})$,

$$\chi_b^{(n,1)} = \begin{cases} MIN\ (N_l \beta_1\ ,\ \ldots\ldots,\ N_l \beta_k\ ,\ \ldots\ldots\ ,\ N_l \beta_n) \\ \qquad\qquad\qquad \cdot \\ MIN\ (N_l \beta_1\ ,\ \ldots\ldots,\ N_l \beta_k,\ \ldots\ldots\ ,\ N_l \beta_n) \\ \qquad\qquad\qquad \cdot \\ MIN\ (N_n \beta_1\ ,\ \ldots\ldots,\ N_n \beta_k\ ,\ \ldots\ldots\ ,\ N_n \beta_n) \end{cases} \tag{2}$$

$$\delta\_B^1(P) = MAX\ (\chi_b^{(n,1)}). \tag{3}$$

Let *band_req* be the value of the requested bandwidth in the RREQ message arriving at node k in path l. $\delta\_B^k(P)$ denotes the bandwidth of the path we are looking for. (3) is solved by using the following process. During the travelling of the RREQ along a path l, each node k calculates MIN (*band_req*, $N_l \beta_k$) and makes a decision to forward the RREQ if the bandwidth requested is available. If no bandwidth is available, the request is discarded. Upon receiving each RREQ, the destination will initialize randomly the field *band_rep* to a value higher than its bandwidth value. Then, a RREP is generated and sent back to the source. With the receipt of an RREP by an intermediary node k on path l, the calculation of MIN (*band_rep*, $N_l \beta_k$) is performed and stored in the node routing table and then set to a newly generated RREP.

The sorting algorithm for computing the maximum bandwidth is activated at the receipt of the first RREP by the source node, and is performed upon receiving other route response packets. δ_B^1 is the maximum bandwidth value among n RREP received. P^1 points out the associated path. We now identify the guaranty interval of the first service class. For that, we extract the second optimal bandwidth δ_B^2 by using (4).

$$\delta\_B^2(P) = MAX\ (\chi_b^{(n,1)} - \delta\_B^1(P)). \tag{4}$$

$GI(B^1) = [\delta\_B^2,\ \delta\_B^1]$ is the first guaranty interval for bandwidth constraint; the optimal path associated is noted P^1. Similarly, we identify the interval of the second class:

$$\delta\_B^3 = MAX\ (\chi_b^{(n,1)} - \delta\_B^2). \tag{5}$$

$GI(B^2) = [\delta\_B^3,\ \delta\_B^2]$ represents the second guaranty interval, with P^2 as the feasible associated path. We can calculate other guaranty intervals as follows:

$$\delta\_B^k = MAX\ (\chi_b^{(n,1)} - \delta\_B^{k-1}). \tag{6}$$

and P^k is the path generated by δ_B^k, it will be the path associated to $GI(B^{k-1})$.

2.3 Guaranty Intervals

The guaranty intervals in AQOPC are defined by both the interval values and their associated feasible paths. They are used by the source node to select the ad hoc (adequate) paths depending on bandwidth, delay, or hop count requirements. The algorithm shown in Fig.1 checks and selects the suitable service class for a flow bandwidth requirement; we have developed other similar algorithms to achieve class selection concerning delay and hop count constraints.

```
  Check availability (B_flow);
    {
selected  =  false  /*initialization  of  variable  indicating  the
result of selection*/
/* we traverse service classes (the number of class is U) */
    while (CS < U)
      {
        if (B_flow)  ∈ ]βcs ,  βcs+1 ]
          { /* we have found the convenient service
            class to B_flow */
            PBcs = PTSELCT(]βcs ,  βcs+1 ])
      /* Activate the optimal path PBcs corresponding */
            ACTV(PBcs)
            Break;
        }}
/* The bandwidth requirement did not find a suitable service class
*/
      if selected = false then
        {
/* select the smallest service class CS where B_flow is inferior
to βcs */
          if exist CS where (( B_flow < βcs))
          { select MIN (CS)
            PBcs = PTSELCT(]βcs ,  βcs+1 ])
            selected = true
          } } }
    /* start the paths discovery if there is no suitable service
class*/
    if selected = false  Start (Paths discovery process)
    }
```

Fig. 1. Path selection from a bandwidth guaranty interval

2.4 Priority Queuing

The prioritization scheme is an important factor for service differentiation in mobile ad hoc network. SWAN model [12] uses priority queuing by limiting the amount of real time traffic in order to treat the lower priority packets. AQOPC implements a separate queue for each class (Fig. 2). Buffer space is shared between service classes: when it is not occupied by one service class, it will be assigned to other classes if there is need. By using priority queuing, the contention prediction and prevention becomes possible at node level: before occupying all buffer space, a node informs its neighbors to slow down traffic.

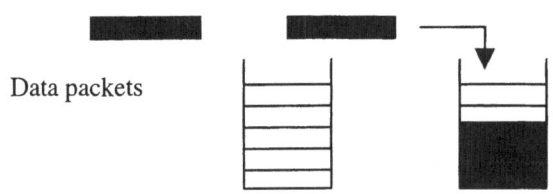

Fig. 2. Priority queuing

3 Detection and Recovery of QoS Violation

3.1 The Detection of QoS Violation

QoS violation detection and recovery mechanisms are of central importance in an ad hoc networks to support multimedia applications. AQOPC proposes a mechanism named *QoS violation by time exceeding* in order to detect QoS violation. QoS violation by time exceeding is caused by congestion as consequence to traffic augmentation in the network and the cumulative treatment nodes. We define:

T_i: the time necessary for generating an RREQ at node i, which has *del_node* as the current delay available at its routing table.

ΔT_T: the estimated packet treatment time at the source node.

ΔT_G: the time between sending two data packets by the application generating traffic.

Now, we can compute the estimated time *TEXC* for receiving a data packet j by a node i.

$$TEXC\ (j) = T_i + \Delta T_T + \sum^{j-1} \Delta T_G \tag{7}$$

Thus, the delay QoS violation can be easily detected at a node by monitoring the delay of the arriving data packets. If a node receives a data packet j whose delay exceeds the maximum time requirement $TEXC(j)$, a QoS violation by time exceeding is detected and the QoS recovery mechanism will be triggered.

3.2 The Recovery of QoS Violation

AQOPC applies a recovery mechanism that can be used to re-establish the route along a new path when a path is broken. The approach adapts the routing paths according to the new network state caused by either nodes mobility or a degraded path state. When a QoS violation is detected at a node x, a neighbor node z will send to the destination D a short update message. Then, D will broadcast back to the source S an update message. The same computations as in the route discovery process in term of admission control will be performed during the traveling of the update message to S. Note that loops are avoided by decreasing the sequence number value of node z before it initiates the recovery process.

Upon receiving the short update message, S activates the service class selection algorithm. Then, the flow will be redirected by the source node according to the new state of the network. The path of a service class which offers a nearly best solution to the flow requirements is automatically selected. Note that the original broken path resources are not held for a long time because resource reservation is released after timeout. The recovery process is performed at the routing layer without adding an excessive load or congestion to other layers.

4   Simulation Results

In the performance assessment experiments, we considered a mobile ad hoc network of 100 nodes. The simulation is generated with the GlomoSim simulator in an area of 5000 by 5000 m. The transmission range of a node is 250m, and its data rate is 2Mb/s.

We implemented a random way point mobility model at each node in the network for the duration of the simulation (600 seconds). We used a constant bit-rate CBR traffic model with fixed data packets of 512 bytes at rate of 20 packets per second. Flows have the required bandwidth B_{min} of 65kb/s and the required delay D_{max} of 0.35s.

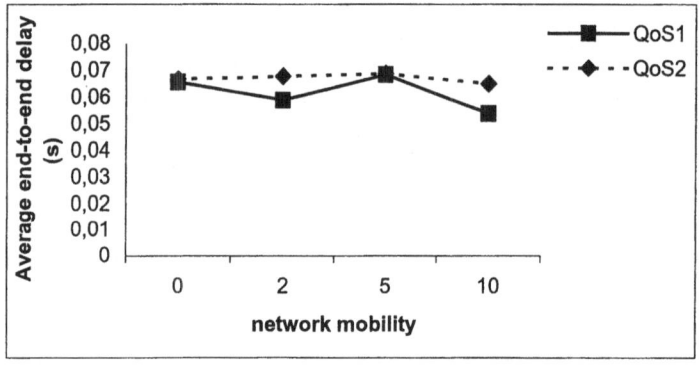

Fig. 3. Average delay versus node mobility

Figure 3 shows the average end-to-end delay versus node mobility. We compare our QoS model represented by QoS1 with the best effort AODV protocol (QoS2). We remark that the average end-to-end delay at the high mobility (above 5 m/s) is smaller than that of the lower mobility because there is less traffic to forward in the network when the mobility becomes high (the traffic delivered to the destination becomes small). We notice that in both QoS1 and QoS2, the stationary scenario (offers slightly better delay than mobility of 5 m/s because at low mobility the paths become longer, therefore the delay of transmitting data from source to destination increases. When the mobility is above 5 m/s, the routes between source and destination become smaller, therefore the average end-to-end delay becomes smaller. AQOPC shows a better result than that of the best effort protocol.

Figure 4 plots the traffic admission ratio versus node mobility. We have considered three types of flow traffics. QoS1, QoS2, and QoS3 assure respectively the traffic of 5, 20, and 50 flows for AQOPC. For the stationary scenario, all the traffic is admitted; about 99.8 % of the traffic is delivered to the destination.

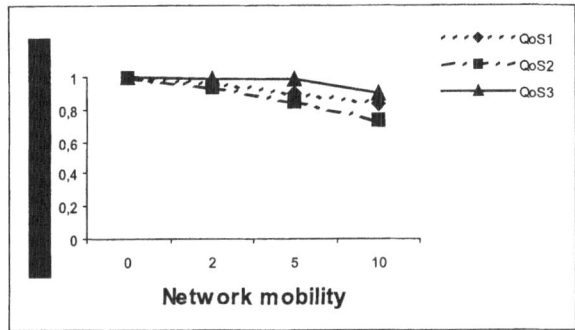

Fig. 4. Traffic admission ratio versus node mobility

We notice that with increased mobility, the traffic admitted by the destination decreases. With high traffic load, less traffic is receipted by the destination. At the mobility 10m/s, the rate is still above 93% for a load of 5 flows, and still above 85% for loads of 20 flows. As the load of the traffic increases, there is more congestion in the network and path errors due to the high mobility, therefore the mechanism of admission control is often called. Therefore, AQOPC routing protocol should not be used in the networks with frequent nodes mobility and fast changing topologies.

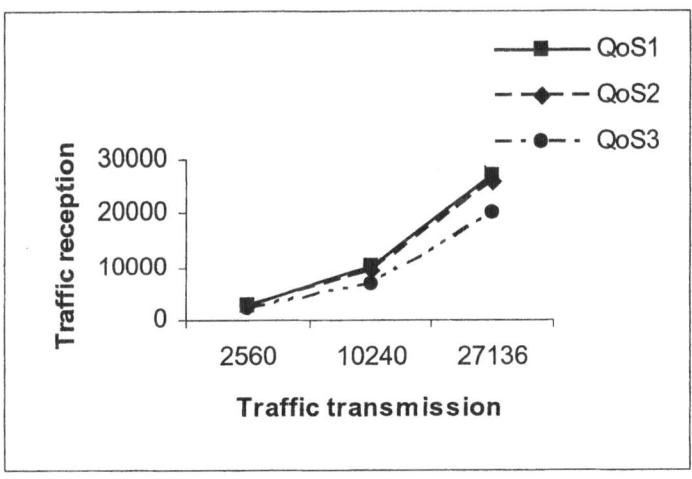

Fig. 5. Packets throughput for mobility = 0, 5, 10 m/s

Figure 5 shows the packet throughput for AQOPC under different traffic loads and node mobility v=0 (QoS1), v=5 (QoS2), v=10 (QoS3). We notice that the scenarios of stationary and mobility 5 m/s present a high performance, and almost all the transmitted packets are received by the destination. When the node speed v increases, the throughput drops. The network throughput is affected by the mobility at both MAC layer by resolving collisions caused by node movements, and at the routing layer by reestablishing, when route break is detected, the new routes after calling the QoS admission control.

5 Conclusion

In this paper we presented a new QoS routing protocol based on multi-service classes that provides per-flow end-to-end QoS support in mobile ad hoc networks. It can build multiple paths between source and destination. AQOPC calculates the feasible paths corresponding to different metrics associated to the application streams. Service classes offer a guaranty for each metric. The priority queuing of AQOPC is an efficient way to prevent a contention in the network. The simulation results show that at lower network mobility, AQOPC produces lower delay than its best effort AODV counterpart. It provides QoS support in mobile ad hoc networks with a high reliability and a low overhead.

References

1. Perkins, C.E., Bhagwat, P.: Highly dynamic destination-sequenced distance-vector routing (DSDV) for mobile computers, Computer Comm.Rev.6 (1994) 234 –244.
2. Perkins, C.E., Royer, E.M.: Ad-hoc on-demand distance vector routing. Proceedings of the Second IEEE Workshop on Mobile Computing Systems and Applications. New Orleans, LA, February 1999, pp.90 –100.
3. Johnson, D., Maltz, D.: Dynamic Source Routing in Ad Hoc Wireless Networks. In T. Imielinski and H. Korth, editor, Mobile Computing. Kluwer Academic Publ., 1996.
4. Haas, Z.J., Pearlman, M.R.: The zone routing protocol (ZRP) for ad hoc networks. Internet Draft,draft-ietf-manet-zone-zrp-02.txt.
5. Qi Xue, Aura Ganz: Ad hoc QoS on-demand routing (AQOR) in mobile ad hoc networks. Journal of Parallel and Distributed, Computing Elsevier Science, USA, 2003.
6. Lee, S., Campbell, A. T.: INSIGNIA: In-band signalling support for QoS in mobile ad hoc networks. In Proc. of the 5th Intl. Workshop on Mobile Multimedia Communication, 1998.
7. Lin, C. R., Liu. J.-S.: QoS Routing in Ad Hoc Wireless Networks. IEEE J. Sel. Areas Commun., JSAC-17 (8): 1426–1438, 1999.
8. Chen, S., Nahrstedt, K : On finding multi-constrained paths, IEEE International conference on communications, June 1998.
9. Chen, S. , Nahrstedt, K.: Distributed Quality-of-Service in AdHoc Networks. IEEE J. Sel. Areas Commun., JSAC-17 (8), 1999.
10. Cansever, D. H., Michelson, A. M, Levesque, A. H.: Quality of Service Support in Mobile ad-hoc IP Networks. In Proc. MILCOM, 1999.
11. Lin, C.-R.: On-Demand QoS Routing in Multihop Mobile Networks. In Proc. of IEEE INFOCOM 2001, pages 1735–1744, April 2001.
12. Ahn, G.H., Campbell, A. T., Veres, A., Sun, L. H.: SWAN: Service Differentiation in Stateless Wireless Ad Hoc Networks, IEEE INFOCOM 2002.
13. Wang, Z., Crowcroft, J.: QoS routing for supporting resource reservation, IEEE J. Sel. Areas Commun., September, 1996.

Integrating Mobile Ad Hoc Network into Mobile IPv6 Network

A. Ali[1], L.A. Latiff[2], and N. Fisal[3]

Faculty of Electrical Engineering,
University Technology Malaysia,
81310 Johor Bahru, Johor Darul Ta'zim, Malaysia
[1] adelali3@lycos.com
[2] liza@citycampus.utm.my
[3] sheila@suria.fke.utm.my

Abstract. In the future more and more devices connected to the Internet will be wireless. Wireless networks can be classified into two types of networks: network with infrastructure (i.e. networks with base stations, gateway and routing support), which is called Mobile IP, and network without infrastructure which is called ad hoc networks. Mobile IP tries to solve the problem of how a mobile node may roam from its network to foreign network and still maintain connectivity to the Internet. Ad hoc network tries to solve the problem if the infrastructure is not available or inconvenient for its use such as in rural environments. Integrating ad hoc into Mobile IP provide new feature for Mobile ad hoc network such as Internet connectivity, streamline communication with another network. This paper presents the development of a test bed for integrating Mobile Ad hoc NETwork (MANET) into Mobile IPv6.

1 Introduction

The Internet Protocol (IP) that is currently used is called IPv4. IPv4 was designed in January 1980 and since its inception, there have been many requests for more addresses and enhanced capabilities due to the phenomenal growth of the Internet. Therefore, IPv6 was developed in 1992. Major changes in IPv6 are the redesign of the header, including the increase of address size from 32 bits to 128 bits. Besides the larger address space IPv6 also provides other new features such as address auto configuration, enhanced mobility support and IP Security (IPSec) integrated into the standard IPv6 protocol stack [3,6,15]. Just having more addresses does not solve the problem of mobility. Because part of the IP address is used for routing purposes, it must be topologically correct. This is where Mobile IP comes in.

Wireless networks are classified into two modes: Infrastructure mode and ad-hoc mode. To provide mobility in infrastructure mode, Mobile IP is sufficient but with ad-hoc mode both the ad-hoc routing protocols as well as the Mobile IP is necessary. The Mobile IP tries to solve the problem of how a mobile node may roam from its network to foreign network and still maintain connectivity to the Internet [1]. Mobile IPv6 has more features compared to Mobile IPv4 such as route optimization and Dynamic Home Agent Address Discovery (DHAAD). Ad hoc network tries to

J.N. de Souza et al. (Eds.): ICT 2004, LNCS 3124, pp. 363–368, 2004.
© Springer-Verlag Berlin Heidelberg 2004

solve the problem if the infrastructure is not available or inconvenient for its use such as in rural environments [7]. Ad hoc networks can be subdivided into two classes: static and mobile [4]. In this paper we will use Mobile ad hoc networks (MANETs).

Integrating MANET with Mobile IPv6 network will extend the capabilities of Mobile IPv6 to MANET which will introduces fast agent discovery, increases cell coverage of access points, monitoring system provision and extend connectivity to other networks or to Internet [1,10,11].

The rest of this paper is organized as follows: Section 2 presents related work in integration of MANET with Mobile IPv6 network. The implementation of our proposal will be described in Section 3 followed by Section 4, which will conclude the paper.

2 Related Work

There are many methods of integrating MANET with Mobile IPv4. The main difference is the routing protocol used in MANET and access point of MANET. One of the method of Integrating Mobile IPv4 with Ad hoc networks [9] is using the Destination-Sequenced Distance Vector (DSDV) routing protocol in the ad hoc network. It extends access to multiple nodes in MANET to create an environment that supports both macro and micro mobility. Another method is MIPMANET: Mobile IPv4 for Mobile Ad hoc Networks [10]. It uses Mobile IP Foreign Agent (FA) as access point to the internet. If any node wants Internet access, it registers in the FA and use its home address for communication. The Ad hoc On-demand Distance Vector (AODV) routing protocol is used to deliver packets between nodes and FA in the ad hoc network. It provides algorithm MIPMANT Cell Switch (MMCS) to determine when mobile node should register with a new FA. Another method of integration is Ad hoc Network and IP Network Capabilities for Mobile Host [13]. It uses AODV as the routing protocol in the ad hoc network and Mobile IPv4 in outside traffic network. It uses Multihomed mobile IPv4 so that any mobile node can register with multiple FAs simultaneously. This connectivity with multiple gateways will enhance performance and reliability. All these methods use mobile IPv4 and they depend of FA. Mobile IPv6, however, does not define foreign agents. To be able to reach the Internet, mobile nodes using Mobile IPv6 need an Internet Gateway.

3 Implementation

Four personal computers identified as HA, MN, CN and Gateway had been configured with IPv6 based on the Linux operating system as shown in Fig .1. Mobile IPv6 which is a distribution from Helsinki University of been setup in all MANET nodes [14]. Kernel AODV is a loadable kernel module Technology had been setup in HA, MN and CN. Kernel AODV which is a distribution from National Institute of Standards and Technology in US had for Linux. It implements AODV routing between computers equipped with WLAN devices.

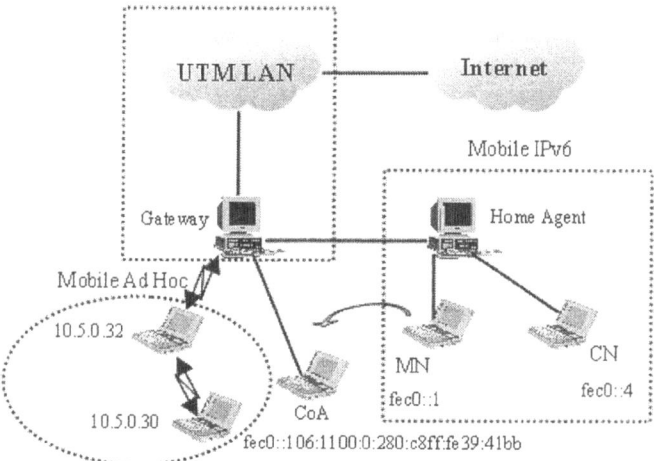

Fig. 1. Network Architecture

```
[root@localhost root]# /etc/rc.d/init.d/./mobile-ip6 restart
Stopping Mobile IPv6:                              [ OK ]
Starting Mobile IPv6:                              [ OK ]
```

Fig. 2. Start Mobile IPv6

NAT-PT has been setup in the gateway to establish communication links between IPv4 (MANET) network and IPv6 (Mobile IPv6) network [5]. The NAT-PT Linux gateway has three Ethernet ports and one WLAN as shown in Fig. 1. All HA, MN and CN are started by running the program as shown in Fig. 2.

3.1   Communication Between MN and CN

When MN starts its Mobile IPv6 program in home network, it sends multicast message (router solicitation) to all routers until HA send router advertisement as shown in Fig. 3. Then, MN will be assigned an address (fec0::280:c8ff:fe39:41bb/64). This will inform MN that it is in the home network. When MN moves to a foreign network, it attaches itself to the gateway, which advertises router advertisement periodically as shown in Fig. 4. After MN receives router advertisement from the gateway, it gets a CoA address (fec0::106:1100:0:280:c8ff:fe39:41bb /64). MN will send a BU to HA and CN. MN will register all sending BU in the BU list as shown in Fig. 4.

```
[root@localhost root]# tcpdump
tcpdump: listening on eth0
18:37:49.565538 fe80::280:c8ff:fe39:41bb > ff02::2: icmp6: router solicitation
18:37:50.671973 fe80::280:adff:fe81:1abc > ff02::1: icmp6: router advertisement
```

Fig. 3. Router Solicitations and Router Advertisements

```
[root@localhost root]#mipdiag -l
Mobile IPv6 Binding update list
Recipient CN:fec0::2
BINDING home address:fec0::1 care-of
address:fec0:106:1100:0:280:c8ff:fe39:41bb
      expires:202 sequence:1 state:1
      delay:1 max delay 256 callback time: 155

Recipient CN:fec0::4
BINDING home address:fec0::1 care-of
address:fec0:106:1100:0:280:c8ff:fe39:41bb
      expires:329 sequence:0 state:1
      delay:1 max delay 256 callback time: 329
```

Fig. 4. Mobile IPv6 Binding update list in MN

```
[root@localhost root]# mipdiag -c
Mobile IPv6 Binding cache in CN
Home Address Care-of Address              Lifetime  Type
fec0::1   fec0:106:1100:0:280:c8ff:fe39:41bb 174     1
Mobile IPv6 Binding cache in HA
Home Address Care-of Address              Lifetime  Type
fec0::1   fec0:106:1100:0:280:c8ff:fe39:41bb 916     2
```

Fig. 5. Mobile IPv6 Binding cache in CN and HA

HA and CN will keep the BU for a certain time (lifetime) in the Binding cache as shown in Fig. 5. Beyond the stipulated lifetime, if no update of BU is received then this shows that MN is unreachable. BU in HA, CN and in MN will be deleted once MN returns home.

3.2 Communication Between MN and MANET Node

Kernel AODV provides route table for all MANET nodes in the MANET coverage area as shown in Fig. 6. It also supports multihop connectivity between MANET nodes. This gateway has a pool of IPv4 address including IPv4 subnet 10.5.0/24. When MN attaches itself to the gateway, it gets a CoA of (fec0:106:1100:0:280:c8ff:fe39 :41bb/64). If MN wants to communicate with MANET node (10.5.0.30), MN creates a packet with the Source Address, SA=fec0:106:1100:0:280:c8ff:fe39:41bb and destination address, DA= PREFIX: 10.5.0.30. The PREFIX is static and any packet originating from an IPv6 node destined to the IPv4 network will contain that PREFIX as a part of the IPv6 destination address.

NAT-PT will translate the IP header including the source and destination address. After translation, the source address will be a one pool address (say 10.5.0.33) and the destination address is 10.5.0.30. The NAT-PT will retain the mapping between 10.5.0.33 and fec0:106:1100:0:280:c8ff: fe39: 41bb until the end of the session. In case of a reverse trip, source address will be 10.5.0.30 and the destination address will be 10.5.0.33. NAT-PT will change the source address to

PREFIX: 10.5.0.30 and destination address to fec0: 106: 1100:0: 280: c8ff: fe39: 41bb. The communication will continue as shown in Fig. 7 and Fig. 8.

```
[root@localhost root]#cat /proc/aodv/route_table

Route Table
_____
 IP    | Seq |Hop Count| Next Hop
_____
 10.5.0.33  1    1     10.5.0.33  Valid sec/msec: 2/892 0
 10.5.0.30  1    0     10.5.0.30  Valid sec/msec: 1976 10239/239 1
 127.0.0.1  1    0     127.0.0.1  Valid sec/msec: 1976 10239/239 1
```

Fig. 6. Routing Table in gateway

```
[root@localhost root]#ping 10.5.0.33
PING 10.5.0.33 (10.5.0.33) from 10.5.0.30 : 56(84) bytes of data.
64 bytes from 10.5.0.33: icmp_seq=1 ttl=64 time=4.84 ms
64 bytes from 10.5.0.33: icmp_seq=1 ttl=63 time=149 ms (DUP!)
64 bytes from 10.5.0.33: icmp_seq=2 ttl=64 time=4.20 ms
64 bytes from 10.5.0.33: icmp_seq=2 ttl=63 time=5.22 ms (DUP!)
```

Fig. 7. Ping one pool address in Ad hoc node

```
localhost root]#tcpdump
tcpdump: listening on eth0
13:01:02.963824 fe80:2e0:7dff:fec7:7450 > fec0:106:1100:0:280:c8ff:fe39:41bb: icmp6:
neighbor sol: who has fec0:106:1100:0:280:c8ff:fe39:41bb
13:01:02.963908 fec0:106:1100:0:280:c8ff:fe39:41bb > fe80:2e0:7dff:fec7:7450: icmp6:
neighbor adv: tgt is fec0:106:1100:0:280:c8ff fe39:41bb
13:01:03.079703 ::10.5.0.30 > fec0:106:1100:0:280:c8ff:fe39:41bb: icmp6: echo request
13:01:03.079800 fec0:106:1100:0:280:c8ff fe39:41bb > ::10.5.0.30: icmp6: echo reply
```

Fig. 8. MN receives ping request from gateway

4   Conclusion

This paper has presented a mechanism for integrating MANET with Mobile IPv6. The IPv6 is based on Linux operating system. Mobile IPv6 test bed with MN, HA and CN functions has been successfully setup and configured in a wired LAN environment. MANET software has been successfully setup and configured in gateway by using Universal Serial Port (USB) Wireless Local Area Network (WLAN) card. However, this Mobile IPv6 test bed is limited to Linux platform because it works under the kernel level of Linux operating system.

NAT-PT software has been successfully setup and configured in the gateway which allows communication between MANET (IPv4) network and mobile IPv6 network.

References

1. M. Ergen and A. Puri, " MEWLANA-Mobile IP enriched wireless local area network architecture", Vehicular Technology Conference, Proceedings, IEEE vol.4, pp. 2449 - 2453, 24-28 Sept.2002
2. Antti J. Tuominen and Henrik Petander, " MIPL Mobile IPv6 for Linux ", Proceedings of Ottawa Linux Symposium 2001, Ottawa, Canada, June 2001
3. Christer Engman,"A study on using IPv6 in home networks", master thesis, Department of Teleinformatics Network Services Royal Institute of Technology Stockholm, Sweden, 1999
4. M. Mauve, J. Widmer and H. Hartebstein, "A Survey on Position-based Routing in Mobile Ad Hoc Network, IEEE Network Magazine, November 2001
5. "Linux- based Userspace NAT-PT", http://www. ipv6.or .kr /english
6. C.E. Perkins and D.B. Johnson, "Mobility support in IPv6." Proc. 2nd Ann., Int'l Conf. Mobile Computing and Networking (Mobicom 96), ACM Press, 1996,pp.27-37
7. Jun-Zhao Sun, "Mobile Ad Hoc Networking: An Essential Technology for Pervasive Computing", International Conferences on Info-tech & Info-net, Beijing, China, 2001
8. Y. Sun, E. M. Belding-Royer, C. E. Perkins, "Internet Connectivity for Ad hoc Mobile Networks" International Journal of Wireless Information Networks, Vol. 9, No. 2, April 2002
9. Yu-Chee Tseng; Chia-Ching Shen; Wen-Tsuen Chen, " Integrating mobile IP with ad hoc networks "; Computer, IEEE Computer Society, Vol. 36, No. 5, pp. 48 –55, May 2003
10. Johnsson, U.; Alriksson, F.; Larsson, T.; Johansson, P.; Maguire, G.Q., Jr.;"MIPMANET-mobile IP for mobile ad hoc networks ", Mobile and Ad Hoc Networking and Computing, MobiHOC. IEEE conference, pp. 75 -85, 11 Aug. 2000
11. C. E. Perkins, R. Wakikawa, A. Nilsson, A. J. Tuominen, "Internet Connectivity for mobile ad hoc Networks" wireless communication and mobile computing, pp. 465-482, 2002
12. M. Frodigh, P. Johansson and P. Larsson," Wireless ad hoc networking-The art of networking without a network", Ericsson Review No. 4, 2000
13. Ahlund, C.; Zaslavsk, A.; " Integration of ad hoc network and IP network capabilities for mobile hosts ", Telecommunications, IEEE conference, Vol. 1, pp. 482 -489, Feb. 23 - Mar. 1, 2003
14. L.Klein-Berndt,"Kernel AODV", National Institute of Standards and Technology, Technology Administration, U.S Department of Commerce. May 2003.
15. J. W. Atwood, Kedar C. Das, and Ibrahim Haddad," NAT-PT: Providing IPv4/IPv6 and IPv6/IPv4 Address Translation", Ericsson IPv6 Activities, a technical report, July 2003

Network Architecture for Scalable Ad Hoc Networks

Jose Costa-Requena, Juan Gutiérrez, Raimo Kantola, Jarrod Creado, and
Nicklas Beijar

Networking Laboratory
Helsinki University of Technology
P.O. Box 3000, 02015 – Finland
{Jose,Raimo,juan,jarrod,nbeijar}@netlab.hut.fi

Abstract. Ad Hoc networking is a technology that is gaining relevance. It is
still an area under development where there are multiple open items such as
routing, addressing, interoperability and service distribution. However, one of
the major problems is the scalability. Most of the protocols designed
specifically for Ad Hoc networks show scalability problems. In this paper, we
propose network architecture for scalable Ad Hoc networks based on node
classification. Moreover, an Ad Hoc framework test bed is implemented for
validating the Ad Hoc nodes taxonomy concept. Preliminary results of the Ad
Hoc framework integrated in PDAs (iPAQ) are presented.

Keywords: Ad Hoc networks, Scalability, Backbone architecture, Routing

1 Introduction

Ad Hoc networks can be rapidly deployed, without preexistence of any fixed
infrastructure. However, scalability is a critical functionality in these networks. The
existing routing protocols in Ad Hoc networks can be either reactive or proactive.
There are many proposals in both categories in order to minimize message overhead
and also to be able to deal with the high dynamics and changing topology of these
networks. The reactive protocols have the advantages that create the route to
destination nodes only when needed. They cache the established routes and remove
them is not needed. In that sense reactive protocols provide a suitable mechanism for
[1]scalable networks. Nonetheless, reactive protocols have the drawback that route
discovery may take excessive time since the route request may traverse the whole
network before the destination node is found and a suitable route is discovered.
Moreover, reactive protocols impose additional processing requirements to all nodes
involved in the route discovery. Proactive protocols are the alternative but it requires
that nodes keep all routing about neighbor links ever if the routes are not used. The
proactive or link state protocols maintain the information about all network topology.
These protocols have additional overhead since they have to periodically exchange
topology information updates with the neighbors. Thus, proactive protocols do not
suffer from the route discovery latency but instead they are less dynamic since the
network changes rapidly and the topology updates may not reflect those changes.

J.N. de Souza et al. (Eds.): ICT 2004, LNCS 3124, pp. 369–380, 2004.
© Springer-Verlag Berlin Heidelberg 2004

There are other alternatives that define hybrid solutions where the nodes have to implement proactive mechanism within a certain range while reactive mechanisms for discovering the routes in a wider range (e.g. ZRP). Other optimization consists of using a proactive mechanism but reducing the updates rates and the topology information depending on the network range (e.g. Fisheye). Alternative solutions keep using proactive protocols but limiting the neighbors involved in the topology updates (e.g. OSLR). However, in all the cases the nodes have to implement the specific mechanism either reactive, proactive or hybrid independently of their capabilities and conditions.

Therefore, we propose to solve the scalability problem in AD Hoc networks using a different perspective. This proposal consists of defining a node taxonomy, which leaves up to each node the decision to identify itself with the contribution or role they want to play in the Ad Hoc network.

The rest of the paper presents the proposed node taxonomy where the nodes are classified in either *smart* or *dummy* nodes. The nodes when joining the network will automatically classify themselves as one of these two types of nodes. This model intends to study the deployment of large networks relying on the capacity provided by these two types of nodes. Section B summarizes the proposed nodes taxonomy and the concept of *smart* and *dummy* nodes. This section includes the algorithm for creating an Ad Hoc backbone formed by *smart* nodes in order to increase the network scalability. Section C describes the test bed implemented in order to support the concept of *dummy* and *smart* nodes. Section D presents the preliminary results of the nodes taxonomy in a real environment using the test bed using real devices (i.e. iPAQ). Finally, section E presents a conclusion and directions for further research.

2 Ad Hoc Node Taxonomy

The actual proposals are either reactive (on demand routing protocols, e.g. AODV [1], DSR [2], TORA [3]), proactive (link state routing protocols, e.g., OLSR, FSR), or combinations of these (e.g. ZRP). There are many proposals in both categories in order to minimize message overhead and also to be able to deal with the high dynamics and changing topology of these networks. The existence of this variety of protocols and solutions proposed for Ad Hoc networks resides in the fact that scalability and reliability is still a critical problem in these networks.

As described in the introduction, the reactive protocols suffer from considerable latency in the route discovery since the route query has to traverse the entire network. Moreover, the reactive protocols create a lot of interferences or noise to the rest of nodes while traversing the entire network during the route discovery.

The proactive proposals maintain a routing table with all the routes to reach all the nodes in the network. They have a quick response time since the routes are already available in the routing table. However, proactive protocol requires a lot of memory for keeping all the route information on their tables. The proactive proposals have to keep the route information up to date, which means that periodically they have to send update messages to all the nodes. This means a constant overhead that is reflected in the network by having constant interferences and noise to all the nodes in the network. Thus, the battery and bandwidth consumption required for keeping the routing tables up to date in the link state routing protocols is a considerable drawback.

Based on the state of the art in Ad Hoc networks and the performance results from existing routing proposals, seems reasonable to have a hybrid solution. This solution will use proactive protocols within a limited area and reactive protocols for a wider area. This approach (e.g. ZRP) will solve the problem of the route discovery latency within the closest area, while the latency remains when the destination nodes reside out of that area. This proposal reduces the size of the routing table since the nodes only keep the route information of all the closest nodes. The noise is also reduced and limited to the neighborhood.

The proposed hybrid solutions consider that all nodes have to implement the proposed hybrid protocol algorithm. Thus, all the nodes have to implement the selected protocol. Thus, we are facing again certain overhead and we are imposing certain requirements to the nodes since all of them have to implement the algorithm defined in the hybrid approach. Thus, when designing a suitable routing proposal we have to keep in mind that Ad Hoc networks require a higher degree of freedom and a fully decentralized architecture. Ad Hoc networks suffer from reliability and scalability problems because the nodes have to act as routers and forward other nodes messages within the network. In addition to this essential packet-forwarding requirement for building Ad Hoc networks, the nodes have to implement the proposed hybrid protocol algorithm. Thus, Ad Hoc nodes may exhaust quickly their resources and became useless.

Therefore, we propose a different perspective for solving the problem. The selected philosophy for this proposal is still based on a hybrid solution that includes both reactive and proactive protocols. The reactive protocols do not require nodes to store link state information, so they are suitable for Ad Hoc networks, where the topology is extremely dynamic. Moreover, the reactive protocols are suitable for nodes with low resources that still can be part and contribute to the packet forwarding required in Ad Hoc network. The proactive reduce the route discovery latency. Therefore, our proposal consists of having a reactive and proactive or hybrid solutions but using them based on the nodes capabilities rather than mandate their usage. We define a node taxonomy [6] that will classify the nodes into *dummy* and *smart* nodes. In the proposed mechanism, according to their capabilities and resources the nodes are classified into *dummy* and *smart* nodes. Moreover, we propose defining a generic mechanism for deploying a network backbone infrastructure for ad hoc networks. Our proposal for extending the lifetime of the Ad Hoc networks is based in a set of core nodes (i.e. *dummy* nodes), which will form a backbone. The backbone constitutes a fully and randomly distributed infrastructure to provide service routing and resource discovery information. This backbone is self-created by the *smart* nodes, thus relieving non-backbone nodes (i.e. *dummy* nodes) from participating on this activity.

Any node within the network can act as *smart* nodes and contribute to maintaining and extending the dimensions of the AD Hoc network. In addition to extending the lifetime of the network, the smart nodes may form a service-provisioning infrastructure, which only requires a registry for storing the service description and the mechanism for discovering that node.

Ad result, the Ad Hoc networks will have a fully decentralized architecture based on the proposed backbone (i.e. group of *smart* nodes). The complexity of routing in Ad hoc networks can be hidden behind the proposed architecture backbone. The node taxonomy differentiates the nodes that are capable of being part of the service backbone, from the nodes that will use it.

3 Ad Hoc Backbone Creation

The proposed model consists of implementing fully distributed backbone architecture for Ad Hoc networks. This approach is based on a node taxonomy that defines two types of nodes [6]; *smart* and *dummy* nodes. The *smart* nodes are considered as nodes with enough resources, that may run multiple protocols simultaneously and they are willing to maintain routing and information about other network resources.

In real networks the AD Hoc backbone can be equivalent to have few *smart* nodes randomly distributed creating the backbone (i.e. Wireless Access Points attached to fixed infrastructure with enough capacity and resources).

3.1 Backbone Creation Algorithm

The "Ad Hoc backbone" is formed with one or many *smart* nodes. In order to implement this model in real networks requires that *smart* nodes implement a new "multiprotocol" architecture. The *dummy* nodes will be devices with limited resources, running single Ad Hoc MANET protocol and s single network interface.

The main advantage of this architecture is that it does not add any requirement to the nodes with limited resources (i.e. *dummy*) that may be running a single IETF MANET protocol (i.e. preferably a reactive protocol such as AODV). The nodes with enough resources (i.e. *smart*) are in charge of creating the backbone and assist the overall routing process in the network. This architecture allows having a huge diversity of nodes with different routing protocols. The *smart* nodes will interact with those native IETF MANET nodes and will cooperate with them in order to extend the network lifetime. The *smart* nodes may become *dummy nodes* at any time along their lifetime because they exhaust their resources.

In order to implement the node taxonomy concept a multiprotocol architecture is required in order to perform a link state routing protocol between the smart nodes and "reactive" or "on demand" routing protocol with the dummy nodes.

3.2 Routing Information Distribution in the Backbone

The smart nodes that implement the backbone maintain the routing information between them using link state protocols (i.e. OSLR). The smart nodes can be organized in clusters. These clusters will use a link state mechanism between them for sharing the routing information. The clusters will use reactive mechanism between them for discovering new routes in different clusters. Thus if any of the smart nodes in the local cluster has received a route request that is not available in its own cluster, the smart node will issue a reactive request to the neighbor clusters asking for the service (i.e. ZRP). If the route is not found a route error response will be returned.

3.3 Routing Information Access

Any node (i.e. *dummy*) can issue a service request. The nodes will not notice the existence of a smart node except in the overall functioning of the Ad Hoc network. The *smart* node will respond to the message query providing the address of the destination node. This route request is broadcasted (TTL=1) to all the neighbors. Any intermediate dummy node that receives a route query and has the address of a smart node on its cache, will unicast the route query to that specific smart node that will attend the query.

Nodes Taxonomy. The proposed distributed backbone requires implementing the node taxonomy composed by *smart* and *dummy* nodes. This section presents the algorithm that each node has to perform.
    "Smart" node:

Network attach procedure is executed every time a *dummy* node becomes a smart node. The smart node sends a broadcast message with TTL=1. In this message the smart node is announcing its capabilities and the available information. If the smart node does not receive any response it means that no other smart node is reachable in the same region either directly or through any local *dummy* node. In case the smart node receives a response from an existing smart node an "Ad Hoc backbone" creation is initiated. The backbone creation consists of checking whether the number of smart nodes reaches an optimum. The optimum consists of a maximum level of smart nodes for avoiding excessive traffic to link state information. If the minimum is not reached the smart node will set a link state relationship among the smart nodes for distributing the routing information within the local region and a clustering infrastructure with other smart nodes for the route discovery in neighbor regions. The responding smart node will indicate to the new node whether it can join the "Ad Hoc backbone" or not depending on the access algorithm (i.e. depending on the number of nodes in the network and the number of nodes that already joined the "Ad Hoc backbone").

Network attach reception occurs when the smart node receives the request from another smart node, which is performing the network attachment. The incoming requests are received either directly from another smart node or indirectly via an intermediate dummy node. The smart node caches the address of the new smart node in the network. The smart node executes the "Ad Hoc backbone" access algorithm and sends back a unicast message to the originating smart node informing whether it can become a smart node and join the backbone or not. If the new node can join the backbone a link state relationship between them is established.

Route request reception occurs when the smart node receives the route request directly from a "dummy" node or from another smart node.
    a) The smart node may receive the route request in a reactive protocol message from a dummy node. The smart node will contain all the routing information about

the local cluster formed by all the smart nodes located in the region nearby. If the smart node does contain the address of the destination node, it will be inserted in the route reply sent back to the dummy node. If the smart node does not contain the address of the requested node, the route request will be forwarded in reactive manner to the neighbor clusters. The smart node times out if a route response from the neighbors cluster is not received within certain period of time. The smart node will respond to the dummy node with route request error or route unavailable message.

b) If the smart node receives the route request from another smart node it means that the neighbor cluster does not contain the destination node address and will issue a reactive route discovery over all the adjacent service clusters. If the contacted smart node contains the requested node, it will be indicated directly via unicast to the dummy node that issued the route request. If the smart node does not contain the address of the destination node, the route request will be forwarded according to the multicast tree formed by the neighbor cluster heads (e.g. ZRP mechanism may be utilized for implementing the service clustering).

"Dummy" node:

Route discovery process is executed when the node needs to find out the address of a destination node. The route discovery process will be implemented by sending a route request as part of the reactive protocol implemented in the dummy nodes (e.g. AODV).

Network attach reception occurs when a dummy nodes receives the broadcast message sent by the smart nodes entering or attaching the network.

a) The dummy receives the message and caches the smart node address. The dummy node will use the cached address of the smart node as gateway address.

b) If the dummy node already has the address of another smart node it will be returned to the smart node that initiated the broadcast.

Route discovery reception occurs when another dummy node is sending a route request broadcast because it did not had the address of any smart node.

a) If the contacted dummy node had the address of any smart node, the route request will be forwarded to that smart node that will respond directly to the dummy node that issued the request with the address of the destination node.

b) If the contacted dummy node does not have the address of any smart node, the route request is broadcasted following the reactive route discovery mechanism.

1) Node self classification:
```
Node_Classification (){
If (start) Then
CheckNodeResournces()
          If (resources>threshold) Then
node = smart
NodeRegistration()
          Else
```

```
node=dummy
          start=False
AttendQueries()
}
```

2) Node enters the AD Hoc network.
```
NodeRegistration(){
If (node==smart) Then
        CreateServiceDescription(message)
        SendInfoBroadcast (TTL=1,message);
        Wait(Timeout)
        If  (response)
          CreateBackbone (response)
        Else
          CreateBackbone(NULL)
}
```

3) Create backbone
```
CreateBackbone(response){
If (response==NULL) Then
        BackboneNeighbours=0
Else if (response)
        SetLinkWithSmartNeighbours()
}
```

4) Attending incoming queries
```
AttendQueries() {
If (smart) Then
        If (query==nodeRegistration) Then
          If (SmartNodes<Optimum)
            SendResponse(ACKresponse)
          Else
              SendResponse(REJECTresponse)
        If (query==routeDiscovery) Then
          If (routeFoundInCache)
              SendResponse (RouteReply)
          Else
              ForwardRequest(RouteRequest,
ToNeighbourCluster)

If (dummy) Then
        If (nodeRegistration) Then
          If (SmartNodeAddressInCache) Then
              SendResponse(SmartNodeAddress)
        If (routeDiscovery) Then
          If (SmartNodeAddressInCache) Then
              ForwardRequest(RouteRequest,
ToSmartNode)
```

```
            Else
                ForwardRequest(RouteRequest,
    Broadcast)
    }
```

4 Ad Hoc Framework as Test Bed for Nodes Taxonomy

Because of the extreme conditions where the Ad Hoc networks should run, it is
envisioned that the specific conditions (i.e. network size, nodes mobility and density,
etc) will determine the suitable routing protocol to be executed. Thus it is rather
difficult to define an optimal protocol that suits for these continuously changing
conditions. In small networks suits either reactive or proactive routing protocols while
in medium to large-scale networks a Hybrid protocol between proactive and reactive
protocols should be used. Therefore, in order to validate the proposed node taxonomy
a test bed has been implemented. This test bed consists of an Ad Hoc framework that
is implemented and integrated in real nodes (i.e. Personal Digital Assistants, PDA).
Fig. 1 shows the main modules implemented in the AD Hoc framework.

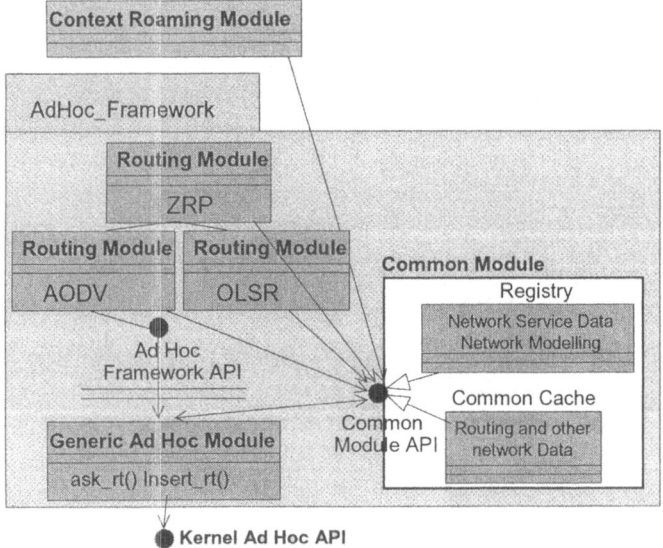

Fig. 1. AD Hoc framework modules

The AD Hoc framework consists of the software package that will be implemented
in the smart nodes. This software package has been designed for supporting multiple
routing protocols running simultaneously. Thus, the smart nodes will use link state
protocol between them while reactive protocol for receiving the route request coming
from dummy nodes, which will only implement a single reactive protocol. Thus, the
Ad Hoc framework is designed as multiprotocol platform.

The Ad Hoc framework we differentiate the component that implements the specific routing protocols (i.e. Protocols Module) and the common module that implements a routing cache. The common cache is the databases for storing all routing or service information provided by the multiple routing protocols running simultaneously in the node.

The Protocols Module is formed by routing protocols such as AODV, OLSR or ZRP protocols, though more protocols can be added. In the actual implementation AODV and OLSR have been used although other ones can be used instead of those.

The Common Module is used to keep common information in the Ad Hoc Network in order to be shared by different protocols. This information is basically routing entries (with their routing information), time stamps, protocols running, collaboration between protocols, etc. Inside this module there are two important parts. The Registry which is an information file where protocol parameters are stored and the Common Cache which is formed by the Common Cache Register Server that is a message center which is listening for register protocols, messages to cache and between protocols and the Cache which is a common routing table where protocols share routes. In this way a node running one protocol can reach other nodes running a different protocol.

The Generic Ad Hoc module is also part of the Ad Hoc framework and provides the interface for accessing the Linux kernel. This module has functions for adding a new route discovered, deleting an old route, updating a route or asking if a route exists. All these operations work with Kernel Routing Table and the route information is updated there.

Fig. 1 shows the **Context Roaming module**, which is not included in the AD Hoc framework but is an important component. This module is responsible of deciding whether the node should become smart or dummy. Thus, the Context Roaming module has to request information from the routing and MAC layer using the API provided by the Common Module. Moreover, this module also request to the Common Module information about the active interfaces and their signal strengths in case the node has to roam into a different radio interface (i.e. from WLAN into Bluetooth or WCDMA). At the moment of performing the test this module is not implemented yet so we base all the tests using a single WLAN interface.

5 Test Environment

The concept of nodes taxonomy is implemented in practical terms as having nodes with multiprotocol architecture (i.e. *smart* nodes) and other nodes having a single protocol; e.g. AODV (i.e. *dummy* nodes). Therefore, the smart nodes implementing a "multiprotocol" architecture exploit the benefits of different routing algorithms running simultaneously in the node and will assist the dummy nodes in the routing process. Moreover, the *smart* nodes should have bigger capacity, memory, resources and battery. Thus, in practical terms the *dummy* nodes will consists of PDA (i.e. iPAQ [7]) while the *smart* nodes will consist of are laptops.

Fig. 2 shows the testing environment composed by 1 *smart* node and 4 *dummy* nodes. The test was done in the Electrical and Communication Building in HUT. The building has a lot of metal and iron doors and other wireless networks, which infers a real environment for the tests. The results are affected by interferences and they had certain changes. Fig. 3 and Fig. 4 represent the average of results in multiple tests. The test was performed using a single smart node (i.e. the central node 1) and four dummy nodes (i.e. the rest of nodes 2,3,4 and 5).

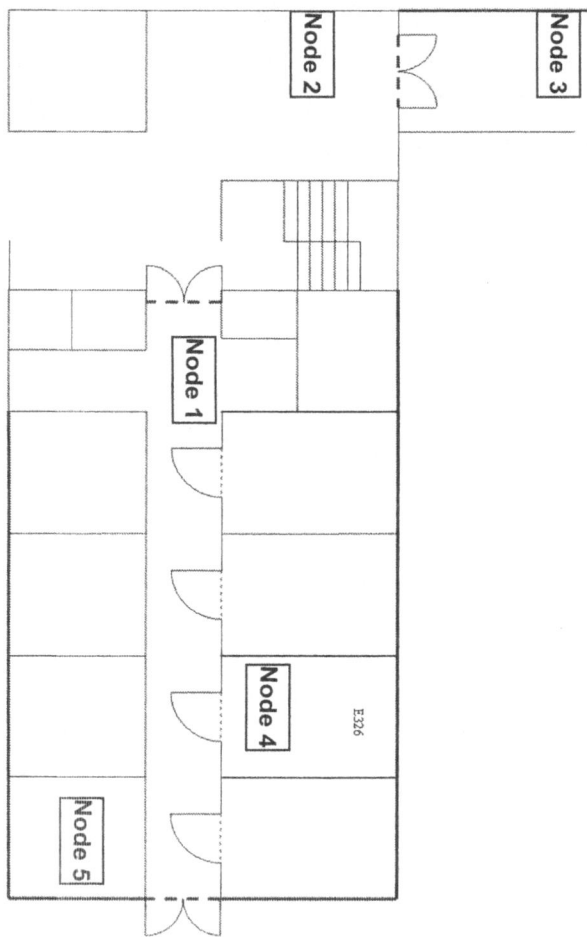

Fig. 2. Test bed environment

The central node was running AODV+OLSR protocol (i.e. smart node is node 1) in a laptop. Two of the other nodes were running AODV (i.e. dummy node running reactive protocol; nodes 2 and 3) and the other two nodes were running OLSR (i.e. dummy node running a proactive protocol; nodes 4 and 5). All nodes were dynamic, that means they were constantly moving and some links were broken during the test.

For the test, ICMP packets with 64 bytes long were used. The test results depicted in following figures represent are the average after running multiple repetitions.

The results of the test shows that including a *smart* node in the AD Hoc network provides remarkable results while the rest of nodes are *dummies* and has lower resources and gain for contributing to the overall network lifetime.

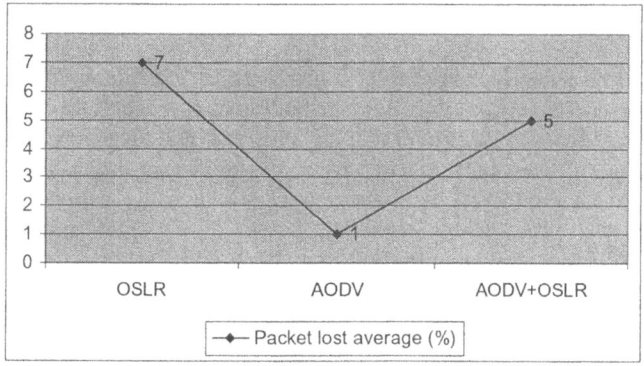

Fig. 3. Packet losses in test bed with Ad Hoc framework.

Fig. 3 shows that packet loss in the test with all the nodes running only OSLR is highest versus the test where all the nodes are running AODV. In high mobility conditions there are bigger packet loss in OSLR because of the latency during the link state update so if the link is broken the packet will be lost until the link information is updated in the routing table. After introducing a single *smart* node the results are much favorable.

Fig. 4 shows that OSLR has lowest average round trip versus AODV because the proactive protocols have available all the routes in their cache at the moment of initiating the packet transmission. Instead the reactive protocols (e.g. AODV) have to discover the route before they are able to initiate the data transmission. The results show that by adding a single smart node in the network the route latency in all the network reach the value equivalent to having all the nodes running proactive protocols even part of the network are running a reactive protocol.

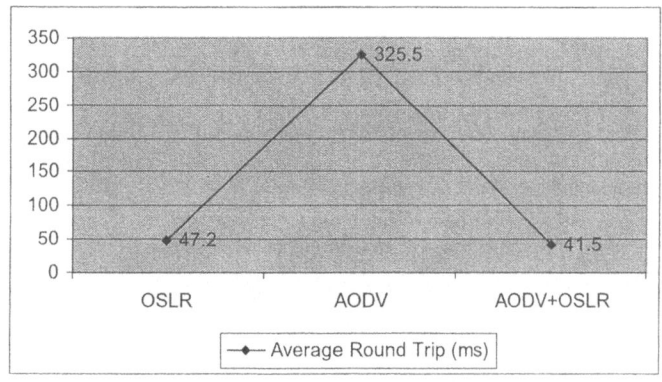

Fig. 4. Round trip test in Ad Hoc framework.

6 Conclusions

The scalability and reliability are two characteristics that are difficult to achieve simultaneously in Ad Hoc networks. This paper proposes a method that does not invent a new routing protocol but instead moves the problem to a different plane. We propose to solve the problem by relying on the node incentive to contribute to the Ad Hoc network.

This method proposes the creation of a non-reliable "Ad Hoc backbone" based on best-effort approach depending on the contribution from many nodes as possible. We classify the nodes in order to differentiate the ones that will contribute to the "Ad Hoc backbone" from the ones that benefit from it. We prove our proposal by real testing using nodes that have extra resources and enhanced capabilities for contributing to the Ad Hoc network. The results show that having those smart nodes the overall performance of the network increases considerably.

Directions of future research include the analysis of the right number of smart nodes required for obtaining the optimal network performance without requiring excessive amount of smart nodes in the network. Moreover, game theory can be applied for analyzing the node incentive to become part of the Ad Hoc backbone and contribute to the network performance. The next step would be to locate the Nash equilibrium for this behavior.

References

[1] C. E. Perkins and E.M. Royer, "Ad-hoc On Demand Distance Vector Routing", Second IEEE Workshop on Mobile Computing Systems and Applications, pp. 90-100, February 1999.
[2] D.b. Johnson and D. A. Maltz, "Dynamic Source Routing in Ad Hoc Wireless Networks", in Mobile Computing, edited by T. Imielinski and H. Korth, chapter 5, pp. 153-181, Kluwer Academic Publishers, 1996.
[3] A. Iwata, C.-C. Chiang, G. Pei, M. gerla and T.-W. Chen, "Scalable Routing Strategies for Ad Hoc Wireless Networks", IEEE Journal on Selected Areas in Communications, Special Issue on Wireless Ad Hoc Networks, vol. 17, No 8, pp. 1369-1379, August 1999.
[4] M.R. Pearlman and Z.J. Haas, "Determining the Optimal Configuration for the Zone Routing Protocol", IEEE Journal on Selected Areas in Communications, Special Issue on Wireless Ad Hoc Networks, vol 17, No 8, pp. 1395-1414, August 1999.
[5] Mingliang Jiang, Jinyang Li and Y.C. Tay," Cluster Based Routing Protocol", IETF Internet Draft (work in progress), 199
(http://community.roxen.com/developers/idocs/drafts/draft-ietf-manet-cbrp-spec-01.txt).
[6] J. Costa-Requena, Raimo Kantola and Nickla Beijar, "Replication of Routing Table for mobility management in Ad Hoc networks", Journal WINET.
[7] Compaq iPAQ Pocket PC H3900 Series
http://www.pocketpccentral.net/pdfs/ipaq3900specs.pdf

Dual Code-Sensing Spread Spectrum Random Access (DCSSRA) Protocol for Ad Hoc Networks*

Chun-Jiang Wang, Wei Quan, Zhen-Hua Deng, Yuan-An Liu, Wen-Bo Wang, and Jin-Chun Gao

Wireless Research Center, Beijing University of Posts and Telecommunications, Beijing 100876, China
`chunj_cn@sohu.com, {deng, yuliu, gjc@bupt.edu.cn}`

Abstract. The DCSSRA protocol is proposed by using receiver-oriented spread spectrum technology and the multiple mini-slots with dual code sensing technology as the controlling method of packet collisions. The Markovian model is proposed for analysis of average Throughput, Delay, Probability of failure to deliver and stability. The numerical results confirm that the high performance and stability of networks is achieved by the new protocol even under a high traffic load condition.

1 Introduction

The focus of our research is generally on Spread Spectrum (SS) Ad Hoc networks, which includes many complex problems. Particularly, this paper focuses on the multiple access protocol. The SS Ad Hoc networks can benefits from the deep research of multiple-user detection technology. E.g., Iltis [5-6] proposed a method of multi-user detection for quasi synchronous CDMA signals using linear decorrelators; more importantly, he [7] proposed a model of QSPNET networks, wherein the GPS-based synchronization, multi-user detection were employed and tell-and-go protocol (a reservation-based connection-oriented communication protocol), the networks achieved a rational performance without power control. From Iltis 's experience, we should separate the research of multiple access protocol from the research of whole networks and assumes the noiseless channel (MAI is ignored), which is credible and helps to the simplification of the problem. If we ignore the error resulted from synchronization or multiple user interference, the packet collision is assumed as only resource of failing transmission, which includes the premier collision resulted from simultaneous arrival of multiple packets (the collided transmission will be reject) and secondary collision resulted from the arrival packets in the transmitting receiver (receiver work in half-duplex).

Many researchers concentrate on the similar area of this research. For distributed SS networks, Sousa and Silvester [1] proposed two Spreading Code Protocols (C-T

---

* The research is supported by the NTT DoCoMo research projects

J.N. de Souza et al. (Eds.): ICT 2004, LNCS 3124, pp. 381–387, 2004.
© Springer-Verlag Berlin Heidelberg 2004

and R-T protocols) that really are multiple access protocols; Chen [2] proposed a series of multiple access protocols with different collision resolution methods; Joa-Ng [3] proposed the MACA/C-T and MACA/R-T protocols combining the Sousa's C-T and R-T protocols and the MACA (Multiple Access with Collision Avoidance) protocol; Bao and Tong [4] proposed a random access protocol using the transmitter-based SS scheme. However, note that most of protocols mentioned above focus on resolution of premier collision, and ignore the secondary collision. In fact, the rejected want-to-transmitting nodes may transit their state from transmitting to receiving, and receive other packets desired to them successfully. It is the main motivation that we propose the DCSSRA protocol. Although, the DCSSRA protocol doesn't resolve the entire problem, its improvement of performance is exciting.

2 DCSSRA Protocol

Firstly, we suppose that the time is synchronized and divided into slots which length is long enough to transmit a data packet and several control packets. As shown in fig.1, the slot time is divided into a data slot and control slot, and control slot further is divided several mini-slots. The data packet is transmitted in data slot and the control packet in one of mini-slots.

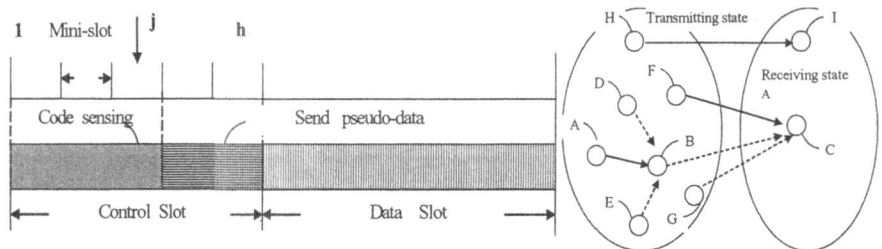

Fig. 1. Structure of Slot Time

Fig. 2. An example of Ad Hoc networks

Secondly, we suppose that each node is equipped with one transceiver and one spreading code, and knows all of spreading code allocated to other nodes. Each source node transmits packets encoded with the receiver's spreading code, which is a receiver-oriented CDMA system. Before the transmission of data packet, source node sends a PSEUDO-DATA encoded by spreading code of destination from the mini-slot selected randomly to the end of control slot. In the other time of control slot, source node keeps sensing the code channel of itself if it hasn't data to send, otherwise, it keeps sensing its spreading code and the spreading code of destination selected by itself (so called dual code-sensing).For understanding the DCSSRA protocol, detail description is taken in an example as shown in Fig 2.

Suppose that nodes C and I are in receiving state, and other nodes are in transmitting state. We take an example with F it wants to send a packet to C. At the same time, B and G also want to send to C. each node independently select a mini-slot as transmitting time of PSEUDO-DATA. If the sequence number of mini-slot of F is less than that of C, B and G, C will receives the PSEUDO-DATA sent by F. At the same time, B and G will sense that code channel of C has been occupied, and release their request of transmission. After receiving of PSEUDO-DATA, C sends an ACK packet to F. Therefore, F can send packet to C without collision in the followed DATA slot. The feedback ACK of the request is sent in end of control slot, which will help to avoid the hidden terminal problem.

Continuing our example, if B fails to contend for the channel (sensing the PSEUDO-DATA sent by F), it transits its state to receiving. At the same time, other nodes A, D and E want to send packets to B. Supposed that the node B accepts the request of node A, which efficiently controls the secondary collision.

Note that each node that wants to transmit (e.g. B) senses dual spreading code (including that of itself and destination). Suppose that before the mini-slot selected by B arrives, and A sends PSEUDO-DATA to A, B will receives the PSEUDO-DATA and accepts the request of A, finally, rejects request of itself.

3   Performance Analysis

We assume the network is in equilibrium and each of the nodes has the same behavior and characteristics. The individual node can represent the characteristics of the whole networks. Based on the theory, we employ two Markovian models. One is based on the number of busy nodes in the current slot, which is defined only by the state in the last slot and is a Markovian chain. The other is based on the M/M/1/k queuing model. The two models are coupled with the idle probability of the queue. The idle probability is defined by the behavior of all of the other nodes and represents the characteristics of the whole networks.

In the first Markovian model, the number of the busy nodes is defined as the state variable, denoted by $N(t)$ in the slot t. For a M-node network, the transition matrix $P=[p_{nk}]$ characterizes the Markov, $P_{nk}=\{N(t+1)=k| N(t)=n\}$, P_{nk} is the probability that the network state goes from n in the slot t to k in the next slot $t+1$. If we assume that the busy probability of each node is independent, and the P_{nk} is given by

$$P_{nk} = \binom{M}{k} \cdot P_a^k(n) \cdot \left(1 - P_a(n)\right)^{M-k}, \text{where } P_a(n) = \left(1 - P_0'(n)\right)\left[P_s(n) + P_r \cdot \left(1 - P_s(n)\right)\right] \qquad (1)$$

P_a is the busy probability of individual node in the next slot, $P_0'(n)$ is the idle probability of the node, $P_s(n)$ is the probability of successful transmission, and P_r is the probability of retransmission. Since the retransmitted packet is always buffered in the head of the queue, the successful transmission indicates the next packet is a new packet. Therefore, if the queue is not idle and the first packet is a retransmitted packet in the queue, the node is busy with probability P_r in the next slot; if the packet is a new packet; the node is busy with probability 1.

To get $P_0'(n)$, we introduce the second Markovian model ,the M/M/1/k queuing model with finite buffer size. The input traffic flow per node is defined as Poisson process with parameter λ and is equilibrium distribution in the networks. The mean service rate is Poisson process with parameter μ.

3.1 Probability of Successful Transmission

Suppose that there are n nodes in the transmitting state in the beginning of a slot. P_s (probability of successful transmission) is given by.

$$P_s(n) = \frac{1}{h} \cdot \sum_{j=1}^{h} P_s(n,j) \tag{2}$$

Where h is the number of mini-slots, $P_s(n,j)$ is the average success probability of transmission when the selected sequence number of mini-slot is j, which is given by following set of linear equations.

$$P_s(n,j) = P_1(n,j) \cdot P_2(n,j) \cdot \left[\frac{M-n}{M-1} + \left[\frac{h-j}{h} + \frac{1}{h} \cdot \sum_{i=1}^{j-1}(1-P_s(n,i)) \right] \cdot \frac{n-1}{M-1} \right] \tag{3}$$

where

$$P_1(n,j) = \sum_{k=0}^{n-1} Q(n-1,k,1/(M-1)) \cdot ((h-j)/h)^k , \ P_2(n,j) = \sum_{k=0}^{n-1} Q(n-1,k,1/(M-1)) \cdot ((h-j+1)/h)^k$$

$$Q(x,y,p) = \binom{x}{y} \cdot p^y \cdot (1-p)^{x-y}$$

Where h is the number of mini-slots, M is the number of nodes; we define the $0^0 = 1$.

3.2 Idle Probability and Block Probability

In the M/M/1/k model, the μ is defined as the average number of packets transmitted successfully during one slot and presented as $1/N_d$, where N_d is the average time taken to transmit a packet successfully. Considering the retransmission probability P_r, and the maximum allowed number of retransmission R_{max}, N_d is given by.

$$N_d = (1-P_s) \cdot \left(\sum_{R=1}^{R_{max}} (1-P_r \cdot P_s)^{R-1} \cdot P_r \cdot P_s \cdot (R+1) + (1-P_r \cdot P_s)^{R_{max}} \cdot (R_{max}+1) \right) + P_s \tag{4}$$

If λ and μ are known, according to the classical queuing theory, the idle probability and block probability are obtained as

$$P_0' = \frac{1-\lambda/\mu}{1-(\lambda/\mu)^{K+1}} , \ P_{block}' = \frac{1-\lambda/\mu}{1-(\lambda/\mu)^{K+1}} \cdot \left(\frac{\lambda}{\mu} \right)^K \tag{5}$$

Substituting (2) (5) into (1), we can get the transition matrix P, and finally get the stationary distribution $\pi = P \cdot \pi$ of the Markov chain.

3.3 Throughput, Delay, and Probability of Failure to Deliver

The throughput (Packets/Slot) is defined as the average number of packets transmitted successfully in a slot and the $P_s(n)$ of the nodes are independent. So, the throughput is given by $\beta(n)=n \cdot P_s(n)$, The average throughput and average delay are respectively

$$\overline{\beta}=\sum_{n=1}^{M}\beta(n)\cdot\pi(n), \overline{D}=\sum_{n=1}^{M}n\cdot\pi(n)\Big/\overline{\beta} \tag{6}$$

The maximum allowed number of retransmission R_{max} and block probability P_{block} define the probability of failure to deliver P_{fd} that is given by

$$P_{fd}(n)=P_{block}+(1-P_{block})\cdot P_{drop} \text{ ,where } P_{drop}=(1-P_s)\cdot(1-P_r\cdot P_s)^{R_{max}} \tag{7}$$

P_{drop} is the probability of drop because of R_{max}. The average probability of failure to deliver is

$$\overline{P}_{fd}=\sum_{n=1}^{M}P_{fd}(n)\cdot\pi(n) \tag{8}$$

4 Numerical Results

Suppose that the number of nodes is 20, the channel is noiseless and the collision is the only resource of error transmission. We show the results in Fig 3, wherein the line called "ideal Ad Hoc network" is for the maximum throughput for Ad Hoc networks. The line called "ideal cellular networks" is for the maximum throughput for cellular CDMA networks, where the time is divided into uplink and downlink (TDD) slots. The line called "J. Q. Bao Tx-based" is for the protocol of Bao in [4], it is assumed that the receiver can receive all of the packets desired to it, the maximum throughput is given by

$$\beta_{Bao}(n)\leq n\cdot\frac{(M-n)}{M-1} \tag{9}$$

In Fig 3, DCSSRA protocol is better than the Bao's protocol when the traffic load (number of transmitting nodes) is heavy, and is better than "idle cellular networks" when traffic load is not heavy. It confirms that DCSSRA protocol can efficiently control both of collision. Besides, the throughput increases as the number of mini-slots increases. Note that throughput of DCSSRA protocol is very good even if the number of mini-slots is very small (such as 3-5).

For "ideal cellular networks", the delay always is 2 (slots/Packet), the probability of failure to deliver always is 0. As shown in Fig 4-5, distributed SS networks used DCSSRA protocol as multiple access protocol achieve lower delay than "ideal cellular networks", at the same time, the DCSSRA protocol keeps low probability of failure to deliver when the traffic load is not very heavy. Besides, performance of DCSSRA protocol is equal to that of R-T proposed by Sousa in [1] when the mini-slots number *h* is 1. When *h* increases, the DCSSRA protocol greatly outperforms both R-T protocol and Bao's protocol.

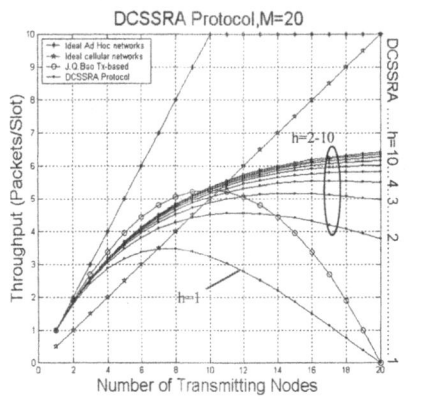

Fig. 3. Comparison with the Various Protocols, the throughput vs number of transmitting nodes, for *M*=20

Fig. 4. the average delay vs average throughput for h from 1 to 10, *M*=20,Qmax=10, Rmax=10 and P_r=0.6

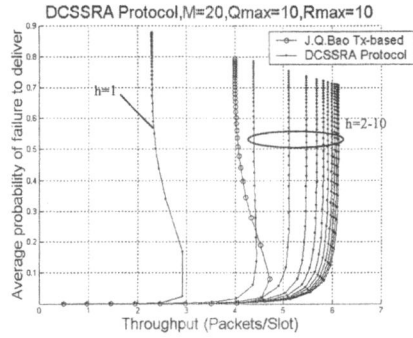

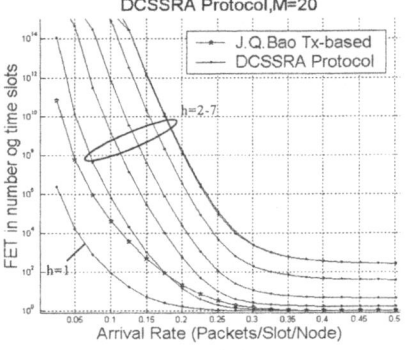

Fig. 5. the average probability of failure to deliver vs average throughput for h from 1 to 10, *M*=20,Qmax=10, Rmax=10 and P_r=0.6

Fig. 6. FET vs average throughput for h from 1 to 10, *M*=20,Qmax=10, Rmax=10 and P_r=0.6

5   Stability Analysis

The stability analysis of DCSSRA protocol follows the FET method in [4]. Given a state threshold n_c, FET T_i is the average time the network state first exceeds n_c assuming at time zero the network is in state i. The T_i is given by following linear equations:

$$T_i = 1 + \sum_{j=0}^{n_c} P_{i,j} \cdot T_j, \qquad i = 0,1,\ldots n_c \tag{10}$$

where, the network state threshold n_c is chosen as the state corresponding to the largest throughput. The results are shown in Fig 6.

The stability of DCSSRA protocol is better than Bao's protocol when mini-slots number h is more than 2. When h is more than 6, the stability of DCSSRA protocol will not increase observably as the increasing of h. Above all, our proposed protocol is very stable and high performance.

6 Conclusion

The DCSSRA protocol creatively utilizes the terminals that fail to transmit as receivers of other transmitting terminals, which dramatically improves the throughput of networks in a high traffic load condition.

References

1. Sousa, E., Silvester, J.: Spreading code protocols for distributed spread-spectrum packet radio networks. IEEE Trans. Commun, Vol. 33, NO.3, pp. 272-281, March 1988
2. Chen, X.H., Oksman, J.: Destructive collision-free protocol for distributed DS/CDMA wireless networks using code-sensing and chip-rate-division techniques. IEE Proceedings Commun., Vol. 143, No. 1, pp. 47-55, Feb. 1996
3. Joa-Ng, M., I-Tai, Lu: Spread spectrum medium access protocol with collision avoidance in mobile ad-hoc wireless network. IEEE INFOCOM '99, Vol. 2, pp. 776-783, Mar. 1999
4. Bao, J.Q., Tong, L.: A Performance Comparison Between Ad Hoc and Centrally Controlled CDMA Wireless LANs. IEEE Trans. Commun., VOL. 1, NO. 4, pp. 829-841, Oct. 2002
5. Iltis, R.A.: Demodulation and code acquisition using decorrelator detectors for QS-CDMA. IEEE Trans. Commun., vol. 44, pp. 1553-1560, Nov. 1996
6. Iltis, R.A., Mailaender, L.: Multiuser detection of quasi-synchronous CDMA signals using linear decorrelators. IEEE Trans. Commun., vol. 44, pp. 1561-1571, Nov. 1996
7. Banerjee, A., Iltis, R.A., Varvarigos, E.A.: Performance evaluation for a quasi-synchronous packet radio network (QSPNET). IEEE/ACM TRANS. ON NETWORKING, VOL. 9, NO. 5, pp. 567-577, OCT. 2001
8. Gross, J., Yellen, J.: Graph theory and its applications, Boca Raton, Fla.: CRC Press, 1999

Exploiting the Small-World Effect to Increase Connectivity in Wireless Ad Hoc Networks

Dave Cavalcanti[1], Dharma Agrawal[1], Judith Kelner[2], and Djamel Sadok[2]

[1] ORB Center for Distributed and Mobile Computing, University of Cincinnati,
Cincinnati, Ohio, USA
{cavalcat, dpa}@ececs.uc.edu
http://www.ececs.uc.edu/~cdmc
[2] Centro de Informatica, Universidade Federal de Pernambuco,
Recife, Brazil
{jk, jamel}@cin.ufpe.br

Abstract. This paper investigates how the *small world* concept can be applied in the context of wireless ad hoc networks. Different from wireless ad hoc networks, small world networks have small characteristic path lengths and are highly clustered. This path length reduction is caused by long-range edges between randomly selected nodes. However, in a wireless ad hoc network there are no such long-range connections. Then, we propose to use a fraction of nodes in the network equipped with two radios with different transmission ranges in order to introduce the long-range shortcuts. We analyze the system from a percolation perspective and show that a small fraction of these "special nodes" can improve connectivity in a significant way. We also study the effects of the special nodes on the process of information diffusion and on network robustness.

1 Introduction

Wireless ad hoc and sensor networks have been extensively studied in the recent years. The efficiency of multihop routing used in these networks depends on the network connectivity, i.e., the existence of a direct or multiple hops path between any two nodes in the network. However, connectivity depends on factors such as, interference, noise and energy constraints and cannot be always assured [5]. Connectivity also affects the network capacity, and as shown in [6], a fundamental condition for scalability is to keep the average distance between any source/destination pair small as the system grows.

Recent results by Watts and Strogatz [8] have shown the existence *Small World* networks that have a small degree of separation between nodes while maintaining highly clustered neighborhoods. The small degree of separation in such networks is obtained through the introduction of some long-range edges that result in faster information propagation. The small word effect has been widely studied in the context of relation graphs, such as, the graph formed by hyperlinks in Web. However, these

J.N. de Souza et al. (Eds.): ICT 2004, LNCS 3124, pp. 388–393, 2004.
© Springer-Verlag Berlin Heidelberg 2004

long-range edges are not possible in wireless ad hoc networks due to the limited the transmission range of the nodes.

In this paper, we propose to introduce some long-range edges in an ad hoc network by using *special nodes* equipped with two radios, one with a short transmission range and another with a longer transmission range. In fact, nodes with multiple interfaces are expected to be common in a near future. For instance, wireless devices can be equipped with cellular and 802.11 interfaces. As we will show, the introduction of the special nodes can improve network connectivity, reduce the route length and consequently, enhance the network capacity and scalability.

The remainder of this work is organized as follows: In Section 2, we present the basic concepts and some related works. Next, in Section 3, we evaluate the effect of the special nodes with multiple radios in the network connectivity, average hop distance between nodes and network robustness. Finally, we present some concluding remarks and future work in Section 4.

2 Background and Related Work

Recently, Watts and Strogatz [6] have studied a class of networks that present a small characteristic path length (L), and are highly clustered, known as *Small World* networks (or small world graphs). Given the network graph G, the characteristic path length can be calculated as the mean number of hops between any two nodes in G. The clustering coefficient C_u of a given node u in G, is the relation between number of edges connecting neighbors of u in G, and the total number of possible edges between the neighbors of u. The overall clustering coefficient C is the average over all C_u for u in G.

The reduction in L is important in various aspects, including the network capacity and the way information propagates through the network. The small world effect has been observed in real networks such as the WWW and overlay peer-to-peer networks. However, these networks form relational graphs in which there is no distance constraint in the edges lengths. Thus, they are not appropriate to model wireless networks, in which the connections are limited by physical distance.

The fixed radius graph $G=G(N, r)$ is a spatial model widely used in studies of wireless ad hoc networks. In this case, given N points placed randomly according to some distribution on the Euclidian plane, G is constructed connecting the nodes whose corresponding points are within a distance r of each other. Clearly, the constraint on the connection range r precludes the existence of long-range shortcuts in G. Nevertheless, Watts suggested in [7] that, if a graph is constructed following the spatial graph concept but using a probability distribution to connect nodes with slow-decay and infinite variance, one might expect that this distribution would generate small world spatial graphs.

In terms of ad hoc and sensor networks, the shortcuts could be interpreted as the global edges serving as the links between the clusters heads which could gather information about cluster members using local connections. However, some practical questions have to be considered: How far a shortcut should be? How many shortcuts are needed to keep the network connected? An attempt is made to answer these questions in the remainder of this paper.

In [3], Helmy has the suggested to use long-range shortcuts to reduce the number of queries during the search for a given target node in a large-scale ad hoc network. In this case, a shortcut is defined as a logical link that a node maintains with a random selected node. Therefore, a logical link between a pair of nodes can eventually correspond to several physical hops, and this logical link is implemented by making each node in the path to keep a route to the long-range contact.

In [2], the authors have used percolation theory to analyze the introduction of fixed base stations to increase connectivity in large-scale ad hoc networks. The main idea is that base stations would allow distant nodes to communicate through a fixed, wired infrastructure. The authors have assumed the base stations transmission ranges are the same as the ones of the wireless nodes. The connectivity level is measured as the fraction of nodes connected to a giant cluster, which is defined as percolation probability. Although analytical modeling and simulations [2] have shown that, in the one-dimensional case, the fixed infrastructure does improve the network connectivity, in the 2-dimensional case, the base stations do not enhance connectivity significantly.

The idea of using base stations is also proposed in [1], where the authors proved that if base stations can communicate at a distance larger than twice the maximum communication distance to the users, a giant connected cluster forms almost surely for large values of the density of users, regardless of the covering algorithm used to place the base stations in the covered region.

3 Long-Range Shortcuts in Ad Hoc Networks

Transforming the network into a small world by simple increasing the transmission range of the nodes is an alternative to reduce the average separation between nodes without imposing restrictions to network traffic as suggested in [4]. However, this approach has practical restrictions, such as power consumption, and interference, and the impact of these constraints on the network performance need to be well understood.

Helmy [3] has also pointed out the possibility of having all nodes equipped with several radios, one for short-range communications and another with a longer transmission range, in order to introduce long-range shortcuts in the network. Indeed, a more practical approach is to have a limited number of *special nodes* equipped with two different radios. A typical short-range radio (r) and a longer range one for introducing the long-range shortcuts (r_f).

We consider that the longer-range radios operate in a different spectrum from the typical short-range radios. Then, two nodes i and j are directly connected in the network graph if they satisfy one of the following conditions: the distance d_{ij} between i and j is smaller than their typical communication range r; or, if i and j are special nodes and $d_{ij} < r_f$

In the remainder of this section we analyze the effect of the introduction of the special nodes on the network connectivity, on the broadcast efficiency and on the network robustness. We measure the connectivity level as the percolation probability, as in [2]. We consider a network in which 1000 nodes are uniformly distributed over a 1kmx1km square area and all results plotted are averaged over 100 runs.

Initially, we considered a disconnected scenario where $r = 35$ m. As shown in Figure 1(a), the percolation probability increases with f for different values of r_f. For $f <$ 0,01, the network remains disconnected regardless of the r_f used. On the other hand, for $f > 0,2$ all values of r_f result in an almost connected network. From this point on, there is not much connectivity gain by adding more special nodes. This is an important result, as it suggests that there is an upper bound on the number of special nodes required to transform a disconnected topology in an almost connected network.

Figure 1(b) shows the percolation probability as a function of the r_f/r relation. We can observe that for all values of f simulated there is a critical value of r_f after which the connectivity has a drastic increase and that, further increase in r_f has no effect on connectivity. This confirms that there is a maximal range r_f after which there is no reduction on L. The reduction in L can be seen in Figure 1(c), for two values of r_f ($10r$ and $3r$). Although we do not show, no significant change was observed in the clustering coefficient. In this case, we have assumed $r=60$, to make sure that the network was initially connected.

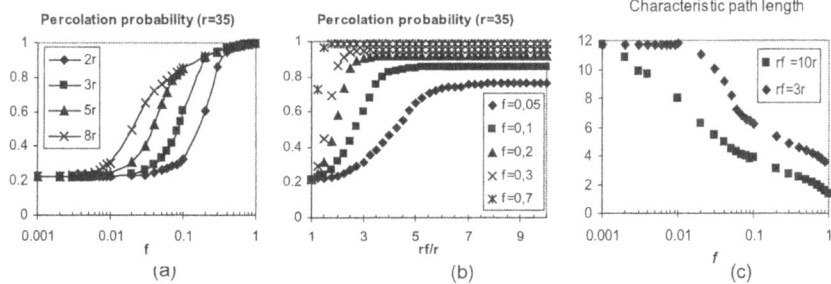

Fig. 1. *(a)* Percolation probability as a function of the fraction of nodes with two radios, *(b)* Percolation probability as a function of r_f, *(c)* Characteristic path length as a function of f

We have also studied the effects of the special nodes on the propagation of a broadcast packet in the network. We randomly select a node to start a broadcast and at each time step, each node that received the packet forwards it to its neighbors with a transmission probability p_{tr}. Then, we computed the fraction of nodes at the end of the simulation that did not receive the message $S(t_{max})$. We call $S(t_{max})$ as the susceptible nodes as an analogy to an epidemic model in which the susceptible nodes represent the nodes that were not infected by the information at the end of the simulation.

The diamonds in Figure 2(a) represent $S(t_{max})$ for the fixed radius model (spatial graph) with $r=60$m. Note that in general, the models with special nodes ($f=0,1$) result in more efficient information diffusion than the simple spatial model for $0.1<p_{tr}<0.3$.

As shown in Figure 2(b), for p_{tr}=0,2 and different values of r_f, $S(t_{max})$ decreases as f increases.

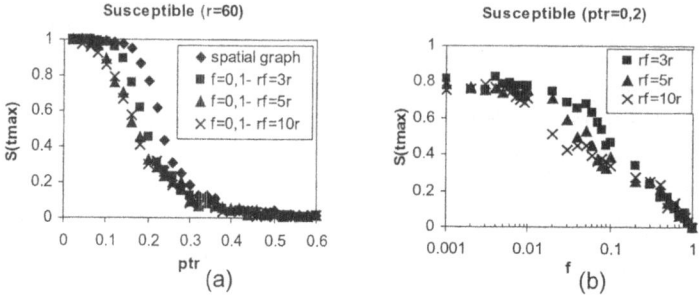

Fig. 2. *(a)* $S(t_{max})$ a function of the transmission probability, *(b)* $S(t_{max})$ as a function of f

Finally, we performed simulations in which at each time step we switch off a node and calculate the percolation probability in the resultant network. In Figure 3(a), we show the effect of r_f on the percolation probability for f=0,1 and f=0. In this case, we can clearly observe a more significant connectivity gain compared to the spatial model (f=0) as r_f increases from $3r$ to $5r$. On the other hand, changing r_f from $5r$ to $10r$ seems to produce no effect on network robustness. This result indicates that there is also a value of r_f after which, increasing r_f there is no effect on network robustness.

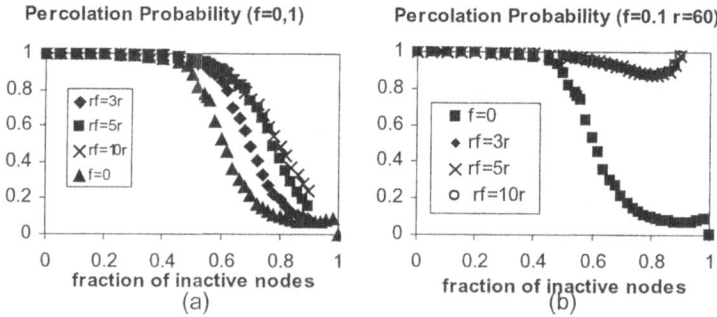

Fig. 3. *(a)* Percolation probability for different r_f, *(b)* Percolation probability without switching off the special nodes

It is important to note that, even with special nodes, the percolation probability decreases as more nodes are switched off. This happens because the connectivity highly relies on the long shortcuts introduced by the special nodes. Therefore, if shortcuts are removed, the network quickly partitions. We have also performed experiments in which no special node is switched off (Figure 3(b)). In this case, the percolation probabilities as the typical nodes became inactive, for f =0,1 with different r_f ($3r$, $5r$ and $10r$), are almost indistinguishable and very high compared to the spatial model (f =0).

This results also suggests that, even in the worst case ($r_f = 3r$), the special nodes form a kind of backbone, which is used by the other nodes in order to maintain the connectivity at high levels.

4 Conclusions

In this paper, we have applied the small world concept to ad hoc networks by adding a fraction f of special nodes equipped with two radios, a short-range radio and a long-range one, operating in different frequency bands. Simulations results have shown that there are critical values of f above which, the network connectivity significantly increases. Also, there is a critical r_f above which there is no connectivity gain. These results have an important practical impact as they limits the power increase required to produce the long-range connections, as well as the number of long-range connections. It is also important to note that to obtain the same connectivity levels without special nodes we needed to increase the transmission ranges of all nodes. Furthermore, the simulation results showed the effectiveness of the special nodes in improving the broadcast efficiency and the network robustness as the nodes are switched off at random. As future work, we will also perform more realistic simulations including the protocols of the physical-link, MAC and network layers to evaluate the capacity of a wireless ad hoc network with the special nodes.

References

1. Booth, L., Bruck, J., Franceschetti, M., Meester, R.: Covering Algorithm, Continuum Percolation and the Geometry of Wireless Networks. Annals of Applied Probability, Vol.13, No. 2, (2003) 722-741.
2. Dousse, O., Thiran, P., Hasler, M.,: Connectivity in Ad-Hoc and Hybrid Networks. Proceedings of INFOCOM'02, (2002).
3. Helmy, A.: Small Large-Scale Wireless Networks: Mobility-Assisted Resource Discovery. ACM Technews, Vol 4, Issue 391, (2002).
4. Li, J., Blake, C., Couto, D., Lee, H., Morris, R.: Capacity of Ad Hoc Wireless Networks. Proceeding of MOBICOM'01 (2001).
5. Ramanathan, R., Rosales-Hain, R.: Topology Control of Multihop Wireless Networks Using Transmit Power Adjustment. Proceedings of the IEEE INFOCOM'00, (2000) 404-413.
6. Watts, D., Strogatz, H.: Collective Dynamics of Small-World Networks. Nature, Vol. 393, (1998) 440-442.
7. Watts, D.: Small Worlds – The Dynamic of Networks Between Order and Randomness. Princeton University Press (1999).

Adaptive Resource Management in Mobile Wireless Cellular Networks

Monir Hossain[1], Mahbub Hassan[2], and Harsha R. Sirisena[3]

[1] School of Computer Science & Engg., The University of New South Wales
NSW 2052, Australia, monirh@cse.unsw.edu.au
[2] School of Computer Science & Engg., The University of New South Wales
NSW 2052, Australia, mahbub@cse.unsw.edu.au
and
National ICT Australia Ltd., Locked bag 9013, NSW 1435, Australia.
[3] Dept. of Electrical & Computer Engineering, University of Canterbury
Christchurch, New Zealand, sirisehr@elec.canterbury.ac.nz

Abstract. Emerging mobile wireless networks are characterized by significant uncertainties in mobile user population and system resource state. Such networks require adaptive resource management that continuously monitor the system and dynamically adjust resource allocations, such as the number of guard channels reserved for handoff calls, for efficient system performance. The main objective of this research is to manage resource such a way so that it can maintain the target call dropping probability as well as utilize the scarce wireless resource efficiently. We propose a simple adaptive resource management scheme which is event based to adjust the number of guard channels for maintaining target call dropping probability. We try to evaluate the performance of the proposed scheme using extensive discrete event simulation. The results indicate that the scheme can guarantee the target call dropping probability under a variety of traffic conditions, and so can utilize the scarce wireless resource efficiently.

1   Introduction

To support a large number of mobile users through frequency reuse, next generation mobile wireless networks are expected to have much smaller cell boundaries (called picocells). One of the direct consequences of smaller cell boundaries is the increased number of handoffs during the lifetime of a mobile call. If there is not enough resources in the new cell when a handoff is being attempted, the call has to be terminated prematurely. Such premature call termination is referred to as *call dropping*. From the user's point of view, call dropping is more annoying than being blocked at call setup time (referred to as *call blocking*). For mobile data transactions, call dropping has more detrimental effect. A data transaction dropped in the middle of processing has to be restarted from the beginning. Therefore, call dropping causes not only user dissatisfaction, but also wastes wireless network bandwidth due to retransmission. Consequently, techniques which can reduce call dropping become increasingly critical in picocell

J.N. de Souza et al. (Eds.): ICT 2004, LNCS 3124, pp. 394–399, 2004.

wireless networks. Ideally, wireless network providers should be able to precisely control the call dropping probability (CDP) in their networks at any time of the day, regardless of the traffic load and user movement patterns.

A basic approach to reduce CDP is to give handoff calls priority over *new* calls. The well known Guard Channel (GC) scheme [1] and its numerous variants [2,3] reserve a fixed number of channels (guard channels) in each base station for serving handoff calls. GC therefore trades off lower CDP with higher call blocking probability (CBP). Determining the optimum number of guard channels for a desirable CDP is the key design goal of GC schemes. FGC schemes, however, are not effective in future networks, where handoff traffic is expected to be highly dynamic. For dynamic environments, the guard channels in a given cell should be adjusted adaptively in response to traffic dynamics.

Schemes which adaptively allocate guard channels are called adaptive guard channel (AGC) schemes. Some AGC schemes require the network to perform additional functions, e.g., prediction of user mobility and call status exchange with neighbouring cells [4,5,6,7]. Because the network overhead of these schemes increases with number of mobile users, they do not provide scalable solutions for future wireless networks. To make it practically employable and scalable, we need localized schemes which do not depend on motion prediction or call status of neighbouring cells.

Recently several methods have been proposed to design AGC schemes which require only local information. Luo et al.[8] proposed a localized AGC where a base station periodically measures the actual handoff rate and uses a M/M/1 queueing model to estimate the number of guard channels. Zhang et al. [9] used Wiener prediction theory to predict the future resource requirements of handoff calls.

We propose a simple scheme which is event based to adjust the number of guard channels for maintaining target CDP. The event is the call departure. When a call is departed from the cell, the scheme calculates the CDP and compares with target CDP, and adjusts the guard channel. The beauty of this scheme is that it is simple, and it can maintain the target CDP very well under a variety of traffic conditions. Only local information (number of handoff call coming and number of handoff call dropped) is used in this scheme. The computational overhead of this scheme is to compute the CDP once an event occurs.

The rest of the paper is organized as follows. Section 2 introduces adaptive resource reservation scheme. The simulation model is presented in section 3. The simulation results are presented and discussed in section 4. Finally, in section 5, we draw our conclusions.

2   Adaptive Resource Reservation Scheme

We consider a cellular network where a mobile station communicates with other mobile stations through a base station. When a mobile station crosses cell boundary in the middle of communication, a handoff occurs. In order to support the handoff calls, some channels are reserved (for example G out of C channels) for handling the handoff calls. The rest of the channels are shared between new calls and handoff calls. The new call is accepted if the number of free channels is greater than G, and handoff call is accepted if there is any free channel in the base station. This is just like the guard channel scheme where the number of guard channels is fixed. In this scheme, we adjust the number of guard channels dynamically for varying traffic conditions and maintain the target CDP. The scheme is based on event which is call departure. When a call is departed, the system calculates the current CDP and compares with the target CDP, and adjusts the number of guard channels in the following way:

1. If the current CDP is higher than the desired CDP then the performance of the scheme is assumed to be worse than the desired one. The reason behind this is that more calls arrived than expected. Therefore, the reserved resource can not accept all of the handoff calls and, a greater number of calls are dropped. Hence the number of guard channels has to be increased in order to accept more handoff calls. Therefore, we increase the number of guard channels by K (Where K is a tuning parameter, and its value can be 1,2,....n) to reduce the dropping probability.
2. If the current CDP is equal to the desired CDP, the performance of the scheme is satisfactory. Therefore, the system will keep the same number of guard channels.
3. If the current CDP is less than the desired CDP, the controller performance is better than that desired. The reason behind this is that the system reserved too much resource for handling the handoff calls. The desired CDP could be satisfied with less number of guard channels. Therefore, the controller must reduce the number of guard channels by K to utilize the resource efficiently.

3   Simulation Model

We have written a discrete event simulation program using C programming language to evaluate the performance of our proposed resource reservation scheme. Each call occupies one channel or one unit of bandwidth. The new calls originate within the cell, and handoff calls come from surrounding cells. The arrival process of new and handoff calls is Poisson with mean arrival rates λ_n and λ_h, respectively. The call holding time and dwell time are exponentially distributed with mean $1/\mu$ and $1/\eta$, respectively.

The performance metrics of our scheme are CDP and CBP. CDP is the ratio of total handoff call dropped to the total handoff call arrived at a cell during the entire simulation time. CBP is the ratio of total new call blocked to the total new call arrived at a cell during the entire simulation time.

The cell capacity is 80. The handoff call arrival rate is exactly half of the new call arrival rate. The mean holding time and dwell time are the same i.e., 112sec. The target CDP is 0.01. The toal simulation time is 180000 sec. Both new and handoff call arrival rates were varied during simulations. The value of K is chosen 1 because it gives the better performance.

4 Results

Simulation results are shown in this section in order to see the performance of the proposed scheme. The predetermined target CDP is 0.01 (1%). The performance of the scheme in maintaining the predetermined CDP is presented in the second paragraph, and the performance in terms of the CBP is presented in the third paragraph. The performance of this scheme in terms of CDP after changing the traffic parameters such as call arrival rate, holding time online is presented in the last paragraph of this section.

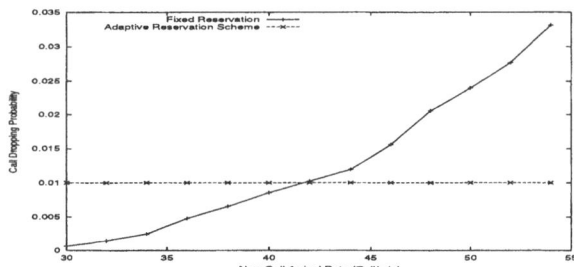

Fig. 1. Call Dropping Probability as a function of traffic load

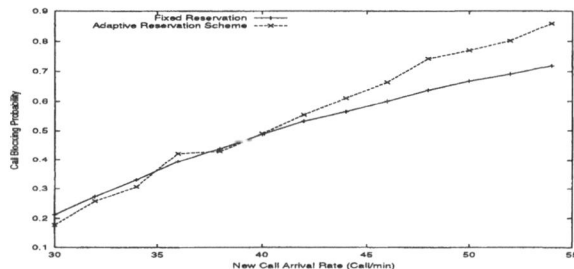

Fig. 2. Call Blocking Probability as a function of traffic load

Figure 1 shows the CDP with the variations of λ_n. The new call arrival rates are varied from 30 calls/min (low load) to 54 calls/min (high load). Results of the fixed reservation scheme with a constant reservation of 3 guard channels are also shown for comparison. The shortcoming of the fixed reservation scheme is immediately apparent. The CDP is too low (below target) for smaller new call arrival rates because then there are too many reserved channels for handoff calls. Conversely, the CDP is too high for higher new call arrival rates because there is smaller number of guard channels for handling the larger number of handoff calls. Our scheme precisely keeps the CDP because it dynamically adjust the number of guard channels depending on the traffic conditions. The scheme adjusts the number of guard channels once a call is departed, so if the CDP is more than the target CDP then it gets an opportunity to adjust the number of guard channels. Therefore, the scheme can keep the target CDP 1% precisely for changing traffic conditions. For low load, our scheme keeps the smaller number of guard channels. On the other hand, a greater number of guard channels are kept for high load for maintaining the target CDP.

Figure 2 shows the CBP with the variations of λ_n. Here we also compare the result of our scheme with a fixed reservation scheme with a constant reservation of 3 guard channels. When the traffic load is low, the CBP is higher for fixed reservation than our scheme because more channels are reserved for handoff calls in fixed scheme. On the other hand, our scheme reserves fewer number of channels, and thus our scheme can accept more number of new calls. Conversely, in fixed reservation, the blocking probability is lower for high load because less channels are reserved for handoff calls, so it suffers more CDP as was seen in Figure 1.

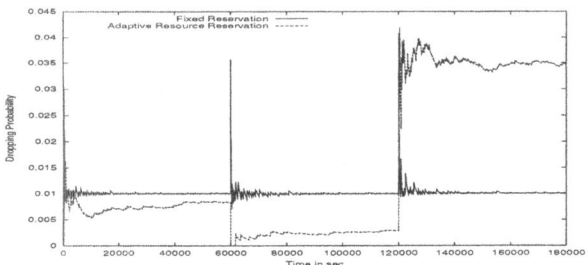

Fig. 3. Dropping Probability under a variety of call arrival rate

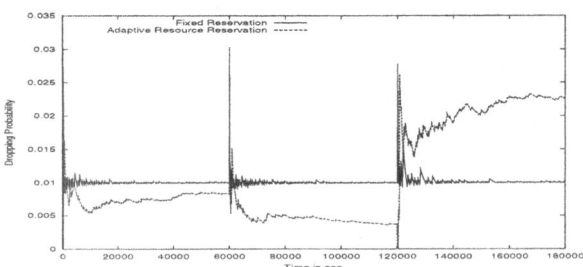

Fig. 4. Dropping Probability user a variety of call holding time

In Figure 3, the call arrival rate (in calls/minute) is changed over time, from 42 (between 0 s to 60000 s) to 32 (between 60001 s and 120000 s) and then to 58 (between 120001 s and 180000 s). Similarly, in Figure 4, the mean call holding time is changed over time from 112 s (between 0 s and 60000 s) to 100 s (from 60001 s to 120000 s) and to 140 s(from 120001 s to 180000 s). Figures 3 and 4 clearly show that the adaptive resource reservation scheme can accurately track the target CDP under a variety of call arrival rates and mean holding times, whereas the CDP varies for the fixed resource reservation scheme with a constant reservation of 3 guard channels.

5   Conclusion

In this paper, we have proposed a simple adaptive resource management to maintain the target call dropping probability for mobile wireless networks. Our scheme can dynamically adjust the number of guard channels depending on the traffic conditions such as arrival rate, holding time. It was shown through simulation that the proposed scheme can maintain the target dropping probability under varying traffic conditions. For future research, we are trying to extend the proposed scheme for multimedia service where different type of services need different amount of bandwidth.

References

1. D. Hong and S. S. Rappaport, *Traffic model and performance analysis for cellular mobile radio telephone systems with prioritized and no-protection handoff procedure*, IEEE Transaction on Vehicular Technology, Vol. 35, No. 3. 1986.
2. R. Ramjee, R. Nagarajan and D. towsley, *On Optimal Call Admission Control in Cellular Networks*, IEEE INFOCOM '96. San Francisco, March 1996.
3. B. Li, C. Lin and S. Chanson, *Analysis of a Hybrid Cutoff Priority Scheme for Multiple Classes of Traffic in Multimedia Wireless Networks*, ACM/Baltzer Journal of Wireless Networks, 1998.
4. Chun-Ting Chou and Kang G. Shin, *Analysis of Combined Adaptive Bandwidth Allocation and Admission Control in Wireless Networks*, INFOCOM 2002.
5. Carlos Oliveira, Jaime Bae Kim, and Tatsuya Suda, *An Adaptive Bandwidth Reservation Scheme for High-Speed Multimedia Wireless Networks*, IEEE Journal On Selected Areas in Communications, Vol. 16, No. 6, August 1998.
6. P. Ramanathan, K. M. Sivalingam, P. Agrawal, and S. Kishore, *Dynamic Resource Allocation Schemes During Handoff for Mobile Multimedia Wireless Networks*, IEEE Journal On Selected Areas in Communications, Vol. 17, No. 7, July 1999.
7. David A. Levine, Ian F. Akylidiz, and Mahmoud Naghshineh, *A Resource Estimation and Call Admission Algorithm for Wireless Multimedia Networks Using the Shadow Cluster Concept*, IEEE/ACM Transactions On Networking, Vol. 5, No. 1. February 1997.
8. X. Luo, I. Thng, W. Zhuang, *A dynamic pre-reservation scheme for handoffs with GoS guarantee in mobile networks*, IEEE International Symposium on Computers and Communications, July 1999.
9. Tao Zhang, J. Chennikara, P. Agrawal, Eric van den Berg, T. Kodama, *Autonomous Predictive Resource Reservation for Handoff in Multimedia Wireless Networks*, Sixth IEEE Symposium on Computers and Communications, 2001.

Comparative Analysis of Ad Hoc Multi-path Update Techniques for Multimedia Applications

Manus Kwan and Kutluyıl Doğançay

University of South Australia, School of Electrical and Information Engineering,
Mawson Lakes, South Australia, Australia

Abstract. Multi-path transmission in ad-hoc networks provides higher bandwidth and better guarantee on packet delivery than the traditional single shortest-path method. This is important to multimedia applications since processing of data packets is sensitive to error and delay. Current ad-hoc protocols are classified as either proactive or reactive. Reactive protocols are more suitable for optimizing path selection than proactive protocols. Among a set of reactive protocols, DSR (Dynamic Source Routing) is the most commonly used protocol for multi-path transmission. However DSR lacks the ability to give a constant update on backup multi-paths. The aim of this comparative study is to investigate existing path updating techniques and incorporate them on backup multi-path update. Multicast, unicast and hello broadcast are compared analytically and tested on ns-2. The result of the simulated test has shown that an extension of AODV's hello broadcast gives the best overall performance.

1 Introduction

The advance of multi-media applications has prompted researchers to undertake the task of finding suitable methods to carry multi-media data through ad-hoc wireless networks. The task of finding suitable methods for this particular application is rather difficult to achieve since ad-hoc networks exhibit highly dynamic link connectivity and are prone to error from radio transmission through the physical environment.

In addition to all of the above, the connectivity of wireless nodes is maintained by a set of network messages. The type of messages depends on the selection of routing protocols. Network messages either perform the task of updating network information periodically (Proactive Routing) or search for routes between source and destination on demand (Reactive Routing). Either way, the link status is maintained by nodes working co-operatively with each other. Co-operation between nodes has to occur because of the fact that there is no centralised body to monitor and control network resources. The absence of centralised network monitor has made the task even more difficult because there is no bandwidth guarantee or negotiation during the process of traffic connection.

The task of finding suitable methods to transport multi-media packets through ad-hoc networks is difficult but not impossible. In [1], [2] and [3] using a set of multi-paths to transport split video streams has been proposed. The original compressed video data is split into multiple streams [10] so that if one or more streams are lost

J.N. de Souza et al. (Eds.): ICT 2004, LNCS 3124, pp. 400–409, 2004.
© Springer-Verlag Berlin Heidelberg 2004

due to link failure or traffic congestion, the receiver can still have video playback with reduced quality.

In general a set of multi-paths provides higher bandwidth and better guarantee on packet delivery than a single shortest path. However the unpredictable nature of wireless link can still produce heavy packet losses. Further investigation has been done on each individual link and a set of links on each path. In [1] K-state Markov process were used to model a wireless link in ad-hoc network. Each K-state in the model corresponds to a certain range of received signal-to-noise ratio. Modeling of packet loss rate of the entire path is done by the accumulation of packet error rate (PER) of each individual link. In [3] and [9] link bandwidth availability of each path is measured by the accumulation of queuing packets in waiting and the total number of nodes.

The work presented in [1], [2], [3] and [9] has used DSR (Dynamic Routing Protocol) as the basic routing protocol to perform analysis on multi-path transmission. Other protocols such as AODV (Ad-hoc On Demand Distance Vector), ZRP (Zone Routing Protocol) and TORA (Temporally Ordered Routing Algorithm) are capable of performing multi-path transport, but DSR is still the most popular choice to implement network adaptation for multi-media traffic.

There are still issues that need clarification regarding the use DSR on multi-path systems. Although DSR has the ability to search and maintain active routes and the modified DSR in [3] has the ability to maintain multiple active routes for a pair of source and destination nodes with minimal network messages, DSR does not have the ability to maintain inactive routes for the purpose of multi-path selection. In this paper, active routes are referred to paths that carry data packets, while inactive routes remain connected but stay idle.

The purpose of this paper is to give a comparative study on various methods for inactive paths maintenance. This paper will not deal with specific problems to do with the physical medium or multiple channels of frequency, time and coding for radio transmission between nodes. The problems of coupled paths in alternate path routing are addressed in [8]. Packets would be blocked if two paths were coupled. Current methodology of multi-path routing would have incoming and outgoing packets to be accumulated in transmit and receive buffers, respectively.

2   Criteria on Multi-path Maintenance

There are similarities and differences between active and inactive route maintenance. In the case of active link failure in reactive routing, error messages are generated by a node located upstream of the broken link to inform the source about the failure [5]. Detection of the link failure is brought about by either (i) the expiry of required time of acknowledgement for the transmitted data packet or (ii) an end-to-end acknowledgement exceeding the required time. The former is performed by any two consecutive nodes in an active path at the data link layer. The latter is performed by the two end nodes of the path at the network layer [6]. If users require further transmission of packets on the network, then Route Discovery [5] is required to search for alternate route. In the case of inactive route, there are no error messages to inform the source since timeout for data packet acknowledgement cannot be used to detect link failure due to absence of data traffic.

Fresh update of link status is required for both active and inactive routes. The source needs to know the status of the path before it can decide which set of paths to use for optimal solution [3]. Proactive routing protocol is effective in constant update of route connectivity for both active and inactive routes. Network messages containing network status are exchanged between nodes periodically. Among a collection of single hop route entries protocols, TORA [7] is the most suitable protocol for multi-path. TORA, like most of proactive protocols, can only store one hop (namely next hop) information in its table entries. In the case of multi-path connection, a node at the intersection of multiple paths can store multiple neighboring hop status for the same destination. This leads to a new problem when the source node needs a set of new multi-paths.

Both [3] and [9] have suggested that feedback data is needed in order to select a set of optimal paths for multimedia traffic. Feedback information includes the accumulation of each individual queue sizes associated with each intermediate nodes and the number of nodes in the path. Although the total number of nodes and total queue size can be accumulated by the propagation of TORA update packets [7], the task has proven to be more difficult at the site of intersection.

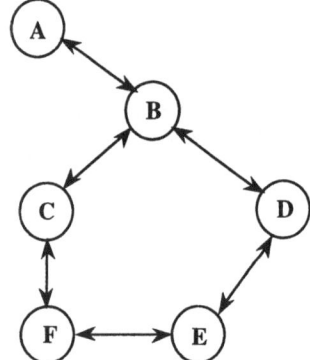

Fig. 1. The topology of path intersection.

In figure 1, when node A requires a route to E, it would generate query messages and flood the network until the destination node has been found or a relay node with the knowledge of next hop towards the destination node in its routing table. Node E in this case generates update messages and propagates back to Node A. If total queue length is produced by accumulation of individual queue size then Node B has two different total queue sizes, one of them from D->E and the other from C->F->E. Node B would have difficulty passing this multi-path information to node A since TORA only requires one hop information. As far as node A is concerned, it only knows that the next hop to E is through B and the number of hops to get there would be three, assuming that the shortest path method has been used.

3   Inactive Multi-path Maintenance Methods

3.1 Forward Route Discovery

As mentioned previously, connectivity update and route finding are the two criteria that inactive paths maintenance requires. In terms of searching for optimal multi-path, proactive protocols that use total node numbers and total queue size along a set of multi-paths prove to be extremely inappropriate. This leaves us with the reactive approach, which uses route discovery to search for a set of multi-paths as well as obtaining a set of useful information for optimal path selection.

Normal DSR route discovery has the ability to search for multi-paths as well as carry useful network information such as total queue sizes as well as total number of nodes. However every time a node initiates route request (RREQ) [5] broadcast, the wireless network tends to be flooded with RREQ packets. Additional flooding of the network is not desirable but is necessary in this case. Periodic flooding of RREQ is one technique that can be used to solve the problem of path finding and eliminate the need of route update. However this would have excessive increase in network overhead. Network overhead can be reduced if periodic routing update is operated on inactive paths. Source node can send network packets along inactive multi-paths to investigate their link status. If the link status of several inactive paths is down then the source node simply deletes it from its route entry. Flooding of the network is not required unless there are not enough spare paths for back up.

3.2 Periodic Neighbor Update

Sending network packets along inactive paths can have repercussions. If sections of inactive and active paths are fused together, then sending network packets to investigate path connectivity can affect data traffic on active paths. This point can be illustrated by observing path branching in figure 2. Assuming there are no further paths branching beyond the scope of the figure, Node D in this case would have two paths leading to destination (not shown), viz., Path D->G->H->... (the inactive path) and D->E->... (the active traffic path). Node B in this case would have two paths accumulated from node D as well as two extra inactive paths through C, which are in turn accumulated from F. If node A decided to send packet updates to investigate its inactive path connectivity, then it has to send three network packets containing routing information on link A<->B. If there is more branching further down the path, then more packets has to be carried by link A<->B. This is a serious problem since link A<->B is also an active link for data traffic.

Two methods of overcoming this update problem shall be described in this paper. In this section we shall describe the neighbor update method. This updating technique is also known as "Hello Messages Updating" in AODV [4]. In AODV when a node has not sent packets to its next hop towards destination node within a hello interval, it broadcasts a hello messages to find out its connectivity amongst its neighbors. The TTL of the hello broadcast is to be set to one so that it would not propagate throughout the network. If a node fails to receive a consecutive "Allowed_Hello_Loss" [5] hello message from its neighbor, then the connectivity around its neighborhood has changed. The advantage of this hello method is to relieve excess load on a set of active links in case of path branching. Nodes among neighbors can simply transmit their identity with minimal information. The second advantage of updating neighboring information is that in multi-user environment, the connectivity status among network neighbors gathering around the intersection of inactive paths is readily available. Even though additional users may join in the network, there would be only a slight increase in the number of hello messages within the intersected region.

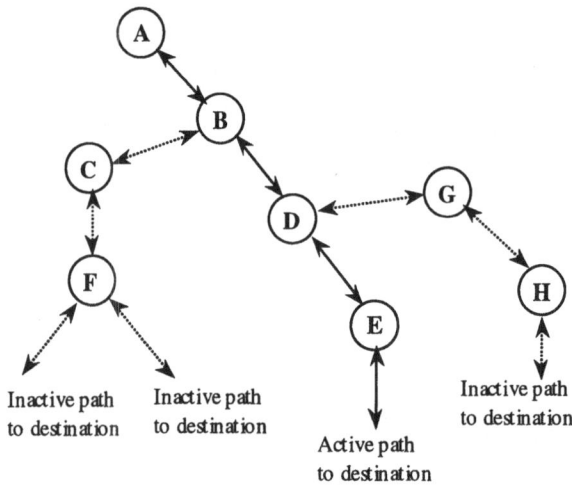

Fig. 2. Topology illustrating inconvenient use of sending network packets along inactive paths.

3.3 Updating with Multicast

The disadvantage of hello message technique is the large overhead when a small number of users are accessing the network. This problem can be overcome by sending update multicast message down both active and inactive paths. From now on, an update message is simply a network message issued by the source to investigate inactive path connectivity. This is not be confused with Update Packets of TORA. Multicast update can be illustrated by referring figure 2 again. If an update multicast packet is sent to node B, then two separate update packets can be replicated for D and C at the split path. This method only transmits one copy of update packet on each link regardless of path branching. This would guarantee that overloading of traffic on active link due to path branching would be eliminated. Link failure can be detected if reply timeout from downstream node has expired.

The disadvantage of this technique is that high number of users in the network can increase the number of network messages. This is due to the fact that sending update packets through known routes only guarantees path connectivity between each source and destination pair.

3.4 Broadcast from Destination

Forward broadcasting need both request and reply messages to discover paths between source and destination. If broadcasting were to be initiated at destination node, then route reply packets can be eliminated. Traditionally DSR [5] route request packets contain path information from source to the received node. The received node need not be the destination node that is specified in Target_Address [6] of route request packet. After running through route logic, routing information from the request packet is copied into the route cache of the received node.

The same mechanism can be utilised if Target_Address is the source node and IP_Source is the destination node. Upon capturing the request packets, route information, which leads to the destination, is to be recorded in the route cache. Route reply packets are not to be issued from the source node, rather it would send updating packet to each inactive routes if necessary. The construction of the reverse path that leads back to the source can be formed by issuing update packets.

4   Combined Path Search and Update for Inactive Path Maintenance

In the previous section, two path finding methods and three updating methods have been introduced to provide suitable inactive path maintenance. Performance on path management can be improved if some of these techniques are integrated together.

Hello messages and forward discovery can be integrated together to produce an combined effort on inactive route finding and updating. Whenever the source node needs updating, it simply broadcasts its hello message to its neighbor. If failed connection has been detected, error message would be propagated back to the source from the node, which detected the failure.

Backward broadcast would produce less overhead than forward discovery. However it alone cannot be combined with hello message updating method. This is due to the fact that relaying nodes only has information on forward paths only. Whenever link failure is located, it cannot send error messages back to upstream sources because of the absence of reverse path. Reverse path knowledge can be restored if the source multicasts update messages to its new discovered paths. The multicast update procedure only has to be initiated once to ensure the reverse path entries of relay nodes have been set. The rest of the update mechanism can be handled by hello messages. Multicast update can be combined with either forward path discovery or backward broadcast.

5   Simulation Studies

The discrete-event network simulator ns-2 was used to simulate three different updating techniques, namely, multicast, unicast and hello broadcast.

5.1 Additional Implementation for Update

The three updating methods are to update as many required multi-paths as possible without too much network overhead. The overhead restriction requires an additional algorithm to select as few nodes as possible to participate in the update process.

To begin with the implementation process, let us introduce three new states to each node. Each node would have its identity as well as a state variable corresponding to either ACTIVE, INACTIVE or NON_USE. As the name suggests, ACTIVE means that the node in question is located in an active path, INACTIVE means that it lies on an INACTIVE path and NON_USE means that it does not lie on any paths.

At the beginning of the updating process, a unicast update must traverse all the active and inactive paths to initialise each state variable within the node to INACTIVE. The state variable can only be set to ACTIVE status by receiving data packets. In hello broadcast, the initial unicast would also update the source table in each node. The source table can be implemented in different ways, but it must contain a state variable and be logically indexed by source node address. When a node receives an initial update, it would locate the source state variable in the source table and switch it to ACTIVE.

In addition to source table, each node must have a neighbor table containing a state variable indexed by the address of the node's neighbor. The state variable of the associated neighbor is set to INACTIVE when the node receives an initial update packet. The information for the identity of each inactive neighbor can be found on an initial update packet due to the fact that the header field of an initial update packet is similar to that of DSR data packet. They both contain addresses of all intermediate nodes along the paths.

After the state variable of a node has been set to INACTIVE, each node would broadcast a hello message to its neighbor on every hello interval. Link breakage is detected if a hello broadcasting node does not receive a more than ALLOWED_HELLO_LOSS [11] consecutive hello messages from one of its neighbors that lies on either an active or inactive paths. When a link breakage has been detected, the detecting node would broadcast error packets containing its identity and the identity of the other node at the other end of the break. The detected node would set its out of range active or inactive neighbor to NON_USE and delete its routing table entries that use the broken link. The detecting node also checks whether there are still any paths that lead to its sources. If all the paths to a source have been deleted, then this source is set to NON_USE and if all the sources in its table entry are set to NON_USE then its node state is set to NON_USE, and periodic hello broadcasts for that node can be terminated.

When a node receives an error packet, it will delete path entries that use the broken link. The received node would only rebroadcast error message packets if it has path entries that use the broken link. The received node also updates its neighbor table entries, source table entries and its node state. When the user terminates its connection, the terminated source node would unicast an end of service packet to all its active and inactive paths. A node that receives the end of service packet performs the same updating procedure that it would after receiving an error packet.

The application of the hello message as described previously is slightly different to AODV's hello message. Although they both concern the connectivity of its active neighborhood, hello in AODV is not concerned with overhead and multi-path update. In section 6.5 of [11] it is mentioned that every node generates a "hello" message on every HELLO_INTERVAL. This could mean that all nodes in the network are broadcasting hello messages to check for connectivity. Reaction to link failure only takes place only when the broken link is an active link. Nevertheless the algorithm that was implemented in this simulation is an extension of AODV's hello broadcast.

Multicast and unicast methods are quite straightforward to implement. In both methods, the source node only has to send update to its active and inactive paths periodically. The detection of link failure can be handled by the data link layer as described in section 3.3.

5.2 Results

A scenario of 18 nodes and a number of up to 8 users were used to generate the simulation results shown in figures 3 and 4. Each user transmits CBR (Continues Bit Rate) data with 0.025 seconds between 500-byte packets. The duration of the simulation is 7.0 seconds. Figure 3 shows the plot of the total number of network packets generated for different number of users. The network packet variable includes other DSR packets generated by route discovery and route reply, thus there is a general increase in the number of total network packets in all three methods including hello broadcast.

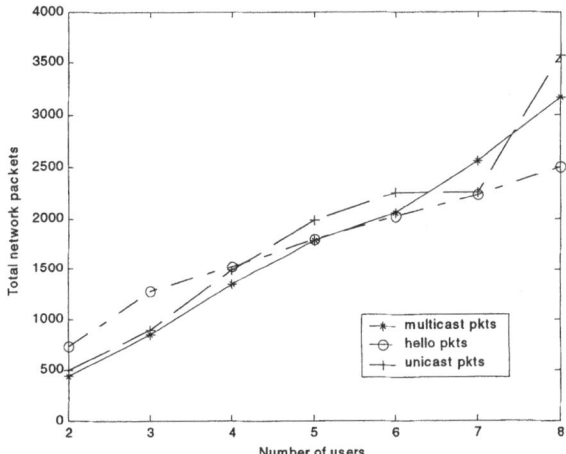

Fig. 3. Number of network packets versus number of users.

As shown in figure 3 both unicast and multicast generate a lower number of network packets than hello broadcast for 2 users. As the number of users increases, both unicast and multicast have a higher rate of increase in the number of total generated network packets. Between 3 users and 4 users, the total number of network packets generated by unicast is greater than that of hello broadcast. Similar observations are made between 5 users and 6 users, but in this case it is between multicast and hello broadcast. Generally the total number of multicast packets is lower than unicast packets, but there is a crossover between 6 users and 7 users and unicast crosses back again between 7 users and 8 users. The unsteady nature of unicast behavior could be caused by variation in source and destination locations. However the general trend of plot in figure 3 is that in most cases unicast has a higher number of network packets than that of multicast.

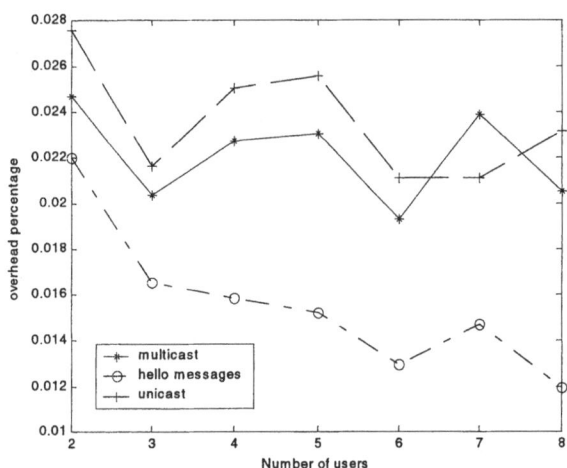

Fig. 4. Overhead percentage versus number of users.

Figure 4 shows the overhead percentage over the number of users. The overhead percentage is the ratio of the total number of bytes generated for network messages to the total number of bytes generated in the simulated period. Hello broadcast is always lower than both unicast and multicast due to its small size update packets. As the number of users increases the overhead percentage of all three methods decrease. In particular hello broadcast has a higher rate of reduction in overhead percentage than both multicast and unicast.

6   Conclusions

A comparative analysis has been presented of different multi-path updating methods based on DSR. The purpose of multi-path update is to ensure that paths from source to destination are valid even though they are not active. Path update in conjunction with path discovery forms a multi-path management system for source traffic that require high bandwidth and low delay.

The DSR route discovery is capable of searching for additional routes between source and destination, but it causes too much overhead if activated too often. Destination broadcasting uses similar path discovery techniques to DSR route discovery, but requires no route reply messages.

Updating techniques are required to refresh connectivity information of all active and inactive paths so that minimum full network broadcasting can be achieved.

Among the three techniques analyzed, the extended AODV's hello broadcast gives the best result in ad hoc networks with a moderate-to-high number of users.

Acknowledgements. This work was partially supported by an Australian Research Council (ARC) Linkage Project Grant and the Motorola Australia Software Centre.

References

[1]   Yao Wang, Shivendra Panwar, Shunan Lin, and Shiwen Mao, Reliable Transmission of Video over Ad-hoc Networks Using Automatic Repeat Request and Multi-path Transport. Proc. IEEE Fall VTC 2001, vol.2, pp.615-619, Atlantic City ,October 8-10, 2001.

[2]   Yao Wang, Shivendra Panwar, Shunan Lin, and Shiwen Mao, Video Transport over Ad-hoc Networks Using Multiple Paths. Invited Paper, Proc. IEEE 2002 International Symposium on Circuit and Systems, Scottdale, Arizona, May 26-29, 2002.

[3]   Manus Kwan, Kutluyıl Do ançay, Lakhmi Jain, Fair Multi-path Selection For Real-Time Video Transmission In Ad-Hoc Networks Using Artificial Intelligence. Proc. HIS'03 conf. on Design and Application of Hybrid Intelligent Systems, pp.830-841, December 2003.

[4]   Charles E Perkins, Elizabeth M. Royer, Ad-hoc On-Demand Distance Vector Routing, MILCOM '97 panel on Ad Hoc Networks, Nov 1997.

[5]   David B. Johnson, David A. Maltz, Dynamic Source Routing in Ad Hoc Wireless Networks, in Mobile Computing (T. Imielinski and H. Korth, eds.), Kluwer Academic Publishers, 1996.

[6]   Josh Broch, David B. Johnson, David A. Maltz, The Dynamic Source Routing Protocol for Mobile Ad-Hoc Networks, IETF MANET Working Group, Internet draft, December 1998(work in progress).

[7]   Vincent D. Park and M. Scott Corson, A Highly Adaptive Distributed Routing Algorithm for Mobile Wireless Networks, Proc. INFOCOM'97, pp 1405-1413, April 1997.

[8]   M.Pearlman, Z.J.Hass, P.Sholander, and S.S.Tabrizi, On the Impact of Alternate Path for Load Balancing in Mobile Ad Hoc Networks, ACM MobiHOC'2000, Boston, MA, August 11, 2000

[9]   Min Sheng, Jiandong Li and Yan Shi, Routing protocol with QoS guarantees for ad-hoc networks, Electronics Letters, Vol. 39 No.1, 9th January 2003

[10]   Dapping Wu, Yiwei Thomas Hou and Ya-Qin Zhang, Scalable Video Coding and Transport over Broad-Band Wireless Networks. Proc. IEEE, vol.89, No.1, January 2001.

[11]   Charles Perkins, Ad Hoc On Demand Distance Vector (AODV) Routing, IETF MANET Working Group, Internet draft, November 1997(work in progress).

PRDS: A Priority Based Route Discovery Strategy for Mobile Ad Hoc Networks

Bosheng Zhou[1], Alan Marshall[1], Jieyi Wu[2], Tsung-Han Lee[1], and Jiakang Liu[1]

[1] Advanced Telecommunication Systems Laboratory,
School of Electrical and Electronic Engineering, Stranmillis Road,
The Queen's University of Belfast, BT9 5AH Belfast, Northern Ireland, UK
{B.Zhou, A.Marshall, Th.Lee}@ee.qub.ac.uk
[2] Research Center of CIMS,
Southeast University, Nanjing China, 210096

Abstract. Most reactive route discovery protocols in MANETs employ a random delay between rebroadcast route requests (RREQ) in order to avoid "broadcast storms". However this can lead to problems such as "next hop racing", and high redundancy in the rebroadcasting. In this paper we propose a Priority-based Route Discovery Strategy (PRDS) to address both problems. PRDS assigns a high rebroadcast priority to "good" candidates to reduce next hop racing; and introduces a competing procedure to keep "bad" candidates from rebroadcast to alleviate "rebroadcast redundancy". PRDS is a general mechanism that can be applied to different reactive routing protocols to improve the routing performance. As an example, a macrobian route discovery strategy using PRDS (PRDS-MR) is introduced. Simulation results show that PRDS-MR outperforms AODV in terms of packet delivery ratio and delay while decreasing routing overhead up to 50%~80%.

1 Introduction

Mobile ad hoc networks (MANETs) are considered as resource limited, e.g., low wireless bandwidth, limited battery capacity and the like. Because of this, the conventional routing protocols used in fixed networks are no longer appropriate for MANETs due to the heavy routing overheads that consume too much bandwidth and energy. Existing MANET routing protocols can be classified into three categories: proactive, reactive, and hybrid. Proactive routing protocols are table-driven and rely on periodical exchange of route information. Each node maintains route entries to all other nodes of the entire network. In large and highly dynamic MANETs, Frequent routing information exchanges are needed to keep routing table up-to-date, and this leads to heavy routing overhead and large memory consumption. Reactive and hybrid routing protocols are designed to address these problems. In reactive routing protocols, each node only maintains active route entries and discovers routes only when needed. In hybrid protocols, a network is partitioned into clusters or zones. Proactive and reactive routing protocols are then deployed in intra-cluster/ intra-zone and inter-cluster/inter-zone, respectively. The major advantage of hybrid routing is improved scalability; however, hierarchical address assignment and zoning/clustering management are complicated and can lead to heavy control overhead in high dynamic networks.

J.N. de Souza et al. (Eds.): ICT 2004, LNCS 3124, pp. 410–416, 2004.
© Springer-Verlag Berlin Heidelberg 2004

The operation of a reactive routing protocol has three basic stages: route discovery, packet delivery, and route maintenance. Different reactive routing protocols are distinguished by different strategies taken in route discovery and route maintenance. In this paper, we focus on route discovery strategies for reactive routing protocols. A source node initiates a route discovery by constructing a Route Request (RREQ) and flooding it to the network. On receiving a new RREQ, an intermediate node usually rebroadcasts it in a random delay, e.g. in AODV [1] and DSR [2], so as to avoid a "broadcast storm" due to synchronization as identified by Ni *et al* [3]. We abbreviate this random rebroadcast delay route discovery approach as RD-random in this paper. Li *et al* [4] argued that RD-random might not find the most desirable route, and Zhou *et al* [5] demonstrated that flooding, which is a broadcasting scheme using random rebroadcast delay, cannot guarantee the least delay.

Fig. 1 illustrates a route discovery scenario. Two of the possible paths from Source *S* to Destination *D* are shown in the figure, i.e. path *S-C-E-F-D* and the shortest path *S-A-B-D*. Two problems exist if RD-random is used in this scenario: 1) Path *S-C-E-F-D* may be selected instead of the shortest path *S-A-B-D* by the destination *D*. This phenomenon was identified as "next-hop racing" problem in [4]. 2) All nodes except for the destination *D* will rebroadcast the RREQ. This is not a serious problem in this scenario; however it will lead to heavy routing overhead and consequent implications such as extra bandwidth and energy consumption in a large scale dynamic network. We identify this phenomenon as "rebroadcast redundancy" problem.

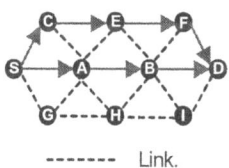

The key motivation of this paper is to address the both these problems. The central idea is that as long as the above two problems are solved; the route discovery strategy expects that a route with better quality may be found while minimising the routing overhead.

------- Link.

Fig. 1. Route discovery: *S* to *D*

The rest of this paper is organized as follows. Details of PRDS are given in Section 2. As an example, a macrobian route strategy based on PRDS is described in Section 3. Simulation results can be found in Section 4. Finally, the paper is concluded in Section 5.

2   PRDS

In this section, we describe the Priority-based Route Discovery Strategy (PRDS). The mechanism of PRDS is quite simple. It assigns a high rebroadcast priority to a "good" candidate for the next hop to solve the "next-hop racing" problem; it uses a competing procedure to prohibit "bad" candidates for the next hop from rebroadcast so as to solve "rebroadcast redundancy" problem.

In PRDS, a priority index (*PI*) indicates how good an intermediate node is for the next hop of the constructing route. *PI* is defined by some node/link/network state parameters according to different route design purposes. For convenience, we restrict the values of *PI* from 0 to 1, i.e. *PI* $\in$ [0, 1]. In the next section, we will define a *PI* for a macrobian route strategy.

Then, the RREQ rebroadcast delay is estimated according to *PI* using the following formula.

$$d = d_{max} \cdot (f(PI) + 0.1*random()) \qquad (1)$$

Where, d is the rebroadcast delay of a RREQ; d_{max} is the upper bound of d. *random*() is a random function whose values is randomly distributed between 0 and 1. This term is used to differentiate rebroadcast delay when nodes have same *PI* value. $f()$ is a function of *PI* that requires the features of (i) a bounded function with upper bound   1 and lower bound   0; (ii) $f()$ decreases as *PI* increases. We define the function $f()$ as follows,

$$f(PI) = \tanh((1.0-PI)/u_0) \qquad (2)$$

where $\tanh(x)$ is a hyperbolic tangent function; u_0 is a constant and the value of 0.3 is appropriate for most cases. $f(PI) \in [0,0.998]$ when $PI \in [0,1]$. $f(PI)$ decreases rapidly when *PI* approaches 1 so as to differentiate rebroadcast delay efficiently between high priority nodes.

When an intermediate node has determined the rebroadcast delay of a RREQ, it enters a competing procedure to rebroadcast the RREQ. In PRDS, each node needs to maintain a Competing State Table (CST). A CST contains three fields:
(i) *RREQ ID* that is used to identify a unique RREQ. It is represented as (*Source ID, broadcast sequence*). (ii) The number of times of receiving the same RREQ - n_h. n_h is initialised to 1 when a node receives a first copy of a RREQ. It is also used to represent the competing state. It is set to 0 when the competition is over. Any following RREQs will be deleted as long as their related n_h equals 0. (iii) The timestamp of receiving the first copy of the RREQ. This field is used to maintain the CST with soft state mechanism, i.e. timeout mechanism.

In PRDS, we define two events: receiving a RREQ (This event is triggered when a node receives a RREQ.), and rebroadcast delay time out (This event is triggered when a rebroadcast delay expires.). We pre-assign the number of times (n_0) that a RREQ may be received. When a rebroadcast delay times out, PRDS checks the related n_h in CST. The node will rebroadcast the RREQ if n_h n_0. Otherwise, the rebroadcast operation will be prohibited. We denote PRDS using different n_0 as PRDS/n_0, e.g. PRDS/1, PRDS/2, and so on. The sequence of operations for PRDS is shown in Fig.2.

3   PRDS-MR

In this section, we design a Macrobian Routing Protocol using PRDS. We term it PRDS-MR. We assume that (i) each node has its own location and mobility knowledge that can be learned from some positioning system; (ii) each node is equipped with an omni-directional antenna that has a transmission range R. PRDS-MR aims at finding the route that has the following features in comparison with RD-random: the lifetime of the route is relatively long; the route length (hops) is not significantly long; routing overhead is minimised.

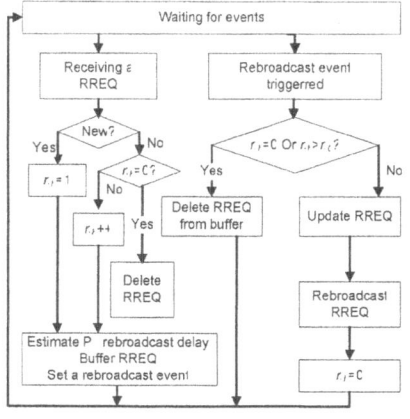

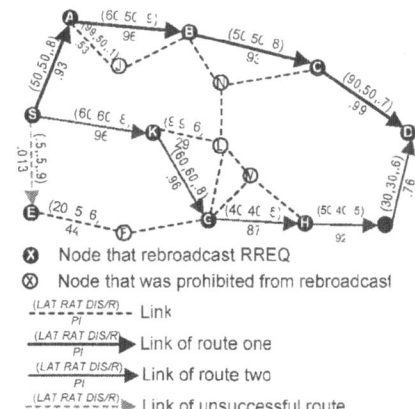

Fig. 2. PRDS flow chart

Fig. 3. A route discovery scenario using PRDS/1-MR

We first define two parameters: Link Alive Time (*LAT*) and Route Alive Time (*RAT*). *LAT* is the amount of remainder time during which two nodes remain connected. *RAT* is the minimum *LAT* of the links along the route from source to destination.

We denote the coordinates and moving speed of Node i as (x_i, y_i, z_i) and (u_i, v_i, w_i), respectively. The distance between Node *1* and Node *2* can then be expressed as,

$$DIS = \sqrt{x_d^2 + y_d^2 + z_d^2}, \tag{3}$$

where $x_d = x_1 - x_2$, $y_d = y_1 - y_2$, $z_d = z_1 - z_2$.

A link exits between Node *1* and Node *2* if *DIS* $\leq$ R, i.e. Node *1* and Node *2* can communicate each other directly. The *LAT* of the link can be estimated as follows,

$$LAT = \frac{-(x_d u_d + y_d v_d + z_d w_d) + \sqrt{A - B}}{u_d^2 + v_d^2 + w_d^2}, \tag{4}$$

where

$$A = (u_d^2 + v_d^2 + w_d^2)R^2,$$

$$B = (u_d y_d - v_d x_d)^2 + (v_d z_d - w_d y_d)^2 + (u_d z_d - w_d x_d)^2,$$

$u_d = u_1 - u_2$, $v_d = v_1 - v_2$, $w_d = w_1 - w_2$,

Now, we define *PI* for PRDS-MR as following,

$$PI = p^{LAT} * p^{DIS} * p^{RAT}, \tag{5}$$

$p^{LAT} = \tanh((LAT/LAT_0)/C_1)$, $p^{DIS} = \tanh((DIS/R)/C_2)$, $p^{RAT} = \tanh((RAT/RAT_0)/C_3)$.

where p^{LAT} is the contribution of *LAT* of the upstream link. It is the main part of *PI*. It guarantees that those links with long lifetime have a higher *PI*. p^{DIS} is the contribution of the length of upstream link. This term prevents very short links from being included in the route. p^{RAT} is the contribution of lifetime of the path from source to the current node. It prevents short lifetime routes from being selected. C_1, C_2, C_3, LAT_0, and RAT_0 are parameters whose values can be determined by experiment. In this paper, we set these parameters as follows,

C_1=0.30; C_2=0.17; C_3=0.05; LAT_0=100 sec; RAT_0=10 sec.

Figure 3 illustrates a route discovery scenario using PRDS-MR. n_0 is set to 1 in this scenario. Node S broadcasts a RREQ to discover a route to Node D. The numbers above links are $(LAT, RAT, DIS/R)$; the number under a link is the PI for the receiving node to rebroadcast the RREQ. For example, numbers (60,50,.9) above the Link A-B mean that LAT of Link A-B is 60 sec; RAT of Route S-A-B is 50 sec; length of Link A-B is 0.9R. The number .96 under Link A-B means the PI for Node B to rebroadcast the RREQ is 0.96. In the figure, Node J is prohibited from broadcast because Link A-J is very short (so p^{DIS} is very small). Node F is prohibited from rebroadcast because the RAT of Path S-E-F is very short (so p^{RAT} is very small). In this example, Path S-A-B-C-D is selected (RAT=50 sec); five nodes are prohibited from rebroadcast.

4   Simulation Results

To evaluate the performance of PRDS-MR, we have implemented PRDS-MR based on AODV in the Global Mobile Simulation (GloMoSim) developing library [6]. In the simulations, IEEE 802.11 Distributed Coordination Function (DCF) is used as the MAC protocol. The random waypoint model is utilized as the mobility model and the pause time is 0 second. The bandwidth of the wireless channel is 2Mbps. The data packet size is 512 bytes. The flow pattern is Constant Bit Rate. The maximum rebroadcast delay d_{max} in Eq.1 is set to 30 ms. Table 1 gives the simulation parameters.

The metrics measured in the simulations are: Packet delivery ratio (PDR), throughput, average end-to-end delay, routing overhead (number of sending times of routing packets in the simulation), and route lifetime.

In the following figures, PRDS/n_0-MR denotes simulation results of the protocols using PRDS/n_0-MR mechanism, where n_0 is the threshold of the RREQ duplicate number as described above. n_0=ALL means each node will rebroadcast every non-duplicated RREQ once.

Table 1. Simulation parameters.

# node	Area(m²)	Simulation time (Sec)	Other pameters
50	1500x300	500	Communication pairs: 10;
100	2200x600	500	Communication load:
200	3500x900	300	4 Pks/source's;
500	4500x1500	300	Maximum speed: 20m/s;
1000	5500x2500	300	Minimum speed: 0m/s.

Fig. 4 illustrates the average route lifetime of each scheme. It shows that route lifetimes of all PRDS/n_0-MR schemes are nearly two times as much as AODV. In AODV, RD-random is used in route discoveries; while in PRDS/n_0-MR, the long lifetime link has the high priority to be selected. The simulation results demonstrate that

PRDS/n_0-MR mechanism is effective. Within PRDS/n_0-MR schemes, PRDS/1-MR and PRDS/2-MR perform better than PRDS/ALL-MR.

The performance of PRDS against route length is shown in Fig. 5. The results show that the average route length of each PRDS/n_0-MR scheme is 5~14% longer than that of AODV. The macrobian route tends to include shorter links than AODV and leads to longer paths. Considering the benefit from route lifetime, the cost in route length is worthy. PRDS/n_0-MR leverages the route length and the route lifetime.

Fig. 6 shows the same variation tendency of routing overhead for all routing strategies. That is, routing overhead increases as network size increases. However it may be seen that the routing overhead of AODV increases much more rapidly than that of PRDS/n_0-MR. The reasons for low routing overhead for PRDS/n_0-MR are 1) macrobian route decreases the number of route discoveries; 2) a large amount of rebroadcasts are avoided in the route discoveries.

Fig. 7 illustrates the packet delivery ratio. As expected, the packet delivery ratio of each scheme decreases as network size increases. However, the packet delivery ratio of AODV decreases more quickly than any of the PRDS/n_0-MR schemes. The difference between them increases as network size increases. The delivery ratio of PRDS/n_0-MR is 2%~5% higher than AODV in 50 node network, while 20%~25% higher in 1000 node network. Among all schemes, PRDS/2-MR performs best; PRDS/1 and PRDS/ALL-MR perform similarly; AODV performs worst. The throughput results of different schemes are of the same features as the packet delivery ratio.

Fig. 8 presents the average end-to-end delay of each scheme. In general, delay of any scheme increases as network size increases. In small networks (network size is less than 100), the delay of AODV is lower than that of PRDS/n_0-MR. Whereas, the delay of AODV increases rapidly as network size increases and exceeds the delay of PRDS/n_0-MR soon.

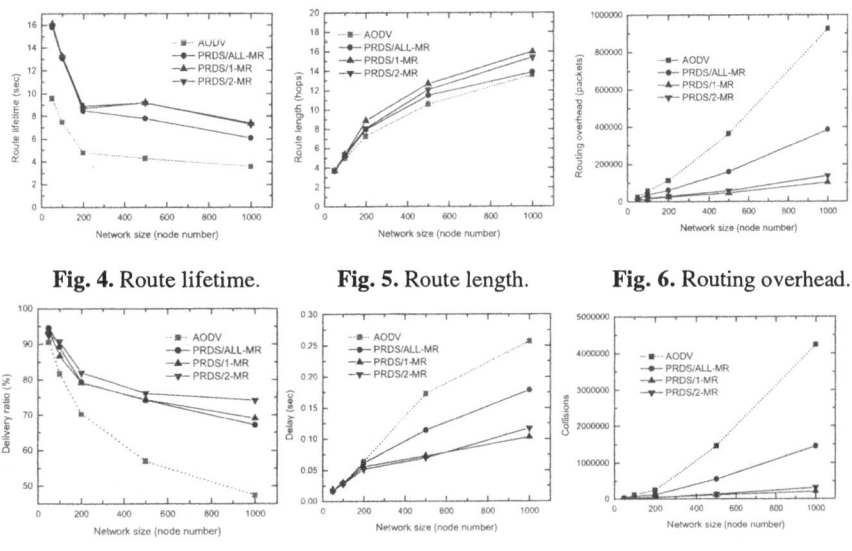

Fig. 4. Route lifetime.　　**Fig. 5.** Route length.　　**Fig. 6.** Routing overhead.

Fig. 7. Delivery ratio　　**Fig. 8.** Average delay.　　**Fig. 9.** Collisions.

Fig. 9 shows the signal collisions measured at MAC layer. If we compare Fig. 9 with Fig. 6, we will find that the curves in the two figures are very similar. This implies that signal collisions are highly correlated with routing overhead. Hence, AODV also consumes more bandwidth and energy than PRDS/n_0-MR.

From the above result analyses, we can conclude that PRDS/n_0-MR outperforms AODV in terms of packet delivery ratio, end-to-end delay, and route lifetime. At the same time, it decreases routing overhead significantly. It saves much network bandwidth and energy.

5 Conclusions

Current reactive route discovery protocols in Mobile Ad-hoc Networks (MANETS) employ random delays between rebroadcasting route requests. This has been shown to lead to "next-hop racing" and "rebroadcast redundancy". In this paper, a novel route discovery strategy termed Priority-based Route Discovery Strategy (PRDS) is described that addresses both these problems. PRDS can be applied to different reactive routing protocols by defining the Priority Index (*PI*) to find better quality route while decreasing routing overhead.

As an example, we have applied PRDS to design a macrobian route discovery strategy (PRDS-MR) and integrate it into AODV. Simulation results show that PRDS-MR outperforms AODV in terms of packet delivery ratio and end-to-end delay while decreasing routing overhead. PRDS-MR has better scalability than AODV.

Acknowledgement. The authors gratefully acknowledge the support and financial assistance provided by the UK EPSRC, under the project GR/S02105/01 "Programmable Routing Strategies for Multi-Hop Wireless Networks".

References

1. Perkins, C.E., Royer, E.M.: Ad-hoc On Demand Distance Vector Routing. In Proc. WMCSA 1999, New Orleans, LA. 90-100.
2. Johnson, D.B., Maltz, D.A.: Dynamic Source Routing in Ad Hoc Wireless Networks. In Mobile computing, edited by Tomasz Imielinski and Hank Korth, Chapter 5, Kluwer Academic Publishers (1996) 153-181.
3. Ni, S., Tseng, Y., Chen, Y., Sheu, J.: The Broadcast Storm Problem in a Mobile Ad Hoc Network. In Proc. ACM/IEEE MobiCom, Seattle, WA, USA (1999) 151-162.
4. Li, J., Mohapatra, P.: A Novel Mechanism for Flooding Based Route Discovery in Ad Hoc Networks. In Proc. IEEE GlobeCom 2003, San Francisco, USA.
5. Zhou, B., Wu, J., Fei, X., Zhao, J.: PCBA: A Priority-based Competitive Broadcasting Algorithm in Mobile Ad Hoc Networks. J. Computer Sci. & Technol., Vol.18 (2003) 598-606.
6. Bajaj, L., Takai, M., Ahuja, R., Tang, K., Bagrodia, R., Gerla, M.: GloMoSim: A Scalable Network Simulation Environment. UCLA Computer Science Department Technical Report 990027 (1999).

Performance Evaluation of AODV Protocol over E-TDMA MAC Protocol for Wireless Ad Hoc Networks

V. Loscrì, F. De Rango, and S. Marano

University of Calabria, D.E.I.S. Department, 87036 Rende (CS), Italy.
{vloscri, derango, marano}@deis.unical.it

Abstract. In this paper, a new protocol for scheduling TDMA transmissions in a mobile ad hoc network has been considered. In this protocol, nodes may reserve time slots for unicast, multicast or broadcast transmissions. The protocol uses contention when generating the schedules, and its operation is distributed and concurrent, hence its operation is not affected by the network size but only by the node density. Consequently it is scalable and can be used in large networks. This protocol is termed E-TDMA (Evolutionary-TDMA). This last one is not based on the contention and there is a very little contention through another protocol for the MAC level that has been considered: Five Phase Reservation Protocol (FPRP). The protocol's performance has been studied via simulation of a routing protocol (AODV) over E-TDMA MAC and the change of MAC parameters in order to evaluate the impact of MAC protocol over routing protocol has been evaluated.

1 Introduction

Wireless ad-hoc networks offer mobile nodes the possibility to share a common wireless channel without any centralized control or network infrastructure. Each node acts as a router forwarding packets to its neighbour nodes. In this environment, routing is a critical issue.

There are frequent unpredictable topological changes in these networks, which makes the task of finding and maintaining routes as difficult. Conventional routing protocols based on distance vector or link state algorithms can not be applied here, since the amount of routing related traffic would waste a large portion of the wireless bandwidth, and such discovered routes would soon become obsolete due to mobility of nodes [1].

Reactive protocols determine the proper route only when required, that is, when a packet needs to be forwarded. In this instance, the node floods the network with a route-request and builds the route on demand from the responses it receives. This technique does not require constant broadcasts and discovery, but on the other hand causes delays since the routes are not already available [6].

It is very interesting to investigate as the changes of the E-TDMA MAC's parameters impact over the network performance. The protocol performances for high speeds, differently by simulations given in literature [2-3] has been evaluated. It is due because we believe that an opportune combination of the permanent colours and those temporary ones can allow to obtain optimal performances also for increasing speeds.

J.N. de Souza et al. (Eds.): ICT 2004, LNCS 3124, pp. 417–424, 2004.
© Springer-Verlag Berlin Heidelberg 2004

Section 2 summarizes the basics of the MAC protocol. Here we consider two algorithms developed for the MAC level, the Five Phase Reservation Protocol (FPRP) and the Evolutionary-TDMA (E-TDMA).

Section 3 describes the Ad hoc On Demand Distance Vector (AODV) protocol that is a routing protocol designed for mobile ad hoc networks. The protocol's algorithm creates routes between nodes only when the routes are requested by the source nodes, giving the network the flexibility to allow nodes to enter and leave the network at will. Section 4 summarises simulation assumptions, and Section 5 gathers numerical results. Conclusions are given in Section 6

2   Medium Access Control (MAC) Protocol

The medium access control and its integration and interaction with routing protocol are a key issue for the deployment of efficient wireless ad hoc networks. Differently by classic studies in literature focusing on 802.11 MAC, this work focuses on a novel MAC protocol and its interaction with a routing protocol. The considered E-TDMA MAC is interesting for its contention phase technique exploited through *Five Phase Reservation Protocol* (FPRP) and for its update schedules mechanism based on two-hop neighbour schedule exchange. FPRP and *Evolutionary Time Division Multiple Access* (E-TDMA) make possible to build a routing protocol with QoS mechanisms [4] in order to exploit services different by classical best-effort service over wireless ad hoc networks. In the following a brief explanation is given.

2.1   Five Phase Reservation Protocol (FPRP)

A node can reserve time slot for data transmissions if it has a particular slot defined temporary colour. A node acquires a temporary colour through a protocol called Five Phase Reservation Protocol (FPRP) [3]. FPRP is a protocol for a node to reserve a broadcast slot with contention using a five-phase message exchange.

The protocol jointly and simultaneously performs the tasks of channel access and node broadcast scheduling. The protocol allows nodes to make reservations within TDMA broadcast schedules. It employs a contention-based mechanism with which nodes compete with each other to acquire broadcast TDMA slots. A node needs no a priori information about the network. A node uses the FPRP to explore its neighborhood and to make nearly *conflict-free* reservations.

A reservation cycle has five phases and to see the specific tasks of these phases to see [3].

2.2   Evolutionary-TDMA

The E-TDMA is a new protocol for scheduling TDMA transmissions in a mobile ad hoc network [2]. In this protocol, nodes may reserve time slots for unicast, multicast or broadcast transmissions.

This protocol is very interesting because it is not centralized and it is independent from a central controller like other developed protocols [4].

The E-TDMA protocol allows nodes to assign TDMA transmission slots among themselves as network composition and bandwidth demands change. Nodes determine who can reserve transmission slot by contending for a permission called *temporary color* (t_c). Many nodes can acquire this permission simultaneously. A temporary color is a permission to reserve new information slots or permanent colors. If a node needs to make new reservation in a control epoch, it first needs to acquire one of these permissions. Its temporary color becomes invalid after this control epoch, and if it wants to make another reservation later it has to contend again.

A node needs a *permanent color* (p_c) in the *ctrl-schedule* for exchanging its scheduling information with its neighbors (but not for making new reservations). Once a node acquires a *permanent color*, it transmits in every slot designated this color as long as its transmission does not collide with others. If a collision occurs due to some topological change, a node will discard its current permanent color and reserve a new one.

The protocol produces two TDMA schedules simultaneously, each used in a different portion of the same channel and for different purpose. The first schedule is a broadcast schedule in which every node is assigned one slot. This broadcast schedule is used for nodes to exchange information in the control frame and is called the *control schedule* (ctrl-schedule). The second schedule carries user generated traffic in the information frame, and is called the *information schedule* (info schedule).

There are two phases in the Control Epoch: In the Contention phase there are N FPRP cycles (where N corresponding to N temporary colour) as shown in fig.1. For every cycle FPRP there is an Allocation (in the Allocation Phase) as shown in fig.2

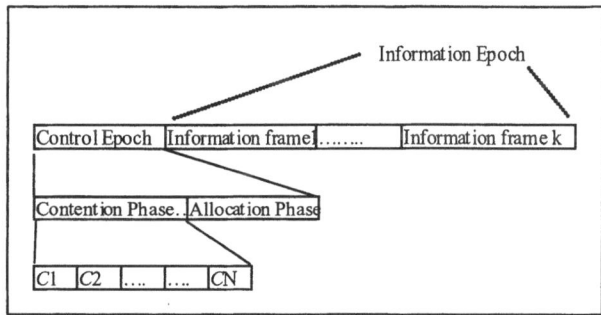

Fig. 1. Control schedule and information schedule structure in E-TDMA MAC protocol.

In the protocol, all nodes participate in the scheduling process on an equal basis. The scheduling process is executed across the entire network at the same time. Nodes do not wait in some particular order to schedule their transmissions.

Every node is responsible for its own transmission schedule. After a transmission is complete, the transmitter releases the slot, which can be reserved for another transmission. The protocol is characterized for a local nature. Every node exchange information with its one-hop neighbors, so it is not sensitive to the network size (Fig.3). The protocol operates within a single TDMA channel. The channel is partitioned into two epochs: a *control epoch* where the schedules are updated by the protocol, and an *information epoch* where user data transmission takes place. The two epochs are interleaved periodically. Every node generates and maintains its own schedules in collaboration with its neighbors. No single node has global information such as the size, the membership or the schedules of the entire network. A node only directly interacts with its one-hop neighbors. Every node has information to other node (two-hop neighbors) indirectly through information to one-hop neighbors.

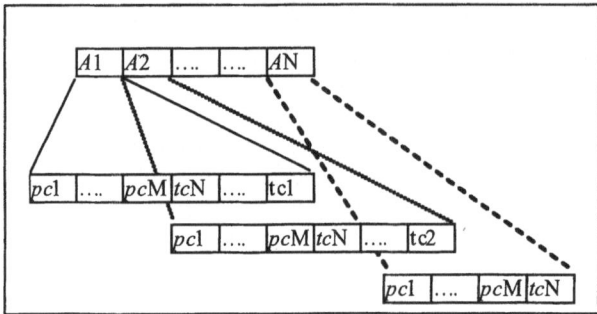

Fig. 2. Allocation phase structure (pc1....pcM are permanent colours and tcN....tc1 are temporary colours)

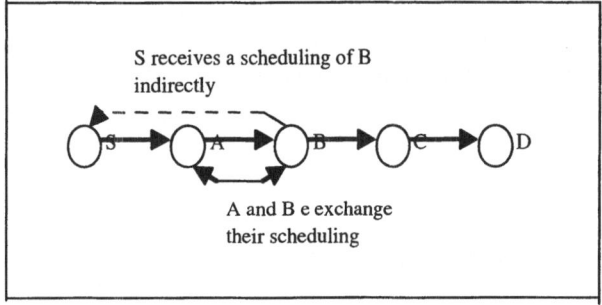

Fig. 3. Schedules updates between neighbour nodes in the control phase of E-TDMA protocol.

If a node S has to communicate with a node D the communication has to be *conflict-free*, in the sense that E-TDMA compute the scheduling above a path and assign the *conflict-free* slot above every link. If a communication is established, at the moment that it is established it's *conflict-free* if there is the E-TDMA protocol at the MAC layer. If the number of temporary' colours is increased the collision probability of the nodes until two hops of distance is reduced, because if two nodes that are to one-hop or two-hop distance contend for a temporary color in the same Control epoch they can receive both a temporary color for same FPRP cycles if there is more than a temporary color

3   AODV Protocol

The Ad hoc On Demand Distance Vector (AODV) [6-7], routing algorithm is a routing protocol designed for ad hoc mobile networks. AODV is capable of both unicast and multicast routing. It is an on demand algorithm, meaning that it builds routes between nodes only as desired by source nodes. It maintains these routes as long as they are needed by the sources.

It is loop-free, self-starting, and scales to large numbers of mobile nodes.

We can see in fig.4 as AODV protocol reacts when a source S establishes a route with a destination D. S sends a RREQ' packet in broadcast and your neighbourhood (nodes A, B and C) receive this request. E will process only the first one packet that it receives. Nodes A, B and C rebroadcast the request packet (RREQ) received from S, because not have any route from the destination. B will not process the RREQ's packet received from C.

When a source node desires a route to a destination for which it does not already have a route, it broadcasts a route request (RREQ) packet across the network. Nodes receiving this packet update their information for the source node and set up backwards pointers to the source node in the route tables.

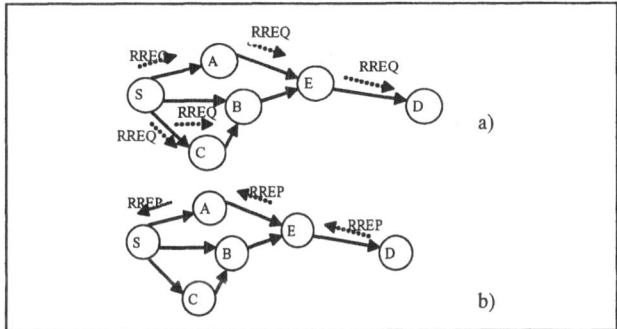

Fig. 4. a) Route request phase b) Route reply phase in AODV routing protocol.

A node receiving the RREQ may send a route reply (RREP) if it is either the destination or if it has a route to the destination with corresponding sequence number greater than or equal to that contained in the RREQ. If this is the case, it unicasts a RREP back to the source, otherwise, it rebroadcasts the RREQ.

As the RREP propagates back to the source, nodes set up forward pointers to the destination. Once the source node receives the RREP, it may begin to forward data packets to the destination. If the source later receives a RREP containing a greater sequence number or contains the same sequence number with a smaller hopcount, it may update its routing information for that destination and begin using the better route.

4   The Simulation Model

The performance of the MAC and routing protocols are studied with simulations realized on Network Simulator [8]. The implementation is based on the AODV module contributed by the MONARCH group, and the MAC protocol E-TDMA is added. The number of nodes in the simulated ad hoc network is 40; they are positioned in a 1000m x 1000m square region and have a transmission range of 250m.

We assume that each node moves with max speed $v_{\max}$ and has a pausing time equal to 10 seconds. The node speed is uniformly distributed in the range $[0, v_{max}]$.

We run our simulations with movement patterns generated for 5 different maximum speed 0.5, 5, 10, 15, 20 m/s. 28 sources have been considered where a CBR source generates packets of 64 bytes (a packet becomes 84 bytes after IP header is added) at a rate of 20 packets per second.

Table 1 summarises the simulation parameters.

Table 1. Simulation Parameters

Input Parameters	
Simulation area	1000x1000 m
Traffic sources	CBR
Number of sources	28
Sending rate	20 packets/s
Size of data packets	64 bytes
Transmission range	250 m
Simulation Time	500 s
Mobility Model	
Mobility Model	Random Way point
Pause time	10 s
Mobility average speed	0.5 ,5,10,15,20 m/s
Traffic pattern	Peer-to-peer
Simulator	
Simulator	NS-2 (version 2.1b8a)
Medium Access Protocol	IEEE 802.11
Link Bandwidth	2 Mbps
Confidence interval	95%
MAC Parameters	
Number of Permanent Color	15, 20
Number of Temporary Color	1,2
Number of bytes for a slot	32
Number of slots for a frame	40
Number of info frames for a cycle	4
Number of FPRP cycles	8

4.1   Numerical Results

In fig.5 the throughput performance of wireless ad hoc networks for different maximum speeds is given. It is possible to observe the decreasing of data packet delivery ratio for high speed. This is due to instability of mobility scenario and to the great number of broken links of mobile nodes. The MAC parameters impact differently on AODV performances. For increasing permanent colours number a light improvement of the throughput (2-3%) is showed. More permanent colours gives more quickly the acquisition permission of channel for exchanging control information. The increase of temporary colours produces a improvement too (10%). More temporary colours are used in the networks, more chances of obtaining wireless channel are given. The improvement given by increasing temporary colours is more effective than increasing permanent colours. This due to greater possibility of giving transmission information slot to mobile node in few FPRP cycles, reducing the mobile nodes attempts to acquire the channels. In this way is reduced the collision probability. The increasing chance to transmit that is given to mobile node lead to decrease the normalized control overhead as shown in fig.6.

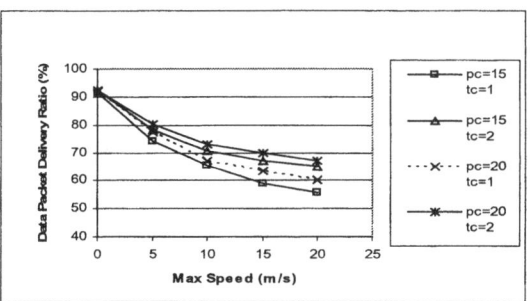

Fig. 5. Number of received packet over dent data for AODV protocol over E-TDMA varying permanent (pc) and temporary (tc) colours

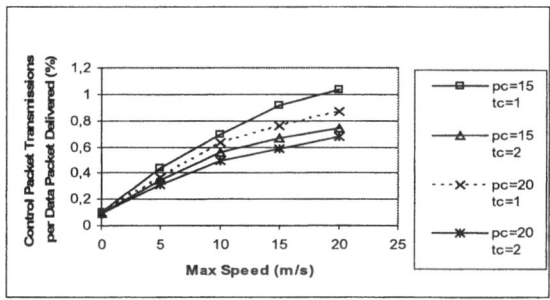

Fig. 6. Control Packet Overhead of AODV over E-TDMA varying permanent (pc) and temporary (tc) colours

The drawbacks to increase the permanent and temporary colours could be to have control epochs longer in terms of time for a fixed number of FPRP cycles (in our case 8 FPRP cycles have been considered). However, the fig.7 shows the improvement in terms of average end-to-end data packet delay. So for 15-20 permanent or 1-2 temporary colours the longer control epoch does not impact so negatively on routing protocol and the greater chance to obtain the info slot given by temporary colours or the better chance to exchange schedule updates given by permanent colours improve also the data packet delay.

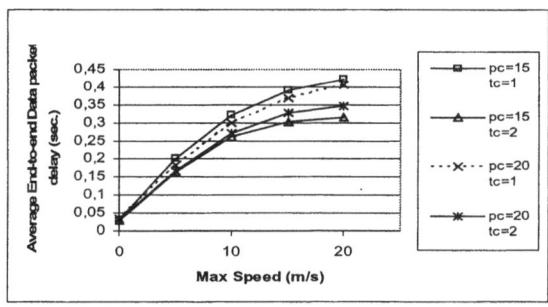

Fig. 7. Average end-to-end data packet delay for AODV protocol over E-TDMA varying permanent (pc) and temporary (tc) colours

5   Conclusions

This paper studies the impact of a novel MAC protocol called E-TDMA over a known routing protocol. The routing protocol considered has been the AODV protocol. The performance of E-TDMA has been evaluated for different permanent and temporary colours and the different impacts over considered routing protocol have been evaluated. This work lead out that to give to the wireless ad hoc networks more permanent and temporary colours can improve the routing protocol performance. This improvement are more effective for increasing temporary colours number than permanent colours ones.

References

1. Elizabeth M. Royer Charles Perkins and Samir R. Das. Quality of Service for Ad Hoc On-Demand Distance Vector Routing. In *Internet-Draft, draft-ietf-manet-aodvqos-00.txt*, Work in Progress, July 2000.
2. Chenxi Zhu. *Medium Access Control and Quality-of-Service Routing for Mobile Ad Hoc Networks*. PhD thesis, Department of Electrical and Computer Engineering, University of Maryland, College Park, MD 20906, 2001.
3. C. Zhu and M. S. Corson. A five phase reservation protocol (FPRP) for mobile ad hoc networks. In *Proc. IEEE INFOCOM*, 1998
4. C. Zhu and M. S. Corson, "QoS Routing for Mobile Ad Hoc Networks," INFOCOM 2002.
5. S. Ramanathan. A Unifed Framework and Algorithm for (T/F/C)DMA Channel Assignment in Wireless Networks. In Proc. INFOCOM, Kobe, Japan, April 1997
6. C. Perkins, E. M. Royer and S. R. Das. Ad Hoc On-Demand Distance Vector routing. In *Internet-Draft, draft-ietf-manet-aodv-06.txt*, July 2000.
7. Elizabeth M. Royer, Chai-Keong Toh, "A review of current routing protocols for ad hoc mobile wireless networks", *IEEE Personal Communications*, no. 2, April 1999 pp. 46-55.
8. K.Fall, K.Varadhan (Editors), "Ucb/lbnl/vint Network Simulator–ns(version2)", http//www-mash.cs.berkeley.edu/ns/".

Mitigating the Hidden Terminal Problem and Improving Spatial Reuse in Wireless Ad Hoc Network*

Jun-Wan Bang[1], Sung-Kwan Youm[1], Chul-Hee Kang[1], and Seung-Jun Seok[2]

[1] Department of Electronics Engineering, Korea University
5-ga, Anam-dong, Sungbuk-gu, Seoul 136-701 Korea
{venezia, skyoum, ssj, Chkang}@widecomm.korea.ac.kr
[2] Dep. of Computer Science and Engineering, Kyungnam University
449 wolyoung-dong, Mssan, 631-701 Korea

Abstract. The 802.11 DCF was designed for the single-hop wireless LAN scenario, where one central control system such as AP controls many nodes and mitigated the hidden terminal problem by exchanging the RTS/CTS message. However, in Mobile Ad Hoc Network (MANET), RTS/CTS exchange mechanism is not sufficient to solve hidden terminal problem and also this mechanism reserved a space more than needs so that spatial reuse is reduced. In this Paper, we proposed new Distributed MAC protocol to mitigate the hidden terminal problem and to increase the spatial reuse. we classify the neighbor nodes into two states – free state and constraint state – based on the interference tolerance of a receiving node.

1 Introduction

A MANET uses the IEEE 802.11 MAC DCF which is used in WLAN to access a wireless medium and transmit a frame [2]. With this scheme, hidden terminal problem is partially solved especially in multi-hop ad hoc networks. Because the 802.11 DCF was devised only for a single AP where all nodes are within the transmitting range of one other, various problems occur in wireless multi-hop environments. Figure (1) shows that the interference range of nodes are varied up to the distance between a sender and a receiver. In figure 1(a), the interference range (R_i) is larger than transmission range (R_{tx}) of a node, node D is out of the transmission range of node B but within the interference range of node B. As a result node B could experience collisions with a reception frame. Also, in figure 1(b), as the interference range of a node B is smaller than the transmission range of the node B, the reserved wireless medium by exchanging the RTS/CTS control message is larger than the interference range of the node. Although node C is out of the interference range of the node B, node C is deferring the transmission of a frame during a NAV time. As a result spatial reuse efficiency is reduced.

---

* This research was supported by S.A.I.T.(Samsung Advance Institute of Technology), under the project "Research on Core Technologies and its Implementation for Service Mobility under Ubiquitous Computing Environment.

J.N. de Souza et al. (Eds.): ICT 2004, LNCS 3124, pp. 425–430, 2004.

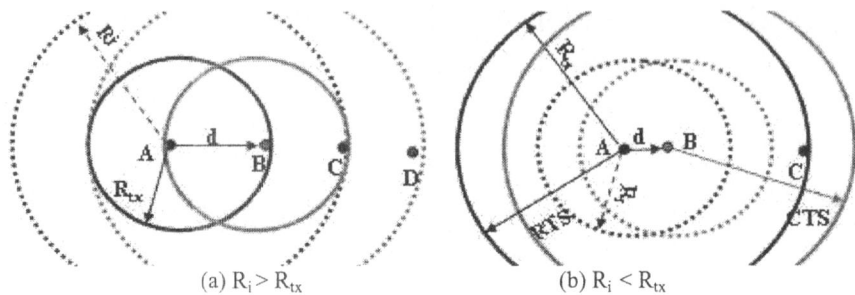

Fig. 1. The relationship between interference range and transmission range up to the distance between nodes

In [1][6], the 802.11 RTS/CTS handshake is analyzed and these are shown that this mechanism is not always effective due to the fact that the power needed for interrupting a packet reception is much lower than that for delivering a packet successfully. Thus the virtual carrier sensing (VCS) implemented by RTS/CTS handshake cannot prevent all interferences. [1] shows that the interference range is varied up to the distance between the transmitter and receiver. The Interference range of the nodes is 1.78 times the distance of two nodes. When the distance of two nodes is greater than 0.56 times R_{tx}, R_i of the nodes is larger than the R_{tx} of the nodes. So, Hidden terminal problem is not completely solved. To solve this problem, a node only replies a CTS frame for a RTS when the receiving power of that RTS frame is larger than the a certain threshold. Up to the distance between sender and receiver, [3] classified three different case with respect to the effectiveness of medium reservation, underactive, moderate, overactive respectively and showed spatial reuse index to evaluate the efficiency of the medium reservation accomplished by 802.11 VCS. [3] also showed that the space reserved by 802.11 for a successful transmission does not always match the space in which interference may occur. In case of overactive RTS/CTS scenario, [3] shows the spatial reuse efficiency is reduced prominently.

In this paper, to solve the hidden terminal problem and improve the spatial reuse, we classify neighbor nodes into two states, free state and constraint state , with respect to the interference tolerance of a receiver. When neighbor nodes are in the free state, they are robust to the interference due to potential hidden terminals and a spatial reuse efficiency is modest. To improve spatial reuse efficiency, these neighbor nodes in the free state could transmit a frame regardless of remaining NAV values which these neighbor nodes have. When neighbor nodes are in the constraint state they are susceptible to the interference caused by potential hidden terminals. To mitigate this problem we propose alert control message and using this message potential hidden terminals can be prevented from transmitting a frame.

2   A Proposed Distributed MAC Protocol

When neighbor nodes overhear RTS or CTS, these neighbor nodes are classified into two state with regard to the interference tolerance of a receiver. To classify these neighbor nodes into "Free state" and "Constraint state", we modify the existing RTS/

CTS messages to the Extended RTS (E_RTS) and Extended CTS (E_CTS) such as figure (2). As a sender transmit the E_RTS the transmitting power (P_{tx_rts}) of the sender is specified in the power filed of the E_RTS header. On receiving the E_RTS, the receiver transmit the E_CTS as a response to it. When sending the E_CTS, the receiver includes its own interference tolerance power ($P_{i\_tolerance}$) and P_{tx_cts}. The $P_{i\_tolerance}$ at the receiver is the margin of the power which receiver can endure and receive a frame without any errors even though the receiver experiences interferences caused by neighbor nodes' transmission. On receiving E_RTS, the sender measure a receiving power (P_{rx}) and evaluate $P_{i\_tolerance}$ using eq.(1)

$$P_{i\_tolerance} = \frac{P_{rx}}{SIR_{\_threshold}} \tag{1}$$

Where the $SIR_{\_threshold}$ is the signal to interference ratio threshold at which receiver can decode the encoded receiving frame without any errors. If experienced interference power at the receiver is greater than $SIR_{\_threshold}$, the receiver can not decoded the frame. P_{rx} should greater than or equal to the $P_{r\_threshold}$ which is a minimum reception power that receiver can receive and decode a frame.

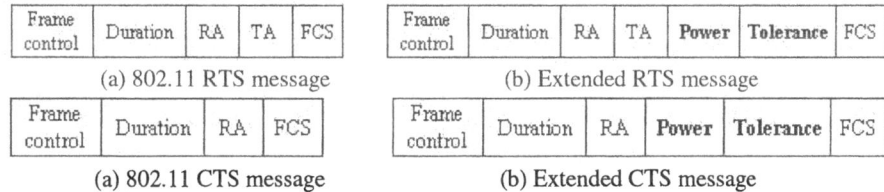

(a) 802.11 RTS message   (b) Extended RTS message

(a) 802.11 CTS message   (b) Extended CTS message

Fig. 2. Modified Extended RTS/ Extended CTS to classify neighbor nodes` states

2.1 A Constraint State

A constraint state: when the nodes in a constraint state defer a transmission during NAV and broadcast the alert message to mitigate the hidden terminal problem. To do this, when neighbor node k overhear the E_RTS or E_CTS from sender i or receiver j respectively, they measure P_{rx} and evaluate the channel gain G_{ik} or G_{jk} using eq.(2) – P_{tx_rts}, P_{tx_cts} is specified in the field of power.

$$G_{jk} = \frac{P_{rx}(receiving\ power)}{P_{tx\_rts,cts}(transmission\ power)} \tag{2}$$

When node k overhearing a E_RTS or a E_CTS transmits a frame with a transmission power of P_{tx_k}, this power appear at node j with the interference $G_{jk} * P_{tx\_k}$. If this interference is greater than $P_{i\_tolerance}$, it could disrupt a current receiving so that the node satisfied eq.(3) is considered oneself constraint state and is deferring a transmission for NAV.

$$P_{i\_tolerance} < G_{jk} * P_{tx\_k} \tag{3}$$

Alert message: as figure (3) the R_i of a node is greater than the R_{tx} of it , the nodes (ex: node D) which is out of the R_{tx} of the receiver (e.g. B) but within the R_i could disrupt the current reception at receiver. To solve this problem, among the nodes which overhear only E_CTS message, only node which satisfy the eq.(3) above and eq.(4) below can broadcast

$$P_r \cong P_{r\_threshold} \tag{4}$$

the alert message. on the reception of this alert message, nodes (e.g. D) should defer their transmission for NAV so that current reception at receiver (e.g. B) couldn't be disrupted by the hidden nodes (ex: D). That is, the node which is in the constraint state and within the transmission range of the receiver j can broadcast the alert message. The wireless channel access procedures for the alert message transmission shows in figure 3(a). A node which meets the eq. (3), (4) can access a wireless channel with random backoff mechanism. the maximum backoff time is 6*SIFS. If nodes which ready to send the alert message receive the alert message, this nodes stop to send and drop it. That is, the only node which access the channel first can broadcast the alert message. If nodes receive the alert message but didn't overhear E_RTS/E_CTS , these node are out of the R_{tx} of both sender and receiver. So these hidden nodes potentially interfere with a reception at the receiver. To prevent this, nodes which receive the alert message but didn't overhear E_RTS/E_CTS should

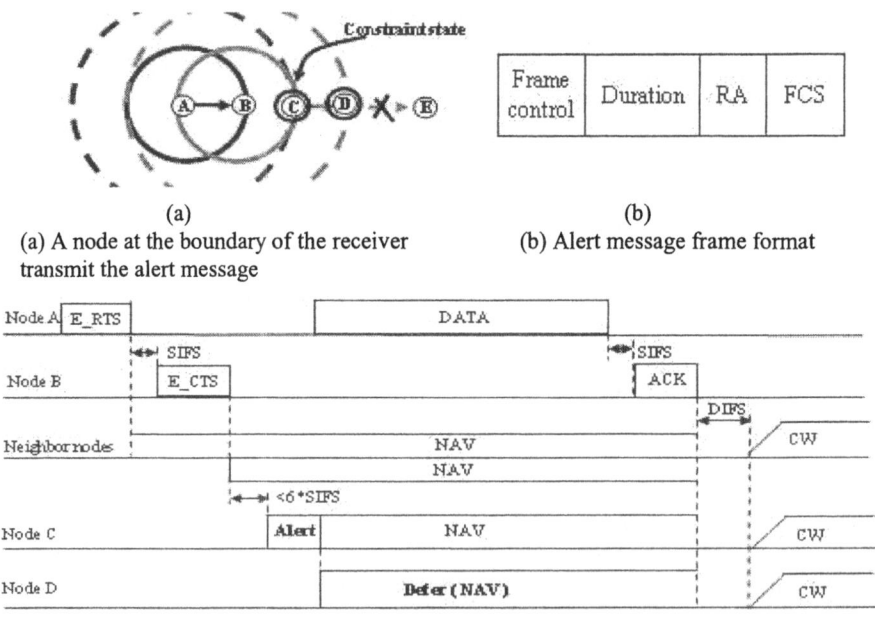

	(a)	(b)
	(a) A node at the boundary of the receiver transmit the alert message	(b) Alert message frame format

(c) Exchange of E_RTS/E_CTS/DATA/ACK

Fig. 3. Alert message to mitigate the hidden terminal problem

defer their transmission during NAV which is included in duration field of alert message. This NAV indicate the duration until sender can receive ACK frame.

2.2 A Free State

If neighbor nodes receive E_RTS or E_CTS or both and satisfy eq.(6), they are considered to be in free state.

$$P_{i\_tolerance} > G_{jk} * P_{tx\_k} \tag{5}$$

That is, as receiving signal power of receiver is even larger than interference noise power, the free state is robust to interference. Therefore, neighbor node that satisfy eq.(5) can transmit frames without interfering with a reception at the current receiver. So, nodes in free state can access to wireless channel through normal CSMA/CA mechanism regardless of remaining there own NAV and transmit frames to another nodes except node that sender ID is same as E_RTS or E_CTS overheard.

3 Simulation

In this paper, the performance of proposed MAC protocol is compared with standard IEEE 802.11 MAC protocol in ad hoc environment using ns-2.26[4]. We use a String topology and analyzed performance with varying distance between nodes. Wireless channel is assumed as symmetric in both direction, and we use the Two-way ground model[5]. Bandwidth of wireless channel is 2 Mbps, and we didn't consider the movement of nodes.

In figure 4, we can see the throughput of both 802.11MAC protocol and proposed MAC protocol are increased as the distance between nodes is far apart. Because the number of nodes that can transmit frames at the same time is increase as the distance between two nodes is increasing.

Proposed MAC protocol mitigate the hidden terminal problem and enlarged the space efficiency, therefore the total throughput is improved as shown by figure (4).

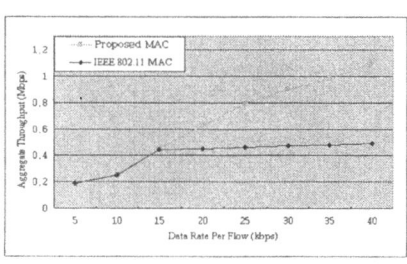
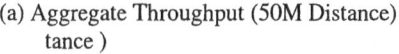
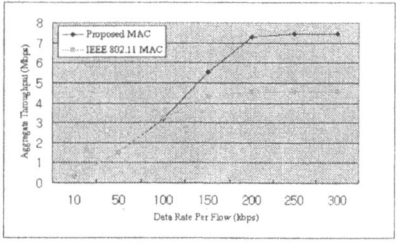

(a) Aggregate Throughput (50M Distance)   (b) Aggregate Throughput (200 M Dis-
    tance)

Fig. 4. The throughput up to the distance between nodes

4 Conclusion

In this paper, we specified neighbor nodes into two states, the free state and the con-straint state, to mitigate the hidden terminal problem and enlarge the space reuse effi-ciency. To determine neighbor node's state, we introduced E_RTS/CTS messages that was modified from original RTS/CTS messages. The case that the neighbor nodes are in the constraint state, they could transmit the new alert message we proposed to solve the hidden terminal problem. And the case that neighbor node is in the free state, transmitting data without physical/virtual carrier sensing to enlarge the space efficiency and increase throughput.

References

[1] S. Xu and T.Saasdawi," Dose the IEEE 802.11 MAC Protocol Work Well in Multihop Wireless Ad Hoc Networks?" IEEE Communications Magazine, vol 39,no 6, pp 130-137, June 2001.
[2] ANSI/IEEE std 802.11: Wireless LAN Medium Access Control(MAC) and PHY Spec, 1999.
[3] Fengji Ye, Su Yi Biplab Sikdar," Improving Spatial Reuse of IEEE 802.11 Based Ad Hoc Network" IEEE GLOBECOM, San Francisco, CA Dec 2003.
[4] CMU Monarch Group, "CMU Mornarch Extensions to ns" <http://www.monarch.cs.cmu.edu/802.11>
[5] T. Rappaort, "Wireless Communications: principle and practice", Prentice Hall, New Jersey, 1996.
[6] K. Xu, M. Gerla, and S. Bae, "How Effective is the IEEE 802.11 RTS/CTS Handshake in Ad Hoc networks?" In Proc. GLOBECOM 2002, Taipei, November 2002.

A New Distributed Power Control Algorithm Based on a Simple Prediction Method

Raimundo A. de O. Neto, Fabiano de S. Chaves, Francisco R.P. Cavalcanti,
and Tarcisio F. Maciel

Wireless Telecommunications Research Group - GTEL
Federal University of Ceará, Fortaleza, Brazil
{neto, fabiano, rod, maciel}@gtel.ufc.br
http://www.gtel.ufc.br

Abstract. Most distributed power control algorithms have been proposed assuming constant interference and constant path gain. These considerations may result in lower performance gains in fast time-varying
channel conditions. The algorithm presented herein addresses this problem efficiently and it is based on a simple prediction method, utilizing
Taylor's Series. In each iteration, the proposed algorithm predicts both
the path gain and the interference and after that, adjust the transmit
power.

1 Introduction

In cellular wireless systems, good communication can be efficiently provided by
ensuring just a minimum signal quality for individual connections. By appropriately adjusting transmission power levels, minimum link quality requirements
can be attained without incurring in unnecessary interference generation. This
technique is called power control.

However, the employment of this technique is not trivial when we strive
with a multipath environment, where fast fading occurs, since SINR depends
on the path gain and the co-channel interference, which are influenced by fast
fading. Fast fading, also called short-term fading or multipath fading, is the
phenomenon that describes the rapid amplitude fluctuations of a radio signal
over a short period of time or travel distance. These rapid fluctuations cause
degradation in the action of power control [1].

Some papers have studied the performance of power control in fast fading
environment. In [2], a new algorithm is derived from the classical Distributed
Power Control algorithm (DPC) [3], considering a time-varying path gain. In
[4], a neural network is used to predict the future channel conditions. Similarly,
[5] and [6] use adaptive filters, with tap weights updated by least-mean-square
(LMS) and recursive least-square (RLS) algorithms, respectively, in order to
predict the future path gain.

In this work, a new distributed power control algorithm is presented, outperforming the classical Distributed Power Control algorithm (DPC) [3] when a

J.N. de Souza et al. (Eds.): ICT 2004, LNCS 3124, pp. 431–436, 2004.
© Springer-Verlag Berlin Heidelberg 2004

time-varying channel is considered. This new algorithm differs from the DPC in that its deduction assumes both path gain and interference to be time-varying functions and it predicts this variations through the Taylor's Series.

2   The New Algorithm

The discrete-time SINR $\rho_i(k)$ of a link is given by:

$$\rho_i(k) = \frac{g_i(k) \cdot p_i(k)}{I_i(k)} \tag{1}$$

So, the instantaneous transmit power necessary to balance this link for $\rho_i(k) = \rho_t$, for all instants k, is such that:

$$\rho_t = \frac{g_i(k) \cdot p_i(k)}{I_i(k)} \Rightarrow p_i(k) = \frac{\rho_t \cdot I_i(k)}{g_i(k)} \tag{2}$$

We do not predispose of values of $I_i(k)$ and $g_i(k)$, because these are instantaneous values. In order to solve this problem, we propose a simple prediction method based on Taylor's Series.

Taylor's Series is used to expand continuous functions $f(x)$ in following form [7]:

$$f(x) = f(x_0) + \sum_{n=1}^{\infty} \frac{f^{(n)}(x_0) \cdot (x - x_0)^n}{n!} \tag{3}$$

where the term $f^{(n)}(x)$ represents the nth derivative of $f(x)$ with respect to x. Due to $(x - x_0)^n$ and $n!$, when x and x_0 are adjacent values, the higher order terms can be neglected. Thus, keeping only the first two terms of the series, we have:

$$f(x) \approx f(x_0) + f'(x_0) \cdot (x - x_0) \tag{4}$$

Now, we transform (4) into a difference equation. For this, we assume that x_0 is the current discrete time instant k and x the next instant $k + 1$. Further, $f'(x_0)$ is approximated by $f(k) - f(k - 1)$. In this way, we obtain:

$$f(k + 1) \approx 2 \cdot f(k) - f(k - 1) \tag{5}$$

Therefore, we can use (5) in order to predict the path gain and interference:

$$\hat{g}_i(k + 1) = 2 \cdot g_i(k) - g_i(k - 1) \tag{6}$$

$$\hat{I}_i(k + 1) = 2 \cdot I_i(k) - I_i(k - 1) \tag{7}$$

So, using (2), (6) and (7), the transmit power at instant $(k + 1)$ is expressed by the following proposed algorithm:

$$p_i(k + 1) = \rho_t \cdot \frac{\hat{I}_i(k + 1)}{\hat{g}_i(k + 1)} = \rho_t \cdot \left[\frac{2 \cdot I_i(k) - I_i(k - 1)}{2 \cdot g_i(k) - g_i(k - 1)} \right] \tag{8}$$

The obtained SINR in discrete time $k + 1$ is:

$$\rho_i(k + 1) = \rho_t \cdot \frac{\widehat{I}_i(k + 1)}{I_i(k + 1)} \cdot \frac{g_i(k + 1)}{\widehat{g}_i(k + 1)} \tag{9}$$

Note that when the estimations tend to correct values, that is, $\hat{g}(k + 1) \approx g(k + 1)$ and $\hat{I}(k + 1) \approx I(k + 1)$, the SINR tends to ρ_t.

3   Simulation Results

We now illustrate the performance of the proposed algorithm by simulations utilizing a simulator consisting of a co-channel set of trisectorized base station in downlink direction. The co-channel sector set size comprises one layer of interferes. Base stations are localized on the corner of sectors. The sector antenna radiation pattern employed is ideal. The main-lobe gain is 0 dBi and the gain outside sector is -200 dBi. A snapshot simulation model is assumed where mobile stations are uniformly distributed over the cell area. In each snapshot, up to 600 iterations of the power control algorithm are performed, in intervals of 1 ms. Other simulation parameters are set as follows. The cell radius is set to 1.5 km. A simplified path loss model is used, where $PL(d) = 120 + 40 log_{10}(d)$ [dB]. The distance d is expressed in kilometers and represents the distance between mobile and base stations. Shadowing standard deviation is assumed 6 dB. Fast fading is implemented following the Jakes' model [8] with two different Doppler spreads: 18.5 Hz and 92.5 Hz.

The target SINR ρ_t for both algorithms is set to 8 dB. Maximum base station transmit power is limited to 35 dBm and the initial transmit power is set to the minimum transmit power (-70 dBm). The noise power is set to -110 dBm.

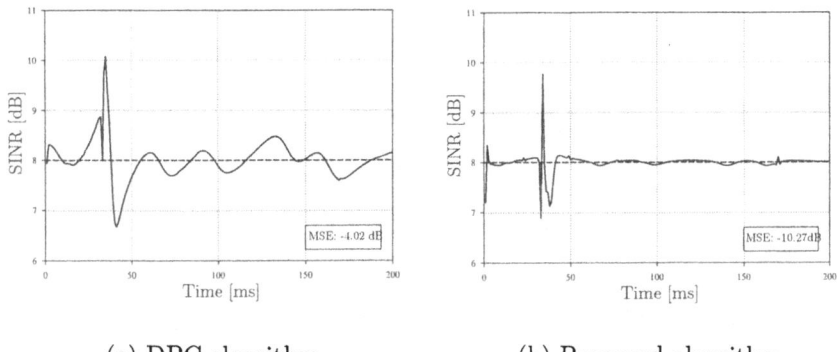

(a) DPC algorithm.   (b) Proposed algorithm.

Fig. 1. Sample of SINR evolution for the evaluated power control algorithms, with Doppler spread 18.5 Hz and reuse pattern 3/9.

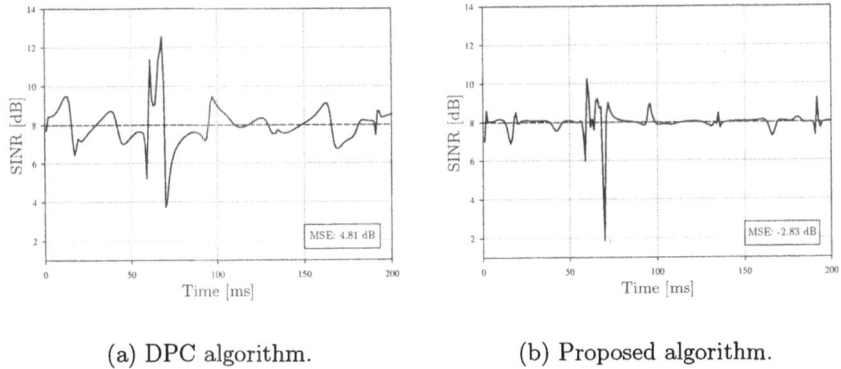

(a) DPC algorithm.   (b) Proposed algorithm.

Fig. 2. Sample of SINR evolution for the evaluated power control algorithms, with Doppler spread 18.5 Hz and reuse pattern 1/3.

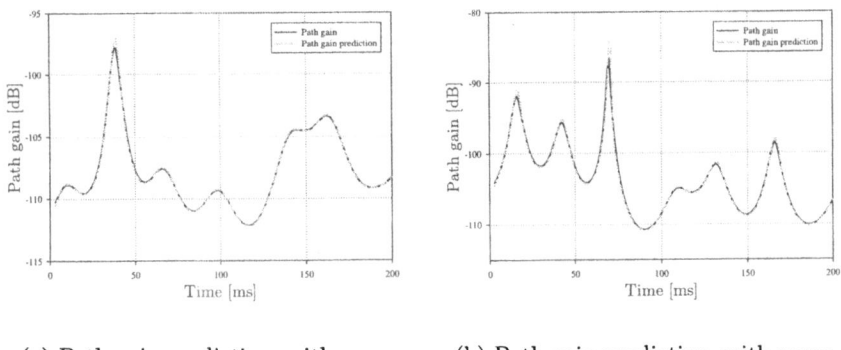

(a) Path gain prediction, with reuse   (b) Path gain prediction, with reuse
3/9   1/3

Fig. 3. Comparison between actual and predicted path gain using the proposed prediction method, for a Doppler spread 18.5 Hz.

Figs. 1 and 2 show a sample of the SINR evolution achieved by a given user in a typical snapshot for DPC and the proposed algorithm. In this case, seven co-channel base stations are considered and the same system configuration and fading realizations are used for both algorithms. The simulated reuse pattern for each figure is 3/9 and 1/3, respectively, with Doppler spread 18.5 Hz.

From figs. 1 and 2, it is clearly observable that the proposed algorithm is able to stabilize the SINR around the target SINR better than DPC algorithm. In other words, the mean squared error (MSE) between the actual and the target SINR is smaller for the proposed algorithm than for DPC. This behavior was observed for all snapshots.

A sample of how the proposed algorithm performs with gain and interference prediction is shown in figs. 3 and 4 for the same snapshot in figs. 1 and 2. Figs.

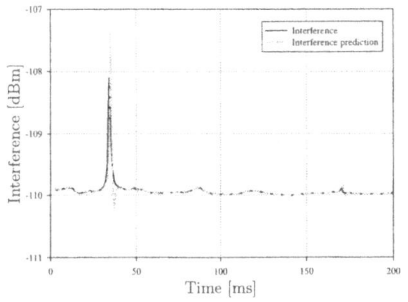
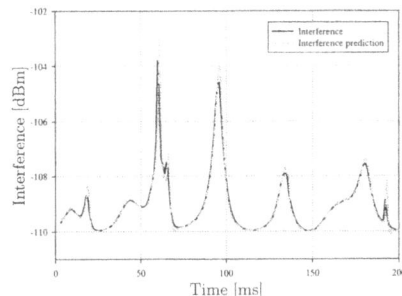

(a) Interf. prediction, with reuse 3/9   (b) Interf. prediction, with reuse 1/3

Fig. 4. Comparison between actual and predicted interference using the proposed prediction method, for a Doppler spread 18.5 Hz.

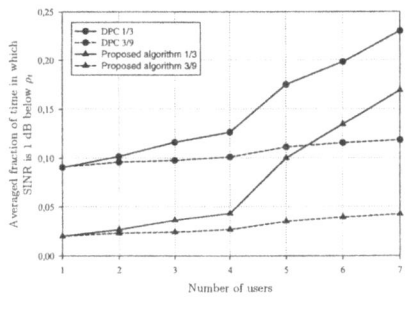
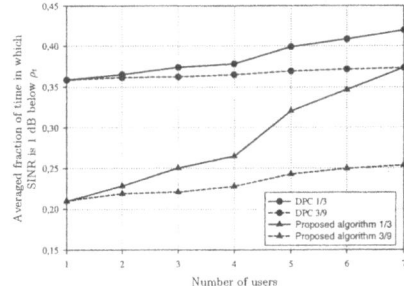

(a) Doppler spread 18.5 Hz   (b) Doppler spread 92.5 Hz

Fig. 5. Averaged fraction of time in which SINR is 1 dB below ρ_t, for two different Doppler spreads.

3(a) and 3(b) present the behavior of the path gain and the tracking performance of the path gain prediction for reuse patterns 3/9 and 1/3, respectively, with the Doppler spread 18.5 Hz. Equivalently, figs. 4(a) and 4(b) show interference and its prediction. It can be observed that prediction based on Taylor's Series achieves good performance for both path gain and interference. The same behavior was observed in all snapshots.

In fig. 5, we illustrate how the superior tracking capability of the proposed algorithm translates into a system-level advantage. In practical systems, it is difficult to keep the SINR exactly at the target value, especially for high speeds [1]. Therefore, we assume an SINR margin below the target SINR in which signal quality is assumed acceptable. We simulated 5000 snapshots for several system loads with reuse pattern 3/9 and 1/3 and calculate the average fraction of time in which the achieved SINR is below the target SINR by a margin of 1 dB.

The simulated maximum load is equivalent to seven co-channel cell, that is, the central cell and an interferer ring with six co-channel cells.

In fig. 5(a), it can be observed that the employment of the new algorithm allows for a significant capacity gain when compared to the DPC, when a Doppler spread of 18.5 Hz is considered. Fig. 5(b) shows the performance of the algorithm with a Doppler spread 92.5 Hz. As the channel variation rate increases, it is expected a performance decrease for both algorithms. However, it can be observed that the proposed algorithm still outperform the DPC algorithm for this Doppler spread.

4   Conclusions

This work has presented a new algorithm for power control in wireless communications systems. The proposed algorithm works well in fast time-varying channels, since they predicts both fast fading and interference variations. The prediction method is based on Taylor's Series and it has low complexity. We demonstrated through simulations that the proposed algorithm is superior to the conventional DPC algorithm, thus resulting in potential capacity gains in mobile communications systems.

Acknowledgments. This work is supported by a grant from Ericsson of Brazil - Research Branch under ERBB/UFC.07 technical cooperation contract.

Fabiano de S. Chaves is supported by CNPq. Tarcisio F. Maciel is supported by FUNCAP-CE.

References

1. Toskala, A. and Holma, H., "WCDMA for UMTS - Radio Access for Third Generation Mobile Communications", *Wiley, England*, 2001.
2. Lee, G. J. and Miljanic, Z., "Distributed Power Control in Fading Channel", *IEE, Elec. Letters*, vol. 38, pp. 653-654, Jun. 2002.
3. Foschini, G. J. and Miljanic, Z., "A Simple Autonomous Power Control Algorithm and its Convergence", *IEEE Trans. Veh. Technol.*, vol. 42, pp. 641-646, Nov.1993.
4. Visweswaran, B. and Kiran, T., "Channel Prediction Based Power Control in W-CDMA Systems", *3G Mobile Communication Technologies*, Conference Publication n° 471, pp. 41-45, 2000.
5. Evans, B. G., Gombachica, H. S. H. and Tafazolli, R., "Predictive Power Control for S-UMTS Based on Least-Mean-Square Algorithm", *3G Mobile Communication Technologies*, Conference Publication n° 489, pp. 128-132, 2002.
6. Lau, F. C. M. and Tam, W. M., "Novel Predictive Power Control in a CDMA Mobile Radio System", *Vehicular Technology Conference*, vol. 3, pp. 1950-1954, May 2000.
7. Apostol, T. M., "Calculus", 2nd ed. *Editorial Reverté*, vol. 1, 1967.
8. Jakes, W. C., "Microwave mobile communications", 2nd ed. *Wiley, New York*, 1974.
9. Lee, J. S. and Miller, L. E., "CDMA Systems Engineering Handbook", *Artech House Publishers*, 1997.

A Heuristic Approach to Energy Saving in Ad Hoc Networks*

R.I. da Silva[1], J.C.B. Leite[1], and M.P. Fernandez[2]

[1] Universidade Federal Fluminense (UFF) - Instituto de Computação (IC)
Rua Passo da Pátria, 156 Bloco E, 3° andar - Niterói – Rio de Janeiro - Brasil
{julius,rsilva}@ic.uff.br
[2] Universidade Estadual do Ceará (UECE) - Lab. Redes de Comunicação e Seg. (LARCES)
Av. Paranjana 1700 – Itaperi - Fortaleza - Ceará – Brasil
marcial@larces.uece.br

Abstract. This paper presents a new algorithm, Extra, for extending the life-time of ad hoc networks. Extra tries to conserve energy by identifying and switching off nodes that are momentarily redundant for message routing in the network. Extra is independent of the underlying routing protocol and uses solely information that is collected locally. Simulation studies conducted have shown promising results.

1 Introduction

In recent years, computer ad hoc networks have been receiving more attention. These are networks without defined topology, whose nodes are moving and they communicate through radio channels. As it does not have a centralized element, all the nodes in network should cooperate to permit messages routing. Several routing protocols in ad hoc networks were proposed in the literature, e.g., DSDV [1], DSR [2], TORA [3] and AODV [4].

An important characteristic of mobile devices in an ad hoc network is the energy consumption, because they usually depend on battery. In that way, besides the traditional metric to evaluate a protocol like packet delay and drop rate, lifetime of the network is important. The nodes in a ad hoc network consumes energy not only when they are transmitting or receiving messages, but also when they listen for any data (idle). That happens because the electronics of the radio can be energized to maintain the capacity to receive messages. Several studies indicate that the necessary power for transmission, reception and listen is, typically, in the order of 1,60W, 1,20W and 1,0W, respectively[5]. However, if the radio goes in sleeping state, the consumption falls to 0,025W. Those values indicate that, to be an effective decrease in the battery consumption the radio should be put in sleeping state by a certain period of time, during the operation of the network.

An interesting comparison on the consumption of energy of several ad hoc network protocols was showed in [6]. In this study, the authors observed the energy

---

* This work was partially supported by FAPERJ, CNPq and CAPES.

J.N. de Souza et al. (Eds.): ICT 2004, LNCS 3124, pp. 437–442, 2004.

consumption of AODV, DSR, DSDV and TORA protocols presents approximate the same value of energy consumption when the idle state consumption is considered. Then, we noticed the importance of minimizing such periods, although we have to guarantee the connectivity in the network or, at least, try to maintain the metrics in acceptable values compared to the value when all nodes are energized.

The BECA (Basic-Energy Conserving Algorithm) algorithm[7], minimize the energy consumption maintaining the radio turned off by the maximum possible period, supporting a higher latency, but having a smaller energy consumption. Another algorithm, GAF (Geography-informed Energy Conservation for Ad Hoc Routing) [6], uses the geographical positioning information (e.g., using a GPS) to support the mechanism of energy conservation.

Several works discussed the conditions what a node should put to sleep and maintaining the network connectivity using only locally information, are all based in heuristics, without a formal proof[8]. In this work, the authors show and demonstrate a theorem of necessary and sufficient conditions for a node to turn off its radio, maintaining the connectivity of the network.

In this paper, we will present the Extra algorithm to deal the problem of the coordination of the entrance in sleeping state of nodes in ad hoc network. This algorithm allows to each node in the network to make decisions in an autonomous way, based only on local information. Additionally, it can work together with any ad hoc routing protocol.

2   The Extra Algorithm

The Extra algorithm implements an energy saving mechanism in ad hoc networks. It can be implemented on any reactive protocol, demanding few changes in routing protocol. Another important point is that the mechanism only uses locally information in each node, and it is not also necessary any communication.

The energy saving procedure is based on put network nodes in sleeping state. In other words, the node radio is turned off to avoid listens in idle mode, that consumes as many energy as the transmission and the reception of messages.

The main characteristic of this energy saving mechanism, and what differs it from other proposals, is the decision to sleep and the duration of sleeping state. In this algorithm, each node will have a probability P, chosen based on heuristics, to start sleeping. In highly populated networks, where there is more active nodes to maintain network connectivity, P is high. When the density decreases, we have to take care to avoid disconnection, reducing the probability.

2.1   Algorithm Operation

During its operation, a node can be in one of three states: Active, Listening and Sleeping, as showed in Figure 1. A node starts operation sending a hello message, reporting its neighbors that it is in the Active state, i.e., the radio is on and ready to transmit. It will stay in this state during Ta seconds. If during this period it receives a data packet (as the destination or is part of a route) or a route response, the counting of Ta time is restarted. Elapsed this time, and not having received any message, the node decide, with probability P, if it will sleep or it will stay active. If the node stays active, another cycle of Ta seconds is started.

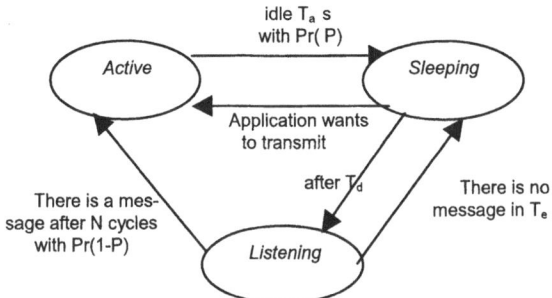

Fig. 1. Node state transition in Extra

A node stays in the Sleeping state for Td seconds. At the end of this time, the node changes its state for Listening. In this situation, the radio is turn on to listen the medium to verify if there is some message destined to him. If it is true, the node restarts the whole cycle, going to the Active state. In case there are no messages during a period of Te seconds, the node goes back to the Sleeping state. This procedure is repeated N times, after that starts the cycle. In order to control neighbors' state, each node has a cache mechanism updated when a packet is received. If, after an interval, no new message be received from a certain neighbor, its information is considered obsolete and it will be withdrawn from the cache.

2.2 Obtaining Probability P

The procedure to calculates the probability P is responsible for the decision to start the energy saving mechanism. The P value is varied to adjust the energy saving mechanism in agreement with the network conditions in the moment. For the first implementation of the algorithm, only information regarding the connectivity and energy level is considered. Simply, P should be higher when the node energy level is high and also when the number of neighbors is high. Based on this comment, some heuristics were developed.

Heuristic 1: its characteristic is to try to preserve the network connectivity. Then, greater value of P and, i.e., high probability to sleeping, is attributed to the nodes in high density area. Smaller value of P is attributed in low density area. The value of P is defined by the product of the number of neighbors by a constant K. However, to avoid that all the nodes in a very populated region go asleep at the same time, the P value is limited by the constant, L. In that way:

P(h1) = K * neighbor_number ()
If (P(h1)> L), then P(h1) = L

where neighbor_number() is the quantity of neighbors of this node. In the tests accomplished for that work the values chosen were K = 0,1 and L = 0,9.

Heuristic 2: its characteristic is also to try to preserve the network connectivity. In this case, the value of P will be the division of the number of active neighbors by the total number of neighbors of the node.

Heuristic 3: the main objective of this heuristics is to save energy when it is insufficient, in spite of the effect that can produce in the connectivity. Here, the value of P is obtained by the relation between level of current energy and the level of initial energy of the node. This heuristics can be seen as a selfish behavior. In that way:
P(h3) = 1 - current_energy() / initial_energy()
where current_energy() is a the amount of energy remaining in the node and initial_energy() is the initial energy of the node.

Heuristics 4: this heuristics is a combination of the heuristics 1 and heuristics 3. In other words, it try to save energy balancing the network connectivity and energy conservation in the node. Its definition is the following:
if (P(h1) = ref) and (P(h3) = ref) => P(h4) = max(P(h1), P(h3))
if (P(h1) = ref) and (P(h3) <ref) => P(h4) = P(h1)
if (P(h1) <ref) and (P(h3) = ref) => P(h4) = P(h3)
if (P(h1) <ref) and (P(h3) <ref) => P(h4) = min(P(h1), P(h3))
where P(h1), P(h3) and P(h4) are the values of P returned by the heuristics 1, 3 and 4, respectively, and ref is an adjustable parameter, whose value attributed in the implementations was 0,5.

Heuristics 5: has similar objective to the heuristics 4. It is the combination of the heuristics 2 and heuristics 3, also looking for save energy where it is in shortage, although trying to maintain the network connectivity. Its definition is the following:
if (P(h2) = ref) and (P(h3) = ref) => P(h5) = max(P(h2), P(h3))
if (P(h2) = ref) and (P(h3) <ref) => P(h5) = P(h2)
if (P(h2) <ref) and (P(h3) = ref) => P(h5) = P(h3)
if (P(h2) <ref) and (P(h3) <ref) => P(h5) = min(P(h2), P(h3))
Finally, in all the accomplished tests, the attributed values N, Ta, Te and Td were, respectively, 15, 4s, 0,05s and 0,5s.

3   Performance Evaluation

The Extra algorithm was evaluated through simulation methodology. The goal was validate the algorithm in a great number of ad hoc network scenarios. To evaluate the performance it was compared with basic AODV protocol, as the reference protocol, and with GAF algorithm with AODV protocol. To validate the simulation model, each metric was measured in 10 different scenarios, and the average value was presented with confidence interval of 95%.

The simulator used in the tests was the Network Simulator - ns-2, version 2.26 [9]. To create the simulation scenario, it was used the BonnMotion version 1.1 [10].

The simulation scenario used 60 nodes, moving in a random way (random waypoint model) in a 1200 m by 600m area. There were defined 4 static nodes on the edges of simulation plane acting as source and sink of traffic. The nodes speed were 0 m/s, a static network used as reference, 1 m/s, equivalent to a person walking, and 10 m/s, a vehicle in urban environment (36 Km/h).

The simulation time was 900 seconds, to permit a comparison with measures accomplished for GAF algorithm in [6]. In all scenarios, we use pause times of 0, 30, 60, 120, 300, 600 and 900 seconds. The radio has 250m range, and the propagation model was the two-ray-ground. For traffic model we choose two sources and two sinks implemented in fixed nodes, with CBR traffic over UDP transport. The packet size was 512 bytes. The evaluated rates were 1 pkt/s, 10 pkts/s and 20 pkts/s, producing in each simulation rates of 2, 20 and 40 pkts/s, respectively.

The energy model chosen is based on the measures from [5] of WaveLAN 2 Mb/s board. The values obtained in this work were 1,6W for transmission, 1,2W for reception, 1,0W in idle (listen) and 0,025W in sleeping state. The initial energy attributed to each node was 500 J, enough to maintain the network working for about 450s with AODV protocol. As the simulation time was 900s, we can observe the behavior of energy saving mechanism.

3.1 Results

In the results presented here none control of energy is considered for AODV protocol and GAF is assumed running with AODV. The first measure indicates the network lifetime obtained by Extra algorithm. In Figure 4 we show a comparison among the performance of AODV, GAF and all Extra heuristics. For this graph it was used a pause time of 0s, the mobile nodes had speed of 1m/s and a rate of 10 pkts/s was maintained for each fonts. On this figure, we can see the benefit of Extra algorithm.

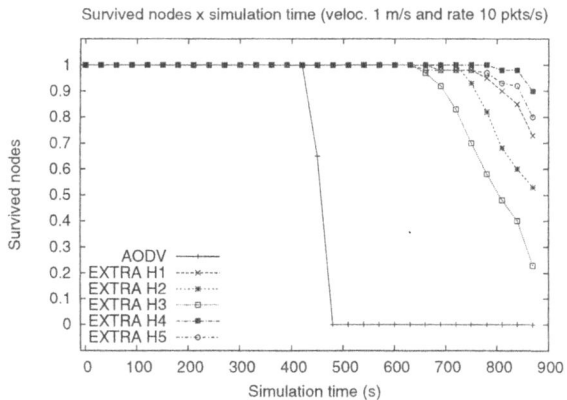

Fig. 2. Lifetime: AODV, GAF and Extra (for several heuristics)

The results of packet drop rate and percentile 90 of the delay with variation in pause time, transmission rate and node speed, for GAF and Extra (using heuristic 4) shows that Extra doesn't produce degradation in QoS metrics.

4 Conclusions and Future Works

This work presents a new algorithm, Extra, for energy conservation of nodes in ad hoc networks. As well several other algorithms, its objective is maximize the lifetime of the network, but trying to maintain the connectivity. Its operation is controlled by a function that defines when the node should enter in sleeping state. Against other algorithms, the entrance in this state happens with a certain probability. This probability is calculated in execution time and heuristics are used to calculate. We tested five different heuristics, all of them with very simple structure. As seen in the work, in all heuristics are used only information obtained locally.

As future works, it can be appraised solutions where the neighboring nodes change information about its current energy and the neighbors' density can be more explored. In that last case, for instance, it can be obtained a notion of who are (or how many are) the neighbors of the neighboring nodes. Obviously, this will cause an increase in control messages, what produces more energy consumption. However, it is possible that this negative effect is not enough to degrade the performance.

In any way, the accomplished tests show the importance of maintaining the simplicity of the control parameters. A larger number of parameters implicates in larger complexity in get information and also in the algorithm structure.

References

1. Perkins, C.E. e Bhagwat, P. Highly Dynamic Destination-Sequenced Distance-Vector Routing (DSDV) for Mobile Computers, ACM Conf. on Communications Architecture, Protocols and Applications, August, p.234-244. (1994)
2. Johnson, D.B. e Maltz, D.A. Dynamic Source Routing in Ad Hoc Wireless Networks, in Mobile Computing, T. Imielinski e H. Korth (eds.), Kluwer, p. 153-181(1996)
3. Park, V. e Corson, S. A Highly Adaptive Distributed Routing Algorithm for Mobile Wireless Networks, INFOCOM, April, p. 1405-1413.(1997)
4. Perkins, C.E. e Royer, E.M. Ad Hoc On-Demand Distance Vector Routing, 2nd IEEE Work. on Mobile Computer Systems and Applications, February, p. 90-100. (1999)
5. Stemm, M. e Katz, R.H. Measuring and Reducing Energy Consumption of Network Interfaces in Hand-held Devices, IEEE Trans. on Communications, E80-B(8), August, p. 1125-1131. (1997)
6. Xu, Y. *et alli*. Geography-informed Energy Conservation for Ad Hoc Routing, 7th Annual ACM/IEEE Int. Conf. on Mobile Computing and Networking, July 16-21, Rome, Italy, p. 70-84. (2001)
7. Xu, Y. *et alli*. Adaptive Energy-Conserving Routing for Multihop Ad Hoc Networks, Technical Report n° 527, USC/ISI, October 12, Los Angeles, USA, 14 p. (2000)
8. Koushanfar, F. et alli. Low Power Coordination in Wireless Ad-hoc Networks, ISLPED'03, 25-27 de August, Seul, Korea, p. 475-480. (2003)
9. ns-2. The Network Simulator - ns-2, http://www.isi.edu/nsnam/ns/. (2003)
10. BonnMotion. A Mobility Scenario Generation and Analysis Tool, http://www.informatik.uni-bonn.de/IV/BonnMotion/ (2003)

Calculating the Maximum Throughput
in Multihop Ad Hoc Networks*

Bernardo A.M. Villela and Otto Carlos M.B. Duarte

GTA/PEE/COPPE-Poli – Universidade Federal do Rio de Janeiro
CP 68504 - 21945-970 - Rio de Janeiro, Brazil.

Abstract. This paper analyzes the communication between two points in wireless ad hoc networks operating at the IEEE 802.11 standard mode. The path from the source to the destination consists of a chain of nodes. We aim at deriving an analytical expression for the maximum throughput. We show that the maximum throughput can be increased, under certain conditions, through the use of two paths: the shortest path and an appropriately chosen alternative path that takes into account the interference problem. The alternative path strategy allows a gain up to 50% of the maximum throughput achieved in the single path strategy.

1 Introduction

Nowadays, the IEEE 802.11 standard is widely deployed in local wireless networks. In the ad hoc mode, it implements a distributed medium access mechanism called DCF (Distributed Coordination Function), which applies the CSMA/CA (Carrier-Sense Multiple Access/Collision Avoidance) access method.

The hidden terminal problem is a classical challenge in wireless networks. The DCF proposes the use of RTS (Request To Send) and CTS (Clear To Send) frames to solve this problem. By means of an RTS frame, the sender shows all his neighbors the intention to transmit and the receiver allows the transmission by sending a CTS frame, showing that its neighborhood is free.

Xu *et al.* [1] showed that the RTS/CTS handshake cannot completely solve the hidden terminal problem, due to the effect of the interference. This mechanism assumes that all nodes that could interfere with the frame reception (which will be called hidden nodes) are able to receive the CTS too. Xu *et al.* [1] derived an expression showing that the interference range is a variable range depending on the distance from the sender to the receiver and the signal to interference ratio at the receiver. Moreover, they showed that when the distance from the sender to the receiver is longer than a threshold value, the interference range becomes greater than the transmission range. As a consequence, we cannot assume that the CTS frame is received by a hidden node.

Gupta *et al.* [2] and Li *et al.* [3] analyzed the capacity of ad hoc networks. They showed that the capacity depends on some local radio parameters, the MAC (Medium Access Control) protocol, the network size, and the traffic patterns. Saadawi *et al.* [4] considered the performance of the IEEE 802.11 MAC protocol in multihop ad hoc networks, through the analysis of a TCP (Transmission Control Protocol) traffic.

---

* This work has been supported by CNPq, CAPES, and FAPERJ.

J.N. de Souza et al. (Eds.): ICT 2004, LNCS 3124, pp. 443–449, 2004.
© Springer-Verlag Berlin Heidelberg 2004

Li *et al.* [3] analyzed the capacity of a chain of nodes considering the interference range. They showed the maximum utilization achievable for a fixed interference range in a specific chain case. In a companion paper [5], we analytically generalized an expression for the maximum utilization of a chain, considering a variable interference range and also considering all the possible relevant distances between two neighbors in the chain. As it was assumed by Li *et al.* [3], we assumed [5] a simplification, considering the carrier sensing range equal to the transmission range. As a matter of fact, this value, which depends on the sender equipment, is typically greater than the transmission range. Accordingly, in this work, we allow the carrier sensing range to assume greater values and we realize that it may have an important impact on the maximum throughput. Although more complex, this model is more complete and precise. Again we generalize an expression for the maximum throughput, now taking into account both the carrier sensing and the interference effects.

We showed in [5] that the maximum utilization can be increased through the simultaneous use of two paths to the destination: the shortest path and an appropriate alternative path that takes into account the interference problem. In the present work, we also generalize an expression for the maximum utilization achievable using an alternative path, now considering both the carrier sensing and the interference effects.

In Section 2, we present the effects of the interference and carrier sensing. In Section 3, we analyze the maximum utilization of a chain of nodes. In Section 4 we analytically show that the simultaneous use of the shortest path and an alternative path can improve, under certain conditions, the throughput. Section 5 presents our conclusions.

2   The Effects of the Interference and the Carrier Sensing

In order to correctly receive a packet, the signal power at the receiver must be strong enough. Hence, the signal to noise plus interference ratio must be greater than the minimum value specified for the receiver equipment. Increasing the distance from the sender to the receiver results in the reduction of the signal power at the receiver, which means that nodes more distant from the receiver can become hidden nodes.

Let d be the distance from the sender to the receiver, r be the distance from the receiver to a third node that might want to transmit, and SIR_{TH} be the minimum value for the signal to interference ratio required for a successful reception. Hence Xu *et al.* [1] showed that every node separated by less than $d\sqrt[4]{SIR_{TH}}$ meters from the receiver can indeed interfere with its reception.

Let R_{Tx} be the transmission range. Xu *et al.* [1] showed that, when the sender and the receiver are more than $\frac{1}{\sqrt[4]{SIR_{TH}}}R_{Tx}$ meters away from each other, the RTS/CTS handshake does not solve the hidden terminal problem.

According to the CSMA/CA mechanism, every node, that wants to transmit some data, before sending it, should sense the medium. This procedure may help avoiding the hidden terminal problem. That happens because the carrier sensing range, whose value is fixed and specified by the sender equipment, is typically greater than the transmission range. Let R_{CS} be the carrier sensing range, Xu *et al.* [1] showed that the hidden terminal problem is completely solved [1] when $R_{CS} = (1 + \sqrt[4]{SIR_{TH}})R_{Tx}$. Increasing R_{CS} beyond this value is not necessary. Moreover, increasing R_{CS} also increases the number of nodes that could transmit without interfering with the initial transmission, but they are prevented from doing it, due to the carrier sensing. Therefore the R_{CS} should not be too high, because this way it reduces the network capacity.

3   The Capacity of a Chain of Nodes

The throughput experienced by a source of traffic is maximized when the destination is in the transmission range of the source, situation called "direct communication", and the medium is free. As an example, Li *et al.* [3] showed that, even in this best case, the IEEE 802.11 at a 2 Mbps rate is only able to manage a maximum throughput of 1.7 Mbps for 1500 bytes of packet size. The data rate is also reduced for the IEEE 802.11 standard operating at others rates, like 11 Mbps or 54 Mbps. The reduction is due to the overhead added by the RTS/CTS/ACK exchange, to the preamble sent before the transmission of each MAC frame, to the interframe spaces, and to the *backoff* mechanism. Furthermore, the maximum throughput depends on the data packet size.

The scenario with only two neighboring nodes gives an upper bound for the maximum throughput, which can be taken as a baseline for comparison. So, in the following analysis, the maximum utilization of a general case is a fraction of this reference value [3, 5]. The general case is a "multihop communication". In this case, the nodes have to cooperate, forwarding the packets from the source to the destination.

Let U_{max} be the maximum utilization. Now we will analyze the maximum utilization of a chain of nodes P_S, P_1, P_2, ..., P_D when the distance d between neighbors is such as $\frac{R_{Tx}}{2} < d \leqslant R_{Tx}$. So it can be written that $d = L \times R_{Tx}$, with $0.5 < L \leqslant 1$. This constraint arises because if $d > R_{Tx}$, then it is impossible reaching the destination and if $d \leqslant \frac{R_{Tx}}{2}$, then the nodes would not send the packets to their closer neighbors, instead they would send them to their neighbors closer to the destination, minimizing the number of hops. Both the carrier sensing and the interference caused by hidden nodes play an essential role on the analysis of the maximum utilization, which will be accomplished in three steps. In the first one, our analysis will consider only the carrier sensing effect, letting R_{CS} assume different values, and it will be called $U_{max_{CS}}$. In the second one, we will calculate $U_{max_{Int}}$, considering only the interference effect [5]. In this case, SIR_{TH} may assume different values. And in the last step, we will find a single value, U_{max}, based on the results of the previous steps. In this paper, the use of the term maximum utilization without any other specification implicitly means that the interference and the carrier sensing range were considered.

Let R_{CS} be the carrier sensing range. It is well known that $R_{CS} \geqslant R_{Tx}$. Hence, $R_{CS} = V \times R_{Tx}$, with $V \geqslant 1$. Let I be the number of nodes that can sense a transmission from the source. Hence, $I = \lfloor \frac{R_{CS}}{d} \rfloor = \lfloor \frac{V}{L} \rfloor$. If we assume that the source is transmitting, then none of its successors until P_I may transmit simultaneously, since the medium is sensed as busy by these nodes. But P_{I+1} (and its successors in the chain) does not sense the transmission from the source and it can send a frame concomitantly. Therefore $U_{max_{CS}} = \frac{1}{I+1}$.

Let R_{Int} be the interference range and let K be the number of nodes located in the interference range of the source. Then $K = \lfloor \frac{R_{Int}}{d} \rfloor = \lfloor \sqrt[4]{SIR_{TH}} \rfloor$. The reception by P_1 of a frame from P_S can be interfered by the transmission of all the successors of P_1

until the node P_{K+1}. Moreover, P_{K+2} and its successors in the chain cannot interfere with the transmission of P_S. Therefore $U_{max_{Int}} = \frac{1}{K+2}$. It only depends on K, which means it only depends on the SIR_{TH}, whose value must be greater than 1.

As both the carrier sensing and the interference effects play an important role on the definition of the maximum utilization, we must consider both of them together. So the maximum utilization assumes the minimum value between $U_{max_{CS}}$ and $U_{max_{Int}}$. This way, U_{max} is given by the dominant effect, depending on R_{CS}, d, and SIR_{TH}.

In the Figures 1(a) and 1(b), it is presented the U_{max} value when $1 < SIR_{TH} < 16$ and $V \in \{1; 2; 3\}$ and when $16 \leqslant SIR_{TH} < 81$ and $V \in \{1; 2; 3; 4\}$, respectively. The maximum value of V in each case was chosen, in order to illustrate a limit case, since Xu et al. [1] showed that when $R_{CS} = (1 + \sqrt[4]{SIR_{TH}})R_{Tx}$, the hidden terminal problem is completely solved and that a greater value of R_{CS} is useless.

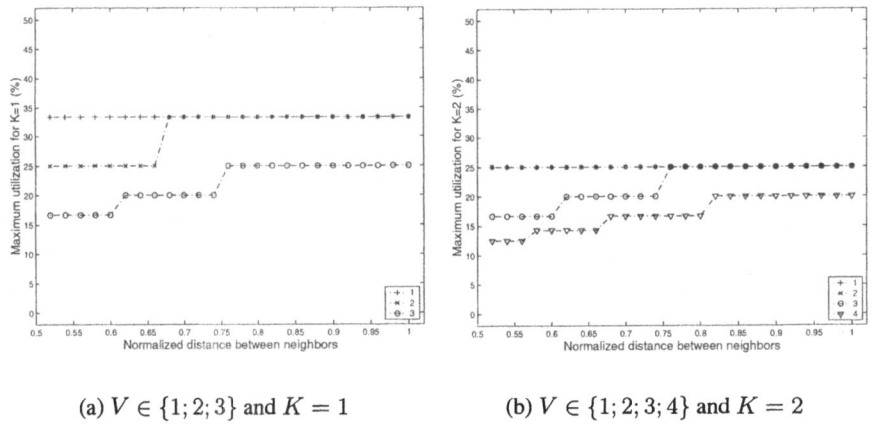

(a) $V \in \{1; 2; 3\}$ and $K = 1$ (b) $V \in \{1; 2; 3; 4\}$ and $K = 2$

Fig. 1. Maximum utilization.

Once defined the SIR_{TH}, $U_{max_{Int}}$ remains constant for $\frac{R_{Tx}}{2} < d \leqslant R_{Tx}$ and it defines the upper bound of $\frac{1}{3}$ and $\frac{1}{4}$ in the Figures 1(a) and 1(b), respectively. The existence of these limits explains why some curves overlap partially, in the Figure 1(a) and in the Figure 1(b) when $U_{max_{CS}} \geqslant U_{max_{Int}}$. The $U_{max_{CS}}$ value depends on the ratio between R_{CS} and d. Increasing this ratio, the carrier sensing effect may become dominant, reducing U_{max} some more, as it can be seen in the Figures 1(a) and 1(b).

4   The Use of an Alternative Path

As presented in the Figure 2, let us suppose there is a second path $(P_S, A_1, A_2, \ldots, P_D)$ to the destination, called alternative path. This path is longer, then a routing protocol, which uses only one path per destination implementing a minimum number of hops metric, will continue using the path $P_S, P_1, P_2, \ldots, P_D$, called primary path. Aiming at increasing the throughput [5], the flow can be split in two parts, as it is shown in the Figure 2, using two different paths: the primary and the alternative path. The source injects frames into the network, alternating each frame into each path.

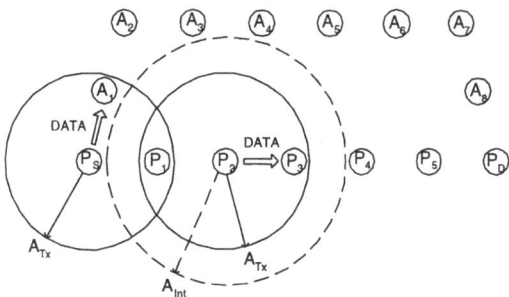

Fig. 2. A primary and an alternative path.

Once defined the carrier sensing range (R_{CS}), if P_S is transmitting, then all the nodes separated from P_S by a distance less than or equal to R_{CS} will be prevented from transmitting, because they sense the medium as busy. That holds true for nodes in the primary path and also for nodes in any other alternative path that may exist. If we define the maximum utilization using an alternative path and considering only the carrier sensing range as $Alt_{max_{CS}}$, then $Alt_{max_{CS}} = U_{max_{CS}} = \frac{1}{I+1}$.

Nevertheless, we showed [5] that the maximum utilization, using an alternative path and considering only the interference effect, now defined as $Alt_{max_{Int}}$, can be increased when compared to $U_{max_{Int}}$. Again, let us assume that the neighboring nodes in each path are separated by a distance d, such as $\frac{R_{Tx}}{2} < d \leqslant R_{Tx}$. The function $dist(P_A, P_B)$ gives the distance from the node P_A to P_B. Again, $K = \lfloor \sqrt[4]{SIR_{TH}} \rfloor$, which is the number of nodes in the interference range of the source. In the chain case, P_S and P_{K+1} cannot transmit simultaneously, since $dist(P_1, P_{K+1}) \leqslant R_{Int}$. Aiming at obtaining a value for $Alt_{max_{Int}}$ greater than $U_{max_{Int}}$, now we allow this concomitant transmission, if P_S can send the frame to a neighbor A_1, such as $dist(A_1, P_{K+1}) > R_{Int}$. An equivalent condition should be respected by P_1 and A_{K+1}. An analogous procedure is recommended at the end of the process, when the two paths get closer to each other. And in the middle of the chain, we recommend that a distance greater than R_{Int} should be guaranteed. This procedure, called $Proc_{Alt}$, is exemplified in the Figure 2, which considers only the interference effect. In the Figure 2, we have $1 < SIR_{TH} < 16$. In this case, P_S can transmit to A_1 simultaneously with the transmission from P_2 to P_3.

We showed in [5] that using only one alternative path with $Proc_{Alt}$ is enough for achieving the optimum throughput. In a general case, $Proc_{Alt}$ allows the simultaneous transmission of P_S and P_{K+1} (and also of P_S and A_{K+1}). So $Alt_{max_{Int}} = \frac{1}{K+1}$ [5].

Let us define the maximum utilization using an alternative path and considering both the carrier sensing and the interference effects as Alt_{max}. Then the value of Alt_{max} is the minimum between $Alt_{max_{CS}}$ and $Alt_{max_{Int}}$. The Figures 3(a) and 3(b) present Alt_{max}, when $1 < SIR_{TH} < 16$ and $V \in \{1; 2; 3\}$ and when $16 \leqslant SIR_{TH} < 81$ and

$V \in \{1; 2; 3; 4\}$, respectively. Once specified the SIR_{TH}, $Alt_{max_{Int}}$ defines the upper bound of $\frac{1}{2}$ and $\frac{1}{3}$ in the Figures 3(a) and 3(b), respectively.

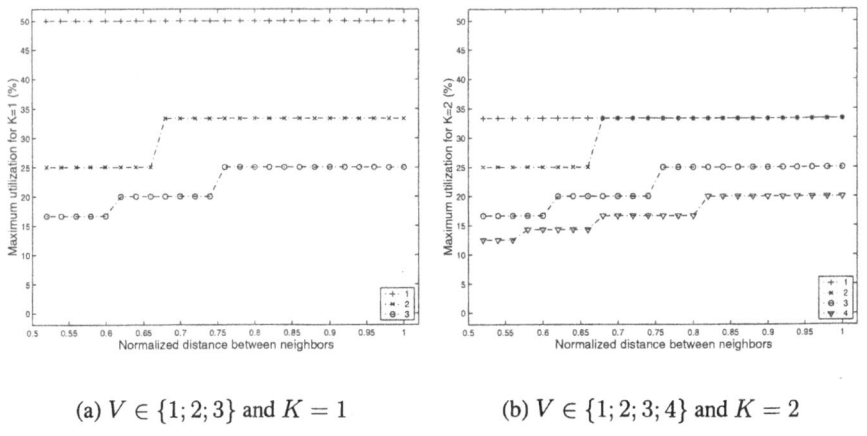

(a) $V \in \{1; 2; 3\}$ and $K = 1$ (b) $V \in \{1; 2; 3; 4\}$ and $K = 2$

Fig. 3. Maximum utilization implementing the alternative path.

The maximum utilization can be increased with the alternative path strategy when, in the chain case, we have $U_{max_{CS}} > U_{max_{Int}}$; otherwise, the carrier sensing defines the maximum utilization and the use of $Proc_{Alt}$ cannot change its value. Therefore we have $Alt_{max} > U_{max}$ when $\lfloor \frac{V}{L} \rfloor < K + 1$. The gain in percentage achieved by the alternative path strategy when compared to the chain case can be generalized as $100 \left(\frac{Alt_{max_{Int}}}{U_{max_{Int}}} - 1 \right) = 100 \left(\frac{K+2}{K+1} - 1 \right)$. When $K = 1$, it is possible achieving the maximum gain, which is 50%. Such gain is achievable for $V = 1$ and any value of d or for $1 \leqslant V < 2$ and values of d, such as $L > \frac{V}{2}$. When $K = 2$, the gain is 33% and it is achievable for $1 \leqslant V \leqslant 1.5$ and any value of d or for $1.5 < V < 3$ and $L > \frac{V}{3}$.

5   Conclusions

Due to the characteristics of the IEEE 802.11 medium access mechanism, the communication between two nodes through a chain of nodes has serious access constraints. The maximum utilization of a chain is $\frac{1}{3}$ in the best case. This situation becomes even more restrictive if the signal to interference ratio necessary for a correct reception (SIR_{TH}) is greater than or equal to 16, due to the hidden terminal problem. The presence of the interference gives an upper bound for the maximum utilization. In addition, we must consider the carrier sensing effect. The ratio between the carrier sensing range and the distance d between neighboring nodes is a relevant parameter. Increasing this ratio, namely increasing R_{CS} or reducing d, the carrier sensing effect may become dominant in the definition of the maximum utilization of the chain, reducing it even more.

We show that it may be possible to increase the maximum throughput, using multiple paths. Therefore we simultaneously use the shortest path and an appropriately chosen alternative path, which takes into account the interference problem. Only one alternative path is enough for achieving the optimum throughput. When $1 < SIR_{TH} < 16$, the use of the alternative path allows a gain of 50% in the maximum utilization when compared to the chain case. When $16 \leqslant SIR_{TH} < 81$, the allowed gain is 33%.

References

1. K. Xu, M. Gerla, and S. Bae, "How effective is the IEEE 802.11 RTS/CTS handshake in ad hoc networks?" In *Proc. IEEE Globecom*, nov 2002.
2. P. Gupta and P. R. Kumar, "The capacity of wireless networks," *IEEE Transactions on Information Theory*, vol. 46, no. 2, Mar. 2000.
3. J. Li, C. Blake, D. S. De Couto, H. I. Lee, and R. Morris, "Capacity of ad hoc wireless networks," In *Proc. ACM MobiCom*, July 2001.
4. S. Xu and T. Saadawi, "Does the IEEE 802.11 MAC protocol work well in multihop wireless ad hoc networks?" *IEEE Communications Magazine*, pp. 130–137, June 2001.
5. B. A. M. Villela and O. C. M. B. Duarte, "Maximum throughput analysis in ad hoc networks," *Proc. of The Third IFIP-TC6 Networking Conference (Networking 2004)*, May 2004.

Throughput of Distributed-MIMO Multi-stage Communication Networks over Ergodic Channels

Mischa Dohler, Athanasios Gkelias, and A. Hamid Aghvami

Centre for Telecommunications Research
King's College London, WC2R 2LS London, UK
http://www.ctr.kcl.ac.uk
mischa@ieee.org

Abstract. Distributed-MIMO multi-stage communication networks are known to yield superior performance over traditional networks which is due to the facilitated MIMO channels as well as gains obtained from relaying. The emphasis of this paper is on a suitable allocation of resources to each terminal so as to maximise the end-to-end throughput, which is proportional to the capacity offered by such network. The contribution of this paper is the derivation of an explicit low-complexity near-optimum resource allocation strategy, the precision and performance of which is also analysed for a two-stage relaying network.

1 Introduction

Link capacity in bits/s/Hz is the fundamental measure of any communication system. With the ever increasing need for mobile information the single link capacity, as derived by Shannon [1], no longer suffices. Multiple-Input-Multiple-Output (MIMO) channels are a promising solution to that problem; these were shown to yield significant capacity benefits over Single-Input-Single-Output (SISO) systems [2,3]. The derived capacities constitute an upper bound on MIMO communication which can be approached utilising appropriate multiplexing and/or space-time coding (STC) techniques.

These techniques perform optimum assuming that a MIMO channel is available and all sub-channels are mutually uncorrelated, which is rarely fulfilled in a real world system. Furthermore, to overcome the disadvantage of having only one antenna in a mobile terminal (MT), spatially adjacent mobile terminals were allowed to communicate with each other and thus form a virtual array of more than one antenna element. This concept was termed Virtual Antenna Array (VAA) [4,5], which is deployed here in a more general context.

Concept. A fairly generic realisation of a distributed-MIMO multi-stage communication network with the utilisation of VAAs is depicted in Figure 1. Here, a source mobile terminal (s-MT) communicates with a target mobile terminal (t-MT) via a number of relaying mobile terminals (r-MTs). Spatially adjacent r-MTs form a VAA, each of which receives data from the previous VAA and relays data to the consecutive VAA until the t-MT is reached.

J.N. de Souza et al. (Eds.): ICT 2004, LNCS 3124, pp. 450–460, 2004.

Each of the terminals may have an arbitrary number of antenna elements; furthermore, the MTs of the same VAA may cooperate among each other. With a proper setup, this clearly allows the deployment of MIMO capacity enhancement techniques.

Background. To the authors' best knowledge, the general concept of distributed-MIMO multi-stage communication systems has been introduced in [4,5]. It based on previous contributions by [6]–[8] on relaying and by [2, 3] on MIMO communication. Independent research has led to similar concepts, see e.g. the excellent contributions by [9]–[11].

The principle of VAA has already been analysed in the context of cellular type networks, e.g. [12]. In [13,14], it has been generalised to multi-hop communication scenarios, where explicit resource allocation strategies for each relaying hop in terms of fractional bandwidth, slot duration and power have been derived for ergodic wireless channels.

Aim of the Paper. In contrast to previous research, the aim of this paper is to maximise the end-to-end data throughput of above-described networks over ergodic fading channels. This is achieved by optimally assigning resources in terms of fractional frame duration, frequency band and transmission power to each of the MTs.

Similar optimisation problems have been analysed in [15]–[18]. The derived solutions, however, require some form of <u>numerical</u> optimisation. It is the aim of this paper to introduce <u>explicit</u> near-optimum resource allocation strategies for regenerative distributed-$\overline{\text{MIMO}}$ multi-stage ad-hoc systems, where transmission from source to sink is constrained by a total power S, bandwidth W and frame duration T. The allocation strategies can then be applied to channel-aware medium access control (MAC) protocols.

Paper Structure. In Section 2, the system model is briefly described. In Section 3, the exact and approximate ergodic MIMO capacities are exposed. In Section 3.2, the fractional resource allocation algorithms for ergodic fading channels are derived. The performance and precision of the developed algorithms is then assessed in Section 5. Final remarks and conclusions are given in Section 6.

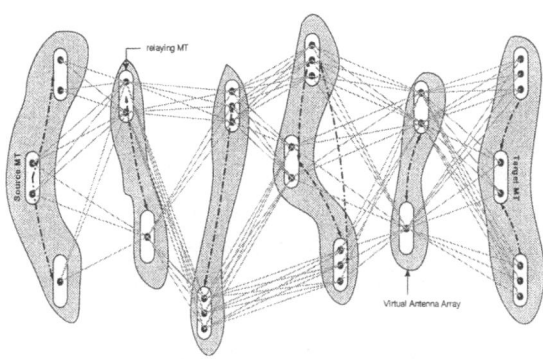

Fig. 1. Distributed-MIMO Multi-Stage Communication Network.

2   System Model

Each MT within the distributed-MIMO multi-stage communication network might be in possession of any number of antenna elements and any number of MTs belonging to the same VAA tier may cooperate among each other. Cooperation is assumed to take place over systems not requiring any optimisation. The correlation among the antenna elements and the various relaying terminals is assumed to be negligible. The channel realisations are assumed to be non-ergodic flat Rayleigh fading, where the sub-channel gains spanned by a MIMO channel of each relaying stage are assumed to be equal.

Transparent or regenerative relaying are the two approaches to accomplish a relaying network. Regenerative relaying allows the deployment of MIMO capacity enhancement technologies at each relaying stage. Therefore, this paper only considers regenerative relaying, where the signal stream is decoded, re-encoded and retransmitted by each relaying stage. The process of terminal selection and data routing is beyond the scope of this paper. For the subsequent analysis, synchronisation is also assumed to be perfect.

Furthermore, it is assumed that orthogonal relaying is deployed, i.e. available resources are allocated such that no interference between the relaying stages occurs. Thus, bandwidth (FDMA)/frame duration (TDMA) has to be divided into non-overlapping frequency-bands/time-slots such that at any time they are used by only one relaying link. The functioning of an FDMA-based relaying system is described in [19] and is briefly summarised for clarity.

Source MT. The s-MT continuously broadcasts the data to the remaining r-MTs in the first relaying VAA tier, utilising negligible power and bandwidth, and possibly not deploying all of its available antenna elements.

First Relaying VAA Tier. After cooperation between the s-MT and the remaining r-MTs of the first relaying VAA tier, the data is space-time encoded according to a given code book. Each MT then transmits only its prior negotiated spatial codewords such that no transmitted codeword is duplicated.

v^{th} **Relaying VAA Tier.** The reception, cooperation, de-coding, re-encoding and re-transmission process is congruent to the proceeding described above. Retransmission from the vth relaying VAA tier is accomplished at frequency band W_v with total power S_v.

V^{th} **Relaying VAA Tier.** The final relaying tier contains the t-MT. Similar to the 1st tier, only cooperative MTs are considered here (no cooperation between the r-MTs and the t-MT would terminate the data flow in the respective r-MTs).

Target MT. After cooperation between the r-MTs and the t-MT, the data is space-time decoded and passed on to the information sink in the t-MT.

For a TDMA-based system, all fractional bandwidths $W_{v \in (1,K)}$ need to be replaced by fractional frame durations $T_{v \in (1,K)}$. Here, K denotes the number of relaying stages and is related to the number of VAA relaying tiers via $K = V - 1$. For any scenario, the total communication duration is denoted by T and the total bandwidth by W. For orthogonal TDMA and FDMA-based relaying systems, $T = \sum_{v=1}^{K} T_v$ and $W = \sum_{v=1}^{K} W_v$ respectively.

3   Exact and Approximated MIMO Capacities

Given K relaying stages, the optimum assignment of resources in terms of fractional bandwidth $\alpha_{v \in (1,K)}$ and transmission power $\beta_{v \in (1,K)}$ per stage requires the evaluation of the MIMO Shannon capacity over ergodic narrowband fading channels. An exact expression for the capacity is given below for traditional and orthogonalised MIMO channels and is shown to be inapplicable to the resource optimisation problem. Therefore, a functional approximation is used to derive an approximate closed form expression to the exact capacity over ergodic channels which is very well suited to the optimisation problem. Note also that it has been demonstrated in [14] that the resource allocation of FDMA and TDMA based systems is equivalent, which is the reason why only FDMA based systems are considered here.

3.1   Exact Fractional Capacity

With reference to the theory exposed in [2], the capacity in [bits/s/Hz] normalised by the bandwidth W of the vth relaying stage for an ergodic flat Rayleigh fading MIMO channel can be expressed as

$$C_v = \alpha_v \cdot \mathrm{E}_{\lambda_v} \left\{ m \log_2 \left(1 + \lambda_v \gamma_v \frac{\beta_v}{\alpha_v} \frac{S}{N} \right) \right\} \tag{1}$$

where α_v is the fractional bandwidth and β_v the fractional power allocated to the vth relaying stage. The probability density function (pdf) of λ_v can be expressed as

$$pdf_{\lambda_v}(\lambda_v) - \frac{1}{m} \sum_{k=0}^{m-1} \frac{k!}{(k+n-m)!} \left[L_k^{n-m}(\lambda_v) \right]^2 \lambda_v^{n-m} e^{-\lambda_v} \tag{2}$$

where $m \triangleq \min\{t_v, r_v\}$, $n \triangleq \max\{t_v, r_v\}$ and $L_k^{n-m}(\lambda_v)$ is the associated Laguerre polynomial of order k. A closed form expression in terms of finite sums for the capacity given in (1) was derived in [19] and is given as

$$C_v = \alpha_v \cdot \sum_{k=0}^{m-1} \frac{k!}{(k+d)!} \left[\sum_{l=0}^{k} A_l^2(k,d)\, \hat{C}_{2l+d}(\tau_v) + \right. \tag{3}$$

$$\left. \sum_{l_1=0}^{k} \sum_{l_2 \neq l_1}^{k} (-1)^{l_1+l_2} A_{l_1}(k,d)\, A_{l_2}(k,d)\, \hat{C}_{l_1+l_2+d}(\tau_v) \right]$$

where $d \triangleq n - m$, $A_l(k,d) \triangleq (k+d)!/[(k-l)!\,(d+l)!\,l!]$ and $\tau_v \triangleq \frac{\beta_v}{\alpha_v} \frac{\gamma_v}{t_v} \frac{S}{N}$. The capacity integral $\hat{C}_\zeta(\tau)$ has been defined in [19] as

$$\hat{C}_\zeta(\tau) \triangleq \int_0^\infty \log_2(1+\tau x)\, x^\zeta\, e^{-x}\, dx \tag{4}$$

Given $\mathrm{Ei}(y) \equiv \int_{-\infty}^{y} \frac{e^t}{t} dt$ to be the exponential integral, eq. (4) can be solved as

$$\hat{C}_\zeta(\tau) = \frac{1}{\log(2)} \sum_{\mu=0}^{\zeta} \frac{\zeta!}{(\zeta - \mu)!} \times \tag{5}$$

$$\left[(-1)^{\zeta-\mu-1}(1/\tau)^{\zeta-\mu} e^{1/\tau} \mathrm{Ei}(-1/\tau) + \right.$$

$$\left. \sum_{k=1}^{\zeta-\mu} (k-1)!(-1/\tau)^{\zeta-\mu-k} \right].$$

3.2   Approximated Fractional Capacity

Clearly, any analytical optimisation on (3) in terms of the fractional bandwidth α_v and power β_v is intractable due to the exponential integral and the fairly evolved expression. To overcome that problem, an approximation to the exact MIMO Shannon capacity has been suggested and assessed in [13,14]. In there, a simple functional approximation is suggested in form of $\log_2(1 + x) \approx \sqrt{x}$. The suggested approximation simplifies (1) to

$$C_v \approx \alpha_v \sqrt{\frac{\beta_v}{\alpha_v}} \sqrt{\gamma_v \frac{S}{N}} \cdot \mathrm{E}_{\lambda_v} \left\{ m\sqrt{\lambda_v/t_v} \right\} \tag{6}$$

The above expression decouples the fractional resources α_v and β_v from the term $\mathrm{E}_{\lambda_v}\{m\sqrt{\lambda/t}\}$, which is associated with the MIMO capacity gain. The expectation with respect to λ_v is evaluated following exactly the same approach as exposed by (1)–(5), to arrive at [20, §7.414.4.1]

$$\mathrm{E}_{\lambda_v} \left\{ m\sqrt{\lambda_v/t_v} \right\} \approx$$

$$\frac{1}{\sqrt{t_v}} \sum_{k=0}^{m-1} \frac{k!}{(k+d)!} \frac{\Gamma^3(d+k+1)\Gamma(d+\frac{3}{2})\Gamma(k-\frac{1}{2})}{(k!)^2 \Gamma(d+1)\Gamma(-\frac{1}{2})} \times$$

$$_3F_2(-k, d+\frac{3}{2}, \frac{3}{2}; d+1, \frac{3}{2}-k; 1) \tag{7}$$

where $\Gamma(\cdot)$ is the complete Gamma function, and $_3F_2(\cdot)$ is the generalised hypergeometric function with three parameters of type 1 and two parameters of type 2.

The term $\mathrm{E}_{\lambda_v}\left\{ m\sqrt{\lambda_v/t_v} \right\}$ is henceforth denoted as $\Lambda(t_v, r_v)$, which allows simplifying the MIMO capacity expression to

$$C_v \approx \alpha_v \sqrt{\frac{\beta_v}{\alpha_v}} \sqrt{\gamma_v \frac{S}{N}} \cdot \Lambda(t_v, r_v). \tag{8}$$

4   Maximum Throughput for Ergodic Channels

Throughput in [bits] is defined as the information delivered from source towards sink, which requires a certain communication time T and frequency band W. Subsequent analysis will therefore refer to the throughput Θ in [bits/s/Hz] normalised by T and W.

In contrast to non-ergodic channels, an ergodic channel can provide a normalised capacity C in [bits/s/Hz] with 100% reliability [2], which allows relating capacity and throughput via $\Theta = C$. Therefore, optimising the throughput Θ is equivalent to optimising capacity C in the case of ergodic fading channels.

As defined by Shannon, capacity relates to error-free transmission. Hence, if a certain capacity was to be provided from source to sink, all channels involved must guarantee error-free transmission. From this it is clear that the end-to-end capacity C is dictated by the capacity of the weakest link [9, page 429].

Since optimisation has to be performed on the weakest link (or one of the equally strong links) at each of the K relaying stages, notation can be simplified further. To this end, it is assumed that the weakest link in the vth relaying stage has t_v antennas acting as transmitters and r_v antennas acting as receivers. The K-stage relaying network is henceforth denoted as $(t_1 \times r_1)/(t_2 \times r_2)/\ldots/(t_K \times r_K)$.

The capacity C_v of the vth relaying stage is hence determined by t_v and r_v, and the occurring channel conditions. It is thus the aim to find for all $v = 1,\ldots,K$ the fractional allocations of bandwidth α_v and power β_v for given channel conditions λ_v and γ_v such as to maximise the minimum capacity C, i.e.

$$C = \sup_{\boldsymbol{\alpha},\boldsymbol{\beta},v\in(1,K)} \left\{ \min\{C_v(\alpha_v,\beta_v,\lambda_v,\gamma_v)\} \right\} \tag{9}$$

where the optimisation is performed over the fractional sets $\boldsymbol{\alpha} \triangleq (\alpha_1,\ldots,\alpha_K)$ and $\boldsymbol{\beta} \triangleq (\beta_1,\ldots,\beta_K)$. Furthermore, $C_v(\alpha_v,\beta_v,\lambda_v,\gamma_v)$ denotes the dependency of the capacity in the vth link on the fractional resource allocations α_v and β_v, and on the channel conditions λ_v and γ_v. Fractional capacity allocation strategies satisfying (9) under applicable constraints are derived below.

With the obvious parameter constraints $\sum_{v=1}^{K} \alpha_v \equiv 1$ and $\sum_{v=1}^{K} \beta_v \equiv 1$, increasing one capacity inevitably requires decreasing the other capacities. The minimum capacity is maximised if all capacities are equated and then maximised [13,14]. The normalised capacity of the vth stage is given as

$$C_v = \alpha_v \cdot \mathrm{E}_{\lambda_v} \left\{ \log_2 \left(1 + \lambda_v \frac{\gamma_v}{t_v} \frac{\beta_v}{\alpha_v} \frac{S}{N} \right) \right\} \tag{10}$$

Thus, α_v is obtained by equating (10) for all $v = 1,\ldots,K$, which can be shown to be [14]

$$\alpha_v = \frac{\prod_{w\neq v} \mathrm{E}_{\lambda_w} \left\{ \log_2 \left(1 + \lambda_w \frac{\gamma_w}{t_w} \frac{\beta_w}{\alpha_w} \frac{S}{N} \right) \right\}}{\sum_{k=1}^{K} \prod_{w\neq k} \mathrm{E}_{\lambda_w} \left\{ \log_2 \left(1 + \lambda_w \frac{\gamma_w}{t_w} \frac{\beta_w}{\alpha_w} \frac{S}{N} \right) \right\}} \tag{11}$$

The end-to-end capacity $C = C_1 = \ldots = C_K$ is obtained by inserting (11) into (10) to arrive at

$$C = \left[\sum_{v=1}^{K} \frac{1}{\mathrm{E}_{\lambda_v} \left\{ \log_2 \left(1 + \lambda_v \frac{\gamma_v}{t_v} \frac{\beta_v}{\alpha_v} \frac{S}{N} \right) \right\}} \right]^{-1} \tag{12}$$

which constitutes a $2K$-dimensional optimisation problem w.r.t. the fractional bandwidth and power allocations α_v and β_v, respectively.

Exact Optimisation via Lagrangian. Using Lagrange's method for maximising (12) under given constraints suggests the Lagrangian

$$\mathcal{L} = \left[\sum_{v=1}^{K} \frac{1}{\mathrm{E}_{\lambda_v} \left\{ \log_2 \left(1 + \lambda_v \frac{\gamma_v}{t_v} \frac{\beta_v}{\alpha_v} \frac{S}{N} \right) \right\}} \right]^{-1} + \iota \left[1 - \sum_{v=1}^{K} \alpha_v \right] + \kappa \left[1 - \sum_{v=1}^{K} \beta_v \right] \tag{13}$$

which is differentiated K times w.r.t. $\alpha_{v \in (1,K)}$ and then w.r.t. $\beta_{v \in (1,K)}$ another K times. The obtained $2K$ equations are equated to zero and the resulting system of equations is resolved in favour of any α_v and β_v, where ι and κ are chosen such as to satisfy the given constraints.

As previously stated, a pdf given in form of (2) leads to $2K$ equations which are not explicitly resolvable in favour of the sought variables. This is the main reason why no explicit resource allocation strategy has been developed to date.

Optimised Bandwidth and Optimised Power. An explicit resource allocation algorithm has been developed, which yields a close to optimum end-to-end capacity. Similar to [14], it is suggested to reduce the $2K$ dimensional optimisation problem (12) to a K-dimensional optimisation problem by optimising along β_v/α_v, the ratio between the fractional power and bandwidth allocation. In [14] it has been shown that $\sum_{v=1}^{K} \beta_v/\alpha_v \approx K$. Utilising this approximation with (8), eq. (12) can be simplified to a $(K-1)$-dimensional optimisation problem w.r.t. β_v/α_v with $v = 2, \ldots, K$. As demonstrated in [14], the $K-1$ equations can be resolved for any β_v/α_v which finally yields

$$\frac{\beta_v}{\alpha_v} \approx K \cdot \frac{\prod_{w \neq v} \sqrt[3]{\gamma_w \cdot \Lambda^2(t_w, r_w)}}{\sum_{k=1}^{K} \prod_{w \neq k} \sqrt[3]{\gamma_w \cdot \Lambda^2(t_w, r_w)}} \tag{14}$$

The fractional bandwidth allocations $\alpha_{v=(1,\ldots,K)}$ can now be obtained by substituting (14) into (11). The fractional power allocations $\beta_{v=(1,\ldots,K)}$ can then be obtained by inserting the prior obtained α_v into (14) and solving for β_v.

Since the derived fractional resource allocation rules are the result of various approximations, the obtained fractional bandwidths and powers not necessarily obey the given constraints. To overcome this problem, it is suggested to derive $K-1$ coefficients α_v and β_v, and then obtain the remaining two α_v and β_v from the constraints $\sum_{v=1}^{K} \alpha_v \equiv 1$ and $\sum_{v=1}^{K} \beta_v \equiv 1$. Since the weakest link will yield the largest fractional bandwidth and power, it is suggest to calculate these last as to guarantee that the constraints hold.

Note finally that due to the approximations deployed, the obtained K capacities $C_{v \in (1,K)}$ do not entirely coincide. The near-optimum end-to-end capacity utilising the above-given technique is hence dictated by the smallest of all C_v.

The case of equal bandwidth and optimised power, as well as the non-optimised case, are easily derived in a similar fashion as above. They have hence been omitted here; however, they are utilised in the next section to compare the performance benefits w.r.t. the optimised case.

5   Performance of Algorithms

The developed fractional resource allocation algorithms for distributed relaying networks with and without resource reuse, as well as with deployed STBCs, are assessed below. The simplest scenario is the 2-stage relaying scenario with only one relaying VAA tier. More relaying stages have not been analysed here due to the lengthy numerical optimisation. The obtained graphs are generally labelled on the parameter p, defined as

$$p \triangleq \left[10 \log_{10} \left(\frac{\gamma_1}{\gamma_1} \right), 10 \log_{10} \left(\frac{\gamma_2}{\gamma_1} \right) \right] \tag{15}$$

which characterises the relative strength in dB of the second relaying stages with respect to the first stage.

a) Single Antenna Element. The derived resource allocation strategies are obviously also applicable to traditional relaying networks with one antenna element per MT. The precision of the developed fractional resource allocation algorithm is assessed in Figure 2. It depicts the optimum end-to-end capacity obtained via numerical optimisation on (13) and the approximate end-to-end capacity obtained from (14), (11) and (10) versus the SNR in the first relaying stage. The graphs are labelled on the parameter p as defined in (15) with $K = 2$, where the second relaying channel is 10dB and 5dB stronger than the first one, equally strong than the first one, and 5dB and 10dB weaker than the first one.

It can be observed that the exact and developed end-to-end capacity almost coincide. The error was found not to exceed 3% for any of the depicted cases. The developed explicit resource allocations are hence a powerful tool in obtaining a near-to-optimum end-to-end capacity without the need for lengthy numerical optimisations. The algorithm is shown to be applicable for channels with attenuations differing by magnitudes.

b) Multiple Antenna Elements. The algorithms were also scrutinised for communication scenarios where the MTs possess multiple but equal number of antenna elements. Although not shown here, the occurring errors between derived allocation strategy and an optimum allocation is below 3%.

The demonstrated performance gains and power savings clearly underline the merit of the developed fractional resource allocation algorithms.

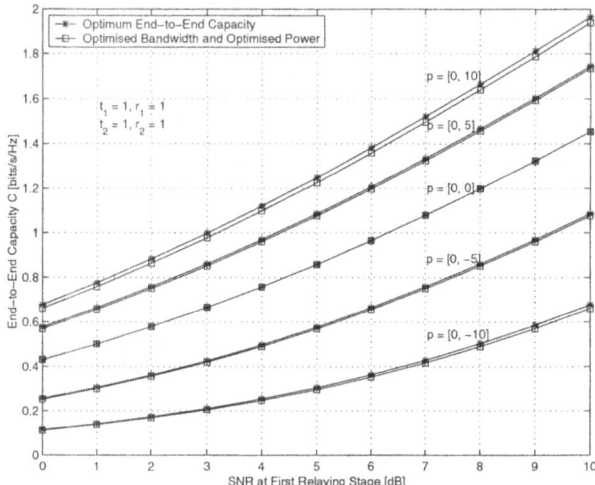

Fig. 2. Comparison between optimum and near-optimum end-to-end capacity for a 2-stage network.

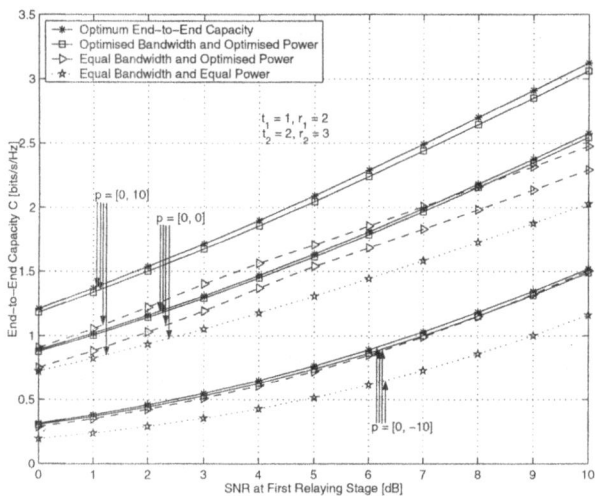

Fig. 3. Achieved end-to-end capacity of various fractional resource allocation strategies for a 2-stage network.

c) Differing Antenna Elements. The importance of the developed strategy, however, becomes apparent when the 2-stage communication scenario is optimised for terminals with a different number of antenna elements. The precision of the fractional resource allocation algorithm, as well as its performance gains when compared to sub-optimal solutions, is exposed in Figures 3. It depicts the case where a s-MT with one antenna element communicates with a t-MT (or t-VAA) with three elements via a relaying stage, which effectively provides two distributed relaying antennas.

The asymmetry of the gains provided by the respective distributed-MIMO relaying stages causes the sub-optimum allocation strategies not to overlap with the optimum one for $p = [0, 0]$dB. Furthermore, non-linearities can be observed in the end-to-end capacity for the case of optimised power only, which is due to the approximation utilised in the derivation of the allocation strategy. In fact, one can observe a breakpoint which divides the zones where one or the other approximate capacity dominates the end-to-end capacity.

However, the optimised fractional bandwidth and power allocation strategy yields close to optimum performance, even if the link attenuations and the created MIMO configurations differ significantly. Additionally, the gains obtained from a bandwidth/power optimised system when compared to a power optimised or non-optimised system increase with the balance between both links decreasing.

For instance, for $p = [0, 10]$dB, the first (1×2) MIMO link is much weaker than the second (2×3) link. At an SNR of 6dB the capacity losses of a power optimised system is then 20% and of a non-optimised system a considerable 40%. Alternatively, the power required to maintain 2 bits/s/Hz is about 55% higher for the power optimised system, whereas a non-optimised system requires 120% more power.

6 Conclusions

A regenerative distributed-MIMO multi-stage relaying network without resource reuse has been introduced. Analysis concentrated on FDMA-based relaying, where fractional bandwidths α and fractional powers β were derived. Explicit fractional resource allocation rules over ergodic channels were derived under the constraint to maximise the end-to-end throughput which, in the ergodic case only, is equivalent to the end-to-end capacity. The derivation of the explicit resource allocation algorithms relied on the exposed exact and approximated capacities for traditional MIMO flat Rayleigh fading. The approximation allowed decoupling the fairly intricate expression related to the MIMO capacity gain from terms requiring optimisation, such as transmission power and bandwidth.

The developed algorithms were tested over a variety of scenarios, i.e. different number of antennas per stage and different channel attenuations. Despite the utilised approximation, the algorithms were found to be robust and yield near-optimum end-to-end throughput.

References

1. C.E. Shannon, "A mathematical theory of communication," *Bell Syst. Tech. J.*, vol. 27, pp. 379-423, 623-656, July-Oct. 1948.
2. E. Telatar, "Capacity of multi-antenna Gaussian channels," *European Trans. on Telecomm.*, vol. 10, no. 6, pp. 585-595, Nov./Dec. 1999.

3. G.J. Foschini and M.J. Gans, "On limits of wireless communications in a fading environment when using multiple antennas", *Wireless Personal Communications*, vol. 6, pp. 311-335, 1998.

4. M. Dohler, F. Said, A. Ghorashi, H. Aghvami, "Improvements in or Relating to Electronic Data Communication Systems", patent publication no. WO 03/003672, priority date 28 June 2001.

5. Mischa Dohler, et al., "VAA for hot-spots with applied STC", *M-VCE Internal Reports* I, II, III and IV, 1999-2002.

6. T. Cover, J.A. Thomas, *Elements of Information Theory*, John Wiley & Sons, Inc., 1991.

7. T. J. Harrold, A. R. Nix, "Capacity Enhancement Using Intelligent Relaying For Future Personal Communications System", *Proceedings of VTC-2000 Fall*, pp. 2115-2120.

8. 3G TR 25.924 V1.0.0 (1999-12) *3rd Generation Partnership Project*, Technical Specification Group Radio Access Network; Opportunity Driven Multiple Access.

9. J.N. Laneman, D.N.C. Tse, G.W. Wornell, "Cooperative diversity in wireless networks: Efficient protocols and outage behaviour," submitted to *IEEE Trans. on Inform. Theory*.

10. A. Sendonaris, E. Erkip, B. Aazhang, "User Cooperation Diversity - Part I: System Description," *IEEE Transactions on Communications*, vol. 51, no. 11, November 2003, pp. 1927-1938.

11. A. Sendonaris, E. Erkip, B. Aazhang, "User Cooperation Diversity - Part II: Implementation Aspects and Performance Analysis," *IEEE Transactions on Communications*, vol. 51, no. 11, November 2003, pp. 1939-1948.

12. Mischa Dohler, E. Lefranc, H. Aghvami, "Virtual Antenna Arrays for Future Wireless Mobile Communication Systems", *IEEE ICT 2002*, Beijing, China, 2002, Conference CD-ROM.

13. M. Dohler, A. Gkelias, H. Aghvami, "2-Hop Distributed MIMO Communication System", *IEE Electronics Letters*, vol. 39, no. 18, Sept. 2003, pp.1350-1351.

14. M. Dohler, A. Gkelias, H. Aghvami, "A Resource Allocation Strategy for Distributed MIMO Multi-Hop Communication Systems", *IEEE Communications Letters*, accepted in September 2003.

15. P. Gupta and P.R. Kumar, "The Capacity of Wireless Networks", IEEE Transactions on Information Theory, vol. 26, no. 2, pp. 388-404, March 2000.

16. S. Toumpis, A.J. Goldsmith, "Capacity Regions for Wireless Ad Hoc Networks", Submitted to the IEEE Transactions on Wireless Communications, September 2001.

17. T.E. Batt and A. Ephremides, "Joint Scheduling and Power Control for Wireless Ad-hoc Networks", IEEE INFOCOM, June 2002, Conference CD-ROM.

18. B. Radunovic and J-Y.L. Boudec, "Joint Scheduling, Power Control and Routing in Symmetric, One-dimensional, Multi-hop Wireless Networks", WiOpt'03, March 2003, Conference CD-ROM.

19. M. Dohler, *Virtual Antenna Arrays*, PhD Thesis, King's College London, 2003.

20. I.S. Gradshteyn, I.M. Ryshik, *Table of Integrals, Series, and Products*, Academia Press, sixth edition, 2000.

Delivering the Benefits of Directional Communications for Ad Hoc Networks Through an Efficient Directional MAC Protocol

Jacir L. Bordim, Tetsuro Ueda, and Shinsuke Tanaka

ATR Adaptive Communications Research Laboratories,
2-2-2 Hikaridai, Keihanna Science City, Kyoto, 619-0288 Japan

Abstract. It is known that wireless ad hoc networks employing omnidirectional communications suffer from poor network throughput and inefficient spatial reuse. Although directional communications are expected to provide significant improvements in this respect, efficient Medium Access Control (MAC) protocols capable of harvesting the benefits of directional communications in the context of ad hoc networks are yet to be seen. In this work we propose a MAC scheme aiming to provide efficient means to explore the transmission range extension and improved spatial division provided by directional antennas. In order to measure the potentiality of our MAC protocol, we have extensively evaluated its performance in terms of average throughput and end-to-end delay under single and multi-hop communication scenarios. The performance evaluation results indicate that our protocol is highly efficient, particularly with increasing number of communications and data rates.

1 Introduction

Stimulated by the availability of faster and cheaper electronic components, the past decade has witnessed enormous advances in wireless communication technologies. These advances have fostered the research in ad hoc networks, which are envisioned as rapidly employable, infrastructure-less networks where each node is equipped with wireless capabilities and act as a mobile router. These characteristics make ad hoc networks suitable to support communications in urgent and temporary tasks, such as business meetings, disaster-and-relief, search-and-rescue, law enforcement, among other special applications. As in the Wireless LAN (WLAN) standard IEEE 802.11 [1], nodes in an ad hoc network are usually assumed to share a common channel and to operate with omnidirectional antennas. Since nodes sufficiently apart from each other can communicate simultaneously, one could expect the throughput to improve with the area they cover. However, the relay load imposed by distant nodes and the inefficient spatial reuse provided by omnidirectional antennas results in poor network throughput [2]. The routing strategy used, and other optimizations may have a positive impact, but only up to a certain extent. In [3], Gupta and Kumar showed that no matter which routing protocol or channel access scheme is used, the throughput

J.N. de Souza et al. (Eds.): ICT 2004, LNCS 3124, pp. 461–470, 2004.
© Springer-Verlag Berlin Heidelberg 2004

obtained per node is $\Theta\left(\frac{W}{\sqrt{n\log n}}\right)$, where n is the number of nodes and W is the data rate.

Aiming to provide better spatial reuse and increase the network capacity, the research community started considering ad hoc networks where the nodes are empowered with directional antennas. The key benefits provided by directional antennas include reduced co-channel interference, transmission range extension, better spatial reuse and signal quality as compared to their omnidirectional counterparts [4]. However, developing efficient directional Medium Access Control (MAC) protocols in the context of ad hoc networks is a challenging task due to the difficulty in coping with mobility and the absence of a centralized control. In spite of this, few MAC protocols tailored for directional communications have been proposed. In [5], Ko *et. al.* proposed a MAC protocol in which each node is assumed to know its physical location (perhaps with the aid of a GPS). Two schemes were proposed: (*i*) RTS packets are sent directionally; (*ii*) RTS packets are send omnidirectionally if none of its antenna elements are blocked. In both schemes, CTS packets are sent omnidirectionally. In the MAC protocol proposed in [6], source and destination identify the location of each other during the omni RTS/CTS exchange. In order to obtain accurate location information without the aid of a GPS, the MAC protocol proposed in [7] associates each neighboring node with an antenna element. RTS/CTS packets are sent directionally, or selective multi-directional, while DATA and ACK packets are sent directionally. The above MAC protocol was refined in [8], where the number of control messages to obtain location information is significantly reduced. Choudhury *et al.* [9,10] proposed MAC protocols aiming to explore the higher gain provided by directional antennas. In [9], RTS packets are send directionally while the destination waits in omnidirectional sensing mode. The antenna element that receives the RTS is then used for subsequent directional transfers. Note that the above approach, in essence, allows for Directional-Omni communications. In [10], a multi-hop RTS mechanism was proposed aiming to provide Directional-Directional communication.

From the above discussion, one can see that those approaches that rely on omni RTS and/or omni CTS exchange provide little improvement in terms of spatial reuse. Furthermore, none of the aforementioned results can efficiently explore the transmission range extension provided by directional antennas. Although the scheme proposed in [10] attempted to explore the transmission range extension, their approach is inefficient since multi-hop RTS wastes bandwidth, increases contention and communication time. In this work we propose a directional MAC scheme, termed DD-MAC, aiming to overcome the aforementioned limitations. More specifically, our MAC protocol centers in exploring the transmission range extension and spatial reuse provided by directional antennas. At the heart of our MAC protocol lies an elegant, yet simple, mechanism to enable packets to be send and received directionally[1]. In order to measure its efficiency, we have successfully implemented DD-MAC in QualNet [12] and extensively evaluated its performance in terms of average throughput and end-to-end delay,

---

[1] Preliminary results of this work have been presented in [11].

under single and multi-hop communication scenarios. Also, unlike other experiments which normally include some upper layer's interactions in its results, we use *static routes* in our simulations to avoid any interferences that might distort the outcome (e.g., with the choice of a particular routing strategy). In addition, to obtain a more realistic evaluation of DD-MAC, the directional beam patterns in our simulations use the real hardware measurements of the ESPAR antenna [13], which is a prototype of a directional antenna tailored for mobile communications. The experimental results indicate that our protocol is highly efficient with increasing number of communications and data rates. For instance, our MAC protocol outperformed the throughput of the IEEE802.11 MAC protocol in more than 150%, on average, for 6 and 10 simultaneous communications.

The remainder of this paper is organized as follows: Section 2 shows a concise discussion on the IEEE 802.11 Distributed Coordination Function (DCF) and presents the antenna model considered in this work. The details or our directional MAC protocol is presented in Section 3. Simulation results are discussed in Section 4, and conclusions are given in Section 5.

2   Preliminaries

2.1   The IEEE 802.11 DCF Operation

The IEEE 802.11 Distributed Coordination Function (DCF) is a contention based MAC protocol designed to operate with omnidirectional antennas [1]. The RTS/CTS exchange is used to inform neighboring nodes about the eminent start of communication. Each of these packets contain the proposed duration of communication and the destination address. Neighboring nodes that overhear any of these packets must themselves defer communication for the proposed duration. This mechanism is known as *Virtual Carrier Sensing* and is implemented through the use of the *Network Allocation Vector* (NAV) variable. DATA and ACK packets follows the successful RTS/CTS exchange. When the MAC protocol at a node S receives a request to transmit a packet, both physical and virtual carrier sensing are performed. If the medium is found idle for an interval of DIFS (DCF Inter Frame Space) time, then S chooses a random back off period for additional deferral. When the back off period expires (*i.e.*, reaches zero), S transmits the packet. If a collision occurs, a new back off interval is selected. A Short Inter Frame Space (SIFS) is used to separate transmissions belonging to the same dialog.

2.2   Directional Antenna Model

There are basically two types of directional antennas: *Switched Beam*, and *Adaptive Arrays*. Switched beam antennas are relatively simple to implement, comprising a number of antenna elements, a basic Radio Frequency (RF) switching function, and a control logic to select a particular beam. The antenna elements are deployed into a number of fixed sectors, among which the one experiencing the highest sinal level is selected to collect the incoming signals. Adaptive array antennas rely on sophisticated digital signal processing algorithms to direct the beam towards the intended user and simultaneously suppress interfering signals (by setting nulls in the direction of interferences). Although adaptive antennas can provide better performance than a switched beam antenna, the engineering cost associated with it is a limiting factor. On the other hand, switched beam antennas are expected to be produced at a much lower cost while being able to provide most of the benefits of a more sophisticated system [14]. Hence, switched beam antennas seems to be a feasible option as a first generation technology to be used in ad hoc networks. Indeed, efforts aiming to enhance mobile devices with directional antennas already exist, an example is the ESPAR (Electronically Steerable Passive Array Radiator) antenna currently being developed at ATR(ACR) [13].

The ESPAR antenna relies on analog RF beamforming, instead of digital beamforming usually employed in array antennas, which drastically reduces the circuit complexity. Since it requires only one receiver chain, the ESPAR antenna provides drastic improvements, both in power dissipation and fabrication costs, by eliminating the need for frequency converters and analog-to-digital converters by the number of array branches. The ESPAR antennas can be used not only as a switched beam antenna but also as an omnidirectional antenna, by selecting the value of reactance for one specific directional beam among multiple directional beam patterns or omnidirectional, without using multiple receiver chains (frequency converters and analog-digital converters). The advantage of using ESPAR antenna as switched beam antenna is that, variable number of beam-pattern is possible with only one receiver chain. Since ESPAR antenna would be a low-cost, low-power, small-sized antenna, it is expected to help reducing power consumption of the user terminals in wireless ad hoc networks and would be able to deliver all the advantages of a switched beam antenna. In this work we assume that each mobile terminal is equipped with a ESPAR antenna. Our MAC protocol, presented in the next section, also works with antenna models that have similar characteristics, like those assumed in [6,7,10].

3 Enabling Directional Communications

We start by showing how we detect the direction of the received signal and then go on to present the details of our directional MAC protocol. For latter reference, we define three types of neighbors: *Omni-Omni* (OO), *Directional-Omni* (DO), and *Directional-Directional* (DD) neighbors. Nodes A and B are DO-neighbors if node B, in omni mode, can receive a directional transmission from node A (OO and DD-neighbors are defined similarly). Following the above terminology, the communication between two DO-neighbors (also OO and DO) is referred to as DO-communication (also OO and DO-communication). A MAC protocol to enable DO-communication was presented in [9]. Using those results, a mechanism to provide DD-communication was later proposed in [10]. The basic idea in [10] is to create a *route* between two DD-neighbors. Such route is composed of DO-neighbors whose task is to forward an RTS request from a DO-neighbor to another (here, the directional RTS is received and decoded in omni mode). On receiving the multi-hop RTS, the destination node then replies with a directional CTS. (Note that the destination has to know the exact location of the source node, an issue not addressed in [10].) Directional communication takes place from this point onwards. Clearly, the above method is inefficient since multi-hop RTS waste bandwidth, increase contention and communication time. In this work we propose a MAC protocol that can realize directional "single-hop" RTS transmission and reception. Furthermore, means to associate each neighboring node to a particular beam of a node is also addressed.

To enable DD-communications, DD-MAC relies on an interesting mechanism to detect the direction of the received signals. As a first solution, whenever a mobile station is idle, it waits in rotational-sector receive-mode. In this mode, a node N rotates its directional antenna sequentially in all directions, covering the entire space (360 degrees) in the form of directional receiver. Each control packet is sent with a preceding tone. By properly selecting the duration of a tone, the receiver is able to identify the best possible direction to receive the signal. In other words, the time to complete one full rotation is less than the tone's duration. Upon hearing a signal that is above the sensing threshold, node N selects the antenna element that maximizes the signal strength to receive the packet that follows the tone. We now show that DD-communication is possible even if idle nodes wait in omni mode. In such case, when a signal is sensed in omni mode, the node switches to directional sensing mode to obtain the direction that maximizes the signal level, as performed above. The omni-sensing and one cycle of the directional rotation are performed within the duration of the preceding tone. Since the antenna gain of an omni antenna is lower than that of a directional antenna, the normal receiving threshold leads to DO-communication. However, if we set the receiver sensitivity to "[(normal receiving threshold) - (gain difference of omni and directional)]", then DD-communication is available even if idle nodes wait in omni mode.

Once the packet is received (using either of the above approaches), node N knows the ID number of the sender and the antenna element associated with it. This information is then appended into a table, named Angle Signal Table (AST). An entry in the AST includes the maximal signal strength of the received signal at a particular antenna element, the direction and node ID of the sender as well as the time the information was received. Note that a node might associate two or more entries of the AST to a particular node. DD-MAC protocol, presented in the next subsection, uses the above mechanism to provide DD-communications.

3.1 DD-MAC Protocol

In our Directional-Directional MAC (DD-MAC), channel reservation is performed using directional RTS/CTS exchange between the sender and the receiver. When idle, each node senses the medium using rotational-sector sensing (or using omni-sensing with the next cycle of directional rotation, as explained above). Suppose that node S wishes to communicate with a neighboring node D. First, node S performs both physical and virtual carrier sensing to determine whether it is safe to transmit towards D or not. If the medium is free, then node S transmits the RTS directionally. If the medium is sensed busy, then S postpones its RTS transmission, similar to the IEEE802.11.

When node D, in rotational sector mode (or using omni-sensing with the next cycle of directional rotation) hears the tone that precedes the directional RTS (DRTS), it locks its antenna in the sector that maximizes the signal. This allows DD-MAC to directionally receive and decode the DRTS, without resorting to multi-hop RTS. Upon receiving the DRTS, node D proceeds by carrier sensing and waiting for SIFS time before sending a directional CTS (DCTS) back to S. The tone and DCTS are sent over the same antenna element from which the DRTS was received. Meanwhile, S remains beamformed towards D to receive the DCTS. Directional DATA and ACK packets follow the successful DRTS/DCTS exchange. If node S does not receive a DCTS within a timeout period, a new DRTS is issued. Again, node S has to go through all the steps of carrier sensing and back-off time before re-transmission.

Neighboring nodes, that overhear the DRTS, DCTS, or both, set their Directional Network Allocation (DNAV) in the direction of the detected signals. Such nodes will defer their own transmissions for the proposed duration of transfer. However, they are allowed to engage in communication as long as the desired direction of communication does not interfere with any ongoing transfers.

4   Performance Evaluation

We have simulated the proposed MAC protocol in QualNet [12] and evaluated its performance in terms of average end-to-end delay and throughput for both single and multi-hop communications. The simulation results of the IEEE802.11 MAC protocol (802.11, for short) are used as benchmark in comparing the results. The simulated environment consists of 50 nodes randomly distributed in a 2D-plane of size 1000×1000 meters. All the nodes have the same transmission power, which is set to 10dBm. In our simulations, we have used the ESPAR antenna in the form of a switched beam antenna with a 30 degrees beamwidth, which is discretely steered to cover the 360 degrees span. Based on hardware measurements of the ESPAR antenna, the duration of the preceding tone in control packets is set to $200\mu s$ in the simulations. The results are drawn from the average of twenty runs with different seeds, where each run lasts for four minutes. We have simulated over 900 scenarios, which accounts to over 60 hours of simulation time, in order to obtain the averaged results.

The performance evaluation consists of two parts. First, we evaluate the performance improvement of DD-MAC in terms of single-hop communications efficiency, where single-hop source-destination pairs are randomly selected for communication. Since a directional antenna receives/transmitts more power towards a specific direction, the *gain* of a directional antenna is usually greater than the gain of an omnidirectional antenna [4]. Hence, using the same transmis-

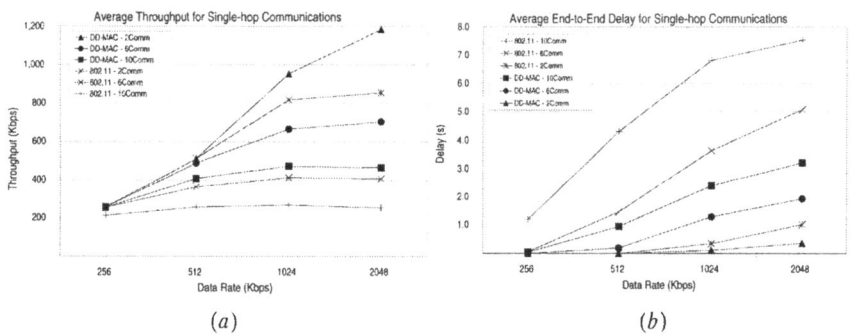

Fig. 1. Comparison of the average throughput (*a*) and delay (*b*) for the IEEE802.11 and DD-MAC at different data rates and number of single-hop communications.

sion power, nodes communicating in directional mode can communicate over a larger distance, as compared to nodes communicating in omnidirectional mode. For example, the transmission range in your simulations is nearly 260 meters, for the 802.11 MAC, and around 500 meters for DD-MAC. Since the farthest distance between any two stations in our simulation is $\approx 1,414$ meters, source and destination are at most 3 hops away in DD-MAC, against 6 hops, in the worst case, for the 802.11 MAC. Thus, in our second set of experiments we focused on the impact of the extended transmission range provided by DD-MAC in a multi-hop environment. Since this work aims in evaluating the performance of MAC-layer protocols, we use static routes in the simulations to prevent packet generation from the upper layers.

Figure 1 shows the average throughput and average end-to-end delay, respectively, for the 802.11 MAC and DD-MAC with 2, 6 and 10 concurrent single-hop communications. The results are shown for four different data rates: 256, 512, 1024 and 2048Kbps (1024Bytes of CBR traffic generated at every 32, 16, 8 and $4ms$ interval, respectively). As can be seen in Figure 1(*a*), the throughput for 802.11 and DD-MAC is comparable at low data rates (256Kbps). This is because the packet injection gives enough time for the MAC protocols to deliver the generated packets without much competition in grabbing the channel. As the number of communications increase though, the gap in throughput widens, even at moderate data rates, as is the case for 10 communications at 512Kbps. With higher data rates, DD-MAC provides an environment of lower contention that the 802.11 with omnidirectional antennas cannot match. In addition, directional communications provide higher gain than omnidirectional communications. Since higher gain reflects in higher SINR (Sinal-to-Interference-and-Noise-Ratio) level, the chances of missing data packets or receiving corrupted data packets are minimized. All these accounts for performance improvements, even when only few nodes are engaged in communication, as is the case for 2 communications at 2048Kbps, where DD-MAC has a throughput improvement or nearly 40% higher than the 802.11 MAC. In comparison with the 802.11 MAC,

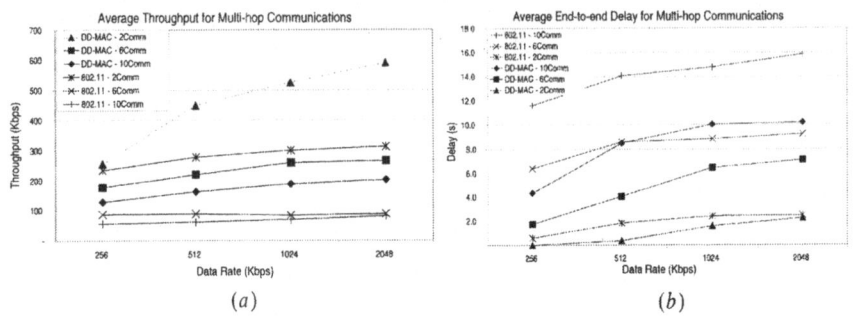

Fig. 2. Comparison of the average throughput (a) and delay (b) for the IEEE802.11 and DD-MAC at different data rates and number of multi-hop communications.

DD-MAC provides a throughput improvement of more than 70%, for 6 communications, and surpasses 80% for 10 communications at 2048Kbps. In addition, as shown in Figure 1(b), when the data rates are higher than 512Kbps, DD-MAC is capable of delivering packets in less than half of the time needed by the 802.11 MAC.

We now focus on performance evaluation of DD-MAC in a multi-hop setting. The static routes for multi-hop communications are computed using the standard shortest-path algorithm. Figure 2 shows the average throughput and average end-to-end delay for multi-hop communications for the 802.11 MAC and DD-MAC with 2, 6 and 10 concurrent communications for four different data rates. With the exception of the results for 2 communications at 256Kbps, all the other results show a wide throughput gap between 802.11 and the DD-MAC, as can be seen in Figure 2(a). On average, DD-MAC outperforms the 802.11 MAC protocol in more than 60% for 2 communications with data rates higher than 256Kbps. The improvements in throughput surpasses 150%, on average, for 6 and 10 simultaneous communications. The higher throughput achieved by DD-MAC comes primarily from the following facts: (i) DD-MAC uses the RTS-CTS packets exchange only to inform neighboring nodes about the eminent start of communication, not to silence them, as is the case of the 802.11. In other words, when neighboring nodes overhear the DRTS and/or DCTS exchange, they set their DNAV in the direction which the incoming signals were detected. These neighboring nodes are allowed to communicate in other directions, provided that the direction towards they wish to communicate is not blocked by the DNAV. Otherwise, they will have to wait for DNAV to expire, similar to the NAV in the IEEE802.11. (ii) Due to the higher gain provided by directional communication, the average number of hops between source-destination pairs is significantly reduced. The average number of hops for DD-MAC and 802.11 in our simulations is 1.6 and 2.6 hops, respectively. Thus, with a reduced number of hops and a large number of nodes competing for the channel, as is the case for 6 − 10 communication pairs, DD-MAC is capable of achieving a much better performance than the 802.11 MAC. The aforementioned facts also allows DD-MAC to minimize the queueing delay, thus reducing the average end-to-end delay, as can be observed in Figure 2(*b*). From the above, one can expect the gap in performance between 802.11 and DD-MAC to be even wider when considering larger multi-hop simulation environments.

5   Conclusions

The main contribution of this work was to present a directional MAC protocol scheme which aims to explore the transmission range extension and improved spatial reuse provided by directional antennas. In order to evaluate the performance of our scheme, termed DD-MAC, we have simulated it in QualNet and measured its performance in terms of average throughput and end-to-end delay for both single and multi-hop communications. The IEEE802.11 MAC protocol is used as benchmark in comparing the results. The simulation results for single-hop communications showed that DD-MAC can provide significant improvement in terms of average throughput and end-to-end delay, even with only few nodes engaged in communication. With an increased number of simultaneous communications and higher data rates, DD-MAC provided throughput improvements in the order of 70% to 80% higher than the IEEE802.11, with 6 and 10 communication pairs, respectively. In our second set of experiments we focused in verifying the impact of the extended transmission range provided by DD-MAC in a multi-hop environment. Here, DD-MAC outperformed the throughput of the IEEE802.11 MAC protocol in more than 150%, on average, for 6 and 10 simultaneous communications. As future work, we plan to verify the performance of DD-MAC under different routing strategies.

Acknowledgements. The authors would like to thank Dr. B. Komiyama for his encouragement and support throughout this work. This research is supported in part by The National Institute of Information and Communications Technology (NICT).

References

1. Institute of Electrical and Electronics Engineers (IEEE), *Wireless LAN Medium Access Control (MAC) and Physical Layer (PHY) Specifications*, IEEE Standard 802.11, June 1999, http://standards.ieee.org.
2. J. Li, C. Blake, et. al., *Capacity of Ad Hoc Wireless Networks*, in Proceedings of ACM International Conference on Mobile Computing and Networking (MOBI-COM), 2001.
3. P. Gupta and P. R. Kumar, *The capacity of wireless networks*, IEEE Transactions on Information Theory, Mar 2000, Volume: 46, Issue: 2, pp. 388-404.
4. T. Rappaport, *Wireless Communications: Principles and Practice*, Prentice Hall PTR, Upper Saddle River, NJ, 2001.
5. Y.-B. Ko, V. Shankarkumar and N. H. Vaidya, *Medium Access Control Protocols Using Directional Antennas in Ad Hoc Networks*, IEEE INFOCOM'2000, March 2000.

6. A. Nasipuri, S. Ye, and R. E. Hiromoto, *A MAC Protocol for Mobile Ad Hoc Networks Using Directional Antennas*, Proceedings of the IEEE Wireless Communications and Networking Conference (WCNC 2000), Chicago, September 2000.

7. S. Bandyopadhyay, K. Hasuike, S. Horisawa, S. Tawara, *An Adaptive MAC Protocol for Wireless Ad Hoc Community Network (WACNet) Using Electronically Steerable Passive Array Radiator Antenna*, Proc of the GLOBECOM 2001, November 25-29, 2001, San Antonio, Texas, USA.

8. T. Ueda, S. Bandyopadhyay, and K. Hasuike, *An approach Towards Improving Quality of Service in Ad Hoc Networks with ESPAR Antenna*, in Proceedings of The 16th International Workshop on Communications Quality & Reliability (CQR) 2002.

9. R. R. Choudhury and N. Vaidya, "Impact of Directional Antennas on Ad Hoc Routing", in Proc. of the 8th Conference on Personal and Wireless Communication (PWC), pp. 590-600, Venice, September 2003.

10. R. R. Choudhury, X. Yang, R. Ramanathan, and N. Vaidya, *Using directional antennas for Medium Access Control in Ad hoc Networks*, ACM International Conference on Mobile Computing and Networking MobiCom, September 2002.

11. J. L. Bordim, T. Ueda, and S. Tanaka, *A Directional MAC Protocol to Support Directional Communications in Ad Hoc Networks*, in Proc. of the First International Conference on Mobile Computing and Ubiquitous Networking (ICMU-2004), pp. 52-57, Jan. 8-9, 2004, Tokyo, Japan.

12. Scalable Networks Technologies, *QualNet - Network Simulator*, http://www.scalable-networks.com/.

13. T. Ohira and K. Gyoda, *Electronically steerable passive array radiator antennas for low-cost analog adaptive beamforming*, in Proceedings of the IEEE International Conference on Phased Array Systems and Technology, 2000. pp: 101-104.

14. J. C. Liberty and T. Rappaport, *Smart Antennas for Wireless Communications: IS-95 and Third Generation CDMA Applications*, Prentice Hall PTR, Upper Saddle River, NJ, 1999.

Synchronization Errors Resistant Detector for OFDM Systems

Yeheskel Bar-Ness[1] and Rubén Solá[2]

[1] Center for Communications and Signal Processing Research, 323 Martin Luther King Jr. Blvd. Newark, NJ 07102-1982, yeheskel.barness@njit.edu
[2] Center for Communications and Signal Processing Research, 323 Martin Luther King Jr. Blvd. Newark, NJ 07102-1982, ruben@confluencia.net

Abstract. An OFDM (Orthogonal Frequency Division Multiplexing) detector both resistant to frequency and time offset is proposed in this paper. The detector combines a modified frequency offset estimation technique based on the one presented by P.H. Moose in [1], with a Minimum Mean Square Error (MMSE) techniques for time offset correction. Results show that the frequency offset estimation can be performed even when the time offset is unknown, within a given interval, followed with correction of the distortion of the received signal caused by the time offset by the means of an adaptive filter.

1 Introduction

Multicarrier systems have been proven as much more sensitive to synchronization errors than single carrier systems [2],[3]. Different studies have been done in the past to estimate error parameters either jointly or individually.

In [1] P. H. Moose derived the Maximum Likelihood Estimator (MLE) for the carrier frequency offset, which is calculated in frequency domain after FFT (Fast Fourier Transform). The method exploits the fact that when two same consecutive OFDM symbols are transmitted, the received symbols at the output of the FFT differ only by a phase term which depends on the frequency offset. Therefore, it was proposed to send two identical symbols and estimate the frequency offset from the differences of phase rotation between them. A MLE function (S. Kay method) was derived, fundamentally averaging the phase differences over all subcarriers. The method needs two training symbols and is limited by the acquisition range, which is ±1/4 the subcarrier spacing, and by the prior knowledge of symbol timing.

Van de Beek et al proposed in [4] a blind estimation method for tracking time and frequency offsets using the guard interval redundancy. A MLE function was derived correlating the samples in time domain for an AWGN (Average White Gaussian Noise) channel. Although the variance is small, when MLE peaks are averaged over a number of symbols, the individual estimations of this method have bigger offsets [5].

Schmidl&Cox proposed in [6] a way for burst detection and carrier frequency offset estimation. The method gives a robust way of estimating frequency offset without the limitation in [4] but presented a *plateau* for the timing metric which leads

J.N. de Souza et al. (Eds.): ICT 2004, LNCS 3124, pp. 471–481, 2004.
© Springer-Verlag Berlin Heidelberg 2004

to some uncertainty for the start of the symbol. The modifications introduced by H.Minn, M Zeng and V. K. Bhargava in [7] for the Schmidl&Cox's method were aimed to reduce this uncertainty.

In this paper a complete scheme for synchronisation parameters correction, rather than just an estimation method, is presented. This study may be considered as a complementary work to the ones previously discussed, wherein the time offset is assumed within a given range, i. e., an initial acquisition for the start of the symbol is assumed using some of the methods previously discussed. As usually these time estimation methods involve correlation calculations over a large number of OFDM symbols, are computationally complex and yet do not provide a perfect estimation of the start of the OFDM symbol, the presented technique may be used for correcting the frequency offset on one hand and the residual timing offset on the other.

2 Problem Definition

Considering the structure presented in figure 1 and following the notation used in previous works [8], the actual OFDM transmitted waveform after translation to higher frequeny can be written as:

$$X_T(t) = \frac{1}{N} \sum_{j=-\infty}^{j=\infty} \sum_{n=0}^{N_c-1} d(n,j) \cdot rect(\frac{t-jT}{T} - \frac{1}{2}) \; e^{j[2\pi f_n(t-jT)+2\pi f_c t+\phi_c]} \tag{1}$$

where N is the number of points used for IFFT and FFT, N_c is the number of sub-carriers, $T=T_g+T_s$ where T_g represents the duration of the guard interval used to avoid ISI and T_s is the duration of the OFDM symbol, $f_n = n/T_s$, $0 \le n \le N_c-1$ are the baseband subcarrier frequencies and f_c, ϕ_c represent the carrier frequency and phase respectively. When assuming that the coherence bandwidth of the channel is larger than the subcarrier bandwidth, then with flat Rayleigh fading over each subcarrier, no guard interval is needed and $T=T_s$. Under this assumption a received symbol $z(l,i)$ is ideally obtained from the original symbol $d(l,i)$ according to:

$$z(l,i) = a_l(i)d(l,i) + n(l,i) \qquad 0 \le l \le N_c - 1 \tag{2}$$

where $a_l(i)$ is the flat fading factor and $n(l,i)$ represents the AWGN component on the lth subcarrier, i^{th} OFDM symbol, respectively. However, in a real system a frequency offset will always exist due either to the limitations on carrier stability (causing a mismatch between the transmitter frequency f_c and the receiver frequency f_c') or to a Doppler shift f_D caused by the movement of the mobile station. In such case, the received OFDM envelope can be written after band translation as:

$$s(t) = S(i-1) + S(i) + \sum_{j \neq i-1,i} S(j) + n(t) \tag{3}$$

where $S(j) = \dfrac{1}{N} \sum_{n=0}^{N-1} a_n(j)d(n,j)rect(\dfrac{t-jT_s}{T_s} - \dfrac{1}{2})e^{j[2\pi(t-jT_s)n/T_s+2\pi\Delta ft+\phi_0]}$ $\Delta f = f_c - f_c' - f_D$, ϕ_0

$= \phi_c - \phi_c'$ and $N=N_c$ has been taken for the sake of simplicity.

Ideally, the sampling at the receiver for the ith OFDM symbol should be performed at time instants $t_k^c = kT_s/N + iT_s$, $0 \le k \le N-1$ thus only recovering the information related to the term $S(i)$, i.e., the current OFDM symbol. When this ideal situation is not in effect and the sampling is performed before the start of the symbol (a similar derivation can be done for a sampling done after the correct timing) at time instants $t_k^0 = (k-D)T_s/N + iT_s$, $0 \le k \le N-1$ where D represents the integer sampling error[1], there will be a contribution from the previous symbol introducing ISI in the system. To illustrate the importance of an accurate sampling at the receiver note that for a conventional OFDM system the minimum integer sampling error ($D=1$) already leads to random detection (BER=0.5). The sampled version of $s(t)$, under the assumption $D<N$, will be given by:

$$\begin{cases} \dfrac{1}{N}\displaystyle\sum_{n=0}^{N-1} a_n(i-1)d(n,i-1)e^{j2\pi[((k-D+N)T_s/N)n/T_s+\gamma k/N]}\cdot e^{j\phi_{cur}(i)} + n(k,i-1)\,,\ 0\le k\le D-1 \\[4mm] \dfrac{1}{N}\displaystyle\sum_{n=0}^{N-1} a_n(i)d(n,i)e^{j2\pi[((k-D)T_s/N)n/T_s+\gamma k/N]}\cdot e^{j\phi_{cur}(i)} + n(k,i-1)\,,\ D\le k\le N-1 \end{cases} \quad (4)$$

where $\phi_{cur}(i) = -2\pi\gamma D/N + 2\pi i\gamma$ with $\gamma = \Delta f T_s$.

After some mathematical manipulation it can be shown that the received symbols at the output of the FFT will yield:

$$z(l,i) = \frac{e^{j2\pi\gamma(-D/N+i)+j\phi_0}}{N}\sum_{n=0}^{N-1} a_n(i-1)d(n,i-1)e^{j2\pi n(N-D)/N}f_1(n-l)$$

$$+\frac{e^{j2\pi\gamma(-D/N+i)+j\phi_0}}{N}\sum_{n=0}^{N-1} a_n(i)d(n,i)e^{-j2\pi nD/N}f_2(n-l)+\xi(l,i) \qquad 0<l<N-1 \quad (5)$$

where $f_1(n-l) = \dfrac{sin(\pi(n-l+\gamma)D/N)}{sin(\pi(n-l+\gamma)/N)}e^{j(D-1)\pi(n-l+\gamma)/N}$, $f_2(n-l) = \dfrac{sin(\pi(n-l+\gamma)(N-D)/N)}{sin(\pi(n-l+\gamma)/N)}e^{j(N+D-1)\pi(n-l+\gamma)/N}$

and $\xi(l,i)$ represents the FFT of the noise samples.

The undesired effects over the received signal are basically: (1) the received constellation incrementally rotates from OFDM symbol to OFDM symbol due to the frequency offset (term $e^{j2\pi\gamma}$) (2) correlation among the information carried by the different subcarriers appears (Inter Carrier Interference-ICI) due to the frequency offset as well as to the sampling error, and (3) the current symbol information is corrupted by the previous (or next) symbol due to the sampling error (Inter Symbol Interference-ISI). When combining the N components of the sequence $z(l,i)$ into a $Nx1$ vector $\mathbf{z}(i)$ the outputs of the FFT can be rewritten in matrix notation as:

$$\mathbf{z}(i) = \mathbf{F}_1^T\mathbf{G}_1(i-1)\mathbf{d}(i-1)+\mathbf{F}_2^T\mathbf{G}_2(i)\mathbf{d}(i)+\boldsymbol{\xi}(i) \qquad (6)$$

where $\boldsymbol{\xi}(i)=[\xi(0,i),\xi(1,i),....\xi(N-1,i)]$ are gaussian noise samples of zero mean and variance σ^2, $\mathbf{d}(i-1)$, $\mathbf{d}(i)$ are $Nx1$ vectors containing the original symbols modulated onto each subcarrier for the $i-1$ and i OFDM symbols respectively,

---

[1] Only D integer is considered, correction of small (fractional) sampling error was studied in for example [9]; by means of differential encoding. The case of integer + fractional time offset is a subject for further research

$$\mathbf{G}_1(i-1) = \frac{e^{j2\pi\gamma(-D/N+i)+j\phi_0}}{N} \; diag(a_0(i-1), a_1(i-1)e^{j2\pi(N-D)/N}, \ldots a_{N-1}(i-1)e^{j2\pi(N-D)(N-1)/N}),$$

$$\mathbf{G}_2(i) = \frac{e^{j2\pi\gamma(-D/N+i)+j\phi_0}}{N} \; diag(a_0(i), a_1(i)e^{-j2\pi D/N}, \ldots, a_{N-1}(i)e^{-j2\pi D(N-1)/N})$$

and
$$\mathbf{F}_i = \begin{bmatrix} f_i(0) & f_i(-1) & \cdots & f_i(-N+1) \\ f_i(1) & f_i(0) & \cdots & f_i(-N+2) \\ \cdots & \cdots & \cdots & \cdots \\ f_i(N-1) & f_i(N-2) & \cdots & f_i(0) \end{bmatrix}$$

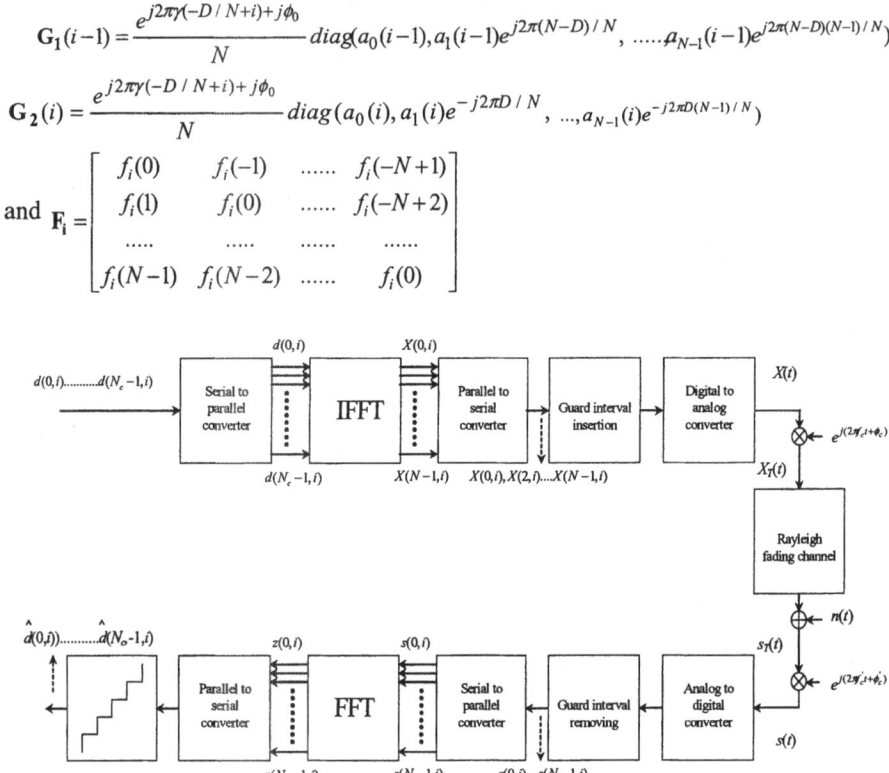

Fig. 1. Conventional OFDM system

3   The Proposed Correction Scheme

When the joint effect of the integer sampling error and the frequency offset are considered, the amount of distortion on the received signal is such that it becomes necessary to include some kind of estimation of the frequency offset for a reasonable detection. In the proposed scheme the frequency offset estimation as well as the current phase estimation (rotation originated by the term $e^{j2\pi\gamma}$) is performed at the beginning of each defined OFDM frame (set of M consecutive OFDM symbols). The estimation is realized using the samples of the training symbol $s_{TR}(k,i)$ at the input of the FFT block. The considered OFDM frame for the proposed scheme is illustrated in figure 2 which depicts samples after the Parallel-to-serial converter at the receiver. It shows the considered frame using two training symbols to estimate the frequency offset and current phase, though it will be shown that it is possible to use only a single training symbol for the estimation. Although the sequences $X(k,i)$ are in general complex sequences, the figure could represent either the real or imaginary parts of $X(k,i)$.

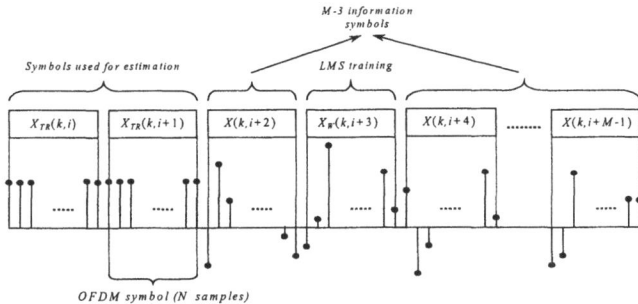

Fig. 2. Considered frame using two symbols for frequency offset estimation

Different methods have been considered to estimate the frequency offset, some of them need computing the timing metric beforehand which represents a large computational load at the receiver. Since it is very important that the amount of computation needed at the receiver be small (this is even more crucial in this case where the estimation is performed for each OFDM frame), a somewhat similar method to that presented by P.H..Moose in [1], which is more computationally simple, is used. However, some modifications are necessary because Moose assumed that there was no sampling error at the receiver. Moreover, the same training symbols used to estimate the frequency offset are also used to estimate the current phase term.

Once the estimation of the frequency offset and the current phase term has been made at the beginning of the OFDM frame the information symbols are corrected as follows:

$$s_c(k,i) = s(k,i)e^{-j2\pi\hat{\gamma}k/N}e^{-j\hat{\phi}_{cur}(i)} \quad 0 \le k \le N-1 \tag{7}$$

where $\hat{\gamma}$ and $\hat{\phi}_{cur}(i)$ stand for the estimate of the frequency offset and current phase respectively. Notice that the frequency offset can be considered constant within an OFDM frame, but it is necessary to keep track of the current phase term. A good estimation of the frequency offset is doubly important because it results in a good estimation of the current phase at the same time. To illustrate this, note that:

$$\phi_{cur}(i+1) = \phi_{cur}(i) + 2\pi\gamma \tag{8}$$

Therefore, the current phase term estimation within one frame has to be updated from OFDM symbol to OFDM symbol as follows:

$$\hat{\phi}_{cur}(i+1) = \hat{\phi}_{cur}(i) + 2\pi\hat{\gamma} \tag{9}$$

At the same time (9) shows that the more recent is the frequency offset estimation, the best is the correction of the OFDM symbol. Assuming a perfect initial acquisition of the phase term ($\hat{\phi}_{cur}(i) = \phi_{cur}(i)$) the current phase estimation for the last symbol of the frame would be:

$$\hat{\phi}_{cur}(i+(M-1)) = \phi_{cur}(i) + (M-1)\cdot 2\pi\hat{\gamma} = \phi_{cur}(i+(M-1)) + \varepsilon(M) \tag{10}$$

i.e., the further from the beginning of the OFDM frame, the worse is the current phase term estimation because $\varepsilon(M)$ is proportional to $M-1$.

Apart from the frequency offset and current phase estimation, a MMSE detector is used to correct the effect of the integer sampling error. A linear filter is applied at the output of the FFT before detection to equalize the samples $z(i) = FFT\{s_C(i)\}$, where $s_C(i)$ is the vector containing the N samples of $s_c(k,i)$, as follows:

$$y(i) = W^H(i)z(i) \tag{11}$$

where $z(i)$ is the Nx1 vector containing the outputs of the FFT, $W^H(i)$ is a NxN matrix containing the MMSE coefficients and $y(i)$ is the Nx1 vector containing the equalized samples. The optimal coefficients for the matrix $W^H(i)$ which minimize the Mean Square Error between the desired and equalized data are given by the well known Wiener-Hopf equation:

$$w_l(i) = R_{zz}^{-1}(i)r_{zd}(l,i) \tag{12}$$

where $w_l(i)$ is the lth column of the matrix $W(i)$, $r_{zd}(l,i)=E\{d(l,i)z^H(i)\}$ and $R_{zz}(i)=E\{z(i)z^H(i)\}$. From (6) and (12) the set of optimal coefficients will be given by:

$$w_l(i) = (F_1^T A(i-1)F_1^* + F_2^T A(i)F_2^* + \sigma^2 I)^{-1} F_2^T g_{2,l}(i) \tag{13}$$

where $A(i)=\text{diag}(a_0^2(i),a_1^2(i),.....,a_{N-1}^2(i))$, $g_{2,l}(i)$ is the lth column of the matrix $G_2(i)$ and σ^2 stands for the noise power.

Note that, since the frequency offset is assumed to be close to zero after frequency offset correction, the optimal coefficients depend basically on D. The value of D is unknown at the receiver and therefore a solution based on the LMS (Least Mean Squares) algorithm is proposed to converge to the optimal coefficients. In each frame a LMS training symbol is sent and the weights of the filter are updated as follows:

$$w_l(i) = = w_l(i-1) - \mu(y_W(l,i) - d_W(l,i))^* z_{WC}(i) \tag{14}$$

where $z_{wc}(i)$ represents the output of the FFT for the training symbol after frequency correction, $y_w(l,i)$ is the output of the filter for the lth subcarrier, $d_w(l,i)$ is the original training data and μ is the scalar parameter determining the rate of convergence and stability.

The LMS training symbol is not inserted right after the frequency estimation symbols which must have certain properties (defined in the next section) not compatible with a proper training for the MMSE filter. Note that the distortion on the LMS training symbol depends on the previous symbol, the values of which should be ideally random, which is not the case for the frequency estimation symbol.

Furthermore, in order to have better correction, the LMS training symbol has to be inserted as close to the beginning of the frame as possible (see equation 10). Thus, the LMS training symbol is always inserted two symbols after the estimation ones. For the simulations, the LMS training sequences are chosen sequentially from a set of 50 randomly generated sequences of length N ($d_w(l,i) =1,-1$) and are known for both the transmitter and receiver.

4 Frequency Estimation Methods

4.1 Two Estimation Symbols Method (TESM)

In [1] Moose derived the MLE of the frequency offset given the realizations of two received symbols at the output of the FFT, $z_{TR}(i)$ and $z_{TR}(i+1)$ in response to identical transmitted data, obtaining the following result:

$$\hat{\gamma} = \frac{1}{2\pi} \text{tg}^{-1} \left\{ \frac{\sum_{l=0}^{l=N-1} \text{Im}\{z_{TR}(l,i) z_{TR}(l,i+1)^*\}}{\sum_{l=0}^{l=N-1} \text{Re}\{z_{TR}(l,i) z_{TR}(l,i+1)^*\}} \right\} \tag{15}$$

Basically, to reduce effect of noise, Moose performs an estimation of $\text{tg}(2\pi\gamma)$ averaging over all the subcarriers. When a sampling error occurs, this method is no longer satisfactory because, at the output of the FFT, the information of the estimation symbol and the information from the previous or next symbol is totally mixed. However, at the input of the FFT, a similar operation with some modifications can be performed if the sampling error D is assumed within a given interval (it is assumed that it occurs in no more than $N/8$ samples away from the correct timing point). Under this assumption the central $N'_{avfreq}=3N/4$ samples of the realizations of two received symbols at the input of the FFT, $s_{TR}(i)$ and $s_{TR}(i+1)$, in response to identical transmitted data, can still be used for averaging.

Furthermore, another drawback of the estimation presented by Moose is that the function arctg(x) used for estimation, returns a value between $[-\pi/2, \pi/2]$ whereas a phase term, $e^{j2\pi\gamma}$, which belongs to the interval $[-\pi, \pi]$, is being measured. This means that when using the arctg(x) function, only a phase term $|2\pi\gamma| < \pi/2$ can be estimated, therefore limiting the frequency offset acquisition to $|\gamma| < 0.25$.

To solve this problem, we use for estimation the following expression instead:

$$\hat{\gamma} = \frac{1}{2\pi} \angle \left\{ \sum_{N/8}^{7N/8-1} \text{Re}\{s_{TR}(k,i) s_{TR}(k,i+1)^*\} + j \sum_{N/8}^{7N/8-1} \text{Im}\{s_{TR}(k,i) s_{TR}(k,i+1)^*\} \right\} \tag{16}$$

One can show that (16) can give, like (15), an estimate for γ. For the simulations the function angle(x) provided in Matlab software is used. One could also use the complementary information provided by $\cos^{-1}(x)$ and $\sin^{-1}(x)$ to achieve the same result and be able to measure phases in the interval $[-\pi, \pi]$. This leads, as discussed earlier, to frequency acquisition range limited to $|\gamma| < 0.5$. If the frequency offset is larger than this value (integer frequency offset), the transmitted sequence appears shifted at the receiver [6] and other techniques become necessary.

Once the frequency offset γ has been estimated, the samples of $s_{TR}(i)$ and $s_{TR}(i+1)$ are corrected to eliminate the different rotation experienced by each sample depen-ding on γ. Additionally, the samples of $\mathbf{s}_{TR}(i)$ are corrected to assure that both symbols are in phase[2] as follows:

$$s_{TC}(k,i) = s_{TR}(k,i)e^{-j2\pi\hat{\gamma}k/N}e^{j2\pi\hat{\gamma}}$$

$$s_{TC}(k,i+1) = s_{TR}(k,i+1)e^{-j2\pi\hat{\gamma}k/N} \qquad (17)$$

Notice that, from (4) and (17), for a sampling performed before the correct point and within the considered interval for D, it is always true that (for simplicity and justification of the used estimation method it is assumed that $\hat{\gamma} = \gamma$):

$$s_{TC}(k,i) = a_n(i)X_{TR}(k-D)e^{j\phi_{cur}(i+1)} + n(k,i) \qquad N/8 \le k \le N-1$$

$$s_{TC}(k,i+1) = a_n(i)X_{TR}(k-D)e^{j\phi_{cur}(i+1)} + n(k,i+1) \qquad 0 \le k \le N-1 \quad (18)$$

where $X_{TR}(k) = \dfrac{1}{N}\displaystyle\sum_{n=0}^{N-1} d_{TR}(n,i)e^{j2\pi kn/N}$ and $X_{TR}(k-D) = X_{TR}(k-D+N)$.

Analogously, for a sampling performed after the correct point, it can be stated:

$$s_{TC}(k,i) = a_n(i)X_{TR}(k+D)e^{j\phi_{cur}(i+1)} + n(k,i) \qquad 0 \le k \le N-1$$

$$s_{TC}(k,i+1) = a_n(i)X_{TR}(k+D)e^{j\phi_{cur}(i+1)} + n(k,i+1) \quad 0 \le k \le 7N/8-1 \quad (19)$$

If the training symbol is such that $\angle\{X_{TR}(k)\}=L, \forall k$, i. e., all the samples of the transmitted OFDM symbol are in phase, an averaging can be performed in order to determine $\phi_{cur}(i+1)$. Under these conditions the estimate of the current phase term may be calculated over the $N^r_{avph}=7N/4$ samples of $s_{TC}(k,i)$ and $s_{TC}(k,i+1)$ as follows:

$$\hat{\phi}_{cur}(i+1) = \angle\left\{ \sum_{k=N/8}^{N-1} s_{TC}(k,i) + \sum_{k=0}^{7N/8-1} s_{TC}(k,i+1) \right\} - L \qquad (20)$$

In order to keep the transmitted power constant and fulfill the previous requirement, $\angle\{X_{TR}(k)\}=L, \forall k$, the training symbol is chosen $X_{TR}(k) = \sqrt{N}\cdot IFFT\{\delta_N(n)\}$ where $\delta_N(n) = [1,0,0.........0]$.

4.2 Single Estimation Symbol Method (SESM)

In order to reduce the number of pilot symbols used to estimate the frequency offset and current phase, therefore increasing the bandwidth efficiency, a similar method can be defined over one single training symbol.

If the transmitted symbol $\mathbf{X}_{TR}(i)$ has two equal halves[3], a similar averaging to that of (16) can be performed over the $N^s_{avfreq}=N/4$ usable samples for each half of $\mathbf{s}_{TR}(i)$ as shown:

$$\hat{\gamma}_s = \frac{1}{\pi}\angle\left\{ \sum_{N/8}^{3N/8-1} Re\{s_{TR}(k,i)s_{TR}(k+N/2,i)^*\} + j \sum_{N/8}^{3N/8-1} Im\{s_{TR}(k,i)s_{TR}(k+N/2,i)^*\} \right\} \qquad (21)$$

---

[2] Note that from (4) two consecutive received symbols at the input of the FFT (sent identical) differ mainly in a phase term $e^{j2\pi\gamma}$ for $D \le k \le N-1$

[3] Under this condition, the property $s_{TR}(k+N/2,i) = s_{TR}(k,i)e^{j\pi\gamma}$, $N/8 \le k \le 3N/8-1$ is fulfilled within the considered margin for D

Analogously, the current phase is estimated using the central $N^s_{avph}=3N/4$ samples of

$$\mathbf{s}_{TC}(i): \qquad \hat{\phi}^s_{cur}(i) = \angle \left\{ \sum_{k=N/8}^{7N/8-1} s_{TC}(k,i) \right\} \approx \pm 2\pi\gamma D/N + 2\pi i\gamma = \phi_{cur}(i) \qquad (22)$$

Notice that in this case the acquisition range for the frequency offset is increased in a 100%. The limitation will be given by $|\pi\gamma|<\pi \Rightarrow |\gamma|<1$. It is also important that, on the other hand, the averaging is performed over less samples for both the frequency offset and current phase, therefore loosing immunity against noise.

5 Simulation Results and Discussion

Simulations were carried over AWGN channel first letting the LMS algorithm converge to the optimal coefficients using the frequency corrected samples $\mathbf{z}_{wc}(i)=\mathrm{FFT}\{\mathbf{s}_{wc}(i)\}$ and applying (14) to update the filter coefficients. Then, the bit error counting is started over the information symbols of the frame. The number of information symbols within one frame is taken 5 and the original symbols are BPSK mapped. In figures 3 and 4 the system performances for TESM and the SESM respectively are depicted for $\gamma=0.4$ and $N=64$. The two schemes perform differently for this number of subcarriers mainly because of the larger noise reduction with the TESM and because the frequency offset and phase estimation are much more accurate. In some applications it might be desirable to loose some spectral efficiency (ratio between the information symbols and the total number of symbols sent) in order to achieve a much better performance.

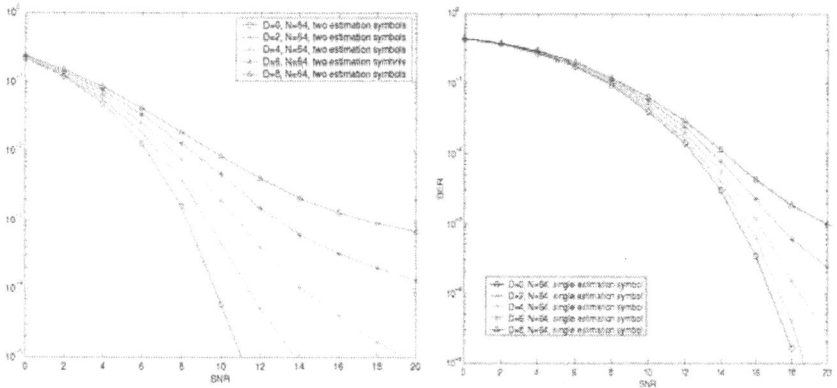

Fig. 3. TESM system performance   **Fig. 4.** SESM system performance

Nevertheless, it has been shown by simulation that for a number of subcarriers larger than 128 both systems became undistinguishable over AWGN channel. It was also noticed that the acquisition range is reduced to the half when using the TESM[4].

---

[4] $N^T_{avfreq}=3N^s_{avfreq}$ and $N^T_{avph}=7N^s_{avph}/3$

Thus it is concluded that for a large number of subcarriers ($N \geq 128$) it is clearly preferable to use the SESM as the performance is almost equal, the acquisition range is doubled and the efficiency is augmented in almost a 10 %.

It was also shown by simulation that the TESM for $N \geq 32$ yields already a very accurate frequency offset estimation when compared to the overall performance of the system. This means that the performance of the system in the absence of frequency offset is equaled; the same statement could be done regarding the SESM for $N \geq 128$. The overall performance of the systems under these conditions is therefore only dependant on the value of the integer sampling error.

6 Conclusions

In this letter a complete scheme for time and frequency offset correction for OFDM systems has been presented. The system uses a frequency offset estimation method for posterior correction combined with a MMSE detector for time offset correction. Two different frequency offset estimation methods have been presented and it has been shown that for a large number of subcarriers is possible to achieve a good estimation using a single pilot symbol. The overall performance of the system remains strongly dependant on the value of the sampling error since this last parameter is blindly corrected instead of estimated as the frequency offset[5].

References

[1]   Moose P.H, "A technique for orthogonal frequency division multiplexing frequency offset correction," *IEEE Trans. Comm.*, Vol. 42, no. 10, pp. 2908-2914, Oct 1994.

[2]   T. Pollet, M. Van Bladel and M. Moeneclaey, "BER Sensitivity of OFDM Systems to Carrier Frequency Offset and Wiener Phase Noise," *IEEE Trans. On Comms.*, Vol. 43, No. 2/3/4, Feb/Mar/Apr 1995, pp. 191-193.

[3]   M. Gudmundson and P. O. Anderson, "Adjacent Channel Interference in an OFDM System," *Proc. Vehicular Tech. Conf,* Atlanta, GA, May 1996, pp. 191-193.

[4]   Van de Beek J-J, Sandell M. and B. P.O. ,"ML Estimation of time and Frequency offset in OFDM system," *IEEE Trans. Sig. Proc.,* vol. 45, no.7, pp. 1800-1805, July 1997.

[5]   Dusan Matic. A.R.J.M and R. Prasad, "OFDM timing synchronisation: Posibilities and Limits to the usage of the Cyclic Prefix for Maximum Likelihood Estimation," *Vehicular Technology Conference (VTC'99),* September 1999., Amsterdam.

[6]   Schmidl T.M. and Cox C.C , "Robust time and frequency synchronization for OFDM," *IEEE Trans. Comm.,* vol 45, no. 12,pp 1612-1621, Dec 1997.

[7]   H. Minn, M. Zeng and V. K. Bhargava, Fellow, IEEE , "On timing Offset Estimation for OFDM systems", *Communication letters. CL 1035.*

[5] This work was partially supported by NSF grant # CCR-9903381. Rubén Solá was invited to the CCSPR as a guest student from the UPC (Universidad Politécnica de Cataluña).

[8]   Dipl.-Ing. Stefan Kaiser, " Multicarrier CDMA Mobile Radio Systems – Analysis and Optimization of Detection, Decoding, and Channel Estimation,", Telecommunications Research Laboratories. München.

[9]   Y. Bar-Ness and R. Solá, " Adaptive Scheme for Joint Frequency and Time Offset Correction over OFDM systems", Accepted paper for *PIMRC'04*, September 2004, Barcelona.

The Discrete Cosine Transform over Prime Finite Fields

M.M.C. de Souza, H.M. de Oliveira, R.M.C. de Souza, and M.M. Vasconcelos

Federal University of Pernambuco - UFPE, C.P. 7800, 50711-970, Recife - PE, Brazil
{hmo, marciam, ricardo}@ufpe.br, mmv@ee.ufpe.br

Abstract. This paper examines finite field trigonometry as a tool to construct trigonometric digital transforms. In particular, by using properties of the k-cosine function over GF(p), the Finite Field Discrete Cosine Transform (FFDCT) is introduced. The FFDCT pair in GF(p) is defined, having blocklengths that are divisors of $(p+1)/2$. A special case is the Mersenne FFDCT, defined when p is a Mersenne prime. In this instance blocklengths that are powers of two are possible and radix-2 fast algorithms can be used to compute the transform.

1 Introduction

The manifold applications of discrete transforms defined over finite or infinite fields are very well known. The Discrete Fourier transform (DFT) has, since long, been playing a decisive role in Engineering. Another powerful discrete transform is the Discrete Cosine Transform (DCT), which became a standard in the field of image compression [1]. Such transforms, despite being discrete in the variable domain, have coefficients with amplitudes belonging to an infinite field. They can therefore be understood as some kind of "analog transforms". In contrast, transforms defined over finite fields are discrete in both variable and transform domain. Their coefficients are from a finite alphabet, so they can be understood as "Digital Transforms".

A very useful transform related to the DFT is the Finite Field Fourier Transform (FFFT), introduced by Pollard in 1971 [2] and applied as a tool to perform discrete convolution using integer arithmetic. Since then several new applications of the FFFT have been found, not only in the fields of digital signal and image processing [3-5], but also in different contexts such as error control coding and cryptography [6-7].

The Finite Field Hartley Transform, which is a digital version of the Discrete Hartley Transform, has been recently introduced [8]. Its applications include the design of digital multiplex systems, multiple access systems and multilevel spread spectrum digital sequences [9-12].

Among the discrete transforms, the DCT is especially attractive for image processing, because it minimises the blocking artefact that results when the boundaries between subimages become visible. It is known that the information packing ability of the DCT is superior to that of the DFT and other transforms [13]. The DCT is now a tool in the international JPEG standard for image processing, but there is no equivalent transform over finite fields. A natural question that arises is whether it is possible to find a DCT over such fields. The first task towards such a new transform is to define the equivalent of the cosine function over a finite field.

J.N. de Souza et al. (Eds.): ICT 2004, LNCS 3124, pp. 482–487, 2004.
© Springer-Verlag Berlin Heidelberg 2004

Based on the trigonometry over finite fields proposed in [8], this paper introduces a new digital transform, the finite field discrete cosine transform (FFDCT). The FFDCT is defined for signals over GF(p), $p\equiv3$ (*mod* 4) and its components are in GF(p).

2 Background and Preliminaries

2.1 Fundamentals on Finite Field Complex Numbers

Definition 1. The set of Gaussian integers over GF(p) is the set GI(p) = {$a + jb, a, b \in$ GF(p)}, where p is a prime such that $j^2 = -1$ is a quadratic nonresidue over GF(p). Only prime $p \equiv 3$ (*mod* 4) meet such a requirement [14].   ❏

The extension field GF(p^2) is isomorphic to the "complex" structure GI(p) [15]. From the above definition, GI(p) elements can be represented in the form $a + jb$ and are referred to as complex finite field numbers.

Definition 2. (unimodular set): The elements $\zeta = (a+jb) \in$ GI(p), such that $a^2+b^2\equiv1$ (*mod p*) are referred to as unimodular elements.   ❏

2.2 Finite Field Trigonometry

This session describes briefly trigonometric functions over a finite field, which hold many properties similar to those of the standard real-valued trigonometric functions [16]. In what follows, the symbol := means *equal by definition*.

Definition 3. Let ζ be a nonzero element of GI(p), where $p \equiv 3$ (*mod* 4). The k-trigonometric functions cosine and sine of $\angle(\zeta^i)$ (arc of the element ζ^i) over GI(p), are

$$\cos_k(\angle\zeta^i) := (2^{-1} \bmod p) \ (\zeta^{ik} + \zeta^{-ik}) \text{ and } \sin_k(\angle\zeta^i) := (2^{-1} \bmod p) \ (\zeta^{ik} - \zeta^{-ik})/j,$$

$i, k = 0, 1,..., N$-1, where ζ has order N.   ❏

For the sake of simplicity, these are denoted by $\cos_k(i)$ and $\sin_k(i)$. Over the field of real numbers, the DCT is defined by the pair

$$C[k] := \sum_{n=0}^{N-1} x[n]\cos[\tfrac{(2n+1)k\pi}{2N}], \quad x[n] = \sum_{k=0}^{N-1}\beta[k]C[k]\cos[\tfrac{(2n+1)k\pi}{2N}],$$

where $\beta[k]$ is the weighting function $\beta[k] = \begin{cases} \frac{1}{2}, & \text{if } k = 0 \\ 1, & \text{if } k = 1 \end{cases}$

The steps leading to the expression for the DCT of the length N sequence $x[n]$, involve replicating $x[n]$ and computing its DFT, from which the DCT coefficients can be obtained. Therefore, to construct a length N DCT, a kernel of order $2N$ is required.

3 The Discrete Cosine Transform in a Finite Field

Let $f = (f_i)$ be a length N vector over GF(p). To define its DCT using k-cosines, the following lemma is required.

Lemma 1 (k-cos lemma): If $\zeta \in$ GI(p) has multiplicative order $2N$, then

$$A = \sum_{k=1}^{N-1} \cos_k(i) = \begin{cases} N-1, & \text{if } i=0 \\ -1, & \text{if } i \text{ is even } (\neq 0) \\ 0, & \text{if } i \text{ is odd} \end{cases}$$

Proof: By definition

$$A = \sum_{k=1}^{N-1} \cos_k(i) = \tfrac{1}{2}\sum_{k=1}^{N-1}(\zeta^{ki} + \zeta^{-ki}),$$

so that, clearly, $A = N-1$ if $i = 0$. Otherwise, $A = \tfrac{1}{2}[\dfrac{\zeta^i(\zeta^{i(N-1)} - 1)}{\zeta^i - 1} + \dfrac{\zeta^{-i}(\zeta^{-i(N-1)} - 1)}{\zeta^{-i} - 1}].$

Since ζ has order $2N$, then $\zeta^N = -1$. Multiplying the second term by $(-\zeta^i)$, yields $A = \tfrac{1}{2}[\dfrac{(-1)^i - \zeta^i}{\zeta^i - 1} + \dfrac{1 - (-1)^i \zeta^i}{\zeta^{-i} - 1}]$. Therefore, for i even, $A = \tfrac{1}{2}[\dfrac{1 - \zeta^i + 1 - \zeta^i}{\zeta^i - 1}] = -1$ and, for i odd,

$A = \tfrac{1}{2}[\dfrac{-1 - \zeta^i + 1 + \zeta^i}{\zeta^i - 1}] = 0.$ □

From this k-cos lemma, it is possible to define a new digital transform, the finite field discrete cosine transform (FFDCT).

Definition 4: If $\zeta \in$ GI(p) has multiplicative order $2N$, then the finite field discrete cosine transform of the sequence $f = (f_i)$, $i = 0,1,... N-1$, $f_i \in$ GF(p), is the sequence $C = (C_k)$, $k = 0,1,... N-1$, $C_k \in$ GI(p), of elements

$$C_k := \sum_{i=0}^{N-1} 2f_i \cos_k\left(\tfrac{2i+1}{2}\right).$$

The inverse FFDCT is given by theorem 1 below.

Teorema 1 (The inversion formula): The inverse FFDCT of the sequence $C = (C_k)$, $k = 0,1,.. N-1$, is the sequence $f = (f_i), i = 0,1,.. N-1$, $f_i \in$ GF(p), where

$$f_i = \tfrac{1}{N}\sum_{k=0}^{N-1} \beta_k C_k \cos_k\left(\tfrac{2i+1}{2}\right),$$

and the weighting function β_k is given by $\beta_k = \begin{cases} (2^{-1} \bmod p), & \text{if } k=0 \\ 1, & \text{if } k \neq 0 \end{cases}$

Proof: To establish the inversion formula, it is sufficient to show that $g_i = f_i$, $i = 0,1,... N-1$, where

$$g_i := \tfrac{1}{N}\sum_{k=0}^{N-1} \beta_k C_k \cos_k\left(\tfrac{2i+1}{2}\right).$$

From definition 4, one may write

$$g_i = \tfrac{1}{N}\sum_{k=0}^{N-1} \beta_k [\sum_{r=0}^{N-1} 2f_r \cos_k\left(\tfrac{2r+1}{2}\right)] \cos_k\left(\tfrac{2i+1}{2}\right),$$

which is the same as

$$g_i = \tfrac{2}{N} \sum_{r=0}^{N-1} f_r [\sum_{k=0}^{N-1} \beta_k \cos_k (\tfrac{2r+1}{2}) \cos_k (\tfrac{2i+1}{2})].$$

Using the addition of arcs formula leads to

$$g_i = \tfrac{2}{N} \sum_{r=0}^{N-1} f_r [\tfrac{1}{2} + \tfrac{1}{2} \sum_{k=1}^{N-1} [\cos_k (r+i+1)] + \tfrac{1}{2} \sum_{k=1}^{N-1} \cos_k (r-i)]].$$

From the k-cos lemma and observing that $(r+i+1)$ is even whenever $(r-i)$ is odd and vice-versa, the evaluation of the above expression requires considering three cases:

i) If $r+i+1 = 0$, then $r = -i-1$, which implies $f_r = 0$. Therefore, in this case, $g_i = 0$.

ii) If $r-i = 0$, then $r = i$. In this case $g_i = \tfrac{2}{N} f_i [\tfrac{1}{2} + \tfrac{1}{2}(0) + \tfrac{1}{2}(N-1)] = f_i$.

iii) If both, $r+i+1$ and $r-i$ are different from zero, considering the parity for these terms it is possible to write

$$g_i = \tfrac{2}{N} \sum_{r=0}^{N-1} f_r [\tfrac{1}{2} + \tfrac{1}{2}(0) + \tfrac{1}{2}(-1)] = 0,$$

so that $g_i = f_i$, $i = 0,1,... N-1$. ❑

The elements of an FFDCT of length 8 over GF(31) are presented in example 1.

Example 1: For $p = 31$, the element $\zeta = (7+j13) \in GL(31)$ has order $(p+1)/2 = 16$. The FFDCT of length $(p+1)/4 = 8$ of the sequence $f = (1,2,3,4,5,6,7,8)$ is the sequence $C = (10,20,0,17,0,12,0,5)$. The transform matrix $\{2\cos_k(\tfrac{2i+1}{2})\}$, $i, k = 0,1,..,7$, is

$$M_{k,i} = \begin{bmatrix} 2 & 2 & 2 & 2 & 2 & 2 & 2 & 2 \\ 27 & 10 & 20 & 22 & 9 & 11 & 21 & 4 \\ 14 & 5 & 26 & 17 & 17 & 26 & 5 & 14 \\ 10 & 9 & 4 & 11 & 20 & 27 & 22 & 21 \\ 8 & 23 & 23 & 8 & 8 & 23 & 23 & 8 \\ 20 & 43 & 22 & 10 & 21 & 9 & 27 & 11 \\ 5 & 17 & 14 & 26 & 26 & 14 & 17 & 5 \\ 22 & 11 & 10 & 4 & 27 & 21 & 20 & 9 \end{bmatrix} \cdot$$

The inverse matrix, which is equal to $\{\tfrac{\beta_k}{N}\cos_k(\tfrac{2i+1}{2})\}$, $i, k = 0, 1,..., 7$, is given by

$$M^{-1}_{k,i} = \begin{bmatrix} 2 & 23 & 28 & 20 & 16 & 9 & 10 & 13 \\ 2 & 20 & 10 & 18 & 15 & 8 & 3 & 22 \\ 2 & 9 & 21 & 8 & 15 & 13 & 28 & 20 \\ 2 & 13 & 3 & 22 & 16 & 20 & 21 & 8 \\ 2 & 18 & 3 & 9 & 16 & 11 & 21 & 23 \\ 2 & 22 & 21 & 23 & 15 & 18 & 28 & 11 \\ 2 & 11 & 10 & 13 & 15 & 23 & 3 & 9 \\ 2 & 8 & 28 & 11 & 16 & 22 & 10 & 18 \end{bmatrix} \cdot$$

It is interesting to observe, at this point, that due to the expressions defining the FFDCT pair, there is a simple relation between the direct and inverse transform matrices. In fact, the elements of these matrices are related by

$$m_{i,k}^{-1} = \begin{cases} (2^{-1} \bmod p) \, m_{k,i}, & \text{if } k = 0 \\ m_{k,i}, & \text{if } k \neq 0 \end{cases} \qquad \square$$

From a practical point of view, an important family of finite field transforms may be obtained from the FFDCT, namely the Mersenne FFDCT. These are defined over GF(p) when $p=2^s-1$, a Mersenne prime. The blocklength is $N=2^{s-2}$, which is attractive since that radix-2 fast algorithms can be used in this case.

In general, the transformed vector (DCT spectrum) lies over the extension field GF(p^2). However, if a unimodular element ζ is used to define the finite field trigonometry, then it can be shown that *cos* and *sin* are real functions [17].

Proposition 1. If $\zeta = a + jb$ is unimodular, then $cos_k(i)$ and $sin_k(i) \in$ GF(p), for any i, k. $\square$

This is the situation illustrated in example 1. In this case ζ is unimodular and has a square root λ that is also unimodular. This λ has order ($p+1$), so that ζ has order ($p+1$)/2, which implies an FFDCT of length $N = (p+1)/4$. Table 1 below lists the parameters of some real FFDCT.

Table 1. Parameters for the FFDCT over a few Finite Fields: Ground field, transform blocklength, unimodular element used to define $cos_k(i)$ and its order over the extension field.

Ground field GF(p)	Blocklength N	Unimod element ζ	Extention field GF(p^2)	Order Ord(ζ)
7*	2	2+j2	GF(49)	8
23	6	4+j10	GF(529)	24
31*	8	2+j11	GF(961)	32
47	12	4+j19	GF(2209)	48
71	18	8+j24	GF(5041)	72
79	20	2+j32	GF(6241)	80
103	26	2+j10	GF(10609)	103
127*	32	2+j39	GF(16129)	128
151	38	2+j65	GF(22801)	152
167	42	4+j73	GF(27889)	168
191	48	6+j27	GF(36481)	192
199	50	2+j14	GF(39601)	200

* Mersenne FFDCT.

4 Conclusions and Suggestions

In this paper the discrete cosine transform in a finite field GF(p) was introduced. The FFDCT uses the k-trigonometric $cos_k(.)$ function over GF(p) as kernel. An important lemma concerning such functions was given, from which the inversion formula was established. The length of the transform is a divisor of $p+1$, where $p \equiv 3$ (*mod* 4). If p is a Mersenne prime, the transform is called Mersenne FFDCT and has a blocklength that is a power of two. Unimodular elements were selected to guarantee real values for the k-trigonometric functions, thus producing real (GF(p)-valued) transforms.

Since there exists many definitions for the classical DCT, it is interesting to investigate alternative definitions for the FFDCT as well. Further Transforms such as discrete sine transform and 2-D FFDCT could also be defined. Possible applications of the FFDCT for multiplex, CDMA schemes and image processing are currently under investigation.

References

1. Transform Coding: Past, Present and Future. IEEE SP Mag., Vol. 18. (2001) 6-93
2. Pollard, J. M.: The Fast Fourier Transform in a Finite Field. Math. Comput., Vol. 25. (1971) 365-374
3. Reed, I.S., Truong, T.K.: The Use of Finite Field to Compute Convolutions. IEEE Trans. Inform. Theory, Vol. IT-21. (1975) 208-213
4. Agarwal, R.C., Burrus, C.S.: Number Theoretic Transforms to Implement Fast Digital Convolution. Proc. IEEE. Vol . 63. (1975) 550-560
5. Reed, I.S., Truong, T.K., Kwoh V.S., Hall, E.L.: Image Processing by Transforms over a Finite Field. IEEE Trans. Comput. Vol.C-26. (1977) 874-881
6. Blahut, R.E.: Transform Techniques for Error-Control Codes. IBM J. Res. Dev. Vol. 23. (1979) 299-315
7. Massey, J.L.: The Discrete Fourier Transform in Coding and Cryptography. IEEE Information Theory Workshop, San Diego, CA, (1998).
8. Campello de Souza, R.M., de Oliveira, H.M., Kauffman, A.N.: Trigonometry in Finite Fields and a New Hartley Transform. Proc. of the IEEE Int. Symp. on Info. Theory. (1998) 293
9. de Oliveira, H.M., Campello de Souza, R.M., Kauffman, A.N.: Efficient Multiplex for Band-Limited Channels. Proc. of the Work. on Coding and Cryptography. (1999) 235 - 241
10. Miranda, J.P.C.L., De Oliveira, H.M.: On Galois-Division Multiple Access Systems: Figures of Merit and Performance Evaluation. Proc. of the 19° Braz. Telecom. Symp., (2001) (in English).
11. de Oliveira, H.M., Miranda, J.P.C.L., Campello de Souza, R.M.: Spread-Spectrum Based on Finite Field Fourier Transforms. Proc. of the ICSECIT - Int. Conf. on Systems Engineering, Communication and Information Technology. Vol. 1. Punta Arenas (2001).
12. de Oliveira, H.M., Campello de Souza, R.M.: Orthogonal Multilevel Spreading Sequence Design. In: Farrell, P.G., Darnell, M., Honary, B. (eds): Coding, Communications and Broadcasting. Research Studies Press / John Wiley, Baldock (2000) 291-303
13. Lim, J.S.: Two-Dimensional Signal and Image Processing. Prentice-Hall, New Jersey (1990)
14. Burton, D.M.: Elementary Number Theory. McGraw Hill, New York (1997)
15. Blahut, R.E.: Fast Algorithms for Digital Signal Processing. Addison-Wesley, Reading, (1985)
16. Campello de Souza, R.M., de Oliveira, H.M., The Complex Hartley Transform over a Finite Field. In: Farrell, P.G., Darnell, M., Honary, B. (eds): Coding, Communications and Broadcasting. Research Studies Press / John Wiley, Baldock (2000) 267-276
17. Campello de Souza, R.M., de Oliveira, H.M., Campello de Souza, M.M.: Hartley Number-Theoretic Transforms. Proceedings of the 2001 IEEE International Symposium on Information Theory. (2001) 210

A Lattice Version of the Multichannel Fast QRD Algorithm Based on *A Posteriori* Backward Errors

António L.L. Ramos and José A. Apolinário Jr.⋆

Instituto Militar de Engenharia, **IME – DE/3**
Praça General Tibúrcio, 80
22290-270, Rio de Janeiro, RJ, Brazil

Abstract. Fast QR decomposition (QRD) RLS algorithms based on backward prediction errors are well known for their good numerical behavior and their low complexity when compared to similar algorithms with forward error update. Although the basic matrix expressions are similar, their application to multiple channel input signals generate more complex equations. This paper presents a lattice version of the multichannel fast QRD algorithm based on *a posteriori* backward errors updating. This new algorithm comprises scalar operations only; its modularity and pipelinability favors its systolic array implementation.

1 Introduction

Digital processing of multichannel signals using adaptive filters has recently found a variety of new applications including color image processing, multispectral remote sensing imagery, biomedicine, channel equalization, stereophonic echo cancellation, multidimensional signal processing, Volterra –type nonlinear system identification, and speech enhancement [1]. This increased number of applications has spawned a renewed interest in efficient multichannel algorithms. One class of algorithms, known as multichannel fast QR decomposition least-squares adaptive algorithms based on backward error updating, has become an attractive option because of fast convergence properties and reduced computational complexity.

In the case of one single channel, a unified formulation for Fast QRD-LS algorithms is available in [2]. In this paper, a new algorithm based on the *a posteriori* backward error updating is developed, using an approach similar to the one used in [3]. Our starting point is the block multichannel algorithm based on the *a posteriori* backward error updating presented in [4]. It is well known that, due to its blocking characteristic, the algorithm of [4] deals with matrix inversions which are potential sources of instability. To overcome this, a new expression is derived; moreover, with the help of order recursive implementations, the resulting new lattice–type algorithm is introduced. This new algorithm uses scalar operations only.

⋆ The authors thank CNPq and CAPES for partial funding of this paper.

J.N. de Souza et al. (Eds.): ICT 2004, LNCS 3124, pp. 488–497, 2004.

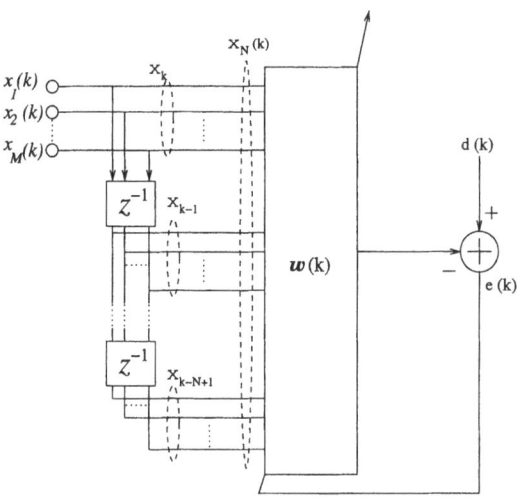

Fig. 1. Multichannel Adaptive Filter.

This paper is organized as follows. In Section 2, we review the basic matrix equations of the Multichannel Fast QR Decomposition Least Squares (**MCFQRD-LS**) algorithms based on backward prediction errors. Section 3 addresses the Multichannel Fast QRD based on the updating of the *a posteriori* error vector (**MCFQRD_POS_B**). In Section 4, the Lattice version of the **MCFQRD_POS_B** algorithm is presented. Simulation results and conclusions are Summarized in Sections 5 and 6, respectively.

2 The Multichannel Fast QRD Algorithm Based on Backward Prediction Errors Updating

The objective function to be minimized according to the least-squares (LS) algorithm is defined as

$$\xi_{LS}(k) = \sum_{i=0}^{k} \lambda^{k-i} e^2(i) = e^T(k)e(k) \tag{1}$$

where $e(k) = \begin{bmatrix} e(k) & \lambda^{1/2}e(k-1) & \cdots & \lambda^{k/2}e(0) \end{bmatrix}^T$ is an error vector and may be represented as follows.

$$e(k) = \begin{bmatrix} d(k) \\ \lambda^{1/2}d(k-1) \\ \vdots \\ \lambda^{k/2}d(0) \end{bmatrix} - \begin{bmatrix} x_N^T(k) \\ \lambda^{1/2}x_N^T(k-1) \\ \vdots \\ \lambda^{k/2}x_N^T(0) \end{bmatrix} w_N(k) = d(k) - X_N(k)w_N(k) \tag{2}$$

where, as seen in Figure 1,

$$x_N^T(k) = \begin{bmatrix} x_k^T & x_{k-1}^T & \cdots & x_{k-N+1}^T \end{bmatrix} \tag{3}$$

and $\boldsymbol{x}_k^T = [x_1(k) \quad x_2(k) \quad \cdots \quad x_M(k)]$ is the input vector at time instant k. Note that N is defined here as the order or the number of filter coefficients per channel, M is the number of input channels, and $\boldsymbol{w}_N(k)$ is the $MN \times 1$ coefficient vector at time instant k.

If $\boldsymbol{U}_N(k)$ is the Cholesky factor of the $(k+1) \times M$ input data matrix $\boldsymbol{X}_N(k)$, obtained through the Givens rotation matrix $\boldsymbol{Q}_N(k)$, then

$$e_q(k) = \boldsymbol{Q}_N(k)e(k) = \begin{bmatrix} e_{q1}(k) \\ e_{q2}(k) \end{bmatrix} = \begin{bmatrix} d_{q1}(k) \\ d_{q2}(k) \end{bmatrix} - \begin{bmatrix} 0 \\ \boldsymbol{U}_N(k) \end{bmatrix} \boldsymbol{w}_N(k) \qquad (4)$$

From the definition of the forward prediction problem in a multichannel scenario, we can write matrix $\boldsymbol{X}_{N+1}(k)$ as follows

$$\boldsymbol{X}_{N+1}(k) = \left[\begin{array}{c|c} \boldsymbol{D}_f(k) & \begin{matrix} \boldsymbol{X}_N(k-1) \\ \boldsymbol{0}^T \end{matrix} \\ \hline \boldsymbol{0}_{(M-1)\times(MN+M)} \end{array} \right] \qquad (5)$$

where $\boldsymbol{D}_f(k) = [\boldsymbol{x}_k \quad \lambda^{1/2}\boldsymbol{x}_{k-1} \cdots \lambda^{k/2}\boldsymbol{x}_0]^T$ is the $(k+1) \times M$ forward reference signal and the subscript $N+1$ corresponds to the $(N+1)$th order problem.

The triangularization process of $\boldsymbol{X}_{N+1}(k)$ leading to $\boldsymbol{U}_{N+1}(k)$ is performed applying Givens rotations to (5) as follows.

$$\begin{bmatrix} \boldsymbol{Q}(k-1) & 0 \\ 0 & \boldsymbol{I}_{M\times M} \end{bmatrix} \left[\begin{array}{c|c} \boldsymbol{D}_f(k) & \begin{matrix} \boldsymbol{X}_N(k-1) \\ \boldsymbol{0}^T \end{matrix} \\ \hline \boldsymbol{0}_{(M-1)\times(MN+M)} \end{array} \right] = \begin{bmatrix} \boldsymbol{E}_{fq1}(k) & 0 \\ \boldsymbol{D}_{fq2}(k) & \boldsymbol{U}_N(k-1) \\ \lambda^{1/2}\boldsymbol{x}_0^T & \boldsymbol{0}^T \\ \boldsymbol{0}_{(M-1)\times(MN+M)} \end{bmatrix} \qquad (6)$$

Now, premultiplying (6) by a series of Givens rotations $\boldsymbol{Q}_f(k)$, that zeroes its first $k - MN$ rows, and by $\boldsymbol{Q}'_f(k)$ that completes the triangularization process, we have

$$\boldsymbol{U}_{N+1}(k) = \boldsymbol{Q}'_{\theta f}(k) \begin{bmatrix} \boldsymbol{D}_{fq2}(k) & \boldsymbol{U}_N(k-1) \\ \boldsymbol{E}_f(k) & 0 \end{bmatrix} \qquad (7)$$

after removing the resulting *null* sections. In (7), $\boldsymbol{Q}'_{\theta f}(k)$ is a fixed order matrix obtained from $\boldsymbol{Q}'_f(k)$. It is clear that the $(MN+M) \times (MN+M)$ matrix $\boldsymbol{Q}'_{\theta f}(k)$ contains the Given rotations that annihilates $\boldsymbol{D}_{fq2}(k)$ against the diagonal of $\boldsymbol{E}_f(k)$, the $M \times M$ Cholesky factor of the forward error covariance matrix.

Based on (7), it is possible to obtain

$$[\boldsymbol{U}_{N+1}(k)]^{-1} = \begin{bmatrix} 0 & \boldsymbol{E}_f^{-1}(k) \\ \boldsymbol{U}_N^{-1}(k-1) & -\boldsymbol{U}_N^{-1}(k-1)\boldsymbol{D}_{fq2}(k)\boldsymbol{E}_f^{-1}(k) \end{bmatrix} \boldsymbol{Q}'^T_{\theta f}(k). \qquad (8)$$

The result in (8) will be used in the next section to derive an expression for the updating of the *a posteriori* backward prediction error vector. Also from (7) we can write

$$\begin{bmatrix} 0 \\ \boldsymbol{E}_f^0(k) \end{bmatrix} = \boldsymbol{Q}'_{\theta f}(k+1) \begin{bmatrix} \boldsymbol{D}_{fq2}(k) \\ \boldsymbol{E}_f(k) \end{bmatrix} \qquad (9)$$

where $E_f^0(k)$ corresponds to the zero order forward error covariance matrix.

Algebraic manipulations based on Givens rotations applied to (6), generate the following equation using fixed order matrix $Q_\theta(k)$ obtained from $Q_N(k)$.

$$\begin{bmatrix} \tilde{e}_{fq1}^T(k+1) \\ D_{fq2}(k+1) \end{bmatrix} = Q_\theta(k) \begin{bmatrix} x_{k+1}^T \\ \lambda^{1/2} D_{fq2}(k) \end{bmatrix} \tag{10}$$

where $\tilde{e}_{fq1}^T(k+1)$ is the first row of $E_{fq1}^T(k+1)$.

Similarly, from (6) it is possible to obtain

$$\begin{bmatrix} 0^T \\ E_f(k+1) \end{bmatrix} = \overline{Q}_f(k+1) \begin{bmatrix} \tilde{e}_{fq1}^T(k+1) \\ \lambda^{1/2} E_f(k) \end{bmatrix} \tag{11}$$

where $\overline{Q}_f(k+1)$ is a fixed order matrix of the orthogonal matrix $Q_f(k+1)$, responsible for annihilate $\tilde{e}_{fq1}^T(k+1)$ against $\lambda^{1/2} E_f(k)$.

Finally, the joint process estimation is performed by the following expressions [1]

$$\begin{bmatrix} e_{q1}(k+1) \\ d_{q2}(k+1) \end{bmatrix} = Q_\theta(k+1) \begin{bmatrix} d(k+1) \\ \lambda^{1/2} d_{q2}(k) \end{bmatrix} \tag{12}$$

$$e'(k) = e_{q1}(k)/\gamma(k) = e(k)/\gamma^2(k) \tag{13}$$

Remark: The expression needed to obtain $Q_\theta(k)$ depends on the type of error to be updated (*a priori* or *a posteriori*) and is provided in the next section for the *a posteriori* case.

3   The Multichannel Fast QRD_POS_B Algorithm

Regardless the type of triangularization applied to $X_N(k)$ to generate $U_N(k)$, matrix $Q_\theta(k)$ can be partitioned as

$$Q_\theta(k) = \begin{bmatrix} \gamma(k) & g_N^T(k) \\ f_N(k) & E_N(k) \end{bmatrix} \tag{14}$$

where $\gamma(k) = \prod_{i=0}^{MN} \cos\theta_i(k)$; $f_N(k)$, $g_N(k)$, and $E_N(k)$ have more complicated expressions and depend on the type of triangularization used (in our case, lower triangularization, corresponding to backward prediction errors).

From (6) and (14), we can write

$$\begin{bmatrix} 0^T \\ U_N(k) \end{bmatrix} = Q_\theta(k) \begin{bmatrix} x_N^T(k) \\ \lambda^{1/2} U_N(k-1) \end{bmatrix} = \begin{bmatrix} \gamma(k) & g_N^T(k) \\ f_N(k) & E_N(k) \end{bmatrix} \begin{bmatrix} x_N^T(k) \\ \lambda^{1/2} U_N(k-1) \end{bmatrix} \tag{15}$$

We know that $Q_\theta(k)$ is orthogonal. Hence,

$$I_{N+1} = Q_\theta(k) Q_\theta^T(k) = \begin{bmatrix} \gamma(k) & g_N^T(k) \\ f_N(k) & E_N(k) \end{bmatrix} \begin{bmatrix} \gamma(k) & f_N^T(k) \\ g_N(k) & E_N^T(k) \end{bmatrix} \tag{16}$$

From (15) and (16), we can obtain the two following relations:

$$f_N(k)x_N^T(k) + \lambda^{1/2}E_N(k)U_N(k-1) = U_N(k) \tag{17}$$

and

$$\gamma(k)f_N(k) + E_N(k)g_N(k) = 0 \tag{18}$$

From (15), we can observe that $U_N(k)$ is Cholesky factor of $[x_N(k) \quad \lambda^{1/2}U_N^T(k-1)]^T$. Hence, we can premultiply (15) by its transpose to have

$$U_N^T(k)U_N(k) = x_N(k)x_N^T(k) + \lambda U_N^T(k-1)U_N(k-1) \tag{19}$$

Premultiplying (17) by $U_N^T(k)$ and comparing to (19) we can write

$$f_N(k) = U_N^{-T}(k)x_N(k) \tag{20}$$
$$E_N(k) = \lambda^{1/2}U_N^{-T}(k-1)U_N^T(k-1) \tag{21}$$

$f_N(k)$ is referred to as the *a posteriori* backward error vector and it is our particular vector of interest in the fast QRD-RLS algorithms based on *a posteriori* backward error updating. By substituting (20) and (21) in (18), it is possible to obtain

$$g_N(k) = -\gamma(k)U_N^{-T}(k-1)x_N(k)/\sqrt{\lambda} \tag{22}$$

which is the quantity of interest in the Fast QRD-RLS algorithms based on the *a priori* backward error updating.

From (20), it is straightforward to see that

$$f_{N+1}(k+1) = U_{N+1}^{-T}(k+1)x_{N+1}(k+1) \tag{23}$$

Now, combining (8) and the expression above, we obtain [5]

$$f_{N+1}(k+1) = Q'_{\theta f}(k+1)\begin{bmatrix} f_N(k) \\ p(k+1) \end{bmatrix} \tag{24}$$

where

$$p(k+1) = E_f^{-T}(k+1)\tilde{e}_f(k+1) \tag{25}$$

with $\tilde{e}_f(k+1)$ –the first line of the multichannel forward error, transposed– being the *a posteriori* forward error vector. Also from [5], we have the following expression, similar to its single dimension counterpart, to update $Q_\theta(k)$.

$$Q_\theta(k+1)\begin{bmatrix} 1 \\ 0 \end{bmatrix} = \begin{bmatrix} \gamma(k+1) \\ f_N(k+1) \end{bmatrix} \tag{26}$$

The expression in (25) requires a matrix inversion operation which can be numerically unstable, leading to stability problems. For calculating $p(k+1)$ in a simpler manner, it is easily shown from (11) that the following equation can be used instead of (25):

$$\overline{Q}_f(k+1)\begin{bmatrix} \gamma(k) \\ 0 \end{bmatrix} = \begin{bmatrix} * \\ p(k+1) \end{bmatrix} \tag{27}$$

Proof. From (11), it is clear that $E_f(k + 1)$ is the Cholesky factor of $[\tilde{e}_{fq1} \quad \lambda^{1/2} E_f^T(k)]^T$ [6]. Consequently, we can write

$$E_f^T(k+1)E_f(k+1) = \begin{bmatrix} \tilde{e}_{fq1}^T(k+1) \\ \lambda^{1/2}E_f(k) \end{bmatrix}^T \begin{bmatrix} \tilde{e}_{fq1}^T(k+1) \\ \lambda^{1/2}E_f(k) \end{bmatrix}$$

$$= \tilde{e}_{fq1}(k+1)\tilde{e}_{fq1}^T(k+1) + \lambda E_f^T(k)E_f(k) \qquad (28)$$

The above equation is the product form of (11). Now, premultiplying and post multiplying (28) by $E_f^{-T}(k+1)\gamma^2(k)$ and $E_f^{-1}(k+1)$, respectively, after some algebraic manipulations, yields

$$\gamma^2(k)I = p(k+1)p^T(k+1) + \Psi \qquad (29)$$

where $\Psi = \lambda\gamma^2(k)E_f^{-T}(k+1)E_f^T(k)E_f(k)E_f^{-1}(k+1)$.

Finally, premultiplying and post multiplying (29) by $p^T(k+1)$ and $p(k+1)$, respectively, it simplifies to

$$\gamma^2(k) = p^T(k+1)p(k+1) + \frac{p^T(k+1)\Psi p(k+1)}{p^T(k+1)p(k+1)} = p^T(k+1)p(k+1) + *^2$$

$$(30)$$

With farther manipulation, it can be shown that the quantity represented by the *asterisk* is known prior to the computation of $p(k+1)$; this knowledge, however, is useless for that purpose because it suffices to know $\gamma(k)$ and $\overline{Q}_f(k+1)$ to obtain $p(k+1)$ in a simple manner.

The expression in (30) is clearly a Cholesky product. Hence, there must exist an orthogonal matrix Q such that

$$\begin{bmatrix} \gamma(k) \\ 0 \end{bmatrix} = Q \begin{bmatrix} * \\ p(k+1) \end{bmatrix} \qquad (31)$$

Now, recalling our starting point in (11), we figure out that Q is related to $\overline{Q}_f(k+1)$. Moreover, from the knowledge of the internal structure of $\overline{Q}_f(k+1)$, we finally realize that $Q = \overline{Q}_f^T(k+1)$ satisfies (31) leading to (27), which concludes the proof.   □

4   The New LATTICE Multichannel Fast QRD_POS_B Algorithm

Because of the blocking nature of the input vector used to derive the equations of the algorithm presented in the previous section, the quantities $D_{fq2}(k)$, $d_{q2}(k)$, and $f_N(k)$ can be split up into N blocks from top to bottom. For the matrix $D_{fq2}(k)$ we have

$$D_{fq2}(k) = \begin{bmatrix} D_{fq2}^{(1)}(k) \\ \vdots \\ D_{fq2}^{(N)}(k) \end{bmatrix} \qquad (32)$$

Table 1. The MCFQRD_POS_B Equations.

Initializations:

$\boldsymbol{f}_N(0) = 0; \qquad \boldsymbol{D}_{fq2}(0) = 0$

$\boldsymbol{d}_{q2}(0) = 0; \qquad \boldsymbol{E}_f(0) = \boldsymbol{I}$

All *cosines* = 1, and all *sines* = 0;

For each k, do

{

  1. Obtaining $\boldsymbol{D}_{fq2}(k+1)$ and $\widetilde{\boldsymbol{e}}_{fq1}(k+1)$

$$\begin{bmatrix} \widetilde{\boldsymbol{e}}_{fq1}^T(k+1) \\ \boldsymbol{D}_{fq2}(k+1) \end{bmatrix} = \boldsymbol{Q}_\theta(k) \begin{bmatrix} \boldsymbol{x}_{k+1}^T \\ \lambda^{1/2} \boldsymbol{D}_{fq2}(k) \end{bmatrix}$$

  2. Obtaining $\boldsymbol{E}_f(k+1)$

$$\begin{bmatrix} \boldsymbol{0}^T \\ \boldsymbol{E}_f(k+1) \end{bmatrix} = \overline{\boldsymbol{Q}}_f(k+1) \begin{bmatrix} \widetilde{\boldsymbol{e}}_{fq1}^T(k+1) \\ \lambda^{1/2} \boldsymbol{E}_f(k) \end{bmatrix}$$

  3. Obtaining $\boldsymbol{p}(k+1)$

$$\begin{bmatrix} * \\ \boldsymbol{p}(k+1) \end{bmatrix} = \overline{\boldsymbol{Q}}_f(k+1) \begin{bmatrix} \gamma(k) \\ \boldsymbol{0} \end{bmatrix}$$

  4. Obtaining $\boldsymbol{Q}'_{\theta f}(k+1)$

$$\begin{bmatrix} \boldsymbol{0} \\ \boldsymbol{E}_f^0(k+1) \end{bmatrix} = \boldsymbol{Q}'_{\theta f}(k+1) \begin{bmatrix} \boldsymbol{D}_{fq2}(k+1) \\ \boldsymbol{E}_f(k+1) \end{bmatrix}$$

  5. Obtaining $\boldsymbol{f}_N(k+1)$

$$\boldsymbol{f}_{N+1}(k+1) = \boldsymbol{Q}'_{\theta f}(k+1) \begin{bmatrix} \boldsymbol{f}_N(k) \\ \boldsymbol{p}(k+1) \end{bmatrix}$$

  6. Obtaining $\boldsymbol{Q}_\theta(k+1)$ and $\gamma(k+1)$

$$\boldsymbol{Q}_\theta(k+1) \begin{bmatrix} 1 \\ 0 \end{bmatrix} = \begin{bmatrix} \gamma(k+1) \\ \boldsymbol{f}_N(k+1) \end{bmatrix}$$

  7. Joint Estimation

$$\begin{bmatrix} e_{q1}(k+1) \\ \boldsymbol{d}_{q2}(k+1) \end{bmatrix} = \boldsymbol{Q}_\theta(k+1) \begin{bmatrix} d(k+1) \\ \lambda^{1/2} \boldsymbol{d}_{q2}(k) \end{bmatrix}$$

  8. Obtaining the *a priori* error

    $e'(k+1) = e_{q1}(k+1)/\gamma(k+1)$

}

---

where $\boldsymbol{D}_{fq2}^{(i)}(k)$ has dimensions $M \times M$. In light of this assumption, (9) can be rewritten as

$$\begin{bmatrix} \boldsymbol{0}_{M(N-i-1) \times M} \\ \boldsymbol{0}_{M(i-1) \times M} \\ \boldsymbol{E}_f^{(i-1)}(k+1) \end{bmatrix} = \boldsymbol{Q}'_{\theta f}{}^{(N-i+1)}(k+1) \begin{bmatrix} \boldsymbol{0}_{M(N-i) \times M} \\ \boldsymbol{D}_{fq2}^{(N-i+1)}(k) \\ \boldsymbol{0}_{M(i-1) \times M} \\ \boldsymbol{E}_f^{(i)}(k+1) \end{bmatrix} \tag{33}$$

for $i = N, N-1, \cdots, 1$, which means a backward execution. From the previous equation, it is easy to see that

$$\boldsymbol{Q}'_{\theta f}(k+1) = \boldsymbol{Q}'_{\theta f}{}^{(N)}(k+1) \boldsymbol{Q}'_{\theta f}{}^{(N-1)}(k+1) \cdots \boldsymbol{Q}'_{\theta f}{}^{(1)}(k+1).$$

Nevertheless, (33) suggests that it can also be performed in a forward manner, that is, for $i = 1, 2, \cdots, N$. This property is the key to derive the lattice version of the algorithm. Now, recalling that $\boldsymbol{Q}'_{\theta f}(k)$ is used to update $\boldsymbol{f}_N(k)$, we can rewrite (24) as

Table 2. The Lattice MCFQRD_POS_B Equations.

Initializations:

$\boldsymbol{f}_N(0) = 0;$ $\boldsymbol{D}_{fq2}(0) = 0;$ $\gamma_0(0) = 1;$

$\boldsymbol{d}_{q2}(0) = 0;$ $\boldsymbol{E}_f^i(0) = \mu\boldsymbol{I},$ $\mu = small\ number$

All *cosines* = 1, and all *sines* = 0;

For each k, do

$\{\ \widetilde{\boldsymbol{e}}_{fq1}^{(0)}{}^T(k+1) = \boldsymbol{x}_{k+1}^T$

   A. Obtaining $\boldsymbol{E}_f^{(0)}(k+1)$ and $\overline{\boldsymbol{Q}}_f^{(0)}(k+1)$

$$\begin{bmatrix} \boldsymbol{0}^T \\ \boldsymbol{E}_f^{(0)}(k+1) \end{bmatrix} = \overline{\boldsymbol{Q}}_f^{(0)}(k+1) \begin{bmatrix} \widetilde{\boldsymbol{e}}_{fq1}^{(0)}{}^T(k+1) \\ \lambda^{1/2}\boldsymbol{E}_f^{(0)}(k) \end{bmatrix}$$

   B. Obtaining $\boldsymbol{p}_0(k+1)$

$$\begin{bmatrix} * \\ \boldsymbol{p}_0(k+1) \end{bmatrix} = \overline{\boldsymbol{Q}}_f^{(0)}(k+1) \begin{bmatrix} \gamma_0(k) \\ 0 \end{bmatrix}$$

$\boldsymbol{f}^{(N+1)}(k+1) = \boldsymbol{p}_0(k+1);$ $\gamma_0(k+1) = 1;$

$e_{q1}(k+1) = d(k+1);$

for $i = 1 : N$

   $\{$ 1. Obtaining $\boldsymbol{D}_{fq2}^{(N-i+1)}(k+1)$ and $e_{fq1}^{(i)}(k+1)$

$$\begin{bmatrix} \widetilde{\boldsymbol{e}}_{fq1}^{(i)}{}^T(k+1) \\ \boldsymbol{D}_{fq2}^{(N-i+1)}(k+1) \end{bmatrix} = \boldsymbol{Q}_\theta^{(i)}(k) \begin{bmatrix} \widetilde{\boldsymbol{e}}_{fq1}^{(i-1)}{}^T(k+1) \\ \lambda^{1/2}\boldsymbol{D}_{fq2}^{(N-i+1)}(k) \end{bmatrix}$$

   2. Obtaining $\boldsymbol{E}_f^{(i)}(k+1)$

$$\begin{bmatrix} \boldsymbol{0}^T \\ \boldsymbol{E}_f^{(i)}(k+1) \end{bmatrix} = \overline{\boldsymbol{Q}}_f^{(i)}(k+1) \begin{bmatrix} \widetilde{\boldsymbol{e}}_{fq1}^{(i)}{}^T(k+1) \\ \lambda^{1/2}\boldsymbol{E}_f^{(i)}(k) \end{bmatrix}$$

   3. Obtaining $\boldsymbol{p}_i(k+1)$

$$\begin{bmatrix} * \\ \boldsymbol{p}_i(k+1) \end{bmatrix} = \overline{\boldsymbol{Q}}_f^{(i)}(k+1) \begin{bmatrix} \gamma_i(k) \\ 0 \end{bmatrix}$$

   4. Obtaining $\boldsymbol{Q}_{\theta f}'{}^{(N-i+1)}(k+1)$

$$\begin{bmatrix} \boldsymbol{0}_{M(N-i-1)\times M} \\ \boldsymbol{0}_{M(i-1)\times M} \\ \boldsymbol{E}_f^{(i-1)}(k+1) \end{bmatrix} = \boldsymbol{Q}_{\theta f}'{}^{(N-i+1)}(k+1) \begin{bmatrix} \boldsymbol{0}_{M(N-i)\times M} \\ \boldsymbol{D}_{fq2}^{(N-i+1)}(k) \\ \boldsymbol{0}_{M(i-1)\times M} \\ \boldsymbol{E}_f^{(i)}(k+1) \end{bmatrix}$$

   5. Obtaining $\boldsymbol{f}^{(N-i+1)}(k+1)$

$$\begin{bmatrix} \boldsymbol{0}_{M(N-i)\times M} \\ \boldsymbol{f}^{(N-i+1)}(k+1) \\ \boldsymbol{0}_{M(i-1)\times M} \\ \boldsymbol{p}_{i-1}(k+1) \end{bmatrix} = \boldsymbol{Q}_{\theta f}'{}^{(N-i+1)}(k+1) \begin{bmatrix} \boldsymbol{0}_{M(N-i)\times M} \\ \boldsymbol{f}^{(N-i+2)}(k) \\ \boldsymbol{0}_{M(i-1)\times M} \\ \boldsymbol{p}_i(k+1) \end{bmatrix}$$

   6. Obtaining $\boldsymbol{Q}_\theta^{(i)}(k+1)$ and $\gamma_i(k+1)$

$$\boldsymbol{Q}_\theta^{(i)}(k+1) \begin{bmatrix} \gamma_{i-1}(k+1) \\ 0 \end{bmatrix} = \begin{bmatrix} \gamma_i(k+1) \\ \boldsymbol{f}^{(N-i+2)}(k+1) \end{bmatrix}$$

   7. Joint Estimation

$$\begin{bmatrix} e_{q1}^{(i)}(k+1) \\ d_{q2}^{(N-i+1)}(k+1) \end{bmatrix} = \boldsymbol{Q}_\theta^{(i)}(k+1) \begin{bmatrix} e_{q1}^{(i-1)}(k+1) \\ \lambda^{1/2}d_{q2}^{(N-i+1)}(k) \end{bmatrix}$$

   $\}$

   8. Obtaining the *a priori* error

   $e'(k+1) = e_{q1}(k+1)/\gamma(k+1)$

$\}$

$$\begin{bmatrix} \boldsymbol{0}_{M(N-i)\times M} \\ \boldsymbol{f}^{(N-i+1)}(k+1) \\ \boldsymbol{0}_{M(i-1)\times M} \\ \boldsymbol{p}_{i-1}(k+1) \end{bmatrix} = \boldsymbol{Q}_{\theta f}'{}^{(N-i+1)}(k+1) \begin{bmatrix} \boldsymbol{0}_{M(N-i)\times M} \\ \boldsymbol{f}^{(N-i+2)}(k) \\ \boldsymbol{0}_{M(i-1)\times M} \\ \boldsymbol{p}_i(k+1) \end{bmatrix} \tag{34}$$

for $i = 1, 2, \cdots, N$. Note that we are taking into account the forward option for both (33) and (34).

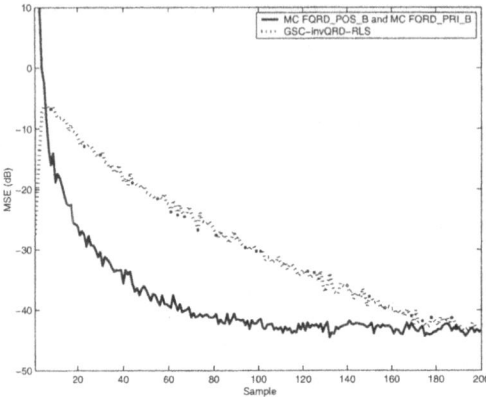

Fig. 2. Convergence in a beamforming scenario using GSC.

From the last two equations, we realize that steps 4 and 5 of the algorithm in Table 1 can now be carried out in a forward manner. The rotation angles $Q_\theta^{(i)}(k+1)$ are obtained through

$$Q_\theta^{(i)}(k+1)\begin{bmatrix}\gamma_{i-1}(k+1)\\0\end{bmatrix}=\begin{bmatrix}\gamma_i(k+1)\\f^{(N-i+2)}(k+1)\end{bmatrix} \tag{35}$$

and the joint estimation is performed according to

$$\begin{bmatrix}e_{q1}^{(i)}(k+1)\\d_{q2}^{(N-i+1)}(k+1)\end{bmatrix}=Q_\theta^{(N-i+1)}(k+1)\begin{bmatrix}e_{q1}^{(i)}(k+1)\\\lambda^{1/2}d_{q2}^{(N-i+1)}(k)\end{bmatrix} \tag{36}$$

In order to adequate the equations of steps 1 to 3 of the algorithm as in Table 1 to this formulation, it suffices to observe that they can be easily split up into $M \times M$ blocks that will be executed recursively as shown in Table 2. It is worth mentioning that, for the sake of simplification due to space constraints, we have used matrix notation in the single loop operations as shown in Table 2. However, when implementing these equations, it is straightforward to reduce the simple Givens rotations matrices into scalar operations.

5   Simulations Results

In this section we perform an evaluation of the Multichannel Fast QRD-RLS in an adaptive beamforming scenario. Although this kind of application requires the use of constrained algorithms, structures like the *Generalized Sidelobe Canceller* (GSC) [7], used here, or the Householder structure [8], make possible the use of unconstrained algorithms to solve constrained problems.

In our adaptive beamforming experiment, we have used a linear array of 7 sensors with a look-direction set to $0°$ and three jammers with incident angles corresponding to $-25°$, $45°$, and $50°$. The signal-to-noise ratio (SNR) was set to

$0dB$ and a jammer-to-noise ratio (JNR) of $30dB$ was used. The forgetting factor (λ) was set to 0.98.

The MSE converging paths (identical) are presented in Figure 2 for both MCFQR_POS_B (introduced here in its lattice version) and the MCFQR_PRI_B of [3]. Both algorithms are of $O(NM^3)$ computational complexity. Nevertheless our proposed algorithm saves $2NM$ multiplications and $2NM$ divisions in steps 3 and 6 when compared with its equivalent counterpart of [3]. After 10 independent runs of a considerably large number of samples (6×10^6), we have observed no sign of divergence, as expected for algorithms of the QRD-LS family.

6   Conclusions

In this paper we have introduced the Lattice version of the Multichannel Fast QRD-RLS algorithm based on the *a posteriori* backward error updating. Its order recursiveness and stability are very attractive features and it can be used in a wide range of applications, many of them in the field of telecommunications.

Although the new algorithm introduced here presents the same converge properties as the one of [3], it is worth mentioning that it saves computational load that makes it particularly attractive as N and M increase.

References

1. N. Kalouptsidis and S. Theodoridis, *Adaptive Systems Identification and Signal Processing Algorithms*, Englewood Cliffs, NJ: Prentice Hall, 1993.
2. J. A. Apolinário Jr., M. G. Siqueira, and P. S. R. Diniz, "Fast QR Algorithms Based on Backward Prediction Errors: A New Implementation and Its Finite Precision Performance," *Birkhäuser Circuits, Systems, and Signal Processing*, vol. 22, no. 4, pp. 335–349, July/August 2003.
3. A. A. Rontogiannis and S.Theodoridis, "Multichannel fast QRD-LS adaptive filtering: New technique and algorithms," *IEEE Transactions on Signal Processing*, vol. 46, pp. 2862–2876, November 1998.
4. C. A. Medina S., J. A. Apolinário Jr., and M. G. Siqueira, "A unified framework for multichannel fast QRD-LS adaptive filters based on backward prediction errors," *IEEE Midwest Symposium on Circuits and Systems*,vol.3, pp. 668–671 , USA, August 2002.
5. J. A. Apolinário Jr., *New algorithms of adaptive filtering: LMS with data-reusing and fast RLS based on QR decomposition*, D.Sc. Thesis, COPPE/Federal University of Rio de Janeiro, Rio de Janeiro, Brazil, 1998.
6. G. H. Golub and C. F. Van Loan, *Matrix Computations*, Baltimore: The Johns Hopkins University Press, 1983.
7. L. J. Griffiths and C. W. Jim, "An alternative approach to linearly constrained adaptive beamforming," *IEEE Transactions on Antennas and Propagation*, vol. AP-30, pp. 27–34, January 1982.
8. M. L. R. de Campos, S. Werner, J. A. Apolinário Jr., and T. I. Laakso, "Constrained adaptation algorithms employing Housholder transformation," *IEEE Transactions on Signal Processing*, vol. 50, no. 9, pp. 2187–2195, September 2002.

New Blind Algorithms Based on Modified "Constant Modulus" Criteria for QAM Constellations

Carlos A.R. Fernandes and João C.M. Mota

Universidade Federal do Ceará (UFC)
Departamento de Engenharia de Teleinformática (DETI)
Campus do Pici, C.P. 6005
Fortaleza–CE, Brazil {carlosalexandre, mota}@deti.ufc.br

Abstract. This work presents a family of new algorithms aiming to perform blind equalization in Quadrature Amplitude Modulation (QAM) signals. The algorithms are based on modified Constant Modulus (CM) criteria and they use decision direction. One of the proposed algorithms is based on the Constant Modulus Algorithm (CMA). Another one is based on the Recursive CMA (RCMA). The other two ones are dual-mode versions for high-level modulations. The first mode is used to avoid a too large number of incorrect decisions. Computer simulations show they outperform the CM based algorithms.

Keywords: Blind equalization, decision-directed, constant modulus, dual-mode.

1 Introduction

This work focuses on blind equalization for high level Quadrature Amplitude Modulations (QAM) signals. The main drawback of blind equalization is the low speed of convergence, which is about one order of magnitude lower than that obtained by the Least Mean Square (LMS) Algorithm [1] in the training mode. In this work, we study algorithms based on the constant modulus (CM) criterion. The CM criterion has been historically used to perform blind equalization and it works very well for use with modulations in which all points of the signal constellation have the same radius, like in Phase Shift Keying (PSK) modulations [2]. However, when the constellation points are allowed to assume multiple radii, the error of the algorithms based on CM criterion never goes to zero, even if the signal is perfectly equalized. This is one of the reasons for the unsatisfactory performance of conventional CM algorithms with QAM signals.

In this paper we will present four novel algorithms for blind equalization based on modified CM criteria. The first one is based on the Constant Modulus Algorithm (CMA). Another one is based on the Recursive Constant Modulus Algorithm (RCMA). The other two ones are dual-mode algorithms that start with the conventional CM algorithms and then switch to the novel ones, aiming to avoid an excessive number of incorrect decisions at initial iterations. In order to achieve better performance on QAM systems we make a modification on the CM cost function. Actually, the proposed modified CM-based algorithms can be viewed as a generalizations of the CM-based classical algorithms for PSK systems.

J.N. de Souza et al. (Eds.): ICT 2004, LNCS 3124, pp. 498–503, 2004.

2   System and Signal Models

A simplified version of the linear system model employed in this work is shown in fig. 1. The transmitted sequence $\{a(n)\}$ can assume the value of any constellation symbol with equal probability. The output sequence of the equalizer $\{y(n)\}$ is given by (1), where $\mathbf{h}_i$ is the impulse response of the channel and $v(n)$ is an additive white Gaussian noise (AWGN) component. The number of taps of the equalizer is M and $\mathbf{w}(n)$ is its tap-weight vector. $\hat{a}(n)$ is the output of the decision device (estimated symbol).

$$y(n) = \sum_{i=1}^{M} \mathbf{w}^{\top}(i)x(n-i), \; where \quad x(n) = \sum_{i=0}^{N} a(n-i)\mathbf{h}_i + v(n). \quad (1)$$

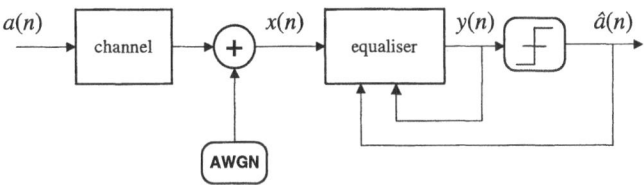

Fig. 1. Simplified structure of the system.

3   Classical Algorithms Based on CM Cost Function

The CMA was developed by Godard [2] and Treichler e al. [3] independently. It is one of the most used algorithms for blind equalization and it has several well-known applications. The equation for updating the tap-weight vector in CMA is given by:

$$\mathbf{w}(n+1) = \mathbf{w}(n) - \mu y(n)(|y(n)|^2 - R)\mathbf{x}^*(n), \, where \quad R = \frac{\mathrm{E}\{|a(n)|^4\}}{\mathrm{E}\{|a(n)|^2\}}, \quad (2)$$

$\mathbf{x}(n) = [x(n)x(n-1)\cdots x(n-M+1)]$ is the vector of filter inputs and μ is the step-size parameter. The CMA is one of the most robust algorithms for the blind equalization, although it may converge to a point of local minimum, due to the cost function expression 2. However, these cases of non-optimal convergence are very unusual in practice [1].

The application of the CMA with QAM signals leads to a non-optimal performance, because even when the signal is perfectly equalized, the adjusted term of the tap weight vector never reaches zero. Higher is the order of the QAM system, higher is the midsadjustment of the algorithm.

Many authors proposed generalized CM cost functions, which must deal with multi-modulus constellations, just like the Multiple Modulus Algorithm (MMA) in [4]. Another alternative for a CM algorithm in multiple radii constellations is the Decision

Adjusted Modulus Algorithm (DAMA), proposed in [4] for real value modulations, and reintroduced in [5] for QAM signals, with the name of Radius Decision Equalization (RDE). The tap-weight vector is updated according to the equation below:

$$\mathbf{w}(n+1) = \mathbf{w}(n) - \mu y(n)\{\min[(|y(n)|^2 - R_i)]\}\mathbf{x}^*(n). \tag{3}$$

where R_i are the squared magnitudes of the constellation points. Clearly, the DAMA cost function goes to zero if the signal is perfectly equalized. However, in many cases, DAMA will not converge because of the large number of incorrect radius decisions during the initial period. The performance of the DAMA can be improved a lot if it works in a dual-mode, just like it is developed in [6], where the authors proposed the CMA - Assisted Decision Adjusted Modulus Algorithm (CADAMA) to improve the stability of the DAMA. The CMA performs the initial adjustment of the tap weight vector and then switches to DAMA.

4   Proposed Modified CM Algorithms

In this section we present new cost functions based on the CM cost function and the respective algorithms. The first new cost function is the Decision-Directed Modulus (DDM) cost function, that is a generalization of the CM cost function and uses the squared magnitude of the decided symbol. The DDM cost function is expressed by:

$$J = \mathrm{E}\{(|y(n)|^2 - |\hat{a}(n)|^2)^2\}. \tag{4}$$

The first proposed algorithm using the DDM criterion is the Decision-Directed Modulus Algorithm (DDMA), which is a version of the CMA with a variable reference modulus. By calculating the stochastic gradient of the DDM cost function we get the DDMA tap-weight vector equation:

$$\mathbf{w}(n+1) = \mathbf{w}(n) - \mu y(n)(|y(n)|^2 - |\hat{a}(n)|^2)\mathbf{x}^*(n). \tag{5}$$

The derivative of $\hat{a}(n)$ relative to $\mathbf{w}(n)$ was assumed to be zero. For PSK modulations, the DDM and the CM cost functions are equivalent. The main advantage of the DDMA algorithm is the great performance in QAM constellations, in terms of speed of convergence and steady-state error. When the perfect equalization is achieved, the DDMA tap-weight vector adaptation term goes to zero while in CMA it never does so.

The performance of the DDMA is also prejudiced if the number of incorrect decisions is too large. To solve this problem the initial adjustment of the tap weight vector can be done by the CMA and when the mean square error goes below a threshold value, the algorithm switches to DDMA. This is the second proposed algorithm, the dual-mode DDMA. It has a great steady-state error and robustness, even in high level modulations.

Many authors proposed cost functions based on the CM cost function, but this work brings something different. We propose not just a cost function based on the CM but another based on the recursive CM. The performance of the DDMA can also be improved at the expense of increased complexity. The second proposed cost function is the Recursive DDM. The RDDM cost function is expressed next:

$$\phi(n) = \sum_{i=0}^{n} \lambda^{n-1}(|y(i)|^2 - |\hat{a}(i)|^2), \tag{6}$$

where $\lambda \leq 1$ is the forgetting factor. The solution to this optimization problem leads us to the Recursive DDMA (RDDMA), which can be described by the following equations:

$$\mathbf{k}(n) = \frac{\mathbf{P}(n-1)\mathbf{s}^*(n)}{\lambda + \mathbf{s}^T(n)\mathbf{P}(n-1)\mathbf{s}(n)}, \quad where \quad s(n) = y_*(n)\mathbf{x}(n)$$

$$\mathbf{P}(n) = \lambda^{-1} \cdot \left[\mathbf{P}(n-1) - \mathbf{k}(n)\mathbf{s}^T(n)\mathbf{P}(n-1)\right]$$

$$\mathbf{w}(n) = \mathbf{w}(n-1) + \mathbf{k}(n) \cdot \left(|y(n)|^2 - |\hat{a}(n)|^2\right)$$

where $\mathbf{P}(n)$ is initialized as $\mathbf{P}(0) = \delta^{-1}\mathbf{I}_N$, and δ is a small positive constant and $\mathbf{I}_N$ is the N-by-N identity matrix. Again, we considered the derivative of $\hat{a}(n)$ relative to $\mathbf{w}(n)$ equal to zero. For PSK modulations, the RDDMA is identical to the Recursive CMA (RCMA), proposed in [7,8]. For QAM constellations, RDDMA has similar performance to that of the DDMA, but with higher speed of convergence.

As well as the DDMA, the performance of the RDDMA can be improved a lot if number of incorrect decisions is not too large. The dual-mode RDDMA starts with RCMA in order to achieve the initial adjustment of the tap weight vector and when the mean square error goes below a threshold value, the algorithm switches to RDDMA. The proposed dual-mode RDDMA also has a great robustness and steady-state error, even in high level modulations.

5   Simulation Results

In this section we present some simulation results concerning the performance of the proposed algorithms, DDMA and RDDMA, in the single-mode and in dual-mode. The channel model used in our computational simulations consists in a telephone line model [9]. Its discrete-time impulse response is expressed by the following equation:

$$h(n) = 0.04\delta(n) - 0.05\delta(n-1) + 0.07\delta(n-2) - 0.21\delta(n-3) - 0.50\delta(n-4)$$

$$+0.72\delta(n-5) + 0.36\delta(n-6) + 0.21\delta(n-8)0.03\delta(n-9) + 0.07\delta(n-10).$$

The modulations used are 16-QAM and 64-QAM. The Signal-to-Noise Ratio (SNR) is 40dB and M = 16. A PLL is used to correct phase shift at the output of the equalizer. The curves are averaged over 50 independent realizations of the experiment.

Fig. 2(a) shows the mean square error (MSE) of the DDMA, DAMA and CMA for a 16-QAM constellation. The green line shows the Wiener optimum square error. Any value of the step-size parameter leads to the convergence of the DAMA, because of the large number of initial wrong decisions, as already mentioned. We can see the proposed DDMA has a better speed of convergence and a little smaller steady-state error than

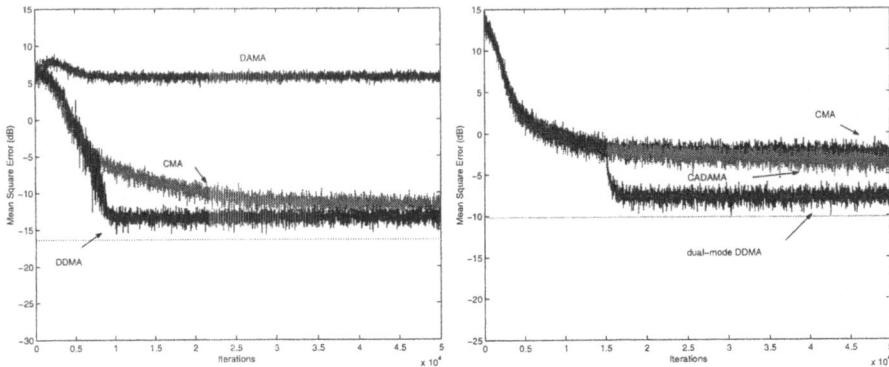

Fig. 2. The learning curve of (a) DDMA, DAMA and CMA for a 16-QAM signal and (b) dual-mode DDMA and DAMA and CMA for a 64-QAM signal .

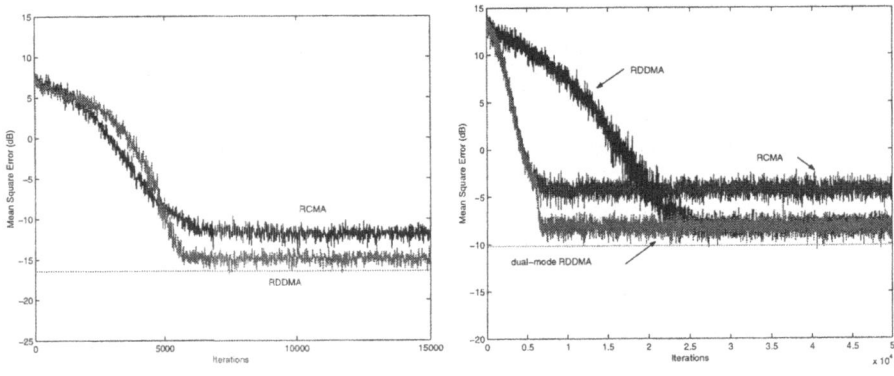

Fig. 3. The learning curve of (a) RDDMA and RCMA for a 16-QAM signal and (b) DDMA, dual-mode DDMA and RCMA for a 64-QAM signal .

CMA. For high QAM constellations, the DDMA and DAMA do not perform very well. The solution used in this case is to use dual-mode algorithms, as mentioned in section 4. Fig. 2(b) shows the learning curves of the CMA, dual-mode DAMA and dual-mode DDMA, for a 64-QAM signal. We can see that the dual-mode DDMA algorithm have much better performance than the all tested CMA-based algorithms.

The following figure presents simulation results relative to the recursive algorithms. Fig. 3(a) shows the MSE of RDDMA and RCMA for a 16-QAM constellation. It should be highlighted that RDDMA presents best steady-state error, as shown in fig. 3(a).The recursive-CMA algorithms, such as the CMA-like ones, do not perform very well for high QAM constellations. Again, we show that the equalization can be improved with dual-mode algorithms. In the fig. 3(b) we can see the much better performance of the dual-mode RDDMA relative to RCMA and RDDMA single-mode for 64-QAM signal.

Based on these results we can point out the some remarks: in signals up to 16-QAM, the DDMA outperforms the CMA and DAMA, and the RDDMA outperforms the RCMA. For high level QAM signals, the dual-mode DDMA has better performance than the others CM based algorithms, and the dual-mode RDDMA has also a much better performance than the others recursive CM based algorithms. The proposed algorithms were tested in others channel and the results were very similar.

6   Conclusions

We have presented a new family of algorithms characterized for some properties very desirable when performing blind equalization in QAM signals. The DDMA outperforms the conventional CMA in terms of speed of convergence and, in some cases, in terms of residual error. The RDDMA performs better than the conventional RCMA in terms of speed of convergence and in terms of residual error. For QAM signals of high modulation levels (up to 32-QAM) these algorithms present a large number of incorrect decisions during the initial phase and this make their performance worse. In these cases, the proposed dual-mode versions of DDMA and RDDMA showed to have very good performances. The next steps of this work include a study of stability and applications in ARMA (Auto Regressive – Moving Average) structures.

References

1. da Rocha, C.A.F.:   Técnicas Preditivas para Equalização Autodidata.   PhD thesis, FEE/UNICAMP, Campinas, Brazil (1996) (*in portuguese*).
2. Godard, D.N.:   Self-recovering equalization and carrier tracking in two dimensional data communication system. IEEE Trans. on Communic. **28** (1980) 1867–1875
3. Treichler, J., Agee, B.: A new approach to multipath correction of constant modulus signals. IEEE Transactions on ASSP **31** (1983) 459–472
4. Sethares, W.A., Rey, G.A., Jr., C.R.J.: Approaches to blind equalization of signal with multiple modulus. IEEE International Conference on Acustic, Speech and Signal Processing (ICASSP) (1989) 972–975
5. Ready, M.J., Gooch, R.P.:   Blind equalization based on radius directed adaptation.   IEEE International Conference on Acoustic, Speech and Signal Processing (ICASSP) (1990) 1699–1702
6. Axford, R.A., et al.: A dual-mode algorithm for blind equalization of QAM signals: CADAMA. Asilomar Conf. Signals, Systems and Computers **1** (1996) 172 –176
7. Pickholtz, R., Elbarbary, K.: The recursive constant modulus algorithm: a new approach for real time array processing. ASILOMAR Conf. on Signals, Systems and Computers (1993) 627–632
8. Hilal, K.: Algorithmes Accélérés d'Egalisation Adaptative Autodidacte. PhD thesis, ENST Paris, France (1993) (*in french*).
9. Proakis, J.G.: Digital Communications. McGraw-Hill Int. Ed. (1989)

Channel Estimation Methods for Space-Time Block Transmission in Frequency-Flat Channels

Nikolai Nefedov[1,2]

[1] Nokia Research Center, P.O.Box 407, 00045 Finland,
nikolai.nefedov@nokia.com
[2] Helsinki University of Technology, Communications Lab., 02015 Finland

Abstract. In this paper we consider different approaches for channel estimation based on semi-blind subspace and training channel estimation for orthogonal and non-orthogonal space-time block codes in HSDPA environment. It is found that application of subspace channel estimation in flat fading channels is rather limited, while estimation based on training symbols provides acceptable performance.

1   Introduction

To achieve 10-20Mbps data rates specified for High Speed Downlink Packet Access (HSDPA) in CDMA, a large fraction of the available downlink channelization codes is assigned to a given user, combined with high order constellations (e.g. 16-QAM or 64-QAM) and high rate (e.g. rate 3/4) turbo codes. On the other hand, instead of high order modulation, high data rates may be sought from transmit diversity and/or multiple-input-multiple-output (MIMO) channels. At the moment the most practical concepts are based on the extensions of space-time block codes (STBC). However, the main limitation of the orthogonal full-diversity, full symbol rate STBC is that it exists only for 2 transmitting antennas, $N_T = 2$ [1]. With increased transmit diversity order, using an orthogonal STBC for 3 or 4 transmit antennas, the maximal symbol rate is reduced to 3/4 [2]. Since the main target is to increase the data rate, such reduction in the symbol rate is not desired.

Solutions addressing the trade-off between diversity order, code rate, and performance are proposed in [2][3][4] where orthogonality is sacrificed in order to keep full symbol rate STBC for $N_T = 4$ transmit antennas [4].

In MIMO channels the number of channel parameters is increased, and it in turn degrades the overall performance of MIMO systems. Efficient channel estimation is needed to overcome performance saturation due to channel estimation errors. One of efficient methods to improve performance is iterative data-aided channel estimation. However, complexity of iterative receiver for CDMA is very high and a more feasible way to improve MIMO channel estimates may be seen in application non-iterative semiblind subspace methods based on second order statistics or/and traditional methods based on training symbols.

J.N. de Souza et al. (Eds.): ICT 2004, LNCS 3124, pp. 504–509, 2004.

In this paper we consider training-based and subspace channel estimation for frequency flat channels[1] with orthogonal STBC (OSTBC) and non-orthogonal STBC (NOSTBC) transmission schemes with/without channel coding. Notation and a brief overview of OSTBC and NOSTBC are given in Section 2. Subspace channel estimation in flat fading channels with STBC is addressed in Section 3. Simulation results for HSDPA with different channel estimation methods are presented in Section 4, with conclusions followed in Section 5.

2   STBC for High Data Rate Systems

Let's consider MIMO transmission with N_T transmitting and N_R receiving antennas. We use bold and capital bold letters for vectors and matrices with real entries, respectively. Vectors and matrices with complex elements are marked with a line on the top, $E\{\}$ and $()^T$ stand for the expectation and matrix transpose. Transmitted signal with STBC may be presented in the following form

$$\bar{\mathbf{X}}(\bar{s}_1, \bar{s}_2, ...\bar{s}_K) \triangleq \sum_{k=1}^{K}[\mathrm{Re}\{\bar{s}_k\}\mathbf{A}_k + i\,\mathrm{Im}\{\bar{s}_k\}\mathbf{B}_k]$$

where K is number of information complex-valued symbols in a transmitted block, $\mathbf{A}_k$, $\mathbf{B}_k$ are $N \times N_T$ modulation matrices, $k = 1, ..., K$; N is number of transmitted coded symbols in the block. Below we consider only minimum delay STBC, i.e. $N = N_T$. In the OSTBC case the modulation matrices satisfy the following conditions:

$$\mathbf{A}_k^T\mathbf{A}_k = \mathbf{I}, \mathbf{B}_k^T\mathbf{B}_k = \mathbf{I} \text{ for } k = 1, ..., K; \mathbf{A}_j^T\mathbf{B}_k = \mathbf{B}_k^T\mathbf{A}_j \text{ for } k, j = 1, ..., K;$$
$$\mathbf{A}_j^T\mathbf{A}_k = -\mathbf{A}_k^T\mathbf{A}_j, \mathbf{B}_j^T\mathbf{B}_k = -\mathbf{B}_k^T\mathbf{B}_j \text{ for } 1 \leq j < k \leq K$$

The received signal $\bar{\mathbf{Y}}$ may be presented in matrix form is $\bar{\mathbf{Y}} = \bar{\mathbf{X}}\bar{\mathbf{H}} + \bar{\mathbf{N}}$, where $\bar{\mathbf{H}}$ and $\bar{\mathbf{N}}$ are matrices describing channel and complex additive white Gaussian noise (AWGN), respectively. In case of one receiving antenna, $N_R = 1$, the received signal vector is $\bar{\mathbf{y}} = \bar{\mathbf{X}}\bar{\mathbf{h}} + \bar{\mathbf{n}}$, where channel vector $\bar{\mathbf{h}} = [\bar{h}_1, ..., \bar{h}_{N_T}]^T$ is formed by complex fading channel coefficients, $\bar{h}_m$, $m = 1, ..., N_T$; $\bar{\mathbf{n}}$ is complex-valued AWGN.

Expanding $N \times 1$ complex-valued received vector $\bar{\mathbf{y}}$ into $2N \times 1$ real vector $\mathbf{y} = [\mathrm{Re}\{\bar{\mathbf{y}}\}, \mathrm{Im}\{\bar{\mathbf{y}}\}]^T$ we get $\mathbf{y} = \mathbf{X}\mathbf{h} + \mathbf{n}$, where $\mathbf{h} \triangleq [\mathrm{Re}\{\bar{\mathbf{h}}\}, \mathrm{Im}\{\bar{\mathbf{h}}\}]^T$; $\mathbf{n} \triangleq [\mathrm{Re}\{\bar{\mathbf{n}}\}, \mathrm{Im}\{\bar{\mathbf{n}}\}]^T$;

$$\mathbf{X} \triangleq \sum_{k=1}^{K}[\mathrm{Re}\{s_k\}\mathbf{P}_k + \mathrm{Im}\{s_k\}\mathbf{Q}_k];$$

$$\mathbf{P}_k \triangleq \begin{pmatrix} \mathbf{A}_k & \mathbf{0}_{N \times N_T} \\ \mathbf{0}_{N \times N_T} & \mathbf{A}_k \end{pmatrix} \qquad \mathbf{Q}_k \triangleq \begin{pmatrix} \mathbf{0}_{N \times N_T} & -\mathbf{B}_k \\ \mathbf{B}_k & \mathbf{0}_{N \times N_T} \end{pmatrix}$$

Note that vectors $\mathbf{P}_k\mathbf{h}$ and $\mathbf{Q}_k\mathbf{h}$ are orthogonal to each other for any expanded channel vector $\mathbf{h}$. It may be shown that for random data ($E\{\mathbf{ss}^T\} = \mathbf{I}$) the autocorrelation matrix of the received vector $\mathbf{y}$ is given by

$$\mathbf{R_y} \triangleq E\{\mathbf{yy}^T\} = E\{\mathbf{Xhh}^T\mathbf{X}^T\} + \frac{\delta^2}{2}\mathbf{I}$$

$$= \sum_{k=1}^{K}(\mathbf{P}_k\mathbf{h})(\mathbf{P}_k\mathbf{h})^T + \sum_{k=1}^{K}(\mathbf{Q}_k\mathbf{h})(\mathbf{Q}_k\mathbf{h})^T + \frac{\delta^2}{2}\mathbf{I} \qquad (1)$$

---

[1] Subspace channel estimation for transmit diversity schemes in multipath MIMO channel are addressed in [8].

Review of non-orthogonal STBC may be found in [5]. As an example of the NOSTBC we consider a scheme known as ABBA ($N_T = 4$, $N_R = 1$, symbol rate 1, i.e. $N = K$) [4][5]. In this case modulation complex-valued matrices are
$\mathbf{A}_1 = \mathbf{T}_1 \otimes \mathbf{T}_1$; $\mathbf{A}_2 = \mathbf{I}_{2 \times 2} \otimes \mathbf{T}_3$; $\mathbf{A}_3 = \mathbf{T}_4 \otimes \mathbf{I}_{2 \times 2}$; $\mathbf{A}_4 = \mathbf{T}_4 \otimes \mathbf{T}_3$; $\mathbf{B}_1 = \mathbf{I}_{2 \times 2} \otimes \mathbf{T}_2$; $\mathbf{B}_2 = \mathbf{I}_{2 \times 2} \otimes \mathbf{T}_4$; $\mathbf{B}_3 = \mathbf{T}_4 \otimes \mathbf{T}_2$; $\mathbf{B}_4 = \mathbf{T}_4 \otimes \mathbf{T}_4$;
where
$$\mathbf{T}_1 = \begin{bmatrix} 1 & 0 \\ 0 & 1 \end{bmatrix} \quad \mathbf{T}_2 = \begin{bmatrix} 1 & 0 \\ 0 & -1 \end{bmatrix} \quad \mathbf{T}_3 = \begin{bmatrix} 0 & 1 \\ -1 & 0 \end{bmatrix} \quad \mathbf{T}_4 = \begin{bmatrix} 0 & 1 \\ 1 & 0 \end{bmatrix};$$
transmitted signal is
$$\bar{\mathbf{X}}(\bar{s}_1, \bar{s}_2, ... \bar{s}_N) \triangleq \sum_{n=1}^{N} [\mathrm{Re}\{\bar{s}_n\}\mathbf{A}_k + i\,\mathrm{Im}\{\bar{s}_n\}\mathbf{B}_n];$$
with the autocorrelation matrix $\mathbf{R}_\mathbf{y}$ similar to (1), where $K = N$.

3   Subspace Channel Estimation in Flat Fading Channels

3.1   OSTBC

In blind subspace methods the observation space is first partitioned into signal subspace and noise subspaces. Then channel parameters are estimated based on orthogonality between subspaces. This method allows to identify channel with transmit diversity up to a right multiplication of an invertible matrix. To resolve the ambiguity a small amount of transmitted symbols (e.g., tailing symbols) should be known at the receiver.

Let's consider first multiple-input-single-output (MISO) flat fading channel channels with $N_R = 1$. Note that oversampling at the receiver side typically used to convert a multipath MISO channel into MIMO channel [9][10] can not be applied for flat fading MISO channels. Hence, other methods should be used to facilitate channel estimation. Below we consider a reduction of the STBC symbol rate to obtain orthogonal subspaces.

Eigenvalue decomposition of the received autocorrelation matrix (1) gives $\mathbf{R}_\mathbf{y} = \mathbf{V}\Lambda\mathbf{V}^T$, where matrices $\mathbf{V}$ and Λ are formed by eigenvectors $\mathbf{v}_i$ and eigenvalues λ_i: $\mathbf{V} \triangleq [\mathbf{v}_1, \mathbf{v}_2, ..., \mathbf{v}_{2N}]$; $\Lambda \triangleq diag[\lambda_1, \lambda_2, ..., \lambda_{2N}]$, $\lambda_1 \geq .... \geq \lambda_{2K} > \lambda_{2K+1} = ... = \lambda_{2N}$. Let's form matrices $\mathbf{V}_S \triangleq [\mathbf{v}_1, \mathbf{v}_2, ..., \mathbf{v}_{2K}]$ and $\mathbf{V}_N = [\mathbf{v}_{2K+1}, ..., \mathbf{v}_{2N}]$ whose columns span signal and noise subspaces of $\mathbf{R}_\mathbf{y}$, respectively. Hence,
$$\mathbf{V}_N^T \mathbf{P}_k \mathbf{h} = 0; \quad \mathbf{V}_N^T \mathbf{Q}_k \mathbf{h} = 0, \text{ for } k = 1, ..., K$$
or
$$\sum_{k=1}^{K} ||\mathbf{V}_N^T \mathbf{P}_k \mathbf{h}||^2 + \sum_{k=1}^{K} ||\mathbf{V}_N^T \mathbf{Q}_k \mathbf{h}||^2 = 0$$
which is equivalent to
$$\mathbf{h}^T \left(\sum_{k=1}^{K} \mathbf{P}_k^T \mathbf{V}_N \mathbf{V}_N^T \mathbf{P}_k + \sum_{k=1}^{K} \mathbf{Q}_k^T \mathbf{V}_N \mathbf{V}_N^T \mathbf{Q}_k \right) \mathbf{h} = 0$$
Then the estimation of the expanded channel vector $\mathbf{h}$ may be found as
$$\hat{\mathbf{h}} = \arg \min_{\mathbf{h}} (\mathbf{h}^T \mathbf{U} \mathbf{h})$$
where $\mathbf{U}$ is $2N_T \times 2N_T$ matrix,
$$\mathbf{U} \triangleq \sum_{k=1}^{K} \mathbf{P}_k^T \mathbf{V}_N \mathbf{V}_N^T \mathbf{P}_k + \sum_{k=1}^{K} \mathbf{Q}_k^T \mathbf{V}_N \mathbf{V}_N^T \mathbf{Q}_k.$$

3.2 Resolving Channel Ambiguity

To identify $\mathbf{h}$ (within a scalar ambiguity) we have to find an unique null-eigenvector of $\mathbf{U}$. However, there exist $2(N - K)$ orthogonal null-eigenvectors of $\mathbf{U}$ that satisfy $\mathbf{h}^T \mathbf{U} \mathbf{h} \to 0$. To resolve the ambiguity problem we have to introduce some redundancy. For example, for complex-valued OSTBC during some part of the slot transmission we may set the real part of the first symbol in some OSTBC blocks to zero, i.e. $\mathrm{Re}\{s_1\} = 0$ [7]. Then for the corresponding blocks the autocorrelation matrix and its eigenvalue decomposition are:

$$\mathbf{R}_{\mathbf{y}}^- = \sum_{k=2}^{K}(\mathbf{P}_k\mathbf{h})(\mathbf{P}_k\mathbf{h})^T + \sum_{k=1}^{K}(\mathbf{Q}_k\mathbf{h})(\mathbf{Q}_k\mathbf{h})^T + \frac{\delta^2}{2}\mathbf{I}$$

where $\mathbf{R}_{\mathbf{y}}^- = \mathbf{V}^- \varLambda (\mathbf{V}^-)^T$, $\mathbf{V}^- \triangleq [\mathbf{v}_1^-, \mathbf{v}_2^-, ..., \mathbf{v}_{2N}^-]$, such that $\mathbf{V}_S^- \triangleq [\mathbf{v}_1^-, \mathbf{v}_2^-, ..., \mathbf{v}_{2K}^-]$ spans signal space of $\mathbf{R}_{\mathbf{y}}^-$.

By construction of the matrices $\mathbf{R}$ and $\mathbf{R}^-$, columns of the matrices $\mathbf{V}_N$ and $\mathbf{V}_S^-$ span orthogonal (i.e. $\mathbf{V}_N(\mathbf{V}_S^-)^T = \mathbf{0}$) subspaces of dimension $2(N - K)$ and $2K - 1$, respectively. Signal subspace of $\mathbf{R}$ (spanned by $\mathbf{V}_S$) has dimension $2K$ and is orthogonal to the subspace spanned by $\mathbf{V}_N$. Hence, a subspace which is orthogonal to both subspaces formed by $\mathbf{V}_N$ and $\mathbf{V}_S^-$ must have dimension $2K - (2K - 1) = 1$ and is spanned by $\mathbf{P}_1\mathbf{h}$. Formally it may be expressed as [7]

$$\mathbf{h}^T \left(\mathbf{P}_1^T\mathbf{V}_N\mathbf{V}_N^T\mathbf{P}_1 + \mathbf{P}_1^T\mathbf{V}_S^-(\mathbf{V}_S^-)^T\mathbf{P}_1\right)\mathbf{h} = 0$$

Then the estimation of the channel vector $\mathbf{h}$ (within a scalar ambiguity) may be found as $\hat{\mathbf{h}} = \arg\min_{\mathbf{h}}(\mathbf{h}^T\mathbf{U}_1\mathbf{h})$, where $\mathbf{U}_1$ is $2N_T \times 2N_T$ matrix:

$$\mathbf{U}_1 \triangleq \mathbf{P}_1^T\mathbf{V}_N\mathbf{V}_N^T\mathbf{P}_1 + \mathbf{P}_1^T\mathbf{V}_S^-(\mathbf{V}_S^-)^T\mathbf{P}_1$$

4 Simulation Results

4.1 OSTBC

It is obvious that for full symbol rate ($N = K$) OSTBC with $N_T = 2$ and $N_R = 1$ no blind subspace channel estimation possible since no orthogonal subspaces ($N - K = 0$) are available. This reveals a trade-off between detector-estimator complexity and OSTBC symbols rate. In case of full symbol rate OSTBC ($N_T = 2$) we have simple linear detectors, but no blind channel estimation possible. To estimate channel we have to sacrifice symbol rate either by reducing the STBC symbol rate or by inserting known training symbols.

Let's consider the OSTBC with $N_T = 4$ and $N_R = 1$. The highest symbol rate for this case is $3/4$ ($K = 3$, $N = 4$) [3][4]. Modulation matrices for this case may be presented as follows [5]

$$\mathbf{A}_1 = \mathbf{T}_1 \otimes \mathbf{T}_1; \quad \mathbf{A}_2 = \mathbf{T}_3 \otimes \mathbf{T}_1; \quad \mathbf{A}_3 = \mathbf{T}_4 \otimes \mathbf{T}_3$$
$$\mathbf{B}_1 = \mathbf{T}_2 \otimes \mathbf{T}_1; \quad \mathbf{B}_2 = \mathbf{T}_4 \otimes \mathbf{T}_2; \quad \mathbf{B}_3 = -\mathbf{T}_4 \otimes \mathbf{T}_4$$

BER performance for $3/4$ rate OSTBC for different estimation methods is presented at Fig.1A. Simulations correspond to a flat block-fading channel without channel coding where each block consists of 50 transmitted matrices made of 200 symbols (4QAM). To facilitate semi-blind channel estimation we set $\mathrm{Re}\{s_1\} = 0$ in a half of the transmitted matrices as described in the Section above (data rate loss is 6.25%). Note that there are no redundant symbols inserted, but less data symbols transmitted. However, one may treat missing transmitted symbols as a data rate loss due to redundancy symbols. As an alternative to the semi-blind method (line with stars at Fig.1A) we may allocate the same amount of redundancy for the explicit training (line with circles at Fig.1A). Our simulations show that is it better to allocate redundancy directly to the training than to have data rate loss (the same as for training) to resolve ambiguities in subspace blind channel estimation (Fig.1A).

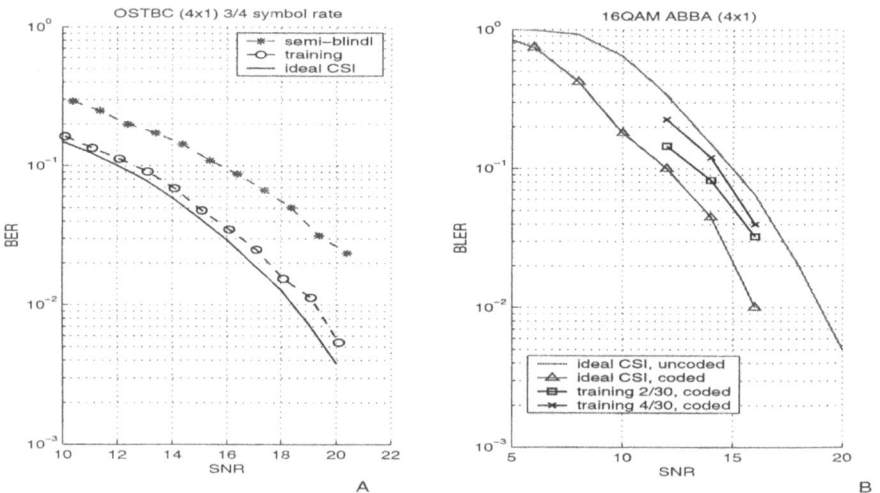

Fig. 1. A) BER performance of channel estimation methods: OSTBC of rate 3/4; B) BLER performance of coded HSDPA, NOSTBC(ABBA), coded block=480bits

4.2 NOSTBC (ABBA)

Despite the fact that ABBA is NOSTBC, it may be shown that *blind subspace channel estimation is not possible if full symbol rate of ABBA is kept* (the proof is not presented due to lack of space, it will appear elsewhere).

For channel estimation we need to allocate some redundancy either in form of direct training or in the way presented in Section 3. However, as it is mentioned above, it is better to allocate redundancy directly for training than to use it only for resolving subspace blind channel estimation ambiguities. To evaluate a performance loss due to channel estimation based on training we simulate the NOSTBC (ABBA) with high channel coding rates based on zigzag codes [11][12].

Simulated block error rate (BLER) performance with 16QAM transmission, channel coding (4/5 rate zigzag codes, block size 480) and channel estimation based on training is presented at Fig.1B. To provide soft decisions needed for channel decoder we used a soft-output lattice detector for MQAM [13]. In each block of 30 matrices we used from 2 to 4 training matrices. As one can see, in coded HSDPA the performance loss due to errors in channel estimation is 1-2 dB (depending on training) w.r.t. to ideal channel state information (CSI). This gap may be further reduced if training is integrated into semi-blind estimation.

5 Conclusions

In this paper we compare training-based and semi-blind subspace channel estimation for different STBC schemes in HSDPA/CDMA environment. Both approaches require to sacrifice the data rate to make channel estimation feasible. However, simulation results show that in STBC it is preferable to allocate redundancy directly to the training than to allow data rate loss for resolving ambiguities in subspace blind channel estimation in flat fading channels. Also it is found that application of subspace channel estimation in flat fading channels is rather limited in general, while estimation based on training symbols provides acceptable performance.

References

1. Alamouti, S.M.: A simple transmit diversity technique for wireless communications. IEEE JSAC **16** October (1998) 1451-1458
2. Tarokh, V., et al: Space-time codes for high data rate wireless communications: performance criteria. In: Proc. IEEE ICC (1997)
3. Jafarkhani, H.: A quasi-orthogonal space-time block code. IEEE Trans. Comm. **49** January (2001) 1-4
4. Tirkkonen, O., Hottinen,A.: Complex space-time block codes for four Tx antennas. In: Proc. IEEE Globecom (2000) 1005-1009
5. Hottinen, A., Tirkkonen, O., Wichman, R.: Multi-antenna Transceiver Techniques for 3G and Beyond. Wiley Ltd (2003)
6. Tirkkonen, O., Hottinen, A.: Square-matrix embeddable space-time block codes for complex signal constellations. In: IEEE Trans. Inform. Theory **48** Feb. (2002) 384-395
7. Psaromiligkos, I., Stamous,A.: On Semi-Blind Channel Estimation of Space-Time Block-Coded Transmission. In: Proc. Conf. ISS (2002)
8. Nefedov, N.: On Subspace Channel Estimation in Multipath SIMO and MIMO Channels. In: this volume
9. Swindlehurst, A., Leus, G.: Blind and Semi-Blind Equalization for Generalized Space-Time Block Codes. IEEE Trans. Sig. Proc. **50** October (2002) 2489-2498
10. Zhou, S., Muquet, B., Giannakis, G.: Subspace-Based (Semi-) Blind Channel Estimation for Block Preceeded Space-Time OFDM. IEEE Trans. Sig. Proc. **50** May (2002) 1215-1228
11. Ping, L., Huang, X., Phamdo, N.: Zigzag Codes and Concatenated Zigzag Codes. IEEE Trans. Inform Theory **47** Feb. (2001) 800–807
12. Nefedov, N.: Evaluation of Low Complexity Concatenated Codes for High Data Rate Transmission. In: Proc. IEEE PIMRC (2003).
13. Ramirez, E., Nefedov, N.: Soft Output Detection using Path Detector for Multiple Antennas. In: Proc. IEEE NorSig (2004)

Faster DTMF Decoding

J.B. Lima, R.M. Campello de Souza, H.M. de Oliveira, and
M.M. Campello de Souza

Departamento de Eletrônica e Sistemas - UFPE, C.P. 7800, 50711-970, Recife-PE, Brasil
flagbros@elogica.com.br, {ricardo, hmo,marciam}@ufpe.br

Abstract. In this paper, a new method for the decoding of DTMF signals is proposed. The approach, which applies the Arithmetic Fourier Transform, is more efficient, in terms of computational complexity, than existing techniques. Theoretical aspects and features that determine the accuracy and the complexity of the proposed method are discussed.

1 Introduction

Computational complexity is a decisive figure of merit in algorithms intended for frequency analysis. Bruns [1] developed a method for computing the coefficients of a Fourier series using the Möbius inversion formula for finite series. The technique, later called the Arithmetic Fourier Transform (AFT) [2], requires mainly trivial multiplications, except for a few scaling factors. Tufts and Sadasiv [3] discovered a very similar algorithm that had the constraint to only deal with even signals. This constraint was later removed by Reed and Tufts [4]. Reed and Shih [5] improved the previous algorithm and proposed the simplified AFT. In this paper, a new method for decoding DTMF (Dual-Tone Multi-Frequency) signals is proposed, which is based on the AFT. The method applies the simplified AFT that has a lower computational complexity than its previous versions. Specifically, the number of multiplications involved in the decoding operation is much lower than that required by an FFT or by the Goertzel algorithm [6]. In the next section basic facts about the AFT are presented. On section 3, the relation between the DFT and the Fourier series is shown. The choice of AFT parameters, such as the sampling rate and transform length, are discussed in section 4. Some numerical results concerning the decoding errors are shown. Section 5 presents the conclusions of the paper.

2 The Arithmetic Fourier Transform

This section briefly reviews a few basic facts concerning the AFT, algorithm that is in the heart of the proposed DTMF decoding method.

Theorem 1 (The Möbius inversion formula for finite series [4]): Assume that n is a positive integer and f_n is a nonzero sequence confined to the interval $1 \leq n \leq N$.

J.N. de Souza et al. (Eds.): ICT 2004, LNCS 3124, pp. 510–515, 2004.

If $g_n = \sum_{k=1}^{\lfloor N/n \rfloor} f_{kn}$, then $f_n = \sum_{m=1}^{\lfloor N/n \rfloor} \mu(m) g_{mn}$, where μ is the Möbius function.

2.1 Reed-Tufts

Let $v(t)$ be a real signal with period T, whose finite Fourier series has the form

$$v(t) = a_0 + \sum_{k=1}^{N} a_k \cos\left(\frac{2\pi k t}{T}\right) + \sum_{k=1}^{N} b_k \sin\left(\frac{2\pi k t}{T}\right), \tag{1}$$

where $f_0 = 1/T$ and a_0 is the mean of $v(t)$. If the dc component is removed from $v(t)$, one obtains $\bar{v}(t)$. Shifting the periodic function $\bar{v}(t)$ by an amount αT, where $|\alpha| < 1$,

yields $\bar{v}(t + \alpha T) = \sum_{k=1}^{N} c_k(\alpha) \cos\left(\frac{2\pi k t}{T}\right) + \sum_{k=1}^{N} d_k(\alpha) \sin\left(\frac{2\pi k t}{T}\right)$, where

$$c_k(\alpha) = a_k \cos(2\pi k\alpha) + b_k \sin(2\pi k\alpha), \ d_k(\alpha) = -a_k \sin(2\pi k\alpha) + b_k \cos(2\pi k\alpha). \tag{2}$$

Definition 1: The kth partial average is

$$S_k(\alpha) = \frac{1}{k} \sum_{m=0}^{k-1} \bar{v}\left(\frac{m}{k}T + \alpha T\right), \text{ where } -1 < \alpha < 1. \tag{3}$$

a_k and b_k can be expressed as a function of $c_k(\alpha)$. Firstly, we relate $c_k(\alpha)$ with the S_k.

Theorem 2: The coefficients $c_k(\alpha)$ can be computed by theorem 1

$$c_k(\alpha) = \sum_{l=1}^{\lfloor N/k \rfloor} \mu(l) S_{lk}(\alpha) \cdot \tag{4}$$

The coefficients a_k and b_k for $k = 2^r(2m+1)$ are computed by

$$a_k = c_k(0), \ b_k = (-1)^m c_k\left(\frac{1}{2^{r+2}}\right), \ k = 1, ..., N \tag{5}$$

where r and m are obtained from the factorization of k.

2.2 Reed-Shih (Simplified AFT)

In this method, the partial averages are redefined according to Bruns [1].

Definition 2 (Bruns averages): The Bruns partial averages, $B_{2k}(\alpha)$, are defined by

$$B_{2k}(\alpha) = \frac{1}{2k} \sum_{m=0}^{2k-1} (-1)^m v\left(m\frac{T}{2k} + \alpha T\right). \tag{6}$$

From the definition of c_k in (2) and using theorem 2 and definition 1, we obtain [5].

Theorem 3: The coefficients $c_k(\alpha)$ are given by the Möbius inversion formula:

$$c_k(\alpha) = \sum_{l=1,3\ldots}^{\lfloor N/k \rfloor} \mu(l) \, B_{2kl}(\alpha). \tag{7}$$

In (2), two conditions can be distinguished: $a_k = c_k(0)$ and $b_k = c_k(1/4k)$. From them and theorem 3, the next result follows.

Theorem 4 (Reed-Shih): The Fourier coefficients a_k and b_k, $k=1,\ldots,N$, are given by

$$a_0 = \frac{1}{T} \int_0^T v(t)\,dt \,, \quad a_k = \sum_{l=1,3\ldots}^{\lfloor N/k \rfloor} \mu(l) \, B_{2kl}(0), \quad b_k = \sum_{l=1,3\ldots}^{\lfloor N/k \rfloor} \mu(l)(-1)^{\frac{(l-1)}{2}} B_{2kl}(1/4kl). \tag{8}$$

3 The DFT and the Coefficients of the Fourier Series

This section shows the relation between the DFT of a sequence and the Fourier coefficients of the continuous version of the signal. This allows to take advantage of the low computational complexity of the AFT for implementing the decoder.

Definition 3 (The Discrete Fourier Transform): Let v be a complex-valued N-dimensional vector. The DFT of v is the vector V, of elements $V[k]$, given by

$$V[k] = \sum_{n=0}^{N-1} v[n] \exp\left(-j\frac{2\pi k n}{N}\right), \quad k = 0,1,\ldots,N-1. \tag{9}$$

The inverse transform is given by

$$v[n] = \sum_{k=0}^{N-1} V[k] \exp\left(j\frac{2\pi k n}{N}\right), \quad n = 0,1,\ldots,N-1. \tag{10}$$

The Fourier series up to the $(N/2)$-th harmonic of $v(t)$ is

$$v(t) = a_0 + \sum_{k=1}^{N/2} a_k \cos\left(\frac{2\pi k t}{T}\right) + \sum_{k=1}^{N/2} b_k \sin\left(\frac{2\pi k t}{T}\right), \tag{11}$$

where N is even and all the other terms of the series are supposed to be negligible. By sampling N equidistant points through a period of $v(t)$, a sequence $v[n]$ is obtained. Then, the discrete version of (11) can be written as

$$v[n] = a_0 + a_{N/2}(-1)^n + \sum_{k=1}^{(N-2)/2} a_k \cos\left(\frac{2\pi n k}{N}\right) + \sum_{k=1}^{(N-2)/2} b_k \sin\left(\frac{2\pi k n}{N}\right). \tag{12}$$

Writing the components $V[k]$ of the DFT in cartesian form $V[k] = \mathrm{Re}\{V[k]\} + j\,\mathrm{Im}\{V[k]\}$ and substituting in (10), leads to

$$v[n] = \frac{V[0]}{N} + \frac{V\left[\frac{N}{2}\right](-1)^n}{N} + \frac{2}{N}\sum_{k=0}^{\frac{N-2}{2}} \mathrm{Re}\{V[k]\}\cos\left(\frac{2\pi k n}{N}\right) - \frac{2}{N}\sum_{k=0}^{\frac{N-2}{2}} \mathrm{Im}\{V[k]\}\sin\left(\frac{2\pi k n}{N}\right). \tag{13}$$

Comparing expressions (12) and (13), we may write $k=1,\ldots,(N-2)/2$.

$$a_0 = \frac{V[0]}{N}, \; a_{N/2} = \frac{V[N/2]}{N}, a_k = \frac{2.\operatorname{Re}\{V[k]\}}{N} \cdot b_k = -\frac{2.\operatorname{Im}\{V[k]\}}{N}, \tag{14}$$

If N is odd,

$$a_0 = \frac{V[0]}{N}, \; a_k = \frac{2\operatorname{Re}\{V[k]\}}{N}, \; b_k = -\frac{2\operatorname{Im}\{V[k]\}}{N}, \text{ for } k=1,...,\frac{(N-1)}{2}, \tag{15}$$

$$\sqrt{a_k^2 + b_k^2} = \frac{2}{N}|V[k]|. \tag{16}$$

4 DTMF Decoding via AFT

In the DTMF system, each key-press generates the sum of two audible tones. When a DTMF signal is received, its frequency content is analysed to identify which digit was transmitted. This analysis can be made by computing its DFT. The magnitudes of the eight components associated to the frequencies nearest the DTMF frequencies are observed. Then, decoding is accomplished by selecting the two higher components. Typically, either a radix-2 FFT algorithm is employed, or the Goertzel algorithm is used when computing only a few components of the DFT.

4.1   The Sampling Frequency and the Length of the Transform

Spectrum analysis of a discrete-time signal by the AFT requires a choice of parameters that are relevant to the accuracy and computational efficiency of the process. The sampling rate of the original continuous signal is one such factor. We set this frequency to 8 kHz, a standard value of PCM-based telephone systems. Another parameter is the transform length N. A suitable value for N, one that allows the detection of the eight DTMF frequencies, should be selected. The relation

$$k = f_k N / f_s, \tag{17}$$

where f_s is the sampling frequency, gives the index k of the coefficient that corresponds to the detected frequency f_k. In general, it returns a noninteger value for k, which is rounded. It is thus possible to define a relative error measure by

$$E_{R,k} = \left| f_k - \tilde{f} \right| / f_k, \tag{18}$$

where $\tilde{f}$ is the rounded frequency. We calculate the mean relative error, $\bar{E}_R$, by averaging out $E_{R,k}$ over the DTMF frequencies. Higher quality detection can be achieved by using the transform length that minimises $\bar{E}_R$. An exhaustive search procedure leads to $N=114$.

4.2   Applying the Arithmetic Fourier Transform

Equation (17) provides the coefficient indexes necessary to decoding. Equation (8) shows the Bruns averages needed to compute such coefficients (table 1). From (6), values of the signal for fractional times must be known, what requires interpolation. The type of interpolation affects on the error and the cost of the algorithm. The decoding was implemented using linear interpolation. The coefficients of the indexes k shown in table 2, were computed for each of the 16 DTMF signals.

Table 1. DTMF frequencies, indexes of coefficients and Bruns averages, $f_s = 8$ KHz, $N = 114$.

f(Hz)	k	Bruns averages	f(Hz)	k	Bruns averages
697	10	$B_{20}, B_{60}, B_{100}, B_{140}, B_{220}$	1209	17	B_{34}, B_{102}, B_{170}
770	11	$B_{22}, B_{66}, B_{110}, B_{154}$	1336	19	B_{38}, B_{114}, B_{190}
852	12	$B_{24}, B_{72}, B_{120}, B_{168}$	1477	21	B_{42}, B_{126}, B_{210}
941	13	$B_{26}, B_{78}, B_{130}, B_{182}$	1633	23	B_{46}, B_{138}

The key point of adopting an AFT-DTMF decoding is its low complexity. The number of floating-point multiplications and the number of additions was computed following [5]. Table 2 presents a comparison between the complexity of the proposed method and Goertzel algorithm [6].

Table 2. Computational Complexity: Simplified AFT, Goertzel algorithm, $f_s = 8$ kHz, $N = 114$.

Operation	Simplified AFT	Goertzel
Multiplication	56	904
Addition	~ 4500	1800

The higher number of additions needed in the simplified AFT is fully compensated for the difference in the number of multiplications. A similar result is observed when comparing the AFT with the Cooley-Tukey FFT. For $N=128$, the FFT requires 712 multiplications and 2504 additions [6].

4.3   An Error Measurement

The interpolations required to compute Bruns averages are responsible for errors in the estimates of a_k and b_k. In order to evaluate this inaccuracy, we use equations (14) and (15) to estimate the tones that identify a digit in the DTMF keypad. For instance, for the signal that represents the digit "1", we estimate $V[10]$ and $V[17]$. The errors in these components are computed, relatively to their exact (DFT) values:

$$E_k = \left\| |V[k]| - |\tilde{V}[k]| \right\| / |V[k]|, \tag{19}$$

where $|\tilde{V}[k]|$ is the estimate of the DFT component of index k, and $|V[k]|$ is its exact value. After computing E_k for each component, we cluster those that correspond to a same frequency and calculate the arithmetic mean, $\bar{E}_k$. Table 3 shows DTMF frequencies, and the associated estimation mean error. The error presented in table 3 significantly varies with the sampling frequency and transform length. The same happens with the complexity of the algorithm. It is thus possible to vary f_s and N, establishing a trade-off between these parameters so as to suit specific project needs.

Table 3. DTMF frequencies, coefficient indexes and estimation mean errors, f_s=8 kHz, N=114.

f (Hz)	697	770	852	941	1209	1336	1477	1633
K	10	11	12	13	17	19	21	23
$\bar{E}_{Rn} \times 10^{-2}$	3.86	6.48	4.19	10.82	4.51	7.30	8.13	5.87

5 Conclusions

This paper offered a new tool for decoding DTMF signals: the Arithmetic Fourier Transform. Some theoretical and practical aspects of the method were discussed, and results concerning the algorithm computational complexity were presented. The AFT decoding presents a low number of floating-point multiplications when compared to existing algorithms for the same application, allowing a faster DTMF decoding. Details for further simplifying the decoding algorithm are currently under investigation, such as, for instance, to deal only with the most significant Bruns partial averages to compute a Fourier coefficient.

References

1. Bruns, H.: Grundlinien des Wissenschaftlichenen Rechnens. Leipzig (1903)
2. Wintner, A.: An Arithmetical Approach to Ordinary Fourier Series. Baltimore, privately monograph, (1947)
3. Tufts, D.W., Sadasiv, G.: The Arithmetic Fourier Transform. IEEE ASSP Mag. (1988) 13-17
4. Reed, I.S., Tufts, D.W., Yu, X., Truong, T., Shih, M.-T., Yin, X.: Fourier Analysis and Signal Processing by Use of the Möbius Inversion Formula. IEEE Transactions on Acoustics, Speech and Signal Processing. Vol. 38. (1990) 459-470
5. Reed, I.S., Shih, M.-T., Truong, T.K., Hendon, E., Tufts, D.W.: A VLSI Architecture for Simplified Arithmetic Fourier Transform Algorithm. IEEE Transactions on Signal Processing. Vol. 40. (1992) 1122-1133
6. Blahut, R.E.: Fast Algorithms for Digital Signal Processing. Addison-Wesley, Reading (1985)

Adaptive Echo Cancellation for Packet-Based Networks

Victoria Stewart, Colin F.N. Cowan, and Sakir Sezer

School of Electrical & Electronic Engineering
Queen's University Belfast
Ashby Building, Stranmillis Road,
Belfast, BT9 5AH, Northern Ireland

Abstract. Echo cancellation is essential if the transmission of voice data is to be of an acceptable quality over networks which introduce a significant end-to-end delay. A significant echo cancellation problem for this type of network is due to the fact that the delay is variable. This means that the canceller needs to track the delay and to reconverge for each change quickly enough to maintain efficiency in its control of the echo. This paper describes a system for dealing with abrupt changes in the network impulse response. The proposed model consists of an arrangement of echo cancellers, each referring to a particular delay, with the most appropriate canceller coefficients being chosen throughout the simulation as the system converges to the changing delay.

Keywords: Echo cancellation, packet-based transmission, adaptive algorithms, adaptive filtering, and variable delay.

1 Introduction

As the Information Age progresses, the need to communicate and to transfer data is quickly increasing. In fact, the rate at which data transfer is growing is much greater than that for any other type of service and already data transmission takes up more than 50% of the network usage since 2000. Though the Public Switched Telephone Network (PSTN) is probably the most wide-spread and dominating communication media at present, new technologies are being established that will support both the current and future communication demands of a range of services, such as voice, video, data, etc.. Packet-based methods of transmission aim to provide a method of communication for all these services, as well as an economic QoS and an efficient use of network resources as expected from an integrated network [1], [2].

Certain packet-switched technologies convey information in fixed length packets [3], maintaining a good balance between the benefits of packet-switching and of circuit switching, e.g. Voice over IP, [4]. By using statistical multiplexing, these packet-based transmissions are capable of making the most efficient use of the available bandwidth, giving a flexibility that circuit switching cannot provide. However, this statistical multiplexing, with fixed or variable packet size, contributes additional delay components that circuit switching does not. These delay components are due to queuing, packet scheduling, packetisation, as well as other general delays across the network. To make matters more complicated still, packet delay variation or

J.N. de Souza et al. (Eds.): ICT 2004, LNCS 3124, pp. 516–525, 2004.

jitter is caused mostly by queuing and packet scheduling at each network node, such as switches and routers.

These delays have a significant impact on the controlling of echo, which is caused by signal reflections at various points in the network. For a telephone call made over the PSTN, i.e. a circuit switched network, the time taken to hear the echo is constant throughout the duration of the call. However, for a call made over a packet-switched network, the echo delay can significantly vary over the duration of a single call. This variation in the delay causes difficulties with the convergence of the adaptive algorithm in the echo canceller and consequently the echo is not necessarily reduced efficiently.

2 Echo Cancellation

The basic principle of echo cancellation is, 'to remove the echo, subtract it!' An estimate of the echo is created through the use of an adaptive filter, Figure 1. This estimate is then subtracted from the actual echo in the system. The error between these is then used to update the coefficients of the filter through the calculation of an adaptive algorithm. As the filter coefficients approach their convergence values, the estimated echo models the actual echo and the error is minimised. Thus the echo in the network is reduced. This is the basic operation of an echo canceller, [5]. There are other features associated with echo cancellation such as non-linear processing, addition of comfort noise, single-talk and double-talk situations but this paper is only concerned with the fundamentals of reducing echo.

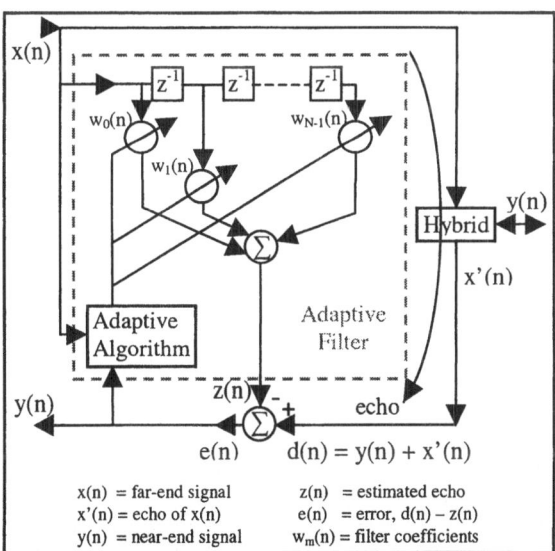

Fig. 1. Simple echo canceller

For a network with a constant delay, the size of the filter is determined by the time taken to hear the echo, i.e. the round-trip delay. For example, if the delay is 64ms and the system is running at a sampling rate of 8kHz, then a 512 tap filter would be needed, [6]. Once the echo canceller is switched on, the adaptive filter will converge to the optimal tap values for the constant delay, giving a good reduction in the echo, [7].

However, for a network with a variable delay, the size of the filter needed is best determined by the maximum delay expected.   Again, once the echo canceller is switched on, the adaptive filter will begin to converge to the optimal tap values for the current delay.   Nevertheless, when the delay changes, the canceller begins to re-converge for the new delay from the tap values obtained for the previous delay.   So unless the delay varies infrequently, the filter coefficients never manage to converge to their optimum and the echo is not efficiently reduced.

3   Proposed Echo Canceller System

The model considered consists of a system of echo cancellers (ECANs) in parallel, Figure 2, [8]. The cancellers are identical; they are each made up of adaptive filters (FIR filter and algorithm) of the same length and use the same adaptive algorithm. The only difference between them is that they are initialised for different delays; the centre canceller is set for a delay time of τ ms., the canceller immediately before this one is set for one time period before τ ms., the canceller after the centre one is set for one time period after τ ms. and so on with each ECAN being separated from its neighbours by one time period.

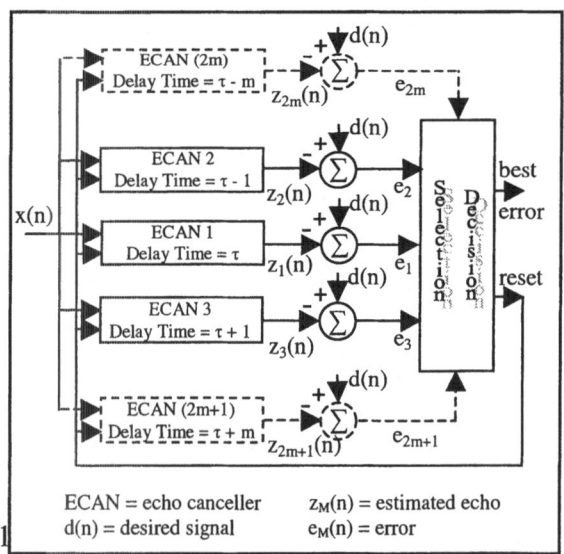

Fig. 2. Adaptive echo canceller system

Each ECAN will be converging towards the same delay (the current network delay) but only one of them will be starting from a excellent coefficient estimate.

Consequently this ECAN will produce the best cancellation at this particular time and subsequently will be considered to be the master ECAN. The master ECAN maintains the 'centre' position with the other ECANs being reset so they are converging towards delays falling on either side of the delay setup of the master. Thus, there is a virtual shifting of position of the ECANs, [9].

To determine which ECAN is currently giving the best reduction in echo and should be considered to be the master, requires a selection process to be performed, [9]. This decision is made based, very generally, on the smallest mean square error (MSE) taken over a specified number of samples, i.e. over a specified number of iterations of the algorithms. The selection process is as follows:

- Determine if a change in delay has occurred.
- Initiate 1st reset of the ECANs.
- Select the ECAN performing best for this new delay.
- Initiate 2nd reset of the ECANs.

The change in delay is determined by comparing the present and immediate past MSE values of the current master ECAN. If the present MSE value is greater than the previous MSE by a pre-determined tolerance level, then this indicates that there has been a disruption to the convergence of the filter coefficients caused by a change in the delay. Using the MSE rather than the straight error value provides a very stable comparison test due to the MSE being an average, thus giving a smoothed error value. This enables an approximate rather than highly precise tolerance level to be used accurately during this delay change determination.

It is necessary to stipulate a condition wherein the delay change determination is not carried out more than once during the same delay period. This is in order to stop an oscillation between master ECANs, particularly during the early iterations after a correct virtual master switch has occurred. This is simply performed by setting a condition that the comparison test cannot be carried out within a specified number of iterations after the delay change has been detected. This ensures that the system does not incorrectly cause an ECAN reset.

The first reset of the ECANs occurs when a change in delay has been established. During this 1st reset, the coefficients are reset to values obtained during an initialisation stage but are shifted to correspond to the current delay (and not the newly established delay). Thus, since at this point, the ECANs do not change their delay set-up, the centre ECAN will not actually be the master (the virtual switch will occur in the 2nd reset). Also during this reset, the error value to all the adaptive algorithms is reset to 0, with the stored error values used in the calculation of the MSE being cleared except for the current error value which is kept. All these procedures enable the ECANs to start converging for the new delay from equally viable positions. However, only one of the ECANs will quickly give a reasonably small error, thus indicating the ECAN that is actually set up with a delay corresponding to the new delay.

The actual selection of the new master ECAN involves a number of different steps and tests to be carried out in order, as well as a determination as to whether the selection can be made directly at the time of the 1st reset or at a slightly later stage. At the 1st reset, the ECAN selection can be made via one of two routes, i.e. if either route is true then the selection can be made and the system can proceed to the 2nd reset.

Route 1 requires a comparison between two basic selection methods. If in each case the same ECAN is chosen then this ECAN will become the new master. The first basic selection chooses the best ECAN based on the smallest difference between the pre-past MSE value and the present MSE value, each relevant to a specific ECAN, where the pre-past value refers to the MSE obtained two iterations previously. The purpose of this calculation is to reduce the effect caused by the delay change which occurred on the previous iteration and thus to produce an indication of which ECAN was set up to converge for the new current delay. The second basic selection chooses the ECAN which is producing the smallest magnitude of error on this iteration, excluding the current master ECAN. The results from these tests need to agree for a new master ECAN to be conclusively selected through Route 1, as the effect on the error and MSE caused by the delay change do not always follow the ideal expectations.

Route 2 compares the similarity between the current errors of the slave ECANs to the previous error from the master ECAN and also to each other slave ECAN. If the errors between the slaves are to similar then no decision on a new master ECAN can be made. Otherwise, the slave ECAN error that is significantly different from the master ECAN's previous error indicates the choice for the new master ECAN. The previous error from the master ECAN refers to when the delay change occurred and shows this by a sudden large increase in value, thus providing a 'worst-case error' for comparison. Calculations are made of the magnitude of the difference between this 'worst-case error' and each of the slave ECAN errors. If all the difference values are less than a specified tolerance level or all the difference values are greater than the specified tolerance level, then the slave ECANs are producing errors that are to similar. Thus no decision on the ECAN selection can be made. Otherwise the slave ECAN with the largest difference is chosen and the 2nd reset is initiated.

If no selection has been made via either Route 1 or Route 2, then the ECANs are allowed to continue converging as they are, for a few more iterations, until a conclusive decision can be made. This allows the ECANs to settle into the convergence and thus it will be more obvious as to which ECAN is producing a relevant smaller error. Again, the similarity between the errors being produced by the slaves is monitored and if the differences between these errors are not above a specified tolerance level then no decision is made. Otherwise, the selection is made simply on the smallest MSE of the slave ECANs.

Upon the selection of a new master ECAN, the system proceeds to the 2nd reset; the ECANs are reset to coefficient values obtained during an initialisation stage but which are shifted to correspond to the relevant new delays for each ECAN. The error value to each of the adaptive algorithms in each of the ECANs is reset to 0.

The system continues to monitor the reduction in echo and performs the procedure described above in anticipation of further changes in delay necessitating a virtual switch of master ECAN.

Note: The initialisation stage enables a set of coefficients to be obtained (for a known delay) that are closer to their optimal values for the network than a set of general tap-values. Thus these coefficients provide a set of best-estimate values that can be used when resetting the ECANs when a delay change and corresponding ECAN switch has occurred.

4 Simulations

The adaptive echo canceller system was simulated in Matlab and Simulink. Each of the separate functions of the system were written in 'C' code and incorporated as S-Functions, thus enabling the system to be built up in Simulink. This provided a graphical interface which could model and simulate the system. For the purpose of this paper, the proposed adaptive echo canceller system consists of only three echo cancellers in parallel. This is the simplest form of the system and will be known as the Tri-ECAN system (or N-ECAN system, N referring to the number of ECANs in parallel). Thus the results included here refer to a Tri-ECAN system.

The input signal, i.e. the 'original speech signal' was generated as a random binary (0,1) sequence. Each of the echo cancellers are assumed to be Finite Impulse Response (FIR) filters, each having 256 coefficients. The first canceller was set to one time period before the initial delay value, the second to the initial delay value and the third to one time period after the initial delay. The initial delay value is chosen as some number of filter taps, obviously within the size of the filters being used (256 taps, in this case). The time period is set to the packetisation delay, as due to the management of packets through the network, the packet delay varies in multiples of the packetisation delay. The echo cancellers all use the Normalised Least Mean Squares (NLMS) algorithm given by Equation (1):

$$\mathbf{W}(n+1) = \mathbf{W}(n) + \frac{2\mu e(n)\mathbf{X}(n)}{\mathbf{X}(n)'\mathbf{X}(n) + \Omega} \tag{1}$$

where $\mathbf{W}(n+1)$ = new coefficient values,
   $\mathbf{W}(n)$ = past coefficient values
   $e(n)$ = error
   $\mathbf{X}(n)$ = input signal
   μ = step size constrained by $0 < \mu < 1$
   Ω = some very small value to prevent a
       divide by zero.

The NLMS is the most common industrially used algorithm for echo cancellation as it performs well for speech signals, gives a stable result and has a fast convergence, [4], [5], [6]. The NLMS algorithm follows a stochastic gradient approach to finding the optimal solution by minimisation of the MSE, with normalisation of the input signal. Speech signals have a varying signal level, i.e. they are non-stationary and normalisation gives a much better performing adaptive algorithm for such signals. Noise was added into the echo path at –40db. The performance of the echo canceller system was measured by calculating the Echo Return Loss Enhancement (ERLE) for the master canceller throughout the simulation, [10], [11]. The ERLE is given by Equation (2):

$$\text{ERLE (db)} = 10\log\frac{P_{echo}}{P_{error}} \tag{2}$$

where P_{echo} = echo power
     P_{error} = error power

The ERLE provides a figure of merit for determining how effective is the echo cancellation process. It assumes that there is always a certain amount of loss incurred by echo and then shows the rate of improvement after echo cancellation. A good echo canceller will output a very large steady state ERLE in a short convergence time.

The system was initially allowed to converge before a delay change occurred, that is the simulation was run for 10,000 iterations before the delay was varied. This enabled the optimal coefficient values to be obtained for a particular delay and these were used when resetting the cancellers throughout the simulation. This meant that the cancellers were starting from a 'best estimate' position each time.

The simulation was run for different random binary sequences and different random delay sequences, through the use of different seeds in Matlab/Simulink. The proposed adaptive echo canceller system was also compared to a system with only one echo canceller. This was simulated in the same way.

5   Results

The results shown, for both the Tri-ECAN system and the single ECAN system, highlight a section of the simulation run, iterations 37,000 to 41,000. Plots showing the whole simulation run tend to have too much detail and it is difficult to make clear observations. However the trends shown in the focused-upon sections published in this paper, are observed throughout the whole simulation.

Figure 3 refers to the Tri-ECAN system and shows the MSEs for each of the three ECANs over the selected section of the simulation. Also shown is how the delay changes over this section.

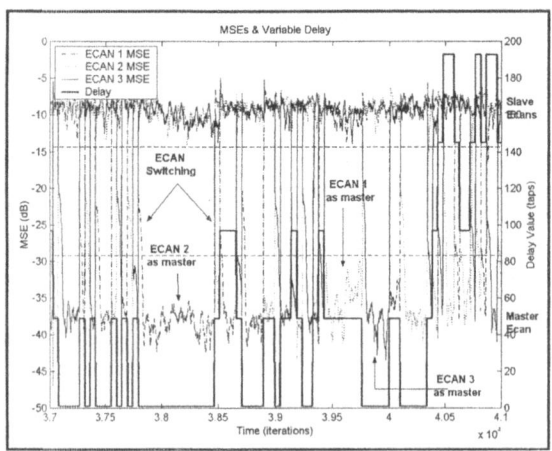

Fig. 3. Tri_ECAN: MSE for all three ECANs.

The upper horizontal MSE refers to the MSEs of the two respective slave ECANs and ranges approximately between −7dB and −12dB. The lower horizontal MSE refers to the MSE of the respective master ECAN and ranges approximately between −34dB to −41dB. Thus there is an obviously distinction between the performance of the slaves and the master with the master providing a significant improvement. The vertical lines on Figure 3 linking the upper and lower lines of the MSE refer to when there is the virtual switching to a new master ECAN. These lines correctly correspond to the change in delay and this fact can be observed in Figure 3. Thus from this figure, it can be seen that the Tri-ECAN system correctly switches and the current master ECAN maintains a low MSE level while the slave ECANs perform poorly.

Figure 4 illustrates the behaviour of the MSE from a single ECAN for the same delay changes as supplied to the Tri-ECAN system. From this figure, it can be observed that during the first 10,000 iterations that the MSE progressively reduces (improves). This section refers to the initialisation period wherein the delay remains constant. Thus it would be expected that the single ECAN would be converging towards optimal coefficient values and a small MSE.

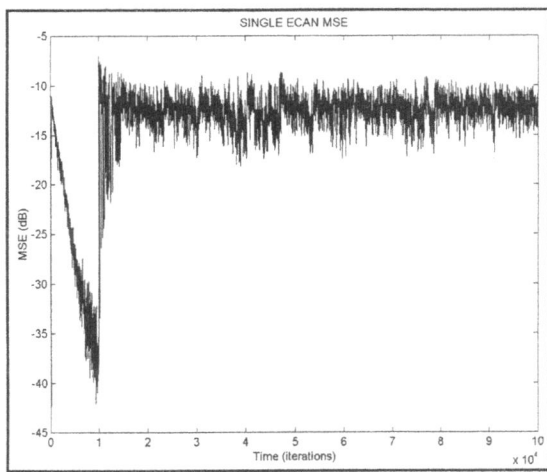

Fig. 4. Single ECAN:MSE over full simulation

This is confirmed by the results for this section shown on Figure 4. When the delay changes, after 10,000 iterations, the MSE for the single ECAN deteriorates as the coefficients are now incorrect for the current delay. This is shown on Figure 4 by the sudden increase in the MSE. As the delay continues to change, the single ECAN never recovers and the MSE is maintained at a very poor level. The single ECAN never 'tracks' the delay and does not respond in any way to the delay changes that occur, as can be observed in Figure 5. There is no pattern to the MSE behaviour as the delay changes, though it does indicate that a single ECAN cannot provide an appropriate reduction in echo for a network where the delay may be variable.

In Figure 6, the upper line with downward spikes refers to the ERLE for the Tri-ECAN system. The spikes refer to the occurrences of the delay changes when there is a sudden sharp deterioration in the echo reduction performance of the system, i.e. before the virtual switch to a new master ECAN. However the Tri-ECAN system recovers immediately and returns to giving an ERLE of approximately 29dB (between

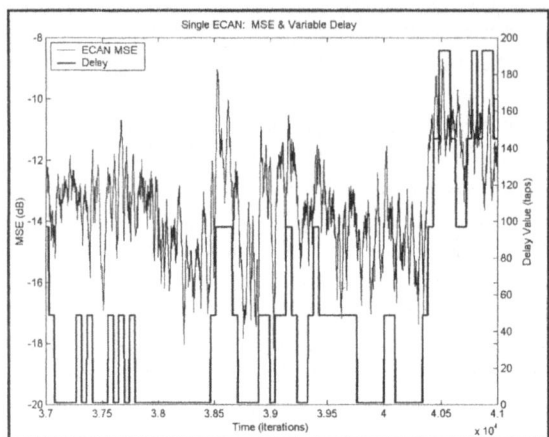

Fig. 5. Single ECAN: MSE & Delay Changes

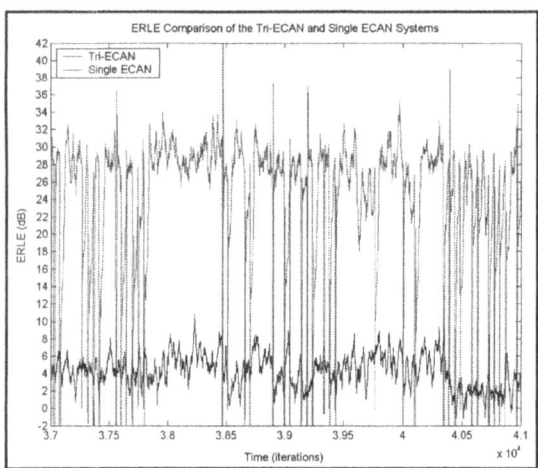

Fig. 6. Performance Comparison.

26dB and 32dB). The ERLE is a comparison of the echoes before and after cancellation and theoretically, for a perfect echo canceller, should increase indefinitely [10], [11], [12].

The performance of the single ECAN is shown as the lower line on Figure 6 and from this it can be observed that it gives an ERLE which is approximately 4dB (between 1dB and 7dB). There is no improvement in the echo reduction performance of the single ECAN throughout the rest of the simulation. So, the ERLE for the single ECAN is, on average, 25dB lower than for the Tri-ECAN system. This is a significant difference and highlights the large improvement in echo reduction that can be made by using the proposed Tri-ECAN system.

6 Conclusions

The results clearly show that for a network where the echo delay is not constant, a classical single echo canceller will not efficiently cancel the echo. The proposed adaptive echo canceller composed of multiple cancellers in parallel is able to adapt to the echo delay variation and achieves an improved echo cancellation. The proposed system is able to maintain a high cancellation performance while sustaining a reasonably constant MSE and ERLE throughout. It has the capability of being able to detect accurately when a change in delay occurs, compared to the classical single canceller which cannot deal with variation of echo delay.

The proposed adaptive echo canceller system is highly capable of efficiently cancelling the echo in a packet switched network where jitter is a significant component of the end-to-end delay.

References

[1]   H. Ahmadi, W.E. Denzel, 'A Survey of Modern high-performance Switching Techniques', IEEE Journal on Selected Areas in Communications, Vol. 7, No. 7 pp 1091-1103, 1989

[2]   B.G. Lee, M. Kang, J. Lee, 'Broadband Telecommunications Technology', Artech House London Publisher 1993, ISBN 089006653X

[3]   J.M. Pitts, J.A. Schormans, 'Intorduction to IP and ATM Deisgn and Performance', Wiley & Sons, 2000, ISBN 047149187

[4]   David J. Wright, 'Voice Over Packet Networks', Wiley & Sons, 2001, ISBN 0471495166

[5]   B. Widrow, S.D. Stearns, 'Adaptive Signal Processing', Prentice-Hall, 1985, ISBN 0130040290

[6]   B. Farhang-Boroujeny, 'Adaptive Filters Theory and Applications', John Wiley & Sons, 1999, ISBN 0471983373

[7]   S. Haykin, 'Adaptive Filter Theory', 3rd edition, Prentice-Hall, 1995, ISBN 013322760x

[8]   V. Stewart, C.F.N. Cowan, S. Sezer, 'Adaptive Echo Cancellation For ATM Networks', 4th World Multiconference on Systemics, Cybernetics and Informatics (SCI 2000) and 6th International Conference on Information Systems Analysis and Synthesis (ISAS2000).

[9]   M.V. Stewart, 'Echo Control For Packet-Switched Digital Telephony', PhD Thesis, Queen's University of Belfast, 2002

[10] F.A. Westall, S.F.A. Ip,'BT Telecommunications Series, Digital Signal Processing in Telecommunications', Chapter 4.3, pp 114-120, Chapman & Hall, 1993, ISBN 0412477602

[11] N. Kalouptsidis, S. Theodoridis, 'Adaptive System Identification and Signal Processing Algorithms', Prentice Hall, 1993, ISBN 0130065455

[12] Youhong Lu, Joel M. Morris, 'Gabor Expansion for Adaptive Echo Cancellation', IEEE Signal Processing Magazine, pp 68-80, March 1999

The Genetic Code Revisited: Inner-to-Outer Map, 2D-Gray Map, and World-Map Genetic Representations*

H.M. de Oliveira[1] and N.S. Santos-Magalhães[2]

Universidade Federal de Pernambuco
[1] Departamento de Eletrônica e Sistemas–Grupo de Processamento de Sinais
Caixa postal 7.800 – CDU, 51.711-970, Recife-Brazil
[2] Departamento de Bioquímica–Laboratório de Imunologia Keizo-Asami–LIKA
Av. Prof. Moraes Rego, 1235, 50.670-901, Recife, Brazil
{hmo,nssm}@ufpe.br

Abstract. How to represent the genetic code? Despite the fact that it is extensively known, the DNA mapping into proteins remains as one of the relevant discoveries of genetics. However, modern genomic signal processing usually requires converting symbolic-DNA strings into complex-valued signals in order to take full advantage of a broad variety of digital processing techniques. The genetic code is revisited in this paper, addressing alternative representations for it, which can be worthy for genomic signal processing. Three original representations are discussed. The inner-to-outer map builds on the unbalanced role of nucleotides of a 'codon' and it seems to be suitable for handling information-theory-based matter. The two-dimensional-Gray map representation is offered as a mathematically structured map that can help interpreting spectrograms or scalograms. Finally, the world-map representation for the genetic code is investigated, which can particularly be valuable for educational purposes –besides furnishing plenty of room for application of distance-based algorithms.

1 Introduction

The advent of molecular genetic comprises a revolution of far-reaching consequences for humankind, which evolved into a specialised branch of the modern-day biochemistry. In the 'postsequencing' era of genetic, the rapid proliferation of this cross-disciplinary field has provided a plethora of applications in the late-twentieth-century. The agenda to find out the information in the human genome was begun in 1986 [1]. Now that the human genome has been sequenced [2], the genomic analysis is becoming the focus of much interest because of its significance to improve the diagnosis of diseases. Motivated by the impact of genes for concrete goals –primary for the pharmaceutical industry– massive efforts have also been dedicated to the discovery of modern drugs. Genetic signal processing (GSP) is being confronted with a redouble amount of data, which leads to some intricacy to extract meaningful information from it [3]. Ironically, as more genetic information becomes available, more the data-mining task becomes higgledy-piggledy. The recognition or comparison of long DNA sequences is often nearly un-come-at-able.

---

* This work was partially supported by the Brazilian National Council for Scientific and Technological Development (CNPq) under research grants N.306180 (HMO) and N.306049 (NSSM). The first author also thanks Prof. G. Battail who decisively influenced his interests

J.N. de Souza et al. (Eds.): ICT 2004, LNCS 3124, pp. 526–531, 2004.

The primary step in order to take advantage of the wide assortment of signal processing algorithms normally concerns converting symbolic-DNA sequences into genomic real-valued (or complex-valued) genomic signals. *How to represent the genetic code?* Instead of using a look-up table as usual (e.g., [4], [5]), a number of different ways for implementing this assignment have been proposed. An interesting mapping from the information theory viewpoint was recently proposed by Battail [6], which takes into account the unbalanced relevance of nucleotides in a 'codon'. Anastassiou applied a practical way of mapping the genetic code on the Argand-Gauss plane [7]. Cristea has proposed an interesting complex map, termed as tetrahedral representation of nucleotides, in which amino acids are mapped on the 'codons' according to the genetic code [8]. This representation is derived by the projection of the nucleotide tetrahedron on a suitable plane. Three further representations for the 3-base 'codons' of the genetic code are outlined in this paper, namely *i*) inner-to-outer map, *ii*) 2D-Gray genetic map, and *iii*) genetic world-chart representations.

2   Process of Mapping DNA into Proteins

Living beings may be considered as information processing system able to properly react to a variety of stimuli, and to store/process information for their accurate self-reproduction. The entire set of information of the DNA is termed as the genome (Greek: *ome*=mass). The DNA plays a significant role in the biochemical dynamics of every cell, and constitutes the genetic fingerprint of living organisms [4]. Proteins –consisting of amino acids– catalyse the majority of biological reactions. The DNA controls the manufacture of proteins (Greek: *protos*=foremost) that make up the majority of the dry mass of beings. The DNA sequence thus contains the instructions that rule how an organism lives, including metabolism, growth, and propensity to diseases. Transcription, which consists of mapping DNA into messenger RNA (*m*-RNA), occurs first. The translation maps then the *m*-RNA into a protein, according to the genetic code [3], [4]. Despite nobody is able to predict the protein 3-*D* structure from the 1-*D* amino acid sequence, the structure of nearly all proteins in the living cell is uniquely predetermined by the linear sequence of the amino acids.

Genomic information of eukariote and prokariote DNA is –in a very real sense– digitally expressed in nature; it is represented as strings of which each element can be one out a finite number of entries. The genetic code, experimentally determined since 60's, is well known [5]. There are only four different nucleic bases so the code uses a 4-symbol alphabet: A, T, C, and G. Actually, the DNA information is transcribed into single-stand RNA –the *m*RNA. Here, thymine (T) is replaced by the uracil (U). The information is transmitted by a start-stop protocol. The genetic source is characterised by the alphabet $N := \{U, C, A, G\}$. The input alphabet N^3 is the set of 'codons' $N^3 := \{n_1, n_2, n_3 \mid n_i \in N, i=1,2,3\}$. The output alphabet A is the set of amino acids including the nonsense 'codons' (stop elements). $A := \{Leu, Pro, Arg, Gln, His, Ser, Phe, Trp, Tyr, Asn, Lys, Ile, Met, Thr, Asp, Cys, Glu, Gly, Ala, Val, Stop\}$. The genetic code consequently maps the 64 3-base 'codons' of the DNA characters into one of the 20 possible amino acids (or into a punctuation mark). In this paper we are barely concerned with the standard genetic code, which is widespread and nearly universal.

The genetic code is a map $GC: N^3 \rightarrow A$ that maps triplets (n_1, n_2, n_3) into one amino acid A_i. For instance, $GC(\text{UAC})=Stop$ and $GC(\text{CUU})=Leu$. In the majority of standard biochemistry textbooks, the genetic code is represented as a table (e.g. [4], [5]). Let $\|.\|$ denote the cardinality of a set. Evaluating the cardinality of the input and the output alphabet, we have, $\|N^3\|=\|N\|^3=4^3=64$ and $\|A\|=21$, showing that the genetic code is a highly degenerated code. In many cases, changing only one base does not automatically change the amino acid sequence of a protein, and changing one amino acid in a protein does not automatically affect its function.

3   The Genetic Code Revisited

Even worthy, ordinary representations for the genetic code can be replaced by the handy descriptions offered in this paper. The first one is the so-called inner-to-outer diagram by Battail, which is suitable when addressing information theory aspects [9]. We present in the sequel a variant of this map by using the notion of the Gray code to systematise the diagram, gathering regions mapped into the same amino acid (figure 1).

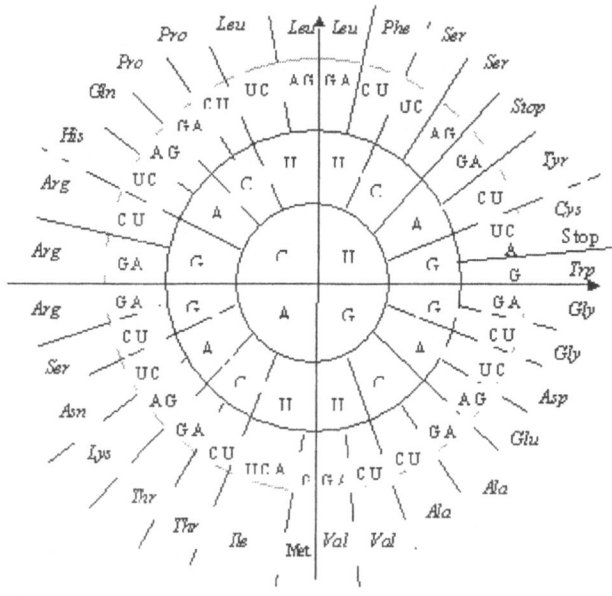

Fig. 1. Variant of Battail's inner-to-outer map for the genetic code [6]. The first nucleotide of a triplet (or 'codon') is indicated in the inner circle, the second one in the region surrounding it, the third one in the outer region, where the abbreviated name of the amino acid corresponding to the 'codon', read from the centre, has also been plotted

Another representation for the genetic code can be derived combining the foundations of Battail's map and the technique for generalized 2-D constellation proposed for high-speed modems [10]. Specifically, the map intended for 64-QAM modulation shown in figure 2 can properly be adapted to the genetic code.

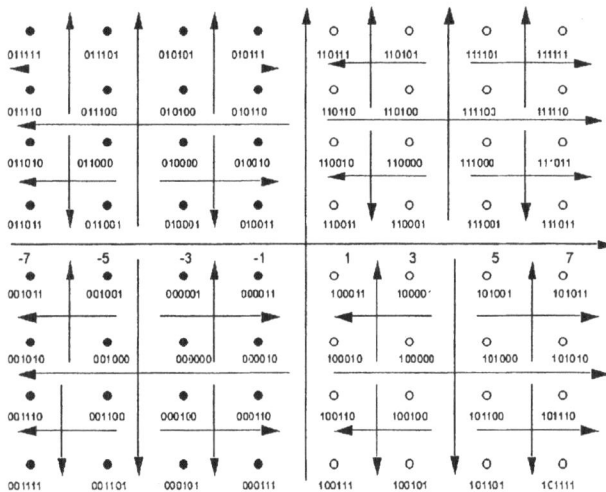

Fig. 2. 2D-Gray bit-to-symbol assignment for the 64-QAM digital modulation [10]

Binary labels are replaced by nucleotides according to the rule ($x \leftarrow y$ denotes the operator "*replace y by x*"): U$\leftarrow$ [11]; A $\leftarrow$ [00]; G $\leftarrow$ [10]; C $\leftarrow$ [01]. The usefulness of this specific labelling can be corroborated by the following argument. The "complementary base pairing" property can be interpreted as a parity check. The DNA-parity can be defined as the sum modulo 2 of all binary coordinates of the nucleotide representations. Labelling a DNA double-strand gives an error-correcting code. Each point of the 64-signal constellation is mapped into a 'codon'. This map (figure 3a) furnishes a way of clustering possible triplets into consistent regions of amino acids (figure 3b). In order to merge the areas mapped into the same amino acid, each of amino acids can be coloured using a distinct colour as in figure 4.

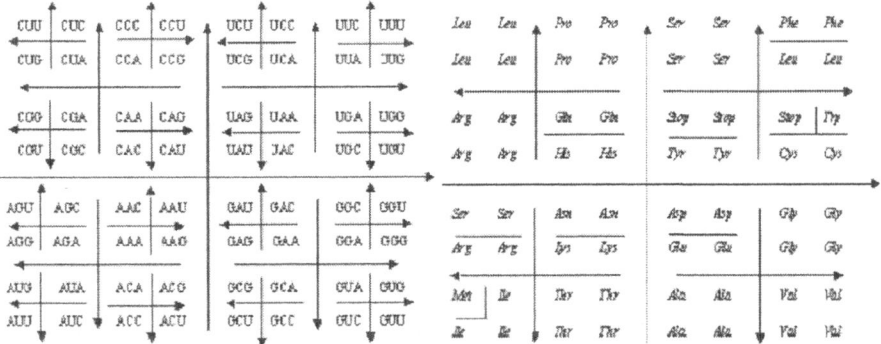

Fig. 3. (a) Genetic code map based on the 2D-Gray map for the 64 possible triplets ('codons'). Each triplet differs to its four closest neighbours by a single nucleotide. The first nucleotide specifies the quadrant regarding the main axis system. The second one provides information about the quadrant in the secondary axis system, and the last nucleotide (the wobble base) identifies the final position of the 'codon'; (b) Genetic code map based on the 2D-Gray genetic map for the 64 possible 'codons' into one of the twenty amino acids (or start/stop)

Val	Ile	Ile	Thr	Thr	Ala	Ala	Val	Val	Ile	Ile
Phe	Leu	Leu	Pro	Pro	Ser	Ser	Phe	Phe	Leu	Leu
Leu	Leu	Leu	Pro	Pro	Ser	Ser	Leu	Leu	Leu	Leu
Trp	Arg	Arg	Gln	Gln	Stop	Stop	Stop	Trp	Arg	Arg
Cys	Arg	Arg	His	His	Tyr	Tyr	Cys	Cys	Arg	Arg
Gly	Ser	Ser	Asn	Asn	Asp	Asp	Gly	Gly	Ser	Ser
Gly	Arg	Arg	Lys	Lys	Glu	Glu	Gly	Gly	Arg	Arg
Val	Met	Ile	Thr	Thr	Ala	Ala	Val	Val	Met	Ile
Val	Ile	Ile	Thr	Thr	Ala	Ala	Val	Val	Ile	Ile
Phe	Leu	Leu	Pro	Pro	Ser	Ser	Phe	Phe	Leu	Leu

Fig. 4. 2D-Gray genetic map for the 64 possible 'codons' into one of the twenty possible amino acids (or punctuation). Each amino acid is shaded with a different colour, defining codification regions on the genetic plane. The structure is supposed be 2D-periodic

Evoking the two-dimensional cyclic structure of the above genetic mapping, it can be folded joining the left-right borders, and the top-bottom frontiers. As a result, the map can be drawn on the surface of a sphere resembling a *world-map*. Eight parallels of latitude are required (four in each hemisphere) as well as four meridians of longitude associated to four corresponding anti-meridians. The Equator line is imaginary, and the tropic circles have 11.25°, 33.75°, 56.25°, and 78.5° (North and south). Starting from a virtual and arbitrary Greenwich meridian, the meridians can be plotted at 22.5°, 67.5°, 112.5°, and 157.5° (East and west). Each triplet is assigned to a single point on the surface that we named as "Nirenberg-Kohama's Earth"[1] (figure 5).

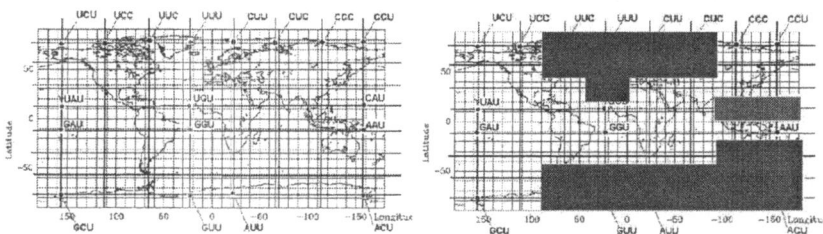

Fig. 5. (a) Nirenberg-Kohama's Earth: Genetic code as a world-map representation. There are four meridians of longitude as well as corresponding anti-meridians comprising four 'codons' each. The eight parallels of latitude (tropics) appear as containing eight 'codons' each. The 'codon' UUU, for instance, has geographic co-ordinates (22.5°W, 78.75°N). The Voronoi region [10] of each triplet can be highlighted according to the associated amino acid colour; (b) Continents of Niremberg-Kohama's Earth: regions of essential amino acid correspond to the land and nonessential amino acids constitutes the ocean. There are two continents (one in each hemisphere), and a single island (the Hystidine island)

If every one of essential amino acids are grouped and said to stand for ground, two continents and a lone island emerge (figure 5b). The remained (nonessential) amino acids materialize the sea. Several kinds of charts can be drawn depending on the criteria used to cluster amino acids [4]. Amino acids can be put together according to

---

[1] In honour to Marshall Nirenberg and M. Gobind Kohama, who independently were the main responsible for cracking the genetic code in three nucleotides ('codons')

their metabolic precursor or coloured by the characteristics of side chains. This approach allows a type of genetic geography. Each one of such representations has idiosyncrasies and can be suitable for analysing specific issues of protein making.

4   Closing Remarks

The innovative representations for the genetic code introduced in this paper are mathematically structured so they can be suitable for implementing computational algorithms. Although unprocessed DNA sequences could be helpful, biologists are typically involved in the higher-level, location-based comments on such strings. Much of signal processing techniques for genomic feature extraction and functional cataloguing have been focused on local oligonucleotide patterns in the linear primary sequences of classes of genomes, searching for noteworthy patterns [3], [7]. GSP techniques provide compelling paraphernalia for describing biologically features embedded in the data. DNA spectrograms and scalograms are among GPS powerful tools [7], [11], *which depend on the choice of the genetic mapping*. The miscellany of maps proposed in this paper supplies further cross-link between telecommunications and biochemistry, and can be beneficial for "deciphering" genomic signals. These maps can also be beneficial for educational purposes, furnishing a much rich reading and visualisation than a simple look-up table.

References

1. Int. Human Genome Sequencing Consortium: Initial Sequencing and Analysis of the Human Genome. Nature **409** (2001) 860-921
2. NIH (2004) National Center for Biotechnology Information, GenBank [on line], Available 05/05/04: http://www.ncbi.nlm.nih.gov/Genomes/index.html
3. Zhang, X-Y., Chen, F., Zhank, Y-T., Agner, S.C., Akay, M., Lu, Z-H., Waye, M.M.Y., Tsui, S.K-W.: Signal Processing Techniques in Genomic Engineering. Proc. of the IEEE **90** (2002) 1822-1833
4. Nelson, D.L., Cox, M.M.: Lehninger Principles of Biochemistry. 3rd ed. Worth Publishers, New York (2000)
5. Alberts, B., Bray, D., Johnson, A., Lewis, J., Raff, M., Roberts, K., Walter, P.: Essential Cell Biology. Garland Pub. New York (1998)
6. Battail, G.: Is Biological Evolution Relevant to Information Theory and Coding? Proc. Int. Symp. on Coding Theory and Applications, ISCTA 2001, Ambleside UK (2001) 343-351
7. Anastassiou, D.: Genomic Signal Processing, IEEE Signal Processing Mag. (2001) 8-20
8. Cristea, P.: Real and Complex Genomic Signals. Int. Conf. on DSP. **2** (2002) 543-546
9. Battail, G.: Does Information Theory Explain Biological Evolution? Europhysics Letters **40** (1097) 343-348
10. de Oliveira, H.M., Battail, G.: Generalized 2-dimensional Cross Constellations and the Opportunistic Secondary Channel, Annales des Télécommunications **47** (1992) 202-213
11. Tsonis, A. A., Kumar, P., Elsner, J.B., Tsonis, P.A.: Wavelet Analysis of DNA Sequences. Physical Review E **53** (1996) 1828-1834

On Subspace Channel Estimation in Multipath SIMO and MIMO Channels

Nikolai Nefedov[1,2]

[1] Nokia Research Center, P.O.Box 407 00045, Finland,
nikolai.nefedov@nokia.com
[2] Helsinki University of Technology, Communications Lab., 02015, Finland

Abstract. Earlier studies showed that iterative processing at receiver may significantly improve performance in wireless communications. In particular, performance of space-time block codes (STBC) extended for frequency selective channels may be notably improved by iterative channel estimation (ICE). On the other hand, the ICE at least doubles the receiver complexity that limits its implementation in handsets. In this paper we address *non-iterative* blind subspace estimation based on the second order statistics (SOS), which is simpler than the iterative receiver. Accuracy and robustness of the SOS (semi-)blind estimators with respect to different channel profiles are evaluated and compared to the ICE. Based on sensitivity analysis appropriate parameters for (semi-)blind estimators are selected. It is found that the SOS semi-blind subspace methods may approach accuracy (in terms of MSE) of the ICE.

1 Introduction

Transmit diversity schemes significantly improve spectral efficiency and currently attract a lot of attention. For example, application of space-time block codes (STBC) extended for frequency selective channels for enhanced GSM evolution (EDGE) may notably improve performance [1][2]. However, most studies addressing transmitting diversity schemes assume perfectly known channel. In practice, the channel is estimated based on known symbols inserted in every transmitted block. On the other hand, the transmit diversity gain is obtained by artificially introduced multipath and it requires more channel parameters to be estimated given the same training sequence. Hence, the transmit diversity gain may not be fully utilized due to degradation of channel estimation accuracy. For example, accuracy of channel estimates based only on GSM training sequences (originally designed for channels without transmit diversity) is not enough in case of the STBC applied for EDGE, which in turn deteriorates the receiver performance. This fact motivated us to consider blind and semi-blind methods where unknown data are used to estimate (identify) the channel.

Blind channel identification is based on channel output signal and some *a priori* knowledge of the statistics of the channel and input signals. These methods may be classified into maximum likelihood (ML) methods and moment-based

J.N. de Souza et al. (Eds.): ICT 2004, LNCS 3124, pp. 532–540, 2004.

methods [6]. The ML methods are based on different versions of the expectation-maximization algorithm that attempts to find the ML solution by iterations between so-called expectation and maximization steps. The iterative receiver for EDGE with/without transmit diversity provides significant gain [3][4], but it comes for the price of extra complexity growing linearly with number of iterations. Another class of blind estimation algorithms, moment-based methods, includes subspace methods (SSM). An attractive property of the SSM (though they are not ML optimal) is that they do not require iterations between a channel estimator and detector/decoder. As a result, a receiver with SSM-based channel estimation is expected to be much simpler than the iterative receiver.

In this paper we address the SSM for channel estimation. After a brief overview of the SOS methods in Section 2, a more detailed description of blind sub-space channel estimation is given in Section 3. Some simulation results are presented in Section 4 followed by conclusions in Section 5.

2 Overview of the SOS Blind Channel Identification

First proposed moment-based blind channel identification methods were based on higher order statistics (HOS). These HOS methods require long statistics collected from a stationary baud-rate (T-sampled) output sequence, and it limits their application in cases where only short data blocks are available. At the same time, it is known that the SOS of T-sampled stationary signals are inadequate for blind identification of non-minimum phase channels. The SOS blind equalization [5] was a breakthrough in the field. Recognizing the existence of excess channel bandwidth, Tong et al [5] suggested fractionally sampled channel identification. By sampling the channel output at higher than baud rate, fractionally spaced samples form a cyclostationary sequence such that its SOS provides channel phase information.

In general, the basic idea of [5] is to introduce additional time or spatial diversity at the receiver. This additional diversity transforms single-input-single-output (SISO) channel estimation problem into single-input-multiple-output (SIMO) problem that may be solved based on the SOS for a large class of channels[1]. An important class of SOS-SIMO algorithms is the SSM. The main idea of the SSM is to exploit the orthogonality of subspaces of certain matrices obtained from (possibly cyclic) SOS of the observations.

Since the SSM introduction, many proposals have been made to improve the estimation both from the computational and statistical point of view (see [6] and references within). In this paper we use the SSM [7] that explicitly utilizes the special structure of channel matrix . At the same time a number attempts has been made to apply blind estimation for wireless communications (e.g.,[11]). However, semi-blind methods, where both known training symbols and unknown data are used together to improve channel estimates [12][13], seem to be a more attractive solution, especially for transmit diversity schemes.

---

[1] There is a class of multipath channels where methods of [5] are unable to correctly identify the CIR [8].

3   Subspace Channel Estimation

3.1   SIMO Signal Model and Subspaces

Subspace blind methods are described in a number of papers (e.g., [5]-[9]). Below the basic SSM method is outlined in a plain way to clarify its implementation. In the following the standard notation is used: small and capital bold letters denote vectors and matrices, $()^H$ and $()^T$ stand for Hermitian and transpose operations; Sylvester (filtering) matrices are presented by calligraphic letters.

Denote signal transmitted at time kT as $s(k)$. In case of SIMO channel multiple outputs may be created either oversampling the received signal with P samples per T, or by P sensors where the received signal is sampled at symbol rate T. Let's consider the oversampling where received $P \times 1$ vector is $\mathbf{y}(k) = \mathbf{x}(k) + \mathbf{w}(k) = \sum_{l=0}^{L} \mathbf{h}_k s(k - l) + \mathbf{w}(k)$.

Stack N observation vectors into $NP \times 1$ vector $\mathbf{Y}(k) = \mathbf{X}(k) + \mathbf{W}(k) = \mathcal{H}_N \mathbf{S}(k) + \mathbf{W}(k)$, where: $\mathbf{X}(k) = [\mathbf{x}(k)^T, \mathbf{x}(k-1)^T, ..., \mathbf{x}(k - N - 1)^T]^T$; $\mathbf{S}(k) = [s(k), s(k - 1), ..., s(k - N - L - 1)]^T$; $\mathbf{W}(k) = [\mathbf{w}(k)^T, \mathbf{w}(k-1)^T, ..., \mathbf{w}(k - N - 1)^T]^T$; $\mathcal{H}_N$ is size $NP \times (L + N)$ convolution Sylvester matrix

$$\mathcal{H}_N = \begin{bmatrix} \mathbf{h}_0 & \mathbf{h}_1 & ... & \mathbf{h}_L & 0 & ... & 0 \\ 0 & \mathbf{h}_0 & \mathbf{h}_1 & ... & \mathbf{h}_L & ... & 0 \\ \vdots & \ddots & \ddots & \ddots & \ddots & \ddots & 0 \\ 0 & ... & 0 & \mathbf{h}_0 & \mathbf{h}_1 & ... & \mathbf{h}_L \end{bmatrix}$$

$\mathbf{h}_k = [h(kP), h(kP + 1), ... h(kP + P - 1)]^T$, $\mathbf{h} = [\mathbf{h}_0^T, \mathbf{h}_1^T, ..., \mathbf{h}_L^T]^T$ is the oversampled channel impulse response (CIR).

Since signal and noise are uncorrelated, the received signal subspace may be separated into signal subspace and noise subspaces, $E\{\mathbf{Y}(k)\mathbf{Y}(k)^H\} = \mathbf{R}_Y = \mathbf{R}_X + \mathbf{R}_W$, such that signal space $\mathbf{R}_X$ is spanned by $(\mathbf{v}_0, \mathbf{v}_{1,...}, \mathbf{v}_r)$, the largest eigenvectors of signal matrix $\mathbf{R}_Y$.

Noise subspace is the orthogonal compliment to the signal subspace. Using the cyclostationarity property of oversampled (or multi sensor) signals the channel may be identified (up to a complex scalar) based on second order statistics $E\{\mathbf{Y}(k)\mathbf{Y}(k)^H\} = \mathbf{R}_Y = \mathcal{H}_N \mathbf{R}_S \mathcal{H}_N^H + \sigma^2 \mathbf{I}$,

where $\mathbf{R}_S = E\{\mathbf{S}(k)\mathbf{S}(k)^H\}$, σ^2 is noise variance.

Channel $\mathbf{h}$ may be identified from $\mathbf{R}_Y$ under conditions that both $\mathcal{H}_N$ and $\mathbf{R}_S$ are full rank. Let $\lambda_0 \geq \lambda_1 \geq ... \geq \lambda_{PN-1}$ be eigenvalues of $\mathbf{R}_Y$. Since $\mathbf{R}_S$ is full rank, then $rank(\mathbf{R}_X) = rank(\mathcal{H}_N \mathbf{R}_S \mathcal{H}_N^H) = L + N$. It follows that $\lambda_i > \sigma^2$ for $i = 0, ..., L+N-1$; and $\lambda_i = \sigma^2$ for $i = L+N, ..., PN-1$. The eigenvectors for signal and for noise subspaces are $(\mathbf{v}_0, \mathbf{v}_{1,...}, \mathbf{v}_{L+N-1})$ and $(\mathbf{g}_0, \mathbf{g}_{1,...}, \mathbf{g}_{PN-L-N-1})$, respectively. Then $\mathbf{R}_Y = \mathbf{V} \, diag(\lambda_0, \lambda_{1,...}, \lambda_{L+N-1})\mathbf{V}^H + \sigma^2 \mathbf{G}\mathbf{G}^H$, where $\mathbf{V} = [\mathbf{v}_0, \mathbf{v}_{1,...}, \mathbf{v}_{L+N-1}]$, $\mathbf{G} = [\mathbf{g}_0, \mathbf{g}_{1,...}, \mathbf{g}_{PN-L-N-1}]$.

By definition, the columns of matrix $\mathbf{V}$ span the signal subspace of dimension $N + L$, while columns of $\mathbf{G}$ span its orthogonal compliment (noise subspace). Columns of $\mathcal{H}_N$ are orthogonal to any vector in noise subspace, $\mathbf{g}_i^H \mathcal{H}_N = 0$ for $0 < i < PN - L - N$. The orthogonality of signal and noise subspaces allows for the identification of the channel vector $\mathbf{h} = [\mathbf{h}_0^T, \mathbf{h}_1^T, ..., \mathbf{h}_L^T]^T$ (up to a complex scalar constant) by minimization of a quadratic form $q(\mathbf{h}) \to min$, where

$$q(\mathbf{h}) \triangleq \sum_{i=0}^{PN-L-N-1} \mathbf{g}_i^H \mathcal{H}_N \mathcal{H}_N^H \mathbf{g}_i = \sum_{i=0}^{PN-L-N-1} |\mathbf{g}_i^H \mathcal{H}_N|^2.$$

To solve this minimization problem the following way may be used:

▶ Partition each of $PN \times 1$ noise eigenvectors into N shorter vectors $\mathbf{g}_i^{(j)}$ ($j = 0, ..., N-1$) of length $P \times 1$, such as $\mathbf{g}_i = [\mathbf{g}_i^{(0)T}, \mathbf{g}_i^{(1)T}, ..., \mathbf{g}_i^{(N-1)T}]^T$, and form $(L+1) \times (L+N)$ filtering matrix $\mathcal{V}_i$, such that

$$\mathcal{V}_i = \begin{pmatrix} \mathbf{g}_i^{(0)} & \mathbf{g}_i^{(1)} & \cdot & \cdot & \mathbf{g}_i^{(N-1)} & 0 & \cdot & \cdot & 0 \\ 0 & \mathbf{g}_i^{(0)} & \mathbf{g}_i^{(1)} & \cdot & \cdot & \mathbf{g}_i^{(N-1)} & 0 & \cdot \\ \cdot & & \cdot & \cdot & \cdot & & & \\ 0 & \cdot & \cdot & 0 & \mathbf{g}_i^{(0)} & \mathbf{g}_i^{(1)} & \cdot & \cdot & \mathbf{g}_i^{(N-1)} \end{pmatrix}$$

▶ Form $P(L+1) \times (L+N)$ matrix $\mathcal{G}_i$ associated with eigenvector $\mathbf{g}_i$ and obtained by stacking P filtering matrices $\mathcal{V}_i$, $i = 0, ..., P-1$; $\mathcal{G}_i = [\mathcal{V}_i^T, \mathcal{V}_i^T, ..., \mathcal{V}_i^T, \mathcal{V}_{P-1}^T]^T$. It may be shown that $\mathbf{g}_i^H \mathcal{H}_N \mathcal{H}_N^H \mathbf{g}_i = \mathbf{h}^H \mathcal{G}_i \mathcal{G}_i \mathbf{h}$ for $i = 1, ..., PN - L - N - 1$ [7]. Hence, the minimization may by obtained as a solution of

$$\mathbf{h}^H \left(\sum_{i=0}^{PN-L-N-1} \mathcal{G}_i \mathcal{G}_i^H \right) \mathbf{h} = \mathbf{h}^H \mathbf{Q} \mathbf{h} \rightarrow \min.$$

▶ To find eigenvalues λ_i of $\mathbf{R}_Y$ and to solve the minimization problems above a convenient way is to use singular value decomposition (SVD).

In derivations above the noise subspace is used such that it is made maximally (in the least square (LS) sense) orthogonal to the estimated signal subspace. Signal subspace method, being a mirror image of the noise subspace method, uses similar linear parameterization of the noise subspace.

In the methods above a plain LS procedure applied such that all terms contributed to the objective function with the same weight. Better estimates may be obtained by solving the overdetermined system in a weighted LS sense [9] where estimation procedures are based on objective functions quadratic in the unknown parameters and are not plagued by existence of local minima.

3.2 Blind Identification of MIMO Channels

Denote transmitted ($M \times 1$) vector over M transmitting antennas at time $t = kT$ as $\mathbf{s}(k) = [s_1(k), ..., s_M(k)]^T$. Received ($P \times 1$) vector at time $t = kT$ is $\mathbf{x}(k) = \sum_{i=0}^{L} \mathbf{H}_i \mathbf{s}(k-i) + \mathbf{w}(k)$, where $\{\mathbf{H}_i\}_{i=0,...L}$ are unknown ($P \times M$) matrices; m^{th} column of $\mathbf{H}_i$ is ($P \times 1$) vector $\mathbf{h}_m^{(i)} = [h_m(iP), h_m(iP+1), ..., h_m(iP+P-1)]^T$; $\mathbf{H}_i = [\mathbf{h}_1^{(i)}; \mathbf{h}_2^{(i)}; ...; \mathbf{h}_M^{(i)}]$.

Stack N observation vectors into $NP \times 1$ vector, such that
$\mathbf{Y}(k) = \mathbf{X}(k) + \mathbf{W}(k) = \mathcal{H}_N(\mathbf{H})\mathbf{S}(k) + \mathbf{W}(k)$,
where $\mathbf{X}(k) = [\mathbf{x}(k)^T, \mathbf{x}(k-1)^T, ..., \mathbf{x}(k-N-1)^T]^T$ is $NP \times 1$ vector;
$\mathbf{S}(k) = [\mathbf{s}(k)^T, \mathbf{s}(k-1)^T, ..., \mathbf{s}(k-N-L-1)^T]^T$ is $(N+L)M \times 1$ vector,
$\mathbf{W}(k) = [\mathbf{w}(k)^T, \mathbf{w}(k-1)^T, ..., \mathbf{w}(k-N-1)^T]^T$ is $NP \times 1$ vector;
$\mathcal{H}_N(\mathbf{H})$ is channel convolution matrix of size $NP \times (L+N)M$,

$$\mathcal{H}_N(\mathbf{H}) = \begin{bmatrix} \mathbf{H}_0 & \mathbf{H}_1 & \dots & \mathbf{H}_L & 0 & \dots & 0 \\ 0 & \mathbf{H}_0 & \mathbf{H}_1 & \dots & \mathbf{H}_L & \dots & 0 \\ \vdots & \ddots & \ddots & \ddots & & \ddots & 0 \\ 0 & \dots & 0 & \mathbf{H}_0 & \mathbf{H}_1 & \dots & \mathbf{H}_L \end{bmatrix}$$

and $\mathbf{H}_i = [\mathbf{h}_1^{(i)}; \mathbf{h}_2^{(i)}; ...; \mathbf{h}_M^{(i)}]$, $\mathbf{H} = [\mathbf{H}_0^T, \mathbf{H}_1^T, ..., \mathbf{H}_L^T]^T$.

The subspace identification is the same as for the SIMO case in Section 3.1. In particular, covariance matrix built from received data

$\mathbf{R}_Y = E\{\mathbf{Y}(k)\mathbf{Y}(k)^H\} = \mathcal{H}_N(\mathbf{H})\mathbf{R}_S\mathcal{H}_N^H(\mathbf{H}) + \sigma^2\mathbf{I}$,

where $\mathbf{R}_S$ is $M(L + N) \times (L + N)M$ signal correlation matrix; $\mathbf{R}_S = E\{\mathbf{S}(k)\mathbf{S}(k)^H\}$; σ^2 is noise variance.

Channel $\mathbf{H}$ may be identified from $\mathbf{R}_Y$ under conditions that both $\mathcal{H}_N(\mathbf{H})$ and $\mathbf{R}_S$ are full column rank. If $\mathcal{H}_N(\mathbf{H})$ has full column rank, then $PN - M(L + N)$ smallest eigenvalues are equals to σ^2 and the corresponding eigenvectors span the noise subspace (left null space of $\mathcal{H}_N(\mathbf{H})$). Let $\mathbf{A}$ be an invertible $M \times M$ matrix. Then the left null spaces of $\mathcal{H}_N(\mathbf{H})$ and $\mathcal{H}_N(\mathbf{HA})$ coincide, so $\mathbf{H}$ may be identified up to right multiplication to an invertible matrix $\mathbf{A}$.

Note that the subspace identification of frequency-selective SIMO and MIMO channels implies that $\mathbf{R}_S$ (and relevant space-time block codes, STBC, if any) is full rank. Application of the SSM for non-orthogonal STBC designs (where $\mathbf{R}_S$ may be not full rank) and for flat-fading MIMO channels is not straightforward and is addressed in [10].

4   Simulation Results

For preliminary performance evaluation of the described blind SSM we use BPSK transmission of data arranged in blocks similar to GSM, but without training symbols. Different channels with minimum and non-minimum phase channel impulse responses (CIR) are considered. The first task of this preliminary simulations is to evaluate accuracy and robustness of the SOS blind estimators with respect to different channel profiles. Another task is to evaluate sensitivity of the blind SSM to different estimator' parameters and find their appropriate values.

The following criteria are used to characterize quality of channel estimation:
- Estimated channel taps averaged over blocks, $\bar{h}_i = E\{||\hat{h}_i||\}$, where the average $E\{\}$ is made over $N_b=500$ simulated blocks;
- Mean square error (MSE) normalized per channel taps, i.e.

$MSE = \frac{1}{(L+1)||\mathbf{h}||} \sum_{i=0}^{L} E\{||h_i - \hat{h}_i||^2\} = \frac{E\{||(\mathbf{h} - \hat{\mathbf{h}})^H(\mathbf{h} - \hat{\mathbf{h}})||\}}{(L + 1)||\mathbf{h}||};$

- Bias for each channel tap $\delta_i = \frac{|E\{||\hat{h}_i||\} - h_i|}{||\mathbf{h}||} = \frac{|\bar{h}_i - h_i|}{\sqrt{\sum_{i=0}^{L} ||h_i||^2}};$

- Bias averaged over all channel taps $\Delta = \frac{1}{L+1} \sum_{i=0}^{L} \delta_i;$

- Standard deviation (STD) for a channel tap $\omega_i = \sqrt{var\{\hat{h}_i\}};$

- The STD averaged over all channel taps $\Omega = \frac{1}{L+1} \sum_{i=0}^{L} \omega_i$

It is known that blind subspace methods inherently have ambiguity (a complex constant in SIMO channels, unitary matrix **A** in MIMO channels). Amplitude ambiguity is usually resolved by energy normalization. To resolve phase ambiguity one needs some side information, e.g. presence of symbol(s) with known phase or an assumption about the phases of some channel taps. In case of SIMO channel we assume that the first channel tap is a real number.

Channel profiles for oversampled ($P=4$) non-minimum phase channel and its corresponding eigenvalues $\lambda_0 \geq \lambda_1 \geq ... \geq \lambda_{PN-1}$ are presented at upper and lower parts of Fig.1, respectively. As it may be seen from Fig.1, at high SNR=10dB the signal subspace may be clearly separated from noise subspace at $(N + L)$ point even without knowledge about the channel order L. However, at low SNR=3dB the signal and noise subspaces separation may be done only provided with a known (or accurately estimated by other means) channel order. The same is valid for minimum phase channels.

The accuracy (in terms of MSE) of the SOS blind estimation may be improved by providing higher time or spatial diversity. For example, higher oversampling P leads to the lower MSE (Fig.2, Fig.3). For comparison, the MSE of channel estimation based on GSM training sequence and iterative channel estimation (ICE) with the LMS update [4] are shown on the same plots. Another reference used here is the LS channel estimate when *all* transmitted data are known at receiver (marked as direct inverse at Fig.2, Fig.3). These figures show that the SSM method may provide accuracy (in MSE terms) close to the direct inverse. According to [4], 2-3 dB improvement in channel estimation accuracy usually results in 1-2dB improvement in BER. Hence, similar improvement in BER may be expected from the SOS subspace estimators.

Note that complexity of the SSM depends on oversampling P factor and stacking depth N (e.g., the SVD complexity grows as $(PN)^3$). On the other hand, for considered channels we did not find advantages in increasing stacking depth above $N=10$. So the complexity-accuracy trade-off in the SSM mainly depends on the available diversity determined by the oversamping P and channel order L. It is also found that the SSM accuracy is about the same both for minimum-phase and non-minimum-phase channels.

It should be noted that while the SSM can approach the accuracy of the direct inverse in the MSE sense, it may include a bias. Our simulations reveal an estimation bias both for min-phase and on non-min-phase channels. As an example, the upper part of Fig.4 presents estimation bias for different taps δ_i (thin lines) and its averaged value Δ (line with circles) for a non-minimum phase channel as a function of SNR. The STD averaged over channel taps, Ω, is depicted at the same figure below. These results also show that asymptotically the SSM estimation is unbiased and consistent, but at low SNR the biasness of the SSM estimates should be taken into account or compensated via semi-blind estimation. Due to lack of space these methods will be addressed elsewhere.

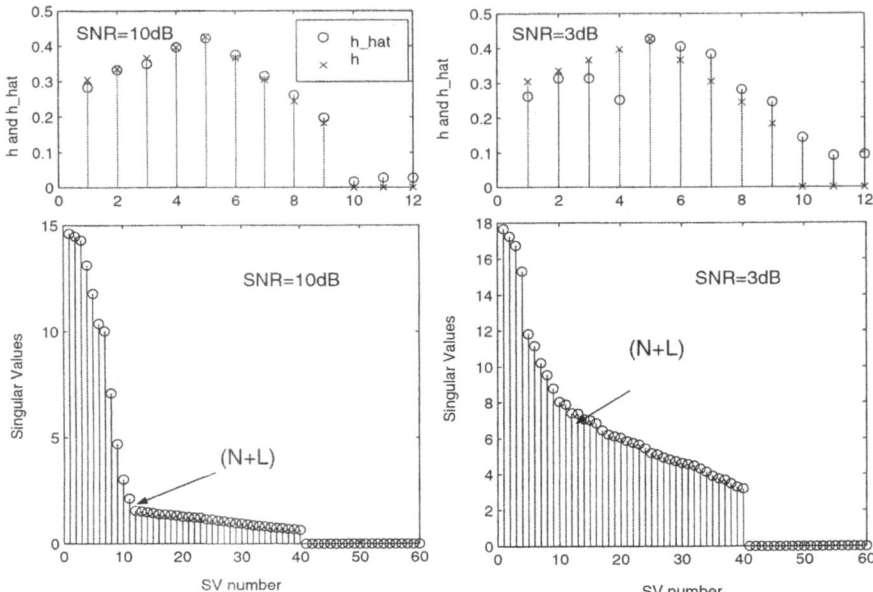

Fig. 1. Oversampled non-minimum phase CIR and its eigenvalues (P=4, N=10)

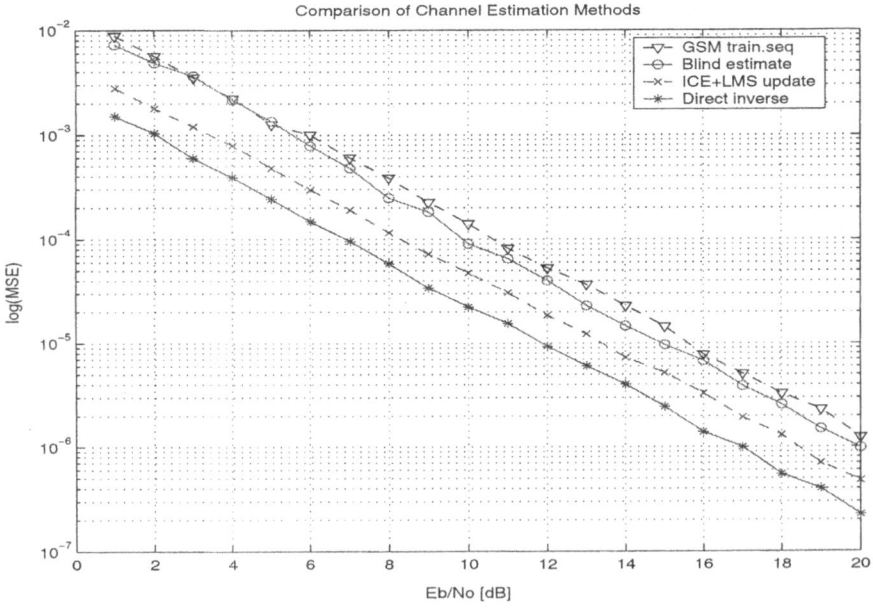

Fig. 2. MSE for minimum phase channel: oversampling P=2; stack length N=10

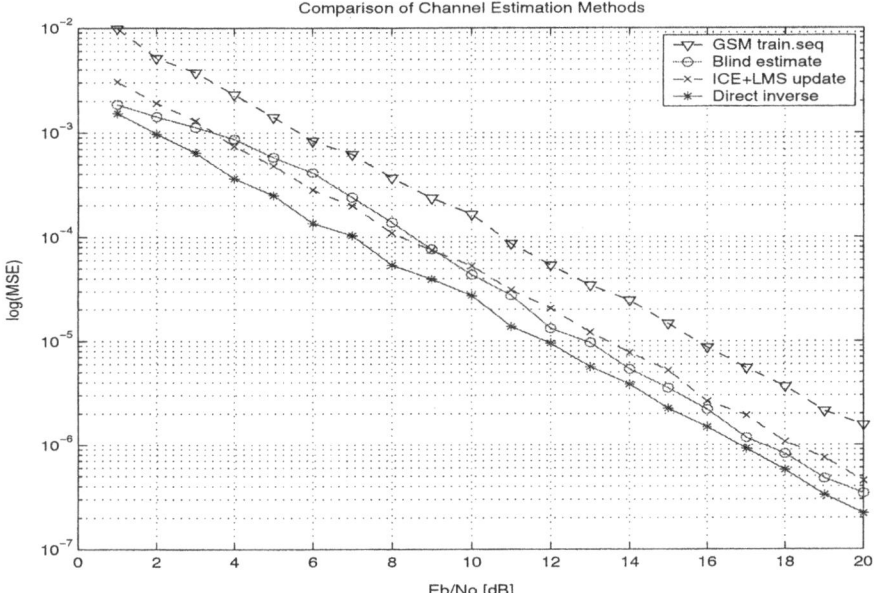

Fig. 3. MSE for non-min-phase channel: oversampling P=4; stack length N=10

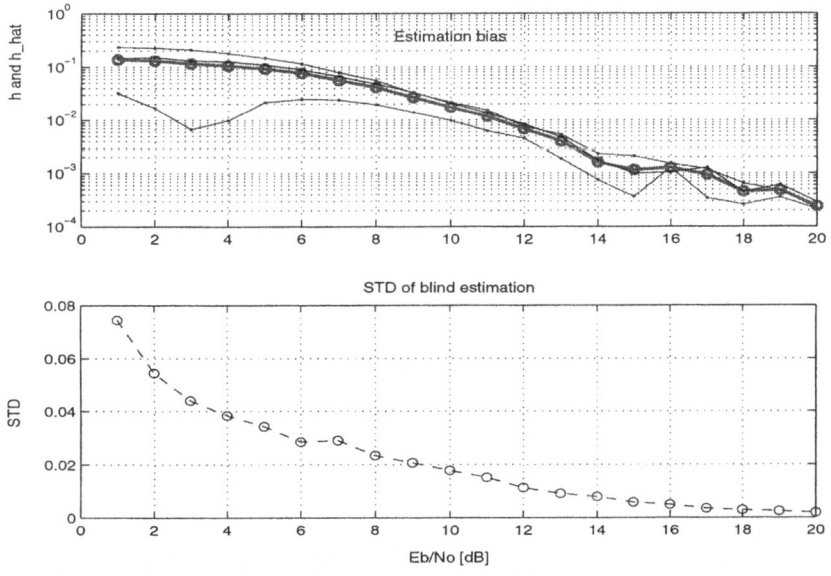

Fig. 4. Estimation bias and the STD. Non-min-phase channel: P=4, N=10

5   Conclusions

The iterative receivers may be a suitable solution for base-stations where complexity and power consumption requirements are not so strict as for mobiles. However, for handsets a less complex solution is highly desirable. In this paper we show that the SOS semi-blind subspace estimators may approach accuracy (in MSE sense) of iterative channel estimation even for relatively short data blocks as used in GSM/EDGE. However, it is found that the channel estimates provided by the considered blind subspace method are biased and it should be taken into account at the detection stage, or compensated via semi-blind estimation. Semi-blind estimators combining data-aided and the SOS blind subspace estimation may be seen as a low complexity solution to improve performance of transmit diversity schemes, especially in handsets. Complexity-performance trade-off of subspace semi-blind methods is the topic of the on-going research.

References

1. Mattellini,G.-P., Kuchi,K., Ranta,P.: Space Time Block Codes for EDGE System. In: Proc. IEEE VTC-Fall (2001) 1235-1239
2. Al-Dhahir,N., Uysal,M., Georghiades,C.N.: Three Space-Time Block-Coding Schemes for Frequency-Selective Fading Channels with Application to EDGE. In: Proc. IEEE VTC-Fall (2001) 1834-1838
3. Nefedov,N., Mattellini,G.-P.: Evaluation of Potential Transmit Diversity Schemes with Iterative Receivers in EDGE. In: Proc. IEEE PIMRC (2002).
4. Nefedov,N., et al: Iterative Data Detection and Channel Estimation for Advanced TDMA Systems. IEEE Trans. on Comm. **51** No.2 (2003) 141-144
5. Tong,L., Xu,G., Kailath,T.: A New Approach to Blind Identification and Equalization of Multipath Channels. In: Proc. IEEE Asilomar Conf.(1991) 856-860
6. Tong,L., Perreau,S.: Multichannel Blind Identification: From Subspace to Maximum Likelihood Methods. Proc. of the IEEE **86** October (1998) 1951-1968
7. Moulines,E. et al: Subspace Method for the Blind Identification of Multichannel FIR filters. IEEE Trans. Sig. Processing **43** Feb. (1995) 516-525
8. Tugnait,J., Huang,B.: Blind Estimation and Equalization of MIMO Channels via Multidelay Whitening. IEEE JSAC **19** August (2001) 1507-1519
9. Abed-Meraim, K. et al: On Subspace Methods for Blind Identification of Single-Input Multiple-Output FIR Systems. IEEE Trans. Sig. Processing **45** January (1997) 42-55
10. Nefedov,N.: Channel Estimation Methods for Space-Time Block Transmission in Frequency-Flat Channels. In: this volume
11. Boss,D., Kammeyer,K.-D., Petermann,T.: Is Blind Channel Estimation Feasible in Mobile Communications Systems. A Study Based on GSM. IEEE JSAC October (1998) 1479-1492
12. Buchoux,V. et al: On Performance of Semi-Blind Subspace-Based Channel Estimation. IEEE Trans. Sig. Processing **48** June (2000) 1750-1759
13. Zhou,S., Muquet,B., Giannakis,G.: Subspace-based (semi-)blind channel estimation for block precoded space-time OFDM. IEEE Trans. Sig. Processing **50** May (2002) 1215 -1228

A Novel Demodulation Technique for Recovering PPM Signals in the Presence of Additive Noise

Paeiz Azmi[1, 3] and Farokh Marvasti[2, 3]

[1] Electrical Engineering Department, Tarbiat Modarres University, Tehran-Iran
[2] Electrical Engineering Department, Sharif University of Technology, Tehran-Iran
[3] Signal Processing and Multimedia Research Lab, Iran Telecom Research Centre (ITRC)
Tehran-Iran

Abstract. In this paper, we propose a novel method for demodulating Pulse Position Modulation (PPM) signals based on Reed-Solomon decoding algorithm. It is shown that the proposed technique outperforms the conventional demodulating method base on Reed-Solomon algorithm in case that the PPM signals are corrupted by Additive White Gaussian Noise (AWGN).

1 Introduction

The basic framework of research into Pulse Position Modulation (PPM) was laid down around 50 years ago but it is only recently that a revival of interest has been experienced with the development of Impulse Radio and Fiber-Optic transmission systems [1], and [2].

In the PPM modulator, constant amplitude pulses are generated at the crossing points of a modulating signal with a sawtooth waveform, so that the PPM signal can be considered as a specific case of non-uniform sampling where the position of the non-uniform samples conveys the signal information. The demodulator of the PPM consists of converting the PPM pulses to a set of non-uniformly spaced samples by multiplying the pulses at the output of matched filter and threshold comparator by the saw-tooth wave; the demodulator then becomes a method for recovering the original base-band signal from this set of irregularly spaced samples.

The easiest way for recovery of the signal from its non-uniform samples is by employing a low pass filter. In this case, in order to achieve satisfactory demodulation, over-sampling has to be introduced, which in turn leads to significant increment of the modulation bandwidth. Reduction of the amount of extra bandwidth can be achieved by using more efficient and sophisticated recovery techniques such as Lagrange, Iterative and Reed-Solomon decoding techniques [3]-[4].

In this paper, we propose a novel method for demodulating PPM signals based on Reed-Solomon decoding technique which outperforms the conventional demodulating method base on Reed-Solomon algorithm [4]-[5] in case that the PPM signals is corrupted by Additive White Gaussian Noise (AWGN).

J.N. de Souza et al. (Eds.): ICT 2004, LNCS 3124, pp. 541–547, 2004.

Our analysis is implemented on the N-dimensional space of complex numbers C^N. In this case, the information signal is a discrete-time signal and is regarded as a block of samples of a continuous signal that is over-sampled at a rate that is the higher than Nyquist rate. In such a system, the PPM pulses correspond to the position of samples that are closest to a discrete sawtooth wave. This extension automatically transforms the aforementioned modulation method to a discrete equivalent. Because the original samples are discrete in time, the time position of the PPM pulses represents a quantized version of the non-uniform sampling amplitudes in the time domain.

In what follows, for this particular discrete modulation method, we propose a new demodulating technique based on the Reed-Solomon to be able to demodulate the PPM signal corrupted by Additive White Gaussian Noise in section 2. In section 3, we will compare the performance of the proposed technique with the conventional one. Finally, section 4 concludes our work.

2 The Proposed Demodulating Technique

As it was stated in the introduction section, the demodulator of PPM signals consists of converting the PPM pulses back to a set of non-uniformly spaced samples by multiplying the pulses by saw-tooth wave and then recovering the original base-band signal from this set of non-uniformly spaced samples. The conversion part of a PPM demodulator is the same in all the demodulation methods. Therefore, we focus on the recovering part.

Let us consider that the discrete transmitted signal $x(n)$ to be a low-pass signal consisting of N real samples which are taken in a frame size of T seconds. Let us also consider that $x(n)$ is a low-pass signal with $2K-1 < N$ nonzero components in the DFT domain, therefore $2K-1$ samples of signal $x(n)$ should be sufficient to determine all the N samples of $x(n)$.

Let $r(n)$ be the signal produced from the received pulse train (original pulse sequence plus added pulses due to channel noise) that has been processed in the receiver stage using hard decision decoding. In this case, $r(n)$ can be decomposed as follows,

$$r(n) = x(n) + e_1(n) - v(n) \tag{1}$$

where $v(n)$ represents the values of $x(n)$ in the position of the samples lost either due to the modulation process or due to the additive channel noise, and $e_1(n)$ represents the noise produced from the false pulses generated due to the additive channel noise. (Note that $r(n)$ is assumed to be zero in the position of missing samples). It should be noted that the value of $v(n)$ is zero when the n-th sample is not lost, and the value of $e_1(n)$ is zero when the n-th sample is lost or when no false pulses is generated due to the channel noise.

Let $r'(n)$ be denoted by,

$$r'(n) = r(n)H\left(e^{j\frac{2\pi}{N}n}\right) \tag{2}$$

where $H\left(e^{j\frac{2\pi}{N}n}\right)$ can be found as follows,

$$H\left(e^{j\frac{2\pi m}{N}}\right) = \prod_{l=1}^{p}\left(e^{j\frac{2\pi m}{N}} - e^{j\frac{2\pi i_l}{N}}\right) \tag{3}$$

and i_l for $\{l=1,2,...,p\}$ denotes the position of the p missing samples either due to the modulation process or due to the channel noise. In this case, we have,

$$r'(n) = \left(x(n) - v(n)\right)H\left(e^{j\frac{2\pi}{N}n}\right) + e_1'(n) \tag{4}$$

where,

$$e_1'(n) = e_1(n)H\left(e^{j\frac{2\pi}{N}n}\right) \tag{5}$$

Furthermore, from (3) it can be seen that $H\left(e^{j\frac{2\pi}{N}n}\right)$ is zero in the position of the p

missing samples. Furthermore, $v(n)$ is nonzero only in the position of missing samples. Thus we have,

$$v(n)H\left(e^{j\frac{2\pi}{N}n}\right) = 0 \qquad \forall n \tag{6}$$

and equation (4) reduces to,

$$r'(n) = x(n)H\left(e^{j\frac{2\pi}{N}n}\right) + e_1'(n) \tag{7}$$

Let R', X', and E_1' be the DFT of the sets r', $x'(n) = x(n)H\left(e^{j\frac{2\pi}{N}n}\right)$, and e_1',

respectively, and let h be found by taking the DFT of H, therefore we have,

$$R' = C' + E_1' \tag{8}$$

where, we have,

$$C' = X' * h \tag{9}$$

In equation (9), "$*$" is the convolution operator.

It is well-known that equation (8) can be represented as follows,

$$r' = c' + e'_1.\tag{10}$$

where,

$$c'(n) = x(n)H\left(e^{j\frac{2\pi}{N}n}\right)\tag{11}$$

and c' is the inverse DFT of C'.

If $r(n)$ has p missing samples, from equation (3), it can be easily shown that we have[1],

$$h(n) = 0 \qquad p+1 \le n \le N-1,\tag{12}$$

On the other hand, $x(n)$ is a real low-pass signal with $2K-1 < N$ nonzero components in the DFT domain, therefore, we have,

$$X'(n) = 0 \qquad K \le n < N-K+1\tag{13}$$

From equations (11), (12), and (13) it can be easily shown that C' has $N - 2K - p + 1$ zeros whose positions are as follows,

$$C'(n) = 0 \qquad K+p \le n \le N-K\tag{14}$$

From equations (8) and (14), it can be seen that the vector E'_1 coincides with the DFT of the vector r'_1 in the positions of the zeros of C'. Based on the theory of complex field Reed-Solomon error-correcting codes [3]-[4], because there are $N - 2K - p + 1$ contiguous zeros in C' (the DFT of c'), the set c' forms a $(N, 2K + p - 1)$ Reed-Solomon code that is capable of removing the added noise to $\left\lfloor \dfrac{N - 2K - p + 1}{2} \right\rfloor$ components of c'.

For removing the added noise to the components of c', using the technique similar to the method discussed in [3]-[4], it can be shown that for the new $(N, 2K + p - 1)$ code we have,

$$\sum_{n=0}^{t} h'(n) \cdot E'(t+r-n) = 0 \qquad r = 0,1,2,...,N-1,\tag{15}$$

where t is the number of errors, $h'(0) = 1$ and $h'(n)$ for $n = 1,2,...,t$ are unknown coefficients. Since $E'(i) = R'(i)$ for i denoting the positions of $N - 2K - p + 1$ consecutive zeros in C', the number of errors and the unknown $e^{j\frac{2\pi}{n}n}$ coefficients ($h'(n)$ for $n = 1,2,...,t$) can be found simultaneously by

---

[1] This is because $H\left(e^{j\frac{2\pi}{n}n}\right)$ is a p order polynomial of $e^{j\frac{2\pi}{N}n}$.

using algorithms such as Levinson-Durbin, and Berlekamp-Massey [6],[7]. Then, the remaining values of E' can be found as follows,

$$\sum_{n=1}^{t} h'(n) \cdot E'(t+r-n) = -E(t+r) \tag{16}$$

In the following, we will apply this algorithm to reduce the degradation in the performance of the proposed technique in the presence of additive channel noise. The procedure of the proposed generalized demodulating algorithm is as follows,

1- Generate $H\left(e^{j\frac{2\pi}{N}n}\right)$ from the positions of missing samples of the received

signal ($r(n)$) and by using equation (3).

2- Form the signal $r'(n)$, by using equation (2).

3- Remove the noise added to the components of c' by using the method discussed in subsection A as follows:

  a- Compute R' the DFT of the signal $r'(n)$.

  b- Compute the number of errors and coefficients $h'(n)$ for $n = 1, 2, ..., t$, based on equation (15) and by applying one of the known algorithms such as Levinson-Durbin or Berlekamp-Massey.

  c- Compute the remaining values of E' by using equation (16).

  d- Remove the error E' form R' to get c'

4- Evaluate the correct values of the recovered pulses at the output of the threshold using equation (11).

5- Recover the lost samples of signal $x(n)$ by using the technique discussed in [4]-[5].

3   Comparison Between the Methods

In this section, we examine the performance of the algorithm proposed in section 2 for demodulating PPM signals under presence of channel noise. We have simulated the proposed Reed-Solomon based algorithm and the conventional technique discussed in [4]-[5] under presence of AWGN. In order to evaluate the performance of the techniques, we have calculated the SNR value for the correlation of 0.95. Table 1

Table 1. SNR values resulting in the correlation value of .95

SNR \ Threshold	.65	.70	0.75	0.80
Conventional Technique	15	15.30	15	15.10
The Proposed Technique	14.30	14.10	14	13.9

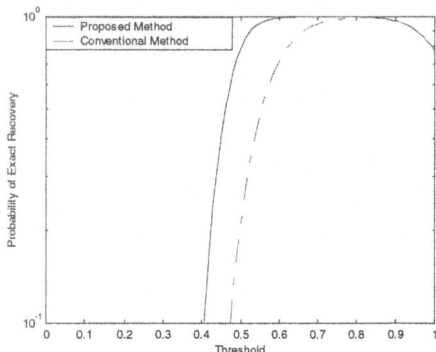

Fig. 1. Probability of recovery for both Reed-Solomon based demodulating algorithms

shows the results. It can be seen that using the proposed technique, the SNR for the correlation of .95 is reduced by about 1dB.

Figure 1 shows the probability of recovery of the transmitted signal for the technique introduced in sections 2 and the conventional technique for the SNR value of 12dB. Our interest is focused on sampling rates of about 1.4 times the equivalent Nyquist rate (N=256 and pm=249). From Figure 1, it can be seen that the proposed technique outperforms the conventional method. But for high threshold values, the probability of generating a false pulse decreases therefore the generalized technique reduces to the technique discussed in [4]-[5].

4 Conclusions

In this paper, we have introduced a demodulating method for PPM signals based on Reed-Solomon decoding technique. We have found out the proposed technique is quite efficient. The simulation results show that the proposed algorithm improves the performance of the conventional method under the presence of additive noise.

References

1. L. Zhao, and A. Haimovich, "Multi-User Capacity of M-ary PPM Ultra-Wideband Communications" *in Proc. IEEE Conference on Ultrawideband Systems and Communications (UWBST) 2002,* Baltimore, Maryland, pp. 175-179, May 2002.
2. B. Wilson, and Z. Ghassemlooy, "Pulse Time Modulation Techniques for Optical Communications: A Review", *IEE Proceeding-J,* vol. 140, no. 6, pp. 346-357, Dec. 1993.
3. F. Marvasti, M. Hasan, M. Echart, S. Talebi, "Efficient Algorithms for Burst Error Recovery Using FFT and Other Transform kernels", *IEEE Trans. on Signal Proc.,* vol. 47, no. 4, pp. 1065-1075, April 1999.
4. F. Marvasti, *Nonuniform Sampling: Theory and Practice,* Kluwer/Plenum Pub. Corp., 2001.

5. D. Meleas, F. Marvasti, 'Signal Recovery from PPM Pulse in the DFT Bandlimited C^N Domain Using Reed-Solomon Decoding", *in Proc. of International Symposium on Telecommunications, IST 2001, Tehran, Iran,* pp.776-779 , September 2001.
6. P. J. S. G. Ferreira, and J. M. N. Vieria, "Locating and Correcting Errors in Images", *in Proc.* IEEE Conference on Image Processing ICIP-97, pp. 691-694, Santa Barbara, USA, Oct 1997.
7. H. M. Zhang, and P. Duhamel, "On the Methods for Solving Yule-Walker Equations", *IEEE Trans on Signal Processing,* vol. 40, no. 12, pp. 2987-3000, Dec. 1992.
8. H. M. Zhang, and P. Duhamel, "On the Methods for Solving Yule-Walker Equations", *IEEE Trans on Signal Processing,* vol. 40, no. 12, pp. 2987-3000, Dec. 1992.

A Novel Iterative Decoding Method for DFT Codes in Erasure Channels

Paeiz Azmi [1,3] and Farokh Marvasti [2,3]

[1] Electrical Engineering Department, Tarbiat Modarres University, Tehran-Iran,
[2] Electrical Engineering Department, Sharif University of Technology, Tehran-Iran
[3] Signal Processing Research Lab, Iran Telecom Research Centre (ITRC), Tehran-Iran

Abstract. One of the categories of decoding techniques for DFT codes in erasure channels is the class of iterative algorithms. Iterative algorithms can be considered as kind of alternating mapping methods using the given information in a repetitive way. In this paper, we propose a new iterative method for decoding DFT codes. It will be shown that the proposed method outperforms the well-known methods such as Wiley/Marvasti, and ADPW methods in the decoding of DFT codes in erasure channels.

1 Introduction

Real/Complex field Reed-Solomon error control codes such as DFT (Discrete Fourier Transform) codes have been considered to recover the missing samples in erasure channels [1]-[5] for several years. For the DFT error control codes, the information vector, K-tuple $\mathbf{u} = \begin{bmatrix} u(1) & u(2) & ... & u(K) \end{bmatrix}^T$, is encoded into an N-tuple $\mathbf{v} = \begin{bmatrix} v(1) & v(2) & \cdots & v(N) \end{bmatrix}^T$, called a codevector (codeword), where we have $N > K$. The encoding procedure of the DFT codes is as follows:

    1- Take the DFT of $\mathbf{u}$ to get a K-tuple $\mathbf{U}$ vector.

    2- Insert N-K consecutive zeros to get an N-tuple $\mathbf{V}$ vector.

    3- Take the inverse DFT to get an N-tuple codeword $\mathbf{v}$.

One of the categories of decoding techniques for DFT codes in erasure channels is the class of iterative algorithms [2]. The block diagram of a conventional iterative algorithm for decoding of DFT codes is shown in Figure 1.

In erasure channels, if $r(n)$ denotes the received signal, we have,

$$r(n) = v(n) + e(n) \tag{1}$$

where $v(n)$ represents the coded signal and the error $e(n)$ is due to lost samples in erasure channels. In the iterative decoding techniques, we have [6]-[7],

$$r_{k+1}(n) = \lambda\left(PSr(n) - PSr_k(n)\right) + r_k(n) \qquad n = 1,2,...,N, \tag{2}$$

or equivalently,

$$r_{k+1}(n) = \lambda PSr(n) + Er_k(n) \qquad n = 1,2,\cdots,N, \tag{3}$$

J.N. de Souza et al. (Eds.): ICT 2004, LNCS 3124, pp. 548–553, 2004.
© Springer-Verlag Berlin Heidelberg 2004

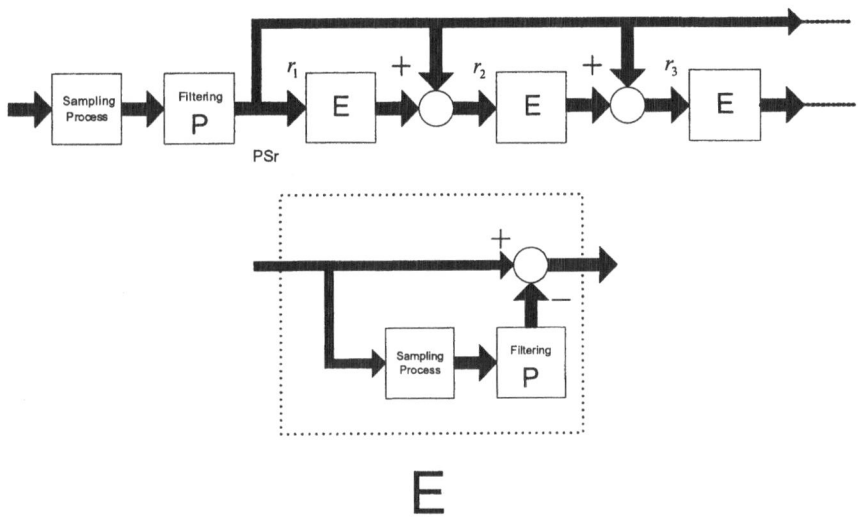

Fig. 1. The block diagram of a conventional iterative algorithm for decoding of DFT codes

where λ and $r_k(n)$ are a convergence constant, and the k-th iteration, respectively. We attempt to set the convergence parameter to a value in $[0,1]$ which would yield the best performance. Furthermore P and S are respectively, the filtering and the sampling and $E = I - \lambda PS$ is the operator taking the input signal to the error signal.

For filtering (F) in the DFT domain, the samples of the received signal in positions in which the zeros were added in the encoding are set to zero. For sampling (S), the various forms of conventional iterative methods use different processes. In the Wiley/Marvasti method, the sampling process is implemented and is given by [1], [6],

$$S_{W/M}r(n) = \sum_{l=1}^{N} r(n_l)\delta(n - n_l),$$
(4)

where

$$S_{W/M}r(n) = \begin{cases} 0 & n = i_l \\ r(n) & n = n_l \end{cases}$$
(5)

where i_l for $\{l = 1,2,...,P\}$ denotes the position of the P missing samples due to erasure channel and n_l for $\{l = 1,2,...,N - P\}$ denotes the position of available samples. It is shown that using each extra iteration, improves the signal-to-noise ratio and potentially the transmitted coded signal can be recovered after infinite iterations [1].

The ADPW method can be considered as a version of Wiley/Marvasti method. In this technique, we have [6],

$$S_{ADPW}x(n) = \sum_{l=1}^{N} w_l x(n_l)\delta(n - n_l),$$
(6)

where,

$$w_l = \frac{n_{l+1} - n_{l-1}}{2}, \tag{7}$$

which may be understood as the length of the Voronoi interval of i. In this case, we have,

$$S_{ADPW} r(n) = \begin{cases} 0 & n = i_l \\ w_l r(n) & n = n_l \end{cases} \tag{8}$$

In the following, we propose a new iterative technique that outperforms both Wiley/Marvasti and ADPW methods. In all the cases, we assume the number of lost samples due to erasure channels is less or equal to the number of consecutive zeros in DFT domain, that is N-K.

2 The Proposed Technique

The block diagram of the proposed algorithm for decoding DFT codes is shown in Figure 2. In this system, we have,

$$r_{k+1}(n) = \lambda \big(FSr(n) - FSr_k(n) \big) + r_k(n) \qquad n = 1,2,...,N, \tag{9}$$

or equivalently,

$$r_{k+1}(n) = \lambda FSr(n) + Er_k(n) \qquad n = 1,2,\cdots,N, \tag{10}$$

where λ, and $r_k(n)$ are a convergence constant and the k-th iteration, respectively. S is the same sampling process which is used in the Wiley/Marvasti method and is

given by $Sr(n) = \begin{cases} 0 & n = i_l \\ r(n) & n = n_l \end{cases}$, and $E = I - FS$ is the operator taking the input

signal to the error signal.

In the proposed method, the linear operator F can be decomposed into the following two linear operators:

1.   A linear operator L that is a $K \times N$ matrix that gives an estimate of the information vector as given below

$$\hat{u} = \arg \Big\{ \underset{u}{\text{Min}} \big(\| Lr - u \|^2 \big) \Big\}, \tag{11}$$

where $\mathbf{r} = \begin{bmatrix} r(1) & r(2) & \cdots & r(N) \end{bmatrix}$.

Because $\mathbf{r} = \mathbf{v} + \mathbf{e} = G\mathbf{u} + \mathbf{e}$, where $\mathbf{e}$ denotes an additive noise which is independent of $\mathbf{r}$, and G is the $N \times K$ generator matrix of the DFT code, we have,

$$L = pinv(G)$$
$$\hat{u} = pinv(G)\mathbf{r} \tag{12}$$

where $pinv(G)$ is the $K \times N$ pseudo-inverse of G.

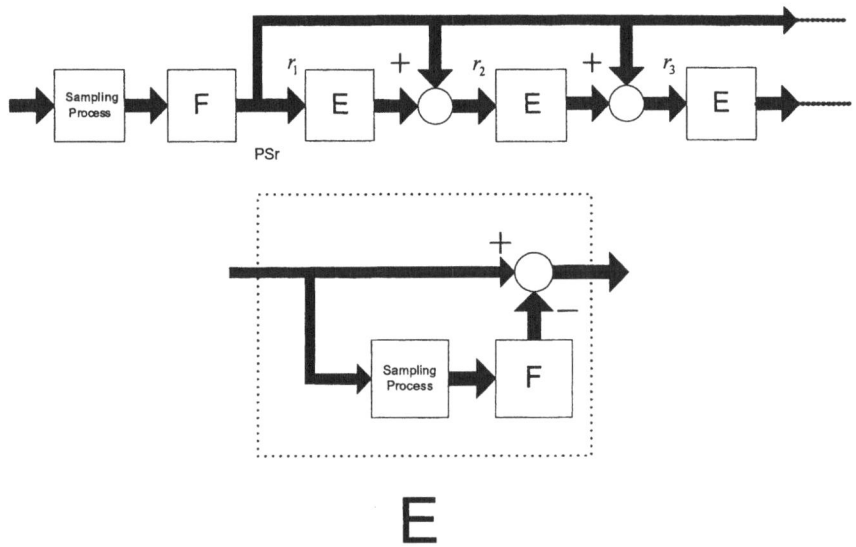

Fig. 2. The block diagram of the proposed iterative algorithm for decoding DFT codes

2.   The linear operator G that is an $N \times K$ matrix that produces an estimate of
    the coded vector $\mathbf{v}$ from $\hat{\mathbf{u}}$ which is given by,

$$\hat{\mathbf{v}} = G\hat{\mathbf{u}} \tag{13}$$

Therefore the linear operator F is an $N \times N$ matrix that can be found as follows,

$$F = G \times pinv(G) \tag{14}$$

It should be noted that we have $pinv(G) \times G = I_K$ where I_K is a $K \times K$
identity matrix. However the multiplication is not commutative and $G \times pinv(G)$ is
an $N \times N$ matrix that gives a proper estimate of $\mathbf{v}$ from the received signal $\mathbf{r}$.

3   Simulation Results

In this section, we examine the performance of the algorithm proposed in section 2
for decoding of DFT codes in erasure channels. To evaluate the proposed algorithm,
as a criterion, we use Mean Square Error (MSE) of the decoded signal, i.e.

$$MSE = E\left\{\|\mathbf{v} - \hat{\mathbf{v}}\|^2\right\} = E\left\{(\mathbf{v} - \hat{\mathbf{v}})^H (\mathbf{v} - \hat{\mathbf{v}})\right\} \tag{14}$$

where $E\{.\}$ denotes the expected value of a random variable. $\mathbf{v}$, and $\hat{\mathbf{v}}$ are the
transmitted coded signal and its estimation after a number of iterations, respectively.
To evaluate the mean value of the square error, its sample-mean is evaluated by using
400 different independent signals.

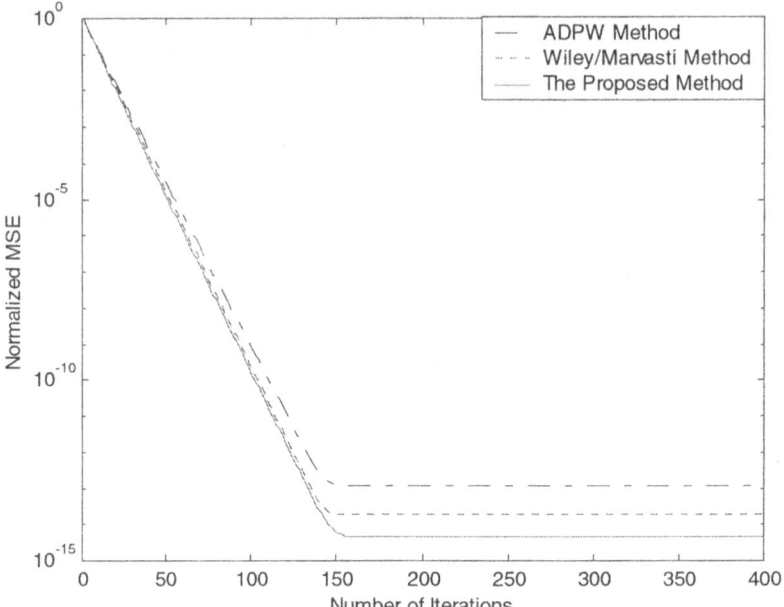

Fig. 3. The Normalized MSE versus the number of Iterations of Wiley/Marvasti, ADPW, and the proposed methods in an erasure channel with 20 missing samples

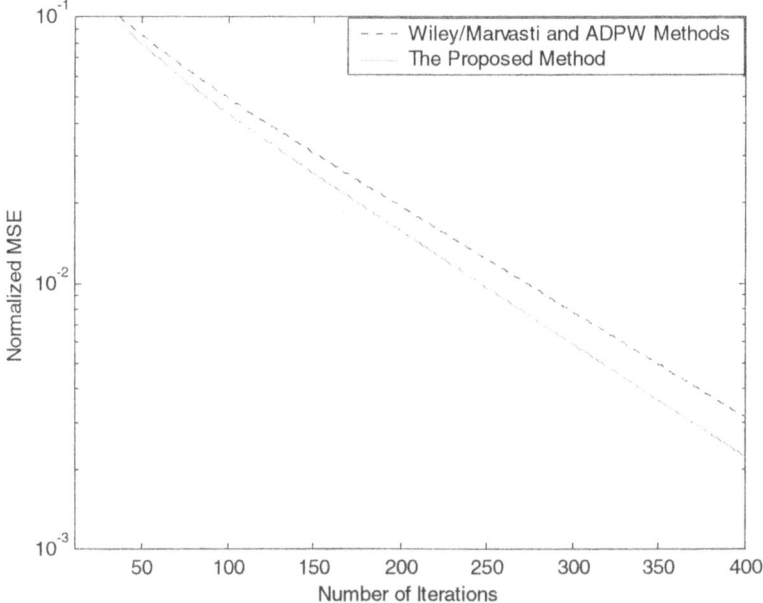

Fig. 4. The Normalized MSE versus the number of Iterations of Wiley/Marvasti, ADPW, and the proposed methods in an erasure channel with 28 missing samples

To compare the proposed decoding methods, we use them to decode a (128,100) DFT code. Figure 3 shows the curves of the Normalized MSE versus the number of Iterations of Wiley/Marvasti, ADPW, and the proposed methods in different scales. In this case, we assume that 20 samples with non-uniformly spaced positions are lost due to erasure channel. It can be seen that the proposed method outperforms both the Wiley/Marvasti and ADPW methods.

In Figure 4, the techniques are compared in decoding of received signals with 28 lost samples in an erasure channel. It can be seen that the proposed method slightly outperforms both Wiley/Marvasti and ADPW methods which have nearly equal performance.

4 Conclusions

In this paper, a new iterative method for decoding DFT codes has been proposed. It has been shown that the proposed method outperforms the well-known methods such as the Wiley/Marvasti, and ADPW methods in the decoding DFT codes in erasure channels.

References

1. F. Marvasti, *Nonuniform Sampling Theory and Practice,* Kluwer Academic/Plenum Publishers, New-York, 2001.
2. F. Marvasti, and M. Nafie, "Using FFT for Error Correction Decoders", *in proc. IEE colloquium* on *DSP Applications on Communication systems",* pp 9/1-9/4, 1993
3. G. Rath, and C. Guillemot, "Performance Analysis and Recursive Syndrome Decoding of DFT codes for Bursty Erasure Recovery", *IEEE Trans. on Signal Processing,* vol. 51, no. 5, pp. 1335-1350, May 2003.
4. F. Marvasti, M. Hassan, M. Echhart, and S. Talebi, "Efficient Algorithms for Burst Error Recovery Using FFT and Other Transform Kernels", *IEEE Trans. on Signal Processing,* vol. 47, no. 4, pp. 1065-1075, April 1999.
5. P. j. S. G. Ferreira, and J. M. N. Vieira, "Locating and Correcting Errors in Images", *in proc. IEEE Conf on Image Processing, ICIP-97,* pp. 691-694,Oct. 1997.
6. Marvasti, M. Analoui, and M. Gamshadzahi, "Recovery of Signals from Non-uniform Samples Using Iterative Methods.", *IEEE Trans on Signal Processing,* vol. 39, no. 4, pp. 872-878, April 1991.
7. S. A. Sauer, and J. P. Allbach, "Iterative Reconstruction of Band-Limited Images from Non-uniformly Spaced Samples.", *IEEE Trans. on Circuit and Systems,* vol. 34, no. 12, Dec. 1987.

Comparison Between Several Methods of PPM Demodulation Based on Iterative Techniques

M. Shariat, M. Ferdosizadeh, M.J. Abdoli, B. Makouei, A. Yazdanpanah, and
F. Marvasti

Multimedia Laboratory, Iran Telecommunication Research Center
Department of Electrical Engineering, Sharif University of Technology

Abstract. In this paper we examine demodulation methods for nonuniform Pulse Position Modulation (PPM), which is generated by the crossings of a modulating signal with a sawtooth wave. Here, we are interested in PPM at the Nyquist rate and illustrate the efficiency of the iterative methods. Several iterative methods such as Wiley/Marvasti, Time-varying, adaptive weight, Zero-order-hold, Voronoi, linear interpolation algorithms are discussed, where the PPM demodulation is converted as a nonuniform sampling reconstruction problem. Also another approach called the inverse system approach is introduced and the performance of the two algorithms is assessed.

1 Introduction

The idea of the PPM modulation was developed around 50 years ago but it is only recently that a revival of interest has been experienced with the development of Impulse Radio and Fiber-Optic transmission systems [4]. The special properties of the PPM modulation make it a suitable candidate for fiber-optic communications. Specially, because there is no information in the amplitude of PPM signals, this modulation technique is immune to the effects of nonlinearity of the fiber-optic devices. Also the PPM modulation is a good choice for the short-range communications in dense multi-path environments [6-7] using the Impulse Radio communications. Theoretically it has been shown that PPM systems are effective when the signals are power-limited rather than bandlimited [5]–[6].

The main idea in the PPM modulation is to send pulses with equal amplitude at the times of intersection between the message signal and a sawtooth wave $r(t)$. when the frequency of the sawtooth signal is very higher then the Nyquist rate lowpass filtering is a suitable method for demodulation of the PPM signal. In the following sections we will show that iterative methods work well even at sampling rates close to the Nyquist rate.The block diagram of a generic iterative algorithm is shown in Fig.1. The block G is the 'distortion block'. In fact the iterative method is a general method for recovery of signals distorted by any kind of distortion. It can be shown that if SNR at the output of the block is greater than one, then the iterative process converges to the original signal at infinity [1].

In the PPM case, the block G can be replaced by various systems. We have classified these systems into 2 groups. The first group treats the PPM demodulation as a nonuniform sampling reconstruction problem while the second group takes conventional PPM

J.N. de Souza et al. (Eds.): ICT 2004, LNCS 3124, pp. 554–559, 2004.
© Springer-Verlag Berlin Heidelberg 2004

modulator for the G block. In the next sections we will present different methods for each group and then compare the results.

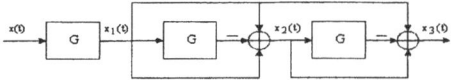

Fig. 1. Block diagram of the proposed iterative method

2   Iterative Approach

In this part, we introduce several approaches for PPM demodulation using iterative methods: the non-uniform sampling approach and the PPM inverse system approach.

2.1   Non-uniform Sampling Approach

In this approach, we consider the G distortion in Fig.1 as a sampling process S followed by a low pass filtering P [2]. Thus we have:

$$x_{k+1}(t) = \lambda(PSx(t) - PSx_k(t)) + x_k(t) \tag{1}$$

The main difference between the various forms of iterative methods is due to the difference of sampling processes [3]. We examine six techniques in this category: Wiley/Marvasti, Time-varying, adaptive weight method, Zero-order-hold, Voronoi and linear interpolation. In the Wiley/Marvasti method the operator S is as follow:

$$Sx(t) = \sum_{i=n}^{N} x(t_i)\delta(t - t_i) \tag{2}$$

and the time-varying method as follow:

$$Sx(t) = \frac{P\left\{\sum_{i=n}^{N} x(t_i)\delta(t - t_i)\right\}}{P\left\{comb(t)\right\}} \tag{3}$$

where $P\{y(t)\}$ is the low pass filtered version of $y(t)$ and $comb(t) = \sum_{i=1}^{N} \delta(t - t_i)$. In the Zero-Order-Hold method we have:

$$Sx(t) = \sum_{i=n}^{N} x(t_i)\phi_i(t) \tag{4}$$

where

$$\phi_i(t) = I(t_i, t_{i+1}) \tag{5}$$

and

$$I(t_i, t_{i+1}) = \begin{cases} 1 & t_1 \leq t \leq t_2 \\ 0 & otherwise \end{cases} \tag{6}$$

The Voronoi method have sampling process similar to Eq. (4) but with different $\Phi_i(t)$. In this method we have:

$$\phi_i(t) = I\left(\frac{t_{i-1} + t_i}{2}, \frac{t_i + t_{i+1}}{2}\right) \tag{7}$$

In the linear interpolation method we can write:

$$Sx(t) = \sum_{i=n}^{N}\left[\frac{x(t_{i+1}) - x(t_i)}{t_{i+1} - t_i}(t - t_i) + x(t_i)\right]I(t_i, t_{i+1}) \tag{8}$$

Adaptive weight method (ADPW) is the next technique that we apply it to our problem. In this technique we multiply each sample by the weight equal to the corresponding Voronoi interval:

$$Sx(t) = \sum_{i=n}^{N}\frac{(t_{i+1} - t_{i-1})}{2}x(t_i)\delta(t - t_i) \tag{9}$$

2.2 PPM Inverse System Approach

Use In this approach the block **G** (in Fig.1) consists of PPM modulator followed by PPM demodulator. Both methods use the same PPM modulator shown in Fig.2, but different in PPM demodulation. This approach is divided into two methods:

Algorithm I: Analyzing the spectrum of a PPM signal, it can be shown [1] that the modulation output has the form of:

$$y(t) = f_s(1 - (\frac{1}{\mu})\frac{d}{dt}x(t))\left(1 + 2\sum_{i=1}^{\infty}\cos(n\,\omega_s(t - \frac{x(t)}{\mu}))\right) \tag{10}$$

Where y(t) is the PPM wave of constant amplitude with pulses positioned at nonuniformly sampled times of:

$$t_k = kT_s + \frac{x(t)}{\mu} \tag{11}$$

and μ is the saw tooth slope. Interpreting Eq.(10) we see that PPM with non-uniform sampling is a combination of linear and exponential carrier modulation. At each harmonic the signal is phase modulated and its derivative is amplitude modulated at the output. We can therefore retrieve the original message by low-pass filtering and integrating with the incorporation of a DC block as shown in Fig.3.

Algorithm II: Since the signal samples at the times k t are equal to the value of the saw tooth signal at these times, we can multiply the PPM pulses (which have unity amplitudes) by the saw tooth to obtain nonuniform samples of the signal. These nonuniform samples are then low pass filtered as shown in Fig.4.

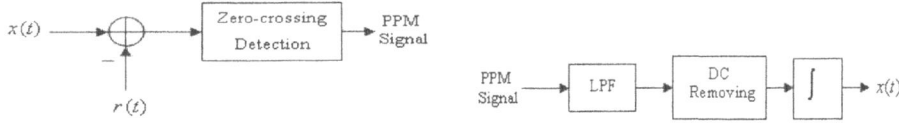

Fig. 2. PPM modulator   **Fig. 3.** Demodulator of the algorithm I

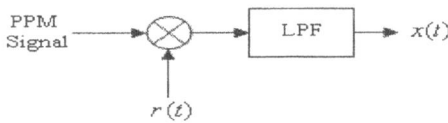

Fig. 4. Demodulator of the algorithm II

3   Simulation Results

The following results have been obtained for the demodulation of a lowpass signal using different iterative techniques described before. The period of the sawtooth signal is chosen so that the average number of PPM pulses in unit time satisfies the Nyquist rate.

In each method we have chosen the optimum λ that was obtained from simulations. We considered the optimality in the sense of minimum mean-square error in the tenth iteration.

Fig.5 shows the mean square error of demodulator (in dB) using different techniques of the first approach (nonuniform sampling) versus the number of iterations. The parameter λ in each method is chosen optimally. In this figure we see that the Voronoi method (symmetric hold) has the highest convergence of all, but its complexity is relatively high and some delay is necessary in its implementation. The mean square error of all methods finally converges to approximately -21.5 dB except the simple zero-order hold method (non symmetric). Another important result is approximately the same performance of the Wiley/Marvasti and ADPW method.

We can observe from the Figure that the method I has better results than the method II. The MSE obtained from the method I in the tenth iteration is about 3dB larger than that is obtained from the method II. Although the method II has better performance than the method I in the sense of MSE, the synchronization is a serious problem for this method. In fact the sawtooth signal required at receiver for the method II must be exactly synchronous with that of transmitter. This critical parameter causes to use the first method widely than the second. Fig.6 shows the MSE versus the number of iterations using different techniques of the second approach (inverse system). The complexities of the first approach methods are approximately the same except in ADPW that we have the additional complexity of the multiplication for each sample. For the algorithms I and II there is an additional complexity because of PPM modulation block in each iteration. In fact this block involves N additions (plus N comparisons). Furthermore, in the algorithm I the demodulation stage requires 2N additions and a filtering (the same as others). But the algorithm II involves N additions and a filtering at the demodulation stage.

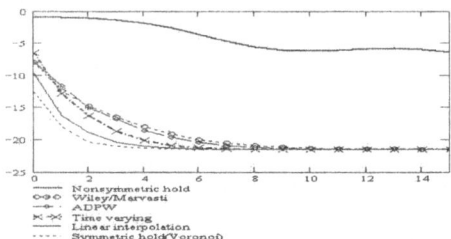

Fig. 5. The mean square error of the first approach methods

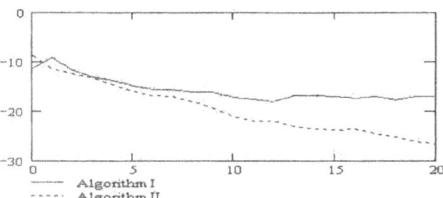

Fig. 6. The mean square error of the second approach methods

The complexities of the first approach methods are approximately the same except in ADPW that we have the additional complexity of the multiplication for each sample. For the algorithms I and II there is an additional complexity because of PPM modulation block in each iteration. In fact this block involves N additions (plus N comparisons). Furthermore, in the algorithm I the demodulation stage requires 2N additions and a filtering (the same as others). But the algorithm II involves N additions and a filtering at the demodulation stage.

4   Conclusion

We introduced several techniques for PPM demodulation and assessed their performances. We can compare these techniques from two points of view: the Mean Square Error (MSE) and the convergence rate. The MSE is in the saturation region of the performance curves depicted in Figs.5-6, which we calculate at the 20th iteration. The convergence rate is a criterion for comparing the MSE in the transition region that is in the lower number of iterations.

As shown in Figures, Algorithm II yields better performance after twenty iterations but the convergence rate is slow, however the convergence rate of the Voronoi technique is the best between one and three iterations.

References

1. F. Marvasti, Nonuniform Sampling Theory and Practice, Kluwer Academic/Plenum Publisher, New York, 2001.

2. F. Marvasti, M. Analouei and M. Gamshadzahi, Recovery of Signals from Samples Using Iterative Methods., IEEE Trans on Signal Processing, vol. 39, no.4, pp. 872-878, April 1991

3. P. Azmi, and F. Marvasti, "Comparison between Several Iterative Methods of Recovering Signals from Nonuniformly Spaced Samples", in Proc. IEEE Conference on Sampling Theory and Applications SampTA2001, Florida-U.S.A., pp. 49-54, 13-17 May 2001.

4. M. Z. Win, and R. A. Scholtz, "Ultra-Wide Bandwidth Time-Hopping Spread-Spectrum Impulse Radio for Wireless Multiple Access Communications" IEEE Trans on Com., vol. 48, no. 4, pp. 679-689, April 2000.

5. L. Zhao, and A. Haimovich, "Multi-User Capacity of M-ary PPM Ultra-Wideband Communications" in Proc. IEEE Conference on Ultrawideband Systems and Communications (UWBST) 2002, Baltimore, Maryland, pp. 175-179, May 2002.

6. B. Wilson, and Z. Ghassemlooy, "Pulse Time Modulation Techniques for Optical Communications: A Review", IEE Proceeding-J, vol. 140, no. 6, pp. 346-357, Dec. 1993.

7. J. R. Pierce, and E. C. Posner, Introduction to Communication Science and System, Plenum, 1976.

A Network Echo Canceler Based on a SRF QRD-LSL Adaptive Algorithm Implemented on Motorola StarCore SC140 DSP

Constantin Paleologu, Andrei Alexandru Enescu, and Silviu Ciochina

University Politehnica of Bucharest, Electronics and Telecommunication Faculty,
Telecommunications Department, 1-3, Iuliu Maniu Blvd., Bucharest, Romania
{pale, aenescu, silviu}@comm.pub.ro

Abstract. A novel network echo canceler based on a square root free QR-decomposition-based least-squares lattice (SRF QRD-LSL) adaptive algorithm is presented. Specific characteristics of this type of algorithms are their high convergence rate, numerical stability and superior robustness to double-talk, which is not present in the case of the normalized least mean squares (NLMS) family of algorithms. This echo canceler is implemented on a powerful fixed-point digital signal processor (DSP), Motorola SC140. The experimental results prove superior performances over the classical echo cancelers based on NLMS algorithms.

1 Introduction

There is need for network echo cancelers for echo paths with long impulse response. Consequently, long FIR adaptive filters ($M \geq 256$) are required. It is well known that longer impulse response implies slower convergence rate, thus rendering traditional algorithms like normalized least mean squares (NLMS) inadequate. In [1], [2] faster converging algorithms called proportionate normalized least mean squares (PNLMS) and PNLMS++, respectively, are proposed. These algorithms achieve higher convergence rate by using the fact that the active part of a network echo path is usually much smaller compared to the whole path that has to be covered. Nevertheless, the experimental results prove that the convergence rate improvement is not significant, particularly for high correlated signals and high length filters.

Based on convergence performance alone, a least squares (LS) algorithm is clearly the algorithm of choice. However, the requirements of an echo canceler are for both rapid convergence and a low computational cost. Thus, a highly desirable algorithm is a low cost (i.e. fast) LS algorithm. On the other hand, another consideration is algorithm stability because it is unacceptable for the algorithm to diverge unexpectedly.

The QR-decomposition-based least-squares lattice (QRD-LSL) algorithm [3] combines the good numerical properties of QR-decomposition and the desirable features of a recursive least-squares lattice. As a disadvantage, the classical QRD-LSL uses square root operations, which are quite complex in a digital signal processor (DSP) implementation. The solutions proposed in [4], [5] are based on some approximations

J.N. de Souza et al. (Eds.): ICT 2004, LNCS 3124, pp. 560–567, 2004.

in order to obtain a square-root free (SRF) version of QRD-LSL algorithm. Two SRF QRD-LSL adaptive algorithms are obtained in this manner. Assuming the use of infinite precision, both versions are mathematically equivalent. However, in a practical situation involving the use of fixed-point arithmetic, these algorithms behave differently. In [5] we proved that only one of them is suitable for a fixed-point implementation. This version is our choice for the implementation of the network echo canceler.

Besides convergence rate and complexity issues, an important aspect of an echo canceler is its performance during „double-talk" (i.e. near-end speech). In the case of NLMS algorithms, the presence of near-end signal considerably disturbs the adaptive process. To eliminate the divergence of echo cancelers the standard procedure is to inhibit the weights updating during the double-talk. The presence of double-talk is detected by a „double-talk detector" (DTD). A number of samples are required by the DTD to detect the double-talk presence. Nevertheless, this very small delay can be enough to generate a considerable perturbation of the echo estimate. Another method to improve overall performance is to allow variability of the global step-size parameter (μ) [6], [7]. However, the algorithm by which the optimal step-size is found is fairly complex and it is difficult to adjust it fast enough when double-talk occurs.

In [8], [9] we have analyzed some versions of QRD-LSL algorithm according to some ITU-T standard requirements concerning echo cancelers. The performed experiments show that the QRD-LSL algorithms fulfills by far the requirements of the ITU-T G.168 recommendation [10] concerning the steady-state echo return loss enhancement and convergence speed. A special interest was given to the problem of the behavior of the algorithms during the double-talk periods. Two distinct effects were identified [8]: the incomplete attenuation of the far-end signal and the unwanted attenuation of the near-end signal as a result of the near-end signal leakage to the output of the adaptive filter through the error signal. Generally, one can assert that this class of algorithms is much more robust to double-talk than the NLMS type algorithms. This suggests that the QRD-LSL algorithms could satisfactorily operate even in the absence of the DTD. Our latest research was focused on the fixed-point DSP implementation of these algorithms [5] and now we are ready to present this network echo canceler based on a SRF QRD-LSL algorithm.

The paper is organized as follows. In section 2 we briefly review the SRF QRD-LSL algorithm. An overview of Motorola SC140 DSP is presented in section 3, followed by experimental results in section 4. Finally, section 5 concludes this work.

2 SRF QRD-LSL Algorithm

The standard versions of QRD-LSL algorithms use Givens rotations for implementing the QR-decomposition, which is basic to their theory. The square-root operations are expensive and awkward to calculate, constituting a bottleneck for overall performance. The solutions proposed in [4], [5] are based on some approximations in order to obtain a square-root free version of QRD-LSL algorithm. Consequently, four versions of QRD-LSL algorithm have been proposed. The last version, called in this paper as SRF QRD-LSL, is the most advantageous from the computational complexity point of view. The simulation results prove that this modified algorithm keeps the fast rate of convergence specific to the family of QRD-LS algorithms. Moreover, the experiments performed in fixed-point arithmetic prove that it offers good numerical properties in finite-precision implementation. The SRF QRD-LSL algorithm is presented below.

1. *Initialization*
For time $n = 0$ and order $m = 1, 2, ..., M$ where M is the final prediction order

$$\overline{\pi}_{f,m-1}(0) = \overline{\pi}_{b,m-i}(0) = 0$$

$$\overline{p}_m(0) = 0$$

$$\overline{J}^f_{m-1}(0) = \overline{J}^b_{m-1}(-1) = \delta^{-1}$$

where δ is a small positive constant
For time $n = 1, 2, ...$ and order $m = 0$

$$\overline{\varepsilon}_{f,0}(n) = \overline{\varepsilon}_{b,0}(n) = x(n)$$

$$\overline{\varepsilon}_0(n) = d(n)$$

$$\overline{\gamma}_0(n) = 1$$

$$\beta_{b,0}(n-1) = \beta_{f,0}(n-1) = 1$$

where $x(n)$ is the input at time n and $d(n)$ is the desired response at time n

2. *Prediction*
For time n and order m

$$t_{b,m-1}(n-1) = \beta_{b,m-1}(n-1)\left|\overline{\varepsilon}_{b,m-1}(n-1)\right|^2 \overline{J}^b_{m-1}(n-2)$$

$$\overline{c}_{b,m-1}(n-1) = \frac{1}{1+t_{b,m-1}(n-1)}$$

$$\overline{s}_{b,m-1}(n-1) = \beta_{b,m-1}(n-1)\overline{\varepsilon}^*_{b,m-1}(n-1)\overline{J}^b_{m-1}(n-2)\overline{c}_{b,m-1}(n-1)$$

$$\overline{J}^b_{m-1}(n-1) = \lambda^{-1}\overline{J}^b_{m-1}(n-2)\overline{c}_{b,m-1}(n-1)$$

$$\overline{\varepsilon}_{f,m}(n) = \overline{\varepsilon}_{f,m-1}(n) - \overline{\varepsilon}_{b,m-1}(n-1)\overline{\pi}^*_{f,m-1}(n-1)$$

$$\overline{\pi}^*_{f,m-1}(n) = \overline{c}_{b,m-1}(n-1)\overline{\pi}^*_{f,m-1}(n-1) + \overline{s}_{b,m-1}(n-1)\overline{\varepsilon}_{f,m-1}(n)$$

$$\overline{\gamma}_m(n-1) = \overline{c}_{b,m-1}(n-1)\overline{\gamma}_{m-1}(n-1)$$

$$t_{f,m-1}(n) = \beta_{f,m-1}(n-1)\left|\overline{\varepsilon}_{f,m-1}(n)\right|^2 \overline{J}^f_{m-1}(n-1)$$

$$\overline{c}_{f,m-1}(n) = \frac{1}{1+t_{f,m-1}(n)}$$

$$\overline{s}_{f,m-1}(n) = \beta_{f,m-1}(n-1)\overline{\varepsilon}^*_{f,m-1}(n)\overline{J}^f_{m-1}(n-1)\overline{c}_{f,m-1}(n)$$

$$\overline{J}^f_{m-1}(n) = \lambda^{-1}\overline{J}^f_{m-1}(n-1)\overline{c}_{f,m-1}(n)$$

$$\overline{\varepsilon}_{b,m}(n) = \overline{\varepsilon}_{b,m-1}(n-1) - \overline{\varepsilon}_{f,m-1}(n)\overline{\pi}^*_{b,m-1}(n-1)$$

$$\overline{\pi}^*_{b,m-1}(n) = \overline{c}_{f,m-1}(n)\overline{\pi}^*_{b,m-1}(n-1) + \overline{s}_{f,m-1}(n)\overline{\varepsilon}_{b,m-1}(n-1)$$

$$\overline{\varepsilon}_{m+1}(n) = \overline{\varepsilon}_m(n) - \overline{\varepsilon}_{b,m}(n)\overline{p}^*_m(n-1)$$

$$\overline{p}^*_m(n) = \overline{c}_{b,m-1}(n-1)\overline{p}^*_m(n-1) + \overline{s}_{b,m-1}(n-1)\overline{\varepsilon}_m(n)$$

$$\beta_{b,m}(n-1) = \overline{c}_{b,m-1}(n-1)\beta_{b,m-1}(n-1)$$

$$\beta_{f,m}(n) = \overline{c}_{f,m-1}(n)\beta_{f,m-1}(n-1)$$

3. *Filtering*
For time n and order M

$$t_{b,M}(n) = \beta_{b,M}(n) \left| \overline{\varepsilon}_{b,M}(n) \right|^2 \overline{J}_M^b(n-1)$$

$$\overline{c}_{b,M}(n) = \frac{1}{1 + t_{b,M}(n)}$$

$$\overline{s}_{b,M}(n) = \beta_{b,M}(n) \overline{\varepsilon}_{b,M}^*(n) \overline{J}_M^b(n-1) \overline{c}_{b,M}(n)$$

$$\overline{J}_M^b(n) = \lambda^{-1} \overline{J}_M^b(n-1) \overline{c}_{b,M}(n)$$

$$\overline{\varepsilon}_{M+1}(n) = \overline{\varepsilon}_M(n) - \overline{\varepsilon}_{b,M}(n) \overrightarrow{p}_M^*(n-1)$$

$$\overrightarrow{p}_M^*(n) = \overline{c}_{b,M}(n) \overrightarrow{p}_M^*(n-1) + \overline{s}_{b,M}(n) \overline{\varepsilon}_M(n)$$

$$\overline{\gamma}_{M+1}(n) = \overline{c}_{b,M}(n) \overline{\gamma}_M(n)$$

$$e_{M+1}(n) = \overline{\gamma}_{M+1}(n) \overline{\varepsilon}_{M+1}(n)$$

The parameters involved in the algorithm are denoted similarly to [3] and according to [5], as follows (for prediction order m):

- $\overline{J}_m^b, \overline{J}_m^f$ - sum of weighted backward/forward prediction error squares;
- $\overline{\varepsilon}_{b,m}, \overline{\varepsilon}_{f,m}$ - modified angle-normalized backward /forward prediction error;
- $\overline{\pi}_{b,m}, \overline{\pi}_{f,m}$ - modified backward/forward prediction auxiliary parameter;
- $\overline{c}_{b,m}, \overline{s}_{b,m}, \overline{c}_{f,m}, \overline{s}_{f,m}$ - modified parameters of the Givens rotations;
- $\overline{\varepsilon}_m$ - modified angle-normalized joint-process estimation error;
- $\overline{p}_m$ - modified joint-process auxiliary parameter;
- $\overline{\gamma}_m$ - modified conversion factor;
- $\beta_{b,m}, \beta_{f,m}$ - new parameters used for square-root free Givens rotations;
- $t_{b,m}, t_{f,m}$ - auxiliary parameters;
- λ - exponential weighting factor.

In Table 1 we present a comparison of the computational complexities of the classical QRD-LSL algorithm and the SRF QRD-LSL algorithm. We observe that the SRF QRD-LSL algorithm is square-root free and have M less multiplication operations per iteration than classical QRD-LSL algorithm. Moreover, the number of division operations for the SRF QRD-LSL algorithm is reduced to $2M+1$ per stage (half division operations comparative with the classical QRD-LSL algorithm), which represent an important issue for any practical implementation.

The reduced number of the division operations is not the only advantage of the SRF QRD-LSL adaptive algorithm. Another important aspect is related to the cost

Table 1. Computational complexities (number of operations per iteration) for the QRD-LSL algorithms

Operations	QRD-LSL	SRF QRD-LSL
multiplications	$25M+11$	$24M+11$
divisions	$4M+2$	$2M+1$
additions/ subtractions	$8M+3$	$8M+3$
square-root	$4M+2$	0

function values ($\overline{J}_m^b, \overline{J}_m^f$). It is well known that in a fixed-point implementation context the absolute values of all involved parameters have to be smaller than unit. In the case of others QRD-LSL algorithms [3], [4] the cost function will asymptotically increase and it will be upper bounded by $1/(1-\lambda)$. When dealing with a value of the exponential weighting factor λ very close to unit (as in the case of an echo canceler [8], [9]) this implies very large values for the cost function. In order to prevent any unwanted overflow phenomenon it is necessary to scale the cost function. As a consequence, the major drawback is the precision loss because of these factors. In the case of SRF QRD-LSL algorithm the update procedure of the cost function is made in a „reverse" manner. The maximum value of these functions will be the initial value and then they will asymptotically decrease to a lower bound. Therefore, the scale factors are less critical leading to an increased precision of representation [5]. Taking these into account, we may conclude that this version of SRF QRD-LSL adaptive algorithm is more suitable for fixed-point D.S.P. implementation.

3 MOTOROLA SC140 DSP Overview

For the practical implementation of the echo canceler it is necessary to choose a powerful processor, with a large number of MIPS (million instructions per second) and a parallel architecture that allows several instructions to be executed simultaneously. This is the main reason for choosing Motorola SC140 processor [11]. In addition, we have structured the algorithm in a way to allow a high complexity algorithm to be run in real-time in a specific application.

The main feature of this DSP is the C compiler and the ability to convert C source code into assembly code. The complexity of SRF QRD-LSL algorithm is quite large and therefore the need for flexibility is important, since programming in C code is much easier than implementing the algorithm direct in assembly code. The C compiler supports ANSI C standard and also intrinsic functions for ITU/ETSI primitives. Assembly code integration is also possible, which optimizes supplementary the code.

One of our main goals is to minimize the number of cycles needed by the algorithm per iteration, in order to lower the computational time per iteration under the sampling time of the CODEC. If we take advantage of the fact that the structure of the algorithm is symmetrical (i.e. similarities between the forward prediction structure and the backward prediction structure) then we can use two identical blocks for each lattice cell, thus we can call twice a function in C language during one iteration. The filtering part is included in backward prediction part and is performed if a flag is set. This flag is set before the backward prediction and reset before the forward prediction. Another optimization technique, accomplished using this procedure is that all the transformations are made in-place, regardless of the iteration (i.e. moment of time), saving a large amount of memory. Choosing an appropriate level of optimization from the C compiler, Code Warrior (0-3), makes further optimization. As well, the proper use of intrinsic functions from C compiler can further reduce the number of cycles.

4 Experimental Results

Let us consider the „interference cancellation" configuration (Fig. 1). The purpose of the scheme is to extract the signal $x(n)$ from the mixture $x(n)+v(n)$. In the case of an echo canceler, $u(n)$ is the far-end signal, $x(n)$ is the near-end signal, H is the echo path equivalent to a FIR filter with the impulse response $\mathbf{h}(n)$ and W is an adaptive filter, having the coefficients $\mathbf{w}(n)$.

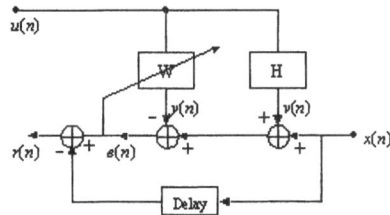

Fig. 1. Configuration for experimental results

In the real case of any adaptive algorithm the coefficients $\mathbf{w}(n)$ depend on the signal $x(n)$. As a consequence, two effects appear [8]:
- $\mathbf{w}(n)$ differs to $\mathbf{h}(n)$ in a certain extent and this may be viewed as a divergence of the algorithm. As a direct consequence, will result a decrease of the echo return loss enhancement (ERLE).
- $y(n)$ will contain a component proportional to $x(n)$, that will be subtracted from the total near-end signal. This phenomenon is in fact a leakage of the $x(n)$ in $y(n)$, through the error signal $e(n)$. The result consists of an undesired attenuation of the near-end signal.
As we demonstrated in [8] the leakage process is important for low λ, where $y(n) \cong v(n)+x(n)$, and is practically absent for $\lambda \cong 1$ where $y(n) \cong v(n)$. As was suggested in [1], to make more apparent in results, it is convenient to subtract out the direct near-end component from the error signal $e(n)$. In this manner, the residual error $r(n)$ cumulates the undesired attenuation of the near-end signal $x(n)$ and the imperfect rejection of the echo path response $v(n)$. In a real application such a subtraction can never be done because the signal $x(n)$ is not available.
The standard ITU-T G.168 [10] recommends certain test procedures for evaluating the performance of echo cancelers. Test signals used are so-called composite source signals (CSS) that have properties similar to those of speech with both voiced and unvoiced sequence as well as pauses. Moreover, we choose a long echo path (64 ms) according to the same above recommendation. The loss of this echo path (ERL-echo return loss) is about 10 dB (averaged over the voice band) and it is considered typical.
The first experiment refers to the convergence rate and echo return loss enhancement. In Fig. 2 are shown the convergence results obtained by the SRF QRD-LSL (with $\lambda = 0.9999$) and NLMS (with $\mu = 1$) adaptive algorithms. For ITU recommendation G.168 testing purposes, the method defined for measuring the level of the signals is a root mean square (RMS) method, using a 35 ms rectangular sliding window.

The measurement device comprises a squaring circuit and an exponential filter (35 ms, 1-pole). The following conclusions are obvious:

- convergence speed and ERLE are superior in the case of SRF QRD-LSL algorithm.

-    test 1 (steady-state and residual echo level test) and the requirements of the recommendation test 2B (convergence speed) are accomplished in the case of SRF QRD-LSL algorithm.

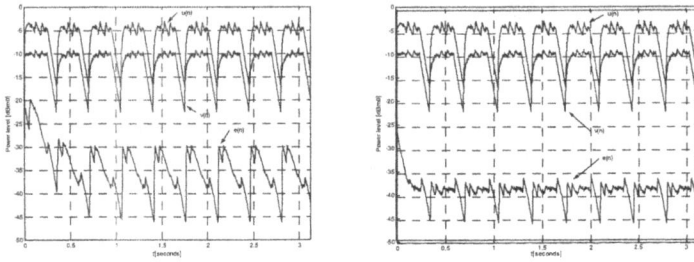

Fig. 2. Power levels [dBm0] for far-end signal $u(n)$ (CSS), echo signal $v(n)$, error signal $e(n)$ for NLMS (left) and SRF QRD-LSL (right).

The second experiment was performed using speech as excitation signals in order to simulate a real-world conversation. It evaluates the performances of the echo canceler for a high level double-talk sequence (similar to test 3B). The double-talk level in this case is about the same as that of the far-end signal. The results are presented in Fig. 3. In the case of SRF QRD-LSL algorithm one can see that the near-end signal $x(n)$ is recovered in $e(n)$ with slight distortions. The NLMS based echo canceler (using DTD) fails in this situation because double-talk appears during initial convergence phase so that the adaptive process is prematurely inhibited. Let us remind that we don't use any DTD in our echo canceler based on SRF QRD-LSL algorithm.

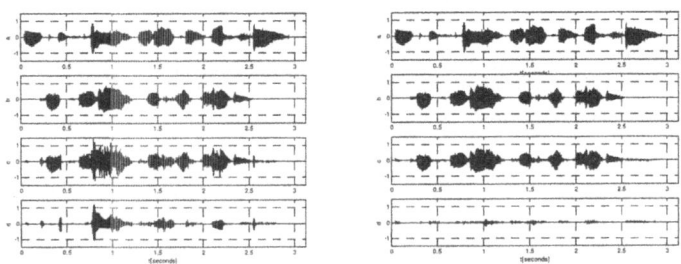

Fig. 3. Performances of NLMS (left) and SRF QRD-LSL (right) algorithms for speech signal during double-talk: (a) far-end signal $u(n)$; (b) near-end signal $x(n)$; (c) recovered near-end signal in $e(n)$; (d) residual error $r(n)$.

Other experiments were performed using more „difficult" echo paths (ERL up to 6dB). It should be noted that 6 dB is a typical worst case value encountered for most networks, and most current networks have typical ERL values better than this. Nevertheless, it was obvious the superior convergence rate and double-talk robustness of the echo canceler based on SRF QRD-LSL adaptive algorithm.

5 Conclusions

The paper presents a novel network echo canceler based on a SRF QRD-LSL adaptive algorithm implemented on Motorola StarCore SC140 DSP, a powerful fixed-point processor. This type of adaptive algorithm is attractive because of its fast convergence rate even for high length filters and high correlated signals and it is much more robust to double-talk than NLMS type algorithms. Moreover, the computational complexity is lower, compared with the classical QRD-LSL algorithm. The modified algorithm is square-root free and has $2M+1$ less division operations per iteration, leading to a feasible fixed-point implementation.

The experimental results show that the requirements of the ITU-T G.168 recommendation concerning the steady-state echo return loss enhancement and convergence speed are fulfilled. The most important issue is that the SRF QRD-LSL algorithm could satisfactorily operate even in the absence of the DTD. This is crucial in the case of a high level of ambient noise corrupting the near-end speech. In this case, a DTD algorithm never enables the adaptive process.

Our future work will focus on optimization techniques used for DSP implementation. Because this adaptive algorithm is known to be complex and because it needs to run in real time, it is interesting to furthermore reduce the number of cycles. The next step is to write special assembly routines within the program, replacing the complex operations.

References

1. Duttweiler D.L.: Proportionate Normalized Least-Mean-Squares Adaptation in Echo Cancelers, IEEE Trans. Speech and Audio Proc., vol. 8, no. 5 (2000) 508-518
2. Gay S.L.: An Efficient, Fast Converging Adaptive Filter for Network Echo Cancellation, Proc. Asilomar, Pacific Grove, CA (1998) 394-398
3. Haykin S.: Adaptive Filter Theory, 4th edn., Prentice Hall, Upper Saddle River, N.J. (2002)
4. Paleologu C., Ciochina S., Enescu A.A.: Modified versions of QRD-LSL Adaptive Algorithm with Lower Computational Complexity, Rev. Roum. Sci. Techn. – Electrotechn. et Energ., vol. 46, no.3 (2001) 333-352
5. Paleologu C., Ciochina S., Enescu A.A.: On the Behaviour of Two Square-Root Free QRD-LSL Adaptive Algorithms in Fixed-Point DSP Implementation, Rev. Roum. Sci. Techn. – Electrotechn. et Energ., vol. 47, nr.4 (2002) 387-399
6. Gänsler T., Gay S.L., Sondhi M.M., Benesty J.: Double-Talk Robust Fast Converging Algorithms for Network Echo Cancellation, IEEE Trans. Speech and Audio Proc, vol. 8, no. 6 (2000) 656-663
7. Liu J.: A Novel Adaptation Scheme in the NLMS Algorithm for Echo Cancellation, IEEE Sign. Proc. Letters, vol. 8, no. 1 (2001) 20-22
8. Ciochina S., Paleologu C.: On the Performances of QRD-LSL Adaptive Algorithm in Echo Cancelling Configuration, Proc. IEEE ICT 2001, Bucharest, Romania, vol.1 (2001) 563-567
9. Paleologu C., Ciochina S., Enescu A.A.: A Simplified QRD-LSL Adaptive Algorithm in Echo Cancelling Configuration, Proc. IEEE ICT 2002, Beijing, China, vol.1 (2002) 240-244
10. ITU-T Recommendation G.168: Digital Network Echo Cancellers, Draft 3 (2000)
11. Motorola SC140 DSP Core Reference Manual, Revised 1, 6 (2000)

Detection of Equalization Errors in Time-Varying Channels

Juraci Ferreira Galdino[1], Ernesto Leite Pinto[1], Marcelo Sampaio de Alencar[2], and Elvino S. Sousa[3]

[1] Instituto Militar de Engenharia, 22290-279, Rio de Janeiro, RJ, Brazil
{galdino, ernesto}@ime.eb.br
[2] Universidade Federal da Paraíba, 58.109-970, Campina Grande, PB, Brazil
{malencar}@dee.ufcg.edu.br
[3] University of Toronto, Toronto, M5S 3G4S, Canada
es.sousa@utoronto.ca

Abstract. This article presents a new procedure to detect equalization errors in time-varying channels (TVC). This procedure has low computational complexity and can be used with different modulation or equalization schemes. In order to detect the occurrence of errors a test of hypothesis is performed using a proper measure of distance between decision-directed channel estimates produced by two adaptive filters with distinct characteristics of robustness to the use of erroneous references. Several results of performance evaluation by simulation are presented and indicate the effectiveness of the proposed scheme.

1 Introduction

The detection of equalization errors is of interest for performance improvement in digital transmission for time-varying frequency-selective channels. In this context, the receiver is usually equipped with decision-directed adaptive filters which may feedback the equalization errors via erroneous adaptations. This feedback mechanism when activated rapidly leads to the complete degradation of the detected data stream. In order to interrupt this catastrophic error events, tools for fast detection of errors in the received symbols are required.

A technique for equalizer error detection was proposed in [1], assuming the time-invariance of the channel. As an attempt to deal with time-varying channels, an evolution of this procedure was proposed in [2] which was based on the use of the recursive least squares (RLS) algorithm for estimating the channel impulse response using the detected symbol stream as reference. In both cases good performance characteristics of equalization error detection have been attained at the price of a computationally intensive algorithm.

A novel and efficient scheme for error detection is addressed in this article which relies on the explicit assumption of a time-varying channel (TVC). This scheme is based on the use of two adaptive filters for channel estimation using detected symbols, which have different characteristics of robustness to errors. In order to detect the occurrence of errors in the detected data stream a test of hypothesis is performed using a proper measure of distance between the channel estimates provided by the adaptive filters. This

J.N. de Souza et al. (Eds.): ICT 2004, LNCS 3124, pp. 568–577, 2004.
© Springer-Verlag Berlin Heidelberg 2004

procedure may be implemented with reduced computational effort and can be used with different modulation or equalization strategies. Results of performance evaluation by simulation were obtained under different conditions of channel variability and signal-to-noise ratio, and indicate that the proposed procedure is an effective tool for the detection of equalization errors in doubly-selective communication channels.

The remaining of the paper is organized as follows. The proposed technique for error detection and some related analytical tools are presented in section 2. Section 3 presents several simulation results of performance evaluation and the concluding remarks appear in Section 4.

2   Proposed Scheme

The adopted communication system model and a block diagram of the proposed procedure for detection of equalization errors are shown in Fig.1. This procedure uses two adaptive filtering algorithms (AFA) to supply estimates of the channel impulse response (CIR), using the detected data stream as reference. These CIR estimates are used to perform a hypothesis test on the occurrence of equalizer errors. It should be noted that the operation of this scheme for equalizer errors detection is independent of the type of equalization and modulation techniques employed.

When two AFA with distinct degrees of robustness to detection errors are used to estimate the CIR on the basis of detected symbols, their behavior may be significantly distinct considering the presence or absence of detection errors. Specifically, they may produce very different estimates in the presence of those errors and, on the other hand, produce similar estimates when operating on the basis of correct symbols.

One example of this property can be observed when using the LMS algorithm and Kalman filter (KF) for estimation purposes [3, 4]. When a correct reference is used for parameter estimation, both algorithms produce low residual errors. Besides, the KF also presents faster convergence. When there are errors in the reference signal the KF can still present good tracking characteristics, particularly if a well-adjusted statistical model is used to describe the parameters to be estimated. On the other hand, the LMS algorithm generally presents poor performance when using an erroneous reference signal, since its operation does not rely on the statistical modeling of the parameters evolution in time.

Therefore, for decision directed channel estimation it is reasonable to expect that the steady-state estimates produced by the LMS and KF algorithms are quite similar when using the detected symbols are correct, while in the presence of equalization errors these algorithms produce significantly different estimates. The procedure here proposed explores this difference of behavior in order to detect the occurrence of equalization errors.

To be effective, this procedure depends on the design of an appropriate statistical test to exploit the differences between the CIR estimates produced by the two AFA.

The hypothesis test here proposed uses a convenient measure of this separation which is denoted by r and given by

$$r = ||\hat{\boldsymbol{h}}_1(k) - \hat{\boldsymbol{h}}_2(k)||^2 = \sum_{i=0}^{L-1} \left| \hat{h}_i^1(k) - \hat{h}_i^2(k) \right|^2, \tag{1}$$

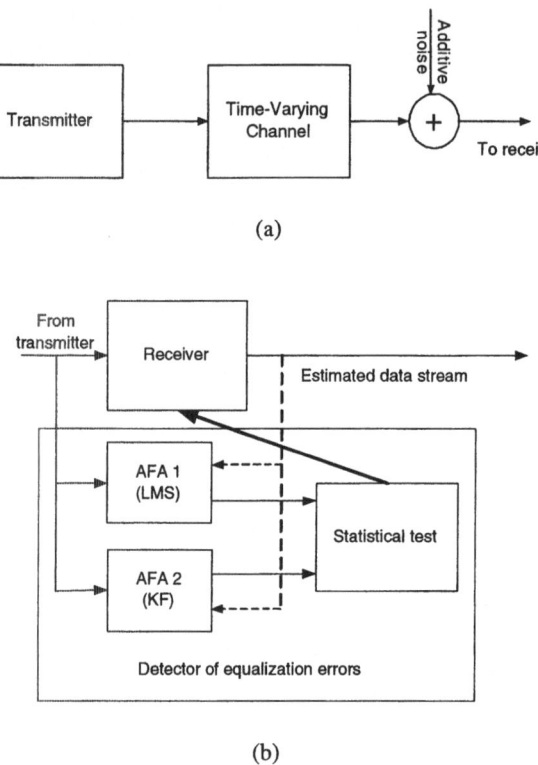

Fig. 1. Communications system model (a) and block diagram of the proposed scheme for the detection of equalizer errors (b).

where $\hat{h}_1(k) = (h_0^1(k), \cdots, h_{L-1}^1(k))'$ and $\hat{h}_2(k) = (h_0^2(k), \cdots, h_{L-1}^2(k))'$ are respectively the CIR estimates obtained with the LMS and KF algorithms at the kth symbol interval, and L is the number of CIR taps.

It is assumed that the estimation is performed in additive white Gaussian noise with zero mean and also that in the absence of detection errors the KF and LMS algorithms produce unbiased and independent CIR estimates which are expressed as

$$\hat{h}_1(k) = h(k) + e_1(k) \text{ and } \hat{h}_2(k) = h(k) + e_2(k), \tag{2}$$

where $e_1(k)$ and $e_2(k)$ are independent random vectors with L components which model the corresponding estimation errors. These vectors are assumed to be Gaussian null mean vector and diagonal covariance matrices denoted by Λ_1 and Λ_2, respectively.

Under these assumptions, the test statistics r may therefore be expressed as [5]

$$r = \sum_{i=0}^{L-1} \alpha_i, \tag{3}$$

where, in the absence of detection errors, the random variables α_i are independent and exponentially distributed, and have means $\overline{\alpha}_i$ given by

$$\overline{\alpha}_i = \Lambda_1(i,i) + \Lambda_2(i,i), \text{ for } i \in [0, L-1]. \tag{4}$$

It can be shown that the conditional probability density function (pdf) of r, given that there are no detection errors, may be expressed as

$$f(r) = \sum_{i=0}^{L-1} \frac{\Pi_i}{\overline{\alpha}_i} \exp\left(-\frac{r}{\overline{\alpha}_i}\right) \text{ for } r \geq 0, \tag{5}$$

where

$$\Pi_k = \prod_{i=0, i \neq k}^{L-1} \frac{\overline{\alpha}_k}{\overline{\alpha}_k - \overline{\alpha}_i}. \tag{6}$$

In the above expression it was assumed that the values of $\overline{\alpha}_i$ are distinct, which is the case of interest here. For identical values of $\overline{\alpha}_i$ a central chi-square distribution is obtained [5].

Using the Neyman-Pearson criterion [6], this pdf can be used to determine the decision threshold, λ, for a pre-established value of the false alarm probability, P_{FA}, which is given by

$$P_{FA} \triangleq \Pr(r \geq \lambda) = \sum_{k=0}^{L-1} \Pi_k \exp\left(-\frac{\lambda}{\overline{\alpha}_k}\right). \tag{7}$$

In order to obtain the value of λ the above equation can be solved numerically. It can also be easily verified that this value fall within the following range

$$\overline{\alpha}_{min} C \leq \lambda \leq \overline{\alpha}_{max} C, \tag{8}$$

where $\overline{\alpha}_{max}$ and $\overline{\alpha}_{min}$ are the maximum and minimum values of $\{\overline{\alpha}_k\}$, respectively, and

$$C = \ln\left[\frac{L \sum_{k=0}^{L-1} \Pi_k}{P_{FA}}\right] \tag{9}$$

where $\ln(\cdot)$ stands for the natural logarithm function.

The bounds in (8) can be used for a fast computation of λ. In particular, the upper bound shown on the right side of (8), implies a conservative design of the test of hypothesis, in the sense that it may be expected that the false alarm probability effectively obtained is smaller than the nominal value of (P_{FA}) used to calculate that bound.

3   Performance Evaluation

Several simulation experiments were conducted for performance evaluation of the proposed procedure. For the sake of generality, no particular reception scheme was simulated and the received data stream was generated by random insertion of errors in the transmitted one.

A *Wide Sense Stationary – Uncorrelated Scattering* (WSS-US) CIR model [7] was assumed, with three T-spaced taps (T is the signaling interval). The Jakes' Doppler power spectrum [8] was adopted and two values of the maximum Doppler shift f_D were considered, which are given by $f_D T = 0.01$ and $f_D T = 0.001$.

For use within the KF algorithm a second-order autoregressive process was adjusted to the Jakes' model by minimizing the prediction error variance [4].

The upper bound of (8) was used for designing the hypothesis test, with $P_{FA} = 0.1$, and the required parameters of the conditional pdf of r in the absence of detection errors were obtained as follows. The analytical results of [9] were used to compute the minimum value of the normalized steady-state mean square error (MSE) produced by the LMS algorithm during the channel estimation, σ_1^2, and the corresponding value of the step-size parameter, in terms of the transmitted signal power, the additive noise variance and the channel autocorrelation function. Considering that $\sigma_{c_i}^2$ is the power of the ith CIR gain, which is specified in the multipath intensity profile, the value of $\Lambda_1(i, i)$ was evaluated as $\Lambda_1(i, i) = \sigma_{c_i}^2 \sigma_1^2$. On the other hand, the value of $\Lambda_2(i, i)$ was supplied by the KF algorithm.

The miss probability $(P_M)^1$ is used as a performance measure of the proposed scheme. Each estimate of (P_M) presented in the following was obtained from 100,000 independent samples of the zero-one random variable associated with the miss event.

The proposed hypothesis test was accomplished after the processing of a block of 42 independent and equiprobable QPSK symbols. The first 32 symbols were supplied to the CIR estimators for training purposes. The other 10 symbols of each block were corrupted by a previously fixed amount of random errors before being used by those estimators.

Several values of the signal to noise ratio (SNR) at the channel output have been considered in the performance evaluation. The SNR was measured as the ratio between the signal energy per bit and the noise power spectral density.

Initially, the analytic approach here adopted for obtain the test variable, whose conditional pdf is given in (5), was evaluated comparing results generated by this pdf with histograms obtained by computational simulation. These histograms were calculated using samples of the test variable obtained in conditions of correct detection of the final 10 symbols in each simulated block.

The results presented in Fig. 2 were obtained with $f_D T = 0.001$ and considering two values of SNR (20 e 40 dB). It can be observed in this figure that, for both values of SNR, the analytic approach produces values of pdf larger than the those obtained by simulation in the range of interest for calculation of the probability of false alarm (tails of the distributions). Therefore, this pdf can be effectively employed for the purpose of setting the decision threshold of the proposed hypotheses test [6].

It should also be noticed that the increase of SNR of 20 for 40 dB produces a significant reduction on the dispersion of the values of the test variable. In fact, the pdf (obtained by analysis and simulation) is more lumped in the second case than in the

---

[1] P_M is the probability that the proposed procedure fails in the detection of a block with erroneous symbols [6].

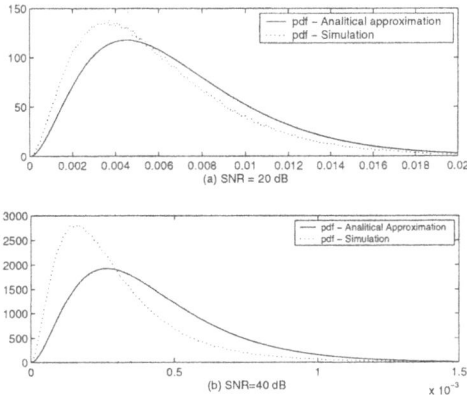

Fig. 2. Conditional pdf of r in the absence of detection errors obtained by analysis and simulation, with $f_D T = 0.001$ and SNR of 20 dB (a) and 40 dB (b).

first one.[2] The results presented in Fig. 3 were obtained with the SNR fixed at 20 dB, for two values of $f_D T$ (0.001 and 0.005). It can be observed again that the analytical approximation produces larger values of the pdf than those obtained by simulation in the range of interest for calculation of the probability of false alarm, confirming its usefulness. With regard to $f_D T$, the results show a small increase in the dispersion of the pdf (both analytical and estimated by simulation) due to the increase in the Doppler spread. This is an expected behavior, since the filtering algorithms face more difficulty in the estimation of channels with faster variations in time. Simulations experiments were also conducted in order to estimate the time variation of the expected value of the test variable r under the occurrence of detection errors.

The transmission of blocks with 100 QPSK symbols was considered, and five random errors were introduced in the 40th, 50th, 60th, 70th and 80th signaling intervals. The value of $f_D T$ was fixed at 0.001 and three values of SNR were used: 20, 30 and 40 dB. The simulation results are exhibited in the Fig. 4 by way of curves of expected value of r versus time (expressed in symbol intervals).

The curves of Fig. 4 show that the mean of the test variable changes abruptly and clearly indicates the instants of occurrence of the detection errors. In addition, it can be observed that the intensity of the variations increase with the increase in the SNR. Other simulations were also performed in order to estimate the conditional pdf of r given the occurrence of detection errors, for different values of the number of detection errors introduced in each block of transmitted symbols. Fig. 5 shows curves of the estimated pdf obtained with $f_D T = 0.001$, SNR=20 dB and three values of the amount of detection errors (2, 5 and 8). For the sake of comparison this figure also exhibits the estimated conditional pdf of r in the absence of errors. It can be verified in Fig. 5 that the conditional pdf r, given the occurrence of detection errors, is significantly different, given

---

[2] It should be observed that the range of values associated to the horizontal axis of parts a and b of Fig. 2 are very different.

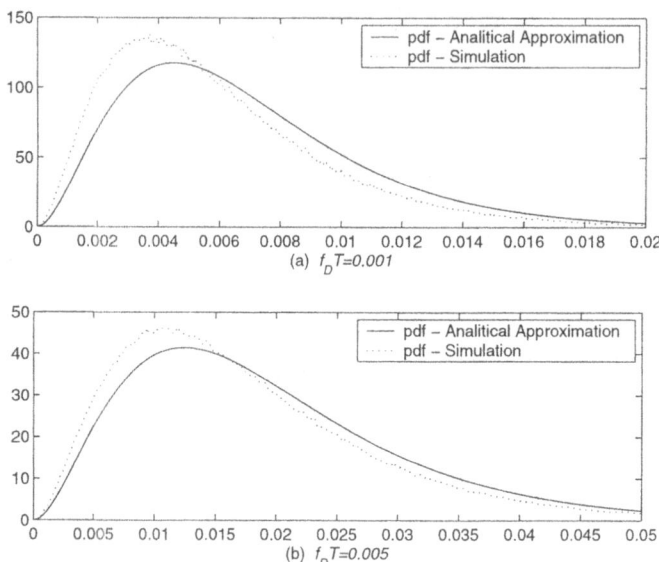

Fig. 3. Conditional pdf of r in the absence of detection errors obtained by analysis and simulation, with SNR of 20 dB and $f_D T = 0.001$ (a) e $f_D T = 0.001$ (b).

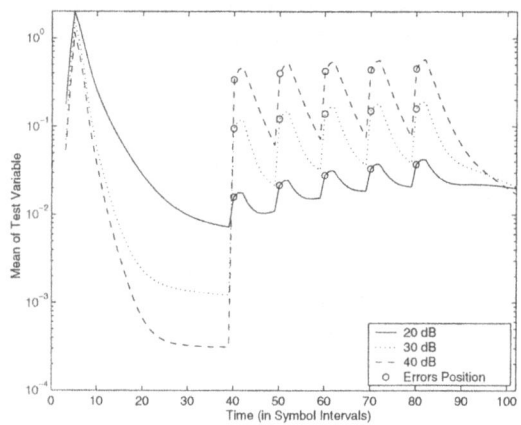

Fig. 4. Mean of the test statistics as a function of time with detection errors in some previously set instants, for $f_D T = 0.001$.

the absence of errors. It is worthy to note that the difference between these pdf's increases with the increase in the amount of detection errors. This is an important evidence that the proposed procedure for error detection is effective and indicates that its performance will with the number of errors.

Fig. 6 shows curves of estimated miss probability of this procedure as a function of the amount of detection errors per block, which were obtained for $f_D T = 0.01$ and

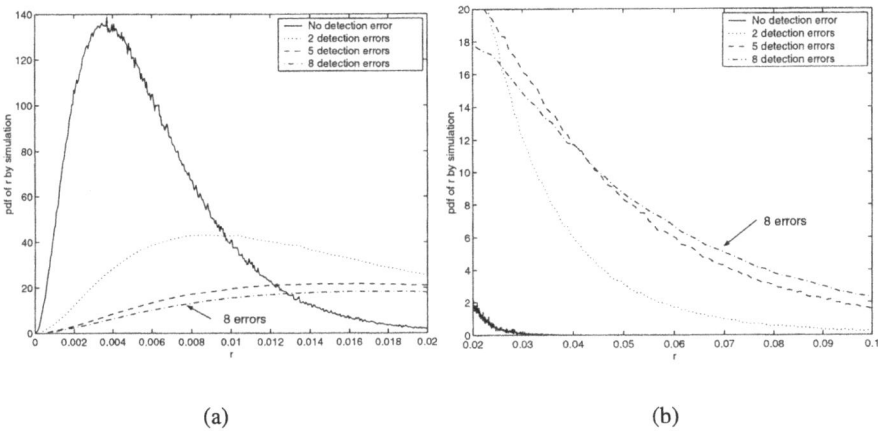

(a)   (b)

Fig. 5. Conditional pdf of r given the occurrence of detection errors, for different values of the amount of errors.

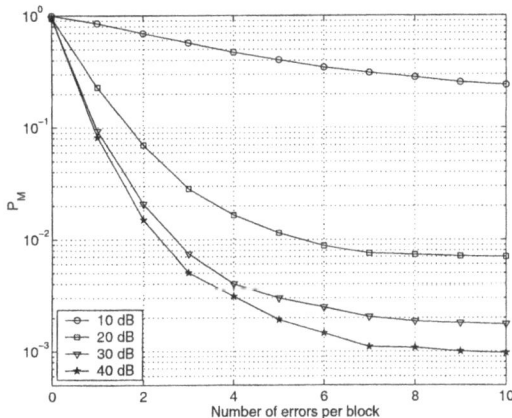

Fig. 6. Miss probability of the proposed scheme for some values of SNR as a function of the amount of errors per block for $f_D T = 0.01$.

four values of SNR. Those curves attest that the error detection ability of this technique increases (P_M decreases) with the amount of errors per block. Besides, they also show that this ability is also improved with the increase in SNR. Fig. 7 shows two curves of miss probability (P_M) versus number of errors per block. which were obtained with SNR=40 dB and $P_{FA} = 0.1$ and two values of the maximum Doppler shift ($f_D T = 0.001$ and $f_D T = 0.01$). It is important to observe that the performance obtained with the smaller value of $f_D T$ is remarkably better. In particular it should be noted that in this case ($f_D T = 0.001$) all the blocks with more than two erroneous symbols were detected.

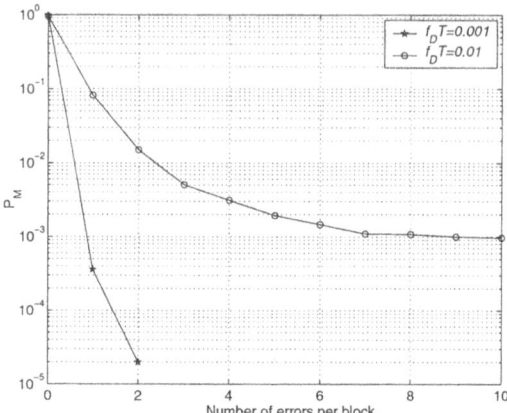

Fig. 7. Miss probability of the proposed scheme as a function of the amount of errors per block, for SNR = 40 dB, $P_{FA} = 0.1$ and two values of $f_D T$.

Several other simulation experiments were conducted under different conditions for $f_D T$, SNR and the number of detection errors and a similar behavior was observed.

Therefore, the performance results obtained so far show that the proposed procedure is an effective technique for the detection of equalization errors in time-varying and frequency-selective communication channels.

4   Conclusion

In this work, a procedure to detect equalization errors has been proposed, which is based on the use of two adaptive filtering algorithms with different characteristics of robustness to errors in the reference used for channel estimation.

The proposed procedure is simple and has excellent performance characteristics, as shown by the simulation results here presented. Besides, this procedure may be applied in association with different receiver structures for time-varying channels.

It is important to mention that in many cases the receiver is originally equipped with an adaptive filter for channel estimation. In such situations the proposed scheme would require only an additional adaptive filter, along with the evaluation of the test statistics and its comparison to a threshold.

In future works the authors intend to investigate alternatives for the choice of the test statistics and evaluate the impact of the proposed procedure on the performance of some equalization schemes for time-varying channels, either using training sequence or adopting blind detection approaches.

Acknowledgments. The authors would like to thank the Brazilian Council for Scientific and Technological Development (CNPq) and the Ministry of Education (CAPES) for the financial support to this work.

References

1. K. Dogançay and R. A. Kennedy. Blind detection of equalization errors in communication systems. *IEEE Transactions on Information Theory*, 43(2):469–482, March 1997.
2. K. Dogançay and V. Krishnamurthy. Blind on-line testing for equalization errors in digital communication systems. *IEEE Transactions on Information Theory*, 44(4):1677–1686, July 1998.
3. P. S. R. Diniz. *Adaptive Filtering Algorithms and Practical Implementation*. Kluwer Academic Publishers, 1997.
4. S. Haykin. *Adaptive Filter Theory*. Prentice Hall, 1991.
5. J. G. Proakis. *Digital Communications*. McGraw-Hill, 1995.
6. H. L. Van Trees. *Detection, Estimation, and Modulation Theory*. Wiley, 1968.
7. J. D. Parsons. *The Mobile Radio Propagation Channel*. John Wiley, 1992.
8. W. C. Jakes. *Microwave Mobile Communications*. John Wiley, 1974.
9. J. F. Galdino, E. L Pinto, and M. S Alencar. Analytical performance of the LMS algorithm on the estimation of wide sense stationary channels. *Submitted to IEEE Transactions on Communications*, 2003.

A Design Technique for Oversampled Modulated Filter Banks and OFDM/QAM Modulations

Didier Pinchon[1], Cyrille Siclet[2], and Pierre Siohan[3]

[1] Université Paul Sabatier, MIP Laboratory, 31062 Toulouse cedex, France
pinchon@mip.ups-tlse.fr
[2] IUT de Saint-Dié, UHP Nancy I - CRAN, 88100 Saint-Dié, France
cyrille.siclet@iutsd.uhp-nancy.fr
[3] France Télécom R&D, DMR/DDH, 35512 Cesson Sévigné, France
pierre.siohan@francetelecom.com

Abstract. This paper describes a new resolution method concerning the orthogonality equations of oversampled complex modulated filter banks, and it is illustrated in the context of oversampled OFDM/QAM modulations. A first simplification is related to the introduction of an equivalent system of orthogonality equation of reduced size. Its solution is expressed using different sets of angular parameters. Furthermore, a compact representation, we recently introduced for OFDM/OQAM modems, is adapted in order to accurately represent the angular parameters of the OFDM/QAM modulation. Both simplifications allow the design of OFDM/QAM modems corresponding to rational oversampling factors and having a high number of carriers.

1 Introduction

The orthogonal multicarrier modulation, OFDM (orthogonal frequency division multiplex), is now integrated in several transmission standards. Its efficiency has particularly been proved for transmission through the radio-mobile channel or over the twisted pair channel. Nevertheless, classical OFDM, using quadrature amplitude modulation without offset, also called OFDM/QAM [1], suffers from several limitations. Indeed, the transmitted symbols have to be extended by a cyclic prefix, which reduces the spectral efficiency. Moreover, the use of a rectangular pulse shape does not allow a good frequency localization. The alternatives to get well-localized pulse shapes are, either, to introduce an offset, which leads to the so-called OFDM/OQAM (Offset QAM) modulation, or an oversampling, which leads to oversampled OFDM/QAM modulation [2,3,4, 5][1]. The design of perfect reconstruction prototypes for OFDM/OQAM modulations, which is equivalent to the design of modified DFT filter banks [8], has been the subject of various publications [9,10]. But, at the contrary, the design of perfect reconstruction optimized prototypes for oversampled OFDM/QAM

---

[1] An extension of both modulations to the biorthogonal case is possible [6,7] but we focus here on the orthogonal case.

J.N. de Souza et al. (Eds.): ICT 2004, LNCS 3124, pp. 578–588, 2004.
© Springer-Verlag Berlin Heidelberg 2004

modulations, which is equivalent to the design of oversampled complex modulated filter bank ([3,4]), has received fewer attention and is much more difficult, particularly if the oversampled ratio is not an integer value.

In this paper, we focus on the design of perfectly orthogonal linear-phase prototype filters in the context of oversampled OFDM/QAM modulation, with a rational oversampling ratio r, such that $1 < r < 2$. In Section 2 we introduce the basics of oversampled OFDM/QAM modulation and give the reduced set of orthogonality equations. In Section 3 we present our parametrization method. Its use to carry out an optimization is presented in Section 4, with design examples for OFDM/QAM systems such that $r = 3/2$ and $5/4$ and more than one thousand of carriers. Finally, the interest of oversampled OFDM/QAM modulations is illustrated with some simulation results presented in Section 5.

2   Oversampled OFDM/QAM Modulations

2.1   Oversampled OFDM/QAM System

We consider an oversampled OFDM/QAM system with M carriers and an oversampling ratio $r = \frac{N}{M} \geq 1$. We denote $p[k]$, $0 \leq k \leq L - 1$, the linear phase prototype filter used on each carrier. It can be shown [6,4] that an oversampled OFDM/QAM modulation system is equivalent to a complex modulated transmultiplexer, as the one depicted in Figure 1 where

$$f_m[k] = p[k]e^{j\frac{2\pi}{M}m\left(k-\frac{L-1}{2}\right)},\ 0 \leq k \leq L-1, \tag{1}$$

$$h_m[k] = p[k]e^{j\frac{2\pi}{M}m\left(k-\frac{L-1}{2}\right)},\ 0 \leq k \leq L-1, \tag{2}$$

and with a and b two integer parameters such that $L-1 = aN-b, 0 \leq b \leq N-1$.

2.2   Orthogonality Conditions

We denote M_0 and N_0 the two coprime integer parameters such that $M_0 N = M N_0$ is the lowest common multiple of M and N. In the following, in order to get simplified expressions, we restrict ourselves to the case where $L = mM_0N$ i.e. $a = mM_0$ and $b = 1$. We also denote $G_l(z)$, $0 \leq l \leq M_0 N - 1$, the $M_0 N$ polyphase components of $P(z)$ defined by $P(z) = \sum_{l=0}^{M_0 N-1} z^{-l} G_l(z^{M_0 N})$. Under these assumptions and using general biorthogonal conditions derived in [7,4], it can be shown that the orthogonal conditions write[2], for $0 \leq l \leq M - 1$ and $1 \leq \lambda \leq M_0$,

$$\sum_{n=0}^{n_{l,\lambda}} G_{nM+l}(z)\tilde{G}_{nM+l+(M_0-\lambda)N}(z) + z^{-1} \sum_{n=n_{l,\lambda}+1}^{N_0-1} G_{nM+l}(z)\tilde{G}_{nM+l-\lambda N_0}(z) = \frac{\delta_{\lambda,M_0}}{M}, \tag{3}$$

---

[2] It is worthwhile mentioning that this corresponds to the orthogonality conditions already derived by R. Hleiss et al [11] and Z. Cvetković et al [12], ignoring the reconstruction delay and the normalization factor.

with $\delta_{m,n}$ the Kronecker operator, $n_{l,\lambda} = \left\lfloor \frac{\lambda N - 1 - l}{M} \right\rfloor$ and $\tilde{\ }$ the paraconjugation operator $(\tilde{G}_l(z) = G(1/z)$ for a real-valued prototype filter).

Now, denoting Δ the greatest common divisor of M and N defined by $M = M_0 \Delta$ and $N = N_0 \Delta$, and setting $H_l^{(d)}(z) = \frac{1}{\sqrt{M}} G_{l\Delta + d}(z)$ and $n'_{l,\lambda} = \left\lfloor \frac{\lambda N_0 - 1 - l}{M_0} \right\rfloor$ we can simplify the system of equations given by (3). Indeed, it can be shown [13] that system (3) is equivalent to a set of Δ independent subsystems of M_0^2 equations given, for $0 \le d \le \Delta - 1$, $0 \le l \le M_0 - 1$ and $1 \le \lambda \le M_0$, by

$$\sum_{n=0}^{n'_{l,\lambda}} H_{nM_0+l}^{(d)}(z) \tilde{H}_{nM_0+l+(M_0-\lambda)N_0}^{(d)}(z) + z^{-1} \sum_{n=n'_{l,\lambda+1}}^{N_0-1} H_{nM_0+l}^{(d)}(z) \tilde{H}_{nM_0+l-\lambda N_0}^{(d)}(z) = \delta_{\lambda,M_0}.$$

(4)

$p[k]$ being linear-phase, this system can be further reduced to $\Delta/2$ subsystems.

3   Resolution Methods of the Orthogonality Conditions

3.1   Parametrization Based on Paraunitary Matrices

For coprime M_0 and N_0, we define $p(k,l)$ and $q(k,l)$ for $0 \le k \le N_0 - 1$, $0 \le l \le M_0 - 1$, by the identity $k + p(k,l)N_0 = l + q(k,l)M_0$, with $0 \le p(k,l) \le M_0 - 1$ and $0 \le q(k,l) \le N_0 - 1$. We also define $e(k,l)$ by $e(k,l) = 1$ if $q(k,0) + q(0,l) - q(k,l) = N_0$ and $e(k,l) = 0$ if $q(k,0) + q(0,l) - q(k,l) = 0$. Then, to a given prototype filter $P(z)$ we can associate a $N_0 \times M_0$ matrix $U(z)$ such that

$$[U]_{k,l}(z) = z^{e(k,l)} G_{k+p(k,l)N_0}(z).$$

(5)

If $U(z)$ may be paraunitary, it has nevertheless some specificities compared to the general paraunitary matrices presented in [14]. Indeed, owing to the $z^{e(k,l)}$ multiplicative factors, all its entries are not necessarily polynomials in z^{-1}. Transformations such as multiplication of lines or columns by a power of z can be attempted [12], but that leads to a constrained representation with leading or trailing coefficients being necessarily zero-valued. Therefore the parametric representation derived in [14, p735] is not used here.

3.2   New Parametrization Method

In this section, we focus on the resolution of the Δ subsystems defined by equation (4). Since these systems are independent, but formally equivalent, we only need to describe the resolution of one of them. Thus, for notational convenience, we now omit the superscript $^{(d)}$ which is present in equation (4) and we, therefore, consider any of the Δ systems (4), in which d is arbitrarily set. Besides, we denote by $S_{M_0,N_0,m}$ the resulting subsystem. Moreover, in order to give a simple illustration of our resolution method, we will consider the example of $S_{2,3,1}$. In this case, for a symmetrical prototype, we have $\Delta/2$ independent sets of $M_0^2 = 4$ equations of the form

$$H_0(z)\tilde{H}_3(z) + H_2(z)\tilde{H}_5(z) + z^{-1}H_4(z)\tilde{H}_1(z) = 0 \tag{6}$$

$$H_1(z)\tilde{H}_4(z) + z^{-1}H_3(z)\tilde{H}_0(z) + z^{-1}H_5(z)\tilde{H}_2(z) = 0 \tag{7}$$

$$H_0(z)\tilde{H}_0(z) + H_2(z)\tilde{H}_2(z) + H_4(z)\tilde{H}_4(z) = 1 \tag{8}$$

$$H_1(z)\tilde{H}_1(z) + H_3(z)\tilde{H}_3(z) + H_5(z)\tilde{H}_5(z) = 1 \tag{9}$$

Admissible Systems. For each $H_i(z), i = 0, \ldots, M_0N_0 - 1$, we denote

$$H_i(z) = \sum_{k=0}^{m-1} \alpha_{i,k} z^{-k} . \tag{10}$$

The number of variables of the system $\mathcal{S}_{M_0,N_0,m}$ is therefore mM_0N_0. For our example, $m = 1$, and $H_i(z) = \alpha_i$, and (6–9) are equivalent to

$$\alpha_0\alpha_3 + \alpha_2\alpha_5 = 0 \tag{11}$$

$$\alpha_1\alpha_4 = 0 \tag{12}$$

$$\alpha_0^2 + \alpha_2^2 + \alpha_4^2 = 1 \tag{13}$$

$$\alpha_1^2 + \alpha_3^2 + \alpha_5^2 = 1 \tag{14}$$

From (4) and (10), we immediately notice that we generally get two kinds of equations. Indeed, they are of the form $\sum_i x_i^2 = 1$ (like (13) and (14)) or such that $\sum_i x_i y_i = 0$ (like equations (11) and (12)), where x_i and y_i correspond to some values of the coefficients $\alpha_{i,k}$. We will call equations of the form $\sum_i x_i^2 = 1$ *square equations*, and equations of the form $\sum_i x_i y_i = 0$ *orthogonal equations*. Square equations are obtained when $\lambda = M_0$, since we have in this case $n_{l,\lambda} = N_0 - 1$ and the corresponding equations write for $0 \leq l \leq M_0 - 1$

$$\sum_{n=0}^{N_0-1} H_{nM_0+l}(z)\tilde{H}_{nM_0+l}(z) = 1, \tag{15}$$

and, for each equation (15), considering the zero degree coefficients, we obtain

$$\sum_{n=0}^{N_0-1} \sum_{k=0}^{m-1} \alpha_{nM_0+l,k}^2 = 1 , \quad l = 0, \ldots, M_0 - 1 . \tag{16}$$

All other equations of system $\mathcal{S}_{M_0,N_0,m}$ are some orthogonal equations. Therefore, there exists a partition $\mathcal{P}_S$ of the set of variables of the system $\mathcal{S}_{M_0,N_0,m}$ into M_0 subsets $S_l = \{\alpha_{nM_0+l,k}, n = 0, \ldots, N_0 - 1, k = 0, \ldots, m - 1\}, l = 0, \ldots, M_0 - 1$ such that each of these subsets is the set of the variables of a square equation of the system (for $\mathcal{S}_{2,3,1}$, $\mathcal{P}_S = \{\{\alpha_0, \alpha_2, \alpha_4\}, \{\alpha_1, \alpha_3, \alpha_5\}\}$). This partition is called *the partition of the squares*. We now introduce a second partition $\mathcal{P}_O$ of the set of variables. Let x and y be two variables, not necessarily different, of the system $\mathcal{S}_{M_0,N_0,m}$. We say that x is relation with y, and we denote $x\mathcal{R}y$, if the product xy appears in one of the equations of the systems. The examination of the equations then shows that the relation $\mathcal{R}$ is an equivalence relation and we denote $\mathcal{P}_O$ the associated partition (for $\mathcal{S}_{2,3,1}$, $\mathcal{P}_O = \{\{\alpha_0, \alpha_2\}, \{\alpha_1, \alpha_4\}, \{\alpha_2, \alpha_5\}\}$). We will say that a system of algebraic equations composed of orthogonal and square equations, and for which there exists a partition $\mathcal{P}_S$ and a partition $\mathcal{P}_O$, is *admissible*.

An admissible system without orthogonal equation is said *trivial*. In this case, the system is composed of n independent systems where n is the cardinal of $\mathcal{P}_S$. Each square equation then admits some solutions that can be represented thanks to $k - 1$ independent angular parameters if k is the number of variables of the equation. If $k = 1$, the equation is of the form $x^2 = 1$ and its solutions are $x = \pm 1$. If $k > 1$, the solution is of the form

$$x_1 = \prod_{i=1}^{k-1} \cos \theta_i , \quad x_k = \sin \theta_{n-1} \prod_{i=n}^{k-1} \cos \theta_i , \ n = 2, \ldots, k .$$

The initial systems, deduced from (4), are admissible. The resolution method consists in replacing an initial system by a set of trivial equivalent systems thanks to a sequence of two type of transformations:

1. The splitting which replaces an admissible system by an equivalent set of systems;
2. The rotation that operates a substitution of variables, depending upon an angular parameter, over an admissible system replacing it by an equivalent system.

The definitions of these operations and the conditions of their application to an admissible system are given in the following paragraphs. By now, the validity of our method is not proved for any system, but many examples show that it is successful for various sets of values of M_0, N_0 and m.

The Splitting Operation. We say that an admissible system can be split if it contains one or several orthogonal equations with a single monomial. If $xy = 0$ is an equation of this type, we say that we split the system when we replace it by the equivalent union of the resulting systems, one with the equation $x = 0$, and the other one with $y = 0$. The cancellation of one variable in an admissible system preserves the property of admissibility. Furthermore the obtained systems can be split as long as it is possible, *i.e.* as long as the obtained systems contain a monomial orthogonal equation. At the end, the union of the obtained systems can be redundant, so we only keep the equations that are strictly necessary. For instance splitting a system with 3 variables, x, y and z and 2 equations such that $xy = 0$, $xz = 0$, leads to 4 redundant subsystems, while the splitting operation will only keep the two independent ones. Each of the systems obtained after a splitting is simpler than the initial one since it contains less variables and less orthogonal equations.

For example, equation (12) shows that the initial system $\mathcal{S}_{2,3,1}$ can be split, leading to the two systems

$$\begin{cases} \alpha_0\alpha_3 + \alpha_2\alpha_5 = 0 \\ \alpha_0^2 + \alpha_2^2 + \alpha_4^2 = 1 \\ \alpha_3^2 + \alpha_5^2 = 1 \end{cases} \text{ and } \begin{cases} \alpha_0\alpha_3 + \alpha_2\alpha_5 = 0 \\ \alpha_0^2 + \alpha_2^2 = 1 \\ \alpha_1^2 + \alpha_3^2 + \alpha_5^2 = 1 \end{cases} \tag{17}$$

The rotation operation. Let $\mathcal{S}$ be an admissible system that cannot be split. Suppose that there exist two distinct subsets O_1 and O_2 of its partition $\mathcal{P}_0$ and a one-to-one correspondence $\phi : O_1 \to O_2$ satisfying the following properties:

1. For all $x \in O_1$, x and $\phi(x)$ belong to the same subset of the partition $\mathcal{P}_S$;
2. For all orthogonal equation containing the monomial xy with $x, y \in O_1$ then, the same equation contains the monomials $\phi(x)\phi(y)$ elements of O_2.

We then say that the system is *regular*. The subsets O_1 and O_2 have therefore the same number of elements, greater or equal to 2. We denote $\{x_1, x_2, \ldots, x_k\}$ the elements of O_1 and $\{y_1, y_2, \ldots, y_k\}$ the elements of O_2 with $y_i = \phi(x_i), i = 1, \ldots, k$. Let θ be an angular parameter. The rotation consists in replacing x_i and y_i by

$$\begin{bmatrix} x_1 \\ y_1 \end{bmatrix} = \begin{bmatrix} r_1 \cos\theta \\ r_1 \sin\theta \end{bmatrix}, \quad \begin{bmatrix} x_i \\ y_i \end{bmatrix} = \begin{bmatrix} \cos\theta & -\sin\theta \\ \sin\theta & \cos\theta \end{bmatrix} \begin{bmatrix} r_i \\ s_i \end{bmatrix}, \tag{18}$$

where $r_1, r_i, s_i, i = 2, \ldots, k$ are the new variables. We denote by $\mathcal{R}$ the resulting system. The sum $x_1^2 + y_1^2$ which occurs in one of the square equations, since x_1 and y_1 belong to the same subset of the square partition, is replaced by r_1^2 and similarly, for $i = 2, \ldots, k$, we have

$$x_i^2 + y_i^2 = r_i^2 + s_i^2 . \tag{19}$$

In the orthogonal equations, we have the groups $x_1 x_i + y_1 y_i, i = 2, \ldots, k$, where $x_i x_j + y_i y_j, 2 \le i, j \le k, i \ne j$. Then, we get:

$$x_1 x_i + y_1 y_i = r_1 r_i, \tag{20}$$

$$x_i x_j + y_i y_j = r_i r_j + s_i s_j . \tag{21}$$

The obtained system is admissible. The $2k$ variables $x_1, \ldots, x_k, y_1, \ldots, y_k$ are replaced by the $2k - 1$ variables $r_1, \ldots, r_k, s_2, \ldots, s_k$ and the partition $\mathcal{P}_0$ is replaced by the partition obtained when replacing the subset O_1 by the subset $\{r_1, \ldots, r_k\}$ and O_2 by $\{s_2, \ldots, s_k\}$.

If one or several orthogonal equations of $\mathcal{S}$ are identical to the left part of (20), we see that the obtained system $\mathcal{R}$ can be split.

Remark. There is no guarantee that the system $\mathcal{R}$ is regular if it cannot be split, nor that the systems obtained after a splitting of $\mathcal{R}$ are regular.

As for $\mathcal{S}_{2,3,1}$, the systems (17) obtained after the first splitting operation are both regular. Considering for example the first one, obtained with $\alpha_1 = 0$, we see that we have the one-to-one correspondence $\alpha_2 = \phi(\alpha_0)$ and $\alpha_5 = \phi(\alpha_3)$. Thus, we make the following variable substitution

$$\begin{bmatrix} \alpha_0 \\ \alpha_2 \end{bmatrix} = \begin{bmatrix} r_0 \cos\theta_0 \\ r_0 \sin\theta_0 \end{bmatrix}, \quad \begin{bmatrix} \alpha_3 \\ \alpha_5 \end{bmatrix} = \begin{bmatrix} \cos\theta_0 & -\sin\theta_0 \\ \sin\theta_0 & \cos\theta_0 \end{bmatrix} \begin{bmatrix} r_1 \\ s_1 \end{bmatrix} \qquad (22)$$

and we get the equivalent system composed by the three equations $r_1^2 + s_1^2 = 1$, $r_0^2 + \alpha_4^2 = 1$ and $r_0 r_1 = 0$. We observe that we get again a system that can be split. At the end, considering the different systems obtained after each splitting, we get the following parametrical solutions

$$\begin{cases} \alpha_0 = \alpha_1 = \alpha_2 = 0 \\ \alpha_3 = \cos\theta_0 \cos\theta_1 - \sin\theta_0 \sin\theta_1 = \cos(\theta_0 + \theta_1) \\ \alpha_4 = \pm 1 \\ \alpha_5 = \sin\theta_0 \cos\theta_1 + \cos\theta_0 \sin\theta_1 = \sin(\theta_0 + \theta_1) \end{cases} \qquad (23)$$

$$\begin{cases} \alpha_0 = \cos\theta_0 \cos\theta_1 \\ \alpha_1 = 0 \\ \alpha_2 = \sin\theta_0 \cos\theta_1 \\ \alpha_3 = \mp \sin\theta_0 \\ \alpha_4 = \sin\theta_1 \\ \alpha_5 = \pm \cos\theta_0 \end{cases} \quad \begin{cases} \alpha_0 = \pm \cos\theta_0 \\ \alpha_1 = \cos\theta_1 \\ \alpha_2 = \pm \sin\theta_0 \\ \alpha_3 = -\sin\theta_0 \sin\theta_1 \\ \alpha_4 = 0 \\ \alpha_5 = \cos\theta_0 \sin\theta_1. \end{cases} \qquad (24)$$

Dimension of a parametrical solution. The $\mathcal{S}_{3,2,1}$ example is very simple since there are only 3 different solutions. But the calculus can rapidly become very heavy. For example, $\mathcal{S}_{4,5,2}$ leads to 13502 solutions. That is why we have to restrict to solutions with the best potential of optimization. We have noticed that solutions of maximal dimensions (that is to say with the highest number of free parameters), provide the best solutions after optimization. In the case of $\mathcal{S}_{2,3,1}$, we immediately see that solutions (24) are of maximal dimension 2.

In general, a direct computation of the dimension is not feasible, but a probabilistic approach can nevertheless provide an exact value of the dimension. Thus, we restrict ourself to the values of the angular parameters θ_i such that $\cos\theta_i$ and $\sin\theta_i$ are rational. The evaluation of the polynomials in $\cos\theta_i$ and $\sin\theta_i$ occurring in the computation of the dimension may then be done exactly. Then, our probabilistic approach consists in a computation of the dimension using a random selection of one or several rational values of $\cos\theta_i, \sin\theta_i$ [13].

4   Design Method and Examples

The design problem consists in finding the coefficients of the prototype filter that satisfy some optimization criterion. Two important points have to be noted concerning the method we propose. First, as the perfect reconstruction conditions are implicitly included in the parametrical representation, the design problem corresponds to an unconstrained optimization. Compared to a design

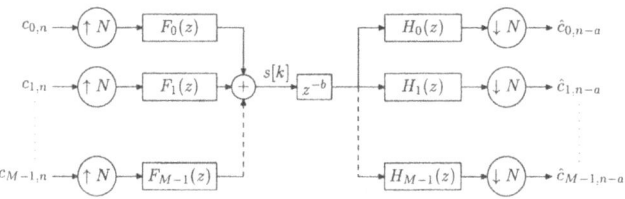

Fig. 1. Transmultiplexer associated to OFDM/QAM modulations.

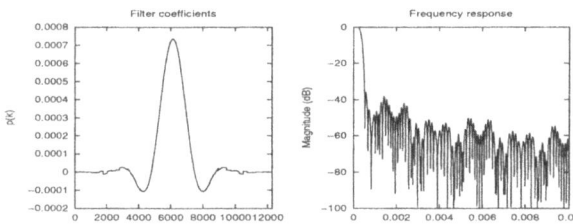

Fig. 2. OFDM/QAM prototype filter minimizing the out-of-band energy, with $M_0 = 2, N_0 = 3, m = 4, M = 1024, L = 12288, E = 3.3976 \times 10^{-4}$.

directly optimizing the $p[k]$'s and taking into account the orthogonality constraints, we get a reduced number of parameters. Second point, but still more important, is the fact that the compact representation method initially proposed for OFDM/OQAM multicarrier modulation [10] can be adapted in the present context [13]. So, let us see now how this method, that can significantly reduce the number of parameters to optimize, can be used for oversampled OFDM/QAM. For some values of M_0, N_0, m, and Δ let us denote $|\theta|$ the number of angular parameters corresponding to the parametrization of each of the $\frac{\Delta}{2}$ subsystems $\mathcal{S}_{M_0,N_0,m}$. Each of these systems can be parameterized with a different solution so that $|\theta|$ is not necessarily the same for each subsystem, but in order not to outrageously complicate the optimization, we will assume that the same parametrical solution is used for each subsystem. The parameter Δ can be very high, for example if $M = 1024$ and the oversampling ratio is equal to $\frac{3}{2}$, $\Delta = 512$, and therefore, the number of parameters to optimize, $\frac{\Delta}{2}|\theta|$ can be prohibitive. But, the compact representation method allows us to considerably reduce this number. Let denote by $\theta_i^{(p)}$, $0 \leq i \leq |\theta| - 1$, $0 \leq p \leq \frac{\Delta}{2} - 1$, the ith angular parameter of the pth subsystem. For a design criterion that is the minimization of the out-of-band energy, the problem is to find the set of $\theta_i^{(p)}$ that minimizes the quantity $E = \int_{\frac{1}{M}}^{\frac{1}{2}} |P(\nu)|^2 d\nu$. When the problem dimension is not too large a direct optimization is possible and, then, it can be checked for this criterion, and for the time-frequency localization criterion [13], that $\theta_i^{(p)}$ is generally a smooth function of p for fixed i. That is why we assume that, at the optimum, each angular parameter leads to a smooth curve that can be easily fitted by a

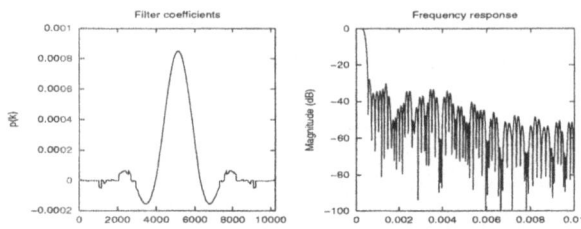

Fig. 3. OFDM/QAM prototype filter minimizing the out-of-band energy, with $M_0 = 4, N_0 = 5, m = 2, M = 1024, L = 10240, E = 3.4092 \times 10^{-3}$.

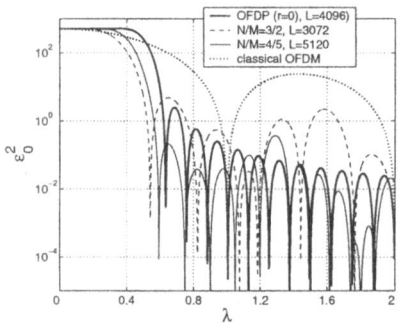

Fig. 4. Comparison of the square error of oversampled OFDM/QAM prototype filters with OFDM (4096 samples, [15]) and classical OFDM, for $M = 512$ carriers.

polynomial. Thus, setting to $K - 1$ the degree of this polynomial, we have

$$\theta_i^{(p)} = \sum_{k=0}^{K-1} x_{i,k} \left(\frac{2p+1}{2M} \right)^k, \ i = 0, \dots, |\theta| - 1. \tag{25}$$

We have then $K|\theta|$ parameters $x_{i,k}$ to optimize instead of $\frac{A}{2}|\theta|$ angular parameters, if we only take advantage of the reduced system (3), and of $L/2$ in a direct approach. Generally, setting $K = 5$ is sufficient to get very accurate results. A first design example in Figure 2, for $r = \frac{3}{2}$, shows that, compared to [3], where the design is limited to $M = 32$ and $m = 2$, *i.e.* a prototype filter of length 192, our approach allows to get results for a very high number of carriers, $M = 1024$, and long prototypes $m = 4$ and $L = 12288$. A second illustration in Figure 3, shows with a design for $r = \frac{5}{4}$, that we are also able to significantly increase the spectral efficiency.

5   Simulation Results

In order to illustrate the possible interest of oversampled OFDM/QAM modulations, we consider, as in [15], the very simple case of a narrowband interference. In this case, the received baseband signal writes

$$r[k] = s[k] + \sigma^2 e^{j \frac{2\pi}{M} \lambda \left(k - \frac{D}{2} \right)}, \tag{26}$$

with $s[k]$ the transmitted signal, σ^2 the power of the interference and λ its frequency (with reduced frequency convention). It can simply be shown [4] that the square error on the mth subband writes

$$\varepsilon_m^2 = \frac{\sigma^2}{\sigma_s^2} \|f\|^2 \left| P\left(\frac{\lambda - m}{M}\right) \right|^2 \frac{M}{N}, \tag{27}$$

with $P(\nu)$ the Fourier transform of the prototype filter and σ_s^2 the variance of the transmitted signal. We have compared in Figure 4 the results obtained for classical OFDM ($M = N$, rectangular prototype), two OFDM/QAM prototype filters optimized to minimize the out-of-band energy and an OFDM/OQAM prototype, called OFDP for optimal finite duration pulse, optimized in [15] in order to minimize the error produced by narrowband interference. Each of the prototype correspond to a 512 carrier system. The curves correspond to the error on the subband number 0. We clearly see that the more we oversample, the lower the square error is, which can compensate the loss of spectral efficiency due to oversampling.

6 Conclusion

In this paper we have presented a new design method for the oversampled OFDM/QAM modulation. We have given, at first, a reduced, but equivalent, form of the perfect reconstruction constraints. Our approach of these equations also allows to characterize a large set of solutions corresponding to rational oversampling ratios, and it has been illustrated in the case of the 3/2 and 5/4 ratios. Furthermore, an adaptation to OFDM/QAM of our compact representation method [10] also permits prototype designs for a very high number of carriers. The interest of oversampled OFDM/QAM in a practical setting has also been illustrated in the case of narrowband interference.

References

1. Le Floch, B., Alard, M., Berrou, C.: Coded orthogonal frequency division multiplex. Proceedings of the IEEE **83** (1995) 982–996
2. Vallet, R., Taieb, K.H.: Fraction spaced multi-carrier modulation. Wireless Personnal Comm. **2** (1995) 97–103
3. Hleiss, R., Duhamel, P., Charbit, M.: Oversampled OFDM systems. In: IEEE DSP'97, Santorini, Greece (1997)
4. Siclet, C.: Application de la th orie des bancs de filtres l'analyse et la conception de modulations multiporteuses orthogonales et biorthogonales. PhD thesis, Universit de Rennes I, Rennes (2002)
5. Lin, Y.P., Phoong, S.M.: ISI-free FIR filterbank transceivers for frequency-selective channels. IEEE Trans. on Signal Processing **49** (2001) 2648–2658
6. Siclet, C., Siohan, P., Pinchon, D.: New results for the design of biorthogonal transmultiplexers with offset QAM. In: IASTED AIC'01, Rhodes, Greece (2001)

7. Siclet, C., Siohan, P., Pinchon, D.: Analysis and design of OFDM/QAM and BFDM/QAM oversampled orthogonal and biorthogonal multicarrier modulations. In: IEEE ICASSP'02, Orlando, USA (2002) Student forum.

8. Siohan, P., Siclet, C., Lacaille, N.: Analysis and design of OFDM/OQAM systems based on filterbank theory. IEEE Trans. on Signal Processing **50** (2002) 1170–1183

9. Bregović, R., Saram ki, T.: A systematic technique for designing prototype filters for perfect reconstruction cosine modulated and modified DFT filter banks. In: IEEE ISCAS'01, Sidney, Australia (2001) 33–36

10. Pinchon, D., Siohan, P., Siclet, C.: Design techniques for orthogonal modulated filter banks based on a compact representation. To appear in IEEE Trans. on Signal Processing (2004)

11. Hleiss, R.: Conception et galisation de nouvelles structures de modulations multi-porteuses. PhD thesis, ENST de Paris (2000)

12. Cvetković, Z., Vetterli, M.: Tight Weyl-Heisenberg frames in $l_2(\mathbf{Z})$. IEEE Trans. on Signal Processing **46** (1998) 1256–1259

13. Pinchon, D., Siclet, C., Siohan, P.: Une m thode de calcul rapide et de g n ra-tion automatis s de modulateurs/d modulateurs OFDM/QAM sur chantillonn s reconstruction parfaite. Technical Report 03-06, CNRS (MIP) (2003)

14. Vaidyanathan, P.P.: Multirate systems and filter banks. Prentice Hall, Englewood Cliffs (1993) ISBN 0-13-605718-7.

15. Pfletschinger, S., Speidel, J.: Optimized impulses for multicarrier Offset-QAM. In: IEEE Globecom'2001, San Antonio, USA (2001)

Nonlinear Bitwise Equalization[*]

Francisco-Javier González Serrano and Manel Martínez Ramón

Departamento de Teoría de la Señal y Comunicaciones
Universidad Carlos III de Madrid. Avda. Universidad 30, 28911 Leganés (SPAIN)
fran@tsc.uc3m.es

Abstract. A new family of nonlinear equalizers that detect Quadrature Amplitude Modulated (QAM) signals by decomposing the process into a binary decision "cascade" is proposed. First, the quadrant of the symbol is determined; second, we determine which quarter of the chosen quadrant contains our symbol, and so on. We have used two nonlinear architectures to approximate the optimal decision boundaries: a Volterra filter and a Generalized Cerebellar Model Arithmetic Computer. Our analysis demonstrates that bitwise equalizers achieve lower Bit Error Rate (BER) than transversal ones.

1 Introduction

Power amplifiers (PA) and physical channels can introduce significant nonlinear distortion effects in digital communication systems. These non-linearities limit the ability of symbol-by-symbol linear equalizers to reduce the error probability. However, the direct application of nonlinear channel models for sequence detection or nonlinear equalizers, such as Volterra filters, Artificial Neural Network (ANN) based structures, presents drawbacks regarding convergence speed, tracking abilities and computational costs.

The present approach is based on a bitwise or staircased classification scheme. Such a methodology consists of solving the symbol decision bit by bit. When a data is introduced into the equalizer, first its (in-phase and quadrature phase) sign is decided, using a specifically trained boundary. Then, the next most significant bit is decided with a boundary selected in function of the previous decision, and so on until the less significant bit is decided.

2 Proposed Scheme

Equalization can be considered as a classification problem in which equalizers define the decision boundaries that allow the received symbols to be assigned to a specific class, which is assumed to be the most likely transmitted symbol. If the objective is to minimize the error probability, the optimum decision boundary between symbols x_i and x_{i+1} will be defined by those received (input) vectors $\mathbf{r}_p$ that satisfy:

---

[*] This work was partially funded by CICYT, TIC2002-03498

J.N. de Souza et al. (Eds.): ICT 2004, LNCS 3124, pp. 589–594, 2004.
© Springer-Verlag Berlin Heidelberg 2004

$$f_\mathbf{r}(\mathbf{r}_P \mid x_i) = f_\mathbf{r}(\mathbf{r}_P \mid x_{i+1}) \tag{1}$$

where $\mathbf{r}_P[k] = (R[k+M_f], ..., R[k], ..., R[k-N_b])^T$ is a vector composed of received samples $R[k]$, M_f is the number of precursor samples, N_f is the number of postcursor samples, $P = M_f + N_b + 1$ is the equalizer order, and $f_\mathbf{r}(\mathbf{r}_P \mid x_i)$ represents a state conditional probability density function for $\mathbf{r}_P$. It is important to notice that even for linear channels, the optimum discriminant functions are nonlinear.

Conventional multilevel classifier design requires an optimization procedure in which all the input samples are used. The objective is to find a set of decision boundaries that minimize some cost with respect to the optimum set defined by Eq (1)). In this work, we suggest splitting the multilevel classification procedure into a binary decision tree. Our aim is taking advantage of binary classifiers of proven quality, and adapting classifiers to the local properties of the input space. Note, for instance, that when nonlinearities are present and saturation effects occur, the shape of each decision boundary varies substantially from one region to another. This situation requires the use of classifiers that can be adapted to the local properties of the input space.

Let us consider a simple example. For an 8-PAM system, the tree has three decision stages: the sign, or the most significant bit (MSB), half of the semiplane which determines if the symbol is in $\{\pm1,\pm3\}$ or in $\{\pm5,\pm7\}$) and, finally, the Least Significant Bit (LSB) determines the symbol value (see Fig. 1). Under linear transmission conditions, the algorithm only requires adding an adjustable bias term to conventional transversal equalizers (except for the tree root), so as to allow each binary decision boundary to deviate from the origin. Taking again the 8-level Pulse Amplitude Modulation (8-PAM) system, initial values for such bias coefficients could be $(\pm2,\pm4,\pm6)$.

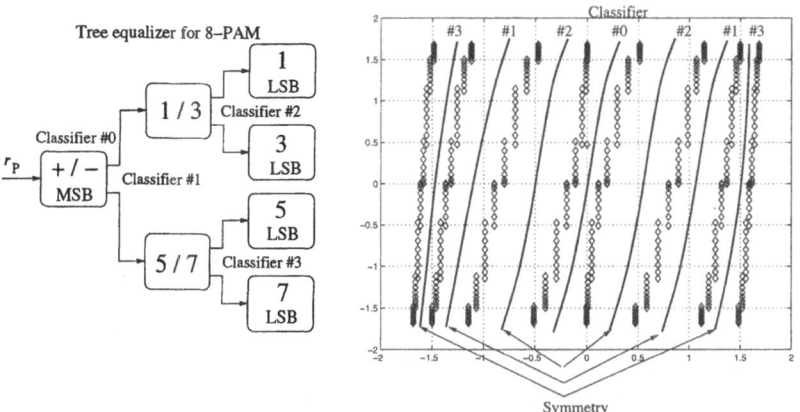

Fig. 1. Bitwise equalizer for a 8-PAM. Symmetries allow to reduce the number of classifiers

3 Training the Bitwise Architecture

We have used networks with a linear-in-the-parameters structure (networks with this property include Radial Basis Functions (RBFs) and the GCMAC or the Volterra Filters (VF)). For these networks, the output can be written as

$$o = \sum_l w_l \Phi_l(\mathbf{r}_P) = \mathbf{w}^T \Phi(\mathbf{r}_P) \tag{2}$$

where $\{w_l\}$ are the parameters and $\{\Phi_l(\cdot)\}$ are some function of the input $\mathbf{r}_P$. Simple instantaneous laws can be used for training, for instance

$$\mathbf{w}[k+1] = \mathbf{w}[k] - \frac{\mu}{2} \nabla_\mathbf{w} J(\varepsilon[k]) \tag{3}$$

where μ is the learning parameter, and $J(\varepsilon)$ is some cost function.

Since each decision stage of the bitwise equalizer outputs a binary result, saturation functions can be employed at the output of any of the nonlinear networks. In particular, the use of the hyperbolic tangent function allows the minimization of cost functions such as the Kullback-Leibler relative entropy:

$$\nabla_{\mathbf{w}_i} J(\varepsilon_{KL}) = 2\lambda (s_i - o_i)^* \Phi_i(\mathbf{r}_P) \tag{4}$$

Since not all of the decisions are made with the same frequency, the information stored in the most frequently updated network, the network that decides the quadrant of samples, can be partially transferred (soft context swapping) to the other networks. We proceed as follows:

$$\mathbf{w}_i[k+1] = \mathbf{w}_i[k] + \beta_i \Delta \mathbf{w}_i[k] + (1-\beta_i) \Delta \mathbf{w}_0[k] \tag{5}$$

$\mathbf{w}_i$ being the weight vector of network i (network 0 decides the most significant bit, i.e. the quadrant, where the symbol is located), $\Delta \mathbf{w}_i$ is the vector that contains the correction and β_i is the mixture parameter ($\beta_0 = 0$; the smaller β_i is, the greater the parallelism in training). In practice, there is a high cross-correlation between weight vectors $\mathbf{w}_i$, so a small value of β_i is taken ($\beta_i \approx 0.05$). This strategy makes the tracking capabilities of the system equal or better than those of a transversal Finite Impulse Response (FIR) filter, as we will see in the simulation section.

4 Nonlinear Networks

As stated above equalization should be performed using nonlinear architectures, which, in addition, should make practical implementation relatively straightforward.

Our approach, which is depicted in Fig. 2, is based on the ensemble averaging method of two experts. Thus, the outputs of the experts are linearly weighted by η_L and $\eta_{NL} = (1 - \eta_L)$. Weights are updated following a first-order law (initial

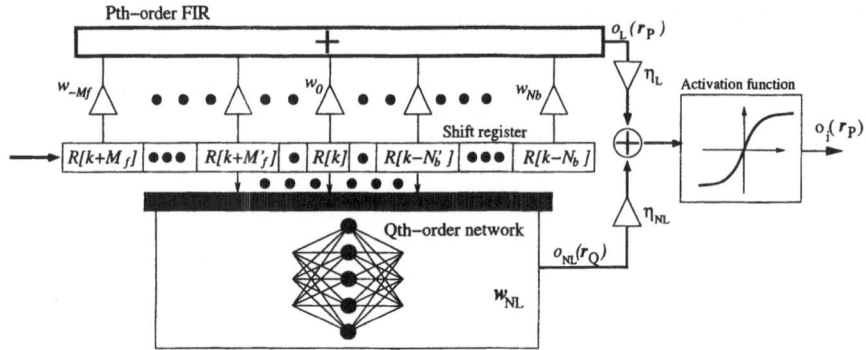

Fig. 2. Committee machine composed by two low complexity experts.

value $\eta_L = 0.7$). We have considered two experts: a P-th order linear FIR filter, and a Q-th order ($Q = N_b' + M_f' + 1$, $Q \leq P$) nonlinear network. The reason why we use the committee machine architecture is that in usual communications channels under multipath most of the distortion energy is produced by the linear impulse response of the channel, whose length, typically, may be about tens of symbol periods. For this reason using a nonlinear structure over the entire duration of the distortion will be a waste of resources. We suggest the use of the GCMAC network [1],[3] as nonlinear expert. The GCMAC can take advantage of the "smoothness" of the optimum decision boundaries to achieve fast equalization for typical nonlinear distortions. In addition, and unlike global approaches, the GCMAC operates locally and as a consequence, it can capture fine details of the decision boundaries, which, in practice, results in better performance.

5   Performance Analysis

In order to compare the performance of the proposed equalizer, we have simulated a 16-QAM system (symbol rate 10 Msymbols per second) with root-raised cosine filters ($\alpha = 0.22$) over a nonlinear land mobile satellite channel. The satellite-to-mobile (or equivalently, mobile-to-satellite) channel is comprised of two cascaded process: the "satellite process", and the "terrestrial process". The "satellite process" is typically described in terms of a memoryless nonlinear function. The "terrestrial process" includes the effects of mobile motion relative to terrestrial scatterers. The nonlinear behavior of the on-board "satellite process" is described by the Saleh' model [4]. For the "terrestrial process", we have used a time-variant discrete multipath model with 2-rays ($h[k] = h_0 \delta[k] + h_1 \delta[k-1]$).

In the first test, we focus on the equalizers' ability for compensating for the channel nonlinearities. We have computed the equivalent SNR degradation caused by the residual nonlinear distortion at a specified BER. The SNR degradation, expressed in dB, is defined as the difference between the SNR required by the equalized system to reach the specified BER at a given output back-off, and the SNR required to obtain the same BER on the Gaussian flat channel. The results for a target BER of 10^{-4} on the Model # 0 channel are shown in Fig. 3(a). All of the equalizers have the same equalization order $P = 3$ ($M_f = N_b = 1$). The equalizers we compare are a linear FIR, and a third-order ($S = 3$) Volterra Filter (VF) and a GCMAC-based ($M'_f = 1, N'_b = 1 \rightarrow Q = 3$, $L = 64$, $\rho_{max} = 8$), equalizer in transversal and bitwise configurations.

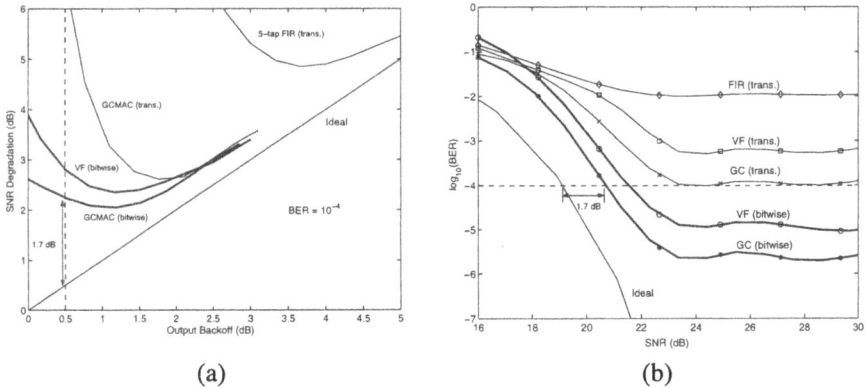

(a)   (b)

Fig. 3. (a) SNR degradation vs. output backoff for a BER=10^{-4}. (b) BER vs. SNR for a backoff of 0.5 dB

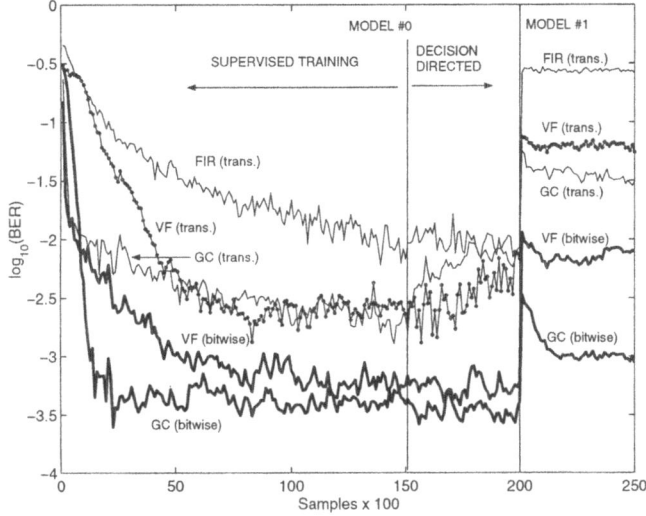

Fig. 4. BER for transversal and bitwise equalizers on a nonlinear time-variant channel.

As the backoff decreases, the equalizers' operation, and especially that of the transversal versions, gets worse. For the case of a saturated PA (0 dB backoff), the Volterra bitwise equalizer requires about 1.5 dB more signal power than the GCMAC equalizer to achieve the same BER. Fig 3.(b) is the complement of the previous figure, showing the BER evolution against SNR for 0.5 dB of output backoff. Again, the transversal filters barely reach the 10^{-4} BER limit, even for extremely high SNR.

The second test is a comparison to determine the tracking abilities of the equalizers on a time-varying channel. Again, the amplifier operates at 0.5 dB of output back-off and the SNR is 20 dB. In this test, all of the equalizers have the same equalization order $P = 4$. Convergence curves, which are shown in Fig. 4, consist of three regions. The first, which spans 15,000 samples, is produced by supervising the equalizers with correct decisions. In the second region, which has a 5,000 samples length, no external reference is used in the training. In the third region, which occurs as a result of 20,000 samples, a sudden change in the channel linear response is considered ($h[k] = \delta[k] \rightarrow h[k] = 0.0995\delta[k] + 0.995e^{j\frac{3\pi}{4}}\delta[k-1]$); again, the training is solely decision-directed. To rank the equalizers behavior, all of them were trained using the learning step that provides the lowest BER.

Fig. 4 represents the smoothed BER evolution. Bold curves are used to highlight the bitwise equalizers behavior. It is observed that in both supervised and decision-directed regions, the bitwise equalizers provide a lower BER than their transversal counterparts. After the change, the linear FIR and the transversal Volterra equalizers are not able to maintain the BER at the levels of the supervised regions.

6 Conclusions

In this paper we have proposed a new structure to equalize nonlinear channels. By means of combining the concepts of equalization and classification, we have introduced a bitwise equalizer that separates the detection of QAM symbols into a binary decision cascade which can easily employ nonlinear schemes. Symmetries of the QAM constellations are exploited to reduce the computational effort.

References

[1] J.S. Albus, A New Approach to Manipulator Control: the Cerebellar Model Articulation Controller, Jnl. on Dynamic Systems, Measurements and Control, 63 (3) (1975) 220-227.
[2] A. R. Figueiras-Vidal, M. Martínez-Ramón, A. Artés-Rodríguez, J. Cid Sueiro, Iterative Decision for Multilevel Equalization, in: Proceedings of the IASTED Int. Conference Expert Systems and Neural Networks, Hawaii, USA, 1996 pp. 357--360.
[3] F. J. González-Serrano, A. Artés-Rodríguez, A. R. Figueiras-Vidal, Generalizing CMAC Architecture and Training, IEEE Transactions on Neural Networks 9 (6) (1998).
[4] A. A. M. Saleh, Frequency-Independent and Frequency-Dependent Nonlinear Models of TWT Amplifiers, IEEE Trans. on Communications COM-29 (11) (1981) 1715-1720.

Modelling and Performance Evaluation of Wireless Networks

G.H.S. Carvalho[1,2], R.M. Rodrigues[1], C.R.L. Francês[1],
J.C.W.A. Costa[1], and S.V. Carvalho[2]

[1] Electrical and Computing Engineering Department,
Universidade Federal do Pará-UFPA,
Belém/PA, Brazil.
{ghsc, rmr, rfrances, jweyl}@ufpa.br
http://www.deec.ufpa.br/lea/index.htm
http://www.deec.ufpa.br/laca/index.php
[2] Laboratory of Computing and Applied Mathematics-LAC,
Instituto Nacional de Pesquisas Espaciais-INPE,
São José dos Campos/SP, Brazil.
{ghsc, solon}@lac.inpe.br
http://www.lac.inpe.br/

Abstract. Integration between voice services and data services brought about new concepts in modelling and performance evaluation of wireless networks. In this paper it is proposed a new approach to deal with that putting together a formal method called Statecharts and a Continuous Time Markov Chain (CTMC)[1]. From the specification of the system made in Statecharts it is automatically generated a CTMC, and then evaluated the performance measures. The advantages of this technique are better representation of the behavior of the system without ambiguity and with consistency.

1   Introduction

The growth of Internet has stimulated an increasing demand for wireless data services. However, circuit-switched data services were not able to satisfy user's and service provider's requirements. This drawback has been overcome by implantation of packet switching based overlays to existing cellular mobile networks, being a evolutive step towards the third-generation (3G) mobile communication systems.

In this paper it is proposed a new approach for modelling and performance evaluation of wireless networks, which combines the Statecharts and a Continuous Time Markov Chain (CTMC) for specifying and solving the resource allocation in Global System for Mobile communications/General Packet Radio Service (GSM/GPRS) networks, respectively. Three resource allocation schemes are modelled and analyzed. Their results are compared to simulation results and it may be seen a good agreement between them.

---

[1] This work was partially supported by CNPq.

J.N. de Souza et al. (Eds.): ICT 2004, LNCS 3124, pp. 595–600, 2004.
© Springer-Verlag Berlin Heidelberg 2004

A lot has been written about GSM/GPRS networks. In [1] it is investigated how many PDCH must be allocated to carry GPRS data traffic. In [2] four resource allocation schemes are studied. It showed that Dynamic Resource Allocation (DCA) schemes outperform Fixed Resource Allocation (FCA) schemes.

The remainder of this paper is organized as follows: Sections 2 and 3 present a overview of the GSM/GPRS network and Statecharts, respectively. Section 4 describes the modelling process of GSM/GPRS resource allocation, while analytical and simulation results are presented in Section 5. Finally, in Section 6, conclusions are draw about the proposed model.

2   GSM/GPRS Network

GPRS is a enhancement of GSM network improving its data capabilities and providing connection with external Packet Data Network (PDN) as X.25 network, Internet, and others [3]. As a packet-switched technology, GPRS only allocates physical channel (PDCH) when there are data transfers over the air interface. Unused GSM radio channels may be allocated as on-demand PDCH. To ensure the integrity of transmitted data packets four Coding Schemes (CS) are defined: CS-1 (9.05 kbit/s), CS-2 (13.4 kbit/s), CS-3 (15.6 kbit/s), and CS-4 (21.4 kbit/s).

3   Statecharts

Statecharts are a visual formalism that was designed to deal with complex reactive systems. These systems are characterized by being large event drive systems and reacting to stimuli received from external and internal medium. Statecharts are the extension of the conventional state diagrams including orthogonality (concurrency), hierarchy (depth), and an appropriated communication mechanism (broadcast communications). Its elements are: states, events, transitions, labels, and expressions. More details see [4].

4   Modelling Resource Allocation in GSM/GPRS Networks

4.1   Resource Allocation Scheme 1

In the first scheme, the radio resource are completely shared between voice and data service. In addition, voice service has preemptive priority over data service, and data packets are accommodated in the buffer while waiting for service. The arrival process of GSM voice calls and GPRS data packets follow a Poisson process with mean λ_v and λ_p, respectively [2,5]. Their service time are exponentially distributed with mean $t_h = 1/\mu_v$ and $t_p = 1/\mu_p$ [2,5]. The number of available radio channels in BTS is N while the buffer size is B_s.

The Statecharts specification that performs these functions is illustrated in Fig. 1. The sub-states of this specification are described bellow:

1. **Source**: Source generator of GSM voice calls and GPRS data packets requests. It has a single sub-state called 'Ready'. *inc_ch* and *inc_bf* associated actions fire new arrivals in the orthogonal components **Voice Channel** and **Buffer**, respectively.
2. **Voice Channel**: It has two single sub-states – 'Free' and 'Busy'. Throughout the *inc_ch* action is fired a new call arrival that may change the status of this component either from 'Free' to 'Busy' or from 'Busy' with v to $v+1$ GSM voice calls.
3. **Buffer**: This template has similar behavior to template **Voice Channel**. The *dec_bf* action decrements one GPRS data packet in the buffer by means of the events *cond* 1 and *cond* 2, while the *inc_bf* action increments one GPRS data packet in the buffer by means of events *cond* 3 and *cond* 4.
4. **Packet Channel**: This template has similar behavior to **Voice Channel** and **Buffer**. The events, which control its behavior (status) are:
   a) Event (Cond 1): It will change the status from 'Free' to 'Busy' if it is true that there is available resource ($v < N$) and there is, at least, one GPRS data packet in the buffer ($Buffer.Busy$).
   b) Event (Cond 2): It will change the status from 'Busy' with p to 'Busy' with $p+1$ GPRS data packets if it is true that there is available resource ($v < N$) and there is, at least, one GPRS data packet in the buffer ($Buffer.Busy$), and the number of GPRS data packets currently in service is less than the available resources ($p < N - v$).
   c) Event (Cond 3): It will change the status from 'Busy' with $p + 1$ to 'Busy' with p GPRS data packets if either a GPRS data packet finishes its service ($p\mu_p(p > 1)$) or it is true that there is not available radio resource ($p = N - v$) and an arrival of GSM voice call takes a place (inc_ch).
   d) Event (Cond 4): It will change the status from 'Busy' with 1 to 'Free', if either the unique GPRS data packet finishes its service (μ_p) or it is true that there is not available resource ($p = N - v$) and the number of GPRS data packets currently in service is 1 and an arrival of GSM voice call takes a place (inc_ch).

The process of generating a CTMC from a Statecharts specification corresponds to construct an infinitesimal generator Q, which contains all transition rates among states. At a given time, a state is defined by the combination of variables (q, v, p) that refer to the current number of GPRS data packets in the buffer, the current number of GSM voice calls in service, and the current number of GPRS data packets in service, respectively. The steady-state probability $\pi = (\pi_0, \pi_1, \cdots, \pi_{max})$ may be evaluated through the matrix equation $\pi.Q = 0$ together with the normalization condition $\sum_i \pi_i = 1$. In this work we have used Gauss-Seidel method to get the steady-state probability.

The blocking probability of a GSM voice call and GPRS data packet are given by (1) and (2), respectively.

$$P_{bv} = \sum_{q=0}^{B_s} \pi_{q,N,F} \; . \tag{1}$$

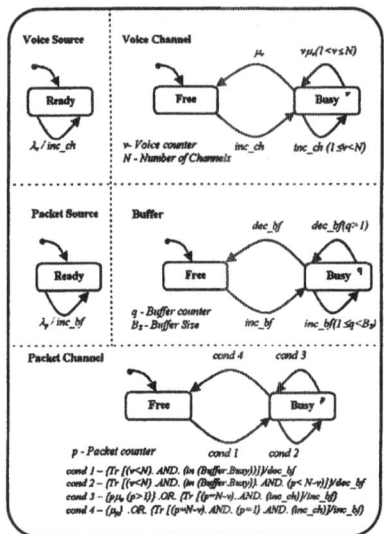

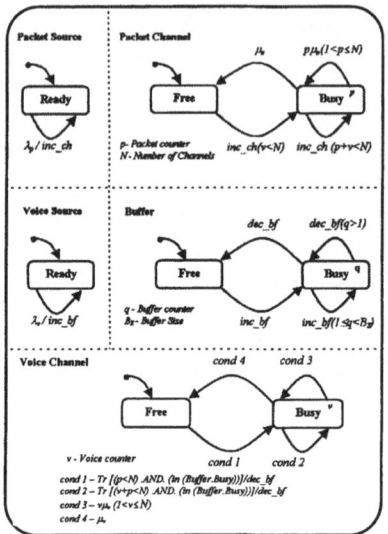

Fig. 1. Statecharts specification of the resource allocation scheme 1

Fig. 2. Statecharts specification of the resource allocation scheme 3

$$P_{bd} = \sum_{q+v+p=N+B_s} \pi_{q,v,p} \; . \tag{2}$$

The Preemption and Dropping Probability of GPRS data packets are given by (3) and (4), respectively.

$$P_{pd} = \frac{1}{\lambda_p(1 - P_{bd})} \sum_{\substack{v+p=N \\ \&p \geq 1}} \lambda_V \pi_{q,v,p} \; . \tag{3}$$

$$P_{dd} = \frac{1}{\lambda_p(1 - P_{bd})} \sum_{\substack{v+p=N \\ \&p \geq 1}} \lambda_V \pi_{B_s,v,p} \; . \tag{4}$$

The mean waiting time of GPRS data packets that is given by (5), while the mean throughput of GPRS data packets is given by (6).

$$W_{qd} = \frac{\sum_{q=1}^{B_s} \sum_{v=0}^{N} \sum_{p=0}^{N-v} q\pi_{q,v,p}}{\lambda_p(1 - P_{bd})[1 + P_{pd}(1 - P_{dd})]} \; . \tag{5}$$

$$T_h = \lambda_p(1 - P_{bd})[(1 - P_{pd}) + P_{pd}(1 - P_{dd})] \; . \tag{6}$$

4.2 Resource Allocation Scheme 2

The only difference between this resource allocation scheme from the one before is the number of available channels dedicated to GPRS (N_p). Therefore, in this model no more than $N - N_p$ voice call will be accepted. So that the performance measures (1)–(6) must be adapted to include it.

4.3   Resource Allocation Scheme 3

This resource allocation scheme is completely different from the others. GSM voice call are taken to the buffer, while GPRS data packets are directly taken to the radio channels. There is not preemptive priority of voice service over data service.

The Statecharts specification that represents this system is showed in Fig. 2.

The blocking probability of GSM voice call and GPRS data packet are given by (7) and (8), respectively. The mean waiting time of GSM voice call is given by (9), while the mean throughput of GPRS data packets is given by (10).

$$P_{bv} = \sum_{q+v+p=N+B_s} \pi_{q,v,p} \ . \tag{7}$$

$$P_{bd} = \sum_{q=0}^{B_s} \sum_{v+p=N} \pi_{q,v,p} \ . \tag{8}$$

$$W_{qv} = \frac{\sum_{q=1}^{B_s} \sum_{v=0}^{N} \sum_{p=0}^{N-v} q\pi_{q,v,p}}{\lambda_v (1 - P_{bv})} \ . \tag{9}$$

$$T_h = \lambda_p (1 - P_{bd}) \ . \tag{10}$$

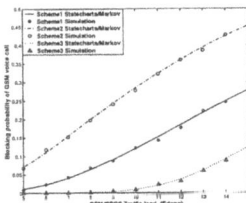

Fig. 3. Blocking probability of GSM voice call vs GSM/GPRS traffic load

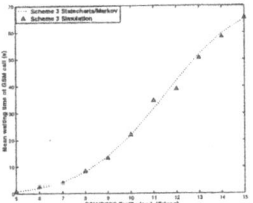

Fig. 4. Mean waiting Time of GSM voice call vs GSM/GPRS traffic load

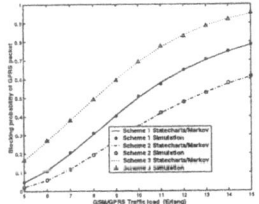

Fig. 5. Blocking probability of GPRS data packet vs GSM/GPRS traffic load

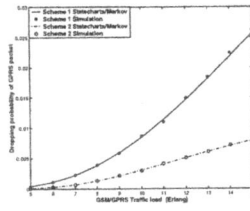

Fig. 6. Dropping probability of GPRS data packet vs GSM/GPRS traffic load

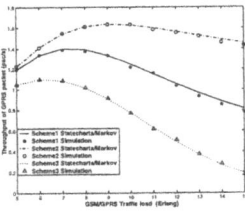

Fig. 7. Mean throughput of GPRS data packet vs GSM/GPRS traffic load

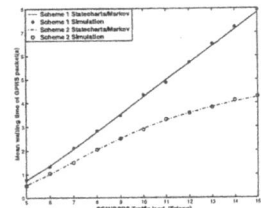

Fig. 8. Mean waiting Time of GPRS data packet vs GSM/GPRS traffic load

5   Performance Evaluation

One cell with 8 radio channels is considered. One radio channel will used for signaling. So seven traffic channels will be available ($N = 7$) [2]. The buffer size is set to be 7. All GPRS data packets have the same priority in addition to carrying their service out in a single slot manner. According to literature the GPRS size packet is exponentially distributed with mean 2×13.4 kbit, resulting in a mean service time of 2 seconds (s) for CS-2, while the mean GSM voice call holding time is set to be 180 s [5]. The GSM/GPRS traffic load varies from 2.5 to 7.5 Erlang each. The confidence level used in the simulation runs was 95 %.

Regarding voice service, it may be seen in Fig. 3 that resource allocation scheme 3 outperforms scheme 1 and 2. Also, scheme 1 has lower blocking probability of GSM voice call than scheme 2 because there are no radio channels dedicated to GPRS. In Fig. 4 it is showed that a GSM voice call has to wait approximately 1 minute to be served for higher values of GSM/GPRS traffic load in scheme 3. Regarding to data service, it may be seen from Fig. 5 to Fig. 8 that the resource allocation 2 outperforms 1 and 3. It happens because there are some radio channels dedicated to carry out data traffic. In addition, resource allocation scheme 3 has the worst performance for data services, because the GPRS data packets are only served if there are available radio channels.

6   Conclusions

In this paper it is proposed a new analytical approach to deal with voice and data integration in wireless network putting together the formal method Statecharts and a CTMC. Regarding GSM/GPRS networks, it must be notice that a trade-off has to be established to satisfy both data service QoS and voice service QoS. Possible enhancement in this model follows the following direction: multiclass queueing system with non exponential service time and inter arrival time of data packets in order to characterize a environment of 3G wireless networks.

References

1. Lindermann, C., and Thummler, A.: Performance Analysis of the General Packet Radio Service, 21st International Conference on Distributed Computing Systems (2001) 673–680
2. Lin, P., Lin., Y.B.: Channel Allocation for GPRS, IEEE Transaction on Vehicular Technology, Vol. 50, No. 2 (2001) 375–387
3. Yacoub, M. D.: Wireless Technology: Protocols, Standards, and Techniques, CRC Press (2002)
4. Harel, D.: Statecharts: a visual formalism for complex systems, Science of Computer Programming, Vol. 8 (1987) 231-274
5. Chen, W.Y., Wu, J.L.C., Lu, L.L.: Performance Comparision of Dynamic Resource Allocation With/Without Channel De-Alloccation in GSM/GPRS Networks, IEEE Communication Letters, Vol. 7, No. 1, (2003) 10–12

Improving End-System Performance with the Use of the QoS Management

Marcio Andrey Teixeira[1] and Jamil Salem Barbar[1]

FACOM - Universidade Federal de Uberlandia (UFU),
Av. Joao Naves de Avila, 2121, 34800-902, Uberlandia - Brasil
marcioandrey@pos.facom.ufu.br, jamil@ufu.br

Abstract. Nowadays, due to the great variety of resources and multimedia applications available in the end system, especially those related to distributed multimedia application , it is been necessary to rely on the Quality of Service management to the improvement of the performance. The present article shows how this can be possible through the use of the management protocol Simple Network Management Protocol and its primitives, by means of the use of adequate management resources and applications.

1 Introduction

Due to the dynamic increase of the computational power of end systems, such as personal computers and networks, the local resource availability and the increasing use of Internet have been leading to the emergence of several distributed multimedia applications, especially the ones related to distributed multimedia applications.

These applications give priority to a closer interaction between users through the massive use of multimedia resources. However, they cause a great increase in the consumption of the net and the end system, which have to be managed to guarantee minimum acceptable Quality of Service (QoS), in order to please the final user end increase the systems' performance.

The most used management protocol nowadays, especially in the monitoring of networks, is the SNMP (*Simple Network Management Protocol*). Thus, this article aims at showing that the use of primitives and the MIB (*Management Information Base*) [1] of the SNMP will be able to be extended to the QoS management in end systems. These procedures, besides being innovative, take into consideration the advantages and the simplicity of the management protocol.

2 QoS and End System

The QoS comprises some stages such as the specification, that consists of obtaining QoS requirements in the application level, the QoS management mechanisms responsible for keeping the defined QoS level and, at last, the so-called provision [2].

J.N. de Souza et al. (Eds.): ICT 2004, LNCS 3124, pp. 601–606, 2004.

The end systems manipulate multimedia streams, such as audio and video streams, proceeding from the multimedia applications. Depending on the existing resources, the end systems are able to generate and show such streams with a certain level of quality, as well as send or receive them through the network.

The distributed multimedia applications use multimedia resources massively; that is the reason why the intense execution of these applications leads to the scarce availability of the end system resources and a performance below of the waited one, being necessary to proceed to its management, in order to guarantee (if possible) the execution of the applications in a satisfactory way. Each multimedia application have quality of service requirements, and such requirements are referred to by QoS parameters. The QoS parameters are defined in the ISO patterns, as being the the collective effect of the performance of a certain service, which determines a service user's satisfaction level [3].

The parameters of QoS specify the amount of resources placed for the services, as well as your discipline of use of the shared resource in one distributed application system, for example, the end-to-end delay parameters determines behavior of the transmission services to the long one of the passage of the media since the origin to the destination, with respect to the scheduling of process (allocation of bandwidth), queuing, (allocation of space), to the o scheduling of tasks (allocation of cpu processing time).

3   QoS Management

The multimedia applications share the need of QoS management to guarantee the meeting of its requirements and, consequently, the keeping of a certain quality (or service) level, having always in mind the user's satisfaction. Memory, CPU, storing Buffer, video, audio, etc. are examples of these requirements.

There are some Frameworks [4] used in the QoS management that define complete sets of management mechanisms. These mechanisms must be inserted in the end system in order to manage the QoS of the multimedia applications in general, provided they are not present. In subsection 3.1 the basic QoS management mechanisms will be described.

3.1   QoS Management Mechanisms

The end system can be seen as an abstraction of several levels or layers, as observed in Fig. 1 [3]. For better agreement, the end system can be divided in three levels: user's level, application level and system level with the audio, video and network device. There is a fourth abstraction level, which is the network level, but the present article focus only on the QoS management of the end system.

This abstraction is used in the understanding of the fundamental mechanism performance in managing the QoS and is part of the proposed management, as it follows:

Throughput Maximization of ARQ Transmission Protocol Employing Adaptive Modulation and Coding

Cristian González[1], Leszek Szczeciński[2], and Sonia Aïssa[2]

[1] Department of Electronics
Universidad Técnica Federico Santa María, Valparaíso, Chile
[2] INRS-EMT
University of Quebec, Montreal, QC, Canada

Abstract. Link adaptation techniques such as adaptive modulation and coding (AMC) aim at maximizing the throughput in wireless networks. In this paper, we propose new schemes allowing for the combining of AMC at the physical layer with automatic repeat request (ARQ) at the link layer. The method we propose formally maximizes the throughput under delay and packet-loss constraints. In addition, we change the basic paradigm of the AMC and propose to vary the modulation/coding levels depending on the state of the ARQ process. Compared to previous approaches, our results show that the proposed AMC scheme enhances the system's throughput.

1 Introduction

Wireless links experience significant temporal and spatial variation in the link quality [1] making it a challenge to optimize the resource utilization and maximize throughput in wireless networks. Link adaptation techniques, such as adaptive modulation and coding (AMC) allow one to cope with the varying nature of the propagation conditions to optimally exploit the available capacity of the channel in order to maximize throughput and provide the required quality of service (QoS) [2]. AMC consists in adapting the transmission parameters, namely, the coding rate and/or modulation levels, depending on the instantaneous channel state, so as to satisfy the QoS requirements, most often expressed in terms of bit or packet error rates (BER, PER), and/or packet delay [2,3]. In flat fading channels, AMC consists in defining a set of signal-to-noise-ratio (SNR) thresholds which are next used to select the most adequate modulation/coding scheme, depending on the instantaneous SNR.

For delay-tolerant applications, AMC can be used along with automatic repeat request (ARQ) to satisfy the QoS requirements. Recently, a truncated selective-repeat ARQ, i.e., with limited number of retransmissions, was considered in [3] where the AMC parameters were chosen so as to guarantee that the PER be lower than a threshold defined by the considered application. Allowing for more retransmissions relaxes the constraints imposed on the individual transmission quality, which in turn allows for a higher spectral efficiency. For instance,

J.N. de Souza et al. (Eds.): ICT 2004, LNCS 3124, pp. 607–615, 2004.
© Springer-Verlag Berlin Heidelberg 2004

in [3], throughput is shown to increase with the number of retransmissions used. However, this increase in throughput is obtained based on a heuristic choice of the transmission and retransmission parameters, leaving space for additional improvement if the parameters were to be optimized.

In this paper, using AMC combined with ARQ, we consider explicit maximization of the throughput, while satisfying the constraints imposed on the transmission delay and on the PER; the SNR thresholds defining the AMC adaptation process being our optimization variables. Considering a Rayleigh fading channel and using a pre-defined set of transmission modes (TMs), defined by modulation-coding pairs, we propose and analyze different approaches to throughput maximization. First, we show how by changing the SNR thresholds only, we are able to increase the system's throughput while satisfying the average PER constraints at no additional cost in complexity. In the example implementing this approach, we limit the number of ARQ retransmissions to one and show that our design outperforms the AMC with two retransmissions based on the heuristic method implemented in [3]. In this way, we are able to increase the performance while maintaining stringent constraints on the transmission delay. Next, we change the paradigm of the AMC design suggested in [3] where the SNR thresholds are defined independently of the state of the ARQ process. We propose an AMC adaptation process in which the TM may be different in the transmission and in the subsequent retransmissions. This approach, referred to as aggressive AMC (A-AMC), improves the throughput because it chooses a more spectrally-efficient TM for the first transmission and switches to the more conservative TM in the retransmissions. According to our analysis, the A-AMC enhances the throughput at the cost of a slight increase in the average transmission delay. The maximum delay, however, is kept within prescribed number of transmissions.

The rest of this paper is organized as follows. The system model is described in section 2. The presentation of the original problem is given in section 3. Our formulation of the problem is then presented in section 4 followed by the numerical results and discussions presented in section 5. Finally, we provide concluding remarks in section 6.

2   System Model

Consider a communications system which employs adaptive modulation and coding at the physical layer, and automatic repeat request at the link layer. AMC is implemented at the transmitter so as to choose the appropriate TM, defined by a modulation level and an FEC coding rate, based on the channel state information (CSI). The latter is obtained from the receiver through the reverse channel or is estimated by the transmitter if the forward and reverse transmissions are performed over the same RF channel such as in time-division duplexing.

At the link layer, selective-repeat truncated ARQ is implemented. Thus, if errors are detected in the received packet (e.g., by means of CRC code), the receiver generates a negative acknowledgement (NACK) and sends it to the transmitter, over the reverse channel, as a retransmission request. Upon reception of the

NACK, the transmitter sends again the packet to the receiver. Retransmissions can continue until the packet is correctly received or that the maximum number of allowed retransmissions is attained; in the latter case the packet is considered lost. We consider hybrid ARQ of type I, i.e., each retransmission carries the same information as the original transmission [4]. Moreover, we assume that no code-combining is performed. Hence, all packets that are received in error are discarded at the receiver.

At the physical layer, data are organized in frames each containing M packets. Each packet within a frame can be transmitted using different TMs if necessary. For further considerations, it is necessary to adopt a suitable analytical expression for the PER. This is done using the following approximation

$$PER_n(\gamma) = \begin{cases} 1 & \text{if} \quad 0 < \gamma < \gamma_{t,n} \\ a_n \exp(-g_n\gamma) & \text{if} \quad \gamma \geq \gamma_{t,n} \end{cases} \quad (1)$$

where $PER_n(\gamma)$ is the PER value corresponding to the n-th TM when the received SNR is γ, and the parameters a_n, g_n, and $\gamma_{t,n}$, defined for $n = 1, \ldots, N$ with N the number of TMs, are obtained through Least-Squares curve fitting with the actual PER vs. SNR curves obtained through simulation [3].

As of the channel model, we assume a Rayleigh quasi-static flat-fading model. The channel is considered invariant within a frame and experiences independent fading from frame to frame. For flat-fading channels, the CSI is uniquely defined by the SNR γ which is a random variable with probability density function

$$p_\gamma(\gamma) = \frac{1}{\overline{\gamma}} \exp\left(-\frac{\gamma}{\overline{\gamma}}\right), \quad (2)$$

where $\overline{\gamma} = E[\gamma]$ denotes the average received SNR.

3   Cross-Layer Design

3.1   AMC Combined with Truncated ARQ

We start by briefly reviewing the main concept of the cross-layer design which consists of choosing the AMC parameters under the constraints imposed by the QoS [3]. The QoS requires that the maximum number of retransmissions, allowed per packet, be no greater than N_{max}, and that the probability of loosing the packet after N_{max} retransmissions be smaller than P_{loss}. Both constraints take into account different QoS requirements defined for the targeted applications (e.g., multimedia) depending on the application's sensitivity to the delay and/or to the packet loss rate.

For AMC with N TMs, the SNR scale has to be partitioned into $N + 1$ intervals, each delimited by two threshold values. Given, this partitioning, the n-th TM, defined by the interval $[\gamma_n, \gamma_{n+1})$, is chosen if the instantaneous SNR γ satisfies

$$\gamma \in [\gamma_n, \gamma_{n+1}), \quad (3)$$

where, by definition, $\gamma_0 = 0$ and $\gamma_{N+1} = +\infty$.

Let $PER_{\max}$ be the maximum instantaneous PER in each transmission. Knowing that $N_{\max} + 1$ independent transmissions are allowed, the cross-layer design consists in determining the SNR thresholds $\{\gamma_n\}_{n=1}^{N}$ under the constraint (called here PER_OBJ) imposed on the PER and given by

$$PER_{\max}^{N_{\max}+1} \leq P_{\text{loss}}$$

$$PER_{\max} \leq P_{\text{loss}}^{\frac{1}{N_{\max}+1}}. \tag{4}$$

Using (1), the SNR thresholds can then be defined as [3]

$$\gamma_n = \gamma_n(PER_{\max}) = \frac{1}{g_n} \ln\left(\frac{a_n}{PER_{\max}}\right), n = 1, 2, \ldots, N, \tag{5}$$

where, for further considerations, we emphasize the functional dependence of the thresholds γ_n on the parameter $PER_{\max}$.

In order to analyze the system's performance both in terms of throughput and PER, we present the PER and the throughput formulae following the development provided in [2] [3]. Given that the n-th TM is chosen with the probability

$$P_n = \int_{\gamma_n}^{\gamma_n+1} p_\gamma(\gamma) d\gamma, \tag{6}$$

and that the average packet error rate for the n-th TM is given by

$$\overline{PER}_n = \frac{1}{P_n} \int_{\gamma_n}^{\gamma_n+1} PER_n(\gamma) p_\gamma(\gamma) d\gamma, \tag{7}$$

the overall average PER is expressed as

$$\overline{PER} = \frac{\sum_{n=1}^{N} P_n \overline{PER}_n}{\sum_{n=1}^{N} P_n}, \tag{8}$$

where the numerator corresponds to the number of incorrectly received packets and the denominator to the total number of transmitted packets.

We assume that if the SNR falls below γ_1 and no transmission is carried out, the idle time is the same as for the first TM. Then, the throughput is given by

$$TH = \frac{\sum_{n=1}^{N} P_n(1 - \overline{PER}_n)}{\sum_{n=1}^{N} P_n/R_n + \left(1 - \sum_{n=1}^{N} P_n\right)/R_1}, \tag{9}$$

where the numerator represents the number of correctly received bits, the denominator indicates the total transmission time, T_{tot}, and R_n is the spectral efficiency of the n-th TM (expressed in bits/symbol). The packet loss rate (PLR) can then be calculated as

$$PLR = (\overline{PER})^{N_{\max}+1}. \tag{10}$$

Note that satisfying the constraints PER_OBJ implies that the PLR is always lower then P_{loss}.

Numerical values of the transmission parameters may be obtained using (2) in (6) and in (7) which leads to the expressions

$$P_n = \exp\left(-\gamma_n/\overline{\gamma}\right) - \exp\left(-\gamma_{n+1}/\overline{\gamma}\right) \tag{11}$$

$$\overline{PER}_n = \frac{a_n}{P_n\overline{\gamma}} \frac{\exp(-x_n\gamma_n) - \exp(-x_n\gamma_{n+1})}{x_n} \tag{12}$$

where $x_n = g_n + 1/\overline{\gamma}$.

3.2 Optimizing the Throughput

The cross-layer design, described in the previous subsection does not directly address the issue of throughput maximization. It rather starts with conditions guaranteeing that the PER constraints are satisfied [cf. (4)] and then by allowing for multiple transmissions increases the spectral efficiency. Note, that both the throughput (9) and the PLR (10) depend on the average SNR $\overline{\gamma}$ and on the thresholds $\{\gamma_n\}$. In other words, $TH \equiv TH(\{\gamma_n\}; \overline{\gamma})$ and $PLR \equiv PLR(\{\gamma_n\}; \overline{\gamma})$. Here we propose to directly maximize the throughput TH under the constraints imposed on the PLR and using the SNR thresholds $\{\gamma_n\}$ as optimization variables. Thus, instead of considering the constraint PER_OBJ in (4), we impose the constraints on the PLR

$$PLR(\{\gamma_n\}; \overline{\gamma}) < P_{\text{loss}} \tag{13}$$

and call these constraints PLR_OBJ. Note that these constraints are less stringent than the PER_OBJ because only the PER averaged over all channel states is constrained while in PLR_OBJ we constrain the PER for *each* packet.

Our optimization problem can then be stated as follows: maximize the throughput for a target SNR $\overline{\gamma}_o$ while ensuring that the PLR is satisfied in the case of different SNRs $\overline{\gamma}_{c,k}$ for $k = 1, \ldots, K$. This is done through the multiple constraints in the following optimization problem:

$$\{\gamma_n\} = \arg_{\{\xi_n\}} \max\{TH(\{\xi_n\}; \overline{\gamma}_o)\} \tag{14}$$

$$PLR(\{\xi_n\}; \overline{\gamma}_{c,k}) < P_{\text{loss}}, \quad k = 1, \ldots, K \tag{15}$$

In this approach, that we refer to as the AMC_PLR solution, packets transmitted over the channel with SNR close to the thresholds $\{\gamma_n\}$ may experience a probability of loss that is higher than P_{loss}. Nevertheless, the PER averaged over the channel states, i.e. the PLR, is ensured to stay within the constraints. The potential problem with the AMC_PLR solution is that the PLR_OBJ constraints are satisfied only if the channel states are independent among the transmissions as it is assumed in the model. In the case of a correlated channel that fades slowly, packets might be transmitted in the conditions with almost the same SNR, that remains close to the threshold value during a long period of time, which can cause undesirable packet-error bursts.

$$PER_{\max,\mathrm{TX}} = \arg_F \max\{TH(\{\gamma_{\mathrm{TX},n}(F)\}, \{\gamma_{\mathrm{RT},n}(P_{\mathrm{loss}}/F)\}; \overline{\gamma}_o)\} \tag{17}$$

$$[\{\gamma_{\mathrm{TX},n}\}, \{\gamma_{\mathrm{RT},n}\}] = \arg_{[\{\xi_n\}, \{\chi_n\}]} \max\{TH(\{\xi_n\}, \{\chi_n\}; \overline{\gamma}_o)\} \tag{18}$$
$$\overline{PLR} \quad (\{\gamma_{\mathrm{TX},n}\}, \{\gamma_{\mathrm{RT},n}\}, \overline{\gamma}_o) \leq P_{\mathrm{loss}}.$$

4   Aggressive Adaptive Modulation and Coding

The AMC considered up to now assumed that the TMs are chosen in the same way for the first transmission and its corresponding retransmissions. We want to change this paradigm noting that the average number of retransmissions is relatively low (usually of order of few percent), so it might be possible to enhance the throughput by increasing the spectral efficiency of the first transmission. Since this would also increase the corresponding PER, retransmissions would have to take care of ensuring the satisfactory packet loss. Given that the TM selection is more "daring" in the first transmission than in the retransmission, we refer to this scheme as *aggressive* AMC (A-AMC). Again, as in the approach presented in the previous section, we consider two different performance constraints PER_OBJ (4) and PLR_OBJ (13) resulting in two design schemes which we respectively call A-AMC and A-AMC_PLR. For the purpose of the demonstration we limit our consideration to the design with only one retransmission, i.e., $N_{\max} = 1$.

In the A-AMC scheme, the SNR thresholds for the first transmission and for the retransmissions may be different, we denote them respectively by $\{\gamma_{\mathrm{TX},n}\}_{n=1}^N$ and $\{\gamma_{\mathrm{RT},n}\}_{n=1}^N$. Now, the throughput and the PLR depend on both sets of the SNR thresholds, i.e., $TH \equiv TH(\{\gamma_{\mathrm{TX},n}\}, \{\gamma_{\mathrm{RT},n}\}; \overline{\gamma})$ and $PLR \equiv PLR(\{\gamma_{\mathrm{TX},n}\}, \{\gamma_{\mathrm{RT},n}\}; \overline{\gamma})$. Denote also, by $PER_{\max,\mathrm{TX}}$ and $PER_{\max,\mathrm{RT}}$ the maximum instantaneous PER for the first transmission and for the retransmission, respectively. In the A-AMC design, the constraint PER_OBJ is given by

$$PER_{\max,\mathrm{TX}} \cdot PER_{\max,\mathrm{RT}} < P_{\mathrm{loss}}. \tag{16}$$

The optimization problem yielding the A-AMC parameters is given by (17) where the constraint (16) is already included, hence only $P_{\max,\mathrm{TX}}$ is the optimization variable. Once its optimal value is found, $P_{\max,\mathrm{RT}}$ is calculated from (16). Both maximum PERs define finally the SNR thresholds through (5).

The A-AMC_PLR design treats both SNR threshold sets $\{\gamma_{\mathrm{TX},n}\}$ and $\{\gamma_{\mathrm{RT},n}\}$ as the optimization variables, so following the idea presented in Section 3.2, the optimization problem (18) is solved.

The A-AMC design implies that packets with different modulation levels may co-exist within the same frame, depending on whether they are transmitted for the first time or correspond to retransmissions. Therefore, the formulae for the throughput and the PER have to be re-derived. The throughput is given by

$$TH = \frac{M_{\mathrm{TX\_OK}} + M_{\mathrm{RT\_OK}}}{T_{\mathrm{tot}}} \tag{19}$$

where $M_{\mathrm{TX\_OK}}$ and $M_{\mathrm{RT\_OK}}$ are the numbers of correctly received packets in the first transmission and the retransmission, respectively, and T_{tot} is the total transmission time. These values are given in (21), (22) and (23), where $P_{m,n}$ represents the probability that the first transmission is done using the n-th TM and the retransmission using the m-th TM, and is given by

$$P_{m,n} = \left(\int_{[\gamma_{TX,n}, \gamma_{TX,n+1}) \cap [\gamma_{RT,m}, \gamma_{RT,m+1})} p(\gamma)d\gamma \right)^2 \qquad (20)$$

The average number of packets in the first transmission and in the retransmission are, respectively $\overline{M}_{TX}$ and $\overline{M}_{RT} = M - \overline{M}_{TX}$ (M being the number of packets per frame). The probability of having the first transmission and not being allowed to retransmit is given by $P_{0,1} = \Pr(\gamma_{TR,1} \leq \gamma \leq \gamma_{TR,2} \wedge \gamma \leq \gamma_{RT,1})$. Finally, replacing in (19) the terms given in (21), (22) and (23), the expanded expression of throughput can be obtained.

$$N_{TX\_OK} = \overline{N}_{TX} \left(\sum_n (1 - \overline{PER}_{TX,n}) P_{TX,n} + (1 - \overline{PER}_{TX,1}) P_{0,1} \right) \qquad (21)$$

$$N_{RT\_OK} = \overline{N}_{RT} \sum_{m,n} \left(1 - \overline{PER}_{RT,m}\right) P_{m,n} PER_{TX,n} \qquad (22)$$

$$T_{tot} = \overline{N}_{TX} \sum_n \frac{P_{TX,n}}{R_n} + \overline{N}_{RT} \sum_{n,m} \frac{P_{m,n}}{R_m} \overline{PER}_{TX,n}$$
$$+ \overline{N}_{RT} P_{0,1} \sum_n \overline{PER}_{TX,n} P_{TX,n} + N \int_0^{\gamma_{1n}} p(\gamma)d\gamma \qquad (23)$$

$$PLR = \frac{\sum_m \overline{N}_{TX} \overline{PER}_{RT,m} P_{RT,m}}{\overline{N}_{TX} \sum_n \frac{P_{TX,n}}{R_n} + \overline{N}_{RT} \sum_{n,m} \frac{P_{m,n}}{R_m} \sum_n \overline{PER}_{TX,n} + \overline{N}_{TX} P_{0,1}} \qquad (24)$$

The PLR can be computed as the ratio of the number of packets incorrectly received in the second transmission, over the number of all transmitted packets. The resulting expression is provided in (24). Note here, that the formula (19) and (24) are valid for $PER_{TX,max} \geq P_{loss}^{1/2}$ which implies that $\gamma_{TX,1} \leq \gamma_{RT,1}$. The average number of retransmitted packets $\overline{M}_{RT}$, employed in (21)-(24), depends on the average number of packets in the first transmission and on its average PER. Therefore,

$$\overline{M}_{RT} = \overline{PER}_{TX} \overline{M}_{TX}, \qquad (25)$$

can further be expressed as

$$\overline{M}_{RT} = \frac{M \overline{PER}_{TX}}{1 + \overline{PER}_{TX}}, \qquad (26)$$

which is used to provide the final expressions for the throughput and the PLR by replacing (26) in (21)-(24).

5   Numerical Comparisons

The numerical results presented in this section were obtained for $P_{loss} = 10^{-4}$, $\overline{\gamma}_o = 20\text{dB}$, $\gamma_{TRc,k} = k; k = 1, \ldots, 20\text{dB}$, and a maximum number of retransmissions N_{max} that varies from 0 to 2 in the AMC case. For the A-AMC scheme, a

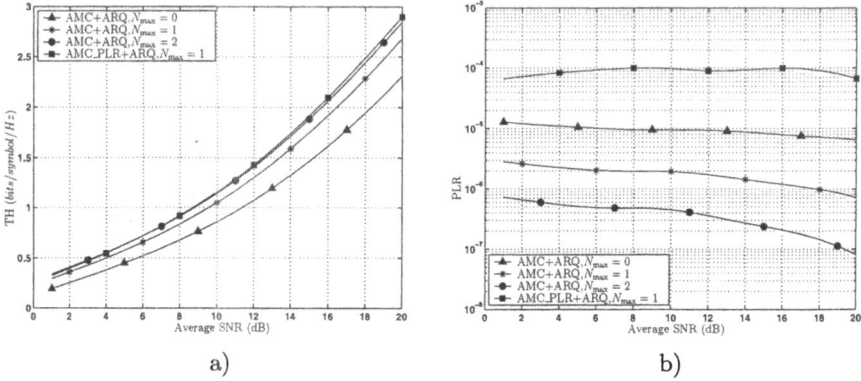

Fig. 1. AMC proposed in [3] and AMC_PLR defined in Section 3.2 a) Throughput, b) Packet Loss Rate.

value of $N_{max} = 1$ was considered. All the optimization procedures were implemented with numerical functions of the Matlab optimization toolbox.

We show in Fig. 1a and Fig. 1b the results of the optimization as described in (14)-(15). In these figures, we also provide results corresponding to the design approach described in subsection 3.1 for comparison purposes. We observe, that allowing for one retransmission increases the throughput of the original AMC design by 0.2-0.4 bit/symbol (depending on the value of $\bar{\gamma}_o$) when compared to one transmission only. Note, that 0.1 bits/symbols translates into 1Mbps of additional bandwidth if the symbol rate is 10Msymbols/s [3]. Compared to the original AMC, our AMC_PLR solution allows us to increase the throughput by 0.1-0.2 bits/symbols while ensuring that the PLR satisfies the constraint of $P_{loss} = 10^{-4}$ (tightly between 7 and 16dB).

The comparison of the throughput obtained by means of the A-AMC and the AMC is provided in Fig. 2a. For the same number of retransmissions allowed, i.e., $N_{max} = 1$, it is shown that the proposed A-AMC outperforms the AMC by 0.1-0.2 bit/symbol in the SNR range 10-20dB. Thus, an additional 1-2Mbps of bandwidth becomes available (symbol rate of 10Msymbols/s). The A-AMC used with one retransmission only outperforms even the AMC scheme used with two retransmissions. The improvement in throughput is obtained without any additional cost for the receiver/transmitter design and not affecting the PLR which is always lower than P_{loss} and in fact, is practically unchanged compared to the one obtained using the original AMC scheme (results not shown for lack of space). Considering the A-AMC_PLR approach, we see in Fig. 2b that an improvement of the throughput with respect to A-AMC (cf. Fig. 2a) is possible; further 0.1 bit/symbol is yielded at the expense of an increase in the PLR which, nevertheless, is kept within the constraint (13) in a similar way as it was already shown in Fig. 1b.

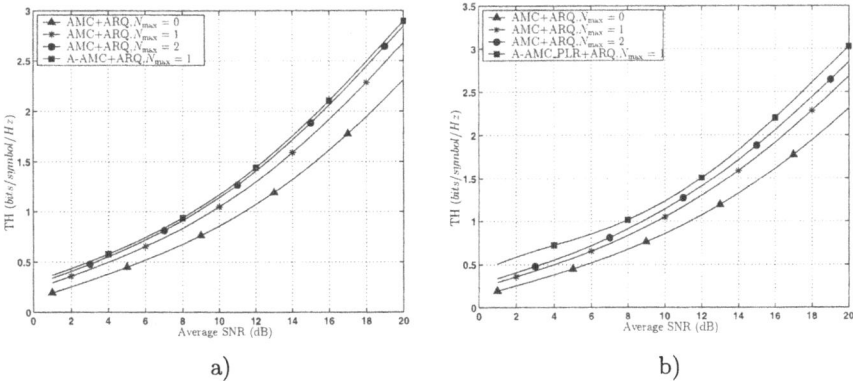

Fig. 2. Aggressive AMC proposed in Section 4 compared with AMC from [3]

6 Conclusions

We presented and analyzed different approaches to throughput maximization when employing adaptive modulation and coding (AMC). We used the constraints defined at the application layer and took into account the improved reliability that can be offered by ARQ at the link layer. The proposed AMC design treats the SNR thresholds as the optimization variables and, unlike the approach in [3], directly maximizes the throughput expression. We considered two types of constraints in the optimization problem we solved. The first imposes the constraints on the individual PER (PER_OBJ) [3] and the second constrains the PLR (PLR_OBJ). We showed that by optimizing the SNR thresholds, throughput can be improved with no necessary modification to the transceiver's design. We also changed the AMC paradigm of [3], by allowing the modulation/coding schemes to change from one transmission to another. The so-called Aggressive-AMC technique is characterized by robustness to channel modelling errors as it imposes the constraints on the probability of error in *each* packet (PER_OBJ) so that the final PER is independent of the actual fading. If the model of the channel is adequate, the A-AMC_PLR solution should be retained as it practically doubles the gain obtained with the heuristic approach presented in [3].

References

1. A. Molisch, *Wideband Wireless Digital Communications*. Prentice Hall, 2001.
2. A. Goldsmith and S.-G. Chua, "Variable-rate variable-power MQAM for fading channels," *IEEE Transactions on Communications*, vol. 45, Oct. 1997.
3. Q. Liu, S. Zhou, and G. B. Giannakis, "Cross-layer combining of adaptive modulation and coding with truncated ARQ over wireless links," *IEEE Transactions on Wireless Communications*, 2004.
4. S. Lin and P. S. Yu, "A hybrid ARQ scheme with parity retransmission for error control of satellite channels," *IEEE Trans. Comm.*, vol. 30, pp. 1701–1719, Jul 1982.

CR-LSP Dynamic Rerouting Implementation for MPLS Network Simulations in Network Simulator

Taumaturgo Antônio Moura Oliveira[1,3], Eduardo Guimarães Nobre[1,2], and Joaquim Celestino Júnior[1]

[1]LARCES – Universidade Estadual do Ceará (UECE)
Av. Paranjana 1700 – Itaperi – 60740-000 – Fortaleza – CE – Brasil
{taumaturgo, eduardo, celestino}@larces.uece.br
[2]Programa de Pós-Graduação em Engenharia Elétrica
Universidade Federal do Ceará (UFC)
Av. Mr. Hull, s/n – Pici – 60455-760 – Fortaleza – CE – Brasil
[3]Centro Federal de Educação Tecnológica do Ceará (CEFET-CE)
Av. Treze de Maio, 2081 – Benfica – 60040-531 – Fortaleza – CE – Brasil

Abstract. This work presents an implementation carried out in Network Simulator that allows the user to simulate and to analyze in a MPLS network the dynamic rerouting of CR-LSPs, in conformity with RFC3214. The addition of the parameter "Actionflag" to the header of the message and the creation and alteration of functions allow the dynamic rerouting of LSPs, using the same identifier, and consequently, a better administration of the resources of the network.

1 Introduction

In the last years, the transmission of multimedia data through the Internet has significantly increased, demanding that the packets have a differentiated treatment. Therefore, the network must guarantee the delivery of information, the delay reduction or even these characteristics combined to make the service qualified for the final user.

We have observed, especially along the last mile traveled by the messages transmitted through the Internet, that difficulties such as network congestion and slowness appear due to the permanence of network that use protocols that don't allow the implementation of "Quality of Service" (QoS). In this scenario appears "Multiprotocol Label Switching" (MPLS) that makes possible the establishment of virtual circuits through several types of networks found along the way. The control over these virtual circuits allows the implementation of QoS and Traffic Engineering concepts.

The Traffic Engineering is concerned with the performance optimization in the networks. In general, it unites technology with scientific principles to measure, model, characterize and control the traffic, and treats the application of such knowledge and

J.N. de Souza et al. (Eds.): ICT 2004, LNCS 3124, pp. 616–621, 2004.
© Springer-Verlag Berlin Heidelberg 2004

techniques to reach specific objectives of performance. It is a process that improves the network utilization by trying to create a uniform and differentiated traffic distribution [4].

2 Network Simulator (NS) and MNSv2

The Network Simulator is a object-oriented simulator of networks, developed at Berkley UC, very used in the academic environment for simulating computer networks. This simulator was developed in C++ and Object Tcl. Its open source code, the many types of networks already implemented and the ease to implement and compile new types of networks are great advantages. Since the object of our study involves MPLS Networks, we used the package "MNS v2.0 for ns-2.1b8a", that compiles on "Network Simulator version 2.1b8a".

2.1 Mapping the Nomenclatures and Concepts of RFCs with the Nomenclatures and Concepts of MNSv2

- LSR and LER: they are declared by the "node" command of NS with a parameter to enable MPLS;
- FTN and ILM: they are implemented in the three structures: 1)LIBEntry (Label Information Base Entry), where the sets of incoming label/interface and the sets of outcoming label/interface are mapped; 2)ERBEntry (Explicit Route Information Base Entry), where there is the FEC and the LSPID mapping with LIBEntry; and 3)PFTEntry (Partial Forwarding Table Entry), where there is the FEC mapping and the data flow ID with LIBEntry.
- NHLFE: In *MNSv2* there is no structure for it. It is implemented along the code. For instance, the get-cr-mapping procedure (in the ns-mpls-ldpagent.tcl file) is a part of this code. The information of next node is stored in LIBEntry.
- Action Flag: variable declared as "actflag" that transports the value "1", that corresponds to the "modify" value cited in [2].

Functions in C++ and Object TCL had to be created and others modified so that the Network Simulator could simulate the dynamic rerouting, as we will see next section.

3 Simulating an Example

In this section we simulate a dynamic rerouting using a network with the following configuration:

- Nodes: Hosts 0 and 7, LSRs 2, 3, 4 and 5, LERs 1 and 6, configured in the same way as hosts of the Fig. 1;
- Links: 10Mbit/s bandwidth, 1ms delay, DropTail queue;

- Flow (1) (UDP, CBR, 1500bytes packet size, 5ms delay);
- Flow (2) (UDP, CBR, 1500bytes packet size, 5ms delay);

Visualizing Rerouting Animations. At first we made a rerouting simulation where a CR-LSP with identifier 1000 was established following route 0-1-2-3-6-7, as is shown in Fig. 1. Then, this CR-LSP had its route modified starting to follow route 0-1-2-3-4-5-6-7, as Fig. 2 shows.

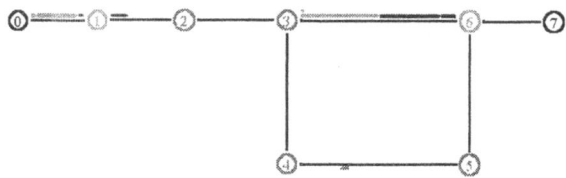

Fig. 1. Network Animator showing two data flows following original CR-LSP and the label mapping packet going through node 5 to 4, mapping the new route.

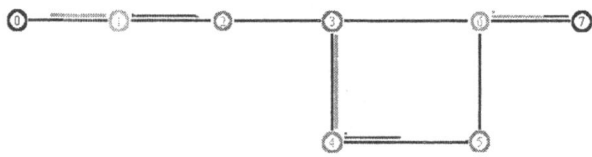

Fig. 2. Network Animator showing data flows following original CR-LSP with modified route.

Initial CR-LSP establishment. The CR-LSP must use the route 0-1-2-3-6-7 initially. It is observed in table 1 the summary of the LIB tables of nodes 1, 2 and 3, with the set of ingress and egress interface/label mappings that characterize CR-LSP. It is also important to observe that node 1 does not have ingress interface (iIface) nor ingress label (iLabel), because it is itself an ingress router, and node 6 does not appear in the LIB tables, because MNSv2 uses Penultimate Hop Popping, where the penultimate LSR of a CR-LSP does not need to store label mappings. It can also be noticed in table 2 that summarizes the tables ERB of nodes 1, 2 and 3, that all lines in the LIBPtr (pointer to LIB, that stores the label mappings of a LSPID) column have the value "0" (zero) because they are pointing to the same CR-LSP. Only the ingress node (1) has the PFT table, shown in table 3, because this table maps the data flows identifiers (Fid) with the LIB entries (LIBPtr). As we created exactly two UDP flows (Fid = "1" and Fid = "2") and send them through the same CR-LSP (LIBPtr = "0"), the PFT table has two lines mapping each Fid to the same LIBPtr.

Table 1. Set of three LIB tables of nodes 1, 2 and 3 for the same CR-LSP.

LIB Tables of nodes 1, 2 and 3							
node:	#	IIface	IIabel	Oiface	Olabel	LIBptr	Linkerror?
1	0	-1	-1	2	1	-1	-1
2	0	1	1	3	1	-1	-1
3	0	2	1	6	0	-1	-1

Table 2. Set of three ERB tables of nodes 1, 2 and 3 for the same CR-LSP 1000.

ERB Tables of nodes 1, 2 and 3									
Node	FEC	LSPid	LIBptr	SLIBptr	QoSid	aPATHptr	ILabel	iIface	Fail Next
1	7	1000	0	-1	-1	-1	-1	-1	*
2	7	1000	0	-1	-1	-1	-1	-1	*
3	7	1000	0	-1	-1	-1	-1	-1	*

Table 3. PFT table of node 1 for the CR-LSP 1000.

PFT Table					
FEC	Fid	Prio	LIBptr	SLIBptr	AlternativePath
7	1	0	0	-1	-1
7	2	0	0	-1	-1

CR-LSP Rerouting Process. The route should be modified to 0-1-2-3-4-5-6-7. It is observed in table 4 that summarizes five LIB tables, that during the process, a new set of labels is created in each node used by the new route, and that in nodes 1, 2 and 3, where the old and new routes coincide, two sets of labels are established for the same CR-LSP. We observe in Table 5, that summarizes five ERB tables of nodes 1, 2, 3, 4 and 5, that the LIBPtr fields of nodes 1, 2 and 3 now points to entry "1" of its respectively LIB table because the value "1" represents the new set of labels (of same CR-LSP but rerouted). Now for nodes 4 and 5, this field points to "0" because in these routers, only a unique set of labels was created for CR-LSP. Table 6, that summarizes PFT tables, remains unaltered because the pointer that attributes UDP flows to LIB entries, was not changed yet as variable LIBPtr shows (it is "0"). Then, the UDP flows remain following the old route.

Table 4. Set of five LIB tables of nodes 1, 2, 3, 4 and 5 for the rerouted CR-LSP.

LIB Tables of nodes 1, 2, 3, 4 and 5							
node:	#	Iiface	iLabel	Olface	oLabel	LIBptr	Linkerror?
1	0	-1	-1	2	1	-1	-1
	1	-1	-1	2	2	-1	-1
2	0	1	1	3	1	-1	-1
	1	1	2	3	3	-1	-1
3	0	2	1	6	0	-1	-1
	1	2	2	4	1	-1	-1
4	0	3	1	5	1	-1	-1
5	0	4	1	6	0	-1	-1

Table 5. Set of five ERB tables of nodes 1, 2, 3, 4 and 5 for the rerouted CR-LSP.

ERB Tables of nodes 1, 2, 3, 4 and 5									
Node	FEC	LSPid	LIBptr	SLIBptr	QoSid	aPATHptr	ILabel	iIface	Fail Next
1	7	1000	1	-1	-1	-1	-1	-1	*
2	7	1000	1	-1	-1	-1	-1	-1	*
3	7	1000	1	-1	-1	-1	-1	-1	*
4	7	1000	0	-1	-1	-1	-1	-1	*
5	7	1000	0	-1	-1	-1	-1	-1	*

Table 6. PFT table of the node 1 for the rerouted CR-LSP.

PFT Table					
FEC	Fid	Prio	LIBptr	SLIBptr	AlternativePath
7	1	0	0	-1	-1
7	2	0	0	-1	-1

Rerouting Finalization and Release of old labels. After the establishment of the new set of labels, a label release message is sent through the old route and, the old labels of this CR-LSP are released. As we can visualize in Table 7, the only sets of labels and interfaces that remain in LIB table are those related with the new route of CR-LSP. ERB table remains unaltered, as we can see in Table 8. After the end of the release operation, we modify the pointer, attributing the UDP flows to the rerouted CR-LSP, as variable LIBPtr of Table 9 shows (it is "1" now).

Table 7. Set of five LIB tables of nodes 1, 2, 3, 4 and 5 for the rerouted CR-LSP without the old labels, already released.

LIB Tables of nodes 1, 2, 3, 4 and 5							
Node:	#	iIface	iLabel	Oiface	oLabel	LIBptr	Linkerror?
1	1	-1	-1	2	2	-1	-1
2	1	1	2	3	2	-1	-1
3	1	2	2	4	1	-1	-1
4	0	3	1	5	1	-1	-1
5	0	4	1	6	0	-1	-1

Table 8. Set of five ERB tables of nodes 1, 2, 3, 4 and 5 for the rerouted CR-LSP without the old labels, already released.

ERB Tables of nodes 1, 2, 3, 4 and 5									
Node	FEC	LSPid	LIBptr	SLIBptr	QoSid	aPATHptr	ILabel	iIface	Fail Next
1	7	1000	1	-1	-1	-1	-1	-1	*
2	7	1000	1	-1	-1	-1	-1	-1	*
3	7	1000	1	-1	-1	-1	-1	-1	*
4	7	1000	0	-1	-1	-1	-1	-1	*
5	7	1000	0	-1	-1	-1	-1	-1	*

Table 9. PFT table of node 1 for the CR-LSP rerouted, without the old labels, already released.

PFT Table					
FEC	Fid	PRio	LIBptr	SLIBptr	AlternativePath
7	1	0	1	-1	-1
7	2	0	1	-1	-1

4 Final Conclusions About the Work

At the end of our work, after having implemented the dynamic rerouting in Network Simulator (NS) as established in [2], we observed that in fact, in the dynamic rerouting of a CR-LSP, there is a great use of the available and used resources, by the fact that the same is liable to being accomplished without making necessary the interruption of the service for the modification of the route of a CR-LSP. However, in the static rerouting, it is first necessary to release CR-LSP and that interrupts the services that use it, and later it establishes the new CR-LSP with parameters identical to the old ones, changing only its route. After, it is necessary to reestablish the services.

References

1. Andersson, L., et al.: LDP Specification - RFC 3036. IETF (2001).
2. Ash, J., et al.: LSP Modification using CR-LDP - RFC 3214. IETF (2002).
3. Ashwood-Smith, P. and Jamoussi, B.: MPLS Tutorial. Nortel Networks. http://www.nanog.org/mtg-9905/ppt/mpls.ppt, (1999).
4. Awduche, D., et al.: Requirements for Traffic Engineering over MPLS - RFC 2702. IETF (1999).
5. Black, U.: MPLS and Label Switching Networks. Prentice-Hall, (2001).
6. IEC, The International Engineering Consortium.: Multiprotocol Label Switching. http://www.iec.org/online/tutorials/mpls/, (2001).
7. Jamoussi, B., et al.: Constraint-Based LSP Setup using LDP - RFC 3212. IETF (2002).
8. MNSv2: MPLS extension for NS2. http://www.isi.edu/nsnam/archive/ns-users/webarch/2001/msg05364.html, (2001).
9. NS2: Network Simulator version 2.1b8a. http://www.isi.edu/nsnam/ns/, (2001).
10. Rosen, E., Tappan, D., et al.: MPLS Label Stack Encoding - RFC 3032. IETF (2001).
11. Rosen, E., Viswanathan. A., et al.: Multiprotocol Label Switching Architecture - RFC 3031. IETF (2001).
12. Stallings, W.: Multiprotocol Label Switching MPLS. Cisco Systems, http://www.cisco.com/, (2001).

FuDyLBA: A Traffic Engineering Load Balance Scheme for MPLS Networks Based on Fuzzy Logic

Joaquim Celestino Júnior, Pablo Rocha Ximenes Ponte,
Antonio Clécio Fontelles Tomaz, and Ana Luíza Bessa de Paula Barros Diniz

Communication Networks and Information Security Laboratory (LARCES) – Universidade Estadual do Ceará (UECE)
Av. Parajana, 1700 – Itaperi – 60.720-020 – Fortaleza – CE – Brasil
{celestino,pablo,clecio,analuiza}@larces.uece.br

Abstract. This article describes a new traffic engineering scheme based on a load balance reactive method for congestion control in MPLS networks that makes use of local search and fuzzy logic techniques. It is an algorithm based on the "First-Improve Dynamic Load Balance Algorithm" (FID) and it works with linguistic values while performing load balance decisions. Computer simulation experiments have shown a better traffic distribution over the links of a network if compared to the behavior of the original version of the algorithm under the same circumstances.

1 Introduction

One of the most interesting applications of MPLS in IP based networks is Traffic Engineering (TE) [2]. The main objective of TE is to optimize the performance of a network through an efficient utilization of network resources. The optimization may include the careful creation of new Label Switched Paths (LSP) through an appropriate path selection mechanism, the re-routing of existing LSPs to decrease network congestion and the splitting of traffic between many parallel LSPs. The main approach to MPLS Traffic Engineering is the so-called Constraint Based Routing (CBR) [4]. In this article we present a new scheme to control the congestion in an MPLS network by using a load balancing mechanism based on fuzzy logic entitled FuDyLBA. FuDyLBA was built upon the ideas of the First-Improve Dynamic Load Balance Algorithm (FID) [4]. The main weakness of FID is that it does not take into account the link utilization intensity to define how many LSPs it will re-route which makes necessary in some cases several iterations of the whole algorithm. The proposed scheme adds a fuzzy controller to the structure of FID, thus eliminating the problems concerning its limitations. The article is organized as follows: Section 2 highlights general aspects about the proposed scheme; Section 3 describes in details the pseudocode of the proposed algorithm; in Section 4 we discuss the experiments performed to analyze FuDyLBA and their results; and finally in Section 5 we make some final considerations and sugestions for possible future works.

J.N. de Souza et al. (Eds.): ICT 2004, LNCS 3124, pp. 622–627, 2004.

2 Overview of FuDyLBA

Our algorithm is named FuDyLBA, because it is an improved version of First-Improve DyLBA (FID) by means of fuzzy logic techniques, or Fuzzified DyLBA (FuDyLBA).

In our method, the core of FID was modified in a way that enabled it to work with intensity levels, thus creating a link unload function. This function gets as input an intensity level that indicates how much a certain link should be released of its load. Such function performs in a way very similar to FID, but it does not stop after finding the first suitable LSP and rerouting it. In fact, it keeps restarting after every finding, accounting the amount of traffic load that has been released. When the released traffic load reaches the level defined by the input parameter, the local search stops and the next link is then examined. That goes on until all the links in the network are examined.

To calculate the input parameter of the link unload function, a fuzzy controller was designed. This controller has the percentage of link utilization as input parameter and the intensity level in which the link should be released of load as an output.

To represent the level of link utilization three linguistic input values were defined, as follows: "low congested", "medium congested" and "high congested". Each one of these values is defined by its own triangular membership function, hereby named input memberships. To understand linguistically the intensity level necessary for the unload function, three non-continuous membership functions were defined, as follows: "low unload", "medium unload", "high unload", hereby named output memberships.

Every link in the network is measured in terms of percentage of link utilization and when a certain measurement fits any of the input memberships the link is considered congested. The unload function is then executed having the output of the fuzzy controller's defuzzification process (center of gravity) as an input.

3 The Pseudocode

The pseudocode of our algorithm is shown in figure 1. Let us define the notation. Three pseudocode blocks are defined. The first block shows the general structure of the algorithm, while the second and third constitute the fuzzified link unload procedure. On the first block the variable *LinkSet* receives the set of all the links in the network, and a loop is started. Inside this loop, one element of *LinkSet* is taken and registered in the variable *link* per iteration. After that, the unload function is called with *link* as input. The loop ends after the variable *LinkSet* has no more elements left.

The unload function is a fusion of the second and third blocks. Actually, this function starts with a fuzzy controller, illustrated in the second block, that feeds the **core** of the unload function with the level which the link should be uloaded, based on the percentage of the link utilization. In the third block, we have the core of the unload function. In this portion of the pseudocode, first the variable change is initialized with zero. This variable will hold the accounting for how much the link was released of load. After that, the first loop is started and it will only stop when the

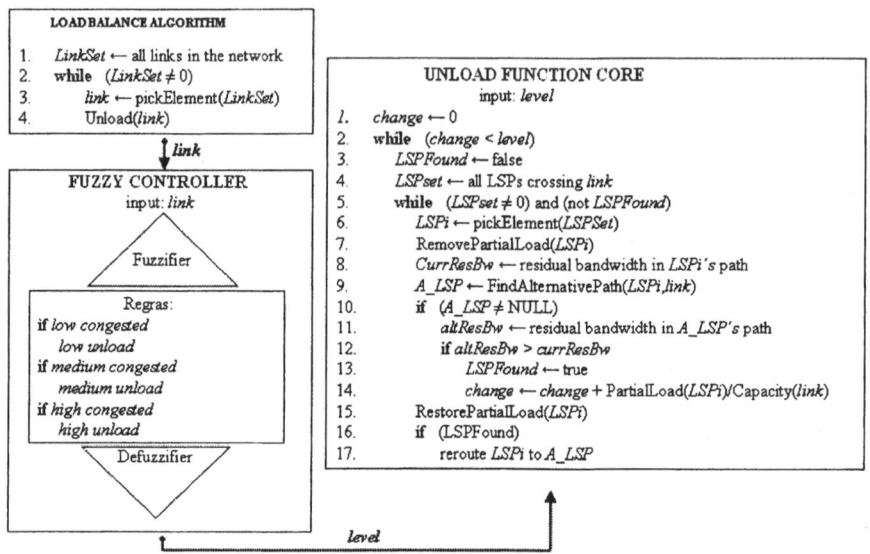

Fig. 1. Pseudocode for FuDyLBA

variable change is greater than or equal to the variable level (which holds the intensity wich the link should be unloaded). Inside this first loop, we initialize the variables LSPFound (with false) and LSPSet (with the set of all the LSPs crossing the link) and a new loop is then started. This second loop will only stop when LSPFound is equal to the logical value "true" or when there is no more elements left in LSPSet. Inside this loop the local search is performed first by taking an element from LSPSet and placing it in the variable LSPi. This is the LSP to be examined. After that, the contribution of the LSP stored in *LSPi* in terms of bandwidth consumption is removed from the network through the function *RemovePatialLoad*. The residual bandwidth for the current LSP is, then, calculated and stored in *currResBw*. After that, the algorithm tries to find a new path that does not include the current link and tries to store it in *A_LSP*. If it succeeds, the residual bandwidth for this alternate path is calculated, stored in *altResBw*, and compared to *currResBw*. If it is greater, *LSPfound* is set to "true" and change is increased with the percentage of link capacity that was released. Finally, the partial load of the LSP stored in *LSPi* is restored to the network and if *LSPFound* is "true" the LSP defined by *LSPi* is rerouted to the path defined by *A_LSP*.

4 Simulations and Results

The objective of the computer simulations was mostly to evaluate the improvements of FuDyLBA over FID, since FID was very well documented and simulated in [3], with comparisons to other load balance techniques that are very much mentioned in the literature, thus making its behavior well known.

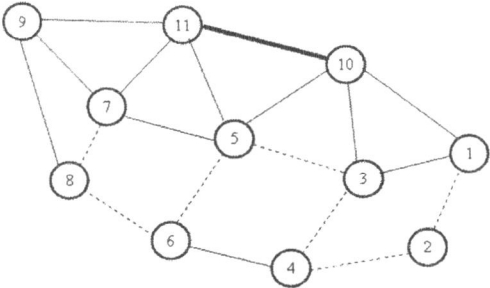

Fig. 2. Topology for the Simuled Network Used in the Experiments

4.1 The Experiments

To accomplish the experiments, the topology defined in the figure 2 was used. The thin traced lines represent links with capacity of 1024 units, the thin non-traced lines represent links with capacity of 2048 units and the thick line represent a link with capacity of 3072 units. The LSP requests are limited to the pair of LSRs 1 and 9, always starting at the LSR 1.

Two experiments were performed, both with the duration of 30 seconds, reproducing the exact same network characteristics for the two algorithms.

The first experiment used low granularity LSPs while the second one used higher granularity LSPs. The LSPs varied in a round-robin style in terms of bandwidth consumption, path and duration while they were being injected in the network at the rate of one at every 0,05 seconds. Six hundred LSPs were injected for both experiments. And the paths used were 1-3-10-11-9 and 1-10-11-9.

In the first experiment the bandwidth consumption used for the injected LSPs varied between 8 and 32 units as their lasting time varied between 3 and 8 seconds.

In the second experiment the bandwidth consumption used for the injected LSP varied between 32 and 128 units and their lasting time was the same used in the first experiment.

4.2 Results

The technique of choice for the evaluation of the quality of the load balance for each algorithm was the calculation of the standard deviation, during the time, for the percentages of residual bandwidth for each link in the simulated network.

The standard deviation is a well known dispersion measurement technique that is very suitable for the quality quantification of any load balance technique.

The standard deviation measures how much the elements of a sample set are apart from each other. Specifically in our case, those elements are the residual bandwidth for the link. A higher standard deviation would mean, in a final analysis, a worse load balance quality.

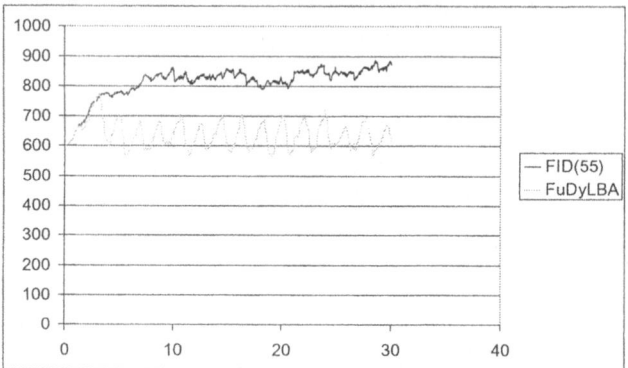

Fig. 3. Standard deviations of the desidual dandwidths for the first experiment, along the time in seconds, respectively.

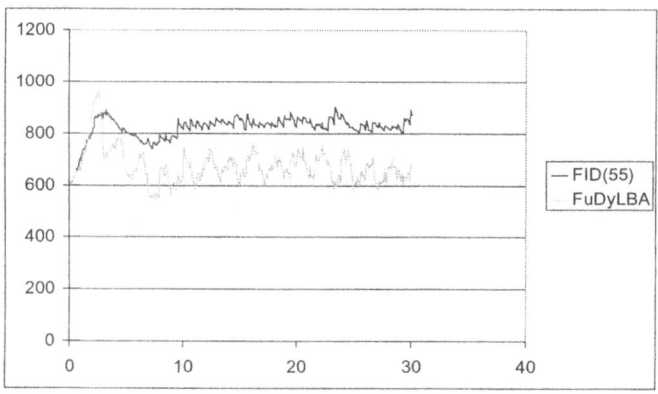

Fig. 4. Standard reviations of the residual bandwidths for the second experiment, along the time in seconds, respectively

In figure 3 we can see the chart for the first experiments. This chart shows the standard deviation of the residual bandwidth for every link in the network along the time for FID and FuDyLBA. The same measurements for the second experiment can be found in figure 4.

Analyzing the presented charts, we can notice that FuDyLBA performs a better load balance than FID, not only as a final result, but for the most part of the experiment, even under attenuating circumstances.

The second experiment intended to show the effects of LSP granularity in FuDyLBA. By the charts it is possible to realize that the curves for both algorithms get closer, thus showing that under the insertion of high granularity LSPs FuDyLBA performs more similar to FID. This is a result of the link based characteristic of FuDyLBA. This does not constitue a downside because the computational complexity follows the same behavior.

5   Conclusions and Future Works

In this paper we presented the benefits of Fuzzy Logic techniques to traffic engineering reactive congestion management methods. Specially, we presented FuDyLBA, a reactive load balance algorithm based on fuzzy logic based that has its origins in the FID algorithm. Computer simulations were able to demonstrate how better FuDyLBA performs if compared to its predecessor. Is was shown as well that attenuating circumstances, like LSP granularity, can suppress the improvements brought by fuzzy logic techniques, making the functionality of the modified algorithm to be similar to the original one. That, in the other hand, does not constitute a downside, because computational complexity follows the same behavior, thus making the fuzzified algorithm, during attenuating circumstances, to behave like its non-fuzzy version with the same computational cost.

Some aspects of the algorithm are yet to be adressed in future works. This includes a better design of the membership functions and the integration of neural networks and genectic algorithms techniques.

References

[1]   Awduche, D.; Chiu, A.; Elwalid, A.; Widjaja, I. e Xiao, X.. (2002) "Overview and Principles of Internet Traffic Engineering.", IETF RFC 3272,
      http://www.faqs.org/rfcs/rfc3272.html, May

[2]   Awduche, D. O. e Jabbari, B. (2002) "Internet Traffic Engineering using Multi-Protocol Label Switching (MPLS)", Computer Networks, (40):p. 111–129, September.

[3]   Battiti, R. e Salvadori, E. (2002) "A Load Balancing Scheme for Congestion Control in MPLS Networks. Technical report", Universitá di Trento, Dipartimento di Informatica e Telecomunicazioni, November.

[4]   Salvadori, E. Sabel, M e. Battiti, R. (2002) "A Reactive Scheme For Traffic Engineering In Mpls Networks. Technical report", Universitá di Trento, Dipartimento di Informatica e Telecomunicazioni, December.

Service Differentiation over GMPLS

Hassine Moungla and Francine Krief

LIPN Laboratory – Paris XIII university, 99 avenue J. Baptiste clement,
93430 Villetaneuse, France
{hassine.moungla, krief}@lipn.univ-paris13.fr

Abstract. This paper presents the service differentiation over Generalized Multi-Protocol Label Switching (GMPLS) networks. A method is proposed to provide necessary bridges between DiffServ and GMPLS network. Policy framework defined by the IETF is also extended with new policies for controlling and managing the life-cycle of Label Switched Paths (LSPs) and for the mapping of DiffServ traffic onto existing LSPs.

1 Introduction

The Multi-Protocol Label Switching (MPLS) technology is a suitable method to provide Traffic Engineering (TE), independent of the underlying layer2 technology [1],[2]. MPLS cannot provide a mapping between a LSP and a service differentiation, which brings up the need to complete it with another technology capable of providing such a feature: DiffServ. DiffServ is becoming prominent in providing scalable network designs supporting multiple classes of services. Currently, in Internet Engineering Task Force (IETF), DiffServ and MPLS, both provide their own benefits [3].

Generalized Multi-Protocol Label Switching (GMPLS) is a new technology proposed by the IETF [4]. GMPLS is a protocol [9] that use advanced network signaling and routing mechanisms to automate set up of end-to-end connections for all types of network traffic: TDM, Packets or Cell-based, Waveband or Wavelength. GMPLS extends MPLS to encompass new technologies (optical networks). We would also like to extend the differentiation of service in this Generalized MPLS networks. In this paper, a new method is proposed to engineer QoS paths within a Diffserv over GMPLS domain. This notion applies a set of rules across the domain (GMPLS domain) to allocate LSP to the traffic based on the Diffserv classes of service and dynamic link metrics.

We will apply policy-based management (PBM) concepts to manage a DiffServ GMPLS network, because we consider this as an appropriate way of dealing with large sets of managed elements. Using PBM for networks and systems is very popular since the early work on policies such as [5],[6],[7].

This article is organized as follows: section 2 describes the GMPLS network and its basic mechanisms, section 3 describes the DiffServ mechanism over MPLS, section 4 presents our proposition for accomplishing the service differentiation in

J.N. de Souza et al. (Eds.): ICT 2004, LNCS 3124, pp. 628–637, 2004.

GMPLS, section 5 and 6 deals with the application for PBM in DiffServ/GMPLS networks. And finally, we conclude in section 7.

2 GMPLS

Traditional MPLS is designed to carry Layer 3 IP traffic using established IP-based paths and associated these paths with arbitrarily assigned labels. These labels can be configured explicitly by a network manager, or be dynamically assigned by means of a protocol such as the Label Distribution Protocol (LDP) or the Resource Reservation Protocol (RSVP).

From MPLS to GPMLS. GMPLS generalizes MPLS in defining labels to switch varying types of Layer 1, Layer 2, or Layer 3 traffic. A functional description of extensions MPLS signaling needed to support the new classes of interfaces and switching is provided in [9]. GMPLS nodes can have links with one or more of the following switching capabilities:

- Fiber-switched capable (FSC),
- Lambda-switched capable (LSC),
- Time division multiplexing (TDM) switched-capable (TSC),
- Packet-switched capable (PSC).

Label-switched paths (LSPs) must start and end on links with the same switching capabilities GMPLS extends the reach of MPLS through a control plane allowing it to reach into other networks and provide centralized control and management of these networks. This will bring greater flexibility to somehow rigid optical networks and provide a centralized management and control.

GMPLS basic mechanisms. GMPLS uses a set of mechanisms to achieve scalability, handle the network heterogeneity regarding switch and forward mechanisms and ease of configuration, besides the requirements for resiliency. Some of these GMPLS specific mechanisms are [10][11]:
- A separation of Control channel from Data Channel,
- A suggested Label to speed up label assignment,
- Use of Forwarding Adjacency (FA) Routing and Signaling.

3 DiffServ over MPLS

In a multi-service network, different applications have varying QoS requirements. The IETF has proposed the DiffServ architecture as a scalable solution to provide Quality of Service (QoS) in IP Networks. In order to provide quantitative guarantees and optimization of transmission resources, DiffServ mechanisms should be complemented with efficient traffic engineering (TE) mechanisms, which operate on an aggregate basis across all classes of service. The MPLS technology is a suitable method to provide TE, independently of the underlying layer2 technology [3].

In recent years there has been active research in the field of MPLS and an increasing number of networks are supporting MPLS. An MPLS network consists of label switched paths (LSPs) and edge/core label switch routers (LERs/LSRs). The LSRs store the label translation tables. Core LSRs provide transit services in the middle of the network while LERs (edge LSRs) provide an interface with external networks. The RFC 3270 gives an answer to a research problem related to the DiffServ model and MPLS, this document [12] presents a solution for the support of DiffServ over MPLS networks. There are two types of MPLS LSPs that support DiffServ extensions the E-LSP and the L-LSP:

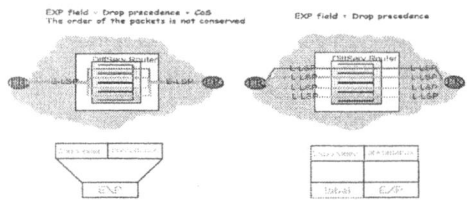

Fig. 1. E-LSP et L-LSP representation

E-LSP (EXP-inferred-PSC LSPs): This type of LSP can support only eight Behavior Aggregates (BA). This is because the three-bit EXP field is used to infer the PHB that is applied to a packet. These LSPs are referred to as "EXP-inferred-PSC LSPs" (E-LSP). The PHB Scheduling Class (PSC) of a packet transported on this LSP type depends on the EXP field value for that packet. With this approach, a single LSP can be used to support one or more Ordered Aggregates (OAs). Here, the mapping of an EXP field with a PHB is either explicitly done during label establishment or pre-configured.

L-LSP (Label-Only-Inferred-PSC LSP): Separate LSPs are established for a single <FEC, OA> pair. With such LSPs, PSC is explicitly signaled at the time of label establishment, so that after label establishment, the LSR can exclusively infer, the PSC to be applied to a labeled packet from the label value. When the Shim Header is used, the Drop Precedence to be applied by the LSR to the labeled packet is conveyed inside the labeled packet MPLS Shim Header using the EXP field. When the Shim Header is not used (e.g., MPLS Over ATM), the Drop Precedence to be applied by the LSR to the labeled packet is conveyed inside the link layer header encapsulation, using link layer specific drop precedence fields (e.g., ATM CLP)[13].

4 DiffServ over GMPLS

As for MPLS, in GMPLS, the setting up of QoS based differentiated services rests on TE and DiffServ, but here E-LSP solution is excluded and only that based on L-LSP can considered. For the simple reason that data routing in the optical nodes is transparent. Indeed, in the optical nodes, the label is quite simply a time slot, a wavelength or a fiber. Thus, here the QoS management mechanisms are gathered around two techniques:

- Traffic Engineering : it rests on in GMPLS an evolved RSVP-TE for the resource reservation, the generalized LSP establishment, on routing protocols adapted to GMPLS for the topology knowledge and on other network characteristics.
- DiffServ L-LSP : it consists in associating the PHB (class of traffic and drop precedence) with a LSP. In fact, neither the class of service nor the drop precedence can be transported in the data.

4.1 Traffic Engineering in GMPLS

The objectives and principles of TE in GMPLS are identical to those of MPLS. On the other hand the mechanisms and protocols evolved in order to adapt to heterogeneous environments and optical networks constraints. In GMPLS, routing and signaling processes are optimized and adapted to the heterogeneous networks. They use an out of band signalization (control channel) established by LMP. LMP provides also effective mechanisms for the checking and the detection of breakdown on the links. We limit ourselves here to describe, in what follows, their differences compared to those of MPLS.The information distribution with the TE routing protocols. The link information included in the Label Switch, which is transmitted by the node is similar to that in MPLS: maximum bandwidth, Maximum reserved Bandwidth, Unreserved bandwidth, Traffic Engineering metric, etc. However, they are related here to interface type and coding information: Package, Ethernet, time slot (SDH/Sonet), lambda (wavelength) or fiber.

Generalized labels distribution and resources reservation by RSVP-TE for GMPLS. The evolutions of RSVP-TE for GMPLS are proposed in RFC 3473 at the IETF [18]. RSVP-TE for GMPLS is an evolution of the RSVP-TE, it allows the establishment of a generalized LSP and the setting up of complex management resource reservation for heterogeneous environments.

4.2 DiffServ L-Lsp

L-LSP (Label only inferred PHB Scheduling Class LSP) is the only solution making it possible to associate a generalized LSP with a couple <FEC, OA>. Indeed, for two points:

- In the optical networks, there is no header added to the packet,
- The optical nodes can not examine header.

For the same reasons, we will not be able to manage either drop precedence within the same class of service in the optical nodes. The PHB will depend only on the value of the label, that amounts establishing a LSP by class of service towards the same recipient. In a given L-LSP, each LSR will use the label for at the same time conveying the packet and to determine which treatment or PHB (class of service) apply to the packets. In the case of classes AF and only in the not-optics nodes, we can use the EXP field of the "shim header" for the drop precedence.

4.3 A Suggested Method

We have modified the work presented in [14], and adapted it to GMPLS networks. In order to support the differentiation of service, we propose that the L-LSP is established according to the priority of the customers, the best link corresponding to the best class according to the Explicit Routing. The cost of the link will depend on:

- the priority of the customer,
- the link utility (appearance frequency in the shortest link),
- the available bandwidth.

The first stage consists in calculating the various links utilities by using Disjktra algorithm. From each peripheral node (Edge Router), we seek with Disjktra algorithm the shortest ways towards all the other peripheral nodes. The link utility is the number of their individual appearances in these shorter ways. Thus each link will have a Utility called **LU** which will be used for the costs calculation.

The second stage consists in calculating the costs of the various links for each class of flow according to a formula which will be detailed further. Lastly, we calculate the best ways for each class by minimizing the total cost given by the following formula:

$$C \text{ (Total)} = \sum_{}^{N} C(\text{ LINK I }) \tag{1}$$

such as : N a number of links in the way and I Link included to the way.

The calculation of the ways for a given class is used for the distribution of labels and construction of the switching table. For the same destination we will have eight different labels corresponding to eight different ways referring each one to a given class (a given priority level). The links cost are calculated using the formulas below. The first corresponds to the first two priority levels, the second concerns the two following levels and so on.

$$CLink = \begin{cases} e^{\frac{1}{B1 \left(\frac{P1 \ x \ Lu}{LU \ max} + \frac{P2 \ x \ BPrest}{BP \ max} \right)}} & 1 \\ \\ e^{\frac{1}{B2 \left(\frac{P1 \ x \ Lu}{LU \ max} + \frac{P2 \ x \ BPrest}{BP \ max} \right)}} & 2 \\ \\ e^{B2 \left(\frac{P1 \ x \ Lu}{LU \ max} + \frac{P2 \ x \ BPrest}{BP \ max} \right)} & 3 \\ \\ e^{B1 \left(\frac{P1 \ x \ Lu}{LU \ max} + \frac{P2 \ x \ BPrest}{BP \ max} \right)} & 4 \end{cases} \tag{2}$$

where:
- **B1** and **B2** are two entities such as $B1 > B2$, chosen by the operator to differentiate the classes.
- **P1** and **P2** with **P2 > P1** and **P1 + P2 = 1,**
- **Lu** (Link Utility): frequency of link appearance in the shortest ways.
- **LUmax** the greatest value of LU in the network.
- **LP max** the max capacity of link bandwidth.
- **BPrest** remaining bandwidth for an arriving flow. This parameter depends on its class.

If, at one moment $Ti,$ a flow of priority J arrives, then:

$$\mathbf{BPrest = BP \ max - \ BPreserved \ (I)} \text{ (I has a greater priority than J)} \tag{3}$$

To homogenize the parameters of our formula, we divide **LU** by **LU max** and **BPresv** by **BPmax.**

5   Policy Based Control Architecture

Policy based QoS management has been widely supported by standardization organizations such as IETF and DMTF to address the need of QoS traffic management. They proposed a PBM framework that defines a set of components to enable policy rules that govern network resources, including conditions and actions, with parameters that determine when the policies are to be implemented in the network. Many works have been carried out in the area of policy-based QoS management for wired networks, but only few works concern GMPLS networks.

Policy scheme is one of the powerful tools to control resources allocation and automate network behavior based on the QoS requirements specified at different levels.

A policy rule is a translation of the Service Level Agreements (SLAs) to a set of actions to be performed on target provided that some conditions are satisfied [6].

Given a different levels of policies that describe users and applications needs and system constraints, a flexible framework, which we describe later, is required to manage these policies, translate them into lower level policies and monitor different events that can trigger one or more policy actions.

A three level architecture. Taking in the consideration the network model (GMPLS) presented in the last section, we suggest the configuration of such networks with policy-based concepts [16]. We extend it with GMPLS networks and service level policies.

Since we need to manage different policies at different levels, we propose a three level architecture for the policy server. The upper level is concerned with the handle of services, the midlevel deals with the network configurations, and the lowest layer configures the network-element. With the three-layered architecture we get a clear separation of different levels. Note that the layering is just logical, each layer may be implemented separately with different tools and deployed independent from the other layers. We map this three level policy on CADENUS architecture, in the purpose of enabling the automation of a number of processes collectively defined as SLA-based service creation for end-user services. The CADENUS Mediation Component architecture proposes the partitioning of system functionalities into three major blocks, called Access Mediator, Service Mediator and Resource Mediator [17].

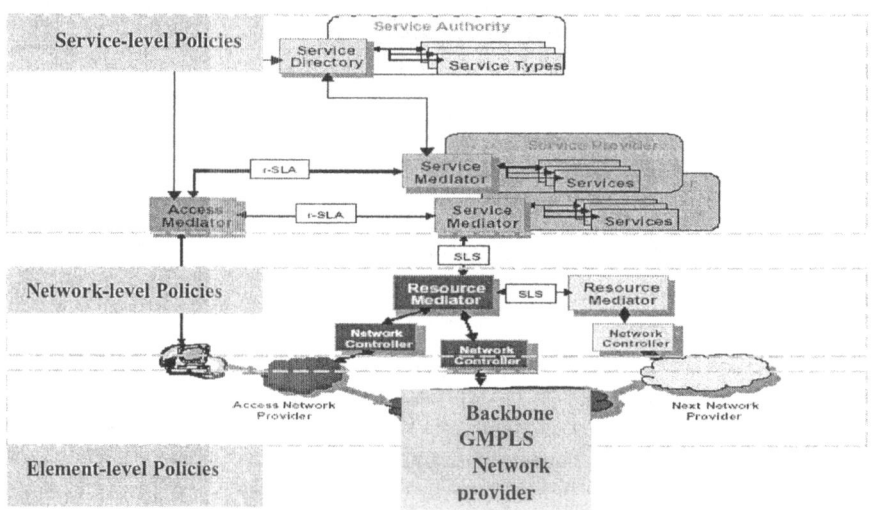

Fig. 2. Logical Three-Level Policy over CADENUS Architecture

The requirements and high-level architecture have been defined in [17]. In the following, we focus on DiffServ over GMPLS specific configurations in backbone network provider and briefly mention DiffServ , because this issue is already covered by the IETF.

6 GMPLS Policies

PBM allows the network control as a whole instead of controlling each individual managed object (network device, interface, queue, LSP) in an independent way. Policies in the GMPLS networks focus on mapping of diffServ traffic on LSPs, for the purpose of traffic engineering and life-cycle management of LSPs.

Configuring DiffServ/GMPLS mainly consists of configuring the LERs. However, the information about routers capabilities is needed by the policy server, in order to take appropriate/device-specific decisions. Information about the capabilities of the LER is needed to decide which configuration applied on the device.

Policies are concerned with LSPs, including the mapping of traffic to LSPs. This LSP is related to a Forward Equivalence Class (FEC), an OA, a traffic profile, a route specification, a role and a resource class.

The <FEC, OA> couple specifies what type of traffic is used by the LSP and the traffic profile specifies a resource profile for the LSP, e.g. delay, bandwidth, etc. The route specification may be used for explicitly set up the route of an LSP or it is used to store the route chosen by the network independently of the policy server. The "role" concept, introduced in PCIM, permits the grouping of the network entities (ex.LSPs) according to the role they play in the network . The role is used to assign a certain property to an LSP from a management point of view. This concept of assigning roles to LSPs allows a network administrator to specify policies which are valid for all LSPs having the same role. The management of a GMPLS network is simplified. The resource class parameter specifies the class from which the LSP may find the resources.

DiffServ mapping in LSP policies. The policy rule set is extended in order to perform the mapping from DiffServ Ordered Aggregates (DOA) to GMPLS paths. DiffServ mapping in LSP policies. The policy rule set is extended in order to perform the mapping from DOA to GMPLS paths. The mapping of traffic to LSPs is generally performed in the LER (edge LSR) of a DiffServ-GMPLS network. However, the traffic mapping is often referred to as <FEC, OA> couple, which forwards a group of packets along the same path with the same per-hop behavior (service differentiation with L-LSP method). The concept of a FEC specifies, what traffic is mapped into a specific LSP. The specification of FECs can be just the destination IP address or a set of filters including IP addresses, ports, DSCPs etc.

In the following example, we are considering a network policy that varies bandwidth allocations based on the time of day and the type of traffic. Traffics in the network are assigned to "roles" in accordance with the Olympic Services Model, i.e. they are marked as "gold", "silver", or "bronze" according to the QoS traffic properties that they carry. A network policy controls the bandwidth allocation associated with this traffic. When the traffic profiles of the gold Traffic, for example, are modified, the "module"(a suggested method in above section) is activated to verify whether the current mapping of traffic to LSPs is still valid. For each modified traffic, the module might also modify the bandwidth of the currently assigned LSPs.

We define an action to modify some parameters of the method proposed in order to calculate a new LSP, and to have an LSP which favorites for example the bandwidth by using, in this case P1 and P2 variables. After the <FEC, OA> mapping LSPs can be established. Example:

- **IF** Time is between 9h00 and 11H00
  THEN *Attribute 2/3 to p2 variable and 1/3 to p1 variable*
  (LSPs will be calculate with the suggested method with new parameters)

This policy can change the manner to calculate the best LSP by minimizing the LSP cost (with the suggested method in section 4.4), where p1 and p2 are weights assigned to the variables used to favorite the utility and bandwidth, respectively.

LSP life-cycle management policies. The control and management of LSP life-cycles include mainly the creation and update of LSPs and the assignment of QoS properties to LSPs. Signaling in GMPLS consists of setting up, releasing and updating an LSP. It should be possible to control the initiation and the execution of these processes using policies. The typical example comprises signaling of an LSP with QoS requirements, Examples:

- The signaled request for a bandwidth X may be permitted or refused, depending on the policies of the network domain.
- **IF** Destination address = 192.247.93.4
  THEN do not choice Y LSR in currant LSP.
  because the LSP setting up requires signaling or configuration mechanisms. Fundamentally two types are possible using an extended hop-by-hop signaling protocol like RSVP-TE, or configuring the LSP over SNMP or COPS.

7 Conclusion

This paper proposes a method and a PBM architecture for GMPLS networks using DiffServ. However the suggested method of differentiation service presented in section 4.4 can be used with CoS (Class of Service) instead of DiffServ. A policy architecture has been introduced with three logical levels. This architecture extends the IETF policy architecture for DiffServ over GMPLS. Particularly the policies rules are used for the LSP life-cycle management, service level request handling, and traffic engineering capabilities. This mainly includes the mapping of DiffServ specific traffic into LSPs on the LERs, and the mapping of LSRs to Per-Hop Behaviors on the core LSRs. This work defines a flexible solution for the support of DiffServ over GMPLS networks.

References

1. D.O. Awduche, J. Malcom, J. Agogbua, M. O'Dell, J. McManus, Requirements for Traffic Engineering over MPLS, IETF RFC 2702, September 1999.
2. D.O. Awduche, L. Berger, D. Gan, T. Li, V. Srinivasan, G. Swallow, RSVP-TE: Extensions to RSVP for LSP Tunnels, IETF RFC 3209, December 2001.
3. I.F. Akyildiza, and all. : "A new traffic engineering manager for DiffServ/MPLS networks: design and implementation on an IP QoS Testbed" Computer Communications 26-pp388–403.2003.
4. T. Li.:"MPLS and the Evolving Internet Architecture", IEEE Communications Magazine, Vol. 37(12), 1999.
5. M. Masullo, S. Calo, "Policy Management: An Architecture and Approach", Proceedings of the First IEEE Intl. Workshop on System Management, LA, USA, April 1993.
6. J. Moffett, M. Sloman, "Policy Hierarchies for Distributed Systems Management", IEEE JSAC Special Issue on Network Management, Vol 11, December 1994.
7. R. Wies: "Policies in Network and Systems Management - Formal Definition and Architecture", Journal of Networks and Systems Management, Vol 2(1), 1994.
8. IETF Policy Framework Working Group, http://www.ietf.org/html.charters/policy-charter.html.
9. L. Berger, "Generalized Multi-Protocol Label Switching (GMPLS) Signaling Functional Description"-RFC 3471- January 2003
10. Ayan Banerjee, John Drake, Jonathan P. Lang, and Brad Turner, Kireeti Kompella, Yakov Rekhter, "Generalized Multiprotocol Label Switching: An Overview of Routing and Management Enhancements"- IEEE Communications Magazine- January 2001.
11. Guillermo Ibáñez, "GMPLS. Towards a Common Control and Management Plane: Generalized Multiprotocol Label Switching in Optical Transport Networks (GMPLS)"-2003.
12. F. Le Faucheur, L. Wu, B. Davie, S. Davari, P. Vaananen, R. Krishnan, P. Cheval, J. Heinanen, "Multi-Protocol Label Switching (MPLS) Support of Differentiated Services". RFC 3270 - May 2002.
13. Ji-Feng Chiu, Zuo-Po Huang, Chi-Wen Lo, Wen-Shyang Hwang and Ce-Kuen Shieh "Supporting End-to-End QoS in DiffServ/MPLS Networks"- 2003 IEEE.
14. Girish Keswani and Miguel A. "Service Differentiation in IP over DWDM Optical Networks, Labrador2002.
15. Leonidas Lymberopoulos, Emil Lupu and Morris Sloman, "An Adaptive Policy Based Framework for Network Services Management", 2002.
16. http://www.ietf.org/
17. G. Cortese, R. Fiutem, D. Matteucci/Editors, CADENUS Deliverable D3.2: "Service Configuration and Provisioning Framework: Requirements, Architecture, Design", CADENUS consortium, 2001.
18. Berger L.:"Generalized Multi-Protocol Label Switching (GMPLS) Signaling Resource ReserVation Protocol (RSVP-TE) extensions", RFC 3473- January 2003

Sliding Mode Queue Management in TCP/AQM Networks

Mahdi Jalili-Kharaajoo

Young Researchers Club, Islamic Azad University, Tehran, Iran
mahdijalili@ece.ut.ac.ir

Abstract. From the viewpoint of control theory, it is rational to regard AQM as a typical regulation system. In this paper, a robust variable structure Sliding Mode Controller (SMC) is applied to Active Queue Management (AQM) in TCP/AQM networks. This type of controller is a robust control strategy, which is insensitive to noise and variance of the parameters, thus it is suitable to time varying network systems. Simulation results show the effectiveness of the proposed SMC in providing satisfactory queue management system.

1 Introduction

Active Queue Management (AQM), as one class of packet dropping/marking mechanism in the router queue, has been recently proposed to support the end-to-end congestion control in the Internet [1-3]. AQM highlights the tradeoff between delay and throughput. By keeping the average queue size small, AQM will have the ability to provide greater capacity to accommodate nature-occurring burst without dropping packets, at the same time, reduce the delays seen by flow, this is very particularly important for real-time interactive applications.

Sliding mode control as a particular type of variable structure control systems is designed to drive and then constrain the system to lie within a neighborhood of a switching function [4,5]. There are two main advantages of this approach. Firstly, the dynamic behavior of the system may be tailored by the particular choice of switching functions. Secondly, the closed-loop response becomes totally insensitive to a particular class of uncertainty. In addition, the ability to specify performance directly makes sliding mode control attractive from the design perspective. This design approach consists of two components. The first, involves the design of a switching function so that the sliding motion satisfies design specifications. The second is concerned with the selection of a control law, which will make the switching function attractive to the system state.

Although PI controller successfully related some limitations of RED, for instance, the queue length and dropping/marking probability are decoupled, whenever the queue length can be easily controlled to the desired value; the system has relatively high stability margin. The shortcomings of PI controller are also obvious. The modification of probability excessively depends on buffer size. As a result, for small buffer the system exhibits sluggishness. Secondly, for small reference queue length, the system tends to performance poorly, which is unfavorable to achieve the goal of

J.N. de Souza et al. (Eds.): ICT 2004, LNCS 3124, pp. 638–643, 2004.
© Springer-Verlag Berlin Heidelberg 2004

AQM because small queue length implies small queue waiting delay. Thirdly, the status of actual network is rapidly changeable, so we believe that it is problematic and unrealistic, at least inaccurate, to take the network as a linear and constant system just like the designing of PI controller. Affirmatively, the algorithm based on this assumption should have limited validity, such as inability against disturbance or noise. We need more robust controller to adapt complex and mutable network environment, which will be our motivation and aim in this study. In the paper, we will apply SMC to design the AQM system for congestion avoidance. The simulation results show the superior performance of the proposed controller in comparison with classic PI controller.

2   TCP Flow Control Model

In [6,7], a nonlinear dynamic model for TCP flow control has been developed based on fluid-flow theory. The state space description of this model can be as follows [1]

$$
\begin{cases}
\dfrac{dx_1}{dt} = x_2 \\[2mm]
\dfrac{dx_2}{dt} = -a_1(t)x_1 - a_2(t)x_2 - b(t) + F(t)
\end{cases}
\tag{1}
$$

$$
a_{1\min} \le a_1 \le a_{1\max}, a_{2\min} \le a_2 \le a_{2\max}, 0 < b_{\min} \le b \le b_{\max}
\tag{2}
$$

where

$$
a_1(t) = \frac{1}{T_1(t)T_2(t)}, a_2(t) = \frac{T_1(t)+T_2(t)}{T_1(t)T_2(t)}, b(t) = \frac{K(t)}{T_1(t)T_2(t)}
$$

$$
F(t) = \frac{d^2}{dt^2}q_o + \frac{T_1(t)+T_2(t)}{T_1(t)T_2(t)}\frac{d}{dt}q_o + \frac{1}{T_1(t)T_2(t)}q_o
\tag{3}
$$

For the purpose of the design of intelligent controller, the varying scope of parameters in TCP/AQM system is assumed as following

$$
N(t) : 1-300, T_p = 0.02\sec, q_o : 0-300 packets, C(t) : 1250-7500 packet/\sec
$$

Therefore,

$$
a_{2\min} = 3.8501, a_{2\max} = 1250, a_{1\min} = 0.0015, a_{1\max} = 60000, b_{\min} = 26042, b_{\max} = 2812500 \quad (4).
$$

3   Sliding Mode Control Design

A Sliding Mode Controller is a Variable Structure Controller (VSC). Basically, a VSC includes several different continuous functions that map plant state to a control surface, and the switching among different functions is determined by plant state that is represented by a switching function. Without lost of generality, consider the design of a sliding mode controller for the following second order system:

$$\ddot{x} = f(x, \dot{x}, t) + bu(t)$$

Here we assume $b > 0$. $u(t)$ is the input to the system. The following is a possible choice of the structure of a sliding mode controller [8]

$$u = -k\,\mathrm{sgn}(s) + u_{eq} \tag{5}$$

where u_{eq} is called equivalent control which is used when the system state is in the sliding mode [8]. k is a constant, representing the maximum controller output. s is called switching function because the control action switches its sign on the two sides of the switching surface $s = 0$. s is defined as

$$s = e + \lambda e \tag{6}$$

where $e = x - x_d$ and x_d is the desired state. λ is a constant. The definition of e here requires that k in (5) be positive. $\mathrm{sgn}(s)$ is a sign function, which is defined as

$$\mathrm{sgn}(s) = \begin{cases} -1 & if \quad s < 0 \\ 1 & if \quad s > 0 \end{cases}$$

The control strategy adopted here will guarantee a system trajectory move toward and stay on the sliding surface $s = 0$ from any initial condition if the following condition meets:

$$\dot{s}s \le -\eta|s|$$

where η is a positive constant that guarantees the system trajectories hit the sliding surface in finite time.

Using a sign function often causes chattering in practice. One solution is to introduce a boundary layer around the switch surface [8]

$$u = -ksat(\frac{s}{\phi}) + u_{eq} \tag{7}$$

where constant factor ϕ defines the thickness of the boundary layer. $sat(\frac{s}{\phi})$ is a saturation function that is defined as

$$sat(\frac{s}{\phi}) = \begin{cases} \dfrac{s}{\phi} & if \quad \left|\dfrac{s}{\phi}\right| \le 1 \\[2mm] \mathrm{sgn}(\dfrac{s}{\phi}) & if \quad \left|\dfrac{s}{\phi}\right| > 1 \end{cases}$$

This controller is actually a continuous approximation of the ideal relay control. The consequence of this control scheme is that invariance property of sliding mode control is lost. The system robustness is a function of the width of the boundary layer. A variation of the above controller structures is to use a hyperbolic tangent function instead of a saturation function [9]

$$u = k \tanh(\frac{s}{\phi}) + u_{eq} \qquad (8)$$

It is proven that if k is large enough, the sliding model controllers of (1), (3) and (4) are guaranteed to be asymptotically stable [8]. For a 2-dimensional system, the controller structure and the corresponding control surface are illustrated in Fig. 1.

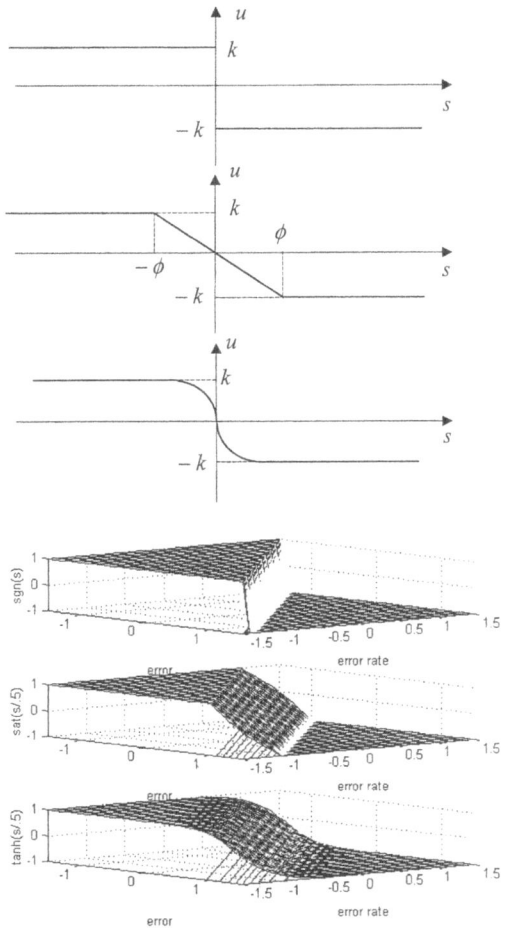

Fig. 1. Various Sliding Mode Controllers and control surfaces

4   Simulation Results

The network topology used for simulation is a FTP+UDP+HTTP network, which has been described in details in [1]. We introduced short-lived HTTP flows and non-responsive UDP services into the router in order to generate a more realistic scenario, because it is very important for a perfect AQM scheme to achieve full bandwidth utilization in the presence of noise and disturbance introduced by these flows [1].

For FTP flows, the instantaneous queue length using the proposed SMC is depicted in Fig. 2. After a very short regulating process, the queue settles down its stable operating point. To investigate the performance of the proposed FLC, we will consider a classic PI controller as

$$p(k) = (a-b)(q(k)-q_o)+b(q(k)-q(k-1))+p(k-1) \qquad (9)$$

The coefficients a and b are fixed at $1.822e^{-5}$ and $1.816e^{-5}$, respectively, the sampling frequency is 500Hz, the control variable p is accumulative [1]. Because the parameter b is very small, and the sample interval is very short, the negative contribution to p made by the second item in the right can be omitted in initial process, then the positive contribution mainly come from the first item. The queue evaluation using PI controller is shown in Fig. 3. As it can be seen SMC acts much better that PI one.

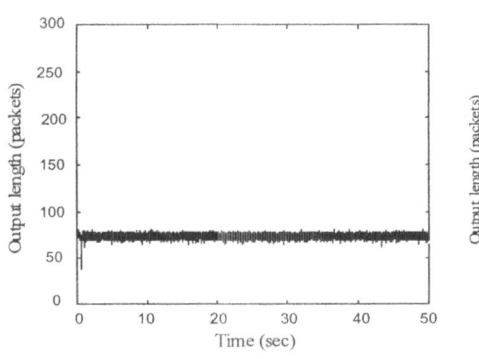

Fig. 2. Queue evaluation (SMC)

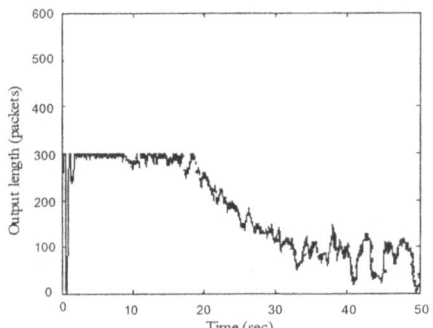

Fig. 3. Queue evaluation (PI)

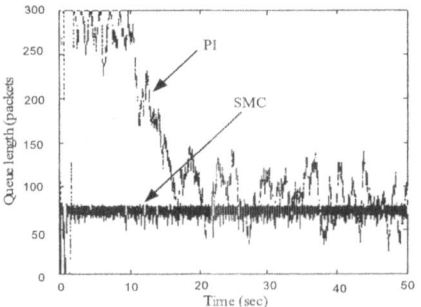

Fig. 4. Queue evaluation (FTP+HTTP)

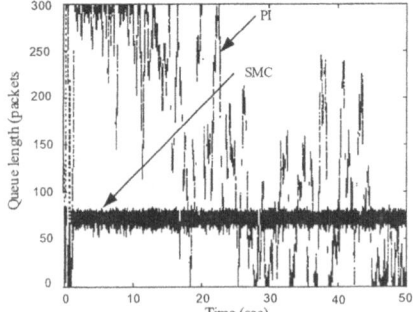

Fig. 5. Queue evaluation (FTP+UDP)

For FTP+HTTP flows, the evaluation of queue size is shown in Fig. 4. As it can be seen, the proposed SMC has better performance than that of PI one. Next, we further investigate performance against the disturbance caused by the non-responsive FTP+UDP flows. Fig. 5 shows the results, obviously, PI is very sensitive to this disturbance, while SMC operates in a relatively stable state. The queue fluctuation increases with introducing the UDP flows, but the variance is too much smaller comparing with PI controller.

5   Conclusion

In this paper, variable structure sliding mode controller wad applied to active queue management in TCP/AQM networks. This kind of controller is insensitive to system dynamic parameters and is capable of being robust against disturbance and noise, which is very suitable for the mutable network environment. The proposed robust controller was very responsive, stable and robust, especially for the small reference queue system. We took a complete comparison between performance of the proposed SMC and classical PI controller under various scenarios. The conclusion was that the integrated performance of SMC was superior to that of PI one.

References

1. Jalili-Kharaajoo, M., Application of robust fuzzy second order sliding mode control to active queue management, *Lecture Notes in Computer Science*, 2957, pp.109-119, 2004.
2. Barden, B. at al., Recommendation on queue management and congestion avoidance in the internet, *REC2309*, April 1998.
3. Firoiu, V. and Borden, M., A study of active queue management for congestion control, *Proc. INFOCOM*, March 2000.
4. Edwards, C. and Spurgeon, S. K. Sliding mode control. *Taylor & Francis Ltd.*, 1998.
5. Bartolini, G. and Pydynowski, P. An improved, chattering free, VSC: scheme for uncertain dynamics systems. *IEEE Transaction on Automatic Control* ,41(8), August, 1996.
6. Misra, V., Gong, W.B. and Towsley, D., Fluid-based analysis of network of AQM routers supporting TCP flows with an application to RED, *Proc. ACM/SIGCOMM*, 2000.
7. Hollot, C., Misra, V., Towsley, D. and Gong, W.B., On designing improved controllers for AQM routers supporting TCP flows, *Proc. INFOCOM*, 2001.
8. J.J. Slotine and W. Li, Applied Nonlinear Control, *Englewood Cliffs, NJ: Prentice-Hall*, 1991.
9. J.S. Glower and J. Munighan, Designing Fuzzy Controllers from a Variable Structures Standpoint, *IEEE Transactions on Fuzzy Systems*, vol.5, no.1, pp.138-144, 1997.

A Distributed Algorithm for Weighted Max-Min Fairness in MPLS Networks*

Fabian Skivée and Guy Leduc

Research Unit in Networking University of Liège (Belgium)
{Fabian.Skivee,Guy.Leduc}@ulg.ac.be

Abstract. We propose a novel distributed algorithm to achieve a weighted max-min sharing of the network capacity. We present the Weight Proportional Max-Min policy (WPMM) that supports a minimal rate requirement and an optional maximal rate constraint and allocates network bandwidth among all aggregates based on a weight associated with each one. Our algorithm achieves this policy for IP/MPLS networks using the RSVP-TE signalling protocol. It uses per-LSP accounting in each node to keep track of the state information of each LSP. It uses a novel explicit bottleneck link strategy and a different control architecture in which we update the control packet in the forward path. Simulations show that these two elements improve substantially the convergence time compared to algorithms designed for ATM networks.

1 Introduction

Traffic engineering aims at using information about traffic entering and leaving the network to optimize network performance. An ISP using MPLS [11] would like to optimize the utilization of its available resources between all the LSPs.

Consider a set of LSPs, each carrying many TCP connections, creating congestion. Without explicit policing, more aggressive LSPs (with more flows) get more that their fair share, independently of their reservations.

The problem of fair sharing of the network bandwidth has been widely treated in the past and especially in the ATM context ([6] [9]). The classical max-min rate allocation has been accepted as an optimal network bandwidth sharing criterion among user flows [1]. Informally, the max-min policy attempts to maximize the network use allocated to the users with the minimum allocation [5].

To provide differential service options, [8] proposes a generic weight proportional network bandwidth sharing policy, also called Weight-Proportional Max-Min (WPMM). The WPMM policy generalizes the classical max-min by associating each flow with a generic weight, which is decoupled from its minimum rate and supports the minimum rate requirement and the peak rate constraint for each flow.

Since a centralized algorithm for the max-min rate allocation requires global network information, which is not scalable to flood, we must develop distributed

---

* This work was supported by the European ATRIUM project (IST-1999-20675) and by the TOTEM project funded by the DGTRE (Walloon region)

J.N. de Souza et al. (Eds.): ICT 2004, LNCS 3124, pp. 644–653, 2004.

algorithms to achieve the same rate allocation in the absence of global knowledge about the network and without synchronisation of different network components. We focus on an edge-to-edge rate-based feedback control scheme, where special control packets are used in both forward and backward paths.

Hou, in [8], provides a good distributed solution for computing the WPMM policy but has two major drawbacks. Firstly, this solution is implemented in the ATM context and uses a lot of Resource Management (RM) ATM cells that create a substantial overhead. And secondly, the convergence is usually slow as illustrated later.

Our proposal solves these two limitations by adding an explicit bottleneck link information in each control packet and by using a different control packet update architecture. These two additions addition change radically the dynamics of the protocol and improve the convergence time by a factor 2 or 3 depending on the precision (see simulation results).

Another contribution is the adaptation of the solution to the actual MPLS architecture using the widespread RSVP-TE protocol [2] instead of RM cells. With our proposed integration in RSVP, we improve the scalability of the protocol by decreasing the overhead added by control packets.

As a possible application of this work, the weighted max-min fair rate allocated to an LSP could be used by a marker at the ingress router to mark the traffic using three colours : green (under reserved rate), yellow (between reserved rate and the fair rate) and red (above the fair rate). In case of congestion, core routers discard the red packets first and possibly, during transient periods, some of the yellow packets by using a WRED policy for example. The objective of the algorithm is to compute the fair rate of each LSP to obtain a network in which only the red packets are discarded and all the green and yellow packets successfully get through.

2   The Generic Weight-Proportional Allocation Policy

A unified definition of the max-min fairness is provided by [10]. Consider a set $\mathcal{X} \subset \mathbb{R}^N$, the definition of the weighted max-min fair vector with respect to set $\mathcal{X}$ is defined as follows :

Definition 1. *A vector $\overrightarrow{x}$ is "weighted max-min fair on set $\mathcal{X}$" if and only if*

$$(\forall \overrightarrow{y} \in \mathcal{X}) \left[(\exists s \in \{1,\ldots,N\}) \; y_s \; > \; x_s \Rightarrow (\exists t \in \{1,\ldots,N\}) \; \frac{y_t}{w_t} < \frac{x_t}{w_t} \leq \frac{x_s}{w_s} \; \right]$$

where w_i is the weight associated with the element i

We can use this theoretical definition to share the free bandwith available in an MPLS network. Our work is based on a previous work in the ATM context provided by [8]. We adapt their definition to the MPLS context.

An MPLS network is a set of IP/MPLS routers interconnected by a set of links $\mathcal{L}$. A set of LSPs $\mathcal{S}$ traverses one or more links in $\mathcal{L}$ and each LSP $s \in \mathcal{S}$

is allocated a fair rate r_s. Denote $\mathcal{S}_l$ the set of LSPs traversing link $l \in \mathcal{L}$. The aggregate allocated rate F_l on link l in $\mathcal{L}$ is

$$F_l = \sum_{s \in \mathcal{S}_l} r_s$$

Let C_l be the capacity (maximum allowable bandwidth) of link l. A link l is saturated or fully utilized if $F_l = C_l$. Denote RR_s and MR_s, the reserved rate requirement and the maximal rate constraint for each LSP $s \in \mathcal{S}$, respectively. For feasibility, we assume that the sum of all LSPs' RR requirements traversing any link does not exceed the link's capacity, i.e. $\sum_{s \in \mathcal{S}_l} RR_s \leq C_l$ for every $l \in \mathcal{L}$. This criterion is used by admission control at LSP setup time to determine whether or not to accept a new LSP on a link.

Definition 2. *We say that a rate vector $r = \{r_s \mid s \in \mathcal{S}\}$ is **MPLS feasible** if the following two constraints are satisfied :*

$$RR_s \leq r_s \leq MR_s \text{ for all } s \in \mathcal{S}$$
$$F_l \leq C_l \text{ for all } l \in \mathcal{L}$$

In the generic weight-proportional max-min (WPMM) policy, we associate each LSP $s \in \mathcal{S}$ with a weight (or priority) w_s. Informally, this policy first allocates to each LSP its RR. Then from the remaining network capacity, it allocates additional bandwidth for each LSP using a proportional version of the max-min policy based on each LSP's weight while satisfying its MR constraint. The final bandwidth for each LSP is its RR plus an additional "weighted" max-min fair share. Formally, this policy is defined as follows and directly derives from Def 1.

Definition 3 (WPMM-feasible vector). *A rate vector r is weight-proportional max-min (WPMM) if it is MPLS-feasible, and for each $s \in \mathcal{S}$ and every MPLS-feasible rate vector r' in which $r'_s > r_s$, there exists some LSP $t \in \mathcal{S}$ such that $\frac{r_s - RR_s}{w_s} \geq \frac{r_t - RR_t}{w_t}$ and $r_t > r'_t$.*

Definition 4 (WPMM-bottleneck link of a LSP). *Given an MPLS-feasible rate vector r, a link $l \in \mathcal{L}$ is a WPMM-bottleneck link with respect to r for a LSP s traversing l if $F_l = C_l$ and $\frac{r_s - RR_s}{w_s} \geq \frac{r_t - RR_t}{w_t}$ for all LSP t traversing link l.*

The following proposition links the relationship between the above WPMM policy and the WPMM-bottleneck link definitions.

Proposition 1 (WPMM vector). *An MPLS-feasible rate vector r is WPMM if and only if each LSP has either a WPMM bottleneck link with respect to r or a rate assignment equal to its MR.*

The centralized Water Filling algorithm computes the fair rate for each LSP according to this policy [5]. This centralized algorithm for the WPMM rate allocation requires global information which is not scalable to flood. It is intended to be used as the network bandwidth sharing optimality criterion for our distributed algorithm, which will be presented in the next section.

3   Proposed Distributed WPMM Algorithm

In this section, we propose an algorithm that converges to the WPMM policy quickly through distributed and asynchronous iterations.

3.1   Basic Algorithm

Our distributed solution uses the RSVP signalling protocol to convey information through the network.

The PATH and RESV packets contain the following parameters :

- RR (Reserved Rate [1]) : provided at the creation of the LSP
- W (Weight [1]) : provided at the creation of the LSP
- ER (Expected Fair Rate) : the fair rate that the network allows for this LSP
- BN (BottleNeck) : id of the LSP's bottleneck link

Periodically, the ingress sends a PATH packet. Each router in the path computes a local fair share for the LSP and updates the ER and BN fields if its local fair rate is less than the fair rate present in the PATH packet. Upon receiving a PATH packet, the egress router sends a RESV packet, initialized with the data obtained in the PATH, to the ingress router.

In the backward path, each router updates its information with the RESV parameters (ER,BN). Upon receiving a RESV packet, the ingress obtains the information about its allowed fair share.

Many prior efforts in ATM networks have been done on the design of ABR algorithms to achieve the classical max-min ([6] [9]). Charny's work [7] was one of the few algorithms that were proven to converge to the max-min in the ATM context. Hou extends Charny's technique to support the minimum rate, peak rate and weight for each flow. Our proposition uses Hou's work but improves the performance by modifying the update mechanism. In Hou's algorithm, the routers update their information in the forward path. So, they cannot have the information computed by the downstream routers. These routers have the correct information in the next cycle. In our approach, we update the router in the backward path, so they have the correct information like the ingress node. Moreover, we add a new parameter (BN) that conveys explicitly the bottleneck link in the path. This information improves considerably the convergence time.

The following are the link parameters and variables used by the algorithm.

- $\mathcal{L}$: Set of links
- $\mathcal{S}_l$: Set of LSPs traversing link l
- FR_l^i : fair rate value of the LSP $i \in \mathcal{S}_l$ as known at link l

---

[1] This information is only needed during the establishment of the LSP or when these values are modified.

- BN_l^i : bottleneck link id of the LSP $i \in \mathcal{S}_l$ as known at link l (refer to Definition 4)
- $\mathcal{B}_l$: Set of LSPs bottlenecked at link l, i.e. $\mathcal{B}_l = \{i \mid i \in \mathcal{S}_l \text{ and } BN_l^i = l\}$
- $\mathcal{U}_l$: Set of LSPs not bottlenecked at link l, i.e. $\mathcal{U}_l = \{i \mid i \in \mathcal{S}_l \text{ and } BN_l^i \neq l\}$
- We have $\mathcal{U}_l \cup \mathcal{B}_l = \mathcal{S}_l$

The fair rate of a LSP i is composed of the reserved rate of the LSP (RR^i) and an additional fair share allocated by the network. This additional fair share is proportional to the weight of the LSP. On a particular link, we can compute a value φ_l that gives the additional fair share per unit of weight for the LSPs bottlenecked on this link. Algorithm2 computes φ_l :

Algorithm 2 : Calculation of φ_l (if $\mathcal{S}_l \neq \emptyset^2$)

$$\text{FB}_l := C_l - \sum_{i \in \mathcal{S}_l} RR^i - \sum_{i \in \mathcal{U}_l} (FR_l^i - RR^i)$$

$$\varphi_l := \left\{ \begin{array}{ll} \frac{\text{FB}_l}{\sum_{i \in \mathcal{B}_l} W^i} & \text{if } \mathcal{B}_l \neq \emptyset \\ \frac{(C_l - \sum_{i \in \mathcal{S}_l} FR_l^i)}{\sum_{i \in \mathcal{S}_l} W^i} + \max_{i \in \mathcal{S}_l} \frac{(FR_l^i - RR^i)}{W^i} & \text{otherwise} \end{array} \right\} \quad (1)$$

The basic case (i.e. $\mathcal{B}_l \neq \emptyset$) occurs when some LSPs are bottlenecked on link l, we compute the free bandwidth on link l (FB_l) by taking the capacity of the link minus the RR of all LSPs traversing l, minus the additional fair share of the LSPs not bottlenecked on l. φ_l is equal to the free bandwidth divided by the sum of the LSP's weights for each LSP bottlenecked on l, i.e. φ_l is the free bandwidth we will allocate to each $i \in \mathcal{B}_l$ per unit of weight.

The second case (i.e. $\mathcal{B}_l = \emptyset$) occurs when all LSPs are bottlenecked at another link than l. In this case, the value of φ_l is chosen as in [8] and [7] to achieve convergence.

The fair rate (FR_l^i) of a LSP i bottlenecked on link l is computed by :

$$(\forall i \in \mathcal{B}_l) \quad FR_l^i := RR^i + \varphi_l * W^i$$

A key element of the algorithm is the strategy to set the BN_l^i information correctly. We use the following definition of "bottleneck consistency" :

Definition 5 (Bottleneck-consistent). *Let $\mathcal{U}_l$ be the set of LSPs not bottlenecked at link $l \in \mathcal{L}$ and φ_l be calculated according to Algorithm2. $\mathcal{U}_l$ is bottleneck-consistent if*

$$(\forall i \in \mathcal{U}_l) \quad FR_l^i \leq RR^i + \varphi_l * W^i$$

This definition derives directly from Definition 4 and means that all LSPs not bottlenecked at a link must have a bottleneck elsewhere or reach their maximal rate, so that they have an allocated fair rate less that the one proposed by the

---

[2] If there is no LSP using link l (i.e. $\mathcal{S}_l \neq \emptyset$), $\varphi_l := \infty$

current link. If that were not the case, some LSPs in $\mathcal{U}_l$ would have to be moved to $\mathcal{B}_l$ (i.e. would be bottlenecked at l).

Our algorithm employs per-LSP accounting at each output port of a node. That is, we maintain a table to keep track of the state information of each traversing LSP. For each LSP i, we keep $FR_l^i, RR^i, W^i, BN_l^i$. Based on this state information, we compute the explicit rate for each LSP to achieve the WPMM rate allocation.

The following is the node algorithm, with each output port link initialized with $\mathcal{S}_l := \emptyset$ and $\varphi_l := \infty$.

Algorithm 3: Node Behaviour

```
Upon receipt of a PATH³ {
    LSPCreationAndTermination();
    updateER();
}
```

```
Upon the receipt of a RESV³ {
    FR_l^i := ER; BN_l^i := BN;
    updateφ_l();
    Forward RESV(i,RR,ER,W,BN)
}
```

```
LSPCreationAndTermination() {
    if LSP termination then {
        S_l := S_l − {i};
        updateφ_l();
    } else if LSP creation then {
        B_l := B_l ∪ {i};
        RR^i := RR; W^i := W;
        updateφ_l();
    } }
```

```
updateER() {
    NER := φ_l * W^i + RR^i;
    ER' := max(min(ER, NER), RR^i);
    if (ER' < ER) then {
        BN := l;
        ER := ER';
    }
    Forward PATH(i,RR,W,ER,BN);
}
```

```
updateφ_l() {
    use Algorithm2 to calculate φ_l;
    if (U_l ≠ ∅) then {
        repeat {
            // stops when bottleneck-consistency (see Def5) is achieved
            φ_l^1 := φ_l;
            p := argmax_{i ∈ U_l} (FR_l^i − RR^i)/W^i);
            if (((FR_l^p − RR^p)/W^p) > φ_l^1) then {
                Move p from U_l to B_l;
                use Algorithm2 to calculate φ_l;
            }
        } until ((((FR_l^p − RR^p)/W^p) ≤ φ_l^1) or (U_l = ∅));
    } }
```

The updateφ_l() procedure will first recompute φ_l based on current values of FR_l^i and BN_l^i for all i. The next step, i.e. the repeat-until loop, will ensure

---
[3] PATH and RESV packets contains the following parameters (i,RR,W,ER,BN)

that, when it terminates, the set $\mathcal{U}_l$ is bottleneck-consistent. To do so, LSPs not satisfying $(FR_l^i - RR^i)/W^i) \leq \varphi_l$ should be moved from $\mathcal{U}_l$ to $\mathcal{B}_l$. At each iteration, one LSP, say p, such that $(FR_l^p - RR^p)/W^p) = \max_{i \in \mathcal{U}_l} (FR_l^i - RR^i)/W^i)$ is removed. The new $\mathcal{B}_l$ and $\mathcal{U}_l$ sets may lead to a new φ_l, which needs to be recalculated. This process ends when either $\mathcal{U}_l$ is empty or $\mathcal{U}_l$ is bottleneck-consistent.

Finally, the edge behaviour is simple. The ingress is responsible for sending PATH packets and for updating the LSP fair rate information upon the reception of a RESV packet. The ingress also computes the fair rate of the first link and so uses Algorithm3.

Upon receiving a PATH packet, the egress sends a RESV packet, initialized with the data obtained in the PATH, to the upstream node.

The structure of Algorithm3 guarantees that for every LSP $i \in \mathcal{S}$, the fair rate (FR_l^i) information in each node is MPLS-feasible, i.e. $RR^i \leq FR_l^i \leq MR^i$.

With our update architecture in the backward path and the explicit bottleneck link information, we decrease the convergence time substantially as shown in the next section.

3.2   Improve Algorithm to Deal with the RSVP Refresh Mechanism

The RSVP Refresh Overhead Reduction Extensions [4] minimize the processing time of the PATH and RESV packets. If two successive PATH (or RESV) packets are the same, the upstream node only sends a refresh PATH. The downstream node refreshes the LSP entry but doesn't process the whole PATH packet.

Our solution can easily be extended to keep this property. We must develop a strategy to determine if a node must send a full new PATH or just a refresh PATH. On each output port, we associate with each LSP i a special bit (NR^i) that is set if we must send a full new PATH packet. When some value of an LSP changes in the output port table, we set this bit for all the LSPs of this table. When we receive a PATH and we refresh the LSP entry, we clear this bit (for more information see [3]).

When the system has converged, only refresh RSVP packets are used. With this improved algorithm, we can keep the advantages of the RSVP refresh mechanism.

The use of RSVP and its refresh mechanism reduce the overhead needed compared to ATM RM cells. In ATM, one RM cell is sent every 32 cells. Therefore, the overhead introduced by ATM depends on the flow rate. On the other hand, our scheme introduces a fixed overhead. Finally, the use of the refresh mechanism reduces the processing needed by the core routers that only need to refresh the LSP entry in their tables.

4   Simulation Results

To compare our solution with Hou's, we have developed a dedicated simulator. We have implemented Hou's solution adapted to the MPLS context (using RSVP mechanism in place of RM cells).

Our simulation process consists of generating a network topology and LSPs on this topology. Next, we add the LSPs one by one and we execute the two algorithms on this topology.

For simulating the RSVP process, we use the concept of iteration and RSVP-cycle. A RSVP-cycle consists of the forwarding of a PATH packet along all the nodes in the path from the ingress to the egress node and the backwarding of a RESV packet along all nodes from the egress to the ingress node. An iteration is the execution of an RSVP-cycle for all the LSPs. At the end of each iteration, we have a vector of LSP fair rates.

We have two possibilities to stop the process. The first is to execute the iterative process until the mean relative error between the last rate vector and the WPMM allocation vector (computed a priori using the Water Filling algorithm [5]) is under a fixed precision (e.g 10^{-4}). Another possibility is to stop the process if the mean relative error between two successive rate vectors is under a fixed precision. The first possibility shows us how our solution reaches the optimum and with which convergence speed. The second approach shows us how our algorithm can be used in practice without any a priori knowledge of the WPMM rate allocation. The simulation in the sequel are based on the first approach.

The simple network configuration we use in our simulation is organized like the olympic symbol. There are five MPLS routers connected with 7 links of 100Mb/s and 5 LSPs traversing the topology Fig 1a.

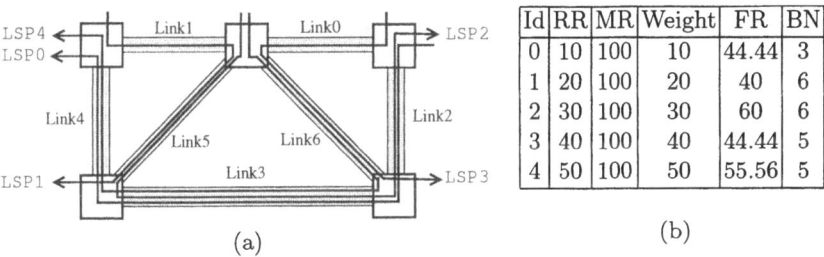

Id	RR	MR	Weight	FR	BN
0	10	100	10	44.44	3
1	20	100	20	40	6
2	30	100	30	60	6
3	40	100	40	44.44	5
4	50	100	50	55.56	5

(a)   (b)

Fig. 1. *(a)* Network topology *(b)* LSP configuration

Fig. 1b lists the LSP parameters : RR requirement, MR constraint, weight and fair rate allocation for each LSP traversing the network configuration. Fig 2a shows the evolution of the fair rates under Hou's distributed algorithm. The rates converge to their optimal WPMM rate listed in Fig 1b. The algorithm takes 41 iterations before getting sufficiently close to the optimal fair share (i.e. euclidean distance under 0.01 %). Fig 2b shows the evolution of the fair rates under our algorithm. Our solution reaches the optimum in 33 iterations.

Extensive Simulation

We made extensive simulations on a large number of complex topologies. We created 63 topologies with 20 to 100 nodes and with 20, 50, 100, 200, 300 and 1000 LSPs. We executed the two solutions with different levels of precision. With a precision of 10^{-4}, our solution is in general 2.86 times faster than Hou's solution. Table 1 presents a short report of the results.

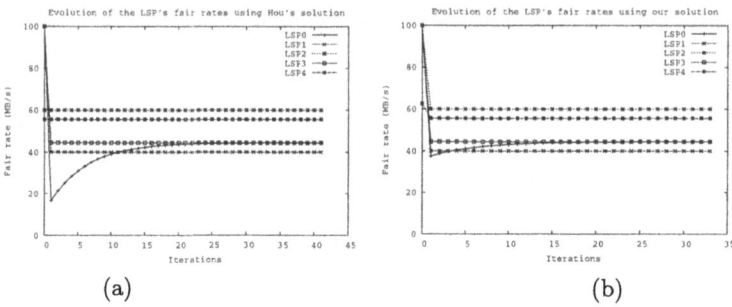

Fig. 2. *(a)* Hou's solution *(b)* our solution on the complex network topology

Table 1. Average number of iterations on 63 topologies

Precision	Hou's algorithm	Our algorithm	Gain
0.01	5.46	2.05	2.67
0.001	18.40	6.81	2.70
0.0001	40.43	14.11	2.86
0.00001	66.86	18.98	3.52

The number of iterations needed by our solution on large topologies is relatively high but if we look at the number of LSPs that reach their fair share, we see that after a few iterations 90% of the LSPs have reached them. So only a few LSPs continue to change. The improved solution adapted to RSVP becomes very useful and will improve hugely the performance and scalability of the algorithm.

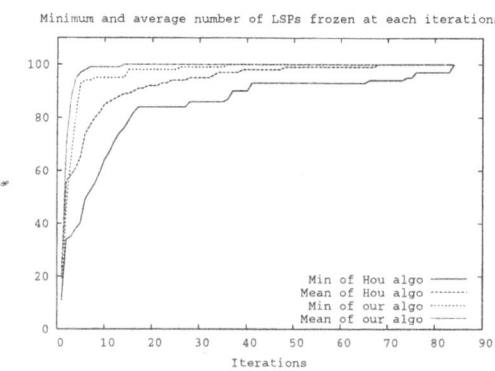

Fig. 3. Minimum and average LSPs frozen at each iterations

We also executed the simulations on 50 topologies with 100 to 1000 LSPs and we have plotted the minimum and average percentage of LSPs that have reached their fair share after each iteration for the two algorithms (Fig 3). Normally, at the last iterations, the min and the average values reach 100 %. Hou's solution takes 84 iterations to converge in the worst case and our solution takes only 36 iterations. We can also see that Hou's solution takes 16 iterations to stabilize 90% of the LSPs while our solution takes only 4.

5   Conclusion

We presented a novel distributed algorithm to achieve the WPMM rate allocations. This solution provides a scalable architecture to share the available bandwidth among all the LSPs according to their weights.

Compared to a direct MPLS adaptation of Hou's solution designed for ATM-ABR, our algorithm improves considerably the performance by using a new update scheme of the control packet and an explicit bottleneck link marking strategy. Our algorithm is compatible with RSVP Refresh extensions [4], thus improving the overhead further compared to ATM networks.

Our future work will focus on the development of a rigourous convergence proof as well as executing our algorithm on even more complex topologies. Another challenging issue is to improve the weight concept by using utility functions to describe the traffic. The objectives become to maximize the global utility of the clients. And finally, we will investigate the integration of this algorithm in a real Diffserv MPLS network.

References

1. ATM Forum Technical Commitee. *Traffic Management Specification - Version 4.0*, February 1996. ATM Forum/95-0013R13.
2. D. Awduche, L. Berger, D. Gan, T. Li, V. Srinivasan, and G. Swallow. *RSVP-TE: Extensions to RSVP for LSP tunnels*, December 2001. RFC 3209.
3. S. Balon, F. Blanchy, O. Delcourt, G. Leduc, L. Mélon, and F. Skivée. Synthesis and recommendation for intra-domain traffic engineering. Technical Report ATRIUM Deliverable D3.3, University of Liège, Novembre 2003.
4. L. Berger, D. Gan, G. Swallow, P. Pan, F. Tommasi, and S. Molendini. *RSVP Refresh Overhead Reduction Extensions*, April 2001. RFC 2961.
5. D. Bertsekas and R. Gallager. *Data Networks*. Prentice Hall, 1992.
6. Y. Chang, N. Golmie, and D. Siu. *A Rate-Based Flow Control Switch Design for ABR Service in an ATM Network*, August 1995. Twelfth International Conference on Computer Communication ICCC'95.
7. A. Charny. An algorithm for rate allocation in a packet-switching network with feedback. Technical Report MIT/LCS/TR-601, 1994.
8. Y. Thomas Hou, Henry Tzeng, Shivendra S. Panwar, and Vijay P. Kumar. A generic weight-proportional bandwidth sharing policy for atm abr service. *Performance Evaluation, vol 38*, pages 21–44, 1999.
9. R. Jain, S. Kalyanaraman, R. Goyal, S. Fahmy, and R. Viswanathan. *ERICA switch algorithm: A complete description*, August 1996. ATM Forum/96-1172.
10. J. Y. Le Boudec and B. Radunovic. A unified framework for max-min and min-max fairness with applications. October 2002.
11. E. Rosen and al. *Multiprotocol label switching architecture*, January 2001. RFC 3031.

Applying Artificial Neural Networks for Fault Prediction in Optical Network Links

Carlos Hairon R. Gonçalves[1], Mauro Oliveira[1],
Rossana M.C. Andrade[2], and Miguel F. de Castro[3*]

[1] Laboratório Multiinstitucional de Redes de Computadores e Sistemas Distribuídos
Centro Federal de Educação Tecnológica do Ceará (LAR/CEFET-CE)
Av. 13 de Maio, 2081, Benfica, Fortaleza, Ceará, Brasil, 60.531-040
{hairon,mauro}@lar.cefet-ce.br
[2] Universidade Federal do Ceará, Departamento de Computação (DC/UFC)
Campus do Pici, Bloco 910, Fortaleza, Ceará, Brasil, 60.455-760
rossana@ufc.br
[3] Institut National des Télécommunications
9 rue Charles Fourier 91011 Evry Cedex - France
miguel.castro@int-evry.fr

Abstract. The IP+GMPLS over DWDM model has been considered a trend for the evolution of optical networks. However, a challenge that has been investigated in this model is how to achieve fast rerouting in case of DWDM failure. Artificial Neural Networks (ANNs) can be used to generate proactive intelligent agents, which are able to detect failure trends in optical network links early and to approximate optical link protection mode from 1:n to 1+1. The main goal of this paper is to present an environment called RENATA2 and its process on how to develop ANNs that can give to the intelligent agents a proactive behavior able to predict failure in optical links.

1 Introduction

At the end of the 90's, ATM was seen as the technology that would become a *de facto* standard for wide, metropolitan, and even for local area networks in few years. Currently. IPv6 and technologies such as MPLS (Multiple Protocol Label Switching) and DiffServ (Differentiated Services), which promise to assure quality of service (QoS) in IP, have emerged to replace ATM. For instance, according to [1], IP + GMPLS (Generalized Multiple Protocol Label Switching) over DWDM (Dense Wavelength-Division Multiplexing) networks have become a strong trend in substitution to the following model of four layers: IP for application and services; ATM for traffic engineering; SONET/SDH for data transport; and DWDM for physical infrastructure.

Independently of the technology used, the Internet management is based on the Agent x Manager paradigm and it uses SNMP (Simple Network Management Protocol), whose manager centers the information processing and, consequently, the decision making process, leaving the agents with limited computational capabilities

---

* Miguel F. de Castro is supported by CAPES-Brazil

J.N. de Souza et al. (Eds.): ICT 2004, LNCS 3124, pp. 654–659, 2004.

and, practically, without autonomy. An alternative for solving this problem is the use of Intelligent Agents, once they have more autonomy and processing power [2][3]. Artificial Intelligence techniques can be used to make possible the proactive characteristics of an agent.

Amongst Artificial Intelligence mechanisms used as an inference machine for intelligent agents, ANNs appear as a viable and motivated technology to several research groups [3][4]. ANNs characteristics, such as learning, generalization, adaptability, robustness and fault tolerance, solves the demands that are imposed by the intelligent agent characteristics as well as by the particular traffic restrictions of different network technologies. In this paper, we present the RENATA 2 environment to support the proactive management for IP, MPLS and GMPLS networks. Traffic control and path allocation are possible tasks to be performed by the agents generated for this environment.

We organize this paper as follows. Next section shows the main issues about IP, MPLS and GMPLS networks. Section 3 presents more details about the proposal. Section 4 describes the simulation specification. Section 5 outlines the obtained results. Finally, Section 6 summarizes our main contributions and future work.

2 Proposal: ANNs for Fault Prediction in Optical Links

One of the problems that limit the use of MPL(ambda)S over DWDM links is rerouting, which is slow due to its reactive restoration for failure [2]. A solution for such problem is the use of 1+1 protection mechanisms, where one link is used as a mirror to provide immediate failure recovery. Such proposal is too expensive, since an exclusive channel must be allocated to offer the necessary infrastructure [6]. The use of ANNs can predict problematic situations in a link what allows taking proactive measures before the problem becomes critical. This solution enables to make possible a 1:n restoration mechanism that can become close to 1+1 models, once the possibility of occurring a failure is detected, the backup channel would start to function as a mirror of the supposedly problematic link.

In particular, fault detection in optical links involves several variables, and the relationship among these variables in order to predict faults is not easily modeled by analytical methods [7]. In this work, we argue that ANNs are able to capture this unknown relationship, making possible to accurately predict physical link layer faults. Some works have already tried to predict faults in different domains using ANN as AI component [8][9]. In order to develop RENATA2, first of all, we investigate the hardware components and monitor equipment with their respective parameters, which can be candidates to input variable for a specific ANN.

After that, RENATA2 is simulated with NS (Network Simulator), which has been chosen for its open feature and the diversity of network technologies it works. However, NS does not provide an appropriate error treatment for the problem solving of generating intelligent agents based on ANNs applied to the failure management. The error treatment considered by NS is imperative, for example, to simulate a link failure between 4 and 6 Label Switching Routers (LSRs) in the instant 4,5s of a MPLS network simulation, we must execute the following command: ***$$ns rtmodel-at 4,5 down $$LSR6 $$LSR4***. In this example, no factor related to the simulation contributes for the link failure. This command reflects only the user input and there is no relationship with the real world. ANNs is trained with a knowledge base that approaches as maximum as possible the real world. Therefore, since the error treatment considered by NS is not adequate for the generation of a repository of examples that is essential to the learning process of the ANN. For this reason, we develop a tool, called Perturbation Generator (PG), which solves this problem for the RENATA2 environment.

Figure 1 shows the functional components of RENATA 2 and how they work together. The process for building intelligent agents is described as follows. An optical link must be modeled and simulated using PG. The physical link behavior description is fed into NS, which models and simulates link and network layers, including MPLS. Simulation logs are generated by NS and treated by DSPM (Data Selection and Preparation Module). The result of this treatment generates a knowledge base to be used in the neural network training process performed by JNNS (Java Neural Network Simulator). The resulting trained ANN must be prepared by the TI (Training Interface) to be inserted into the Intelligent Agent Generator Module (IAGM). In this last module, the trained ANN is encapsulated into an intelligent agent and deployed to the managed object.

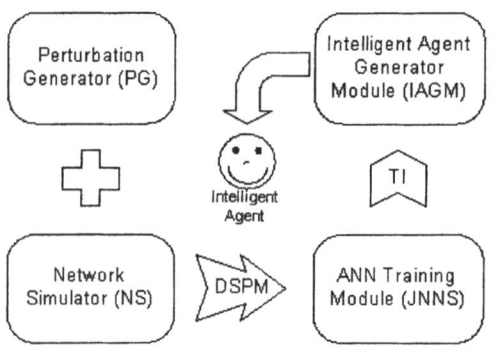

Fig. 1. RENATA 2 Functional Components

The PG simulates the behavior of the main hardware devices of a optical fiber network, taking in consideration the physical variables (transmitter temperature and voltage) that dictate the behavior of such devices. The simulated equipments for the Perturbation Generator (PG) are Optical Transmitter, Optical Fiber and Optical Receiver. These devices are considered essential in our process, for being mandatory in optical fiber networks either directly in the own equipment or indirectly in a transmitter that is part of a DWDM multiplexer, for example. The PG was developed using pure Java, which allows this tool to become multiplatform. Finally, a solution is implemented with intelligent agents for the fast re-routing problem after a failure occurs in a MPLS link transmission, when the link is used over optical networks (MPλS or GMPLS). These agents are based on ANNs and they can foresee failures in IP+GMPLS links over DWDM guaranteeing more ability in the rerouting on these networks [10].

3   Simulation Specification

As stated in Section 2, an example set is needed in order to perform ANN training. In RENATA2, several simulations of physical, link and network layers are performed with this goal.

In the physical layer, a full-duplex optical link was modeled and simulated together with a semiconductor laser transmitter in its input and a semiconductor photoreceptor in its output. This link does not comprise either signal amplifiers or regenerators. The configuration parameters employed to this equipment were chosen using as basis a typical DWDM system. The following parameters are used: Optical fiber: Attenuation Coeficient = 0,25 db/km, Length = 44km; Transmitter: Output Power (max.) = -3dBm, Insertion Loss = 6dBm; and Receiver: Input Power (min.) = -26dBm, Insertion Loss = 6dBm.

Optical channel throughput was defined as 10 Gbps, with a maximum operational wavelength of 1550 nm for transmitter and receiver. Such values are based on Alcatel 1640 OADM DWDM system, and they are also in conformance to Scientific Atlanta Mux/Demux DWDM system. Different specifications for DWDM systems are employed not because of divergences, but because of the lack of a complete documentation for each system. Although the simulated system is hypothetic, it represents a good approximation for a real DWDM system.

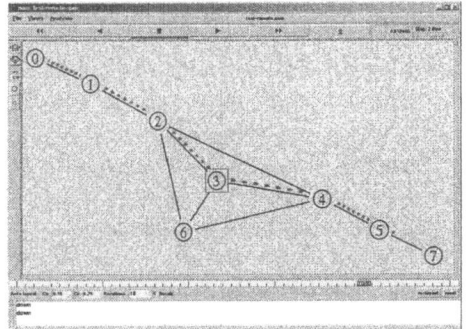

Fig. 2. Simulated Network Topology

Figure 2 shows a snapshot of the Network Animator tool showing the simulated network topology fed into Network Simulator. IP and MPLS are chosen for simulation. MPLS NS (MNS), which is an extension for NS in order to simulate MPLS networks, is used. We do not use any specific extension to simulate GMPLS. However, this fact does not interfere on simulation integrity, once the optical link is simulated in the GPD module separately, and GMPLS is only the generalized form of MPLS to be employed on different networks types. Hence, MNS extension can also be representative to GMPLS simulation performance.

In the simulated topology illustrated in Figure 2, nodes 0 and 7 are IP routers, while all the others are LSR, in particular, nodes 1 and 5 are Edge LSR. Two Label Switching Paths (LSPs) are established: LSP ID 1000 with pathway 1-2-6-4-5 and LSP ID 2000 using path 1-2-3-4-5. LSP 1000 is the main path, while LSP 2000 is the alternative one that is negotiated in case of LSP 1000 failure. LSR 6 node takes the role of traffic protector, indicating the possible need for rerouting generated by link 6-4 failure. The link modeled in Section 3.1 refers to the LSR 4 and 6, which means that PG communicates directly with NS to specify this link behavior.

In order to adapt simulations of layers 2 and 3 to layer 1 simulation, the bandwidth supported by the modeled link is defined as 10 Gbps, while the other links support more than 10 Gbps. This bandwidth definition was needed to create a network bottleneck in which problematic situations will be created. This differentiation must teach the ANN to distinguish link failures from network congestion. The IP router 0 (node 0) also takes the role of traffic generator, while node 7, which also employs IP protocol, is the sink for traffic generated by node 0. The traffic generated in the simulation is described as CBR UDP, with 64 KB packets and 5 MB bursts, in which throughput varies from 5 to 12 Gbps.

4 Results and Discussion

The knowledge base that is necessary to ANN training was generated based on the log files created by NS. The DSPM is responsible for gathering information from log files and transforming them into a knowledge base (i.e., example set) used by the learning algorithm to train and validate the ANN. Neural network training and validation were performed by the JNNS tool.

In order to predict link failures, the ANN must query former parameter values and associate current ones, and build a conclusion about the next BER. Simulated data gathering granularity was adjusted to 0.1 second. Two different types of ANNs were modeled. The first one (Type 1) uses input parameters from physical equipments, while the second one (Type 2) employs input parameters from optical link traffic.

The ANN topology for Type 1 is a 20-20-1 feed forward network with back propagation learning algorithm. Input parameters are extracted from former values of BER on intervals of 2s, i.e., for each input value, 19 former BER readings and the current one are taken into account. The ANN output presents the BER value for the next tenth part of second. Hence, values of BER must be collected in the following instants: t-0.19, t-0.18, t-0.17, ..., t-0.3, t-0.2, t-0.1 and t. They must then be fed as ANN input. The ANN output is the BER value for instant t+0.1. The Mean Square Error (MSE) for neural network validation is 0.0455. When only peak values from the knowledge base were used in training, i.e., only in situations where failure states are imminent, the validation MSE is 0.1020. For this type of neural network, the information's majority for assembly the knowledge base comes from PG simulations. In real world is not possible calculate the BER of physical layer directly, when the system (computer network) is operating. An alternative for this case will be to amplify the optical signal before this one arrive at the photo detector and calculate the BER using formula 3.1 as follows.

$$BER = EXP\left(-\frac{2\overline{P}_{rec}}{hvB}\right)\bigg/2 \qquad (3.1)$$

where $\overline{P}_{rec}$ is the receptor input medium power, hv is the photon energy and B is the bit rate in seconds. This work proposes to deduce the BER through traffic performance of a network computer, the ANNs type 2 are responsible for this.

The ANN topology for Type 2 has the same topology and learning algorithm as ANN Type 1. The input parameters adopted are: Link Status, the amount of used bandwidth (UsedBand), received packets (PktRec) and lost packets (PktLost). ANN output gives BER on physical layer. Candidate variables to the ANN, such as wavelength, maximum power for each wavelength, and OSNR for each DWDM channel, are more adequate to ANN Type 1. Besides this, these variables have not been considered because of the lack of available laboratorial infrastructure. Input variables for ANN Type 2 were collected on instants: t-0.4, t-0.3, t-0.2, t-0.1 e t, where t is given on tenth of seconds. The ANN output indicates BER parameter for instant t. The inferred value of BER for the current time t presents MSE of 0.0015 and when taking into account only the peak points, the MSE is 0.0003. To this type of ANN, where network traffic is used to infer BER, the main goal is to predict BER value on future time t+0.1. Although MSE for overall knowledge base is satisfactory (0.0536), the MSE for training using only critical peak points is not sufficiently good (0.2997).

5 Conclusions

This article applies intelligent agents based on ANNs to predict failures and to guarantee proactive fast re-routing of optical networks. We have developed two neural network experiments. The first one (ANN Type 1) tries to predict BER value using as input former readings of BER parameter. In the second experiment (ANN Type 2), a neural network is fed with other parameters that describe actual operation, such as wavelength, maximum power for each wavelength, and OSNR for each DWDM channel. As ANN Type 2 infers BER value for the current time and ANN Type 1 predicts BER based on current and past BER values, the first one could be the input for the second one. However, this arrangement is not possible because the system propagates the error from layer to layer on both ANNs, which compromises the final BER prediction. As a future work, we will concentrate on developing the intelligent agent signaling and its interaction with the communicating system as well as its modeling process.

References

[1]   Banerjee, A. et al., "Generalized Multiprotocol Label Switching: An Overview of Routing and Management Enhancements". *IEEE Comm. Mag.*, vol.39, num.1, pp. 145-150, Jan, 2001.

[2]   Franklin, M.F. *Redes Neurais na Estimativa da Capacidade Requerida em Comutadores ATM*. Master's Thesis, Universidade Federal do Ceará, Fortaleza, Apr, 1999.

[3]   Nascimento, A.S. *Desenvolvendo Agentes Inteligentes para a Gerência Pró-Ativa de Redes ATM*. Master's Thesis, Universidade Federal do Ceará, Fortaleza, 1999.

[4]   Soto, C.P. *Redes Neurais Temporais para o tratamento de Sistemas Variantes no Tempo*. Master's Thesis, PUC/RJ, Rio de Janeiro, 1999.

[5]   Doverspike, R., Yates, J. "Challenges for MPLS in Optical Network Restoration". *IEEE Communications Magazine*, vol.39, num.2, pp. 89-96, Feb, 2001.

[6]   Kartalopoulos, S.V. *Fault Detectability in DWDM – Toward Higher Signal Quality & System Reliability*. New York: IEEE Press, 2001.

[7]   Mas, C., Thiran, P. "An Efficient Algorithm for Locating Soft and Hard Failures in WDM Networks". *IEEE JSAC*, vol.18, num.10, pp. 1900-1991, Oct, 2000.

[8]   Wong, P.K.C., Ryan, H.M., Tindle, J. "Power System Fault Prediction Using Artificial Neural Networks", In *Proc. of International Conference on Neural Information Processing*, Hong Kong, September 1996.

[9]   Abraham, A., "A Soft Computing Approach for Fault Prediction of Electronic Systems", In *Proc of. The Second International Conference on Computers In Industry (ICCI 2000)*, Published by The Bahrain Society of Engineers, Majeed A Karim *et al.* (Eds.), Bahrain, pp. 83-91, 2000.

[10]   Lawrence, J. "Designing Multiprotocol Label Switching Networks". *IEEE Comm. Mag.*, vol.39, num.7, pp. 134-142. Julho de 2001.

Capacity Allocation for Voice over IP Networks Using Maximum Waiting Time Models*

Sanaa Sharafeddine[1], Nattha Kongtong[1], and Zaher Dawy[2]

[1] Institute of Communication Networks, [2] Institute for Communications Engineering
Munich University of Technology, Arcisstr. 21,
80290 Munich, Germany
{sharafeddine, kongtong, dawy}@mytum.de

Abstract. As voice services impose stringent quality of service (QoS) guarantees to perform well over IP networks, large network resources should be allocated to their traffic class. It gets unaffordable when hard guarantees are required as in deterministic-based mechanisms such as the guaranteed services model of the integrated services (IntServ) architecture. However, the amount of network resources could be drastically decreased if only a small number of all voice connections are allowed to be negatively affected. In this work, a new capacity allocation method based on the maximum waiting time model is explored. It is established from the following concept: by providing statistical quality guarantees to those packets that experience the maximum waiting time among all packets of the active voice connections, all other packets are implicitly protected from excess delay and, thus, from service degradation. This method is investigated and mathematically analyzed for the voice service class in converged IP networks.

1 Introduction

The ongoing convergence of information technology and telecommunications places a high demand on IP networks to support new services having strict and different requirements such as interactive voice and video. Over the past years, several technologies and quality of service (QoS) mechanisms have been developed to realize such multi-service IP networks. One fundamental prerequisite of these networks is the capability to differentiate between packets of different traffic classes and to forward the packets appropriately, each with the desired QoS. To do so, each traffic class should be provided with sufficient share of the total link capacity; otherwise, it experiences quality degradation.

The guaranteed services (GS) model [1] provides hard guarantees per traffic flow with a deterministic upper bound on the network delay [2]. To provide hard QoS guarantees to the voice traffic class, worst-case considerations should be accounted for: packets of active voice connections that are routed to the same outgoing link of a

---

* This work is funded by Siemens AG, ICN EN.

J.N. de Souza et al. (Eds.): ICT 2004, LNCS 3124, pp. 660–670, 2004.
© Springer-Verlag Berlin Heidelberg 2004

network node arrive simultaneously at their corresponding queue and just miss their service turn which is proceeded to service another queue containing a maximum-size packet. The GS model then assures that the voice packet that happened to be queued last is fully transmitted within the defined delay bound. To achieve this, extremely high capacity values are required [3, 4]. In [5, 6], deterministic delay bounds for DiffServ networks are derived where it is also shown in [6] that high speed links are needed to assure these bounds. However, if a slight violation of the delay bounds is tolerated, large capacity resources can be saved.

The contribution of this work is two-fold: 1) we perform a statistical study of the distribution of delay bounds (maximum waiting time) for voice traffic in DiffServ networks and 2) we investigate a new capacity allocation method for voice traffic class based on the derived distribution so as to provide statistical performance guarantees to interactive voice services over IP networks. Previous work of the authors presents simulation results that show the advantage of applying this method for capacity allocation for voice traffic over IP networks [4]. This fact motivates further development and mathematical study of this method as carried out in this work.

In Section 2, we describe our network model along with the assumptions and present the capacity allocation method proposed for voice over IP (VoIP). In Section 3, we mathematically analyze the method for a single network node by providing ways of computing the maximum waiting time distribution and interpolating the needed capacity value, subject to a given set of QoS parameters. In addition, we extend the analysis to include converged IP networks that support other traffic classes that compete for the available link capacity. Finally, Section 4 concludes the paper.

2 A New Method for Capacity Allocation of VoIP

2.1 General Network Model

In this work, we consider a general IP network model consisting of a number of nodes and links, through which traffic with different characteristics could be flowing. Information transmission between a pair of nodes forms communication paths, each of which consists of a succession of hops in tandem.

Should we be conservative, we consider a pessimistic scenario in which traffic interfering with a target flow is injected at each node independently from other nodes [3, 7, 8]. Consequently, the delay caused by the interfering traffic is independent at each node, allowing us to investigate the per-hop delay contributions separately.

2.2 Single-Hop Model

Based on the assumption of independent hops, we extract a single hop from the source-destination path of the target flow and analyze the multiplexing process of several incoming lines onto one outgoing port of the associated node as shown in Figure 1. We assume that K active voice connections are carried by those incoming lines and they are destined to the same output link with capacity C. To obtain a high-level performance of the traffic classes supported in the network node considered, sufficient amount of capacity C should be provided to be shared among all classes. As voice traffic demands strict QoS requirements, its share of the link capacity should be explicitly evaluated. Later in this work, we present and investigate a new method to compute the voice share of the link capacity C.

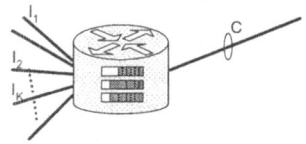

Fig. 1. Single-hop model (the unlabeled incoming lines do not carry voice connections)

For simplification, we assume that the K voice traffic flows of the active voice connections are carried by different incoming lines. We need to make sure via proper capacity allocation that the queuing delay of voice packets is bounded within a delay threshold D in order to attain a smooth and interactive communication between the users [9]. Therefore, we can represent our problem as a queuing system whose performance depends on the arrival and service processes of the packets. The arrival process depends on the pattern in which traffic flows are generated by the voice sources and the service process depends on the traffic characteristics in addition to the scheduling policy employed in a multi-class-network node.

Voice sources transmit fixed-size packets at regular intervals if voice activity detection (VAD) is not employed. In case VAD is activated in the voice encoders of the sources, it is expected that the resulting waiting time distribution is bounded by the distribution of non-VAD-traffic (i.e. constant bit rate (CBR) traffic) and so if the latter distribution is used for capacity allocation then additional overdimensioning for voice traffic class is induced. According to [7, 10], it is shown, however, that very slight overdimensioning is required since the waiting time distribution tails of both VAD- and non-VAD-traffic match very closely. As a result, we consider non-VAD-traffic in the analysis; however, the results apply also to networks with VAD-traffic. In this work, we assume that each voice source generates a packet of fixed size L in the same periodicity of T seconds, producing traffic at rate $r = L/T$. The arrival times of packets belonging to the different K voice connections are assumed mutually independent and uniformly distributed over the time interval $[0, T)$.

It is worth mentioning that we are interested in evaluating the minimum link capacity that is required for voice traffic only. The presented method solely is not enough for computing the total link capacity needed to serve a set of traffic classes. Currently, it is still not possible to devise a single method for IP networks that can evaluate the total capacity value required by all available traffic classes having different properties and quality constraints. To do so, other methods should be considered to evaluate the capacity required per each class and then the capacity values are combined in a certain way to obtain the total link capacity. One direct approach of combining the individual capacity requirements is to simply sum them up. Such approach might, however, lead to underutilized links since no multiplexing gain is considered.

Subsequently, in regards to the service process, the employed scheduling policy determines the way in which packets are serviced from their corresponding queues. In our network model, we consider priority queuing (PQ) as the employed scheduling policy. PQ has been proposed as an adequate scheduling for the expedited forwarding per-hop behavior EF-PHB [11], which grants premium service to a defined aggregate of traffic in the DiffServ model [12] and has been introduced to support critical real-time traffic such as voice. With PQ, however, it is possible that traffic classes with the high priority take up the whole bandwidth and push out lower-priority traffic unless some sort of traffic policers are employed to control the high-priority traffic. An alternative to PQ is class-based weighted fair queuing (CB-WFQ), which allocates a weight to each class/queue and shares the link capacity among the busy queues in direct proportion to their weights. Note that both scheduling schemes are work-conserving and so idle periods of one traffic class can be used by another class. In this work, however, we will only consider PQ and grant voice traffic the highest priority while we expect a straightforward extension to the case of CB-WFQ.

2.3 Method Description

To properly allocate capacity for VoIP traffic, it is important to study the waiting time introduced in IP networks that should be restricted to a defined bound if high performance is desired. The allocated capacity is fully utilized to service voice packets within the delay bound at times of pessimistic arrival scenarios which cause maximum waiting times. Therefore, if we know the maximum waiting time then we are able to determine the needed capacity.

In Figure 1, all packets of the K connections might arrive simultaneously at their queue as a worst-case scenario; the GS model (or any other deterministic-based mechanisms [6, 13]) then makes sure that the voice packet that happened to be queued last is definitely served within a defined per-hop delay limit (i.e. required service rate = $K \cdot L/D$). This way, extremely high capacity would be needed to provide hard QoS guarantees. Therefore, if we are aware of the packets that experience the maximum delay among all packets, we can practically protect all other packets of the active voice connections from extra delay, and, thus, from quality degradation. This notion forms the basis of our proposed capacity allocation method. Slightly softened guarantees that allow the maximum waiting time among packets of certain arrival pattern to be exceeded if the arrival patterns occur very rarely lead to significant reduction in the capacity requirements. The proposed method offers such guarantees and could be summarized by the following steps:

1. Calculate the *maximum* waiting time distribution of voice traffic at a node.
2. Define a delay threshold D of the waiting time at the specific node.
3. Define an outage probability P that represents the frequency of arrival patterns, in which the maximum waiting time is allowed to exceed D.
4. Compute the link capacity such that the voice packet with the maximum waiting time at a given node undergoes a delay that exceeds D in at most a probability P.

Note that the allocation of one capacity value for a number of connections makes this method applicable to DiffServ networks, which service behavior aggregates [12].

3   Mathematical Analysis

To study the method mathematically, we start with a simplified model so as to obtain a feasible mathematical solution. Afterwards, we attempt to generalize the model gradually to account for new issues and extend the results accordingly.

3.1 Voice-Only IP Networks

To start with, we consider a single-hop of an IP network that runs voice traffic only. By doing so, we study certain aspects of this traffic in isolation to the effect introduced by other interfering traffic that compete for the available capacity. In voice-only networks, the previously defined queuing system is reduced to a queue with periodic arrivals and constant service times. The resulting scenario can be modeled as an N*D/D/1 queuing system, which models a superposition of N independent equiperiodic sources serviced in constant service time. The output process is considered as an equal-slotted transmission channel that can transmit exactly one packet per timeslot (i.e. the length of a timeslot is equal to the service time per packet $T_s = L/C$, where C, in this case, is fully allocated for voice traffic).

To compute the necessary capacity for VoIP service using the proposed method, the maximum waiting time model of the traffic is required as a first step. The virtual waiting time distribution (unfinished work) of the N*D/D/1 model had been extensively studied in the literature [14, 15, 16], while the *maximum* waiting time distribution is briefly handled and is not widely applied for practical purposes. The former, however, if used for capacity allocation, results in underestimated capacity values since it accounts for zero unfinished work (when the queue is empty). The latter, on the other hand, is shown by simulations in [4] to be more appropriate for capacity allocation purposes when QoS guarantees are required as in VoIP.

To obtain the *maximum* waiting time distribution of the N*D/D/1 model, we return to methods given by Ott and Shantikumar in [17] for a discrete time model and generalized by Hajek in [18] for the continuous case. The virtual waiting time process of an N*D/D/1 queue is stationary and periodic with period T and it has the same distribution as the virtual waiting time of an M/D/1 queue with the condition that K customers arrive during the interval $[0, T)$ and the server is idle at time T [18].

In our model, K active voice connections are routed over the channel. Each connection sends a packet once every M timeslots (i.e. $M = \lfloor T/T_s \rfloor$). The probability that the maximum waiting time W_{max} exceeds a constant c, where $0 \le c \le K \cdot T_s$, is given by [18]

$$P\{W_{max} > c\} = \begin{cases} 1 - \dfrac{\displaystyle\sum_{j=1}^{K} r_K^{*j}(c) f\left(\lambda\left(1-\frac{K}{M}\right), j\right)}{\left(1-\frac{K}{M}\right) f(\lambda, K)} & 1 \leq K \leq M-1, \\[4ex] 1 - \dfrac{r_K(c)\lambda}{f(\lambda, K)} & K = M, \end{cases} \tag{1}$$

where $r_K(c)$ is the probability that during a typical busy period of an M/D/1 queue (arrival rate $= \lambda = M/(1\sec)$, service time $= (1\sec)/M$), exactly K customers arrive and are served, and the virtual waiting time remains less than or equal to c throughout the busy period. The factor $r_K^{*j}(c)$ denotes the j-fold convolution of $r_K(c)$ and $f(\mu, i) = \exp(-\mu) \cdot \mu^i / i!$. To compute the probability $r_K(c)$ and the convolution $r_K^{*j}(c)$ using the methods of [17], we approximate the delay bound c to integer multiples of T_s, i.e. $c = \lfloor c/T_s \rfloor \cdot T_s + \varepsilon \approx n \cdot T_s$, where $n = \lfloor c/T_s \rfloor$ is an integer and $0 \leq \varepsilon < T_s$.

Let $\Pi(n)$ be the $n \times n$ matrix:

$$\Pi(n) = \begin{bmatrix} \pi_1 & \pi_2 & \cdots & \pi_n \\ \pi_0 & \pi_1 & \cdots & \pi_{n-1} \\ 0 & \pi_0 & \cdots & \pi_{n-2} \\ \vdots & \vdots & \vdots & \vdots \\ \cdots & 0 & \pi_0 & \pi_1 \end{bmatrix}, \tag{2}$$

where $\pi_i = \exp(-1)/i!$, $i = 0, 1, \ldots, n$, and $\pi_1^{(k)}(n)$ is the top leftmost element of the kth power of $\Pi(n)$, $(\Pi(n))^k$, for $k = 0, 1, \ldots, K$. Hence, $r_k(c)$ can be closely approximated as:

$$r_k(c) \approx \begin{cases} 0, & k \leq 0 \\ \exp(-1)\pi_1^{(k-1)}(n-1), & k = 1, 2, \ldots, K \end{cases} \tag{3}$$

and the convolution $r_K^{*j}(c)$, for any integer $j > 0$, can be iteratively calculated as follows:

$$r_K^{*0}(c) = \delta(K, 0),$$
$$r_K^{*j}(c) = \sum_{k=1}^{K} r_k(c) r_{K-k}^{*(j-1)}(c). \tag{4}$$

where $\delta(x, y) = 1$ for $x = y$ and $\delta(x, y) = 0$ otherwise. Now that all components of (1) are computable, we can form the maximum waiting time distribution of the N*D/D/1 system. Given K active voice connections and a delay threshold $D = c$, we allow an outage probability P. Finally, we can solve iteratively for C using (1).

Figure 2 plots the computed complementary cumulative distribution function (CCDF) of W_{max} and compares it to Monte-Carlo simulation results. We observe that the two plots almost overlap for different system parameters verifying the mathematical calculation of the distribution. In the simulation, the single-hop model presented in Section 2.2 is implemented in voice-only networks, where we generate T-periodic arrivals of K voice connections with uniformly distributed arrival time over $[0, T)$ for each connection. For each new set of arrivals, we compute the maximum waiting time encountered among all packets.

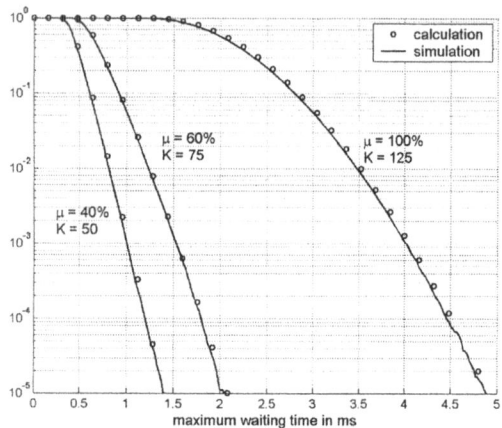

Fig. 2. Calculation and simulation results of W_{max} CCDF. $C = 10$ Mbps, $r = 80$ kbps ($L = 200$ byte, $T = 20$ ms), $K = 50$, 75, and 125 users corresponding to link utilization $\mu = 40\%$, 60%, and 100%, respectively

3.2 Converged IP Networks

In general, the existence of other queues/classes in a network sharing part of the available capacity negatively affects the performance of the voice service by introducing extra delay to voice packets that wait their turn of service among the other queues. As a pessimistic assumption, packets of other classes are assumed with maximum transfer unit (MTU) size and are constantly filling up their queues.

If PQ is the scheduling policy used, voice traffic is assigned to a separate FIFO queue with the highest priority. When the preemptive version of PQ is applied, voice traffic class is strongly protected against other traffic classes as if it is the only class available in the network. So to speak, the associated maximum waiting time maintains the same distribution as in voice-only networks and the same results presented in Section 3.1 still apply to this case.

However, when the non-preemptive version of PQ is used, deterioration in the voice class performance is observed due to the interference caused by other traffic classes. The waiting time incurred on voice packets is affected by the residual transmission time of lower priority traffic. As a result, the voice traffic periodicity is negatively influenced when a variable residual time caused by transmission of packets of other classes is added. In this section, we examine the resulting voice queuing model of a single-hop network and aim to extend the results of the previous section to include the residual transmission time of lower priority packets.

Figure 3 shows the CCDF of W_{max} corresponding to a set of Monte-Carlo simulations with various *MTU*-size packets. It is obvious that the introduction of another traffic class in the network causes W_{max} distribution of voice traffic to deviate from its initial distribution in voice-only networks. It is realized that the deviation distance (calculated in percentage of one service time of the corresponding *MTU*, T_{MTU} = *MTU/C*) does not exceed T_{MTU} for any value of *MTU* and it increases as the *MTU* size increases. Such observation is logical as the inclusion of *MTU*-size traffic in addition to voice traffic affects the maximum waiting time of voice traffic by at most T_{MTU}.

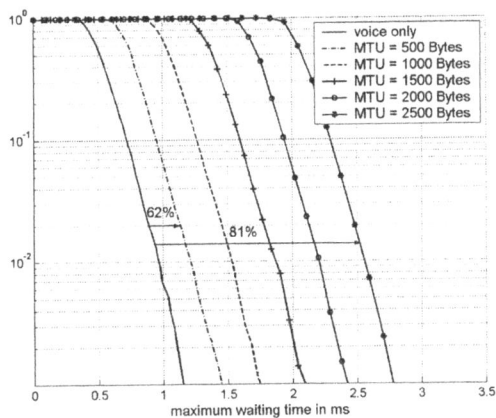

Fig. 3. W_{max} CCDF for 60 active voice connections. $C = 10$ Mbps and $r = 80$ kbps

The maximum waiting time among all voice packets can be formulated as

$$W_{max} = \max(w_{voice} + w_{residue}),\tag{5}$$

where w_{voice} is the waiting time resulted from queuing behind other voice packets in the same queue, and $w_{residue}$ is the residual transmission time of lower priority packets. The distribution of $w_{residue}$ is typically assumed to be uniform in the interval $[0, T_{MTU}]$ [8]. A lower bound on W_{max} is achieved when no lower priority packets are being serviced when a voice packet arrives to an empty queue. This scenario yields similar results to the case of voice-only networks, where no other traffic class affects the transmission of voice packets. Therefore

$$W_{max} = \max\left(w_{voice} + w_{residue}\right) \geq \max(w_{voice}) = \left(W_{max}\right)_{voice-only}.\tag{6}$$

On the other hand, an upper bound on W_{max} is achieved when the specific packet that experiences the maximum waiting time in voice-only networks arrives to the queue exactly when a full *MTU*-size packet has just been placed in service, that is

$$W_{max} \leq \max\left(w_{voice}\right) + \max\left(w_{residue}\right) \leq \max\left(w_{voice}\right) + T_{MTU},\tag{7}$$

where w_{voice} and $w_{residue}$ are assumed mutually independent. The distributions resulting from the upper and lower bounds are computed and plotted in Figure 4, where they are compared to Monte-Carlo simulation results that fall in between the two bounds. The actual W_{max} CCDF can be well approximated by calculating W_{max} as

$$W_{max} \approx \max\left(w_{voice}\right) + w_{residue} \tag{8}$$

at quantile probabilities, since the influence of voice traffic, with enough number of connections, increases at low probabilities and becomes the dominant factor of the delay percentile as compared to the residual transmission time of only one *MTU* packet. So, whether (8) or (5) is used, very close W_{max} CCDFs are resulted as illustrated in Figure 4. The calculation curve in the figure falls below the simulation curve at high probability values, where the actual W_{max} is the time when the maximum residual transmission time occurs, while the approximate calculated W_{max} adds a uniformly distributed residual transmission time. The simulation and calculation curves cross at lower probabilities when the influence of voice traffic increases. For higher voice loads, the two curves cross at higher probability values, since voice traffic gets to have more effective influence on W_{max}.

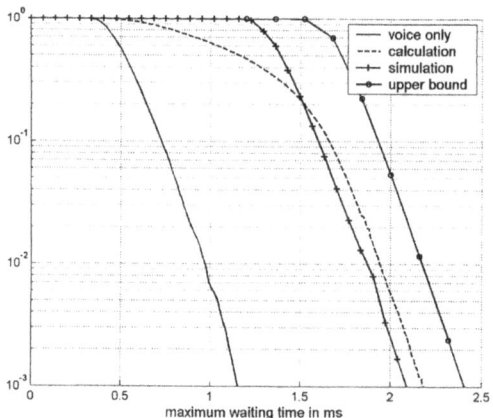

Fig. 4. W_{max} CCDF in converged IP networks. $K = 60$ users, $C = 10$ Mbps, $MTU = 1500$ bytes and $r = 80$ kbps

4   Conclusion

In this paper, we mathematically analyze and elaborate a new capacity allocation method for voice over IP. We investigate this method per network hop analytically in voice-only networks and extend our analytical study to converged networks supporting multiple traffic classes with the condition that voice traffic is given a separate queue and being serviced using priority queuing. Knowing that the assumed queuing model still yields accurate results in cases of low link utilization [14], this method can be generalized to end-to-end capacity allocation where the maximum waiting time distribution can be directly obtained by simple convolution over the network hops.

This method is shown to provide a tradeoff between allocated network resources and the probability of quality guarantees and also shown to be suitable for the Diff-Serv model where it makes use of the multiplexing gain in order to reduce the needed resources. This method will be utilized in a network planning tool that evaluates the minimum capacity required for voice traffic in large-scale IP networks supporting interactive services.

References

1. S.Shenker, C. Partridge and R. Guerin. Specification of Guaranteed Quality of Service. Request for Comments RFC 2212, IETF, September 1997.
2. A.K. Parekh, R.G. Gallager. "A Generalized Processor Sharing Approach to Flow Control in Integrated Services Networks: The Multiple Node Case," IEEE/ACM Transactions on Networking, April 1994.
3. M. Mandjes, K. van der Wal, R. Kooij, and H. Bastiaansen. "End-to-End Delay Models for Interactive Services on a Large-Scale IP Network," in Proceedings of the 7th IFIP Workshop on Modeling and Evaluation of ATM / IP Networks, June 1999.
4. S. Sharafeddine, A. Riedl, and J. Totzke. "A Dimensioning Strategy for Almost Guaranteed Quality of Service in Voice over IP Networks" in Proceedings of IEEE High Speed Networks and Multimedia Communications (HSNMC 2003), July 2003.
5. Charny and J.-Y. Le Boudec. "Delay Bounds in a Network with Aggregate Scheduling," in Proceedings of Quality of future Internet Services Workshop (QofIS 2000), September 2000.
6. J. Schmitt, P. Hurley, M. Hollick, and R. Steinmetz. "Per-Flow Guarantees under Class-Based Priority Queuing" in Proceedings of Globecom 2003, December 2003.
7. M. Karam, F. Tobagi. "Analysis of the Delay and Jitter of Voice Traffic Over the Internet," in Proceedings of Infocom'01, April 2001.
8. L. Kleinrock. Queueing Systems: Computer Applications, Vol. 2, John Wiley and Sons, New York, 1975.
9. ITU-T Recommendation G.108: Application of the E-Model: A Planning Guide, September 1999.
10. K. Sriram and W. Whitt. "Characterizing Superposition Arrival Processes in Packet Multiplexers for Voice and Data," IEEE Journal on Selected Areas in Communications, Vol. SAC-4, No. 6, pp. 833-846, Sept. 1986.
11. V. Jacobson, K. Nichols, K. Poduri. The Virtual Wire 'Per-Domain Behavior': Analysis and Extensions. Internet Engineering Task Force, July 2000.
12. S. Blake, D. Black, M. Carlson. An Architecture for Differentiated Services. Request for Comments RFC 2475, IETF, December 1998.
13. W.-J. Jia, H.-X. Wang, J.-C. Fang, and W.Zhao. "Delay Guarantees for Real-Time Traffic Flows with High Rate," in Proceedings of ITC18, August/September 2003.
14. Eckberg. "The Single Server Queue with Periodic Arrival Process and Deterministic Service Times," IEEE Transactions on Communications, Vol. Com-27, No. 3, pp. 556-62, March 1979.
15. P. Humblet, A. Bhargava, and M. Hluchyj. "Ballot Theorems Applied to the Transient Analysis of nD/D/1 Queues," IEEE/ACM Transactions on Networking. Vol. 1, No. 1, pp. 81-95, February 1993.

16. J.W. Roberts and J.T. Virtamo, "The Superposition of Periodic Cell Arrival Processes in an ATM Multiplexer," IEEE Transactions on Communications, Vol. 39, pp. 298-303, February 1991.
17. T.J. Ott and J.G. Shantikumar. "On a Buffer Problem for Packetized Voice with N-Periodic Strongly Interchangeable Input Processes," Journal on Applied Probability, pp. 630-646, 1991.
18. Hajek. "A Queue with Periodic Arrivals and Constant Service Rate," Probability, Statistics and Optimization -- A Tribute to Peter Whittle, F.P. Kelly ed., John Wiley and Sons, pp. 147-158, 1994.

SCTP Mobility Highly Coupled with Mobile IP

Jin-Woo Jung[1], Youn-Kwan Kim[2], and Hyun-Kook Kahng[1]

[1] Department of Electronics Information Engineering, Korea University
♯208 Suchang-dong Chochiwon Chungnam, Korea
[2] LG Gangnam Tower, 679, Yeoksam-dong, Gangnam-gu, Seoul, Korea 135-080
jjw@korea.ac.kr

Abstract. This paper presents results, which have obtained by extensive simulations for Mobile IP (MIP) and Stream Control Transmission
Protocol (SCTP) from the perspective of mobility support in wireless Internet. After illustrating the problem in the existing Mobile SCTP and
SCTP over Mobile IP for diverse cases, we propose SCTP highly coupled with MIP for realizing seamless mobility. In this thesis we use ns2
with SCTP module to compare the throughput and end-to-end delay
for packet transmission associated with the Mobile SCTP, SCTP over
Mobile IP, and the proposed approach. Simulation results show that our
proposed approaches achieve better performance than other proposals.

1 Introduction

In this paper, we examine a recently standardized transport protocol - Stream
Control Transmission Protocol (SCTP) and the Mobile IPv4 (MIP) standard
from the perspective of mobility support in wireless Internet [1][2].
The SCTP is a new standardized transport protocol operating on top of the
Internet Protocol (IP). Although it shares many characteristic with TCP, it has
many significant and interesting differences. The distinct differences to TCP are
multi-homing and the concept of several streams within a connection. In this
paper we focus on multi-homing feature and flow control mechanism from the
view point of mobility support in SCTP. Even though the original SCTP protocol
did not consider the mobility of the end nodes, there have been ongoing research
efforts to support mobility in the current SCTP protocol [3][4][6].
The basic concept of the proposed approach is that the SCTP-mobility approach
does not exclude the MIP-based approach, it may work to complement based on
the kind of application. That is, the solution suggested in this paper is to use MIP
for non-SCTP based applications (e.g. telnet, ftp, tftp, http, and etc) but to use
the proposed MIP-based SCTP for SCTP-based applications. Unlike standard
MIP, however, the proposed approach limits tunneling to SCTP associations
that are active during a movement.
In this paper we present performance figures, which have been obtained by extensive simulations for SCTP over MIP, SCTP-based mobility and the proposed

This research was supported by University IT Research Center Project.

J.N. de Souza et al. (Eds.): ICT 2004, LNCS 3124, pp. 671–677, 2004.
© Springer-Verlag Berlin Heidelberg 2004

approach. The simulation results show that our proposed approach can achieve better performance than both SCTP over mobile IP and SCTP-based mobility, and overcome the drawbacks of each protocol.

2   Related Works

2.1   SCTP-Based Mobility

Mobile SCTP is proposed as a transport layer mobility management by using multi-homing feature in SCTP [3][5][6]. The most significant limitation is the absence of a mobility management for the existing TCP/IP suite (TCP and UDP) or non-SCTP based applications. Second, a Mobile SCTP by itself can not support roaming. If there are no additional entity or devices for location management, the INIT chunk for a new association with MN's home address cannot be correctly routed to the new location of MN after a mobile node moves a new network. Third, it can cause a significant disruption time if association update is not completed while the mobile node is in the overlapped area.

2.2   Solutions Related to Improve TCP Performance in Mobile Environments

To understand the aims of this paper, it is useful to investigate several solutions, proposed to overcome the problems of standard TCP operation over wireless networks. Even if a single packet is dropped for any reason, the current standard of TCP(and SCTP) assumes that the loss was due to congestion and throttles the transmission by bringing the congestion window down to the minimum size. For disconnections due to handoff, this coupled with the TCP(and SCTP)'s slow-start mechanism means that the successive TCP timeouts then increase the TCP(and SCTP) timeouts interval such that, even though handoff processing has completed. TCP will not immediately resume the communication. Many papers have been written for solving the performance degradation of TCP over lossy wireless links(such as Indirect-TCP, Wireless-TCP, Mobile-TCP, Snoop, Forward Error Correction, and Freeze-TCP, etc)[8][9].

3   Integration of MIP and SCTP-Mobility

3.1   Location Management

Fig. 1 shows the time diagram of handoff procedure for the proposed approach. When a mobile node enters a new network, it detects movement and obtains an initial CoA. At the same time, MN's MIP must contact the DHCP(or FA) server to obtain a globally unique CCoA and inform SCTP instance of its movement. It then sends a registration request through the foreign agent to the home agent requesting mobility binding update for a period of time. The HA updates its mobility binding table and sends a registration reply back to the MN through

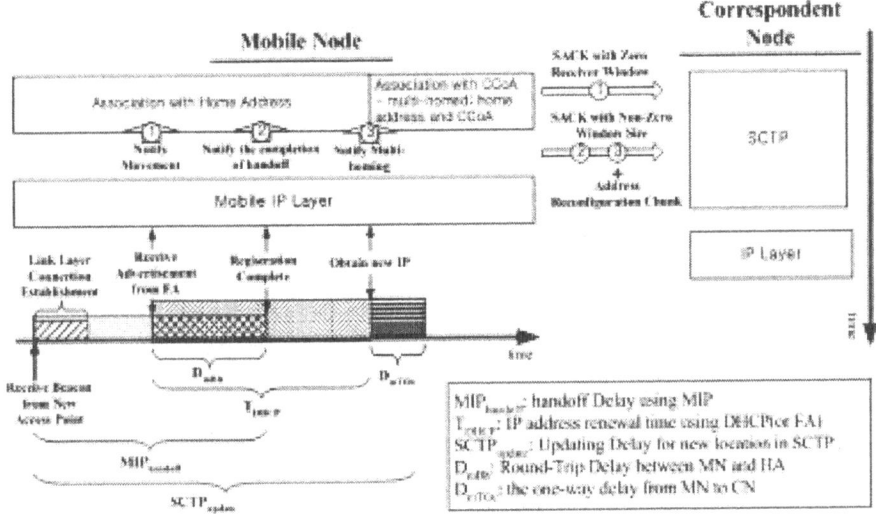

Fig. 1. Operation of the proposed approach

the FA allowing or denying that registration. Once MN acquire a new IP address (CCOA), it notifies the completion of address renewal to SCTP instance in MN. By using this approach, if a MN cannot acquire the CCoA, the mobility for SCTP-based application is supported by MIP.

3.2   Standstill SCTP

In this section, we suggest an explicit notification method for confirming the completion of handoff. This causes the sender to immediately transmit data chunks, which eliminates the waiting period. The proposed Standstill SCTP is similar mechanism to Freeze-TCP and M-TCP using zero window advertisement for improving TCP performance in mobile environments [8][9]. Unlike a Freeze-TCP, it is able to obviously detect long disconnection since a Standstill SCTP is depends on mobile IP for detecting an impending handoff. For an association from CN to MH, when a MN notices its movement, its SCTP instance sends SACK chunk with zero window size to CN. As soon as a CN receives this SACK chunk, it freezes the cwnd, ssthresh and the retransmission timer, and prevents further data chunk. To resume a association, the MN sends immediately SACK with non-zero window size + Heartbeat chunk to the CN as soon as it identifies whether the completion of handoff from MIP or the finishing of IP renewal from DHCP. When the CN receives this packet, it recovers cwnd, ssthresh and the retransmission timer, and starts data chunk transmission.

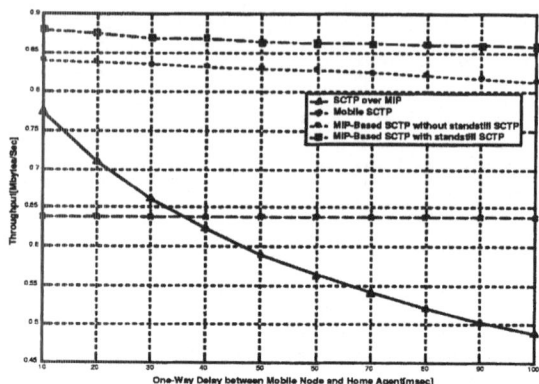

Fig. 2. SCTP throughput vs. delay between the new FA and the old FA

4   Performance Evaluation

4.1   Simulation Model

Simulation model of SCTP over MIP is based on the library modules provided by ns2 [7] with SCTP module, while the SCTP-mobility simulation module is implemented based on the description provided in [3]. MNs are connected to access point(APs) using the ns-2 carrier sense multiple access with collision avoidance(CSMA/CA) wireless link model. And, the bandwidth and the link delay between AP and MN are set to 10Mbps and 2ms, respectively. The SCTP association from CN to MN performs bulk file transfer using MSS equal to 1448 bytes(that is, there is no IP fragmentation due to MIP's tunneling. If the MSS is larger than maximum transmission unit(MTU) of Ethernet, the SCTP performance over MIP is more degraded). Also, we exploit the default values that are recommended in RFC 2960 suitable for SCTP deployments in the public Internet(minimum retransmission timeout value: RTOmin= 1sec, initial retransmission timeout value:RTOinit = 3 sec, SACK delay = 200ms, and so on)[1].

4.2   Simulation Results

In this paper, we consider two kinds of factor which affects the handoff completion delay and the SCTP performance over wireless networks:
• DnTOo: the one-way delay to exchange packets between the new FA and the old FA
• DmnTOcn: the one way delay from MN to CN or vice versa
Fig. 2 illustrates the throughput at the MN as the DnTOo delay increases. Here the DmnTOcn is set to be 30ms. For showing distinct results, we measure the average SCTP throughput of each approach during a 25 second around handoff point. Obviously, the SCTP throughput using Mobile SCTP approach becomes better than that of SCTP over MIP as the direct path between the MN and CN

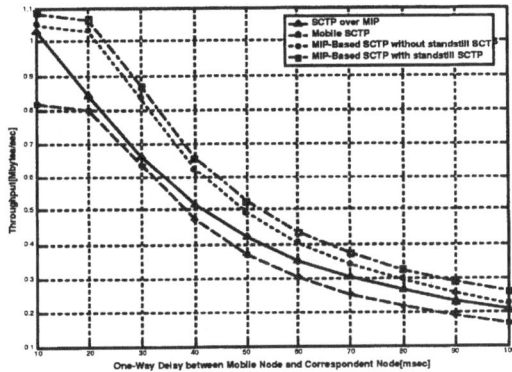

Fig. 3. SCTP throughput vs. delay between MN and CN

is shorter than the distance go through FA and HA. Also, the Mobile SCTP is independent of the distance between MN and HA, its performance is fixed about 0.642 Mbytes/sec. For the proposed MIP-based SCTP, it shows more effective results. If the handoff completion delay of MIP is shorter than that of Mobile SCTP, the packets destined for MN is tunneled to FA through HA until the handoff of Mobile SCTP is completed. Also, Fig. 2 shows that the MIP-based SCTP using Standstill algorithm can more improve SCTP performance than that of MIP-based SCTP by reducing redundant waiting time due to the successive timeouts after the handoff of MIP or the handoff of SCTP is completed.

Fig. 3 shows the SCTP throughput as the delay DmnTOcn increases. Here the DnTOo is assumed to be 30ms. Since all methods including MIP and Mobile SCTP, and the proposed approach are depends the End-to-End delay between CN and MN, the average throughput for each approach is degraded as the delay DmnTOcn increases. However, as remarked above, since the MIP-based approach uses the fast handoff method between MIP and SCTP, it shows more improved throughput than other two approaches. The reason for a drop in throughput near 20 ms for Fig. 3 is that a longer End-to-End delay (over 25ms) results in a second timeout during the handoff.

Fig. 4 and Fig. 5, respectively, illustrate the SCTP throughput and variation of congestion window size in worst scenario for Mobile SCTP. In Fig. 4, at time t = 9.5s the MN switches its old BS to a new BS in a new domain and establishes link layer connection. Since the MN is no longer attached at its previous address, all of the further segments destined for previous address are lost. In Fig. 4 Mobile SCTP approach experiences third timeout because it requires additional time to obtain a new IP address. That is, even though its handoff is completed at t = 12.5s and MN can receive further data, the CN does not immediately restart data transmissions because the CN is still waiting for SACKs for the lost segments. Fig. 4 and Fig. 5 also show that the Standstill SCTP can successfully deal with the redundant waiting time due to the successive timeouts when the DmnTOc delay is slightly longer.

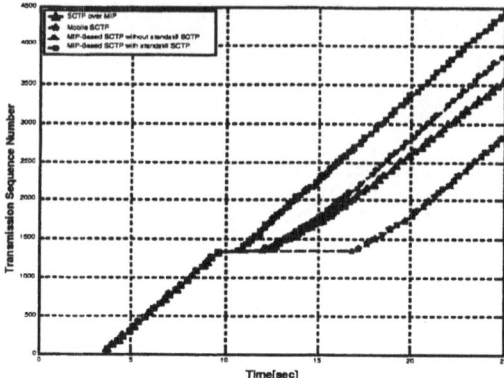

Fig. 4. SCTP throughput at DmnTOcn= 100ms and DnTOo= 30ms

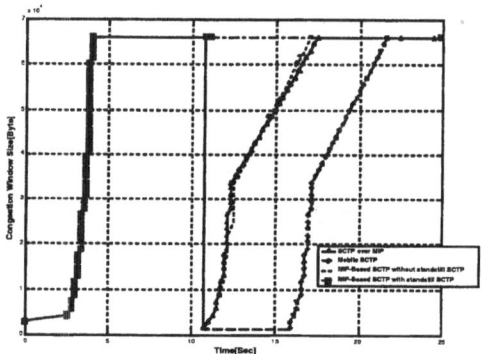

Fig. 5. cwnd at DmnTOcn= 100ms and DnTOo= 30ms

As a result, we conclude that the proposed methods can achieve significant improvements during handoff and the best performance in various environments.

5   Conclusion

The proposed mobility management method has two main features: First, the proposed scheme uses SCTP's packet delivering coupled with a location management scheme derived from the MIP's location management scheme to provide a seamless mobility. Second, to alleviate the performance degradation of SCTP during disconnections due to handoff, we propose a Standstill SCTP highly coupled with events of MIP. The simulation results show that the proposed methods can achieve more improved performance than Mobile SCTP and SCTP over MIP. The best performance in various environments was achieved when the two proposed schemes were combined. However, the implementation of the proposed approach is needed in real world since the proposed approach is simulated under some assumptions and the processing complexity is ignored (e.g., IP renewal delay is one second, Move Detection delay in MIP is 0.5 second, respectively).

References

[1] R. Stewart, Q. Xie, et al., "Stream Control Transmission Protocol", RFC 2960 in IETF, Oct. 2000.

[2] C. Perkins, "IP Mobility Support", RFC 3344 in IETF, August 2002

[3] M. Riegel, M. Tuexen, "Mobile SCTP", draft-riegel-tuexen-mobile-sctp-01.txt, Feb. 2002.

[4] J. Noonan, P. Perry, J. Murphy, "A Study of SCTP Services in a Mobile-IP Network", Proc. of IT&T Annual Conference, 30-21 Oct. 2002.

[5] R. Stewart and Q. Xie. "SCTP extensions for Dynamic Reconfiguration of IP Addresses.", draft-ietf-tsvwg-addip-sctp-07.txt",work in progress, Feb. 2003

[6] Shinichi Baba, Nobuyasu Nakajima, Hnning Schulzrinne, "M-SCTP: Design And Prototypical Implementation Of An End-To-End Mobility Concept", In Proc. Of 5th International Workshop, Berlin, Germany, Oct 2002.

[7] UC Berkeley, LBL, USC/ISI, and Xerox Parc, version 2.1b8, 2001, http://www.isi.edu/nsnam/ns

[8] Goff, T., Moronski, J., Phatak, D.S., Gupta, V.,"Freeze-TCP: A true end-to-end TCP enhancement mechanisms for mobile environments", INFOCOM 2000, 19th Annual Joint Conference of the IEEE Computer and Communications Societies, Proceedings. IEEE, Volume:3, Mar 2000.

[9] K. Brown and S. Singh, "M-TCP:TCP for Mobile Cellular Networks", ACM Computer Communications Review(CCR), vol. 27, no. 5, 1997.

Migration to the New Internet – Supporting Inter Operability Between IPv4 and IPv6 Networks

Teresa Vazão[1,2,3], Luís Raposo[1], and João Santos[1]

[1] Instituto Superior Técnico, Lisboa, Portugal,
teresa.vazao@tagus.ist.utl.pt
[2] INESC-ID, Lisboa, Portugal
[3] INOV - Inesc Inovação, Lisboa, Portugal

Abstract. The evolution of communications towards a single unified network based on IP, brings new issues to the Internet, as a new business paradigm appears and an explosion in the number of users will continue. IPv6 will support this new Internet, but, in spite of its advantages, it will be expectable that it will coexist with IPv4 for a long time, due to the huge size of the Internet today.

The migration phase will play a central role in this process and special care has to be taken concerning inter-operability issues. From the existing set of inter-operability solutions, NAPT-PT is the most promising for the nearest future, as it efficiently uses network resources, without wasting the spare IPv4 addressing space.

This paper presents an Inter Working Unit based on NAPT-PT, which have been developed in GNU/Linux Operating System. Considerations about the IWU design are taken and functional and performance tests are described.

1 Introduction

The convergence of data and telecommunication networks around a single network based on IP, brings an unpredictable business importance to the Internet, leading to a continuous and exponential increasing in the number of users. Wireless and broadband communications will contribute to this expansion, bringing to the network non-traditional costumers, new applications and services.

Although the advantages of having an unified network, based on a widely deployed technology, IPv4 addressing space is reaching its limits, due to the 32 bit size of address' field. Internet Engineering Task Force (IETF), being aware of this problem, provided the Network Address Translation (NAT) – a short term solution –, already supported in most of the existing enterprise networks to face the lack of available IP public addresses.

In the meanwhile, IETF have developed a long term solution, which is a new version of IP protocol – IPv6 – targeted to sustain the main issues of the new generation Internet[1]. It solves the scalability's problem of IPv4 addressing

J.N. de Souza et al. (Eds.): ICT 2004, LNCS 3124, pp. 678–687, 2004.

space, by providing longer IP addresses; it has native Mobility and Quality of Service support and, due to its new header structure, faster processing may be achieved and future evolution is easily supported.

The 3rd generation UMTS networks will lead this unified approach to the limit, by putting together data, voice and cellular networks, traditional applications and multimedia ones, as well as service and content providers. IPv6 will play a central role in the core network, but, in the meanwhile both IPv4 and IPv6 will coexist.

Nowadays, a significant effort is being doing in the dissemination of IPv6 technology and several task forces [2] [3] and pilot networks [4] are being developed around the world. The main idea is not only to develop the technology, in order to achieve enough maturity, but also to promote it in the industry and users. Although IPv6 have started in the 90's , Internet Service Providers (ISPs), Consumers and Enterprises will have different timescales to adopt it and the migration will not start simultaneously. One of the main problems is that many ISPs currently are hesitating to invest in major new activities, due to the actual economic situation. Although new business is welcome, deploying IPv6 does not automatically imply an increasing of revenues. Indeed, in the short term, costs may increase, due to the dual IPv4/IPv6 infrastructure that is needed, during migration phase.

From a Consumer or Enterprise point of view, IPv6 will bring important advantages to their infrastructure – seamless mobility, broadband access across different technologies and quality of service – that might justify their investment, even because the new protocol is already supported in most of existing operating systems, like Windows XP and Linux.

An effective migration solution is essential to assure the success of the transition phase. While both technologies coexist, IPv6 users must be transparently connected to IPv4 users, and vice-versa, with a minimum overhead, great flexibility and total transparency.

This paper addresses this issue and proposes an **Inter Working Unit** (IWU) that supports communication between IPv6 and IPv4 protocols. The IWU is based on a flexible inter operability solution and have been developed and tested in GNU/Linux Operating System.

The remaining part of the paper is organised as follows: section 2, presents and evaluates the existing inter operability scenarios; section 3, presents related work; section 4, describes the proposed architecture; section 5, presents the implementation; section 6, shows the test results and, finally, in section 6 some conclusions are drawn and future work is presented.

2   Inter Operability

2.1   Migration Scenario

During the migration phase from IPv4 to IPv6 it is expectable that both type of networks coexist and must cooperate to transparently provide end-to-end service to their costumers.

The migration from the present situation, where the Internet is based on IPv4, to the final one, where it is only supported on IPv6, will comprise two main phases. In the beginning most of the networks use IPv4, but, as long as IPv6 has success, IPv4 networks are progressively replaced by IPv6 and the situation progressively reverses. Probably, this migration phases will take a long time, as it will be very difficult to replace the entire set of IPv4 islands, due to the huge size of the Internet. Therefore, even when two IPv6 costumers communicate, inter operability may be needed, as the communication process will probably cross IPv4 domains.

Several inter operability solutions have been proposed by IETF and will be presented in the next sections.

2.2   Dual Stack Approach

The **Dual Stack** approach, defined by IETF in RFC 2893 [5], provides a very simple inter operability scenario, which basically consists in having complete IPv4 stacks in every Network Element that supports IPv6. If a dual stack node needs to communicate to IPv4 uses IPv4 datagram, while if it needs to communicate to IPv6 nodes it uses IPv6 datagram. When the communication process passes through IPv4 and IPv6 domains, header translation is performed in the edge.

This is a simple approach, which presents two important disadvantages: first, every dual stack node needs an IPv4 address and an IPv6 address and, IPv4 addressing space is sparse; second, when header translation is performed from IPv6 to IPv4 the new header fields of IPv6, like flow label, will be lost forever, as they are no corresponding ones in IPv4.

2.3   Tunnelling

The other solution proposed in the same RFC – **Tunnelling** – solves both problems described before. Dual stack nodes exist at the edge of IPv4 and IPv6 domains. When an IPv6 datagram needs to cross an IPv4 domain to reach its destination, at the egress of the IPv6 domain, the IPv6 datagram is encapsulated into an IPv4 datagram and put into a tunnel; after passing through the IPv4 domain, at the ingress of the other IPv6 domain, the tunnel is released and the IPv4 header removed in order to recover the original IPv6 datagram.

Besides the overhead introduced by the tunnel, this approach implies that both costumers uses the same type of IP protocol, as encapsulation implies the reverse procedure.

2.4   Translator

A more complex solution – **Network Address Translation - Protocol Translation** (NAT-PT) – has been proposed also by IETF in [6], to support communication between IPv6 and IPv4 stations. It merges the protocol translation proposed in Stateless IP/ICMP Translation Algorithm (SIIT) with the dynamic addressing procedures defined in NAT.

When an IPv6 station triggers a communication process to an IPv4 station, it uses a **Translator** to require an IPv4 address and to perform header conversion. This address – **IPv4-translated address** – is obtained from a pool of addresses and a session is created, in order to share the address among all datagrams associated to the same communication process.

NAT-PT has a drawback: when the pool of addresses reaches its limit, no more communication processes between IPv6 and IPv4 may be established.

If port translation is used, as proposed in [6] **Network Address Port Translation - Protocol Translation** (NAPT-PT), this limitation is solved, because ports are also mapped and associated to the session. However, in such a case, and due to the use of port number, it is not possible to have the same service in different servers of the same network. In spite of this, NAPT-PT is the most adequate solution for Small Office, Home Office (SOHO) environments, and it will be the chosen solution.

3   Related Work

Today, there are several projects aimed to promote the deployment of IPv6 networks, through the development of trial network platforms and applications. In the European community, the Information Society Technologies (IST) program promoted more than 10 projects, which have been identified in *Euro6IX project* [7]. Also different countries, around the world, have their own national IPv6 research networks, which joins together academic, industry and network operators. Euro6IX provides a Pan-European network, connecting its own network to other networks, such as 6Bone and 6Net [8].

As far as inter operability is concerned, a few projects are being developed. The *Microsoft Research IPv6*, had developed an IPv6 stack, containing a translator, based on NAT-PT approach. Recently, *Windows XP* has provides Dual Stack implementation and also supports inter operability, by using a Tunnelling variant called Teredo. The project 6Talk [9], developed by South Korean organisations is aimed to support Dual Stack, Tunnelling, SIIT and NAT-PT inter operability solutions.

4   Architecture

As depicted in figure 1, the IWU uses NAPT-PT to support inter operability among several IPv6 and IPv4 domains. The Network Element that contains the IWU must interfaces both type of domains and all the communications established between them must use the IWU.

According to NAPT-PT characteristics, the IWU supports TCP and UDP port translation and IPv6/ IPv4 and ICMPv6/ICMP protocol translation. Specific **Application Layer Gateways** (ALG) should be developed to support

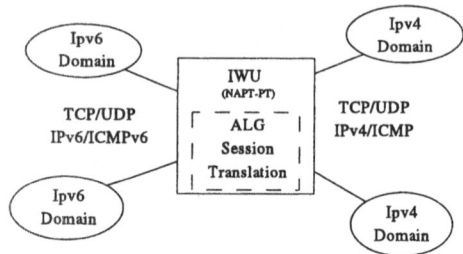

Fig. 1. IWU Architecture

the applications that have some particularities, which do not allow transparent translation.

Concerning IP address/port translation, it occurs only when a communication process is initiated. To support this, the architecture defines a special entity – **Session** –, that keeps all the information needed to translate incoming packets, from whose the translation had been previously done. As far as TCP or UPD traffic is considered, the entity Session is characterised by the tuple (Protocol, Source Address, Destination Address, Source Port, Destination Port); if ICMP is considered, the tuple is (Protocol, Source Address, Destination Address, Identifier).

During a communication process the following steps are executed:

- **Packet association to a Session** – all UDP packets and ICMPv6 Echo Request may trigger the creation of a new session, if there is not any one that matches its tuple. Concerning TCP, only connection establishment packets lead to the creation of a session; all the packets that match an existing tuple are associated to the corresponding session entity; all the others are discarded.
- **Packet translation** – when a packet is associated to a session, the session's information is used to translate it from IPv6 to IPv4, or vice-versa, depending on the direction of the communication. Thus, the IPv6/IPv4 header is mapped into the corresponding IPv4/IPv6 one, and UPD or TCP ports are mapped.
- **Session termination** – all UDP and ICMP sessions end by timeout, as they are connectionless protocols. Concerning TCP, the session is finalised when the TCP connection is terminated, or by timeout, if some problem occurs and the FYN segment is lost.

5   Implementation

The IWU has been designed and implemented in GNU/Linux Operating Systems, in user space mode and C language.

According to the architecture depicted in figure 2, the IWU comprises three main functional blocks:

- **Capture** – that is responsible for retrieving a packet from the Kernel, when it enters the IWU, and for sending it to the Translator.
- **Translation** – that converts an IPv6 packet into the corresponding IPv4 packet, and vice-versa.
- **Output** – that receives a packet from the Translator and sends it to the Kernel, in order to be transmitted.

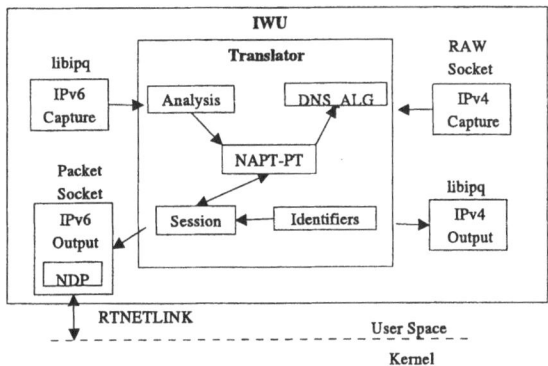

Fig. 2. IWU Implementation scheme

One of the first design options that have been taken is related to the type of mechanism used to support the communication between the IWU and the Kernel of Linux. At a first glance, Linux provides different mechanisms to perform this communication, but several precautions had been taken, in order to assure an adequate functionality.

The mechanisms used to capture a packet must assure that the packet is retrieved from the Kernel, in stead of copied, in order to avoid duplications. Thus, some of the existing mechanisms, such as *RAW Sockets*, *Linux Packet Filters* and *libpcap*, could not be used, as none of them completely intercept the packet. From the two available options – *Divert Sockets* and *lipipq* – the former was quit, because it was only available at the older Linux Kernel and does not provide support to IPv6.

libipq is a library that gives access to *Netlink* through an easy API. Netlink allows the IWU to interact with the Kernel, in order to completely capture an IPv6 or IPv4 packet, using an Inter-Process Communication (IPC) similar to a socket.

In the Output block, the basic problem is related to the fact that, to translate packet the IP header must be modified. The **RAW Sockets** may be used to send IPV4 packets, using the option (IP-HDRINCL) that includes the IP header in the message that is sent in the socket.

Concerning IPv6, there is limitation of RAW sockets in this protocol that forbids its use: it is not possible to change the IP addresses and to use an address different to the interface's address. Thus, a lower layer communication mechanisms was selected – **Packet Sockets** – , which capture packets at the *Logical Link Layer*. No limitations arise from its use, but the protocol responsible for the conversion between IP and MAC addresses had to be developed. In IPv6, this is the **Neighbour Discover Protocol** (NDP). The **RTNetLink** is used by NDP to look at, or insert new entries, at the cache table that contains the IP/MAC addresses information, which is located at the Kernel.

The Translator uses a **thread** from IPv6 to IPv4 translation and another one, for the reverse direction.

When an incoming packet enters the Translator, it is validated by the **Analysis component**, which decides if the translation is to continue, or if the packet must be rejected. If the translation goes on, the **NAPT-PT component** controls it: it triggers the creation of new sessions, associates packets to existing sessions, translates the protocol and triggers the session's termination. Session management is done by the **Session component**, with the help of the **Identifier component**, which is used to manage identifiers for ICMP packets.

The **DNS ALG** is a special component used to translate DNS queries that IPv6 station makes to name servers, existing in IPv4 domains. AAAA Queries are converted into A Queries and the server replies are also converted.

6   Trial Results

The IWU was tested using the trial platform represented in figure 3. The IWU runs on a PC, with *Red Hat Linux* , Kernel 2.4.20-18.9, compiled to support the module *ip6 queue*. IPv6 Stations and IPv4 Servers both use *Red Hat Linux 9* and *MS Windows XP*. The IWU uses Ethernet interfaces at 10/100 Mb/s. Some IPv4 servers are accessed through the Internet, while others lie in the Local Area Network.

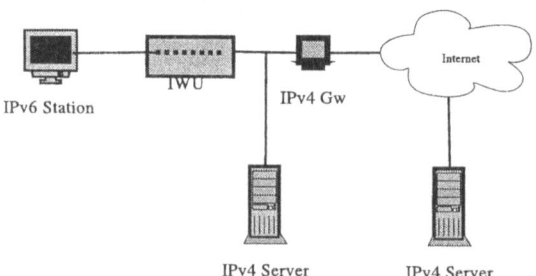

Fig. 3. IWU Platform

Functional tests have been performed using different applications and the results are described in table 1.

Almost all the applications that does not require other ALGs than DNS have performed well. The only exception appears in the DNS application, if it uses an *MS Windows XP* Station, and is due to the fact that this Operating Systems did not contain a DNS resolver for IPv6, at the time the tests were performed.

Concerning, TFPT the problem lies on the fact that no special ALG have been yet develop for this protocol, which can not be not natively supported by NAPT-PT. TFTP uses a port in the reply different to the one used in the query, and so, when the reply returns, no session matches the reply packet.

Table 1. Functional Tests

Application	Client	Server	Result
HTTP	Mozilla 1.2.1	micro httpd	Ok
HTTP	Mozilla Firebird 0.7	Apache	Ok
HTTP	Internet Explorer 6	IIS	Ok
SSH	OpenSSH 3.5p1		Ok
DNS	Linux	Any	Ok
DNS	Windows XP	Any	NOk
TFPT	Linux	Linux	NOk
ICMP	Linux	Linux	Ok

Also performance tests have been done, regarding the Latency introduced by the IWU. The PC used in the tests is a Pentium II, at 333 MHz, with 224MB RAM, Fast Ethernet 10/100Mbps, with chipset RTL8139.

To evaluate the Latency introduced by the IWU, tests have been done using a network with two Stations connected through an IWU and another network where the IWU was replaced by a router. In such a case, the communication comprises only IPv6 or IPv4 Stations. The Round Trip Time (RTT) has been measured, using ICMP packets, with different sizes. For each test 30000 ICMP requests are done, at rate of 100 packets/sec.

The average RTT is represented in figure 4 for the three situations tested: IPv4 communication, IPv6 communication and IPv6 to IPv4 communication, using the IWU. The Latency introduced by the IWU, also represented, is evaluated by the difference between the RTT of the IWU and the RTT achieved with IPv6.

It is possible to conclude that, as expectable, the RTT increases with the packet size; concerning the Latency of the IWU, it is kept almost constant, in the order to magnitude of 200 msec, which is an acceptable value due to the performance characteristics of the test machine. Even considering the situation of a communication process across the Internet, this is an acceptable value.

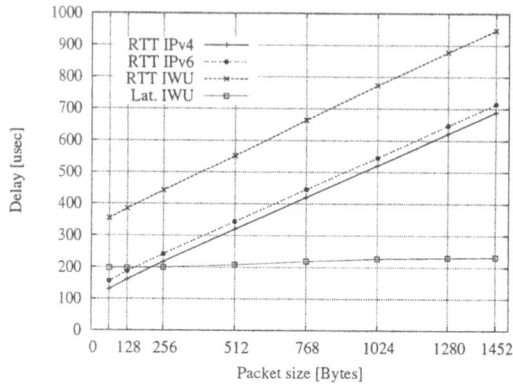

Fig. 4. IWU Platform

7 Conclusions

This paper proposes and InterWorking Unit used to support the inter operability between IPv6 and IPv4, during the migration phase, which is expectable to be long, due to the huge size of the Internet today and the financial investments needed to support the transition.

From the existing set of inter-operability solutions - Dual Stack, Tunneling, NAT-PT and NAPT-PT, the later one is the most promising for the new future, as it efficiently uses network resources, without wasting the spare IPv4 addressing space.

The IWU presented in this paper is based on NAPT-PT approach and was implemented at user space mode, in GNU/Linux. It represents a good solution for SOHO environments, as it does imposes significant requirements.

A specific ALG block have been developed, to support name resolution. Several design approaches have been identified, regarding packet capture and output at both IPv6 and IPv4 sides. Due to the different maturity and implementation options that have been taken in Linux's IPv4 and IPv6 stacks, an asymmetrical design of the IWU has chosen.

Functional tests have shown that the IWU performs well using different applications, developed for different Operating Systems, if they do not require other special ALGs. Performance tests have shown that, in spite of being developed at user space, the IWU introduces a small Latency in the communication process, in the order of magnitude of 200 usec, which is kept almost constant, independently of the traffic load.

Future work comprises developing specific ALGs for other applications, not yet supported and implementing the IWU as a Kernel Loadable Module. Support of redundancy mechanism and load balancing among different IWUs must also be addressed.

References

[1] Deering, S., Hinden R.: Internet Protocol Version 6 (IPv6) Specification. IETF RFC 2460. December (1988)

[2] Steering Committee of IPv6 Taskforce.: IP Overall Status. Deliverable 3.3,, October (2003)

[3] Portugal - IPv6 Taskforce.: Home Page of Portugal IPv6 Taskforce. http://www.ipv6-tf.com/home.pt, (2004)

[4] IST IPv6 Cluster.: European IPv6 Research and Development Projects. http://www.ist-ipv6.org, (2004)

[5] Gilligan, R., Nordmark, E.: Transition Mechanisms for IPv6 Hosts and Routers. IETF RFC 2893, August (2000)

[6] Tsirtsis, G. and P. Srisuresh.: Network Address Translation - Protocol Translation (NAT-PT). IETF RFC 2766, February (2000)

[7] Rao, S. et al.: Dissemination and Use Plan. Euro6IX, Deliverable 5.1, March (2002)

[8] Ralli, C. et al.: Specification of the Backbone Network Architecture. Euro6IX, Deliverable 2.2, December (2002)

[9] 6TALK.: IPv6 Translator for Krv6. http://www.6talk.net, (2004)

Smart Profile: A New Method for Minimising SIP Messages

Ahmed Meddahi[1], Gilles Vanwormhoudt[1], and Hossam Afifi[2]

[1] ENIC Telecom Lille 1, 59658 Villeneuve d'Ascq Cedex France
{ahmed.meddahi, gilles.vanwormhoudt}@enic.fr
[2] Institut National des Télécommunications, 91011 Evry Cedex France
hossam.afifi@int-evry.fr

Abstract. SIP (Session Initiation Protocol) is a text-based protocol engineered for high data rate links. As a result, SIP messages size have not been optimised. This is problematic, particularly with the planned usage of this protocol in wireless handsets as part of 2.5G and 3G cellular networks. With low bit rate IP connectivity or signalling channel, the transmission delays for call setup and feature invocation are significant. In this paper, we propose a new approach called "Smart Profile" where SIP messages are minimised in the access network for performance optimisation and completely reconstructed at the edge. The message reconstruction is based on SIP call profile that contains session attributes extracted from the user and terminal profiles. We discuss about Smart Profile implementation and focus on measuring its performance. Results and comparison with compression method show that Smart Profile solution is globally better.

1 Introduction

The Session Initiation Protocol [RFC 3261] is intended for initiating multimedia sessions but also for managing these sessions. SIP is a text based protocol engineered for bandwidth rich links. As a result, the messages have not been optimised in term of size. For example, typical SIP messages range from a few hundred bytes up to several kilo bytes [1]. When SIP is used in wireless handsets as part of 2.5G and 3G cellular networks, where the bandwidth and energy represent high cost resources, with a potential high packet loss and collision rates (compare to wired networks), the large message size and the need to handle high message/transaction rates is problematic [1,2]. Generally, the maximum bandwidth has to be dedicated to the media flow and call set-up delay must be under a certain threshold [3]. Taking into account retransmissions and the multiplicity of messages that are required in some flows, call set-up and feature invocation are dramatically affected.

To cope with SIP performance issues, IETF proposes to use a compression method (SigComp [RFC 3320]) at the application level (SIP level). This approach generally needs a negotiation protocol for capability exchange leading to some interoperability issues (e.g. compression support, type of compressors). It also

J.N. de Souza et al. (Eds.): ICT 2004, LNCS 3124, pp. 688–697, 2004.
© Springer-Verlag Berlin Heidelberg 2004

requires CPU and memory resources which are quite scarce in some conditions (PDA, terminal or mobile handsets).

We propose a new approach called Smart Profile (that could replace or be associated with SigComp, in some specific conditions) where the major part of the complexity is moved from the terminal toward the edge SIP proxy. This method is transparent to existing SIP servers. The paper is organised as follows: section 2 describes SIP performance issues in term of message size and transmission delays. Section 3 presents compression method while "Smart Profile" and its implementation are described in section 4. Section 5 presents our Benchmark and gives results. Finally, we comment on results and give some perspectives.

2   SIP Performance Issues

To date, it appears that there are few published studies about SIP performance or benchmarks. Some measures of set-up delays obtained from simulation are given in [4] for the context of IP telephony. There also exists a methodology to measure server performance [5]. In [6], we have presented a SIP benchmark that provides results of SIP performance for multiple user accesses (LAN, Wireless LAN 802.11, Modem, Cable Modem) and compared these results with ITU-T recommendations [3] which specify optimal delays for real-time connections (particularly voice connection in "circuit" networks). This benchmark reflects that SIP does not achieve good performance for low bit rate and WiFi accesses. The set-up delay can reach more than 20 seconds which is far from ITU-T recommendations for local (3 seconds) and international connection (8 seconds).

Analysis of results obtained from our benchmark identifies at least three factors that have an impact on SIP performance and introduce overhead.

First, SIP messages are quite large, due to their textual format and structure. SIP messages size, including lower layers (UDP, IP, ...) goes from about 500 up to 2000 bytes[1] if we consider large SDP bodies, obtained using Ethereal tool on typical UAs (Ubiquity Software Agent). As experienced by our benchmark, the typical size of SIP messages constitutes an issue for low bit rate IP connectivity. For example, if we consider a 9.6 Kbs bandwidth channel [2] each byte takes 0.8 ms (serialisation delay) and if we assume a 500 bytes SIP message, the serialisation delay represents by itself an unacceptable value of 0.5 sec and 2 sec for a 2000 bytes message. With the propagation delay in a 3G network, the total call set-up delay can reach approximately 7.9 sec. [1]. For comparison, a typical GSM call set-up is about 3.6 sec. [1].

Secondly, processing time, due to SIP messages parsing process and Proxy-servers along the signalling path with store and forward behaviour add delays. This delay increases dramatically if SIP servers logic complexity is high, resulting in high message/transaction handling rates. Finally, in wireless networks, characterised by high packet loss and collision rates, the radio network frame sizes

---

[1] By comparison, when ITU-T H323 protocol is used for IP multimedia applications, call signalling messages are more compactly encoded with a maximum size of 300 bytes.

[2] This is the basic channel for a 3G typical CDMA radio link, 8 kbs for GERAN.

add constraints meaning that signalling message should fit into certain amount of radio layer frames. Given this, general techniques to reduce call set-up delays while saving bandwidth could be (1) increasing the bit rate per user which lead to reduce the number of supported users; (2) decreasing the RTT (in Wireless networks) which is not be trivial with fundamental delay factor like interleaving and FEC; (3) stripping the protocol which would break transparency and end to end semantics; (4) use IP header compression such as RObust Header Compression [ROHC]. Nevertheless, today, there is a major trend to privilege Minimisation techniques dedicated to SIP. We dedicate the next section to deal with current compression techniques.

3   Minimising SIP Messages: Signalling Compression

Compression is an active area within the IETF and the need for SIP compression is clear [RFC 3322]. We consider SigComp [RFC 3320] as one of the main contributions. Sigcomp defines an architecture for compressing messages generated by application level protocols such as SIP (or RTSP). One of the main component of this architecture is a Universal Decompressor Virtual Machine (UDVM) specified at application layer (SIP). This virtual Machine has a limited and simple instructions set (33 instructions) that is optimised and targeted at implementing decompression and manage states related to the compression/decompression process. The compressor algorithm bytecode (commonly used compressors are textual ones such as DEFLATE, LZ77, LZW[3], ...) is uploaded to the UDVM for being executed when each compressed message is received for decoding. So the choice of algorithm becomes a local decision at the compressor side. SigComp protocol has only one message and uses UTF-8 coding, allowing mixing of compressed and normal messages (a SigComp message begins with an illegal bitstring).The SigComp message is composed of several fields, containing: a state identifier, the state identifier length or the bytecode, following by the location (memory address) of the bytecode algorithm, possible feedback request or data and the compressed message. TCP or UDP transport protocol is used for signalling and depends on the required reliability. Generally, compression based approach is not transparent to existing SIP servers and needs high processing and storage capabilities in terminals (for the compression/decompression context and dictionary). It also requires a capacity negotiation protocol before exchanging any SIP compressed messages, with the fact that multiple compression algorithms (standard or proprietary) exist, leading to potential interoperability issues. These constraints require high complexity on the terminal side. They are incompatible with terminals characterised by limited resources (CPU, memory) such as PDAs and mobile handsets. This is more critical with smart cards if we consider that SIP could be embedded and used for security/authentication purposes [7].

---

[3] Compression ratio depends on the type of content. If we considered a large text book, it is about 3 for LZ77, 2.5 for LZW and 1.5 for Compact.

4   Minimising SIP Messages: "Smart Profile"

We propose in this section a new approach where the complexity is more located on the edge SIP proxy instead of the terminal (this leads to an asymmetric scheme). We call it Smart Profile for minimising SIP messages length. It can be considered as a trade-off between compression and caching techniques.

4.1   Principles

"Smart Profile" uses the Virtual Home Environment (VHE) [8] or Personal Service Environment (PSE) [9] concepts for providing services adapted to the user or its environment. It is based on the user, service, network or terminal profiles but also on the mobility profile for location based services. These service personalisation models (VHE, PSE) are mainly defined for being supported in wireless networks (e.g. 3G, GPRS, WLAN, ...). Hence, profile related information such as: terminal identifier, users preferences, network, terminal or service characteristics and user location are considered to be known and stored in various databases. Based on these information, it is easy to guess the capabilities of the terminal, network or service and hence to build and select a SIP oriented call profile, which could be dynamically updated from the VHE or PSE environment (mobile environment).

The basic concept of Smart Profile is to exploit SIP oriented call profiles to minimise and reconstruct messages based on stored information. Figure 1 shows this concept, with a typical call scenario. The first complete request is used by the edge proxy to construct the call profile (stored also in the terminal for minimisation in both direction, for requests and responses). For next requests, the SIP UA client will send an INVITE message containing a minimal set of information such as terminal, destination or source identifier. The SIP edge Proxy server will then add all remaining mandatory or optional fields (including SDP protocol) based on the stored call profile. By this way, minimised messages are reconstructed before being sent in the core network [4].

So, considering Smart Profile approach for SIP protocol, most of the optional header fields and SDP body of a SIP message sent by a User Agent describe the sessions attributes the user wants to establish. The sessions attributes are inherited from the terminal, network, user capabilities and preferences. We consider these attributes as static data but they could be updated dynamically, for example in case of mobile environment. This suggests that a call profile representing the sessions attributes is created and refreshed periodically [5].

Below, we show a typical SIP message (INVITE) of about 2 Kbytes used in our testbed. Lines starting with "**" represent all information that will not be sent in subsequent minimised messages, as they will be extracted from the SIP call profile in the Proxy. For SIP messages between the User Agent and the Proxy, we remove the fields that will not changed or changed in a predictive manner

---

[4] The bottleneck is generally located more on the access network but this does not preclude the use of Smart Profile in the core network on a hop by hop basis.

[5] Register and re-Invite method could be used for updating SIP call profile.

during a call session, that is : (1) static header fields (e.g. Contact); (2) header fields which are body related (e.g. Content-Disposition, Content-Type) ; (3) the SDP body which can be retrieved from call profiles. Other fields, although they can be static fields (e.g. Max-Forwards, From) are mandatory in SIP messages and thus must be kept. Given this, the SIP call profile will be build based on these considerations.

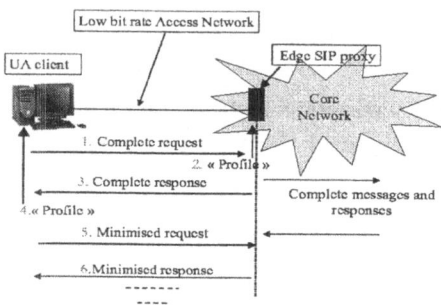

Fig. 1. Call scenario using "Smart Profile"

```
INVITE sip:SmartProfile@193.48.251.169;transport=udp SIP/2.0
CSeq: 1 INVITE
Call-Id: 8cf4bd79f89bc5c15b8ab93088dc7c6f@193.48.251.66
From: "Caller" <sip:rodin@193.48.251.66:5060>;tag=6789-1234
To: "Callee" <sip:durer@192.168.246.225>;tag=12345-6789
Via: SIP/2.0/UDP 193.48.251.66:5060
Via: SIP/2.0/UDP 80.248.33.37:5070;branch=z9hG4bK50F8212513CE00F4B525D73
Contact: "Caller" <sip:rodin@193.48.251.66:5060>;expires="Sun, 30 Mar 2003 05:18:32 GMT"
** Organization: E.N.I.C. Telecom Lille 1
** Subject: Testing SIP protocol
** Accept-Encoding: gzip
** Accept:text/HTML
** Accept-Language: French
** Date: Sun, 29 June 2003 05:18:32 GMT
** Content-Type: application/sdp
Max-Forwards:70
** Content-Length: 1737 (this field equal 0 if minimised)

** v=0
** o= enic +33 20 33 55 6  IN IP4  192.48.251.107
** c=IN IP4 193.48.251.169                  // SDP body of
** m=audio 1032 RTP/AVP 0 8 4 3 96    // 1.7 Kb length
** a=rtpmap:0 PCMU/8000
** a=rtpmap:8 PCMA/8000
   ..
```

Minimising SIP messages could be obtained by creating SIP call profiles, related to the user, network or services characteristics, while maintaining and updating them in a database. Such a database could be located in the edge of the network (edge SIP proxy), but also in the core network or in the access network (depending on performance considerations). After having created these call profiles and making them easily accessible for being updated by their associated users, user agents can then send their messages with the minimum set of header fields and body to the SIP proxy server maintaining the call profiles, which in turn rebuilds the message using the call profile and forwards the complete message to the intended destination. Maintaining call profiles by the outbound SIP proxy server of a low bit rate IP connectivity will allow user agents, on the access network to send minimised SIP messages to the proxy server, where the messages will be reconstructed and forwarded in their complete form inside the core network.

The main advantages of Smart Profile approach are to decrease the transaction delay while saving bandwidth consumption. In some conditions Smart Profile achieves a minimisation ratio of up to 85 percent (see figure 5). Smart Profile is transparent to existing SIP proxies while compression oriented methods are not. Tests through an external SIP proxy (sipcenter.com) which does not implement Smart Profile, shows that minimised message are forwarded while not compressed messages, meaning that existing SIP proxies (Smart Profile not supported) are transparent to Smart Profile messages. Notice that Smart Profile could be associated with compression method, as these two approaches could be considered as complementary and not antagonistic, allowing better performance: minimised messages can be compressed; both Smart Profile and Compression oriented Proxys can co-exist on the call path. Also, in order to increase performance, the SIP proxy may ask through the SIP Record-Route mechanism to be in the signalling path for all future requests that belong to the dialog, allowing the minimisation of all subsequent messages and not only the first messages (INVITE, TRYING, . . .).

4.2   Implementation

The Smart Profile algorithm (figure 2) which is implemented in the SIP proxy, performs the following main tasks:

- Creating and managing call profiles, by getting the information needed to create these call profiles and storing them as one call profile object per user in a data structure. There is a unique profile for each user-terminal combination representing a set of session attributes.
- Receiving and sending minimised SIP messages on the access network, reconstruct them at the edge of the core network and forward them to the intended user (based on the call profiles). This is an existing SIP proxy process that forwards SIP messages to the intended destination with an extra job, consisting in retrieving the associated call profile, building the complete message, for being forwarded afterward.

To provide the Smart Profile proxy with the information needed to build call profiles, the user agents has to send a non-minimised or complete SIP request message before they can send any minimised ones to the edge proxy server (figure 2). The non-minimised message will give all the required information needed to reconstruct subsequent messages in their complete format. Once a non-minimised SIP message is sent by a user agent client, the associated call profile is created on the proxy server. If the call profile is created successfully, the user agent can send

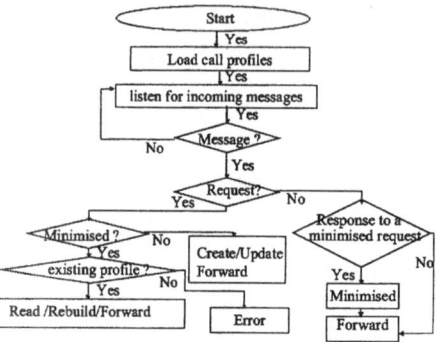

Fig. 2. "Smart Profile" algorithm

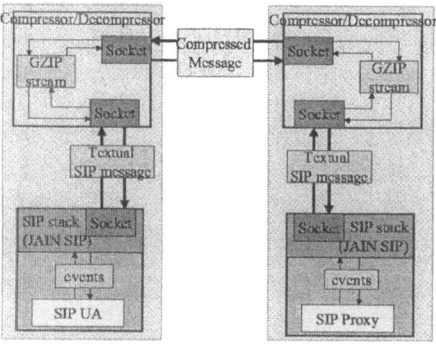

Fig. 3. "Compression" Architecture

minimised messages. If a user agent client, that has already created a call profile in the proxy server, sends a non-minimised SIP message, it will see his call profile updated based on the new content of this message. To experiment and validate Smart Profile, we have developed a JAVA implementation. This implementation uses an existing stack compliant with the JAIN-SIP specification proposed by SUN [10]. This is the implementation that we used to build up the benchmark described in the next section.

5   Benchmarking

In the previous part, we described the compression and Smart Profile approaches for minimising SIP messages. To evaluate efficiency of both approaches and compare their performance, we designed a benchmark. This benchmark provides call-set-up delays under specific conditions. In particular, we take care to measure these delays for low bit rate access.

To build up this benchmark, we have developed our own implementation of the compression approach. This implementation compresses/decompresses SIP message using the Lempel-Ziv algorithm which is defined as one of the main compression algorithms for being used in SigComp. The compression/decompression process is achieved by a module running on the same host as the SIP User Agent and located between this User Agent and the network (see figure 3). More precisely, the SIP UA sends textual SIP messages to the UDP port corresponding to the compressor/decompressor module. Once the module receives a SIP message, it compresses/decompresses and sends it to the intended host at the port number where the remote decompressor/compressor is listening. Then, the decompressor/compressor extracts the compressed message and sends it to the local SIP UA or Proxy module for processing the SIP level.

5.1 Architecture and Tests Descriptions

The benchmark architecture (figure 4) is composed of the following components:

SIP User Agent Clients: These components are host systems. Each of these clients includes the same SIP User Agent to generate SIP requests and record call set-up delays. Low bit rate IP connectivity is reflected by using loaded WiFi (802.11b) and modem links (V90) at variable bite rate.

SIP Servers: The SIP server is a host system from our Intranet located at the "sip.enic.fr" public address. This server is a Pentium III machine with an integrated location database. It implements Smart Profile compression methods (server part). Call profiles are also stored in this server.

SIP User Agent Server: This component is also a host system from our Intranet that includes a SIP User Agent simulating called destinations in our targeted architecture (both users and services). The task of this agent consists in automatically returning provisional and non-provisional responses.

The benchmark currently consists of a scenario that is based on a single SIP Proxy server. In this scenario, each SIP client sends an INVITE request to our Intranet SIP server, which implement Smart Profile and compression

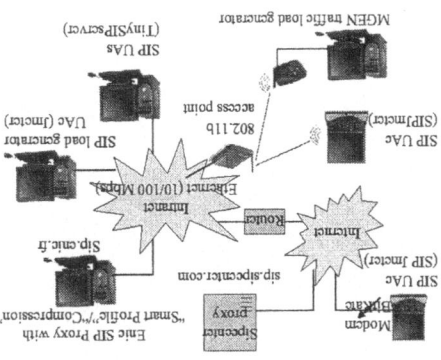

Fig. 4. Benchmark Architecture

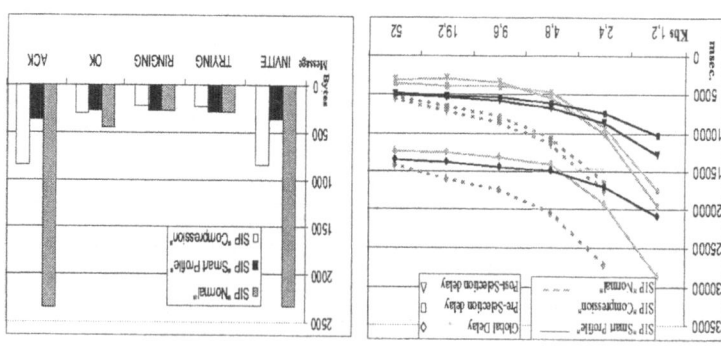

Fig. 5. Messages size and call set-up delays with "Smart Profile" and "Compression"

methods. This configuration causes the SIP server to process the SIP messages[6] and forward them to the targeted UAs. We also use an INVITE request with a SDP content that gives a large 2000 bytes SIP message length. Results collected are classical call set-up delays [5]. We measure delays and messages size related to Smart Profile, Compression and Normal SIP method. "Pre-selection" delay is the time between INVITE request and TRYING response. "Post-selection" delay is the time between INVITE request and RINGING response. "Connection" delay is the time between INVITE request and OK response.

5.2 Results and Discussion

Even if call set-up delays are still important, results show that performance are globally better for Smart Profile. Figure 5 shows the message sizes related to the three methods. Smart Profile method gives a maximum minimisation ratio for large messages (Invite and Ack messages) including large SDP content. The ratio (INVITE message) is about 85 percent for the Smart Profile while it is 65 percent for compression. Minimisation and compression ratio are equivalent and much more less significant for short messages (350 Bytes).

Figure 5 also gives the modem call set-up delays for the three methods with a modem link at variable bit rate. If we consider Smart Profile, the post-selection delay at 1.2 Kbs is 13 sec. and 20 sec. for SIP compression while we get request timed out for Normal SIP protocol (respectively 8, 10 and 17sec. at 2.4 Kbs). The gap becomes less important when bandwidth increases. After 4.8 Kbs compression performance are slightly better than Smart profile. These results could be improved if we keep only few main messages required to establish the connection or if SIP Record-Route mechanism is used. Figure 5 shows also that SIP requests are timed out when serialisation delay is too high due to the large message size (SIP Normal at 1.2 Kbs) or when the external SIP proxy (sipcenter.com) is crossed. Also regarding these figures, the protocol complexity and

---

6 For this scenario, we use UDP for transport protocol of SIP message. SIP over TCP will be analysed in a next study, but we assume that delays will be superior, due to TCP "overhead", flow control and congestion mechanisms.

minimisation ratio obtained with Smart Profile, has to be compared to the cost of CPU and memory capacity needed to execute compression algorithms on the terminals. In our proposal with Smart Profile, minimisation should not face this cost since it is mainly implemented in the edge SIP server.

6 Conclusion

Although SIP protocol brings many improvements and simplifications, it still needs some additional optimisation to become a high performance scalable protocol, particularly for low bit rate IP connectivity and for bandwidth consumption saving. To deal with this optimisation, we propose a new contribution called Smart Profile which can be considered as an alternative or complementary solution to compression approach. As previous results shown, Smart Profile is globally better than compression, allowing call set-up delay optimisation and bandwidth saving, while introducing a minimal complexity (CPU, memory resources) inside the terminal. Also, Smart Profile is transparent to non Smart Profile proxy servers, allowing compatibility and interoperability with existing SIP servers. Thus, Smart Profile can be seriously envisioned in networks where resources are limited or bandwidth saving is required (e.g. Wlan, 3G networks). In term of perspectives, we will investigate two directions: First direction is related to mobility : when a UA moves from one access network to the other, its local edge proxy will change. How does the new edge proxy acquire state/profile for an ongoing calls/sessions? Second one is related to cost performance, we will characterise overheads at the edge proxy for state maintenance of on-going calls and overhead for converting "minimised" SIP messages to expanded SIP messages.

References

1. Nortel Networks, A comparison between GERAN Packet-Switched call Setup using SIP and GSM Circuit-Switched call Setup.
2. A. Meddahi, H. Afifi, Improving SIP performance and scalability in a PSTN/3G framework, IEEE Networks 2002 ; Atlanta USA, Aug. 2002; pages: 36-44.
3. ITU-T recommendation E.721: Network grade of service parameters and target values for circuit-switched services in the evolving ISDN.
4. T. Eyers and H. Schulzrinne, Predicting Internet Telephony Call Setup Delay, in Proc. of First IP Telephony Workshop, Berlin, April 2000.
5. H. Shulzrinne, Narayanan, Lennox, SIPstone: Benchmarking SIP server performance, Apr. 2002.
6. A. Meddahi, G. Vanwormhoudt, SIP for e-learning services, IEEE ICT 2003, Papeete. Feb 2003.
7. B. Zouari, G. Pujolle, H. Afifi, H. Labiod, P. Urien, A Novel Authentication Model Based on Secured IP Smart Cards, IEEE ICC 2003, Anchorage. May 2003.
8. The Virtual Home Environment, 3GPP TS 22.121, 2001.
9. M.M. Lankhorst, H. van Kranenburg, A. Salden, A. Peddemors, Enabling technology for personalizing mobile services, HICSS 2002, Hawaii. Pages: 1107-1114.
10. JSR-000032 JAIN SIP Specification, Java Community Process.

Lecture Notes in Computer Science 3124

Commenced Publication in 1973
Founding and Former Series Editors:
Gerhard Goos, Juris Hartmanis, and Jan van Leeuwen

Editorial Board

José Neuman de Souza   Petre Dini
Pascal Lorenz (Eds.)

Telecommunications and Networking – ICT 2004

11th International Conference on Telecommunications
Fortaleza, Brazil, August 1-6, 2004
Proceedings

 Springer

Volume Editors

José Neuman de Souza
Universida de Federal do Ceará, Departamento de Computação
Campus do Pici - Bloco 910, 60455-760 Fortaleza, Ceará, Brazil
E-mail: neuman@ufc.br

Petre Dini
Cisco Systems, Inc., USA and Concordia University, Canada
170 West Tasman Drive, San Jose, CA 95134, USA
E-mail: pdini@cisco.com

Pascal Lorenz
University of Haute Alsace
34 rue du Grillenbreit, 68008 Colmar, France
E-mail: lorenz@ieee.org

Library of Congress Control Number: 2004109602

CR Subject Classification (1998): C.2, H.4, H.3, H.5.1, K.4.4, K.8.1, D.2

ISSN 0302-9743

ISBN 978-3-540-22571-3   ISBN 978-3-540-27824-5 (eBook)
DOI 10.1007/978-3-540-27824-5

springeronline.com

© Springer-Verlag Berlin Heidelberg 2004

Originally published by Springer-Verlag Berlin Heidelberg New York in 2004.

Typesetting: Camera-ready by author, data conversion by PTP-Berlin, Protago-TeX-Production GmbH
Printed on acid-free paper   SPIN: 11306153   06/3142   5 4 3 2 1 0

Preface

Welcome to the 11th International Conference on Telecommunications (ICT 2004) hosted by the city of Fortaleza (Brazil).

As with other ICT events in the past, this professional meeting continues to be highly competitive and very well perceived by the international networking community, attracting excellent contributions and active participation. This year, a total of 430 papers from 36 countries were submitted, from which 188 were accepted. Each paper was reviewed by several members of the ICT 2004 Technical Program Committee. We were very pleased to receive a large percentage of top-quality contributions.

The topics of submitted papers covered a wide spectrum from photonic techniques, signal processing, cellular networks, and wireless networks, to ad hoc networks. We believe the ICT 2004 papers offer a wide range of solutions to key problems in telecommunications, and describe challenging avenues for industrial research and development.

In addition to the conference regular sessions, seven tutorials and a workshop were organized. The tutorials focused on special topics dealing with next-generation networks. The workshop focused on particular problems and solutions in heavily distributed and shareable environments.

We would like to thank the ICT 2004 Technical Program Committee members and referees. Without their support, the creation of such a broad conference program would not be possible. We also thank all the authors who made a particular effort to contribute to ICT 2004. We truly believe that due to all these efforts the final conference program consisted of top-quality contributions.

We are also indebted to many individuals and organizations that made this conference possible. In particular, we would like to thank the members of the ICT 2004 Organizing Committee for their help in all aspects of the organization of this professional meeting.

The 11th International Conference on Telecommunications was an outstanding international forum for the exchange of ideas and results between academia and industry, and provided a baseline for further progress in the telecommunications area.

We hope that the attendees enjoyed their stay in Brazil and were able to visit various points of interest in this lovely country.

June 2004

José Neuman De Souza
Petre Dini
Pascal Lorenz

IEEE

ICT 2004 General Chair

José Neuman de Souza, UFC, Brazil

ICT Steering Committee

Hamid Aghvami, King's College, London, UK
Tulin Atmaca, Institut National des Télécommunications, France
Petre Dini, Cisco Systems, Inc., USA, and Concordia University, Canada
Pascal Lorenz, University of Haute Alsace, France
Farokh Marvasti, King's College, London, UK

ICT 2004 Technical Program Co-chairs

Petre Dini, Cisco Systems, Inc., USA and Concordia University, Canada
Pascal Lorenz, University of Haute Alsace, France

ICT 2004 International Scientific Committee

Aghvami, Hamid, King's College, London, UK
Aguiar, Rui L., Universidade de Aveiro, Portugal
Agoulmine, Nazim, Université d'Évry, France
Aissa, Sonia, INRS-Telecommunications, Canada
Alencar, Marcelo Sampaio, UFCG, Brazil
Alfa, Attahiru Sule, University of Manitoba, Canada
Andrade, Rossana, UFC, Brazil
Atmaca, Tulin, Institut National des Télécommunications, France
Atwood, Bill, Concordia University, Montreal, Canada
Baier, Paul Walter, University of Kaiserslautern, Germany
Barbeau, Michel, Carleton University, Canada
Brito, Jose Marcos Câmara, National Institute of Telecommunications, Brazil
Casaca, Augusto, INESC, Portugal
Celestino, Júnior Joaquim, UECE, Brazil
Cherkaoui, Soumaya, University of Sherbrooke, Canada
Chemouil, Prosper, France Telecom R&D, France
Chen, Xiaodong, University of London, UK
Cooklev, Todor, San Francisco State University, USA
Correia, Luis M., Technical University of Lisbon, Portugal
Dini, Petre, Cisco Systems, Inc., USA and Concordia University, Canada
Ferreira, Afonso, CNRS-INRIA, France
Galis, Alex, University College London, UK
Kabacinski, Wojciech, Poznan University of Technology, Poland

Kahng, Hyun-Kook, Korea University, Korea
Lee, Mike Myung-Ok, Dongshin University, Republic of Korea
Logrippo, Luigi, University of Ottawa, Canada
Lorenz, Pascal, University of Haute Alsace, France
Marshall, Alan, Queen's University of Belfast, UK
Marvasti, Farokh, King's College, London, UK
Mohr, Werner, Siemens, Germany
Molinaro, Antonella, University of Messina, Italy
Mota, João César Moura, UFC, Brazil
Mynbaev, Djafar, City University of New York, USA
Nefedov, Nikolai, Nokia Research Center, Finland
Oliveira, Mauro, CEFET-CE, Brazil
Pujolle, Guy, LIP6, France
Sestini, Fabrizio, European Commission DG Information Society, Belgium
Sezer, Sakir, Queen's University Belfast, UK
Soulhi, Said, Ericsson Research, Canada
Souza, José Neuman de, UFC, Brazil
Stasiak, Maciej, Poznan University of Technology, Poland
Takasaki, Yoshitaka, Tokyo University, Japan
Wenbo, Wang, Beijing University of Posts and Telecommunications, China
Westphall, Carlos Becker, UFSC, Brazil

Reviewer List

Aghvami, Hamid, King's College, London, UK
Agoulmine, Nazim, LIP6, France
Aguiar, Rui Luis, Universidade de Aveiro, Portugal
Aissa, Sonia, INRS-EMT, Canada
Alencar, Marcelo Sampaio de, UFCG, Brazil
Alfa, Attahiru Sule, University of Manitoba, Canada
Andrade Rossana, Maria de Castro, UFC, Brazil
Atmaca, Tulin, INT-Evry, France
Atwood, John William, Concordia University, Montreal, Canada
Baier, Paul Walter, University of Kaiserslautern, Germany
Barbeau, Michel, School of Computer Science, Carleton University, Canada
Barreto, Guilherme de Alencar, UFC, Brazil
Brito, Jose Marcos Camara, INATEL, Brazil
Casaca, Augusto, INESC, Portugal
Cavalcanti, Francisco Rodrigo P., UFC, Brazil
Celestino, Júnior Joaquim, UECE, Brazil
Chemouil, Prosper, France
Chen, Xiaodong, University of London, UK
Cherkaoui, Soumaya, Université de Sherbrooke, Canada
Coelho, Pedro Henrique Gouvêa, Brazil

Cooklev, Todor, San Francisco State University, USA
Correia, Luis M., Technical University of Lisbon, Portugal
Costa, Max, UNICAMP, Brazil
Dini, Petre, Cisco Systems, Inc., USA and Concordia University, Canada
Farrell, Paddy, UK
Ferreira, Afonso, CNRS/INRIA, France
Galdino, Juraci Ferreira, Brazil
Galis, Alex, University College London, UK
Gameiro, Atílio, Portugal
Godoy, Júnior Walter, CEFET-PR, Brazil
Granville, Lisandro Zambenedetti, UFRGS, Brazil
Hauck, Franz J., Universität Ulm, Germany
Kabacinski, Wojciech, Poznan University of Technology, Poland
Kahng, Hyun-Kook, Korea University, South Korea
Kassler, Andreas J., SCE, NTU, Singapore
Lee, Mike Myung-Ok, Dongshin University, South Korea
Logrippo, Luigi, Université du Québec en Outaouais, Canada
Lorenz, Pascal, University of Haute Alsace, France
Madeiro, Francisco, UNICAP, Brazil
Marshall, Alan, Queen's University Belfast, UK
Martins, Filho Joaquim F., UFPE, Brazil
Marvasti, Farokh, King's College, London, UK
Medeiros, Álvaro Augusto Machado de, UNICAMP, Brazil
Mobilon, Eduardo, CPqD, Brazil
Mohr, Werner, Siemens, Germany
Molinaro, Antonella, University of Messina, Italy
Monteiro, Edmundo, Portugal
Mota, João César Moura, UFC, Brazil
Mynbaev, Djafar, City University of New York, USA
Nefedov, Nikolai, Nokia Research Center, Finland
Oliveira, Hélio Magalhães de, UFPE, Brazil
Oliveira, Mauro, CEFET-CE, Brazil
Olveira, Antonio Jeronimo Belfort de, UFPE, Brazil
Peixeiro, Custodio, Portugal
Pelaes, Evaldo, UFPA, Brazil
Pereira, Jorge, European Commission, Belgium
Pimentel, Cecilio, UFPE, Brazil
Pinto, Ernesto Leite, IME, Brazil
Pujolle, Guy, LIP6, France
Queiroz, Wamberto José Lira de, UFPB, Brazil
Ramos, Rubens Viana, UFC, Brazil
Rocha Jr., Valdemar C. da, UFPE, Brazil
Rocha, Mônica de Lacerda, CPqD, Brazil
Sabbaghian, Maryam, Canada
Salvador, Marcos Rogério, CPqD, Brazil

Table of Contents

Multimedia Services

Enhancing 3G Cellular Systems With User Profiling Techniques
for Multimedia Application QoS Adaptation 1
   G. Araniti, P. De Meo, A. Iera, D. Ursino

Delivery of Streaming Video in Next-Generation Mobile Networks 7
   S. Bates

Use of the 2500-2690 MHz Band
for Multimedia Broadcast Multicast Service 16
   A.R. Oliveira, A. Correia

Throughput Fairness and Capacity Improvement
of Packet-Switched Networks Through Space-Time Scheduling 22
   E.B. Silva, L.S. Cardoso, T.F. Maciel, F.R.P. Cavalcanti,
   Y.C.B. Silva

Antennas

Performance of MIMO Antenna Systems With Hybrids of Transmit
Diversity and Spatial Multiplexing Using Soft-Output Decoding 28
   W.C. Freitas Jr., A.L.F. de Almeida, J.C.M. Mota,
   F.R.P. Cavalcanti, R.L. de Lacerda Neto

Simple Model to Determine the Gap Effect on the Input Impedance
of a Cylindrical Dipole. Application to Standard Antennas 38
   A. Kazemipour, X. Begaud, D. Allal

An Interference Avoidance Technique
for Ad Hoc Networks Employing Array Antennas 43
   T. Hunziker, J.L. Bordim, T. Ohira, S. Tanaka

Analysis of a Cylindrical Microstrip Antenna
by the Nonorthogonal FDTD Method and Parallel Processing 53
   R.O. dos Santos, R.M.S. de Oliveira, F.J.B. Barros,
   C.L. da S.S. Sobrinho

Transmission Technologies and Wireless Networks

Unified Analysis of the Local Scatters Channel Model 60
   W.J.L. Queiroz, F.G.S. Silva, M.S. Alencar

Spacial Analysis of the Local Scatters Channel Model 67
    W.J.L. Queiroz, F.G.S. Silva, M.S. Alencar

Propagation Prediction Based on Measurement at 5.8GHz
for Fixed Wireless Access 74
    T.M. Keen, T.A. Rahman

Dimensioning of Wireless Links Sharing Voice and Data 82
    M. de Oliveira Marques, I.S. Bonatti

Communication Theory (I)

An Iterative Matrix-Based Procedure
for Finding the Shannon Cover for Constrained Sequences 88
    D.P.B. Chaves, C. Pimentel, B.F. Uchôa-Filho

Truncated Importance Sampling Simulation
Applied to a Turbo Coded Communication System 94
    B.B. Albert, F.M. de Assis

Zero-Error Capacity of a Quantum Channel 100
    R.A.C. Medeiros, F.M. de Assis

Communication Theory (II)

Communication System Recognition by Modulation Recognition 106
    A.R. Attar, A. Sheikhi, A. Zamani

An Extension to Quasi-orthogonal Sequences 114
    F. Vanhaverbeke, M. Moeneclaey

Trellis Code Construction for the 2-User Binary Adder Channel 122
    V.C. da Rocha Jr., M.L.M.G. Alcoforado

Rayleigh Fading Multiple Access Channel
Without Channel State Information 128
    N. Marina

Communication Theory (III)

Adaptive Decision Feedback Multiuser Detectors
With Recurrent Neural Networks for DS-CDMA in Fading Channels 134
    R.C. de Lamare, R. Sampaio-Neto

Comparison of Two Filterbank Optimization Criteria
for VDSL Transmission 142
    J. Louveaux, C. Siclet, D. Pinchon, P. Siohan

Phase Estimation and Phase Ambiguity Resolution by Message Passing 150
    J. Dauwels, H. Wymeersch, H.-A. Loeliger, M. Moeneclaey

An Expurgated Union Bound for Space-Time Code Systems 156
V.-D. Ngo, H.-W. Choi, S.-C. Park

Achieving Channel Capacity With Low Complexity RS-BTC
Using QPSK over AWGN Channel.................................... 163
R. Zhou, A. Picart, R. Pyndiah, A. Goalic

Telecommunication Pricing and Billing

Design of Sub-session Based Accounting System
for Different Service Level of Mobile IP Roaming User 171
B. Lee, H. Kim, K. Chung

Priority Telephony System With Pricing Alternatives 183
S. Yaipairoj, F. Harmantzis

Mobile Telephony Industry: Price Discrimination Strategies
for Interconnected Networks.. 192
L. Cricelli, F. Di Pillo, N. Levialdi, M. Gastaldi

Network Performance and Telecommunication Services

A Novel ECN-Based Congestion Control and Avoidance Algorithm
With Forecasting and Verifying 199
H.-s. Liu, K. Xu, M.-w. Xu

Competitive Neural Networks for Fault Detection and Diagnosis
in 3G Cellular Systems 207
*G.A. Barreto, J.C.M. Mota, L.G.M. Souza, R.A. Frota, I. Aguayo,
J.S. Yamamoto, P.E.O. Macedo*

Performance Analysis of an Optical MAN Ring
for Asynchronous Variable Length Packets 214
H. Castel, G. Hébuterne

Enabling Value-Generating Telecommunications Services
upon Integrated Multiservice Networks 221
D.X. Adamopoulos, C.A. Papandreou

Active Networks and Mobile Agents

An Architecture for Publishing and Distributing Service Components
in Active Networks 227
N. Dragios, C. Harbilas, K.P. Tsoukatos, G. Karetsos

Performance Comparison of Active Network-Based and Non Active
Network-Based Single-Rate Multicast Congestion Control Protocols 234
Y. Darmaputra, R.F. Sari

MAMI: Mobile Agent Based System for Mobile Internet 241
    M.A. Haq, M. Matsumoto

Design and Implementation of an ANTS-Based Test Bed
for Collecting Data in Active Framework 251
    V. Damasceno Matos, J.L. de Castro e Silva, J.C. Machado,
    R.M. de Castro Andrade, J.N. de Souza

Adaptive QoS Management for Regulation
of Fairness Among Internet Applications 257
    M.F. de Castro, D. Gaïti, A. M'hamed

Optical Photonic Technologies (I)

Black Box Model of Erbium-Doped Fiber Amplifiers in C and L Bands 267
    A. Teixeira, D. Pereira, S. Junior, M. Lima, P. André, R. Nogueira,
    J. da Rocha, H. Fernandes

Optical Packet Switching Access Networks
Using Time and Space Contention Resolution Schemes 272
    L.H. Bonani, F.J.L. Pádua, E. Moschim, F.R. Barbosa

Full-Optical Spectrum Analyzer Design
Using EIT Based Fabry-Perot Interferometer 282
    A. Rostami

A Powerful Tool Based on Finite Element Method
for Designing Photonic Crystal Devices 287
    A. Cerqueira Jr., K.Z. Nobrega, F. Di Pasquale,
    H.E. Hernandez-Figueroa

Optical Photonic Technologies (II)

Analysis of Quantum Light Memory in Atomic Systems 296
    A. Rostami

Wavelength Conversion With 2R-Regeneration
by UL-SOA Induced Chirp Filtering 304
    C. de Mello Gallep, E. Conforti

Performance Analysis of Multidimensional PPM Codes
for Optical CDMA Systems in the Presence of Noise 312
    J.S.G. Panaro, C. de Almeida

Smart Strategies for Single-Photon Detection 322
    J.B.R. Silva, R.V. Ramos, E.C. Giraudo

Optical Networks (I)

Inter-arrival Planning for Sub-graph Routing Protection
in WDM Networks 328
   D.A.A. Mello, J.U. Pelegrini, M.S. Savasini, G.S. Pavani,
   H. Waldman

An Adaptive Routing Algorithm
for Intelligent and Transparent Optical Networks 336
   R.G. Dante, F. Pádua, E. Moschim, J.F. Martins-Filho

Novel Wavelength Assignment Algorithm for Intelligent Optical Networks
Based on Hops Counts and Relative Capacity Loss 342
   R.G. Dante, J.F. Martins-Filho, E. Moschim

Ad Hoc Networks (I)

Assigning Codes in a Random Wireless Network 348
   F. Djerourou, C. Lavault, G. Paillard, V. Ravelomanana

Flexible QoS Routing Protocol for Mobile Ad Hoc Network 354
   L. Khoukhi, S. Cherkaoui

Integrating Mobile Ad Hoc Network into Mobile IPv6 Network 363
   A. Ali, L.A. Latiff, N. Fisal

Network Architecture for Scalable Ad Hoc Networks 369
   J. Costa-Requena, J. Gutiérrez, R. Kantola, J. Creado, N. Beijar

Ad Hoc Networks (II)

Dual Code-Sensing Spread Spectrum Random Access (DCSSRA) Protocol
for Ad Hoc Networks 381
   C.-J. Wang, W. Quan, Z.-H. Deng, Y.-A. Liu, W.-B. Wang, J.-C. Gao

Exploiting the Small-World Effect to Increase Connectivity
in Wireless Ad Hoc Networks 388
   D. Cavalcanti, D. Agrawal, J. Kelner, D. Sadok

Adaptive Resource Management in Mobile Wireless Cellular Networks 394
   M. Hossain, M. Hassan, H.R. Sirisena

Comparative Analysis of Ad Hoc Multi-path Update Techniques
for Multimedia Applications 400
   M. Kwan, K. Doğançay

Ad Hoc Networks (III)

PRDS: A Priority Based Route Discovery Strategy
for Mobile Ad Hoc Networks 410
   B. Zhou, A. Marshall, J. Wu, T.-H. Lee, J. Liu

Performance Evaluation of AODV Protocol over E-TDMA MAC Protocol
for Wireless Ad Hoc Networks 417
   V. Loscrì, F. De Rango, S. Marano

Mitigating the Hidden Terminal Problem
and Improving Spatial Reuse in Wireless Ad Hoc Network 425
   J.-W. Bang, S.-K. Youm, C.-H. Kang, S.-J. Seok

A New Distributed Power Control Algorithm
Based on a Simple Prediction Method 431
   R.A.O. Neto, F.S. Chaves, F.R.P. Cavalcanti, T.F. Maciel

Ad Hoc Networks (IV)

A Heuristic Approach to Energy Saving in Ad Hoc Networks 437
   R.I. da Silva, J.C.B. Leite, M.P. Fernandez

Calculating the Maximum Throughput in Multihop Ad Hoc Networks 443
   B.A.M. Villela, O.C.M.B. Duarte

Throughput of Distributed-MIMO Multi-stage Communication Networks
over Ergodic Channels 450
   M. Dohler, A. Gkelias, A.H. Aghvami

Delivering the Benefits of Directional Communications
for Ad Hoc Networks Through an Efficient Directional MAC Protocol 461
   J.L. Bordim, T. Ueda, S. Tanaka

Signal Processing (I)

Synchronization Errors Resistant Detector for OFDM Systems 471
   Y. Bar-Ness, R. Solá

The Discrete Cosine Transform over Prime Finite Fields 482
   M.M.C. de Souza, H.M. de Oliveira, R.M.C. de Souza,
   M.M. Vasconcelos

A Lattice Version of the Multichannel Fast QRD Algorithm
Based on *A Posteriori* Backward Errors 488
   A.L.L. Ramos, J.A. Apolinário Jr.

New Blind Algorithms Based on Modified "Constant Modulus" Criteria
for QAM Constellations 498
   C.A.R. Fernandes, J.C.M. Mota

Signal Processing (II)

Channel Estimation Methods for Space-Time Block Transmission
in Frequency-Flat Channels 504
   N. Nefedov

Faster DTMF Decoding 510
   J.B. Lima, R.M. Campello de Souza, H.M. de Oliveira,
   M.M. Campello de Souza

Adaptive Echo Cancellation for Packet-Based Networks 516
   V. Stewart, C.F.N. Cowan, S. Sezer

The Genetic Code Revisited: Inner-to-Outer Map, 2D-Gray Map,
and World-Map Genetic Representations 526
   H.M. de Oliveira, N.S. Santos-Magalhães

Signal Processing (III)

On Subspace Channel Estimation in Multipath SIMO
and MIMO Channels.. 532
   N. Nefedov

A Novel Demodulation Technique for Recovering PPM Signals
in the Presence of Additive Noise 541
   P. Azmi, F. Marvasti

A Novel Iterative Decoding Method for DFT Codes in Erasure Channels 548
   P. Azmi, F. Marvasti

Comparison Between Several Methods of PPM Demodulation
Based on Iterative Techniques 554
   M. Shariat, M. Ferdosizadeh, M.J. Abdoli, B. Makouei,
   A. Yazdanpanah, F. Marvasti

Signal Processing (IV)

A Network Echo Canceler Based on a SRF QRD-LSL Adaptive
Algorithm Implemented on Motorola StarCore SC140 DSP 560
   C. Paleologu, A.A. Enescu, S. Ciochina

Detection of Equalization Errors in Time-Varying Channels 568
   J.F. Galdino, E.L. Pinto, M.S. Alencar, E.S. Sousa

A Design Technique for Oversampled Modulated Filter Banks
and OFDM/QAM Modulations 578
   D. Pinchon, C. Siclet, P. Siohan

Nonlinear Bitwise Equalization 589
   F.-J. González Serrano, M. Martínez Ramón

Network Performance and MPLS

Modelling and Performance Evaluation of Wireless Networks 595
    G.H.S. Carvalho, R.M. Rodrigues, C.R.L. Francês, J.C.W.A. Costa,
    S.V. Carvalho

Improving End-System Performance With the Use of the QoS Management 601
    M.A. Teixeira, J.S. Barbar

Throughput Maximization of ARQ Transmission Protocol
Employing Adaptive Modulation and Coding 607
    C. González, L. Szczeciński, S. Aïssa

CR-LSP Dynamic Rerouting Implementation
for MPLS Network Simulations in Network Simulator 616
    T.A. Moura Oliveira, E. Guimarães Nobre, J. Celestino Jr.

Traffic Engineering (I)

FuDyLBA: A Traffic Engineering Load Balance Scheme for MPLS
Networks Based on Fuzzy Logic 622
    J. Celestino Jr., P.R.X. Ponte, A.C.F. Tomaz, A.L.B.P.B. Diniz

Service Differentiation over GMPLS 628
    H. Moungla, F. Krief

Sliding Mode Queue Management in TCP/AQM Networks 638
    M. Jalili-Kharaajoo

A Distributed Algorithm for Weighted Max-Min Fairness in MPLS Networks ... 644
    F. Skivée, G. Leduc

Traffic Engineering (II) and Internet (I)

Applying Artificial Neural Networks for Fault Prediction
in Optical Network Links 654
    C.H.R. Gonçalves, M. Oliveira, R.M.C. Andrade, M.F. de Castro

Capacity Allocation for Voice over IP Networks
Using Maximum Waiting Time Models 660
    S. Sharafeddine, N. Kongtong, Z. Dawy

SCTP Mobility Highly Coupled With Mobile IP 671
    J.-W. Jung, Y.-K. Kim, H.-K. Kahng

Migration to the New Internet – Supporting Inter Operability
Between IPv4 and IPv6 Networks 678
    T. Vazão, L. Raposo, J. Santos

SIP, QoS (I), and Switches

Smart Profile: A New Method for Minimising SIP Messages 688
A. Meddahi, G. Vanwormhoudt, H. Afifi

A QoS Provisioned CIOQ ATM Switch With m Internal Links 698
C.R. dos Santos, S. Motoyama

Multi-constrained Least Cost QoS Routing Algorithm 704
H. Jiang, P.-l. Yan, J.-g. Zhou, L.-j. Chen, M. Wu

The New Packet Scheduling Algorithms for VOQ Switches 711
A. Baranowska, W. Kabaciński

Optical Networks (II)

An Algorithm for the Parameters Extraction
of a Semiconductor Optical Amplifier 717
R.P. Vivacqua, C.M. Gallep, E. Conforti

All-Optical Gain Controlled EDFA: Design and System Impact 727
J.C.R.F. Oliveira, J.B. Rosolem, A.C. Bordonalli

Iterative Optimization in VTD to Maximize the Open Capacity
of WDM Networks 735
K.D.R. Assis, M.S. Savasini, H. Waldman

Link Management Protocol (LMP) Evaluation for SDH/Sonet 743
P. Uria Recio, P. Rauch, K. Espinosa

Optical Networks (III) and Network Operation and Management (I)

Experimental Investigation of WDM Transmission Properties
of Optical Labeled Signals Using Orthogonal IM/FSK Modulation Format 753
P.V. Holm-Nielsen, J. Zhang, J.J. Vegas Olmos, I. Tafur Monroy,
C. Peucheret, V. Polo, P. Jeppesen, A.M.J. Koonen, J. Prat

Performance Assessment of Optical Burst Switched Degree-Four
Chordal Ring Networks 760
J.J.P.C. Rodrigues, M.M. Freire, P. Lorenz

All-Optical Routing Limitations Due to Semiconductor Optical
Amplifiers Dynamics 766
A. Teixeira, P. André, R. Nogueira, P. Monteiro, J. da Rocha

The Hurst Parameter for Digital Signature of Network Segment 772
M. Lemes Proença Jr., C. Coppelmans, M. Bottoli, A. Alberti,
L.S. Mendes

Network Operation and Management (II)

High Performance Cluster Management Based on SNMP: Experiences on
Integration Between Network Patterns and Cluster Management Concepts 782
  R. Sanger Alves, C. Cassales Marquezan, L. Zambenedetti Granville,
  P.O.A. Navaux

A Web-Based Pro-active Fault
and Performance Network Management Architecture 792
  A.S. Ramos, A. Salles Garcia, R. da Silva Villaça, R.B. Drago

A CIM Extension for Peer-to-Peer Network and Service Management 801
  G. Doyen, O. Festor, E. Nataf

A Generic Event-Driven System for Managing SNMP-Enabled
Communication Networks 811
  A.P. Braga, R. Rios, R. Andrade, J.C. Machado, J.N. de Souza

Network Management Theory and Telecommunications Networks

Algorithms for Distributed Fault Management
in Telecommunications Networks.................................... 820
  E. Fabre, A. Benveniste, S. Haar, C. Jard, A. Aghasaryan

Fault Identification by Passive Testing 826
  X.H. Guo, B.H. Zhao, L. Qian

Script MIB Extension for Resource Limitation
in SNMP Distributed Management Environments 835
  A. da Rocha, C.A. da Rocha, J.N. de Souza

UML Specification of a Generic Model for Fault Diagnosis
of Telecommunication Networks 841
  A. Aghasaryan, C. Jard, J. Thomas

Mobility and Broadband Wireless

IEEE 802.11 Inter-WLAN Mobility Control With Broadband Supported
Distribution System Integrating WLAN and WAN 848
  M. Rahman, F. Harmantzis

A Cost-Effective Local Positioning System Architecture Based on TDoA 858
  R.M. Abreu, M.J.A. de Sousa, M.R. Santos

Indoor Geolocation With Received Signal Strength Fingerprinting Technique
and Neural Networks 866
  C. Nerguizian, C. Despins, S. Affès

Implementation of a Novel Credit Based SCFQ Scheduler
for Broadband Wireless Access 876
  E. Garcia-Palacios, S. Sezer, C. Toal, S. Dawson

Cellular System Evolution (I)

BER for CMOS Analog Decoder With Different Working Points 885
  K. Ruttik

BiCMOS Variable Gain LNA at C-Band
With Ultra Low Power onsumption for WLAN 891
  F. Ellinger, C. Carta, L. Rodoni, G. von Büren, D. Barras,
  M. Schmatz, H. Jäckel

Joint MIMO and MAI Suppression for the HSDPA 900
  M. Marques da Silva, A. Correia

A Joint Precoding Scheme and Selective Transmit Diversity 908
  M. Marques da Silva, A. Correia

Cellular System Evolution (II)

Application of a Joint Source-Channel Decoding Technique
to UMTS Channel Codes and OFDM Modulation 914
  M. Jeanne, I. Siaud, O. Seller, P. Siohan

Transmit Selection Diversity for TDD-CDMA
With Asymmetric Modulation in Duplex Channel 924
  I. Jeong, M. Nakagawa

Planning the Base Station Layout in UMTS Urban Scenarios:
A Simulation Approach to Coverage and Capacity Estimation 932
  E. Zola, F. Barceló

A General Traffic and Queueing Delay Model
for 3G Wireless Packet Networks 942
  G. Aniba, S. Aïssa

Personal Communication, Terrestrial Radio Systems, and Satellites

MobiS: A Solution for the Development of Secure Applications
for Mobile Devices 950
  W. Viana, B. Filho, K. Magalhães, C. Giovano, J. de Castro,
  R. Andrade

Compensating Nonlinear Amplifier Effects on a DVB-T Link 958
  V. Vale do Nascimento, J.E.P. de Farias

Controlled Load Services in IP QoS Geostationary Satellite Networks 964
    F. De Rango, M. Tropea, S. Marano

Mobility Management

Fast Authentication for Inter-domain Handover 973
    H. Wang, A.R. Prasad

Session and Service Mobility in Service Specific Label Switched
Wireless Networks 983
    P. Maruthi, G. Sridhar, V. Sridhar

Mobility Management for the Next Generation of Mobile Cellular Systems 991
    M. Carli, F. Cappabianca, A. Tenca, A. Neri

Handoff Delay Performance Comparisons
of IP Mobility Management Schemes Using SIP and MIP 997
    H.-s. Kim, C.H. Kim, B.-h. Roh, S.W. Yoo

Multimedia Information, Network Reliability, EMC in Communications, and Multicast

A Video Compression Tools Comparison for Videoconferencing
and Streaming Applications for 3.5G Environments 1007
    N. Martins, A. Marquet, A. Correia

Fairness and Protection Behavior of Resilient Packet Ring Nodes
Using Network Processors 1013
    A. Kirstädter, A. Hof, W. Meyer, E. Wolf

Markov Chain Simulation of Biological Effects
on Nerve Cells Ionic Channels Due to Electromagnetic Fields
Used in Communication Systems 1023
    D.C. Uchôa, F.M. de Assis, F.A.F. Tejo

Multiobjective Multicast Routing Algorithm 1029
    J. Crichigno, B. Barán

Image Processing, ATM, and Web Services

A Comparison of Filters for Ultrasound Images 1035
    P.B. Calíope, F.N.S. Medeiros, R.C.P. Marques, R.C.S. Costa

Filtering Effects on SAR Images Segmentation 1041
    R.C.P. Marques, E.A. Carvalho, R.C.S. Costa, F.N.S. Medeiros

The Implementation of Scalable ATM Frame Delineation Circuits 1047
    C. Toal, S. Sezer

Web Based Service Provision
A Case Study: Electronic DesignAutomation 1057
    S. Dawson, S. Sezer

Communication, Security, and QoS (II)

Identification of LOS/NLOS States Using TOA Filtered Estimates 1067
    A.G. Guimarães, M.A. Grivet

A Hybrid Protocol for Quantum Authentication of Classical Messages 1077
    R.A.C. Medeiros, F.M. de Assis

Attack Evidence Detection, Recovery, and Signature Extraction
With ADenoIdS 1083
    F.S. de Paula, P.L. de Geus

QoS-Differentiated Secure Charging in Ad-Hoc Environments 1093
    J. Girão, J.P. Barraca, B. Lamparter, D. Westhoff, R. Aguiar

Switching and Routing

Multi-rate Model of the Group of Separated Transmission Links
of Various Capacities 1101
    M. Głąbowski, M. Stasiak

A Multicast Routing Algorithm Using Multiobjective Optimization 1107
    J. Crichigno, B. Barán

A Transsignaling Strategy for QoS Support in Heterogeneous Networks 1114
    D. Gomes, P. Gonçalves, R.L. Aguiar

Evaluation of Fixed Thresholds for Allocation and Management
of Dedicated Channels Transmission Power in WCDMA Networks 1122
    C.H.M. de Lima, E.B. Rodrigues, V.A. de Sousa Jr.,
    F.R.P. Cavalcanti, A.R. Braga

Next Generation Systems (I)

Efficient Alternatives to Bi-directional Tunnelling for Moving Networks 1128
    L. Burness, P. Eardley, J. Eisl, R. Hancock, E. Hepworth,
    A. Mihailovic

Group Messaging in IP Multimedia Subsystem of UMTS 1136
    I. Miladinovic

The Effect of a Realistic Urban Scenario on the Performance
of Algorithms for Handover and Call Management
in Hierarchical Cellular Systems 1143
    E. Natalizio, A. Molinaro, S. Marano

Providing Quality of Service for Clock Synchronization 1151
    A.C. Callado, J. Kelner, A.C. Frery, D.F.H. Sadok

Next Generation Systems (II) and Traffic Management (I)

Application of Predictive Control Algorithm to Congestion Control
in Differentiated Service Networks. 1157
    M. Jalili-Kharaajoo, B.N. Araabi

Design of a Manageable WLAN Access Point 1163
    *T. Vanhatupa, A. Koivisto, J. Sikiö, M. Hännikäinen,
    T.D. Hämäläinen*

Performance Evaluation of Circuit Emulation Service
in a Metropolitan Optical Ring Architecture 1173
    *V.H. Nguyen, M. Ben Mamoun, T. Atmaca, D. Popa, N. Le Sauze,
    L. Ciavaglia*

Expedited Forwarding End to End Delay Variations 1183
    H. Alshaer, E. Horlait

Traffic Management (II)

Internet Quality of Service Measurement Tool
for Both Users and Providers 1195
    A. Ferro, F. Liberal, E. Ibarrola, A. Muñoz, C. Perfecto

A XML Policy-Based Approach for RSVP 1204
    E. Toktar, E. Jamhour, C. Maziero

SRBQ and RSVPRAgg: A Comparative Study 1210
    R. Prior, S. Sargento, P. Brandão, S. Crisóstomo

TCP Based Layered Multicast Network Congestion Control 1218
    A.A. Al Naamani, A.M. Al Naamany, H. Bourdoucen

A Novel Detection Methodology of Network Attack Symptoms
at Aggregate Traffic Level on Highspeed Internet Backbone Links 1226
    B.-h. Roh, S.W. Yoo

Wireless Access (I)

Precision Time Protocol Prototype on Wireless LAN 1236
    J. Kannisto, T. Vanhatupa, M. Hännikäinen, T.D. Hämäläinen

Moving Telecom Outside Plant Systems
Towards a Standard Interoperable Environment 1246
    G.M. Weiss, E.Z.V. Dias

Analysis and Contrast Between STC and Spatial Diversity Techniques
for OFDM WLAN With Channel Estimation 1252
  E.R. de Lima, S.J. Flores, V. Almenar, M.J. Canet

Fair Time Sharing Protocol: A Solution for IEEE 802.11b Hot Spots 1261
  A. Munaretto, M. Fonseca, K. Al Agha, G. Pujolle

Wireless LANs (II)

An Algorithm for Dynamic Priority Assignment
in 802.11e WLAN MAC Protocols................................... 1267
  A. Iera, G. Ruggeri, D. Tripodi

DHCP-Based Authentication for Mobile Users/Terminals
in a Wireless Access Network 1274
  L. Veltri, A. Molinaro, O. Marullo

Performance Analysis of Multi-pattern Frequency Hopping Wireless LANs 1282
  D. Chen, A.K. Elhakeem, X. Wang

A Delayed-ACK Scheme for MAC-Level Performance Enhancement
of Wireless LANs 1289
  D. Kliazovich, F. Granelli

Wireless LANs (III) and Network Planning and Optimization

The VDSL Deployments and Their Associated Crosstalk
With xDSL Systems 1296
  K.-A. Han, J.-J. Lee, J.-C. Ryou

Design of PON Using VQ-Based Fiber Optimization 1303
  P. Shah, N. Roy, A. Roy, K. Basu, S.K. Das

IPv6 Deployment Support Using an IPv6 Transitioning Architecture
– The Site Transitioning Architecture (STA) 1310
  M. Mackay, C. Edwards

An Intelligent Network Simulation Platform Embedded
With Multi-agents Systems for Next Generation Internet 1317
  M.F. de Castro, H. Lecarpentier, L. Merghem, D. Gaïti

Internet (II)

Reduced-State SARSA With Channel Reassignment
for Dynamic Channel Allocation in Cellular Mobile Networks 1327
  N. Lilith, K. Doğançay

Optimal Bus and Buffer Allocation for a Set
of Leaky-Bucket-Controlled Streams 1337
  E. den Boef, J. Korst, W.F.J. Verhaegh

Scalable End-to-End Multicast Tree Fault Isolation 1347
     T. Friedman, D. Towsley, J. Kurose

Internet Service Pricing Based on User and Service Profiles 1359
     Y. Khene, M. Fonseca, N. Agoulmine, G. Pujolle

Design and Performance of Asymmetric Turbo Coded Hybrid-ARQ 1369
     K. Oteng-Amoako, S. Nooshabadi, J. Yuan

ICT 2004 Poster Papers 1382

Author Index 1385

A QoS Provisioned CIOQ ATM Switch with *m* Internal Links

Carlos Roberto dos Santos[1] and Shusaburo Motoyama[2]

[1]DEE-INATEL, Santa Rita do Sapucaí MG, Brazil, Phone: +55-3471-9221
carlos@inatel.br
[2]DT-FEEC-UNICAMP, Campinas SP, Brazil, Phone:+55-19-3788-3765
motoyama@dt.fee.unicamp.br

Abstract. A QoS provisioned CIOQ ATM switch with *m* internal links is proposed in this paper. In the proposed switch the incoming cells at each input port are discriminated into service classes. For each service class, at each input port, a single buffer is provided whereas each output port needs *m* buffers, where *m* is the number of internal links (or channels) that connect each input port to each output port. The proposed switch uses physical internal links instead of time division, thus no speedup is required. The results of simulation of proposed switch using simple priority schedulers show that for $m \geq 3$ the number of cell waiting at input queue is small, less than 0,1 cells in average, independent of service classes. The proposed switch has also the feature that facilitates the choice of scheduler in order to satisfy the QoS of each class of service.

1 Introduction

The intense increasing in Internet traffic is demanding high capacity packet switches. One of the approaches to obtain these switches is combining the switching speed of layer 2 with the routing features of layer 3. Such kind of switch is called IP switching. Another approach is based on ATM switches, which have the design characteristics appropriate to get high capacity switching. ATM switches have been used as backbone in local area networks (LANs). To become the backbone much more effective higher switching capacity ATM switches are still required. Most of ATM switches for LANs have the structures based on input buffering. Thus, many input-buffering switching structures have been proposed and its performance analyzed in the literature [1], [2], [3], [4].

Recently two new schemes have received attention: Virtual Output Queuing (VOQ) and Combined Input-Output Queuing (CIOQ). VOQ is a queuing scheme used to solve the HOL blocking problem associated with FIFO input queuing or to provide facility to satisfy quality of service (QoS) in input queuing. In this technique each input port maintains a virtual separate queue for each output port. CIOQ is a queuing scheme that combines input and output queuing.

A CIOQ structure named High-speed Statistical Retry Switch (HSR switch) is presented in [5]. This structure uses an internal speedup factor of *m*, i.e., cells are transmitted between input and output buffers *m* times the input line speed (1<*m*<N,

J.N. de Souza et al. (Eds.): ICT 2004, LNCS 3124, pp. 698–703, 2004.
© Springer-Verlag Berlin Heidelberg 2004

where N is the switch size). With ever-increasing line rates, this kind of scheme are becoming increasingly difficulty to implement. The algorithm used in the HSR switch, is not fair, because first m input ports have always priorities over the rest. Thus the last input-buffer will have more cells waiting than the first m input-buffers. The advantage of the HSR switch is the simplicity of the arbitration circuit at the crosspoints.

Another structure presented in [6], uses a combination of VOQ and CIOQ. At input port N^2 buffers are provided and a two-stage approach for scheduling purpose is implemented. The arbiters at the input side perform input contention resolution, while the output-buffered switch element performs classical output contention resolution. The two main bottlenecks in this switch structure are the shared memory interconnection complexity (each input and each output must be able to reach each memory location) and the output queue bandwidth (up to N+1 accesses per queue per packet cycle).

We have proposed in [7] and [8], by using the combined input-output queued structure an efficient way to transfer the cells from input buffers to output buffers, without the need of internal speedup. The proposed switch is based on crossbar type of structure, but vertical lines of this structure have each one a bunch of m lines instead only one line. This simple provisioning permits to the switch a parallel transfer of cells, thus giving a very high throughput at the output port. The proposed structure avoids internal speedup as is required in [5], and its structure is simpler than presented in [6]. In this paper, the proposed structure presented in [7] is modified in order to handle QoS in cell traffic.

The paper is organized as follows. In section 2 a QoS provisioned ATM switch is presented. In following section the results of performance analysis of proposed switch are shown. Finally, the conclusions are presented in last section.

2   QOS Provisioned ATM Switch Structure

In the proposed structure the incoming cells at the input ports are discriminated into class of services as is shown in Fig. 1. The figure shows that five FIFO buffers for each ATM service class (CBR, rtVBR, nrtVBR, ABR and UBR) are provided at each input port. Two sets of REQ and ACK lines, connecting each input port to each scheduler to transport request and acknowledge information are also provided. Each output needs 5xm buffers, where m is the number of internal links. It is important to note that each internal link needs a physical buffer, but service class separation can be done by a logical buffer (shared memory). The figure shows that a scheduler (SCH) for each output port is also provided. Any kind of scheduling algorithm can be used at the scheduler.

2.1   Operation Principle

The operation of proposed switch depends on the scheduling algorithm used at the scheduler (SCH). There are many scheduling algorithm proposed in the literature for different applications. For illustration purpose, the simple non-preemptive priority algorithm will be adopted.

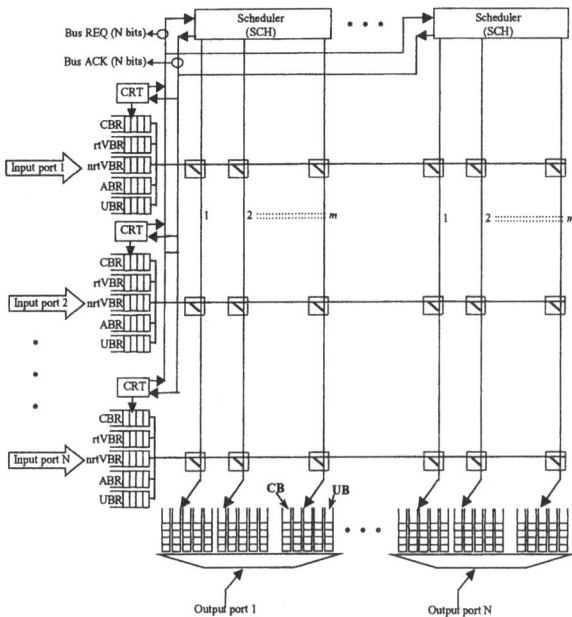

Fig. 1. QoS provisioned ATM switch structure.

In the priority scheme, the incoming cells in each input port are discriminated and stored to the appropriate service class queues. The service class CBR has highest priority followed by rtVBR whereas UBR is the lowest service class priority. Thus, each control unit (CRT) first examines if CBR class of service has any cell to transmit. If any cell is waiting in the CBR queue, the CRT sends a CBR code signal (3 bits, for instance, 111) using the REQ line. If there are more than m requests from the CBR service classes in the same time slot, each SCH selects m inputs on a Round-Robin basis and sends ACK signals (only a bit for each ACK) to the chosen CRTs through the reserved lines (ACK bus in Fig. 1). If there is no CBR class of service cell, the CRT then examines rtVBR class of service buffer. If any cell is waiting an rtVBR code signal (for instance, 110) is sent to the SCH in REQ line. In the same way, if there are more than m requests, in the same time slot, only m inputs will be served, chosen on a Round-Robin basis. This sequence is performed at each input port, thus the service class UBR will be served only if there are no others service classes to be served. If a SCH receives more than m request from different service classes in the same time slot, it will serve only m inputs, obeying the service class priority discipline.

3   Queue Length Performance Analysis

3.1   Simulation Model

To study the performance of proposed switch in relation to cell queuing lengths at input and output buffers a simulation was carried out. Since the switch structure is symmetric the simulation model can be simplified for only one generic output as is shown in Fig. 2.

The model considers that 5 input buffers are provided for each class of service. In each service class buffer the cells arrive from N inputs at a rate λ_i. Assuming that the cell arrivals at the inputs obey independent and identical Bernoulli process, the probability of a cell arrival in a slot is p, the traffic percentage of service class i is S_i ($\Sigma_i S_i = 1$) and the probability that a cell in an input port will be sent to the particular output port is 1/N, then the rate λ_i at each service class buffer is $NpS_i 1/N$. In each time slot, the server j looks first for a cell to transmit in the CBR buffer that is the highest priority class and can serves until *m* cells simultaneously. If there is no cell or there are less than *m* cells in CBR buffer the server goes to the next buffer or buffers (rtVBR, nrtVBR, ABR and UBR buffers, in this sequence) until it completes *m* cells. The output buffer at each of *m* internal links is divided into 5 service classes and the cells are transmitted to the output j in a priority order. First all CBR cells from *m* output buffers are transmitted, then all rtVBR cells from *m* output buffers are transmitted, and so on. The simulation was carried out in MatLab software.

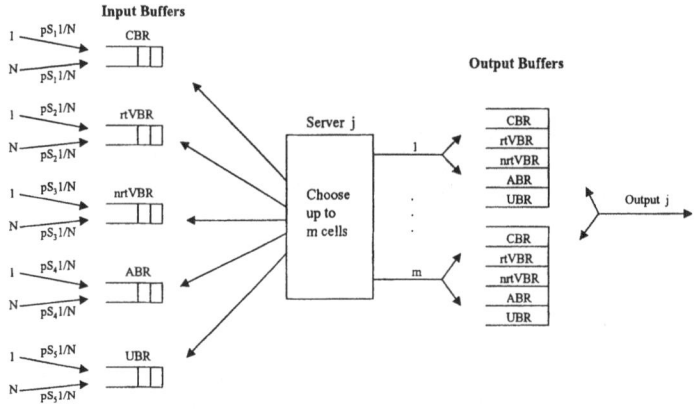

Fig. 2. Simulation model for proposed switch.

3.2 Performance at Input Buffer

The average number of cells in queue for each class of service in function of internal link *m* obtained by simulation at input buffer is shown in Fig. 3, considering switch size N=16 and offered load p=0.9. The percentages of service class are 40, 20, 20, 10 and 10% for Class1, Class2, Class3 Class4 and Class5, respectively. As it was expected, simulation results show that the number of cells waiting at each service class input buffer, in average, is small. For m 3 there is no practically cell waiting in the input buffer independent of service class. Even for m=2, the lowest priority class has a small number of cells in queue, less than one cell in average.

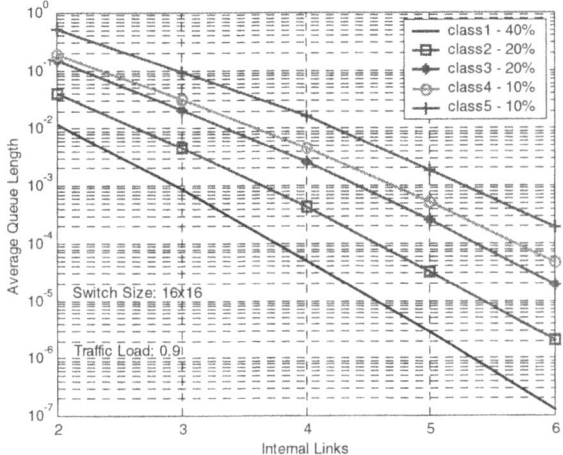

Fig. 3. Average queue length at each service class input buffer.

3.3 Performance at Output Buffer

At the output buffer the performance of average number of cells for each service class is evaluated in terms of internal links m. The simulation results are shown in Fig. 4, considering switch size N=16 and offered load p=0.9.

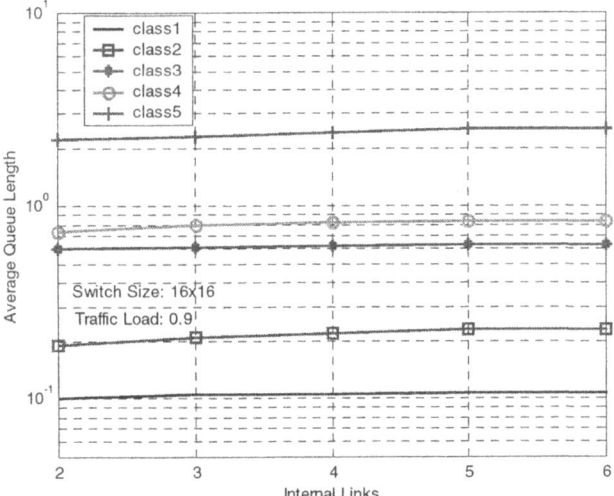

Fig. 4. Average queue length for each service class in function of internal links m.

As it can be observed, results show that only the service Class5 has a number of cells waiting greater than one cell, about 2.5 cells in average. Note that increasing the number of internal links m, there is no practically increase in the number of cells waiting in buffer. For the classes 1 to 4, the number of cells at each output buffer is small, less than one cell in average. It can be concluded that, for this example, about *m*=3 internal links are enough for rapid cell transfer and transmission.

4 Conclusion

A QoS provisioned CIOQ ATM switch with *m* internal links was proposed in this paper. To provide QoS the proposed structure discriminates the incoming cells into service classes at input ports. The discriminated cells at input buffers are transferred to the output buffers by means of *m*x*N* internal links. Since no speedup is required the proposed structure is suited for high-speed switches.

To verify the switch performance, simulation was carried out using a simple priority-scheduling scheme. The simulation results showed that only 3 internal links for each output port are sufficient for quick cell transfer from input buffers to the output buffers. Furthermore, the results have shown that the cell average queuing length at output buffers can be kept small for any service class in the simulation conditions used (N = 16 and p = 0,9).

References

1. Motoyama, S., Petr, D. W. and Frost, V. S.: "Input – queued switch based on a scheduling algorithm", Electronics Letters, 31(14), July 1995, 1127-1128.
2. McKeown, N., Varaiya, P., Walrand, J.: "Scheduling cells in an Input-Queued Switch", Electronics Letter, 3(11), December 1993, 323-325.
3. Motoyama, S.; Ono, L. M., Mavigno, M.C.: "An iterative cell scheduling algorithm for ATM input-queued switch with service class priority", IEEE communications Letters, 3(11), 1999, 323-325.
4. Anderson, T. E., Owicki, S.S., Saxe, J.B and Tacker, C. P.: "High Speed Switch Scheduling for Local Area Networks", Proc. Fifth International Conference on Architectural Support for Programming Languages and Operating Systems, Oct. 1992, 98-110.
5. Genda, K., Yamanaka N., Doi, Y.: "A High-Speed ATM Switch that uses a Simple Retray Algorithm and Small Input Buffers", IEICE Trans. Commun. Vol. E76-B. No. 7, July 1993.
6. Minkenberg, C., Engbersen, T.: "A combined Input and Output Queued Packet-Switched System Based on PRIZMA Switch-on-a-Chip Technology", IEEE Commun. Magazine, 38(12), December 2000, 70-77.
7. Motoyama, S. Santos, C. R.: "A Combined Input-Output Queuing ATM Switch With m Internal Links for Cell Transfer", International Telecommunications Symposium, September 2002, 142-146.
8. Motoyama, S. Santos, C. R.: "Performance Analysis of a CIOQ ATM Switch With m Internal Links for Cell Transfer", 10th International Conference on Software, Telecommunications & Computer Networks – SoftCom 2002, November 2002, 636 – 640.

Multi-constrained Least Cost QoS Routing Algorithm

Hao Jiang, Pu-liu Yan, Jian-guo Zhou, Li-jia Chen, and Ming Wu

School of Electronic Information, Wuhan University, Wuhan 430079
Hubei, China
jianghaow@263.net

Abstract. To obtain multi-constrained QoS path with minimal cost, a least-cost QoS routing algorithm—LBPSA(Lagrangian Branch-and-bound path selection algorithm) is presented. This algorithm uses Lagrangian relaxation to obtain the lower bound and the upper bound of the least cost of multi-constrained QoS path. And finds the multi-constrained QoS path with least cost by branch-and-bound method. In solving the Lagrangian multiplier problem, it introduced a new iteration method. Through recursion from destination to source based on branch-and-bound scheme, the least-cost multi-constrained path is obtained. The proposed algorithm has a pseudopolynomial running time. The simulation shows that the algorithm can find the path with high success ratio and low complexity. To the topology in reality, the success ratio of the algorithm is very high.

1 Introduction

Future networks are expected to support various application, specially multimedia application such as VoIP, VoD. End-to-End quality of service (QoS) has attracted much attention in both academia and industry. An important method of End-to-End quality of service is QoS routing[1, 2].

Recently, many papers provide solutions for QoS routing. The recent works in QoS routing have some features. Some papers propose QoS routing method[3] . But in our opinion, the end user or ISP hope that their data can go through the networks with quality of service and spend little money.

In this paper, we want to find the path that not only satisfy the multi-constrains, but also with least cost. The multi-constrained path is a NP-complete problem, the least cost multi-constrained path is NP-hard problem[4].

To the user, the QoS path spending least money is very important. In this paper, we presented the least-cost QoS routing algorithm--LBPSA. To this problem, we use Lagrangian relaxation to find the feasible paths that accommodate the multi-constrains and the lower bound, upper bound of cost, then select the least cost QoS path by branch-and-bound method. In paper of Jutter and Szviatovski[5], they also use Lagrangian relaxation, but they only solve one constrains QoS routing, not multi-constrains routing, and they do not take care for least cost.

J.N. de Souza et al. (Eds.): ICT 2004, LNCS 3124, pp. 704–710, 2004.

The simulation shows that LBPSA can easily obtain the least cost multi-constrained path through few iterations. We believe that the result of this paper provides an important method to find the least QoS path for ISP. Our results can be used in a variety of practical applications. For example, in MPLS networks, LBPSA can be used to generate LSP that pay little money.

The rest of this paper is structured as follows. Section 2 introduces and motivates the models and problems. Section 3 provides a description of the algorithm. The results of simulation is presented in Section 4. Finally, conclusions and future work are discussed in Section 5.

2 Model and Problems

We represent by a directed graph $G(V, E)$. V is the set of nodes, there are n nodes in the network. E is the set of links. There is a single source s and a single destination t. The user QoS request have k *additive* constrains, so every $e(u, v)$ offers k QoS metrics. $w_i(e)$,$1 < i < k$. $w(e) = \{w_1(e), w_2(e), \dots, w_k(e)\}$ represents the metrics of the link e. p is the path from s to t. $w_i(p) = \sum w_i(e)$, $e \in p$, $1 < i < k$ is the metrics of path p. Every e has the link cost $c(e)$, the cost is money or the other. The cost of p is $c(p) = \sum c(e)$, $e \in p$. M_i is the i_{th} QoS additive constrain. If $w_i(p) < M_i$, p is the feasible path. Our purpose is to find the path p_o that $w_i(p_o) < M_i, 1 < i < k$ and $c(p_o) = \min(c(p) \mid w_i(p) < M_i, 1 < i < k)$.

The programming formula of this problem is:

$$\min c(p)$$
$$s.t. \ w_1(p) \le M_1$$
$$w_2(p) \le M_2$$
$$\vdots \tag{1}$$
$$w_k(p) \le M_k$$

(1) is a NP-hard problem. To this problem, we propose Lagrangian relaxation and branch-and-bound search to resolve it.

3 Routing Algorithm

3.1 Lagrangian Relaxation and Iteration Method

Lagrangian relaxation[6] is a general solution strategy for solving mathematical programs. The Lagrangian relaxation of (1) is:

$$L(\mu) = \min[c(p) + \mu_1(w_1(p) - M_1) + \cdots + \mu_k(w_k(p) - M_k)] \tag{2}$$

The maximum value of $L(\mu)$ is the low bound that is much close the optimal result.

$$L(\mu) \le \max{}_{\mu} L(\mu) \le z^*$$

z^* is the least cost of the feasible paths. $L^* = \max{}_{\mu} L(\mu)$.

The (2) is equal to the path form of the Lagrangian relaxation.

$$L(\mu) = \min(c(p) + \mu_1(w_1(p) - M_1) + \cdots + \mu_k(w_k(p) - M_k))$$
$$= \min(c(p) + \mu_1 w_1(p) + \cdots + \mu_k w_k(p) - \mu_1 M_1 - \cdots - \mu_k M_k)$$
So the $L(\mu)$ is the length

of the shortest path under the aggregated metrics.

In this section, our goal is to obtain the lower bound of z^* and the upper bound of z^* which are used in the branch-and-bound search for the least cost multi-constrained QoS path.

The $\max{}_{\mu} L(\mu)$ or the $L(\mu)$ close to $\max{}_{\mu} L(\mu)$ can be the lower bound. We propose that the cost of a feasible path be the upper bound of the least cost of feasible paths. So we want to search the $\max{}_{\mu} L(\mu)$ and get a feasible path at the same time.

Claim 1: The $\max{}_{\mu} L(\mu)$ is a point, in the $L(\mu)$s that form the $\max{}_{\mu} L(\mu)$, there are $L(\mu)$s which is correspond the feasible paths.

In μ iteration process, if $L(\mu)$ increase to $\max{}_{\mu} L(\mu)$, we easily find a feasible path, and then obtain lower & upper bound. So we need a method that make $L(\mu)$ increase through changing μ and find a feasible path. In this paper, we propose a new iteration method that can find a feasible path quickly near $\max{}_{\mu} L(\mu)$ in networks structure ---- maximum μ_i iteration.

To the least cost QoS routing, the $L(\mu)$ is a discrete unlike the other nonlinear function. And these $L(\mu)$s intersect to a point $\max{}_{\mu} L(\mu)$. In the $L(\mu)$s that form the $\max{}_{\mu} L(\mu)$, there is a $L(\mu)$ that correspond to a path which satisfy the QoS constraints, it is descript in Claim 2.

For example, when there are two constrains, $L(\mu)$ is a plane. And the $L(\mu)$s form a polyhedron.

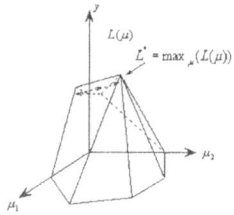

Fig. 1. example of $L(\mu)$s

Our iteration method is the maximum direction iteration. Lagrangian multiplier μ moves in the maximum direction every iteration. It means that every time only change the maximum μ_i. The μ updates from 0, every time we only update the μ_i

that correspond to $w_i(p)-M_i=\max(w_j(p)-M_j,1\le j\le k)$, and the other μ_j s do not change. This iteration method makes the μ move around on the space surrounded by the $L(\mu)$ s.

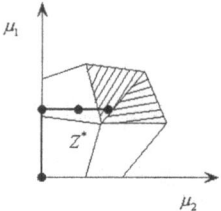

Fig. 2. Iteration process of μ (black line represents the μ)

At first, μ is 0. In the iteration, if $\mu_i < 0$, we set $\mu_i = 0$. The step size is $\theta=\lambda_k[UB-L(\mu)]/(w_i(p)-M_i)^2$, in practice, it is the popular heuristic for selecting the step size.

  UB is the upper bound, at first it is a estimated value large enough, in the iteration we use the cost of feasible path to replace the initial value, if $c(p_f) < UB$, $UB = c(p_f)$, p_f is feasible path. In general, $\lambda_k = 2$, if $L(\mu)$ don't increase in some steps, $\lambda_{k+1} = \lambda_k / 2$. If λ_k is small enough or $L(\mu)$ don't increase, we stop the iteration. To reduce the complexity, we also stop the iteration while find the feasible path.

3.2 Branch-and-Bound

From Claim 1, we know that $L^* \le c(p_o) \le UB$, p_o is the least-cost multi-constrained QoS path that satisfy the multi-constrains.

  According to the properties of the least-cost multi-constrained path, there are three conditions for the branch-and-bound search .

1. $L^* \le c(p_o) \le UB$
2. $w_i(p_o) - M_i \le 0, \quad 1 \le i \le k$ (3)
3. $L^* \le c(p_o) + \sum_{i=1}^{k} \mu_i(w_i(p_o) - M_i) \le UB$

Because it is difficult to obtain L^* and μ^*, we adopt an approximate scheme in fact. We use the maximum $L(\mu)$ in the iterations $\max_I L(\mu)$ to replace the L^*.

  So the least-cost multi-constrained QoS path must satisfy these approximate conditions.

  1) $\max_I L(\mu) \le c(p_o) \le UB$

2) $w_i(p_o) - M_i \le 0, \quad 1 \le i \le k$ (4)

3) $L(\mu^f) \le c(p_o) + \sum_{i=1}^{k} \mu_i^f (w_i(p_o) - M_i) \le UB$

We use these condition in the branch-and-bound search. The branch-and-bound[7] search in this paper do not start from the source node, but from the destination node. When $\mu = \mu^f$, the $L(\mu^f)$ must correspond to the feasible path that is found in the iteration procedure . We find the feasible path by Dijkstra algorithm. So Dijkstra algorithm will find the minimum span tree, the label of the node in the span tree is the distance from the source to this node, the distance is the combined distance, $dist(p) = c(p) + \sum_{i=1}^{k} \mu_i^f w(p)$. If node j on the path that satisfy condition 3, then $dist(p_{s \to j}) + dist(p_{j \to t}) \le UB + \sum_{i=1}^{k} \mu_i^f M_i$. To the metric of cost, there is a minimum span tree found by Dijkstra. If $c(p_{s \to j}) + c(p_{j \to t}) \le UB$, then condition 1 is satisfied, if $w(p_{s \to j}) + w(p_{j \to t}) - M \le 0$, condition 2 is satisfied. When condition 1, 2 and 3 are all satisfied, record the node j and $dist(p_{j \to t})$, $c(p_{j \to t})$, the algorithm keep on searching the neighbor node of j untill the source node. When found source, compare whether $L(\mu^f) + \sum_{i=1}^{k} \mu_i^f \le dist(p_{s \to t})$ and $\max L(\mu) \le c(p_{s \to t})$, and the min cost path that satisfy the condition is the least-cost multi-constrained QoS path.

3.3 Computational Complexity of Algorithm

In the simulation, we find that the iteration times is related to the number of feasible paths. In this section, we define that $f(d, k+1)$ is the number of feasible paths add $k+1$---the times of the shortest path calculation. The worst time of branch-and-bound search is $O(d\, n)$, so the time complexity of the algorithm in this paper is $O(n \lg n + m + n)$.

4 Simulation

The goal of the simulation is to compare the performance of LBPSA and H_MCOP for the Traint-Stub topologies that is close to real world network topologies. One Traint-Stub topology has 150 nodes, other Traint-Stub has 250 nodes. In the simulation, the w_i and *cost* of the links are randomly generated in the range of [0,

100], for $1 \le i \le k$. In each topology, we select source-destination node pair 100 times, and the minimum hop between the source and destination node is more than two.

We use two methods to evaluate the LSPSA :1) Iteration times, that indicates LSPSA's iteration times in get feasible path, 2) Success ratio is defined as the ratio of the number of requests satisfied by using the algorithms and the total number of requests generated.

In QoS routing, the success ratio is related to the feasbile paths in the network. The feasible paths is determined by the Constrains. In this paper, the constrains of the weight is defined as $M_i = shortest\_path(w_i) + random(100)$. $shortest\_path(w_i)$ is the length of shortest path for the weight w_i. $random(100)$ is a random number ranged in [0, 100], it is used to relax the constrains to get more feasible paths.

4.1 Results of Different Topology and Same Constrains

This figure is the success ratio and iteration times of TS topology 150 nodes and 250 nodes with 4, 10 constrains.

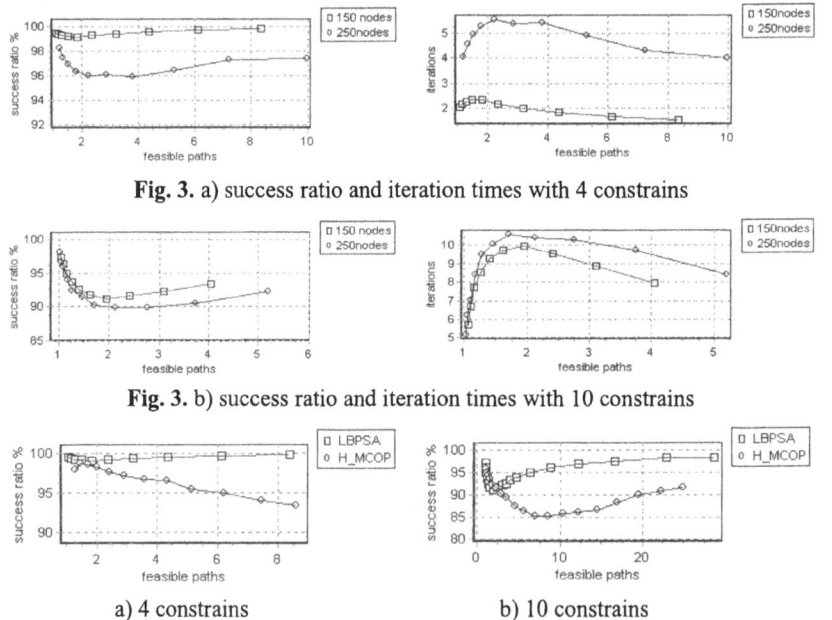

Fig. 3. a) success ratio and iteration times with 4 constrains

Fig. 3. b) success ratio and iteration times with 10 constrains

a) 4 constrains   b) 10 constrains

Fig. 4. The success ratio of LBPSA and H_MCOP with 4 and 10 constrains for 150 nodes topology

From the result, we find that the success ratio is very high when there are much few feasible paths in the networks, and the selection process through only little iteration.

4.2 Comparison the Success Ratio of the Algorithm and H_MCOP

We compare LBPSA with H_MCOP[8], which is based on Dijkstra algorithm.

Fig 4 compares the success ratio of the LBPSA and H_MCOP for 150 nodes TS topology with 4 constrains and 10 constrains. We find that LBPSA is more better than H_MCOP , its success ratio is higher than H_MCOP.

5 Conclusion

This paper presents a least-cost QoS routing algorithm. This algorithm uses Lagrangian relaxation to obtain the lower bound and the upper bound of the least cost of multi-constrained QoS path. And finds the multi-constrained QoS path with least cost by branch-and-bound method. In solving the Lagrangian multiplier problem, it introduced new iteration method: Lagrangian multiplier moves in the maximum direction. The simulation shows that the algorithm can find the path with high success ratio and low complexity.

References

[1]   E Crawley, R Nair, B Rajagopalan, *et al.* A framework for QoS-based routing in the Internet. RFC 2386, 1998.
[2]   Wang Z, Crowcroft J . QoS routing for supporting resource reservation[J]. IEEE Journal on Selected Areas in Communications, 1996, 14(7): 1228~1234.
[3]   Ma Q, Steenkiste P. Quality-of-Service routing with performance guarantees. Proceedings of the 4th International IFIP Workshop on Quality of Service, New York: IFIP WG6.1, 1997.
[4]   Zheng Wang, Jon Crowcroft. Quality-of-service routing for supporting multimedia applications[J]. IEEE Journal on Selected Areas in Communications, 1996, 14(7): 1228~1234.
[5]   Juttner. A, Szviatovski, Mecs, B.I. *et al.* Lagrange relaxation based method for the QoS routing problem. Proceedings of the IEEE INFOCOM 2001. IEEE Communication Society, 2001. 859~868.
[6]   K. Ahuja, L. Magnanti. Network Flows: Theory, Algorithms, and Applications[B]. Prentice Hall, Upper Saddle River, New Jersey, 1993.
[7]   Thomas H, Cormen Charles E, Leiserson Ronald L, *et al.* Introduction to Algorithms. MIT Press, 2001.
[8]   T. Korkmaz, M. Krunz, Multi-Constrained Optimal Path Selection. In Proceedings of IEEE INFOCOM, 2001, pp:834-843.

The New Packet Scheduling Algorithms for VOQ Switches

Anna Baranowska and Wojciech Kabaciński

Institute of Electronics and Telecommunications,
Poznań University of Technology,
ul. Piotrowo 3a, 60-965 Poznań, Poland
{Anna.Baranowska, Wojciech.Kabacinski}@et.put.poznan.pl

Abstract. We present two new scheduling algorithms, the *Maximal Matching with Random Selection* (MMRS) and the *Maximal Matching with Round-Robin Selection* (MMRRS). These algorithms provide high throughput and fair access in a packet switch that uses multiple input queues. They are based on finding a maximal matching between input and output ports without iterations. Simulations performed show that for large load these algorithms perform better than other iterative algorithms.

1 Introduction

In fixed-length switching technology, which is widely accepted as an approach to achieve high switching efficiency for high-speed packet switches, variable-length packets are segmented into fixed-length packets, called cells, at inputs and reassembled at the outputs. Each cell occupy a time interval called a time slot. In the further part of this paper the fixed-length cell will be called a packet. Depending on the traffic pattern, in a given time slot there may be more than one cells directed to one output port, which can accept only one cell in one time slot. Therefore, buffering is used to resolve this output contention. Buffer memories can be placed at inputs, outputs, crosspoints, or may be distributed between different parts of the switch.

The Input Queuing (IQ) architecture is desirable for building high-speed and large capacity ATM switches and IP routers, since the buffer memory speed is equal to the line speed. The best performance (100% throughput, low mean time delay) is achieved in the Output Queuing (OQ) packet switches, but the switch fabric and the buffer memory speed must operate at N times link speed if packet loss is to be prevented. The main drawback of IQ switches is the head-of-line (HOL) blocking phenomena, which limits the throughput to 58,6% for uniformly distributed traffic [1]. Virtual Output Queuing (VOQ) was proposed to overcome this problem and combine the advantages of IQ and OQ architectures [2]. In the VOQ switch each input port maintains N queues, one for each output port. Each arriving packet is classified and then queued in the appropriate VOQ according to its determined destination output port. However, since each input

J.N. de Souza et al. (Eds.): ICT 2004, LNCS 3124, pp. 711–716, 2004.
© Springer-Verlag Berlin Heidelberg 2004

port can transmit only one packet in a time slot and one output port can receive also only one packet in a time slot, a proper scheduling algorithm is needed for choosing queued cells for transmission. Such a scheduling algorithm must resolve output contention swiftly, provide high throughput, satisfy quality-of-service (QoS) requirements, and should be easily implemented in hardware.

The scheduler must perform a one-to-one matching of non-empty VOQs during a period of one time slot, which is getting smaller when line speed is greater. The throughput of the switch is then a function of the number of matches made in each time slot. A matching can be maximum or maximal [3], [4]. A maximum matching is one that matches the maximum number of inputs and outputs, while a maximal matching is a matching where no more matches can be made without modifying the existing matches. One class of scheduling algorithms is based on finding a maximal matching, for example Parallel Interactive Matching (PIM) [2], Iterative Round-Robin Matching (RRM) [5], iSLIP [6], Dual Round-Robin Matching (DRRM) [7]. Depending on the implementation they may use one or multiple iterations to converge on a maximal matching. A maximum matching may result in instability for some schedulable flows, i.e. they cannot be carried by the switch, resulting in forever increasing queue length [8]. Maximum weight matching algorithms were also proposed to improve the switch performance but they are not implemented in hardware [9].

In this paper two algorithms are proposed. They are based on finding a maximal matching between inputs and outputs, and are performed in one iteration. The paper is organized as follows. In section 2 the proposed algorithms are described. In the next section performance evaluation of the proposed algorithms is given and compared with other algorithms, followed by conclusions.

2   Scheduling Algorithms

2.1   The Switch Architecture

The switch considered in the paper is the VOQ switch. Each input prot contains the input buffer divided into N separate queues, one for each output port. These queues are denoted by $VOQ(i, j)$, where i is the input port number and j is the output port number. Packets transmitted through the switching fabric are fixed-length packets, and the switching fabric is nonblocking. One packet occupies one time slot. In one input port only one packet can arrive in one time slot. Similarly, only one packet can be transmitted through the switching fabric from the given input port to the given output port in one time slot (there is no speed-up in the switching fabric). Any output port can accept a packet from at most one input port at one time slot. In each time slot the scheduler selects packets for transmission in the next time slot, i.e. it finds one-to-one matching of non-empty VOQs. The centralized scheduler is used to resolve output contention. The architecture of the scheduler and scheduling algorithms will be considered in the next subsection.

2.2   The Arbitration Schemes

The proposed scheduler consists of N logical circuits called counters and one circuit called the selector, as it is shown in Fig. 1. Each VOQ is connected to the scheduler by means of two control lines. One line is used for sending the request signal from the VOQ, and the second line is used for sending the grant signal to the VOQ. This control lines are shown in Fig. 1 by solid and dashed arrows, respectively. In the scheduler, control lines are connected to the counters in such a way, that each counter is connected to one VOQ from each input port, and VOQs connected to the same counter contains packets directed to different output ports (one-to-one matching). In this way, when all VOQs have HOL packets, these N counters check N perfect matchings between inputs and outputs. The example of such perfect matchings for 4×4 switch is shown in Fig. 2. The selector chooses the set of requests which will be realized in the successive time slot. The algorithm is performed in two steps: *request* and *grant*.

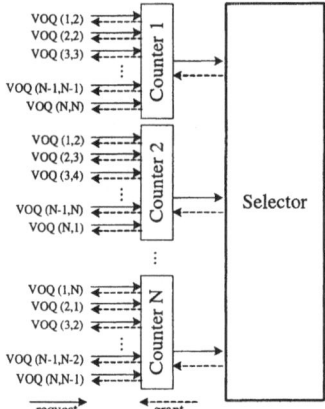

VOQ	Output	VOQ	Output
(1,1) •----------• 1		(1,2) • • 1	
(2,2) •----------• 2		(2,3) • • 2	
(3,3) •----------• 3		(3,4) • • 3	
(4,4) •----------• 4		(4,1) • • 4	
Counter 1		Counter 2	

VOQ	Output	VOQ	Output
(1,3) • • 1		(1,4) • • 1	
(2,4) • • 2		(2,1) • • 2	
(3,1) • • 3		(3,2) • • 3	
(4,2) • • 4		(4,3) • • 4	
Counter 3		Counter 4	

Fig. 1. The general architecture of the scheduler

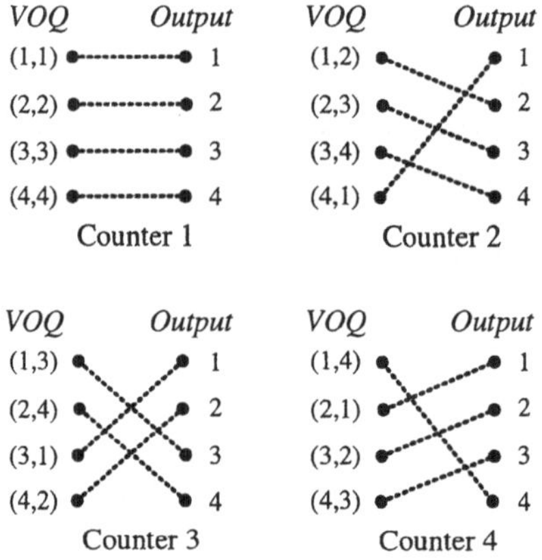

Fig. 2. Matchings being checked in counters 1 to 4 in 4×4 switch

- *Request*
  Each non-empty VOQ sends the request signal to the scheduler. Counters determine the number of request signals received and pass it to the selector.
- *Grant*
  The selector selects the counter with the biggest number of requests and sends the grant signal to the selected counter. This signal is further passed by the counter to the respective VOQs.

In the *request* step each non-empty VOQ sends the request signal to the arbiter. In the arbiter, these signals are grouped into N groups of N signals in such a way that each group constitutes a one-to-one matching. Signals from group i are then counted in the counter i, $1 \leq i \leq N$. Each counter determines the number of request signals received, and sends it to the selector. The selector selects the counter with the biggest number of requests. In the *grant* step the selector sends the grant signal to the selected counter and this counter sends grant signals to respective VOQs which sent request signals. When there are more than one counter with the same number of request signals, the selector chooses one of these counters either in random or using the round-robin algorithm. In the former case the algorithm will be called the *Maximal Matching with Random Selection* (MMRS). In the later case, the selector selects the first counter with the biggest number of request signals starting from the next counter to those selected in the previous time slot. We will call this algorithm the *Maximal Matching with Round-Robin Selection* (MMRRS). In both cases, the selection is done only between counters with the same and the biggest number of request signals.

An example of the MMRRS algorithm execution is shown in Fig. 3. States of HOL cells in time slots 1, 2 and 3 are given in Fig. 3a, where the black square denotes that the given HOL cell has the packet for transmission. At the beginning of time slot 1 we assume that in the previous time slot counter 3 was selected and the pointer in the selector is set to 4. In the first step VOQs with HOL packets send request signals to the arbiter (bold arrows in Fig. 3b). Each counter counts the number of received request signals and passes it to the selector. In time slot 1, counter 2 has the biggest number of requests (4) and it receives the grant signal (bold and dashed arrow). The pointer is set to 3. In the next two time slots each counter receives four requests so, in time slots 2 and 3, counters 3 and 4 are selected, respectively (Fig. 3c and d).

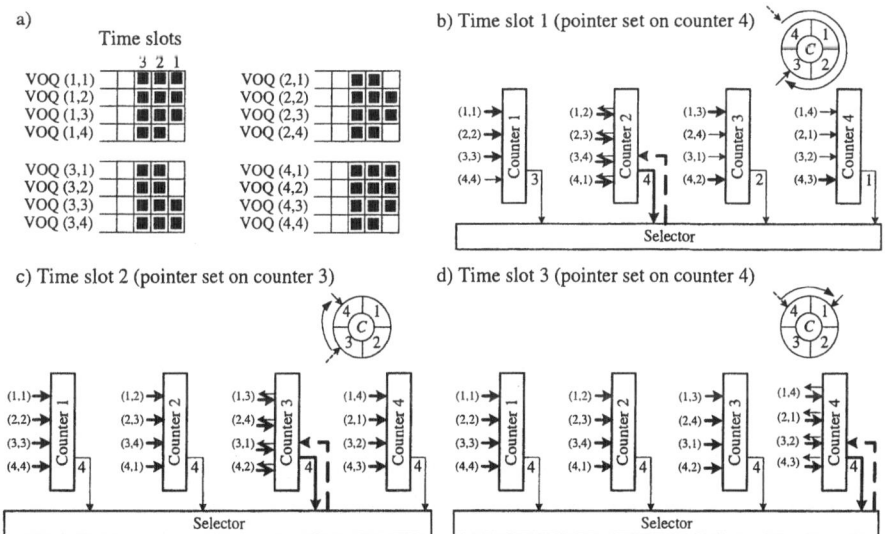

Fig. 3. An example of the MMRRS execution in 4×4 switch; HOL cells in time slots 1, 2, and 3 (a), and arbiter's states after these slots (b, c, and d, respectively)

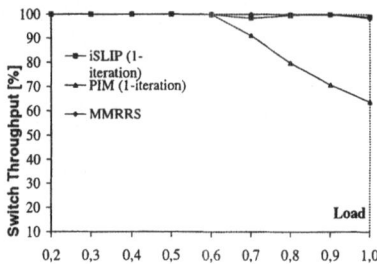
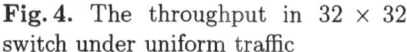
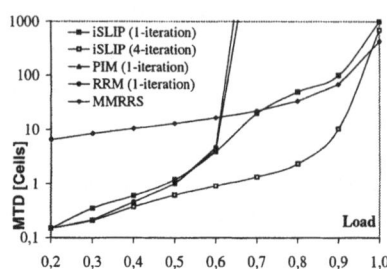

Fig. 4. The throughput in 32×32 switch under uniform traffic

Fig. 5. The mean time delay in 16×16 switch under uniform traffic

3   Simulation Results

We evaluated the performance of the proposed algorithms under uniform and non-uniform traffics for switches of different sizes and compared them with other algorithms using simulation. In all cases we assumed that a packet size is 1 time slot and that it may arrive at the input with probability p.

In the uniform traffic each packet can be destined to the given output port with probability $1/N$. Fig. 4 compares the throughput of 32×32 switch, while Fig. 5 shows the mean time delay in 16×16 switch. The results obtained for MMRS and MMRRS are very similar so only the curve for MMRRS is shown in the figures. The throughput is almost 100% and is slightly better than that of iSLIP. The mean time delay for the load greater than 0.7 MMRS and MMRRS are better than iSLIP with one iteration. For the high load, close to 1, our algorithms are also better than iSLIP with four iterations. In general, for large traffic, the mean time delay obtained in our algorithms is much better than that of other iterative algorithms when one iteration is used.

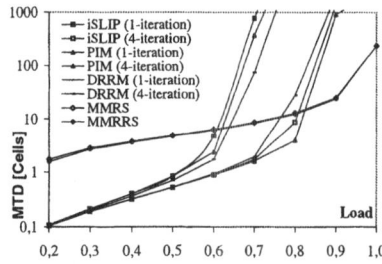
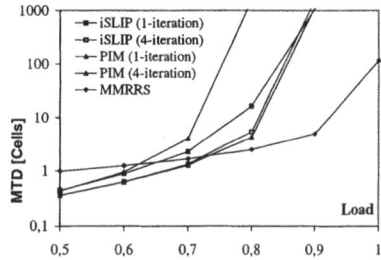

Fig. 6. The mean time delay in 8×8 switch under non-uniform traffic

Fig. 7. The mean time delay in 32×32 switch under diagonal traffic

We consider also a non-uniform traffic, in which a packet at input i is directed to output i with greater probability than to other outputs (1/2 and 1/14 for 8×8

switch, respectively). The simulation results are shown in Fig. 6. It can be seen that our algorithms are in this case much better than other iterative algorithms (with one and four iterations performed) for the load greater than about 0.8. For the low load, our algorithm introduce a mean time delay of about 1 cell.

The proposed algorithms were also simulated under the diagonal traffic, where a packet from input i can be only directed to ouputs i or $i+1$ (N and 1 when $i = N$). Simulation results are shown in Fig. 7. The MMRRS is only shown since for MMRS we obtained similar results. Similarly as for uniform and non-uniform traffics for higher loads our algorithms ensure lower mean time delay than other iterative algorithms with one and four iterations.

4   Conclusion

In this paper we proposed MMRRS and MMRS scheduling algorithms for VOQ switches. The performance of these algorithms was evaluated by simulation for uniform, non-uniform, and diagonal traffics. Simulation results show that these algorithms perform better than other iterative algorithms, while they are executed using one iteration only. It is also known, that iterative algorithms suffer form instability, and some VOQs may be blocked (i.e. due to the pointers set-up, some of VOQs will not be matched for longer time and queues will built-up) [8]. Up till now we observed this phenomena for iterative algorithms simulated, but we do not obtained any state of instability when our algorithms were used.

In future works we will perform more simulation experiments for switches of greater capacity to check if the proposed algorithm reach such instability states.

References

1. Karol, M.K., Hluchyj, M., Morgan, S.: Input versus output queuing on a space-division packet switch. IEEE Trans. Commun. **35** (1987) 1347–1356
2. Anderson, T., et al.: High-speed switch scheduling for local-area networks. ACM Trans. Comp. Syst. **11** (1993) 319–352
3. Chao, H.J., Lam, C.H., Oki, E.: Broadband Packet Switching Technologies: A Practical Guide to ATM Switches in IP Routers. Willey, New York (2001)
4. Chao,H.J.: Next generation routers. Proc. of the IEEE **90** (2002) 1510–1558
5. McKeown, N., Varaiya, P., Warland, J.: Scheduling cells in an input-queued switch. IEE Electronics Letters **29** (1993) 2174–2175
6. McKeown, N.: The iSLIP scheduling algorithm for input-queued switches. IEEE/ACM Trans. Network. **7**, (1999) 188–200
7. Chao, H.J.: Saturn: A terabit packet switch using dual round-robin. IEEE Commun. Mag. **38** (2002) 78–84
8. Yoshigoe, K., Christensen, K.J.: An evolution to crossbar switches with virtual ouptut queuing and buffered cross points. IEEE Network **17** (2003) 48–56
9. McKeown, N., Mekkittikul, A., Anantharam, V., Walrand, J.: Achieving 100% throughput in input-queued switches. IEEE Trans. Commun. **47** (1999) 1260–1267

An Algorithm for the Parameters Extraction of a Semiconductor Optical Amplifier

Rafael P. Vivacqua[1], Cristiano M. Gallep[2], and Evandro Conforti[1]

[1] DMO-FEEC-UNICAMP, Caixa Postal 6101,
13083-970 Campinas, SP, Brazil
conforti@dmo.fee.unicamp.br
http://www.ifi.unicamp.br/photon/conforti-en.htm
[2] Div. Tecnologia em Telecomunicações - CEFET-UNICAMP,
R. Paschoal Marmo, 1888, CEP 13484-370 Limeira, SP, Brazil

Abstract. The parameters of an encapsulated semiconductor optical amplifier (SOA) are obtained by a proper extraction algorithm using a technique based both on measurements and modeling. The theoretical model is based on rate equations and the experiments use continuous wave optical gain curves measured for different injected currents and optical input powers. The algorithm analyses the derivatives of the gain curves in relation to the longitudinal SOA dimension. After the parameters extraction procedure, a good accuracy is achieved for the SOA operating as a high frequency carrier modulated optical amplifier.

1 Introduction

The semiconductor optical amplifier (SOA) is a promising device for optical processing in optical fiber networks, operating as a linear or a non-linear device. It can be an alternative to costly wavelength-flattened Erbium-doped optical fiber amplifiers in wavelength-division multiplexing (WDM) metropolitan networks [1] or be used in optical processing as wavelength switching [2]. In addition, data erase on a downstream signal and data rewrite on the upstream has been proposed in order to reuse the WDM carrier wavelength [3].

However, present day SOAs are expensive and the availability of a reliable and flexible computer-aided design programs is an important task to cut designing costs and also to predict the performance of very high speed based data links [4]. Theoretical models with different complexity levels were being applied to predict SOA-based functionalities, and complex models can provide more accurate knowledge of what happens inside the semiconductor optical active region. Alternatively, a simple model might be good enough in many practical situations of system applications. In any case, however, the modeling parameters should be estimated basing on experimental data.

A simple algorithm, based on the analysis of the continuous wave (CW) optical gain curve, is presented to optimize the SOA modeling. The algorithm was developed using the Agrawal and Olson formulation [5], where a single differential equation is used to predict the optical gain behavior [6]. Using a commercial available SOA, the

J.N. de Souza et al. (Eds.): ICT 2004, LNCS 3124, pp. 717–726, 2004.

experimental optical gain data, for different injected currents and optical input powers, was obtained. With those values, the modeling parameters are extracted. Then, the SOA model parameters are optimized, including to the gain coefficient, the transparency current, the saturation optical power and the loss coefficients. The optimization is based on the minimization of a proper error function. Within the optimized parameters, the simple model predicted the differential gain, from CW up to 6 GHz, with good accuracy. Papers about the extraction of diode lasers parameters have been presented in literature [7], and this paper introduces a similar work for the SOA case. The model can predict both continuous wave and high frequency modulated optical carrier SOA operation.

2 Basic Modeling

For a complete modeling description and applications, see [8, 9]. In short description, the SOA active region optical gain (g), has a time differential equation.

$$\frac{\partial g(z,t)}{\partial t} = \frac{g_0 - g(z,t)}{\tau_c} - \frac{g(z,t)P(z,t)}{E_s} . \tag{1}$$

In equation (1), E_s is the active region saturation energy, τ_c is the carrier lifetime, P is the optical power level in the active region and g_0 the small-signal gain. Due to constraints of software running time, the amplitude spontaneous noise (ASE) has not been considered here. Therefore, the simulation of very small levels of optical signal to be amplified is not adequate. In order to obtain a further reduction in computational time, an analytical integration has been done to avoid the SOA longitudinal discretization. Regarding the simplicity of the model used here, is interesting to note that a true one-dimensional model has been implemented in order to include carrier transport in the longitudinal (optical propagation) direction and also multi-wavelength operation with ASE [10], but it needs many big matrices, much more computer memory and runs slower. Following the approach of Agrawal and Olson to exclude the longitudinal discretization, the optical gain is integrated over the SOA active region longitudinal length L. In this way, the optical output power is simply related to the input by the SOA optical total gain factor $h(t)$, where

$$h(t) = \int_0^L g(z,t)\, dz . \tag{2}$$

Therefore, the h evolution can be evaluated by a single differential equation [5,6]

$$\frac{\partial h(t)}{\partial t} = \frac{g_0 L - h(t)}{\tau_c} - \frac{P_{in}(t)}{E_s}\left(e^{h(t)} - 1\right) . \tag{3}$$

In addition, the carrier lifetime dependence with the carrier concentration is introduced by [6,11]

$$\tau_c = 1 \Big/ \Big(A + B\, N + C\, N^2 \Big).$$ (4)

In equation (4), the parameters A, B and C are, respectively, trapping, spontaneous and Auger recombination coefficients (see Table 1), and N is the carrier density in the SOA active region. The saturation energy E_s is related to the SOA dimensions and parameters by $E_s = h\, \omega_0\, w\, d / a\, \Gamma$ where: h is the Plank constant; ω_0 is the laser angular frequency; a is the transversal section gain; w and d are the active cavity width and height; Γ is the optical confinement factor. The SOA parameters shown in Table 1 are typical values for a bulk SOA operating at 1550 nm. The cavity length L of 0.677 mm has been experimentally obtained by measuring the distance between two adjacent lines of the spectral spontaneous emission noise.

Table 1. SOA modeling parameters

Parameter	Definition	Value
A	trap recombination factor	$5 . 10^7$ /s
B	spontaneous recombination factor	$5 . 10^{-10}$ cm^3/s
C	Auger recombination factor	$7.5 . 10^{-29}$ cm^6/s
w	cavity width	1.4 µm
d	cavity height	0.2 µm
L	cavity length	0.677 mm
a	transversal section gain	$1.0\ 10^{-16}$ cm^2
N_0	transparency carrier density	$1.3\ 10^{18}$ cm^{-3}
R	facet reflection	0.0001
n_{ef}	effective refractive index	3.4
E_s	saturation energy	1.2 pJ
Γ	confinement factor	0.4

3 Algorithm for the Parameters Extraction and Optimization

The parameters extraction are based on the analysis of the continuous wave (CW) optical gain versus the bias injected current curve ($G \times I$) of a commercial bulk-type SOA (*E-TEK*, HSOA) operating in three different optical input power levels. The main purpose is to find optimized values for simple and composed parameters necessary to the best SOA CW behavior fitting. The parameters are: the transparency current I_0; the SOA saturation power P_s; and a factor k given by $k = \Gamma a\, N_0$ where N_0 is the SOA transparency carrier density. The relation between the saturation energy and power is given by $P_s = E_s / \tau_c$. In addition, the parameter α_{int} is introduced here to consider the semiconductor attenuation in the SOA active region.

The first step in the algorithm is the determination of P_S and α_{int}. In order to do that, it is necessary to find the SOA operating currents I_X where the active region gain is compensated by its internal losses for three levels of optical input power. Therefore, assuming the device under these transparency conditions (bias current enabling an optical gain equivalent to the total cavity attenuation), the optical power P in the active region remains constant in the longitudinal direction, and assuming the optical carrier as a slow varying amplitude modulated carrier with optical power P it is possible to write [5,6,8]

$$\frac{\partial P}{\partial z} = (g - \alpha_{int}) P = 0 \ .$$

(5)

This behavior occur for a particular current $I = I_X$. In this case, the gain coefficient $g(z,t)$ is also constant in the z longitudinal direction, and equation (1) leads to

$$g(I,P) = k\left[(I/I_0)-1\right] \Big/ 1+(P/P_S) \ .$$

(6)

The output power inside the optical fiber is written as

$$P_{out} = \gamma_{cpl}^2 \, P_{in} \exp\left[(g - \alpha_{int})L\right] \ .$$

(7)

In the above equation, the parameter γ_{cpl} includes the coupling loss between the optical fiber end and the SOA active region. Also, the input and output SOA coupling losses are considered to have the same values. In addition, using (6) in (7), and taking the SOA overall gain in a dB scale,

$$G(I) = 20\log(\gamma_{cpl}) + 4.34 \, L\left\{\left[k\left[(I/I_0)-1\right]\Big/ 1+(P_{in}\,\gamma_{cpl}/P_S)\right] - \alpha_{int}\right\} \ .$$

(8)

Taking the first derivative of (8) in relation to the bias current,

$$\frac{\partial G}{\partial I} = 4.34 \, k \, L \Big/ I_0\left[1+(P_{in}\,\gamma_{cpl}/P_S)\right] \ .$$

(9)

The above expression is valid only for $I = I_x$. The SOA under test presents different values of I_x, for different values of the optical input power P_{in}. Considering the input power as a parameter for each curve $G \times I$, the values of I_x are taken from the experimental data by finding the point where $G(I_x) = 20\log(\gamma_{cpl})$, using a trial value for γ_{cpl}. Thus, it is possible to re-write (9) for the three different I_x values

$$G'_1\big|_{I=I_{x1}} = 4.34 \, k \, L \Big/ I_0\left(1+(P_{in1}\,\gamma_{cpl}/P_S)\right) \ .$$

(10.1)

$$G'_2\big|_{I=I_{x2}} = 4.34 \, k \, L \Big/ I_0\left(1+(P_{in2}\,\gamma_{cpl}/P_S)\right) \ .$$

(10.2)

$$G'_3\big|_{I=I_{x3}} = 4.34 \, k \, L \Big/ I_0\left(1+(P_{in3}\,\gamma_{cpl}/P_S)\right) \ .$$

(10.3)

The derivative terms in the left side of equations (10.1, 10.2 and 10.3) can also be evaluated numerically with the experimental data. Dividing (10.1) by (10.2), dividing

(10.2) by (10.3), and (10.3) by (10.1), three new set of equations are formed, allowing the calculation of the SOA saturation power P_s. In this way,

$$P_{s12} = \gamma_{cpl} \frac{P_{in2}\,G'_2 - P_{in1}\,G'_1}{G'_1 - G'_2} \,. \tag{11.1}$$

$$P_{s23} = \gamma_{cpl} \frac{P_{in3}\,G'_3 - P_{in2}\,G'_2}{G'_2 - G'_3} \,. \tag{11.2}$$

$$P_{s31} = \gamma_{cpl} \frac{P_{in1}\,G'_1 - P_{in3}\,G'_3}{G'_3 - G'_1} \,. \tag{11.3}$$

The calculation depends on a trial value for γ_{cpl}. The graphic on Figure 1 shows values of P_s calculated for trial values of $20 \log(\gamma_{cpl})$ within the range of -3 to -6 dB. It is clear to identify the convergence point (-4.85 dB and 1.16 dBm), as shown in Fig.1, which gives $\gamma_{cpl} = 0.57$ and $P_s = 1.3$ mW.

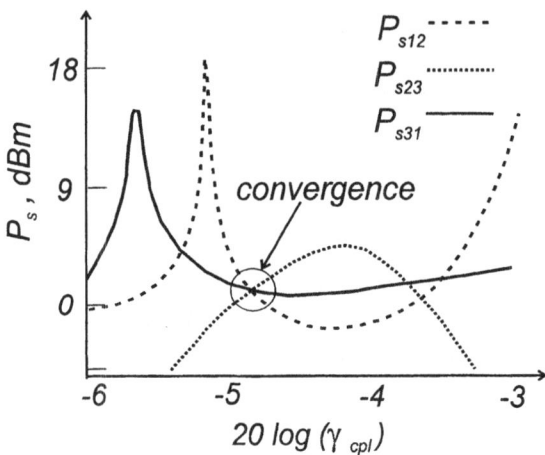

Fig. 1. The convergence point of the SOA power saturation versus 20 log of the coupling loss

Therefore, using equation (6) together with the condition $g(I_x, P) = \alpha_{int}$, leads to

$$I_0 = \frac{I_x}{1 + \dfrac{\alpha_{int}}{k}\left(1 + \dfrac{\gamma_{cpl}\,P_{in}}{P_s}\right)} \,. \tag{12}$$

This equation can also be evaluated for each of the three curves $G \times I$, and a convergence point of I_0 be searched using trial values for the factor α_{int} / k. For an average SOA, this factor is below 0.4 . Here, the value $\alpha_{int} / k = 0.2$ was obtained for the particular SOA used in the experiments. In addition, (9) can be expressed by

$$k\,L = \frac{G'(I_x)\,I_0\left(1+\dfrac{\gamma_{cpl}\,P_{in}}{P_s}\right)}{4.34}. \tag{13}$$

The value of k can be obtained by using I_0 from (12), the measured value of L, and taking the mean value of the results of (13) for the three curves $G \times I$. The internal loss coefficient was found directly using $\alpha_{int} = 0.2\,k$. Assuming the measured value of L, the above procedure gives good results for the parameters α_{int} and γ_{cpl}. To improve the accuracy of the remaining parameters, the simulated results was fitted with experimental data using the gradient method based on the minimization of a square error function of the difference between the experimental and the simulated SOA gain. In this way, a better estimation of k, I_0 and P_s was obtained. The results of the SOA parameters optimization are shown in Table 2.

Table 2. SOA optimized parameters

Parameter	Definition	Initial Value	Optimized Value
k	$k = \Gamma a\,N_0$	52 cm^{-1}	53.0 cm^{-1}
I_0	transparency current	30 mA	15.7 mA
P_s	saturation power	1 mW	1.28 mW
α_{int}	semiconductor attenuation	20 cm^{-1}	10.6 cm^{-1}
γ_{cpl}	coupling loss	0.5	0.57

4   Experimental Setup

The set-up to obtain the experimental data for the SOA parameters optimization is illustrated in Fig. 2. Only the CW SOA optical gain is necessary to obtain the optimized parameters of Table 2. However, the set up of Fig. 2 has been also used to obtain the SOA high frequency differential gain. The measurements were performed using an optical power meter (Anritsu MI910A; photo-detector MA302A) for the CW optical SOA gain. An Optical Network Analyzer (HP-8702B, with matched HP laser/detector) was used to measure the SOA high frequency differential gain, with signal sweep from 3 MHz to 6 GHz. The differential gain data was collected to a personal computer by GPIB interface and communication software. In addition, the SOA temperature was maintained around 25 ± 0.01 Celsius and the bias was provided using the internal Network Analyzer test set facility.

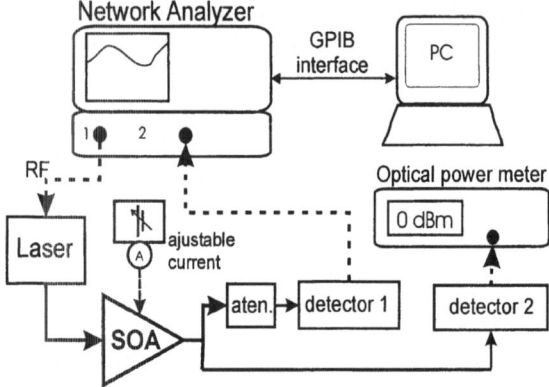

Fig. 2. Experimental setup for the CW and the high frequency differential SOA optical gain

5 Results

Figure 3 presents the experimental (points) data and simulated results for optimized parameters (lines) for the SOA CW gain, using three optical powers at the SOA input port: 0.63 mW, 1.6 mW and 2.4 mW. The good matching demonstrates the ability of the function error minimization process used here. It is interesting to note that the theory predicts the behavior at low values of the polarization current, where the SOA acts as an attenuator.

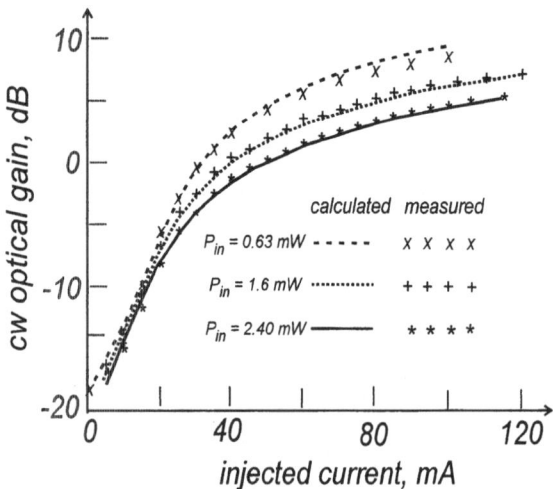

Fig. 3. Experimental and simulated data for the SOA CW optical gain

After the described optimization, the simulator [8] was used to calculate the high frequency differential gain behavior in order to compare with experimental data. The measured results are presented in Figure 4 and the theoretical results with the SOA parameters of Table 2 are presented in Fig. 5. Reasonable agreement can be found from 3 MHz until 6 GHz

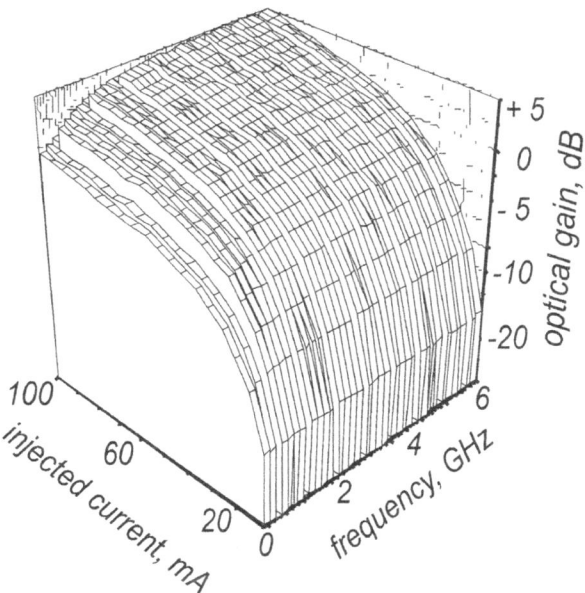

Fig. 4. Measured high frequency SOA differential gain for an optical input power of 2.4 mW

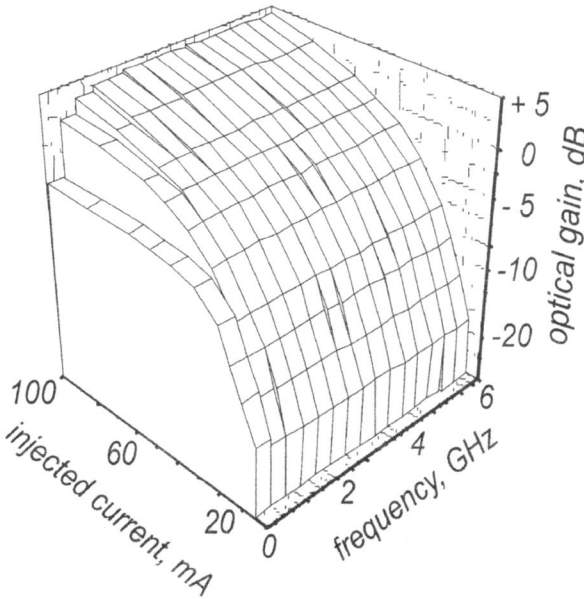

Fig. 5. Calculated high frequency SOA differential gain for an optical input power of 2.4 mW

The comparison of Figures 4 and 5 shows that the final error between corrected and measured value of the SOA differential gain are within 1 dB, with the exception of the results for very low modulation frequencies or very low polarization currents, where a higher discrepancy can be found. In those cases the SOA can even act as an attenuator. The discrepancy may be a consequence of the simple SOA model used here, since it does not take the amplitude spontaneous emission noise (ASE) in consideration. In the region of low values of optical power (or very low values of the SOA differential gain), ASE tends to deteriorate the optical signal to noise ratio and also the SOA gain. In the SOA used here, the measured noise figure at 1550 nm was 7.6 dB for a bias current of 100 mA.

Results have also been obtained for other carrier average powers. However, when the average optical power is low, the agreement between theory and experiments are not so good, due to the ASE effect commented above. Those effects can be computed using more sophisticated models that take ASE in consideration [10,12]. However, those models are difficult to implement and the parameters extraction involves a complex procedure.

6 Conclusions

A simple technique for a SOA modeling parameters estimation was presented based on the derivatives of the continuous wave gain curves in relation to the SOA longitudinal dimension. The gain curves are experimentally obtained for three levels of optical input power and for various values the polarization current. After the calibration, the model predicted with good accuracy the SOA continuous wave and the high frequency differential gain behavior, notwithstanding the modeling simplifications.

The authors acknowledge the partial financial support of CEPOF – Center in Photonics, FAPESP (Fundação de Amparo a Pesquisa do Estado de São Paulo), CNPq (Conselho Nacional de Desenvolvimento Científico e Tecnológico), CAPES (Coordenação do Aperfeiçoamento do Pessoal Docente) and MCT (Ministério de Ciência e Tecnologia).

References

1. Sun, Y., Srivastava, A.K., Banerjee, S., Sulhoff, J.W., Pan, R., Kantor, K., Jopson, R.M., Chraplyvy, A.R.: Error-free transmission of 32 x 2.5Gbit/s DWDM channels over 125km using cascaded in-line semiconductor optical amplifiers. Electron. Lett., Vol. 35 (1999) 1863-1865
2. van Thourhout, D., Bernasconi, P., Miller, B., Yang, W., Zhang, L., Sauer, N., Stulz, L., Cabot, S.: Novel Geometry for an Integrated Channel Selector. IEEE J. Select. Topics Quantum Electron., Vol. 8 (2002) 1211-1214
3. Takesue, H., Sugie, T.: Wavelength Channel Data Rewrite Using Saturated SOA Modulator for WDM Networks With Centralized Light Sources. J. Lightwave Technol., Vol. 21 (2003) 2546-2556
4. Morikuni, J.J., Kang, S.-M.: Computer-Aided Design of Optoelectronic Integrated Circuits and Systems. SPIE Optical Engineering Press (1997)

5. Agrawal, G.P., Olsson, N.A.: Self-Phase Modulation and Spectral Broadening of Optical Pulses in Semiconductor Laser Amplifiers. J. Quantum Electr., Vol.25 (1989) 2297-2306
6. Agrawal, G.P., Dutta, N.K.: Semiconductor Lasers, Van Nostrand Reinhold (1993)
7. Cartledge, J.C., Srinivasan, R.C.: Extraction of DFB Laser Rate Equation Parameters for System Simulation Purposes. J. Lightwave Technol., Vol. 15 (1997) 852-860
8. Gallep, C.M., Conforti, E., Custodio, J.B.P., Bordonalli, A.C., Ho, S.H., Kang, S.-M.: SOASim: a Simulator for Semiconductor Optical Amplifier with Feed Gain Control. Proceedings Microwave and Optoelectronics Conference, 1999. SBMO/IEEE MTT-S, APS and LEOS - IMOC '99, 9-12 Aug. (1999) 386-390
9. Conforti, E., Gallep, C.M., Bordonalli, A.C., Kang, S.-M.: Optical Regeneration Using Feed-Forward Semiconductor Optical Amplifier with Chirp Controlled Filtering. Microwave and Optical Technology Letters, Vol. 30 (2001) 438-442
10. Gallep, C. M., Conforti, E.: Simulations on Picosecond Non-Linear Electro-Optic Switching Using an ASE-calibrated Optical Amplifier Model. Optics Communication, Vol. 236/1-3 (2004) 131-139
11. Ghafouri-Shiraz, H.: Fundamentals of Laser Diode Amplifiers, John Willey & Sons (1996)
12. Conforti, E., Gallep, C.M., Bordonalli, A.C.: Decreasing Electro-Optic Switching Time in Semiconductor Optical Amplifiers by Using Pre-Pulse Induced Chirp Filtering. In: Mørk, J., Srivastava, A. (eds.): TOPS 92, Optical Amplifiers and Their Applications 2003, Optical Society of America, Vol. 92 (2003) 111-116

All-Optical Gain Controlled EDFA: Design and System Impact

J.C.R.F. Oliveira[1, 2], J.B. Rosolem[2], and A.C. Bordonalli[1]

[1] School of Electrical and Computer Engineering, State University of Campinas, P. O. Box
6101, 13083-970, Campinas, SP, Brazil
{oliveira, aldario}@dmo.fee.unicamp.br;
[2] CPqD, Rodovia Campinas-Mogi Mirim, km 118,5, Campinas, SP, 13086-902, Brazil
{julioc, rosolem}@cpqd.com.br

Abstract. An erbium doped fiber amplifier with automatic gain control was designed using the all-optical gain control technique. The amplifier performance
for different control channel wavelengths and feedback loop attenuation values
was experimentally investigated under static and dynamic conditions by measuring the bit error rate of a 2.5-Gb/s 8-channel DWDM system.

1 Introduction

In dense wavelength division multiplexing (DWDM) networks, the number of channels coupled into Erbium doped fiber amplifiers (EDFAs) can vary due to optical path
reconfiguration or network failure [1,2], affecting the gain response of the already
saturated amplifiers. The power variation resulting from the addition or removal of
channels at the EDFA input leads to oscillations in the gain levels that create power
transients with overshoots and undershoots in the amplified surviving channels, compromising system performance. To avoid system penalties, EDFA gain control techniques can be applied [3-7]. Compared to other techniques, the all-optical gain control presents several advantages such as optical domain operation, simple hardware,
and low cost. Gain variations are drastically minimized with all-optical gain controlled EDFAs (AOGC-EDFAs), but severe transient power fluctuations and gain
reduction are still possible if the control channel wavelength is incorrectly allocated
and the feedback loop attenuation is too high [5-6]. Therefore, careful gain control
design is required as both parameters are responsible for the control efficiency and
gain suppression in this type of amplifiers. In this work, an experimental investigation
of the system impact caused by the use of an AOGC-EDFA is presented. Firstly, an
EDFA with optical gain control was designed and characterized in terms of the feedback loop attenuation and control channel wavelengths. Then, the system impact
caused by the insertion of this gain controlled EDFA in a 2.5-Gb/s 8-channel DWDM
link was experimentally analyzed and quantified in terms of BER penalties. The BER
was measured during static and dynamic EDFA operation conditions, considering
different loop attenuation values and control channel wavelengths.

J.N. de Souza et al. (Eds.): ICT 2004, LNCS 3124, pp. 727–734, 2004.

2 All-Optical Gain Control Technique

Figure 1 shows the schematic diagram of an AOGC-EDFA [6-7]. The technique is based on a positive feedback loop where part of the EDFA output passes through a tunable optical filter and a variable attenuator before being coupled back into its input by coupler 1. This signal, along with the transmitted channels, is then amplified by the EDFA and once again sampled by the coupler 2 for a new feedback cycle. After several cycles, this control signal (control channel) acquires enough power to overcome losses and compete with the transmitted channels for the EDFA gain. As a result, assuming an appropriate loop attenuation level, the control channel power responds in accordance to the power fluctuations of the transmitted channels, maintaining the overall EDFA gain practically constant. Finally, by properly setting the optical filter center wavelength, the control channel can be formed from the EDFA amplified spontaneous emission (ASE) and tuned within the EDFA bandwidth.

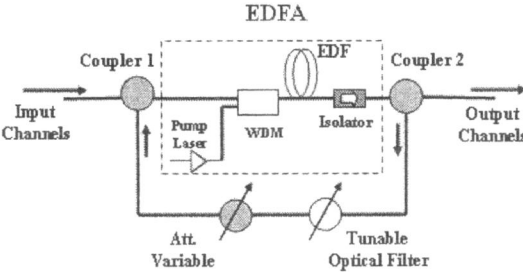

Fig. 1. Schematic diagram of an all-optical gain controlled EDFA. EDF: erbium doped fiber

An ordinary EDFA will present a deviation from a level of gain to another if its input power is dynamically changed. In an AOGC-EDFA, the control technique is designed to maintain the EDFA gain constant even during input power transitions. However, the amplifier is unable to respond instantaneously to these transitions due to the power transfers among the transmitted and control channels. Consequently, this leads to a fluctuation in the carrier population of the erbium metastable level, causing power oscillations (relaxation oscillations) in all channels during a period of time called transient time. In EDFAs, the transient time is of the order of hundreds of microseconds, decreasing as more amplifiers are added to the link.

Unfortunately, the relaxation oscillations are an intrinsic characteristic and main limiting factor of the all-optical gain controlled technique. The power variation caused by these oscillations can compromise receiver sensibility during high bit rate transmissions or induce photodetector saturation. Yet, the effect of the relaxation oscillations can be suppressed by an appropriate choice of the control channel wavelength and loop attenuation level. Another limitation factor in all-optic gain controlled EDFA is the spectral hole burning (SHB) [8], which can produce cross-gain modulation among closely spaced channels.

3   The Experimental Results

Figure 2 shows the experimental set-up schematic diagram. The all-optical gain control technique was incorporated to a commercial amplifier through a loop composed by two optical couplers, a tunable optical filter, and a variable attenuator. The transmitter was built with eight DFB lasers equally spaced from 1537.4 to 1558.5 nm. To maximize the effect of power transients in the EDFA, Channel 8 (1558.5 nm) was selected as the only surviving channel during add/drop experiments. Also, it was directly modulated by a 2.5 Gb/s pattern generator using a 2^{23}-1 pseudo-random binary sequence. The other lasers were externally modulated by the same generator and combined with Channel 8 by an optical coupler.

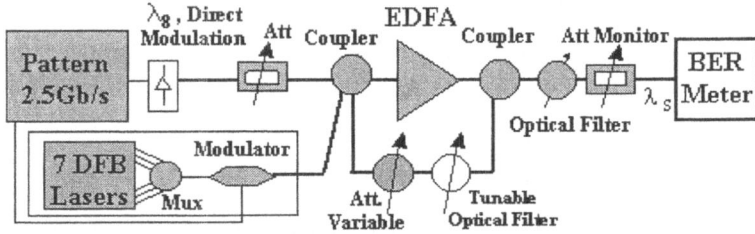

Fig. 2. Schematic diagram of the experimental set-up used in the all-optical gain controlled EDFA analysis

The impact of the use of the AOGC-EDFA over the performance of Fig. 2 set-up was statically and dynamically analyzed. To investigate the static case, no add/drop was induced and all channels were kept on. Sensitivity measurements were then conducted for four different control channel wavelengths (1535, 1548, 1554, and 1557 nm), with the loop attenuation set at 10 dB. For the dynamic case, only Channel 8 remained on to simulate the most severe seven-channel add/drop situation. Its optical power at the receiver was set to produce a 10^{-10} BER. After six BER measurements, the other seven channels were turned on and six new BER values were recorded for Channel 8. By using this procedure, it was possible to evaluate the penalties introduced in the surviving channel by the addition of channels. The measurements were repeated for each one of the control-channel wavelengths listed above and for different loop attenuation values to verify the effects caused by the control channel position and the strength of the optical feedback. For the channel-dropping situation, a similar measurement method can be adopted, where all channels are initially on before the Channel 8 power is set to produce a 10^{-10} BER at the receiver.

In order to illustrate the operation of the AOGC-EDFA, Fig. 3 shows the amplitude of the surviving channel photodetected signal for different values of loop attenuation, as the remaining 7 channels are added/dropped in intervals of 1 ms. The measured surviving channel power fluctuation was equivalent to 6.4 dB for operation close to the "off" gain-control situation (30 dB attenuation). However, when the loop attenuation was set to 1 dB (attenuator intrinsic loss), the channel power was controlled within 0.1 dB. Although a considerable decrease in power fluctuation was observed for lower attenuation levels, it is important to point out that this condition produces a stronger control channel, which tends to greatly reduce the gain of the transmitted channels. Therefore, the use of this gain control technique requires a compromise between the gain control precision and the transmitted channel gain.

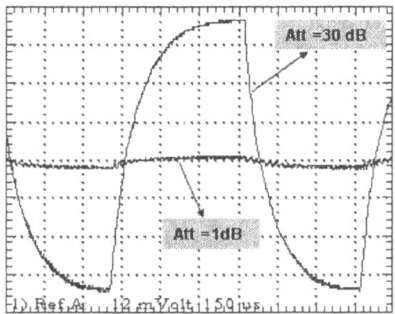

Fig. 3. Surviving channel photodetected signal amplitude during channel adding/dropping, having the loop attenuation as a parameter

Figures 4 (a) and (b) show the surviving channel gain and the surviving channel gain variation after channel dropping as a function of the loop attenuation, considering a -20 dBm/channel input power. In a system with either 4 or 8 channels, Fig. 4 (a) and (b), respectively, the surviving channel gain grows and reaches a constant value as the loop attenuation increases. This behavior is expected, since the control channel power (gain control) is gradually reduced and the EDFA can more efficiently amplify the transmitted channels. However, in case of channel dropping, Fig. 4 also shows that the surviving channel gain becomes progressively more susceptible to variations due to poor gain control. For a greater number of transmitted channels, Fig. 4 (b), the surviving channel gain suffers an even more distinct variation. In this case, the surviving channel gain before dropping is lower than that of the 4-channel example, suggesting that the amplifier is operating in a deeper saturation condition. Thus, as the control channel effect vanishes, the surviving channel gain varies more significantly with channel dropping due to greater EDFA gain sensitivity to power changes.

As mentioned before, the all-optical gain control technique requires a compromise between the gain control precision and the transmitted channel gain. In fact, Fig. 4 demonstrates that the loop attenuation plays a fundamental role in the gain performance of the EDFA. In this sense, the attenuation has to be set to values that allow high overall amplifier gain (high gain for each transmitted channel) and, at same time, a considerable reduction in the transmitted channel power variations. Unfortunately, it is impossible to ideally satisfy both conditions and the loop attenuation level must be determined based on the particular design requirements for each EDFA system application. For instance, Fig. 4 (a) shows that the surviving channel gain is 30.5 dB and its variation is lower than 1 dB for 17.5 dB loop attenuation. In order to obtain the same gain variation for the 8-channel example, Fig. 4 (b), the loop attenuation has to be reduced to 13 dB and the surviving channel gain drops to around 28 dB.

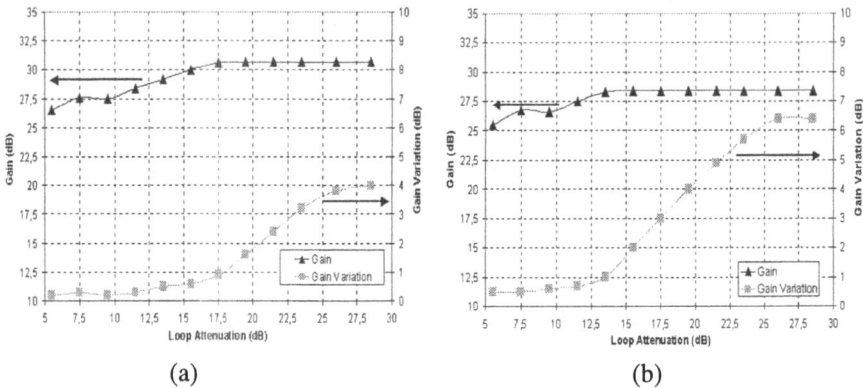

(a)   (b)

Fig. 4. Surviving channel gain and surviving channel gain variation as a function of the loop attenuation, considering (a) 4 and (b) 8 channels and an input power of -20 dBm/channel

By considering the static case analysis, Fig. 5 shows the receiver sensitivity measured for different wavelength control channels. It can be seen that an increase in the control channel wavelength causes a growth in the BER penalty. The penalty increases rapidly for wavelengths over 1554 nm. Synchronism is completely lost when the control channel reaches 1557 nm (close to the surviving channel wavelength). Therefore, the control channel wavelength must be carefully designed to avoid system penalties. In this particular example, the surviving channel was chosen to be Channel 8. However, Fig. 5 would no longer be valid if the surviving channel were set to be, for instance, Channel 1 (1537 nm). Further measurements showed that it is necessary at least a 10 nm distance between control and surviving channel wavelengths to keep low power penalties (around 1 dB) at 10^{-9} BER. These same measurements indicated that good overall gain control performance could also be observed if the control channel wavelength is positioned near the amplifier ASE peak (around 1532 nm) while the DWDM channels operation is concentrated within the flat gain region of the EDFA.

To illustrate the distortions caused by the control channel wavelength sweeping from 1535 nm to 1557 nm, Fig. 6 (a) to (d) present, respectively, the surviving channel received bits for a low transmission rate (52 Mb/s). For a 1535 nm operation, no significant oscillation is present. The same happened for the 1548-nm control channel operation, however, with bit amplitude decrease. The bit amplitude was even lower for 1554-nm operation and oscillations started to become apparent [9]. Finally, for a 1557-nm control wavelength, the bits presented oscillations and distortions that compromised proper receiver decisions, in a way similar to that verified during the sensitivity measurements at 2.5 Gb/s. The amplitude suppression in Fig. 6 (b) to (d) is related to a grater surviving channel gain suppression due to the proximity with the control channel wavelength. Fig. 7 (a) to (d) show the dynamic analysis results for the surviving channel, considering different control channel wavelengths and loop attenuation values.

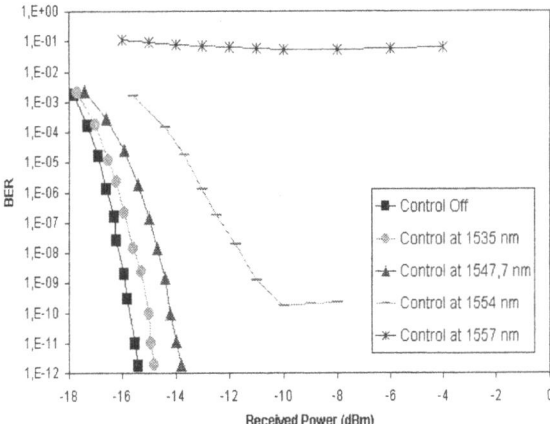

Fig. 5. BER versus received optical power, having the control channel wavelength as a parameter. Loop attenuation: 10 dB

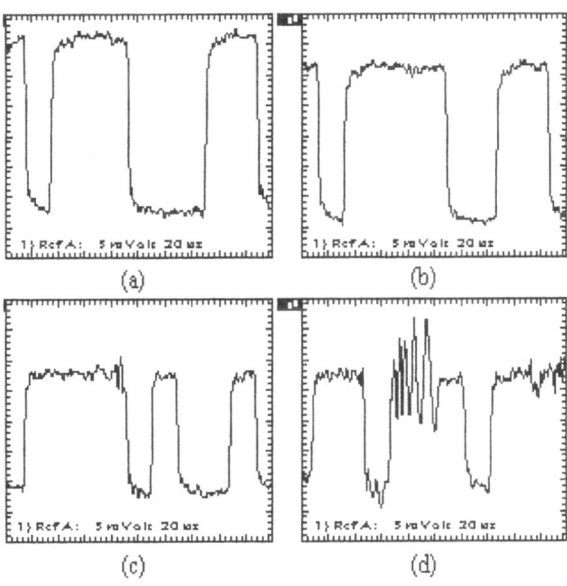

Fig. 6. Surviving channel received bits for a 52 Mb/s transmission, assuming different control channel wavelengths: (a) 1535, (b) 1548, (c) 1554, and (d) 1557 nm. Loop attenuation: 10 dB

In Fig. 7 (a) and (b), it is possible to observe the gain control efficiency by measuring the BER variation during channel adding with no gain control (*loop off*) and active control (*loop on*). Both plots show that, after channel adding, there is a reduction of around nine orders in the BER magnitude when gain control with 1-dB loop attenuation is active. With gain control operation, a surviving channel gain reduction

of 6 dB was measured when the loop attenuation changed from 10 to 1 dB after channel adding. Fortunately, the equivalent system impact was considerably low (BER
variation of around one order of magnitude), indicating that the loop attenuation can
be adjusted to provide low gain reduction with a small system impact. By operating
the control channel at 1554 nm and 1557 nm, Fig. 7 (c) and (d), respectively, it was
not possible to align the BER at 10^{-10}, as already suggested by Fig. 5. Similar results
and conclusions were found after a similar analysis for channel dropping.

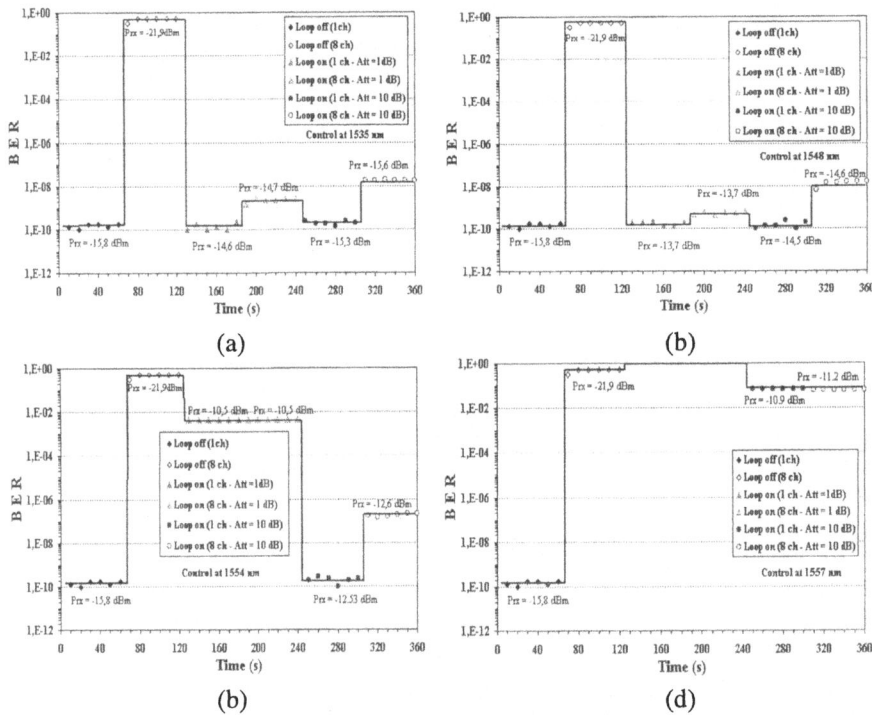

Fig. 7. Dynamic analysis of the BER when the gain control is off, on with 1 dB loop attenuation, and on with 10 dB loop attenuation with control channel at 1535 nm.(a), 1548 nm (b),
1554 nm (c) and 1557 nm (d). Prx: optical power at receiver.

Figure 7 (a) shows that channel adding without loop gain control implies on a ten-
order-of-magnitude BER penalty. If gain control is active and loop attenuation is low
(1 dB), the penalty is reduced to one order of magnitude, but 6 dB gain suppression is
imposed to the EDFA in comparison to the gain for 10 dB loop attenuation. However,
by setting the loop attenuation at 10 dB, the BER penalty increases by two orders of
magnitude. Thus, the loop attenuation level must be properly chosen to guarantee
both a high stabilized amplifier gain and low system penalties. Similar conclusions
can be obtained from Fig. 7 (b). Above 1548 nm, Fig. 7 (c) and (d), the proximity
between control and surviving channels severely compromises the system performance.

4   Conclusions

In this work, the requirements for the design of an AOGC-EDFA were investigated and the system impact caused by its insertion in a 2.5-Gb/s 8-channel DWDM system was analyzed for different control channel wavelengths and loop attenuation values. It was observed that the all-optical gain control technique imposes a compromise between the gain control precision and the transmitted channel gain, forcing the loop attenuation and the control channel wavelength to be set in accordance with the design requirements for a particular EDFA system application. In addition, it was verified that the control channel wavelength position must be at least 10 nm apart from that of the surviving channel to guarantee low system penalties. Further measurements also showed that good gain control performance could be achieved if the control channel wavelength is located near the EDFA ASE peak while the DWDM channels operate within the flat gain region of the EDFA. This work was supported by CAPES, CNPq, CePOF/FAPESP FAEP/UNICAMP, and CPqD Telecom & IT Solutions, Brazil.

References

1. Kuwano S.,Wematsu H.: Two-fiber unidirectional OADM ring system for L-band. Proceedings NFOEC (2000) 74-85
2. Ono H., Shimizu M.: Analysis of gain dynamics of Erbium-doped fiber amplifiers for wavelength-division-multiplexing networks. IEEE J. Quantum Electron. 39 (2003) 541-547
3. Jolley N., Davis F., Mun J.: Out-of-band electronic gain clamping for a variable gain and output power EDFA with low dynamic gain tilt. Proceedings OFC (1997) 134-135
4. Park S. Y., Kim H. K., Lyu G. Y., Kang S. M., Shin S-Y.: Dynamic gain and output power control in gain-flattened Erbium-doped fiber amplifier. IEEE Photon. Technol. Lett. 10 (1998) 787-789
5. Luo G., Zyskind J. L., Sun Y., Srivastava A. K., Sulhoff J. W., Ali M. A.: Relaxation oscillations and spectral hole burning in laser automatic gain control of EDFA's. Proceedings OFC (1997), WF4.
6. Chung J., Kim S. Y., Chae C. J.: All-optical gain-clamped EDFAs with different feedback wavelengths for use in multiwavelength optical networks. Electron. Lett. 32 (1996) 2159-2161
7. Richards D., Jackel J., Ali M.: A theoretical investigation of dynamic all-optical automatic gain control in multichannel EDFAs and EDFAs cascades. IEEE J. of Select. Topics Quantum Electron. 3 (1997) 1027-1036
8. Luo G., Zyskind J. L., Nagel J. A., Ali M. A.: Experimental and theoretical analysis of relaxation-oscillations and spectral hole burning effects in all-optical gain-clamped EDFA's for WDM networks. J. Lightwave Technol. 16 (1998) 527-533
9. Luo G., Zyskind J. L., Sun Y., Srivastava A. K., Sulhoff J. W., Wolf C., Ali M. A.: Performance degradation of all-optical gain-clamped EDFA's due to relaxation-oscillations and spectral-hole burning in amplified WDM networks. IEEE Photon. Technol. Lett. 9 (1997) 1346-1348

Iterative Optimization in VTD to Maximize the Open Capacity of WDM Networks

Karcius D.R. Assis, Marcio S. Savasini, and Helio Waldman

DECOM/FEEC/UNICAMP, CP. 6101, 13083-970 Campinas, SP-BRAZIL
Tel: +55-19-37883793, FAX: +55-19-32891395
{karcius, savasini, waldman}@decom.fee.unicamp.br

Abstract. We propose an iterative algorithm for the Virtual Topology Design (VTD) in optical networks. The algorithm eliminates lighter traffic lightpaths and re-arrange the traffic through the remaining lightpaths. This tries to preserve the open capacity for the accommodation of future unknown demands. The results suggest that it is feasible to preserve enough open capacity to avoid blocking of future requests in networks with scarce resources; this is, maximize the traffic scaling.

1   Introduction

We are witnessing a new evolution of WDM networks today, as a consequence of a big change in the application scenario. Data traffic is going to overcome traditional telephone traffic in volume: statically modeling of network load has to be modified to describe a new reality with less regular flows, more and more independent from geographical distances. The change is also reflected by the evolution of WDM protocol standardization. The simple static Optical Transport Network (OTN) is already well-defined by the main standard bodies, while the new model know as Automatic Switched Optical Network (ASON) is currently under development. Its main feature is the ability to accommodate on-line connection request issued to the network operating system, which is responsible of the activation of new lightpaths in real time [1], [2].

It is however likely that the evolution from OTN to ASON is going to happen as a gradual process in order to preserve the investments of the network operators. In this transition phase static and dynamic traffic will co-exist and share the same WDM network infrastructure [1].

In general, the network design problem in static phase can be formulated as an optimization problem aimed at maximizing network throughput or other performances measures of interest. Typically, the exact solution can be shown to be NP-hard, and heuristic approaches are needed to find realistic good solutions. For this purpose, the problem can be decomposed into two subproblems [5]. The first is to decide what virtual topology to embed on a given physical topology, that is, what are the lightpaths to be implemented, as seen from the client layer: this is the virtual topology design (VTD) or lightpath topology design (LTD) problem. The second is the routing–and-wavelength assignment (RWA) for these lightpaths at the physical layer.

J.N. de Souza et al. (Eds.): ICT 2004, LNCS 3124, pp. 735–742, 2004.

The routing of packet traffic on the lighpaths is also usually seen to be a part of the VTD problem, since its objective function is usually some parameter function of the traffic routing.

Ideally in a network with N nodes, we would like to set up lightpaths between all the $N.(N-1)$ pairs. However this is usually not possible because of two reasons. First, the number of wavelengths available impose a limit on how many lightpaths can be set up. (This is also a function of the traffic distribution). Secondly, each node can be source and sink of only a limited number of lightpaths Δ. This is determined by amount of optical hardware that can be provided (transmitters and receivers) and by the amount of information the node can handle in total [4].

This paper studies a particular situation in which the new on-line connection requests are generated as an expansion of the original static traffic. The key point is the iterative formulation of the well-known VTD problem for the elimination of the least congested lightpaths λ_{min} in the virtual topology, allowing a degradation of the objective function in VTD until one predefined bound C. In other words, we have used a iterative (modified) VTD algorithm to configure the lightpaths, which is oriented towards preservation of open capacity for the accommodation of future unknown demands [6], [7].

The maximum congestion λ_{max} determine the viability of the solution in iterative routing VTD. If after eliminating the lightpath with traffic λ_{min} and re-arrange the traffic through the remaining lightpaths the λ_{max} is supported by the system, i.e., $\lambda_{max} < C$, then the solution is feasible. Whenever this happens, the rollback to the routing problem of the virtual topology asks for the next best solution, thus implying objective function degradation.

Section 2 describes the VTD and its main variables. In section 3 we show the iterative VTD proposed and simulations for a small network. Section 4 shows simulations for a hypothetical large network over Brazilian territory and in section 5 we finish the paper with some conclusions.

2   Static Problem Statement

Although lightpaths underlie SDH networks in a natural way, cell- and packet-switching client networks, like ATM and IP, they would be better served by more packet-oriented WDM layer mechanisms and protocols. However, current optical packet-switching technologies do not yet deliver the same performance that is possible in electronic networks. For this reason, lightpath provisioning is still the best available service the optical layer can now offer to its client layers, including those that would probably benefit from optical packet switching, such as IP. However, as IP becomes the dominant client in most environments, the integration between the control planes of the client (IP) and server (optical) layers becomes attractive, generating the so called peer-to-peer framework.

2.1 Physical Topology

A physical topology (G_p) is a graph representing the physical interconnection of the wavelength routing nodes by means of fiber-optic cables. Fig.1.(a) shows a physical topology of a six-node wide-area network. The wavelength routing nodes are numbered from 0 to 5. We consider an edge in the physical topology to represent a pair of fibers, one in each direction.

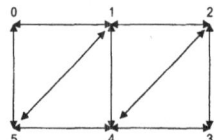   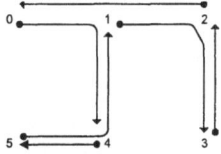

Fig. 1. (a) physical topology (Gp)   (b) virtual topology (Gv)

2.2 Virtual Topology

Let $T = (\lambda^{sd})$ be the traffic matrix, i.e., λ^{sd} is the arrival rate of packets (or Gb/s) at s that are destined for d. We try to create a virtual topology G_v and route the given traffic in this G_v minimizing $\lambda_{max} = \max_{ij} \lambda_{ij}$ where λ_{ij} denotes the offered load on link (i,j) of the virtual topology. λ_{max} is the maximum offered load to a virtual link and is called the *congestion*. Let G_p be the given physical topology of the network, Δ the degree of the virtual topology, and W the number of wavelengths available. An informal description of the virtual topology design problem is as follow (a precise definition as a mixed-integer linear program (MILP) is given in [5]):

$$\text{Min } \lambda_{max}$$

Such that:

- Each virtual link in G_v corresponds to a lightpath and two lightpaths that share an edge in the physical topology are assigned different wavelengths.
- The total number of wavelengths used is at most W.
- Every node in G_v has Δ incoming edges and Δ outgoing edges.
- Traffic is routed so that flow of traffic from each source-detination pair is conserved at each node.

Note that the topology design problem includes routing as a subproblem. The set of all unidirectional lightpaths set up among the access nodes is the virtual topology G_v or lightpath topology. For example, Fig.1.(b) shows a possible virtual interconnection with $\Delta = 1$, in which arrows are bent so as to show their physical routing. Notice, however, that physical routing of the lightpaths is actually not visible in the virtual or lightpath topology. There is an edge in the virtual topology between node 2 and node 0 when the data or packets from node 2 to node 0 traverse the optical network in the optical domain only, i.e., undergo no electronic conversion in the intermediate wavelength routing nodes. Edges in a virtual topology are called virtual links, and are defined only by their source and destination nodes (i.e., they are shapeless, or straight, arrows, unlike in Fig.1.(b). Note that to send a packet from node 2 to node 4 we would have to use two virtual links (or lightpaths) *2-0* and *0-4*. The logical connection would then use two virtual hops.

3   Iterative Virtual Topology Design

In this section we propose a loose topology, i.e., a multi-client physical topology that must accommodate both a static and a dynamic traffic demands. An iterative formulation of the well-known VTD problem is used for the elimination of the least congested lightpaths, i.e., with   traffic λ_{min} in the virtual topology, allowing a degradation of the objective function in VTD until one predefined bound C. After, all available wavelengths may be used to solve the static RWA problem. The wavelengths are then re-used to set up lightpaths adaptively to dynamic traffic demands by dynamic RWA. Blocking probability of dynamic path requests is to be minimized while allowing a degradation of the objective function in to the static traffic demand. The objective function degradation in VTD until system supported capacity orients towards preservation of open capacity for the accommodation of future unknown demands.

3.1 Heuristic Algorithm for Iterative VTD Problem

Here we present a heuristic algorithm based on VTD design:
Step 1: Given a static traffic matrix and the system capacity C, find the virtual links (original G_v) ;
Step 2: Route the given static traffic on the virtual links and find λ_{max};
Step 3: (If $\lambda_{max} \geq C$ and it's the first iteration, then stop. Or if $\lambda_{max} \geq C$ and any virtual links has already been removed, go to step 6). Else continue;
Step 4: Remove the least congested virtual link (λ_{min});
Step 5: If all nodes remain virtually connected, return to the step 2. If any node becomes disconnected, continue.
Step 6: Re-add the last removed virtual link, so that the last network state is recovered, and continue;
Step 7: Solve the RWA problem;
Step 8: Find the blocking probability for a future dynamic traffic.

Table. 1. Traffic matrix for 6-node network, [5]

-	0,537	0,524	0,710	0,803	0,974
0,391	-	0,203	0,234	0,141	0,831
0,06	0,453	-	0,645	0,204	0,106
0,508	0,660	0,494	-	0,426	0,682
0,480	0,174	0,522	0,879	-	0,241
0,950	0,406	0,175	0,656	0,193	-

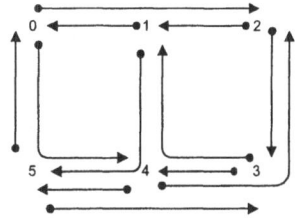

Fig. 2. Original virtual topology

3.1.1 Example

We solve the above heuristic for the 6-node network shown in Fig.1.(a), for a $\Delta=2$ with linear programming, using the routine CPLEX in static phase. The traffic matrix used is shown in table 1 and is the same as in [5].

With C=5 Gb/s, there are 5 possible virtual topologies {(a),(b),(c),(d) and (e)}. The table (a) is the original virtual topology (Fig.2) with λ_{max} =2,04 and λ_{min}=1,4 in virtual link 5-0. The (e) is optimized topology (Fig.3) with λ_{max} =4,96 and λ_{min} =2,28. Note that (f) is not viable topology, because virtual link 0-4 would have λ_{max} =7,04 and the system capacity is 5, besides one more elimination would lead to a node disconnection. After, the static RWA problem was solved for original and optimized topologies with shortest path routing and link load minimization.

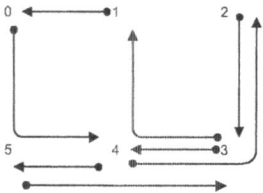

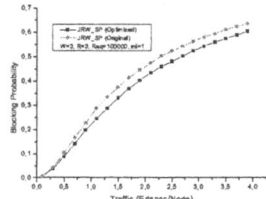

Fig. 3. Optimized virtual topology   **Fig. 4.** Blocking probability for future demands

-		2,04		2,04	
1,53	-				1,85
	1,77	-	1,87		
	2,04		-	1,51	
		2,04		-	2,04
1,4			2,04		-

(a) Original Topology

-		2,07		1,78	
2,69	-				**1,38**
	1,81	-	2,5		
	2,69		-	2,15	
		2,69		-	1,76
			2,69		-

(b) 1 lightpath eliminated

-		2,35		2,61	
3,8	-				
	1,94	-	3,79		
	2,32		-	3,8	
		3,8		-	3,14
			2,68		-

(c) 2 lightpaths eliminated

-		**1,44**		3,52	
3,79	-				
		-	4,23		
	4,23		-	3,1	
		3,2		-	3,9
			3,4		-

(d) 3 lightpaths eliminated

-				4,96	
3,8	-				
		-	3,4		
	4,2		-	**2,28**	
		3,88		-	3,89
			3,43		-

(e) Optimized Topology

-				7,24	
6,08					
				4,48	
	6,5				
		4,9			2,8
			2,4		

(f) Inviable Topology

Fig. 5. Tables (Iterative VTD)

3.1.2 Dynamic Phase

In dynamic phase, several first-fit algorithms have been studied and compared in the literature [3, Chap. 8]. The simplest one uses an *a priori* wavelength list: the algorithm will then look up the list and pick the first wavelength under which the path can be accommodate. This will be called the *fixed priority algorithm* (FP). Other algorithms favor the use of the wavelength that is being most used in the network at assignment time.

A good performance has been observed with the MaxSum (MS) algorithm, which chooses the wavelength which minimizes the number of routes that will become blocked with the wavelength assignment.

Combining the decision rules of MS with route allocation algorithms leads to a JRW (*Joint Routing and Wavelength Assignment*) algorithm, hence in this case the algorithm must compare, according to some criterion, all the *route-wavelength* pairs.

In this example, for dynamic environments, JRW algorithms were used in which a shortest path route is always chosen, i.e., JRW_SP. In Fig. 6, note that in dynamic phase the optimized topology rises the open capacity of the network in function the heuristic strategy.

4   Large Optical Networks

In order to study of a network with large number of nodes approaching known data, the 12-node hypothetical network of Fig. 6 was considered. Each node corresponds to one of 12 states in Brazil, chosen for their economic regional importance.

A plausible mesh was assumed for the physical links. In this network each edge represents a pair of directed edges. The distances among nodes were based on the geographical positions of the State capitals. The traffic matrix (Table 2) has to be released over the Brazilian Network and was obtained through the following expression (1) following (next page).

The utilization factor F_i is some measure of the use of Telecommunications in state i. In the network under study, it will be the level of services of telecommunications requested by a state. Observe that equation (1) implies a symmetrical matrix (Table 2).

We studied the network with $W=2$ or $W=3$. A certain virtual degree $\Delta=2$ is initially assigned for the static configuration, (MILP formulation [5]) and $R=2$ is the remaining virtual degree could be used for a dynamic traffic.

Using the proposed algorithm with system capacity $C=10$Gb/s, there are 7 possible virtual topologies, the original with $\lambda_{max}=6,5$ and the last (optimized) with $\lambda_{max}=9,955$.

In Fig. 7, note that for $W=2$ and $R=2$ using the JRW_SP in dynamic phase the best topology in relation to the open capacity is the optimized one. It generates less blocking probability for dynamic traffic. For heavier traffic the optimization gain tends to be smaller, because traffic will be the key factor for blocking. For lighter traffic the used resources in the static phase play a major role for blocking.

However, if we increase the number of available wavelengths in the network to $W=3$, the blocking probabilities for the optimized and the original topologies get similar (Fig.8) for lighter or heavier traffic demands.

This suggests that optimization in the static phase of the VTD is required if the system has scarce resources (for example, only a few wavelengths).

$$T_{ij} = k\frac{P_i F_i P_j F_j}{d_{ij}} \quad (1)$$

Where :

P_i	Population of State i
P_j	Population of State j
F_i	Utilization factor in State i
F_j	Utilization factor in State j
d_{ij}	Distance between States i e j
K	Proportionality factor

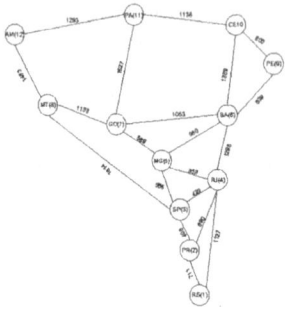

Fig. 6. Hypothetical network

Table 2. Matrix Traffic for Brazilian Network (Gb/s)

.	0.37	1.00	0.39	0.32	0.08	0.08	0.03	0.02	0.03	0.01	0.04
0.37	.	2.56	0.60	0.51	0.11	0.09	0.04	0.03	0.04	0.01	0.03
1.00	2.56	.	4.08	3.72	0.65	0.36	0.16	0.12	0.16	0.05	0.14
0.39	0.60	4.08	.	2.42	0.22	0.22	0.07	0.05	0.06	0.02	0.04
0.32	0.51	3.72	2.42	.	0.54	0.35	0.12	0.10	0.08	0.04	0.08
0.08	0.11	0.65	0.22	0.54	.	0.10	0.04	0.03	0.03	0.02	0.03
0.08	0.09	0.36	0.22	0.35	0.10	.	0.10	0.04	0.04	0.01	0.01
0.03	0.04	0.16	0.07	0.12	0.04	0.10	.	0.04	0.02	0.01	0.01
0.02	0.03	0.12	0.05	0.10	0.03	0.04	0.04	.	0.04	0.01	0.01
0.03	0.04	0.16	0.06	0.08	0.03	0.04	0.02	0.04	.	0.02	0.01
0.01	0.01	0.05	0.02	0.04	0.02	0.01	0.01	0.01	0.02	.	0.01
0.04	0.03	0.14	0.04	0.08	0.03	0.01	0.01	0.01	0.01	0.01	.;

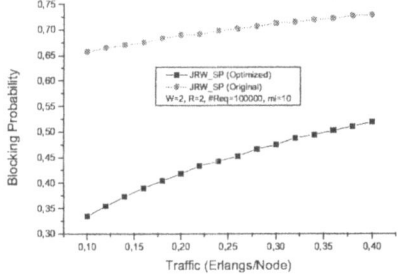

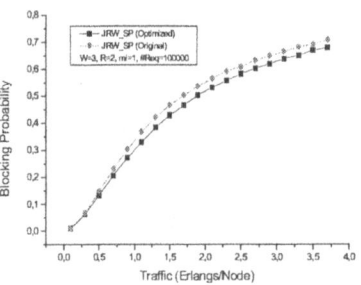

Fig. 7. Blocking probability for future demands

Fig. 8. Blocking probability for future demands

4.1 Some Comments

In static phase, Integer Linear Programming models are popular in the literature as they provide formal descriptions of the problem. In practice, however, scalability to networks with at least 10´s of nodes, with 100´s of demands is required. In many cases all but trivial instances of theses ILP´s are computationally difficult with current state-of-the-art software. In our 12-node mesh network, our strategy took about 2 hours using the optimization software CPLEX© on an Intel Pentium IV/1.6Ghz to find the original topology in the static phase. However, each iteration took around one second, because it should only re-route the traffic through the remaining lightpaths. Only linear constraints had to be handled.

5   Conclusions

The issue we intended to investigate was whether the set of resources left idle after an iterative optimization procedure on the network for a particular static traffic can be used to accommodate a traffic expansion. The approach proposed a maximization of open capacity for future online demands.

The results suggest that it is feasible to preserve enough open capacity to avoid blocking of future requests in networks with scarce resources; this is, maximize the traffic scaling. Some resources like transmitters and receivers can save on initial costs (network implementation), since the optimization eliminates some lightpaths, reducing the virtual degree at some nodes for the static phase.

Acknowledgments. This work was supported by FAPESP, CNPq and by the Research and Development Center, Ericsson Telecomunicações S.A., Brazil.

References

[1]   L. Barbado, G. Maier, "Maximum Traffic Scaling in WDM Networks Optimized for an Inicial Static Load", ONDM 2003, Vol.1pp. 41-60. February 3-5, 2003. Budapest, Hungary.

[2]   E. Ozdalgar, D.P. Bertsekas, "Routing and Wavelength Assisnment in Optical Networks IEEE/ACM transactions on networking, vol.11, n.2. april 2003.

[3]   R. Ramaswami and K.N. Sivarajan , "Optical Networks: A Practical Perspective" Morgan Kaufmann Publishers, 1998.

[4]   R. Ramaswami and K.N. Sivarajan, "Routing and Wavelength Assignment in All-Optical Networks", *IEEE/ACM Transactions on Networking*, pp. 489-500, Oct. 1995.

[5]   R. Ramaswami and K.N.Sivarajan, "Design of logical topologies for wavelength-Routed All Optical Networks", IEEE/JSAC, vol. 14, pp. 840-851, june 1996.

[6]   M. Kodialam, T. V. Laksshman, "Integrated Dynamic IP and Wavelength Routing in IP over WDM Networks", in Proc. IEEE INFOCOM , 2001.

[7]   K.D.R. Assis, H. Waldman, " Approaches to Maximize the Open Capacity of Optical Networks", the 8th IFIP Working Conference on Optical Network Design & Modelling (ONDM'04), vol.1 pp. 599-618.

Link Management Protocol (LMP) Evaluation for SDH/Sonet

Pedro Uria Recio[1], Peter Rauch[2], and Koldo Espinosa[3]

[1] Lucent Technologies Network System GmbH (Diploma Thesis)
Camino de Iturrigorri 21 2B 48002 Bilbao, Spain
uriarecio@gmx.net

[2] Lucent Technologies Network Systems GmbH,
Thurn-und-Taxis-Strasse 10, D-90411 Nuremberg, Germany
prauch@lucent.com

[3] University of the Basque Country. Dpto. de Electrónica y Telecomunicaciones
Alda. Urquijo s/n - 48013 – Bilbao, Spain
jtpesacj@bi.ehu.es

Abstract. Next generation optical networks will operate under a common control plane, called Generalized Multiprotocol Label Switching (GMPLS), which simplifies network management and operation of equipment that switches in either packet, time, wavelength or fiber domain. Link Management Protocol (LMP) is a new internet-draft protocol and represents a fundamental member of the GMPLS protocol family. LMP runs between adjacent nodes and eases the provisioning of bearer links. An LMP protocol machine has been created in order to evaluate the LMP protocol itself as well as to extract performance conclusions in network scenarios. This LMP protocol machine was designed for the Lucent Technologies LambdaUnite® MultiService Switch (MSS). The application is distributed on a bi-processor host according to the Half Object Plus Protocol (HOPP) design pattern. A critic about several improvable aspects of the LMP specifications has been made. This paper is exclusively focused on LMP from the SDH/Sonet point of view.

1 Introduction

In next-generation optical networking panorama a wide range of devices of very different technologies interact with each other. Packet routers, switches, add-drop multiplexers (ADM), digital crossconnects (DXC), Dense Wavelength Division Multiplexed (DWDM) systems or even newer equipment such as optical crossconnects (OXC) compose complex multi-vendor networks throughout the world, in which both standard and proprietary protocols are deployed.

Automating resource management and end-to-end provisioning based on traffic engineering (TE) metrics for all these devices would reduce configuration costs by several orders of magnitude. Furthermore it would make it possible to provide new services such as bandwidth on demand, load balancing or Quality of Service (QoS) for sophisticated applications.

J.N. de Souza et al. (Eds.): ICT 2004, LNCS 3124, pp. 743–752, 2004.
© Springer-Verlag Berlin Heidelberg 2004

To achieve these goals a new standardized common control plane for all equipment is necessary, no matter which switching technology is involved: Packet Switching (ATM, IP), Time Division Multiplex (TDM) like SDH/Sonet, DWDM/Lambda Switching or Fiber Switching. Only under a common control plane can the concept of intelligent optical network come true.

To deal with this challenge, the Internet Engineering Task Force (IETF) is standardizing the Generalized Multiprotocol Label Switching (GMPLS) [1] protocol family. GMPLS consists of enhancements to the Open Shortest Path First / Intermediate System to Intermediate System (OSPF/ISIS) routing protocols, extensions to the Resource Reservation Protocol / Routing-Label Distributed Protocol (RSVP/LDP) and finally a new Link Management Protocol (LMP) [2].

GMPLS addresses issues concerning the switching and forwarding diversity of the network, fast auto-configuration, scalability and efficient resource usage, reliability based on protection and restoration techniques and finally data routing according to the availability of resources, the current and the expected traffic

On the other hand the International Telecommunication Union (ITU) develops the Architecture for Automatically Switched Optical Networks (ASON) [3]. ASON is not a protocol collection but an architecture that defines control plane components and the interactions between them.

The Optical Internetworking Forum (OIF) works on the User Network Interface (UNI) [4], which allows client devices (IP routers or ATM switches) to request services from the optical core network. UNI is a fusion of high priority ASON requirements with a profile of GMPLS protocols.

2 LMP Features

LMP provides an IP Control Channel that is allowed to be physically diverse from the bearer links. This is important for transparent photonic switches, which cannot monitor or inject signals in the bearer links.

Next-generation devices will be interconnected by thousands of bearer links (Data Links). This involves a serious scalability problem. To address this issue, an LMP procedure called Link Property Correlation (LPC) matches link parameters between adjacent nodes and aggregates Data Links with similar attributes into a single bundle called Traffic Engineering (TE) Link. This process is called Link Bundling. TE Links, along with their aggregated attributes, are advertised into the routing protocols. As a result, the size of the link-sate database is reduced by a large factor as well as the number of messages exchanged in signaling or routing protocols. However the increased abstraction level results in loss of granularity in the network resources.

LMP Link Connectivity Verification (LCV) checks the physical connectivity of Data Links by sending in-band Test messages over them.

LMP Fault Management (FM) localizes link failures and inhibits unnecessary alarms in either transparent or opaque networks. However LMP cannot compete with the 50 ms protection switch times of SDH/Sonet.

LMP has been extended by the OIF to provide UNI Neighbor and Service Discovery between UNI clients (UNI-C) and network (UNI-N) devices.

Other extensions [5] enable information exchanges between LMP nodes and DWDM Optical Line Systems (OLS) attached to them.

LMP enhancements to fit SDH/Sonet legacy equipment [6] ease Link Connectivity Verification by allowing sending out-of-band Test messages over the Control Channel. The method is based on the correlation of certain SDH/Sonet bytes (e.g. J0) with patterns transmitted in the test messages.

A Management Information Base (MIB), compliant to the Structure of Management Information (SMIv2), describes LMP managed objects for use with Simple Network Management Protocol (SMNP) [7].

Finally, LMP procedures can be interpreted in the context of the ASON terminology [8] promoting a common understanding between the ITU and the IETF.

3 LMP System Architecture

An LMP protocol machine has been designed for the Lucent Technologies LambdaUnite® MultiService Switch (MSS). From the point of view of Software Development for the LMP application, two processors are relevant: A and B.

LMP Control Channels run over Local Area Networks (LAN) or SDH/Sonet Data Control Channels (DCC), whose interfaces are available in A. B provides persistent memory services and a Transcription Language 1 (TL1) [9] management interface.

The Software architecture is organized in three layers: Operative System, a General Protocol Framework, which is universal for any protocol but specific for the target platform, and the LMP Hierarchy, which is LMP specific.

3.1 Design Patterns

The Half Object Plus Protocol (HOPP) design pattern [10] is concerned with systems that are forced to be implemented across two address spaces. System components are divided into two independent half-objects with a protocol between them.

A Real-Associated Duality has been proposed to determine the objects' behavior in the LMP system. While Real Objects reside in processor A, processor B contains Associated Objects. Real Objects implement the internal operation of a system component and Associated Objects are the interface of that component for management and persistence purposes.

Between each Real Object and each Associated Object there is a one-to-one partnership. A synchronization protocol is necessary so that a Real Object and its partner Associated Object share the same state. This protocol is based on full-duplex data unit updates between partner Real and Associated Objects. To assure the absence of inconsistencies, data units updated in different directions must be disjoint (semantically simplex protocol).

A second pattern is used: the Bridge pattern [11] decouples an abstraction from its implementation and provides flexibility to implement a component in different ways depending on the Real-Associated Duality or on the platform.

3.2 General Protocol Framework

A Generalized Socket System that integrates UDP and TCP has been designed for this LMP protocol machine. LMP communications are peer-to-peer, no client or server exists but equals of the same category. Therefore a new Active-Passive paradigm has been proposed to suit LMP. Each node has one passive organ to receive messages and connections, and several active organs to send messages and start connections. Inside a node signaling mechanisms between the active and the passive organs are needed.

A TL1 interface provides remote management and performance monitoring capabilities. The system uses a limited syntax. TL1 has standard messages for all management procedures that suit big optical networks. SMNP lacks those procedures.

The system also provides of an encoding/decoding module, (binary for LMP and ASCII for TL1), a queue system, a tidy termination system, an error handling system, a highly scalable and reliable timing server and a persistence server to store system information that needs to be ready after a failure or restart.

Since all services are not available in both processors an inter-processor communication mechanism is necessary: HOPP. The HOPP System is based on a hermaphrodite Client-Server design and its goal is to make the physical distribution transparent.

One process, composed of several threads, is running on each processor. Using hierarchical allocation of resources [12] prevents deadlocks.

Several threads, called dispatchers, process LMP messages but only one dispatcher processes TL1 messages.

3.3 LMP Hierarchy

The LMP Hierarchy is not mere MIB but contains the LMP internal operation logics (Real LMP Hierarchy in A) as well as the LMP management interfaces (Associated LMP Hierarchy in B).

The LMP Local instance stores local configuration. Each LMP Neighbor is represented in the LMP hierarchy and contains several Control Channel Management Finite State Machines (FSM) and only one UNI Service Discovery FSM.

Both Data Links and TE Links connecting the local node with each neighbor are generically represented by Abstract Link instances, which perform Link Property Correlation and Connectivity Verification. Pointers denote the Link Bundling relationships between TE and Data Links. Fault Management is not required at this stage.

Moreover, a reporting system is available to inform other (non-LMP) system applications of the successes or failures of LMP procedures.

4 LMP System Operation

LMP messages (Fig. 1) are received by the Generalized Passive Socket in A and left in the queue. LMP Dispatchers take messages from the queue and process them by making use of the logics contained in the Real LMP Hierarchy. Timing services may be needed. HOPP communications with B may take place after an LMP event to update the Associated LMP Hierarchy for persistence, management or observability purposes. Eventually an LMP answer may be sent.

TL1 Input messages are processed by the TL1 Dispatcher in B and may be addressed to any component in B or in A (via an appropriate Associated Object). Fig. 1 illustrates the case that the TL1 message is addressed to the LMP Hierarchy. The TL1 Dispatcher may need timing, persistence or HOPP services to process it. Afterwards, a TL1 Response message is always sent.

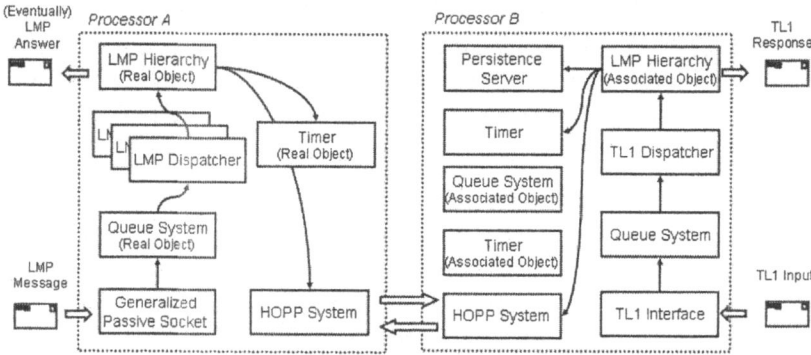

Fig. 1. LMP system operation

5   LMP Simulation Environment

A simulation environment was designed to create an arbitrary LMP network in a computer cluster.

TL1 commands allow changing the LMP nodes' configuration on the fly as well as stimulating the network. The stimulation can be controlled manually by an operator or totally automatic. LMP nodes react sending statistical data in TL1 Response messages. This information is integrated, interpreted and stored.

5.1   Scenarios

The following scenarios were simulated (Table 1):

5.2   Performance

LMP discovers services, correlates and verifies links in tens of seconds. In contrast, manual configuration, as traditionally in SDH/Sonet, would take weeks or months.

In this implementation, LMP Nodes managing more than 8 Control Channels or more than 256 Data Links can be considered under very high load conditions.

Many small bundles of Data Links lead to worse LMP performance than fewer TE Links composed of a larger number of Data Links.

A node can be modeled as an M/M/1 queue where the effective service rate is:

µeff = Number of LMP Dispatchers / LMP Message Service Time .   (1)

The system performance can be optimized with 2 or 3 LMP Dispatchers (Fig. 2): when several LMP dispatchers must compete with each other, the service times increase exponentially but LMP messages are removed faster from the queue. This leads to a maximum in the effective service rate and to a minimum in the waiting time and queue length. After this point is reached additional LMP dispatchers degrade the performance.

Table 1. Simulated scenarios

Scenario	Description
Scenario 1	Two LMP nodes with an arbitrary number of Control Channels.
Scenario 2	Two LMP nodes with 4 Control Channels, 1 UNI neighborhood and different combinations of Data Links and TE Links.
Scenario 3	Totally meshed networks of several nodes, with 1 Control Channel and 1 UNI neighborhood between each pair of nodes.
Scenario 4	Meshed networks of different topologies making use of different combinations of the LMP procedures.

Scheduling desynchronizations between LMP nodes lead to message retransmissions. As a result, an LMP network's behavior is characterized by relatively random oscillations around one or several resonance points (depending on the topology) in which the network is most stable. Each resonance point is defined by a certain value for each observed parameter.

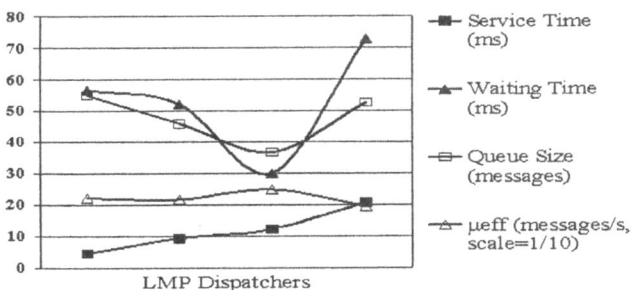

Fig. 2. Optimizing performance

6 Link Bundling Criteria

According to [13], all Data Links in the same TE Link must begin and end on the same pair of Label Switching Router (LSR), have the same point-to-point or multicast nature, the same switching capability (packet, time, lambda, or fiber) and the same Traffic Engineering (TE) metric at each end of the link.

A suitable TE metric facilitates reliable and efficient network operations and optimizes network resource utilization and traffic performance.

The TE metric recommended for SDH/Sonet in this evaluation is composed of the following parameters (Table 2).

Table 2. Link Bundling Parameters

Parameter	Description
Shared Risk Link Group (SRLG)	The SRLG is a vector of SRLG IDs. Each SRLG ID indicates links that share a certain risk (e.g. links sharing the same fiber or duct). SRLG is used for protected connections, especially those demanded by a UNI Client, which is not allowed to know the core network's topology.
Virtual Private Network (VPN) ID [14]	The VPN ID identifies ports that have been contracted to carry traffic for a specific customer (e.g. Ethernet over SDH/Sonet).
Maximum/Minimum Switching Granularity	The signal rate range that a port can multiplex and/or demultiplex at. (e.g. sts1, vc4/au4).
Propagation Delay	Essential to guarantee QoS.
Administrative Mode	A link may be under normal operation, out of service for maintenance purposes or being phased out for administrative deletion.
Administrative Cost	Routing protocols select the least-cost path.

These properties are coded for each Data Link in an LMP object called Data Link Sub-object and exchanged in Link Summary messages. SRLG is the only one of them which is already considered in the LMP specifications [5]. For the others, additional Data Link Sub-objects have been defined in this evaluation.

7 LMP Protocol Specifications Evaluation

Some LMP aspects are discussible or improvable:

7.1 Link Summary Message and TCP

The Link Summary message's length depends on the number of Data Links belonging to a particular TE Link and properties associated to each Data Link.

LMP messages are transmitted via UDP [2]. The maximum size of a UDP datagram (64 KB) allows bundling up to 600 Data Links using the previously defined TE metric. However, UDP implementations hardly ever allow reaching this limit [15]. Thus, most applications avoid exceeding 8KB (75 Data Links).

In the future, providers will deploy hundreds of parallel fibers between each pair of nodes, each carrying hundreds of lambdas. The scalability of these networks must not be limited by LMP.

This problem can be solved by allowing both TCP and UDP Control Channels. LMP messages must be processed in the same way independently of the nature of the channel over which they where transmitted.

7.2 Control Channel Selection

Allowing LMP control channels to be transmitted over different physical media (LAN, DCC, a separate wavelength or fiber or a tunnel through a separate management network) and to use different transport protocols involves providing a mechanism so that LMP neighbors can agree on the same Control Channel [16].

This requirement has not still been covered by LMP. Additional information must be added to the LMP Config object for this purpose and transmitted in the Config / Config Nack messages.

7.3 Test Status Failure Message

TE Links in SDH/Sonet are correlated by sending one Test message, containing a J0 byte trace, over the Control Channel for each Data Link [6]. For each Data Link this trace is compared with the J0 pattern received over the Data Link. If they match, a Test Status Success message containing the Local Interface ID is sent. Otherwise a Test Status Failure, which does not contain any Interface ID [2], is sent.

If Test messages are sent simultaneously for several Data Links, it is impossible to know which Data Link did not work after the process for the whole TE Link is over. As a result Test, Test Status Failure and Test Status Ack messages are retransmitted unnecessarily. This problem can be solved by adding the Local Interface ID to the Test Status Failure messages.

7.4 Deadlock in LMP FSMs

An LMP message requiring acknowledgement is retransmitted a maximum number of times. After reaching the limit, the FSM should return to the initial state to avoid deadlocks (transition 1). In the same way, the time an LMP node waits for a particular message should also be limited and the FSM should also return to the initial state after this time expires (transition 2).

None of these transitions are described in [2] for the following messages: Config (Control Channel), Link Summary (LPC) and Begin Verify / End Verify (LCV). Due to the Verify Interval only transition 1 is needed for the Test Status Success / Failure messages (LCV).

7.5 Missing FSMs for LCV

Link Connectivity Verification (LCV) involves tasks associated to both Data and TE Links. The Active and Passive LMP Data Links FSM defined in [2] only implement the operation logic associated to Data Links. LCV does also require active and passive FSMs for TE Links, which are not yet defined and must not be considered mere implementation details.

The following FSM (Fig. 3) deals with this issue. The PreTesting state is involved in exchanging Begin Verify messages. End Verify messages are exchanged in the PostTesting state. The Testing state initializes the LPC Data Link FSMs, coordinates and executes them until all of them deliver positive or negative results.

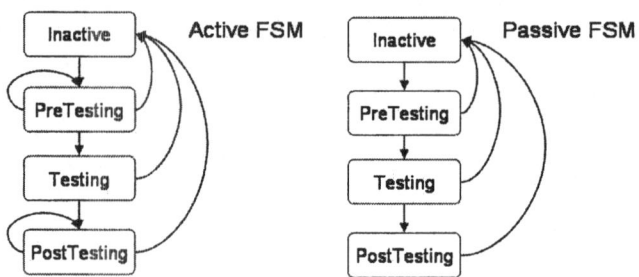

Fig. 3. LCV Missing FSMs for TE Links

7.6 UNI Service Config Object

UNI Client and UNI network devices exchange attributes about signaling, data channels and network services in Service Config messages. These attributes are wrapped in Service Config objects. According to [4], a Service Config message is allowed to contain only one Service Config object. This contradicts the idea expressed in the same document of sending Network Transparency properties and Network Diversity & Tandem Connection Monitoring (TCM) attributes in the same Service Config message (type 3).

Indeed, Service Config messages are required to contain several Service Config Objects. Doing so also improves the protocol performance reducing the number of Service Config messages of type 2 exchanged in a Service Discovery process.

7.7 UNI Client Port-Level Service Attribute

In GMPLS every Data Link is assigned an ID at each end. These IDs may be IPv4, IPv6 addresses or unnumbered IDs (due to the scarcity of IP addresses and the management burden of assigning an IP address to each fiber, lambda or TDM channel)

The 4-byte Local Interface ID field of the Client Port-Level Service Attribute defined in [2] does not indicate the ID type and its size is not IPv6-compatible.

We propose replacing this field by a 4-byte Interface Type and a variable length Interface ID, which is 4 bytes long for IPv4 and unnumbered IDs, and 16 bytes for IPv6.

8 Conclusions

In the very long term GMPLS control plane enables traditional four-layer networks (IP, ATM, SDH/Sonet, and DWDM) to bypass the SDH/Sonet and ATM layers. This results in a more efficient and faster network [17]. However it is worth deploying LMP in today's SDH/Sonet equipment since it allows network providers to set out for a cost-effective migration into next generation networking when the market emerges.

Open IP tools like LMP reduce costs and provide compatibility among vendors but are inherently insecure. Thus, security models in today's equipment are of paramount importance, above all in the UNI interface as UNI clients should not be trusted.

LMP is still an Internet draft protocol in its infancy. Some aspects in the LMP specifications involve potential ambiguities and inconsistencies. LMP needs refining in order to meet the real needs of the industry.

References

1. E. Mannie, "Generalized Multi-Protocol Label Switching Architecture", draft-ietf-ccamp-gmpls-architecture-07, March 2003
2. J. Lang, "Link Management Protocol (LMP)", draft-ietf-ccamp-lmp-10, October 03.
3. "Architecture for the Automatically Switched Optical Network (ASON)", ITU-T Rec. G.8080/Y.1304
4. "User Network Interface (UNI) 1.0 Signaling Specification", Optical Networking Forum, October 2001
5. A. Fredette, et al., "Link Management Protocol (LMP) for Dense Wavelength Division Multiplexing (DWDM) Optical Line Systems", draft-ietf-ccamp-lmp-wdm-02, March 2003.
6. J. Lang, D. Papadimitriou, "SONET/SDH Encoding for Link Management Protocol (LMP) Test messages", draft-ietf-ccamp-lmp-test-sonet-sdh-03, May 2003
7. M. Dubuc, "Link Management Protocol Management Information Base", draft-ietf-ccamp-lmp-mib-07, October 2003
8. O. Aboul-Magd et al., "A Transport Network View to LMP", draft-aboulmagd-ccamp-transport-lmp-01, June, 2003
9. "Operations Application Messages - Language for Operations Application Messages", Telcordia, GR-831
10. L. Rising et al.,"Design Patterns in Communication Software", Cambridge University Press, UK, 2001
11. E. Gamma, "Design Patterns: Elements of Reusable Object-Oriented Software", Addison-Wesley, Holland, 1998
12. N.A. Lynch, "Distributed Algorithms", Morgan Kaufmann Publishers, Inc., CA 1997.
13. K. Kompella et al., "Link Bundling in MPLS Traffic Engineering", draft-ietf-mpls-bundle-04, January 2003.
14. B. Fox et al., "Virtual Private Networks Identifier", Internet RFC 2685, September 99
15. W. R. Stevens, "Maximum UDP Datagram Size", TCP/IP Illustrated, Volume 1, Addison-Wesley, 1994
16. N. Jerram, A. Farrel, "MPLS in optical networks", page 33, Data Connection, Enfield, UK, October 2001
17. A. Banerjee et al. "Generalized Multiprotocol Label Switching: An overview of Routing and Management Enhancements", IEEE Communications Magazine, January 2001.

Experimental Investigation of WDM Transmission Properties of Optical Labeled Signals Using Orthogonal IM/FSK Modulation Format

P.V. Holm-Nielsen[1], Jianfeng Zhang[1], J.J. Vegas Olmos[2], I. Tafur Monroy[2], C. Peucheret[1], V. Polo[3], P. Jeppesen[1], A.M.J. Koonen[2], and J. Prat[3]

[1] Research Center COM, Technical University of Denmark, Building 345V, DK-2800 Kgs. Lyngby, Denmark
Tel:+45 4525 3635, Fax:+45 4593 6581, vhn@com.dtu.dk
[2] COBRA Research Institute, Faculty of Electrical Engineering, Eindhoven University of Technology, The Netherlands.
[3] GCO Optical Communications Group, Technical University of Catalonia (UPC), Barcelona, Spain.

Abstract. In this paper we report on WDM transmission of optical labeled signals using the orthogonal IM/FSK modulation format aimed for optical packet/burst switching. WDM transmission in both a point-to-point transmission system and a demonstration network with a label-swapping node are implemented.

1 Introduction

With the continuing growth of the Internet and the introduction of high-bit rate WDM connections in metro and backbone networks, the need for effective and flexible switching solutions will become more apparent.

One proposed solution is optical label switching (OLS), that enables the implementation of packet routing and forwarding functions in IP-over-WDM [1]. OLS supports high bit rates for payload data transmission while employing low speed electronics in the core nodes for label processing. Labels are received and swapped at every node, while the payload information is transparently forwarded with possible wavelength conversion [2].

Several approaches have been studied for labeling optical packets [2,3]. Among those the combined intensity modulation/frequency-shift keying (IM/FSK) [4,5] is one of the promising methods due to its in-band character and simplicity of implementation. IM/FSK optical labeling [6] uses frequency modulation for the label orthogonally superimposed onto the intensity modulated payload at the same wavelength. Orthogonally in this context means that the label information can be detected independently from the payload.

This technique enables transmitting payload information at high bit rates, while allowing label information to be easily extracted from the bit stream. IM/FSK optical labeling offers a larger tolerance to the laser linewidth than other labeling techniques, which makes it easier to implement [4]. However, it requires careful fiber dispersion compensation due to its relatively larger optical spectrum.

J.N. de Souza et al. (Eds.): ICT 2004, LNCS 3124, pp. 753–759, 2004.
© Springer-Verlag Berlin Heidelberg 2004

This paper presents the latest experimental investigation on wavelength division multiplexing (WDM) transmission of IM/FSK modulated signals. It is organized as follows: Section 2 presents the results of point-to-point WDM transmission. Two systems are implemented. First, 8 IM channels where only one of them has a superimposed FSK label, then three fully labeled IM/FSK signals are transmitted. In section 3, an optical label-swapping node is implemented for the first time on a WDM signal and error-free transmission performance is verified. Finally, conclusions are given in section 4.

Two methods of generating the labeled IM/FSK signal are used in these experiments. As reported in [7], a DFB/EA laser can be used for FSK modulation by directly modulating the electrical current of a DFB laser with the data of the label signal (at 156 Mbps), while feeding the inverted data electrical signal into a subsequent electro-absorption modulator (EAM). The payload of the signal is then imposed (at 10 Gbps) on the intensity of this signal by a second modulator. The modulation depth is chosen to generate carrier frequency tones spaced apart by 20 GHz.

A grating assisted co-directional coupler with sampled reflector (GCSR) laser can also be used to generate a FSK signal by modulating the phase current, usually used for fine wavelength tuning [8,9].

At the receiver, the labeled signal is split equally. One part is used for directly detect the payload. In the other arm, two optical filter stages are used to filter only a single tone of the FSK labeled signal.

2   Point-to-Point Transmission of WDM IM/FSK Signals

In this section, experimental results on an eight-channel WDM system with 200 GHz channel spacing and a three-channel system with 100 GHz channel spacing are reported.

2.1   One FSK/IM Channel with Seven IM Channels

As shown in Fig 1, the FSK signal generated by a DFB/EA laser is multiplexed with the outputs of seven CW DFB lasers in an 8-channel arrayed waveguide grating (AWG). The payload data is generated by a PRBS source with a sequence length of 2^{31}-1, then is imposed on the eight channels by a chirp-free intensity modulator with a 6 dB IM extinction ratio. After the modulator, the signals are transmitted through a dispersion-compensated 80 km standard fiber link. At the receiving end, another AWG is used to demultiplex the wavelength channels. Finally, the demultiplexed labeled channel is detected for BER measurement.

The eight channels cover the wavelength range between 1549.3 nm and 1560.6 nm, with 1.6 nm (200 GHz) channel spacing. The IM/FSK signal is modulated on the first channel at a center wavelength around 1549.3 nm. The effective channel suppression ratio at the output of the wavelength demultiplexer is more than 40 dB.

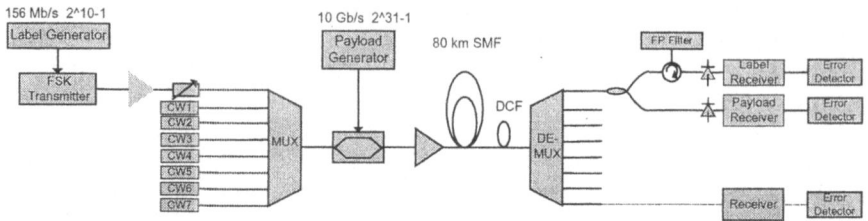

Fig. 1. A WDM transmission system consisting of one IM/FSK channel and seven IM channels.

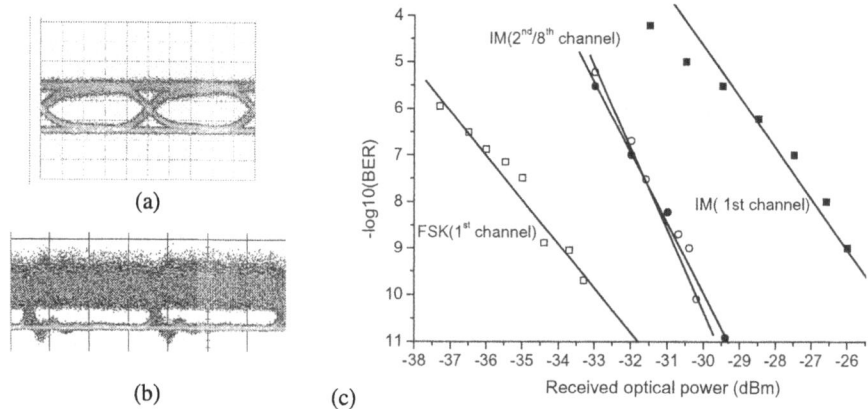

Fig. 2. Eye diagrams of the received (a) IM payload and (b) FSK label of the labeled channel and (c) measured BER curves of different wavelength channels.

As the channel spacing is very large compared to the spectral bandwidth of the signal, the wavelength multiplexing and demultiplexing processes add little degradation to the whole system. The transmission over 80 km fiber also induces negligible degradations to the system performance. Fig. 2 shows the eye-diagrams of the IM payload and demodulated FSK label after transmission, while Fig. 3 shows the BER curves of selected wavelength channels.

2.2 Three-Channel IM/FSK Signal Transmission

A three-channel IM/FSK signal transmission system is implemented, as shown in Fig. 1. Two GCSR-DBR lasers and one DFB/EA laser are used to generate multi-wavelength FSK/IM signals. With 100 GHz (0.8 nm) spacing, the wavelength channels are at 1548.5 nm (GCSR), 1549.3 nm (DFB/EA) and 1550.1 nm (GCSR) respectively.

Due to the non-uniform frequency modulation response of the GCSR-DBR lasers, 8B10B encoding is applied to the label data (PRBS 2^9-1) [11] before it modulates the two GCSR-DBR lasers, while the DFB/EA laser is frequency-modulated by a PRBS data with a length of 2^{10}-1. The outputs of the three lasers are multiplexed using two optical couplers, and then intensity modulated in a chirp-free Mach-Zehnder modulator with a 6 dB extinction ratio. The generated multi-wavelength IM/FSK signals are amplified and then input into 80 km of dispersion compensated SMF with an average power of 10 dBm.

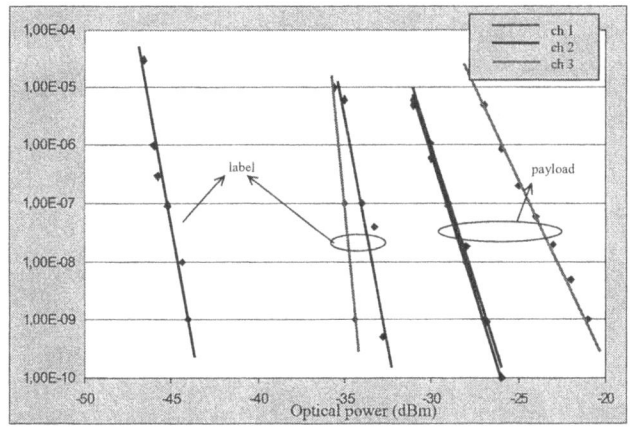

Fig. 3. Measured BER curves for the three wavelength channels after transmission.

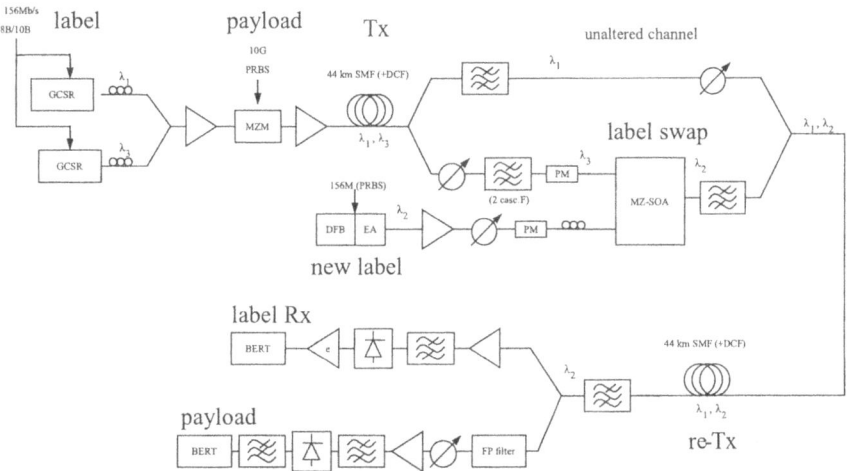

Fig. 4. A WDM transmission system with a label-swapping node.

At the receiver end, a tunable optical filter is used to demultiplex the wavelength channels. The demultiplexed channels are finally detected for BER measurements. The suppression ratio of the demultiplexed signal is around -30 dB. All the three channels show an error-free performance. As indicated in Fig. 3, there is no obvious degradation introduced by transmission effects. Among the three wavelength channels, channel 2 (generated with the DFB/EA laser) has the best FSK modulation performance due to its high FM efficiency; The other two channels have better IM performance as the GCSR lasers generate much less residual intensity ripples when they are frequency modulated.

3   Performance of IM/FSK Signals in a WDM Transmission System with a Label-Swapping Node

As shown in Fig. 4, a three-channel WDM transmission system with a label-swapping node is implemented using a Mach-Zehnder-interferometer semiconductor optical amplifier (MZI-SOA) wavelength converter.

The three channels are at 1548.5 nm, 1549.3 nm and 1550.1 nm respectively. IM/FSK signals are generated at the transmitter node at wavelengths corresponding to channel 1 (1548.5 nm) and channel 3 (1550.1 nm) and are transmitted over a dispersion compensated 44 km standard fiber span before reaching the label-swapping node. At the swapping node, the IM/FSK signal at channel 3 is label-swapped, and its wavelength is changed to that of channel 2 (1549.3 nm). The IM/FSK signals at channel 1 and channel 2 are transmitted again over another 44 km standard fiber span before being detected.

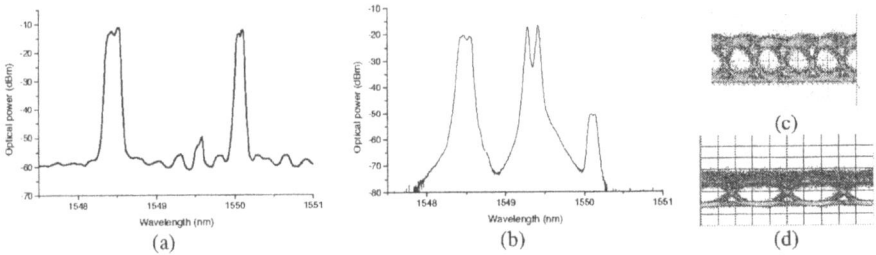

Fig. 5. Spectra of the IM/FSK signals at the (a) transmitting and (b) swapping nodes. Eye-diagrams of (c) IM payload and (b) FSK label of channel 2 at the receiving node.

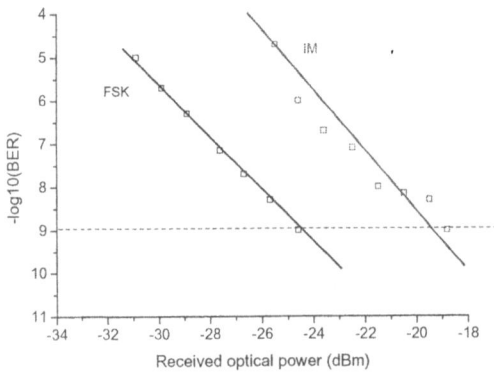

Fig. 6. Measured BER curves of channel 2 after 2 hop transmission including label swapping

Fig. 5 shows the measured spectra of IM/FSK signals at different locations. In all cases, the signals have a sufficiently large optical-signal-to-noise-ratio (OSNR) to ensure acceptable BER performance. Fig. 6 shows the measured eye diagrams of channel 2 at the receiving node. In the entire transmission link, channel 1 is kept unaltered while channel 3 is terminated at the swapping node. Only channel 2 experiences both label-swapping and WDM transmission process, so channel 2 is given a detailed analysis through BER measurements.

As shown in Fig. 4, the FSK label and IM payload of channel 2 are detected error-free. It can be seen that WDM transmission and label-swapping introduce a rather large power penalty that could be alleviated by transmitter optimization, having FSK transmitters designed specifically to the working wavelength with a low IM residual modulation.

4 Conclusion

Some basic issues in IM/FSK WDM transmission systems including the impact of wavelength channel spacing, and multi-wavelength IM/FSK signal generation have been presented. The performance of IM/FSK signals in a point-to-point WDM transmission system has been experimentally investigated, showing that IM/FSK channels can be added to pure IM channels to upgrade legacy WDM networks.

A three-channel IM/FSK transmission system was implemented with a channel spacing of 100 GHz and the error-free transmission performance was verified. Finally, a WDM transmission system with a label-swapping node has been demonstrated. The experimental results show that IM/FSK can be a feasible scheme to upgrade metro-distance WDM transmission networks, which strengthen the feasibility of using the orthogonal IM/FSK labeling as a technique for future IP-over-WDM networks.

Acknowledgement. This work was performed in the framework of the STOLAS project, which is partly funded by the IST Program of the European Commission.

References

[1]   Chunming Qiao, "Labeled Optical Burst Switching for IP-over-WDM Integration", IEEE Communications Magazine, vol. 38, no. 9, pp. 104-114, 2000.
[2]   D.J. Blumenthal et al, "All-optical label swapping networks and technologies" Journal of Lightwave Technology. vol. 18, no. 12, pp. 2058-2075, 2000.
[3]   I.T.Monroy, et al, "Techniques for labeling of optical signals in burst switched networks," Proc of WOBS'03, pp. 1-11, 2003.
[4]   T. Koonen et al, "Optical packet switching in IP-over-WDM networks deploying two-level optical labeling", Proc. of ECOC'01, paper Th.L.2.1, pp. 608-609, 2001.
[5]   K. G. Vlachos et al, "STOLAS: Switching Technologies for Optically Labeled Signals", IEEE Communications Magazine, vol. 41, no. 11, pp. 43-49, 2003.
[6]   T. Koonen et al, "Optical Labeling of Packets in IP-over-WDM Networks", Proc. of ECOC'02, paper 5.5.2, 2002.
[7]   J.Zhang et al, "An optical FSK Transmitter Based on an Integrated DFB Laser-EA Modulator and Its Application in Optical Labeling", IEEE Photonics Technology Letters, vol. 15, no. 7, pp. 984-986, 2003.

[8]   Saavedra et al, "Amplitude and frequency modulation characteristics of widely tunable GCSR lasers", IEEE Photonics Technology Letters, vol. 10, no. 10, pp. 1383-1385, 1998.

[9]   Vegas Olmos et al, "High bit-rate combined FSK/IM modulated optical signal generation by using GCSR tunable laser sources", Optics Express, vol. 11, no. 23, pp. 3136-3140, 2003.

[10] P.-J. Rigole et al, "Quasi-continuous tuning range from 1560 to 1520 nm in a GCSR laser, with high power and low tuning currents", Electronics Letters, vol. 32, no. 25, pp. 2352-2354, 1996.

[11] S.P. Gangopadhyay et al, "Performance of linecoded optical heterodyne FSK systems with nonuniform laser FM response", Journal of Lightwave Technology, vol.13, no.4 , pp. 628-638.

Performance Assessment of Optical Burst Switched Degree-Four Chordal Ring Networks

Joel J.P.C. Rodrigues[1], Mário M. Freire[1], and Pascal Lorenz[2]

[1] Department of Informatics, University of Beira Interior,
Rua Marquês d'Ávila e Bolama,
6201-001 Covilhã, Portugal
{joel, mario}@di.ubi.pt
[2] IUT, University of Haute Alsace
34, rue du Grillenbreit, 68008 Colmar, France
lorenz@ieee.org

Abstract. This paper presents a performance analysis of optical burst switching (OBS) networks with degree-three and degree-four chordal ring topologies. The analysis considers just-in-time (JIT), Jumpstart, JIT$^+$, just-enough-time (JET) and Horizon signaling protocols. For a network with 20 nodes and for the considered traffic loads, it is shown that the nodal degree gain due to the increase of nodal degree from two (ring) to three (degree-three chordal ring) is between one and two orders of magnitude in the last hop of each topology, whereas the nodal degree gain due to the increase of nodal degree from three (degree-three chordal ring) to four (degree-four chordal ring) is between two and three orders of magnitude. It is also shown that the performance of the five signaling protocols above referred is very similar.

1 Introduction

Optical burst switching (OBS) [1]-[8] has been proposed to overcome the technical limitations of optical packet switching, namely the lack of optical random access memory and to the problems with synchronization. OBS is a technical compromise between wavelength routing and optical packet switching, since it does not require optical buffering or packet-level processing and is more efficient than circuit switching if the traffic volume does not require a full wavelength channel. In OBS networks, IP (Internet Protocol) packets are assembled into very large size packets called data bursts. These bursts are transmitted after a burst header packet, with a delay of some offset time. Each burst header packet contains routing and scheduling information and is processed at the electronic level, before the arrival of the corresponding data burst. Several signaling protocols have been proposed for optical burst switching networks. In this paper, we concentrate on Just-In-Time (JIT) [3], JumpStart [4]-[6], JIT$^+$ [7], Just-Enough-Time (JET) [1], and Horizon [2] signaling protocols.

Although JIT is a conceptually simple protocol, some previous studies have shown that JIT has a performance worse than either JET or Horizon. On the other hand, JET

J.N. de Souza et al. (Eds.): ICT 2004, LNCS 3124, pp. 760–765, 2004.
© Springer-Verlag Berlin Heidelberg 2004

and Horizon require complex scheduling and void filling algorithms. However, most of the existing studies ignore many important parameters such as the offset length, the processing time of setup messages, and the optical switch configuration time, which have significant impact on burst loss probability. Therefore, there is a need for more detailed studies in order to explore in depth the differences among the various protocols. In this paper, we use accurate models for an OBS mesh network operating under the JIT, JumpStart, JIT$^+$, JET, and Horizon signaling protocols.

A major concern in OBS networks is the contention and burst loss. The two main sources of burst loss are related with the contention on the outgoing data burst channels and on the outgoing control channel. In this paper, we consider bufferless networks and we concentrate on the loss of data bursts in OBS networks with chordal ring topologies. For comparison purposes, ring topologies are also considered.

After this introductory section, section 2 describes the model of the OBS network under study and presents a performance analysis of OBS networks with chordal ring topologies. Section 3 concludes the paper.

2 Performance Assessment

In this session, we present a performance assessment of JIT, JumpStart, JIT$^+$, JET, and Horizon signaling protocols in OBS networks with ring and degree-three and degree-four chordal ring topologies for N=20 nodes.

Chordal rings are a well-known family of regular degree three topologies proposed by Arden and Lee in early eighties for interconnection of multi-computer systems [9]. A chordal ring is basically a bi-directional ring network, in which each node has an additional bi-directional link, called a chord. The number of nodes in a chordal ring is assumed to be even, and nodes are indexed as 0, 1, 2, ..., N-1 around the N-node ring. It is also assumed that each odd-numbered node i (i=1, 3, ..., N-1) is connected to a node $(i+w) \bmod N$, where w is the chord length, which is assumed to be positive odd. According to [10] it is assumed that each link is replaced by a chord and, instead of a topology with nodal degree of 3, we have a topology with a nodal degree of n, where n is a positive integer, and instead of having 3 chords we have n chords. Then, using this notation, a general degree n topology is represented by DnT($w_1,w_2,...,w_n$), and a chordal ring family with a chord length of w_3 is represented by D3T(1,N-1,w_3), a chordal ring family with a chord length of w_4 is represented by D4T(1,N-1,w_3,w_4), and a bi-directional ring is represented by D2T(1,N-1).

We consider that each node of the OBS network supports F+1 wavelength channels per unidirectional link. One wavelength is used for signaling (carries setup messages) and the other F wavelengths carry data bursts. It is assumed also that each OXC consists of non-blocking space-division switch fabric, with full conversion capability, but without optical buffers.

In this study, it is assumed that [8]: T_{OXC} = 10 ms, T_{setup}(JIT)=12.5 μs, T_{setup}(JumpStart)=12.5 μs, T_{setup}(JIT$^+$)=12.5 μs, T_{setup}(JET)=50 μs, and

$T_{setup}(Horizon)$=25 μs. The mean burst size, $1/\mu$, was set to 50 ms, and the burst arrival rate λ, is such that λ/μ=38.4.

In chordal ring topologies, different chord lengths can lead to different network diameters, and, therefore, to a different number of hops. One interesting result that we found is concerned with the diameters of the D3T(w_1,w_2,w_3) families, for which w_2=(w_1+2)mod N or w_2=(w_1-2)mod N. Each family of this kind, i.e. D3T(w_1,(w_1+2)mod N, w_3) or D3T(w_1,(w_1-2)mod N, w_3), with $1{\le}w_1{\le}19$ and $w_1{\ne}w_2{\ne}w_3$, has a diameter which is a shifted version (with respect to w_3) of the diameter of the chordal ring family (D3T(1, N-1, w_3)). For this reason, we concentrate the analysis on chordal ring networks.

Fig. 1 shows the network diameter as a function of chord length for D3T(1,19,$w3$), D4T(1,19,3,$w4$), and D4T(1,19,5,$w4$). As may be seen, for N=20 nodes, the maximum and minimum diameter of the degree-three chordal ring family are 6 and 4, respectively, while, the maximum and minimum diameter of the degree-four chordal ring family are 4 and 3, respectively.

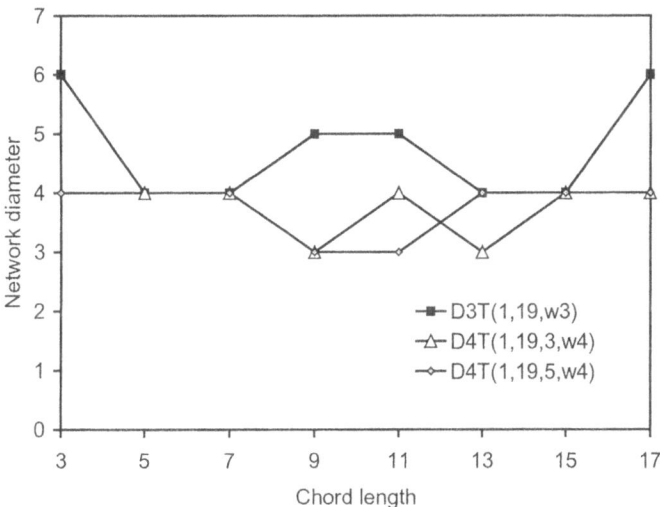

Fig. 1. Network diameter, as a function of chord length for D3T(1, 19, $w3$), D4T(1, 19, 3, $w4$), and D4T(1, 19, 5, $w4$).

Fig. 2 shows the burst blocking probability in the last hop of D2T(1,19), D3T(1,19,5), and D4T(1, 19, 3, 9), for JIT, JumpStart, JIT$^+$, JET and Horizon signaling protocols. As may be seen in Fig. 2, chordal rings clearly have better performance that rings, and, in terms of chordal rings, degree-four topologies have better performance than degree-free.

Fig. 3 shows the burst blocking probability, for N=20 nodes, for D2T(1, 19), D3T(1, 19, 5), and D4T(1, 19, 3, 9) as a function of the number of hops, for JIT, JumpStart, JIT$^+$, JET and Horizon. This figure clearly confirms the results of Fig. 2, i.

e., degree-three chordal rings clearly have better performance than rings, and degree-four chordal rings clearly performs better than degree-three chordal rings.

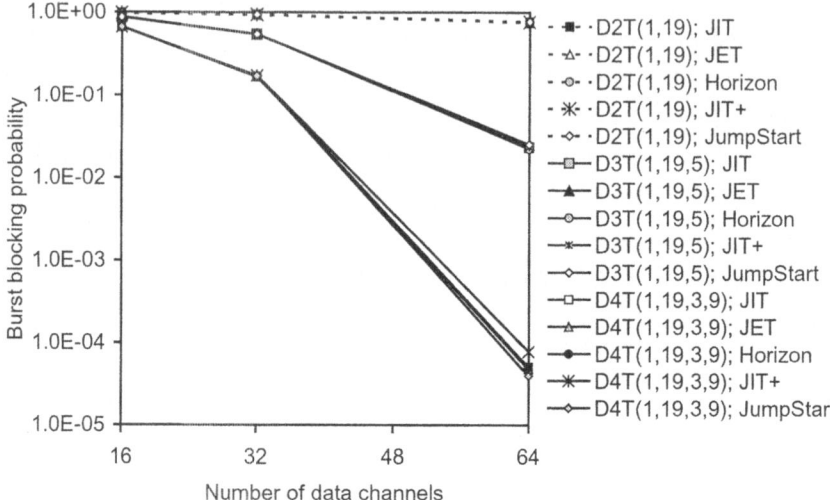

Fig. 2. Burst blocking probability in the last hop of each topology versus number of data channels for D2T(1, 19), D3T(1, 19, 5), D4T(1, 19, 3, 9); $N=20$.

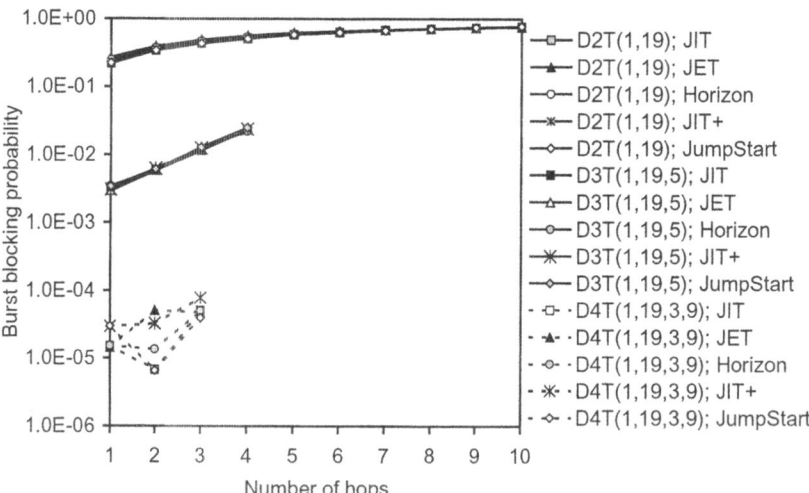

Fig. 3. Burst blocking probability versus number of hops for D2T(1, 19), D3T(1, 19, 5), D4T(1, 19, 3, 9); $N=20$, $F=64$.

In order to quantify the benefits due to the increase of nodal degree from 2 to 3 and from 3 to 4, we introduce a new performance metric: the nodal degree gain, $G_{x,y}(i,j)$. The nodal degree gain, $G_{x,y}(i,j)$, is defined as:

$$G_{x,y}(i,j) = \frac{P_i\big(D_{(n-1)}T(w_1, w_2,..., w_{n-1})\big)}{P_j\big(D_nT(w_1^*, w_2^*,..., w_n^*)\big)}. \tag{1}$$

where $P_i(D_{(n-1)}T(w_1, w_2, ..., w_{n-1}))$ is the burst blocking probability for the i-th hop of the $D_{(n-1)}T(w_1, w_2, ..., w_{n-1})$, and $P_j(D_nT(w_1^*, w_2^*, ..., w_n^*))$ is the burst blocking probability for the j-th hop of the $D_nT(w_1^*, w_2^*, ..., w_n^*)$, for the same network conditions (same number of data wavelengths per link, same number of nodes, etc) and for the same signaling protocol. We use $G_{x,y}$ where x represents the nodal-degree of the $P_i(D_{(n-1)}T(w_1, w_2, ..., w_{n-1}))$ topology and y represents the nodal-degree of the $P_j(D_nT(w_1^*, w_2^*, ..., w_n^*))$ topology.

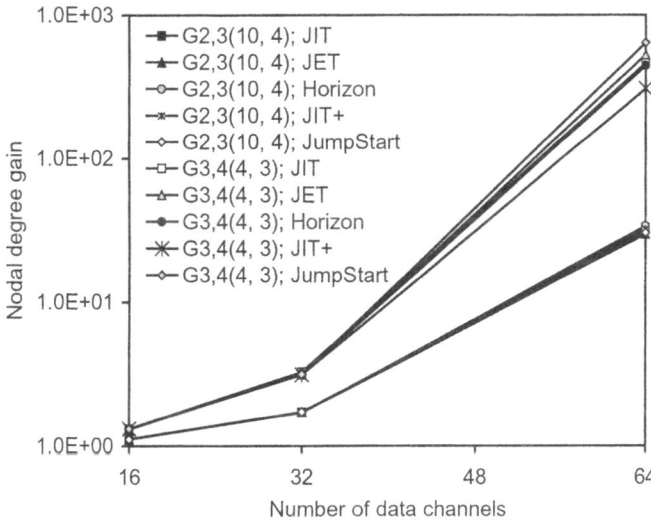

Fig. 4. Nodal degree gain, in the last hop of each topology, due to the increase of the nodal degree from 2 (D2T(1, 19)) to 3 (D3T(1, 19, 5)) and from 3 (D3T(1, 19, 5)) to 4 D4T(1, 19, 3, 9); N=20, F=64.

Fig. 4 shows the nodal degree gain, $G_{x,y}(i,j)$ in the last hop of each topology, due to the increase of nodal degree from 2 (D2T(1, 19)) to 3 (D3T(1, 19, 5)), and from 3 (D3T(1, 19, 5)) to 4 (D4T(1, 19, 3, 9)) for JIT, JumpStart, JIT⁺, JET, and Horizon (F=64). For F=16 and F=32 the nodal degree gain is very small due to the high burst blocking probability. However, when the number of data channels per link increases to 64, a nodal degree gain between 1 and 2 orders of magnitude is observed for degree-three chordal rings. On other hand, the increase of nodal degree from 3 (D3T(1,19,5)) to 4 (D4T(1, 19, 3, 9)) leads to a performance improvement between 2 and 3 orders of magnitude. Another important observation that can be made from this figure is that the five signaling protocols under study lead to very similar nodal degree gains both when the nodal degree increases from 2 to 3 $(G_{2,3})$ and when the nodal degree increases from 3 to 4 $(G_{3,4})$.

3 Conclusions

A performance analysis of OBS degree-three and degree-four chordal ring networks was presented for JIT, JIT$^+$, JumpStart, JET and Horizon protocols. For comparison purposes, ring topologies are also considered. It was shown that, for a network with 20 nodes, the nodal degree gain due to the increase of nodal degree from two (ring) to three (degree-three chordal ring) is between one and two orders of magnitude in the last hop of each topology. It was also shown that the nodal degree gain due to the increase of nodal degree from three (degree-three chordal ring) to four (degree-four chordal ring) is between two and three orders of magnitude. It was observed in both cases that the performance of the five signaling protocols under study is very similar.

References

1. Qiao, C., Yoo, M.: Optical burst switching (OBS)-A New Paradigm for an Optical Internet. In Journal of High Speed Networks, Vol. 8, No. 1 (1999) 69-84
2. Turner, J.S.: Terabit Burst Switching. In Journal of High Speed Networks, Vol. 8, No. 1 (1999) 3-16
3. Wei, J.Y., McFarland, R.I.: Just-in-time signaling for WDM optical burst switching networks. In Journal of Lightwave Technology, Vol. 18, No. 12 (2000) 2019-2037
4. Baldine, I., Rouskas, G.N., Perros, H.G., Stevenson, D.: JumpStart: A just-in-time signaling architecture for WDM burst-switched networks. In IEEE Communications Magazine, Vol. 40, No. 2 (2002) 82-89
5. Zaim, A.H., Baldine, I., Cassada, M., Rouskas, G.N., Perros, H.G., Stevenson, D.: The JumpStart Just-In-Time Signaling Protocol: A Formal Description Using EFSM. In Optical Engineering, Vol. 42, No. 2, February (2003) 568-585
6. Baldine, I., Rouskas, G.N., Perros, H.G., Stevenson, D.: Signaling Support for Multicast and QoS within the JumpStart WDM Burst Switching Architecture. In Optical Networks, Vol. 4, No. 6, November/December (2003)
7. Teng, J., Rouskas, G. N.: A Detailed Analysis and Performance Comparison of Wavelength Reservation Schemes for Optical Burst Switched Networks, submitted for publication.
8. Teng, J., Rouskas, G.N.: A Comparison of the JIT, JET, and Horizon Wavelength Reservation Schemes on A Single OBS Node. In The First International Workshop on Optical Burst Switching, Dallas Texas, USA, October 16 (2003)
9. Arden, B.W., Lee, H.: Analysis of Chordal Ring Networks. In IEEE Transactions on Computers, Vol. C-30, No. 4 (1981) 291-295
10. Rodrigues, J.J.P.C., Freire, M.M., Lorenz, P.: Performance Comparison of Optical Burst Switching Ring and Chordal Ring Networks Using Just-in-Time and Just-Enough-Time Signaling Protocols. In Proceedings of the Third International Conference on Networking (ICN'04), Gosier, Guadeloupe, French Caribean, March 1-4 (2004) 65-69

All-Optical Routing Limitations Due to Semiconductor Optical Amplifiers Dynamics

António Teixeira[1,2], Paulo André[2,3], Rogério Nogueira[2,3], Paulo Monteiro[1,2,4], and José da Rocha[1,2]

[1] Departamento de Telecomunicações, Universidade de Aveiro,
3810-193 Aveiro, Portugal
teixeira@ua.pt, frocha@ieee.org, pandre@av.it.pt,
rnogueira@fis.ua.pt, paulo.monteiro@siemens.com
http://www.ua.pt
[2] Instituto de Telecomunicações, Campus Universitário Santiago,
3810-193 Aveiro, Portugal
[3] Departamento de Física da Universidade de Aveiro,
3810-193 Aveiro, Portugal
[4] IC/WOW, Siemens S.A., Rua Irmãos Siemens,
2720-093 Alfragide, Portugal

Abstract. In this work we describe a technique which associates the already matured code generation and detection technique of Optical Code Division Multiplexing with the all-optical routing needs and discuss its advantages and limitations due to one of the current optical gating techniques, based on the gain saturation of a Semiconductor Optical Amplifier.

1 Introduction

The next generation of optical networks will demand high scalability and fine granularity in addition to high bandwidths. In present times, techniques like D-WDM (Dense-Wavelength Division Multiplexing) have been able to produce quite high bandwidths and make it available at least in core networks, where, in combined total rates, one can find more than 1 Tbit/s. However, as regards the granularity, the WDM light-path network is quite coarse. Anyway, efforts are being driven to increase the network light-path granularity, as it is the case of the IP (Internet Protocol) over GMPLS (Generalized Multi-Protocol Label Switching) over WDM technology which provides mechanisms for finer granularity [1]. However, in this case, due to electronic header processing, memory access for header analysis and all the involved processing tends to be the bottleneck of this next generation of networks. In order to overcome this limitation, an all-optical header processing technique can be the solution for the needed high speed processing. Several approaches to this technique are being developed nowadays, showing the great interest on the field. Many of them are based on OCDM (Optical Code Division Multiplexing) [2]. To achieve such goal

J.N. de Souza et al. (Eds.): ICT 2004, LNCS 3124, pp. 766–771, 2004.

many building blocks have to be matured first, however, studies and demonstrators are already available, where an optical label processing unit is the key for achieving Optical Packet Switching (OPS) or other [3],[4.]. The main building blocks of this technology are an all-optical header generation and detection technique, which can be achieved by FBG (Fiber Bragg Gratings), PLC (Planar Lightwave Circuits) or other techniques [3], [4] and a gating device, which nowadays are currently achieved by optical amplifier structures.

In this work we will address the OCDM technique based on FBG, section 2, following the block diagram of the router will be addressed, section 3, and finally we will address the gating device in the router, which we considered in this work, to be a SOA (Semiconductor Optical Amplifier), section 4. The work finishes by drawing some conclusions, section 5.

2   Optical Code Division Multiplexing Based on Fiber Bragg Gratings

In order to achieve all-optical routing, in this work, we aim to have high power gating pulse which will be the starting point to initiate some kind of optical gating, either by saturation or by any other linear or nonlinear process in a gating device. So, we want to design a technique which allows us to generate these pulses recurring to an all-optical process or scheme. The elected technique to perform this work, is based on the OCDM code properties. This technique allows high power output pulses when there is good correlation between the generated code and the correlator, and low power pulses when there is low or absent correlation. A schematic of the technique to generate and detect the code is presented in Fig. 1. The represented schematic is valid for time/wavelength codes, where the pulses are, in the source, scrambled in time and wavelength by the correlator in order to achieve a unique combination suitable for high correlation at the correlating node. There are other dimensions and codes which can and are being used like for example the time/phase codes [4], time-polarization, wavelength-polarization, and even more complex codes with higher dimension mixing combinations of more than two of the properties which we can control in light (time, wavelength, phase, polarization, space, and others). In this work, aiming simplicity in the codes and low cost in the correlators/code generators we used Time-Wavelength codes, since they are quite easy to process with a cheap off-shelf technology like the FBG's.

The FBG generation of this kind of codes is addressed with the block diagram shown in Fig. 2. The FBG is a reflective device, so it acts as rejection band filter on the transmission and as a band-pass filter in the reflection. Therefore, by sending white noise into an array of such filters, considering that they are not all centered at the same wavelength, the result in reflection will be a comb of lines centered at the wavelengths in which each FBG was written and in transmission is a sequence of deeps complementary to the reflected spectrum.

Considering the referred code generation method and presented in Fig. 2a), one can conclude that this kind of code generation is quite simple to implement. A comb of multi-wavelength pulses arrives to a circulator which forwards it to the FBG's which are placed in space according to the time sequence that we want to generate and with the correctly addressed wavelengths, the reflected wave will be a time spread of the pulses in time and wavelength in accordance to the coder. At the output of the FBG array there are only the remaining pulses unneeded to generate the wanted code. The spatial position of the gratings is calculated such that a round trip from the circulator to the FBG is the value of the time spread which that wavelength should have, so the far the FBG is, higher is the delay with respect to the first reflected pulse.

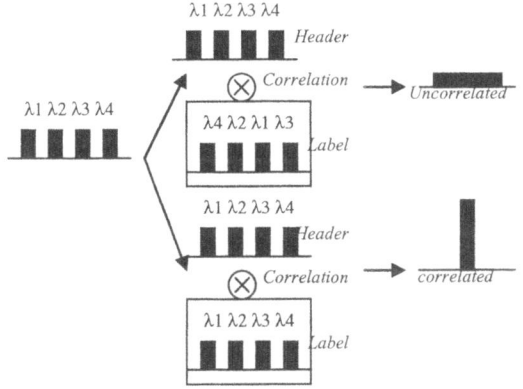

Fig. 1. All-optical label Processing based on two dimensional time/ wavelength codes.

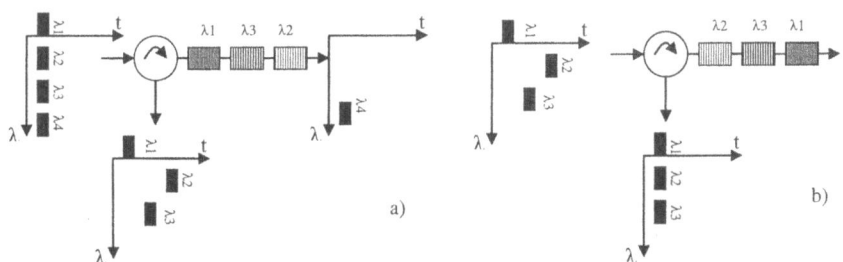

Fig. 2. Code generation and correlation based on Fiber Bragg Grating Structures.

To conceive the decoder the principle is the same and also the structure, since we only need to reverse the time delays to have the input comb again. So, the correlator is the same structure only the FBG array is placed in the reverse position, which can be implemented very easily since these devices are bidirectional. The block diagram of the correlator is shown in Fig. 2b).

As it can be observed in the schematic in Fig. 2b), at the output of the correlator, in case of matched correlation between the optical code and the label held by the FBG's, the output power in all the wavelengths is high since all pulses are aligned in time. If there is no or small correlation between the code and the correlator the power at decision time chip would be much smaller and depending on the code used.

3 All-Optical Routing Architecture

In core routers the number of routes which come out of the device are in very low number, since only transfer traffic goes by, and only a small part is dropped there. For this kind of devices, we should mainly have quite fast label processing since the aggregate data rates are very high. A small number of route addresses is needed, based on what was said, and therefore optical if we use the above described technique, OCDM, to label the routes of the packets we need a very small number of codes, which makes the process much simpler, due the already known limitations of the OCDM codes. The problem becomes even more restrictive if we use several of these codes in the same time and wavelength window in order to route several packets in parallel due to the multiple access interference.

In this work we considered an architecture, not very effective in terms of throughput, however it allows simple all optical routing implementation, which, in conjunction with other management schemes can be clearly improved.

The base block diagram of the generating node and routing device is presented in Fig. 3.

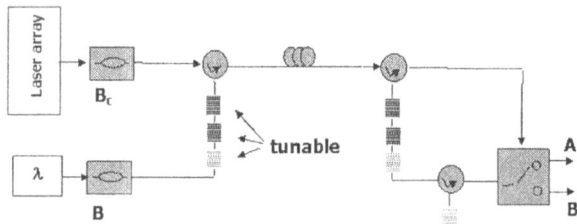

Fig. 3. Block diagram of the proposed router architecture.

At the source we are considering a laser array which will be modulated at a rate which is the Bc (Chip rate). This rate, Bc, is much smaller than the rate of the packet since all traffic will pass in one time chip (1/Bc) at the router. The wavelength bands for the label and for the packets are different in order to allow multiplexing of either the labels or the packets and simultaneously allow time overlappi ng ofmultiple packets, which will turn out to reduce the packet transit time and increase throughput. At the coder, the packet data is modulated in bursts at bit rate B in a wavelength which can flow through the header encoder FBG's. The tuneable FBG's allow the transmitter to address the several ports of the gate at the router, by matching or not the code, as explained in section 2. If the header matches the receiver coded label there will be a high power pulse available, if not there will be a low power time spread signal available to control the gate. Based on the power available the gate should make the decision.

4 Gating with Semiconductor Optical Amplifiers

One of the most commonly used gating devices is the SOA. This device has very peculiar properties which allow several mechanisms to be highly efficient inside the device. It is the case of the gain dependence on the input power, which is quite sharp, near saturation, which can be clearly observed in Fig. 4. Typically, the gain varies quite significantly as we vary the input saturating power, as it can be noticed form Fig. 4, where we depict the SOA gain versus the input power for several values of the internal device differential gain. So, if we consider the pulse which comes out of the correlation and send it simultaneously with the packet through the same SOA device, if there is high correlation between the label and the header there will be an high power state available, if not there will be a low power state available. Entering the device while there will be a high power state the gate will block the signal due to low overall gain, on the other hand, if there is a low power state the gate will let the packet through. In Fig. 4a) it is presented the time sequence of the two packets where one was blocked and the other passed through.

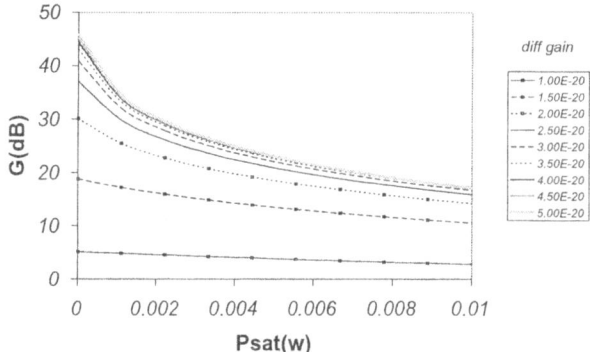

Fig. 4. Gain dependence of a Semiconductor Optical Amplifier as a function of the saturating power, for different internal differential gains.

The results of Fig. 5a), show the effects of the internal dynamics of the SOA, evidencing the oscillation and patterning in the flat tops of the non-blocked packets, however it can be observed clearly the blocking effect of the routing device, which is incomplete due to the gain characteristics of the device. Also, the slow dynamics of the SOA can also be observed, especially in the fall times due to carrier recombination times inside the device, which depend on the length of the pattern of the data packet.

In Fig. 5b) we represent the relative eye opening in dB, which is defined here as the ratio between the maximum inner opened portion of the eye inside eye opening and the total peak to peak power of the eye, as a function of the correlated header pulse extinction ratio, both in dB. We define this last parameter as the ratio between the levels at the state of high correlation with the level of the state of low correlation. As it can be seen if the code does not result in a very high extinction ratio when correlated the eye closes and the blocking does not occur with the needed quality. Also, it can be noticed that the tendency of the curve denotes the penalty imposed by the saturation properties of the SOA since, the increase in the extinction ratio of the pulse does not bring any improvement after the about 28dB of Extinction ratio.

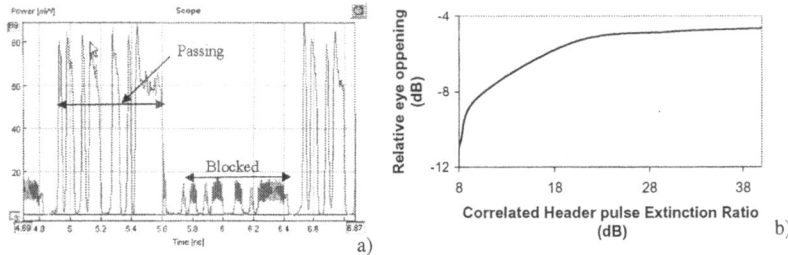

Fig. 5. Optical packet switching time curves for a 40Gbit/s packet data rate and a 1.25GBit/s chip header rate. b) Relative eye opening as a function of the output pulse of the header correlation extinction ratio

5 Conclusions

In this work we have suggested a technique for achieving all-optical high data rate routing. This technique is based on simple and cheap off-the-shelf components, the fibre bragg gratings, circulators and semiconductor optical amplifiers. The architecture allows reasonable eye opening of the transit packet as long as the chosen codes allow an extinction ration between the correlated and uncorrelated decoded pulses on the order of the 18dB.

Acknowledgments. We would like to acknowledge the projects Wo-Net (POSI/2002/CPS40009), FEDER and Siemens SA for the support through the ADI CONVOCO project.

References

1. K. Sato, et al., IEEE Commun. Magazine, Vol. 40, No. 3, (2002), 96-101
2. N. Wada, et al. J. Lightwave Technol., Vol17, No10, (1999). 1758-1765
3. A. Teixeira, et al., Proceedings of LEOS annual meeting, (2003)
4. K. Sato, et al. Proceedings of ONDM, Ghent, (2003), 411-423

The Hurst Parameter for Digital Signature of Network Segment*

Mario Lemes Proença Jr.[1], Camiel Coppelmans[1],
Mauricio Bottoli[2], A. Alberti[2], and Leonardo S. Mendes[2]

[1]State University of Londrina (UEL) – Computer Science Department – Londrina, PR.
{proenca, camiel} @uel.br
[2]State University of Campinas (UNICAMP) – Communications Department – Campinas, SP
{Bottoli, alberti, lmendes} @decom.fee.unicamp.br

Abstract. This paper presents results of the Hurst parameter for digital signature of network segments. It's also presented a model for digital signature automatic generation which aims at the characterization of traffic in network segments. The use of the digital signature allows the manager to: identify limitations and crucial points of the network; establish the real use of network resources; better control the use of resources and the establishment of thresholds for the generation of more accurate and intelligent alarms which suit the real network characteristics. The obtained results validate the experiment and show in practice significant advantages in networks management.

1 Introduction

The digital signature of network segment (DSNS) can be defined as the set of basic information that shows the traffic profile in a segment of the network, through minimum and maximum thresholds about volume of traffic, quantity of errors, types of protocols and services that flow through this segment along the day [1]. It can also be defined as a baseline of the network segment.

The forecast of a determined instant, about the characteristics of the traffic of the segments that make up the network backbone, make the management decisions on anomalies that might be happening, more reliable and safer [2][3].

The use of the DSNS can help the network manager to identify limitations and control the use of resources that are critical for services that are latency-sensitive such as Voice over IP and video transport, because they can't take retransmission or even network congestion. Besides improving the resources control, its use also facilitates the planning on the network increase, because it clearly identifies the real use of resources and the critical points along the backbone, avoiding problems of performance and faults that might happen.

The use of the DSNS also offers the network manager advantages related to performance management, by means of the previous knowledge of the maximum and

* This work was made with support of CNPq

J.N. de Souza et al. (Eds.): ICT 2004, LNCS 3124, pp. 772–781, 2004.

minimum quantities of traffic in the segment along the day. This enables the establishment of more effective and functional alarms and controls, because they are using limits that suit the DSNS, respecting the variations of traffic along the day instead of using the linear limits, normally found in the networks management systems (NMS) that exist at present [4]. Deviations in relation to what are being monitored real-time and what the DSNS expresses must be observed and analyzed carefully, and can or can not be considered as problems. In order to do that, the use of an alarm system integrated to the DSNS and to the real-time monitoring will deal with these problems, warning the network manager when it is necessary.

As for security management, the use of the DSNS can offer information related to the analysis of the users behavior, because the previous knowledge of the behavior and the traffic characteristics of a determined segment is directly related to the profile users manipulation, using this as information to prevent intrusion aspects or even network attacks, by means of the intrusion detection software [5][6].

Another use for the DSNS is related to the monitoring of a network segment which is normally performed manually by means of visual control, using only the empirical knowledge with the network acquired by the manager. An example of this can be seen with the utilization of tools like GBA (Automatic Backbone Management) [7] and MRTG (Multi Router Traffic Grapher) [8] that generate graphs with statistical analysis which consist of averages along a determined period of time about an analyzed segment or object. However, the simple use of these graphs establishes limitations for the network manager concerning discovery and solution of problems. The limitations are caused especially by the non-automation of this task, where the monitoring of these graphs is performed visually, depending exclusively on the empirical knowledge about the functioning of the network acquired by the manager and due to the large quantity of graphs that have to be analyzed continuously. It manages to detect the problems and unusual situations in a reactive way.

Networks with a great number of segments turn their management more complex, considering the great quantity of graphs to be analyzed [3]. The graphs usually present information on the volume of input and output traffic of a certain segment, not aggregating information that could help the manager more efficiently in his decision-making with the purpose of solving problems that might be happening or that might have already happened.

There is a lot of work done in traffic characterization and traffic measurement that is related to the proposal in this work [9][10]. Traffic characterization and traffic measurement are important aspects that have to be considered for network management and control. In [9][10] is presented a survey of the main research done for traffic characterization in telecommunication networks. However these models intend to traffic modeling in a generic way, while the proposal presented in this paper intends to a traffic characterization generated from collected real data of each segment of analyzed network. The aim of this characterization is to create a particular profile for each monitored segment, which we call a baseline or digital signature of the network segments (DSNS).

Another important area that is related to work presented in this paper is anomaly detection [2][3]. Thottan et al [2], presents a review about anomaly detection methods and a statistical signal processing technique based on abrupt change detection that uses analysis of SNMP MIB variables for anomaly detection. In [2] is used a 15s sampling frequency, and it assumes, like an open issue, that there exist some changes in MIB data that don't correspond to network anomalies. The use of an effective and real baseline can help to solve this problem for knowing the real behavior of the traffic.

In the rest of this paper it will be presented a description about the model used for the construction of the digital signature of network segments (DSNS), which we also refer as baseline; the results of the Hurst parameter calculation used for avail DSNS, some results that show in practice the gains with its use for the networks management and at last, conclusions and suggestions for future works.

2 DSNS Implementation

The main purpose to be achieved with the construction of the DSNS is the characterization of the traffic of the segment it refers to. This characterization should reflect initially the profile expected for the traffic along the day and as well as other existing characteristics such as: types of protocols, types of applications, types of services. These characteristics are used to create a profile of the users. The DSNS was initially developed to analyze the quantity of input and output of octets stored in the *ifInOctets* and *ifOutOctets* objects which belong to the *Interface* group of the MIB-II [11].

The use of the GBA tool (Automatic Backbone Management) was chosen as a platform for the development of the DSNS due to the great quantity of historical information related to monitoring carried out along the last years in the main network segments of UEL. The GBA was initially developed to help with the networks management with ATM backbone and it performed its duty as it became a platform of learning and development, helping with the management as well as with the understanding about the networks functioning. Further information on the GBA can be found at http://proenca.uel.br/gba or in [12].

As for the tests and validation of the model, the data gathered by the GBA were used since 2002 up to the present. The use of the data from the last two years was considered an important sample, characterized by periods of winter and summer vacations as well as holidays which contributed to the tests and validations of the ideas presented in this work. The analyzed data is related to the network segments with traffic TCP/IP based on Ethernet and ATM with LAN Emulation. The tests of the proposed model were carried out in three segments, which are described below:

1. The first one which is called segment S_1 is responsible for interconnecting the ATM router to the other backbone segments of State University of Londrina (UEL) networks; it gathers a traffic of approximately 2500 computers;
2. The second one which is called S_2 interconnects the office for undergraduate studies of academic affairs in UEL; it gathers a traffic of 50 computers;
3. The third one which is called S_3 interconnects State University of Campinas (UNICAMP) network to academic network at São Paulo (ANSP), it gathers a traffic of all UNICAMP (about 5000 computers) to Internet.

For the generation of the DSNS a model was developed based on statistical analyses that we call BLGBA. The analyses were carried out for each second of the day, each day of the week. Figure 2.1 illustrates the operational diagram used in the

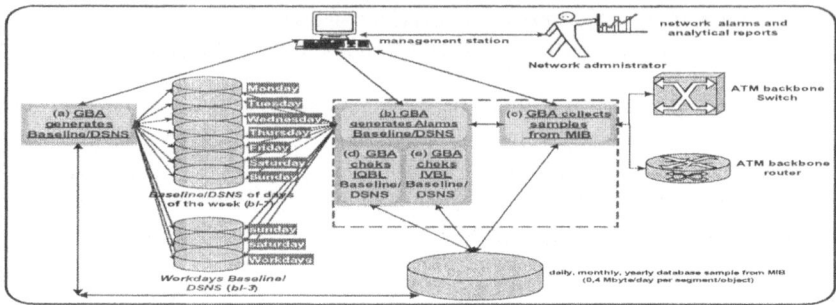

Fig. 2.1. Operational functioning diagram for the generation of DSNS and alarms.

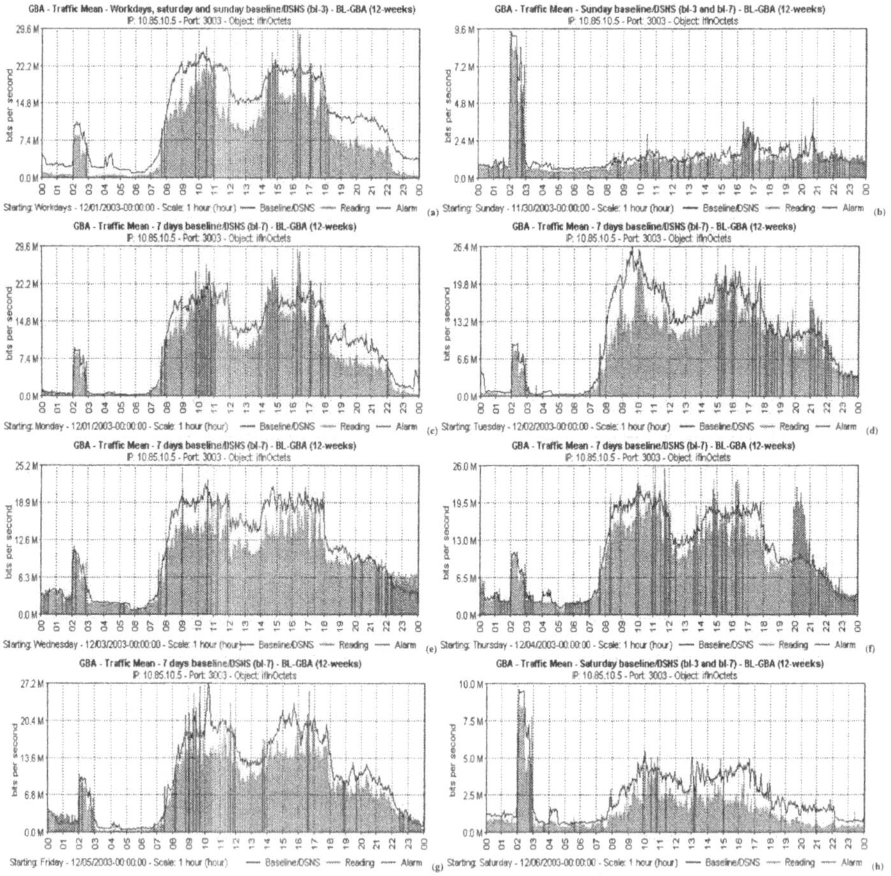

Fig. 2.2. DSNS and the daily movement for S_i segment analyze.

implementation of the DSNS, which is carried out by the <u>GBA generated Baseline/DSNS</u> module. This module reads information from the database, with data gathered daily and generates the DSNS based on a period requested by the network manager.

Two types of DSNS were created, one called *bl-7* which consists of seven DSNS files, one for each day of the week, and the other one called *bl-3* which consists of three DSNS files, one for the workdays from Monday to Friday, one for Saturday and another one for Sunday, as shown in Figure 2.1. The choice for generating the DSNS separating the workdays of the week from Saturday and Sunday, was in order to minimize the margin of error in the final result, concerning the alterations in the volume of traffic that occur between the workdays and the other days. The results showed that it was the right choice, because the variation that was found in the volume of traffic between the workdays was of 10% and over 200% comparing workdays and weekends, as can be seen in figure 2.2.

The model for DSNS generation proposed and presented in this work, performs statistical analysis of the collected values, respecting the exact moment of the collection, second by second for twenty-four hours, preserving the characteristics of the traffic based on the time variations along the day. For the generation of the DSNS, the holidays were also excluded due to the non-use of the network on these days. Moreover, the process of DSNS generation also considered faults in the collected samples which occur along the day, eliminating these faults from the calculations for the DSNS generation.

The GBA makes collections at each second at the MIBs of the network equipments. Along each day, 86400 samples are expected. Problems usually occur and may affect some of these samples due to the loss of package or congesting in the network. In this case, for the generation of the DSNS, the exclusion of these samples was chosen in the calculation of the DSNS related to that second. This problem occurs in less than 0.05% a day, for the analyzed samples.

The processing for the DSNS generation is done initially in batch aiming at its creation through data related to a pre-established period. The DSNS is generated second by second for a period of days represented by N which makes up the set n_j (j = 1, 2, 3, 4, ..., N); with the daily gathering there is a set of samples of the day represented by a_i (i = 0, 1, 2, ..., 86399). Then the bi-dimensional matrix is built with 86400 lines and N columns which must be previously sorted and that will be represented by M_{ij}.

The algorithm used for the calculation of the DSNS (BLGBA) is based on a variation in the calculation of *mode*, which takes the frequencies of the underlying classes as well as the frequency of the modal class into consideration. The calculation takes the distribution of the elements in frequencies, based on the difference between the greatest G_{aj} and the smallest S_{aj} element of the sample, using only 5 classes. This difference divided by five, forms the amplitude h between the classes, $h = (G_{aj} - S_{aj})/5$. Then the limits of each L_{Ck} class are obtained. They are calculated by $L_{Ck} = S_{aj} + h.k$, where Ck represents the k class (k = 1...5).

The proposal for the calculation of the DSNS of each Bl_i second has the purpose of obtaining the element that represents 80% of the analyzed samples. The Bl_i will be defined as the greatest element inserted in class with accumulated frequency equal or greater than 80 %. The purpose is to obtain the element that would be above most samples, respecting the limit of 80%. This process is used for the generation of DSNS models *bl-7* and *bl-3*.

The BLGBA model used for the calculation of the DSNS was chosen after the performance of tests with other statistical models based on the *mean, octile, decile average* and on the *mode*. The choice for the BLGBA model was based on:

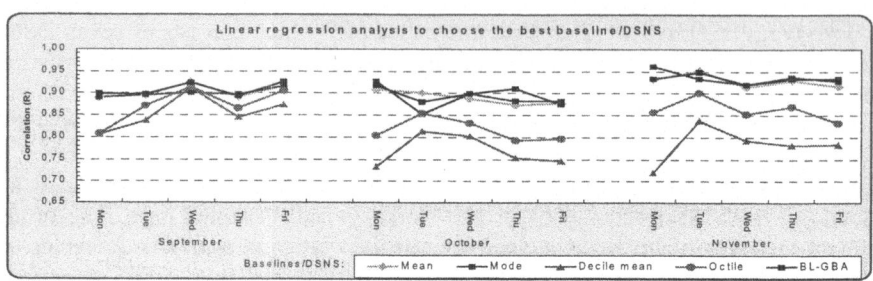

Fig. 2.3. Linear regression analysis aiming at evaluating which is the best method for DSNS generation.

1.   Visual analysis of graphics containing the DSNS and its respective daily movement, as illustrated in figure 2.2;

2.   Deviation analysis proposed by Bland and Altman [13], which takes into consideration the differences between the predicted and observed movements. Such differences must lie between an interval defined by $\overline{d} \pm 2 * s$, where $\overline{d}$ is the differences mean and s is the standard deviation of these differences. With this an upper and lower limit are set where the deviation must be contained. The model that presented better adjustment was the BLGBA, with 95% of the differences in these limits;

3.   Residual analysis – the model which showed less residual index between the predicted and the occurred movement was the BLGBA;

4.   Linear regression [14][15] between the models aimed at evaluating which one showed a better correlation coefficient between the DSNS and the daily movement. Figure 2.3 shows the result of the correlation tests for the segment S_1 related to the months of September to November 2003. In this figure it is possible to notice that the BLGBA shows a better correlation coefficient between the daily movement and the DSNS.

We created an index with the purpose of evaluating the coefficient of variation of the DSNS of one month in relation to the other. This index is called Index of Variation of the Baseline (IVBL). The IVBL is calculated based on the difference between one DSNS and the other, as shown in equation (1). The IVBL was used to calculate the variation of DSNS generated from n weeks and compared to a DSNS of $(n - 1)$ weeks, and in the comparison between the DSNS of 1 week with the DSNS of n weeks. These calculations using weekly DSNS were carried out with the purpose of evaluating and demonstrating the minimum quantity of samples necessary for the formation of the DSNS. With the comparison of the DSNS of n weeks with the one of $(n - 1)$, during 24 weeks, it was observed that the percentage of variation tends to stabilize from the 12th week on, and not being significant for the formation of the DSNS. And when a DSNS of 1 week was established and a comparison was carried out for 24 weeks, it was also noticed that, from the 12th week on, the percentage of variation tends to stabilize around 20%, showing no more significant variations that could be added to the DSNS from this point on. The IVBL test shows that would be necessary at least 4 and no more than 12 weeks for the formation of the DSNS.

$$IVBL = \left(\sum_{i=1}^{86400} BL\,'_i - BL\,''_i \right) \Big/ 86400 \qquad (1)$$

Where *IVBL* = variation index of a *baseline* in relation to another

The obtained results show the validity of the model for the generation of the DSNS, bearing in mind the performed analyses and the comparison with the real movement that occurred. An example of that can be seen in figure 2.2 that illustrate in the form of a histogram, the daily movement of the segment S_1 and their respective DSNS. In these figures some graphs are shown, concerning the first week of December 2003, with the DSNS in blue and the real movement that occurred on the day in green. We came to the following conclusions with the results shown in figures 2.2:

1. Clear peaks of traffic in the DSNS everyday between 2:00 and 3:00 o'clock in the segment S_1 that are related to the backup performed in this period in the network server;

2. The DSNS is influenced by time factors which, in this case, are related to the working day that starts at 8:00 a.m. and finishes at 10:00 p.m.

3. Periods in which the traffic of the day becomes higher than the DSNS. In this case, its color is changed from green to red, which means a peak of traffic above the DSNS, and this could or could not be interpreted as an alarm;

4. The profile of traffic for the workdays, figures 2.2 (a), generated by the *bl-3* model and 2.2 (c), (d), (e), (f) and (g), generated by the *bl-7* model, is quite similar with a strong time dependence along the day which, in this case, is related to the working day hours of the university where the tests were performed. In the case of Saturdays and Sundays, the DSNS generated for these days are exactly the same for *bl-3* and *bl-7* models, figures 2.2 (b), (h), show this results;

5. Not only the DSNS generated for the workdays *bl-3* but also the one generated for all the days of the week *bl-7*, showed to be suitable for the characterization of the traffic. The *bl-7* is a model of DSNS to be used in cases in which there is the need to respect individual particularities which occur in each day of the week, such as backup days, whereas the *bl-3* is the most suitable for the cases where this is not necessary, that is, all the workdays can be dealt with in a single DSNS, leaving the decision on what model to be used to the network manager;

6.   The generated DSNS fulfill their main objective which is the characterization of the traffic in the analyzed segments.

2.1 Hurst Parameter for DSNS

Besides the visual evaluation of the results, other analytical tests have been carried out aiming to evaluate the reliability of the DSNS generated by the BLGBA in relation to the real movement. The tests were been carried out from January to December of 2003 using:

I. Linear Regression [14][15]: The results demonstrate a high correlation and adjustment between the movement that occurred in the days in relation to its DSNS;

II.   Test purposed by Bland & Altman [13]: Refer to the deviations analysis that occurs between the DSNS and the real movement. 95 % of the deviations/errors observed during all days from January to December 2003, in segments S_1, S_2 and S_3, are between the required limits of $\overline{d} \pm 2 * s$, where $\overline{d}$ is the mean and S is the standard deviation of the differences between the DSNS and the real movement, confirm the reliability of the model;

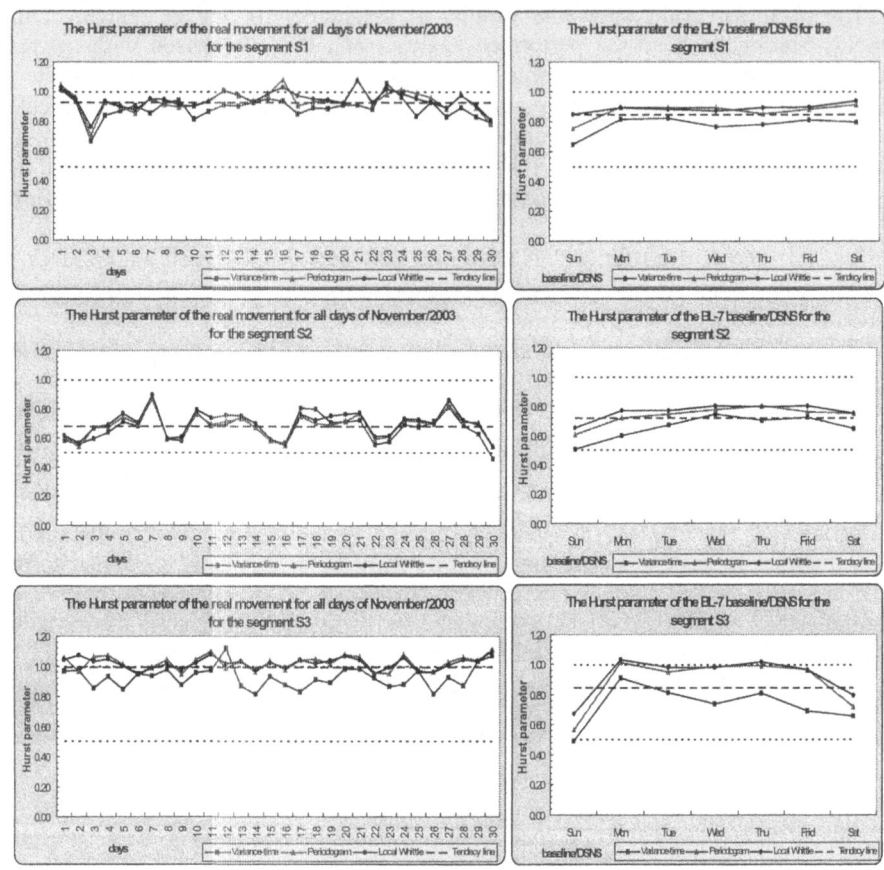

Fig. 2.4. The Hurst Parameter (H) for the real movement and its DSNS of S1, S2 and S3 segments in November 2003.

III. Hurst parameter (*H*): Tests carried out with the real movement and the DSNS generated by the BLGBA, using the statistical methods Variance-time, Local Whittle and Periodogram [16], generate the Hurst parameter *H*. The analysis confirms that the traffic is self-similar and the DSNS is also self-similar, however presenting a lower Hurst parameter. Figure 2.4 illustrates an example of these calculations for real movement and its DSNS (BL-7) for S_1, S_2 and S_3 segments during November 2003. In most of the cases, these tests also allow us to notice that in segments with lower number of computers like S_2, the Hurst parameter presents a lower rate, between 0.6 and 0.7, in segments with great aggregated traffic like the S_1, and S_3 it presents a rate between 0.8 and 1.0. The Hurst parameter evaluation was made using the samples collected second by second with the GBA tool. Calculations were been made for each day between 8:00 and 18:00 hours, the period when the traffic is more similar with to a stationary stochastic process [16]. Its utilization makes possible the evaluation of the DSNS quality in segments of different burstiness. Indicating that the greater the burstiness of the segment the bigger the Hurst parameter and the better the characterization shown by the DSNS. And the lower the burstiness of the segment, the smaller the Hurst parameter and worse the results shown by the DSNS. These results are corroborated by the other tests utilized to validate the DSNS that also indicate an increase of the DSNS quality in segments with a higher burstiness.

3 Conclusion

This work presents a contribution related to the automatic generation of Digital Signature of Network Segment (DSNS), which constitutes itself into an important mechanism for the characterization of the traffic of the analyzed segment, through thresholds that reflect the real expectation of the volume of traffic respecting the time characteristics along the day and the week. This enables the network manager to identify the limitations and the crucial points in the network, control the use of the network resources, establish the real use of the resources, besides contributing to the planning about the needs and demands along the backbone.

The use of an alarm system integrated to the DSNS as well as with the monitoring performed real time by the GBA, figure 2.1 (b) and (c), make it possible for the network manager to be informed through messages, at the exact moment a difference related to the expected traffic and the DSNS, was found out. This possibility is fundamental for the segments or crucial points of the networks that demand perfect control and pro-active management in order to avoid the unavailability of the services rendered.

The use of graphs such as the ones shown in figures 2.2 with information about the digital signature of network segment (DSNS) and about the daily movement, makes a better control over the segments possible.

It could be noticed that the behavior of the traffic of the Ethernet networks is random, self-similar and extremely influenced by the quantity of bursts, which intensify as the number of hosts connected to the segment increase, as shown in [16]. It also showed that the model chosen for the DSNS, presented in this work, is viable for the characterization of the traffic in backbone segments that concentrate the traffic of a great number of hosts, as shown in the examples of section 2.

Tests were also realized with DSNS from other MIB objects, like ipInReceives, icmpInMsgs, udpInDatagrams. The results have been satisfactory and demonstrate that the BLGBA model can be used for other MIB objects, however more tests must be done aiming to evaluate this possibility.

Despite the tests performed at the networks of UEL and in the Communications Department of the Electric Engineering Faculty of UNICAMP, with results validating the model presented in this work, tests with different types of networks, such as factories, large Internet providers and industries shall be performed, aiming to evaluate and perfect the model proposed for generation of DSNS.

Another future work that can be made is related with establishment of criteria for generation of alarms based on presuppositions that establish classes of thresholds in differentiated levels, which would indicate specific conditions to customizable problems to the network.

References

[1]   Paul Barford, J. Kline, D. Plonka, A. Ron; A signal analysis of network traffic anomalies, Internet Measurement Workshop; Proceedings of the second ACM SIGCOMM Workshop on Internet measurement, Marseille, 2002, ISBN:1-58113-603-X.

[2]   Thottan, M.; Chuanyi Ji; Anomaly detection in IP networks, Signal Processing, IEEE Transactions on Volume: 51, Issue: 8, Aug. 2003.

[3]   Papavassiliou, S.; Pace, M.; Zawadzki, A.; Ho, L.; Implementing enhanced network maintenance for transaction access services: tools and applications, Communications, 2000. ICC 2000. IEEE International Conference on, Volume: 1, June 2000.

[4]   Hajji, H.; Baselining network traffic and online faults detection; Communications, 2003. ICC '03. IEEE International Conference on, Volume: 1, 11-15 May 2003.

[5]   NORTHCUTT, Stephen, NOVAK Judy. Network Intrusion Detection, Third Edition, New Riders, 2002.

[6]   Xinzhou Qin; Wenke Lee; Lewis, L.; Cabrera, J.B.D.; Integrating intrusion detection and network management, Network Operations and Management Symposium, 2002. NOMS 2002. 2002 IEEE/IFIP, 15-19 April 2002.

[7]   Ferramenta para Auxílio no Gerenciamento *Backbone* Automatizado, Available by Web in http://proenca.uel.br/gba/ (28/04/2004).

[8]   The Multi Router Traffic Grapher (MRTG), Disponível por WWW em 26/10/2002 no endereço: http://people.ee.ethz.ch/~oetiker/webtools/mrtg/.

[9]   Rueda, A.; Kinsner; A survey of traffic characterization techniques in telecommunication networks, Electrical and Computer Engineering, 1996. Canadian Conference on, Vol.2, Iss., 26-29 May 1996, Pages:830-833 vol.2.

[10] Adas, A.; Traffic models in broadband networks, Communications Magazine, IEEE, Vol.35, Iss.7, Jul 1997, Pages:82-89.

[11] INTERNET ENGINEERING TASK FORCE (IETF). Management Information Base for Network Management of TCP/IP-based internets: MIB-II, RFC 1213, mar.1991.

[12] PROENÇA, Mario Lemes, Jr. "Uma Experiência de Gerenciamento de Rede com *Backbone* ATM através da Ferramenta GBA", XIX Simpósio Brasileiro de Telecomunicações – SBrT 2001, Fortaleza 03-06/09/2001.

[13] Bland J. Martin and Altman Douglas G., Statistical Methods For Assessing Agreement Between Two Methods of Clinical Measurement, The LANCET i:307-310, 1986.

[14] Bussab, Wilton O.; Morettin Pedro A. Estatística Básica, Editora Saraiva, 5a edição 2003.

[15] PAPOULIS, Athanasios, Pillai S. Unnikrishna. Probability, Random Variables and Stochastic Processes, Fourth Edition, McGraw-Hill, 2002.

[16] Leland Will E., Taqqu M. S., Willinger W., Wilson D. V., On the Self-Similar Nature of Ethernet Traffic (Extended Version), IEEE/ACM Transactions on Networking, volume 2, No 1, February 1994.

High Performance Cluster Management Based on SNMP: Experiences on Integration Between Network Patterns and Cluster Management Concepts

Rodrigo Sanger Alves, Clarissa Cassales Marquezan,
Lisandro Zambenedetti Granville, and Phillippe Olivier Alexandre Navaux

Federal University of Rio Grande do Sul - Institute of Informatics
Av. Bento Gonçalves, 9500 - Bloco IV - Porto Alegre, RS - Brazil
{sanger, clarissa, granville, navaux}@inf.ufrgs.br

Abstract. High performance clusters, like any network element, require management. Cluster management area has no standard protocols or *de facto* management tools like network management. For this reason the interaction with other tools is so hard. To obtain interoperability, cluster management particularities should be adapted to the management architecture used in the network. This work presents the experiences obtained with SNMP-based cluster management. Moreover, a cluster management tool based on SNMP is proposed.

1 Introduction

High performance clusters are foundation for building computing grids that spread along computer networks [1]. In order to manage grids, clusters need to be managed first. Maintaining a cluster infrastructure requires administrative interventions executed by the cluster administrator and sometimes this administrator accumulates both tasks of managing the cluster and the network.

Considering this an integration problem arises: the tools to manage networks and the tools to manage clusters are different. In this case, the network/cluster administrator, willing to provide proper grid services, is forced to use one set of tools to manage networks, and another set of tools to manage clusters.

Since the widely accepted solution for traditional, TPC/IP-based network management is the SNMP (Simple Network Management Protocol), integration with cluster management could be achieved through the introduction of SNMP support in the cluster infrastructures. In this paper we present an SNMP-based cluster management system developed to manage two clusters of our main campus network. The system is composed by an SNMP agent and a Web-based management station. The SNMP agent provides management information of a cluster MIB and interacts with the cluster infrastructure in order to expose information to the Web-based station. It is also possible to access the agent via SNMP set-request messages to configure internal cluster resources. We believe that the main contribution of this work comes from the fact it shows that cluster and network management integration can be reached through

J.N. de Souza et al. (Eds.): ICT 2004, LNCS 3124, pp. 782–791, 2004.

SNMP, since the developed SNMP agent allows, even for our particular management necessities, the real integration of cluster management in the SNMP framework.

The remainder of this paper is organized as follows. Section 2 presents related work concerning common cluster and network management issues. In Section 3 we introduce the cluster MIB defined and supported by the SNMP cluster agent. Section 4 shows a case study where the Web-based cluster management station is presented. Finally, we finish this paper with conclusions and future work in Section 5.

2 Related Work

From a network management perspective, a cluster is nothing more than a network resource that needs to be managed in order to be properly accessed. However, from a cluster perspective, the cluster management requires intervention on every single internal cluster element (e.g. hosts, CPUs and processes). Although each element needs to be accessed (which is a cluster management requirement), the whole cluster should expose only one management interface to be seen as a unique resource (which is a network management requirement). In this case, the management infrastructure of a cluster should provide a single point of management where internal cluster information could be retrieved from.

Today, differently from what happens in the network management area, there is no consolidated and widely accepted notion of what really *cluster management* means. Indeed, a large number of different concepts are used to try to define it, but such concepts are often confusing. In this scenario, we try to organize the current cluster management tools in three broad and general classes described below.

Cluster monitoring tools are used to verify the internal status and utilization of the cluster resources. Ganglia [2], the most spread monitoring tool, is a distributed monitoring system able to monitor inter-cluster and intra-cluster information. Inter-cluster monitoring is accomplished through a multicast-based listen/announce protocol, while intra-cluster monitoring is based on point-to-point connections. Another tool is SIMONE (SNMP-based Monitoring System for Network Computing) [3]. It requires SNMP agents installed in each cluster node, which is not adequate if we want a unique management interface for the whole cluster. Basically, SIMONE implements most of the management information Ganglia does.

Users' tasks management tools are tools designed for job scheduling. Such tools allow the definition of cluster fragmentation in smaller parts to help the allocation of nodes. The most spread tool here is PBS (Portable Batch System) [4]. The basic PBS operation is implemented as a FIFO queue.

Administration management tools automate administrative tasks such as node image replication and parallel commands. An example is the tools from the Oscar project [5], which is a set of softwares to automate administrators' interventions.

From a network management perspective, all these cluster tools may not always be identified as management tools in the strict word mean, but in cluster terms it is not rare to find such denomination. A more critical problem is the fact that these tools can not be integrated with network management without complex adaptation work. Integration is not achieved today due to the lack of a common management interface in the cluster management tools. Our developed work uses a common management interface through the SNMP.

3 The SNMP Cluster Agent

Since the main management tasks required in our two clusters were previously executed through *cluster monitoring tools* and *users' tasks management tools,* a developed SNMP agent is mainly concerned about providing information about nodes features (e.g. number of processors) and nodes utilization (e.g. CPU and memory usage). Additionally, the agent supports a simplified node allocation mechanism.

The design of the agent allows a single instance of the agent per cluster, thus providing a single point of management for each cluster. However, the information exposed by each agent allows the management of the cluster internal information. Fig. 1 presents our cluster management scenario where a NMS (Network Management System) accesses the clusters A and B through SNMP. In this case, each cluster is managed via its SNMP agent located in the cluster front-end.

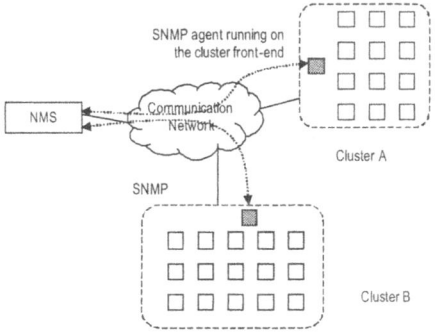

Fig. 1. Cluster management scenario

3.1 The Cluster MIB

The design of the cluster MIB was based on the necessities of our two clusters. Particularly, we were interested in (a) characterizing each cluster node, (b) monitoring cluster nodes CPUs and memory, (c) monitoring and controlling users processes and (d) reserving cluster nodes. Also, we wanted a clear differentiation between the management information related to the cluster front-end, where the SNMP cluster agent is located, and the information related to the cluster nodes. Based on these requirements, the *CLUSTER-MIB*, presented in Fig. 2, was defined.

The *clServer* group exposes data about the cluster front-end. *clServerName* and *clServerDescr* describe the front-end. *clServerTotalCPUs* gives the number of CPUs the cluster front-end has. Additionally, the *clServerCPUsFreq* gives the clock of these CPUs. We assume, in this case, that all CPUs in the front-end are identical. Finally, the *clServerUserCPUs* provides the percentage of use of the total CPUs. When in 100%, it indicates that all CPUs are used. *clServerTotalMem* and *clServerFreeMem* gives, respectively, the total amount of memory in the cluster front-end, and how much of this memory is free to be used.

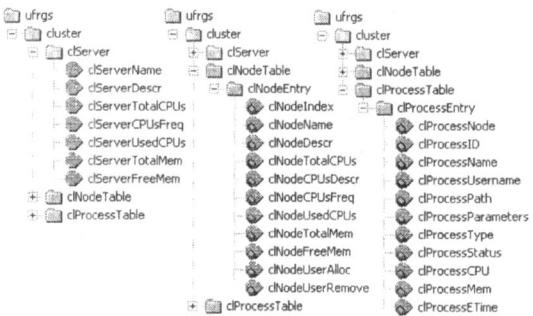

Fig. 2. The Cluster MIB

In order to manage the nodes of a cluster, the manager use the *clNodeTable*. It provides objects either to monitor the nodes or to allocate/remove specific users to/from such nodes. Each row in the table addresses one single node. The *clNodeIndex* indexes every node in the table, while the *clNodeName* and *clNodeDescr* describe the nodes. Similarly with the *clServer* group objects, the *clNodeTotalCPUs*, *clNodeCPUsDescr*, *clNodeCPUsFreq* and *clNodeUsedCPUs* objects allow the monitoring of the CPUs. Here again we suppose that all the CPUs in a node are equal. The *clNodeTotalMem* and *clNodeFreeMem* objects, on their turn, provide information about the memory of each node.

Node allocation is supported through the *clNodeUserAlloc* and *clNodeUserRemove* objects. Once a node is required to be allocated, the manager should set the *clNodeUserAlloc* with the username of the user to have the node allocated to. Although allocation is executed just after the object is set, the user will not have full access to the node until all previous users leave the node and their processes finish. However, after allocation, no other user, except the one that received the node, is able to log into that node. To deallocate a node (grant access back to every user) the administrator has to use an empty username in setting the *clNodeUserAlloc* object.

Even with the node allocation processes above, a user may have to wait indefinitely to have full access to the allocated node, since previous running processes are not killed by an allocation request. In critical situation, however, a node should have to be totally freed. The *clNodeUserRemove* object is designed to support process removal. Once the manager needs to remove all processes of a specified user from a node, the *clNodeUserRemove* object must be set to the username of the user to be removed. Furthermore, if all users of a node are have to be removed, the manager can use the special string "allusers" to indicate that every user process must be killed, except those started by the user indicated in the *clNodeUserAlloc*.

Finally, cluster process management is accomplished through the *clProcessTable*. This table is deeply based on the *hrSWRunTable* and *hrSWRunPerfTable* tables of the HostResources MIB [6]. However, since such MIB was designed to provide information about processes of a specific host, it does not fit in the approach we took to our solution, where a single point of management provides information about the whole cluster. In our Cluster MIB the pair *clProcessNode* and *clProcessID* indexes a process within a cluster. Since two different cluster nodes may have processes with the same identifier (PID), the differentiation from one process to the other is reached with the identification of the nodes that hold each process. With this mechanism it is possible to uniquely identify every process running in the cluster.

The *clProcessName* and *clProcessUserName* provides, respectively, the name of the process and its owner. *clProcessPath* indicates the execution path of the process, while the *clProcessParameters* provides the execution parameters. Each running process is of one kind: (1) unknown, (2) operating system, (3) device driver or (4) application. The process type is retrieved through the *clProcessType* object. Also, each process has a status: (1) running, (2) runnable, when waiting for a system resource such as CPU or memory, (3) not runnable or sleeping, (4) invalid, (5) stopped, when debugging a process or (6) zombie. The *clProcessStatus* provides the status of a process. The probably more "strange" status is invalid. Actually, this is not a status, but a special value to allow the process killing via SNMP. If a process needs to be killed, one can do that setting the *clProcessStatus* object to invalid. The last three objects of the process table are intended to be used as performance attributes. The *clProcessCPU* gives the amount of CPU used by a process. The *clProcessMem* provides the amount of memory used by a process, and the *clProcessETime* informs the elapsed time since the process was started.

3.2 The Agent Implementation

The cluster agent was implemented as an extension of the NET-SNMP agent [7]. Currently, the agent runs in a Linux environment, since this is the operating system used to implement the front-end of our clusters. The cluster agent operations executed in response to SNMP requests are carry out basically through bash scripts. Such scripts generally scans over the results of shell commands and operating system files to retrieve the information to be send back to the management station. The scripts are executed either local and remotely in order to retrieve information from both frontend and cluster nodes. Accordingly to these operations we will have two different kinds of agent interaction: front-end and node interaction.

In the *front-end interaction* the cluster agent communicates directly with the frontend system. Since the agent is also running inside the front-end, only front-end internal communication is observed. In this kind of interaction the system files provide important information. As shown in Fig. 3, when an SNMP request arrives to the cluster front-end, the SNMP agent executes a bash script and searches inside the results until the requested information is found. Then the agent manipulates the information to assemble an SNMP reply sent back to the management station.

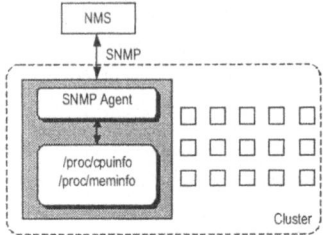

Fig. 3. Local Information Acquision

For the *clServer* group the *hostname* command is used to provide the *clServer-Name*. The system file */proc/cpuinfo* provides information related to the *clServer-Name*, *clServerDescr*, *clServerTotalCPUs*, *clServerCPUsDescr*, *clServerCPUsFreq* and *clServerUsedCPUs* objects. Memory information described in *clServer* group is provided by the */proc/meminfo* file, which informs the total and free memory on the front-end for the *clServerTotalMem* and *clServerUsedMem* objects.

In the *node interaction*, the cluster agent communicates with a cluster node, thus introducing management traffic within the cluster. The nodes information required by the cluster agent is found in a configuration file. Thus, among other information, the configuration file has a list of the nodes the cluster agent is responsible to access.

Node interactions are required to retrieve information to fill data in the *clNode-Table* and *clUserProcessTable*. Fig. 4 presents how information is retrieved from the cluster nodes: a retrieving script parses the configuration file and executes an RSH command for each registered node. The results are stored in a temporary file that is analyzed by the cluster agent to remove unwanted information. The remaining information is assembled in SNMP replies send back to the management station.

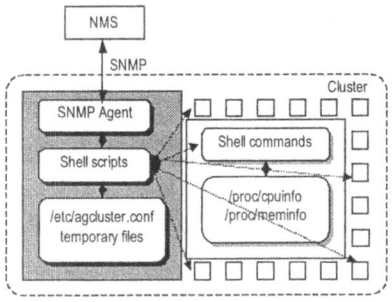

Fig. 4. Remote Information Acquision

In the *clNodeTable*, the *clNodeName* object is obtained with the *hostname* command. The *clNodeDescr*, *clNodeTotalCPUs*, *clNodeCPUsDescr*, *clNodeCPUsFreq*, *clNodeUsedCPUs*, *clNodeTotalMem* and *clNodeUsedMem* have the same behavior of *clServer* group objects presented before, except that the information is retrieved through a remote command execution. Complementary, in order to verify which processes (and associated information) are being executed in the cluster nodes, the *clUserProcessTable* and its objects are filled through the use of the *ps* command and its options executed in each cluster node.

The *clNodeUserAlloc* and *clNodeUserRemove* objects use a different strategy than the one used by previous objects. In the case of a cluster based on Linux, the */etc/security/access.conf* file is manipulated. Such file describes who may access a cluster node. If a user name is passed to *clNodeUserAlloc* the configuration file is manipulated to allow the access only to the user informed (see the line below).

    -:ALL EXCEPT root username

When setting the *clNodeUserAlloc* object with "allusers" the line above is remove from *access.conf*, allowing every one to have access again.

A user node reservation does not mean the user will have full control of the node. Processes of other users can remain active in indefinitely, then *clNodeUserRemove* is used. A process is killed through a combination of the basic shell commands.

4   The SNMP-Based Management System

In this section we present the developed system that communicates with the cluster agent to allow the administrator to take advantages over agent features. The management system is developed as a set of PHP4 [8] scripts that provides a Web-based interface throughout the administrator can communicate with the cluster agent.

Based on our none automated management experiences, we can point three critical cluster management tasks: checking the load of the front-end and nodes; checking the nodes utilization by users; checking and proceeding with node allocation. Thus, the development of the management system was driven to support such tasks.

4.1   Ranking the Cluster

Checking the load of the cluster front-end and nodes is needed in order to discover if the front-end is overloaded and to check the most used nodes. This allows the cluster administrator, to decide which nodes use to execute a demanding task.

The management system provides support for checking the load of the cluster front-end and nodes through two tables. The first and simplest (Fig. 5) shows the information related to the front-end, where the administrator can check the front-end load in terms of its number of CPUs, CPUs clock and CPUs usage, and in terms of its total, available and free memory.

The node table (Fig. 5) typically has several rows, one row per node. Although the columns of the node table are identical to the ones from the front-end table, here we have the option to rank the rows based on the values for a particular column. For example, supposing that the administrator intends to select the two nodes with larger available memory, then he or she can click on the "Free Memory" column and the same table would be presented again with rows with larger free memory first.

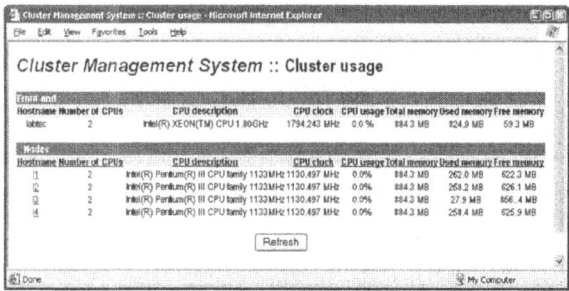

Fig. 5. Front-end and node management tables

Internally, every time the PHP4 scripts that generate the front-end and node table are accessed, several SNMP requests are sent to the cluster agent in order to retrieve the current values of the managed cluster.

4.2 Is the Cluster Policy Being Respected?

Once the cluster is available to users, it is expected that such users consume the resources in a proper way. A "proper way" depends on the goals of the cluster; and that goals are accomplished through the definition of the cluster policy of use. A cluster policy is an abstract concept understood by the users and by the administrator. One of the key tasks is to verify if the cluster users are consuming the available resources in a coherent way, consistent with the cluster policy. "Bad users" are those that do not respect the cluster policy; "good users" are those that do.

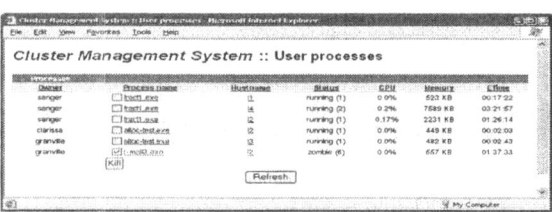

Fig. 6. User processes management table

The problem here is to verify how the cluster manager identifies good/bad users. The management system is able to reveal user's activities through the user process table. This table lists the user processes in order to let the administrator verifies which users are respecting the cluster policy of use. Again, different strategies to show this processes can be selected by clicking in the columns of the table (Fig. 6). The default strategy is to show the processes grouped by users, but the administrator may chose, for example, to group the same processes by nodes.

4.3   Controlling the Nodes Allocation

Node allocation is required to grant special access to special users. The determination of who is these special users depends on the goals of the cluster and the intention each used has when using the cluster resources. This can not be determined by the cluster itself, but by the cluster administrator. In the end, the administrator has to have mechanisms to allocate/deallocate cluster nodes to the users.

To control the allocation of cluster nodes, the management system provides the allocation table. This table complements the tables previously presented and allows the cluster administrator manage the node allocation strategy. The table is composed by a column to identify each cluster node and two additional columns (owner and users) that present, respectively, the user which has to node allocated to and the list of active users that have processes still running on the node (Fig. 7).

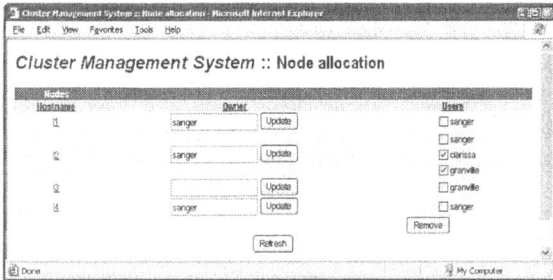

Fig. 7. Node allocation management table

Allocation can be partial or total. In partial allocation, the administrator determines the user that will receive the node but does not remove the previous users. In this case, newer processes can only be created by the new node owner, but previous running processes created by other users will still remain. In the total allocation, however, the administrator also determines the new node owner and additionally all the previous processes that were not started by the new owner are removed. Partial and total allocation is performed by the manipulation, in the system Web-interface, of the owner and users columns of the allocation table. Same interface facilities are provided in order to help the administrator in proceedings with partial and total allocations. For example, once a node is allocated to a user, the user interface already presents the users of this node, except the owner, with a checked box ready to remove such users. Some other facilities as this one are provided along the PHP4 scripts.

5   Conclusions and Future Work

With the SNMP agent and the management system developed the main management functionalities required in our research clusters were achieved. With the developed solution we are able to proceed with the most critical and frequent tasks of the cluster administration, namely the cluster nodes load verification, the user utilization of the cluster resources and node allocation.

From the development of the cluster agent we conclude that a single cluster management interface is an interesting feature because the whole coordination to retrieved the cluster information is executed in the front-end agent, freeing the management system of this coordination task. However, an scalability drawback may be present when the number of cluster nodes increases. In this situation the cluster agent will interact with several nodes and the performance of the agent will surely degrade. In this point a future work is to proceed with scalability checks to verify the agent behavior with different numbers ofcluster nodes.

From the development of the management system we conclude that the automation of cluster management tasks can be easily achieved. We have provided in the management system a first set of management functionalities that helped to carry the most common tasks of management of our test cluster. Further automation may be developed to improve the system. An immediate work is the development of a support to describe the cluster policy of use and an automatic way to check if such policy is being accomplished by the users. Today, as explained in the previous section, this policy checking is manually executed observing the user processes. An automatic checking would easy even more the execution of this task.

A final work to be executed is the investigation of the management traffic introduced in the standard network when the cluster management solution developed is used. Complementary, we also need to verify the amount of cluster internal management traffic in order to guarantee that such traffic does not overloads the cluster decreasing its performance.

References

1. Foster, I.: The Grid: Blueprint for a New Computing Infrastructure, Morgan Kaufmann, San Francisco (1999)
2. Massie, M.: Ganglia - Distributed Monitoring and Execution System, http://ganglia.sourceforge.net, May (2003).
3. Subramanyan, R., Miguel-Alonso, J., Fortes, J. A. B.: Design and Evaluation of a SNMP-based Monitoring System for Heterogeneous, Distributed Computing, School of Electrical and Computer Eng, Purdue University, July (2000).
4. PBS: Portable Batch System, http://www.openpbs.org, May (2003).
5. OSCAR: Open Source Cluster Application Resources, http://www.csm.ornl.gov/oscar, May (2003).
6. Waldbusser, S., Grillo, P.: Host Resources MIB, IETF RFC 2790, March (2000).
7. Hardaker, W.: The NET-SNMP Project, http://netsnmp.sourceforge.net, May (2003).
8. PHP4: http://www.php.net, May (2003).

A Web-Based Pro-active Fault and Performance Network Management Architecture

Andrea Silva Ramos, Anilton Salles Garcia, Rodolfo da Silva Villaça, and
Rodrigo Bonfá Drago

Universidade Federal do Espirito Santo – LPRM – Av. Fernando Ferrari, s/n – Campus de
Goiabeiras – CT VII – Informática – Vitória – ES – Brasil – CEP 29060-970.
{dearamos, anilton, bonfa}@inf.ufes.br,
rodolfo@maxima-ti.com.br

Abstract. This paper proposes a Service Level Management architecture for an
IT (Information Technology) environment based on the concept of the creation
of a Remote Network Operations Center through the Internet and Web interface
under safe connections, viewing cost reduction and simplification of the work
underlying network management. Additionally, a methodology for fault and
performance management is exposed and a laboratory case is carried out in
order to validate the proposal.

1 Introduction

The most common profile of network managers and administrators can be defined by
the reactive management in which first the problems occur and only then they are
analyzed and fixed. None or very few pro-active actions are taken with the objective
of anticipating and preventing problems.

The management activity consists in observing and controlling the most important
events during the daily operations of an IT environment. Through the manipulation of
the data obtained by management tools it is possible to fully comprehend the whole
environment, seeking tendencies and reducing the time between isolating and fixing
the faults, and, if possible, foreseeing them.

There are many solutions elaborated and developed to solve this question.
However, the fast evolvement of technology asks for a flexible computational
methodology which is able to gather procedures and tools in order to constantly
reduce the cost of systems management.

Bearing in mind the integration of all management activities in a unique
environment which is able to produce cost reductions and to be as flexible as possible,
this work has the purpose of specifying and generating a prototype of a hardware and
software tool to help IT management integrated in a remote NOC (Network
Operations Center). Also, an adequate pro-active methodology for fault and
performance environment is proposed.

Firstly, a brief explanation of the main concepts on network management involved
in this paper is given . It covers tools, standards and architectures. Secondly, a proactive methodology for fault and performance management is exposed. This has been
adapted from the original found in [Monteiro, 2000]. The NOC architecture to support

J.N. de Souza et al. (Eds.): ICT 2004, LNCS 3124, pp. 792–800, 2004.
© Springer-Verlag Berlin Heidelberg 2004

this methodology is proposed and specified. A case study finalizes the work, showing the current stage of development and the results previously obtained.

1.1 Network Management Protocols, Standards, and Architectures

Among the management protocols specifically developed for this activity, the SNMP (Simple Network Management Protocol), is currently the standard for Internet management and it is based on the Manager/Agent/MIB (Management Information Base) architecture to designate the management functions in the network. See [Stallings, 1999] and [Feit, 1995] for more detailed information on SNMP architecture. The RMON (Remote Monitoring) [Stallings, 1999] standard was developed so as to complement the SNMP features for network traffic management information.

The fast growth of the WWW (World Wide Web) made it possible to develop a wide range of web based network management architectures. According to [Tsai, 1998], there are three advantages of using web based network management: 1) there is no need of specific software to manage and configure the devices; 2) there are no longer problems of version in the management stations; 3) the management framework and its location are independent.

Basically there are three NMS Architectures (Network Management System Architectures) as classified by [Leinwand, 2000] and [Kahani, 1997]: centralized, hierarchical, and distributed. Most NMS are centralized, in other words, a unique host keeps an eye on all IT systems. Although simple, it shows problems such as a high concentration of fault probability on a single element and low scalability.

In the hierarchical architecture the MOM (Managers of Managers) and the domain manager concepts are applied. Each domain manager has to monitor its own domain only. The highest level domain manager make periodical polls to the lowest level domain managers. Distributed architectures are like peer-to-peer, where multiples managers, each managing its own domain, communicate with the others on a peer-system. A manager from any domain can poll information to another manager from another domain.

1.2 Pro-active Performance and Faults Management Methodology

Pro-active management is the one in which the network manager searches continuously for information that can help it to foresee network problems. Statistical resources and daily event monitoring are used to follow up behavior changes and to predict faults and loss of performance in the IT environment.

The OSI (Open Systems Interconnection) research group, through the 7498-4 document, divided network management into five functional areas: fault, configuration, security, accounting and performance management.

Fault management is a process that consists in finding and fixing problems or faults in a network. Fault management is probably the most important task in the network management process. Basically three steps are involved: 1) locating the fault; 2) isolating its cause; 3) fixing it, if possible. The simple use of a fault management tool in this process can optimize this task, especially tasks 1 and 2. There are two ways to locate network faults: polling and traps [Leinwand, 2000].

According to [Stallings, 1999], networks are made by grouping shared resources in which each network demands attention to respect the limits of performance by the services that made use of those resources. This is the scenario in performance management. There are two central tasks involved in performance management: monitoring (performance analysis) and control (capacity planning). Monitoring consists in collecting information about traffic (error rates, throughput, loss, utilization, collisions, volume, matrix), services (protocols, overhead, matrix, response time) and resources (CPU utilization, memory available, disk utilization, ports) and comparing this information with the normal and/or desirable values for each. Controlling consists in taking actions to plan or modify network configurations and capacities to meet the desirable performance.

[Monteiro, 2000] proposes a pro-active methodology for performance management of computer systems. This methodology was extended so as to include the fault management. The methodology proposed divides the action of the manager into four steps placed in ranking order according to the needs of investigating and solving performance problems as they occur. Figure 1 displays the four steps.

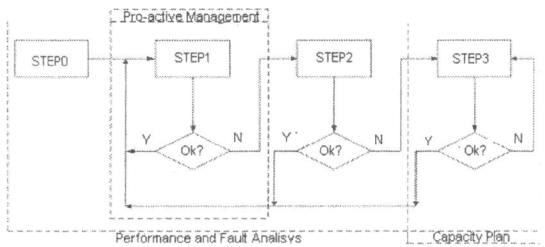

Fig. 1. Pro-active Fault and Performance Methodology. Source: [Monteiro, 2000]

The first step (Step 0) has to establish the initial conditions for the methodology, in other words, it must define the SLA (Service Level Agreement) for each service and resource monitored by pollings and traps.

In the next step (Step 1) information about the indicators of each resource monitored is generated. These indicators can be obtained directly or calculated by manipulating other indicators. This represents the SLM (Service Level Management) implementation in the managed environment. If an SLA is not being accomplished, the methodology goes to the next step (Step 2) which is basically the pro-active part of the methodology and it must be repeated as fast as the changes that occurred in the managed environment.

If in Step 1 problems are detected through performance indicators, Step 2 starts investigating them, and, if possible, it isolates and identifies them as well as generate suggestions to repair them. The NOC architecture intends to develop a tool to relate all (or some) fault and performance events based on a problem diagnosis and symptoms database, as proposed in [Lopes, 2000].

Step 3 involves capacity planning, where the current and the previous state of the environment is explored to simulate and predict responses for all selected indicators in the future. These activities are not included in this version of this work.

To make the IT environment management easier, this paper also suggests a four-layer division as illustrated in Figure 2.

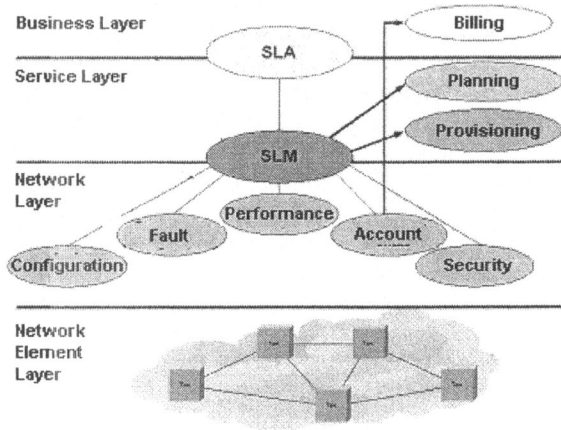

Fig. 2. Management Layers

- Network Element Layer: devices and servers and their own performance and fault indicators separately.
- *Network Layer:* integrates the management of the network elements. Here, data collected at Layer 1 is consolidated so as to produce fault and performance indicators related to the SLA. Layer 1 and 2 correspond to Step 1 and 2 of the methodology.
- *Service Layer:* manages the availability and well-functioning of the services that support the applications. The SLM database produced at the lower layers helps the managers to plan the capacity and provisioning of the services and applications of the managed environment. It corresponds to Steps 2 (SLM share) and 3 of the methodology proposed.
- *Business Layer:* aims to increase the satisfaction of the users and the productivity by defining the SLA (Step 0) that meets the requirements of productivity. The stricter the SLA definition is, the better the management and planning of the IT environment has to be, what implies costs. These costs need to be equivalent to the return of the investment in management (Billing).

1.3 The NOC Architecture

A hybrid architecture which incorporates hierarchical and distributed characteristics was chosen to meet the requirements of the four layers described in the last section of this paper. At NOC the Service and Business layers are managed, producing charts and reports to the highest level administrators by summarizing and consolidating data collected from the management server (MBOX).

The main motivation underlying this work is the creation of a remote Network Operations Center (NOC). The NOC is composed of a central network management server and a database that keeps historical statistics about the networks managed. It can be web-accessed safely, based on the concept of virtual private networks (VPN). The other elements involved in NOC are an event and information database containing fault and performance data collected (through SNMP protocol) and a management server (MBOX), located in each network monitored by NOC as shown in Figure 3.

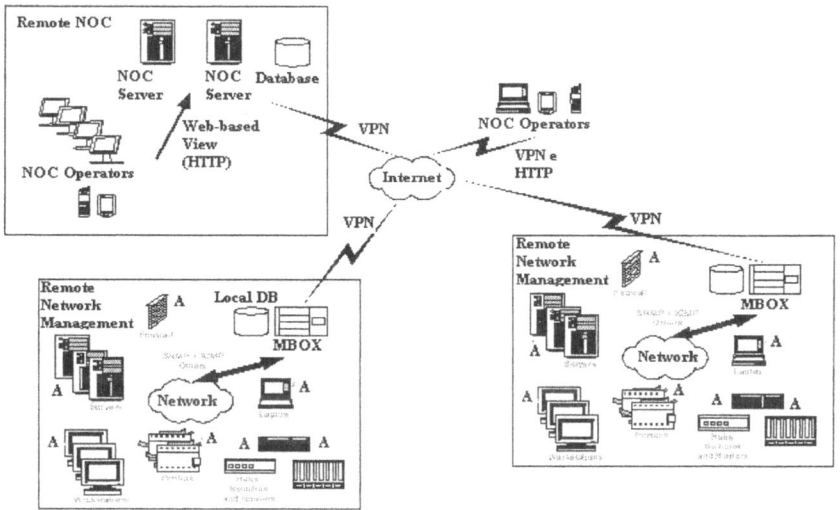

Fig. 3. NOC Architecture sketch

In Figure 3 some important elements can be pointed out:

- *MBOX* – Low cost connectivity hardware and software platform based on Linux operating system. The MBOX connects NOC to the managed networks using ADSL (Asymmetric Digital Subscriber Line) even if it is based on a dynamic IP solution offering a minimum of 95% of availability to the link. MBOX monitors managed objects, stores data information collected at its Local Database, generates events (traps) for the NOC operators, communicates with NOC to store SLM data at the historical database and relates the events generated in order to try to find and anticipate fault and performance problems;

- *Agents (A)* – Makes information about the managed objects available in the device in which it resides. To make this work compatible with the most common management protocol (SNMP), it supports MIB-2, RMON MIB and private MIBs. By using the SNMP Proxy concept, this work intends to establish a unique interface to make it possible to manage any device, even the SNMP non-compatible elements;

- *NOC Server*: central element in this architecture, it provides the NOC operators a web-based interface that is capable to supply them with all information needed to manage the monitored networks. A simple or more detailed interface can be chosen, depending on the NOC operator's needs. The NOC database has to maintain historical and statistical fault and performance management indicators to generate charts and reports about the behavior of the managed environment.

The SLA management model, in which a user can establish the desirable quality of service level to be supplied by its provider is the model adopted here. Penalties and taxes can be established previously for when the quality of the service is not reached.

The migration to this model of work is not simple and can generate high initial costs, even technically, because establishing good SLAs relies on having very good knowledge of the managed environment. Its availability, traffic shape, and fault tolerance needs must be measured. In spite of these problems,

SLA definitions depend on which indicators will be monitored and what values are adequate to each case. Some indicators are listed below, according to [Azambuja, 2001] and [Wise 1997].

Technology and Services independent indicators:

- *Availability percent:* given a monitored period, indicates the portion of the total time that the service was available to its users;
- *MTTR (Mean Time To Repair):* mean time needed to fix an unavailable service;
- *MTPS (Mean Time to Provide Service):* mean time needed to install and make a new service required available.
- *Response Time:* The amount of time that elapses between the user requesting a service and the service provider response, accomplishing the request.

Technology and Service- Dependent indicators:

- *Link Layer:* throughput (Bytes/s, Packets/s), error rates, collision rates, latency;
- *Servers: CPU* utilization, memory available, disk utilization.

1.4 Tool Presentation (Current Stage of Development)

The solution to safe connectivity between NOC and managers is already in use through the first version of MBOX. Currently, a tool prototype to collect SNMP based fault and performance data, based on [RAMOS, 2004] and [DRAGO, 2001] is working. This tool communicates with a Local Database to store information about the collected data. A web interface, which now simulates the NOC Server, is used to produce charts and reports about the management performed in a particular period and device. Up to the present moment, not all the performance and fault indicators have been collected to produce the SLM data.

The SLA input, monitoring and an event correlation tool are still in progress and will generate a Computer Science graduation project, due on July 2004. This implementation is based on the specification work of [DRAGO, 2004]. The NOC Server and the complete NOC architecture are also in progress and will be an Electrical Engineer Master Dissertation (in Portuguese).

To bring scalability to this project the research group decided to use Java language in the development of all components of the architecture. The web interface is totally based on servlets and HTML language. The database management system chosen is PostgreSQL because it is free software and it is robust enough to support the amount of data expected. To generate charts and reports the Java software named JFreeChart and JFreeReport were chosen.

2 Case Study

In this section some results obtained by the group with the NOC architecture will be shown. The managed environment used is the Multimedia and Networks Research Lab (LPRM), located at the Federal University of Espírito Santo, Brazil. This experience was carried from 01/15/2004 to 01/23/2004.

The objective of this case study will be taken as satisfactory if the performance and fault indicators can be collected and able to generate SLM data. Thus the SLM data will have to be available to the NOC Operators using a web interface and a safe internet connection.

Corresponding to Step 0 of the methodology exposed at the last sections, the following fault and performance indicators were chosen to be monitored by the collect tool based on SNMP MIB-2 and private MIB agents: CPU utilization, amount of memory available, read and write disk utilization, network throughput, error rates and system availability.

The acceptable level of performance and fault indicators (defined in Step 0) weren't analyzed in their absolute values. Only the mean, maximum, minimum, standard deviation and coefficient of variation values were calculated. The decision to not establish SLA presently was due to the fact that its input and monitoring tool was still being developed when this study was carried.

On Step 1 of the methodology, as mentioned in the previous paragraphs, an SNMP based collect tool was developed. This tool is based on pollings to the agents and their managed objects, corresponding to the fault and performance indicators chosen in Step 0. The data collected is stored in a local database. Everything in Step 1 resides at MBOX. The manager can choose a period of time to monitor based on the critical activity of the managed environment. The monitoring of fault and performance indicators is available to the manager to each element managed by this tool.

The web interface provides daily and monthly reports based on the monitoring periods chosen, the devices and the indicators selected by the user. Selecting the daily summary option the system will show you an on-line chart where all the values collected for each indicator will be piloted. The monthly summary will bring a chart with the daily mean values per day and a report containing statistical values for each indicator selected.

In this case study the MBOX has been used as the network manager of LPRM. SNMP performance agents were installed at the Lab Server and configured to be used. A managed switch device was also used for monitoring.

In this first experience the following fault and performance indicators were monitored: the Lab Server network interface availability, CPU utilization, amount of memory available, free disk space and read/write utilization and network throughput. Remotely, using a web interface, charts and a statistical table together with summary are presented.

Further, these values (which compose the SLM data) will be constantly monitored and compared to the desired SLA. This consolidated data will also be transferred to the NOC Database that will create a historical database to generate the normal shape and a baseline for each indicator.

Presently, the results obtained have met the expectations and all NOC components which are already in an advanced stage of development have worked satisfactorily.

The fault and performance indicators selected at Step 0 of the proposed methodology were all collected and analyzed by the developed tools.

3 Conclusion

Although it was not possible to thoroughly test the proposed architecture because it is still in development, the previous results show that it is very adequate to perform remote pro-active management on IT environments.

The main advantages of this architecture are the management database distribution and the centralization of the management operations at a remote and specialized NOC, which means that it can be applied from large to small environments. To solve the NOC Sever dependence on centralization, dual links and database replication can be used when necessary. Even when a link failure occurs between NOC and the managed network, the MBOX continues to collect, manipulate and monitor the SLA, generating reports and traps when it fails, which means that an alternative solution can be used to deliver the trap and reports to the NOC Operators. For instance, a Wireless Cellular Network can be used to do this work. In this case, a link failure will only temporarily stop the local and NOC database data transfer.

The use of free software has proven to be totally adequate to the objective of this project, providing NOC with an excellent way to manage scalability and flexibility, since the project evolves as the technologies involved in it evolve, too . Some free management software were analyzed by the research group, such as Nagios and NET-SNMP, and contributed to the work inspiring and supplying data during the early tests. The final work which will complement the NOC specification intends to include a standardized interface in which any data collected via SNMP through any collecting tool will be able to be inserted in the MBOX local database and, when ready, these can generate the SLM data and be incorporated to the proposed methodology.

As a further step, the research group involved with NOC development plans to turn automatic reports as simple as possible so as to simplify the analysis of the management results.

References

Azambuja, Marcelo C. PSWeM: Development and Implementation of a Web-based Tool for Service Level Management. Masters dissertation (in portuguese). Electrical Engineer Post-Graduate Program. Pontifícia Universidade Católica from Rio Grande do Sul. Porto Alegre: 2001.

Drago, Ádrian B. A computational tool to help practical Computer Network Management. Masters dissertation (in Portuguese). Electrical Engineer Post-Graduate Program. Federal University of Espírito Santo - UFES. Vitória: 2001.

Drago, Rodrigo B. An IT environment Management Tool specification. Graduation Project on Computer Engineer. Federal University of Espírito Santo - UFES. Vitória: 2004.

Feit, Sidnie. SNMP A Guide to Network Management. McGraw-Hill, 1995.

Kahani, Mohsen.; BEADLE, H. W. Peter. Decentralized Approaches for Network Management. Computer Communications Review. ACM SIGCOMM, Vol. 27 N.3. July (1997).

Leinwand, Allan.; FANG, Karen. Network Management: A practical perspective. 2 ed. Massachusetts: Addison Wesley, 1995.

Lopes, Raquel V. Best Practices for Computer Networks Management. Masters dissertation (in Portuguese). Computer Science Post-Graduate Course. Federal University of Paraíba. Campina Grande: 2002.

Monteiro, Maxwell E. Methodology for Data Communications Network performance evaluation. Masters dissertation (in Portuguese). Computer Science Post-Graduate Program. Federal University of Espírito Santo - UFES. Vitória: 2000.

Monteiro, Maxwell E. DRAGO, Ádrian B. GARCIA, Anilton S. A Methodology for Network Performance Management. Paper accepted for presentation in the 25th Annual IEEE Conference on Local Computer Networks. Tampa, Flórida: 2000.

Ramos, Andrea S. Methodology for Data Communications Network performance evaluation. Masters dissertation (in portuguese). Computer Science Post-Graduate Program. Federal University of Espírito Santo UFES. Vitória: 2004.

Stallings, William. SNMPv1, SNMPv2, SNMPv3 e RMON 1 e 2. 3. ed. Massachusetts: Addison Wesley, 1999.

Tsai, Ching W.;CHANG, Ruay-Shiung. SNMP through WWW. International Journal of Network Management. John Wiley & Sons, 1998.

Wise, Sid. Client/Server Performance Tuning. New York: McGraw Hill, 1997.

A CIM Extension for Peer-to-Peer Network and Service Management

Guillaume Doyen, Olivier Festor, and Emmanuel Nataf

The Madynes Research Team, LORIA
615 rue du Jardin Botanique
54602 Villers-lès-Nancy, France
Guillaume.Doyen@loria.fr

Abstract. Peer-to-peer (P2P) networks and services are increasingly present in the networking world. Emerging P2P based applications targeting enterprise solutions require an open approach to their management that ensures that the operation of any service is in agreement with QoS parameters. Our contribution deals with the design of a management framework for P2P networks and services. As part of this effort, we have designed a management information model that captures the different functional, organizational and topological aspects of the P2P model. It is built on the standard CIM model and is the focus of this paper.

1   Introduction

Peer-to-peer (P2P) networking is built on a distributed model where peers are software entities which both play the role of client and server. Today, the most famous application domain of this model concerns the file sharing with applications like E-mule, Napster, Gnutella and Kazaa among others. However, the P2P model also covers many additional domains [1] like distributed computing (Seti@Home [2], the Globus Project ([3])), collaborative work (Groove (www.groove.net), Magi (www.endtech.com)) and instant messaging (JIM [4]). To provide common grounds to all these applications, some middleware infrastructures propose generic frameworks for the development of P2P services (Jxta (www.jxta.org), Anthill [5], FLAPPS (flapps.cs.ucla.edu)).

The P2P model enables valuable service usage by aggregating and orchestrating individual shared resources [6]. The use of existing infrastructures that belong to different owners reduces the costs of maintenance and ownership. The decentralized topology increases fault tolerance by suppressing any central point of failure, and improves both load balancing and scalability. At last, the distributed nature of algorithms and some embedded mechanisms allow participating peers to keep a great level of anonymity.

While some applications use built-in incentives as a minimal self-management feature, advanced management services are required for enterprise oriented P2P environment. These services are the focus of our attention and we work on the design of a generic management framework for them.

J.N. de Souza et al. (Eds.): ICT 2004, LNCS 3124, pp. 801–810, 2004.
© Springer-Verlag Berlin Heidelberg 2004

The major step toward this objective consists in designing a generic management information model for P2P networks and services that can be used by any management application as a primary abstraction. We have chosen CIM as the framework for the design of our model because it uses the object oriented model and because it provides a large set of classes covering several domains of computing that can be easily reused and extended. The P2P management information model we have designed is described in this paper.

Therefore, the organization of this paper is the following: section 2 presents the diverse contributions that address P2P management. Then, section 3 presents our extensions of CIM for P2P networks and services. Finally, section 4 gives some conclusions and draws some directions for our future works.

2 Related Works

A lot of effort was invested in detecting and monitoring P2P traffic in order to constrain or prohibit their service [7]. Thus, enterprise management products provide solutions to integrate P2P traffic among the other networking applications. Recently, managing P2P networks and services in order to ensure their efficient operation is emerging. For example, in the file sharing domain, some incentive approaches embedded in applications tend to improve the global QoS[1]. For example, some applications use a reputation system that allow peers respecting some constraints to obtain better results in their file discovery or downloading. These constraints can be the total size of the shared files, or the constancy of the connection.

The main research initiative that uses incentives is MMAPPS (`www.mmapps-.org`). This project aims at managing P2P services by using market management techniques. Three application domains are targeted by this project. First, the file sharing with the establishment of resource share incentive mechanisms [8], then the wireless communication [9] and telemedecine. Nonetheless, this project is market management oriented and doesn't aim at providing a generic P2P management framework.

Finally, Jxta as a framework provides a P2P abstract model. While some of its concepts are common with our model like the peer and community notion or the pipe one. Nevertheless, Jxta doesn't aim at providing a management information model and so doesn't provide the abstraction for the full monitoring of the functional, organizational and topological aspects of its peers.

3 The P2P Extension

This section presents our CIM extension for P2P networks and services. The DMTF Common Information Model (CIM) is an approach to the management of systems and networks that applies the basic structuring and conceptualization techniques of the object-oriented paradigm [10], [11].

---

[1] Quality of Service

3.1   General Overview

The proposed extension for the P2P networks and services we have designed aims at providing a general management information model, that is topology-oriented, for such a type of application. This way, it allows any P2P application to subclass it in order to provide dedicated classes that will represent the specific application features at best. Instances of these classes will provide a distributed MIB[2] that a management application will use to administrate the application.

The model is composed of five main parts that are:

- **The peer and community model:** models the notion of peers, community, virtual topology and links this abstraction to the current Core classes.
- **The communication topology model:** provides information about the way peers communicate in term of protocols and communication medias.
- **The resource model:** informs about the resources a peer shares with another in order to contribute to the smooth running of a service.
- **The service model:** represents the services that can be provided and consumed in the context of P2P networking.
- **The routing model:** models the routing and forwarding services and the routing tables hosted by peers.

3.2   Peer and Community

In the P2P domain, a peer represents a piece of a distributed application built over the P2P model that is executed on a computer system. Such applications can be of several types ranging from file sharing to distributed computing. The value of provided and consumed service comes mainly from the aggregation of the peers' ressources and their collaboration. In order to represent the gathering of peers around a common interest, we introduce the notion of community. A community is a set of peers that participate to a common service.

Given this definition, we have addressed several use cases. First, a peer can belong to more than one community if, for example, it runs more than one P2P application. Secondly, a community can host sub-communities. It may be the case of an Instant Messaging application that offers different topics. Finally, a peer always belongs to at least one community. If it doesn't manage to connect an existing one it will create a personal default community.

The peer and community model. The peer and community model, represented on Figure 1 is the root of our model. All other classes of the extension we have designed are built around it.

A peer is modeled by the *P2P_Peer* class. It inherits from the *CIM_Enabled-LogicalElement* class of the core model. Its association with the *CIM_Computer-System* class models its hosting in a particular computer system. The community a peer belongs to, is represented by the *P2P_Community* class. It is a subclass of the *CIM_AdminDomain* class since the latter, defined in the Network Common

---

[2] Management Information Base

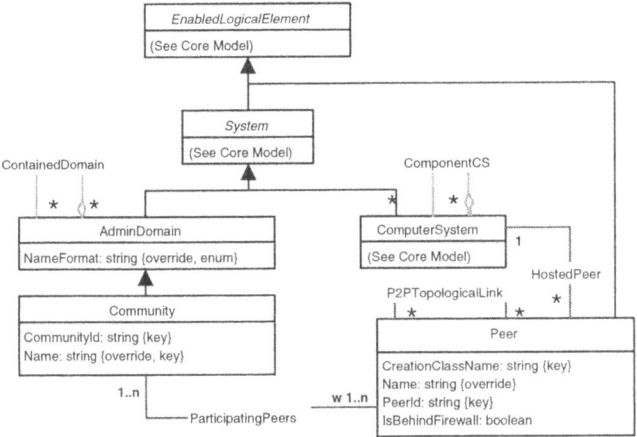

Fig. 1. The peer and community model

Model [12], represents any set of elements (devices, services, ...) that belongs
to a common domain. Peers and communities are linked together through the
P2P_ParticipatingPeer class. One may note that its cardinality means that a
peer can belong to one or more communities and that a particular community
must contain at least one peer. Moreover, the *CIM_ContainedDomain* association
class allows us to define sub-communities as presented above.

The virtual topology. A P2P network is one of application level overlay.
Establishing the topology of such network can rely on different criteria. For
example, it can be based on knowledge, routing, interest or technological considerations. This is why, we have chosen to integrate the virtual topology aspect
of the P2P model but without associating any particular semantic. We capture this notion in our model through the *P2P_P2PVirtualTopology* association
class. It enables us to represent an association that two peers present and thus,
to build a virtual topology. A description property adds semantic information
to the association.

3.3   Communication

In a P2P application, peers communicate together via an application level protocol. The latter binds two underlying transports endpoints. CIM captures this
notion through the pipe notion. A pipe represents a virtual communication channel that links two entities and allows them to communicate.

Our model is built around the *CIM_NetworkPipe* class. A pipe is bound to
two protocol endpoints with the *CIM_EndpointOfNetworkPipe* association class.
Considering the pipe modeling in a P2P context, we have designed two association classes that bind pipes with peers and communities. First, the *P2P_Peer-
UsesPipe* association class links the *CIM_NetworkPipe* with the *P2P_Peer* class.

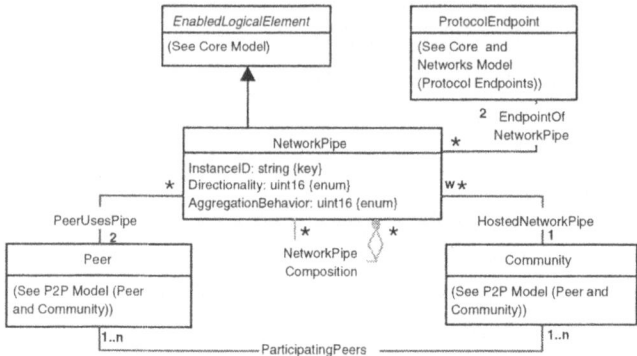

Fig. 2. The virtual topology model

The cardinalities mean that a peer may use zero or several pipes but a pipe is attached to two peers. Then, the *P2P_HostedNetworkPipe* association class links the *CIM_NetworkPipe* with the *P2P_Community* class. Indeed, we have chosen to define pipes in the context of a community. Thus a pipe is unique in a particular community and a community can host zero or several pipes.

3.4   Resources

In order to participate in the service delivery, a peer may provide part of its resources to the community it belongs to. We have considered different types of resources that can be shared by a peer.

First it can be some processing power. For example, in case of a distributed computing application, peers will receive a chunk of a global computation, process it and return the result to another peer responsible for the aggregation of the different results. Then, it can be storage capacity. A peer can share empty space of its hard drive and thus allow someone to store some files in it. Moreover, files that are explicitly shared by a peer are considered as a resource too. Finally, it can be bandwidth. Indeed, the latter can be used in order to ensure a good service operation and not just for the local service usage.

In order to model all types of resources described previously, we have designed the *P2P_PeerResource* abstract class and linked it to the *P2P_Peer* class through the *P2P_PeerSharesResource* association class. Concerning the cardinalities, we consider that a peer can share several resources but a particular resource can only be shared by zero or one peer. In order to model the use of resources in the context of a service, we have linked the *P2P_PeerResource* with the *P2P_P2PService* with the *P2P_ServiceUsesResource* association class; a service can use zero or several resources and a resource can be used by zero or several services.

Given the *P2P_PeerResource* abstract class, we have modeled the four concrete resources that can be shared by a peer. First, the *P2P_SharedProcessing-Power* class represents the different way for sharing processing power. Then,

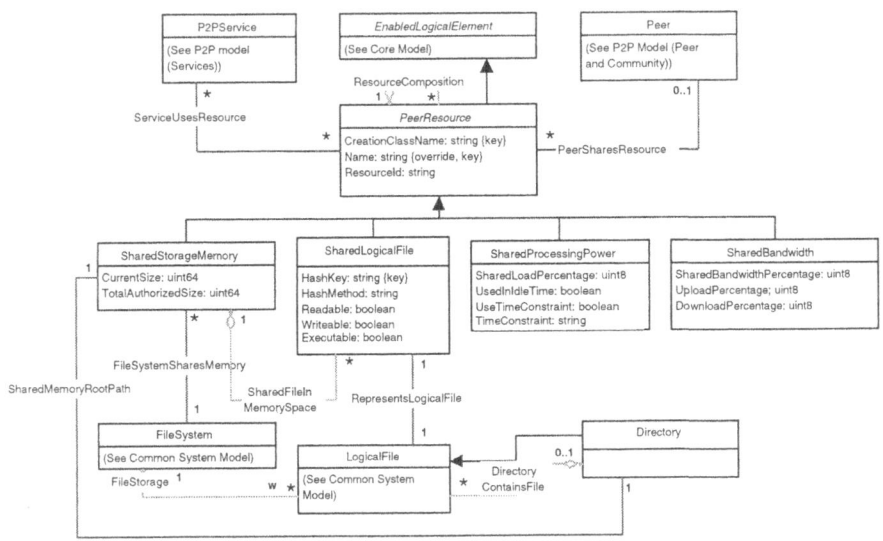

Fig. 3. The resource model

through the *P2P_SharedBandwidth* class, we have modeled the fact that a peer may share some of its communication resources. This share can be defined in terms of upload or download bandwidth.

Thirdly, we have represented the shared storage memory. The *P2P_Shared-StorageMemory* class is linked to the underlying file system through the *P2P_-FileSystemSharesMemory* dependency class. This association allows us to obtain information about the properties of the file system. The *P2P_SharedStorage-Memory* class is linked to the *CIM_Directory* too. Thus we have concrete information about the directory that is the root of the shared space.

Lastly, we have modeled different files that enter shared storage memory. This is done through the *P2P_SharedLogicalFile* class. A shared file is attached to its storing shared memory through the *P2P_SharedFileInMemorySpace* aggregation class: a file can just be stored in a shared location but the latter can host zero or several shared files.

3.5   Service

According to the CIM standard specifications, a service is a functionality provided by a device or a software feature. For example, a computer system may run a large variety of services like a word processing, an email client or a print service. Some of them are local while others are remote and use networks facilities. In this way, CIM models the configuration, operational data and management of functions. Moreover, for a particular service, CIM represents its access point as the way in which a service is invoked. To continue the previous example, the email client service access point is the protocol endpoint used to pull email from a server.

The translation of these concepts in the CIM Core model has led to the
design of the *CIM_Service* and *CIM_ServiceAccessPoint* abstract classes.

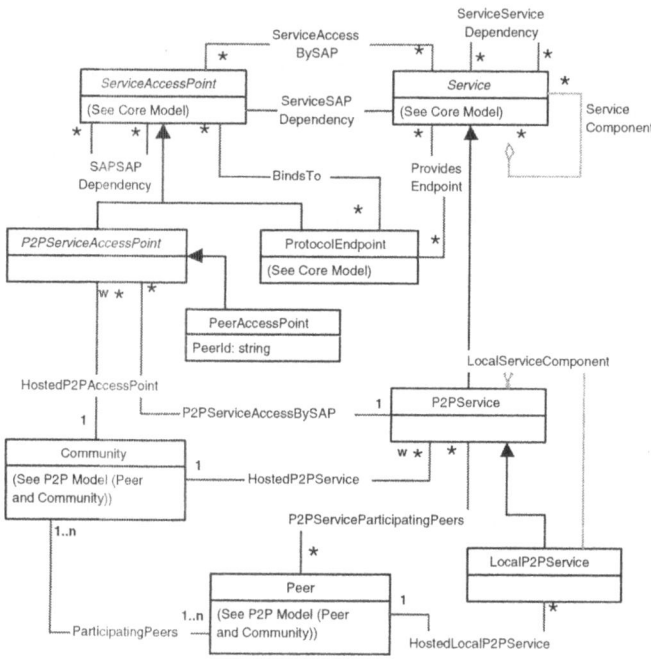

Fig. 4. Service

The P2P service model. In the P2P context we have reconsidered the definition of a service and defined the following one: *a P2P service is a software
functionality provided and/or consumed by a set of peers belonging to the same
community.* Contrary to the definition given in the CIM Core model, P2P services present a distributed aspect. Thus we have followed the cluster service
mode defined in the Common System Model [13]. Concerning their service access point, we consider that a P2P service access point is the identifier of any
subset of peers that participates in the service operating.

Given these concepts, we have designed the class diagram shown in Figure
4. First, the *P2P_P2PService* abstract class, that inherits from the *CIM_Service*
class represents a P2P service. Then, the *P2P_PeerServiceAccessPoint* class inherits from the *P2P_P2PServiceAccessPoint* one and defines a new attribute
that is the identifier of a peer operating as an access point for a considered service. The *P2P_P2PService* and *P2P_P2PServiceAccessPoint* classes are linked
together by the *P2P_P2PServiceAccessBySAP* dependency class: a P2P service
owns one or more access points and an access point can operate on one or more
services.

Now considering the links between services, peers and their communities, we have designed association classes that represent the hosting of a service in a particular community (*P2P_HostedP2PService*) and the fact that just a subset of peers can participate in the service operation (*P2P_P2PServiceParticipating-Peers*).

Local service vs global service view. In case of a standard networking model, services are local to a computer system, like a word processing one, or provided by a server through a remote connection, like a mail service. Their common characteristic is that, in this context, a service is provided by a single entity or in case of a service composition by the aggregation of single entities. However, in a P2P context, a service is a distributed function that is consumed and provided by a set of peers. Thus, global services features depend on the different peers that contribute to the service and they are not bound to a particular entity. Moreover, each peer that participates in a service delivery presents a local view of the service.

To model these two views of a service we have chosen to use a service composition association. The *P2P_LocalServiceComponent* association class allows the definition of global P2P services that are the aggregation of local services. About the cardinalities, a global service may be composed of zero or several local services, but a local service can just belong to zero or one global service.

3.6   Routing, Forwarding, and Routes

Routing in P2P overlay networks is closely bound to the infrastructure. Many P2P applications propose different ways of routing. For example, Gnutella [14] uses a routing process that lies on message identifiers; CAN [15] uses a content routing method.

Our model enables the monitoring of routing and forwarding services as well as routes hosted by peers. According to our P2P service definition, the route calculation and forwarding services are considered as distributed P2P services that are composed of a set of local services.

Now, concerning the routes, the current routes model of the Common Network Model [12] fits our requirements. Figure 5 shows it. The *CIM_NextHop-Route* represents an entry in a routing table and contains the destination address of the route. Moreover, the *P2P_P2PHostedRoute* association class links a set of routing entries to the peer that hosts them.

4   Conclusion and Future Works

In this article, we have presented a generic management information model for P2P networks and services based on CIM. Our model is organization, functionality and topology oriented and covers several aspects of the P2P domain. First it deals with the notion of peer and its belonging to one or several communities. The mutual knowledge of peers allows the establishment of a virtual topology

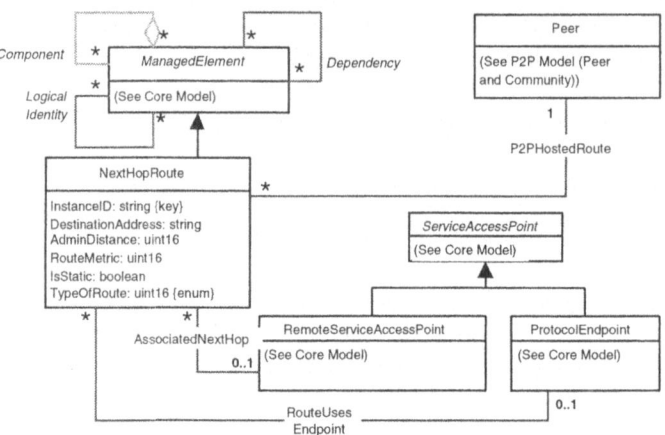

Fig. 5. The routes model

that we represent through pipes. Then our model features the resources available in a community and especially the ones shared by its composing peers. Resources are consumed or provided in the context of services that is our fourth model aspect. Indeed, a P2P service is a basic functionality that is distributed among a set of participating peers. Finally, particular basic services offered by P2P frameworks concern the routing of packets in a P2P domain that is an overlay network. Thus we have modeled such services and the routing tables they generate or use. In this way, our CIM extension for P2P networks and services is a major step toward the design of a generic P2P management framework that can have an abstract view of a P2P network located in a manageable environment.

The use of CIM as a formalism makes our model easily extensible to deal with particular areas of the management. For example, security concerns as well as QoS or fault tolerance aspects could be added by extending classes of our model or by using existing dedicated CIM classes.

In order to prove the validity of the model, it has been instantiated on the Chord P2P infrastructure. The instantiation showed that with our model we are able to monitor a Chord community, its different participating peers, the discovery service among others and the routing tables. Moreover, we are now working on the instantiation of our model on the Jxta framework. This framework, as a common middleware over which dedicated applications can be built, is very interesting in our management context. In a longer time perspective, we will focus our work on the performance management of P2P networks. We plan to refine our generic model for the performance evaluation of such a type of virtual networks.

References

1. Oram, A., ed.: Peer-to-peer: Harnessing the Power of Disruptive Technologies. O'Reilly & Associates, Inc. (2001)
2. Anderson, D.: SETI@Home. Number 5. In: (in [1]) 67–76
3. Foster, I., Kesselman, C.: Globus: A metacomputing infrastructure toolkit. The International Journal of Supercomputer Applications and High Performance Computing **11** (1997) 115–128
4. Doyen, G., Festor, O., Nataf, E.: Management of peer-to-peer services applied to instant messaging. In Marshall, A., Agoulmine, N., eds.: Management of Multimedia Networks and Services. Number 2839 in LNCS (2003) 449–461 The End-to-End Monitoring Workshop 2003.
5. Babaoglu, O., Meling, H., Montresor, A.: Anthill: A framework for the development of agent-based peer-to-peer systems. In: The 22th International Conference on Distributed Computing Systems (ICDCS '02), IEEE Computer Society (2002)
6. Milojicic, D., Kalogeraki, V., Lukose, R., Nagaraja, K., Pruyne, J., Richard, B., Rollins, S., Xu, Z.: Peer-to-peer computing. HP Laboratories HPL-2002-57 (2002)
7. Kim, M., Kang, H., Hong, J.: Towards peer-to-peer traffic analysis using flows. In Brunner, M., Kekker, A., eds.: Self-Managing Distributed Systems, DSOM 2003. Number 2867 in LNCS, International Federation for Information Processing (2003) 55–67
8. Hausheer, D., Liebau, N.C., Mauthe, A., Steinmetz, R., Stiller, B.: Token-based accounting and distributed pricing to introduce market mechanisms in a peer-to-peer file sharing scenario. In: 3rd IEEE International Conference on Peer-to-Peer Computing, Linkoping, Sweden (2003)
9. Antoniadis, P., Courcoubetis, C., Efstathiou, E., Polyzos, G., Strulo, B.: Peer-to-peer wireless lan consortia: Modelling and architecture. In: 3rd IEEE International Conference on Peer-to-Peer Computing, Linkoping, Sweden (2003)
10. Bumpus, W., Sweitzer, J.W., Thompson, P., R., W.A., Williams, R.C.: Common Information Model. Wiley (2000)
11. Distributed Management Task Force: CIM Concepts White Paper, CIM versions 2.4+. www.dmtf.org/standards/documents/CIM/DSP0144.pdf (2003)
12. Distributed Management Task Force: CIM Network Model White Paper, CIM version 2.7. www.dmtf.org/standards/published_documents/DSP0152.pdf (2003)
13. Distributed Management Task Force: CIM System Model White Paper, CIM version 2.7. www.dmtf.org/standards/documents/CIM/DSP0150.pdf (2003)
14. Kan, G., Gnutella, GoneSilent.com: Gnutella. Number 8. In: (in [1]) 94–122
15. Ratnasamy, S., Francis, P., Handley, M., Karp, R., Shenker, S.: A scalable content addressable network. In: Proceedings of ACM SIGCOMM 2001. (2001)

A Generic Event-Driven System for Managing SNMP-Enabled Communication Networks

Aécio P. Braga [1], Riverson Rios [2], Rossana Andrade [2], Javam C. Machado [2], and José Neuman de Souza [2]

[1] Núcleo de Processamento de Dados – Universidade Federal do Ceará (UFC)
Campus do Pici, Bloco 901 – 60.455-770 – Fortaleza – CE – Brasil, aecio@ufc.br
[2] Departamento de Computação – Universidade Federal do Ceará (UFC)
Campus do Pici, Bloco 910 – 60.455-770 – Fortaleza – CE – Brasil
{riverson, rossana, javam, neuman}@ufc.br

Abstract. In the area of monitoring communication networks, GEDSystem is a tool for supporting the development of programs driven to the management of domains that may show up when networks are monitored by SNMP. This tool is supported by an automatic modeling of data and events and automates the grouping of MIB objects from several SNMP agents, producing data structures that represent Domains and Events. The structures are automatically transformed into Tables and Views of a relational database. The views implement the perception mechanisms of the Events defined in the tool. The GEDSystem also provides a Monitoring Agent that automatically recognizes, collects and stores information about the Domains.

1 Introduction

After an analysis of several works in the area of network management, it is possible to notice that, in general, the used approaches try to identify and notice Events as a means for detection, diagnosis and correction of anomalies on the network. They make use, for instance, of probabilistic calculations and graphs as well as case- and rule-based reasoning. The methods are used in the following way:

- Probabilistic approaches are used in the control of the Quality of Service (QoS) and of the resource reservations in ATM networks [4], in the production of fault prediction alarms [11], in the identification of "signatures" from the network traffic and from the behavior of the managed objects [7], and in the anticipation of potential problems in a web server [9].
- Causality and dependency graphs are used in the determination of the simple causes of a chain of alarms or events [5] and [12].
- In [3], rules are formulated as thresholds whose definitions are based on time series of states of the objects in a network.
- Case-based reasoning is applied in maintenance support systems aided by a Knowledgebase of Events [1], in the integration of heterogeneous networks and in the correction of flaws on the nodes of a network [8]. The technique is also integrated into a Trouble Tickets Systems (TTS) architecture so that solutions to flaws based on past episodes that have occurred in a computer network can be formulated [6].

J.N. de Souza et al. (Eds.): ICT 2004, LNCS 3124, pp. 811–819, 2004.
© Springer-Verlag Berlin Heidelberg 2004

Briefly, these applications are fed by alarms or logs files. Starting from events, they attempt to foresee tendencies and anomalies, to detect, to isolate and to correct the original causes of the events and to map the physical observations with the states of the managed network objects. To accomplish the task, such techniques as thresholds comparisons, statistical analyses, graphs and historical information are used. On the other hand, these techniques employ, for instance, more algorithms in the routers, complex mathematical and statistical models, more layers in the management models etc., which increase computing and message overhead.

There is nowadays a great amount of network management applications whose computational characteristics vary tremendously. Unfortunately, the development of these applications does not take into account the management domains in general [2], in the sense that those approach lacks of a generic and automatic mechanism for preparing data models capable to represent these domains in such a way to facilitate the perception of events.

The main objective of this work is to specify and implement a Generic Event-Driven System for Monitoring Communication Networks (*GEDSystem*). *GEDSystem* is a tool for supporting the construction of Monitoring Systems for Communication Networks that provides:

1. an automatic way to define management domains;
2. an abstraction of the SNMP protocol in the non-reactive phases of the management process (communication, data structures etc.);
3. an automation of the processes of collecting, dating and storing information;
4. an automation of the perception of Events; and
5. an operational Web interface.

This paper is organized as follows. Section 2 depicts the event-driven network management; section 3 describes a generic management model; section 4 presents the GEDSystem tool; section 5 outlines some GEDSystem's functionalities and finally the section 6 has the conclusion and further works.

2 Event-Driven Network Management

The main operations of network management consist of tracing, interpreting and manipulating events. In general, an event is defined as an anomalous condition observed in the operation of a network. Usually, these are problems that happen in the hardware and/or software of the nodes [12]. Regarding the SNMP protocol, the events or conditions can be noticed through the variations that happen in the management information stored in the MIBs of the Agents and/or probes of the RMON. When provided by an Agent, this information is obtained as regular instantaneous snapshots. In the case of RMON probes, it is obtained as groups of information collected along a period of time. In both cases, the Manager application is entrusted of periodically requesting the information that needs to analyze, and of executing some control action on the managed nodes.

The compilation of the information from the RMON and RMON2 probes is accomplished through package selection together with the scanning of their contents. The aim is to generate statistics and/or events in conformity with the definitions of events from the probes themselves. These events can trigger traps that send information from the RMON/RMON2's MIBs to some Manager.

As mentioned before, systems managed by SNMP are fed with information from the MIBs of the Agents and/or from the RMON probes. Each MIB is a general collection of information that change over time. In other words, they should be grouped, based on some supposed relationships, so that a Management Domain can be defined. The most immediate relationship than one can notice among them is time.

According to [10], the RMON and RMON2 probes use circular buffers. Consequently, the Manager application must have a more accurate control on the polling so that the buffers do not overflow, thereby avoiding the loss of previously compiled information. The probes, too, are computationally burdened with the execution of processes that generate statistics, events and traps. The buffers and processes could be left under the responsibility of another processing entity. This way, they would be freed from tasks other than "listening" to the interfaces and scanning packages.

3 A Generic Monitoring Model

Up to now, this paper has touched upon the data sources for management applications and the diversity of domains managed by the applications. This work proposes a tool to support the development of programs capable of managing domains deriving from SNMP-enabled communication networks, based on events.

The tool allows the gathering of information from the several nodes of the networks with SNMP agents. This operation results in the creation of a database to keep this information in a transparent way. In other words, it provides the modeling of data structures that represent management domains, so as to feed network management processes.

Dating attributes are also automatically included into the data models, indicating time of requisition and time of reception of the information from the modeled domains. This information can help, for instance, performance analyses, historical behavior and statistics procedures.

Rules that evaluate the possible variations of the grouped information can be defined and eventually used in the formulation of expressions that can represent several types of events. Additionally, Views of the repositories, based on the expressions, can be dynamically defined in such a way to provide a perception mechanism of the events. The feasible management processes can consult these Views in order to detect, diagnose and correct any anomaly in the monitored management domain.

The built tool provides a monitoring agent that automatically recognizes, monitors and stores the information from the management domains. This way, the applications can be exclusively dedicated to the management procedures, freeing themselves from doing polls. Their decisions are based only on the perception of the events implemented as database Views or on other analyses done on the information contained in the repositories. Small countless applications, therefore, can manage their own domain, facilitating the development of a Network Management System. The tool is based on such technologies as Domain Modeling, Event Modeling, Database Modeling, Web Modeling and Client/Server Modeling.

4 The GEDSystem Tool

A Generic Event-Driven System for Monitoring Communication Networks (*GEDSystem*) was born from the main issues involving a monitoring model explained in Section 3. *GEDSystem* is a tool composed of technological components produced by recent progresses in the areas of operating systems, web systems, databases systems, graphical user interfaces, programming languages, communication protocols and network management. *GEDSystem*'s architecture is composed of one or more Clients in the first layer, a Monitoring Server in the second layer and one or several Information Management Servers (SNMP agents) in the third layer, as can be seen in Figure 1.

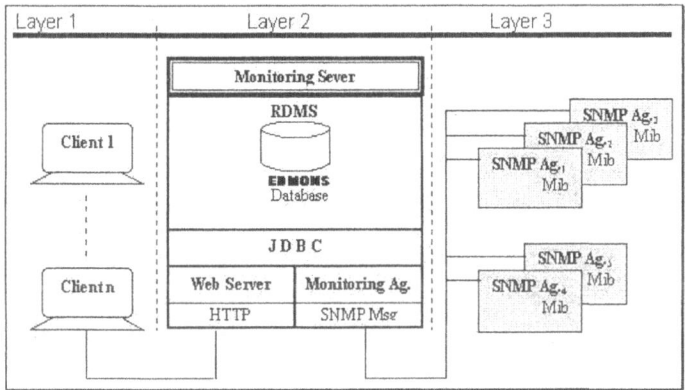

Fig. 1. GEDSystem 's Client/Server Architecture

The First Layer – The Client Layer. A Client can be any computer whose Internet browser can access to a Java Virtual Machine (JVM). Such a browser is enough to provide the communication between the first and second layers of the architecture, additionally offering a graphical user interface that makes it possible for the network administrator to easily interact with the Monitoring and the Database Servers.

The Second Layer – The Monitoring System. The Monitoring Server is a computer running any operating system with a Web Server that supports Java Servlets and JDBC (Java Database Connectivity). The support for any RDBMS is also mandatory. To implement a prototype of GEDSystem, the Apache Server's Tomcat, AB's MySQL database server and Java servlets / JDBC and AdventNet APIs were adopted. In addition, the tool provides a Monitoring Agent implemented in the Java language. The functional base of the process will be shown later in this paper along with the definition of the Generic Event-Driven Data Model (MDGE) for monitoring communication networks. Through the MDGE, the Monitoring Agent accomplishes the following tasks: Collection of information of the managed elements' MIBs; Grouping of information in domains; Storage of information in a database.

The Third Layer – The Layer of the Management Information Servers. The third layer of the GEDSystem 's architecture is composed of the nodes of the network with SNMP Agents. This way, the agents are Management Information Servers, which provide the values of the managed objects that are found in the MIBs. In other words, the second layer's server takes the role of a management Information Client.

GEDSystem's Database is based on the Generic Event-Driven Data Model (MDGE). It is composed by the supporting structures of the monitoring process, by the information obtained by the monitoring agent, and by the mechanisms used for noticing Events. In the database, data and historical events coming from SNMP Agents are stored as time series. By doing this, the database can help in the production of management reports about the behavior of the network.

The Generic Event-Driven Data Model (MDGE) represents a group of entities in the form of Tables whose attributes and relationships(Figure 2) specify which, where, when and how the managed Objects are collected and stored by the Monitoring Agent. These specifications represent the administrator's monitoring needs. Besides, MDGE in an event-driven model and, thus, allows the perception of the variations of the values collected at any moment by means of a certain restrictive rule of those values.

Some entities of the Model are created in a transparent and dynamic way, depending on the configuration of the monitoring process previously defined by the Administrator. The entities are the repositories of the values of the managed SNMP Objects.

MDGE is composed by the following entities: *Community, Domain, PrimitiveState, CompositeState, Index, Formula* and *GeneralTable*. In addition, an uncertain number of entities named *Repositoryes* are also part of the Model.

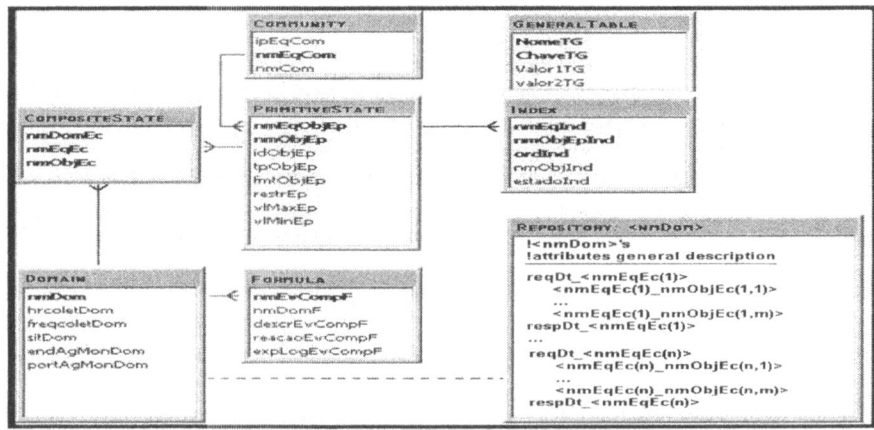

Fig. 2. General view of MDGE's Entity-Relationship Model

Figure 3 shows the relationship among the entities that provide the bases for the automatic construction of the repositories. In the upper part of the illustration, one can see MDGE's entities whose relationships represent the definition of Management Domains. The lower part shows this relationship in the form of connections between tables that implement the entities presented in the upper part.

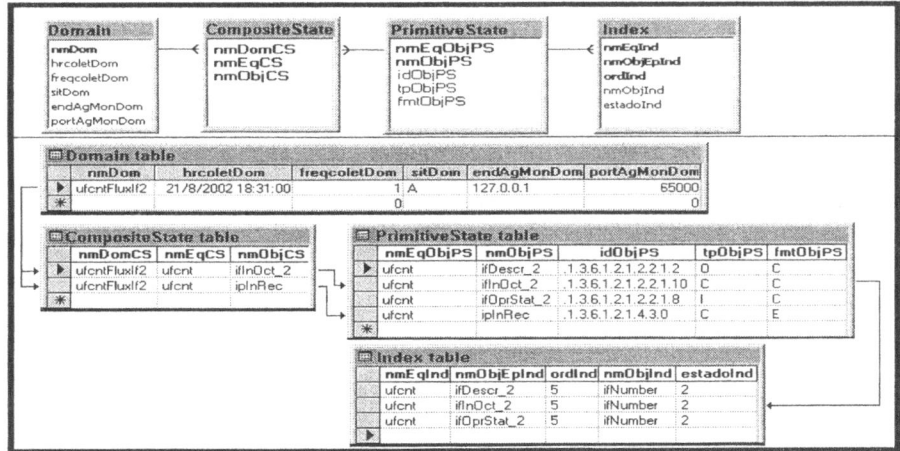

Fig. 3.

Figure 3 illustrates the definition of a domain called *ufcntFluxIf2*. This domain contains the objects *ifInOct_2* (columnar) and *ipInRec* (scalar), both of a node called *ufcnt*. This definition leads to the creation of a table, the repository of the domain being treated, in a database called *R_ufcntFluxIf2* with the following lay-out:

Attribute Name	Data Type
reqDt_ufcnt	Date
ufcnt_ifInOct_2	Double
ufcnt_ipInRec	Double
respDt_ufcnt	Date

R_ufcntFluxIf2 : repository table

Fig. 4.

Figure 4 illustrates, in the upper part, the entities whose relationship defines the possible Events of a Domain. The lower part shows this relationship in the form of connections between tables that implement the entities presented in the upper part.

Figure 5 illustrates the definition of an Event called *ufcntFluxSuspIf2*. The Rule that characterizes it is the same one that limits the ´where´ part of the SQL clause that creates the view of the database. This can be clearly seen in the contents of the *expLogEvCompF* attribute shown in Figure 4.

5   GEDSystem's Functionalities

The tool is composed by three functional groups of services that allow *GEDSystem* to reach its objectives. The first group configures the monitoring process of the network nodes and defines Events. The second one defines and dynamically generates management reports from the data stored in its repositories. The last group, formed only by the Monitoring Agent, executes the monitoring process itself.

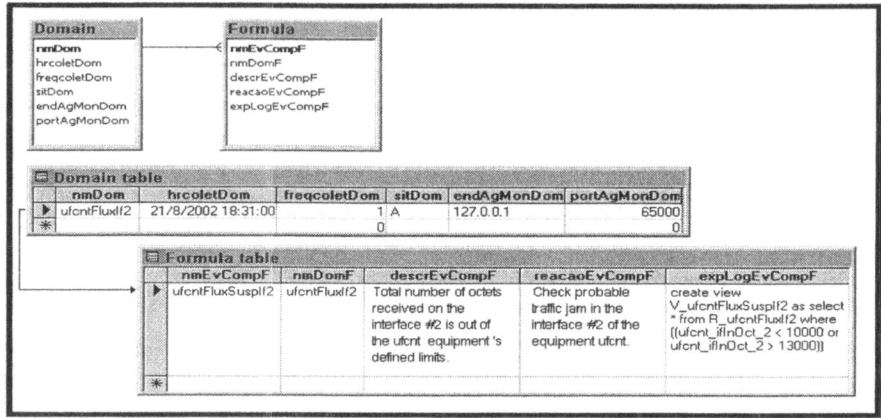

Fig. 5.

The configuration of the monitoring process allows the Administrator to define the SNMP Objects that he/she wants to monitor. These definitions are made by the services: Communities Registration, Primitive States Registration, Indexing Registration, Domains Registration, Collection Control Registration and Events Registration.

The Communities Registration service maintains the information that allows the Monitoring Agent to request data from the SNMP Agents; the Primitive States Registrations service defines the SNMP Objects to which the network administrator's attentions are directed. The service also allows the establishment of restrictions or Rules that evaluate the possible variations of the instances of the Objects. Similarly, in the case of the registration of columnar Objects, the Indexation Registration service allows the definition of its indices; the Domains Registration service contains the SNMP Objects or Primitive States that, according to the Administrator's opinion, have some kind of logical relationship. Each group forms one Management Domain or simply a Domain, and the values of its components are stored in a repository. One repository is built for each Domain; the Collection Control Registration service fires a monitoring process for each of the defined Domains, at the same time that creates their repositories; the Events Registration service defines formulas, based on the Rules established for the Primitive States that lead to the definition and creation of Views on the repositories of the previously registered Domains. The views constitute the perception mechanisms of the Events.

The Definition and the Generation of Management Reports provide the construction of the repositories of the data obtained by the monitoring processes. This way, the definition and report generation services can generate HTML forms already filled with a list of attributes of the repositories that can be easily selected to compose the reports. The Administrator can also specify time filters for the reports. And all is done in a very friendly way. Besides the definitions of the structures of the repositories, MDGE also comprises the definitions of the Views. Thus, the transactions Definition and Generation of Management Reports can consult the Views to list the events that happened in the network in a given period of time.

The Monitoring Process is executed by one of GEDSystem´s component called the Monitoring Agent. In order to have a total independence of the hardware and the operating system the Monitoring Agent is implemented in the Java language. It continuously verifies the current situation of the Domains, enabling/disabling their monitoring as requested by the Administrator. Different process threads make the surveillance of each Domain individually.

6 Future Work

As future work, the possibility of implementing the idea of Domain-based polling and the concept of Events based on database views directly in the SNMP protocol and in the RMON/RMON2 probes could be verified. The use of Triggers and Stored Procedures, in the automation of reactive procedures, could also be investigated.

In the conventional so much as in the active networks, one enhancement could be the implementation of an database where the definitions of the Domains that the Administrator wants to monitor would be contained. This way, the protocol could obtain the data of the domains locally and store them directly into the specified database. In active networks context, the active packages will constitute of data collectors of the monitored Domains. In both cases one could expect a decrease on the traffic of SNMP messages in the network, the non-use of the UDP protocol and the continuity of the support for the construction of management applications fed by relational databases.

References

1. Burgess, John, Guillermo, Ray. (2000)."Raising Network Fault Management Intelligence". In *IEEE/IFIP Network Operations and Management Seminar*.
2. Hariri, Salim, Kim, Yoonhee. (2000). "Design and Analysis of a Proactive Application Management System (PAMS)". In *IEEE/IFIP Network Operations and Management Seminar*.
3. Ho L.Lawrence, Cavuto, David.J, Papavassiliou, Symeon [et al] (2000). "Adaptive and Automated Detection of Network/Service Anomalies in Transaction-Oriented WAN´s:Network Analysis, Algorithms Implementation and Deployment". In *IEEE Networks Journal;* v. 18, n. 5.
4. LI, Jung-Shian (2000)."Mesurement and in-service monitoring for QoS violations and spare capacity estimations in ATM network". *Computer Communications*; v.23, p162-170.
5. Lo, Chi-Chum, Chen Shing-Hong.(1998) "Robust Event Correlation Scheme for Fault Identification in Communications Network". IEEE.
6. Melchiors, C., Tarouco, L.M.R. (2000)."Troubleshooting Network Faults Using Past Experience". In *IEEE/IFIP Network Operations and Management Seminar*.
7. Papavassiliou, Symeon., SAVANT, V.S., TUPINO, J.J. [et al] (1998). Enhanced Network Management for Online Services. In *Proc. IEEE International Conference on Computer Communications and Networks* IC3N'98, Louisian, Oct.
8. Penido, G., Nogueira, J.M, Machado, C. (1999). "An automatic fault diagnosis and correction system for telecommunications management".In Integrated Network Management VI.
9. Shen, Dongxu, Hellerstein, Joseph.(2000). "Predictive Models for Proactive Network Management: Application to a Production Web Server". In *NOMS 2000 IEEE/IFIP Network Operations and Management Seminar*.

10. Stallings,William. *SNMP, SNMPv2, SNMPv3, and RMON1 and 2*.Massachusetts: Addison Wesley, c1999. 619p.

11. Thottan, M., JI, C. (2000). "Fault Prediction at the Network Layer using Intelligent Agents". In *Proc. of Sixth IFIP/IEEE International Symposium on Integrated Network Management IM'99, Boston, May.*

12. Yemini, Shaula Alexander, KLIGER, Shmuel, MOZES, Eyal [et al] (1996). "High Speed and Robust Event Correlation.

Algorithms for Distributed Fault Management in Telecommunications Networks*

Eric Fabre, Albert Benveniste, Stefan Haar[1], Claude Jard[2], and Armen Aghasaryan[3]

[1] IRISA/INRIA, Campus de Beaulieu, 35042 Rennes cedex, France
[2] IRISA/ENS-Cachan Campus de Ker-Lann; 35042 Rennes cedex, France
[3] Alcatel Research & Innovation, Next Generation Network and Service Management Project, Alcatel R&I, Route de Nozay, Marcoussis, 91461 France

Abstract. Distributed *architectures* for network management have been the subject of a large research effort, but distributed *algorithms* that implement the corresponding functions have been much less investigated. In this paper we describe novel algorithms for model-based *distributed* fault diagnosis.

1 Introduction

Distributed self-management is a key objective in operating large scale infrastructures. Fault management is one of the five classical components of management, and is the subject of our work. In this paper, we consider a distributed architecture in which each supervisor is in charge of its own domain, and the different supervisors cooperate at constructing a set of *coherent* local views for their respective domains. By coherent, we mean that the different views for the different supervisors agree on the interfaces of their respective domains. Of course, to ensure that the approach can properly scale up, the corresponding global view should never be actually computed.

Fault diagnosis has been addressed by different means. Historically, the first approach was by means of rule-based expert systems [10,11]. Such systems are generally limited to simple correlation rules, with small scope, and hardly capture all the complexity of the network reaction to a failure. Connexionist and related learning techniques from Artificial Intelligence area have been investigated to avoid the burden of providing detailed and explicit knowledge on the supervised domain [5]. But again, such methods are not scalable, and difficult to update when the network evolves. To avoid these drawbacks, the main trend nowadays is in favor of model based approaches [8,7,9]. Most of them rely on topological information, both physical and logical (this information is easy to get by scanning Management Information Bases). Using the topology of interactions, successive correlations with an observed symptom can be traced back to their possible alternative causes.

* This work was supported by the French National RNRT project MAGDA2, funded by the Ministère de la Recherche.

J.N. de Souza et al. (Eds.): ICT 2004, LNCS 3124, pp. 820–825, 2004.

The present approach goes further in this direction. First of all in the size and details of the model (see fig. 1): the physical and logical topologies (Network Elements + connections) are modeled as a graph of interconnected Managed Objects (MO), and each MO is itself described as a small *dynamic* system. In the management plane, MOs are equiped with their own fault management function and are responsible of the production of alarms. Alarms are emitted when a local failure is detected, or in reaction to the failure of a neighboring MO that was necessary to guarantee a correct service. This second phenomenon is at the origin of failure propagations in the network. We refer the reader to [1] for details on the modeling, and for the description of a tool that automatically builds the model. Second key feature of our approach: fault diagnosis is performed in a *distributed* way. Specifically, we assume alarms are not sent to a global supervisor, but are rather collected by local supervisors, in charge of part of the system. In this scenario, a local supervisor only needs to know the local model of its domain (for example one NE in fig. 1), and cooperates asynchronously with supervisors of neighboring domains. We believe this is the key to cross-domain fault management.

For space reasons, we only give a glance at the theory behind distributed diagnosis algorithms, by means of a toy example (see [2,3,4] for details).

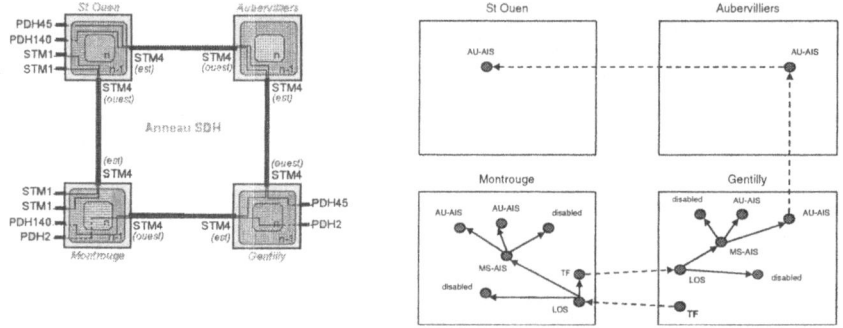

Fig. 1. Left: Part of the Paris area SDH/SONET ring (topological view, at the Network Element level). In the model of this network, each NE is further decomposed into Managed Objects, corresponding to the SDH hierarchy (SPI, RS, MS, etc.). Right: a failure scenario on this model, with propagation of the failure, entailing correlation in the alarms raised.

2   A Structure to Represent Runs with Concurrency

A central feature in large distributed dynamic systems is that many components run in parallel, so several events can occur at the same time, *i.e.* "concurrently." It is therefore crucial to represent runs of such systems in a framework that captures this independence, and sequences of events are definitely inappropriate for that. Since the components of our network are finite state machines, this suggests to use safe Petri nets to model components, and to represent runs of the network (*i.e.* failure propagations) with so-called "true concurrency semantics." We illustrate this on a toy example, given in fig. 2.

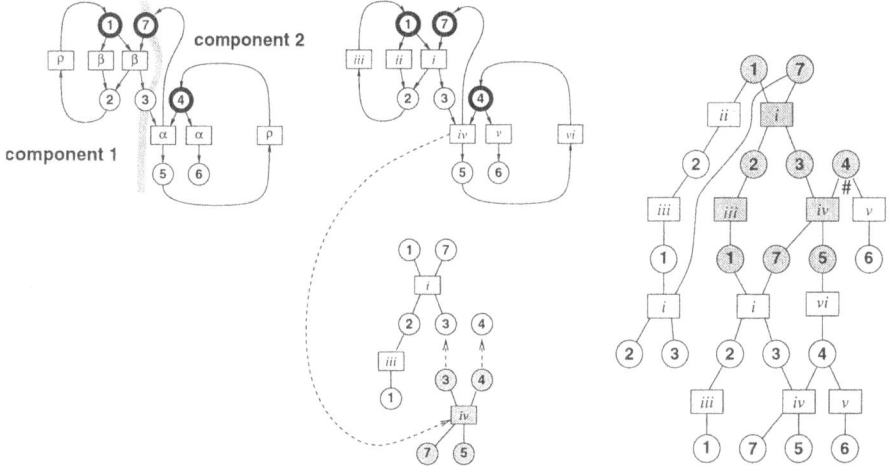

Fig. 2. Running example in the form of a Petri Net, viewed as two components interacting through shared places 3 and 7 (left). Thick places indicate the initial marking. Right: representation of runs of this system under the form of a branching process. "Time" goes from top to bottom.

On the example, places 1 and 2 represent states *ok* or *down* of component 1 (with a possibility of self-repair by transition *iii*). Component 1 can thus go down in two ways, through *ii* or *i*. The second possibility fills place 3 which models a propagation of the failure to component 2: place 4 corresponds to state *ok*, place 5 to a temporary failure (self-repair possible by *vi*), and place 6 to a definitive *down* state. Notice that, on the leftmost right picture of the model, transitions are labeled by α, β, ρ which corresponds to the *alarms* produced when they fire.

The mechanism of constructing a run of the Petri net $\mathcal{P}$ in the form of a partial order is illustrated in the 2nd and 3rd diagrams. Initialize any run of $\mathcal{P}$ with the three conditions labeled by the initial marking $(1, 7, 4)$. Append to the pair $(1, 7)$ a copy of the transition $(1, 7) \rightarrow i \rightarrow (2, 3)$. Append to the new place labeled 2 a copy of the transition $(2) \rightarrow iii \rightarrow (1)$. Append, to the pair $(3, 4)$, a copy of the transition $(3, 4) \rightarrow iv \rightarrow (7, 5)$ (this is the step shown). We have constructed (the prefix of) a run of $\mathcal{P}$, where concurrency is explicitly displayed: the ordering of events labeled *iii* and *iv* is left unspecified, we only know they occur after (or as a consequence of) *i*.

Now, all runs of $\mathcal{P}$ can be constructed in this way. Different runs can share some prefix. The rightmost diagram shows a *branching process* (BP) of $\mathcal{P}$, obtained by superimposing the shared parts of different runs. The gray part is a copy of the run shown in the middle. The alternative run on the extreme left of this diagram (involving successive transitions ii, iii, i) shares only its initial places with the run in gray. On the other hand, replacing, in the gray run, the transition labeled iv by the one labeled v yields another run. We say there is a *conflict* at place 4: a choice must be made between firing iv or v. A conflict takes place each time a place is branching in this diagram. Branching processes thus encodes sets of executions of a PN in a compact manner: a run of the PN corresponds to selecting part of the BP which is both conflict free and causaly closed. Observe that by nature a BP has no oriented cycle. The maximal BP of a Petri net $\mathcal{P}$ is called its *unfolding*: $\mathcal{U}_\mathcal{P}$. Places and transitions of a BP are rather called *conditions* and *events*, and are labeled by places and transitions of the original net $\mathcal{P}$.

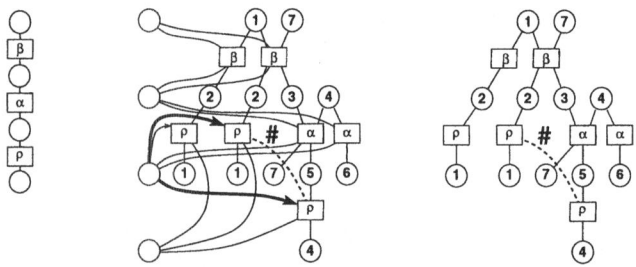

Fig. 3. Left: an alarm pattern $\mathcal{A}$. Center: the product $\mathcal{A} \wedge \mathcal{U}_\mathcal{P}$. right: same net without the conditions of $\mathcal{A}$.

Assume the system modeled by $\mathcal{P}$ runs and that labels (alarms) produced by its transitions are collected by a sensor. We represent the collected alarm pattern under the form of a net $\mathcal{A}$ (left diagram on fig. 3). The (centralized) diagnosis problem can be formalized in the following manner: compute all runs of $\mathcal{P}$ that could explain the observed alarm pattern $\mathcal{A}$. Deciding what failure actually occured is then a post-processing on this set of possible runs. Formally, this amounts to computing all runs of the product net $\mathcal{A} \times \mathcal{P}$, which is $\mathcal{P}$ *constrained* by observations $\mathcal{A}$. Equivalently, we want $\mathcal{U}_{\mathcal{A} \times \mathcal{P}} = \mathcal{A} \wedge \mathcal{U}_\mathcal{P}$, where $\wedge$ is a special product on branching processes, designed to avoid oriented cycles. The result is depicted on fig. 3, and can be built recursively, in an asynchronous manner, just like branching processes of $\mathcal{P}$. Observe that an extra conflict relation is created (dashed line labeled with $\#$) to capture the fact that alarm ρ can be explained either by $(2) \rightarrow iii \rightarrow (1)$ or by $(5) \rightarrow iii \rightarrow (4)$. Possible explanations to $\mathcal{A}$ correspond to maximal runs of $\mathcal{A} \wedge \mathcal{U}_\mathcal{P}$ that are labeled by *all* alarms of $\mathcal{A}$ (some runs of $\mathcal{A} \wedge \mathcal{U}_\mathcal{P}$ may only explain the first alarms and then stop).

3   Distributed Diagnosis with Two Supervisors

This is the main novelty of this paper. Assume a local sensor collects alarms produced component i, under the form of a local alarm pattern $\mathcal{A}_i$, $i = 1, 2$. The objective is to build a *distributed* view of $\mathcal{U}_\mathcal{P} \wedge \mathcal{A}_1 \wedge \mathcal{A}_2$. Alarm patterns are processed by two local supervisors communicating asynchronously.

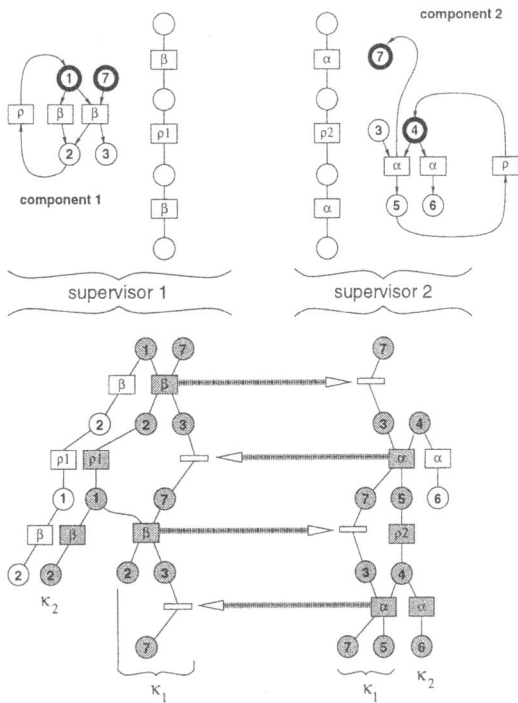

Fig. 4. Distributed diagnosis: two coherent local views of the unfolding $\mathcal{U}_{\mathcal{P}} \wedge \mathcal{A}_1 \wedge \mathcal{A}_2$ are built by two supervisors cooperating asynchronously.

Fig. 4 (top) displays the information available to supervisor i: $\mathcal{A}_i$ together with the local model of component i. The bottom diagrams represent the local views of the diagnosis. They correspond to projections on events of each component of $\mathcal{U}_{\mathcal{P}} \wedge \mathcal{A}_1 \wedge \mathcal{A}_2$. Specifically, on the side of component 1, one has transitions of component 1 (explaining local alarms $\mathcal{A}_1$) plus abstractions of the action of component 2 on shared places 7 and 3. These events appear as empty rectangles. To recover a complete trajectory of $\mathcal{P}$, one has to glue matching trajectories on each side. This corresponds for example to gluing events indicated by arrows on the picture, to obtain the grey trajectory of $\mathcal{P}$. Observe that this distributed representation of the diagnosis is more efficient than the centralized one: to be able to glue two local runs, it is enough that they have maching behaviours *only on shared places*. So, for a given history of these shared places, m trajectories of component 1, and n trajectories of component 2 actually correspond to $m \times n$ trajectories of $\mathcal{P}$.

As for the centralized case, the distributed view of the diagnosis can be obtained recursively, but requires communication between supervisors. Each arrow actually corresponds to the communication of a new event to append. However, the cooperation between the two supervisors need only asynchronous communication. Each supervisor can simply "emit and forget." Diagnosis can progress concurrently and asynchronously at each supervisor. For example, supervisor 1 can construct the branch $[(1) \rightarrow \beta \rightarrow (2) \rightarrow \rho_1 \rightarrow (1) \rightarrow \beta \rightarrow (2)]$ as soon as the corresponding local alarms are collected, without ever synchronizing with supervisor 2.

4 Conclusion

We have proposed an unfolding approach to the distributed diagnosis of concurrent and asynchronous discrete event dynamical systems. Our presentation was informal, based on a toy illustrative example. The related mathematical material is found in [3]. A prototype software implementing this method was developed at IRISA. This software was subsequently deployed at Alcatel on a truly distributed architecture, no modification was necessary to perform this deployment. Model-based approaches suffer from the burden of getting the model. We have developed a tool [1] that automatically generates the needed model. This work is also presented at this conference.

References

1. A. Aghasaryan, C. Jard and J. Thomas: UML specification of a genericmodel for fault diagnosis of telecommunications networks. ICT 2004.
2. A. Benveniste, E. Fabre, C. Jard and S. Haar: Diagnosis of asynchronous discrete event systems, a net unfolding approach. IEEE Trans. on Automatic Control, 48(5), 714–727, may 2003.
3. A. Benveniste, S. Haar, E. Fabre and C. Jard: Distributed monitoring of concurrent and asynchronous systems. In *Proc. of CONCUR'2003*, Sept. 2003. See also IRISA report No 1540, http://www.irisa.fr/bibli/publi/pi/2003/1540/1540.html
4. E. Fabre. Monitoring distributed systems with distributed algorithms. In Proc of the 2002 IEEE Conf. on Decision and Control, 411–416, Dec. 2002, Las Vegas, 2002.
5. Bennani Y. and Bossaert F: Modular Connectionist Modeling and Classification Approaches for Local Diagnosis in TelecommunicationTraffic Management. *Int. J. of Computational Intelligence and Applications,* World Scientific Publishing Company.
6. J. Engelfriet: Branching Processes of Petri Nets. Acta Informatica 28, 1991, pp 575–591.
7. B. Gruschke: Integrated Event Management: Event Correlation using Dependency Graphs. In Proceedings of the 9th IFIP/IEEE International Workshop on Distributed Systems: Operations & Management 1998, DSOM 98, Newark, DE, USA, October, 1998.
8. S. Kaetker and K. Geihs: A Generic Model for Fault Isolation in Integrated Management System. Journal of Network and Systems Management, Special Issue on Fault Management in Communication Networks, Vol. 5, No. 2, June 1997.
9. S. Kliger, S. Yemini, Y. Yemini, D Ohsie, and S. Stolfo. A coding approach to event correlation. In IFIP/IEEE Int. Symp. on Integrated Network Management IV, 1995, 266-277.
10. J. Liebowitz, editor. Expert System Applications to Telecommunications. John Wiley and Sons, New York, NY, USA 1988.
11. Y. A. Nygate: Event correlation using rule and object based techniques. In IFIP/IEEE Int. Symp. on Integrated Network Management IV, 1995, 278-289.

Fault Identification by Passive Testing*

X.H. Guo[1], B.H. Zhao[1], and L. Qian[1]

Department of Computer Science
Univ. of Sci&Tech of China,Hefei,Anhui,230027
gxh@mail.ustc.edu.cn

Abstract. In this paper, fault identification in passive testing is investigated. First, the fault detection algorithm and an existing fault identification algorithm in passive testing are introduced. Then, two more
efficient fault identification algorithms which can give out fewer possible faults than the existing one are proposed. One combines passive
testing with active testing, another adopts further passive testing. An
experiment on a practical routing protocol BGP is done to compare the
improved algorithms with the original one.

1 Introduction

Recent advances in communication softwares have enlarged and complicated
communication protocols. To increase the reliability of such protocols, testing
techniques ensuring that a protocol implementation conforms to its specification
are necessary. The traditional testing method is that testing system actively
communicates with IUT (Implementation Under Testing), then based on the
observation of IUT's responses to judge whether IUT is consistent with the
specification. This method is called as "Active Testing". "Passive Testing" is
a different testing method: It doesn't communicate with IUT, it just passively
collects the mutual messages between IUT and other systems, then analyzing
these messages to find faults of IUT. As it doesn't interfere with the normal
running of network systems, it's the best method to find the potential faults on
the real network.

In the early 70's there existed some researches on passive testing to test integrated circuit, but not until recently the need of network management changes
its research into a hot topic. In 1996, Lee proposed some basic principles and
methods of testing finite state machines using passive testing in [8] and presented
a fault detection algorithm in [7] in the next year; Miller adopted the model of
CFSM (Communication Finite State Machine) for passive testing in [3] and did
some research on the problem of fault location based on CFSM in [2] in 2000.
Then in 2001, Miller presented a fault identification algorithm base on DFSM
(Deterministic Finite State Machine) in [1]; Arisha and Miller applied these theoretical results to the practical network management systems in [6], [7]; All of

---

* Part of the research was supported by National Natural Science Foundation of China
  grants 90104010 and 60241004 and National Major Basic Research of China grants
  2003CB314801.

J.N. de Souza et al. (Eds.): ICT 2004, LNCS 3124, pp. 826–834, 2004.
© Springer-Verlag Berlin Heidelberg 2004

the former researches focus on the control flow, recently Lee researched on the data flow in [5].

Based on the significant research of Miller [1], two improved fault identification algorithms, which can give fewer possible faults and should be helpful to diagnose a network system, are proposed in this paper. The remainder of this paper is organized as follows. In Section 2, the model of DFSM, some necessary assumptions and a corresponding fault model are introduced. Section 3 includes all the details of Miller's approach for fault identification and the results of our analyzing on it. In Section 4 an algorithm combining passive testing with active testing is presented. Section 5 proposes another improved algorithm. In Section 6 an experiment on the practical routing protocol BGP to verify the effectiveness of our algorithms is included. Finally, Section 7 concludes the paper and points for future research.

2   The Specification Model

2.1   The DFSM Model

Definition 1. *A deterministic finite state machine (DFSM) M is a six-tuple:* $M = (I, O, S, s_0, \delta, \lambda)$ *where:*

- *I, O and S are finite non-empty sets of input symbols, output symbols, and states respectively.*
- *s_0 is a designated initial state.*
- *$\delta : S \times I \rightarrow S$ is the state transition function.*
- *$\lambda : S \times I \rightarrow O$ is the output function.*

When the machine is in state $s \in S$ and receives an input $a \in I$, it moves to the next state specified by $\delta(s, a)$ and produces an output given by $\lambda(s, a)$.

2.2   Assumptions for the DFSM Model

The algorithms in this paper are based on the following two assumptions:

- **Single Fault** Assume that there exists only one fault during a test cycle. This assumption is effective considering that the real systems won't produce faults frequently and the test cycle is not too long to allow multiple faults occur. What's more, if taking multiple faults into count, the faults maybe hide each other and cause fault identification by passive testing more complicated or even impossible.
- **Complete Machines** Assume that a machine will response to any inputs in spite of its current state. If not so, for the unspecified transitions import an error state F and an error output symbol "f" first and change these unspecified transitions to the imported state F with the output of "f". After these two steps the machine will be complete defined and it will be helpful to the fault identification algorithm. But this error state F doesn't exist in the actual specification, so if in a test cycle the machine reaches this state that means some faults happening.

2.3   Fault Model

The following two types of faults are considered in this paper:

- **Output Fault.** A transition has an output fault if, for the corresponding state and received input, the implementation provides an output different from the one specified by the output function.
- **Tail State Fault.** A transition has a tail state fault if, for the corresponding state and received input, the implementation enters a different state than specified by the state transition function.

3   Fault Detection and Identification

3.1   Fault Detection Overview

Lee gives out a fault detection algorithm for passive testing in [4]. It assumes that there exist a specification machine A and an implementation machine B. We can passively observe the input/output sequences of machine B. And we can judge that B is "faulty" if its behavior is different than that of A under the same input/output sequences.

When start observing there's no information to judge the current state of B. So we assume that B could be in any state of A at the beginning. Let L^0 designate this initial set of possible states. As each input/output pair i_j/o_j is observed, a new set L^j is produced from the set L^{j-1} such that L^j contains all states that are tail states of transitions with head states being in L^{j-1} with input/output i_j/o_j . If a set L^k becomes empty, this indicates a fault in the implementation. The sequence $L^0, L^1, \cdots, L^k$ is called as forward trace.

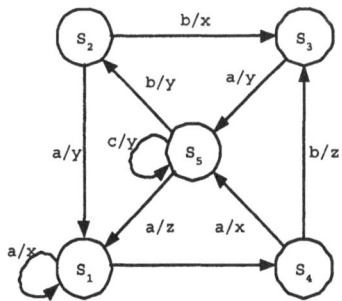

Fig. 1. An example FSM

For example, assuming that the observed input/output sequence is {a/y, c/y, b/z, a/x}, then the forward trace is: $L^0 = \{S_1, S_2, S_3, S_4, S_5\}, L^1 = \{S_1, S_5\}, L^2 = \{S_4, S_5\}, L^3 = \{S_3\}, L^4 = \emptyset$. That means a fault is detected at the fourth step.

3.2   Fault Identification Overview

After detecting a fault by passive testing, how to identify the possible faults in an implementation? Miller gives out a fault identification algorithm in [1] to solve this problem. First, he proposes a process to produce a sequence of sets of states called as backward trace.

1. Assume that a fault is detected at the step k, let $(L^k)^R$ be the set of all states of A.
2. Form set $(L^{j-1})^R$ from $(L^j)^R$ as follows: $(L^{j-1})^R$ contains all states that are head states of transitions with input/output i_j/o_j with tail states being members of $(L^j)^R$.

Then he proves the following three theorems:

Theorem 1. *The corresponding backward and forward sets are disjoint, i.e,*
$L^j \cap (L^j)^R = \emptyset.$

Theorem 2. *If L^j has a state s_p that under i_{j+1} has an output $\neq o_{j+1}$ and $(L^{j+1})^R$ has $\delta(s_p, i_{j+1})$ as an element, then the output fault $s_p \xrightarrow{i_{j+1}/o_{j+1}} \delta(s_p, i_{j+1})$ could have occurred.*

Theorem 3. *If L^j has a state s_p with transition $s_p \xrightarrow{i_{j+1}/o_{j+1}} s_q$ and there is a $s_r \in (L^{j+1})^R$, then $s_p \xrightarrow{i_{j+1}/o_{j+1}} s_r$ is a tail state fault could have occurred.*

Finally the forward/backward crossover algorithm is presented as follows:

1. Do the forward trace analysis for the observed input/output sequence, let k be the least k such that $L^k = \emptyset$.
2. Do the backward trace analysis for the observed input/output sequence.
3. Add crossover arrows by applying theorem 2 and 3 to find the faults between L^j and $(L^{j+1})^R$.

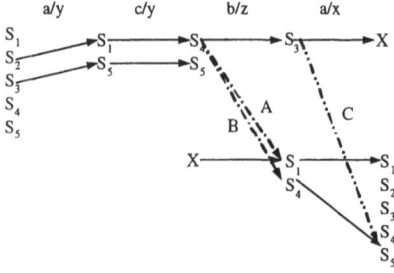

Fig. 2. Example of the forward and backward trace

To the FSM shown in Fig.1, assume that the observed input/output sequence is: $\{a/y, c/y, b/z, a/x\}$. Then the forward trace is: $L^0 = \{S_1, S_2, S_3, S_4, S_5\}$, $L^1 = \{S_1, S_5\}$, $L^2 = \{S_4, S_5\}$, $L^3 = \{S_3\}$, $L^4 = \emptyset$. The backward trace is: $(L^4)^R = \{S_1, S_2, S_3, S_4, S_5\}$, $(L^3)^R = \{S_1, S_4\}$, $(L^2)^R = \emptyset$. As shown in Fig.2 we can get three possible faults that would cause the implementation to produce the observed input/output sequence by application of theorem 2 and 3: Applying theorem 3, the crossover "A" depicts a tail state fault $S_4 \xrightarrow{b/z} S_1$ and the crossover "B" depicts a tail state fault $S_4 \xrightarrow{b/z} S_4$. Applying theorem 2, the crossover "C" depicts an output fault $S_3 \xrightarrow{a/x} S_5$.

3.3   Thinking About Fault Identification Algorithm

Corollary 1. *The maximal number of faults that the above algorithm may find is* $\sum_{i=0}^{k-1} |L^i| * |(L^{i+1})^R|$

Proof. During the run step 3 of the algorithm, the maximal number of crossover arrows is $|L^i| * |(L^{i+1})^R|$ if $|L^i| \neq 0$ and $|(L^{i+1})^R| \neq 0$. According to theorem 1 we can get $|(L^0)^R| \neq 0$, then we can find the minimal j such that $|(L^j)^R| \neq 0$. So the maximal number of faults is $\sum_{i=j}^{k-1} |L^i| * |(L^{i+1})^R|$. As for all $i < j$ we have that $|(L^i)^R| = 0$, $\sum_{i=0}^{k-1} |L^i| * |(L^{i+1})^R| = \sum_{i=j}^{k-1} |L^i| * |(L^{i+1})^R|$. That's the conclusion.   □

The implementation machine would keep running even though it is known to be faulty, and thus further testing method (continuing passive testing or active testing) might be possible and useful for gaining more information. Reconsider the example shown in Figure 2. In this example three possible faults are found: two tail state faults "A" and "B", one output fault "C". If the fault is "A", the final state of the implementation machine should be the state 1. While if the fault is "B" or "C", the final state of the implementation machine should be the state 5. This information is very useful as if we can find some methods to distinguish the current state of a system. If so, we can discard the faults that cannot cause the system to the current state. Based on this thought we can define an equivalent relationship of the faults that the algorithm [1] finds.

Definition 2. *To the same observed input/output sequence if two faults cause the system to the same final state, then the two faults are equivalent.*

4   Combined with Active Testing

Assume that in the FSM shown in Fig.3 an observed input/output sequence is $\{a/0, b/1, a/1, b/0\}$ and a fault is detected. Using Miller's algorithm in [1] we can get two possible faults: one tail state fault $S_2 \xrightarrow{a/1} S_3$ and another output fault $S_2 \xrightarrow{b/1} S_3$. Obviously if the fault is $S_2 \xrightarrow{a/1} S_3$, the current state of this FSM is S_1 . If the fault is $S_2 \xrightarrow{b/1} S_3$, the current state of this FSM is S_3 . According to

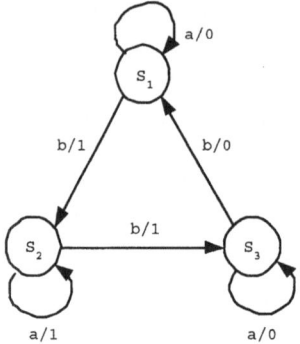

Fig. 3. Another FSM

Definition 1 we can get two equivalent sets $\{S_2 \xrightarrow{a/1} S_3\}$ and $\{S_2 \xrightarrow{b/1} S_3\}$. In [9] we know that Distinguishing Sequence or Adaptive Distinguishing Sequence can be used to judge the current state of a FSM. To the FSM shown in Fig.3 there is a Distinguishing Sequence $\{ab\}$. After detecting a fault, we actively inject the input sequence $\{ab\}$. If the observed output sequence is $\{01\}$, it means that the current state of FSM is S_1, then $\{S_2 \xrightarrow{a/1} S_3\}$ is the only possible fault. If the observed output sequence is $\{00\}$,then the current state of FSM is S_3 and $\{S_2 \xrightarrow{b/1} S_3\}$ is the only possible fault. So we can combine the algorithms in [1], [9] to improve fault identification as follows:

1. Using Miller's algorithm in [1] to get the original possible faults set E.
2. According to Definition 1, make an equivalent partition to the set E.
3. Using the algorithm in [9] to judge whether the specification FSM has a Distinguishing Sequence or Adaptive Distinguishing Sequence. If the sequence doesn't exist the set E is the final output faults set and this algorithm ends.
4. If the sequence exists, use it to judge the current state of FSM and output the subset of E corresponding to the current state as the final output faults set.

5 Further Passive Testing

In the upper section an algorithm combined with active testing method is presented. But the algorithm has a limitation: not every FSM has Distinguishing Sequence or Adaptive Distinguishing Sequence. While further passive testing method still might be useful for gaining more information. Reconsider the example shown in Fig.2. If an input/output sequence $\{a/y, c/y, b/z, a/x\}$ is observed, we detect a fault happened and keep observing. Assume that the next observed input/output is a/x. Now $\{a/y, c/y, b/z, a/x, a/x\}$ is the total input/output sequence. According to the construction of backward trace we can get that $(L^5)^R = \{S_1, S_2, S_3, S_4, S_5\}$, $(L^4)^R = \{S_1, S_4\}$, $(L^2)^R = \{S_1\}$, $(L^2)^R = \emptyset$. Now in the fourth step (we detect the fault here), the implementation should be in state 1 or 4 (originally it's the set of all states). Still using Miller's algorithm we can get only one possible fault, a tail state fault $S_4 \xrightarrow{b/z} S_1$. Compared with the three faults found by the original algorithm of Miller's, our method can give out less possible faults that will be very useful to fault management. Now summarize our method as follows:

1. First run fault detection to get the forward trace and the observed input/output sequence, assume that in the step k we detect a fault, which means $L^k = \emptyset$.
2. Keep observing for m steps (The number m should be appropriate, for that if m is too large that will cause our single fault assumption failed, while if m is too small it may be not efficient to narrow the possible faults set).
3. Construct the backward trace and find the first nonempty element, assume the corresponding subscript number is j.
4. For $i = j - 1$ to $k - 1$ apply theorem 2 and 3 to find the faults between L^i and $(L^{i+1})^R$.

There's a problem to be solved using the algorithm: how to select the number m? In the following section we'll present an experiment simulation method to get this number.

6   Experiment on a Routing Protocol BGP

6.1   Simulation of BGP

BGP is a widely used routing protocol. It includes three major parts: Interconnection, Handling Errors and Route Updating. The FSM model of the Interconnection part of BGP(version 4) is shown in Fig.4. It includes four states and twenty-two transitions.

Using the algorithm in [9] we can get an Adaptive Distinguishing Sequence of the FSM model for BGPv4. This sequence is shown in Figure 5.Our simulation steps are listed as follows:

1. Generate two random numbers N_1 and N_2. To insure the fault injected to be detected we should have $N_1 < N_2$.
2. Randomly select the possible transition and execute the transition, which is called one step. Run for N_1 steps from the initial state and start fault detection algorithm at step N_1.
3. At step N_2 randomly inject a possible output or tail state fault.
4. Assume that at step N_3 we detect a fault, run the Miller's original algorithm to get a fault set.
5. Save the current information of the simulation. Run the algorithm listed in Section 4 to get a fault set.
6. Load the saved information. Run the algorithm listed in Section 5 to get a fault set.
7. Repeat step 1 to 6 for 5000 times and compare the three algorithms.

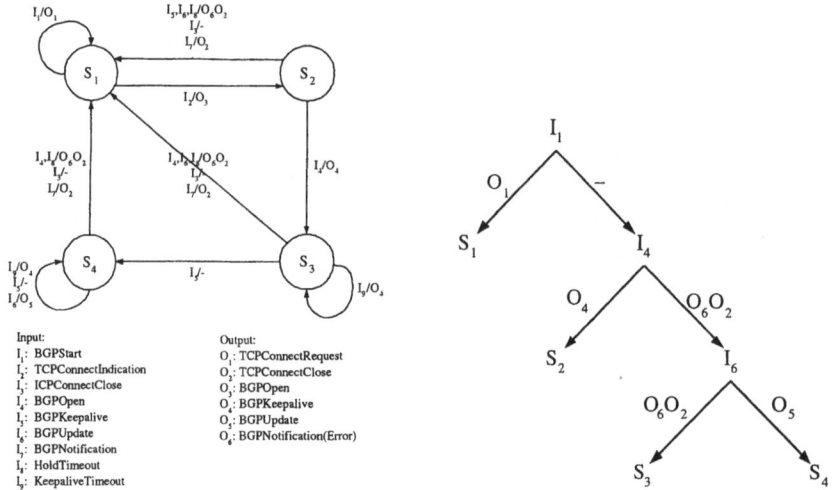

Fig. 4. Model for BGPv4

Fig. 5. Adaptive Distinguishing Sequence for BGP

6.2 Experimental Results

During the experiment we get 5,000 random input/output sequences to the FSM shown in Fig. 4. Figure 6 shows the distribution of the sequence length:

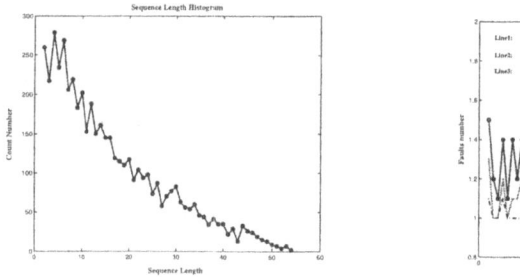

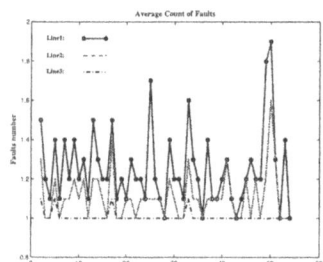

Fig. 6. Histogram of test sequence length **Fig. 7.** Average Count of Identified Faults

It can be seen that most of the sequence lengths lie between 2 and 20 and the machine does not take many steps (one input/output symbol in passive testing) to detect the injected fault. Figure 7 shows the average count of identified faults by three algorithms: The Line1 (solid line with blue color) shows the average count of identified faults found by Miller's algorithm. The Line2 (dashed line with purple color) shows the average count of identified faults found by our algorithm of observing one more step. The Line3 (dash-dot line with black color) shows the average count of identified faults found by our algorithm combined with active testing. According to Fig.7 we can see that:

- Even only observing one more step in passive testing, fewer identified faults may be found.
- Using the algorithm combined with active testing we can get the smallest identified faults, to this example the fairly constant count of faults (similarly to 1) is a very good result. Since it shows how efficient we can determine the faults that may occur.

Continuing the simulation we find that even if further observing step is greater than 1, we get the equable average count of faults with observing only one more step. That means $m = 1$ to this example.

7   Conclusion and Future Extensions

In this paper two improved fault identification algorithms are proposed. Under the assumption of single fault, these two algorithms can give out fewer possible faults that are very useful in fault management. According to the experimental result we advocate that under the condition that the specification has a Distinguishing Sequence or Adaptive Distinguishing Sequence and active testing is applicable, we should use the algorithm in Section 4 to identify faults.

Much remains to be done. Such as: Considering the relationship of different faults to detect multiple faults and the recurrent faults; Extending our fault identification approach for FSM to a network specified by the CFSM model.

References

[1] Miller, R.E.; Arisha, K.A. "Fault identification in networks by passive testing". Proceedings of the 34th Annual Simulation Symposium, 2001 , Page(s): 277 -284
[2] Miller, R.E.; Arisha, K.A. "On fault location in networks by passive testing". Proceedings of IEEE IPCCC 2000, Page(s): 281 -287
[3] Miller, R.E. "Passive testing of networks using a CFSM specification". Proceedings of IEEE IPCCC 1998, Page(s): 111 -116
[4] Lee, D.; Netravali, A.N.; Sabnani, K.K.; Sugla, B.; John, A. "Passive testing and applications to network management". Proceedings of ICNP 1997, Page(s): 113 -122
[5] Lee, D.; Dongluo Chen; Ruibing Hao; Miller, R.E.; Jianping Wu; Xia Yin. "A formal approach for passive testing of protocol data portions". Proceedings of ICNP 2002, Page(s): 122 -131
[6] Arisha, K.A. "Fault management in avionics telecommunication using passive testing". The 20th Conference of Digital Avionics Systems, 2001, Page(s): 1C7/1 - 1C7/13 vol.1
[7] Miller, R.E.; Arisha, K.A. "Fault management using passive testing for mobile IPv6 networks". IEEE Global Telecommunications Conference, 2001, Page(s): 1923 -1927 vol.3
[8] D.Lee; M. Yannakakis. "Principles and methods of testing finite state machines - A survey". Proc. Of the IEEE, vol.84, Aug 1996
[9] D.Lee; M.Yannakakis. "Testing finite state machines: state identification and verification". IEEE Trans. Computers, 1994,vol.43, no.3, pp.306-320.

Script MIB Extension for Resource Limitation in SNMP Distributed Management Environments

Atslands da Rocha[1], Cris Amon da Rocha[2], and J. Neuman de Souza[1]

[1] Universidade Federal do Ceará (UFC) - Dept. de TeleInformática (DETI)
Campus do Pici - Fortaleza - Brazil
atslands@deti.ufc.br, neuman@ufc.br
[2] Faculdade 7 de Setembro (FA7)
Alm. Maximiano da Fonseca 1395, Luciano Cavalcante - Fortaleza - Brazil
crisamon@fa7.edu.br

Abstract. Resource limitation on management scripts has been an important requirement for distributed management environments. This article proposes a Script MIB extension with new objects able to control the usage of specific resources (physical memory, processing cycles, among others) for each script launched inside the distributed environment.

1 Introduction

A widespread dissemination of TCP/IP networks has caused an increasing search for new and more efficient management environments. Following this purpose the Internet Engineering Task Force (IETF) has created the Simple Network Management Protocol (SNMP), which supplies a functional management framework.

However, the traditional SNMP management model is becoming a problem because of processing bottlenecks and bandwidth saturation. In large networks this kind of management environment is impracticable and opens room for new frameworks like distributed management model.

Development of distributed management standards for SNMP environments is a task for DISMAN-WG [1] that has released a set of new Management Information Bases (MIB) and distributed management framework documents to improve the SNMP model. The main purpose is to avoid changes in the protocol and make use of the actual devices already installed.

One of these proposals uses management by delegation [2], where a high level manager is responsible for submanagers. This proposal defines management functions by means of scripts and makes use of the Script MIB to provide mechanisms for the transfer, execution, administration and control of management scripts.

With this kind of environment, problems of bottlenecks and bandwidth saturation are solved by distribution of management tasks. Problems like scalability and fault tolerance are covered too. In spite of these advantages, it's necessary and important to analyze tradeoffs for submanagers. These elements are network devices with different functions and, therefore, different resources (Processor, memory and bandwidth). These functions cannot be damaged by scripts.

J.N. de Souza et al. (Eds.): ICT 2004, LNCS 3124, pp. 835–840, 2004.

Aware of the problem cited above, DISMAN advices hardware resource limitation in management scripts execution, but anything else is specified. Thus, this article proposes a Script MIB extension in order to deploy this requirement. New management objects are added in order to control hardware resources in distributed managers.

2   IETF Script MIB

The IETF Script MIB provides means of delegating and calling management scripts for/in distributed managers. According to RFC 3165 [3], a distributed manager is a processor entity able to provide network management functions. This distributed manager can be broken up in two elements. The first one is a SNMP entity, which implements the Script MIB, and the other is an execution environment responsible for execution of scripts.

A standard interface is defined by the SNMP management architecture for delegation of scripts. In short, the Script MIB provides mechanisms to: transfer management scripts to distributed manager; start, suspend, resume and finish management scripts; send arguments to management scripts; monitor/control active management scripts and bring management scripts execution results back.

Scripts can be written in any programming language supported by the MIB implementation. Nothing is said about the format of management scripts during transfers. The most part of programming languages is registered by IANA and other specific languages can be registered in enterprise subtrees.

The Script MIB allows the execution of various instances of the same scripts. A running script can be suspended, resumed or aborted. Active scripts timeouts are used to avoid problems such as infinite loops. A notification with an exit code, an error description and a timestamp is generated when a script ends in an abnormal way.

To start a script is needed to create an entry in smLaunchTable table, which store all script parameters, its maximum time to live and its owner. An abstract launch button is responsible for starting the script and this action can be done through a single SetRequest primitive. So many instances of the same script can be executed, but different parameters and permissions must be applied.

The smRunTable table lists all scripts that are running at the moment or finished one second before. A manager can retrieve script states, control their execution and retrieve results of finished scripts through objects from this table. The agent stores script results until maximum time to live.

However, the Script MIB specification leaves open the resource limitation problem. This issue is addressed just one time when suggestions of mapping the owner Script MIB object to a local operating system (or another execution environment) user are done.

3 Script MIB Extension

Mapping of the owner MIB object to a local user in the execution environment
is a great suggestion to solve the resource limitation problem, however this approach is followed by many tradeoffs to be considered. The lack of user based
policies in simple execution environments or no related information about users
and their pre-defined resources are some examples. A user in an general purpose operating system nothing says about constraints such as limit of processing
cycles.

Therefore, this article proposes the creation of a new table called smResourceTable, which provides new objects able to specify and control resources
during management script executions. Some computational resources are very
important to the right working of devices. Four essential features were chosen
to control script execution: Processing time; Physical memory; Number of open
files; File size. The Script MIB extended with the new table is shown at Fig. 1.

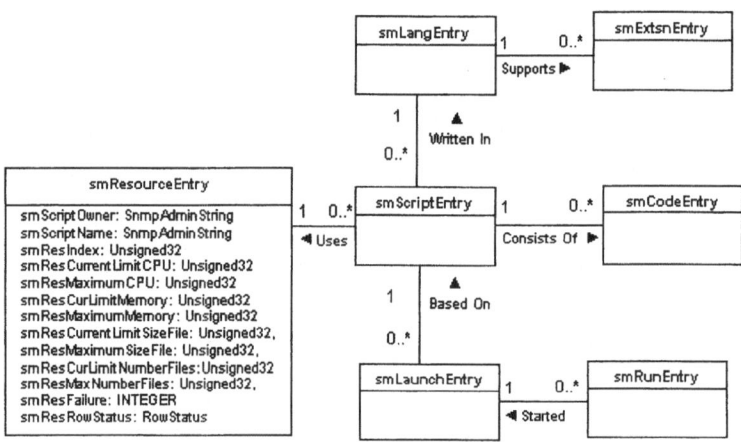

Fig. 1. Extended Script MIB diagram

An entry is associated with a management script through the smScriptOwner
and smScriptName objects. smResourceIndex object associates a specific resource limitation configuration with a script execution. All atributtes are optional.

smResourceCurrentLimitCPU object stores the maximum of processing time,
in seconds, that a script can make use. smResourceCurrentLimitMemory stores
the maximum physical memory available to its execution. Management scripts
cannot exceed these limits and the execution environment needs certify this
constraint.

Maximum limits that a user can setup to the current limit of processing
time and physical memory are presented at smResourceMaximumCPU and

smResourceMaximumMemory objects. Moreover, they cannot be changed by the script owner because are pre-defined by the SNMP agent.

The smResourceCurrentLimitFileSize object shows the maximum size, in bytes, of files written by management scripts and its maximum possible value is found at smResourceMaximumFileSize. In the same way, the smResource-CurrentLimitNumberFiles and smResourceMaximumNumberFiles objects store the maximum number of open files that a specific script can handle and the maximum possible values. Management scripts cannot exceed these values.

The smResourceFailure object points out that the related script has reached one of the pre-defined limits. This objects value is related with the last incident. Possible values are: sigxcpu (Processing time limit has been reached); notAllocMemory (Script has tried to exceed the maximum physical memory limit); sigxfsz (Script has tried to write to a file bigger than the maximum file size value defined); noOpenFile (Script has tried to open more files that the value defined); noFail (No limit has been reached yet).

When a script reaches a pre-defined limit for any resource or tries to exceed this limit, a notification is generated in order to identify which limit was responsible for the exception.

Even with definitions for each object cited above, there are so many ways to deploy them. Their behavior is left open for each environment that implements the proposed extension. However, two situations need to be considered:

The first one concerns about objects that store maximum possible values. Management scripts can belong to different owners, which can have different permissions and privileges in the system, so the agent responsible for the Script MIB can specify different values in order to give different priorities to the owners.

The second one is about penalties for management scripts that tried to exceed limits pre-defined at current objects. Notifications send to the manager are just warning messages and not penalties.

A complete Script MIB specification extended by the proposal cited in this article can be found at the project website [4].

4   Script MIB Extension Implementation

4.1   JASMIN – A Java Script MIB Implementation

The Script MIB has been implemented in Java programming language at Technical University of Braunschweig and C&C Research Laboratories of NEC Europe Ltd. at Berlin inside a project called JAva Script Mib ImplementatioN (JASMIN) [5]. The final projects purpose is to experiment the IETF Script MIB functionalities and share results with DISMAN working group. Figure 2 details the Jasmin internal structure.

The left side shows the master agent process, where Jasmin makes use of NET-SNMP toolkit . Subagents are loaded inside the master agent in runtime mode through dynamically loaded subagents modules.

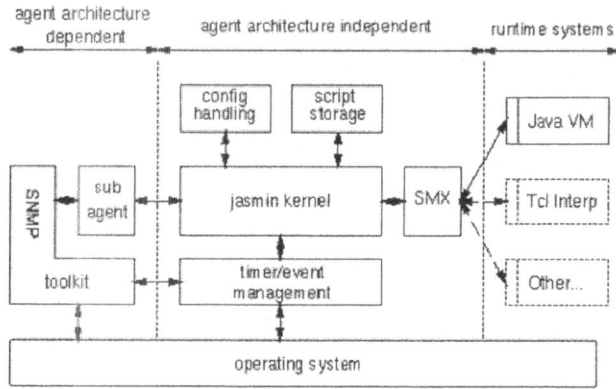

Fig. 2. JASMIN internal structure

Jasmin core shows the autonomous kernel and its auxiliaries mechanisms. This kernel changes information with execution systems via Script Mib eXtensibility (SMX) [6], which eases the adding of new execution systems. The Java virtual machine is used as the main Jasmin execution system and provides support for execution systems in Tcl and Perl. The Jasmin kernel starts execution systems and a local security service certifies the system authentication process.

Jasmin uses security profiles in order to define what an active script can do. The first one is the operating system security profile, which specifies a set of operating system services that can be used by the script. The second one is the runtime system security profile, which specifies a set of services that can be used at some well-defined moments during script execution.

4.2   The SmResourceTable Implementation

Jasmin is an open source SNMP agent built over a modular architecture that provides extensibility and scalability. Besides, it provides support for resource limitation in distributed managers.

Based on these remarks, Jasmin project has been chosen as a framework to the Script MIB extension implementation, following suggestions found at own Jasmin documentation of mapping script credentials to Jasmin security profiles.

Jasmin has three essential files: a MIB file that defines the module itself; a C header file with all function prototypes and C implementation files. The MIB file was generated according to RFC 2592 [7] and the C header file is called jasmin.h.

To add the new proposed objects to Jasmin environment it was needed change the local Script MIB file in order to extend it. Jasmin.h file was changed with new object definitions and their handle SNMP functions. Two files called resource.c and write_resourcetab.c have been added and concerns about read-only and read-write access methods.

Initial object values were specified in jasmin.conf configuration file, which has been changed with new smResourceTable objects. Another configuration file named snmpd_jasmin.conf was changed to support access to smResourceTable.

All new objects were implemented, but smResourceCurrentLimitMemory and smResourceMaximumMemory objects will not be instrumented because Jasmin haven't any limitation rule for them.

Object instrumentation was done at C language because of NET-SNMP toolkit. However, the resources limitation mechanism has been developed in Java. Warnings and JasminSecurityException exceptions are sent when a script tries to access an unauthorized resource. This implementation has a main purpose of validating the proposal presented in this article.

5 Final Remarks

Script MIB has many advantages in distributed network management. However, the possibility of execute management scripts in remote devices can bring many problems as controlling of submanagers; monitoring of scripts and their instances and controlling of arguments and scripts results. The computational processing to realize all these tasks cited above can damage essential distributed manager functions.

In order to define a model for delegation of credentials over the Script MIB, allowing resource limitation per script, this article presented a Script MIB extension, which add new objects to monitor and control essential system resources (processing cycles, physical memory and file handle).

The implementation has been finished and was deployed inside the JASMIN architecture, which has saved days of work. Some experiments will be done next step in order to certify viability and efficiency.

As future works, our group will try to port this extension for other execution environments such as Tcl and Perl, which couldn't be supported at this first moment. Besides, new experiments and improvements will be done in order to reach an IETF draft style as a RFC 3165 addendum.

References

1. DISMAN - The Distributed Management Working Group. Available at: http://www.ietf.org/html.charters/disman-charter.html/.
2. J. Schönwälder. Network Management by Delegation - From Research Prototypes Towards Standards. Computer Networks and ISDN Systems, Nov 1997.
3. D. Levi and J. Schönwälder. Definitions of Managed Objects for the Delegation of Management Scripts. RFC 3165, Nortel Networks, TU Braunschweig, Aug 2001.
4. XScript MIB Project. Available at: http://www.deti.ufc.br/~atslands/xscriptmib.
5. Jasmin - JAva Script Mib ImplementatioN. Available at: http://www.ibr.cs.tu.bs.de/projects/jasmin.
6. J. Schönwälder, M. Bolz, S. Mertens. SMX Script MIB Extensibility Protocol Version 1.0. RFC 2593, Nortel Networks, TU Braunschweig, Aug 2001.
7. D. Levi and J. Schönwälder. Definitions of Managed Objects for the Delegation of Management Scripts. RFC 2592, Nortel Networks, TU Braunschweig, May 1999.

UML Specification of a Generic Model
for Fault Diagnosis of Telecommunication Networks*

Armen Aghasaryan[1], Claude Jard[2], and Julien Thomas[3]

[1] Alcatel Research & Innovation, Route de Nozay, 91461 Marcoussis, France
Armen.Aghasaryan@alcatel.fr
[2] IRISA/ENS Cachan, Campus de Ker-Lann, 35170 Bruz, France
Claude.Jard@irisa.fr
[3] IRISA/INRIA, Campus de Beaulieu, 35042 Rennes, France

Abstract. This document presents a generic model capturing the essential structural and behavioral characteristics of network components in the light of fault management. The generic model is described by means of UML notations, and can be compiled to obtain rules for a Viterbi distributed diagnoser.

1   Introduction

This paper presents the results of the continued efforts on generic modeling initiated within the Magda projects [1] and [2]. The generic model captures the essential structural (generic components and their relations) and behavioral (interactions between the generic components) characteristics of telecommunications network components in the light of their utilization in fault management tools. The generic model covers both circuit-based and packet-based networks, despite of divergent approaches adopted by the respective standardization bodies. The generic model is described by means of UML notations, namely, class diagrams, sequence diagrams and instance diagrams. These diagrams are intended to be used in 1/ derivation of the technology-specific models and 2/ generation of rules on generic component instances. Although the targeted management applications work with the derived specific models, but in certain cases they can directly apply the rules defined on generic components. Thus, the effort necessary for deriving a specific model is significantly reduced. For the diagnosis application we have considered, we proved that the generic model can be compiled to obtain generic rules.

2   Basic Concepts

2.1   Structure

The basic concepts of our generic model are guided by the ITU-T standard Generic Functional Architecture of Transport Networks [6]. The layering concept in this architecture introduces client-server relations between the adjacent layer networks,

---

* The presented results have been funded by the French national RNRT Magda2 project on fault management.

J.N. de Souza et al. (Eds.): ICT 2004, LNCS 3124, pp. 841–847, 2004.
© Springer-Verlag Berlin Heidelberg 2004

while the partitioning concept allows the decomposition of a layer network into sub-networks and links. The end-to-end connectivity in the server layer, network connection, is obtained by concatenation of link connections. A *trail* in the server layer provides the communication service between two neighboring nodes of the client layer network, see Fig. 1a. In the case of multiplexing, it carries several *link connections* of a client network in the server network; we speak also of a *containment* relation between the respective Connection Termination Point (CTP) and Trail Termination Point (TTP) managed entities of the neighboring network layers. The CTP/TTP entities of the same network layer are connected, as shown in the figure, through *matrices* and subnetworks.

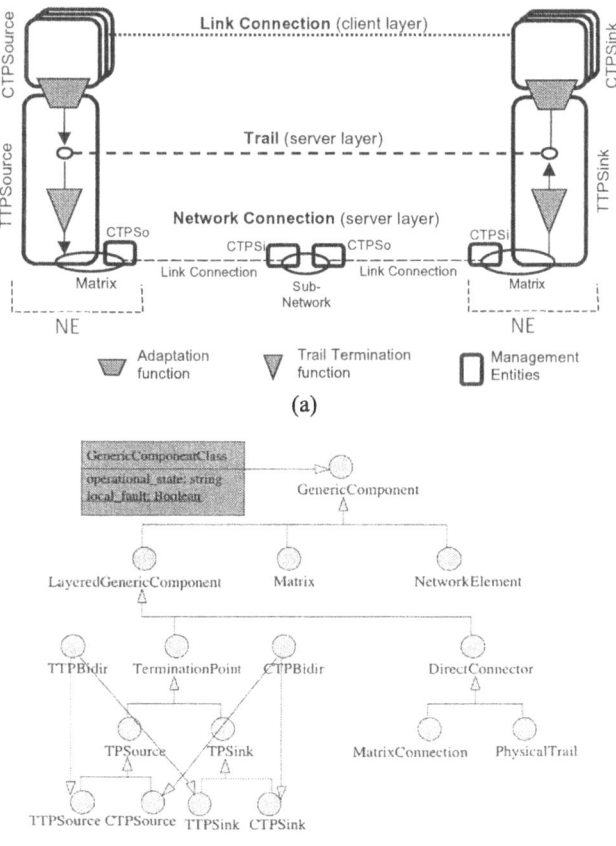

(a)

(b)

Fig. 1. a. Generic functions and managed entities, **b.** Generic Components.

These concepts are largely adopted in circuit-based technologies and they are represented in object-oriented management information models [5]. In packet-based technologies however these generic concepts are not explicitly expressed and the information models are often based on SNMP table structures. Nevertheless, it can be shown that these concepts remain pertinent also for packet-based connection-oriented networks, and the respective objects can be extracted from table-based information models. In MPLS [8], the FEC (Forwarding Equivalence Class) aggregation approximates the termination function and can be seen as a TTP, while the label assignment is related to the adaptation function and can be represented by a CTP.

2.2   Fault Behavior

Due to their physical nature and the associated monitoring mechanisms, the main role in fault propagation across the network is played by transmission failures. A transmission failure is physically propagated through the network: horizontally, along a network connection at a given layer network, and vertically, through the higher layer networks. In addition, the detected faults can be propagated via *monitoring signals*. So, once a failure/degradation is detected at a TTPSink, failure indications are sent to the corresponding CTPSink objects on the client layer. Further, this information can be forwarded downstream along the network connection in the client layer until the respective TTPSink is reached (*Forward Defect Indication*). Note that the propagation can be interrupted whenever one of the objects on the propagation path is in a "non-communicating" (e.g. Disabled) state (*alarm masking*).

3   Structural Relations

3.1   Generic Components Package

We use an abstract notion of *GenericComponent* to represent any network component that can be faulty and/or can participate in a fault propagation, see Fig.1b. The generic components are interrelated by means of peering and containment relations. In order to allow working with a programming language (e.g. Java) which does not support multiple inheritance, all the generic components are declared as Interfaces. However, the state variables can not be declared as attributes of an Interface, this is why we define them in an additional class GenericComponentClass and impose (informally) that in a specific model any class implementing a generic component must inherit from this class to obtain the uniform definition of state variables. We identify 4 main classes of generic components : *NetworkElement* represents a physical network element (device); *Matrix* represents a logical or physical component that regroups a set of cross-connections (matrix connections); *TerminationPoint* represents a logical or physical component used in transmission; *DirectConnector* represents a matrix connection and physical link components, the latter is a trail at the lowest layer network.

3.2   Layer Network (Partitioning) Package

The diagram of Fig. 2. defines the horizontal (or peering) relations between the generic components. *MatrixConnection* class represents a cross-connection between two termination points on the same network element. *LinkConnection* class represents a logical (*indirect*) connection between two CTP objects on two neighboring network elements. *Link* class represents a list of link connections grouped together for usage in routing algorithms. *Trail* class represents a logical (*indirect*) connection between two

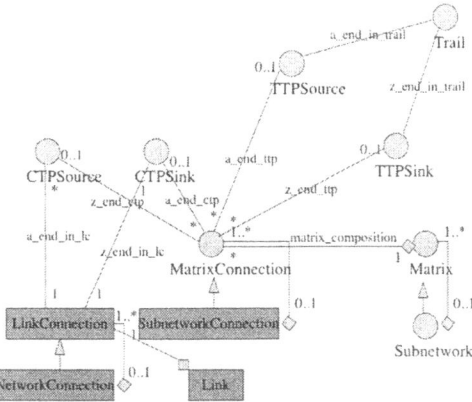

Fig. 2. Peering relations : layer network (horizontal) partitioning.

TTP objects on two distant network elements, it symbolizes a transport service provided by a given layer network to its client layer network. The *Client-Server Relations package* is used to define the client-server (vertical) layering as containment relations between *CTPSource* and *TTPSource* (source aggregation), and between *CTPSink* and *TTPSink*. The *Physical Location package* introduces another type of containment relations: containment by physical location. A network element contains a matrix, and a number of physical ports. The *NE view package* is aimed at presenting a local view of the network hierarchy. It proposes a structure of recursively embedded *Layers* where each layer assembles the local termination points belonging to the given layer network. This view is important if one aims at automatic model discovery from the information available in the network devices. In that case, a complete model instance is constructed having as an input the entities and dependencies described in this package.

4   Fault Dynamics

In order to describe the fault-related behavior of model components the notion of tile was introduced in [4]. The tiles are composed of a pre-condition part - conditions on attribute values and reception of messages; an action part - sending of messages, possibly under some conditions, and a post-condition part - new attribute values. The tiles can be easily described with a rule script and can be directly called in fault management applications that make use of a rule engine. We introduce an equivalent UML compliant description mechanism which allows to associate fault-related behaviors with the generic components and to transform them into a rule script.

4.1   Rule Description with UML Diagrams: The Concept

The main idea is to describe a rule by means of one sequence diagram and one instance diagram. The sequence diagram represents the *message exchange* between the class instances and defines the *conditions* on attributes in the associated comment blocks. On the other hand, the instance diagram specifies the *structural relations* between the class instances. The instance diagram will therefore introduce new conditions to be verified in the rule.Fig. 3. shows such a pair of sequence and instance diagrams. The corresponding rule script (in Ilog JRules syntax) generated with default conventions is shown on the right.

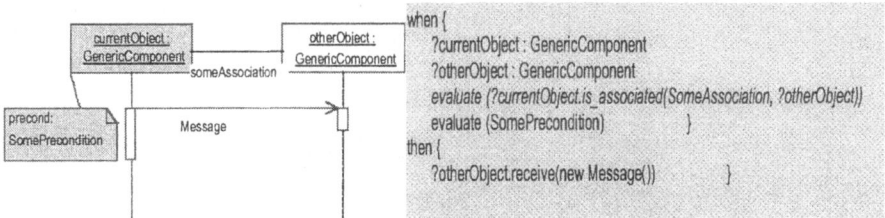

Fig. 3. A sequence diagram with an instance diagram describes a rule.

We use this mechanism for defining *generic tiles*, i.e. behaviors associated with generic components such that the corresponding rule script is readily applicable, by inheritance, to the respective components in a specific model. Of course, the same mechanism can be applied during the definition of the specific model in order to describe supplementary specific fault behaviors.

4.2   Horizontal and Vertical Propagation

Propagations between the components of the same network layer can be in *forward* or *backward* directions. In either direction, they can follow *direct connections* (matrix connections) or *indirect connections* (link connections, trails). For the sake of briefness, we chose to present the case FDI-1 of direct connections.

The sequence diagram of Fig. 4 defines a behavior where a generic component with *operational_state = Disabled* communicates a message FDI to another generic component. The right part summarizes the 5 cases where such a behavior can happen: via matrix connections or physical trails. The concerned objects are outlined with blue rectangles, the associations between them are indicated with red lines, the *currentObject* is indicated with a circle where an arrow originates. Propagation and/or Masking is a behavior inherent to all the layered generic components and can be easily modeled by pre-conditioning the forwarding of messages from an upstream object

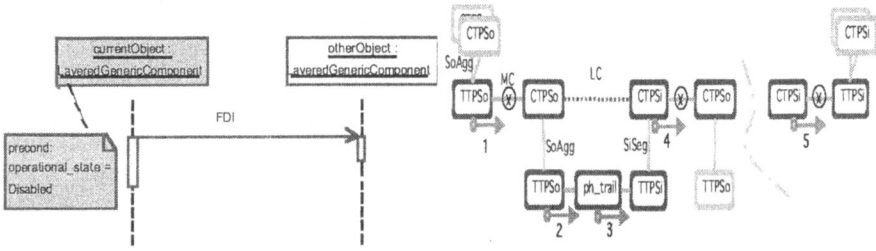

Fig. 4. Forward Propagation via a direct connection.

to an downstream object, by the test of the *operational_state* of the intermediate object. Vertical propagation is modeled as a message exchange between two generic components with a precondition on the operational state of the object sending the message. The idea behind it is that any generic component in a faulty state will automatically propagate its state to all the components it contains. So, this behavior is essentially based on the containment relation.

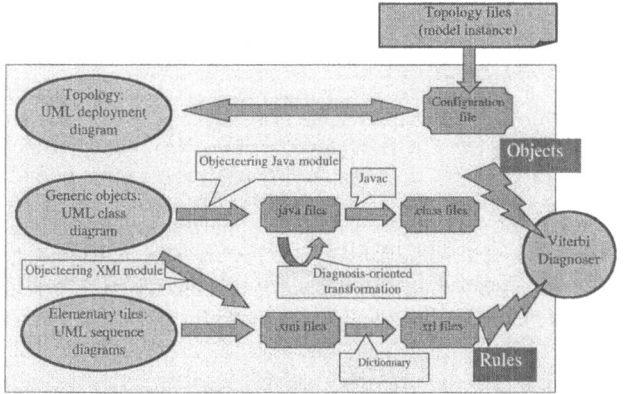

Fig. 5. The functional view of the OSCAR model-compiler

5   Model Compiler

Using the class and deployment diagrams, our tool builds the actual objects and their associated links, in order to obtain the model instance of the specific network to be supervised. These objects will be used by the diagnosis application. On the other hand, the sequence diagrams are compiled to produce the rules (expressed in Jrules) needed by the diagnoser. The OSCAR prototype has been connected to the Objecteering case tool using the XMI interface. Precisely, OSCAR takes in input a UML model and produces a set of Java classes, a set of rules (an XML file describing a set of condition/action) and a configuration file, as illustrated in Fig. 5. Another step of facilitation of the modeling task consists in the discovery of network topology and the respective instantiation of the correlation model used by diagnosis modules. The automatic instantiation can be applied to the structural part of the model by inspecting the supervised network elements and by extracting the connectivity information contained in the SNMP MIB tables. This information then is mapped onto an object structure of a specific model for MPLS networks derived from our generic model which in its turn is compliant with ITU-T G.805 Recommendations. As a result of this mapping one generates a set of XML files representing the logical and physical topology of the supervised network in the terms comprehensible for diagnoser modules. The same schema can be used in order to take into account the network reconfigurations that may happen after the initial model instance was communicated to diagnoser modules. This functionality is integrated within a generic Topology Manager tool developed in Alcatel R&I Lab. In our experiment, the distributed alarm correlation task is performed by a collection of Viterbi Diagnoser (VD) modules.

Each of these modules is in charge of a limited part of the network, typically one network element . The task of a VD is to collect alarms produced by the region it supervises, and to recover all behaviors of the supervised region that could explain these alarms [3]. The whole chain of our model-based approach to diagnosis have been demonstrated on the Alcatel management platform ALMAP.

References

[1] MAGDA2-GMPLS Architecture, A. Aghasaryan, F. Touré, A. Benveniste, S. Haar, E. Fabre, F. Krief, Magda2 project deliverable MAGDA2/HET/LIV/1, November 30, 2002.

[2] Modeling Fault Propagation in Telecommunications Networks for Diagnosis Purposes, A. Aghasaryan, C. Dousson, E. Fabre, A. Osmani, and Y. Pencolé, XVIII World Telecommunications Congress, Paris, 22-27 September 2002.

[3] Algorithms for Distributed Fault Management in Telecommunications Networks, E. Fabre, A. Benveniste, S. Haar, C. Jard, A. Aghasaryan, This conference.

[4] Fault detection and diagnosis in distributed systems : an approach by partially stochastic Petri nets, A. Aghasaryan, E. Fabre, A. Benveniste, R. Boubour, and C. Jard, Journal of Discrete Event Dynamic Systems, Kluwer Academic Publishers, Boston, Vol.8, no.2, June 1998.

[5] Generalized Multiprotocol Label Switching: An Overview of Signaling Enhancements and Recovery Techniques, Ayan Banerjee, John Drake, Jonathan Lang, Daniel Awduche, Lou Berger, Kireeti Kompella, and Yakov Rekhter, IEEE Communications Magazine, July 2001.

[6] ITU-T G.805 Generic Functional Architecture of Transport Networks, March 2000.

[7] ITU-T G.782 Types and general characteristics of synchronous digital hierarchy (SDH) equipment, 1994.

[8] Multiprotocol Label Switching (MPLS) FEC-To-NHLFE (FTN) Management Information Base, Internet Draft, draft-ietf-mpls-ftn-mib-09.txt, October 2003.

IEEE 802.11 Inter-WLAN Mobility Control with Broadband Supported Distribution System Integrating WLAN and WAN

Moshiur Rahman[1] and Fotios Harmantzis[2]

[1] AT&T Labs, 200 Laurel Avenue, Middletown, NJ 07748, U.S.A.
moshiurrahman@att.com
[2] Stevens Institute of Technology, Castle Point on Hudson, Hoboken, NJ 07030, U.S.A.
fharmant@stevens.edu

Abstract. The integration of wireless local area networks (WLANs) and the high speed wide area network (WAN) using 'always on' broadband access that could provide an efficient and cost effective inter-WLAN mobility management requires the current WLAN internetworking protocol enhanced. The Inter Access Point Protocol (IAPP) defines methods for access point coordination over a distribution system to support WLAN interworking without any central intelligence to support mobility. The implementation of mobility management with any external central control in a distribution system, including broadband such as DSL, is not specified in any WLAN, including 802.11. This paper, first, presents an implementation of a distribution system with broadband DSL accessed central network control for IEEE 802.11 inter-WLAN mobility. Then the modification to the IAPP protocol messaging required for the broadband supported mobility management for inter-WLAN is presented. Preliminary simulation results demonstrating the effectiveness of the proposed distribution and its signaling are also provided.

1 Introduction

To take advantage of the broadband supported distribution system, the WLAN interworking protocol needs to be enhanced. The distribution system is formed by a distribution system medium (DSM) and distribution system service function in each access point and central server. The implementation of mobility management with any external central control in a distribution system, including broadband such as DSL, is not specified in any current WLAN. The coverage and the performance of an inter-WLAN mobility would largely depend on the implementation of the distribution system and the control messaging. [1] proposes three options of the implementation of the distribution system, including a server controlled WLANs using IP layer, to forward messages between stations in a IEEE 802.11 infrastructure network.

Our paper concentrates on mobility management in addition to the broadband distribution system implementation network topology. Thus our work differs from [1]

J.N. de Souza et al. (Eds.): ICT 2004, LNCS 3124, pp. 848–857, 2004.

in two important ways: first our distribution implementation architecture option uses DSL accessed network server residing in high speed backbone data network for inter WLAN communications and its mobility management. DSM such as Ethernet provides limited range in local area but the broadband DSM will provide wider range with efficient connectivity. This, integration of WLANs and the WAN (high speed data network) using broadband access provides an efficient and cost effective mobility management for the WLANs.   Secondly, this work leverages IAPP and proposes supplementary protocol procedures for handover and specifies required additional signaling messages for the inter-WLAN mobility management. Cellular network mobility management protocol [2] does not seem to be a good starting for the connectionless packet networks, such as 802.11 WLAN.

The IEEE 802.11 Wireless LAN standard [3] specifies which messages are exchanged between an access point and a mobile terminal in intra-WLAN. The Inter Access Point Protocol (IAPP) [4] defines methods for access point coordination over a distribution system to support APs interworking without any central control and it is not adequate to support the proposed distribution architecture.   The 802.11F [5] that makes use of IAPP and specifies the information to be exchanged between APs among themselves and a higher-layer management entity residing in the AP, is also not adequate for the Inter-WLAN mobility with DSL accessed network server architecture as it does not address any network-based external server provided control mobility management. Typically, this management entity is a main operational program of AP which is AP vendor specific and its implementation is not defined. The 802.11F and IAAP are geared towards the multi-vendors' APs interworking and not sufficient for the Inter-WLAN mobility management. The notion of our proposed broadband accessed distribution system for inter-WLAN goes beyond the traditional WLAN mobility.

This paper addresses the network server control mobility management based on the IAAP preliminary work to support the mobility for the 802.11 WLANs. If and how the IEEE 802.11F features could be leveraged for the proposed architecture is for future study. The proposed distribution system integrating WLAN and high speed WAN with a central control would obviate the multicasting and improve the overall efficiency. IAPP protocol implemented on top of UDP/IP uses IP multicast function to link all access points. The IAAP IP multicasting introduces signaling traffic overhead and affects the QoS. It is not capable of supporting mobility in the proposed broadband supported distribution system. In this work we are mainly interested in IEEE 802.11 inter-WLAN mobility, not Intra-WLAN mobility.

The rest of the paper is organised as follows: Section II provides a description the DSL broadband supported distribution system medium with central network server control for inter-WLAN mobility. Section II also gives a summary on mobility management approaches using mobile IP. Section III describes the proposed HO procedure and the signalling messages required for the HO in broadband supported distribution system and central network control.   Section IV provides the simulation results. Finally, Section V summarizes our conclusions and contributions.

2   Broadband-Based DSM and Inter-WLAN Mobility Implementation

Figure 1 shows a network diagram that illustrates the various network entities and the functions that may comprise such an end-to-end distribution network over DSL that supports WLAN mobility. DSL is an effective encoding technology-over which higher layer encapsulations like ATM, protocols, including IP, and higher order services such as Web access or multicasting are deployed.

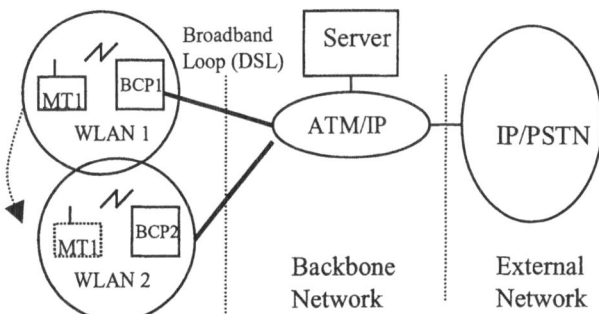

Fig. 1. Broadband-supported DSM and Network Control WLAN Mobility

Our implementation architecture is about the inter-WLAN mobility, i.e., inter-access points (APs) messaging via the wired part of the "always on" DSL loop accessed backbone network to support mobility. With an appropriate protocol architecture mechanism, it is possible to detect the loss of the connectivity to the control point of the first WLAN and establishing a connection to the next available control point of a different WLAN. This control point in an infrastructure network of IEEE 802.11 is referred to as an AP. In this paper this access point will be termed as broadband control point (BCP). The mobile station will be referred to as mobile terminal (MT).

In order to support mobility in an inter-WLAN, the network infrastructure needs to have a set of network entities and functions, that, when functioning together, are able to provide the necessary services for the application. First, a wireless access point with DSL interface capability to extend transport service to the network-based server. Second, a signaling protocol that is capable of connection establishment and seamless handoffs when the mobile terminal is migrating across the WLAN boundaries is required. Finally, a wireless control mechanism is needed for management of radio resources, handoff etc. between BCP and mobile terminals.

In the proposed signaling protocol architecture, handover signaling interactions between a terminal and a BCP take place over the WLAN air interface but the inter-BCPs signaling communications of different WLANs will take place via the always on DSL access network. The network-based server coordinating the required handover will be accessed from all the provisioned BCPs of the respective WLAN via the DSL access. A typical IEEE 802.11 WLAN consists of a collection of WLAN adapters and APs interconnected via a distribution system. In IEEE 802.11 WLAN standard the coverage area of a single Access Point is referred to as a basic service set (BSS); to extend this, multiple BSSs are connected through a distribution system (usually the wired network) to form an extended service set (ESS). In typical WLAN all the access points are connected via Ethernet as DSM that gives limited range. The use of DSL access will provide extended environment. DSL access technology will provide the wired network to build the ESS with wider coverage for different and large WLANs. This is the uniqueness in inter-WLAN communication where always on DSL access loop integrates WLAN with backbone network server for inter-WLAN mobility. As shown in Figure 1 the WLANs are connected to the backbone network via a broadband, always on, access network, such as DSL. In this distribution architecture, each WLAN is managed by the respective BCP and can communicate with any MT of the different WLAN through the backbone network. MTs in the same WLAN communicates with each other via the local BCP without going through the backbone network. A mobile MT (voice or high speed data) connected to the backbone via any one of these WLANs, moves away from current WLAN service area to another WLAN service area, the connection with the backbone will be maintained under the control of the backbone-based network server. In this network arrangement, the link between the DSL Access Multiplexing and the server could be ATM employing IP. The BCP maintains a permanent virtual connection with the network server over DSL access employing ATM. With ATM, the connection-oriented technology, the interaction between the BCP and the network server is permanent virtual circuit (PVC) or switched-virtual circuit (SVC).

Mobility management enables telecommunication networks to locate roaming terminals for call delivery and to maintain connections, as the terminal is moving into a new service area. The task of the mobility management in a data network is to route the incoming packets towards mobile nodes. The 802.11 standard does not specify how the distribution system will transfer a message from the source AP or central server to the destination AP or server over broadband access integrating LAN and WAN. In order to manage the mobility between the WLANs connected via the DSL-accessed network server, we need a mobility manager, i.e., an internetworking strategy. There are several alternatives, including Mobile IP to interconnect the two networks. While mobile IP is widely used in wireless WANs, it is not known how well it performs in WLAN environment. Mobile IP is intended to enable nodes to move from one IP subnet to another. [6, 7, 8, 9] describe mobile IP application for specific network inter-working and integration but provide similar conclusion that Mobile IP approach may not be suitable for handover in all packet services because of latency, delay, and packet loss. This approach suffers from the triangle routing between networks if mobile IP does not support route optimization. Network based central server would provide both home and foreign agents like capabilities allowing mobile user to keep the same IP address when changes its point of attachment. This would be simple, cost effective and efficient.

3 Proposed Signaling Protocol

There are two protocols involved in Inter-WLAN mobility for IEEE 802.11: Announce Protocol and the Handover Protocol. The IAPP Announce Procedure needs to be modified to make it work for the proposed architecture. More specifically, Anounce.request message must be sent to the central server only and that warrants procedure changes. This Announce protocol requirements are beyond the scope of this paper. Here, we concentrate only on the Handover protocol.Two parts in inter-WLAN end-to-end signaling needed in Handover procedure (Fig. 2a): 1) Wireless part – between BCP and MT 2) Wired part – between BCP and network server.

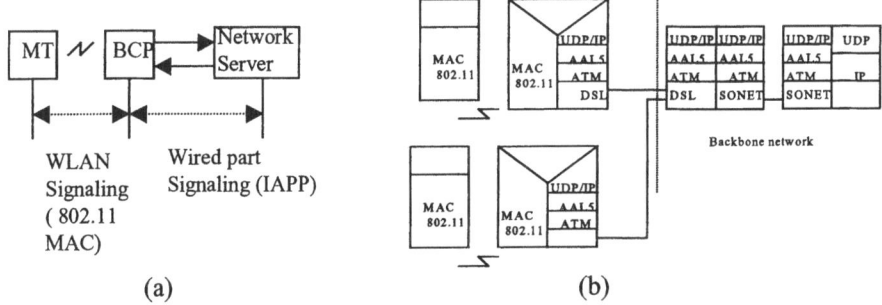

Fig. 2. a) Two parts of end-to-end signaling and b) End-to-end protocol stacks

Here the idea is to leverage IAPP as much as possible. The wireless part would be IEEE 802.11 interface as described in [3] but the wired part would be modified for implementing the proposed broadband assisted inter-WLAN mobility in IEEE 802.11. In order to identify required signaling in wired part of the above-mentioned signaling, a handoff procedure needs to be defined. Figure 3 shows protocol stacks using IP over ATM in DSL access. This allows location management and forwarding packets over an IP network composed of various WLANs interconnected by a central control server. A routing signaling message from one BCP or central server is not broadcast to all other BCPs and to the central server. The IP addresses of all BCPs belonging to the ATM over DSL supported virtual private network need to be known to the server only unlike [1]. With this broadband supported distribution, all these WLANs will appear as if they are in same subnet under the control of a single server. Where the WLANs are off different broadband access network, it is possible that all the BCPs are served by the same central server. This could be achieved via some intelligence signaling technique. This signaling interconnecting technique using IP and PSTN SS7 signaling is beyond the scope of this paper. More than one distribution system can be connected via any backbone mechanism, such as IP, to form a single distribution system that would cover wider geographic area. This arrangement would obviate Mobile IP like management approach. Note that with the typical distribution system, such as Ethernet, the inter-WLAN mobility range is limited. But with this broadband supported distribution system, we can have broader inter-WALN mobility range with high quality for both data and voice. In this distribution system, the network server will maintain an address-mapping table for recording address of the MTs and the associated BCPs. These addresses in the address-mapping table could be pre-provisioned and could be populated in real time dynamically. Each BCP can inform the server of the MAC address of the MT's in its WLAN using a message. Note that in IEEE 802.11, Association function provides that capability. After receiving this Association information from the MT, the BCP transfers this information to the server over the ATM PVC. In this broadband distribution arrangement, BCP does not have to maintain any address mapping table. All information related to routing and addressing are handled by the network server. When a MT has moved out into another WLAN, the new BCP responsible for the new WLAN must inform the server about the change. The server connects all the WLANs (i.e. BSS) to form a virtual private network, with WLANs distributed in different building interconnected by the internet and the voice network.

3.1 Proposed Handoff

The intention of the modified HO protocol is not only to inform the old BCP about the move but also provide the required mobility control messages. Our HO procedure described is inspired from the connection oriented mobility management concept. The existing IAPP Handover protocol only informs the old access point that support for the MT has been assumed by another access point and the old access point takes some local action such releasing the resources, update its filter table, and discard or forward buffered frames for the MT. Table 1 shows modified IAPP protocol summary and protocol requirements.

Table 1. Modified IAPP: Protocol Summary and Protocol Requirements

Modified IAPP: Protocol Summary	Modified IAPP: Protocol Requirements
1. Inform the central control about the intent of the MT's re-association with new access point 2. Central control determines the availability of the bandwidth for the QoS required for the proposed service 3. Central control find out best possible access point for the migrating MT 4. Inform the old access point that support for the MT has been assumed by another access point (allowing it to release the resources, update its filter table, discard/forward buffered frames) 5. Update filter tables of intermediate MAC-bridges	1. An access point must transmit a HANOVER.request to acentral control when it receives a Reassociation-request from the IEEE 802.11 MAC protocol (this an attempt of handoff by a Mobile station) 2. Central control receiving HANOVER.request, must respond to it with a RESRC_REQ (this RESRC_REQ must include the messageID from the associated HANDOVER.request) 3. Access point receiving RESRC_REQ, must send RESRC_RES indicating the available bandwidth. 4. An access point, receiving no response after having transmitted a HANOVER.request, should retransmit the HANOVER.request at least once 5. An access point, receiving no response after having retransmitted a HANDOVER.request, may retransmit additional HANDOVER.requests. These additional retransmissions may be at increasing intervals

3.2 Proposed Messaging

Additional messages are needed in IAPP to support the mobility management using the proposed architecture. The flowing messages, referred to as Protocol Data Units (PDUs) in IEEE 802.11, defined in the modified IAPP: HO_ACK, RESRC_REQ, and RESRC_RES. Figure 3a shows the required signaling message flow for the proposed handoff procedure. As the MT migrates away from old BCP and towards the WLAN coverage area of new BCP, it listens to the beacon signals from both BCPs. Based on its measurements of the BCP beacons, the MT can suggest when to initiate handoff from old BCP to new BCP based on its measurements of the BCP beacons. Note that these beacons measurement and HO decision is specific vendor implementation dependent and beyond the scope of this work. After the measurement is done, the MT sends Reassociation Request (Reasso_REQ), an existing IEEE 802.11 MAC message, to the new BCP using the broadcast radio channel, specifying all its audible frequencies and their corresponding signal to interference ratio (SIR) status.

Upon receiving this HO indication message (Reasso_REQ), the new BCP sends Hanover Request (HO_REQ), an existing IAPP message to the network-based server over the DSL/ATM PVC interface. Based on this message (user device ID, potential handoff BCP (new BCP) and the call connection ID of the existing call) from the new BCP, the server establishes a co-relation between the existing call and the old BCP that is serving the MT. The BCPs maintain a permanent virtual circuit with the network server through the DSL access. The network server would maintain a connection table of all the existing calls (data or voice) using BCPs served by the server. The server then sends a query Resource message, RSRC_REQ, to the new BCP asking about its resource availability. This would be a first new message to IAPP. In response, the new BCP sends RSRC_RES (second new message) informing about the resource availability (in this message flow it is assumed that the resource is available). The network server could use an inter-BCP (WLAN) protocol to find out the best possible new BCP (WLAN) for the migrating remote mobile terminal. After receiving RSRSC_RES from new BCP, the network server sends an acknowledgement, HO_ACK(third new message) to the new BCP. This is a second new message to the existing IAPP. Assured of the establishment of the data path to MT via the new BCP, the sever sends a HO_REQ message to the old BCP. The old_BCP sends HO_RES to server and in turn the server sends this HO_RES to the new BCP. Then the new BCP sends this Reasso_RES to the MT to complete the handover initiation phase. After receiving the Reasso_RES from the new BCP, the MT changes its operating frequency and starts communicating through new BCP. Now the resource allocation tables both in the BCPs and the network server are updated. The routing table in the server is also updated. Figure 3b shows modified HO exchange (retransmission).

4   Simulation Results

Figures 4-5 demonstrate the effectiveness of the proposed distribution architecture and its signalling scheme over the current WLAN-interworking system. Figure 4a compares the performance of the proposed broadband accessed network-based server mobility solution with current non-coordinated system solution in worst case when the systems are in highly interference environment with no recovery from the degraded links. Figure 4b depicts the performance comparison when the systems are in highly interference mode with an inference combat strategy by switching to better channel under the control of the network server.

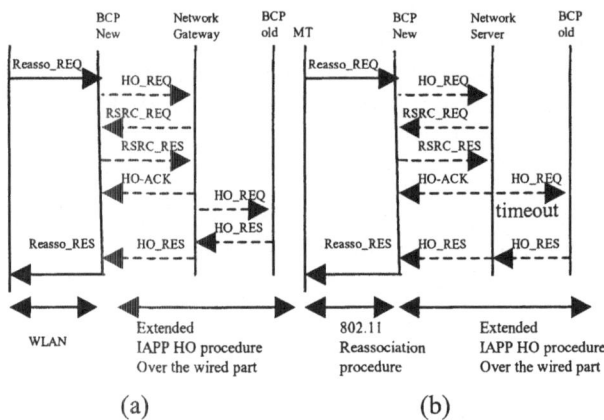

Fig. 3. a.) Modified HO Exchange (normal) and b) Modified HO Exchange (re-transmission)

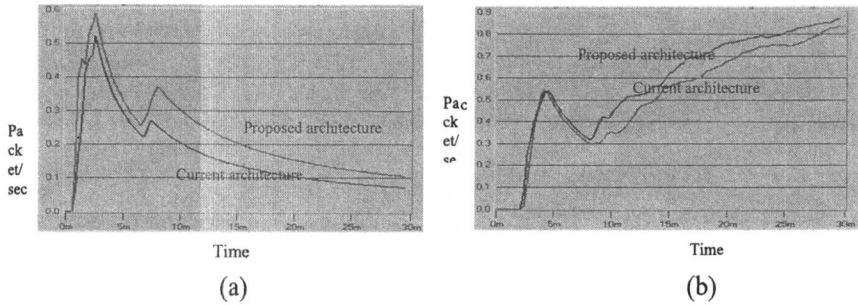

Fig. 4. Traffic received at MT for current vs. proposed system with high interference: a) with no combat strategy and b) with combat strategy

Figure 5a shows the comparison of the traffic received in the WLAN when the system has less interference than the previous two cases shown above. In all scenarios, our architecture and its signalling scheme always performs better because of broadband high speed accessed central control mobility management scheme. In the current WLAN interworking architecture, the traffic loss is higher due the fact that all the incoming traffic to the mobile node has to go to the mobile node's home WLAN and then to the present WLAN where the mobile happens to roam to. With our architecture this tunnelling is avoided by using the server's intelligence. Figure 5b shows the delay comparison between the proposed signalling scheme and the current WLAN interworking scheme. As seen in this figure, even though the proposed signalling scheme has more controlling messages to be transported than the current WLAN interworking scheme, the delay for our case is lower than the current architecture and its signaling scheme.

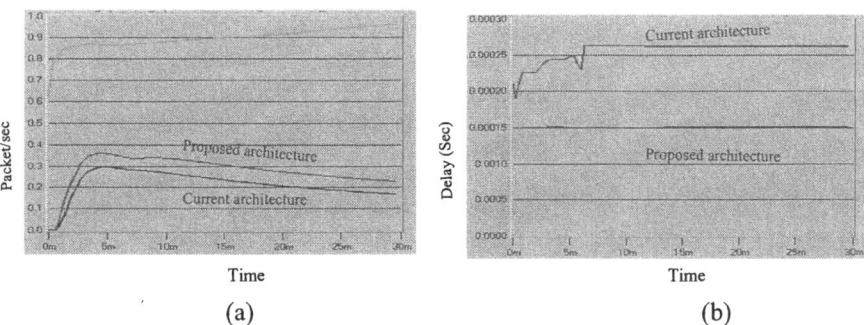

(a)

Fig. 5. a) Traffic received at MT with current vs. proposed system (with less interference than in Fig 4a and 4b and no combat strategy), and b) Comparison of delay with the proposed system vs. the current system

5   Conclusions

An implementation of a distribution system with broadband DSL accessed central network control for IEEE 802.11 inter-WLAN mobility is presented. Our implementation architecture has changed the traditional notion of WLAN extended infrastructure network and its wired distribution system. This paper has provided a survey result on the mobility management approaches including Mobile IP to assess and develop a mobility management strategy for the broadband supported distribution system for IEEE802.11 inter-WLAN mobility. With this broadband supported distribution, different WLANs will belong to the same subnet under the control of a single server. The IAAP is not adequate for the Inter-WLAN mobility with DSL accessed network server architecture, as it does not have any central intelligence to manage the mobility with QoS. The IAAP/IEEE 802.11F specification is developed for the interworking between the APs over the local distribution system. and does not support network-based mobility management like the proposed architecture and its signaling scheme. Another deficiency of the IAAP is IP multicasting introducing signaling traffic overhead and affecting the QoS. This paper addresses the network server control mobility management based on the IAAP without IP multicasting to support the mobility. We have proposed a handover procedure and the modification to the IAPP protocol and identified required messaging for the broadband supported network server control inter-WLAN mobility. This paper presents simulation results demonstrating the better efficiency of the proposed distribution architecture and its signaling scheme over the current WLANs interworking architecture and the scheme. This mobility management approach using broadband based DSM would highly benefit the service providers in offering a wide range of voice and data services in WLANs with minimum efforts and cost.

References

1. El-Hoiydi, A.: Implementation Options for the Distribution Systems in the 802.11 Wireless LAN Infrastructure network. In: IEEE International Conference on Communications (2000), Vol. 1, 164-169
2. Kou, W.K., Chan, C.Y., Chen, K.C.: Time Bounded Services and Mobility Management in IEEE 802.11 WLANs. In: ICPWC (1997), Mubai, India, 157-161
3. IEEE Standard for Wireless LAN MAC and PHY spec, IEEE Standard 802.11,1997, June 26, 1997
4. Moelard, H., Trompower, M.: Inter Access Point Protocol. In: Internet Draft Standard , March 1998
5. IEEE Standard 802.11F, 2003
6. Tsao, S., Lin, C.: Design and Evaluation of UMTS-WLAN Interworking Strategies. In: Proceedings 56th IEEE Vehicular Technology Conference (2002), Vol. 2, 777-781
7. Stemm, M., Katz, R.H.: Vertical Handoffs in Wireless Overlay Networks. In: Mobile Networks and Applications (1998), Vol. 3, No. 4, 335-350
8. Honkasola, H., Pehkonen, K., Neimim, T., Leino, A.T.: WCDMA and WLAN for 3g and beyond. In: IEEE Wireless Communications (2002), Vol. 9, Issue. 2, 14-18
9. Morand, L., Tessir, S.: Global mobility approach with mobile IP in "All IP" networks. In: IEEE International Conference on Communications (2002), Vol. 4, 2075-2079

A Cost-Effective Local Positioning System Architecture Based on TDoA

Ricardo Matos Abreu, Miguel Jorge Alves de Sousa, and Mário Rui Santos

IT - Instituto de Telecomunicações, Aveiro site.

Abstract. Joining the capabilities of two different worlds, location/positioning systems and RFID systems, is the aim of the Real Time Location System (RTLS) described in the present paper. Here we propose a cost-effective architecture for such a system based on Time Difference of Arrival (TDoA). Delay measurement of very small time intervals is also approached. Trial experiences in controlled medium results were made and shown the effectiveness of such methods.

1 Introduction

Location and positioning systems are in widely use these days. Perhaps the most known among them is the Global Positioning System (GPS), although several others (Decca, Omega, Loran, Glonass, to name a few) have/had their own importance [1]. These positioning systems typically allow a user to know where he is. The location system here described has a different focus: it enables us to locate things in a restrict area, as products in a warehouse, vehicles in a parking, people in a building, and so on.

Also, in the last years we have seen an ever increasing demand for RFID (Radio Frequency Identification) systems throughout several markets [2]. RFID tags are being put everywhere (products, machines, animals) providing several kinds of information, such as provenience, destiny, date of manufacturing, owner, and so on. Then RFID devices are used as part of information systems, which have proven to be invaluable tools for numerous tasks such as warehouse and fleet management. A LPS (Local Positioning System) would be able to give the answer to the eternal question "Where is it?" for any tagged item.

Joining location systems with RFID would thus be highly desirable for businesses such as management and logistics.

2 Proposed System

2.1 Motivation

In a scenario of a large parking yard, for example, a car factory parking for the new cars, or a large parking for buses, it is of the most interest to know where each vehicle is and, if possible, to get information about it, including a description/identification of it, data from sensors, mileage, oil levels, or even a link to the vehicle's central computer to check any malfunctioning, flag, alarm, etc.

J.N. de Souza et al. (Eds.): ICT 2004, LNCS 3124, pp. 858–865, 2004.

For the LPS to be usable on these and other scenarios it is desirable a competitive price especially for the expectable high number of tags.

2.2 Guidelines

Our proposed LPS is composed of two clearly distinct parts: mobile subsystem (the tags to be attached to appropriate items) and fixed subsystem (radio base stations, control computer, wiring). The tags should emit a somewhat strong radio signal in order to be located. Hence, unlike some proximity RFID systems, they should be active devices, i.e., to have their own power supply (e.g., a battery).

Let us here expose some main guidelines a LPS system should obey:
- Information retention;
- Radio modem capabilities;
- About 1m location error;
- Cost-effective system;
- Outdoor operation;
- Tag reading range around 100m ;

Data transfer through radio is an easy task nowadays. The main problem in a LPS is exactly location. For that reason, the paper is mainly about location.

2.3 Referential Problems

Within a positioning system, we must be able to measure/inference the distance from the item being localized to several different known locations (our referential). Once those distances determined, finding the item's position is a matter of algebra.

When using radio signals, it is obvious that we can measure propagation times and determine those distances knowing $\Delta x = c \times \Delta t$, where c is the speed of light.

One of the imperative requirements for a local positioning system is its cost-effectiveness. As can be easily realized, precision reference sources such as atomic clocks, GPS-like, are not viable [3]. Conventional quartz clocks, even with very tight tolerance crystals, have too much clock skew. So, we soon realized that having clocks on every device synchronized to a common time frame was not the way to do it. One must then choose another common reference to our system.

The solution came out to be simple: one of the known fixed points (our "base stations") will be used as a common reference. That particular base station will act as a mirror for the radio signal. The tag's signal will be received and retransmitted on a different frequency. So, it happens every other base station will receive two radio signals: the direct one from the tag and the reflected one from the mirror base station. These two radio signals have, in fact, the same content. The difference relies on the distance traveled by each one. Hence, a fundamental principle can be stated: the distance information does not lie on the signal itself, but rather on the delay between the two received signals.

Fig. 1 illustrates this principle. Base station 1 (BS1) is acting like a mirror. Let us take base station 2 (BS2) as an example: this particular base station receives the tag's signal directly from the tag delayed by a time amount of t2, and the reflected one delayed by t1+t2'. Consequently, BS2 is able to estimate the time difference between the signals. Since BS1 and BS2 are at known fixed positions, t2' is constant. Therefore, we can in fact compute the time difference t2-t1.

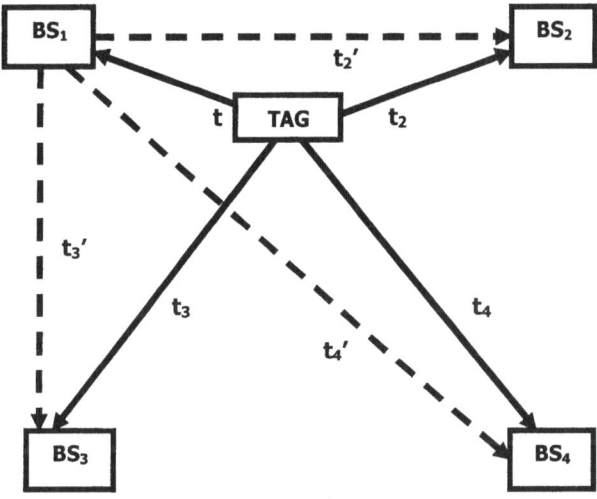

Fig. 1. Propagation delays with a mirror base station

The above leads us to the following conclusion: in our system there is no absolute time frame. We can only make relative measures between signals. Being so, unlike GPS, we never know t1 or t2 (or t3 or t4) but the difference t2-t1 (or t3-t1 or t4-t1). This time difference is directly converted to path difference by the means of the speed of light, c.

2.4 Position Finding Method

Unlike GPS, the tag's location can not be found as the intersection of spherical surfaces centered at known points, since those surfaces' radii are not known [3].

For the sake of clarity, we will constrain ourselves to two dimensions from now on. The generalization to three dimensions is straightforward [4], [5]. Hence, the following assumes we are working on a plane surface.

In a plane, the locus of all points whose distance difference to two fixed points is constant is a hyperbola. Those fixed points are named focus (plural foci). This clearly fits our system: the tag's location given a time difference lies on a hyperbola. That hyperbola's foci are the base stations involved in the measurement of the given time difference (for example: BS1 and BS2 for t2-t1).

The point of intersection of two or more hyperbolae is the seeked tag's location. Fig. 2 illustrates the situation. The black square represents the tag whereas the base stations are indicated by triangles. When the tag sends a signal, several time difference measurements can be made, one by each pair of base stations. The dotted, dashed and full lines represent the hyperbolae due to the time measurements made by BS2 and BS3, BS1 and BS2, BS1 and BS3, respectively. Each hyperbola has two branches (thin line and thick line). As one can see, there are two distinct points where the lines intersect. However, the sign of the time difference (positive or negative) tells us which branch should we use.

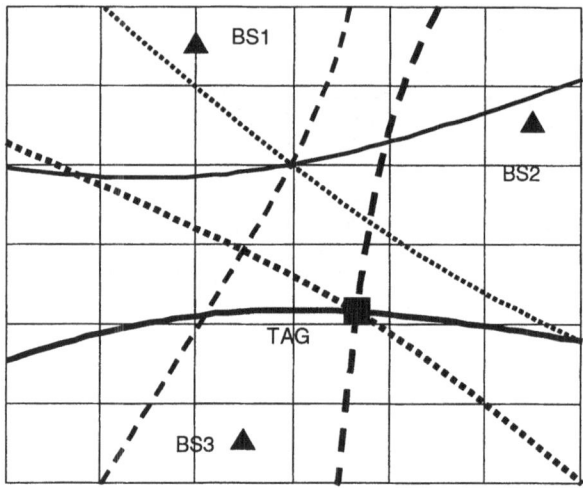

Fig. 2. Hyperbolae intersection

It is clear we need at least two hyperbolae in order to find the tag's position. Hence, at least three base-stations are needed for the two-dimensional case.

2.5 Delay Measuring Method

The speed of light is in the order of 3×10^8ms-1. Easily we find that it takes about 3.3ns for radio signals to travel 1 meter. We must target a smaller tolerance in order to achieve a final accuracy of 1 meter. Targeting 10cm measurement accuracy, we have to deal with time intervals as small as 0.33ns. Such a measurement, of course, cannot be made with a conventional chronometer system, "start/stop" fashion. In fact, a GPS like correlation mechanism is required [3].

Our proposal is to use a pseudo-noise sequence. A pure white noise signal, as we know, has an auto-correlation peak at zero lag and a near zero value at any other lag. In fact, pseudo-noise sequences are periodic, and so it is its auto-correlation function. This periodicity enables us to finely reconstruct the signal using a coarse (low rate) sampling. Take Fig. 3 as an example: we are sampling a periodic signal (a sawtooth signal) for which we do know the period. Of course, we also do know the sampling period.

Now we sample the waveform. In the first period we get the samples represented by squares. In the second period we also do get samples but, relative to the previous period, these last samples have a little time offset. Same happens with the last period. After this, the samples positions within a period start to repeat themselves. As should be apparent, this repeating period equals the least common multiple of the sampling period and the waveform's period. Since we know a priori the sampled signal's period, we may find the position of the second and third period's samples within the first

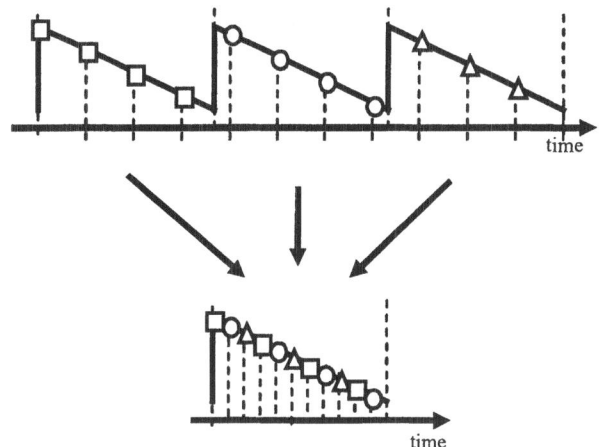

Fig. 3. Sampling resolution enhancement

period, thus reconstructing the signal with a much finer resolution. Judicious choice of the signal and sampling periods can lead us to the desired resolution. Of course, frequency tolerances are of concern. Tight frequency drift must be accomplished to guarantee proper performance.

2.6 Measurement Trials

Some trials were made in order to validate the described delay measurement method. The propagation medium was not the air but RG-58U coaxial cable. We fed several different length cables with a pseudo-noise sequence and sampled the signal at both ends. The cables were driven by a 50Ω output impedance generator and correctly terminated with a 50Ω load. This arrangement enables us to test the delay measurement method without multiple reflections or multiple paths, which are undesirable effects at this development stage. Some trial parameters are summarized at Table 1. A cross-correlation graph example is shown at Fig. 4 (the time axis is normalized to the system's resolution).

Those trial results are shown at Fig. 5 and undoubtedly validate our measurement architecture. The linear dependence of the measured time delay with the cable length is apparent.

The best fit line parameters given by least-squares fitting are presented at Table 2. The tolerances shown are for 66% confidence intervals.

Table 1. Measurement trials parameters

Signal type	7-bit pseudo-noise sequence
Signal frequency	210.650 KHz
Sampling frequency	20 MHz
Time resolution	0.4 ns

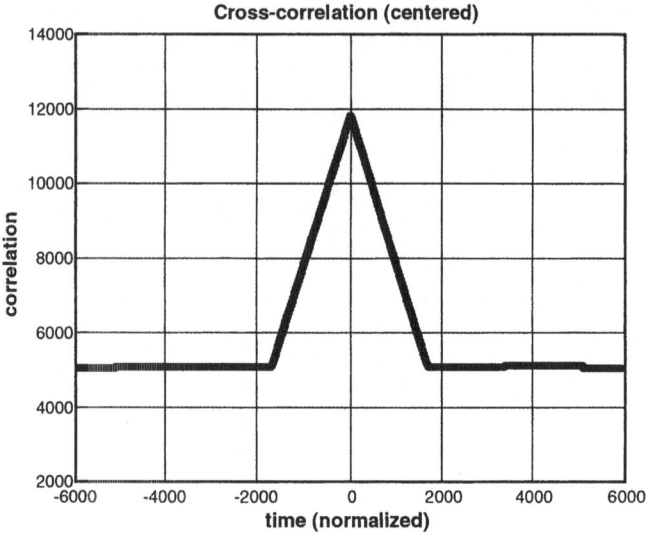

Fig. 4. Example of cross-correlation between signals

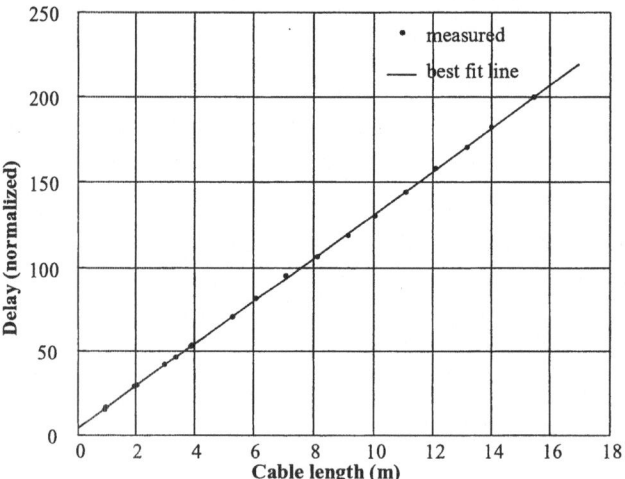

Fig. 5. Time delay versus cable length

Table 2. Best fit line's parameters

Line's slope	$(5.089 \pm 0.034) \times 10^{-9} [s/m]$
Line's y-intercept	$(1.51 \pm 0.12) \times 10^{-9} [s]$
Correlation coefficient	0.9998
Cable's propagation speed (inverted slope)	$(1.965 \pm 0.013) \times 10^{8} [m/s]$

According to several different manufacturers, the propagation speed on RG-58U cables is 66% of the speed of light, about $1.978 \times 10^8 \mathrm{ms}^{-1}$. Our result is accurate with an error under 0.66% of this value and every measurement made had an offset relative to the best fit line under 0.4ns.

These measurements were made using a rather inexpensive acquisition board designed for this purpose, providing two input channels, one bit resolution each, and 8 Kbytes on-board memory. An 8-bit micro controller is employed to interface it with a PC. The sampling clock is provided by a 2.5ppm TCXO. The total cost is under €75.00 for the prototype unit.

3 Conclusions

This paper proposes a LPS architecture that could be adopted for RF and microwave communication systems. This architecture targets a low-cost, high accuracy system providing modem and location capabilities. The TDoA method used for localization solves the clock distribution problems inherent to GPS-like approaches.

Fine time resolutions were achieved on the sub-nanosecond range with inexpensive hardware, thanks to a proper signal choice and appropriate mathematical treatment. Preliminary tests made on cables have shown delay measurement accuracies on the order of 0.4ns, which translates to 8cm on RG-58U cable or 12cm on air.

Future work comprises delay measurements with air interface at the 2.4GHz ISM band, where multipath and fading may compromise final accuracy. Once that issue is solved, we can validate this proposal as a high-potential LPS architecture for a large number of applications.

Acknowledgment. The authors gratefully acknowledge the contributions of Davide Azevedo and Jorge Amaral for their work and cooperation on this project.

References

[1] Laurie Tetley, David Calcutt, "Electronic navigation systems," Butterworth-Heinemann, 2001;
[2] Klaus Finkenzeller, "RFID handbook," John Wiley & Sons, 2003.
[3] B. Hofmann-Wellenhof, H. Lichtenegger, J. Collins, "GPS – Theory and Practice," Springer-Verlag, 4th edition, 1997.
[4] James J. Caffery Jr., "Wireless location in CDMA cellular radio systems," Kluwer Academic Publishers, 2000.
[5] Bertrand F. Fang, "Simple Solutions for Hyperbolic and Related Position Fixes," IEEE Trans. Aerospace and Electronic Systems, vol. 26, pp. 748-753, Sept. 1990.

Biographies

Ricardo Matos Abreu was born in Porto, Portugal, on January 25, 1978. He studied electronics and telecommunications engineering and graduated at the University of Aveiro, Portugal.

His experience is on instrumentation, measurement and automation. Previous works include instrument design for industrial control, assistant at the Dept. of Physics, University of Aveiro and research on semiconductor device modeling.

Currently he is with IT - Instituto de Telecomunicações, Aveiro site.

Miguel Jorge Alves de Sousa was born in 1975, at Lourenço Marques (now Maputo), Mozambique. He graduated on electronics and telecommunications engineering in 2002, and worked as a researcher on satellite-earth link propagation experiments, developing radio-frequency and microwave electronics as well as instrumentation systems. Since April 2003 he has been developing real time location systems.

Mário Rui Santos was born in 1973 at Lubango, Angola, at that time a Portuguese colony. He graduated on electronics and telecommunications engineering in 1996 at University of Aveiro, Portugal.

He immediately joined IT - Instituto de Telecomunicações, where he worked on satellite television and broadband wireless communication systems development. He got his MSc on Telecommunications at University of Aveiro in 2001. Since then he has been developing RFID systems, GPS based information systems and local positioning systems.

Indoor Geolocation with Received Signal Strength Fingerprinting Technique and Neural Networks

Chahé Nerguizian[1], Charles Despins[2,3], and Sofiène Affès[3]

[1] École Polytechnique de Montréal,
2500 Chemin de Polytechnique
Montréal, (Qc) Canada, H3T 1J4
chahe.nerguizian@polymtl.ca
[2] Prompt-Québec,
1010 Sherbrooke ouest, bureau 1800
Montréal, (Qc) Canada, H3A 2R7
cdespins@promptquebec.com
[3] INRS-EMT,
800 de la Gauchetière ouest, suite 6900
Montréal, (Qc) Canada, H5A 1K6
affes@inrs-emt.uquebec.ca

Abstract. The location of people, mobile terminals and equipments is highly desirable for operational enhancements and safety reasons in indoor environments. In an in-building environment, the multipath caused by reflection and diffraction, and the obstruction and/or the blockage of the shortest path between transmitter and receiver are the main sources of range measurement errors. Due to the harsh indoor environment, unreliable measurements of location metrics such as RSS, AOA and TOA/TDOA result in the deterioration of the positioning performance. Hence, alternatives to the traditional parametric geolocation techniques have to be considered. In this paper, we present a method for mobile station location using narrowband channel measurement results applied to an artificial neural network (ANN). The proposed system learns off-line the location 'signatures' from the extracted location-dependent features of the measured data for LOS and NLOS situations. It then matches on-line the observation received from a mobile station against the learned set of 'signatures' to accurately locate its position. The location precision of the proposed system, applied in an in-building environment, has been found to be 0.5 meter for 90% of trained data and about 5 meters for 45% of untrained data.

1 Introduction

A problem of growing importance in indoor environments is the location of people, mobile terminals and equipments. In in-building environments, geolocation with good performance is essential in order to improve operational efficiency,

J.N. de Souza et al. (Eds.): ICT 2004, LNCS 3124, pp. 866–875, 2004.

worker's safety and remote control of mobile equipments. In indoor environments where conditions of signal propagation are severe (multipath, NLOS), the traditional parametric indoor geolocation techniques (RSS, AOA, TOA/TDOA) or their combinations (TDOA with RSS) fail to provide adequate location accuracy. For these techniques, all the paths used for triangulation must have a LOS to ensure an acceptable accuracy, a condition that is not always met in an indoor environment. An improvement of the accuracy may be obtained by using the location fingerprinting technique in which the effect of multipath is used as constructive information.

This paper provides a method for mobile station location using a fingerprinting technique based on narrowband channel measurement results in conjunction with an artificial neural network (ANN). For the studied in-building environment, results show a distance location accuracy of 0.5 meter for 90% of trained data and about 5 meters for 45% of untrained patterns. In section 2, we discuss the various wireless fingerprinting geolocation techniques used in outdoor and indoor environments. In section 3, we present our proposed system and give the position location results by applying the measured indoor data to an artificial neural network. Finally, we close this paper with a conclusion in section 4.

2   Wireless Fingerprinting Geolocation Techniques

2.1   Fingerprinting Geolocation Technique

The process of geolocation based on the received signals' fingerprint is composed by two phases: a phase of data collection (off-line phase) and a phase of locating a user in real-time (real-time phase). The first phase consists of recording a set of fingerprints (in a database) as a function of the user's location, covering the entire zone of interest. During the second phase, a fingerprint or a 'signature' pattern is measured and compared with the recorded fingerprints of the database. A pattern-matching algorithm is then used to identify the closest recorded fingerprint to the measured one and hence to infer the corresponding user's location (Fig. 1).

To constitute a fingerprint or a 'signature pattern', several types of information [1] can be used such as received signal strengths (RSS), angular power profile (APP) and power delay profile (PDP). Moreover, several types of pattern-matching algorithms may be employed which have the objective to give the position of the mobile station with the weakest location error. Among the commonly used algorithms, one can find algorithms based on the measure of proximity, on the cross correlation of signals and on artificial neural networks. Due to physical constraints of indoor environments, the database containing the set of fingerprint information may not contain all the necessary fingerprints to cover the entire zone of interest. Hence, the pattern-matching algorithm must be robust and respect the generalization property against perturbations and lack of fingerprint data, respectively. Since an artificial neural network respects these properties, an architecture based on neural networks has been used in the proposed geolocation system as the pattern-matching algorithm.

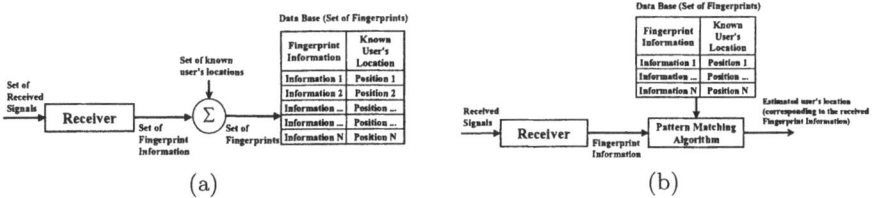

Fig. 1. Process of geolocation using received signal's fingerprint, a) off-line phase, b) real-time phase.

2.2   Wireless Geolocation Systems Using The Fingerprinting Technique

Several geolocation systems, using the fingerprinting technique, have been recently deployed in outdoor and indoor environments. The main differences between these systems are the types of fingerprint information and pattern-matching algorithms. RADAR [2,3] is an RF-based system for locating and tracking users inside buildings. It uses received signal strength (RSS) information gathered at multiple receiver locations to triangulate the user's coordinates. The system, operating with WLAN technology, has three access points (fixed stations) and covers the entire zone of interest. A pattern-matching algorithm, which consists of the nearest neighbor(s) in signal space, is used to estimate the user's location. Another system similar to RADAR, EKAHAU [4], uses signal strength information gathered at multiple receiver locations to perform an indoor positioning using a WLAN infrastructure. In the framework of the project WILMA [5], RSS fingerprint information has been used to estimate user's location in a building equipped with a WLAN technology. The pattern-matching algorithm employed has been an MLP type artificial neural network [6] to achieve the generalization needed when confronted with new data, not present in the training set. RadioCamera [7], DCM [8,9,10], and a third system found in [11] use fingerprinting techniques to locate and track mobile units in metropolitan outdoor environments. RadioCamera, operating with cellular technology, uses multipath angular profile information (APP) gathered at one receiver to locate the user's coordinates (one-antenna array per cell). A measure of proximity is used as the pattern-matching algorithm [12]. DCM, operating with cellular GSM and UMTS technologies, uses RSS and channel impulse response (CIR) measured fingerprint information with a cross-correlation metric algorithm to do the localization process. As a measure of performance, the median resolution of the location estimation for indoor and outdoor geolocation systems, using fingerprinting techniques, is reported to be in the range of 2 to 3 meters and 20 to 150 meters, respectively.

Although RSS type of information (RADAR, EKAHAU and WILMA used for indoor) requires the involvement of several fixed stations to compute the user's location, its implementation is simple and non expensive because of the use of narrowband receivers. On the other hand, the pattern-matching algorithm used in RADAR and DCM systems may show a lack of generalization (an algorithm that gives an incorrect output for an unseen input), a lack of robustness

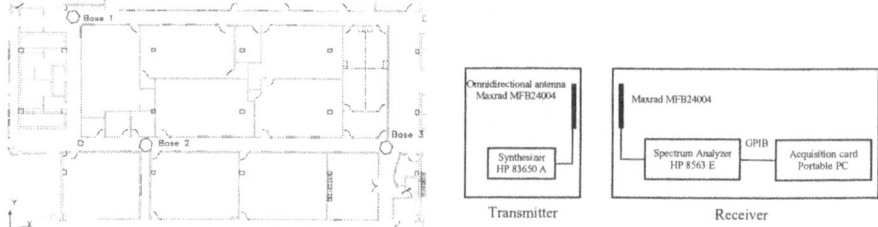

Fig. 2. Map of the in-building environment. **Fig. 3.** Schematic diagram for narrowband measurement system.

against noise and interference, and a long search time needed for the localization (real-time localization) especially when the size of the environment or the database is large. Hence, the use of an artificial neural network (ANN) as the pattern-matching algorithm is essential since ANN is robust against noise and interference, has a good generalization property and the time of localization during the real-time phase is almost instantaneous [1,6]. Accordingly, it has been decided to choose location-dependent RSS data (measured at three narrowband receivers) in conjunction with an artificial neural network for the geolocation of mobile units in the considered in-building environment.

3 Geolocation in an In-Building Environment Using the Fingerprinting Technique

3.1 Collection of Fingerprint Information (RSS)

Narrowband measurements were conducted in the mid part of the 5th floor within the Decelles building at École Polytechnique of Montreal. Figure 2 illustrates the map of the indoor measurement area, which consists of a rectangular central bloc (composed of laboratories and meeting rooms), two corridors and restrooms within an islet at the left end of the central bloc. The measurement area stretches over a length of 56.6 meters with a width and a height of 18.8 and 2.7 meters, respectively (no measurements have been performed in the laboratories and rooms situated in the lower side of the second corridor). The islet containing the restrooms is made of 18.5 cm width of concrete walls, whereas the walls of the central bloc are composed of 10 cm width of gypsum. The laboratories and the meeting rooms of the central bloc have at least one glass region (a window or a vitreous section). Moreover, some walls, separating the laboratories, contain also a glass region. Hence, the existence of line of sight (LOS), obstructed line of sight (OLOS) and non-line of sight (NLOS) propagations is noted.

A central frequency of 2.4 GHz has been used throughout the measurements in order to have a compatibility with WLAN systems, which may be used for data, voice and video communications as well as for radiolocation purposes. Figure 3 gives the schematic diagram of the setup for narrowband measurements [13]. The synthesizer HP83650A has been used as the source for the transmitter (indicated as the Base in Figure 2). Its frequency range was between 10MHz and 50 GHz and the maximum power at the operation frequency of 2.4 GHz was 12 dBm. Omni directional antennas of type MFB24004 from Maxrad were used for both transmission and reception. The central frequency of the antennas was 2.4 GHz with a gain of 4 dBi. The zero span mode of the spectrum analyzer HP8563B, acting as an envelope detector, has been used as the receiver. The bandwidth of the analyzer's filters (band pass and low pass) was set to 100 Hz. A set of 601 points of measurements, for each location, have been registered by the analyzer, corresponding to a sweep time of 1.2 seconds and a sampling frequency of 500 Hz (number of points divided by the sweep time), respecting the Nyquist criteria. Finally, the National Instrument acquisition card installed in a portable computer with the Labview software was used for data acquisition. It has to be noted that no synchronization cable, between transmitter and receiver, was needed since the 2.4 GHz frequency of operation was in the operating range of the spectrum analyzer. The maximum operating frequency being 26 GHz, no mixing operation was involved. Hence the problem of synchronization between local oscillators of the transmitter and the receptor was non- existent.

On the other hand, because of the radiolocation purpose (fingerprinting technique), the experimental procedures given in this article are different from those encountered in previous works. Three sets of measurements were taken, with the transmitter, i.e. the synthesizer and the transmit antenna, placed at a different location for each set. The receiver, consisting of the spectrum analyzer, the PC and the receive antenna, was moved to a new position for each power measurement. The position of the transmitter for the three sets was (x=8.9m, y=28.5m) for TX1, (x=18.9m, y=11.6m) for TX2 and (x=52m, y=11.6m) for TX3 with respect to the predefined referential (x=0, y=0) of Fig.2. As for the mobile receiver RX, it covered the entire indoor measurement area (situations with LOS, OLOS and NLOS) by varying its position by 1 meter widthwise and lengthwise in the corridors and by 2 meters widthwise and lengthwise in the rest of the measured area. As a result, 480 narrowband location measurements have been performed for each transmitter location. During the measurements, transmit and receive antennas were both mounted on carts at a height of 1.8 meters.

3.2   ANN-Based Pattern-Matching Algorithm

A trained artificial neural network can perform complex tasks such as classification, optimization, control and function approximation. The pattern-matching algorithm of the proposed geolocation system can be viewed as a function approximation problem (nonlinear regression) consisting of a nonlinear mapping from a set of input variables containing information about the three received signal strengths (power levels) onto a set of two output variables representing the two dimensional location (x, y) of the mobile station. The feed-forward artificial neural networks that can be used as function approximation are of two types, Multi-Layer Perceptron (MLP) networks and Radial Basis Function (RBF) networks. Either type of the two networks can approximate any nonlinear mapping

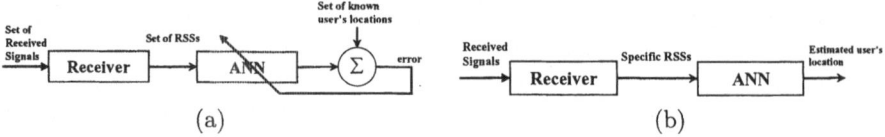

(a)   (b)

Fig. 4. Operation of the proposed system, a) learning phase (off-line phase), b) recalling phase (real-time phase).

to an arbitrary degree of precision provided the right network complexity is selected [14]. A specific learning algorithm is associated for each type of the two networks, which has the role of adjusting the internal weights and biases of the network based on the minimization of an error function, and defines the training of the network. MLP networks can reach globally any nonlinear continuous function due to the sigmoid basis functions present in the network, which are nonzero over an infinitely large region of the input space; accordingly, they are capable of doing a generalization in regions where no training data are available (generalization property). On the other hand, RBF networks can reach the given nonlinear continuous function only locally because the basis functions involved cover only small, localized regions. However, the design of a RBF network is easier, and the learning is faster compared to the MLP network. A generalized regression neural network (GRNN), which is an RBF-type network with a slightly different output layer, and an MLP-type network have been tested for the proposed geolocation system. The GRNN network showed a lower location error, compared to the MLP, during the memorization of the data set and the generalization phase of the network. Accordingly, the GRNN-type network has been chosen for the pattern-matching algorithm used in the proposed geolocation system. The GRNN-type ANN, used in the proposed system, consisted of two phases: a supervised learning phase (training of the network) and a recalling (testing) phase. During the off-line phase, the GRNN network is trained to form a set of fingerprints as a function of user's location and acts as a function's approximation (nonlinear regression). Each fingerprint is applied to the input of the network and corresponds to the three RSSs measured data at the fixed receiver stations. This phase, where the weights and biases are iteratively adjusted to minimize the network performance function, is equivalent to the formation of the database (recording of the set of fingerprints as a function of user's location) seen with other fingerprinting systems. During the real-time phase, the aforementioned fingerprint (three RSSs) from a specific mobile station is applied to the input of the artificial neural network (acting as a pattern-matching algorithm). The output of the ANN gives the estimated value of the user's location (Fig. 4).

The used GRNN architecture consisted of three inputs corresponding to the three RSS measured data, one hidden layer (radial basis layer) and an output layer (special linear layer) with two neurons, corresponding to (x, y) location of the user (Fig. 5). A radial basis type of transfer function (Gaussian function) has been associated for neurons in the hidden layer and a linear one for the output layer.

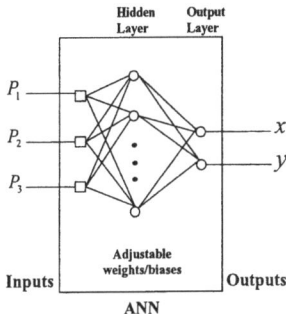

Fig. 5. Proposed pattern-matching ANN.

3.3   Location Estimation Results

The proposed neural network architecture has been designed using the function newgrnn.m of the Neural Network Toolbox of Matlab [15]. The simulation results showed that a spread constant value of 0.8 was adequate to do the required regression with a good generalization property. In the learning phase, the set of the three measured RSS data and the measured true mobile positions have been used as the input and as the target of the ANN, respectively. From the 480 measured data, 360 patterns have been employed to train the network. For recalling phase, as a first step, the same 360 patterns have been applied to the pattern-matching neural network to obtain the location of the mobile station (validation of the memorization property). The location errors as well as their cumulative density functions (CDF) have been computed for analysis purposes. The plots of the corresponding location errors and CDFs of location errors are given in Figures 6 and 7. It has to be noted that the localization error has been calculated as the difference between the exact position of the user and the winning position estimate given by the localization algorithm, and hence represents the RMS position location error. Moreover, by analogy with FCC requirements [16], the CDF of location error has been used as the performance of the system.

In the training set of data, it can be seen (Fig. 6) that the location error in x varies between -6.1 meters and 6.8 meters, the location error in y varies between -5.1 meters and 5.1 meters and the maximum error in Euclidean distance, between the estimated and the true positions, is equal to 8.2 meters. Moreover, it can be seen, from Figure 7, that a distance location accuracy of 0.5 meter is found for 90% of the trained patterns. The pattern-matching algorithm in [2], which consists of the nearest neighbor(s) in signal space, gives a location accuracy of 5 meters for about 75% of the empirical data. The comparison of these two results shows the advantage of using a neural network as a pattern-matching algorithm in the fingerprinting technique. As a second step, the remaining 120 non-trained patterns have been applied to the network to verify the generalization property of the proposed geolocation system. The location errors as well as their cumulative density functions (CDFs) have been computed and plotted (Figs. 8 and 9).

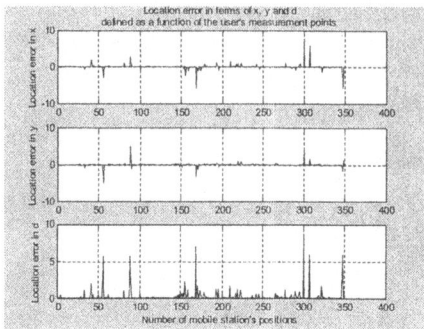

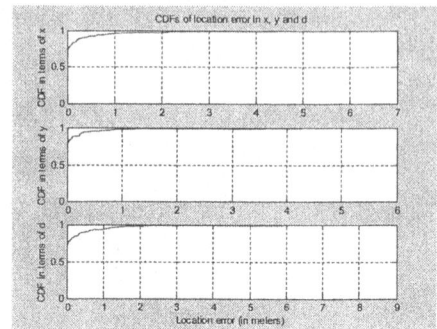

Fig. 6. Location errors in x, y and Euclidean distance (d), with inputs corresponding to the training set of data defined by the number of positions of the mobile station.

Fig. 7. Cumulative distribution functions (CDFs) of location errors in x, y and d, with inputs corresponding to the training set data defined by the number of positions of the mobile station.

For the untrained set of data, it can be seen (Fig. 8) that the location error in x varies between -32 meters and 43.3 meters, the location error in y varies between -15.7 meters and 13.9 meters and the maximum error in Euclidean distance, between the estimated and the true positions, is equal to 43.2 meters. Moreover, it can been seen, from Figure 9, that a distance location accuracy of 5 meters is achieved for about 45% of the untrained patterns. It has to be noted that the accuracy of the position estimate depends on the resolution of the map, which in turn depends on the distance threshold used in the map building process. After localization has been achieved, the theoretical error between the actual and estimated position (localization error) should therefore vary between zero and the distance threshold. When the size of the grid is 1 meter widthwise and 1 meter lengthwise (case of the corridors), the geolocation accuracy that one may expect with the proposed fingerprinting technique, should be between 0 and 1.4 meters (distance threshold) in terms of the Euclidean distance. When the size of the grid is 2 meters by 2 meters (case of the rest of the measured area), the distance threshold is equal to 2.8 meters.

4 Conclusions

This paper has shown that the RSS fingerprinting technique using an artificial neural network can give an accurate mobile station location in the studied in-building environment. The results showed that distance location accuracies of 0.5 meter and 5 meters have been found for 90% and 45% of the trained and untrained patterns, respectively. Moreover, since the studied radio channel represents a typical in-building environment, other indoor situations with similar characteristics may give similar results.

The use of an artificial neural network as a pattern-matching algorithm for the proposed system is a new approach that has the advantage to give a robust

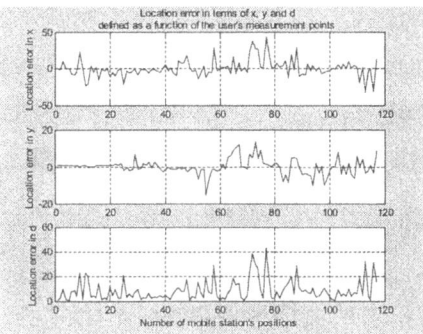

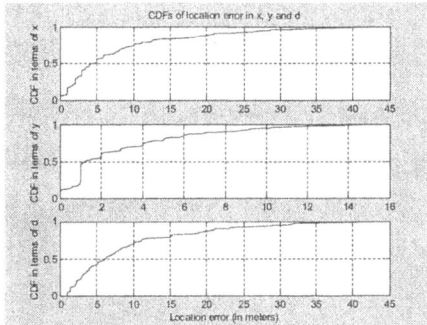

Fig. 8. Location errors in x, y and Euclidean distance (d), with inputs corresponding to the untrained set of data defined by the number of positions of the mobile station.

Fig. 9. Cumulative distribution functions (CDFs) of location errors in x, y and d, with inputs corresponding to the untrained set of data defined by the number of positions of the mobile station.

response with a good generalization property (the location fingerprint does not have to be in the fingerprint database). Moreover, since the training of the ANN is off-line, there is no convergence and stability problems that some control (real-time) applications encounter.

The transposition of the system from two to three dimensions is easy (addition of a third neuron in the ANN's output layer corresponding to the z position of the user) and constitutes an advantage of the ANN. On the other hand, the RSS fingerprinting technique has some deficiencies. First, due to the fading environment, the estimated set of RSS information may have poor reproducibility and uniqueness properties, which will deteriorate considerably the user's location accuracy. Channel impulse response information (via wideband measurements) may be used in order to resolve this first deficiency [17]. Secondly, the fingerprinting technique needs the digital map of the environment and is not well suited for dynamic areas (variations of the channel with time). Therefore, updates of database's information (a new training of the neural network) must be performed when important changes of the channel's characteristics occur.

It is interesting to note that this RSS fingerprinting technique may also be applicable to any other indoor applications (shopping centers, campuses, office buildings). In addition, some advanced simulation programs may be used to generate RSS information as a function of user's location (for the training set of data of the neural network) instead of getting these RSS values via narrowband measurements. This approach will reduce the database generation time and will act in favor of the proposed system's implementation.

Finally, the radio access technology used for an effective implementation of the proposed system may employ different types of system such as WLAN, impulse radio (UWB) and mobile radio.

Acknowledgment. The authors wish to thank M. Gilbert Lefebvre and M. Mourad Djadel for their precious collaboration in the narrowband measurement campaign.

References

1. Nerguizian, C., Despins, C., Affès, S.: A Framework for Indoor Geolocation using an Intelligent System. 3rd IEEE Workshop on WLANs, Boston, USA, September 2001. http://www.wlan01.wpi.edu/proceedings/wlan44d.pdf
2. Bahl, P., Padmanabhan, V.N.: RADAR : An In-Building RF-based User Location and Tracking System. Proceedings of IEEE INFOCOM 2000, Tel Aviv, Israel, March 2000.
3. Microsoft Corporation. http://www.microsoft.com
4. Ekahau Inc. http://www.ekahau.com
5. Wireless Internet and Location Management Architecture. http://www.wilmaproject.org
6. Battiti, R., Villani, A., Nhat, T.L.: Neural Network Models for Intelligent Network : Deriving the Location from Signal Patterns. Autonomous Intelligent Networks and Systems, UCLA, Los Angeles, USA, May 2002.
7. US Wireless Corporation. http://www.uswcorp.com
8. VTT Technical Research Centre. http://www.vtt.fi
9. Laitinen, H., Nordström, T., Lähteenmäki, J.: Location of GSM Terminals using a Database of Signal Strength Measurements. URSI XXV National Convention on Radio Science, Helsinki, Finland, September 2000.
10. Ahonen, S., Lähteenmäki, J., Laitinen, H., Horsmanheimo, S.: Usage of Mobile Location Techniques for UMTS Network Planning in Urban Environment. IST Mobile and Wireless Telecommunications Summit 2002, Thessaloniki, Greece, June 2002.
11. Nypan, T., Gade, K., Maseng, T.: Location using Estimated Impulse Responses in a Mobile Communication System. 4th Nordic Signal Processing Symposium (NORSIG 2001), Trondheim, Norway, October 2001.
12. Wax, M., Hilsenrath, O.: Signature Matching for Location Determination in Wireless Communication Sustems. U.S. Patent 6,112,095.
13. Djadel, M., Despins, C., Affès, S.: Narrowband Propagation Characteristics at 2.45 and 18 GHz in Underground Mining Environments. Proceedings IEEE GLOBE-COM 2002, Taïpai, Taïwan, November 2002.
14. S. Haykin, Neural Network, a Comprehensive Foundation, MacMillan, 1994.
15. Demuth, H., Beale, M.: Neural Network Toolbox for use with Matlab (User's Guide). The MathWorks Inc., 1998.
16. Caffery Jr., J.J., Stüber, G.L.: Overview of Radiolocation in CDMA Cellular Systems. IEEE Communications Magazine, April 1998.
17. Nerguizian, C., Despins, C., Affès, S.: Geolocation in Mines with an Impulse Response Fingerprinting Technique and Neural Networks. submitted to the IEEE Transactions on Wireless Communications, December 2003.

Implementation of a Novel Credit Based SCFQ Scheduler for Broadband Wireless Access

Emiliano Garcia-Palacios, Sakir Sezer, Ciaran Toal, and Stephen Dawson

School of Electrical and Electronic Engineering, Stranmillis Rd,
Belfast, BT9 5AH, Northern Ireland, U.K
{E.Garcia, S.Sezer, C.Toal}@ee.qub.ac.uk
http://www.ee.qub.ac.uk

Abstract. A novel tag computation circuit for a credit based Self-Clocked Fair Queuing (SCFQ) Scheduler is presented in this paper. The scheduler combines Weighted Fair Queuing (WFQ) with a credit based bandwidth reallocation scheme. The proposed architecture is able to reallocate bandwidth on the fly if particular links suffer from channel quality degradation .The hardware architecture is parallel and pipelined enabling an aggregated throughput rate of 180 million tag computations per second. The throughput performance is ideal for Broadband Wireless Access applications, allowing room for relatively complex computations in QoS aware adaptive scheduling. The high-level system breakdown is described and synthesis results for Altera Stratix II FPGA technology are presented.

1 Introduction

Broadband Wireless Access (BWA) is proving to be the next generation wireless technology that will provide access to broadband multimedia services. The system is ideal for SME customers, Small Office Home Office (SOHO) environments and in the future (when the technology becomes inexpensive) for the residential customer.

The IEEE 802.16 standardises a fixed broadband wireless access alternative to existing cable and DSL, employing a point-to-multipoint architecture, which sets the basis for next generation broadband access technology. Quality of Service (QoS) and the support of delay sensitive interactive services have been addressed with the specification of a connection-oriented Medium Access Control (MAC). All services, including inherently connectionless services, are mapped to a virtual connection. This connection oriented nature provides a mechanism for requesting bandwidth, associating QoS and traffic parameters. IEEE 802.16 uses the concept of "service flows" to define unidirectional transport of packets on either downlink or uplink. Service flows are characterized by a set of QoS parameters such as latency, jitter and throughput which are uniquely identified by a 32bit Service Flow ID (SFID).

J.N. de Souza et al. (Eds.): ICT 2004, LNCS 3124, pp. 876–884, 2004.

The Base Station (BS) and the Subscriber Station (SS) must reserve resources to comply with those QoS parameters. The principal resource to be reserved is bandwidth. The BS is in control and can allocate the required bandwidth to a downlink connection according to its SFID field. Each connection in the uplink direction is mapped onto 1 of 4 existing types of uplink scheduling services [6]. Some of these scheduling services prioritise the uplink access to the medium for a given service, for example by using unsolicited bandwidth grants. Others use polling mechanisms for real time and non-real time services or even pure random access for non real time Best Effort (BE) services [6].

Scheduling algorithms in the BS and SS may be very different since the SS may use bandwidth in a way that is unforeseen by the BS. The BS sees requests on a per connection basis and based on this, grants bandwidth to the SS while trying to maintain QoS and fairness. However, according to the standard, the SS may be granted an aggregated amount of bandwidth (Grant Per Subscriber Station- GPSS) rather than bandwidth on a connection basis (Grant Per Connection-GPC); in this case the SS scheduler has to maintain QoS among its connections and is responsible for sharing the bandwidth among the connections (maintaining QoS and fairness). Since different connections have to provide different QoS levels, the importance of building a QoS aware scheduler for SSs is paramount.

In the case of the BS, the scheduler not only needs to be QoS aware but must also be extremely fast. The BS scheduler needs to manage bandwidth for a large amount of downlink connections. In many commercial systems each BS usually serves up to 256 SSs. Since each of these SSs may support a SOHO network at the other end, the number of connections may easily increase to several thousand (each of them with associated QoS constraints). Furthermore, interactive services such as Voice over IP (VoIP) and video conferencing introduce IP traffic with relatively small packet sizes and critical QoS constraints.

The aggregated traffic of diverse services from different subscribers and applications and the need for throughput rates beyond gigabits enforces the need for fast processing QoS aware BWA scheduler architectures in the BS.

Furthermore, wireless channels are receptive to atmospheric effects such as harsh weather conditions and also to man made electromagnetic noise. Despite the Forward Error Correction (FEC) and other packet loss combat mechanisms, QoS degradation due to noisy wireless channels are expected. If channel quality and subsequently QoS begins to degrade, immediate resource reallocation is essential for the maintenance of QoS of premium links. Resource reallocation mechanisms incorporating credit based schemes for resource scheduling can temporarily resolve the degradation of QoS especially for premium services.

In this paper we introduce an extended Self Clocked Fair Queuing (SCFQ) algorithm specifically targeted for packet scheduling in BWA. In this algorithm, the weight determining the bandwidth allocation in the traditional SCFQ is extended with an additional weight parameter that can be enabled by specifying a credit for a finite number of packets. This additional feature enables a configurable on demand bandwidth reallocation mechanism for broadband wireless access.

A highly parallel, pipelined finishing tag computation architecture for the proposed credit based SCFQ scheduling algorithm and its hardware implementation using FPGA technology is presented.

2 Weighted Fair Queuing

Weighted Fair Queuing (WFQ) algorithms have proved ideal for scheduling various sized packets, such as Internet Protocol [2]. It allows an arbitrary number of end-to-end connections having a fair access to a link. It is complex in its computation and so causes a significant implementation problem for high throughput rates. Equation (1) shows finishing tag commutation for WFQ [1].

$$F_k^i = \frac{L_k^i}{r_k} + \max[F_k^{i-1}, v(a_k^i)] \tag{1}$$

where

L_k^i = Packet Length of the ith packet of flow/class k

r_k = Weight assigned to flow/class k

a_k^i = Arrival time of ith Packet of flow/class k

$v(a_k^i)$ = Virtual Time

WFQ based packet scheduling involves the computation of finishing tags for each packet that is scheduled to be serviced and the subsequent servicing of these packets in accordance with their finishing tag values. Therefore, the fairness and the programmability of a WFQ based scheduling scheme is determined by the accurate computation of the finishing tag value. Specific algorithms such as Self Clocked Fair Queuing, (SCFQ) [7], Worst-case Fair Weighted Fair Queuing WF²Q [8] and others [2], [3] have been proposed and implemented. These algorithms are a deviation of the classical WFQ algorithm with various trade-offs between the finishing tag computation complexity and the fairness of the scheduling scheme.

The SCFQ is an approximation of WFQ. The virtual time used in SCFQ; $v(t)$ is a measure of the progress of the system itself. Whenever the system changes state from busy to idle, $v(t)$ resets to zero. In fact, SCFQ computation of virtual time is much simpler than that of WFQ, resulting in SCFQ being a much more practical solution.

3 Credit Based SCFQ

In the case of a Credit based SCFQ (C-SQFQ), two additional parameters are introduced.

wr_k = Wireless weight to be assigned to k'th flow.

c_k = Credit assigned to k'th flow (number of packets that will be serviced with the added wireless weight)

$$
F_k^i = \begin{cases} \dfrac{L_k^i}{r_k + wr_k} + \max[F_k^{i-1}, v(a_k^i)] & for\, c_k > 0 \\[3mm] \dfrac{L_k^i}{r_k} + \max[F_k^{i-1}, v(a_k^i)] & for\, c_k = 0 \end{cases} \tag{2}
$$

Based on the available channel quality and the best achieved packet loss probability, the BS traffic management is able to reallocate bandwidth to the effected premium services to maintain their QoS at the cost of the non-premium services. This is accomplished in the form of credit by increasing the weights of the k'th flow with an additional wireless weight value, wr_k, for c_k number of packets. The credit based bandwidth reallocation simplifies the control overhead of the BS traffic management. The value r_k represents the weight allocated by the BS connection admission and wr_k for c_k packets represents the fine adjustment and dynamic reallocation of bandwidth by the traffic management.

3.1 C-SCFQ Architecture

Figure 1 illustrates a typical BWA Basestation incorporating a fair queuing scheduler. At the ingress data path, IP packets arriving from the high speed fixed network are classified based on their traffic type and physical destination address.

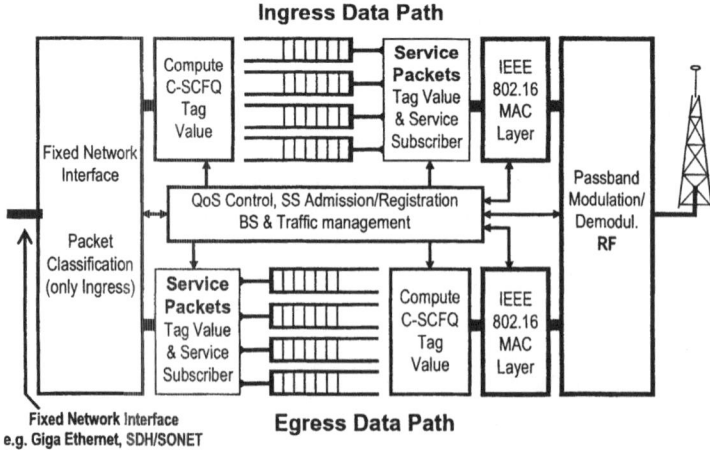

Fig. 1. IEEE 802.16 Base Station

The C-SCFQ scheduler is a part of the MAC layer and schedules IP packets based on their service subscriber ID, priority, packet length and allocated weight (bandwidth). The scheduling process is composed of two phases; finishing tag computation and packet servicing. The packets are serviced from the smallest tag value to the largest (fair queuing algorithm).

Figure 2 illustrates the high-level breakdown of the C-SCFQ scheduler architecture. The tag computation block is responsible for assigning finishing tag values to every data packet according to its properties. The finishing tag computation and assignment circuit is discussed in more detail in Section IV.

The Packet Buffer Write Control is responsible for storing the arriving packets in the shared memory and updating the "finishing tag / packet pointer" vector in the Content Addressable Memory (CAM).

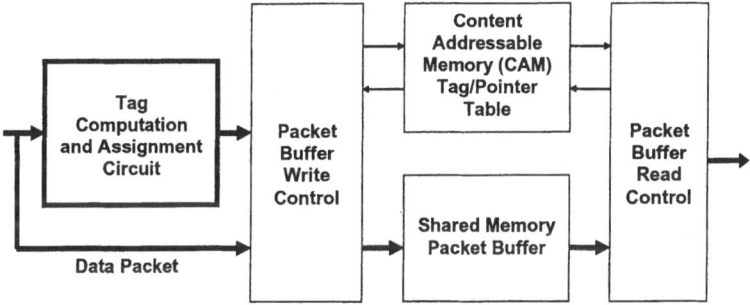

Fig. 2. C-SCFQ Scheduler Architecture

The CAM accommodates the "finishing tag / packet pointer" values in a number of Hash tables. The finishing tag value represents the content subsequently the address of the packet pointer. CAM is ideal for such an implementation where data content is used to address the memory location. The smallest finishing tag values in the CAM will be at the head of the line ensuring that packets with the lowest finishing tag values are transmitted first.

The Shared Memory stores the packets before they can be serviced by the packet Buffer Read Control. Address pointers identify the location and size of each packet within the shared memory.

4   Tag Computation Architecture

The tag computation plays a vital role in the scheduling procedure of many fair queuing scheduling schemes. The method for computing finishing tags determines not only the fairness of the scheduling process but also the throughput rate i.e. how many finishing tags can be computed per second. In this research, technology optimised architectures for Field Programmable Gate Array (FPGA) implementations are investigated. For example, the use of on chip distributed RAMs for the storage of r_k,

wr_k, c_k and finishing tag values. The presented finishing tag computation circuit implementation is highly parallel and pipelined incorporating a range of distributed on chip memories.

The *Traffic Address* (physical address tag) and *Traffic Class* are used to identify a particular IP flow or traffic class while the *Packet Length* input parameter is required for the finishing tag calculation.

Figure 3 shows the block level description of the finishing tag computation block.

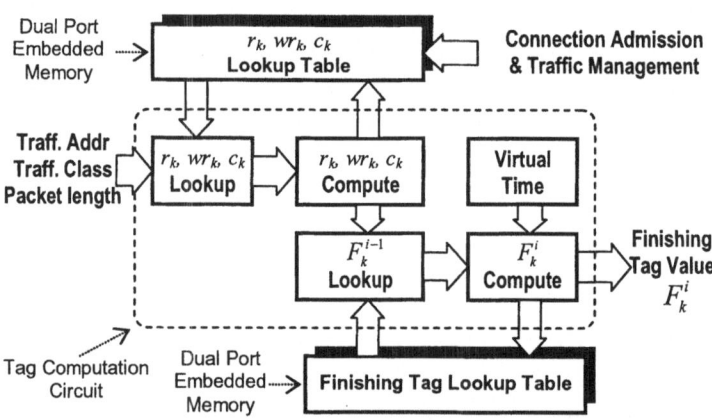

Fig. 3. Block Level Description

The finishing tag computation circuit is composed of 5 main blocks; r_k, wr_k, c_k lookup, rk, wr_k, c_k compute, i-1 finishing tag lookup, virtual time generator, and finishing tag compute circuit. Four lookup tables, for r_k, wr_k, c_k and finishing tag data lookup, are implemented using Altera Stratix II embedded M4K RAM blocks[4].

4.1 r_k, wr_k, c_k Lookup and Weight Calculator

An IP packet i belonging to a flow k is identified by an internal address assigned by the IP classifier circuit at the BS input. This internal address could be a tag identifying the *Traffic Class* or *Traffic Flow*. An internal MUX translates this address into a singe *Traffic Address* allowing the support of per-class and per-flow based queuing. The internal *Traffic Address* is used to lookup the flow specific r_k, wr_k, c_k values.

The r_k, wr_k and c_k lookup table is composed of 64 Stratix II M4K dual port memory blocks of 4kbits per block accommodating up to 4096 lookup entries. The lookup table translates the traffic address into an equivalent r_k, wr_k and c_k value which will be used to calculate the individual finishing tag for each IP packet. The lookup table size determines the number of flows that can be simultaneously supported by the scheduler. The presented implementation supports up to 4096 flows i.e. 256 SS supporting

16 flows per SS. The lookup table size can be increased by incorporating a larger embedded or an external memory.

As stated in previous sections, bandwidth allocation for an individual traffic flow k is achieved by assigning it a specific r_k value. If particular flows experience packet loss and thus sub sequential QoS degradation, bandwidth adjustment can be accomplished by assigning a wireless weight wr_k in the form of credit for c_k number of IP packets. The r_k, wr_k, c_k entries are updated by the BS connection admission and traffic management software via the uP interface. The calculation processes involves, firstly the summation of the bandwidth value r_k and the corresponding wireless weight wr_k, if $c_k>0$ and secondly decrementing the value of c_k. The new weight value and the internal traffic address are both forwarded to the second computation phase.

4.2 Virtual Time

The Virtual Time block is a counter circuit producing an integer count value which emulates the motion of time for the SCFQ scheduler. Even if the system has been running for an infinite period of time, the *Current Virtual Time* (CVT) will always indicate finite value. A mechanism must be put in place to ensure smooth operation of the CVT along with the finishing tag assignment.

The Virtual Time block is composed of a 14 bit binary counter producing a *CVT* value between 0 and 16383, and a 2 bit counter *CTV-Window* (CVT register flag) creating 4 time windows. The *CTV-Window* increments every time the *CVT* count reaches 16383. The finishing tag values range between 0 and 65536. It is assumed that the finishing tag difference within an IP flow of two consecutive packets will always be smaller than 16383. Having four time windows of 16384 clock cycles and a maximum consecutive finishing time smaller than the virtual time interval allows the development of a circular virtual time generator. It is important that the virtual time operates in a circular manner to prevent infinite CVT values. The use of a window mechanism prevents smaller or higher priority packets from misusing their network connection whilst also ensuring that larger or lower priority packets are not delayed infinitely.

4.3 Finishing Tag Lookup and Calculator

The finishing tag lookup and calculator block is similar to the r_k and wr_k lookup block. It retrieves the F_k^{i-1} for the current flow k and calculates F_k^i. The lookup table is composed of 16 Stratix II M4K dual port memory blocks accommodating up to 4096 finishing tag lookup entries. Similar to the r_k and wr_k lookup block, the finishing tag lookup is addressed by the internal *Traffic Address* and can support both per class and per flow queuing. The lookup table size can also be increased using either embedded or external memory. The tag lookup table is updated with the new finishing tag value by the tag calculation block with every new computed tag value. If a previous finishing tag value has been read for the next tag computation then it will also be replaced after one clock cycle by the new finishing tag value. This is a necessary requirement for a pipelined tag computation procedure as a consecutive IP packet may belong to the same flow.

According to the SCFQ algorithm (equation (2)) the largest value of the current CVT and F_k^{i-1} is used to compute the finishing tag. This is determined with the use of a comparator which compares the *CVT*, *CVT-Window* and F_k^{i-1} and selects the maximum value. Each IP flow can be composed of both very large data and very small acknowledgment packets resulting in different finishing tag calculations for the same IP flow. The comparison between CVT and F_k^{i-1} ensures that each packet within a specific flow is transmitted in the correct order independent of its size.

The final computation involves the addition of the maximum virtual time and the result of the binary division. This new finishing tag value is then assigned to the processed IP packet and forwarded to the CAM to be scheduled according to its value. In parallel, the finishing tag lookup table is updated with the new value.

5 Synthesis and Circuit Study

The presented architecture of the finishing tag computation circuit has been implemented using hardware description language VHDL, synthesised and targeted to a Altera Startix II [4] FPGA using Synplicity and Altera Quartus II tools. Speed and area performances are examined. The post-layout synthesis results are included in Table 1.

Table 1. Post-layout Synthesis results

Tag Assign Circuit Post-layout Synthesis					
Device	Clock Speed	Registers	Adaptive logic Modules	ALUTs	Embedded Memory
EP2S30	181.36	1644	1457 10.7%	2412 7.1%	80 M4K RAM Blocks 55%

The speed optimised architecture achieves a clock frequency of up to 181 MHz. Due to the parallel pipelined nature of the architecture, the circuit is able to compute one finishing tag value every clock cycle, achieving an estimated peak rate of up to 181 million tag computation per second.

6 Conclusions

In this paper the architecture and implementation of a tag computation circuit for a Broadband Wireless Access Base Station using a Credit based SCFQ scheduler is presented. The implementation has demonstrated that a custom parallel processing architecture with a relatively small hardware cost is able to perform extremely fast tag computation. The presented implementation occupies less than 11% of the logic resources and 55% of the embedded memory resources of a medium size Altera Stratix II (EP2S30) device, leaving plenty of logic resources for a full C-SCFQ scheduler implementation. Synthesis and circuit analysis proved that the presented implementation is able to operate at 181 MHz achieving a maximum finishing tag computation rate of 181 million per second. Assuming a minimum IP packet length of 100 bytes, the tag computation circuit is able to service a link with a throughput rate of up to 150 Gbps. Hence, this architecture offers enormous potential for next generation BWA systems operating at 28 GHz and 32 GHz frequencies.

For current commercial BWA systems operating in the 2 GHz to 5 GHz frequency region, which target lower link speeds (150 Mbps), a fast architecture such as the one presented, leaves plenty of room in the proposed scheduler architecture to include the necessary QoS parameters in order to achieve a fully QoS aware scheduler in our future work. Further research envisaged in this area also contemplates the implementation of intelligent schedulers that can adapt scheduling algorithms to traffic profiles, existing network loads and wireless link states [9].

The presented work has also demonstrated that typical FPGA architectures with a range of embedded peripherals and memories are an ideal platform for high throughput programmable network processing allowing the implementation of next generation fixed-wireless networks.

References

1. Mark W. Garrett, Bellcore. *"A Service Architecture for ATM: From Application to Scheduling"*, IEEE Network Magazine, Vol. 10, No. 3, pp 6-14, May/June 1996
2. A. Varma, D. Stiliadis, *"Hardware Implementation of Fair Queuing Algorithms for ATM Networks"*, IEEE Comm. Magazine, pp 54-68, Dec. 1997.
3. Donpaul Stephens, Jon Bennett, Hui Zhang, *"Implementing Scheduling Algorithms in High Speed Networks"*. IEEE JSAC Journal on Selected Areas in Communications, Vol. 17, pp.1145-1158, September 1999
4. Altera Stratix II Handbook , Altera Corporation, 101 Innovation Drive, San Jose, California 95134, USA, http://www.altera.com/literature/hb/stx2/stratix2_handbook.pdf.
5. IEEE 802.16-2001, *"IEEE Standard for Local and Metropolitan Area Networks — Part 16: Air Interface for Fixed Broadband Wireless Access Systems,"* Apr. 8, 2002.
6. C Eklund, R Marks, et Al, *"IEEE Standard 802.16: A Technical Overview of the Wireless-MAN™ Air Interface for Broadband Wireless Access"*, IEEE Communications Magazine, June 2002, pp. 98-107.
7. S. J. Golestani, *"A Self-Clocking Fair Queuing Scheme for Broadband Applications"*, INFOCOM'94 Proceedings, pp 636-646, 1994.
8. J. C. R. Bennett, H. Zhang, *"WF^{2}Q: Worst-case Fair Weighted Fair Queuing"*, INFOCOM'96 Proceedings, pp 120-128, 1996.
9. Walsh S., Garcia-Palacios E, Sezer S "A Framework for Traffic-profile based Wireless Fair Queuing for Next Generation Wireless networks", Irish Signals and Systems Conference, ISSC'04, Belfast, June.04.

BER for CMOS Analog Decoder with Different Working Points*

Kalle Ruttik

Helsinki University of Technology, Communications Laboratory,
P.O. Box 2300, FIN-02015 HUT, Finland
Kalle.Ruttik@hut.fi

Abstract. In this work we investigate how to select the working point for a CMOS based analog turbo decoder. We study the performance of such decoder by modelling mismatch error of a transistor for wide range of drain currencies: the transistor is in weak, medium, strong inversion. The main building block of an analog decoder is modified Gilbert multiplier. The CMOS process based multiplier performs ideal multiplication as long as transistors operate in weak inversion. Based on our simulation we can conclude that the decoder is capable to decode even when the multiplication is not ideal.

The dominating source of error in an analog decoder is not nonideality of multiplication but mismatch between the transistors. However, even without special minimisation of the mismatch error the analog decoder was able to show BER performance within 0.5 dB from an ideal decoder.

1 Introduction

The turbo decoder can be interpreted as an instance of the sum product algorithm [1]. The crucial operation for this algorithm is multiplication of probabilities. In 1998 two research groups [2] [3] noticed that the required multiplication can easily be carried out by an analog circuit - modified Gilbert multiplier. That approach is very promising the test chips have achieved decoding speed around half a gigabit per second with power consumption less than 1 W [5].

For proper multiplication the transistors in Gilbert multiplier should have exponential dependency between base voltage and source current. The first analog decoder circuits were using bipolar transistors where such dependency is readily given [5] [6]. Development of digital technology has significantly reduced the price of CMOS process. The price advantage compared to bipolar processes has increased the interest in implementations of analog decoders on CMOS [7]. The MOS transistors provide the required exponential dependency only when operated in subthreshold. However, in subthreshold the small current density hinders the speed of circuit.

The published studies of CMOS based analog decoders are dominantly simulations or implementations of small simple turbo decoders. In this work we are

---

* This work is supported by the National Technology Agency (TEKES) of Finland

J.N. de Souza et al. (Eds.): ICT 2004, LNCS 3124, pp. 885–890, 2004.
© Springer-Verlag Berlin Heidelberg 2004

pursuing a different approach: by using a detailed transitor model we study the impact of selection of different parameters on performance of large analog decoder circuit. Our interest has been to identify the multiplier imperfections and process nonidealities contribution to the decoding error

The paper is organised as follows. In section 2 we describe the structure of the used decoder. In section 3 we proceed to the transistor level block description and introduce transistor and error models used in simulations. The simulation results are covered in section 4. The section 5 concludes the paper.

2   Decoder Description

The main module of turbo decoder, SISO block, calculates marginal probability distribution $P(u_i|c)$ of each transmitted bit u_i, given the received sequence c. The detiles of the algorithm can be found in the literature [4].

The decoder is best visualised by using a code trellis. The parts of the SISO algorithm: "forward-backward" recursion and bit probability evaluations can be described by connection of trellis sections. The operations in the trellis section can directly be mapped to a modified Gilbert multiplier module [6]. In such module the multiplication and summation are performed in parallel.

As an example in this work we use parallel concatenated convolutional code with random uniform interleaver. The systematic 1/3 rate code is created by concatenation of two codes created by the same rules: tail-biting code derived from a convolutional codes with recursive generation polynome [5 7]. For performance studies we simulate block size with 100 information bits. Properties of such code are well studied and can be found in the literature [8].

Utilisation of Gilbert multipliers allows us to create a simple decoder structure. The amount of different type of modules depends on how many different types of sections the code has. As shown in Fig. 1 in practice the decoder can be constructed by using only few different module types.

Each block in Fig. 1 corresponds to one modified Gilbert multiplier circuit. The $X1$ and $X2$ type modules compute the transition probabilities in the trellis. The blocks L calculate extrinsic information for the coded and information bits. The S type module calculates probabilities of states during "forward" and "backward" recursion in the trellis.

3   Transistor Model

In Fig. 2 is shown the transistor level description of the S block. In this circuit the input currents are interpreted as input probabilities and output currents are interpreted as the output probabilities [7].

We are interested how error in individual transistors contributes to the bit error ratio (BER) performance of the whole decoder. In SPICE like simulators evaluation of BER of large analog circuits is impractical. For faster simulations we have build C++ based transistor level simulation program.

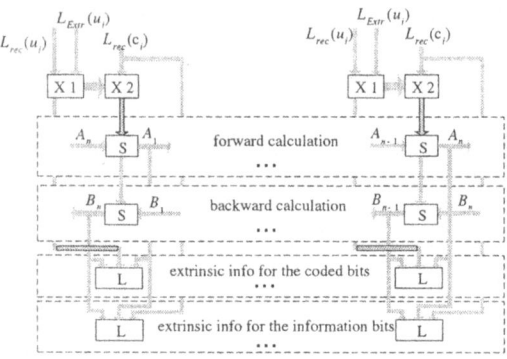

Fig. 1. Block diagram of the decoder

The transistors are modeled accordingly to the model proposed in [9]. In our applications the transistor nonidealities are the mismatch error and the "non-exponential" characteristic of the transistor.

In weak inversion the CMOS transistors drain current, I_D, is small. I_D and gate voltage, V_G, have exponential relationship and the Gilbert multiplier operates as the ideal multiplier.

The exponential relation is satisfied when the transistor is in saturation, the drain base voltage V_D is much higher than source voltage V_S, $V_{DD} \gg V_S$.

$$I_D = I_{D0}e^{\left[\frac{V_G-nV_S}{nU_T}\right]} \tag{1}$$

where $I_{D0} = I_S e^{\left[\frac{-V_{T0}}{n \cdot U_T}\right]}$, $I_S = 2n\beta U_T^2$, $\beta = \mu_n C_{ox}\frac{W}{L}$ Here V_{T0} is the threshold voltage, n is subthreshold slope, U_T is thermal voltage, $\mu_n C_{ox}$ is current factor parameter, W and L are the width and length of the transistor gate.

When the current trough the transistor is increased the transistor starts to operate in strong inversion. In strong inversion the base voltage and drain current relation can be characterised by square law. Between the strong and weak inversion the transistor behaviour can be characterised by transition from exponential to square law behaviour. The drain current can be approximated over a wide range of currents by the simple interpolation function [9].

$$I_D = I_S \left[\log\left(1 + \exp\left[\frac{V_G - V_S}{2U_T}\right]\right)\right]^2 . \tag{2}$$

Whether transistor is in strong or weak inversion can be selected by limiting the amount of current through it. In the scheme on Fig. 2 this is done by selecting the value for the current source I_{ref}. The value of the current source defines the maximum current that can flow to any branch of the multiplier.

Due to the physical properties of the producing process two identical designed transistors have random time independent differences in their operation [10].

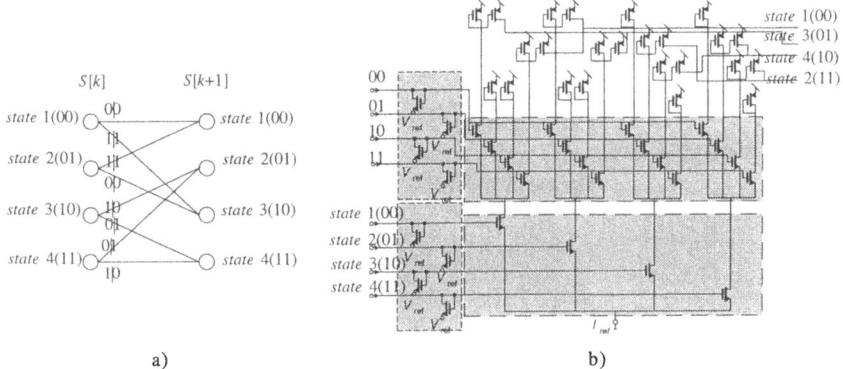

Fig. 2. The block S a) trellis transition. b) transistor level schematic

This mismatch is characterised by random variation of threshold voltage V_{T0}, current factor β, and body factor γ. If transistors are located close to each other the last one of these is negligible and we can model only errors of β and V_{T0}. In the following we describe these errors as ε_I and ε_V correspondingly.

Together with error component the modified (1) for the drain current is

$$I_{D0} = (I_S + \varepsilon_I)\, e^{-\frac{V_{T0}+\varepsilon_V}{nU_T}} = I_S \left(1 + \frac{\varepsilon_I}{I_S}\right) e^{-\frac{\varepsilon_V}{nU_T}} \Rightarrow \left(1 + \frac{\varepsilon_I}{I_S}\right) e^{-\frac{\varepsilon_V}{nU_T}}. \quad (3)$$

Here we extracted the part that contains only error terms.

The variance of body factor and current factor mismatch are process dependent. We are using in our simulations values values proposed in the [10], where the variances are calculated as

$$\sigma^2\left(V_{T0}\right) = \frac{A_{VT0}^2}{2WL}, \quad \left(\frac{\sigma\left(\beta\right)}{\beta}\right)^2 = \frac{A_\beta^2}{2WL}. \quad (4)$$

For 0.7 μm process [10] suggests to use following values: for nMOS $A_{VT0}^2 = 13\ mV\,\mu m$, $A_\beta^2 = 1.9\,\%\mu m$; and for pMOS $A_{VT0}^2 = 22\ mV\,\mu m$, $A_{VT0}^2 = 22\ mV\,\mu m$.

4   Simulations

We characterise the decoder BER performance with the transistors being in weak, medium, or strong inversion. The working point is selected accordingly to the inversion coefficient (IC) [11].

In (1) we can notice that the transistor drain current I_D depends on the process dependent current I_S and on the gate voltage. For particular realisation of the decoder the I_S is fixed and the inversion coefficient can be expressed as $IC = \frac{I_D}{I_S(W/L)}$. For IC values less than 0.1 the transistor is in weak, for $0.1 > IC > 10$ in moderate and for $IC > 10$ in strong inversion.

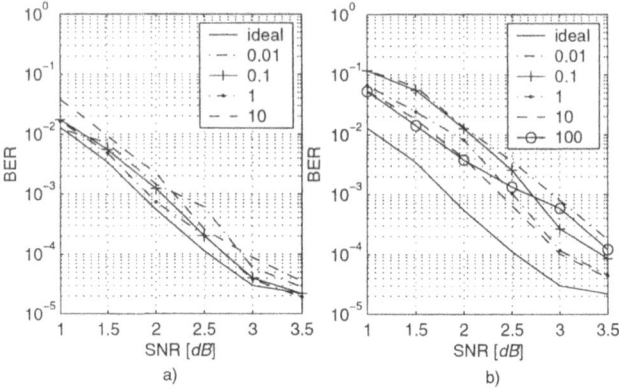

Fig. 3. BER with the transistor model a) no mismatch error b) with mismatch error

The currents in any of the Gilbert multipier brances are less or equal to I_{ref}. That means all the transistors are operating in inversion conditions that is less or equal to the $\frac{I_{ref}}{I_S(W/L)}$. The maximal inversion of a transistor controls much the Gilbert multiplier deviates from the ideal multiplier.

In our simulations we fixed the width and length to be same $0.35\,\mu m$ for both N and P type transistors. In all the simulations we are using 3.3 V power supply.

In first set of simulations we investigate the impact of the IC on the decoder performance (Fig. 3 a). Without mismatch error the analog decoder demonstrates performance near to the ideal digital decoder with floating point calculations . The trade-off between the dynamic range and behaviour of the Gilbert multiplier can explain the difference of the performance for different working points.

For working point $IC = 0.01$ the dynamic range limits the behaviour of the decoder. The dynamic range is comparable to the clipping level in digital signals. Difference between the maximal and minimal value is not sufficient. For $IC = 100$ the error is dominated by the nonideality of the multiplication. In the simulations the best trade off is at the working point 10.

The simulations in Fig. 3 indicate that the error in the decoder is mainly dominated by the mismatch error, not by nonideality of the multiplier or by its range.

5   Conclusions

In this paper we have investigated how the working point of the transistor and the mismatch error impact the BER performance of an analog turbo decoder. This study is made by crating a transistor level simulator of the decoder in C++ based program.

We have used the model that characterises the transistor in weak, medium and strong inversion. The inversion level characterises the working conditions of Gilbert multiplier, the main building block of the decoder.

The simulations show that despite the error in individual blocks the total system is stable. The simulated BER at the best is within 0.5 dB from the ideal decoder.

For different current values the CMOS transistor based multiplier is dominated by different type of errors. The simulations results reported in this work suggest that the best BER performance is achieved when the error due to the dynamic range and due to the nonideality of the multiplication operation is balanced. In our simulation it was achieved for inversion coefficient 10.

The main concern for designer of an analog decoder is the mismatch error. Without it the decoder has BER performance close to ideal. It should be noted that with careful selection of transistor size the impact of mismatch can be reduced.

References

1. F.R. Kschischang, B.J. Frey, H-A. Loeliger, " Factor graphs and the sum-product algorithm," *IEEE Trans. Inform. Theory,* vol. 47, pp. 498–519, Feb. 2001.
2. J. Hagenauer and M. Winklhofer, "The analog decoder," *Proc. 1998 IEEE Int. Symp. On Inform. Theory,* pp. 145, Aug. 1998.
3. H-A. Loeliger, F. Lustenberger, M. Helfenstein, F. Tarköy, "Probability propagation and decoding in analog VLSI," *Proc. 1998 IEEE Int. Symp. On Inform. Theory,* pp. 146, Aug. 1998.
4. C. Berrou,A. Glavieux, P. Thitimajshima, "Near Shannon limit error-correcting coding and decoding: Turbo-codes. 1" *Proc. ICC'93,* pp. 1064 - 1070, May. 1993.
5. M. Moerz, T. Gabara, R. Yan, J. Hagenauer, "An analog 0.25 mm BICMOS tailbiting MAP decoder," *IEEE International Solid-State Circuit Conference,* pp. 356–357, 2000.
6. F. Lustenberger, "On the design of analog VLSI iterative decoders," *Diss ETH,* No 13879, Nov. 2000.
7. A. Xotta, D. Vogrig, A. Gerosa, A. Neviani, A. Amat, G. Montorsi, "An all-analog CMOS implementation of a turbo decoder for hard-disc drive read channels," *IEEE Int. Symp. on Circuits and Systems, ISCAS 2002,* Vol. 5, pp. 69–72, May 2002.
8. C. Weiß, C. Bettstetter, S. Riedel, "Code construction and decoding of parallel concatenated tail-biting codes," *IEEE Trans. Inform. Theory,* Vol 42., No 1., pp. 366–386, Jan. 2001.
9. C. Enz, E.A. Vittoz, 'CMOS low-power analog circuit design," *Designing Low Power Digital Systems, Emerging Technologies (1996)* , pp. 96–133, 1996.
10. P. Kinget, M. Steyaert, "Analog VLSI integration of massive parallel signal processing systems," *Dordrecht: Kluver,* 1997.
11. D.M. Binkley, M. Bucher, D. Foty, "Design-oriented characteristization of CMOS over the continuum of inversion level and channel length," *Proc. 7 IEEE Int.Conf.on Electronics, Circuit & Systems ICECS'2k,* pp. 161–164, 2000.

BiCMOS Variable Gain LNA at C-Band
with Ultra Low Power Consumption for WLAN

Frank Ellinger[1,4], Corrado Carta[2,4], Lucio Rodoni[1,4], George von Büren[1,4],
David Barras[1,4], Martin Schmatz[3], and Heinz Jäckel[1]

[1] Electronics Laboratory, Swiss Federal Institute of Technology (ETH) Zurich,
CH-8092 Zurich, Switzerland, ellinger@ife.ee.ethz.ch
[2] Laboratory for Electromagnetic Fields and Microwave Electronics, ETH Zurich
[3] IBM Research, Zurich Research Laboratory, 8803 Rüschlikon, Switzerland
[4] IBM/ETH Center for Advanced Silicon Electronics, Zurich, 8803 Rüschlikon, Switzerland

Abstract. An ultra low power consuming low noise amplifier (LNA) at C-band with variable gain for adaptive antenna combining is presented in this paper. The microwave monolithic integrated circuit (MMIC) was fabricated using commercial 0.25 µm bipolar complementary metal oxide semi-conductor (BiCMOS) technology. At 5.2 GHz, a supply voltage of 1.2 V and a current consumption of only 1 mA, a maximum gain of 12.7 dB, a noise figure of 2.4 dB and a third order intercept point at the output (OIP3) of 0 dBm were measured. A large amplitude control range of 36 dB was achieved. To the knowledge of the authors, the obtained gain/supply power (S_{21}/P_{dc}) figure of merit of 11 dB/mW is by far the highest ever reported for silicon based C-band LNAs. The characteristics of different bias methods for amplitude control of the cascode circuit are elaborately discussed. A bias control method is proposed to significantly decrease the transmission phase variations versus gain.

1 Introduction

In comparison to the declining mobile phone market, a strong market growth is predicted for wireless local area networks (WLANs) [1]. In this context, the high performance LAN (HIPERLAN) II and the IEEE 802.11a standards at C-band will play an important role. Adaptive antenna combining offers a high potential to improve the performance of those WLAN systems. Especially for adaptive antenna receivers there is the need to lower the power consumption of the components. In comparison to standard RF frontends with single antenna, the required supply currents are significantly higher, because several active antenna paths have to be fed with current. Fig. 1 shows an example of such an adaptive antenna receiver with n active antenna paths. To maximize the quality of the signal available in the base band, the weighting vector of each active antenna path is adjusted. The weighting vectors are functions of phase and amplitude. The amplified and weighted analog RF signals are combined and mixed down to the IF. Then they are digitally coded and finally processed in the base band, where the required weighting vectors are calculated.

J.N. de Souza et al. (Eds.): ICT 2004, LNCS 3124, pp. 891–899, 2004.

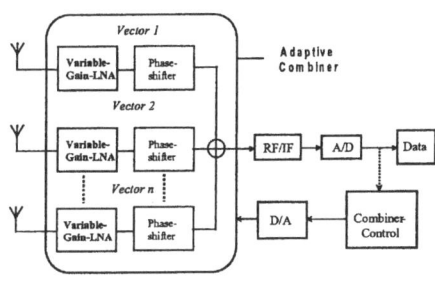

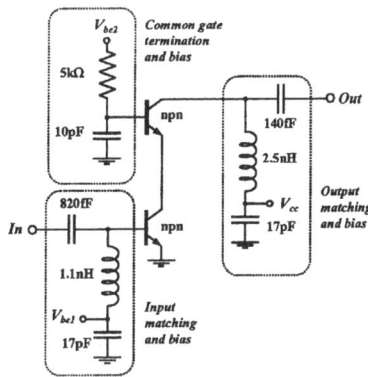

Fig. 1. Adaptive antenna receiver with n active antenna paths. LNAs with variable gain can be used to weight the amplitude of each antenna path, RF: Radio frequency, IF: Intermediate frequency, A/D: Analog to digital converter, D/A: Digital to analog converter.

Fig. 2. Simplified schematics of the VGLNA

Variable gain LNAs (VGLNAs) can be used to adjust the amplitude of each antenna path. This has the advantage that no additional attenuator or variable gain amplifier is required. Thus, power consumption, chip size and cost can be minimized at the same time. Today, there is the trend to lower the supply voltage down to the voltage of one elementary battery cell, which is below 1.4 V. Batteries still consumes a major part of system size and weight. A minimization of the number of battery cells significantly decreases the system size and weight, thus allowing the launch of new applications.

The circuit was particularly optimized for low power consumption. For information concerning typical requirements for WLAN systems it is referred to literature [2]. A challenge was the high amplitude control range, which is defined as maximum power gain minus maximum power attenuation. Due to the strong substrate coupling, silicon based circuits have a limited attenuation range. Ground shields are used for inductors and pads to decrease the substrate coupling. The performances of different bias methods for gain control of the cascode circuit are elaborately discussed. A bias control technique is proposed to significantly decrease the transmission phase variations versus gain, thus simplifying the control complexity of adaptive antenna systems. Several low power consuming MMIC LNAs with excellent RF performances have been reported in the past. A summary is listed in TABLE II comparing their key performances [3-16]. Most of them consider no amplitude control. Some of them allow switching from maximum gain to maximum attenuation [5,8].

A S_{21}/P_{dc} figure of merit of 0.6 dB/mW and an amplitude control of 12.6 dB have been reported for a 0.18 μm CMOS VGLNA [10]. This relative small amplitude control range limits the potential of adaptive antenna systems. Good performances with a S_{21}/P_{dc} figure of merit of 1.7 dB/mW and an amplitude control range of 35 dB has been reported using 0.6 μm GaAs MESFET technology [7]. Unfortunately, in mass fabrication, III/V based products are much more expensive than the silicon based counterparts. Thus, their commercial competitiveness is limited.

For this work, a commercial 0.25 μm silicon germanium (SiGe) BiCMOS technology is used, which combines the good RF performances of hetero bipolar transistors (HBTs) with the excellent performances of CMOS devices for the digital baseband, enabling low cost single chip solutions. Today, cost differences between CMOS and BiCMOS technologies are small. To the knowledge of the authors, the achieved S_{21}/P_{dc} figure of merit of 11 dB/mW is by far the highest ever reported for silicon based C-band LNAs. This result is even slightly better than the best reported for III/V technologies [8]. Furthermore, the authors believe that this is the first study elaborately discussing the properties of a HBT based VGLNA.

Table 1. Comparison with state-of-the-art.

Ref.	Technology	Amplitude Control	S_{21}	NF	OIP3	V_{dc}	I_{dc}	S_{21}/P_{dc}
III/V based technologies								
3	0.15μ PHEMT	No	16dB	0.8dB	n.a.	3V	40.3mA	0.13dB/mW
4	GaAs HBT	No	23.4dB	1.6dB	22dBm	5V	12.3mA	0.38dB/mW
5	InP HBT	20dB (S)	23dB	2.3dB	15dBm	3V	9.3mA	0.82dB/mW
6	0.6μ MESFET	No	11dB	1.9dB	16dBm	3V	4.4mA	0.83dB/mW
7	0.6μ MESFET	33dB	15dB	1.7dB	10dBm	3V	3mA	1.7dB/mW
8	0.6μ MESFET	45dB (S)	12.3dB	2.4dB	-0.8dBm	1V	1.2mA	10dB/mW
Silicon based technologies								
9	SiGe HBT	No	17dB	1.6dB	n.a.	4.5V	7.5mA	0.51dB/mW
10	0.18μ CMOS	No	13.2dB	2.5dB	n.a.	1V	22.2mA	0.6dB/mW
11	0.25μ SiGe BiCMOS	12.6dB	11dB	4.4dB	9dBm	3.75V	4.3mA	0.7dB/mW
12	0.35μ CMOS	No	19.3dB	2.5dB	14dBm	3.3V	8mA	0.73dB/mW
13	0.18μ CMOS	No	14.2dB	0.9dB	15dBm	1V	16mA	0.89dB/mW
14	0.25μ CMOS	No	11dB	2.2dB	11dBm	2V	5mA	1.1dB/mW
15	SiGe HBT	No	15dB	1.65dB	5dBm	1V	13mA	1.15dB/mW
16	0.35μ SiGe BiCMOS	No	17dB	2.5dB	10dBm	3.3V	4mA	1.3dB/mW
This work	0.25μ SiGe BiCMOS	36dB	12.7dB	2.4dB	0dBm	1.2V	1mA	11dB/mW

2   Low Noise Amplifier

The circuit was fabricated with the commercial IBM 6HP SiGe BiCMOS process. At optimum bias current, the HBTs yield transit frequencies f_t of up to 47 GHz and minimum noise figures f_{min} of around 1.2 dB at 5 GHz. Furthermore, the process features inductors with quality factors up to 19 at 5 GHz, metal insulator metal capacitors with specific capacitances per area of 0.7 fF/μm^2 and poly resistors with specific resistances up to 3600 Ω/square. For more information concerning the MMIC process it is referred to [17, 18]. Fig. 2 shows the simplified circuit schematics of the VGLNA. A cascode configuration was used since cascode topologies offer a high amplitude control range. Compared to topologies with single transistor, cascode circuits have a higher attenuation range and a higher maximum gain without increasing the supply current. The higher maximum attenuation range in the attenuation mode is obtained by the lower parasitic input to output capacitance. This parasitic capacitance can be approximated by series connection of the feedback capacitances of the common emitter stage and the common base stage. The resulting parasitic capacitance is smaller than the one of a single transistor topology.

An important design goal was the minimization of the chip size and the corresponding chip costs. The chip size is mainly determined by the size of the bulky inductors. The number of inductors was minimized by using the same inductors for matching and biasing. Therefore, only two inductors were required for the whole circuit. Ground shields were used for inductors and pads to maximize the quality factor of the elements and to decrease the substrate coupling. Substrate coupling limits the maximum attenuation in the attenuation mode. The input was reactively noise matched to 50 Ω. The output was reactively gain matched to 50 Ω. The base of the common base circuit was terminated by a RF shunt capacitor and biased by a high ohmic resistor. Equal transistors were used for the common emitter and the common base stage. The collector emitter voltages of the common emitter stage and common base stage are approximately 0.6 V.

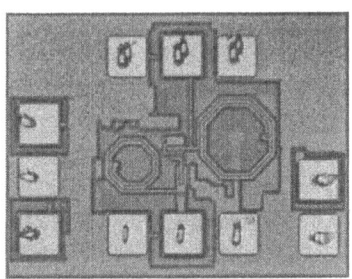

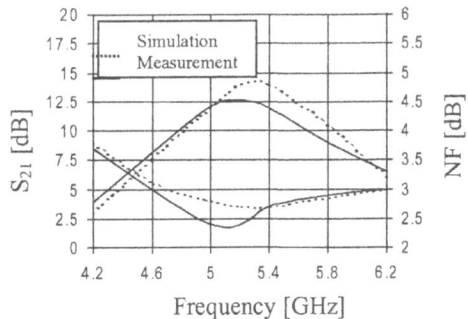

Fig. 3. Photograph of the VGLNA, overal chip size is 0.7 mm x 0.9 mm.

Fig. 4. Measured and simulated gain (S_{21}) and noise figure (NF) versus frequency, $V_{cc} = 1.2$ V, $I_{cc} = 1$ mA.

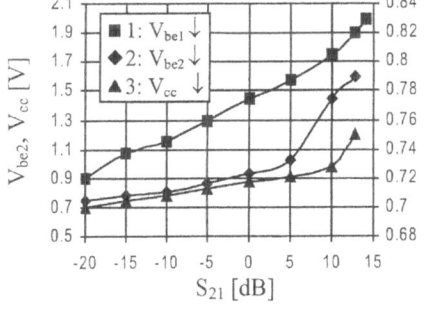

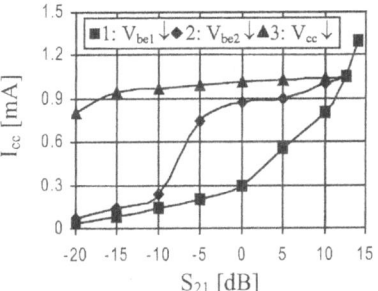

Fig. 5. Measured bias control linearity at 5.2 GHz, decreasing of gain by decreasing of one bias voltage while keeping the two others constant.

Fig. 6. Measured supply current versus gain at 5.2 GHz.

Low power consuming HBTs with small emitter area were chosen to scale down the supply current. The decrease of the emitter area is limited by impedance matching constraints, since the input and output impedances of the transistors are scaling with the emitter area. Inductors with large inductances would be required. Unfortunately, this increases the resistive losses and the noise of the circuit. Thus, an optimum tradeoff has to be found. Fig. 3 shows a photo of the compact MMIC chip, which has an overall size of 0.9 mm x 0.7 mm. The effective circuit area is less than 0.2 mm^2. The main reason for that achievement is the minimization of the number of inductors. The measured and simulated gain and noise figure versus frequency are shown in Fig. 4. The device models of the design kit were used for the simulations. As expected for a commercial and well-established IC technology, good agreement between measurements and simulations were obtained.

At 5.2 GHz, a supply voltage of 1.2 V and a supply current of 1 mA, a gain of 12.7 dB and a noise figure of 2.4 dB were measured. The measured input and output return losses are 7 dB and 9.6 dB, respectively. An OIP3 of 0 dBm was reached.

3   Variable Gain Characteristics

The performances of different bias techniques of the cascode circuit versus gain are compared and discussed. Three different bias modes to decrease the gain from maximum to minimum are investigated:

Mode 1: Decreasing of the base emitter voltage of the common emitter stage V_{be1}. The two other bias voltages (V_{be2}, V_{cc}) are kept constant. Gain is decreased since the transconductance of the transistors is decreasing with the bias current.

Mode 2: Decreasing of the base voltage of the common base stage V_{be2}. The two other bias voltages (V_{be1}, V_{cc}) are kept constant. Gain is decreased since the collector emitter voltage of the common emitter stage $V_{ce, CES} = V_{be2} - V_{be1}$ is driven into the resistive region.

Mode 3: Decreasing of the supply voltage V_{cc}. The two other bias voltages (V_{be1}, V_{be2}) are kept constant. Gain is decreased since the collector emitter voltage of the common base stage $V_{ce, CBS} = V_{cc} - V_{be2} + V_{be1}$ is driven into the resistive region. Generally, this bias mode is not recommended since a relatively high control current has to be provided. Bias modes 1 and 2 have the advantage that their control current (base current) is very low. For comparison of the different bias modes, the most important characteristics were measured versus gain and are discussed in sub-sections a)-f). Fortunately, for most adaptive antenna system, the requirements for b)-e) decrease with decreasing gain. In most cases, only those antenna paths have to be attenuated which have a strong signal amplitude at the antenna input, thus originally having a good signal to noise ratio.

a) Control linearity: A high control linearity minimizes the resolution required for the D/A converter, which converts the digital control voltages calculated in the baseband. As shown in Fig. 5, the best linearity was reached for bias mode 1.

b) Current consumption: The supply current is decreasing with decreasing bias. This is advantageous since it lowers the power consumption. The corresponding performances are depicted in Fig. 6. As expected, the highest decrease of the supply current was obtained by bias mode 1 since the decreasing of V_{be1} significantly lowers the supply current.

c) Noise figure: The noise figure is increasing with decreasing bias. To minimize the supply current, the circuit is operated below the supply current required for minimum noise. The measured performances are compared in Fig. 7. The noise is significantly increased with bias mode 1 since the decrease of V_{be1} moves the bias current far away

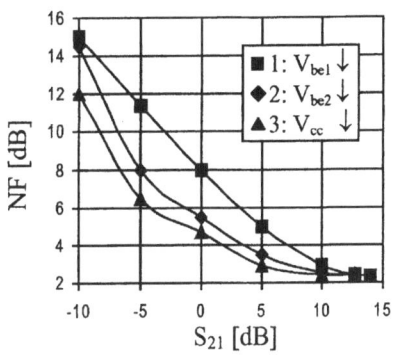

Fig. 7. Measured noise characteristics versus gain at 5.2 GHz.

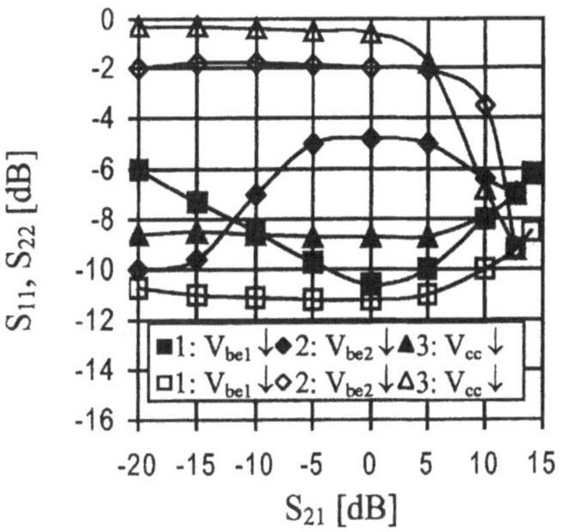

Fig. 8. Measured input return loss (filled symbols) and output return loss (symbols not filled) versus gain at 5.2 GHz.

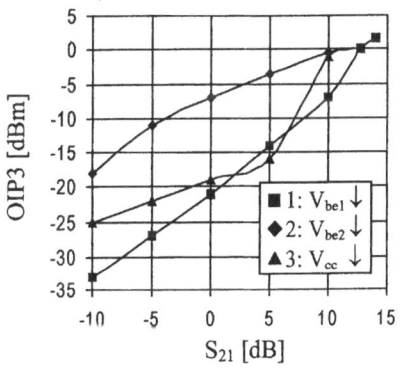

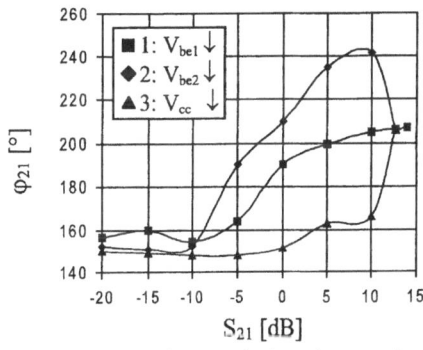

Fig. 9. Measured third order intercept point at the output versus gain at 5.2 GHz.

Fig. 10. Measured transmission phase variations versus gain at 5.2 GHz.

from the current for minimum noise. A medium noise increase was measured for bias mode 2. Noise is increased since the common emitter stage is driven into the resistive region. The lowest increase of the noise figure was obtained for bias mode 3. The noise of the common base stage (output stage) is increased because it is driven into the resistive region, whereas the noise increase and gain decrease of the common emitter (input stage) is weak. According to the formula of Friis, the system noise is dominated by the input stage. Thus, the total noise increase versus gain is weak.

d) Return losses: Decreasing of the return losses could generate undesired signal reflections which may degrade the performance of the system. As illustrated in Fig. 8, the best results with relatively high return losses within a wide gain control range were obtained for bias mode 1. It is noted that the matching was optimized for bias mode 1. A strong degradation of the output return losses is observed for bias mode 3. The output impedance is getting very high ohmic since the output stage is driven in the resistive region.

e) OIP3: The OIP3 is decreasing with bias since the supply current and the supply voltages are decreased, thus lowering the amplitude of the signal at the fundamental frequency and increasing non-linear effects. For maximum gain, the circuit is operated in class AB. As depicted in Fig. 9, good performance is reached for bias mode 2, since the input stage acts as a variable attenuator at the input. The non-linearities of this resistive transistor are relatively weak. A strong degradation of the OIP3 is observed for bias mode 1. The lowering of V_{be1} drives the amplifier from class-AB to class-C operation, where the transistor generates high non-linearities.

The strongest decrease of the OIP3 was measured for bias mode 3. The reason is that within a wide gain control range, the gain of the input stage is staying high. Thus, compared to bias mode 2, the output stage is sooner driven into compression.

f) Transmission phase: Unfortunately, the transmission phase (phase of S_{21}) is varying with the gain and the related bias point. Phase variations are generated by variations of RC time constants. A constant phase versus gain can be mandatory for some types of adaptive antenna systems. Variations of the transmission phase can be compensated by the phase shifters as shown in Fig. 1. However, in this case, amplitude and phase can not be controlled independently. This would require a feedback control loop, which would significantly increase the control complexity. Thus, a VGLNA with constant phase versus gain is highly preferred. A smart bias method reaching that goal is proposed in the next section. The measured performances are shown in Fig. 10. The lowest phase variation was reached by using bias mode 1. High phase variations were measured with bias mode 2 and 3. These phase variations are generated by the strong resistive and capacitive variations occuring around the saturation region (transistion from the forward active to the resistive region).

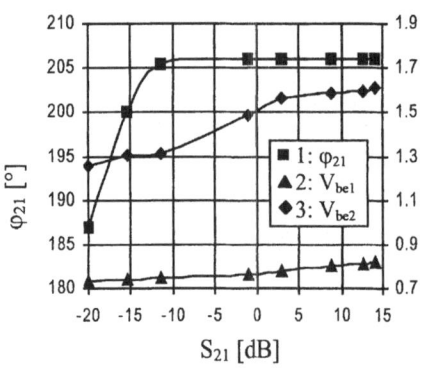
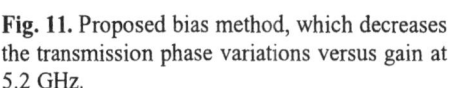
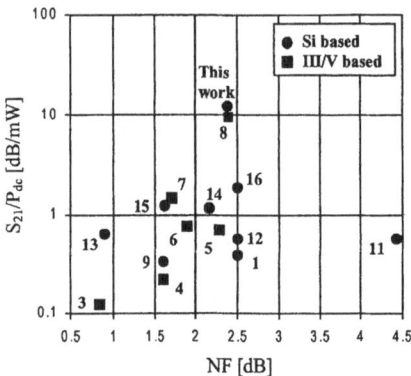

Fig. 11. Proposed bias method, which decreases the transmission phase variations versus gain at 5.2 GHz.

Fig. 12. Gain per supply power figure of merit versus noise figure for state-of-the-art C-band LNAs.

4 Bias Technique for Constant Transmission Phase

As demonstrated in the previous section, good results were obtained with bias mode 1. This bias mode is well suited for short to medium range systems. However, the transmission phase variations versus amplitude control can be a limitation for low cost systems requiring a low control complexity of the antenna array. In this section, a bias method is proposed which significantly decreases these transmission phase variations. The basic idea is that within a given gain range, bias mode 1 and bias mode 2 have opposite phase characteristics. Decreasing of V_{be1} decreases the phase, whereas decreasing of V_{be2} increases the phase. This has been illustrated in Fig. 10. Thus, within a certain amplitude control range, the phase variations can compensate each other. The measured results are shown in Fig. 11. The phase can be kept constant for an amplitude range of approximately 25 dB. This amplitude control range is sufficient for demanding applications.

5 Conclusions

An ultra low power consuming C-band VGLNA has been presented. The compact MMIC has been fabricated using low cost 0.25 µm SiGe BiCMOS technology. The cascode circuit has a high gain control range making it well suited for adaptive antenna combining. The characteristics of different bias methods for amplitude control have been elaborately compared and discussed. Furthermore, a bias control technique has been proposed to significantly decrease the transmission phase variations versus gain. To the knowledge of the authors, the presented VGLNA has by far the lowest power consumption ever reported for silicon based C-band LNAs, while still providing adequate gain, large signal and noise performance. A comparison with other works by means of the gain per supply figure of merit versus noise figure is summarized in Fig. 12. The reached figure of merit is even slightly higher than the highest reported for III/V based technologies. This MMIC is an excellent candidate for low cost WLAN applications operating in accordance to the 802.11a and HIPERLAN standards.

References

1. CommWeb, "Falling costs driving WLAN penetration", *www.commweb.com*, 16. April 2003.
2. T. H. Lee, H. Smavati and H. R. Rategh, "5-GHz CMOS wireless LANs", *IEEE Transactions on Microwave Theory and Techniques*, Vol. 50, No. 1, pp. 268-280, Jan. 2002.
3. B. G. Choi, Y. S. Lee, C. S. Park and K. S. Yoon, "Super low noise C-band PHEMT MMIC low noise amplifier with minimum input matching network", *Electronics Letters*, Vol. 36, No. 19, Sept. 2000.
4. K. W. Kobayashi, L. T. Tran and M. D. Lammert, T. R. Block, P. C. Grossmann, A. K. Oki, D. C. Streit, "Sub 1.3 dB noise figure direct coupled MMIC LNAs using a high current gain 1 µm GaAs HBT technology", *GaAs IC Symposium*, pp. 240-243, Nov. 1997.

5. Y. Aoki, N. Hayama, M. Fujii and H. Hida, "A 23/3-dB dual-gain low-noise amplifier for 5-GHz-band wireless applications", *Gallium Arsenide Integrated Circuit Symposium*, pp. 197 -200, Nov. 2002.

6. S. Yoo, D. Heo, J. Laskar and S. S. Taylor, "A C-band low power high dynamic range GaAs MESFET low noise amplifier", *IEEE Radio and Wireless Conference*, Denver, pp. 191-194, May 1999.

7. F. Ellinger, U. Lott and W. Bächtold, "A 5.2 GHz variable gain LNA MMIC for adaptive antenna combining", *IEEE MTT-S International Microwave Symposium*, Vol. 2, page 87-89, June 1999.

8. F. Ellinger, U. Lott and W. Bächtold, "Ultra low power GaAs MMIC low noise amplifier for smart antenna combining at 5.2 GHz", *IEEE Radio Frequency Integrated Circuit Symposium*, pp. 157-159, June 2000.

9. U. Erben, H. Schumacher, A. Schuppen and J. Arndt, "Application of SiGe heterojunction bipolar transistors in 5.8 and 10 GHz low-noise amplifiers", *Electronics Letters*, Vol. 3, No. 1, pp. 1498 -150, July 1998.

10. T. K. K. Tsang and M. N. El-Gamal, "Gain and frequency controllable sub-1 V 5.8 GHz CMOS LNA", *IEEE International Symposium on Circuits and Systems*, Vol. 4, pp. IV-795 -IV-798 Vol. 4, 2002.

11. S. Chakraborty, S. K. Reynolds, T. Beukema, H. Ainspan and J. Laskar, "Architectural trade-offs for SiGe BiCMOS direct conversion receiver front-ends for IEEE802.11a", *Gallium Arsenide Integrated Circuit Symposium*, pp. 120 -12, Nov. 2002.

12. Choong-Yul Cha and Sang-Gug Lee, "A 5.2-GHz LNA in 0.35 μm CMOS utilizing inter-stage series resonance and optimizing the substrate resistance", IEEE Journal of Solid-State Circuits, Vol. 38, No. 4, pp. 669-672, April 2003.

13. D. J. Cassan and J. R. Long, "A 1-V transformer-feedback low-noise amplifier for 5-GHz wireless LAN in 0.18μm CMOS", *IEEE Journal of Solid-State Circuits*, Vol. 38 No. 3 pp. 427 -435, March 2003.

14. Hong-Wei Chiu and Shey-Shi Lu, "A 2.17 dB NF, 5 GHz band monolithic CMOS LNA with 10 mW DC power consumption", *IEEE Symposium on VLSI Circuits*, pp. 226 -229, 2002.

15. M. Soyuer and J-O. Plouchart, H. Ainspan, J. Burghartz, "A 5.8 GHz 1 V low noise amplifier in SiGe bipolar technology", *1997 IEEE Radio Frequency Integrated Symposium*, pp. 19-22, June 1997.

16. J. Sadowy, I. Telliez, J. Graffeuil, E. Tournier, L. Escotte and R. Plana, "Low noise, high linearity, wide bandwidth amplifier using a 0.35 μm SiGe BiCMOS for WLAN applications, *IEEE Radio Frequency Integrated Circuits Symposium*, 2002 IEEE , pp. 217-220, June 2002.

17. S. A. St. Onge, et al., "A 0.24 μm SiGe BiCMOS mixed-signal RF production technology featuring a 47 GHz f_t HBT and 0.18 μm L_{eff} CMOS" *Bipolar/BiCMOS Circuits and Technology Meeting*, pp. 170-120, Sept. 1999.

18. www.mosis.com.

Joint MIMO and MAI Suppression for the HSDPA

Mário Marques da Silva[1] and Américo Correia[2]

[1] Institute for Telecommunications,
[2] ADETTI
IST, Torre Norte 11.10, Av. Rovisco Pais, 1049-001 Lisboa, Portugal
http://www.it.pt
marques.silva@engenheiros.pt, americo.correia@iscte.pt

Abstract. We consider the use of a specific type of Clipped Soft Decision (CSD) – Parallel Interference Cancellation (PIC), optimized for Multiple-Input-Multiple-Output (MIMO) operation. It is shown that the use of the MIMO combined with the proposed PIC scheme can achieve a high performance improvement over the correspondent schemes, and hence, the High Speed Downlink Packet Access (HSDPA) is able to increase the data rate, also with a performance improvement in the Bit Error Rate (BER).

1 Introduction

The main purpose in UMTS is to allow high data rates, low delays, high capacity and flexibility in services. Direct Sequence Code Division Multiple Access (DS-CDMA) technology is the main key to reach this convergence, once we develop systems that combat the interferences.

To combat the fading and also taking advantage of fading in order to provide diversity, the current work considers a MIMO scheme. Furthermore, once all spectrum is available to all users at the same time, there is MAI, which is caused because several spreading codes are used by several users and they are not received by one reference user perfectly orthogonal. This happens because, even with the use of orthogonal spreading codes and for a synchronous network, the multipath environment breaks the orthogonality, generating MAI. Thus, in addition to the MIMO scheme, we also consider a specific type of PIC to cancel the MAI.

We consider the use of a 2X2 MIMO scheme based on the Alamouti Scheme, where both transmitter and receiver have 2 antennas. However, even if only one user is considered, the several parallel physical channels (PPC) create the appearance of MAI between them. These are the interfering signals that need to be cancelled to achieve higher data rate users with low probability of BER. Thus, once the several PPC present the same received power and once the receiver has knowledge about the several spreading sequences, it is considered a CSD–PIC, optimized for MIMO operation.

This paper is structured as follows: section 2 presents the system model for the 2X2 MIMO scheme; section 3 presents the system model for the CSD-PIC, optimized for MIMO operation; section 4 presents the performance results and analysis for the proposed schemes; the key findings of this paper are then summarized in Section 5.

J.N. de Souza et al. (Eds.): ICT 2004, LNCS 3124, pp. 900–907, 2004.

2   System Model for the 2X2 MIMO

The 2X2 MIMO is spectral efficient and resistant to fading, where the BS uses $M=2$ transmit antennas and the MS also uses $M=2$ receive antenna [3]. As long as the antennas are spaced sufficiently far apart, the transmission and receiving signals from each antenna undergoes independent fading. Fig. 1 shows a dual-antenna transmitter and receiver. The encoding performed by the proposed MIMO scheme is the same as applied by 2X1 Space-Time Transmit Diversity (STTD) initially proposed by Alamouti. The lowpass equivalent transmitted signals at the two antennas are:

$$s_1(t) = \sum_{n=-\infty}^{+\infty} \left[\sqrt{E_c} \cdot \sum_{k=1}^{K} b'(k, \lfloor n/N \rfloor) \cdot S_d(k,n) \right] \cdot g(t - nT_c)$$

$$s_2(t) = \sum_{n=-\infty}^{+\infty} \left[\sqrt{E_c} \cdot \sum_{k=1}^{K} b''(k, \lfloor n/N \rfloor) \cdot S_d(k,n) \right] \cdot g(t - nT_c)$$

(1)

where E_c is the chip energy of data channel. The operation $\lfloor \cdot \rfloor$ stands for the integer part of operand. $S_d(k,n)$ means the data channel spreading sequence of the kth user. N is the spreading factor (SF). T_c is the chip interval and $g(t)$ is the chip waveform. $b'(k, \lfloor n/N \rfloor)$ and $b''(k, \lfloor n/N \rfloor)$ are the encoded data symbols and are given by:

$$b'(k, \lfloor n/N \rfloor) = \begin{cases} b_1(k, \lfloor n/N \rfloor) & \lfloor n/N \rfloor \text{ is odd} \\ b_2(k, \lfloor n/N \rfloor) & \lfloor n/N \rfloor \text{ is even} \end{cases}$$

$$b''(k, \lfloor n/N \rfloor) = \begin{cases} -b_2^*(k, \lfloor n/N \rfloor) & \lfloor n/N \rfloor \text{ is odd} \\ b_1^*(k, \lfloor n/N \rfloor) & \lfloor n/N \rfloor \text{ is even} \end{cases}$$

(2)

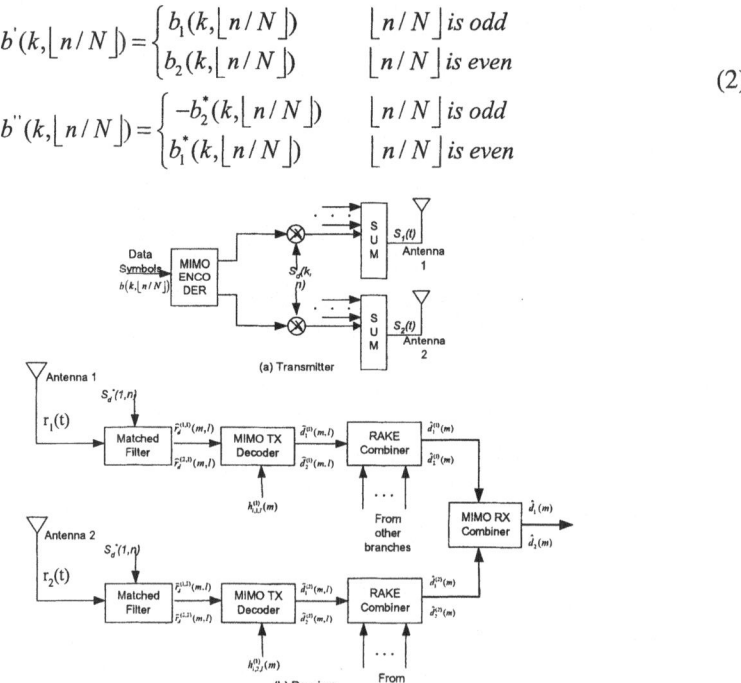

Fig. 1 Scheme of a 2X2 MIMO Scheme (a) Transmitter (encoder) and (b) Receiver (decoder)

We consider a discrete tap-delay-line channel model where the channel from ith transmit antenna to the jth receive antenna comprises discrete resolvable paths, expressed through the channel coefficients. The four sets of temporal multipaths corresponding to paths between the two transmit antennas and the two receive antennas experience independent and identical distributed (i.i.d.) Rayleigh fading. Therefore, the channel coefficients $h_{i,j,l}$ are complex-valued Gaussian random variables with zero-mean and same variance σ^2 in both real and imaginary parts. The lowpass equivalent complex impulse response of the channel between ith transmit antenna and jth receive antenna is given by $h_{i,j}(t) = \sum_{l=1}^{L} h_{i,j,l} \cdot \delta(t - \tau_l)$, where τ_l is the delay of the lth resolvable path.

For a given temporal path index l, the fading coefficients $h_{1,l}$ and $h_{2,l}$ are Rayleigh distributed and independent from each other. Additionally, $h_{i,j,l}$ is assumed invariant over at least one MIMO block that refers to every two consecutive symbols. For any mobile station, downlink signals from K synchronous data channels experience the same frequency selective fading and reach the receiver as:

$$r_j(t) = \sum_{l=1}^{L} \left[h_{1,j,l} \cdot s_1(t - \tau_l) + h_{2,j,l} \cdot s_2(t - \tau_l) \right] + n_j(t)$$

(3)

where $n_j(t)$ represents the additive white Gaussian noise (AWGN) of the jth receive antenna, with double-sided power spectral density of $N_0/2$.

The signal of each path can be resolved by a matched filter (MF) with the local delayed spreading sequence. It is assumed that multipath delays are approximately a few chips in duration and smaller than the symbol period so that Inter-Symbol Interference (ISI) can be neglected. Without loss of generality, let us focus on the mth MIMO block. By sampling the output of the pulse-matching filter, the received signal in the first and second symbol interval of the mth MIMO block, before MIMO TX decoding, can be, respectively, given by:

$$u_j^{(1)}(m,n) = \sum_{k=1}^{K} \sum_{l=1}^{L} \sqrt{E_c} \cdot S_d\left(k, 2mN + n - \lfloor \tau_l / T_c \rfloor\right)$$
$$\cdot \left[b_1(k,m) \cdot h_{1,j,l} - b_2^*(k,m) \cdot h_{2,j,l} \right] + n_j^{(1)}(m,n)$$

(4)

$$u_j^{(2)}(m,n) = \sum_{k=1}^{K} \sum_{l=1}^{L} \sqrt{E_c} \cdot S_d\left(k, (2m+1) \cdot N + n - \lfloor \tau_l / T_c \rfloor\right)$$
$$\cdot \left[b_2(k,m) \cdot h_{1,j,l} + b_1^*(k,m) \cdot h_{2,j,l} \right] + n_j^{(2)}(m,n)$$

where and $n=0,1,\ldots,N-1$ and $b_j(k,m)(j=1,2)$ stands for the jth data bit of the kth active user transmitted in the mth MIMO block. $n_j^{(i)}(m,n)$, $(i=1,2)$, is the sampled AWGN in the ith symbol period, j^{th} receive antenna, of the mth MIMO block.

The first user ($k=1$) is assumed to be the desired user. Assuming perfect chip timing synchronization and that the local despreading sequence is locked to the lth resolvable path, the data channel of the first user at the lth path despread during the first and second symbol period of the mth MIMO block, before TX MIMO decoding, comes:

$$
\begin{aligned}
r_d^{(1,j)}(m,\hat{l}) &= \sum_{n=0}^{N-1} u^{(1)}(m,n)\cdot S_d^*\left[1,2mN+n-\lfloor \tau_j/T_c\rfloor\right] \\
&= \sqrt{E_c}\cdot N\cdot\left[b_1(1,m)\cdot h_{1,j,\hat{l}} - b_2^*(1,m)\cdot h_{2,j,\hat{l}}\right] + n_j^{(1)}(m,\hat{l}) \\
&\quad + \sum_{k=1}^{K}\sum_{\substack{l=1\\l\neq\hat{l}}}^{L}\sqrt{E_c}\cdot R_{d,d}^{(1)}(m,k;l,\hat{l})\cdot\left[b_1(k,m)\cdot h_{1,j,l} - b_2^*(k,m)\cdot h_{2,j,l}\right]
\end{aligned}
$$

$$
\begin{aligned}
r_d^{(2,j)}(m,\hat{l}) &= \sum_{n=0}^{N-1} u^{(2)}(m,n)\cdot S_d^*\left[1,(2m+1)N+n-\lfloor \tau_j/T_c\rfloor\right] \\
&= \sqrt{E_c}\cdot N\cdot\left[b_2(1,m)\cdot h_{1,j,\hat{l}} + b_1^*(1,m)\cdot h_{2,j,\hat{l}}\right] + n_j^{(2)}(m,\hat{l}) \\
&\quad + \sum_{k=1}^{K}\sum_{\substack{l=1\\l\neq\hat{l}}}^{L}\sqrt{E_c}\cdot R_{d,d}^{(2)}(m,k;l,\hat{l})\cdot\left[b_2(k,m)\cdot h_{1,j,l} + b_1^*(k,m)\cdot h_{2,j,l}\right]
\end{aligned}
$$

(5)

where the superscript $*$ means the complex conjugate operation $n_j^{(i)}(m,\hat{l})$, $i=1,2$, $j=1,2$ represents the coloured noise components. $R_{d,d}^{(i)}(m,k;l,\hat{l})$, $i=1,2$, is the N-length discrete aperiodic correlation function of two time delayed data channel spreading sequences used in the ith symbol period of the mth MIMO block.

Introducing the proper simplified matrix-vector representation we may define $\mathbf{y}=\mathbf{Hb}+\mathbf{n}$, where $\mathbf{H}=\begin{bmatrix} h_{1,j} & -h_{2,j} \\ h_{2,j}^* & h_{1,j}^* \end{bmatrix}$, $\mathbf{b}=\begin{bmatrix} b_1 \\ b_2^* \end{bmatrix}$, $\mathbf{n}=\begin{bmatrix} n_{1,j} \\ n_{2,j} \end{bmatrix}$ and, $\mathbf{y}=\begin{bmatrix} r_{1,j} \\ r_{2,j} \end{bmatrix}$, and where j^{th} means the receive antenna. This is equivalent to write $\mathbf{r}=\mathbf{Bh}+\bar{\mathbf{n}}$ where $\mathbf{B}=\begin{bmatrix} b_1 & -b_2^* \\ b_2 & b_1^* \end{bmatrix}$, $\mathbf{h}=\begin{bmatrix} h_{1,j} \\ h_{2,j}^* \end{bmatrix}$ and $\bar{\mathbf{n}}=\begin{bmatrix} n_{1,j} \\ n_{2,j}^* \end{bmatrix}$, which is the function possible for real implementation.

Assuming that the ideal CSI $h_{i,j,\hat{l}}$ is available at the jth MS, and thus, the random variable of the first estimated data bit in the mth MIMO block is constructed through MIMO decoding as follows:

$$
d_1^{(j)}(m,\hat{l}) = r_d^{(1,j)}(m,\hat{l})\cdot h_{1,j,\hat{l}}^* + r_d^{(2,j)*}(m,\hat{l})\cdot h_{2,j,\hat{l}}
$$

$$
d_2^{(j)}(m,\hat{l}) = r_d^{(2,j)}(m,\hat{l})\cdot h_{1,j,\hat{l}}^* - r_d^{(1,j)*}(m,\hat{l})\cdot h_{2,j,\hat{l}}
$$

(6)

Or, using the simplified matrix-vector representation as $\hat{\mathbf{b}}=\mathbf{d}=\underset{(a)}{\underbrace{\dfrac{1}{h^2}}}\mathbf{H}^H\mathbf{y}$, where (a)

stands for the gain introduced by the MIMO transmit diversity, which is expressed as $h^2=\sum_{l=1}^{L}\left|h_{1,j,l}\right|^2+\sum_{l=1}^{L}\left|h_{2,j,l}\right|^2$. Each transmitted signal, after demodulation, is multiplied

by the square of the channel coefficients. So, if the channel has L multipaths between each transmit antenna and the each receive antenna then the transmitter diversity is able to get a diversity order of $2L$ (2 is the number of transmit antennas).

In the equal gain combining receiver, the signals of first L_c arriving paths among total L resolvable paths are selected and combined. Assuming that the fading of each path is independent and the random variables in distinct branches of 2D-RAKE are independent from each other, the output of the 2D-RAKE receiver can be represented as $d_{i,j}(m) = \sum_{l=1}^{L_c} d_{i,j}(m,l)$. This expression means the signal at the output of each of the two 2D-RAKE receivers. Thus, considering an equal gain combiner, the signal at the output of the MIMO receiver combiner comes $d_i(m) = \sum_{j=1}^{2} d_{i,j}(m)$, where 2 stands for the number of receive antennas. The order of diversity is now expressed by $4L$, where L is the number of resolvable paths.

When we consider a MIMO scheme of order higher than two, it was shown in [4], for the STTD, that there is IIS from only one symbol in the decoding of any other symbol, for a modulation scheme higher than BPSK, which can also be extrapolated for the Alamouti based MIMO scheme with more than two transmit antennas.

3   System Model for the Parallel Interference Cancellation

There are two different kinds of subtractive sub-optimal Multi-User Detectors: Successive Interference Cancellation (SIC) and Parallel Interference Cancellation (PIC). These kinds of detectors make the estimation and subtraction of the MAI seen by each user. However, they present the disadvantage of introducing a symbol period delay for each level of cancellation. By level of cancellation we mean the partial cancellation of each user (SIC) or the partial cancellation of all users in parallel (PIC). The term partial is used because it is not possible to make the total cancellation as it depends on the precision of the estimation of the transmitted symbols and this estimation can vary in accordance with several conditions, such as the noise level (E_b/N_0), the amount of ISI present, the estimation of the CSI and the amount of MAI seen by each user. To improve the liability of the MAI cancellation, this can be done in several levels of cancellation, increasing the value of the subtractive coefficient with the increase of level of cancellation, as the estimation of the transmitted symbols tends to improve. In an environment with near-far problem (or without power control) we tend to obtain a better performance with the SIC, rather than with a conventional PIC [2]. In the presence of power control, the conventional PIC tends to perform better, although, it is more complex. The scheme shown in Fig. 2 for the proposed PIC is generic, although most of the processing for different users is the same, except the spreading sequences, since it is considered the downlink situation. MIMO Encoder and MIMO Decoder of Fig. 2 are described in more detail in Fig. 1. The resulting signal for the kth user, after subtracting MAI is $r_k^{(m)}(t) = r(t) - \sum_{i=1,i \neq k}^{K} \hat{r}_i^{(m)}(t - \tau_i)$, being m the cancellation order, A_i the received amplitudes of the several users ($i=1$ to K, $i \neq k$) and s_i the spreading sequences of the ith interfering PPC.

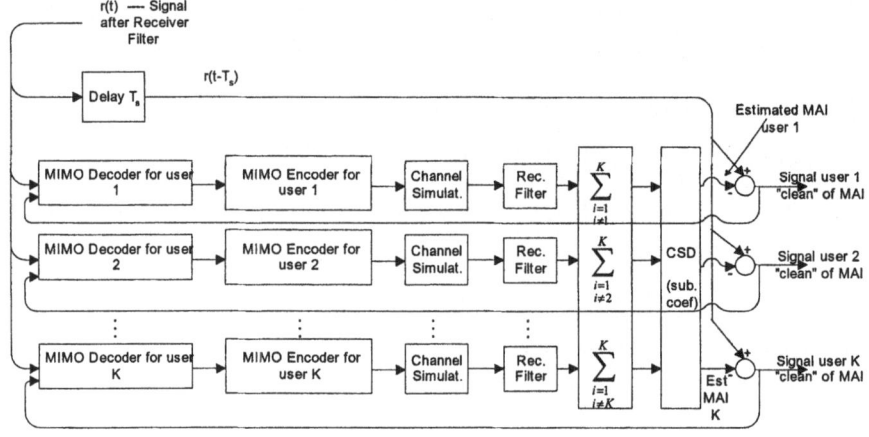

Fig. 2. Scheme of the modified PIC Detector

A problem arises when we are dealing with the optimization of the subtractive co-efficients. The solution to that problem is as follows: based on MMSE error considerations and convenient Gaussian approximations, it is possible to derive an optimal tentative decision function for the intermediate estimates [6]. For QPSK modulation the function is a hyperbolic tangent of the matched filtered statistics applied independently to the in-phase and quadrature components. The piecewise Linear Clipped soft Decision (CSD) is a good approximation of the hyperbolic tangent. The CSD function retains the advantages of both the linear and the hard-decision functions, while avoiding some of their obvious shortcomings. Small correlator outputs lead to unreliable data estimates, and thus, only a small amount of signal cancellation should be carried out. The inverse also applies.

4   Results and Discussion

All the simulation results presented in this section considered the QPSK modulation in a frequency selective Rayleigh fading channel, namely the Vehicular A propagation models of 3GPP. The mentioned propagation model was selected to evaluate the effect of IPI in the performance of 2X2 MIMO and PIC. The average power profile of Vehicular A is modeled as [0.4874 0.3134 0.0905 0.0446 0.0397 0.0090 0.0055 0.0099], with a maximum delay spread correspondent to seven chip periods. The Walsh-Hadamard spreading sequences were considered in the simulation, with a spreading factor of 16. The double of the binary debit was considered for the results with the 2X2 MIMO (with and without PIC), relating the Single-Input-Single-Output (SISO). Additionally, the RAKE receiver was always considered, namely the 1D_RAKE for the curves entitled SISO and PIC, and two modules of 2D_RAKE in case of a 2X2 MIMO (with and without PIC). By number of parallel physical channels (PPC) in the several results we mean the number of PPC in the downlink of HSDPA, corresponding, in our case, to only one user. However, it is similar to considering several (PPC) co-sited users.

When we refer to HSDPA in the results, it means that the binary debit of 384 kb/s is increased by a number corresponding to PPC, for the case of SISO and PIC, and 2xPPC, for the case of 2X2 MIMO and 2X2 MIMO + PIC. CSD-PIC with three levels of cancellation was selected in our simulations because it corresponds to a good balance between performance and delay in the signal.

In Fig. 3a) it is shown the performance obtained with 15 PPC. As can be seen, due to the presence of higher level of MAI, the use of PIC corresponds to an appreciated performance improvement over the SISO. When we increase the order of diversity to $4L$ (where L corresponds to the number of multipaths) with the use of 2X2 MIMO, we can observe that, even with the double of the binary debit, the performance is much improved. However, due to the presence of the MAI, we can see that the improvement of performance with the increase of E_b/N_o is low. This was the main motivation to add the PIC to the 2X2 MIMO, whose results correspond to a much-appreciated gain over the 2X2 MIMO alone. The gain obtained with the 2X2 MIMO + PIC over the 2X2 MIMO is much higher than the gain obtained with the PIC over the SISO. This confirms that the 2X2 MIMO and the PIC fit very well for the HSDPA.

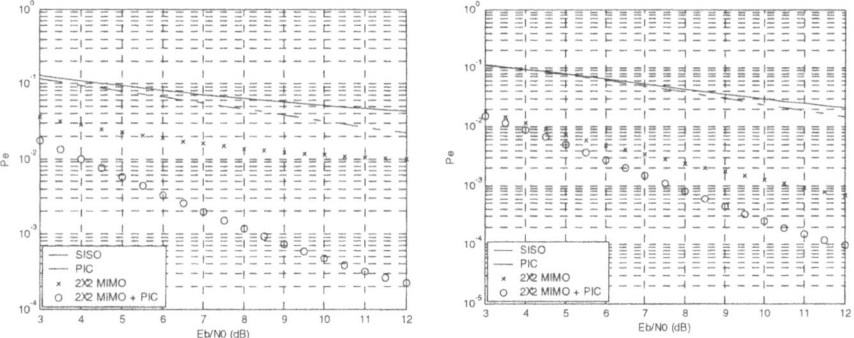

Fig. 3. BER performance for Vehicular A propagation of 3GPP with a) 15 PPC and b) 4 PPC

In Fig 3b) it is shown the performance obtained with 4 PPC, which corresponds to lower level of MAI, when compared to Fig. 3a). Due to lower level of MAI, the gain obtained with the PIC (relating to SISO) is lower than that of Fig 3a).

The performance improvement of the 2X2 MIMO + PIC relating to 2X2 MIMO is much higher than the performance improvement of the PIC relating to the SISO. This confirms that the system considering the 2X2 MIMO with the PIC is able to achieve a performance with very low level of bit errors. For 4 PPC and E_b/N_o=12 dB, the performance obtained for the 2X2 MIMO + PIC is 10^{-4}, which is almost the same performance obtained for the same configuration but with 15 PPC ($P_e=10^{-3.9}$). The same does not happen when we consider only the 2X2 MIMO. Hence, we may say that the system composed of the combination of 2X2 MIMO and PIC is almost insensitive to an increase in the number of PPC.

5   Conclusions

In this paper we have studied the downlink of HSDPA with the combination of two specific schemes, namely the 2X2 MIMO and the CSD-PIC for the Vehicular A propagation model of 3GPP.

It was shown that the 2X2 MIMO combined with the CSD-PIC tends to reach a performance improvement over the corresponding schemes isolated. When the level of MAI is moderate to high we observed that, even with the use of the 2X2 MIMO, the performance obtained was limited. This was the main motivation to add the PIC to the 2X2 MIMO scheme, whose results correspond to a much-appreciated gain over the 2X2 MIMO alone. The gain obtained with the 2X2 MIMO + PIC over the 2X2 MIMO is much higher than the gain obtained with the PIC over the SISO. This confirms that the combination of 2X2 MIMO and the PIC fit very well for the downlink of HSDPA. We have also shown that the system considering the 2X2 MIMO combined with the PIC is almost insensitive to an increase in the number of PPC.

Acknowledgements. This work has been partially funded by the IST-2003-507607 EU _Broadcasting and Multicasting Over Enhanced UMTS Mobile Broadband Networks_ project.

References

1. A. Correia, "Optimized Complex Constellations for Transmission Diversity", Wireless Personal Communications 20, Journal of Kluwer Academic Pub, Mar. 2002, pp. 267-284;
2. M. M. Silva, A. Correia, "Joint Multi-User Detection and Inter-symbol Interference Cancellation for WCDMA Satellite UMTS", Internation Jounal of Satellite Communications and Networking – Special Issue on Interference Cancellation – John Willey and Sons, Ltd, n.° 21, 2003, pp. 93-117;
3. S. M. Alamouti, "A Simple Transmitter Diversity Scheme for Wireless Communications", IEEE JSAC, Oct., 1998, pp. 1451-1458;
4. M. M. Silva, A. Correia, "Space Time Block Coding for 4 antennas with coding rate 1", Proc. of IEEE ISSSTA'02, Prague, Check Republic, Sep. 2002, pp. 318-322;
5. M. M. Silva., A. Correia, "Parallel Interference Cancellation with Commutation Signaling", Proc. IEEE ICC 2000, New Orleans, USA, Spring 2000;
6. P. Rooyen, M. Lötter, D. Wyk, Space-Time Processing for CDMA Mobile Communications – Kluwer Academic Publishers, Boston, 2000;

A Joint Precoding Scheme and Selective Transmit Diversity

Mário Marques da Silva[1] and Américo Correia[2]

[1] Institute for Telecommunications,
[2] ADETTI
IST, Torre Norte 11.10, Av. Rovisco Pais, 1049-001 Lisboa, Portugal
http://www.it.pt
marques.silva@engenheiros.pt, americo.correia@iscte.pt

Abstract. We propose the use of the precoding (PC) scheme, which can be used to combat both the Multiple Access Interference (MAI) and Inter-Path Interference (IPI), combined with the Selective Transmit Diversity (STD) for high data rate transmissions. The proposed scheme is considered over frequency selective Rayleigh fading channels jointly with a RAKE in the receiver. With the PC, the increase in performance is achieved with a small increase in power processing in the BS, avoiding any need to increase complexity in the Mobile Station (MS) [1,2,3]. It is shown that the use of the proposed PC scheme, alone or combined with the STD achieves a performance improvement over the corresponding schemes without PC.

1 Introduction

Much research has been undertaken in the area of the Multi-User Detectors [4] for Direct Sequence Code Division Multiple Access (DS-CDMA) technology. It can be mainly used by Base Stations (BS) where there is enough power processing capability and where it is easier to know/estimate the uplink Channel Impulse Responses (CIR) and the spreading sequences of interfering users. However, the trend of the telecommunications is to become more and more asynchronous and with higher data rates in the downlink than in the uplink. This is useful for services like Web browsing, Data Base Access, Multimedia, etc. To obtain this, it is important to transfer the complexity as much as possible to the BS, where the power processing and electrical power available is higher, so as to combat the several sources of interference.

With the aid of pre-distorting the signals to be transmitted by the BS, the orthogonality between the signals seen by the different users can be improved. The proposed PC scheme is done taking into account the spreading sequences, transmitted symbols and the CIR of the several users present in the cell. The current approach considers a RAKE receiver in the MS. Moreover, we consider a combination of PC with the STD.

This paper is structured as follows: section 2 presents the system model for the proposed Precoding; section 3 presents the system model for the Selective Transmit Diversity; section 4 presents the performance results and analysis for the proposed schemes; the key findings of this paper are then summarized in section 5.

J.N. de Souza et al. (Eds.): ICT 2004, LNCS 3124, pp. 908–913, 2004.

2   System Model for the Precoding

Multi-user detection has been investigated mostly from the viewpoint of receiver optimization. However, in the downlink situation, the MS usually does not have access to information needed to perform the Multi-user Detection (MUD). Such information consists of spreading sequences, CIR and the symbols transmitted by different users. Additionally, the processing capability and power energy at the MS is limited. For these reasons, we propose to transfer the complexity from the MS to the BS, and hence, we implement the PC scheme. The transmitted signal is $x(t) = s^T(t)\mathbf{TAb}$, where $\mathbf{T}$ is a *(2M+1)K X (2M+1)K* matrix to be chosen according to some optimality criterion. Therefore, with PC, the vector of MF output is given by $\mathbf{y} = \mathbf{RTAb} + \mathbf{z}$.

Thus, the optimum PC transformation $\mathbf{T}$, which minimizes the bit error rate (BER), is derived in [1] as $\mathbf{T} = \mathbf{R}^{-1}$. This makes $\mathbf{y} = \mathbf{Ab} + \mathbf{z}$.

Assuming that the signal of each user is subject to frequency selective multipath fading, the received signal at the pth receiver can be expressed as [1]:

$$r_p(t) = S_p(t,b) + n_p(t) \tag{1}$$

$$S_p(t,b) = \sum_{i=-M}^{M} \sum_{k=1}^{K} \sum_{l=1}^{L-1} b_k(i) A_k(i) \alpha_p^l(i) s_k(t - iT_b - lT_c) \tag{2}$$

being T_c the chip duration, $b_k(i)$ the transmitted symbol of the kth user, $k=1,...,K$, $i=-M,...,M$, $A_k(i)$ is the amplitude of the kth signal in the ith symbol interval. $\alpha_p^l(i)$ represents the gain of the channel of the lth multipath component during the ith symbol interval over the channel between the transmitter and the pth receiver. When we consider $(L-1)T_c < T_b$ the window should have size 3, i.e., $M = [-1,0,1]$. Thus, the output of the RAKE receiver comes:

$$y_p(i) = \sum_{n=0}^{L-1} \int_{iT_b + nT_c}^{(i+1)T_b + nT_c} \alpha_p^n(i) r_p(t) s_p(t - iT_b - nT_c) dt \tag{3}$$

To make this notation compact, we define a $(2M+1)K \times (2M+1)K$ $\mathbf{R}(l-n)$ matrix as in [1]:

$$\mathbf{R}(l-n) = \begin{bmatrix} \mathbf{R}(0,l-n) & \mathbf{R}(-1,l-n) & \cdot & & \cdot & & \cdot \\ \mathbf{R}(1,l-n) & \mathbf{R}(0,l-n) & \mathbf{R}(-1,l-n) & \cdot & & \cdot \\ \cdot & \mathbf{R}(1,l-n) & \mathbf{R}(0,l-n) & \mathbf{R}(-1,l-n) & \cdot & & \cdot \\ \cdot & & \mathbf{R}(1,l-n) & \mathbf{R}(0,l-n) & \cdot & & \cdot \\ \cdot & & \cdot & \mathbf{R}(1,l-n) & \cdot & & \cdot \\ \cdot & & \cdot & & \cdot & & \mathbf{R}(-1,l-n) \\ \cdot & & \cdot & & \cdot & \mathbf{R}(1,l-n) & \mathbf{R}(0,l-n) \end{bmatrix} \tag{4}$$

where, due to the size window 3, $\{\mathbf{R}(l-n)\}_{i,j} = 0$ for every i,j, such that $|i-j| > 1$. Additionally, we define $\mathbf{R}(m,l-n)$ as a $K \times K$ matrix whose values are:

$$R_{p,k}(m,l) \int_{-\infty}^{+\infty} s_p(t) s_k(t - mT_b - lT_c) dt \qquad (5)$$

Assuming that multipath intensity profiles do not change over the block length, we define [1]:

$$\mathbf{A} = diag\left[A_1(-M),..., A_K(-M),..., A_1(M),..., A_K(M) \right]_{(2M+1)K \times (2M+1)K}$$

$$\alpha(l) = diag\left[\alpha_1^l(-M),..., \alpha_K^l(-M),..., \alpha_1^l(M),..., \alpha_K^l(M) \right] \qquad (6)$$

and $K(2M+1) \times 1$ vectors such that:

$$\mathbf{b} = \left[b_1(-M)^T,..., b_K(-M)^T,..., b_1(M)^T,..., b_K(M)^T \right]^T \qquad (7)$$

$$\mathbf{z} = \left[z_1(-M)^T,..., z_K(-M)^T,..., z_1(M)^T,..., z_K(M)^T \right]^T$$

Hence, we express the vector of the RAKE receiver outputs at all K sites as:

$$\hat{\mathbf{y}} = \left[\mathbf{y}(-M)^T,..., \mathbf{y}(M)^T \right]^T \qquad (8)$$

$$= \sum_{n=0}^{L-1} \sum_{l=0}^{L-1} \alpha(n) \alpha(l) \mathbf{R}(l-n) \mathbf{A} \mathbf{b} + \mathbf{z}$$

where $\mathbf{y} = \{y_1,..., y_K\}^T$.

Making $\mathbf{R} = \sum_{n=0}^{L-1} \sum_{l=0}^{L-1} \alpha(n) \alpha(l) \mathbf{R}(l-n)$ the matrix $\mathbf{R}$ can, now, be interpreted as the cross-correlation matrix for an equivalent synchronous AWGN problem where the whole transmitted sequence is considered to result from $(2M+1) \times K$ users, during one transmission interval of duration $T_c = (2M+1) \times T_b$. The matrix $\mathbf{R}$ is not symmetric anymore. The vector of decision statistics becomes now $\mathbf{y} = \mathbf{RTAb} + \mathbf{z}$. Now, a multiuser detection problem in multipath channels is reduced to K decoupled single user detection problems.

3   System Model for the Selective Transmit Diversity

The STD scheme has a low rate (750 Hz-3 kHz) feedback link from the receiver (MS) telling the transmitter (BS) which antenna should be used in transmission. There is a common or dedicated pilot sequence, which is transmitted. Different antennas with specific pilot patterns/codes enable antenna selection. Then, the transmitter has a link quality information about the M (number of antennas) links and it transmits a single symbol stream over the best antenna. The receiver is supposed to re-acquire the carrier phase after every switch between antennas. The Antenna Switch has the capability to switch every slot duration, i.e., 0.667 ms (rate 1.5 kHz). After using STD from

M antennas over the L-paths, each with equal average energy, then we have changed the probability density function for each link. The BER analytical expression is described in [5]. The decision for the antenna selection of the STD combined with the RAKE receiver (maximize multipath diversity) is done in order to select the antenna $i \in \lfloor 1,...,M \rfloor$ which maximizes the multipath diversity, i.e., maximizes the sum of the square of the absolute values of the paths gains, which corresponds to maximize $\max \left(\sum_{l=1}^{L} |h_{i,l}|^2 \right)$.

4   Results and Discussion

All the simulation results presented in this section considered the QPSK modulation in a frequency selective Rayleigh fading channel, namely the Vehicular A propagation model of 3GPP. This propagation models was selected to evaluate the effect of IPI in the performance of the proposed PC scheme. Thus, the Vehicular A propagation model presents a high number of paths, whose average power profile is modelled as [0.4874 0.3134 0.0905 0.0446 0.0397 0.0090 0.0055 0.0099], with a maximum delay spread correspondent to seven chip periods.

The Walsh-Hadamard spreading sequences were considered in the simulation, with spreading factor 16 and 64. Positions of interfering MSs were defined randomly using a uniform distribution inside a cell, as well as the correspondent CIRs for different users. If we consider all users co-located with the reference user, i.e., same CIRs for all users, performances are better than those presented in this paper. As can be seen inFig. 1 for SF=64 and 15 users, an increase in the order of diversity to $2L$ (where L corresponds to the number of multipaths) with the use of the STD, corresponds to an improvement of performance. However, due to the presence of MAI, we can see that the improvement of performance with the increase of E_b/N_o is low. This was the main motivation to add the proposed PC scheme to the STD, whose results correspond to a gain over the STD alone. The use of the PC corresponds to an appreciated performance improvement over the "Only RAKE", except for low values of E_b/N_o, where the noise enhancement generated by the decorrelator type precoding reaches a high value. The PC is able to partly "clean" the MAI which was generated due to the presence of the multipath environment. Additionally, the best performance is achieved with the combined scheme composed of STD and PC.

The results of Fig. 2a) are similar to the previous one, except that we now consider SF=16. Similar results were obtained, with the exception that, in this case, the performance obtained with STD+PC only outperforms the STD for values of E_b/N_o below 11 dB. There are two steps of performance: the first corresponds to the system without any kind of PC (with and without STD), and hence, with a high level of MAI; the second corresponds to the system in which the PC was used to remove part of the MAI, which originates a step in the performance.

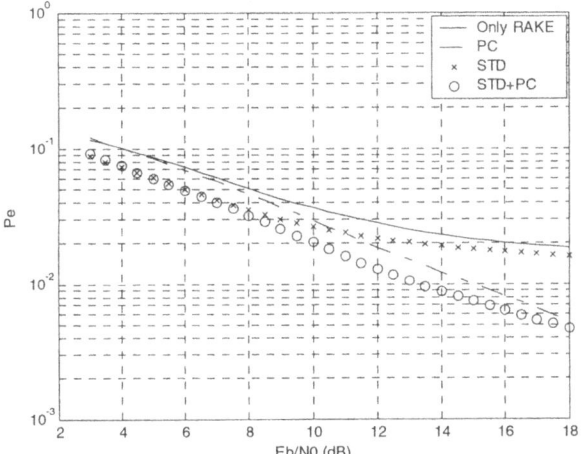

Fig. 1. BER performance for 15 users, SF=64

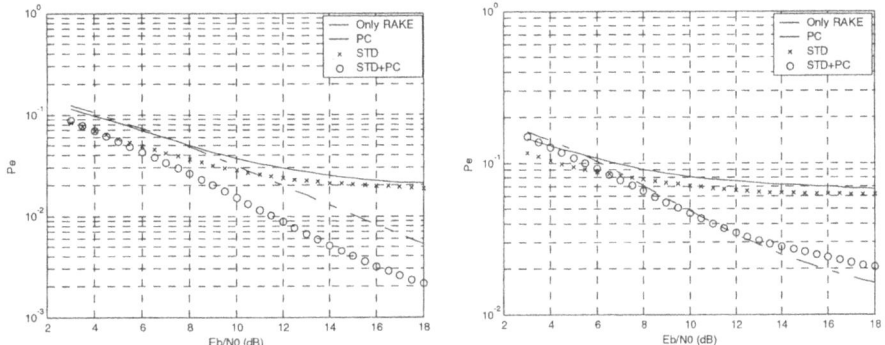

Fig. 2. BER performance for SF=16, with a) 15 users and b) 4 users

Due to lower level of MAI in Fig. 2b), the gain obtained with the PC (relating to "Only RAKE") is lower than that of Fig. 2a). The performance improvement of the STD+PC relating to STD is much higher than the performance improvement of the PC relating to the "Only RAKE". This confirms that the system considering the STD+PC is able to achieve a performance with very low level of bit errors. With four users, the gain obtained with STD+PC relating to "Only RAKE" is much higher than with 15 users. This happens once the system is "cleaner" of MAI and, hence, the STD is able to reach a better performance.

5   Conclusions

In this paper we have studied the combination of two specific schemes, namely the precoding scheme with the selective transmit diversity for the downlink, through the Vehicular A propagation models of 3GPP. When the level of MAI was more intense we observed that, even with the use of the STD, the performance obtained was limited. This was the main motivation to add the PC to the STD scheme. It was shown that the use of the PC combined with the STD tends to reach a performance improvement over the corresponding schemes isolated. This tends to happen because the PC "cleans" the MAI and, hence, the diversity provided by the use of the STD is able to achieve better results. This confirms that the combination of STD and the PC fit very well for the downlink 3GPP.

Our work considered the vehicular A propagation model which presents higher number/level of multipath diversity, and consequently, originates higher level of MAI. However, once most of propagation models of 3GPP present an order/level of multipath diversity similar to the Vehicular A propagation model, we should conclude that the combination of STD with PC tends to achieve better results in several propagation environments.

Acknowledgements. This work has been partially funded by the IST-2003-507607 EU _Broadcasting and Multicasting Over Enhanced UMTS Mobile Broadband Networks_ project.

References

1. B. R. Vojcic, W. M. Jang, "Transmitter Precoding in Synchronous Multiuser Communications", IEEE Transactions on Communications, Vol. 46, n.°10, October 1998, pp. 1346-1355.
2. J. Choi, "Interference Mitigation Using Transmitter Filters in CDMA Systems", IEEE Transactions on Vehicular Technology, n°4-Vol.51, July 2002, pp. 657-666.
3. M. M. Silva, A. Correia, "Interference Cancellation with combined Pre-distortion filtering and Transmit Diversity", 14th IEEE International Symposium on Personal Indoor and Mobile Radio Communications, Beijing, China, 7-10 September 2003.
4. M. M. Silva, A. Correia, "Joint Multi-User Detection and Intersymbol Interference Cancellation for WCDMA Satellite UMTS", Internation Jounal of Satellite Communications and Networking – Special Issue on Interference Cancellation – John Willey and Sons, Ltd, n.° 21, 2003, pp. 93-117.
5. A. Correia, "Optimized Complex Constellations for Transmission Diversity", Wireless Personal Communications 20, Journal of Kluwer Academic Pub, Mar. 2002, pp. 267-284.

Application of a Joint Source-Channel Decoding Technique to UMTS Channel Codes and OFDM Modulation*

Marion Jeanne, Isabelle Siaud, Olivier Seller, and Pierre Siohan

France Télécom R&D, DMR/DDH, Site de Rennes, 4 rue du Clos Courtel,
B.P. 59, 35512 Cesson-Sévigné Cedex, France
{firstname.lastname@francetelecom.com}

Abstract. This paper describes the application of a joint source channel decoding technique (JSCD) of variable length codes (VLCs), presented at first by Guivarch et al., in the context of future public land mobile telecommunication systems (FPLMTS) involving UMTS channel codes and OFDM modulation. OFDM parameter sets have been adjusted both to the UMTS chip rate and to the selectivity of the urban multipath propagation channel. The application of JSCD to UMTS codes may lead to signal to noise improvements greater than 3 dB. A methodology is also proposed to get soft error patterns attached to the transmission system, including the OFDM modulation and the multipath channel. Thus, several JSCD schemes can be accurately evaluated thanks to an equivalent simplified model.

1 Introduction

One of the main objectives of FPLMTS is to provide high capacity mobile radio systems allowing the access to various types of high data services, as for instance video. However, to be implemented at a user level, the computational complexity and delay involved by these new systems and services have to be relatively low. In the last few years, it has been shown that JSCD techniques could be the right approach to improve the transmission performance given these complexity and delay constraints, see for instance a tutorial introduction by Van Dyck and Miller [1]. In this field, the most recent studies have focused on JSCD methods that could handle the widely spread case of a variable length encoding of the source, as it exists for instance in video encoders. However, most of these techniques have been evaluated with a simple additive white gaussian noise (AWGN) channel [2], [3], and, in some cases on a Rayleigh fading channel [4]. In this paper, the first objective is to evaluate the JSCD proposed in [3] through a mobile radio channel involving UMTS channel codes and an OFDM modulation scheme. Indeed, recent 3GPP investigations on the OFDM modulation for upcoming downlink UMTS-standards on adjacent UMTS-bands [5] have shown its relevance to face the problem of multipath propagation in urban areas. Thus, from a wideband selectivity parameters analysis carried out on simulated outdoor ITU-R models [6] and previous France Telecom measurements [7], we have selected the appropriate ITU-R propagation models. In accordance with

---

* This work was supported in part by the French ministry of research under the contract named COSOCATI (http://www.telecom.gouv.fr/rnrt/index_net.html).

J.N. de Souza et al. (Eds.): ICT 2004, LNCS 3124, pp. 914–923, 2004.
© Springer-Verlag Berlin Heidelberg 2004

the UMTS-chip rate, two low-complexity OFDM dimensionings fit with micro and small cell areas, respectively, are defined afterwards. A second objective of our work is to provide a methodology to get soft error patterns files attached to the transmission system. Consequently, a unique soft error pattern can be used to test several JSCD schemes without having to simulate the modulation/demodulation set-up. Our paper is organized as follows. In section 2 we present the OFDM transmission system. The generation and validation of the soft error pattern are detailed in section 3. Finally, section 4 reports JSCD performance with convolutional codes and turbo codes.

2   The Transmission System

2.1   Propagation Modelling

Wideband tapped delay line models consist in a discrete version of the scattering function of the propagation channel under wide sense stationary uncorrelated scatterers (WSSUS) assumptions. Hence, the scattering function $P_s(\nu, \tau) = |S(\nu, \tau)|^2$ is a simplified version of the correlation function $R_s(\nu, \nu', \tau, \tau')$ where $S(\nu, \tau)$ is the Doppler-variant channel impulse response of the channel [8] (ν: Doppler frequency, τ: delay)

$$P_s(\nu, \tau) = \sum_{i,j} |S_{i,j}(\nu, \tau)|^2 \delta(\tau - \tau_i)\delta(\nu - \nu_j). \tag{1}$$

The average power delay profile (APDP), $P(\tau)$, also depends on the delayed and attenuated taps (τ_i, A_i):

$$P(\tau) = \sum_i \delta(\tau - \tau_i) \sum_j |S_{i,j}(\nu_j, \tau)|^2. \tag{2}$$

The excursion of the Doppler power spectrum $P_s(\nu, \tau_i)$ is identical for each tap and is given by the Clarke model [8]:

$$P_s(\nu, \tau_i) = \frac{A_i}{4\pi\nu_{max}} \frac{1}{\sqrt{1 - (\frac{\nu}{\nu_{max}})^2}}. \tag{3}$$

The ITU-R tapped delay line model [6] selection representative of an urban area (micro and small cells) is based on a comparison between urban measured propagation channels and filtered tapped delay line models (equivalent band for the comparison). A micro-cell (MC) environment is typical of a dense urban area with the base station (BS) height below the average rooftop-height of buildings while a small cell (SC) corresponds to BS height slightly higher than the rooftop-height of buildings [9]. We have evaluated the delay spread and the delay window set to 95% (W95%), after a pre-filtering processing (in a 25 MHz bandwith and a roll-off equal to 0.61) applied on ITU-R models. The coherence bandwidth (B_c−0.5) has been deduced from the OFDM dimensioning (section 2.2) used to sample the average frequency correlation function. Table 1 provides wideband selectivity parameters of outdoor ITU-R tapped delay line models and France Telecom R&D measurements carried out in MC and SC environments.

   SC(1) and SC(2) scenarios are connected to SC cases with a distance range inferior to 500 m. while MC(1) and MC(2) describe MC cases covering up a 200-300 m distance range. Results show that the frequency selectivity significantly increases with the BS

Table 1. Wideband selectivity parameters in urban area

1-ITU-channel	Outdoor to Indoor and Pedestrians		Vehicular test Environment	
	A	B	A	B
delay spread (μs)	0.045	0.633	0.371	4.00
delay window 95% (μs)	0.20	2.35	1.16	12.94
($B_c - 1/2$) MHz	0.781	0.234	0.351	0.390
2-FT R&D channels	SC(1)	SC(2)	MC(1)	MC(2)
delay spread (μs)	0.44	0.673	0.186	0.271
delay window 95% (μs)	1.533	1.803	0.696	1.163

height. The delay spread for SC is twice as large as in MC. Table 1 and [7] show that the pedestrian A is related to large indoor areas and vehicular B to urban MC environments. Table 1 also confirms that the vehicular test environment A is rather attached to MC [7], while pedestrian B shall be related to either very selective macro-cell or SC environments. Hence, in the following, we select the vehicular test environment channel A to simulate a typical MC case and the pedestrian channel B environment to simulate a typical SC case. Relative delays and attenuations are reported in Table 2.

Table 2. Test Environments

Vehicular Test Environment (channel A)						Pedestrian Test Environment (channel B)					
Tap	Delay (μs)	Power (dB)	Tap	Delay (μs)	Power (dB)	Tap	Delay (μs)	Power (dB)	Tap	Delay (μs)	Power (dB)
1	0.0	0.0	4	1.09	-10.0	1	0.0	0.0	4	1.2	-8.0
2	0.31	-1.0	5	1.73	-15.0	2	0.20	-0.9	5	2.30	-7.8
3	0.71	-9.0	6	2.51	-20.0	3	0.80	-4.9	6	3.70	-23.9

2.2 OFDM System Parameters

The conventional OFDM modulation set-up results in N_u overlapped sub-carriers with a transfer function of the k-th sub-carrier given by:

$$G_k(f) = T_u \delta(f - f_k) e^{j2\pi f T_u} \text{sinc}(\pi f T_u) \qquad (4)$$

where T_u denotes the useful symbol duration. As shown in Fig. 1, transmitted QPSK data symbols are serial to parallel (S/P) mapped through N_u branches modulating N_u sub-carriers. The OFDM signal is obtained by an inverse fast Fourier transform (FFT) of length N_{fft}, with $N_{fft} > N_u$. A cyclic prefix (CP) of duration Δ is inserted at the beginning of each OFDM symbol to cancel intersymbol interference between successive OFDM symbols. The time-variant multipath propagation channel filters data. The received OFDM signal is the sum of a filtered version of the OFDM transmitted signal and of an additional Gaussian noise.

After removal of the CP, the OFDM demodulation, operated by an FFT, exploits the orthogonality between sub-carriers to estimate incoming data symbols. A perfect channel estimation is assumed. The OFDM parameter definition has then consisted in adjusting Δ to the largest delay excursion trying also to minimize N_{fft}. Denoting by T_e the sampling period we, firstly, impose that: $T_u = p \cdot \Delta = N_{fft} T_e$.

Secondly, by refining the parameters p and N_u, we have adjusted the bandwidth B_w to the UMTS channelization in order to get a net bit rate (D_u) similar to the UMTS chip rate. At the end, in order to limit the spectrum efficiency loss (SEL), we have got $p = 8$. Denoting by R_c the code rate, corresponding to the channel code used for JSCD, and m the number of bits per subcarrier, we get

$$B_w = \frac{N_u}{T_u} = \frac{N_u}{p\Delta}, SEL = \frac{p}{p+1}, D_u = \frac{mR_cN_u}{(p+1)\Delta}. \tag{5}$$

Table 3 summarizes the OFDM parameter sets. They are slightly different for both environments along an accurate CP refinement intended to reduce the FFT size for each scenario. The parameter SEL is kept constant in order to allow a bit error rate (BER) comparison between MC and SC environments.

Table 3. System parameters

	Vehicular A	Pedestrian B
N_u: # of sub-carriers	112	145
ν_{max}: Doppler excursion (Hz)	89.6	89.6
$T_s = T_u + \Delta$: OFDM symb. duration (μs)	28.8	37.65
$\Delta = T_u/p(\mu s)$: Cyclic prefix duration	3.2	4.2
B_w: trans. bandwidth (MHz)	4.37	4.33
F_e: Sampling rate (MHz)/N_{fft}: FFT size	5.62/144	5.71/191
T: Interleaver depth (ms)	14.45	14.45
D_u: Net bit rate (Mbps) ($R_c = 1/2, 1/3$)	(3.89, 2.59)	(3.85, 2.56)
$-SEL$ (dB)	0.51	0.51
Sub-carrier modulation	QPSK	QPSK

To test our JSCD technique, convolutional codes (CC) and turbo codes (TC) have been implemented. A pseudo-random binary interleaving (Π) has been performed to cope with fast fading. The interleaving depth T, is much larger than the inverse positive Doppler excursion. An additional inner interleaving (frequency interleaving, not reported in Fig. 1) has also been applied within every OFDM symbol.

2.3 The JSCD Technique

In [3], [10], [11], the authors propose a method to provide a low complexity JSCD of VLCs used in conjunction with convolutional or turbo codes. The key idea is to improve the channel decoding by using, at a bit level, an *a priori* source information. Assuming source symbols probabilities are known at the receiver, it is then possible to compute, for each bit associated with a VLC tree branch, a source bit probability. By keeping a one-to-one correspondence between the VLC tree and the channel code trellis, these *a priori* bit probabilities can be inserted in the metrics of the channel decoding algorithm to improve the receiver performance. In the case where CC are used, the channel decoding is based on a sequence *maximum a posteriori* (MAP) algorithm. When TC are employed, it is a bit to bit MAP decoder that we are looking for. However, in this last case, in order to keep the complexity low when the *a priori* source information is taken into account, a SUBMAP [12] is used instead of a BCJR [13]. As in most JSCD techniques applied on VLC, packetization is carried out to limit the error propagation phenomena at some point. In [3], [10], [11], JSCD performances are given on the AWGN channel for different source statistics and different channel codes. Here, we also evaluate its performances over a time-variant multipath channel with an OFDM modulation.

3   Soft Error Pattern Principle

The proposed JSCD method may only involve a slight modification in the computation of the channel code metrics at the decoder side. That means no additional delay and only a small extra computational cost. However its simulation over a multipath channel using OFDM modulation may be time consuming. Hence, it may be convenient to get the soft encoded metrics corresponding to long sequences of transmitted bits over the overall channel, including modulation. Then different configurations of source coding (with VLCs) and channel coding can be tested simply using the resulting soft error pattern file. Such procedures, leading to hard error patterns, are well known in the case of hard input-hard output channels. Here, we propose an extension leading to the concept of soft error pattern file, *i.e.* a file that contains real values representing all components between the channel coding output and the channel decoding input. Thus, for a given modulation scheme, a soft error pattern describes symbol decision errors on the encoded bits due to multipath degradations. In the following, the generation method, its use and validation, are illustrated in the case of two test environments, named vehicular A and pedestrian B (see Table 2).

3.1   Generation and Use

Soft error pattern generation is deduced from the global transmission scheme depicted in Fig. 1. For each carrier-to-noise ratio (C/N), a soft error pattern file shall be generated and fed to various codec structures.

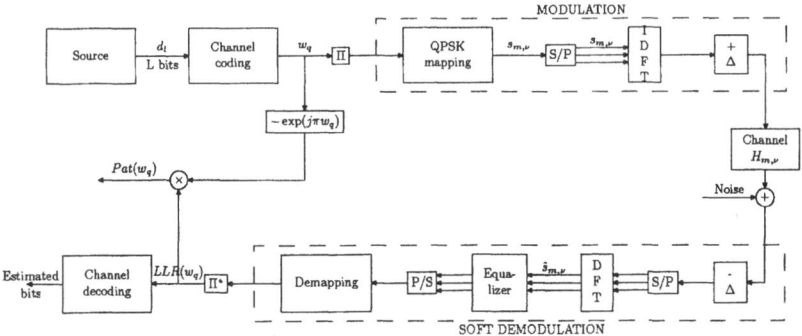

Fig. 1. OFDM transmission scheme through a time-variant multipath channel with soft error pattern generation.

The log likelihood ratios (LLRs) at the deinterleaver (Π^*) output can be modelled by a sequence of real random variables (RVs). As a QPSK mapping is used, each coded bit after interleaving undergoes the same channel degradation. Then, the RVs can be viewed as identically distributed. We note LLR for one RV of this sequence. For another mapping, one should consider as many RVs as classes of LLRs with identical probability mass function (pmf). For example, for the 16 quadrature amplitude modulation (QAM) mapping, one should consider 2 RVs. We denote by w a uniformly distributed discrete RV with alphabet $\{0, 1\}$. It represents a bit at the channel coding output. As the QPSK mapping and the channel are symmetric, the pmf can be written as:

$$P(LLR|w = 0) = P(-LLR|w = 1), P(LLR|w = 1) = P(-LLR|w = 0). \quad (6)$$

Similarly to the LLRs, the soft error pattern can be modelled by a sequence of real and identically distributed RVs. We note Pat as one of them, and we define it, as shown in Fig. 1, by: $Pat = -LLR \exp(j\pi w)$. Then, its pmf is such that: $P(Pat|w = 0) = P(-LLR|w = 0)$ and $P(Pat|w = 1) = P(LLR|w = 1)$. With equation (6), we get $P(Pat|w = 0) = P(Pat|w = 1) = P(Pat)$. Hence, the random variable Pat is independent of the transmitted bit. As illustrated in Fig. 2, to use the soft error pattern with a new source, we set: $LLR' = -Pat \exp(j\pi w')$. w' is a RV associated with a bit at the new channel code output and LLR' is the RV associated with the new LLR.

Then, the pmf of the RV LLR' is: $P(LLR'|w' = 0) = P(-Pat)$ and $P(LLR'|w' = 1) = P(Pat)$, since Pat is independent of w'.

Finally we get:

$$P(LLR'|w' = 0) = P(LLR|w = 0), P(LLR'|w' = 1) = P(LLR|w = 1). \quad (7)$$

These two last equations mean that for the two considered transmission schemes, the one in which all the components between channel coding and decoding are present and the one where they are replaced by the pattern, the LLRs have an identical pmf. Hence, these two schemes are equivalent. A soft error pattern is then a suitable way to represent the OFDM modulation and demodulation and the channel degradation. It is associated to a number of simulated bits, a C/N ratio, a given interleaver and a mapping. Its statistics do not depend on the channel and source codes used to generate it.

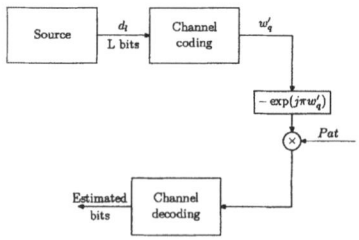

Fig. 2. Application of the soft error pattern.

3.2 Validation

To validate the soft error pattern process, we have compared the BER of the two systems: the overall transmission scheme presented in Fig. 1 and the simplified one proposed in

Fig. 2, for several ratios of the useful bits energy over noise (E_b/N_0). The simulations have been run on the vehicular A channel using an *i.i.d* source and with CC of constraint length K equal to 9 for the UMTS channel codes of rate 1/2 and 1/3 [14], and to $K = 5$ for another CC. The octal representations of these CC are $(561, 753)_o$ and $(557, 663, 711)_o$ for the UMTS codes and $(31, 27)_o$ for the third CC. In Fig. 3 a) and b) the labels indicate, in first position, the channel code being used and in second position, when applicable, the CC used for generating the soft pattern files. This second indication also means that the results have been obtained running the simplified scheme with this CC. Otherwise the BER results from a simulation carried out using the overall OFDM transmission scheme.

In Fig. 3 a) it can be seen that we get similar results with the overall scheme and the simplified one, independently of the code rate used to generate patterns. Another validation is proposed in Fig. 3 b), this time with a pattern built using a UMTS CC ($K = 9, R_c = 1/2$) different from the CC ($K = 5, R_c = 1/2$) used for encoding the *i.i.d* source. Again the curves are superimposed showing that the pattern files accurately represent the vehicular A channel. A similar result, not reported here, also holds for the pedestrian B channel.

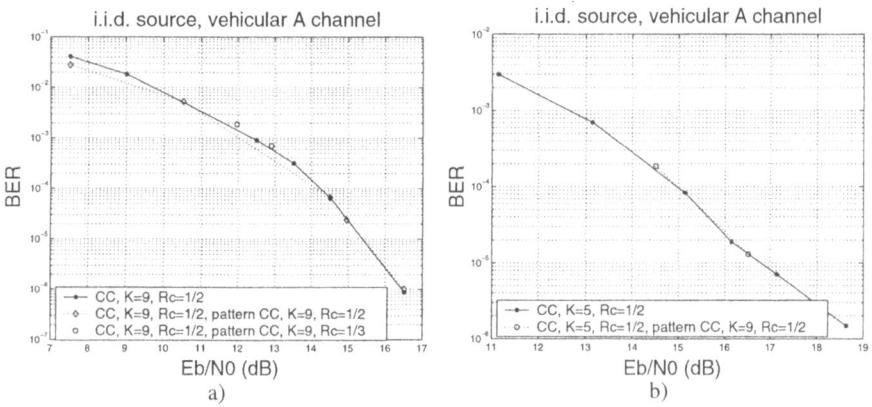

Fig. 3. "Soft error pattern" validation.

4   Results

4.1   JSCD Performance with a Convolutional Code

In this section, the JSCD algorithm [3] recalled in section 2.3 is evaluated using the soft pattern files generated by the multipath OFDM radio transmission depicted in Fig. 1. We used the Huffman encoded first order Markov source proposed in [2]. The stationary and transition probabilities of this 3-symbol source are given in [11], its residual redundancy, defined by the difference between the VLC encoder's code rate and the entropy rate of the source, is 0.67. To limit the error propagation at a certain point, we have used resynchronization, packetizing data into blocks of 256 symbols. Our simulations correspond to the transmission of 5×10^7 coded bits, which is also the size of the soft error pattern file.

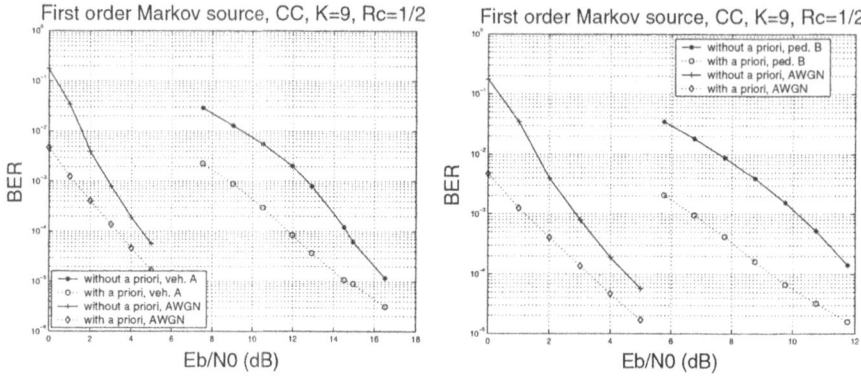

Fig. 4. BER with and without *a priori* source information for the first order Markov source of Murad and Fuja and a convolutional code $K = 9$, $R_c = 1/2$. Vehicular A channel on the left, pedestrian B channel on the right.

In a classical, or tandem, scheme the channel decoder does not use any *a priori* source information, while, as explained in subsection 2.3, for JSCD the *a priori* used are the VLC tree structure and the source statistics. In Fig. 4 these two approaches are compared, in terms of BER as a function of the E_b/N_0, for the vehicular A and pedestrian B environments. These displays also include, as a matter of reference, our simulation results on the AWGN channel. The pedestrian B channel presents better performance than the vehicular A channel with a gain near 2 dB although the pedestrian B channel is the most frequency selective channel. This result shows that the OFDM dimensioning exploits the frequency selectivity of the propagation channel in an efficient way. For a BER of 10^{-4}, a gain of 3 dB, by using the JSCD technique, can be observed on both the vehicular A channel and the pedestrian B channel. Both displays also illustrate the fact that, compared to a tandem decoding, the impact of JSCD is still more important for mobile radio channels than for AWGN ones.

4.2   JSCD Performance with Turbo Codes

In a second round of simulations we have used a UMTS turbo code [14]. This TC of rate 1/3 corresponds to a parallel concatenation of two systematic recursive coders, with generators $(13, 15)_o$. Contrary to the previous case, packet size is now given in bits and not in symbols. As we work with VLCs, it means that a packet does not obviously correspond to an integer number of symbols, so some extra bits are added to fill the packet. Here the packet size is chosen as 4096 bits, it is also the size of the internal line-column turbo code interleaver.

In Fig. 5 we show the results obtained at the third iteration of this turbo decoder for the vehicular A and pedestrian B channels. Results obtained with an AWGN transmission are also presented on the two corresponding displays. For a BER fixed to 10^{-4}, the improvement (in dB) obtained when using the *a priori* source information is around 2.4 dB on the pedestrian B channel and around 2.2 dB for the vehicular A channel.

Another set of experiments has been carried out puncturing the TC output in order to get a code rate $R_c = \frac{1}{2}$. In Fig. 6 we can compare JSCD and tandem decoding for

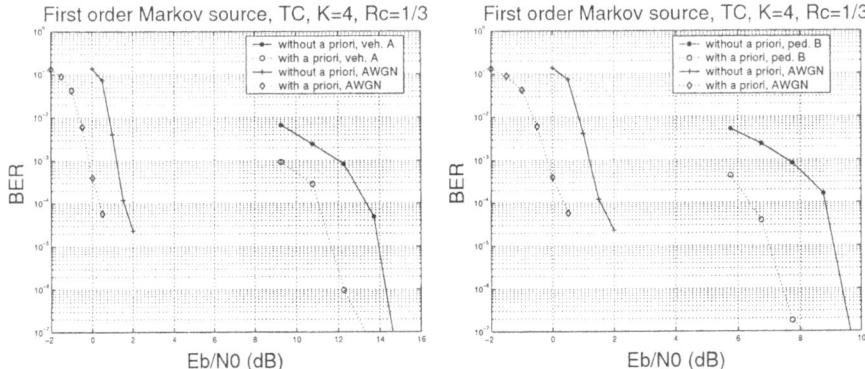

Fig. 5. BER with and without *a priori* source information for the first order Markov source of Murad and Fuja and a turbo code (3rd iteration, $K = 4$, $R_c = 1/3$). Vehicular A channel on the left, pedestrian B channel on the right.

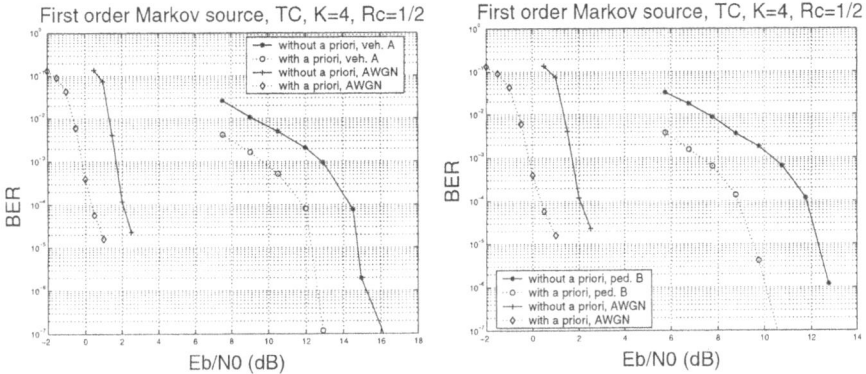

Fig. 6. BER with and without *a priori* source information for the first order Markov source of Murad and Fuja and a turbo code (3rd iteration, $K = 4$, $R_c = 1/2$). Vehicular A channel on the left, pedestrian B channel on the right.

the two types of mobile environment and also in the case of the AWGN channel. For a BER fixed to 10^{-4}, the JSCD improvements in E_b/N_0 are 2.5 dB and 2.9 dB for the vehicular A channel and the pedestrian B channel, respectively. So, it can be noted, at least in this case, that the positive impact of JSCD becomes stronger for lower channel code rates. Note also that, for both TC rates, the JSCD technique still yields more benefit in mobile environments than with the AWGN channel.

5   Conclusion

We have presented an application, in the context of FPLMTS, of a JSCD method for VLCs [3], [10], [11]. The channel modellization is based on ITU-R models [6] and France Telecom measurements [7], it also assumes an OFDM modulation with which dimensioning has been carried out to fit to micro and small-cell areas. Results exhibit an efficient OFDM implementation in favour of a selective propagation channel attached to a small-cell area that should be translated into an extended radio coverage in the face of a micro-cell deployment. A soft error pattern generation is proposed that greatly simplifies the simulation of various JSCD techniques. When using convolutional codes or turbo codes borrowed from the UMTS standard [14], it is shown, with a Markov source [2] that, thanks to our JSCD technique transmission, gains between 2 and 3 dB are obtained.

References

1. Van Dyck, R.E., Miller, D.J.: Transport of wireless video using separate, concatenated, and joint source-channel coding. Proceedings of the IEEE **87** (1999) 1734–1749
2. Murad, A.H., Fuja, T.E.: Joint source-channel decoding of variable length encoded sources. In: Proc. of ITW, Killarney, Ireland (1998) 94–95
3. Guivarch, L., Carlach, J., Siohan, P.: Joint source-channel soft decoding of Huffman sources with Turbo-codes. In: Proc. of DCC, Snowbird, Utah, USA (2000) 83–92
4. Bauer, R., Hagenauer, J.: Iterative source/channel-decoding using reversible variable length codes. In: Proc. of DCC, Snowbird, UT, USA (2000) 93–102
5. Javaudin, J.P., Lacroix, D., Rouxel, A.: Pilot-aided channel estimation for OFDM/OQAM. In: Proc. of VTC, Jeju, Korea (2003) 1581–1585
6. International Telecommunication Union-Radio Study groups: Guidelines for evaluation of radio transmission technologies for imt-2000/fplmts. FPLMTS.REVAL, document 2/29-E (1996)
7. Yuan-Wu, Y., Siaud, I., Duponteil, D.: Radio performance evaluation of the DECT for WPCN based on ITU and recorded channel models. In: Proc. of VTC, Ottawa, Canada (1998)
8. Parsons, J.D.: The mobile radio propagation channel. Pentech-Press (1992)
9. Siaud, I.: A mobile propagation channel model with frequency hopping based on a digital signal processing and statistical analysis of wideband measurements applied in micro and small cells at 2.2 ghz. In: Proc. of VTC, Phoenix, Arizona, USA (1997)
10. Guivarch, L., Siohan, P., Carlach, J.: Low complexity soft decoding of Huffman encoded Markov sources using Turbo-codes. In: Proc. of ICT, Acapulco, Mexico (2000) 872–876
11. Jeanne, M., Carlach, J.C., Siohan, P., Guivarch, L.: Source and joint source-channel decoding of variable length codes. In: Proc. of ICC. Volume 2., New York, USA (2002)
12. Robertson, P., Villebrun, E., Hoeher, P.: A comparaison of optimal and sub-optimal MAP decoding algorithms operating in the log-domain. In: Proc. of ICC, Seattle, USA (1995) 1009–1013
13. Bahl, L.R., Cocke, J., Jelinek, F., Raviv, J.: Optimal decoding of linear codes for minimizing symbol error rate. IEEE Transactions on Information Theory (1974) 284–287
14. 3rd Generation Partnership Project (3GPP): Technical Specification Group (TSG), Radio Access Network (RAN), Working Group 1 (WG1), Multiplexing and Channel Coding (FDD). TS 25 212 (1999)

Transmit Selection Diversity for TDD-CDMA with Asymmetric Modulation in Duplex Channel

Incheol Jeong[1] and Masao Nakagawa[2]

[1] Div. of Computer and Information Science, Sungkonghoe University, 1-1
Hang-dong, Guro-gu, Seoul 152-716, Korea,
jeong@mail.skhu.ac.kr
[2] Dept. of Information and Computer Science, Keio University, 3-14-1 Hiyoshi,
Kohoku-ku, Yokohoma-shi, Kanagawa 223–8522 Japan,
nakagawa@nkgw.ics.keio.ac.jp

Abstract. This paper presents a new transmit antenna diversity technique for TDD-CDMA systems which consists of different modulation schemes between the forward and the reverse link where the system utilizes single carrier modulation for the reverse link and multicarrier modulation for the forward link, respectively. A transmit antenna diversity system which needs a simple detection method at the mobile station is proposed and the system performance is evaluated by computer simulation in frequency selective fading.

1 Introduction

Recently, to achieve robust system performance in frequency selective fading channel multicarrier (MC) modulation has attracted a great deal of attention in various application of communication systems [1]-[4]. On the other hand, antenna diversity is open used to improve the system performance in mobile communications. Even though space diversity gain is achieved by employing diversity antennas, it is difficult to equip multiple antennas at the mobile unit since it needs large space. In order to address this problem, we have proposed a transmit diversity system for single carrier TDD-CDMA in [5]. However, it is hard to apply the transmit antenna diversity directly to the system presented in [6] because the modulation scheme is different on the forward and the reverse link. In [7] a transmit diversity method have been proposed to exploit the transmit diversity in the base station. The system performance has been improved as the number of antennas increased. Nevertheless, for the large number of users the performance degraded rapidly.

In this paper, we propose a new transmit antenna diversity technique for TDD-CDMA with asymmetric modulation addressing the above the problem. To evaluate the proposed system performance computer simulation is carried out in multipath fading channel and the BER results are compared with that of the conventional systems.

---

[1] This work has been done at Keio University

J.N. de Souza et al. (Eds.): ICT 2004, LNCS 3124, pp. 924–931, 2004.
© Springer-Verlag Berlin Heidelberg 2004

2 System Model

The structure of the proposed system is shown in Fig.1. From the figure, we can
see that the modulation scheme is different between the forward and the reverse
link. In the system single carrier modulation is utilized for the revere link and
MC modulation for the forward link, respectively. In both links CDMA is used as
a multiple access method, and multiple antennas are equipped at the base station
to obtain a space diversity gain. In order to estimate the fading channel, pilot
symbol is inserted between the data symbols. Fig.2 show the block diagram of
the mobile station (MS). In the forward link the MS detects the received signals
by using the pilot symbol.

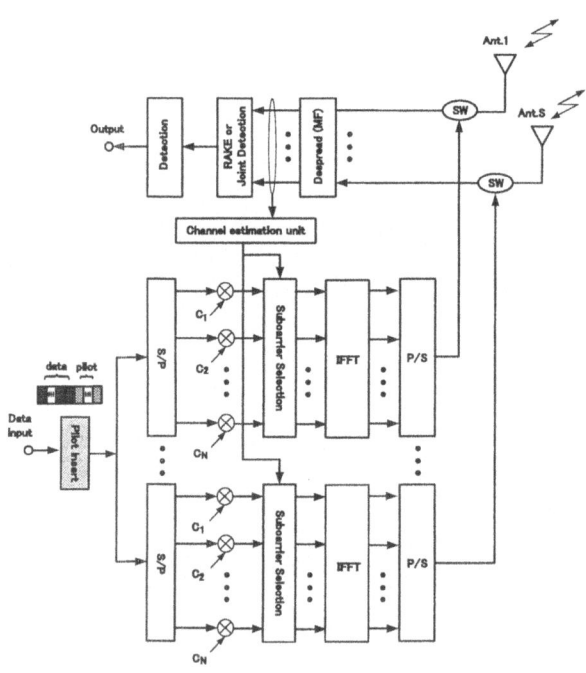

Fig. 1. Base station

2.1 Reverse Link

In the reverse link, the received multipath signals are separated and combined
by a RAKE receiver for each antenna. Parameters including the phase and delays are measured and estimated by channel estimation unit for each antenna.
The channel impulse response of the forward link can be estimated at the base
station (BS) by measuring the received signal in the reverse link period since
the reciprocity of TDD channel is utilized in the system.

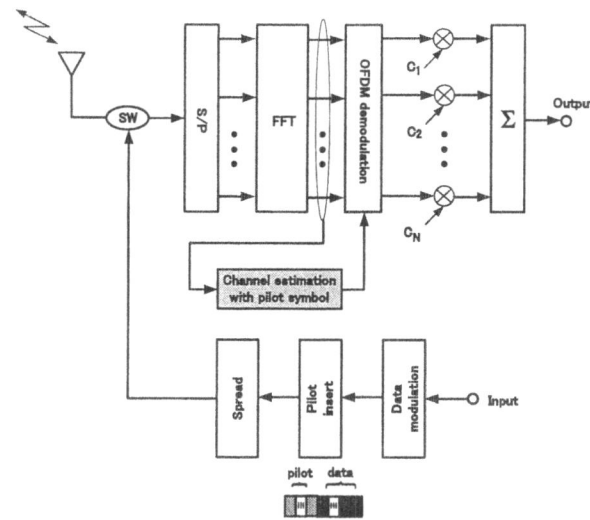

Fig. 2. Mobile station

Then, the channel impulse response is analyzed in the frequency domain by performing a fast Fourier transformation (FFT) for each antenna. It is noticed that the signal of frequency domain can be obtained from the channel impulse response of time domain. This process is also carried out at the channel estimation unit.

Fig.3 shows the principle of the proposed transmit diversity scheme with 2 antennas. Each channel impulse response received by corresponding antenna is Fourier transformed by FFT. By performing this operation, we can obtain the signal level of frequency domain for each antenna. For each antenna the sub-carriers whose signal level is higher than the other antenna's are selected and utilized in the forward link transmission.

2.2   Channel Model

For an impartial comparison, the channel condition is assumed to be equal for all systems. The reverse channels are assumed to be statistically independent for all users. The complex low-pass impulse response of the channel of user k is given for the s th antenna by

$$h_k^s(t) = \sum_{l=0}^{L-1} \beta_{k,l}^s \exp\left(j\gamma_{k,l}^s\right)\delta(t - lT_c) \tag{1}$$

where L is the number of channel paths, the path gains $\beta_{k,l}^s$ are independent identically distributed (i.i.d.) Rayleigh random variables (r.v.'s) for all s, k, and l, the angles $\gamma_{k,l}^s$ are i.i.d. uniformly distributed in $[0, 2\pi)$ and T_c is the spreading code chip duration for the reverse link.

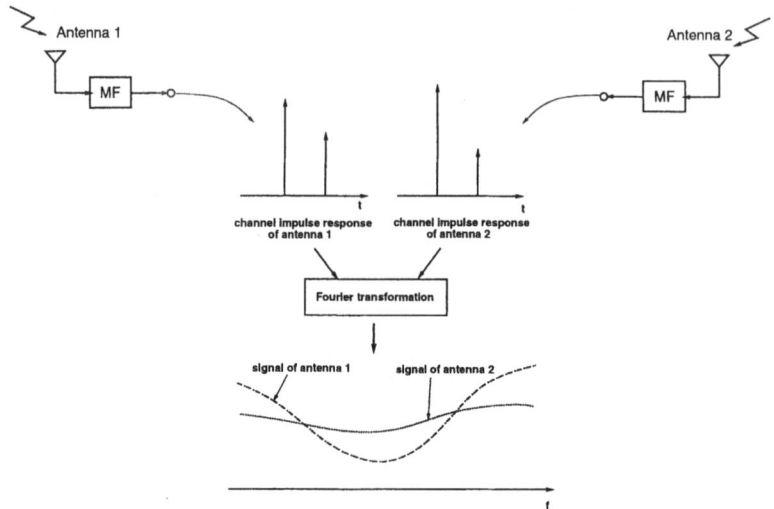

Fig. 3. Frequency analyzing process for each received signal by multiple antennas

2.3   Antenna Selection for Subcarriers

Given a frequency selective fading channel for CDMA with multipath intensity profile $\phi_c(\tau)$, we can find spaced-frequency correlation function $\phi_C(\Delta f)$ of the time-invariant channel by taking the Fourier transform of $\phi_c(\tau)$ [8], where the multipath intensity profile can be obtained by the channel impulse response expressed in Eq.1. Using this characteristic to the proposed system, we can estimate the correlation of fading among subcarriers in the forward link. Taking the Fourier transform of the multipath intensity profile, the frequency domain signal of the s th antenna for user k is given by

$$H_k^s(f) = |H_k^s(f)| \cdot \exp\left[j\theta(f)\right] \tag{2}$$

Assuming that $f_m = f_1 - (m+1)\frac{1}{T}$ where $m = 1, 2, \cdots, M$, $H_k^s(f_m)$ can be write as follows for simplicity

$$H_k^s(f_m) = \rho_{k,m}^s \tag{3}$$

where T is a OFDM symbol duration in the forward link and $\rho_{k,m}^s$ means the path gains of the s th antenna correlated for different m and same k. The fading levels of subcarriers are known from Eq.3 so that the antenna with the highest power can be chosen for each subcarrier by comparing the path gains of the same subcarrier. Thus, we can choose the antennas which transmit the signal of subcarrier m for user k as follows.

$$\rho_{k,m}^{\hat{s}} = \max_{1 \leq s \leq S}\{|\rho_{k,m}^s|\} \tag{4}$$

From Eq.4, the $\hat{s}$ th antenna which has the maximum value is selected for subcarrier m of user k and utilized for the forward link transmission.

2.4   Forward Link

In the forward link, the input data is copied in serial to parallel and then spread in frequency domain with orthogonal code for each antenna. The output of each inverse fast Fourier transform (IFFT), which means multicarrier modulated signal, is then converted in parallel to serial.

The output of IFFT for antenna s is given by

$$s^s(t) = \sum_{k=1}^{K} s_k^s(t) = \sum_{k=1}^{K} \sqrt{2P} b_k(t) \sum_{m=1}^{M} \lambda_{k,m}(s) c_m^k \exp j\omega_m t \tag{5}$$

where

$$\lambda_{k,m}(s) = \begin{cases} 1, & s = \hat{s} \\ 0, & \text{otherwise} \end{cases}$$

Here, P represents the transmitted power per carrier and $b_k(t)$ is the data stream for user k consisting of a train of i.i.d. data bits with duration T which takes the value of ± 1 with equal probability. The data symbol is spread with a spreading code c_m^k. We assume that c_m^k is Walsh Hadamard (WH) code, which is the user specific sequence, where $(k = 1, \cdots, K, \; m = 1, \cdots, M)$, and ω_m is the mth carrier frequency. The orthogonal frequencies ω_m are related by

$$\omega_m = \omega_1 + (m-1)\frac{2\pi}{T}, \text{ where } m = 1, 2, \cdots, M. \tag{6}$$

In the reverse link, several peaks which show multipaths at the output of the despreader (MF) give the phases, amplitudes, and delays. These values are utilized for the antenna selection in forward link transmission. Guard intervals are inserted for the output of the IFFT to prevent inter symbol interference and inter carrier interference. Finally, the combined signals are transmitted to the mobile unit through multiple antennas simultaneously.

For the MC-CDMA system with K users, the received signal at user i during the forward link is given by

$$r_i(t) = \text{Re}\left[\sum_{s=1}^{S} \sum_{j=0}^{L-1} \beta_{i,j}^s s^s(t - jT_c) \exp\left(j\gamma_{i,j}^s\right) \right] + \eta(t)$$

$$= \sum_{k=1}^{K} \sum_{s=1}^{S} \sum_{j=0}^{L-1} \sqrt{2P} \beta_{i,j}^s \cdot b_k(t - jT_c) \cdot \sum_{m=1}^{M} \lambda_{k,m}(s) c_m^k$$

$$\cdot \cos\left(\omega_m(t - jT_c)\right) \cos\left(\gamma_{i,j}^s\right) + \eta(t) \tag{7}$$

where $\eta(t)$ is the zero mean AWGN with two sided power spectral density $N_0/2$. Assuming that user 1 is the desired user, the desired signal at the front of FFT is given by

$$r_1(t) = \sum_{s=1}^{S} \sum_{j=0}^{L-1} \sqrt{2P} \beta_{1,j}^s b_1(t - jT_c)$$

$$\sum_{m=1}^{M} \lambda_{1,m}(s) c_m^1 \cos\left(\omega_m(t - jT_c)\right) \cdot \cos\left(\gamma_{1,j}^s\right). \tag{8}$$

From Eq.8, we can see that the phases of each subcarrier's signal distorted by fading can be adjusted by using the pilot symbol received by different antennas. The detection process can be carried out in frequency domain after FFT. With this process, the mobile unit receives the transmitted signal and also achieves the path and the antenna diversity effect.

As shown in Eq.7, orthogonality between the codes can be maintained since the phases are identical for all users in each subcarrier. Therefore, if we apply an appropriate detection method, we can achieve a good performance by using the proposed system.

3   Simulation Results

Computer simulations is carried out to evaluate the BER performance of the proposed system in the forward link. We assume that the bandwidth is equal on both the forward and the reverse link. The simulation conditions are shown in Table 1. 64 subcarriers are used for the multicarrier modulation. The burst length used in both links is 12 bits which means 0.75 msec long in a 16 kbit/sec TDD. The maximum Doppler frequency f_d is set to 32 Hz.

Table 1. Simulation parameters

Transmission data rate	$16[Kb/s]$
Spreading code	64 Walsh Hadamard
No. of sub-carrier	64
Modulation	BPSK
Detection method	EGC
Maximum delay spread (τ)	977 ns
Maximum Doppler shift	32 Hz
Burst length	12 bits (0.75 ms)
No. of antennas	1, 2

3.1   Multiple Users Environment

Fig.4 shows the BER vs. E_b/N_o for $1 \sim 32$ users case under the condition of 2 paths. Equal Gain Combining (EGC) is used as a detection method. In the figure, we find that the performance does not degrade up to 32 users for both 1 and 2 antennas case (proposed system) and sufficient diversity effect is achieved for the proposed system. From this result it is found that the diversity gain increases as the number of antennas increases and the orthogonality of WH code can be kept up to around this number of users.

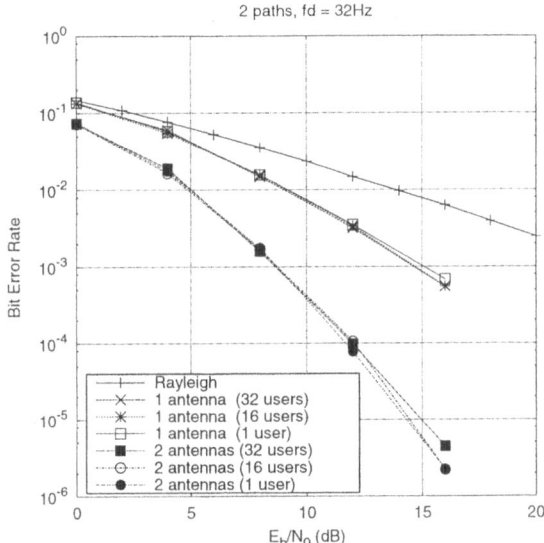

Fig. 4. BER performances for the proposed system under multiple users environment (2 paths)

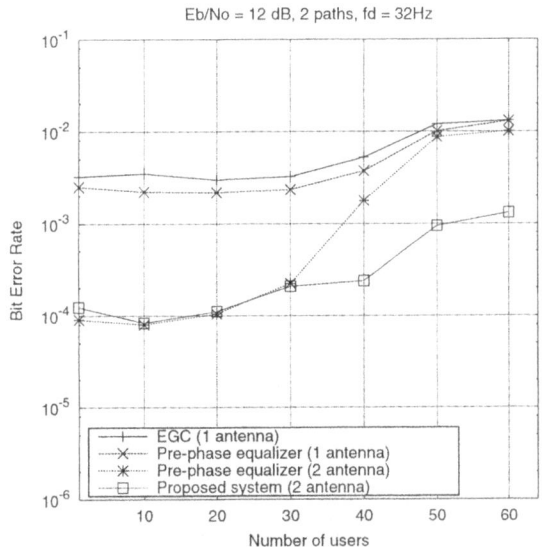

Fig. 5. BER performances for various number of users (2 paths, $E_b/N_0 = 12dB$)

3.2   Various Number of Users

The system performance is also evaluated for various number of users. Fig.5 shows the BER result for the case of 2 paths and E_b/N_0=12dB. As a reference, we show the BER performance for 1 antenna system with EGC detection and the BER performance for the 2 antenna system using pre-phase equalizer shown in [7]. From the figure, it is found that the BER performance of the proposed system is greatly improved for 2 antennas overall number of users and outperforms the other conventional systems.

4   Conclusion

We have proposed a transmit antenna diversity system for TDD-CDMA with asymmetric modulation where the fading channel is estimated by pilot symbol. In spite of the asymmetric structure, the proposed system achieves space diversity gain without needing any reception antenna at the mobile unit. From the simulation results it is found that the BER performance of the proposed system is greatly improved for 2 antennas overall number of users and outperforms the other conventional systems.

References

1. K. Fazel and G. P. Fettweis, eds., Multi-carrier spread-spectrum, Kluwer Academic Publishers, 1997.
2. R. Prasad and S. Hara, "An overview of multi-carrier CDMA," Proc. IEEE ISSSTA'96, pp.107–113, Mainz, Germany, Sept. 1996.
3. E. A. Sourour and M Nakagawa, "Performance of orthogonal multicarrier CDMA in a multipath fading channel," IEEE Trans. Commun., vol.COM-44, no.3, pp.356–367, Mar. 1996.
4. N. Yee, J. Linnartz, and G. Fettweis, "Multi-carrier CDMA in indoor wireless radio networks," Proc. IEEE PIMRC'93, pp.D1.3.1–D1.3.5, Pacifico Yokohama, Yokohama, Japan, 1993.
5. I. Jeong and M. Nakagawa, "A Novel transmission diversity system in TDD-CDMA," IEICE, Trans. commun., vol.E81-B, no.7, pp.1409–1416, July 1998.
6. I. Jeong and M. Nakagawa, "A Time division duplex CDMA system using asymmetric modulation scheme in duplex channel," IEICE Trans. Commun., vol.E82-B, no.12, pp.1956–1963, Dec. 1999.
7. I. Jeong and M. Nakagawa, "Transmit antenna diversity for TDD-CDMA with asymmetric modulation in duplex channel", Proc. IEEE PIMRC'00, vol.1, pp.378–382, London, UK, Sept. 2000.
8. J. G. Proakis, Digital Communications, 3rd Ed., New york: McGraw-Hill, 1995.

Planning the Base Station Layout in UMTS Urban Scenarios: A Simulation Approach to Coverage and Capacity Estimation[*]

Enrica Zola and Francisco Barceló

Wireless Networks Group
Entel Dept., Technical University of Catalonia
c/ Jordi Girona 1-3, 08034 Barcelona (Spain)
enrica@entel.upc.es, barcelo@mat.upc.es

Abstract. This paper analyzes the performance of a base station layout for a UMTS network in a densely populated city. The study is carried out using snapshot simulations of an actual city (Barcelona, Spain) with specific traffic and propagation profiles. A first layout is proposed in order to guarantee good coverage with a minimum number of base stations. This layout is approached taking into account the link budget analytical calculations and the first simulation results for a single-cell environment. A second layout is then presented in an attempt to solve the problems that arise in the first (e.g. shadows, blind zones, etc.). This is carried out by finely tuning parameters from the first layout, adding new base stations and changing the locations of the existing ones.

1 Introduction

A new feature of UMTS is higher user bit rates: 384 kbps on Circuit-Switched (CS) connections, and up to 2 Mbps on Packet-Switched (PS) connections. At the beginning of the UMTS era, most of the traffic will be voice, but the share of data will grow progressively. It is necessary to study the 3G radio network planning carefully, in order to fulfill the requirements for coverage, capacity and quality of service [1].

Planning the Base Station (BS) layout in a city is a complex task since there are many constraints that make it difficult to locate antennas in the desired positions. This paper presents an approach to the layout problem by positioning BSs according to the Link Budget (LB) calculations carried out for the city of Barcelona. In a second step, the specific problems that arise from the first plan are taken into account. Only a long process of trial-and-error, which is beyond of the scope of this paper, can lead to an optimum trade-off between full coverage and low investment in BSs.

The goal of this study is to show the basics of the aforementioned trial-and-error process. Although only two steps are carried out in the paper due to space constraints, this two-steps-only approach gives good results. With regard to capacity planning, simulations show how a slight improvement can be obtained by finely balancing the

---

[*] This research has been funded by the Spanish Ministry of Science and Technology through CICYT project RUBI 2003-01748.

J.N. de Souza et al. (Eds.): ICT 2004, LNCS 3124, pp. 932–941, 2004.

power share in the downlink (DL). However, the benefit obtained in terms of coverage does not carry through in terms of capacity.

The paper is structured as follows. In Section 2, we describe the LB evaluation process. Section 3 introduces the snapshot simulation tool that was used to analyze the network. The simulator is used to verify the coverage and capacity results presented in Section 2. According to the results, the first BSs location plan for the city of Barcelona is drawn in Section 4. Slight modifications to the first plan improve the overall coverage greatly, while capacity problems relating to DL power sharing and interference cannot be solved.

2 Analytical Evaluation of the Link Budget

This section presents UMTS radio network planning including capacity and coverage planning [2, 3]. In the dimensioning phase, the number of BS sites is estimated on the basis of the operator's requirements for coverage, capacity and quality of service. Capacity and coverage are closely related in W-CDMA networks [4] and must be considered simultaneously in the study.

2.1 Coverage Plan

In order to guarantee sufficient coverage and to limit costs by using the lowest possible number of BSs, it is important to position BSs at an optimal distance from each other. One can estimate this distance by computing the maximum losses that the UMTS signal suffers. This first approach to radio network planning is based on the analytical estimation of radio signal propagation. This process is known as the *link budget* [5, 6] and is independent from the simulation results presented in the following sections. Note that the simulation tool carries out its own calculations by means of propagation models that are better suited to the environment analyzed, and are hence more complex. The LB estimation is carried out using the Okumura-Hata propagation model, in which the propagation losses (L_{OH}) in an urban scenario are computed as:

$$L_{OH} = 133.76 + 34.79 \log d, \qquad (1)$$

where d is the distance in kilometers. More details can be found in [7, 8].

Table 1 shows the cell coverage ranges (i.e. distance in meters) obtained from the uplink (UL) budget for an urban environment. A similar calculation is obtained (although it is not displayed in this paper) for different system loads and for different environments, since both have an impact on the cell range. Other parameters are set by the standard, e.g. the maximum power a Mobile Station (MS) can transmit depends on its class. The standard classifies four types of MSs, but Class 4, for which the maximum power is limited to 21 dBm, will be most used. Class 3 will be used for high data rate transmissions (384 kbps) only. Moreover, the bit rate of the communication directly settles the SIR target at the receiver and, consequently, has an impact on the sensibility of the BS. The *fading margin* groups a set of margins: log-normal and Rayleigh fading, soft handover and diversity gain, power-control, etc.

It can be observed that, when considering a service with a higher data rate, the cell coverage decreases, unless a higher transmission power is used as in CS384. Depending on the services to be provided, the operator must establish minimum cell coverage (i.e. the coverage for the most constraining service) required in order to guarantee good coverage to every user. In the DL, the transmission power of the BS (20 W) is shared between all users. As higher bit rate communications need more power than low rate ones, the coverage range is highly sensitive to the cell load: the UL is coverage-limited (i.e. distance from the BS) while the DL is capacity-limited.

Table 1. Results from the link budget in an urban environment (uplink)

		Voice	CS64	CS144	CS384
Total system load: 60% **Urban environment** **Pedestrian user**	MS tx power (P_{tx}) [dBm]	21	21	21	24
	BS noise power [dBm]	-101.2	-101.2	-101.2	-101.2
	Bit rate [kbps]	12.2	64	144	384
	Sensibility (S) [dBm]	-122.5	-115.6	-112.7	-109.6
	Total losses (L) [dB]	4	1	1	-1
	Fading margin (fm) [dB]	3	3	3	3
	Propagation losses (P_{tx}-S-L-fm) [dB]	136.46	132.6	129.7	131.58
	Cell coverage range [m]	1196	924	764	866

2.2 Capacity Dimensioning

W-CDMA systems are interference-limited: the link performance depends on the ability of the receiver to discern signal in presence of interference. The LB displayed in Table 1 is valid for 60% load only: in general, the coverage range depends on the cell load. It represents the capacity of the system to carry a specific traffic load (depending on the density of traffic and on the kind of services considered). In [3], the authors present simulations with different traffic densities for a single BS environment that show how coverage decreases when load increases.

The second phase of dimensioning is estimating the amount of traffic supported per BS: we have to find an agreement between the maximum load inside a cell (i.e. the maximum number of simultaneous users for each service) and the number of traffic channels allocated to the BS [9]. In this process, only CS traffic has been taken into account, since PS services do not need circuit reservation. By estimating the traffic carried with a given GoS, which is represented here by the blocking probability, one could visualize PS services being provided during idle periods of CS traffic.

Cell ranges for each environment and load are estimated in the UL budget. We must ensure that the BSs DL power can be shared between all users and that this leaves a sufficient margin for efficient performance.

3 The Simulation Tool

The simulation program performs coverage analysis from a set of antennas in a specific traffic and propagation environment, from which it is possible to determine the coverage area of the layout under study. It also performs analysis of the traffic carried on maps that can be shaped by the user (i.e. the kind of services offered and traffic density for each service). By analyzing iterative snapshots of the system, the tool evaluates the average percentage of calls that the system can handle simultaneously.

3.1 Coverage Analysis

By working with propagation models that can be molded to the scenario, the simulator estimates the losses and generates a coverage map. In this way, it is possible to compute the BS coverage ranges by simulation [10]. In [3], the authors presented the propagation model used for these simulations. It is an adaptation of the classic Okumura-Hata model in which the parameters are tuned to the specific scenario and topography characteristics of Barcelona. By performing propagation analysis, the simulator can draw maps in which colors represent the signal strength around the BS.

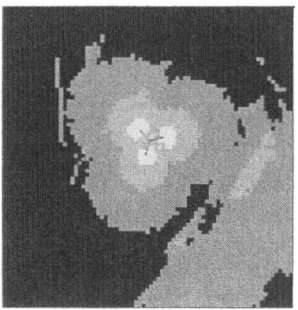

Fig. 1. Coverage Map for a Tri-sectorial BS

Fig. 1 shows the coverage map around a tri-sectorial BS in the city of Barcelona. Black represents the signal of a power strength under -90 dBm. This corresponds to the out-of-coverage area for this BS. Dark grey is where the strength remains between -90 and -80 dBm. This area extends to the bottom right-hand side, where the sea begins. As the signal moves away from the BS, it finds obstacles which can weaken its strength. We must determine the limits of the power strength in order to receive the communication correctly (i.e. *signal-to-interference target*).

3.2 Traffic Analysis

The tool simulates a real environment in which each BS serves a specific number of users that request different services. It is possible to work with traffic maps in which the user specifies the Erlang quantity for each service and area. Table 2 shows the values that are commonly assumed in this kind of analysis.

Table 2. Traffic density for different environments

Environment	Traffic density in the busy hour (E/km^2)
Dense urban	10 – 90
Urban	1 – 30
Suburban	0.5 - 5
Vehicular	It depends on the environment
Rural	0 – 0.3

3.3 Monte Carlo Simulation

The simulator, based on the Monte Carlo approach, relies on the analysis of different snapshots of the system for a variety of conditions. For each snapshot, a number of users with various characteristics are planned in the system and randomly allocated throughout the area analyzed (with a uniform space distribution). Once the system has reached the desired convergence in terms of blocking probability (i.e. the number of dropped calls in the actual iteration converges to the average of all iterations), statistics and pictures describing the system are generated, since it is assumed that enough snapshots have been run in order to give a good statistical representation [2].

4 BS Layout in the City of Barcelona

When planning the BSs layout in a real environment, it is important to know the coverage that a BS can provide in order to establish the optimum distances between sites. The authors first ran simulations in a single tri-sectorial site in order to check that the results from the analytical LB of Section 2 were accurate. The results were presented in [3]. There is a good agreement between the cell range simulation and the values listed in Table 1. The multiple service case (i.e. voice and data users at all bit rates) was also analyzed. An important result is that high data rate services increase cell load and tend to decrease the GoS. In this paper, we present the results obtained when simulating a complete BS layout in Barcelona. Coverage problems can be solved by changing the layout, while capacity problems remain.

4.1 The First Layout

In the first plan, a regular pattern was followed when positioning BSs: the underlying is hexagonal with some necessary adjustments. Fig. 2 displays the position of the BSs in the city. Only two services, voice and CS64, were considered. The latter has a smaller coverage range, so this distance (i.e. the one for the most constraining service) was used in the plan. The area analyzed corresponds to the central area of the city of Barcelona, in which two different environments can be found: dense urban, corresponding to the center of the city, and urban, corresponding to the residential areas. In the first zone, we placed a greater number of BSs in order to carry a higher density of traffic. The related coverage map in Fig. 3 shows the signal strength for each point of the city. The lower right-hand side is where the sea begins (see Fig. 2), which is why there is good coverage even though there are no sites.

Fig. 2. First BSs layout plan for Barcelona

Fig. 3. Coverage map for the first BSs plan

The next step in the study is to analyze the system when giving service to voice users only. Table 3 shows the parameters used in this first simulation. The number of traffic channels is computed according to Erlang-B [11], in order to limit the GoS to 2%. An extra 30% of the channels is allocated for handover purposes only.

Table 3. Parameters for the first simulation

Environment	Dense urban	Urban
Users	Voice	Voice
Offered traffic	40 E/km^2	20 E/km^2
Total area	11.28 km^2	9.45 km^2
Traffic channels	65 for each BS	

The report provided by the simulator in Table 4 shows that 97.7% of users can be served. 99.9% of all dropped users (i.e. 2.3% of the *users attempted*) is due to the *Mobile ERP Limit* (Equivalent Radiated Power). These users are concentrated in areas where coverage is not good. A negligible quantity of users cannot be served due to the *CPICH Power Limit* (Common PIlot CHannel). These users cannot receive the beacon sequences transmitted by the BSs correctly. In this first analysis, the limited power of the BSs can be efficiently shared between all users (i.e. 0% is dropped due to *DPCH Power Limit*, Dedicated Physical CHannel). The simulator allows the user to set the power limits for both the CPICH and DPCH, according to the recommendations given by the operator (in Section 4.2, an example of tuning these values is provided).

The simulator analyzes snapshots of the system. By evaluating the power levels at each point, it can estimate the percentage of users that remain in one of the different handover (HO) situations. A high proportion of users (48.7%) is in a soft-HO with three different BSs (*soft-soft HO*); 29.2% is not in an HO, while 20.7% is in an HO with two BSs (*soft HO*). Small percentages of users remain in other soft-HO situations: 0.3% in *softer HO*, which is an HO involving two sectors of the same BS; 0.9% and 0.1% are in HO with two sectors of the same BS and another BS.

The next case study consists in the addition of 10% of CS64 data traffic. The parameters are shown in Table 5.

Table 4. Report for voice traffic (first layout)

		Reason	No. of users	% of users (of the 2.3% / of the total)
Dropped user statistics		Mobile ERP limit	14.4	99.9 / 2.3
		Noise rise limit	0	0
		Primary CE limit	0	0
		CPICH power limit	0	0.1 / 0
		DPCH power limit	0	0
		Code limit	0	0
HO statistics		**Type of HO**	**No. of users**	**% of users**
		Not in HO	182.1	29.2
		Softer HO	2.0	0.3
		Soft HO	128.8	20.7
		Soft-soft HO	303.8	48.7
		Softer-soft HO	6.6	1.1

User type: Voice at 12.2 kbps
Users attempted: 637.9; Users served: 623.4 (97.7%); Total dropped users: 2.3%

Table 5. Parameters for the mixed-traffic case study

Environment	Dense urban		Urban	
Users	Voice	CS 64	Voice	CS 64
Offered traffic	40 E/km^2	4 E/km^2	20 E/km^2	2 E/km^2
Total area	11.28 km^2		9.45 km^2	
Traffic channels	78 for each BS			

Table 6. Report for mixed-traffic (first layout)

		Reason	No. of users	% of users (of the 6.9% / of the total)
Dropped user statistics		Mobile ERP limit	32.9	68.4 / 4.72
		Noise rise limit	0	0
		Primary CE limit	0	0
		CPICH power limit	2.2	4.5 / 0.3
		DPCH power limit	13.1	27.2 / 1.88
		Code limit	0	0
HO statistics		**Type of HO**	**No. of users**	**% of users**
		Not in HO	187.8	28.7
		Softer HO	1.9	0.3
		Soft HO	131.2	20.1
		Soft-soft HO	324.8	49.7
		Softer-soft HO	8.3	1.2

User type: Voice and CS data at 64 kbps
Users attempted: 702.1; Users served: 654 (93.1%); Total dropped users: 6.9%

Table 6 reports a reduction to 93.1% of calls attended. The percentage of non-served calls due to coverage problems increases to the 4.7% of the *users attempted*. By simply adding a small percentage of data service, problems that are not directly related to the coverage range arise. About 1.88% of the *users attempted* fail due to the *DPCH Power Limit*, while 0.3% cannot be attended due to the *CPICH Power Limit*. A large proportion of users (49.7%) stays in an HO involving three BSs; 28.7% are not in an HO, and 20.1% remain in a soft-HO with two different BSs.

4.2 The Second Layout

The second layout (see Fig. 4) is an attempt to improve the system's performance when combining voice and CS64 data traffic. The coverage problems in the first plan were solved by tuning down-tilt and orientation of antennas, and by adding new BSs when necessary.

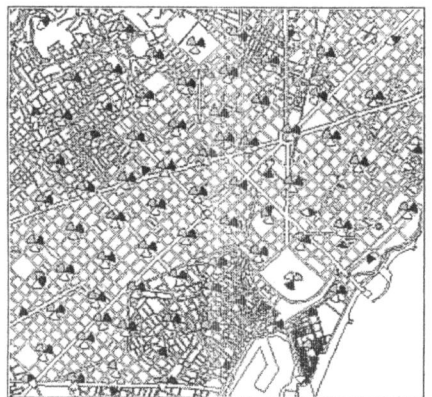

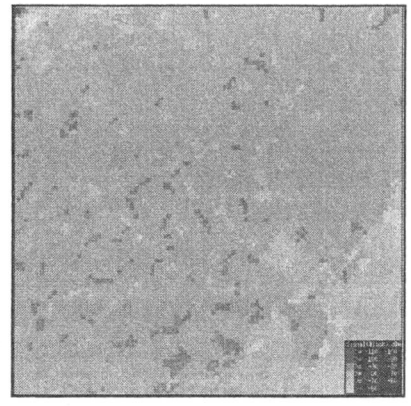

Fig. 4. Second BSs layout plan

Fig. 5. Coverage map for the second BSs layout plan

Fig. 5 shows that signal strength improves with respect to Fig. 3 in most of the city area. With this new layout, voice and CS64 data services can be provided to 94.9% of the users (Table 7). The simulator reports that coverage problems were reduced (dropped calls due to the *Mobile ERP Limit* are now 0.4% of the dropped users, which in turn represent 5.1% of all attempted users), while problems with power sharing in the DL cannot be solved by merely changing the parameters of the antennas. Table 7 reports that 4.5% of users (88.8% of all dropped users) are dropped due to the *DPCH power limit*. This simulation also shows that a high proportion of users (55.1%) are in a soft-HO with three different BSs; 24.1% are not in an HO, while 18.7% are in a soft-HO with two BSs.

Table 7. Report for mixed-traffic (second layout)

		User type: Voice and CS data at 64 kbps	
		Users attempted: 699.5; Users served: 663.8 (94.9%); Total dropped users: 5.1%	
		Reason	**No. of users**	**% of users (of the 5.1% / of the total)**
Dropped user statistics		Mobile ERP limit	0.1	0.4 / 0.02
		Noise rise limit	0	0
		Primary CE limit	0	0
		CPICH power limit	3.9	10.8 / 0.55
		DPCH power limit	31.7	88.8 / 4.53
		Code limit	0	0
HO statistics		**Type of HO**	**No. of users**	**% of users**
		Not in HO	160.3	24.1
		Softer HO	2.2	0.3
		Soft HO	124.4	18.7
		Soft-soft HO	365.6	55.1
		Softer-soft HO	11.2	1.8

Table 8. Simulation report with changes in the power limits of DPCH and CPICH

		User type: Voice and CS data at 64 kbps	
		Users attempted: 700.9; Users served: 676.8 (96.6%); Total dropped users: 3.4%	
		Reason	**No. of users**	**% of users (of the 3.4% / of the total)**
Dropped user statistics		Mobile ERP limit	0.2	0.8 / 0.03
		Noise rise limit	0	0
		Primary CE limit	0	0
		CPICH power limit	5.2	21.8 / 0.74
		DPCH power limit	18.6	77.4 / 2.63
		Code limit	0	0
HO statistics		**Type of HO**	**No. of users**	**% of users**
		Not in HO	165.5	24.5
		Softer HO	2.2	0.3
		Soft HO	131.1	19.4
		Soft-soft HO	366.0	54.1
		Softer-soft HO	12.0	1.8

While it was not possible to solve the power sharing problem by changing the layout of the BSs, the authors tried to perform new simulations after changing the limits of the power dedicated to DPCH and CPICH. As reported in Table 8, it is possible to partially improve the performances: the system can now attend 96.6% of the users. We could try to tune these power limits further, but this would not be enough to solve the problems inherent to sharing a common resource between too many users and load. While providing the possibility for one user's signal to be recovered from different receivers in the UL, macrodiversity creates an overload in the DL since more than one BS has to provide a part of its limited power to just one user.

5 Conclusions

The problem of distributing UMTS base stations in a densely populated area was addressed through simulation analysis in a realistic scenario (the city of Barcelona, Spain). The LB was computed and checked against the single-cell case with excellent results. This is useful to validate the simulator against the model and to gain confidence in the simulation results that follow.

To illustrate the trial-and-error process needed in this kind of planning, a first distribution of BSs following the classical hexagonal pattern was evaluated. Several problems in coverage (i.e. shadows) and capacity were observed. The second layout introduces slight changes to the first and improves coverage. However, capacity results are very difficult to improve in this process. It is foreseen that this trend will not improve even if more trials are carried out.

References

1. Holma, H., Toskala, A.: WCDMA for UMTS Radio Access for 3rd Generation Mobile Communications. John Wiley & Sons (2002)
2. Dehghan, S., Lister, D., Owen, R., Jones, P.: W-CDMA Capacity and Planning Issues. Electronics and Communication Engineering Journal, Vol. 12, Issue 3 (2000) 101-118
3. Zola, E., Barceló, F., Martín, I.: Simulation Analysis of Cell Coverage and Capacity Planning in a UMTS Environment: A Case Study. IASTED International Conference Communication Systems and Networks (2003) 342-346
4. Veeravalli, V.V., Sendonaris, A.: The Coverage-Capacity Tradeoff in Cellular CDMA Systems. IEEE Trans. on Vehicular Technology, Vol. 48, No 5 (1999) 1443-1450
5. Coinchon, M., Salovaara, A., Wagen, JF.: The Impact of Radio Propagation Predictions on Urban UMTS Planning. International Zurich Seminar on Access, Transmission and Networking (2002) 32-1 - 32-6
6. Schroder, B., Liesenfeld, B., Weller, A., Leibnitz, K., Staehle, D., Phuoc Tran-Gia: An Analytical Approach for Determining Coverage Probabilities in Large UMTS Networks. IEEE 54th Vehicular Technology Conference, VTC 01 Fall (2001) 1750 -1754
7. Hata, M.: Empirical Formula for Propagation Loss in Land Mobile Radio Service. IEEE Trans. on Vehicular Technology, Vol. 29, No. 3 (1980)
8. Laiho, J., Wacker, A., Novosad, T.: Radio Network Planning and Optimisation for UMTS. John Wiley and Sons, 2001
9. Owen, R., Burley, S., Jones, P., Messer, V.: Considerations for UMTS Capacity and Range Estimation. IEEE Colloquium on Capacity and Range Enhancement Techniques for the Third Generation Mobile Communications and Beyond, Ref. No. 2000/003 (2000) 5/1-5/6
10. Pattuelli, R., Zingarelli, V.: Precision of the estimation of area coverage by planning tools in cellular systems. IEEE Personal Communications, Vol. 7, Issue 3 (2000) 50 -53
11. Garg, V., Yellapantula, R.: A Tool to Calculate Erlang Capacity of a BTS Supporting 3G UMTS System. IEEE International Conference on Personal Wireless Communications (2000) 173 – 177

A General Traffic and Queueing Delay Model for 3G Wireless Packet Networks*

Ghassane Aniba and Sonia Aïssa

INRS-EMT
University of Quebec
Montreal, QC, Canada
{ghassane, aissa}@inrs-emt.uquebec.ca

Abstract. This paper considers traffic modelling and queueing delay estimation of different 3G packet network services. First, a general traffic model for conversational (voice) and streaming (video) services is presented. This model is subdivided into three levels: session level, burst level, and packet level. Based on the proposed model, the statistical behavior of the queueing delay is studied. It is shown that the queueing delay corresponding to voice and video streaming services follows an exponential distribution. Analytical modelling of the probability density function (PDF) of the queueing delay being untractable, we resort to simulated data and provide simple mathematical formulation of the different parameters that characterize the density functions of the different services. Indeed, we present useful equations which could be utilized, directly in network dimensioning, as a reference to satisfy a certain quality of service (QoS) and in the design of radio resource management algorithms.

1 Introduction

In third generation (3G) wireless networks [1], packet-based services have different requirements in terms of quality of service (QoS) and tolerance to delay. Thus, modelling the traffic and the queueing delay of these services is of extreme importance. Indeed, such models are needed in order to dimension practical networks and design efficient radio resource management strategies and communication protocols for these networks. In this context, the focus of this paper is to model the queueing delay of different packet services. For this purpose, a general traffic model is provided and used to derive the parameters that specify the queueing delay PDF (probability density function) corresponding to the services under consideration. Taking into account studies that have previously been made, our focus is on conversational and streaming services.

In 3G wireless packet networks, traffic modelling of the different classes of services is not finalized yet. In [2], a traffic model is provided for the web-browsing service only. In this paper, we generalize the model presented therein [2] to the other classes of traffic, namely, conversational service (voice), and streaming

---

* Work supported by the Canadian NSERC Research Grants Program.

J.N. de Souza et al. (Eds.): ICT 2004, LNCS 3124, pp. 942–949, 2004.

service (video streaming). This model is subdivided into three levels: session level, burst level, and packet level.

On the other hand, the air interface throughput in such networks is shared between different users. In order to increase the capacity of these systems, the capability of the network to carry packet-switched traffic is used along with multiplexing the traffics of different users. However, loading the network causes delay which, in turn, can cause degradation in the QoS of delay-sensitive applications and in the overall performance of the network. In order to avoid degradation, it is essential to model the queueing delay of the different services and use these models for the purpose of developing the communication strategies that ensure the increased capacity. Because of the complexity of the packet traffics involved, modelling the queueing delay in a packet-switched network through analytical computation is untractable. In [3], model fitting based on simulated data is used to model the queueing delay PDF for web browsing applications. In this work, with the use of the general traffic model that is proposed herein, we provide a general delay model for voice and video streaming traffic classes following the approach in [3] for web browsing. For each service, we provide the different parameters that characterize the queueing delay PDF as a function of the air throughput and the loading of the network. The latter being measured through a percentage utilization of this throughput.

The remainder of this paper is organized as follows. Section 2 describes the general traffic model with the parameters associated to each service under consideration. Our queueing delay modelling and results corresponding to these services are provided in Section 3. Finally, conclusions are drawn in Section 4.

2   Traffic Modelling

Based on the definition provided in [2] for the web-browsing service, the general traffic model, proposed herein, is divided into three levels:

- *Session level:* user sessions are modelled at this level. We suppose the arrival of each session to follow a Poisson distribution with the assumption that each user has only one session during the busy hour.
- *Burst level:* each packet session is formed by one or many bursts of packets. The distribution of the inter-arrival of these bursts depends on the class of traffic as will be presented later.
- *Packet level:* each packet burst is composed of a number of data packets. The distribution of the packet size and that of the inter-arrival time are specified at this level.

2.1   Voice Traffic Model

The voice service traffic model follows the representation shown in Figure 1. When a user is in an active voice session, its traffic generation pattern is modelled via two activity states which are called *High* and *Low* states [4]. Each state has its own packet generation characteristics with state durations assumed to be exponentially distributed. In Table 1, we provide the different parameters and the values considered herein [4]. The choice of $3s$, as an average value for the *High* and *Low* state durations, is motivated by a speech activity of 50% [1].

For the packet level, we use a deterministic model, with constant packet size and packet inter-arrival time. The packet size for the *High* state and the *Low* state correspond to a source rate of $16kbps$ and $2.8kbps$, respectively. The *Low* state represents the transmission control when no speech is transmitted. The mean traffic of each voice session is approximatively equal to $9.6kbps$.

2.2   Video Streaming Model

Due to the lack of detailed models for video streaming over wireless packet networks, we propose to adapt the traffic model used in fixed networks to the three-level model of communication (session, burst, packet). The adaptation of the traffic flow to the three-level model is shown in Figure 2. A video session consists of several packet calls, each composed of a number data blocks. Each block consists of a group of pictures (GOP) that is structured according to the MPEG standard [5]. Each GOP is composed of three types of frames: Intra-coded (I), Predictive (P), and Bidirectional (B). The I frames are those for which intra-frame coding is used (without motion estimation), the P frames are those for which inter-frame coding is used (with motion estimation), and the B frames specify the frames that can be predicted using forward and backward prediction.

Previous studies and measurements showed that the frame length follows a log-normal distribution (1) at the output of the encoder. In [6], measurements made show that the distribution of video traffic does not change, even after many wireless nodes. We therefore adopt the log-normal distribution in our modelling:

$$f(x) = \begin{cases} \frac{1}{x\sigma\sqrt{2\pi}} \exp\left[\frac{-(\log x - \mu')^2}{2\sigma'^2}\right] & x > 0 \\ 0 & \text{otherwise}, \end{cases} \tag{1}$$

$$\text{with} \quad \sigma'^2 = \log\left(1 + \frac{\sigma^2}{\mu^2}\right), \quad \text{and} \quad \mu' = \log\left(\frac{\mu^2}{\sqrt{\sigma^2 + \mu^2}}\right).$$

Video sessions are assumed to arrive as independent Poisson processes and their durations are assumed to be exponentially distributed. In the multiplexing of video sessions, we use direct multiplexing of the video frames (I, B and P) because of the relatively high air-throughput values (1Mbps or higher) in 3G wireless networks. The parameters corresponding to the proposed video traffic model are summarized in Table 2 along with the associated values considered in our simulations. These values are deduced from [5] and [7].

3   Delay Modelling: Results and Analysis

The previously described traffic models are now applied to model the queueing delay of the services under consideration, namely, voice and video streaming.

Table 1. Voice traffic parameters

Session arrival process	Poisson
Voice session duration	120s (exponential r.v)
High state duration	3s (exponential r.v)
Low state duration	3s (exponential r.v)
Packet inter-arrival time	20ms
Packet size: High state	40 bytes
Packet size: Low state	13 bytes

Table 2. Video traffic parameters

Session arrival process	Poisson
Video session duration	10mn (exponential r.v)
GOP size	12 frames
Inter-arrival time	
between frames	40ms (25f/s)
Frame I: (μ, σ)	(3000 bytes, 42 bytes)
Frame P: (μ, σ)	(1000 bytes, 25 bytes)
Frame B: (μ, σ)	(500 bytes, 17 bytes)

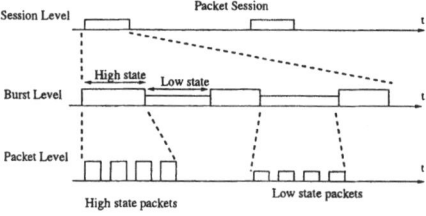

Fig. 1. Voice packet service session

Fig. 2. Video packet service session

Specifically for each service, we provide the different parameters that characterize the queueing delay PDF as a function of the air throughput and the loading of the network. The latter is measured through the percentage utilization of the throughput during the busy hour [3]. For each class of traffic, the number of users is chosen so that the sum of the loaded traffic equals an average percentage U of the air throughput R, given for the busy hour.

For each traffic case, a number of simulation runs, ranging between 30 and 100 runs were performed in order to provide precision to our delay modelling. In the same vein, unlike the approach adopted in [3] where delay is represented by average values computed on a per-second basis, our measurement of the delay was performed by collecting, for each packet in a session, the corresponding value in the queueing buffer. From our simulations, and for the voice and video streaming traffics, we found that the queueing delay follows a weighted exponential distribution, similar to the web browsing traffic as shown in [3], according to:

$$f_\tau(\tau) = p.\delta(\tau) + (1-p)(\frac{1}{\mu}\exp(-\frac{\tau}{\mu})), \qquad (2)$$

where p is the probability that the queueing buffer is empty, and μ is the mean queueing delay when delay is non-zero:

$$E(\tau/\tau > 0) = \mu. \qquad (3)$$

The mean queueing delay is hence given by

$$E(\tau) = \int_0^\infty \tau \, f_\tau(\tau) \, d\tau = (1-p)\,\mu. \qquad (4)$$

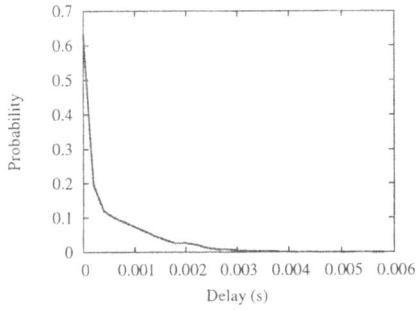
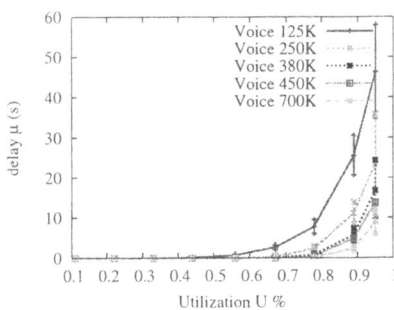

Fig. 3. Delay PDF of voice traffic with 44% utilization of the air throughput R of value 380kbps

Fig. 4. The relationship between μ and U for voice traffic

In the following, we present a mathematical representation of μ and p, as a function of throughput R, in order to deduce the mean queueing delay from (4) for the voice and video streaming services.

3.1   Voice Traffic

The PDF of the delay corresponding to the voice traffic model described in Section 2.1 is represented in Figure 3. Analysis of these results along with the corresponding cumulative distribution shows that the delay PDF can be modelled using the weighted exponential distribution (2). In Figure 4, we show the variation of μ as a function of the utilization percentage U for different values of the throughput R.

As for the probability of no-delay, $p = P_\tau(\tau = 0)$, we represent its variation with respect to U in Figure 5 for different values of R. The bars in these figures represent standard deviations about the means.

Following an extensive comparison of our simulation's results to different models (exponential, polynomial, logarithmic, linear and fractional), we found that the most suitable mathematical representation for μ and p can be expressed as

$$\mu = \frac{c}{(1 - U)} \quad , \quad p = 1 - U^b \tag{5}$$

where c and b are constants that depend on R.

Indeed, using the results shown in Figures 6 and 7 representing b and c for specific values of R, and performing comparisons, in terms of the correlation coefficient, with other mathematical models, we found that the parameter c is inversely proportional to the throughput R, and that the parameter b is a linear function of R. Based on our analysis, we provide in the following the most suitable mathematical formulation we found for these parameters:

$$b = 9.48.10^{-4} * R + 0.967 \quad , \quad c = \frac{283.11}{R} \tag{6}$$

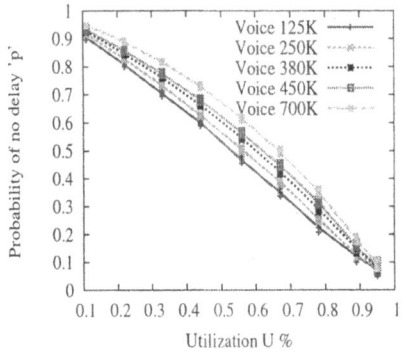

Fig. 5. The relationship between p and U for voice traffic

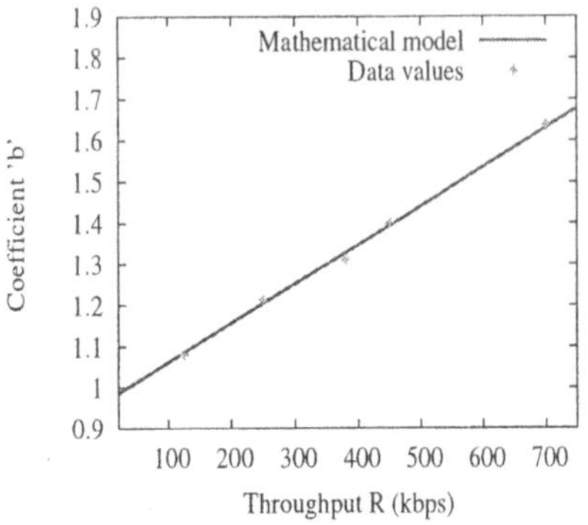

Fig. 6. Voice traffic: variation of parameter b as a function of throughput R

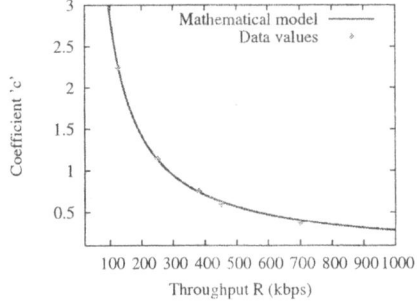

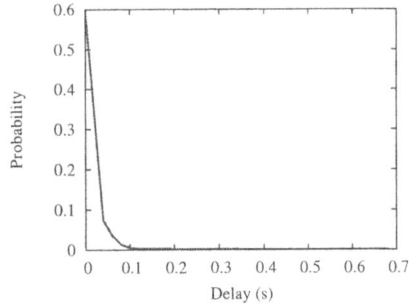

Fig. 7. Voice traffic: variation of parameter c a as function of throughput R

Fig. 8. Delay PDF of video streaming traffic with 33% utilization of the air throughput R of value 1000 kbps

Representation of these parameters as a function of R, respectively in Figure 6 and Figure 7, and comparisons of these curves with the data values (represented in marks on these figures), shows the validity of our modelling.

3.2   Video Streaming Traffic

In this case, simulation results also show that the queueing delay can be modelled through an exponential distribution as for the voice traffic case. This can be seen in Figure 8 showing the queueing delay PDF corresponding to the video streaming traffic as modelled in Section 2.2.

Similarly to the previously studied traffic, we represent in Figures 9 and 10, the evolution of μ and p with the utilization U for different values of the throughput R. As can be seen, the variation of μ is not similar to that observed for the voice traffic, and therefore, finding a simple equation to model this parameter is not straightforward. Following an analysis of the results we obtained using different models (exponential, polynomial, logarithmic, linear and fractional), we found that a suitable function for the mean μ is a polynomial function of the second degree. In Table 3, we provide the expression of μ, as a second-degree polynomial in U, corresponding to the values we considered for the throughput R. For other values of R, the corresponding value of μ can be found through interpolation.

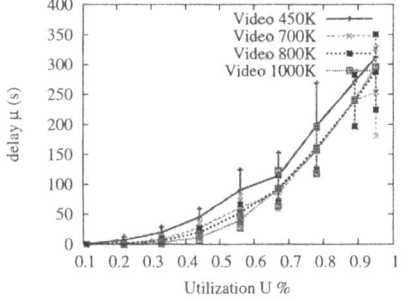

Fig. 9. The relationship between μ and U for video streaming traffic

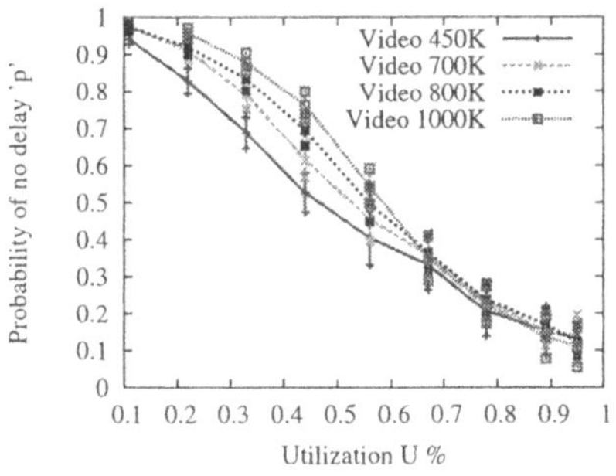

Fig. 10. The relationship between p and U for video streaming traffic

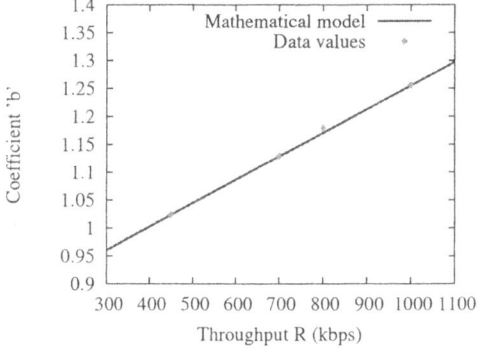

Fig. 11. Video traffic: variation of the parameter b as a function of throughput R

Table 3. The expression of μ, as a second-degree polynomial in U for video streaming traffic

R (kbps)	Mean μ (s)
450	$583.68*U^2$-287*U+31.14
700	$483.38*U^2$-201.45*U+19.56
800	$511.33*U^2$-175.3166*U+18.14
1000	$721.1*U^2$-444.7*U+66.14

For the modelling of the variation of the probability of no-delay p as a function of U (Fig. 10), our analysis showed that the most correlated function to the simulation's results is the linear function

$$p = b(1 - U) \tag{7}$$

which, as can be seen, is not as simple as for the voice traffic case.

Next, we need to express the parameter b as a function of R. Representation of the values of b, obtained from simulated data is provided in Figure 11. These results show that the parameter b is a linear function of the throughput R. In this case, we found that the model that provides the best fit is given by

$$b = 4.21.10^{-4} * R + 0.834. \tag{8}$$

Based on the results provided in this section, we can directly use the curves to get the suitable combination of throughput value R and utilization U in order to get a desired average-delay value for a given class of traffic. For example, consider the voice traffic case, a pair $(R, U) = (125kbps, 80\%)$ yields an average delay value $\mu \simeq 10s$. Hence, if we want to increase the number of users, the results of Figure 4 show that a choice of $(R, U) = (250kbps, 88\%)$ could be made to increase the number of users while keeping the same delay value $\mu \simeq 10s$. The proposed mathematical formulation can also be used to derive the exact combination of parameters that could be used to get a required delay value or a given QoS for a given class of traffic.

4 Conclusions

In the first part of this paper, we generalized the web-browsing traffic model presented in [2] to two classes of traffic: voice and video streaming. For each case, the traffic model is subdivided into three levels, namely, session level, burst level, and packet level. Then, based on the traffic models proposed, we provided a queueing delay model and showed that the queueing delay of these traffics is exponentially distributed as for the web browsing traffic considered in [3]. For each of the two studied traffics, we presented a mathematical formulation for the different parameters that characterize the queueing delay. The proposed formulae can be used to deduce different density functions of queueing delay as a function of the traffic parameters and of the air throughput. In addition, direct use of our curves can provide the combinations of values (throughput, utilization) that allow to obtain a required value for the mean queueing delay.

Acknowledgment. We wish to thank M. Shafi and K. Butterworth for discussions related to [3].

References

1. TS 122 105 V5.2.0, "UMTS Services & service capabilities, (3GPP TS 22.105 version 5.2.0 Release 5)", ETSI 2002.
2. TR 101 112 V3.2.0, "Selection procedures for the choice of radio transmission technologies of the UMTS", ETSI 1998.
3. K. Butterworth, M. Shafi and P. Smith, "A Model for Estimating the Queuing Delay in a Wireless Packet Network", Private Communication.
4. A. Garcia and al., "Quality of Service Support in the UMTS Terrestrial Radio Access Network", Proc. HPOVUA'2002, 11-13 June 2002.
5. E. He, F. Du, X. Dong, L.M. Ni and H.D. Hughes, "Video Traffic Modeling Over Wireless Networks", IEEE ICC'2000, vol. 1, pp. 536–542, 18-22 June 2000.
6. Frank H. P. Fitzek and M. Reisslein, "MPEG-4 and H.263 Video Traces for Network Performance Evaluation", IEEE Network, vol. 15, no. 6, pp. 40–54, Nov-Dec 2001.
7. http://www-tkn.ee.tu-berlin.de/research/trace/trace.html

MobiS: A Solution for the Development of Secure Applications for Mobile Devices[*]

Windson Viana[1], Bringel Filho[1], Katy Magalhães[3], Carlo Giovano[2],
Javam de Castro[1], and Rossana Andrade[1]

[1] Universidade Federal do Ceará (UFC), Mestrado em Ciência da Computação (MCC),
Campus do Pici, Bloco 910, Fortaleza, Ceará, Brasil, 60.455-760
{bringel, javam, windson,rossana}@lia.ufc.br
http://great.lia.ufc.br

[2] Instituto Atlântico, Chico Lemos 956, Cidade dos Funcionários, Fortaleza-CE,
Brasil, 60.822-780
cgiovano@atlantico.com.br

[3] Universidade Federal do Ceará (UFC), Programa de Pós-Graduação em Engenharia Elétrica
(PPGEE), Campus do Pici, Caixa Postal 6007, CEP 60.755-640
katy@deti.ufc.br

Abstract. With the mobile computing evolution, portable devices have emerged in the market with the possibility of connection to IP networks using wireless technologies. The migration from fixed terminals to wireless mobile devices and the increasing availability of information for such devices impose new challenges in the development of applications. The security and the portability are examples of these challenges that constitute the motivation for this work. Our main goal is to introduce a solution, called MobiS (Mobility and Security), to the development of secure applications for data transmission using wireless technologies.

1 Introduction

A great diversity of mobile devices has appeared in the market with the possibility of IP network connection using wireless medium, for example, cellular phones with WAP support, GPRS, i-Mode (in Japan), as well as notebooks, Palms and Pocket PCs with Bluetooth and IEEE 802.11. The possibility of changing from fixed access terminals to mobile devices imposes new challenges in the development of applications. The mobility and storage capabilities, the low power processing, the compact size and the possibility of wireless communication with other devices connected to the Internet are examples of these challenges. Moreover, the increasing availability of information for such devices causes serious risks to the security. It is worth mentioning that these devices offer security mechanisms less powerful than the ones offered in the fixed networks.

---

[*] This work is supported by Instituto Atlântico, FUNCAP, CAPES and CNPq, which are research agencies from Brazil

J.N. de Souza et al. (Eds.): ICT 2004, LNCS 3124, pp. 950–957, 2004.

In this context, the main focus of this paper is to present an approach, called Mo-biS (Mobility and Security), that allows the development of secure applications for mobile devices. As a case study, we describe the development of an application named MobileMulta, which makes the transmission of infraction tickets in a secure way, from the mobile device to an application server located in the Internet.

The rest of this paper is divided as follows. In Section 2, we present a generic view of applications for mobile devices as well as risks and current security strategies to emphasize the motivation of this work. In Section 3, we introduce MobiS for the development of secure applications for mobile devices. In Section 4, we describe the development of a MobiS case study, the MobileMulta application. Finally, in Section 5, we present our conclusions and future works.

2 Applications for Mobile Devices: Risks and Solutions

Corporative Applications developed for mobile devices are usually inserted in a sce-nario like the one illustrated in Fig. 1. These applications, in general, are executed in an off-line state that allows insertion and other operations in a local database (called LDB) of a mobile device. At any time, they can establish a connection using a wire-less technology and send data to the application server that treats these data with its business rules. When the application server sends the confirmation that the transmit-ted information had been received successfully, the mobile device application deletes the LDB registers. The connection is then released by the application that returns to the off-line state.

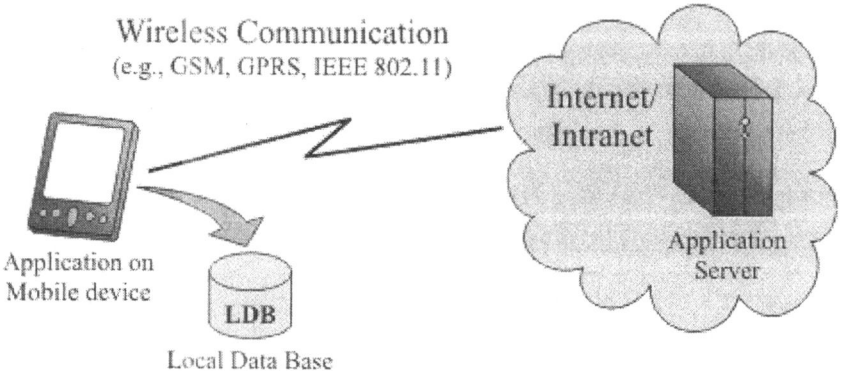

Fig. 1. A generic scenario for mobile devices transmitting information to an application server

In this generic scenario, it is necessary to worry about risks caused by, for example: non-authorized access to the application in the mobile device; limitations of the wireless communication medium; and external attacks to the application server. We present more details about these risks and current solutions to overcome them as follows.

First, we should consider that mobile devices can be lost and non-authorized ac-cesses to the application as well as to the stored information in the device can happen.

A complete security solution against this access must combine the protection against non-authorized accesses to the device itself with the protection to the application using password.

Second, when developing applications for mobile devices, we should pay attention to the wireless communication media that not only have limitations of bandwidth but also are vulnerable to signal variations (e.g., external interferences). Moreover, it is difficulty to protect them against passive and active attacks that put in risk the confidentiality, integrity and authenticity of the transmitted information. Then, considering also computational limitations of the mobile devices and the security demanded by each application, adaptations in the existing security protocols to transmit information over these media have been presented in the literature. For example, [3] has identified and reported security failures in WEP (Wired Equivalent Privacy). Cellular systems, such as GSM/GPRS (Global System for Mobile Communications/General Packet Radio Service), also need additional mechanisms for security. In short, the existing solutions make impracticable their use for applications in mobile devices either due to computational power demanded (e.g., VPN – Virtual Private Network [4] and SSL/TLS – Secure Socket Layer/Transport Layer Security [5]) or due to the non-availability of resources in the applications (e.g., TCP for applying KSSL [6]). However, mobile device applications can get an independent protection from the wireless communication medium used (e.g., IEEE 802,11, GSM/GPRS and Bluetooth) if developers apply cryptography in the application layer.

Finally, an application server that is connected to the Internet is exposed to external attacks that can cause service interruptions. These attacks are called DoS (i.e., Denial of Service [7]). According to [8], the implementation of a firewall can protect an application server against certain attacks.

3   Developing Secure Applications for Mobile Devices

As a result of our research in the literature, [1] is the only work directly related to our proposed approach. The authors present an end-to-end security solution in the application layer for applications developed in J2ME platform. This security solution uses Java components to provide data authentication and confidentiality between a customer implemented in J2ME platform and a server implemented in J2EE platform. They apply Rijndael´s AES algorithm, which is a version implemented by the Legion of the Bouncy Castle [2], for the user authentication process and for the cryptography of data to be sent. However, this solution does not guarantee the integrity of the transmitted information that can be modified during the transmission. On the other hand, our approach, called MobiS, guarantees authenticity, confidentiality and integrity of the transmitted information.

Fig. 2 shows the MobiS architecture that focuses on the mobile device access, the wireless communication medium and the application server. MobiS applies the solutions detailed in Section 2, for example, the user access to the application is protected by login and password, which have their digest stored in the device during the application installation. The digest is calculated using a hash function that brings more security to the user authentication. Although there are reported failures in some current hash functions (for example, in [9], a failure in MD4 is presented), the choice of the hash function to be used in our approach must have criteria, taking into account algorithm performance and security.

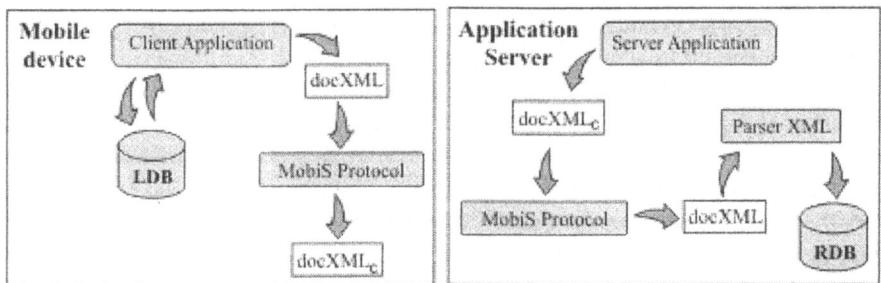

Fig. 2. MobiS architecture for secure applications

In Fig. 2, a mobile device is used to have access to the application information stored in a local database (LDB). When the user sends the information stored in the LDB, the application creates a XML document (Extensible Markup Language) that contains all stored data. The XML Document (called docXML) is created to facilitate the data exchange between the mobile device and the application server.

In our approach, the security mechanism used to protect LDB accesses depends on the storage resources used by the application. For instance, depending on the mobile device, the operational system and the development platform, an application can neither have access to the data files of the operational system (e.g., pdb in the Palm OS) nor have access to the mobile users of database. The LDB security, therefore, is dependent of the application, as presented in the case study implementation (see Section 4).

Section 2 also introduces solutions, such as VPNs, SSL/TLS and KSSL, for protecting application data transmissions using wireless communication technologies. When there is a need to guarantee the portability of the application, however, the use of these solutions are not suitable since some devices have limitations (e.g., processing power) imposed by hardware that make impracticable their use. Although KSSL presents acceptable performance, it needs TCP support in the devices, what is not always available. Thus, for this last security problem, we propose the specification and the implementation of a security protocol in the application layer that is independent of the communication infrastructure used to transmit data. This protocol, which is the MobiS protocol shown in Fig. 2, assures confidentiality, integrity and authenticity of the transmitted information. The MobiS protocol is based on adaptations of WEP, SSL and PGP (Pretty Good Privacy) presented in the literature. These adaptations take in consideration hardware limitations, which make the original protocols not suitable for mobile devices, and the possibility of bringing more portability to the application.

The docXML document is ciphered using the MobiS protocol (e.g., a Symmetrical MobiS protocol as illustrated in Fig. 3). The $docXML_c$ transmission is executed using HTTP (HiperText Transfer Protocol). When the application server receives the docXML$_c$, it deciphers the message using also the MobiS protocol that verifies its authenticity and its integrity. If the verification is successful, then the XML parser is instantiated and data are recovered from the docXML and inserted in the remote database (RDB).

The MobiS protocol includes three different specifications: Symmetrical MobiS, Anti-symmetrical MobiS and Hybrid MobiS. These specifications refer to the type of cryptographic algorithms used. For example, Symmetrical MobiS uses symmetrical cryptography algorithms, Anti-symmetrical MobiS algorithms apply anti-symmetrical cryptography and, finally, Hybrid MobiS combines the use of these two types of algorithms. For this paper, we present only the symmetrical implementation. The other versions have been specified, but we are still working on implementation and tests.

The symmetrical MobiS protocol is based on the WEP version used in IEEE 802.11b networks to provide data confidentiality. Despite the reported failures, WEP presents a strong security scheme. As mentioned in [3], the protocol failures are related to the RC4 cryptography algorithm and to the key size used. Fig. 3 illustrates the symmetrical MobiS protocol. In I), we describe the operations applied by the protocol before sending a message. In II), we describe the operations applied by the protocol when receiving the message.

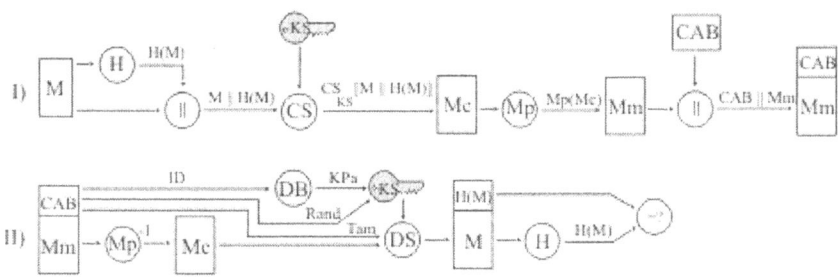

Fig. 3. Symmetrical MobiS Protocol

The sending process, presented in Fig. 3.I, is detailed as follows. A digest, called H(M), is calculated using a hash function, H, in the message to be sent, M. The digest is then concatenated to the message and the resultant text is generated, M ‖ H(M). The resultant text is ciphered using a symmetrical cryptography algorithm block cipher CS (e.g., Triple DES) using the KS session key, thus, Mc = CSKS [M ‖ H(M)]. The session key can be generated in the following ways: concatenating the symmetrical key KPa present in the device with a Rand calculated for each transmission (KS = KPa ‖ Rand) or running an operation Xor (or exclusive) between KPa and the generated Rand (KS = KPa $\oplus$ Rand).

After ciphering the message, a conversion algorithm, called Mp, is applied over Mc (e.g., Base64) transforming invalid characters generated in the cryptography into acceptable characters for the transmission (Mm = Mp(Mc)). This mapping process must be executed since the cryptography algorithms generate characters that are not transmitted correctly using certain protocols, such as MIME and HTTP. Mapping algorithms transform bytes (256 possible characters) into a table of fewer characters. Base64 is one of these algorithms that transform byte in alphanumeric characters according to a table using only 64 characters. The final message increases around 30%.

A header, named CAB, is added to the Mm and contains the size of the original message (TAM), the device identification (ID), the identification of algorithm combinations (ICA) (e.g., an identification for the Triple DES + MD5 + HexEncoder combination) and the generated Rand. The transmission of TAM in the CAB is necessary since the block cipher symmetrical cryptography algorithms require that the input size is multiple of the block size that the algorithm works (e.g., the Rijndael works with blocks of 16 bytes). Thus, before cryptography, a padding to the end of the message is executed to make its size multiple of the block size of the algorithm.

The receiving process, shown in Fig. 3.II, is detailed as follows. After receiving a message, the CAB is read and the inverse mapping process (Mp-1) is done over the rest of the message. It is possible to recover the KPa symmetrical key in the key database (DB) using ID. KPa is concatenated (or, as explained previously, a Xor is executed) with the sent Rand, generating KS that is used to decipher the message (DS). The hash function is applied in the deciphered data, generating a digest that is compared to the sent digest. If they are equal, message integrity is assured.

One of the problems in the symmetrical cryptography is the vulnerability of known ciphered text attacks. In the symmetrical MobiS, this problem is solved using a random number, Rand, in the composition of the session key KS. Symmetrical MobiS guarantees data confidentiality, since only the parts that communicate (i.e., mobile device and application server) in the architecture can recover the original data. Data integrity is assured using the hash function. In case of modified data, the digest calculated in the receiving process does not match to the digest sent. Finally, data production is prevented using symmetrical cryptography; therefore, only the ones who possess the key are able to produce a valid message. Reply attacks can still happen and they should be solved by the application. A possible solution would be the addition of an incremental code that identifies only the sent data. Thus, the server is able to discard the received data with the same code.

Another existing problem in the use of symmetrical cryptography is related to the key distribution. However, the MobiS protocol does not worry about it, during the development of the application, the software engineer is responsible for choosing one of the existing key distribution mechanisms. In the next section, we detail the key distribution mechanism used in our case study.

Thus, the robustness of the symmetrical MobiS protocol is directly related to the algorithm combinations as well as to the size of the cryptographic key.

4   MobileMulta Application: A Case Study

We develop an application, called MobileMulta, to register traffic infraction tickets as a case study. The system user (i.e., the traffic warden) uses a mobile device (e.g., Palm) to register infractions in the LDB. Data transmission to the server (i.e., the sending process of the infractions to the traffic central office) is executed using an infrared (IrDA) connection with a cellular phone, which can use GSM/GPRS systems.

The sending process of infraction tickets, which are registered in the LDB, to the application server using a wireless communication medium requires data confidentiality, integrity and authenticity. Thus, the use of a MobiS protocol is necessary in the MobileMulta application to guarantee these security requirements.

For the implementation of the MobiS protocol in the MobileMulta, we use the following combination of symmetrical algorithms: AES (Advanced Standard Encryption) block cipher, which is a cryptography algorithm based on the Rijndael [11], with 128 bits key, SHA1 hash function (digest of 160 bits) and the Base64 mapping algorithm. The KS session key is generated from the concatenation of KPa and Rand (KS = KPa || Rand). KPa contains 96 bits and Rand is 32 bits, composing 128 bits that are necessary to constitute a key for the AES algorithm. The use of Rand and KS composition makes impracticable the application of crypto analysis techniques such as known cipher text (see more details in Section 3, receiving process).

The key distribution occurs as follows: in the MobileMulta installation, KPa key is generated in the server and synchronized with the mobile device using a data cable. KPa copy is stored in the key database (DB) located in the server and it is associated to the device identification ID. Therefore, the key is protected against eavesdropping. KPa is exchanged every week, period where the traffic warden goes to the central office for system maintenance. This period is determined according to the application business rules, since it is possible force brute attacks as presented in [12].

For the implementation of the MobileMulta application, we choose J2ME CLDC/MIDP platform for its portability to cellular systems (with support to MIDP 1.0) and Palm OS e Windows systems CE [13]. We also use Sun One Studio IDE for development [14] integrated to Wireless ToolKit 1.0 [13], which are widely used free tools for the J2ME development.

J2ME CLDC/MIDP 1.0 platform does not contain APIs for the treatment of security aspects (e.g., support to SSL). A development group, called Legion of the Bouncy Castle [2], maintains APIs J2ME freely available that implements the most known cryptography algorithms. We use, therefore, AES, SHA1 and Base64 implementations of this group to compose the symmetrical MobiS protocol.

MobileMulta has been tested using the Sun virtual machine in the M130 and M515 Palms (both 33 Mhz processors) with 8MB and 16MB of memory, respectively. We use Siemens S45 cellular phone to allow the communication with the server using a GSM dialed connection. We have run another test using the Motorola A388 cellular phone (2 Mb memory, 10 Mhz processor and support to J2ME MIDP 1.0) using GPRS system to transmit infraction. In the simulation tests, we have got a satisfactory performance. The use of symmetrical MobiS protocol by the application caused a 250 ms delay, in average, for sending each infraction.

5   Conclusion and Future Work

This paper presents the MobiS approach that is a solution for the development of secure applications for mobile devices. We introduce MobiS architecture and apply a case study, called MobileMulta, to validate the architecture. MobiS guarantees authenticity, confidentiality and integrity of the transmitted information in contrast to [1], a related work that presents a solution only for integrity of the transmitted information. In [1], information can be still modified during the transmission. Mobile-Multa application uses the Symmetrical MobiS protocol to provide confidentiality, integrity and authenticity of the transmitted infractions.

As future work, we can point out: implementation and tests of the Asymmetrical and Hybrid MobiS protocol versions; performance evaluation of the Symmetrical MobiS protocol compared to other combinations of algorithms (e.g., TripleDES + MD5 + HexCodec) and tests of the solution using other transmission medium (e.g., IEEE 802.11b).

References

1. W. Itani and A. Kayssi, "J2ME End-to-End Security for M-Commerce", in Proc. IEEE Wireless Communications and Networking Conference (WCNC 2003), p. 2015 - 2020, March 2003, New Orleans, Louisiana.
2. The Legion of Bouncy Castle. Cryptography API para Java. Disponível em <www.bouncycastle.org> Access in: Feb 2003.
3. Borisov, N.: Goldberg, I.; Wagner, D.; Intercepting Mobile Communications: The Insecurity of 802.11. p. 1-9, 2000.
4. Nunes, Bruno, A. A.; Moraes, Luís F. M. Avaliando a Sobrecarga Introduzida nas Redes 802.11 pelos Mecanismos de Segurança WEP e VPN/IPSec. Workshop de segurança - WSEG. In: Anais SBRC 2003.
5. RFC 2246. "The TLS Protocol version 1.0" Available at <http://www.faqs.org/rfcs/rfc2246.html>. Access in: 15 Aug. 2003.
6. Gupta, V.; Gupta, S.; Experiments in Wireless Internet Security. Wireless Communications and Networking Conference – WCNC 2002. v. 2, p. 17-21, Mar, 2002.
7. Stallings, William. Network security essentials: applications and standards / William Stallings. ISBN: 0-13-016093-8.
8. Murthy U.; Bukhres O.; Winn W.; Vanderdez E.; Firewalls for Security in Wireless Networks. p. 1-9, 2000.
9. Kasselman, P.R. A fast attack on the MD4 hash function. Communications and Signal Processing (COMSIG97). South African Symposium Proceedings, p.147-150, sept. 1997.
10. An Open Specification for Pretty Good Privacy. Available at: <http://www.ietf.org/html.charters/openpgp-charter.html>. Access in: Jan 2003.
11. AES Advanced Encryption Standard. Disponível em: <http://csrc.nist.gov/CryptoToolkit/aes/>. Access in: Jan 2003.
12. Gilmore, John. Cracking DES: Secrets of Encryption Research, Wiretap Politics & Chip Design. 1998. 272p
13. J2ME – Java 2 Platform, Micro Edition. Available at <http://java.sun.com/j2me>. Access in: Sept 2002.
14. Sun[TM] One Studio 5, Standard Edition. Available at: <http://wwws.sun.com/software/sundev/jde/buy/index.html>. Access in: Sept 2002.

Compensating Nonlinear Amplifier Effects on a DVB-T Link

Vagner Vale do Nascimento and José Ewerton P. de Farias

Federal University of Campina Grande, Department of Electrical Engineering,
Campina Grande, Paraíba, Brazil
{vagnervale, ewerton}@dee.ufcg.edu.br

Abstract. Due to its ability to combat multipath effects, COFDM is now part of some well known digital broadcasting standards. However, COFDM can be seriously affected by nonlinear distortion caused, mainly, by the power amplifier at the transmitter. In this paper, an investigation on the nonlinear effects in a DVB-T link is reported. To minimize the nonlinear effects on link performance, a linearization subsystem is applied before the power amplifier. Simulation results for bit error rate, signal constellations, and power spectral densities for the received signal are included.

1 Introduction

The DVB-T (*Digital Video Broadcasting - Terrestrial*) standard is specified by ETSI EN 300 744 v1.4.1 recommendation [1] which describes the channel coding/modulation system intended for digital multi-programme terrestrial services.

The DVB-T system digital transmission is based on the multicarrier modulation principle called COFDM (*Coded Orthogonal Frequency Division Multiplexing*). It is a modulation system well suited for VHF and UHF terrestrial broadcasting in a frequency selective response propagation environment. Typically hundreds (or even thousands) of carriers are used, each one carrying a small part of the coded data stream. The frequency spacing is carefully defined to ensure orthogonality among the subcarriers.

The small data rate per subcarrier decreases the multipath susceptibility. A guard time interval is used to increase the COFDM signal robustness to the multipath. Channel coding and data interleaving are essential processes that contribute to a good performance of this technique.

The OFDM signal resembles a gaussian random process. This characteristic implies a large dynamic range, therefore making it vulnerable to nonlinear distortions.

To improve energy efficiency, the power amplifier is required to operate as close to its saturation point as possible. This condition makes the system vulnerable to nonlinear distortion.

J.N. de Souza et al. (Eds.): ICT 2004, LNCS 3124, pp. 958–963, 2004.

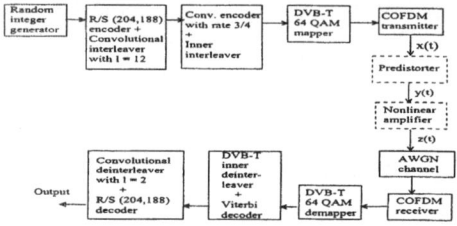

Fig. 1. DVB-T transmission: baseband model.

2 System Model

A block diagram of the transmission system used in the simulations is illustrated in Figure 1. The nonlinear amplifier and the predistorter blocks are dashed to indicate they are removable components during the simulations. The abbreviation R/S stands for Reed/Solomon and l=12 is the block length used by the interleaver. The transmitter works in the "2k" mode, that is, 1705 subcarriers are used for the transmitted signal. The receiver consists of a 64-QAM demapper that performs smooth decisions, resulting in a set of six real numbers for each complex number in the input. These six numbers represent smooth decisions over the real and imaginary components of the first, second and third bits. The Viterbi decoder works together with a punctured convolutional encoder of rate 3/4, that is important for applications with limited bandwidth availability, as in digital television transmission [2].

3 Power Amplifier Model

Using baseband representation for the signals involved ([3], Chapter 4), let $y(t)$ represent the complex envelope of the waveform at the power amplifier input, with r_y and θ_y being the time-varying magnitude and argument, respectively. The signal complex envelope, $z(t)$, at the power amplifier output can be written as

$$z(t) = y(t)G[r_y(t)] \qquad (1)$$

where

$$G[r_y(t)] \triangleq \frac{A[r_y(t)]\exp\{j\Phi[r_y(t)]\}}{r_y(t)}. \qquad (2)$$

Functions $\mathcal{A}[\cdot]$ and $\Phi[\cdot]$ represent AM/AM and AM/PM conversions [4], respectively. According to the Saleh model [5] for the travelling wave tube amplifier (*TWTA*), the functions $\mathcal{A}[r]$ and $\Phi[r]$ are:

$$\begin{cases} \mathcal{A}[r] = \frac{\nu\,r}{(1+\beta_a\,r^2)} \\ \Phi[r] = \frac{\alpha_\phi\,r^2}{(1+\beta_\phi\,r^2)} \end{cases} \qquad (3)$$

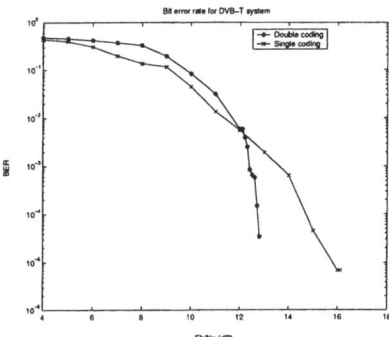

Fig. 2. Bit error rate versus E_b/N_o for the DVB-T system with single and double coding layers in a AWGN channel.

where ν is the small-signal gain, and $\mathcal{A}_{sat} = 1/\sqrt{\beta_\alpha}$ is the input saturation voltage. The maximum output amplitude is $\mathcal{A}_0 = m\acute{a}x_r\{A[r]\} = \nu\,\mathcal{A}_{sat}/2$. In addition, $\phi_\infty = \alpha_\phi/\beta_\phi$ is the maximum phase displacement that can occur in the amplified signal. A set of values for the above parameters are:

$$\nu = 2.00 \quad \beta_\alpha = 1.00 \quad \alpha_\phi = \frac{\pi}{3} \quad \beta_\phi = 1.00 \ . \tag{4}$$

Nonlinear distortions depend on the *output backoff* (OBO), see reference [4].

4   Power Amplifier Linearization

The power amplifier linearization method used in the system model is based on that described in reference [6].

According to equation (1), and considering the values given in (4), the magnitude and argument of the complex envelope at the power amplifier output can be written as:

$$\begin{cases} r_z = \mathcal{A}(r_y) \\ \theta_z = \theta_y + \Phi(r_y) \end{cases} \tag{5}$$

where dependence upon time has been dropped for simplicity. Recalling that (1) is a function of the squared input amplitude r_y^2, it is convenient to express (5) as follows:

$$\begin{cases} r_z^2 = \mathcal{A}_2(r_y) \\ \theta_z = \theta_y + \Phi(r_y) \end{cases} \tag{6}$$

where $\mathcal{A}_2(r_y) \overset{\Delta}{=} \mathcal{A}^2(r_y)$.

Let $\mathcal{A}_2^{-1}$ denotes inverse transformation of $\mathcal{A}_2$, and r_x and θ_x represent magnitude and argument, respectively, of the complex envelope x at the predistorter input. To obtain an exact linearization of the power amplifier, the input-output

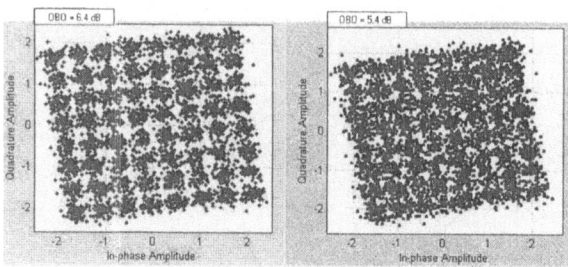

Fig. 3. 64-QAM signal constellation for OBO equal to a) 6.4 dB and b) 5.4 dB over an ideal channel.

response of the predistorter should be equal to the inverse response of the power amplifier. Then, the ideal predistorter output is:

$$y = \mathcal{A}_2^{-1}(r_x^2) \exp j[\theta_x - \Phi(r_y)]$$
$$= x\mathcal{B}(r_x^2) \exp[-j\Theta(r_x^2)] \qquad (7)$$

where

$$\begin{cases} \mathcal{B}(r_x^2) \triangleq \dfrac{\mathcal{A}_2^{-1}(r_x^2)}{r_x} \\ \Theta(r_x^2) \triangleq \Phi[\mathcal{A}_2^{-1}(r_x^2)] \ . \end{cases} \qquad (8)$$

In practical realization, the real-valued functions $\mathcal{B}(r^2)$ and $\Theta(r^2)$ can be approximated by polynomials over the interval $r \in (r', r'')$, as follows:

$$\begin{cases} \mathcal{B}_{N_B}(r^2) \triangleq \sum_{i=0}^{N_B} \alpha_i r^{2i} \\ \Theta_{N_\Theta}(r^2) \triangleq \sum_{i=0}^{N_\Theta} \beta_i r^{2i} \ . \end{cases} \qquad (9)$$

The coefficients of the above polynomials are determined by considering a finite number N_s of uniformly-spaced samples $\mathcal{A}(r_i)$ and $\Phi(r_i)$, $i = 1, 2, ..., N_s$ of the functions $\mathcal{A}(r)$ and $\Phi(r)$, over the interval $r \in (r', r'')$. Reference [6] has the details on how to obtain these coefficients.

Assuming $N_B = N_\Theta = 2$ and using $N_s = 20$ samples of responses in (3) with uniform spacing in $r \in (0,1)$, for 64-level QAM, (9) yields

$$\begin{cases} \mathcal{B}_2(r^2) = 0.2813 \cdot r^4 + 0.0207 \cdot r^2 + 0.5078 \\ \Theta_2(r^2) = 0.1995 \cdot r^4 + 0.2065 \cdot r^2 + 3.2 \cdot 10^{-3} \ . \end{cases} \qquad (10)$$

5   Simulation Results

A program in MATLAB© to interact with the DVB-T system simulator of Figure 1, has been developed. The program executes the simulator 28 times. For each iteration a different E_b/N_o value for the AWGN channel is taken, and the corresponding value of the bit error rate is evaluated, both at the link end and between the inner and outer coding layers. The results are presented in Figure 2.

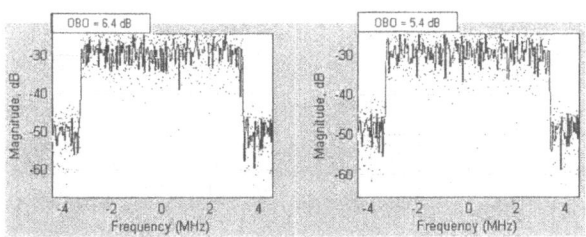

Fig. 4. Power spectra of the COFDM signal for OBO equal to a) 6.4 dB and b) 5.4 dB over an ideal channel.

A sizable difference between the performances for these two situations has been found. In the case of a single layer of coding, the link only achieves a reasonable bit error rate for $E_b/N_o = 16$ dB. When using a double layer of coding, a quasi-error-free transmission can be reached with no more than approximately $E_b/N_o = 13,5$ dB. At this point the curve tends to a 90° slope as seen in Figure 2.

In order to isolate the effects of the power amplifier, an ideal channel has been used. Due to the insertion of the amplifier model described in section 4, large amplitude and phase distortions have been observed in the COFDM signal. This is illustrated in Figure 3. A bit error rate of 10^{-1}, much greater than the maximum tolerable practical value [7], has been obtained. Setting the amplifier operating point closer to saturation, that is, using $OBO = 5.4$ dB (Figure 3b) instead of $OBO = 6.4$ dB (Figure 3a), a greater distortion is verified in the detected signal. Thus, the signal distortion must be traded off against the power efficiency of the amplifier.

Now, concerning the COFDM signal spectrum, an out-of-band amplification has been observed as in any practical amplifier. In Figure 4 the spectra for two values of output backoff ($OBO = 6.4$ dB and $OBO = 5.4$ dB) are presented. This undesirable amplification gives rise to a difference between the in-band and out-of-band power levels, for $OBO = 6.4$ dB and $OBO = 5.4$ dB, of about 23 dB and 20 dB, respectively.

To alleviate the nonlinear power amplifier effects, the predistortion model described in section 4 is inserted before the power amplifier block, as shown in Figure 1. For the worst case studied ($OBO = 5.4$ dB), the application of the predistortion technique resulted in a quasi-distortion-free constellation, as illustrated in Figure 5a. This confirms the usefulness of the linearization method.

Another benefit of the predistortion technique is the decrease of adjacent channel interference caused by intermodulation products generated in the power amplifier. The out-of-band amplification is considerably reduced, as can be seen in Figure 5b. The spectrum has been plotted for an output backoff equal to 5.4 dB.

6   Conclusion

In this paper, the importance of using two layers of channel coding, instead of just one, has been confirmed. Transmission quality is either high (quasi-error-

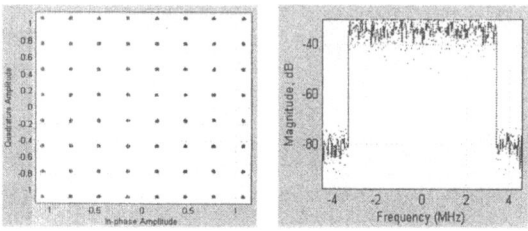

Fig. 5. a) 64-QAM signal constellation and b) Power spectrum of the COFDM signal, both using the predistortion technique for $OBO = 5.4\ dB$.

free transmission for E_b/N_o approximately greater than or equal to 13.5 dB) or very low (high bit error rate for E_b/N_o approximately less than or equal to 12.5 dB).

A considerable amplitude and phase distortion has been observed in the received COFDM signal. An undesirable out-of-band amplification is also present. This effect may originate adjacent channel interference on others stations. By choosing an amplifier operating point closer to saturation, that is, using $OBO = 5.4\ dB$ instead of $OBO = 6.4\ dB$, a greater distortion is verified in the detected signal. Thus, signal distortion must be traded off against amplifier power efficiency.

A predistortion technique has been used to mitigate the transmission impairments generated by the power amplifier. The linearization method employed is efficient, for the amplitude and phase distortions have been almost completely eliminated. The predistorter also plays the role of an RF filter, as the out-of-band amplification is now almost negligible. Finally, we can conclude that, according to our simulations, the chosen predistortion technique is an adequate approach to mitigate the nonlinear power amplifier effects on a DVB-T link.

References

1. ETSI EN 300 744 V1.4.1, Digital Video Broadcasting (DVB); Framing structure, channel coding and modulation for digital terrestrial television, Sophia Antipolis, France, 2001.
2. G. Begin e D. Haccoun, "High-rate punctured convolucional codes: structure properties and construction technique", IEEE Trans. on Communications, vol. 37, no. 12, December 1989, pp. 1381-1385.
3. J. G. Proakis, Digital Communications, 4ed., McGraw-Hill, 2000.
4. R. M. Gagliardi, Satellite Communications, Van Nostrand Reinhold, 2nd ed., 1991.
5. A. A. M. Saleh, "Frequency-Independent and Frequency-Dependent Nonlinear Models of TWT Amplifiers", IEEE Trans. on Communications, vol. COM-29, No 11, November 1981, pp. 1715-1720.
6. A. N. D'Andrea, V. Lottici e R. Reggiannini, "RF Power Amplifier Linearization Through Amplitude and Phase Distortion", IEEE Trans. on Communications, vol. 44, No 11, November 1996, pp. 1477-1484.
7. Gerald W. Collins, Fundamentals of Digital Television Transmission, Wiley, 2000.

Controlled Load Services in IP QoS Geostationary Satellite Networks

F. De Rango, M. Tropea, and S. Marano

University of Calabria, D.E.I.S. Department, 87036 Rende (CS), Italy.
{derango,marano}@deis.unical.it
mauro.tropea@tin.it

Abstract. In this paper a multimedia satellite platform called EuroSkyWay and the IntServ Controlled Load Services (CLS) have been considered. The behavior of the system loaded with Controlled Load (CLS) traffic and Guaranteed (GS) traffic has been evaluated. The management of CLS is important in order to manage the category the traffic called "better than best-effort" and to guarantee a full interoperation with the terrestrial network. In particular, Call Admission Control phase for CLS in the satellite network has been studied. Different simulations have been outlined and the conformance to QoS parameters for the approaches have been verified. Furthermore, the satellite utilization and the average end-to-end delay bound for different values of bandwidth levels declared in satellite admission phase have been evaluated.

1 Introduction

The recent development in telecommunication networks has revealed two important tendencies. The first one is the huge drive in Internet traffic. Many applications like browsing the World Wide Web (WWW) or Electronic Mail (Email) have become a part of everyday life. Secondly, the demand for wireless services has caused wireless networks to extend at an enormous rate. With the introduction of satellite system for wireless networks, operating with the global terrestrial system, network providers aim at closing the gap between these two trends.

The new applications which make use of enormous amounts of bandwidth it has pushed the community to study new architecture for adapting the current internet technologies to multimedia services, since it does not provide adjustable quality of service (QoS).

Several concepts for QoS at the IP level have been proposed, in particular the Internet Engine Task Force (IETF) has been proposed integrated services (IntServ) [1] and differentiated services (DiffServ) architectures [2].

In our work we have considered a QoS architecture inside a geostationary satellite network [7]. The satellite architecture considered is EuroSkyWay platform in which a GEO satellite in Ka band with On Board Processing (OBP) payload can interface with heterogeneous broadband networks [5].

We have considered the QoS architecture defined by IETF: Integrated Services (IntServ) and we have study in particular the class of service Controlled Load and the

J.N. de Souza et al. (Eds.): ICT 2004, LNCS 3124, pp. 964–972, 2004.
© Springer-Verlag Berlin Heidelberg 2004

behaviour of the system loaded with Controlled Load (CLS) traffic and Guaranteed (GS) traffic [4].

The rest of this article is structured as follows. Section II describes the reference EuroSkyWay [5] satellite system chosen as an example of next generation broadband satellite platform. Section III reports Guaranteed and Controlled Load services[3].; in section IV the performance evaluations have been considered and at then in section V, the conclusions have been summarized

2   Satellite Platform

The GEO satellite system architecture we take as a reference in this work is Euro Sky Way (ESW) [5], which is an enhanced satellite platform for multimedia applications. The satellite system, with the its main elements, is presented in fig.1. It is composed mainly of the following parts:
- a multi-spot beam geostationary Ka-band satellite with multi-frequency TDMA (MF-TDMA) access on uplink and onboard processing (OBP) capability;
- a gateway earth station which interfaces the terrestrial network through protocol adapting interfaces (ISDN, ATM, PSTN, etc.) and inter-working functionality;
- Satellite user Terminals (SaT) of different types (SaT-A, SaT-B, SaT-C);
- a Master Control Station (MCS) which is responsible for the overall ESW system super visioning and control. In particular, the MCS is responsible for Call Admission Control (CAC). The reference system uses statistical CAC to increase satellite resources utilization.

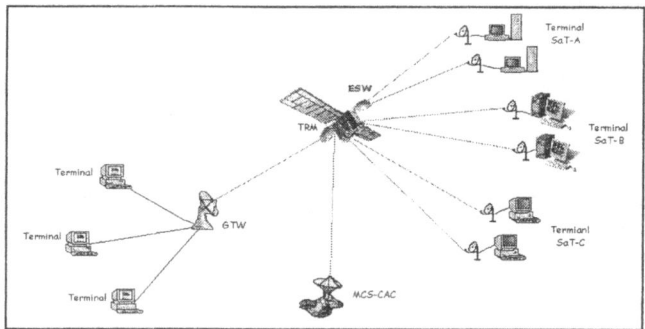

Fig. 1. Geostazionary Satellite Reference Network

The satellite has OBP capability and implements traffic and resource management functions (TRM). Functionality of traffic and resource management is distributed in different modules of the overall system. Call Admission Control and resource management functions are distributed in the satellite user terminal and earth stations. The satellite network can be seen as an underlying network, aiming to interface a wide user segment based on different protocols, such as IP, asynchronous transfer mode (ATM), X.25, frame relay, narrowband integrated services digital network (N-ISDN), and MPEG-based ones (so called overlying networks, OLNs).

Two channel assignment modes are provided in the satellite network: permanent and semi-permanent connections. According to the permanent mode, resources are assigned for the whole duration of the connection and released only following the reception of an explicit connection-end notification. In the case of semi-permanent connections, the TRM module on-board allocates satellite resources for a fixed time interval (multiple of frame period). In other words, available resources are dynamically allocate by the TRM to semi-permanent connections each time the relevant sources generate a new information burst.

2.1   Call Admission Control and Resource Management

Call Admission Control and resource management functions are distributed in the satellite user terminal and earth stations [6].

Call Admission Control (CAC) represents a module of Network Operation Center (NOC) disposed on a terrestrial station. Its task is to regulate the access to satellite segment. It permits a flexible handling of the bandwidth and avoids the *a priori* partitioning of the resources among different types of service. The CAC algorithm has been designed also to fulfil the objectives of minimizing the signalling exchange between the on-board and on-earth segments of the system. In order to reduce delays due to the processing of the call requests on board, the parameters relevant to the processed calls are stored and elaborated within the ground segment.

The Traffic Resource Manager (TRM) is the satellite entity which manages transmission resources. It is equipped with different databases and with a calculator for an optimum resource management. The calculator, for each frame time (26.5 ms), memorizes the arrived requests. In the next frame time, it analyses the requests and checks the possibility of satisfying the requests through the consulting of the occupation resources table. If it is possible to satisfy the requests, TRM sends a resource assignment message to the terminal requesting satellite channels. For more details see [6].

3   IntServ on Satellite Network

The Integrated Services (IntServ) have the advantage of providing quantitative guarantees on traffic. This architecture [7] is suitable for a satellite network capable of offering guaranteed services with control on the flow basis. Reference is made to the Euro Sky Way (ESW) network [5] where it is possible to use permanent and semi-permanent connections and it is important to define the mapping of IntServ classes on satellite connections.

3.1   Guaranteed Services

Guaranteed service (GS) guarantees that datagrams will arrive within the guaranteed delivery time and will not be discarded due to queue overflows, provided the flow's traffic stays within its specified traffic parameters. This service is intended for applications which need a firm guarantee that a datagram will arrive no later than a certain time after it was transmitted by its source.

The flow's level of service is characterized at each network element by a bandwidth (or service rate) R and a buffer size *B*. *R* represents the share of the link's bandwidth the flow is entitled to and *B* represents the buffer space in the network element that the flow may consume [4].

In [7] the rate *R* requested by the receivers for permanent and semi-permanent connections has been computed. R is determined by starting from the value of the requested end-to-end delay bound (DB) which is:

$$DB = \frac{(b-M)}{R} \cdot \frac{p-R}{p-r} + \frac{M+C_{tot}}{R} + D_{tot} \qquad \text{if } p > R \geq r \qquad (1)$$

$$DB = \frac{M+C_{tot}}{R} + D_{tot} \qquad \text{if } R \geq p \geq r \qquad (2)$$

In (1) e (2) C_{tot} and D_{tot} are two error terms, which represent how the network elements' implementation of Guaranteed Services (GS) deviates from the fluid model [7].

The two connection modes present different *D* values which impact terrestrial network in terms of a different requested rate in order to obtain the same end-to-end maximum delay. So we have:

- D_{perm} = 26.5 ms (frame delay) for permanent connections.
- $D_{semiperm}$ = 250ms (round-trip delay) + 26,5ms (maximum processing time onboard the spacecraft) + 26,5ms (frame duration) = 303ms for semi-permanent connections. For C_{tot} values see previous works [6-7].

Only semi-permanent connections have been considered in this work.

3.2   Controlled Load Services

The end-to-end behavior provided to an application by a series of network elements providing controlled-load service tightly approximates the behavior visible to applications receiving best-effort service "under unloaded conditions" from the same series of network elements. Assuming the network is functioning correctly, these applications may assume that:

- A very high percentage of transmitted packets will be successfully delivered by the network to the receiving end-nodes.
- The transit delay experienced by a very high percentage of the delivered packets will not greatly exceed the minimum transmit delay experienced by any successfully delivered packet.

To ensure that these conditions are met, clients requesting controlled-load service provide the intermediate network elements with a estimation of the data traffic they will generate; the *TSpec*.

It guarantees a QoS similar to those achievable by best effort traffic in an unloaded network, with low queuing delay and low dropping probability.

The controlled load service is intended to support a broad class of applications which have been developed for use in today's Internet, but are highly sensitive to overloaded conditions. Important members of this class are the "adaptive real-time applications" currently offered by a number of vendors and researchers. These applications have been shown to work well on unloaded nets, but to degrade quickly under overloaded conditions. A service which mimics unloaded nets serves these applications well.

The controlled-load service does not accept or make use of specific target values for control parameters such as delay or loss. Instead, acceptance of a request for controlled-load service is defined to imply a commitment by the network element to provide the requestor with service closely equivalent to that provided to uncontrolled (best-effort) traffic under lightly loaded conditions.

Differently from [8] in which the CLS services are served to bandwidth level corresponding to the peak value p, now it is proposed the reduction of bandwidth more CLS traffic because it can be more tolerant to maximum end-to-end delay. So the CLS traffic is discriminated not only on TRM trough the queue priority, but is discriminated also in admission phase. The CAC module requests two parameters for admitting a call: burstiness value for regulating the statistical multiplexing of traffic sources and a bandwidth requests for assigning a certain number of satellite channels [6]. For CLS it is given a bandwidth level in this way:

$$P_{smoothed} = k * p \qquad (3)$$

where k is a reduction factor called *smoothing factor* and p is the token bucket peak value. k can vary between 0 and 1. Reducing k, less satellite channels are given to CLS services an this means more time consumed to serve CLS traffic but more traffic managed on satellite connections.

4   Performance Evaluation

In order to evaluate the performance of the satellite system introducing the smoothing factor, different simulation trials have been lead out. Only semi-permanent satellite connections have been evaluated. In Table 1 are summarized the most important considered and fixed parameters in the simulation scenario.

The parameters considered for performance evaluation are:

- Total Satellite Utilization: r/R where r is the average token bucket data rate and R represents the bandwidth requested by the satellite receiver for GS service.
- CLS Satellite Utilization: r/R where r is the average token bucket data rate and R represents the bandwidth requested by the satellite receiver for CLS service.
- Average CLS End-to-end Delay bound: it is the average delay bound registered by satellite receivers.

The simulation campaign has been conducted considering different traffic conditions. In Scenario 1 there are high requests of CLS traffic and low requests of GS traffic. In Scenario 2, instead, the inverse situation has been evaluated. The different percentage have been simulated through the introduction a traffic load percentage. There is the assumptions that for each traffic source there is a satellite receiver that requests a service to the traffic source considered. In Table 2 the percentage values of the traffic load have been showed.

Table 1. Simulation Parameters

Source Parameters	
Traffic Sources	Real time variable bit rate
Number of terrestrial sources	256
Burstiness β	4
Bucket size b	512
Peak rate p	$\beta* r$
Rate r	128
Receiver parameters	
Number of satellite receivers	256
Satellite receiver type	SaT-C
Delay Bound (DB) requested	500, 600, 700, 800, 900, 1000 ms
Smoothing Factor k	1, 0.9, .08, 0.7, 0.6
Satellite parameters	
Round Trip Time	540 ms
Medium Access Protocol	MF-TDMA
Timeout n_{GS} for GS requests in OBP	1 frame time (26.5 ms)
Timeout n_{CLS} for CLS requests in OBP	13 frames time(344.5 ms)
DB$_{semiperm}$	477 ms
Satellite Link Bandwidth	16 Mbit/s
Atomic satellite channel	16 Kbit/s
Target burst loss probability (ε)	0.01

Table 2. Traffic Load

	GS Traffic Percentage	CLS Traffic Percentage	
Scenario 1	80%	20%	High CLS traffic
Scenario 2	20%	80%	Low CLS traffic

4.1   Scenario 1

Fig. 3 demonstrates the better utilization of satellite link for CLS class of service. This is due to increasing number of CLS calls for lower values of smoothing factor. The simulations have showed the reduction of GS utilization (not presented here) but this reduction is minimal in respect of the improvement of CLS utilization. This produces a better total satellite utilization as shown in fig.2

The simulation in fig.2-fig.3 shown how the smoothing factor can give improvement in terms of satellite link utilization for different DB requested by satellite GS receivers.

So, it is suitable to use a reduction of bandwidth requested by CLS calls through the smoothing factor k, in order to admit more calls CLS that can tolerate more delay before being served on TRM module. For k=1, the satellite receivers request a bandwidth corresponding to the token bucket peak value of traffic source [8]. This values is the maximum bandwidth request that can be given to CLS requests. The simulation shows that for different DB and burstiness values β (only the case of β=4 is reported in this paper) it is possible to declare in admission phase a bandwidth for CLS traffic proportional to smoothing factor k and peak value p, preserving the EDP of 1% for both GS and CLS traffic classes (the EDP is not showed in this work for space limitations but it is verified).

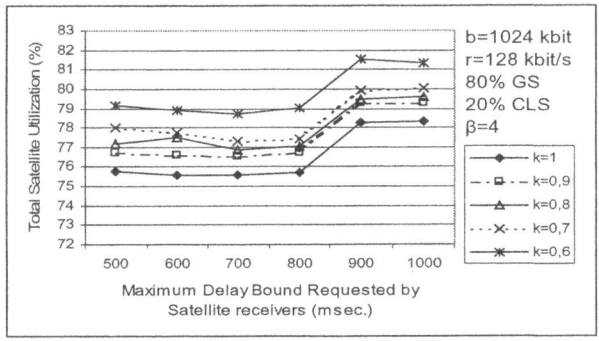

Fig. 2. Satellite Utilization versus smoothing factor for β=4

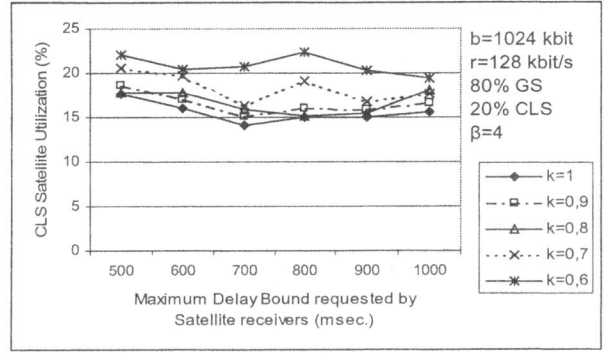

Fig. 3. CLS Satellite Utilization versus smoothing factor for β=4.

In fig.4 the average end-to-end delay requested by satellite receivers is showed. It is possible to observe as for increasing values of smoothing factor k, the average delay increase. This is due the increasing number of CLS admitted calls that can be managed on TRM module on satellite system. For k=0,6 the maximum delay bound is greater than 500ms but the average delay bound is limited to 500ms. This is important for CLS services because also if the maximum delay bound is specified only by GS services, the CLS can not wait in the queues indefinitely.

4.2   Scenario 2

In fig.5 and fig.6, as for the scenario 1 case, the reduction of smoothing factor k reduces the satellite bandwidth improves the bandwidth utilization of satellite system

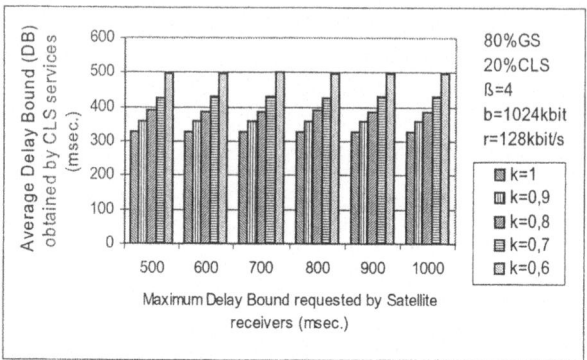

Fig. 4. Average Delay Bound requested by satellite receivers versus smoothing factor for β=4

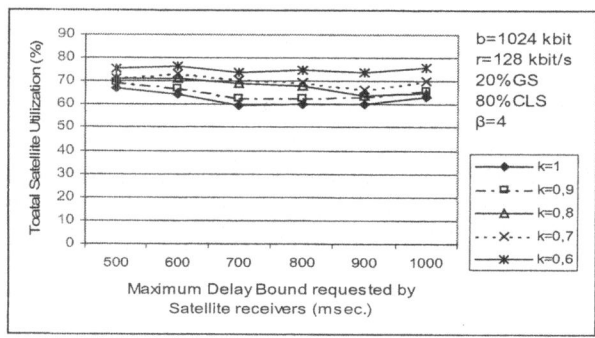

Fig. 5. Total Satellite Utilization for GS and CLS classes of traffic versus smoothing factor for β=4.

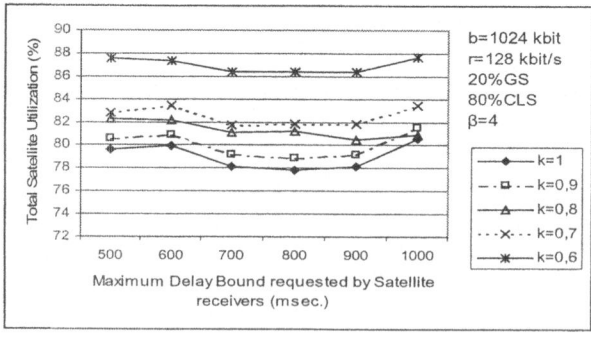

Fig. 6. Total Satellite Utilization for CLS class of service versus smoothing factor. for β=4.

and CLS traffic. Fig.5 shows that the improvement of overall satellite system is more effective (8-10%).

The improvement of CLS utilization is more effective because the high traffic of CLS sources use the satellite channels for more time and the admitted CLS sources are maximized. In scenario 1, instead, reducing the bandwidth request of GS services the queue on TRM module is not enough fill up offering a lower overall utilization.

The average delay bound requested by satellite receivers is not reported for scenario 2 because its behavior is similar to that reported in fig.4.

5   Conclusions

In this paper the management of Controlled Load services has been addressed. A policy based to priority queues on TRM satellite module has been considered and a reduction of bandwidth for CLS services has been proposed. The simulations show the improvement given by smoothing factor for high CLS traffic but they give also an idea of how the smoothing factor impact on overall system performance. The light traffic of CLS sources does not give effectiveness to bandwidth level reduction. The average end-to-end delay bound is limited to 500ms for different smoothing factor values preserving the DB requested by satellite receivers. So reduction of bandwidth levels for CLS services improves the total satellite utilization without delaying indefinitely the CLS services.

References

[1]   P.White, "RSVP and Integrated Services in the Internet: A Tutorial," *IEEE Communications Magazine*, May 1997.
[2]   Blake, S., Black, D., Carlson, M., Davies, E., Wang, Z. and W. Weiss, "An Architecture for Differentiated Services," *RFC 2475*, December 1998.
[3]   J. Wroclawski, "Specification of the Controlled-Load Network Element Service", *RFC 2211*, Sept. 1997.
[4]   Shenker, S., Partridge, C., and R Guerin, "Specification of Guaranteed Quality of Service", *RFC 2212*September 1997.
[5]   G. Losquadro, M. Marinelli "The EuroSkyWay system for interactive multimedia and the relevant Traffic Management," *Proc. of the third Ka-band Utilisation Conference*, pp.17-24, Sep. 1997
[6]   A. Iera, A. Molinaro, S. Marano, "Call admission control and resource management issues for rel-time VBR traffic in ATM-satellite networks," *IEEE J.Selct Areas Commun.*, vol.18, pp.2393-2403, Nov.2000.
[7]   A. Iera, A. Molinaro, S. Marano, "IP with QoS guarantees via GEO satellite channels: Performance issues," *IEEE Personal Communications*, no. 3, June 2001 pp. 14-19.
[8]   F.De Rango, M.Tropea, S.Marano, "Controlled Load Service Management in Int-Serv Satellite Access Network", to appear on *CCECE 2004 Int. Conference*, Ontario, Canada, May 2004.

Fast Authentication for Inter-domain Handover

Hu Wang and Anand R. Prasad

DoCoMo Communications Laboratories Europe GmbH
Landsberger Str. 308-312, 80687, Munich, Germany
{wang, prasad}@docomolab-euro.com

Abstract. This paper proposes a method of fast authentication for inter administrative domain handover between two foreign mobile or wireless communication network domains. It is based on mutual trust and agreement between the two foreign domains and can be used as a complement of full authentication to reduce delay in handover phase.

1 Introduction

The IMT-2000 (International Mobile Telecommunications) system, usually called 3rd Generation (3G) mobile communication system, provides high bandwidth data communication over large areas. With technologies like Wireless Local Area Network (WLAN), much higher data rate can be achieved in relatively small spots. In the future, the Beyond 3rd Generation (B3G) mobile communication systems will encompass heterogeneous radio and network technologies to provide user seamless access to network and services, even when the networks are owned and managed by different operators. To support seamless mobility, inter-domain and vertical handover should be performed without much delay and packet loss.

Inter-domain handover in this paper means inter-administrative-domain, i.e. an active communication session changes its attachment point from one network domain to another, in which the two domains are managed by different operators. As part of most public networks' security policy, Mobile Node (MN) must be authenticated when it handovers from other administrative domain. On the other side, MN also requires authentication of the network in order to avoid connecting to a roguish Base Station or Access Point (AP).

When inter-domain handover happens between foreign administrative domains, a full mutual authentication as performed for 3G roaming can hardly fulfil the requirement of short delay [16]. A study on Mobile IP (MIP) authentication performance shows that the delay of a full authentication for inter-domain MIP handover is one order of magnitude longer than that for intra-domain handover [4]. Though MIP handover is different from the handover concept of B3G systems, some of the analyses also apply to the inter-domain handover studied in this paper. The reasons of the long delay are: full authentication involves MN's Home Network (HN), and the end-to-end delay between foreign network and HN becomes the main part of the total delay.

This paper proposes a fast authentication method for seamless inter-domain handover to reduce the delay. In Section 2 the handover domain model and security re-

J.N. de Souza et al. (Eds.): ICT 2004, LNCS 3124, pp. 973–982, 2004.
© Springer-Verlag Berlin Heidelberg 2004

quirements are studied. The proposed method is described in Section 3, and a proposal to combine the authentication method with handover is given in Section 4.

2 Background

This section describes background study of inter-domain handover, including its domain model, security requirements on authentication and related works.

2.1 Domain Model of Inter-domain Handover

Domain model of inter-domain handover is studied in [16] and given in Fig. 1. The involved network domains are MN, HN, SN (Serving Network), TN (Target Network) and CN (Corresponding Node) (see Table 1). An administrative organization controls one or many networks, thus forms an administrative domain. In this paper the said network domains are assumed to be controlled by different administrative organizations, which may be service providers, network operators, or persons. Therefore, this paper is studying handover between two foreign domains. The reference points are identified in Fig. 1, which represent various relation and interaction (e.g. service provision, signalling) between a pair of domains.

Table 1. Administrative domains involved in inter-domain handover

MN	it is user domain including the mobile terminal
HN	the network domain in charge of user subscription and other supporting services, like billing and authentication
SN	the network domain that serves MN before handover
TN	the network domain that serves MN after handover
CN	the node that MN is talking with

2.2 Security Requirements on Authentication

Security requirements on authentication of inter-domain handover are studied in [16] and summarized here:

- It must be a mutual authentication between MN and TN.
- Fast authentication should be built on minimal trusts between SN and TN. To be specific, the old security keys shared by MN and SN must not be known by TN, and the new security keys shared by MN and TN must not be known by SN. However, depending on TN's security policy, part of the information may be disclosed to SN in order to smooth the handover (see Section 4).
- MN's identity must be protected.
- Authentication must be resistant to various attacks, such as man in the middle, denial of service, session hijacking and replay attack.

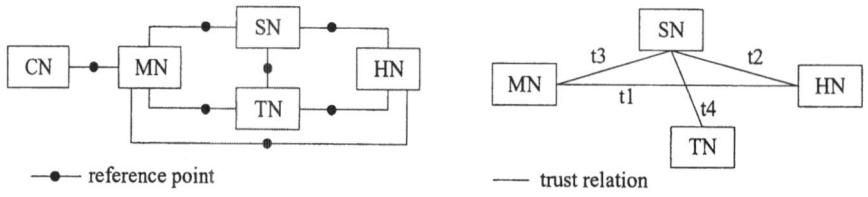

Fig. 1. Administrative domain model for inter-domain handover [16]

Fig. 2. Trust relations existing before inter-domain handover

2.3 Related Works

Inter-domain handover is not well studied in both academia and industry due to a very low demand among mobile users. Besides few proprietary solutions, it is not supported in most mobile networks. In 3GPP's latest technical specification of 3G mobile system, inter-domain handover remains to be an optional feature with only the basic requirements identified [9] – there it is called inter-PLMN (Public Land Mobile Network) handover. On the other side, the present distribution of WLANs is so sparse that inter-domain handover can hardly make any sense. It is a technology for future mobile communication systems.

IETF Seamoby Working Group is working on seamless IP mobility within one network domain. A technique to smooth network layer handover is context transfer: to transfer context of QoS, AAA, security, header compression, etc., from previous access router to new router in order to avoid the lengthy process of setting up the states from scratch [10-11]. The ideas of reusing existing state and processing in a distributed way inspired this proposal of fast authentication. Besides, some security issues of context transfer are also relevant [8].

IEEE 802.11f study group proposed a protocol called IAPP (Inter-Access Point Protocol) to support interoperability between WLAN APs. Transferring and sharing context parameters to APs to which mobile station might handover is standardized in [5] to support Layer 2 fast handover. Though their scope is limited to WLANs of one network operator, some ideas like proactive key distribution [12] are valuable to inter-domain handover, too.

MIP is usually regarded as a macro-mobility solution for IP networks [1-2]. The authentication of MIP handover is not a fast authentication. However, the MIP architecture is similar to the inter-domain handover model given in Fig. 1. Therefore, an evaluation of MIP authentication performance [4] is adopted for the analysis of handover delay, as discussed in Section 1.

The general authentication mechanisms used for mobile and wireless networks are discussed in many references, like [13-15]. However, only with them the inter-domain handover cannot be seamless.

3   Fast Authentication for Inter-domain Handover

In B3G systems, seamless mobility of mobile terminal requires short delay, low or no packet loss of handover between access networks, even for inter-domain vertical handover. A full authentication that involves MN's HN usually implies a long delay, especially when handover happens between two foreign networks; thus, communicating with HN should be avoided in the time critical parts of the handover. On the other hand, mutual authentication of MN and TN must be performed for every inter-domain handover. To achieve the goal of seamlessness, a third party is preferable, which can play the role of HN in the authentication, but introduces much less delay to the handover. In this proposal SN is selected as the third party. In brief, the proposal is to setup a temporary trust relation between MN and TN in the phase of inter-domain handover based on the existing trust relations between SN and MN, and SN and TN, instead of the trust to MN's HN. This section discusses prerequisites of the proposed method at first. Then the authentication method is described with some remarks.

3.1   Prerequisites and Trusts

The proposed authentication method for inter-domain handover between two foreign domains has a few prerequisites that should be sufficed before start the handover.

- SN and TN provide similar level securities in their networks, e.g., mutual authentication and secure communication.[1]
- MN and SN are mutually authenticated and their bilateral trust relations are ensured by security context of the session, such as a session key. The authentication might have been performed with help of HN or a trusted third party; however, this is out of the scope of the inter-domain handover.
- SN and TN have the trust relations and a secure channel established beforehand.
- To be independent of access technology, a universal naming scheme of User / MN should be adopted or methods to relate different formats are supported. Candidates include Network Access Identifier and International Mobile Subscriber Identity.
  The trust relations existing before handover are given in Fig. 2:
- t1. Trust between MN and HN is a long-term relation established when subscriber signed a contract to HN.
- t2. Trust between HN and SN could be a long-term relation or be dynamically setup, e.g. with help of third parties like AAA brokers. It is established at latest before the handover happens.
- t3. Trust between MN and SN is a short-term trust relation (usually valid for the ongoing session) and is established with (either direct or indirect) help of the HN.
- t4. Because SN and TN are close to or overlap each other, they could have the trusts built beforehand in order to support inter-domain handover of a roaming MN. The trusts may be: TN trust that SN handover only the MNs belonging to a pre-defined set of HNs; and SN trust that TN will treat those MNs according to the agreement.

---

[1] To handover a MN from a hotspot WLAN not supporting mutual authentication to a 3G network with mutual authentication enforced is dangerous and insecure. Technologies like IEEE 802.1X could solve the problem of WLAN to some extent.

3.2 Fast Authentication Method

In this proposal, MN's HN is excluded from the authentication in order to reduce delay. The security context built between MN and SN are used to generate cryptographic keying material for the authentication. The trust between SN and the TN is employed, too. However, there is no violation of any domain's security policy, e.g. the session keys used for the SN are not known to the TN and vice versa. The method is described here (see Fig. 3).

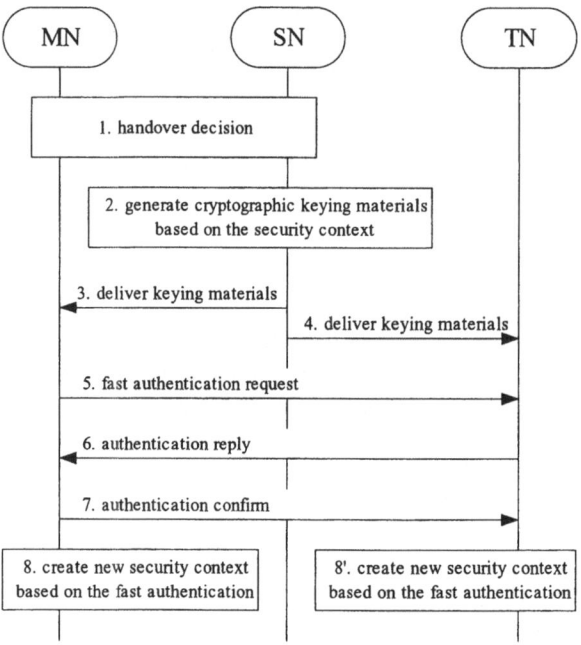

Fig. 3. Fast authentication for inter-domain handover (1)

1. The method starts after handover decision is made by MN, SN or both of them co-operatively.
2. SN generates a cryptographic keying material k_M for MN and TN. It must be unpredictable to any others including MN and TN, and should be only used for the upcoming handover authentication. An exemplary construction is:

```
k_M := r0 XOR k_S
```

   in which k_S is a session key used between MN and SN, and $r0$ is a random number generated by SN to mask k_S. XOR is bit operation of logic exclusive OR.
3. SN sends MN the identity of TN – ID_{TN}, which could be its Mobile Network Code, domain name or other format; the temporary identity of MN – TID_{MN}, which is only used in the handover in order to hide the identity to eavesdroppers; and the keying material k_M. The message is encrypted with the shared secret key of MN and SN: k_SM.

$$\text{SN} \rightarrow \text{MN}: \{ \text{ID}_{TN} \mid \text{TID}_{MN} \mid k_M \}_{k\_SM}$$

4. SN sends MN's temporary and real identities and the keying material to TN. The message is encrypted by the shared secret key of SN and TN: k_ST.

$$\text{SN} \rightarrow \text{TN}: \{ \text{TID}_{MN} \mid \text{ID}_{MN} \mid k_M \}_{k\_ST}$$

Based on this information, TN can create a quaternion (ID_{SN}, TID_{MN}, ID_{MN}, k_M), in which ID_{SN} is SN's identity, and store it in a list 'inbound-HO-waiting'. ID_{SN} is known to TN since they trust each other.

5. MN generates a random number $r1$, creates a piece of message, and sends the message with a keyed hash digest to TN as fast authentication request. Hash function like Secure Hash Algorithm can be used here.

```
msg1 := IDSN | TIDMN | r1
MN → TN: msg1 | Hash( msg1 | kM )
```

6. Based on the received message, TN looks up in its inbound-HO-waiting list. If it matches one entry and the hash digest is verified, the MN is then identified; otherwise the authentication is failed. If success, TN generates a random number $r2$, creates a piece of message, and sends it together with a keyed hash digest to MN.

```
msg2 := IDSN | TIDMN | r1 | r2
TN → MN: msg2 | Hash( msg2 | kM )
```

7. From the received message, MN can authenticate TN by verifying the hash digest. If success, MN sends a confirmation message to TN.

```
msg3 := IDSN | TIDMN | r2
MN → TN: msg3 | Hash( msg3 | kM )
```

8, 8'. Based on the received message, TN can authenticate MN by verifying the hash digest. If success, a new security context for MN and TN can be created and data transportation can be resumed through TN.

It can happen that MN loses its connection to SN before it decides to handover. In this case the keying material generated by SN cannot be delivered to MN. A slight revision of the method in Fig. 3 can solve the problem (see Fig. 4).

1. SN generates cryptographic keying material k_M before handover decision.
2. SN sends a temporary identity together with the keying material to MN:

$$\text{SN} \rightarrow \text{MN}: \{ \text{TID}_{MN} \mid k_M \}_{k\_SM}$$

3. MN confirms SN the reception of the message.
4. In a later time MN decides to start an inter-domain handover to TN. However, if handover does not happen within the lifetime of the keying material k_M, SN will regenerate and send to MN again.
5. MN generates a random number $r1$, creates a message, and sends the message with a keyed hash digest to TN as fast authentication request.

```
msg1 := IDSN | TIDMN | r1
MN → TN: msg1 | Hash( msg1 | kM )
```

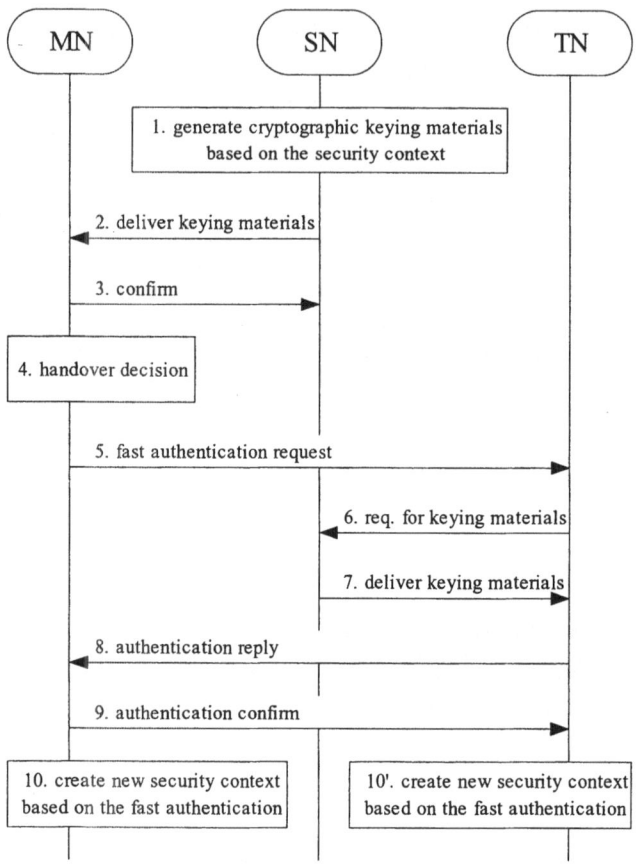

Fig. 4. Fast authentication for inter-domain handover (2)

6. TN cannot find the entry in its inbound-HO-waiting list. The message is forwarded to SN requesting for necessary information:

   TN $\rightarrow$ SN: { msg1 | Hash(msg1 | k_M) }$_{k\_ST}$

7. SN can verify the hash digest. If failed, SN should inform TN about it. If success, SN will provide MN's identity and the keying material to TN:

   SN $\rightarrow$ TN: { TID_{MN} | ID_{MN} | k_M }$_{k\_ST}$

   The steps 8 – 10 and 10' are similar to those steps 6 – 8 and 8' in Fig. 3.

3.3 Remarks

As part of the method's objective, the random numbers $r1$ and $r2$ can be used as input to the Diffie-Hellman key agreement [7], to agree on a temporary session key between MN and TN without disclosing to SN. The key agreement provides perfect forward secrecy; however, if it is not required in TN's security policy, other key agreement can be used, too. The random numbers generated in the method should only be used for the present inter-domain handover.

SN, as a trusted third party, generates the keying material based on the security context between MN and itself. Other ways of generating keying material are not excluded, supposing that TN can not deduce the original security context from the generated material, because it compromises MN and SN's security policies. The keying material generated by SN must have a short lifetime because it is only used for the handover and must be invalidated after the temporary security context is created. The lifetime could be specified by SN and enforced by MN and TN.

TN may cover a large geographic area that requires multiple inbound-HO-waiting lists distributed in its network. However, SN can provide TN the location information of MN in order to ease TN's decision of which list to use.

In Fig. 3 the message 4 and 5 must be synchronized otherwise it may fail to authenticate a legal MN. A possible way of synchronization is to let TN to send a request to SN when MN's request can not match any entry in the waiting list.

When the fast mutual authentication is successfully done, MN and TN have established trusts between themselves. TN can then start to provide service to MN and MN can accept the TN as new SN. The trust relations existing after fast authentication are shown in Fig. 5. The new trust relation established after the fast authentication is t5. It is based on the trusts t3 and t4 and is not comparable to t3, because:

- t3 depends on t1 (long term trust) and t2 (could be long term trust)
- t5 depends on t3 (short term trust) and t4 (could be long term trust)

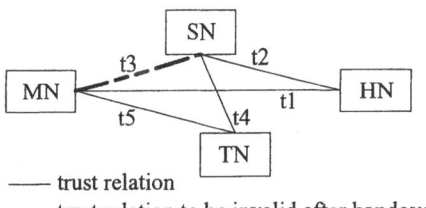

—— trust relation
‒‒‒ trust relation to be invalid after handover

Fig. 5. Trust relations existing after fast authentication

According to the agreement between SN and TN, connection of MN will be resumed when the fast authentication succeeds. In order to keep a same or similar security level after inter-domain handover, a full mutual authentication between MN and TN, with involvement of HN must be performed. However, it can be done after the handover, and the end-to-end transmission delay between HN and TN is avoided during the inter-domain handover phase.

- After the inter-domain handover is complete, any logical / physical connections between MN and SN are disrupted. Therefore, the trust relation t3 is invalid.
- Trust relation between TN and HN is not necessary to be setup during the fast authentication, this is the design objective: do not involve HN during the authentication in order to reduce delay. This trust could be setup either dynamically or statically; it is out of the scope of this paper.

4 Inter-domain Handover

One form of integrating fast authentication to inter-domain handover is given in Fig. 6. Before handover, the communication between MN and CN is going through the route a-b. Then MN and SN decide to handover (1). The secrecy is generated by SN and distributed to MN (2) and TN (3). A fast authentication is performed between TN and MN (4). If successful, SN will relay the data (5), which then goes through a-c-d. A full authentication (6) and MIPv6 Binding Update (7) are then performed. If successful, the communication will be rerouted to e-d. Then the handover is completed. The full authentication (6) could be performed in parallel and start after the procedure 4. Nevertheless, the influence of full authentication delay is avoided by the fast authentication method.

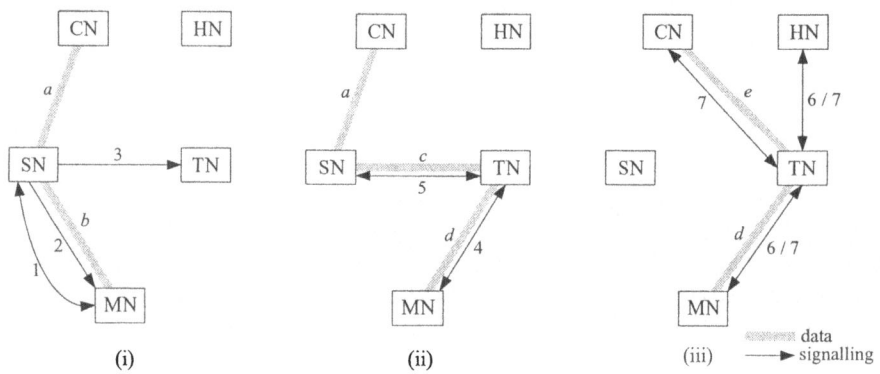

Fig. 6. Integration fast authentication method with inter-domain handover.

In this proposal, data is relayed by SN after fast authentication (Fig. 6-ii). Thus the perfect forward secrecy provided by Diffie-Hellman agreement may be unnecessary. Nevertheless, the security context built after a full authentication of MN and TN must not be disclosed to SN (Fig. 6-iii).

5 Conclusions

This paper presents a fast authentication method for inter-domain handover. Trust between two adjacent network domains, SN and TN, is used to establish a temporary trust between MN and TN. SN – as a trusted third party – generates keying materials for MN and TN, based on which a mutual authentication can be performed without involving MN's HN. The exclusion of HN in time critical part of handover can reduce delay dramatically.

The method is fast, secure, and does not compromise security policy of any administrative domain. It is flexible and allows to be used for heterogeneous networks. However, the proposal is far from mature. Issues like prerequisites, integration with B3G system's authentication and authorization architecture, adaptation to heterogeneous technologies, performance evaluation and synchronization, etc. need to be studied in the future.

References

[1]   R. Koodli (editor), et al., "Fast Handovers for Mobile IPv6", Internet Draft, work in pro-
      gress, draft-ietf-mobileip-fast-mipv6-07.txt, September 2003.
[2]   M. Laurent-Maknavicius, and F. Dupont, "Inter-Domain Security for Mobile IPv6", in
      Proc. of 2nd European Conference on Universal Multiservice Networks (ECUMN'02),
      Colmar, France, April 2002.
[3]   Hasan, J. Jähnert, S. Zander, and B. Stiller, "Authentication, Authorization, Accounting,
      and Charging for the Mobile Internet", TIK Report Nr. 114, ETH Zürich, TIK, June
      2001.
[4]   Hess, and G. Schäfer, "Performance Evaluation of AAA / Mobile IP Authentication", in
      Proc. of the 2nd Polish-German Teletraffic Symposium (PGTS'02), Gdansk, Poland, Sep-
      tember 2002.
[5]   IEEE, "IEEE Trial-Use Recommended Practice for Multi-Vendor Access Point
      Interoperability via an Inter-Access Point Protocol Across Distribution Systems Sup-
      porting IEEE 802.11™ Operation", IEEE standard 802.11F-2003, July 2003.
[6]   H. Wang, and A. R. Prasad, "Security Context Transfer in Vertical Handover", in Proc.
      of the 14th International Symposium on Personal, Indoor, Mobile Radio Communication
      (PIMRC 2003), Beijing, China, September 2003.
[7]   Menezes, P. van Oorschot, and S. Vanstone, Handbook of Applied Cryptography, CRC
      Press, p.515-520, 1996.
[8]   Aboba, and T. Moore, "A Model for Context Transfer in IEEE 802", Internet Draft, ex-
      pired, draft-aboba-802-context-02.txt, April 2002.
[9]   3GPP, "Handover Requirements between UTRAN and GERAN or other Radio Sys-
      tems", Tech. spec., 3GPP TS 22.129 v.6.0.0, March 2003.
[10] J. Kempf (editor), "Problem Description: Reasons For Performing Context Transfers
      Between Nodes in an IP Access Network", Internet RFC 3374, September 2002.
[11] J. Loughney (editor), M. Nakhjiri, C. Perkins, and R. Koodli, "Context Transfer Proto-
      col", Internet draft, work in progress, draft-ietf-seamoby-ctp-05.txt, October 2003.
[12] Mishra, M. Shin, and W. Arbaugh, "Proactive Key Distribution to support fast and secure
      roaming", IEEE working document, IEEE 802.11-03/084r1, January 2003.
[13] 3GPP, "3G Security; Security architecture", Tech. spec. 3GPP TS 33.102 v6.0.0, Sep-
      tember 2003.
[14] G. Schäfer, H. Karl, and A. Festag, "Current Approaches to Authentication in Wireless
      and Mobile Communications Networks", Technical Report TKN-01-002, Telecommuni-
      cation Networks Group, Technische Universität Berlin, March 2001.
[15] K. Boman, G. Horn, P. Howard, and V. Niemi, "UMTS security", Electronics &
      Communication Engineering Journal, vol.14, issue 5, p.191-204, October 2002.
[16] H. Wang, A. R. Prasad, and P. Schoo, "Research Issues for Fast Authentication in Inter-
      Domain Handover", in Proc. of the 8th Wireless World Research Forum meeting
      (WWRF#8), Beijing, China, February 2004.

Session and Service Mobility in Service Specific Label Switched Wireless Networks

P. Maruthi, G. Sridhar, and V. Sridhar

Applied Research Group, Satyam Computer Services Ltd.,
14 Langford Avenue, Lalbagh Road, Bangalore 560 025 INDIA
{sridhar_gangadharpalli, sridhar, uday_golwelkar }@satyam.com

Abstract. In this paper, we have extended our earlier works related to service-specific label switching in the context of campus wide wireless networks. The extension is to support session and service mobility. The session mobility is exemplified by the mobile users carrying out their work while on move and the service mobility is illustrated by multiple servers (mirror servers) offering the same service. We have provided a brief description of the various algorithms related to session and service mobility.

1 Introduction

Session continuity in wireless networks during handoff is increasingly gaining importance with a wide verity of services being provided in these networks. In a scenario such as that in a university campus, which consists of multiple WLANs, typically users of specific services at any given time are spread across different WLANs while the service providing servers are attached to a particular WLAN. In this scenario, both the micro and macro mobility have specific effects with respect to the session continuity. For the discussions in this paper, we consider a campus network, which has multiple WLANs and are inter-connected by WLAN routers. We focus on the session continuity during macro mobility, that is, when a mobile terminal (MT) travels across the boundary from one WLAN to other neighboring WLAN.

Session continuity requires that an effective session transfer mechanisms to be in place. A typical session transfer involves (i) identification of the router to which the session is to be transferred; (ii) establishing the path between the new router and the application server; and (iii) provisioning for the resources across the selected path for the session.

We consider that the identification of the router to which the session is to be transferred is quite crucial for fast and effective session transfer. In this paper, we consider this to be a mobile assisted identification based on the relative signal strengths received by the MS. The path establishment and resource provisioning are proposed based on label switching mechanism. Further, we also present an approach for session transfer between multiple mirror servers that simultaneously offer the same service.

J.N. de Souza et al. (Eds.): ICT 2004, LNCS 3124, pp. 983–990, 2004.

2 Related Work

Ensuring the session continuity during the mobility of an MT across the WLAN boundaries is a challenging task. Efficient transfer of sessions across the network with minimum disruption of QoS has been the subject of many recently reported research activities. Session transfer across heterogeneous platforms is addressed in [4] and the work focuses on application session handoff between heterogeneous platforms. It considers a typical scenario where multiple users are running typical applications continuously across client platforms which communicate with application server acting as a data repository. These clients could be any devices including desktops, laptops, or handheld devices such as PDAs. A session handoff mechanism is proposed which facilitates user to move an application's session seamlessly from one machine to another running the same application.

One of the significant experiences a mobile terminal will have during migration is in terms of resource availability. Therefore, a terminal needs to monitor the resource availability continuously during the migration and if necessary, need to switch from obsolete resources to new resources to maintain continuity in service provision. This calls for customized resource discovery and selection mechanisms as requirements are application specific. Ohata *et.al* [2] propose an adaptive terminal middleware that performs policy-based dynamic resource selection and host based session management to hide session failures and resource changes from applications and users. The proposed middleware realizes a virtual socket, on top of a real socket, which employs a seamless session handoff mechanism for resource changes and a resume and retransmission mechanism against disconnection of the wireless link.

Our present work is in the context of management of session continuity during mobility between different WLANs in a campus scenario. [1] considers a similar network scenario and proposes a wireless mobility framework for a university campus. The objective of this framework is to let a mobile user, equipped with a PDA device, experiences real wireless IP mobility while moving on a large spatial scale. It is proposed that this is achieved by means of middleware that leverages from three main wireless access technologies: Bluetooth, WiFi, and GPRS. These are managed as a hierarchy of spatially overlapping access domains. During the mobility of the user the client side of the middleware triggers smart switching between the currently best available and more appropriate wireless access path. This further facilitates the continuity of sessions during the mobility of the mobile stations between disjunct WiFi access domains by temporarily relaying on GPRS access domains.

A scheme for seamless handoff of mobile terminals in an MPLS-based Internet is proposed in [3]. This scheme consists of a mobile label switched tree architecture that is used for dynamic location tracking of a mobile terminal and adaptive re-routing of traffic to the mobile terminal during the bi-directional packet forwarding over MPLS paths. The current BS with which mobile terminal is associated and the neighboring BSs, which are potential handoff BSs, form the members of this tree. Further, the scheme consists of a Robust Fast Reservation Protocol to support per-flow reservation along the MPLS paths.

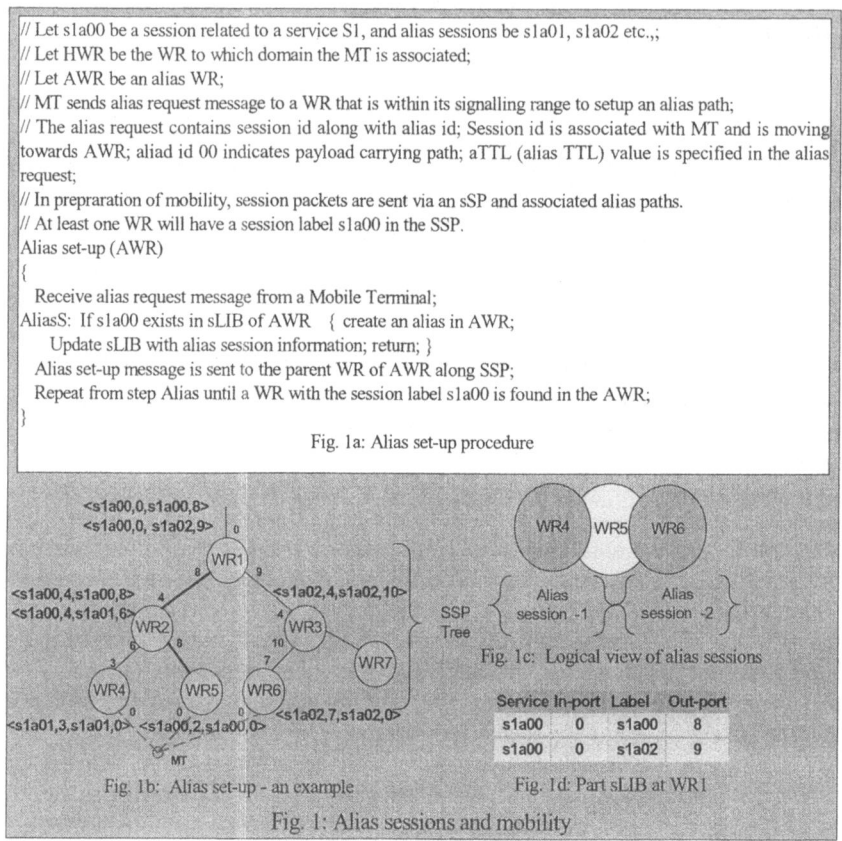

// Let s1a00 be a session related to a service S1, and alias sessions be s1a01, s1a02 etc.,;
// Let HWR be the WR to which domain the MT is associated;
// Let AWR be an alias WR;
// MT sends alias request message to a WR that is within its signalling range to setup an alias path;
// The alias request contains session id along with alias id; Session id is associated with MT and is moving towards AWR; aliad id 00 indicates payload carrying path; aTTL (alias TTL) value is specified in the alias request;
// In prepraration of mobility, session packets are sent via an sSP and associated alias paths.
// At least one WR will have a session label s1a00 in the SSP.
Alias set-up (AWR)
{
    Receive alias request message from a Mobile Terminal;
AliasS: If s1a00 exists in sLIB of AWR   { create an alias in AWR;
    Update sLIB with alias session information; return; }
    Alias set-up message is sent to the parent WR of AWR along SSP;
    Repeat from step Alias until a WR with the session label s1a00 is found in the AWR;
}

Fig. 1a: Alias set-up procedure

Fig. 1c: Logical view of alias sessions

Service	In-port	Label	Out-port
s1a00	0	s1a00	8
s1a00	0	s1a02	9

Fig. 1b: Alias set-up - an example

Fig. 1d: Part sLIB at WR1

Fig. 1: Alias sessions and mobility

Fig. 1. Alias sessions and mobility

In a recent work [6], we have proposed a scheme for service-specific label switching for managing data traffic due to various services in a campus network. And the same is extended in [5] to support service-specific QoS and is achieved by identifying an SSP Tree (a collection of Service Specific Paths (SSPs)) rooted at the WLAN router connected to the server hosting the service. Further, the service QoS parameters are distributed across the SSP Tree for network resource reservations purposes. This reserved SSP Tree is used to set up bi-directional QoS supported session-specific paths. Refer to Appendix –A for additional details.

3   Session Mobility

Let a mobile terminal MT be associated with a WLAN router WR1. When a mobile terminal moves, there is a possibility of the terminal coming within the range of additional WRs. Subsequent mobility could cause the terminal to go out of range of

Session Handover
{
   Source: HWR; Target: AWR; Mobile terminal: MT;
   Mobile assisted hanoff initiated by MT;
   MT sends a handoff request to HWR and AWR;
   HWR: On receiving a handoff request from MT:
   { Determines alias ID that is taking over as HWR;
      Alias point between HWR and AWR is determined as APWR;
      APWR is requested to swap Alias ID and 00 in its sLIB;
      This request is propagated downwards along both the paths and labels are changed appropriately
      // If WR has 00 as alias id, then it is made Aliash ID and similarly, if WR has Alias ID as alias id, then it is set
00;
   }
   AWR: On receiving handoff request form MT:
   { Sets aTTL to a large value so that buffered data is not dropped;
      Waits for label swap to get completed;
      Establishes session with MT;
      Receive info about last packet received by MT via HWR;
      Avoid duplicate delivery of packets and deliver the remaining packets in the buffer;
      Establishes normal session with MT;
   }
// On session handover, in a typical case when MT is still within the range of both AWR and HWR, the original live
session becomes an alias session and an alias session becomes a live session;
}

Fig. 2a: Session handover procedure.

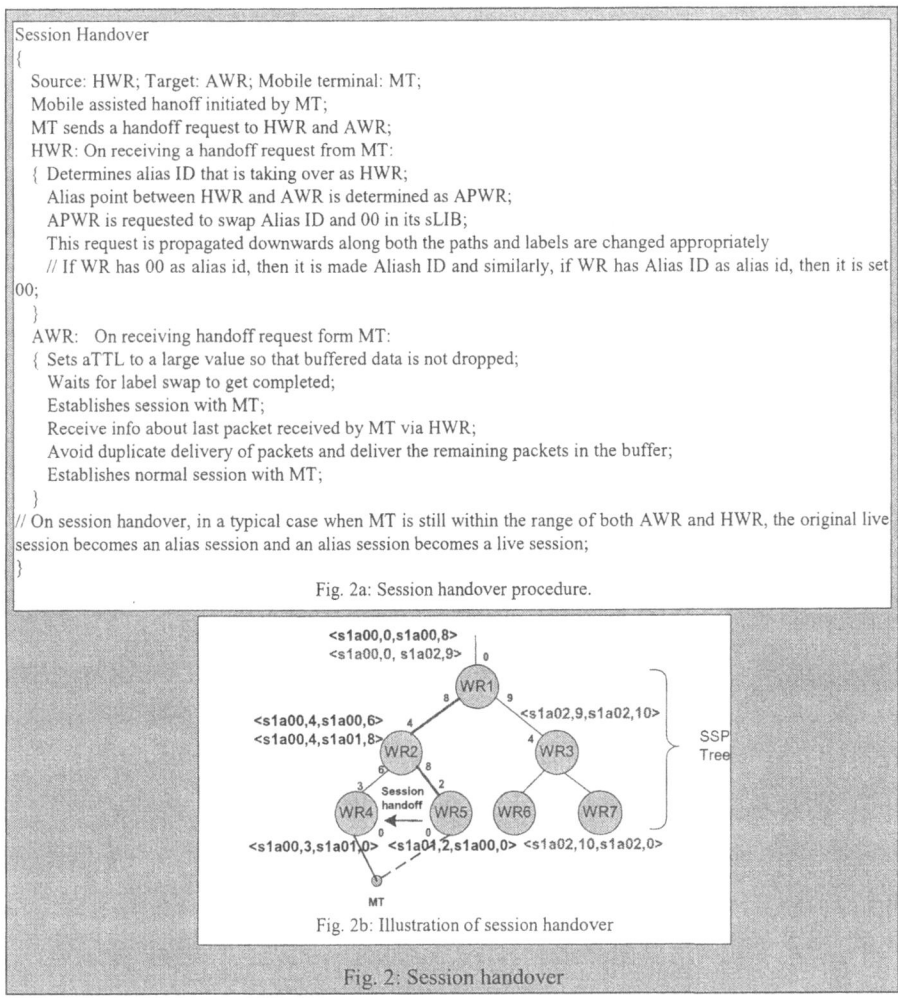

Fig. 2b: Illustration of session handover

Fig. 2: Session handover

Fig. 2. Session handover

WR1. Under this condition, it is necessary to support continued interaction with the server without any interruption. Furthermore, mobility should not lead to any loss of data. In order to achieve these objectives, we suggest that mobility of a terminal needs to be tracked and whenever there is a possibility of the terminal moving away from its associated WR into a different WR, some prior preparation should reduce the handover latency. An approach to achieve this is depicted in Fig. 1. Observe that whenever a terminal is within the range of a WR, an alias path is setup to deliver packets simultaneously with the delivery to Home WR. As described in the figure, Home WR is the WR to which the terminal is presently associated to interact with the service and alias WRs are those WRs to one of which there is a possibility of the terminal getting associated with. The packets received by alias routers are buffered in anticipation and are discarded once aTTL (alias TTL) expires. The appropriate value for aTTL depends on the typical handoff latency value and can be configured. Fig. 1a provides a brief overview of the alias setup procedure and Fig. 1b illustrates the same with an example. Observe that, in Fig. 1b, two alias paths are indicated: WR1, WR2, WR4, and WR1, WR3, WR6 with alias ids 01 and 02 respectively. The sLIBs at these WRs are suitably altered to enable multicasting at WR1 and WR2 as shown in Fig. 1b. This updation of sLIBs is done during alias path setup. Fig. 1c provides a logical view of alias sessions and Fig. 1d provides a portion of sLIB at an intermediate router When a mobile terminal is within the range of multiple WRs, it is required to determine at what point handoff is initiated. An obvious mandatory handoff occurs when the terminal looses connection with its Home WR. In a related paper [7], we have discussed various possible handoff and how point of handoff can be used to achieve certain objectives such as QoS. On handoff, it is required to establish a session from the new WR and this is achieved as follows: Observe that, on service launch, every WR that is part of CAN would have established an SSP and these collections of SSPs form an SSP Tree [5]; further, on account of alias path, a handoff candidate WR would already have been part of an ongoing session. Based on this observation, a brief procedure for handoff is provided in Fig. 2a while an illustration is provided in Fig. 2b. Observe that the handoff procedure discussed is an example of a mobile assisted handoff and in the illustration, it is assumed that while MT stays within the range of WR4 while moving closer to WR5, it is no longer within the range of WR6. As a consequence, the sLIB entries in WR6, WR3, and WR1 are suitably updated to eliminate the alias path. Further, WR2, WR5, and WR4 are suitably updated to reflect the alias id swap (this is indicated in Fig. 2b in blue color while the dropped entries are shown in red color).

AWR receives packets along the alias path and buffers them. On timer, the aTTL field of the buffered packets are updated and expired packets are dropped. This is done as alias sessions are set up in anticipation of handoff and, in reality, handoff may not take place at all. Hence, a suitable value for aTTL is chosen so that it does not consume too much of AWR's resources.

Typically, the life of an alias session is determined based on mobile terminal handoff latency. Hence, as soon as a mobile terminal goes out of range of a particular AWR, both MT and AWR initiate a procedure to remove the information related to the alias session. AWR clears the alias buffer, updates its sLIB, and communicates alias clear message to its parent until Alias Point WR is reached. Under certain situations, however, it is useful to keep the alias session alive even after the corresponding mobile terminal has moved out of the range of AWR. This helps, for example, to suitably address ping-pong effects. In this case, a pre-defined time elapses before the alias information is removed by both MT and AWR.

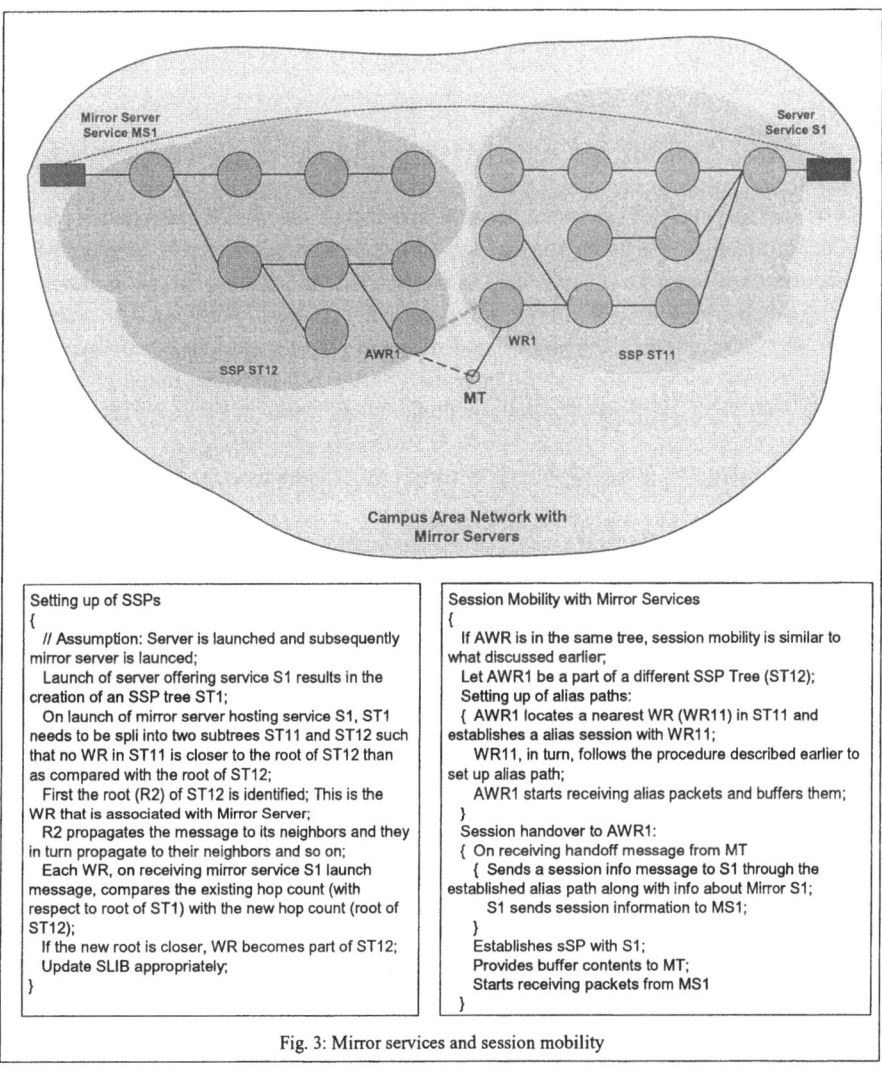

Fig. 3: Mirror services and session mobility

Fig. 3. Mirror services and session mobility

4   Service Mobility

In a campus like environment, thousands of students and faculty members access various services such as Library information services. Accessing a service with more number of hops results in a degraded quality of service and may even unnecessarily load the network. In such cases, it is appropriate to locate mirror servers at various points on the campus so that load can be distributed across these multiple mirror servers. This also provides a benefit of offering uninterrupted availability of a service even if one or more servers hosting the service fail. In this section, we discuss how

SSPs are set up when there are multiple mirror servers and address the issues related to session mobility.

Mirror Services and Session Mobility

Launching of a service results in the identification of an SSP Tree that covers all the WRs of the CAN [5]. If the same service is launched from multiple (say, K) mirror servers, K different SSP trees are formed. In one approach, each WR is made part of only of the SSP Trees. In other words, launch of K mirror services results in splitting the original SSP tree into K subtrees as described in Fig. 3. Alternatively, each WR can be part of multiple SSP trees related to mirror servers and at any point in time, gets associated with only one of the servers. Observe that this latter approach is advantageous when multiple servers are launched and brought down quite frequently. Note that, in either of the approaches, a WR always obtains service from a nearest server there by reducing the latency and distributing the load on the network.

Fig. 3 provides a brief overview of session mobility when there are multiple mirror servers. Note that whenever a WR supports an alias path, there is a possibility of that WR becoming part of two SSP trees as indicated in Fig. 3. While AWR1 receives alias packets from S1, if MT gets associated with AWR1 on handoff, it establishes a session on behalf of MT with MS1 and receives further packets from MS1 instead of from S1. As a consequence, most of the packets get delivered from a nearest server. Also, the temporary link between AWR1 and WR1 is released once AWR1 establishes a session with MS1.

5   Simulation Setup

We are currently simulating the proposed scheme as part of an ongoing project that studies various aspects of Service Specific Label Switched Wireless Networks. We are in the process of building a larger simulation framework representing the campus network considered in some of our recent work [5,6]. The simulation setup is being implemented in *ns2* and consists of about 20 WLANs each with a WLAN Router. Typically, we consider about 25 to 50 mobile users associated with each WLAN at any given time. A typical scenario we have considered for simulation is as follows:

- 10 services are launched simultaneously from 10 servers;
- 5 of the above services have varying number of mirror servers

One of objectives of the ongoing simulations is to study the service specific traffic flow in the network from the point of view of network and service throughput.

6   Conclusion

In this paper, we have extended our earlier works related to service-specific label switching in the context of campus wide wireless networks. The extension is to support session and service mobilities. The session mobility is exemplified by the mobile users carrying out their work while on move and the service mobility is illustrated by multiple servers (mirror servers) offering the same service. We have provided a brief description of the various algorithms related to session and service mobilities, and are presently working towards setting up a simulation environment to achieve (a) defining multiple interconnected WLANs; (b) multiple, mobile and stationary users distributed over the campus; (c) label manager to distribute unique service and session labels; (d) setting up and tearing down of service specific and session specific paths; (e) setting up and tearing down of alias paths; (f) setting up of multiple service specific trees to account for multiple mirror servers; and (g) simulation of session and service mobilities.

References

1. Calvagna A., G. Morabito and A. la Corte, "WiFi Bridge: wireless mobility framework supporting session continuity," *Proceedings of Percom'03,* 2003.
2. Ohata K., T. Yoshikawa, T. Nakagawa, "Adaptive terminal middleware for session mobility," *Proceedings of ICDCSW'03,* 2003.
3. Oliver T.W.Yu , "Next generation MPLS-based wireless mobile networks," *Proceedings WCNC 2000,* Sept. 2000.
4. Phan T., R. Guy., J. Gu, and R. Bagrodia, "A New TWIST on Mobile Computing: Two-Way Interactive Session Transfer," *Proceedings of WIAPP'01,* July 2004.
5. Maruthi P, G. Sridhar and V. Sridhar, "QoS Management in Service Specific Label Switched Wireless Networks." Submitted to *Wireless Telecommunication Symposium 2004 (WTS 2004),* May 2004.
6. Sridhar, G., and V. Sridhar, "Service-Specific Label Switching in Wireless Networks." Submitted to *Networks 2004,* June 2004.
7. Sridhar, G., and V. Sridhar, "QoS management by selective handoff." Submitted to *WNET 2004,* July 2004.

Mobility Management for the Next Generation of Mobile Cellular Systems

Marco Carli, Fabio Cappabianca, Andrea Tenca, and Alessandro Neri

Applied Electronics Dep., University of ROMA TRE, Via della Vasca Navale 84,
I-00146 Rome, ITALY
{carli, neri}@uniroma3.it
http://www.comlab.ele.uniroma3.it

Abstract. In this paper, a performance comparison among three main procedures for mobility management (MM) in a cellular framework for the 3.5 generation and beyond is presented. The MM procedures integrate and improve two mobility protocols: Mobile IP for macro mobility and Cellular IP for micro mobility. Specifically version 4, version 4 with routing optimization, and version 6 of the said protocols have been compared by means of simulations performed with Network Simulator II. The experimental results show that version 4 with routing optimization is effective in terms of Quality of Service (QoS), since it offers optimized routing, faster handoff procedures, and a very low packet loss rate. In addition, it supports real time communications.

1 Introduction

In many fields of telecommunications, Internet has become the main infrastructure. The reason of this success is its flexibility that allows network integration among systems with different characteristics of lower layers, and transport of flows generated by a variety of applications with different requirements over a common infrastructure. One of the most challenging aspects in designing tomorrow's Internet-based communication systems is an access networks that can guarantee an acceptable Quality of Service (QoS) in an environment requiring great capacity, capillary distribution and global accessibility.

Global access can be obtained thanks to wireless LAN and mobile cellular networks. In this framework, the mobility management (MM) is one of the most important functionalities to be provided in the next generation of Internet. At this aim, the Internet Engineering Task Force (IETF) has introduced a reference network architecture based on access domains representing different sub-networks, managed by Mobile IP [1]. This architecture is based on the concept that most of the mobility can be managed locally within one domain without overloading the core network [1]-[8].

Presently, the working version Mobile IP v.4 (MIPv4) is based on IP v.4 with some modifications to adapt a wired born protocol to the mobility needs, namely routing optimization. Nevertheless, during the handoff, a certain amount of packets can be lost. The consequent latency affects the QoS. To effectively reduce this delay, it is necessary to improve the existing mobility protocols by ad hoc extensions supporting seamless delivery of voice, video, and data with high quality.

J.N. de Souza et al. (Eds.): ICT 2004, LNCS 3124, pp. 991–996, 2004.

The main problem is to preserve the service continuity while a user moves from one cell to another. At this aim in [9] the authors proposed a solution for managing a Cellular Intranet (see Fig. 1) based on the integration of the Mobile IP (MIP) v.4 and the Cellular IP (CIP) v.4 protocols. A Cellular Intranet is a set of subnets able to supply to a mobile host (MH) a framework similar to the one available in cellular systems (GSM, DCS, and PCS). Recently, the IETF has defined the next version of MIP, denoted as Mobile IP v.6 (MIPv6), specifically designed to support MM. Being based on IPv6, it will also solve the problems associated with the size of the address space, and consequently, to the limitation on the number of possible Internet users. The diffusion of this protocol is directly related to the effective spread of IP v6.

The aim of this paper is to compare the performances of MIP v.4, MIP with routing optimization (MIPRO) and MPI v.6 with respect to handoff.

The paper is organized as follows. In Section II the Cellular Intranet is described, in Section III a brief description of the simulation framework is reported. Section IV describes the simulation results and finally conclusions are drawn in Section V.

2 The Cellular Intranet

A cellular intranet is a system that integrates two MM levels: the local and the global component. This architecture uses the standard IP for the core network. MIP is used as an inter-subnet mobility protocol for macro MM; while CIP is employed for the intra subnet mobility as support to the micro-mobility and paging management (see [9]-[16]). Compared to version 4 of the aforementioned protocols, CIPv6 introduces control packets that are localized inside the access network, so that updating of the IP Router is not required. As shown in Fig. 2, the third layer ensures the internal cellular mobility and Internet visibility. The interface between the wireless and the wired world is constituted by the gateway (GW/FA) node.

By operating as a Foreign Agent (FA), the gateway filters the micro-mobility traffic within the cellular IP network. In macro MM management the MH directly exchanges datagram with its FA and is indirectly engaged with the Home Agent (HA)

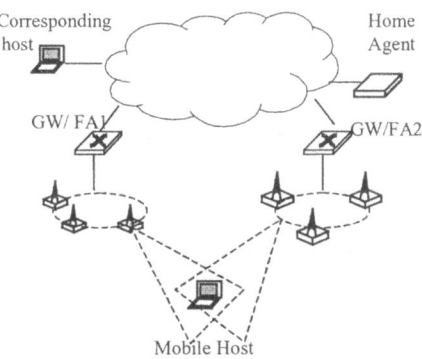

Fig. 1. Wireless-Wired Scenario

via FA. In this configuration, the MH is globally visible only if it is registered with the HA (macro-mobility) and the route cache maps of the cellular IP nodes serving the MH are updated. Simulation results validated our choice to separate the two protocols to efficiently utilize the pair CIP/MIP simultaneously: it is equivalent to consider an intranet with heterogeneous cellular access subnets.

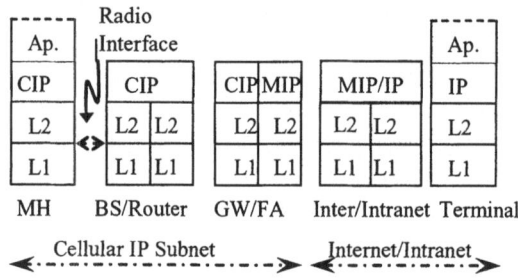

Fig. 2. MIP/CIP protocol stack architecture

Since the CIP architecture is centralized, the MH macro-MM can be left within GW operations. In this situation the MH utilizes only CIP and the GW performs the FA function with some modifications in the access network interface. The GW/FA also represents the interface between CIP and MIP. IP packets within the cellular network are processed by the layer 3 entities of GW/FA depending on their IP address.

Generally, the GW/FA can manage two states of the mobile station: the active state with consequent priority on operation management on the core and then access network, and the idle state without priority needs for the MH.

3 The Simulated Framework

To compare the performances of the three protocols, a cellular network has been simulated by means of the Network Simulator version 2.1b7a. In Fig. 3, a scheme of the simulated network is presented. In the figure each link is labeled with the corresponding delay (in milliseconds). The considered scenario is composed by a core Internet network based on wired links, (composing of 8 routers, one corresponding host, and one home agent) and two cellular networks.

When the simulation starts, the corresponding host sends packets to the MH trough the Internet network and the Cellular network #1. The MH is already registered in the Foreign Router 1 (FR1) that manages the cellular subnet CIPv.6) and the Home Agent has the *bind cache* updated and knows the present Care of Address (CoA) of the MH. The session starts at $t=0$ sec when the bind cache, paging cache and route cache timer start. At $t=2$ sec the corresponding node (CN) starts sending packets to MH with a bit rate of 160 Kb/s. At time $t=7$ sec, the mobile host moves towards the cellular network number two, and after $t=15.7569$ sec. it detects the new Foreign Agent (FA2), by receiving the Router Advertisement sent from the FA2: at this time the handoff procedure starts.

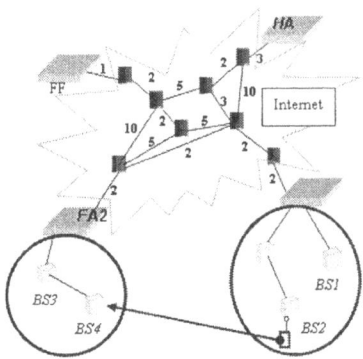

Fig. 3. The cellular network used in the experiments.

4 Experimental Results

In the following, we present the simulation results of the performance assessment of the mobility protocols MIP, MIPRO, MIPv6, CIP, and CIPv6.. In this simulation, we assumed a transmission length of 30 secs, a packet emission frequency from the CN of 100 packets/sec, and a packet length of 520 bytes, including IP header. In addition, the DSDV ad-hoc routing protocol has been employed, while the time needed for the execution of Duplicate Address Detection procedure has been set equal to zero. In Fig. 4 for each packet the transmission time for the three different versions of the MM protocols, for a maximum core network reserved link capacity of 1Mb/s, is reported. The MIP protocol loses 12 packets, starting from the number 9 and ending to 21. When MIP-RO is employed the number of lost packets is equal to 5. The two packets (15, 16) are received out of sequence and then throw out in absence of a procedure to recover the packets order. The delay is due to the packet encapsulation during the forward procedure among the old and the new GW/FA.

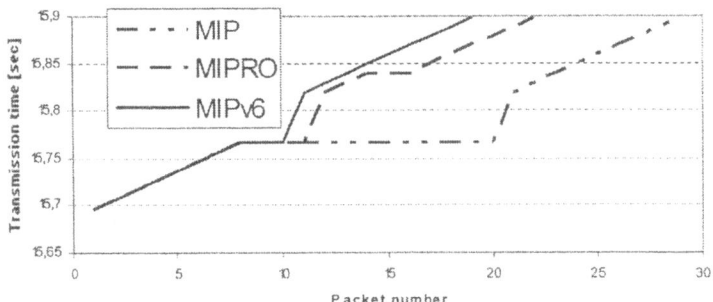

Fig. 4. Packet loss, bandwidth: 1 Mb/s.

MIPv6 presents the best performances showing only a loss of 2 packets. With respect to the IPv.4 based protocols (MIP, MIPRO), the MIPv6 protocol does not require to encapsulate packets: in fact, the presence of a co-located CoA allows processing the packets directly to the MH than to the network Foreign Agents.

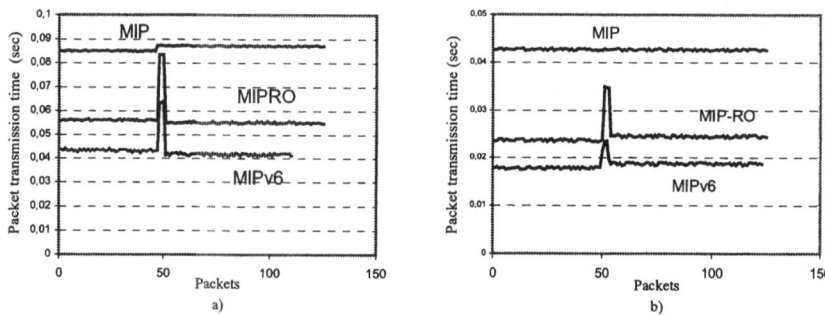

Fig. 5. Packet delay during handoff. a) Bandwidth=1 Mb/s. b) Bandwidth=100 Mb/s.

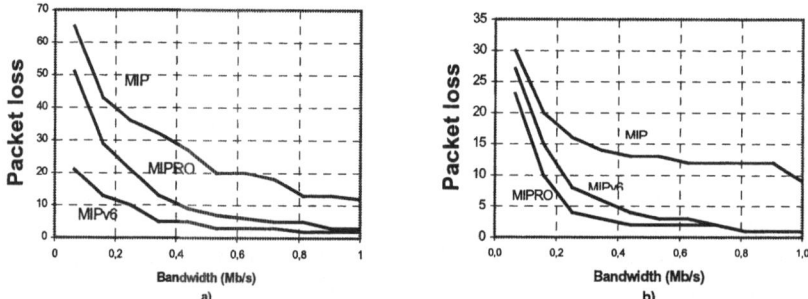

Fig. 6. Packet loss vs. bandwidth. *a)* Bidirectional traffic, *b)*Unidirectional traffic.

By increasing the internet bandwidth to 100 Mb/s, there is an improvement for MIP (10 packets lost) and for MIPRO (3 packets lost), while the performances do not change for the MIPv6. This fact confirms that by increasing the network congestion, the number of packets lost during the hand off procedure also increases.

In the Fig. 5 the delay associated to the three protocols in the case of 1Mb/s and 100Mb/s Internet bandwidth is reported. In the 1Mb/s case, the MIP transmission delay is approximately constant with an average value of 87 msec. With MIPRO the average delay reduces to 56 msec, presenting a peak in correspondence of the handoff, due to encapsulating procedure performed by FA1. For MIPv6 the average transmission delay is about 43 msec, which is lower than in the other cases, because it does not tunnel the packets towards the new network.

In Fig. 6.a we report how the available bandwidth affects the packet loss with a source throughput of 160 kb/s. The advantage of the MIPv6 with respect to the previous version of protocols is evident. Nevertheless, by increasing the bandwidth, the gap between MIPv6 and MIPRO is reduced, while it remains noticeable if the network congestion increases. This behavior is due to the piggyback feature used in MIPv6 to transmit control messages inside data packets, limiting network congestion.

In Fig. 6.b, we present the behavior of a unidirectional transmission from CN to the MH (sink transmission). In this case, MIPv6 performance decreases faster than MIPRO by decreasing the bandwidth. This is due to control packets whose size is bigger than in IPv4.

5 Conclusions

As illustrated in the previous section, although MIPv6 outperforms MIPRO, in presence of heavy symmetric traffic, thanks to the piggybacking feature, the two protocols present almost the same performances when the network congestion decreases. Nevertheless the improvements introduced by MIPRO are rather effective, so that it appears a viable solution if we consider the actual possibility of IPv6 to impose itself as a standard. On the other hand MIPRO outperforms MIPv6, especially at low bit rates when the traffic is asymmetric and the source bit rate is constant (i.e. ftp transfer), independently from the congestion rate.

References

1. C. Perkins, "IP Mobility Support", *IETF, RFC 2002*, October 1996.
2. T. F. La Porta, L. Salgarelli, G. T. Foster, "Mobile IP and Wide Area Wireless Data", *Proc. of IEEE Wireless Communications and Networking Conf.*, pp.1528-1532 vol.3, Sept. 1999.
3. R.Ramjee, T. La Porta, L. Saltarelli, S Thuel, K. Varadhan, L Li. "IP-based Access Network Infrastructure for Next Generation Wireless Data Networks", *IEEE Personal Communications*, Vol. 7 Issue: 4, pp. 34-41, Aug. 2000.
4. C. Perkins, D. B. Johnson, "Route Optimization in Mobile IP", *IETF Draft. Draft-ietf-mobileip-optim-09*, 15 February 2000.
5. C. Castelluccia. "A Hierarchical Mobile IPv6 Proposal", *Tech. Rep., INRIA*, Nov. 1998.
6. C. E Perkins, K.Y. Wang, "Optimized Smooth Handoffs in Mobile IP", *Proceedings IEEE Int. Symposium on Computers and Communications*, pp. 340 –346, 1999.
7. C. Perkins. "Mobile IP", *IEEE Communications Mag.*, Vol. 35, May 1997, pp. 84 -99
8. I. Gronback. "Cellular and Mobile IP: overview and enhancements", *Project 1 Paper, March 1999.* http://pi.nta.no/users/inge/cellular.pdf.
9. M. Carli, A. Neri, A. Rem Picci, "Mobile IP and Cellular IP Integration for Inter Access Networks Handoff" – *Proc. IEEE ICC2001*, 11-14 June 2001 Helsinki, Finland.
10. A. T. Campbell, S. Kim, J. Gomez, C-Y. Wan, Z. Turanyi, A. Valko. "Cellular IP". *IETF Draft, draft-ietf-mobileip-cellularip-00, Mobile IP Working Group*, December 1999.
11. A. Valkò. "Design and Analysis of Cellular Mobile Data Networks", *Ph. D. Dissertation*, Technical University of Budapest, Budapest 1999.
12. A.Valkò, J. Gomez, S. Kim, A. Campbell. "Performance of Cellular IP Access Networks", *Sixth IFIP Int. Workshop on Protocols For High-Speed Networks* (PfHSN '99), 1999.
13. A. T. Campbell, S. Kim, J. Gomez, C-Y Wan, Z. Turanyi, A. Valkò. "Cellular IP Performance", IETF *Draft, Draft-gomez-cellularip-perf-00*, October 1999.
14. The Cellular IP Project at Columbia University. Http://www.comet.columbia.edu/cellularip.
15. A.T. Campbell, J. Gomez, S. Kim, A. G. Valkó, and C.Y. Wan, Zoltán R. Turányi, J. Postel, "Design, Implementation, and Evaluation of Cellular IP", *IEEE Personal Communications*, pp. 42-49, August 2000.
16. Y.C. Tay, K.C. Chua. "On the Performance of a Mobile Internet Protocol and Wireless CSMA Cell", Centre for Wireless Computing, National Univ. of Singapore, 1997.

Handoff Delay Performance Comparisons of IP Mobility Management Schemes Using SIP and MIP

Hyun-soo Kim, C.H. Kim, Byeong-hee Roh, and S.W. Yoo

Graduate School of Information and Communication, Ajou University,
San 5 Wonchon-dong, Youngtong-Gu, Suwon, 443-749, Korea
{bhroh,swyoo}@ajou.ac.kr

Abstract. In this paper, we analyze and compare the handoff delays for several architectures such as MIP-only, SIP-only, MIP/CIP and SIP/CIP, which are expected to be most promising solutions to support IP mobility management. For the analysis, we make the detailed message flows of those architectures in handoffs, and then make an analytical model to compute the handoff delay for each scheme. Numerical results showed that the SIP/CIP outperforms other schemes in the viewpoints of the handoff delay.

1 Introduction

Mobility support is one of the most important functions to provide seamless data transfer of real-time multimedia traffic in wireless networks. For the mobility support, as a standard track, Mobile IP (MIP) has been proposed[1]. However, it has the triangle routing problem that all packets sent to the mobile host(MH) should be delivered through its home agent(HA). To overcome the triangle routing problem of MIP, routing optimization solutions (MIP-RO) have been proposed[2]. It is known that the MIP-based approaches show inefficient performances when handoffs occur within an administrative domain. For reducing the complexity when mobility occurs within an administrative domain, various micro-mobility support schemes have been proposed. Cellular IP(CIP) is one of those schemes, and operates at layer-2 [2][8]. Since MIP is designed operating at network layer, it can provide transparent communications regardless of all applications above the network layer. However, if the mobility solution were to be implemented at a higher layer, it would be inefficient. Session Initiation Protocol(SIP) is an application layer protocol for signaling and controlling a session consisting of multiple streams[3], and can support mobilities at application layer using redirect server. An architecture for supporting terminal, personal, session and service mobilities using SIP has been proposed[4]. Unlike other MIP-based schemes, SIP does not require the triangle routing and the overhead due to IP encapsulation as in MIP. Kwon et al.[5] analytically computed and compared the handoff delays for MIP and SIP-based approaches without adopting any micro-mobility schemes. According to their result, MIP outperforms the SIP approach in most situations.

J.N. de Souza et al. (Eds.): ICT 2004, LNCS 3124, pp. 997–1006, 2004.
© Springer-Verlag Berlin Heidelberg 2004

The results are mainly due to the IP address assignment from DHCP in SIP approach[6][10]. Gatzounas et al.[9] proposed an mobility management architecture with the integration of SIP and CIP. However, they did not provide detailed performances of the architecture. Likewise, in order to support IP mobility management, much of schemes have been proposed. It is a question that among those, which scheme can be a most useful solution for delivering real-time multimedia traffic in the hybrid of wired and wireless environments. In this paper, we analyze and compare handoff delay performances for various promising mobility support architectures such as MIP-only, MIP/CIP, SIP-only and SIP/CIP. MIP-only and SIP-only are schemes using MIP with route optimization and SIP only, respectively, without adopting any micro-mobility scheme, while MIP/CIP and SIP/CIP denote those adopting CIP as a micro-mobility scheme. For the handoff delay analysis, we make explicit message flows for those promising mobility support schemes. With comparisons between MIP-only and SIP-only, we can find the intrinsic properties of MIP and SIP for the mobility support. While, with comparisons between MIP/CIP and SIP/CIP, we can recognize the role of MIP and SIP for macro-mobility support.

The rest of the paper is organized as follows. In Section 2, we discuss on four comparable architectures such as MIP-only, SIP-only, MIP/CIP and SIP/CIP, and give message flows to deal with handoff for those architectures. In Section 3, we analyze handoff delays for those architectures. Section 4 gives some experimental results, and finally, we conclude the paper in Section 5.

2   Mobility Architectures

2.1   Mobile IP with Route Optimization (MIP-RO)

MIP[1] allows a MH to move between IP subnets, while keeping communications with its corresponding host(CH). In MIP, home agent(HA) and foreign agent(FA) located in home network(HN) and foreign network(FN), respectively, play a key role in supporting the mobility. When a MH moves to a FN, it gets a temporary IP address called a care of address(CoA) from a DHCP server, and informs the HA of its CoA. After this step, with the route optimization, packets are re-routed from CH to MH directly without going to HA first[2]. In Fig.1(a), it is shown a message flow for MIP-only architecture when macro-mobility occurs. After a MH acknowledges that the domain has been changed by receiving a beacon signal, it gets a CoA from the new FA [flow ① to ③]. Then, the binding tables maintained in HA and old FA are updated [flow ④ to ⑧]. Finally, by sending re-INVITE message, the path between MH and CH are re-established[flow ⑨ to ⑩]. Message flows of ⑤, ⑥ and ⑦ are not required when intra-domain handoff occurs.

2.2   SIP-Based Mobility Support (SIP-Only)

SIP is an application layer protocol for signaling and controlling a session consisting of multiple streams[3]. SIP basically supports not only personal mobility

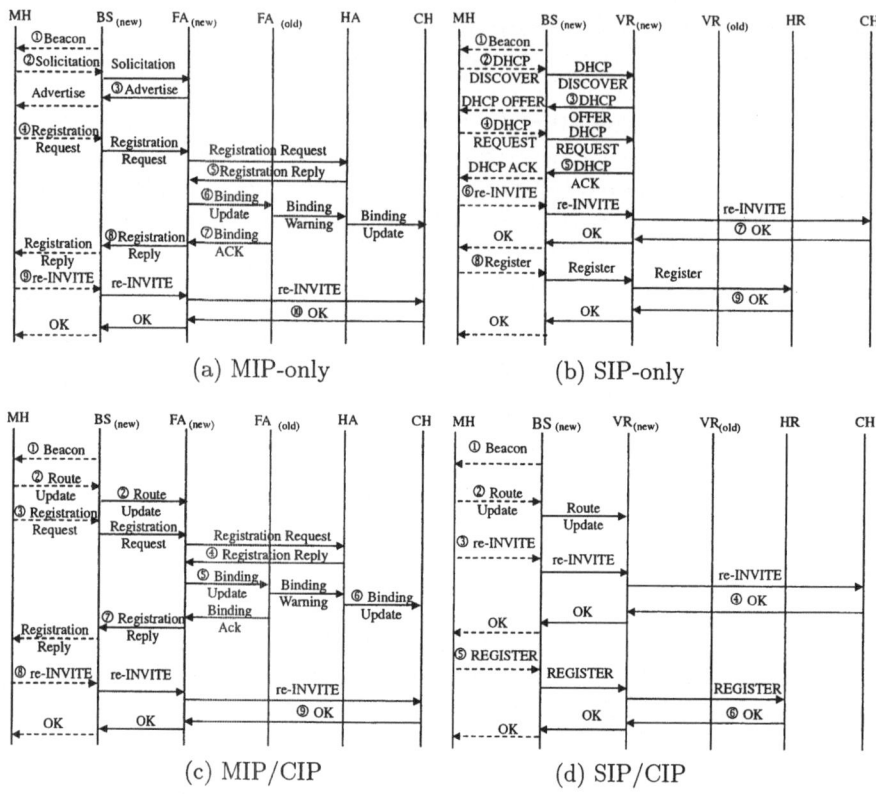

Fig. 1. Message flows to handle the inter-domain handoff

but also the terminal mobility using home redirect(HR) and visited redirect(VR) servers. Such ability of SIP can be used for finding and updating the location of MH. The location information of MH is registered in SIP redirect server instead of HA as in MIP, and a SIP proxy server provides functions as similar as FA in MIP. In Fig.1(b), a SIP message flow in macro-mobility is shown. After a MH acknowledges that the domain has been changed by receiving a beacon signal, it gets a CoA using DHCP mechanism [flow ① to ⑤]. Then, MH re-establishes a session with CH by sending re-INVITE message through new VR[flow ⑥ to ⑦]. After this, MH registers its modified location information to HR for the next time calls.

2.3   MIP-RO Interworking with CIP (MIP/CIP)

Cellular IP(CIP)[7] supports micro-mobility at layer-2. In CIP, when handoff occurs in an administrative domain, MHs are not required to register their CoAs in the HA. Instead, the decision to reroute packets are performed by using routing and paging caches that each CIP node maintains. For handoffs between CIP domains, normal MIP procedures are used. Any MH in CIP networks does not need to use DHCP to obtain a temporary IP address. So the integration of MIP and CIP can reduce the delay overhead for the frequent registrations. Fig.1(c) depicts a message flow to support macro-mobility in MIP/CIP. In MIP/CIP, functions of HA and FA are implemented in CIP gateways on the home and foreign networks, respectively. MH receiving a beacon signal sends a Route Update message to CIP gateway through BS, and a new routing path along the path delivering the message is established [flow ① to ②]. After this, MH updates its location information in HA, and the path between new FA and CH is established [flow ③ to ⑦]. After the registration is successfully performed, MH setup call to CH again by sending re-INVITE message [flow ⑧ to ⑨]. In case of micro-mobility, MH can re-establish a connection to CH without registration process such as ③, ④, ⑤, ⑥, and ⑦ because the MH's CoA is not changed though the routing path is changed.

2.4   SIP Interworking with CIP (SIP/CIP)

To integrate SIP with CIP, SIP proxy server functions, such as home redirect(HR) and visited redirect(VR) servers, can be implemented within CIP gateways. By adding a new type of control information in CIP route/paging-update packets, an email-like SIP user identifier of the form "user@home" can be used in CIP domain[9]. SIP HR server maintains the binding information of SIP URL with CIP gateway address, and it provides SIP proxy server with the information for the next calls to the MH. In SIP INVITE message, there exists a Contact field containing the current location information, and the field can be used for supporting handoff. By referencing the field, CH can send packets to MH directly. A possible mobility support scenario for inter-domain handoff is shown in Fig.1(d). MH receiving a beacon signal sends a Route Update message to CIP gateway through BS, and a new routing path along the path delivering the message is established [flow ① to ②]. Then, the session between MH and CH are updated by sending re-INVITE message to the CH via the new VR server [flow ③ to ④]. For next calls, MH registers its new location information in HR [flow ⑤ to ⑥]. It is noted that in MIP/CIP, the registration process must be completed before MH sends a re-INVITE message to CH in order to eliminate the triangle routing problem. On the contrary, in SIP/CIP, since the location information of MH is included in the Contact field in SIP re-INVITE message, the handoff can be completed by sending the message. The registration process after this is only for next calls, which are regardless of the current session. From the facts, we can intuitively perceive that the routing update by SIP/CIP can be done much faster than that by MIP/CIP.

3   Handoff Delay Analysis

In this Section, we give an analytical model to compute the handoff delay, which is the delay for re-establishing a new path in a handoff. For the analysis, we make a simple network model as shown in Fig.2, which is as similar as in Kwon

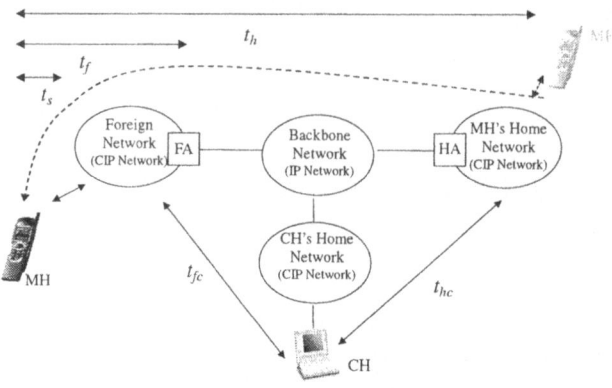

Fig. 2. A simple network model for handoff delay analysis

et al.s'[5]. While Kwon et al. assumed that all domain and core networks use MIP-only or SIP-only, we extend it to the hybrid use of CIP for micro-mobility within each domain. In the network model, only message flow is considered. That is, security and authentication mechanisms are not considered. We assume that the CH is located in the CN that is the home network of the CH.

Let t_s be the delay for delivering a message through the wireless link between MH and BS. The delay corresponds to the delay for MH to receive a beacon signal through the wireless link. Let t_f and t_h be the delay between MH and FN and between MH and HN, respectively. Let t_{hc} be the delay between CH and HN, and t_{fc} be the delay between CH and FN. Let us define t_{up} the delay between FAs. Let $T_{mip-inter}$, $T_{sip-inter}$, $T_{mip/cip-inter}$ and $T_{sip/cip-inter}$ be the inter-domain handoff delays for MIP-only, SIP-only, MIP/CIP and SIP/CIP, respectively.

We can obtain $T_{mip-inter}$ from Fig.1(a). It takes t_s for MH to receive a beacon message, $2t_f$ for MH to send Solicitation message to FA_{new} and to receive Advertise message that contains a new IP address, t_h for MH to send a Registration Request to HA, and t_h-t_f for HA to send Registration Reply to FA_{new}. It takes $2t_{up}$ for FA_{new} to send Binding update message to FA_{old} and to receive ACK message from FA_{old}. It takes t_h-t_f and t_{hc} requiring for that FA_{old} sends Binding Warning to HA, and that HA sends Binding update message to CH, respectively. Then, it takes t_f in which HA sends Registration Reply to FA_{new}. At this time handoff procedure is completed. Then, it additionally takes t_f+t_{fc} for MH to send re-INVITE message to CH via FA, and t_f+t_{fc} for CH to reply OK message to MH via FA for call setup. To sum up the above delay times, we can easily obtain $T_{mip-inter}$ given by

$$T_{mip-inter} = t_s + 3t_h + 3t_f + 2t_{up} + t_{hc} + 2t_{fc} \qquad (1)$$

From Fig.1(b), $T_{sip-inter}$ can be calculated. It takes t_s for receiving a beacon message, $2t_f$ for exchanging DHCP DISCOVER and DHCP OFFER messages between MH and VR_{new}. Then, t_f for MH to select one DHCP server and to send DHCP REQUEST to the selected server, t_f for the DHCP server sends DHCP

ACK to confirm the assignment of the address to MH. Then, it additionally takes $2(t_f+t_{fc})$ for MH to send re-INVITE message to CH via VR, and then for CH to reply OK message to MH via FA for call setup. It takes $2t_h$ to send a REGISTER message to HR and to reply a OK message to MH. Then, we have

$$T_{sip-inter} = t_s + 2t_h + 6t_f + 2t_{fc} \tag{2}$$

Similarly, $T_{mip/cip-inter}$ can be obtained from Fig.1(c). It takes t_s for MH to receive a beacon message, t_f for MH to send a Route Update to FA_{new} according to CIP mechanism, t_h for MH to send a Registration Request to HA, and t_h - t_f for HA to send Registration Reply to FA_{new}. It takes $2t_{up}$ for FA_{new} to send Binding update message to FA_{old} and to receive ACK message from FA_{old}. It takes t_h-t_f and t_{hc} requiring for that FA_{old} sends Binding Warning to HA, and that HA sends Binding update message to CH, respectively. Then, it takes t_f in which HA sends Registration Reply to FA_{new}. At this time handoff procedure is completed. Then, it additionally takes $2(t_f+t_{fc})$ for MH to send re-INVITE message to CH via FA, and then for CH to reply OK message to MH via FA for call setup. To sum up the above delay times, we can easily obtain $T_{mip/cip-inter}$ given by

$$T_{mip/cip-inter} = t_s + 3t_h + 2t_f + 2t_{up} + t_{hc} + 2t_{fc} \tag{3}$$

Finally, we can compute the handoff delay in SIP/CIP, $T_{sip/cip-inter}$, according to the message flow shown in Fig.1(d). It takes t_s for MH to receive a beacon signal, t_f for sending a Route Update according to CIP mechanism. It also takes $2(t_f+t_{fc})$ for delivering re-INVITE message from MH to CH via VR, and then for replying OK message from CH to MH via VR_{new}. By doing so, the actual handoff procedure is completed. For next calls, it takes $2t_f$ for exchanging REGISTER and OK messages between MH and HR. Then, we have

$$T_{sip/cip-inter} = t_s + 2t_h + 3t_f + 2t_{fc} \tag{4}$$

For intra-domain handoffs, since a registration with HA or HR is not needed, we can obtain the intra-domain handoff delays $T_{mip-intra}$, $T_{sip-intra}$, $T_{mip/cip-intra}$ and $T_{sip/cip-intra}$ for MIP-only, SIP-only, MIP/CIP and SIP/CIP, respectively, as following

$$T_{mip-intra} = t_s + 6t_f + 2t_{fc} \tag{5}$$

$$T_{sip-intra} = t_s + 8t_f + 2t_{fc} \tag{6}$$

$$T_{mip/cip-intra} = t_s + 3t_f + 2t_{fc} \tag{7}$$

$$T_{sip/cip-intra} = t_s + 3t_f + 2t_{fc} \tag{8}$$

4   Experimental Results

4.1   Handoff Delay Performance

For the analysis of the model described in Section 3, we used the values given in [5] for the delay time parameters. That is, 10ms for t_s, 12ms for t_f, and 17ms for t_h. And, we assumed that the delay within each network is constant at 5ms, and that MH is moving from HN to FN.

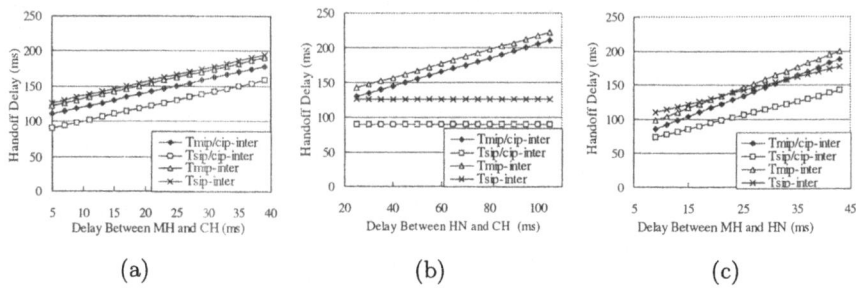

Fig. 3. Inter-domain handoff delay performances varying (a) the delay between MH and CH, (b) the delay between HN and CH, and (c) delay between MH and HN

Inter-domain handoff delay performance: The inter-domain handoff delay performances varying the delay between MH and CH, t_{fc}, the delay between HN and CH, t_{hc}, and the delay between MH and HN, t_h, are depicted in Fig.3(a), (b), and (c), respectively. It is noted that the increase of delays between those nodes and networks means the increase of the distance or the decrease of available bandwidth between them. It is also noted that the delay within a network can be caused by the decrease of available bandwidth due to the increase of traffic within the network.

As shown in Fig.3(a), schemes with CIP such as SIP/CIP and MIP/CIP show better delay performances than MIP-only and SIP-only. This is because MIP-only and SIP-only have some complicate steps to get a temporary IP address in a new FN. Without CIP, MIP-only outperforms SIP-only because MIP-only uses an advertise message only or an advertise with a solicitation message, but SIP-only requires four additional DHCP messages to get a temporary IP address as shown in Fig.1(b). In the whole, SIP/CIP outperforms other schemes. As illustrated in Fig.3(b), as the delay t_h increases, the handoff delays of MIP-only and MIP/CIP increase while those of SIP-only and SIP/CIP keep at a constant level. This is because that the SIP-based handoff delay is independent of t_{hc} as illustrated in Section 3, while the MIP-based is not. From Fig.3(c), we can see that the schemes adopting CIP have better handoff delay performances than those schemes without CIP, and that SIP/CIP outperforms MIP/CIP.

In general, SIP-only shows larger inter-domain handoff delay than MIP-based schemes. However, by integrating CIP with SIP, it can get much better handoff delay performances than other schemes including that of SIP-only.

Intra-domain handoff delay performance: In Fig.4, the intra-domain handoff delay performances are shown. As shown in Fig.4(a), the handoff delay keeps a constant level regardless of the variance of the delay between FN and HN. This is because they do not need to register with a HA or a HR. While, in Fig.4(b), the handoff delays tend to increase as the delay between FN and CH increases, since there exists message exchanges between MH and CH though in intra-handoffs. In overall, SIP-only shows the largest intra-domain handoff delay,

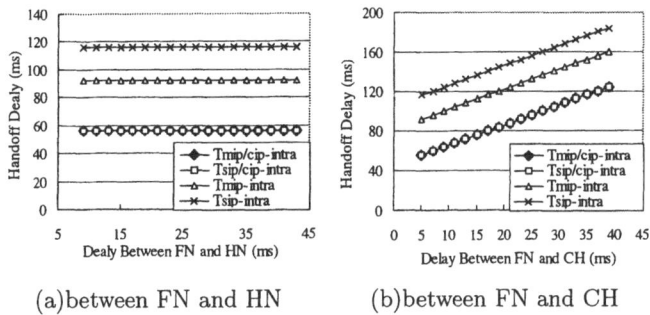

(a)between FN and HN   (b)between FN and CH

Fig. 4. Intra-domain handoff delay performances varying delays

and the handoff delay of MIP-only is larger than both MIP/CIP and SIP/CIP. This is because SIP-only has very complicate temporary IP address assignment mechanism using DHCP, and MIP-only has a next complexity for it as shown in Fig.1(b). The delay of MIP/CIP and SIP/CIP show same values because they use the same CIP operation at the intra-domain handoff.

4.2 Real-Time Data Delivery Performance

In the previous subsection, we show that the architectures adopting CIP outperforms those without CIP. Here, for MIP/CIP and SIP/CIP, we carried out the following simulation using ns-2[11], in order to show how the handoff delay performances evaluated by the analysis model can affect the actual delivery of real-time data. Fig.5 depicts the network model for the simulation as similar as in [12]. In the network model, there are three separate wireless networks such as HN, FN1 and FN2 with different administrative policy each other, and those wireless networks are interconnected through wired IP network. HN is the home network for MH. It is assumed that the link delays between networks are same as 5 ms and their bandwidths are 5Mbps. We also assumed that the size of each wireless network is 75m, and there is no overlap between the wireless networks. We let CH send packets at constant rate of 500 Kbytes/sec by using UDP. And, we let MH start from HN, then move to FN2 through FN1, so two handoffs occurs.

Fig.6 plots each packet's end-to-end delay from CH to MH when MH's moving speed is 6m/sec and the cells are very closed to each other but not overlapped. Handoffs from HN to FA1 and from FA1 to FA2 occurred at about 24 and 50 second, respectively. Fig.6(a) corresponds to MIP/CIP case, and Fig.6(b) to SIP/CIP case. As we can see, handoff delays of SIP/CIP are much smaller than those of MIP/CIP. We have shown the similar analytical results in the previous subsection. In addition, we can see the severe delays of packets just after handoffs are completed in MIP/CIP, while the phenomenon is not appeared in SIP/CIP. Fig.7 shows the average packet loss ratio varying the distance between wireless network cells. The average packet loss ratio is defined as the number of lost

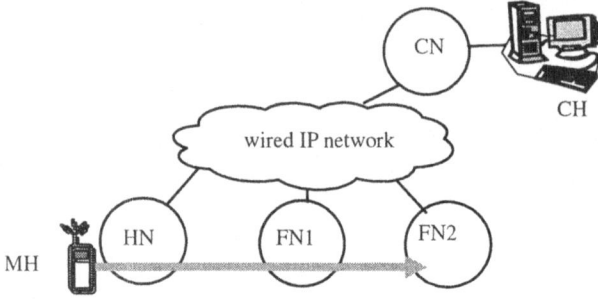

Fig. 5. Network model for evaluating real-time data delivery performance

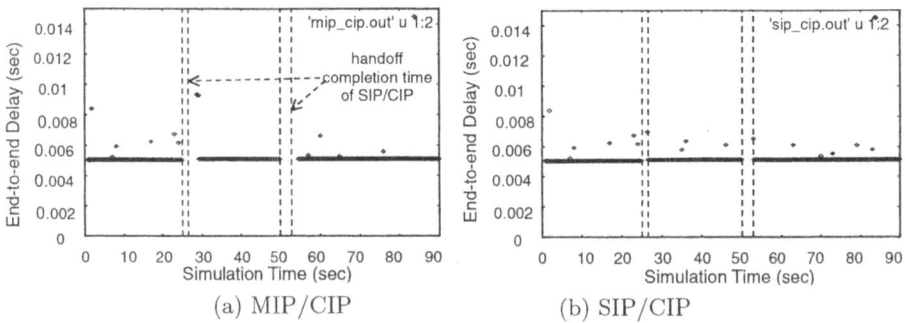

(a) MIP/CIP   (b) SIP/CIP

Fig. 6. End-to-end delay characteristics (moving speed of MH=6m/sec)

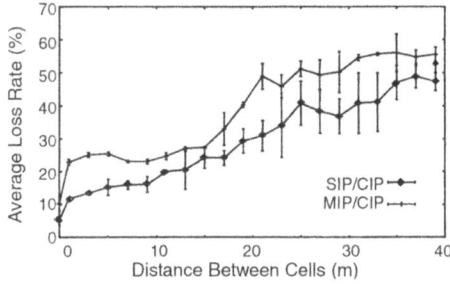

Fig. 7. Average packet loss rate varying the distance between cells

packets over the number of total packets sent. It is noted that the environment of handoff is getting worse as the distance between wireless networks increases. From Fig.7, it is shown that as the distance between cells increases, the average packet loss ratio also increases. However, the average packet loss ratio of SIP/CIP is much lower than that of MIP/CIP. This is because the handoff delay by SIP/CIP is much smaller than MIP/CIP as shown in Fig.6.

5   Conclusion

In this paper, we analyzed and compared handoff delay performances for several mobility support architectures such as MIP-only, SIP-only, MIP/CIP and SIP/CIP. For the analysis, we described the detailed message flows of those architectures in handoffs, and then made an analytical model to compute the handoff delay for each scheme. Numerical results showed that SIP/CIP can reduce the handoff delay significantly compared to other mobility support architectures. We also showed that SIP/CIP can achieve much improved loss and throughput performances in delivering real-time multimedia traffic by using simulation.

Much of works have been done for supporting IP mobility management. It is questionable that among those, which scheme can be a most useful solution for delivering real-time multimedia traffic in the hybrid of wired and wireless environments. The schemes considered in this paper such as MIP-only, SIP-only, MIP/CIP and SIP/CIP are expected to be most promising solutions to support IP mobility. From our experimental results, we showed that SIP/CIP can be a good solution to support both micro- and macro-mobilities. In addition, SIP is getting widely accepted as a signaling protocol for real-time multimedia services in both the Internet and wireless networks. We think that if SIP can be used for a signaling protocol to support IP mobility management as well, the control of real-time services can be done at one framework, from the start of a session to the end of the session though handoff occurs, which makes it easier and more efficient to manage application services as well as networks.

References

1. Perkins C.: IP Mobility Support, IETF RFC 2002, (1996)
2. Perkins .E, Johnson D.: Route optimization in Mobile IP, IETF Draft,<draft-ietf-mobileip-optim-10.txt> (2000)
3. Rosenberg J., Schulzrinne H., Camarillo G., Johnston A., Peterson J., Sparks R., Handley M., Schooler E.: Session Initiation Protocol, IETF RFC 3261 (2002)
4. Schulzrinne H., Wedlund E.: Application-layer mobility using SIP, ACM Mobile Computing and Commun. Rev., **4** (2000)
5. Kwon T., Gerla M., Das S., Das S.: Mobility Management for VoIP Service: MIP vs. SIP, IEEE Wireless Communications, **9** (2002)
6. Schulzrinne H.: DHCP Option for SIP Servers, IETF draft <draft-ietf-sip-dhcp-05.txt> (2001)
7. Campbell A., et al.: Cellular IP, IETF draft <draft-ietf-mobileip-cellularip-00.txt> (2000)
8. Carli M., Neri A., Picci A.: Mobile IP and Cellular IP Integration for inter Access Network Handoff, IEEE ICC'2001 (2001)
9. Gatzounas D., Theofilatos D., Dagiuklas T.: Transparent Internet Mobility using SIP/Cellular IP Inte-gration, IPCN 2002, Paris France (2002)
10. Wong K., Wei H., Dutta A., Young K., Schulzrinne H.: Performance of IP Micro-Mobility Management Scheme using Host Based Routing, WPMC'01 (2001)
11. Network Simulator - ns (version 2), http://www.isi.edu/nsnam/ns
12. Chen, H.: Simulation of Route Optimization in Mobile IP, WLN'2002 (2002)

A Video Compression Tools Comparison for Videoconferencing and Streaming Applications for 3.5G Environments

N. Martins, A. Marquet, and A. Correia

ADETTI, Edifico ISCTE, Avenida das Forças Armadas, 1600-082 Lisboa, Portugal
{nuno.martins, andre.marquet, americo.correia}@iscte.pt

Abstract. The purpose of this paper is to evaluate the state of the art in standard video coding technologies in what regards to its use in the B-BONE[1] project. The B-BONE project aims to define the technological approaches, such as the usage of Multimedia Broadcast Multicast Services (MBMS), highly efficient modulation and spectrum allocation schemes, necessary to the delivery of high quality video streams to Enhanced Universal Mobile Telecommunications System (E-UMTS) end-users. With this paper the authors intend to present the MPEG-4 Simple Profile and the H.264/ AVC Baseline standard and examine both in terms of rate-distortion efficiency *via* objective quality metrics assessment.

1 Introduction

Presently, MPEG-4 is considered the state of the art standard for video compression. MPEG-4 is an ISO/ IEC standard developed by Moving Pictures Expert Group (MPEG). The Advanced Video Coding[2] (AVC) is the outcome of Joint Video Team (JVT) and the standard final draft for International Organization for Standardization (ISO) approval was submitted only as recently as July 2003 [1]. These standards maintain the same motion compensated approach of previous video standards such as H.261, H.263 and MPEG-2 [2].

Enhanced-Universal Mobile Telecommunications System (E-UMTS), in the scope of B-BONE, will allow bit-rates in the order of megabits *per* second. Novel approaches will be adopted to enhance end-user applications and minimize network resources consumption thus allowing the delivery of Multimedia Broadcast Multicast Services (MBMS) if suitable video coding and proper distribution technologies are adopted.

The intent of this paper is to give a comparative rate-distortion analysis between these two standards and is organized as follows. Section 2 presents a brief overview

---

[1] This work was supported by the EU through the Sixth Framework Programme, http://fp6.cordis.lu/fp6/home.cfm.
[2] In this paper it shall be used the ITU-T designation, H.264, AVC is the common term used by the MPEG community.

J.N. de Souza et al. (Eds.): ICT 2004, LNCS 3124, pp. 1007–1012, 2004.
© Springer-Verlag Berlin Heidelberg 2004

of the encoding features for MPEG-4 Simple Profile (SP) and H.264 Baseline (BL). The rate-distortion efficiency is discussed in Sect. 3.

2 Standards Overview

Lossy video coding standards intend, at cost of some visual information loss, to reduce the amount of bits required to represent the input signal. Although the block-based hybrid motion compensated approach for video coding is conceptually the same for MPEG-4 and H.264, different coding methods are employed, particularly in the motion prediction, transform and entropy coding. Being specifically designed for CPU power constrained devices, MPEG-4 SP [3] and H.264 BL [4], were considered for the trial because the expected target platforms for B-BONE are low end computer devices such as cell phones. The Simple Profile in MPEG-4 uses rectangular frames based on 16x16 samples macroblocks, supporting half-pixel accuracy with bilinear interpolation. A Discrete Cosine Transform (DCT) is applied to 8x8 blocks, followed by scalar quantization and variable length entropy coding [5]. The H.264 Baseline uses an integer transform with similar proprieties as DCT, applied to different block sizes intended to reduce the prediction error signal. H.264 BL quantization is made using scalar quantization followed by context-based adaptive variable length entropy coding [6].

3 MPEG-4 and H.264 Visual Performance

A metrics evaluation for MPEG-4 SP and H.264 BL is presented in this paper in terms of rate-distortion efficiency. The comparison is performed in the YUV colour space. The study focuses the analysis on the luminance component with 8 bit *per* sample signals, thus assuming 255 as the peak signal.The visual quality is indicated through the Peak-to-Signal-Noise Ratio (PSNR). The PSNR measures the cumulative square error between reference and reproduction frame in relation to peak signal. The image distortion is presented in terms of average PSNR, that is, the PSNR *per* frame. The formula comes as,

$$PSNRperframe(x,y) = \left(\sum_{I=0}^{N-1} 20 \log_{10} \left(\frac{255}{\sqrt{\frac{1}{nm} \sum_{i=0}^{n} \sum_{j=0}^{m} \left[V_i(x,y) - V_j(x,y) \right]^2}} \right) \right) \Big/ N \qquad (1)$$

$V_k(x,y),$ *luminance values for each pel [0, 255]*

$m,$ *horizontal pels*

$n,$ *vertical pels*

$N,$ *number of frames*

The rate-distortion analysis is oriented for videoconferencing and video streaming scenarios. Key issues in the evaluation were setting the input image sequences and video coder bit-rate control. Appropriate content was selected to represent each scenario, *vide* Table 3, and video coder bit-rate control was achieved tuning the quantization parameter using variable bit-rate control mechanism.

3.1 Videoconferencing Scenario

Low resolutions are typically used in videoconferencing applications, thus Quarter-Common Intermediate Format (QCIF) at 10 Hz was adopted for the assessment. An Intra period of ten was adopted for both MPEG-4 and H.264 encodings. Some reference bit-rates were specified according to UMTS downlink common channels capacity [7]. The metric was computed for PSNR *per* frame as in Eq. (1). The rate-distortion curves were attained modelling a power function. The resulting rate-distortion curves are depicted in Fig.1, from which it is possible to observe that H.264 BL outperforms MPEG-4 SP in all selected sequences.

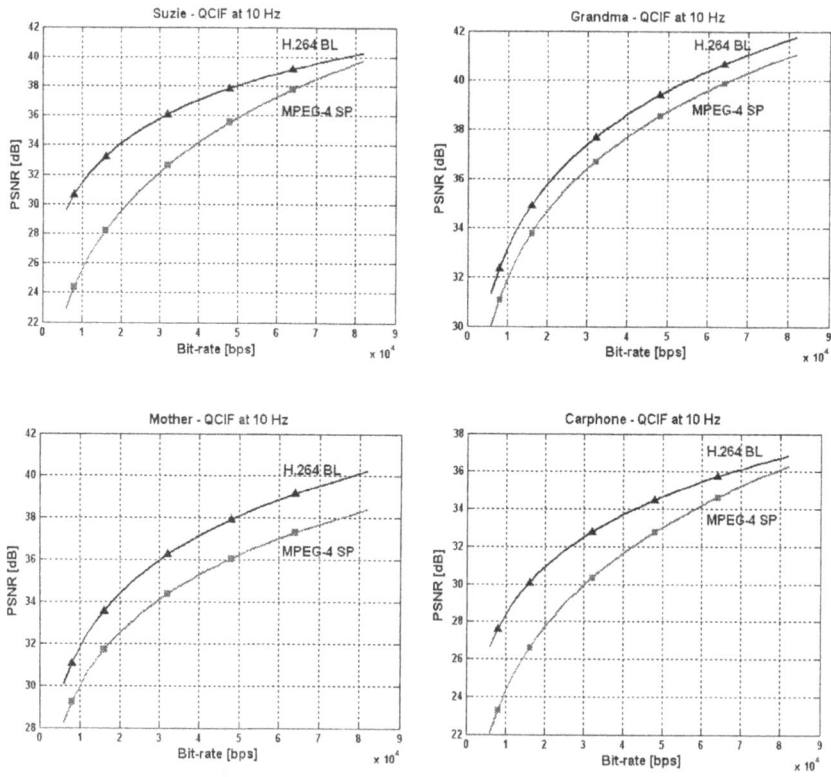

Fig. 1. Rate-distortion curves for videoconferencing scenario.

The rate-distortion curves allowed computing PSNR values for the target reference bit-rates, summarized in Table 1. An intra scenario analysis indicates that the disparity between PSNR values for H.264 BL and MPEG-4 SP diminishes when bit-rate increases. For the 8 kbps, the H.264 BL has an average PSNR 3.4 dB higher than MPEG-4 SP. The average gain of H.264 BL relatively to MPEG-4 SP continuously decreases along the range of reference bit-rate values, until 1.3 dB for 64 kbps. The overall average gain of H.264 BL facing MPEG-4 SP is 2.3 dB. It should be noted that MPEG-4 SP has a higher PSNR standard deviation for the bit-rate interval than H.264 BL, being 4.2 dB and 3.4 dB respectively.

Table 1. MPEG-4 and H.264 PSNR for the videoconferencing reference bit-rates.

Sequence	8 kbps		16 kbps		32 kbps		48 kbps		64 kbps	
	MP4	H.264	MP4	H.264	MP4	H.264	MP4	H.264	MP4	H.264
QCIF, 10 Hz										
Suzie	24.4	30.6	28.2	33.2	32.6	36.1	35.5	37.8	37.7	39.1
Mother	29.2	31.1	31.7	33.5	34.4	36.2	36.1	37.9	37.3	39.2
Grandma	31.0	32.3	33.8	34.9	36.7	37.7	38.5	39.4	39.9	40.7
Carphone	23.3	27.6	26.6	30.1	30.3	32.8	32.8	34.5	34.6	35.7

3.2 Video Streaming Scenario

For the video streaming scenario Common Intermediate Format (CIF) sequences at 30 Hz were used. Encodings were made with an Intra period of 30. Rate-distortion curves were fitted in order to represent the PSNR in terms of bit-rate. As for videoconferencing scenario rate-distortion curves, the power model function revealed to be the best fit with the exception of H.264 BL rate-distortion curve for Akiyo that was modelled as a linear interpolation. These curves are depicted in Fig. 2.

Again H.264 BL achieved better rate-distortion efficiency than MPEG-4 SP in all tested sequences. In average terms PSNR gain for H.264 BL is 3.8 dB compared to MPEG-4 SP. This gain relation has a decreasing behaviour, attaining the maximum value at 128 kbps, 4.9 dB, and the minimum value at 768 kbps, 3.1 dB. The PSNR standard deviation is higher for MPEG-4 than for H.264, 2.0 dB and 1.3 dB respectively. The Table 2 shows the results for the PSNR for the reference bit-rates specified according to expected bit-rates in UMTS downlink shared channels [8].

Table 2. MPEG-4 and H.264 PSNR for the video streaming reference bit-rates.

Sequence	128 kbps		256 kbps		384 kbps		512 kbps		768 kbps	
	MP4	H.264	MP4	H.264	MP4	H.264	MP4	H.264	MP4	H.264
CIF, 30 Hz										
Hall	33.9	36.2	35.4	37.0	36.3	37.5	37.1	37.9	38.0	38.4
Mobile	21.5	30.0	23.7	30.9	25.1	31.4	26.1	31.8	27.7	32.4
Akiyo	39.0	41.2	40.5	43.0	41.4	44.2	42.1	45.5	43.1	48.0
Tempete	25.0	31.4	27.2	32.3	28.7	32.9	29.7	33.3	31.2	33.9

---

[3] A composite analysis for a specific application.

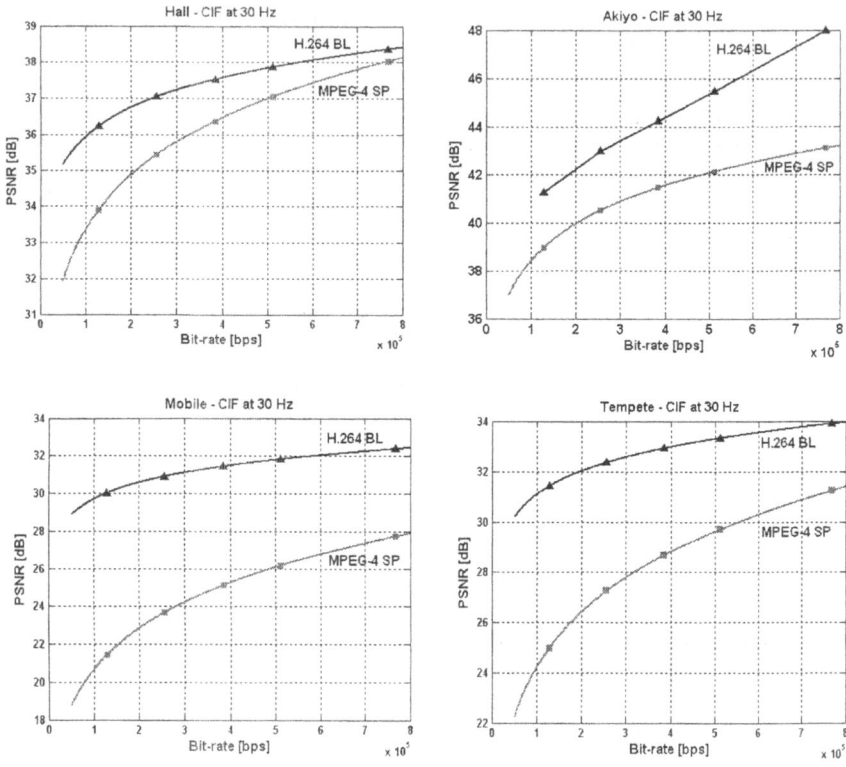

Fig. 2. Rate-distortion curves for video streaming scenario.

Table 3. Test sequences for videoconferencing and video streaming scenarios.

Name	Resolution	Frames	Duration	Description
Carphone	QCIF	129	12.9 sec.	Fast head movements with some content background motion
Grandma	QCIF	290	29 sec.	Still camera with low movement 'talking head'
Suzie	QCIF	50	5 sec.	Still camera with rapid head movements
Mother	QCIF	320	32 sec.	An adult and a child talking with head and hand movements
Akiyo	CIF	300	10 sec.	Still camera newscast with synthetic background
Tempete	CIF	260	8.67 sec.	Fast random motion with detailed background and camera zoom-out
Mobile	CIF	300	10 sec.	High picture detail with camera movements and complex motion
Hall	CIF	300	10 sec.	Static camera, lighting change and localized motion

4 Conclusions

The results reveal that H.264 BL achieved better rate-distortion than MPEG-4 SP for the analyzed application scenarios. Therefore H.264 BL appears like an important candidate for near future multimedia applications. The results turn evident the theoretical premises that point to H.264 BL as an improved coding approach due to the embracement of novel coding algorithms and a built-in deblocking filter.

When comparing the obtained standard deviation values for both standards H.264 BL performs particularly better than MPEG-4 SP for lower bit-rates in the two analysed application scenarios.

Forthcoming this analysis it would be of particular interest to conduct some subjective quality assessment on the same test sequences in order to correlate with the results that are presented in this paper. This is a very time consuming and resource demanding task which is already being conducted and results are expected in the near future.

Furthermore in wireless video communications, the ability of the video standard to provide robust error resilience is of utmost importance. Since H.264 BL proved to have higher rate-distortion efficiency for the studied application scenarios, one can question whether H.264, low redundancy bitstream, will not be specially affected in error-prone channel conditions. Performing quality metrics tests to infer the level of the H.264 bitstream robustness to the expected channel operation bit error rates for E-UMTS scenarios, in relation to MPEG-4, could be decisive on which video standard should be considered for trial and field tests within the scope of B-BONE objectives.

References

1. Draft ITU-T Recommendation and Final Draft International Standard of Joint Video Specification (ITU-T Rec. H.264 | ISO/IEC 14496-10 AVC)
2. Hallsall, F., Multimedia Communications, Pearson Education, England, 2001
3. Coding of Audio-Visual Objects—Part 2: Visual, ISO/IEC JTC1,ISO/IEC 14 496-2 (MPEG-4 visual version 1), 1999.
4. T. Wiegand, G. J. Sullivan, and A. Luthra, "Draft ITU-T Recommendation H.264 and Final Draft International Standard 14496-10 Advanced Video Coding," Joint Video Team of ISO/IEC JTC1/SC29/WG11 and ITU-T SG16/Q.6, Doc. JVT-G050rl, Geneva, Switzerland, May 2003
5. The MPEG-4 Book, Pereira, F., Ebrahimi, T., Prentice Hall PTR, 2002
6. H.264 and MPEG-4, Richardson, I., Wiley, 2003
7. 3GPP TR 25.803 S-CCPCH performance for MBMS (Release 6), 3GPP, 2004
8. 3GPP TS 25.308 v6.0.0 HSDPA Overall Description Stage 2, 3GPP, 2003

Fairness and Protection Behavior of Resilient Packet Ring Nodes Using Network Processors

Andreas Kirstädter[1], Axel Hof[1], Walter Meyer[2], and Erwin Wolf[2]

[1]Siemens AG, Corporate Technology
Information and Communications
Otto-Hahn-Ring 6
81730 München, Germany
{Andreas.Kirstaedter, Axel.Hof}@siemens.com
[2]Siemens AG, ICN
Hofmannstr. 51
81379 München, Germany
{Walter.Meyer, Erwin.Wolf}@siemens.com

Abstract. The Resilient Packet Ring IEEE 802.17 is an evolving standard for the construction of Local and Metropolitan Area Networks. The RPR protocol scales to the demands of future packet networks and includes sophisticated resilience mechanisms that allow the reduction of equipment costs. Network processors are a new opportunity for the implementation of network nodes offering a high flexibility and a reduced time to markets. This paper describes the implementation of a Resilient Packet Ring line card for a SDH/Sonet add-drop-multiplexer using the Motorola C-5 network processor. We show the novel system architecture of the ring node influenced by the use of the network processor. System simulations and field-trial measurements verify the performance of the implemented protection and fairness mechanisms. Even with the usage of a protection steering mechanism implemented on flexible network processor hardware we were able to achieve reconfiguration times well below 50 milliseconds.

1 Introduction

The Resilient Packet Ring (RPR) is a draft standard to transport data traffic over ring-based media with link data rates scalable up to many gigabits per second in Local or Metropolitan Area Networks. The RPR standardization (IEEE 802.17) working group of the Institute of Electrical and Electronic Engineers (IEEE) started to work on the specification in December 2000 with the intention to create a new Media Access Control layer for RPR.

Two counter-rotating buffer-insertion rings build up an RPR [1,2], as shown in Figure 1. Adjacent nodes are interconnected via a (fiber) link pair. The link bit-rate of an RPR can take values in the range from 155 Mbit/s up to 10 Gbit/s [2].

J.N. de Souza et al. (Eds.): ICT 2004, LNCS 3124, pp. 1013–1022, 2004.
© Springer-Verlag Berlin Heidelberg 2004

Among many other deployment areas, RPR rings are especially attractive for the use within SDH/Sonet Add-Drop Multiplexers in Metropolitan Area networks. Here SDH/Sonet paths constitute the links between the RPR nodes.

The RPR line card described in this paper offers on the tributary-interface side the choice between 10/100 Mbps and 1 Gbps Ethernet. On the (SDH) ring side, either VC-4 paths or VC-4-4v paths can be supported. To achieve this flexibility a network processor (NP) was selected for the task of data processing [3]. The network processor C-5 from C-Port/Motorola proved to be appropriate for this kind of application [4].

In principle, several solutions exist for protecting RPR ring networks. These solutions differ in their protection speed and bandwidth efficiency. In our implementation we selected a steering mechanism for the ring protection and implemented it in the software of the NP and its controlling General Purpose Processor (GPP).

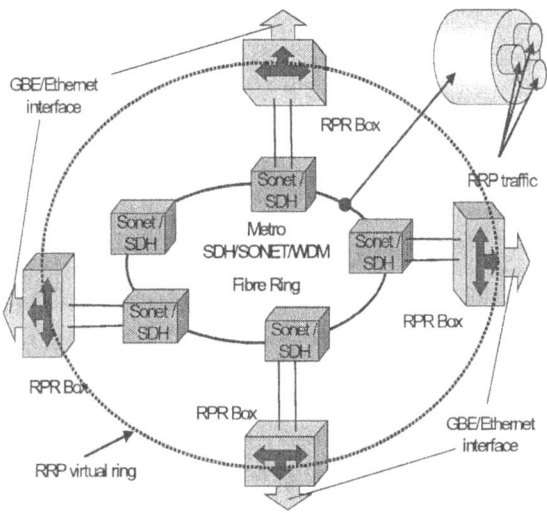

Fig. 1. RPR topology on the basis of SDH links

The rest of the paper is organized as follows: Chapter 2 describes the architecture of the RPR card and the SDH add-drop multiplexer it is connected to. Chapter 3 gives a small overview of the C-5 network processor and shows the main features of this processor. Ring protection and the steering mechanism are presented in chapter 4 together with the measurement results from a field trial with a lead customer. Additionally, we carried out some system simulations of the RPR ring. Chapter 5 describes the simulator and presents various results of system measurements and simulation.

2   RPR Card and SDH Add-Drop Multiplexer

The SDH/Sonet add-drop multiplexer is a multi-service system that is configured in a rack with multiple flavors of line cards. A SDH/Sonet back plane provides the interworking among the cards across a switch fabric. A control processor card manages the operation of the system. The line cards run with OC-3, OC-12 and OC-48.

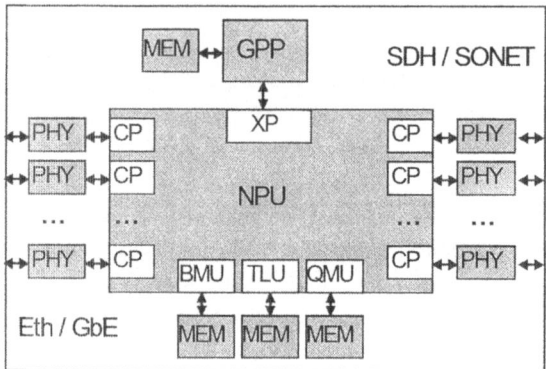

Fig. 2. Architecture of the RPR line card

As shown in Figure 2 the RPR card mainly consists of the NP, a GPP, and some interface and memory chips. The GPP controls the network processor and consists of a Power-PC processor connected via a PCI bridge to the network processor. The GPP takes care about the generation of the routing table, the alarm handling and bandwidth reservation. Via the PCI bridge the GPP can access a part of the data memory of the NP, e.g. for downloading routing tables to the NP or reading of some statistical data. A control processor core (XP) within the NP handles the access of the GPP to the data memory on the NP chip.

The RPR implementation supports stream and best effort traffic. A fairness algorithm on top of the ring guaranties the reserved bandwidth for the stream traffic and distributes the remaining bandwidth between the ring nodes in a fair manner for the best effort traffic. An input rate control at the tributary Ethernet interfaces regulates the throughput of added packets on the ring. A feedback mechanism of the fairness algorithm influences the settings of the input rate control and can back press the Ethernet packets in case of ring congestion.

3 Network Processor Architecture

The C-5 network processor from Motorola contains 16 parallel channel processors (CP). Each of them consists of a RISC core together with a Serial Data Processor (SDP) for the bit and byte processing [5]. Additionally to the XP block mentioned above, there are also four other special-purpose units on the C-5 for the buffering (BMU), queuing (QMU), table lookups (TLU), interconnection to a switch fabric (FP), as shown in Figure 2.

The 16 parallel channel processors (CP) are ordered into four clusters of four processors each. The four processors in one cluster can run the same application and share an instruction memory of 24 kByte that also can be subdivided so that each CP gets a dedicated 6kByte sub-array. Three independent data buses (Figure 3) provide internal communication paths between the different internal processors.

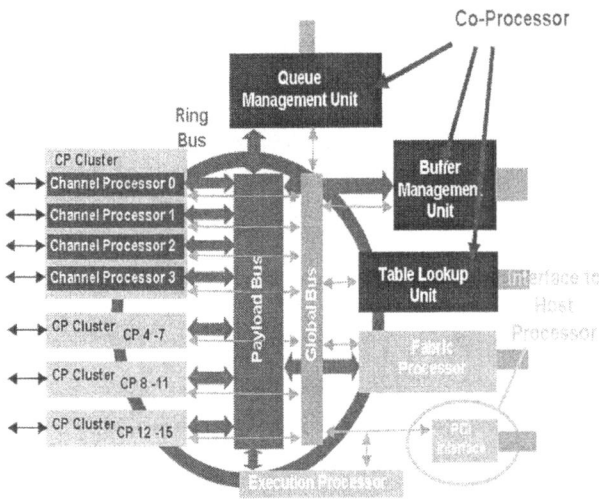

Fig. 3. C-5 network-processor architecture

Each of the sixteen CPs contains a Reduced Instruction Set Computer (RISC) Core controlling cell and packet processing in its channel via the execution of a MIPS ™ 1 instruction set (excluding multiply, divide, floating point).

Packet buffering and queuing in the C-5 is handled as follows: The payload of the incoming packet is stored in the external memory, which is controlled by the Buffer Management Unit (BMU). The BMU controls the storage of the payload and returns a descriptor of the memory block for the payload storage to the CP. After the lookup at the Table Lookup Unit (TLU) the CP sends the descriptor of the payload buffer to the Queue Management Unit (QMU) and enqueues it into the queue of the transmitting CP.

For the programming and configuration of the special entities like the BMU, QMU and TLU exists a library of service functions [6], which is part of the C-Ware Software Toolset (CST). The XP, being the only processor with no linkage to the data path, controls the operation of the other processors and downloads the configuration onto them and the special units. During runtime the XP generates control messages or table entries within the TLU.

4   Ring Protection Measurement and Field Trial

As mentioned in the introduction several alternatives exist for the protection of RPR rings. A pure protection on the SDH level - below the RPR protocol - is surely the fastest way but occupies a lot of protection bandwidth and does not cover failures of the packet node or on the Ethernet level.

The IEEE 802.17 protocol itself will support wrapping and steering for ring protection, which allows the spatial re-use of bandwidth.

The faster alternative is wrapping being less bandwidth effective due to the wrapping loops. Wrapping occurs locally and requires two nodes to perform protection switching. As shown in Figure 4 the two nodes neighbouring the failed span have to loop the traffic onto the other ring. The dashed line is the original traffic flow, whereas the solid line symbolizes the protection path. Fast wrapping generates the lowest packet loss on the cost of higher bandwidth consumption.

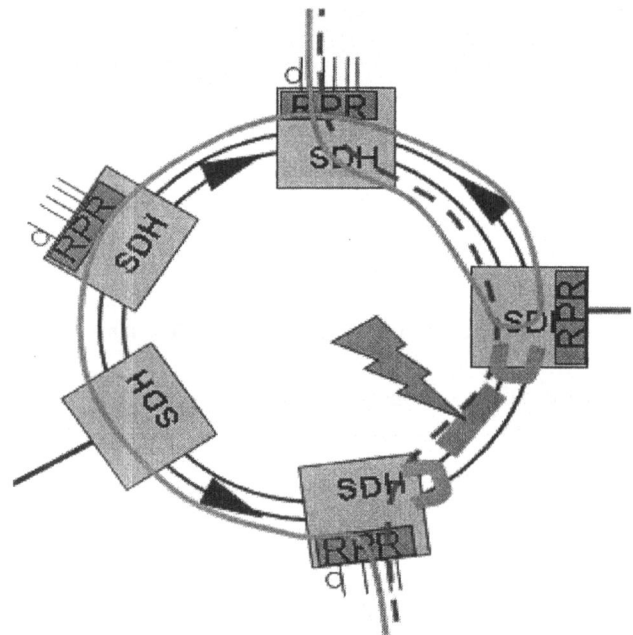

Fig. 4. Wrapping for ring protection

Steering reacts to the failure by modifying the routing tables in all nodes. Therefore it is more bandwidth efficient but also slower due to the messaging and the generation of the tables. In the RPR layer the two neighboring nodes would signal to other nodes span-status changes via control messages carried on opposite ring. Instead of wrapping the ring each node then independently reroutes the traffic it is sourcing onto the ring using the updated topology (see Figure 5).

The RPR foresees a recovery time of 50 milliseconds in event of fibre/node failure on the ring. Steering will be the lowest common denominator when both steering and wrapping nodes are on the ring.

For the wrapping mechanism the outgoing node has to store a high number of packets in case of a switch back to the original link after the failure recovery to avoid packet disorder. The original link is shorter then the protection link and therefore the transmission from the incoming node to the outgoing node includes more hops. The total number of stored packets in a C-5 NP is limited to 16000. The number of packet descriptors is needed for the proper operation of the fairness algorithm. Due to the limited number of packet descriptors we selected a steering mechanism and implemented it in the GPP and NP software.

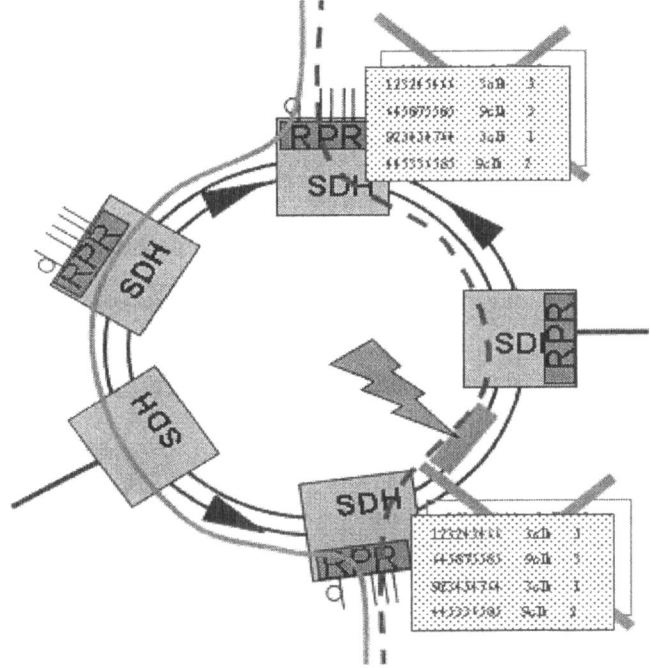

Fig. 5. Steering for ring protection

Link failures are detected by SDH alarming in the SDH overhead or frame. The physical layer device (PHY in Fig. 2) for SDH analyzes the SDH overhead and the SDH framer within the C-5 NP controls the frame errors. The GPP collects all failure alarms and generates an alarm message. The nodes neighboring the failure inform all other nodes via alarm messaging on the RPR level. To reduce the transmission time of the alarm messages, the control packets got the highest priority (above stream traffic).

In the GPP of each single node the routing tables are recalculated according the collected routing information in the received alarm message and then downloaded into the NP via the PCI Bridge. Due to the limited instruction memory in the XP in the C-5 NP the routing generation is part of the GPP software. In [7] the authors describe a solution for the speed up of the interconnection between a NP and GPP, which will lead to a shorter rerouting time. For the C-5e a faster rerouting would be feasible by the inclusion of the routing table generation code into the XP, since the C-5e has twice the instruction memory capacity.

System measurements with different SDH failures verified ring protection times well within 50 milliseconds. Table 1 presents the results for the failure insertion and the failure removal on a 1200 km ring with 12 nodes. In all cases of failures removal the protection switching time stays below 20 milliseconds. All error detections have no integration time to keep the delay as low as possible.

Table 1. L2 protection times for different SDH faults

Failure	L2 Protection Switching time (msecs.)	
	Failure Insertion	Failure Removal
LOS	44	15
AU4-AIS	44	15
UNEQ	20	15
LOM	20	15

The Loss of Signal (LOS) alarm is raised when the synchronous signal (STM-N) level drops below the threshold at which a BER of 1 in 10^3 is predicted. This could be due to a cable cut, excessive attenuation of the signal, or equipment fault. The LOS state will be cleared as soon as two consecutive framing patterns are received and no new LOS condition is detected.

The Alarm Indication Signal (AIS) for STS-3c (AU4) is an all-ONES characteristic or adapted information signal. It is generated to replace the normal traffic signal when it contains a defect condition in order to prevent consequential downstream failures being declared or alarms being raised.

The Loss of Multi-frame (LOM) state occurs when the incorrect H4 values for 8 frames indicate lost alignment.

The unequipped (UNEQ) alarm is raised when z consecutive frames contain the all-ZEROS activation pattern in the unequipped overhead.

After the system integration followed a field trial at a lead customer side. Delay measurements in a 12 nodes ring in Austria (Vienna, Salzburg, Klagenfurt) make up the main part of the trial. For the delay and packet loss measurements we used frame-sizes according to RFC 2544. We observed no packet loss and delays between 6.33 and 6.98 milliseconds depending on the frame size.

5 Simulations During System Development

The C-5 tool environment includes a cycle-accurate simulator together with a per-formance analyzer. Via this tool we made a first rough estimation of the workload on the network processor and were able to trace internal components of the C-5 like the buses and special units like the QMU.

Additionally to the cycle-accurate simulations the overall system behavior had to be verified. The cycle-accurate simulations deliver a very detailed picture of the internal operation of the C-5 running the RPR protocol. But for exact statements on the protocol behavior itself a separate simulator had to be developed since the system of several RPR nodes had to be observed for larger time intervals. Running these simulations with the cycle-accurate simulator was not possible due to the CPU time requirements: The maximum number of packets that could be observed during reasonable CPU times is around 100 to 1000. This short time frame was not sufficient to check the system behavior of a complete RPR ring concerning fairness and delay behavior.

The system simulator is programmed in C++; the libraries of CNCL (Communication Networks Class Library [8]) were used. The simulator is event based and is built in a very modular manner. The protocol and also the simulator are specially adapted to the behavior of the C-5 NP. To simulate different scenarios it is possible to use different sources with different distributions of packet length and destination addresses. Meters can be attached to points of interest in the investigated network to accomplish the behavior and performance investigation of the protocol by collecting data while the simulation is running.

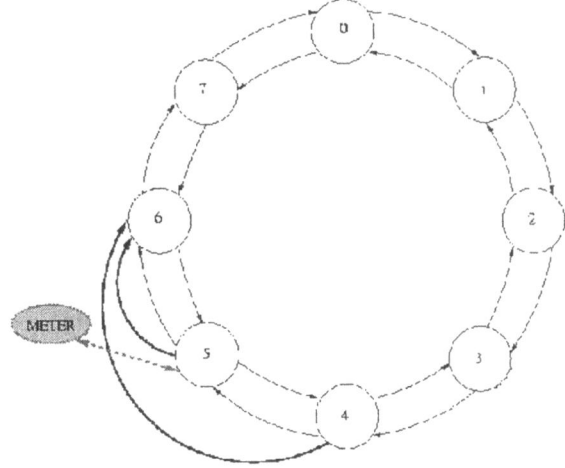

Fig. 6. Eight node network with traffic meter

As an example the following drawings show the priority handling in an eight-node topology where first node #4 at time=0sec sources 100 Mbps of low-priority traffic onto the ring (link capacity: 150 Mbps) for forwarding further downstream towards node #6 (see Figure 6). At the time=0.5sec node #5 sources 100 Mbps of high-priority traffic also destined to node #6.

For the simulations we used different traffic sources. The destination addresses are fix or have a uniform distribution. The packet length is fix or distributed negative exponential or like a so called bath tube (30% 64 bytes, 60% 1540 bytes). The sources are switched on and off after a duty cycle of 0.5sec. In this example the destination address is fix and the source are sending with a constant bit rate.

Figure 7 shows the resulting throughput in the form of forwarded low-priority traffic originated by node #4. Figure 8 shows the amount of high-priority traffic sourced by node #5.

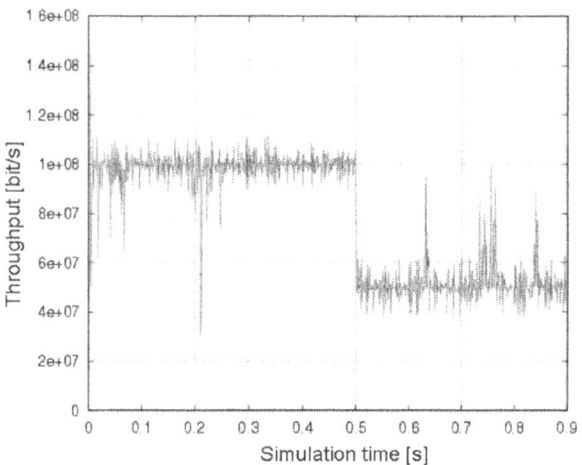

Fig. 7. Throughput from node #4 over time

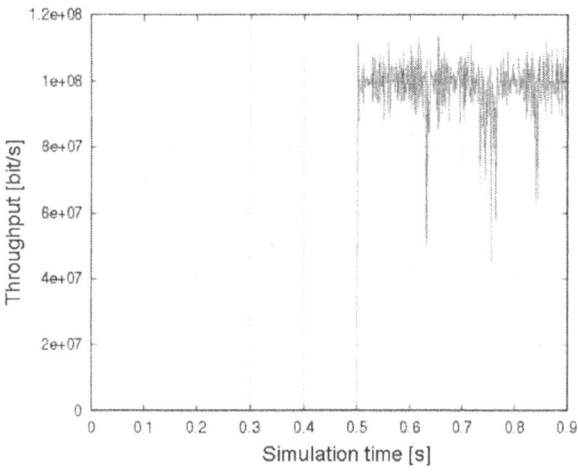

Fig. 8. Throughput from node #5 over time

When the high priority traffic is switched on the throughput of the low-priority traffic is reduced to remaining ring capacity of 50 Mbps

As it can be seen from the diagrams above the ring fairness protocol preserves the strict priority between high and low-priority traffic. Exactly the same result was also measured in the experimental setup.

We repeated comparable simulations with different traffic sources and different ring topologies. During the system test we verified the fair distribution of the ring capacity with similar test scenarios.

In conjunction with the steering mechanism the fairness algorithm guarantees a proper behaviour also in case of ring protection. The additional sourced packets in case of ring protection is controlled and the input of low-priority traffic reduced to keep the guaranties of the high-priority traffic.

6 Conclusion

During the system development phase it was very helpful to have both the cycle-accurate and the system-level simulators at hand. Especially the system-level simulations delivered important details on the operation and optimization of protocol features that otherwise could not be verified in advance.

The system tests verified the fast protection switching with the usage of SDH alarms despite the steering mechanism. The generation of the rerouted tables in the GPP and the PCI transfer into the NP showed to be sufficiently fast without the need any additional special hardware. Additionally the fairness algorithm guaranties the appropriate subdivision of the link bandwidth between the single flows even in protection state.

The field trial provided us with long-time measurements and asserted the smooth system behavior of the RPR line card. Since some months the system is delivered to customers.

References

[1] IEEE 802.17 Resilient Packet Ring Working Group Website, http://www.ieee802.org/rprsg/.
[2] H.R. van As, "Overview of the Evolving Standard IEEE 802.17 Resilient Packet Ring," 7th European Conference on Networks & Optical Communications (NOC), Darmstadt, Germany, June 18-21, 2002.
[3] T. Wolf, "Design of an Instruction Set for Modular Network Processors," IBM Research Report, RC 21865, October 27, 2000
[4] N. Shah, "Understanding Network Processors," Master's Thesis, Dept of Electrical Engineering and Computer Science, Univ. of California, Berkeley, 2001
[5] C-5e Network Processor Architecture Guide Silicon Revision A0, Motorola, http://e-www.motorola.com/brdata/PDFDB/docs/C-5EC3EARCH-RM.pdf
[6] C-5e Application Documentation, Motorola, http://e-www.motorola.com/webapp/sps/site/prod_summary.jsp?code=C-5E#applications
[7] F.T. Hady, T. Bock, "Platform Level Support for High Throughput Edge Applications: The Twin Cities Prototype," IEEE Network Magazin, July/August 2003
[8] RWTH Aachen, http://www.comnets.rwth-aachen.de/doc/cncl/index.html

Markov Chain Simulation of Biological Effects on Nerve Cells Ionic Channels Due to Electromagnetic Fields Used in Communication Systems

Daniel C. Uchôa, Francisco M. de Assis, and Francisco A.F. Tejo

Universidade Federal de Campina Grande
Department of Electrical Engineering
Av. Aprigio Veloso 882, Campina Grande - PB, Brazil

Abstract. The indiscriminate use of mobile communications technology has given rise to a quarrel about possible effects on the population health. The transmembrane ionic channels have been considered as a crucial site of electromagnetic (EM) fields interaction with the exposed living tissue. An overview in the literature was made and a membrane channel model was simulated. The model response to sinusoidal EM field in the ELF range, for a voltage-dependent channel, was analyzed and variations at around 10 % in the open probability have been observed. This study is important to evaluate hereafter the possible impacts due to the use of new technologies.

1 Introduction

Since the beginning of bioelectromagnetism (BEM) studies the cellular membrane has been considered as the primary site of interaction. Experimental observations have shown some effects at this target and different examples can be found in the literature [1]. This is reasonable, since the physiological - biophysical and biochemical-equilibria are managed at cell membrane level.

The cellular membrane is composed by several pores called *ionic channels*. The bioelectrical cell membrane phenomenon is regulated by these channels, so they are extremely important for the vital functions of the living organism.

The EM field generated by some radiation sources used in mobile communications can be interpreted as an additive perturbation to the membrane potential [2], modifying the channels' conductivity. This may even affect the cell's integrity.

Several ionic channels interaction mechanism models have been proposed. We discuss here, the biophysical Hodgkin-Huxley (HH) membrane channel model, associated to a Markovian stochastic process. We have used this model to simulate possible EM effects at cell level.

J.N. de Souza et al. (Eds.): ICT 2004, LNCS 3124, pp. 1023–1028, 2004.

2   Modeling

The ionic flow through the membrane channels (*transmembrane current*) gives rise to a potential gradient across the membrane. This potential difference is called *transmembrane potential*.

From analysis of an experimental single-channel current pattern, obtained by a *patch-clamp* measurement technique, we can observe two well-defined states of the channel: a permissive state - in which the current flow is allowed - that is, the channel is *open*; and a non-permissive state - the channel is *closed* (there is no current flow). What we understand by *state* here is the conformational situation of the channel (the structural shape assumed by the channel protein due to the EM field distribution produced by the transmembrane voltage). Measurements reveal that the states transition, and the time during which the channel remains in a given state, present a stochastic behavior. Most channels can actually exist in three states, namely closed, open and inactivated. Hence, the channel can be considered as a non-deterministic state-machine. Therefore, we can define two random variables [4]:

1. the *occupancy*, that defines the probability of finding the channel in a given state, at a certain instant of time;
2. the *dwell-time*,T_d, that means the time a channel remains in one of the two states - open or closed.

The states are randomly generated and the interevent time between each generation is also random. We must then choose a stochastic model to represent correctly the channel behavior. From the probabilies measurements, we can obtain an accurate information about the electrophysiological features of the channel.

The transition rates among the channel states was studied by Hodgkin and Huxley (HH) [3]. They have obtained a successful equation systems that describe the access rate of states (the *gating*). These equations show that the gating depends on the transmembrane voltage and temperature.

Experimental data have demonstrated that the channel kinetics holds a short range memory step, that is, it depends on its actual state, no matter which was its prior state [5]. A suitable model, encompassing this memoryless property, and that can be governed by the HH transfer rates, could be represented by a continuous-time Markov chain (CTMC) [6]. This mathematical model has a one-step memory: given the present state at a specified time, and the state at a previous instant of time, the system is completely defined.

To explain their model, Hodgkin and Huxley speculated on the ion conductance mechanism. They supposed the existence of "charged" particles which allow the ions to flow when they occupy particular sites in the membrane [3], determining its conformational state. Thus, we need to insert in our model a counting process of these particles in order to determine the current state.

A classical counting process is the Poisson process. Keeping with this we choose a sort of Markov chain called *regenerative structure* [6].

This structure has an embedded uniform Markov chain associated to a homogenous Poisson process (HPP). The HPP is called the *clock* of the chain. If a Markov chain is a regenerative structure, it follows that:

1. $\{X_n\}_{n\geq0}$ is a discrete-time HMC with a transition probability matrix $P = \{p_{ij}\}$ given by:
   $p_{ij} = \frac{q_{ij}}{q_i}$, *if* $q_i > 0$ *and* $j \neq i$, *where the transition rate* $q_i = -q_{ii}$;
2. Given $\{X_n\}_{n\geq0}$, the interevents time sequence $\{\tau_{n+1} - \tau_n\}_{n\geq0}$ is independent; furthermore, $\forall n \geq 0$ and $\forall T_d \in \Re_+$,

$$P[(\tau_{n+1} - \tau_n) \leq T_d | \{X_k\}_{k\geq0}] = 1 - e^{(-q_{X_n} T_d)} \tag{1}$$

The most important ionic channels was simulated: potassium, with a two-states CTMC; sodium and calcium, both with three-states CTMC. The HH equations for the transition rates to the potassium and sodium channels can be found in [3], whereas the calcium transition rates was taken from [4]. The results obtained show that this modeling can be used to describe the channel kinetics, in a voltage clamp situation without exposure condition (see Fig. 1(a)-1(c)).

3   Implemented Simulation Technique

In order to get the occupancies for each state at each instant of time, it is necessary to find the dwell-time distribution in a voltage-clamp situation. As Equation 1 represents a cumulative function distribution (cfd), we can generate the desired distribution, with a uniform random number r, by using the inverse transform method, resulting in:

$$T_d = -\frac{1}{q_{ij}} ln(r) \qquad 0 < r < 1 \tag{2}$$

where q_{ij} is the transition rate from state i to j. For each state generated by the Markov chain, the dwell-time in that state is decided through Equation 2. Each time a transition occurs, the system configuration - the transmembrane voltage and temperature - is considered, the transition rate values are updated, resulting in a pattern which simulate the single-channel record.

As in experimental measurements, we obtain the ionic membrane current, in a voltage clamp situation, by the ensemble average of each patch-clamp record. Then, repeating the simulation N times, we build up a random process composed of N realizations of that event. Summing up the records of each event and dividing by N we obtain an ensemble average. Since the opening probability is linked to the channel conductance, we can obtain the current response of the simulated channel.

4   The Electromagnetic Stimulation

The external EM field penetrates into the surroundings of biological tissues, until it reaches the ionic channels of a particular cell. That EM field, then, induces a coherent voltage signal on the cell's membrane, which could be treated as an additive noise to the transmembrane voltage [2].

We have then an oscillating voltage signal superimposed to the usual voltage level associated to normal electrophysiological conditions. The channel's protein will respond to this external stimuli, gating the channel at the wrong instants of time. This could result in a complete modification of the whole living organism, for long term exposures to EM fields [7].

The existing studies on membrane relaxation times lead us to consider EM fields in the extremely low frequency range (ELF) [7]. As a result, the membrane is able to register and react to the stimulus, almost instantaneously.

A quantitative analysis of the EM coupling to cells has been presented in the specialized literature [1]. In our simulation we use the voltage variation induced on a spherical model of cell membrane (radius $R = 40\mu m$, relaxation time $\tau = 1ps$), for a sinusoidal signal stimulus [1]. This voltage variation will modify the transition rates, altering the opening or closing time, therefore causing an abnormal behavior of the cell.

We can evaluate the degree of perturbation imposed to the cell through a parameter called "relative effect factor" [4], defined as the rate between the variation of the open probability under and without exposure, in a voltage clamp situation, and the opening probability without exposure, in the same voltage clamp situation, that is:

$$\text{Effect} = \frac{p_{o_{exp}} - p_o}{p_o} \tag{3}$$

5   Results

As we are just interested in the voltage dependence, in the following we have fixed the membrane temperature value at $6^0 C$ [3].

A time response to the potassium channel can be seen in Fig. 1(d). The channel was voltage clamped at -40mV for 5ms, so a voltage step of 80mV is applied - clamping the channel at 40mV (for simplicity, we made $V_r = 0mV$). The response to a voltage step in the sodium and calcium channels is a little bit different, since the potassium channel does not have an inactivated state [3].

After a transient period, the open-state probability shows a short fluctuation around an average value, for the three kinds of channels. That means the channel reached a steady state. Thus we can calculate the steady state probability by ensemble averaging for several voltage-clamps (see Figs. 1(a) - 1(c)). These waveforms were compared to the experimental voltage-current results. The experimental data were obtained in [4].

Since our results without EM exposure have shown such a good agreement with the experimental data, we are able to use our model to predict biological effects under EM exposure as well. We simulated a 100Hz external continuous wave (CW) sinusoidal EM signal with a very short amplitude of the order of 20mV/m (with an incident angle $\delta = 0$). This amplitude field is capable to induce an oscillating voltage signal of $1.2\mu V$ at membrane surface (Fig. 1(e)). By

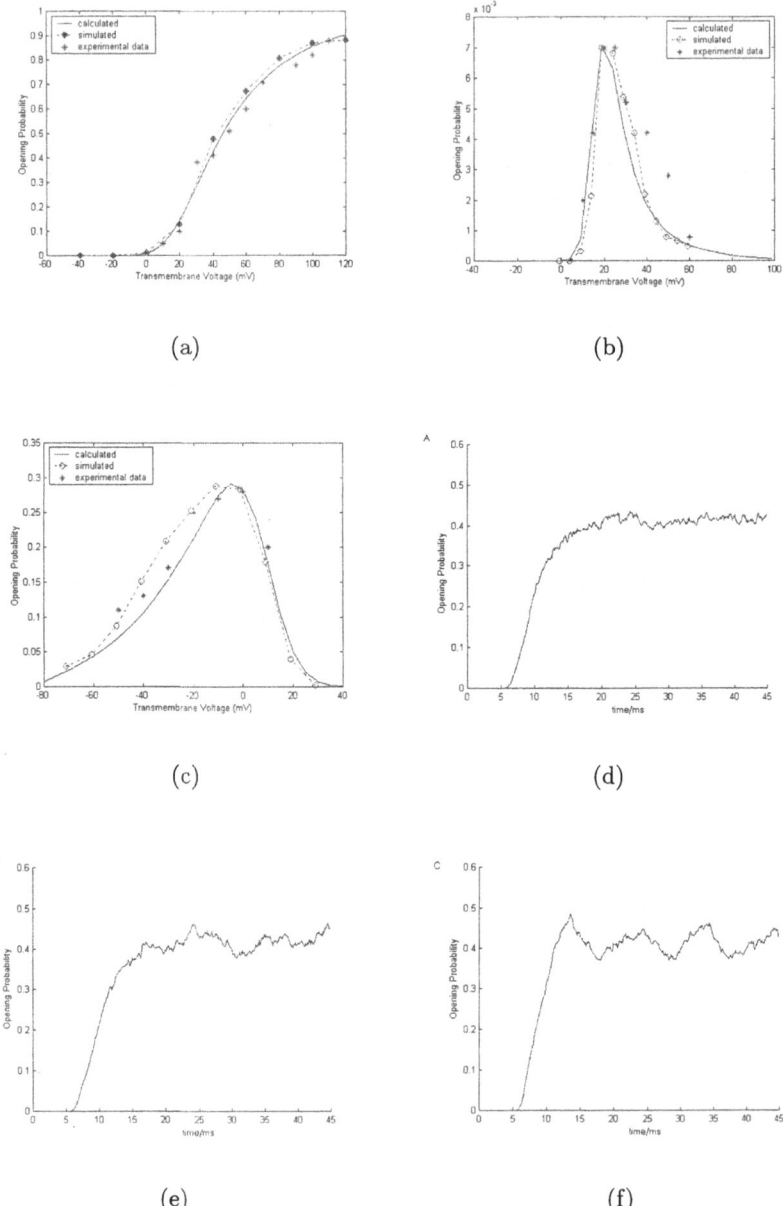

Fig. 1. Results: (a) The open steady state occupancy versus transmembrane voltage clamp values for potassium channel; (b) Sodium channels occupancy versus transmembrane voltage; (c) Calcium channels occupancy versus transmembrane voltage; (d) The instantaneous open-state probability in a potassium channel simulation without exposure; (e) Potassium channels response to a 100Hz, 20mV external voltage signal; (d) Potassium channels response to a 100Hz, 200V external voltage signal.

increasing the external amplitude of the EM field to 200V/m (capable to induce a $12mV$ transmembrane voltage signal) we can see that the channel follows the EM induced signal (Fig. 1(f)). Similar results were obtained to both sodium an calcium channels.

6   Conclusions

Markov chain models provide a real-time evaluation of the single-channel's conductance of voltage-dependent membrane channels under both unexposed and EM field exposure conditions, in a voltage clamp situation. Moreover, due to the instantaneous sensitivity of the HH transition rates to transmembrane voltage and temperature changes, this model can be used to study the effect of EM perturbations at the level of an individual channel.

The simulation reveals a really important feature of the ionic channels: the existence of a proportional link between the current flowing through the channel and the opening channel probability.

We have shown here that even external EM signals with very small amplitudes (the so called non-thermal EM fields) are able to induce a variation of about 10 % on the normal channel conductance. This probability variation produces transitions among the channel states at instants in which it was not supposed to happen (Fig.1(e)), and can result in drastic changes in the cell's functionality. We also noted that the opening channel probability follows the imposed external signal. Examples of several modes of synchronism between an imposed ELF field and neuronal patterns were experimentally demonstrated [1], but we are showing here that this kind of synchronism occurs at channel's level.

References

1. Polk, C., and Postow, E.(1986): "Handbook of Biological Effects of Electromagnetic Fields", CRC Press, Boca Raton, FL.
2. DeFelice, L.J., and Clay, J.R.(1981): "Introduction to Membrane Noise", New York, Plenum.
3. Hodgkin, A.L., and Huxley, A.F.(1952): "A quantitative description of membrane current and its application to conduction and excitation in nerve". J. Physiol. (Lond.) 117:500-44.
4. D'Izeo G., Pisa, S., and Torricone, L.(1993): "Ionic channels gating under EM exposure: A stochastic model", Bioelectron. Bioeng. J., no 29, pp. 290-304.
5. Colquhoun, D., and Hawkes, A.G.(1982): "On the stochastic properties of single ion channel openings and clusters of bursts". Phil. Trans. R. Soc. Ser. B 300, 159.
6. Brémaund, P.(1994):"Markov Chains - Gibbs fields, Monte Carlo simulation, and queues", Ed Springer.
7. Panagopoulos, D.J. and Margaritis, L.H.(2002): "Mechanism for Biological Effects of Oscillating Electromagnetic Fields", 2nd International Workshop in Biological Effects of Electromagnetic Fields, 7-11 October 2002.

Multiobjective Multicast Routing Algorithm

Jorge Crichigno and Benjamín Barán

P. O. Box 1439 - National University of Asuncion
Asunción – Paraguay. Tel/Fax: (+595-21) 585619
{jcrichigno, bbaran}@cnc.una.py
http://www.una.py

Abstract. This paper presents a new multiobjective multicast routing algorithm (*MMA*) based on the Strength Pareto Evolutionary Algorithm (*SPEA*), which simultaneously optimizes the cost of the tree, the maximum end-to-end delay, the average delay and the maximum link utilization. In this way, a set of optimal solutions, known as Pareto set, is calculated in only one run, without a priori restrictions. Simulation results show that *MMA* is able to find Pareto optimal solutions. They also show that for the constrained end-to-end delay problem in which the traffic demands arrive one by one, *MMA* outperforms the shortest path algorithm in maximum link utilization and total cost metrics.

1   Introduction

Multicast consists of concurrently data transmission from a source to a subset of all possible destinations in a computer network [1]. In recent years, multicast routing algorithms have become more important due to the increased use of new point to multipoint applications, such as radio and TV, on-demand video and teleconferences. Such applications need an important quality-of-service (QoS) parameter, which is the end-to-end delay along the individual paths from the source to each destination.

Another consideration in multicast routing is the cost of the tree. It is given by the sum of the costs of its links. A particular case is given with unitary cost. In this case, a multicast tree with minimum number of links is preferred, such that bandwidth consumption is minimized. To improve the network resource utilization and to reduce hot spots, it is also important for a multicast routing algorithm to be able to balance traffic. In order to improve load balancing, minimization of the maximum link utilization is proposed [2].

Most algorithms deal with two metrics: cost of the tree and end-to-end delay. They address the multicast routing as a mono-objective optimization problem, minimizing the cost subjected to a maximum end-to-end delay restriction. In [3], Kompella et al. present an algorithm (*KPP*) based on dynamic programming that minimizes the cost of the tree with a bounded end-to-end delay to each destination. For the same problem, Ravikumar et al. [4] present a method based on a simple genetic algorithm. This work was improved in turn by Zhengying et al. [5] and Araujo et al. [6]. The main disadvantage with this approach is the necessity of an *a priory* upper bound for the end-to-end delay that may discard good solutions.

J.N. de Souza et al. (Eds.): ICT 2004, LNCS 3124, pp. 1029–1034, 2004.
© Springer-Verlag Berlin Heidelberg 2004

Lee et al. [2] present a multicast routing algorithm which finds a multicast tree minimizing the maximum link utilization subject to a hop-count constraint.

In contrast to the traditional mono-objective algorithms, a *MultiObjective Evolutionary Algorithm* (*MOEA*) simultaneously optimizes several objective functions; therefore, they can consider the maximum end-to-end delay, the average delay, the cost of the tree and the maximum link utilization as simultaneous objective functions. *MOEAs* provide a way to solve a multiobjective problem (*MOP*), finding a whole set of Pareto solutions in only one run [7]. This paper presents a Multiobjective Multicast Routing Algorithm (*MMA*), a new approach to solve the multicast routing problem based on a *MOEA* with an external population of Pareto Optimal solutions, called the *Strength Pareto Evolutionary Algorithm* (*SPEA*) [7].

The remainder of this paper is organized as follow. A general definition of a multiobjective optimization problem is presented in Section 2. The problem formulation and the objective functions are given in Section 3. The proposed algorithm is explained in Section 4. Experimental results are shown in Section 5. Finally, the conclusions are presented in Section 6.

2   Multiobjective Optimization Problem

A general Multiobjective Optimization Problem (*MOP*) includes a set of n decision variables, k objective functions, and m restrictions. Objective functions and restrictions are functions of decision variables. This can be expressed as:

Optimize $\mathbf{y} = \mathbf{f}(\mathbf{x}) = (f_1(\mathbf{x}), f_2(\mathbf{x}), ..., f_k(\mathbf{x}))$

Subject to $\mathbf{e}(\mathbf{x}) = (e_1(\mathbf{x}), e_2(\mathbf{x}), ... , e_m(\mathbf{x})) \geq \mathbf{0}$

where $\mathbf{x} = (x_1, x_2, ..., x_n) \in \mathbf{X}$ is the decision vector, and $\mathbf{y} = (y_1, y_2, ... , y_k) \in \mathbf{Y}$ is the objective vector. $\mathbf{X}$ denotes the decision space while the objective space is denoted by $\mathbf{Y}$. The set of restrictions $\mathbf{e}(\mathbf{x}) \geq \mathbf{0}$ determines the set of feasible solutions $\mathbf{X}_f \subseteq \mathbf{X}$ and its corresponding set of objective vectors $\mathbf{Y}_f \subseteq \mathbf{Y}$ The problem consists in finding $\mathbf{x}$ that optimizes $\mathbf{f}(\mathbf{x})$. In general, there is no unique "best" solution but a set of solutions. Thus, a new concept of optimality should be established for *MOPs*. In the minimization context of the present work, given two decision vectors $\mathbf{u}, \mathbf{v} \in \mathbf{X}_f$,

$\mathbf{f}(\mathbf{u}) = \mathbf{f}(\mathbf{v})$ iff $\forall i \in \{1,2,...,k\}$: $f_i(\mathbf{u}) = f_i(\mathbf{v})$;

$\mathbf{f}(\mathbf{u}) \leq \mathbf{f}(\mathbf{v})$ iff $\forall i \in \{1,2,...,k\}$: $f_i(\mathbf{u}) \leq f_i(\mathbf{v})$;

$\mathbf{f}(\mathbf{u}) < \mathbf{f}(\mathbf{v})$ iff $\mathbf{f}(\mathbf{u}) \leq \mathbf{f}(\mathbf{v}) \wedge \mathbf{f}(\mathbf{u}) \neq \mathbf{f}(\mathbf{v})$.

Then, they comply with one of three conditions: (i) $\mathbf{u}$ dominates $\mathbf{v}$ iff $\mathbf{f}(\mathbf{u}) < \mathbf{f}(\mathbf{v})$; (ii) $\mathbf{u}$ and $\mathbf{v}$ are non-comparable iff $\mathbf{f}(\mathbf{u}) \not< \mathbf{f}(\mathbf{v}) \wedge \mathbf{f}(\mathbf{v}) \not< \mathbf{f}(\mathbf{u})$; or (iii) $\mathbf{v}$ dominates $\mathbf{u}$ iff $\mathbf{f}(\mathbf{v}) < \mathbf{f}(\mathbf{u})$. $\mathbf{u} \succcurlyeq \mathbf{v}$ denotes that $\mathbf{u}$ dominates or is equal to $\mathbf{v}$. A decision vector $\mathbf{x} \in \mathbf{X}_f$ is non-dominated with respect to a set $\mathbf{V} \subseteq \mathbf{X}_f$ iff: $\mathbf{x}$ dominates $\mathbf{v}$ or they are non-comparables, $\forall \mathbf{v} \in \mathbf{V}$. When $\mathbf{x}$ is non-dominated with respect to the whole set $\mathbf{X}_f$, it is called an optimal Pareto solution. The Pareto optimal set $\mathbf{X}_{true}$ may be defined as $\mathbf{X}_{true} = \{\mathbf{x} \in \mathbf{X}_f \mid \mathbf{x}$ is non-dominated with respect to $\mathbf{X}_f\}$. The corresponding set of objective vectors $\mathbf{Y}_{true} = \mathbf{f}(\mathbf{X}_{true})$ constitutes the Optimal Pareto Front.

3   Problem Formulation

A network is modeled as a direct graph $G = (V, E)$, where V is the set of nodes and E is the set of links. Let $(i, j) \in E$ be the link from node i to node j. For each link (i,j), let $z(i,j)$, $c(i,j)$, $d(i,j)$ and $t(i,j)$ be its capacity, cost per bps, delay and current traffic, respectively. Let $s \in V$ denote a source, $N \subseteq V - \{s\}$ denote the set of destinations, and $\phi \in R^+$ the traffic demand (in bps) of a current multicast request. Let $T(s,N)$ represent a multicast tree with s as source node and N as destination set. At the same time, let $p_T(s,n)$ denote a path that connects the source node s with a destination node $n \in N$. Clearly, $p_T(s,n)$ is a subset of $T(s,N)$. The multicast routing problem may be stated as a *MOP* that tries to find a multicast tree that minimizes:

1- Maximum Delay:
$$D_M = \underset{n \in N}{Max} \left\{ \sum_{(i,j) \in p_T(s,n)} d(i, j) \right\} . \tag{1}$$

2- Cost of the tree:
$$C = \sum_{(i,j) \in T} c(i, j) . \tag{2}$$

3- Maximum link utilization:
$$\alpha_T = \underset{(i,j) \in T}{Max} \left\{ \frac{\phi + t(i, j)}{z(i, j)} \right\} . \tag{3}$$

4- Average delay:
$$D_A = \frac{1}{|N|} \sum_{n \in N} \left[\sum_{(i,j) \in p_T(s,n)} d(i, j) \right] . \tag{4}$$

subject to
$$\phi + t(i, j) \le z(i, j) , \forall (i, j) \in T . \tag{5}$$

Example 1. Figure 1(a) shows the NSF network, with $d(i,j)$ in ms, $c(i,j)$, and $t(i,j)$ in Mbps. The capacity of the links is 1.5 Mbps. Suppose a traffic request arriving with ϕ=0.2 Mbps, s=5, and N={0, 4, 8, 9, 13}. (b) shows the tree constructed with *KPP* [3], subject to a maximum delay of 40 ms, (c) shows a tree would not be found by *KPP* or other algorithms based on restrictions if a bound delay lower than 40 ms were a priori established, even though it is a good option. If α_T is the most important metric, solution (d) would be the best alternative.

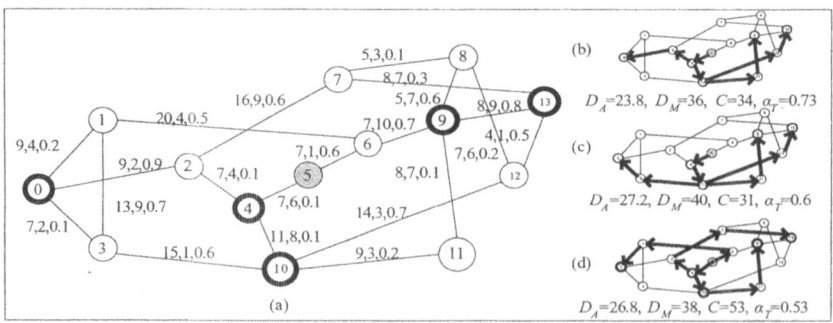

Fig. 1. The NSF Net. $d(i,j)$, $c(i,j)$ and $t(i,j)$ are shown over each (i,j) link. Different alternative trees for the multicast request with s=5, N={0, 4, 9, 10, 13} and ϕ=0.2 Mbps

4   Proposed Algorithm

The proposed algorithm holds an evolutionary population P and an external Pareto solution set P_{nd}. Starting with a random population P, the individuals evolve to optimal solutions to be included in P_{nd}. The algorithm is shown in Figure 2(a).

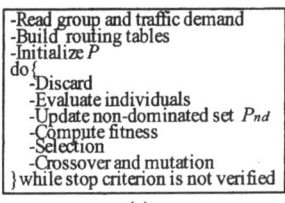   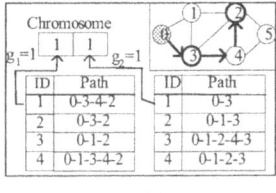

(a)   (b)

Fig. 2. (a) The proposed algorithm. (b) Relation between the chromosome, genes and routing tables. The chromosome represents the tree shown in the same Figure

Build routing tables: Let $N = \{n_1, n_2, ..., n_{|N|}\}$. For each $n_i \in N$, a routing table is built. It consists of the R shortest, R cheapest and R least used paths, where the use of a path is defined as the maximum link utilization along the path. R is a parameter of the algorithm. Yen's algorithm [10] was used for this task. A chromosome is represented by a string of length $|N|$ in which the element (gene) g_i represents a path between s and n_i. See Figure 2(b) to see the chromosome that represents the tree in the Figure.

Discard individuals: In P, there may be duplicated chromosomes. Thus, duplicated chromosomes are replaced by new randomly generated individuals as proposed in [9].

Evaluate: The individuals of P are evaluated using the objective functions. Then, non-dominated individuals of P are compared with the individuals in P_{nd} to *update the non-dominated set*, removing from P_{nd} dominated individuals.

Compute fitness: Fitness is computed for each individual, using *SPEA* procedure [7].

Selection: A roulette selection operator [8] is applied over the set $P_{nd} \cup P$ to generate the next evolutionary population P.

Crossover and mutation: In this work, two-point crossover operator is applied over each selected pair of individuals. Then, some genes in each chromosome of the new population are randomly changed (mutated) with probability P_{mut} [8].

5   Results

Simulation experiments were performed for Example 1. An exhaustive search method was used to find the optimal Pareto set of 16 solutions. The running time of the exhaustive search method was approximately 3 hours.

One hundred runs were done using *MMA* with $|P| = 50$, $P_{mut} = 0.3$, $R = 25$ for 500 generations. The minimum, maximum and average Pareto optimal solutions found by the runs using *MMA* were 16, 10 and 12.72 respectively. The mean running time was 270 ms and its maximum was 300 ms. Clearly *MMA* has a good performance finding at least 62.5% of the Pareto Front.

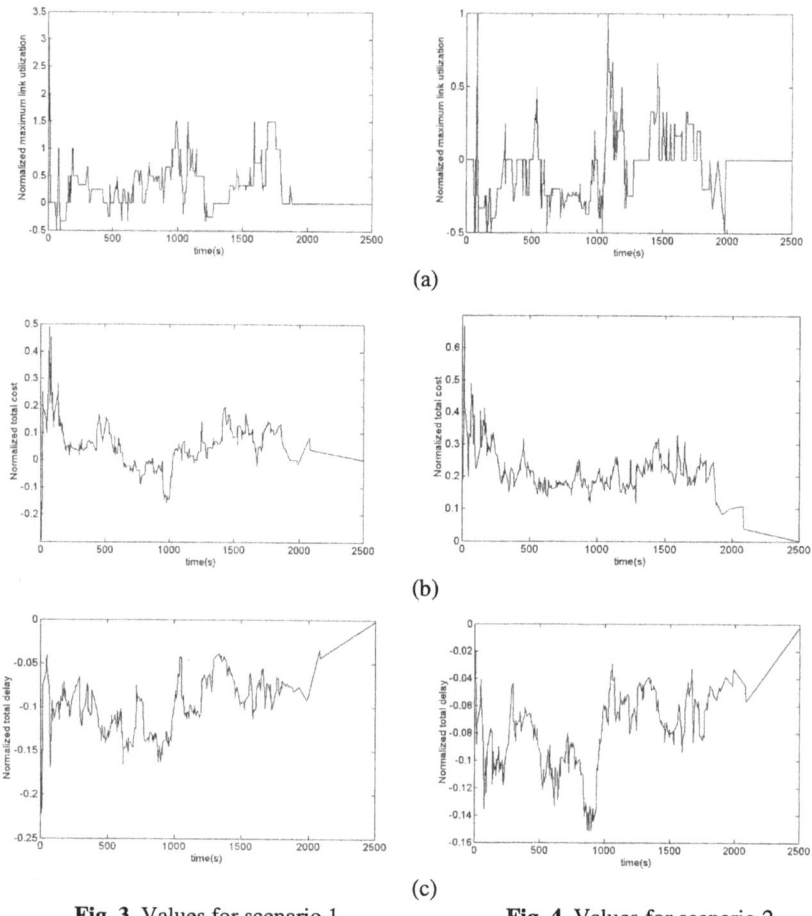

Fig. 3. Values for scenario 1   **Fig. 4.** Values for scenario 2

MMA was also compared against the delay shortest path (*SP*). Two hundred random requests of traffic demands of 0.1 Mbps were generated. The multicast group was randomly selected with a size between 4 and 7. The duration of each traffic demand was exponentially distributed (with an average of 120 s) and the inter-arrival time randomly distributed between 0 and 30 minutes. The maximum end-to-end delay for a group was set to 1.25 times the maximum end-to-end delay of the tree constructed with *SP*. *MMA* was set to $|P|=100$, $P_{mut}=0.3$ and $R = 30$. The mean time consumed to construct a multicast tree was 270 ms. Given that *MMA* may provide several solutions, two different scenarios were simulated: firstly, the trees with minimum α_r subject to end-to-end delay restriction; secondly, trees with minimum C satisfying the restriction. To compare performance, normalized values of maximum link utilization of the network (α), total cost, which is calculated as the sum of the tree costs of the multicast groups already in the net, and the total delay, which is given by the sum of the total delay of the multicast groups already in the net, were calculated. For example, normalized total cost was given by $C_N = (C_{SP} - C_{MMA})/C_{MMA}$. Figure 3(a) shows that *MMA* leads to better α than *SP*. Note that the maximum link utilization using *SP* is sometimes 150 times greater than using *MMA*. Besides, from Figure 3(b), it can be seen that at almost all time, *SP* total cost is more expensive than *MMA*. Since *SP* produces optimal delay trees, total delay using *SP* was lower than *MMA*. Figures 4(a) to 4(c) show normalized values for the second scenario where the cost difference between *SP* and *MMA* has increased since the select criterion of *MMA* is the cost. Note that at almost all time, *SP* total cost is at least 10% more expensive than *MMA*.

6   Conclusion

This paper presents a new multiobjective multicast routing algorithm (*MMA*) to solve the multicast routing problem. This new algorithm minimizes simultaneously four objective functions: 1- maximum end-to-end delay, 2- cost of a tree, 3- maximum link utilization and 4- average delay. *MMA* has a purely multiobjective approach, based on *SPEA*. This approach calculates an optimal Pareto set of solutions in only one run, without *a priori* restrictions, an important feature of MMA.

Experimental results show that *MMA* was able to found Pareto optimal solutions. They also show that α and the total cost of *MMA* were lower than those of the shortest path algorithm. As future work, we plan to consider a traffic engineering scheme using different distribution trees over larger problems.

References

1. A. Tanenbaum, Computer Networks, Prentice Hall, 2003.
2. Y. Lee, Y. Seok, and Y. Choi, "Explicit Multicast Routing Algorithm for Constrained Traffic Engineering," Proc. of 7th Int. Symp. on Computer and Comm. (ISCC'02), 2002.
3. V. Kompella, and A. Gupta, "Multicast routing in multimedia communication," IEEE/ACM Transactions on Networking, Vol. 1 No. 3, 1993, pp. 286-291.
4. C. P. Ravikumar, and R. Bajpai, "Source-based delay bounded multicasting in multimedia networks," Computer Communications, Vol. 21, 1998, pp. 126-132.
5. W. Zhengying, S. Bingxin, and Z. Erdun, "Bandwidth-delay-constraint least-cost multicast routing based on heuristic genetic algorithm," Computer Communications, Vol. 24, 2001, pp. 685-692.
6. P. T. de Araujo, and G. M. Barbosa, "Multicast Routing with Quality of Service and Traffic Engineering Requirements in the Internet, Based On Genetic Algorithm, "Proceedings of the VII Brazilian Symposium on Neural Networks (SBRN'02), 2002.
7. E. Zitzler, and L. Thiele, "Multiobjective Evolutionary Algorithms: A comparative Case Study and the Strength Pareto Approach," IEEE Trans. Evolutionary Computation, Vol. 3, No. 4, 1999, pp. 257-271.
8. D. Goldberg, Genetic Algorithm is Search, Optimization & Machine Learning, Addison Wesley, 1989.
9. R.H. Hwang, W.Y. Do, and S.C. Yang, "Multicast Routing Based on Genetic Algorithms," Journal of Information Science and Engineering, Vol. 16, 2000, pp. 885-901.
10. J. Yen, "Finding the k shortest loopless path in a network," Management Science, 17:712-716, 1971.

A Comparison of Filters for Ultrasound Images

Priscila B. Calíope, Fátima N.S. Medeiros,
Régis C.P. Marques, and Rodrigo C.S. Costa

Federal University of Ceara
Campus do PICI, Bloco 705, C. P. 6007
60455-760 – Fortaleza, CE, Brasil
{priscila,fsombra,rodcosta,regismarques}@deti.ufc.br
http://www.gpi.deti.ufc.br/index.html

Abstract. The coherent nature of ultrasonic waves, that provides information for ultrasound image formation, results in the appearance of speckle noise. The development of speckle noise filtering methods for ultrasound B-scan images is very important for accurate detection of targets and boundaries. Many filters have been proposed in the literature for speckle reduction, most of them in remote sensing applications. This paper presents a comparison of filters for ultrasound images. The tests were performed in images artificially contaminated with speckle noise supposed to be Rayleigh distributed and a phantom ultrasound image.

1 Introduction

For some than two decades, ultrasonography has been considered as one of the most powerful techniques for imaging organs and soft tissue structures in the human body [1]. It is often preferred over other medical imaging modalities because it is noninvasive, portable, and versatile, it does not use ionizing radiations, and it is relatively low-cost. However, the main disadvantage of medical ultrasonography is the poor quality of images, which are affected by speckle noise. Speckle is a troublesome noise, that disturbes image interpretation and target classification. Speckle suppression in ultrasonic imaging is usually done by techniques that are applied directly to the original image domain like Lee and Frost filters, adaptive filters [2,3,4], or image processing techniques in the wavelet domain [1].

In [4] an adaptive algorithm called aggressive region growing filtering (ARGF) was proposed for speckle reduction. It selects a filtering region in the image using an appropriately estimated homogeneity threshold for region growing. The adaptive homogeneity threshold describes the statistical specifications of the homogeneous regions and it is adaptively determined. Whether or not a new region is homogeneous it is determined by comparing its local homogeneity to this adaptive homogeneity threshold. Thus, edges pixels, with higher homogeneity values, are preserved using a nonlinear median filter. Homogeneous regions, with smaller homogeneity values, are smoothed applying an arithmetic mean filter. The algorithm was compared with the

J.N. de Souza et al. (Eds.): ICT 2004, LNCS 3124, pp. 1035–1040, 2004.

adaptive weighted median filter (AWMF) [2] and homogeneous region growing mean filter (HRGMF) [3] methods.

Jin et al. reviewed and applied in [5] some speckle filters (mean, median, Lee, Frost and maximum a posteriori (MAP) filters) in ultrasound B-scan images, along with a non-linear technique based on computing the median on the binary slices of the data. The performance of these filters was assessed both quantitatively and qualitatively [5] and the data was better processed by the Frost filter.

A novel speckle suppression method for medical ultrasound images was developed by Achim et al. [1]. Firstly, the logarithmic transform of the original image was applied to the multiscale wavelet domain. The subbands of the image presented significantly non-Gaussian statistics and thus were described for an alpha-stable distribution. To exploit these statistics was designed a Bayesian estimator and developed a blind speckle-suppression processor that performed a nonlinear operation on the data.

In this paper we present a comparison between speckle filters applied to ultrasound images. We evaluate the modified MAP filter proposed by Medeiros et al. [6] and compare its performance with the mean and median filters [7], the classical Frost [8] and MAP filters [9] and another one based on Daubechies wavelet [10].

2 Medical Ultrasound Speckle Patterns

The nature of the speckle pattern can be categorized into one of three classes according to the number of scatterers per resolution cell or scatterer number density (SND). When many fine randomly distributed scattering sites exist within the resolution cell, the amplitude of the backscattered signal can be modeled as a Rayleigh distributed random variable with a constant SNR of 1.92 [11]. When a spatially variant coherent structure is present within the random scatterer region the signal backscatter can be modeled by the K-distribution [11]. Due to the correlation between scatterers, the effective number of scatterers is finite (SND<10). This type is associated with SNR below 1.92. It can also be modeled by the Nakagami distribution. When a spatially invariant coherent structure is present, the probability density function (PDF) of the backscattered signals becomes close to the Rician distribution [11]. This class is associated with SNR above 1.92.

3 The Filtering Methods Applied to the Test Images

In this article we use the modified MAP algorithm proposed by Medeiros et al. [6]. It applies the MAP estimator in the current adaptive window that is controlled by a measure of homogeneity in the area around the noisy pixel [12]. Among the elements in the current window w_{ij}, only the samples in the boundary of this window are used for the decision of the next window size to save the computational load [12].

In this paper MAP Gauss designates the MAP filter that assigns the a *priori* Gaussian distribution to the original image; MAP Pearlman Gaussian designates the adaptive MAP Gauss filter [6].

4 Filtering Assessment Methodology

The great challenge of speckle filtering algorithms is to filter speckle and to preserve edges. For quantitative assessment, it is measured the mean-square error (*MSE*) that is defined as [1]:

$$MSE = \frac{1}{K}\sum_{i=1}^{K}(\hat{S}_i - S_i)^2 \; . \tag{1}$$

where S and $\hat{S}$ are the original and denoised images, respectively. K is the image size. A common way to evaluate the noise suppression in case of multiplicative noise in coherent imaging is to calculate the signal to *MSE* (*S/MSE*) ratio that is defined as [1,13]:

$$S/MSE = 10\log_{10}\left(\sum_{i=1}^{K}(S_i)^2 / \sum_{i=1}^{K}(\hat{S}_i - S_i)^2\right) . \tag{2}$$

This measure corresponds to the classical SNR in case of additive noise. Thus, it was applied the logarithmic in the contaminated and original images to realize the *S/MSE*. In addition to the above quantitative performance measures, we also consider one qualitative measure for edge preservation, defined in [1,14] as β:

$$\beta = \frac{\Gamma\left(\Delta S - \overline{\Delta S}, \widehat{\Delta S} - \overline{\widehat{\Delta S}}\right)}{\sqrt{\Gamma\left(\Delta S - \overline{\Delta S}, \Delta S - \overline{\Delta S}\right).\Gamma\left(\widehat{\Delta S} - \overline{\widehat{\Delta S}}, \widehat{\Delta S} - \overline{\widehat{\Delta S}}\right)}} \; . \tag{3}$$

where ΔS and $\widehat{\Delta S}$ are the high-pass filtered versions of S and $\hat{S}$, respectively, obtained with a 3x3 pixel standard approximation of the Laplacian operator and the function $\Gamma(S_1, S_2)$ is described by:

$$\Gamma(S_1, S_2) = \sum_{i=1}^{K} S_{1i}.S_{2i} \; . \tag{4}$$

The correlation measure, β should be close to unity for an optimal edge preservation effect.

One measure used in this work to determine the amount of speckle within an image is the contrast to speckle noise ratio *CRS* [15], introduced as an attempt to quantify the ability of an observer to perceive anechoic areas against a background of speckle. This measure evaluates the contrast enhancement and is given by:

$$CRS = \frac{x_0 - x_i}{\sqrt{\sigma_0^2 + \sigma_i^2}} \; . \tag{5}$$

where x_i and σ_i are the average signal value and the variance inside the void, respectively, x_0 and σ_0 are those outside the region.

5 Experimental Results

The filter based on Daubechies wavelet using Db4 basis presented the best measures results to the artificially noisy images and Db1 basis to the phantom image. The mean filter was applied to the details coefficients and these filters were named as MDb4 and MDb1.

Fig. 1(a) displays the original Block image (256x256 pixels) and its artificially contaminated version is showed in Fig. 1(b). The obtained *MSE, S/MSE,* , and *CRS* values are shown in Table 1. From Table 1, it is observed that the MAP Pearlman Gauss filter presented the best speckle suppression results compared with the other ones. In terms of *MSE* and *S/MSE* values the mean, Frost and Map Gauss filters presented similar performance. Related to details preservation the MAP Pearlman Gaussian filter outperformed the others.

The MAP, Frost, wavelet and median filters were good at preserving details. Among the evaluated filters, the best ratio *CRS* values were attained by the MAP Pearlman Gauss, mean, Frost, MAP Gauss, median and MDb4, in decreasing order.

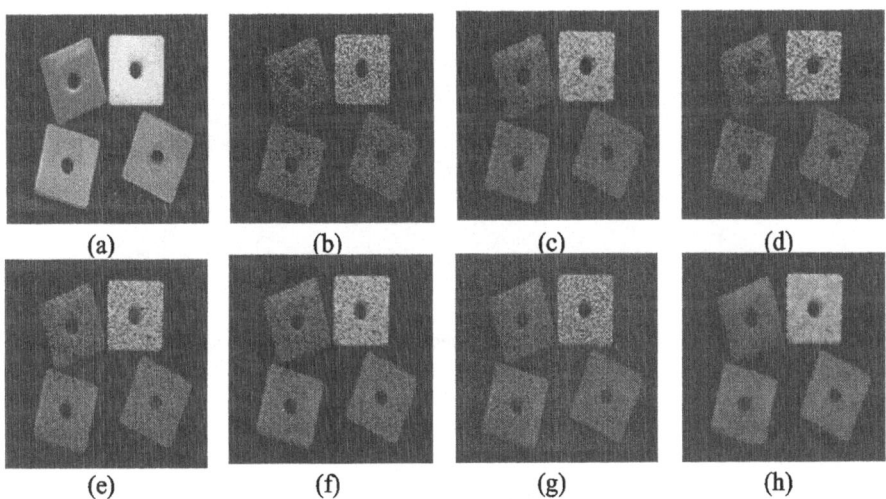

Fig. 1. (a) Original Block image, (b) its speckled version processed by the filters, (c) mean, (d) median, (e) Frost, (f) MAP Gauss, (g) MDb4 and (h) MAP Pearlman Gaussian

Fig. 2(b) presents the filtered versions of the image shown in Fig. 2(a). The best *CRS* values for this image were presented by the filters, in decreasing order: MAP Pearlman Gaussian, MAP Gauss, mean, Frost, median and MDb1. Their respective values obtained were: 0.7288, 0.6757, 0.6924, 1.0398, 1.1502 and 0.6253. According to the *CRS* ratio, the MAP Pearlman Gauss outperformed the others.

Table 1. Image enhancement measures to assess filtering methods applied on the artificially contaminated image. The *S/MSE* is given in dB

Filter	*MSE*	*S/MSE*	*B*	*CRS*
Mean	425.0271	24.6367	0.3382	4.9193
Median	852.6871	21.8664	0.3642	2.9658
MDb4	1029.9761	20.84	0.3241	2.7886
Frost	546.7341	24.0187	0.3959	4.8576
MAP Gauss	450.3533	24.4720	0.3866	4.5149
MAP Pearlman Gauss	237.5794	26.1586	0.5201	13.5959

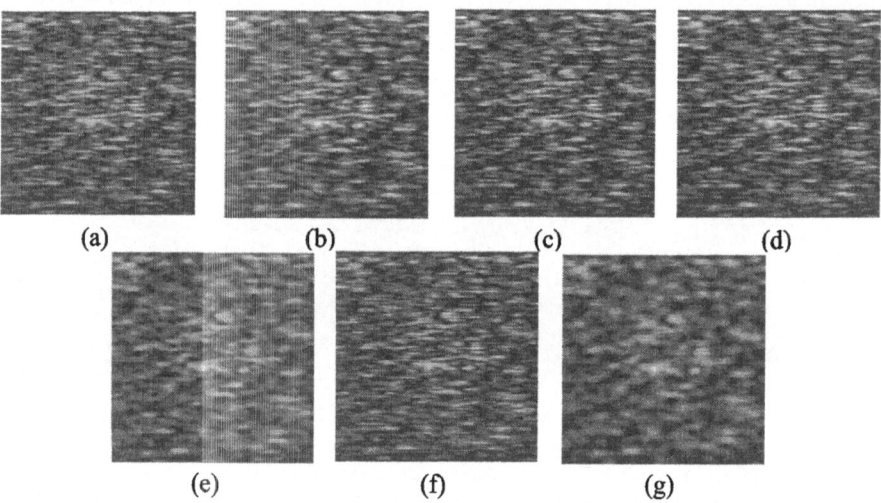

Fig. 2. (a) The phantom image and the processed versions by the filters, (b) mean, (c) median, (d) Frost, (e) MAP Gauss, (f) MDb1 and (g) MAP Pearlman Gaussian

6   Conclusions and Future Work

In this paper we presented a comparison and evaluation of filters for ultrasound images. The data at hand were better processed by the MAP Pearlman Gaussian filter, from both quantitatively and qualitatively criteria. The strategy of looking for a local statistically homogeneous area near pixel to compute the local statistics tends to improve the adaptive filtering. This suggests, as further developments, the application of this adaptive filtering in other distributions associated to the multiplicative model such as Rice and K distributions.

Acknowledgments. The authors are grateful to FUNCAP and CNPq for financial help. We are indebted to Constantine Kotropoulos for providing test data.

References

1. Achim, A., Bezerianos, A.: Novel bayesian multiscale method for speckle removal in medical ultrasound images. IEEE Transactions on Medical Imaging, Vol. 20. (8) (2001) 772–783
2. Loupas, T., McDicken, W.N., Allan, P.L.: Adaptive weighted median filter for speckle suppression in medical ultrasonic images. IEEE Transactions on Circuits and Systems, Vol. 36. (1989) 129–135
3. Koo, J. I., Park, S. B.: Speckle reduction with edge preservation in medical ultrasonic images using a homogeneous region growing mean filter (HRGMF). Ultrasonic Imaging, Vol. 13. (1991) 211–237
4. Chen, Y., Yin, R., Flynn, P., Broschat, S.: Aggressive region growing for speckle reduction in ultrasound images. Pattern Recognition Letters, Vol. 24. (2003) 677–691
5. Jin, J.Y., Silva, G. T., Frery, A.C.: SAR despeckling filters in ultrasound imaging. Latin American Applied Research, Vol. 34. (2004) 49-54
6. Medeiros, F.N.S., Mascarenhas, N.D.A., Marques, R.C.P., Laprano, C.M.: Evaluating an adaptive windowing scheme in speckle noise map filtering. Proceedings of the XV Brazilian Symposium on Computer Graphics and Image Processing. Fortaleza October (2002) 69 -75
7. Gonzalez, R.C., Woods, R. E.: Digital Image Processing. Addison-Wesley Massachusetts (1992)
8. Frost, V. S. et al.: A Model for Radar Image and its Application to Adaptive Digital Filtering of Multiplicative Noise, IEEE Transactions on Geoscience and Remote Sensing, Vol. 4 (2). (1982) 157-166
9. Medeiros, F.N.S.: Filtragem adaptativa de imagens de radar de abertura sintética utilizando a abordagem maximum a posteriori. Tese (Doutorado em Ciências) – Instituto de Física de São Carlos. São Paulo (1999)
10. Daubechies, I.: The wavelet transform, time-frequency localization and signal analysis. IEEE Transactions Information Theory, Vol. 36. (1990) 961-1005
11. Abd-Elmoniem, K.Z., Youssef, A. M., Kadah, Y. M.: Real-time speckle reduction and coherence enhancement in ultrasound imaging via nonlinear anisotropic diffusion. IEEE Transactions on Biomedical Engineering, Vol.49. (9) (2002) 997-1014
12. Park, J.M., Song, W.J., Pearlman, W.A.: Speckle filtering of SAR images based on adaptive windowing. IEE Proc. Vis. Image Processing, Vol. 146. August (1999) 191-197
13. Gagnon, L., Jouan, A.: Speckle filtering of SAR images – a comparative study between complex-wavelet-based and standard filters. SPIE Proc, Vol. 3169. (1997) 80–91
14. Sattar, F., Floreby, L., Salomonsson, G., Lovstrom, B.: Image enhancement based on nonlinear multiscale method. IEEE Transactions on Image Processing, Vol. 6. (6) (1997) 889-895
15. Kang, S.C., Hong, S. H.: A speckle reduction filter using wavelet-based methods for medical imaging application. IEEE 14th International Conference on Digital Signal Processing, Vol. 2. Santorini July (2002) 1169 – 1172

Filtering Effects on SAR Images Segmentation

Régis C.P. Marques, Eduardo A. Carvalho,
Rodrigo C.S. Costa, and Fátima N.S. Medeiros

UFC-DETI, Campus do PICI, Bloco 705
Zip 60455-760 - Fortaleza, CE, Brazil
{regismarques,eduardo,rodcosta,fsombra}@deti.ufc.br
http://www.gpi.deti.ufc.br

Abstract. This paper evaluates filtering effects on SAR image segmentation testing four types of speckle reduction algorithms. A primary goal of these filters is to provide a large amount of speckle noise reduction in homogeneous areas and to preserve edges and details. To assess the effects produced by the filters in the posterior segmentation task, some quality measures are calculated from the processed images and used to indicate the filtering ability for features preservation.

1 Introduction

Speckle noise is a common phenomenon in all coherent imaging systems like laser, acoustic and Synthetic Aperture Radar (SAR). The source of this noise is attributed to random interferences between the coherent returns. Speckle degrades fine details in the image (targets and edges) and makes the segmentation of such corrupted images a complex task. Therefore, filtering is a common pre-processing stage in SAR images applications that carries improvements on class discrimination and image interpretation. Nevertheless, filtering effects like blurring and details degradation must be considered.

Capstick and Harris [1] evaluated the effects of six speckle reduction filters in agricultural applications and data classification. The filters were evaluated in the context of identifying the growth of potatoes fields in the United Kingdom using ERS -2 Earth observation data.

In our approach speckled images are submitted to speckle reduction by applying Lee [2], Kuan [3], Frost [4] and adaptive window MAP [5] filters. After the filtering step the images are segmented by different algorithms: a) thresholding [6], (b) k-means clustering [7], c) global k-means clustering [8] and d) region growing [6]. To evaluate the filtering effects over SAR images segmentation, some quality measures [9, 10] are applied to the processed images. Furthermore, the proposed approach evaluates the filter's ability of reducing speckle noise.

The segmentation algorithms used in this paper are modified versions of the classical ones. In order to improve these algorithms, segmentation parameters such as

J.N. de Souza et al. (Eds.): ICT 2004, LNCS 3124, pp. 1041–1046, 2004.

threshold values, clusters number and seed pixels are estimated automatically by a multiscale wavelet analysis of the image histogram.

In Section 2 the speckle filters used in this paper are described. Section 3 and Section 4 present a brief description of the segmentation algorithms and the assessment measures, respectively. Section 5 exhibits and discusses the experimental results. Finally, Section 6 provides the concluding remarks.

2   Speckle Filters

Lee [2] and Kuan [3] algorithms are locally adaptive filters based on the multiplicative speckle model. These filters are similar in relation to the estimate operator, which is a minimum mean square error estimative for the central pixel within a fixed size window. The Frost filter [4] is a Wiener filter based also on the local statistics of speckle noise multiplicative model. All these filters have the desirable characteristics of edge preservation and speckle reduction in homogeneous areas.

Medeiros et al. [5] proposed an adaptive window MAP filtering inspired in [11]. This method combines a MAP estimator and the adaptive window scheme proposed by Park et al. [11]. This is an interactive process, in which an analysis of the local statistics around the noisy pixel indicates the maximum homogeneous neighborhood and the adequate window size for noise filtering.

3   Segmentation Methods

Segmentation addresses the extraction of important objects of an image by means of isolating and separating them through the analysis of features like gray value, color, texture or edge evidence [11]. Thresholding is the simplest segmentation technique, most times successful in bimodal histogram images. However, when an image presents a multimodal histogram this method tends to be computationally expensive and inaccurate.

The k-means algorithm [7] is one the most widely used clustering methods. In its simplest form, image pixels are assigned to different groups. The k parameter indicates the number of clusters supposed to be present in the image. The centroids are updated iteratively and the clustering process stops when there are no more back and forth movements of pixels from one group to another. This algorithm is a local search procedure and it is also well known that it suffers from a serious drawback: its performance strongly depends on the initial starting conditions [12]. Likas et al. [8] proposed the global k-means algorithm to treat this problem, in which the k clusters problem is solved in an incremental way.

Region growing is one of the conceptually simplest approaches to image segmentation, where neighboring pixels of similar amplitude are grouped together to form a segmented region [6]. The idea is that neighboring pixels of a seed pixel are clustered into the same region, according to some similarity rule.

One major problem in cluster analysis is the determination of the number of clusters [13]. To improve the segmentation algorithms, we estimate the clusters centroids using an undecimated wavelet decomposition of the filtered image histogram. This proposed approach identifies peaks in several scales leading to dominant clusters of the image.

4   Assessment of Segmentation Algorithms

In this paper, some measures [9, 10, 14] are used to assess speckle filtering effects over segmentation results. The results produced by the segmentation algorithms are dependent on the speckle filtering quality.

The gray level uniformity (*GU*) is a goodness measure that defines the amount of homogeneity at certain areas of an image (i.e., the gray level uniformity over a region evaluated from the pixels belonging to that region). It is believed that an adequate segmentation produces regions having higher intra-region uniformity, according to [9]. *GU* is computed on the basis of the variance of the gray levels of the filtered images. A good segmentation produces homogeneous regions, therefore reducing the value of *GU*.

The normalized pixel distance error (*ND*) is a discrepancy measure that takes into account the number and position of mis-segmented pixels [9]. A good segmentation scheme should produce a restricted amount of misclassified pixels. An M by N reference image is segmented and those misclassified pixels are counted. For each misclassified pixel $p_{k,mc}$, one must calculate the Euclidian distance to the nearest pixel that is actually of the misclassified class $p_{k,cc}$. The square root of the summation of all these squared distances, divided by the number of pixels in the image, yields the normalized distance *ND*.

The relative ultimate measurement accuracy (*RUMA*), as proposed by Zhang [9], is an accuracy measure to compare different segmentation algorithms. A segmented image has the highest quality if the object features extracted from it precisely match the features in the reference segmentation, thus diminishing the value of *RUMA*. In this paper the area A_i of the objects is employed in the calculus of *RUMA*.

In Delves et al. [14] is described a methodology to compare SAR segmentation algorithms, based on the fitting of individual regions of segmented images. The *fitness* discrepancy embodies the comparison of regions $\{m_1, m_2, ..., m_q\}$ of a segmented image with those present in the reference image $\{r_1, r_2, ..., r_k\}$, i.e, comparing pairs of elements (r_i, m_j). The *fitness* informs the quality of the segmentation method based on the position on the X-Y plan, average pixel intensity computed from the original image, size of the regions and their shape [10]. Once quantified the fitness discrepancy, it is also possible to compare segmentation algorithms.

5   Experimental Results and Evaluation

The experiment consisted on segmenting filtered images by the prior described methods. In this section we present and discuss the obtained results.

5.1   Considerations on the Filters

Fig. 1(a) illustrates an example of a test image. Fig. 1(b) is the speckled version (speckle is simulated as 1 look, with unitary mean). Fig.1(c-d) are some filtering results produced by the Kuan and AWMAP filters, respectively.

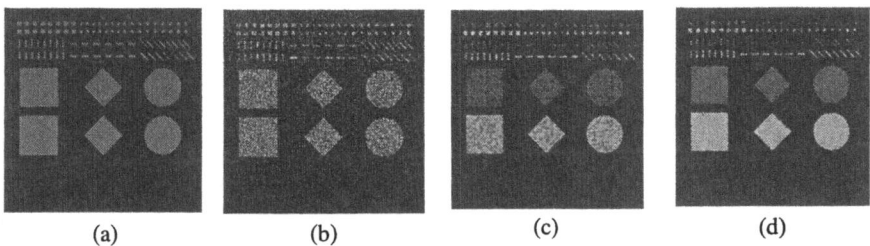

(a)   (b)   (c)   (d)

Fig. 1. (a) Test image, (b) noisy image and the results produced by (c) Kuan and (d) AWMAP

A visual comparison of images in Fig. 1(c) and Fig. 1(d) leads to saying that the AWMAP filtered image was the best result. On the other hand, Lee, Kuan and Frost filters applied to the test image presented similar performance. The measure used to assess these filters was the standard-deviation to mean ratio (β) in homogenous areas [5]. The obtained β measures were 0.2076, 0.2027, 0.2934 and 0.0382, for Lee, Kuan, Frost and AWMAP filters, respectively. By analyzing these β values, we concluded that the AWMAP filter was more effective on suppressing speckle than the others.

5.2   Considerations on the Segmentation Quality Measures

Fig. 2 presents the segmentation quality measures for: (a) *GU*, (b) *fitness*, (c) *RUMA* and (d) *ND*. *GU* measures the intra-class homogeneity, not considering spatial information or pixel connection. For this reason, the best results were presented by the *k*-means algorithm (S2), due to the class concentration around the centroids. The same occurred for the global *k*-means (S3), excluding the segmentation result over the Frost filtered image, because global *k*-means is more sensitive to remnant noise.

Despite the segmentation method, the *fitness* measure performed similar when applied to AWMAP filtered images.

Considering *RUMA*, the best values were attributed to the region growing method, when compared to the others. This measure compares segmented areas with a reference counterpart. Segmentation methods that produce a considerable amount of fragments inside the object tend to produce higher values for this measure (near unity). This was the case of thresholding, k-means and global k-means algorithms.

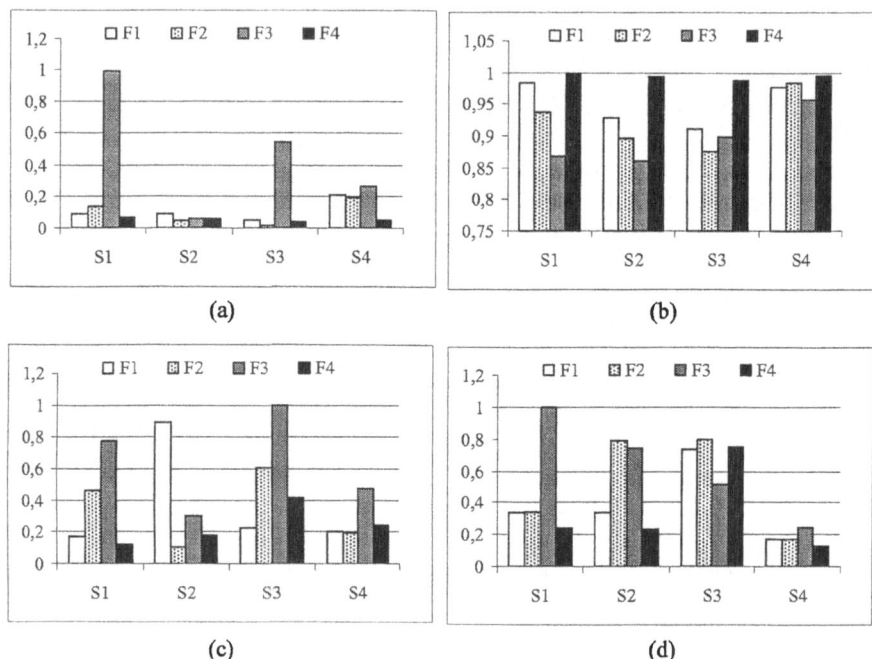

Fig. 2. Measures for segmentation quality assessment (a) goodness gray level uniformity (*GU*), (b) *fitness*, (c) *RUMA* and (d) normalized distance (*ND*). F1, F2, F3, F4 represent Lee, Kuan, Frost and AWMAP filters, respectively. S1, S2, S3 and S4 represent thresholding, k-means, global k-means and region growing segmentation algorithms, in this order.

The normalized distance (*ND*) measures indirectly the amount of misclassified pixels based on a sum of distances. Thus, this measure is dependent on the segmentation technique and its classification capacity, as well as the filtering results. Region growing provided the smaller values for this measure.

6   Concluding Remarks

From the results, we have concluded that the performance of the thresholding method depended on the ability of the speckle filtering scheme. The k-means and global k-means algorithms were less dependent on it, however the latter was more sensitive to the remnant noise. This fact can be perceived through the *GU* measure calculated on the segmented image filtered by the Frost scheme. Region growing produced the smallest values for the *ND* measure due to the smaller amount of misclassified pixels. This method worked well, independently of the filtering process adopted. For *RUMA* and *fitness* measures, this segmentation method presented the best results. Based on the measures, there was an indication that the AWMAP filter presented the best results among the others.

References

1. Capstick, D., Harris, R.: The Effects of Speckle Reduction on Classification of ERS SAR Data, International Journal of Remote Sensing, Vol. 22 (18). (2001) 3627-3641
2. Lee, J. S.: Speckle Analysis and Smoothing of Synthetic Aperture Radar Images', Computer Graphics and Image Process, Vol. 17. (1981) 24-32
3. Kuan, D. T. et al.: Adaptive noise smoothing filter for images with signal dependent noise, IEEE Trans. on Pattern Analysis and Machine Intelligence, Vol. 7 (2). (1985) 165-177
4. Frost, V. S. et al.: A Model for Radar Image and its Application to Adaptive Digital Filtering of Multiplicative Noise, IEEE Transactions on Geoscience and Remote Sensing, Vol. 4 (2). (1982) 157-166
5. Medeiros, F. N. S. et al.: Evaluating an Adaptive Windowing Scheme in Speckle Noise MAP Filtering, SIBIGRAP2002, Fortaleza Brazil (2002) 281-285
6. Pratt, W. K.: Digital Image Processing. 1st edn. John Wiley & Sons, New York (1991)
7. Webb, A.: Statistical Pattern Recognition. 2nd edn. John Wiley & Sons, England (2003)
8. Likas, A., Vlassisb, N., Verbeekb, J.: The Global K-Means Clustering Algorithm, Pattern Recognition, Vol. 36(2). (2003) 451-461
9. Zhang, Y. J.: Evaluation and comparison of different segmentation algorithms, Pattern Recognition Letters, Vol. 18. (1997) 963-974
10. Lucca, E. V. D.: Avaliação e comparação de algoritmos de segmentação de imagens de radar de abertura sintética. M. Sc. Dissertation. INPE, São José dos Campos, Brazil (1998)
11. Park, J. M., Song, W. J., Pearlman, W. A.: Speckle Filtering of SAR Images Based on Adaptive Windowing, IEE Proc. Vis. Image Processing, Vol. 146 (4). (1999) 191-197
12. Peña, J. M., Lozano, J. A., Larrañaga, P.: An Empirical Comparison of Four Initialization Methods for the k-Means Algorithm, Pattern Recognition Letter, Vol. 20 (10). (1999) 1027-1040
13. Guo, P., Chen, C. L. P., Lyu, M. R.: Cluster Number Selection for a Small Set of Samples Using the Bayesian Ying-Yang Model, IEEE Trans. Neural Networks, Vol. 13 (3). (2002) 757-763
14. Delves, L. M. et al.: Comparing the performance of SAR image segmentation algorithms, International Journal of Remote Sensing, Vol. 13 (11). (1992) 2121-2149

The Implementation of Scalable ATM Frame Delineation Circuits

Ciaran Toal and Sakir Sezer

School of Electrical and Electronic Engineering, Queens University Belfast,
Riddle Hall, Belfast BT9 5EE
185 Stranmillis Road, UK
`Ciaran.Toal@ee.qub.ac.uk`

Abstract. This paper presents the design and study of various HEC hunt architectures for ATM frame delineation and explores the trade-offs between the data-path (parallelism) and the hardware cost. A bit-serial, 4-bit, 8-bit, 32-bit and a 64-bit HEC hunt circuit has been implemented and analysed in terms of hardware cost, speed, and data throughput rate. The performances of the bit-parallel architectures have been improved by further pipelining the computation circuit. In the case of the 64-bit data-path architecture, the data throughput capability increased by 63% with an area penalty of only 20%. Post layout results are presented for Altera Stratix FPGA technology.

1 Introduction

In communication networks, the physical layer is responsible for the transmission of raw bit streams between a source and a destination. Framing is an essential process of the data-link layer to provide a mechanism for packet boundary recognition. Usually packets such as the Internet Protocol (IP) do not have a mechanism in place that will indicate the start and end of a packet within streamed data.

Frame delineation is a key function of the framing process of data-link layer protocols, such as Ethernet, PPP, GFP, HDLC, SDLC and ATM. A number of frame delineation mechanisms have been adopted by the standard. Many of these mechanisms are based on unique bit patterns indicating the start and end of a frame [3]. ATM and emerging link layer protocols, such as Generic Frame Procedure (GFP), use cyclic coding for Header Error Check (HEC) and frame delineation. Cyclic code based frame delineation requires a complex Cyclic Redundancy Check (CRC) computation circuit for error and frame boundary detection. The advantage of this technique is that the frame payload does not need to be modified before transmission, unlike HDLC or PPP which must escape their frame delineation pattern.

ATM Cell delineation is specified by the ITU-T in recommendation I.432 [6]. The HEC field is the fifth byte of the ATM cell header. The HEC field is calculated from the first 4 bytes of the header. When an ATM cell is received the HEC is again calculated from the first 4 header bytes and compared with the HEC field. In the absence of errors, both values are identical and the cell boundary is located.

J.N. de Souza et al. (Eds.): ICT 2004, LNCS 3124, pp. 1047–1056, 2004.
© Springer-Verlag Berlin Heidelberg 2004

As a further safety mechanism, the HEC computation starts with a preset HEC value of "01010101" at both the transmitter and the receiver. Without the preset value an ATM header composed of four zero bytes would yield a valid HEC. The HEC field is calculated as a remainder of the modulo-2 division of the first 4 header bytes with the CRC generator polynomial $G(x) = 1+x+x^2+x^8$.

ATM cell synchronisation is a sequential process in accordance to the state graph in Fig. 1. The receiver initially operates in the HUNT state and assumes no knowledge of the next incoming frame boundary. The streamed data is passed through the CRC computation circuit. Once 4 bytes have been processed by the CRC circuit, the receiver checks if the computed 8-bit CRC value is equal to the next incoming 8 bits i.e. the HEC field in the frame header. If there is a match the system enters the PRESYNC state, otherwise it continues checking incoming data bit-by-bit. The comparison of the computed CRC value with a possible HEC field must be carried out for each bit entering the computation circuit. Potentially every received bit could be the final HEC bit indicating the start of the cell payload.

If a correct HEC pattern has been detected, the synchronisation state machine moves to PRESYNC state and checks subsequent cells for matching HEC fields. If it receives δ consecutive correct HEC fields it enters the SYNC state. During the PRESYNCH phase, the synchronisation circuit will return back to HUNT state if a single incorrect HEC is found. Once in the SYNC state, the system can only return to HUNT if α consecutive incorrect HEC fields are received.

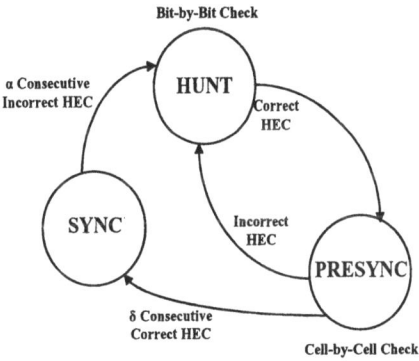

Fig. 1. ATM Cell Delineation State Diagram

The parameters α and δ are to be chosen to make the cell delineation process as robust and secure as possible while satisfying the performance specified by ITU-T. In the ITU-T recommendation I.432, for the SDH-based physical layer, values of α =7 and δ =6 are suggested. For the cell-based physical layer, values of α =7 and δ =8 are suggested.

In this paper we investigate a number of architectures and explore the trade-offs between the data-path (parallelism) and the hardware cost and throughput rate. The design, implementation and the circuit analysis of a bit-serial, 4-bit, 8-bit, 32-bit and a 64-bit HEC hunt are presented in terms of hardware cost, speed, and data throughput rate in sections 3, 4 and 5.

2   HEC Hunt Architectures and Related Work

The implementation of the bit-serial and parallel ATM HEC check architectures have been presented by G.E. Griffith et al [4], Suh Chung-Wook et al [1] and Ng. Leong Seong et al [5]. Chung-Wook's investigation is based on a HEC check implementation for a 16-bit data path targeting a throughput rate of 622 Mbps for ATM over SONET. Leong Seong's investigation explores an 8, 16 and 32-bit CRC computation architecture for the ATM HEC hunt. Both investigations emphasise mainly the CRC computation of the HEC hunt circuit and target a solution only for octet based cell transmission (SONET/SDH).

In our investigation, we consider these designs and also cover the scenario that the ATM cell may not be aligned to an octet boundary and present parallel implementations of a bit-by-bit HEC hunt for parallel circuits ranging from 4-bits to 64-bits. Furthermore, speed limitations due to the complex CRC computation have been significantly reduced by further pipelining the architecture.

2.1   Serial ATM Cell Delineation

Fig. 2 shows the architecture of the CRC computation circuit [2]. It is designed such that the HEC value is available as soon as the first 32 bits of the header has been streamed into the circuit. The architecture is simple and cost-effective.

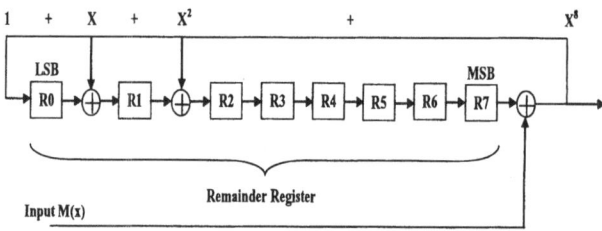

Fig. 2. Serial HEC Compute Circuit

2.2   Parallel ATM Cell Delineation

ATM HEC hunt for transmission rates beyond gigabits can only be achieved with parallel processing architectures. Leong Seong et al [5] demonstrated that, with FPGA technology in 1996, parallel implementation of a HEC hunt circuit for a 16-bit data-path could achieve a speed up by a factor of 6 over a bit-serial circuit.

The degree of computation performed by the parallel circuits is much greater due to the fact that a complete bit-by-bit check is required. A 32-bit parallel HEC hunt circuit requires 32 CRC calculations every clock cycle in order to check for every possible HEC location. Fig. 3 gives an example of a 32-bit parallel HEC hunt architecture. The circuit is composed of 32 32-bit-In/8-bit-Out CRC circuits and 32 compare units. Upon the first HEC match detected by one of the compare units, the data requires realignment so that bit location 0 of the output is mapped with the located cell boundary.

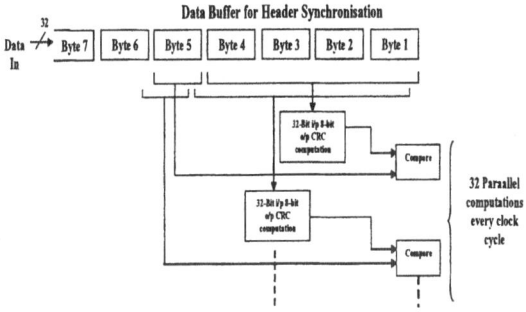

Fig. 3. 32-Bit HEC Hunt Process

Fig. 4. Parallel HEC HUNT Procedure

Fig. 4 illustrates the hunt procedure of parallel HEC hunt circuits.

First of all, an *n*-bit data word is latched into an *n*-bit register. Every possible *n*-bit combination is entered into individual CRC computational units. The output of each CRC circuit is XOR'd with the pattern "01010101", before being compared with the next 8 data bits. If there is a match, the data is aligned to the location 0 of the output data-path of the HEC hunt circuit.

3   Hardware Implementation

We designed 5 different area optimised parallel circuit architectures. In the second phase, the architectures were further pipelined to increase the operational clock frequency to meet throughput rates beyond 6 Gbps. The HEC hunt is based on a CRC8 computation. This fact has a significant impact to the scalability of the parallel processing architecture. 5 different scenarios are investigated:

Case 1: bit-serial architecture. A new data bit is available every clock cycle. The HEC is simply located 40 clock cycles after the first header bit is received.

Case 2: bit-parallel architecture data-path = 4-bit. The 4-bit architecture is the only circuit designed that has for each CRC unit, an input port size that is smaller than the CRC size. The architecture is very different from all the other implementations. A feed back loop is required within the CRC unit to ensure correct calculation. This presents a speed bottleneck that is not present in the other bit-parallel circuits. At least 10 cycles of data must be scanned in before the HEC is located.

Case 3: bit-parallel architecture data-path = 8-bit. This is the fastest operating bit parallel architecture. CRC calculation is extremely fast as it is made up of a straight forward 8-bit in/8-bit out architecture. At least 5 clock cycles of data must be scanned in before the HEC can be located.

Case 4: bit-parallel architecture data-path > 8-bit and < 40-bit. The 16-bit and 32-bit architectures are similar to the 8-bit architecture. For the 16-bit circuit, each CRC unit consists of the two 8-bit CRC circuits staggered as shown in Fig. 5. One CRC unit is for the 8 LSB and the other is for the 8 MSB. The 32-bit circuit requires four 8-bit CRC units structured in the same fashion. The 16-bit circuit will locate the HEC at least 3 clocks after the first header bit is scanned in. The 32-bit circuit will require a minimum of 2 clock cycles.

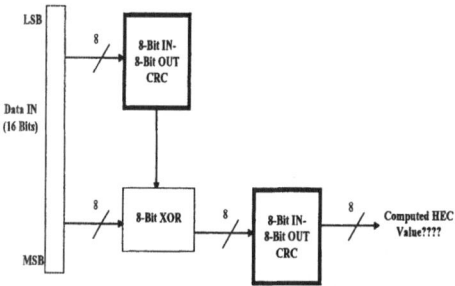

Fig. 5. Parallel HEC HUNT Procedure

Case 5: bit-parallel architecture data-path > 40-bit. In this case the input data is larger than the ATM header of 40 bits. With our 64-bit implementation, it is possible that the cell boundary might be located after only 1 clock cycle. The same CRC calculation unit that is used with the 32-bit circuit is used since no more than 32 bits are involved in calculating the HEC.

3.1 Serial Implementation

Fig. 6 shows a diagram of the serial ATM HEC hunt architecture. It consists of a 40-bit shift register, an 8- bit CRC calculator and an 8-bit comparator. Bit 8 of the shift register feeds into the CRC calculator. The CRC remainder value is XOR'd with the pattern "01010101" and compared with bits 0 to 7 of the shift register. If there is a match then bit 40 will be assumed as the cell boundary.

3.2 Parallel Implementations

For the parallel HEC hunt circuit, a generic parallel CRC core has been developed, based on the work presented in [9], which synthesises optimised CRC computation matrices. There are four different CRC computation units required for the different circuits. Therefore it was desirable to develop an IP core that based on the CRC port size and the polynomial value, synthesises a highly optimised and parallel matrix that performs the CRC computation. Each circuit developed consists of very different CRC computational unit architectures.

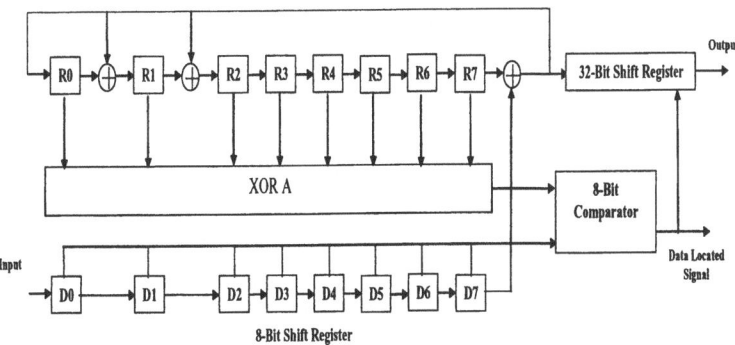

Fig. 6. Bit-Serial HEC HUNT Architecture

The parallel implementations are much more complex than the serial implementation. The serial system computes only one CRC value and performs a single 8-bit comparison for every new bit received. However, considering the 4-bit system, only 4 new bits are received every clock cycle but 32 CRC computations and four 8-bit comparisons must be performed.

In the case of the 4-bit circuit, the frame boundary will occur on 1 of the 4 bit locations. However, since the HEC is calculated from the first four header bytes i.e. 8 clock cycles of incoming data, the circuit must always simultaneously be computing 8 CRC remainder values for each bit location. Therefore 32 CRC computation blocks are required within the design.

Four compare units are also required. Only 4 of the 32 CRC circuits can output a possible correct HEC since they will have processed 32-bits of incoming data before being reset. Therefore only 4 8-bit comparison units are required to see if there is a match between any of the CRC remainder values and the corresponding next 8 bits buffered in the system. Fig. 7 shows a diagram of the fully pipelined 4-bit architecture.

Fig. 8 shows a diagram of the fully pipelined 32-bit architecture. With the 32-bit system, each CRC unit is reset every clock cycle. This is unlike the 4-bit system which has each CRC unit reset after 8 clock cycles, every 4 clock cycles with the 8-bit system and every 2 clock cycles with the 16-bit system.

The 4-bit, 8-bit, 16-bit and 32-bit circuits all require 32 CRC circuits, but different numbers of 8-bit comparison circuits. Each circuit requires one comparator for each incoming bit. The 64-bit circuit contains 64 CRC and comparator units.

4 Synthesis Results and Circuit Analysis

Each HEC Hunt circuit was synthesised and targeted to an Altera Stratix EPIS10C5 FPGA [8] using Synplify Pro and Quartus II tools. Speed and area performance is examined. The post-layout synthesis results are presented in Tables 1 and 2.

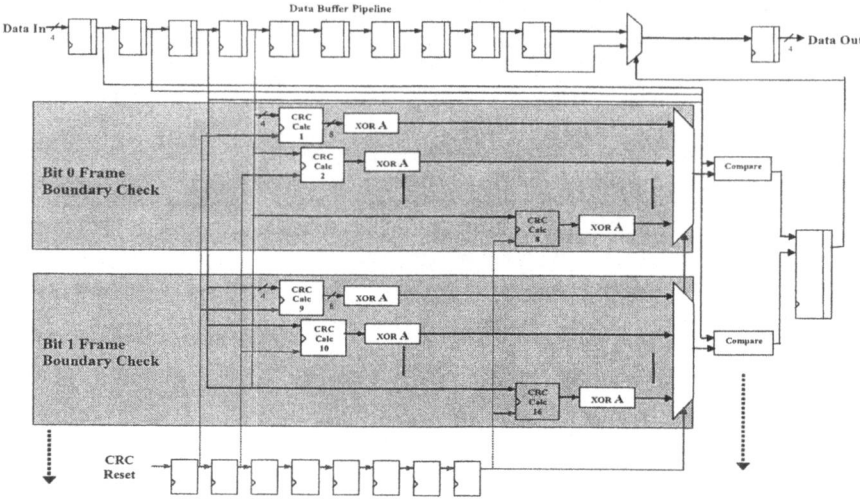

Fig. 7. 4-Bit ATM Bit-by-Bit HEC Hunt Circuit

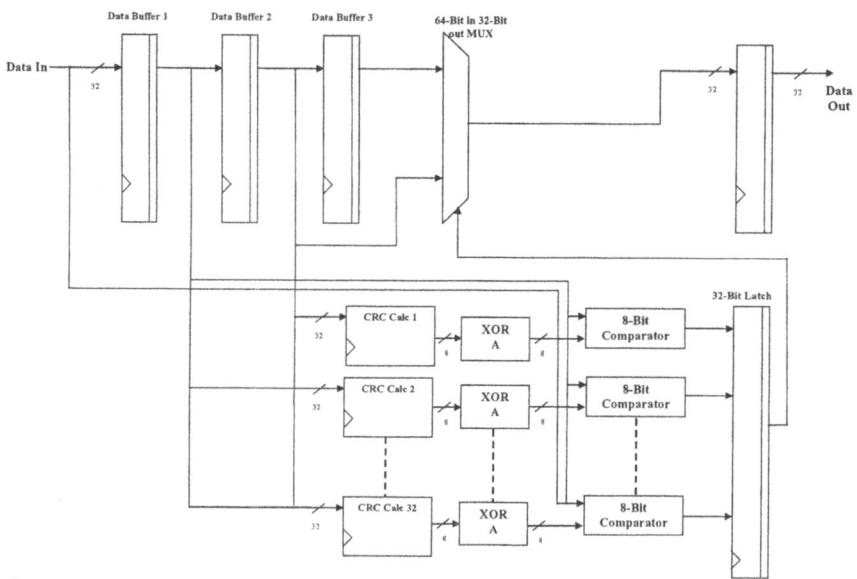

Fig. 8. 32-Bit ATM Bit-by-Bit HEC Hunt Circuit

From Table 1, we can see that before pipelining, the 4-bit circuit has a 23% smaller data throughput capability than the serial circuit. The extra complexity involved requires ten times as many logic cells which drastically slows the circuit down.

Table 1. Post-Layout Synthesis Results for Non-Pipelined Architectures for Altera Stratix Technology

	Serial	4-Bit	8-Bit	16-Bit	32-Bit	64-Bit
Logic Cells	49	491	865	1136	1256	2386
Registers	48	170	308	338	259	518
Clock Frequency (MHz)	422.12	81.23	120.9	93.78	81.53	65.1
Data Throughput (Mbps)	422.12	324.92	967.2	1500.48	2608.96	4166.4

Another unexpected point to note is that the 4-bit circuit operates at close to the same frequency of the 32-bit circuit and a great deal slower than the 8-bit and 16-bit circuits. This is due to the CRC core. Since the 4-bit circuit is the only implementation that requires CRC computation units each with a port size less than the CRC polynomial size, the CRC units are slower due to the extra feedback required within each matrix.

Table 2. Post-Layout Synthesis Results for Pipelined Architectures for Altera Stratix Technology

	4-Bit	8-Bit	16-Bit	32-Bit	64-Bit
Logic Cells	402	955	1159	1274	2856
Registers	202	348	370	386	706
Clock Frequency (MHz)	171.97	204.42	160.41	127.94	106.39
Data Throughput (Mbps)	687.88	1635.36	2566.56	4098.08	6808.96

Inserting an extra pipeline into each parallel circuit after the Compare operation allowed the operational frequency of each circuit to be increase substantially.

Figures 9, 10 and 11 show clearly how the circuit area in terms of logic cells and the speed in terms of operational frequency and data throughput rates vary with data-path parallelism for both pipelined and non-pipelined implementations.

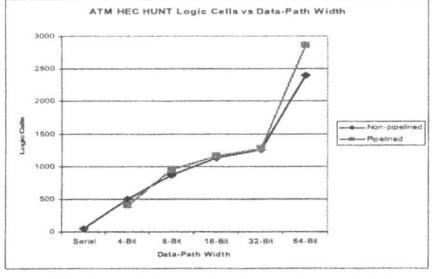

Fig. 9. Logic Cell Count vs Data-Path Width

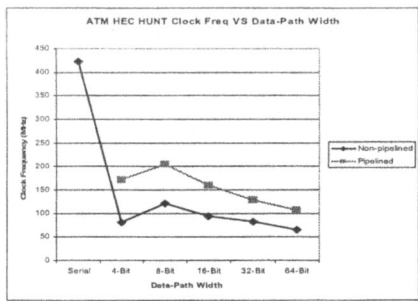

Fig. 10. Clock Frequency vs Data-Path Width

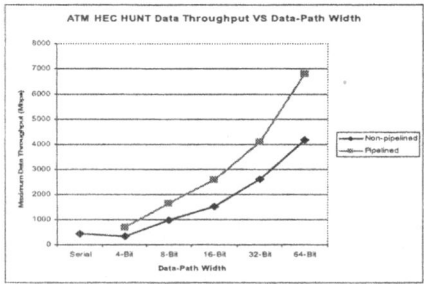

Fig. 11. Data Throughput Rate vs Data-Path Width

With the pipelined architecture, the 16-bit HEC hunt circuit can feed a 2.5 Gbit ATM receiver with complete bit-by-bit HEC check. It is also advantageous to increase the data-path width further. There is a greater increase in data throughput capability compared to area increase. In fact the area increases quite steadily except for the migration from a 32-bit data-path to a 64-bit data-path. This is because the 4-bit, 8-bit, 16-bit and 32-bit systems all require 32 CRC computation units. The 64-bit system however, requires 64 CRC computation circuits and 64 8-bit comparator units. This explains the large area resource increase. This is not a huge disadvantage considering the data throughput increase of more than 2.7 Gbps. The 64-bit circuit performance of 6.8 Gbps data throughput with complete bit-by-bit HEC check is highly impressive for an implementation on established FPGA technology.

5 Conclusions

In this paper, we investigated the architecture and implementation of a range of parallel ATM HEC hunt circuits and analysed them in terms of speed, area and scalability. The architectural limitations of parallel HEC hunt processing are discussed. Introducing an additional pipeline stage significantly improved the overall throughput rate for all HEC hunt circuits. For example, in the case of the 64-bit data-path architecture, the data throughput capability is increased by 63% with an area penalty of only 20%.

Comparisons with referred work [4] in literature revealed that the presented serial architecture on an FPGA outperforms a 0.18µm ASIC technology implementation by a factor of 2.6. Our 8-bit implementation is 27% faster than the 8-bit circuit presented in [4]. This is even more significant considering we include a complete bit-by-bit check and not just a simpler byte-by-byte check as with [4].

The research has demonstrated the advantages of pipelining HEC hunt circuits and the trade-off limitations of increasing the parallelism to achieve higher throughput rates. Circuit study and synthesis has revealed that a Stratix FPGA can achieve 6.8 Gbps complete ATM bit-by-bit HEC check.

It is envisaged that an implementation on the new Stratix 2 FPGA family or standard cell implementation will achieve much higher data rates and achieve throughput rates beyond 10Gbps.

References

1. C. W. Suh, K. S. Kim, "High-speed HEC algorithm for ATM", 1ˢᵗ International Conference on Information, Communications and Signal Processing, 1997
2. Comtel Systems ,"Implementing ATM Header Error Control". A Practical Approach. 1995. White paper, http://www.cmtl.com/pdf/hec_design.pdf
3. C Toal, S Sezer, "A 32-BitSoPC Implementation of a P⁵", International Symposium on Computers and Communications 2003
4. G.E. Griffith, T. Arslan, A. T. Erdogan. "Asynchronous Transfer Mode Cell Delineator Implementations" SoC 2003
5. L.S. Ng, Bill Dewar, "Parallel realization of the ATM cell header CRC", Computer Communications, 1996
6. ITU TS Rec. I.432
7. Stratix FPGA Family Data Sheet, DS-STXFAMLY-3.0, Altera Corporation.
8. T.-Bi-Pei, C. Zukowski, "High-speed parallel CRC circuits in VLSI," IEEE Transaction on Communication, vol. 40, pp. 653-657, April 1992

Web Based Service Provision
A Case Study: Electronic Design Automation

Stephen Dawson and Sakir Sezer

School of Electrical and Electronic Engineering,
Queens University Belfast,
Riddle Hall, Belfast BT9 5EE
185 Stranmillis Road, UK
stephen.dawson@ee.qub.ac.uk

Abstract. As inter enterprise cooperation on the Internet becomes more dynamic and heterogeneous, web services have been widely declared as the foundation for a new Internet service structure. In the future the Web will be populated with many interconnecting self-describing services. Organisations will not only provide their own dedicated services via the web, but they will also provide services to interact and cooperate with other organisations services. To promote an increase in design and intellectual property reuse there is a need for more efficient management of Electronic Design Automation (EDA) tools and computing resources. This makes this area an ideal case study for web based service provision. In this paper we investigate current web based design tools. We then propose two possible business models for companies wishing to offer such a service. Taking these models we devised an architecture specifically designed for offering a web based Electronic Design Automation service. This framework is subsequently implemented and a demonstration system is presented.

1 Introduction

As internet usage increases so does the users' expectations of the services provided by it. No longer is it seen as a medium for personal web pages but as a truly dynamic business platform for both Business to Business and Business to Consumer markets. Services have moved on from the traditional hard coded web pages provided by many online shopping sites and now provide higher levels of interaction for example travel route planners and virtual hotel tours. Although these are very useful and effective services, companies are looking to take these proprietary services and convert them to universal feature rich applications which can be consumed by third parties with no knowledge of the underlying application logic. The technologies grouped as Web services have been widely declared as the foundation for this new service orientated architecture.

Web services are an ideal opportunity to address many of the major difficulties facing EDA tool vendors and their users. The complex and computation intense nature of such tools dictate not only the specification of the system required to run the tools but also the need for constant updates. These updates can often be difficult to

J.N. de Souza et al. (Eds.): ICT 2004, LNCS 3124, pp. 1057–1066, 2004.
© Springer-Verlag Berlin Heidelberg 2004

install and costly for the vendors to supply. Using web services the customer can be assured that they are using the most up to date version and that it is properly installed. The consumer will benefit from the 'pay as you go' nature of the service as they will be billed only for what they use as opposed to the currently expensive system of software licenses. The advantage to the vendor is they no longer have to send out expensive update packages to every user. Although it seems unlikely both the consumer and vendor can gain from a 'pay as you go' system, it is due to lack of security of traditional software licensing. Therefore this system will provide the vendor with the added benefit of preventing software piracy as only the vendor will have the software and only registered users will be billed for each use. "Reducing software piracy by just 10 percentage points worldwide would generate 1.5 million jobs and add \$400 billion to the world economy" [1].

In this paper we present a framework for web based EDA tooling and propose a novel web service architecture that is specially targeted for EDA applications. Finally we developed a demonstrator for Web based ASIC design including Synopsys and ModelSim tools.

2 WEB Services

The web service architecture is built on the Internet standards Hypertext Transfer Protocol (HTTP) [2], Extensible Markup Language (XML) [3] and Simple Object Access Protocol (SOAP) [4] and introduces new concepts such as Service Orientated Architecture (SOA), Universal Description, Discovery and Integration (UDDI) [5] and the Web Services Description Language (WSDL) [6]. Web services are a collection of functions packed into individual, self-contained entities, which are published in a repository for other services to use. This model permits the creation of loosely coupled distributed applications, which use open standards for communication.

There are three roles that are fundamental to the web services architecture. The first is the service provider, their duty is to create a service and publish its interface using the WSDL to the service broker. It is the service brokers' responsibility to list this service description in a repository. Finally the service consumer who has the responsibility to consume the service according to the interface defined in the WSDL. Web services extend the component models key features i.e. providing functionality as a 'black box' with described and published interfaces. Web services provide much more than just a platform for legacy integration as they provide a service, which is platform independent, ad hoc in nature and removed from any single programming paradigm [7].

3 Resource Management

Web based computing platforms enable the sharing, selection, and aggregation of geographically distributed heterogeneous resources, such as computers and data sources. The resource owners and resource users have different objectives, strategies, and requirements. Tight control of all interactions and resources is vital as resource owners are far more likely to allow others to utilize their resources if they have tight control [2]. Effective management of resources can improve software reuse, communication and ultimately productivity [3]. Some of the possible resources that could benefit from computer assisted management are:

- Staff
- Documents
- Computer processing time
- Intellectual Property Cores
- Software Licenses
- Storage Resources

An efficient system is needed for the resource allocation to regulate supply and demand of the available resources listed above. An economic-based framework offers an incentive to resource owners for contributing and sharing resources, and motivates resource users to think about tradeoffs between the processing time (e.g., deadline) and computational cost (e.g., budget), depending on their QoS requirements. This approach is essential for promoting such web based resource sharing as a mainstream computing paradigm [4]

4 EDA Tools

EDA tools seem like an ideal application for such a service orientated resource management tool due to the high design complexity and tool costs. Small hardware design companies would like to test their designs but may not be able to afford the high initial outlay needed for high end tools e.g. Cadence. Start-up tool vendors would also like to offer their tools to a wider audience and hopefully gain interest without high advertising costs. An example solution to this is provided by E*ECAD which offers third-party EDA tools on an hourly or monthly basis. The problem with this implementation is firstly that it does not have the major vendors backing e.g. Synopsys, Cadence and ModelTec. Another and fundamental flaw is that the user is still required to install the software on their system and so this is really just a license managing service for vendors. Xilinx currently offers its ISE WebPACK over the web which is a subset of their Foundation design tools providing instant access to the ISE tools at no cost. This is not really an online service in terms of actual online interaction but rather a simple free download of a basic package which is bundled with another piece of software which allows for checking of Xilinx component costs on the web.

5 Business Model

During our investigations two business models were examined as possible implementation architectures for this case study. These models are not only crucial to the operational behaviour of the system but also to the business strategy of the provider. The first model shown in Fig.1 Virtual Service Provider (VSP) assumes that the user interface and all of its underlying logic components are developed and offered by a third party i.e. a web development company. Using this assumption the VSP developer can utilise various web services provided by different vendors to offer one or more complete services. The developer effectively uses an aggregation of the services offered to create their own unique service.

The advantage of this model is that the user interface can be designed and updated by a web developer with only knowledge of the WSDL describing the service to be consumed. This means that the presentation layer can be left to the relevant experts i.e. web developers and the application layer can be left to the relevant experts i.e. EDA tool vendors. Implementation of this model has the capacity to provide both added revenue to large companies who permit their services to be consumed by VSPs as well as a whole new source of revenue for smaller companies to explore.

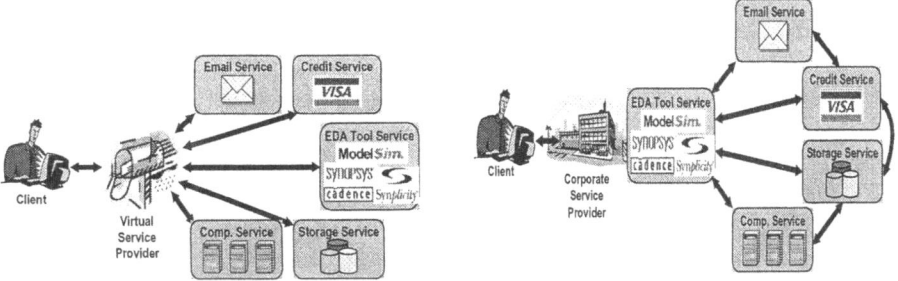

Fig. 1. Virtual Service Provider   **Fig. 2.** Corporate Service Provider

The second of these models, shown in Fig. 2 Corporate Service Provider (CSP) is based on an existing business-to-business model in which many companies and their affiliates operate currently. This system is based on tightly coupled companies, which provides and use services from other trusted companies for example a credit facility for payment. This model would be more attractive to large companies as it gives them complete control over business processes and in doing so provides a higher level of security.

The economic value of this model for an EDA vendor would come from the effective use and billing of resources. This model is also effective for the end user as they can always be confident that they are using the most recent tools and libraries. To both the public user and the provider this model provides the highest level of assurance that sensitive data and resources will be secure.

These two models are the focus of this paper because they show the value added services, which can be provided by two diverse organizations using the web services paradigm. The implementations of both these services are very similar but the manner in which each provider creates revenue from these services is fundamentally different. In the case of the VSP, amalgamating available services to create one service, which can be charged for, creates revenue. On the other hand the CSP is creating revenue from billing for the use of resources that would otherwise be inefficiently utilized.

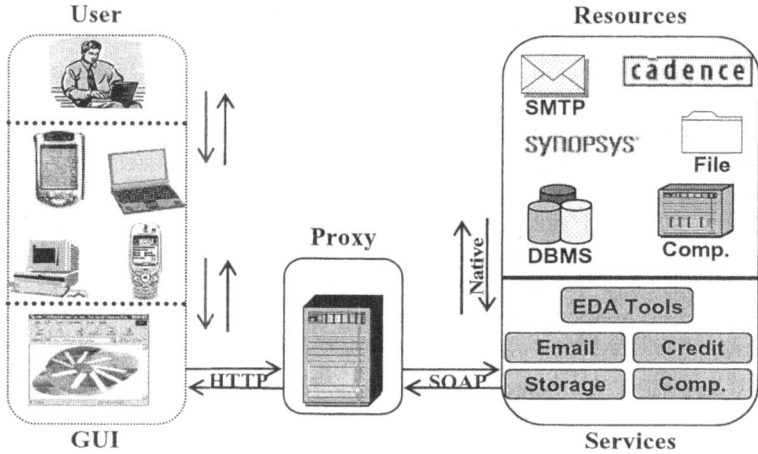

Fig. 3. System Architecture

6 System Architecture

Based on our investigations into various EDA tools used for the design of digital systems, VLSI and System on a Chip (SoC), we propose a generic architecture using XML web services as shown in Fig. 3. This system can incorporate existing and future services including various EDA tools and still provide a web interface to non-Web based legacy systems on all common platforms. This offers providers a low risk and minimal implementation option for adopting a web based hardware design tool. The architecture is viable for both business models discussed in Sect. 5.

The decision to make the system portable and universally available, dictates that a laptop, PC, PDA and modern mobile phone are all essential target devices. The level of interactivity offered by a service to each of these devices must be tailored to fit the devices features and limitations. A mobile phone and a PDA obviously offer the greatest mobility due to their size and lack of dependence on hardwired networks but are limited by relatively low processing power and resources. Therefore the service may report status to the phone via SMS and to the PDA via an email. Due to their relatively high processing power both the PC and laptop would be offered a service with a high degree of interactivity for example an interface for the design of complex chips.

The user interface provides the user with a graphical interface to the services. This interface would ideally be highly portable and available and so for this architecture the natural choice would be a Web based interface since the Web and HTTP are already universally supported. Therefore any device that supports a Web browser could access these services.

As the technology in small devices continues to improve, it may be possible to enable web service accessibility directly, however in order to help facilitate rapid progress in creating a ubiquitous web services infrastructure including mobile devices, a proxy architecture is necessary as a starting point. We utilized the proxy server architecture in order to avoid the high overhead involved in parsing, creating and sending SOAP messages. In the VSP model this will be the layer in which all of the services to be incorporated will be amalgamated into one user interface. This effectively provides access to many linked services from one interface or website. In the CSP model this layer will just act as a bridge between the user and the underlying hardware design service.

The service layer contains numerous XML services, written in various programming languages, which are offered by many organisations. Each service is defined independently using WSDL and provides its own individual service. The services that will be offered are:

- Email
  - o Allows the user to send an email by passing in the sender and destination address as well as the subject and body text and any attachments
- File storage
  - o Permits the user to store their files on a server and transfer them to other servers
- Credit card
  - o Provides a method of billing for services and can be used to either debit or credit a user's card
- Hardware Design Tools
  - o Allows a user to run design tools on a remote server
- High Performance Computing
  - o Allows a user to execute computational intensive simulation and synthesis tasks on high-performance/grid computers.

All the services will have the same interface for both models but will have differing internal logic.

Essentially the difference between the flows of both business models is the central hub of communication. In the VSP model the web interface coordinates all interaction and effectively manages the system whereas in the CSP model this duty is performed by the hardware design service itself. For this case study we assume that the user has already registered and set up an account with the relevant service.

7 Implementation

The implemented architecture illustrated in Fig. 3 was chosen to show many of the integration choices available to architects. This model was developed using the two most popular web service development platforms i.e. .net [11] and Java 2 Enterprise Edition (J2EE) [12] and their respective programming languages C# and Java. To illustrate that both of the models discussed in Sect. 5 can be easily integrated into a real world scenario both have been developed.

7.1 User Interface

This layer contains the user, their desired medium and a web browser. To maximise the usability and availability of the system the choice was taken to develop a web based graphical interface using Macromedia Flash as it is available for all of the desired user mediums and design time is relatively short for web developers. The interface runs on each of the corresponding mediums' web browser with a flash player installed. With all of these options the user can monitor the progress of their synthesis and also initiate further simulations from virtually any conceivable location or scenario.

The quality of the UI has a profound effect on productivity. Schwab estimates that an interactive UI is four times as productive as a request/response Web page UI [13]. The reasons are well-understood. If system response times are between one second and 10 seconds, users will begin to lose focus on the task at hand. As they slow beyond 10 seconds, users will want to move to another task while waiting. In any real world application, users will routinely initiate transactions that take several seconds to complete. Therefore it was necessary to incorporate an asynchronous interaction model that lets them initiate and monitor multiple tasks and data feeds simultaneously, and immediately alerts them when a transaction completes or a critical event occurs. This is performed by spawning a processing thread for each request and having well defined exception returns.

7.2 Proxy Server

Our proxy clients still use the same Internet web services but instead of creating and transmitting the SOAP requests directly, the client sends requests by method calls to the server over HTTP. The workstation is running a proxy server application that takes our method call and converts the request into a SOAP request. This acts as a web services client to perform the SOAP transaction. Requests are sent to the proxy server in the form:

```
[service address]?methodName = [value] & parametersList
[[parameterName] = [parameters]].
```

The proxy server sends the result to our proxy client via the return value of the method call Fig. 3 illustrates the proxy setup.

Table 1. Service Implementation

	Services			
	Email	**File**	**Design**	**Credit**
Programming Language	C#	Java	Java	C#
Operating System	Windows XP	Windows XP	Red Hat Linux	Windows XP
Web Server Host	IIS	Apache	Apache	IIS
Database Driver	ADO	JDBC	JDBC	ADO

7.3 Web Services

In this layer both of the discussed business models are supported. In the case of the VSP model there is no inter-service communication as all the services are managed by the Servlet. All of the implementation technologies used for each service and their subsequent hosts are listed in table 1.

7.4 System Resources

This layer contains five very common resources that a mobile user would like to access and utilise via the Web, namely a database, hardware design tools, an email server, a high performance computer and a remote file system. Communication between the web services and the resources that they encapsulate is bridged using the native libraries of the language that is used to provide the service e.g. Simple Mail Transfer Protocol (SMTP) server is used by the E-Mail service and so the interface to it is bridged using ActiveX Data Objects (ADO) which is provided with C#.

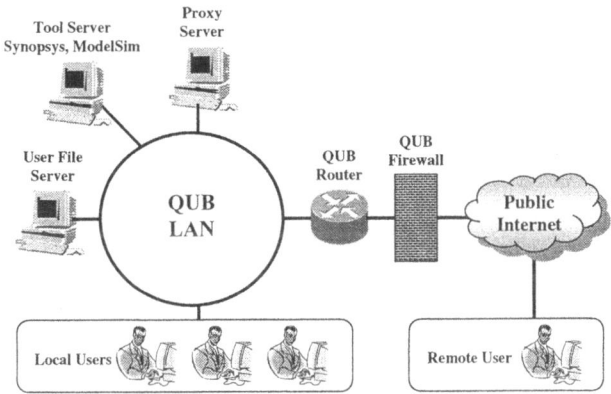

Fig. 4. Demonstrator Configuration

8 Demonstrator System

As part of our system investigation and concept validation, we applied the proposed framework and developed a working demonstrator for web based EDA tool access. The demonstrator is based on the proposed Web service architecture and incorporates all of the design features we have discussed in Sect. 7. Fig. 4 depicts the demonstrator configuration and shows access to the services by both remote and local users. This access is gained through the firewall and via the router.

Fig.5 shows both the flash user interface for Synopsys and the traditional SUN-Solaris interface. The current interface supports the majority of the Synopsys functions including setup, constraint specification, synthesis and design analysis. Interactive schematic view is currently under development. Fig. 6 illustrates a basic demonstration using ModelSim and Synopsys tools for the design of a UART circuit using VHDL. This process has four main stages, which are: logging in i.e. Access Control; downloading and modifying of their existing design; starting and loading the design into their target tool; running synthesis using selected parameters. This design scenario illustrates how the demonstrator test bed can be used for simulation and synthesis of complex circuitry. Any additional services or EDA tools could be seamlessly integrated into the system due to the modular nature of the interfaces and components.

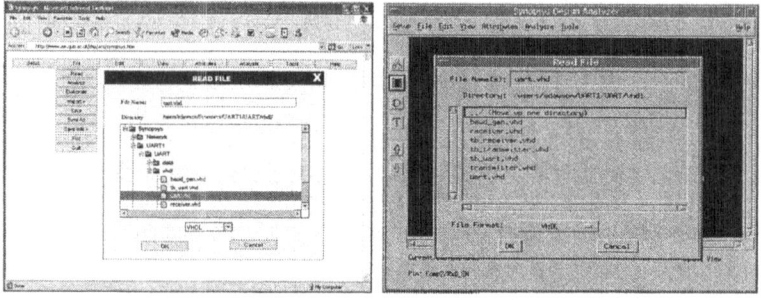

Synopsys *Flash* Interface   Synopsys *SUN-Solaris* Interface

Fig. 5. Synopsys web based Flash user interface versus SUN-Solaris user interface

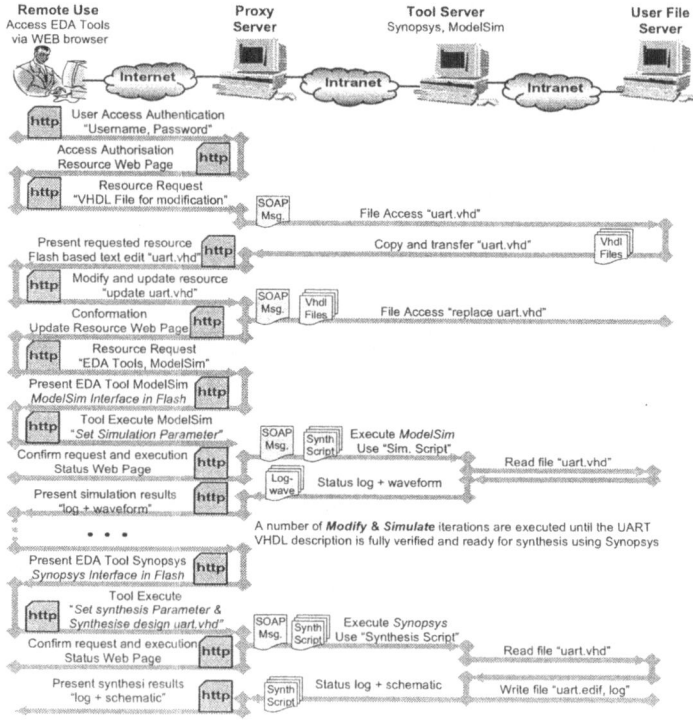

Fig. 6. Demonstrator operation, simulation and synthesis example

9 Conclusions

The aim of the work presented in this paper was to investigate and implement a web service for EDA. Our research encompassed the integration of many different programming languages, servers, operating systems and database drivers in order to provide five individual and unique services, which can be amalgamated into one feature rich service. The two business models implemented show the manner in which different specialist companies can take advantage of this new architecture. This architecture addresses many of the current limitations of EDA tool access and so provides an attractive option for both types of organizations discussed. This highly scalable architecture allows additional services or resources to be 'plugged in' with ease using a clearly defined interface. The architecture presented shows a significant step forward in the provision and use of EDA tools.

References

1. Expanding Global Economies: The Benefits of Reducing Software Piracy. Available from: http://global.bsa.org/idcstudy/pdfs/White_Paper.pdf
2. R Fielding, J Gettys, J Mogul, H Frystyk, T Berners-Lee, "Hypertext transfer protocol – http/1.1", Internet Engineering Task Force, RFC 2616, 1997. Available from: http://www.ietf.org/rfc/rfc2616.txt
3. "Extensible Markup Language 1.0 (XML)", Second Edition, W3C Recommendation, October 2000. Available from: http://www.w3.org/TR/REC-XML.pdf [Accessed Sept 5, 2002]
4. "Simple Object Access Protocol 1.1", W3C Note. Available from:http://www.w3.org/TR/SOAP [Accessed Sept 10, 2002]
5. "UDDI Version 3.0", Published Specification, Universal Description, Discovery and Integration Organisation, UDDI.org, 19 July 2002. Available from: http://uddi.org/pubs/uddi_v3.htm
6. "Web Services Description Language 1.1", W3C Note, March 2001. Available from: http://www.w3.org/TR/WSDL.html [Accessed Sept 12,2002]
7. J Coco, "Maximizing the Potential of Web Services", XML-Journal. SYS-CON Media, Inc. (2001). Available from: www.sys-con.com/xml.
8. "R. Figueiredo, P. A. Dinda, J. A. B. Fortes, A Case for Grid Computing on Virtual Machines, Proc. International Conference on Distributed Computing Systems (ICDCS), May 2003.",
9. Economic-based Distributed Resource Management and Scheduling for Grid Computing Rajkumar Buyya PhD Thesis
10. R. Buyya, D. Abramson, J. Giddy, and H. Stockinger, Economic Models for Resource Management and Scheduling in Grid Computing, The Journal of Concurrency and Computation: Practice and Experience (CCPE), Wiley Press, May 2002
11. Microsoft, .net Framework Available from : http://www.microsoft.com/net/
12. Sun Microsystems Inc. Java 2 Platform, Enterprise Edition. Available from http://java.sun.com/j2ee/
13. "Grid Application Design", John Blair, EAI Journal, June 2003

Identification of LOS/NLOS States Using TOA Filtered Estimates

Alberto Gaspar Guimarães and Marco Antonio Grivet

Pontifical Catholic University - CETUC, Rio de Janeiro, Brazil
{agaspar,mgrivet}@cetuc.puc-rio.br

Abstract. One of the major difficulties to finding accurate mobile terminal location is the lack of line-of-sight propagation (NLOS) caused by blocking of the signal's direct path by obstacles. A tracking approach based on the Kalman recursive filtering of time-of-arrival (TOA) measurement has been successfully employed to mitigate NLOS error, but it strongly depends on proper identification of NLOS/LOS states. In this paper we propose a composite hypothesis testing to detect LOS⇆NLOS transitions. The decision thresholds are set to minimize false alarm and miss probabilities considering statistical parameters of the location scenario. Restricted assumptions such as uncorrelation of TOA measurements (when NLOS corrupted) and knowledge of NLOS error statistics are not required. Simulation results show the validity of the proposal when compared with the technique of identifying NLOS/LOS states by standard deviation computation.

1 Introduction

The E-911 Report issued by the US Federal Communications Commission (FCC) required that all wireless operators provide location information to Emergency 911 Public Safety Answering Points (PSAPs) in USA. Besides the regulatory aspect, other location based applications are of great interest to service providers such as location sensitive billing, vehicle fleet management and data source to cell coverage design.

One of the conventional methods for wireless positioning is based on the measurement of time-of-arrival (TOA) of signals transmitted among mobile station (MS) and a sufficient number of base stations (BS). However, the characteristics of the wireless mobile channel poses several challenges to make such measurements with the required location estimate accuracy. Non-line-of-sight (NLOS) propagation, which occurs when the LOS path between the MS and BS is blocked by some structure, has been identified as one of the primary factors that limits the location systems efficiency.

Several methods have been proposed to mitigate NLOS errors from TOA measurements. The tracking approach by using Kalman filters to recursively estimate TOA data seems to be a promising alternative [1][2][3]. Besides smoothing the noisy timing measurements, a Kalman tracker is able to cancel the NLOS

J.N. de Souza et al. (Eds.): ICT 2004, LNCS 3124, pp. 1067–1076, 2004.

bias by (artificially) increasing the diagonal elements of the noise covariance matrix when the channel is under NLOS. In this way the Kalman filter disregards the TOA data, which is NLOS corrupted, and produces an output based on prior estimates obtained under LOS conditions.

A crucial task in this technique is to identify whether the TOA measurements are NLOS corrupted or not. Once the NLOS state is correctly detected, the algorithm can efficiently remove the NLOS error. On the other hand, if the NLOS or LOS states are not properly recognized, the NLOS error is not mitigated and Kalman algorithm can loose tracking, as can be checked by simulation.

In [1] and [3] similar methods for detecting the NLOS/LOS condition were developed. In both cases, the decision criterion is based on the comparison between the standard deviation of TOA measurements (periodically computed) and a given threshold empirically chosen. If the measured standard deviation is greater than the threshold then the NLOS state is assumed. The serious drawback of this approach is that the measurements *must be* mutually uncorrelated when NLOS corrupted, otherwise the standard deviation estimation by time averages does not correspond to the actual standard deviation of TOA measurements. Unfortunately, for most practical channels, there is no evidence of such property[1]. Furthermore, the statistics of NLOS error should be known to properly set the threshold value of the decision rule.

In this paper we formulate a binary hypothesis test to detect LOS⇄NLOS transition based solely on the difference between prior Kalman estimates and TOA measurements. The test is applied on a sample by sample basis and at each time instant it is decided whether the previous state was modified or kept unaltered. There is no need to assume that the measured data are uncorrelated and the test can be constructed without previous knowledge of the NLOS error statistics.

2   Hypothesis Test for LOS→NLOS Identification

Measured TOA is corrupted by several factors related to the propagation environment and the transmission system. The widely accepted model [1][3][5] for the TOA measurement (τ_n^{meas}) between MS and a given BS at time instant t_n is

$$\tau_n^{meas} = \tau_n + \alpha_n b_n + v_n \ , \tag{1}$$

where τ_n is the true TOA, b_n is the NLOS error and v_n is the noise measurement, usually considered a zero-mean white gaussian process. The parameter α_n defines the channel state and it models the random interruption of the LOS path ($\alpha_n = 1$ for NLOS ; $\alpha_n = 0$ for LOS). All the quantities in the right-hand

---

[1] Timing offsets due to NLOS propagation are determined by the geometrical distribution of scatterers around the MS and BS [4]. As the MS moves the surrounding environment changes, but it is not plausible to assume lack of correlation between scatterer positions over short periods of time.

side of equation (1) except τ_n are modelled as random variables associated to independent stochastic processes observed at t_n.

We assume that a Kalman filter is used for recursive estimation of TOA at each time sample. Let the random variable $\Delta\tau$ be given by

$$\Delta\tau = \tau_n^{meas} - \hat{\tau}_{n,n-1} \ , \tag{2}$$

where τ_n^{meas} is the measured TOA at t_n and $\hat{\tau}_{n,n-1}$ is the Kalman filter prediction.

Assuming that at t_{n-1} the channel is in LOS state ($\alpha_{n-1} = 0$), there are two possible alternatives at t_n: i) state transition to NLOS ($\alpha_n = 1$), referred to as hypothesis $H1$, or; ii) state sojourn in LOS ($\alpha_n = 0$), referred to as hypothesis $H0$. In each case the variable $\Delta\tau$ has distinct representations.

Assuming that $H1$ is true, $\Delta\tau$ is given by

$$H1: \quad \Delta\tau = \tau_n - \hat{\tau}_{n,n-1} + b_n + v_n \ , \tag{3}$$

and under the hypothesis $H0$, $\Delta\tau$ becomes

$$H0: \quad \Delta\tau = \tau_n - \hat{\tau}_{n,n-1} + v_n. \tag{4}$$

The Kalman prior estimate can be written as

$$\hat{\tau}_{n,n-1} = \tau_n + s_{n,n-1} \ , \tag{5}$$

where $s_{n,n-1}$ is the *prior error*, modelled as a zero-mean random variable (assuming that $\hat{\tau}_{n,n-1}$ is unbiased) whose variance can be evaluated by the Kalman filter.

In order to detect the state changing LOS→NLOS, the following binary hypothesis test is formulated:

$$\begin{aligned} H1 &: \Delta\tau \sim p_{\Delta\tau|H1} \\ H0 &: \Delta\tau \sim p_{\Delta\tau|H0} \ , \end{aligned} \tag{6}$$

where $p_{\Delta\tau|H1}$ and $p_{\Delta\tau|H0}$ are the probability density functions (pdf) of $\Delta\tau$ under the hypotheses $H1$ and $H0$, respectively. From (3), (4) and (5) we may state

$$\begin{aligned} H1 &: \Delta\tau = v_n' + b_n \\ H0 &: \Delta\tau = v_n' \ , \end{aligned} \tag{7}$$

where

$$v_n' = v_n - s_{n,n-1} \ . \tag{8}$$

Considering that the measurement noise is a stationary zero-mean white Gaussian process, v_n is statistically independent from v_{n-1} and consequently from $s_{n,n-1}$. Moreover, $s_{n,n-1}$ has also a Gaussian distribution. Hence $v_n' \sim \mathcal{N}(0, \sigma_0)$, where $\sigma_0^2 = \text{Var}[v_n] + \text{Var}[s_{n,n-1}]$. Throughout this paper $x \sim \mathcal{N}(\mu, \sigma)$ denotes that x is Gaussian distributed with mean μ and standard deviation σ, and Var[.] denotes the variance of a random variable. In practice both variances can be estimated, hence we consider in the following that σ_0 is known. The bias b_n by its turn is non-negative and may have known or unknown statistics.

In a realistic scenario the prior probabilities of both hypotheses are not precisely known. Therefore it is convenient to use the Neyman-Pearson (NP) approach which maximizes the probability of detection for a given probability of false alarm [6]. This approach gives rise to the following likelihood ratio test (LRT):

$$l(\Delta T) = \frac{p_{\Delta\tau|H0}(\Delta T|H0)}{p_{\Delta\tau|H1}(\Delta T|H1)} \underset{H_1}{\overset{H_0}{\gtrless}} \gamma \ , \tag{9}$$

where $l(.)$ is the *likelihood function*, ΔT is an outcome of the random variable $\Delta\tau$ and γ is a threshold to be determined.

Since b_n is non-negative, that is, has a one-sided pdf, the NP test for the hypotheses presented in (7) can be formulated conditioning the LRT to b_n:

$$\frac{p_{\Delta\tau|H0}(\Delta T)}{p_{\Delta\tau|H1,b_n}(\Delta T)} = \frac{(\sqrt{2\pi}\sigma_0)^{-1}\exp\left[-\frac{(\Delta T)^2}{2\sigma_0^2}\right]}{(\sqrt{2\pi}\sigma_0)^{-1}\exp\left[-\frac{(\Delta T - b_n)^2}{2\sigma_0^2}\right]} \underset{H_1}{\overset{H_0}{\gtrless}} \gamma. \tag{10}$$

This is in fact a *Uniformly Most Powerful* (UMP) test, which has a performance equivalent to the case of known b_n [6][7]. The existence of a UMP test is a rather interesting result, since it outperforms any other composite test and can be applied whether the NLOS error statistics is known or not.

Simplifying the LRT in (10) we arrive at the following decision rule

$$\Delta T \underset{H_0}{\overset{H_1}{\gtrless}} \gamma_{01} \ , \tag{11}$$

in which the threshold γ_{01} depends on the maximum false alarm desired to NLOS detection (β). Under $H0$, $\Delta\tau \sim \mathcal{N}(0, \sigma_0)$, yielding

$$P_F = \Pr(\Delta\tau > \gamma_{01}|H0) = 1 - Q(\gamma_{01}/\sigma_0) = \beta \ , \tag{12}$$

where $Q(.)$ is the cumulative distribution function of a unit variance and zero mean Gaussian variable. Thus,

$$\gamma_{01} = \sigma_0 Q^{-1}(1 - \beta) \ . \tag{13}$$

For a given β, which defines γ_{01}, the corresponding *miss probability* (P_M) depends on the pdf associated to b_n:

$$P_M = \Pr(\Delta\tau < \gamma_{01}|H1) = \int_{-\infty}^{\gamma_{01}} p_z(Z)dZ, \tag{14}$$

where $z = v_n' + b_n$. Hence,

$$P_M = \int_{-\infty}^{\gamma_{01}} \int_0^\infty p_{b_n}(B)g(Z - B, 0, \sigma_0)dBdZ = \int_0^\infty p_{b_n}(B)Q\left(\frac{\gamma_{01} - B}{\sigma_0}\right)dB \ , \tag{15}$$

in which p_{b_n} is the pdf of the NLOS error b_n and $g(x, \mu, \sigma)$ denotes a Gaussian pdf with argument x and parameters (μ, σ).

Previous works have considered different probability density functions for b_n: one-sided Gaussian [8], Exponential [9], Uniform [3] and others [2][10]. Figure 1 shows P_F and P_M plotted (continuous line) against the threshold γ_{01}, for $\sigma_0 = 100/c$ sec and assuming that b_n is uniformly distributed in the interval $[0 \ 1000/c]$ sec, where c is the speed of light. Since P_F and P_M are conflicting objectives, it is necessary to find a trade-off solution for the threshold values. Assigning identical costs for both error types, the best threshold choice would be $\gamma_{01} \approx 120$, which yields $P_F \approx P_M \approx 13\%$

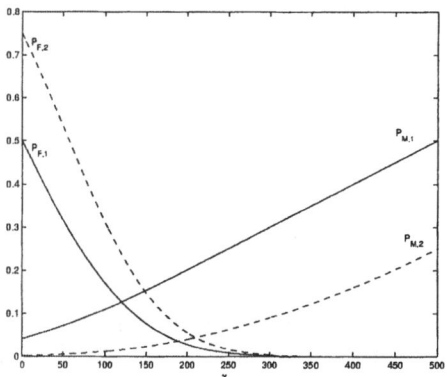

Fig. 1. False alarm and miss probabilities for b_n uniformly distributed in $[0 \ 1000/c]$ sec and $\sigma_0 = 100/c$. $P_{F,1}$ and $P_{M,1}$ (continuous line plot) were obtained from (12) and (15) and $P_{F,2}$ and $P_{M,2}$ (dashed plot) from (19) and (18).

3   Hypothesis Test of Higher Performance

3.1   Double Hypothesis Test

An alternative test routine is proposed in the following in order to obtain smaller values of P_F and P_M. The decision is based on the comparison between two consecutive samples of TOA, namely τ_n^{meas} and τ_{n+1}^{meas}, and the Kalman prior estimate at time t_n $(\hat{\tau}_{n,n-1})$. The following random variables are defined:

$$\Delta\tau_1 = \tau_n^{meas} - \hat{\tau}_{n,n-1} \quad \text{and} \quad \Delta\tau_2 = \tau_{n+1}^{meas} - \hat{\tau}_{n,n-1} \qquad (16)$$

and two hypothesis tests are set as follows:

$$H1_1 : \Delta\tau_1 \sim p_{\Delta\tau|H1,b_n} \qquad \text{and} \qquad H1_2 : \Delta\tau_2 \sim p_{\Delta\tau|H1,b_n} \qquad (17)$$
$$H0_1 : \Delta\tau_1 \sim p_{\Delta\tau|H0} \qquad\qquad\qquad H0_2 : \Delta\tau_2 \sim p_{\Delta\tau|H0} \ ,$$

where $p_{\Delta\tau|H1,b_n}$ and $p_{\Delta\tau|H0}$ have the same characterization showed in (10). Assuming that at t_{n-1} the channel state is LOS $(\alpha_{n-1} = 0)$, the decision for the state change at t_n is taken if it is decided in favor of $H1_1$ and/or $H1_2$ in (17). Otherwise it is decided that the channel remains in LOS state at t_n. Table 1 summarizes the decision criterion according to the results obtained in tests 1 and 2.

In order to analyze the performance associated to this proposal it is necessary to identify the possible combinations of states at t_n and t_{n+1}, recalling that the channel state is LOS at t_{n-1}. Table 2 shows the possible hypotheses, where the sequence NLOS LOS was not considered since it is assumed that the mean time of the channel in NLOS state is much greater than the sampling interval.

Table 1. Decision criterion for the hypotheses $H1$ (LOS$\rightarrow$ NLOS) or $H0$ (LOS$\rightarrow$ LOS) at t_n.

Test 1	Test 2	Final decision
$H1_1$	$H1_2$	$H1$
$H1_1$	$H0_2$	$H1$
$H0_1$	$H1_2$	$H1$
$H0_1$	$H0_2$	$H0$

Table 2. Possible combination of states at t_n and t_{n+1} (assuming LOS at t_{n-1}) .

	t_n	t_{n+1}
a	NLOS	NLOS
b	LOS	LOS
c	LOS	NLOS

Letting $P_{M,1}$ be the miss probability for the single hypothesis testing, the resulting miss probability ($P_{M,2}$) obtained with the criterion of Table 1 is given by

$$P_{M,2} = \Pr(\hat{H}0|\text{``a'' occurs}) = P_{M,1}^2 , \tag{18}$$

and the new false alarm probability is

$$P_{F,2} = \Pr(\hat{H}1|\text{``b'' occurs}) = 1 - \Pr(\hat{H}0|\text{``b'' occurs}) = P_{F,1}(2 - P_{F,1}) . \tag{19}$$

We may observe that $P_{M,2} \leq P_{M,1}$ and $P_{F,2} \geq P_{F,1}$ (equalities hold when $P_{M,1}$, $P_{F,1} = 0$ or 1), that is, in this test the probability of detection is increased at the expense of increasing the false alarm probability. However a performance comparison of the single and double tests indicates that for the latter threshold values may be found resulting in lower probabilities for both error types. In other words, the trade between false alarm and detection is worthy in this case.

As a basis of comparison, Figure 1 also shows $P_{F,2}$ and $P_{M,2}$ plotting (dashed curves) considering the same scenario as in the single test case. As we can observe the double testing leads to a better performance since it is possible to set a threshold to simultaneously obtain smaller P_F and P_M.

It's worthy to mention that when the NLOS error statistics is not known, the double test approach becomes even more important since it significantly reduces the unknown P_M for the range of threshold values typically chosen. Hence, the consequences of a threshold choice regarding P_M are less critical.

The decision probability for $H1$ when "c" occurs is $\Pr(\hat{H}1|\text{``c'' occurs}) = 1 - P_{M,1}(1-P_{F,1})$. When this decision is taken the channel state at t_n is erroneously considered as NLOS. That is, the detection would happen at a time instant prior to the instant when the state transition actually takes place. Nevertheless this situation has a negligible impact on the whole performance of TOA Kalman estimates, as can be verified by simulation.

3.2   Computing Prior Probabilities

Under the NP approach the threshold is chosen considering the *conditional* probability $P_F = \Pr(\hat{H}1|H0)$. However, the "true" false alarm probability must be evaluated by $P_F = \Pr(\hat{H}1|H0)\Pr(H0)$, and similarly the actual miss probability is given by $P_M = \Pr(\hat{H}0|H1)\Pr(H1)$.

Even though the hypothesis test can be implemented without the knowledge of the prior probabilities $\Pr(H0)$ and $\Pr(H1)$, this information may be important to appropriately choose a threshold value according to the desired P_F, mainly in the case where $\Pr(H0)$ is significantly different from $\Pr(H1)$.

It is possible to derive an approximate expression for the ratio $\Pr(H0)/\Pr(H1)$ based on some parameters of the underlying propagation environment and terminal mobility. Assuming that the LOS/NLOS states are modelled as a two-state continuous time Markov chain, with transition probability rate of LOS→NLOS given by λ_{01} s^{-1}, the ratio between the prior probabilities may be given approximately by $\frac{\Pr(H1)}{\Pr(H0)} \approx \lambda_{01} \cdot \delta_t$, where δ_t is the time interval between TOA measurements.

This result was determined recalling that λ_{01}^{-1} is the mean time of LOS state, which can be derived from [5] resulting in $\lambda_{01} = \frac{p_1\lambda_{10}}{1-p_1} = \frac{p_1}{1-p_1}\frac{v}{<L>}$, where p_1 is the probability of NLOS state, $<L>$ is the average size of the projection of NLOS shadow onto mobile's route and v is the mobile speed. The parameter $<L>$ can be obtained from typical widths of building and roads in urban or suburban areas. The quantity p_1 was defined in [4] according to the nature of terrain (urban, suburban or rural). Finally, v can be determined from mobile position estimates.

4   Hypothesis Test for NLOS→LOS Identification

Now assuming that the channel state at t_{n-1} is NLOS ($\alpha_{n-1} = 1$), the two possible situations at t_n are: *i*) state transition to LOS ($\alpha_n = 0$), referred to as hypothesis $H1$, or; *ii*) state sojourn in NLOS ($\alpha_n = 1$), referred to as hypothesis $H0$.

Under normal operation and NLOS condition, the Kalman algorithm filters out the v_n and b_n variations and tends towards the bias produced by the NLOS error[2]. In this case the Kalman predictions may be given by

$$\hat{\tau}_{n,n-1} \approx \tau_n + s_{n,n-1} + \bar{b} \ , \tag{20}$$

where $\bar{b}$ is the bias of Kalman output due to the NLOS error. Given the above result we may observe that under hypothesis $H1$ we have

$$H1: \quad \Delta\tau = v'_n - \bar{b} \ , \tag{21}$$

---

[2] This is referred as "unbiased filtering" (as in [3]).

where the variables $\Delta\tau$ and v'_n are defined as in equations (2) and (8), respectively. On the other hand, if hypothesis $H0$ occurs we have

$$H0: \quad \Delta\tau = v'_n + b_n - \bar{b} \ . \tag{22}$$

Thus,

$$E[\Delta\tau|H0] = E[b_n - \bar{b}] \approx 0 \ , \tag{23}$$

where $E[.]$ denotes the expectation of a random variable, and

$$\mathrm{Var}[\Delta\tau|H0] = \mathrm{Var}[v_n] + \mathrm{Var}[s_{n,n-1}] + E[(b_n - \bar{b})^2] \ . \tag{24}$$

The last quantity on the right-hand side of the above equation depends on the correlation between samples of b_n. Its maximum value is $\mathrm{Var}[b_n]$, when the samples are mutually uncorrelated, and its minimum value is 0, when b_n has a unique value during the NLOS state.

We can thus consider the following hypotheses to detect NLOS→LOS transition

$$\begin{aligned} H1 &: \Delta\tau = v'_n - \bar{b} \\ H0 &: \Delta\tau = v'_n + \varepsilon_b \ , \end{aligned} \tag{25}$$

where ε_b is a zero-mean variable with variance in the interval $(0 \ , \ \mathrm{Var}[b_n])$.

The UMP test does not exist for the above hypotheses. There are some approaches which may be applied (Bayes, GLRT)[6][7], depending whether the b_n statistics is known or not. For the sake of simplicity we further assume that $\varepsilon_b = 0$. Although this solution might not seem prudent (the detection is subjected to a greater number of false alarms), it will be checked by simulation that the resulting performance is not seriously degraded.

When $\varepsilon_b = 0$ the UMP test may be applied to (25) giving rise to

$$l(\Delta T) = \Delta T \overset{H_0}{\underset{H_1}{\gtrless}} \gamma_{10} \ . \tag{26}$$

The threshold γ_{10} will be negative and can be determined following the same approaches developed in Sections II and III. It should be noted, however, that the ratio between prior probabilities in this case is $\Pr(H1)/\Pr(H0) \approx \lambda_{10}.\delta_t$, where λ_{10}^{-1} is the mean time of NLOS state and may be evaluated by $\lambda_{10} = v/<L>$.

5   Simulation Results

The performance of the proposed method to detect LOS/NLOS states is evaluated against the standard deviation checking used in previous works. A location system is implemented using Kalman filters for TOA estimation. Once detected a NLOS environment for a given BS, the corresponding element of the measurement covariance matrix is weighted by 10^3 (experimentally chosen) to cancel the NLOS error. The Kalman outputs are fed into a LS algorithm [10] to produce the MS position estimation.

Most of simulation conditions follow the scenario presented in [3]. TOA measurements are collected over a total of 200 sec at intervals of 0.2 sec, the MS describes a straight trajectory with constant speed of 30 m/sec and 3 BS's are considered with coordinates at: BS1(-3000,-2000) , BS2(3000,5000) and BS3(6000,2000). For each one of the three TOA sequences, the measurement noise is simulated as a zero-mean white noise gaussian process with $\mathrm{Var}[v_n] = (100/c)^2$ sec^2, and the NLOS/LOS states are considered as a two-state Markov process, with mean time in NLOS and LOS given respectively by $< L > /v$ and $(1 - p_1) < L > /(vp_1)$, where $p_1 = 0.3$ (suburban area) and $< L >= 100$ m.

Regarding the NLOS error (b_n), the sequences are generated considering three different degrees of correlation between samples. In the first case b_n is defined as a white process, i.e, with null autocorrelation coefficient $(\rho = 0)$. In the second a unique value is assigned for b_n during the NLOS period $(\rho = 1)$, and, in the third case, b_n is represented as an autoregressive process of order 1, in which the autocorrelation coefficient has an exponential-like decay with $\rho_{0.5}^{-1} = 70$, where $\rho_{0.5}^{-1}$ denotes the lag between samples for which $\rho = 0.5$. The first order distribution of b_n in all cases is uniform in the interval $[0 \ \ 1000/c]$ sec .

In the hypothesis testing based on standard deviation calculation the interval for repeatedly checking LOS/NLOS condition and the length for sample calculation are the same: 15 samples. The threshold was experimentally chosen as $1.5 \ \sigma_0$. For the method proposed in this paper, two test routines are applied for LOS$\rightleftharpoons$NLOS detection. The first follows the UMP test presented in Section II while the second applies the UMP test with the approach presented in Section III. The threshold values $\{\gamma_{01}, \gamma_{10}\}$ used in each case are $\gamma^{(1)} \equiv \{120, -120\}$ and $\gamma^{(2)} \equiv \{320, -294\}$, respectively, determined by setting $P_F = P_M$ and considering $\mathrm{Var}[v_n] \gg \mathrm{Var}[s_{n,n-1}]$ with the parameters of the location scenario given above.

Table 3 shows the mean location error in meters obtained after 30 simulation runs in the same trajectory and neglecting the 100 initial points in each run. In the table line labeled "$\hat{\sigma}_\tau \gtrless \kappa\sigma_0$"the location system performance using standard deviation comparison is presented. It can be seen, as expected, that this alternative is valid only if the uncorrelation between samples is assumed. The error magnitude verified when b_n samples are correlated is comparable to the results of "Unbiased Filter" (table line 'UnbF"), which shows that the NLOS error mitigation by TOA Kalman filtering becomes useless when the LOS/NLOS states can not be properly recognized. The mean location error obtained using the proposed methods for LOS/NLOS detection are shown in the table lines named "$\Delta\tau \gtrless \gamma^{(1),(2)}$". When the pair $\gamma^{(1)}$ is used the method does not perform well, but on the other hand setting $\gamma^{(2)}$ under the approach developed in Section III the location error is substantially decreased. Comparing both results the importance of the formulation presented in Section III becomes clear. Considering $\gamma^{(2)}$, the best performance is achieved when NLOS error samples are totally correlated $(\rho = 1)$. This behavior is expected since we have neglected the variable ε_b in (25). Nevertheless even for the worst scenario $(\rho = 0)$ the resulting mean error is comparable to the best result obtained by standard deviation checking approach. Anyway, the robustness of the algorithm might be increased by proper characterization of the variable ε_b in NLOS$\rightarrow$LOS detection.

Table 3. Mean location error (in m)

	$\rho = 0$	$\rho_{0.5}^{-1} = 70$	$\rho = 1$
UnbF	280.3	293.7	298.0
$\hat{\sigma}_\tau \gtrless \kappa\sigma_0$	104.6	300.0	319.3
$\Delta\tau \gtrless \gamma^{(1)}$	155.3	259.7	272.1
$\Delta\tau \gtrless \gamma^{(2)}$	115.7	108.8	84.1

$\gamma^{(1)} \equiv \{120, -120\}$, $\gamma^{(2)} \equiv \{320, -294\}$

6 Conclusions

We have developed a suitable hypothesis testing for LOS/NLOS identification comparing TOA data with Kalman prior estimates. It was shown that an UMP test may be applied to detect state transitions with no previous knowledge of NLOS error statistics and under different degrees of correlation between TOA measurements. Good results were obtained taking into account the hypotheses prior probabilities and implementing double comparisons for a sequence of measured TOA's. Even though the number of false alarms in NLOS$\rightarrow$LOS detection tends to increase when there is low correlation between TOA samples, the whole test performance is still acceptable applying the proposed scheme.

References

1. Thomas, N., Cruickshank, D., Laurenson, D.: A robust location estimation architecture with biased Kalman filtering of TOA data for wireless systems. In: Proc. IEEE Spread Spectrum Techniques and Applications Symposium. (2000) 296–300
2. Thomas, N., Cruickshank, D., Laurenson, D.: Performance of a TDOA-AOA hybrid mobile location system. In: Proc. IEE 3G Mobile Comm. Techn. (2001) 216–220
3. Le, B.L., Ahmed, K., Tsuji, H.: Mobile location estimator with NLOS mitigation using Kalman filtering. In: Proc. IEEE Wir. Comm. and Net. (2003) 1969–1973
4. Thomas, N., Cruickshank, D., Laurenson, D.: Channel model implementation for evaluation of location services. In: Proc. IEE 3G Mobile Comm. Techn. (2000) 446–450
5. Wylie-Green, M.P., Wang, S.S.: Robust range estimation in the presence of the non-line-of-sight error. In: Proc. IEEE Vehic. Techn. Conf.-Fall. (2001) 101–105
6. Van Trees, H.L.: Detection, Estimation, and Modulation Theory. John Wiley & Sons, USA (1968)
7. Kay, S.M.: Fundamentals of Statistical Signal Processing (Volume II-Detection Theory). Prentice Hall, New Jersey-USA (1998)
8. Silventoinen, M., Rantalainen, T.: Mobile station emergency locating in GSM. In: Proc. IEEE Personal Wireless Communications Conference. (1996) 232–238
9. Cong, L., Zhuang, W.: Non-line-of-sight error mitigation in TDOA mobile location. In: Proc. IEEE Global Telecomm. Conf. (Globecom). (2001) 680–684
10. Caffery, Jr., J.J.: A new approach to the geometry of TOA location. In: Proc. IEEE Vehic. Techn. Conf.-Fall. (2000) 1943–1949

A Hybrid Protocol for Quantum Authentication of Classical Messages

Rex A.C. Medeiros* and Francisco M. de Assis

Federal University of Campina Grande
Department of Electrical Engineering
Av. Aprígio Veloso, 882, Bodocongó,
58109-970 Campina Grande-PB, Brazil
{rex,fmarcos}@dee.ufcg.edu.br

Abstract. Quantum authentication of classical messages is discussed. We propose a non-interactive hybrid protocol reaching informationtheoretical security, even when an eavesdropper possesses infinite quantum and classical computer power. We show that, under certain conditions, a quantum computer can only distinguish a sequence of pseudo random bits from a truly sequence of random bits with an exponentially small probability. This suggests the use of such generator together with hash functions in order to provide an authentication scheme reaching a desirable level of security.

1 Introduction

Until the last decade, the expression "quantum cryptography" referred basically to protocols for quantum key distribution (QKD) [1]. Recently, several researches have been made in the sense of applying quantum mechanics resources in the resolution of others problems related to data security. Curty and Santos [2] proposed a protocol to quantum authentication of unitary-length classical messages (bit). As for the secret key, they use a maximally entangled EPR pair previously shared between Alice and Bob. For the types of attacks discussed, the probability P_d that Eve deceives Bob was $0.5 \leq P_d < 1$, depending on the choice of an unitary operator used to create the tag. Later, the same authors proposed a protocol to quantum authentication of unitary-length quantum messages (qubit) [3].

More recently, Barnum *et al.* [4] described a protocol to authenticate quantum messages of length m. They propose a scheme that both enables Alice to encrypt and authenticate (with unconditional security) an m qubit message by using a stabilizer code to encode the message into $m+s$ qubits, where the probability of failure decreases exponentially in the security parameter s. Such scheme requires a private key of size $2m + O(s)$ to be shared between Alice and Bob.

All protocols discussed above require unfeasible quantum resources according with state-of-art or near-future technology. In this paper we address the problem of authenticating classical messages of arbitrary length transmitted over a

---

* Graduate Program in Electrical Engineering – Master's degree student.

J.N. de Souza et al. (Eds.): ICT 2004, LNCS 3124, pp. 1077–1082, 2004.

noiseless quantum channel. We propose a non-interactive scheme that just requires preparation of quantum states into orthornormal bases, transmission and measurements of these states in the same bases. Our hybrid protocol extend the concept of computational security hash authentication introduced by Brassard [5].

2   Computationally Secure Hash Authentication

There exist several Message Authentication Codes with different levels of security [6]. The first MAC scheme providing informational security was proposed by Wegman and Carter [7]. The protocol makes use of strongly universal-2 classes of hash functions so that an eavesdropper even with infinite computing power can not forge or modify a message without detection. This allows Bob to be certain that the received message is authentic. Unfortunately, their scheme requires a long secret key to be shared previously between Alice and Bob, even in the case where Alice wishes to send Bob more than one message. More precisely, if Alice and Bob need to exchange n authentic messages, the length of the key must be $n \log(1/P_{WC})$, where P_{WC} is the probability of failure they are willing to tolerate. These restrictions make prohibitive the use of such scheme in real systems.

The first computationally secure scheme for authentication based on hash functions was proposed by Brassard [5]. In his work, Brassard suggests the use of cryptographically strong sequences of pseudo random bits to reduce drastically the length of the key when the participants wish to exchange a large number of messages. The protocol works as follows. Let P_{WC} be the acceptable probability of failure and k an integer greater than $\log(1/P_{WC})$; let M be the space of messages, and let B be the set of all bit strings of length k. Let $\mathcal{H}$ be a strongly universal-2 class of functions from M to B. Now, the secret key that Alice and Bob must share consists of a particular hash function $h \in \mathcal{H}$ and a seed x_0 for the pseudo random bits generator. For the n-th message m exchanged between them, the authentication tag is simply

$$a(m, n) = h(m) \oplus x_0(n), \tag{1}$$

where $x_0(n) = x_0[(n-1)k+1, \ldots, nk]$ denote the bits of inclusive rank $(n-1)k+1$ to nk generated from seed x_0. The effect is to provide "pseudo one-time-pad" cryptography to hide the value of the secret hash function.

The security of the scheme above depends on how secure pseudo random bit generators are. Generally, the security of generators is based on the intractability of some problems of number theory, including prime factorization, discrete logarithm problem and so on. Indeed, all these problems can be reduced to the index finding problem. Because an eavesdropper can copy and manipulate classical information without detection, we can conclude that this scheme is not secure in a quantum setting. Quantum computers can solve the index problem in polynomial time [8].

2.1   The Blum-Micali Generator

The first probably secure pseudo random bits generator (PRBG), known as BM generator, was described by Blum and Micali [9]. In their work, Blum and Micali showed that the BM-PRBG is unpredictable in polynomial time assuming intractability of the discrete logarithm problem. Let p an odd prime and g a generator for the group $G = \mathcal{Z}_p^*$. The secret seed x_0 is an element randomly chosen from $\mathcal{Z}_{p-1}$. The BM generator works as follows. For the i-th bit b_i, starting with $i = 1$, let

$$x_i = g^{x_{i-1}} \mod p, \tag{2}$$
$$b_i = \delta_{x_i > (p-1)/2}. \tag{3}$$

The latter means that $b_i = 1$ if and only if $x_i > (p-1)/2$. To demonstrate that this generator is unpredictable, the authors showed that any procedure able to predict the previous bit of a given piece of a sequence is also able to calculate the logarithm discrete efficiently. Equivalently, the problem of inferring the generator is reduced to the logarithm discrete problem.

3   A Protocol for Quantum Authentication of Classical Messages

The protocol we describe here is an extension of the scheme proposed by Brassard. Let P_{WC}, k, M, B, and x_0 be defined like in Sec. 2. Remember that the Brassard's protocol make use of two secret keys, a particular hash function $h \in \mathcal{H}$ and a seed x_0. Our protocol requires another key, a new seed we call y_0. When Alice wants to send Bob a certified message, she makes all steps of the Brassard's protocol. For the n-th message m exchanged, Alice has a tag $a(m, n)$ of k bits given by Eq. (1). Next, we have the quantum round.

Assume that Alice and Bob agree on two orthonormal bases for the 2-dimension Hilbert space, $\mathcal{Z} = \{|0\rangle, |1\rangle\}$ and $\mathcal{X} = \{|+\rangle = \frac{1}{\sqrt{2}}(|0\rangle + |1\rangle), |-\rangle = \frac{1}{\sqrt{2}}(|0\rangle - |1\rangle)\}$. These are the same bases used to create the four quantum states in the BB84 protocol. For each bit of $a(m, n)$, Alice prepares a non-entangled quantum state $|\psi_{n_j}\rangle$ based on the corresponding bit generated from the seed y_0. Then, if the j-th bit of $y_0(n)$ is 0, Alice prepares $|\psi_{n_j}\rangle$ using $\mathcal{Z}$ basis, such that

$$|\psi_{n_j}\rangle = \begin{cases} |0\rangle & \text{if the } j\text{-th bit of } a(m, n) \text{ is } 0 \\ |1\rangle & \text{if the } j\text{-th bit of } a(m, n) \text{ is } 1. \end{cases} \tag{4}$$

Similarly, if the j-th bit of $y_0(n)$ is 1, Alice prepares $|\psi_{n_j}\rangle$ using $\mathcal{X}$ basis, such that

$$|\psi_{n_j}\rangle = \begin{cases} |+\rangle & \text{if the } j\text{-th bit of } a(m, n) \text{ is } 0 \\ |-\rangle & \text{if the } j\text{-th bit of } a(m, n) \text{ is } 1. \end{cases} \tag{5}$$

After the qubits generation, Alice sends the state $|\psi_{n_j}\rangle^{\otimes k}$ to Bob through the noiseless quantum channel, and the message m using either an unauthentic quantum or a classical channel.

At the reception, Bob makes measurements to obtain a sequence $a_B(m, n)$ of k bits. For the j-th received qubit, Bob measures it using the basis $\mathcal{Z}$ or $\mathcal{X}$, depending on the j-th bit of $y_0(n)$ is 0 or 1, respectively. Because the quantum channel is perfect, Bob recognizes that the message is authentic if $a_B(m.n) \oplus x_0(n) = h(m_B)$, where m_B is the message received from the unauthentic channel. Otherwise, Bob assumes that Eve tried to send him an unauthentic message. He then discards the received message.

4   Security Analysis

We analyze the security of our scheme considering enemies possessing infinite quantum and classical resources. Our arguments for proving unconditional security are based on quantum algorithms for solving problems from number theory, and our strategy for creating the quantum states.

The Wegman and Carter's protocol makes use of truly random sequence in order to perform one-time-pad cryptography on the tag. Their protocol achieves informational security, even when quantum resources are available to enemies. The security of our protocol is based on the security of the pseudo random bits generator.

4.1   Quantum Information Processing

We analyze the case where Eve intercepts the quantum transmission, possibly store the states sent by Alice, and try to process them using a quantum computer. Note the states Eve possesses are non-orthogonal so quantum computers can not perfectly distinguish them. On the other hand, Eve knows nothing about the hash function used by Alice and Bob. We conclude that quantum information processing of the quantum states do not aid Eve with her task of cheating Bob. Eve should perform measurements in order to use her quantum and classical computer power.

4.2   Measurement Attack

It is clear at this point that the security of our scheme depends on how secure BM-PRBG is when used in a quantum setting. We should investigate how Eve can make use of quantum resources to predict such generator. First, we prove the following result concerning the Blum-Micali PRBG with parameters p, g, and x_0, where p is an odd prime such that $p \equiv 3 \mod 4$, g is a primitive element of $\mathcal{Z}_p^*$ and x_0 is a secret seed chosen from $\mathcal{Z}_{p-1}$, having a l-bit representation.

Lemma 1. *Let $b_{i+1} \ldots b_{i+s}$ be a piece of s bits from a sequence generated by the PRBG described above. The best probabilistic algorithm $A_{BM}(g, b_{i+1}, \ldots, b_{i+s})$ to predict the entire sequence backward (and therefore forward) needs at least $s = l$ bits for which*

$$Prob[A_{BM}(g, b_{i+1}, \ldots, b_{i+s}) = b_i] = 1. \tag{6}$$

Proof. We prove the Lemma by observing that this problem is equivalent to compute the hard-core bit[1], and the latter can be reduced to the discrete logarithm problem. Let $A_{DL}(g, x^{i+1})$ be an algorithm running on a quantum computer to find the discrete logarithm x^i, where $b_i = \delta_{x_i > (p-1)/2}$.

The result follows by contradiction. Suppose that such algorithm $A_{BM}(\cdot, \cdot)$ exists. Then, it should exist a function $f(b_{i+1}, \ldots, b_{i+s})$ such that

$$x^{i+1} = f(b_{i+1}, \ldots, b_{i+s}), \qquad s < l, \tag{7}$$

and

$$x^i = A_{DL}(g, x^{i+1}), \tag{8}$$

$$b_i = \delta_{x_i > (p-1)/2}. \tag{9}$$

But such function does not exist, since the cardinality of its domain (2^s) is smaller than the cardinality of its codomain (2^l).

Eve initially has no information about the keys x_0, y_0 and $h \in \mathcal{H}$ shared between Alice and Bob. The seed x_0 is associated with the tag $a(n, m) = h(m) \oplus x_0(n)$ and y_0 with bases choices. Eve may choose a seed y_{E_0} to indicate bases for measurements or she can choose bases randomly. The former choice is successfully with probability 2^{-l} so we consider the latter.

When Eve measures the state $|\psi_{n_j}\rangle^{\otimes k}$ using k random bases, it is clear that she obtains each bit of $a(m, n)$ correct with probability $p_b = 3/4$. Then,

Lemma 2. *When Eve chooses bases at random to measure the quantum states sent by Alice, the bit sequence generated by the Blum-Micali generator with seed x_0 appears to be a truly random sequence with probability*

$$P_r \geq 1 - \left(\frac{3}{4}\right)^l, \tag{10}$$

where l is the number of bits used to represent the seed x_0.

Proof. By Lemma 1, it is required at least l consecutive bits, beginning with b_{i+1}, for which the previous bit b_i is calculated with certainty, when the best procedure $A_{BM}(\cdot, \cdot)$ is used. Then, a procedure $A'_{BM}(\cdot, \cdot)$ to predict the BM-PRBG from $a(m, n)$ needs evidently at least l bits. But Eve has only access to bits from measurements. Considering that each bit is sent independently, the result follows.

Since Eve can not distinguish the pseudo random sequence from a truly random sequence, our scheme reduces to the Wegman and Carter's scheme. Now we are ready to present our main result.

Theorem 1. *The protocol described in Sec. 3 achieves information-theoretical security such as in the Wegman and Carter's scheme with probability P_{WC}, even when an eavesdropper possesses both infinite classical and quantum computer power.*

---

[1] The hard-core bit problem [9] consists in find a bit $b_i = \delta_{x_i > (p-1)/2}$ from the value $x_{i+1} = g^{x_i} \mod p$.

Proof. Let P_{WC} be the probability of undetected forgery that Alice and Bob are willing to tolerate in a Wegman and Carter's scheme. We simply make $(3/4)^l$ smaller than P_{WC} by increasing l so that the security level is always reached.

5   Conclusion

In this work we presented a non-interactive scheme for quantum authentication of classical messages. According to quantum mechanics theory, and considering the approach adopted in the creation of quantum states, we showed that our protocol presents unconditional security, even against an eavesdropper that possesses infinite quantum and classical computer power. This is because an eavesdropper can only distinguish between a pseudo random sequence of bits and a truly random sequence with an exponentially small probability. The scheme allows for authentication of a large number of messages by making use of shorter secret keys.

Acknowledgments. The authors thank the Brazilian National Council for Scientific and Technological Development (CNPq) for support (CT-INFO Quanta, grants # 552254/02-9).

References

1. C. H. Bennett and G. Brassard. Quantum cryptography: public-key distribution and coin tossing. In *Proceedings of IEEE International Conference on Computers, Systems, and Signal Processing, Bangalore, India, 1984*, pages 175–179. IEEE Press, New York, 1984.
2. M. Curty and D. J. Santos. Quantum authentication of classical messages. *Phys. Rev. A*, 64:062309, 2001.
3. M. Curty, D. J. Santos, E. Pérez, and P. Garcia-Fernandez. Qubit authentication. *Phys. Rev. A*, 66:022301, 2002.
4. H. Barnum, C. Crépeau, D. Gottesman, A. Smith, and A. Tapp. Authentication of quantum messages. *e-print quant-ph/0205128*.
5. G. Brassard. On computationally secure authentication tags requiring short secret shared keys. *Proc. of Crypto'82*, pages 79–86, 1983.
6. W. Stallings. *Cryptography and Network Security: Principles and Practice*. Prentice-Hall, Inc., New Jersey, 2 edition, 1998.
7. M. N. Wegman and J. L. Carter. New hash functions and their use in authentication and set equality. *J. Comput. Syst. Sci.*, 22:265–279, 1981.
8. M. A. Nielsen and I. L. Chuang. *Quantum Computation and Quantum Information*. Cambridge University Press, Cambridge, 2000.
9. M. Blum and S. Micali. How to generate cryptographically strong sequences of pseudo-random bits. *SIAM J. Comput.*, 13(4):850–864, 1984.

Attack Evidence Detection, Recovery, and Signature Extraction with ADENOIDS

Fabrício Sérgio de Paula[1,2] and Paulo Lício de Geus[2]⋆

[1] State University of Mato Grosso do Sul (UEMS), Dourados, MS, Brazil
[2] Computing Institute, State University of Campinas, Campinas, SP, Brazil
{Fabricio, Paulo}@las.ic.unicamp.br

Abstract. This paper presents the ADENOIDS intrusion detection system (IDS). ADENOIDS takes some architectural inspiration from the human immune system and automates intrusion recovery and attack signature extraction. These features are enabled through attack evidence detection. This IDS is initially designed to deal with application attacks, extracting signature for remote buffer overflow attacks. ADENOIDS is described in this paper and experimental results are also presented. These results show that ADENOIDS can discard false-positives and extract signatures which match the attacks.

1 Introduction

The Internet was designed to be an open and distributed environment with mutual trust among users. Security issues are rarely given high priority by software developers, vendors, network managers or consumers. As a result, a considerable number of vulnerabilities raises constantly. Once explored by an attacker, these vulnerabilities put government, businesses, and individual users at risk [1,2].

Intrusion detection systems (IDSs) are useful tools to improve the security of a computer system and, because of their importance, they have become an integral part of modern network security technology. An IDS acts by monitoring events in a computer system or network, analyzing them for signs of security problems [3]. Several techniques are used to achieve intrusion detection such as expert systems, state transition approaches, statistical analysis, and neural networks [3]. More recently, several approaches based on the immune system were proposed [4,5,6]. Most of these approaches concentrate on building models and algorithms for behavior-based detection.

This paper presents the ADENOIDS IDS which is intended to mimic, mainly at the architectural level, several human immune system features, some of them little explored in other works. Examples of these features are intrusion tolerance, attack evidence detection, automated attack signature extraction and system recovery mechanisms.

One of the most important aspects of this IDS is its assumption that successful attacks are inevitable, and its strongest feature is its ability to deal with such

---

⋆ The authors would like to thank the FAPESP agency for supporting this research.

situation. Note that this is also the case with the vertebrate immune system. Some disease-causing agents are successful in invading the organism and causing harm to it before the immune system can eliminate them. After that, the immune system learns to cope with this type of agent, and some repair strategy is taken to recover the damaged parts. In this way, this IDS is more related to a research in virus identification [7] than previous work in intrusion detection.

ADENOIDS was developed to detect attack evidences in running applications, restore the system after an attack using a file system undo mechanism, and extract the attack signature for remote buffer overflow attacks.

In fact, applications that provide publicly available services have been the most intended targets of attack in the last years [8]. Among several techniques employed to exploit application vulnerabilities, buffer overflow has been one of the most explored [8].

ADENOIDS was tested against two datasets and the experimental results are encouraging. The proposed signature extraction algorithm can find the attack signature and discard candidate signatures which do not correspond to an attack.

In this paper will not be discussed the immune system features and its analogies with security systems, and the reader is referred to [9] for an introduction to these issues.

This paper is organized as follows. Sect. 2 presents an overview of the ADENOIDS IDS. Sect. 3 describes the main implementation aspects of this IDS and experimental results are shown in Sect. 4. Sect. 5 concludes the paper.

2   ADENOIDS Overview

ADENOIDS was designed to monitor a single computer in such way to detect application-level attacks and automate signature extraction for remote buffer overflow attacks. The attack evidences are detected in running process at the system call level and the attack signatures are extracted at the network level. Fig. 1 illustrates the ADENOIDS modules and the communication flow between them. In this figure, a short name for some modules is indicated inside parenthesis.

Some of these modules are well-known intrusion detection building blocks such as the Console, the Data Source and the Behavior-Based Detector. The role of each module is as follows:

- Data Source: is not a module by itself but represents the source of all information needed for the correct IDS working.
- ADCON: is an interface between the IDS and the system administrator.
- ADEID: is responsible for monitoring the computer in search for events that indicate a successful attack. Although all modules are designed to work concurrently, ADEID can initiate an automated response (by activating ADIRA) and the signature extraction process (by activating ADSIG).
- ADIRA: is responsible for restoring the computer system after an attack.
- ADBID: performs anomaly detection by monitoring incoming network traffic in search for candidate attack signatures.

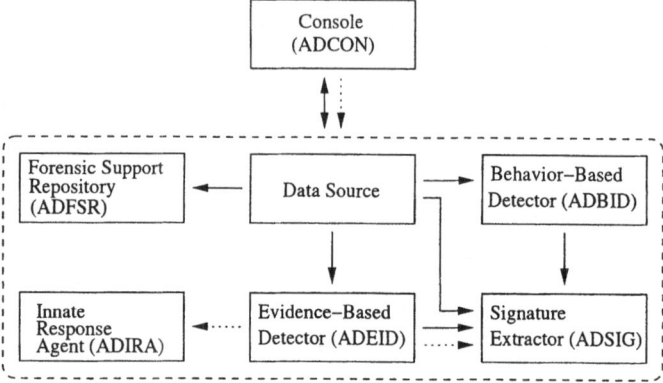

Fig. 1. The ADENOIDS modules. Each module is represented by a solid line rectangle. Solid directed lines indicate information flow and dotted directed lines show control flow. Each flow occurs between two modules or between one module and all other modules of the grouping (represented by a dotted line rectangle)

- ADSIG: analyzes the collected candidate signatures in attempt to: 1) discard false-positives; and 2) extract signatures which match the attack.
- ADFSR: it is only modeled to provide support for manual forensic analysis by preserving data that cannot be corrupted even after a system restore.

The purpose of the signature extraction process is to enable a more efficient detection of this attack in the future by using a signature-based IDS like Snort-inline [10]. This process involves the ADBID and ADSIG modules and a general algorithm for the signature extraction problem is proposed in Sect. 2.1.

2.1 The Signature Extraction Algorithm

This section proposes a signature extraction algorithm which takes inspiration from the negative selection process of the human immune system and it is suitable for general attacks. The algorithm divides the signature extraction into two phases: the search for candidate signatures and the maturation of the candidates.

Unlike other works [4,6] which generate candidate detectors randomly, the proposed algorithm takes advantage from the evidence detection of ADEID and selects anomalous events prior the attack to be the candidates.

The proposed approach seems to be more appropriate for searching good candidates than randomly generation. In fact, the most appropriate use of the negative selection can be as a filter for invalid detectors, and not for the generation of effective detectors [11].

The proposed algorithm is as follows. The input is composed of a real number $p \in \;]0; 1]$, a set E of events prior the evidence detection and a set N of events generated by the computer system during normal working, where $N \cap E = \emptyset$. The output is a set $C \subseteq E$ of events, which are the extracted attack signatures with estimated probability less than p of false-positives occurring during further detection. The steps of this algorithm are as follows:

1. Restore the computer system to a safe state.
2. Select a set C of events to be the candidate signatures, where $C \subseteq E$.
3. $progress \leftarrow 0$.
4. While $progress < \left\lceil \frac{|C|}{p} \right\rceil$ do:
    4.1. Get a new event $n \in N$ during the normal computer system working.
    4.2. For all $c_i \in C$, if c_i **matches** n, then $C \leftarrow C \setminus \{c_i\}$.
    4.3. $progress \leftarrow progress + 1$.
5. Return each signature in C. If $|C| = 0$, return null.

Step 2 involves the search for candidates and Step 4 performs the maturation of the candidates. The system restoration (Step 1) is provided by ADENOIDS through the ADIRA module. The set E comprehends the incoming network traffic prior to ADEID detection and the initial set C is built by collecting the anomalous traffic detected by ADBID from E. Because ADENOIDS focuses overflow attacks, ADBID works by detecting large requests[1] in the network traffic. The set N is built by collecting related incoming network traffic after the system restoration. The matching criterion of Step 4.2 takes into account the size of requests into the network traffic. If a new attack evidence is found during or soon after the signature extraction process, the algorithm is restarted with the initial set C, because this new attack can discard relevant events of the prior attack.

By considering the matching operation to be dominant and m to be the size of the initial set C, the running time of this algorithm is, in the worst case, $O\left(\frac{m^2}{p}\right)$. However, it should be noted that the real execution time is also dependent upon the generation rate of normal events. Therefore, this process may be long and it is not intended to provide a response in real-time.

3   ADENOIDS Implementation

ADENOIDS was implemented in C over the Linux kernel version 2.4.19. All information required by this IDS are distributed in two levels: system calls and network traffic.

The ADCON module is provided through a set of configuration files and a set of log files. The remaining modules are described as follows.

3.1   ADEID

The ADEID module monitors running applications in the search for events which violate pre-specified access policies. Each access policy specifies a set of operations which can be performed by an specific process. The events analyzed by ADEID are the following:

---

[1] The term "request" is was adopted to refer to both application-level command and response.

- Files, directories and links: open, creation, erasing, renaming, truncation and attribute changing (owner, group and permissions).
- Process: creation and execution.
- Kernel modules: creation and deletion.
- Communication: signal sending, TCP connection creation and acceptance, and UDP datagram sending and receiving.

The monitoring policies must be specified obeying the following structure:

```
policy_name[/fully/qualified/program/pathname]
{
    fs_acl { list of pathnames and access permissions }
    can_exec { list of programs which can be executed }
    max_children = maximum number of child process
    can_send_signal = yes | no
    can_manip_modules { list of kernel modules which can be created and deleted }
    connect_using_tcp = yes | no
    send_using_udp = yes | no
    accept_conn_on_ports { list of port ranges which can be used to accept connections }
}
```

Although the system call policies proposed in [12] can be more powerful, the ADEID policies make the specification a simpler task. For building a good monitoring policy it is necessary to know about the Linux file system hierarchy and the main purpose of the application intended to be monitored. ADEID has been used to monitor named, wu-ftpd, amd, imapd and httpd applications for two months. It has demonstrated to be very efficient to detect attacks, being free from false-positives and false-negatives during the tests.

ADEID is implemented as a kernel patch by rewriting some system calls which deal with the monitored events. The policies are read from disk and loaded into kernel memory. Whenever a new process is executed the related monitoring policy is attached to the process, if any is defined.

By detecting attack evidences, ADEID analyzes only successful system calls. This feature—which characterizes the evidence detection, unlike [12]—also helps to reduce the false-positive rate because unauthorized actions will not be analyzed.

Whenever ADEID detects some attack evidence the ADIRA module becomes active by calling a kernel procedure and, after, a SIGUSR1 is sent to ADSIG and information about the attack—current process and violated policy—are also delivered. For testing purposes an user can disable these activation mechanisms.

Preliminary results show that the performance cost imposed by ADEID is imperceptible for users. A general benchmark for the most expensive operation—opening and reading cached files—showed that this cost is, on average, lower than 5% in an Athlon XP 1900+ with 512MB RAM.

3.2 ADBID

ADBID captures packets through pcap library and delivers them to application-specific procedures. An application-specific procedure decodes the related application-level protocol delivering the request to be analyzed.

Because ADENOIDS focuses on signature extraction for remote buffer overflow attacks, ADBID works by building an statistical profile to detect requests whose length are less probable to be found during normal operation and are found in overflow attacks. Actually ADBID detects requests whose length is greater than $\mu + 2s$, where μ is the arithmetic mean of requests length and s is the standard deviation of requests length. The detection procedure and parameters can also be easily changed.

3.3   ADIRA and ADFSR

The ADIRA module is implemented as a kernel patch and works by restoring the computer system after an attack. This restoration is done through the following steps: 1) block all user processes; 2) restore the file system; 3) restart the monitored applications; 4) kill attacked process; and 5) unblock all blocked processes.

The file system restoration is implemented by applying *undo* techniques into any file system which can support both, reading and writing data. This mechanism is activated before the following operations over files, directories and links can be done: creation, erasing, renaming, writing, truncation and attribute changing. For each operation is created an specific undo log. This log holds the necessary information in such way that the operation can be reversed in the future. Redo logs are created by operations done during the undo process. A kernel procedure can be called to request a file system undo or redo up to a defined checkpoint. Actually the checkpoints are inserted automatically during the system startup. A configuration file states what directories are covered by this mechanism.

The undo performance depends on the operation to be done. Appending bytes to a file adds only a fixed-size log. File truncation requires to read the bytes to be truncated and to write them into the log file. File erasing and the overwrite operation are also expensive. It should be noted that the default configuration file includes vital directories which are rarely modified and, therefore, the imposed cost is very acceptable.

The ADFSR module implements an interface to provide step-by-step redo by calling redo procedures after a system reboot. In this way, it is enabled the manual analysis of all file system events done during an attack, if a system administrator or forensics specialist want to do that.

3.4   ADSIG

This module works exactly as proposed in Sect. 2.1. Once activated by ADEID, ADSIG reads the delivered information about the violated policy (policy name and related process) and begins a new signature extraction by loading the candidate signatures into a proper data structure. By default, the set E is built by selecting requests for the last 24 hours prior to ADEID detection. Therefore the initial set C contains the candidates for the last 24 hours.

A matching criterion was chosen to discard the candidates which are most probable to be found during normal operation. In the actual ADSIG implementation all candidates whose length is lower than or equal to a normal request length are discarded. This works well for buffer overflow attacks because if a request is normal—and probably does not overflow a buffer size—a candidate whose length is at most equal the request length probably will not overflow this buffer size too and can also be considered normal[2]. At the end of the signature maturation process ADSIG outputs the extracted signatures.

3.5 ADenoIdS Self-Protection

It is implemented a simple self-protection mechanism, which consists of denying access to ADenoIdS modules, data and configuration files from the processes being monitored. In this way ADenoIdS considers that all possible attack target—usually server applications—must be monitored.

4   Experimental Results

This section presents experimental results obtained by testing the ADenoIdS IDS. The main objectives of the tests were to evaluate the ability of evidence-based detection, behavior-based detection and signature extraction mechanisms.

The test system were customized from a Red Hat Linux 6.2 to provide vulnerable named, wu-ftpd, amd and imapd applications over the Linux kernel 2.4.19. All these applications can be successfully attacked through buffer overflow exploits collected around the world and these attacks are launched from an external machine.

The ADEID, UNDOFS and ADFSR modules have been used for two months whereas the ADBID, ADIRA and ADSIG modules were tested for two weeks.

Each ADEID monitoring policy was built in two steps by observing the reported violations:

1. Initial policy establishment. This step spent about half an hour of intensive work.
2. Policy refinement. This step spent about two days of sparse work.

After these steps, the ADEID demonstrated to be very efficient to detect attacks, being free from false-positives and false-negatives during the tests.

The complete ADenoIdS IDS was tested against the 1999 Darpa Offline Intrusion Detection Evaluation dataset—available at http://www.ll.mit.edu/IST/ideval/index.html—and against a dataset collected at our research laboratory (LAS dataset).

The 1999 Darpa Offline IDS Evaluation dataset is composed of training and test datasets which include network traffic data, event logs and other audited

---

[2] A more complete analysis should consider a different buffer size for each request type.

data. Several attack types are present in this evaluation, including buffer overflow attacks. ADENOIDS was tested only against named buffer overflow attacks because this dataset does not provide training data for the imapd overflow and the vulnerable sendmail daemon was not available. To compensate this, some buffer overflow attacks in the test data for the wu-ftpd daemon were inserted. Because ADENOIDS analyzes only events produced by one host the test was done considering the network traffic destined to hosts separately.

The LAS dataset was collected under normal conditions at our external DNS server during 43 days. This dataset was chosen by two factors: 1) named is a very important application and often vulnerable; and 2) DNS queries can be replayed easily. This dataset was first analyzed before the test phase and was verified to be free of attacks.

Table 1 summarizes the results for the 1999 Darpa Offline IDS Evaluation and the LAS datasets. An appropriate label is placed before the beginning of each dataset results. The first column describes the target daemon being considered and the second column shows the average number of requests per day to the considered target host in the whole dataset. The third column presents the number of requests spent in the ADBID training. The fourth column shows the number of requests prior to the attack which were captured in the last 24 hours ($|E|$). Each test was performed by considering an exclusive set of prior events. For the LAS dataset, it was used a fixed number of 10000 requests prior to the attack, exceeding the average number of requests per day. The fifth column shows the number of ADBID candidate signatures extracted (the initial $|C|$) from each of these sets. The sixth column presents the number of requests required by the complete signature extraction process. In some tests the final ADSIG output can be known by using only 1000 normal events in the maturation process, but to satisfy the p parameter (indicated inside parenthesis) the process must be continued. The seventh column indicates the number of requests outputted by ADSIG at the end of the maturation process. Some attacks can present more than one attack signature and the eighth column indicates if the main overflow request is found by ADSIG. The last column shows the number of false-positives after the signature extraction process.

The ADBID module was very efficient to found the candidate signatures. Its detection was capable of selecting fewer candidates and the main buffer overflow request was always inside the candidates' set.

The ADSIG module has also demonstrated to be very appropriate to discard erroneous candidates. The overflow requests were the only candidates at the end of the signature extraction process in seventeen out of twenty one attacks analyzed. The wu-ftpd false-positives were probably produced due to a fewer number of requests in the dataset and, consequently, in the maturation process. The first named false-positive was not also an overflow, but it looks like a malformed host name query.

Although some false-positives can happen, the signature generation algorithm claims that extracted signatures which are valid requests will be probabilistically rare events in further detection.

Table 1. Experimental results for the Darpa and LAS datasets

Target Daemon	Average # of Reqs/Day	# Reqs Training (24 hours)	# Reqs (24 hours)	# Candidates (24 hours)	Required # of Normal Events	# Outputted Requests	Signature Found?	# False-Positives
Experimental results for the 1999 Darpa Offline IDS Evalutaion dataset								
named	174559	50000	58942	6	10000 ($p = 0.0001$)	1	yes	0
named	174559	50000	267336	11	10000 ($p = 0.0001$)	1	yes	0
named	174559	50000	266995	8	10000 ($p = 0.0001$)	1	yes	0
wu-ftpd	1575	2000	1873	29	4342 ($p = 0.003$)	13	yes	1
wu-ftpd	1575	2000	1603	36	4008 ($p = 0.003$)	12	yes	0
wu-ftpd	827	2000	918	22	3340 ($p = 0.003$)	10	yes	0
wu-ftpd	827	2000	795	22	3674 ($p = 0.003$)	11	yes	1
wu-ftpd	761	2000	670	24	4008 ($p = 0.003$)	12	yes	0
wu-ftpd	761	2000	1097	38	4008 ($p = 0.003$)	12	yes	0
Experimental results for the LAS dataset								
named	8590	40922	10000	25	10000 ($p = 0.0001$)	1	yes	0
named	8590	40922	10000	7	10000 ($p = 0.0001$)	1	yes	0
named	8590	40922	10000	14	21099 ($p = 0.0001$)	2	yes	1
named	8590	40922	10000	20	10000 ($p = 0.0001$)	1	yes	0
named	8590	40922	10000	18	10000 ($p = 0.0001$)	1	yes	0
named	8590	40922	10000	15	10000 ($p = 0.0001$)	1	yes	0
named	8590	40922	10000	10	10000 ($p = 0.0001$)	1	yes	0
named	8590	40922	10000	18	10000 ($p = 0.0001$)	1	yes	0
named	8590	40922	10000	20	20000 ($p = 0.0001$)	2	yes	1
named	8590	40922	10000	25	10000 ($p = 0.0001$)	1	yes	0
named	8590	40922	10000	20	11640 ($p = 0.0001$)	1	yes	0
named	8590	40922	10000	23	30955 ($p = 0.0001$)	1	yes	0

5   Conclusions and Future Works

This paper presents the ADENOIDS IDS which takes inspiration from the human immune system. This IDS is originally intended to deal with application attacks, extracting attack signatures for remote buffer overflow attacks.

ADENOIDS was tested against the Darpa 1999 Offline IDS Evaluation dataset and against another collected dataset. The experimental results presented were very encouraging. The proposed signature extraction algorithm can find the attack signatures and discard candidate signatures that would only produce false-positives.

Future work includes new tests considering other vulnerable applications, correlation of subsequent attacks, and an study about ADENOIDS generalization capability.

Although the ADENOIDS signature extraction mechanism covers only buffer overflow attacks it is extensible to other classes of attacks. The ideas described here can also have straight applications in other areas, such as honeypot automation and forensic analysis.

References

1. Garfinkel S., Spafford G.: Practical UNIX & Internet Security. 2nd edn. O'Reilly and Associates (1996)
2. Pethia, R.: Computer Security. Cert Coordidantion Center. Available on the web at http://www.cert.org/congressional_testimony/Pethia_testimony_Mar9.html (2000)
3. Bace, R.: Intrusion Detection. 1st edn. Macmillan Technical Publishing (2000)
4. Hofmeyr S., Forrest, S.: Architecture for an Artificial Immune System. Evolutionary Computation, Vol. 8 (2000) 443-473
5. Dasgupta, D.: Immunity-Based Intrusion Detection System: A General Framework. Proceedings of the 22nd National Information System Security Conference (1999) 147-160
6. Kim, J., Bentley, P.: An Artificial Immune Model for Network Intrusion Detection. Proceedings of the 7th European Congress on Intelligent Techniques and Soft Computing (1999)
7. Kephart, J.: A Biologically Inspired Immune System for Computers. Artificial Life IV: Proceedings of the Fourth International Workshop on the Synthesis and Simulation of Living Systems (1994) 130-139
8. CERT Coordination Center: CERT Summaries 1995-2003. Available on the web at http://www.cert.org/summaries (2004)
9. de Castro, L.N., Timmis, J.: Artificial Immune Systems: A New Computational Intelligence Approach. 1st edn. Springer-Verlag (2002)
10. Haile, J., McMillen, R.: Snort-inline tool. Available on the web at http://project.honeynet.org/papers/honeynet/tools (2004)
11. Kim, J., Bentley, P.: Evaluating Negative Selection in an Artificial Immune System for Network Intrusion Detection. Proceedings of the Genetic and Evolutionary Computation Conference (2001) 1330-1337
12. Provos, N.: Improving Host Security with System Call Policies. Proceedings of the 12th USENIX Security Symposium (2003)

QoS-Differentiated Secure Charging in Ad-Hoc Environments*

João Girão[1,2], João P. Barraca[1,2], Bernd Lamparter[1], Dirk Westhoff[1], and Rui Aguiar[2]

[1]NEC Europe, Networking Labs, Germany
{joao.girao, bernd.lamparter, dirk.westhoff}@netlab.nec.de
[2]Universidade de Aveiro, D.E.T.-I.T., Portugal
{jpbarraca@av.it.pt, ruilaa@det.ua.pt}

Abstract. In order to keep up with new networking needs, it is necessary to adopt mechanisms for charging network usage in a universal way. The Secure Charging Protocol (SCP) aims at answering this complex authentication, authorization, accounting and charging (AAAC) problem. SCP fits business models especially adequate for ad-hoc networks. This document discusses SCP as a possible solution to the AAAC problems in MANETs and presents the improvements made to this protocol in terms of Quality of Service (QoS). An implementation of this protocol on PDAs and the results achieved are discussed.

1 Introduction

Wireless technologies have broadened the concept of networks to shared uncontrolled mediums where all are responsible for a little part of the whole. Words like "hotspot" and "mesh" have become popular in marketing and, step by step, the average user becomes aware of the new possibilities new wireless technologies offer.

Many researchers have exploited the advantages of this evolution by considering decentralised scenarios, where any node is not only a client but also a server. The concept of Mobile Ad-hoc networks (MANET) [1] became a major research area under this framework. MANETs can range from networks made only of nodes that forward packets for each other - and therefore have no need for a centralised structure - until networks with special static points of access to a main network or even the Internet. In every case, MANETs are always highly volatile networks which support high mobility, decentralised routing algorithms, and wireless technologies. It is not surprising thus that MANETs have provided new opportunities for research on technologies and business (and AAAC) models. AAAC architecture concepts [2] can be applied to MANETs with slight modifications: distributed accounting must be implemented, as all nodes are possible accounting and data collection points; and authentication and authorization must be considered case by case.

MANETs flexibility is attracting network service providers to incorporate ad-hoc technology support in their products. MANETs seem to provide mechanisms to

---

* This work is in part supported by the EU Framework Programme 6 for Research and Development Daidalos (IST-2202-506997).

J.N. de Souza et al. (Eds.): ICT 2004, LNCS 3124, pp. 1093–1100, 2004.

extend connectivity range, reduce node's power consumption and increase both node's mobility and failure tolerance - all without a significant investment from the service provider. There are two problems, though. Ad-hoc technologies require a critical mass of well-behaved nodes willing to forward other's traffic. Moreover, the ISP also expects to charge the users accessing the network, in a parallel way to traditional connections. Both problems create the need to develop new charging paradigms and business models, capable of securely charging over "trust-no-one" environments and motivating node's participation in ad-hoc networks.

This document introduces a business model adequate to traffic charging in ad-hoc networks, while increasing the participation level by motivating nodes to forward traffic. A protocol appropriate to this business model, the SCP protocol, and a novel implementation supporting traffic differentiation, will be presented. We will discuss issues concerning security and key sizes and present experimental results obtained. Finally, in the end, we will present the conclusions reached.

2   The SCP Protocol

Charging is clearly associated with business models, and ad-hoc environments present an opportunity for the appearance of novel business models. In the future, ideally ISP revenues will be maximised by increasing cooperation and promoting communication between users, instead of the traditional Access services.

The implementation of this type of business concept, where the operator will benefit by the users' willingness to promote communication (peer-to-peer services), has some challenges. Especially in the cases of small devices in ad-hoc environments, it is not usually to the owner's advantage to loose processing and bandwidth by forwarding other people's packets. Therefore, a user will want his packets forwarded, but not to forward other's traffic. By providing "incentives", novel business models [3] intend to reward users with money for the (forwarding) service they provide. The more packets a node forwards the more money it will receive at the end. This money can then be reused in its own usage of the cooperative service, to have his packets forward by other nodes, or can become a net profit at the end of the contract.

Two price factors, P+ and P-, are defined, to reflect the ratio between the money a user has to pay for sending packets and the amount he receives for forwarding. This concept is the basis for a service an ISP can provide anywhere a MANET can be deployed, using the ISP access points. The actual protocol that has been developed to support this approach is the Secure Charging Protocol [3]. SCP is adequate for the Business Models suggested, where the service provided is the inter-cooperation, the forwarding of traffic by multiple nodes. Notice that other alternative solutions based on incentives have been proposed (e.g. [4, 5]), with different trade-offs in terms of scenarios and required security levels.

SCP is an AAAC protocol that focuses on securely retrieving accounting information for traffic flows, while retaining relevant information associated with the routes that were taken by each of the packets. The protocol aims to handle situations where the receiving node (a paying node) cannot be trusted to report the correct traffic information, and provide per-packet assurances, if required.

To solve the questions associated with AAAC problems in this environment, a set of cryptographic primitives like signing, verifying and keyed hash chains, were used within the protocol to create digital signatures. These primitives are essential to guarantee the authenticity of the accounting information provided. This accounting information is sent by the last forwarding node, which can accurately determine which packets were received by the receiving end. This node has the interest of providing as much information as possible, as it will benefit (economically) of this (it is the last node in the communication path that will forward data). SCP further defines an Access Router, which is trust-worthy, and will collect all relevant information. Note that this "Access Router" will often be the router interconnecting the ad-hoc network to a physical infrastructure (hotspot), but it is not necessarily so.

The protocol operation (Fig. 1) is divided into three phases: registration, forwarding and charging phases.

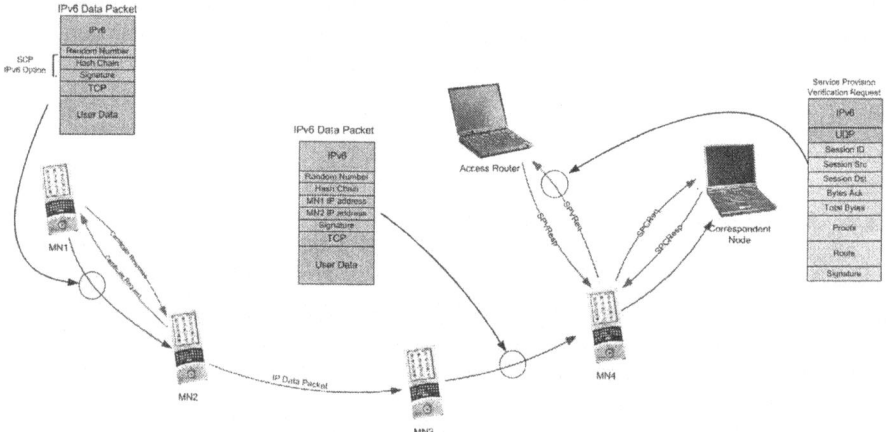

Fig. 1. Diagram showing operation of SCP.

a) During the registration phase, nodes acquire their "identity" within the domain. An Access Request is sent to the Access Router containing relevant information on the node's domain, plus a username and password. In response, the Access Router returns over an encryption tunnel the node's certificate, the node's private key, a shared key with the access router and charging information. This information can be used to authenticate node-related information.

b) During data transfer, when a node needs to send out data, it must sign it with its own private key so that the information can be verified as originating from that particular node (this is of particular importance to avoid charging wrong people). Together with the payload, a hash chain is initiated using a random number and a shared key so that the route can be confirmed at Access Router (AR) level.

Each time the packet hops through a node, the data is confirmed as being correct and authentic and the hash chain is further increased, by hashing the last value with the shared key of each node, before the packet is forwarded. This way the node is not wasting processing time on invalid packets, and introduces a confirmation of his activities (forwarding) inside the packet. The freshness of this hash is guaranteed by a random number seed introduced by the sender.

c) The last forwarding node is responsible for the accounting. It keeps database information on packets, routes and hash chains and periodically sends this information to the Access Router when reachable, during the accounting phase.

The accounting information is confirmed with the receiving node and both his answer and the information held by the last forwarding node are sent to the accounting server. By reproducing the order of hashing, together with the shared key, in the AR, the end result calculated in the server is compared to the value provided by the network, and thus provides a path confirmation mechanism. Having verified the path reported, the AR is then able to derive credits and charges to issue to all nodes in the communication path.

These concepts have been extended to support QoS. Making sure that a node complies with given QoS parameters is usually difficult, and is especially hard in MANETs, where nodes have different capabilities, and links have varying performance. In this context, we handle QoS simply as class differentiation: each node supports a finite number of classes (queues) that receive different processing times. The definition and parameterization of these classes is an issue for the operator.

A packet is transmitted associated with a class. The QoS class information is marked in the flow label of the IPv6 packet and is confirmed because it is part of the signed data of each packet. This creates a register of the QoS class provided for the packet by the nodes. A queue exists for each class. Service differentiation is provided by sorting packets to different QoS classes (different queues). The heaviest processing functions (in this case, the signature verification) are then used within a weighted round robin (WRR) scheme to distribute processing time unevenly, providing differentiated services to the QoS classes. Note that more advanced scheduling disciplines could be chosen, at the expense of larger implementation complexity.

During the charging phase, the AR derives the credits and charges taking in consideration the QoS reported information coming from the last forwarding node.

3 Implementation

SCP was implemented in a very common PDA, the Sharp Zaurus, running the Linux operating system. The implementation was performed in such a way that it can run on most Linux platforms and on different architectures (currently i386 and StrongArm). One of the main objectives of the implementation was to have a threaded modular prototype that could easily be extended, having pieces of code replaced without affecting the overall performance and public interface. The modular approach allows for any extension to be both simple and less time consuming because only a particular piece of the code has to be changed. Threading was implemented in order to provide the SCP implementation with flexibility, non-blocking asynchronous tasks and the possibility of service differentiation on a per-packet basis.

The SCP implementation consists of two layers: the SCP finite state machine, in the high level, and the network interface: the Packet Handler (Fig. 2). The Packet Handler is the basis for SCP. All SCP verification functions and packet differentiation are done at this level. Functions are then called depending on the packet type (class) from the finite state machine which characterizes SCP. There are three main threads in the Packet Handler. Two belong to the Packet Manager which interfaces directly with the network. One is for packet reading (referenced as Read in the schematic) and the other for sending (represented as Write). These processes are asynchronous to the rest of the program; and depend only on the state of the packet. The other thread is the WRR mechanism that distributes the processing power by calling the verification function as many times for each queue as it has been configured to (the Scheduler). Naturally, empty queues' timeslots will be used by pending packets in other queues.

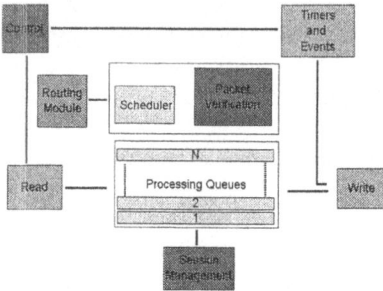

Fig. 2. Packet Handler Implementation and Key Global Functions

The SCP state machine implementation is divided into two classes that share some of the code and belong independently to the Node or the Access Router. (Note that this state machine is quite complex, and thus is not completely represented on the figure). At this level, all the functions regarding SCP verification functions, storing of data and cryptographic methods are grouped so they can be easily accessed from the underlying layer. This is represented as the Verification module in Fig. 3.

At this same level is another thread, the Timer thread, which is responsible for timed events such as a packet being re-sent or the cleaning of some stale data.

The implementation also includes Session Management capabilities. It includes a monitoring server that reads on a specified UNIX domain socket for information requests from other programs in the same machine mainly for debugging and demonstration purposes. After a program connects to the socket, it can request information about general variables, traffic statistics, and also details about the active sessions and known nodes. Information like traffic statistics or general information can be important for the end user to access, while the details about the active sessions and the known nodes are somewhat more important for debugging and testing purposes of the protocol.

A graphical interface running in both ARM and x86 architectures was developed to exploit this monitoring facility. This graphical interface is implemented using the QT libraries distributed by Trolltech Corporation [8] allowing a seamless integration both with the OPIE environment of the OpenZaurus distributions, and the QT based graphical environment distributed with the majority of the Linux distributions.

Notice that the modularity characteristics of this software allows for different verification or scheduling algorithms to be supported with minimal effort. Furthermore, the interfacing with the ad-hoc routing module is quite flexible, and different ad-hoc routing mechanisms can be easily supported. Moreover, run-time change of the routing protocol (because of node movement across different networks, e.g.) is supported. Currently the implementation supports AODV6 and static routing.

The choice of the signature algorithm will greatly influence the overall performance of any SCP implementation. Moreover, characteristics of ad-hoc networks like multi-hopping, reduced bandwidth, processing power and battery, and high distrust on the intentions of the nodes, impose some restrictions on the level of security and, therefore, in the signature algorithm to be used. Three key points needed to be evaluated: security level, processing power and network overhead. Security level should be high enough to avoid impersonation and guaranty non-repudiation of the network traffic. These requirements are related to the key and algorithm to be used and also to the duration of the keys. If the ISP generates a new certificate every hour, key sizes can be very small. This may be not generally appropriate to all ad-hoc networks because the network overhead to exchange certificates could be too high in some networks. We supposed that certificates can have a high duration, and were to be secure enough to comply with the security requirements.

Different keying algorithms will have different signing and verifying times (Table 1). Specially, RSA and ECDSA are very different in terms of timings. The following table compares the time, in milliseconds, needed to perform a sign or verify operation using RSA or ECDSA. These tests [6] were executed in a PDA equipped with a StrongARM 206Mhz processor using the standard OpenSSL library [9].

Table 1. Comparison between ECDSA and RSA.

Algorithm	RSA		ECDSA	
Key size	704	1024	131	163
Sign (ms)	32.4	78	22.1	28.8
Verify (ms)	2.5	4.3	43.4	55.9

According to [8] if we consider an average number of hops equal or lesser than 5 and a level of security equivalent to an RSA-1024, ECDSA proves to be the most appropriate choice. However if the key length is below the previous value, RSA may be a better choice to maximise implementation performance.

RSA and ECDSA key sizes are very different. ECDSA provides a much more compact representation of the key regarding to the same security level, and thus is the most appropriate choice for the signature algorithm in order to minimise network overhead. A smaller key size also implies smaller certificates and consequently less memory occupied at each node by other nodes certificates.

After comparing RSA with ECDSA, ECDSA was found to provide better security, performance and overhead ratios than RSA, when considering the target environments of our SCP implementation. Final choices were of a security level related to an RSA key with 1024 bits or a security-equivalent 163 bits ECDSA key [7].

4   Implementation Results

We used a highly optimised implementation [6] of ECDSA which is capable of performing signing and verifying operations about 3.2 times faster than the OpenSSL implementation as in the CVS snapshot 20021202.

Our test scenario consisted in four nodes creating a bidirectional flux of packets with different packet rates, different packet sizes, different rates of packet verification at each node, and using a key size of 163 bits. The intermediate nodes are Zaurus PDAs and edge nodes are Pentium 500Mhz processors. With these scenarios we could effectively evaluate the protocol's impact in the round-trip delay of a bidirectional connection like a VoIP communication, and QoS differentiation behaviour.

QoS differentiation was quite simple to demonstrate, and performed according with the expected WRR performance. Processing power was scheduled asymmetrically by the four queues implemented according to the weights allocated.

Other measures determined the benefit of verifying only a percentage of the packets forwarded. Fig. 3 shows the round-trip delay without SCP, SCP with verification of all packets crossing a node, and SCP with verification of half of the packets crossing the node. This last approach should not introduce many security problems, since control packets are all verified and, statistically all packets have a high probability of being verified along the route.

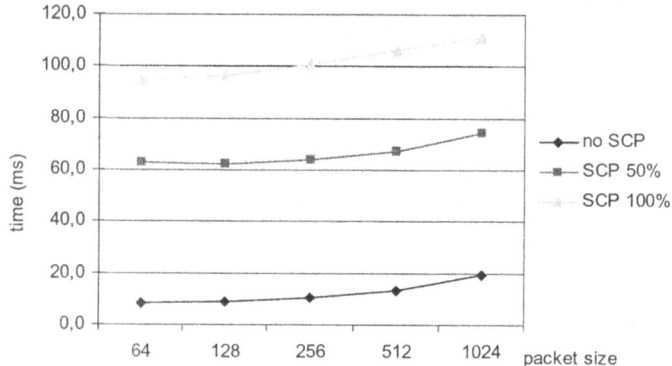

Fig. 3. Round trip delays with different verification ratios

The round-trip delay results show a constant increment on the measured time in function of the packet size. The delay difference between packets without using SCP in the network is the same, only added by an additional delay. This delay is mainly caused by the cryptographic functions verify and sign. (Measurements shown were taken for a single QoS class).

We identified three major functions introducing delay: SCP processing, cryptographic verify and cryptographic sign. Taking the case of a packet with 256 bytes, and doing signature verification in all nodes, the round-trip delay increased from ~10ms to ~100ms. Considering that nodes perform 6 verifications and 2 signatures in a round-trip, Table 2 shows the impact of each component in the overall performance of the protocol, along the different nodes.

It is clear that verification is the key function in this implementation. The choice of a key length appropriate to the security requirements from a specific scenario plays a major role in maximising protocol performance – and in our case we have a 163-bit ECDSA, usually regarded as life-time security. Also, the use of dedicated cryptographic hardware integrated in the mobile nodes, a trend already present in many wireless devices, could push the security of SCP even further by allowing these very secure keys without significant penalty loss, and minimising battery consumption.

Table 2. Percentage of CPU time at each node by function

	MT	**FW Node**	**Third**	**CN**
SCP	0,8%	1,6%	1,6%	0,8%
Sign	6,0%	0,0%	0,0%	6,0%
Verify	13,9%	27,8%	27,8%	13,9%

5 Conclusion

The SCP protocol can perform secure charging in ad-hoc networks without relying on any centralised infrastructure to account traffic, and considering that the nodes are not trusted. This allows the implementation of novel business approaches, promoting users' willingness to cooperate and benefiting from peer-to-peer trends.

The implementation described offers a secure and modular implementation of the Secure Charging Protocol, able to evaluate the current mechanisms offered by the protocol specification, and including service differentiation. It also provides a stable base for future development, evaluation and testing of new features and proposals. In these regards, the SCP implementation has proved to outperform the expectations.

The key length is very important to determine the security achieved and also the performance penalty imposed with the charging process. The use of dedicated cryptography hardware, like FPGA technology, could greatly improve the results obtained and reduce dramatically packet delay in the case of very long key lengths.

References

[1] S. Corson and J. Macker, "Mobile Ad hoc Networking (MANET): Routing Protocol Performance Issues and Evaluation Considerations", RFC 2501, January 1999

[2] B. Aboba, et al., "Criteria for Evaluating AAA Protocols for Network Access", RFC 2989, November 2000

[3] B. Lamparter, K. Paul, D. Westhoff, "Charging Support for Ad Hoc Stub Networks", In Elsevier Journal of Computer Communications Special Issue on Internet Pricing and Charging: Algorithms, Technology and Applications, Elsevier Science, Aug. 2003.

[4] Sheng Zhong, Jiang Chen, Yang Richard Yang "Sprite: A Simple, Cheat-Proof, Credit-Based System for Mobile Ad-Hoc Networks", In Proceedings of IEEE Infocom 2003, San Francisco, USA

[5] Naouel Ben Salem, et al, "A Charging and Rewarding Scheme for Packet Forwarding in Multi-hop Cellular Networks", ACM MobiHoc '03, 2003.

[6] Ingo Riedel "Security in Ad-hoc Networks: Protocols and Elliptic Curve Cryptography on an Embedded Platform", Diploma Thesis, Ruhr-Universität Bochum, March 2003.

[7] A.J. Menezes, P. C. van Oorschot and S. A. Vanston, "Handbook of Applied Cryptography", CRC Press, 1996

[8] Web: Trolltech Creator of QT, Trolltech corp, http://www.trolltech.com, as in 14-10-2003.

[9] Web: OpenSSL Project, http://www.openssl.org, as in 14-10-2003.

Multi-rate Model of the Group of Separated Transmission Links of Various Capacities*

Mariusz Głąbowski and Maciej Stasiak

Institute of Electronics and Telecommunications
Poznań University of Technology
ul. Piotrowo 3A, 60-965 Poznań
Telephone: +48 61 8750801, Fax: +48 61 6652572
{mglabows,stasiak}@et.put.poznan.pl

Abstract. In this paper it is proposed an approximate recurrent method of calculations of the occupancy distribution and the blocking probability in the *generalised model of the limited-availability group*, i.e. the group which consists of separated transmission links of various capacities. The group is offered a mixture of different multi-rate traffic streams. The results of the analytical calculations of blocking probability are compared with the results of the simulation of the limited-availability groups.

1 Introduction

The analysis of Broadband Integrated Services Digital Networks (B-ISDN) with virtual-channel connections, which guarantee Quality of Service parameters for calls of many different traffic classes, indicate the necessity of elaboration of effective methods of analysis and designing of links and nodes of broadband networks. One of the most effective methods of analysis of such systems are the methods which use the concept of the so-called *equivalent bandwidth* [1]. The assignment of several equivalent bandwidths to the Variable Bit Rate sources enables the evaluation of traffic characteristics of links and nodes in B-ISDN by means of multi-rate models worked out for the multi-rate circuit switching [1,2].

The basic link model in the traffic theory is the *full-availability group* which is a discrete link model that uses complete sharing policy [3,4]. However, from the viewpoint of realisation of groups of links of broadband networks the *limited-availability groups* (LAG) are more important than the full-availability ones. LAG is a discrete model of a system of separated transmission links. The limited-availability groups were the subject of many professional analyses, e.g. [5,6,7,8,9]. However, in all the known published papers only the so-called *basic model of the limited-availability group* (BMLAG), i.e. the group consisting of several links of identical capacities has been analysed. This model is appropriate for narrowband networks in which groups of PCM links are considered. In broadband networks, groups usually consist of links of various capacities (e.g. physical links in ATM networks include a number of virtual paths of various bandwidths).

---

* This work was supported by the Polish Committee for Scientific Research Grant 4 T11D 020 22 under Contract No 1551/T11/2002/22

J.N. de Souza et al. (Eds.): ICT 2004, LNCS 3124, pp. 1101–1106, 2004.
© Springer-Verlag Berlin Heidelberg 2004

In the paper a new method of blocking probability calculation in the so-called *generalised model of the limited-availability group* (GMLAG) with multi-rate traffic is proposed. The proposed method is a generalisation of the method quoted in [9] for blocking probability calculation in BMLAG. Following the proposed conception, calculations of blocking probability in GMLAG consists in approximating a multi-dimensional service process (occurring within the system) by one-dimensional Markov chain which is characterised by a product form solution and can be described by the generalised Kaufman-Roberts recursion [9,10].

The remaining part of this paper is organised as follows. In Section 2 BMLAG is analysed. In Section 3 the proposed method of blocking probability calculation in GMLAG is described. In Section 4 the results of analytical calculations of limited-availability groups are compared with the results of simulation. Section 5 concludes the paper.

2   Basic Model of the Limited-Availability Group

Let us consider the basic model of the limited-availability group, i.e. the system composed of k identical separated transmission links (Fig. 1). Let us assume that the system services call demands having an integer number of the so-called Basic Bandwidth Units (BBU)[1] and that each of links of the group has the capacity equal to f BBU's. Thus, the total capacity of the system is equal to $V = kf$. The system services a call – only when this call can be entirely carried by the resources of an arbitrary single link. The group is offered M independent classes of Poisson traffic streams having the intensities: $\lambda_1, \ldots, \lambda_M$. A class i call requires t_i BBU's to set up a connection. The holding time for calls of particular classes has an exponential distribution with the parameters: $\mu_1, \ldots, \mu_M$. Thus, the mean traffic offered to the system by the class i traffic stream is equal to $a_i = \lambda_i / \mu_i$.

In [9] an approximate method of blocking probability calculation in BMLAG was proposed. According to this method, the occupancy distribution in the system is determined by the generalised Kaufman-Roberts (KR) recursion [10,11,12]:

$$nP(n) = \sum_{i=1}^{M} a_i t_i \sigma_i (n - t_i) P (n - t_i), \tag{1}$$

where $P(n)$ is the state probability, i.e. the probability of an event that there is n busy BBU's in the system and $\sigma_i(n)$ is the conditional probability of passing between the adjacent states of the process. The probability $\sigma_i(n)$ is calculated on assumption that in the considered group n BBU's are busy [9].

Let us designate the number of all the possible arrangements of n busy BBU's in the group by symbol $\alpha(n)$. According to [9], the problem of determination of the number $\alpha(n)$ in LAG can be reduced to determination of the number of arrangements of $x = V - n$ free BBU's in k links, each of which has the capacity

---

[1] It is assumed that BBU is the greatest common divisor of the equivalent bandwidths of all call streams offered to the system [1].

limited to f BBU's. The number of the arrangements can be calculated by means of combinatorial function $F(x, k, z)$ which let us determine the number of arrangements of x free BBU's in k links which capacities are limited to z BBU's:

$$F(x, k, z) = \sum_{i=0}^{\lfloor \frac{x}{z+1} \rfloor} (-1)^i \binom{k}{i} \binom{x + k - 1 - i\,(z + 1)}{k - 1} . \tag{2}$$

On the basis of (2), the parameter $\alpha(n)$ can be be written as follows:

$$\alpha(n) = F(V - n, k, f) . \tag{3}$$

Subsequently, let us designate the number of events for which the blocking state occurs (the state in which each link of the group contains less than t_i free BBU's for servicing class i call) by symbol $\beta_i(n)$. The number of such events can be calculated as the number of arrangements of $V - n$ free BBU's in k links, each of which has the capacity limited to $t_i - 1$ BBU's. According to (2), we obtain:

$$\beta_i(n) = F(V - n, k, t_i - 1) . \tag{4}$$

Thus, the conditional probability of blocking for class i traffic stream in the limited-availability group in which n BBU's are busy can be expressed as follows:

$$\gamma_i(n) = \beta_i(n)/\alpha(n) . \tag{5}$$

At last the conditional probability of passing $\sigma_i(n) = 1 - \gamma_i(n)$ can be determined:

$$\sigma_i(n) = [F(V - n, k, f) - F(V - n, k, t_i - 1)] / F(V - n, k, f) . \tag{6}$$

Having determined all the probabilities $\sigma_i(n)$, we can calculate the occupancy distribution $P(n)$ and subsequently blocking probability for class i calls [9]:

$$b(i) = \sum_{n=V-k(t_i-1)}^{V} P(n)[1 - \sigma_i(n)] . \tag{7}$$

3   Generalised Model of the Limited-Availability Group

Let us consider now GMLAG, i.e. the system composed of links of q types. Each link type is explicitly featured by the following parameters: k_q – the number of links of type q; f_q – capacity of links of type q (Fig. 2). Total capacity of the group is equal to $V = \sum_{s=1}^{q} k_s f_s$. The considered system services a call only when this call can be entirely carried by the resources of an arbitrary single link.

In this paper it is assumed that the occupancy distribution in GMLAG is determined by the generalised KR recursion (1) in which probabilities $\sigma_i(n)$ are calculated according to the method proposed below.

Firstly let us consider the limited-availability group composed of links of two types ($q = 2$). In the considered case, the total capacity of the group V can

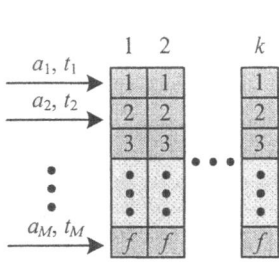

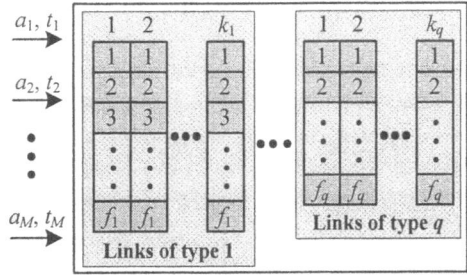

Fig. 1. Basic model of the limited-availability group

Fig. 2. Generalised model of the limited-availability group

be expressed as a sum of capacities of links of the first and the second type: $V = V_1 + V_2$, where $V_1 = k_1 f_1$, $V_2 = k_2 f_2$. The problem of determination of all the possible arrangements of x free BBU's in the group can be considered in two stages. In the first stage we determine the number of all the possible ways of division of x BBU's into x_1 BBU's in links of the first type and $x - x_1$ BBU's in links of the second type whereas in the second stage we determine the number of arrangements of a given number of BBU's in links of the same type, i.e. the number of arrangements of x_1 BBU's in links of the first type and $x - x_1$ BBU's in links of the second type. The number of arrangements in the second stage can be calculated by (2). So, the number of all the possible arrangements of x free BBU's in the group composed of links of two types can be expressed as follows:

$$F(x, k_1, k_2, f_1, f_2) = \sum_{x_1=0}^{x} F(x_1, k_1, f_1) F(x - x_1, k_2, f_2) \ . \tag{8}$$

Continuing our considerations we can determine the number of all the possible arrangements of x free BBU's in the group composed of links of q types:

$$F(x, k_1, \ldots, k_q, f_1, \ldots, f_q) = \sum_{x_1=0}^{x} \sum_{x_2=0}^{x-x_1} \cdots \sum_{x_{q-1}=0}^{x-\sum_{r=1}^{q-2} x_r} F(x_1, k_1, f_1) \cdot F(x_2, k_2, f_2) \cdot$$

$$\cdots \cdot F(x_{q-1}, k_{q-1}, f_{q-1}) \cdot F(x - \sum_{r=1}^{q-1} x_r, k_q, f_q) \ . \tag{9}$$

At last, $\sigma_i(n)$ in GMLAG can be calculated by the modified (6):

$$\sigma_i(n) = 1 - [F(V - n, k_1 \ldots k_q, t_i - 1)/F(V - n, k_1 \ldots k_q, f_1 \ldots f_q)] \ . \tag{10}$$

Taking into account all the blocking states in the limited-availability group composed of links of q types, the blocking probability for class i calls can be calculated:

$$b(i) = \sum_{n=V-\sum_{s=1}^{q} k_s(t_i-1)}^{V} P(n)[1 - \sigma_i(n)] \ . \tag{11}$$

4   Numerical Results

In order to confirm the adopted assumptions accepted in the proposed method, the analytical results of blocking probability have been compared with the simulation results. In this section, selected numerical results are presented for BM-LAG ($q = 1$, $k_1 = 4$, $f_1 = 30$, $V = 120$) and GMLAG ($q = 3$, $k_1 = 4$, $f_1 = 5$, $k_2 = 7$, $f_2 = 10$, $k_3 = 2$, $f_3 = 15$, $V = 120$). The systems were offered three traffic classes in the following proportion: $a_1t_1 : a_2t_2 : a_3t_3 = 1 : 1 : 1$. The number of BBU's demanded for calls of particular traffic classes is equal to $t_1 = 1$, $t_2 = 2$, $t_3 = 6$, respectively. The results are presented in Fig. 3(a) and Fig. 3(b) in relation to the value of traffic offered to a single BBU: $a = \left(\sum_{i=1}^{M} a_i t_i \right) / V$. The simulation results are shown in the charts in the form of marks with 95% confidence intervals that were calculated after the t–Student distribution for the five series with 1,000,000 calls of this traffic class that generates the lowest number of calls.

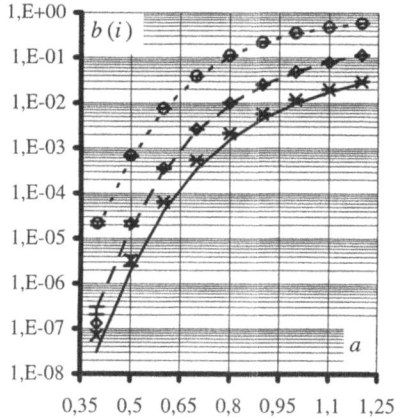

(a) Blocking probability in BMLAG

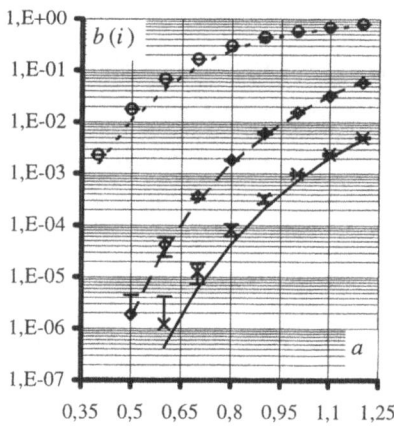

(b) Blocking probability in GMLAG

Fig. 3. Calculations: —— class 1; – – – class 2; - - - class 3. Simulations: × class 1; ◇ class 2; ○ class 3.

Figure 3(a) shows the results of blocking probability in the limited-availability group composed of links of one type. The obtained results allow us to compare accuracy of the method worked out in [9] for BMLAG with the accuracy of the calculation method proposed for GMLAG. In Figure 3(b) the results of blocking probability in the limited-availability group composed of links of various capacities are presented. The presented figure show the power of the proposed method of blocking probability calculation. For each value of traffic, calculation results are characterised by fair accuracy.

5   Conclusions

In this paper a new approximate model of a multi-rate system (a group of transmission links) has been proposed. The model is based on transforming multi-dimensional service processes in the system into a one-dimensional discrete Markov chain characterised by a product form solution. The proposed model, called the generalised model of the limited-availability group, has been used for blocking probability calculation in the group of separated transmission links of various capacities. The obtained results have been compared with the data of digital simulation. This research has confirmed a great accuracy of the proposed model, which is comparable to the accuracy of the method worked out for the groups composed of identical links (basic model of the limited-availability group). A lot of simulation experiments carried out by the authors indicate that fair accuracy is obtained regardless of the number of type of links, link capacities and the number of traffic classes. It confirms all the adopted theoretical assumptions accepted in the proposed method.

References

1. Roberts, J., Mocci, V., Virtamo, I., eds.: Broadband Network Teletraffic, Final Report of Action COST 242. Springer Verlag, Berlin, Germany (1996)
2. Reiman, M., Schmitt, J.: Performance models of multirate traffic in various network implementations. In: Proc. 14th ITC. Vol. 1b., Antibes Juan-les-Pins, France, Elsevier Science B.V. (1994) 1217–1227
3. Kaufman, J.: Blocking in a shared resource environment. IEEE Transactions on Communications **29** (1981) 1474–1481
4. Roberts, J.: A service system with heterogeneous user requirements — application to multi-service telecommunications systems. In Pujolle, G., ed.: Proceedings of Performance of Data Communications Systems and their Applications, Amsterdam, Holland, North Holland (1981) 423–431
5. Conradt, J., Buchheister, A.: Considerations on loss probability of multi-slot connections. In: Proc. 11th ITC, Kyoto, Japan (1985) 4.4B–2.1
6. Karlsson, J.: Loss performance in trunk groups with different capacity demands. In: Proc. 13th ITC. Vol.Discussion Circles, Copenhagen, Denmark (1991) 201–212
7. Lutton, J., Roberts, J.: Traffic performance of multi-slot call routing strategies in an integrated services digital network. In: Proc. ISS'84, Florence, Italy (1984)
8. Ramaswami, V., Rao, K.: Flexible time slot assignment — a performance study for the integrated services digital network. In: Proc. 11th ITC. (1985)
9. Stasiak, M.: Blocking probability in a limited-availability group carrying mixture of different multichannel traffic streams. Ann. des Télécomm. **48** (1993) 71–76
10. Beshai, M., Manfield, D.: Multichannel services performance of switching networks. In: Proc. 12th ITC, Torino, Italy (1988) p.5.1A.7
11. Stasiak, M.: An approximate model of a switching network carrying mixture of different multichannel traffic streams. IEEE Trans. on Commun. **41** (1993) 836–840
12. Ross, K.: Multiservice Loss Models for Broadband Telecommunication Network. Springer Verlag, London, UK (1995)

A Multicast Routing Algorithm Using Multiobjective Optimization

Jorge Crichigno and Benjamín Barán

Universidad Católica "Ntra. Sra. de la Asunción"
Asunción –Paraguay. Tel/Fax: (+595-21) 334650
{jcrichigno, bbaran}@cnc.una.py
http://www.uca.edu.py

Abstract. Multicast routing problem in computer networks, with more than one objective to consider, like cost and delay, is usually treated as a mono-objective Optimization Problem, where the cost of the tree is minimized subject to a priori restrictions on the delays from the source to each destination. This paper presents a new multicast algorithm based on the Strength Pareto Evolutionary Algorithm (*SPEA*), which simultaneously optimizes the cost of the tree, the maximum end-to-end delay and the average delay from the source node to each destination node. Simulation results show that the proposed algorithm is able to find Pareto optimal solutions. In addition, they show that for the problem of minimum cost with constrained end-to-end delay, the proposed algorithm provides better solutions than other well-known alternatives as *Shortest Path* and *KPP* algorithms.

1 Introduction

Multicast consists of concurrent data transmission from a source to a subset of all possible destinations in a computer network [1]. In recent years, multicast routing algorithms have become more important due the increased use of new point to multipoint applications, such as radio and TV transmission, on-demand video, teleconferences and e-learning. Such applications have an important quality-of-services (QoS) parameter, which is the end-to-end delay along the individual paths from the source to each destination. Another important consideration in multicast routing is the cost of a tree. It is given by the sum of the costs of its links. Most algorithms dealing with cost of a tree and delay from source to each destination, address multicast routing as a mono-objective optimization problem, minimizing the cost subjected to an end-to-end delay restriction. In [2], Kompella et al. present an algorithm (*KPP*) based on dynamic programming that minimizes the cost of the tree with a bounded end-to-end delay. For the same problem, Ravikumar et al. [3] present a method based on a simple genetic algorithm. This work was improved in turn by Zhengying et al. [4] and Araujo et al. [5]. The main disadvantage with this approach is the necessity of an a priory upper bound for the delay that may discard solutions of very low cost with a delay only slightly larger than a predefined upper bound. In contrast to the mono-objective algorithms, a *MultiObjective Evolutionary Algorithm*

J.N. de Souza et al. (Eds.): ICT 2004, LNCS 3124, pp. 1107–1113, 2004.

(*MOEA*) simultaneously optimizes several objective functions; therefore, they can consider end-to-end delay as a new objective function. Multiobjective Evolutionary Algorithms provide a way to solve a multiobjective problem (*MOP*), finding a whole set of Pareto solutions in only one run. This paper presents a new approach to solve the multicast routing problem based on a *MOEA* called the *Strength Pareto Evolutionary Algorithm* (*SPEA*) [6].

The remainder of this paper is organized as follow. A general definition of a multiobjective optimization problem is presented in Section 2. The problem formulation and the objective functions are given in Section 3. The proposed algorithm is explained in Section 4. Experimental results are shown in Section 5. Finally, the conclusions are presented in Section 6.

2 Multiobjective Optimization Problem

A general Multiobjective Optimization Problem (*MOP*) includes a set of n decision variables, k objective functions, and m restrictions. Objective functions and restrictions are functions of decision variables. This can be expressed as:

Optimize $\mathbf{y} = \mathbf{f}(\mathbf{x}) = (f_1(\mathbf{x}), f_2(\mathbf{x}), ..., f_k(\mathbf{x}))$

Subject to $\mathbf{e}(\mathbf{x}) = (e_1(\mathbf{x}), e_2(\mathbf{x}), ... , e_m(\mathbf{x})) \geq \mathbf{0}$

where $\mathbf{x} = (x_1, x_2, ..., x_n) \in \mathbf{X}$ is the decision vector, and $\mathbf{y} = (y_1, y_2, ... , y_k) \in \mathbf{Y}$ is the objective vector. $\mathbf{X}$ denotes the decision space while the objective space is denoted by $\mathbf{Y}$. The set of restrictions $\mathbf{e}(\mathbf{x}) \geq \mathbf{0}$ determines the set of feasible solutions $\mathbf{X_f} \subseteq \mathbf{X}$ and its corresponding set of objective vectors $\mathbf{Y_f} \subseteq \mathbf{Y}$. The problem consists in finding $\mathbf{x}$ that optimizes $\mathbf{f}(\mathbf{x})$. In general, there is no unique "best" solution but a set of solutions. Thus, a new concept of optimality is established for *MOP*s. In the minimization context of this work, given $\mathbf{u}, \mathbf{v} \in \mathbf{X}$,

$\mathbf{f}(\mathbf{u}) = \mathbf{f}(\mathbf{v})$ iff $\forall i \in \{1,2,...,k\}$: $f_i(\mathbf{u}) = f_i(\mathbf{v})$;

$\mathbf{f}(\mathbf{u}) \leq \mathbf{f}(\mathbf{v})$ iff $\forall i \in \{1,2,...,k\}$: $f_i(\mathbf{u}) \leq f_i(\mathbf{v})$;

$\mathbf{f}(\mathbf{u}) < \mathbf{f}(\mathbf{v})$ iff $\mathbf{f}(\mathbf{u}) \leq \mathbf{f}(\mathbf{v}) \wedge \mathbf{f}(\mathbf{u}) \neq \mathbf{f}(\mathbf{v})$.

Then, they comply with one of three conditions: (i) $\mathbf{u}$ dominates $\mathbf{v}$ iff $\mathbf{f}(\mathbf{u}) < \mathbf{f}(\mathbf{v})$; (ii) $\mathbf{u}$ and $\mathbf{v}$ are non-comparable iff $\mathbf{f}(\mathbf{u}) \nless \mathbf{f}(\mathbf{v}) \wedge \mathbf{f}(\mathbf{v}) \nless \mathbf{f}(\mathbf{u})$; or (iii) $\mathbf{v}$ dominates $\mathbf{u}$ iff $\mathbf{f}(\mathbf{v}) < \mathbf{f}(\mathbf{u})$. $\mathbf{u} \succcurlyeq \mathbf{v}$ will denote that $\mathbf{u}$ dominates or is equal to $\mathbf{v}$. A decision vector $\mathbf{x} \in \mathbf{X_f}$ is non-dominated with respect to a set $V \subseteq \mathbf{X_f}$ iff: $\mathbf{x}$ dominates $\mathbf{v}$ or they are non-comparables, $\forall \mathbf{v} \in V$. The set $\mathbf{X_{true}} = \{\mathbf{x} \in \mathbf{X_f} \mid \mathbf{x}$ is non-dominated with respect to $\mathbf{X_f}\}$ is known as Optimal Pareto set, while the corresponding set of objective vectors constitutes the Optimal Pareto Front.

3 Problem Formulation

For this work, a network is modeled as a direct graph $G = (V, E)$, where V is the set of nodes and E is the set of links. Let $(i, j) \in E$ be the link from node i to node j. For each link (i,j), let c_{ij} and d_{ij} its cost and delay. Let $s \in V$ denote a source and $N \subseteq V - \{s\}$ denote a set of destination nodes of a multicast group. Let $T(s,N)$ represent a multicast tree with s as source node and N as destination set. Let $p_T(s,n)$ be the subset of $T(s,N)$ that connects the source node s with a destination $n \in N$. The multicast routing problem may be stated as a *MOP* that tries to find a tree minimizing the maximum delay (D_M), the cost of the tree (C) and the average delay (D_A):

$$D_M = \underset{n \in N}{Max} \left\{ \sum_{(i,j) \in p_T(s,n)} d_{ij} \right\} \cdot \qquad (1)$$

$$C = \sum_{(i,j) \in T} c_{ij} \cdot \qquad (2)$$

$$D_A = \frac{1}{|N|} \sum_{n \in N} \left[\sum_{(i,j) \in p_T(s,n)} d_{ij} \right] \cdot \qquad (3)$$

Example 1. Given the multicast group shown in Figure 1, a tree with an end-to-end delay less than 40 ms is a priori chosen. (a) shows the Shortest Path Tree (*SPT*). (b) shows the tree constructed with *KPP* [3], that minimizes C subject to the bound delay of 40 ms. (c) shows a tree that would not be found by *KPP* or other algorithms based on restrictions if an a priori restriction of 40 ms were given. This alternative may be a good option since it has lower cost and a bound delay only slightly larger than the predefined bound.

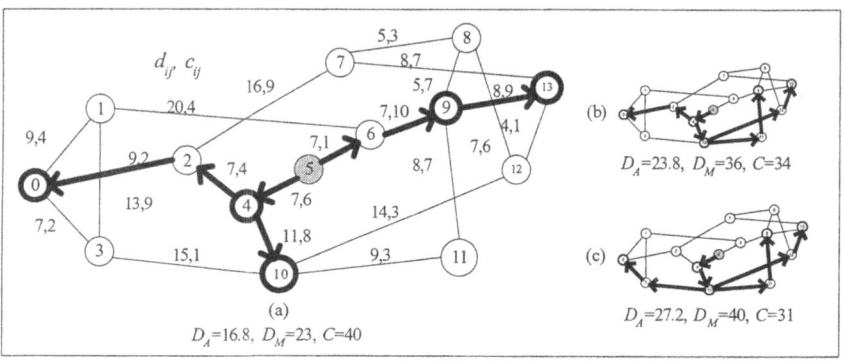

(a)
D_A=16.8, D_M=23, C=40

(b)
D_A=23.8, D_M=36, C=34

(c)
D_A=27.2, D_M=40, C=31

Fig. 1. National Science Foundation (NSF) Net. Each link with its delay and cost assigned

It is important to note from the mathematical formulation that the three objective functions are treated independently and should be minimized simultaneously. Therefore, a set of optimal solution, like the Fig. 1 shows, is provided in one run.

4   Proposed Algorithm

The proposed algorithm holds an evolutionary population P and a Pareto set P_{nd}. Starting with a random population P, individuals evolve to optimal solutions, and these are included in P_{nd}. The algorithm, shown in Figure 2(a), is briefly explained. *Construct routing tables.* Let $N = \{n_1, n_2, ..., n_{|N|}\}$. For each $n_i \in N$, a routing table is built. It consists of the R shortest and R cheapest paths. R is a parameter of the algorithm. Yen's algorithm [9] was used for this task. A chromosome is represented by a string of length $|N|$ in which the element (gene) g_i represents a path between s and n_i. The relation between a chromosome, genes and routing tables is shown in Figure 2(b). The chromosome represents the tree in the same Figure.

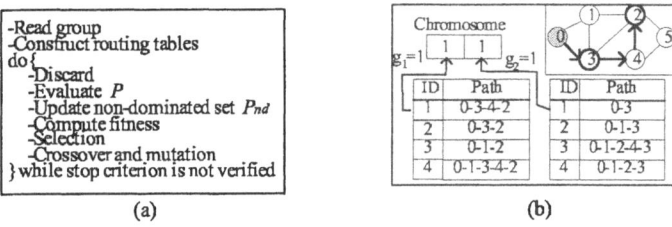

(a)   (b)

Fig. 2. (a) Proposed algorithm. (b) Relation between the chromosome, genes and routing tables

Discard. In P, there may be duplicated chromosomes. Applying genetic operations like crossover on two of them will yield the same chromosome. Therefore, the searching ability could be reduced. Duplicated individuals are replaced by new randomly generated ones, as recommended in [8].

Evaluate P. Evaluate P computes the objective vector of each individual in P, using the objective functions defined in Section 3.

Update non-dominated set P_{nd}. Each non-dominated individuals of P is compared with the individuals in P_{nd}. If that in P is not dominated by anyone of P_{nd}, then it is copied to P_{nd}. Besides, if an individual in P_{nd} is dominated by someone in P, it is removed from the external set.

Compute fitness. Fitness of every individual is computed using *SPEA* procedure [6].

Selection. The selection operator is applied on each generation over the union set of P_{nd} and P, to select good individuals to generate the next population P. The roulette procedure has been implemented as a selection operator [7].

Crossover and mutation. Two point crossover operator is applied over each selected pair of individuals. Then, some genes in each chromosome of the new population are randomly changed (mutated) with probability P_{mut} [7].

5   Results

The proposed algorithm has been tested in different network topologies. Firstly, simulation experiments for the Example 1, labeled as P1, were performed. Besides, two test problems from [5] were used. They were labeled as P2 and P3. The parameters of the proposed algorithm were set to $|P|=100$, $P_{mut}=0.3$ and $R = 25$. An exhaustive search method, which finds the Pareto optimal solutions, was used to compare the results. This algorithm simply calculates all possible chromosomes and picks up the non-dominated individuals. The run time was approximately 3 hours for P1 and 10 minutes for P2 and P3. For each of the last two problems, 50 runs were done using the proposed algorithm with $|P|=50$, $P_{mut}=0.3$ and $R=25$. The runs were stop when no new non-dominated solution was found for 250 successive generations.

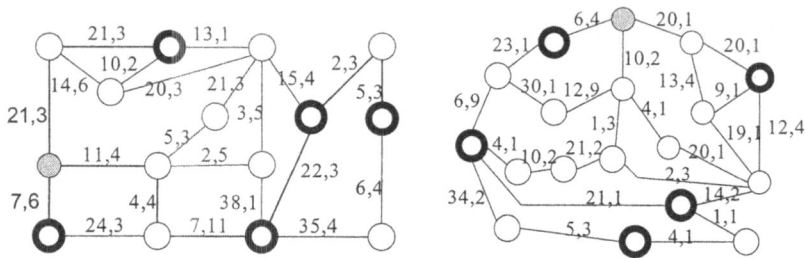

Fig. 3. Test problems P2 and P3

Table 1 presents the average (t_a) and maximum (t_{max}) running time in ms; the number of Pareto optimal solutions (S); the minimum (SF_{min}), maximum (SF_{max}) and average (SF_a) number of theoretical optimal solutions found in a run. Note that even in the worst case 66% of the theoretical Pareto set was found (4/6 for P3). Furthermore, the lower ratio SF_a/S was at least 0.857 (6/7, for P2).

Table 1. Results of problems P1, P2 and P3

	t_a	t_{max}	S	SF_{min}	SF_{max}	SF_a
P1	210	215	8	7	8	7.8
P2	350	390	7	5	7	6
P3	350	380	6	4	6	5.2

Besides the above test problems, the proposed algorithm was compared against the *SPT* and *KPP* [2] to find the minimum cost tree subject to an end-to-end delay restriction. The proposed algorithm was tested using the NSF Net. In order to measure the performance of the algorithm, average normalized cost and delay were computed:

$$C_N = (1/Y)\sum_{i=1}^{Y}\left(C_H^i - C_{MMA}^i\right)/C_{MMA}^i. \tag{4}$$

$$D_N = (1/Y)\sum_{i=1}^{Y}\left(D_H^i - D_{SPT}^i\right)/D_{SPT}^i. \tag{5}$$

where
Y : Number of runs with the same bound delay and size of multicast group.
C_H^i : Cost of the tree using H (H=*SPT* or H=*KPP*) on run i.
C_{MMA}^i : Cost of the tree using the proposed Multicast Multiobjective Algorithm (*MMA*) on run i.
D_H^i : Average delay of the tree using H (H=*MMA* or H=*KPP*) on run i.
D_{MMA}^i : Average delay of the *SPT*.

One hundred runs for each of four different multicast group sizes and for each of four different bound delays were done. Thus, 1600 runs with different multicast groups were tested. Costs of the links were generated at random and uniformly from the set $\{3, 4, \ldots, 10\}$ for each run. *MMA* was set to $|P|=100$, $P_{mut}=0.3$ and $R=30$. The runs stopped when no new non-dominated solutions were found for 100 generations. Given that *MMA* provides more than one solution, the one showing the minimum cost

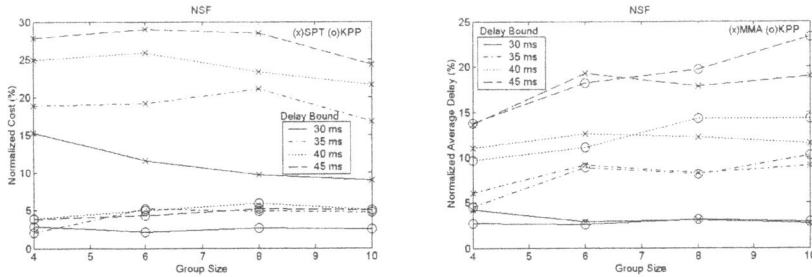

Fig. 4. Normalized cost and average delay

subject to the end-to-end delay restriction was picked out. Figure 4 summarize the results. They show that *MMA* constructs lower cost trees than *KPP* and *SPT*. The normalized *SPT* costs are between 10 and 30 %, while those of *KPP* are between 2 and 5 %. The effect of increasing the delay constraint is clear: *KPP* and *MMA* increase their average delay compared against *SPT*, while the cost of their trees are lowered. This implies a notorious tradeoff between both metrics. Clearly an approach that can find Pareto solutions is much more suitable for this type of problems.

6 Conclusions

This paper presents a new multiobjective approach to solve the multicast routing problem. To solve this problem, a multiobjective multicast routing algorithm was proposed. This algorithm optimizes simultaneously three objective functions: 1-maximum end-to-end delay, 2- cost of a tree, 3- average delay. The proposed evolutionary algorithm has a purely multiobjective approach, based on *SPEA*. This approach calculates not only one solution, but also an optimal Pareto set of solutions, in only one run. This last feature is of special importance, since the most adequate solution for each particular case can be chosen without a priori restrictions.

The proposed algorithm was evaluated with three test problems. Even in the worst case, it was able to find the 66% of the real Pareto set. Next, the proposed algorithm was compared against *SPT* and *KPP* to solve the problem of minimum cost tree subject to end-to-end delay restriction. Besides constructing the lowest cost tree, the proposed algorithm produces solutions with lower average delay than *KPP* in several cases (this is, cheaper trees with lower average delay), proving it is able to find better solutions including several theoretical Pareto optimal ones.

As future work, we will consider other objective functions as maximum link utilization and larger networks.

References

1. A. Tanenbaum, Computer Networks, Prentice Hall, 2003.
2. V. Kompella, and A. Gupta, "Multicast routing in multimedia communication," IEEE/ACM Transactions on Networking, Vol. 1 No. 3, 1993, pp. 286-291.

3. C. P. Ravikumar, and R. Bajpai, "Source-based delay bounded multicasting in multimedia networks," Computer Communications, Vol. 21, 1998, pp. 126-132.
4. W. Zhengying, S. Bingxin, and Z. Erdun, "Bandwidth-delay-constraint least-cost multicast routing based on heuristic genetic algorithm," Computer Communications, Vol. 24, 2001, pp. 685-692.
5. P. T. de Araujo, and G. M. Barbosa, "Multicast Routing with Quality of Service and Traffic Engineering Requirements in the Internet, Based On Genetic Algorithm," Proceedings of the VII Brazilian Symposium on Neural Networks (SBRN'02), 2002.
6. E. Zitzler, and L. Thiele, "Multiobjective Evolutionary Algorithms: A comparative Case Study and the Strength Pareto Approach," IEEE Trans. Evolutionary Computation, Vol. 3, No. 4, 1999, pp. 257-271.
7. D. Goldberg, Genetic Algorithm is Search, Optimization & Machine Learning, Addison Wesley, 1989.
8. R.H. Hwang, W.Y. Do, and S.C. Yang, "Multicast Routing Based on Genetic Algorithms," Journal of Information Science and Engineering, Vol. 16, 2000, pp. 885-901.
9. J. Yen, "Finding the k shortest loopless path in a network," Management Science, 17:712-716, 1971.

A Transsignaling Strategy for QoS Support in Heterogeneous Networks

Diogo Gomes, Pedro Gonçalves, and Rui L. Aguiar

Universidade de Aveiro/Instituto de Telecomunicações, 3800-193 AVEIRO, Portugal
{dgomes, pasg}@av.it.pt, ruilaa@det.ua.pt

Abstract. The increasing usage of multiple signalling mechanisms, with associated QoS extensions, creates several problems to commercial data networks. New and scalable approaches are required for the network operator to support this diversity. This paper discusses a highly flexible, scalable architecture for processing QoS Admission Control in public networks. The architecture relies on the cooperation of two different entities, an agent and a manager, with fully distributed implementation, and able to perform the required signalling, authorization, and admission control decisions. If required, the agent is capable of interfacing with different signalling mechanisms. Early implementation conclusions are also presented. This architecture is capable of operating with multiple QoS frameworks, with minimal added overhead.

1 Introduction

Internet traffic is increasing at an unprecedented rate as Internet-driven service demand grows. New applications require higher bandwidths and are often quality sensitive. Differentiated traffic treatment is expected for better management of available resources. For this, the Internet Engineering Task Force (IETF) proposed models to support QoS requirements, such as the Differentiated Services (DiffServ) and the Integrated Services (IntServ)/RSVP frameworks. Furthermore, other QoS-related signaling proposals appeared, associated with protocols oriented towards multimedia communications [1].

The IntServ architecture [2] was proposed in order to give QoS guarantees to a specific flow between a source and a destiny. Unfortunately, it presents a severe scalability problem in large networks. On the other hand, the DiffServ [3] framework solves the scaling problem aggregating the traffic flows with similar QoS requirement in Classes of Service (CoS) but does not provide, by itself, end-to-end QoS guarantees, and just provides different treatments to this CoS-marked traffic. In DiffServ scenarios, end-to-end QoS guarantees can be achieved by more complex control strategies: adding a Bandwidth Broker [4] to DiffServ networks; or using hybrid networks, with an access network supporting IntServ, and a transport network DiffServ-aware. Implementations [5][6][7] already exist for these scenarios but they suffer from various problems: lack of flexibility (not able to adapt to new signalling protocols); scalability problems (centralizing the QoSBroker functions in one machine); or

J.N. de Souza et al. (Eds.): ICT 2004, LNCS 3124, pp. 1114–1121, 2004.

extreme inefficiency (carrying IntServ information over the DiffServ network, without any impact on network control).

For an ISP, this QoS multiplicity poses severe difficulties, which are compounded by the extensive requirements of the "new multimedia services". The operator network will have to adapt to the fact that some applications do not explicitly signal QoS requirements, although they require QoS assurances for proper behaviour, while other applications use complex protocols to negotiate their QoS-related needs. Furthermore, an operator usually has to handle customers with quite different relevance, and it would be preferable to provide differentiated service according to the type of user. This is the target environment for future networks [8], where QoS support will be widespread, and provided in a diversity of situations. In this "4G wireless world" context, the classic Internet paradigm ("keeping all intelligence in the edges") cannot be maintained, as operators will aim to control network usage. Notice also that the recent security concerns already forces operators to intervene and monitor network traffic.

The next section describes a proposed network architecture for supporting QoS, able to handle this multiplicity of requirements, using a transsignalling unit at the edges. Some potential application scenarios of this architecture are presented in "Transisgnalling usage", and key conclusions, based on current implementation, are presented in the final section.

2 An Advanced QoS Control Architecture

For scalability reasons QoS support at the network level will require DiffServ enabled core networks. Our improved QoS architecture (Fig. 1), based on [9], assumes such a network, and defines specific entities for signalling and control. Three entities are defined, an AAAC server responsible for contract level QoS issues, an AQMUA (Advanced QoS Manager of Universidade de Aveiro) which is mainly a QoS Broker [4] with added functionalities, and SPAAQE (Signalling Processing, Access Authorization and QoS Enforcement) units, an advanced entity lying in access routers, that provides advanced signalling and QoS processing.

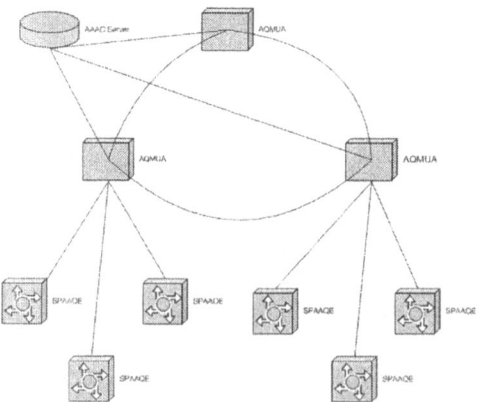

Fig. 1. Proposed QoS-Network Architecture

Each AAAC defines an administrative region for providing a common set of QoS services. Each AQMUA defines an "autonomous" region, independent in its capability of allocating resources, and controlling its associated edge devices (SPAAQE entities). The multiple AQMUAs act as a distributed overlay network for internal signalling, capable of interchanging QoS related information in a fast and simple way.

2.1 AQMUA

The Advanced QoS Manager of Universidade de Aveiro is the architecture element responsible for managing end-to-end QoS in its "autonomous" region. The AQMUA is responsible for performing global network management, keeping QoS levels per class; making admission control decisions; keeping information on network topology, as well as information about the neighbor network architectures; exchanging network information with SPAAQE entities; supporting SPAAQE in traffic conditioning decisions, feeding proper queue parameters to that entity and keeping accounting information of user network usage. For doing this, each AQMUA has detailed knowledge of its domain topology and receives reports from its associated SPAAQE entities.

Fig. 2 shows AQMUA internals. It contains four major components (the service interface with AAAC is not represented):

- A QoS Broker that performs control admission decisions, and overall network management tasks;
- A SPAAQE driver, that interfaces with SPAAQE entities from its network;
- An AQMUA driver, used to communicate with other AQMUA entities, in order to implement a distributed environment.
- A configuration and control graphical user interface.

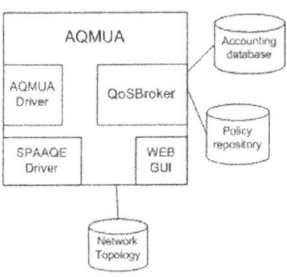

Fig. 2. AQMUA internals

In order to store all the information it needs, the AQMUA has three databases: an accounting database where it registers the network usage; a policy database, where it has service descriptions in terms of QoS parameters; and a network topology database, where it keeps the network element description and utilization, as well as the network architecture of neighbor areas. Furthermore, AQMUA can be configured using a Web GUI, where network elements can be listed into the database and their characteristics can be defined.

The QoSBroker is the key element in AQMUA. It is the QoSBroker that makes AQMUA resource management decisions. After a resource request, the QoSBroker examines the network resources availability, decides upon the acceptance of that traffic flow, and informs the SPAAQE. Information related with the authorized traffic flows is inserted in the accounting database, and can be used later for admission control decisions (deriving trends) or for accounting purposes.

2.2 SPAAQE

In this distributed architecture, the SPAAQE is the element positioned at the interface with the access network. The SPAAQE is responsible for detecting; classifying and processing data flow's going from and into the DiffServ network core. Typically this entity would act as a DiffServ Edge Router [3], but the added functionalities and individual processing capabilities implemented in SPAAQE lead us to a significantly different entity. In the SPAAQE the flow's are not only treated for QoS, but further processed in terms of signalling, accounting, destination address translation and any other kind of processing the Service Provider might want to add through the use of plug-ins. All decisions made by SPAAQE are local and relative to flow processing done by means of information received from the AQMUA: the SPAAQE does not take decisions by itself. However, once configured by the AQMUA, the SPAAQE is capable of handling localized decisions based on the information exchanged with the AQMUA.

One of the SPAAQE's major improvements is the capability to handle micro-flows identifiable by selectable mechanisms. These flows represent a greater granularity in the way the DiffServ network can differentiate handled traffic. For instance, they can represent specific User/Application traffic pairs that can provide the network administrator new ways of customizing available services to each specific client. This provides an IntServ-granularity applicable to user applications even when they are not IntServ-aware. Furthermore, these features are provided without any constrains to the final user. This is a consequence of SPAAQE main objective: a flexible signalling interface for the network provider, totally transparent to the end-user.

The SPAAQE's architecture (Fig. 3) consists of four main elements, a Classifier, a Scheduler, an Accounting module and an Advanced QoS Agent (ASQA). The Classifier is capable of classifying incoming packets based on information provided by the ASQA. The Scheduler manages queues, and is responsible for enforcing quality of service on packets going through SPAAQE; it further includes dropping capabilities (policing). The Accounting module interfaces with the accounting system deployed in the network (either Diameter or Radius servers) providing information on network usage. This module interfaces with the scheduler and ASQA through pluggable interfaces.

Fig. 3 also shows a simplified vision of the information flow inside the SPAAQE. Packets going through the Classifier are marked according to the installed configuration. Packets that don't match any configuration directive are considered unmarked and routed to the ASQA for further processing. The Classifier also routes packets that require advanced processing to the ASQA, such as packets it has no information on which action to take.

The ASQA is the SPAAQE's main element, where all control tasks are processed and treated. The ASQA is not only responsible for configuring the Classifier and Scheduler but is the responsible for the advanced functionalities that differentiate the SPAAQE from a common Edge Router.

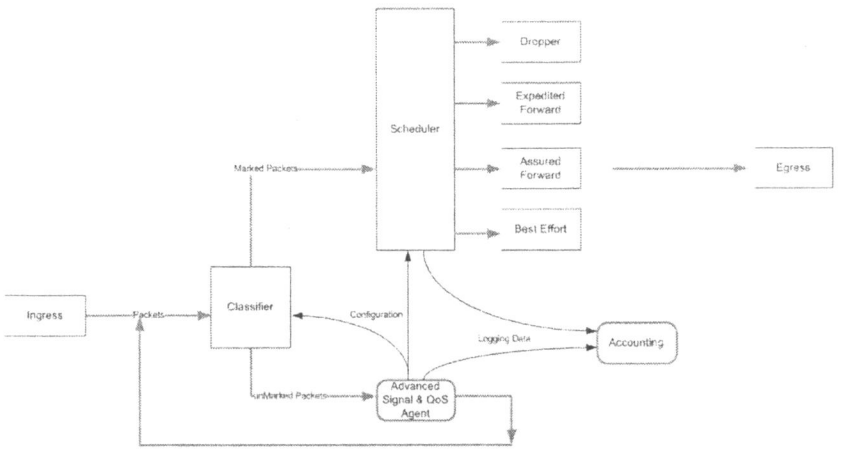

Fig. 3. SPAAQE Architecture and information flow

The ASQA is made of three functions; a Flow Detection unit capable of identifying micro-flows; a Flow Processing unit that treats the packet flows according to adequate configuration and a Flow Management process, which controls the SPAAQE. The ASQA is also the element that interfaces with the AQMUA, requesting instructions, providing information and receiving commands by means of COPS messages. Thus the AQMUA keeps a permanent knowledge of the current state of the network.

The most powerful feature of SPAAQE is the capability of the Flow Management process in the ASQA to process signalling messages, such as RSVP packets or SIP messages. This feature provides the means of achieving the desired network functionality for providing a homogeneous signalling to the core entities, while fully supporting any type of signalling in the access network. For instance, in terms of RSVP messages, the Flow Management receives the RSVP packets and according to information exchanged with the AQMUA enforces adequate QoS to the flow packets. The Flow Management process is not only capable of interpreting the RSVP packets and further translate IntServ into DiffServ, but is also capable of generating RSVP packets – which implies translating DiffServ-marked packets back into a IntServ flow - and issuing the proper messages to the applications.

This advanced feature is modular, in the sense that new modules capable of processing other signalling protocols can be incorporated into the SPAAQE, even during run-time, enabling it to translate any signalling protocol that the operator defines.

This *transsignalling* capability separates user-related signalling from network internal mechanisms. The transsignalling provided by the ASQA enables precise means of authorizing and enforcing QoS upon applications running on the operator Network. These features constitute a set of mechanisms that provide additional intelligence to the routers, enabling the deployment of advanced services (as per Application QoS) and support for legacy heterogeneous applications - while keeping a single signalling framework in the operator's core network. The classifier could also enable mechanisms of tracing and stopping Internet traffic; for instance, virus could be detected and stopped from spreading through the core network.

3   TransSignalling Usage

TransSignalling capabilities can potentially have multiple applications in future operator networks, especially when multiple heterogeneous environments are considered to be seamlessly supported – the usual network assumption for 4G scenarios [9].

3.1   Signaling Heterogeneity

Handling signaling in a heterogeneous network can be a very complex task, as different frameworks have their own signaling mechanisms (e.g. the multiple QoS frameworks supported by IETF). Furthermore, several signaling schemes would optimally require proper integration with network support, in terms of QoS. For example, SIP messages should trigger also network level reservation [1]. This implies that QoS-related signaling has to be generated from non-QoS signaling, or for non-QoS aware applications.

When multiple types of clients or applications are connected to the same access point, the access router has to handle these diverse types of signaling. Pushing this intelligence to the Broker (AQMUA) would lead to low performance implementations. Furthermore, these signaling protocols would all need to understand how the network is internally managed. All these problems disappear using the SPAAQE. The interface between SPAAQE and the AQMUA is uniform, COPS-based. The applications have their own signaling mechanisms, and the SPAAQE has the required intelligence to hide this signaling from the network, translating the application messages (e.g. a SIP message) into QoS requests, and simplifying the overall network QoS management.

A similar situation happens with cellular wireless access networks (e.g. UMTS). These technologies have very specific QoS mechanisms and supporting QoS at the physical layer may require a complex set of messages exchanged between the mobile node and the access point/base station (which incorporates the access router in our model [6]). The SPAAQE is able to provide this physical layer signaling adaptation.

This system can be applied even for QoS unaware applications (such as common Internet gaming applications). The cost of changing all the network client applications is very high, so legacy applications have to be supported in future networks. In those cases, the SPAAQE can identify the application (or the user) and query AQMUA in order to know the QoS profile that should be used by this traffic. AQMUA answers can depend on the application that generated the traffic, allowing (for example) the operator to provide different QoS services to email and web browsing traffics.

3.2   Integration of Different QoS Architectures

Integration of different QoS architectures (namely IntServ and DiffServ) is a good illustrative example of this type of capabilities, and this module is already operational in our current SPAAQE implementation. This implementation is based in proven Linux API's such as Netfilter [10], TC [11] and L7 Filters [12]. These API's have been extended in functionalities, but maintaining their stability and scalability features as well as good performance. DiffServ approaches rely on DSCP marking associated to each packet, while IntServ relies on RSVP signaling. The integration of IntServ traffic (common on the access) and DiffServ networks (common on the operator network) has some problems, as these two frameworks use two different signaling strategies.

Our QoS system is able to decide the signal adaptation that should be made in order to provide the correct QoS signaling to the next network to be visited. The SPAAQE is a network stateful entity capable to change QoS signaling, in both directions. The AQMUA knows the network topology as well as the network QoS architecture. When traffic comes from a network border, SPAAQE queries AQMUA about the next network technology. In the case of the QoS architecture still being the same, SPAAQE does nothing, except routing the flows. But if the next network uses a different QoS framework, SPAAQE requests to the next network valid QoS profiles and "formats" the traffic with the correct QoS parameters. In a typical end-to-end communication, RSVP reservations are "logically" propagated along the AQMUA interoperation network, and restored at the end access link by the SPAAQE, while traffic is transmitted under a DiffServ QoS-framework.

Fig. 4 depicts this process. Upon the request for a new RSVP flow (1, a PATH message) the SPAAQE that lays on the Access Network requests its AQMUA for access authorization and associated QoS Profile for the requested flow. This request triggers an overall end-to-end resource availability evaluation. This AQMUA requests information on whether resources are available or not in the networks the flow must traverse by conducting several requests in chain to the AQMUAs managing the successive QoS domains (3, 4, 5, 6, COPS messages). As soon as each AQMUA gets a acknowledgement that the flow can go through, it issues a COPS-PR message to its border SPAAQE's (7, 8, 9). On the destination Access Network the SPAAQE hides the whole process from the application by engaging in an appropriate RSVP negotiation process with the destination (12 – PATH, and 13 - RESV).

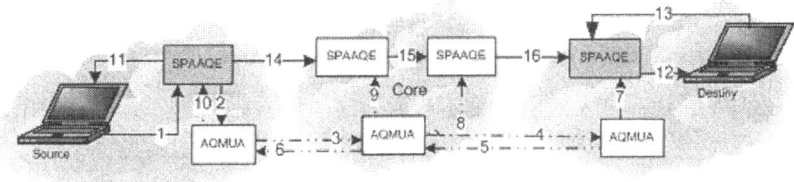

Fig. 4. Architecture Deployment example: Intserv-Diffserv interoperation

Finally the initial COPS Request (2) is answered (10) and the "RSVP" negotiation is completed (11 - RESV). When the communication flows through the network (14, 15, 16) the traffic presents the desired QoS performance, based on the previously provisioned QoS information, and an optimal mapping to the DiffServ network classes. In this process, the AQMUA processing occupies a minimal percentage of time: path transversal is the dominant delay.

4 Conclusions

We have proposed a flexible heterogeneous QoS architecture that can be overlaid in any operator IP-based QoS network. This architecture can handle most problems created by QoS signaling heterogeneity and by traffic flowing between different QoS environments, through the interoperation between an AQMUA and complex SPAQQE units.

The proposed architecture enables an end-to-end QoS support independently of the QoS architectures used in the access networks. Traffic QoS requests are always "adapted" to the existing architecture in the next neighbor network, enabling the reutilization of the existing network elements and architectures. Furthermore, integration of network QoS signaling with non-QoS related signaling schemes (e.g. SIP) is also supported simply by adding a new processing module.

Current tests with IntServ - DiffServ signaling adaptation show that the transsignalling unit can process hundreds of flows per second in current hardware, for practical cases of RSVP applications.

References

1. L. Veltri, S. Salsano, D. Papalilo, "SIP Extensions for QoS support", draft-veltri-sip-qsip-01, October 2002.
2. R.Braden, D.Clark, S. Shenker, "Integrated Services in the Internet Architecture: An Overview," RFC 1633, July 1994.
3. S. Blake, D. Clarke et all, "An Architecture for Differentiated Services", RFC 2475, December 1998.
4. M Gilmer; T. Braun, "Evaluation of bandwidth broker signalling", ICNP '99, 7th Int. Conf. on Network Protocols, 1999.
5. Eunkyu Lee, Sang-Ick Byun, Myungchul Kim, "A Translator between Integrated Service/RSVP and Differentiated Service for End-to-End QoS", ICT 2003, 10th Int. Conf. on Telecommunications, Vol. 2, 2003.
6. Werner Almesberger, Silvia Giordano, et all, "A Prototype Implementation for IntServ Operation over DiffServ Networks", IEEE Globecom, S. Francisco, Dec. 2000.
7. H. Bai, M. Atiquzzaman, W. Ivancic, "Running Integrated Services over Differentiated Service Networks: Quantitative Performance Measurements", Quality of Service over Next-Generation Internet, Boston, Aug. 2002.
8. H. Einsiedler, Rui L. Aguiar, et all, "The Moby Dick Project: a Mobile Heterogeneous All-IP Architecture", Advanced Technologies, Applications and Market Strategies for 3G - ATAMS 2001, ISBN 3-88309-20-X, pp. 164-171, 2001
9. Victor Marques, Rui L. Aguiar, et all, "An IP-based QoS architecture for 4G operator scenarios", IEEE Wireless Communications Magazine, June 2003.
10. "Netfilter", http://www.netfilter.org/
11. "Linux Advanced Routing & Traffic Control", http://lartc.org/
12. "Application Layer Packet Classifier for Linux", http://l7-filter.sourceforge.net/

Evaluation of Fixed Thresholds for Allocation and Management of Dedicated Channels Transmission Power in WCDMA Networks*

Carlos H.M. de Lima**, Emanuel B. Rodrigues***, Vicente A. de Sousa Jr.†, Francisco R.P. Cavalcanti, and Andre R. Braga

Wireless Telecommunications Research Group – GTEL
Federal University of Ceará, Fortaleza, Brazil
Phone/Fax: +55-85-2889470 - http://www.gtel.ufc.br
{carlos, emanuel, vicente, rod, andrerb}@gtel.ufc.br

Abstract. The major goal of this contribution is to evaluate strategies for allocation and management of the Dedicated Traffic Channels (DCH) transmission power in a WCDMA radio access network. The power allocation and control solution is tested aiming power consumption saving while the system QoS requirements are kept in acceptable levels. We focus on multi-cellular dynamic system-level simulations of the WCDMA radio access network in the forward link for the conversational service class. From these investigations we address the feasibility of our solution.

1 Introduction

Recently, some investigations have been carried out in order to develop efficient power management algorithms providing capacity and Quality of Service (QoS) gains for the 3G WCDMA system [1]. Such system is power restricted, that is, the base and mobile station power are a hard limiting resource. A limited base station power configures a scenario where interference and user blocking are the most important limiting factors. Thus, with the growth of wireless consumers, coverage and capacity problems arise.

In this scenario, the transmission power therefore becomes an essential resource and requires careful planning and suitable control if optimal system performance and maximal user capacity are to be achieved. Considering this fact, we aim to study a method of easy application and capable of enhance WCDMA downlink performance through efficient power allocation and control.

This investigation addresses the influence of fixed thresholds for allocation and management of the DCH transmission power regarding voice users. This technique envisages a WCDMA system on the downlink direction where base

---

* This work was supported by a grant from Ericsson of Brazil - Research Branch under ERBB/UFC.07 Technical Cooperation Contract.
** Graduate scholarship support by FUNCAP – Brazil.
*** Graduate scholarship support by CAPES - Brazil.
† Doctorate scholarship support by FUNCAP – Brazil.

J.N. de Souza et al. (Eds.): ICT 2004, LNCS 3124, pp. 1122–1127, 2004.
© Springer-Verlag Berlin Heidelberg 2004

stations are not allowed to transmit to the mobile station with a power higher than a pre-determined threshold. We intend to study how the employment of our proposed algorithm can impact on the system performance considering multi-cellular dynamic system-level simulations in the forward link. Case studies will present how system performance changes with this power allocation scheme and which threshold is better suited for the system.

The organization of this paper is as follows. In section 2, our strategies of allocation and management of radio resources are introduced. The Call Admission Control (CAC), Power Control (PC) and our proposed fixed thresholds for Power Utilization strategies are described. Section 3 describes the QoS requirement and the test scenarios which are used to evaluate our proposal. Section 4 draws the performance results and in section 5, we summarize the conclusions and envisage some perspectives.

2 Solution Conception

WCDMA cellular systems are essentially interference limited [2]. Therefore, the efficient power resource management enhances the network performance. Traditionally, the transmission power resources which are made available in the forward direction must be shared between all connected users according to their necessities, i.e., in order to compensate for the radio channel impairments and the interference levels as well [3,4].

The power control algorithm dynamically adjusts the transmited power in accordance with its default procedure, which imposes no restriction over the transmitted power levels per traffic channels on the forward direction. Then, some users can require higher transmission power values generating more interference and compromising the connection of other users. In this way, the utilization of the power resources can be restricted to few users in detriment of the others.

In this way, we can reduce the amount of interference generated in the network by limiting the maximum transmission power which is allowed per traffic channel [5]. The unconstrained power control approach was addressed in many works, e.g., [6,7]. Additionally, the investigations conducted in this contribution consider the feasibility of limiting the maximum transmission power per traffic channel in the forward link of WCDMA radio networks. Furthermore, our investigations intends to capture the maximum transmission power limit impact over the WCDMA network dynamics. This superior limit is defined higher enough in order to avoid coverage problems. In addition, the call admission control is also influenced by the superior limitation.

2.1 Definition of the Power Utilization Strategies

In the downlink direction, the transmission power is a scarce resource which all users are sharing. This fact imposes that WCDMA system capacity becomes strongly dependent on the downlink power utilization strategy. In this way, we study three distinct strategies to distribute and manage the power resource to WCDMA dedicated channels: **Basic Power Utilization**, **Infinity-Resource Power Utilization** and **Fixed Threshold Power Utilization**.

The **Basic Power Utilization** algorithm shares the power resource until total BS power reaches its specified maximum output power. The BS transmission power constraint is guaranteed all the time throughout the system dynamics due to our Call Admission Policy.

The **Infinity-Resource Power Utilization** is a theoretical reference case, where there is no resource limitation and all user power request will be assured.

Due to mobility and channel condition variability, a user could reach situations with poor or even no radio coverage. In such situations it is not reasonable to ensure the negotiated requirements for such users, because it would require too much resources and hence the QoS of all other users would get worse. The **Fixed Threshold Power Utilization** algorithm is in charge of balance the power resource utilization. Differently of the Basic Power Utilization, this scheme manages the Call Admission Control and Power Control algorithms in order to avoid excessive consumption from one individual link, i.e., a maximum pre-determined power threshold (P_{TH}) is defined for all active links of the system.

Considering the Call Admission Control, if the requisited power by the user at the beginning of the connection is higher than the upper limit it is set to this superior limit (P_{TH}). Regarding the Power Control, if during the call a user requires more power, our algorithm adjusts the user power obeying the upper power limit (P_{TH}). Thus, the system will react in a controlled way, trying to satisfy the requirements of as many users as possible. This limit can be dependent on the radio network planning and is defined in the radio link budget stage.

3   Performance Indicators and Test Scenarios

We assume two main criteria to define the Quality of Service (QoS) for circuit-switched oriented conversational services: user blocking rate and Frame Erasure Probability (FEP) which must be lower than 3% in order to satisfy the quality requirements [8]. The FEP must be kept below a tolerable level in order to guarantee a reasonable service quality, which is characterized for a good voice intelligibility.

FEP is obtained from a mapping of the Signal-to-Interference plus Noise Ratio (SINR) averaged over a voice frame. The user blocking rate is the ratio between the number of blocked users and the number of users that required a connection to the system. We do not evaluate the user dropping rate explicitly. Alternatively, we capture the effect of outage periods into the final user quality.

From our Power Utilization Strategies defined in the section 2.1, we present our simulation results by means of a performance comparison of five test scenarios shown in the table 1. Note that the power constraints for the **Fixed Threshold Power Utilization** (scenarios II, III and IV) strategies are based on the link-budget power estimation.

Table 1. Test Scenarios for Power Utilization Investigation.

Algorithm	Scenario	BS Power Constraint	DCH Power Constraint
Basic Power Utilization with Limited BS Power	Scenario I	43 dBm	Without Constraint
Fixed Threshold Power Utilization with Limited BS Power	Scenario II Scenario III Scenario IV	43 dBm	27.62 dBm (587.94 mW) 30.62 dBm (1157.88 mW) 33.62 dBm (2315.76 mW)
Infinity-Resource Power Utilization	Scenario V	Without Constraint	Without Constraint

4   Performance Results

The load is presented in terms of Poisson rate in Calls/Seconds. The solution efficiency is measured according to a user satisfaction metric and BS power utilization. The values of capacity and quality metrics presented in the next graphics should not be regarded as absolute performance indicators. Indeed, the reader should focus on the relative comparisons presented.

Figs. 1(a) and 1(b) show the 95th percentile of the FEP and user blocking rate for different offered loads and all proposed test scenarios, respectively (see table 1). A joint evaluation of these figures can determine the spectral efficiency of the system for each proposed power utilization solution.

Considering the scenario V, where the BS power resource is unlimited, the lowest FEP and user blocking rate values are found for all tested loads. Consequently, this scenario presents the best spectral efficiency. However, it is an unreal system configuration spending excessive power resource, as it will be present afterward.

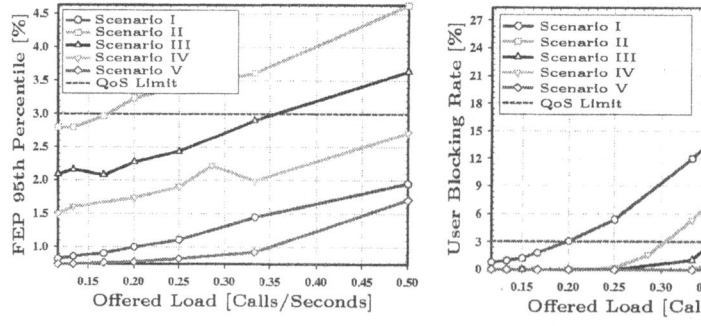

(a) FEP 95th Percentile vs. System Offered Load ($P_{TH} = 24.61$ dBm).

(b) User Blocking Rate vs. System Offered Load ($P_{TH} = 24.61$ dBm).

Fig. 1. Definition of system capacity.

Differently, the basic scenario (Scenario I) offers an adequate service quality, but the user blocking rate is high. This can be explained by our rigorous call admission control. This fact restricts the system capacity resulting a low spectral efficiency.

Now, we will discuss scenarios II, III and IV, where the maximum power of the link is limited. In this scenarios, the system capacity and service quality depend on the threshold applied to the link power. One can verify that the lower the link power limit, the higher the FEP and the lower the user blocking rate. But there is an power limit, where we can establish a tradeoff between service quality and system capacity. The Scenario II, where the maximum link power is two-fold the required power at the cell border, presents the worst results. In this scenario, in spite of the user blocking probability becomes lower, the FEP presents a high value, unsuitable to guarantee the pre-determined user QoS levels. A dissimilar behavior is perceived in the Scenario IV. Here, the FEP remains in accepted level, but the user blocking rate is compromised.

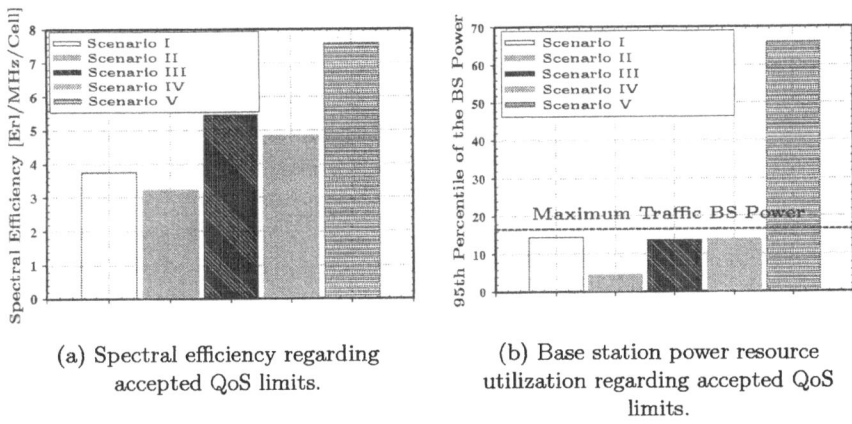

(a) Spectral efficiency regarding accepted QoS limits.

(b) Base station power resource utilization regarding accepted QoS limits.

Fig. 2. System capacity and transmission power resource utilization.

Until now, we assessed the system behavior based on system blocking and Frame Erasure Probability. However, it is important to show a BS power utilization study. A careful observation of the Figs. 2(a) and 2(b) can build a parallel evaluation between spectral efficiency and BS power utilization. Fig. 2(a) draws the spectral efficiency while Fig. 2(b) presents the 95th percentile of the total BS power measured at QoS limit (FEP or user blocking rate). One can perceive that our best technique (Scenario III) has BS power consumption lower than Basic Power Utilization. This power saving can be used to enable the application of resource reservation techniques such as power reservation for handover calls [9]. We can also notice an increased power utilization of the Scenario V. This fact was expected, it helps us to understand the influence of BS power utilization constraints on the spectral efficiency and assess the feasibility of practical solutions.

5 Conclusions

We analyzed the performance of a power utilization strategy that manages the call admission control and power control algorithms. In this way, we can point out that the application of the Fixed Threshold Power Utilization presents a tradeoff between system capacity and service quality, i.e., user blocking rate and FEP under accepted levels providing better spectral efficiency. Our research on resource management will be extended to formulate an adaptive framework, which is capable to set the link power thresholds according to the actual system load and user propagation conditions. Another interesting investigation will include the employment of our power utilization solution in a multiple service scenario, where both conversational and data service are provided. This scenario is more dependent on the BS power and therefore more crucial to manage in an efficient manner.

References

1. A. Furuskär, "Radio resource sharing and bearer service allocation for multi-bearer service, multi-access wireless networks - methods to improve capacity", Ph.D. dissertation, Royal Institute of Technologie, May 2003.
2. J.S. Lee and L.E. Miller, *CDMA Systems Engineering Handbook*, Artech House, 1 edition, October 1998.
3. H. Holma and A. Toskala, *WCDMA for UMTS - Radio Access for Third Generation Mobile Communications*, John Wiley & Sons, Ltd, 2001.
4. J. Laiho, A. Wacker, and T. Novosad, *Radio Network Planning and Optimisation for UMTS*, John Wiley & Sons, Ltd, 2002.
5. D. Kim, "A Simple Algorithm for Adjusting Cell-Site Transmitter Power in CDMA Cellular Systems," *IEEE Transactions on Vehicular Technology*, vol. 48, pp. 1092–1098, July 1999.
6. M. Almgren, H. Andersson, and K. Wallstedt, "Power Control in a Cellular Systems," *IEEE 44th Vehicular Technology Conference*, vol. 2, pp. 833–837, June 1994.
7. G. Foschini and Z. Miljanic, "A Simple Distributed Autonomous Power Control Algorithm and its Convergence," *IEEE 44th Vehicular Technology Conference*, vol. 42, pp. 641–646, November 1993.
8. UMTS, "Selection procedures for the choice of radio transmission technologies of the UMTS," Tech. Rep., Technical Report - Universal Mobile Telecommunications System (UMTS) - TR 101 112 V3.2.0 - UMTS 30.03, April 1998.
9. Emanuel B. Rodrigues, Carlos H. M. de Lima, Vicente A. de Sousa Jr., Francisco R. P. Cavalcanti, and Andre R. Braga, "Qos and load management via admission control in UMTS forward link," *IEEE Vehicular Technology Conference - VTC Spring*, May 2004.

Efficient Alternatives to Bi-directional Tunnelling for Moving Networks

Louise Burness[1], Philip Eardley[1], Jochen Eisl[2], Robert Hancock[3], Eleanor Hepworth[3], and Andrej Mihailovic[4]

[1]BT Exact, UK
{louise.burness, philip.eardley}@bt.com
[2]Siemens AG, Germany
jochen.eisl@siemens.com
[3]Siemens/Roke Manor Research
{robert.hancock, eleanor.hepworth}@roke.co.uk
[4]UK King's College London UK
andrej.mihailovic@kcl.ac.uk

Abstract. The following paper discusses ways to provide reachability to devices residing within a moving network. A device is reachable when an external node can successfully contact it. In this paper we introduce two novel solutions to support reachability in moving networks – one based on a proxy signalling agent and the other on the use of temporary names. We analyse aspects of these proposals compared to existing solutions, namely the bi-directional tunnelling solution proposed within the IETF's NEMO (Network Mobility) working group.

1 Introduction

The area of moving networks is gaining popularity both as a research topic and as a potential source of revenue for WLAN (Wireless Local Area Network) hotspot operators. A moving network (or vehicular network -VN) is a wireless or fixed LAN deployed on a vehicle that offers network services to passengers. A number of companies are now participating in trials of this technology with a view to offering Internet access to passengers on vehicles, such as trains [1] and planes [2].

The following paper presents some of the conclusions of a joint in-house project between Siemens, BT Exact and King's College London. The ASCOT (Architecture for Self Contained mOving neTworks) project investigated the problems associated with moving networks to support product innovation and broaden technical expertise.

To provide a seamless moving network service, the VN must provide session continuity, connectivity and reachability to user devices on-board the vehicle, including when the VN hands over between fixed networks. Whilst numerous different approaches to providing mobility solutions for moving networks have been investigated, the issue of reachability maintenance has not received such attention. The following paper discusses the issue of reachability for moving networks in greater detail, first providing justification for why such research is useful, before introducing two novel approaches for reachability maintenance in moving networks.

J.N. de Souza et al. (Eds.): ICT 2004, LNCS 3124, pp. 1128–1135, 2004.

1.1 Background

Figure 1 illustrates an example of a moving network architecture. The key component of the system is the Mobile Router (MR), which allocates IP addresses to user devices attaching to the network within the vehicle. It also provides the connectivity between the on-board vehicle and the ground network (the network that is providing the backhaul wireless link between the vehicle and the Internet).

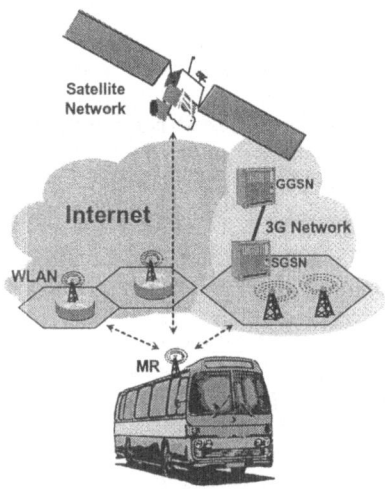

Fig. 1. Moving Network Scenario

It is useful to highlight a number of key characteristics of VNs that need to be considered when developing mobility solutions. Firstly, the link between a VN and the Internet is wireless, comparatively expensive, and resource limited e.g. cellular or satellite link. The utilisation of these backhaul links must be efficient, with the majority of bandwidth dedicated to user data (i.e. minimal signalling).

Secondly, the ground network must handle VN movement, and associated IP address changes. For example, intermittent satellite coverage along a railway track [3] may require handover to alternative technologies such as WLAN or cellular, resulting in IP address changes. Ideally, this movement should not impact on-going user sessions.

In addition, users will access the network for long periods of time (the duration of the journey), and are likely to want to be reachable by correspondent nodes in external networks, for example, to receive incoming connection requests.

2 NEMO Base Solution

Work has been carried out within the IETF NEMO working group [4] to develop a moving network solution. This solution, described in [5], is summarised below:

The VN contains a mobile router (MR) acting as a gateway between devices in the VN and the Internet. The MR is identified by a home address, which is advertised by a mobility agent in the home network of the MR. The MR is also delegated a prefix that is used to allocate addresses to user devices within the VN.

When traffic arrives at the home network destined for the prefix of the devices in the VN, the home agent intercepts and tunnels the data to the current care-of-address of the MR. It is the responsibility of the MR to update care-of-address information as the VN changes point of attachment to the network.

In the reverse direction, traffic generated by the user devices is reverse tunnelled to the home network before being routed to the correspondent device. Effectively, the VN and its home network form a single logical network, connected via a bi-directional tunnel.

This solution has advantages by eliminating signalling across the air interface, by hiding mobility events from user devices. Therefore, they will not generate, for example, Mobile IP binding updates [8], or Dynamic DNS (Domain Name System) updates [7] as the VN moves. In addition, the NEMO base solution inherently supports reachability, since the addresses allocated to the user devices are already advertised by the MR home network.

However, the base solution does have a number of drawbacks. Firstly, the use of a bidirectional tunnel introduces a routing inefficiency and more single points of failure along the data path. There is also additional tunnel encapsulation overhead that may be of concern to network operators. This is perhaps a significant drawback for operators who may have to pay additional fees for backhauling user data along the extended route, and who also have to consider continuity of service on home agent failure (or alternatively the deployment of potentially complex failover mechanisms). Secondly, this solution is heavily IPR encumbered, which may prevent its wide adoption.

Ideally, the reachability information for a user device should refer to an address that is topologically close to the device, resulting in fewer network nodes having to maintain state related to the device and less inefficient routing to and from the moving network. The following section of this paper introduces two reachability solutions for moving network scenarios that have the above desirable attributes.

3   Alternative Solutions

There are a number of alternative moving network solutions that make different assumptions about which part of the network is responsible for managing the address space from which addresses are allocated to the user device. These include:

- Option 1: Address from Ground Network
  The ground network delivers an address prefix or set of addresses to the VN, which the MR can then allocate to attaching user devices. The user device then appears as a node within the ground network and has a globally reachable IP address (for the purposes of this discussion, we have assumed that the ground network does not deploy any NAT - Network Address Translation functionality). This option provides the most efficient routing support, with little or no support required from the network. However, VN movement is visible to the user devices.

- Option 2: Address from MR Private Network
  The MR provides addresses to user devices from a pre-configured private address space. This option implies the use of a NAT (Network Address Translator) at the

MR, to translate between the device's private address (used within the VN) and a public (globally reachable) address allocated by the ground network. Note that a NAT swaps IP addresses in network and transport headers, but not in a payload. This option provides efficient routing support and conservation of public address space. Mobility events are transparent to the user devices.

[6] provides further details the various addressing options and their associated advantages and disadvantages.

Whist routing to user devices and managing device mobility using the above addressing options is well understood, the ways of maintaining reachability are limited. The following sections discuss novel techniques to handle this flooding issue, and an alternative reachability solution that can be used in moving network environments.

3.1 Proxy Signalling Agent

In this method a node in the fixed network acts as a signalling agent on behalf of the user device. The main goal of this technique is to reduce signalling across the bandwidth limited backhaul link by eliminating the need for individual reachability update messages to be transmitted by individual user devices.

The primary motivation for developing proxy signalling functionality for user devices in moving networks is maintenance of reachability while the vehicles move between different points of attachment to the network. At the same time, use of a signalling agent (SigAG in Figure 2) reduces inevitable overhead in situations where all user devices perform independent updating of corresponding states in the Internet and relaxes handover notification procedures in moving networks that would be associated with these. The signalling floods occur when addressing Option 1 is used since user devices are "virtually" connected to ground IP infrastructure and need to update their corresponding states in the Internet upon handover. When Option 2 is used as addressing solution, the need for management of signalling becomes a functional necessity since user devices are unaware of the mobility of moving network and hence require additional support to maintain the correct reachability information in the corresponding states in the Internet.

The scheme operates as follows; when the MR is initialized, it registers with a Signalling Agent (SigAg). On user attach, the user device is allocated an IP address (either from the ground network, or by the MR) to which traffic can be routed. The presence of the SigAg is discovered by the user devices (e.g. from advertisements made by the MR), and each device within the network requests the SigAg to send Mobile IP binding updates to its home agent on its behalf when the VN moves. The user device may also provide some secret or authorization information so that the home agent knows to trust the binding updates from a third party. This binding update contains the new address mapping to reach the new VN location.

When the VN moves, the MR informs the SigAg of this event, and includes the new addresses that have been assigned to user devices (in fact, this could just be the new prefix if the host part of the address is always consistent). The SigAg generates a series of binding updates towards the appropriate home agents for each user device. These messages can be secured using the information established as part of the user device registration phase. When the user device leaves the VN, the SigAg is informed that it should no longer send binding updates on behalf of the user device.

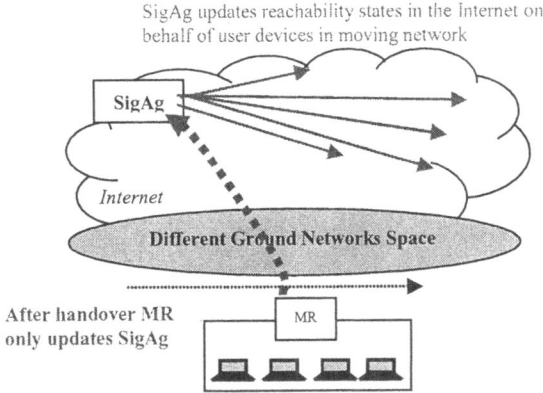

Fig. 2. Proxy Signalling Agent Operation

Analysis of the advantages of this solution depends on the addressing option chosen. For Option 1, the proxy signalling agent provides performance advantages by minimizing signaling overhead across resource limited backhaul links. Option 2 does not support reachability by default, and so our solution allows user devices on the moving network to be reachable whilst the operator keeps the benefit of preservation of globally routable addresses.

There are several possibilities for realising the Signalling Agent's functionality requiring different levels of complexity within different parts of the network. One example is the route optimisation requirement where users devices in moving network can directly update their corresponding nodes (not only their HAs) via Signalling Agent. In this scenario, validity of security associations and shared secrets becomes an important issue resulting in more complexity in the Signalling Agent for overcoming some of the limitations present in current mobility protocols related to surrogate registrations of current points-of-attachments.

3.2 Temporary Names

An alternative approach to maintain reachability for network nodes attached to a moving network is accomplished by the concept of temporary names. This temporary name (CNAME in DNS terminology) belongs to a name space managed by the VN operator. For example, whilst attached to a VN operated by "vn.co.uk", the user temporarily becomes node1.vn.co.uk. This temporary name is used to redirect name queries from a mobile user's home domain name server to a name server provided by the VN operator. Note that the VN name server is located in a fixed network domain. Consequently as the IP address of the moving network changes, only the entries in the VN name server need to be updated.

This solution makes use of the new Dynamic DNS capabilities being developed within the IETF that allow secure, low latency updates of domain name mapping information. Suitably selected DNS time-to-live information prevents over aggressive caching of rapidly changing DNS information, as described in [7].

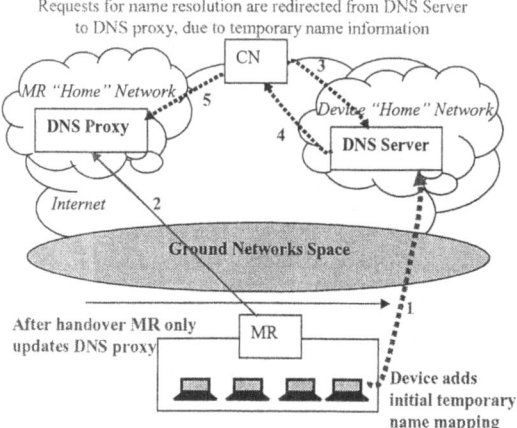

Fig. 3. Temporary Name Operation

A more detailed description of the reachability solution for moving networks is as follows; when a user device joins the VN, it is allocated an address (e.g. a private address from the MR). It also gets a DNS name which it can use as an alias (CNAME) in the DNS system. It registers (Dynamic DNS UPDATE) the temporary name with its home DNS server. The MR device updates the proxy DNS server with the user device's CNAME to address mapping. It is important to note that the new component in the naming system – the "proxy name server" acts in behalf of the VN and resides somewhere in the fixed network, but external to the VN.

When a correspondent node (CN) wishes to contact a user device, it looks up the user device's permanent name. The query is directed to the device's home DNS server which in turn re-directs the query to the local DNS server in the VN home network, as a result of the information embedded in the CNAME. The local DNS server resolves the query and returns the current IP address of the user device to the CN.

When the VN moves, the MR updates the information in the proxy DNS server with the new addressing information related to the user devices, thereby maintaining their reachability as the moving network changes point of attachment to the network. The address updating in the proxy DNS server in this example is based on a single message containing the new address prefix. Resolving of names is possible since all user devices connected to the VN keep the host specific part of their IP address. The message flow for temporary names concept is visualized in figure 2.

The user device may obtain an address from the ground network – option 1 (as in figure 3), or from an address space behind a NAT – option 2. This solution is applicable to both addressing options as long as there is a single routable IP address that uniquely identifies the user device.

The key feature of this solution is that the optimised signalling operation does not compromise the security relationships defined by Dynamic DNS; i.e. only the user device is allowed to update DNS entries in its "home" DNS server, whilst only the MR is allowed to update entries in the "local" DNS server. Once the initial DNS update has been generated, all dynamic DNS signalling across the air interface is eliminated whilst the user device remains within the VN.

In both the addressing option scenarios, signalling on VN handover is minimised, however, this solution does require updates to the user device software. Firstly, the user device must know how to parse and register the temporary name, and secondly, in addressing option 1 where MR prefix changes are visible, must know not to issue a Dynamic DNS update on IP address change. The latter issue is not a major one, since the user device could be configured to only be concerned about temporary name changes, not IP address changes. For addressing option 2, IP address change of the moving network is transparent to the user devices.

4 Discussion and Conclusion

We have presented two novel approaches to managing reachability for nodes in a mobile network. Our key concern has been efficient use of the wireless link between the VN and the rest of the network - because this link is likely to be comparatively expensive and low bandwidth, e.g. cellular or satellite. Both solutions minimise this signalling load. In particular when the VN hands over, global reachability is maintained for all devices with a single message (rather than a burst of signalling from all the user devices). The solutions are also IP-version agnostic, as user devices with an IPv4 or IPv6 protocol stack can be supported.

The first approach based upon a signalling agent was described just in relation to mobile IP binding updates to home agents. However, it is applicable to many different reachability management types - it can simultaneously provide support for SIP (Session Initiation Protocol), DNS, Mobile IP (including route optimisation with correspondent nodes) and even new techniques such as DCCP (Datagram Congestion Control Protocol) session re-direction. However it does require several messages to establish the initial state in the signalling agent and there are some security issues to resolve (delegation to the signalling agent of the right to send mobility management updates on behalf of the user devices). The solution requires no support in the user device, but requires additional functionality within the network. In addition, a more complex security model is required to support the generation of signalling towards the user device home network by a third party.

The second approach, based upon temporary names supports reachability of user devices in mobile networks over NAT-based addressing models. By using names rather than IP addresses, the user device and its DNS server do not need IPv6 support, even though it is likely that the visited networks, such as 3G networks will run IPv6. The main drawback of this solution is that it requires support from the user device, including Dynamic DNS extensions and enhancements to handle temporary names.

These solutions provide alternative strategies for managing device reachability in moving networks to those proposed within NEMO. They have an additional benefit of supporting more efficient routing towards the VN, not requiring state management within devices in the middle of the data path. There is a historic trend to push networking complexity up the network stack and out towards the edges of the network. We feel that our proposed solutions conform to this trend, providing more scaleable network architectures and more resilient network operation.

Whilst our alternative solutions do not handle mobility and reachability together, as is the case with NEMO, it is worth considering frequency of IP address change on moving network handover versus the overhead of managing this event. Typical backhaul technologies will initially consist of satellite or cellular networks. In the former case, IP address changes will occur infrequently, satellite coverage is extensive. In the latter case, the cellular network handles mobility on behalf of the mobile device, allowing handovers to occur between many radio access networks without the need to change IP address.

However, if multiple backhaul technologies are to be used simultaneously, with intelligent load balancing and failover as links appear and disappear, then it will be necessary for data sessions to handle mobility and reachability updates. The NEMO solution allows the mobility to be hidden from the sessions, but the details associated with supporting multihomed operation in the network still require refinement.

A full comparison of the solutions requires simulations and/or test-bed experiments for a variety of scenarios to provide quantitative results of the expected signalling load on the radio link between the VN and the ground network. The signalling load reduction becomes more apparent for moving networks containing a large number of users. This is also the scenario within which the tunnelling overhead of the NEMO base solution may become prohibitive to operators. In summary, the motivation behind this paper was to illustrate that there are promising alternatives to NEMO to support the reachability of user devices moving networks.

References

[1] "GNER to spend £1m on Wi-Fi trains", The Register, electricnews.net, 15/08/2003, http://www.theregister.co.uk/content/69/32351.html
[2] "Lufthansa Begins Trial of In-flight Wireless and Wired LAN Service", Converge Network Digest, 15/01/2003, http://www.convergedigest.com/WiFi/wlanarticle.asp?ID=6078
[3] "Specialised satellite long haul links for WLANs on trains", BNSC S@tcom, 3rd round, http://www.bnsc.gov.uk/
[4] IETF NEMO working group http://www.ietf.org/html.charters/nemo-charter.html
[5] "Network Mobility (NEMO) Basic Support Protocol", http://www.ietf.org/internet-drafts/draft-ietf-nemo-basic-support-02.txt
[6] "Examining a key topic in the ASCOT project – address management in moving hotspots", K. Boukis, et al. Submitted to European Wireless 2003
[7] Domain names – concepts and facilities (RFC 1034), implementation and specification (RFC 1035), P. Mockapetris, Nov 1987, and subsequent updates.
[8] IP Mobility Support for IPv4 (RFC 3344), C. Perkins (editor), Aug 2002; Mobility Support in IPv6 (work in progress), D. Johnson, C. Perkins, J. Arkko, June 2003

Group Messaging in IP Multimedia Subsystem of UMTS

Igor Miladinovic

Telecommunications Research Center Vienna (ftw.)
Donau-City-Strasse 1, 1220 Vienna, Austria
Tel. +43-1-5052830-54
miladinovic@ftw.at

Abstract. Messaging service is an always-on service widely used in second generation (2G) networks today. In third generation (3G) networks it is expected that the messaging service will play even more important role. It will alow not only exchange of text messages, but also exchange of any multimedia content, including images, audio, and video in near real time fashion. Messaging service offers a great opportunity for group communications, since a message can be sent to a group of recipients. This paper investigates possibilities to deploy group messaging service in the Universal Mobile Telecommunication System (UMTS), as the most important representative of 3G networks. Given that the content of messages is transported by the signaling protocol, traffic generated by the messaging service interfere with the traffic for session signaling. This can cause increase of signaling delay, or even the loss of signaling information. Especially for group messaging, an efficient mechanism is necessary to distribute the message to each recipient without wasting the network resources. This work propose multi-recipient (MR) messages for distribution of group messages in the Internet Protocol (IP) Multimedia Subsystem (IMS) that is responsible for signaling in UMTS (Release 5 and above).

Keywords: Messaging, IMS, SIP, Multi-Recipient Messages, Group communication, UMTS

1 Introduction

Different kinds of messaging service have gained on importance in mobile networks in the last several years. They offer users the possibility to exchange messages, or more generally any kind of multimedia content, in near real-time. The short messaging services (SMS) [11], for example, is a widely used messaging service in the Global System for Mobile communication (GSM) network that is the most important representative of second generation (2G) networks. SMS, first introduced in 1996, supports only text messages of up to 160 characters. Despite this limitation, SMS has become extremely popular [7]. Third generation (3G) networks will provide users with considerably more bandwidth than 2G networks. This gives possibility to enable multimedia messaging service

J.N. de Souza et al. (Eds.): ICT 2004, LNCS 3124, pp. 1136–1142, 2004.

(MMS) capable of exchanging any type of multimedia content between users, including any combination of text, animations, digital images, and audio and video download or streaming.

The Universal Mobile Telecommunication System (UMTS) is expected to become the most important representative of the 3G networks, similarly to GMS in 2G networks. It has been standardizing by the 3rd Generation Partnership Project (3GPP) and the current specification draft is referred to as Release 6. The UMTS architecture has changed importantly with Release 5, since it has introduced the Internet Protocol (IP) Multimedia Subsystem (IMS) [2] as an extension to the existing Circuit Switched (CS) and Packet Switched (PS) domains. IMS is an significant step towards the all-IP networks architecture [14]. It is responsible for traditionally telephony services and new multimedia services. As the signaling protocol in IMS, 3GPP has chosen the Session Initiation Protocol (SIP) [12][13]. SIP is an application layer signaling protocol able to initialize, modify, and terminate any kind of multimedia session in IP-based networks. With its extension for instant messaging [4], SIP is capable of distributing content of any Multipurpose Internet Mail Extensions (MIME) [6] type between users in near real time. 3GPP is currently specifying messaging services in IMS based on this SIP extension.

This paper investigates possibilities to deploy the messaging service in IMS with the particular focus on group messaging. In this context, we are speaking about group messaging when a message is sent to at least 2 recipients. The only way to distribute such a message in SIP is to sent a separate message to each recipient in turn. This behavior not only wastes the bandwidth but also increments the end-to-end delay of the message when the number of recipients is large, as discussed in [8]. To overcome this problem, 3GPP proposes the use of an application server (AS), which distributes the message to the recipients. In this way, the message is sent only once from the sender to AS. However, between AS and recipients the message is still distributed by replicated sending to each recipient. After receiving the message, each recipient replies with a response that confirm the arrival of the message. These responses generate additional signaling traffic. This paper proposes a special type of SIP messages, called multi-recipient (MR) messages [10], for group messaging in IMS. These messages are tailored to cover the needs of multiparty communication in SIP and to optimize SIP traffic in group communications. The paper proposes a slight adoption of MR messages for IMS and describes the new message flow.

2   Messaging Service in IMS

As mentioned in Section 1, messaging service provides exchange of any type of multimedia content between users in near real time fashion. In order to support this services in IMS, 3GPP is considering SIP capabilities for instant messaging. The current IMS specification [2] proposes to use the SIP extension for instant messaging [4]. This extension introduces a new type of SIP request, called MESSAGE request, that can be exchanged between users without initialization any session. Similar to other SIP requests, the MESSAGE request is composed of a message header and an optional message body. Given that the message body is capable to carry any MIME [6] type content, it is used to carry the multimedia content of the message. IMS messaging specification [3] differentiates between three types of messaging: immediate messaging, deferred delivery messaging, and session-based messaging.

The main focus of this paper is immediate messaging and deferred delivery messaging, because they use SIP for message delivery. Session-based messaging is not fully specified so far, and it is not clear how messages will be exchanged. The current specification of the IETF proposes SIP for the session initialization and the Message Session Relay Protocol (MSRP) [5] for the exchange of messages within a session in order to reduce SIP traffic. However, MSRP can be used only within an established session. For immediate messaging and deferred delivery messaging, when only a few message are to be exchanged, it is not efficient to use it, because the initialization and termination of a SIP session require at least 5 messages.

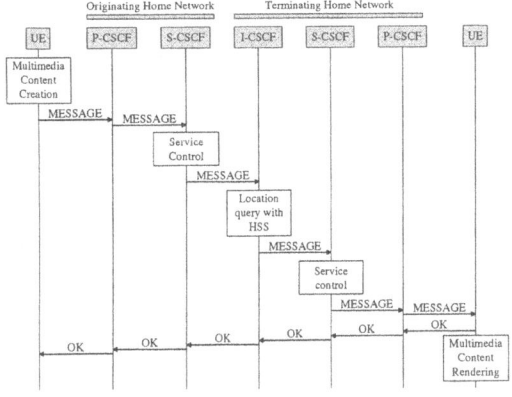

Fig. 1. Immediate Messaging in IMS

Figure 1 shows the message flow for the delivery of an immediate message in IMS. The sender's User Equipment (UE) generates the MESSAGE request with the multimedia content and sends it to the corresponding Proxy Call State Control Function (P-CSCF) that forwards this message to the Serving-CSCF (S-CSCF). Now, S-CSCF can apply any service control on this message. For example, it can interact with an AS responsible for processing this request. After that, S-CSCF forwards the request to I-CSCF of the terminating home network. The Interrogating-CSCF (I-CSCF) first interacts with the Home Subscriber Server (HSS) in order to determines the address of the corresponding S-CSCF and then forwards the request to this S-CSCF. S-CSCF applies any kind of service control required by the terminating side. Finally, S-CSCF forwards the request to the recipient's UE over P-CSCF. This UE replies immediately with a 200 (OK) response and notifies the user about the received message. Note that this response is merely a confirmation that the request has been received by the other side, and not that the user has seen the message. The response takes the reverse path of the request until it reaches the sender's UE. In the case that the recipient is not available, the message is stored at S-CSCF in the terminating network. This recipient will receive the message after registration, if the message has not expired.

In the current IMS specification, there are two possibilities to sent a single message to several recipients. Both of them uses an AS, contacted by the S-CSCF in the originating home network. The first possibility requires the creation of a new group with an unique identifier on the appropriate AS [1]. The messages are addressed to the group identifier and it is the responsibility of AS to distribute messages to each group member. Users have to subscribe to this group in order to receive messages addressed to it. The second possibility does not require any group management. The sender simply includes addresses of each recipient as a part of the message content. The corresponding AS interprets this part of the message content and sends messages addressed to each intended recipient.

3 Optimized Messaging Service

The basic SIP specification does not provide any possibility to distribute a message to several recipients efficiently. The only way is to sent a single message to each recipient in turn. This is because SIP was originally designed for signaling of large multiparty conferences over multicast. Nowadays, the main focus of SIP has moved to signaling of two-party session. The current SIP specification discourages from using multicast for transport of SIP messages because of security reasons [12].

MR messages [10] are designed to optimize SIP traffic in group communications. In SIP there is two types of messages, requests and responses. Since both of them can be MR enabled, we can differentiate between MR requests and responses. A MR request carries addresses of all its recipients in its header. MR enabled network servers (for example, proxy servers) use this information to route the request to each of its destinations. MR request not only reduce the amount of signaling traffic compared with regular SIP requests, but also improve the end-to-end delay.

After receiving a MR request, each recipient replies with a response. In the case of the MESSAGE request, the response is sent immediately by the recipient's terminal. Given that a MR request reaches several recipients, there are several responses that are sent to the sender of the request. These responses can be collected by the network servers and sent as a single MR response [9], reducing the amount of signaling traffic.

This work propose the use of MR messages for messaging service in IMS. As stated in Section 2, in order to reduce traffic between the sender and first S-CSCF (see Figure 1), immediate messages to several recipients include recipients' addresses in the multimedia content of the message. This S-CSCF, after interaction with the corresponding AS, distributes the message by replicated sending to each recipient. This behavior reduces the amount of traffic between the sender and first S-CSCF, but traffic between this S-CSCF and recipients is not optimized, since the same message is sent to each recipient separately. Furthermore, IMS proposal does not provide any optimization of traffic generated by responses. Each recipient sends a response which is routed to the sender of request as usual. Therefore, the sender receives a separate response from each recipient.

Using MR messages in IMS, a single request containing addresses of all recipients is sent to the S-CSCF, similarly as proposed in IMS. However, the recipients' addresses are placed in a special header field of the request. Compared with the IMS proposal, the benefit is that this information, which is actually needed for the routing, is placed in the request header, together with other routing information and not together with the message content. After receiving a MR request, the S-CSCF of the sender, called originating S-CSCF, determines the next hop CSCF for each recipient. If two or more recipients have the same next hop CSCF, the request is sent as a MR request there. Otherwise, if a recipient has a next hop CSCF different from all others, the S-CSCF sends the request as a regular requests there. All other CSCFs behave similarly as the sender's S-CSCF. Figure 2 shows an IMS topology and the flow of MR requests in the scenario where the sender (S) sends an immediate message to recipients 1-6. As we can see, the message is sent never more the once between any two CSCFs. In contrast to that, according to the current IMS proposal there are some links on which the message must be transmitted several times. For example, on all the links between the first S-CSCF (II) and I-CSCF (III, IV, and V) the message is transmitted twice.

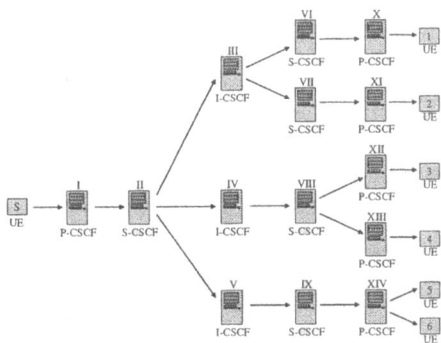

Fig. 2. Example IMS messaging scenario

MR requests require MR enabled CSCFs in order to route requests. On the other side, the IMS proposal requires S-CSCF to interacts with the corresponding AS. An AS does not need to be contacted when MR requests are used. This reduces the number of interactions and therefore the end-to-end delay of the request. Because the additional routing information is placed in the header of a MR request, routing of MR requests is ensured even if some CSCFs are not MR enabled. In this case, such a CSCF would replay with a 420 (Bad Extension) response. After that, the request can be sent again but this time as several regular requests to each recipient separately [10].

Besides of MR requests, MR responses can be additionally used to further reduce signaling traffic [9]. This work proposes a special use of MR responses. The goal is to reduce response traffic without introducing significant changes in the behavior of the IMS components. Therefore, this paper proposes MR responses only between the originating S-CSCF and the sender. In this way, only S-CSCFs should be enabled to collect responses. However, if a S-CSCF has not this feature, responses are sent as usual. The distance between the originating S-CSCF and the sender includes the most critical link in this scenario – the radio interface of the sender. MR responses can reduce traffic on this link, since only a single response is transmitted over that link instead of a separate response of each recipient. A precondition for applying MR responses is that at least two recipient replies with the same type of response, for example with a 200 (OK) response. A MR response includes information about recipients that have sent this response, such as recipients' addresses.

4   Conclusion

Messaging service has become very popular in 2G networks over the last several years. Given that 3G networks provide more access bandwidth, it is expected that this service will be used even more in 3G networks, in particular in UMTS. Combination of text, voice, video, and other data in a single message gives a new dimension to the messaging service. On the other side, since the content of message is carried by the signaling protocol, messaging traffic interfere with session initialization traffic. This can cause increasing of the call setup delay in IMS. Therefore, the amount of traffic generated by the messaging service should be kept as low as possible. Messaging service is often used for the group communication, what means that the same message is sent to several recipients. Replicated sending the message to each recipient is not acceptable since it wastes network resources, in particular the bandwidth of the sender's radio interface. To overcome this problem in IMS, 3GPP proposes the use of an AS. The content of the message include the addresses of each recipient, and the message is sent to the AS that is able to interpret this part of the message and to distribute the message to each recipient.

This paper propose the MR messages to be used for group messaging service in IMS. In this way, it is not necessary to communicate with an AS. However, an AS still can be contacted in order to apply some value added services, operator specific services or both. MR messages requires only application layer network components (CSCFs) to support them. Even if some of them are not MR enabled, the message can be delivered to all of its destinations. Concluding, this work has shown that the combination of MR requests and responses can be used in IMS to optimize traffic in group messaging communications.

References

1. 3GPP TSG SSA. IP Multimedia Subsystem (IMS) group management - Stage 1 (Release 6). TS 22.250 v 6.0.0, 3rd Generation Partnership Project, 2002.
2. 3GPP TSG SSA. IP Multimedia Subsystem (IMS) - Stage 2 (Release 6). TS 23.228 v 5.9.0, 3rd Generation Partnership Project, 2003.
3. 3GPP TSG SSA. IP Multimedia System (IMS) messaging - Stage 1 (Release 6). TS 22.340 v 6.1.0, 3rd Generation Partnership Project, 2003.
4. B. Campbell, J. Rosenberg, H. Schulzrinne, C. Huitema, and D. Gurle. Session Initiation Protocol Extension for Instant Messaging. RFC 3428, Internet Engineering Task Force, December 2002.
5. B. Campbell, J. Rosenberg, R. Sparks, and P. Kyzivat. The Message Session Relay Protocol. Internet draft, work in progress, Internet Engineering Task Force, October 2003.
6. N. Freed and N. S. Borenstein. Multipurpose Internet Mail Extensions (MIME) Part One: Format of Internet Message Bodies. RFC 2045, Internet Engineering Task Force, November 1996.
7. A. J. Huber and J. F. Huber. *UMTS and mobile computing*. Artech House, Inc., 2002.
8. I. Miladinovic and J. Stadler. Improving End-to-End Signaling Delay in Multiparty Communications. In *Proceedings of the ICT 2003, Bangkok, Thailand*, pages 197–206, April 2003.
9. I. Miladinovic and J. Stadler. Instant Messaging Traffic Reduction in the Session Initiation Protocol. In *Proceedings of the 3rd IASTEDInternational Conference on Wireless and Optical Communications WOC 2003, Banff, Canada*, Juli 2003.
10. I. Miladinovic and J. Stadler. Multi-Recipient Requests in the Session Initiation Protocol. In *Proceedings of the 21st IASTED International Conference AI PDCN 2003, Innsbruck, Austria*, pages 802–806, Februar 2003.
11. G. Peersman, P. Griffiths, H. Spear, S. Cvetkovic, and C. Smythe. A Tutorial Overview of the Short Message Service within GSM. *IEEE Computing and Control Engineering Journal*, 11(2):79–89, April 2000.
12. J. Rosenberg, H. Schulzrinne, G. Camarillo, A. Johnston, J. Peterson, R. Sparks, M. Handley, and E. Schooler. SIP: Session Initiation Protocol. RFC 3261, Internet Engineering Task Force, June 2002.
13. H. Schulzrinne and J. Rosenberg. The Session Initiation Protocol: Internet-Centric Signaling. *IEEE Communications Magazine*, 38(10):134–141, 2000.
14. J. Yang and I. Kriaras. Migration to all-IP Based UMTS Networks. In *Proceedings of the First International Conference on 3G Mobile Communication Technologies*, pages 19–23, March 2000.

The Effect of a Realistic Urban Scenario on the Performance of Algorithms for Handover and Call Management in Hierarchical Cellular Systems

Enrico Natalizio, Antonella Molinaro, and Salvatore Marano

DEIS - University of Calabria, I-87036 Arcavacata di Rende (CS), ITALY
{enatalizio,molinaro,marano}@deis.unical.it

Abstract. In this paper we show how the use of real data on vehicular traffic characterization affects the performance of call and handoff management algorithms conceived for hierarchical cellular systems. The main contribution of this work is the scenario where these algorithms are analysed. Studies and statistics on traffic in several cities centres are exploited to derive a realistic statistical distribution of the user velocity. The estimation of the user velocity is the key parameter which all the algorithms analysed in this paper are based on, in order to effectively exploit the presence of overlapping coverage layers (macrocell and microcell) to increase the system capacity and to minimise the number of handoff events, which can be detrimental to the offered service quality.

1 Introduction

A consolidated solution for the rising demand for mobile communications consists in deploying *multi-tier* cellular systems, which offer hierarchical overlapping coverage layers {1, 2}. The multitier coverage gives the possibility to transfer active connections from one coverage layer to another during the user roaming, when resources are not available in the current layer. User velocity estimation allows regulating the traffic flows among layers. In order to minimise the signalling load, due to unnecessary handovers, and to optimise the resource assignment, each layer may take care of the users of a specific class of velocity [3, 4].

In this paper, we investigate into the performance of three algorithms for handover and call management, in a scenario consisting of two tiers (microcells and macrocell) and realistic vehicular traffic data. The chosen algorithms are reported in [5-7]; they all share some common features:

- reject the classic assumption of constant user speed during a call;
- transfer the task of velocity classification from the network to the mobile terminal;
- classify the terminal velocity also when the terminal is idle.

The paper is organized as follows. In section 2 we illustrate the simulated scenario, focusing on the features that make it "realistic". In section 3 we briefly describe the algorithms chosen for performance comparison. Section 4 reports simulation results. Concluding remarks are summarised in Section 5.

J.N. de Souza et al. (Eds.): ICT 2004, LNCS 3124, pp. 1143–1150, 2004.
© Springer-Verlag Berlin Heidelberg 2004

2 The "Realistic" Urban Context

The main aim of our work is to simulate "realistic" vehicular traffic and teletraffic distributions within a typical city centre in order to assess the performance of some call management algorithms. To this aim we have developed a composite simulative tool whose main inter-operating modules are the *system* and the *traffic* simulator. The former implements the radio coverage and the algorithms operation. The latter generates vehicle mobility profiles in a realistic traffic jam situation.

Our reference scenario is a two-tier cellular system placed in a city's centre. The focus is on a one-way, two-lanes street segment, characterised by the presence of crossing points and traffic lights, whose number, position, and distance can be selected as input parameters of the *system* simulator.

Pedestrian and vehicular users travel along the street setting up and releasing voice connections. Pedestrian users are assumed standing along the street margins; vehicles can move accounting for stopping at traffic lights, overtaking or queuing other cars, and, in any case, respecting safety rules.

The *traffic* simulator generates vehicular traffic and teletraffic flows according to realistic data about vehicle characteristics (maximum acceleration/deceleration, intentional speed), driving rules (safety distance, overtaking) and physical laws (braking distance, adaptation distance). Vehicles characteristics refer to typical values for cars currently available on the market. Data on the typical vehicles velocity in a city centre are drawn from [8], which contains a study on the users velocity measured in several European big cities' centres during different hours of the day and different days of the week. In Figure 1 we show a graph taken from these measurements. As we can see the most recurring measured velocity (32%) is 20 km/h and the range between 20-40 km/h represents the 85% of total observations. In traffic jam situations the average velocity is 13 km/h, while in fluent traffic condition it varies from 26 to 32 km/h.

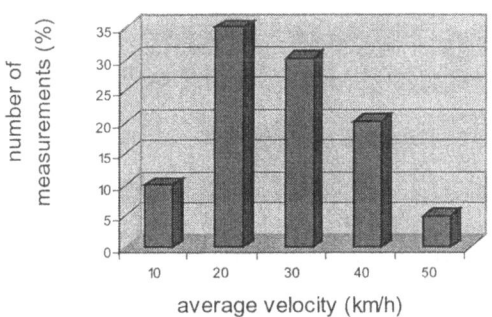

Fig. 1. Measured average velocity distribution.

The *traffic* simulator can take as an input the profile of the *intentional velocity* for a vehicle. This is the velocity the user would reach if traffic conditions allow doing it. So we approximated the measured distribution in Fig. 1 with a normalised Erlang distribution that is reported in Fig. 2 and is described by the following function:

$$f(v) = (1/14!)v^{14}e^{-v}. \tag{1}$$

This distribution has been given in input to the traffic simulator as the intentional velocity profile. A further issue concerns the calculus of the arrival rate of new cars in the simulated street segment. We calculated it starting from a relationship found in [9] and reported in Fig. 3 (interpolated from measured values) between the average velocity a car can maintain travelling in an urban centre and the density of cars on the same road. From this plot, we calculated the product of car density and velocity and obtained the other key parameter for our traffic simulator, that is the *number of vehicles on the road per hour*. Within the range of velocities of our interest we found a minimal (0.33-*light*) and a maximum (0.55-*heavy*) value of vehicles per second. We used them as the mean values of a Poisson distribution describing the car inter-arrival time.

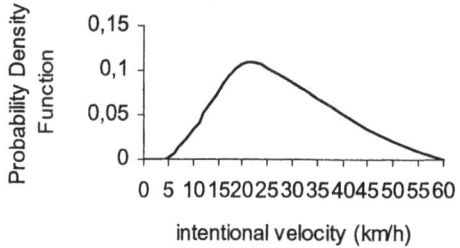

Fig. 2. Approximated pdf distribution for the vehicles' intentional velocities.

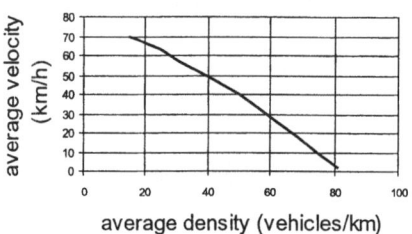

Fig. 3. Relationship between average car velocity and measured density of cars

3 Handover Management Algorithms

In the next subsections we present the basic ideas of the handover and call management algorithms chosen for performance comparison. They are named *idle-bonus* [6], *Y&N* [5], and *Max_vel* [7]. The reader is encouraged to refer to the cited documents for more extensive descriptions. They all remove the classic unrealistic assumption of constant user speed during a call and they all transfer the user-speed classification task from the network to the mobile terminal. Their technique to estimate the user velocity is different. *Idle-bonus* combines the dwell time spent in the last microcell and a "bonus" it gets from the network. *Y&N* and *Max_vel* use the dwell time spent in a given number of crossed microcells. *Y&N* relies on the estimation of the *average* mobile speed, *Max_vel* relies on the *maximum* user speed during the microcell crossings.

3.1 Idle-Bonus Algorithm

The distinguishing feature of this algorithm, which we proposed in [6], is the concept of *bonus* used by the user-speed classification mechanism, and assigned to a mobile terminal according to the street and traffic conditions.

The terminal recognizes itself as slow or fast by taking the cell border crossing times into account. At the connection set-up, the mobile terminal communicates its estimated slow/fast status to the network, which, thereafter, is able to assign it to the right hierarchical layer. The terminal keeps on classifying itself during the call.

If the terminal classifies itself as "fast" (its dwell time is smaller than a dwell threshold time) it receives a *bonus*, consisting in an amount of time to be added with the dwell threshold time. It can be spent during the crossing of a certain number of subsequent microcells. The bonus has the purpose of counterbalancing temporary slowdowns due to traffic jam or traffic lights.

The usage of the bonus makes the system classifying more users as fast, and for this reason it mainly exploits the macrocell tier. As in the algorithm designers' objective, the heavier the macrocell usage, the smaller the handover rate per call.

3.2 Y&N Algorithm

This algorithm, proposed in [5], enriches the literature with at least two more new elements. First it uses an *exponential* averaging of the past dwell times samples collected in a given number of crossed microcells, and compares the results with a dwell threshold time. The main assumption for using exponential averaging is that the average user speed (and then the dwell time) is slowly varying.

Second, Y&N establishes a relationship between the teletraffic load and the threshold velocity. The optimal threshold velocity v_o is calculated from the user velocity distribution $f(v)$, and by fixing the macrocell load corresponding to a target blocking probability (using the B-Erlang formula). When the traffic load increases, the threshold v_o increases as well, so that more mobile terminals can be assigned to the microcells and the system capacity increases. This adaptation is made by the network.

Y&N obtained a relationship [5] between the traffic load and threshold velocity.

3.3 Max_vel Algorithm

In [7] we demonstrated that Y&N algorithm does not perform as in authors' intentions, because in an urban context the average user velocity is not slowly varying, as they suppose. Thus we proposed to implement the same kind of exponential averaging as Y&N does, but using a sequence of *virtual* dwell times collected in the past crossed microcells instead of *measured* dwell times. The virtual dwell time is calculated by assuming that the users cross the entire microcell by travelling at the *peak velocity* they were able to reach in that cell. The result of the exponential averaging does not present aberrations or excessive deviations from its mean value. In [7] we demonstrated that *Max_vel* exploits the virtues of both algorithms: it is successful in trading off the handoff rate per call (idle_bonus) and the system GoS (Y&N).

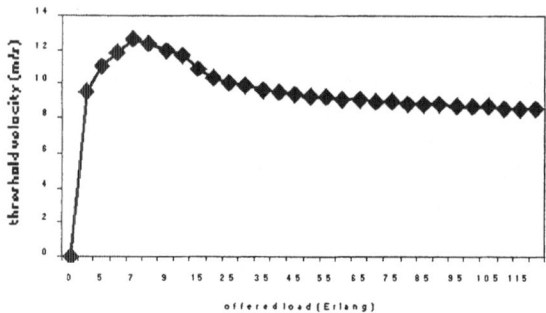

Fig. 4. New relationship between threshold velocity and offered teletraffic load.

Table 1. Main simulation parameters and assumptions.

Environment	3 km one-way two-lanes street
N° of traffic-lights	6
Distance between traffic-lights	about 500 m
Traffic light times	cycle 60–80 s, red 30–35 s
Radio coverage	10 microcells + 1 macrocell
Microcell shape	Circular shape; radius = 200 m
Distance between microcells	300 m
Microcell overlapping length	100 m
Microcell threshold time (idle-bonus)	40, 50, 60 s
Bonus value (idle bonus)	40 s
N° of successive cells to spend the bonus in (idle bonus)	2
N° of past cell sojourn times (Y&N)	5
N°. of radio channels in the macrocell	8
N°. of radio channels in the microcells	12
Car arrivals	Poisson distribution
Mean vehicle generation rate	0.33 (light traffic) - 0.55 (heavy traffic) vehicles/s
Call duration	negative exponential distribution, mean 120 s
Call inter-arrival time per user	negative exponential distribution
Average N° of users (pedestrian+ vehicular)	800
Maximum car acceleration	$1.5 \sim 3.5$ m/s^2
Maximum car deceleration	$(-7) \sim (-10)$ m/s^2
Intentional speed	Erlang distribution (Figure 2)

4 Numerical Results

In this paper, we extend the algorithms comparisons reported in [7] by introducing the more realistic assumptions described in Section 2. Specifically, we recalculate the relationship between the threshold velocity and the offered teletraffic load by considering the new user velocity distribution illustrated in Fig. 1. Following the mathematical calculus explained in [5], we obtained the new relationship plotted in Fig. 4.

Table 1 reports the main simulation parameters and assumption.

Fig. 5 reports the handover rate per call at *light* (0.33 vehicles/s) and *heavy* (0.55 vehicles/s) traffic conditions. In the new realistic urban scenario *Max_vel* performs still better than all other algorithms, even at low offered teletraffic and heavy car traffic loads. It shows even better behaviour when tested under lighter car traffic load.

Numerically, *Max_vel* on average performs 58% and 40% fewer handovers for light and heavy traffic, respectively. This means that the user classification method used by *Max_vel* is successful in assigning fast users to the macrocell layer. The worse performance under heavy traffic is due to the fact that in this situation vehicles rarely reach their intentional velocity due to the traffic jam.

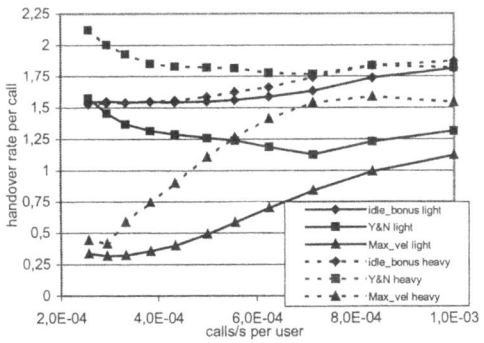

Fig. 5. Handover rate per call for light and heavy car traffic

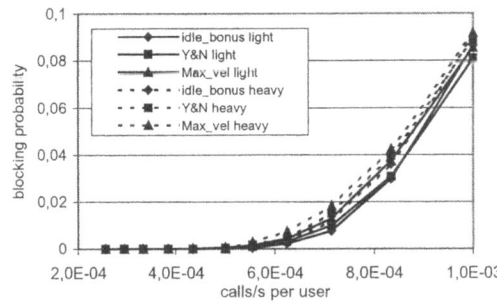

Fig. 6. Blocking probability for light and heavy car traffic

The reduced number of handover in *Max_vel* is not paid in terms of a worse exploitation of the radio resources. The evidences are in Figures 6 and 7 that show new call blocking and handover dropping probabilities. The performances regarding both parameters are very similar for the three algorithms, especially at low call rate per user (under the target blocking rate of 0.02).

Fig. 8 shows the teletraffic (Erlang) conveyed on the macrocell layer. *Max_vel* carries more traffic than the other algorithms as we demonstrated in Fig. 5. It depends on the velocity classification method. In fact, *Y&N* tends to consider as fast only a user who is fast "on average". This means that it is not enough for a user to instantaneously reach a velocity higher than the threshold to be classified as fast, but the average velocity must be higher than the threshold.

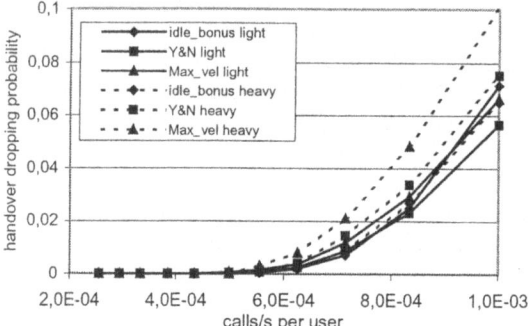

Fig. 7. Handover dropping probability for light and heavy car traffic

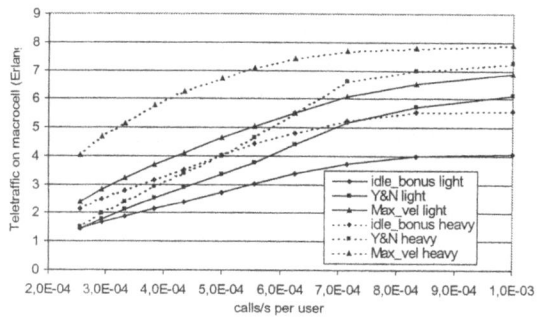

Fig. 8. Teletraffic conveyed in the macrocell layer for light and heavy car traffic

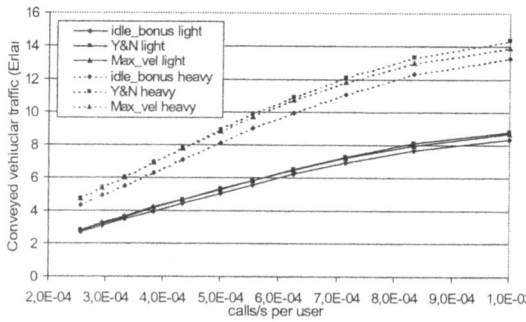

Fig. 9. Teletraffic successfully conveyed in the system for light and heavy car traffic

Finally, we show in Fig. 9 the throughput expressed in terms of teletraffic (Erlang) that is successfully conveyed by the overall system. The results are referred to vehicular users only (pedestrian are always classified as slow by all the algorithms) The low performance of idle_bonus in terms of handover rate in the curves shown up to this point can be explained by the fact that it uses a fixed velocity threshold. On the contrary, Y&N and Max_vel dynamically adapt the threshold according to the teletraffic load (Figure 6). For this reason we show two other curves demonstrating how the idle_bonus performance is dependent on the choice of the threshold. These also testify the possibility for an optimal threshold calibration that could be achieved in order to perform better than the other analysed algorithms.

5 Conclusions

In this paper we have analysed the effect of a realistic urban context on the behaviours of three algorithms for call and handover management in a two-tier cellular system. Differently from previous studies on this topic we have focused on a very realistic urban scenario, trying to simulate not only the high speed variability a user can experience in a city centre, but also a noteworthy users velocity distribution and a significant relationship between the average velocity and density of cars in the observed system. Simulation results have shown it is possible to optimize the resources usage, providing users with a good grade of service together with a very low number of handovers.

References

[1] X. Lagrange, "Multitier cell design," *IEEE Commun. Mag.*, pp. 60–61, Aug. 1997
[2] A. Ganz, C. M. Krishna, D. Tang, and Z. J. Haas, "On optimal design of multitier wireless cellular systems," *IEEE Commun. Mag.*, vol. 35, pp. 88–93, Feb. 1997.
[3] L. C.Wang, G. L. Stuber, and C. T. Lea, "Architecture design, frequency planning, and performance analysis for a microcell/macrocell overlaying system," *IEEE Trans. Veh. Technol.*, vol. 46, pp. 836–848, Nov. 1997.
[4] Y. Chung, D.-J. Lee, D.-H. Cho, and B.-C. Shin, "Macrocell/Microcell Selection Schemes Based on a New Velocity Estimation in Multitier Cellular System", *IEEE Trans. Veh. Technol.*, vol. 51, pp. 893-903, September 2002.
[5] K. L. Yeung and S. Nanda, "Channel management in microcell/macrocell cellular radio systems," *IEEE Trans. Veh. Tech.*, vol. 45, pp. 601–612, Nov. 1996.
[6] A. Iera, A. Molinaro, Marano, "Handoff management with mobility estimation in hierarchical systems", *IEEE Trans. Veh. Technol.*, vol. 51, pp. 915-934, September 2002.
[7] A. Iera, A. Molinaro, E. Natalizio and S. Marano, "Call management based on the mobile terminal-peak velocity: virtues and limitations in a two-tier cellular system", *IEEE Trans. Veh. Technol.*, vol. 52, pp. 794-813, July 2003.
[8] M. Andre, U. Hammarstrom, "Driving speeds in Europe for pollutant emissions estimation", *Transportation Research*, part D 5, Elsevier, 2000.
[9] P. Ferrari, F. Giannini, *Geometria e progetto di strade*, (Geometry and design of roads) vol. 2, ISEDI (in Italian).

Providing Quality of Service for Clock Synchronization

Arthur C. Callado, Judith Kelner, Alejandro C. Frery, and Djamel F.H. Sadok

Centro de Informática – Universidade Federal de Pernambuco
Caixa Postal 7851 – 50732-970 – Recife – Brazil
{acc2, jk, frery, jamel}@cin.ufpe.br

Abstract. The Network Time Protocol, an over 2-decade old and always improving algorithm for synchronizing networked computer clocks, still finds problems for its efficient operation. Many applications need a trustable time system to function correctly (e.g., banking and distributed database servers). With the advent of Quality of Service in computer networks, this problem can be elegantly approached and solved. This article suggests a framework to dealing with clock synchronization on DiffServ Domains, introduces a novel treatment to packets and validates this proposal on a case-study done with live application metrics in a network emulation environment.

1 Introduction

Clock synchronization in the Internet is fundamental to give integrity guarantees to time-sensitive services. This way, the accuracy of clocks depends on the algorithms used and on the network delays and jitter. Guaranteeing minimum and stable network delay is a way of ensuring better clock precision. The best synchronization protocol for IP, the Network Time Protocol (NTP), has been accepted as an Internet standard.

Statistical techniques are used to yield an acceptable solution, but there had been no proposal based on quality of service (QoS). With QoS it is possible to offer guarantees about delay and jitter, improving synchronization. In this work, we analyze the benefits of QoS for clock synchronization.

2 Clock Synchronization in Computer Networks

Nowadays, a huge number of time-servers [8] around the world use the *Network Time Protocol* (NTP) [7][8] to replicate their time. It is the "de facto" standard.

NTP uses a hierarchical and distributed algorithm to share time [7]. Physical clocks [6] are used as references and are attached to *stratum 1* [8] computers. These serve a very trustable time, but at a very high price and require special skills to install and configure. Any computer that synchronizes from a stratum 1 computer has an *association* with it and is called *stratum 2*. It has a local clock to keep time and may serve time to *stratum 3* ones and so on up to *stratum 15*, forming a synchronization tree. The stochastic process of clock synchronization is modeled in [5] and [1].

J.N. de Souza et al. (Eds.): ICT 2004, LNCS 3124, pp. 1151–1156, 2004.

Some interesting metrics can be used for time synchronization [7]. The *stability* of a clock is how well it can maintain a constant frequency, while *accuracy* is how well its frequency compares to time standards and *precision* is how these quantities can be maintained on a system (maximum error estimation). The *offset* of two clocks (here, an NTP client and a server) is the time difference between them. The *skew* represents the frequency difference between them (computed as the first derivative of offset with time) and the *drift* is the variation of skew (second derivative of offset with time).

The synchronization process requires periodic message exchanges between client and server, named T_{1i} to T_{4i}, for each message exchange i. The timestamps are used to compute estimations for clock offset δ and roundtrip time θ, as follows:

$$\delta_i = (T_{4i} - T_{1i}) - (T_{3i} - T_{2i}) \quad \theta_i = \frac{(T_{2i} - T_{1i}) + (T_{3i} - T_{4i})}{2} \tag{1}$$

Based on δ and θ, the clock discipline algorithm corrects the local clock, compensates for its intrinsic frequency error and dynamically adjusts various parameters in response to network jitter and oscillator frequency stability.

According to Mills [6] NTP works better when the network load is not high and even that is not a problem since network links generally spend very little time in high loads. But during high and even low load periods, NTP loses synchrony and the machine depends solely on its local clock. Synchronization depends on the reference clock quality, local clock stability and network delay and jitter. Therefore, by guaranteeing minimum and stable delay for request packets one guarantees permanently good synchronization for computers that use NTP. There has been no work analyzing synchronization algorithms on the light of quality of service.

3   The Differentiated Services Architecture

Diffserv was designed to have a smooth implementation in today's Internet by being able to work with legacy applications and network topologies. Only router update is actually required and can be done on a per-network basis.

The IPv4 header has a byte-long field named *"Type of Service"* (TOS), designed to select different treatment to packets marked by applications. The main idea in Diffserv is to map each configuration (a *Diffserv Code Point* - DSCP) of this byte (called *"DS Field"*) to a different packet forwarding treatment, named *Per Hop Behavior* (PHB). Some PHBs have been standardized, as explained below.

The *Best Effort* (BE) PHB represents the treatment used on the Internet. It delivers fairness (among its packets) without guarantees. It was not standardized for Diffserv, and its definition from [9] is still valid. The *Assured Forwarding* (AF) [2] is a *Behavior Aggregate* (BA) that offers discard priority guarantees within a class. There are 4 AF classes, each with 3 different drop precedence values. It can be used by an application with different flows or by applications with differentiated importance.

Expedited Forwarding (EF) [3] guarantees no loss and delays close to the minimum. EF traffic should be independent of any traffic on the router.

4   QoSYNC Framework Proposal

We propose a configuration scheme for improving quality of clock synchronization (QoSYNC) with of NTP and the Diffserv architecture by using either of two PHBs: the Expedited Forwarding and the proposed Hot-Potato Forwarding, described next.

4.1   The Expedited Forwarding PHB

The EF PHB was proposed for providing minimum delay and no loss, and using it for NTP would have good stability. But NTP does not need the loss guarantees of EF and the delay guarantees could be a little improved. With simultaneous arrivals and small bursts it is common to have a little queuing of EF traffic, hurting synchronization.

In order to configure the routers to transport NTP traffic, there should be a good estimation of NTP traffic load. The EF configuration proposed *must* be done by adding to the reserved EF bandwidth the one that will be used by NTP. However, the average bandwidth is not equal to the peak bandwidth, since routers do not receive uniformly sparse requests. Practice shows that on reasonably loaded NTP servers traffic peaks are rarely more than twice the average. Therefore, system administrators *must* reserve twice the average bandwidth for NTP traffic. If EF is implemented in such a manner that starves other traffic [3], routers *must* also limit (token bucket) the rate of EF at the reserved bandwidth. Without a Bandwidth Broker, periodic measurements on Edge Routers can estimate the amount of NTP traffic to reserve.

4.2   The Hot-Potato Forwarding PHB

This work introduces the Hot-Potato Forwarding (HPF) PHB, intended to be the ideal treatment to packets with strong delay requirements and no delivery requirements. This PHB can be used to construct a lossy, very low latency, very low jitter service. Applications with time-sensitive information, like the exchange of high-resolution timestamps or the monitoring of physical links delay can make good use of it.

The HPF PHB provides forwarding treatment for a particular DiffServ aggregate called HPF traffic. HPF packets arriving at any Diffserv router *should* be immediately forwarded or dropped. No HPF packet should wait longer than a packet time (the time to send a packet in the outgoing network interface) to be sent. This assures a limited minimum jitter, based on the number of nodes and on the packet serialization delays.

No bandwidth reservations are necessary or even desired, although a maximum bandwidth can prevent that misuse of HPF will affect other PHBs, notably EF. HPF traffic *should not* break guarantees of other PHBs. The decision of discarding or not an HPF packet must be based on these guarantees. A router implementing HPF *should not* use traffic shaping of HPF traffic other than discarding nonconforming packets.

Two scheduling mechanisms accomplish the guarantees: *priority queue* and *class based queuing*. A suggestion is to use a queue for HPF and a queue for EF, and the HPF queue with a higher priority and be sized to a packet (if measured in packets) or to the size of the biggest packet expected (bytes). If an HPF packet arrives when another is being sent and there is an EF packet waiting, the HPF should be dropped.

HPF can be employed on any DiffServ router that implements any other PHBs, as long as the HPF specification is respected. If two DS Domains exchange HPF traffic, they should agree on the DSCP for HPF or should do packet remarking. Even though no reliability guarantees are made, a router should only discard an HPF packet if it really has to, to avoid starvation of HPF. No bandwidth is reserved, but the fact that HPF must have the highest priority makes it interfere with other traffic jitter.

5 Case-Study: Evaluation of QoSYNC with Network Emulation

In order to validate the QoSYNC framework, we decided to use a network emulation environment to check how much it improves clock synchronization quality with NTP.

Two NTP machines (a client and a server) were needed. The server had to be synchronized to international standards and could be connected to an atomic physical clock. A simple Diffserv network should separate the machines, with two routers. The routers should be connected through a single link with 40ms delay, shared with other Internet flows. Here, the applications mark their and no remarking is needed.

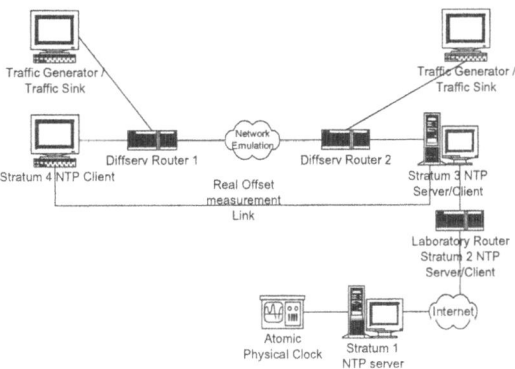

Fig. 1. Implemented Network Topology (with network emulation)

On the emulated topology, a few items were added or replaced (Figure 1). In replacement to external networks, traffic generators (for self-similar traffic) were used. A dedicated link between client and server was added to measure real clock offset. A network emulation machine was used as a link between routers. It delayed all packets in any direction by exactly 40ms, with no artificial jitter introduction (besides queuing). Without a physical clock, an external time reference was used. The laboratory router was used as an NTP gateway between our server (now stratum 3) and a remote stratum 1. To make sure our server furnished a trustable time, all servers were configured with a short polling interval ($P = 32$ seconds), and run for several days before the starting the experiment to let corrections stabilize.

A preliminary test was run to decide the number of replications needed to achieve statistical guarantees [4], based on the maximum error acceptable at the confidence level of 90%. Table 1 compares the results for each variable in each scenario. As a result, the test was made with 120 replications (hours), which resulted in 5 days running each of the three scenarios. Though we gathered real and estimated offsets, this work also analyzed the difference between them, the offset estimation error.

Table 1. Determining Number of Replications

Analyzed Variable	Scenario	Sampled Average	Max. Error	Replications
Real Offset	BE	0.009817	15%	117.8108
Real Offset	EF	0.004889	15%	49.98287
Real Offset	HPF	0.003376	15%	32.98568
Estimation Error	BE	0.002501	15%	110.1464
Estimation Error	EF	0.002707	15%	42.46485
Estimation Error	HPF	0.002421	15%	33.31118

Table 2. Maximum Possible Jitter Estimate

Type of Traffic	Observed Jitter
Best Effort	1.5 ms
Expedited Forwarding	0.275 ms
Hot-Potato Forwarding	0.095 ms

The maximum jitter (Table 2) was measured in each scenario for the type of traffic of NTP with simple "ping" (changing the DS Field). After the run of each experiment, 1000 ping packets were sent and half of the difference between the maximum and minimum observed RTT (round-trip time) delay was taken to represent the one-way jitter. This difference is an estimate of the maximum possible jitter, though the jitter itself is the difference observed between two adjacent packets of a microflow.

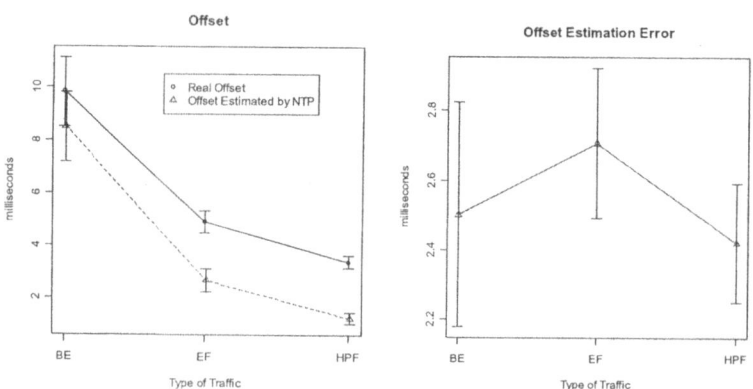

Fig. 2. Offset (a) and Offset Estimation Error (b) Comparison

The jitter observed on HPF still a little higher than what should be, but it is due to the use of general-purpose computers (with network activities treated by software) to act as routers. In all types of traffic, the jitter is limited by the use of only two routers, and on a given network topology the number of routers and the configuration of the queues affect observed jitter. The three scenarios did not show any significant change in the loss rate of the background traffic, and the number of total lost packets did not differ significantly (0.499% for all tests). The goal of the background traffic was to imitate Internet behavior for NTP, not to analyze how NTP would affect other traffic.

On Figure 2 (a) we see a comparison between the resulting clock offsets for each type of traffic, computed as absolute values. The vertical bars represent the standard deviation of the data weighted by the trust interval coefficient [4], based on the confidence level of 90%. NTP performs systematically better as EF or HPF traffic than it does as BE traffic. As expected, HPF traffic represents a little improvement over EF. Figure 2 (b) shows the offset estimation error, computed as absolute values. Changing the type of traffic did not significantly affect its offset estimate. Though the average has increased a little for EF, its variation is smaller, resulting in a very small aggravation of this estimate. The jitter caused by EF is small, but it is unpredictable, since some packets receive minimum delay and others receive higher ones. HPF traffic, however, showed a little improvement on estimation mean and variation.

6 Conclusions and Future Work

This work has focused the distribution of time for clock synchronization in networked computers and proposed some network configurations based on the use of Quality of Service for dealing with it. It was shown that the use of the Expedited Forwarding PHB or the proposed Hot-Potato Forwarding PHB can help ensure permanent synchronization of a computer with NTP and also improve the quality of such synchronization. There was no need to alter the behavior of NTP, just QoS markings.

Many networks can benefit from the use of this proposal, which can decrease the overall use of NTP traffic on the Internet (it is well known that in order to recover from its synchronization losses NTP raises its bandwidth) and improve its accuracy.

We devised two future works: a broader study of the peak-to-mean relation for NTP traffic, allowing better estimation of peak EF bandwidth reservation; and a study of the effect of HPF traffic on TCP connection jitter (not using HPF *for* TCP).

References

1. Callado, A., "Clock Synchronization in Computer Networks with Quality of Service", master in science dissertation, Federal University from Pernambuco, Sep. 2003.
2. Heinanen, J., BAKER, F., WEISS, W. et WROCLAWSKI, J.. "Assured Forwarding PHB Group", RFC 2597, June 1999.
3. Jacobson, V., NICHOLS, K. et PODURI, K. "An Expedited Forwarding PHB", RFC 2598, June 1999.
4. Jain, Raj. "The Art of Computer Systems Performance Analysis: Techniques for Experimental Design, Measurement, Simulation, and Modeling". John Wiley & Sons. 1991.
5. Mills, David L., "Adaptive Hybrid Clock Discipline Algorithm for the Network Time Protocol", IEEE Trans. on Net., volume 6, number 5, Oct. 1998.
6. Mills, David L., "Experiments in Network Clock Synchronization", RFC 957, Sep. 1985.
7. Mills, David L., "Network Time Protocol (Version 3) Specification, Implementation and Analysis", RFC 1305, March 1992.
8. Mills, David L., "Simple Network Time Protocol (SNTP) Version 4 for IPv4 IPv6 and OSI", RFC 2030, Oct. 1996.
9. Postel, J. "Internet Protocol", RFC 791, Sep. 1981.

Application of Predictive Control Algorithm to Congestion Control in Differentiated Service Networks

Mahdi Jalili-Kharaajoo[1,2] and Babak N. Araabi[2]

[1] Young Researchers Club
Islamic Azad University, Tehran
[2] Control and Intelligent processing Center of Excellence
ECE Department, University of Tehran, Iran
mahdijalili@ece.ut.ac.ir, araabi@ut.ac.ir

Abstract. The Differentiated Service (Diff-Serv) architectures are proposed to deliver Quality of Service (QoS) in TCP/IP networks. The aim of this paper is to design an active queue management system to secure high utilization, bounded delay and loss, while the network complies with the demands each traffic class sets. To this end, predictive control strategy is used to design the congestion controller. This control strategy is suitable for plants with time delay, so the effects of round trip time delay can be reduced suing predictive control algorithm in comparison with the other exciting control algorithms. Simulation results of the proposed control action for the system with and without round trip time delay, demonstrate the effectiveness of the controller in providing queue management system.

1 Introduction

The rapid growth of the Internet and increased demand to use the Internet for voice and video applications necessitate the design and utilization of new Internet architectures with effective congestion control algorithms. As a result, the Differentiated Service (Diff-Serv) architectures were proposed to deliver Quality of Service (QoS) in TCP/IP networks. Diff-Serv architecture tries to provide QoS by using differentiated services aware congestion control algorithms. Recently several attempts have been made to develop congestion controllers [1,2], mostly using linear control theory. In this paper, the traffic of the network is divided into three types: Premium, Ordinary and Best Effort Traffic Services [3]. For very important people, there are VIPs passes. VIP passes get preferential treatment. This category is likened to our premium traffic Service. For ordinary people, there are common passes. To purchase these tickets, people may have to queue to get the best possible seats, and there is no preferential treatment, unless different prices are introduced for better seats. This category may be likened to our Ordinary Traffic Service. For reasons of economy, another pass may be offered, at a discount price for the opportunists at the door (Best Effort Traffic Service).

In this paper, we will make use of predictive control strategy [4-7] to congestion control in differentiated services networks. Using the proposed control action, congestion control in Premium and Ordinary classes is performed. Best effort class is no-controlled. Some computer simulations are provided to illustrate the effectiveness of the proposed sliding mode controller.

J.N. de Souza et al. (Eds.): ICT 2004, LNCS 3124, pp. 1157–1162, 2004.
© Springer-Verlag Berlin Heidelberg 2004

2 Dynamic Network Model

In this section, a state space equation for M/M/1 queue is presented. The model has been extended to consider traffic delays and includes modeling uncertainties then three classes of traffic services are introduced in a Diff-Serv network.

2.1 Fluid Flow Model

A diagram of a sample queue is depicted in Fig.1. Let $x(t)$ be a state variable denoting the ensemble average number in the system in an arbitrary queuing model at time t. Furthermore, let $f_{in}(t)$ and $f_{out}(t)$ be ensemble averages of the flow entering and exiting the system, respectively. $\dot{x}(t) = dx(t)/dt$ can be written as

$$\dot{x}(t) = f_{in}(t) - f_{out}(t) \tag{1}$$

Equation of this kind of model has been used in the literature, and is commonly referred to as fluid flow equation [8,9]. To use this equation in a queuing system, C and λ have been defined as the queue server capacity and average arrival rate respectively. Assuming that the queue capacity is unlimited, $f_{in}(t)$ is just the arrival rate λ. The flow out of the system, $f_{out}(t)$, can be related to the ensemble average utilization of the queue, $\rho(t)$, by $f_{out}(t)=\rho(t)C$. It is assumed that the utilization of the link, ρ, can be approximated by the function $G(x(t))$, which represents the ensemble average utilization of the link at time t as a function of the state variable. Hence, queue model can be represented by the following nonlinear differential equation [3,8]

$$\dot{x}(t) = -CG(x(t)) + \lambda \tag{2}$$

In this model input and service rates both have Poisson distribution function. For M/M/1 the state space equation would be [9]

$$\dot{x}(t) = -C\frac{x(t)}{1+x(t)} + \lambda \tag{3}$$

The validity of this model has been verified by a number of researchers [3,8,10].

2.2 System Structure and Controller Mechanism

Consider a router of K input and L output ports handling three differentiated traffic classes mentioned above (Fig. 2). The incoming traffic to the input node includes different classes of traffic. The input node then separates each class according to their class identifier tags and forwards the packets to the proper queue. The output port could transmit packets at maximum rate of C_{server} to destination where

$$C_{server} < C_p + C_r + C_b \tag{4}$$

2.3 Premium Control Strategy

Premium traffic flows needs strict guarantees of delivery. Delay, jitter and packet drops should be kept as small as possible. The queue dynamic model can be as

$$\dot{x}_p(t) = -C_p(t)\frac{x_p(t)}{1+x_p(t)} + \lambda_p(t) \qquad (5)$$

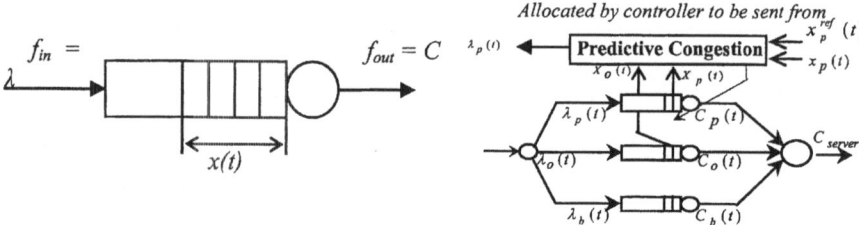

Fig. 1. Diagram of sample queue

Fig. 2. Control strategy at each switch output port

The control goal here is to determine $C_p(t)$ at any time and for any arrival rate $\lambda_p(t)$ in which the queue length, $x_p(t)$ is kept close to a reference value, $x_p^{ref}(t)$, specified by the operator or designer. So in (5), $x_p(t)$ would be the state to be tracked, $C_p(t)$ is the control signal determined by the congestion controller and $\lambda_p(t)$ is the disturbance. Note that we are confined to control signals as

$$0 < C_p(t) < C_{server} \qquad (6)$$

2.4 Ordinary Control Strategy

In the case of ordinary traffic flows, there is no limitation on delay and we assume that the sources sending ordinary packets over the network are capable to adjust their rates to the value specified by the bottleneck controller. The queue dynamic model is as follows

$$\dot{x}_0(t) = -\frac{x_0(t)}{1+x_0(t)}C_0(t) + \lambda_0(t-\tau) \qquad (7)$$

where, τ denotes the round-trip delay from bottleneck router to ordinary sources and back to the router. The control goal here is to determine $\lambda_o(t)$ at any time and for any allocated capacity $C_o(t)$ so that $x_o(t)$ be close to a reference value $x_o^{ref}(t)$ given by the operator or designer. There are two important points that must be considered, first, $C_o(t)$ is the remaining capacity, $C_o(t)=C_{server}-C_p(t)$ and would be considered as disturbance which could be measured from the premium queue. In our controller scheme we would try to decouple the affect of $C_o(t)$ on the state variable $x_o(t)$, and the another point is that λ_o is limited to a maximum value λ_{max} and no negative λ_o is allowed i.e.

$$0 <= \lambda_o(t) <= \lambda_{max} <= C_{max}$$

2.5 Best-Effort Traffic

As mentioned in the previous section, best effort traffic has the lowest priority and therefore could only use the left capacity not used by Premium and Ordinary traffic flows. So, this class of service is no-controlled.

3 Predictive Congestion Controller Design

A predictive control anticipates the plant response for a sequence of control actions in future time interval, which is known as prediction horizon [4]. The control action in this prediction horizon should be determined by an optimization method to minimize the difference between set point and predicted response. Predictive control belongs to the class of model based controller design concepts. That is, a model of the process is explicitly used to design the controller, as is illustrated in Fig. 2. Usually, predictive controllers are used in discrete time. Supposed the current time is denoted by sample k, $u(k)$, $y(k)$ and $w(k)$ denote the controller output, the process output and the desired process output at sample k, respectively. More details about this strategy can be found in [4-7].

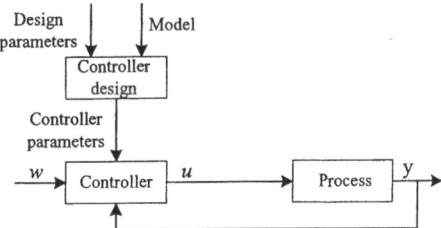

Fig. 3. Scheme of model based control

To design the controller, we have made the following assumptions for controller design throughout this paper
$C_{max} = 300000$ Packets Per Second

$\lambda_{max} = 150000$ Packets Per Second

In addition at first is assumed there is not any delay in system ($\tau = 0$).

The simulation results are depicted in Figs. 4, 5 and 6 for premium traffic, and in Figs. 7, 8 and 9 for ordinary traffic. Figs. 4 and 7 show $x(t)$ with $x_{ref}(t)$ for Premium and Ordinary classes, respectively where good behavior for rising and settling of $x(t)$ is clear. The input and output rates of Premium buffer are shown in Figs. 5 and 6, respectively. Figs. 8 and 9 shows the input and output rates for the Ordinary buffers as well. To investigate the robustness of proposed controller, the round trip time delay and uncertainty is applied to the system as follows:

$$G(x(t)) = (1 + \frac{10}{100}) \frac{x(t)}{1 + x(t)}, \tau = 3 \ m \sec \qquad (8)$$

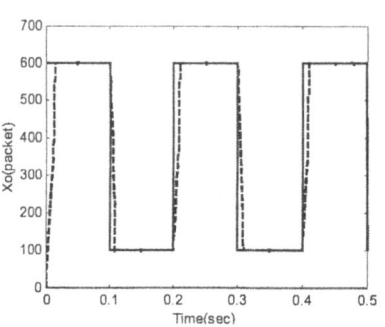

Fig. 4. $x_p^{ref}(t)$ and $x_p(t)$.

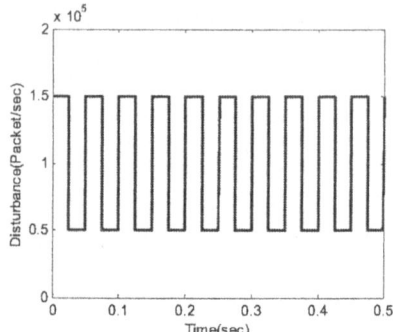

Fig. 5. Input rate of Premium's buffer

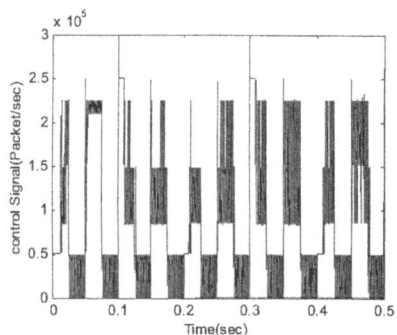

Fig. 6. Output rate of Premium's buffer

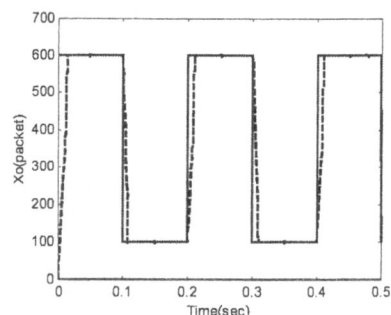

Fig. 7. $x_o^{ref}(t)$ and $x_o(t)$

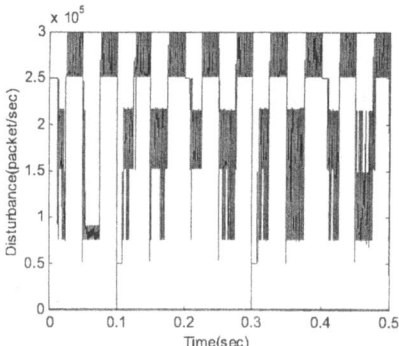

Fig. 8. Input rate of Ordinary's buffer

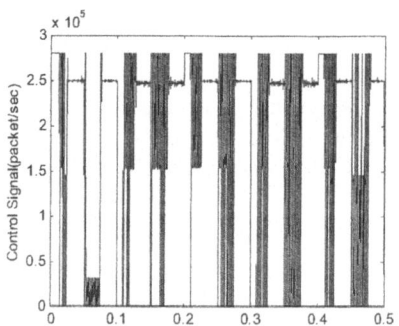

Fig. 9. Output rate of Premium's buffer

Figs. 10 and 11 shows the set point tracking behavior of $x_o(t)$ and $x_p(t)$, respectively with above conditions. It is evident that the performance of $x_p(t)$ with the proposed control action does not vary much; so the above uncertainty does not effect on the closed-loop system very much. The performance of $x_o(t)$ is a little worst than the case of without delay. It means that our proposed robust controller still needs to be improved to compensate the effect of round trip time delay.

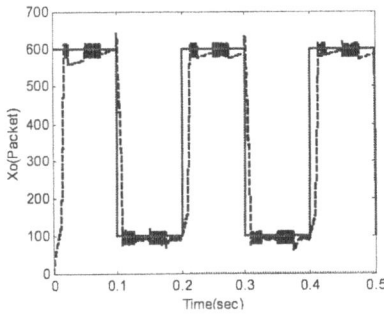   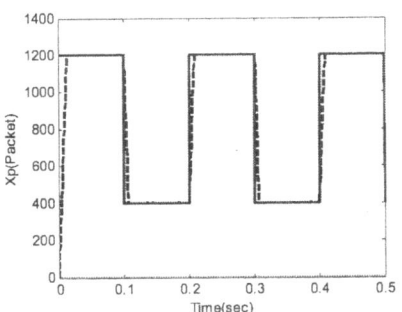

Fig. 10. $x_p^{ref}(t)$ and $x_p(t)$ with delay   **Fig. 11.** $x_p^{ref}(t)$ and $x_p(t)$ with delay

4 Conclusion

In this paper, predictive Controller was applied to congestion control in Differentiated-Services networks. A differentiated-services network framework was assumed and the control strategy was formulated for three types of services: Premium Service, Ordinary Service, and Best Effort Service. The proposed control action demonstrated robust performance against round trip time delay. Some computer simulations showed good and satisfactory performance for the proposed controller.

References

1. Kolarov and Ramamurthy G., A control theoretic approach to the design of an explicit rate controller for ABR service, *IEEE/ACM Transactions on Networking*, October 1999.
2. Pitsillides and Lambert J., Adaptive congestion control in ATM based networks: quality of service with high utilization, *Journal of Computer Comm.*, 20, 1997, pp. 1239-1258.
3. Pitsillides A.and Ioannou P., Non-linear Controllers for Congestion Control in Differentiated Services Networks, *TR-99-1, Dept. CS, University of Cyprus*, 2001.
4. Camacho, E.F. *Model predictive control*, Springer Verlag, 1998.
5. Garcia, C.E., Prett, D.M., and Morari, M. Model predictive control: theory and practice- a survey, *Automatica*, 25(3), pp.335-348, 1989.
6. Parker, R.S., Gatzke E.P., Mahadevan, R., Meadows, E.S., and Doyle, F.J. Nonlinear model predictive control: issues and applications, *In Nonlinear predictive control theory and practice, Kouvaritakis, B, Cannon, M (Eds.)*, IEE Control Series, pp.34-57, 2001.
7. Jalili-Kharaajoo, M. and Araabi, B.N. Neural network control of a heat exchanger pilot plant, *to appear in IU Journal of Electrical and Electronics Engineering*, 2004.
8. Sharma, S., D. Tipper, Approximate models for the study of nonstationary queues and their applications to communication networks, IEEE ICC 93, May 1993.
9. Tipper D., Sandareshan M. K., Numerical Methods for modeling Computer Networks Under Non-stationary Conditions, *IEEE Journal SAC*, Dec. 1990.
10. Rossides L., Pitsillides A. and Ioannou P., Non-linear Congestion control: Comparison of a fluid flow based model with OPNET simulated ATM switch model, TR-99-1, Dept. Computer Science, University of Cyprus, 1999.

Design of a Manageable WLAN Access Point

Timo Vanhatupa, Antti Koivisto, Janne Sikiö, Marko Hännikäinen,
and Timo D. Hämäläinen

Institute of Digital and Computer Systems, Tampere University of Technology
Korkeakoulunkatu 1, FIN 33720 Tampere, Finland
Tel. +358 3 3115 2111, Fax +358 3 3115 3095
{timo.vanhatupa, marko.hannikainen}@tut.fi

Abstract. This paper presents the design and prototype implementation of a manageable WLAN Access Point (mAP). mAP has been developed for managing WLAN Quality of Service (QoS), frequency selection, client configuration, and for collecting a wide range of management information. The prototype is implemented using a Linux platform. With the presented architecture, the mAP functionality can be easily extended by adding new management functions and automated services.

1 Introduction

The number of Wireless Local Area Network (WLAN) installations has increased rapidly during the past few years. This growth has been fuelled by user mobility and the easier installation of WLANs compared to wired LANs. Most of the currently used networks meet the IEEE 802.11b, 802.11g, and 802.11a WLAN standards, providing the nominal bandwidth from 11 MBit/s up to 54 MBit/s [1][2].

Increasing WLAN usage has revealed problems for managing these networks. The requirements set for traditional wired networks for configuration, fault, performance, security and accounting management apply also for WLANs [3]. Special requirements for WLAN management are emphasized as the number of parallel networks is increased and service requirements of applications become more demanding. A natural location for the management is WLAN Access Point (AP), which connects a wired backbone and WLAN [5].

Recent research work related to WLAN management has concentrated on developing features such as firewall technologies [8], algorithms for load balancing [9], and mathematical models for allocating radio channels in WLANs [4]. A significant development work is also the Inter Access Point Protocol (IAPP) specified by the IEEE 802.11f standard [10]. Its target is to provide tools for roaming between APs.

In the infrastructure WLAN, AP has a significant responsibility when reaching for efficient and autonomous management. In this paper, we present our own concept for a manageable WLAN AP that we call mAP. The target operating environment for mAP is presented in Fig. 1. In the presented network, a Network Management System (NMS) [3] is responsible for managing a network containing mAPs, legacy WLAN APs and a gateway router, providing access to the Internet.

J.N. de Souza et al. (Eds.): ICT 2004, LNCS 3124, pp. 1163–1172, 2004.

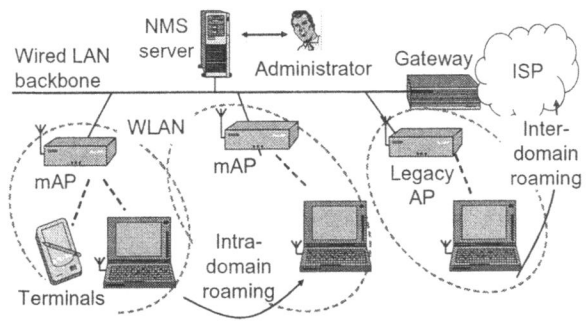

Fig. 1. The operating environment of mAP.

The network management functionality in the presented architecture is divided between the NMS server and mAPs. All management functions in mAP have corresponding functions in the NMS server. They are responsible for the network management in the scope of the whole LAN, thus the architecture of NMS is centralised [11].

This paper is organized as follows. Sect. 0 defines the requirements for mAP. The functional architecture of mAP is presented in Sect. 0 and the prototype implementation in Sect. 0. As a proof of mAP functionality, Sect. 0 presents the test results of two management operations. Finally, Sect. 0 gives conclusions and presents future work.

2   Requirements for a Manageable AP

Traditional network management focus on the whole network. This consequently leads to the requirements placed on a single managed network node. We have divided the functional requirements for mAP into five groups. The groups are the *QoS, radio, terminal, bridge* and *security* management, and the *non-functional* requirements.

For enabling usable QoS, WLAN must provide a negotiated share of bandwidth to applications [6]. This is possible only if application traffic flows (e.g. TCP sessions and UDP streams) are identified and controlled. Consequently, mAP is required to implement traffic classifying and conditioning functions. Traffic can be controlled either flow basis as in the Integrated Services (IntServ) [12] or class basis as in the Differentiated Services (DiffServ) [13]. If a link level QoS mechanism, such as IEEE 802.11e [25] or 802.1D priorities [16], is used in network, mAP should provide mapping from traffic classes used in wireless link to those used in backbone network.

One of the main factors affecting WLAN throughput is the interference caused by other WLANs located on the same geographical area [4]. In the presented network architecture, NMS is responsible for the overall network configuration. To facilitate frequency planning, NMS needs information about AP signal propagation as well as neighbour APs on the area. Traditionally, this information has been collected by manual on-site measurements. However, they can not be used to collect up-to-date information because of frequent changes in the network topology and environment.

The terminal management includes collecting terminal information and controlling of their operations. Statistical information on terminal behaviour needs to be available for developing the network infrastructure [7]. The terminal control relates to load balancing between APs, terminal roaming, and consequently it affects QoS.

The load balancing algorithms control which terminals are allowed to associate with mAP. This is also necessary for minimizing the negative effects of uncontrolled handovers on QoS due to reassociation and possible routing changes [14]. Roaming between APs located in different management domains requires inter-domain management signalling. A good alternative for interoperability is to implement the IAPP standard in mAP.

The bridge management monitors the state of mAP, including CPU loading and traffic load in each network interface. NMS needs information about load peaks and traffic trends to enable load balancing, capacity planning, and network monitoring [6].

Security should be considered on the scope of the whole network. Thus, requirements for a specific node depend on the selected approach for implementing the security functionality. WLAN security has mainly been pursued using centralized security architectures [1]. However, several requirements can be identified specifically for mAP. First, mAP should collect behaviour and position information about neighbour APs and terminals in order to detect possible rogue APs and misbehaving/unauthorized terminals. Second, mAP should contain a firewall for dividing a network into different security zones. Third, mAP should be able to force the disassociation of misbehaving or unwanted terminals.

Manual configuration should be minimized by automating the management functionality in mAP. As the complexity of managed systems increase, the data centric approach used by the Simple Network Management Protocol (SNMP) comes more and more difficult to handle by an administrator and NMS. It is more effective when administrators view the managed nodes in task oriented manner, and not as a set of configurable attributes [15]. The task oriented approach avoids semantic mismatch between mAP and the administrator view.

To provide a task oriented view, mAP should contain a message based management interface that consists of all accepted management operations. Additionally, mAP should also support traditional data centric monitoring with SNMP. For cost-efficiency, it is required that implementation has low footprint, processing, and memory requirements. Accurate clock synchronization is required for traffic analysis and QoS measurement purposes.

3 Functional Architecture of mAP

The functional architecture of mAP is presented in Fig. 2. The main parts are the *mAP adaptor* and the *legacy AP components*. Existing legacy AP components are WLAN and Ethernet interface, bridge, and Management Information Base (MIB) module. The mAP adaptor is an execution environment for *management functions* and *services*. The mAP adaptor, management functions, and services are developed for mAP.

The mAP adaptor is able to dynamically load and execute management functions. This enables one to easily add new services and functions, also remotely. The mAP adaptor also provides the basic inter-communication for management functions.

Management functions contain the intelligence in mAP. Currently, six management functions have been designed and implemented. These are the *terminal, AP,* and *flow monitors,* and *QoS, traffic,* and *frequency managers.*

The purpose of executing management functions in mAP is to reduce the amount of management network traffic and the workload of the NMS server. This is possible because a considerable part of the management decisions can be done locally. However, the mAP architecture preserves a single point of configuration because the management functions in mAP receive their configuration from the corresponding functions in the NMS server.

In mAP, services are utilized by management functions. Services execute conceptually low level management tasks and tasks that can be shared by several management functions. A service implements an Application Programming Interface (API) for a certain purpose but does not contain management intelligence. Examples of low level tasks are the setting of the current frequency channel, transmission power, and radio scanning interval.

3.1 Services

Currently, the services defined for mAP are the *AP control, flow meter, mMIB* and *traffic control.* The AP control service provides an access to legacy AP components. It keeps an up-to-date record of neighbour APs, their network names, used frequency channels, and signal strengths. The service also monitors associated terminals and contains an access list for defining the terminals allowed to associate. The access list is a set of Medium Access Control (MAC) addresses. The list is used primarily for load balancing purposes, while its security value is low due to such concerns as stealing the MAC address [1].

The flow meter service analyses the Internet Protocol (IP), TCP, and UDP headers of received and transmitted packets. The service assigns timestamps to the packets and measures their queuing delays. The packet measurement data, such as transmitted

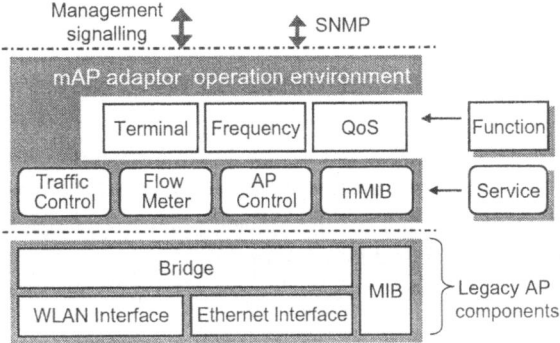

Fig. 2. Functional mAP architecture.

bytes, is summarized for a detected flow. This information is used by the flow monitor function, which is responsible for supervising the actual flow characteristics and informing NMS.

The traffic control service is responsible for classifying and conditioning traffic flows in network interfaces. Traffic is classified using filters that identify data packets belonging to a certain flow. The flow identification is based on source and destination IP addresses and port numbers, and optionally protocol and Type of Service (TOS) bits.

In traffic control, bandwidth for each interface is divided into *dedicated* and *statistical* traffic. The difference between these traffic types is that mAP reserves bandwidth for each dedicated flow separately, whereas each statistical traffic class may contain several flows.

The parameters configured for each dedicated flow and statistical class are the average and maximum data rate, the size of the burst allowed to exceed the average rate, and priority. The priority for dedicated flows can be used for controlling delays, while in statistical traffic a priority value defines a traffic class priority compared to other classes. Seven statistical traffic classes are supported in mAP. These are the network control, voice, video, controlled load, excellent effort, best effort, and background, as defined in [16].

The traffic control service is an IP level solution for QoS management and it controls both wireless and wired interfaces, and works in the same way for uplink and downlink traffic. The service is also capable for supporting link level QoS mechanisms. For example, each traffic class of IEEE 802.11e can be directed to a separate queue inside mAP.

The mMIB service supports SNMP based monitoring with NMS and traditional network management tools. The service also provides API for management functions for storing and retrieving local management information.

3.2   Functions

The frequency manager function is responsible for monitoring the networks, channels and signal strengths of the neighbour APs. This information is stored using the AP control service. Neighbour APs are scanned with specified intervals (e.g. 10 s), thus the data transfer is interrupted as little as possible. If neighbour APs appear, vanish, or their signal strength changes significantly, the changed information is sent to NMS. Consequently, NMS is capable of calculating the optimal channel for mAPs in the network.

Neighbour APs are also monitored by the AP monitor function that is responsible for providing an overview of the WLANs in the area to NMS. The AP monitor utilizes same scanning information as the frequency manager but is implemented as a separate function because of its different purpose.

The traffic manager function observes traffic loads of the network interfaces in mAP. When capacity limit is exceeded, the function informs NMS that mAP is overloaded. Thus, NMS can assign terminals to less congested mAPs.

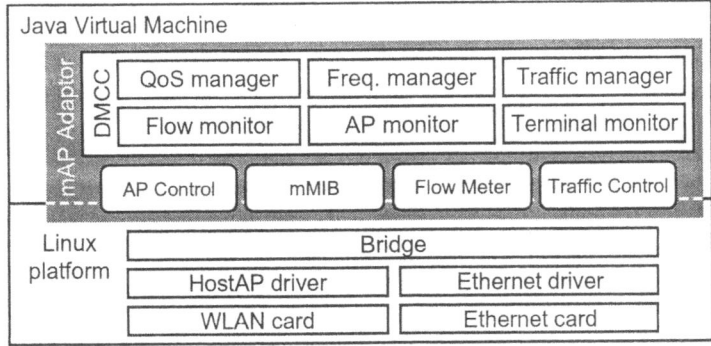

Fig. 3. mAP implementation.

Preventing a terminal from using a congested mAP can be done with the mAP access list but finding an alternative mAP and guiding the terminal there is more complicated. Several load balancing algorithms are being evaluated, including the use of position of a terminal, allowing a terminal to associate only with a single defined mAP, and using client software to control the association.

The terminal monitor function collects information about associated terminals, including MAC addresses and signal strength levels of received packets. This information is used by NMS for providing up-to-date view of the network usage.

The QoS manager and flow monitor functions are responsible for controlling traffic conditioning and measuring that reserved bandwidth is provided. The QoS manager receives input QoS parameters from NMS and reserves transfer capacity using the traffic control service. The flow monitor provides flow characteristics to NMS for further processing. NMS combines traffic characteristics from several mAPs and provides a view in the scope of the whole WLAN. NMS is also responsible for synchronizing mAPs to provide comparable timestamp values from multiple sources.

4   Prototype Implementation

The prototype was implemented based on the Linux operating system because of its rich variety of existing services such as *Traffic Control* [17], *HostAP* device driver [18] and *Bridge* [19]. The physical platform is a laptop PC. The main component, the mAP adaptor, is implemented jointly on top of the Java Virtual Machine (JVM) and Linux. The Java programming language was selected for rapid development and platform independence.

For the mAP prototype, all management functions, the AP control, mMIB, and main parts of the flow meter service are implemented with Java. However, due to the performance requirements, some services are implemented using C++. For example, lower levels of the flow meter service are implemented with C++, containing time stamping, protocol header analysis, and the merging of packet measurements into flow measurements. This gives reasonable performance for testing. For large scale flow monitoring, critical parts such as timestamping should be implemented closer to

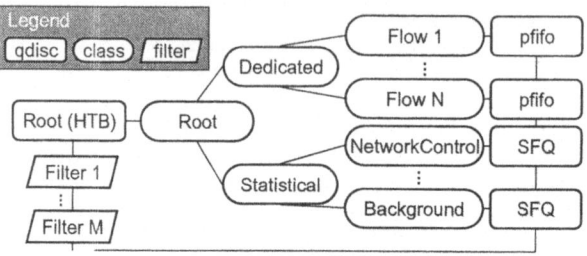

Fig. 4. The traffic control service implementation.

the hardware. The execution environment for management functions in mAP is named as Distributed Management Code Container (DMCC). Fig. 3 presents the prototype implementation of mAP.

The traffic control service is implemented using a Linux Traffic Control module with an added management API for controlling it. The Linux traffic control module implements several queuing disciplines, classes, and filters [17][20]. Traffic is controlled in output interfaces, where bandwidth is divided into classes using a Hierarchical Token Bucket (HTB) queuing discipline, as presented in Fig. 4. HTB supports several attributes on which the average and maximum data rate, burst size, and priority are used.

Statistical and dedicated traffic utilize different queuing disciplines. For statistical traffic, a Stochastic Fairness Queueing (SFQ) discipline is used for each traffic class to guarantee fairness between the flows inside a class. A default queuing discipline, pfifo, is used with the dedicated flows, since there is no need for additional queuing policies within a flow.

The HostAP driver is used for controlling the WLAN adapter card. The driver implements the basic AP control functionality, such as frequency and power control, frequency scanning, terminal association control, and the collection of traffic statistics. It supports wireless LAN cards based on the Prism2/2.5/3 chipset [21].

The NMS server contains a Java Message Service (JMS) [22] implementation called OpenJMS [23] that is responsible for the communication between management functions. JMS supports two communication models queues (send – receive) and topics (publish – subscribe) that are both used in the implementation. Each management function has own queue where it receives messages targeted only to that particular function. Additionally, NMS utilizes several topics that are listened by management functions.

One example of using topics is roaming inside a management domain. When a terminal changes its association from one mAP to another, the destination mAP sends a message to the terminal roaming topic listened by the source mAP. The HostAP driver keeps a record of associated terminals where the terminals are removed after a fixed inactivity period. This time period is necessary for preventing the oscillation when a terminal is soon re-associated. However, by informing other mAPs about terminal roaming, mAP can react faster to the permanent disassociation of the terminal.

In the prototype, the NMS server implementation is similar to mAP. The server also contains DMCC and management functions. WLAN is controlled using the NMS management functionality instead of configuring each mAP individually. The NMS

server is implemented with Java. It contains a WWW server for loading a user interface that is implemented as a Java applet [11].

5 Prototype Measurements

This section presents two measurement cases of mAP management functions. These
are QoS management, and flow monitoring. The test cases cover the main WLAN
management tasks for verifying the mAP functionality.

Fig. 5 presents the test setup for the QoS management measurements. In the test,
terminals A and B, associated in the same mAP, receive data from a gateway server
acting as a traffic source. A test application used for the traffic generation and measurement is a UDP sender/receiver utility RUDE [24]. The receiver utility logs all
packets received by the terminal. The throughput graph seen in Fig. 5 was created by
combining the logs from the terminals A and B.

The sender utility creates four data flows, from which two are sent to the terminal
A and two to the terminal B. All flows have a constant 2.2 Mbit/s send rate, but use
different statistical traffic classes. The reserved capacities for the classes used by
flows 1, 2, 3 and 4 are 2 Mbit/s, 600 kbit/s, 400 kbit/s, and 200 kbit/s, respectively. At
the beginning of the measurement, traffic is not controlled in mAP and throughput for
all flows is about 1.2 Mbit/s, as can be seen from Fig. 5. The traffic control is activated at 20 s from the start, and after a short stabilization period all flows received
configured throughputs. As can be seen, the QoS management is able to provide controlled throughput for application traffic flows.

The previous example covers the downlink from AP to the terminal. QoS management works in the same way also for uplink traffic. However, uplink UDP traffic
is more difficult to control because the terminal has no rate control mechanism (as
opposed to the slow-start mechanism of the TCP). Thus, a misbehaving terminal is
able to send UDP traffic at any rate, possibly using more than its configured bandwidth. The means for mAP to prevent this is forcing the terminal to disassociate.

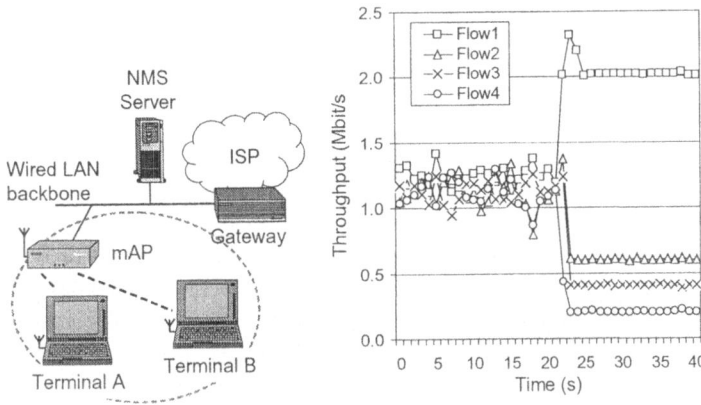

Fig. 5. The setup for QoS management (left), and QoS management test results (right).

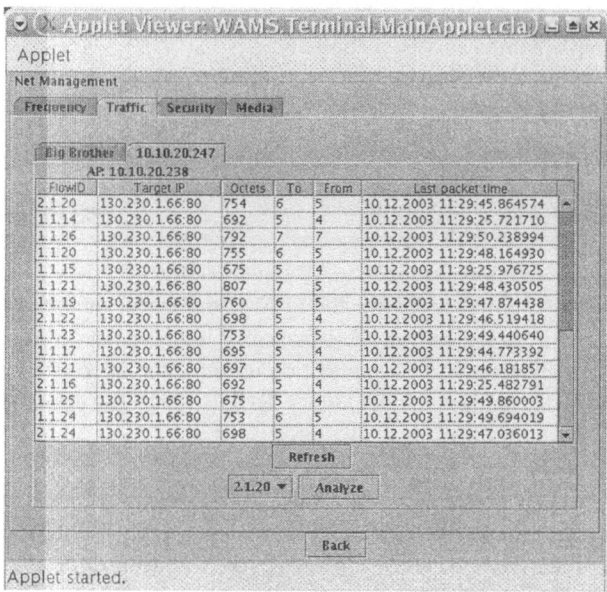

Fig. 6. Flow monitor user interface view.

Consequently, without either a link level QoS mechanism or a traffic control extension at the terminal, traffic differentiation cannot be guaranteed for uplink traffic.

In the second test, flows were monitored using the tools provided by NMS and mAP. The flow meter service in mAP analyzed the headers of transmitted packets and summarized the measurement data for detected flows. This information about flows is given to the flow monitor for further analysis. The flow monitoring view generated by NMS in Fig. 6 presents the total transmitted octets, number of received and transmitted packets, target IP address and port number, and time the last packet for each flow in the selected terminal. This data can be further analyzed by NMS to calculate delays in the route through the management domain.

6 Conclusions

The functional requirements for mAP, the designed architecture and the implementation of a mAP prototype were presented in this paper. mAP has been developed for managing WLAN Quality of Service (QoS), frequency selection, client configuration, and for collecting a wide range of management information. The functionality of mAP can be easily extended using the presented architecture, by adding new management functions and services. Two tests of WLAN management operations were presented for verifying the prototype functionality. Future work will concentrate on developing new automated management functionality for mAP and for the scope of the whole WLAN.

References

1. Park, J.S. et al.: WLAN Security:Current and Future. IEEE Internet Computing, Vol. 7 (2003) 60-65
2. IEEE 802.11 working group, available at: http://grouper.ieee.org/groups/802/11
3. Subramanian, M.: Network Management principles and practice. Addison-Wesley (2000) 40-44
4. Youngseok, L. et al.: Optimization of AP placement and channel assignment in wireless LANs. IEEE Conference on Local Computer Networks (LCN'2002) 831-836
5. Kuorilehto, M. et al.: Implementation of wireless LAN access point with quality of service support. IECON (2002) 2333-2338
6. Balachandran, A. et al.: Wireless hotspots: current challenges and future directions. WMASH (2003) 1-9
7. Balachandran, A. et al.: Characterizing user behavior and network performance in a public wireless LAN. SIGMETRICS (2002) 195-205
8. Haverinen H.: Improving user privacy with firewall techniques on the wireless LAN access point. PIMRC (2002) 987 -991
9. Balachandran A. et al.: Hot-spot congestion relief in public-area wireless networks. Mobile Computing Systems and Applications. (2002) 70-80
10. IEEE Trial-Use Recommended Practice for Multi-Vendor Access Point Interoperability via an Inter-Access Point Protocol Across Distribution Systems Supporting IEEE 802.11™ Operation, IEEE Standard 802.11F (2003)
11. Rantanen, T., et al.: Design of a Management System for Wireless Home Area Networking. Euro-Par, (2003), 1141-1147
12. Braden, R., et al.: Integrated Services in the Internet Architecture: an Overview. RFC 1633 (1994)
13. Blake, S., et al.: An Architecture for Differentiated Services. RFC 2475 (1998)
14. Mishra, M., et al.: An Empirical Analysis of the IEEE 802.11 MAC Layer Handoff Process. University of Maryland Tech Rep. UMIACS-TR-2002-75 (2002)
15. Schoenwaelder, J.: Overview of the 2002 IAB Network Management Workshop. RFC 3535 (2003)
16. IEEE Common Specifications – Part 3: Media Access Control (MAC) Bridges, IEEE Standard 802.1D (1998)
17. Almesberger, W.: Linux Network Traffic Control -- Implementation Overview. Annual Linux Expo, Raleigh, NC (1999) 153-164
18. Malinen, J.: Host AP driver for Intersil Prism2/2.5/3, http://hostap.epitest.fi/
19. Linux IEEE 802.1d ethernet bridging,: http://sourceforge.net/projects/bridge/
20. Hubert B.: Linux Advanced Routing & Traffic Control, http://lartc.org/howto/
21. GlobespanVirata, PRISM WLAN, http://www.globespanvirata.com/prism.html
22. Sun Microsystems,: Java Message Service Version 1.0.2b, http://java.sun.com/products/jms/docs.html
23. OpenJMS, Open source JMS implementation, http://openjms.sourceforge.net
24. Laine, J., et al.: RUDE homepages, http://www.atm.tut.fi/rude/
25. IEEE 802.11e draft/D8.0: Part 11: Wireless Medium Access Control (MAC) and Physical Layer (PHY) specifications: Medium Access Control (MAC) Quality of Service (QoS) Enhancements. (2004)

Performance Evaluation of Circuit Emulation Service in a Metropolitan Optical Ring Architecture

V.H. Nguyen[1], M. Ben Mamoun[2], T. Atmaca[1], D. Popa[1], N. Le Sauze[3], and
L. Ciavaglia[3]

[1] Institut National Des Télécommunications - 9, Rue Charles Fourier, 91011 Evry, France
{viet_hung.nguyen, tulin.atmaca, daniel.popa}@int-evry.fr
[2] Université Mohammed V, B.P 1014, Rabat, Maroc
ben_mamoun@fsr.ac.ma
[3] Alcatel R&I, Route de Nozay, F-91460 Marcoussis, France
{nicolas.le_sauze, laurent.ciavaglia}@alcatel.fr

Abstract. Circuit Emulation Service is a new technology which allows the transport of TDM service such as PDH (E1/T1/E3/T3) as well as SONET/SDH circuit over a packet switched network. This paper presents the simulation based performance evaluation and QoS guarantee of circuit emulation service over a peculiar metropolitan optical ring network.

Keywords. Circuit Emulation Service, TDM, Quality of Service, Metropolitan Networks, Optical Ethernet Ring, Performance Evaluation.

1 Introduction

Traditionally, voice has been carried over Time Division Multiplexed (TDM) based networks. TDM based networks such as SONET/SDH offer high reliability and survivability for connections and predictable delays for voice samples, thus providing a superior quality. Today, the volume of data traffic in the world's telecommunication networks has outstripped the volume of voice traffic. This growth of data traffic is primarily due to the growth of internet and the increase in multimedia traffic. This growth has also lead to the ubiquitous development and deployment of packet switched networks (PSN). With the increase in access transport network and the evolution of packet switched networks, service providers have benefit to transport TDM (not only voice but also video) traffic and data traffic over the same packet switched network architecture. This convergence of TDM and data traffic in an existent PSN architecture could save considerable equipment and installation cost. Thus packet switched networks, which were initially designed for only transporting data traffic, are now facing the challenge of carrying TDM service.

Circuit Emulation Service (CES) is a technology allowing the transport of TDM service such as PDH (E1/T1/E3/T3) as well as SONET/SDH circuit over a packet switched network. Circuit emulation originally comes from Asynchronous Transfer Mode (ATM) world [1]. The idea has been taken up in the packet switched world by a number of organisms, including the Internet Engineering Task Force (IETF), the Metro Ethernet Forum (MEF) and the Multi-Protocol Label Switching (MPLS) forum. The main CES standards are being set by the Pseudo-Wire Emulation Edge to

J.N. de Souza et al. (Eds.): ICT 2004, LNCS 3124, pp. 1173–1182, 2004.

Edge (PWE3) working group in the IETF [2, 3]. This group is chartered to develop methods to carry Layer-1 and Layer-2 services across a packet switched network (principally IP or MPLS). Hence the group is looking at TDM circuit emulation, and also carriage of Layer-2 service such as ATM, Frame Relay and Ethernet across the PSN. The Metro Ethernet Forum [4] is looking to extend the work of the PWE3 group to make it applicable to a metropolitan Ethernet context. Similarly, the MPLS Forum [5] is also looking at the same standards from the point of view of an MPLS network.

In this paper, we study the Circuit Emulation Service on a singular metropolitan optical Ethernet ring network. The rest of this paper is organized as following. Section 2 describes the ring architecture and its main features. Section 3 presents the studied CES model, the CES packet format, the TDM frame segmentation mechanisms and the QoS definition for CES. Section 4 discusses the considered performance parameters and the QoS requirements for CES over a Metro Ethernet network. Section 5 provides some simulation results on the CES performance. Finally, section 6 gives some conclusions and future researches.

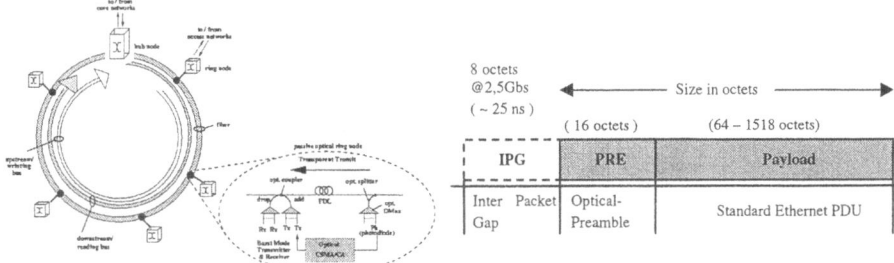

Fig. 1. Metropolitan optical ring architecture   **Fig. 2.** Optical Ethernet frame structure

2   Optical Ring Architecture and Main Features

The network considered in this work is a metropolitan optical Ethernet ring network, called DBORN - Dual Bus Optical Ring Network [6]. DBORN uses a ring topology with a spectral separation of upstream/writing (from ring nodes to the hub node) and downstream/reading buses (from the hub to ring nodes) (Figure 1). This spectral separation allows the use of a simple passive structure for the optical part of ring nodes. At the optical level, ring nodes first use an optical coupler to separate an incoming signal into two identical signals: the main transit signal and its copy used for the control. A Fiber Delay Line (FDL) creates a fixed delay on the transit path between the control and the add/drop function. This FDL, which is larger than the MTU (Maximum Transmission Unit) of the communication protocol used, allows avoiding collisions. The packet insertion on the upstream bus is insured by a Carrier Sense Multiple Access with Collision Avoidance (CSMA/CA) medium access control (MAC) protocol, which is based on void detection.

To have a simple and flexible architecture, Ethernet is used as the convergence layer for the data plane. The structure of an optical Ethernet frame (OEF) is shown in Figure 2. The frame size is kept unchanged (from 64 bytes to 1518 bytes). The existing standard extensions (IEEE 802.1Q/802.1p /802.3ad) are still applicable, and only the preamble field is adapted for an asynchronous optical transport layer. In this study, we use a traffic control mechanism called TCARD (Traffic Control Architecture using Remote Descriptors) [7] that has been designed in the framework of DBORN to face the unfairness issues on the upstream/writing bus due to the ring topology. Indeed, several ring nodes share the upstream bus bandwidth. Hence the upstream nodes may grab all the bandwidth and block the emission of downstream nodes. TCARD is a preventive mechanism that guarantees the access to the resource for downstream ring nodes. With TCARD, free bandwidth is preserved by a ring node according to traffic requirements of downstream nodes (e.g. based on Service Level Agreements - SLA). This is completed by the generation of anti-tokens that forbid the node emission, and hence preserving voids for downstream nodes. More details about the TCARD mechanism and the DBORN architecture can be founded in [6, 7].

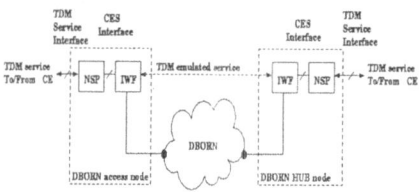

Fig. 3. CES reference model

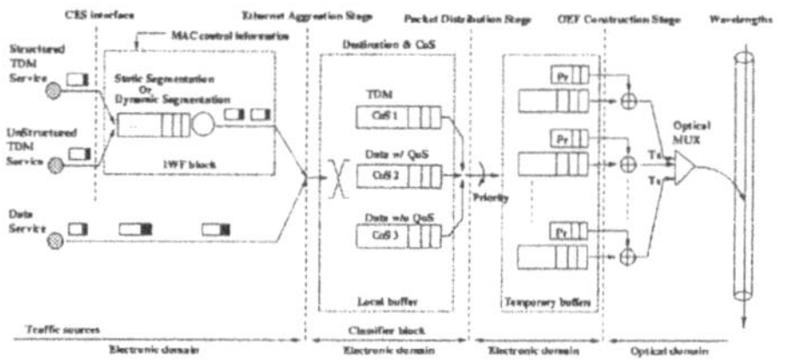

Fig. 4. CES logical DBORN access node model

3   CES Model and QoS Definition

3.1   CES Reference Model

The global model for circuit emulation presented in PWE3 draft [2] is adopted as the reference CES model for our work. Figure 3 presents the general model of CES on DBORN. We have two TDM customers' edges (CE) communicating via DBORN. One CE is connected to a ring access node; the other CE is connected to the ring HUB node (this study is limited on DBORN upstream bus only). TDM service generated by ingress TDM customer edge is transported/emulated by DBORN to egress TDM customer edge. The emulated TDM service between the two CEs is managed by two inter-working functions (IWF) implemented at appropriate nodes.

CES has two principle modes of functioning. In the first one, called "unstructured" emulation mode, the entire TDM service bandwidth is emulated transparently. The frame structure of TDM service is ignored. The ingress bit stream is encapsulated into an emulated TDM flow (called also CES flow) and is reproduced at the egress destination. The second mode, called "structured" emulation, requires the knowledge of TDM frame structure being emulated. Individual TDM frames are visible and are byte aligned in order to preserve the frame structure. "Structured" mode allows frame-by-frame treatment, permitting overhead stripping, flow multiplexing/demultiplexing. In the reference model of CES, the Native Service Processor (NSP) block performs some necessary operations (in TDM domain) on native TDM service such as overhead treatment or flow multiplexing/demultiplexing, terminating the native TDM service coming/going from/to CE.

The main functions of an Inter-Working Function are to encapsulate TDM service in DBORN packets (i.e., Ethernet packets in our case), to perform TDM service synchronization, sequencing, signaling, and also to monitor performance parameters of emulated TDM service. Each TDM emulated service requires a pair of IWF installed respectively at the ingress and egress provider edges (PE). The aim of our study is to evaluate the logical performance of CES on DBORN. Hence, we ignore some operations and functional blocks in the CES reference model, which are outside the scope of this paper, such as the synchronization and signaling aspect of the IWF block, and the functioning of the NSP block.

3.2   CES Logical Model at DBORN Node

A logical architecture for a DBORN access node (the CES ingress side), which supports the emulation of TDM service, is shown in Figure 4. Each access node is composed of an electronic part and an optical part. The incoming TDM service (treated as "structured" or "unstructured") is mapped into Ethernet packets thanks to the IWF block. A segmentation mechanism can be applied on large TDM frames in order to fit them to Ethernet packet. We will discuss the static and dynamic segmentation policies in the next section. Ethernet packets transporting data service and Ethernet packets transporting TDM service are aggregated into an electronic local buffer. Here all packets are classified, according to their destinations and classes of service (CoS), into three separated buffers corresponding to three CoS. A scheduler, taking into account the CoS priority, distributes all packets from local buffer to temporary sending electronic buffers. The optical Ethernet frames (OEF) are built by adding an optical preamble (Pr) to each electronic packet, and then are sent on appropriate wavelength.

At the egress side (the HUB node), the same architecture with some modifications concerning the IWF block is used. A jitter buffer (or playout buffer) is introduced in IWF block. On one hand, the size of the jitter buffer should be large enough in order to accommodate the expected TDM frame jitter. On the other hand, the jitter buffer should not introduce excessive delay in the emulated circuit. The jitter buffer size is a local parameter of the IWF block and can be variable according to the statistics of the frame delay variation. In this work, we focus on the performance measure of CES on DBORN. The main function of the egress IWF is to measure the CES performance (e.g., delay, jitter, and loss) based on delivered Ethernet packets transporting TDM service. The other aspects of IWF, such as jitter buffer dimensioning and reconstruction of native TDM frames, are not considered.

3.3   CES Ethernet Packet Format

We adopted the CES packet format proposed in the PWE3 draft [3] (Figure 5). A CES control word is added to each TDM payload. The main functions of the CES control word are to differentiate the network outage and the emulated service outage, to signal problems detected at the IWF egress to the IWF ingress, to save bandwidth by not transferring invalid data, and to perform packet sequencing if RTP header is not used . An RTP header may be added to the resulting packet for the synchronization and sequencing. The new resulting packet is encapsulated in the CES Ethernet packet by adding Ethernet and multiplexing header. All details about the structure of CES control word and RTP header are described in [3].

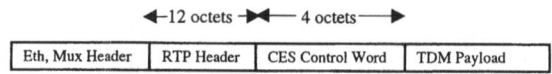

Fig. 5. CES Ethernet packet format

3.4   Segmentation Mechanism for TDM Frames

As we explained above, in order to perform CES on DBORN, TDM frames are encapsulated into Ethernet packets. A TDM frame would ideally be relayed across the emulated TDM service as a single unit. However, when the combined size of TDM frame and its associated header exceeds the maximum transfer unit (MTU) size of DBORN, a segmentation and re-assembly process should be performed in order to deliver TDM service over DBORN.

We have proposed two segmentation mechanisms. The first one, called *dynamic segmentation*, fragments a TDM frame into small segments according to void size detected on the medium (wavelength) by the MAC unit. This approach promises a good use of wavelength bandwidth, but technically it is complex to implement. The second one, called *static segmentation*, segments the TDM packet according to a predefined threshold. This technique is simple to implement, and it provides resulting TDM segments with predictable size. Thus current TDM monitoring methods could be reused, simplifying the management of CES. In the framework of this study, we used the static segmentation method to evaluate the performance of CES on DBORN.

Segmentation threshold is a parameter that we have to determine during this work. The PWE3 draft [3] has recommended some rules to determine the segmentation threshold. First, the segmentation threshold should be either an integer multiple or an integer divisor of the TDM payload size. For example, for all unstructured SONET/SDH services, the segmentation threshold could be an integer multiple of STS-1 or STM-0 frame of 810 bytes. Second, for unstructured E1 and DS1 services, the segmentation threshold for E1 could be 256 bytes (i.e., multiplexing of 8 native E1 frames), and for DS1 could be 193 bytes (i.e., multiplexing of 8 native DS1 frames).

3.5 QoS Definition

The QoS definition for all types of service in DBORN is given in the Table 1. As the TDM service requires high quality of service, it is given the highest priority. The medium class (data service with guarantee of QoS) can be considered as real-time data traffic (e.g., video streaming, and voice over IP). Real-time traffic generates packets whose size is distributed between 100 bytes and 300 bytes. For the reason of simplicity, we suppose the mean packet size of 250 bytes for CoS2. The CoS3 or Best-Effort (BE) service is sporadic internet traffic, which has no guarantee of QoS. Generally, BE packets have size varying between 50 bytes and 1500 bytes .

Table 1. QoS definition

CoS Type	QoS			
	Priority	PLR	Delay	Jitter
CoS1 (TDM)	High	10^9	Strictly limited	Strictly limited
CoS2 (data with QoS)	Medium	10^{-6}	Limited	Limited
CoS3 (data without QoS)	Low	No guarantee	No guarantee	No guarantee

4 Performance Parameters

Many performance parameters must be met in order to support circuit emulation on a packet switched network such as DBORN. Based on the MEN requirements [4], we focused on three main parameters for CES below.

The CES Ethernet Frame Loss (FL) is defined as the ratio of lost Ethernet frames carrying TDM service among total sent Ethernet frames carrying TDM service. The CES Ethernet end-to-end Frame Delay (FD) is the maximum delay measured for a percentile P (superior to 95%) of successfully delivered Ethernet frames carrying TDM service over a measured interval T. The CES Ethernet Frame Jitter (FJ) is derived from the FD measured over the same measurement interval T and percentile P. FJ is obtained by subtraction of the lowest frame delay from FD. FJ is used typically to size the Jitter buffer at the egress side. All these parameters must meet the MEN requirements for CES given in [4]. Concretely, the FL and FD shall be kept to a minimum, and the FJ shall not exceed *10 ms*.

5 Simulation Parameters and Results

We used simulations to evaluate the performance parameters of CES. In our study we consider an upstream/writing bus that consists of 8 access nodes and one HUB node. The bus starts at the first access node and ends at the HUB node. At each access node, we set the CoS1 buffer capacity to 100 Kbytes, the CoS2 buffer capacity to 100 Kbytes and the CoS3 capacity to 250 Kbytes. One or more wavelengths running at 2.5 Gbs may be used on the upstream bus.

To model the packet arrival process at each access node, we used three types of traffic source. The Constant Bit Rate (CBR) source is used to model TDM traffic. The exponential source with packet size of 250 bytes is used to model CoS2 traffic. The CoS3 traffic is modeled by Interrupted Poisson Process (IPP) source. The packet size is assumed of 50 bytes (small), of 500 bytes (medium) or of 1500 bytes (big). The packet length repartition is 10% of small size packets, 40% of medium size packets and 50% of big size packets. This distribution of packet length is inspired from the internet traffic statistic in [8]. We suppose that the offered traffic on the upstream bus is repartitioned uniformly. This means that each access node is fed with the same arrival processes, and the average arrival rate at each access node is identical. The TCARD mechanism is enabled at each access ring node, which allocates for downstream nodes (i.e., all access nodes closer to the HUB node than the current access node) the mean bandwidth equal to their mean bit rate.

5.1 Segmentation Threshold Choice

We first analyze the impact of segmentation threshold on the performance of circuit emulation. We suppose that all access nodes on the upstream bus share two wavelengths running at 2.5 Gbs. Each access node is fed with one unstructured STS-1 service, which corresponds to frame of size 810 bytes generated each 125 microseconds. The volume distribution of the offered traffic at access node is 10% of TDM traffic, 20% of CoS2 traffic and 70% of CoS3 traffic. We set the average arrival rate at each access node to 0.5 Gbs. Thus, the average offered load of upstream bus is 80%.

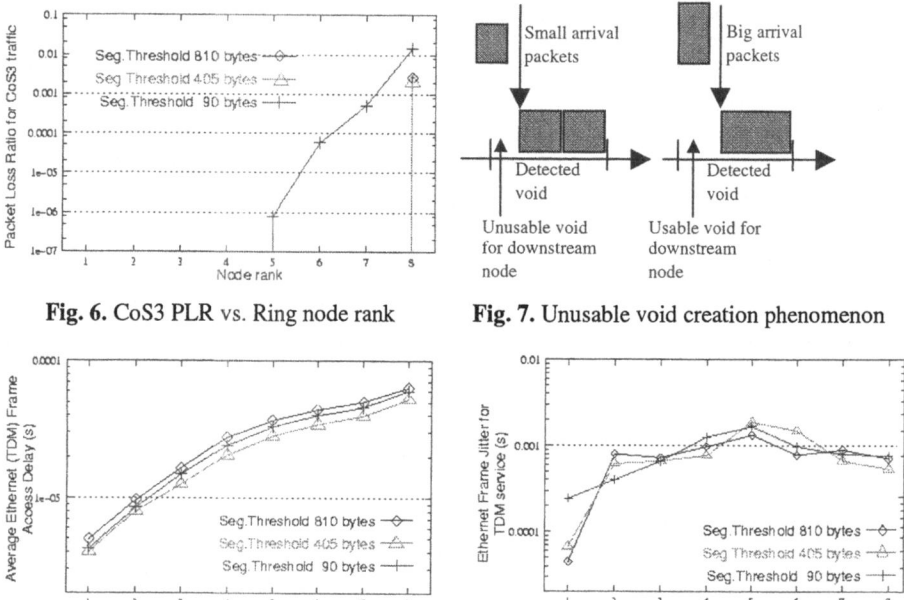

Fig. 6. CoS3 PLR vs. Ring node rank   **Fig. 7.** Unusable void creation phenomenon

Fig. 8. TDM AAD vs. Ring node rank   **Fig. 9.** TDM FJ vs. Ring node rank

We consider three performance measures: Packet Loss Ratio (PLR), average access delay (AAD) and jitter. Figure 6 shows the CoS3 traffic PLR in the case where the segmentation threshold for TDM frames is 810 bytes, 450 bytes or 90 bytes. In our experiments, we observe no loss for high priority traffic such as TDM and CoS2. Thus the FL for TDM service is zero. We observe in the Figure 6 that the segmentation threshold of 90 bytes provides a highest PLR for CoS3 traffic, followed by the segmentation threshold of 810 bytes and 405 bytes. In addition, the segmentation threshold of 90 bytes causes losses at several access nodes, while the segmentation threshold of 810 bytes and 405 bytes cause losses at the last access node only.

We give the explanation of these losses. On the upstream bus, all access nodes share the bus bandwidth. The first access node, which starts the bus, has no problem to access bandwidth. Moving forward on the bus, the downstream nodes encounter the difficulty to access bandwidth. This explains the CoS3 loss at the last node in all cases. The highest PLR obtained with the segmentation threshold of 90 bytes is due to the overhead added to each segment. Indeed, for each segment of 90 bytes, we have to add 40 bytes of overhead (4 bytes of CES header, 12 bytes of RTP header and 24 bytes of optical overhead). This leads to the waste of available bandwidth, which penalizes the emission of downstream nodes. Moreover, introducing small packets on the upstream bus may create small voids that are not exploitable for downstream nodes, evoking bandwidth loss. Figure 7 shows an example illustrating this phenomenon. When a void is detected, packets are inserted on this void. In the case of small packet, only two packets can be inserted and this leaves a small (unusable) void for downstream nodes. In the case of big packet, only one packet can be inserted, and this leaves a bigger (usable) void for downstream nodes.

Let us now look for the AAD of TDM service. We define the access delay for an arrival packet as the waiting time from the moment when it is inserted in the electronic buffer of the node, until it is transmitted successfully on the medium. Figure 8 plots the TDM service AAD versus ring node rank. The segmentation threshold of 810 bytes gives the highest AAD, followed by the segmentation threshold of 90 bytes and 405 bytes. We observe that the AAD increases when the ring node rank increases, due to the difficulty of downstream nodes to find voids to emit packets. Obviously, it is more difficult to find appropriate void to emit a segment of 810 bytes, than emitting a smaller segment of size 405 bytes. However, as we explain above, inserting small segments of 90 bytes on the upstream bus leads to an inefficient use of bandwidth, thus it increases the AAD.

Figure 9 shows the FJ for TDM service versus ring node rank. Recall that this jitter is computed, according to the MEF definition [4], as the difference between the maximum and minimum end-to-end delay value measured for a set of successfully delivered Ethernet packets transporting TDM service. We notice in Figure 9 that the FJ at each access node is fluctuating according to different segmentation thresholds, because it is very related to the variation of maximum and minimum value of end-to-end delay. We observe also that for all segmentation thresholds, the FJ of all access nodes are largely lower than 10 ms, satisfying the MEN requirements in [4].

Overall, all segmentation thresholds of 810 bytes, 405 bytes and 90 bytes provide satisfying performance for TDM service in terms of PLR, access delay and FJ. The small segmentation of 90 bytes gives the worst performance in terms of PLR for CoS3 traffic. Since segmentation process consumes network resource, it should be avoided whenever possible. Thus we think that the segmentation threshold of 810 bytes is preferable for all SONET/SDH services with frame size larger than 810 bytes (e.g., STS-3, STS-12...). For lower rate services such as PDH service (E1, E3...), there is no need to segment.

5.2 Volume of TDM Traffic

We now study the behavior of the network and the performance of CES when the volume of TDM service in offered traffic changes. We suppose that all access nodes on the upstream bus share one wavelength running at 2.5 Gbs. For each simulation scenario, we set the average arrival rate at each access node to 0.25 Gbs. Thus, the average offered load of upstream bus is 80%. The offered traffic at each access node is described in the Table 2. No segmentation is performed for all TDM services in these scenarios.

Table 2. Scenarios for TDM volume study

TDM:CoS2:CoS3	TDM service	Number of TDM service flows per ring node	TDM flow bit rate (Mbs)	TDM flow Frame size (bytes)
10%:18%:72%	Unstructured E-1	12	2.048	32
20%:20%:60%	Unstructured STS-1	1	51.84	810
48%:18%:34%	Unstructured E-3	2	34.368	537
	Unstructured STS-1	1	51.84	810

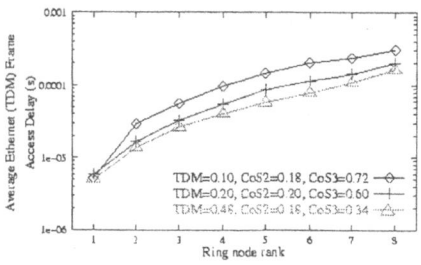

Fig. 10. TDM AAD vs. Ring node rank

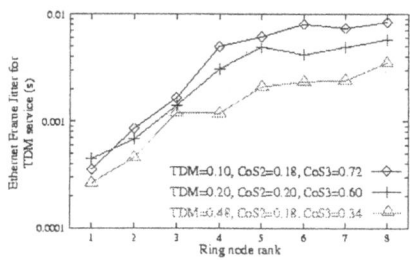

Fig. 11. TDM FJ vs. Ring node rank

We observe no loss for high priority traffic such as TDM and CoS2. The CoS3 traffic PLR (not shown here due to limited space) is highest when the volume of TDM service is 10%, followed by the TDM volume of 20% and 48%. The behavior of these losses is due to the fact that, when the TDM volume is low, the CoS3 volume is high, and this leads to the high loss of CoS3 traffic. When the TDM volume increases, the CoS3 volume decreases, and this decreases the CoS3 PLR.

Figure 10 shows the AAD of TDM service versus the ring node rank. At the first access node, the AAD for TDM service in all cases is almost the same, due to the facility of bandwidth access. At all downstream nodes, the case of 10% TDM volume gives the highest access delay for TDM service. When the TDM volume increases, the AAD for TDM service at downstream nodes decreases.

We now explain the behavior of the curves in Figure 10. The CES performance depends on the relation between different traffic profiles inserted on the sharing medium. In our simulation, with a small volume of TDM service, we have a big volume of bursty CoS3 traffic (e.g., 10% TDM against 72% CoS3). As the TDM service has the highest priority, the CoS3 traffic does not influence the performance of TDM service at the same access node. However, CoS3 traffic inserted by upstream nodes disturbs and consumes free bandwidth for premium service at downstream nodes. Thus it can influence the performance of TDM service at downstream nodes. Concretely it increases the AAD for TDM service at downstream nodes. Therefor, as the volume of CoS3 traffic at upstream nodes decreases, it leaves more usable bandwidth for downstream nodes, which causes the AAD for TDM service to decrease.

Figure 11 shows the FJ for TDM service versus ring node rank. We observe the same trend as in the case of AAD. Moreover, in all configurations, the FJ is always lower than 10 ms, satisfying the MEN requirements.

From the simulation, we observe that there is no superior boundary for the volume of TDM service in realizing CES on DBORN. Indeed, we have performed simulation with 100% TDM traffic on the network, and we have obtained good performance of CES. The other important parameter is that the volume of bursty traffic inserted at upstream nodes can influence the performance of TDM service at downstream nodes.

6   Conclusion and Future Research

In this paper we have provided performance evaluations of circuit emulation service on a peculiar metro Ethernet optical ring architecture - DBORN. By the simulation, we have concluded that CES can be realized in DBORN, even if the bandwidth reservation mechanism used is asynchronous (TCARD). The static segmentation method with threshold of 810 bytes (STS-1 frame size) is proved appropriate to emulate unstructured TDM service. An obstacle that limits the performance of CES on DBORN is the impact of bursty traffic inserted by upstream nodes on the performance of CES at downstream nodes. Further study should be done in order to avoid this drawback. In the future work, we think improving the current access protocol MAC to eliminate a fragmentation of free bandwidth into small voids due to insertion of bursty traffic at upstream nodes. Or we could also imagine a new complicated access protocol that ensures the synchronous transmission for TDM service, and at the same time performs the asynchronous transmission for data service.

References

1. ATM Forum, *Circuit Emulation Service Interoperability Specification 2.0*, http://www.atmforum.com, January 1997.
2. PWE3 - IETF working document, *PWE3 Architecture*, http://www.ietf.org/internet-drafts/draft-ietf-pwe3-arch-04.txt , June 2003.
3. PWE3 - IETF working document, *TDM Circuit Emulation over Packet Switched Network (CESoPSN)*, http://www.ietf.org/internet-drafts/draft-vainshtein-cesopsn-06.txt, June 2003
4. MEF - Metro Ethernet Forum, MEF working document, Requirements for Circuit Emulation Services in Metro Ethernet Networks, revision 3.0, April 2003.
5. MPLS Forum, http://www.mplsforum.org.
6. N.Le Sauze, E. Dotaro, A. Dupas and al., DBORN: A Shared WDM Ethernet Bus Architecture for Optical Packet Metropolitan Network, Photonic in Switching, July 2002.
7. N. Bouabdallah, N. Le Sauze, E. Dotaro, L. Ciavaglia, *TCARD*, OPTICOM 2003, Dallas-Texas USA, October 2003.
8. *IP packet length distribution*, http://www.caida.org/analysis/AIX/plein_hist, June 2002.

Expedited Forwarding End to End Delay Variations

Hamada Alshaer[1] and Eric Horlait[1]

Lip6,UPMC, 8 rue Capitaine Scott,
75015 Paris, France
{hamada.alshaer,eric.horlait}@lip6.fr
http://www-rp.lip6.fr

Abstract. An end to end (e2e) packet delay variations (jitter) has a negative impact on the offered QoS in IP networks. Therefore, in this paper we clarify this passive impact, and discuss the delay jitter that is based on the analysis done in [1]. However, here we focus on the expedited forwarding (EF) class in the differentiated services network (DiffServ). EF flows are represented by renewal periodic ON-OFF flows, and the background (BG) flows by Poisson process. We analyze the jitter effects of these BG flows on EF flows patterns when they are serviced by a single class scheduling discipline, such as FIFO, and a multiclass scheduling discipline, such as static priority service discipline (SPS). Thus, we have simulated a DiffServ network, where different users were provided with different service classes. Consequently, along the simulations different scenarios were formed to see the impact of BG flows and their characteristics on EF flows. As a result, we have found out from these simulations that the EF Per-Hop Behaviors (PHBs) configuration according to RFC 2598 can't stand alone in guaranteeing the EF flows delay jitter. Therefore, playout buffers must be added to the DiffServ network for handling the EF delay jitter problem.

1 Introduction

Nowadays the networks with guaranteed quality of service (QoS) are greatly paid attention. These networks will offer alternatives to the existent ones, which offer a single service class called best effort. An example of such networks is the Internet network, where the end-users have no guarantee on the quality of their required services. Therefore, service models such as asynchronous transfer mode (ATM), integrated service (IntServ), and DiffServ have been developed to support new service classes with varying traffic characteristics and QoS requirements. Recently, the efforts in the world have been intensified on redefining, and improving the design of the DiffServ, so that it supports better the QoS of the supported service classes. Hence, the new demand of QoS (i.e, stringent delay, delay variations, and loss) required for real time applications, such as video streaming, audio, and IP telephony are abstracted in the real time service classes offered by these service models. Furthermore, new management algorithms have

J.N. de Souza et al. (Eds.): ICT 2004, LNCS 3124, pp. 1183–1194, 2004.

been embedded in the control mechanisms in these service models, to manage efficiently the dramatic increase in the network capacity.

In spite of these control mechanisms, still at a given time instant, the cumulative peak rates of the connections supported in the previous service models exceed the network capacity, where serious congestions may occur, resulting in QoS degradation. Even though, network congestion can be controlled by source traffic shaping or regulating in order to reduce the traffic burstiness, the series of traffic connections multiplexing in the network core recreate once again the traffic burstiness causing traffic distortion.

In DiffServ network as shown in Figure 1; EF flows are statistically multiplexed with the BE flows. Thereby, packets belong to the EF flows experience different individual delays at the servers along their paths toward their destinations, which cause distortion in the timing sequence of the real time applications flows serviced as EF flows, arising from their packets delay variations. This can be measured at each node "n" traversed by EF flow through computing the difference between the inter departure and inter arrival times of two consecutive packets belong to the EF flow.

$$\gamma_{j_n} = |(d_i - d_{i-1}) - (a_i - a_{i-1})| \tag{1}$$

Also, packets delay variations can be measured by another parameter, rate jitter, which is the rate of difference between the minimum inter arrival and maximum inter arrival times. This parameter is very convenient for measuring video broadcast over the DiffServ domain, since a slight deviation of rate is translated to only a small paly out delay[6].

$$\delta_{playout} = \lim_{t->\infty} \frac{|(a_i - a_{i-1})^{max} - (a_j - a_{j-1})^{min}|}{t} \tag{2}$$

Thus, playout buffers at the end users are needed to reorganize the time sequence of the interactive applications flows. These de-jitter buffers have been considered the main solution for compensating the packets delay variations in their flows. However, their design still causes a major challenge in dealing with the delay jitter. This challenge has been represented in two points: the first point is the playout buffer mechanism choice, so that if it is a static playout buffer, then the playout instants are chosen according to the first packet arrival in its flow, however if it is adaptive playout buffer, then the playout instants change according to the packets arrivals instants. The second point in the challenge is the playout buffer depth; for example if it is small, then this increases the packets loss probability. However, if it is long, then this adds a significant delay to the flows delay budget, which is not tolerated by the real time flows. Therefore, an optimal choice for the playout buffer's depth, and play out buffer mechanism are required. Nevertheless, the choice is an application dependent [5,2,3,4].

The rest of this paper is organized as follows: In section 2, we discuss some of the most important works that have been done around the EF jitter, and we clarify the problem to be treated in this paper. In section 3, and 3.1 we describe the simulation environment, different traffic classes, and its characteristics. In

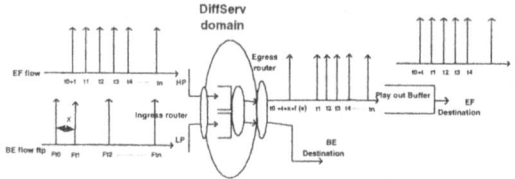

Fig. 1. Delay jitter in Differentiated Services Network.

section 4, and 5, we analyze the EF jitter in a DiffServ network serves aggregates through a single and multiple class service discipline. In section 6, we interpret the delay jitter results of the different EF flows in the DiffServ network shown in Figure 2, and finally, section 7 will include a conclusion to this paper.

2   Related Work and Problem Statement

A relationship between a customer and an IP telephony Operator works on the DiffServ network is controlled in two directions. In one direction the customer is committed to inject his traffic within specific traffic bounds, such as, peak rate and burst size. And, the network operator is designed to guarantee the signed QoS service in a contract; such as, delay, delay variation, and loss rate. Furthermore, the network operator supports different connections with different characteristics, where service overlapping can be occurred at any node in the network core causing jitter for the user flows as shown in Figure 1. Therefore, some regulators are installed at each node in the network core in order to hold a packet in a buffer until its eligibility time for transmission to the next node. In [6] Y.Mansour, and B.Patt-Shamir propose to install a bounded buffer size at each core node to regulate the input traffic to the node, and thus eliminating the jitter at its output. Furthermore, in [8] R.Landry, and I.Stavrakakis introduce a scheduling service discipline called peak output rate enforcement(PORE), which recognizes at the network core nodes the traffic profiles carried out at the edge routers, for each flow, or each traffic class. Then, according to these profiles, it guarantees the periodicity of the flows through keeping their packets at minimum space of time units at each output core node, which results in keeping the flows time sequences organized along their traversals toward their destinations.

Nevertheless, packets inter arrivals and inter departures might be randomly changed due to the statistical multiplexing with other different flows. But, this random change can be characterized by a stochastic process, where certain statistical functions are derived to characterize the distribution of this randomness. This represents in turn the jitter distribution of the flows to which the packets belong. The jitter distribution function might be based on markov chain as it is introduced in [8], by forming a markov chain of the points representing the eligibility times and service times of the tagged packets. Further statistical analysis around the jitter is done in [9], which derives the first order statistical function, that provides information about the jitter length of the different flows in a traffic aggregate, and packets that constitute these flows. Furthermore, it derives the second order statistical functions characterized with the autocorrelation function (ACF), Which provides information about the neighboring packets jitters in their periodic flows.

DiffServ network is an example of multiple traffic classes' networks. It supports three traffic classes: expedited forwarding(EF), assured forwarding "AF", and best effort "BE". The EF class is the key ingredient in the DiffServ for providing a low delay, jitter, and loss. According to RFC 2598 [13], EF is defined as a forwarding treatment for a DiffServ aggregate, where its aggregate's departure rate from any DiffServ node must equal or exceed the configured rate of this node. However, the coexistence of multiple traffic classes in the network makes the different traffic flows to compete for the limited resources, which creates the randomness in the interarrivals and interdeparture of the packets belong to the EF traffic class. Therefore, EF flows suffer from delay jitter in the network. This problem is analyzed in [10] by Jean C.R.Bennett, and J.Yves Le Boudec through analyzing a method to reorganize the structure of the Diffserv network traffic flows, such that EF flows suffer from very limited minimum jitter. However, we see that this approach is hard to be realized. Because, the arrivals and departures of the different flows belonging to the different traffic classes at the different network nodes are random. Furthermore, their routes intersections are totally random. As well as, we see the rearrangement of the EF flows in a tree structure as it is introduced in [10] is nearly impossible. Therefore, we consider the EF delay jitter still needs more clarifications, and in this paper we analyze this problem based on the previous analysis done on the delay jitter in ATM networks [1]. Then, we analyze it through real simulations done on the DiffServ network shown in Figure 2.

3   Network Topology

In order to investigate the e2e delay variations of EF flows, and the effect of other classes properties on this class, a simulation environment of DiffServ network has been developed using NS-2 simulator [11]. The sources and destinations are connected through multiple paths, where a circular traffic can occur. For example, if we follow the arrows in Figure 2, we can notice that: the traffic flows generated from source 2,3 form quasi circular traffic. Furthermore, this network can support more users by connecting them through edges connected to core D,C shown in Figure 2. All the routers output links have the same capacity"C_0", except the circled links have a capacity equivalent to $\lfloor \frac{3}{4} * C_0 \rfloor$. Since, they have been considered the network bottleneck, where the DiffServ network shows its real functionality, and at which we have analyzed the different queuing systems in order to choose the one that meets our study goal.

3.1   Network Traffic

Network traffic is divided into three classes $\{C_{EF}, C_{AF}, C_{BE}\}$, where C_{EF}, and C_{AF} contains a determined number of flows $\{1, 2.., N_{EF}\}$, and $\{1, 2.., N_{AF}\}$ respectively. However, the number of flows in class C_{BE} is kept unknown, and they

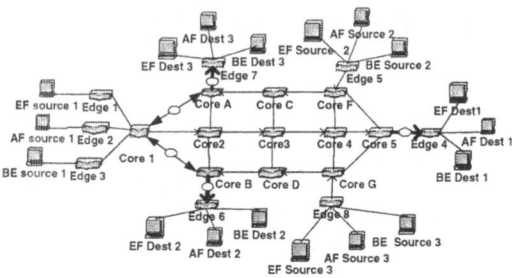

Fig. 2. Differentiated Services Topology.

don't receive a special treatment by the DiffServ nodes. Therefore, this type of traffic can be injected by any network user at any time. Nevertheless, we have kept it under control along the simulations, to see their effects on the EF flows. Along our analysis we focus on the EF class under the varying effects of BE characteristics, burstiness and peak rate on this class. The AF class hasn't been paid attention to, but it is remained as a future point to guarantee its bandwidth requirement.

Nine users divided into three groups according to their Edge routers have been allowed to inject these different traffic classes in the DiffServ domain as shown in Figure 2. A static priority scheduler serves these different classes at the routers output links. Therefore, each class has its own priority queue, and reserved bandwidth. A limited controlled number of tagged flows belong to EF class enter the network through the ingress router 1,5 and 8 as shown in figure 2, until they reach their EF destinations 1,2 and 3 respectively, as shown in the same figure. The tagged flows are those whose e2e delay variations have been monitored, and calculated from their sources until their destinations. The flows that belong to AF and EF class have the same characteristics, peak rate ρ, burst σ, and packet size. While, the flows belong to BE class has different characteristics, and we have changed these from worst case until the optimal case in order to see the impact of this class on e2e delay variations of EF flows. For example, we see clearly from Figure 3, how the EF e2e delay jitter changes in term of the BE packet size.

4   Jitter in Single Class Service Discipline

Scheduling service disciplines(S.S.D) have been considered the main tool in reducing, and improving the delay variations of the traffic flows. Therefore, some S.S.D's have been supplied with further equipments, such as clocks for stamping the different packets. Then, the packets stamps are used by next S.S.D's along their paths to hold the delayed or earliest packets in their regulators to become eligible for transmission. This keeps in turn the flows sequence time of the different flows, which eliminates the jitter of these flows. However, there is a big

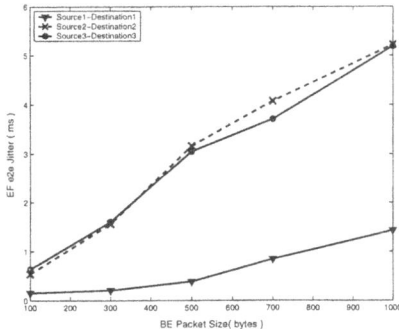

Fig. 3. Effect Of BE Packet Size On The EF e2e Delay Jitter Bound.

gap between what have been theoretically discussed about the scheduling service disciplines, and what is really implemented in the real networks.

FIFO scheduling service discipline is mostly used in the actual networks. This service discipline is used due to its simplicity and availability. However, it causes distortion for a real time traffic class, whenever it supports this class with another one. This distortion is represented in the delay variations of packets belonging to the real traffic flows, which is explained deeply in [1]. However, in this paper we adopt the results of this important reference for analyzing the delay jitter in DiffServ network that support the EF traffic class and others through FIFO S.S.D.

Therefore, if we replace the static priority S.S.D's by FIFO S.S.D's, and disable the AF traffic sources in the network shown in Figure 2, two traffic aggregates compete for the service on the link capacity at the nodes outputs: real time periodic traffic flows (EF), and background traffic "BG"(BE). Furthermore, EF packets are given a priority higher than BG packets. Thus, EF packets are served a head of BG packets that arrive at the same time. Therefore, the delay jitter of EF packets at m^{th} node equals to the difference of BG backlog times measured at two EF consecutive packets instants, if we consider the fluid queuing system. However, in this paper, we consider the discrete queuing time model, where FIFO server serves a packet each time unit. consequently, EF delay jitter is the difference of backlogged BG packets numbers seen by two EF consecutive packets $(Q^n_{BG_{n+1},m^{th}}, Q^{n+1}_{BG_n,m^{th}})$, and their inter arrival time is $I_{n,n+1}$

$$J^{m^{th}}_{EF} = Q^{n+1}_{BG_{n+1},m^{th}} - Q^n_{BG_n,m^{th}} + I_{n,n+1} \qquad (3)$$

Henceforth, delay jitter analysis is coincided with the analysis of the variations of the queue size at the arrival instants of EF packets. From [1], The queue size in Z-domain can be expressed at the arrival instants of EF packets as follows:

$$Q(z) = G'_{EF}(1)(1 - \rho_{total}) \frac{(1 - z^{-1})(B(z)/z)^k}{(1 - zG_{EF}(B(z)/z))}$$

$$\times \frac{\pi_{k=1}^K [(z/B(z)) - (r_k/B(r_k))]}{\pi_{k=1}^K [1 - (r_k/B(r_k))]} \tag{4}$$

Where, $G'_{EF}(1) = (\partial G_{EF}(z)/z)_{z=1}$. $k = G_{max} - 2$, $\rho_{total} = \rho_{BG} + \rho_{EF}$ is the total load at the server. The inter arrival time (I) of EF flows has finite supports; G_{min}, and G_{max}, where $1 \leq G_{min}, \leq G_{max} \prec \infty$. $B(z)$ is the probability generating function (p.g.f) of the random variable corresponding to the BG batch size. And, $G_{EF}(z)$ is the (p.g.f) of the integer random variable I, which can be expressed as follows :

$$G_{EF}(z) = \sum_{i=G_{min}}^{G_{max}} Pr(I = i)z^i = \sum_{i=G_{min}}^{G_{max}} g_i z^i \tag{5}$$

$$J_{EF}^{k^{th}}(Z) = \sum_{i=G_{min}}^{G_{max}} g_i J_i(z) \tag{6}$$

where,

$$J_i(z) = z(B(z))^i + (B(z))^{i-1}(z - 1) \sum_{k=1}^{i-1} (B(z)/z)^{-k} \phi(z^{-1}; k) \tag{7}$$

$$\phi(z^{-1}; k) = \sum_{L=0}^{k-1} z^L \pi_k(0; L) Pr(Q = L) \qquad \forall 1 \leq k \leq i - 1 \tag{8}$$

where, $\pi_k(0; L)$ is the probability that the queue is empty at the Kth time instant following the arrival of EF packet, after all the new arrivals at this time instant have been counted. It is immediate that $\pi_k(0; L) = 0$ for $1 \leq k \leq min(L, G_{max} - 1)$. Hence, $\pi_k(0; L) = Pr(Q(k) + B = 0|Q(0) = L)$.

5   Jitter in Multiclass Service Discipline

Multiclass S.S.D's are often employed in multiclass networks to isolate the different traffic classes from one to another, and manage the output link capacity among the backlogged traffic flows belonging to the different traffic classes according to their QoS requirements. However, the real time traffic flows are subjected at any node in a network to be delayed or their sequence times to be distorted due to other traffic classes' characteristics.

In the network shown in Figure 2 the static priority scheduling (SPS) has been chosen to manage the priorities, and to serve the different traffic classes in the DiffServ domain. We have changed the SPS parameters values such as the EF traffic flows will have absolute priority over the background (BE)flows. Nevertheless, EF flows still suffer from the problem of delay jitter as we are going to see in the simulation section. This problem can be analyzed or interpreted in the following three scenarios:

First, when the packets belong to EF class 1 in the DiffServ network shown in Figure 2 arrive to the server of core router 1, while it is busy in servicing a packet belongs to BE transmitted by BE source 1 or 2. Then, as long as this server doesn't finish servicing the BE packet, other EF packets belong to other EF sources probably arrive, where two sources of delay variations can be underlined in this situation: one due to the BE packet that is being currently serviced by the server. Which, results in a worst delay jitter equivalent to $max \left[\frac{L_{min}}{C_0 - C_{EF}}, \frac{L_{max}}{C_0 - C_{EF}}\right]$. This jitter source forms the second scenario which is similar to that we have explained in section 4. However, in this case the back ground for EF flows generated by EF source 1 are other EF packets generated by other EF sources 2,3 in the DiffServ shown in Figure 2. Nevertheless, we can apply the same analysis of the previous section, but we have to change the values of G_{max}, G_{min} according to the following tow equations respectively: $G_{max} = max \left[\frac{L_{min}}{C_0 - C_{EF}}, \frac{L_{max}}{C_0 - C_{EF}}\right]$, and $G_{min} = min \left[\frac{L_{min}}{C_0 - C_{EF}}, \frac{L_{max}}{C_0 - C_{EF}}\right]$.

Third, when the EF packets transmitted by EF source 1 arrive to the server of core router 2 in Figure 2, where the probability of BG packets arrival rate is around zero. Then, the BG distribution can be characterized by the Taylor series around $\rho_{BE} = 0$. Hence, the EF queue size $Q(z, \rho_{BG})$ at the EF packets arrival instants can be characterized through the expression clarified in the theorem 1, [1,12]:

Theorem 1 (Light Back Ground Traffic). *Given* $B(z, \rho) = 1 + \rho(a(z) - 1) + O(\rho)^2$, *and* $a(z) = \sum_{i=0}^{A_{max}} a_i z^i$, *then if* $A_{max} \prec G_{min} \Rightarrow Q(z, \rho) = 1 + \rho \left[\frac{(a(z) - z)}{z - 1}\right] + O(\rho^2)$

Where, A_{max} is the upper bound on the minimum spacing of arriving EF packets. Which, is determined by the characteristics of the BG traffic. Then, the EF flows jitter can be characterized through the expression clarified in the theorem 2, [12]:

Theorem 2 (EF Flows Jitter). *If* $A_{max} \prec G_{min} \Rightarrow J(z) = G(z) \left[1 + \frac{\rho}{2}(f_1(z) - 1)\right] + O(\rho^2).where, f_1(z) = \frac{a(z) - 1}{z - 1}$

In our simulations, we have used FTP application, whose Poisson distribution as a back ground traffic. Z-transform of Poisson distributed batches can be described as follows:

$$B(z) = e^{\rho(z-1)} \tag{9}$$

Then, by using taylor series we can expand this as follows :

$$B(z) = 1 + \rho(z - 1) + \rho^2 \frac{(z - 1)^2}{2} + O(\rho^3) \tag{10}$$

Then from theorem 1, and background taylor expansion 10, we get a(z) = z, then from theorem 2, f(z) = 1, thus

$$J_{EF}(z) = G(z) + O(\rho^2) \ if \ G_{min} \succ 1 \qquad (11)$$

where, we see no effect for the first order term appeared in expression 10.

So fare, we have focused on a single node analysis. Nevertheless, the previous results form the base in the analysis of jitter in a multiple nodes. Since, if we approximate the departure process of the EF traffic flows from any node in the DiffServ domain shown in Figure2 as a renewal process, the marginal distribution of the departure process of EF flows from node k is approximated by a renewal process with the inter arrival time distribution identical to $J_k(i)$. Denote the sequence $G_{k+1}(i)$ as the probability of inter arrival time of EF flows entering node $k+1$ to be i time service units.

$$G_{k+1}(i) = J_k(i) \quad 1 \leq k \leq N - 1 \qquad (12)$$

and in the Z-domain, we have

$$G_{k+1}(z) = J_k(z) \quad 1 \leq k \leq N - 1 \qquad (13)$$

Consequently, once the EF arrival process at the first node of the network is periodic, then we simply have $G_1(z) = Z^T$. Therefore, if we track the EF flows' jitter from their sources to destinations, then their exact marginal jitter distribution can be obtained.

6   Simulation and Results

In this section, some simulations are provided on the e2e delay jitter of the different EF flows generated by the different EF sources in the network shown in Figure 2. The EF traffic sources are periodic ON-OFF. The Background traffic generated by the different BE traffic sources are identical. Furthermore, they are characterized by the same statistical distribution described in the expression 9. The BE (BG) , AF, and EF sources are controlled at their traffic generation starting times in order to form different scenarios, where different traffic flows generated from the different sources arrive at the same time at the network core nodes. Afterward, we focused on analyzing the EF flows jitters due to other traffic flows' existence through measuring their delay jitter using the expression 1.

Among the different scenarios that have been realized, we focus on only four scenarios due to space limitation. In the first scenario, all the traffic sources, AF,BE and EF are allowed to generate traffic at the same time. In the second scenario, we disabled the AF source, and allowed the BE and EF to generate traffic at the same time. In the third scenario, we changed the burstiness of the BE flows, and we allowed only the BE and EF sources to generate traffic in the network, while keeping the AF sources disabled. In the fourth scenario, we disabled the AF and BE sources, and allowed only the EF sources to generate traffic.

The behaviors of the different EF flows e2e delay jitters along the previous Four scenarios are described in the Figures shown in 4, 5, 6, 7,and 8. From Figure 4, we see that the EF flows from source 1 keep their jitter bounds within small limits, however when they arrived at core router 1, a spike is formed, which represents the delay jitter of these EF flows, due to other traffic coming from source 2, and 3. Afterward, their delay jitter increases as they traverse more nodes, but with decreasing probability, since the background traffic load rate is decreasing. Furthermore, we see that when we increase the BG burstiness in the third scenario, the probability of EF delay jitter is increased. From Figure 5, we see that EF flows from source 1 suffer from a very small jitter at the core router 1 when they meet other EF flows coming from source 2, 3, however, after their delay jitter almost approach zero in the network core.

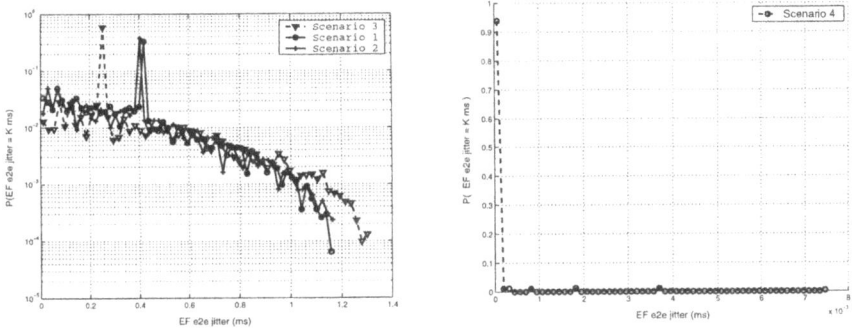

Fig. 4. EF e2e jitter distribution of the EF flows generated by EF source 1.

Fig. 5. EF e2e jitter distribution of the EF flows generated by EF source 1.

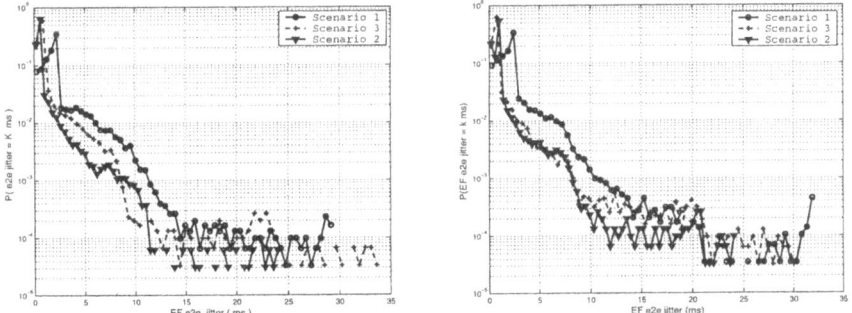

Fig. 6. EF e2e jitter distribution of the EF flows generated by EF source 2.

Fig. 7. EF e2e jitter distribution of the EF flows generated by EF source 3.

From the figures shown in 6, and 7 we see that the behaviors of EF flows of source 2, and 3 along the three scenarios are almost similar. However, if we look carefully in these two figures, we find out that the probabilities of delay jitter of EF flows of source 3 are greater than those of source 2. This can be referred to

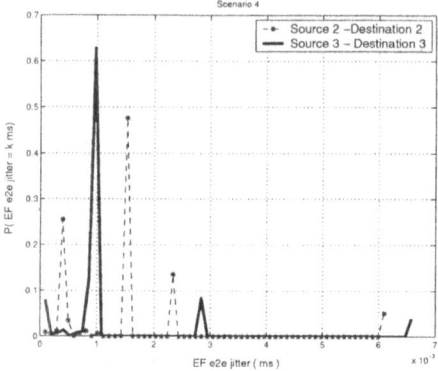

Fig. 8. EF e2e jitter distribution of the EF flows generated by EF source 2 and 3.

that the EF source 2 started to generate traffic before the EF source 3. Therefore, the EF flows from source 2 are mostly served before the ones generated from source 3 at core B, core 1, and core A , where both of these EF flows, and other BE flows compete for the service at these servers. However, the spikes in these two figures, which represent the delay jitter of the EF flows are formed due to their meeting at the links that connect core B and core 1, and core 1 and core A. After that the probability of delay jitter decreases rapidly until they arrive around the probability of delay jitter steady state, then their delay jitter stays within certain jitter bounds. However, From figure shown in 8, the delay jitter of EF flows of sources 2,3 approach almost zero, because the back ground load rate is decreased to arrive around zero. Nevertheless, a number of spikes are formed, because both of EF flows of source 2, and 3 compete for the common resources along their path.

7   Conclusion

In this paper, we analyzed the EF e2e delay jitter in the DiffServ network shown in Figure 2. This network was designed according to the DiffServ norms defined in ns-2, and further the EF aggregate departure rate was configured according to RFC 2598[13]. Therefore, we thought that the EF flows would not suffer from any delay jitter, since through this configuration they will be isolated from other traffic classes. However, we found out from the simulations carried out on the network that they are not isolated, and they are affected by other traffic class characteristics as we have seen in Figure 3, the effect of BE packet size change on the e2e delay jitter of EF flows. Furthermore, the different Figures in section 5 show us clearly that the EF flows are affected by the existence of other traffic classes through their traversals to their destinations. Therefore, the DiffServ norms in ns-2, and RFC 2598 configuration can't stand alone in guaranteeing the QoS demanded by real time traffic offered the EF service class. Thus, playout buffers as shown in Figure 1 must be added to the DiffServ network for compensation the delay jitter of EF flows. Furthermore, we found out that delay jitter of EF flows depend on the back ground traffic intensity in the network. In this, we have seen in Figures 4, 6, 7, that when the different traffic flows meet at the core router 1 some spikes are formed in the three scenarios. Therefore, the background traffic intensity must be controlled to guarantee the EF delay jitter.

Consequently, there are two points that require further studies:

- From the different simulations results we see that the DiffServ network requires de-jitter buffers to improve the QoS offered through Expedited service class to the real time applications.
- Adding a control mechanism that controls the best effort (BG) traffic intensity in the DiffServ network to guarantee the QoS offered to EF flows.

References

1. Matragi,W.,Bisdikian,C.,Sohraby,K.: Jitter Calculus in ATM networks: Single Node Case. INFOCOM. 1994.
2. Sreenan,C.,Chen,J.,Agrawa,P.,Narendran,B.: Delay reduction techniques for playout buffering. IEEE Trans. Multimedia 2(2). (2000)88-100.
3. Ramjee,R.,Kurose,J.,Towsley,D.,Schulzrinne,H.: Adaptive delay mechanisms for packetized audio applications in wide-area networks. Proc. of the IEEE infocom. Toronto, Canada. June 1994,pp.680-688.
4. Jacobson,V.: Congestion avoidance and control. Proc. ACM SIGGCOMM, August. 1998,pp.314-329.
5. Fujimoto,K.,Ata,S.,Murata,M.. Adaptive Playout Buffer Algorithm for Enhancing Perceived Quality of Streaming Applications. To appear in Telecommunication Systems. January 2004.
6. Mansour,Y.,Shamir,B.P.: Jitter Control in QoS Networks. IEEE/ACM Trans. on Networking. 9(4):492-502, 2000.
7. Belenki,S.: An Enforced Inter-Admission Delay Performance- Driven connection Admission Control Algorithm. ACM SIGCOM,Computer communication review. Vol.32, No.2, April 2002.
8. Landry,R.,Stavrakakis,I.: Study delay Jitter with and without peak rate enforcement. IEEE/ACM Trans. on Networking. Vol.5, No.4, August 1997.
9. Fulton,C.A.,Li,S.Q.: Delay Jitter First Order and Second Order Statistical Functions of General Traffic on High Speed Multimedia Networks. IEEE/ACM Trans. on Networking. Vol.6, No.2, April 1998.
10. Bennett,J.C.R.,Benson,K.,Charny,A.,Courtney,W.F,Le Boudec,J.Y.: Delay Jitter Bounds and Packet Scale Rate Guarnatee for Expedited Forwarding. IEEE/ACM Trans. on Networking. Vol.10, No.4,Auguust 2002.
11. Altman,E,Jiménez,T.: NS simulator for beginners. In *Lecture notes*. Automn 2002.
12. Matragi,W,Bisdikian,C,Sohraby,K.: Light Traffic Analysis of jitter in ATM multiplexers. IBM Research Report, RC 19413, 1993.
13. Jacaobson, V., Nicohols, K.,Poduri,K.: Expedited Forwarding PHB. IETF RFC 2598. June 1999.

Internet Quality of Service Measurement Tool for Both Users and Providers

Armando Ferro, Fidel Liberal, Eva Ibarrola, Alejandro Muñoz, and Cristina Perfecto

Departamento de Electrónica y Telecomunicaciones – Área de Ingeniería Telemática
Escuela Superior de Ingenieros de Bilbao
Universidad del País Vasco / Euskal Herriko Unibertsitatea
Alameda de Urquijo s/n – 48013 Bilbao (Spain)
Tel: +34 94 601 39 00   Fax: +34 94 601 42 59
{jtpfevaa, jtplimaf, jtpibara, jtpmumaa, jtppeamc}@bi.ehu.es

Abstract. This paper offers an approach to the definition of access speed measurements. An ever growing increase of Internet Service Providers (ISP) and different types of technologies in the Internet access, makes it difficult for users to decide which is the best or most suitable connection to satisfy their needs. From these considerations, the possibility for Internet users of obtaining a real and neutral measurement of their access may help them to decide if it is covering their demands or, on the contrary, the provider is not complying with the terms of the contract. To perform this measurement, an Internet speed test has been defined, as well as the way to obtain a measure which can comply with the objective of giving users a proper idea of how their Internet access is working. This project is included in the Quality of Service (QoS) investigation area in our investigation group.

1 Introduction

The growth of the Internet over the last years along with the spread of new access technologies, have given rise to new situations. Internet services end users are becoming aware of the importance of a sensible choice when it comes to selecting the ISP and the technology that best fulfil their connection needs.

Knowing if an ISP is complying with the terms of agreement, according to time and access speed to end services, is becoming more and more important. Moreover, among end users there is a real interest for being able to perform their own speed measurements since, this way, they can obtain actual results about the status of his connection whenever the ISP seems to fail to provide the agreed quality. Similarly It is also useful to objectively compare the degree of quality obtained against other users with the same characteristics.

On one hand, ISPs are conscious of the competition they must face [1] and know that the QoS offered to customers is critical to preserve their market share[2]. ISPs also know that advertised speeds are no longer considered an objective factor for users. Not only ISPs, but also their customers, would take advantage of a measurement service developed by a neutral entity. This would allow users to evaluate the speed of their connection at any time, and ISPs to compare their service to the one offered by

J.N. de Souza et al. (Eds.): ICT 2004, LNCS 3124, pp. 1195–1203, 2004.

competitors. Other working groups have developed other measurement systems[3], however, most of them only present static results. None of them manages to carry out statistical analysis with the data obtained from different tests. Nonetheless, this is a main interest focus in our approach.

On the other hand, within the present environment of Internet accesses that are offered by ISPs, we are involved in an era of great changes and new developments, specially concerning traditional access technologies such as POTS and ISDN. Among those new trends, ADSL and cable-modem and newly arrived PLC can be considered outstanding. All these provide users with much faster speeds and are remarkable because they are billed as a real flat-rate as well. Speeds higher than 256 kbps can be achieved along with permanent connectivity. The average user no longer looks only for an economical access, but also for one that warrants the QoS required.

Internet service providers offering these services are frequently limited by factors outside their own network infrastructure. Sometimes they are forced to hire services to higher level providers or carriers that usually offer Internet access services as well.

A customer who purchases one of the fastest connections (i.e. basic ADSL) expects at least a download rate near 256 kbps. It is obvious that he must be very keen on knowing the actual status of his connection at times when downloading files or web surfing takes much longer than usual.

Last of all, it is important to describe the special situation that takes place in Spain. The whole access network belongs to the same provider, and so does the local loop too. In the particular case of ADSL technology –currently the most important in Spain–, this ISP has a market share of near 85 per cent. In addition, Spain has one of the higher connection fees in Europe, taking into account Spain's population average income and the data transfer rate the users get. In general, the access speed to these networks is lower that the hired one and the customer service is very deficient. This fact get worse due to the absence of any law, because the ISP only commit themselves to give users an access speed of up to 10 per cent of the highest hired!.

2   Pursued Objectives

The objective of this paper is to present the system developed within our research group to estimate access connection speed achieved by an ISP. To do so it is important to analyse both download and upload channels and be able to check whether the performance achieved fulfils the agreement.

The platform www.velocimetro.org pretends to be a point of reference in the area of QoS measurements, offering a simple and fast way of knowing the real speed of our Internet access. Besides, the tool owns others functionalities that allow users to maintain a repository with all the tests that permits examine the evolution of their connections speed. The measurements service also allows the storage of resulting data from the tests to perform further filtering and analysis that led to the publication of periodical reports in our site.

Being so, users can have an estimate of their Internet access speed so they could decide if the service provided complies with ISP´s contract terms and if it is covering their demands regarding a good service quality.

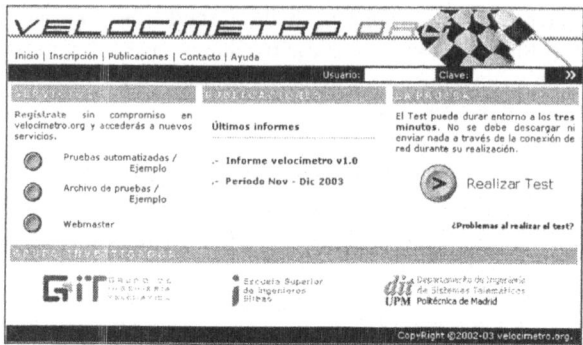

Fig. 1. Screenshot of the start of the test

As a global goal, the results of the test are stored in the database to analyse and draw conclusions, carrying out comparisons and generating interesting statistics through data mining.

3   System Description

Figure 1 shows the initial screen of the application. It can be reached through the following URL: http://www.velocimetro.org.

From an ISP customer's point of view, this system allows him to estimate its connection speed in both download and upload channel. He is also informed about certain relevant aspects of the system itself: estimation of the type of hired access, date and hour of the test, number of tests carried out so far, etc. and other type of information or computations of exclusive interest for the user of the service.

From ISPs' point of view, this is a very useful tool to estimate which is the final speed their customers are obtaining [4]. It also allows them to compare the speeds that other ISPs offer to their customers. Because of the special system architecture, all kinds of statistics can be obtained from the data stored in every step of the test. Some examples of statistics could be a graphic representation of a connection's speed at any given time slot, most used connections to access the Internet, and so on.

Once welcome page has been loaded, the test begins by clicking on the start button. Then, an initial small sample of data is downloaded in order to estimate the connection speed by measuring that time slot. The results of this initial part of the test are not shown to the user since they are not very accurate and are only useful to get an approximation of the user's connection type. After doing so, test parameters are customized to best suit each connection's characteristics; mainly the size of file to communicate with the remote servers.

While the test is being performed a form is presented to the user. By means of it, some interesting information can be interactively collected from the user, an information that will help in later result processing and study; Requested information includes ZIP code –only for connections from Spain– and technology used in the access to the Internet, where individual and LAN connections are distinguished. All this information is verified in order to guarantee the validity of data.

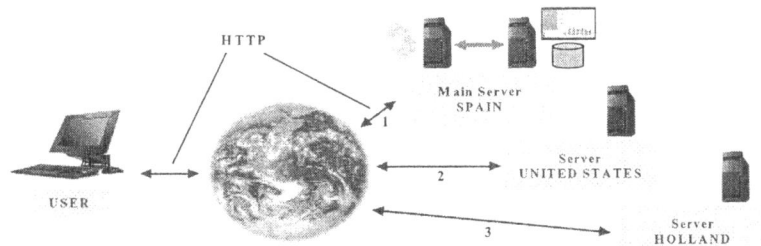

Fig. 2. Geographical Distribution of hosts

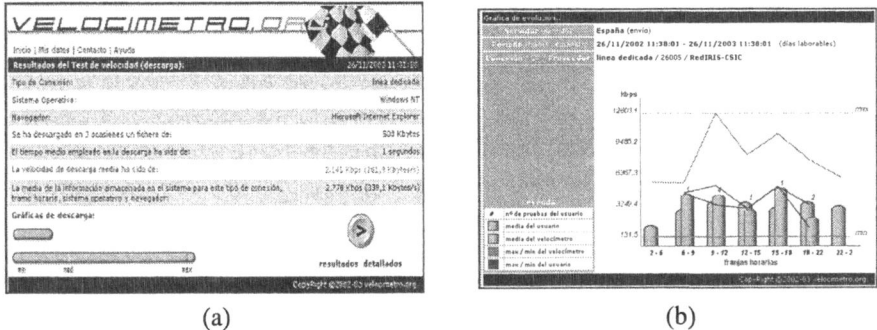

(a)   (b)

Fig. 3. Examples of results: (a) basic test, and (b) automatic programmable tests

As a result of the test, a measurement of the connection speed is displayed (Figure 3a). That final result is the average value of measured times against already mentioned destinations. Besides, a comparison is showed between the measured speed and the average of all the results obtained so far for that connection type in the same time slot. Results are also graphically compared to the maximum and minimum measured speeds obtained in previous tests developed in identical conditions If the user wants to obtain further information, he can access to a detailed results section. where, all the partial upload and download speed measurement tests against the three destinations,are shown.

Registered users are entitled to gain access to some advanced features, such as automatic test scheduling or personal results repository where time-evolution graphics of their connection are displayed).Due to the considerable amount of tests presently stored in the database, our system can provide reliable information about this test, arranging then according toconnection types, user role, upload and download channels, time zone, destination, etc.

4   Architecture and Methodology

Next, the architecture of our web based speed test is described. Although some aspects will be detailed later, we would like to insist on the separation of the different modules not only from the users' point of view, but also from the server side.

4.1 Platform

The system consists of a group of hosts located in strategically disperse locations with a suitable and optimum connection and the basic software to perform the measurements. This way, results are more reliable and independent from a specific location.

Moreover, so that the test can give coherent data and provide statistics classified by zone, the test platform must be located in independent and separated hostings. These should have a good and warranted connectivity.

Figure 2 shows the distribution of the hosts. The first one is located in Spain, in a neutral point, called HISPANIX, where all ISPs exchange their data; it is the main one, so that it also holds the database and management module. The second host is located in the USA and the third one is in Holland. The locations have been chosen taking into account that most of web traffic exchanged by the main target audience, Spanish users, is due to Spain, USA and Europe.

The main server, which contains the whole software suite to perform all the actions, is situated in a dedicated machine. The rest of the remote locations (Holland and the United States) are hired in a virtual-host way. There is only a nominal part of the software in those hosts: the upload and download related modules (agent and manager). These modules directly communicate with the result manager module.

In particular, the main site consists of two servers in a master-slave configuration. The master is responsible for collecting requests from clients, data processing and returning the suitable responses to the users. The second machine communicates with the other one and executes operations that require higher data processing needs. In this way, the slave generates statistical graphics and holds the database. All these operations take place in execution time.

Each host contains a web server, a Java Virtual Machine and a Web Container (for running Servlets and JSPs). The central host also includes a DataBase Management System (DBMS). The basic software is Tomcat and PostgreSQL. All the software used in the platform is opensource and runs on a Solaris platform. Finally, Extensible Markup Language (XML) is internally used for generating and formatting dynamic contents from static data.

4.2 System Internal Architecture and Technology

A clear need for portability between different operating systems and architectures lead us to an architecture based on the use of several modules of the Java 2 Enterprise Edition platform (J2EE). These elements are Servlets, JSP pages and JavaBeans objects. These three elements make up the model-view-controller design pattern.

First of all, Servlets are responsible for controlling the application operation; they include the processing logic and all the methods to interact with the other elements of the global architecture. Java Server Pages implement the interface management and they are in charge of presenting information to the user. Finally, JavaBeans objects contain the logic and the facilities to represent the designed data model.

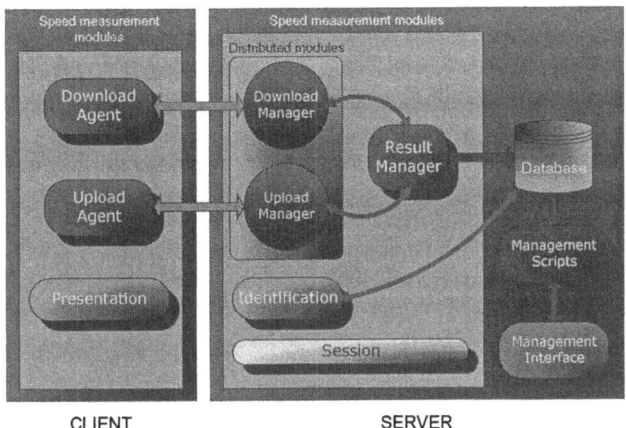

CLIENT   SERVER

Fig. 4. System internal architecture

The database is a basic element of the system architecture. Not only it stores information about results, users and configuration data, but also collects data to enhance statistical information access. It is essential to choose an efficient DBMS and design the database properly, since it has a direct impact on the overall system performance. Because of that, a specific data model was designed to represent the information in the database.

There are numerous potential sources of error in the system. For example, a user could just close his Internet browser before finishing the test or the connection could be lost. This would lead the system to gather incomplete or inaccurate results. This incorrect or incomplete information should be deleted to prevent statistics from being corrupted.

To achieve this, a management module consisting of some maintenance scripts and a web interface has been incorporated. By means of the web interface, the system manager can easily program maintenance tasks and check the status of the system.

Concerning to the operation of the system, it is based on a client-server model. It involves client processes requesting service from server processes.

4.3   Methodology

The measuring process is divided into three individual estimations. These match up with servers located in Spain, Holland and USA. When the user launches the test, his host establishes a TCP connection and executes an HTTP request to the first server. In this moment, the host begins to download a packet of a certain size. The size is the result of an initial estimation carried out when the user enters to the site and it is different for each type of connection considered in the speed test.

After that, the host returns the downloaded packet in the opposite direction. Each server is responsible for measuring the time needed for downloading and uploading. Following this, the user repeats these operations this time connecting to the servers located in Holland and the United States.

Due to the special configuration of the platform, the test collects a huge amount of data for each destination. The main server processes the results in order to return to the user just most representative data of his connection capacity. System stores all the data introduced by the user and partial results of the test. It is important to guarantee the correct maintenance of the information in the database because it is all used to generate periodical reports and they are essential to provide added value services.

Being so, the speed test evaluates the real capacity of the users to surf in the Internet, so that the results refer to World Wide Web browsing (HTTP protocol over TCP/IP stack).

4.4   Server Site

The main tool runs on a Solaris platform; that fact gives the platform a high stability. The server is intended to keep the interface of the test, gather the data and statistics generated, and store and manage this information properly. In order to perform all these actions, Java Servlets and JSP pages are used (based on J2EE platform). These two elements are responsible for controlling system operation, executing the suitable actions and presenting results to users. The modules involved in the test that reside in the server are:

- *Identification module.* It is responsible for collecting all available information about a client, such as IP address, browser type or the Internet Service Provider. It is also in charge of logging in users' access. Every user is uniquely identified; thus, cookies are used to store or retrieve the their identification. In this way, each client uses the same identification every time he connects, which would allow detailed statistics per user to be offered periodically.

- *Upload manager.* It is in charge of managing the actions that the upload agent carries out. These actions are related to the initial estimates of the user's speed. Once gathered the data from the agent, it transfers it to the result manager.

- *Download manager.* As the upload manager does, this module manages the actions that the download agent must carry out; It collects the download speed estimates from the agent and transfers them to the result manager. It is also responsible for discriminating the optimum size of the packet to transfer according to user's connection type.

- *Result manager.* It acts as the interface between the rest of the modules and the database. It stores all the information from download/upload managers into the database. It also computes some cumulative statistics and stores them in the appropriate context, so that further statistics are quickly available. This reduces the need to continuously access the database when showing test results to the user.

- *Database.* The database manager module interacts between the result manager and the database itself. It receives the results of the test in the proper way and maintains them in the database. It also holds system configuration data and users information.

4.5   Client Site

Client site is one of the most active elements. The platform takes advantage of the characteristics of the Internet browser and HTTP protocol to download samples and files of a certain size. In the same way, web technologies, such as POST method, are used to send files back to server.

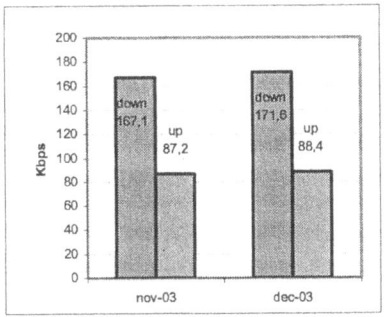 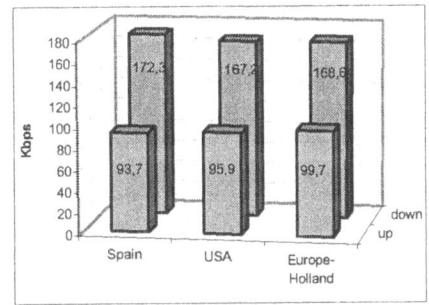

Fig. 5. Graphics of a report representing the results obtained for ADSL 256 Kbps technology

The use of HTTP methods involves the need of an adapted control logic over this operations. The programming language adopted to solve these questions is Javascript. The main reasons for choosing this client-side scripting language was its simplicity, lightness and the availability of an interpreter in almost every modern browser. On the other hand, it is powerful enough to carry out the needed operations (time capturing, file transferring, automatic redirections, etc.).

The client site is also responsible for maintaining the upload and download agents. These modules are in charge of transferring data between client and server site.

5   Future Research

Research on this project will go on by introducing some improvements, analysing results and also developing new tools to make a good use of them. Last 28th of October 2003 the new version of the application was released. It incorporates all the functionalities described in this article.

Nowadays, new services are been developed. Most of them are result of the unselfish aid our users. It is planned to publish periodical reports with the results of the tests, classified by region, user role, type of connection and so on. These reports are available online. It is also planned to translate the whole application into English and Basque (local language in the Basque Country). Some other particular improvements that will be carried out include:

- Improving testing methodology.
- Automation of maintenance activities.
- Developing of new tools for accurate handling of tests results.
- New web interfaces and tools will be defined so that results statistics can be extended to different entities of the Internet community.

6 Conclusions

Through this system users can obtain an estimation of the effective speed they have available to surf the Internet. This kind of measurement is not valid to properly estimate the real raw bandwidth, since TCP, a non stateless protocol, is used. However, it is a valid service to grant users a real impression of the maximum speed they can achieve.

Our measurement system gives multiple benefits not only for users, but also for ISPs and companies associated to the Wold Wide Web. In conjunction with single individual test results, resulting comparatives are used to extract conclusions about the real state of the users´ connections to the Internet. This information is extremely valuable to ensure whether ISPs are complying with the terms of agreement, according to time and speed of access to end services.

It is very important to remark that our investigation working group is a neutral entity. That is, all the results collected are handled the right way, guarantying their integrity and never helping out private providers. Moreover, home users are handed full results completely out of charge.

Last of all, the work developed in the frame of this project has contributed to turn the Telematics Engineering Group into a point of reference in Spain in the area of quality of service measurements. Presently several other QoS related measurement tools and methodologies are under development both for Spain's Ministry of Science and Technology and also to audit European Union's Community R&D Information Service (CORDIS) External Monitoring and user feedback.

References

1. Susan O'Keefe; "A New Era in Networking"; Telecommunications Online, January 1999
2. Web of the QoS Forum. http://www.qosforum.com
3. Other Internet speed tests:
   http://www.telefonica-data.es, http://extranet.iies.es/CGI/speed.cgi
   http://pcpitstop.com/internet, http://www.dslreports.com
4. CAIDA Measurement Tool Taxonomy. *"Internet measurements tool Survey"*.
   http://www.caida.org/tools/taxonomy

A XML Policy-Based Approach for RSVP

Emir Toktar, Edgard Jamhour, and Carlos Maziero

Pontifical Catholic University of Paraná, PUCPR, PPGIA
{toktar,jamhor,maziero}@ppgia.pucpr.br

Abstract. This work proposes a XML-based framework for distributing and enforcing RSVP access control policies, for RSVP-aware application servers. Policies are represented by extending XACML, the general purpose access control language proposed by OASIS. Because RSVP is a specific application domain, it is not directly supported by the XACML standard. Hence, this work defines the XACML extensions required for representing and transporting the RSVP access control policy information.

1 Introduction

Policy based network management (PBNM) is an important trend for IP-based networks. Recent works developed by IETF have defined a standard model for representing policies on different areas of network management. Besides IETF, others organizations are proposing standard policy models for PBNM. The OASIS (Organization for the Advancement of Structured Information Standards) proposed a language for representing access control policies, on general purpose, denominated XACML (eXtensible Access Control Markup Language). This work proposes the use of the XACML for modeling and distributing RSVP access control policies for RSVP-aware application servers. Because RSVP is a specific application domain, it is not directly supported by the XACML standard. Hence, this work defines the XACML extensions required for representing and transporting the RSVP access control policy information. This proposal also supply the information for defining the Tspec and Rspec parameters transported in the PATH and RESV messages, defined by RSVP. Therefore, the XACML policy also provides the information used for "admission control" by the network elements along the path between the transmitter and the receiver.

This paper is structured as follows. Section 2 presents an analysis of the models that can be employed for describing RSVP access control policies, and the strategies for distributing and enforcing those policies. The section 3 describes how the XACML can be used for describing RSVP policies, and presents the required extensions for adapting XACML to the RSVP issue. The section 4 illustrates the modeling of RSVP policies through a case study. Finally, the conclusion reviews the principal aspects of this study and indicates the future works.

J.N. de Souza et al. (Eds.): ICT 2004, LNCS 3124, pp. 1204–1209, 2004.
© Springer-Verlag Berlin Heidelberg 2004

2 RSVP Policy Control Strategies

In this paper, the strategy for representing, distributing and enforcing RSVP access control policies follows Policy Based Network Management (PBNM) approach. The concept of PBNM is already widely adopted by organizations that propose Internet standards, such as IETF [11] and the OASIS [7]. Although the definitions for PBNM could diverge according to the organization, the main concepts are relatively universal. The basic idea for PBNM is to offer a strategy for configuring policy on different network elements (nodes) using a common management framework, composed by a policy server, denominated PDP (Policy Decision Point) and various policy clients, denominated PEPs (Policy Enforcement Points) [12]. The PDP is the entity responsible for storing and distributing the policies to the diverse nodes in the network. A PEP is, usually, a network node component responsible for interpreting and applying the policies received from the PDP. IETF defines as well a standard protocol for supporting the communication between the PEP and the PDP. This protocol is denominated COPS (Common Open Policy Service). The basic structure of the COPS protocol is described in the RFC 2748 [1]. The PBNM approach can be applied in various aspects of network management. This section will explore how this approach can be applied for managing access control policies in RSVP server (sender) applications.

The IETF already published various works concerning the use of PBNM approach for RSVP policy control. The works cover the definition of a framework for admission control [11] and the utilization of COPS in outsourcing (COPS-RSVP) [4] and provisioning (COPS-PR) models. The provisioning approach is still under development, being necessary additional definitions for its complete specification.

The XACML proposal from OASIS also follows the PDP/PEP approach. However, OASIS does not make a distinction between the outsourcing and provisioning models, neither defines a standard protocol for supporting the communication between the PEP and the PDP. An analysis of the XACML indicates, however, that it was primarily conceived for supporting the outsourcing approach. An important difference between the approaches adopted by OASIS and IETF relates to how policies are represented and stored. OASIS proposes XACML as a particular model for access control, represented and stored as XML documents. On the other side, IETF defines PCIM as a generic model, independent from the way the policies will be represented and stored. The PCIM model is abstract, and needs to be extended in order to support particular areas of management, such as QoS [10]. IETF indicates strategies for mapping the information models to LDAP (Lightweight Directory Access Protocol) schemas, but this form of storage requires a supplementary effort by developers.

A work describing the implementation and performance evaluation of a PBNM framework, using COPS in outsourcing model with RSVP (COPS-RSVP) was presented by Ponnappan [8]. The QoS policies were represented using QPIM (QoS Policy Information model), an IETF PCIM extension described by Snir [10]. The policies were represented and stored using LDAP .This work uses CORBA (Common Object Request Broker Architecture) for supporting the interaction between the application components.

3 Proposal

This paper proposes a XACML-based framework for distributing and enforcing access control policies to RSVP-aware application servers. The PEP element represents a component of the server application, responsible for requesting policy decisions to the PDP and interacting with the RSVP daemon in the host computer. The code of the PEP must be integrated with the application server. In our proposal, the PEP is responsible for all interaction with the RSVP daemon, releasing the application from the task of any QoS negotiation. This interaction includes retrieving the traffic information for building PATH messages and granting or not the reservation request on receiving the RESV message. This approach can be implemented in any system that supports the RSVP APIs described in the RFC 2205. The sequence of events and messages exchanged during the establishment of a RSVP reservation, using the proposed framework, is described as follows:

---

1. A RSVP client requests a connection to a multimedia server for obtaining services with QoS.
2. In the multimedia server, the application calls the PEP for evaluating the request. Then, the PEP sends to the PDP a XACML request context message informing a "Target" containing its IP address (Resource), the IP address of the client (Subject) and the requested operation (Action).
3. The PDP evaluates the policy defined in XACML for the supplied target, and returns to the PEP a XACML response context message having, besides the result (permit or deny), the information of traffic specification (*Tspec*, supplied through the Obligations structure).
4. In case of positive decision, the PEP calls its RSVP daemon, informing the *Tspec* parameters. The RSVP daemon, then, sends a RSVP PATH message to the receiver (i.e., the RSVP client). The *Tspec* parameters are stored in the PEP for further analysis (see step 6).
5. The RSVP client, on receiving a RSVP PATH message, calls its RSVP daemon, which obtains the traffic parameters from the PATH message and formats a RESV RSVP message, returning it to the sender (i.e., the PEP).
6. On receiving the RESV message from the client, the RSVP daemon of the server triggers an event to the PEP forwarding the *Tspec* information. The PEP compares the *Tspec* information received from the client with the *Tspec* information saved in step 4. If the *Tspec* parameters are identical or smaller than those saved in step 4, the PEP confirms the reservation to the RSVP daemon. In this step, the RSVP daemon also verifies if it has enough resources to satisfy the request (admission control).

Fig. 1. The steps 1 to 6 refer to a well-succeeded scenario of reservation, and exception treatment was omitted.

The strategy adopted in this work for describing a RSVP policy in terms of XACML is illustrated in Fig 2. The reason for defining a RSVP policy in terms of a <PolicySet> and not in terms of a single <Policy> element is related to the XACML definition. In XACML version 1.1 the <Obligations> element is mapped to <Policy> or <PolicySet>, but it can't be mapped to Rules, i.e., all <Rules> in a policy defines the same <Obligations>. Therefore, distinct service levels can't be represented in a single policy. Another important point is to define where the users and services information is located. If we consider the PCIM approach, defined by IETF, a logical approach would consist in representing users and services through CIM (Common Information Model objects). The XACML definition permits to define all the information concerning the policy (subjects, resources and actions) in the same XML document, as defined by the *"xacml policy scheme"*. However, OASIS points that it will be possible to write XACML policies that refers to information elements stored in a LDAP repository. The 1.0 specification does not define how it can be done. However, because *xacml* is based on standard xml definitions, a possible solution would be create references in a policy to external documents using the *XML Pointer Language (XPointer)* strategy [6]. There are some references about the use of *XPointer* in the 1.0 OASIS specification, however, its use is limited to request documents (i.e., context scheme) and its use in policy documents is not supported. However, using *XPointer* for creating policies with reusable subjects and services information is a logical extension for future XACML versions. Hence, this work adopts the use of XPointer for defining policies with reusable subjects (users) and services information.

`<PolicySet PolicySetId="RSVP_Aware_Server_Application">`
  `<Target>`
    `<!—Defines the services (resources) to which the policy applies →`
  `</Target>`
  `<Policy PolicyId="Service Level 1"> <!—e.g. GOLD →`
    `<Rule>`
      `<Target> <!—Subjects to which the policy applies → </Target>`
        `<Condition> <!-- Time and client's IP addresses restrictions -->`
        `</Condition>`
    `</Rule>`
    `<Obligations> <!—TSpec specification for service level 1 →`
    `</Obligations>`
  `</Policy>`
  `<Policy PolicyId="Service Level 2"> ... </Policy> <!—e.g. SILVER →`
  `<Policy PolicyId="Service Level N"> ... </Policy> <!—e.g. BRONZE →`
  `<Policy PolicyId="Default Policy"> <!—usually denies all → </Policy>`
`</PolicySet>`
	A `<PolicySet>` represents the QoS service levels policies offered by na RSVP-aware application.
	
	The `<Policy>` elements in the `<PolicySet>` are used for defining distinct QoS service levels offered by the same application. For example, "GOLD", "SILVER", etc.
	
	The `<Rule>` defines the users (subjects) that have authorization to receive the service level.
	
	`<Obligations>` element describes the Tspec parameters.

Fig. 2. In the proposed strategy each "RSVP-aware" server application (or group of applications) is mapped to a XACML `<PolicySet>`.

4 Case Study

In order to illustrate the use of the XACML approach for describing RSVP policies, the following scenario was considered: A set of "video streaming" servers in a university campus offers "tutorials" to registered and unregistered students (visitors). The policy adopted for having access to the video streaming is defined as follows: a) Registered students have permission to access any server in the campus offering a *"TutorialVideoStreaming"* service without time restrictions. If a student connects to a server using a client host from inside the campus, he will receive a "GOLD" or "SILVER" service level. b)Unregistered students can have access to the *"TutorialVideoStreaming"* service only from the internal network and not in business-time. They can receive only the "SILVER" service level.

The service level are described by an external XML document as illustrated in Fig 3. The RSVP policy is defined in terms of a XACML `<PolicySet>`, as described in section 3. Fig 4 illustrates the structure of a policy in the `<PoliceSet>`. The `<Action>` element, in the `<Target><Rule>`, defines that a PEP requesting a RSVP policy for a specific application must specify a *getResourceQoS* action. The `<Subject>` element defines that the policy applies only to registered students. The xpath-node-match function is a proposed XACML extension permitting to create XPointer references to external documents. In the example, the information concerning the student status is represented in the external document *"resource.xml"* not shown in this paper. The `<Condition>` of the `<Rule>` determines that the policy applies only to requests where the client host is located inside the university campus. Finally, the `<Obligation>`

element uses *XPointer* references to the Service information file in order to retrieve the QoS parameters.

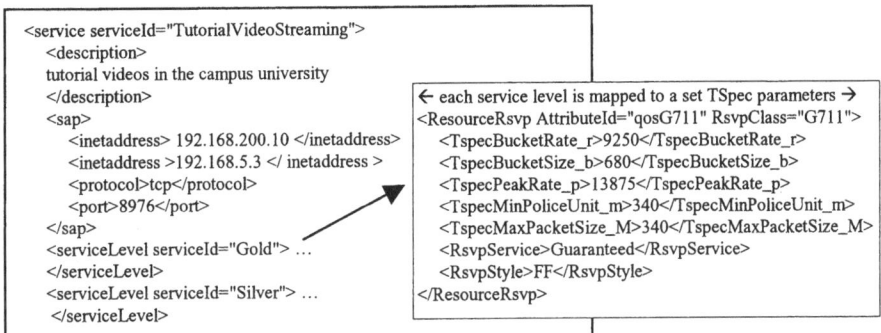

Fig. 3. The figure illustrates how a service information is represented. Note that the <SAP> structure defines the services in the campus that are subjected to the policy.

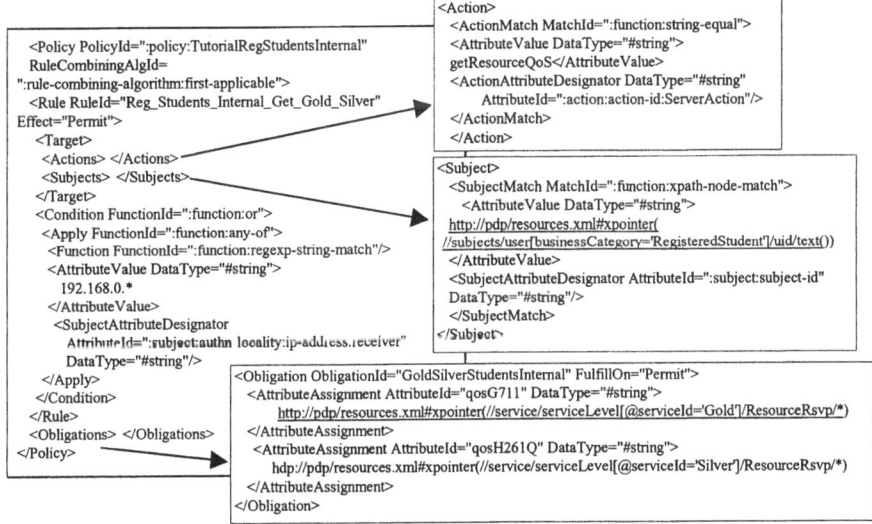

Fig. 4. Policy Structure for Registered Students in Internal Network

5 Conclusion

In this work, XACML use was extended beyond the access control functionalities, because the decisions generated by the PDP include the *Tspec* parameters necessary for building the PATH messages. The capacity of returning configuration parameters through PDP decisions is an important feature for many PBNM scenarios. This feature, easily supported in IETF PCIM-based models, is quite difficult to implement in XACML. To support the RSVP scenario, modifications in the <Obligations> structure were required. The 1.0 XACML specification is deficient in returning results that are not simple deny or permit decisions. In the proposed work, some features have been added to the XACML framework without modifying its scheme: <Obligations> are dynamically processed and *XPointer* references to external documents are used for creating policies with reusable resources and subjects.

Some modifications on XACML scheme, however, would be useful. First, we suggest a more flexible way of mapping conditional <Obligations> to policies. Mapping <Obligations> to <Rules> would permit to define different service levels in a single policy. This modification would certainly be useful for other application domains. Another suggested modification is to formalize the use of XPointer references in the XACML scheme.

References

1. Boyle, J.; Cohen, R.; Durham, D.; Herzog, S.; Rajan, R.; Sastry, A. The COPS (Common Open Policy Service) Protocol, RFC2748, Jan. 2000.
2. Braden, R.; Zhang, L.; Berson, S.; Herzog, S.; Jamin, S. Resource Reservation Protocol (RSVP) Version 1 Functional Specification, RFC2205, Sep. 1997.
3. Chan K.; Seligson, J.; Durham, D.; Gai, S.; McCloghrie, K.; Herzog, S.; Reichmeyer, F.; Yavatkar, R.; Smith, A. COPS Usage for Policy Provisioning (COPS-PR), RFC3084, Mar. 2001.
4. Herzog, S.; Rajan, R.; Sastry, A. COPS usage for RSVP, RFC2749, Jan. 2000.
5. Moore, B.; Ellesson, E.; Strassner, J.; Westerinen, A. Policy Core Information Model - Version 1 Specification, RFC3060, Feb. 2001.
6. W3C, XPointer Framework, W3C Recommendation, 25 March 2003.
7. OASIS, eXtensible Access Control Markup Language (XACML) Version 1.0. OASIS, Feb. 2003.
8. Ponnappan, A.; Yang, L.; Pillai, R.; Braun, P. "A Policy Based QoS Management System for the IntServ/DiffServ Based Internet". Proceedings of the Third International Workshop on Policies for Distributed Systems and Networks (POLICY.02). IEEE, 2002 .
9. Shenker, S.; Wroclawski, J. General Characteri-zation Parameters for Integrated Service Network Elements, RFC 2215, Sep. 1997.
10. Snir, Y.; Ramberg, Y.; Strassner, J.; Cohen, R. "Policy QoS Information Model, work in progress, draft-ietf-policy-qos-info-model-05.txt". IETF, May. 2003.
11. Yavatkar, R., Pendarakis, D.; Guerin, R. A Framework for Policy-Based Admission Control, RFC2753, Jan. 2000.
12. Westerinen, A. et. al. Terminology for Policy Based Management. RFC3198, Nov. 2001.
13. Wroclawski, J. RSVP with INTSERV, RFC 2210, Sep. 1997.

SRBQ and RSVPRAgg: A Comparative Study

Rui Prior[1], Susana Sargento[1,2], Pedro Brandão[1], and Sérgio Crisóstomo[1]

[1] DCC-FC & LIACC, University of Porto, Portugal
[2] Institute of Telecommunications, University of Aveiro, Portugal
{rprior, ssargento, pbrandao, slc}@ncc.up.pt

Abstract. This paper presents a comparative evaluation of the Scalable Reservation-Based QoS (SRBQ) and the RSVP Reservation Aggregation (RSVPRAgg) architectures, both designed to provide QoS levels similar to RSVP/IntServ without the scalability concerns that prevent its usage in high-speed core networks. The comparative analysis, based on simulation results, shows that SRBQ provides the same QoS guarantees of RSVPRAgg, with significantly increased network resource utilisation and a small penalty in signalling processing overhead. This stems from the fact that although based on end-to-end reservations, SRBQ makes use of techniques and algorithms that reduce the computational complexity of signalling processing, increasing its scalability.

1 Introduction

With the goal of benefiting from the virtues of both IntServ [1] and DiffServ [2] and mitigating their respective problems, several architectures have been proposed in the literature. None of them, however, ensures simultaneously the strict and differentiated QoS support and the maximisation of the usage of network resources without scalability concerns. One of the most promising architectures [3] is based on the aggregation of per-flow reservations, where the RSVP is extended to allow RSVP signalling messages to be hidden inside an aggregate. In the simplest case, reservations of aggregate bandwidth are performed between ingress and egress routers of a network domain; these reservations are updated in bulks much larger than the individual flow's bandwidth. Whenever a flow requests admission in an aggregate region, the edge routers of the region check if there is enough bandwidth to accept the flow on the aggregate. If enough resources are available, the flow is accepted without any need for signalling the core routers. Otherwise, the core routers will be signalled in an attempt to increase the aggregate's bandwidth. If this attempt succeeds, the flow will be admitted; otherwise, it will be rejected. This architecture benefits from the fact that signalling messages are only exchanged when the aggregate's bandwidth needs to be updated. Unfortunately, the decrease in signalling rate is accompanied by a decrease in resource utilisation.

In order to address the requirements of end-to-end QoS support without resource utilisation and scalability concerns, we developed a new architecture, Scalable Reservation-Based QoS (SRBQ) [4]. The underlying architecture of SRBQ is based on Diff-Serv, with the addition of signalling-based reservations subject to admission control. The network is partitioned into domains, consisting of core and edge routers; access domains contain also access routers. Flows are aggregated according to service classes,

J.N. de Souza et al. (Eds.): ICT 2004, LNCS 3124, pp. 1210–1217, 2004.

mapped to DiffServ PHBs (Per-Hop Behaviors), and packet classification and scheduling are based on the DS field of the packet headers. Besides Best-Effort (BE), SRBQ supports a Guaranteed Service (GS) class, providing strict QoS guarantees, and one or more Controlled Load (CL) classes, emulating lightly-loaded BE networks, based on the Assured Forwarding (AF) PHB. SRBQ's queuing model is compatible with DiffServ; the two models may coexist in the same network. The main scheduler is priority-based: the highest priority belongs to the GS class, which must be shaped by a token-bucket; below, there is a class for signalling, which must be rate-controlled; the CL class(es) come next, with optional rate-control; finally, at the bottom priority is the BE class.

In SRBQ, all nodes perform signalling and support the previously described queuing model. Access routers perform per-flow policing for the CL class and per-flow ingress shaping for the GS class. Edge routers perform aggregate policing and DSCP remarking. Core routers perform no policing.

Flows are subject to admission control, performed at every node. GS flows are characterised by token-buckets; CL flows are characterised by 3 rate water-marks, corresponding to different drop priorities. A scalable hop-by-hop signalling protocol was developed to perform unidirectional, sender initiated, soft-state reservations. Several techniques and algorithms have been developed aiming at the minimisation of the computational complexity and, therefore, the improvement of the signalling scalability. More specifically, a label switching mechanism, was developed to allow direct access to the reservation structures, avoiding expensive lookups in flow reservation tables. The labels are installed at reservation setup time, and all subsequent signalling messages use them. Moreover, a scalable implementation of expiration timers for soft reservations, with a complexity that is low and independent from the number of flows, was also developed. In terms of QoS guarantees, [5] has shown that our architecture is able to support strict and soft QoS guarantees to each flow, irrespectively of the behaviour of the other flows in the same and in different classes, with resource utilisation similar to that of IntServ, but increased scalability.

The rest of the paper is orgnanised as follows. In section 2 we compare both architectures in terms of QoS guarantees and resource utilisation, evaluating their relative merits and shortcomings, as well as their suitability to replace the reference RSVP/IntServ architecture, which suffers from scalability problems that disallow its usage in high traffic core networks. The results indicate that both the Scalable Reservation-Based QoS (SRBQ) [4] and the RSVP Reservation Aggregation (RSVPRAgg) [3] models provide adequate QoS levels and may, therefore, be used in place of RSVP/IntServ. Section 3 presents the most important conclusions and points out some topics for future work.

2   Comparison Between SRBQ and RSVPRAgg

Both the SRBQ and RSRVPRAgg models aim at providing QoS levels comparable to RSVP/IntServ, but in a scalable manner. Both of them make use of flow aggregation in order to achieve scalability in packet classification and scheduling, using the DSCP field of the IP header to this end, in a DiffServ-like approach. The main differences between these architectures stem from the different approaches to signalling. In RSVPRAgg, reservations at the core are performed in an aggregate basis and their bandwidth is updated in bulk quantities, reducing the amount of state stored and the number of signalling messages processed. The unused bandwidth of all aggregates, however, adds up, leading to poor resource utilisation. Large bulk sizes aggravate this problem, but are needed for signalling to be scalable. Additionally, at edge routers of transit domains, where traffic is still very high, per-flow signalling (and, in some cases, classification and scheduling) is performed, imposing a scalability limitation. In SRBQ, the end-to-end character of reservation signalling is preserved, and scalability is achieved by using highly efficient techniques (like label switching and efficient timers). The amount of state stored is not really a problem [5], and resource usage is always optimal. Overall, the signalling processing overhead of these models is expected to be comparable.

Both architectures were implemented in the ns-2 network simulator. Although not scalable, an existing implementation of the RSVP/IntServ model is also used as a reference for QoS results. It is important to keep in mind that ns-2 has some limitations, the most significant of which is the inability to simulate and measure processing delays.

These models have some adjustable parameters. In RSVP and RSVPRAgg, the R parameter (average refresh period) used was the default of 30 s. The reservation expiration timer in SRBQ was chosen so that refreshes would be sent every 32 s, the closest value. Controlled Load (CL) flows in SRBQ are characterised by 3 rate water-marks; their target utilisation values were adjusted to 0.999, 1.0 and 3.0 times the bandwidth assigned to the CL class in order to ensure that, using the reservation parameters given below for each set of simulations, admission control would be performed based on the second water-mark. In the RSVPRAgg model there are two tunable parameters related to aggregate bandwidth management [6]: we used a value of 15 s for the bulk release delay timer, related to hysteresis, and a value of 5 s for the hold timer which prevents repeated failed attempts at increasing the bandwidth of a given aggregate. Simulations with the RSVPRAgg model were performed with two different bulk sizes: 300 kbps and 600 kbps. The admission control used in all models is parameter-based (PBAC).

A mapping between the QoS architectures needs to be performed for an accurate comparison. The aggregation regions of RSVPRAgg and the non-aggregated RSVP regions correspond to the core and access domains of SRBQ, respectively; the aggregators and deaggregators in RSVPRAgg correspond to edge routers in SRBQ. Given this mapping, the simulations used the same topology, depicted in figure 1. It includes 1 transit (TD) and 6 access (AD) domains. Each terminal simulates a set of terminals. The bandwidth of the connections in the transit domain, and in the interconnections between the transit and the access domains, is 10 Mbps. The propagation delay is 2 ms in the transit domain connections and 1 ms in the interconnections between the access and the transit domain.

The simulated scenario contains a class for signalling traffic, CL and BE classes. At each link, the bandwidth assigned to signalling is 1 Mbps. Note that, although this seems very high, the unused signalling bandwidth is used for BE traffic. The bandwidth assigned to the CL class is 7 Mbps. The remaining bandwidth, as well as unused CL and signalling bandwidth, is used for BE traffic.

Each terminal of the access domains on the left side generates a set of flows belonging to the CL and BE classes. The destination of each flow is randomly chosen among the terminals in the right side access domains. With 3 source and 3 destination edge routers in the core domain, the number of required end-to-end aggregates in the domain is 9. Traffic belonging to the CL class is a mixture of different types of flows: CBR, exponential on-off and Pareto on-off. These flows are initiated according to a Poisson process with a certain mean time interval between calls (MTBC), and flows' durations are exponentially distributed. Filler traffic in the BE class is composed by on-off Pareto and FTP flows.

All simulations presented in this paper are run for 5,400 simulation seconds, and data for the first 1,800 seconds is discarded. All values presented are an average of, at least, 5 simulation runs with different random seeds. The simulation results are presented in the next sub-sections.

2.1 QoS, Utilisation, and Signalling

In the first set of simulations we used, in the CL class, only 64 kbps constant bit-rate (CBR) flows with a packet size of 500 bytes. This is done to completely avoid the unfairness originated by disproportionate rejection rates between flows of different types in different models. The average flow duration is 120 s, and the mean time between calls is adjusted to vary the offered load between 0.8 and 1.2 times the bandwidth allocated to the CL class at the core link. The results from this set of simulations are presented in figure 2.

In all models, the mean delay is not much higher than the sum of transmission and propagation delays (12.08 ms), meaning that the time spent in queues is low. Nevertheless it is lower in SRBQ, as is jitter (not shown). These results are probably due to the use of WFQ in RSVP and in RSVPRAgg outside the aggregation domain. In all models presented there are no losses, since the reserved bandwidth of the CBR flows is equal to the maximum required bandwidth, being sufficient to accommodate the accepted flows.

Regarding the utilisation of bandwidth allocated to the CL class, it is much higher in SRBQ (similar to RSVP) than in RSVPRAgg. In RSVPRAgg, the utilisation is very noticeably lower with a bulk size of 600 kbps than with a bulk size of 300 kbps. Notice that a bulk size of 600 kbps is less than a factor of 10 higher than the flow rates. This suggests that the use of larger bulk sizes in order to increase the scalability would lead to very poor network resource utilisation. Corresponding to the lower utilisation figures, the blocked bandwidth in the RSVPRAgg model is higher than in SRBQ, and is higher when using larger bulk sizes. In SRBQ the blocked bandwidth figures are similar to regular RSVP, since in both end-to-end reservations are accepted up to the bandwidth

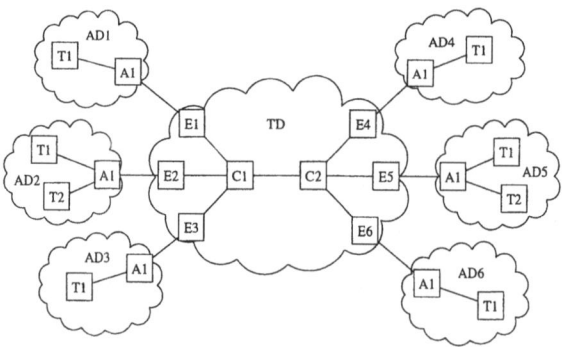

Fig. 1. Simulation topology

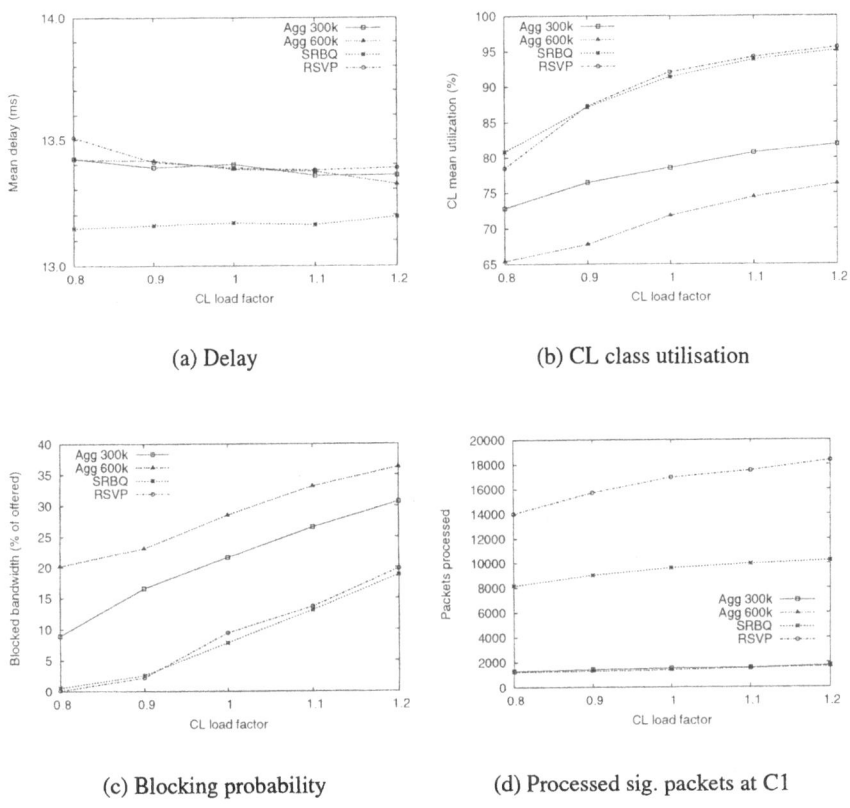

(a) Delay   (b) CL class utilisation

(c) Blocking probability   (d) Processed sig. packets at C1

Fig. 2. Performance of SRBQ, RSVPRAgg and RSVP with a single type of flows

reserved for the CL class, contrasting to the RSVPRAgg model in which end-to-end reservations are only accepted up to the reserved rate of the corresponding aggregate.

We also evaluated the number of signalling messages processed at core node 1 (see figure 1). This number is much lower in RSVPRAgg (about 1500 packets on average during the 3600 useful simulation seconds) than in SRBQ or RSVP (respectively, about 9000 and 16000 packets processed under the same conditions). This is an obvious result, since at interior nodes only aggregate messages are processed in RSVPRAgg. The almost twofold difference between SRBQ and RSVP is due to the fact that in RSVP both Path and Resv refreshes are needed. From the number of processed messages only, the RSVPRAgg model would be the clear winner in terms of signalling processing scalability. Keep in mind, though, that one big strength of the SRBQ model is the use of low complexity, highly efficient algorithms (labels, timers, etc.), which translate in much less CPU time used to process each message.

We performed a similar set of simulations using a mixture of different flow types (CBR, exponential on-off and Pareto on-off). The results of these simulations have shown that both models provide adequate QoS to the different flow types, except for

Table 1. Flow characteristics for the isolation test

Type	Avg. rate (kbps)	Pkt. size (Bytes)	On (ms)	Off (ms)	Pk. rate (kbps)	Token Bucket		Watermarks (kbps)			MTBC (s)	Avg. dur. (s)
						R (kbps)	B (Bytes)	1	2	3		
cbr64cl	64	500				64	1500	64	64.064	72	11	120
exp1cl	48	500	200	200	96	64	15000	32	64	96	11	120
exp2cl	48	500	var	var	var	64	15000	32	64	96	11	120

Pareto on-off flows which suffer higher delay and very significant losses in RSVPRAgg (about 10%, compared to about 0.003% in SRBQ). The higher delay is inflicted by WFQ outside the aggregation domain, while the packet losses occur mostly at the aggregator due to policing. Both of these problems stem from the fact that, having a heavy-tailed distribution with infinite variance, Pareto flows are not well suited for the token-bucket characterisation. SRBQ's rate water-marks characterisation is more tolerant of this type of flow. The per-class utilisation curves were similar to the previous ones, but lowered by about 10%.

2.2   Flow Isolation

With this set of simulations we evaluate the behaviour of the different models in the presence of misbehaved flows, that is, flows that transmit at rates much higher than they reserved for considerable periods of time. We measure not only the quality of service received by these bursty flows but also the impact in the other flows. Three flow types were used in this test (table 1): (1) CBR flows (cbr64cl) that are considered well behaved flows; (2) on-off exponential flows (exp1cl) with a burstiness of 50% (average busy and idle times of 200 ms) and a peak rate of 96 kbps, that are considered nearly well behaved flows, since they send at a rate a little higher than the reserved; and (3) on-off exponential flows (exp2cl) with varying burstiness and peak rate that are considered misbehaved flows, since they send at a rate much larger than the reserved one for considerable periods of time. Their burstiness is variable, from 50% to 12.5%, varying their peak rate between 96 kbps (average busy and idle times of 200 ms) and 384 kbps (average busy and idle times of 50 ms and 350 ms, respectively). Notice that the sum of the average idle and busy times remains constant (400 ms), as does the average rate. It is the high mismatch between the requested rate and the peak transmission rate that turns exp2cl flows into misbehaved ones.

Each of the 4 transmitting terminals generates flows with these parameters, in the CL class. The total mean offered load at the core link is, therefore, 120% of the bandwidth allocated to CL, in terms of reserved rate.

Figure 3 shows some results from this set of simulations. As observed, the mean delay for misbehaved flows is not affected by their burstiness in SRBQ (where it is mostly the sum of transmission and propagation delays), contrary to the other models. This is due to the fact that in the SRBQ model all CL data packets share the same queue. Notice however, that in this model, the traffic is policed before entering the domain. On the other hand, we may observe that in all models highly bursty flows have no noticeable impact in the delay of the low burstiness flows; the delay of CBR flows (not shown) is not affected either. The same kind of results are obtained for jitter, not shown here due to space limitations, but the difference is even higher: contrasting to the approximately constant value of 0.6 ms in SRBQ, jitter for the aggregation model varies from more

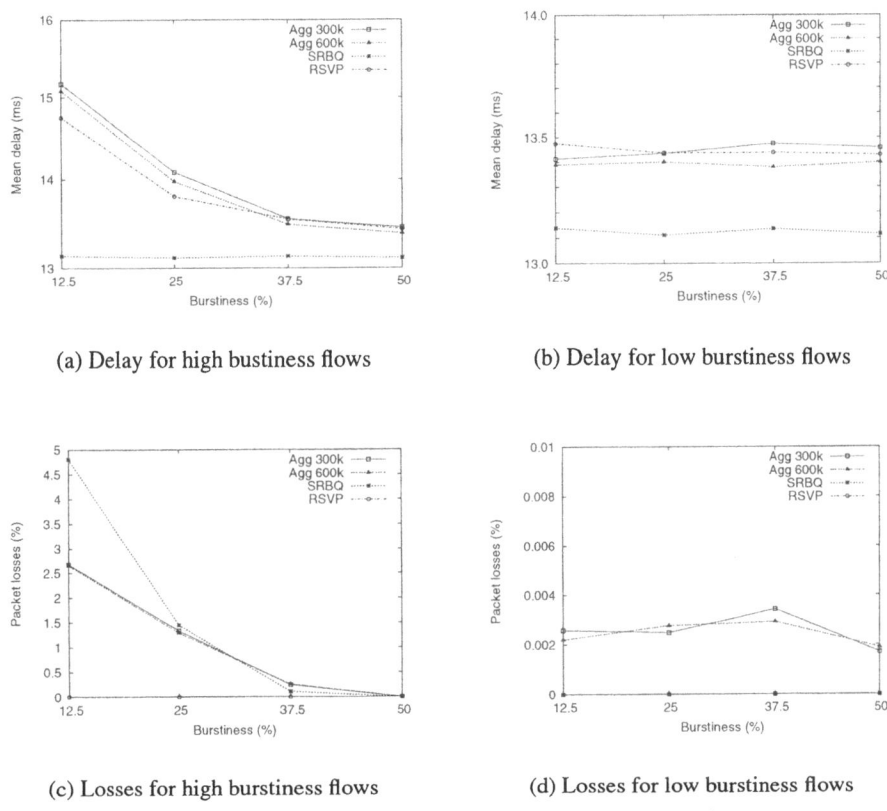

(a) Delay for high bustiness flows

(b) Delay for low burstiness flows

(c) Losses for high burstiness flows

(d) Losses for low burstiness flows

Fig. 3. QoS results for low and high burstiness exponential flows

than 10 ms with a burstiness of 12.5% to little more than 1 ms with a burstiness of 50%. Jitter for low burstiness flows is constant in all models, being about 0.6 ms in SRBQ and just above 1 ms in RSVPRAgg.

Misbehaved flows are heavily penalised in terms of packet losses. In SRBQ, packet losses for these flows almost reach 5% with a burstiness of 12.5%, while they are about 2.6% in RSVPRAgg. The high loss values are due to the fact that these flows transmit at rates much higher than they reserved for considerable periods of time. A relatively large bucket size absorbs these bursts up some level in RSVPRAgg, but in SRBQ the reservations for CL traffic have no bucket parameter, only 3 rate water-marks, of which even the third one is much lower (96 kbps) than the peak transmission rate (384 kbps). Clearly, these flows are violating their contracts, so measures must be taken against them to prevent QoS degradation for other flows. SRBQ is inflicting higher penalisations, better protecting other flows. Loss figures for other flows are not affected by the misbehaved ones. They are very low in RSVPRAgg and null in SRBQ for low burstiness exponential flows and are null in all models for CBR flows (not shown).

These results show that all models are able to provide adequate QoS levels to flows respecting their traffic contracts even in presence of misbehaved flows. The service of these flows is degraded in order to protect the well behaved flows: in SRBQ this degradation is only in terms of packet losses; in RSVPRAgg they are less penalised in terms of losses, but also have increased delay and jitter.

3   Conclusions and Future Work

In this paper we performed a comparative evaluation of two QoS architectures, SRBQ and RSVPRAgg, aimed at providing QoS levels similar to those provided by the well-known RSVP/IntServ architecture, but which are scalable enough for use in high traffic core networks. Several sets of simulations were performed for both models in order to evaluate different relevant aspects of the architectures (QoS parameters, flow isolation and resource utilisation). From the simulation results discussed in the previous section we may state that both the RSVPRAgg and SRBQ models provide adequate QoS levels and flow isolation in the CL class and are, therefore, real alternatives to RSVP/IntServ. Regarding scalability, SRBQ and RSVPRAgg are similar in terms of packet classification and scheduling procedures at the core, both using a scalable DiffServ-like approach. In terms of the raw number signalling messages processed at core nodes, RSVPRAgg wins by a wide margin. This is due to the fact that signalling scalability in SRBQ is not obtained by performing it at an aggregate level with bulk updates; instead, SRBQ makes use of highly efficient techniques and algorithms while keeping the end-to-end character of signalling. Due to these different approaches, utilisation figures are significantly better in SRBQ: there is no trade-off with signalling scalability like in RSVPRAgg, and resource utilisation is optimal under all conditions.

As future work, we plan to evaluate the possibilities for interoperability between SRBQ and other QoS architectures. We also plan to perform simulations with less synthetic, more realistic generation of flows, based on flow data collected in a real network. Finally, we expect to implement prototypes of both architectures in order to evaluate performance parameters which are not provided by ns-2, namely those regarding processing power required.

References

1. Braden, R., Clarck, D., Shenker, S.: Integrated Services in the Internet Architecture: an Overview. RFC 1633, Internet Engineering Task Force (1994)
2. Blake, S., Blake, D., Carlson, M., Davies, E., Wang, Z., Weiss, W.: An Architecture for Differentiated Services. RFC 2475, Internet Engineering Task Force (1998)
3. Baker, F., Iturralde, C., Faucheur, F.L., Davie, B.: Aggregation of RSVP for IPv4 and IPv6 Reservations. RFC 3175, Internet Engineering Task Force (2001)
4. Prior, R., Sargento, S., Cris stomo, S., Brand o, P.: End-to-end Quality of Service with Scalable Reservations. In: Proceedings of the 11th International Conference on Telecommunication System, Modeling and Analysis. (2003)
5. Prior, R., Sargento, S., Brand o, P., Cris stomo, S.: Efficient Reservation-Based QoS Architecture. In: Interactive Multimedia on Next Generation Networks. Volume 2899 of Lecture Notes in Computer Science., Springer-Verlag (2003) 161–181
6. Prior, R., Sargento, S., Brand o, P., Cris stomo, S.: Performance Evaluation of the RSVP Reservation Aggregation Model. In: Proceedings of the 7th IEEE International Conference on High Speed Networks and Multimedia Communications (to appear). (2004)

TCP Based Layered Multicast Network Congestion Control

Ahmed A. Al Naamani[1], Ahmed M. Al Naamany[2], and Hadj Bourdoucen[2]

[1] Petroleum Development Oman
Ahmed.A.Naamani@pdo.co.om
[2] Sultan Qaboos University
Muscat Oman
{Naamany,Hadj}@squ.edu.om

Abstract. The rapid increase of the internet and intranet usage has significantly increased the data sent to networks and in most cases this has resulted in congestion. Several techniques are used nowadays in networks to control network congestion. Most of these techniques are used for a unicast transmission. This paper is proposing a new technique to alleviate the multicast network congestion. This technique is also, trying to use the available network resources while reducing packets drops.

1 Background of Congestion Control

The following example offer good introduction of network congestion as shown in network diagram figure1. In this diagram, three senders transmit data to 6 receivers. The available bandwidth is different for each receiver as shown on the diagram. The intermediary devices like router 2 is receiving data at 4Kbps and transmitting it on 3Kbps due to capacity of line hence 1Kbps is stored on its buffer. The same will be noted for the routers 4, 5 and 6. They will all stores data on their buffers because of the difference between the receiving and the transmission rates. With the increase of the data storage on the routers, the time required for the data to reach the destination increases. This cause a delay on receiving data and also a delay of data flow of other transmissions from senders 2 and 3. When the buffers are filled with data, the router drops all extra packets. TCP will ensure reduction of sending rates and the dropped data is retransmitted. This further increases complexity of network congestion. [1,2,3]

The congestion caused by over transmissions results in inefficient utilization of the network and specifically bandwidth resources. The congestion control reduces this inefficiency by controlling the transmission levels to the actual throughputs. This benefits both methods of transmissions: multicast and unicast transmissions. [1,2,3]

J.N. de Souza et al. (Eds.): ICT 2004, LNCS 3124, pp. 1218–1225, 2004.

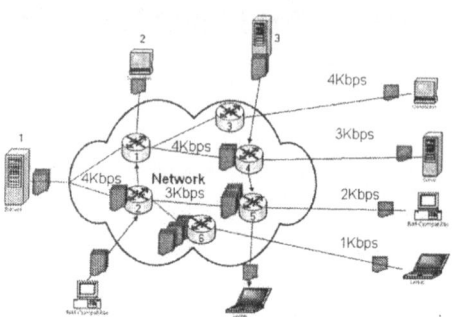

Fig. 1. Congestion example

2 Network Congestion Control Methods

Recently there has been attempt to control network congestion by controlling the transmission rate. These techniques are used for unicast and multicast transmission. The technique presently used in TCP/IP is Additive increase multiplicative decrease (AIMD)[4,5,6]algorithm which is simply controlled by a congestion window which is halves the transmission rate for every window of data containing a packet drop, and increased by roughly one packet per window of data otherwise.

A good attempt to break off the AIMD method is TCP Friendly Rate Control (TFRC) for Unicast [7] and multicast [8]which developed equation that can be used to control network congestion. In Unicast TFRC the receiver measures the packet loss rate (p) and feed it back to the sender. The sender calculates the round trip time RTT (R) to the receiver and uses an equation to identify the new transmission rate (T) which is adjusted accordingly.

For Multicast TFRC[7] uses receivers to measure packet loss rate, round trip time and calculate the new sending rate that is suitable for them and send it back to the sender. The sender adjusts its sending rate to the lowest one among the received ones from the receivers. This is despite of the fact that some of the receivers are capable of receiving at higher rates.

3 Layered Multicast Network Congestion Control

This proposed method strives to make use of the available network resources and to control network congestion more efficiently. This method is a based on layered transmission rates which makes use of the multitasking capabilities of computer systems. Network congestion is controlled by setting different transmission rates for different groups of receivers. These transmission rates levels are selected based on the receiver's throughput capabilities see Figure 2 below.

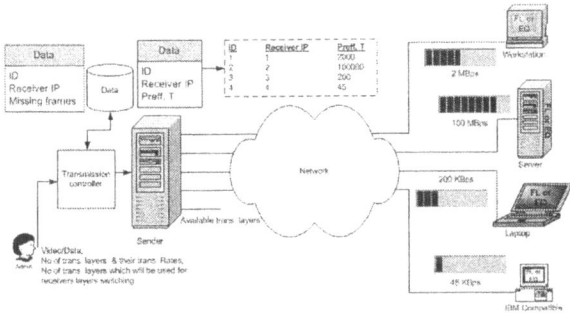

Fig. 2. Layered Multicast Network Congestion Control

Layered Multicast congestion control is based in the following algorithm:

1. Sender sets the initial transmission file type, the number of transmission groups and their transmission rates based on the receivers bandwidth on the network and he allocates the number of processes to start new transmissions.
2. Receivers calculate the preferred transmission rate using the equation used in the equation based congestion control [7] hence, selecting their preferred transmission group layer. This information is then feed back to the sender.
3. Sender store the information in its database and starts to send data on different transmission rates based on the layers groups.

3.1 Transmission Rate Change

The system is flexible since it allows receiver to change its transmission rate level if their throughput is changed during transmission. But live data are treated a bit different than the non-live data. This is due to the fact that live data has to be received by all receivers at the same time.

3.2 Changing to Lower Transmission Layer

If receiver's calculated transmission rates are significantly lower (due to packet drops which translates to mean that present sending rates are high), receiver sends a feedback to the transmission controller in the sender. When the sender receives a request from the receiver to switch to a lower transmission layer, sender checks if it can multitask new process i.e. a new transmission layer if it can if will start transmitting at this new rate. But if the allocated sender capabilities do not allow starting new transmission layer, the receiver should wait for the others receivers in the lower transmission layer to reach its synchronization and then continue with them.

3.3 Changing to Higher Transmission Layer

In cases where receiver is receiving at lower rates than what is capable of receiving as far as it calculated transmission rates are concerned, it sends a feedback to the sender requesting a change in transmission rate. Sender moves receiver directly to the higher transmission layer. This means the receiver will miss some data because of the un-synchronization in the transmission. To recover this missing data, the controller check for the sender capability to make new transmission layer to transmit it in parallel with the higher transmission layer connection transmission. This means the receiver will not be allowed to move to the higher transmission layer unless it can receive the data in the upper transmission layer in parallel with the missing data from the new started transmission layer. If the sender is not able to starts new transmission layer, the missing data position will stored in a database so that they will be sent later when it is possible to start new transmission layer.

4 Sender and Receiver Capabilities

To be able to carry out Layered Multicast Congestion Control, sender should be able to sends simultaneously data in different independent transmission layers with different transmission rates without affecting any transmission layer. Receiver is required to be able to receive data from more than two sources at the same time. It should be able to combine the received segments from different layered transmissions into a uniform file. Before, the receiver is allowed to move to a higher layer transmission rate it should be able to receive data at least into two transmission rates. One is the higher transmission rate and the other one is at least the lowest transmission rate so that it recovers the missing data because of the un-synchronization between different levels.

4.1 Congestion Control Overheads

Layered Multicast Congestion Control like other congestion control methods has limited benefits in cases of small size transmissions. Using such methods to sends small size files would not help to improve the transmission rate because the time spent recognizing and defining transmission rates can be used for the real transmission. However with the increase of the transmission file size the performance of this method increase comparing to the conventional transmission methods.

4.2 Benefits of Layered Multicast Congestion Control

The Layered Multicast Congestion Control improves multicast transmissions. Benefits of this method are measurable by the following three network parameters: Packet Drops, Throughput and Usage of available network resources.

It can be seen in figure 1 that the sender is trying to sends 12 Kb to 4 receivers which can receives data in different transmission rates (4,3,2 and 1 Kbps).With the standard TCP congestion control method, the data will be sent to all receivers with transmission rate of 4 Kbps. This transmission rate would create network congestion on routers 2, 4, 5 and 6 which will affect the data flow on that network as explained earlier.

This behavior will happen at the beginning of the transmission but then the TCP congestion control will starts to work and will send the data on the slowest transmission rate of the receivers i.e. 1Kbps.

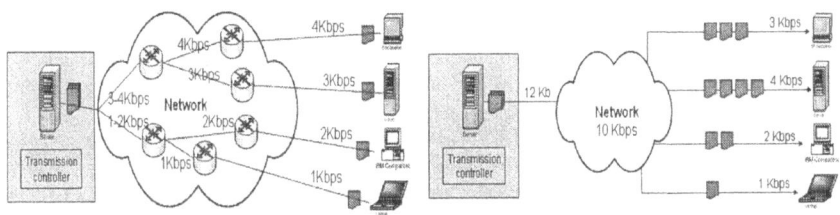

Fig. 3. Layered multicast transmission and Data transmission in the proposed methods

However, in Layered Multicast Congestion Control the sender will send transmission information to all receivers which will select the preferred transmission layer based on their available throughput. Then the sender will send the data in different transmission layers (1, 2, 3 and 4 Kbps). So in this case the data will not be sent through any route unless it is expected that the route is capable to handle the data transmission without losing any data. The sender will transmit bits at a rate of 4 Kbps in the route which passes through routers 1 & 3, 3 Kbps in route which passes through routers 1 & 4, 2Kbps in route which passes through routers 2 & 5 and 1 Kbps in route which passes through routers 2 & 6. The sender will continue sending data very smoothly and in a way that uses maximum of available network resources and after making sure that each route is capable of handling that transmission rate as explained above. With such a way of transmission, the number of dropped packets is sharply reduced. This example shows that the root of each connection goes through only two routers but in the real life applications it might go through many several routers, hubs and switches which should increase the packet lost problem.

Layered Multicast Congestion Control method at stable sending rates improves the overall network throughput. Figure 4 shows example where the sender is transmitting 12Kb to four receivers using conventional method at the rates shown.

Once the transmission become stable in the standard method, the data will be sent via the lowest receiver throughput to all receivers. This means that the sender is sending only at the lowest possible rate.

In the Layered Multicast Congestion Control, the sender should send the data to the four receivers in different transmission rates based on their throughputs. This does not mean that the new method is going to overload the network with data but it is going to use all available resources to finish the transmission as soon as possible, hence, freeing the bandwidth for other transmission. In other words, if we have a 10 lanes road,

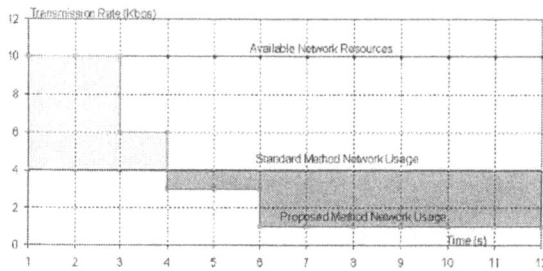

Fig. 4. Comparison between standard and the new method on the number of bits send to the network

the old method is going to use only 4 lanes of them while the new method is going to use all available lanes. The total amount of data sent to the network in the new method will continue going down once each group of transmission layer finish receiving their data.

The first area on the chart represents resources that are not used by the standard method and used by the new method. The second lower area on the chart represents resources that are freed in the Layered Multicast Congestion Control method to be used for other transmission. Above example also indicates that Layered Multicast Congestion Control method offer better benefit in cases where the multicast group is having many high throughput receivers and few low throughput receivers.

5 Network Simulation and Results

The proposed method is simulated in a network simulation tool called Opnet. The network used in this simulation is selected in a way that contains several bottlenecks. Also, it contains several nodes with several connections capabilities to the internet. The aim is to generate the congestion and to show that the new proposed system is able to help in controlling it.

The network covers six subnets which belong to three independent organizations networks. To ensure that the system is as realistic as possible real organizations with their actual configurations were simulated. The organizations were PDO(4 subnets which include 314 workstations), Omantel and SQU(71 workstations). Omantel is the internet service provider (isp) which has several users who connects via 56K fax/modem, ADSL and ISDN connections. The other organizations are connected to Omantel via several technologies among them, E1 standards with bandwidth of 2Mbps. IDF's in PDO are connected to the backbone via 100Mbps to each other except Fahud area is connected by bandwidth of 10Mbps. The simulation is run for 10 minutes and several network parameters are used for the simulation comparison to shows the existence of the congestion. These parameters are the packet drop, network delays and download response time.

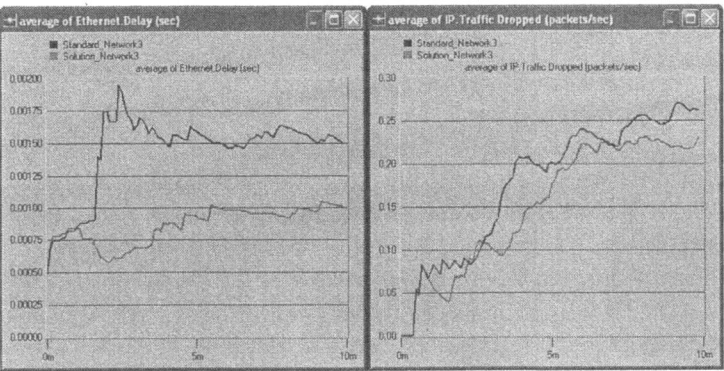

Fig. 5. Average of ethernet delay (sec) and average of IP traffic dropped

Figure 5 shows the average IP traffic dropped in both scenarios during the data transmission. Packet drop rate is the number of packets that are lost and did not reach the destination or the receiver during the transmission period. As it is clear, during most of the transmission period the packet dropped for the Layered Multicast Congestion Control method is better than the one for the standard method. Figure 5 also shows the average Ethernet delay which represents the delay of the packets to reach the destination or the receiver. At the beginning the delay in both methods are similar. This is due to the fact that the Layered Multicast Congestion Control method sends most of the data at the beginning of the transmission at higher transmission rates than the standard method.

The standard method delay time increases quickly and become much higher than the new method. That is because of two reasons. The first one is the new method has already finished the transmission to group of receivers and it is not sending data to them again. So less data is sent to network and so less delay. The second reason, the standard method at this stage already filled the network with packets and it start to handle these packets which cause the network loosing some of it resources for such tasks. Layered Multicast Congestion Control method responds faster because the network is not loaded with a lot of data and it is not overloaded. This means that the congestion is controlled better in the new proposed method. In addition, the network resources are used better and not wasted in saving and getting rid of lost packets.

5.1 Benefit from the Layered Multicast Network Control Method

This method offer different benefits for different parts of the network these are summarized as follows:

Sender: With the reduction of the packet dropped the request to resend packet should be reduced as well. This reduction should reduce the amount of tasks required to be undertaken by the sender.

Receiver: Receivers with high capabilities and good connection will receive data much faster than the slow ones. They do not need to wait for other receivers which are slower or which are in slower connections.

Network: Network Components will not be overloaded and will not be tasked in getting rid of overflowed buffers. This indicates that the expensive connection to the internet and WAN resources will be used in more efficient way. Also, in this method the data will be sent once to the network and resending due to packet drop would be reduced sharply.

6 Conclusion

Simulation results shows that the new method has less packet drop, faster download response time, and less Ethernet delays than the standard method. This demonstrates that the congestion is less in the new Layered Multicast Congestion Control method.

Implementation of Layered Multicast Congestion Control method adds more control of data flow on the network which offers efficient utilization of the resources of the sender, receiver and the in-between network (especially ISP's).

This method is effective mainly for multicast congestion control and for cases where the non-live data files must not be received by receivers at the same time. Similarly, this method is most effective when used to transmit large files.

References

1. S. Floyd, "A report on Recent Development in TCP congestion control", IEEE communication magazine, April 2001.
2. H. Balakrishnan et al., "TCP Behavior of Busy Web Server: Analysis and Improvement", IEEE Infocom, Mar 1998.
3. Sally Floyd and Kevin Fall, "Promoting the use of end-to-end congestion control in the Internet" in IEEE/ACM Transactions on Networking, May 1999.
   http://www.icir.org/floyd/papers/collapse.may99.pdf
4. Ahmed M. Al-Naamany and Hadj Bourdoucen, "Fuzzy Logic Based TCP Congestion Control System", Network Control and Engineering for QoS, security and mobility II in Kluwer Academic publishers, Oct 2003.
5. Chris Palmer: Basics of Networking
   http://www.ardenstone.com/projects/seniorsem/reports/Basics.html
6. Yang Richard Yang, Simon S. Lam, "General AIMD Congestion Control" in proceedings of ICNP 2000, Osaka, Japan, May 9, 2000
   ftp://ftp.cs.utexas.edu/pub/lam/ToN/Pubs/YangLam00tr.pdf
7. Sally Floyd, Mark Handley, Jitendra Padhye, and Jorg Widmer, "Equation based congestion control for unicast applications", In Proceedings of ACM SIGCOMM 2000, USA, August 2000. http://www.icir.org/tfrc/tcp-friendly.pdf
8. Jorg Widmer and Mark Handley, "Extending Equation-Based Congestion Control to Multicast Applications", In Proceedings of ACM SIGCOMM 2000, USA , August 2001.

A Novel Detection Methodology of Network Attack Symptoms at Aggregate Traffic Level on Highspeed Internet Backbone Links*

Byeong-hee Roh and S.W. Yoo

Graduate School of Information and Communication, Ajou University,
San 5 Wonchon-dong, Youngtong-Gu, Suwon, 443-749, Korea
{bhroh,swyoo}@ajou.ac.kr

Abstract. In this paper, we investigate the network attack traffic patterns appeared on Internet backbone links. Then, we derive two efficient measures for representing the network attack symptoms at aggregate traffic level. The two measures are the power spectrum and the packet count-to-traffic volume ratio of the aggregate traffic. And, we propose a new methodology to detect networks attack symptoms by measuring those traffic measures. Unlike existing methods based on individual packets or flows, since the proposed method is operated on the aggregate traffic level, the computational complexity can be significantly reduced and applicable to high-speed Internet backbone links.

1 Introduction

Recently, we have experienced several troubles of the Internet services due to various network attacks. Attacks on the Internet infrastructure can lead to enormous destructions, since different infrastructure components of the Internet have implicit trust relationship with each other.

Houle and Weaver[1] surveyed the trends in deployment, use, and impact of variety of denial of service (DoS) attacks. In their report, most of attack tools alter packets' major attributes such as source/destination IP addresses and port numbers for different purposes of attacks. Some of representative network attack types classified regarding the degree of alteration of those attributes are listed in Table.1 [1][2], which includes IP spoofing attacks with varied destination port number (vspoof) as well as fixed one (fspoof), host scanning attacks (hostscan), and port scanning attacks (portscan).

Much of works on of network attacks and corresponding solutions[3] have been focused on detecting and reacting to network attacks at individual end networks for their own safety. However, since global-scale network attacks are far more defined and visible in the backbone before it spreads out towards targets, it might be more efficient that to detect and react the symptoms of those global-scale network attacks are done in backbone domain. Some works for detecting

---

* This work was supported by grant (No. R05-2004-000-10824-0) from Ministry of Science & Technology

J.N. de Souza et al. (Eds.): ICT 2004, LNCS 3124, pp. 1226–1235, 2004.

Table 1. Classification Of DoS Attack Types

attributes attack types	source IP address	destination IP address	destination port number
vspoof	varied	fixed	varied
fspoof	varied	fixed	fixed
hostscan	fixed	varied	fixed
portscan	fixed	fixed	varied

network attacks on backbone links, e.g. [2], have been carried out. However, those works can not be applied to all the attacks since attack mechanisms and tools continue to improve and evolve. One promising direction to react those evolving various attack patterns is to develop a global defense infrastructure[4]. However, it is only suggestion, but not provide any actual mechanism or architecture.

In this paper, we propose a new methodology for traffic measurement based detection of network attack symptoms on high-speed Internet backbone links. The proposed method deals with the network attacks from the view points of the aggregate traffic level different from the existing works, in which network attacks were dealt with on the individual packet or flow basis. The reasons why we focus on network attacks from the aggregate traffic's viewpoints are as follows: First, it can provide a robust mechanism even though the advent of new attack types. This is because the pattern of the normal traffic can be known and forecastable by conventional traffic modeling or measurement techniques, so that if new network attack types are appeared, the traffic pattern affected by those attacks can be easily detected since the aggregate traffic patterns may be significantly changed different from the normal traffic patterns. Second, to detect network attacks at individual packet or flow levels require much higher computational complexities, while the complexity of the works at the aggregate traffic level can be significantly reduced. Finally, since the traffic monitoring and management at each network segment are currently working individually and it can be extended for global Internet management, if the detect mechanism at aggregate traffic level are cooperated with the global network management systems, it can possible to develop a global defense infrastructure.

The rest of the paper is organized as follows. In Section 2, we show some characteristics of network attack traffic, and how those network attacks can affect the normal traffic pattern. And, the proposed method for detecting network attacks at aggregate traffic level and experimental results are presented in Section 3 and Section 4, respectively. Finally, we conclude the paper in Section 5.

2   Characteristics of Network Attack Traffic

In this Section, we investigate how network attack traffics are appeared on backbone links and how those attacks affect the normal traffic pattern. For doing so, first, we captured packets on two trans-pacific T-3 links connecting the U.S. and a Korean Internet Exchange with the help of National Computerization Agency,

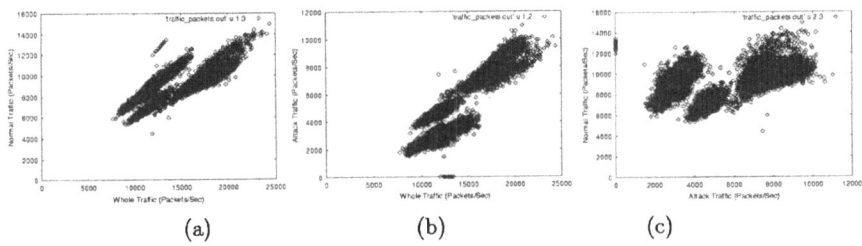

Fig. 1. Relationship in packet count (packets/sec) between traffic types (a)aggregate and normal (b) aggregate and attack (c)attack and normal

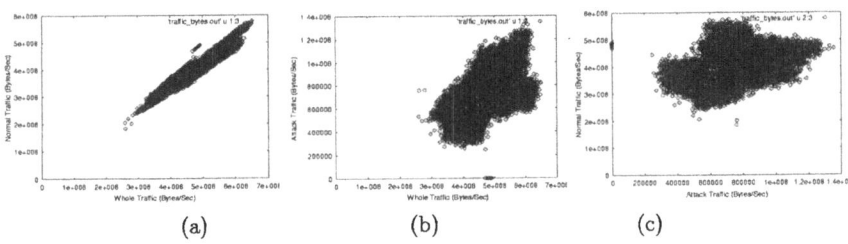

Fig. 2. Relationship in traffic volumes (bytes/sec) between traffic types (a)aggregate and normal (b) aggregate and attack (c)attack and normal

Korea. Then, we classified network attack packets from the captured data by using the method proposed in [2] and [7], which are based on the phenomenon of attack types as shown in Table. 1. We define *the attack traffic* as the sequence of packets classified for network attacks among captured packets, *the normal traffic* as the sequence with remaining packets, and *the aggregate traffic* as the sequence of whole packets.

In Fig.1, it is shown how each traffic type is mutually correlated with other types in the packet counts. Each point in the figure represents the number of generated packets during the same 1 second period from the two comparative traffic types. Fig.1 shows strong relationships not only between the aggregate and the normal, but also between the aggregate and the attack, for their packet counts. While, no relationships between the normal and the attack can be seen. Fig.2 also depicts the relationship in traffic volume between each traffic types. The traffic volume of the aggregate traffic is highly correlated to that of the normal traffic, while we can not see any relationships between other traffic types. As we can see from Fig.1 and Fig.2, the packet counts of the aggregate traffic are governed by both the normal and the attack traffic, but the variant of aggregate traffic volume is mainly affected by the normal traffic.

In order to show the reason why the results of Fig.1 and Fig.2 are appeared, we depict the cumulative density function (cdf) for packet sizes of each traffic type in Fig.3. As we can see, most of the attack packets have much smaller bytes than the normal traffic. That is, more than 90% of the attack packets show sizes below 80 bytes. While, the normal traffic shows more spread pattern around the

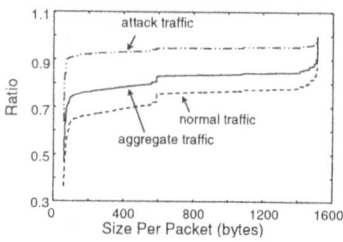

Fig. 3. Cumulative density function for packet sizes

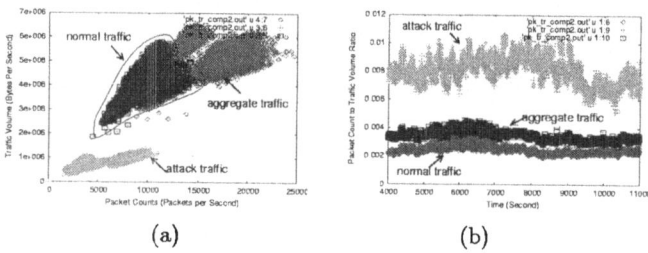

Fig. 4. Relative statistics:(a)between packet counts and traffic volumes (b)packet counts-to-traffic volumes ratio(CVR)

whole packet sizes from 60 to 1500 bytes, but there are some sizes where much of packets have, such as below 70 bytes, around 590 bytes, and above 1450 bytes.

To view how the statistics shown in Fig.3 affects the aggregate traffic flow from a different standpoint, the relationships between the traffic volume and the packet counts are depicted in Fig.4(a), in which all values are measured in 1 second period. From Fig.4(a), we can see that when the attacks are added, the degree of packet count variation in the aggregate traffic becomes much higher than that of traffic volume variation. Fig.4(b) shows the ratio of the packet count to the traffic volume in a certain time period, in which more attacks are detected than in other time period. As we can expect, the packet count-to-the traffic volume ratio(CVR) of the attack traffic is much higher than that of the normal traffic. Accordingly, the ratio of the aggregate traffic is getting increase by adding the attack traffic as shown in Fig.4(b).

It has been well known that the Ethernet traffic is statistically self-similar[5]. The self-similar nature of a traffic flow can affect the development of network congestion control schemes as well as the source traffic characterization. It is noted that the higher the self-similarity of the traffic is, the more the network performances such as link utilization, throughput, and so on, are affected. In Fig.5, it is shown the variance-time-plot (VTP) [5] for estimating Hurst parameters for each traffic type. We can see that the attack traffic shows higher self-similarity than the normal traffic, and that the self-similarity of the aggregate traffic is increased by adding the attack traffic. This indicates that since the addition of the attack traffic increases not only the traffic volume in networks, but also the self-similarity of the aggregate traffic, it can significantly affect the

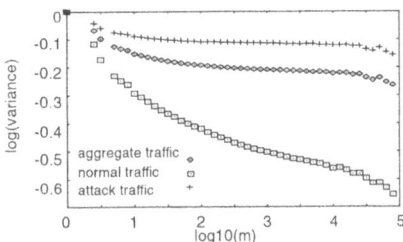

Fig. 5. Variance time plot (VTP)

network performances more than in the situation that the same amount of the normal traffic is added.

3   Detection of Network Attack Symptoms Based on Traffic Measurement

In the previous section, we used the VTP method for finding the Hurst parameter. However, the Hurst parameter can be estimated by using periodogram method[6]. For the discrete-time sequence with l samples X_0, X_1,...,X_{l-1}, the periodogram is defined as

$$S(\omega) = \frac{1}{2\pi N}\left|\sum_{k=0}^{l-1} X_k e^{jk\omega}\right|^2 \tag{1}$$

The relationships between S(ω) and ω can be written as

$$\log_{\omega\to 0} S(\omega) = a_0 \log|\omega| + a_1 \tag{2}$$

When S(ω)is plotted against ω on a log-log plot, it can be approximated by a straight line with a slope a_0. Then, the Hurst parameter is given by $H = (1 - a_0)/2$. It is noted that Eq.(1) is based on the discrete-time Fourier transform (DFT). Li et al[8] showed that the queuing performances in networks can be dominated by the average power spectrum obtained by DFT of the input traffic. From the researches related to [8] and the results of the previous section, we use the average power spectrum as one of the measures for detecting the network attack symptoms. The average power spectrum is defined as follows. Let us assume that the time is divided into a constant period of Δ, then Δ is a basic unit for traffic measurement. Let c_n and v_n (n=0,1,2,...) be the packet count and the traffic volume of the aggregate traffic measured during n-th Δ duration, respectively. Let $L\Delta$ be the detection period, which is the time duration that the detection algorithm is applied to. That is, each detection period consists of unoverlapped L consecutive Δs. And, we define the packet counts and traffic volumes measured during m-th detection period as the vectors $c_m = [c_{mL}, c_{mL+1}, ..., c_{(m+1)L-1}]$ and $v_m = [v_{mL}, v_{mL+1}, ..., v_{(m+1)L-1}]$ (m=0,1,2,...), respectively. Then, we have the average power spectrum for the vector c_m as following

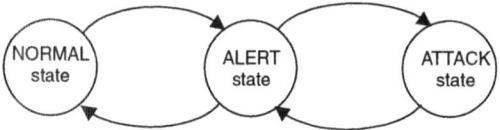

Fig. 6. Transition between states for the detection of the attack symptom

$$\overline{P}(m) = \sum_{k=0}^{L-1} \phi_{mk} \tag{3}$$

where $\Psi_m = [\phi_{m0}, \phi_{m1}, ..., \phi_{m(L-1)}]$ is obtained through DFT of c_m. That is, $\Psi_m = L^{-2}|DFT(c_m)|^2$.

It is noted that the average power is a measure for reflecting the effect of the self-similarity due to the network attacks. Besides the self-similarity, we showed several characteristics of the packet count and the traffic volume statistics, and they can be another good measure for detecting network attack symptoms. Especially, it is shown that the CVR is significantly changed when the network attacks are added. In this paper, we use the ratio as another measure for detecting the network attack symptoms. The CVR is given by

$$\overline{R}(m) = \frac{c_m \cdot e}{c_m \cdot e} \tag{4}$$

where $e = [1, 1, ..., 1]^T$, and $[\bullet]^T$ indicates a transpose matrix.

The proposed method to detect the network attack symptoms considering two measures given in Eq.(3) and Eq.(4) is as follows. Let $x_p(m)$ and $x_r(m)$ be the weighted averages of the average power and the CVR measured at m-th detection period, respectively, and given by

$$x_p(m+1) = \alpha_p x_p(m) + (1 - \alpha_p)\overline{P}(m) \qquad , m = 0, 1, 2, ... \tag{5}$$

$$x_r(m+1) = \alpha_r x_r(m) + (1 - \alpha_r)\overline{R}(m) \qquad , m = 0, 1, 2, ... \tag{6}$$

where α_p and α_r are the constant values between 0 and 1.

In general, it is known that the normal traffic flows in a stationary state vary within a forecastable range. Let m_p and δ_p be the average and the tolerance for the weighted average of the average power of the normal traffic within a certain stationary state, respectively. The m_p is calculated by the conventional way to obtain the arithmetic mean, and the δ_p is determined by an administrative policy of the network operators. Similarly, we define m_r and δ_r as the average and the tolerance for the weighted average of the CVR of the normal traffic within a certain stationary state, respectively. It is assumed that the tolerances δ_p and δ_r are determined based on the normal traffic flows prior to the actual measurement for the detection. The case when the measured $x_p(m)$ and/or $x_r(m)$ exceed the tolerances δ_p and δ_r, we can assume that it may be a symptom of the network attacks. In order to determine the situation that the network infrastructure is currently being attacked, we define the three states such as NORMAL, ALERT, and ATTACK. The NORMAL state is the state that there is no attack. The ALERT state is the one that the attack symptom is in question but the decision of the attack symptom is not completed. And, in the ATTACK state, it is inferred that the network resources are being attacked. The transition between those states is shown in Fig.6, and the main algorithm to detect the attack symptom considering those states is illustrated as follows.

<variables>
  $attack\_count$: counter for representing the degree of attack
  $Alert\_Threshold$: threshold value to change between ALERT and ATTACK states
  $Attack\_Threshold$: maximum value of $attack\_count$
  $state$: current state of the algorithm

<main algorithm>
  at the end of every detection period, update $x_p(\cdot)$ and $x_r(\cdot)$ by using (5) and (6)
  considering $x_p(\cdot)$ and $x_r(\cdot)$, the state at the detect period is determined by
  the following sequence.

```
if ( state == NORMAL )
    if ( ( x_p(·) > δ_p AND x_r(·) ≤ δ_r ) OR ( x_p(·) ≤ δ_p AND x_r(·) > δ_r ) )
        state = ALERT;
        attack_count += 1;
    elseif ( x_p(·) > δ_p AND x_r(·) > δ_r )
        state = ALERT;
        attack_count += 2;
    endif
elseif ( state == ALERT )
    if ( ( x_p(·) > δ_p AND x_r(·) ≤ δ_r ) OR ( x_p(·) ≤ δ_p AND x_r(·) > δ_r ) )
        attack_count += 1;
    elseif ( x_p(·) > δ_p AND x_r(·) > δ_r) )
        attack_count += 2;
    elseif ( x_p(·) ≤ δ_p AND x_r(·) ≤ δ_r )
        attack_count -= 2;
    else
        attack_count -= 1;
    endif
    if ( attack_count > Alert_Threshold )
        state = ATTACK;
    elseif ( attack_count ≤ 0 )
        state = NORMAL;
        attack_count = 0;
    endif
elseif ( state == ATTACK )
    if ( x_p(·) > δ_p AND x_r(·) > δ_r) )
        attack_count = MIN ( attack_count+1, Attack_Threshold);
    elseif ( x_p(·) ≤ δ_p AND x_r(·) ≤ δ_r )
        attack_count -= 2;
    else
        attack_count -= 1;
    endif
    if ( attack_count ≤ Alert_Threshold )
        state = ALERT ;
    endif
endif
```

---

4   Experimental Results

For the experiment, we used the self-similar traffic generation method proposed in [10]. Traffic parameters used for the experiment were obtained from our captured traffic data explained in Section 2. That is, for the parameters of the normal traffic, we used 0.9 for the Hurst parameter and 9080.97 packets per second for the average packet count. For the attack traffic, we used 0.99 for the Hurst parameter and 5292.8 packets per second for the average packet count. More detail about the statistics for those parameters is illustrated in [9]. We had the packet size of each generated packet followed the distribution shown in Fig.3. And, we had the attack traffic kept in a certain period as shown in Fig.7.

Fig.8 shows the variation of the weighed averages $x_p(m)$ and $x_r(m)$ for the normal traffic and the aggregate traffic, respectively, when $\Delta=10$msec and $L=10$. We can see from Fig.8 that the two traffic measures such as the average power and the CVR can be appropriately used for detecting the attack symptoms at the aggregate traffic level. That is, those weighted average values in the normal traffic are varied within the forecastable range during the time interval in a stationary sate. On the other hand, when the irregular traffic patterns such as network attacks are added, those weighted average values in the aggregate traffic deviate from those in the normal traffic pattern. From this point of view, the two measures reflect the phenomenon of the network symptoms very well.

For testing the effectiveness of the proposed method, we define the following performance measures

$$Exactness = (T_{detected}/T_{actual}) \times 100(\%) \tag{7}$$

$$ErrorRatio = (T_{error}/T_{actual}) \times 100(\%) \tag{8}$$

$$DetectDelay = T_{detected} - T_{actual} \tag{9}$$

where T_{actual} and $T_{detected}$ are the time duration that the attack traffic is appeared in the test sequence used for the experiment and the time duration that the attack is detected by the proposed algorithm, respectively. T_{error} is the time duration that the detection is not correct, which consists of the duration classified into ATTACK state though it is not the actual attack duration and the duration that is not classified into ATTACK state though it is within the actual attack duration. And, T_{actual} and $T_{detected}$ are the start time of the attack in the test sequence and the time that the attack is firstly detected, respectively.

Table.2 shows the experimental results for the performance measures varying the unit time Δ and the detection period L. The format of A/B/C in each cell in Table.2 represents the performance measures such as Exactness/ Error-Ratio/ DetectDelay. The parameters for obtaining Table.2 are as follows: $\alpha_p = \alpha_r = 0.9$, $Alert\_Threshold=5$, $Attack\_Threshold=10$. For δ_p and δ_r, the maximum weighted average values obtained from the normal traffic sequence used for the experiment by using Eq(5) and Eq(6), respectively, are used. As we can imagine intuitively, the smaller the values of Δ is, the better the performance measures such as Exactness and ErrorRatio are. While the bigger the value of L is, the larger the DetectDelay is. However, it is noted that as Δ and L are decreased, the computational complexity will be increased, and it will give the more load to the system. For any treatable parameters, the exactness of the detection is greater than 97%.

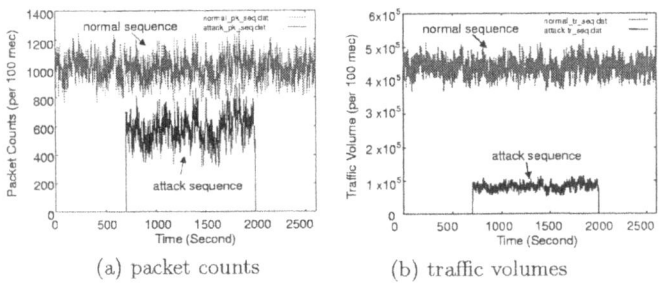

(a) packet counts   (b) traffic volumes

Fig. 7. Traffic sequence used for the experiment

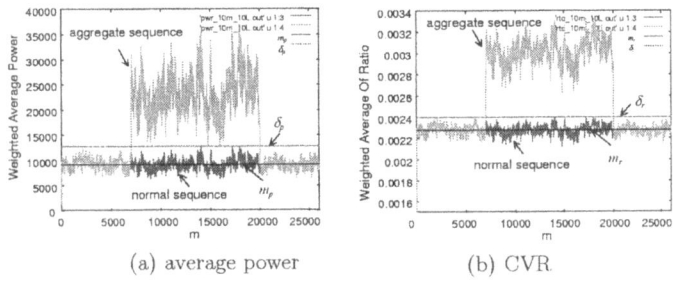

(a) average power   (b) CVR

Fig. 8. Variation of weighted averages (Δ=10msec and L=10)

5   Conclusion

In this paper, we investigated the nature of network attacks from the viewpoint of the aggregate traffic flow. We showed that the network attacks alter the pattern of the normal traffic significantly and the degree of the alteration can be detected by two measures such as the power spectrum of the packet count and the CVR. Then, we proposed a method for traffic measurement based detection of network attack symptoms at the aggregate traffic level, and showed the effectiveness and the applicability of the proposed method by experiments.

Detecting the network attack symptom based on individual packet or flow level requires for much higher computational complexities. In addition, finding the detection mechanisms to deal with each attack mechanism that is continuously and diversely evolving will get to a limitation. However, if the detection is done at the aggregate traffic level as in the proposed method, the required computational complexity can be significantly reduced, and it is possible to provide a robust detection mechanism regardless of the attack types. This is because the basic nature of the normal traffic flows can be known and forecastable by conventional traffic modeling or measurement techniques, so that if new network attack types are appeared, the altered traffic pattern affected by those attacks can be easily detected.

Table 2. Experimental Results

$L \setminus \Delta$	10 msec	100 msec	1 sec
10	99.98%/0.001/300msec	99.77%/0.016/3sec	97.71%/0.275/30sec
100	99.77%/0.016/3sec	97.71%/0.274/30sec	N/A
1000	97.71%/0.274/30sec	N/A	N/A

Backbone networks have their network management systems to monitor and maintain the networks. The traffic measurement based detection methodology can cooperate with those network management systems, so that it can possible to develop a global defense infrastructure. This paper focused on the issue to detect the attack symptoms on individual backbone links, so it requires further study for the proposed method to be extended to and used for the global defense infrastructure. In addition, we derived the measures for detecting attack symptoms by using captured packet data empirically. So, to investigate the traffic model for reflecting the variety of network attacks and to develop more practical control method to detect and deal with those attacks requires further study.

References

1. Houle K., Weaver J.: Trends in Denial of Service Attack Technology, CERT Coordination Center (2001)
2. Kim H., et al.: Fast Classification, Calibration, and Visualization of Network Attacks on Backbone Links, Tech. Report, http://net.korea.ac.kr (2003)
3. Chakrabarti A., Manimaran G.: Internet Infrastructure Security: A Taxonomy, IEEE Networks, vol.16, no.6 (2002)
4. Chang R.: Defending Against Flooding-Based Distributed Denial-of-Service Attacks: A tutorial, IEEE Communications Magazine, Oct. (2002) 42-51
5. Leland W.E., Taqqu M.S., Willinger W., Wilson D.V.: On the Self-Similar Nature of Ethernet Traffic (extended version), IEEE/ACM Tr. Networking, vol.2, no.1, Feb.(1994) 1-15
6. Beran J., Sherman R., Taqqu M.S., Willinger W.: Long-Range Dependence in Variable-Bit-Rate Video Traffic, IEEE Tr. Commun., vol.43, no. 2/3/4, (1995)
7. Chung E.: Effective Detecting DoS Attack and Scanning At Internet Backbone Using Bloom Filter, Master Degree Thesis, Ajou University, Feb.(2004)
8. Li S., Hwang C.L.: Queue Response to Input Correlation Functions: Discrete Spectral Analysis, IEEE/ACM Tr. Networking, vol.1, no.5, Oct.(1993) 522-533
9. Jeon Y., Roh B., Kim J.: Traffic Characterization For Network Attack Flows On the Internet Backbone Links," Internet Computing 2004, Las Vegas, June (2004)
10. Paxon V.: Fast, Approximate Synthesis of Fractional Gaussian Noise for Generating Self-Similar Network Traffic, ACM SIGCOMM Computer Communication Review, vol.27, no.5, (1997)

Precision Time Protocol Prototype on Wireless LAN

Juha Kannisto, Timo Vanhatupa, Marko Hännikäinen, and Timo D. Hämäläinen

Institute of Digital and Computer Systems, Tampere University of Technology
Korkeakoulunkatu 1, FIN 33720 Tampere, Finland
Tel. +358 3 3115 2111, Fax +358 3 3115 3095
{juha.kannisto,timo.vanhatupa}@tut.fi

Abstract. IEEE 1588 is a new standard for precise clock synchronization for networked measurement and control systems in LAN environment. This paper presents the design and implementation of an IEEE 1588 PC software prototype for Wireless LANs (WLAN). Accuracy is improved using two new developed methods for outbound latency estimation. In addition, an algorithm for adjusting the local clock is presented. The achieved accuracy is measured and compared between WLAN and fixed LAN environments. The results show that 2.8 µs average clock offset can be reached on WLAN, while wired Ethernet connection enables 2.5 µs.

1 Introduction

Clock synchronization is needed in various home, office, and industrial automation applications. As computer clocks in general have limited accuracy, synchronization protocols are used to precisely synchronize independent clocks throughout a distributed system. Synchronization allows transactions between distributed systems to be controlled on timely basis.

Basically, synchronization accuracy can be improved using two methods [1]. First, computer hardware can be extended with a more accurate clock. Second, computer clocks can be synchronized to one accurate external clock. This paper focuses on the second method, which is considered more cost efficient and scalable.

Network Time Protocol (NTP) [2] is a wide-spread protocol for time synchronization. It is targeted for wide area networking (Internet scale) and for maintaining the correct time and date in personal workstations. Consequently, its accuracy requirements are within a millisecond scale [3]. This is inadequate for time critical automation processes.

Applications in home, office and industrial environments are increasingly built using Local Area Network (LAN) technologies. IEEE 1588 is a new standard for precise clock synchronization for networked measurement and control systems in the LAN environment [4]. The standard defines a Precision Time Protocol (PTP) developed for synchronizing independent clocks running on separate network nodes. The standard aims for sub-microsecond accuracy, while higher accuracy is targeted by defining hardware implemented extensions.

Wireless LANs (WLAN) are increasingly used to extend wired networks due to easier installation and freedom of movement. This paper presents the design and im-

J.N. de Souza et al. (Eds.): ICT 2004, LNCS 3124, pp. 1236–1245, 2004.

plementation of a software based prototype for implementing the IEEE 1588 standard on IEEE 802.11 WLAN [5]. The accuracy of the implementation is improved with two developed methods for outbound latency estimation. Additionally, an algorithm for adjusting the local clock is presented.

The prototype has been implemented on Windows PC platform. Measurement results on switched and shared Ethernet are compared to the results on WLAN. Additionally, the effect of outbound latency methods is evaluated by measurements in WLAN environment.

The paper is organised in the following way. Sect. 2 presents related work on time synchronization. Sect. 3 gives an overview of the IEEE 1588 standard. Next, Sect. 4 presents the prototype design and implementation. Sect. 5 defines the measurements arrangements and gives the measurement results. Finally, Sect. 6 concludes the paper.

2 Related Research

The use of hardware extensions for improving NTP accuracy has been proposed in [6]. The results show significant improvement in accuracy as the average offset between two computer clocks is reduced to 90 ns. Without extensions the average offset is within the range of few milliseconds [3].

In the proposal, special hardware used in synchronization is added to Network Interface Card (NIC). Hardware recognizes NTP packets, corrects their timestamps, and calculates a new checksum for a packet. The operation is done just before transmitting the Ethernet frame. The delay between the timestamping and the actual start of the transmission is practically a constant. When NIC receives a NTP packet, it triggers a timestamp latch, which a NIC device driver can read. Timestamps are generated from a time register on NIC. The value in the time register is the time of the clock that is synchronized.

In NTP, the clock adjustment algorithm [2] is based on Phase Locked Loop (PLL), Frequency Locked Loop (FLL), and a loop filter. PLL and FLL analyse the phase and frequency errors of the local clock based on NTP messages. The analysis results are combined and filtered using a loop filter. The filter output is used to adjust the clock frequency.

Agilent Technologies has implemented a hardware-based prototype of the IEEE 1588 standard. The accuracy of the implementation was measured on a 10 Mbit/s repeater (hub) and a gigabit Ethernet switch. According to the measurements, it has 32 ns average offset with 6400 ns^2 variance on a repeater network and 49 ns average offset with 19600 ns^2 variance on a switched network [7].

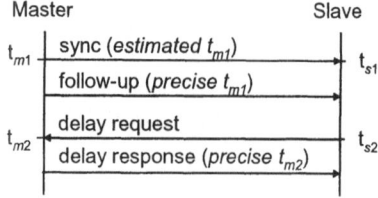

Fig. 1. PTP messaging.

Although good results have been achieved in both proposals, they require special hardware. Hardware-based implementation can be used in special applications but it is neither cost-efficient nor feasible approach for synchronizing the existing computer systems. Our goal is to reach the best possible accuracy without special hardware.

3 IEEE 1588 Overview

PTP divides the topology of a distributed system into network segments enabling direct communication between PTP clocks. These segments, denoted as communication paths, may contain repeaters and switches connecting same LAN technology. Devices connecting communication paths, such as routers, introduce possibly asymmetric and variable delay to the communication, and are therefore treated separately. PTP defines two types of clocks, ordinary clocks and boundary clocks. An ordinary clock connects to only one communication path while a boundary clock has connections to two or more communication paths. The boundary clock is responsible for synchronization between communication paths.

On each communication path a single clock is selected as a master clock while others are slave clocks synchronizing to it. The selection is done using the *best master clock algorithm*. The algorithm calculates the status of the local clock independently by listening synchronization messages and no negotiation between clocks is necessary. If two or more precise clocks are available, such as atomic clocks or clocks coupled with Global Positioning System (GPS), the algorithm makes only one active at a time.

PTP messaging between a master clock and a slave clock is presented in Fig 1. A master clock sends a *synchronization (sync) message* once in every two seconds in a default configuration. The message contains information about the clock and an estimated timestamp t_{m1} of the message transmission time. The clock information contains the identification and the accuracy of the master clock. When a slave clock receives the sync message, it stores a timestamp t_{s1} of the reception time. As it may be difficult to timestamp a sync message with an exact transmission time, the master clock can send a *follow-up message*, which contains a more precise value for the timestamp t_{m1}.

A slave clock sends periodically a *delay request message* and stores its transmission time with a timestamp t_{s2}. When a master clock receives the message, it sends a *delay response message*, which contains the timestamp t_{m2} of the reception time of the corresponding request message.

The slave clock calculates the *master to slave delay* d_{ms} and the *slave to master delay* d_{sm} according to these timestamps as

$$d_{ms} = t_{s1} - t_{m1},\tag{1}$$

$$d_{sm} = t_{m2} - t_{s2}.\tag{2}$$

The slave clock calculates the estimation of the one way delay d_w and the *offset from master* o_{fm} using the results from (1) and (2) as

$$d_w = \frac{d_{ms} + d_{sm}}{2},\tag{3}$$

$$o_{fm} = d_{ms} - d_w. \tag{4}$$

The offset from master is used to adjust the computer clock frequency and/or time.

All PTP messages described above use a multicast address, thus every clock on the communication path receives the messages and the discovering of other clocks is not necessary. Due to multicasting, it is necessary that PTP messages are not propagated between communication paths. The Ethernet implementation, defined by an annex of the standard uses a Time To Live (TTL) value in the Internet Protocol (IP) packet header to limit the propagation distance. On top of IP, User Datagram Protocol (UDP) is used as a transport protocol.

The accuracy of a generated timestamp is an important factor of the overall PTP system accuracy. It is affected by the delay fluctuation in the protocol stack thus the closer the timestamp is taken from the transmission of the message, the better the accuracy is. The timestamp can be generated either on hardware, driver, or application layer, as presented in Fig. 2. The delay between the timestamping point and the sending of message bits to the medium is called outbound latency while inbound latency is corresponding delay for received messages. Regardless of the location, the implementation is required to correct the values of reported timestamps with the estimation of the corresponding latencies. The estimation method is outside the standard scope, but averaging is advised to be used.

4 Prototype Implementation

Our prototype is an implementation of the IEEE 1588 standard for Windows platform. The remote management is not implemented and the number of network connections is limited to one.

The prototype topology and test arrangements are presented in Fig. 3. The prototype consists of a master and slave clock, a reference pulse generator, and a connecting LAN technology.

The reference pulse generator is connected to a Clear-To-Send signal (CTS) in a serial port of the both clocks. Each time CTS is set, an interrupt is generated and the implementation creates a reference timestamp. Reference timestamps are values of the local clock at the times of reference pulses. Thus, they can be used for calculating the clock offsets. The generator creates approximately one pulse per second independ

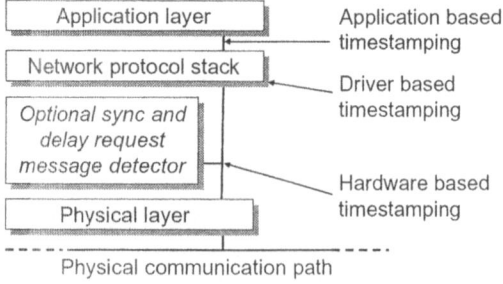

Fig. 2. Possible timestamping methods.

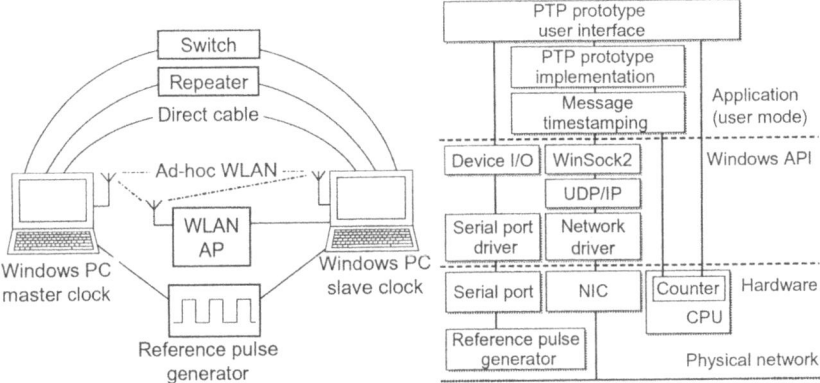

Fig. 3. Prototype topology (left) and implementation (right).

ently from the synchronization process. The used LAN technologies are a direct cable, repeater, switch, ad-hoc WLAN, and WLAN Access Point (AP).

Fig. 3 also shows the implementation architecture. The prototype uses UDP/IP with Windows Sockets 2 (WinSock2) for communication [8]. It is implemented as a user mode application consisting of three modules: the user interface module, the protocol implementation module, and the message timestamping module.

When the message timestamping module is loaded, it creates three threads. Two of the threads are listening UDP sockets and when a socket receives a message, the listening thread attaches a timestamp to it. Threads have a high priority in Windows for achieving as precise timestamping as possible. The third thread is used for transmission. It sets the transmission timestamp field of the sync and delay request messages. The module provides an Application Programming Interface (API) for the protocol implementation module.

The protocol implementation module contains the state machine of the protocol. Possible inputs for the state machine are timer events and received PTP messages from the timestamping module. The protocol has two types of timer events. First is for issuing a sync or a delay request message while the second is for reconfiguration timeouts. When sync messages are not received from the master clock, the timeout is used to initiate automatic selection of new master clock.

The user interface module receives status information from the protocol implementation module and shows it on display as presented in Fig. 4. The information contains the state of the state machine, the estimated one way delay, and the scaling factor of the timestamp counter of the processor. The received messages are also displayed.

The user interface sends administrative messages and the initial configuration to the implementation module. The reference timestamps are created by the user interface module, which also shows them on the display.

We have developed two methods for estimating the outbound latency that are used in the prototype. They are called *local* and *external echo* methods. The inbound latency is assumed constant.

The local echo method estimates the outbound latency as follows. When a clock sends a PTP message, the protocol stack echoes the message back to the sender, because PTP uses a multicast address that is listened also by the sender. The elapsed time between the timestamping of the message and reception of the echo is called an *echo delay* in the implementation. The measured echo delay is decreased by the calculated average echo delay and used as an estimate of the outbound latency. According to the test measurements, the changes in actual outbound latency are followed by the echo delay value. Thus estimating the outbound latency with the local echo method is more accurate than simply using a constant estimate.

The external echo method can only be used when both master and slave clock are using WLAN as a LAN technology with a single WLAN AP. With this setup, the second echo of each multicast transmission can be received when AP broadcasts the message. The sender and receiver receive the broadcasted message at the same time and the reception time is used as an estimate of the precise transmission time. The variation in inbound latency is smaller than in outbound latency because no waiting due to access contention is involved before accessing the medium. Additionally, the effect of the delay variation in WLAN AP is completely removed. This extension to the standard increases the accuracy but requires that both clocks use the same method.

The prototype uses averaging when calculating the delays from master to slave and slave to master. When a slave clock receives a sync message, the estimated master to slave delay is calculated and filtered using an 11-tap median filter. The filter output is a median of the eleven latest input values. When the slave clock receives a delay response message, the slave to master delay is calculated and filtered respectively. The master to slave and the slave to master delays are used to calculate the offset from a master.

The offset from a master is used to adjust the local clock. The clock adjustment method is not defined in the IEEE 1588 standard. The general time unit of the Windows system clock is a millisecond, which is too long for synchronization purposes. Instead, the implementation uses the timestamp counter c_p of Central Processing Unit (CPU) as a clock source. The implementation reads the counter value directly using assembler commands.

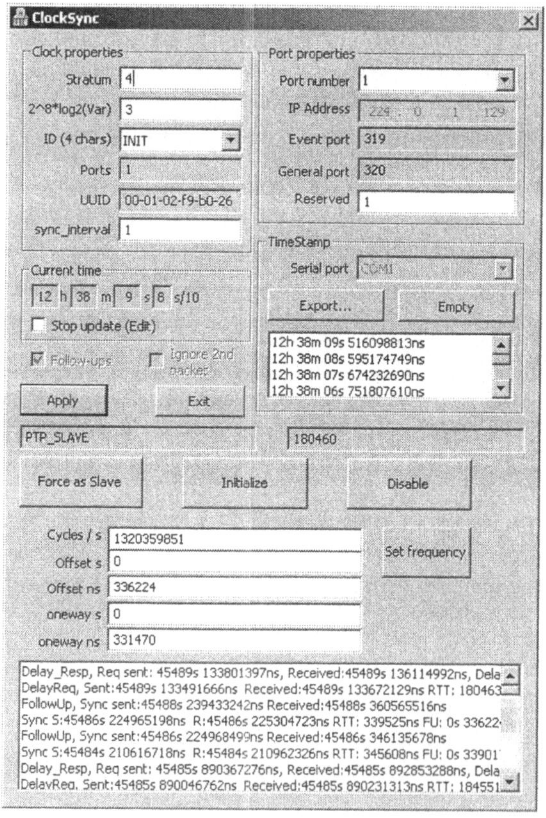

Fig. 4. Screen capture of the application.

The time unit in PTP is one nanosecond while c_p is increased once in a clock cycle. The time of the clock t_{ns} (in nanoseconds) is calculated from c_p with a linear function

$$t_{ns} = \frac{10^9}{f_p} c_p + t_{base},$$ (5)

where t_{base} is the time of the clock (in nanoseconds) when the processor was started, since c_p starts from zero, and f_p is the clock scaling factor that converts clock cycles to nanoseconds.

At the beginning of the synchronization, the values of f_p and t_{base} are estimated and set in order to achieve a rough estimate of t_{ns}. Later, t_{ns} is changed by adjusting f_p with a small value af_p instead of directly changing the time value. This is necessary because time must be strictly increasing, and it guarantees that the clock does not run backwards. When f_p is changed, the value of t_{base} is changed accordingly to keep t_{ns} constant in the adjustment point.

The adjustment is based on the offset from master and the first derivate of it. The adjustment a is calculated

$$a = \frac{A o_{fm} + B(o_{fm} - o'_{fm})}{10^9 t_d c_s},$$ (6)

where o_{fm} is the calculated offset from master in nanoseconds, o'_{fm} is its previous value, t_d is the time elapsed from previous adjustment in seconds, and c_s is a sequence number of the adjustment. Values $A=1.5$ and $B=5.0$ are scaling factors chosen based on test measurements. The divisor c_s is used for enabling faster adjustment at the beginning. The implementation restarts synchronization when c_s increases too much. Synchronization is also restarted when the absolute value of a reaches the value 1, thus slave clock is too far from the master for gradual adjustment.

5   Measurement Results

Three laptop PCs and one desktop PC were used in the measurements. Test equipment details are presented in Table 1. All computers were equipped with 10/100 Mbit/s Ethernet NICs and in addition, laptops were equipped with 11 Mbit/s 802.11b WLAN adapters. In all measurements, the 500 MHz laptop was used as a master clock and the 700 MHz laptop as a slave clock unless otherwise stated. The local echo method was selected for the outbound latency estimation.

In each test run, the clocks are first set to different times. The length of each test is about 15 minutes. At first, about 5 minutes is needed before the slave clock is stabilized, depending on the original clock offset. The reference timestamps generated before the slave clock was stabilized are ignored. The rest of the corresponding timestamps are compared and the difference of each timestamp pair is calculated. The clock offset for a single test run is defined as the average of the timestamp differences. Each test run is repeated 15 times. The average offset for a set of test runs is the average value of the test run offsets and the variance is the variance of the test run offsets.

Six measurement cases were used to analyze the accuracy of the prototype. Summary of the results is presented in Fig. 1.

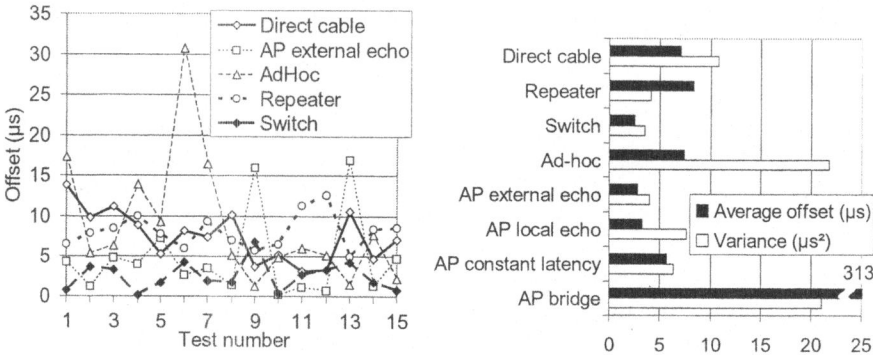

Fig. 5. Overview of the measurements (left) and summary (right).

1) The best achieved accuracy with each LAN technology was evaluated according to measurement arrangements presented in Fig. 3. The results of each test run are presented in Fig. 1. Outbound latency estimation methods used in this measurement were the external echo method for WLAN AP and the local echo method for the others. The results show that the average offset reached using a direct cable connection was 7.0 µs while the variance was 10.8 µs². A repeater network is almost as accurate as the direct cable connection with 8.3 µs average offset and 4.1 µs² variance. A switched network enables 2.5 µs average offset with 3.5 µs² variance. In WLAN, the results show that ad-hoc enables 7.4 µs average offset with 22 µs² variance. The WLAN AP enables the average offset of 2.8 µs with 4.0 µs² variance.

The most accurate setup was the 100 Mbit/s switched network, which supports full-duplex operation. The full-duplex operation probably improves the accuracy because transmission medium can be used without a multiple access algorithm.

With WLAN AP, the usage of the external echo method improves the accuracy close to the switched network although the switched network is more than ten times faster than WLAN. With the ad-hoc setup WLAN can achieve accuracy close to 10 Mbit/s repeater and direct cable setups.

2) The effect of the computer speed was measured with the direct cable setup. By exchanging the master and slave clock the average offset was 9.5 µs and the variance was 9.0 µs². This is close to original values of 7.0 µs offset and 10.8 µs² variance. Us-

Table 1. Test equipment.

Equipment	Description
Laptop PC 1	Intel Pentium III 500 MHz WinXP
Laptop PC 2	Intel Pentium III 700 MHz WinXP
Laptop PC 3	Intel Pentium III 1.0 GHz Win2000
Desktop PC	AMD Athlon 1.33 GHz Win2000
Direct cable	10 Mbit/s half-duplex
Repeater	10 Mbit/s MIL-4710H half-duplex
Switch	100 Mbit/s Cisco Catalyst 3500 Series XL full-duplex
WLAN AP	11 Mbit/s Nokia A020 802.11b WLAN AP
WLAN adapter	11 Mbit/s Nokia C110 802.11b

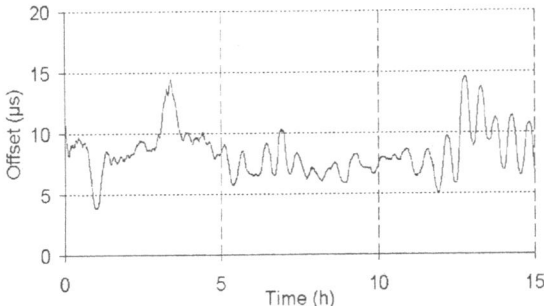

Fig. 6. Fifteen hour measurement results.

ing 1.33 GHz desktop computer as a slave clock slightly increased the accuracy. The average offset was 6.6 µs with a variance of 2.9 µs^2.

3) The accuracy of the prototype in a longer timescale was evaluated with a fifteen hour measurement using the repeater network. The offsets are shown in Fig. 5. The offset changed between 0.8 µs and 14.6 µs with the average of 8.3 µs.

4) The effect of the selected outbound latency estimation method was analyzed with WLAN AP measurements. First, outbound latency was assumed a constant and the average offset was 5.7 µs with 6.3 µs^2 variance. With the local echo method, the average offset was 3.3 µs with 7.6 µs^2 variance. Finally, with the external echo method the average offset was 2.8 µs with 4.0 µs^2 variance.

5) The need for a boundary clock in a WLAN AP bridge was analyzed with the following measurement. A master clock connected with WLAN and a slave clock connected with Ethernet were bridged by WLAN AP. According to the standard this setup requires a boundary clock, which was verified by the results. The average offset was 310 µs with 21 µs^2 variance. The result is 30 to 100 times worse than with other setups.

6) The effect of background network traffic on synchronization accuracy was estimated using a third station, the 1.0 GHz laptop, transmitting 1500 byte UDP packets to a station that was not involved with the measurements. The station transmitted with the loading rates of 1.2, 2.4, 3.6, and 4.8 Mbit/s on the repeater network and with rates of 0.6, 1.2, 1.8, and 2.4 Mbit/s on the ad-hoc WLAN. The results are shown in Fig. 7. for the repeater and for WLAN. On the repeater, the average offset was decreased

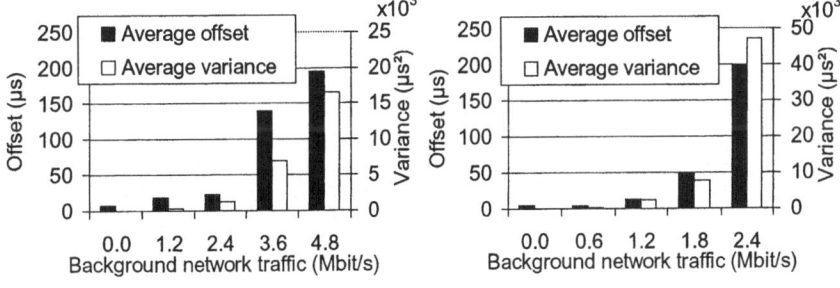

Fig. 7. Offset on loaded repeated (left) and ad-hoc (right) network.

from 8.3 µs to 18.9 µs at 1.2 Mbit/s and to 195 µs at 4.8 Mbit/s. On WLAN the decrease on average offset was from 7.4 µs to 12.6 µs at 1.2 Mbit/s and to 198 µs at 2.4 Mbit/s. Thus increased background traffic decreases the accuracy considerably. Increasing background traffic from 1.2 to 4.8 Mbit/s with a repeater network increases the average offset by a decade.

6 Conclusions

The prototype of PTP implementation is done and measurements are carried out to find out the precision of software based clock synchronization for WLAN environment.

According to the results, the average offset less than 10 µs can be reached with software implementation on fixed LANs and WLANs. Consequently, the 1588 standard can be implemented with application based timetamping and used with WLANs. The accuracy is increased considerably using the local and external echo methods presented in this paper. To achieve better stability, the clock adjustment algorithms used with NTP could be adopted. In future, better accuracy will be targeted by timestamping the messages closer to the transmission medium. A driver and hardware aided timestamping will be evaluated. Additionally, PTP will be implemented in the WLAN AP and used as a boundary clock.

References

1. Sun-Mi Jun, Dong-Hui Yu, Young-Ho Kim and Soon-Yong Seong: A Time Synchronization Method for NTP. Real-Time Computing Systems and Applications RTCSA '99, Hong Kong, (1999) 466-473
2. Mills D.: Network Time Protocol Ver3 Specification, Implementation and Analysis. RFC-1305 (1992)
3. Mills D.: Measured Performance of the Network Time Protocol in the Internet System. RFC1128, (1989)
4. IEEE std. 1588-2002: IEEE Standard for a Precision Clock Synchronization Protocol for Networked Measurement and Control Systems. (2002)
5. IEEE Std 802.11-1997: Wireless Lan Medium Access Control (MAC) And Physical Layer (PHY) Specifications. (1997)
6. Butner S. E., Vahey S: Nanosecond-scale Event Synchronization over Local-area Networks. Local Computer Networks (2002) 261-269
7. Agilent Technologies, IEEE-1588 prototype,:
    http://tycho.usno.navy.mil/ ptti/ptti2002/paper21.pdf
8. Microsoft: Platform SDK Windows Sockets 2 (2003)

Moving Telecom Outside Plant Systems Towards a Standard Interoperable Environment

Gerson Mizuta Weiss and Eliane Zambon Victorelli Dias

CPqD Telecom & IT Solutions – Rodovia Campinas-Mogi Mirim km 118,5 13086-902, Campinas, São Paulo, Brazil
{weiss;eliane}@cpqd.com.br

Abstract. There is a growing need for integrating telecommunication systems. This paper introduces the Telecommunication Outside Plant Markup Language (TOPML), an OpenGis GML (Geographic Markup Language) application schema designed to describe telecommunication outside plant data in which geographical information is an important issue. The use of TOPML can be a key factor for cost reduction on data gathering, conversion process of georeferenced data and system integration. By using TOPML, telecom systems can now communicate in different ways such as Web Services. Additionally, with TOPML and the definition of services following the OpenGIS specifications and the NGOSS component based architecture, the telecom outside plant systems can move towards a standard interoperable environment. The results of this paper are being implemented in the CPqD Outside Plant System, a geographical information system that automates the telecommunications outside plant management.

1 Introduction

Technology innovations and business drivers are in constant shift in the telecommunication area. This constant evolution leads to alterations in telecom outside plant management systems internal properties such as the addition of new equipments, services or business rules [4]. These new pieces of information also have to be exchanged among other systems, usually resulting in new communication interfaces.

In order to decrease costs concerned with the alteration of application programming interfaces (APIs) and integration mechanisms, it is necessary to use generic APIs and data importation and exportation mechanisms. The construction of these generic mechanisms and APIs can be accomplished with the use of XML [8], by means of which the schema of the elements can be shared among the communicating systems.

Telecom outside plant management systems has been used in the automation of many operation carriers processes. These systems should be integrated with many other systems, like workflow, customer care and ERP (Enterprise Resource Planning)

J.N. de Souza et al. (Eds.): ICT 2004, LNCS 3124, pp. 1246–1251, 2004.

systems. In order to facilitate data handling and data exchange, it is required to describe these data in XML during data importation and exportation.

Georeferenced data is an important characteristic in the representation of telecom outside plant elements, such as the basic urban mapping elements and telecom network elements like underground duct placement, connectors, cables, poles and terminal boxes.

Thus, the goal of this paper is to present an XML encoding to describe the several types of existing elements in an outside plant management system. The GML [1] language were used since it is an OpenGIS Consortium (OGC) standard that act as an XML encoding for the transport and storage of geographic features. The next stage of our work is to define services and interfaces following the OpenGIS and the NGOSS (New Generation Operations Systems and Software) specifications [3][6].

The rest of this paper is organized as follows. Section 2 is an introduction on telecom outside plant and on the CPqD Outside Plant (OSP) system, used as the basis to gather data about outside plant elements. Section 3 presents the TOPML language. Section 4 describes the possible contributions for the integration of telecommunication systems. Finally, section 5 presents the conclusions of the paper.

2 Telecom Outside Plant

A telecom outside plant or telecommunication network is a combination of several elements required in order to support telecom services (such as voice and data), both in local and long-distance scenarios. A telecom outside plant is the basis for the whole telecommunication activity, as it connects the final user to the world through the use of several technologies such as: copper pairs, coaxial cable, optical fiber or wireless, microwaves and satellite. Telecommunications outside plant is the name given to the set of cables and equipments external to the central office, responsible to connect the customer to the central office, and interconnect different central offices.

An important characteristic of a telecom network management system is the use of georeferenced data to precisely locate the several existing elements in an outside plant, like poles and distribution boxes, and elements of the basic urban mapping, such as streets and blocks. So, an outside plant management system should be a geographical information system (GIS) that adds geographic data to database elements, and functions that allow specific uses in outside plant management. The use of GIS can provide the following benefits:

- Establish a standard representation for the network elements;
- Calculate project costs, based on distances and length of cables and fibers in the network;
- Search for paths among connected network elements;
- Search automatically for the best paths to serve a specific subscriber.

Currently, many operation companies still have all their outside plant usage recorded on paper notebooks and have their network management systems in a non-integrated way, greatly increasing the difficulties to keep the network data up to date. Fortunately, the advances in technology also brought new tools to help with the required revolution in the operating companies environment. Next section describes a system that automates the process related to outside plant network.

2.1 CPqD Outside Plant System

The CPqD Outside Plant (OSP) is a system comprised of several modules developed by CPqD Telecom & IT Solutions for the telecom outside plant management. It was developed to automate the process related to outside plant, from the planning, design and construction phases, and continuing with the network operation and maintenance.

CPqD OSP puts the power of GIS technology to work for the network infrastructure operator [4]. Assignments and management of existing facilities, capacity planning for future requirements, and engineering of new facility implementation can be conducted in concert, sharing consistent views of the operating environment, service delivery facilities, and spatial and non-spatial representations of customer locations and service demands. This system has been used in the process automation several operating companies.

The CPqD OSP has approximately 400 objects described in a proprietary format. Many of these objects have geographic properties such all objects related to the basic urban mapping. However, this format has some disadvantages: lack of mechanisms to organize the relationship of the outside plant element properties; mechanisms for reading and validation are complex and hard to maintain; reuse of data types is impossible; difficult communication with external systems.

In order to prevent these problems, a new format was defined. The use of GML was chosen to represent the outside plant elements. The following sections present the new format.

3   GML and TOPML Schema

The Open GIS Consortium, an international industry consortium of more than 230 companies, government agencies and universities participating in a consensus process to develop publicly available interface specifications, adopted GML [1] as the standard for description of geographic content.

Since GML documents are both human-readable and machine-parsable, they are easier to understand and maintain than proprietary formats. GML brings an alternative to expensive proprietary software, making possible for many smaller companies to beat industry barriers and deploy GML in a variety of applications, like the telecom industry.

We developed TOPML [7], an XML encoding that uses GML to describe the telecom outside plant data. Figure 1 describes the structure of the XML Schema that was created for the TOPML application schema showed above.

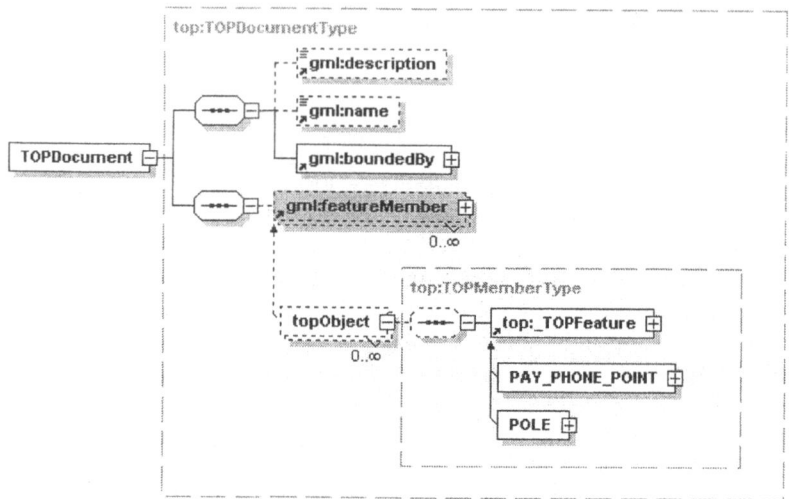

Fig. 1. TOPML schema structure

Consider, for instance, a pole object, where there is a geographic property indicating its location and a set of attributes describing it, such as for instance, its identification number and height. The definition of a type in the XML Schema, according to TOPML schema would be as follows:

```
<complexType name="POLE_TYPE">
<complexContent>
<extension base="gml:AbstractFeatureType">
  <sequence>
    <element ref="gml:location"/>
        <element name="number" type="string" minOccurs="1"/>
        <element name="height" type="float" minOccurs="0"/>
        < element name="situation" type="integer" minOccurs="1"/>
  </sequence>
</extension>
</complexContent>
</complexType>
```

Notice that in the case of a pole, its geographic property is a point, represented by a *location* element. The *location* type is one of the geographic properties defined by GML. Other properties are lines, polygons and several combinations of these properties [1].

For the construction of the TOPML language, the concepts applied in the CPqD OSP system were applied as the basis for the collection of data about outside plant. Therefore, mapping rules were defined to map the CPqD OSP proprietary format schema into the TOPML schema.

The use of GML may standardize the representation of the telecom network objects that have spatial information, making even more easy its handling and dissemination.

4   Towards a Standard Interoperable Environment

Currently, most operation carriers have all registers from the outside plant in hard copy or have non-integrated computer systems, showing several inconsistent and scattered databases throughout many company areas. This implies the possibility of errors due to these databases being outdated, which may lead to errors in the network facility designation.

In this context, the great challenges to the widespread use of GIS in telecom is the reduction of costs concerned to data gathering and to system integration solutions.

Some activities were developed aiming at the establishment of a proper infrastructure to integrate a GIS for telecom outside plant management with other existing systems in the telecommunication area. Standard models are intended to be used for GIS interoperability, such as the Service Architecture defined by OpenGIS consortium [3]. The OpenGis Web Feature Server (WFS) and Web Map Server (WMS) were used in prototypes to allow data exchange in the Internet. The WFS can also be used in data export mechanisms.

APIs and mechanisms were developed in order to make the most requested integrations to the CPqD OSP system feasible. This way, some of the APIs that were developed to meet the integration of GIS for outside plant management with some telecom carriers legacy systems were implemented. Some examples include:

- Service Order Management System for Privative Lines,
- Feasibility Analysis Systems for the Installation of ADSL Services.
- Service Order Management Systems for Pay Phones.
- Supervision Systems for Pay Phones.
- Geographic Positioning Simulation Systems.

The TOPML language is being applied to the defined integration APIs and to the data importation and exportation mechanisms. The built import mechanism performs validation and reading of TOPML documents and its data insertion into a database. This mechanism is configurable through a metadata file also described in XML. This file defines which objects will be imported, their format and type, in addition to their database schema. So, to import data of other objects, change an object data format, or import data in different database tables, it is enough to change the metadata file.

These results are very important for the GIS integration in the telecommunication domain and enabled to keep the CPqD OSP in line with the market expectations, giving greater agility in the product deployment and decreasing maintenance and development costs of specific APIs.

The integration infrastructure will be consolidated through the use of the results achieved by the international efforts in standardization of telecom operation system interfaces [5][6] and geographic information systems interoperability [3].

By using TOPML, telecom systems can now communicate in different ways such as Web Services. The TOPML can be transformed into a SVG (Scalable Vector Graphics) format, allowing geographic data visualization in the Internet [2]. The integration mechanisms based on XML can be more stable and easier to maintain, modify and extend. A standard format defined along with the use of GML can make the information exchange easier, which also improves data mapping and extraction process.

The next goal is define interfaces using the NGOSS architecture model [6]. The NGOSS specifies a component based architecture where each component interacts with all other components in a deterministic fashion. The adherence to NGOSS will enable the CPQD OSP system to interoperate with other NGOSS complaint systems.

5 Conclusions

The definition of the TOPML language may facilitate system integration as other systems can work with the proposed language, avoiding the creation of specific solutions, typically based on the creation of text documents with proprietary formats. Thanks to GML standard, TOPML can act as the basis to a standard format to describe telecommunications outside plant data.

TOPML is part of a project aimed at the use of the OpenGIS specifications in order to define an open architecture to build interoperable telecom outside plant systems. This architecture will enable interoperable data services through interface standardization.

Additionally, with TOPML and the definition of services following the OpenGIS specifications and the adherence to the NGOSS architecture, the telecom outside plant systems can move towards a standard interoperable environment.

References

1. Cox, S., et al.: OpenGIS® Geography Markup Language (GML) Implementation Specification, version 2.1.1. Available in http://www.opengis.org/techno/specs/02-009/GML2-11.html.
2. Kolbowicz, C et. Al.: Web-based Information System using GML. In GML Dev Days 2002. Vancouver – Canada (2002).
3. OGC - Open GIS Consortium: The OpenGIS Abstract Specification Topic 12: OpenGIS Service Architecture Version 4.3. Available in http://www.opengis.org/techno/abstract/02-112.pdf.
4. Previdelli, C. A. and Magalhães, G. C.: Moving Geospatial Applications Towards a Mission Critical Scenario. In: 25th Annual Congress of the GeoSpatial Information Technology Association (GITA), Tampa/USA (2002).
5. TeleManagement Forum (TMForum).: ETOM Bussiness Process Framework. Available in http://www.tmforum.org.
6. TeleManagement Forum (TMForum).: The NGOSS Technology Neutral Architecture - TMF053 v3.5. July 2003. www.tmforum.org.
7. Weiss, G. M. and DIAS, E. Z. V.: Integrating Telecom Outside Plant Systems Through GML. In GML Developer Days 2003, Vancouver – Canada (2003)
8. Zaslavsky, I. et al.: XML-based Spatial Data Mediation Infrastructure for Global Interoperability. In: 4th Global Spatial Data Infrastructure Conference, Cape Town, South Africa (2000).

Analysis and Contrast Between STC and Spatial Diversity Techniques for OFDM WLAN with Channel Estimation

Eduardo R. de Lima[1], Santiago J. Flores[1], Vicenç Almenar[1], and María J. Canet[2]

Universidad Politécnica de Valencia, EPS de Gandia, [1]Dep. Comunicaciones, and
[2]Ing. Electrónica
Carretera Nazaret-Oliva s/n, 46730 Gandia – Valencia - Spain
http://www.gised.upv.es
{edrodde, macasu}@doctor.upv.es,
{sflores, valmenar}@dcom.upv.es

Abstract. This paper regards an evaluation of different spatial diversity techniques contrasted to a Space-Time Code (STC) technique called Alamouti-scheme, applied to OFDM WLAN, using channel estimation based in the least squares criteria and with a limited preamble. The results here presented are based on IEEE802.11a Physical Layer simulations, but could be straightforward extended to HIPERLAN/2. The implications of using such a recent approach as STC, and conventional diversity schemes in existent WLAN standards are addressed, as well as the implications of using channel estimation.

1 Introduction

Wireless transmissions through multipath fading channels have been always a challenge that researchers and developers have to face when designing or implementing wireless communication systems that will work over that scenario.

Demand for higher data rates has enforced the rising of new wireless technologies to support communication with high and low mobility. These new wireless data technologies appear to force the wireless systems towards Shannon's frontier. Nevertheless, more issues arise. With the aim of mitigate these problems, several solutions have been proposed on the last years: channel codification, interleaving, more robust modulations, diversity, multiple access technologies in different flavors, "Smart Antennas", adaptive equalization, power control, turbo coding and, more recently, Space-Time Code (STC) and the use of multiple antennas at both end of the radio link (MIMO). These techniques can reduce drastically the inconvenient behavior of the wireless channel. Some of them improving average throughput and others the signal-to-interference ratio. But MIMO has a breakthrough concept that is to exploit the rich-scattering nature of wireless channel, under no-line-of-sight (NLOS) conditions, to give a diversity gain lineal with the number of antennas. A core idea behind MIMO is to complement time signal processing with the spatial dimension inherent in the use of multiple antennas.

J.N. de Souza et al. (Eds.): ICT 2004, LNCS 3124, pp. 1252–1260, 2004.

Space-Time Code is the set of schemes aimed at realizing joint encoding of multiple TX antennas. Transmission schemes in MIMO are typically Spatial Multiplexing, Space-Time Trellis Code (STTC) and Space Time Block Codes (STBC). Spatial multiplex consists in split the data stream in several independent streams as the number of Tx antennas and focus on increasing the average capacity. STTC consists in the joint coding of the independent streams, created by multiplexing, in order to maximize diversity gain and/or code gain. STBC has reached a strong penetration in standards due to the use of a simple linear decoder, and after its discovery, the popularity of STC has risen in importance [1].

A special STBC implementation that reduces the receiver complexity was proposed by Alamouti [2] and is currently part of the standards cdma2000, UMTS, IEEE802.16a. Taking into account its simplicity and advantages presented in the literature, we are going to consider the evaluation of the basic Alamouti's scheme in MISO case, i.e. 2Tx and 1Rx antenna, as an initial step towards more complex MIMO systems. This scheme is applied in an OFDM WLAN system and compared with other more conventional diversity schemes.

This paper is organized as follows: the section two presents an overview concerning the physical layers of two WLAN standards: IEEE802.11a and HIPERLAN/2 (HL/2). The section three presents the diversity and the STC schemes. Section four is dedicated to channel estimation. Section five is addressed to simulation results for receive diversity techniques and STC applied to OFDM modulation, using channel estimation as well as for perfect channel knowledge. Finally, in section six we present a summary and the conclusion.

2   WLAN Physical Layer Overview

HL/2 and IEEE802.11a, thanks to joint efforts of IEEE and ETSI, have the same characteristics in the physical layer (PHY), with a modulation scheme based on OFDM, due to its good performance on highly dispersive channels. Table 1 shows a summary of their characteristics and Figure 1 presents their block diagram.

In both cases, baseband signal is built using a 64-FFT, and a 16 samples cyclic prefix is added to avoid multipath effects. Since the sampler frequency is 20 MHz, each symbol is 4 µs (80 samples) long, and the guard interval is 800 ns long. In order to facilitate implementation of filters and to achieve sufficient adjacent channel suppression, only 52 subcarriers are used: 48 data carriers and 4 pilots for phase tracking. One main difference between HIPERLAN/2 and IEEE802.11a is the multiple access technology adopted by each one: TDMA for the former and CSMA for the last.

The preamble structure of both standards have small differences, but its use for training purposes (synchronization, frequency and channel estimation) stands for both of them. Figure 2 presents the training structure for IEEE802.11a, applied in this work. The preamble T was used for channel estimation and the system was supposed perfectly synchronized.

PHY

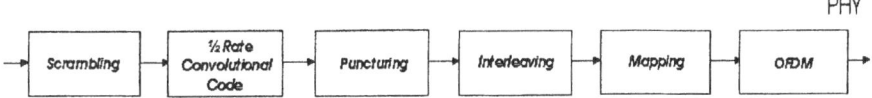

Fig. 1. HIPERLAN/2 and IEEE802.11a Physical Layer (PHY)

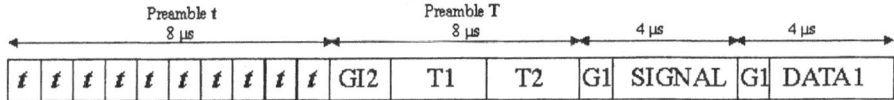

Fig. 2. Training structure for IEEE802.11a standard

HIPERLAN/2 has 7 transmission modes while IEEE802.11a has 8. Modes 1 to 6 are mandatory for HIPERLAN/2, and modes 1, 3 and 5 for IEEE802.11a.

Simulations carried out in this work are based on the block diagram presented in figure 1, except for the scrambling, code and puncturing, which were not considered. The simulator was set up to mode 6 therefore the results could be straightforward extrapolated d to HL/2.

Table 1. Main Physical Layer (PHY) Parameters for IEEE802.11a and HIPERLAN/2 Standards

MODE	MODULATION		CODE RATE		BIT RATE (Mbps)	
	HIPERLAN/2	IEEE 802.11a	HIPERLAN/2	IEEE 802.11a	HIPERLAN/2	IEEE 802.11a
1	BPSK	BPSK	1/2	1/2	6	6
2	BPSK	BPSK	3/4	3/4	9	9
3	QPSK	QPSK	1/2	1/2	12	12
4	QPSK	QPSK	3/4	3/4	18	18
5	16QAM	16QAM	9/16	1/2	27	24
6	16QAM	16QAM	3/4	3/4	36	36
7	64QAM	64QAM	3/4	2/3	54	48
8	-	64QAM	-	3/4	-	54

3   Diversity and Space-Time Code Scheme

In this section we present a short explanation regarding the spatial diversity techniques applied to receiver and transmitter, and also the space-time code scheme.

3.1   Diversity Techniques

Usually, when different multipath components fade independently, diversity is the chosen method. The reason is that if p is the probability that one of the paths is below a detection threshold, then p^L (a value considerable smaller than p) is the probability that all L paths are below the threshold. The cost of diversity is an additional complexity due to path tracking and additional components processing.

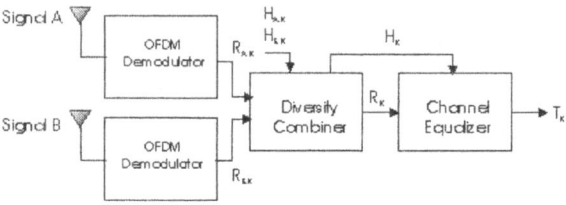

Fig. 3. Spatial Receiver Diversity Block Diagram

3.1.1 Receiver Diversity Techniques

Habitually, most of wireless communications systems design, either with or without mobility, use spatial receiver diversity. But this diversity technique is generally applied to the Base Station (BS) side, while the other end of the link uses just a single antenna. In a very simple way one can say that spatial diversity techniques are usually applied when space to put the extra antennas is not a problem and when it causes less economic losses for the network vendor.

Figure 3 presents the block diagram of spatial receiver diversity techniques implemented in this work.

3.1.1.1 Maximal Ratio Rx Combining (MRRC)

Subcarriers in both antennas are phase aligned and weighted by their power. The output of the combiner is given by $R_k = R_{A,k}(H_{A,k})^* + R_{B,k}(H_{B,k})^*$. So that, the values to be compensated by the equalizer are given by the equation $|H_{A,k}|^2 + |H_{B,k}|^2$, for all k. These operations are shown in figure 5.

3.1.1.2 Rx Subcarrier Selection Combining (RSSC)

This combiner selects the subcarrier with highest magnitude response. That is, R_k output is either $R_{A,k}$ or $R_{B,k}$ for each k, depending on $|H_{A,k}|$ is greater or not than $|H_{B,k}|$. So, for each subcarrier the equalizer compensates the channel response at the subcarrier frequency of the selected entry.

3.1.2   Transmission Diversity Techniques

Traditionally, transmission diversity techniques were not chosen. But recently it has received a lot of attention, especially due to its improvements in high data rate dedicated systems. The transmission diversity techniques here presented are dual of those presented for reception in the previous section. One point to highlight here is that, in spite of its gain, the use of such transmission diversities brings more complexity to the transmitter and a lost of throughput in the reverse link, for systems that do no use TDMA, due to the necessity of channel state feedback. Figure 5 presents the block diagram of spatial transmitter diversity.

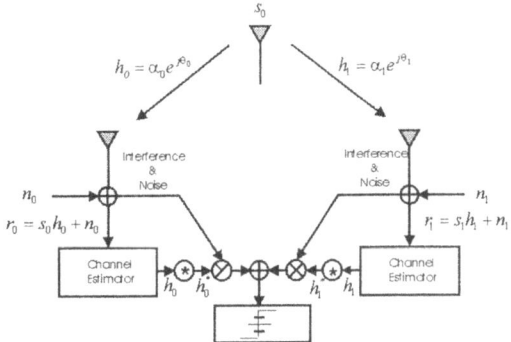

Fig. 4. Maximal Ratio Receiver Combining block diagram

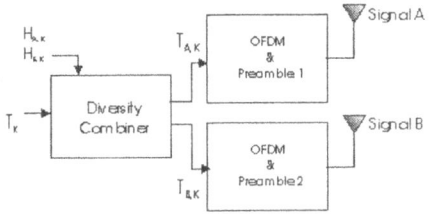

Fig. 5. Spatial Transmitter Diversity Block Diagram

3.1.2.1 Tx Subcarrier Selection Combining (TSSC)

The combiner selects the subcarrier with highest magnitude response. So, output is either $T_{A,k}$ or $T_{B,k}$ for each k, depending on $|H_{A,k}|$ is greater or not than $|H_{B,k}|$. Also, for each subcarrier the equalizer compensates the channel frequency response.

3.1.2.2 Maximal Ratio Tx Combining (MRTC)

In this technique, subcarriers are rotated, so that they are aligned at the receiver, weighted by their power, and transmitted on each antenna, for all k,

$$T_{A,k} = T_k \left(H_{A,k} \right)^* / \left(\left| H_{A,k} \right| \sqrt{2} \right) \tag{1}$$

$$T_{B,k} = T_k \left(H_{B,k} \right)^* / \left(\left| H_{B,k} \right| \sqrt{2} \right) \tag{2}$$

3.2 Space-Time Code (STC)

Space-Time Codes were introduced as a method of providing diversity in wireless fading channels using multiple transmit antennas [3]. Until then, multipath fading effects in multiple antennas wireless system were mitigated by means of time, frequency, and antenna diversity. Receive antenna diversity was the most commonly applied technique. For cost reasons, multiple antennas are preferably located at base station (BS), so transmit diversity schemes for BS are increasing in popularity.

Alamouti's scheme is a particular case of STBC, and consequently STC, that minimizes the receiver complexity and reaches a diversity gain similar to MRC, but using diversity at transmitter side instead of the receiver.

3.2.1 Alamouti's Space Time Scheme

Alamouti has shown that a scheme using two Tx and one Rx antenna provides the same diversity order as MRRC with one Tx antenna, and two Rx antennas [2]. This scheme does not require bandwidth expansion, any feedback from the receiver to transmitter, and its complexity is similar to MRRC. Figure 6 illustrates it.

The receiver combiner performs the following operation,

$$\tilde{s}_0 = \left(\alpha_0^2 + \alpha_1^2\right)s_0 + h_0^* n_0 + h_1 n_1^* \tag{3}$$

$$\tilde{s}_1 = \left(\alpha_0^2 + \alpha_1^2\right)s_1 + h_0 n_1^* + h_1^* n_0 \tag{4}$$

This scheme may be easily generalized to 2 Tx and M Rx antennas to provide a diversity order of 2M [2]. The proposed scheme support maximum likelihood detection and it is as complex as Maximal Ratio Combining (MRC). Even when one receive chains fails, the combiner works as well as in case of no diversity (soft failure).

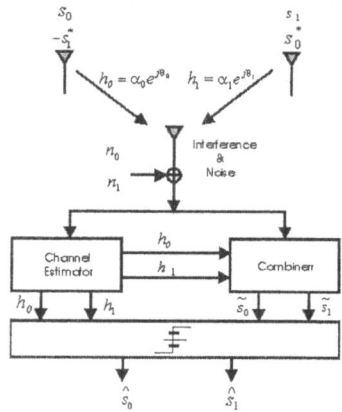

Fig. 6. Alamouti's scheme for 2 Tx and 1 Rx antennas.

4   Least Square Channel Estimation Based on Limited Preamble

Here we present the Least Square (LS) estimator used to implement channel estimation with limited preamble in IEEE802.11a, (preamble T shown in figure 2).

Given a training sequence at the receiver, the LS criteria applied to channel estimation in the frequency domain will lead us to equation 5.

$$\hat{h}_{LS} = X^{-1}y = \left[\frac{x_0}{y_0} \frac{x_1}{y_1} \frac{x_2}{y_2} \ldots \frac{x_{N-1}}{y_{N-1}} \right] \tag{5}$$

Operating (5) over an average of preambles T1 and T2, we obtain the frequency response LS-estimated. Figure 7 shows the block diagram for an OFDM system and the transmitted and received symbols in frequency domain.

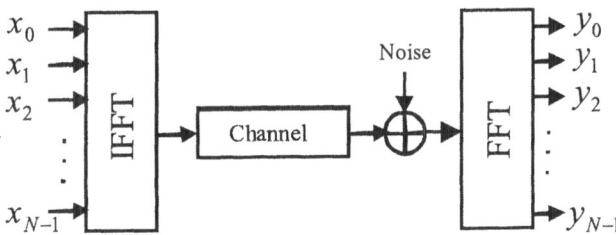

Fig. 7. Block diagram of the OFDM modulation and demodulation. The figure shows the transmitted and received symbols in frequency domain as well

5   Results and Discussion

Following we will present the simulation results for Maximal Ratio Receiver Combining (1Tx,2Rx), Maximal Ratio Transmitter Combining (2Tx,1Rx), Space Time Block Code in its Alamouti's version (2Tx,1Rx), Space Time Block Code with Soft Failure(SF) (2Tx,1Rx), Transmitted Subcarrier Selection (2Tx,1Rx) and Receiver Subcarrier Selection Combining (1Tx,2Rx) for perfect knowledge of the channel(CSI) and also for channel estimation based in a limited preamble.

Channel model was based in the taped delay type A presented in [4] and it was supposed to be invariant. So, Figure 8 presents the performance using channel estimation with LS criteria, and figure 9 shows the results with perfect CSI. As one can see, the best performance was reached by the MRRC with and without channel estimation. Next better performance was reached by RSSC, with quite simpler implementation than MRRC, and next one is MRTC, with similar performance as STBC. Next, we have TSSC that, like its dual in Rx, has a quite simple implementation; nevertheless, it needs to feed the transmitter combiner with channel information. Alamouti's scheme with soft failure performance presents a better result than zero forcing case when using channel estimation. Nevertheless, for perfect knowledge of the channel, it presents similar performance than zero forcing. It is necessary to clarify that, in Alamouti scheme with perfect channel knowledge, signal power is 3dB greater than in the channel estimation case. This does not happen with STBC, so this difference must be compensated in the C/N ratio.

Each proposed technique has its drawbacks. Depending on the application, one might considerer if the implementation premises are stronger than the improvement of a specific technique. For instance, transmitter diversity must be considered when the number of antennas is a problem at the receiver side, and receiver combining techniques must be considered in order to avoid increase complexity at the transmitters.

Receiver Subcarrier Selection is the technique with simplest implementation and it has a good performance. Nevertheless it is necessary the use of channel estimation in transmission. This is not the case of STBC that does not need such information. Nevertheless for some cases the premise that channel does not vary within two OFDM symbol could not be true and the channel could lost its orthogonality what could be an issue for STBC in its Alamouti's version.

For implement all these diversity techniques is necessary to take some changes in the preamble structure of the standards as well as the addition of a new preamble for each antenna for channel estimation purposes. For transmission diversity, except for the STBC, is necessary to have information about the channel at the transmitter.

Using transmission diversity and channel estimation, we need to estimate the channel response of the channel, by means of the training structure, before to start transmitting the data.

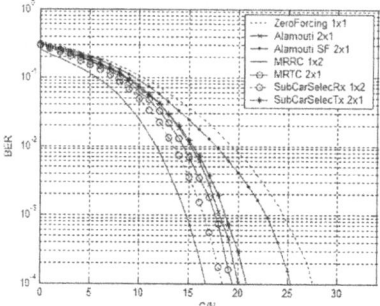

Fig. 8. Performance of IEEE802.11a and HIPERLAN/2 using transmission and receiver diversities techniques and least squares channel estimation, based in a limited preamble

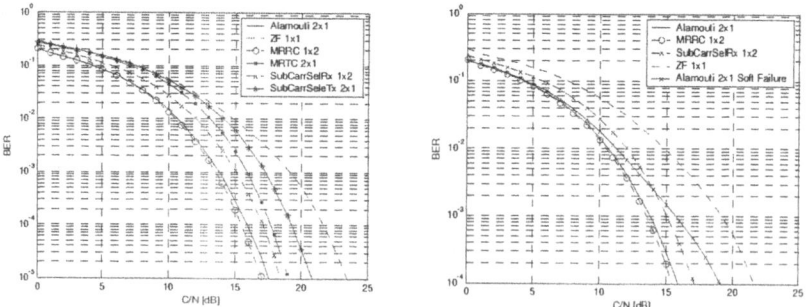

Fig. 9. Performance of IEEE802.11a and HIPERLAN/2 using transmission and receiver diversities techniques (left) and STBC with and without Soft Failure (right), with perfect CSI.

6   Conclusions

In this paper we present the performance of physical layer simulations for OFDM applied to WLAN standards IEEE802.11a and HIPERLAN/2. We have applied spatial diversity techniques at the transmitter and receiver sides as well as STBC in its Alamouti's version. We have assumed perfect channel knowledge and channel estimation based in a limited preamble using the least squares criteria in frequency domain. Among those results here presented the MRRC presents the best performance while the RSSC/TSSC presents the smallest complexity. The Alamouti's scheme when applied in time domain could not keep its performance when in presence of fast channel variations (within one OFDM symbol). Nevertheless its implementation in frequency domain (Space Frequency Block Code-SFBC) is a choice to fix that drawback. The Alamouti scheme when contrasted to MRTC has the advantage of no need of channel information at the transmitter and has the diversity order that MRRC/MRTC. Channel estimation in OFDM has an important impact in the performance of the system and stands for new preamble structure when using diversity techniques.

Acknowledgements. This work was supported in part by the Spanish Ministerio de Ciencia y Tecnología under Research Project TIC2001-2688-C03-01.

References

1. D. Gesbert, M. Shafi, Da-shan, P. J.Smith and A. Naguib.: From Theory to Practice: An Overview of MIMO Space-Time Coded Wireless Systems. IEEE Journal on Selected Areas in Communications, Vol.21 No.3, April 2003.
2. S. M. Alamouti: A simple Transmit Diversity Technique for Wireless Communications. IEEE Journal on Select Areas in Communications, Vol. 16 Number 8, October 199, pp. 1451-1458.
3. V. Tarokh, N. Seshadri, R. Calderbank: Space-Time Codes for High Data Rate Wireless Communication: Performance Criterion and Code Construction. IEEE Transactions on Information Theory, Volume 44, Number 2, Mar 1998, pp. 744-765.
4. ETSI/BRAN: document no. 30701F, 1998BRAN WG3 PHY Subgroup. Criteria for Comparison.

Fair Time Sharing Protocol: A Solution for IEEE 802.11b Hot Spots

Anelise Munaretto[1,2], Mauro Fonseca[1,4], Khaldoun Al Agha[2,3], and Guy Pujolle[1]

[1] LIP6 Laboratory, University of Paris VI, 75015 Paris, France
[2] LRI Laboratory, University of Paris XI, 91405 Orsay, France
[3] INRIA Laboratory, Rocquencourt, 78153 Le Chesnay, France
[4] PPGIA, PUC-PR, 80215-901 Curitiba, Parana, Brazil

Abstract. To adapt the data rate in accordance with the quality of the link, the IEEE 802.11b standard proposes the variable rate shifting functionality. This intrinsic functionality of the 802.11b products progressively degrades the bit rate when a host detects unsuccessful frame transmissions. Furthermore, the basic CSMA/CA channel access method guarantees that the long-term channel access probability is equal for all hosts. When one host captures the channel for a long time because its bit rate is low, it penalizes other hosts that use the higher rate, inciting a performance anomaly. This paper aims at avoiding this performance anomaly and the consequent waste of bandwidth. We propose the Fair Time Sharing (FTS) approach to perform real fair sharing among the active hosts in the hot spot, thus avoiding the performance degradation caused by one or more slow hosts. This paper presents the FTS architecture and its performance evaluation, showing the improvement achieved.

1 Introduction

The economic feasibility and an easy deployment of IEEE 802.11b for wireless local area networks, leading to the wireless mobile Internet, incite its widely adoption as future access networks. The raise of the bit rate in the wireless networks technologies offers to these networks a comparable performance with respect to wired local area networks. However, the wireless nature of the radio networks introduces many problems that were unknown in the wired networks. These are mainly due to the transmission of a signal over the radio medium. Furthermore, there are various effects due to fading, interference from other users, and shadowing from objects, all them degrading the channel performance [JA94]. Measurements of a particular radio LAN appear in [DU92], showing that the packet-error rates critically depend on the distance between the transmitter and receiver, and surprisingly, not monotonically increasing with the distance. Because of such constraints, transmission in radio networks is not reliable enough to allow the construction of a control based access mechanism for the network management. Likewise, several performance evaluation studies of the IEEE 802.11 DCF [CH04] [CR97] [WE97] [BI00] show that performance is very sensitive to the number of competing stations on the channel, especially when the Basic Access mode is employed. To adapt the data rate in accordance with the quality of the link, the IEEE 802.11 standard [ST99] proposes the variable rate shifting functionality.

J.N. de Souza et al. (Eds.): ICT 2004, LNCS 3124, pp. 1261–1266, 2004.
© Springer-Verlag Berlin Heidelberg 2004

This intrinsic functionality of the 802.11b products progressively degrades the bit rate from the nominal 11 Mbps to 5.5, 2, or 1 Mbps when a host detects unsuccessful frame transmissions. The multiple data transfer rate capability performs dynamic rate switching with the objective of improving performance. Nevertheless, to ensure coexistence and interoperability among multirate-capable stations, the standard defines a set of rules that shall be followed by all stations. A performance anomaly causing a performance degradation in a hot spot due to the influence of a host with lower bit rate, was first analyzed in [HE03]. The authors examine the performance of 802.11b showing that the throughput is much smaller than the normal bit rate. They also analyze how a host with a lower bit rate influences the throughput of the other hosts that share the same radio channel. This results in a performance anomaly perceived by all hosts. The authors solely analyze the performance anomaly without introducing any solution. Our paper introduces the new Fair Time Sharing approach to avoid the performance anomaly in a dynamic and adaptive manner. Such an approach is based on the attribution of time slot for each host, to perform a real fair sharing among the active heterogeneous hosts in the hot spot.

2   Fair Time Sharing (FTS)

The fair access to the channel provided by CSMA/CA causes a slow host transmitting at 1 Mbps to capture the channel for a period eleven times longer than hosts transmitting at 11 Mbps [HE03]. Such a procedure induces, at the long-term, a non equitable time sharing, because the active hosts degrade their bit rate to the smaller bit rate in the hot spot, e.g. around 1 Mbps. This characterizes a performance anomaly. We introduce the *Fair Time Sharing* approach to avoid the performance anomaly. In our approach we fairly share the access time destined for each host according to the configured bit rate. We attribute a specified time slot to each slow host, shaping their maximum output traffic to prevent the unfair share due to the data rate degradation. The assessment of the time slot is based on the number of competing hosts within the hot spot and the useful throughput. The Fair Time Sharing (FTS) architecture proposes a simple architecture composed by three components over the UDP protocol: FTS-Server, FTS-Client, and Traffic shaper. To achieve a better bandwidth optimization in the hot spot, two classes are implemented: Slow class and Fast class. In the slow class, all client stations that can compromise the hot spot performance are included, i.e. client stations having a slow data rate. In such a class, the resources are shared based on the number of clients and a resource reservation is performed. In the fast class, all client stations work in fast data rate. The FTS-Server is located on the access point. The access point logs the number of connected users in the hot spot and their corresponding bit rate. The FTS-Server broadcasts this number of connected users and sets the fast class threshold. It works over UDP to avoid connection state management. The FTS-Client is located on the client station, it receives the FTS-Message and implements the traffic shaper. It works over UDP remaining simple and connectionless. The Traffic shaper is located on the client station and responsible for adapting the outgoing traffic to network bandwidth availability. This traffic shaping policy is implemented using the Traffic Control (TC) program from the "iproute2" package, detailed in [HOTC]. It works at network level.

Each FTS-Client enforces the calculated allowed time slot using the traffic shaper element. The FTS-Client does not limit the time slot when it belongs to the fast class, using as much bandwidth as possible. This means that a FTS-Client in fast class uses the default fair sharing method, allowing this class of clients to also consume the idle bandwidth left by slow class clients. In order to optimize the bandwidth utilization, the FTS-Message sets a threshold for the fast class. For instance, when there is no host working at 11 Mbps, the threshold for the fast class is set to the highest data rate in the hot spot. Therefore, this field allows changing the maximum data rate used in the fast class in a dynamic manner. Then, each node i maintains a local control of the bandwidth in setting its traffic shaper.

3   Performance Measurements

In this section, we present the performance measurements of the proposed FTS protocol. The measurements are performed using the **iperf** tool [IPER]. The **iperf** represents a TCP/UDP bandwidth measurement tool. We set up a testbed platform to measure the throughput that hosts can obtain when sharing a 11 Mbps 802.11b wireless hot spot. We have used three notebooks (host1, host2, and host3) with 802.11b cards (Lucent Orinoco). The wired part of the network is connected by the access point. We implement the FTS protocol comparing the results with a classical hot spot. The first step is to measure the real useful throughput in data rate degradation. These measurements depicted in Table 1, are stored in the host to gauge its allotted time slot and are then used to set the traffic shaper. Table 1 presents the measurements of the useful throughput through our performance evaluation results.

Table 1. The useful throughput obtained for one single host in a hot spot with UDP traffic.

Bit Rate	Useful throughput	Proportion of useful throughput
11 Mbps	6.05 Mbps	55%
5.5 Mbps	3.81 Mbps	69.27%
2 Mbps	1.68 Mbps	84%
1 Mbps	0.891 Mbps	89.1%

The FTS protocol uses a bandwidth value to set the traffic shaper to avoid the degradation of the average throughput. This bandwidth is calculated according to the number of connected clients in the hot spot and the maximum useful throughput, presented in Table 1. In the measurement environment, the choice of the destination nodes for the flows is somewhat arbitrary and any destination could have been chosen for each flow without affecting the results. Unless otherwise specified, the following assumptions are made: Each flow is active throughout the duration of the experimental. All packets on all flows contain 1534 bytes. The time duration of our implementation is of 400 s. The results of the first experimentation are depicted in Fig. 1(a). Observe that this experi-

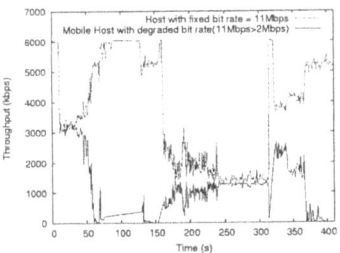
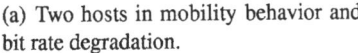
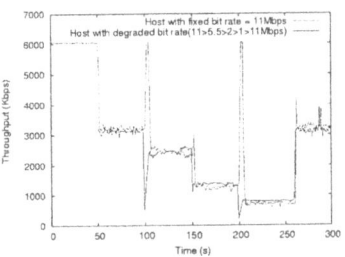

(a) Two hosts in mobility behavior and bit rate degradation.

(b) The throughput degradation according to the bit rate in two hosts.

Fig. 1. Measurements results

mentation shows the effect perceived by improving the mobility of one host in the hot spot.

In time 0 s, only one station (host1) is active in the hot spot with 11 Mbps as bit rate and approximately 6 Mbps as useful throughput. After approximately 10 s, another station (host2) also with 11 Mbps as bit rate arrives in the hot spot. The useful throughput decreases to 3 Mbps for the two hosts. Afterwards, the host2 starts to move away from the Access Point, decreasing its useful throughput, although it keeps the same bit rate. At the other hand, we can see the throughput increase of host1. From 50 s until 150 s, the useful throughput perceived by host2 is strongly degraded. At about 160 s, we set the bit rate to 2 Mbps only for the host2. In this instant, the slower host increases its useful throughput to about 500 kbps. Thus, the fast host decreases drastically its useful throughput to about 2.5 Mbps. In such a case, we can verify that the degradation of the bit rate in one mobile host imposes the degradation in the other host. However, as the slower host gets closer the Access Point, and consequently near the fast host, the useful throughput becomes the same. For the two hosts this stabilization is perceived between time 240 s and time 310 s, when the two stations achieve approximately the same throughput. Afterwards, about the time 320 s, we set the bit rate to 11 Mbps for the host2. In such an instant, the faster host increases its useful throughput to about 4 Mbps and the slower host to about 2.5 Mbps. Then, the useful throughput of the host2 starts again to decrease because of the mobility. Next experimentation measures the throughput of the competing hosts if there are 2 heterogeneous hosts in the hot spot. The results presented in Fig. 1(b), show the final throughput according to the bit rate degradation.

At the beginning, only one host is active in the hot spot, thus achieving 6 Mbps as useful throughput. At 50 s, a second host arrives in the hot spot, fairly sharing the useful throughput. At 100 s, the bit rate of the first host is degraded to 5.5 Mbps. Note that the degradation of the 2 hosts reaches nearly the same throughput, about 2.5 Mbps, despite that only one host has the bit rate degraded. The same anomaly is depicted each time that the bit rate of a single host is degraded. At 260 s, the bit rate of the slower host is once again set to 11 Mbps. Then, we can verify that the useful throughput perceived by the two hosts is again of 3 Mbps. The effectiveness of the proposed FTS approach is

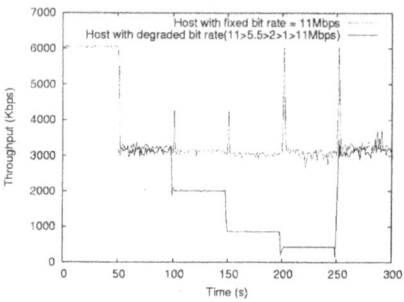

Fig. 2. The Fair Time Sharing improvement avoiding the throughput degradation.

demonstrated in Fig. 2. This scenario demonstrates that our proposed technique is able to track abrupt variations in the network state, while keeping a very high level of accuracy.

In these experimentations, our FTS-Server broadcasts each 5 s the FTS-Message that represents the number of connected clients (2 clients) and the fast class threshold (11 Mbps). At 100 s, the bit rate of the first host, is degraded to 5.5 Mbps. In this point, the FTS-Client checks its bit rate and the corresponding useful throughput according to Table 1. In such a case, as we have 5.5 Mbps as bit rate configuration, we use the 3.81 Mbps value. Then, the FTS-Client divides this value with the last received number of connected clients cad 1.905 Mbps. This value is applied in the traffic shaper limiting the output bandwidth as shown in Fig. 2. We verify that the FTS approach prevents the degradation in faster host's throughput. Only the hosts belonging to a slow class have their throughput degraded. Likewise, the same improvement occurs when the bit rate of the slower host is degraded to 2 Mbps. The FTS-Client checks its bit rate and the corresponding useful throughput according to Table 1. As we have 2 Mbps as bit rate configuration, we use the 1.68 Mbps value. Then, the FTS-Client divides this value with the last received number of connected clients, that is equal in this case to 0.840 Mbps. Therefore, this value will be applied in the traffic shaper limiting the bandwidth output as stated in Fig. 2. The FTS-Client checks its bit rate and the corresponding useful throughput according to Table 1. In such a case, as we have 1 Mbps as bit rate configuration, we use the 0.891 Mbps value. Then, the FTS-Client divides this value with the last received number of connected clients, that is equal in this case to 0.445 Mbps. Therefore, this value will be applied in the traffic shaper limiting the output bandwidth as depicted in Fig. 2. After 250 s, the bit rate of the slower host is once again set to 11 Mbps. Then, we can verify that the useful throughput perceived by the two hosts returns to 3 Mbps. A drawback of the FTS architecture is the need to deploy a FTS-Server application in the Access Point in order to broadcast the number of clients and a FTS-Client application in every host to set the shaper. Therefore, we assume that all stations in the hot spot support the FTS architecture. Our findings show the improvement achieved by our architecture: the faster host keeps its throughput unchanged regardless of the presence of a slower competing host.

4   Conclusion

In this work, we have analyzed the performance of a 802.11b network. This performance is very sensitive to the number of competing stations on the channel. Through our analysis, we verify that a lower bit rate host influences the throughput of other hosts that share the same radio channel. This paper has discussed the impact of bit rate degradation in an 802.11b wireless LAN that characterizes a performance anomaly. We have then introduced a new approach, Fair Time Sharing to avoid the performance anomaly. The proposed FTS approach implements a real fair sharing scheme among the active heterogeneous hosts in the hot spot, avoiding the performance degradation caused by one or more slow hosts. We attribute a specified time slot to each slower host, shaping their maximum traffic during the degradation period. The FTS approach presents the cure for the performance anomaly of 802.11b.

References

[JA94]   William C. Jakes, "Microwave Mobile Communications", Wiley-IEEE Press., May, 1994.

[DU92]   Dan Duchamp and Neil F. Reynolds, "Measured Performance of a Wireless LAN", In Proc. of 17th Conf. on Local Computer Networks, Minneapolis, 1992.

[CH04]   H. Chaouchi and A. Munaretto, "Adaptive QoS Management for IEEE 802.11 future Wireless ISPs", ACM Wireless Networks Journal, Kluwer, Vol.10, No. 4, July, 2004.

[CR97]   B. P. Crow, I. Widjaja, J. G. Kim, and P. T. Sakai, "IEEE 802.11 Wireless Local Area Networks", IEEE Communications Magazine, September, 1997.

[WE97]   J. Weinmiller, M. Schlager, A. Festag, and A. Wolisz, "Performance Study of Access Control in Wireless LANs IEEE 802.11 DFWMAC and ETSI RES 10 HIPERLAN", Mobile Networks and Applications, Vol.2, 1997.

[BI00]   G. Bianchi, "Performance Analysis of the IEEE 802.11 Distributed Coordination Function", IEEE Journal of Selected Areas in Telecommunications, Wireless Series, Vol.18, March, 2000.

[ST99]   IEEE 802.11: "Wireless LAN medium access control (MAC) and physical layer (PHY) specifications", 1999.

[HE03]   M. Heusse, F. Rousseau, G. Berger-Sabbatel, and A. Duda., "Performance anomaly of 802.11b", In Proceedings of IEEE INFOCOM 2003, San Francisco, USA, 2003.

[IPER]   http://dast.nlanr.net/Projects/Iperf/

[HOTC]   http://tldp.org/HOWTO/Adv-Routing-HOWTO/

An Algorithm for Dynamic Priority Assignment in 802.11e WLAN MAC Protocols

Antonio Iera, Giuseppe Ruggeri, and Domenico Tripodi

Department D.I.M.E.T.- Università *"Mediterranea"* di Reggio Calabria,
Via Graziella, Località feo di Vito,
89060 Reggio Calabria Italy.
{iera, ruggeri, tripodi}@ing.unirc.it

Abstract. Nowadays, the need for communications systems able to contemporary support best-effort and "QoS sensitive" (real-time, for example) traffic, is constantly increasing. The urge is particularly felt in environments experiencing a dramatic growth, such as the Wireless LANs. Unfortunately, 802.11, which is the WLAN standard with the widest diffusion, does not seem to be able to satisfy those requirements. Many research efforts have been put to fill up this gap both by academic and standardization groups. They brought to the formalization of the IEEE 802.11e standard, representing a MAC layer enhancements of the former IEEE 802.11. This paper proposes an novel algorithm specifically tailored to *dynamically assign the priorities* to the packets, depending on application level QoS requirements.

1 Introduction

Wireless LAN (WLAN) are clearly emerging as a wireless and mobile access methodology that is complementary/alternative to the 3G and beyond mobile radio systems. Therefore, the role of the WLAN as a mere alternative to the fixed LAN in private office environment belongs to the past. The current telecommunications scenario is, in fact, characterized by *global* systems fast converging towards the idea of *virtual home/office environment*. Consequently, the user while moving feels the exigency of being continuously end efficiently connected to multimedia information systems and databases scattered across a fixed interconnected backbone (mainly characterized by the Internet technology). The way he/she wants to access particular services is expected to be "personalized" and "differentiated".

Unfortunately, IEEE 802.11 [1] does not offer any support to the Quality of Service (QoS) differentiation. To this aim, an ad-hoc conceived task group is currently going towards the standardization of an enhanced version of the 802.11 MAC, the IEEE 802.11e [2]. This "annex e", which is still a draft, defines a new operational mode for the MAC, called the Hybrid Coordination Function (HCF). HFC couples a contention based medium access modality, the Enhanced Distributed Channel Access (EDCA), with a polling based medium access procedure controlled by the Hybrid Coordinator (HC) that will be typically located within the AP in "infrastructured" WLANs.

EDCA adds a *prioritization* mechanism to the legacy DCF by allowing up to four different Access Categories (ACs). Each AC has a different queue, different Arbitration

J.N. de Souza et al. (Eds.): ICT 2004, LNCS 3124, pp. 1267–1273, 2004.

Inter-Frame Spaces and contention window parameters. The HCF polling mechanism essentially is an improved version of the legacy Point Coordination Function (PCF).

The authors of [3,4,5] highlight that the EDCA modality is able to support acceptable QoS levels under high load conditions even when it is not coupled to any polling modality. We agree with this approach but in our opinion still some issues need to be investigated to make this latter choice *actually effective* in any condition. As an example, up untill now criteria have not been formulated to effectively associate the EDCA priorities to traffic flows with different QoS exigencies. Solutions reported in literature assume a heuristic association between traffic category and priority [2].

Two main drawbacks are implied by such a *static* priority association. On the one hand, the system has no certainty that the priority assigned to a given service is actually able to always match its exigencies in terms of QoS guarantees. On the other hand, no mechanism is available to dynamically adapt each associated priority to the current congestion level of the network; this strongly affects the resulting application QoS. Therefore, *automatically adapting* the assigned priorities to the current system conditions intuitively could greatly enhance the QoS levels guaranteed to each application traffic.

This paper proposes an effective algorithm specifically tailored to dynamically assign the EDCA priorities, depending not only on the application level QoS requirements, but also on the medium loading conditions.

The paper is structured as follows. Section 2 describes the proposed algorithm. The description of the performance evaluation campaign is the topic addressed in section 3. Concluding remarks are given in section 4.

2   The Proposed Algorithm

The methodology for *dynamically assigning priorities to the applications* is specified by following a design approach based on two steps:

- first, the access time, associated to each priority, is constantly evaluated,
- then, for each packet, a transmission priority is dynamically chosen according to the applications requirements and the measured access times.

As for the second step, it is worth highlighting that each packet has a "delay budget" associated, this implying that the packet has to be transmitted within this delay budget. The delay budget is assigned according to the specific requirements of the application (which are derived from the SLA the user subscribes) and is associated to each packet by means of a suitable label put into the TOS field. In the following subsections more details are given, which allow the reader to have a more detailed idea on the functional behavior of the proposed methodology.

2.1   Dynamically Updating the Transmission Delay Estimation: The Virtual MAC

The transmission delay D_i associated to a given priority is evaluated on the basis of the delay experienced by each packet that is transmitted with such a priority. More specifically, the time interval ΔT_i^j from the instant in which the *j-th* packet with the *i-th* priority is inserted into the *i-th* priority queue to the instant when it is sent to the radio interface for transmission is evaluated. This is done each time a packet is scheduled from the *i-th* priority queue and the result contributes to the update of the estimated average delay for the transmission of packets with the *i-th* priority as follows:

$$D_i^j = (1 - \alpha)D_i^{j-1} + \alpha\Delta T_i^j \tag{1}$$

where α represents a suitable smoothing factor, whose best value has been computed to be 0.8 following a comprehensive simulation campaign under different traffic and system conditions.

As long as the WSTA sends traffic at the *i-th* priority, the D_i relevant to the *i-th* priority is continuously updated. It clearly emerges that, if no traffic at the *i-th* priority is sent for a given time interval, then the estimation relevant to this priority soon turns to be obsolete. Therefore, finding a method to update the delay estimations relevant to those priorities we call *"temporary inactive"* (without introducing further traffic, such as probe traffic, into the network) is vital for the correct behavior of our algorithm.

In [6] the concept of Virtual MAC (VMAC) has been introduced, which provides the WSTA with a MAC operating on *"virtual packets"* (i.e. packets not actually sent). By means of these virtual packets, a suitable functionality estimates the traffic currently loading the network and the quality of service which would likely be perceived by the WSTA at the moment of the *"real packet"* transmission.

In the present work the concept of *virtual*, or *dummy*, packets is also exploited. Specifically, each time the D_i estimation, relevant to the *i-th* priority, turns to be obsolete, a dummy packet is inserted into the *i-th* priority queue. It will be processed as a normal packet *until the instant just preceding its transmission over the air interface*. Logically, the packet *will not be actually transmitted*.

According to what proposed in [6], a collision is assumed whenever a different station chooses the same slot to perform the transmission. If a collision is detected, the backoff algorithm is repeated. Eventually, the instant in which the packet is *"virtually"* passed to the physical interface is exploited for the computation of ΔT_i^j. This way, D_i can be estimated according to equation (1).

In our proposal, following a comprehensive parameter tuning campaign, it has been decided that a dummy packet with the *i-th* priority need to be sent each time the D_i estimation is not updated for more than $250ms$ and no packets at the *i-th* priority are in the relevant queue.

2.2   The Decision Criterion

Last point to address is the criterion used to assign a given priority to a newly originated packet. Each time a given WSTA has a packet ready to be transmitted, it compares the packet $Delay\,Budget$ with the D_i values associated to the different priorities and chooses for the packet transmission the value of priority for which the following condition is verified:

$$Delay\,Budget - T_{Physical} >$$
$$\begin{cases} D_i & if\,queue_i\,is\,empty \\ max(D_i, TLT_i) & otherwise \end{cases} \tag{2}$$

Where TLT_i represents the time from the last transmission at priority i. $T_{physical}$ is computed as $\frac{PS}{R}$, being PS the packet size and R the physical interface rate, and represents the difference between the time instant in which a packet gains the access to the medium and the instant in which the last bit of the packet is completely delivered. The second of the two alternatives in equation 2 is added to take into account the possibility that the priority queue is blocked and the last estimation of D_i is too old and far from the actual value of the experienced delay.

3   Results

Several simulation campaigns have been performed by means of a proprietary simulation tool implemented by exploiting the Network Simulator NS-2 [7]. Our studies mainly aim at testing the capacity of the proposed algorithm in effectively adapting the priorities to the traffic conditions and the consequent performance achievable by the cited procedure. During the simulations, according to [8], the channel bandwidth is assumed to be 11Mb/s, while the parameters related to the backoff process are set as suggested in [2].

We modelled the video traffic as reported in [9], speech sources have been assumed to be CBR coded at 64kb/sec according to ITU-T Rec. G.711[10], and, finally, for bulk traffic, we simply adopted the *ns-2* pre-existing models. Finally let us remark that, according to [2], in our notation priority 0 is the highest one while priority 3 is the lowest one.

3.1   The Dynamic Priority

Early studies conducted aim at gathering a deeper knowledge on the capacity of the proposed algorithm to dynamically choose the priorities according to the traffic loading the system. For this study the reference scenario is that in figure 1.a). The node tagged

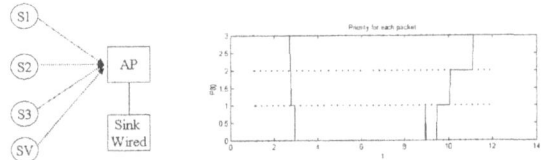

a) Simulative Scenario b) Priority experienced for each packet

Fig. 1.

with SV is equipped with the adaptive priority mechanism proposed, while nodes S1-S3 originate interfering traffic flows. The observed flow is a speech flow. Due to the high delay sensitivity of Voice flows [11] it has a delay budget of 10 ms associated.

To the aim of modelling an interfering traffic at different levels of "aggressiveness", a priority 1, 2, 3 is associated to the S1, S2, S3 source respectively. In the envisaged scenario, the adaptive source starts its transmission at the instant $t=1$ sec, while the interfering sources behave as follows: S3 starts at *2 sec*, S2 starts at *2,5 sec*, and S1 starts at *3 sec*, then S1 ends at *8 sec*, S2 ends at *9 sec* and S3 ends at *10 sec*. Figure 1.b) shows the priority at which packets are transmitted vs. the simulated time (in seconds). It shall be pointed out that in figure 1.b) the lines represent the measured values relevant to the *"real"* packets, while the dotted values refer to the *"dummy"* packets. From the observation of figure 1.b) it can be noticed how the proposed algorithm, following the activation of the interfering sources, adapts the priority of the transmission priority of the speech flow so to keep the transmission latency. As the interfering sources begin their transmission, the proposed algorithm forces the speech source to transmit at a higher priority.

3.2 Overall System Performance

During the second campaign of simulations, we investigated the advantages, in terms of system performance, achievable when the adopted algorithm is employed. To this aim, we considered to use the proposed algorithm for the transmission of a high quality video (average bandwidth 3,65 Mb/s). This latter is the flow under examination during the test campaigns. Two scenarios with different loading levels are considered: in the *"light loaded scenario"* the interfering traffic is represented by ftp flows while in the *"heavy loaded scenario"* the interference is caused by an additional video flow and a variable number of ftp flows. The second video flow is characterized by the same traffic parameters as the flow under examination; furthermore, according to [2], static priority of 1 is assigned to it. Having the ftp interfering flows a typical best effort nature, they have a priority 3 assigned. Some results of the simulation campaign are shown in fig. 2 More specifically, the graphics in the first row refer to the *"light load scenario"* while that on the second one are relevant to the *"heavy loaded scenario"*. Let us start to analyze the results in figures 2.a and 2.b. These figures show the mean priority assigned to the packets of the flow under examination when varying the *delay budget* assigned to them and the number of interfering ftp flows.

From 2.a and 2.b we can observe that in both scenarios the priority assigned to each packet increases (i.e. tends to zero) as the *"delay budget"* is reduced; moreover, we can observe that while in the *"light loaded scenario"* the mean priority assigned to the packets can be also kept low (i.e. values greater than 1), in the other scenario it shall be kept almost always below 1. This means that the adaptive flow, when it is required, i.e. under high traffic load, will assume a priority greater than the other one.

The second experimental results refer to the throughput evaluation of the adaptive flow. In order to have a clear idea on this issue, we compare our results with the reference "static priority assignment" case (always priority 1 assigned to the Video). In figures 2.c and 2.d we plotted the ratio between the throughput obtained in the adaptive case and the throughput obtained in the static case, against the *delay budget* assigned to the adaptive flow. Also in this case we observed a different behavior in the two scenarios. In the *light loaded* one we get ratios very close to one (i.e. the same throughput with adaptive and non adaptive algorithm). This is because both in the adaptive and in the static case the observed video flow does not suffers from the presence of the interfering traffic and it get all the bandwidth it needs. Please note that, in the case of adaptive priority assignment, as shown in figure 2.a, packets are, on the average, transmitted with a priority lower than

Light Loaded Scenario

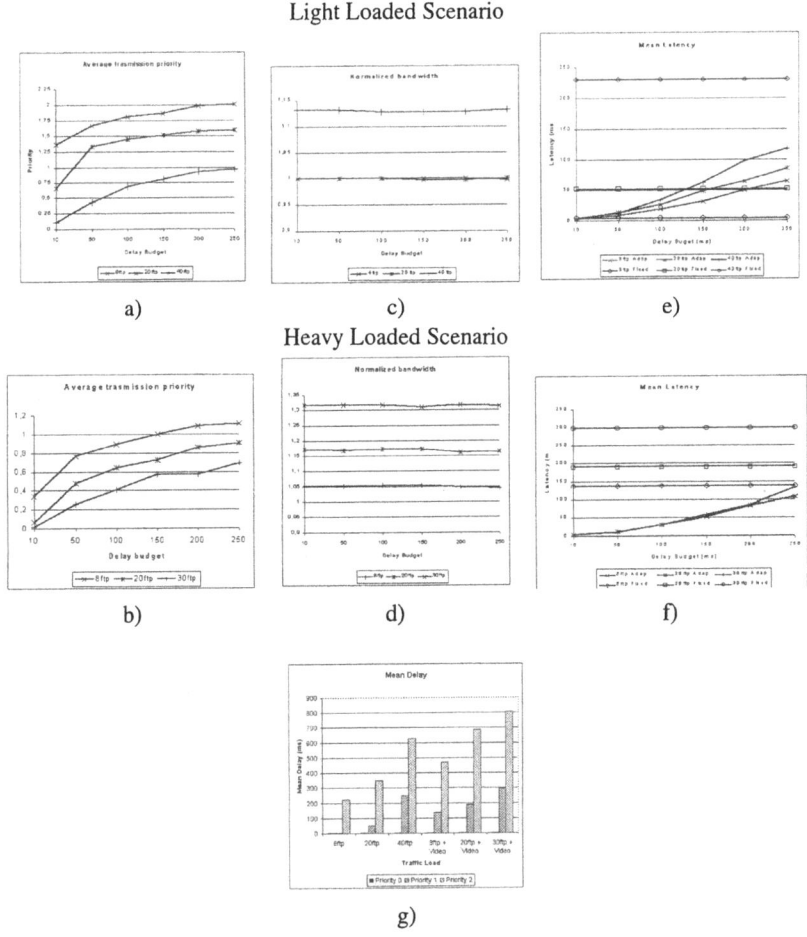

Heavy Loaded Scenario

Fig. 2. System behavior when varying the delay budget and the system load

that used in the reference case (i.e. always 1). This means that in the adaptive case even if we get similar throughput results than in the static case, this happens *without allowing the observed source to abuse of the system resources as it does when a static priority 1 is associated to it*. As a consequence, also an improvement in the performance of the lower priority traffic is expected.

In the *"heavy loaded scenario"* we get ratios up to 1.3, which means up to 30% throughput gain of the adaptive priority assignment approach against the static one. The improvement in the gain of the adaptive sources at heavy loads, and the contemporary improvement in the behavior of the lower priority sources at light loads, are the beneficial effects of an adaptive priority based approach.

In figures 2.e and 2.f, we reported the latency experienced during the transmission of the packets against the *delay budget* committed. At a first glance it is clear that the latency we obtain is frequently *well below* the one we committed to the system (given by the delay budget). This can appear as an inefficiency of our algorithm, as is seems that the systems increases too much the priority of the dynamic sources. Nevertheless, we should remember that we can choose only between a finite set of priorities and very often, the difference in the latency associated to each priority level is high (see figure 2.g); therefore, in order to transmit a packet within a certain *delay budget* we must choose to adopt a priority with a latency significatively lower than the *delay budget* itself. In other words, this is just a mere *granularity* problem, which can be easily overcome by foreseeing a greater set of priorities.

4 Conclusions

In this paper a mechanism for the adaptive allocation of the access priorities to the application traffics loading a IEEE 802.11e WLAN is proposed. As emerges from the simulation campaign the throughput of the traffic sources exploiting this dynamic priority association (usually real-time applications) can be improved when compared to the static allocation case. Furthermore, the dynamic association of priority to the transmission of high exacting traffic allows for an expected contemporary improvement in the QoS of the low priority traffics.

References

1. "Wireless LAN Medium Access Control(MAC) and Physical Layer (PHY) Specification," Tech. Rep. 802.11, IEEE, 1999.
2. "Medium Access Control(MAC) Enhancements for Quality of Service (QoS)," Tech. Rep. IEEE std 802.11e/D5.0, IEEE, July 2003.
3. S. Mangold, S. Choi, P. May, O. klein, G. Hiertz, and L. Stibor, "IEEE 802.11e Wireless LAN for Quality of Service," *European Wireless'02, Florence Italy*, February 2002.
4. Truong Vannuccini Ibm, "The ieee 802.11e mac for quality of service in wireless lans," IBM research Zurich Laboratory.
5. Y. Xiao, "Enhanced DCF of IEEE 802.11e to Support QoS," March 2003.
6. A. T Campbel A. Veres, M. Barry, and Li-Hsiang Sun, "Service Differentiation in Wireless Packet Networks Using Distributed Control," *IEEE Jornal on Selacted Areas in Communications (JSAC)*, vol. 19, no. 10, October 2001.
7. "Network Simulator- ns (version 2)," available from http://www.isi.edu/nsnam/ns/.
8. "Part 11:Wireless LAN Medium Access Control(MAC) and Physical Layer (PHY) Specification: High-speed Physical Layer in the 2.4GHz Band," Tech. Rep. 802.11b, IEEE, September 1999.
9. Michael Frey and Son Nguyen-Quang, "A Gamma-Based Framework for Modeling Variable-Rate MPEG Video Sources: The GOP GBAR Model," *IEEE/ACM Transaction on Networking*, vol. 8, no. 6, December 2000.
10. ITU-T, "Rec. G.711 Pulse code modulation (PCM) of voice frequencies," Tech. Rep., ITU-T, November 1988.
11. F. Beritelli, S.Casale, and G. Ruggeri, "New Speech Processing Issues in IP Telephony," *ICCT2000, Beijing (China)*, August 2000.

DHCP-Based Authentication for Mobile Users/Terminals in a Wireless Access Network

L. Veltri[1], A. Molinaro[2], and O. Marullo[2]

[1] Dip. Ingegneria dell'Informazione - Università degli Studi di Parma – Italy
luca.veltri@unipr.it
[2] DEIS - Università degli Studi della Calabria - Italy
molinaro@deis.unical.it

Abstract. In this paper we present a simple solution to control the wireless users/terminals access to the services offered by a wireless access provider and ultimately to the global Internet. We enhance the DHCP configuration procedure with authentication capability, and add further filtering control at the access router interfacing with the global Internet. This required adding some functionality in the DHCP server; while the DHCP client is unchanged. This feature allows personalizing the offered service based on user's credentials and profile.

1 Introduction

Wireless local area networks (WLANs) have experienced unusual growth in the last few years, especially encouraged by the success of the Internet and the availability of low-priced laptop computers and PDA with WLAN cards.

Although WLANs were originally conceived for enlarging the range of wired LAN in corporate environments, they are gaining popularity as a means for providing IP connectivity in office, residential and campus environments as well as in public hotspots, such as airports, hotels, and convention centers.

Authentication and authorization is a vital requirement for providing access to the Internet or other services via a WLAN system. Several mechanisms are currently implemented to provide user authentication and access control, but no standard mechanism is available yet. Existing solutions range from link-level authentication, which provides security mechanism at the MAC layer, to application-level authentication. The IEEE 802.11 [1] provides at data link level a shared key authentication scheme based on challenge-response with a shared. However the authentication is not mutual and the security mechanism is very weak [3]. The new IEEE 802.11i standard tries to enhance the 802.11 security by the use of IEEE 802.1X (port-based access control) as the authentication framework. IEEE 802.1X in turn uses the Extensible Authentication Protocol (EAP) that allows for end-to-end mutual authentication between a terminal and an Authentication Server (AS) through an intermediate authenticator node (e.g. the AP). At the contrary, upper level authentication mechanisms performs authentication and authorisation directly at IP or application level layers and

J.N. de Souza et al. (Eds.): ICT 2004, LNCS 3124, pp. 1274–1281, 2004.
© Springer-Verlag Berlin Heidelberg 2004

they are very implementation dependent. Usually an authentication gateway is placed in front of the wireless network controlling network and services connectivity, and forcing users to authenticate against it before using the network. The gateway/access server is responsible for opening and closing firewall rules and thus allowing users to reach the services provided by the network. Such access mechanisms are often referred as *Captive Portals*, and are becoming a popular way for community WiFi infrastructure and hotspot operators [4]. A rather common approach is to use simple web based interfaces between the user and the access system. The authentication procedure may begin when an un-authenticated user starts his/her web browser attempting to browse to any web page. At this point, the HTTP request is redirected to a new HTTPS URL inviting the user to enter his/her credentials (e.g. login/password).

Besides many advantages, these approaches have some drawbacks; for example the link-layer dependency of IEEE 802.x schmes, the use of implementation-dependent mechanisms of *Captive Portal*, and more in general, the lack of personalization of the terminal configuration parameters as function of the authentication result and user credentials.

In this paper we present an authentication mechanisms based on the DHCP protocol that tries to overcome some of these limitations. The basic idea is to enhance the existing DHCP configuration procedure with authentication and service personalization capability. In section 2 a brief overview of the standard DHCP protocol is given. In section 3 an extension of the DHCP configuration procedure with authentication and authorization functionality is presented, while in section 4 an implementation of the authentication mechanism is described. Conclusive remarks are given in section 5.

2 Dynamic Host Configuration Protocol

DHCP [5] is an UDP-based client/server protocol, designed to dynamically assign IP addresses and to bind several network parameters to fixed or mobile hosts. The entities involved are: i) *DHCP client* - a node that initiates request for network parameters from one or more DHCP servers, on a local access network, ii) *DHCP server* - a node that responds to the client's requests; it may be on the same link of the client or, alternatively it may be attached to a different subnet, iii) *DHCP relay* - a node that acts as an intermediate entity between DHCP client and server; a DHCP relay is always located on the same link of the client and acts as relay for DHCP messages exchanged between a client and a server. The basic mechanism for the dynamic allocation of IP addresses is is summarized in Fig. 1.

- The client broadcasts a DHCP DISCOVER message on its local physical subnet. Relay agents may pass the message to DHCP servers not on the same physical subnet.
- Each server responds with a DHCP OFFER message that includes an available IP address and other configuration
- The client receives DHCP OFFER message(s) from one or more servers. The client chooses one and broadcasts a DHCP REQUEST including the 'server identifier'.

- The server selected in the DHCP REQUEST message commits the binding for the client to persistent storage and responds with a DHCP ACK message, containing the configuration parameters.
- The client receives the DHCP ACK message with configuration parameters and re-configures the network layer.

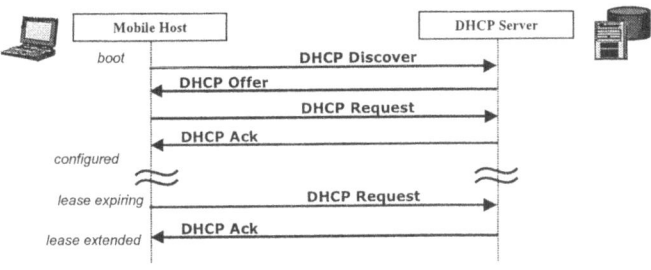

Fig. 1. Basic DHCP message exchange

The DHCP protocol can be also extended with authentication functionality. In RFC 3118 [6] it is defined a new DHCP option that can provide both entity authentication and message authentication. Through this mechanism newly attached hosts with proper authorization can be automatically configured from an authenticated DHCP server (and vice-versa).

3   A DHCP Extension for User/Terminal Authentication

The basic idea behind our proposed architecture is quite simple. To control the access of wireless terminals into a local site and to the Internet we provide the DHCP server with authentication and authorization capability and provide further packet filtering at the access router. The main strength of our proposal is that it can be implemented in each wireless provider's site by only requiring the use of a common protocol like DHCP and very few changes in the software of its *server* component. We name our modified DHCP server as DHCP+. In addition, our authentication mechanism is transparent to either the specific link layer protocol (e.g. IEEE 802.11) or the design options of the wireless terminal's manufacturer. The only concern for the local site administrator is to provide the authorized wireless terminals in its domain with an *authentication client* application. Fig. 2 illustrates the main components of our reference scenario.

In the wireless MT two processes are running: 1) a *DHCP client*, compliant to the original DHCP specifications in RFC 2131 [5]; normally this daemon is running as part of the terminal OS; 2) an *authentication client* (AC), which is supplied by the site administrator to authorized users only. AC exchanges messages with the DHCP server on a given UDP client port; this process does not require being a module of the OS and can simply run at application level. The AP should simply behave like a hub/bridge between the MT and the rest of the wired LAN.

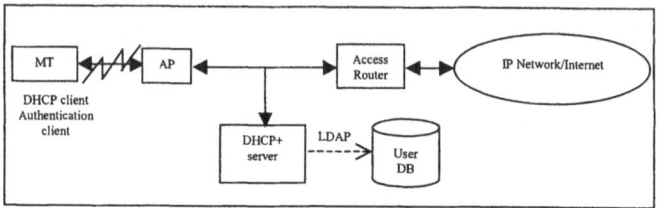

Fig. 2. Proposed reference scenario

The DHCP+ server is a standard DHCP server (RFC 2131) extended to support authentication functionality through additional messages exchange with the AC. These messages allow for user authentication or for mutual authentication of both the user and the DHCP server. As addressed in section 3, RFC 3118 also introduces authentication capability in the DHCP protocol by defining a new authentication option to be conveyed in the DHCP messages. Nevertheless, this solution requires modifications within the OS of the clients and could not be advantageous in our context.

The DHCP+ server is in charge of authorizing the user by checking his/her profile stored in a *User Database* (e.g. accessed through LDAP). The database contains the subscribers' credentials and profiles (username and password, MAC address, type of service, user class, lease attributes, etc.). The site administrator is responsible for the database creation and management policy. An administrator uses a user/terminal identifier as the key attribute to identify the authorized wireless user/terminal. This identifier can be the MAC address of the user's terminal, or a NAI (Network Access Identifier) [7]. In case of static MAC address's associations, MAC address filtering can also be implemented by the AP. However, in a more dynamic scenario, it is the access router, opportunely configured through the DHCP+ server, to perform both MAC address based and IP address based packet filtering. This approach has the advantages to be manufacturer-independent and to scale with the number of authorized users/terminals.

Another advantage of our architecture is that it does not require the unauthenticated user to start a specific application session to gain IP connectivity (browser pages/applets or pop-up windows). The authentication process is, in fact, triggered by any request for an IP address originated from the wireless terminal. The procedure starts when the DHCP client broadcasts a DHCP DISCOVER message, according to the normal DHCP operation. On receiving the DISCOVER message, the DHCP+ server issues an authentication request (*challenge*). This message is captured by the authentication client running in the wireless terminal and listening at a predetermined UDP port. The AC replies with a *response* message that includes all authentication information. This information contains the user identifier and the hash of the user credentials combined with the server challenge. If the authentication is successful, the DHCP+ server authenticates the client and issues the DHCP OFFER message followed by the client's DHCP REQUEST, and server's DHCP ACK (Fig. 3). The authentication is valid for a pre-configured time, and the DHCP+ server reissues periodically the request for user/terminal (re-)authentication. This can avoid the fraudulent use of the credentials of an authenticated user after he/she moved from the network.

If mutual authentication is supported, the AC can independently perform network authentication, by issuing an authentication request to the DHCP+ server is the same.

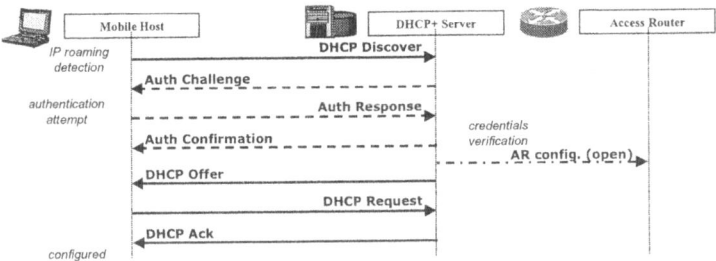

Fig. 3. DHCP-based authentication

In a more general architecture, the authentication could be implemented through a separate backend AS, implemented through a standard AAA server. In this case, the DHCP+ server acts as intermediate authenticator, i.e. as a pass-through agent forwarding authentication messages back and forth between the AC and the backend AS. Such messages are exchanged encapsulated within the proper AAA protocol (e.g. RADIUS or Diameter). This is illustrated in Fig. 4. Separation of the authenticator from the backend authentication server simplifies credentials management and policy decision making.

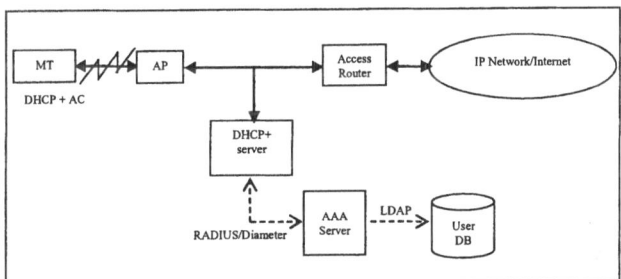

Fig. 4. Architecture with the AAA server

To address situations where un-authorized users try to configure manually their IP address or may be misconfigured through the use of bogus DHCP servers, the access router should perform proper traffic control and packet filtering, opportunely configured by the authenticator DHCP+ server. The basic idea is that the DHCP+ server, after assigning an IP address to an authenticated terminal, communicates to the access router a new filtering rule entry for it. The access router is responsible for opening and closing firewall rules and thus allowing only authenticated users the access to the global Internet or to particular, per-user, services. This can be done according to implementation specific requirements and using the same mechanism as for captive portals. Furthermore, the DHCP configuration parameters for the user/terminal (such as TOS value, DSCP, etc.) allow the access router to implement QoS mechanism and differentiated traffic handling policy.

The main distinguishing feature of our architecture is the possibility for the DHCP+ server to *personalize* the configuration parameters according to the users' credentials and profiles stored in the User database. This possibility is not available in solution like captive portals or when the MT directly communicates with the access router.

With DHCP+, service personalization can be performed at various levels, e.g.:

- *type of lease*: some attributes could count for the time-of-day in which the access can be granted according to the user class (e.g. certain user can be allowed accessing only at a certain time of the day);
- *IP address*: according to the user credentials the DHCP+ server could assign public or private IP addresses to control IP connectivity;
- *Default router and static routes*: they could be settled according to the user class and privileges;
- *Type of Service*: the access router could use information to establish different QoS control mechanisms over the data packets.
- *IP Telephony*: the DHCP+ can configure the MT with a default outbound SIP proxy server and/or with a default media gateway, or the access router (acting as media gateway) can enable the media relay to/from external ISDN/PSTN/IP networks.

4 Authentication Implementation

In this section, a possible implementation of the proposed authentication mechanism is presented and a testbed realization is described.

The main actors in the DHCP-based authentication mechanism are: i) the *Authentication Client (AC)*, ii) the *Authenticator*, i.e. the DHCP+ server, iii) the *Backend Authentication Server* (AS), eventually co-located with the DHCP+ server.

All authentication messages exchanged between the AC and the Authenticator, and between the Authenticator and the AS are carried by EAP [8] messages, encapsulated within UDP between the AC and the authenticator and within the proper AAA protocol between the authenticator and the AS. The advantage of using EAP as a transport mechanism is its flexibility; in fact, it supports multiple authentication protocols (e.g. CHAP, OTP, TLS, etc.) without having to pre-select a particular one. Moreover, EAP provides directly support for the use of a backend AS.

EAP uses four different types of messages: 1-*Request*, 2-*Response*, 3-*Success*, 4-*Failure*, which are distinguished by the first EAP field (the 'Code' field).

The *Request* messages are sent by the authenticator to the AC. Each *Request* has a 'Type' field, which indicates what type of authentication procedure is used. *Responses* are bound to the corresponding request through the 'Identifier' field. The contents of the 'Data' field depend on the *Request* type. The AC sends a *Response* messages in reply to a valid *Request* packet. The format of the EAP *Request/Response* messages is shown in Fig.5.a. As authentication method, we use the challenge-response authentication protocol (CHAP) [9]. CHAP implements the three-way challenge-response authentication procedure described in the previous section.

Since the same procedure applies for both re-authentications and the initial authentication (i.e. also before the DHCP procedure is completed), there must be a mechanism to delivery requests to ACs that do not have yet a valid IP address. For this reason, the broadcast IP address is used as destination address for Request messages (and for *Success/Failure* messages), while the address 0.0.0.0 is used as source address for *Response* messages (a TTL equal to 1 is recommended).

In order to map the EAP messages to the target AC, a new EAP over UDP (EAPoU) encapsulation protocol has been defined. The EAPoU messages are encapsulated into UDP datagrams and are simply formed by the EAP packet leaded by an EAPoU header. The new header is only composed of two fields: the 'Length' (1 byte) and the 'Client-Identifier' (variable size). The 'Client-Identifier' is a unique identifier used by the authenticator to address the AC, while the 'Length' field indicates the size of the identifier (Fig.5.b). In our implementation the 'Client-Identifier' is the client hardware address, copied by the *chaddr* DHCP field [5].

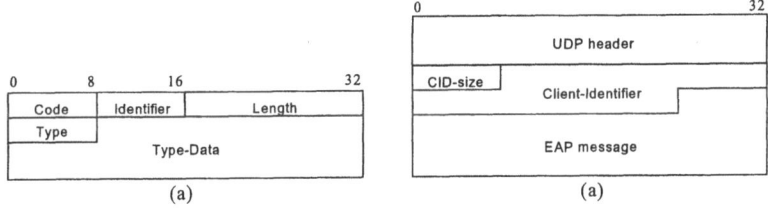

Fig. 5. (a) EAP message format; (b) EAP over UDP encapsulation

It is important to note that although in our implementation one-way CHAP authentication procedure has been utilized, the described architecture can be used with any standard authentication algorithms implemented in the AC and the Authenticator (or the Backend authentication server), such as OTP, TLC, UMTS SIM-based, etc. Moreover mutual authentication is also possible.

The wireless access authentication scenario has been implemented into a demonstration tesbed where an 802.11b MT (a laptop PC with a WiFi card) tries to access to an IP network. For simplicity the DHCP+ server has been located directly on the access router, implemented on a Linux PC (with kernel 2.4), with packet filtering enabled (IPTables/NetFilter). The 802.11b AP has been connected directly to the internal interface of the access router. When the MT starts the DHCP procedure sending a DHCP DISCOVER message to the DHCP server, this issues a new CHAP authentication request to a default AC's UDP port. If the challenge-response authentication succeeds, the DHCP+ server notifies the AC with an EAP *Success* message and proceeds with the DHCP configuration procedure (OFFER/REQUEST/ACK). When the new IP address has been correctly assigned to the authenticating MT, the DHCP+ server proceeds to configure the IPTables/NetFilter rules (i.e. the Linux firewall) in order to grant the access to the external IP network. Periodically the DHCP+ server repeats the authentication procedure in order to renew the access grant to the mobile terminal.

5 Conclusions

In this paper we presented a quite simple solution to control the Internet access of mobile wireless terminals roaming in a WLAN access network. The main design goal was to give the local provider simple mechanisms to authenticate the roaming users in order to enable basic IP connectivity and application-level services. Authentication is performed by the DHCP server in a link-layer and application-independent manner. This required adding some functionality in the software of the DHCP server; but the DHCP client side is kept unchanged in the wireless terminal population. A distinguishing feature of our approach is the personalization of the services offered to the wireless users. This is possible by enabling the DHCP server to assign configuration parameters according to the user credential and profile, and to communicate these parameters to an access router connecting to the global Internet.

References

1. LAN MAN Standards of IEEE Comp. Soc.: Wireless LAN Medium Access Control (MAC) and Physical Layer (PHY) Specification (1999)
2. Supplements to 802.11: IEEE 802.11b-1999: Higher speed Physical Layer (PHY) extension in the 2.4 GHz band; IEEE 802.11a-1999: High-speed Physical Layer in the 5 GHz band; IEEE 802.11g-2003: Further Higher-Speed Physical Layer Extension in the 2.4 GHz Band
3. Arbaugh, W. A., Shankar, N., Wan, Y. C. J., Zhang, K.: Your 802.11 Wireless Network has no Clothes, IEEE Wireless Comm. (2002) 44-51
4. Captive Portals, NoCat, http://nocat.net, Opengate, http://www.cc.saga-u.ac.jp/opengate/index-e.html, WiCap , http://www.geekspeed.net/wicap/, StockholmOpen , http://software.stockholmopen.net/index.shtml, OpenSplash, http://opensplash.qalab.com/
5. R. Droms - "Dynamic Host Configuration Protocol", IETF Request For Comments, RFC 2131, March 1997.
6. Droms, R., Arbaugh, W.: Authentication for DHCP Messages, IETF Request For Comments, RFC 3118 (2001)
7. Aboba, B., Beadles, M.: The Network Access Identifier, IETF Request For Comments, RFC 2486 (1999)
8. Blunk, L., Vollbrecht, J., Aboba, B., Carlson, J., Levkowetz, H.: Extensible Authentication Protocol (EAP), IETF draft-ietf-eap-rfc2284bis-07 (2003)
9. Simpson, W.: PPP Challenge Handshake Authentication Protocol (CHAP), IETF Request For Comments, RFC 1994 (1996)

Performance Analysis of Multi-pattern Frequency Hopping Wireless LANs

Danyan Chen, A.K. Elhakeem, and Xiaofeng Wang

Electrical and Computer Engineering Department, Concordia University
1455 de Maisonneuve Blvd. W., Montreal, Quebec H3G 1M8, Canada
{d_chen, ahmed, xfwang}@ece.concordia.ca

Abstract. This paper proposes the use of multiple patterns (MP) to increase the capacity of Frequency Hopping (FH) Wireless Local Area Networks (WLANs). The use of MP in WLANs can significantly reduce the probability of collision on the Medium Access Control (MAC) layer. However, it also introduces more interference on the physical layer to the neighboring WLANs. We analyze the combined effects of the MAC and physical layers and evaluate the performance in terms of packet error probability, normalized throughput and transmission delay of FH MP WLANs under saturated condition. Results show that the use of multiple patterns can significantly improve the performance.

1   Introduction

With the rapid growth of Wireless Local Area Networks (WLANs) services, it becomes harder to meet the traffic demands. There are few ways to improve network capacity, such as cell splitting. However, the coverage of WLANs is already small compared to that of macro-cellular systems. Decreasing the cell size even more will cause considerable capacity loss due to more frequent handoffs.

In this paper we consider means of improving the capacity of frequency hopping (FH) WLANS due to its superior capability in combating interference over other physical layer schemes such as direct spread spectrum. Instead of being assigned one FH pattern as stated in the IEEE 802.11 standard [1], each BSS would be assigned multiple patterns (MP). All the patterns used by the same BSS are perfectly coordinated, i.e., synchronized. It is clear that the more coordinated the FH patterns each BSS uses, the fewer users contend on each pattern and the less collision probability. However, since it is difficult to keep the collocated networks of different service providers synchronized, or coordinated, more than one pattern may hop onto the same FH band occasionally resulting in the so-called *FH hit interference*. The more FH pattern each BSS is allowed to use, the more interference it causes to other neighboring patterns. Therefore, a tradeoff study is necessary to evaluate the combined effects of increased hopping patterns on network capacity.

WLAN protocols cover both the Medium Access Control (MAC) sub-layer and the physical layer of the open system interconnection reference model. References [2], [3], and [4] discuss the performance of WLAN on either MAC or physical layers alone. Both layers are considered in this paper, since both of them affect the overall performance. In

J.N. de Souza et al. (Eds.): ICT 2004, LNCS 3124, pp. 1282–1288, 2004.

this paper, we conduct a detailed performance analysis for MP WLANs with saturated traffic loads. Numerical results show that the impairment of the increase in interference can be outweighed by the benefit of the reduced probability of collision.

2 System Model

The IEEE802.11 WLAN [1] defines Basic Service Sets (BSSs) as small wireless local area networks interconnected by a Distribution System (DS) to form an Extended Service Set (ESS). Each BSS consists of a group of stations. We will use the term "user" to denote a station.

Carrier Sense Multiple Access with Collision Avoidance (CSMA/CA) is the MAC layer protocol used by WLANs to mediate the access to the shared medium. In this paper, we will evaluate the performances of the MP WLAN with saturated traffic load, i.e., the transmission queue of each station is assumed to be always nonempty. In the protocol, the time, when the channel is idle, is slotted, and a station is allowed to transmit only at the beginning of each slot time. To reduce the probability of collision, a backoff scheme is adopted where the backoff window is doubled after each unsuccessful transmission attempt, up to a maximum value $CW_{\max}$.

For a packet to be received correctly, it has to be transmitted successfully on both the MAC and physical layers, i.e.,

$$P_p = P_{MAC} \cdot P_{phy} \tag{1}$$

where P_{MAC} and P_{phy} are packet correct reception probabilities on the MAC and physical layers, respectively.

When the system is saturated, the transmission process using a hopping pattern can be described by a discrete time Markov chain [4]. Using this model, two limiting parameters: the probability of collision, P_{MAC}, and the probability that each station transmits in a randomly chosen slot time, τ, can be readily found [4]. Let n denote the number of users contending on a pattern in a MP WLAN. A simple extension of the results in [4] leads to

$$P_{MAC} = (1 - \tau)^{n-1} \tag{2}$$

$$\tau = \frac{2(2P_p - 1)}{(2P_p - 1) + W(P_p - 2(1 - P_p)^{m+1})} \tag{3}$$

where W is the minimum backoff window size and m is the maximum allowable backoff stages, i.e., $CW_{\max} = 2^m W$.

As the state of art indicates [7], [8] and [9], most of the recent research efforts are concentrated on the MAC layer issues. The issue on the physical layer, FH hit, becomes more relevant in the MP WLAN. In order to evaluate the system performance precisely, we need to take both the MAC and physical layers into account as indicated by (1). In the sequel, we will first evaluate the system performance in the term of the probability of correct demodulated packet, P_p. Then the normalized throughput, G, and packet transmission delay, D, will be derived.

3   Probability of Correct Packet Reception

In this section, we derive the probability of the correct reception of a packet. According to (1), we need to first find the probability of the correct reception of a packet that is received without collision. Thus only physical-layer impairments including noise, fading, and interference are considered.

We assume Non-coherent MFSK modulation and Ricean fading channel with a dominant signal component of power A_s^2 and a diffuse component of power $2\sigma_s^2$ [5]. In the absence of the diffuse component, e.g., $\sigma_s^2 = 0$, the only impairment in the channel is additive white Gaussian noise (AWGN). The fading is assumed to be slow such that the channel can be regarded as constant during each hop. Below, we first find the probabilities of correct reception of a hop with and without interference.

Assume that the channel undergoes independent fading in different frequency bands. In the absence of interference, the probability of correct reception of a hop with N_S symbols is

$$P_H = \int_0^\infty (1 - P_s(e|r_s))^{N_s} p(r_s) dr_s \ , \tag{4}$$

where $r_s^2 = A_s^2 + 2\sigma_s^2$ is the total power of the signal and $P_s(e|r_s)$ and $p(r_s)$ are the symbol error probability conditioned on r_s and the probability density function of r_s, respectively. They can be written as [6]

$$p(r_s) = \frac{r_s}{\sigma_s^2} \exp\left(-\frac{r_s^2 + A_s^2}{2\sigma_s^2}\right) I_0\left(\frac{r_s A_s}{\sigma_s^2}\right), r_s > 0, A_s > 0 \tag{5}$$

and

$$P_H = \int_0^\infty (1 - P_s(e|r_s))^{N_s} p(r_s) dr_s \ , \tag{6}$$

When interference is present, we assume that the interference has the same power as the signal of interest. Under this assumption, the symbol error probability can be written as

$$PP_s = \frac{1}{M} PP_s(e|signal\ branch\ is\ hit) + \frac{M-1}{M} PP_s(e|signal\ branch\ is\ not\ hit) \tag{7}$$

With some manipulation, the symbol error probability when the signal branch is hit, $PP_s(e|signal\ branch\ is\ hit)$, can be found as

$$PP_s(e|signal\ branch\ is\ hit) = \sum_{m=1}^{M-1} \binom{M-1}{m} \frac{(-1)^{m+1}}{m+1} \exp\left[-\frac{r_s^2 + r_1^2}{\sigma^2} \frac{m}{m+1}\right]$$
$$\cdot I_0\left[\frac{2mr_s r_1}{(m+1)\sigma^2}\right] \tag{8}$$

where $I_0()$ is the modified Bessel function of the first kind and order zero. On the other hand, if the signal and interference energy is far larger than that of the AWGN,

$$PP_s(e|signal\ branch\ is\ not\ hit) \approx Pr(r_s^2 > r_1^2) = 0.5 \tag{9}$$

One can get PP_s by substituting (8) and (9) into (7), and then get the probability of correct reception of a hop, PP_H, by removing the conditioning on r_s and r_I, in (6), i.e.,

$$P_H = \int_0^\infty \int_0^\infty (1 - PP_s)^{N_s} \cdot p(r_s)p(r_1)dr_s dr_1 . \tag{10}$$

Having obtained the probabilities of correct reception of a hop, we are now ready to find the probability of correct reception of a packet. With (4) and (10), the probability that a packet of N_h hops being hit at h hops but still received correctly can be expressed as

$$P_h = PP_H^h \cdot P_H^{N_h - h} . \tag{11}$$

It follows that

$$P_{phy} = \sum_{h=0}^{N_h} [P(h) \cdot P_h] \tag{12}$$

where

$$P(h) = \sum_{s=0}^{S} \left[P(h, s|s) \cdot P(s) \right] , \tag{13}$$

is the probability that h out of the N_h hops of a packet are hit, $P(h, s|s)$ is the probability that h out of the N hops of a packet are hit given that there are s busy interfering patterns, and $P(s)$ is the probability of s out of the S interfering patterns being busy according to MAC protocol activities. $P(s)$ depends on both the traffic load n (assume every BSS has the same number of users) and probability of those users transmitting on the interfering FH patterns, i.e.,

$$P(s) = \binom{S}{s} \left[1 - (1 - \tau)^n \right]^s \left[(1 - \tau)^n \right]^{S-s} , \tag{14}$$

The conditional probability $P(h, s|s)$ in (13) can be found as

$$P(h, s|s) == \binom{N}{h} \cdot \left[P_h(s) \right]^h \cdot \left[1 - P_h(s) \right]^{N-h} , \tag{15}$$

where

$$P_h(s) = 1 - \left(1 - \frac{1}{H} \right)^s$$

is the probability that at least one of the s busy uncoordinated patterns hop to the same FH band as the subject pattern in each hop.

Substituting P_h and $P(h)$ into (12), one obtains the performance of physical layer, P_{phy}, and together with (1), (2), and (3), one can obtain the final expression of the probability of packet correct reception.

4   Throughput and Delay

In this section, we evaluate system performance in terms of throughput and transmission delay.

4.1   Normalized Saturated Throughput

P_p obtained in section 3 is conditioned on the fact that at least one packet is transmitted. It is the performance seen by a transmitted packet. From the viewpoint of the entire system, normalized throughput, G, is defined as the fraction of time the channel is used to successfully transmit payload bits. It can be expressed as the portion of the time spent in successful packet transmission, i.e.,

$$G = \frac{P_{suc} \cdot T_{PA}}{\overline{N_\sigma} \cdot T_\sigma + T_{PA}}, \tag{16}$$

where T_{PA} is packet transmission time or the time needed to transmit a packet, P_{suc} is the probability that a transmission occurring on the channel is successful, T_σ is a slot time, which is a constant, and N_σ is the number of slots during each idle period. N_σ is a random variable having geometric distribution related to the traffic load. $\overline{N_\sigma}$ is the mean value of N_σ, i.e.,

$$\Pr(N_\sigma = n_\sigma) = (1 - P_{tr})^n_\sigma P_{tr}, \qquad \overline{N_\sigma} = \frac{1}{P_{tr}} - 1, \tag{17}$$

where $P_{tr} = 1 - (1 - \tau)^{n_M}$ is the probability that there is at least one transmission on the intended FH pattern.

The probability P_{suc} that a transmission occurring on the channel is successful is given by the probability that exactly one station transmits on the channel, conditioned on the fact that at least on station transmission, i.e.,

$$P_{suc} = \frac{n_M \tau (1 - \tau)^{n_M - 1}}{P_{tr}} \cdot P_{phy}, \tag{18}$$

4.2   Transmission Delay

Another important network performance parameter transmission delay, D, which is the time it takes a packet to be transmitted, including backoff and retransmissions. For network applications, especially the delay sensitive applications, D is a critical parameter.

Let N_t be the number of packet retransmission, and N_b be the number of backoff slots. Both N_t and N_b are random variables with geometric distribution related to P_p and $\tau_{1,1}$ and with means $\overline{N_t}$ and $\overline{N_b}$, respectively, i.e.,

$$\Pr(N_t = n_t) = (1 - P_p)^{n_t} P_p, \qquad \overline{N_t} = \frac{1}{P_p} - 1, \tag{19}$$

and

$$\Pr(N_b = n_b) = (1 - \tau)^{n_M} \tau, \quad \overline{N_b} = \frac{1}{\tau} - 1 . \tag{20}$$

Thus, we have

$$D = (\overline{N_t} + 1) \cdot (\overline{N_b} \cdot T_\sigma + T_{PA}) . \tag{21}$$

5 Numerical Results

The performance of WLANs using 1 pattern, 2 patterns and 3 patterns are compared
in Fig. 1. We assume that each WLAN has one additional collocated interfering BSS,
i.e., for the IEEE 802.11 single pattern case, we assume one interfering pattern, i.e., S
in (13) and (14) is 1. And the MP WLANs using 2 and 3 patterns have 2 ($S = 2$) and
3 ($S = 3$) interfering patterns, respectively. All the parameters are fixed and same for
these three WLANs, such as $M = 4$, $R_b = 1$ Mbits/sec, $R_h = 500$ Hz, $L_{pac} = 2000$
bits/packet, so the packet transmission time, T_{PA}, is 2 msec. SNR is 20 dB. All the BSSs
are assumed to have the same traffic loads, which are perfectly balanced on the patterns
used by the same BSS.

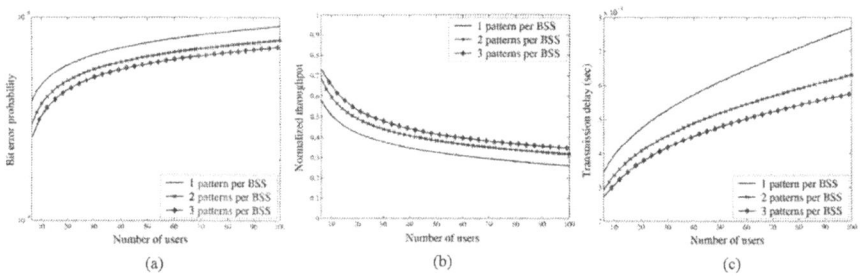

Fig. 1. Performance of MP WLANs, (a) bit error probabilities, (b) normalized throughputs, (c)
transmission delays. With 1 interfering BSS, each has 50 users. $M = 4$, $R_b = 1$ Mbps, $R_h = 1.5$
KHz, $L_{pac} = 2000$ bits/packet, $E_b/N_o = 10$ dB, and slot time = $10\mu s$.

Fig. 1(a) shows that the average bit error probability, P_b, is improved by using
multiple patterns. We take $P_b = 5 \times 10^{-3}$ as the minimum uncoded performance
requirement the networks have to provide. The IEEE 802.11 WLAN using 1 pattern can
support a maximum of 21 users (Fig. 1(a)). The MP WLAN using 2 patterns can support
40 users (Fig. 1(b)), and the MP WLAN using 3 patterns can support up to 53 users (Fig.
1(c)). The capacity in terms of the maximum number of users the networks can support is
improved. However, the efficiency of each patter, the number of users supported by one
pattern, becomes lower is due to the increased FH hit on the physical layer. The capacity
improvements achieved by using MP WLANs are much more than the one made by
adopting cell splitting, or multiple cells, since there is not considerable capacity loss

due to handoff, and the FH interference hit is minimized by the synchronization of the patterns used by the same BSS.

The normalized network throughputs given by (16) and the packet transmission delays given by (21) of 3 networks are shown in Fig. 1(b) and 1(c), respectively. Since all the users contend on the only pattern, the WLAN using only 1 pattern faces the most failure due to collision on the MAC layer, and hence, has the lowest throughput and highest transmission delay. The shortened transmission time makes MP WLANs amenable to supporting more delay sensitive applications, which cannot be supported by the IEEE 802.11 WLAN due to its high delay.

6 Conclusion

In this paper, we propose the use of multiple patterns to improve the capacity of FH WLANs. The network performance is investigated considering both the MAC and physical layers with saturation condition. The probabilities of a packet transmission failure, normalized throughput and the packet network delay are taken as the metrics of the performance evaluation. With the overall effect of collisions and interference, all the performances are improved.

References

[1] IEEE 802.11, 1999 Edition (ISO/IEC 8802-11: 1999), IEEE Standards for Information Technology – Part 11: Wireless LAN Medium Access Control (MAC) and Physical Layer (PHY) Specifications
[2] H. S. Chhaya and S. Gupta, "Performance modeling of asynchronous data transfer methods of IEEE 802.11 MAC protocol," *Wireless Network*, vol. 3, pp. 217–234, 1997.
[3] G. Bianchi, L. Fratta, and M. Oliveri, "Performance analysis of IEEE 802.11 CSMA/CA medium access control protocol," in *Proc. IEEE PIMRC*, Taipei, Taiwan, Oct, 1996, pp. 407–411.
[4] G. Bianchi, "Performance analysis of the IEEE 802.11 distributed coordination function," *IEEE JSAC*, vol. 18, No. 3, March 2000.
[5] Theodore S. Rappaport, "Wireless communications, principles and practice", Prentice Hall PTR, 1996.
[6] David L. Nicholson, "Spread spectrum signal design, LP and AJ system," Computer Science Press, 1988.
[7] H. Fattah and C. Leung, "An overview of scheduling algorithms in wireless multimedia networks," *IEEE Wireless*, vol. 9, no.5, Oct. 2002, pp. 76–83.
[8] W. Pattara-Atikom and P. Krishnamurthy, S. Banerjee, "Distributed mechanisms for Quality of service in wireless LANs," *IEEE Wireless Commun.*, June 2003.
[9] J. Deng and R. S. Chang, "A priority scheme for IEEE 802.11 DCF access method," *IEICE Trans. Commun.*, vol. E82-B, no. 1, 1999, pp. 96–102.
[10] Danyan Chen, A.K. Elhakeem and Xiaofeng Wang, "Performance of unsaturated multi-pattern frequency hopping WLANs," SPECT 2004.

A Delayed-ACK Scheme for MAC-Level Performance Enhancement of Wireless LANs*

Dzmitry Kliazovich and Fabrizio Granelli

DIT - University of Trento
Via Sommarive 14, I-38050 Trento (Italy)
{klezovic,granelli}@dit.unitn.it

Abstract. The IEEE 802.11 MAC protocol provides reliable link layer data transmission using the well-known Stop & Wait ARQ. The cost for high reliability is the overhead due to acknowledgement packets in the direction opposite to the actual data flow, which decreases transmission performance on the wireless link. In this paper, the design of a new protocol (DAWL) as an enhancement of IEEE 802.11 is proposed, with the aim of reducing supplementary traffic overhead and increasing the bandwidth available for actual data transmission. The performance is evaluated through simulations, underlining significant advantages against existing solutions and confirming the value and potentiality of the approach.

1 Introduction

In the past few years, wireless communications gained worldwide importance. In such framework, IEEE 802.11 standard [1] represents the leading MAC (Medium Access Control) protocol for wireless local area networks (WLAN). In wireless networks, packets can be lost due to errors, collisions and hidden nodes. 802.11 provides reliable link layer transmission by handling the packet delivery problems using a Stop&Wait ARQ (Automatic Repeat Request) scheme. This means that in IEEE 802.11 the receiver of a packet must reply with positive acknowledgement (PACK frame) to sender in order to enable continuation of the packet flow. The reception of this acknowledgement indicates successful frame transmission. If either the packet or its acknowledgement is lost, the sender of the packet will not receive any acknowledgement and it will retransmit the packet after a given timeout period.

However, even if this scheme provides relevant advantages (high reliability of data delivery and ease of implementation), such ARQ scheme is inefficient as any other Stop & Wait scheme due to the idle time spent in waiting for the receiver acknowledgement after each transmission [2]. An experimental study of the IEEE 802.11 ARQ scheme is described in [3], where its weak points are underlined.

Several studies were performed for improvement of the performance of 802.11 logical link control via modification of its ARQ scheme. The alternative local area network protocol proposed in [4] and Enhanced Retransmission Scheme [5] were designed to reduce the number of control frames used for single-packet delivery. A new SSCOP-based protocol was proposed in [6] for the improvement of the acknow-

---

* This work is partially funded by the Province of Trento in the framework of the DIPLODOC project (http://dit.unitn.it/~diplodoc/)

J.N. de Souza et al. (Eds.): ICT 2004, LNCS 3124, pp. 1289–1295, 2004.

ledgement scheme, aimed at the reduction of the overhead due to acknowledgements. This becomes possible by collecting the acknowledgement information at the receiver side and then sending it after being polled by transmitter by using a single control frame. This technique was already implemented in ATM networks and known as Service Specific Connection Oriented Protocol (SSCOP).

This paper presents an additional step forward in the reduction of the acknowledgement overhead for improvement of throughput in wireless networks. The proposed protocol, that can be considered as a modification of the original 802.11 standard, exploits the main concepts of the TCP Delayed-ACK scheme as well as a negative acknowledgement technique.

The paper is organized as follows: the proposed scheme is described in sections 2 and 3 (main concepts and modifications required to 802.11); performance evaluation and conclusions are presented in sections 4 and 5, respectively.

2   Description of the Proposed Method

2.1   General Description

The proposed protocol – Delayed-ACK for Wireless LANs (DAWL) – is a combination of the TCP Delayed-ACK scheme and of the SSCOP concepts. The main concept behind DAWL is that the receiver does not immediately answer to packets delivery, but it delays their acknowledgement. Assuming to have data going in the opposite direction, the acknowledgement can be sent together with data packets for increasing usage of the wireless medium and, as a consequence, decreasing the overall packet delivery time. To this aim, Positive ACKnowledgements (PACK) are used to acknowledge the data packet delivery and Negative ACKnowledgements (NACK) to request retransmission of missing packets.

In the following, for ease of presentation and without losing generality, we assume that only one station is transmitting in a given time interval. In case more than one station is working at the same moment transmission will incur in packet drops due to collisions. On the basis of such assumption, all transmitted packets are going continuously one-by-one. Then, it is possible to detect packet losses by analyzing the order of sequence numbers of the received packets. When a missing packet is detected, the receiver sends a NACK message, mentioning the sequence number and the amount of missed packets immediately to let the sender retransmit the missing packets. The frames for positive and negative acknowledgement are control frames at the MAC layer and they are to be transmitted after the Short Inter Frame Space (SIFS) time interval.

Fig. 1 shows an example of the basic operations of DAWL protocol. The RTS – CTS framework is omitted for convenience of presentation only. The data frames Data (1.1) and Data (1.2) from Node 1 are received by Node 2 without any MAC-level acknowledgement, until there is a data frame Data (2.1) going in the opposite direction (i.e. from Node 2 to Node 1). PACK (1.1-1.2) is transmitted within such data frame in order to acknowledge the previously received frames, without requiring specific channel allocation for a standalone control frame. Then, Node 2 has no additional data to send, acknowledgements to be sent to Node 1 are collected and, when

PACKDelay timeout expires, Node 2 reports to Node 1 the reception of frames (1.3 - 1.7) through the transmission of PACK(1.3 - 1.7).

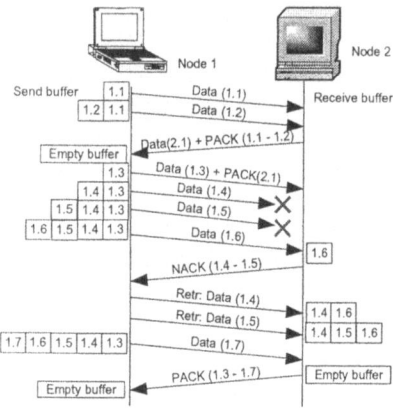

Fig. 1. Basic operation, error recovery and buffer management in DAWL.

The main difference with respect to the IEEE 802.11 MAC protocol is the same as in the case of the SSCOP-based protocol – elimination of ACK timeout and reduction of the medium-busy time due to data packet acknowledgement. The difference with the SSCOP-based protocol is the elimination of control packets transmission (such as STAT, required from the trasmitter after sending a specified amount of data frames). The STAT frame is a control frame of variable length which is to be sent after SIFS without medium reservation by Network Allocation Vector (NAV). Therefore, its elimination will decrease the probability of collision in the medium. In the framework of DAWL, in most of the cases data frames are present in both directions and acknowledgements go along with data frames.

2.2 Error Recovery in the DAWL Protocol

DAWL provides a fast error recovery mechanism that enables keeping total throughput at a high rate. When the receiver detects the loss of data frames Data (1.4) and Data (1.5) (see Fig. 1), it informs the sender by sending a retransmission request for the missed frames. It is possible to request more than one packet because NACK contains a sequence number and the amount of packets to be retransmitted. After the reception of NACK, the sender must retransmit the requested frames. In case there is no retransmission caused by NACK, but there is a continuation of data flow, NACK request must be repeated. After successful retransmission of the lost frames, transmitted data flow continues (with Data 1.7 frame). If the sender has not received positive acknowledgement for the transmitted data within the POLL timeout time, it will poll the receiver through transmission of POLL frame. Upon reception of the POLL frame, the receiver must immediately respond with PACK.

The main differences with the SSCOP-based protocol [6] are:
- the possibility of requesting retransmission of more than one frame;
- the sender is not freezing the data flow while waiting for the acknowledgement to sent packets;

– the retransmission of POLL using SSCOP-based protocol takes much more channel resources than retransmission of POLL frame in the proposed protocol, because in the latter case it is just a control frame.

When frame retransmission is required, a Retry Counter is assigned to the packet to be retransmitted. The Retry Counter is increased with each unsuccessful delivery of the packet or reset to zero for successful delivery of the packet or in case of packet discard (if the number of retransmissions of the packet exceeds the Retry Count).

2.3 Timeouts

DAWL provides reliable link layer data transmission. However, some data frames and acknowledgements can be lost. The POLL timeout is used to handle the situation of missing acknowledgements and its expiration causes the generation of POLL frame by sender to get the status information from the receiver. This is the only time when a POLL frame can appear in packet exchange.

One more timer, PACKDelay timeout, is similar as in the case of TCP Delayed-ACK scheme. During this time, a node deliberately delays sending PACK, assuming it can send the positive acknowledgement along with data. In case there is no data to send, it will send a standalone PACK.

Such timers should be carefully configured in order to trigger retransmission before a higher layer timeout occurs. The POLL timeout value must be bigger than PACKDelay timeout in order to avoid transmission of un-necessary POLL frames. The optimal value of the PACKDelay timeout depends from the channel error rate. As a guideline, the timeout value should be enough for the transmission of 3-4 transport layer frames each consisting of 1K bytes. This value was set empirically on the basis of a series of experiments that were carried out.

2.4 Buffer Management

All data frames to be sent must be placed in buffer before transmission. They can be deleted from the buffer after their successful acknowledgement was received.

On the receiver side, when a frame loss is detected, all frames from this time are to be put in the buffer until missing frames are retransmitted and successfully received. As it is shown in Fig. 1, frames (1.1) and (1.2) are buffered at the transmitter until PACK received within data frame (2.1). The receiver detects a frame loss when frame (1.6) is received and puts it into the buffer as well as retransmitted frames (1.4) and (1.5). When frame (1.7) is received, the receiver releases the buffer. Finally, upon reception of PACK standalone frame, the sender deletes successfully transmitted frames from the buffer.

2.5 Packet Delivery Time and Sender Notification

One of the most important parameters of a protocol functionality is the packet delivery time, especially in case control frames accompany data packet delivery.

Figure 2a shows the packet delivery time of DAWL protocol in comparison with 802.11 MAC for a standalone higher layer packet. Upon the DATA packet reception

and its checksum verification, the receiver is able to decide about the successful reception of data and can pass received data to the upper layer of the protocol stack.
Since the timing of the RTS – CTS – DATA exchange is exactly the same in both
cases, we can conclude that packet delivery time of DAWL and IEEE 802.11 is exactly the same. As a result, we can infer that the modifications proposed in the DAWL
protocol are completely transparent to higher layer protocols operation, such as TCP
with its Round Trip Time (RTT) calculation algorithms. The final stage of the packet
delivery is the sender notification of the delivery status. The PACKDelay timeout in
DAWL is bigger than SIFS in 802.11 MAC. This generates a longer delay for a
standalone DATA packet transmission (see Fig. 2a), but it does not imply a significant impact on data flow, since notification events are related to the release of the resources allocated for transmission and for data flow control mechanisms.

3 Proposed Modifications of IEEE 802.11 MAC Protocol

• **NAV modification.** The elimination of ACK frames in data frame transmission
scheme of 802.11 requires a modification in the NAV value calculation. Modified
NAV will not include SIFS interval and ACK frame durations as it is shown on Fig.
2b, where the difference between 802.11 standard and DAWL in the calculation of
NAV is underlined. POLL frame includes NAV for SIFS and PACK answer durations. The contention phase is still exactly the same as in 802.11.

• **Sequence number management.** IEEE 802.11 assigns a sequence number counter
(SNC) to each data frame, that is increased modulo 4096.

Frame loss detection and further retransmission of the lost frames in the DAWL
protocol are based on the analysis of the order of frame sequence numbers. In order to
have an uninterrupted increase of sequence number in data exchange between two
nodes, DAWL protocol needs to modify the original 802.11 scheme: each node of the
network should have one SNC for broadcast packets and one SNC for each node to
which the current node is exchanging data. This means that when a node needs to
send data to any other node it needs to allocate a new SNC.

• **Frame formats.** In the DAWL protocol, 5 new frame types are introduced:
  – Data + PACK frame to carry acknowledgement information;
  – PACK frame as a standalone packet;
  – NACK frame to inform the sender about a frame loss;
  – NACK + PACK frame as a combination of ACK and NACK frames;
  – POLL frame to request receiver status information.

All the designed frame types are compatible with the MAC-layer frame format
specified in the IEEE 802.11 standard. For the definition of the new frames presented
by DAWL, the reader should refer to [8].

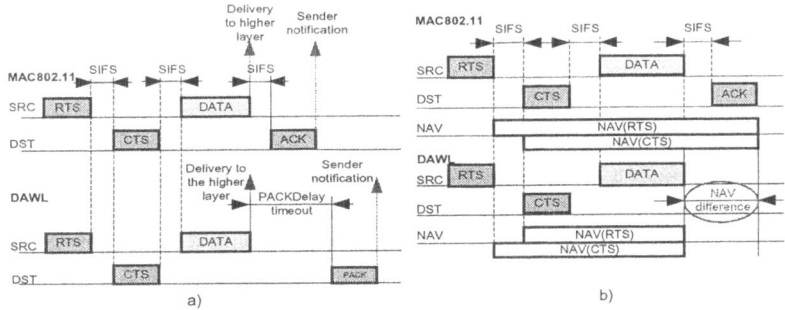

Fig. 2. a) Packet delivery and sender notification events; b) the difference between NAV calculation in the original 802.11 and DAWL protocols.

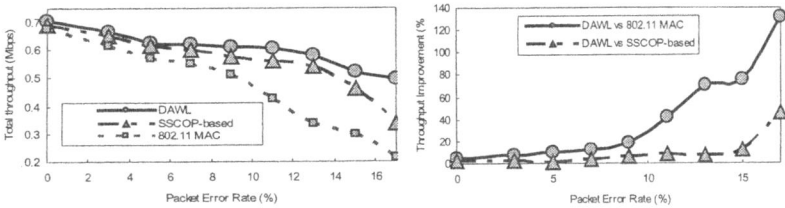

Fig. 3. Performance comparison among 802.11, SSCOP-based and DAWL protocols.

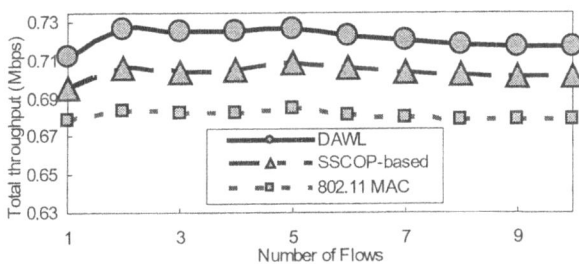

Fig. 4. Throughput performance of 802.11, SSCOP-based and DAWL protocols in a multi-flow environment.

4   Performance Evaluation

The performance of the proposed protocol is analyzed by using ns-2 [7]. In order to be able to evaluate a performance bound for the proposed scheme, it is important to consider only a single TCP flow (data + ACKs) between nodes, by disabling routing, ARP protocol, connection establishment and slow start phase of TCP. Channel data rate is set to 1 Mbps, data packet size to 1Kbytes.

Fig. 3 shows the simulation results. As the packet error rate (PER) increases, DAWL protocol has an increasing advantage comparing to 802.11 MAC and SSCOP-based protocols. The reason why the proposed protocol achieves higher throughput in comparison with SSCOP-based protocol is the elimination of transmission of control frames, the improved retransmission algorithm and the absence of delay in data transmission while waiting for the acknowledgement of sent data frames. The difference with IEEE 802.11 is mainly due to the significant improvement in the acknowledgement scheme, which leads to faster data exchange between nodes.

The average throughput improvement of DAWL protocol (Fig. 3) in the interval of PER from 0 to 10 percent is 3.128% comparing with SSCOP-based protocol and 10.70% for IEEE 802.11 MAC protocol. In case of higher error rates, DAWL provides more advantages. With a PER ranging from 10 to 17 percent, in fact, the throughput improvements are in average 18.97% for SSCOP-based protocol and 79.95% for IEEE 802.11 MAC.

The DAWL protocol operation was also tested in a multi-flow environment where there are N TCP flows produced by 2*N nodes on the same medium (N = 1 ÷ 10). This scenario showed conceptual similarity with the two-node scenario (Fig. 4).

5   Conclusions

This paper presented a new protocol, DAWL, as an enhancement of the IEEE 802.11 MAC, proposing a combination of the Delayed-ACK and negative acknowledgement techniques as a new alternative ARQ scheme. The performance is compared with IEEE 802.11 MAC as well as with a SSCOP-based protocol. Results underline significant advantages of DAWL, especially in bi-directional traffic exchange scenarios.

References

[1]   "Wireless LAN Medium Access Control (MAC) and Physical Layer (PHY) Specifications," IEEE 802.11 standard, 1997.
[2]   Medeiros de Lima, H. and Duarte O.C.M.B., "An Effective Tselective repeat ARQ strategy for high speed point-to-multipoint communications", TGLOBECOM, 1996.
[3]   Chung Ho Nam, Soung C. Liew, Cheng Peng Fu, T"An Experimental Study of ARQ Protocol in 802.11b Wireless LAN,"T http://www.broadband.ie.cuhk.edu.hk/ research/paper/Namicc2002.pdf.
[4]   B.E. Mullins, N.J. Davis, S.F. Midkiff, "A wireless local area network protocol that improves throughput via adaptive control", ICC, Vol. 3 , pp. 1427-1431, 8-12 June 1997.
[5]   H.-L. Wang, J. Miao, J. Morris Chang, "An Enhanced IEEE 802.11 Retransmission Scheme", WCNC, Vol. 1 , pp. 66-71, 16-20 March 2003.
[6]   H.-L. Wang, A. Velayutham, "An SSCOP-based Link Layer Protocol for Wireless LANs", GLOBCOM '03, Vol 1, p. 453 -457.
[7]   NS-2 simulator tool home page. TU http://www.isi.edu/nsnam/ns/UT, 2000.
[8]   Dzmitry Kliazovich and Fabrizio Granelli, "A Delayed-ACK Scheme for Performance Enhancement of Wireless LANs", Technical Report #DIT-03-042, July 2003, http://dit.unitn.it/research/publications/techRepDit?anno=DIT&lang=en.

The VDSL Deployments and Their Associated Crosstalk with xDSL Systems

Kyung-Ah Han [1], Jae-Jin Lee [1], and Jae-Cheol Ryou [2]

[1] Korea Telecom Technology Laboratory
463-1 Jeonmin-dong, Yusung-gu, Daejeon, 305-348, Korea
{kahan, jaejin}@kt.co.kr
[2] Division of Electrical and Computer Engineering, Chungnam National University
220 Gung-dong, Yusung-gu, Daejeon, 305 - 764, Korea
jcryou@home.cnu.ac.kr

Abstract. In Korea, VDSL deployment has been started recently. This is because VDSL can overcome the effects ADSL relatively asymmetric and slow data rate. Though it seems certain that VDSL is a promising technology for ILECs who possess enormous copper cables to implement Full-Service Broadband Access Networks, its broad frequency occupancy used for transmission over existing copper pairs impose several obstacles in massive service deployment. Especially Crosstalk interference produced by between xDSL services provided in the same cable binder is critical issue in VDSL service deployment.

In this paper, we present our experience encountered during VDSL deployment in the multi-services existence environment on the same cable and the experimental results caused by crosstalk interference between services. We describe expectable crosstalk problems in the VDSL environment and analyze the influence of crosstalk.

1 Introduction

With the eager desire of customers for faster data services, the broadband access services have been explosively spread in Korea through last a few years. Among several high-speed access technologies, ADSL service is already most popular these days. Then as the result of saturated ADSL market, VDSL deployment has been started recently. This is because VDSL can overcome the effects ADSL relatively asymmetric and slow data rate. And it can accommodate a discriminated and luxurious service to customers.

Though it seems certain that VDSL is a promising technology for ILECs who possess enormous copper cables to implement Full-Service Broadband Access Networks, its broad frequency occupancy used for transmission over existing copper pairs impose several obstacles in massive service deployment. Especially Crosstalk interference produced by between xDSL services provided in the same cable binder is critical issue in VDSL service deployment.

The issues of crosstalk are more severely revealed by the VDSL deployment.
-   VDSL uses relatively high frequency compared with ADSL.
-   Standard incompliant VDSL system is introduced because the development

J.N. de Souza et al. (Eds.): ICT 2004, LNCS 3124, pp. 1296–1302, 2004.

- of standard compliant VDSL is slowed in progress.
- The variation of CO & CPE distance by deployment scenario increase the affection of FEXT.

This paper presents our experience encountered during VDSL deployment in the multi-services existence environment on the same cable and the experimental results caused by crosstalk interference between services. The representative systems what should be considered crosstalk interference in this paper are such things like standard compliant VDSL, standard incompliant VDSL and ADSL mainly.. In here, standard compliant VDSL system means that a system follows on of frequency band plans specified on ITU-T recommendation G.993.1. We describe expectable crosstalk problems in the VDSL environment and analyze the influence of crosstalk. Also, we present the issues of standard to minimize the influence of crosstalk and their general effects briefly.

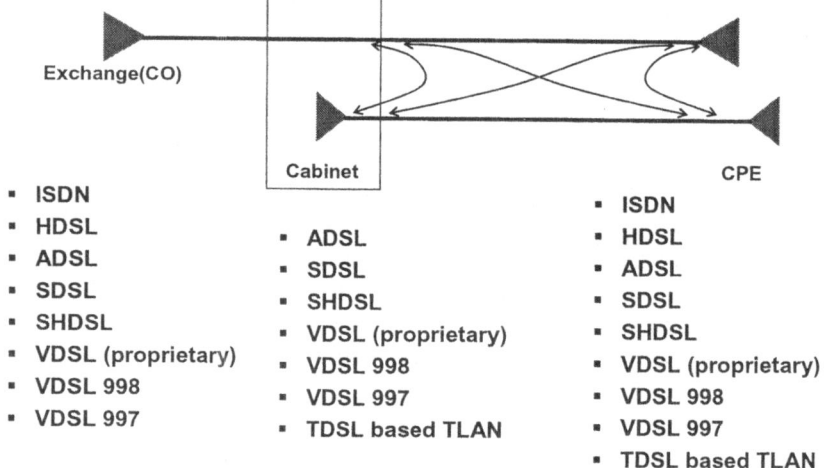

Exchange(CO)

Cabinet   **CPE**

- ISDN
- HDSL
- ADSL
- SDSL
- SHDSL
- VDSL (proprietary)
- VDSL 998
- VDSL 997

- ADSL
- SDSL
- SHDSL
- VDSL (proprietary)
- VDSL 998
- VDSL 997
- TDSL based TLAN

- ISDN
- HDSL
- ADSL
- SDSL
- SHDSL
- VDSL (proprietary)
- VDSL 998
- VDSL 997
- TDSL based TLAN

Fig. 1. xDSL Network Configuration

2   xDSL Network Configuration and Their Crosstalk

Crosstalk is a noise introduced by the pairs within same binder group. It is distinguished into NEXT and FEXT by the source of noise. The figure 2 shows NEXT and FEXT. NEXT is defined as the crosstalk effect between a receiving path and a transmitting path of DSL transceivers at the same end of two different subscriber loops within the same binder. FEXT is defined as the crosstalk effect between a receiving path and a transmitting path of DSL tranceivers at opposite ends of two different subscriber loops within the same binder.

self-NEXT and self-FEXT is initiated from adjacent transmitters of the same type of DSL system. When self-NEXT is absent(through the arrangement of Frequency Division Multiplex(FDM) systems between transmissions in opposite directions, for example), the effect of self-FEXT could dominate that of background noise.

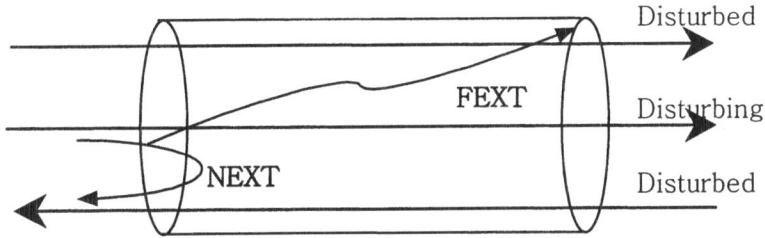

Fig. 2. NEXT and FEXT

Due to capacitive and inductive coupling there is crosstalk between each copper pair in the same binder group even though pairs are well insulated at DC.
The main causes of crosstalk in same binder group are as follows.

- the overlap of the bandwidth between upstream and downstream
- the difference of PSD templates between xDSL

In FDM mode, the bandwidth of upstream and downstream is not overlapped. Usually, NEXT is more intrusive than FEXT in ADSL environment. There is no self-NEXT in FDM theoretically and self-FEXT can be neglected within same distances in ADSL. But self-FEXT is mainly considered in VDSL environment because VDSL use high frequency (~12MHz) and high speed.

3 The Characteristics of Variable Access Technologies

Table 1. xDSL technologies deployed in Korea

		ADSL	VDSL (proprietary)	VDSL 998	VDSL 997	T-LAN	HDSL 2B1Q	HDSL CAP
Performance (DN/UP, bps)		8M/1M	10~25M/ 10~16M	40-50M/ 8~15M	40~45M/ 20~23M	2M/2M	E1,T1	E1,T1
coding		DMT	QAM	QAM,DMT	QAM,DMT	TDSL	2B1Q	CAP
Duplex		FDD	FDD	FDD	FDD	TDD	EC	EC
Reach		About 3km	1~1.3km	About 1.5km	About 1.5km	1km	3km	3km
Frequency	UP	25K~138K	120K~3.9M	3.75M~5.2M 8.5M~12M	3.0M~5.1M 7.05M~12M	20K~2.5M	0~196K(T1) 0~292K(E1)	21K~160K (T1)
	DN	138K~1.1M	3.1~8.0M	138K~3.75M 5.2M~8.5M	138K~3.0M 5.2M~7.05M			21K~256K (E1)
PSD (dBm/Hz)	UP	-38	FTTCab M1 standard PSD (-60dBm/Hz) *Uplink PBO			-40 (peak)	-37	-40
	DN	-40						

4 The VDSL Deployments and Issues Concerning Crosstalk

Major broadband carriers are now seeking to set forth its vision for the next generation of the Internet by exploring implications of the ubiquitous network and digital convergence. Carriers in Korea are attempting to determine how this vision could be practically implemented, by launching the VDSL service, expected to be the driving factor for the next phase of the broadband boom.

KT has deployed VDSL systems since end of last year, 2002, with 13Mbps link speed symmetrically and in near future over 50Mbps link rate will be served. Other ILECs are starting to deploy VDSL services. We will present expectable crosstalk issues attendant on VDSL deployments in this section.

4.1 The Band Plan Issues in VDSL Systems

When the service providers deploy xDSL services, the management of band plan is very important issues concerning crosstalk. The VDSL systems can be divided into two main classes; standard compliant VDSL systems and standard incompliant(proprietary) VDSL system.

Because the development of standard compliant VDSL systems is slowed in progress, many 2-band VDSL systems which do not comply with standard band plan(997 or 998) was deployed. The band plan issue is one of the most important issues in Korea and further study about band plan is needed to solve problems due to the crosstalk interference between the different VDSL systems.

The next figure presents the diversities of frequency that various VDSL systems use.

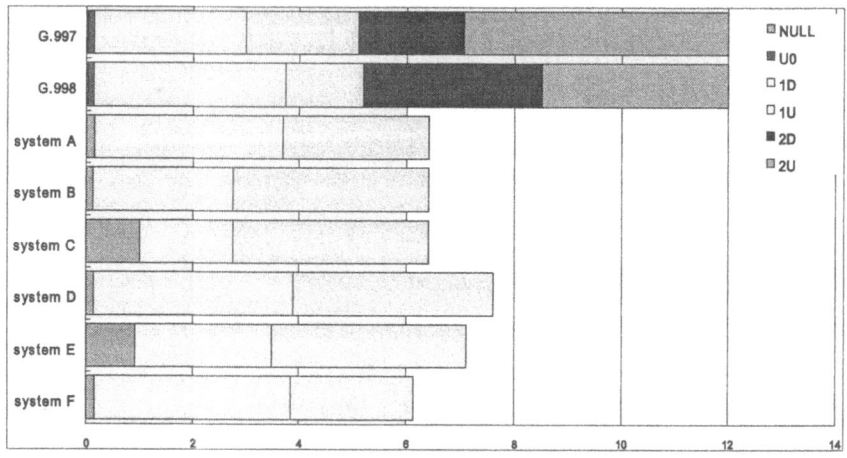

Fig. 3. Standard compliant and incompliant VDSL systems

Figure 3 shows that the overlapping of frequency can occur very severe NEXT interference. We need to minimize NEXT interference between standard compliant and incompliant VDSL systems.

To solve this problem, we are investigating the possible approaches like frequency limitation, power adjustment and pair selections to reduce crosstalk interferences.

4.2 The Self Crosstalk Issues in VDSL Systems

There is no self-NEXT in FDM theoretically and self-FEXT can be neglected within same distances in ADSL environment, because bandwidth of upstream and downstream is not overlapped in FDM mode. But self-FEXT is mainly considered in VDSL environment because VDSL use high frequency (~12MHz) and high speed.

The implementation of VDSL system is very important in terms of minimizing the self-FEXT. All VDSL ports in the same binder should provide the same quality of VDSL service simultaneously. In the case of unstable VDSL systems, some ports can affect or be affected with another port. The performance of each port in the same binder has very large deviation.

We analyzed the degree of self-xtalk in the VDSL frequency bandwidth. At first, we measured the FEXT & NEXT power sum loss and line loss of CPEV 0.5mm cable. And then, the SNR range of no self-disturber could be compared with the SNR range of 24 self-disturbers(NEXT, FEXT).

- Disturbers: 24 self-disturbers FEXT and NEXT noise
- Test loop: FEXT & NEXT power sum loss and line loss of CPEV 0.5mm cable (0.3km,1km)
- Background noise: -140dBm/Hz
- Average PSD: -60dBm/Hz

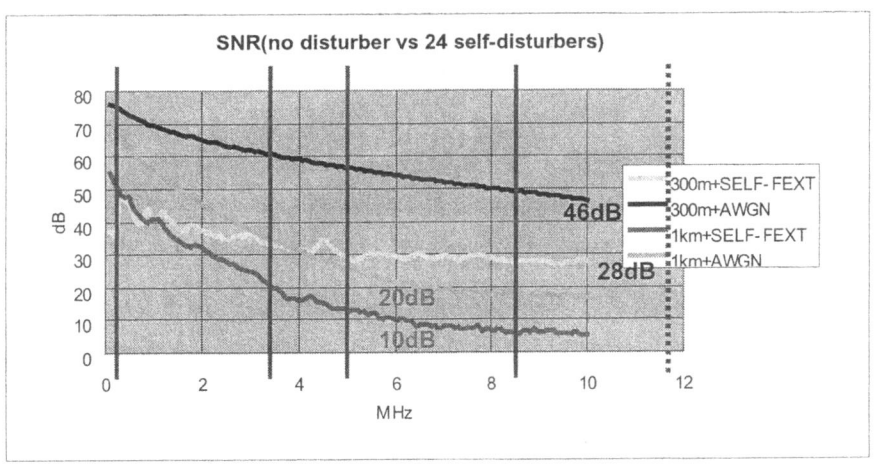

Fig. 4. The Result of SELF-FEXT

If we use 998 4B, the effect of SELF-FEXT follows;
- 0.3km in U2 bandwidth, 46dB(at least, 2048 QAM) -> 28dB(32~64 QAM)
- 1km in D2 bandwidth, 20dB(8QAM) -> 10dB(?)

If we use 998 4B and system A in the same binder(the overlap of the bandwidth between upstream and downstream in D2), the effect of SELF-NEXT follows;
- 0.3km in U2 bandwidth, 50dB or more -> 10dB and less
- 1km in D2 bandwidth, 20dB -> 0dB

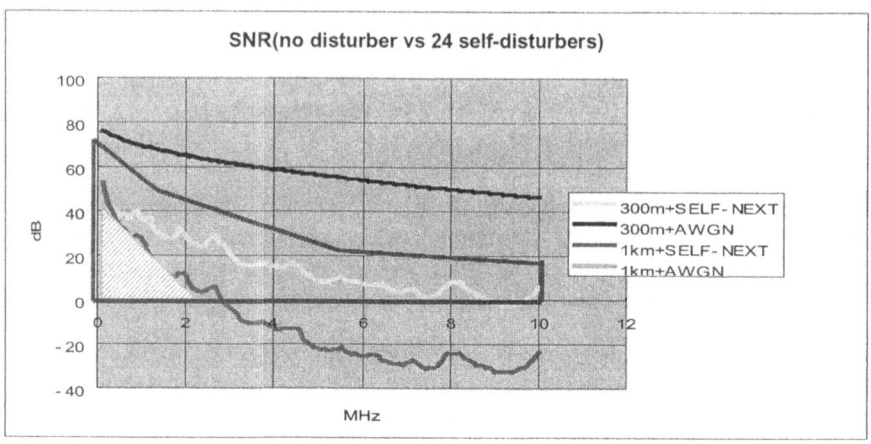

Fig. 5(a). The Result of SELF-NEXT

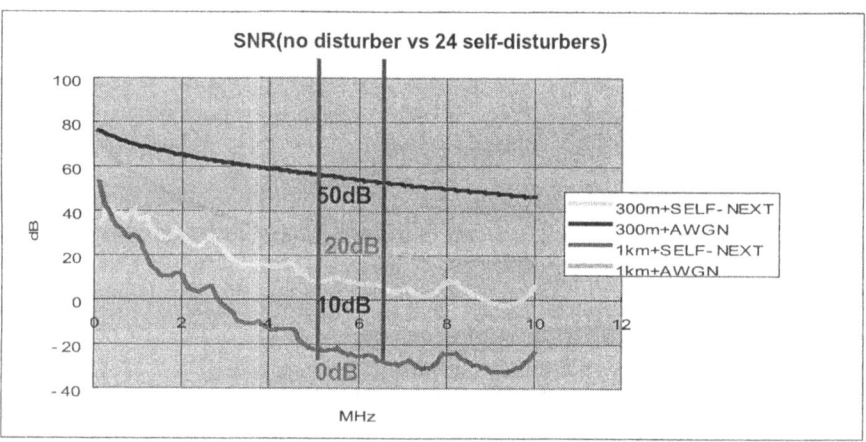

Fig. 5(b). The Result of SELF-NEXT

When the standard compliant and in compliant VDSL systems is accommodated in the same binder, self-NEXT is severely affected with each other. It results in performance degradation.

The difference of CO or CPE distance can occur another kind of self-FEXT by the difference of signal power. The strong signal can affect weak signal in relatively long distance.

We can use the UPBO or DPBO to solve this kind of self-FEXT interference. ITU-T recommendation G.993.1 regulates that UPBO is mandatory function and DPBO is optional function. The most VDSL systems in Korea implemented UPBO function.

4.3   The Crosstalk Issues with ADSL Technologies

Table 1 describes the characterstics of xDSL systems deployed in Korea. In this environment, we need to consider the problem of electromagnetic interference among xDSL systems. Because many subscribers use ADSL and VDSL technologies in Korea, we are considering mainly the effect of crosstalk in ADSL and VDSL environment.

As tested in our laboratory and fields, the degradation of exchange-based ADSL transmission rate due to the interference of other xDSL systems appears as the line loss gets higher. The performance of exchange-based ADSL systems is degraded by the cabinet-based VDSL systems, about 3Mbps in 3km. We are investigating the strategies to minimize the performance degradation of exchange-based ADSL. At first, when exchange-based ADSL and cabinet-based VDSL systems are mixed in the same region, we try to change exchange-based ADSL subscribers to cabinet-based VDSL.

Also, in the VDSL systems, the PSD reduction function in the frequency region below 1.104MHz(ADSL) use to minimize the influence of VDSL systems. Operators determine whether the function is used or not.

5 Conclusion

In Korea, VDSL deployment has been started recently. This is because VDSL can overcome the effects ADSL relatively asymmetric and slow data rate. Though it seems certain that VDSL is a promising technology for ILECs who possess enormous copper cables to implement Full-Service Broadband Access Networks, its broad frequency occupancy used for transmission over existing copper pairs impose several obstacles in massive service deployment. Especially Crosstalk interference produced by between xDSL services provided in the same cable binder is critical issue in VDSL service deployment.

As described in this paper, we knew the performance degradation of VDSL or ADSL due to the interference of massive VDSL deployment. We described the issues of crosstalk by deployment of VDSL in detail and proposed possible approaches like frequency limitation, power adjustment and pair selections to reduce crosstalk interferences. Also, the study for spectrum management to minimize interference by VDSL deployment is considering continuously in Korea..

References

1. Transmission Systems for Communications, Fifth Ed. Holmodel, NJ:Bell Laboratories, 1982.
2. D.Rauschmayer, "ADSL/VDSL Principles", MTP, 1999.
3. Walter Y.Chen, "DSL simulation techniques", MTP, 1998.
4. Bingham, "ADSL, VDSL, and Multicarrier", Wiley, 2000.

Design of PON Using VQ-Based Fiber Optimization

Prateek Shah, Nirmalya Roy, Abhishek Roy, Kalyan Basu, and Sajal K. Das

CReWMaN, CSE Dept., Univ. of Texas at Arlington, TX 76019, USA
{prshah, niroy, aroy, basu, das}@cse.uta.edu

Abstract. The rapid development of broadband technologies has already set the stage for next generation access networks with optical fibers proliferating from the long haul backbone into the local area networks (LANs). However, the bottleneck of bandwidth availability still remains at the end-user. Passive optical network (PON) has the potential to solve this "last mile problem". In this paper we have investigated into the major research issues regarding the deployment of the PON technology. An estimate of effective bandwidth consumption by residential and business users is obtained to maximize the bandwidth utilization. A new vector quantization (VQ) based location determination algorithm is formulated to determine the optimal location of splitters, thus minimizing the fiber layout cost. Inherent drawbacks of the standard ranging scheme is also discussed. Simulation results corroborate the maximization of bandwidth utilization, minimization of fiber layout cost and points out the necessity of an efficient ranging scheme to reduce the ranging overhead time.

1 Introduction

While the capacity of the backbone networks has been keeping pace with the tremendous growth of traffic, there has been little progress in the access technology, thus depriving the subscribers from the real benefits of broadband services. The "last mile" (or "first mile"), which connects the service provider's central offices to users, still remains a major bottleneck. Although the currently deployed broadband solutions, like digital subscriber line (DSL) and cable modem (CM) provide an improvement over the 56 Kbps dial-up lines, they are unable to provide enough bandwidth for real-time services. Passive optical networks (PON) possess the inherent capability of delivering voice, data and video services with, high reliability over long distances, low cost and simple maintenance [8, 9]. However, deployment of PON posses significant challenges, like bandwidth assignment, minimizing fiber cost and collision free upstream data transmission.

In this paper, we address the basic design issues in deploying PON as the access network for residential and business users. Section 2 highlights the major components and functionalities of the PON architecture. A Pareto-distributed traffic model for estimating effective bandwidth is discussed in Section 3. This gives an optimal combination of the two classes of users, maximizing the bandwidth utilization. In order to minimize the fiber layout cost (amount of fiber), the

J.N. de Souza et al. (Eds.): ICT 2004, LNCS 3124, pp. 1303–1309, 2004.
© Springer-Verlag Berlin Heidelberg 2004

distances between the network components should be reduced. Minimizing distance has a perfect analogy with minimization of distortion in VQ terminology. With this motivation, in Section 4, we develop a VQ-based near-optimal location determination algorithm to obtain optimal locations for splitters and OLT. Ranging is a process to avoid upstream traffic collision by making the ONUs transmit with some additional delay, if required. Section 5 discusses the ranging scheme to calculate the delay offset for each ONU to synchronize the ONUs. Simulation results in Section 6 corroborates the effective bandwidth requirements, optimal fiber cost and the inefficiency of the existing ranging scheme when ONUs are almost equidistant from the splitter. Section 7 concludes the paper.

2   Architecture of PON

PON is a point-to-multipoint (P2MP) optical network with no active components in the signal's path from source to destination. Among several PON topologies like tree, ring or bus, the tree topology has been widely accepted as the standard for deployment and possible performance evaluation purposes [7, 8]. Figure 1 shows a tree based PON consisting of Optical Line Terminal (OLT), splitters, Optical Network Units (ONUs) and users. All transfers occur via the OLT and no direct traffic exists between the ONUs. PON can be used to implement a FTTx (fiber-to-the-home/building/curb) access network [3,5]. Each splitter can handle as many as 32–64 ONUs at a maximum distance of 20 Km. In the upstream direction, PON is a multipoint-to-point network where data from multiple ONUs may collide at the OLT. Channel separation mechanism like wavelength or time division multiplexing (WDM, TDM) can be employed to avoid collisions and fairly share the channel capacity. In a WDM solution, different wavelengths are assigned to different ONUs and the need of a tunable receiver at the OLT makes it really expensive. However in a TDM solution, single wavelength is shared by all ONUs, making it the preferred upstream access solution. Hence, our discussions are made in respect to a tree based TDM PON.

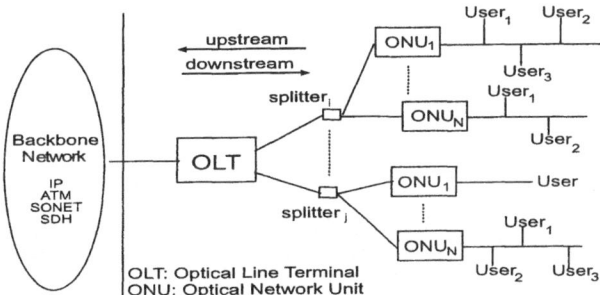

Fig. 1. Architecture of PON

3 Traffic Modeling

It is well known that Poisson processes fail to successfully capture the Internet traffic dynamics [16], especially when the time scale is smaller than user-session arrivals. This fact motivates us to model the self-similar, long-range-dependent traffic of PON by multiplexing several *Pareto-distributed* [6] "on/off" sources. This traffic model is required to estimate the bandwidth distribution among residential and business users. The expected (mean) value of this pareto distribution is estimated by: $E[x] = \frac{\alpha b}{\alpha-1}[1 - p^{\frac{\alpha-1}{\alpha}}]$ where α is the shape parameter, b is the minimum value from a Pareto distribution and p is the smallest non-zero uniformly distributed value. Denoting transmitted mean packet size as $\mathcal{K}$, the ith 'on' and i^{th} 'off' periods as τ_{1i} and τ_{2i} respectively, the load of a particular source can be calculated as $\mathcal{L}_i = \frac{\mathcal{K}}{\mathcal{K}+\tau_{2i}/\tau_{1i}}$. If N denotes the maximum number of ONUs/splitter then we have:

Result 1 *Total load* $(\mathcal{L})$ *from a splitter to the OLT, is an aggregation of* M *streams (ONUs), where* $M \in N$, *is estimated by* $\mathcal{L} = \sum_{i=1}^{M} \mathcal{L}_i = \sum_{i=1}^{M} \frac{\mathcal{K}}{\mathcal{K}+\tau_{2i}/\tau_{1i}}$.

The empirical work of Kelly [2] helps us to estimate the effective bandwidth (β_{eff}) corresponding to our design as:

$$\beta_{eff} = \begin{cases} am[1 + 3z(1 - m/h)], & \text{if } 3z < min(3, h/m) \\ am[1 + 3z^2(1 - m/h)], & \text{else } 3 < 3z^2 < h/m \\ ah, & \text{otherwise} \end{cases} \tag{1}$$

where $a = 1 - \frac{\log_{10} P_{loss}}{50}$; $z = -2\frac{\log_{10} P_{loss}}{(c/h)}$; $P_{loss} \approx 10^{-9}$ is the loss probability, m, h and c respectively denote mean, peak and channel rates. This formulation of effective bandwidth helps to estimate the number of ONUs under every splitter.

Result 2 *If* β *represents the available line rate for upstream traffic,* η_r *and* η_b *denote the number of residential and business ONUs under a splitter,* β_{r_i} *and* β_{b_j} *represent the effective bandwidth consumed by the* i^{th} *residential and* j^{th} *business ONU respectively, then number of ONUs per splitter is estimated by the following two conditions:* $\sum_{i=1}^{\eta_r} \beta_{r_i} + \sum_{j=1}^{\eta_b} \beta_{b_j} \leq \beta$ *and* $\eta_r + \eta_b \leq N$. $\qquad$ (2)

4 Fiber Layout Optimization

One prime objective behind designing of PON lies in the minimization of overall fiber layout cost. The fiber layout cost is estimated by connecting the splitters with the respective ONUs and the OLT in turn with the splitters. Hence, in order to minimize this cost one needs an optimal placement for these splitters and the OLT. The objective is to minimize the distance between the splitters and ONUs. Intuitively the objective has a nice analogy with vector-quantization strategy, where a set of points is represented by a code-vector, such that, the overall distortion (measured in terms of distance) is minimized. Motivated by this basic analogy, we have formulated a new Vector Quantization (VQ)-based algorithm [11] for obtaining such near-optimal locations of splitters (with respect to ONUs) and OLT (with respect to the splitters). Thus, the set of ONUs actually represent the output points and the splitters represent the code-vectors. The distance between a splitter and the corresponding ONUs provide the estimate of the distortion generated by such a quantization scheme. Our newly proposed fiber layout optimization algorithm works on the concept of modified *LBZ* or *Loyd's* algorithm [11], available for VQ.

The algorithm starts with the set of source output points (ONUs) and obtains an *initial code-vector* $\bar{V}_0^0$ representing the *center-of-mass* for this entire set of output points. It now perturbs the existing vector(s) by using a *perturbation vector* to get a new set of vectors $\bar{V}_0^0 \cup \bar{\vartheta}_0^0$ having twice cardinality. The entire search space is now divided into quantization regions (R) such that source outputs nearer to a particular code-vector belong to a particular region. This forms the basis of grouping and quantization scheme, where the corresponding code-vector is the representative of the set of output points. The total distortion of all the points are computed. This distortion actually provides a measure of distance between the ONUs and the splitters. The algorithm now loops back to start computing the new code-vectors representing the *center of masses* of the newly formed regions. A similar approach for perturbing the set of new code-vectors and forming new quantization regions follows. Subsequently, the new value of distortion measure is estimated. This same procedure is repeated at every iteration, until the change in distortion is less than a predefined threshold (ϵ). For practical situations, splitting of overloaded and merging of under-utilized regions have been considered. Now, in order to connect the set of ONUs with the respective splitter, a Minimum Cost Spanning Tree (MCST) is formed with splitter as the source and the set of ONUs as the leaf nodes. We have chosen Prim's algorithm [14] for the formation of this MCST. The set of all splitters needs to be connected with the OLT. In order to obtain the optimal location of the OLT, the same technique is applied to get the center-of-mass of the set of splitters. Figure 2 gives a pseudo-code of this approach, where $f(X)$ is the Euclidean distance function required for computing the center-of-mass.

5   Ranging

Once the PON is deployed with optimal splitter-locations, the upstream synchronization problem needs to be solved to avoid any data collisions. Although ONUs transmit data in their assigned slots, variation in their propagation delays may result in collision. Since the round trip propagation time is linearly proportional to the ONU-OLT distance, the OLT can assign appropriate slots to respective ONUs, provided it knows ONUs' distance. *Ranging* is a process where all ONUs are made to be at the same logical distance by inserting appropriate delays to their actual roundtrip propagation time. Hence, ONU transmits with a delay offset to form an upstream frame without collision.

As described in [4], $T_{response}$ is estimated as sufficient signal processing

1. Initialize $k := 0$;
2. Get the vector(s)

$$\{\bar{V}_i^k\}_{i=1}^k = \frac{\int_{\forall X \in R^k} X f(X) dX}{\int f(X) dX};$$

3. Perturb $\{\bar{V}_i^k\}_{i=1}^k$ to get new vectors $\{\bar{\vartheta}_i^k\}_{i=1}^k$;
4. Combine the existing and new vectors to get: $\{\bar{V}_i^k\}_{i=1}^k \cup \{\bar{\vartheta}_i^k\}_{i=1}^k$;
5. Find the quantized regions:
   $R_i^k := \{X : d(X, \bar{V}_i) \le d(X, \bar{V}_j)$
   $\forall (i, j) \in [1 \dots k], j \ne i\}$;
6. Get distortion
   $$d_n^k := \sum_{i=1}^k \int_{\forall k} |X - \bar{V}_i^k|^2 f(X) dX;$$
7. If $(\frac{d_n^k - d_n^{k-1}}{d_n^k} \le \epsilon)$ goto step 10;
8. Increment k by 1 and go to step 2;
9. If (no. of ONUs in a region > N)
10. Split the regions in two or more groups;
11. If (no of ONUs in a region < $(0.1 \times N)$)
12. Merge two or more regions;

1. OLT broadcasts a *MUTE* control message: all ONUs suspend their operations;
2. OLT starts an internal timer when it sends a *RANGE* message.
3. *RANGE* message arrives at ONU_i after time $\frac{T_{process}}{2} + T_{pd}$. ONU_i sends a *REPLY* immediately. Actual round trip time is calculated by the timer in OLT as $T_{rt} = \frac{T_{process}}{2} + T_{pd} + T_{response}$
   $+ T_{pd} + \frac{T_{process}}{2}$;
4. If *REPLY* messages collide, OLT sends a *Serial_number_mask*. ONUs with a match sends a *REPLY* back.
   (binary tree mechanism [4]);
5. Goto Step 2 till all ONUs are ranged;
6. OLT calculates T_δ for ONU_i;
7. OLT lets all ONU know about its T_δ;

Fig. 2. Location Determination Algorithm   **Fig. 3.** Ranging Algorithm

time in an ONU, for instance $3136 - 4032$ bits for 156 Mbps line speed. Let the equalized round trip time be T_{eqd}, the ONU's actual propagation time be T_{pd} and the signal processing time at the OLT be $T_{process}$. Hence, $T_{eqd} = 2 \times T_{pd} + T_{response} + T_\delta + T_{process}$. So, for a constant T_{eqd}, variation of T_{pd} and $T_{response}$ results in different equalization delay (T_δ). The actual ranging procedure is shown in Figure 3. In the worst case, OLT has to range each ONU individually to know their round trip propagation time if their *REPLY* messages collide. Our results show that the number of iterations (*Range+Reply* message tuple) required to range all ONUs is significantly large when ONUs are equidistant from the OLT.

6 Simulation Experiments

We consider data and voice traffic as per specifications in [13] for a population size of $\mathcal{P} = 10^4$ (80% residential, 20% business users), distributed randomly over $A = 20 \times 20$ sq. Km with 156Mbps as the available upstream line rate (β). Any splitter can support a maximum of 32 ONUs, each having $1 - 3$ users. Assuming channel rate (c) as 5 and 25 Mbps, m and h is obtained as 0.333, 0.667 and 6.475, 12.95 Mbps for residential and business users respectively [13]. Using these results in Equation (1), we obtain the effective bandwidth consumption as 0.786 and 15.281 Mbps respectively. Figure 4 delineates the effective bandwidth distribution for different, valid combinations of residential and business users, according to Equation (2). The minimum bandwidth wastage is observed for a ratio of residential to business ≈ 2.55. Assuming the cost of unit length fiber and each splitter be C and $10 \times C$ respectively, Figure 5 obtained from the algorithm in Figure 2 demonstrates the minimum cost achieved by using 257 splitters, supporting 4772 ONUs. The corresponding total fiber and splitter cost is $\sim 4704C$ units, using $2,134.26$ Km fiber length.

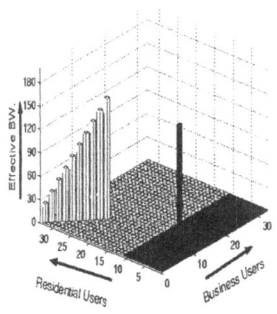 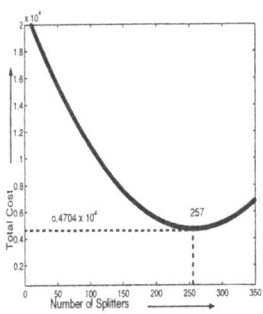 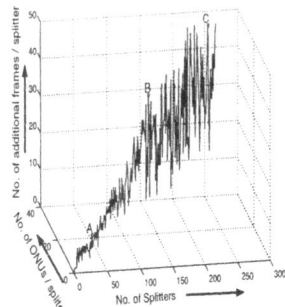

Fig. 4. Effective Bandwidth

Fig. 5. Total Cost

Fig. 6. Additional Frames Required for Ranging

We calculate the time in terms of the number of frames required to obtain the equalization delay (T_δ) for each ONU. Figure 6 shows the plot for the number of additional frames (*Serial_Number_Mask* and *Reply*) required to range all ONUs for every splitter. Point 'A' indicates the 21*st* splitter supporting only 7 ONUs, requires 0 additional frames. 'B' indicates the 153*rd* splitter supporting 21 ONUs requires 24 additional frames while 'C' indicates the 245*th* splitter supporting the maximum number of 32 ONUs requires 50 additional frames to range all ONUs. We observe that with optimal splitter-location, roundtrip propagation time for all ONUs under a given splitter belong to a concentrated time interval, leading to repeated collisions while being ranged. Hence the total ranging duration adds a considerable amount of overhead time before actual data transmission takes place and it is more prominent when ONUs are equidistant from the OLT.

7   Conclusion

In this paper we have investigated into the research issues behind the deployment of PON, an attractive solution for the last mile problem. A traffic model for residential and business users is developed to obtain an estimate of effective bandwidth consumption. Subsequently, a fair combination of these two classes of users is obtained to maximize the effective bandwidth utilization. The location of splitters and OLT are determined by a new VQ-based region forming algorithm, which aims at minimizing the amount and cost of fiber layout. In order to avoid collision and synchronize the ONUs during upstream transmission, a ranging scheme is also discussed. Our research results indicate that although the ranging is capable of avoiding collisions, ranging time increases appreciably when ONUs are almost equidistant from the OLT. Our future work lies in the improvement of the ranging scheme to reduce the overhead time.

References

1. A. Erramilli, O. Narayan, and W. Willinger, "Experimental Queueing Analysis with Long-Range Dependent Traffic", *IEEE/ACM Trans. on Networking*, pp. 209–223, April 1996.
2. F. Kelly, "Notes on Effective Bandwidths", *Stochastic Networks: Theory and Applications*, Oxford Univ. Press, 1996, pp. 141-168.
3. F. J. Effenberger, H. Ichibangase and H. Yamashita, "Advances in Broadband Passive Optical Networking Technologies", *IEEE Comm. Magazine*, Vol. 39, Issue 12, Dec. 2001, pp. 118-124.
4. "Broadband optical access systems based on Passive Optical Networks (PON)", ITU-T, G.983.1.
5. G. Pesavento and G. Kelsey, "PONs for Broadband local loop", *Lightwave*, Vol. 16, No. 10, September. 1999, pp. 68-74.
6. G. Kramer, B. Mukherjee and G. Pesavento, "IPACT: A Dynamic Protocol for an Ethernet PON (EPON)", *IEEE Communications Magazine*, Vol. 40, Issue 2, February 2002, pp. 74-80.
7. G. Kramer and G. Pesavento, "Ethernet Passive Optical Network (EPON): Building a Next-Generation Optical Access Network", *IEEE Communications Magazine*, Vol. 40, Issue 2, February 2002, pp. 66-73.
8. G. Kramer and B. Mukherjee, "Design and Analysis of an Access Network based on PON Technology", *UCD Student Workshop on Computing*, September 2000.
9. H. Ueda et-al, *"Deployment Status and Common Technical Specifications for a B-PON System"*, *IEEE Communications Magazine*, Vol. 39, Issue 12, December 2001, pp. 134-141.
10. IEEE 802.3ah Ethernet in the First Mile Task Force, "http://ieee802.org/3/efm/baseline/p2mpbaseline.html".
11. K. Sayood, *"Introduction to Data Compression"*, *Morgan Kaufmann Publishers*, 1996.
12. P. Shah, "A Game Theoretic Approach for Dynamic Bandwidth Assignment in Ethernet Passive Optical Network(EPON)", Masters Thesis, Univ. of Texas at Arlington, (expected) June 2004.
13. *"Speech Coding with Fixed and Variable Bit Rate"*, ITU-T specifications, G.729.
14. T. H. Cormen, C. E. Leiserson, R. L. Rivest, *"Introduction to Algorithms"*, Prentice-Hall Publishers, 2001.
15. T. M. Cover and J. A. Thomas, *"Elements of Information Theory"*, J. Wiley & Sons Publishers.
16. V. Paxcon and S. Flyod, "Wide-Area Traffic: The Failure of Poisson Modeling", *IEEE/ACM Transactions on Networking*, Vol. 3, Issue 3, June 1995, pp. 226-244.

IPv6 Deployment Support Using an IPv6 Transitioning Architecture – The Site Transitioning Architecture (STA)

Michael Mackay and Christopher Edwards

Computing Department, Lancaster University, Bailrigg, Lancaster, LA1 4YW, UK
{m.mackay,ce}@comp.lancs.ac.uk

Abstract. With IPv6 being increasingly regarded as 'ready' for deployment, developers are now addressing the operational issues related to IPv6. One key aspect of this is IPv6 transitioning and more specifically it's management and use within networks. This paper puts forward an IPv6 transitioning architecture that offers a managed approach to the deployment and usage of transitioning mechanisms within medium-to-large scale networks. The aim of this is to provide a simple, robust and secure architecture for IPv6 deployment regardless of the type of network used.

1 Introduction

As IPv6 nears readiness, there are still certain aspects of it that require further development to avoid delaying its uptake within the wider community. One such aspect is IPv6 transitioning and more specifically transitioning deployment and management architectures. It has been found that while the necessary functionality to support IPv6 deployment exists, there is still a need for mechanisms to administer and manage the transitioning infrastructure and this is the key purpose of transitioning architectures. This obviously implies both an operational and network management aspect to the architecture, the later of which probably handled via existing IPv6 network management methods. This paper will put forward and present one such transitioning management architecture, the STA [1], and describe its design and possible usage within large network environments.

The STA is essentially an administrative framework designed specifically to support the IPv6 migration process, unifying the sites' transitioning functionality within a single entity. This will provide a highly manageable transitioning environment for the duration of the IPv6 deployment process in addition to introducing extra functionality and customisability that would be far harder to implement otherwise. The STA provides an open framework into which any combination of suitable transitioning mechanisms can be integrated and where their performance and functionality can be customised from high-level site-wide permissions to low-level device control.

Section 2 gives a brief overview of the development of a transitioning architecture and its management and section 3 introduces the STA itself, describing its design. Section 4 will outline a number of STA deployment examples and finally, section 5 will discuss implementation.

J.N. de Souza et al. (Eds.): ICT 2004, LNCS 3124, pp. 1310–1316, 2004.

2 Problem Analysis

IPv6 Transitioning Architectures. At present, there has been little work done on developing a transitioning architecture, certainly within the larger bodies such as the IETF, however, at least one draft is now in circulation [2]. In it, there is an attempt to establish some design guidelines for a common transitioning approach and discuss how transitioning mechanisms fit into it given the current trends in IPv6 transitioning deployment scenarios [3]. When considering a transitioning architecture there are several key design principles that must be considered, as a basis it must for example be robust, secure and simple.

The architecture must be both robust and secure in order to cope with hostile network conditions and guard against misuse but also simple in terms of its deployment and operation. One other element that must be considered is that of service provision, this is important as the type and requirements of the traffic to be handled will affect quite radically the transitioning resources that must be deployed.

Of course, the actual transitioning infrastructure deployed will depend on a number of factors, both internal and external. Internal factors might include the 'starting point of the network (IPv4-only as today or Dual stack or IPv6 only as may be the case in the future), its size and transitioning requirements while external factors include the service available from its provider and developments related to IPv6 deployment within the Internet as a whole. Because it is expected to accommodate this variety of requirements implies a final, very important aspect, flexibility.

Transitioning Management. Since 1998, IPv6 has been modelled within MIB II and therefore supported by SNMP (though this is still in progress [4]) and recently IPv6 transport support was added to certain SNMP implementations [5]. However, while SNMP is the *de facto* protocol for fault management and monitoring, various protocols are used for other network management tasks. Configuration management for example, uses protocols like SNMPConf, while Authentication, Authorisation and Accounting (AAA) management might use DAIMETER, RADIUS or Kerberos V. However, to some extent most management protocols now support IPv6 [6].

Developing support for transitioning mechanisms in protocols such as SNMP depends on two factors, its availability across a transitioning network i.e. support for both IPv4 and IPv6 which has now been overcome, and a system for modelling transitioning management information within a common structure such as the MIB. Here we focus on the later as it has yet to be effectively addressed. In order to do this, it is important to establish what should be modelled within the MIB and how.

When considering the information to be held within the MIB, it will obviously be important to hold information on all transitioning mechanisms supported or in use within the deployment community. A discussion of the exact information modelled within the transitioning MIB is obviously beyond the scope of this paper however it is reasonable to expect configuration and performance information for each mechanism.

The second aim is to establish a structure for a transitioning MIB and in order to do this, transitioning elements can be placed within three basic groups based roughly on their operation; tunnelling, translator and dual stack as is shown in Fig 1. Using this model, mechanism-specific information is held in their respective tables with another 'general' table to handle non-specific transitioning information.

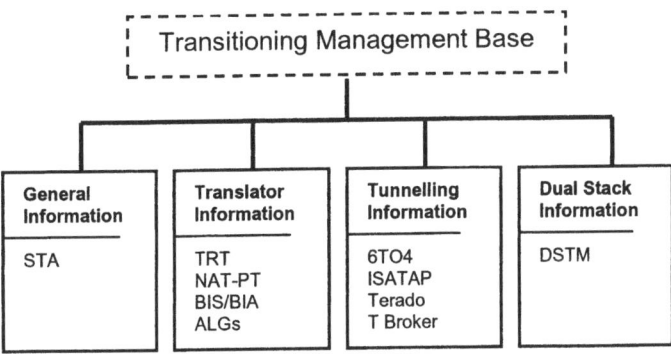

Fig. 1. Provisional grouping for a transitioning MIB

3 Introducing the STA

Having outlined the basic aspects of a transitioning architecture we now move on to introduce our solution, the STA. The primary objective of the STA is to simplify the deployment of and migration to IPv6 during the transitioning process. By deploying a flexible service-oriented framework within which transitioning can take place, the STA can simplify the process of IPv6 deployment in a network over the course of the transitioning process from initial IPv6 deployment to a prolonged co-existence period and beyond. The key objectives of the STA are: to provide a unified framework suitable for the whole transitioning period, to exhibit simplified management properties and to introduce a service-based framework to the transitioning process.

3.1 Design

While the actual functionality, and therefore management components, deployed within the STA differ according to the transitioning support required in an actual deployment, it is important to first identify the core components and topology of the architecture. The STA operates on two abstract planes, operational and management. The operational plane defines the actual transitioning configuration deployed within the site while the management plane dictates how they are controlled and operated. In terms of the operation plane, the framework consists of 3 key transitioning components; hosts, servers and border devices each offering specific operational functionality to the framework and describing the layout of transitioning mechanisms used within the site. The management plane utilises a client/server model that provides the administrator with remote access to the operational components deployed within the framework. The layout of the STA is shown in Fig 2.

3.2 Operational Plane

While the predefined components are sufficient to describe most transitioning infrastructures, the STA will also be capable of incorporating additional components as deemed necessary.

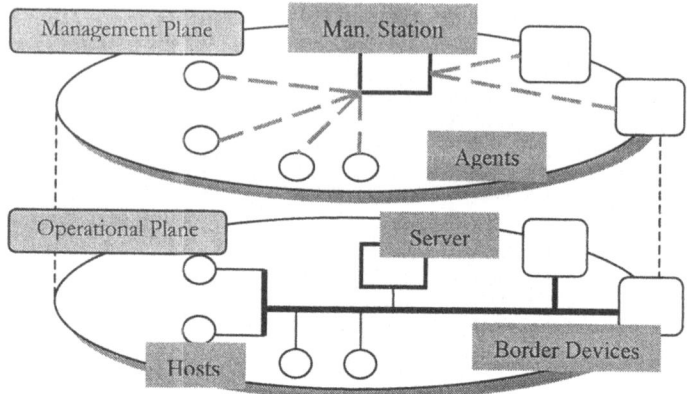

Fig. 2. Layout of the STA architecture

Border Devices. Border devices define devices located at the IP border of the network, and as such they have a major role in most transitioning mechanisms. In the STA, while the transitioning complexity is still in the network, i.e. in the border devices, the administrative complexity is removed and unified inside the network, making the border devices 'dumb'. This will also improve the scalability of these transitioning devices making the deployment of more complex architectures possible.

STA Server. The STA server will be the only persistent component of the architecture. The core of this component contains the actual STA server entity and the management station used to control the rest of the STA architecture.

Additionally, the STA server component may also include mechanisms which implement a server component such as the tunnel broker. Such server devices may be collocated with the actual STA server or distributed to improve scalability. This may include core network services such as DHCPv6 (when providing DSTM functionality for example).

Hosts. To reduce management overheads, hosts will not be specifically modelled in the STA unless their operation is specified as part of a mechanism, as in the case of DSTM for example. Where host management is necessary, a 'group policy' may be applied thereby relieving the issue significantly.

3.3 Management Plane

The management plane is a key feature of the STA as it allows administrators additional control over the operational plane. It will conform to an SNMP structure with agents on managed devices accessing the transitioning MIB. The management station allows the administrator to control the devices through their agents as shown in Fig 3.

In addition to SNMP, the STA management plane will also include configuration management capabilities to define transitioning policies defining how the architecture

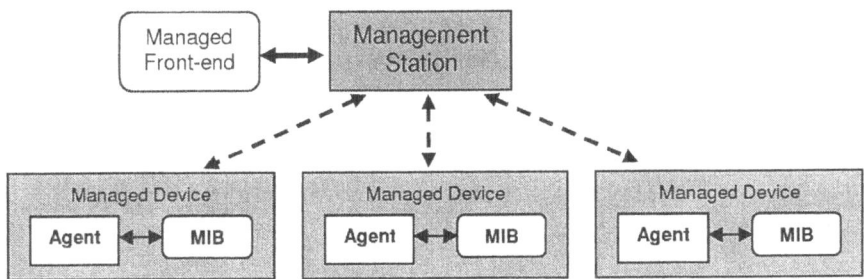

Fig. 3. Layout of the STA management framework

will behave according to certain profiles. This will be achieved via the SNMPConf protocol [7] that introduces policy MIBs and enforcement points to the base SNMP architecture. These policies will allow transitioning devices to offer a service-based infrastructure tailored to supporting QoS or end-to-end security for example.

4   STA Deployment Examples

While the STA design describes it's the core components, it is important to show how this can then be applied to suit transitioning scenarios. It must be stressed that in each case the STA does not change, rather it is applied to meet specfic transitioning requirements.

Case 1. Enterprise Deployment. Consider a University campus deployment where IPv6 deployment requires a mix of interoperation and tunnelling functionality. Internal tunnelling may be provided by ISATAP or, in the case where labs or residence networks are connected via NAT, Terado. Initial IPv6 deployments may also be based extensively on dual stack. Whilst any dual stack deployment inherently impacts all the devices in the 'dual stacked' path, it is realistic to consider this within the campus backbone where the impact will be more limited. The backbone can then support further IPv6 networks as they appear. Over time, tunnelling mechanisms will be removed as native IPv6 becomes available and interoperation mechanisms may be needed where IPv4-only and IPv6-only nodes have to interact. The mechanisms will also be required to scale accordingly over time.

Case 2. IPv6-only Deployment. This case considers a sizeable IPv6-only deployment within a large corporate network. The system administrators have decided that the upgraded infrastructure is entirely IPv6-only with external connectivity to the wider corporate network being via dual stack. In the absence of dual stack internally a significant transitioning deployment will be necessary to support IPv6 interoperation. In this case, a variety of mechanisms may be employed including protocol relays (TRT), IP translators (NAT-PT) or dual stack tunnelling mechanisms such as DSTM. The decision as to which tools to use will be made based on the size of the deployment and the functionality to be provided. In this case, the majority of the interoperation complexity will be deployed at the subnet boundary and will move outwards and be extended over time as the rest of the corporate network deploys IPv6.

Case 3. ISP Deployment. Consider an ISP that wishes to transition to IPv6 with the aim to provide IPv6 connectivity, eventually natively, to its customers. Such a deployment may involve upgrading both the core and access network and has the potential to be both costly and time consuming. Therefore, initial IPv6 connectivity may be provided by configured tunnels or via a tunnel broker. Native IPv6 connectivity can then be rolled-out either by upgrading the existing components or deploying a new IPv6-only infrastructure alongside the existing network. In addition to providing IPv6 connectivity, an ISP may choose to provide additional transitioning services to its customers in the form of interoperation. This would require a considerable deployment based on translators and protocol relays but as existing IPv4 resources are gradually exhausted, it is a service that may be increasingly utilised.

5 Implementation

An implementation of the STA is currently underway, based loosely on the interoperation example (case 2). DSTM and NAT-PT will be deployed, here, the aim being to demonstrate how the STA can provide an interoperation-specific managed infrastructure supporting IPv6-only operation. In this case NAT-PT will be used as the common interoperation mechanism with DSTM providing limited 'extended' functionality via dual stack on a restricted basis. The implementation is based on a Linux platform using existing implementations. The implementation was carried out in parallel transitioning and management 'threads', discussed below.

Transitioning Development. The basic modifications necessary to update transitioning mechanisms for use in the STA are limited to implementing an interface through which it will communicate with the STA server (which should be part of the configuration procedure). In this case however, we have decided to further modify the NAT-PT mechanism to share address-allocation functionality with DSTM via the server. This optimises IPv4 address usage and improved manageability giving a number of DSTM-enabled hosts, a central server, and a number of DSTM TEP and NAT-PT border devices. Such a configuration also exhibits improved scalability characteristics. The implementation uses the ENST DSTM and the KAME NAT-PT mechanisms that have been modified to support STA and DSTM/NAT-PT integration. Finally, it is necessary to implement the STA server and control device.

Management Development. Implementation of the management plane has been based on the Net-SNMP project which is one of the implementations to support IPv6 transport. This provides us with an extensible SNMPv3 agent and a collection of applications for sending SNMP commands which have been integrated to create a management station. One missing piece of this is the actual transitioning MIB that has been developed from scratch and incorporates NAT-PT and DSTM support based on the mechanisms used. This is because one has yet to be formally discussed or developed and so will also be used to stimulate progress in this area.

6 Conclusions

It is positive to note that the management infrastructure for IPv6 deployment is for the most part either in place or under development. The transitioning management aspects however are still in need of significant development. This paper puts forward a design for an IPv6 transitioning architecture, describes its design and operation and shows how it can be used as part of a managed network deployment. This paper outlines the design of the STA, a transitioning management architecture capable of supporting IPv6 transitioning deployments in a site throughout the migration process.

Acknowledgments. This paper was written with support from the 6NET project, a Pan-European IPv6 research project.

References

[1] M. Mackay, C Edwards, "IPv6 migration implications for Network Management – Introducing the Site Transitioning Framework (STF)", IEEE IPOM2003 workshop, October 2003

[2] P. Savola, "A View on IPv6 Transition Architecture", draft-savola-v6ops-transarch-03.txt, January 2004, work in progress

[3] M. Mackay, C. Edwards, M. Dunmore, T. Chown, G. Carvalho, "A Scenario Based Review of IPv6 Transition Tools", IEEE Internet Computing Special Edition on IPv6, June 2003

[4] Shawn A. Routhier (Ed), "Management Information Base for the Internet Protocol (IP)", draft-ietf-ipv6-rfc2011-update-07.txt, February 2004, work in progress

[5] Net-SNMP home page, http://www.net-snmp.org

[6] Isabelle Astic, Olivier Festor, "Current Status of IPv6 Management", http://www.inria.fr/rrrt/rt-0274.html, December 2002

[7] M. MacFaden, D. Partain, J. Saperia, W. Tackabury, "Configuring Networks and Devices with Simple Network Management Protocol (SNMP)", April 2003, RFC 3512

An Intelligent Network Simulation Platform Embedded with Multi-agents Systems for Next Generation Internet

Miguel Franklin de Castro[1], Hugues Lecarpentier[2],
Leïla Merghem[2], and Dominique Gaïti[3]

[1] GET/Institut National des Télécommunications
9 rue Charles Fourier - 91011 Evry Cedex, France
miguel.castro@int-evry.fr
[2] LIP6 - Université de Paris 6
8 rue du Capitaine Scott - 75015 Paris, France
{hugues.lecarpentier;leila.merghem}@lip6.fr
[3] LM2S - Université de Technologie de Troyes
12 rue Marie Curie - 10010 Troyes Cedex, France
dominique.gaiti@utt.fr

Abstract. Next Generation Networks (NGN) impose new challenges to network management. In particular, QoS management presents a big complexity to an efficient network operation, requiring real-time decisions to keep QoS commitments. In order to solve these problems, there is a growing interest in including intelligent mechanisms inside the network, enabling adaptive behavioral management. This article presents *Corail-Sim*, a new network simulation platform enabled with Multi-Agent Systems (MAS) to develop intelligent management services to NGN. This new tool is intended to design, model and test intelligent behaviors over a number of technologies such as DiffServ, IntServ, MPLS, Wireless LAN, etc.

1 Introduction

Several different network infrastructures of nowadays are running towards convergence into a single IP-based infrastructure. With this fusion, different services will coexist and compete for the same – sometimes scarce – resources. Furthermore, mixing services inherited from circuit switching networks with packet-based services changes abruptly the dynamic of these networks, turning network modeling task still more complicated.

In this new environment, network management, which had already a crucial role in traditional networks, becomes still more important. Management systems must now deal with several types of media, each one potentially with its own priorities and requirements.

This adaptation to this new reality on management goes sometimes beyond the simple *on-the-fly* monitoring and control, and must be extended to the early phases of network deployment, also adapting the way networks are designed,

J.N. de Souza et al. (Eds.): ICT 2004, LNCS 3124, pp. 1317–1326, 2004.
© Springer-Verlag Berlin Heidelberg 2004

modeled and tested. In this early phase of network deployment, network simulation seems to be the more practical solution. However, traditional network simulation does not address the inclusion of artificial intelligence mechanisms as management helpers inside the network. The lack of such capabilities in traditional network simulation tools argues against development of intelligent management services. In network simulators such as NS-2 [1] and OPNET^{TM} [2], the deployment of intelligent management services requires a deep comprehension of specific model libraries, and intelligent components (such as Multi-Agent Systems) must be implemented from the earliest stage.

The objective of this article is to propose *Corail-Sim*, an intelligent network simulation platform for new generation of Internet. This tool employs Multi-Agent Systems (MAS) as intelligent component to help improving network management, especially for QoS (Quality of Service) support.

This article is organized as follows. Section 2 outlines some existent network simulation tools, evaluating the required capabilities to build up intelligent and adaptive management. In Section 3, we present the Multi-Agent Systems capabilities to be built into Corail-Sim. A case study where adaptive management based on MAS improves network performance. Finally, Section 6 presents our conclusions for this work, and our outlooks.

2   Network Simulation

Network simulation is a very important tool to network development. Testing new services, protocols and algorithms can be performed with low costs and high flexibility. There are several options of network simulator tools. They are characterized by – among others – programming language, graphical interface, performance issues, model library, portability and extension flexibility.

In particular, extension flexibility is a very important feature for modeling new generation of Internet. As new intelligent services are being widely deployed on Internet, network modeling and simulation tools must follow this trend by offering new facilities to test such services as well as it is desirable to embed intelligent systems inside the network simulation core and model library.

The following subsections outline some popular network management tools and briefly show how complex would be to model and test intelligent management services with these tools.

2.1   Network Simulator (NS-2)/VINT

NS is a discrete event simulator targeted at networking research. NS provides substantial support for simulation of TCP, routing, and multicast protocols over wired and wireless (local and satellite) networks [1].

The simulator is written in C++, and it uses OTcl as a command and configuration interface. NS v2 has three substantial changes from NS v1: (1) come of the complex objects in NS v1 have been decomposed into simpler components for greater flexibility and composability; (2) the configuration interface is now

OTcl, an object oriented version of Tcl; and (3) the interface code to the OTcl interpreter is separate from the main simulator [1].

We have seen lately a growing interest of the research community in this tool. Its open source license is the most attractive characteristic for researchers that want to test new protocols and services. However, there is no existent support for the inclusion of intelligent algorithms like Multi-Agent Systems inside the simulation core and its model library. The implementation of such feature would require a deep knowledge of its existent model library as well as it would need to make deep changes in its source code in order to accommodate such new potentials.

Moreover, there are still some characteristics that have not yet been well developed, like graphical user interface (GUI). Modeling a network topology with NS-2 requires OTcl script programming, while it would be desirable to have a friendly interface to facilitate modeling task. Still, there are still limitations about OS this tool can run over.

2.2   J-Sim

J-Sim is a component-based simulation environment, built upon the notion of the autonomous component architecture and developed entirely in Java [3]. The behavior of J-Sim components are defined in terms of contracts and can be individually designed, implemented, tested, and incrementally deployed in a software system. A system can be composed of individual components like a hardware module is composed of IC chips. Moreover, components can be plugged into a software system, even during execution. This makes J-Sim a truly platform-neutral, extensible, and reusable environment.

This component-based architecture makes it possible to compose different simulation scenarios – for different network architectures – from a set of basic components and classes proposed by J-Sim and/or defined by the user.

INET Node Structure

On the top of the component-based architecture, J-Sim proposes an implementation of an abstract network model "INET". The internal node structure (router or host) is a two layers structure: the Upper Protocol Layer "UPL" and the Core Service Layer "CSL" (see Figure 1(a)). The UPL contains transport, signaling, and application protocol modules. The CSL encapsulate the network, the link and the physical layers and provides a set of well-defined services to modules of the UPL.

The Core Service Layer

As we have seen in the previous section, the CSL provides many services to UPL modules: packet forwarding/delivery services, identity services, routing table services, interfaces/neighbor service, and packet filter configuration services.

These services are defined in terms of contracts. A protocol module that uses the core services may own a port that is bounded to the specific service contract. As shown in Figure 1(a), a typical INET node is composed of the core service layer and one or more application and transport layer protocols.

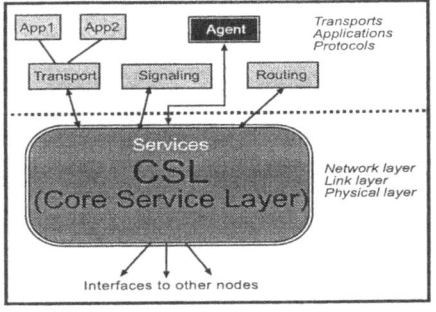
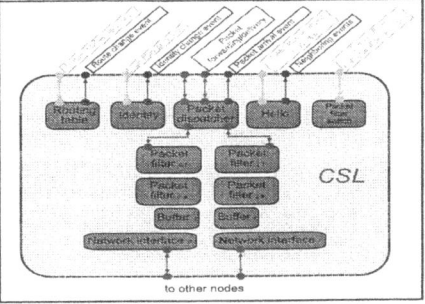

(a) INET internal node structure   (b) CSL decomposition

Fig. 1. J-Sim Structure.

As shown in Figure 1(b), the CSL is composed of many components which
provide services to the UPL module and forward events. The *Identity* component
keeps the identities of the node and provides the identity services. The *Routing
Table* component maintains the routing table entries and provides the routing
services. The *Packet Dispatcher* component provides the data sending/delivery
services. It forwards incoming packets to an interface according to the routing
table. The *Hello* component keeps the interface and neighbor information of
the node and provides the interface/neighbor services. The Packet Filter Switch
component enables UPL module to switch between different packet filters.

We have presented in this section two network simulators. We have chosen
to use the J-Sim tool for designing Corail-Sim. We propose to introduce multi-
agent system in the INET network model to ensure a global and generic control.
Multi-agent systems are presented in the next section.

3   Multi-agent Approach

A multi-agent system is composed of a set of agents which solve problems that
are beyond their individual capabilities. An agent is able of acting in its environment. It is able to communicate directly with other agents, possesses its own
resources, perceives its environment (but to a limited extent), has only a partial
representation of its environment (and perhaps none at all), and has a behavior
which tends towards satisfying its objectives – taking account of the resources
and skills available to it and depending on its perception –, its representation
and the communications it receives [4].

Multi-agent systems are well suited to control distributed systems. Telecommunication networks are good examples of such distributed systems. That is
why there is a considerable contribution to introduce agents in this area. The
aim was to resolve a particular problem or a set of problems in networks like:
the discovery of topology in dynamic networks by mobile agents [5], the opti-

mization of routing process in a satellite constellation [6], the fault location by ant agents [7], and even the maximization of channel assignment in a cellular network [8], etc. However, there is no contribution that uses agents to handle all the problems at once. That is, agents are in charge of some network's control but there must still have a human assistance to deal with the other aspects of network management.

In this section, we have presented multi-agent systems and their applications in telecommunication network. The agent architecture we use in Corail-Sim is introduced in the next section.

4 Architecture

Many software agents architectures are described in the literature. Our goal is to provide a network control dedicated architecture that can be use in many case. Usually, agent architectures are classified in four groups: Deliberative, Reactive, Hybrid and Layered architectures. Deliberative architectures are designed with an explicit word representation. They use symbolic reasoning. They are also called cognitive architecture. Reactive architectures are the simplest ones, based on a perception/action model with a rule base. Hybrid architectures represent a combination of deliberative and reactive architectures. They are the most flexible architectures. Finally, Layered architectures are based on a hierarchic organization.

As in many time dependent domains, the most interesting architectures for the Corail-sim design are hybrid architectures. These architectures allow agent to briefly react in situation that requires quick decision-making process. However, they allow agent to react in a reasoned manner if situations do not requires a quick decision-making process, but require an efficient decision.

In the following subsections, a generic agent structure for telecommunication network inspired from [9] is presented. We first detail the way it controls the node and collects events in Corail-Sim, and then agent components are presented. In addition, we describe an illustration of architecture capabilities.

4.1 Controlling the Node

As shown on Figure 1(a), our agents are components that work as protocols to control the node. They use services provided by the CSL and so can affect routing, filter, and identity. One particular agent must be plugged to services it needs to use and events it needs to know. Agents also use the data sending/delivery service to communicate with other agents. They use agent@node style addresses to send messages to other agents in the network.

4.2 Internal Structure of Corail-Sim Agent

The chosen Corail-Sim agent structure is composed of five main modules: the event monitor, the house keeping, the message interface, the task interface, and the manager (see Figure 2). All the modules are modeled as a J-sim component and then communicate through ports.

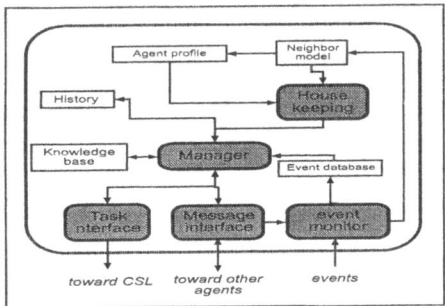

Fig. 2. Corail-Sim Agent Structure.

The Event Monitor is responsible of internal and external events detection and management. The Monitor gets all the events from the CSL: incoming packet, packet drop, routing table modification, identities modification, packet filters modification, buffer overflow, interface failure, etc. The Monitor also gets external events like alert messages from administrator and messages from other agents. This component classifies events in a database, which is used by the Manager. It also maintains the neighbor model and the agent profile. The agent profile is a set of attributes such as name, type, role, capacities, and goals of the agent. These attributes are initialized during agent design and are updated later by the event monitor. They guide the agent behavior during tasks planning, during decision-making, or during interaction with the agent neighborhood.

The House Keeping module fuses the events, task derivatives, agent profile, and neighbor model to provide the manager a consistent set of information. It 'keeps the house' avoiding redundant and useless information. This makes the manager temporal and situational aware. This component is primordial to ensure an optimum software agent working. House keeping work is very important: too many information will make the decision making hard, too few information will provide less efficient decisions.

The Message Interface allows sending and receiving messages. These messages are exchanged between software agents. Agents can use messages from the other agents to update their world representation, but can also alert, inform, or request others. Exchanged messages use the Agent Communication Language from FIPA (Foundation for Physical and Intelligent Agents) [10]. The FIPA ACL language is a standard for messages and provides coding and semantic of the messages. This component accomplishes control operations determined by the manager. In fact, it acts directly on the CSL layer components and can control routing table, packet filters, and identities in this way. This control is executed via services provided by the CSL layer.

The Manager takes decisions and plans agent actions. It uses a knowledge base for. This knowledge base is composed of rules, priorities, capacities, and all useful information for decision making and plans production. The Manager can in this way decide how and when to execute a task. It also manages a history where it stores observations and activated behaviors. It also uses an event database to verify his actions efficiency and to know node potential reactions.

At the creation of an agent, the designer supplies the agent with its profile, its model of neighboring agents, and a set of task directives. The agent profile is a set of attributes such type, role, behavior, etc. These attributes are used in guiding the behavior of the agent in task planning, decision-making, and interactions with neighboring agents. The neighbor model includes a list of agents known to the agent. The task directives is a list which describe the responsibilities

The event monitor module interfaces with the node environments and logs events in the event database. The house keeping module fuses the events, tasks directives, agent profile, and the neighbor model together to provide the Manager module a modified but consistent set of information, making the Manager temporal and situation aware.

The message interface can send and receive messages. The task interface performs the task. The Manager module is responsible for the decision making and planning of the agent, according to its knowledge base.

In this section, we have presented the Corail-Sim agent architecture. We next present simulations that have been made with another tool: oRis [11]. These simulations have been made before the Corail-Sim design, but Corail-Sim follows the same concepts and processes than oRis. First Corail-Sim tests have been done to ensure that oRis and Corail-Sim simulation model and simulation results are always coherent.

5   Case Study

In this section, we will demonstrate the importance of agents in the adaptation of the node management mechanisms. In fact, we argue that by using simple rules and by observing the appropriate parameters, the agents can optimize the node performance by finding a good compromise between the different Quality of Service (QoS) parameters (delay, jitter, loss).

In order to illustrate these ideas, we performed a multi-agent simulation of a DiffServ network. The simulator we chose is oRis [11], a generic multi-agent simulator which is not dedicated to network simulation. That is, we implemented all the structures concerning a network. We also defined and implemented the agents responsible for the network control. The Corail-sim and oRis tools have been validated to ensure coherent results. Both we use same simulation models and concept.

The network topology used in our simulations is composed of six routers each of which has four clients (sources). Each router has a queue of 100 places. A packet is generated by a client and has another client as a destination, which is chosen randomly. The packets generation is in mode ON/OFF following a poissonian process, while the size of each flow of packets follows an exponential process. The parameters of the two processes (generation and size) depend on the class of the generated packets. For each router, one client sends only Best Effort packets, the second one only Premium packets, while the two last ones send only Olympic packets. We simulate the proportion of the different classes of packets as follows: 25% of Best Effort packets, 50% of Olympic packets and 25% of Premium packets. These proportions are not realistic but the aim of such parameters is to show that good performance are guaranteed even when the proportion of prioritized packets (Premium and Olympic) is very high.

In the following, we give some simulation results showing the contribution of agents when handling adaptive scheduling. Results concerning other tasks like routing and queue management, for example, can be found in [12].

5.1   Adaptive Scheduling Case Study

In this example, the management mechanism we are interested in is the scheduling and we aim to demonstrate the performance improvement resulting of the adaptive scheduling. Simulations begin with the activation of Round Robin and each class of traffic has its own queue managed by FIFO mechanism. The size of each queue is as follows: Premium: 25 packets, Olympic: 50 packets and Best Effort: 25 packets. Two scenarios are compared. In the first one, there is no adaptation of the scheduling mechanisms, while in the second one the agent owns rules (Scheduling_Rule1, Scheduling_Rule2) able to realize such an adaptation.

```
Scheduling_Rule 1:
  if (Premium loss is >0.0001% or
        Olympic Loss is >3%) then
    deactivate Round Robin
    activate Priority Queuing
Scheduling_Rule 2:
  if load on queue Q =0 since T''(ms) then
    distribute reserved resources from
                queue Q to the others
```

In Figure 3(b), we can see that when the rules are applied, performance is improved. Indeed, Olympic packets loss becomes almost insignificant from the moment of the rules execution. The same improvement is observed with Olympic delay (Figure 3(d)) which decreases in a very important fashion with the application of these two rules. This proves that by carefully choosing the rules, one can optimize many parameters at the same time (Scheduling_Rule1 observes the loss and by executing it, we improved, in addition to loss, delay and jitter).

5.2   Discussion

We have given an example that shows the efficiency of agents and their ability to guarantee better performance by adapting scheduling mechanisms to the current

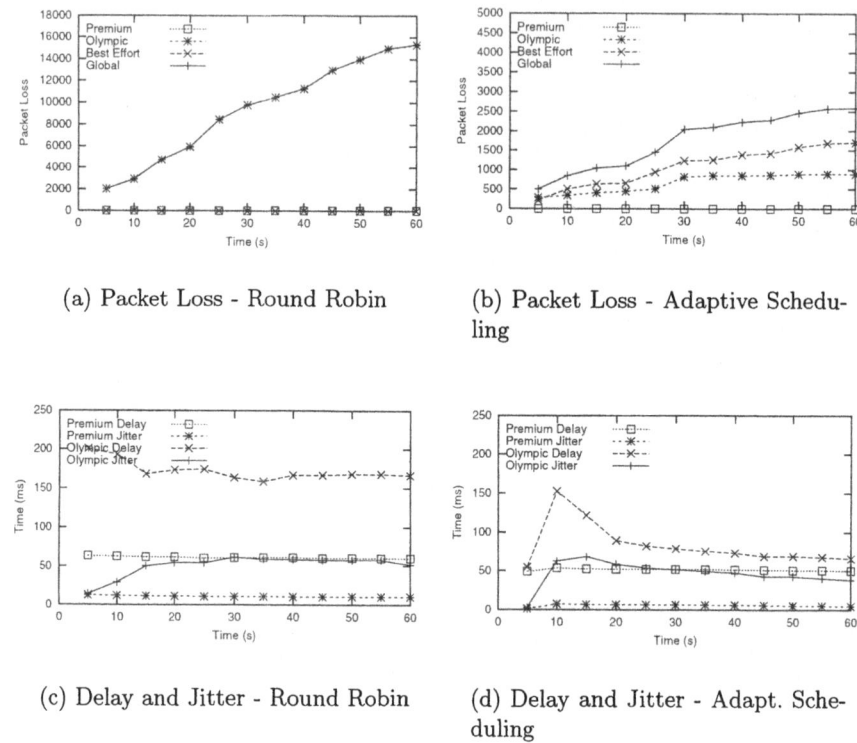

(a) Packet Loss - Round Robin

(b) Packet Loss - Adaptive Scheduling

(c) Delay and Jitter - Round Robin

(d) Delay and Jitter - Adapt. Scheduling

Fig. 3. Packet loss and Delay in Round Robin and Adaptive Scheduling.

traffic conditions. This was possible thanks to the use of simple rules the agent owns and activates when event monitor notices that their triggering conditions are met. In the performed simulations, we have seen that adaptation is realized in order to give the best QoS to Premium packets and to Olympic ones.

In order to realize these simulations and other ones reported in [12], we were forced to implement all the components and functionalities of a DiffServ network, since oRis does not provide these functionalities. This is one of our main motivatations to choose another tool, J-Sim, which will avoid us usefulness effort. The work presented here will be largely reused in "Corail-Sim" because the principles and the aims are the same. Moreover, we have validated the simulation model on both the two platform. The generic Corail-sim model allows us to simply develop new kind of control in different type of network, that would require a difficult implementation in oRis.

6   Conclusions and Outlook

This article presented *Corail-Sim*, a new network modeling and simulation platform embedded with Multi-Agent Systems. This new tool enables the interaction between components from different technologies such as DiffServ, MPLS and wireless networks to communicate and be managed by an intelligent management system, based on MAS. The simulation model and agent architecture used in Corail-Sim have yet shown their efficiency in another simulation tool: oRis. This one offer good performances but Corail-Sim is more generic and many network concepts are already developed.

We have seen by the results shown in the case study that Multi-Agent Systems can bring good enhancements to network management, more specifically to QoS management, which requires intelligent decisions and reactions in real-time. Hence, network simulation must take into account the inclusion of intelligent capabilities into network management. However, we observe that the lack of support of intelligent mechanisms inside network simulation tools argues against the development of intelligent services, like adaptive QoS management. Hence, Corail-Sim represents the solution for developing and testing intelligent services based on MAS for next generation networks (NGN).

Acknowledgement. Miguel F. de Castro is supported by CAPES-Brasil.

References

1. NS-2: The Network Simulator (NS-2) Homepage. (2004) http://www.isi.edu/nsnam/ns/.
2. OPNET: OPNET Technologies Inc. Homepage. (2004) http://www.opnet.com.
3. J-Sim: DRCL J-Sim Homepage. (2004) http://www.j-sim.org.
4. Ferber, J.: Multi-Agent Systems: An Introduction to Distributed Artificial Intelligence. Addison Wesley Longman (1999)
5. Roychoudhuri, R., Bandyopadhyay, S., Paul, K.: Topology discovery ub ad-hoc wireless networks using mobile agents. In: Proc. of 2nd. Intl. Workshop MATA'2000, Paris, France (2000) 1–15
6. Sigel, E., Denby, B., Hégarat-Mascle, S.L.: Application of ant colony optimization to adaptive routing in LEO telecommunications satellite network. Annals of Telecommunications **57** (2002) 520–39
7. White, T., Pagurek, B.: Distributed fault location in networks using learning mobile agents. In: 2nd. Pacific Rim Intl. Workshop on Multi-Agents (PRIMA'99), Kyoto, Japan (1999) 182–96
8. Bodanese, E.L., Cuthbert, L.G.: A multi-agent channel allocation scheme for cellular mobile networks. In: Proc. of ICMAS'2000, Boston, USA (2000) 63–70
9. Tsatsoulis, C., Soh, L.: Intelligent Agents in Telecommunication Networks. In: Computational Intelligence in Telecommunications Networks. CRC Press (2000) 479–504
10. FIPA: FIPA Communicative Act Library Specification. (2004) http://www.fipa.org/specs/fipa00037/.
11. oRis: oRis Homepage. (2004) http://www.enib.fr/~harrouet/oris.html.
12. Merghem, L.: Une Approche Comportementale pour la Modélisation et la Simulation des Réseaux de Télécommunications. PhD thesis, Université de Paris 6 (2003)

Reduced-State SARSA with Channel Reassignment for Dynamic Channel Allocation in Cellular Mobile Networks

Nimrod Lilith and Kutluyıl Doğançay

School of Electrical and Information Engineering
University of South Australia
Mawson Lakes, Australia
Nimrod.Lilith@postgrads.unisa.edu.au,
Kutluyil.Dogancay@unisa.edu.au

Abstract. This paper proposes a novel solution to the dynamic channel allocation problem in cellular telecommunication networks featuring user mobility and call handoffs. We investigate the performance of a number of reinforcement learning algorithms including Q-learning and SARSA, and show via simulations that a reduced-state version of SARSA incorporating a limited channel reassignment mechanism provides superior performance in terms of new call and handoff blocking probability and a significant reduction in memory requirements.

1 Introduction

Cellular systems organise a geographical area into a number of regularly sized cells, each with its own base station. The available bandwidth is divided into a number of channels, which may be time slots or frequencies, each of which may be assigned to a call. Using a cellular system allows a given channel to be assigned simultaneously to multiple calls, as long as each assigning cell is at least a given distance apart, in order to avoid co-channel interference. This distance is termed the 'reuse distance'. Most modern mobile communication systems use a Fixed Channel Assignment (FCA) strategy, whereby channels are pre-allocated to given cells according to a regular pattern that minimises the distance between co-channel cells, i.e. cells that may assign the same channel to a call, whilst not violating the channel reuse distance constraint.

In contrast to FCA, Dynamic Channel Assignment (DCA) strategies do not permanently pre-allocate given channels to particular cells. Instead channels are assigned to cells as they are required, as long as these assignments do not violate the channel reuse constraint. This flexibility in channel assignment allows a cellular system to take advantage of possible stochastic variations in offered call traffic over a given area. A number of DCA schemes have been devised [5] which follow certain procedures in order to attempt to maximise the total traffic carried in a cellular area. This paper proposes a new state reduction and limited channel reassignment mechanism to improve solution performance and reduce memory requirements, and examines the performance of the reduced-state reinforcement learning solutions to

J.N. de Souza et al. (Eds.): ICT 2004, LNCS 3124, pp. 1327–1336, 2004.

DCA in mobile cellular networks with handoffs. The proposed method of state reduction has been motivated by the weak link between one of the states and the blocking probability performance. The rest of the paper is organised as follows. Firstly, a brief introduction to reinforcement learning techniques is included in section 2. Section 3 details the specifics of the problem formulation employed, followed by simulation methods and results in section 4. Lastly conclusions are drawn in section 5.

2 Reinforcement Learning

2.1 Introduction to Reinforcement Learning

Assume $X = \{x_1, x_2, ..., x_N\}$ is the set of possible states an environment may be in, and $A = \{a_1, a_2, ..., a_N\}$ is the set of possible actions a learning agent may take. The learning agent attempts to find an optimal policy $\pi^*(x) \in A$ for all x which maximizes the total expected discounted reward over time. The action-value function for a given policy π, which is defined as the expected reward of taking action a in state x at time t and then following policy π thereafter, may then be expressed as:

$$Q^\pi(x,a) = E\{r(x,a) + \sum_{k=1}^{\infty} \gamma^k r(x_{t+k}, \pi(x_{t+k})) \mid x_t = x, a_t = a\} \tag{1}$$

where $r_t = r(x_t, a)$ is the reward received at time t when taking action a in state x_t and γ is a discount factor, $0 \leq \gamma \leq 1$. This formulation allows modified policies to be evaluated in the search for an optimal policy π^*, which will have an associated optimal state-action value function:

$$Q^*(x,a) = \max_\pi Q^\pi(x,a), \forall x \in X \text{ and } a \in A, \tag{2}$$

The state-action values of all admissible state-action pairs may be either represented in a table form or approximated via function approximation architecture, such as an artificial neural network, to save memory [3,7,12]. These state-action values are then updated as the learning agent interacts with the environment and obtains sample rewards from taking certain actions in certain states. The update rule for the state-action values is:

$$Q_{t+1}(x,a) = \begin{cases} Q_t(x,a) + \alpha \Delta Q_t(x,a), & \text{if } x = x_t \text{ and } a = a_t \\ Q_t(x,a) & \text{otherwise} \end{cases} \tag{3}$$

where α is the learning rate, $0 < \alpha < 1$, and

$$\Delta Q_t = \{r_t + \gamma \max_a Q_t(x_{t+1}, a_{t+1})\} - Q_t(x,a) \tag{4}$$

for Watkins' Q-Learning algorithm [13]. If α is reduced to zero in a suitable way and each admissible state-action pair is encountered infinitely often, then $Q_t(x,a)$ converges to $Q^*(x,a)$ as $t \to \infty$ with probability 1 [13].

The Q-Learning algorithm described previously is an off-policy technique as it uses different policies to perform the functions of prediction and control. If a strictly greedy policy was used for both prediction and control from $t = 0$, then no guarantee could be made regarding convergence. On-policy reinforcement learning methods, such as SARSA [12], differ from off-policy methods, such as Q-Learning, in that the update rule uses the same policy for its estimate of the value of the next state-action pair as for its choice of action to take at time t, that is for prediction and control. SARSA converges to an optimal policy with probability 1 if all admissible state-action pairs are visited infinitely often and its policy converges to a greedy policy. This can be achieved for example by using an ε-greedy policy with $\varepsilon \to 0$ as $t \to \infty$ for both the estimation of the value of the next state-action pair and the decision of action to take at time t.

3 Problem Formulation and Proposed Solution

3.1 Problem Structure

We consider a 7-by-7 cellular array with a total of 70 channels available for assignment to calls as the environment in which the learning agent attempts to find an optimal channel allocation policy. This is the same system simulated in [7-9]. Denoting the number of cells by N, and the number of channels by M, the state at time t is defined as $x_t = (n,m)_t$ where $n \in \{1,2,...,N\}$ is the cell index of the event occurring at time t, and $m \in \{0,1,...,M\}$ is the number of channels already allocated in cell n at time t. This state formulation has been traditionally employed in reinforcement learning solutions to the DCA problem and is similar to that featured in [8] and [10], although our formulation defines m as the number of allocated channels, rather than the number of available channels as in those works. It was decided to define m as a direct measure of a given cell's currently allocated share of the total channels in the interference region centered on that cell which will enable extension of this work to incorporate call admission control.

Admissible actions are restricted to assigning an available channel or blocking a call if no channels are available for assignment, i.e. a call will always be accepted and a channel assigned to it if there is at least one channel available in the originating cell. Actions are defined as $a = h, h \in \{1,2,...,M\}$, the set of available channels that may be assigned without violating the channel reuse constraint. Thus a state-action pair consists of three components, with a total of $(M+1) \times N \times M$ ($71 \times 49 \times 70$) possible state-action pairs, which yields 243,530 distinct state-action pairs. In implementation this reduces to $M \times N \times M$ as a cell with 70 channels already allocated cannot accept a new call and thus action values for these states are not required for learning, resulting in 240,100 admissible state-action pairs.

The reward associated with action a_t in state x_t, $r(x_t, a_t)$, is defined as c, the total number of ongoing calls in the system immediately after action a_t is taken, an approach similar to that in [11]. Handoff calls are treated as new calls by the cell being entered, and as such any channel allocated is rewarded and incorporated in the learning procedure.

3.2 Problem Characteristics

An initial examination into the characteristics of this problem shows that an optimal or near-optimal solution will effectively take the form of preferred 'cell-channel pairings'. Moreover the optimality of these cell-channel pairings is not due to a pre-existing absolute allocation scheme of particular channels to particular cells, but is instead dependent on the location of a channel's allocation relative to its other current or likely future allocations. As the system continues to allocate channels to calls any successful learning solution will tend to favour certain channels for allocation in certain cells, and it is the speed with which these associations can be formed that will help determine its efficiency, especially in a dynamically changing environment.

The states that will be evaluated by our reinforcement learning algorithm consist of two components n and m, the cell index and number of channels already allocated respectively. Given that we are effectively using reinforcement learning to search for optimal cell-channel pairings it is conceivable that the number of channels currently allocated in cell n is of secondary importance in the search for an efficient solution. For example if the allocation of channel p in cell n at time t is optimal given there are already m channels allocated there, it is likely the allocation of channel p will be optimal or near optimal if there are $m+1$ channels allocated in cell n, but if this state has not been encountered yet, i.e. this is the first time cell n has had $m+1$ channels allocated, this potentially useful knowledge will not be used.

Another property of the state component m is that even though its theoretical range is 0-70 it is extremely unlikely that this entire range will be required given there exist at least 6 interfering cells for any cell. A value of 70 for m would equate to every available channel being allocated in a single cell! It is very likely that over a short period the maximum value of m encountered system-wide will be a fraction of its maximum possible value. Of course the probability of a maximal m value is non-zero and thus it cannot be assumed it will never occur, but given its low probability it is conceivable that it will occur with a relative frequency so minute that if it were not included for learning, i.e. no provision were made in the state-action value representation mechanism for an m value of 70, that it would not affect performance of the system significantly. This principle can also theoretically be extended not only for the case of a maximal m, but for all values of m unlikely to occur sufficiently frequently to affect performance.

A property of implementing a limited state value range for a state component using a table-based state-action value representation is that the size of the table required to hold all possible state-action values will be reduced. For large M and N values the size of the table required to store the value function could be prohibitively large. A large state space may also impair convergence towards an optimal policy.

Optimal cell-channel pairings may also be used upon call termination and handoff events in an attempt to improve the system-wide channel allocation pattern. Assuming that for a call arrival event in a non-empty system there exists an optimal or a number of equally optimal possible channel assignment(s), then for a call termination event there will also exist an optimal or a number of optimal channel release(s). If a call which is currently assigned an optimal channel is subject to a termination event, whilst a less optimal channel assignment still carries an ongoing call in that cell, then an intra-cell handoff or channel reassignment may lead to an improved channel allocation pattern. If desired, this reassignment may in turn provoke channel reassignments in neighbouring cells, which may then result in reassignments propagating further throughout the system. Such a scheme introduces issues such as increased computational complexity and signaling overhead. Channel reassignment techniques are not new [5], and may be extended to call setup events as well, whereby a number of channel reassignments may be undertaken in order to free channels in a cell's interference region with the aim of accepting a call that would otherwise be blocked.

3.3 Proposed Solution

Given the above memory issues in the implementation of our reinforcement learning solution we reduced the number of admissible state-action pairs using two different techniques. Firstly we reduced the number of possible state-action pairs by aggregating rarely encountered states. Specifically, the magnitude of the variable representing the number of channels currently allocated was limited to 30, rather than the original range of $0 - 70$, (i.e., roughly half the number of channels available). Simulations of this problem over a similar system have shown that over a 24 hour period a given cell will rarely, if ever, have more than a fraction of the total available channels allocated in it at any one time [6]. Consequently, the states associated with these configurations may rarely, if ever, be visited over this time period. By reducing the range for the number of channels allocated to 0-30, we are effectively trimming the size of the table required from 240,100 ($70 \times 49 \times 70$) to 106,330 ($31 \times 49 \times 70$), a reduction of over fifty percent in memory requirements. Secondly, we have deliberately excluded the state variable representing the number of channels currently allocated, m. This resulted in reducing the number of state-action variables to two and bringing the number of possible state-action pairs from 240,100 ($N \times M \times M$) to 3,430 ($N \times M$), a reduction of over 98%. An important advantage of a reduced-dimension state space for reinforcement learning is that it may lead to a faster learning or convergence rate because of the increased rate of exploration, despite discarding potentially useful information.

Both of these memory saving techniques were simulated and compared to not only a full-table-based reinforcement learning solution, but also to a fixed channel allocation algorithm and a random channel allocation algorithm.

A simple channel reassignment mechanism was also included, whereby upon encountering a termination or handoff event the reinforcement learning agent considered a single channel reassignment.

4 Simulation Methods and Results

4.1 Simulation Methods

New call arrivals were modeled as Poisson processes with a uniform distribution pattern, with mean call arrival rates λ between 100 to 200 calls/hour. New call durations obeyed an exponential distribution with a mean call duration $1/\mu$ of 3 minutes. New calls that were blocked were cleared. A proportion of calls (15%) were simulated as handoff traffic [9]. When a call was about to terminate a check was made to see if the termination was due to the call entering a new cell, i.e. if it fired a handoff event. The entered cell for handoff traffic was chosen randomly using a uniform distribution with all neighbouring cells as eligible. All handoff traffic was simulated according to an exponential distribution with a mean call duration $1/\mu$ of 1 minute in the new cell.

All simulations were initialised with no ongoing calls, and the algorithms were assessed according to both their new call blocking probability and handoff blocking probability.

Both Q-Learning and SARSA were implemented and simulated. The Q-Learning algorithm simulated employed a different reward scheme, classifying channels according to their usage patterns with respect to an individual cell. The simulated Q-Learning algorithm also used an alternate state representation, which was a combination of the cell index and the number of channels currently available in the cell. Channel assignment actions were selected using an ε-greedy algorithm, with ε being diminished over time, and state-action values were stored using a table-based representation. An additional two variations of SARSA were also simulated, a trimmed-table SARSA (TT-SARSA) and reduced-state SARSA (RS-SARSA), both incorporating table-based state-action value storage and the same ε-greedy reduction mechanism. Versions of the SARSA algorithm which included channel reassignment functionality were also simulated.

TT-SARSA uses an identical state-action formulation as the SARSA algorithm but limits the value of the state variable representing the number of channels currently allocated in a cell to thirty, reducing the number of admissible state-action pairs. RS-SARSA further reduces the number of state-action pairs by eliminating the state variable representing the number of channels currently allocated in a cell. Both of these techniques drastically reduce the memory required for table-based storage of the state-action value function, as discussed in Section 3.

In addition to Q-Learning and the three SARSA algorithms, a fixed channel allocation algorithm and a random channel allocation algorithm, where a randomly selected channel from the set of those available was assigned to an incoming call, were also simulated for the purposes of comparison. Events were scheduled according to a system clock and were implemented via a dynamic singly linked list.

4.2 Call Blocking Probabilities

Firstly the system was simulated with a uniform offered traffic load over all cells of 100 to 200 calls/per hour, corresponding to 5 to 10 Erlangs. All algorithms were

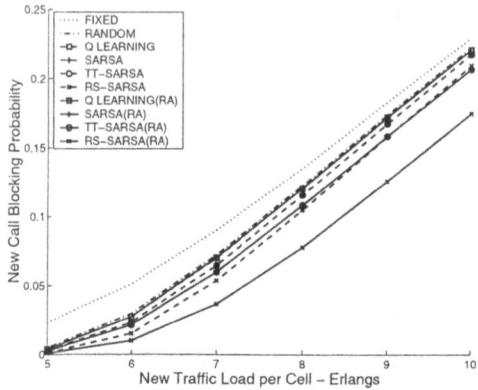

Fig. 1. New call blocking probabilities.

simulated in an online manner, i.e. there was no initial learning period where Q values could be learnt via simulation prior to implementation. After 24 simulated hours the respective new call and handoff blocking probabilities were calculated (Fig. 1 & 2). As can be expected, the fixed channel allocation algorithm performs the worst over the entire range of traffic loads, whilst all the reinforcement learning algorithms exhibit superior performance in terms of both blocking probabilities.

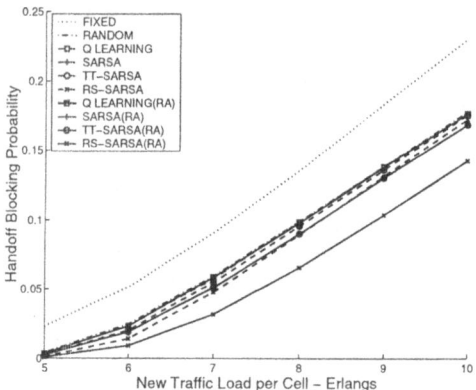

Fig. 2. Handoff blocking probabilities.

The channel reassignment functionality, denoted (RA), has a marked effect in decreasing the blocking probabilities of all of the SARSA algorithms, particularly RS-SARSA(RA) which has a much better new call and handoff blocking probability than all other assignment strategies simulated. Furthermore, TT-SARSA and TT-SARSA(RA) have identical performances to SARSA and SARSA(RA) respectively whilst requiring less than half the memory for state-action value representation, indicating that the aggregated states are rarely if ever visited and their aggregation has no effective bearing on the algorithm's performance. The fact that TT-SARSA can achieve the same levels of performance as SARSA using the full range of possible values of the state component m shows that the states that possess an m value of greater than 30 are not visited often enough that their aggregation negatively impacts call blocking performance. Lastly, not only does RS-SARSA require only a fraction of the memory of SARSA to represent its state-action values, it actually performs better in terms of both the other reinforcement learning algorithms' blocking probabilities. This may seem counterintuitive initially as it has access to less state information to learn from. Whilst true, as was pointed out in Section III, this reduction in total state space can lead to an improvement in learning performance.

4.3 Convergence

The convergences and long term blocking probabilities over time of all three SARSA algorithms featuring channel reassignment, as these were the three best performing algorithms, were simulated for a uniform call arrival rate of 180 calls/hour and the results are shown in Fig. 3. The learning rate parameter α was chosen so as to minimize the long-term blocking probabilities of all the simulated algorithms, enabling a fair comparison of convergence. Fig. 3 shows the blocking probabilities over a 24 hour period. SARSA(RA) and TT-SARSA(RA) have identical call blocking probabilities over the entirety of this period, whilst RS-SARSA(RA) consistently has a lower blocking probability over this period.

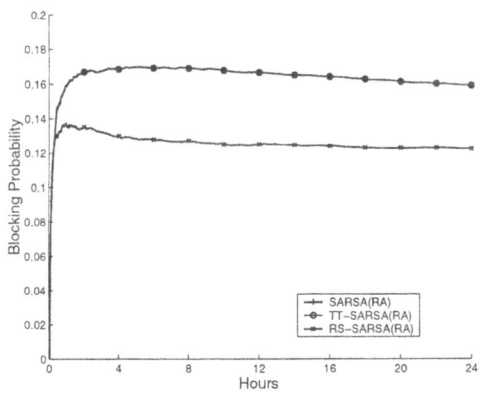

Fig. 3. Convergence of SARSA algorithms over 24 hours.

4.4 Channel Failures

Simulations were also carried out where a number of channels were unavailable for allocation for a given portion of simulation time. A fraction of the total channels for a cellular system may become unavailable for a period of time in the event of

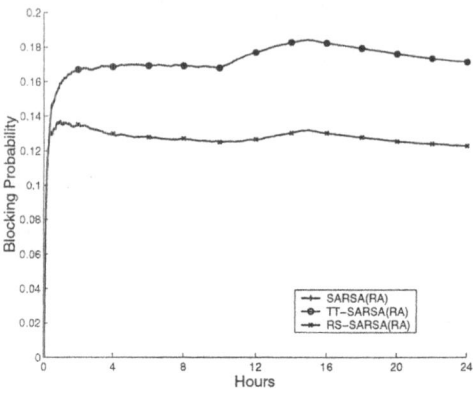

Fig. 4. New call blocking probability with 7 channel failures.

equipment failure, or even jamming in military communications systems. The three SARSA algorithms were tested over a period of 24 simulated hours with a uniform call arrival rate of 180 calls/hour. Between the hours of 10 o'clock and 15 o'clock a number of channels were made unavailable for use, as in [8]. Fig. 4 shows the blocking probabilities for SARSA(RA), TT-SARSA(RA), and RS-SARSA(RA) over a 24 hour period in which 7 channels become unavailable for allocation between the times of 10 o'clock and 15 o'clock. SARSA(RA) and TT-SARSA(RA) exhibit a similar performance when confronted with channel failure, while RS-SARSA(RA) displays both a greater resistance to the channel failures and a superior call blocking probability over the entirety of the simulation.

5 Conclusions

We have proposed a novel reduced state SARSA algorithm featuring channel reassignment for dynamic channel allocation in mobile cellular networks with handoffs. By carefully reducing the state space the problem of dynamic channel allocation in cellular mobile networks with handoffs can be solved, gaining significant savings in terms of memory requirements compared to currently available table-based reinforcement learning schemes. Furthermore, this cost reduction can be achieved with effectively no performance penalty.

Additionally, by eliminating a state component of secondary importance we achieved a further reduction in memory requirements of over 98% in total (Section 3.3). The reduced-state representation incorporating channel reassignment functionality also achieved greatly superior performance in terms of both new call and handoff blocking probability over all other reinforcement learning solutions simulated. The reduced-state SARSA algorithm with channel reassignment also exhibited much lower blocking probability during periods of channel unavailability simulating periods of equipment failure or jamming.

Further issues for investigation are the extension of this work to include call admission functionality and multiple call classes. This technique may also be applicable to other problems exhibiting the same characteristics of disparate state component values and optimal policy actions dependent on initial arbitrary decisions taken.

Acknowledgements. This work was supported by an Australian Research Council Linkage Project Grant and the Motorola Australia Software Centre.

References

1. Bertsekas D.: Dynamic Programming and Optimal Control, Vol. 1. Athena Scientific, Belmont Mass. 1995.
2. Bertsekas D.: Dynamic Programming and Optimal Control, Vol. 2. Athena Scientific, Belmont Mass. 1995.
3. Bertsekas D., Tsitsiklis J.: Neuro-Dynamic Programming, Athena Scientific, Belmont Mass. 1996.
4. Freeman R. L.: Telecommunication System Engineering, 3rd ed. Wiley, New York 1996.
5. Jordan S.: Resource allocation in wireless networks. Journal of High Speed Networks, Vol. 5 no.1, pp. 23-34, 1996.
6. Lilith N., Dogançay K.: Dynamic channel allocation for mobile cellular traffic using reduced-state reinforcement learning. To appear in Proc. WCNC 2004, March 2004.
7. Mitchell T.: Machine Learning, McGraw-Hill, New York, 1997.
8. Nie J., Haykin S.: A Q-learning-based dynamic channel assignment technique for mobile communication systems. IEEE Transactions on Vehicular Technology, Vol. 48 iss. 5, pp. 1676 –1687, Sep. 1999.
9. Rajaratnam M., Takawira F.: Handoff traffic characterization in cellular networks under nonclassical arrivals and service time distributions. IEEE Transactions on Vehicular Technology, Vol. 50 iss. 4, pp. 954-970, Jul. 2001.
10. Senouci S.-M., Pujolle G.: Dynamic channel assignment in cellular networks: A reinforcement learning solution. ICT 2003, 10th International Conference on Telecommunications, Vol.1, pp. 302-309.
11. Singh S., Bertsekas D.: Reinforcement learning for dynamic channel allocation in cellular telephone systems. Advances in NIPS 9, MIT Press, pp. 974-980, 1997.
12. Sutton R., Barto A.: Reinforcement Learning: An Introduction. MIT Press 1998.
13. Watkins C. J. C. H., Dayan P.: Q-learning. Machine Learning, Vol. 8, pp. 279-292, 1992.

Optimal Bus and Buffer Allocation for a Set of Leaky-Bucket-Controlled Streams

Edgar den Boef[1,2], Jan Korst[1], and Wim F.J. Verhaegh[1]

[1] Philips Research Laboratories, Prof. Holstlaan 4, 5656 AA Eindhoven, The Netherlands
[2] Technische Universiteit Eindhoven, P.O. Box 513, 5600 MB Eindhoven, The Netherlands
denboef@natlab.research.philips.com
{jan.korst;wim.verhaegh}@philips.com

Abstract. In an in-home digital network (IHDN) it may be expected that several variable-bit-rate streams (audio, video) run simultaneously over a shared communication device, e.g. a bus. The data supply and demand of most of these streams will not be exactly known in advance, but only a coarse traffic characterization will be available. In this paper we assume that data streams are characterized by a concave function f, which gives a bound on the amount of data supplied for each length of a time window. We show how allocations of the bandwidth of a single bus and of buffers connected to the bus can be obtained for all streams, such that for each stream a feasible transmission strategy exists. For this, we show that a feasible solution for a data supply that exactly follows the function f is sufficient. This problem can then be solved by repeatedly solving single-stream problems for which we present efficient methods.

Keywords: Leaky-bucket-traffic characterization, resource management, in-home digital network, smoothing variable-bit-rate streams

1 Introduction

In the future home, all digital devices will be interconnected and streams (audio, video) will run over an in-home network between the different devices. This gives rise to a number of resource management problems, such as handling the communication efficiently. When several people at the same time access multimedia over the home network, then multiple data streams have to be transmitted simultaneously, and consequently the bandwidth of one or more communication links has to be shared. For variable-bit-rate streams this requires an efficient resource allocation strategy.

In this paper we concentrate on a network consisting of a single bus with finite transmission capacity to which several nodes, i.e., devices, are connected. Furthermore, a buffer of finite size is located between each node and the bus; see also Figure 2. In an in-home network typically only a few streams, e.g. 5–10 streams, use a communication link simultaneously. Because of this low number

J.N. de Souza et al. (Eds.): ICT 2004, LNCS 3124, pp. 1337–1346, 2004.

of streams, we may not hope that peaks in the amount of supplied data are cancelled out over the total set of streams. Therefore, a deterministic approach of admission test is required.

In [1], Den Boef, Verhaegh, and Korst show how optimal bandwidth and buffer shares can be determined for streams of which the exact supply and demand is known in advance. While normally the supply and demand are known for pre-recorded streams, they are generally not known for live video streams. For live streams, deterministic traffic characterizations such as the (σ, ρ)-model [2] or the D-BIND model [3] can be used. These characterizations give a worst-case bound on the amount of data that is supplied during any time window. More specifically, let τ be the size of a time window, and let $P(t)$ denote the supply of data up to time t. Then a traffic characterization described by a function f bounds the supply of data as follows. For all t,

$$P(t + \tau) - P(t) \leq f(\tau) . \tag{1}$$

When f is a piecewise-linear, concave function, the traffic characterization corresponds to a stream controlled by one or more leaky buckets [4].

In this paper we assume the data supply of each stream to be characterized by a concave function f. We shall refer to these streams as *concavely-upper-bounded streams*. We also consider the special case in which each stream is controlled by one or more (σ, ρ)-leaky buckets which shape the stream before it enters the network. Figure 1 gives an example of a (σ, ρ)-leaky bucket. When a stream leaves the network, it is deshaped into its original form. Figure 2 shows an example network. The network is placed between the shaping and the deshaping of the stream, which introduces an extra delay of the stream, viz. the time between the moment data enters the network and the moment the same data leaves the network. The problem that we consider is to determine for each stream a reserved share of the bus capacity and a reserved share of the corresponding buffers, as well as a transmission strategy indicating how the data is to be transmitted over time, such that the total bus and buffer capacities are never exceeded and buffers neither underflow nor overflow. We do not consider the actual shaping and deshaping of the streams. Any resources such as buffers that are required for this, are assumed to be allocated outside the network. The solution of the above described problem can be used to decide whether for a given set of streams all streams can be admitted service or whether a new stream can be admitted.

For the (σ, ρ)-characterization an extensive calculus has been developed, originally by Rene Cruz [2], [5], which was later extended by several other authors. For a clear description and overview of this calculus we refer to the book by Cheng-Shang Chang [6]. The developed calculus can be used to determine resource allocations for streams that are controlled by (σ, ρ)-leaky buckets. However, we could not use it to model the situation in which data needs to be buffered at the receiving side, if it arrives before it is demanded.

In Section 2 we first give our assumptions and define the problem. Next, in Section 3 we give a linear programming (LP) model for the problem, which can be decomposed into a master LP problem, and several single-stream problems.

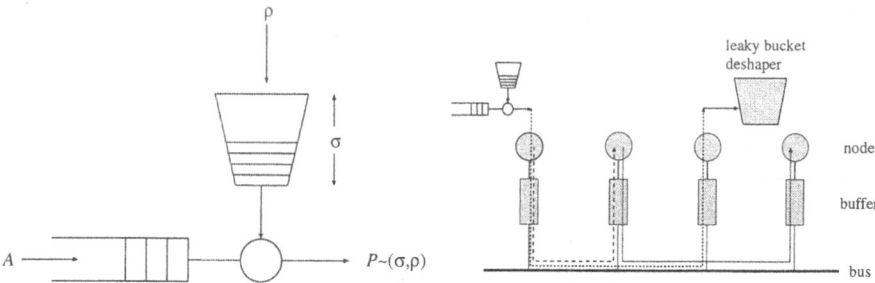

Fig. 1. An example of a (σ, ρ)-leaky bucket controller. It consists of a bucket of size σ in which tokens are generated at rate ρ. Data packets arrive at the controller according to an arrival process A. Each packet needs a token before it may proceed. The resulting data output P is characterized by (σ, ρ), also denoted as $P \sim (\sigma, \rho)$, i.e., $f(t) = \sigma + \rho t$. Here, ρ can be seen as the maximum sustainable rate of data, while σ gives the maximum burst size of data.

Fig. 2. An example network with four nodes and three streams running. The streams are shaped by one or more leaky buckets before they enter the network. After they leave the network they are de-shaped into their original form.

Furthermore, we show that f can be used as the actual supply scheme to obtain a solution. In Section 4 we show how a solution for the single-stream problems can be obtained and propose a greedy transmission strategy.

2   Problem Formulation

We start with giving our assumptions in Section 2.1. Next, we introduce our notation and we define the problem in Section 2.2.

2.1   Assumptions

We split up the time axis into a finite set of identical time units. We assume that during a time unit, data of more than one stream can be transmitted over the bus, and that the switching time between transmissions for different streams is either negligible or subtracted from the available bandwidth beforehand. Furthermore, we assume that there is no loss of data during transmission.

All streams that enter the bus are concavely-upper-bounded streams. An application supplies data of a stream at the stream's sending node. The demand of data takes place at the stream's receiving node by an application after a given delay, i.e., the amount of data that is supplied during a time unit at the sending node, is the exact amount of data that is demanded at the receiving node after the delay. We assume that during the initial delay the demand is zero.

With respect to the amount of data to be buffered, we assume that data is supplied at the beginning of a time unit, and then buffered at the sending node before it is transmitted. After transmission somewhere within a time unit, it is buffered at the receiving node before it is demanded at the end of a time unit. However, other buffering assumptions will only lead to small modifications in the constraints for the transmission strategy of a stream.

2.2 Notation

Let $\mathcal{T} = \{1, 2, \ldots, T\}$ denote the set of time units. Let B represent the maximum amount of data that can be transmitted over the bus during a time unit. Furthermore, let $\mathcal{N}$ denote the set of nodes that are connected to the bus. The capacity of the buffer between node $n \in \mathcal{N}$ and the bus is given by M_n. $\mathcal{S}$ denotes the set of data streams with each stream $i \in \mathcal{S}$ transmitting data from a sending node $s_i \in \mathcal{N}$ to a receiving node $r_i \in \mathcal{N}$. For each stream $i \in \mathcal{S}$ the supply of data during a time window of t time units, $t = 1, \ldots, T$, is bounded by $f_i(t)$ with f_i concave on $[0, T]$. The delay is given by d_i for each stream $i \in \mathcal{S}$. As buffer underflow and overflow depend on the actual supply and demand, we introduce for all $t \in \mathcal{T}$, $p_i(t)$ as the actual supply and $c_i(t)$ as the actual demand. Since demand equals the supply with delay d_i, it follows that for all $t \in \mathcal{T}$,

$$c_i(t) = p_i(t - d_i) . \tag{2}$$

For each stream i the decision variables are the reserved bus bandwidth b_i, the reserved buffer sizes $m_{s_i, i}$ and $m_{r_i, i}$, and the transmission strategy $x_i(t)$, $t \in \mathcal{T}$, which gives for each t the amount of data that should be transmitted depending on the actual amount of data supplied. This gives the following problem.

Problem (Multiple Leaky-Bucket Streams Smoothing Problem)
Given $\mathcal{S}$, $\mathcal{N}$, $\mathcal{T}$, B, M_n for all $n \in \mathcal{N}$, $f_i(t)$ for all $t \in \mathcal{T}$ and for all $i \in \mathcal{S}$, and d_i for all $i \in \mathcal{S}$, determine values for b_i, $m_{s_i, i}$, and $m_{r_i, i} \geq 0$ for all $i \in \mathcal{S}$, and a transmission strategy $x_i(t)$ for all $t \in \mathcal{T}$, such that the total bus and buffer capacities B and M_n are not exceeded and for all supply p_i bounded by f_i buffer shares neither overflow nor underflow.

3 Model

We first define for all $t \in \mathcal{T}$, $P_i(t) = \sum_{k=1}^{t} p_i(k)$, $C_i(t) = \sum_{k=1}^{t} c_i(k)$, and $X_i(t) = \sum_{k=1}^{t} x_i(k)$, as the cumulative supply scheme, cumulative demand scheme, and cumulative transmission strategy, respectively. To avoid tedious formulations concerning function values at the boundaries of their domains, we assume that for all functions g on $\mathcal{T}$ used in this paper, $g(t) = 0$ if $t \leq 0$, and $g(t) = g(T)$ if $t \geq T$.

3.1 Multiple Streams

The Multiple Leaky-Bucket Streams Smoothing Problem contains the following constraints.

The total bandwidth and total buffer sizes reserved may not exceed the bus and buffer capacities, i.e.,

$$\sum_{i \in S} b_i \leq B , \tag{3}$$

and for all $n \in \mathcal{N}$,

$$\sum_{i \in S : s_i = n} m_{n,i} + \sum_{i \in S : r_i = n} m_{n,i} \leq M_n . \tag{4}$$

The amount of data transmitted for stream i during time unit t may not exceed the reserved bandwidth of stream i, i.e., for all $i \in S$ and all $t \in \mathcal{T}$,

$$x_i(t) \leq b_i . \tag{5}$$

Furthermore, the reserved buffer sizes at the sending and receiving node of each stream may not underflow nor overflow, i.e., for all $i \in S$ and all $t \in \mathcal{T}$, we must have

$$P_i(t) - X_i(t) \geq 0 , \tag{6}$$
$$P_i(t) - X_i(t - 1) \leq m_{s_i, i} , \tag{7}$$

and for all $i \in S$ and all $t \in \mathcal{T}$, we must have

$$X_i(t) - C_i(t) \geq 0 , \tag{8}$$
$$X_i(t) - C_i(t - 1) \leq m_{r_i, i} . \tag{9}$$

The bus and buffer shares, and transmission strategies must meet these constraints for all actual supply functions p_i that satisfy (1), and for all actual demand functions c_i that satisfy (2).

3.2   Single Stream

We now consider the constraints that concern only one stream per constraint, i.e., constraints (5)–(9). For ease of notation we omit the subscript i in the rest of this paper when it is clear only one stream is considered. We call a solution b, m_s, and m_r for a single stream *feasible w.r.t.* (5)–(9), if for all p and c that satisfy (1) and (2), there exists a transmission strategy for which (5)–(9) are satisfied. Instead of considering all p that satisfy (1) and c that satisfy (2), we now show that it is sufficient to consider only a worst-case supply which is given by f directly, i.e., let for all $t \in \mathcal{T}$,

$$P(t) = f(t) , \tag{10}$$
$$C(t) = f(t - d) . \tag{11}$$

It can be easily verified that this function P satisfies (1) by the concavity of f.

The feasible solution area for one stream is then given by the following constraints. For all $t \in \mathcal{T}$,

$$x(t) \leq b \, , \tag{12}$$

$$f(t) - X(t) \geq 0 \, , \tag{13}$$

$$f(t) - X(t-1) \leq m_s \, , \tag{14}$$

$$X(t) - f(t-d) \geq 0 \, , \tag{15}$$

$$X(t) - f(t-d-1) \leq m_r \, . \tag{16}$$

We call a solution b, m_s, and m_r feasible w.r.t. (12)–(16), if there exists a transmission strategy for which (12)–(16) are satisfied. The following theorem states that a solution b, m_s, and m_r is feasible w.r.t. (12)–(16), if and only if it is feasible w.r.t. (5)–(9). For space reasons we leave the proof to a future, more elaborate paper.

Theorem 1. *A solution b, m_s, and m_r is feasible w.r.t. (5)–(9), if and only if it is feasible w.r.t. (12)–(16).*

Both solution areas are thus equivalent for b, m_s, and m_r. Therefore, we may use f as the actual supply and we can use (12)–(16) instead of (5)–(9).

3.3   Model Decomposition

The problem now becomes to find for each stream $i \in S$ a feasible solution w.r.t. (12)–(16) such that the total bus and buffer capacities are not exceeded, given by (3) and (4). As (12)–(16) concern each $t \in \mathcal{T}$, this leads to a large number of constraints. We can efficiently deal with these constraints as (12)–(16) only concern one stream per constraint. We use this fact to decompose the multiple-stream problem into single-stream problems. This gives for each stream $i \in S$,

$$\begin{aligned} \text{minimize} \quad & c_b b_i + c_{s_i} m_{s_i,i} + c_{r_i} m_{r_i,i} + c_e \\ \text{subject to} \quad & (12)\text{–}(16) \, , \end{aligned} \tag{17}$$

where c_b, c_s, c_r, and c_e are cost coefficients. This decomposition is based on the Dantzig-Wolfe decomposition algorithm for linear programs [7]. We refer to a previous paper [8] for an elaborate description of this decomposition to the described model.

4   Single Stream Problems

In this section we show how a solution for (17) can be obtained. Although each single-stream problem (17) is in fact an LP, we have shown in [1] how they can be solved more efficiently. Here we show how we can use the concavity and piecewise linearity of f to obtain a solution even more efficiently for leaky-bucket-controlled streams. Constraints (12)–(16) include the transmission strategy $x(t)$, with which no costs are associated. However, we can exploit the specific properties of f to derive four necessary and sufficient constraints on b, m_s, and m_r, that do not involve the transmission strategy. In Section 4.2 we describe how an optimal solution can be obtained efficiently for the single-stream problems.

4.1   Necessary and Sufficient Constraints

We derive four necessary constraints using the fact that the allocated bandwidth share b should be large enough to avoid buffer underflow at the receiving buffer and buffer overflow at the sending buffer. To avoid buffer underflow at the receiving buffer, the bandwidth share should be large enough such that for each time unit the cumulative amount of data demanded could have been transmitted, i.e., such that for all $t \in \mathcal{T}$,

$$f(t) \leq (t+d)b .\tag{18}$$

Now to avoid buffer overflow at the sending buffer, the bandwidth share should be large enough such that for each time unit the difference between the cumulative amount of data supplied and the cumulative amount of data that could have been transmitted, is not larger than the sending buffer share, i.e., such that for all $t \in \mathcal{T}$,

$$f(t) - (t-1)b \leq m_s .\tag{19}$$

Constraints (18) and (19) are derived with the assumption that data can be transmitted using the full bandwidth share. However, all data that is transmitted needs to be buffered at the receiving side. When the buffer share at the receiving side is completely filled, no more data can be transmitted until data is demanded from the receiving buffer. A larger bandwidth share may then be needed to transmit data at a later time such that buffer underflow at the receiving side and buffer overflow at the sending side are avoided.

There are $d+1$ time units in which data can be transmitted before data is demanded from the receiving buffer. The amount of data that can be transmitted during these $d+1$ time units is thus bounded by the receiving buffer share m_r. The bandwidth share b should be large enough such that after the first $d+1$ time units enough data can be transmitted to avoid buffer underflow at the receiving side, i.e., such that for all $t \in \mathcal{T}$,

$$f(t) - m_r \leq (t-1)b .\tag{20}$$

Furthermore, the bandwidth share should be large enough such that after the first $d+1$ time units enough data can be transmitted to avoid buffer overflow at the sending side, i.e., such that for all $t > d+1$, $t \in \mathcal{T}$,

$$f(t) - m_r - (t-d-2)b \leq m_s .\tag{21}$$

Notice that (20) and (21) follow from (18) and (19), respectively, by decreasing the amount of data to be transmitted, given by $f(t)$, with m_r, and by decreasing the number of time units during which data can be transmitted, given directly in front of b, with $d+1$.

The next theorem states that (18)–(21) are not only necessary but also sufficient for any feasible solution of (17). We leave the proof to a future, more elaborate paper.

Theorem 2. *Any solution b, m_s, and m_r to (18)–(21), is also a feasible solution of the single-stream problem (17).*

By rewriting (18)–(21) we can obtain the following constraints on b given values of m_s and m_r. For all $t \in \mathcal{T}$,

$$b \geq f(t - d)/t ,\qquad(22)$$
$$b \geq (f(t + 1) - m_s)/t ,\qquad(23)$$
$$b \geq (f(t + 1) - m_r)/t ,\qquad(24)$$
$$b \geq (f(t + d + 2) - m_r - m_s)/t .\qquad(25)$$

Likewise, we can obtain the following constraints on m_s given values of b and m_r, and on m_r given values of b and m_s. For all $t \in \mathcal{T}$,

$$m_s \geq f(t) - (t - 1)b ,\qquad(26)$$
$$m_s \geq f(t + d + 1) - m_r - (t - 1)b ,\qquad(27)$$
$$m_r \geq f(t) - (t - 1)b ,\qquad(28)$$
$$m_r \geq f(t + d + 1) - m_s - (t - 1)b .\qquad(29)$$

Constraints (27) and (29) are equivalent, however, this gives a clear presentation of the constraints on m_s and m_r separately. This constraint can also be rewritten into the following constraint on the total buffer space required given a value of b. For all $t \in \mathcal{T}$,

$$m_s + m_r \geq f(t + d + 1) - (t - 1)b .\qquad(30)$$

In the next section we show how these constraints can be used to find an optimal solution.

4.2   Solution Methods

To solve a single-stream problem we need to minimize the weighted sum of the bandwidth and buffer shares subject to the existence of a feasible transmission strategy. As the cost coefficients can be either positive or non-positive, the bandwidth and buffer shares themselves either need to be minimized or can be set to their maximum. To find the optimum we can therefore distinguish eight cases in which each of the cost coefficients is either positive or non-positive. Furthermore, we assume that a feasible solution exists, i.e., $b = B$, $m_s = M_s$, and $m_r = M_r$ satisfy (18)–(21). For all cases a feasible transmission strategy is given by the greedy strategy $x(t) = \min\{b, f(t) - X(t - 1), m_r + f(t - d - 1) - X(t - 1)\}$.

Single Resource Minimization. If only one cost coefficient is positive, an optimal value for the corresponding bus or buffer share can be obtained by determining its minimum value that satisfies either (22)–(25) in case of the bus share, or (26) and (27) in case of the sending buffer share, or (28) and (29) in case of the receiving buffer share. As f is a piecewise-linear and concave function, these minimum values occur at one of the bending points of f.

Two-Buffer Minimization. The buffer shares m_s and m_r have to satisfy (26)–(30). For a given value of b, we define $m_1(b) = \max_{t \in \mathcal{T}}(f(t) - (t-1)b)$ and $m_2(b) = \max_{t \in \mathcal{T}}(f(t+d+1)-(t-1)b)$. If $2m_1(b) \geq m_2(b)$ holds, then $m_s = m_1(b)$ and $m_r = m_1(b)$ are the optimal values as then (26) and (28) are tight and (30) has slack. Otherwise, their sum $m_s + m_r = 2m_1 < m_2$ is not guaranteed to satisfy (30). Hence, at least one of them must be increased: the cheapest one first and then, if necessary, the expensive one. Formally, if $c_s \leq c_r$, then $m_s = \min\{m_2(b) - m_1(b), M_s\}$ and $m_r = m_2(b) - m_s$. Notice that $m_r \geq m_1(b)$. If $c_s > c_r$, then $m_r = \min\{m_2(b) - m_1(b), M_r\}$ and $m_s = m_2(b) - m_r$. The optimal values of the buffer shares can now be determined with $b = B$.

Bandwidth-Buffer Trade-off. When the cost coefficient of the bandwidth and of at least one buffer is positive, a trade-off has to be made between bandwidth and buffer space. We assume both c_s and c_r are positive, and that $c_s \leq c_r$. If $c_s > c_r$ or either $c_s \leq 0$ or $c_r \leq 0$, i.e., a trade-off between the bandwidth and just one buffer, the optimal solution can be obtained in an analogous manner.

To perform the trade-off we start with an initial solution with minimum required bandwidth, which is obtained as follows. First, we take $m_s = M_s$ and $m_r = M_r$. The minimum bandwidth share $\tilde{b}$ of this initial solution is then given by the minimum value that satisfies (22)–(25). Next we minimize m_s and m_r given $b = \tilde{b}$ using (26)–(30).

We now define $t_1(b), t_2(b) \in \mathcal{T}$ such that $f(t_1(b)) - (t_1(b) - 1)b = m_1(b)$ and $f(t_2(b) + d + 1) - (t_2(b) - 1)b = m_2(b)$. Then $t_2(b) = t_1(b) - d - 1$ if $t_1(b) > d + 1$, and $t_2(b) = 1$ if $t_1(b) \leq d + 1$. Furthermore, as f is piecewise-linear and concave, $t_1(b)$ is a bending point of f.

The trade-off is now performed by increasing the value of b starting from $\tilde{b}$ by an amount Δ. For a given b, let $\rho(b)$ denote the slope of the line segment of f that ends at $t_1(b)$, i.e., the smallest slope of a line segment of f larger than b. If we increase b to $\rho(b)$, i.e., by an amount $\Delta = \rho(b) - b$, the values of $m_1(b)$ and $m_2(b)$ are attained at the previous bending point of f, i.e., the bending point at the start of the line segment with slope ρ_i. The values of $m_1(b)$ and $m_2(b)$ then decrease by $(\rho(b) - b)(t_1(b) - 1)$ and $(\rho(b) - b)(t_2(b) - 1)$, respectively.

The new values of m_s and m_r can be determined from m_1 and m_2 as described previously. Initially, either $m_2 < 2m_1$ or $m_2 \geq 2m_1$. As $t_2 \leq t_1$, the decrease of $2m_1$ is larger than the decrease of m_2 when b is increased. Therefore, while increasing b, as soon as $m_2(b) \geq 2m_1(b)$ holds, it will hold for any further increase of b. When $m_2(b) < 2m_1(b)$, this happens when b is increased by $\Delta = \frac{2m_1(b)-m_2(b)}{2t_1(b)-t_2(b)-1}$ which can be derived from $m_2(b + \Delta) = 2m_1(b + \Delta)$. Note that this only holds when $\Delta < \rho(b) - b$. Furthermore, when $m_2 \geq 2m_1$, the cheap buffer m_s can maximally increase to M_s. If $m_s < M_s$ it increases by the increase in $m_2 - m_1$. Therefore, $m_s = M_s$ when b is increased by $\Delta = \frac{M_s - (m_2(b)-m_1(b))}{t_1(b)-t_2(b)}$. Again, note that this only holds when $\Delta < \rho(b) - b$.

Finally, we consider the change in costs. If $m_2(b) < 2m_1(b)$, both the change in m_s and the change in m_r are equal to the change in m_1. The change in costs is then equal to $c_b \Delta - c_s \Delta(t_1(b) - 1) - c_r \Delta(t_1(b) - 1)$. If $m_2(b) \geq 2m_1(b)$

and $m_s < M_s$, the change in m_s is given by the change in $m_2(b) - m_1(b)$. The decrease of m_r is to $\Delta(t_1(b) - 1)$. The total change in costs is now equal to $c_b\Delta - c_s\Delta(t_2(b) - t_1(b)) - c_r\Delta(t_1(b) - 1)$ which can be rewritten into $c_b\Delta - c_s\Delta(t_2(b) - 1) - (c_r - c_s)\Delta(t_1(b) - 1)$. When $m_s = M_s$, the decrease of m_r is equal to the decrease of $m_2(b)$. The total change in costs is then equal to $c_b\Delta - c_r\Delta(t_2(b) - 1)$. As t_1 and t_2 only decrease with an increase of b, the change in costs only increases when b increases, and thus it is optimal to stop as soon as the change of costs is non-negative.

To conclude, we remark that the time complexity of the algorithms described in this section depends only linearly on the number of bending points of the function f. As the number of bending points is generally very small, the algorithms are very efficient when dealing with leaky-bucket-controlled streams.

5   Conclusion

In this paper we considered the problem to allocate fixed bandwidth and buffer shares for leaky-bucket-controlled streams using a single bus, such that a feasible transmission strategy exists for each stream. We have shown that if the traffic characterization function f is concave, this function can be used as supply function to determine feasible bandwidth and buffer shares for a stream. This resulted in a linear programming model suitable for decomposition into sub-problems concerning only one stream each. We have derived four sufficient and necessary constraints for a feasible solution to such a single-stream problem. Furthermore, we have shown how these constraints can be used to obtain an optimal solution efficiently.

References

1. den Boef, E., Verhaegh, W.F., Korst, J.: Smoothing streams in an in-home digital network: Optimization of bus and buffer usage. Telecommunication Systems **23** (2003) 273–295
2. Cruz, R.L.: A calculus for network delay, part I: Network elements in isolation. IEEE Transactions on Information Theory **37** (1991) 114–131
3. Knightly, E.W., Zhang, H.: D-BIND: An accurate traffic model for providing QoS guarantees to VBR traffic. IEEE/ACM Transactions on Networking **5** (1997) 219–231
4. Turner, J.S.: New directions in communications (or which way to the information age?). IEEE Communications Magazine **24** (1986) 8–15
5. Cruz, R.L.: A calculus for network delay, part II: Network analysis. IEEE Transactions on Information Theory **37** (1991) 132–141
6. Chang, C.S.: Performance Guarantees in Communication Networks. Springer-Verlag (2000)
7. Dantzig, G., Wolfe, P.: The decomposition algorithm for linear programming. Econometrica **29** (1961) 767–778
8. den Boef, E., Verhaegh, W.F., Korst, J.: Bus and buffer usage in in-home digital networks: Applying the Dantzig-Wolfe decomposition. Journal of Scheduling **7** (2004) 119–131

Scalable End-to-End Multicast Tree Fault Isolation

Timur Friedman[1], Don Towsley[2], and Jim Kurose[2]*

[1] Université Pierre et Marie Curie
Laboratoire LiP6-CNRS
8 rue du Capitaine Scott
75015 Paris, France
timur.friedman@lip6.fr
[2] University of Massachusetts Amherst
Department of Computer Science
140 Governors Drive
Amherst, MA 01003, USA
{towsley, kurose}@cs.umass.edu

Abstract. We present a novel protocol, M3L, for multicast tree fault isolation based purely upon end-to-end information. Here, a fault is a link with a loss rate exceeding a specified threshold. Each receiver collects a trace of end-to-end loss measurements between the sender and itself. Correlations of loss events across receivers provide the basis for participants to infer both multicast tree topology and loss rates along links within the tree. Not all receiver traces are needed to infer the links within the network that exceed the loss threshold. M3L targets the minimal set of receiver traces needed to identify those loss-exceeding links.

While multicast inference of network characteristics (MINC) is well understood, the novelty of M3L lies in the manner in which only a subset of the receiver traces is used. We model this as a problem of establishing agreement among distributed agents, each acting upon incomplete and imperfect information. Considering bandwidth to be a limited resource, we define an optimal agreement to be based upon the smallest possible receiver set. M3L performs well compared to the optimal.

1 Introduction

We know that multicast can serve as an active measurement tool, revealing network topology otherwise hidden from an end-to-end perspective, and allowing inference of loss rates and delays along internal links. Little, however, is known about how such measurement techniques might scale. The process of gathering receiver traces into one place in order to perform inference would seem to require a quantity of measurement-related traffic that is linear in the number of receivers. This paper shows how, by restricting the measurement problem to one of identifying just the lossiest links, the traffic can be made to scale sublinearly. The paper proposes the Multicast Lossy Link Location protocol, M3L, which scales according to a power law with a positive exponent less than one.

M3L is designed to leverage the newly-introduced standards track reporting extensions for the RTP control protocol. RTCP XR [1] is a mechanism that can be used

---

* This work was supported in part by the NSF under grant ITR-0085848.

uniformly across all types of multimedia sessions for providing highly detailed reports on, among other things, packet losses. Data packets from the RTP media stream constitute the active probes upon which measurement is based. If a session participant enables the sending of RTCP XR Loss RLE Report Blocks [1, Sec. 4.1] then it gathers traces specifying, for each sequence number, whether or not a packet was received. These loss traces are compressed by run-length encoding (RLE), and, typically, multicast along with the standard RTCP Report Blocks.

In a multicast RTP session in which all receivers originate RTCP XR Loss RLE Blocks, each participant potentially obtains a full set of detailed loss traces. Thus armed, the participant can use MINC (Multicast Inference of Network Characteristics) [2] to deduce elements in the structure of the multicast tree by discerning branching points between which some loss has occurred, and by estimating the loss rates between those points. Knowledge of tree structure and loss rates is valuable in reliable multicast protocols that promote local recovery of lost packets, as pointed out by Ratnasamy and McCanne [3]. A participant that is a network monitor could also provide information that would allow a network administrator to perceive and respond to problems in multicast data diffusion. For both the reliable multicast participant and the network monitor, one aim is fault isolation: identifying links in the tree that experience high loss.

The literature reveals a number of proposals for multicast tree fault isolation. Reddy, Govindan, and Estrin describe a method [4] that uses mtrace [5] and router "subcast" to identify lossy links. Zappala describes a technique [6] that multicast receivers could use to request alternate routing paths if they detect, with mtrace, that they lie behind a bad link. Saraç and Almeroth's Multicast Routing Monitor (MRM) [7] employs agents within the network to send test traffic among themselves for the purpose of localizing faults. Walz and Levine's Hierarchical Passive Multicast Monitor (HPMM) [8] is made up of a set of daemons located at routers and designed to accomplish the same goal through passive monitoring.

All of these protocols scale well, but in order to achieve their good scaling properties they rely upon the active support of routers or other agents inside the network. M3L is fundamentally different because it is based upon the purely end-to-end mechanisms of RTCP XR and MINC.

Strictly end-to-end methods other than M3L, such as Floyd et al.'s Scalable Reliable Multicast (SRM) [9] or Xu et al.'s Structure-Oriented Resilient Multicast (STORM) [10], do not aim at fault isolation per se, so much as forming multicast receivers into topologically related groups. These groups provide the basis for scalable signalling and loss repair. The end-to-end mechanisms at work in SRM and STORM are very different from that used by M3L. Rather than loss-based inference, SRM uses IPv4 multicast packet time-to-live (TTL) scoping to determine topology. The STORM work reveals a possible weakness in the SRM approach, in that measurements of multicast packets reveal only two TTL values (either 64 or 128) actually being used in practice, which would make scoping on that basis difficult. STORM thus enhances the TTL locality information by adding round-trip-time (RTT) measurements.

One protocol that does use loss-based inference is Ratnasamy and McCanne's Group Formation Protocol (GFP) [11]. By creating a separate multicast group corresponding to each topological group, and by designating a single receiver to send its traces, or "lossprints," within each group, GFP reduces overall traffic. However, during initial tree formation each receiver is liable to send its trace to a common control group. The GFP work does not include an analysis of how rapidly receivers might peel off from the common control group, and whether this process might prevent trace traffic from being linear in the number of receivers. Work in the present paper indicates that waiting for trace information to arrive in a random order can result in a nearly linear quantity of trace traffic when the goal is fault isolation.

There is prior work by this paper's first author, along with others, on scaling MINC inference for loss traces shared through RTCP XR. Oury and Friedman showed [12] that these traces could be compressed by up to a factor of five, and Cáceres, Duffield, and Friedman examined [13] the effects of thinning the traces to accommodate bandwidth constraints. However, neither of these techniques brings better than linear scaling in the number of receivers.

2   Protocol Overview

The key to achieving better than linear scaling in M3L is the insight that not all receivers' traces contribute equally when the task is specifically to identify the lossiest links. Similarly-situated receivers report essentially redundant information. Traces from receivers on low-loss paths might not be necessary at all. M3L employs a heuristic that prioritizes reporting from certain receivers over other receivers. The traces from those receivers are sent using RTCP XR packets, and MINC inference is performed, just as in the prior work just cited. But by prioritizing, we speed up fault isolation for bandwidth-constrained situations.

The protocol works by progressively refining what we call a picture until all lossy links have been identified. Refinement takes place over a series of rounds. In each round, receivers that have not yet contributed traces determine whether their data might be useful or not. If so, they become candidates. One candidate is selected each round at random through probabilistic polling of the sort described by Nonnenmacher and Biersack [14, 15,16]. The protocol terminates after at most three rounds in which no receiver sends a trace.

The essence of the M3L protocol is the process by which a receiver r decides whether it is a candidate. It requires the traces from the set of receivers S that contributed in prior rounds. Though it does not know the underlying tree $\mathcal{T}$, with MINC it can infer a picture $\mathcal{T}(S)$ of part of that tree. It can also infer a second picture by combining its own data with that from S. This is the picture $\mathcal{T}(S \cup \{r\})$. Since r alone can infer this picture, we call it r's "private picture." By contrast, since all receivers can infer $\mathcal{T}(S)$, we refer to it as the "public picture." It is by comparing the public picture to its own private picture that a receiver makes its decision.

Fig. 1 shows an example of a tree $\mathcal{T}$, a public picture $\mathcal{T}(S)$ inferred from the traces of the set of nodes $S = \{4, 11, 14, 15\}$, and the private picture $\mathcal{T}(S \cup \{7\})$ seen by receiver 7. Why might receiver 7 consider itself a candidate for transmitting its data? There are two possibilities. First, receiver 7 might infer that the link $(2, 7)$ is a lossy link. (Though it does not know the identity of node 2, receiver 7 infers its presence. For simplicity, in this text we refer to node 2 and other internal nodes by their labels.) Since

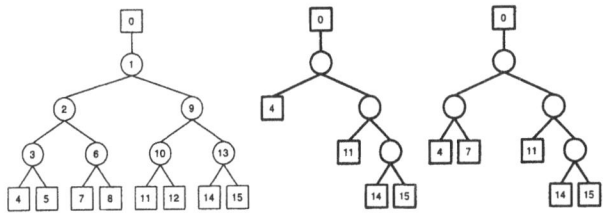

Fig. 1. Tree $\mathcal{T}$, public picture $\mathcal{T}(S)$, and private picture $\mathcal{T}(S \cup \{7\})$

this link appears only in its private picture, not the public picture, receiver 7 has valuable information to contribute. Second, if the public picture shows the link $(1, 4)$ to be a lossy link, receiver 7's private picture shows that that that link can be divided in two. One of the resulting links, $(1, 2)$ or $(2, 4)$, might prove to be a lossy link, in which case receiver 7's data has helped to isolate it. Or receiver 7's data might show that neither is lossy, in which case the data has helped eliminate what would otherwise be a false positive.

In M3L, rounds alternate between each of the possibilities just described. There is a round, called an ADD round, in which data indicating new lossy links is solicited. Then there is a CUT round, in which data indicating the subdivision of existing lossy links is solicited. When there is an ADD round for which there are no candidates followed by a CUT round for which there are no candidates, the protocol halts. The resultant public picture ought to isolate all of the tree's lossy links.

To evaluate the performance of the M3L protocol, this paper compares it to the performance of a hypothetical optimal protocol that would always select the smallest possible set of receivers necessary to isolate the lossy links. This paper also compares M3L against a protocol that demonstrates what might be accomplished if no special knowledge or heuristic were to be employed by choosing receivers at random.

3   The M3L Protocol

This section describes the M3L protocol in formal terms, by showing how to simulate its operation under ideal circumstances (such as perfect MINC inference). The setting is that of a multicast tree $\mathcal{T} = (V, L)$, with nodes V (including root node ρ), and links $L \subset V^2$. The protocol functions over a series of rounds, numbered $i = 0, \dots, n$. The set of receivers in the multicast tree is $R \subset V$. A set, $S_i \subseteq R$, is called the set of "in-picture receivers" for the start of round i, with $S_0 = \emptyset$. Let $C_i \subseteq R \setminus S_i$ designate a set called the "candidates" for a round. Each "out-of-picture receiver" $r \in R \setminus S_i$ makes an independent decision regarding whether it is a candidate or not. The terms of this decision differ depending upon whether the round is an ADD round (i is even) or a CUT round (i is odd), as we now describe.

A receiver makes its decision based upon the picture, $\mathcal{T}(S_i) = (V(S_i), L(S_i))$, that is formed by the set of in-picture receivers. $\mathcal{T}(S_i)$ is itself a tree, made up of a subset of the nodes from $\mathcal{T}$, and with links that follow the paths of links in $\mathcal{T}$. Using $j \prec k$ to indicate that a node j is descended from a node k in $\mathcal{T}$, we define the set of picture nodes

```
1    σ^H(T,α*) {
2        i ← 0
3        S_0 ← ∅
4        do
5            quiescent ← true
6            C_i ← C_ADD(S_i, α*)
7            if C_i ≠ ∅
8                we arbitrarily choose one r ∈ C_i
9                S_{i+1} ← S_i ∪ {r}
10               quiescent ← false
11           i ← i + 1
12           C_i ← C_CUT(S_i, α*)
13           if C_i ≠ ∅
14               we arbitrarily choose one r ∈ C_i
15               S_{i+1} ← S_i ∪ {r}
16               quiescent ← false
17           i ← i + 1
18       while ¬quiescent
19       return S_i
20   }
```

Fig. 2. Algorithm to simulate the M3L protocol

to be $V(S_i) = \{\rho\} \cup \{v \in V : \exists s_1, s_2 \in S_i, s_1 \neq s_2, s_1 \prec v, s_2 \prec v\} \cup S_i$. The set of picture links is $L(S_i) = \{(k,j) \in V(S_i) \times V(S_i) : j \prec k, \nexists v \in V(S_i) : j \prec v \prec k\}$. A picture defines a set of "endogenous nodes," $V_{\text{endo}}(S_i) = \{v \in V : \exists (j,k) \in L(S_i), j \prec v \prec k\}$, and a set of "exogenous links,"

$$L_{\text{exo}}(S_i) = \{(k,r) \in (V(S_i) \cup V_{\text{endo}}(S_i)) \times (R \setminus S_i),$$
$$r \prec k, \nexists v \in V(S_i) \cup V_{\text{endo}}(S_i) : r \prec v \prec k\}.$$

A receiver uses MINC to estimate loss rates on links in the picture, and along the exogenous link that the receiver terminates. We assume a Bernoulli link loss model. Packets are independent and each packet is successfully transmitted across link (k,j) with "passage probability" $\alpha((k,j))$. A threshold value α^* determines whether a link is lossy or not. The set of candidates for an ADD round are

$$C_{\text{ADD}}(S_i, \alpha^*) = \{r \in R \setminus S_i\} : (v,r) \in L_{\text{exo}}(S_i), \alpha(v,r) \leqslant \alpha^*,$$

and the set of candidates for a CUT round are

$$C_{\text{CUT}}(S_i, \alpha^*) = \{r \in R \setminus S_i\} : \exists (k,j) \in L(S_i),$$
$$\alpha(k,j) \leqslant \alpha^*, (v,r) \in L_{\text{exo}}(S_i), j \prec v \prec k.$$

Fig. 2 defines a function $\sigma^H()$ that returns the set $S \subseteq R$ of receivers that results from application of the M3L protocol. This function can be applied to simulate the functioning of the M3L protocol. In this function, quiescence is determined by a variable, "quiescent," which is true if there are no candidates for an ADD round and no candidates in the immediately following CUT round. The function loops through ADD and CUT rounds until quiescence results.

The function $\sigma^H()$ returns what we call a "solution" to the lossy link location problem. This is a member of the set of all possible solutions, defined as follows:

$$S^\star(\mathcal{T}, \alpha^\star) = \{S \subseteq R : \alpha_{\min}(L_{\mathrm{exo}}(S)) > \alpha^\star,$$
$$(k, j) \in L(S) \wedge \alpha(k, j) \leqslant \alpha^\star \Rightarrow \nexists v \in V_{\mathrm{endo}}(S) : j \prec v \prec k\}$$

The proof that a solution indeed isolates all of the lossy links in the tree can be found in the first author's thesis [17]. The thesis also describes possible enhancements to the ADD and CUT rounds.

4 Comparison Protocols

The previous section described the M3L protocol, which employs a heuristic for use by receivers operating on-line and with limited information. To evaluate M3L, we compare its performance to an optimal protocol and a random protocol.

The optimal protocol represents the best that could be done operating off-line and with omniscience. This protocol returns a set $S = \sigma(\mathcal{T}(k), \alpha^\star)$ that is a minimal cardinality solution:

$$S^\star_{\min C}(\mathcal{T}, \alpha^\star) = \{S \in S^\star(\mathcal{T}, \alpha^\star) : |S| = \min_{X \in S^\star(\mathcal{T}, \alpha^\star)} |X|\}.$$

Space does not permit a detailed treatment of an efficient algorithm for calculating the results of the optimal protocol. The details can be found in the first author's thesis [17, Sec. 3.2].

The random protocol consists of a series of rounds, $i = 1, \ldots$. At the beginning of round 0, the set S_0 of in-picture receivers is the empty set: $S_0 = \emptyset$. In each round i, one out-of-picture receiver $r \in R \setminus S_i$ is chosen at random and added to the set of in-picture receivers: $S_{i+1} = S_i \cup \{r\}$. If the set S_{i+1} constitutes a solution for the tree $\mathcal{T}$, that is if $S_{i+1} \in S^\star(\mathcal{T}, \alpha^\star)$, then the protocol halts, and S_{i+1} is returned. Let $\sigma^R(\mathcal{T}, \alpha^\star)$ be the function that returns a set arrived at through simulation of the random protocol.

5 Empirical Evaluation

This section describes the empirical evaluation of M3L. This evaluation is conducted in two stages. In the first stage, we compare M3L against the optimal protocol and the random protocol. These comparisons assume that MINC inference returns perfectly accurate estimates. By studying how well M3L performs under the assumption of perfect MINC inference, we can establish a bound on how well the heuristic can potentially perform. Also, it allows us to experiment upon larger topologies than are feasible when we must pay the computational costs of MINC inference. In the second stage, we introduce the possibility of inaccuracies arising from the MINC inference. In both stages we focus on a special case of the lossy link location problem: locating the single lossiest link in the multicast tree, chosen to facilitate comparisons.

5.1 Experimental with Perfect Inference

The first stage of experiments was performed upon simulated trees of constant fanout, with fanout values of either two, three, or four, and depths of two, three, or four. If computational resources permitted, trees of greater depth were also simulated, to a depth of eight. A time limit of five minutes was placed upon simulation for any given topology. In the case of trees of fanout two, this permitted a depth of eight; for trees of fanout three, a depth of seven; and for trees of fanout four, a depth of five. Increasing the time limit to several hours did not permit the generation of further data points.

Link passage probabilities for each link in each tree were chosen independently from a uniform distribution on the interval $(0, 1)$. For each topology, ten thousand trees were simulated. For each tree $\mathcal{T}$, the threshold α^* was set to be the passage probability of the lossiest link: $\alpha^* = \alpha_{\min}(L)$. Then, for that tree, an optimal set $\sigma(\mathcal{T}, \alpha^*)$ and a heuristic set $\sigma^H(\mathcal{T}, \alpha^*)$ were determined, as well as heuristic variants with enhanced ADD and CUT rounds, and finally a random protocol set $\sigma^R(\mathcal{T}, \alpha^*)$. For each set, the cardinality was recorded, leading to a calculation of the mean.

Results. In Fig. 3(a) we see the results for trees of fanout two. This graph is in log-log scale. The horizontal axis indicates the number of leaves, which is to say receivers, in the tree. Since the depth varies from two to eight, the values reported upon are: 4, 8, 16, 32, 64, 128, and 256. In Fig. 3(b) we see the results for trees of fanout three. This graph is similar to the preceding graph. As the depth of trees varies from two to seven, the numbers of receivers in the topologies reported upon are: 9, 27, 81, 243, 729, and 2187. In Fig. 3(c) we see the results for trees of fanout four. This graph is similar to the two preceding graphs. The depth of trees varies from two to five, so the numbers of receivers are: 16, 64, 256, and 1024.

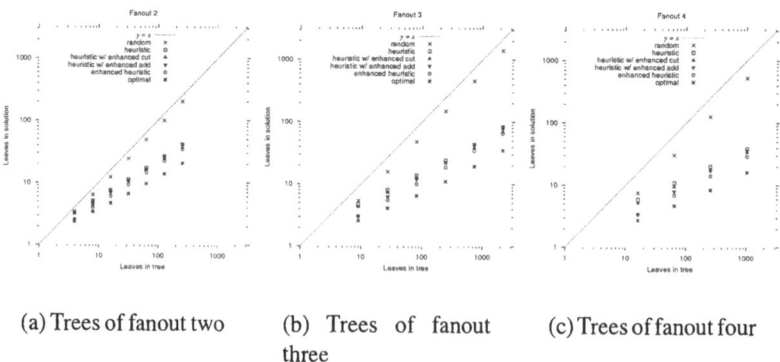

(a) Trees of fanout two   (b) Trees of fanout three   (c) Trees of fanout four

Fig. 3. Number of receivers in solution

The vertical axis indicates the mean number of receivers in a solution for each of the protocols described above. The straight line $y = x$ depicts the theoretical worst case, in which a solution is only arrived at when every receiver has been entered into the picture. The lower the mean cardinality of a solution, the further below the line $y = x$ it appears. The highest set of values is for the random protocol, and the lowest for the optimal protocol. Values M3L with for the various heuristics lie in between. The results for every protocol tested are well fitted by straight lines on a log-log graph, indicating power law scaling (that is, we can fit the points to a curve of the form $y = bx^a$).

Discussion. We see that the random protocol performs relatively poorly, the number of receivers in a solution increasing approximately linearly with the number of receivers in the tree (estimates $\hat{a} = 0.98, 1.01$, and 1.02). The results for the optimal protocol demonstrate that in the best of cases the growth in the number of receivers in a solution could be distinctly sub-linear ($\hat{a} = 0.52, 0.47$, and 0.43). M3L's behavior more closely resembles the optimal than the random. When employing the basic heuristic, we obtain estimates for the exponent of $\hat{a} = 0.61, 0.52$, and 0.45.

Although with M3L the number of receivers sending traces scales sub-linearly, is this definitively better than employing network tomography with data from all receivers? One reason why it might not be is that, if all receivers are to be heard from, they could simply unicast their traces to a single site. With M3L, the traces that are sent are multicast, reaching all receivers. For those hosts that do not need to perform inference for some application-related purpose, M3L relieves some load (they don't necessarily have to send their traces) while creating additional load (the receiving and processing of others' traces).

There is thus a trade-off. In general, the redistribution of a high load from a single point to a considerably lower load more widely shared might be viewed as worthwhile, and in keeping with the motivation behind multicast itself. However, a special concern arises when the load is being placed on receivers that are behind lossy links. Does M3L, by sending them additional traffic, not exacerbate their situation?

The extent of the problem depends upon how much additional traffic is generated. If the RTP control protocol is being used for sharing traces, as in prior work [13], then overhead is limited to 5% of session bandwidth, which may be a sufficient limitation. However, if this additional traffic is genuinely a problem, a hybrid system could be adopted. A receiver that finds itself in such a situation could unicast its trace to another session member. This does not create any load on the receiver beyond what would be necessary for tomography with full data. Then the other session member could act as an M3L proxy on the receiver's behalf, while the receiver drops out of the multicast group on which the traces are being sent.

5.2   Experiment with MINC Inference

The second stage of experiments was also performed upon simulated trees of constant fanout, with fanout values of either two, three, or four, and depths of two, three, or four. Within this range, the computational costs of MINC inference limited us to trees with a maximum of 27 receivers.

A narrower range of link passage probabilities was used for this experiment, to better focus on the problems of incorrect inference. The passage probability for each link in each tree was chosen independently from a uniform distribution on the interval

(0.900001, 1.0). Then one link in the tree was chosen at random, and reassigned a lower passage probability, chosen from a uniform distribution on the interval [0.85, 0.90]. The threshold passage probability for determining that a link was a lossy link was $\alpha^\star = 0.90$. Pseudo-random numbers were generated in the same manner as for the first experiment.

For each topology, a sufficient number of trees was simulated to construct 95% confidence intervals for the statistics that were collected. On each tree, the sending of 8,192 probe packets was simulated. This value is sufficient to obtain relatively good MINC inference. The heuristic sets $\sigma^H(\mathcal{T}, \alpha^\star)$ were generated, employing MINC to create the pictures. This inference was variously based upon the outcomes at the receivers from one probe, two probes, four probes, etc. ..., up to 8,192 probes. Thus, fourteen different heuristic sets were generated for each tree.

Based upon a given heuristic set, each receiver (whether in the set or not) either identified the lossy link correctly, or it did not. This fact was recorded. A correct identification is scored as follows. As we observing from off-line know the true topology, we can identify the set of receivers $R(k)$ that lie below the lossy link k. Each receiver j, in its inference based upon the MINC data from the set of receivers $S \cup \{j\}$, identifies a certain number of lossy links $k_1, k_2, \ldots$. Each set $R(k_i) \cap (S \cup \{j\})$ is compared against the set $R(k) \cap (S \cup \{j\})$. If the two sets are identical, then a correct identification is scored.

For each heuristic set, each receiver might also misidentify a number of links as lossy links. The number of such false positives was also recorded.

In addition, for each given number of probes, MINC inference was conducted using the entire set of receivers R. As for the heuristic set, it was recorded for each receiver whether it made a correct identification of the lossy link, as well as the number of false positives.

Results. The results for each topology were very similar. We show results here for trees of depth three and fanout three, having 27 receivers. In each of the three graphs shown here, the independent variable is the number of probes that were employed in MINC inference. This is plotted in log scale on the horizontal axis, and it varies from 1 to 8, 192. All confidence intervals for the dependent variables are at the 95% level or better.

In Fig. 4(a), the dependent variable is the mean number of receivers in a heuristic set. This number, plotted in linear scale on the vertical axis, ranges from a low of 0.8987 to a high of 16.0984. The horizontal line labelled "ideal" represents the mean number of receivers in the heuristic set if MINC inference were perfectly accurate. (As the loss rates are different from the first experiment, these numbers were recalculated, and in this case the number is 14.4491.)

In Fig. 4(b), the dependent variable is the mean number of unidentified lossy links. This value is plotted in linear scale on the interval [0, 1]. A value of zero is the best. Values are plotted based upon the set of receivers returned by the heuristic, and based upon use of all the receivers.

In Fig. 4(c), the dependent variable is the mean number of false positives. This value, plotted in linear scale on the vertical axis, ranges from a low of 0 to a high of 9.5036. As in the previous graph, values are plotted based upon the set of receivers returned by the heuristic, and based upon use of all the receivers.

The confidence intervals for these results are too narrow to plot in these figures.

Discussion. These experiments confirm the validity of using the heuristic even in the face of inaccuracies introduced in the course of MINC inference. In Fig. 4(a), we see that the number of receivers that results while using MINC inference is very nearly the same as if MINC inference were perfectly accurate. This is so once the number of probes is eight or more.

Of course, many more than eight probes are required in order for the resulting inference to be accurate. In Fig. 4(b), we see that once the number of probes enters the hundreds, the lossiest link is correctly identified most of the time, on average and this improves to 90% once the number of probes exceeds a thousand. In Fig. 4(c), we see that a few thousand probes are required before the number of false positives drops below one. What is striking about both of these graphs is the fact that, after a couple of hundred probes, there is almost no perceptible diminution in performance that results from using MINC traces from M3L's heuristic set of receivers rather than MINC traces from the entire receiver set. This final result confirms the value of M3L in reducing the bandwidth requirements for lossy link identification.

One might ask what happens if loss rates should change over the course of an inference. This is a question inherent to network tomography that M3L cannot by itself

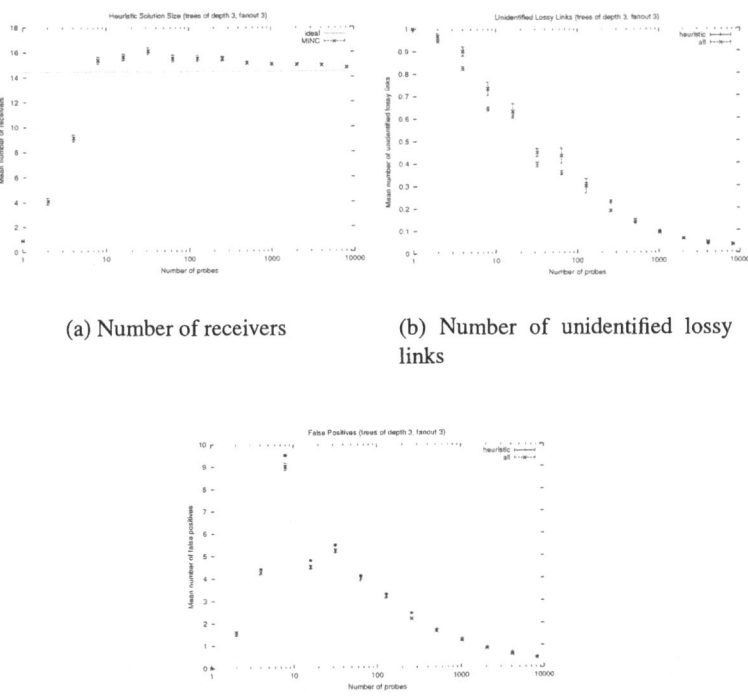

(a) Number of receivers

(b) Number of unidentified lossy links

(c) Number of false positives

Fig. 4. Results with MINC inference

solve. However, M3L by speeding up the process, can help. And what should happen if loss rates are not spatially or temporally independent? Again, this is a general problem for tomography. Whether such dependencies create specific biases for M3L remains a topic for future work.

6   Related and Future Work

A number of works on fault isolation were mentioned in the introduction. This paper is the first to both make use of the end-to-end fault isolation capability provided by MINC inference and reduce the overall amount of data required for that inference.

Although this paper does not make use of delay-based inference, it appears possible to adapt the techniques described in this paper to identify the high-delay links in a multicast tree. Multicast-based inference of network-internal delay characteristics is described by Lo Presti et al. (including a co-author of this paper) [18]. The combination of both loss and delay measurements for improved topology inference is described by Duffield et al. [19].

The strong correlations in outcomes for multicast packets makes multicast an effective tool for end-to-end inference of behavior inside of a network. However, multicast is often not available, and where it is available it may be of limited use for predicting unicast behavior. An alternative measurement tool is to send closely-spaced unicast packets to different receivers. These packets should also show correlated behavior. Coates and Nowack [20] have applied this principle to loss inference. It is not necessary to actively send unicast packets for measurement purposes, as abundant unicast traffic exists that can be passively monitored. Tsang, Coates and Nowack show [21] how TCP flows can be monitored to opportunistically take advantage of such closely-spaced packets as do appear. The use of striped unicast packets for delay inference is described by Duffield et al. (including a co-author on this paper) [22], and by Coates and Nowack [23,24]. The M3L protocol could be adapted for unicast-based inference so long as multicast were available for trace sharing.

Future work specifically building upon M3L will include study of a wider variety of scenarios. How does M3L perform in isolating multiple lossy links in a tree as compared to the case of a single lossy link studied in this paper, for instance? We are also interested in applying M3L to delay inference and to other forms of tomography. Finally, we plan to deploy M3L in the Internet, to study the effects of such things as correlated losses and the loss of trace-bearing packets.

References

[1] T. Friedman (ed.), R. Caceres (ed.), A. Clark (ed.), K. Almeroth, R. G. Cole, N. Duffield, K. Hedayat, K. Sarac, and M. Westerlund, "RTP control protocol extended reports (RTCP XR)," RFC 3611, Internet Engineering Task Force, Nov. 2003.
[2] A. Adams, T. Bu, R. Cáceres, N. Duffield, T. Friedman, J. Horowitz, F. Lo Presti, S.B. Moon, V. Paxson, and D. Towsley, "The use of end-to-end multicast measurements for characterizing internal network behavior," *IEEE Communications Magazine*, May 2000.
[3] S. Ratnasamy and S. McCanne, "Inference of multicast routing trees and bottleneck bandwidths using end-to-end measurements," in *Proc. Infocom '99*.

[4] A. Reddy, R. Govindan, and D. Estrin, "Fault isolation in multicast trees," in *Proc. SIG-COMM 2000*.

[5] B. Fenner, "mtrace (multicast traceroute)," Available from ftp://ftp.parc.xerox.com/pub/net-research/ipmulti/.

[6] D. Zappala, "Alternate path routing for multicast," in *Proc. Infocom 2000*.

[7] K. Saraç and K. C. Almeroth, "Monitoring reachability in the global multicast infrastructure," in *Proc. ICNP 2000*.

[8] J. Walz and B. N. Levine, "A hierarchical multicast monitoring scheme," in *Proc. NGC 2000*.

[9] S. Floyd, V. Jacobson, C.-G. Liu, S. McCanne, and L. Zhang, "A reliable multicast framework for light-weight sessions and application level framing," *IEEE/ACM Trans. on Networking*, vol. 5, no. 6, pp. 784–803, Dec. 1997.

[10] X. Xu, A. Myers, H. Zhang, and R. Yavatkar, "Resilient multicast support for continuous-media applications," in *Proc. NOSSDAV*, 1997.

[11] S. Ratnasamy and S. McCanne, "Scaling end-to-end multicast transports with a topologically sensitive group formation protocol," in *Proc. ICNP '99*.

[12] N. Oury and T. Friedman, "Compression des traces de perte de paquets multicasts sur internet," in *Proceedings of Journ es Doctorales Informatique et R seaux (JDIR)*, Nov. 2000.

[13] R. Cáceres, N. Duffield, and T. Friedman, "Impromptu measurement infrastructures using RTP," in *Proc. Infocom 2002*.

[14] J. Nonnenmacher, *Reliable Multicast Transport to Large Groups*, Ph.D. thesis, Institut Eurécom, 1998.

[15] J. Nonnenmacher and E. Biersack, "Optimal multicast feedback," in *Proc. Infocom '98*.

[16] J. Nonnenmacher and E. Biersack, "Scalable feedback for large groups," *IEEE/ACM Trans. on Networking*, vol. 7, no. 3, pp. 375–386, June 1999.

[17] T. Friedman, *Scalable Estimation of Multicast Characteristics*, Ph.D. thesis, UMass Amherst, May 2002.

[18] F. Lo Presti, N.G. Duffield, J. Horowitz, and D. Towsley, "Multicast-based inference of network-internal delay distributions," *IEEE/ACM Trans. on Networking*, vol. 10, no. 6, pp. 761–775, Dec. 2002.

[19] N.G. Duffield, J. Horowitz, and F. Lo Presti, "Adaptive multicast topology inference," in *Proc. Infocom 2001*.

[20] M. Coates and R. Nowak, "Network loss inference using unicast end-to-end measurement," in *Proc. ITC Conf. on IP Traffic, Modeling and Management*, 2000.

[21] Y. Tsang, M. Coates, and R. Nowak, "Passive network tomography using EM algorithms," in *Proc. IEEE Intl. Conf. on Acoustics, Speech, and Signal Processing*, May 2001.

[22] N.G. Duffield, F. Lo Presti, V. Paxson, and D. Towsley, "Inferring link loss using striped unicast probes," in *Proc. Infocom 2001*.

[23] M. Coates and R. Nowak, "Network delay distribution inference from end-to-end unicast measurement," in *Proc. IEEE Intl. Conf. on Acoustics, Speech, and Signal Processing*, May 2001.

[24] M. Coates and R. Nowak, "Sequential monte carlo inference of internal delays in non-stationary communication networks," *IEEE Trans. Signal Processing*, vol. 50, no. 2, pp. 366–376, Feb. 2002.

Internet Service Pricing Based on User and Service Profiles

Youcef Khene[1], Mauro Fonseca[1], Nazim Agoulmine[2,1], and Guy Pujolle[1]

[1]LIP6 Laboratory 1 University of Paris 6 8, rue du Capitaine Scott 75015 – Paris – France
[2]LSC Laboratory University of Évry – IUP Val d'Essonne 8, rue du Pelvoux 91025
Évry France
{Youcef.Khene; Mauro.Fonseca; Nazim.Agoulmine;
Guy.Pujolle}@lip6.fr

Abstract. Today, the ISPs are looked-for offering differentiated services with an adequate QoS for their end user. This is due to the fact that end users are not expecting the same QoS depending on their activities and consequently are not willing to pay the same price for these services. In this context, this paper proposes a new pricing mechanism that takes into account the customer behaviour to calculate the price of Internet services usage. The idea behind this approach is to optimize the network resources usage as well as the ISP profit by adapting, for each customer, the price of each service according to his profile. Hence, the proposed pricing mechanism is integrated into a dynamic SLA negotiation process. This process is divided into two steps: Pre-service SLA, and On-service SLA. The impact of our pricing model in over-provisioned networks has been analyzes analytically. The obtained results demonstrate that our mechanism diminishes considerably the cost of the over-provisioning. This is obtained thanks to a good balance of the traffic load during a day-scale period.

1 Introduction

The emergence of highly bandwidth consuming applications in the Internet such as multimedia applications or peer to peer applications are raising important problems to the Internet Service Providers. In fact, with the increasing number of customers using these applications the quality of service is deteriorating rapidly. Despite the existence of protocol level quality of service differentiation such as DiffServ [1], it is clear that the main differentiation factor is the price. So the questions are how should the ISP distribute its resources of different classes of service between the end-users? and how should the ISP bill these services?

An effective and complete solution to manage and control a network resources in order to offer adequate QoS to the users must be accompanied by an adequate pricing mechanism. Such a mechanism will enable one to major resources usage and adapt its price which will affect the future behaviour of the end users. However, this pricing mechanism should encourage users to use more network services.

J.N. de Souza et al. (Eds.): ICT 2004, LNCS 3124, pp. 1359–1368, 2004.
© Springer-Verlag Berlin Heidelberg 2004

In this paper, the objective is to propose new pricing mechanisms in the context of DiffServ networks. The choice of this architecture is motivated by the fact that is based on multi-service classes that facilitate the building of a competitive allocation resource among users by applying dynamic pricing mechanism.

The remainder of the paper is organized as follows: section 2 describes the background concepts for the purpose of this work. Section 3 presents the objectives of this work. The next section presents the proposed pricing solution using unified profile with describing in detail our strategy pricing. In the Section 5, we study the impact of our pricing model in an over-provisioning network use case and finally section 6 concludes the paper by working at some future works.

2 Background

Billing and pricing internet services are studied mainly in relation to offering different Quality of Service (QoS) classes. In fact, it is essential to link a choice of QoS to a financial incentive. In the absence of such an incentive, users would always select the highest level, thus obstructing a guarantee and the utilization of several QoS classes. There are many pricing models that have been proposed for implementation in DiffServ architecture: The Flat Rate Pricing model employs a fixed charging scheme where the user is charged a fixed amount irrespective of his usage of the service. The Usage based pricing refers to charging the consumers according to parameters such as the connection time or the volume of exchanged data. In the Content Based Pricing, the user is charged according to the type of traffic that is generated. Separate charges can be applied for data, voice or video. In the case of Congestion Based Pricing, user charging is based on the severity of congestion state. It means that the price may vary according to the load state of the network. In another approach called Schemes Based On User Preferences Pricing, the users have the possibility to choose the price of their requested service. For more information about these different models, the reader can refer to the following articles [2, 3] in which the authors give a detailed and comparative study between different pricing models.

All these pricing models have advantages and limitations but they are all introducing increment benefits of the offered services using optimizations model that do not take into account the customer's behaviours and profile which are important aspect as the end-users is the main concerned actor.

In the Internet, the billing is usually divided into connection price and usage services price with a total price equal to the addition of these costs. The Connection cost represents the cost to access the service, the setting up the user's account and the provision of software or hardware required for the connection, the connection cost has a fixed value and it is identical for all customers of a particular service provider. The Usage Service Cost represents the usage cost of each service. it usually varies according to the amount of resources consumed during the service usage.

This vision of two costs is related to economic as well as social concerns. A connection cost is always fixed; this is to make service access socially equitable to every layer of the society, otherwise, the companies will be suspected of social unfairness (poor and rich). The service cost varies according to the customers request and current provider offers.

One crucial problem the providers are facing is the accurate billing of the provided services [4]. The process of price setting is difficult because the price should reflect the value of the service as perceived by the consumer, and not the real cost of operating the service [5]. Therefore it is important to obtain some indications of this value, e.g. by means of identifying the demand patterns and what we're suggesting here the customers profiles. and which will be detailed in the following sections.

3 Objective

The main idea of this work is to introduce users and service profiles into the pricing mechanism. This approach is based on the user characterisation that helps the operators to customize the negotiation process based on the historical behaviour of the customer and also based on service characterisation which is managed by the operator itself.
Secondly, we propose a new dynamic SLA negotiation process which is negotiated between the user and the ISP. This fulfillment of an SLA (Service Level Agreement) in a particular domain is divided in two steps: Pre-service SLA, and On-service SLA. Finally, will be studied the effect of our pricing mechanism in a context of an over-provisioning network.

4 Proposed Solution

This paper introduces an technique that allows an ISP to determine and negotiate in an intelligent manner with its customers the price of it network services with users. We introduce in this work the concept of unified profile. The unified profile is an information model about the customers' behaviour regarding to the utilization of the provider services. The idea for the provider is to adapt its decisions according to the changing behaviour of the customer over a period of time. A dynamic users profile aims to capture the users' behaviour and to catalogue in a useful way, information related to his fidelity and preferences. This profile is integrated in the pricing process so that the provider can define differentiated pricing for each services among customers according to their profiles. Thus, we have defined a unified profile database composed of two information models related to the user profile and the service profile (as presented in figure 1).

4.1 User Profile

Three classes of customers' profile have been defined to reflect their behaviour: Irregular customer, regular customer and intensive customer. These behaviours concern the frequency of a particular service usage during a period of time.

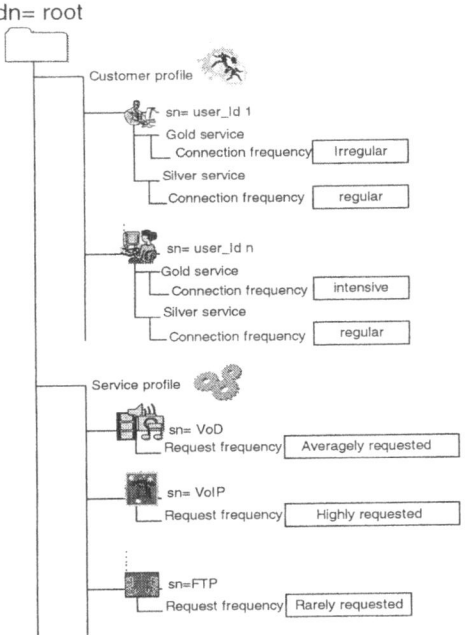

Fig. 1. Unified profile database

This information is collected and maintained by the User Classification Agent (UCA) in the unified profile database of the ISP management system. The UC agent makes the decision to change the class of a particular user if his service usage frequency increases or decreases over predefined thresholds and period of time. When a profile is captured for a particular customer, the pricing of provided services is fixed according to this profile. Table 1 shows an example of price affectation to each user profile. For example, an irregular user will not be offered some special discount prices from the ISP. This group of customers will be asked to pay the full price (100%) for used services. However, as the end user become a more regular customer of the provider services his profile is upgraded accordingly and thus he is allowed to benefit from a more attractive tariffs.

Table 1. Example of benefit customer classification

	Services class	
Profile	Gold	Silver
Irregular	100%	100%
Regular	75%	95%
Intensify	50%	75%

4.2 Service Profile

Similarly to the user profile, the system aims to establish a classification of the provided service in term of interest to the customers. The ISPs must classify the network services into a set of categories according to their usage frequency: highly requested services, average requested services and finally services that are rarely requested. This classification allows the ISP to affect the right price to the service usage depending on how often the service is used. The provider has to identify what the most interesting services for the customers i.e. the services that are mostly requested during a period of time. Based on this information, the system maintains a service profile database which will be used to fix the price of each service. From nowadays experience, we know that the most interesting services are not necessarily the ones that need the highest QoS.

This classification process is performed by a specific agent called Service Classification Agent (SCA). The SC Agent maintains historical information about the request frequency of each provided service to perform this classification. Figure 2 shows the variation of the user and service profiles according to the user behaviour, i.e connection frequency and service request frequency.

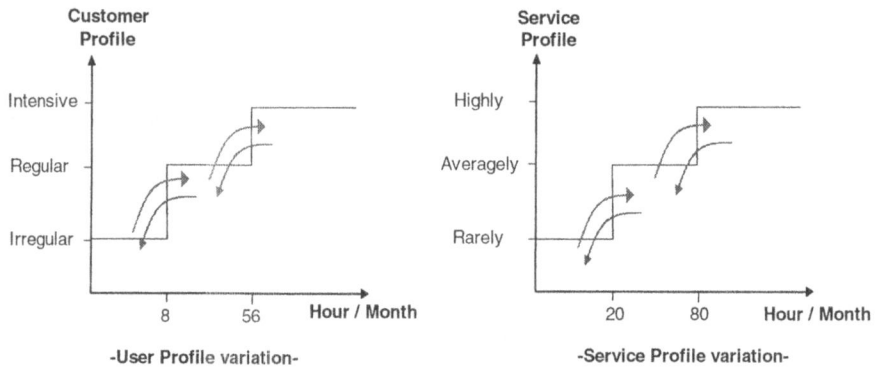

Fig. 2. User and Service Profiles Variation

4.3 Pricing Profile Strategy

Currently, we notice that large companies try to understand the behaviour of their customers in order to customise theirs offer to respond to the demand as well to establish a more accurate pricing strategy. Moreover, customizing the offers to the customers profiles will permit to respond to a real demand and increase the satisfaction of the customers while attracting new ones. Consequently, customers profiles are very important information for the business strategy of the companies. The profile becomes a central point of the overall service usage billing architecture. In this remains true in the telecommunication area. As Internet services providers offer more and more added value services, they have to change their billing strategy to integrate the user's profile in the price negotiation process. Therefore, we aim here to introduce a new pricing function and associated algorithm that take into account customers and services profile when calculating the service usage.

We introduce in the price calculation a function called $\Re(C\_Id, S\_Id)$ which takes as parameters the customer's profile identifier (C_Id) and the service's profile identifier (S_Id). Hence, we have defined the price determination function as a combination of the cost related to connection setup and maintenance P_C^j and the cost related to the usage of a particular service P_U^j. While the connection cost is considered here as independent from the customer profile (this constraint can be released in the future with work in wireless environment for example), the cost related to the service usage is fully dependent on the customer's profile and the type of service that is used (service profile).The total price determination function of jth class P_T^j is as follow:

$$P_T^j(t) = P_C^j + P_U^j * (\alpha.T + \beta.V) * \Re(C\_Id, S\_Id) \tag{1}$$

The usage cost calculation in the formula (1) can depend on the duration of the service usage $\alpha.P_U^j$, the volume of information exchanged during the service usage $\beta.P_U^j$ or both. We can notice that an operator can balance this cost between the parameters α or β to reflect a more sensitive pricing to duration or to volume. Our objective in this paper is to focus more on the benefit of introducing user and service profile in the pricing formula, however for more details about usage based pricing, you'll find information in [6].

In order to specify $\Re(C\_Id, S\_Id)$, we have based our approach on the results of Tianshu and al. [7]. In this work, the authors have implemented a call admission control based on pricing mechanism. They have adopted an exponential growth function that increases the price value when demand D_i^j or a current load for class j exceed its targeted capacity T_j.

$$P_j(t) = \begin{cases} P_{base}^j & if \ D_j(t) \le T_j \\ P_{base}^j * e^{\alpha^j \left[\frac{D_j}{T_j} - 1 \right]} & otherwise \end{cases}$$

where $P_j(t)$ denotes the price for class j at time t, and P_{base}^j is the base price for service class j and α^j is the convergence rate factor. However, in their case, the base price P_{base}^j and the targeted capacity T_j for each service class j are fixed value and usually they are pre-calculated based on the operator business strategy. For this reason we can classify this approach in the non-competitive market price and thus it can not affect the behaviour of the end users.

Based on this observation, we propose a $\Re(C\_Id, S\_Id)$ function that depends on the connection frequency of a particular customer and on his request frequency for each provided services. The R function is as follows:

$$\Re(C\_Id,\ S\_Id) = \begin{cases} \dfrac{1}{g^{\,j}(t)} * \dfrac{1}{f_i^{\,j}(t)} + \dfrac{D_i^{\,j}}{C_{av}^{\,j}} & if\ C_{av}^{\,j} - D_i^{\,j} \geq \lambda\% * C_{max}^{\,j} \\[3ex] g^{\,j}(t) * \dfrac{1}{f_i^{\,j}(t)} + e^{\frac{D_i^{\,j}}{C_{av}^{\,j}}} & otherwise \end{cases} \tag{2}$$

where:

$g^{\,j}$: utilization frequency of jth service.

$f_i^{\,j}$: utilization frequency of jth service by the ith user.

$D_i^{\,j}$: requested bandwidth for j^{th} service from the ith user.

$C_{av}^{\,j}$: current available bandwidth for j^{th} service.

The term $D_i^{\,j}/C_{av}^{\,j}$ represents the effect of the user i traffic on the global traffic increase. This term varies according to the user service usage and the type of service provided by the operator. Figure 3 illustrates our pricing strategy in general. When the load for particular service j is lower than $C_{av}^{\,j} - \lambda\% * C_{max}^{\,j}$, the price of this service decreases and converges towards the connection price $P_C^{\,j}$ proportionally to user and service profiles. Otherwise, this price will rapidly increase according to an exponential function.

We notice that the based usage price $P_U^{\,j}$ and λ are fixed and calculated according to some businesses consideration for all services.

We can see that our proposed pricing strategy gives a competitive market price aspect in the case of a non-congestion state or a congestion state of the network as opposite to a congestion based pricing or a flat pricing mechanism. Our mechanism encourages users to highly use available services in order to get more interesting prices (lower prices) for the requested services.

In the non-congestion state i.e. $D_i^{\,j} \lll C_{av}^{\,j} \Rightarrow D_i^{\,j}/C_{av}^{\,j} \approx 0$

(1) became depended only of the user and service request frequency.

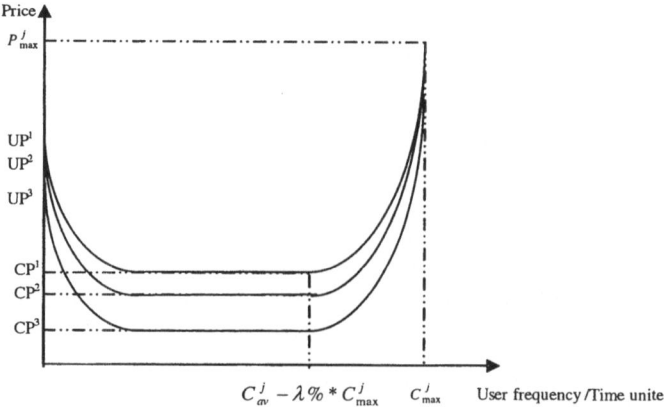

Fig. 3. General pricing strategy

$$\Rightarrow P_T^j(t) = P_C^i + P_U^j * (\alpha T + \beta V) * \frac{1}{g^j(t)} * \frac{1}{f_i^j(t)}$$

It also encourages users to use more intensively network resources in the case of non-congestion network. When the network is congested, the price of service usage increases rapidly and discourages customers to use the provided services. However, this price increases proportionally to user and service request frequency.

4.4 Dynamic SLA Negotiation Process

In this section we describe how to integrate a proposed pricing mechanism into the SLA negotiation architecture which is proposed in [8]. For this purpose, we suppose that All users have a representative mobile agent, which permits to negotiate requested service parameters. The SLA negotiation process is divided into two steps: Pre-service SLA, and On-service SLA (Figure 4). This format aims to facilitate a service negotiation between the customer and the ISP agent.

The price negotiation of requested service parameters between an ISP and a user is introduced at the SLA negotiation level, where the connection price and usage price are negotiated in Per-Service SLA and On-Service SLA respectively.

In *Pre-Service SLA* a customer and the ISP want to make a common agreement that contains the necessary information to identify the parties, the service, the time validity period, the access points, the connection cost, etc. For QoS agreement, Monitoring agreement and cost agreement, we have defined an *On-Service SLA* where the customer and the ISP negotiate the different agreements. The customer specifies the maximum and minimum boundaries for the negotiated SLA parameters as well as a priority in the negotiation process .

Generally, the time validity of *Pre-Service SLA* is long enough (month, year) while the time validity of *On-Service SLA* is short (hour, service duration, day).We note that, for all requested services a *Pre-Service SLA* step is not necessary. For a negotiation of locally requested services, a customer agent negotiates directly an SLA subscription with its ISP.

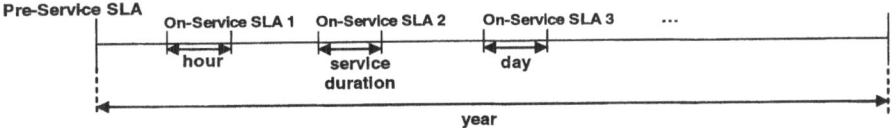

Fig. 4. Stage of SLA negotiation process

5   Over-Provisioning Versus Profile Based Pricing

The main objective of any ISP when he applies a new pricing mechanism is to optimize its resource usage while increasing its benefit and its customers satisfaction.

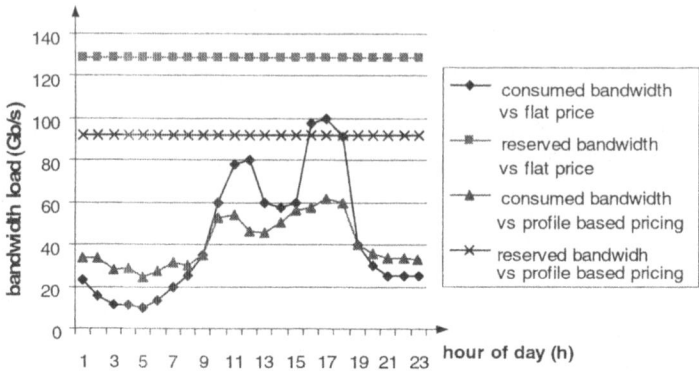

Fig. 5. Over-provisioning vs. Pricing mechanism

Nowadays, an important part of ISP investment is lost because of the over-provisioning strategy of network resources. In fact, the ISPs use this method because of the low cost of adding bandwidth and to enhance the satisfaction of their customers in order to not loose them. However, first we can not say that adding bandwidth is not expensive and also it is important to notice that this method is only efficient when the network usage is high. That means that even if over provisioning is interesting there is different manner to achieve it. The nature of the traffic generated by users application (Figure 5) is very variable and always push the providers towards adding more resources.

In this section, we show how our proposed pricing model affects the behaviour of the users to incite them to use more resources in the non-congestion period of the network. Figure 5 highlights different bandwidth load reserved according to the applied pricing mechanism. We can observe that, with the profile pricing model, the customers are interested to be informed about low charged period of service usage.

The ISPs benefit function $b_j(t)$ of jth class of the service is the maximization of the difference between the total users price and cost of class j reservation.

$$b_j(t) = Max\left[\sum_i P_i^j(t) - f_j(C)\right] \tag{3}$$

If we applied flat pricing, it is clear that ISPs would have increased their base price in order to cover the cost of the global reservation of bandwidth and to increase the benefit. However, this method doesn't affect user's behaviour. But if we apply our proposed pricing mechanism that affects directly the user's behaviour, the ISP increases its benefit saving Δf_j of the overall offers capacity of jth class, (equation 4).

This economy allows to the ISP to reinvest its financial resources to reduce its connection price and base usage price to become more competitive in the telecommunication market.

$$b_j(t) = Max\left[\sum_i P_i^j(t) - (f_j(C) - \Delta f_j)\right] \tag{4}$$

we can see that customers prefer to use resources when the price is the lowest. As a consequence, we will have a good balance between the users traffic load and the reserved volume capacity. The main idea of our approach is to move from a flat pricing strategy and a dynamic traffic to a dynamic profile based pricing strategy and flat traffic. If the expected behaviour of the customers is achieved, then the operator will benefit from a more predictable traffic pattern while proposing a novel pricing schema to the end customers.

6 Conclusion

In this paper, we have proposed a competitive pricing mechanism based on user and service profiles for internet services and we have integrated this pricing model into a dynamic SLA negotiation process. We have also studied the impact of our pricing model in the case of an over-provisioned network. The analytical evaluation carried out that our mechanism diminishes considerably the cost of the over-provisioning. This is obtained thanks to a good balance of the traffic load during a day-scale period. The concept called Unified Profile has been introduced. It aims to model User Profile and Service Profile. These two profiles helps the ISP to adapt its pricing strategy to the customer's behaviour and service demand. This approach allows him to increase its revenues while increasing the users satisfaction.

References

1. S. Blake et al, "Rfc 2474 - An Architecture for Differentiated Services," Network Working Group – IETF, Dec, 1998.
2. M. Falkner and al., "An Overview of Pricing Concepts for Broadband IP Networks," IEEE Communications Surveys, Sept. 2000.
3. L. DaSilva., "Pricing for QoS-Enabled Networks: A Survey," IEEE Communications Surveys, Sept. 2000.
4. F. Hartanto and G. Carle. Policy-based Billing Architecture for Differentiated Services. In Proceeding of IFIP Fifth International Conference on Broadband Communications (BC'99), Sep. 1999.
5. C. Gadecki. "The Price Point," Tele. Com Magazine. Nov. 1997.
6. C. A. Courcoubetis, and al., "A study of simple usage-based charging schemes for broadband networks", Telecommunication Systems, 15(3-4):323-343, 2000.
7. T. Li, and al., "Tariff-based pricing and admission control for DiffServ networks", IFIP/IEEE International symposium on integrated Management (IM'03), USA, March, 2003.
8. Y. Khene, and al., "Unified profile and agent based negotiation service to facilitate end-to-end SLA management", In Proc. 10th HP-OVUA Workshop, Switzerland, July, 2003.

Design and Performance of Asymmetric Turbo Coded Hybrid-ARQ

Kingsley Oteng-Amoako, Saeid Nooshabadi, and Jinhong Yuan

University of New South Wales, Sydney,
NSW 2052, Australia
k.oteng@student.unsw.edu.au

Abstract. The paper presents asymmetric turbo hybrid automatic-repeat-request (ATH-ARQ) schemes that employ component code selection for enhanced performance in Gaussian and fading channels. The resulting system provides for an efficient set of component code pairs from low to high SNR. In addition, a novel low-rate encoder structure that enables standard and asymmetric component code selection by puncturing. The paper illustrates that in certain cases the use of identical component codes yields better performance.

Keywords: Concatenated codes, fading channels

1 Introduction

Automatic-Repeat-reQuest (ARQ) schemes employing turbo codes have been considered in literature and shown to improve throughput performance in burst-error channels by providing rate flexibility [1]. An extension to ARQ, Hybrid forward-error-correction (FEC)/ARQ have been established as a means of exploiting the rate flexibility of ARQ and the preemptive error correction capabilities of FEC [2]. Hybrid-ARQ schemes with rate-compatible-punctured-turbo (RCPT) codes were shown to offer spectrally efficient performance given channel conditions [3]. Turbo hybrid-ARQ have been considered widely in literature due to the close to capacity performance when turbo codes are employed as a FEC scheme [4] [5] [6].

The original turbo encoder structure [7], employed two identical parallel concatenated recursive systematic convolutional (RSC) codes separated by a random interleaver. The component codes chosen for the encoders were typically fixed and selected so as to provide acceptable performance given channel state information [8]. It was shown that the performance at high-SNR required the use of component codes with a correspondingly high the effective free distance (d_e) [9].

In some cases, the combination of low-memory and high-memory order component codes, termed asymmetric turbo codes, can result in an improved performance for given channel conditions [10]. Asymmetric turbo codes, can be considered as a special case of multiple turbo codes [11,12]. Asymmetric turbo codes provide improved performance particularly in the waterfall region of a turbo code. A possible application for asymmetric turbo Hybrid-ARQ follows

J.N. de Souza et al. (Eds.): ICT 2004, LNCS 3124, pp. 1369–1381, 2004.

that of the current use of turbo hybrid-ARQ in a channel [13]. Ideally, a flexible encoder structure that enables component code selection based on channel requirements would further improve performance. Further by combining ARQ successfully with component code for adaptation of transmitted codewords, additional performance improvements can be obtained.

In designing turbo hybrid-ARQ, the selection of a component code required an analysis based on a single metric, usually d_e [14] [15]. Further, in turbo hybrid-ARQ, the selection of a component code was a trade-off between good performance at either low or high SNR. In designing asymmetric turbo hybrid-ARQ, the target error rate required by a component code pair must be selected such that the combined d_e results in the target FER being achieved. Indeed, if the combined free distance of a disparate component code pair is not lower than a required d_e, the minimum error rate constraint is not achieved. In addition, asymmetrically selecting component codes may not always yield a better performance than using identical component code pairs. It will be shown in this paper, that there are conditions where the use of identical component codes are beneficial, particular at extreme points of an SNR range.

This paper is organized as follows. In Section II, a summary is presented on component code convergence and performance at various SNR. In addition, a variety of decoding metrics are also discussed. In section III, a set of selection criteria are presented for a variety of SNR's based on the design metrics of section II. In section IV, a novel encoder structure is presented that enables both code rate and component code selection. In section V, numerical results are provided including a discussion on the results. We conclude in section VI.

2   Component Code Convergence

The error rate performance of a codeword is based on the ability of the codeword to converge during decoding. The convergence is a function of the component codes employed and the effect of the a priori received probabilities [16], [17].

In a SISO decoder, the a priori probabilities effect the convergence characteristics of the trellis decoder. However, given that a logarithm based expression is employed in both SOVA and log-MAP algorithms, large magnitude variations have a correspondingly minimal effect in the log-likelihood ratio [18]. Correct decoding in SISO is dependent on the nature in which the trellis responds to codewords as determined by the component employed.

The ability of a received sequence to travel the maximum-likelihood (ML) path in the trellis corresponds to the selection of an error-free branch at each stage of decoding. In the trellis, a large branch metric results in an inflexible decoder and minimizes gains in iterative decoding. Large branch metrics are analogous to codewords with a high proportion of larger output weight codewords. Hence, good performance of component codes at low SNR and the capability to regularly converge in a trellis is dependent on the ability to regularly generate low weight codewords [19].

Recently, the use of a pair of disparate component codes in turbo coding, termed asymmetric turbo codes, was shown to provide superior code performance over conventional turbo codes [20]. The coding gain in asymmetric codes is attributable to the combined performance of aided decoding in the waterfall region by a "weak" component code and the improved performance in the error-floor region by the "strong" component code [21]. Whilst previous work on asymmetric turbo codes has focused on the use of a single component code pair across an entire SNR range, we focus on the selection of component code pairs based on the SNR. Further, we do not limit our selection of component codes to asymmetric codes, opting to consider identical component codes in the event they offer superior performance.

2.1   Effective Free Distance

It is highlighted in literature that the convergence of turbo codes at high SNR is based primarily on d_e [22]. The d_e is a function of a weight-2 input sequence that commences and terminates in the zero-state, thus a low-weight output caused by the shortest possible path through a trellis based on weight-2 input sequences. The performance of a component code is thus highly dependent on the weight-2 input, and to a lesser extent, the weight-3 input sequence. The effective free-distance, of a turbo code based on identical parallel RSC codes is bounded as,

$$d_e = 2 + 2d_2 \tag{1}$$

where d_2 is the output weight of the shortest weight-2 input through a trellis and the "2" corresponds to the weight of the input sequence. Correspondingly, for an asymmetric turbo encoder, d_e is given by

$$d_e = 2 + d_{2,[1]} + d_{2,[2]} \tag{2}$$

where $d_{2,[1]}$ is the output-weight due to the first component code and $d_{2,[2]}$ the output-weight due to the second component code. Note, if $d_{2,[1]}$ and $d_{2,[2]}$ are identical the expression reduces to the standard turbo code case of equ. 1.

2.2   Minimum Weight Code Sequences

It was proposed in [19], that the analysis of [23] be extended to input sequences of weight-i, where $(d_i > 2)$, in order to get a more accurate analysis of codeword performance. The effect of $(d_i > 2)$ input weights becomes increasingly relevant where optimized interleavers are employed to break low weight inputs.

Component codes with competing weight-i characteristics were discriminated to find the best performing code, based on the ability of a code to minimize BER given a uniform interleaver [24]. The code design criteria of [19] was found to be a more accurate reflection of code performance than [23].

2.3   Irreducible and Primitive Polynomials

In [9], it was shown by simulation that the use of primitive polynomials as the feedback in a component code maximizes the output weight. The selection

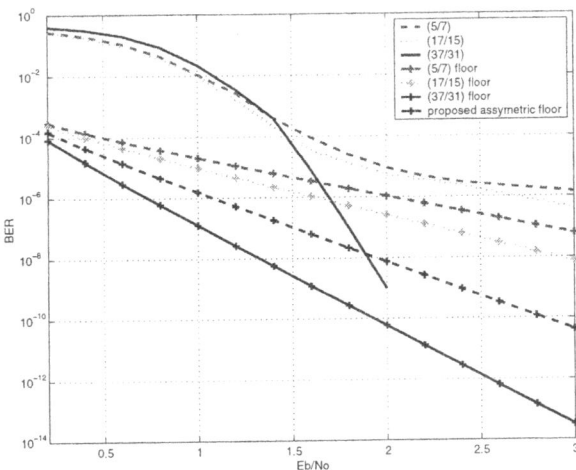

Fig. 1. Performance and error floor of selected component codes and proposed asymmetric component code: error floors denoted by '+'

of primitive polynomials, of memory order 1 to 4, as feedback correspondingly maximizes the free-distance and the performance of the turbo code at high SNR. The use of a primitive polynomial results in a "strong" code that is less likely to break the low-weight codewords during interleaving. Thus the use of primitive polynomials as component codes restrict the performance of turbo codes at low SNR.

2.4 Decoding Complexity

The decoding complexity of a component code is given by the memory-order, v, of the code. At low SNR, the performance of small order component codes outperforms that of larger order component codes due to "weak" codes having low-weight codewords readily broken during interleaving [19]. The performance of component codes as a function of memory order is presented in fig. 1. In addition, at high SNR the error floor achieved by larger component codes is lower than that of smaller order component codes [19]. The improved performance of larger memory order component codes is a result of component codes producing a larger d_e and an increased multiplicity in the minimum weight output sequences of weight-i, for $i > 2$.

3 Component Code Selection Criteria

3.1 Low-SNR Code Search Criterion

The design criterion for component codes at low SNR, can be accurately generalized by employing "weak" codes and minimizing the output weight sequence as detailed in section 2.2. The criteria can be summarized in terms of the metrics of decoding complexity, irreducible polynomial and primitive polynomial. The following are a set of proposed search criteria for component codes for targeting the FER region based on the metrics of section II,

1. limit candidate component code to irreducible and non-primitive feedback polynomials.
2. limitcomponent codes to $v \leq 3$ order polynomials
3. determine d_i for , $(2 \leq i \leq 6)$ for the set of component codes and select the component code pair that maximizes overall d_e of the asymmetric pair and aids performance at high-SNR

3.2   High-SNR Code Search Criterion

The BER performance floor of a codeword of length n, in an AWGN channel given d_e, can be approximated as [25], [14]

$$P_K(e) \simeq \frac{N_f w_f}{2n} erfc\left(\sqrt{d_e R \gamma}\right) \qquad (3)$$

where N_f is the number of codewords capable of generating the effective free-distance d_e, w_f is the average weight of the free-distance information word, γ is the SNR and R is the code rate. Correspondingly, the FER performance floor of the turbo code is approximated as,

$$P_K(e) \simeq erfc\left(\sqrt{d_e R \gamma}\right) \qquad (4)$$

Thus it is ibserved that the error-floor at high SNR is lowered by employing a high d_e. The design of a component code for targeting FER performance at high SNR can be summarized as;

1. Limit candidate component codes to irreducible and primitive polynomials
2. limit component codes to $v > 2$ order polynomials
3. determine d_i, $(2 \leq i \leq 6)$, of the set of component codes
4. limit selection to component codes that maximize the overall d_e of the asymmetric pair
5. select the a component code that maximizes the output weight spectra for weight-i, $d_i > 2$.

3.3   Medium-SNR/Waterfall Code Search Criterion

The design criterion for component codes at medium SNR, termed the *waterfall region*, requires that the component code remain "weak" enough to break low output weight codewords and provide flexibility during trellis decoding to ensure codewords achieve the BER constraint. We propose the following code search criteria for targeting the frame-error rate in the waterfall region.

1. determine d_i, $(2 \leq i \leq 6)$, of the set of component codes

2. select a component code based on the criteria of section 3.2.
3. select a second component code based on the criteria of 3.1 and also simultaneously maximize the overall d_e. This is expressed as

$$d_{e,[2]} \leq 2d_{e,[\phi]} - d_{e,[1]} \tag{5}$$

where $d_{e,[2]}$ is the effective free distance of the first component code and $d_{e,[\phi]}$ is the required effective free distance of the component code pair based on the BER constraint of the system.

Table 1. 4-state, 8-state and 16-state component codes and their effective free distances for rate 1/3 encoder structures

	v	component code $[g_0, g_1]$	d_e
A	1	[3,2] [9]	4
B	2	[7,5] [9]	10
C	2	[5,7]	8
D	3	[13,11] [9] [13]	14
E	3	[15,17] [9]	14
F	3	[17,15]	10
G	3	[13,15] [13]	14
H	3	[11,13]	8
I	4	[31,33] [9]	22
J	4	[37,21] [7] [9]	10
K	4	[23,35]	22
L	4	[23,25]	22
M	4	[31,27]	22
N	4	[37,23]	10
O	4	[37,31] [9] [26]	22
P	4	[25,37]	10

4   Novel ATH-ARQ Encoder Structure

In fig. 2, a rate 1/7 turbo encoder based on the parallel concatenation of non-systematic component codes, of $v = (2, 3, 4)$ is shown. The encoder structure is constructed based on the optimal component codes for low SNR, high SNR and the waterfall region.

It is observed from the figure that the rate 1/7 turbo encoder structure is based on parity bits of six rate 1/2 RSC encoders and the systematic bits of a

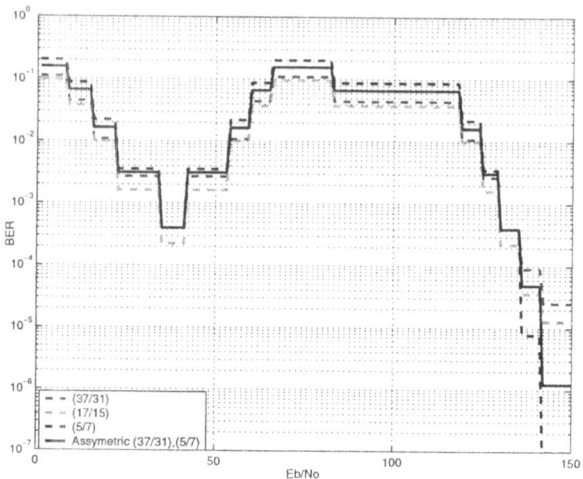

(a) Performance at $f_d = 2$ Hz and $\bar{\gamma} = 0.5$ dB

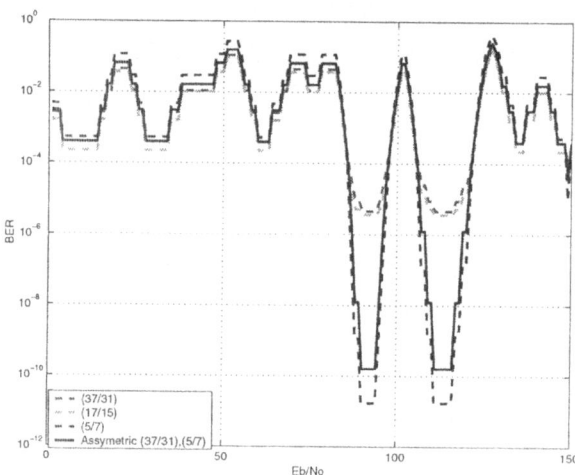

(b) Performance at $f_d = 6$ Hz and $\bar{\gamma} = 0.5$ dB

Fig. 2. Performance of various rate 1/3 component codes during Rayleigh distributed random walk in various fade rates

single encoder. This results in a systematic stream and six parity streams each of length ρ encoded bits. Thus the total encoded frame size prior to puncturing is a $(7 \times \rho)$ matrix where each row corresponds to either a systematic or parity stream and each column corresponds to a puncture period.

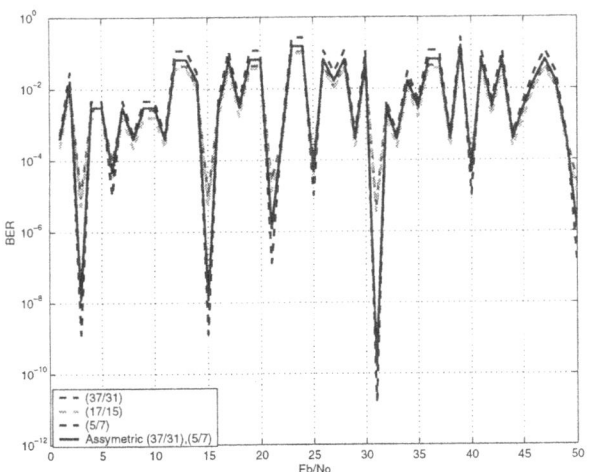

(a) Performance at $f_d = 60$ Hz and $\overline{\gamma} = 0.5$ dB

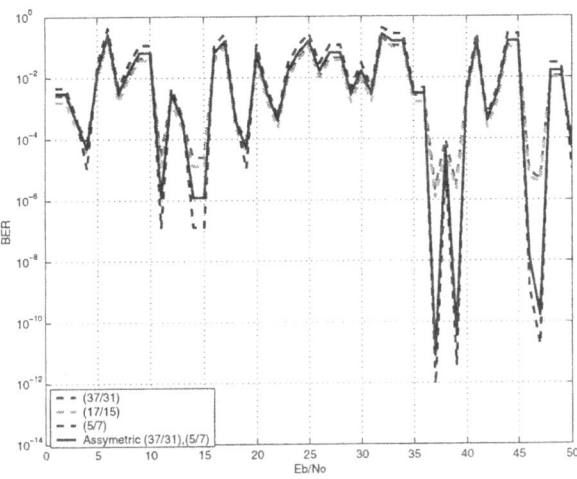

(b) Performance at $f_d = 120$ Hz and $\overline{\gamma} = 0.5$ dB

Fig. 3. Performance of various rate 1/3 component codes during Rayleigh distributed random walk in various fade rates

The puncturing of the encoded blocks in the proposed structure, amounts to sending a specific component code and code rate based on CSI. Thus, the component codes available are $\{(5, 7), (17, 15), (37, 31)\}$ and the code rates available are $\{(1/3), (2/5), (1/2), (2/3)\}$. Consequently S-P$P$ (Systematic-puncture periods) are employed to select code rate and component code pairs. The encoder employs two-stages of puncturing, the first stage is for component code selection and the second stage enables rate-matching. Given that three distinct component codes are employed in the mothercode, the first stage puncturing employs $(3!+1)$ distinct puncturing schemes and thus SP-6 puncturing at the first stage. The puncture patterns of the first stage puncturing are denoted by **P6** prefix. The selection of standard rate 1/3 encoders as proposed by [7], is given by puncturing the mothercode based on matrices $\{\mathbf{P6}(0), \mathbf{P6}(1), \mathbf{P6}(2), \mathbf{P6}(3)\}$,

$$
\begin{array}{cccc}
\mathbf{P6}(0) & \mathbf{P6}(1) & \mathbf{P6}(2) & \mathbf{P6}(3) \\
\begin{bmatrix} 1111 \\ 0000 \\ 0000 \\ 0000 \\ 0000 \\ 0000 \\ 0000 \end{bmatrix} &
\begin{bmatrix} 1111 \\ 1111 \\ 0000 \\ 0000 \\ 1111 \\ 0000 \\ 0000 \end{bmatrix} &
\begin{bmatrix} 1111 \\ 0000 \\ 1111 \\ 0000 \\ 0000 \\ 1111 \\ 0000 \end{bmatrix} &
\begin{bmatrix} 1111 \\ 0000 \\ 0000 \\ 1111 \\ 0000 \\ 0000 \\ 1111 \end{bmatrix}
\end{array}
$$

In addition, the selection of asymmetric codes can be obtained from the mothercode based on the following puncturing schemes $\{\mathbf{P6}(4), \mathbf{P6}(5), \mathbf{P6}(6)\}$

$$
\begin{array}{ccc}
\mathbf{P6}(4) & \mathbf{P6}(5) & \mathbf{P6}(6) \\
\begin{bmatrix} 1111 \\ 1111 \\ 0000 \\ 0000 \\ 0000 \\ 1111 \\ 0000 \end{bmatrix} &
\begin{bmatrix} 1111 \\ 0000 \\ 1111 \\ 0000 \\ 0000 \\ 0000 \\ 1111 \end{bmatrix} &
\begin{bmatrix} 1111 \\ 0000 \\ 0000 \\ 1111 \\ 1111 \\ 0000 \\ 0000 \end{bmatrix}
\end{array}
$$

The second stage puncturing enables rate matching based on the observed channel conditions. Given, the for the second stage three streams (one systematic and two parity) of subblocks of length ρ exist, SP-4 puncturing is required. The puncture patterns of the second stage are SP-4. Thus the puncturing matrices of $\{\mathbf{P4}(0), \mathbf{P4}(1), \mathbf{P4}(2), \mathbf{P4}(3)\}$ are employed,

$$
\begin{array}{cccc}
\mathbf{P4}(0) & \mathbf{P4}(1) & \mathbf{P4}(2) & \mathbf{P4}(3) \\
\begin{bmatrix} 1111 \\ 1111 \\ 1111 \end{bmatrix} &
\begin{bmatrix} 1111 \\ 1110 \\ 0111 \end{bmatrix} &
\begin{bmatrix} 1111 \\ 1010 \\ 0101 \end{bmatrix} &
\begin{bmatrix} 1111 \\ 0001 \\ 0100 \end{bmatrix}
\end{array}
$$

As an example, suppose an encoder based on the asymmetry of $v = 2$ and $v = 4$ component codes is required at rate 2/5, **P6**(6) and **P4**(1) would be employed sequentially to achieve the required puncture rate and assymetry.

5   Numerical Results

5.1   Channel Assumptions

We assume coherent signalling over an AWGN and fading channel where γ corresponds to E_s/N_o, where E_s is the energy-per-symbol and N_o is the single-sided power spectral density of the noise. The output of the channel, therefore, is characterized at a given time as, $y = \alpha x + n$, where α is a Rayleigh distributed fading variable, antipodal modulated (i.e. $x \in \{+1, -1\}$) and n is the zero-mean Gaussian random variable of standard deviation $\sigma = \sqrt{N_o/2E_s}$. A SR-ARQ scheme is considered with sufficiently larger buffers. An error-free low-capacity feedback channel is considered over which positive acknowledgements (ACK) and negative acknowledgements (NACK) can be sent. The channel conditions are assumed known at all times, thus channel state information (CSI) is available at both the transmitter and receiver sides.

The source encoder employs a binary turbo encoder with signal mapping to a QPSK modulator considering an encoded sequence of k information bits and a total length of n bits, such that the rate of the code is represented as $R = k/n$ with $n - k$ parity bits.

The analysis presented is based primarily on AWGN channel and extended to a flat fading channel. It is assumed that the channel remains constant over the entire transmit period of a frame such that the packet can be estimated and assumed known at the receiver.

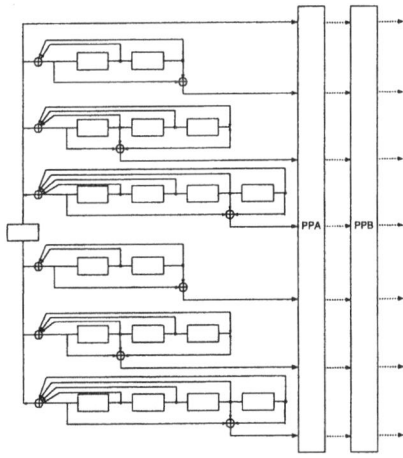

Fig. 4. Novel ATH-ARQ encoder implementation structure

5.2   Code Search Example

We present here a code search example based on the criterion proposed in section 3.1, 3.3 and 3.2 with consideration of asymmetric codes. In Table I, the d_e of a variety of component codes considered in literature, of order $2 \leq v \leq 4$ are presented. In addition, the distance weight spectra $(2 \leq d_i \leq 6)$ of each code is employed in the analysis criteria. The set of distance weight spectra available in [14,27], can be used in the analysis provided.

In Section 2, it was established that at low-SNR optimum performance required "weak" component codes. The selection indicated a requirement for low memory order and non-primitive component codes. Both code A, B and C are low memory order component codes $(v \leq 2)$, however code B has a primitive polynomial that decreases the occurrence of breaking low-weight input sequences during interleaving given criterion (2) of section 3.1. However, code B clearly generates the larger d_e, as observed in Table 1. Considering the combined metrics for selection of component codes at low-SNR, code B appears to be the best component code. Given that the combined effect of employing two "weak" component codes results in the best performance at low-SNR, parallel encoders consisting of two code B generators are suggested.

The selection of a component code at high-SNR is based solely on its distance spectra characteristics; initially maximization of the d_e and in addition, consideration of distance weight spectra characteristics. Based on the selection criteria in section 3.2, codes I, K, L, M and O offer $d_e = 22$, and are all considered high memory order (i.e. $v \geq 3$). However, observation of $Z, (1 \geq Z \geq 6)$ [21], highlights that code L produces the lower output weight codewords, generating $Z = 8$ for $W = 2$. Code O also employs a reducible feedback polynomial. Thus codes I, K and M can be considered as candidate component codes for high-SNR. In order to obtain maximum performance at high-SNR, an identical component code pair of code O would be required to maximise d_e.

It was established that at medium-SNR, a "weak" component code is required for performance at low SNR and another d_e component code for maximization of the distance spectra. Given that code B has been suggested as the optimal component code at low-SNR, it can be employed as the "weak" code in a component code pair. However, due to the relatively low d_e of code B, the use of this component code for situations where the γ changes rapidly (i.e. fading channels) would limit performance. In addition, the low d_e would limit performance in systems requiring a low BER constraint. Thus code D is suggested as an alternate "weak" component code based on the criteria of section 3.3. Code O, can be employed as the "strong" component code for operation at medium-SNR. Therefore an asymmetric structures, consisting of code O and code B for AWGN channels and code O and code D for fading channels, would yield the best throughput.

Based on an application of the design metrics of section II, it has been demonstrated that a turbo encoder structure based on identical component code pairs offer the best performance at low-SNR and high-SNR. In order to achieve optimal performance in fast-fading and non-coherent channels, an asymmetric turbo encoder structure based on the detailed design metrics offer the best performance.

The performance of the asymmetric code is compared to various component codes based on a random walk in a Rayleigh fading channel in fig. 3. The simulation is based on code B (5/7), code F (17/15), code O (37/31) and asymmetric component code based on code B (5/7) and code O (37/31). It is observed in fig. 2(a) and fig. 2(b), that in slow-fade scenarios of low-SNR component code O offers the best performance. It is observed in fig. 3(a) and fig. 3(b), that the asymmetric code offers better performance than code O at low-SNR and good performance at high-SNR.

6   Conclusion

The design criteria for throughput maximization of turbo codes at low, medium and high SNR was proposed based on complexity and distance spectra . The above criteria considered the use of component codes in an asymmetric structure. The optimal component codes were obtained based on the proposed criteria. It was shown that by using an asymmetric turbo encoder structure performance improvements in the dynamic SNR range is achieved over standard turbo encoder structures, particularly at low and medium SNR. A novel low-rate encoder structure was proposed based on the optimal component code pairs.

References

1. S. Lin and J. Costello, *Error Control Coding Fundamentals and Application*, Prentice-Hall, Englewood Cliffs, NJ, 1983.
2. S. Kallel, "Analysis of memory and incremental redudancy arq schemes over non-stationary channels," vol. 40, pp. 1474–1480, 1992.
3. R. H. Deng, "Hybrid arq schemes employing coded modulation and sequence combining," *Electron. Lett.*, vol. 42, pp. 2239–2245, May 1994.
4. P. Jung, J. Pleiochinger, M. Foestsh, and F. M. Berens, "A pragmatic approach to rate compatible punctured turbo codes for mobile radio applications," in *Proc. IEEE International Conference on Advances in Comm. And Control*, Nagoya, Japan, June 1997.
5. D. N. Rowitch and L. B. Milstein, "Rate compatible puncture turbo (rcpt) codes in hybrid fec/arq," in *Proc. IEEE International Conference on Comm. Theory (Globecomm'97)*, Nov. 1997, pp. 55–59.
6. L. Lin, R. Yates, and P. Spasojevic, "Discrete adaptive transmission for fading channels," in *Proc. IEEE International Conference on Communications (ICC'01)*, Nagoya, Japan, Jan. 1997, pp. 290–294.
7. C. Berroux and A. Glavieux, "Near optimum error correcting coding and decoding: Turbo-codes," vol. 1, pp. 77–79, May 1997.
8. A. S. Barbulescu and S. S. Pietrobon, "Rate compatible punctured turbo codes," vol. 44, pp. 591–600, May 1996.
9. S. Benedetto and G. Montorsi, "Design of parallel concatenated convolutional codes," *Electronic Letters*, vol. 31, pp. 534–535, Mar. 1995.
10. O. Y. Takeshita, O. M. Collins, P. C. Massey, and D. J. Costello, "A note on assymetric turbo codes," Mar. 1999, vol. 3, pp. 69–71.

11. P. C. Massey and D. J. Costello Jr, "New low-complexity turbo-like codes," in *Proc.IEEE Information Theory Workshop (ISIT'98)*.

12. C. He, A. Banerjee, P. C. Massey, and D. J. Costello Jr, "On the performance of low complexity multiple turbo codes," in *Proc.IEEE of the 40th Annual Allerton Conference on Communication, Control, and Computing, (Monticello, IL)*.

13. "Physical layer aspects of utra high speed downlink packet access," White Paper, May 2000.

14. S. Benedetto and G. Montorsi, "Unveiling turbo codes: Some results on parallel concatenated coding schemes," vol. 42, pp. 409–428, Mar. 1996.

15. S. Benedetto and G. Montorsi, "Average performance of parallel concatenated block codes," *Electronic Letters*, vol. 31, pp. 156–158, Feb. 1995.

16. J. Hagenauer, E. Offer, and L. Papke, "Iterative decoidng of binary block and convolutional codes," vol. 42, pp. 429–445, Mar. 1996.

17. J. Hagenauer and P. Hoeher, "A viterbi algorithm with soft-decision outputs and its application," in *Proc. IEEE of Global Communications (Globecom'89)*, Nov. 1989, pp. 1680–1686.

18. J. Hagenaur, E. Offer, and L. Papke, "Iterative decoding of binary and convolutional codes," vol. 42, pp. 429–445, Mar. 1996.

19. S. Benedetto and R. Garello, "A search for good convolutional codes to be used in convolutional codes to be used in the construction of turbo codes," vol. 46, pp. 1101–1115, Sept. 1998.

20. P. Massey, O.Y. Takeshita, O.M.Collins, and D. Costello, "Assymetric turbo codes," in *Proc.IEEE International Symposium Inforamtion Technology (ISIT'98)*.

21. S. Benedetto and R. Garello, "A search for good convolutional codes to be used in convolutional codes to be used in the construction of turbo codes," vol. 46, pp. 1101–1115, Sept. 1998.

22. D. Divsalar and R. J. McEliece, "Effective free distance of turbo codes," vol. 30, pp. 1701–1719, July 1982.

23. D. Divsalar and F. Pollara, "On the design of turbo codes," Tech. Rep. TDA Progress Report 42-123, Digital Equipment Corporation, MA, Nov. 1995.

24. H. R. Sadjadpour, N. J. A. Sloane, M. Salehi, and G. Nebe, "Interleaver design for turbo codes," .

25. K. R. Narayanan and G. L. Stuber, "A novel arq technique using the turbo coding principle," vol. 1, pp. 49–51, Mar. 1997.

26. D. Divsalar, S. Dolinar, and F. Pollara, "Proposal for ccsds turbo codes deep sapce and new earth," Tech. Rep. JPL, NASA, MA, Nov. 1996.

27. S. Benedetto and G. Montorsi, "Performance evaluation of turbo-codes," *Electronic Letters*, vol. 31, pp. 163–168, Feb. 1995.

ICT2004 Poster Papers List

Improving Power Delay Profiles Estimates with Wavelet Based De Noising
G.L. Siqueira, M.H.C. Dias

Performance Evaluation of Hardware Cryptographic Algorithms for Ad Hoc Network
A.A. Pires, A.C.P. Pedrosa, A.C.C. Vieira, A.C.M. Filho

Burst-Like Packet Loss Effects over the Quantisation of Speech Lsf Paremeters
A. Alcaim, F.D. Backx, R.C. Lamare

Study of a Microstrip Antenna with Pbg Considering the Substrate Thickness Variation
J.F. Almeida

A New Proposal for Mobility Extension to H.323 Using Sdl
S.W.K. Chung, W.C. Borelli

A New Packet Fec Approach with High Packet Recovering Capability
E. Liu, G. Shen, L. Gui, S. Jin, X. Xu

Design of Tx Block of Antenna System for Satellite Communication
H.S. Noh, K.H. Lee, S.H. Son, S.I. Jeon, U.H. Park

A Multicast-Linked Lightweight Solution for Mobile Sctp-Based Ip Mobility
L. Wang, X. Ke, X. Ming-Wei

Performance Analysis of Dft-Based Method for Clipping Noise Suppression in Ofdm Systems over Fading Channels
P. Azmi, R. Ali-Hemmati

Performance Comparison Between Bpsk and Bppm Th-Uwb Radio Systems in the Presence of Narrow-Band Interference Using Proper Waveform Design
H. Khanee, P. Azmi

Interference Suppression Consisting of Pre-distortion Filtering and Beamforming with Joint Transmit Diversity
A. Correia, M.P.G.M. Silva

Frequency Offset Estimation in Ofdm Systems
C.C. Chiu, C.W. Tung

Heterogeneous Polarization Transmission Diversity Based on Alamouti Coded Ofdm
C.J. Ahn, H. Harada, S. Takahashi, Y. Kamio

Synthesis of Freestanding Frequency Selective Surfaces by Using Neural Networks
A.M. Martins, A.L.P.S. Campos

Double Pass Broadband Edfa: Design and Characterization
A.A. Juriollo, J.B. Rosolem, M.A. Romero, R. Arradi

Design and Characteristics of Optical Filters Using Silicon Pbg Nanostructures
A.A.P. Pohl, P.T. Neves

End-To-End Qos Provisioning in an Ip-Based Umts Terrestrial Radio Access Network
A. Molinaro, A. Iera, G. Araniti, S. Pulitano

A Novel Configuration of Optical Cross-Connects Based on 1 X 2 Switch Elements
C.J.A.B. Filho, E.A.J. Arantes, J.F.M. Filho, S.C. Oliveira

Adaptive Identification of Feedback Systems Using an Artificial Immune Network
F.J.V. Zuben, J.M.T. Romano, L.N. Castro, L.T. Duarte, R.R.F. Attux

Infinite Sequences over Finite Fields and New Digital Transforms
H.M. Oliveira, M.M.C. Souza, M.M. Vasconcelos, R.M.C. Souza

A Linear Equalizer Based on Generalized Orthonormal Bases
G. Favier, J.C. Mota, R.R. Araujo

Ultra-Wideband Pulse Design Approach for Multiple Narrowband Interference
Suppression
H.Gao, Y. Liu, Z. Luo

Multicast in the Diffserv Domain: A Case Study
K.B. Carbonaro, P.R. Guardieiro

Experimental Analysis of Forward Error Correction Improvements in the Presence of
Fwm and Dispersion
A.C. Bordonalli, E. Mobilon, M.R.X. Barros

Performance and Dimensioning Analysis of Optical Packet Switching Access Networks
with Variable Traffic Demands
E. Moschim, F.R. Barbosa, L.H.B. Nascimento

Space-Time Block Codes in Frequency Selective Channels
M.M. Hashemi, M. Chavoshi

2 X 40 Gbit/S Wdm Soliton Transmission Improvement by Initial Time Delay Technique
C.J.A.B. Filho, F.W.B. Rech, J.F.M. Filho, L.P. Salles, T.F. Vieira

Internet Aggreta Traffic: An Analysis at Flow Level
L. Rodrigues, P.R. Guardieiro

Improving End-System' Performance with the Use of the Qos Management
J.S. Barbar, M.A. Teixeira

A Bluetooth Scheduling Algorithm Using Channel State Information
J.H. Kleinschmidt, L.A.P.L. Júnior, M.E. Pellenz

Statistics of Duration and Number of Events of Attenuation in Slant-Path Links at the
Ku-Band in Equatorial Brazil
E.C. Miranda, L.A.R S. Mello, M.S. Pontes

A Wsdl Extension for Protecting Web Services with Ipsec
C. Maziero, C. Ditzel, E. Jamhour

Considerations on the Performance of a Hybrid Resource Allocation Scheme
J.R.B. Marca, L.C. Mello

On Error Rate of Wavelet Coded Psk System over Awgn Channels
E.L. Pinto, F.M. Assis, L.F.Q. Silveira

Optimizing a Multi Agent Based Architecture to Distribution of Internet Services
A.C.F. Thomaz, A.M.B. Oliveira, A. Serra, F. Jackson, G. Cordeiro, M. Franklin

Performance Analysis of Adaptive Modulation in Wcdma/Hsdpa
J.C.B. Brandão, R.D. Vieira, R.J.A. Corrêa

Fast God: A New Version of Godzuk Cryptographic Algorithm for Wireless Network Applications
A.C. Pinho, A.C.C. Vieira, M.V. Fernandes, S.L.C. Salomão

Optical Multiplexing Through Fiber Four-Wave Mixing: Generation and Transmisson of Ask-4 Signals
E. Moschim, E.A.M. Fagotto, I.E. Fonseca, M.L.F. Abbade, R.S. Braga

A New Look at Space-Time Coding: Performance Improvement by Using Rotated Constellations and Stack Algorithm
F. Madeiro, M.S. Alencar, W.T.A. Lopes

Multilayers Ebg Structures Design by Fast Wave Concept Iterative Procedure
G. Fontgalland, H. Baudrand, N. Raveu, P.I.L. Ferreira, R. Garcia, T.P. Vuong

Author Index

Abdoli, M.J.   554
Abreu, R.M.   858
Adamopoulos, D.X.   221
Affès, S.   866
Afifi, H.   688
Aghasaryan, A.   820, 841
Aghvami, A.H.   450
Agoulmine, N.   1359
Agrawal, D.   388
Aguayo, L.   207
Aguiar, R.L.   1093, 1114
Aïssa, S.   607, 942
Al Agha, K.   1261
Albert, B.B.   94
Alberti, A.   772
Alcoforado, M.L.M.G.   122
Alencar, M.S.   60, 67, 568
Ali, A.   363
Allal, D.   38
Almeida, A.L.F. de   28
Almeida, C. de   312
Almenar, V.   1252
Alshaer, H.   1183
Al Naamani, A.A.   1218
Al Naamany, A.M.   1218
Andrade, R.   811, 950
Andrade, R.M.C.   654
André, P.   267, 766
Aniba, G.   942
Apolinário Jr., J.A.   488
Araabi, B.N.   1157
Araniti, G.   1
Assis, F.M. de   94, 100, 1023, 1077
Assis, K.D.R.   735
Atmaca, T.   1173
Attar, A.R.   106
Azmi, P.   541, 548

Bang, J.-W.   425
Bar-Ness, Y.   471
Barán, B.   1029, 1107
Baranowska, A.   711
Barbar, J.S.   601
Barbosa, F.R.   272
Barceló, F.   932

Barraca, J.P.   1093
Barras, D.   891
Barreto, G.A.   207
Barros, F.J.B.   53
Basu, K.   1303
Bates, S.   7
Begaud, X.   38
Beijar, N.   369
Benveniste, A.   820
Ben Mamoun, M.   1173
Boef, E. den   1337
Bonani, L.H.   272
Bonatti, I.S.   82
Bordim, J.L.   43, 461
Bordonalli, A.C.   727
Bottoli, M.   772
Bourdoucen, H.   1218
Braga, A.P.   811
Braga, A.R.   1122
Brandão, P.   1210
Büren, G. von   891
Burness, L.   1128

Callado, A.C.   1151
Calíope, P.B.   1035
Campello de Souza, M.M.   510
Campello de Souza, R.M.   510
Canet, M.J.   1252
Cappabianca, F.   991
Cardoso, L.S.   22
Carli, M.   991
Carta, C.   891
Carvalho, E.A.   1041
Carvalho, G.H.S.   595
Carvalho, S.V.   595
Cassales Marquezan, C.   782
Castel, H.   214
Castro, J. de   950
Castro, M.F. de   257, 654, 1317
Castro Andrade, R.M. de   251
Castro e Silva, J.L. de   251
Cavalcanti, D.   388
Cavalcanti, F.R.P.   22, 28, 431, 1122
Celestino Jr., J.   616, 622
Cerqueira Jr., A.   287

Cerqueira Jr., A.   287
Chaves, D.P.B.   88
Chaves, F.S.   431
Chen, D.   1282
Chen, L.-j.   704
Cherkaoui, S.   354
Choi, H.-W.   156
Chung, K.   171
Ciavaglia, L.   1173
Ciochina, S.   560
Conforti, E.   304, 717
Coppelmans, C.   772
Correia, A.   16, 900, 908, 1007
Costa, J.C.W.A.   595
Costa, R.C.S.   1035, 1041
Costa-Requena, J.   369
Cowan, C.F.N.   516
Creado, J.   369
Cricelli, L.   192
Crichigno, J.   1029, 1107
Crisóstomo, S.   1210

Damasceno Matos, V.   251
Dante, R.G.   336, 342
Darmaputra, Y.   234
Das, S.K.   1303
Dauwels, J.   150
Dawson, S.   876, 1057
Dawy, Z.   660
Deng, Z.-H.   381
Despins, C.   866
Dias, E.Z.V.   1246
Diniz, A.L.B.P.B.   622
Djerourou, F.   348
Doğançay, K.   400, 1327
Dohler, M.   450
Doyen, G.   801
Dragios, N.   227
Drago, R.B.   792
Duarte, O.C.M.B.   443

Eardley, P.   1128
Edwards, C.   1310
Eisl, J.   1128
Elhakeem, A.K.   1282
Ellinger, F.   891
Enescu, A.A.   560
Espinosa, K.   743

Fabre, E.   820

Farias, J.E.P. de   958
Ferdosizadeh, M.   554
Fernandes, C.A.R.   498
Fernandes, H.   267
Fernandez, M.P.   437
Ferro, A.   1195
Festor, O.   801
Filho, B.   950
Fisal, N.   363
Flores, S.J.   1252
Fonseca, M.   1261, 1359
Francês, C.R.L.   595
Freire, M.M.   760
Freitas Jr., W.C.   28
Frery, A.C.   1151
Friedman, T.   1347
Frota, R.A.   207

Gaïti, D.   257, 1317
Galdino, J.F.   568
Gallep, C.M.   717
Gao, J.-C.   381
Garcia-Palacios, E.   876
Gastaldi, M.   192
Geus, P.L. de   1083
Giovano, C.   950
Girão, J.   1093
Giraudo, E.C.   322
Gkelias, A.   450
Głąbowski, M.   1101
Goalic, A.   163
Gomes, D.   1114
Gonçalves, C.H.R.   654
Gonçalves, P.   1114
González, C.   607
González Serrano, F.-J.   589
Granelli, F.   1289
Grivet, M.A.   1067
Guimarães, A.G.   1067
Guimarães Nobre, E.   616
Guo, X.H.   826
Gutiérrez, J.   369

Haar, S.   820
Hämäläinen, T.D.   1163, 1236
Hännikäinen, M.   1163, 1236
Han, K.-A.   1296
Hancock, R.   1128
Haq, M.A.   241
Harbilas, C.   227

Harmantzis, F.   183, 848
Hassan, M.   394
Hébuterne, G.   214
Hepworth, E.   1128
Hernandez-Figueroa, H.E.   287
Hof, A.   1013
Holm-Nielsen, P.V.   753
Horlait, E.   1183
Hossain, M.   394
Hunziker, T.   43

Ibarrola, E.   1195
Iera, A.   1, 1267

Jäckel, H.   891
Jalili-Kharaajoo, M.   638, 1157
Jamhour, E.   1204
Jard, C.   820, 841
Jeanne, M.   914
Jeong, I.   924
Jeppesen, P.   753
Jiang, H.   704
Jung, J.-W.   671
Junior, S.   267

Kabaciński, W.   711
Kahng, H.-K.   671
Kang, C.-H.   425
Kannisto, J.   1236
Kantola, R.   369
Karetsos, G.   227
Kazemipour, A.   38
Keen, T.M.   74
Kelner, J.   388, 1151
Khene, Y.   1359
Khoukhi, L.   354
Kim, C.H.   997
Kim, H.   171
Kim, H.-s.   997
Kim, Y.-K.   671
Kirstädter, A.   1013
Kliazovich, D.   1289
Koivisto, A.   1163
Kongtong, N.   660
Koonen, A.M.J.   753
Korst, J.   1337
Krief, F.   628
Kurose, J.   1347
Kwan, M.   400

Lacerda Neto, R.L. de   28
Lamare, R.C. de   134
Lamparter, B.   1093
Latiff, L.A.   363
Lavault, C.   348
Lecarpentier, H.   1317
Leduc, G.   644
Lee, B.   171
Lee, J.-J.   1296
Lee, T.-H.   410
Leite, J.C.B.   437
Lemes Proença Jr., M.   772
Levialdi, N.   192
Liberal, F.   1195
Lilith, N.   1327
Lima, C.H.M. de   1122
Lima, E.R. de   1252
Lima, J.B.   510
Lima, M.   267
Liu, H.-s.   199
Liu, J.   410
Liu, Y.-A.   381
Loeliger, H.-A.   150
Lorenz, P.   760
Loscrì, V.   417
Louveaux, J.   142

M'hamed, A.   257
Macedo, P.E.O.   207
Machado, J.C.   251, 811
Maciel, T.F.   22, 431
Mackay, M.   1310
Magalhães, K.   950
Makouei, B.   554
Marano, S.   417, 964, 1143
Marina, N.   128
Marques, R.C.P.   1035, 1041
Marques da Silva, M.   900, 908
Marquet, A.   1007
Marshall, A.   410
Martínez Ramón, M.   589
Martins, N.   1007
Martins-Filho, J.F.   336, 342
Marullo, O.   1274
Maruthi, P.   983
Marvasti, F.   541, 548, 554
Matsumoto, M.   241
Maziero, C.   1204
Meddahi, A.   688
Medeiros, F.N.S.   1035, 1041

Medeiros, R.A.C.   100, 1077
Mello, D.A.A.   328
Mello Gallep, C. de   304
Mendes, L.S.   772
Meo, P. De   1
Merghem, L.   1317
Meyer, W.   1013
Mihailovic, A.   1128
Miladinovic, I.   1136
Moeneclaey, M.   114, 150
Molinaro, A.   1143, 1274
Monteiro, P.   766
Moschim, E.   272, 336, 342
Mota, J.C.M.   28, 207, 498
Motoyama, S.   698
Moungla, H.   628
Moura Oliveira, T.A.   616
Munaretto, A.   1261
Muñoz, A.   1195

Nakagawa, M.   924
Nataf, E.   801
Natalizio, E.   1143
Navaux, P.O.A.   782
Nefedov, N.   504, 532
Nerguizian, C.   866
Neri, A.   991
Neto, R.A.O.   431
Ngo, V.-D.   156
Nguyen, V.H.   1173
Nobrega, K.Z.   287
Nogueira, R.   267, 766
Nooshabadi, S.   1369

Ohira, T.   43
Oliveira, A.R.   16
Oliveira, H.M. de   482, 510, 526
Oliveira, J.C.R.F.   727
Oliveira, M.   654
Oliveira, R.M.S. de   53
Oliveira Marques, M. de   82
Oteng-Amoako, K.   1369

Pádua, F.J.L.   272, 336
Paillard, G.   348
Paleologu, C.   560
Panaro, J.S.G.   312
Papandreou, C.A.   221
Park, S.-C.   156
Pasquale, F. Di   287

Paula, F.S. de   1083
Pavani, G.S.   328
Pelegrini, J.U.   328
Pereira, D.   267
Perfecto, C.   1195
Peucheret, C.   753
Picart, A.   163
Pillo, F. Di   192
Pimentel, C.   88
Pinchon, D.   142, 578
Pinto, E.L.   568
Polo, V.   753
Ponte, P.R.X.   622
Popa, D.   1173
Prasad, A.R.   973
Prat, J.   753
Prior, R.   1210
Pujolle, G.   1261, 1359
Pyndiah, R.   163

Qian, L.   826
Quan, W.   381
Queiroz, W.J.L.   60, 67

Rahman, M.   848
Rahman, T.A.   74
Ramos, A.L.L.   488
Ramos, A.S.   792
Ramos, R.V.   322
Rango, F. De   417, 964
Raposo, L.   678
Rauch, P.   743
Ravelomanana, V.   348
Rios, R.   811
Rocha, C.A. da   835
Rocha, J. da   267, 766
Rocha Jr., V.C. da   122
Rodoni, L.   891
Rodrigues, E.B.   1122
Rodrigues, J.J.P.C.   760
Rodrigues, R.M.   595
Roh, B.-h.   997, 1226
Rosolem, J.B.   727
Rostami, A.   282, 296
Roy, A.   1303
Roy, N.   1303
Ruggeri, G.   1267
Ruttik, K.   885
Ryou, J.-C.   1296

Sadok, D.F.H.   388, 1151

Salles Garcia, A.   792
Sampaio-Neto, R.   134
Sanger Alves, R.   782
Santos, C.R. dos   698
Santos, J.   678
Santos, M.R.   858
Santos, R.O. dos   53
Santos-Magalhães, N.S.   526
Sargento, S.   1210
Sari, R.F.   234
Sauze, N. Le   1173
Savasini, M.S.   328, 735
Schmatz, M.   891
Seller, O.   914
Seok, S.-J.   425
Sezer, S.   516, 876, 1047, 1057
Shah, P.   1303
Sharafeddine, S.   660
Shariat, M.   554
Sheikhi, A.   106
Siaud, I.   914
Siclet, C.   142, 578
Sikiö, J.   1163
Silva, E.B.   22
Silva, F.G.S.   60, 67
Silva, J.B.R.   322
Silva, R.I. da   437
Silva, Y.C.B.   22
Silva Villaça, R. da   792
Siohan, P.   142, 578, 914
Sirisena, H.R.   394
Skivée, F.   644
Sobrinho, C.L. da S.S.   53
Solá, R.   471
Sousa, E.S.   568
Sousa, M.J.A. de   858
Sousa Jr., V.A. de   1122
Souza, J.N. de   251, 811, 835
Souza, L.G.M.   207
Souza, M.M.C. de   482
Souza, R.M.C. de   482
Sridhar, G.   983
Sridhar, V.   983
Stasiak, M.   1101
Stewart, V.   516
Szczeciński, L.   607

Tafur Monroy, I.   753
Tanaka, S.   43, 461
Teixeira, A.   267, 766

Teixeira, M.A.   601
Tejo, F.A.F.   1023
Tenca, A.   991
Thomas, J.   841
Toal, C.   876, 1047
Toktar, E.   1204
Tomaz, A.C.F.   622
Towsley, D.   1347
Tripodi, D.   1267
Tropea, M.   964
Tsoukatos, K.P.   227

Uchôa, D.C.   1023
Uchôa-Filho, B.F.   88
Ueda, T.   461
Uria Recio, P.   743
Ursino, D.   1

Vale do Nascimento, V.   958
Vanhatupa, T.   1163, 1236
Vanhaverbeke, F.   114
Vanwormhoudt, G.   688
Vasconcelos, M.M.   482
Vazão, T.   678
Vegas Olmos, J.J.   753
Veltri, L.   1274
Verhaegh, W.F.J.   1337
Viana, W.   950
Villela, B.A.M.   443
Vivacqua, R.P.   717

Waldman, H.   328, 735
Wang, C.-J.   381
Wang, H.   973
Wang, W.-B.   381
Wang, X.   1282
Weiss, G.M.   1246
Westhoff, D.   1093
Wolf, E.   1013
Wu, J.   410
Wu, M.   704
Wymeersch, H.   150

Xu, K.   199
Xu, M.-w.   199

Yaipairoj, S.   183
Yamamoto, J.S.   207
Yan, P.-l.   704
Yazdanpanah, A.   554

Yoo, S.W.   997, 1226
Youm, S.-K.   425
Yuan, J.   1369

Zamani, A.   106
Zambenedetti Granville, L.   782

Zhang, J.   753
Zhao, B.H.   826
Zhou, B.   410
Zhou, J.-g.   704
Zhou, R.   163
Zola, E.   932

Lecture Notes in Computer Science

For information about Vols. 1–3044

please contact your bookseller or Springer-Verlag

Vol. 3158: M. Barbeau, E. Kranakis, I. Nikolaidis (Eds.), Ad-Hoc, Mobile, and Wireless Networks. IX, 344 pages. 2004.

Vol. 3146: P. Érdi, A. Esposito, M. Marinaro, S. Scarpetta (Eds.), Computational Neuroscience: Cortical Dynamics. XI, 161 pages. 2004.

Vol. 3143: W. Liu, Y. Shi, L. Qing (Eds.), Advances in Web-Based Learning – ICWL 2004. XIV, 459 pages. 2004.

Vol. 3140: N. Koch, P. Fraternali, M. Wirsing (Eds.), Web Engineering. XXI, 623 pages. 2004.

Vol. 3139: F. Iida, R. Pfeifer, L. Steels, Y. Kuniyoshi (Eds.), Embodied Artificial Intelligence. IX, 331 pages. 2004. (Subseries LNAI).

Vol. 3138: A. Fred, T. Caelli, R.P.W. Duin, A. Campilho, D.d. Ridder (Eds.), Structural, Syntactic, and Statistical Pattern Recognition. XXII, 1168 pages. 2004.

Vol. 3133: A.D. Pimentel, S. Vassiliadis (Eds.), Computer Systems: Architectures, Modeling, and Simulation. XIII, 562 pages. 2004.

Vol. 3131: V. Torra, Y. Narukawa (Eds.), Modeling Decisions for Artificial Intelligence. XI, 327 pages. 2004. (Subseries LNAI).

Vol. 3129: L. Qing, G. Wang, L. Feng (Eds.), Advances in Web-Age Information Management. XVII, 753 pages. 2004.

Vol. 3128: D. Asonov (Ed.), Querying Databases Privately. IX, 115 pages. 2004.

Vol. 3127: K.E. Wolff, H.D. Pfeiffer, H.S. Delugach (Eds.), Conceptual Structures at Work. XI, 403 pages. 2004. (Subseries LNAI).

Vol. 3126: P. Dini, P. Lorenz, J.N. de Souza (Eds.), Service Assurance with Partial and Intermittent Resources. XI, 312 pages. 2004.

Vol. 3125: D. Kozen (Ed.), Mathematics of Program Construction. X, 401 pages. 2004.

Vol. 3124: J.N. de Souza, P. Dini, P. Lorenz (Eds.), Telecommunications and Networking - ICT 2004. XXVI, 1390 pages. 2004.

Vol. 3123: A. Belz, R. Evans, P. Piwek (Eds.), Natural Language Generation. X, 219 pages. 2004. (Subseries LNAI).

Vol. 3121: S. Nikoletseas, J.D.P. Rolim (Eds.), Algorithmic Aspects of Wireless Sensor Networks. X, 201 pages. 2004.

Vol. 3120: J. Shawe-Taylor, Y. Singer (Eds.), Learning Theory. X, 648 pages. 2004. (Subseries LNAI).

Vol. 3118: K. Miesenberger, J. Klaus, W. Zagler, D. Burger (Eds.), Computer Helping People with Special Needs. XXIII, 1191 pages. 2004.

Vol. 3116: C. Rattray, S. Maharaj, C. Shankland (Eds.), Algebraic Methodology and Software Technology. XI, 569 pages. 2004.

Vol. 3115: P. Enser, Y. Kompatsiaris, N.E. O'Connor, A.F. Smeaton, A.W.M. Smeulders (Eds.), Image and Video Retrieval. XVII, 679 pages. 2004.

Vol. 3114: R. Alur, D.A. Peled (Eds.), Computer Aided Verification. XII, 536 pages. 2004.

Vol. 3113: J. Karhumäki, H. Maurer, G. Paun, G. Rozenberg (Eds.), Theory Is Forever. X, 283 pages. 2004.

Vol. 3112: H. Williams, L. MacKinnon (Eds.), Key Technologies for Data Management. XII, 265 pages. 2004.

Vol. 3111: T. Hagerup, J. Katajainen (Eds.), Algorithm Theory - SWAT 2004. XI, 506 pages. 2004.

Vol. 3110: A. Juels (Ed.), Financial Cryptography. XI, 281 pages. 2004.

Vol. 3109: S.C. Sahinalp, S. Muthukrishnan, U. Dogrusoz (Eds.), Combinatorial Pattern Matching. XII, 486 pages. 2004.

Vol. 3108: H. Wang, J. Pieprzyk, V. Varadharajan (Eds.), Information Security and Privacy. XII, 494 pages. 2004.

Vol. 3107: J. Bosch, C. Krueger (Eds.), Software Reuse: Methods, Techniques and Tools. XI, 339 pages. 2004.

Vol. 3105: S. Göbel, U. Spierling, A. Hoffmann, I. Iurgel, O. Schneider, J. Dechau, A. Feix (Eds.), Technologies for Interactive Digital Storytelling and Entertainment. XVI, 304 pages. 2004.

Vol. 3104: R. Kralovic, O. Sykora (Eds.), Structural Information and Communication Complexity. X, 303 pages. 2004.

Vol. 3103: K. Deb, e. al. (Eds.), Genetic and Evolutionary Computation – GECCO 2004. XLIX, 1439 pages. 2004.

Vol. 3102: K. Deb, e. al. (Eds.), Genetic and Evolutionary Computation – GECCO 2004. L, 1445 pages. 2004.

Vol. 3101: M. Masoodian, S. Jones, B. Rogers (Eds.), Computer Human Interaction. XIV, 694 pages. 2004.

Vol. 3100: J.F. Peters, A. Skowron, J.W. Grzymała-Busse, B. Kostek, R.W. Świniarski, M.S. Szczuka (Eds.), Transactions on Rough Sets I. X, 405 pages. 2004.

Vol. 3099: J. Cortadella, W. Reisig (Eds.), Applications and Theory of Petri Nets 2004. XI, 505 pages. 2004.

Vol. 3098: J. Desel, W. Reisig, G. Rozenberg (Eds.), Lectures on Concurrency and Petri Nets. VIII, 849 pages. 2004.

Vol. 3097: D. Basin, M. Rusinowitch (Eds.), Automated Reasoning. XII, 493 pages. 2004. (Subseries LNAI).

Vol. 3096: G. Melnik, H. Holz (Eds.), Advances in Learning Software Organizations. X, 173 pages. 2004.

Vol. 3095: C. Bussler, D. Fensel, M.E. Orlowska, J. Yang (Eds.), Web Services, E-Business, and the Semantic Web. X, 147 pages. 2004.

Vol. 3094: A. Nürnberger, M. Detyniecki (Eds.), Adaptive Multimedia Retrieval. VIII, 229 pages. 2004.

Vol. 3093: S.K. Katsikas, S. Gritzalis, J. Lopez (Eds.), Public Key Infrastructure. XIII, 380 pages. 2004.

Vol. 3092: J. Eckstein, H. Baumeister (Eds.), Extreme Programming and Agile Processes in Software Engineering. XVI, 358 pages. 2004.

Vol. 3091: V. van Oostrom (Ed.), Rewriting Techniques and Applications. X, 313 pages. 2004.

Vol. 3089: M. Jakobsson, M. Yung, J. Zhou (Eds.), Applied Cryptography and Network Security. XIV, 510 pages. 2004.

Vol. 3087: D. Maltoni, A.K. Jain (Eds.), Biometric Authentication. XIII, 343 pages. 2004.

Vol. 3086: M. Odersky (Ed.), ECOOP 2004 – Object-Oriented Programming. XIII, 611 pages. 2004.

Vol. 3085: S. Berardi, M. Coppo, F. Damiani (Eds.), Types for Proofs and Programs. X, 409 pages. 2004.

Vol. 3084: A. Persson, J. Stirna (Eds.), Advanced Information Systems Engineering. XIV, 596 pages. 2004.

Vol. 3083: W. Emmerich, A.L. Wolf (Eds.), Component Deployment. X, 249 pages. 2004.

Vol. 3080: J. Desel, B. Pernici, M. Weske (Eds.), Business Process Management. X, 307 pages. 2004.

Vol. 3079: Z. Mammeri, P. Lorenz (Eds.), High Speed Networks and Multimedia Communications. XVIII, 1103 pages. 2004.

Vol. 3078: S. Cotin, D.N. Metaxas (Eds.), Medical Simulation. XVI, 296 pages. 2004.

Vol. 3077: F. Roli, J. Kittler, T. Windeatt (Eds.), Multiple Classifier Systems. XII, 386 pages. 2004.

Vol. 3076: D. Buell (Ed.), Algorithmic Number Theory. XI, 451 pages. 2004.

Vol. 3074: B. Kuijpers, P. Revesz (Eds.), Constraint Databases and Applications. XII, 181 pages. 2004.

Vol. 3073: H. Chen, R. Moore, D.D. Zeng, J. Leavitt (Eds.), Intelligence and Security Informatics. XV, 536 pages. 2004.

Vol. 3072: D. Zhang, A.K. Jain (Eds.), Biometric Authentication. XVII, 800 pages. 2004.

Vol. 3071: A. Omicini, P. Petta, J. Pitt (Eds.), Engineering Societies in the Agents World. XIII, 409 pages. 2004. (Subseries LNAI).

Vol. 3070: L. Rutkowski, J. Siekmann, R. Tadeusiewicz, L.A. Zadeh (Eds.), Artificial Intelligence and Soft Computing - ICAISC 2004. XXV, 1208 pages. 2004. (Subseries LNAI).

Vol. 3068: E. André, L. Dybkjær, W. Minker, P. Heisterkamp (Eds.), Affective Dialogue Systems. XII, 324 pages. 2004. (Subseries LNAI).

Vol. 3067: M. Dastani, J. Dix, A. El Fallah-Seghrouchni (Eds.), Programming Multi-Agent Systems. X, 221 pages. 2004. (Subseries LNAI).

Vol. 3066: S. Tsumoto, R. Słowiński, J. Komorowski, J.W. Grzymała-Busse (Eds.), Rough Sets and Current Trends in Computing. XX, 853 pages. 2004. (Subseries LNAI).

Vol. 3065: A. Lomuscio, D. Nute (Eds.), Deontic Logic in Computer Science. X, 275 pages. 2004. (Subseries LNAI).

Vol. 3064: D. Bienstock, G. Nemhauser (Eds.), Integer Programming and Combinatorial Optimization. XI, 445 pages. 2004.

Vol. 3063: A. Llamosí, A. Strohmeier (Eds.), Reliable Software Technologies - Ada-Europe 2004. XIII, 333 pages. 2004.

Vol. 3062: J.L. Pfaltz, M. Nagl, B. Böhlen (Eds.), Applications of Graph Transformations with Industrial Relevance. XV, 500 pages. 2004.

Vol. 3061: F.F. Ramos, H. Unger, V. Larios (Eds.), Advanced Distributed Systems. VIII, 285 pages. 2004.

Vol. 3060: A.Y. Tawfik, S.D. Goodwin (Eds.), Advances in Artificial Intelligence. XIII, 582 pages. 2004. (Subseries LNAI).

Vol. 3059: C.C. Ribeiro, S.L. Martins (Eds.), Experimental and Efficient Algorithms. X, 586 pages. 2004.

Vol. 3058: N. Sebe, M.S. Lew, T.S. Huang (Eds.), Computer Vision in Human-Computer Interaction. X, 233 pages. 2004.

Vol. 3057: B. Jayaraman (Ed.), Practical Aspects of Declarative Languages. VIII, 255 pages. 2004.

Vol. 3056: H. Dai, R. Srikant, C. Zhang (Eds.), Advances in Knowledge Discovery and Data Mining. XIX, 713 pages. 2004. (Subseries LNAI).

Vol. 3055: H. Christiansen, M.-S. Hacid, T. Andreasen, H.L. Larsen (Eds.), Flexible Query Answering Systems. X, 500 pages. 2004. (Subseries LNAI).

Vol. 3054: I. Crnkovic, J.A. Stafford, H.W. Schmidt, K. Wallnau (Eds.), Component-Based Software Engineering. XI, 311 pages. 2004.

Vol. 3053: C. Bussler, J. Davies, D. Fensel, R. Studer (Eds.), The Semantic Web: Research and Applications. XIII, 490 pages. 2004.

Vol. 3052: W. Zimmermann, B. Thalheim (Eds.), Abstract State Machines 2004. Advances in Theory and Practice. XII, 235 pages. 2004.

Vol. 3051: R. Berghammer, B. Möller, G. Struth (Eds.), Relational and Kleene-Algebraic Methods in Computer Science. X, 279 pages. 2004.

Vol. 3050: J. Domingo-Ferrer, V. Torra (Eds.), Privacy in Statistical Databases. IX, 367 pages. 2004.

Vol. 3049: M. Bruynooghe, K.-K. Lau (Eds.), Program Development in Computational Logic. VIII, 539 pages. 2004.

Vol. 3047: F. Oquendo, B. Warboys, R. Morrison (Eds.), Software Architecture. X, 279 pages. 2004.

Vol. 3046: A. Laganà, M.L. Gavrilova, V. Kumar, Y. Mun, C.J.K. Tan, O. Gervasi (Eds.), Computational Science and Its Applications – ICCSA 2004. LIII, 1016 pages. 2004..

Vol. 3045: A. Laganà, M.L. Gavrilova, V. Kumar, Y. Mun, C.J.K. Tan, O. Gervasi (Eds.), Computational Science and Its Applications – ICCSA 2004. LIII, 1040 pages. 2004.